Diagram Symbols

The entries in this table are given roughly in order of appearance in the textbook, but consistency has been maintained in later chapters as well.

Mechanics

Symbol	Meaning
	displacement vector, displacement component
	velocity vector, velocity component
	acceleration vector
	force vector, force component
	momentum vector
	gravitational field vector
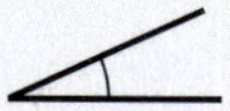	angle marking
	rotational motion

Thermodynamics

Symbol	Meaning
	energy transferred as heat
	energy transferred as work
	cycle or process

Waves and Electromagnetism

Symbol	Meaning
	ray (light or sound)
	positive charge
	negative charge
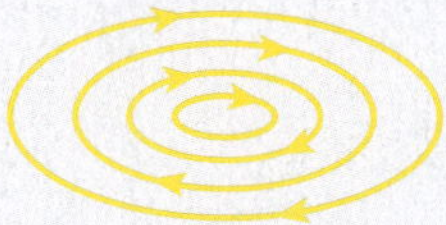	electric field lines
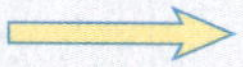	electric field vector
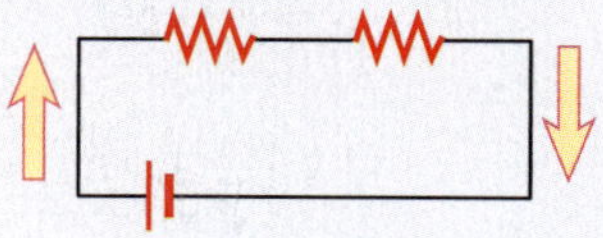	electric current
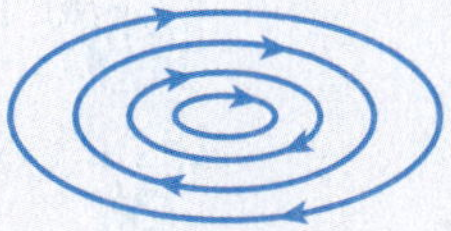	magnetic field lines
	magnetic field vector (into page, out of page)

HOLT

Physics

TEACHER'S EDITION

Raymond A. Serway, Ph.D.
Jerry S. Faughn, Ph.D.

Teacher's Edition Walk-Through

Student Edition Contents in Brief

HOLT, RINEHART AND WINSTON
A Harcourt Education Company
Orlando • Austin • New York • San Diego • London

A Dual Approach to Physics:

Balances conceptual study with problem solving

Holt Physics is the only text that offers a conceptual foundation and a mathematically-based presentation of physics. Written by Raymond Serway and Jerry Faughn specifically for your college-bound high school students, ***Holt Physics*** covers the core physics content. Your students' comprehension will be further extended with the application of multiple print and technology resources.

Why we wrote this book:

As a high school teacher, you face challenges in preparing your students to understand the world around them. You also want to make your class as inviting, interesting, and inclusive as possible. We wanted to write the book that was both "user friendly" and one that would help you and your students achieve these goals.

Get the Physics Right First and foremost, we wanted to give you a book that was technically correct and one that provided good preparation for college. Our previous experience writing *College Physics* gave us the background we needed to write an authoritative, accurate, and up-to-date text that is appropriate for today's students.

Link Concepts and Problem-Solving Students need clear conceptual devopment and plenty of practice working with both fundamental physical concepts ad problem-solving skills. We wanted this book to help students with both.

Focus on the Diagram Learning how to prepare an accurate and informate diagram for a situation is a crucial step that identifies the connection beteen the concrete world and the world of physics. We wanted to provide an abundace of support in preparing and interpreting such diagrams to sharpen students'kills.

Relate to the Student The best way to ensure learning that lasts is throug practical applications and concrete examples that students can relate to ad appreciate. Therefore, we wanted a book filled with examples—from the tt presentation to questions, problems, and other features.

Without a doubt, the most important elements in any learning environmenare you, the instructor, and effective communication between you and your stlents. If you are excited, knowledgeable, and interested in what you teach, and nvey this effectively, you will be very successful in the classroom. We applaud ur contributions to the world and to our future, and we wish you and your stuents much success.

Regards,

Raymond A. Serway and Jerry S. Faughn

LEVELED ACTIVITIES HELP YOU CUSTOMIZE YOUR COURSE TO MEET THE NEEDS OF ALL STUDENTS.

- All activities are leveled **Basic, General,** or **Advanced** to help you effectively plan your lessons.

STUDENTS WILL EXTEND THEIR UNDERSTANDING OF PHYSICS CONCEPTS WITH A FLEXIBLE LAB PROGRAM.

- The extensive laboratory program includes in-text chapter labs, **Quick Labs,** and modified labs in the **Appendix** and on the ***Teacher's One-Stop Planner®***, as well as **Discovery Labs, Invention Labs,** and **CBL™ Labs.**
- The variety of chapter lab types—**Skills Practice, Inquiry** and **CBL™**—meets the needs of your curriculum.
- ***Holt Physics Virtual Investigations*** includes simulations and detailed exploration activities that allow students to interact with the science on screen.

VERSATILE PROBLEM-SOLVING SUPPORT INTEGRATES QUANTITATIVE SKILLS WITH PHYSICS CONCEPTS.

- **Sample Problems** use a four step method—**Define, Plan, Calculate,** and **Evaluate**—to strengthen students' analytical skills. **Tips** and **Practice Problems** are also provided to reinforce problem-solving skills.
- **Problem Guide** in the teacher's wrap assists you in assigning appropriate problems for targeted practice.

COMPREHENSIVE AND CUSTOMIZABLE ASSESSMENTS HELP YOU CHECK STUDENTS' UNDERSTANDING.

- Section assessment—from **Section Reviews** to **Section Quizzes**—helps you check students' understanding along the way.
- Chapter assessments test students' overall knowledge of the chapter objectives with a variety of questions including standardized test prep questions.
- The **ExamView® Version 6 Assessment Suite** on the ***Teacher's One-Stop Planner®*** gives you the power to customize your own assessments, and post them to **Holt Online Assessment** for automatic grading and to track student progress. Algorithmic assessments generate random variables for problems.

PLUS . . . INTEGRATED TECHNOLOGY AND ONLINE RESOURCES REINFORCE STUDENTS' COMPREHENSION OF PHYSICS.

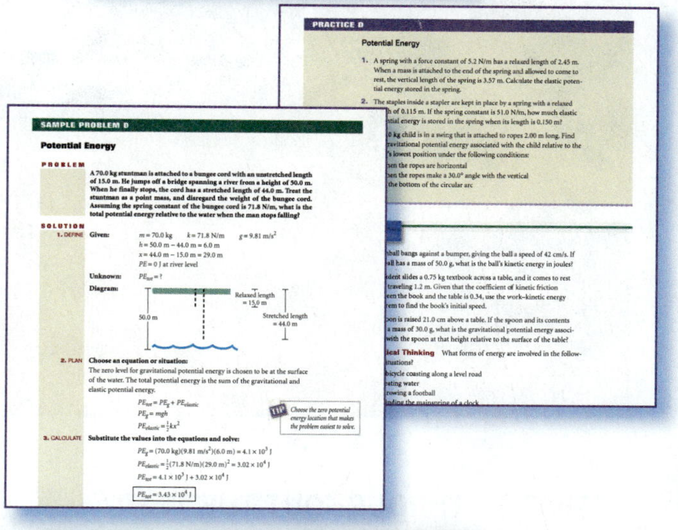

A text that fosters comprehension of physics concepts and mathematics

ENGAGING AND WELL-ORGANIZED STUDENT PAGES MAKE PHYSICS RELEVANT.

Vivid photos accompanied by smaller schematics tie the image's action to the chapter content.

What to Expect and **Why it Matters** encourage students' interest in the chapter.

Chapter Preview prepares students for what they will learn.

ACTIVITIES THROUGHOUT THE TEXT EXTEND STUDENTS' LEARNING.

Did you know? presents interesting facts related to chapter content.

Did you know?

The word *physics* comes from the ancient Greek word for "nature." According to Aristotle, who assigned the name, physics is the study of natural events. Aristotle believed that the study of motion was the basis of physics. Galileo developed the foundations for the modern study of motion using mathematics. In 1632, Galileo published the first mathematical treatment of motion.

Why it Matters: Conceptual Challenge is a point-of-use check for students to verify their understanding of the text by applying physics concepts to real-world applications.

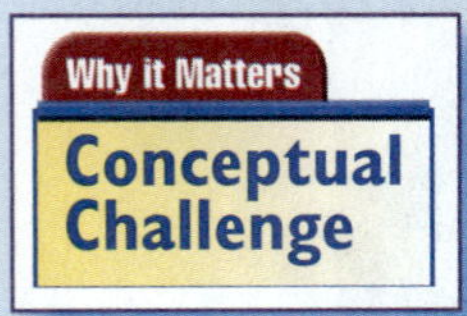

extension

Integrating Health

Visit go.hrw.com for the activity "Energy Costs of Walking and Running."

Keyword HF6WRKX

Sample Problems and **Practice Problems** reinforce students' problem-solving skills.

Advanced Topics refer students to the **Appendix** for more detail on special topics.

Quick Labs give students hands-on learning when time is limited. Some **Quick Labs** are suitable as homework assignments and are labeled as such in the *Teacher's Edition.*

In-text labs—**Skills Practice, Inquiry,** and **CBL™**—provide students the opportunity to apply analytical skills while reinforcing physics concepts.

Extension directs students to **go.hrw.com** for related activities that demonstrate chapter concepts.

Student Edition

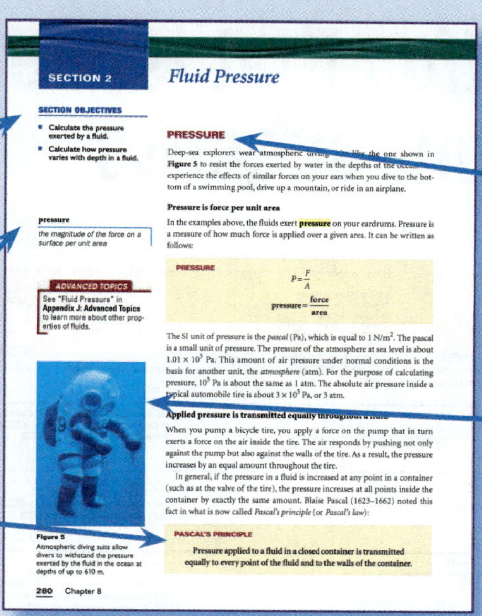
SECTION 2 **Fluid Pressure**

SECTION OBJECTIVES

- Calculate the pressure exerted by a fluid.
- Calculate how pressure varies with depth in a fluid.

PRESSURE

Deep-sea explorers wear atmospheric diving suits like the one shown in **Figure 5** to resist the forces exerted by water in the depths of the ocean. You experience the effects of similar forces on your ears when you dive to the bottom of a swimming pool, drive up a mountain, or ride in an airplane.

Pressure is force per unit area

In the examples above, the fluids exert **pressure** on your eardrums. Pressure is a measure of how much force is applied over a given area. It can be written as follows:

PRESSURE

$$P = \frac{F}{A}$$

$$\text{pressure} = \frac{\text{force}}{\text{area}}$$

The SI unit of pressure is the *pascal* (Pa), which is equal to 1 N/m². The pascal is a small unit of pressure. The pressure of the atmosphere at sea level is about 1.01×10^5 Pa. This amount of air pressure under normal conditions is the basis for another unit, the *atmosphere* (atm). For the purpose of calculating pressure, 10^5 Pa is about the same as 1 atm. The absolute air pressure inside a typical automobile tire is about 3×10^5 Pa, or 3 atm.

Applied pressure is transmitted equally throughout a fluid

When you pump a bicycle tire, you apply a force on the pump that in turn exerts a force on the air inside the tire. The air responds by pushing not only against the pump but also against the walls of the tire. As a result, the pressure increases by an equal amount throughout the tire.

In general, if the pressure in a fluid is increased at any point in a container (such as at the valve of the tire), the pressure increases at all points inside the container by exactly the same amount. Blaise Pascal (1623–1662) noted this fact in what is now called *Pascal's principle* (or *Pascal's law*):

PASCAL'S PRINCIPLE

Pressure applied to a fluid in a closed container is transmitted equally to every point of the fluid and to the walls of the container.

pressure
the magnitude of the force on a surface per unit area

ADVANCED TOPICS
See "Fluid Pressure" in **Appendix J: Advanced Topics** to learn more about other properties of fluids.

Figure 5
Atmospheric diving suits allow divers to withstand the pressure exerted by the fluid in the ocean at depths of up to 610 m.

280 Chapter 8

Section Objectives outline what students will learn.

Key Terms are highlighted in yellow and defined in the margin to develop students' vocabulary skills.

Important physics principles and equations are highlighted prominently in the text for easy reference.

The pages are easy to navigate with outline-style headings and content grouped into chunks.

Visuals are engaging and closely related to the text. You'll also find detailed schematics explaining physics phenomena and relevant photos of students in action.

RELEVANT FEATURES RELATE PHYSICS TO THE WORLD.

Why it Matters, a magazine style feature, explains the physics behind familiar objects, such as seatbelts and refrigerators.

Physics Careers focuses on professionals who apply physics in their everyday work.

Science • Technology • Society explores how science and technological innovation impact society.

Physics and Its World provides timelines that outline key events in the history of physics in the context of other historical events.

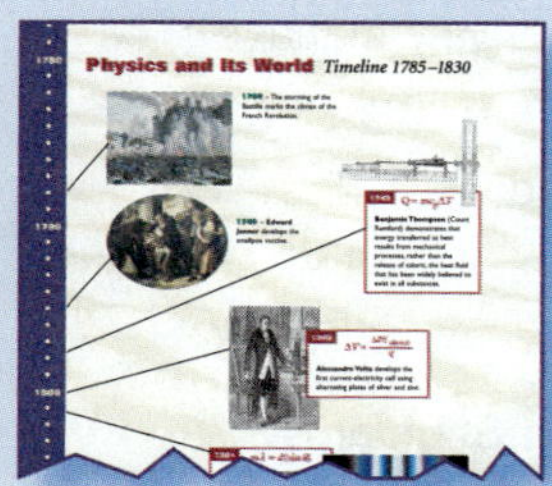

COMPREHENSIVE TESTING SUPPORT PREPARES STUDENTS FOR SUCCESS.

Section Review checks students' understanding of the section objectives.

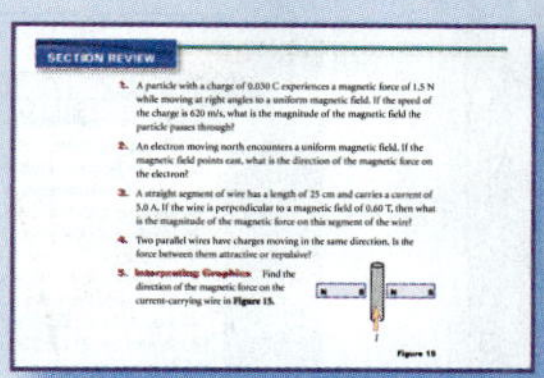

Chapter Highlights is a study tool that provides a summary of **Key Terms, Key Ideas,** and **Variable Symbols.**

Chapter Review is a comprehensive assessment of the chapter objectives that includes concept review questions, practice problems, mixed review, graphing calculator computations, and alternative assessments.

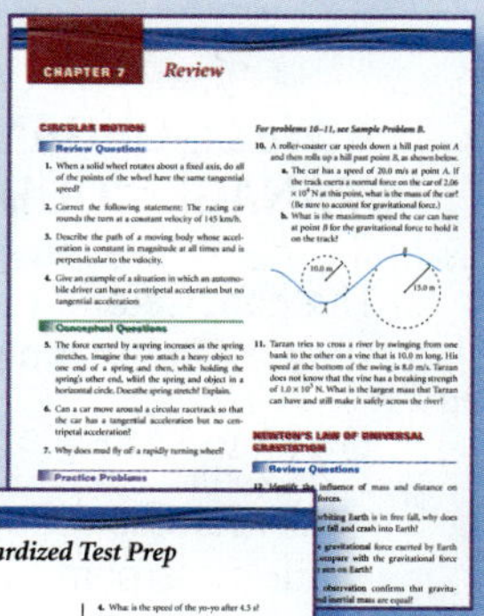

Standardized Test Prep prepares students for standardized tests with multiple-choice, short-response, and extended-response questions. **Test Tip** provides students with helpful hints for testing success.

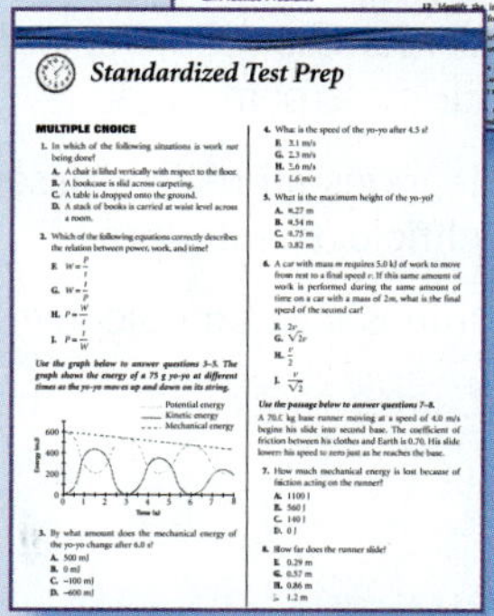

A *Teacher's Edition* that is the foundation of your success

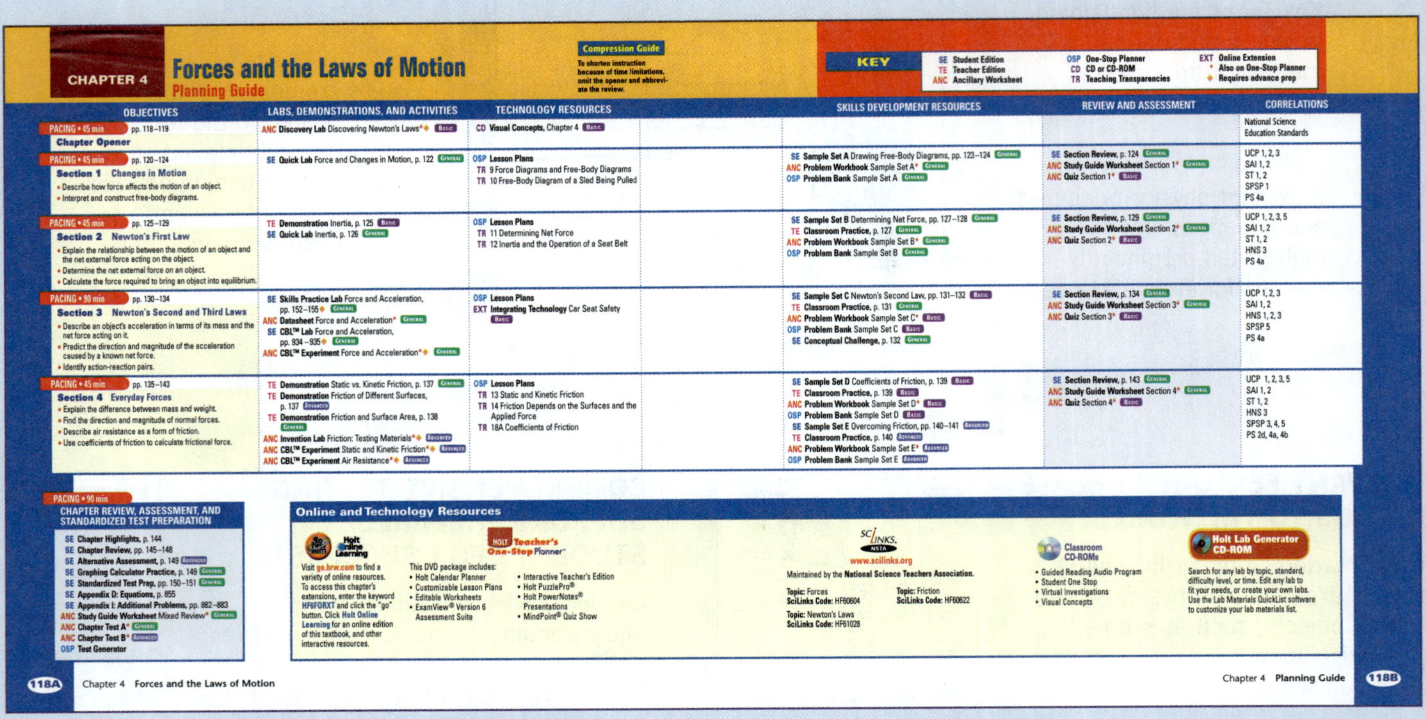

CHAPTER 4 **Forces and the Laws of Motion**
Planning Guide

Compression Guide
To shorten instruction because of time limitations, omit the opener and abbreviate the review.

KEY
SE Student Edition; TE Teacher Edition; ANC Ancillary Worksheet; OSP One-Stop Planner; CD CD or CD-ROM; TR Teaching Transparencies; EXT Online Extension; * Also on One-Stop Planner; ◆ Requires advance prep

OBJECTIVES	LABS, DEMONSTRATIONS, AND ACTIVITIES	TECHNOLOGY RESOURCES	SKILLS DEVELOPMENT RESOURCES	REVIEW AND ASSESSMENT	CORRELATIONS
PACING • 45 min pp. 118–119 **Chapter Opener**	ANC **Discovery Lab** Discovering Newton's Laws*◆ BASIC	CD **Visual Concepts,** Chapter 4 BASIC			National Science Education Standards
PACING • 45 min pp. 120–124 **Section 1 Changes in Motion** • Describe how force affects the motion of an object. • Interpret and construct free-body diagrams.	SE **Quick Lab** Force and Changes in Motion, p. 122 GENERAL	OSP **Lesson Plans** TR 9 Force Diagrams and Free-Body Diagrams TR 10 Free-Body Diagram of a Sled Being Pulled	SE **Sample Set A** Drawing Free-Body Diagrams, pp. 123–124 GENERAL ANC **Problem Workbook** Sample Set A* GENERAL OSP **Problem Bank** Sample Set A GENERAL	SE **Section Review,** p. 124 GENERAL ANC **Study Guide Worksheet** Section 1* GENERAL ANC **Quiz** Section 1* BASIC	UCP 1, 2, 3 SAI 1, 2 ST 1, 2 SPSP 1 PS 4a
PACING • 45 min pp. 125–129 **Section 2 Newton's First Law** • Explain the relationship between the motion of an object and the net external force acting on the object. • Determine the net external force on an object. • Calculate the force required to bring an object into equilibrium.	TE **Demonstration** Inertia, p. 125 BASIC SE **Quick Lab** Inertia, p. 126 GENERAL	OSP **Lesson Plans** TR 11 Determining Net Force TR 12 Inertia and the Operation of a Seat Belt	SE **Sample Set B** Determining Net Force, pp. 127–128 GENERAL TE **Classroom Practice,** p. 127 GENERAL ANC **Problem Workbook** Sample Set B* GENERAL OSP **Problem Bank** Sample Set B GENERAL	SE **Section Review,** p. 129 GENERAL ANC **Study Guide Worksheet** Section 2* GENERAL ANC **Quiz** Section 2* BASIC	UCP 1, 2, 3, 5 SAI 1, 2 ST 1, 2 HNS 3 PS 4a
PACING • 90 min pp. 130–134 **Section 3 Newton's Second and Third Laws** • Describe an object's acceleration in terms of its mass and the net force acting on it. • Predict the direction and magnitude of the acceleration caused by a known net force. • Identify action-reaction pairs.	SE **Skills Practice Lab** Force and Acceleration, pp. 152–155◆ GENERAL ANC **Datasheet** Force and Acceleration* GENERAL SE **CBL™ Lab** Force and Acceleration, pp. 934–935◆ GENERAL ANC **CBL™ Experiment** Force and Acceleration*◆ GENERAL	OSP **Lesson Plans** EXT **Integrating Technology** Car Seat Safety BASIC	SE **Sample Set C** Newton's Second Law, pp. 131–132 BASIC TE **Classroom Practice,** p. 131 GENERAL ANC **Problem Workbook** Sample Set C* BASIC OSP **Problem Bank** Sample Set C BASIC SE **Conceptual Challenge,** p. 132 GENERAL	SE **Section Review,** p. 134 GENERAL ANC **Study Guide Worksheet** Section 3* GENERAL ANC **Quiz** Section 3* BASIC	UCP 1, 2, 3 SAI 1, 2 HNS 1, 2, 3 SPSP 5 PS 4a
PACING • 45 min pp. 135–143 **Section 4 Everyday Forces** • Explain the difference between mass and weight. • Find the direction and magnitude of normal forces. • Describe air resistance as a form of friction. • Use coefficients of friction to calculate frictional force.	TE **Demonstration** Static vs. Kinetic Friction, p. 137 GENERAL TE **Demonstration** Friction of Different Surfaces, p. 137 ADVANCED TE **Demonstration** Friction and Surface Area, p. 138 GENERAL ANC **Invention Lab** Friction: Testing Materials*◆ ADVANCED ANC **CBL™ Experiment** Static and Kinetic Friction*◆ ADVANCED ANC **CBL™ Experiment** Air Resistance*◆ ADVANCED	OSP **Lesson Plans** TR 13 Static and Kinetic Friction TR 14 Friction Depends on the Surfaces and the Applied Force TR 18A Coefficients of Friction	SE **Sample Set D** Coefficients of Friction, p. 139 BASIC TE **Classroom Practice,** p. 139 BASIC ANC **Problem Workbook** Sample Set D* BASIC OSP **Problem Bank** Sample Set D BASIC SE **Sample Set E** Overcoming Friction, pp. 140–141 ADVANCED TE **Classroom Practice,** p. 140 ADVANCED ANC **Problem Workbook** Sample Set E* ADVANCED OSP **Problem Bank** Sample Set E ADVANCED	SE **Section Review,** p. 143 GENERAL ANC **Study Guide Worksheet** Section 4* GENERAL ANC **Quiz** Section 4* BASIC	UCP 1, 2, 3, 5 SAI 1, 2 ST 1, 2 HNS 3 SPSP 3, 4, 5 PS 2d, 4a, 4b

PACING • 90 min
CHAPTER REVIEW, ASSESSMENT, AND STANDARDIZED TEST PREPARATION

- SE **Chapter Highlights,** p. 144
- SE **Chapter Review,** pp. 145–148
- SE **Alternative Assessment,** p. 149 ADVANCED
- SE **Graphing Calculator Practice,** p. 149 GENERAL
- SE **Standardized Test Prep,** pp. 150–151 GENERAL
- SE **Appendix D: Equations,** p. 855
- SE **Appendix I: Additional Problems,** pp. 882–883
- ANC **Study Guide Worksheet** Mixed Review* GENERAL
- ANC **Chapter Test A*** GENERAL
- ANC **Chapter Test B*** ADVANCED
- OSP **Test Generator**

Online and Technology Resources

Holt Online Learning
Visit go.hrw.com to find a variety of online resources. To access this chapter's extensions, enter the keyword HF6FORXT and click the "go" button. Click Holt Online Learning for an online edition of this textbook, and other interactive resources.

Holt Teacher's One-Stop Planner®
This DVD package includes:
- Holt Calendar Planner
- Customizable Lesson Plans
- Editable Worksheets
- ExamView® Version 6 Assessment Suite
- Interactive Teacher's Edition
- Holt PuzzlePro®
- Holt PowerNotes® Presentations
- MindPoint® Quiz Show

SciLinks. NSTA
www.scilinks.org
Maintained by the **National Science Teachers Association.**
Topic: Forces **SciLinks Code:** HF60604
Topic: Newton's Laws **SciLinks Code:** HF61028
Topic: Friction **SciLinks Code:** HF60622

Classroom CD-ROMs
- Guided Reading Audio Program
- Student One Stop
- Virtual Investigations
- Visual Concepts

Holt Lab Generator CD-ROM
Search for any lab by topic, standard, difficulty level, or time. Edit any lab to fit your needs, or create your own labs. Use the Lab Materials QuickList software to customize your lab materials list.

118A Chapter 4 Forces and the Laws of Motion
Chapter 4 Planning Guide 118B

A WELL-ORGANIZED PLANNING GUIDE HELPS YOU EFFECTIVELY MANAGE YOUR CLASSROOM TIME.

The concise and complete **Planning Guide** makes preparing lessons simple. The **Planning Guide**

- is an easy-to-use visual that outlines all your teaching resources.
- incorporates a **Pacing and Compression Guide** to help you effectively manage your classroom time.
- integrates all review, assessments, labs, skills development, and technology resources.
- outlines section objectives to help you keep lessons focused.
- includes section correlations to the **National Science Education Standards.**

Teacher's Edition

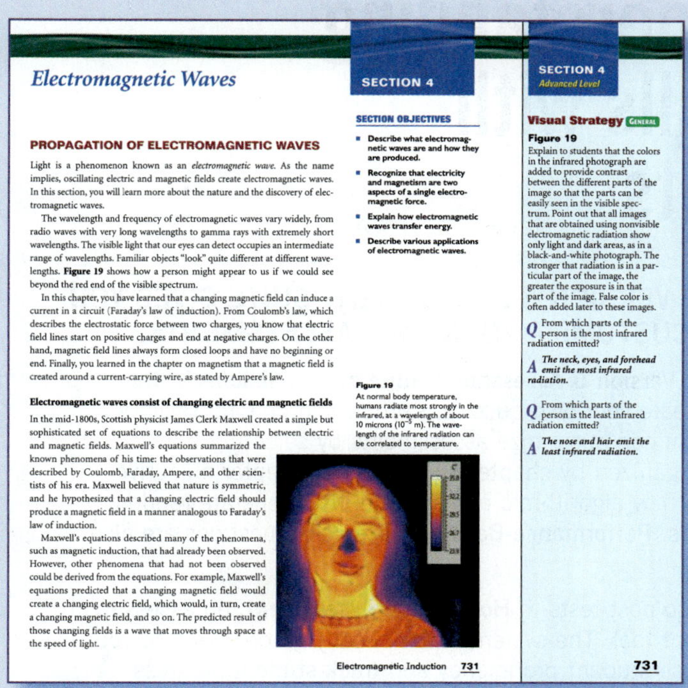
Electromagnetic Waves

SECTION 4

PROPAGATION OF ELECTROMAGNETIC WAVES

Light is a phenomenon known as an *electromagnetic wave.* As the name implies, oscillating electric and magnetic fields create electromagnetic waves. In this section, you will learn more about the nature and the discovery of electromagnetic waves.

The wavelength and frequency of electromagnetic waves vary widely, from radio waves with very long wavelengths to gamma rays with extremely short wavelengths. The visible light that our eyes can detect occupies an intermediate range of wavelengths. Familiar objects "look" quite different at different wavelengths. **Figure 19** shows how a person might appear to us if we could see beyond the red end of the visible spectrum.

In this chapter, you have learned that a changing magnetic field can induce a current in a circuit (Faraday's law of induction). From Coulomb's law, which describes the electrostatic force between two charges, you know that electric field lines start on positive charges and end at negative charges. On the other hand, magnetic field lines always form closed loops and have no beginning or end. Finally, you learned in the chapter on magnetism that a magnetic field is created around a current-carrying wire, as stated by Ampere's law.

Electromagnetic waves consist of changing electric and magnetic fields

In the mid-1800s, Scottish physicist James Clerk Maxwell created a simple but sophisticated set of equations to describe the relationship between electric and magnetic fields. Maxwell's equations summarized the known phenomena of his time: the observations that were described by Coulomb, Faraday, Ampere, and other scientists of his era. Maxwell believed that nature is symmetric, and he hypothesized that a changing electric field should produce a magnetic field in a manner analogous to Faraday's law of induction.

Maxwell's equations described many of the phenomena, such as magnetic induction, that had already been observed. However, other phenomena that had not been observed could be derived from the equations. For example, Maxwell's equations predicted that a changing magnetic field would create a changing electric field, which would, in turn, create a changing magnetic field, and so on. The predicted result of those changing fields is a wave that moves through space at the speed of light.

SECTION OBJECTIVES

- Describe what electromagnetic waves are and how they are produced.
- Recognize that electricity and magnetism are two aspects of a single electromagnetic force.
- Explain how electromagnetic waves transfer energy.
- Describe various applications of electromagnetic waves.

Figure 19
At normal body temperature, humans radiate most strongly in the infrared, at a wavelength of about 10 microns (10^{-5} m). The wavelength of the infrared radiation can be correlated to temperature.

Electromagnetic Induction 731

SECTION 4
Advanced Level

Visual Strategy GENERAL

Figure 19
Explain to students that the colors in the infrared photograph are added to provide contrast between the different parts of the image so that the parts can be easily seen in the visible spectrum. Point out that all images that are obtained using nonvisible electromagnetic radiation show only light and dark areas, as in a black-and-white photograph. The stronger that radiation is in a particular part of the image, the greater the exposure is in that part of the image. False color is often added later to these images.

Q From which parts of the person is the most infrared radiation emitted?

A *The neck, eyes, and forehead emit the most infrared radiation.*

Q From which parts of the person is the least infrared radiation emitted?

A *The nose and hair emit the least infrared radiation.*

731

DIFFERENTIATED INSTRUCTION HELPS YOU MEET ALL YOUR STUDENTS' NEEDS.

Sections and activities in the teacher's wrap are labeled by ability level—**Basic, General,** and **Advanced**—to help you choose appropriate assignments for each student.

- **Basic** activities are designed for all students and address core skills.
- **General** activities are appropriate for most students and require more critical-thinking skills than **Basic** activities.
- **Advanced** activities are more challenging and require more higher-order thinking skills than **General** activities.

Visual Strategy guides your use of diagrams and illustrations from the text to clarify concepts covered. A question and answer format promotes classroom participation.

INTRIGUING ACTIVITIES AND DEMONSTRATIONS ENHANCE YOUR LESSONS.

Demonstrations add engaging visuals to classroom presentations.

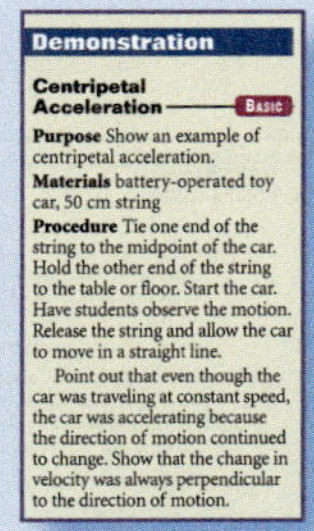
Demonstration

Centripetal Acceleration BASIC

Purpose Show an example of centripetal acceleration.

Materials battery-operated toy car, 50 cm string

Procedure Tie one end of the string to the midpoint of the car. Hold the other end of the string to the table or floor. Start the car. Have students observe the motion. Release the string and allow the car to move in a straight line.

Point out that even though the car was traveling at constant speed, the car was accelerating because the direction of motion continued to change. Show that the change in velocity was always perpendicular to the direction of motion.

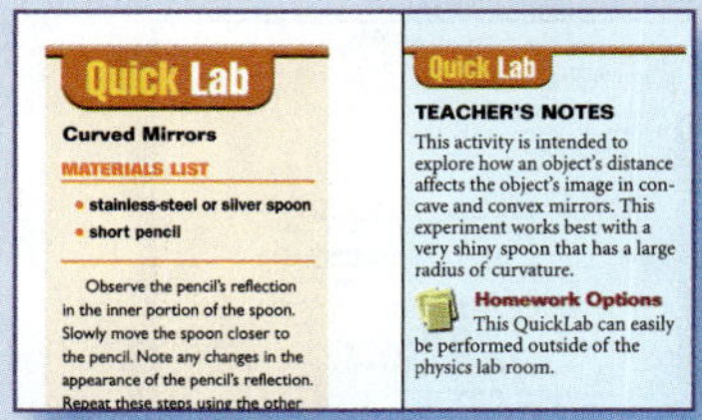
Quick Lab

Curved Mirrors

MATERIALS LIST

- stainless-steel or silver spoon
- short pencil

Observe the pencil's reflection in the inner portion of the spoon. Slowly move the spoon closer to the pencil. Note any changes in the appearance of the pencil's reflection. Repeat these steps using the other

Quick Lab

TEACHER'S NOTES

This activity is intended to explore how an object's distance affects the object's image in concave and convex mirrors. This experiment works best with a very shiny spoon that has a large radius of curvature.

Homework Options
This QuickLab can easily be performed outside of the physics lab room.

Quick Lab notes detail how to use the mini labs effectively. **Quick Labs** labeled **Homework Options** can safely be performed outside the classroom.

Extension directs students to **go.hrw.com** for activities that demonstrate physics concepts.

Classroom Practice is presented along with **Sample Problems** from the student page and supplies students with additional problem-solving practice.

Teaching Tips outline strategies for you to effectively communicate key concepts in different ways.

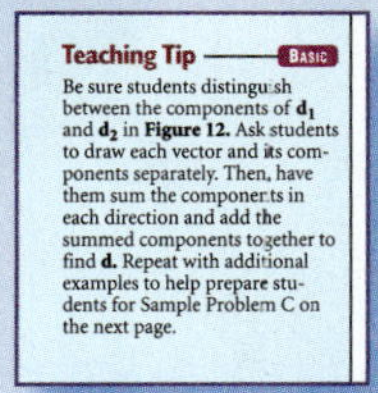
Teaching Tip BASIC

Be sure students distinguish between the components of $\mathbf{d}_1$ and $\mathbf{d}_2$ in **Figure 12.** Ask students to draw each vector and its components separately. Then, have them sum the components in each direction and add the summed components together to find **d.** Repeat with additional examples to help prepare students for Sample Problem C on the next page.

Misconception Alert provides tips on correcting and dispelling false understandings of topics.

Tapping Prior Knowledge is a springboard to building upon topics students have learned. It outlines **Knowledge to Review** and **Items to Probe.**

Problem Guide summarizes your problem-solving practice options from the *Student Edition*, ***Problem Workbook,*** and **Problem Bank** on the ***Teacher's One-Stop Planner®.***

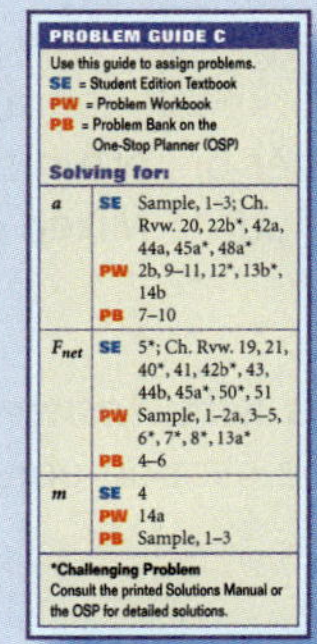
PROBLEM GUIDE C

Use this guide to assign problems.
SE = Student Edition Textbook
PW = Problem Workbook
PB = Problem Bank on the One-Stop Planner (OSP)

Solving for:

a	SE	Sample, 1–3; Ch. Rvw. 20, 22b*, 42a, 44a, 45a*, 48a*
	PW	2b, 9–11, 12*, 13b*, 14b
	PB	7–10
F_{net}	SE	5*; Ch. Rvw. 19, 21, 40*, 41, 42b*, 43, 44b, 45a*, 50*, 51
	PW	Sample, 1–2a, 3–5, 6*, 7*, 8*, 13a*
	PB	4–6
m	SE	4
	PW	14a
	PB	Sample, 1–3

*Challenging Problem
Consult the printed Solutions Manual or the OSP for detailed solutions.

Teacher's Edition

Measure conceptual understanding and problem-solving skills with *Holt Physics* assessments

THE EXAMVIEW® VERSION 6 ASSESSMENT SUITE GIVES YOU THE POWER TO CUSTOMIZE YOUR OWN ASSESSMENTS.

With the **ExamView® Version 6 Assessment Suite** on the ***Teacher's One-Stop Planner®*** you can create your own quizzes, section and chapter reviews, and chapter tests. You can customize assessments by selecting from a bank of questions, organized by chapter and linked to chapter objectives. ExamView® includes new algorithmic assessments that generate random variables for problems. **Performance-Based Assessment** questions are also included.

- You can also post tests to **Holt Online Assessment,** an assessment management tool. The system automatically grades tests so you can diagnose student proficiency and track student progress.
- ***MindPoint® Quiz Show*** provides questions from ExamView® in an engaging, interactive quiz show format.
- **Online Intervention** administers a pretest and diagnoses which concepts from the chapter need further study by students.

SECTION ASSESSMENTS HELP YOU CHECK STUDENTS' UNDERSTANDING ALONG THE WAY.

In-text **Section Reviews** help students confirm their comprehension of the section and include critical-thinking questions.

The ***Study Guide*** workbook contains additional **Section Reviews,** which provide another opportunity for test preparation. These worksheets also include a **Mixed Review.** The answers are in the ***Solutions Manual.***

The ***Section Quizzes with Answer Key*** workbook assesses students' understanding of the section objectives.

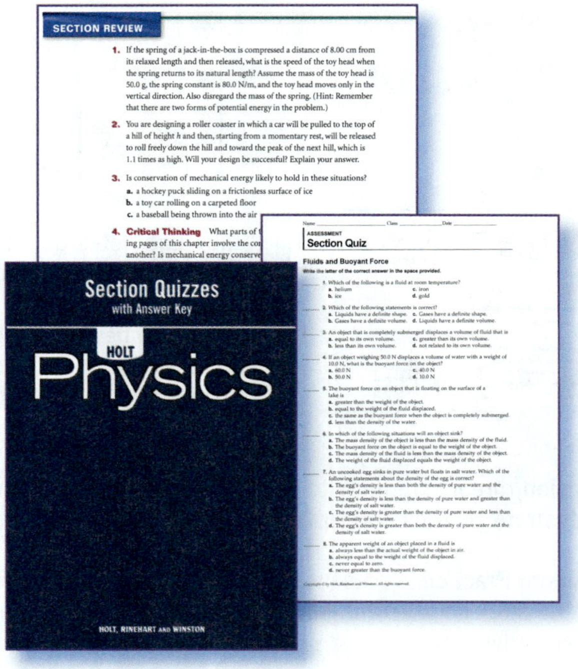

CHAPTER ASSESSMENTS TEST STUDENTS' OVERALL KNOWLEDGE OF THE CHAPTER OBJECTIVES.

In-text **Chapter Reviews** improve students' study and review habits with Review Questions, Conceptual Questions, Practice Problems, and a Mixed Review.

- **Review Questions** focus on key physics facts and relationships.
- **Conceptual Questions** encourage students to apply their understanding of physics.
- **Practice Problems** test the problem-solving techniques introduced in the **Sample Problems.**

Alternative Assessments at the end of each Chapter Review give you a different evaluation option to ensure a thorough assessment.

In-text **Standardized Test Preps** prepare students for testing with multiple-choice, short-response, and extended-response questions. There are two full-pages of test preparation in each chapter of the *Student Edition.* **Test Tip** highlights useful strategies to ensure student success.

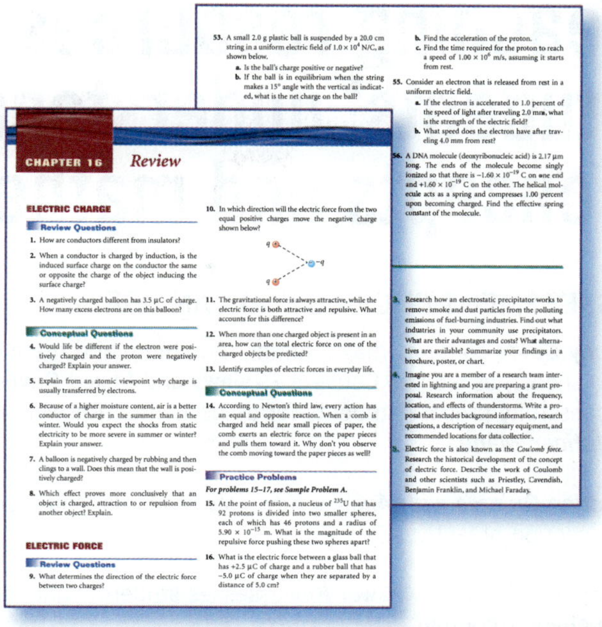

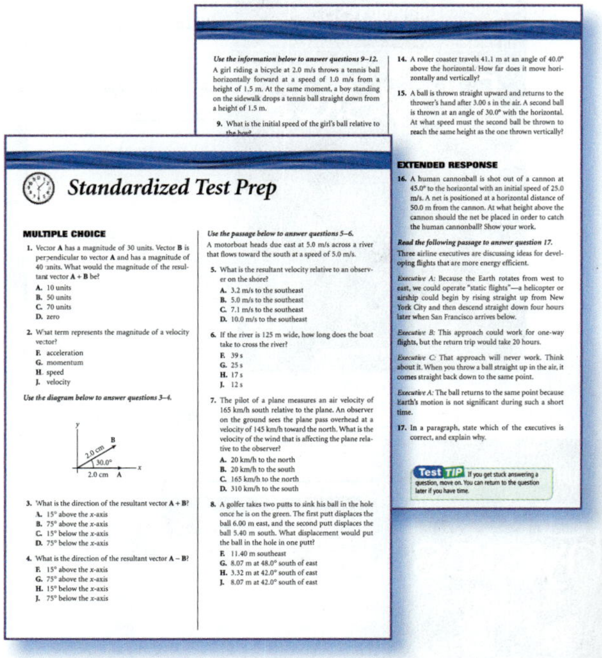

The *SAT® Bellringer Workbook* provides standardized test practice with physics content.

The ***Chapter Tests with Answer Key*** workbook evaluates students' comprehension of the chapter content.

Teaching resources that extend students' understanding of physics concepts and mathematics

ASSESSMENT CHOICES YOU CAN USE.

Study Guide contains diagramming, graphing, and math skills worksheets, as well as **Concept Reviews** and **Mixed Reviews** that successfully prepare students for assessments. Answers are in the ***Solutions Manual.***

Section Quizzes with Answer Key tests students' comprehension of the section objectives.

Chapter Tests with Answer Key assesses students' understanding of the chapter content.

PROBLEM-SOLVING WORKBOOKS HELP STUDENTS IMPROVE THEIR QUANTITATIVE SKILLS.

Problem Workbook contains additional **Sample Problems** and **Practice Problems** that correlate to the text.

Solutions Manual outlines detailed solutions to math problems in the **Section** and **Chapter Reviews, Practice Problems,** the **Additional Problems** appendix in the student text, and the ***Problem Workbook*** worksheets. It also includes answers to the ***Study Guide*** workbook.

LAB BOOKS PROVIDE HANDS-ON LEARNING WHILE STRENGTHENING YOUR STUDENTS' INQUIRY SKILLS.

Datasheets for In-Text Labs are blackline worksheets that deliver the entire text of the end-of-chapter lab, along with space for students' data and answers.

Laboratory Experiments are additional labs that reinforce concept development and encourage exploration. Students will apply inquiry skills by designing their own labs.

Laboratory Experiments, Teacher Edition includes all the student pages with answers plus teacher's notes about safety, equipment, clean-up, and disposal.

CBL™ Experiments provide additional labs that integrate technology and physics concepts. Students will practice calculator and data-probe techniques as they work through labs adapted for probe-ware.

CBL™ Experiments, Teacher Edition includes the answers and student pages plus useful teacher's notes that help you effectively lead experiments.

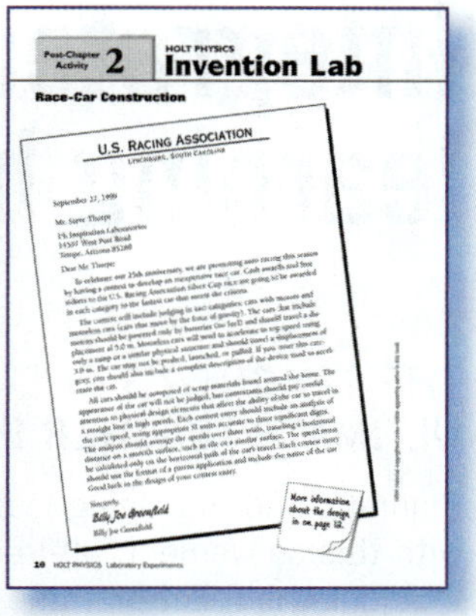

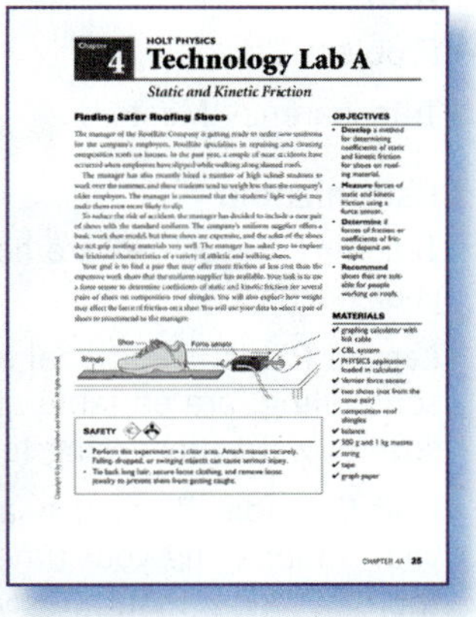

ADDITIONAL RESOURCES PROVIDE CONTINUED SUPPORT.

Teaching Transparencies are full-color visual teaching aids of the key illustrations and diagrams from the text that help you motivate class discussion.

Holt Science: Skills Workshop: Reading in the Content Area contains exercises to target reading skills specific to the comprehension of science concepts. Students learn to analyze text structures, recognize patterns, and organize information in ways that help them construct meaning.

Holt Science: Laboratory Manager's Professional Reference is a teacher reference that explains risk management and the hazards that can occur in the classroom. It gives you knowledge to ensure safety in the classroom.

HOLT Teacher's One-Stop Planner®

All these resources are available in a printable format on your convenient ***Teacher's One-Stop Planner®***.

Integrated technology expands learning beyond the classroom

THE *TEACHER'S ONE-STOP PLANNER*® IS THE ULTIMATE TEACHER PLANNING TOOL.

Planning and managing lessons has never been easier than with this convenient, all-in-one DVD package that includes the following time-saving features:

Printable

- Problem Bank
- Transparency Masters

Customizable

- Teaching Resources are editable to fit your needs.
- **Lesson Plans,** traditional and block-scheduling, are editable and available in several word-processing formats.
- **Holt Calendar Planner** is a tool that assists you in managing your time and resources by the day, week, month, or year.
- **PowerNotes® Presentations** help you to design your own lectures with **PowerNotes®** presentations, an image bank, and standardized test preparation resources.

Powerful

- **ExamView® Version 6 Assessment Suite** assists you in creating quizzes and tests from a bank of hundreds of editable questions. Algorithmic assessments generate random variables for problems. You can post test questions to **Holt Online Assessment** for automatic grading.
- **Alternative Assessments** from the **Comprehensive Conceptual Curriculum for Physics (C^{3}P)** project, a **National Science Foundation** funded effort, are provided to give you more assessment options.
- **Holt PuzzlePro®** helps you create crossword puzzles and word searches that make learning vocabulary words fun.
- **Interactive Teacher's Edition** makes your planning easy with the entire teacher text linked to related resources.
- **MindPoint® Quiz Show** provides questions from ExamView® in an engaging, interactive quiz show format.

CD-ROMs BROADEN STUDENTS' REALM OF LEARNING.

Student One Stop includes the complete *Student Edition*, MP3 files with audio summaries of the chapters, and a variety of student worksheets.

Visual Concepts CD-ROM demonstrates key chapter content with engaging graphics, animations, and movie clips and can be used by students as a study guide or by teachers for presentations.

Holt Physics Virtual Investigations includes simulations and detailed exploration activities that allow students to interact with the science on screen.

Holt Physics Lab Generator features a customizable database of physics labs that are searchable by topic, difficulty level, duration, or standard. Also included is an integrated **Lab Materials QuickList** that allows you to customize science kits to fit your classroom needs.

Online Resources

Connect to the world with current online resources

HOLT BUILDS AND MAINTAINS THE BEST ONLINE RESOURCES.

The *Interactive Online Edition* of ***Holt Physics*** brings the textbook to life. Students will build computer skills while learning physics concepts and problem-solving skills. Your *Interactive Online Edition* includes:

- All *Student Edition* pages
- **Visual Concepts**—multimedia presentations of core concepts
- Interactive activities, such as **Concept Maps**
- Helpful tools, such as **Science Glossary, Periodic Table,** and **Grapher**
- Web links to **go.hrw.com** and **SciLinks®**
- **Classroom Manager** and the ***Teacher's One-Stop Planner®*** so you can easily create lessons and manage resources
- **Holt Online Assessment,** which saves you time by automatically grading tests so you can focus on improving student proficiency
- **Online Intervention,** which administers a pretest, diagnoses which concepts from a chapter need further study, and provides remediation

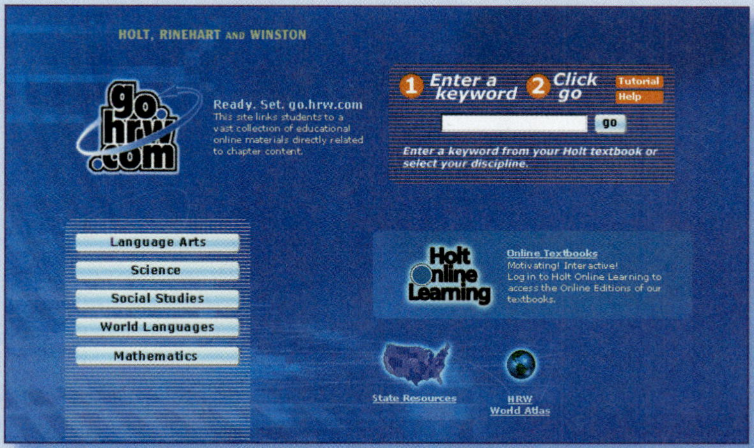

go.hrw.com broadens student learning with resources organized by chapter. The Web site includes worksheets for review and practice and exercises that enrich and extend study.

To integrate technology successfully in the classroom you'll find graphing calculator support at **go.hrw.com.**

- Free lab software
- Free keystroking programs
- Keystroking guides
- System requirements

HOLT WORKS WITH NSTA TO BRING YOU EVEN MORE ONLINE EDUCATIONAL RESOURCES.

SciLinks®, a web service developed and maintained by the **National Science Teachers Association (NSTA),** contains a large collection of prescreened links that include current information directly related to chapter topics.

- **SciLinks** boxes throughout each chapter provide topics and codes so students can investigate further.
- Each topic code leads to multiple links.
- Prescreening saves you valuable time searching for relevant and up-to-date Web sites and screening out inappropriate ones.
- Sites are reviewed by science-content experts and educators.

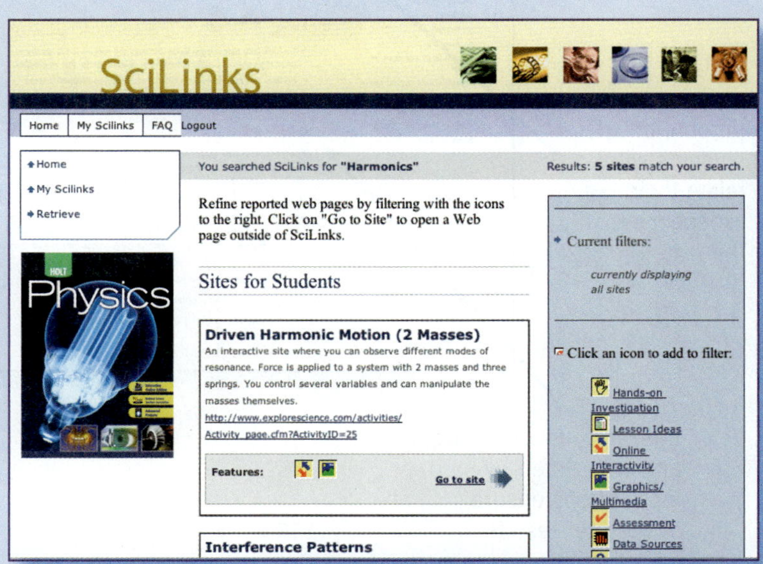

Labs and Activities

A flexible lab program that gives you options

STUDENTS WILL REINFORCE THEIR UNDERSTANDING OF PHYSICS CONCEPTS WITH IN-TEXT LABS.

Students apply scientific methods with chapter experiments, such as **Skills Practice, Inquiry,** and **CBL™ Labs.**

- The ***Teacher's One-Stop Planner®*** includes optional **Skills Practice Labs** adapted from all the **Inquiry Labs.**
- The **Appendix** includes alternative **CBL™** procedures of selected **Skills Practice Labs.**

Analysis and **Conclusions** provide students a structure to apply critical-thinking skills.

Clear, detailed photos show apparatus set-up so students can verify they are on task.

The objectives are outlined to prepare students.

A materials list and safety symbols help students set-up.

A clear description and purpose ensures students will relate their experience to the text.

Easy-to-follow headings and clear, step-by-step instructions ensure successful results.

CHAPTER 3

Inquiry Lab

Velocity of a Projectile

Design Your Own

OBJECTIVES

- **Measure** the velocity of projectiles in terms of the horizontal displacement during free fall.
- **Compare** the velocity and acceleration of projectiles accelerated down different inclined planes.

MATERIALS LIST

- aluminum sheet, edges covered with heavy tape
- C-clamp
- cardboard box
- cord
- inclined plane
- masking tape
- meterstick
- packing tape
- several large sheets white paper
- several sheets carbon paper
- small metal ball
- small metal washer
- support stand and clamp
- towel or cloth

When a ball rolls down an inclined plane, then rolls off the edge of a table, the ball becomes a projectile with some positive horizontal velocity and an initial vertical velocity of zero. However, the length of time that the projectile stays in the air depends not on the horizontal velocity but on the height of the table above the ground. The horizontal velocity determines how far the projectile travels during the time it is in the air.

In this lab, you will roll a ball down an inclined plane, off the edge of a table, and onto a piece of carbon paper on the floor. You will design your own experiment by deciding the details of the setup and the procedure, including how many trials to perform, the angles of the inclined plane, and how high up the plane you will release the ball. Your experiment should include trials with the plane inclined to different angles and multiple trials at different heights along the plane at each angle of inclination. Your procedure should include steps to measure the height from which the ball is released, the length of the ball's travel along the plane, and the horizontal displacement of the ball after it leaves the table.

SAFETY

- **Tie back long hair, secure loose clothing, and remove loose jewelry to prevent its getting caught in moving or rotating parts. Put on goggles.**
- **Perform this experiment in a clear area. Falling or dropped masses can cause serious injury.**

PROCEDURE

1. Study the materials provided, and read the Analysis and Conclusions questions. Design an experiment using the provided materials to meet the goals stated above and to allow you to answer the questions.
2. Write out your lab procedure, including a detailed description of the measurements to take during each step and the number of trials to perform. Create a data table to record your measurements for each trial. You may use **Figure 1** as a guide to one possible setup. Your setup should include a box to catch the ball at the end of each trial.
3. Ask your teacher to approve your procedure.
4. Follow all steps of your procedure.
5. Clean up your work area as directed by your teacher.

116 Chapter 3

ANALYSIS

1. **Organizing Data** Find the time interval for the ball's motion from the edge of the table to the floor using the equation for the vertical motion of a projectile. In those equations, Δy is the vertical displacement of the ball after it leaves the table. The result is the time interval for each trial.
2. **Organizing Data** Using the time interval from item 1 and the value for *Displacement* Δx, calculate the average horizontal velocity for each trial during the ball's motion from the edge of the table to the floor.
3. **Constructing Graphs** Plot a graph of average horizontal velocity versus height of release. You may use graph paper, a computer, or a graphing calculator.
4. **Constructing Graphs** Plot a graph of average horizontal velocity versus length of travel along the plane. You may use graph paper, a computer, or a graphing calculator.

CONCLUSIONS

5. **Drawing Conclusions** What is the relationship between the height of the inclined plane and the horizontal velocity of the ball? Explain.
6. **Drawing Conclusions** What is the relationship between the length of the inclined plane and the horizontal velocity of the ball? Explain.
7. **Evaluating Methods** Why might using the vertical displacement to calculate the time interval be more reliable than using a stopwatch for each trial?
8. **Applying Conclusions** In which trials would the total velocity of the ball when it hits the ground be the greatest?

EXTENSION

9. **Designing Experiments** Design an experiment to test the assumption that the time the ball is in the air is independent of the horizontal velocity of the ball. If you have time and your teacher approves your plan, carry out the experiment.

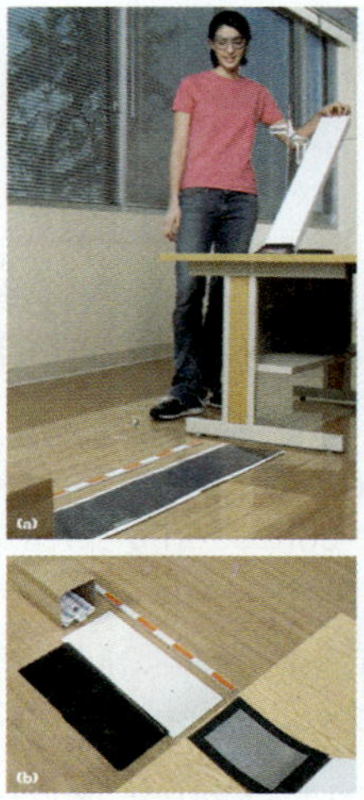

Figure 1
- Use tape to cover the sharp edges of the aluminum sheet before taping it to the end of the plane. The aluminum keeps the ball from bouncing as it rolls onto the table.
- Use a washer hanging from a string to find the zero-displacement point directly under the edge of the table.
- Use a box lined with a soft cloth to catch the ball after it lands.

Two-Dimensional Motion and Vectors 117

RELEVANT ACTIVITIES AND QUICK LABS KEEP STUDENTS INTERESTED.

Quick Lab is a short activity that requires few materials.

Demonstration helps you exemplify physics concepts to help students grasp difficult topics.

Extension refers students to **go.hrw.com** for activities that integrate other subject matter areas and for extra practice problems related to the text.

A materials list is provided for smooth implementation.

Safety Caution is addressed.

Precise directions are easy-to-follow.

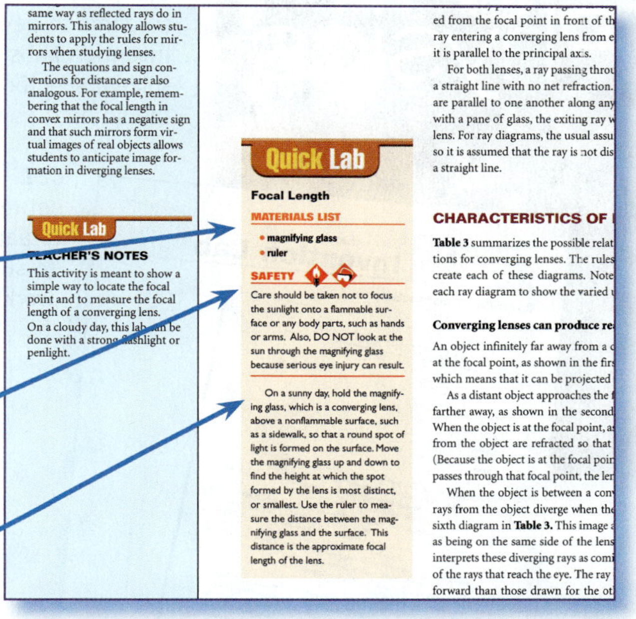
same way as reflected rays do in mirrors. This analogy allows students to apply the rules for mirrors when studying lenses.

The equations and sign conventions for distances are also analogous. For example, remembering that the focal length in convex mirrors has a negative sign and that such mirrors form virtual images of real objects allows students to anticipate image formation in diverging lenses.

Quick Lab

TEACHER'S NOTES

This activity is meant to show a simple way to locate the focal point and to measure the focal length of a converging lens.

On a cloudy day, this lab can be done with a strong flashlight or penlight.

Quick Lab

Focal Length

MATERIALS LIST

- magnifying glass
- ruler

SAFETY

Care should be taken not to focus the sunlight onto a flammable surface or any body parts, such as hands or arms. Also, DO NOT look at the sun through the magnifying glass because serious eye injury can result.

On a sunny day, hold the magnifying glass, which is a converging lens, above a nonflammable surface, such as a sidewalk, so that a round spot of light is formed on the surface. Move the magnifying glass up and down to find the height at which the spot formed by the lens is most distinct, or smallest. Use the ruler to measure the distance between the magnifying glass and the surface. This distance is the approximate focal length of the lens.

ed from the focal point in front of th ray entering a converging lens from e it is parallel to the principal axis.

For both lenses, a ray passing throu a straight line with no net refraction. are parallel to one another along any with a pane of glass, the exiting ray w lens. For ray diagrams, the usual assu so it is assumed that the ray is not dis a straight line.

CHARACTERISTICS OF

Table 3 summarizes the possible relat tions for converging lenses. The rules create each of these diagrams. Note each ray diagram to show the varied

Converging lenses can produce re

An object infinitely far away from a at the focal point, as shown in the firs which means that it can be projected

As a distant object approaches the farther away, as shown in the second When the object is at the focal point, a from the object are refracted so that (Because the object is at the focal poin passes through that focal point, the le

When the object is between a con rays from the object diverge when th sixth diagram in Table 3. This image as being on the same side of the lens interprets these diverging rays as com of the rays that reach the eye. The ray forward than those drawn for the ot

The teacher's wrap identifies **Quick Labs** that are appropriate homework assignments.

YOU RECEIVE STRONG SUPPORT FOR CHAPTER LABS.

The teacher's wrap provides the essential information you need to lead the experiment. You'll be prepared with time requirements, classroom organization, demonstration techniques, checkpoints, and additional notes.

Datasheets for In-Text Labs are organized worksheets with the complete lab text that eliminate the need for textbooks in the lab.

Laboratory Manager's Professional Reference will help you ensure safety in the classroom.

Ordering materials is made simple with the **Lab Materials QuickList Software** on the ***Holt Physics Lab Generator.***

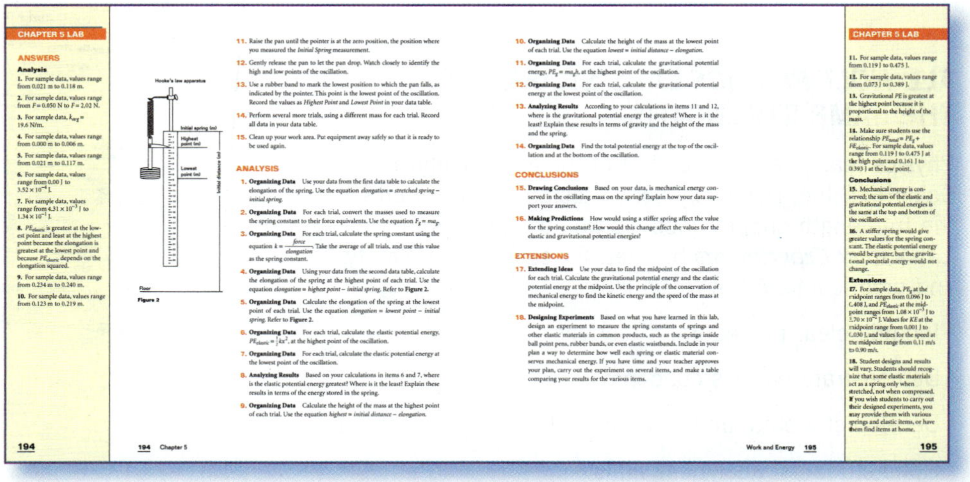

Labs and Activities

STUDENTS WILL EXTEND THEIR UNDERSTANDING OF PHYSICS CONCEPTS WITH THESE ADDITIONAL LAB BOOKS.

Laboratory Experiments include **Discovery** and **Invention Labs.** The **Discovery Labs** require no prior knowledge of a topic, encouraging students to develop their own conceptual understanding of the topic. **Invention Labs** challenge students to create a device or test materials to solve a specific problem. The ***Teacher Edition*** of this workbook includes the student pages plus the answers and useful lab notes.

CBL™ Experiments encourage students to make the connection between the physics principles studied and real-world situations. Students use graphing calculators and data probes to successfully complete these labs. The ***Teacher Edition*** of this workbook includes the student datasheet plus the answers and helpful lab notes.

ALL *HOLT PHYSICS* LABS SHARE THE SAME STRENGTHS.

Physics teachers and the staff of Sargent-Welch/CENCO Scientific, Inc. rigorously bench-tested all procedures for the full-length chapter labs, as well as the labs in the ***Laboratory Experiments*** booklet, for ease of use, safety, and practicality. All of these labs:

- Include clear procedural directions
- Demonstrate physics concepts
- Develop students' understanding of scientific methods
- Hone students' analytical skills
- Have been reviewed by teachers for reliability, safety, and efficiency

Materials ordering made easy with the Holt Physics Lab Generator

- Search for labs by topic, difficulty level, lab duration, National Science Education Standard, or state standard.
- Edit labs to fit classroom needs.
- Develop new labs from our easy-to-use formatted template.
- Customize your own science kits and save time ordering materials using the **Lab Materials QuickList software.**

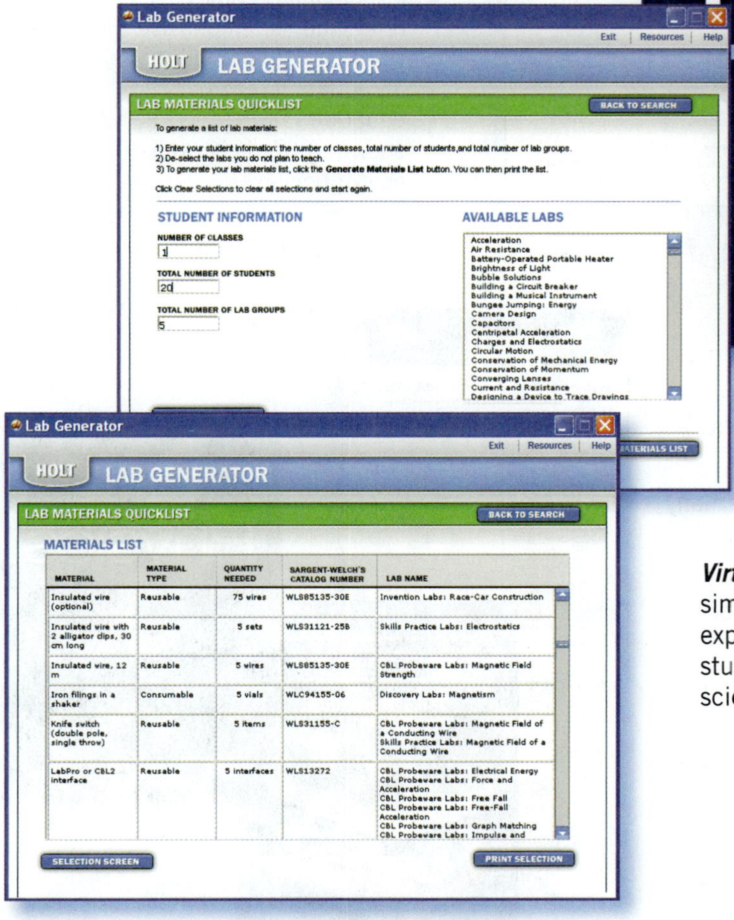

Virtual Investigations includes simulations and detailed exploration activities that allow students to interact with the science in a virtual environment.

Reading Skills

Support for critical reading skills

TEXT FEATURES FOSTER COMPREHENSION OF PHYSICS CONCEPTS.

Section Objectives focus students' attention on the material they will read.

Headline Organization includes topic and summary heads that break the section's contents into manageable chunks. Summary heads clearly state important concepts and can be used to facilitate review before a test.

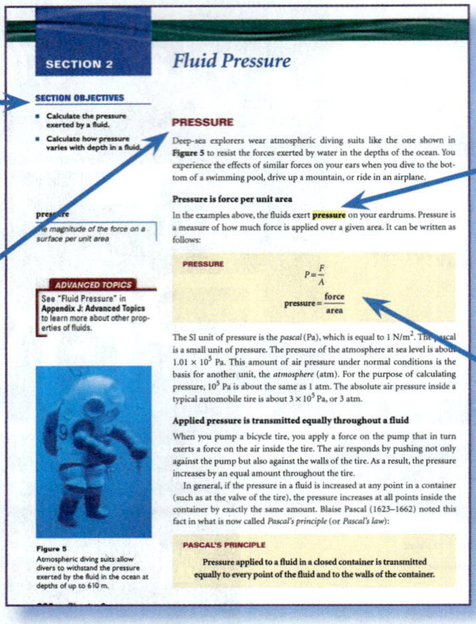

Vocabulary Words are highlighted in the text and defined in the margin when they are first used. A glossary at the back of the textbook includes all **Key Terms.**

Key Equations, Principles, and **Laws** are highlighted for easy reference. To emphasize the concepts behind an equation, each key equation is titled and is shown in words as well as symbols.

ADDITIONAL READING RESOURCES PROVIDE ANOTHER OPTION FOR SUPPORT.

Your *Teacher's Edition* provides strategies for improving students' reading skills.

- **Language of Physics** explains the origins and proper usage of many physics terms to help students improve their vocabulary skills.
- **Teaching Tip** helps you explain physics concepts in an understandable manner to reinforce students' understanding of chapter content.
- **Visual Strategy** provides questions related to the images in the text so students can conceptualize physics concepts.

Section Reviews in the ***Study Guide*** reinforce fundamental knowledge from a section of the text. They also include a **Mixed Review.** Answers are in the ***Solutions Manual.***

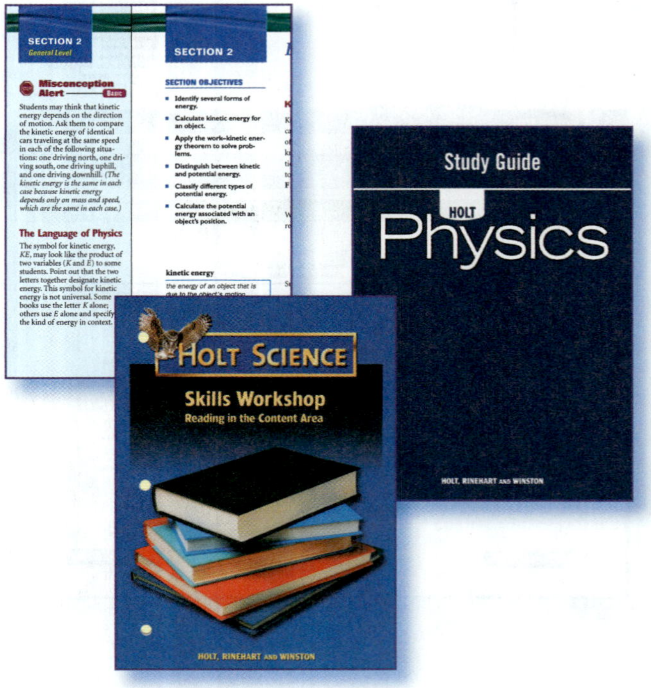

Holt Science Skills Workshop: Reading in the Content Area targets the reading skills specific to the comprehension of science texts. The ***Teacher's Edition*** comes with transparencies to lead the class.

Reinforce essential math skills

MATHEMATICS IS INTEGRATED THROUGHOUT THE TEXT.

Sample Problem provides a step-by-step approach to problem solving. Students can see the logical steps needed when approaching a problem. A **Tip** guides students in the right direction.

Section Review provides additional problems to test student understanding of material presented in the lesson.

Chapter Review presents even more opportunities for students to practice their math skills with **Practice Problems** and **Mixed Review.**

Practice follows each **Sample Problem,** allowing students to immediately apply their new problem-solving skills.

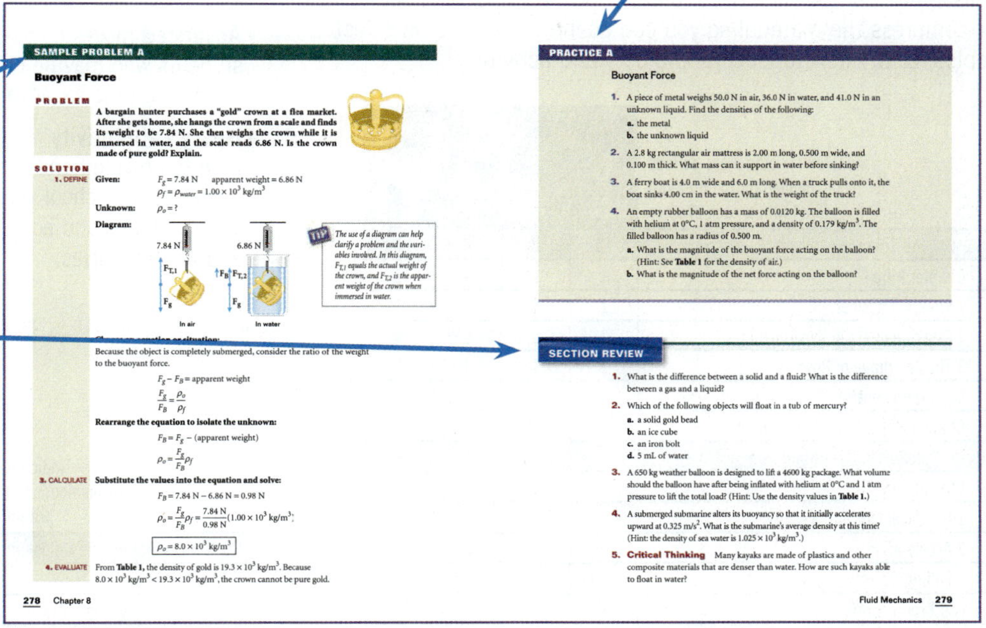

SAMPLE PROBLEM A

Buoyant Force

PROBLEM

A bargain hunter purchases a "gold" crown at a flea market. After she gets home, she hangs the crown from a scale and finds its weight to be 7.84 N. She then weighs the crown while it is immersed in water, and the scale reads 6.86 N. Is the crown made of pure gold? Explain.

SOLUTION

1. DEFINE **Given:** $F_g = 7.84$ N apparent weight = 6.86 N
$\rho_f = \rho_{water} = 1.00 \times 10^3$ kg/m^3

Unknown: $\rho_o = ?$

Diagram:

TIP *The use of a diagram can help clarify a problem and the variables involved. In this diagram, $F_{T,1}$ equals the actual weight of the crown, and $F_{T,2}$ is the apparent weight of the crown when immersed in water.*

Because the object is completely submerged, consider the ratio of the weight to the buoyant force.

$$F_g - F_B = \text{apparent weight}$$

$$\frac{F_g}{F_B} = \frac{\rho_o}{\rho_f}$$

Rearrange the equation to isolate the unknown:

$$F_B = F_g - (\text{apparent weight})$$

$$\rho_o = \frac{F_g}{F_B}\rho_f$$

3. CALCULATE **Substitute the values into the equation and solve:**

$$F_B = 7.84 \text{ N} - 6.86 \text{ N} = 0.98 \text{ N}$$

$$\rho_o = \frac{F_g}{F_B}\rho_f = \frac{7.84 \text{ N}}{0.98 \text{ N}}(1.00 \times 10^3 \text{ kg/m}^3)$$

$$\boxed{\rho_o = 8.0 \times 10^3 \text{ kg/m}^3}$$

4. EVALUATE From **Table 1,** the density of gold is 19.3×10^3 kg/m^3. Because 8.0×10^3 kg/m^3 < 19.3×10^3 kg/m^3, the crown cannot be pure gold.

278 Chapter 8

PRACTICE A

Buoyant Force

1. A piece of metal weighs 50.0 N in air, 36.0 N in water, and 41.0 N in an unknown liquid. Find the densities of the following:
   a. the metal
   b. the unknown liquid
2. A 2.8 kg rectangular air mattress is 2.00 m long, 0.500 m wide, and 0.100 m thick. What mass can it support in water before sinking?
3. A ferry boat is 4.0 m wide and 6.0 m long. When a truck pulls onto it, the boat sinks 4.00 cm in the water. What is the weight of the truck?
4. An empty rubber balloon has a mass of 0.0120 kg. The balloon is filled with helium at 0°C, 1 atm pressure, and a density of 0.179 kg/m^3. The filled balloon has a radius of 0.500 m.
   a. What is the magnitude of the buoyant force acting on the balloon? (Hint: See **Table 1** for the density of air.)
   b. What is the magnitude of the net force acting on the balloon?

SECTION REVIEW

1. What is the difference between a solid and a fluid? What is the difference between a gas and a liquid?
2. Which of the following objects will float in a tub of mercury?
   a. a solid gold bead
   b. an ice cube
   c. an iron bolt
   d. 5 mL of water
3. A 650 kg weather balloon is designed to lift a 4600 kg package. What volume should the balloon have after being inflated with helium at 0°C and 1 atm pressure to lift the total load? (Hint: Use the density values in **Table 1.**)
4. A submerged submarine alters its buoyancy so that it initially accelerates upward at 0.325 m/s^2. What is the submarine's average density at this time? (Hint: the density of sea water is 1.025×10^3 kg/m^3.)
5. **Critical Thinking** Many kayaks are made of plastics and other composite materials that are denser than water. How are such kayaks able to float in water?

Fluid Mechanics 279

Mathematical Review and additional practice problems are found in the **Appendix.**

ADDITIONAL MATH SUPPORT ENSURES STUDENTS' SUCCESS.

Your *Teacher's Edition* offers additional math problems and problem-solving hints.

- **Alternative Problem-Solving Approach** includes strategies to advance students' math skills.
- **Classroom Practice** includes additional math problems that are based on the **Sample Problem.**

Problem Workbook and ***Problem Bank*** on the ***Teacher's One-Stop-Planner®*** are musts for students who need extra help, reteaching or more practice. The workbook includes individual sample problems with accompanying practice problems so students can practice and demonstrate their mastery of problem-solving skills.

SIGNIFICANT FIGURES

Holt Physics provides a clear but rigorous approach to significant figures in Chapter 1. This approach is maintained consistently throughout the rest of the program.

Several valid approaches to significant figures are used by different physics teachers. The discussion of significant figures in ***Holt Physics*** makes it clear that students should consult you, the teacher, for confirmation of which rules you believe are appropriate for your class.

Pacing Guide

Today's physics classroom often requires a more flexible curriculum. ***Holt Physics*** can help you meet a variety of needs and challenges you and your students face in the classroom. The **Pacing Guide** below shows a number of ways to adapt the program to your teaching schedule.

This **Guide** can be further adapted, allowing you to mix and match or compress the material so you can spend more time on select topics, or to allow for special projects and activities.

- **Basic** gives more time for the foundations of physics, especially mathematical problem-solving, with less emphasis on some advanced topics from later in the course.
- **General** provides the recommended course of study as indicated in the *Teacher's Edition,* found in the individual chapter guides preceding each chapter.
- **Advanced** moves quickly through foundations of physics for students who may be comfortable with the basics, to provide additional time for advanced topics.
- **Heavy Lab/Activity** indicates ways to streamline "lecture" time to provide hands-on experience for more than a third of the blocks in the school year. (Note: even this approach does not cover all of the labs and activities that are available with ***Holt Physics*** and its ancillaries.

Numbers indicate class periods recommended for the material within each chapter	Basic	General	Advanced	Heavy Lab/Activity
Ch. 1 The Science of Physics	**10**	**8**	**6**	**8**
Chapter Intro	1	1	–	1
1 What is Physics?	1	1	1	1
2 Measurements in Experiments	3	2	2	2
3 The Language of Physics	2	1	1	1
Lab Experiment(s)	1	1	1	2
Chapter Review and Assessment	2	2	1	1
Ch. 2 Motion in One Dimension	**11**	**8**	**7**	**8**
Chapter Intro	1	1	–	1
1 Displacement and Velocity	2	1	1	1
2 Acceleration	3	2	3	2
3 Falling Objects	2	1	1	1
Lab Experiment(s)	1	1	1	2
Chapter Review and Assessment	2	2	1	1
Ch. 3 Two-Dimensional Motion and Vectors	**10**	**9**	**9**	**9**
Chapter Intro	1	1	–	1
1 Introduction to Vectors	2	1	1	1
2 Vector Operations	2	1	2	1
3 Projectile Motion	2	2	2	2
4 Relative Motion	–	1	2	1
Lab Experiment(s)	1	1	1	2
Chapter Review and Assessment	2	2	1	1
Ch. 4 Forces and the Laws of Motion	**11**	**8**	**7**	**8**
Chapter Intro	1	1	–	1
1 Changes in Motion	2	1	1	1
2 Newton's First Law	2	1	1	1
3 Newton's Second and Third Laws	2	1	1	1
4 Everyday Forces	1	1	1	1
Lab Experiment(s)	1	1	2	2
Chapter Review and Assessment	2	2	1	1
Ch. 5 Work and Energy	**13**	**9**	**8**	**9**
Chapter Intro	1	1	–	1
1 Work	2	1	1	1
2 Energy	3	2	2	2
3 Conservation of Energy	2	1	2	1
4 Power	2	1	1	1
Lab Experiment(s)	1	1	1	2
Chapter Review and Assessment	2	2	1	1

Numbers indicate class periods recommended for the material within each chapter	Basic	General	Advanced	Heavy Lab/Activity
Ch. 6 Momentum and Collisions	**9**	**8**	**8**	**7**
Chapter Intro	1	1	–	1
1 Momentum and Impulse	3	2	2	2
2 Conservation of Momentum	2	1	2	1
3 Elastic and Inelastic Collisions	–	1	2	1
Lab Experiment(s)	1	1	1	1
Chapter Review and Assessment	2	2	1	1
Ch. 7 Circular Motion and Gravitation	**8**	**8**	**8**	**9**
Chapter Intro	1	1	–	1
1 Circular Motion	2	1	2	1
2 Newton's Law of Universal Gravitation	2	1	1	1
3 Motion in Space	1	1	1	2
4 Torque and Simple Machines	–	1	2	1
Lab Experiment(s)	–	1	1	2
Chapter Review and Assessment	2	2	1	1
Ch. 8 Fluid Mechanics	**0**	**6**	**7**	**5**
Chapter Intro	–	1	–	1
1 Fluids and Buoyant Force	–	1	2	1
2 Fluid Pressure	–	1	2	1
3 Fluids in Motion	–	1	2	1
Chapter Review and Assessment	–	2	1	1
Ch. 9 Heat	**7**	**8**	**7**	**9**
Chapter Intro	1	1	–	1
1 Temperature and Thermal Equilibrium	3	2	2	2
2 Defining Heat	1	1	1	2
3 Changes in Temperature and Phase	–	1	2	1
Lab Experiment(s)	–	1	1	2
Chapter Review and Assessment	2	2	1	1

Numbers indicate class periods recommended for the material within each chapter

	Basic	General	Advanced	Heavy Lab/Activity
Ch. 10 Thermodynamics	**5**	**6**	**7**	**5**
Chapter Intro	1	1	–	1
1 Relationships Between Heat and Work	2	1	2	1
2 The First Law of Thermodynamics	–	1	2	1
3 The Second Law of Thermodynamics	–	1	2	1
Chapter Review and Assessment	2	2	1	1
Ch. 11 Vibrations and Waves	**12**	**9**	**10**	**9**
Chapter Intro	1	1	–	1
1 Simple Harmonic Motion	2	1	1	1
2 Measuring Simple Harmonic Motion	2	1	2	1
3 Properties of Waves	3	2	3	2
4 Wave Interactions	1	1	2	1
Lab Experiment(s)	1	1	1	2
Chapter Review and Assessment	2	2	1	1
Ch. 12 Sound	**7**	**7**	**7**	**8**
Chapter Intro	1	1	–	1
1 Sound Waves	2	1	1	1
2 Sound Intensity and Resonance	1	1	2	1
3 Harmonics	–	1	2	2
Lab Experiment(s)	1	1	1	2
Chapter Review and Assessment	2	2	1	1
Ch. 13 Light and Reflection	**12**	**9**	**9**	**10**
Chapter Intro	1	1	–	1
1 Characteristics of Light	2	1	1	1
2 Flat Mirrors	2	1	1	1
3 Curved Mirrors	3	2	2	3
4 Color and Polarization	1	1	2	1
Lab Experiment(s)	1	1	2	2
Chapter Review and Assessment	2	2	1	1
Ch. 14 Refraction	**9**	**8**	**8**	**9**
Chapter Intro	1	1	–	1
1 Refraction	2	1	1	1
2 Thin Lenses	3	2	2	3
3 Optical Phenomena	–	1	2	1
Lab Experiment(s)	1	1	2	2
Chapter Review and Assessment	2	2	1	1
Ch. 15 Interference and Diffraction	**7**	**8**	**8**	**7**
Chapter Intro	1	1	–	1
1 Interference	1	1	2	1
2 Diffraction	2	2	2	2
3 Lasers	–	1	2	1
Lab Experiment(s)	1	1	1	1
Chapter Review and Assessment	2	2	1	1
Ch. 16 Electric Forces and Fields	**8**	**8**	**7**	**8**
Chapter Intro	1	1	–	1
1 Electric Charge	2	1	1	1
2 Electric Force	2	2	2	2
3 The Electric Field	–	1	2	1
Lab Experiment(s)	1	1	1	2
Chapter Review and Assessment	2	2	1	1

Numbers indicate class periods recommended for the material within each chapter

	Basic	General	Advanced	Heavy Lab/Activity
Ch. 17 Electrical Energy and Current	**12**	**9**	**9**	**9**
Chapter Intro	1	1	–	1
1 Electric Potential	2	1	2	1
2 Capacitance	1	1	1	1
3 Current and Resistance	3	2	3	2
4 Electric Power	2	1	1	1
Lab Experiment(s)	1	1	1	2
Chapter Review and Assessment	2	2	1	1
Ch. 18 Circuits and Circuit Elements	**9**	**8**	**8**	**9**
Chapter Intro	1	1	–	1
1 Schematic Diagrams and Circuits	2	1	1	2
2 Resistors in Series or in Parallel	3	2	2	2
3 Complex Resistor Combinations	–	1	2	1
Lab Experiment(s)	1	1	2	2
Chapter Review and Assessment	2	2	1	1
Ch. 19 Magnetism	**9**	**8**	**8**	**8**
Chapter Intro	1	1	–	1
1 Magnets and Magnetic Fields	2	1	1	1
2 Magnetism from Electricity	1	1	2	1
3 Magnetic Force	2	2	2	2
Lab Experiment(s)	1	1	2	2
Chapter Review and Assessment	2	2	1	1
Ch. 20 Electromagnetic Induction	**7**	**9**	**11**	**9**
Chapter Intro	1	1	–	1
1 Electricity from Magnetism	2	2	2	2
2 Generators, Motors, and Mutual Inductance	1	1	2	1
3 AC Circuits and Transformers	–	1	2	1
4 Electromagnetic Waves	–	1	2	1
Lab Experiment(s)	1	1	2	2
Chapter Review and Assessment	2	2	1	1
Ch. 21 Atomic Physics	**0**	**7**	**8**	**6**
Chapter Intro	–	1	–	1
1 Quantization of Energy	–	1	2	1
2 Models of the Atom	–	1	2	1
3 Quantum Mechanics	–	1	2	1
Lab Experiment(s)	–	1	1	1
Chapter Review and Assessment	–	2	1	1
Ch. 22 Subatomic Physics	**0**	**8**	**9**	**7**
Chapter Intro	–	1	–	1
1 The Nucleus	–	1	1	1
2 Nuclear Decay	–	1	2	1
3 Nuclear Reactions	–	1	2	1
4 Particle Physics	–	1	2	1
Lab Experiment(s)	–	1	1	1
Chapter Review and Assessment	–	2	1	1
TOTAL	**176**	**176**	**176**	**176**

National Science Education Standards Correlation

The following lists show the chapter correlation of ***Holt Physics*** with the ***National Science Education Standards.***

The chapter correlations for the Physical Science Content Standards begin on page T25.

Unifying Concepts and Processes	
Standard	**Code**
Systems, order, and organization	UCP 1
Evidence, models, and explanation	UCP 2
Change, consistency, and measurements	UCP 3
Evolution and equilibrium	UCP 4
Form and function	UCP 5
Science as Inquiry	
Standard	**Code**
Abilities to do scientific inquiry	SAI 1
Understanding about scientific inquiry	SAI 2
Science and Technology	
Standard	**Code**
Abilities of technological design	ST 1
Understanding about science and technology	ST 2
History and Nature of Science	
Standard	**Code**
Science as a human endeavor	HNS 1
Nature of science	HNS 2
History of science	HNS 3
Science in Personal and Social Perspectives	
Standard	**Code**
Personal health	SPSP 1
Populations, resources, and environments	SPSP 2
Natural hazards	SPSP 3
Risks and benefits	SPSP 4
Science and technology in society	SPSP 5

Physical Science Content Standards

Structure of Atoms

Standard	Code	Chapter Correlation
Matter is made of minute particles called atoms, and atoms are composed of even smaller components. These components have measurable properties, such as mass and electrical charge. Each atom has a positively charged nucleus surrounded by negatively charged electrons. The electric force between the nucleus and electrons holds the atom together.	**PS 1a**	**Chapter 16** **Chapter 21**
The atom's nucleus is composed of protons and neutrons, which are much more massive than electrons. When an element has atoms that differ in the number of neutrons, these atoms are called different isotopes of the element.	**PS 1b**	**Chapter 22**
The nuclear forces that hold the nucleus of an atom together, at nuclear distances, are usually stronger than the electric forces that would make it fly apart. Nuclear reactions convert a fraction of the mass of interacting particles into energy, and they can release much greater amounts of energy than atomic interactions. Fission is the splitting of a large nucleus into smaller pieces. Fusion is the joining of two nuclei at extremely high temperature and pressure, and is the process responsible for the energy of the sun and other stars.	**PS 1c**	**Chapter 22**
Radioactive isotopes are unstable and undergo spontaneous nuclear reactions, emitting particles and/or wavelike radiation. The decay of any one nucleus cannot be predicted, but a large group of identical nuclei decay at a predictable rate. This predictability can be used to estimate the age of materials that contain radioactive isotopes.	**PS 1d**	**Chapter 22**

Motion and Forces

Standard	Code	Chapter Correlation
Objects change their motion only when a net force is applied. Laws of motion are used to calculate precisely the effects of forces on the motion of objects. The magnitude of the change in motion can be calculated using the relationship $F=ma$, which is independent of the nature of the force. Whenever one object exerts force on another, a force equal in magnitude and opposite in direction is exerted on the first object.	**PS 4a**	**Chapter 4** **Chapter 7** **Chapter 11** **Chapter 16** **Chapter 19**

Motion and Forces *(continued)*		
Standard	**Code**	**Chapter Correlation**
Gravitation is a universal force that each mass exerts on any other mass. The strength of the gravitational attractive force between two masses is proportional to the masses and inversely proportional to the square of the distance between them.	**PS 4b**	**Chapter 4** **Chapter 7**
The electric force is a universal force that exists between any two charged objects. Opposite charges attract while like charges repel. The strength of the force is proportional to the charges, and, as with gravitation, inversely proportional to the square of the distance between them.	**PS 4c**	**Chapter 16**
Between any two charged particles, electric force is vastly greater than the gravitational force. Most observable forces such as those exerted by a coiled spring or friction may be traced to electric forces acting between atoms and molecules.	**PS 4d**	**Chapter 4** **Chapter 16**
Electricity and magnetism are two aspects of a single electromagnetic force. Moving electric charges produce magnetic forces, and moving magnets produce electric forces. These effects help students to understand electric motors and generators.	**PS 4e**	**Chapter 19** **Chapter 20**

Conservation of Energy and the Increase in Disorder			
Standard	**Code**	**Chapter Correlation**	
The total energy of the universe is constant. Energy can be transferred by collisions in chemical and nuclear reactions, by light waves and other radiations, and in many other ways. However, it can never be destroyed. As these transfers occur, the matter involved becomes steadily less ordered.	**PS 5a**	**Chapter 5** **Chapter 6** **Chapter 9** **Chapter 10**	**Chapter 20** **Chapter 22**
All energy can be considered to be either kinetic energy, which is the energy of motion; potential energy, which depends on relative position; or energy contained by a field, such as electromagnetic waves.	**PS 5b**	**Chapter 5** **Chapter 7** **Chapter 9**	**Chapter 11** **Chapter 17** **Chapter 20**
Heat consists of random motion and the vibrations of atoms, molecules, and ions. The higher the temperature, the greater the atomic or molecular motion.	**PS 5c**	**Chapter 9** **Chapter 10**	
Everything tends to become less organized and less orderly over time. Thus, in all energy transfers, the overall effect is that the energy is spread out uniformly. Examples are the transfer of energy from hotter to cooler objects by conduction, radiation, or convection and the warming of our surroundings when we burn fuels.	**PS 5d**	**Chapter 9** **Chapter 10**	

Interaction of Energy and Matter		
Standard	**Code**	**Chapter Correlation**
Waves, including sound and seismic waves, waves on water, and light waves, have energy and can transfer energy when they interact with matter.	**PS 6a**	**Chapter 11** **Chapter 12** **Chapter 13** **Chapter 20**
Electromagnetic waves result when a charged object is accelerated or decelerated. Electromagnetic waves include radio waves (the longest wavelength), microwaves, infrared radiation (radiant heat), visible light, ultraviolet radiation, x-rays, and gamma rays. The energy of electromagnetic waves is carried in packets whose magnitude is inversely proportional to the wavelength.	**PS 6b**	**Chapter 13** **Chapter 20** **Chapter 21**
Each kind of atom or molecule can gain or lose energy only in particular discrete amounts and thus can absorb and emit light only at wavelengths corresponding to these amounts. These wavelengths can be used to identify the substance.	**PS 6c**	**Chapter 21**
In some materials, such as metals, electrons flow easily, whereas in insulating materials such as glass they can hardly flow at all. Semiconducting materials have intermediate behavior. At low temperatures some materials become superconductors and offer no resistance to the flow of electrons.	**PS 6d**	**Chapter 17**

Safety in your laboratory

Risk Assesssment

MAKING YOUR LABORATORY A SAFE PLACE TO WORK AND LEARN

Concern for safety must begin before any activity in the classroom and before students enter the lab. A careful review of the facilities should be a basic part of preparation for each school term. You should investigate the physical environment, identify any safety risks, and inspect your work areas for compliance with safety regulations.

The review of the lab should be thorough, and all safety issues must be addressed immediately. Keep a file of your review, and add to the list each year. This will allow you to continue to raise the standard of safety in your lab and classroom.

Many classroom experiments, demonstrations, and other activities are classics that have been used for years. This familiarity may lead to a comfort that can obscure inherent safety concerns. Review all experiments, demonstrations, and activities for safety concerns before presenting them to the class. Identify and eliminate potential safety hazards.

1. **Identify the Risks**

   Before introducing any activity, demonstration, or experiment to the class, analyze it and consider what could possibly go wrong. Carefully review the list of materials to make sure they are safe. Inspect the equipment in your lab or classroom to make sure it is in good working order. Read the procedures to make sure they are safe. Record any hazards or concerns you identify.

2. **Evaluate the Risks**

   Minimize the risks you identified in the last step without sacrificing learning. Remember that no activity you perform in the lab or classroom is worth risking injury. Thus, extremely hazardous activities, or those that violate your school's policies, must be eliminated. For activities that present smaller risks, analyze each risk carefully to determine its likelihood. If the pedagogical value of the activity does not outweigh the risks, the activity must be eliminated.

3. **Select Controls to Address Risks**

   Even low-risk activities require controls to eliminate or minimize the risks. Make sure that in devising controls you do not substitute an equally or more hazardous alternative. Some control methods include the following:

   - Explicit verbal and written warnings may be added or posted.
   - Equipment may be rebuilt or relocated, parts may be replaced, or equipment may be replaced entirely by safer alternatives.
   - Risky procedures may be eliminated.
   - Activities may be changed from student activities to teacher demonstrations.

4. **Implement and Review Selected Controls**

   Controls do not help if they are forgotten or not enforced. The implementation and review of controls should be as systematic and thorough as the initial analysis of safety concerns in the lab and laboratory activities.

SOME SAFETY RISKS AND PREVENTATIVE CONTROLS

The following list describes several possible safety hazards and controls that can be implemented to resolve them. This list is not complete, but it can be used as a starting point to identify hazards in your laboratory.

Identified risk	Preventative control
Facilities and Equipment	
Lab tables are in disrepair, room is poorly lighted and ventilated, faucets and electrical outlets do not work or are difficult to use because of their location.	Work surfaces should be level and stable. There should be adequate lighting and ventilation. Water supplies, drains, and electrical outlets should be in good working order. Any equipment in a dangerous location should not be used; it should be relocated or rendered inoperable.
Wiring, plumbing, and air circulation systems do not work or do not meet current specifications.	Specifications should be kept on file. Conduct a periodic review of all equipment, and document compliance. Damaged fixtures must be labeled as such and must be repaired as soon as possible.
Eyewash fountains and safety showers are present, but no one knows anything about their specifications.	Ensure that eyewash fountains and safety showers meet the requirements of the ANSI standard (Z358.1).
Eyewash fountains are checked and cleaned once at the beginning of each school year. No records are kept of routine checks and maintenance on the safety showers and eyewash fountains.	Flush eyewash fountains for 5 minutes every month to remove any bacteria or other organisms from pipes. Test safety showers (measure flow in gallons per min) and eyewash fountains every 6 months and keep records of the test results.
Labs are conducted in multipurpose rooms, and equipment from other courses remains accessible.	Only the items necessary for a given activity should be available to students. All equipment should be locked away when not in use.
Students are permitted to enter or work in the lab without teacher supervision.	Lock all laboratory rooms whenever a teacher is not present. Supervising teachers must be trained in lab safety and emergency procedures.
Safety equipment and emergency procedures	
Fire and other emergency drills are infrequent, and no records or measurements are made of the results of the drills.	Always carry out critical reviews of fire or other emergency drills. Be sure that plans include alternate routes. Don't wait until an emergency to find the flaws in your plans.
Emergency evacuation plans do not include instructions for securing the lab in the event of an evacuation during a lab activity.	Plan actions in case of emergency: establish what devices should be turned off, which escape route to use, and where to meet outside the building.
Fire extinguishers are in out-of-the-way locations, not on the escape route.	Place fire extinguishers near escape routes so that they will be of use during an emergency.
Fire extinguishers are not maintained. Teachers are not trained to use them.	Document regular maintenance of fire extinguishers. Train supervisory personnel in the proper use of extinguishers. Instruct students not to use an extinguisher but to call for a teacher.
Teachers in labs and neighboring classrooms are not trained in CPR or first aid.	Teachers should receive training from the local chapter of the American Red Cross. Certifications should be kept current with frequent refresher courses.

Identified risk	Preventative control
Safety equipment and emergency procedures *(continued)*	
Teachers are not aware of their legal responsibilities in case of an injury or accident.	Review your faculty handbook for your responsibilities regarding safety in the classroom and laboratory. Contact the legal counsel for your school district to find out the extent of their support and any rules, regulations, or procedures you must follow.
Emergency procedures are not posted. Emergency numbers are kept only at the switchboard or main office. Instructions are given verbally only at the beginning of the year.	Emergency procedures should be posted at all exits and near all safety equipment. Emergency numbers should be posted at all phones, and a script should be provided for the caller to use. Emergency procedures must be reviewed periodically, and students should be reminded of them at the beginning of each activity.
Spills are handled on a case-by-case basis and are cleaned up with whatever materials happen to be on hand.	Have the appropriate equipment and materials available for cleaning up; replace them before expiration dates. Make sure students know to alert you to spilled chemicals, blood, and broken glass.
Work habits and environment	
Safety wear is only used for activities involving chemicals or hot plates.	Aprons and goggles should be worn in the lab at all times. Long hair, loose clothing, and loose jewelry should be secured.
There is no dress code established for the laboratory; students are allowed to wear sandals or open-toed shoes.	Open-toed shoes should never be worn in the laboratory. Do not allow any footwear in the lab that does not cover feet completely.
Students are required to wear safety gear but teachers and visitors are not.	Always wear safety gear in the lab. Keep extra equipment on hand for visitors.
Safety is emphasized at the beginning of the term but is not mentioned later in the year.	Safety must be the first priority in all lab work. Students should be warned of risks and instructed in emergency procedures for each activity.
There is no assessment of students' knowledge and attitudes regarding safety.	Conduct frequent safety quizzes. Only students with perfect scores should be allowed to work in the lab.
You work alone during your preparation period to organize the day's labs.	Never work alone in a science laboratory or a storage area.
Safety inspections are conducted irregularly and are not documented. Teachers and administrators are unaware of what documentation will be necessary in case of a lawsuit.	Safety reviews should be frequent and regular. All reviews should be documented, and improvements must be implemented immediately. Contact legal counsel for your district to make sure your procedures will protect you in case of a lawsuit.

Identified risk	Preventative control
Purchasing, storing, and using chemicals	
The storeroom is too crowded, so you decide to keep some equipment on the lab benches.	Do not store reagents or equipment on lab benches. Keep shelves organized. Never place reactive chemicals (in bottles, beakers, flasks, wash bottles, etc.) near the edges of a lab bench.
You purchase plenty of chemicals to be sure that you won't run out or to save money.	Purchase chemicals in class-size quantities. Do not purchase or have on hand more than one year's supply of each chemical.
You don't generally read labels on chemicals when preparing solutions for a lab, because you already know about a chemical.	Read each label to be sure it states the hazards and describes the precautions and first aid procedures (when appropriate) that apply to the contents in case someone else has to deal with that chemical in an emergency.
You never read the Material Safety Data Sheets (MSDSs) that come with your chemicals.	Always read the Material Safety Data Sheet (MSDS) for a chemical before using it. Follow the precautions described in the MSDS. File and organize MSDSs for all chemicals where they can be found easily in case of an emergency.
The main stockroom contains chemicals that haven't been used for years.	Do not leave bottles of chemicals unused on the shelves of the lab for more than one week or unused in the main stockroom for more than one year. Dispose of or use up any leftover chemicals.
No extra precautions are taken when flammable liquids are dispensed from their containers.	When transferring flammable liquids from bulk containers, ground the container; before transferring flammable liquids to a smaller metal container, ground both containers.
Students are told to put their broken glass and solid chemical wastes in the trash can.	Have separate containers for trash, for broken glass, and for different categories of hazardous chemical wastes.
You store chemicals alphabetically instead of by hazard class. Chemicals are stored without consideration of possible emergencies (fire, earthquake, flood, etc.), which could compound the hazard.	Use MSDSs to determine which chemicals are incompatible. Store chemicals by the hazard class indicated on the MSDS. Store chemicals that are incompatible with common fire-fighting media like water (such as alkali metals) or carbon dioxide (such as alkali and alkaline-earth metals) under conditions that eliminate the possibility of a reaction with water or carbon dioxide if it is necessary to fight a fire in the storage area.
Corrosives are kept above eye level, out of reach from any unauthorized person.	Always store corrosive chemicals on shelves below eye level. Remember, fumes from many corrosives can destroy metal cabinets and shelving.
Chemicals are kept on the stockroom floor on the days that they will be used so that they are easy to find.	Never store chemicals or other materials on floors or in the aisles of the laboratory or storeroom, even for a few minutes.

Safety Symbols

The following safety symbols will appear in this text when students are asked to perform a procedure requiring extra precautions. The numbered rules on the previous pages apply to all laboratory work.

EYE PROTECTION

- Wear safety goggles when working around chemicals, acids, bases, flames or heating devices. Contents under pressure may become projectiles and cause serious injury.
- Never look directly at the sun through any optical device or use direct sunlight to illuminate a microscope.
- Avoid wearing contact lenses in the lab.
- If any substance gets into your eyes, notify your instructor immediately and flush your eyes with running water for at least 15 minutes.

CLOTHING PROTECTION

- Secure loose clothing and remove dangling jewelry. Do not wear open-toed shoes or sandals in the lab.
- Wear an apron or lab coat to protect your clothing when you are working with chemicals.
- If a spill gets on your clothing, rinse it off immediately with water for at least 5 minutes while notifying your instructor.

CHEMICAL SAFETY

- Always use caution when working with chemicals.
- Always wear appropriate protective equipment. Always wear eye goggles, gloves, and a lab apron or lab coat when you are working with any chemical or chemical solution.
- Never mix chemicals unless your instructor directs you to do so.
- Never taste, touch, or smell chemicals unless your instructor directs you to do so.
- If a chemical gets on your skin, on your clothing, or in your eyes, rinse it immediately and alert your instructor.
- If a chemical is spilled on the floor or lab bench, alert your instructor, but do not clean it up yourself unless your instructor directs you to do so.
- Add an acid or base to water; never add water to an acid or base.
- Never return an unused chemical to its original container.
- Never transfer substances by sucking on a pipet or straw; use a suction bulb.
- Follow instructions for proper disposal.
- Do not allow radioactive materials to come into contact with your skin, hair, clothing, or personal belongings. Although the materials used in this lab are not hazardous when used properly, radioactive materials can cause serious illness and may have permanent effects.

ELECTRICAL SAFETY

- Do not place electrical cords in walking areas or let cords hang over a table edge in a way that could cause equipment to fall if the cord is accidentally pulled.
- Do not use equipment that has frayed electrical cords or loose plugs.
- Be sure that equipment is in the "off" position before you plug it in.
- Never use an electrical appliance around water or with wet hands or clothing.
- Be sure to turn off and unplug electrical equipment when you are finished using it.
- Never close a circuit until it has been approved by your teacher. Never rewire or adjust any element of a closed circuit.
- If the pointer on any kind of meter moves off scale, open the circuit immediately by opening the switch.
- Do not work with any batteries, electrical devices, or magnets other than those provided by your teacher.

HEATING SAFETY

- Avoid wearing hair spray or hair gel on lab days.
- Whenever possible, use an electric hot plate instead of an open flame as a heat source.
- When heating materials in a test tube, always angle the test tube away from yourself and others.
- Glass containers used for heating should be made of heat-resistant glass.
- Wire coils may heat up rapidly. If heating occurs, open the switch immediately, and handle the equipment with a heat-resistant glove.
- Know the location of laboratory fire extinguishers and fire-safety blankets.
- Know your school's fire-evacuation routes.

SHARP OBJECTS

- Use knives and other sharp instruments with extreme care.
- Never cut objects while holding them in your hands. Place objects on a suitable work surface for cutting.
- Never use a double edged razor in the lab.

HAND SAFETY

- To avoid burns, wear heat-resistant gloves whenever instructed to do so.
- Always wear protective gloves when working with an open flame, chemicals, solutions, or wild or unknown plants.
- If you do not know whether an object is hot, do not touch it.
- Use tongs when heating test tubes. Never hold a test tube in your hand to heat the test tube.
- Perform this experiment in a clear area. Attach masses securely. Falling, dropped, or swinging objects can cause serious injury.
- Use a hot mitt to handle resistors, light sources, and other equipment that may be hot. Allow all equipment to cool before storing it.

GAS SAFETY

- Do not inhale any gas or vapor unless your instructor directs you to do so. Do not breathe pure gases.
- Handle materials prone to emit vapors or gases in a well-ventilated area. This work should be done in an approved chemical fume hood.

GLASSWARE SAFETY

- Check the condition of glassware before and after using it. Inform your teacher of any broken, chipped, or cracked glassware, because it should not be used.
- Do not pick up broken glass with your bare hands. Place broken glass in a specially designated disposal container.
- If a bulb breaks, notify your teacher immediately. Do not remove broken bulbs from sockets.

WASTE DISPOSAL

- Clean and decontaminate all work surfaces and personal protective equipment as directed by your instructor.
- Dispose of all broken glass, contaminated sharp objects, and other contaminated materials (biological and chemical) in special containers as directed by your instructor.

HYGIENIC CARE/CLEAN HANDS

- Keep your hands away from your face and mouth.
- Always wash your hands thoroughly when you have finished with an experiment.

The Holt Physics Lab Program

Introduction

Holt Physics offers you and your students several options for classroom labs. Each chapter in the textbook is followed by a full-scale laboratory investigation—either a **Skills Practice Lab** or an **Inquiry Lab. Skills Practice Labs** are investigations of physics principles in a traditional, multi-step format. In contrast, **Inquiry Labs** provide students with a predetermined set of materials and a stated goal and challenge students to design an experiment to satisfy the goal. All answers in the *Teacher Edition* are based on sample data, which is on the **One-Stop Planner CD-ROM.** The workbook ***Datasheets for In-Text Labs*** contains reproducible masters of the complete procedure for each lab. These datasheets provide space for students to record their data and answers. This workbook also contains a **Skills Practice** version of each textbook-based **Inquiry Lab.** The fully-articulated steps in these versions offer you the option to use a more traditional approach.

Appendix K in this textbook contains some end-of-chapter labs modified for use with CBL 2 equipment. The workbook ***CBL™ Experiments*** contains reproducible masters of the complete procedure for each lab in **Appendix K.** It also contains CBL 2-modified versions of some end-of-chapter labs and extra, scenario-based CBL 2 labs that correspond to most book chapters. Alternate versions of the labs using first-generation CBL (rather than CBL 2) equipment can be found at **go.hrw.com.**

The textbook also contains many **Quick Labs,** which are brief demonstrations that require few materials. Some **Quick Labs** are appropriate as homework assignments and are designated as such in the *Teacher Edition.*

A separate ***Laboratory Experiments*** workbook contains two types of additional full-scale labs. **Discovery Labs** allow students to explore physics phenomena before they study the corresponding chapter. **Invention Labs** are open-ended inquiry-based labs couched in terms of an engaging, real-world scenario. Students are challenged to test a material or invent a device to solve a specific problem. In lieu of a traditional lab report, students submit a patent application for their invention or process.

The following pages contain preparation notes and teaching tips for the chapter labs and **Appendix K** labs found in this textbook.

Teaching Tips for Inquiry Labs

Prior to starting an **Inquiry Lab,** each student or group should turn in a detailed procedure for approval. You may wish to start this process a few days before the lab period, to allow yourself time to review the procedures and to allow the students time to make necessary revisions.

Compare student procedures to the **Sample Procedures** found on

the following pages. Use the **Checkpoints** as guidelines for things to look for in the procedures and to identify areas where students may need extra help. Students may use a procedure that differs from the sample procedure only if the alternate procedure meets the following conditions:

- The procedure is safe.
- The procedure can be done in the allotted time.
- All necessary materials are available.
- The procedure will prepare the students to answer the questions at the end of the lab when they are finished.

Before starting the procedure, students should revise their procedures according to your comments. You may want to look over revised procedures before granting final approval.

Evaluate student procedures according to organization, clarity, completeness, safety, how well they follow the steps of the scientific method, and how well they prepare students to answer the questions at the end of the lab. Return procedures with your comments to students before they begin the lab.

Additional Support for In-Text Labs

Ch.1 Skills Practice Lab: Physics and Measurement

Planning

Recommended time: 1–2 lab periods

Classroom organization: Each lab group should have two students. Students should alternate duties so that each student performs all steps. For the second part of the lab, each group needs a large, clear area to work in. If possible, consider conducting this part of the lab in a large, open space, such as outdoors or in a gymnasium.

Materials (for each lab group)
- 3 wooden blocks
- alarm stopwatch, 12- or 24-hour
- balance: portable, electronic balance or triple-beam balance with weight
- meterstick
- metric ruler, 15 cm long

Required Precautions: Wear eye protection and other required safety equipment when cutting wood blocks. Follow all instructions and safety guidelines for the equipment. Do not work with power tools without other people present.

Materials Preparation: The wood blocks should be cut from standard 2 in. by 4 in. lumber. For future labs, each lab group should have 4 blocks, cut into lengths of approximately 10 cm, 15 cm, 20 cm, and 25 cm.

Ch.2 Skills Practice Lab: Free-Fall Acceleration

Planning

Recommended time: 1 lab period
Calibrating the recording timer may be done separately, or it may be skipped entirely if the period of the timer is known.

Classroom organization: Each lab group should have two students. Each group needs a level work surface at least 0.5 m above the floor with an edge to clamp the stand base onto. Each group needs clear floor space of at least 1.0 m^2. For the recording timer calibration, each group needs enough open space so the student can walk away from the timer in a straight line for 3.0 s.

Materials (for each lab group)
- 1-position support base and rod, 1.3 cm × 91 cm
- balance: portable, electronic balance or triple-beam balance with weight
- C-clamp
- meterstick

Additional Materials (for chapter lab)
- alarm stopwatch, 12- or 24-hour
- metric hooked mass set
- recording timer: acceleration timer, tabletop acceleration timer, or compact spark timer
- replacement paper tape, 13 mm
- replacement carbon disks

Additional Materials (for Appendix K lab)
- 3 wooden blocks of different masses
- LabPro® or CBL 2™ interface
- roll of masking tape
- thin foam pad
- TI graphing calculator with link cable
- V-clamp
- Vernier motion detector

Required Precautions: Wear eye protection and other required safety equipment when cutting wood blocks. Follow all instructions and safety guidelines for the equipment. Do not work with power tools without other people present.

Materials Preparation: The wood blocks for the **Appendix K** procedure should be cut from standard 2 in. by 4 in. lumber. Each lab group should have 4 blocks cut into lengths of approximately 10 cm, 15 cm, 20 cm, and 25 cm.

Ch.3 Inquiry Lab: Velocity of a Projectile

Planning

Recommended time: 1 lab period

Classroom organization: Each group must have at least two students. Each group needs a level surface at least 0.5 m above the floor, with at least 2.0 m of space in front of the surface.

Materials (for each lab group)
- 1-position support base and rod, 1.3 cm × 91 cm
- aluminum sheet, 12.5 cm × 25 cm, 0.0010 in. thick
- C-clamp
- carbon paper, 4 sheets
- cardboard box
- inclined plane
- meterstick
- right-angle clamp for 1.3 cm rods
- roll of black nylon cord
- roll of adhesive packing tape
- roll of adhesive tape, 0.5 in. wide
- roll of masking tape

(Ch.3: Velocity of a Projectile, cont.)

- solid steel ball, 25 mm diameter
- stand rod
- steel washer, 26 mm diameter
- towel or washcloth
- white copier paper, 4 sheets

Sample Procedure

1. Prepare a data table with five columns and nine rows. In the first row, label the columns *Trial, Height of Ramp (m), Length of Ramp (m), Displacement* Δx *(m),* and *Displacement* Δy *(m).* In the first column, label the second through ninth rows *1, 2, 3, 4, 5, 6, 7,* and *8.*
2. Set up the inclined plane at an angle as shown in Figure 1. Tape the aluminum to the end of the plane and to the table. Leave at least 5 cm between the bottom end of the inclined plane and the edge of the table.
3. In front of the table, place the box to catch the ball after it bounces. Perform a practice trial to find the correct placement of the box. Use masking tape to secure the box to the floor. Cover the floor with white paper. Cover the white paper with the carbon paper, carbon side down. Tape the paper down.
4. Use a piece of cord and tape to hang a washer from the edge of the table so that the washer hangs a few centimeters above the floor. Mark the floor directly beneath the washer with tape. Move the washer to another point on the table edge, and repeat. Connect the two marks on the floor with masking tape.
5. Use tape to mark a starting line near the top of the inclined plane. Measure the height of the ramp from the tabletop to the tape mark. Measure the length along the ramp from the tape mark to the tabletop. Measure the distance from the top of the tabletop to the floor. Enter these values in the data table for *Trial 1* as *Height of Ramp (m), Length of Ramp (m),* and *Displacement* Δy *(m).*
6. Place the metal ball on the inclined plane at the tape mark. Keep the area around the table clear of people and obstructions. Release the ball from rest so that it rolls down the inclined plane, off the table onto the carbon paper, and bounces into the box.
7. Lift the carbon paper. There should be a carbon mark on the white paper where the ball landed. Label the mark with the trial number.
8. Replace the carbon paper and repeat this procedure as *Trial 2.*
9. With the inclined plane in the same position, place another tape mark about halfway down the inclined plane. Measure and record the height and length of the inclined plane from this mark. Use this mark as the starting point for *Trial 3* and *Trial 4.* Record all data.
10. Raise or lower the inclined plane and repeat the procedure. Perform two trials for each tape mark as *Trials 5, 6, 7,* and *8.* For each trial, measure the distance from the carbon mark to the tapeline. Record this distance as *Displacement* Δx *(m).*
11. Clean up the work area. Put equipment away safely so it is ready to be used again. Recycle or dispose of used lab materials.

Checkpoints

Step 2: Make sure the plane is secure. Make sure all edges on the aluminum sheet are taped.

Step 3: Remind students to watch out for balls on the floor. The box should be taped to the floor and filled with a folded towel or cloth.

Step 4: Students should be able to explain why the distance must be measured from the tapeline.

Step 9: For Trials 3 and 4, start the ball from the lower tape mark without moving the plane.

Step 10: Trials 5 and 6: release the ball from the higher tape mark; Trials 7 and 8: release from the lower mark. Students must measure the distances before lifting the paper from the floor.

Ch.4 Skills Practice Lab: Force and Acceleration

Planning

Recommended time: 1-2 lab periods
For a 2-period lab, add the Extension exercise at the end of the lab.

Classroom organization: Each group needs a level work surface that is at least 2.0 m long and 0.5 m above the floor, with an end to clamp the pulley to. Each group must have at least two students and no more than four. With larger groups, make sure all students are involved.

Materials (for each lab group)

- 1 dynamics cart
- 1-position support base and rod, 1.3 cm × 91 cm, or C-clamp
- adjustable table-clamp pulley
- balance: portable, electronic balance or triple-beam balance with weight
- meterstick
- metric hooked mass, 1000 g
- metric slotted mass set, 1 g–500 g, and holder
- roll of adhesive tape, 0.5 in. wide
- roll of thick, white, braided cord

Additional Materials (for chapter lab)

- recording timer: acceleration timer, tabletop acceleration timer, or compact spark timer
- replacement carbon disks
- replacement paper tape, 13 cm

Additional Materials (for Appendix K lab)
- LabPro® or CBL 2™ interface
- rod and parallel clamp
- square of poster board, 25 cm × 25 cm
- V-clamp
- Vernier dual-range force sensor
- Vernier motion detector

Materials Preparation: Set up a sample apparatus in the laboratory for students to refer to as they set up their equipment.

Ch.5 Skills Practice Lab: Conservation of Mechanical Energy

Planning

Recommended time: 1 lab period

Classroom organization: Each group needs a level work surface. Each lab group should have two students.

Materials (for each lab group)
- lattice rod, 1.3 cm × 30 cm
- meterstick
- metric slotted mass set, 1 g–500 g, and holder
- right-angle clamp for 1.3 cm rod
- support base and rod: 1-position support base and rod, 1.3 cm × 91 cm
- table clamp or tripod base and support rod

Ch.6 Inquiry Lab: Conservation of Momentum

Planning

Recommended time: 1 lab period
For a 2-period lab, have students repeat the experiment to find the value of an unknown mass on one of the carts.

Classroom organization: Each group needs a level work surface with clear table space at least 2.0 m long. Each lab group should have at least two students.

Materials (for each lab group)
- 1 right-angle clamp for 1.3 cm rods
- 1 stand rod, 1.3 cm × 60 cm
- 1-position support base and rod, 1.3 cm × 91 cm
- 2 lattice rods, 1.3 cm × 30 cm
- 2 rubber bands
- balance: portable, electronic balance or triple-beam balance with weight
- C-clamp
- dynamics cart set with two carts
- meterstick
- metric hooked mass, 1000 g
- metric slotted mass set, 1 g–500 g
- recording timer: acceleration timer, tabletop acceleration timer, or compact spark timer
- replacement carbon disks
- replacement paper tape, 13 mm
- roll of adhesive tape, 0.5 in. wide

Materials Preparation: Set up a sample apparatus in the laboratory for students to refer to when setting up their equipment.

Sample Procedure

1. Prepare a data table with seven columns and four rows. In the first row, label the first through seventh columns *Trial*, m_1 *(kg)*, m_2 *(kg)*, *Cart 1 Distance (m)*, *Cart 2 Distance (m)*, *Cart 1 Time Interval (s)*, and *Cart 2 Time Interval (s)*. In the first column, label the second through fourth rows *1*, *2*, and *3*.
2. Choose a location where both carts will be able to move at least 1.0 m without any obstacles.
3. Set up the apparatus as shown in Figure 1. Do not plug in the timer until the teacher approves the setup. Do not close the switch.
4. Calibrate the recording timer with the stopwatch, or use the previously determined value for the timer's period.
5. Record the value for the timer's period on a line near the data table.
6. Measure the mass of one cart, and record it in the data table. Add a 1.0 kg mass to the second cart, and record the mass of the cart plus the 1.0 kg mass.
7. Fasten a timing tape to one end of each cart. Because both tapes pass through the same timer, place two carbon paper disks back to back between the paper tapes.
8. Compress the spring and position the carts. When the teacher approves the setup, plug the recording timer into the wall outlet. Start the timer and release the spring simultaneously.
9. Catch the carts before they reach the edge of the table and then stop the timer. Do not let the carts fall off the table. Remove the tapes. Label each tape so that it corresponds to the cart to which it was attached.
10. On each tape, find a portion where the distance between dots is fairly constant. Use the metric ruler to measure three distances between successive dots.
11. Find the average of these three values and record the average as the distance for that cart in the data table. Record the period of the timer as the time interval in the data table.
12. Using different masses, repeat the experiment two more times for *Trial 2* and *Trial 3*.
13. Clean up the work area. Put equipment away safely so that it is ready to be used again.

(Ch.6: Conservation of Momentum, cont.)

Checkpoints

Step 4: If students have not used the recording timer before, refer to the lab "Free-Fall Acceleration" in the chapter "Motion in One Dimension" for instructions.

Step 8: Students should leave enough room for both carts to move about 1.0 m. Remind students that they will have to catch both carts at about the same time.

Step 9: Make sure students are careful not to let carts fall off the table.

Step 10: Help students choose 3 dots on each tape where the velocity is fairly constant. Students should be able to explain how the dots represent the motion of the carts.

Ch.7 Inquiry Lab: Machines and Efficiency

Planning

Recommended time: 1 lab period

Classroom organization: This lab may be performed by one or more students. Half the class may work with pulleys while the other half works with the inclined plane.

Materials (for each lab group)

- 1-position support base and rod, 1.3 cm × 91 cm
- balance: portable, electronic balance or triple-beam balance with weight
- ball-bearing tandem pulleys, two double and two triple
- ball-bearing tandem pulleys, two single
- C-clamp
- Hall's carriage
- inclined plane
- mass hanger
- mass set: metric slotted mass set or metric hooked mass set
- meterstick
- right-angle clamp
- roll of thick, white, braided cord
- ruler
- suspension clamp

Materials Preparation: Each group will need one cord for the inclined-plane procedure and another cord for the pulleys. The length required will depend upon the height of the plane and the pulley configuration, but most setups will require 1–2 m of cord. You may want to provide several lengths or allow students to cut their own. Students should be supervised when working with scissors or other sharp objects. If desired, attach about 1 m of cord to each dynamics cart before the lab. The classroom may be divided into two parts, with inclined planes set up at half the lab stations and pulley configurations set up at the other half. Half the groups will perform one part of the lab while the other half performs the other part of the lab. Then the groups can switch tasks.

Sample Procedure

1. Prepare a data table with six columns and seven rows. In the first row, label the first through sixth columns *Trial, Machine, Mass*$_1$ *(kg)*, Δh *(m)*, *Mass*$_2$ *(kg)*, and Δd *(m)*. In the first column, label the second through seventh rows *1, 2, 3, 4, 5,* and *6.*

Inclined Plane

2. Set up the inclined plane as shown in Figure 1. Set the incline securely to an angle. Keep the angle constant during this part of the experiment. Place the inclined plane away from the edge of the table, or clamp its base to the edge of the table.
3. Measure the mass of the cart. Attach a piece of cord through the hole on the body of the cart. The cord should be long enough so that the other end of the cord reaches the tabletop before the cart reaches the top of the incline. Place the cart on the plane and run the cord over the pulley at the top of the plane. Attach a mass hanger to the free end of the cord.
4. Place a 200 g mass in the cart. Record the total mass of the cart and its contents as *Mass*$_1$. Attach masses to the mass hanger to find the lowest mass that will allow the cart to move up the plane with a constant velocity. Stop the cart before it reaches the top of the incline. Record the mass of the mass hanger plus the added mass as *Mass*$_2$ in the data table.
5. Measure the distances, and record them. Δh is the vertical distance the cart moves, while the mass hanger on the cord moves the distance Δd.
6. Repeat steps 4 and 5 several times, increasing the mass in the cart by 100 g and finding the mass that will allow the cart to move with a constant velocity each time. Record all data for each trial in your data table.

Pulley

7. Set up a pulley system. For the first trial, use five pulleys. Keep the area beneath the pulley system clear throughout the experiment. Measure the mass of the bottom set of pulleys before including them in the setup. Attach a 500 g mass to the bottom set of pulleys. Record the total mass of the 500 g mass plus the bottom set of pulleys as *Mass*$_1$ in the data table.
8. Starting with 50 g, add enough mass to the mass hanger to prevent the pulleys from moving when released. Place the mass hanger just below the 500 g mass, and measure the initial positions of both masses to the nearest millimeter by measuring the height of each mass above the base.
9. Add masses to the mass hanger to find the mass that will make the 500 g mass move up with constant velocity once it has been started. Record the mass of the mass hanger plus the added mass as *Mass*$_2$ in the data table.
10. Measure the final positions of both masses, and record the distances (final position – initial position) in the data table. Δh is the vertical distance through which the mass on the pulley is raised, while the mass on the mass hanger moves down through the distance Δd.

11. Using the same 500 g $Mass_1$, perform two more trials using different pulley systems (four pulleys, six pulleys, and so on). Record all data. Include the mass of the bottom set of pulleys in the total mass that is raised in each trial.

12. Clean up the work area. Put equipment away safely so that it is ready to be used again.

Checkpoints

Step 2: Make sure the inclined plane is secure. The apparatus should be away from the edge of the table; if not, use a C-clamp to secure the base of the stand to the table. For best results the top of the plane should be at least 20 cm above the table.
Step 3: The cord should be attached through the hole in the body of the cart. Make sure the masses are attached securely. Remind students to include the mass hanger in the total suspended mass.
Step 4: Students should demonstrate that they are using the smallest mass that will allow the cart to move up the plane and that the cart has constant velocity. The hanging mass should hit the table before the cart reaches the top of the plane.
Step 7: Students may need help threading the pulley systems.
Step 8: Measure the position from the bottom of each mass.
Step 9: Students should demonstrate that the added mass is the smallest that will cause the pulley set to move and that the pulley set has a constant velocity. Remind students to include the mass hanger in the total recorded mass.

Ch.9 Skills Practice Lab: Specific Heat Capacity

Planning

Recommended time: 1 lab period for 3 samples; additional samples may require more time.

Classroom organization: This lab presents several safety hazards. Remind students never to leave a hot plate unattended while it is on. Make sure all students wear the appropriate safety gear at all times. Students must take extra care not to break thermometers in this lab. Each group must have at least two students. Groups must perform more than one step at a time, so groups may be larger than usual. With larger groups, make sure all students are involved and paying attention to safety. Bringing water to a boil takes 10–20 min; other tasks can be completed as the water heats. Each group must have a level work surface large enough so that students are not too close to the hot plates. Students should place wet metal shot in designated containers at the end of the lab to be dried and used again.

Materials (for each lab group)

- 2 low-form beakers, 600 mL
- balance: portable, electronic balance, 2000 g capacity or triple-beam balance with weight
- double-wall metal calorimeter and stirring rod with known specific heat capacity
- hot plate
- ice
- metal shot:
  100 g tin
  100 g aluminum
  100 g copper
  100 g zinc
- safety goggles or spectacles
- safety hot mitt
- small plastic dish
- steam generator (metal heating vessel, including metal heating dipper)
- water

Additional Materials (for chapter lab)

- dual magnifier
- fractional thermometer, –5°C to 50°C
- red liquid thermometer, –20°C to 110°C

Additional Materials (for Appendix K lab)

- LabPro® or CBL 2™ interface
- stainless steel temperature probe
- TI graphing calculator with link cable

Required Precautions: Use a hot mitt and wear safety goggles or spectacles when drying the samples in a lab oven.

Materials Preparation: Metal shot should be clean and dry. Between classes, dry the metal shot by placing it in a laboratory oven at low heat. Cool samples before class begins. Each type of metal should be labeled with a number, not the name of the metal. Record the sample numbers and their corresponding metals to check students' results. Metal samples should be placed in large beakers with spatulas for students to measure out. For best results, use crushed ice or clean snow.

Ch.11 Inquiry Lab: Simple Harmonic Motion of a Pendulum

Planning

Recommended time: 1 lab period

Classroom organization: Each group must have at least two students.

Materials (for each lab group)

- balance: portable, electronic balance or triple-beam balance with weight
- roll of pendulum suspension cord
- meterstick
- drilled-ball set (provides pendulum bobs for two workstations)
- pendulum clamp
- protractor
- 12- or 24-hour alarm stopwatch
- 1-position support base and rod, 1.3 cm × 91 cm

(Ch.11: Simple Harmonic Motion of a Pendulum, cont.)

Sample Procedure

1. Prepare a data table with four columns and seven rows. In the first row, label the first through fourth columns *Trial, Mass (kg), Length (m),* and *Time (s).* In the first column, label the second through seventh rows *1, 2, 3, 4, 5,* and *6.*

Constant Mass with Varying Length

2. Measure the mass of the bob to the nearest gram. Record it in the data table.
3. Set up a ring stand with a clamp to serve as the support. Place the ring stand away from the table's edge. Use cord to tie the bob to the clamp on the ring stand, securing it at the desired length (about 75 cm). Choose a location away from other groups, where the pendulum can swing freely.
4. Measure the length of the pendulum to the center of the bob.
5. Use the protractor to find the position of the pendulum where the amplitude is equal to 10°. Lift the bob to this position. Make sure that the path is free of obstructions, and release the bob so that it swings freely.
6. Measure and record the time required for 20 cycles of the pendulum bob. Keep the amplitude in all trials between 5° and 15°. Do not exceed 15°.
7. Perform two more trials using the same mass but different cord lengths.

Constant Length with Varying Mass

8. Replace the bob with a different mass, and repeat steps 2–6. Record the mass and length.
9. Perform two more trials using the same length but different masses.
10. Clean up the work area. Put equipment away safely so that it is ready to be used again.

Checkpoints

Step 5: Make sure students hold the cord straight when lifting the bob. The angle should be 10° to 15°. If the swing traces a circle instead of an arc, the angle may be too large.
Step 7: To change the length of the cord, students should loosen the clamp, move the cord up or down, and tighten the clamp securely.

Ch.12 Skills Practice Lab: Speed of Sound

Planning

Recommended time: 2 lab periods

Classroom organization: Each group must have at least two students. Each group must have a level work surface at least 0.5 m above the floor. The CBL procedure uses a different method to find the speed of sound; be aware of these differences if using them in the same class.

Materials (for each lab group; for chapter lab)

- Erlenmeyer flask, 1000 mL
- food coloring (optional)
- red liquid thermometer, –20°C to 110°C
- resonance apparatus with double 45° clamp
- set of four tuning forks, physical pitch
- soft rubber hammer
- water

Materials (for each lab group; for Appendix K lab)

- cardboard tube
- LabPro® or CBL2™ interface
- meterstick
- roll of masking tape
- stainless steel temperature probe
- support stand with clamp
- TI graphing calculator with link cable

Materials Preparation: For better visibility, water for the resonance apparatus procedure may be tinted with food coloring. Mix a large quantity of colored water from which students may fill their apparatus, or set out the food coloring for students to use.

Ch.13 Skills Practice Lab: Brightness of Light

Planning

Recommended time: 1 lab period

Classroom organization: Each lab group must have at least two students. Because of the complexity of the exercise, lab groups may have more than two students. Each lab group needs a level work surface that is near an electrical outlet and away from any sources of water. Each work area must be at least 1.5 m long.

Materials (for each lab group)

- 1-position support base and rod, 1.3 cm × 91 cm
- battery eliminator with alligator clips, 6 V/0.5 A
- black construction paper
- blackened card tube for bulb shield
- meterstick, plain wood, 1 m long and metal supports
- miniature bulb and base
- replacement bulb, 6.3 V/0.3 A
- roll of adhesive tape, 0.5 in. wide
- round-jaw symmetrical clamp with holder

Required Precautions: The greatest danger in this activity is presented by exposed electrical connections. All bulb sockets should have enclosed contacts and insulated connectors should be used on all wire connectors.

Materials Preparation: Set up a sample apparatus in the laboratory for students to refer to as they set up their equipment.

The blackened card tube to be used as a bulb shield may be constructed using a cardboard toilet paper tube with a hole at the level of the bulb filament. This hole should be adjusted to line up with the probe or meter. For the light meter, use another cardboard tube as a shield for the detector. Use the lens support to hold the tube in place during the lab.

Ch.14 Converging Lenses

Planning

Recommended time: 1 lab period

Classroom organization: This lab can be performed by students working alone, but it is better with groups of two or more students. Each lab group needs a level work surface that is near an electrical outlet and away from any sources of water. Each work area must be at least 1.5 m long.

Materials (for each lab group)
- battery eliminator with alligator clips, 6 V/0.5 A
- miniature bulb and base
- meterstick optical bench set includes:
  - card screen with mm scale (set of 5)
  - lens or mirror support, for 4 cm lenses
  - marker/object riders
  - meterstick, plain wood, 1 m long
  - pair of metal meterstick supports
  - screen support riders
- metric ruler, 15 cm long
- double convex lens
  - 38 mm diam. $f = 10$ cm, or
  - 38 mm diam. $f = 15$ cm, or
  - 38 mm diam. $f = 20$ cm
- illuminated object screen
- pkg. of 10 replacement lamps, 6.2 V/0.5 A
- roll of insulated copper wire, 18 awg, 30 m

Required Precautions: The greatest danger in this activity is presented by exposed electrical connections. All bulb sockets should have enclosed contacts and insulated connectors should be used on all wire connections.

Materials Preparation: Set up a sample apparatus in the laboratory for students to refer to as they set up their equipment. If you are not using complete meterstick optical bench sets, you will need all the equipment listed under the *optical bench* heading for each lab group.

Ch.15 Skills Practice Lab: Diffraction

Planning

Recommended time: 1 lab period

Classroom organization: Each lab group needs a level work surface that is near an electrical outlet and away from any sources of water. Each work area must be at least 1.5 m long. Each lab group should have two or more students.

Materials (for each lab group)
- battery eliminator with alligator clips, 6 V/0.5 A
- black card, 1 sheet
- meterstick optical bench set includes:
  - card screen with mm scale (set of 5)
  - lens or mirror support, for 4 cm lenses
  - marker/object riders
  - meterstick, plain wood, 1 m long
  - pair of metal meterstick supports
  - screen support riders
- miniature bulb and base
- mounted film transmission grating
- paper clips
- roll of self-adhesive tape, 0.5 in. wide
- scale and slit

Materials Preparation: Set up a sample apparatus in the laboratory for students to refer to as they set up their equipment. If you are not using complete meterstick optical bench sets, you will need all the equipment listed under the *optical bench* heading for each lab group. Bent paper clips can be used as riders to mark the positions of the images on the meterstick scale. To make the riders, pull the small part of the paperclip forward. The bent paperclips can be hung over the edge of the meterstick at the position of the images.

Ch.16 Skills Practice Lab: Electrostatics

Planning

Recommended time: 1 lab period Alternatively, each section of the lab could be performed for part of a period on different days.

Classroom organization: Each lab group needs a level work surface. Students may perform this lab alone or in groups of two or more students.

Materials (for each lab group)
- 1-position support base and rod, 1.3 cm × 91 cm
- electroscope, aluminum leaf
- exciting pad, animal fur
- exciting pad, silk
- exciting pad, wool felt
- friction rod, borosilicate glass (hollow)
- friction rod, flint glass (solid)
- friction rod, hard rubber (solid)
- friction rod, polystyrene (solid)
- meterstick, wooden
- roll of adhesive tape, 0.5 in. wide
- roll of fine black nylon cord, 9 m
- round-jaw symmetrical clamp, with holder
- static electricity tube, PVC
- suspension stirrup for rods

(Ch.16: Electrostatics, cont.)

Materials Preparation: Electrostatic laboratories are difficult if undesired charge cannot be eliminated. A common ground consisting of a wire connected to a water pipe will allow students to eliminate this charge.

Ch.17 Skills Practice Lab: Current and Resistance

Planning

Recommended time: 1 lab period

Classroom organization: Each lab group should have a level surface to work on that is near an electrical outlet and away from any sources of water. Each lab group should have two students.

Materials (for each lab group)

- mounted resistance coils
- patch cord, black, insulated alligator clip/stacking banana plug
- patch cord, red, insulated alligator clip/stacking banana plug
- power supply:
  4-output dc regulated power supply, or
  CENCO 6 V ac/dc low-voltage power supply, or
  CENCO universal power supply
- switch: contact key, or knife switch (single pole/single throw)
- wire leads with alligator clips (60 cm)
- One of the following:
  2 basic digital multimeters; 2 student multimeters;
  2 general purpose multimeters; or 1 CENCO standard movement, triple range dc ammeter (50 mA, 500 mA, 5 A) **and** 1 CENCO standard movement, triple range dc voltmeter (0–3 V, 10 V, 15 V)

Materials Preparation: To use multimeters to measure current and potential difference (voltage), follow this setup procedure.

Current: On the front of the multimeter, set the range switch pointer to *DC A* to read direct current in amperes. Set the pointer to the *200 mA* mark. Use a piece of tape to label this meter *A.* Make connections to the sockets labeled *mA* and *COM.*

Potential difference (voltage): On the front of the multimeter, set the range switch pointer to *DC V* to read direct current in volts. Set the pointer to the *20 V* mark. Use a piece of tape to label this meter *V.* Make connections to the sockets labeled *V*Ω and *COM.*

When checking students' circuits, always make sure voltage meters are connected in parallel and current meters are connected in series. Also make sure leads are connected to the correct terminals of the meters.

Ch.18 Inquiry Lab: Resistors in Series and Parallel

Planning

Recommended time: 1 lab period

Classroom organization: Each lab group should have a level surface to work on that is near an electrical outlet and away from any sources of water. Each lab group should have two students.

Materials (for each lab group)

- 150 Ω resistor, 0.5 W
- 68 Ω resistor, 0.5 W
- pair of alligator clip adapters
- patch cord, 30 cm, black, with banana plugs
- patch cord, 30 cm, red, with banana plugs
- power supply: 1.5 V/3 V battery eliminator, or
  20 V dc/500 mA regulated power supply
- switch: contact key or knife switch (single pole/single throw)
- One of the following:
  2 basic digital multimeters; 2 student multimeters;
  2 general purpose multimeters; or 1 CENCO standard movement, triple range dc ammeter (50 mA, 500 mA, 5 A) **and** 1 CENCO standard movement, triple range dc voltmeter (0–3 V, 10 V, 15 V)

Materials Preparation: See the teacher's notes for the lab "Current and Resistance" for instructions on using multimeters to measure current and voltage.

Sample Procedure

1. Prepare a data table with six columns and three rows. In the first row, label the second through sixth columns ΔV_T (V), ΔV_1 (V), I_1 (A), ΔV_2 (V), and I_2 (A). In the first column, label the second row *Series* and the third row *Parallel.*

Resistors in Series

2. Construct a circuit that includes a battery, a switch, and two unequal resistors in series. Do not close the switch until the teacher approves the circuit.
3. With the switch open, connect the current meter in series and the voltage meter in parallel with one of the resistors. Do not close the switch.
4. After the teacher approves the circuit, close the switch. Measure the current in and the potential difference across the resistor. Record the information in the data table. Open the switch.
5. Carefully remove the meters from the first resistor. Rewire the meters to measure the potential difference across and the current in the second resistor. Do not close the switch.
6. After the teacher approves the circuit, close the switch. Measure the current in and potential difference across the resistor, and record the information in the data table. Open the switch.
7. Leave the current meter in place, and carefully rewire the voltage meter to measure the potential difference across both resistors. Record this value in the data table as ΔV_T.

Resistors in Parallel

8. Construct a circuit containing a power supply, a switch, and the two resistors wired in parallel. Do not close the switch.

9. With the switch open, connect the current meter in series and the voltage meter in parallel with one of the resistors. Do not close the switch.

10. After the teacher approves the circuit, close the switch. Measure the current in and potential difference across the resistor, and record the information in the data table. Open the switch.

11. Rewire the meters to measure the potential difference across and current in the second resistor. Do not close the switch.

12. After the teacher approves the circuit, close the switch. Measure the current in and potential difference across the resistor, and record the information in the data table. Open the switch.

14. Leave the current meter in place, and carefully rewire the voltage meter to measure the potential difference across the power supply. Record this value in the data table as ΔV_T.

Checkpoints

IMPORTANT: Always check every circuit before the students close the switch. Make sure the configuration is correct, that all connections are secure, and that power supplies and meters are set to the proper levels. Remind students that they MUST wait for you to check circuits when the circuits are first set up and when any changes are made. Also remind students to keep switches closed only long enough to take readings.

Step 4: Make sure the power supply, resistors, and meters are connected properly and are set at the proper settings. Students should be able to demonstrate that the resistors are in series. Check to make sure that students are taking both measurements for the same resistor.

Step 6: Make sure students have correctly rewired the circuit to measure voltage and current for the second resistor. Remind students to close the switch only long enough to take their readings.

Step 10: Students should be able to demonstrate that the resistors are in parallel. Check to make sure that students are taking both measurements for the same resistor.

Ch. 19 Skills Practice Lab: Magnetic Field of a Conducting Wire

Planning

Recommended time: 1 lab period

Classroom organization: Each lab group should have two students.

Materials (for each lab group)

- 1 Ω resistor, 10 W
- bare copper wire, 1 m
- patch cord, black, insulated alligator clip/stacking banana plug
- patch cord, red, insulated alligator clip/stacking banana plug
- power supply:
  CENCO universal power supply, or CENCO 6 V ac/dc low-voltage power supply, or 20 V dc/500 mA regulated power supply, or 4-output dc regulated power supply
- roll of self-adhesive tape, 0.5 in. wide
- tangent galvanometer kit
- wire leads with alligator clips (pkg. of 10)

Additional Materials (for chapter lab)

- 16 mm compass (12/pkg.)
- knife switch (double pole/single throw)
- multimeter or ammeter:
  basic digital multimeter; student multimeter; or CENCO standard movement, triple range dc ammeter (50 mA, 500 mA, 5 A)

Additional Materials (for Appendix K lab)

- alligator clips
- knife switch (double pole/double throw)
- LabPro® or CBL 2™ interface
- roll of masking tape
- support stand and buret clamp
- TI graphing calculator with link cable
- Vernier magnetic field sensor
- voltage probe

Materials Preparation: See the teacher's notes for the lab "Current and Resistance" for instructions on using multimeters to measure current and voltage.

Ch. 20 Skills Practice Lab: Electromagnetic Induction

Planning

Recommended time: 1 lab period

Classroom organization: Each lab group should have two or more students.

Materials (for each lab group)

- contact key
- cylindrical magnet, Alnico V
- galvanometer:
  student galvanometer, –500 mA to +500 mA, or economical galvanometer, or CENCO 6 in 1 galvanometer
- 2 patch cords, red, insulated alligator clip/stacking banana plug
- power supply:
  1.5 V/3 V battery eliminator, or 20 V dc/500 mA regulated power supply, or dry cells
- primary and secondary coils, or economy primary and secondary coils, or CENCO primary and secondary coils
- rheostat: 10 Ω, or potentiometer

Ch.21 Skills Practice Lab: The Photoelectric Effect

Planning

Recommended time: 1 lab period

Classroom organization: You may conduct this as a demonstration for the entire class, or groups of two or more may perform the lab independently.

Materials (for each lab group)

- lens or mirror support, for 4 cm lenses
- lens support, 7.5 cm
- meterstick, plain wood, 1 m long
- One of the following:
  basic digital multimeter; student multimeter; general purpose multimeter; or CENCO standard movement, triple range, dc voltmeter (0–3 V, 10 V, 15 V)
- black construction paper
- pair of metal meterstick supports
- patch cord, black, 60 cm, with banana clips
- patch cord, red, 60 cm, with banana clips
- photoelectric-effect device with amplifier and filters (red, green, blue)
- power supply:
  CENCO universal power supply, or CENCO 6 V ac/dc low-voltage power supply, 20 V dc/500 mA regulated power supply, or 4-output dc regulated power supply
- lamp base with connecting leads
- roll of self-adhesive tape, 0.5 in. wide
- set of wood blocks, 2 in. × 4 in. lumber, 4 different lengths, 10–20 cm
- xenon light bulb (4.8 V, 850 mA)

Materials Preparation: See the teacher's notes for the lab "Physics and Measurement" for instructions on preparation of wood blocks. For this lab, only one block may be needed, to serve as a support for the apparatus.

See the teacher's notes for the lab "Current and Resistance" for instructions on using multimeters to measure current and voltage. Use the black construction paper to make a tube between the lamp and the photoelectric-effect device. This will yield better results and will also protect students' eyes during the lab. Use the xenon light source for the blue and green filters, and a standard tungsten light for the red filter. The tungsten light will need to be placed adjacent to the red filter to give good results.

Ch.22 Skills Practice Lab: Half-Life

Planning

Recommended time: 1 lab period

Classroom organization: You may conduct this as a demonstration for the entire class, or groups of two or more may perform this lab independently. If students will be performing this exercise independently, be sure the isogenerator for each group is properly charged before each use.

Materials (for each lab group)

- basic nuclear-lab station
- disposable culture dish (small)
- isogenerator set
- round-jaw symmetrical clamp, with holder
- support base with rod, 46 cm long × 0.8 cm diameter

Materials Preparation: The Cs-137/Ba-137 isogenerator set does not present a contamination problem, because the solution has a very short half-life, no longer than an hour. However, be sure you know the cleanup and disposal procedures required by your district and local and state governments.

HOLT

Physics

Raymond A. Serway

Jerry S. Faughn

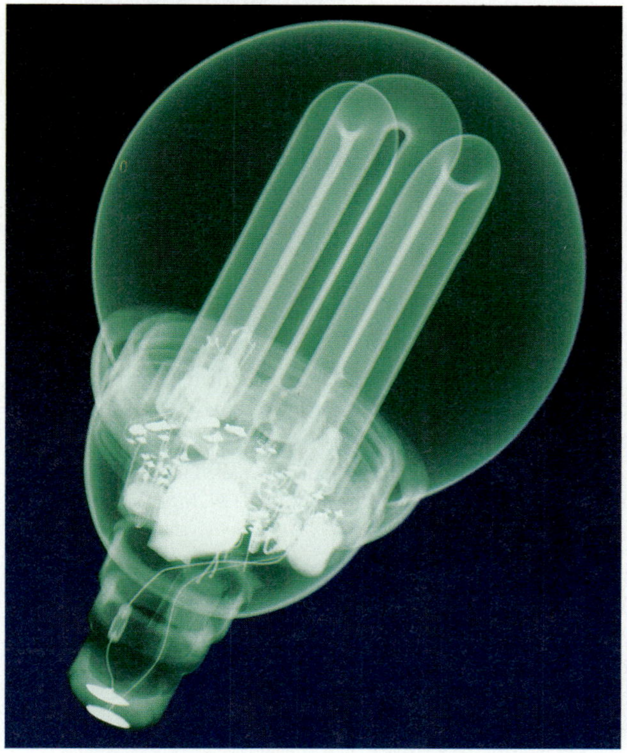

HOLT, RINEHART AND WINSTON

A Harcourt Education Company

Orlando • **Austin** • New York • San Diego • London

TEACHER'S EDITION

Authors

Raymond A. Serway, Ph.D.

Professor Emeritus
North Carolina State University

Jerry S. Faughn, Ph.D.

Professor Emeritus
Eastern Kentucky University

On the cover: The large image is an X ray of an energy-saving lightbulb. The leftmost, small image is a computer model of a torus-shaped magnet that is holding a hot plasma within its magnetic field, shown here as circular loops. The central small image is of a human eye overlying the visible light portion of the electromagnetic spectrum. The image on the right is of a worker inspecting the coating on a large turbine.

Printed in the United States of America

ISBN-13: 978-0-03-036817-2
ISBN-10: 0-03-036817-0

5 6 7 0868 11
4500285240

Acknowledgments

Contributing Writers

Robert W. Avakian
Instructor
Trinity School
Midland, Texas

David Bethel
Science Writer
San Lorenzo, New Mexico

David Bradford
Science Writer
Austin, Texas

Robert Davisson
Science Writer
Delaware, Ohio

John Jewett Jr., Ph.D.
Professor of Physics
California State Polytechnic University
Pomona, California

Jim Metzner
Seth Madej
Pulse of the Planet radio series
Jim Metzner Productions, Inc.
Yorktown Heights, New York

John M. Stokes
Science Writer
Socorro, New Mexico

Salvatore Tocci
Science Writer
East Hampton, New York

Lab Reviewers

Christopher Barnett
Richard DeCoster
Elizabeth Ramsayer
Joseph Serpico
Niles West High School
Niles, Illinois

Mary L. Brake, Ph.D.
Physics Teacher
Mercy High School
Farmington Hills, Michigan

Gregory Puskar
Laboratory Manager
Physics Department
West Virginia University
Morgantown,West Virginia

Richard Sorensen
Vernier Software & Technology
Beaverton, Oregon

Martin Taylor
Sargent-Welch/VWR
Buffalo Grove, Illinois

Academic Reviewers

Mary L. Brake, Ph.D.
Physics Teacher
Mercy High School
Farmington Hills, Michigan

James C. Brown, Jr., Ph.D.
Adjunct Assistant Professor of Physics
Austin Community College
Austin, Texas

Anil R Chourasia, Ph.D.
Associate Professor
Department of Physics
Texas A&M University—Commerce
Commerce, Texas

David S. Coco, Ph.D.
Senior Research Physicist
Applied Research Laboratories
The University of Texas at Austin
Austin, Texas

Thomas Joseph Connolly, Ph.D.
Assistant Professor
Department of Mechanical Engineering and Biomechanics
The University of Texas at San Antonio
San Antonio, Texas

Brad de Young
Professor
Department of Physics and Physical Oceanography
Memorial University
St. John's, Newfoundland, Canada

Bill Deutschmann, Ph.D.
President
Oregon Laser Consultants
Klamath Falls, Oregon

Arthur A. Few
Professor of Space Physics and Environmental Science
Rice University
Houston, Texas

Scott Fricke, Ph.D.
Schlumberger Oilfield Services
Sugarland, Texas

Simonetta Fritelli
Associate Professor of Physics
Duquesne University
Pittsburgh, Pennsylvania

David S. Hall, Ph.D.
Assistant Professor of Physics
Amherst College
Amherst, Massachusetts

Roy W. Hann, Jr., Ph.D.
Professor of Civil Engineering
Texas A & M University
College Station, Texas

Sally Hicks, Ph.D.
Professor
Department of Physics
University of Dallas
Irving, Texas

Robert C. Hudson
Associate Professor Emeritus
Physics Department
Roanoke College
Salem, Virginia

William Ingham, Ph.D.
Professor of Physics
James Madison University
Harrisonburg, Virginia

Karen B. Kwitter, Ph.D.
Professor of Astronomy
Williams College
Williamstown, Massachusetts

Phillip LaRoe
Professor of Physics
Helena College of Technology
Helena, Montana

Joseph A. McClure, Ph.D.
Associate Professor Emeritus
Department of Physics
Georgetown University
Washington, DC

Ralph McGrew
Associate Professor
Engineering Science Department
Broome Community College
Binghamton, New York

Clement J. Moses, Ph.D.
Associate Professor of Physics
Utica College
Utica, New York

Acknowledgments, continued

Alvin M. Saperstein, Ph.D.
Professor of Physics; Fellow of Center for Peace and Conflict Studies
Department of Physics and Astronomy
Wayne State University
Detroit, Michigan

Donald E. Simanek, Ph.D.
Emeritus Professor of Physics
Lock Haven University
Lock Haven, Pennsylvania

H. Michael Sommermann, Ph.D.
Professor of Physics
Westmont College
Santa Barbara, California

Jack B. Swift, Ph.D.
Professor
Department of Physics
The University of Texas at Austin
Austin, Texas

Thomas H. Troland, Ph.D.
Physics Department
University of Kentucky
Lexington, Kentucky

Mary L. White
Coastal Ecology Institute
Louisiana State University
Baton Rouge, Louisiana

Jerome Williams M.S.
Professor Emeritus
Oceanography Department
US Naval Academy
Annapolis, MD

Carol J. Zimmerman, Ph.D.
Exxon Exploration Company
Houston, Texas

Teacher Reviewers

John Adamowski
Chairperson of Science Department
Fenton High School
Bensenville, Illinois

John Ahlquist, M.S.
Anoka High School
Anoka, Minnesota

Maurice Belanger
Science Department Head
Nashua High School
Nashua, New Hampshire

Larry G. Brown
Morgan Park Academy
Chicago, Illinois

William K. Conway, Ph.D.
Lake Forest High School
Lake Forest, Illinois

Jack Cooper
Ennis High School
Ennis, Texas

William D. Ellis
Chairman of Science Department
Butler Senior High School
Butler, Pennsylvania

Diego Enciso
Troy, Michigan

Ron Esman
Plano Senior High School
Plano, Texas

Bruce Esser
Marian High School
Omaha, Nebraska

Curtis Goehring
Palm Springs High School
Palm Springs, California

Herbert H. Gottlieb
Science Education Department
City College of New York
New York City, New York

David J. Hamilton, Ed.D.
Benjamin Franklin High School
Portland, Oregon

J. Philip Holden, Ph.D.
Physics Education Consultant
Michigan Dept. of Education
Lansing, Michigan

Joseph Hutchinson
Wichita High School East
Wichita, Kansas

Douglas C. Jenkins
Chairman, Science Department
Warren Central High School
Bowling Green, Kentucky

David S. Jones
Miami Sunset Senior High School
Miami, Florida

Roger Kassebaum
Millard North High School
Omaha, Nebraska

Mervin W. Koehlinger, M.S.
Concordia Lutheran High School
Fort Wayne, Indiana

Phillip LaRoe
Central Community College
Grand Island, Nebraska

William Lash
Westwood High School
Round Rock, Texas

Norman A. Mankins
Science Curriculum Specialist
Canton City Schools
Canton, Ohio

John McGehee
Palos Verdes Peninsula High School
Rolling Hills Estates, California

Debra Schell
Austintown Fitch High School
Austintown, Ohio

Edward Schweber
Solomon Schechter Day School
West Orange, New Jersey

Larry Stookey, P.E. Science
Antigo High School
Antigo, Wisconsin

Joseph A. Taylor
Middletown Area High School
Middletown, Pennsylvania

Leonard L. Thompson
North Allegheny Senior High School
Wexford, Pennsylvania

Keith C. Tipton
Lubbock, Texas

John T. Vieira
Science Department Head
B.M.C. Durfee High School
Fall River, Massachusetts

Virginia Wood
Richmond High School
Richmond, Michigan

Tim Wright
Stevens Point Area Senior High School
Stevens Point, Wisconsin

Mary R. Yeomans
Hopewell Valley Central High School
Pennington, New Jersey

G. Patrick Zober
Science Curriculum Coordinator
Yough Senior High School
Herminie, Pennsylvania

Patricia J. Zober
Ringgold High School
Monongahela, Pennsylvania

continued on page 973

Contents

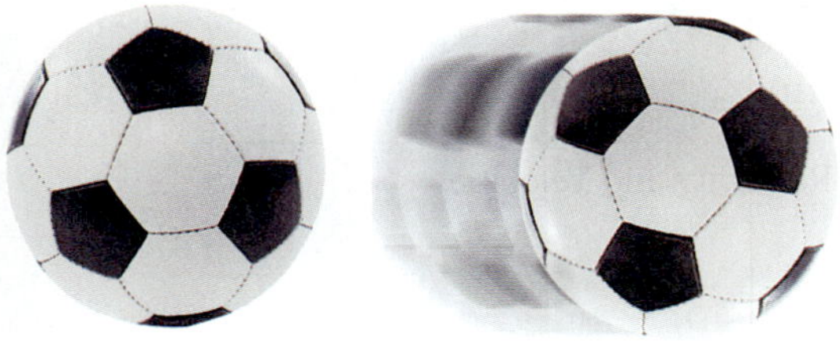

ASTROLAND
CONEY ISLAND

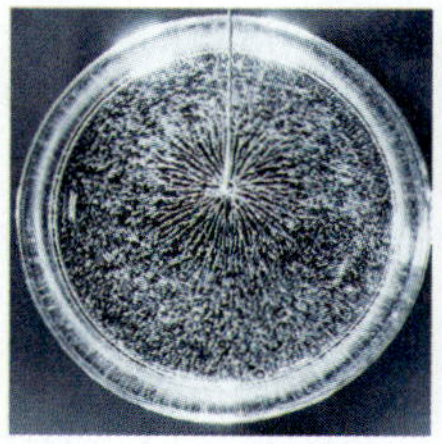

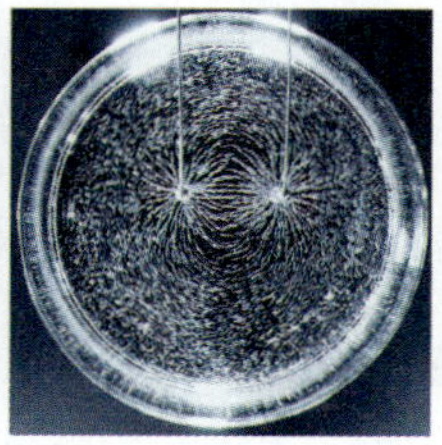

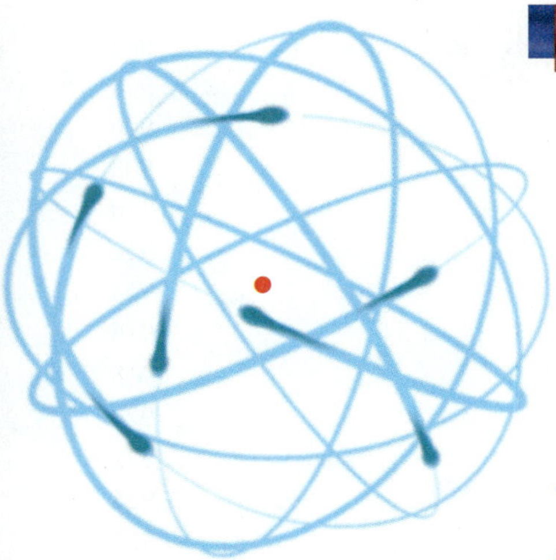

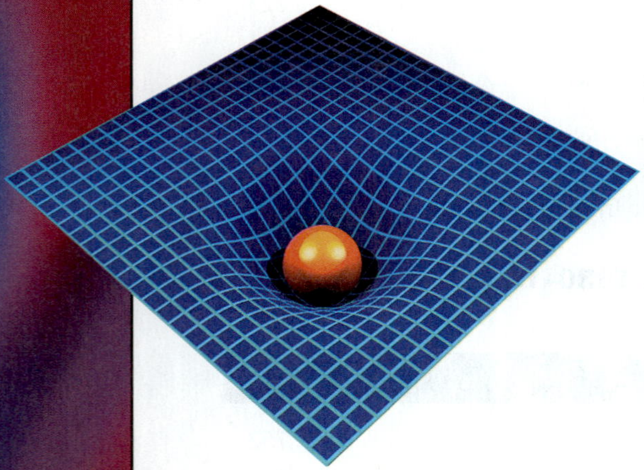

Feature Articles

Why it Matters

Timelines

PHYSICS CAREERS

Science • Technology • Society

Labs

Skills Practice Labs

Inquiry Labs

CBL™ Labs

Quick Labs

Safety Symbols

Remember that the safety symbols shown here apply to a specific activity, but the numbered rules on the following pages apply to all laboratory work.

Eye Protection

- Wear safety goggles when working around chemicals, acids, bases, flames or heating devices. Contents under pressure may become projectiles and cause serious injury.
- Never look directly at the sun through any optical device or use direct sunlight to illuminate a microscope.

Clothing Protection

- Secure loose clothing and remove dangling jewelry. Do not wear open-toed shoes or sandals in the lab.
- Wear an apron or lab coat to protect your clothing when you are working with chemicals.

Chemical Safety

- Always wear appropriate protective equipment. Always wear eye goggles, gloves, and a lab apron or lab coat when you are working with any chemical or chemical solution.
- Never taste, touch, or smell chemicals unless your instructor directs you to do so.
- Do not allow radioactive materials to come into contact with your skin, hair, clothing, or personal belongings. Although the materials used in this lab are not hazardous when used properly, radioactive materials can cause serious illness and may have permanent effects.

Electrical Safety

- Do not place electrical cords in walking areas or let cords hang over a table edge in a way that could cause equipment to fall if the cord is accidentally pulled.
- Do not use equipment that has frayed electrical cords or loose plugs.
- Be sure that equipment is in the "off" position before you plug it in.
- Never use an electrical appliance around water or with wet hands or clothing.
- Be sure to turn off and unplug electrical equipment when you are finished using it.
- Never close a circuit until it has been approved by your teacher. Never rewire or adjust any element of a closed circuit.
- If the pointer on any kind of meter moves off scale, open the circuit immediately by opening the switch.
- Do not work with any batteries, electrical devices, or magnets other than those provided by your teacher.

Heating Safety

- Avoid wearing hair spray or hair gel on lab days.
- Whenever possible, use an electric hot plate instead of an open flame as a heat source.
- When heating materials in a test tube, always angle the test tube away from yourself and others.
- Glass containers used for heating should be made of heat-resistant glass.

Sharp Object Safety

- Use knives and other sharp instruments with extreme care.

Hand Safety

- Perform this experiment in a clear area. Attach masses securely. Falling, dropped, or swinging objects can cause serious injury.
- Use a hot mitt to handle resistors, light sources, and other equipment that may be hot. Allow all equipment to cool before storing it.
- To avoid burns, wear heat-resistant gloves whenever instructed to do so.
- Always wear protective gloves when working with an open flame, chemicals, solutions, or wild or unknown plants.
- If you do not know whether an object is hot, do not touch it.
- Use tongs when heating test tubes. Never hold a test tube in your hand to heat the test tube.

Glassware Safety

- Check the condition of glassware before and after using it. Inform your teacher of any broken, chipped, or cracked glassware, because it should not be used.
- Do not pick up broken glass with your bare hands. Place broken glass in a specially designated disposal container.

Waste Disposal

- Clean and decontaminate all work surfaces and personal protective equipment as directed by your instructor.
- Dispose of all broken glass, contaminated sharp objects, and other contaminated materials (biological and chemical) in special containers as directed by your instructor.

Safety In The Physics Laboratory

Systematic, careful lab work is an essential part of any science program because lab work is the key to progress in science. In this class, you will practice some of the same fundamental laboratory procedures and techniques that experimental physicists use to pursue new knowledge.

The equipment and apparatus you will use involve various safety hazards, just as they do for working physicists. You must be aware of these hazards. Your teacher will guide you in properly using the equipment and carrying out the experiments, but you must also take responsibility for your part in this process. With the active involvement of you and your teacher, these risks can be minimized so that working in the physics laboratory can be a safe, enjoyable process of discovery.

These safety rules always apply in the lab:

1. **Always wear a lab apron and safety goggles.**
   Wear these safety devices whenever you are in the lab, not just when you are working on an experiment.

2. **No contact lenses in the lab.**
   Contact lenses should not be worn during any investigations using chemicals (even if you are wearing goggles). In the event of an accident, chemicals can get behind contact lenses and cause serious damage before the lenses can be removed. If your doctor requires that you wear contact lenses instead of glasses, you should wear eye-cup safety goggles in the lab. Ask your doctor or your teacher how to use this very important and special eye protection.

3. **Personal apparel should be appropriate for laboratory work.**
   On lab days avoid wearing long necklaces, dangling bracelets, bulky jewelry, and bulky or loose-fitting clothing. Loose, flopping, or dangling items may get caught in moving parts, accidentally contact electrical connections, or interfere with the investigation in some potentially hazardous manner. In addition, chemical fumes may react with some jewelry, such as pearl jewelry, and ruin them. Cotton clothing is preferable to clothes made of wool, nylon, or polyester. Tie back long hair. Wear shoes that will protect your feet from chemical spills and falling objects. Do not wear open-toed shoes or sandals or shoes with woven leather straps.

4. **NEVER work alone in the laboratory.**
   Work in the lab only while under the supervision of your teacher. Do not leave equipment unattended while it is in operation.

5. **Only books and notebooks needed for the experiment should be in the lab.**
   Only the lab notebook and perhaps the textbook should be in the lab. Keep other books, backpacks, purses, and similar items in your desk, locker, or designated storage area.

6. **Read the entire experiment before entering the lab.**
   Your teacher will review any applicable safety precautions before the lab. If you are not sure of something, ask your teacher.

7. **Heed all safety symbols and cautions written in the experimental investigations and handouts, posted in the room, and given verbally by your teacher.**
They are provided for a reason: YOUR SAFETY.

8. **Know the proper fire-drill procedures and the locations of fire exits and emergency equipment.**
Make sure you know the procedures to follow in case of a fire or emergency.

9. **If your clothing catches on fire, do not run; WALK to the safety shower, stand under it, and turn it on.**
Call to your teacher while you do this.

10. **Report all accidents to the teacher immediately, no matter how minor.**
In addition, if you get a headache, feel sick to your stomach, or feel dizzy, tell your teacher immediately.

11. **Report all spills to your teacher immediately.**
Call your teacher rather than trying to clean a spill yourself. Your teacher will tell you if it is safe for you to clean up the spill; if not, your teacher will know how the spill should be cleaned up safely.

12. **Student-designed inquiry investigations, such as the Invention Labs in the *Laboratory Experiments* manual, must be approved by the teacher before being attempted by the student.**

13. **DO NOT perform unauthorized experiments or use equipment and apparatus in a manner for which they are not intended.**
Use only materials and equipment listed in the activity equipment list or authorized by your teacher. Steps in a procedure should only be performed as described in the book or lab manual or as approved by your teacher.

14. **Stay alert in the lab, and proceed with caution.**
Be aware of others near you or your equipment when you are about to do something in the lab. If you are not sure of how to proceed, ask your teacher.

15. **Horseplay and fooling around in the lab are very dangerous.**
Laboratory equipment and apparatus are not toys; never play in the lab or use lab time or equipment for anything other than their intended purpose.

16. **Food, beverages, chewing gum, and tobacco products are NEVER permitted in the laboratory.**

17. **NEVER taste chemicals. Do not touch chemicals or allow them to contact areas of bare skin.**

18. **Use extreme CAUTION when working with hot plates or other heating devices.**
Keep your head, hands, hair, and clothing away from the flame or heating area, and turn the devices off when they are not in use. Remember that metal surfaces connected to the heated area will become hot by conduction. Gas burners should only be lit with a spark lighter. Make sure all heating devices and gas valves are turned off before leaving the laboratory. Never leave a hot plate or other heating device unattended when it is in use. Remember that many metal, ceramic, and glass items do not always look hot when they are hot. Allow all items to cool before storing.

19. **Exercise caution when working with electrical equipment.**
Do not use electrical equipment with frayed or twisted wires. Be sure your hands are dry before using electrical equipment. Do not let electrical cords dangle from work stations; dangling cords can cause tripping or electrical shocks.

20. **Keep work areas and apparatus clean and neat.**
Always clean up any clutter made during the course of lab work, rearrange apparatus in an orderly manner, and report any damaged or missing items.

21. **Always thoroughly wash your hands with soap and water at the conclusion of each investigation.**

How to Use this Textbook

Your Roadmap for Success with *Holt Physics*

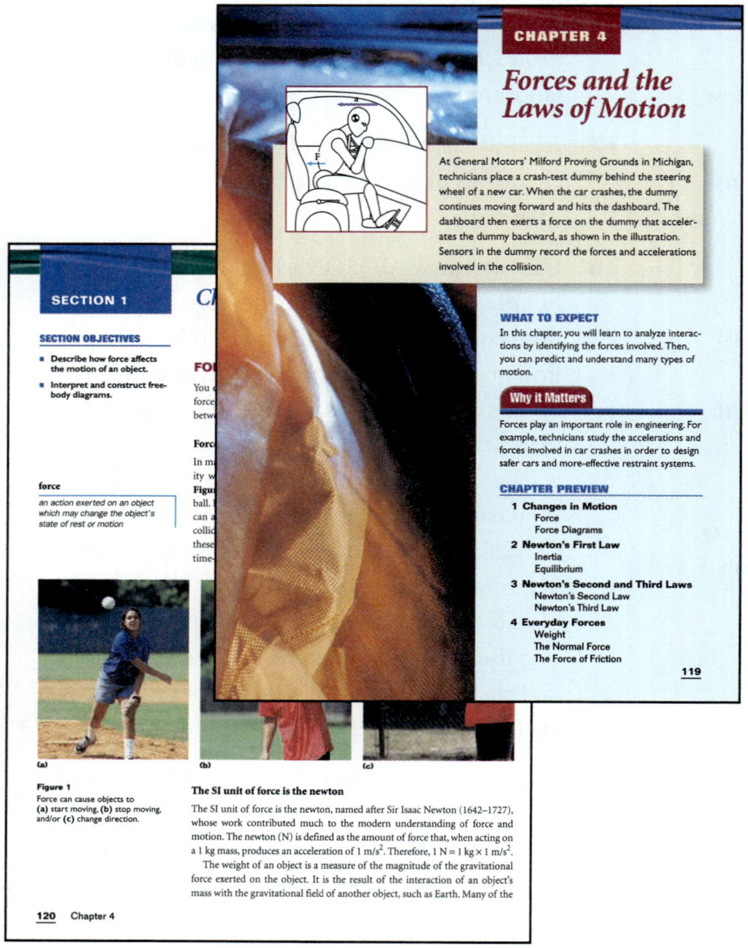

Get Organized

Read **What to Expect** and **Why It Matters** at the beginning of each chapter to understand what you will learn in the chapter and how it applies to real situations and systems.

STUDY TIP Use the **Chapter Preview** outline at the beginning of the chapter to organize your notes on the chapter content in a way that you understand.

Read for Meaning

Read the **Section Objectives** at the beginning of each section because they will tell you what you'll need to learn. Each **Key Term** is highlighted in the text and defined in the margin. After reading each chapter, turn to the **Chapter Highlights** page and review the Key Terms and the **Key Ideas,** which are brief summaries of the chapter's main concepts. You may want to do this even before you read the chapter.

Use the charts at the bottom of the Chapter Highlights page to review important variable symbols and diagram symbols introduced in the chapter.

STUDY TIP If you don't understand a definition, reread the page on which the term is introduced. The surrounding text should help make the definition easier to understand.

Be Resourceful, Use the Web

SCILINKS® NSTA

SciLinks boxes in your textbook take you to resources that you can use for science projects, reports, and research papers. Go to **www.scilinks.org**, and type in the SciLinks code to get information on a topic.

go.hrw.com

Visit go.hrw.com
Find resources and reference materials that go with your textbook. Visit **go.hrw.com**, and enter the keywords found in your textbook to access the available resources.

Work the Problems

Sample Problems, followed by associated **Practice** problems, build your reasoning and problem-solving skills by guiding you through explicit example problems.

Prepare for Tests

Section Reviews and **Chapter Reviews** test your knowledge of the main points of the chapter. Critical Thinking items challenge you to think about the material in different ways and in greater depth. The **Standardized Test Prep** that is located after each Chapter Review helps you sharpen your test-taking abilities.

STUDY TIP Reread the Section Objectives and Chapter Highlights when studying for a test to be sure you know the material.

Use the Appendix

Your **Appendix** contains a variety of resources designed to enhance your learning experience. A **Mathematical Review** sharpens your math skills. The appendices **Symbols, Equations, SI Units,** and **Useful Tables** summarize essential problem-solving information. **Additional Problems** provides more practice in math and problem-solving skills. **Advanced Topics** allows you to delve deeper into areas of physics that lie beyond material presented in the chapters.

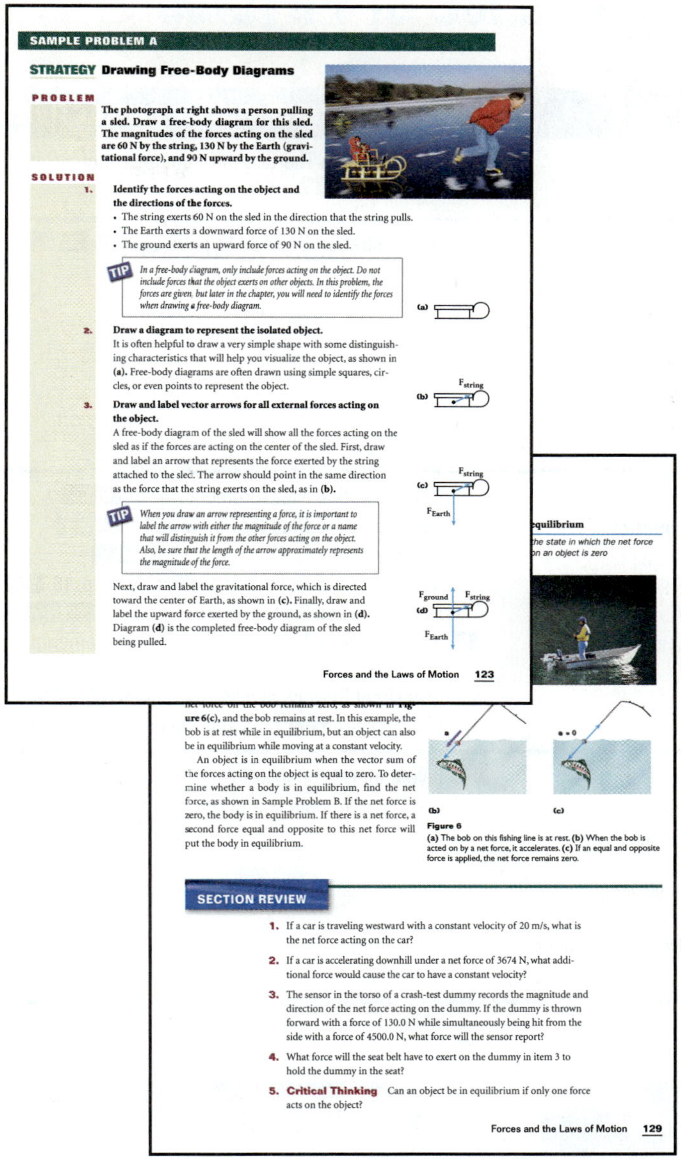

SAMPLE PROBLEM A

STRATEGY Drawing Free-Body Diagrams

PROBLEM

The photograph at right shows a person pulling a sled. Draw a free-body diagram for this sled. The magnitudes of the forces acting on the sled are 60 N by the string, 130 N by the Earth (gravitational force), and 90 N upward by the ground.

SOLUTION

1. **Identify the forces acting on the object and the directions of the forces.**
   - The string exerts 60 N on the sled in the direction that the string pulls.
   - The Earth exerts a downward force of 130 N on the sled.
   - The ground exerts an upward force of 90 N on the sled.

TIP *In a free-body diagram, only include forces acting on the object. Do not include forces that the object exerts on other objects. In this problem, the forces are given, but later in the chapter, you will need to identify the forces when drawing a free-body diagram.*

2. **Draw a diagram to represent the isolated object.**
   It is often helpful to draw a very simple shape with some distinguishing characteristics that will help you visualize the object, as shown in **(a).** Free-body diagrams are often drawn using simple squares, circles, or even points to represent the object.

3. **Draw and label vector arrows for all external forces acting on the object.**
   A free-body diagram of the sled will show all the forces acting on the sled as if the forces are acting on the center of the sled. First, draw and label an arrow that represents the force exerted by the string attached to the sled. The arrow should point in the same direction as the force that the string exerts on the sled, as in **(b).**

TIP *When you draw an arrow representing a force, it is important to label the arrow with either the magnitude of the force or a name that will distinguish it from the other forces acting on the object. Also, be sure that the length of the arrow approximately represents the magnitude of the force.*

Next, draw and label the gravitational force, which is directed toward the center of Earth, as shown in **(c).** Finally, draw and label the upward force exerted by the ground, as shown in **(d).** Diagram **(d)** is the completed free-body diagram of the sled being pulled.

Forces and the Laws of Motion 123

equilibrium
the state in which the net force on an object is zero

ure 6(c), and the bob remains at rest. In this example, the bob is at rest while in equilibrium, but an object can also be in equilibrium while moving at a constant velocity.

An object is in equilibrium when the vector sum of the forces acting on the object is equal to zero. To determine whether a body is in equilibrium, find the net force, as shown in Sample Problem B. If the net force is zero, the body is in equilibrium. If there is a net force, a second force equal and opposite to this net force will put the body in equilibrium.

(b) (c)

Figure 6
(a) The bob on this fishing line is at rest. **(b)** When the bob is acted on by a net force, it accelerates. **(c)** If an equal and opposite force is applied, the net force remains zero.

SECTION REVIEW

1. If a car is traveling westward with a constant velocity of 20 m/s, what is the net force acting on the car?
2. If a car is accelerating downhill under a net force of 3674 N, what additional force would cause the car to have a constant velocity?
3. The sensor in the torso of a crash-test dummy records the magnitude and direction of the net force acting on the dummy. If the dummy is thrown forward with a force of 130.0 N while simultaneously being hit from the side with a force of 4500.0 N, what force will the sensor report?
4. What force will the seat belt have to exert on the dummy in item 3 to hold the dummy in the seat?
5. **Critical Thinking** Can an object be in equilibrium if only one force acts on the object?

Forces and the Laws of Motion 129

Visit Holt Online Learning

If your teacher gives you a special password to log onto the **Holt Online Learning** site, you'll find your complete textbook on the Web. In addition, you'll find some great learning tools and practice quizzes. You'll be able to see how well you know the material from your textbook

CHAPTER 1

The Science of Physics

Planning Guide

Compression Guide

To shorten instruction because of time limitations, omit the opener and abbreviate the review.

OBJECTIVES	LABS, DEMONSTRATIONS, AND ACTIVITIES	TECHNOLOGY RESOURCES
PACING • 45 min pp. 2–3 **Chapter Opener**	ANC **Discovery Lab** The Circumference-Diameter Ratio of a Circle* ◆ BASIC	CD **Visual Concepts,** Chapter 1 BASIC
PACING • 45 min pp. 4–9 **Section 1 What Is Physics?** • Identify activities and fields that involve the major areas within physics. • Describe the processes of the scientific method. • Describe the role of models and diagrams in physics.	TE **Demonstration** Galileo's Hypothesis, p. 8 BASIC	OSP **Lesson Plans** EXT **Integrating Biology** Serendipity and Science BASIC TR 1 Physics and Automobiles TR 2 Physics Analysis of a Thrown Basketball TR 3 Galileo's Thought Experiment TR 1A Areas Within Physics TR 2A Activities of the Scientific Method
PACING • 135 min pp. 10–20 **Section 2 Measurements in Experiments** • List basic SI units and the quantities they describe. • Convert measurements into scientific notation. • Distinguish between *accuracy* and *precision.* • Use significant figures in measurements and calculations.	TE **Demonstration** Measurements, p. 10 BASIC TE **Demonstration** Standard Units, p. 11 GENERAL SE **Quick Lab** Metric Prefixes, p. 12 GENERAL TE **Demonstration** Accuracy and Precision, p. 16 GENERAL TE **Demonstration** Significant Figures, p. 16 GENERAL SE **Skills Practice Lab** Physics and Measurement, pp. 34–37 ◆ GENERAL ANC **Datasheet** Physics and Measurement* GENERAL ANC **CBL™ Experiment** Time and Measurement* ◆ GENERAL	OSP **Lesson Plans** TR 4 Estimation in Measurement TR 3A SI Standards TR 4A Metric Prefixes for Powers of Ten TR 5A Determining Whether Zeros Are Significant Figures TR 6A Rules for Calculating with Significant Figures TR 7A Rules for Rounding
PACING • 45 min pp. 21–25 **Section 3 The Language of Physics** • Interpret data in tables and graphs, and recognize equations that summarize data. • Distinguish between conventions for abbreviating units and quantities. • Use dimensional analysis to check the validity of equations. • Perform order-of-magnitude calculations.	ANC **Invention Lab** Bubble Solutions* ◆ ADVANCED ANC **CBL™ Experiment** Graph Matching* ◆ ADVANCED	OSP **Lesson Plans** EXT **Integrating Chemistry** Dependent and Independent Variables BASIC TR 8A Data from Dropped-Ball Experiment TR 9A Graph of Data from Dropped-Ball Experiment TR 10A Abbreviations for Variables and Units

PACING • 90 min

CHAPTER REVIEW, ASSESSMENT, AND STANDARDIZED TEST PREPARATION

- SE **Chapter Highlights,** p. 26
- SE **Chapter Review,** pp. 27–31
- SE **Graphing Calculator Practice,** p. 30 GENERAL
- SE **Alternative Assessment,** p. 31 ADVANCED
- SE **Standardized Test Prep,** pp. 32–33 GENERAL
- SE **Appendix I: Additional Problems,** p. 880
- ANC **Study Guide Worksheet** Mixed Review* GENERAL
- ANC **Chapter Test A*** GENERAL
- ANC **Chapter Test B*** ADVANCED
- OSP **Test Generator**

Online and Technology Resources

go.hrw.com Holt Online Learning

Visit go.hrw.com to find a variety of online resources. To access this chapter's extensions, enter the keyword **HF6SOPXT** and click the "go" button. Click **Holt Online Learning** for an online edition of this textbook, and other interactive resources.

HOLT Teacher's One-Stop Planner®

This DVD package includes:

- Holt Calendar Planner
- Customizable Lesson Plans
- Editable Worksheets
- ExamView® Version 6 Assessment Suite
- Interactive Teacher's Edition
- Holt PuzzlePro®
- Holt PowerNotes® Presentations
- MindPoint® Quiz Show

KEY

SE	Student Edition	OSP	One-Stop Planner	EXT	Online Extension
TE	Teacher Edition	CD	CD or CD-ROM	*	Also on One-Stop Planner
ANC	Ancillary Worksheet	TR	Teaching Transparencies	◆	Requires advance prep

SKILLS DEVELOPMENT RESOURCES	REVIEW AND ASSESSMENT	CORRELATIONS
		National Science Education Standards
	SE **Section Review,** p. 9 GENERAL ANC **Study Guide Worksheet** Section 1* GENERAL ANC **Quiz** Section 1* BASIC	UCP 2 SAI 1, 2 HNS 1, 2
SE **Sample Set A** Metric Prefixes, p. 15 BASIC TE **Classroom Practice,** p. 15 BASIC ANC **Problem Workbook** Sample Set A* BASIC OSP **Problem Bank** Sample Set A BASIC	SE **Section Review,** p. 20 GENERAL ANC **Study Guide Worksheet** Section 2* GENERAL ANC **Quiz** Section 2* BASIC	UCP 1, 2, 3 SAI 1, 2 HNS 2, 3
	SE **Section Review,** p. 25 GENERAL ANC **Study Guide Worksheet** Section 3* GENERAL ANC **Quiz** Section 3* BASIC	UCP 1, 2, 3 SAI 1, 2 HNS 2

SCILINKS. NSTA

www.scilinks.org

Maintained by the **National Science Teachers Association.**

Topic: Models in Physics
SciLinks Code: HF60977

Topic: SI Units
SciLinks Code: HF61390

Topic: Graphing
SciLinks Code: HF60686

Topic: Orders of Magnitude
SciLinks Code: HF61074

Classroom CD-ROMs

- Guided Reading Audio Program
- Student One Stop
- Virtual Investigations
- Visual Concepts

Holt Lab Generator CD-ROM

Search for any lab by topic, standard, difficulty level, or time. Edit any lab to fit your needs, or create your own labs. Use the Lab Materials QuickList software to customize your lab materials list.

CHAPTER 1
Overview

Section 1 describes the nature of physics and its related fields and activities, introduces the scientific method of inquiry, and discusses the role of models in science.

Section 2 introduces the basic SI units, scientific notation, and significant digits, and distinguishes between precision and accuracy.

Section 3 presents various ways of summarizing data, including tables, graphs, and equations; uses dimensional analysis to check the validity of expressions; and introduces estimation procedures.

About the Illustration

The technology shown in this photograph is known as *motion capture.* In addition to sports research, this technology is used to create synthetic characters in movies based on the motion of human actors. Examples include Jar Jar Binks from *Star Wars I* and Gollum from the *Lord of the Rings* trilogy.

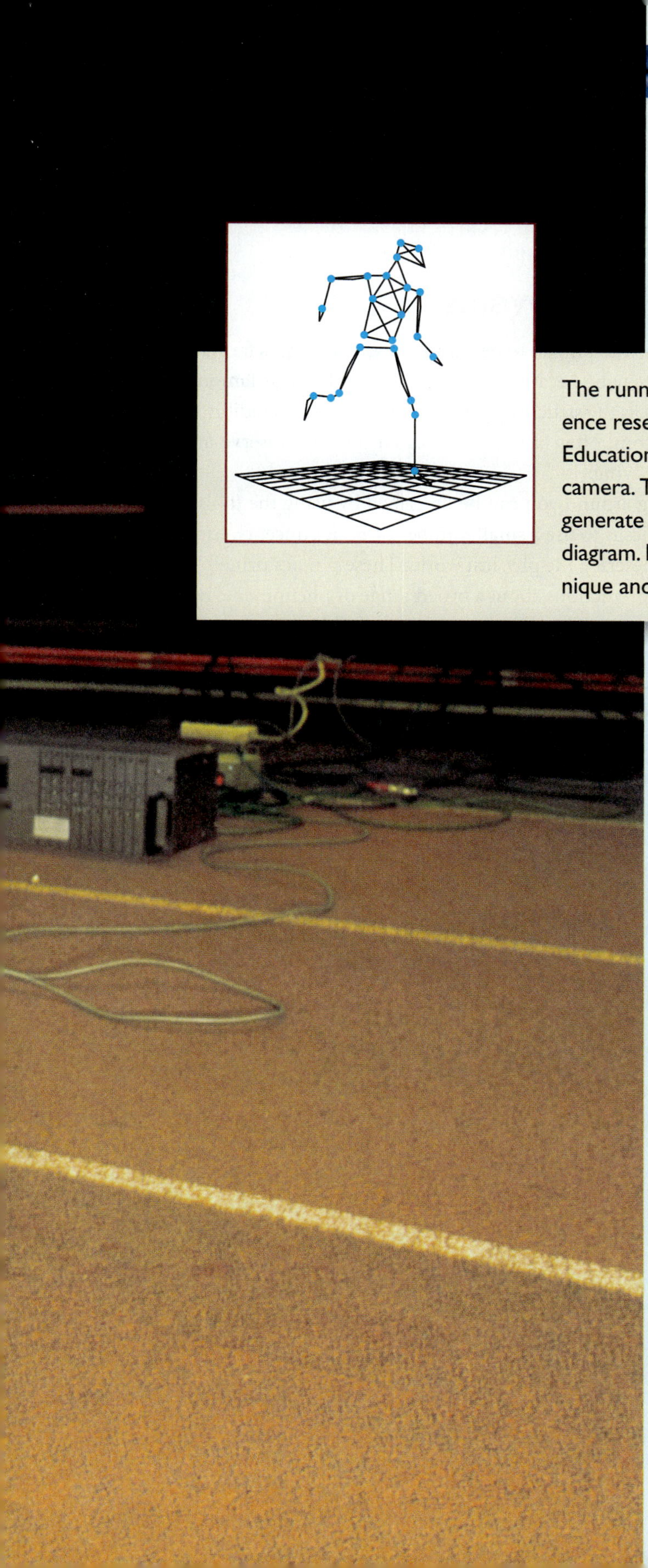

The runner in this photograph is participating in sports science research at the National Institute of Sport and Physical Education in France. The athlete is being filmed by a video camera. The white reflective patches enable researchers to generate a computer model from the video, similar to the diagram. Researchers use the model to analyze his technique and to help him improve his performance.

CHAPTER 1

The Science of Physics

WHAT TO EXPECT

In this chapter, you will learn about the branches of physics, the scientific method, and the use of models in physics. You will also learn some useful tools for working with measurements and data.

Why it Matters

Physics develops powerful models that can be used to describe many things in the physical world, including the movements of an athlete in training.

CHAPTER PREVIEW

Tapping Prior Knowledge

Items to Probe

✔ Ability to work with decimal numbers and powers of ten, and to draw, read, and interpret graphs: Ask students to calculate $(3 \times 10^{-5})(5 \times 10^{8})$.

SECTION 1
General Level

Teaching Tip — GENERAL

Exposure to science, both formal and informal, shapes students' current views of physics. Students should be encouraged to express their perceptions and attitudes in a brainstorming session. Students need to realize that applied science, which is aimed at creating technology, interacts with basic science, which strives to develop fundamental ideas to explain natural phenomena. Point out that physics is a very broad field of study that presents an organized way of modeling and interpreting nature. The different areas of basic science specialize in interpreting different aspects of nature (living things, materials, celestial objects, and so forth). Sometimes, the boundaries between these areas are not clear, but the fundamental ideas of physics underlie all basic and applied sciences.

SECTION 1

What Is Physics?

SECTION OBJECTIVES

- **Identify activities and fields that involve the major areas within physics.**
- **Describe the processes of the scientific method.**
- **Describe the role of models and diagrams in physics.**

THE TOPICS OF PHYSICS

Many people consider physics to be a difficult science that is far removed from their lives. This may be because many of the world's most famous physicists study topics such as the structure of the universe or the incredibly small particles within an atom, often using complicated tools to observe and measure what they are studying.

But everything around you can be described by using the tools of physics. The goal of physics is to use a small number of basic concepts, equations, and assumptions to describe the physical world. These physics principles can then be used to make predictions about a broad range of phenomena. For example, the same physics principles that are used to describe the interaction between two planets can be used to describe the motion of a satellite orbiting Earth.

Many physicists study the laws of nature simply to satisfy their curiosity about the world we live in. Learning the laws of physics can be rewarding just for its own sake. Also, many of the inventions, appliances, tools, and buildings we live with today are made possible by the application of physics principles. Physics discoveries often turn out to have unexpected practical applications, and advances in technology can in turn lead to new physics discoveries. **Figure 1** indicates how the areas of physics apply to building and operating a car.

Figure 1
Without knowledge of many of the areas of physics, making cars would be impossible.

Physics is everywhere

We are surrounded by principles of physics in our everyday lives. In fact, most people know much more about physics than they realize. For example, when you buy a carton of ice cream at the store and put it in the freezer at home, you do so because from past experience you know enough about the laws of physics to know that the ice cream will melt if you leave it on the counter.

Any problem that deals with temperature, size, motion, position, shape, or color involves physics. Physicists categorize the topics they study in a number of different ways. **Table 1** shows some of the major areas of physics that will be described in this book.

People who design, build, and operate sailboats, such as the ones shown in **Figure 2,** need a working knowledge of the principles of physics. Designers figure out the best shape for the boat's hull so that it remains stable and floating yet quick-moving and maneuverable. This design requires knowledge of the physics of fluids. Determining the most efficient shapes for the sails and how to arrange them requires an understanding of the science of motion and its causes. Balancing loads in the construction of a sailboat requires knowledge of mechanics. Some of the same physics principles can also explain how the keel keeps the boat moving in one direction even when the wind is from a slightly different direction.

Figure 2
Sailboat designers rely on knowledge from many branches of physics.

Table 1 Areas Within Physics

Name	Subjects	Examples
Mechanics	motion and its causes, interactions between objects	falling objects, friction, weight, spinning objects
Thermodynamics	heat and temperature	melting and freezing processes, engines, refrigerators
Vibrations and wave phenomena	specific types of repetitive motions	springs, pendulums, sound
Optics	light	mirrors, lenses, color, astronomy
Electromagnetism	electricity, magnetism, and light	electrical charge, circuitry, permanent magnets, electromagnets
Relativity	particles moving at any speed, including very high speeds	particle collisions, particle accelerators, nuclear energy
Quantum mechanics	behavior of submicroscopic particles	the atom and its parts

Teaching Tip — GENERAL

Ask students to list the careers that they are interested in pursuing. Point out how a physics background may contribute to their effectiveness in those careers. Above all, a good understanding of science makes it possible to critically examine scientific theories and make educated decisions about science-related issues facing our society.

SECTION 1

Key Models and Analogies — GENERAL

An investigation provides a good analogy to help students see the scientific method at work. Have students link the procedures of an investigation of a car accident with the stages of the scientific method.

- Observation/collection of data: The investigator examines the crime scene and fills out a report.
- Hypotheses: The investigator imagines several likely scenarios that might have caused the accident. Maybe the driver was intoxicated, fell asleep, or was speeding; maybe the car had mechanical failure; or maybe weather conditions affected the car's traction or the driver's ability to see the road well.
- Experiments/tests: The investigator might order a blood-alcohol-level test, check the car parts, test-drive the car in different weather conditions, or try to reproduce the skid marks left by the car.
- Interpret/revise hypothesis: The investigator must reexamine evidence and possibly revise his hypothesis. The evidence may be inconclusive.
- Conclusions: The investigator goes to court, reexamines the evidence, and defends his theory of how the accident occurred.

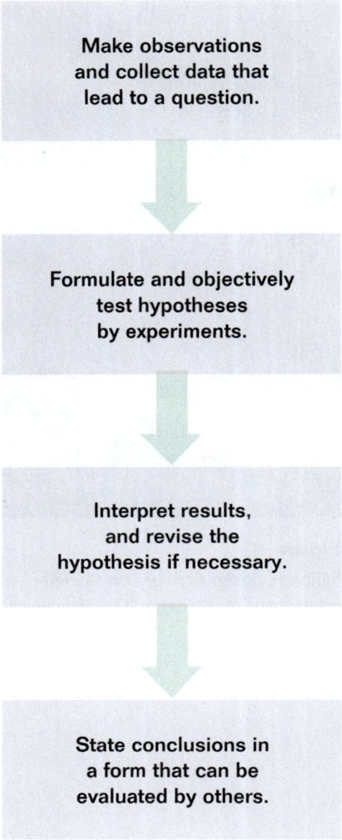

Figure 3
Physics, like all other sciences, is based on the scientific method.

THE SCIENTIFIC METHOD

When scientists look at the world, they see a network of rules and relationships that determine what will happen in a given situation. Everything you will study in this course was learned because someone looked out at the world and asked questions about how things work.

There is no single procedure that scientists follow in their work. However, there are certain steps common to all good scientific investigations. These steps, called the *scientific method,* are summarized in **Figure 3.** This simple chart is easy to understand; but, in reality, most scientific work is not so easily separated. Sometimes, exploratory experiments are performed as a part of the first step in order to generate observations that can lead to a focused question. A revised hypothesis may require more experiments.

Physics uses models that describe phenomena

Although the physical world is very complex, physicists often use **models** to explain the most fundamental features of various phenomena. Physics has developed powerful models that have been very successful in describing nature. Many of the models currently used in physics are mathematical models. Simple models are usually developed first. It is often easier to study and model parts of a system or phenomenon one at a time. These simple models can then be synthesized into more-comprehensive models.

When developing a model, physicists must decide which parts of the phenomenon are relevant and which parts can be disregarded. For example, let's say you wish to study the motion of the ball shown in **Figure 4.** Many observations

model

a pattern, plan, representation, or description designed to show the structure or workings of an object, system, or concept

Figure 4
This basketball game involves great complexity.

can be made about the situation, including the ball's surroundings, size, spin, weight, color, time in the air, speed, and sound when hitting the ground. The first step toward simplifying this complicated situation is to decide what to study, that is, to define the **system.** Typically, a single object and the items that immediately affect it are the focus of attention. For instance, suppose you decide to study the ball's motion in the air (before it potentially reaches any of the other players), as shown in **Figure 5(a).** To study this situation, you can eliminate everything except information that affects the ball's motion.

system

a set of particles or interacting components considered to be a distinct physical entity for the purpose of study

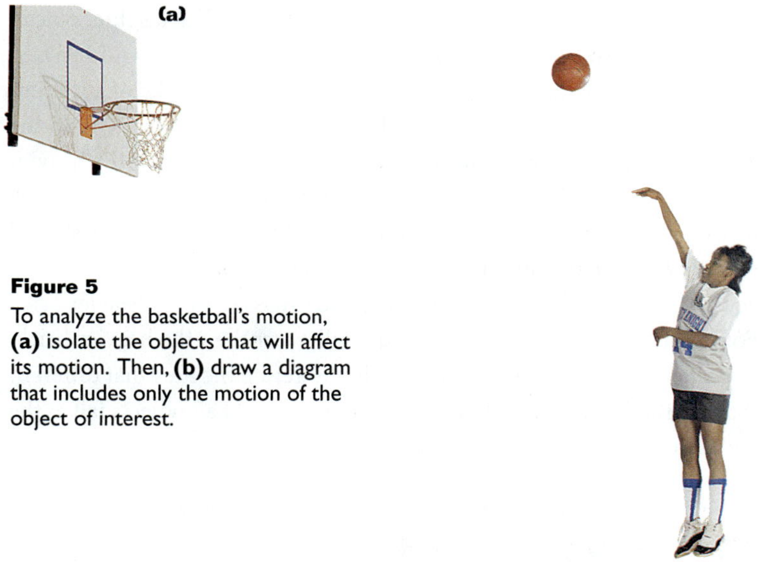

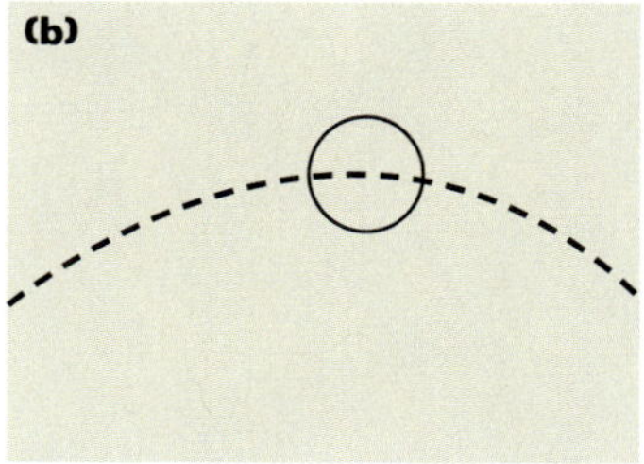

Figure 5
To analyze the basketball's motion, **(a)** isolate the objects that will affect its motion. Then, **(b)** draw a diagram that includes only the motion of the object of interest.

You can disregard characteristics of the ball that have little or no effect on its motion, such as the ball's color. In some studies of motion, even the ball's spin and size are disregarded, and the change in the ball's position will be the only quantity investigated, as shown in **Figure 5(b).**

In effect, the physicist studies the motion of a ball by first creating a simple model of the ball and its motion. Unlike the real ball, the model object is isolated; it has no color, spin, or size, and it makes no noise on impact. Frequently, a model can be summarized with a diagram, like the one in **Figure 5(b).** Another way to summarize these models is to build a computer simulation or small-scale replica of the situation.

Without models to simplify matters, situations such as building a car or sailing a boat would be too complex to study. For instance, analyzing the motion of a sailboat is made easier by imagining that the push on the boat from the wind is steady and consistent. The boat is also treated as an object with a certain mass being pushed through the water. In other words, the color of the boat, the model of the boat, and the details of its shape are left out of the analysis. Furthermore, the water the boat moves through is treated as if it were a perfectly smooth-flowing liquid with no internal friction. In spite of these simplifications, the analysis can still make useful predictions of how the sailboat will move.

extension

Integrating Biology

Visit go.hrw.com for the activity "Serendipity and Science."

Keyword HF6SOPX

Visual Strategy GENERAL

Figure 5

Ask students to compare the photograph in **Figure 5(a)** with the photograph in **Figure 4.** Point out that the details relevant to the study of the ball's motion are the direction that the ball was thrown; the forces on the ball; the ball's spin; the ball's location at every instant; the ball's shape, size, and mass; and the air surrounding the ball.

Q Which of these details are ignored in the diagram of the ball's motion in **Figure 5(b)** for the purpose of simplification?

A ***Size, spin, and air resistance normally affect the ball's flight. However, as a first approximation, the study focuses on the trajectory of a model ball and assumes that the effects of air resistance and spin are negligible.***

Demonstration

Galileo's Hypothesis BASIC

Purpose Illustrate the arguments for and against Galileo's thought experiment.

Materials pennies, tape, coffee filters, deep and clear container, water

Procedure Stack 10 pennies and tape them together. Hold one penny in one hand, and hold the stack of pennies in the other hand; drop them simultaneously. Tell students to listen carefully and decide which object hits the ground first. (It should be difficult to detect any difference in the falling time.)

Hold one coffee filter in one hand and a stack of 10 filters in the other hand; drop them simultaneously. This time, it should be easy to see that 10 filters fall faster than one. How could Galileo defend his hypothesis in the face of such evidence? *(He identified the role of air resistance.)*

Fill the clear container with water, and drop the single penny and the 10-penny stack into the water simultaneously. This time, the single penny will land on the bottom last. Have students use these experiments to defend or criticize the argument that heavy bodies fall faster than lighter ones.

hypothesis

an explanation that is based on prior scientific research or observations and that can be tested

Models can help build hypotheses

A scientific **hypothesis** is a reasonable explanation for observations—one that can be tested with additional experiments. The process of simplifying and modeling a situation can help you determine the relevant variables and identify a hypothesis for testing.

Consider the example of Galileo's "thought experiment," in which he modeled the behavior of falling objects in order to develop a hypothesis about how objects fell. At the time Galileo published his work on falling objects, in 1638, scientists believed that a heavy object would fall faster than a lighter object.

Galileo imagined two objects of different masses tied together and released at the same time from the same height, such as the two bricks of different masses shown in **Figure 6.** Suppose that the heavier brick falls faster than the lighter brick when they are separate, as in **(a).** When tied together, the heavier brick will speed up the fall of the lighter brick somewhat, and the lighter brick will slow the fall of the heavier brick somewhat. Thus, the tied bricks should fall at a rate *in between* that of either brick alone, as in **(b).**

However, the two bricks together have a greater mass than the heavier brick alone. For this reason, the tied bricks should fall *faster* than the heavier brick, as in **(c).** Galileo used this logical contradiction to refute the idea that different masses fall at different rates. He hypothesized instead that all objects fall at the same rate in the absence of air resistance, as in **(d).**

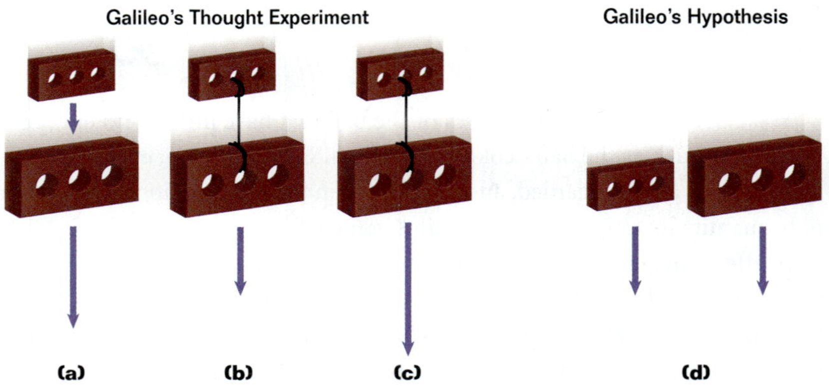

Figure 6
If heavier objects fell faster than slower ones, would two bricks of different masses tied together fall slower **(b)** or faster **(c)** than the heavy brick alone **(a)**? Because of this contradiction, Galileo hypothesized instead that all objects fall at the same rate, as in **(d).**

Models help guide experimental design

Galileo performed many experiments to test his hypothesis. To be certain he was observing differences due to weight, he kept all other variables the same: the objects he tested had the same size (but different weights) and were measured falling from the same point.

The measuring devices at that time were not precise enough to measure the motion of objects falling in air. So, Galileo used the motion of a ball rolling down a ramp as a model of the motion of a falling ball. The steeper the ramp, the closer the model came to representing a falling object. These ramp experiments provided data that matched the predictions Galileo made in his hypothesis.

SCILINKS®
www.scilinks.org
Topic: Models in Physics
SciLinks Code: HF60977

Like Galileo's hypothesis, any hypothesis must be tested in a **controlled experiment.** In an experiment to test a hypothesis, you must change one variable at a time to determine what influences the phenomenon you are observing. Galileo performed a series of experiments using balls of different weights on one ramp before determining the time they took to roll down a steeper ramp.

controlled experiment

an experiment that tests only one factor at a time by using a comparison of a control group with an experimental group

The best physics models can make predictions in new situations

Until the invention of the air pump, it was not possible to perform direct tests of Galileo's model by observing objects falling in the absence of air resistance. But even though it was not completely testable, Galileo's model was used to make reasonably accurate predictions about the motion of many objects, from raindrops to boulders (even though they all experience air resistance).

Even if some experiments produce results that support a certain model, at any time another experiment may produce results that do not support the model. When this occurs, scientists repeat the experiment until they are sure that the results are not in error. If the unexpected results are confirmed, the model must be abandoned or revised. That is why the last step of the scientific method is so important. A conclusion is valid only if it can be verified by other people.

Did you know?

In addition to conducting experiments to test their hypotheses, scientists also research the work of other scientists. The steps of this type of research include

- identifying reliable sources
- searching the sources to find references
- checking for opposing views
- documenting sources
- presenting findings to other scientists for review and discussion

SECTION REVIEW

1. Name the major areas of physics.
2. Identify the area of physics that is most relevant to each of the following situations. Explain your reasoning.
   a. a high school football game
   b. food preparation for the prom
   c. playing in the school band
   d. lightning in a thunderstorm
   e. wearing a pair of sunglasses outside in the sun
3. What are the activities involved in the scientific method?
4. Give two examples of ways that physicists model the physical world.
5. **Critical Thinking** Identify the area of physics involved in each of the following tests of a lightweight metal alloy proposed for use in sailboat hulls:
   a. testing the effects of a collision on the alloy
   b. testing the effects of extreme heat and cold on the alloy
   c. testing whether the alloy can affect a magnetic compass needle

Misconception Alert — GENERAL

Point out that even when the results of an inquiry appear convincing, the conclusions must be formulated carefully by specifying the circumstances under which the experiment was performed and the reasoning that led to accepting or refuting the evidence. Ask students which variable was manipulated when vacuum pumps made it possible to test Galileo's hypothesis *(air resistance).*

SECTION REVIEW ANSWERS

1. mechanics, thermodynamics, vibrations and wave phenomena, optics, electromagnetism, relativity, and quantum mechanics
2. **a.** mechanics; because the ball is a moving, spinning object
   b. thermodynamics; because cooking concerns changes in temperature
   c. vibrations and wave phenomena; because music is a type of sound
   d. electromagnetism; because lightning is a form of electricity
   e. optics; because the lenses decrease the light intensity that reaches your eye
3. observing and collecting data, formulating and testing hypotheses, interpreting results and revising the hypothesis, stating conclusions
4. Answers could include diagrams, equations, and computer simulations.
5. **a.** mechanics
   b. thermodynamics
   c. electromagnetism

SECTION 2
General Level

Demonstration

Measurements — BASIC

Purpose Give students an overview of measurement concepts.

Materials assortment of measuring instruments, including rulers, measuring tapes, various clocks, various metronomes, various bathroom scales, pan balance with standard masses, electronic balance, beam balance, thermometers of various sizes and ranges, ammeters, protractors, graduated cylinders, and measuring cups

Procedure Display the various instruments. Ask students to describe what instrument they would use to measure properties of objects found in the classroom. Let students walk around the exhibit and take notes about the different methods and units of measure. Whenever possible, have students perform measurements and report their results in correct units.

SECTION 2

Measurements in Experiments

SECTION OBJECTIVES

- **List basic SI units and the quantities they describe.**
- **Convert measurements into scientific notation.**
- **Distinguish between *accuracy* and *precision*.**
- **Use significant figures in measurements and calculations.**

NUMBERS AS MEASUREMENTS

Physicists perform experiments to test hypotheses about how changing one variable in a situation affects another variable. An accurate analysis of such experiments requires numerical measurements.

Numerical measurements are different from the numbers used in a mathematics class. In mathematics, a number like 7 can stand alone and be used in equations. In science, measurements are more than just a number. For example, a measurement reported as 7 leads to several questions. What physical quantity is being measured—length, mass, time, or something else? If it is length that is being measured, what units were used for the measurement—meters, feet, inches, miles, or light-years?

The description of *what kind* of physical quantity is represented by a certain measurement is called *dimension.* In the next several chapters, you will encounter three basic dimensions: length, mass, and time. Many other measurements can be expressed in terms of these three dimensions. For example, physical quantities such as force, velocity, energy, volume, and acceleration can all be described as combinations of length, mass, and time. In later chapters, we will need to add two other dimensions to our list, for temperature and for electric current.

The description of *how much* of a physical quantity is represented by a certain numerical measurement depends on the *units* with which the quantity is measured. For example, small distances are more easily measured in millimeters than in kilometers or light-years.

SI is the standard measurement system for science

When scientists do research, they must communicate the results of their experiments with each other and agree on a system of units for their measurements. In 1960, an international committee agreed on a system of standards, such as the standard shown in **Figure 7.** They also agreed on designations for the fundamental quantities needed for measurements. This system of units is called the *Système International d'Unités* (SI). In SI, there are only seven base units. Each base unit describes a single dimension, such as length, mass, or time.

Figure 7
The kilogram is the only SI unit that is defined by a material object. The platinum-iridium cylinder shown here is the primary kilogram standard for the United States.

Table 2 SI Standards

Unit	Original standard	Current standard
meter (length)	$\frac{1}{10\,000\,000}$ distance from equator to North Pole	the distance traveled by light in a vacuum in $3.33564095 \times 10^{-9}$ s
kilogram (mass)	mass of 0.001 cubic meters of water	the mass of a specific platinum-iridium alloy cylinder
second (time)	$\left(\frac{1}{60}\right)\left(\frac{1}{60}\right)\left(\frac{1}{24}\right) =$ 0.000 011 574 average solar days	9 192 631 770 times the period of a radio wave emitted from a cesium-133 atom

The base units of length, mass, and time are the meter, kilogram, and second, respectively. In most measurements, these units will be abbreviated as m, kg, and s, respectively.

These units are defined by the standards described in **Table 2** and are reproduced so that every meterstick, kilogram mass, and clock in the world is calibrated to give consistent results. We will use SI units throughout this book because they are almost universally accepted in science and industry.

Not every observation can be described using one of these units, but the units can be combined to form derived units. Derived units are formed by combining the seven base units with multiplication or division. For example, speeds are typically expressed in units of meters per second (m/s).

In other cases, it may appear that a new unit that is not one of the base units is being introduced, but often these new units merely serve as shorthand ways to refer to combinations of units. For example, forces and weights are typically measured in units of newtons (N), but a newton is defined as being exactly equivalent to one kilogram multiplied by meters per second squared ($1 \text{kg} \bullet \text{m/s}^2$). Derived units, such as newtons, will be explained throughout this book as they are introduced.

SI uses prefixes to accommodate extremes

Physics is a science that describes a broad range of topics and requires a wide range of measurements, from very large to very small. For example, distance measurements can range from the distances between stars (about 100 000 000 000 000 000 m) to the distances between atoms in a solid (0.000 000 001 m). Because these numbers can be extremely difficult to read and write, they are often expressed in powers of 10, such as 1×10^{17} m or 1×10^{-9} m.

Another approach commonly used in SI is to combine the units with prefixes that symbolize certain powers of 10, as illustrated in **Figure 8.**

SCILINKS.
www.scilinks.org
Topic: SI Units
SciLinks Code: HF61390

Did you know?

NIST-F1, an atomic clock at the National Institute of Standards and Technology in Colorado, is one of the most accurate timing devices in the world. NIST-F1 is so accurate that it will not gain or lose a second in nearly 20 million years. As a public service, the Institute broadcasts the time given by NIST-F1 through the Internet, radio stations WWV and WWVB, and satellite signals.

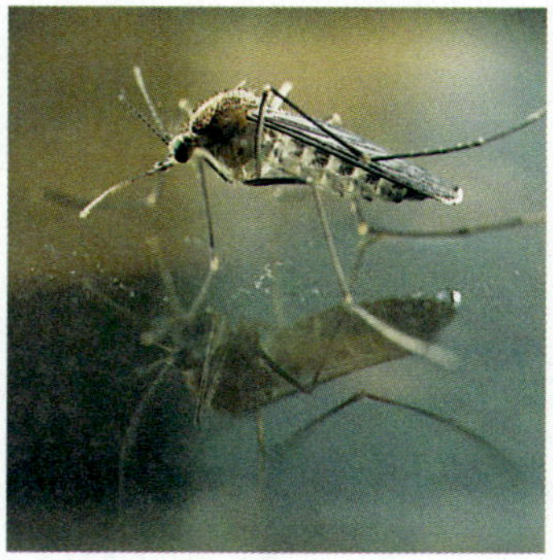

Figure 8
The mass of this mosquito can be expressed several different ways: 1×10^{-5} kg, 0.01 g, or 10 mg.

Demonstration

Standard Units — GENERAL

Purpose Show that standard units of measurement are based on arbitrary convention.

Materials ribbon strips about 1 cm wide but of unequal lengths

Procedure Explain that the class will define its own standard unit of length with a ribbon strip. Have each group measure the length of the classroom in "ribbon units." Each group will have different results (10.5 ribbons, 9.75 ribbons, etc.). Have the groups compare their ribbons and discover that the lengths were not exactly equal. Let students decide which ribbon should be the standard. Ask them to compare their reasons for selecting a particular ribbon. There is no particular reason to prefer any one. Mark two points about 80 cm apart on a ribbon. Tape the ribbon on the chalkboard, and tell the class to calibrate their ribbon units against this new standard. Have them repeat their measurements. Results should be about the same number of standard ribbon units. Point out that inches, meters, and miles were defined by similar conventions. Scientific standards of measure are universally established through a similar but more rigorous process.

The Language of Physics

Derived units describe quantities, such as density, that are calculated from measurements in base units.

Teaching Tip — BASIC

In order to make the terminology relevant to students, have them propose examples of quantities they know to be measured in *milli-*, *micro-*, *kilo-*, etc. For example, the frequency of their favorite radio station could be 94.7 MHz, or the dosage of some vitamins may be 300 mg. Ask students to restate these quantities in powers of 10.

Quick Lab

TEACHER'S NOTES

This activity is intended to have students explore the use of prefixes for powers of 10 with metric units.

This lab works best when the papers are folded to fit on the pan of the balance. Students should record their measurements as given by the scale and try to convert them to smaller or larger units using different prefixes listed in **Table 3.** Let them realize that in this case the most convenient way to represent the measurements is to use grams and decimal numbers.

Teaching Tip — GENERAL

The prefix *kilo* stands for 1000, *mega* stands for 1 million, and *giga* stands for 1 billion. Ask students to write their computer's RAM and hard-drive memory in bytes. Ask how many watts a 20-megawatt power station supplies (*20 million*). How many watts does a 300 milliwatt bulb use? (*0.3 W*)

Table 3
Some Prefixes for Powers of 10 Used with Metric Units

Power	Prefix	Abbreviation	Power	Prefix	Abbreviation
10^{-18}	atto-	a	10^{-1}	deci-	d
10^{-15}	femto-	f	10^{1}	deka-	da
10^{-12}	pico-	p	10^{3}	kilo-	k
10^{-9}	nano-	n	10^{6}	mega-	M
10^{-6}	micro-	μ (Greek letter *mu*)	10^{9}	giga-	G
10^{-3}	milli-	m	10^{12}	tera-	T
10^{-2}	centi-	c	10^{15}	peta-	P
			10^{18}	exa-	E

Quick Lab

Metric Prefixes

MATERIALS LIST

- **balance (0.01 g precision or better)**
- **50 sheets of loose-leaf paper**

Record the following measurements (with appropriate units and metric prefixes):

- the mass of a single sheet of paper
- the mass of exactly 10 sheets of paper
- the mass of exactly 50 sheets of paper

Use each of these measurements to determine the mass of a single sheet of paper. How many different ways can you express each of these measurements? Use your results to estimate the mass of one ream (500 sheets) of paper. How many ways can you express this mass? Which is the most practical approach? Give reasons for your answer.

The most common prefixes and their symbols are shown in **Table 3.** For example, the length of a housefly, 5×10^{-3} m, is equivalent to 5 millimeters (mm), and the distance of a satellite 8.25×10^{5} m from Earth's surface can be expressed as 825 kilometers (km). A year, which is about 3.2×10^{7} s, can also be expressed as 32 megaseconds (Ms).

Converting a measurement from its prefix form is easy to do. You can build conversion factors from any equivalent relationship, including those in **Table 3.** Just put the quantity on one side of the equation in the numerator and the quantity on the other side in the denominator, as shown below for the case of the conversion 1 mm = 1×10^{-3} m. Because these two quantities are equal, the following equations are also true:

$$\frac{1 \text{ mm}}{10^{-3} \text{ m}} = 1 \quad \text{and} \quad \frac{10^{-3} \text{ m}}{1 \text{ mm}} = 1$$

Thus, any measurement multiplied by either one of these fractions will be multiplied by 1. The number and the unit will change, but the quantity described by the measurement will stay the same.

To convert measurements, use the conversion factor that will cancel with the units you are given to provide the units you need, as shown in the example below. Typically, the units to which you are converting should be placed in the numerator. It is useful to cross out units that cancel to help keep track of them.

$$\text{Units } \textit{don't} \text{ cancel: } 37.2 \text{ mm} \times \frac{1 \text{ mm}}{10^{-3} \text{ m}} = 3.72 \times 10^{4} \ \frac{\text{mm}^2}{\text{m}}$$

$$\text{Units } \textit{do} \text{ cancel: } 37.2 \ \cancel{\text{mm}} \times \frac{10^{-3} \text{ m}}{1 \ \cancel{\text{mm}}} = 3.72 \times 10^{-2} \text{ m}$$

Why it Matters

The Mars Climate Orbiter Mission

The $125 million Mars Orbiter mission failed because of a miscommunication about units of measurement.

The *Mars Climate Orbiter* was a NASA spacecraft designed to take pictures of the Martian surface, generate daily weather maps, and analyze the Martian atmosphere from an orbit about 80 km (50 mi) above Mars. It was also supposed to relay signals from its companion, the *Mars Polar Lander,* which was scheduled to land near the edge of the southern polar cap of Mars shortly after the orbiter arrived.

The orbiter was launched from Cape Canaveral, Florida, on December 11, 1998. Its thrusters were fired several times along the way to direct it along its path. The orbiter reached Mars nine and a half months later, on September 23, 1999. A signal was sent to the orbiter to fire the thrusters a final time in order to push the spacecraft into orbit around the planet. However, the orbiter did not respond to this final signal. NASA soon determined that the orbiter had passed closer to the planet than intended, as close as 60 km (36 mi). The orbiter most likely overheated because of friction in the Martian atmosphere and then passed beyond the planet into space, fatally damaged.

The *Mars Climate Orbiter* was built by Lockheed Martin in Denver, Colorado, while the mission was run by a NASA flight control team at Jet Propulsion Laboratory in Pasadena, California. Review of the failed mission revealed that engineers at Lockheed Martin sent thrust specifications to the flight control team in English units of pounds of force, while the flight control team assumed that the thrust specifications were in newtons, the SI unit for force. Such a problem normally would be caught by others checking and double-checking specifications, but somehow the error escaped notice until it was too late.

Unfortunately, communication with the *Mars Polar Lander* was also lost as the lander entered the Martian atmosphere on December 3, 1999. The failure of these and other space exploration missions reveals the inherent difficulty in sending complex technology into the distant, harsh, and often unknown conditions in space and on other planets. However, NASA has had many more successes than failures. A later Mars mission, the Exploration Rover mission, successfully placed two rovers named *Spirit* and *Opportunity* on the surface of Mars, where they collected a wide range of data. Among other things, the rovers found convincing evidence that liquid water once flowed on the surface of Mars. Thus, it is possible that Mars supported life sometime in the past.

The *Spirit* and *Opportunity* rovers have explored the surface of Mars with a variety of scientific instruments, including cameras, spectrometers, magnets, and a rock-grinding tool.

Why it Matters

The Mars Climate Orbiter Mission

Although SI units are widely used in science, they are not universal. Scientists in different fields often use different systems of units depending on what is most convenient and appropriate to their areas of study.

The lesson of the failure of the Mars Climate Orbiter mission is not necessarily that all scientists should use a single system of units but that communication between different groups working on a project is essential. Furthermore, any project, especially a very large or complex one, should have a systematic plan for checking and double-checking all plans and calculations, including the consistent use of units.

Extension — GENERAL

Have students determine the distance from their home to school in miles. Then, have them calculate how far they would fall short of the school if they traveled the same number of kilometers along the same route.

Visual Strategy GENERAL

Figure 9

Point out that all measurements could have been converted to cm.

Q What would be the area in cm^2?

A *$(2035\ cm)(1250\ cm) = 2\,540\,000\ cm^2 = 2.54 \times 10^6\ cm^2$*

Q This number is much greater than the number of square meters. Does it indicate an error?

A ***No, the same-sized area contains a small number of large units (large tiles) or a large number of small units (small tiles), just as 20 one-dollar bills have the same value as 2000 pennies.***

(a)

(b)

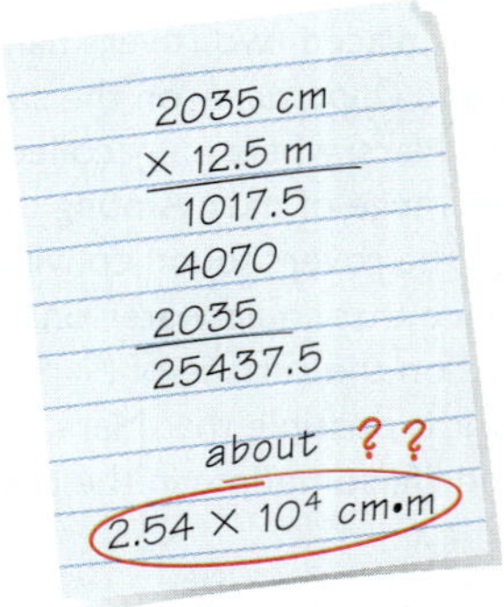

(c)

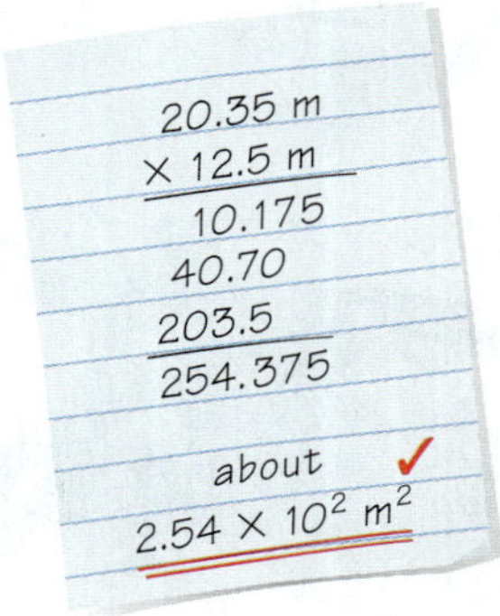

Figure 9
When determining area by multiplying measurements of length and width, be sure the measurements are expressed in the same units.

Both dimension and units must agree

Measurements of physical quantities must be expressed in units that match the dimensions of that quantity. For example, measurements of length cannot be expressed in units of kilograms because units of kilograms describe the dimension of mass. It is very important to be certain that a measurement is expressed in units that refer to the correct dimension. One good technique for avoiding errors in physics is to check the units in an answer to be certain they are appropriate for the dimension of the physical quantity that is being sought in a problem or calculation.

In addition to having the correct dimension, measurements used in calculations should also have the same units. As an example, consider **Figure 9(a),** which shows two people measuring a room to determine the room's area. Suppose one person measures the length in meters and the other person measures the width in centimeters. When the numbers are multiplied to find the area, they will give a difficult-to-interpret answer in units of cm•m, as shown in **Figure 9(b).** On the other hand, if both measurements are made using the same units, the calculated area is much easier to interpret because it is expressed in units of m^2, as shown in **Figure 9(c).** Even if the measurements were made in different units, as in the example above, one unit can be easily converted to the other because centimeters and meters are both units of length. It is also necessary to convert one unit to another when working with units from two different systems, such as meters and feet. In order to avoid confusion, it is better to make the conversion to the same units before doing any more arithmetic.

SAMPLE PROBLEM A

Metric Prefixes

PROBLEM

A typical bacterium has a mass of about 2.0 fg. Express this measurement in terms of grams and kilograms.

SOLUTION

Given: mass = 2.0 fg

Unknown: mass = ? g mass = ? kg

Build conversion factors from the relationships given in **Table 3.** Two possibilities are shown below.

$$\frac{1 \times 10^{-15}\ \text{g}}{1\ \text{fg}} \text{ and } \frac{1\ \text{fg}}{1 \times 10^{-15}\ \text{g}}$$

Only the first one will cancel the units of femtograms to give units of grams.

$$(2.0\ \cancel{\text{fg}})\left(\frac{1 \times 10^{-15}\ \text{g}}{1\ \cancel{\text{fg}}}\right) = \boxed{2.0 \times 10^{-15}\ \text{g}}$$

Then, take this answer and use a similar process to cancel the units of grams to give units of kilograms.

$$(2.0 \times 10^{-15}\ \cancel{\text{g}})\left(\frac{1\ \text{kg}}{1 \times 10^{3}\ \cancel{\text{g}}}\right) = \boxed{2.0 \times 10^{-18}\ \text{kg}}$$

PRACTICE A

Metric Prefixes

1. A human hair is approximately 50 µm in diameter. Express this diameter in meters.
2. If a radio wave has a period of 1 µs, what is the wave's period in seconds?
3. A hydrogen atom has a diameter of about 10 nm.
   a. Express this diameter in meters.
   b. Express this diameter in millimeters.
   c. Express this diameter in micrometers.
4. The distance between the sun and Earth is about 1.5×10^{11} m. Express this distance with an SI prefix and in kilometers.
5. The average mass of an automobile in the United States is about 1.440×10^{6} g. Express this mass in kilograms.

Classroom Practice

Metric Prefixes
If a woman has a mass of 60 000 000 mg, what is her mass in grams and kilograms?

Answer
60 000 g, 60 kg

PROBLEM GUIDE A

Use this guide to assign problems.
SE = Student Edition Textbook
PW = Problem Workbook
PB = Problem Bank on the One-Stop Planner (OSP)

Solving for:

SI units	**SE** Sample, 1–5; Ch. Rvw. 11–14 **PW** Sample, 1–8 **PB** Sample, 1–10

***Challenging Problem**
Consult the printed Solutions Manual or the OSP for detailed solutions.

ANSWERS

Practice A

1. 5×10^{-5} m
2. 1×10^{-6} s
3. a. 1×10^{-8} m
   b. 1×10^{-5} mm
   c. 1×10^{-2} µm
4. 0.15 Tm, 1.5×10^{8} km
5. 1.440×10^{3} kg

Demonstration

Accuracy and Precision GENERAL

Purpose Show the difference between accuracy and precision in measuring length.

Materials shoe, metersticks

Procedure Have three students successively measure the length of a long table with a shoe. Ask students to compare the accuracy of each measurement and the precision of the measuring tool. *(The measurement has little accuracy; i.e., it is difficult to estimate a fractional part of a shoe. The tool lacks precision; i.e., repeated measurements produce little agreement.)* Have three students repeat the measurement with metersticks and compare their results. Are there still differences even though the meterstick is more precise?

Demonstration

Significant Figures GENERAL

Purpose Clarify the meaning of significant figures.

Materials pan balance, small paper cup, salt or sugar

Procedure Pour salt into the cup on one pan until it is balanced with 183 g. Slowly add small amounts of salt, and have students note that the reading remains 183 g until the balance tilts. Try to restore balance by adding 1 g masses (they will probably be too heavy or too light). Ask students to guess the weight of the salt. *(The answer is actually between 183 g and 184 g.)*

ACCURACY AND PRECISION

accuracy

a description of how close a measurement is to the correct or accepted value of the quantity measured

precision

the degree of exactness of a measurement

Because theories are based on observation and experiment, careful measurements are very important in physics. But no measurement is perfect. In describing the imperfection, one must consider both a measurement's **accuracy** and a measurement's **precision.** Although these terms are often used interchangeably in everyday speech, they have specific meanings in a scientific discussion. A numeric measure of confidence in a measurement or result is known as *uncertainty.* A lower uncertainty indicates greater confidence. Uncertainties are usually expressed by using statistical methods.

Error in experiments must be minimized

Experimental work is never free of error, but it is important to minimize error in order to obtain accurate results. An error can occur, for example, if a mistake is made in reading an instrument or recording the results. One way to minimize error from human oversight or carelessness is to take repeated measurements to be certain they are consistent.

If some measurements are taken using one method and some are taken using a different method, a type of error called *method error* will result. Method error can be greatly reduced by standardizing the method of taking measurements. For example, when measuring a length with a meterstick, choose a line of sight directly over what is being measured, as shown in **Figure 10(a).** If you are too far to one side, you are likely to overestimate or underestimate the measurement, as shown in **Figure 10(b)** and **(c).**

Another type of error is *instrument error.* If a meterstick or balance is not in good working order, this will introduce error into any measurements made with the device. For this reason, it is important to be careful with lab equipment. Rough handling can damage balances. If a wooden meterstick gets wet, it can warp, making accurate measurements difficult.

Figure 10

If you measure this window by keeping your line of sight directly over the measurement **(a),** you will find that it is 165.2 cm long. If you do not keep your eye directly above the mark, as in **(b)** and **(c),** you may report a measurement with significant error.

Because the ends of a meterstick can be easily damaged or worn, it is best to minimize instrument error by making measurements with a portion of the scale that is in the middle of the meterstick. Instead of measuring from the end (0 cm), try measuring from the 10 cm line.

Precision describes the limitations of the measuring instrument

Poor accuracy involves errors that can often be corrected. On the other hand, precision describes how exact a measurement can possibly be. For example, a measurement of 1.325 m is more precise than a measurement of 1.3 m. A lack of precision is typically due to limitations of the measuring instrument and is not the result of human error or lack of calibration. For example, if a meterstick is divided only into centimeters, it will be difficult to measure something only a few millimeters thick with it.

In many situations, you can improve the precision of a measurement. This can be done by making a reasonable estimation of where the mark on the instrument would have been. Suppose that in a laboratory experiment you are asked to measure the length of a pencil with a meterstick marked in centimeters, as shown in **Figure 11.** The end of the pencil lies somewhere between 18 cm and 18.5 cm. The length you have actually measured is slightly more than 18 cm. You can make a reasonable estimation of how far between the two marks the end of the pencil is and add a digit to the end of the actual measurement. In this case, the end of the pencil seems to be less than halfway between the two marks, so you would report the measurement as 18.2 cm.

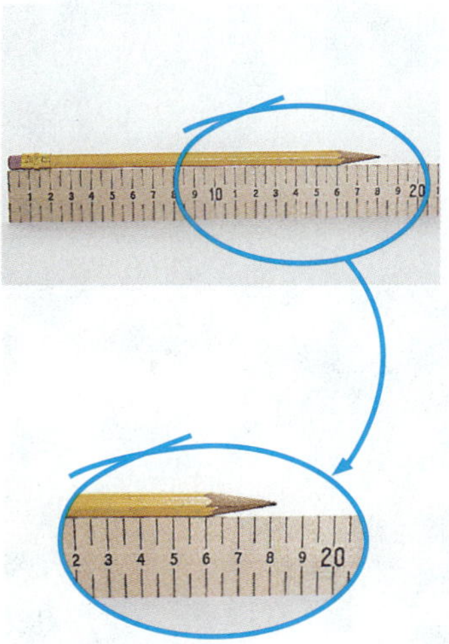

Figure 11
Even though this ruler is marked in only centimeters and half-centimeters, if you estimate, you can use it to report measurements to a precision of a millimeter.

Significant figures help keep track of imprecision

It is important to record the precision of your measurements so that other people can understand and interpret your results. A common convention used in science to indicate precision is known as **significant figures.**

significant figures

those digits in a measurement that are known with certainty plus the first digit that is uncertain

In the case of the measurement of the pencil as about 18.2 cm, the measurement has three significant figures. The significant figures of a measurement include all the digits that are actually measured (18 cm), plus one *estimated* digit. Note that the number of significant figures is determined by the precision of the markings on the measuring scale.

The last digit is reported as a 0.2 (for the estimated 0.2 cm past the 18 cm mark). Because this digit is an estimate, the true value for the measurement is actually somewhere between 18.15 cm and 18.25 cm.

When the last digit in a recorded measurement is a zero, it is difficult to tell whether the zero is there as a place holder or as a significant digit. For example, if a length is recorded as 230 mm, it is impossible to tell whether this number has two or three significant digits. In other words, it can be difficult to know whether the measurement of 230 mm means the measurement is known to be between 225 mm and 235 mm or is known more precisely to be between 229.5 mm and 230.5 mm.

Teaching Tip — ADVANCED

Draw a rectangle on the board, and tell students that the rectangle represents a pool that is 50 m long. Ask students how accurately they think a real 50 m pool can be constructed. Explain to students that in pouring the concrete for pools, there may be a slight variation in length. Might the length of a 50 m pool be 50.1 m or 49.9 m? Might it be 50.01 m or 49.99 m? Might it be 50.001 m or 49.999 m? Show students lengths of 10 cm, 1 cm, and 1 mm on a ruler. Then show students a distance of approximately 50 m.

Have students assume that the pool is 50.00 ± 0.01 m. That would mean that one swimming pool may be 1 cm shorter than 50 m and a second pool in another location may be 1 cm longer than 50 m. Because a 1500 m race requires a swimmer to swim 30 laps, a swimmer may swim 30 cm less in one pool and 30 cm more in a second pool while completing a 1500 m race. Could this 60 cm difference in the lengths of the races explain why a new record might be set in the first pool? Before we can answer this question, we have to also consider the precision of the time measurement of the race.

SECTION 2

Teaching Tip — ADVANCED

Have students assume that the precision of the time measurement in the 1500 m race from the previous Teaching Tip is 0.01 s. Therefore, if the possible 60 cm variation in the length is to be ignored between any two competitions, a swimmer would have to swim this difference in less than 0.01 s. Show that at a rate of 60 cm/0.01 s, a swimmer would have to swim 60 m (more than 30 body lengths) in 1 s. Thus, since it takes longer than 0.01 s to swim 60 cm, then there can be no valid comparisons of time that differ by less than 0.01 s from one 1500 m race to the next. The possible difference in the times may not be due to the ability of the individual swimmers; it may be due to a difference in the lengths of the two pools. Why do Olympic judges measure swimming times to the 0.1 s, 0.01 s, or even 0.001 s? Students can either explain this or have something new to think about as they watch or engage in sports activities.

Visual Strategy GENERAL

Table 4

Students may need to practice with examples in applying the rules shown in **Table 4.**

Q Write each of the examples in scientific notation. Check the number of significant figures.

A *5.03×10^1 m; 3.0025 s; 8.92×10^{-1} kg; 8×10^{-4} ms; 5.700×10^1 g; 2.000 000 kg; 1×10^3 m; 2×10^1 m*

Figure 12

If a mountain's height is known with an uncertainty of 5 m, the addition of 0.20 m of rocks will not appreciably change the height.

One way to solve such problems is to report all values using scientific notation. In scientific notation, the measurement is recorded to a power of 10, and all of the figures given are significant. For example, if the length of 230 cm has two significant figures, it would be recorded in scientific notation as 2.3×10^2 cm. If it has three significant figures, it would be recorded as 2.30×10^2 cm.

Scientific notation is also helpful when the zero in a recorded measurement appears in front of the measured digits. For example, a measurement such as 0.000 15 cm should be expressed in scientific notation as 1.5×10^{-4} cm if it has two significant figures. The three zeros between the decimal point and the digit 1 are not counted as significant figures because they are present only to locate the decimal point and to indicate the order of magnitude. The rules for determining how many significant figures are in a measurement that includes zeros are shown in **Table 4.**

Significant figures in calculations require special rules

In calculations, the number of significant figures in your result depends on the number of significant figures in each measurement. For example, if someone reports that the height of a mountaintop, like the one shown in **Figure 12,** is 1710 m, that implies that its actual height is between 1705 and 1715 m. If another person builds a pile of rocks 0.20 m high on top of the mountain, that would not suddenly make the mountain's new height known accurately enough to be measured as 1710.20 m. The final reported height cannot be more precise than the least precise measurement used to find the answer. Therefore, the reported height should be rounded off to 1710 m even if the pile of rocks is included.

Table 4 Rules for Determining Whether Zeros Are Significant Figures

Rule	Examples
1. Zeros between other nonzero digits are significant.	**a.** 50.3 m has three significant figures. **b.** 3.0025 s has five significant figures.
2. Zeros in front of nonzero digits are not significant.	**a.** 0.892 kg has three significant figures. **b.** 0.0008 ms has one significant figure.
3. Zeros that are at the end of a number and also to the right of the decimal are significant.	**a.** 57.00 g has four significant figures. **b.** 2.000 000 kg has seven significant figures.
4. Zeros at the end of a number but to the left of a decimal are significant if they have been measured or are the first estimated digit; otherwise, they are *not* significant. In this book, they will be treated as *not* significant. (Some books place a bar over a zero at the end of a number to indicate that it is significant. This textbook will use scientific notation for these cases instead.)	**a.** 1000 m may contain from one to four significant figures, depending on the precision of the measurement, but in this book it will be assumed that measurements like this have one significant figure. **b.** 20 m may contain one or two significant figures, but in this book it will be assumed to have one significant figure.

Similar rules apply to multiplication. Suppose that you calculate the area of a room by multiplying the width and length. If the room's dimensions are 4.6 m by 6.7 m, the product of these values would be 30.82 m^2. However, this answer contains four significant figures, which implies that it is more precise than the measurements of the length and width. Because the room could be as small as 4.55 m by 6.65 m or as large as 4.65 m by 6.75 m, the area of the room is known only to be between 30.26 m^2 and 31.39 m^2. The area of the room can have only two significant figures because each measurement has only two. So, the area must be rounded off to 31 m^2. **Table 5** summarizes the two basic rules for determining significant figures when you are performing calculations.

Table 5 Rules for Calculating with Significant Figures

Type of calculation	Rule	Example
addition or subtraction	Given that addition and subtraction take place in columns, round the final answer to the *first column from the left containing an estimated digit.*	97.3 + 5.85 = 103.15 $\xrightarrow{\text{round off}}$ 103.2
multiplication or division	The final answer has the same number of significant figures as the measurement having the *smallest* number of *significant figures.*	123 × 5.35 = 658.05 $\xrightarrow{\text{round off}}$ 658

Calculators do not pay attention to significant figures

When you use a calculator to analyze problems or measurements, you may be able to save time because the calculator can compute faster than you can. However, the calculator does not keep track of significant figures.

Calculators often exaggerate the precision of your final results by returning answers with as many digits as the display can show. To reinforce the correct approach, the answers to the sample problems in this book will always show only the number of significant figures that the measurements justify.

Providing answers with the correct number of significant figures often requires rounding the results of a calculation. The rules listed in **Table 6** on the next page will be used in this book for rounding, and the results of a calculation will be rounded after each type of mathematical operation. For example, the result of a series of multiplications should be rounded using the multiplication/division rule before it is added to another number. Similarly, the sum of several numbers should be rounded according to the addition/subtraction rule before the sum is multiplied by another number. Multiple roundings can increase the error in a calculation, but with this method there is no ambiguity about which rule to apply. You should consult your teacher to find out whether to round this way or to delay rounding until the end of all calculations.

The Language of Physics — GENERAL

In mathematics, we use 97.3, 97.30, 97.300, etc., to represent the same number with a single value (e.g., if $x = 97.3$ and $y = 97.300$, then $x = y$). When numbers represent the results of measurements, they are not pure mathematical quantities: they contain additional information about precision. Measured values actually stand for a range of possible values. This range is defined by the significant digits. Here, x could be 97.25 cm, 97.26 cm, or 97.34 cm. But y could be 97.295 cm, 97.296 cm, or 97.304 cm. There may be a difference between x and y, although it may not be measurable.

STOP Misconception Alert — BASIC

Sometimes, students are reluctant to round off a product so that it has no decimal digits. They might even think they misapplied a rule. An example may familiarize them with such counterintuitive results. Ask them to multiply 53.5824 s by 2.14 m/s and round off. The result, 114.666 336 m, needs to be reduced to three significant digits, that is, 115 m.

Teaching Tip

When results are rounded after each type of operation, the order of operations used to solve a problem sometimes affects the answer. For this reason, students' answers could differ slightly from those in this Teacher's Edition and still be correct. If you ask students to round at the end of a calculation instead of after each type of operation, there will be even more variation.

SECTION 2

Misconception Alert GENERAL

The *symmetric rounding rules* used when the last digit is a five may appear odd. Students may not be clear about their reasons. Explain that when the last digit is five, by convention we round to the *nearest even digit.* Ask them to apply this to rounding 12.05, 12.15, 12.25, 12.35, 12.45 and so on until 12.95. *(12.0, 12.2, 12.2, 12.4, 12.4, etc.)* Students will realize that they alternately round up and down. Point out that this convention allows them to reduce the average error because the numbers are sometimes increased and sometimes decreased.

SECTION REVIEW ANSWERS

1. a. meters
   b. kilograms
   c. seconds
2. a. 6.20×10^{-6} kg
   b. 3×10^{-6} ms
   c. 8.80×10^{4} m
3. a. 0.67
   b. 14
   c. 778.92
   d. 797.5
4. a. accurate and precise
   b. precise
   c. neither

Table 6 Rules for Rounding in Calculations

What to do	When to do it	Examples
round down	• whenever the digit following the last significant figure is a 0, 1, 2, 3, or 4	30.24 becomes 30.2
	• if the last significant figure is an even number and the next digit is a 5, with no other nonzero digits	32.25 becomes 32.2 32.65000 becomes 32.6
round up	• whenever the digit following the last significant figure is a 6, 7, 8, or 9	22.49 becomes 22.5
	• if the digit following the last significant figure is a 5 followed by a nonzero digit	54.7511 becomes 54.8
	• if the last significant figure is an odd number and the next digit is a 5, with no other nonzero digits	54.75 becomes 54.8 79.3500 becomes 79.4

SECTION REVIEW

1. Which SI units would you use for the following measurements?
   a. the length of a swimming pool
   b. the mass of the water in the pool
   c. the time it takes a swimmer to swim a lap
2. Express the following measurements as indicated.
   a. 6.20 mg in kilograms
   b. 3×10^{-9} s in milliseconds
   c. 88.0 km in meters
3. Perform these calculations, following the rules for significant figures.
   a. $26 \times 0.02584 = ?$
   b. $15.3 \div 1.1 = ?$
   c. $782.45 - 3.5328 = ?$
   d. $63.258 + 734.2 = ?$
4. **Critical Thinking** The following students measure the density of a piece of lead three times. The density of lead is actually 11.34 g/cm^3. Considering all of the results, which person's results were accurate? Which were precise? Were any both accurate and precise?
   a. Rachel: 11.32 g/cm^3, 11.35 g/cm^3, 11.33 g/cm^3
   b. Daniel: 11.43 g/cm^3, 11.44 g/cm^3, 11.42 g/cm^3
   c. Leah: 11.55 g/cm^3, 11.34 g/cm^3, 11.04 g/cm^3

The Language of Physics

SECTION OBJECTIVES

- **Interpret data in tables and graphs, and recognize equations that summarize data.**
- **Distinguish between conventions for abbreviating units and quantities.**
- **Use dimensional analysis to check the validity of equations.**
- **Perform order-of-magnitude calculations.**

MATHEMATICS AND PHYSICS

Just as physicists create simplified models to better understand the real world, they use the tools of mathematics to analyze and summarize their observations. Then they can use the mathematical relationships among physical quantities to help predict what will happen in new situations.

Tables, graphs, and equations can make data easier to understand

There are many ways to organize data. Consider the experiment shown in **Figure 13,** which tests Galileo's hypothesis that all objects fall at the same rate in the absence of air resistance (see Section 2). In this experiment, a table-tennis ball and a golf ball are dropped in a vacuum. The results are recorded as a set of numbers corresponding to the times of the fall and the distance each ball falls. A convenient way to organize the data is to form a table like **Table 7.** A clear trend can be seen in the data. The more time that passes after each ball is dropped, the farther the ball falls.

Table 7 Data from Dropped-Ball Experiment

Time (s)	Distance golf ball falls (cm)	Distance table-tennis ball falls (cm)
0.067	2.20	2.20
0.133	8.67	8.67
0.200	19.60	19.59
0.267	34.93	34.92
0.333	54.34	54.33
0.400	78.40	78.39

One method for analyzing the data in **Table 7** is to construct a graph of the distance the balls have fallen versus the elapsed time since they were released. This graph is shown in **Figure 14** on the next page. Because the graph shows an obvious pattern, we can draw a smooth curve through the data points to make estimations for times when we have no data. The shape of the graph also provides information about the relationship between time and distance.

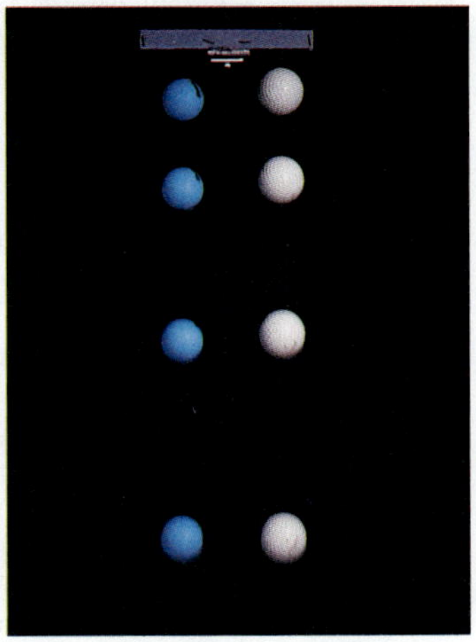

Figure 13
This experiment tests Galileo's hypothesis by having two balls with different masses dropped simultaneously in a vacuum.

Teaching Tip

Point out that although the time intervals shown in **Table 7** cannot be measured by direct observation, these data can be recorded using advanced technological tools, including a computer connected to a probe.

Visual Strategy GENERAL

Table 7

In order to process the information contained in **Table 7,** students need to look at the numbers closely.

Q Were the data collected at regular time intervals?

A *yes, every $\frac{1}{15}$ of a second*

Q According to the distances displayed in the table, did one of the balls appear to fall faster than the other one?

A *No, the measured distances vary by only one digit in the last significant figure.*

Q According to the distances displayed in the table, did the golf ball travel the same distance during every time interval?

A *No, the distance traveled increases with each time interval.*

Visual Strategy GENERAL

Figure 14
Have students verify that the points on the graph correspond to the numbers listed in **Table 7.**

Q How far have the balls fallen after a time of 0.225 s?

A ***26 cm (as read from the graph)***

Q Was the point (0, 0) listed in the data table? How would you justify it?

A ***No, we assume that the clock started exactly at the instant the balls were dropped and that distances were measured from the dropping point.***

The Language of Physics

In the first equation on this page, the term *change in position* stands for the vertical distance from the point where the balls were dropped. Likewise, *time of fall* specifies the number of seconds since the balls were dropped.

We use the symbols Δy and Δt to represent the vertical distance between two points and the time interval, respectively. These symbols are used in the second equation shown on the page.

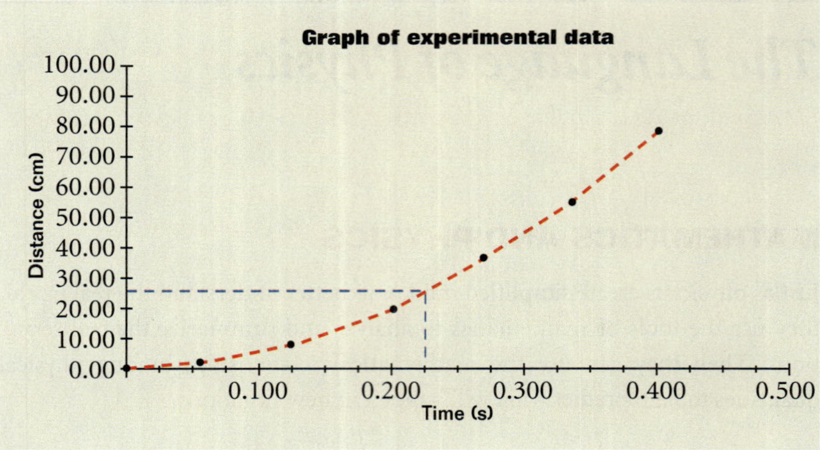

Figure 14
The graph of these data provides a convenient way to summarize the data and indicate the relationship between the time an object has been falling and the distance it has fallen.

We can also use the following equation to describe the relationship between the variables in the experiment:

$$(\text{change in position in meters}) = 4.9 \times (\text{time of fall in seconds})^2$$

This equation allows you to reproduce the graph and make predictions about the change in position for any arbitrary time during the fall.

Physics equations describe relationships

While mathematicians use equations to describe relationships between variables, physicists use the tools of mathematics to describe measured or predicted relationships between physical quantities in a situation. For example, one or more variables may affect the outcome of an experiment. In the case of a prediction, the physical equation is a compact statement based on a model of the situation. It shows how two or more variables are thought to be related. Many of the equations in physics represent a simple description of the relationship between physical quantities.

To make expressions as simple as possible, physicists often use letters to describe specific quantities in an equation. For example, the letter v is used to denote speed. Sometimes, Greek letters are used to describe mathematical operations. For example, the Greek letter Δ (delta) is often used to mean "difference or change in," and the Greek letter Σ (sigma) is used to mean "sum" or "total."

With these conventions, the word equation above can be written as follows:

$$\Delta y = 4.9(\Delta t)^2$$

The abbreviation Δy indicates the vertical change in a ball's position from its starting point, and Δt indicates the time elapsed.

As you saw in Section 2, the units in which these quantities are measured are also often abbreviated with symbols consisting of a letter or two. Most physics books provide some clues to help you keep track of which letters refer to quantities and variables and which letters are used to indicate units. Typically, variables and other specific quantities are abbreviated with letters that are

boldfaced or *italicized.* (You will learn the difference between the two in the chapter "Two-Dimensional Motion and Vectors.") Units are abbreviated with regular letters (sometimes called roman letters). Some examples of variable symbols and the abbreviations for the units that measure them are shown in **Table 8.**

As you continue to study physics, carefully note the introduction of new variable quantities, and recognize which units go with them. The tables provided in Appendices C–E can help you keep track of these abbreviations.

extension

Integrating Chemistry
Visit go.hrw.com for the activity "Dependent and Independent Variables."

Keyword HF6SOPX

Table 8 Abbreviations for Variables and Units

Quantity	Symbol	Units	Unit abbreviations
change in vertical position	Δy	meters	m
time interval	Δt	seconds	s
mass	m	kilograms	kg

EVALUATING PHYSICS EQUATIONS

Like most models physicists build to describe the world around them, physics equations are valid only if they can be used to make correct predictions about situations. Although an experiment is the ultimate way to check the validity of a physics equation, several techniques can be used to evaluate whether an equation or result can possibly be valid.

Dimensional analysis can weed out invalid equations

Suppose a car, such as the one in **Figure 15,** is moving at a speed of 88 km/h and you want to know how much time it will take it to travel 725 km. How can you decide a good way to solve the problem?

You can use a powerful procedure called *dimensional analysis.* Dimensional analysis makes use of the fact that *dimensions can be treated as algebraic quantities.* For example, quantities can be added or subtracted only if they have the same dimensions, and the two sides of any given equation must have the same dimensions.

Let us apply this technique to the problem of the car moving at a speed of 88 km/h. This measurement is given in dimensions of length over time. The total distance traveled has the dimension of length. Multiplying these numbers together gives the dimensions indicated below. Clearly, the result of this calculation does not have the dimensions of time, which is what you are trying to calculate. That is,

$$\frac{\text{length}}{\text{time}} \times \text{length} = \frac{\text{length}^2}{\text{time}} \text{ or } \frac{88 \text{ km}}{1.0 \text{ h}} \times 725 \text{ km} = \frac{6.4 \times 10^4 \text{ km}^2}{1.0 \text{ h}}$$

Figure 15
Dimensional analysis can be a useful check for many types of problems, including those involving how much time it would take for this car to travel 725 km if it moves with a speed of 88 km/h.

Misconception Alert GENERAL

Students may think that algebraic symbols representing variable quantities must always be the same letters shown in the table. Point out that physicists tend to *choose* different symbols for each dimension in different contexts and that there are few fixed rules. For instance, as shown in **Table 8,** one symbol for change in position is Δy; change in position can also by symbolized by Δx, d, s, ℓ, or h.

Teaching Tip GENERAL

Point out that dimensional analysis is a good tool for checking whether the equation you are using gives the kind of quantity you want to calculate. For example, when you want to find volume (V) based on density (ρ) and mass (m), you may find that you erroneously set $V = \frac{\rho}{m}$ before doing the calculations. Simply replace each variable with its dimensions and see whether the equation balances.

$$V = \frac{\rho}{m}$$

$$\text{cm}^3 = \frac{\frac{\text{g}}{\text{cm}^3}}{\text{g}} = \frac{\text{g}}{\text{cm}^3}\frac{1}{\text{g}} = \frac{1}{\text{cm}^3}$$

This mismatch reveals that the equation $V = \frac{\rho}{m}$ must be invalid.

Teaching Tip

The order-of-magnitude estimate mentioned in the "Did You Know?" feature on this page is posed to students in the Chapter Review, and one possible solution is given in the Solutions Manual.

To calculate an answer that will have the dimension of time, you should take the distance and *divide* it by the speed of the car, as follows:

$$\frac{\text{length}}{\text{length/time}} = \frac{\text{length} \times \text{time}}{\text{length}} = \text{time} \qquad \frac{725\ \text{km} \times 1.0\ \text{h}}{88\ \text{km}} = 8.2\ \text{h}$$

In a simple example like this one, you might be able to identify the invalid equation without dimensional analysis. But with more-complicated problems, it is a good idea to check your final equation with dimensional analysis. This step will prevent you from wasting time computing an invalid equation.

Did you know?

The physicist Enrico Fermi made the first nuclear reactor at the University of Chicago in 1942. Fermi was also well known for his ability to make quick order-of-magnitude calculations, such as estimating the number of piano tuners in New York City.

Order-of-magnitude estimations check answers

Because the scope of physics is so wide and the numbers may be astronomically large or subatomically small, it is often useful to estimate an answer to a problem before trying to solve the problem exactly. This kind of estimate is called an *order-of-magnitude* calculation, which means determining the power of 10 that is closest to the actual numerical value of the quantity. Once you have done this, you will be in a position to judge whether the answer you get from a more exacting procedure is correct.

For example, consider the car trip described in the discussion of dimensional analysis. We must divide the distance by the speed to find the time. The distance, 725 km, is closer to 10^3 km (or 1000 km) than to 10^2 km (or 100 km), so we use 10^3 km. The speed, 88 km/h, is about 10^2 km/h (or 100 km/h).

$$\frac{10^3\ \text{km}}{10^2\ \text{km/h}} = 10\ \text{h}$$

This estimate indicates that the answer should be closer to 10 than to 1 or to 100 (or 10^2). The correct answer (8.2 h) certainly fits this range.

Order-of-magnitude estimates can also be used to estimate numbers in situations in which little information is given. For example, how could you estimate how many gallons of gasoline are used annually by all of the cars in the United States?

First, consider that the United States has almost 300 million people. Assuming that each family of about five people has two cars, an estimate of the number of cars in the country is 120 million.

Next, decide the order of magnitude of the average distance each car travels every year. Some cars travel as few as 1000 mi per year, while others travel more than 100 000 mi per year. The appropriate order of magnitude to include in the estimate is 10 000 mi, or 10^4 mi, per year.

If we assume that cars average 20 mi for every gallon of gas, each car needs about 500 gal per year.

$$\left(\frac{10\,000\ \text{mi}}{1\ \text{year}}\right)\left(\frac{1\ \text{gal}}{20\ \text{mi}}\right) = 500\ \text{gal/year for each car}$$

SCILINKS.
www.scilinks.org
Topic: Orders of Magnitude
SciLinks Code: HF61074

Multiplying this by the estimate of the total number of cars in the United States gives an annual consumption of 6×10^{10} gal.

$$(12 \times 10^7 \text{ cars})\left(\frac{500 \text{ gal}}{1 \text{ car}}\right) = 6 \times 10^{10} \text{ gal}$$

Note that this estimate depends on the assumptions made about the average household size, the number of cars per household, the distance traveled, and the average gas mileage.

SECTION REVIEW

1. Indicate which of the following physics symbols denote units and which denote variables or quantities.

a. C **b.** *c* **c.** *C* **d.** *t* **e.** T **f.** *T*

2. Determine the units of the quantity described by each of the following combinations of units:

a. kg (m/s) (1/s) **b.** (kg/s) (m/s^2)
c. (kg/s) $(m/s)^2$ **d.** (kg/s) (m/s)

3. Which of the following is the best order-of-magnitude estimate in meters of the height of a mountain?

a. 1 m **b.** 10 m **c.** 100 m **d.** 1000 m

4. Interpreting Graphics Which graph best matches the data?

Volume of air (m^3)	Mass of air (kg)
0.50	0.644
1.50	1.936
2.25	2.899
4.00	5.159
5.50	7.096

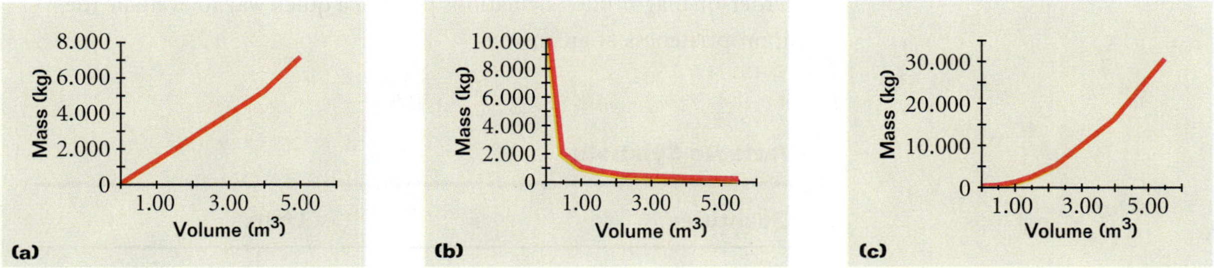

Figure 16

5. Critical Thinking Which of the following equations best matches the data from item 4?

a. $(\text{mass})^2 = 1.29$ (volume) **b.** (mass)(volume) = 1.29
c. mass = 1.29 (volume) **d.** mass = $1.29\ (\text{volume})^2$

SECTION REVIEW ANSWERS

1. units: a, e
variables/quantities: b, c, d, f
2. a. kg•m/s^2
b. kg•m/s^3
c. kg•m^2/s^3
d. kg•m/s^2
3. d
4. a
5. c

CHAPTER 1

Highlights

Teaching Tip

Tell students that they will apply the material in this chapter as they proceed through this book. What they have just learned will make more sense as they study different physics topics. They should remember to refer to this chapter as needed as they study motion, light, electricity, atomic physics, or any other subject.

CHAPTER 1

Highlights

KEY TERMS

model (p. 6)

system (p. 7)

hypothesis (p. 8)

controlled experiment (p. 9)

accuracy (p. 16)

precision (p. 16)

significant figures (p. 17)

PROBLEM SOLVING

If you need more problem-solving practice, see **Appendix I: Additional Problems.**

KEY IDEAS

Section 1 What Is Physics?

- Physics is the study of the physical world, from motion and energy to light and electricity.
- Physics uses the scientific method to discover general laws that can be used to make predictions about a variety of situations.
- A common technique in physics for analyzing a complex situation is to disregard irrelevant factors and create a model that describes the essence of a system or situation.

Section 2 Measurements in Experiments

- Physics measurements are typically made and expressed in SI, a system that uses a set of base units and prefixes to describe measurements of physical quantities.
- *Accuracy* describes how close a measurement is to reality. *Precision* results from the limitations of the measuring device used.
- The significant figures of a measurement include all of the digits that are actually measured plus one estimated digit.
- Significant-figure rules provide a means to ensure that calculations do not report results that are more precise than the data used to make them.

Section 3 The Language of Physics

- Physicists make their work easier by summarizing data in tables and graphs and by abbreviating quantities in equations.
- Dimensional analysis can help identify whether a physics equation is invalid.
- Order-of-magnitude calculations provide a quick way to evaluate the appropriateness of an answer.

Variable Symbols

Quantities		Units	
Δy	change in vertical position	m	meters
Δt	time interval	s	seconds
m	mass	kg	kilograms

Review

THE SCIENCE OF PHYSICS

Review Questions

1. Refer to **Table 1** of this chapter to identify at least two areas of physics involved in the following:

- **a.** building a louder stereo system in your car
- **b.** bungee jumping
- **c.** judging how hot an electric stove burner is by looking at it
- **d.** cooling off on a hot day by diving into a swimming pool

2. Which of the following scenarios fit the approach of the scientific method?

- **a.** An auto mechanic listens to how a car runs and comes up with an idea of what might be wrong. The mechanic tests the idea by adjusting the idle speed. Then the mechanic decides his idea was wrong based on this evidence. Finally, the mechanic decides the only other problem could be the fuel pump, and he consults with the shop's other mechanics about his conclusion.
- **b.** Because of a difference of opinions about where to take the class trip, the class president holds an election. The majority of the students decide to go to the amusement park instead of to the shore.
- **c.** Your school's basketball team has advanced to the regional playoffs. A friend from another school says their team will win because their players want to win more than your school's team does.
- **d.** A water fountain does not squirt high enough. The handle on the fountain seems loose, so you try to push the handle in as you turn it. When you do this, the water squirts high enough that you can get a drink. You make sure to tell all your friends how you did it.

3. You have decided to select a new car by using the scientific method. How might you proceed?

4. Consider the phrase, "The quick brown fox jumped over the lazy dog." Which details of this situation would a physicist who is modeling the path of a fox ignore?

SI UNITS

Review Questions

5. List an appropriate SI base unit (with a prefix as needed) for measuring the following:

- **a.** the time it takes to play a CD in your stereo
- **b.** the mass of a sports car
- **c.** the length of a soccer field
- **d.** the diameter of a large pizza
- **e.** the mass of a single slice of pepperoni
- **f.** a semester at your school
- **g.** the distance from your home to your school
- **h.** your mass
- **i.** the length of your physics lab room
- **j.** your height

6. If you square the speed expressed in meters per second, in what units will the answer be expressed?

7. If you divide a force measured in newtons (1 newton = 1 kg•m/s^2) by a speed expressed in meters per second, in what units will the answer be expressed?

Conceptual Questions

8. The height of a horse is sometimes given in units of "hands." Why was this a poor standard of length before it was redefined to refer to exactly 4 in.?

9. Explain the advantages in having the meter officially defined in terms of the distance light travels in a given time rather than as the length of a specific metal bar.

10. Einstein's famous equation indicates that $E = mc^2$, where c is the speed of light and m is the object's mass. Given this, what is the SI unit for E?

Review

ANSWERS

1. **a.** vibrations and wave phenomena, electromagnetism
b. mechanics, vibrations and wave phenomena
c. optics, thermodynamics
d. thermodynamics, mechanics

2. a, d

3. Collect information on different cars, hypothesize about which car best suits your needs, test your hypothesis with test drives, interpret the results of your test drives, rethink your needs, and choose the best car to buy.

4. the fact that the fox is brown and the dog is lazy

5. Answers may vary.
a. s or das
b. kg or Mg
c. m
d. cm or dm
e. g or dag
f. Ms
g. km
h. kg
i. m or dam
j. cm, dm, or m

6. m^2/s^2

7. kg/s

8. The size of the unit varied depending on who was doing the measuring.

9. The size of the unit is standardized and reproducible.

10. kg•m^2/s^2

11. a. 2×10^2 mm
b. 7.8×10^3 s
c. 1.6×10^7 µg
d. 7.5×10^4 cm
e. 6.75×10^{-4} g
f. 4.62×10^{-2} cm
g. 9.7 m/s

12. a. 1 dekaration
b. 2 kilomockingbirds
c. 1 microphone
d. 1 nanogoat
e. 1 examiner

13. 1.08×10^9 km

14. 11 people

15. Yes, a set of measurements could be precise (have a high number of significant figures) but not close to the true value due to an error with the measuring instrument.

16. a. 1
b. 3
c. 5
d. 4
e. 4
f. 6

17. No, the number of significant figures is not consistent between the two values in each example.

18. a. 3.00×10^8 m/s
b. 2.9979×10^8 m/s
c. $2.997\,925 \times 10^8$ m/s

19. a. 3
b. 4
c. 3
d. 2

20. a. 797 g
b. 0.90 m/s
c. 17.8 mm
d. 23.7 s

21. 228.8 cm

22. 115.9 m

Practice Problems

For problems 11–14, see Sample Problem A.

11. Express each of the following as indicated:

a. 2 dm expressed in millimeters
b. 2 h 10 min expressed in seconds
c. 16 g expressed in micrograms
d. 0.75 km expressed in centimeters
e. 0.675 mg expressed in grams
f. 462 µm expressed in centimeters
g. 35 km/h expressed in meters per second

12. Use the SI prefixes in **Table 3** of this chapter to convert these *hypothetical* units of measure into appropriate quantities:

a. 10 rations
b. 2000 mockingbirds
c. 10^{-6} phones
d. 10^{-9} goats
e. 10^{18} miners

13. Use the fact that the speed of light in a vacuum is about 3.00×10^8 m/s to determine how many kilometers a pulse from a laser beam travels in exactly one hour.

14. If a metric ton is 1.000×10^3 kg, how many 85 kg people can safely occupy an elevator that can hold a maximum mass of exactly 1 metric ton?

ACCURACY, PRECISION, AND SIGNIFICANT FIGURES

Review Questions

15. Can a set of measurements be precise but not accurate? Explain.

16. How many significant figures are in the following measurements?

a. 300 000 000 m/s
b. 3.00×10^8 m/s
c. 25.030°C
d. 0.006 070°C
e. 1.004 J
f. 1.305 20 MHz

17. The photographs below show unit conversions on the labels of some grocery-store items. Check the accuracy of these conversions. Are the manufacturers using significant figures correctly?

(a) (b) (c) (d)

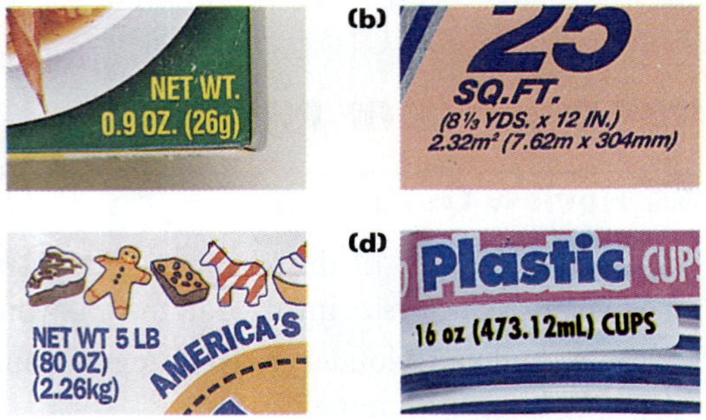

18. The value of the speed of light is now known to be $2.997\,924\,58 \times 10^8$ m/s. Express the speed of light in the following ways:

a. with three significant figures
b. with five significant figures
c. with seven significant figures

19. How many significant figures are there in the following measurements?

a. 78.9 ± 0.2 m
b. 3.788×10^9 s
c. 2.46×10^6 kg
d. 0.0032 mm

20. Carry out the following arithmetic operations:

a. find the sum of the measurements 756 g, 37.2 g, 0.83 g, and 2.5 g
b. find the quotient of 3.2 m/3.563 s
c. find the product of 5.67 mm × π
d. find the difference of 27.54 s and 3.8 s

21. A fisherman catches two sturgeons. The smaller of the two has a measured length of 93.46 cm (two decimal places and four significant figures), and the larger fish has a measured length of 135.3 cm (one decimal place and four significant figures). What is the total length of the two fish?

22. A farmer measures the distance around a rectangular field. The length of each long side of the rectangle is found to be 38.44 m, and the length of each short side is found to be 19.5 m. What is the total distance around the field?

DIMENSIONAL ANALYSIS AND ORDER-OF-MAGNITUDE ESTIMATES

Note: In developing answers to order-of-magnitude calculations, you should state your important assumptions, including the numerical values assigned to parameters used in the solution. Since only order-of-magnitude results are expected, do not be surprised if your results differ from those of other students.

Review Questions

23. Suppose that two quantities, *A* and *B*, have different dimensions. Which of the following arithmetic operations *could* be physically meaningful?

a. $A + B$
b. A/B
c. $A \times B$
d. $A - B$

24. Estimate the order of magnitude of the length in meters of each of the following:

a. a ladybug
b. your leg
c. your school building
d. a giraffe
e. a city block

25. If an equation is dimensionally correct, does this mean that the equation is true?

26. The radius of a circle inscribed in any triangle whose sides are *a, b,* and *c* is given by the following equation, in which *s* is an abbreviation for $(a + b + c) \div 2$. Check this formula for dimensional consistency.

$$r = \sqrt{\frac{(s-a)(s-b)(s-c)}{s}}$$

27. The period of a simple pendulum, defined as the time necessary for one complete oscillation, is measured in time units and is given by the equation

$$T = 2\pi\sqrt{\frac{L}{a_g}}$$

where *L* is the length of the pendulum and a_g is the acceleration due to gravity, which has units of length divided by time squared. Check this equation for dimensional consistency.

Conceptual Questions

28. In a desperate attempt to come up with an equation to solve a problem during an examination, a student tries the following: (velocity in m/s)^2 = (acceleration in m/s^2) × (time in s). Use dimensional analysis to determine whether this equation might be valid.

29. Estimate the number of breaths taken by a person during 70 years.

30. Estimate the number of times your heart beats in an average day.

31. Estimate the magnitude of your age, as measured in units of seconds.

32. An automobile tire is rated to last for 50 000 mi. Estimate the number of revolutions the tire will make in its lifetime.

33. Imagine that you are the equipment manager of a professional baseball team. One of your jobs is to keep a supply of baseballs for games in your home ballpark. Balls are sometimes lost when players hit them into the stands as either home runs or foul balls. Estimate how many baseballs you have to buy per season in order to make up for such losses. Assume your team plays an 81-game home schedule in a season.

34. A chain of hamburger restaurants advertises that it has sold more than 50 billion hamburgers over the years. Estimate how many pounds of hamburger meat must have been used by the restaurant chain to make 50 billion hamburgers and how many head of cattle were required to furnish the meat for these hamburgers.

35. Estimate the number of piano tuners living in New York City. (The population of New York City is approximately 8 million.) This problem was first proposed by the physicist Enrico Fermi, who was well known for his ability to quickly make order-of-magnitude calculations.

36. Estimate the number of table-tennis balls that would fit (without being crushed) into a room that is 4 m long, 4 m wide, and 3 m high. Assume that the diameter of a ball is 3.8 cm.

1 REVIEW

23. b, c
24. **a.** 10^{-2} m
b. 10^0 m
c. 10^1 m to 10^2 m
d. 10^1 m
e. 10^2 m
25. No, dimensional consistency indicates that the equation *may* be valid, but it is not *necessarily* valid.
26. The dimensions are consistent.
27. The dimensions are consistent.
28. The equation is not valid.

(Note: Because the following are estimates, student answers may vary from those shown here.)

29. 4×10^8 breaths
30. 9×10^4 beats
31. 5.4×10^8 s
32. 4×10^7 revolutions
33. 2×10^3 balls
34. 1×10^{10} lb, 2×10^7 head of cattle
35. 7×10^2 tuners
36. 8×10^5 balls

37. **a.** 22 cm; 38 cm^2
b. 29.2 cm; 67.9 cm^2

38. Take the $5000 because it would take you 272 years to count out the $5 billion in single dollar bills.

39. 9.818×10^{-2} m

40. 1.79×10^{-9} m

extension

Graphing Calculator Practice

Visit go.hrw.com for answers to this Graphing Calculator activity.

Keyword HF6SOPXT

MIXED REVIEW

37. Calculate the circumference and area for the following circles. (Use the following formulas: circumference = $2\pi r$ and area = πr^2.)

a. a circle of radius 3.5 cm
b. a circle of radius 4.65 cm

38. A billionaire offers to give you (1) $5 billion if you will count out the amount in $1 bills or (2) a lump sum of $5000. Which offer should you accept? Explain your answer. (Assume that you can count at an average rate of one bill per second, and be sure to allow for the fact that you need about 10 hours a day for sleeping and eating. Your answer does not need to be limited to one significant figure.)

39. Exactly 1 quart of ice cream is to be made in the form of a cube. What should be the length of one side in meters for the container to have the appropriate volume? (Use the following conversion: 4 qt = 3.786×10^{-3} m^3.)

40. You can obtain a rough estimate of the size of a molecule with the following simple experiment: Let a droplet of oil spread out on a fairly large but smooth water surface. The resulting "oil slick" that forms on the surface of the water will be approximately one molecule thick. Given an oil droplet with a mass of 9.00×10^{-7} kg and a density of 918 kg/m^3 that spreads out to form a circle with a radius of 41.8 cm on the water surface, what is the approximate diameter of an oil molecule?

Graphing Calculator *Practice*

Mass Versus Length

What is the relationship between the mass and length of three wires, each of which is made of a different substance? All three wires have the same diameter. Because the wires have the same diameter, their cross-sectional areas are the same. The cross-sectional area of any circle is equal to πr^2. Consider a wire with a diameter of 0.50 cm and a density of 8.96 g/cm^3. The following equation describes the mass of the wire as a function of the length:

$$Y_1 = 8.96X^* \pi(0.25)^2$$

In this equation, Y_1 represents the mass of the wire in grams, and X represents the length of the wire in centimeters. Each of the three wires is made of a different substance, so each wire has a different density and a different relationship between its mass and length.

In this graphing calculator activity, you will

- use dimensional analysis
- observe the relationship between a mathematical function and a graph
- determine values from a graph
- gain a better conceptual understanding of density

Visit go.hrw.com and type in the keyword **HF6SOPX** to find this graphing calculator activity. Refer to **Appendix B** for instructions on downloading the program for this activity.

41. An ancient unit of length called the cubit was equal to approximately 50 centimeters, which is, of course, approximately 0.50 meters. It has been said that Noah's ark was 300 cubits long, 50 cubits wide, and 30 cubits high. Estimate the volume of the ark in cubic meters. Also estimate the volume of a typical home, and compare it with the ark's volume.

42. If one micrometeorite (a sphere with a diameter of 1.0×10^{-6} m) struck each square meter of the moon each second, it would take many years to cover the moon with micrometeorites to a depth of 1.0 m. Consider a cubic box, 1.0 m on a side, on the moon. Estimate how long it would take to completely fill the box with micrometeorites.

43. One cubic centimeter ($1.0\ cm^3$) of water has a mass of 1.0×10^{-3} kg at 25°C. Determine the mass of $1.0\ m^3$ of water at 25°C.

44. Assuming biological substances are 90 percent water and the density of water is $1.0 \times 10^3\ kg/m^3$, estimate the masses (density multiplied by volume) of the following:

a. a spherical cell with a diameter of 1.0 µm (volume $= \frac{4}{3}\pi r^3$)

b. a fly, which can be approximated by a cylinder 4.0 mm long and 2.0 mm in diameter (volume $= \ell\pi r^2$)

45. The radius of the planet Saturn is 6.03×10^7 m, and its mass is 5.68×10^{26} kg.

a. Find the density of Saturn (its mass divided by its volume) in grams per cubic centimeter. (The volume of a sphere is given by $\frac{4}{3}\pi r^3$.)

b. Find the surface area of Saturn in square meters. (The surface area of a sphere is given by $4\pi r^2$.)

Alternative Assessment

1. Imagine that you are a member of your state's highway board. In order to comply with a bill passed in the state legislature, all of your state's highway signs must show distances in miles and kilometers. Two plans are before you. One plan suggests adding metric equivalents to all highway signs as follows: Dallas 300 mi (483 km). Proponents of the other plan say that the first plan makes the metric system seem more cumbersome, so they propose replacing the old signs with new signs every 50 km as follows: Dallas 300 km (186 mi). Participate in a class debate about which plan should be followed.

2. Can you measure the mass of a five-cent coin with a bathroom scale? Record the mass in grams displayed by your scale as you place coins on the scale, one at a time. Then, divide each measurement by the number of coins to determine the approximate mass of a single five-cent coin, but remember to follow the rules for significant figures in calculations. Which estimate do you think is the most accurate? Which is the most precise?

3. Find out who were the Nobel laureates for physics last year, and research their work. Alternatively, explore the history of the Nobel Prizes. Who founded the awards? Why? Who delivers the award? Where? Document your sources and present your findings in a brochure, poster, or presentation.

4. You have a clock with a second hand, a ruler marked in millimeters, a graduated cylinder marked in milliliters, and scales sensitive to 1 mg. How would you measure the mass of a drop of water? How would you measure the period of a swing? How would you measure the volume of a paper clip? How can you improve the accuracy of your measurements? Write the procedures clearly so that a partner can follow them and obtain reasonable results.

5. Create a poster or other presentation depicting the possible ranges of measurement for a dimension, such as distance, time, temperature, speed, or mass. Depict examples ranging from the very large to the very small. Include several examples that are typical of your own experiences.

1 REVIEW

41. The ark ($6 \times 10^4\ m^3$) was about 100 times as large as a typical house ($6 \times 10^2\ m^3$).

42. 3.2×10^{10} years

43. 1.0×10^3 kg

44. **a.** 5×10^{-16} kg
b. 1×10^{-5} kg

45. **a.** $0.618\ g/cm^3$
b. $4.57 \times 10^{16}\ m^2$

Alternative Assessment ANSWERS

1. Students should recognize that preferences for one scheme or the other have nothing to do with the accuracy of the measurements.

2. Students should recognize that their approximation of the mass of a single coin becomes both more accurate and more precise as they increase the number of coins.

3. Students' presentations will vary depending on which scientists they choose. Alfred Nobel (1833–1896), inventor of dynamite, left a fund for the establishment of annual prizes in several fields.

4. Students' procedures should be safe and thoroughly explained. In some cases, they will need to measure several objects together.

5. Students' presentations will vary but should include references for sources of information. Check that all examples are within a reasonable order of magnitude.

Standardized Test Prep

ANSWERS

1. B
2. J
3. A
4. H
5. C
6. G
7. C
8. J
9. B

Standardized Test Prep

MULTIPLE CHOICE

1. What area of physics deals with the subjects of heat and temperature?

A. mechanics
B. thermodynamics
C. electrodynamics
D. quantum mechanics

2. What area of physics deals with the behavior of subatomic particles?

F. mechanics
G. thermodynamics
H. electrodynamics
J. quantum mechanics

3. What term describes a set of particles or interacting components considered to be a distinct physical entity for the purpose of study?

A. system
B. model
C. hypothesis
D. controlled experiment

4. What is the SI base unit for length?

F. inch
G. foot
H. meter
J. kilometer

5. A light-year (ly) is a unit of distance defined as the distance light travels in one year. Numerically, 1 ly = 9 500 000 000 000 km. How many meters are in a light-year?

A. 9.5×10^{10} m
B. 9.5×10^{12} m
C. 9.5×10^{15} m
D. 9.5×10^{18} m

6. If you do not keep your line of sight directly over a length measurement, how will your measurement most likely be affected?

F. Your measurement will be less precise.
G. Your measurement will be less accurate.
H. Your measurement will have fewer significant figures.
J. Your measurement will suffer from instrument error.

7. If you measured the length of a pencil by using the meterstick shown in the figure below and you report your measurement in centimeters, how many significant figures should your reported measurement have?

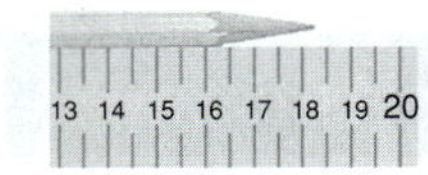

A. one
B. two
C. three
D. four

8. A room is measured to be 3.6 m by 5.8 m. What is the area of the room? (Keep significant figures in mind.)

F. 20.88 m^2
G. $2 \times 10^1 \text{ m}^2$
H. $2.0 \times 10^1 \text{ m}^2$
J. 21 m^2

9. What technique can help you determine the power of 10 closest to the actual numerical value of a quantity?

A. rounding
B. order-of-magnitude estimation
C. dimensional analysis
D. graphical analysis

10. Which of the following statements is true of *any* valid physical equation?
   F. Both sides have the same dimensions.
   G. Both sides have the same variables.
   H. There are variables but no numbers.
   J. There are numbers but no variables.

The graph below shows the relationship between time and distance for a ball dropped vertically from rest. Use the graph to answer questions 11–12.

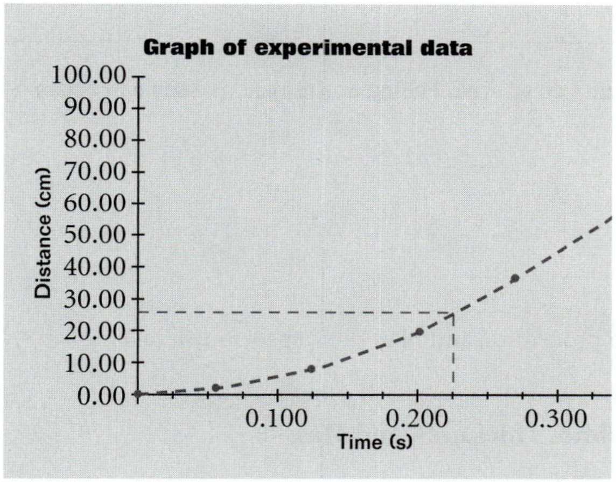

11. About how far has the ball fallen after 0.200 s?
   A. 5.00 cm
   B. 10.00 cm
   C. 20.00 cm
   D. 30.00 cm

12. Which of the following statements best describes the relationship between the variables?
   F. For equal time intervals, the change in position is increasing.
   G. For equal time intervals, the change in position is decreasing.
   H. For equal time intervals, the change in position is constant.
   J. There is no clear relationship between time and change in position.

SHORT RESPONSE

13. Determine the number of significant figures in each of the following measurements.
   A. 0.0057 kg
   B. 5.70 g
   C. 6070 m
   D. 6.070×10^3 m

14. Calculate the following sum, and express the answer in meters. Follow the rules for significant figures.
   (25.873 km) + (1024 m) + (3.0 cm)

15. Demonstrate how dimensional analysis can be used to find the dimensions that result from dividing distance by speed.

EXTENDED RESPONSE

16. You have decided to test the effects of four different garden fertilizers by applying them to four separate rows of vegetables. What factors should you control? How could you measure the results?

17. In a paragraph, describe how you could estimate the number of blades of grass on a football field.

Test TIP If more than one answer to a multiple-choice question seems to be correct, pick the answer that is *most* correct or that most directly answers the question.

10. F

11. C

12. F

13. **a.** 2
b. 3
c. 3
d. 4

14. 26 897 m

15. $\text{distance} \div \frac{\text{distance}}{\text{time}} =$

$$\frac{\cancel{\text{distance}} \times \text{time}}{\cancel{\text{distance}}} = \text{time}$$

16. Sample answer: Because the type of fertilizer is the variable being tested, all other factors should be controlled, including the type of vegetable, the amount of water, and the amount of sunshine. A fifth row with no fertilizer could be used as the control group. Results could be measured by size, quantity, appearance, and taste.

17. Paragraphs should describe a process similar to the following: First, you could count the number of blades of grass in a small area, such as a 10 cm by 10 cm square. You would round this to the nearest order of magnitude, then multiply by the number of such squares along the length of the field, and then multiply again by the approximate number of such squares along the width of the field.

Lab Planning

Beginning on page T34 are preparation notes and teaching tips to assist you in planning.

Blank data tables (as well as some sample data) appear on the **One-Stop Planner.**

No Books in the Lab?

See the ***Datasheets for In-Text Labs*** workbook for a reproducible master copy of this experiment.

CBL™ Option

A CBL™ version of this lab appears in the ***CBL™ Experiments*** workbook.

Safety Caution

Remind students to be aware of other groups' activities. Falling objects can cause serious injury.

CHAPTER 1

Skills Practice Lab: Physics and Measurement

OBJECTIVES

- **Measure** accurately using typical laboratory equipment.
- **Measure** length and mass in SI units.
- **Determine** the appropriate number of significant figures for various measurements and calculations.
- **Examine** the relationships between measured physical quantities by using graphs and data analysis.

MATERIALS LIST

- **2 rectangular wooden blocks**
- **15 cm metric ruler**
- **balance**
- **meterstick**
- **rectangular wooden block**
- **stopwatch**

In this laboratory exercise, you will gain experience making measurements as a physicist does. All measurements will be made using units to the precision allowed by your instruments.

SAFETY

- **Perform this lab in a clear area. Falling or dropped masses can cause serious injury.**

PROCEDURE

Preparation

1. Read the entire lab procedure, and plan the steps you will take.

Measuring Length, Width, Thickness, and Mass

2. If you are not using a datasheet provided by your teacher, prepare a data table in your lab notebook with seven columns and five rows, as shown below. In the first row, label the second through seventh columns *Trial 1, Trial 2, Trial 3, Trial 4, Trial 5,* and *Trial 6.* In the first column, label the second through fifth rows *Length (cm), Width (cm), Thickness (cm),* and *Mass (kg).*

	Trial 1	Trial 2	Trial 3	Trial 4	Trial 5	Trial 6
Length (cm)						
Width (cm)						
Thickness (cm)						
Mass (kg)						

3. Use a meterstick to measure the length of the wooden block. Record all measured digits plus one estimated digit.

4. Follow the same procedure to measure the width and thickness of the block. Repeat all measurements two more times. Record your data.

5. Carefully adjust the balance to obtain an average zero reading when there is no mass on it. Your teacher will show you how to adjust the balances in your classroom to obtain an average zero reading. Use the balance to find the mass of the block, as shown in **Figure 1.** Record the measurement in your data table.

6. Repeat the mass measurement two more times, and record the values in your data table. Each time, turn the block so that it rests on a different side.

7. For trials 4–6, repeat steps 3 through 6 with the second wooden block.

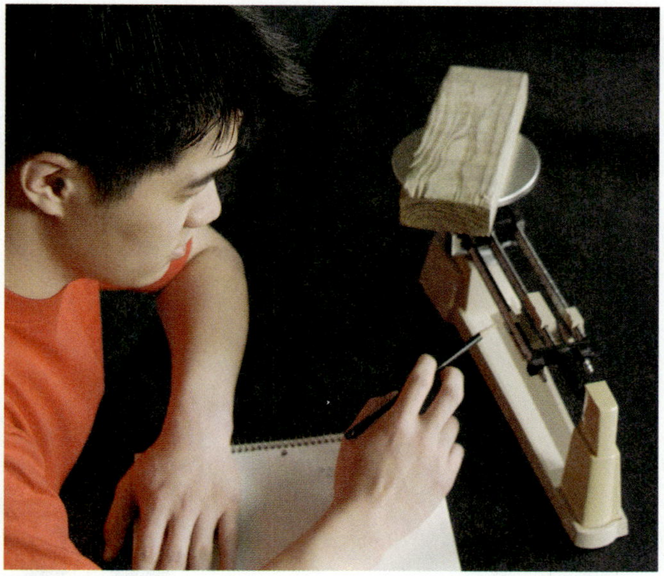

Figure 1

Step 5: Make sure you know how to use the balances in your classroom. The balance should read zero when there is no mass on it. The number of significant figures in your measurement will be determined by your instrument, the object being measured, and the purpose of your measurement.

Measuring Time and Distance

8. If you are not using a datasheet provided by your teacher, prepare a second data table in your lab notebook with three columns and seven rows, as shown below. In the first row, label the columns *Trial, Distance (m),* and *Time (s)*. Label the second through seventh rows *1, 2, 3, 4, 5,* and *6.*

Trial	Distance (m)	Time (s)
1		
2		
3		
4		
5		
6		

Tips and Tricks

- Remind students to record all measured digits plus one estimated digit.
- Students may repeat all length measurements using the 15 cm ruler.
- Students should practice timing the falling block before recording falling-block data.
- Better results that are easier to reproduce will be obtained with a greater falling distance.
- Show students the proper way to hold a meterstick when taking a measurement.
- Show students how to adjust the balances to zero. If you are using triple-beam balances, demonstrate handling the balances and moving the masses to take a measurement.
- Demonstrate how to hold the block between your hands and release the block by pulling your hands straight out to the sides. Show how this method prevents the block from turning, whereas other methods cause the block to turn.
- Make sure students know how to operate and read a stopwatch.

✓ Checkpoints

Step 3: Make sure all students perform all measurements. Students should be able to explain how they assign significant figures to their measurements.

Step 5: Students may need help using the balance for the first time. For some balances, instrument drift may prevent a continuous zero reading, so an average zero reading will be the goal.

CHAPTER 1 LAB

Step 10: Students should hold the block at about shoulder height for each trial. In step 13, the second student should hold the block at his or her own shoulder height.

Step 11: Before recording data, students should practice until they can demonstrate that the timer starts at the moment the block is released and stops when the block hits the floor.

ANSWERS

Analysis

1. Answers will vary, depending on the blocks used. Make sure answers have the right number of significant figures. For sample data, block 1 Trial 1 has a volume of 1.60×10^2 cm^3.

2. Answers will vary.
  a. For the sample data block 1, the answer is 0.10 cm.
  b. For the sample data block 1, the answer is 12 cm^3.
  c. Students should recognize that the answer becomes less precise as several values are multiplied and that the difference between the highest and lowest answers becomes greater.

3. Student answers should state that the block always falls from the same height in the same amount of time. If students answer the question by finding the ratio between distance and time, they will find different values for different heights.

4. The graph should show one point for each height from which the block fell.

Figure 2
Step 10: Hold the block between your hands.
Step 11: Release the block by pulling both hands straight out to the sides. It may take some practice to release the block so that it falls straight down without turning.

9. Perform this exercise with a partner. One partner will drop the wooden block from a measured height, and the other partner will measure the time it takes the block to fall to the floor. Perform this in a clear area away from other groups.

10. One student should hold the wooden block straight out in front of him or her at shoulder height. Hold the block between your hands, as shown in **Figure 2.** Use the meterstick to measure the height of the wooden block above the floor. Record this distance in your data table.

11. Use the stopwatch to time the fall of the block. Make sure the area is clear, and inform nearby groups that you are about to begin. The student holding the block should release it by pulling both hands straight out to the sides. The student with the stopwatch should begin timing the instant the block is released and stop timing as soon as the block hits the floor. In your data table, record the time required for the block to fall.

12. Repeat for two more trials, recording all data in your data table. Try to drop the block from exactly the same height each time.

13. Switch roles, and repeat steps 10 through 12. Perform three trials. Record all data in your data table.

ANALYSIS

1. **Organizing Data** Using your data from the first data table, calculate the volume of the wooden block for each trial. The equation for the volume of a rectangular block is *volume* = *length* × *width* × *thickness.*

2. **Analyzing Data** Use your data from the first table and your results from item 1 above to answer the following questions.

   a. For each block, what is the difference between the smallest length measurement and the largest length measurement?

   b. For each block, what is the difference between the smallest calculated volume and the largest calculated volume?

   c. Based on your answers to (a) and (b), how does multiplying several length measurements together to find the volume affect the precision of the result?

3. **Analyzing Data** Did the block always fall from the same height in the same amount of time? Explain how you found the answer to this question.

4. **Constructing Graphs** Using the data from all trials, make a scatter plot of the distance versus the time of the block's fall. Use a graphing calculator, computer, or graph paper.

CONCLUSIONS

5. **Drawing Conclusions** For each trial in the first data table, find the ratio between the mass and the volume. Based on your data, what is the relationship between the mass and volume?

6. **Evaluating Methods** For each type of measurement you made, explain how error could have affected your results. Consider method error and instrument error. How could you find out whether error had a significant effect on your results for each part of the lab? Explain the role of human reaction time in your measurements.

EXTENSION

7. **Evaluating Data** If there is time and your teacher approves, conduct the following experiment. Have one student drop the wooden block from shoulder height while all other class members time the fall. Perform three trials. Compare results each time. What does this exercise suggest about accuracy and precision in the laboratory?

Conclusions

5. Answers should state that both wooden blocks have the same mass-to-volume ratio, even though their masses and volumes are not the same. Some students may realize that this value is that of the density of the wood.

6. Student answers will vary but should include an analysis of error in the laboratory. Students should understand that reaction time is part of the error in the block experiment.

Extension

7. Typically, a range of values is reported as different class members measure the time of the same event. This result should lead students to realize that for short times, error can be a significant factor.

CHAPTER 2 Motion in One Dimension

Planning Guide

Compression Guide

To shorten instruction because of time limitations, omit the opener and abbreviate the review.

OBJECTIVES	LABS, DEMONSTRATIONS, AND ACTIVITIES	TECHNOLOGY RESOURCES
PACING • 45 min pp. 38–39 **Chapter Opener**	**ANC** **Discovery Lab** Motion*◆ BASIC	**CD** **Visual Concepts,** Chapter 2 BASIC
PACING • 45 min pp. 40–47 **Section 1 Displacement and Velocity** • Describe motion in terms of frame of reference, displacement, time, and velocity. • Calculate the displacement of an object traveling at a known velocity for a specific time interval. • Construct and interpret graphs of position versus time.	**TE** **Demonstration** Displacement, p. 42 BASIC	**OSP** **Lesson Plans** **EXT** **Integrating Technology** Methods of Transportation BASIC **TR** 5 Positive and Negative Displacements
PACING • 90 min pp. 48–59 **Section 2 Acceleration** • Describe motion in terms of changing velocity. • Compare graphical representations of accelerated and non-accelerated motions. • Apply kinematic equations to calculate distance, time, or velocity under conditions of constant acceleration.	**TE** **Demonstration** Acceleration, p. 48 GENERAL **TE** **Demonstration** Constant Acceleration, p. 51 GENERAL **ANC** **Invention Lab** Race-Car Construction*◆ ADVANCED **ANC** **CBL™ Experiment** Acceleration*◆ ADVANCED	**OSP** **Lesson Plans** **TR** 11A Signs of Velocity and Acceleration **TR** 12A Constant Acceleration and Average Velocity **TR** 13A Equations for Constantly Accelerated Straight-Line Motion
PACING • 90 min pp. 60–65 **Section 3 Falling Objects** • Relate the motion of a freely falling body to motion with constant acceleration. • Calculate displacement, velocity, and time at various points in the motion of a freely falling object. • Compare the motions of different objects in free fall.	**SE** **Quick Lab** Time Interval of Free Fall, p. 62 GENERAL **SE** **Skills Practice Lab** Free-Fall Acceleration, pp. 76–79◆ GENERAL **ANC** **Datasheet** Free-Fall Acceleration* GENERAL **SE** **CBL™ Lab** Free-Fall Acceleration, pp. 932–933◆ **ANC** **CBL™ Experiment** Free-Fall Acceleration*◆ GENERAL **ANC** **CBL™ Experiment** Free Fall*◆ ADVANCED	**OSP** **Lesson Plans**

PACING • 90 min

CHAPTER REVIEW, ASSESSMENT, AND STANDARDIZED TEST PREPARATION

- **SE** **Chapter Highlights,** p. 67
- **SE** **Chapter Review,** pp. 68–73
- **SE** **Graphing Calculator Practice,** p. 73 GENERAL
- **SE** **Alternative Assessment,** p. 73 ADVANCED
- **SE** **Standardized Test Prep,** pp. 74–75 GENERAL
- **SE** **Appendix D: Equations,** p. 854
- **SE** **Appendix I: Additional Problems,** pp. 880–881
- **ANC** **Study Guide Worksheet** Mixed Review* GENERAL
- **ANC** **Chapter Test A*** GENERAL
- **ANC** **Chapter Test B*** ADVANCED
- **OSP** **Test Generator**

Online and Technology Resources

go.hrw.com Holt Online Learning

Visit **go.hrw.com** to find a variety of online resources. To access this chapter's extensions, enter the keyword **HF6MODXT** and click the "go" button. Click **Holt Online Learning** for an online edition of this textbook, and other interactive resources.

HOLT Teacher's One-Stop Planner®

This DVD package includes:
- Holt Calendar Planner
- Customizable Lesson Plans
- Editable Worksheets
- ExamView® Version 6 Assessment Suite
- Interactive Teacher's Edition
- Holt PuzzlePro®
- Holt PowerNotes® Presentations
- MindPoint® Quiz Show

KEY

SE Student Edition	**OSP** One-Stop Planner	**EXT** Online Extension
TE Teacher Edition	**CD** CD or CD-ROM	* Also on One-Stop Planner
ANC Ancillary Worksheet	**TR** Teaching Transparencies	◆ Requires advance prep

SKILLS DEVELOPMENT RESOURCES	REVIEW AND ASSESSMENT	CORRELATIONS
		National Science Education Standards
SE Conceptual Challenge, p. 41 GENERAL **SE Sample Set A** Average Velocity and Displacement, p. 44 BASIC **TE Classroom Practice,** p. 44 BASIC **ANC Problem Workbook*** and **OSP Problem Bank** Sample Set A BASIC **SE Conceptual Challenge,** p. 45 GENERAL	**SE Section Review,** p. 47 GENERAL **ANC Study Guide Worksheet** Section 1* GENERAL **ANC Quiz** Section 1* BASIC	UCP 2, 3 SAI 1, 2
SE Sample Set B Average Acceleration, p. 49 BASIC **TE Classroom Practice,** p. 49 BASIC **ANC Problem Workbook*** and **OSP Problem Bank** Sample Set B BASIC **SE Conceptual Challenge,** p. 50 GENERAL **SE Sample Set C** Displacement with Constant Acceleration, p. 53 BASIC **TE Classroom Practice,** p. 53 BASIC **ANC Problem Workbook*** and **OSP Problem Bank** Sample Set C BASIC **SE Sample Set D** Velocity and Displacement, p. 55 GENERAL **ANC Problem Workbook*** and **OSP Problem Bank** Sample Set D GENERAL **SE Sample Set E** Final Velocity After Any Displacement, pp. 57–58 GENERAL **ANC Problem Workbook*** and **OSP Problem Bank** Sample Set E GENERAL **SE Advanced Topics** Angular Kinematics, pp. 898–901 ADVANCED **SE Advanced Topics** Special Relativity and Time Dilation, pp. 914–915 ADVANCED	**SE Section Review,** p. 59 GENERAL **ANC Study Guide Worksheet** Section 2* GENERAL **ANC Quiz** Section 2* BASIC	UCP 2, 3 SAI 1, 2 HNS 3
SE Sample Set F Falling Object, pp. 63–64 GENERAL **TE Classroom Practice,** p. 63 GENERAL **ANC Problem Workbook*** and **OSP Problem Bank** Sample Set F GENERAL	**SE Section Review,** p. 65 GENERAL **ANC Study Guide Worksheet** Section 3* GENERAL **ANC Quiz** Section 3* BASIC	UCP 2, 3 SAI 1, 2

SCILINKS. NSTA
www.scilinks.org

Maintained by the **National Science Teachers Association.**

Topic: Motion
SciLinks Code: HF60996

Topic: Acceleration
SciLinks Code: HF60007

Topic: Galileo
SciLinks Code: HF60633

Topic: Free Fall
SciLinks Code: HF60620

Classroom CD-ROMs

- Guided Reading Audio Program
- Student One Stop
- Virtual Investigations
- Visual Concepts

Holt Lab Generator CD-ROM

Search for any lab by topic, standard, difficulty level, or time. Edit any lab to fit your needs, or create your own labs. Use the Lab Materials QuickList software to customize your lab materials list.

CHAPTER 2
Overview

Section 1 introduces the concepts and relationships between displacement, time, speed, and velocity.

Section 2 describes the difference between accelerated motion and nonaccelerated motion and introduces the kinematic equations for constant acceleration.

Section 3 explores freely falling bodies as examples of motion with constant acceleration.

About the Illustration

This high-speed train is emerging from the Box Tunnel in western England. When this 2939 m long tunnel opened in 1841 as part of the Great Western Railway, it was the longest railroad tunnel in the world.

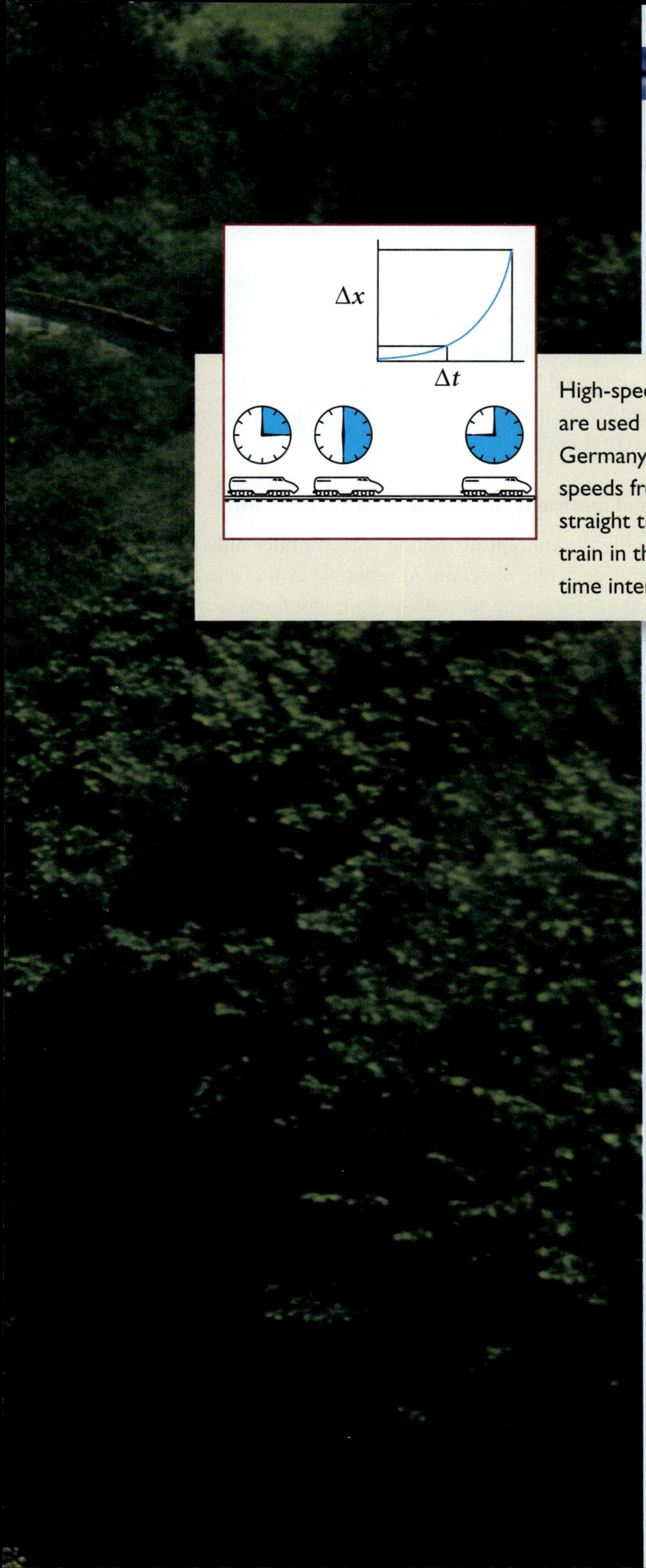

CHAPTER 2

Motion in One Dimension

High-speed passenger trains such as the one shown here are used in many countries, including Japan, France, England, Germany, and South Korea. These trains have operational speeds from 200 to 300 km/h. A train moving along a straight track is an example of *one-dimensional motion.* The train in the diagram is covering greater distances in equal time intervals—in other words, it is accelerating.

WHAT TO EXPECT

In this chapter, you will learn how to analyze one-dimensional motion in terms of displacement, time, speed, and velocity. You will also learn how to distinguish between accelerated and nonaccelerated motion.

Why it Matters

Velocity and acceleration are involved in many aspects of everyday life, from riding a bicycle to driving a car to traveling on a high-speed train. The definitions and equations you will study in this chapter allow you to make predictions about these aspects of motion, given certain initial conditions.

CHAPTER PREVIEW

Tapping Prior Knowledge

Knowledge to Review

- ✔ The number of significant figures in a number equals the number of reliably known digits plus one estimated digit.
- ✔ SI measurements of motion are based on a set of three standards: m, kg, and s.
- ✔ Graphs and tables provide a way to organize data and recognize relationships.

Items to Probe

- ✔ Unit conversions: Ask the students to convert from m/s to km/h.
- ✔ Significant figures: Ask the students to identify the number of significant digits in a set of numbers on the board. Make sure that they understand that 50 m/s has only one significant digit.

SECTION 1
General Level

The Language of Physics

Although this chapter discusses displacement, velocity, and acceleration, the concept of vectors is not introduced here. For the purposes of this chapter, it is sufficient to describe the direction of a quantity with a positive or negative sign because the focus is on motion in one dimension. The transition to two-dimensional motion and vectors is made in the first section of the chapter "Two-Dimensional Motion and Vectors."

SECTION 1

Displacement and Velocity

SECTION OBJECTIVES

- **Describe motion in terms of frame of reference, displacement, time, and velocity.**
- **Calculate the displacement of an object traveling at a known velocity for a specific time interval.**
- **Construct and interpret graphs of position versus time.**

frame of reference

a system for specifying the precise location of objects in space and time

MOTION

Motion happens all around us. Every day, we see objects such as cars, people, and soccer balls move in different directions with different speeds. We are so familiar with the idea of motion that it requires a special effort to analyze motion as a physicist does.

One-dimensional motion is the simplest form of motion

One way to simplify the concept of motion is to consider only the kinds of motion that take place in one direction. An example of this one-dimensional motion is the motion of a commuter train on a straight track, as in **Figure 1.**

In this one-dimensional motion, the train can move either forward or backward along the tracks. It cannot move left and right or up and down. This chapter deals only with one-dimensional motion. In later chapters, you will learn how to describe more complicated motions such as the motion of thrown baseballs and other projectiles.

Motion takes place over time and depends upon the frame of reference

It seems simple to describe the motion of the train. As the train in **Figure 1** begins its route, it is at the first station. Later, it will be at another station farther down the tracks. But Earth is spinning on its axis, so the train, stations, and the tracks are also moving around the axis. At the same time, Earth is moving around the sun. The sun and the rest of the solar system are moving through our galaxy. This galaxy is traveling through space as well.

When faced with a complex situation like this, physicists break it down into simpler parts. One key approach is to choose a **frame of reference** against which you can measure changes in position. In the case of the train, any of the stations along its route could serve as a convenient frame of reference. When you select a reference frame, note that it remains fixed for the problem in question and has an origin, or starting point, from which the motion is measured.

Figure 1
The motion of a commuter train traveling along a straight route is an example of one-dimensional motion. Each train can move only forward and backward along the track.

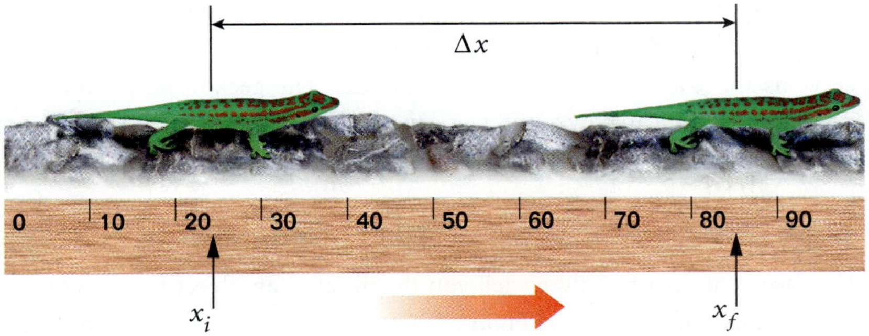

Figure 2
A gecko moving along the x-axis from x_i to x_f undergoes a displacement of $\Delta x = x_f - x_i$.

If an object is at rest (not moving), its position does not change with respect to a fixed frame of reference. For example, the benches on the platform of one subway station never move down the tracks to another station.

In physics, any frame of reference can be chosen as long as it is used consistently. If you are consistent, you will get the same results, no matter which frame of reference you choose. But some frames of reference can make explaining things easier than other frames of reference.

For example, when considering the motion of the gecko in **Figure 2,** it is useful to imagine a stick marked in centimeters placed under the gecko's feet to define the frame of reference. The measuring stick serves as an x-axis. You can use it to identify the gecko's initial position and its final position.

DISPLACEMENT

As any object moves from one position to another, the length of the straight line drawn from its initial position to the object's final position is called the **displacement** of the object.

displacement

the change in position of an object

Displacement is a change in position

The gecko in **Figure 2** moves from left to right along the x-axis from an initial position, x_i, to a final position, x_f. The gecko's displacement is the difference between its final and initial coordinates, or $x_f - x_i$. In this case, the displacement is about 61 cm (85 cm − 24 cm). The Greek letter delta (Δ) before the x denotes a *change* in the position of an object.

DISPLACEMENT

$$\Delta x = x_f - x_i$$

displacement = change in position = final position − initial position

TIP *A change in any quantity, indicated by the Greek symbol delta (Δ), is equal to the final value minus the initial value. When calculating displacement, always be sure to subtract the initial position from the final position so that your answer has the correct sign.*

Why it Matters

Conceptual Challenge

1. Space Shuttle

A space shuttle takes off from Florida and circles Earth several times, finally landing in California. While the shuttle is in flight, a photographer flies from Florida to California to take pictures of the astronauts when they step off the shuttle. Who undergoes the greater displacement, the photographer or the astronauts?

2. Roundtrip

What is the difference between the displacement of the photographer flying from Florida to California and the displacement of the astronauts flying from California back to Florida?

Visual Strategy GENERAL

Figure 2

Point out to students that the choice of a reference point for the coordinate system is arbitrary, but once chosen, the same point must be used throughout the problem.

Q What would the displacement of the gecko be if the zero end of the meterstick had been lined up with the gecko's first position?

A ***the same (because where we put the meterstick does not affect how far the gecko moved)***

The Language of Physics

In this book, Δx refers to a change in position along the x-axis of whatever coordinate system is chosen and Δy refers to a change in position along the y-axis. In later chapters, these two variables refer to the x and y components of a displacement vector.

STOP Misconception Alert GENERAL

Many students may have difficulty understanding that the magnitude of a displacement is the length of the straight-line path between two points rather than the distance traveled. Point out that although the odometer on a car shows that it has been driven 5 mi, the displacement may have been 0 mi.

ANSWERS

Conceptual Challenge

1. neither (because the displacements are the same)
2. The difference between these two displacements is in their direction—one is positive and the other negative.

Demonstration

Displacement — BASIC

Purpose Demonstrate the importance of direction in reference to displacement.

Materials one meterstick, 3 pieces of modeling clay, one toothpick or paper clip, one toy car

Procedure Place the meterstick on edge so that the 0 mark is to the students' left and the students can see the numbers. Put the toothpick in one of the pieces of modeling clay to represent the initial position.

For positive displacement, place the initial position marker and car somewhere between 0 and 10 cm. Roll the car down the meterstick to some point past the 50 cm mark, and place the final position marker (the second piece of modeling clay) at the car's new position. Ask the students to calculate the displacement of the car. *(It should be a positive number.)*

For negative displacement, move the initial position marker to the car, and roll the car back toward the zero end of the meterstick. Stop the car and place the third position marker. Ask the students to calculate the displacement for the second leg of the trip. *(It should be a negative number.)*

Now have them calculate the car's total displacement.

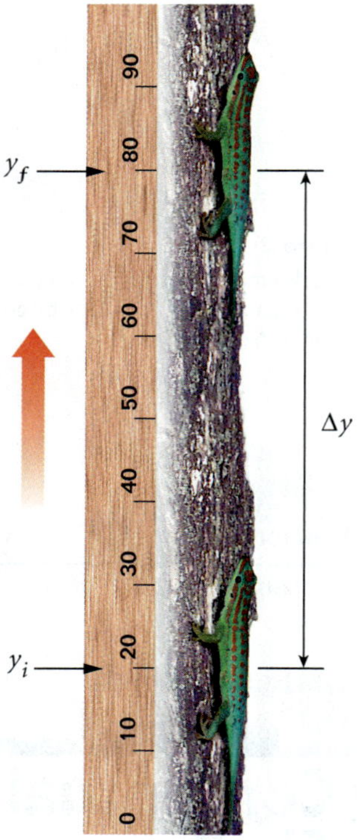

Figure 3
When the gecko is climbing a tree, the displacement is measured on the y-axis. Again, the gecko's position is determined by the position of the same point on its body.

Now suppose the gecko runs up a tree, as shown in **Figure 3.** In this case, we place the measuring stick parallel to the tree. The measuring stick can serve as the y-axis of our coordinate system. The gecko's initial and final positions are indicated by y_i and y_f, respectively, and the gecko's displacement is denoted as Δy.

Displacement is not always equal to the distance traveled

Displacement does not always tell you the distance an object has moved. For example, what if the gecko in **Figure 3** runs up the tree from the 20 cm marker (its initial position) to the 80 cm marker. After that, it retreats down the tree to the 50 cm marker (its final position). It has traveled a total distance of 90 cm. However, its displacement is only 30 cm ($y_f - y_i = 50\text{ cm} - 20\text{ cm} = 30\text{ cm}$). If the gecko were to return to its starting point, its displacement would be zero because its initial position and final position would be the same.

Displacement can be positive or negative

Displacement also includes a description of the direction of motion. In one-dimensional motion, there are only two directions in which an object can move, and these directions can be described as positive or negative.

In this book, unless otherwise stated, the right (or east) will be considered the positive direction and the left (or west) will be considered the negative direction. Similarly, upward (or north) will be considered positive and downward (or south) will be considered negative. **Table 1** gives examples of determining displacements for a variety of situations.

Table 1 Positive and Negative Displacements

Positive

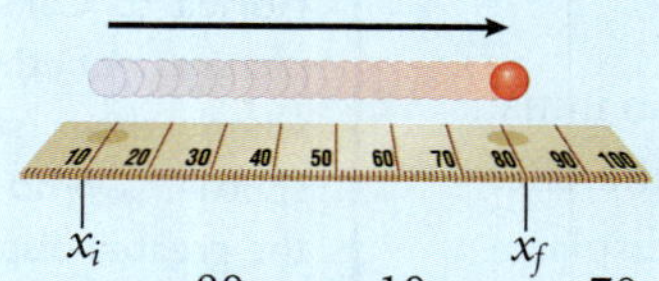

$$\Delta x = x_f - x_i = 80\text{ cm} - 10\text{ cm} = +70\text{ cm}$$

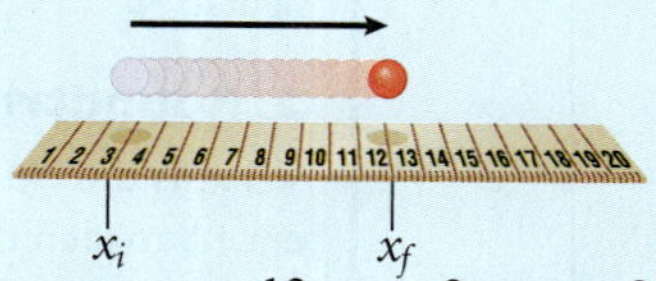

$$\Delta x = x_f - x_i = 12\text{ cm} - 3\text{ cm} = +9\text{ cm}$$

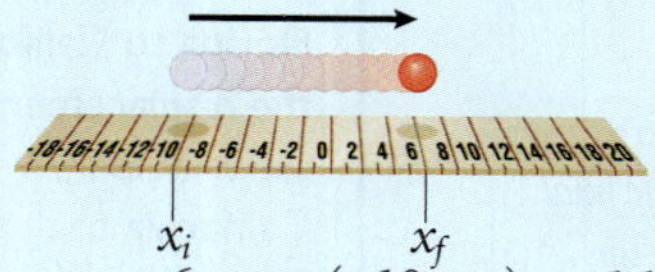

$$\Delta x = x_f - x_i = 6\text{ cm} - (-10\text{ cm}) = +16\text{ cm}$$

Negative

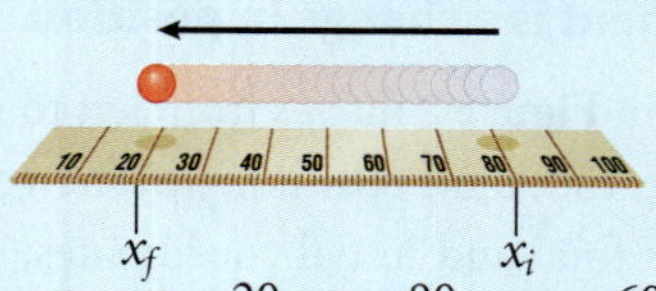

$$\Delta x = x_f - x_i = 20\text{ cm} - 80\text{ cm} = -60\text{ cm}$$

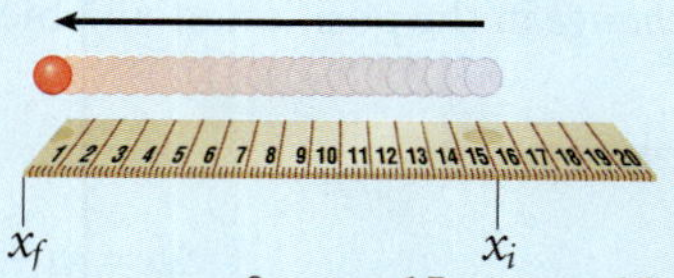

$$\Delta x = x_f - x_i = 0\text{ cm} - 15\text{ cm} = -15\text{ cm}$$

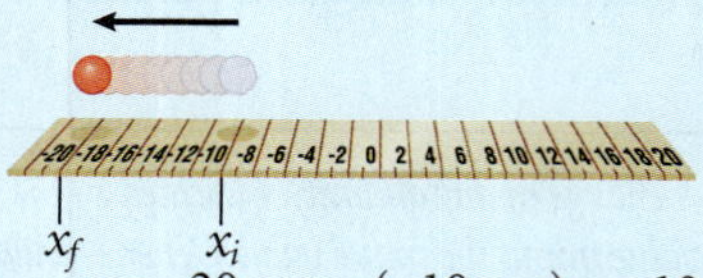

$$\Delta x = x_f - x_i = -20\text{ cm} - (-10\text{ cm}) = -10\text{ cm}$$

VELOCITY

Where an object started and where it stopped does not completely describe the motion of the object. For example, the ground that you're standing on may move 8.0 cm to the left. This motion could take several years and be a sign of the normal slow movement of Earth's tectonic plates. If this motion takes place in just a second, however, you may be experiencing an earthquake or a landslide. Knowing the speed is important when evaluating motion.

Average velocity is displacement divided by the time interval

Consider the car in **Figure 4.** The car is moving along a highway in a straight line (the x-axis). Suppose that the positions of the car are x_i at time t_i and x_f at time t_f. In the time interval $\Delta t = t_f - t_i$, the displacement of the car is $\Delta x = x_f - x_i$. The **average velocity,** v_{avg}, is defined as the displacement divided by the time interval during which the displacement occurred. In SI, the unit of velocity is meters per second, abbreviated as m/s.

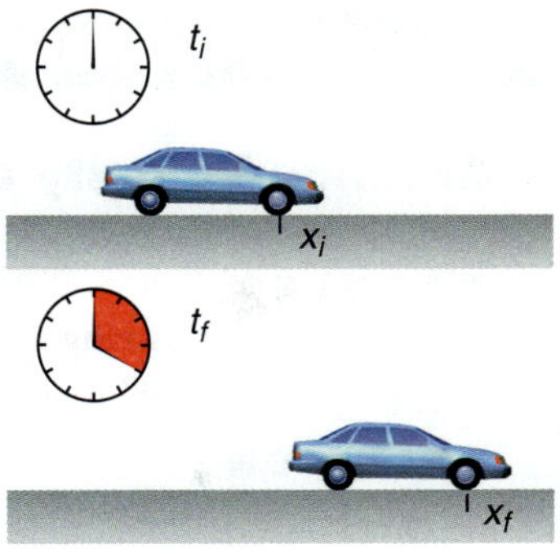

Figure 4
The average velocity of this car tells you how fast and in which direction it is moving.

average velocity

the total displacement divided by the time interval during which the displacement occurred

AVERAGE VELOCITY

$$v_{avg} = \frac{\Delta x}{\Delta t} = \frac{x_f - x_i}{t_f - t_i}$$

$$\text{average velocity} = \frac{\text{change in position}}{\text{change in time}} = \frac{\text{displacement}}{\text{time interval}}$$

The average velocity of an object can be positive or negative, depending on the sign of the displacement. (The time interval is always positive.) As an example, consider a car trip to a friend's house 370 km to the west (the negative direction) along a straight highway. If you left your house at 10 A.M. and arrived at your friend's house at 3 P.M., your average velocity would be as follows:

$$v_{avg} = \frac{\Delta x}{\Delta t} = \frac{-370\ \text{km}}{5.0\ \text{h}} = -74\ \text{km/h} = 74\ \text{km/h west}$$

This value is an average. You probably did not travel exactly 74 km/h at every moment. You may have stopped to buy gas or have lunch. At other times, you may have traveled more slowly as a result of heavy traffic. To make up for such delays, when you were traveling slower than 74 km/h, there must also have been other times when you traveled faster than 74 km/h.

The average velocity is equal to the constant velocity needed to cover the given displacement in a given time interval. In the example above, if you left your house and maintained a velocity of 74 km/h to the west at every moment, it would take you 5.0 h to travel 370 km.

TIP *Average velocity is not always equal to the average of the initial and final velocities. For instance, if you drive first at 40 km/h west and later at 60 km/h west, your average velocity is not necessarily 50 km/h west.*

SCILINKS.
www.scilinks.org
Topic: Motion
Code: HF60996

Did you know?

The branch of physics concerned with motion and forces is called *mechanics*. The subset of mechanics that describes motion without regard to its causes is called *kinematics*.

Misconception Alert — GENERAL

Many students believe that the average speed is always the average of the starting and ending speeds, as discussed in the Tip on this student page. (Note that the student Tip discusses average velocity rather than average speed because average speed has not yet been introduced.) Use counterexamples to address this misconception.

Example: A car travels from city A to city B (100 km). If the first half of the distance is driven at 50 km/h and the second half is driven at 100 km/h, the average speed is given by the following relation.

$$\frac{100\ \text{km}}{\frac{50\ \text{km}}{50\ \text{km/h}} + \frac{50\ \text{km}}{100\ \text{km/h}}} = 67\ \text{km/h}$$

The average speed would be 75 km/h if the car spent equal *time* at 50 km/h and 100 km/h.

Classroom Practice

Average Velocity and Displacement
A doctor travels to the east from city A to city B (75 km) in 1.0 h. What is the doctor's average velocity?

Answer
75 km/h to the east

PROBLEM GUIDE A

Use this guide to assign problems.
SE = Student Edition Textbook
PW = Problem Workbook
PB = Problem Bank on the One-Stop Planner (OSP)

Solving for:

Δx	**SE** Sample, 1–3; Ch. Rvw. 7–10, 11*, 38b **PW** 3b **PB** 8, 10
Δt	**SE** 4–6a; Ch. Rvw. 34, 35*, 38a **PW** Sample, 1–3a, 4a, 8b **PB** 7, 9
v_{avg}	**SE** 6b; Ch. Rvw. 9c–d, 10a, 51* **PW** 4b, 5–8b **PB** Sample, 1–6

***Challenging Problem**
Consult the printed Solutions Manual or the OSP for detailed solutions.

ANSWERS

Practice A

1. 2.0 km to the east
2. 3.1 km
3. 680 m to the north
4. 3.00 h
5. 0.43 h
6. a. 6.4 h
   b. 77 km/h to the south

SAMPLE PROBLEM A

Average Velocity and Displacement

PROBLEM

During a race on level ground, Andra runs with an average velocity of 6.02 m/s to the east. What is Andra's displacement after 137 s?

SOLUTION

Given: $v_{avg} = 6.02$ m/s
$\Delta t = 137$ s

Unknown: $\Delta x = ?$

Rearrange the average velocity equation to solve for displacement.

$$v_{avg} = \frac{\Delta x}{\Delta t}$$

$$\Delta x = v_{avg}\Delta t$$

$$\Delta x = v_{avg}\Delta t = (6.02 \text{ m/s})(137 \text{ s}) = \boxed{825 \text{ m to the east}}$$

CALCULATOR SOLUTION

The calculator answer is 824.74 m, but both the values for velocity and time have three significant figures, so the displacement must be reported as 825 m.

PRACTICE A

Average Velocity and Displacement

1. Heather and Matthew walk with an average velocity of 0.98 m/s eastward. If it takes them 34 min to walk to the store, what is their displacement?
2. If Joe rides his bicycle in a straight line for 15 min with an average velocity of 12.5 km/h south, how far has he ridden?
3. It takes you 9.5 min to walk with an average velocity of 1.2 m/s to the north from the bus stop to the museum entrance. What is your displacement?
4. Simpson drives his car with an average velocity of 48.0 km/h to the east. How long will it take him to drive 144 km on a straight highway?
5. Look back at item 4. How much time would Simpson save by increasing his average velocity to 56.0 km/h to the east?
6. A bus travels 280 km south along a straight path with an average velocity of 88 km/h to the south. The bus stops for 24 min. Then, it travels 210 km south with an average velocity of 75 km/h to the south.
   a. How long does the total trip last?
   b. What is the average velocity for the total trip?

Velocity is not the same as speed

In everyday language, the terms *speed* and *velocity* are used interchangeably. In physics, however, there is an important distinction between these two terms. As we have seen, velocity describes motion with both a direction and a numerical value (a magnitude) indicating how fast something moves. However, speed has no direction, only magnitude. An object's average speed is equal to the distance traveled divided by the time interval for the motion.

$$\text{average speed} = \frac{\text{distance traveled}}{\text{time of travel}}$$

Velocity can be interpreted graphically

The velocity of an object can be determined if the object's position is known at specific times along its path. One way to determine this is to make a graph of the motion. **Figure 5** represents such a graph. Notice that time is plotted on the horizontal axis and position is plotted on the vertical axis.

The object moves 4.0 m in the time interval between $t = 0$ s and $t = 4.0$ s. Likewise, the object moves an additional 4.0 m in the time interval between $t = 4.0$ s and $t = 8.0$ s. From these data, we see that the average velocity for each of these time intervals is +1.0 m/s (because $v_{avg} = \Delta x/\Delta t = 4.0 \text{ m}/4.0 \text{ s}$). Because the average velocity does not change, the object is moving with a constant velocity of +1.0 m/s, and its motion is represented by a straight line on the position-time graph.

For any position-time graph, we can also determine the average velocity by drawing a straight line between any two points on the graph. The slope of this line indicates the average velocity between the positions and times represented by these points. To better understand this concept, compare the equation for the slope of the line with the equation for the average velocity.

Slope of a Line

$$\text{slope} = \frac{\text{rise}}{\text{run}} = \frac{\text{change in vertical coordinates}}{\text{change in horizontal coordinates}}$$

Average Velocity

$$v_{avg} = \frac{\Delta x}{\Delta t} = \frac{x_f - x_i}{t_f - t_i}$$

extension

Integrating Technology
Visit go.hrw.com for the activity "Methods of Transportation."

Keyword HF6MODX

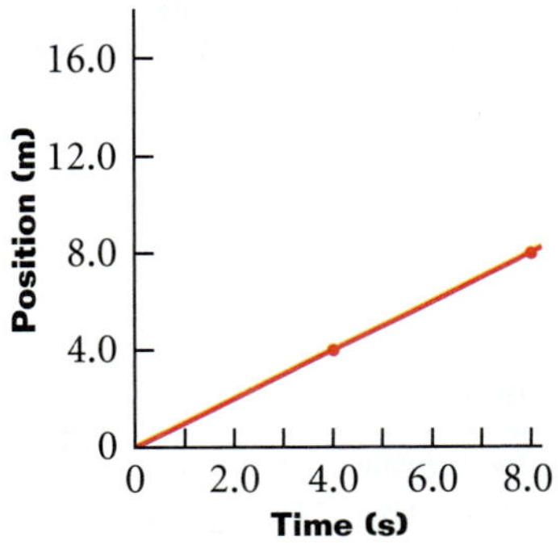

Figure 5
The motion of an object moving with constant velocity will provide a straight-line graph of position versus time. The slope of this graph indicates the velocity.

Why it Matters

Conceptual Challenge

1. Book on a Table A book is moved once around the edge of a tabletop with dimensions 1.75 m × 2.25 m. If the book ends up at its initial position, what is its displacement? If it completes its motion in 23 s, what is its average velocity? What is its average speed?

2. Travel Car A travels from New York to Miami at a speed of 25 m/s. Car B travels from New York to Chicago, also at a speed of 25 m/s. Are the velocities of the cars equal? Explain.

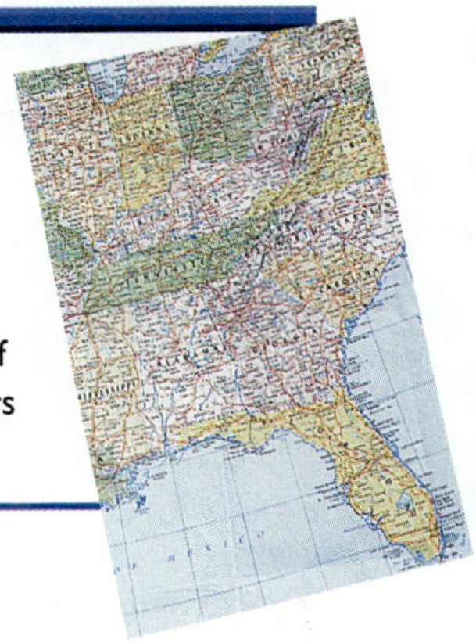

SECTION 1

Key Models and Analogies — GENERAL

The idea of a car race, such as the Indianapolis 500, may help students understand the difference between speed and velocity. Cars have *speedometers,* not "velocitometers." Indy cars spend hours going very fast (high speed) but getting nowhere (zero velocity). In order to have a nonzero velocity, the car must finish at some point other than the starting point.

Teaching Tip — BASIC

Probe your students' understanding of the concept of *slope.* Many students may remember the concept from math class. Students will most likely remember slope from algebra as described by the phrase *rise over run.*

ANSWERS

Conceptual Challenge

1. The book's displacement is zero, its average velocity is zero, and its average speed is 0.35 m/s.
2. The velocity of car A does not equal the velocity of car B even though their speeds are the same, because they are traveling in different directions.

Misconception Alert GENERAL

Some students will use raw data points to calculate the slope of a line. Students should be shown how to draw a "best-fit" line and should be cautioned to use points on the best-fit line, not the raw data points, which may or may not be on the line.

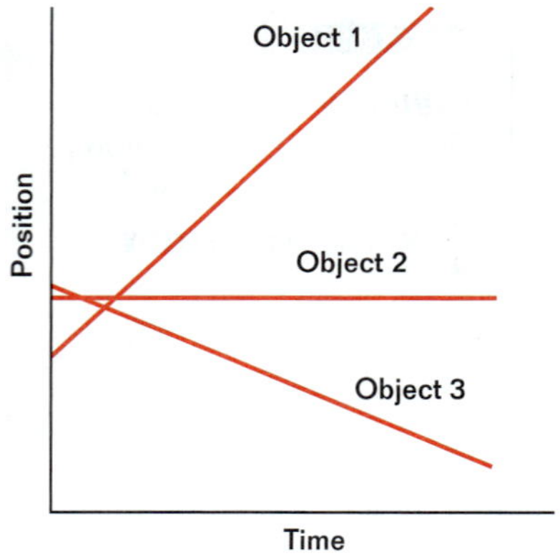

Figure 6
These position-versus-time graphs show that object I moves with a constant positive velocity. Object 2 is at rest. Object 3 moves with a constant negative velocity.

Figure 6 represents straight-line graphs of position-versus-time for three different objects. Object 1 has a constant positive velocity because its position increases uniformly with time. Thus, the slope of this line is positive. Object 2 has zero velocity because its position does not change (the object is at rest). Hence, the slope of this line is zero. Object 3 has a constant negative velocity because its position decreases with time. As a result, the slope of this line is negative.

Instantaneous velocity may not be the same as average velocity

Now consider an object whose position versus time graph is not a straight line, but a curve, as in **Figure 7.** The object moves through larger and larger displacements as each second passes. Thus, its velocity increases with time.

For example, between $t = 0$ s and $t = 2.0$ s, the object moves 8.0 m, and its average velocity in this time interval is 4.0 m/s (because $v_{avg} = 8.0$ m/2.0 s). However, between $t = 0$ s and $t = 4.0$ s, it moves 32 m, so its average velocity in this time interval is 8.0 m/s (because $v_{avg} = 32$ m/4.0 s). We obtain different average velocities, depending on the time interval we choose. But how can we find the velocity at an instant of time?

To determine the velocity at some instant, such as $t = 3.0$ s, we study a small time interval near that instant. As the intervals become smaller and smaller, the average velocity over that interval approaches the exact velocity at $t = 3.0$ s. This is called the **instantaneous velocity.**

instantaneous velocity

the velocity of an object at some instant or at a specific point in the object's path

One way to determine the instantaneous velocity is to construct a straight line that is *tangent* to the position-versus-time graph at that instant. The slope of this tangent line is equal to the value of the instantaneous velocity at that point. For example, the instantaneous velocity of the object in **Figure 7** at $t = 3.0$ s is 12 m/s. **Table 2** lists the instantaneous velocities of the object described by the graph in **Figure 7.** You can verify some of these values by carefully measuring the slope of the curve.

Table 2
Velocity-Time Data

t (s)	v (m/s)
0.0	0.0
1.0	4.0
2.0	8.0
3.0	12.0
4.0	16.0

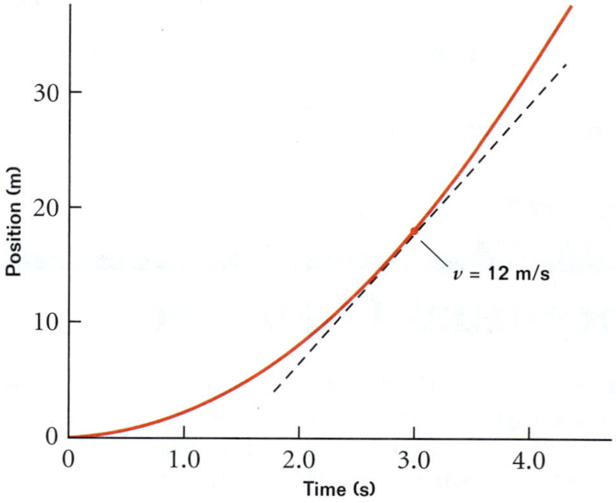

Figure 7
The instantaneous velocity at a given time can be determined by measuring the slope of the line that is tangent to that point on the position-versus-time graph.

SECTION REVIEW

1. What is the shortest possible time in which a bacterium could travel a distance of 8.4 cm across a Petri dish at a constant speed of 3.5 mm/s?

2. A child is pushing a shopping cart at a speed of 1.5 m/s. How long will it take this child to push the cart down an aisle with a length of 9.3 m?

3. An athlete swims from the north end to the south end of a 50.0 m pool in 20.0 s and makes the return trip to the starting position in 22.0 s.
   a. What is the average velocity for the first half of the swim?
   b. What is the average velocity for the second half of the swim?
   c. What is the average velocity for the roundtrip?

4. Two students walk in the same direction along a straight path, at a constant speed—one at 0.90 m/s and the other at 1.90 m/s.
   a. Assuming that they start at the same point and the same time, how much sooner does the faster student arrive at a destination 780 m away?
   b. How far would the students have to walk so that the faster student arrives 5.50 min before the slower student?

5. **Critical Thinking** Does knowing the distance between two objects give you enough information to locate the objects? Explain.

6. **Interpreting Graphics** **Figure 8** shows position-time graphs of the straight-line movement of two brown bears in a wildlife preserve. Which bear has the greater average velocity over the entire period? Which bear has the greater velocity at $t = 8.0$ min? Is the velocity of bear A always positive? Is the velocity of bear B ever negative?

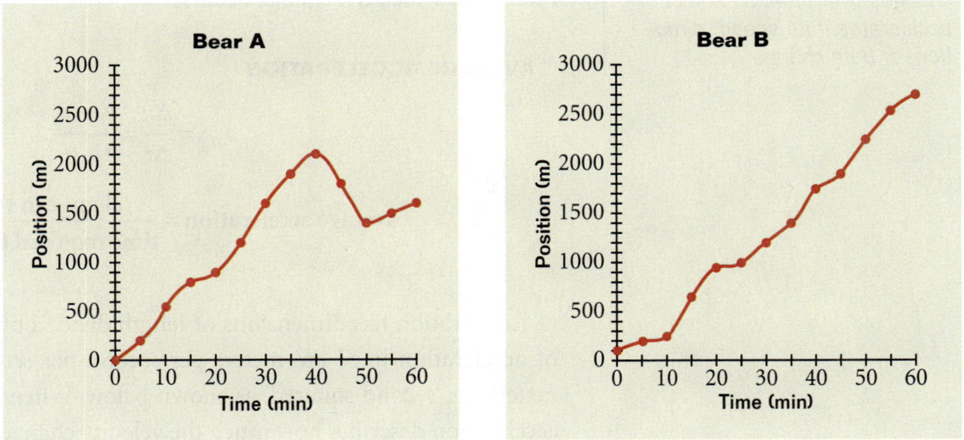

Figure 8

SECTION 1

SECTION REVIEW ANSWERS

1. 24 s
2. 6.2 s
3. **a.** 2.50 m/s to the south
   b. 2.27 m/s to the north
   c. 0.0 m/s
4. **a.** 460 s
   b. 570 m
5. No, because a single distance could correspond to a variety of different positions of the objects
6. bear B; bear A; no; no

SECTION 2
General Level

Demonstration

Acceleration GENERAL

Purpose Visually demonstrate acceleration.

Materials marble, inclined plane, tape, stopwatch, meterstick, protractor, metal or glass cup

Procedure Ask students to observe the motion of the marble. Release the marble from rest at the top of the inclined plane. When the marble reaches the end, ask students to describe the motion. Ask students to identify the locations of lowest speed (*top*) and highest speed (*bottom*). Repeat the demonstration at a different angle, and have students compare the two trials.

After students have learned the kinematic equations, you can add quantitative values by timing the runs. Place a tape marker near the top of the plane to serve as a start line. Use a metal or glass cup at the end of the plane as a sound cue for stopping the stopwatch. Measure the length of the plane. Have students record the time and angle measurements for several trials and then calculate the acceleration for each trial with the equation $\Delta x = \frac{1}{2}a(\Delta t)^2$. Each trial should have a different angle and therefore a different acceleration. Have students create a graph of acceleration versus angle.

SECTION 2

Acceleration

SECTION OBJECTIVES

- **Describe motion in terms of changing velocity.**
- **Compare graphical representations of accelerated and nonaccelerated motions.**
- **Apply kinematic equations to calculate distance, time, or velocity under conditions of constant acceleration.**

SCILINKS.
www.scilinks.org
Topic: Acceleration
Code: HF60007

CHANGES IN VELOCITY

Many bullet trains have a top speed of about 300 km/h. Because a train stops to load and unload passengers, it does not always travel at that top speed. For some of the time the train is in motion, its velocity is either increasing or decreasing. It loses speed as it slows down to stop and gains speed as it pulls away and heads for the next station.

Acceleration is the rate of change of velocity with respect to time

Similarly, when a shuttle bus approaches a stop, the driver begins to apply the brakes to slow down 5.0 s before actually reaching the stop. The speed changes from 9.0 m/s to 0 m/s over a time interval of 5.0 s. Sometimes, however, the shuttle stops much more quickly. For example, if the driver slams on the brakes to avoid hitting a dog, the bus slows from 9.0 m/s to 0 m/s in just 1.5 s.

Clearly, these two stops are very different, even though the shuttle's velocity changes by the same amount in both cases. What is different in these two examples is the time interval during which the change in velocity occurs. As you can imagine, this difference has a great effect on the motion of the bus, as well as on the comfort and safety of the passengers. A sudden change in velocity feels very different from a slow, gradual change.

The quantity that describes the rate of change of velocity in a given time interval is called **acceleration.** The magnitude of the average acceleration is calculated by dividing the total change in an object's velocity by the time interval in which the change occurs.

acceleration

the rate at which velocity changes over time; an object accelerates if its speed, direction, or both change

AVERAGE ACCELERATION

$$a_{avg} = \frac{\Delta v}{\Delta t} = \frac{v_f - v_i}{t_f - t_i}$$

$$\text{average acceleration} = \frac{\text{change in velocity}}{\text{time required for change}}$$

Acceleration has dimensions of length divided by time squared. The units of acceleration in SI are meters per second per second, which is written as meters per second squared, as shown below. When measured in these units, acceleration describes how much the velocity changes in each second.

$$\frac{(\text{m/s})}{\text{s}} = \frac{\text{m}}{\text{s}} \times \frac{1}{\text{s}} = \frac{\text{m}}{\text{s}^2}$$

SAMPLE PROBLEM B

Average Acceleration

PROBLEM

A shuttle bus slows down with an average acceleration of -1.8 m/s^2. How long does it take the bus to slow from 9.0 m/s to a complete stop?

SOLUTION

Given: $v_i = 9.0$ m/s
$v_f = 0$ m/s
$a_{avg} = -1.8$ m/s^2

Unknown: $\Delta t = ?$

TIP *Watch for implied data in problem statements, such as "starts at rest" ($v_i = 0$ m/s) or "comes to rest" ($v_f = 0$ m/s).*

Rearrange the average acceleration equation to solve for the time interval.

$$a_{avg} = \frac{\Delta v}{\Delta t}$$

$$\Delta t = \frac{\Delta v}{a_{avg}} = \frac{v_f - v_i}{a_{avg}} = \frac{0 \text{ m/s} - 9.0 \text{ m/s}}{-1.8 \text{ m/s}^2}$$

$$\boxed{\Delta t = 5.0 \text{ s}}$$

PRACTICE B

Average Acceleration

1. As the shuttle bus comes to a sudden stop to avoid hitting a dog, it accelerates uniformly at -4.1 m/s^2 as it slows from 9.0 m/s to 0.0 m/s. Find the time interval of acceleration for the bus.
2. A car traveling at 7.0 m/s accelerates uniformly at 2.5 m/s^2 to reach a speed of 12.0 m/s. How long does it take for this acceleration to occur?
3. With an average acceleration of -1.2 m/s^2, how long will it take a cyclist to bring a bicycle with an initial speed of 6.5 m/s to a complete stop?
4. Turner's treadmill runs with a velocity of -1.2 m/s and speeds up at regular intervals during a half-hour workout. After 25 min, the treadmill has a velocity of -6.5 m/s. What is the average acceleration of the treadmill during this period?
5. Suppose a treadmill has an average acceleration of 4.7×10^{-3} m/s^2.
   a. How much does its speed change after 5.0 min?
   b. If the treadmill's initial speed is 1.7 m/s, what will its final speed be?

Classroom Practice

Average Acceleration
Find the acceleration of an amusement park ride that falls from rest to a speed of 28 m/s in 3.0 s.

Answer
9.3 m/s^2

PROBLEM GUIDE B

Use this guide to assign problems.
SE = Student Edition Textbook
PW = Problem Workbook
PB = Problem Bank on the One-Stop Planner (OSP)

Solving for:

Δt	**SE** Sample, 1–3; Ch. Rvw. 16 **PW** 7a, 8–9 **PB** 6, 8, 9
Δv	**SE** 5 **PW** Sample, 1–4a, 5a, 8b **PB** 7, 10
a_{avg}	**SE** 4; Ch. Rvw. 17, 45 **PW** 4b, 5b, 6, 7b, 10 **PB** Sample, 1–5

***Challenging Problem**
Consult the printed Solutions Manual or the OSP for detailed solutions.

ANSWERS

Practice B

1. 2.2 s
2. 2.0 s
3. 5.4 s
4. -3.5×10^{-3} m/s^2
5. a. 1.4 m/s
   b. 3.1 m/s

SECTION 2

Visual Strategy – BASIC

Figure 10
Point out to students that the same analysis that is done for positive velocity in the graph can be done for negative velocity.

Q If the graph continued, the velocity would become negative. What does this indicate about the object?

A ***The object has slowed down, stopped (zero velocity), and turned around (negative velocity). It is moving back toward the origin and is speeding up.***

ANSWERS

Conceptual Challenge

1. No; The ball could be speeding up from rest.
2. slowing down
3. Jennifer's acceleration could be positive if she is moving in the negative direction; in this case, the direction of the bike's velocity is opposite the direction of the bike's acceleration, so the bike slows down. In mathematical terms, $a = \frac{\Delta v}{\Delta t}$. If v_f and v_i are negative, and if the magnitude of v_f is less than the magnitude of v_i, then Δv and thus a are positive.

Figure 9
High-speed trains such as this one can travel at speeds of about 300 km/h (186 mi/h).

Acceleration has direction and magnitude

Figure 9 shows a high-speed train leaving a station. Imagine that the train is moving to the right so that the displacement and the velocity are positive. The velocity increases in magnitude as the train picks up speed. Therefore, the final velocity will be greater than the initial velocity, and Δv will be positive. When Δv is positive, the acceleration is positive.

On long trips with no stops, the train may travel for a while at a constant velocity. In this situation, because the velocity is not changing, $\Delta v = 0$ m/s. When the velocity is constant, the acceleration is equal to zero.

Imagine that the train, still traveling in the positive direction, slows down as it approaches the next station. In this case, the velocity is still positive, but the initial velocity is larger than the final velocity, so Δv will be negative. When Δv is negative, the acceleration is negative.

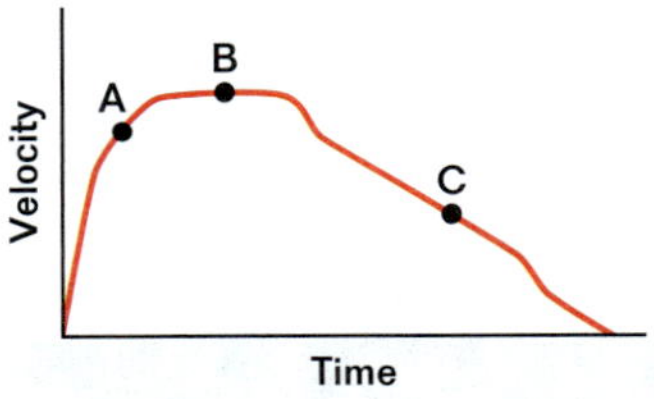

Figure 10
When the velocity in the positive direction is increasing, the acceleration is positive, as at point A. When the velocity is constant, there is no acceleration, as at point B. When the velocity in the positive direction is decreasing, the acceleration is negative, as at point C.

The slope and shape of the graph describe the object's motion

As with all motion graphs, the slope and shape of the velocity-time graph in **Figure 10** allow a detailed analysis of the train's motion over time. When the train leaves the station, its speed is increasing over time. The line on the graph plotting this motion slopes up and to the right, as at point **A** on the graph.

When the train moves with a constant velocity, the line on the graph continues to the right, but it is horizontal, with a slope equal to zero. This indicates that the train's velocity is constant, as at point **B** on the graph.

Finally, as the train approaches the station, its velocity decreases over time. The graph segment representing this motion slopes down to the right, as at point **C** on the graph. This downward slope indicates that the velocity is decreasing over time.

A negative value for the acceleration does not always indicate a decrease in speed. For example, if the train were moving in the negative direction, the acceleration would be negative when the train gained speed to leave a station and positive when the train lost speed to enter a station.

Why it Matters

Conceptual Challenge

1. Fly Ball If a baseball has zero velocity at some instant, is the acceleration of the baseball necessarily zero at that instant? Explain, and give examples.

2. Runaway Train If a passenger train is traveling on a straight track with a negative velocity and a positive acceleration, is it speeding up or slowing down?

3. Hike-and-Bike Trail When Jennifer is out for a ride, she slows down on her bike as she approaches a group of hikers on a trail. Explain how her acceleration can be positive even though her speed is decreasing.

Table 3 shows how the signs of the velocity and acceleration can be combined to give a description of an object's motion. From this table, you can see that a negative acceleration can describe an object that is speeding up (when the velocity is negative) or an object that is slowing down (when the velocity is positive). Use this table to check your answers to problems involving acceleration.

For example, in **Figure 10** the initial velocity v_i of the train is positive. At point **A** on the graph, the train's velocity is still increasing, so its acceleration is positive as well. The first entry in **Table 3** shows that in this situation, the train is speeding up. At point **C,** the velocity is still positive, but it is decreasing, so the train's acceleration is negative. **Table 3** tells you that in this case, the train is slowing down.

Table 3 Velocity and Acceleration

v_i	a	Motion
+	+	speeding up
−	−	speeding up
+	−	slowing down
−	+	slowing down
− or +	0	constant velocity
0	− or +	speeding up from rest
0	0	remaining at rest

MOTION WITH CONSTANT ACCELERATION

Figure 11 is a strobe photograph of a ball moving in a straight line with constant acceleration. While the ball was moving, its image was captured ten times in one second, so the time interval between successive images is 0.10 s. As the ball's velocity increases, the ball travels a greater distance during each time interval. In this example, the velocity increases by exactly the same amount during each time interval. Thus, the acceleration is constant. Because the velocity increases for each time interval, the successive change in displacement for each time interval increases. You can see this in the photograph by noting that the distance between images increases while the time interval between images remains constant. The relationships between displacement, velocity, and constant acceleration are expressed by equations that apply to any object moving with constant acceleration.

Figure 11
The motion in this picture took place in about 1.00 s. In this short time interval, your eyes could only detect a blur. This photo shows what really happens within that time.

Demonstration

Constant Acceleration GENERAL

Purpose Give several visual examples of constant acceleration.

Materials metronome (optional), tile floor (or tape)

Procedure Tell students you are going to demonstrate constant velocity and then constant acceleration. Use a metronome (or have students clap at regular intervals) to show time intervals. For the first part (constant velocity), walk in a straight line at a rate of one tile per time interval. (If you do not have a tile floor, use the tape to mark regular intervals on the floor.)

Now show a constant acceleration of one tile per interval for each interval. Walk a distance of one tile in the first interval, two tiles in the second interval, three in the third, and so on.

Finally, show students a constant negative acceleration. Start walking at a rate of four or five tiles per interval, decreasing by one tile per interval with each interval. When you get to zero tiles per interval, you may want to continue by walking backward one tile per interval, then two tiles per interval, and so on. Explain that the acceleration was still present at the velocity of zero tiles per interval, so your velocity continued to change.

Teaching Tip ADVANCED

Point out that because the acceleration is constant, the distance that the ball travels in each time interval is equal to the distance it traveled in the previous interval, plus a constant distance. Tell students to make a chart with calculations that demonstrate this fact, given an acceleration of 3 m/s^2.

Teaching Tip — GENERAL

Show students that the area under the curve in a graph of velocity versus time equals the displacement during that time interval. Use the simplest case in **Figure 12,** where v_i equals zero, to illustrate this point. Choose a point on the graph, and draw a vertical line from the x-axis to the point and a horizontal line from the y-axis to the point, as shown below.

Use the corresponding velocity and time interval values to find the area of the rectangle ($A = v_f t_f$). Point out that the line in the graph bisects the box; thus, the area under the line equals $\frac{1}{2}v_f t_f$, which is the equation for the displacement of a constantly accelerated object that begins at rest.

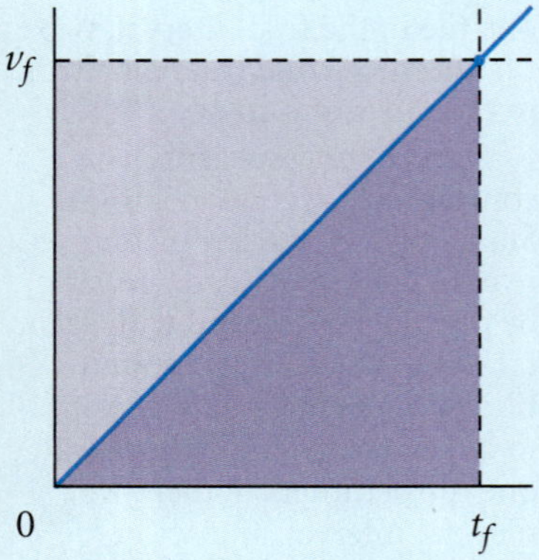

$Area = \frac{1}{2}\,v_f t_f = \frac{1}{2}\,v_f \Delta t = \Delta x$

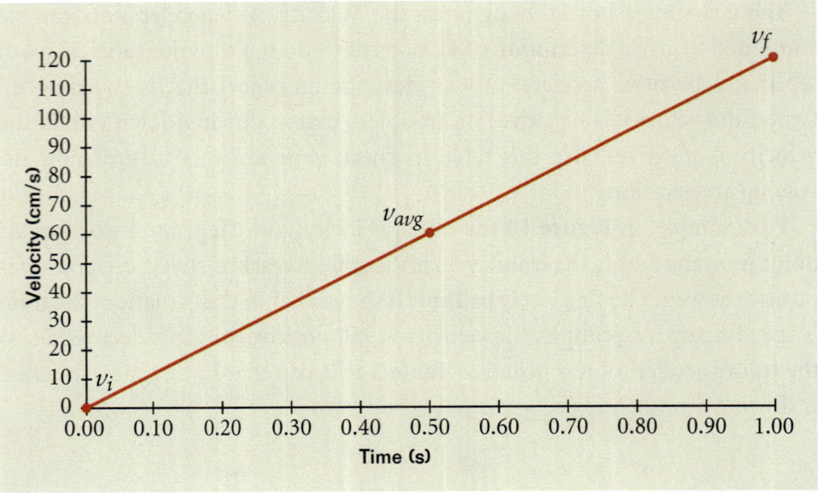

Figure 12
If a ball moved for the same time with a constant velocity equal to v_{avg}, it would have the same displacement as the ball in **Figure 11** moving with constant acceleration.

Did you know?

Decreases in speed are sometimes called *decelerations*. Despite the sound of the name, decelerations are really a special case of acceleration in which the magnitude of the velocity—and thus the speed—decreases with time.

Displacement depends on acceleration, initial velocity, and time

Figure 12 is a graph of the ball's velocity plotted against time. The initial, final, and average velocities are marked on the graph. We know that the average velocity is equal to displacement divided by the time interval.

$$v_{avg} = \frac{\Delta x}{\Delta t}$$

For an object moving with constant acceleration, the average velocity is equal to the average of the initial velocity and the final velocity.

$$v_{avg} = \frac{v_i + v_f}{2} \qquad \text{average velocity} = \frac{\text{initial velocity} + \text{final velocity}}{2}$$

To find an expression for the displacement in terms of the initial and final velocity, we can set the expressions for average velocity equal to each other.

$$\frac{\Delta x}{\Delta t} = v_{avg} = \frac{v_i + v_f}{2}$$

$$\frac{\text{displacement}}{\text{time interval}} = \frac{\text{initial velocity} + \text{final velocity}}{2}$$

Multiplying both sides of the equation by Δt gives us an expression for the displacement as a function of time. This equation can be used to find the displacement of any object moving with constant acceleration.

DISPLACEMENT WITH CONSTANT ACCELERATION

$$\Delta x = \tfrac{1}{2}(v_i + v_f)\Delta t$$

displacement = $\frac{1}{2}$(initial velocity + final velocity)(time interval)

ADVANCED TOPICS

See "Angular Kinematics" in **Appendix J: Advanced Topics** to learn how displacement, velocity, and acceleration can be used to describe circular motion.

SAMPLE PROBLEM C

Displacement with Constant Acceleration

PROBLEM

A racing car reaches a speed of 42 m/s. It then begins a uniform negative acceleration, using its parachute and braking system, and comes to rest 5.5 s later. Find the distance that the car travels during braking.

SOLUTION

Given: $v_i = 42$ m/s $\quad v_f = 0$ m/s

$\Delta t = 5.5$ s

Unknown: $\Delta x = ?$

Use the equation that relates displacement, initial and final velocities, and the time interval.

$$\Delta x = \frac{1}{2}(v_i + v_f)\Delta t$$

$$\Delta x = \frac{1}{2}(42 \text{ m/s} + 0 \text{ m/s})\ (5.5 \text{ s})$$

$$\boxed{\Delta x = 120 \text{ m}}$$

CALCULATOR SOLUTION

The calculator answer is 115.5. However, the velocity and time values have only two significant figures each, so the answer must be reported as 120 m.

TIP

Remember that this equation applies only when acceleration is constant. In this problem, you know that acceleration is constant by the phrase "uniform negative acceleration." All of the kinematic equations introduced in this chapter are valid only for constant acceleration.

PRACTICE C

Displacement with Constant Acceleration

1. A car accelerates uniformly from rest to a speed of 6.6 m/s in 6.5 s. Find the distance the car travels during this time.
2. When Maggie applies the brakes of her car, the car slows uniformly from 15.0 m/s to 0.0 m/s in 2.50 s. How many meters before a stop sign must she apply her brakes in order to stop at the sign?
3. A driver in a car traveling at a speed of 21.8 m/s sees a cat 101 m away on the road. How long will it take for the car to accelerate uniformly to a stop in exactly 99 m?
4. A car enters the freeway with a speed of 6.4 m/s and accelerates uniformly for 3.2 km in 3.5 min. How fast (in m/s) is the car moving after this time?

Classroom Practice

Displacement with Constant Acceleration

A bicyclist accelerates from 5.0 m/s to 16 m/s in 8.0 s. Assuming uniform acceleration, what distance does the bicyclist travel during this time interval?

Answer

84 m

PROBLEM GUIDE C

Use this guide to assign problems.
SE = Student Edition Textbook
PW = Problem Workbook
PB = Problem Bank on the One-Stop Planner (OSP)

Solving for:

Δx	**SE** Sample, 1–2; Ch. Rvw. 18, 19 **PW** 4–6, 9 **PB** 8
Δt	**SE** 3; Ch. Rvw. 41, 43* **PW** Sample, 1–3 **PB** 7, 9–10
Δv	**SE** 4 **PW** 7–8 **PB** Sample, 1–6

*Challenging Problem
Consult the printed Solutions Manual or the OSP for detailed solutions.

ANSWERS

Practice C

1. 21 m
2. 18.8 m
3. 9.1 s
4. 24 m/s

SECTION 2

The Language of Physics

Some texts use the term v_0 to represent the initial velocity (v_i) of the object and the term v to represent the final velocity (v_f) of the object.

Teaching Tip — GENERAL

You may want to present an interesting geometrical interpretation of this equation:

$$\Delta x = v_i \Delta t + \frac{1}{2}a(\Delta t)^2$$

Draw a velocity-versus-time graph representing the equation $v_f = v_i + a\Delta t$, as below.

The area beneath this curve has two parts. The lower part is a rectangle of area $v_i\Delta t$. The upper part is a triangle of area $\frac{1}{2}a(\Delta t)^2$. Therefore, the total area under the straight-line graph is equal to the displacement, Δx, of the object.

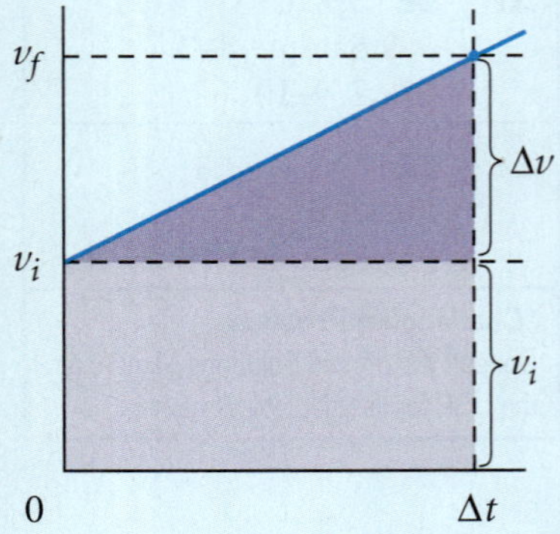

ADVANCED TOPICS

See "Special Relativity and Time Dilation" in **Appendix J: Advanced Topics** to learn about how time intervals are interpreted in Einstein's special theory of relativity.

Final velocity depends on initial velocity, acceleration, and time

What if the final velocity of the ball is not known but we still want to calculate the displacement? If we know the initial velocity, the acceleration, and the elapsed time, we can find the final velocity. We can then use this value for the final velocity to find the total displacement of the ball.

By rearranging the equation for acceleration, we can find a value for the final velocity.

$$a = \frac{\Delta v}{\Delta t} = \frac{v_f - v_i}{\Delta t}$$

$$a\Delta t = v_f - v_i$$

By adding the initial velocity to both sides of the equation, we get an equation for the final velocity of the ball.

$$a\Delta t + v_i = v_f$$

VELOCITY WITH CONSTANT ACCELERATION

$$v_f = v_i + a\Delta t$$

final velocity = initial velocity + (acceleration × time interval)

You can use this equation to find the final velocity of an object after it has accelerated at a constant rate for any time interval.

If you want to know the displacement of an object moving with constant acceleration over some certain time interval, you can obtain another useful expression for displacement by substituting the expression for v_f into the expression for Δx.

$$\Delta x = \frac{1}{2}(v_i + v_f)\Delta t$$

$$\Delta x = \frac{1}{2}(v_i + v_i + a\Delta t)\Delta t$$

$$\Delta x = \frac{1}{2}[2v_i\Delta t + a(\Delta t)^2]$$

DISPLACEMENT WITH CONSTANT ACCELERATION

$$\Delta x = v_i\Delta t + \frac{1}{2}a(\Delta t)^2$$

displacement = (initial velocity × time interval) + $\frac{1}{2}$ acceleration × (time interval)^2

This equation is useful not only for finding the displacement of an object moving with constant acceleration but also for finding the displacement required for an object to reach a certain speed or to come to a stop. For the latter situation, you need to use both this equation and the equation given above.

SAMPLE PROBLEM D

Velocity and Displacement with Constant Acceleration

PROBLEM

A plane starting at rest at one end of a runway undergoes a uniform acceleration of 4.8 m/s^2 for 15 s before takeoff. What is its speed at takeoff? How long must the runway be for the plane to be able to take off?

SOLUTION

Given: $v_i = 0$ m/s $\quad a = 4.8$ m/s^2 $\quad \Delta t = 15$ s

Unknowns: $v_f = ?$ $\quad \Delta x = ?$

First, use the equation for the velocity of a uniformly accelerated object.

$$v_f = v_i + a\Delta t$$
$$v_f = 0 \text{ m/s} + (4.8 \text{ m/s}^2)(15 \text{ s})$$

$$\boxed{v_f = 72 \text{ m/s}}$$

Then, use the displacement equation that contains the given variables.

$$\Delta x = v_i \Delta t + \tfrac{1}{2}a(\Delta t)^2$$
$$\Delta x = (0 \text{ m/s})(15 \text{ s}) + \tfrac{1}{2}(4.8 \text{ m/s}^2)(15 \text{ s})^2$$

$$\boxed{\Delta x = 540 \text{ m}}$$

TIP *Because you know now v_f, you could also use the equation $\Delta x = \frac{1}{2}(v_i + v_f)(\Delta t)$, or $\Delta x = \frac{1}{2}(72 \text{ m/s})(15 \text{ s}) = 540$ m.*

PRACTICE D

Velocity and Displacement with Constant Acceleration

1. A car with an initial speed of 6.5 m/s accelerates at a uniform rate of 0.92 m/s^2 for 3.6 s. Find the final speed and the displacement of the car during this time.
2. An automobile with an initial speed of 4.30 m/s accelerates uniformly at the rate of 3.00 m/s^2. Find the final speed and the displacement after 5.00 s.
3. A car starts from rest and travels for 5.0 s with a constant acceleration of −1.5 m/s^2. What is the final velocity of the car? How far does the car travel in this time interval?
4. A driver of a car traveling at 15.0 m/s applies the brakes, causing a uniform acceleration of −2.0 m/s^2. How long does it take the car to accelerate to a final speed of 10.0 m/s? How far has the car moved during the braking period?

PROBLEM GUIDE D

Use this guide to assign problems.
SE = Student Edition Textbook
PW = Problem Workbook
PB = Problem Bank on the One-Stop Planner (OSP)

Solving for:

Δv	**SE** Sample, 1–3; Ch. Rvw. 20–21a, 22*, 49c* **PW** 10–12 **PB** 8
Δx	**SE** Sample, 1–4; Ch. Rvw. 21b, 22, 23, 42b*, 49b* **PW** 6*, 7–9 **PB** 9b, 10b
Δt	**SE** 4; Ch. Rvw. 42a*, 49a* **PW** Sample, 1–3, 4*, 5a*, 9 **PB** 9a, 10a
a	**PW** 5b, 7, 13–15 **PB** Sample, 1–7

***Challenging Problem**
Consult the printed Solutions Manual or the OSP for detailed solutions.

ANSWERS

Practice D

1. 9.8 m/s; 29 m
2. 19.3 m/s; 59.0 m
3. −7.5 m/s; 19 m
4. 2.5 s; 32 m

Teaching Tip — BASIC

Make sure students are aware that all the kinematic equations are related. Show them that there is more than one way to solve kinematics problems. In general, one of the equations will be easier to use than the others, but it is possible to use different combinations of the equations to reach the same answer.

Did you know?

The word *physics* comes from the ancient Greek word for "nature." According to Aristotle, who assigned the name, physics is the study of natural events. Aristotle believed that the study of motion was the basis of physics. Galileo developed the foundations for the modern study of motion using mathematics. In 1632, Galileo published the first mathematical treatment of motion.

SCILINKS.
www.scilinks.org
Topic: Galileo
Code: HF60633

Final velocity depends on initial velocity, acceleration, and displacement

So far, all of the equations for motion under uniform acceleration have required knowing the time interval. We can also obtain an expression that relates displacement, velocity, and acceleration without using the time interval. This method involves rearranging one equation to solve for Δt and substituting that expression in another equation, making it possible to find the final velocity of a uniformly accelerated object without knowing how long it has been accelerating. Start with the following equation for displacement:

$$\Delta x = \frac{1}{2}(v_i + v_f)\Delta t$$ Now, multiply both sides by 2.

$$2\Delta x = (v_i + v_f)\Delta t$$ Next, divide both sides by $(v_i + v_f)$ to solve for Δt.

$$\left(\frac{2\Delta x}{v_i + v_f}\right) = \Delta t$$

Now that we have an expression for Δt, we can substitute this expression into the equation for the final velocity.

$$v_f = v_i + a(\Delta t)$$

$$v_f = v_i + a\left(\frac{2\Delta x}{v_i + v_f}\right)$$

In its present form, this equation is not very helpful because v_f appears on both sides. To solve for v_f, first subtract v_i from both sides of the equation.

$$v_f - v_i = a\left(\frac{2\Delta x}{v_i + v_f}\right)$$

Next, multiply both sides by $(v_i + v_f)$ to get all the velocities on the same side of the equation.

$$(v_f - v_i)(v_f + v_i) = 2a\Delta x = v_f^2 - v_i^2$$

Add v_i^2 to both sides to solve for v_f^2.

FINAL VELOCITY AFTER ANY DISPLACEMENT

$$v_f^2 = v_i^2 + 2a\Delta x$$

(final velocity)^2 = (initial velocity)^2 + 2(acceleration)(displacement)

When using this equation, you must take the square root of the right side of the equation to find the final velocity. Remember that the square root may be either positive or negative. If you have been consistent in your use of the sign convention, you will be able to determine which value is the right answer by reasoning based on the direction of the motion.

SAMPLE PROBLEM E

Final Velocity After Any Displacement

PROBLEM

A person pushing a stroller starts from rest, uniformly accelerating at a rate of 0.500 m/s^2. What is the velocity of the stroller after it has traveled 4.75 m?

SOLUTION

1. DEFINE **Given:** $v_i = 0$ m/s $a = 0.500$ m/s^2
$\Delta x = 4.75$ m

Unknown: $v_f = ?$

Diagram:

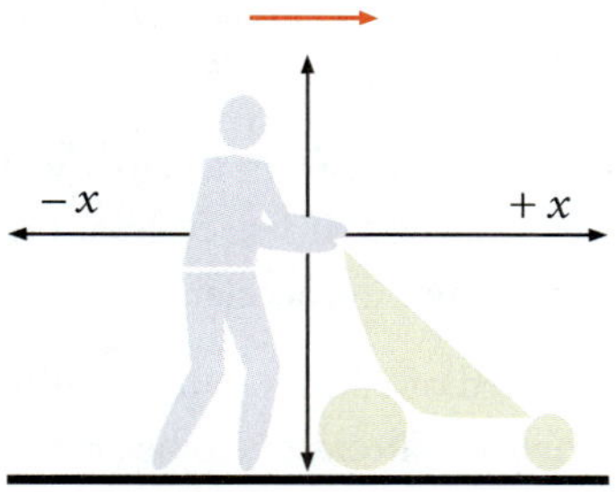

Choose a coordinate system. The most convenient one has an origin at the initial location of the stroller. The positive direction is to the right.

2. PLAN **Choose an equation or situation:**
Because the initial velocity, acceleration, and displacement are known, the final velocity can be found using the following equation:

$$v_f^2 = v_i^2 + 2a\Delta x$$

Rearrange the equation to isolate the unknown:
Take the square root of both sides to isolate v_f.

$$v_f = \pm\sqrt{(v_i)^2 + 2a\Delta x}$$

3. CALCULATE **Substitute the values into the equation and solve:**

$$v_f = \pm\sqrt{(0 \text{ m/s})^2 + 2(0.500 \text{ m/s}^2)(4.75 \text{ m})}$$

$$\boxed{v_f = +2.18 \text{ m/s}}$$

TIP *Think about the physical situation to determine whether to keep the positive or negative answer from the square root. In this case, the stroller is speeding up because it starts from rest and ends with a speed of 2.18 m/s. An object that is speeding up and has a positive acceleration must have a positive velocity, as shown in* ***Table 3.*** *So, the final velocity must be positive.*

4. EVALUATE The stroller's velocity after accelerating for 4.75 m is 2.18 m/s to the right.

Classroom Practice

Final Velocity After Any Displacement
An aircraft has a landing speed of 83.9 m/s. The landing area of an aircraft carrier is 195 m long. What is the minimum uniform acceleration required for a safe landing?

Answer
-18.0 m/s^2

An electron is accelerated uniformly from rest in an accelerator at 4.5×10^7 m/s^2 over a distance of 95 km. Assuming constant acceleration, what is the final velocity of the electron?

Answer
2.9×10^6 m/s

Alternative Problem-Solving Approach

Because acceleration is uniform and $v_i = 0$, we know that $v_f = 2v_{avg}$.

$$v_{avg} = \frac{\Delta x}{\Delta t}$$

$$v_f = 2\frac{\Delta x}{\Delta t}$$

Because $\Delta x = \frac{1}{2}a(\Delta t)^2$,

$$\Delta t = \sqrt{\frac{2\Delta x}{a}}$$

Substitute this expression for Δt into the expression for v_f.

$$v_f = \sqrt{2a\Delta x} = +2.18 \text{ m/s}$$

PROBLEM GUIDE E

Use this guide to assign problems.
SE = Student Edition Textbook
PW = Problem Workbook
PB = Problem Bank on the One-Stop Planner (OSP)

Solving for:

Δv	**SE**	Sample, 1, 2a–2c, 3a; Ch. Rvw. 24a, 25, 30, 39
	PW	4–6, 7a, 7b
	PB	7, 10
Δx	**SE**	4, 6; Ch. Rvw. 39, 44a*
	PW	Sample, 1–3
	PB	6, 8–9
a	**SE**	5
	PW	8, 9
	PB	Sample, 1–5

***Challenging Problem**
Consult the printed Solutions Manual or the OSP for detailed solutions.

ANSWERS

Practice E

1. +2.51 m/s
2. a. +21 m/s
   b. +16 m/s
   c. +13 m/s
3. a. 16 m/s
   b. 7.0 s
4. 7.4 m
5. $+2.3\ \text{m/s}^2$
6. 88 m

PRACTICE E

Final Velocity After Any Displacement

1. Find the velocity after the stroller in Sample Problem E has traveled 6.32 m.

2. A car traveling initially at +7.0 m/s accelerates uniformly at the rate of $+0.80\ \text{m/s}^2$ for a distance of 245 m.

a. What is its velocity at the end of the acceleration?
b. What is its velocity after it accelerates for 125 m?
c. What is its velocity after it accelerates for 67 m?

3. A car accelerates uniformly in a straight line from rest at the rate of $2.3\ \text{m/s}^2$.

a. What is the speed of the car after it has traveled 55 m?
b. How long does it take the car to travel 55 m?

4. A motorboat accelerates uniformly from a velocity of 6.5 m/s to the west to a velocity of 1.5 m/s to the west. If its acceleration was $2.7\ \text{m/s}^2$ to the east, how far did it travel during the acceleration?

5. An aircraft has a liftoff speed of 33 m/s. What minimum constant acceleration does this require if the aircraft is to be airborne after a take-off run of 240 m?

6. A certain car is capable of accelerating at a uniform rate of $0.85\ \text{m/s}^2$. What is the magnitude of the car's displacement as it accelerates uniformly from a speed of 83 km/h to one of 94 km/h?

With the four equations presented in this section, it is possible to solve any problem involving one-dimensional motion with uniform acceleration. For your convenience, the equations that are used most often are listed in **Table 4.** The first column of the table gives the equations in their standard form. For an object initially at rest, $v_i = 0$. Using this value for v_i in the equations in the first column will result in the equations in the second column. It is not necessary to memorize the equations in the second column. If $v_i = 0$ in any problem, you will naturally derive this form of the equation. Referring back to the sample problems in this chapter will guide you through using these equations to solve many problems.

Table 4
Equations for Constantly Accelerated Straight-Line Motion

Form to use when accelerating object has an initial velocity	Form to use when accelerating object starts from rest
$\Delta x = \frac{1}{2}(v_i + v_f)\Delta t$	$\Delta x = \frac{1}{2}v_f\Delta t$
$v_f = v_i + a\Delta t$	$v_f = a\Delta t$
$\Delta x = v_i\Delta t + \frac{1}{2}a(\Delta t)^2$	$\Delta x = \frac{1}{2}a(\Delta t)^2$
$v_f^2 = v_i^2 + 2a\Delta x$	$v_f^2 = 2a\Delta x$

SECTION REVIEW

1. Marissa's car accelerates uniformly at a rate of $+2.60\ m/s^2$. How long does it take for Marissa's car to accelerate from a speed of 24.6 m/s to a speed of 26.8 m/s?

2. A bowling ball with a negative initial velocity slows down as it rolls down the lane toward the pins. Is the bowling ball's acceleration positive or negative as it rolls toward the pins?

3. Nathan accelerates his skateboard uniformly along a straight path from rest to 12.5 m/s in 2.5 s.
   a. What is Nathan's acceleration?
   b. What is Nathan's displacement during this time interval?
   c. What is Nathan's average velocity during this time interval?

4. **Critical Thinking** Two cars are moving in the same direction in parallel lanes along a highway. At some instant, the instantaneous velocity of car A exceeds the instantaneous velocity of car B. Does this mean that car A's acceleration is greater than car B's? Explain, and use examples.

5. **Interpreting Graphics** The velocity-versus-time graph for a shuttle bus moving along a straight path is shown in **Figure 13.**
   a. Identify the time intervals during which the velocity of the shuttle bus is constant.
   b. Identify the time intervals during which the acceleration of the shuttle bus is constant.
   c. Find the value for the average velocity of the shuttle bus during each time interval identified in **b.**
   d. Find the acceleration of the shuttle bus during each time interval identified in **b.**
   e. Identify the times at which the velocity of the shuttle bus is zero.
   f. Identify the times at which the acceleration of the shuttle bus is zero.
   g. Explain what the slope of the graph reveals about the acceleration in each time interval.

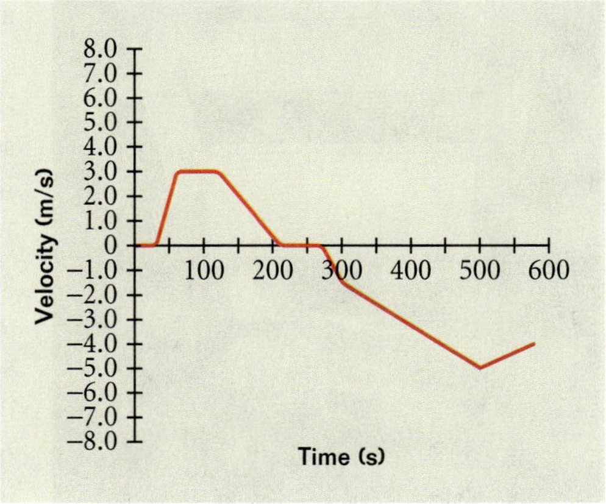

Figure 13

6. **Interpreting Graphics** Is the shuttle bus in item 5 always moving in the same direction? Explain, and refer to the time intervals shown on the graph.

SECTION 2

SECTION REVIEW ANSWERS

1. 0.85 s
2. positive
3. **a.** $+5.0\ m/s^2$
   b. +16 m
   c. +6.4 m/s
4. No, car A's acceleration is not necessarily greater than car B's acceleration; If the two cars are moving in the positive direction, car A could be slowing down (negative acceleration) while car B is speeding up (positive acceleration), even though car A's velocity is greater than car B's velocity.
5. **a.** 0 s to 30 s; 60 s to 125 s; 210 s to 275 s
   b. 0 s to 30 s; 30 s to 60 s; 60 s to 125 s; 125 s to 210 s; 210 s to 275 s; 275 s to 300 s; 300 s to 520 s; 520 s to 580 s
   c. 0 m/s; 1.5 m/s; 0 m/s; 1.5 m/s; 0 m/s; −0.75 m/s; −3.25 m/s; −4.5 m/s
   d. $0\ m/s^2$; $0.1\ m/s^2$; $0\ m/s^2$; $-0.04\ m/s^2$; $0\ m/s^2$; $-0.06\ m/s^2$; $-0.02\ m/s^2$; $0.02\ m/s^2$
   e. 0 to 30 s; 210 to 275 s
   f. 0 s to 30s; 60 s to 125 s; 210 s to 275 s
   g. When the graph slopes upward, acceleration is positive. When it slopes downward, acceleration is negative.
6. No; The bus is moving in the positive direction from 30 s to 210 s (when velocity is positive) and in the negative direction from 275 s to 600 s (when velocity is negative).

SECTION 3
General Level

Misconception Alert

Be sure to emphasize that g is an *acceleration,* specifically the acceleration due to gravity near the surface of Earth. Also, make sure students realize that the acceleration is the same regardless of the height from which an object falls. Some students do not realize that two objects dropped from different heights will experience the same acceleration.

Visual Strategy GENERAL

Figure 14

Point out that the feather and apple will fall at the same rate regardless of the value of a_g.

Q If the feather-and-apple experiment were performed on the moon, where free-fall acceleration is approximately one-sixth the value of free-fall acceleration on Earth, how would the picture compare with **Figure 14**?

A ***The feather and apple would still fall with equal accelerations (1/6 g); because the acceleration is less, the displacement in each time interval would be less.***

SECTION 3

Falling Objects

SECTION OBJECTIVES

- **Relate the motion of a freely falling body to motion with constant acceleration.**
- **Calculate displacement, velocity, and time at various points in the motion of a freely falling object.**
- **Compare the motions of different objects in free fall.**

free fall

the motion of a body when only the force due to gravity is acting on the body

FREE FALL

On August 2, 1971, a demonstration was conducted on the moon by astronaut David Scott. He simultaneously released a hammer and a feather from the same height above the moon's surface. The hammer and the feather both fell straight down and landed on the lunar surface at exactly the same moment. Although the hammer is more massive than the feather, both objects fell at the same rate. That is, they traveled the same displacement in the same amount of time.

Freely falling bodies undergo constant acceleration

In **Figure 14,** a feather and an apple are released from rest in a vacuum chamber. The two objects fell at exactly the same rate, as indicated by the horizontal alignment of the multiple images.

The amount of time that passed between the first and second images is equal to the amount of time that passed between the fifth and sixth images. The picture, however, shows that the displacement in each time interval did not remain constant. Therefore, the velocity was not constant. The apple and the feather were accelerating.

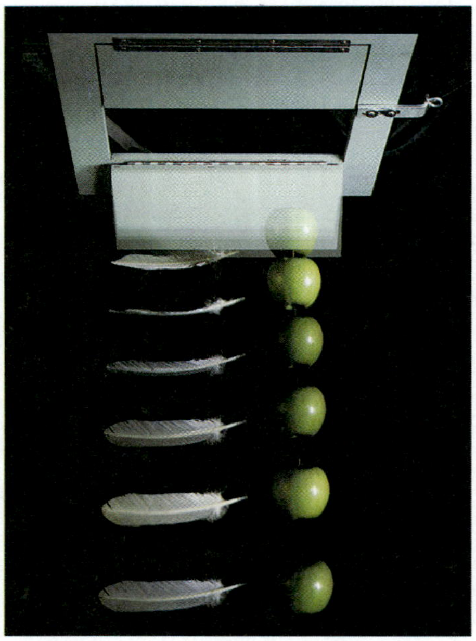

Figure 14
When there is no air resistance, all objects fall with the same acceleration regardless of their masses.

Compare the displacement between the first and second images to the displacement between the second and third images. As you can see, within each time interval the displacement of the feather increased by the same amount as the displacement of the apple. Because the time intervals are the same, we know that the velocity of each object is increasing by the same amount in each time interval. In other words, the apple and the feather are falling with the same constant acceleration.

If air resistance is disregarded, all objects dropped near the surface of a planet fall with the same constant acceleration. This acceleration is due to gravitational force, and the motion is referred to as **free fall.** The acceleration due to gravity is denoted with the symbols a_g (generally) or g (on Earth's surface). The magnitude of g is about 9.81 m/s^2, or 32 ft/s^2. Unless stated otherwise, this book will use the value 9.81 m/s^2 for calculations. This acceleration is directed downward, toward the center of the Earth. In our usual choice of coordinates, the downward direction is negative. Thus, the acceleration of objects in free fall near the surface of the Earth is $a_g = -g = -9.81$ m/s^2. Because an object in free fall is acted on only by gravity, a_g is also known as free-fall acceleration.

Acceleration is constant during upward and downward motion

Figure 15 is a strobe photograph of a ball thrown up into the air with an initial upward velocity of +10.5 m/s. The photo on the left shows the ball moving up from its release toward the top of its path, and the photo on the right shows the ball falling back down. Everyday experience shows that when we throw an object up in the air, it will continue to move upward for some time, stop momentarily at the peak, and then change direction and begin to fall. Because the object changes direction, it may seem that the velocity and acceleration are both changing. Actually, objects thrown into the air have a downward acceleration as soon as they are released.

In the photograph on the left, the upward displacement of the ball between each successive image is smaller and smaller until the ball stops and finally begins to move with an increasing downward velocity, as shown on the right. As soon as the ball is released with an initial upward velocity of +10.5 m/s, it has an acceleration of -9.81 m/s^2. After 1.0 s ($\Delta t = 1.0$ s), the ball's velocity will change by −9.81 m/s to 0.69 m/s upward. After 2.0 s ($\Delta t = 2.0$ s), the ball's velocity will again change by −9.81 m/s, to −9.12 m/s.

The graph in **Figure 16** shows the velocity of the ball plotted against time. As you can see, there is an instant when the velocity of the ball is equal to 0 m/s. This happens at the instant when the ball reaches the peak of its upward motion and is about to begin moving downward. Although the velocity is zero at the instant the ball reaches the peak, the acceleration is equal to -9.81 m/s^2 at every instant regardless of the magnitude or direction of the velocity. It is important to note that the acceleration is -9.81 m/s^2 even at the peak where the velocity is zero. The straight-line slope of the graph indicates that the acceleration is constant at every moment.

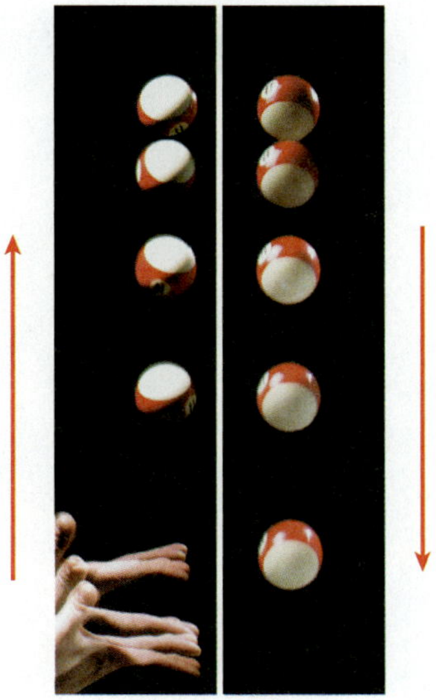

Figure 15
At the very top of its path, the ball's velocity is zero, but the ball's acceleration is -9.81 m/s^2 at every point—both when it is moving up **(a)** and when it is moving down **(b).**

SCILINKS.
www.scilinks.org
Topic: Free Fall
Code: HF60620

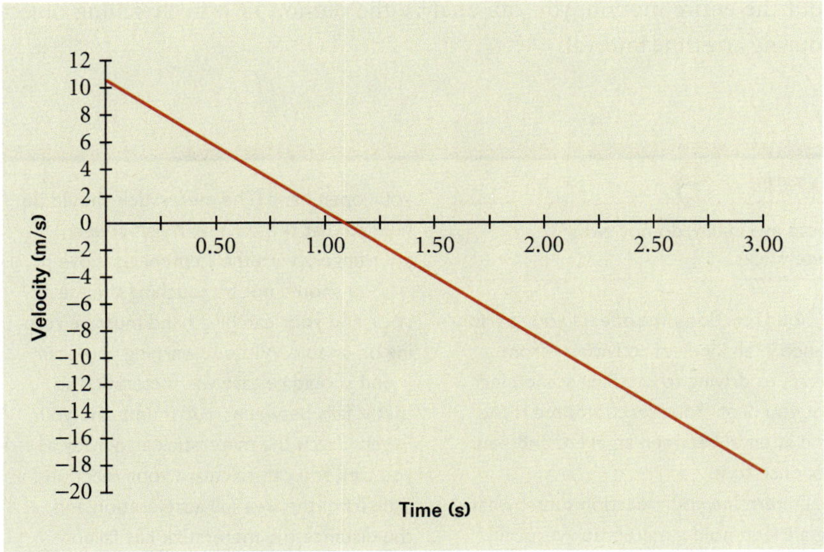

Figure 16
On this velocity-time graph, the slope of the line, which is equal to the ball's acceleration, is constant from the moment the ball is released ($t = 0.00$) and throughout its motion.

SECTION 3

Visual Strategy GENERAL

Figure 15
Because the ball's acceleration is constant, the ball's speed is the same going up as it is going down at any point in the ball's path.

Q If the initial velocity of the ball is 10.5 m/s when it is thrown from a height of 1.5 m, what is the velocity of the ball when it passes this point on the way down?

A *−10.5 m/s*

Misconception Alert

Many students find it very difficult to grasp the idea that an object can have a nonzero acceleration at the top of its flight when the velocity in the y direction is zero. Point out that there is a change in the direction of the velocity at the instant the velocity is zero.

Quick Lab

TEACHER'S NOTES

In order to reduce error caused by anticipating the drop, have students look at the end of the meterstick between their fingers. If time allows, have each student test several times and use the mean distance in the calculation of reaction time.

Homework Options

This QuickLab can easily be performed outside of the physics lab room.

Freely falling objects always have the same downward acceleration

It may seem a little confusing to think of something that is moving upward, like the ball in the example, as having a downward acceleration. Thinking of this motion as motion with a positive velocity and a negative acceleration may help. The downward acceleration is the same when an object is moving up, when it is at rest at the top of its path, and when it is moving down. The only things changing are the position and the magnitude and direction of the velocity.

When an object is thrown up in the air, it has a positive velocity and a negative acceleration. From **Table 3** in Section 2, we see that this means the object is slowing down as it rises in the air. From the example of the ball and from everyday experience, we know that this makes sense. The object continues to move upward but with a smaller and smaller speed. In the photograph of the ball, this decrease in speed is shown by the smaller and smaller displacements as the ball moves up to the top of its path.

At the top of its path, the object's velocity has decreased until it is zero. Although it is impossible to see this because it happens so quickly, the object is actually at rest at the instant it reaches its peak position. Even though the velocity is zero at this instant, the acceleration is still $-9.81\ m/s^2$.

When the object begins moving down, it has a negative velocity and its acceleration is still negative. From **Table 3,** we see that a negative acceleration and a negative velocity indicate an object that is speeding up. In fact, this is what happens when objects undergo free-fall acceleration. Objects that are falling toward Earth move faster and faster as they fall. In the photograph of the ball in **Figure 15** (on the previous page), this increase in speed is shown by the greater and greater displacements between the images as the ball falls.

Knowing the free-fall acceleration makes it easy to calculate the velocity, time, and displacement of many different motions using the equations for constantly accelerated motion. Because the acceleration is the same throughout the entire motion, you can analyze the motion of a freely falling object during any time interval.

Quick Lab

Time Interval of Free Fall

MATERIALS LIST

- meterstick or ruler

SAFETY

Avoid eye injury; do not swing metersticks.

Your reaction time affects your performance in all kinds of activities—from sports to driving to catching something that you drop. Your reaction time is the time interval between an event and your response to it.

Determine your reaction time by having a friend hold a meterstick vertically between the thumb and index finger of your open hand. The meterstick should be held so that the zero mark is between your fingers with the 1 cm mark above it.

You should not be touching the meterstick, and your catching hand must be resting on a table. Without warning you, your friend should release the meterstick so that it falls between your thumb and your finger. Catch the meterstick as quickly as you can. You can calculate your reaction time from the free-fall acceleration and the distance the meterstick has fallen through your grasp.

SAMPLE PROBLEM F

Falling Object

PROBLEM

Jason hits a volleyball so that it moves with an initial velocity of 6.0 m/s straight upward. If the volleyball starts from 2.0 m above the floor, how long will it be in the air before it strikes the floor?

SOLUTION

1. DEFINE

Given: $v_i = +6.0$ m/s $\quad a = -g = -9.81$ m/s^2

$\Delta y = -2.0$ m

Unknown: $\Delta t = ?$

Diagram: Place the origin at the starting point of the ball ($y_i = 0$ at $t_i = 0$).

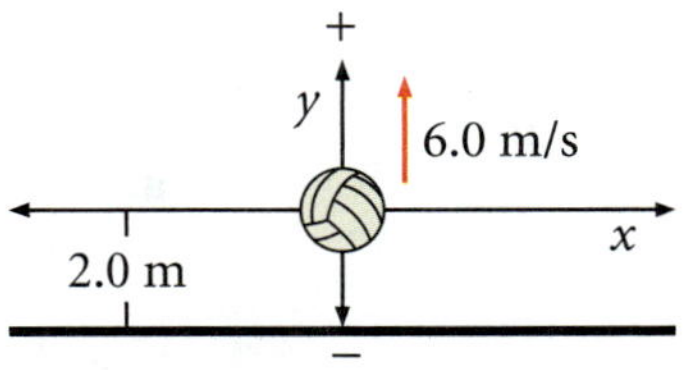

2. PLAN

Choose an equation or situation:

Both Δt and v_f are unknown. Therefore, first solve for v_f using the equation that does not require time. Then, the equation for v_f that does involve time can be used to solve for Δt.

$$v_f^2 = v_i^2 + 2a\Delta y \qquad\qquad v_f = v_i + a\Delta t$$

Rearrange the equations to isolate the unknown:

Take the square root of the first equation to isolate v_f. The second equation must be rearranged to solve for Δt.

$$v_f = \pm\sqrt{v_i^2 + 2a\Delta y} \qquad\qquad \Delta t = \frac{v_f - v_i}{a}$$

3. CALCULATE

Substitute the values into the equations and solve:

First find the velocity of the ball at the moment that it hits the floor.

$$v_f = \pm\sqrt{v_i^2 + 2a\Delta y} = \pm\sqrt{(6.0 \text{ m/s})^2 + 2(-9.81 \text{ m/s}^2)(-2.0 \text{ m})}$$

$$v_f = \pm\sqrt{36 \text{ m}^2/\text{s}^2 + 39 \text{ m}^2/\text{s}^2} = \pm\sqrt{75 \text{ m}^2/\text{s}^2} = -8.7 \text{ m/s}$$

> **TIP** *When you take the square root to find v_f, select the negative answer because the ball will be moving toward the floor, in the negative direction.*

Next, use this value of v_f in the second equation to solve for Δt.

$$\Delta t = \frac{v_f - v_i}{a} = \frac{-8.7 \text{ m/s} - 6.0 \text{ m/s}}{-9.81 \text{ m/s}^2} = \frac{-14.7 \text{ m/s}}{-9.81 \text{ m/s}^2}$$

$$\boxed{\Delta t = 1.50 \text{ s}}$$

4. EVALUATE The solution, 1.50 s, is a reasonable amount of time for the ball to be in the air.

Classroom Practice

Falling Object

A ball is thrown straight up into the air at an initial velocity of 25.0 m/s. Create a table showing the ball's position, velocity, and acceleration each second for the first 5.00 s of its motion.

Answer

t (s)	**y (m)**	**v (m/s)**	**a (m/s^2)**
1.00	20.1	+15.2	−9.81
2.00	30.4	+5.4	−9.81
3.00	30.9	−4.4	−9.81
4.00	21.6	−14.2	−9.81
5.00	2.50	−24.0	−9.81

PROBLEM GUIDE F

Use this guide to assign problems.
SE = Student Edition Textbook
PW = Problem Workbook
PB = Problem Bank on the One-Stop Planner (OSP)

Solving for:

Δt	**SE** Sample, 1a, 2b, 3b; Ch. Rvw. 31, 32, 36a*, 46a, 48b, 48c **PW** 6, 8–9 **PB** 8
Δv	**SE** 1b, 2a, 3a; Ch. Rvw. 36b*, 39, 40a, 46b, 47b*, 47c* **PW** 1–5 **PB** 7, 9–10
Δy	**SE** 4; Ch. Rvw. 36c*, 37*, 40b, 48a **PW** 7 **PB** Sample, 1–6

***Challenging Problem**
Consult the printed Solutions Manual or the OSP for detailed solutions.

ANSWERS

Practice F

1. a. −42 m/s
   b. 11 s
2. a. 22.1 m/s
   b. 2.25 s
3. a. 8.0 m/s
   b. 1.63 s
4. 1.8 m

Why it Matters

Sky Diving

You may use this introduction to air resistance and terminal velocity to explain to students that objects in the real world do not always follow idealized physical principles, such as the principle that free-fall acceleration is the same for all objects near Earth's surface. A feather and a cannonball dropped from the same height will not hit the ground at the same time because the feather quickly reaches a terminal velocity, but the cannonball does not.

You may wish to return to this feature after students have studied air resistance as a force, in the chapter "Forces and the Laws of Motion." Terminal velocity is reached when the force of air resistance equals the force due to gravity (but acts in the opposite direction). The balanced forces result in a constant velocity, in accordance with Newton's first law of motion.

PRACTICE F

Falling Object

1. A robot probe drops a camera off the rim of a 239 m high cliff on Mars, where the free-fall acceleration is $-3.7\ m/s^2$.
   a. Find the velocity with which the camera hits the ground.
   b. Find the time required for it to hit the ground.
2. A flowerpot falls from a windowsill 25.0 m above the sidewalk.
   a. How fast is the flowerpot moving when it strikes the ground?
   b. How much time does a passerby on the sidewalk below have to move out of the way before the flowerpot hits the ground?
3. A tennis ball is thrown vertically upward with an initial velocity of +8.0 m/s.
   a. What will the ball's speed be when it returns to its starting point?
   b. How long will the ball take to reach its starting point?
4. Calculate the displacement of the volleyball in Sample Problem F when the volleyball's final velocity is 1.1 m/s upward.

Why it Matters

Sky Diving

When these sky divers jump from an airplane, they plummet toward the ground. If Earth had no atmosphere, the sky divers would accelerate with the free-fall acceleration, *g*, equal to $9.81\ m/s^2$. They would not slow down even after opening their parachutes. Fortunately, Earth does have an atmosphere, and the acceleration of the sky divers does not remain constant. Instead, because of air resistance, the acceleration decreases as they fall. After a few seconds, the acceleration drops to zero and the speed becomes constant. The constant speed an object reaches when falling through a resisting medium is called *terminal velocity.*

The terminal velocity of an object depends on the object's mass, shape, and size. When a sky diver is spread out horizontally to the ground, the sky diver's terminal velocity is typically about 55 m/s (123 mi/h). If the sky diver curls into a ball, the terminal velocity may increase to close to 90 m/s (200 mi/h). When the sky diver opens the parachute, air resistance increases, and the sky diver decelerates to a new, slower terminal velocity. For a sky diver with an open parachute, the terminal velocity is typically about 5 m/s (11 mi/h).

SECTION REVIEW

1. A coin is tossed vertically upward.

a. What happens to its velocity while it is in the air?

b. Does its acceleration increase, decrease, or remain constant while it is in the air?

2. A pebble is dropped down a well and hits the water 1.5 s later. Using the equations for motion with constant acceleration, determine the distance from the edge of the well to the water's surface.

3. A ball is thrown vertically upward. What are its velocity and acceleration when it reaches its maximum altitude? What is its acceleration just before it hits the ground?

4. Two children are bouncing small rubber balls. One child simply drops a ball. At the same time, the second child throws a ball downward so that it has an initial speed of 10 m/s. What is the acceleration of each ball while in motion?

5. Critical Thinking A gymnast practices two dismounts from the high bar on the uneven parallel bars. During one dismount, she swings up off the bar with an initial upward velocity of + 4.0 m/s. In the second, she releases from the same height but with an initial downward velocity of −3.0 m/s. What is her acceleration in each case? How do the final velocities of the gymnast as she reaches the ground differ?

6. Interpreting Graphics **Figure 17** is a position-time graph of the motion of a basketball thrown straight up. Use the graph to sketch the path of the basketball and to sketch a velocity-time graph of the basketball's motion.

a. Is the velocity of the basketball constant?

b. Is the acceleration of the basketball constant?

c. What is the initial velocity of the basketball?

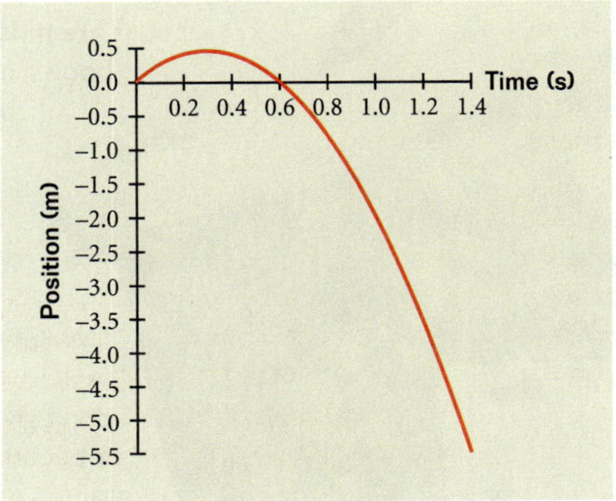

Figure 17

SECTION 3

SECTION REVIEW ANSWERS

1. a. The coin's velocity decreases, becomes zero at its maximum height, and then increases in the negative direction until the coin hits the ground.

b. The coin's acceleration remains constant.

2. 11 m

3. At maximum altitude, $v = 0$ and $a = -9.81$ m/s^2; Just before the ball hits the ground, $a = -9.81$ m/s^2.

4. -9.81 m/s^2 for each ball

5. The gymnast's acceleration (-9.81 m/s^2) will be the same in each case. Her final velocities will be determined by the equation $v_f^2 = v_i^2 + 2a\Delta y$. Because the acceleration and displacement are the same, the final velocity is greater for the larger initial velocity, +4.0 m/s.

6. The ball goes up 0.5 m, returns to its original position, and then falls another 5.5 m. The time axis of students' graphs should follow that in **Figure 17,** and the velocity coordinates should roughly correspond to $\frac{y}{x}$ in **Figure 17** at each point.

a. no

b. yes

c. about 3 m/s

Science Writer
According to Janice VanCleave, working on her science books never stops. Students may enjoy the following anecdote about one of her vacations:

> I cannot say my free time is for vacations; no matter where I visit, I am not a tourist; I am a researcher. Once, for example, while standing on a glacier in Alaska, others in the tourist group were questioning how deep a hole in the ice was. For fun, I dropped a stone (from rest) into the hole and counted until the stone hit the water below. Using free-fall acceleration ($g = 9.8\ \text{m/s}^2$) and the displacement equation ($\Delta x = \frac{1}{2}a\Delta t^2$), I calculated that the hole was about 21 m (70 ft) deep. We were all interested in knowing how deep the hole was, and I was able to tell them.

PHYSICS CAREERS

Science Writer

Science writers explain science to their readers in a clear and entertaining way. To learn more about science writing as a career, read the interview with Janice VanCleave, author of *Janice VanCleave's A+ Projects in Physics* and more than 50 other books about science.

Janice VanCleave writes, on average, three new science books every year.

What does a science writer do?

After the topic and basic format of a book are determined, I start researching. My personal library contains about 1,000 science books. For more information, I use online sources to find current printed research. But most important are science consultants in each specific field, such as a NASA astronomer, a research chemist, and physics professors.

After I write the original manuscript, it is edited five different times. I have to make sure any changes do not affect its scientific correctness and that the art represents the text. On average, I write three new books each year between doing reviews on the books written in previous years.

What sort of training do you have?

I taught science for 27 years, and this was the only training I had for my writing career.

Mine is a Cinderella story. A publisher saw an ad for an elementary science enrichment program that I designed and taught. She sent a letter asking if I was interested in writing a science book for young kids. The answer was a big YES!

It didn't take long to realize that although I had skills to write experiments for my class, I didn't have a clue about how to write a book. Thankfully, the publisher really wanted the book, so I was given a great deal of personal instruction. I learned by trial and error. In fact, even after writing 50 books, I am still learning how to better write a book.

What is your favorite part of your work?

I love writing because, as in teaching, I learn so much. Now, instead of just writing for my classroom, I am able to share my ideas with students and teachers around the world. A downside is that I have less time with students than when I taught.

What advice do you have for students who are interested in science writing?

Study the market, and know what kinds of books are most wanted. Writing a book may be your smallest problem; getting it published can be a big obstacle. If you are a great writer but not a good salesperson, I suggest hiring an agent. Write about something that you have a passion for, and don't give up if your work is not accepted. Try, try, and try again.

Highlights

KEY IDEAS

Section 1 Displacement and Velocity

- Displacement is a change of position in a certain direction, not the total distance traveled.
- The average velocity of an object during some time interval is equal to the displacement of the object divided by the time interval. Like displacement, velocity has both a magnitude (called speed) and a direction.
- The average velocity is equal to the slope of the straight line connecting the initial and final points on a graph of the position of the object versus time.

Section 2 Acceleration

- The average acceleration of an object during a certain time interval is equal to the change in the object's velocity divided by the time interval. Acceleration has both magnitude and direction.
- The direction of the acceleration is not always the same as the direction of the velocity. The direction of the acceleration depends on the direction of the motion and on whether the velocity is increasing or decreasing.
- The average acceleration is equal to the slope of the straight line connecting the initial and final points on the graph of the velocity of the object versus time.
- The equations in **Table 4** are valid whenever acceleration is constant.

Section 3 Falling Objects

- An object thrown or dropped in the presence of Earth's gravity experiences a constant acceleration directed toward the center of Earth. This acceleration is called the free-fall acceleration, or the acceleration due to gravity.
- Free-fall acceleration is the same for all objects, regardless of mass.
- The value for free-fall acceleration on Earth's surface used in this book is $a_g = -g = -9.81$ m/s^2. The direction of the free-fall acceleration is considered to be negative because the object accelerates toward Earth.

Variable Symbols

Quantities	Units	Quantities	Units
x position	m meters	y position	m meters
Δx displacement	m meters	Δy displacement	m meters
v velocity	m/s meters per second	a acceleration	m/s^2 meters per second^2

KEY TERMS

frame of reference (p. 40)

displacement (p. 41)

average velocity (p. 43)

instantaneous velocity (p. 46)

acceleration (p. 48)

free fall (p. 60)

PROBLEM SOLVING

See **Appendix D: Equations** for a summary of the equations introduced in this chapter. If you need more problem-solving practice, see **Appendix I: Additional Problems.**

CHAPTER 2

Highlights

Teaching Tip

Many parents measure their child's height every year by marking notches on a door post. Point out that in this situation the child's growth each year is the displacement. The child's growth rate (velocity) is that displacement divided by the time (1 year), and the change in growth rate per unit of time (acceleration) is the difference in the growth rate from one year to the next.

Review

ANSWERS

1. 5.0 m; +5.0 m
2. the slope of the line at each point
3. t_1: negative; t_2: positive; t_3: positive; t_4: negative; t_5: zero
4. a. 0 to t_1; t_1 to t_2; t_3 to t_4; t_4 to t_5
   b. t_1 to t_2; t_2 to t_3; t_3 to t_4
5. The duck's displacement must also be zero.
6. The time interval is always positive because time can only move in one direction (forward).
7. 10.1 km to the east
8. 4.22×10^1 km
9. a. +70.0 m
   b. +140.0 m
   c. +14 m/s
   d. +28 m/s

CHAPTER 2 Review

DISPLACEMENT AND VELOCITY

Review Questions

1. On the graph below, what is the total distance traveled during the recorded time interval? What is the displacement?

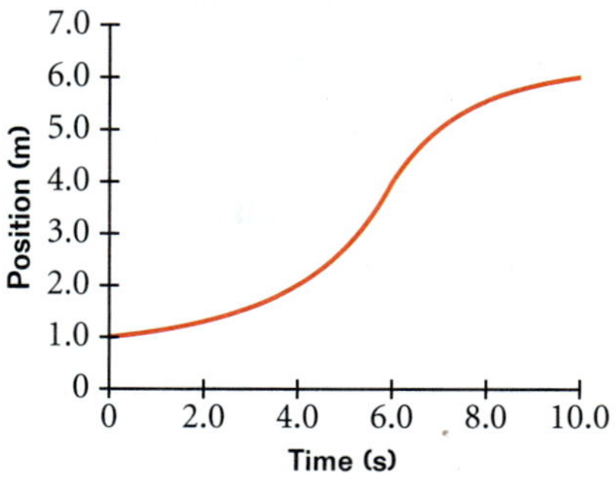

2. On a position-time graph such as the one above, what represents the instantaneous velocity?

3. The position-time graph for a bug crawling along a line is shown in item 4 below. Determine whether the velocity is positive, negative, or zero at each of the times marked on the graph.

4. Use the position-time graph below to answer the following questions:
   a. During which time interval(s) is the velocity negative?
   b. During which time interval(s) is the velocity positive?

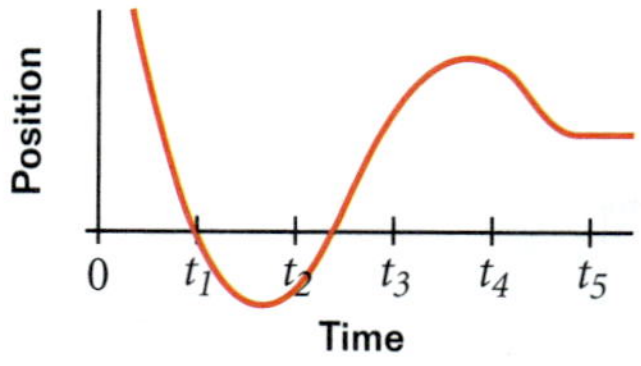

Conceptual Questions

5. If the average velocity of a duck is zero in a given time interval, what can you say about the displacement of the duck for that interval?

6. Velocity can be either positive or negative, depending on the direction of the displacement. The time interval, Δt, is always positive. Why?

Practice Problems

For problems 7–11, see Sample Problem A.

7. A school bus takes 0.530 h to reach the school from your house. If the average velocity of the bus is 19.0 km/h to the east, what is the displacement?

8. The Olympic record for the marathon is 2.00 h, 9.00 min, 21.0 s. If the average speed of a runner achieving this record is 5.436 m/s, what is the marathon distance?

9. Two cars are traveling on a desert road, as shown below. After 5.0 s, they are side by side at the next telephone pole. The distance between the poles is 70.0 m. Identify the following quantities:
   a. the displacement of car A after 5.0 s
   b. the displacement of car B after 5.0 s
   c. the average velocity of car A during 5.0 s
   d. the average velocity of car B during 5.0 s

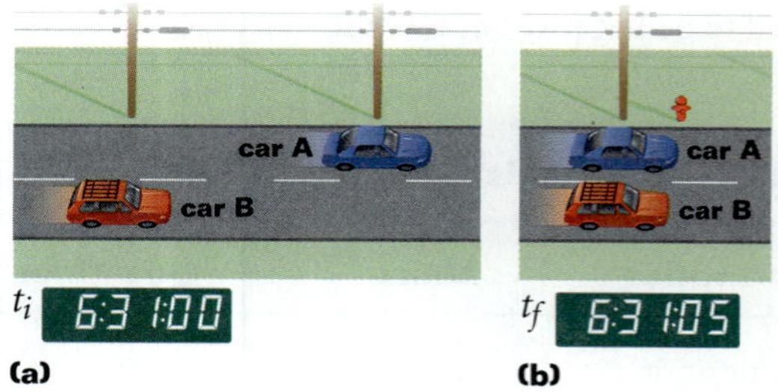

10. Sally travels by car from one city to another. She drives for 30.0 min at 80.0 km/h, 12.0 min at 105 km/h, and 45.0 min at 40.0 km/h, and she spends 15.0 min eating lunch and buying gas.

a. Determine the average speed for the trip.

b. Determine the total distance traveled.

11. Runner A is initially 6.0 km west of a flagpole and is running with a constant velocity of 9.0 km/h due east. Runner B is initially 5.0 km east of the flagpole and is running with a constant velocity of 8.0 km/h due west. What will be the distance of the two runners from the flagpole when their paths cross? (It is not necessary to convert your answer from kilometers to meters for this problem. You may leave it in kilometers.)

ACCELERATION

Review Questions

12. What would be the acceleration of a turtle that is moving with a constant velocity of 0.25 m/s to the right?

13. Sketch the velocity-time graphs for the following motions.

a. a city bus that is moving with a constant velocity

b. a wheelbarrow that is speeding up at a uniform rate of acceleration while moving in the positive direction

c. a tiger that is speeding up at a uniform rate of acceleration while moving in the negative direction

d. an iguana that is slowing down at a uniform rate of acceleration while moving in the positive direction

e. a camel that is slowing down at a uniform rate of acceleration while moving in the negative direction

Conceptual Questions

14. If a car is traveling eastward, can its acceleration be westward? Explain your answer, and use an example in your explanation.

15. The strobe photographs below show a disk moving from left to right under different conditions. The time interval between images is constant. Assuming that the direction to the right is positive, identify the following types of motion in each photograph. (Some may have more than one type of motion.)

a. the acceleration is positive

b. the acceleration is negative

c. the velocity is constant

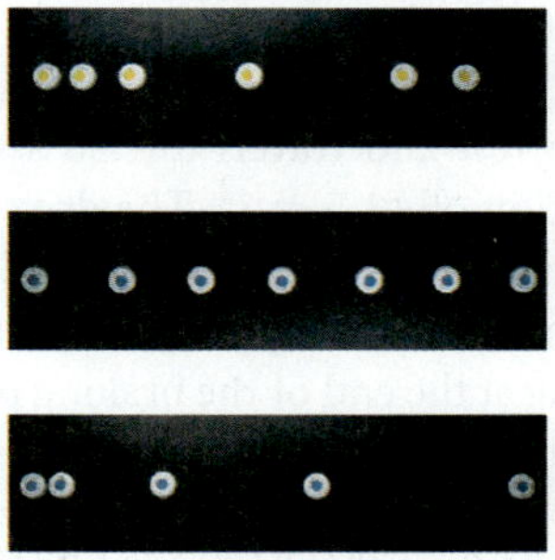

Practice Problems

For problems 16–17, see Sample Problem B.

16. A car traveling in a straight line has a velocity of +5.0 m/s. After an acceleration of 0.75 m/s^2, the car's velocity is +8.0 m/s. In what time interval did the acceleration occur?

17. The velocity-time graph for an object moving along a straight path is shown below. Find the average accelerations during the time intervals 0.0 s to 5.0 s, 5.0 s to 15.0 s, and 0.0 s to 20.0 s.

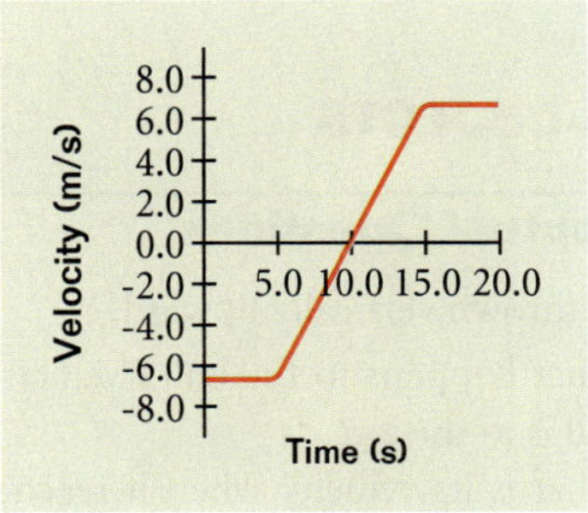

For problems 18–19, see Sample Problem C.

18. A bus slows down uniformly from 75.0 km/h (21 m/s) to 0 km/h in 21 s. How far does it travel before stopping?

2 REVIEW

10. **a.** 53.5 km/h
b. 91.0 km

11. 0.2 km west of the flagpole

12. 0.00 m/s^2

13. **a.** slope is zero
b. slope is positive
c. slope is negative
d. slope is negative
e. slope is positive

14. Yes; A car traveling eastward and slowing down has a westward acceleration.

15. **a.** left half of top photo, bottom photo
b. right half of top photo
c. middle photo

16. 4.0 s

17. 0.0 m/s^2; +1.36 m/s^2; +0.680 m/s^2

18. 2.2×10^2 m

19. 110 m
20. +8.6 m/s
21. a. −15 m/s
b. −38 m
22. +1.5 m/s; +32 m
23. 17.5 m
24. a. 11 m/s
b. 52 s
25. 0.99 m/s
26. a. The ball's velocity decreases, becomes zero at its maximum altitude, then increases in the negative direction.
b. At maximum altitude, the ball's velocity is zero.
c. $-9.81\ m/s^2$
d. $-9.81\ m/s^2$
e. The ball's acceleration remains constant.
27. For each ball, the velocity can be analyzed as the distance between images, and the acceleration can be seen by comparing the distances. Both balls are accelerating; they start with approximately the same acceleration, but the smaller ball's acceleration gradually decreases because of air resistance.
28. The two pins have the same acceleration ($-9.81\ m/s^2$).
29. a. yes
b. yes
c. yes
30. −39.6 m/s
31. 3.94 s
32. 0.60 s
33. 1.51 h

19. A car accelerates uniformly from rest to a speed of 65 km/h (18 m/s) in 12 s. Find the distance the car travels during this time.

For problems 20–23, see Sample Problem D.

20. A car traveling at +7.0 m/s accelerates at the rate of $+0.80\ m/s^2$ for an interval of 2.0 s. Find v_f.

21. A car accelerates from rest at $-3.00\ m/s^2$.

a. What is the velocity at the end of 5.0 s?

b. What is the displacement after 5.0 s?

22. A car starts from rest and travels for 5.0 s with a uniform acceleration of $+1.5\ m/s^2$. The driver then applies the brakes, causing a uniform acceleration of $-2.0\ m/s^2$. If the brakes are applied for 3.0 s, how fast is the car going at the end of the braking period, and how far has it gone from its start?

23. A boy sledding down a hill accelerates at $1.40\ m/s^2$. If he started from rest, in what distance would he reach a speed of 7.00 m/s?

For problems 24–25, see Sample Problem E.

24. A sailboat starts from rest and accelerates at a rate of $0.21\ m/s^2$ over a distance of 280 m.

a. Find the magnitude of the boat's final velocity.

b. Find the time it takes the boat to travel this distance.

25. An elevator is moving upward at 1.20 m/s when it experiences an acceleration of $0.31\ m/s^2$ downward, over a distance of 0.75 m. What will its final velocity be?

FALLING OBJECTS

Conceptual Questions

26. A ball is thrown vertically upward.

a. What happens to the ball's velocity while the ball is in the air?

b. What is its velocity when it reaches its maximum altitude?

c. What is its acceleration when it reaches its maximum altitude?

d. What is its acceleration just before it hits the ground?

e. Does its acceleration increase, decrease, or remain constant?

27. The image at right is a strobe photograph of two falling balls released simultaneously. (This motion does not take place in a vacuum.) The ball on the left side is solid, and the ball on the right side is a hollow table-tennis ball. Analyze the motion of both balls in terms of velocity and acceleration.

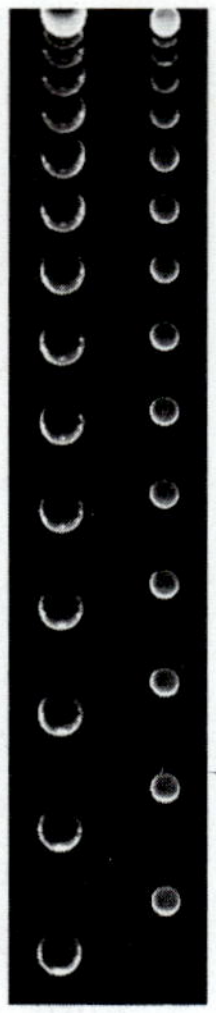

28. A juggler throws a bowling pin into the air with an initial velocity v_i. Another juggler drops a pin at the same instant. Compare the accelerations of the two pins while they are in the air.

29. A bouquet is thrown upward.

a. Will the value for the bouquet's displacement be the same no matter where you place the origin of the coordinate system?

b. Will the value for the bouquet's velocity be the same?

c. Will the value for the bouquet's acceleration be the same?

Practice Problems

For problems 30–32, see Sample Problem F.

30. A worker drops a wrench from the top of a tower 80.0 m tall. What is the velocity when the wrench strikes the ground?

31. A peregrine falcon dives at a pigeon. The falcon starts downward from rest with free-fall acceleration. If the pigeon is 76.0 m below the initial position of the falcon, how long does the falcon take to reach the pigeon? Assume that the pigeon remains at rest.

32. A ball is thrown upward from the ground with an initial speed of 25 m/s; at the same instant, a ball is dropped from rest from a building 15 m high. After how long will the balls be at the same height?

MIXED REVIEW

33. If the average speed of an orbiting space shuttle is 27 800 km/h, determine the time required for it to circle Earth. Assume that the shuttle is orbiting about 320.0 km above Earth's surface, and that Earth's radius is 6380 km.

34. A ball is thrown directly upward into the air. The graph below shows the vertical position of the ball with respect to time.

a. How much time does the ball take to reach its maximum height?

b. How much time does the ball take to reach one-half its maximum height?

c. Estimate the slope of $\Delta y/\Delta t$ at $t = 0.05$ s, $t = 0.10$ s, $t = 0.15$ s, and $t = 0.20$ s. On your paper, draw a coordinate system with velocity (v) on the y-axis and time (t) on the x-axis. Plot your velocity estimates against time.

d. From your graph, determine what the acceleration on the ball is.

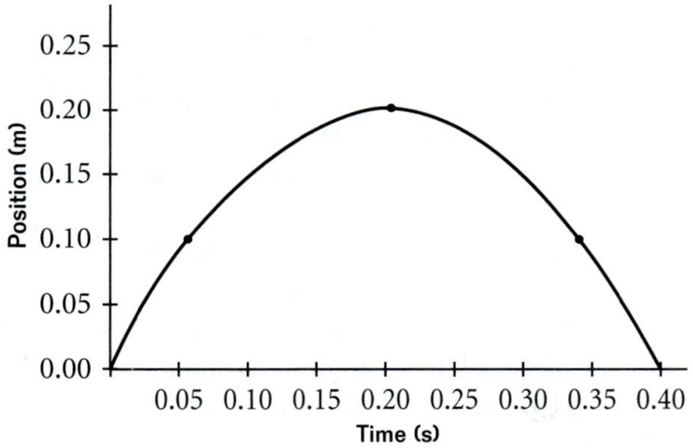

35. A train travels between stations 1 and 2, as shown below. The engineer of the train is instructed to start from rest at station 1 and accelerate uniformly between points A and B, then coast with a uniform velocity between points B and C, and finally accelerate uniformly between points C and D until the train stops at station 2. The distances AB, BC, and CD are all equal, and it takes 5.00 min to travel between the two stations. Assume that the uniform accelerations have the same magnitude, even when they are opposite in direction.

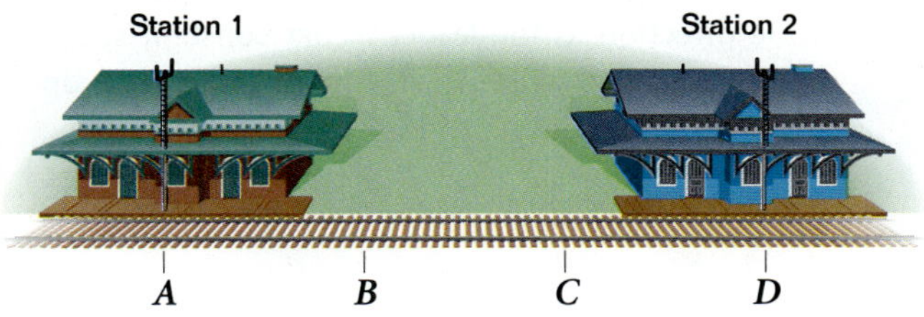

a. How much of this 5.00 min period does the train spend between points A and B?

b. How much of this 5.00 min period does the train spend between points B and C?

c. How much of this 5.00 min period does the train spend between points C and D?

36. Two students are on a balcony 19.6 m above the street. One student throws a ball vertically downward at 14.7 m/s. At the same instant, the other student throws a ball vertically upward at the same speed. The second ball just misses the balcony on the way down.

a. What is the difference in the time the balls spend in the air?

b. What is the velocity of each ball as it strikes the ground?

c. How far apart are the balls 0.800 s after they are thrown?

37. A rocket moves upward, starting from rest with an acceleration of $+29.4 \text{ m/s}^2$ for 3.98 s. It runs out of fuel at the end of the 3.98 s but does not stop. How high does it rise above the ground?

38. Two cars travel westward along a straight highway, one at a constant velocity of 85 km/h, and the other at a constant velocity of 115 km/h.

a. Assuming that both cars start at the same point, how much sooner does the faster car arrive at a destination 16 km away?

b. How far must the cars travel for the faster car to arrive 15 min before the slower car?

39. A small first-aid kit is dropped by a rock climber who is descending steadily at 1.3 m/s. After 2.5 s, what is the velocity of the first-aid kit, and how far is the kit below the climber?

40. A small fish is dropped by a pelican that is rising steadily at 0.50 m/s.

a. After 2.5 s, what is the velocity of the fish?

b. How far below the pelican is the fish after 2.5 s?

41. A ranger in a national park is driving at 56 km/h when a deer jumps onto the road 65 m ahead of the vehicle. After a reaction time of t s, the ranger applies the brakes to produce an acceleration of -3.0 m/s^2. What is the maximum reaction time allowed if the ranger is to avoid hitting the deer?

2 REVIEW

34. a. 0.20 s
b. 0.06 s and 0.34 s
c. Answers may vary slightly but should be close to the following:
$t = 0.05$ s, $v = 2$ m/s
$t = 0.10$ s, $v = 1$ m/s
$t = 0.15$ s, $v = 0.5$ m/s
$t = 0.20$ s, $v = 0$ m/s
d. Answers will vary but should be near -10 m/s^2.

35. a. 2.00 min
b. 1.00 min
c. 2.00 min

36. a. 3.0 s
b. −24.5 m/s for each
c. 23.6 m

37. 931 m

38. a. 0.05 h
b. 81 km

39. −26 m/s; 31 m

40. a. −24 m/s
b. 31 m

41. 1.6 s

42. a. 24.6 s
b. 738 m

43. 5 s; 85 s; +60 m/s

44. a. 6100 m
b. 9 s

45. -1.5×10^3 m/s^2

46. a. 2.33 s
b. −32.9 m/s

47. a. 3.40 s
b. −9.2 m/s
c. −31.4 m/s; −33 m/s

48. a. 310 m
b. 8.5 s
c. 16.4 s

49. a. 4.6 s after stock car starts
b. 38 m
c. +17 m/s (stock car), +21 m/s (race car)

50. a. 13 s
b. −2.9 m/s^2
c. 12 s

51. 4.44 m/s

42. A speeder passes a parked police car at 30.0 m/s. The police car starts from rest with a uniform acceleration of 2.44 m/s^2.

a. How much time passes before the speeder is overtaken by the police car?

b. How far does the speeder get before being overtaken by the police car?

43. An ice sled powered by a rocket engine starts from rest on a large frozen lake and accelerates at +13.0 m/s^2. At t_1 the rocket engine is shut down and the sled moves with constant velocity v until t_2. The total distance traveled by the sled is 5.30×10^3 m and the total time is 90.0 s. Find t_1, t_2, and v.
(See Appendix A for hints on solving quadratic equations.)

44. At the 5800 m mark, the sled in the previous question begins to accelerate at −7.0 m/s^2. Use your answers from item 43 to answer the following questions.

a. What is the final position of the sled when it comes to rest?

b. How long does it take for the sled to come to rest?

45. A tennis ball with a velocity of +10.0 m/s to the right is thrown perpendicularly at a wall. After striking the wall, the ball rebounds in the opposite direction with a velocity of −8.0 m/s to the left. If the ball is in contact with the wall for 0.012 s, what is the average acceleration of the ball while it is in contact with the wall?

46. A parachutist descending at a speed of 10.0 m/s loses a shoe at an altitude of 50.0 m.

a. When does the shoe reach the ground?

b. What is the velocity of the shoe just before it hits the ground?

47. A mountain climber stands at the top of a 50.0 m cliff hanging over a calm pool of water. The climber throws two stones vertically 1.0 s apart and observes that they cause a single splash when they hit the water. The first stone has an initial velocity of +2.0 m/s.

a. How long after release of the first stone will the two stones hit the water?

b. What is the initial velocity of the second stone when it is thrown?

c. What will the velocity of each stone be at the instant both stones hit the water?

48. A model rocket is launched straight upward with an initial speed of 50.0 m/s. It accelerates with a constant upward acceleration of 2.00 m/s^2 until its engines stop at an altitude of 150 m.

a. What is the maximum height reached by the rocket?

b. When does the rocket reach maximum height?

c. How long is the rocket in the air?

49. A professional race-car driver buys a car that can accelerate at +5.9 m/s^2. The racer decides to race against another driver in a souped-up stock car. Both start from rest, but the stock-car driver leaves 1.0 s before the driver of the race car. The stock car moves with a constant acceleration of +3.6 m/s^2.

a. Find the time it takes the race-car driver to overtake the stock-car driver.

b. Find the distance the two drivers travel before they are side by side.

c. Find the velocities of both cars at the instant they are side by side.

50. Two cars are traveling along a straight line in the same direction, the lead car at 25 m/s and the other car at 35 m/s. At the moment the cars are 45 m apart, the lead driver applies the brakes, causing the car to have an acceleration of −2.0 m/s^2.

a. How long does it take for the lead car to stop?

b. Assume that the driver of the chasing car applies the brakes at the same time as the driver of the lead car. What must the chasing car's minimum negative acceleration be to avoid hitting the lead car?

c. How long does it take the chasing car to stop?

51. One swimmer in a relay race has a 0.50 s lead and is swimming at a constant speed of 4.00 m/s. The swimmer has 20.0 m to swim before reaching the end of the pool. A second swimmer moves in the same direction as the leader. What constant speed must the second swimmer have in order to catch up to the leader at the end of the pool?

Alternative Assessment

1. Can a boat moving eastward accelerate to the west? What happens to the boat's velocity? Name other examples of objects accelerating in the direction opposite their motion, including one with numerical values. Create diagrams and graphs.
2. The next time you are a passenger in a car, record the numbers displayed on the clock, the odometer, and the speedometer every 15 s for about 5 min. Create different representations of the car's motion, including maps, charts, and graphs. Exchange your representations with someone who made a different trip, and attempt to reconstruct that trip based on his or her report.
3. Two stones are thrown from a cliff at the same time with the same speed, one upward and one downward. Which stone, if either, hits the ground first? Which, if either, hits with the higher speed? In a group discussion, make your best argument for each possible prediction. Set up numerical examples and solve them to test your prediction.
4. Research typical values for velocities and acceleration of various objects. Include many examples, such as different animals, means of transportation, sports, continental drift, light, subatomic particles, and planets. Organize your findings for display on a poster or some other form.
5. Research Galileo's work on falling bodies. What did he want to demonstrate? What opinions or theories was he trying to refute? What arguments did he use to persuade others that he was right? Did he depend on experiments, logic, findings of other scientists, or other approaches?
6. The study of various motions in nature requires devices for measuring periods of time. Prepare a presentation on a specific type of clock, such as water clocks, sand clocks, pendulum clocks, wind-up clocks, atomic clocks, or biological clocks. Who invented or discovered the clock? What scale of time does it measure? What are the principles or phenomena behind each clock? Can they be calibrated?

Graphing Calculator Practice

Motion in One Dimension

At what speed does a falling hailstone travel? Does the speed depend on the distance that the hailstone falls? In this graphing calculator activity, you will have the opportunity to answer these questions.

Your calculator will display two graphs: one for displacement (distance fallen) versus time and the other for speed versus time. These two graphs correspond to the following two equations:

$$Y_1 = 4.9X^2$$
$$Y_2 = 9.8X$$

You should be able to use **Table 4** of this chapter to correlate these equations with those for an accelerating object that starts from rest.

Visit go.hrw.com and type in the keyword **HF6MODX** to find this graphing calculator activity. Refer to **Appendix B** for instructions on downloading the program for this activity.

Alternative Assessment ANSWERS

1. All examples should involve decreasing speeds. The graphs should show velocity decreasing.
2. Students should be able to reconstruct a trip based on information about direction on a map and about speed from the speedometer.
3. The stone thrown downward hits the ground first, but both have the same speed when they hit the ground.
4. Students should record their values with appropriate units. Check the order of magnitude of their examples.
5. Galileo used both logic and detailed experiments to refute the Aristotelian view that bodies naturally tend to move downward and heavy bodies fall faster.
6. Student answers could include Galileo's use of his pulse to measure equal amounts of time and Huygens' clock, the first clock that could measure with the precision of 1 s.

extension

Graphing Calculator Practice

Visit go.hrw.com for answers to this Graphing Calculator activity.

Keyword HF6MODXT

Standardized Test Prep

ANSWERS

1. B
2. F
3. D
4. H
5. B
6. G
7. C
8. G

Standardized Test Prep

MULTIPLE CHOICE

Use the graphs below to answer questions 1–3.

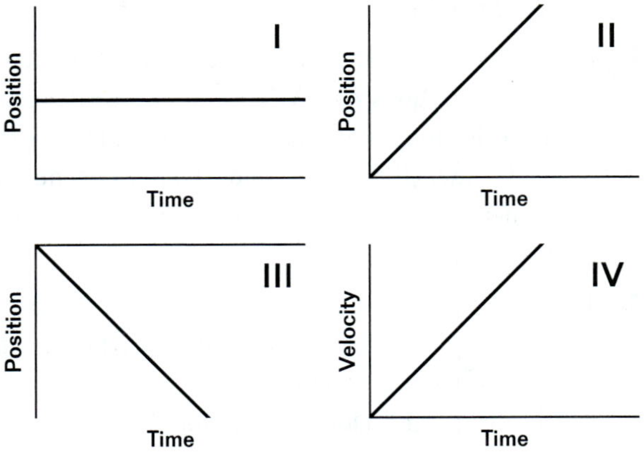

1. Which graph represents an object moving with a constant positive velocity?

A. I
B. II
C. III
D. IV

2. Which graph represents an object at rest?

F. I
G. II
H. III
J. IV

3. Which graph represents an object moving with constant positive acceleration?

A. I
B. II
C. III
D. IV

4. A bus travels from El Paso, Texas, to Chihuahua, Mexico, in 5.2 h with an average velocity of 73 km/h to the south. What is the bus's displacement?

F. 73 km to the south
G. 370 km to the south
H. 380 km to the south
J. 14 km/h to the south

Use the following position-time graph of a squirrel running along a clothesline to answer questions 5–6.

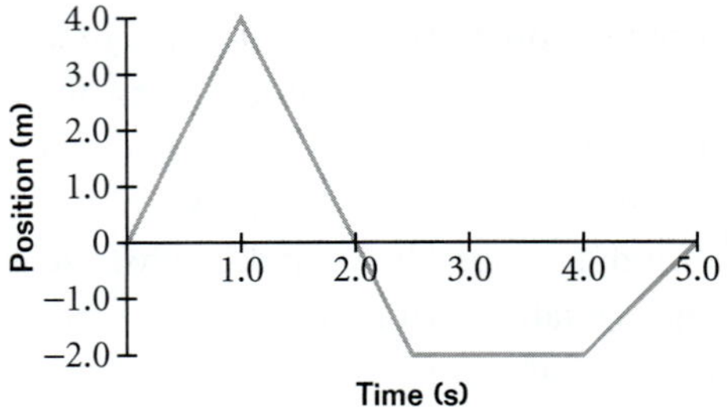

5. What is the squirrel's displacement at time $t = 3.0$ s?

A. −6.0 m
B. −2.0 m
C. +0.8 m
D. +2.0 m

6. What is the squirrel's average velocity during the time interval between 0.0 s and 3.0 s?

F. −2.0 m/s
G. −0.67 m/s
H. 0.0 m/s
J. +0.53 m/s

7. Which of the following statements is true of acceleration?

A. Acceleration always has the same sign as displacement.
B. Acceleration always has the same sign as velocity.
C. The sign of acceleration depends on both the direction of motion and how the velocity is changing.
D. Acceleration always has a positive sign.

8. A ball initially at rest rolls down a hill and has an acceleration of 3.3 m/s^2. If it accelerates for 7.5 s, how far will it move during this time?

F. 12 m
G. 93 m
H. 120 m
J. 190 m

9. Which of the following statements is true for a ball thrown vertically upward?
   - **A.** The ball has a negative acceleration on the way up and a positive acceleration on the way down.
   - **B.** The ball has a positive acceleration on the way up and a negative acceleration on the way down.
   - **C.** The ball has zero acceleration on the way up and a positive acceleration on the way down.
   - **D.** The ball has a constant acceleration throughout its flight.

SHORT RESPONSE

10. In one or two sentences, explain the difference between *displacement* and *distance traveled.*

11. The graph below shows the position of a runner at different times during a run. Use the graph to determine the runner's displacement and average velocity:
   - **a.** for the time interval from $t = 0.0$ min to $t = 10.0$ min
   - **b.** for the time interval from $t = 10.0$ min to $t = 20.0$ min
   - **c.** for the time interval from $t = 20.0$ min to $t = 30.0$ min
   - **d.** for the entire run

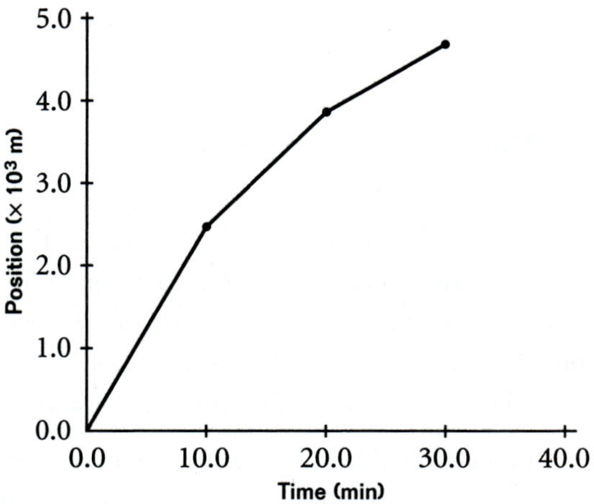

12. For an object moving with constant negative acceleration, draw the following:
   - **a.** a graph of position vs. time
   - **b.** a graph of velocity vs. time

   For both graphs, assume the object starts with a positive velocity and a positive displacement from the origin.

13. A snowmobile travels in a straight line. The snowmobile's initial velocity is +3.0 m/s.
   - **a.** If the snowmobile accelerates at a rate of $+0.50\ m/s^2$ for 7.0 s, what is its final velocity?
   - **b.** If the snowmobile accelerates at the rate of $-0.60\ m/s^2$ from its initial velocity of +3.0 m/s, how long will it take to reach a complete stop?

EXTENDED RESPONSE

14. A car moving eastward along a straight road increases its speed uniformly from 16 m/s to 32 m/s in 10.0 s.
   - **a.** What is the car's average acceleration?
   - **b.** What is the car's average velocity?
   - **c.** How far did the car move while accelerating?

   Show all of your work for these calculations.

15. A ball is thrown vertically upward with a speed of 25.0 m/s from a height of 2.0 m.
   - **a.** How long does it take the ball to reach its highest point?
   - **b.** How long is the ball in the air?

   Show all of your work for these calculations.

Test TIP When filling in your answers on an answer sheet, always check to make sure you are filling in the answer for the right question. If you have to change an answer, be sure to completely erase your previous answer.

9. D

10. Displacement measures only the net change in position from starting point to end point. The distance traveled is the total length of the path followed from starting point to end point and may be greater than or equal to the displacement.

11. Answers will vary depending on graph estimations but should be approximately as follows:
   - **a.** $\Delta x_1 = +2400$ m; $v_1 = +4.0$ m/s
   - **b.** $\Delta x_2 = +1500$ m; $v_2 = +2.5$ m/s
   - **c.** $\Delta x_3 = +900$ m; $v_3 = +2$ m/s
   - **d.** $\Delta x_{tot} = +4800$ m; $v_{avg} = +2.7$ m/s

12. **a.**

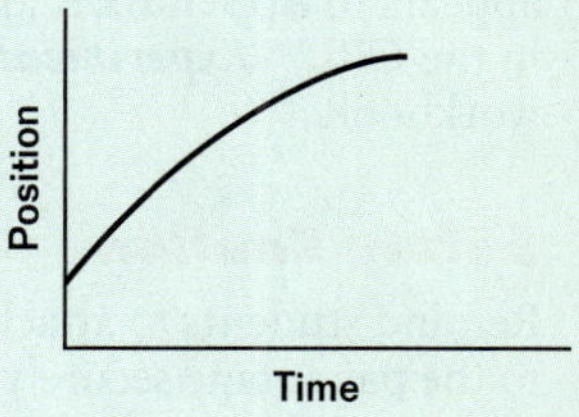

b.

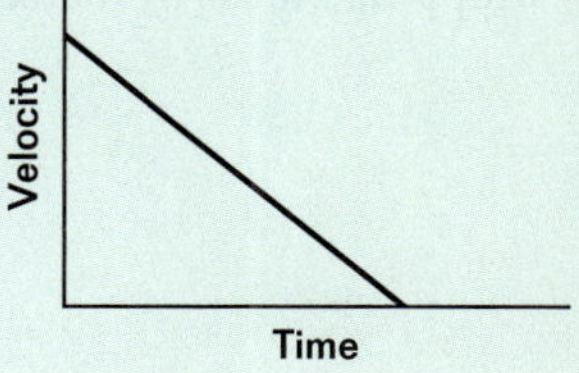

13. **a.** +6.5 m/s
   b. 5.0 s

14. **a.** $1.6\ m/s^2$ eastward
   b. 24 m/s eastward
   c. 240 m
   (See the Solutions Manual or the One-Stop Planner for the full solution.)

15. **a.** 2.55 s
   b. 5.18 s
   (See the Solutions Manual or the One-Stop Planner for the full solution.)

Lab Planning

Beginning on page T34 are preparation notes and teaching tips to assist you in planning.

Blank data tables (as well as some sample data) appear on the **One-Stop Planner.**

No Books in the Lab?

See the ***Datasheets for In-Text Labs*** workbook for a reproducible master copy of this experiment.

CBL™ Option

A **CBL™** version of this lab appears in **Appendix K** and in the ***CBL™ Experiments*** workbook.

Safety Caution

Remind students to attach masses to the paper tape securely and make sure the area is clear before dropping blocks or masses.

CHAPTER 2

Skills Practice Lab — Free-Fall Acceleration

OBJECTIVES

- **Measure** motion in terms of the change in distance during the period of a recording timer.
- **Compare** the speed and acceleration of different falling masses at different stages of free fall.
- **Compare** the experimental value for the average acceleration to the accepted value for free-fall acceleration.

MATERIALS LIST

- **C-clamp**
- **masking tape**
- **meterstick**
- **recording timer and paper tape**
- **ring stand**
- **set of hooked masses**
- **stopwatch**

In the laboratory, you can use a recording timer to determine the velocity and acceleration of bodies moving in one dimension. A recording timer measures the time it takes an object to move a short distance by making marks at regular time intervals on a strip of paper attached to the moving object. As the paper tape is pulled through the timer, the distance between two dots on the tape is the distance the tape moved during one back-and-forth vibration of the clapper. The time required for one back-and-forth motion of the clapper is called the *period* of the timer.

In this experiment, you will first calibrate a recording timer by determining an average value for its period. You will then use the recording timer to determine the average velocity and average acceleration of falling bodies of different masses.

SAFETY

- **Tie back long hair, secure loose clothing, and remove loose jewelry to prevent their getting caught in moving or rotating parts. Put on goggles.**
- **Attach masses securely. Falling or dropped masses can cause serious injury.**

PROCEDURE

Preparation

1. Read the entire lab procedure, and plan the steps you will take.
2. If you are not using a datasheet provided by your teacher, prepare a data table in your lab notebook with six columns and five rows. In the first row, label the first two columns *Trial* and *Mass (kg)*. The space for the third through sixth columns should be labeled *Distance (m)*. Under this common label, columns 3–6 should be labeled *A–B, C–D, E–F,* and *G–H*. In the first column, label the second through fifth rows *1, 2, 3,* and *4*.
3. If you are not using a datasheet provided by your teacher, prepare a second data table with three columns and five rows in your lab notebook. Label this table *Calibration*. In the first row, label the columns *Trial, Time (s),* and *Number of Dots*. Fill in the first column by labeling the second through fifth rows *1, 2, 3,* and *4* for the number of trials.

4. Clamp the recording timer to the ring stand to hold the timer in place. Choose a location that will allow you to pull a long section of paper tape through the timer in a straight line without hitting any obstacles. ***Do not plug in the timer until your teacher approves your setup.***

5. Insert a strip of paper tape about 2.0 m long into the timer so that the paper can move freely and will be marked as it moves. Lay the tape flat behind the timer. As shown in **Figure 1,** one student should hold the end of the tape in front of the timer.

6. When your teacher approves your setup, plug the timer into the wall socket.

7. One student should start the timer and the stopwatch at the same time that the other student holding the free end of the tape begins pulling the tape through the timer at a steady pace by walking away from the timer.

   a. After exactly 3.0 s, the first student should turn off the timer and stop the watch, just as the second student with the tape stops walking. Mark the first and last dots on the tape. Tear or cut the dotted strip of tape from the roll and label it with the trial number and the time interval as measured by the stopwatch.

   b. Repeat this procedure three more times. Label all tapes.

8. Count the number of dots for each trial, starting with the second dot. Record this number in your Calibration data table.

   a. Compute the period of the timer for each trial by dividing the 3.0 s time interval by the number of dots recorded in the table.

   b. Find the average value for the period of the recording timer. Use this value for all your calculations.

Figure 1
Step 4: Put the ring stand at the edge of the table so the tape can be pulled through parallel to the floor for the calibration step. If your timer will not mount on a ring stand as shown, clamp it to the table instead.
Step 5: Thread the tape through the timer and make sure the paper tape is *under* the carbon disk.

Tips and Tricks

- If 110 V recording timers are used, calibration may be omitted to save time; the average period of these timers is 0.017 s (1/60 s).
- If possible, timers should be mounted on support rods as shown.
- Show students how to thread the paper tape through the recording timer guides under the carbon disk.

✓ Checkpoints

Step 6: Check each setup before the timer is plugged in. Make sure the tape is inserted properly and all clamps are tight and positioned where they will not protrude and cause injury.

Step 8: Students should use the spacing of the dots on the tape to show that they walked at a steady pace. They should explain how they find the period of the timer.

Step 9: All clamps must be tight, and the recording timer must be positioned so the masses can fall without hitting any obstacles.

Step 11: The mass should be dropped from the same level for all trials. Students should hold the timer tape to suspend the mass in the starting position.

Step 12: Students should be able to explain how the dots on the tape represent the motion of the falling mass. They should be able to explain in what order the dots were made.

ANSWERS

Analysis

1. Student answers will vary. Answers should show that the velocity increases throughout the motion. Typical values will range from 0.210 m/s to 2.56 m/s.

2. Student answers will vary. Make sure students use the relationship $a_{avg} = \frac{\Delta v}{\Delta t}$.
For sample data, calculated values range from 4.78 m/s^2 to 9.63 m/s^2.

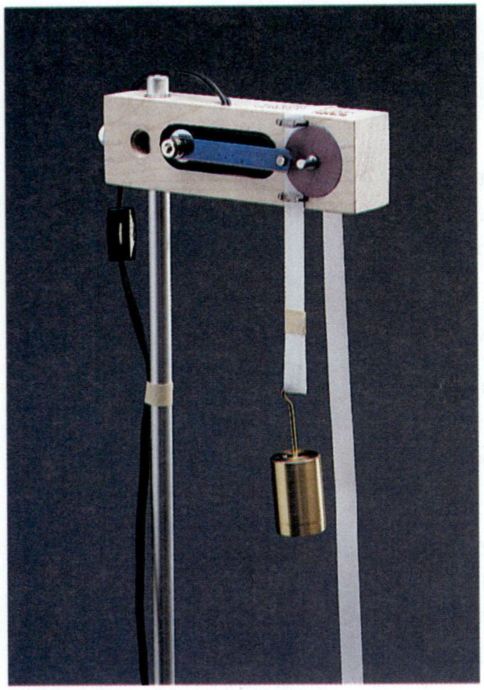

Figure 2

Step 9: When the timer is mounted on the ring stand, tape the cord to the stand to keep it out of the way.

Step 11: Use a piece of tape to mark the stand at the starting point, and start the mass from the same point for all trials.

Speed and Acceleration of a Falling Object

9. Set up the apparatus as shown in **Figure 2.** If the timer cannot be mounted on the stand, clamp the timer to the edge of the table.
10. Cut a length of paper tape that is at least 20 cm longer than the distance between the timer and the floor. Thread the end of the tape through the timer.
11. Fold the end of the paper tape and fasten it with masking tape to make a loop. Hook a 200 g mass through the looped end of the paper tape, as shown.
12. Position the mass at a convenient level near the timer, as shown. Hold the mass in place by holding the tape behind the timer. Make sure the area is clear of people and objects. Simultaneously, start the timer and release the tape so the mass falls to the floor. Stop the timer when the mass hits the floor.
13. Label the tape with the mass used. Label the second and third dots *A* and *B*, respectively. Count four dots from *B* and label the seventh and eighth dots *C* and *D*, respectively. Label the twelfth and thirteenth dots *E* and *F*, and label the seventeenth and eighteenth dots *G* and *H*.
14. Repeat this procedure using different, larger masses, such as 300 g and 400 g masses. Drop each mass from the same level in each trial. Label all tapes, and record all data.
15. On each tape, measure the distance between *A* and *B*, between *C* and *D*, and so on. Record the distance in meters in your data table.
16. Clean up your work area. Put equipment away safely so that it is ready to be used again.

ANALYSIS

1. **Organizing Data** For each trial with the falling mass, find the magnitude of the average velocity, v_{avg}. Divide the distance *A–B* by the average period of the timer. Repeat this calculation for the other marked distances for each trial.
2. **Organizing Data** Using the results from item 1, calculate the average acceleration. Find the change of speed between the distance *A–B* and the distance *C–D*, between the distance *C–D* and the distance *E–F*, and so on. (Hint: Remember to use the total time interval for each calculation. For example, for the first calculation, use the time interval from *A* to *D*.)

3. **Constructing Graphs** Use your data to plot the following graphs for each trial. On each graph, label the axes and indicate the trial number. Use a graphing calculator, computer, or graph paper.

   a. position versus time
   b. velocity versus time
   c. acceleration versus time

4. **Organizing Data** Use the values for the average acceleration for all four trials to find the average value.

5. **Evaluating Results** Use the accepted value for the free-fall acceleration given in the text and the average of your results from item 4.

   a. Determine the absolute error of your results using the following equation:

   $$\text{absolute error} = |\text{experimental} - \text{accepted}|$$

   b. Determine the relative error of your results using the following equation:

   $$\text{relative error} = \frac{(\text{experimental} - \text{accepted})}{\text{accepted}}$$

CONCLUSIONS

6. **Making Predictions** Based on your results, how long would it take a 1000 kg mass to reach the floor if it were dropped from the same height as the masses in this experiment?

7. **Analyzing Graphs** Calculate the slope of each velocity-time graph from item 3b.

8. **Evaluating Results** Find the average value for the slope of the velocity-time graphs. What is the relationship between this value and the values you found for the average accelerations of the masses?

EXTENSION

9. **Designing Experiments** Devise a plan to perform this experiment to study the motion of an object thrown straight up into the air. Make sure you take into account any special safety requirements or equipment you might need to use. If there is time and your teacher approves your plan, perform the experiment. Use your data to plot graphs of the position, velocity, and acceleration versus time.

3. a. Student graphs should show a parabolic curve pointing up and to the right.
b. Student graphs should show a straight line pointing up and to the right.
c. Student graphs should show a straight line parallel to the *x*-axis.

4. all trials: $a_{avg} = 6.92\ m/s^2$ (using sample data)

5. Accepted value = $9.81\ m/s^2$;
a. absolute error = $2.89\ m/s^2$,
b. relative error = −0.295 (using sample data)

Conclusions

6. All blocks will fall with the same acceleration; therefore, the 1000 kg block will fall in the same time as all the other blocks.

7. Answers should be similar to acceleration values.

8. They are the same or very close.

Extension

9. Student plans should be safe and complete, including a list of equipment, measurements, and calculations required. Plans should recognize that the lab is essentially the same while also providing for the safety hazards and technical difficulties of studying an object that changes direction.

CHAPTER 3

Two-Dimensional Motion and Vectors
Planning Guide

Compression Guide
To shorten instruction because of time limitations, omit the opener and Section 4 and abbreviate the review.

OBJECTIVES	LABS, DEMONSTRATIONS, AND ACTIVITIES	TECHNOLOGY RESOURCES
PACING • 45 min pp. 80–81 **Chapter Opener**	ANC **Discovery Lab** Vector Treasure Hunt*◆ BASIC	CD **Visual Concepts,** Chapter 3 BASIC
PACING • 45 min pp. 82–85 **Section 1 Introduction to Vectors** • Distinguish between a scalar and a vector. • Add and subtract vectors by using the graphical method. • Multiply and divide vectors by scalars.	TE **Demonstration** Vector Addition, p. 83 GENERAL	OSP **Lesson Plans** TR 6 Graphical Method of Vector Addition TR 14A Commutative Property of Vectors
PACING • 45 min pp. 86–94 **Section 2 Vector Operations** • Identify appropriate coordinate systems for solving problems with vectors. • Apply the Pythagorean theorem and tangent function to calculate the magnitude and direction of a resultant vector. • Resolve vectors into components using the sine and cosine functions. • Add vectors that are not perpendicular.		OSP **Lesson Plans** TR 15A The Pythagorean Theorem and the Tangent Function TR 16A Adding Vectors That Are Not Perpendicular TR 17A Adding Vectors Algebraically
PACING • 135 min pp. 95–101 **Section 3 Projectile Motion** • Recognize examples of projectile motion. • Describe the path of a projectile as a parabola. • Resolve vectors into their components and apply the kinematic equations to solve problems involving projectile motion.	TE **Demonstration** Air Resistance, p. 96 GENERAL SE **Quick Lab** Projectile Motion, p. 97 GENERAL SE **Inquiry Lab** Velocity of a Projectile, pp. 116–117◆ ANC **Datasheet** Inquiry Lab, Velocity of a Projectile* GENERAL ANC **Datasheet** Skills Practice Lab, Velocity of a Projectile* ANC **Invention Lab** The Path of a Human Cannonball*◆ ANC **CBL™ Experiment** Projectile Motion*◆ ADVANCED	OSP **Lesson Plans** TR 7 Vertical Motion of a Projectile
PACING • 45 min pp. 102–105 *Advanced Level* **Section 4 Relative Motion** • Describe situations in terms of frame of reference. • Solve problems involving relative velocity.		OSP **Lesson Plans** TR 8 Frames of Reference

PACING • 90 min

CHAPTER REVIEW, ASSESSMENT, AND STANDARDIZED TEST PREPARATION

SE **Chapter Highlights,** p. 107
SE **Chapter Review,** pp. 108–113
SE **Graphing Calculator Practice,** p. 113 GENERAL
SE **Alternative Assessment,** p. 113 ADVANCED
SE **Standardized Test Prep,** pp. 114–115 GENERAL
SE **Appendix D: Equations,** pp. 854–855
SE **Appendix I: Additional Problems,** pp. 881–882
ANC **Study Guide Worksheet** Mixed Review* GENERAL
ANC **Chapter Test A*** GENERAL
ANC **Chapter Test B*** ADVANCED
OSP **Test Generator**

Online and Technology Resources

go.hrw.com Holt Online Learning

Visit go.hrw.com to find a variety of online resources. To access this chapter's extensions, enter the keyword **HF6TDMXT** and click the "go" button. Click **Holt Online Learning** for an online edition of this textbook, and other interactive resources.

HOLT Teacher's One-Stop Planner®

This DVD package includes:
- Holt Calendar Planner
- Customizable Lesson Plans
- Editable Worksheets
- ExamView® Version 6 Assessment Suite
- Interactive Teacher's Edition
- Holt PuzzlePro®
- Holt PowerNotes® Presentations
- MindPoint® Quiz Show

KEY

SE Student Edition
TE Teacher Edition
ANC Ancillary Worksheet
OSP One-Stop Planner
CD CD or CD-ROM
TR Teaching Transparencies
EXT Online Extension
* Also on One-Stop Planner
◆ Requires advance prep

SKILLS DEVELOPMENT RESOURCES	REVIEW AND ASSESSMENT	CORRELATIONS
		National Science Education Standards
	SE **Section Review,** p. 85 GENERAL ANC **Study Guide Worksheet** Section 1* GENERAL ANC **Quiz** Section 1* BASIC	UCP 2, 3 SAI 1, 2
SE **Sample Set A** Finding Resultant Magnitude, pp. 88–89 GENERAL TE **Classroom Practice,** p. 88 GENERAL ANC **Problem Workbook*** and OSP **Problem Bank** Sample Set A GENERAL SE **Sample Set B** Resolving Vectors, pp. 91–92 GENERAL TE **Classroom Practice,** p. 91 GENERAL ANC **Problem Workbook*** and OSP **Problem Bank** Sample Set B GENERAL SE **Sample Set C** Adding Vectors Algebraically, pp. 93–94 GENERAL TE **Classroom Practice,** p. 93 GENERAL ANC **Problem Workbook*** and OSP **Problem Bank** Sample Set C GENERAL	SE **Section Review,** p. 94 GENERAL ANC **Study Guide Worksheet** Section 2* GENERAL ANC **Quiz** Section 2* BASIC	UCP 2, 3
SE **Sample Set D** Projectiles Launched Horizontally, pp. 98–99 GENERAL TE **Classroom Practice,** p. 98 GENERAL ANC **Problem Workbook*** and OSP **Problem Bank** Sample Set D GENERAL SE **Sample Set E** Projectiles Launched at an Angle, pp. 100–101 GENERAL TE **Classroom Practice,** p. 100 GENERAL ANC **Problem Workbook*** and OSP **Problem Bank** Sample Set E GENERAL	SE **Section Review,** p. 101 GENERAL ANC **Study Guide Worksheet** Section 3* GENERAL ANC **Quiz** Section 3* BASIC	UCP 2, 3 SAI 1, 2
SE **Conceptual Challenge,** p. 103 ADVANCED SE **Sample Set F** Relative Velocity, pp. 104–105 GENERAL TE **Classroom Practice,** p. 104 GENERAL ANC **Problem Workbook*** and OSP **Problem Bank** Sample Set F GENERAL SE **Appendix J: Advanced Topics** Special Relativity and Velocities, pp. 916–917 ADVANCED	SE **Section Review,** p. 105 ADVANCED ANC **Study Guide Worksheet** Section 4* GENERAL ANC **Quiz** Section 4* GENERAL	UCP 2, 3

SCILINKS. NSTA
www.scilinks.org

Maintained by the **National Science Teachers Association.**

Topic: Vectors
SciLinks Code: HF61597

Topic: Projectile Motion
SciLinks Code: HF61223

Classroom CD-ROMs

- Guided Reading Audio Program
- Student One Stop
- Virtual Investigations
- Visual Concepts

Holt Lab Generator CD-ROM

Search for any lab by topic, standard, difficulty level, or time. Edit any lab to fit your needs, or create your own labs. Use the Lab Materials QuickList software to customize your lab materials list.

CHAPTER 3
Overview

Section 1 discusses scalar and vector quantities and graphical vector addition.

Section 2 explains the use of the Pythagorean theorem and trigonometric functions to find resultant vectors and vector components.

Section 3 explores projectile motion, neglecting air resistance.

Section 4 describes relative motion in terms of vector operations.

About the Illustration

This fountain is located in Bayfront Park in Sarasota, Florida. Point out that each drop of water emerging from the fountain moves as a projectile and obeys the rules for projectile motion described in this chapter. Likewise, the dolphins move as projectiles as they sail through the air.

CHAPTER 3

Two-Dimensional Motion and Vectors

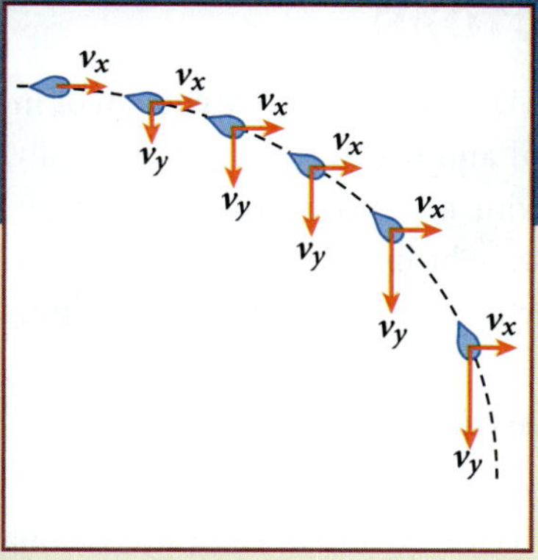

Without air resistance, any object that is thrown or launched into the air and that is subject to gravitational force will follow a parabolic path. The water droplets in this fountain are one example. The velocity of any object in two-dimensional motion—such as one of these water droplets—can be separated into horizontal and vertical components, as shown in the diagram.

WHAT TO EXPECT

In this chapter, you will use vectors to analyze two-dimensional motion and to solve problems in which objects are projected into the air.

Why It Matters

After you know how to analyze two-dimensional motion, you can predict where a falling object will land based on its initial velocity and position.

CHAPTER PREVIEW

Tapping Prior Knowledge

Knowledge to Review

- ✔ Displacement is a change in location relative to a reference point.
- ✔ Velocity includes speed and direction.
- ✔ Acceleration is the rate of change of velocity with respect to time. It has both magnitude and direction.

Items to Probe

- ✔ Displacement versus distance: Have students decide displacement and distance values for scenarios listed on the board, such as round trips versus one-way trips, or by walking from one place to another in the classroom.
- ✔ Acceleration and velocity: Ask students to describe the effects of the four possible combinations of acceleration and velocity in one dimension: $a > 0, v > 0$; $a < 0, v > 0$; $a > 0, v < 0$; $a < 0, v < 0$.

SECTION 1
General Level

Teaching Tip

Carry an arrow with you to class, and use it often to describe the direction of the vector.

The Language of Physics

Establish a convention for distinguishing vectors from scalars for use in your classroom. Using an arrow above the variable is a good substitute for bold type, which would be hard to reproduce on a chalkboard. Make sure to be consistent when using this convention. For instance, when you use an arrow above the symbol to designate a vector, be sure to signify the *direction* of the vector as well as its magnitude and units.

Visual Strategy BASIC

Figure 1

Tell students that the arrows in the figure represent velocity, that is, both speed and direction.

Q Why do we consider velocity instead of speed in order to decide which player reaches the ball first?

A ***Direction is important because a player may run very fast but not toward the ball.***

SECTION 1

Introduction to Vectors

SECTION OBJECTIVES

- **Distinguish between a scalar and a vector.**
- **Add and subtract vectors by using the graphical method.**
- **Multiply and divide vectors by scalars.**

SCALARS AND VECTORS

In the chapter "Motion in One Dimension," our discussion of motion was limited to two directions, forward and backward. Mathematically, we described these directions of motion with a positive or negative sign. That method works only for motion in a straight line. This chapter explains a method of describing the motion of objects that do not travel along a straight line.

Vectors indicate direction; scalars do not

Each of the physical quantities encountered in this book can be categorized as either a scalar quantity or a vector quantity. A **scalar** is a quantity that has magnitude but no direction. Examples of scalar quantities are speed, volume, and the number of pages in this textbook. A **vector** is a physical quantity that has both direction and magnitude.

scalar

a physical quantity that has magnitude but no direction

vector

a physical quantity that has both magnitude and direction

As we look back to the chapter "Motion in One Dimension," we can see that displacement is an example of a vector quantity. An airline pilot planning a trip must know exactly how far and which way to fly. Velocity is also a vector quantity. If we wish to describe the velocity of a bird, we must specify both its speed (say, 3.5 m/s) and the direction in which the bird is flying (say, northeast). Another example of a vector quantity is acceleration.

Figure 1
The lengths of the vector arrows represent the magnitudes of these two soccer players' velocities.

Vectors are represented by boldface symbols

In physics, quantities are often represented by symbols, such as t for time. To help you keep track of which symbols represent vector quantities and which are used to indicate scalar quantities, this book will use **boldface** type to indicate vector quantities. Scalar quantities will be in *italics.* For example, the speed of a bird is written as $v = 3.5$ m/s. But a velocity, which includes a direction, is written as $\mathbf{v} = 3.5$ m/s to the northeast. When writing a vector on your paper, you can distinguish it from a scalar by drawing an arrow above the abbreviation for a quantity, such as $\vec{v} = 3.5$ m/s to the northeast.

One way to keep track of vectors and their directions is to use diagrams. In diagrams, vectors are shown as arrows that point in the direction of the vector. The length of a vector arrow in a diagram is proportional to the vector's magnitude. For example, in **Figure 1** the arrows represent the velocities of the two soccer players running toward the soccer ball.

A resultant vector represents the sum of two or more vectors

When adding vectors, you must make certain that they have the same units and describe similar quantities. For example, it would be meaningless to add a velocity vector to a displacement vector because they describe different physical quantities. Similarly, it would be meaningless, as well as incorrect, to add two displacement vectors that are not expressed in the same units. For example, you cannot add meters and feet together.

Section 1 of the chapter "Motion in One Dimension" covered vector addition and subtraction in one dimension. Think back to the example of the gecko that ran up a tree from a 20 cm marker to an 80 cm marker. Then the gecko reversed direction and ran back to the 50 cm marker. Because the two parts of this displacement are opposite, they can be added together to give a total displacement of 30 cm. The answer found by adding two vectors in this way is called the **resultant.**

Did you know?

The word *vector* is also used by airline pilots and navigators. In this context, a vector is the particular path followed or to be followed, given as a compass heading.

resultant

a vector that represents the sum of two or more vectors

Vectors can be added graphically

Consider a student walking 1600 m to a friend's house and then 1600 m to school, as shown in **Figure 2.** The student's total displacement during his walk to school is in a direction from his house to the school, as shown by the dotted line. This direct path is the *vector sum* of the student's displacement from his house to his friend's house and his displacement from the friend's house to school. How can this resultant displacement be found?

One way to find the magnitude and direction of the student's total displacement is to draw the situation to scale on paper. Use a reasonable scale, such as 50 m on land equals 1 cm on paper. First draw the vector representing the student's displacement from his house to his friend's house, giving the proper direction and scaled magnitude. Then draw the vector representing his walk to the school, starting with the tail at the head of the first vector. Again give its scaled magnitude and the right direction. The magnitude of the resultant vector can then be determined by using a ruler. Measure the length of the vector pointing from the tail of the first vector to the head of the second vector. The length of that vector can then be multiplied by 50 (or whatever scale you have chosen) to get the actual magnitude of the student's total displacement in meters.

The direction of the resultant vector may be determined by using a protractor to measure the angle between the resultant and the first vector or between the resultant and any chosen reference line.

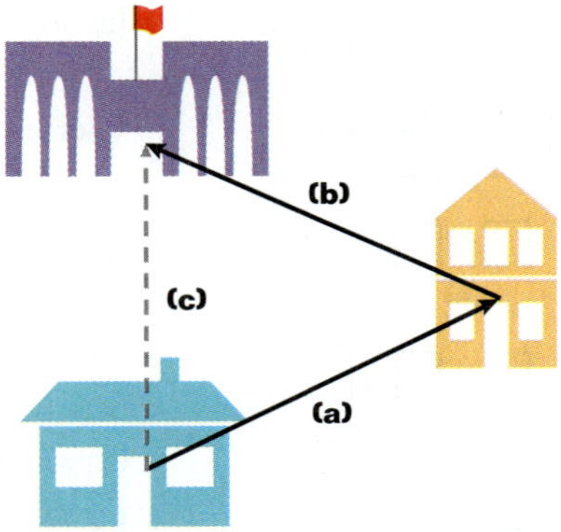

Figure 2
A student walks from his house to his friend's house **(a),** then from his friend's house to the school **(b).** The student's resultant displacement **(c)** can be found by using a ruler and a protractor.

SECTION 1

Demonstration

Vector Addition — GENERAL

Purpose Preview force as a vector quantity to demonstrate vector addition.

Materials large, empty box

Procedure Ask for a student volunteer. Explain to the class that you will push the box in one direction while the student will push the box in a second, perpendicular direction at the same time. Ask students to predict the motion of the box.

Have the student volunteer take a practice push alone, and emphasize that the box moves in the direction of the push. Take a practice push by yourself on the box, and again emphasize that the box moves in the direction of the push. Return the box to its original location, and both you and the student push the box perpendicular to each other at the count of three. Ask a student to explain why the box moved along the diagonal. *(The box moved in a direction between the directions of the two pushes.)*

Show the vector addition for the demonstration qualitatively on the chalkboard. This should lay the groundwork for a discussion of why vector addition is a valuable activity.

Misconception Alert

Students may have difficulty visualizing the movement of the toy car on the walkway. Illustrate the situation on the board, and emphasize that the motion represented in **Figure 3** is drawn as viewed by a stationary observer above the walkway. Frame of reference will be covered in Section 4.

The Language of Physics

The triangle method of vector addition is also called the *polygon method,* the *head-to-tail method*, or the *tip-to-tail method.*

Teaching Tip — GENERAL

Students will better understand the commutative property of vector addition if they work several examples on paper with a ruler and protractor.

Visual Strategy BASIC

Figure 4

Reinforce students' understanding of the difference between distance and displacement by pointing out that if the runner starts and ends at the positions shown in **Figure 4,** the runner's displacement will be the same regardless of the path the runner chooses to follow.

Q How can the runner increase distance but begin and end at the same points?

A ***The runner can take a different, longer path but will have to double back somewhere along the path to end up at the same point.***

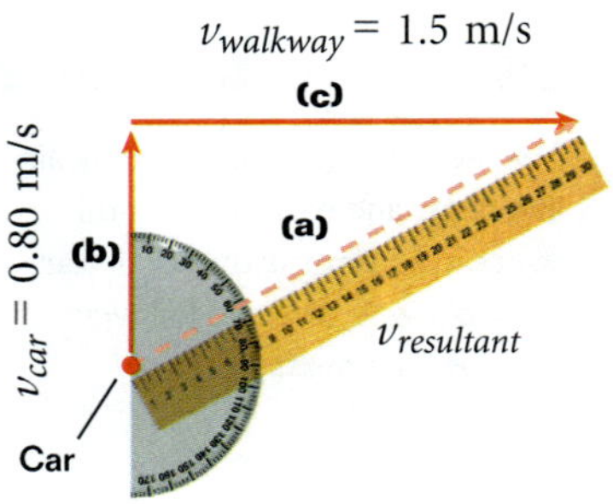

Figure 3
The resultant velocity **(a)** of a toy car moving at a velocity of 0.80 m/s **(b)** across a moving walkway with a velocity of 1.5 m/s **(c)** can be found using a ruler and a protractor.

PROPERTIES OF VECTORS

Now consider a case in which two or more vectors act at the same point. When this occurs, it is possible to find a resultant vector that has the same net effect as the combination of the individual vectors. Imagine looking down from the second level of an airport at a toy car moving at 0.80 m/s across a walkway that moves at 1.5 m/s. How can you determine what the car's resultant velocity will look like from your view point?

Vectors can be moved parallel to themselves in a diagram

Note that the car's resultant velocity while moving from one side of the walkway to the other will be the combination of two independent motions. Thus, the moving car can be thought of as traveling first at 0.80 m/s across the walkway and then at 1.5 m/s down the walkway. In this way, we can draw a given vector anywhere in the diagram as long as the vector is parallel to its previous alignment (so that it still points in the same direction). Thus, you can draw one vector with its tail starting at the tip of the other as long as the size and direction of each vector do not change. This process is illustrated in **Figure 3.** Although both vectors act on the car at the same point, the horizontal vector has been moved up so that its tail begins at the tip of the vertical vector. The resultant vector can then be drawn from the tail of the first vector to the tip of the last vector. This method is known as the *triangle* (or *polygon*) *method of addition.*

Again, the magnitude of the resultant vector can be measured using a ruler, and the angle can be measured with a protractor. In the next section, we will develop a technique for adding vectors that is less time-consuming because it involves a calculator instead of a ruler and protractor.

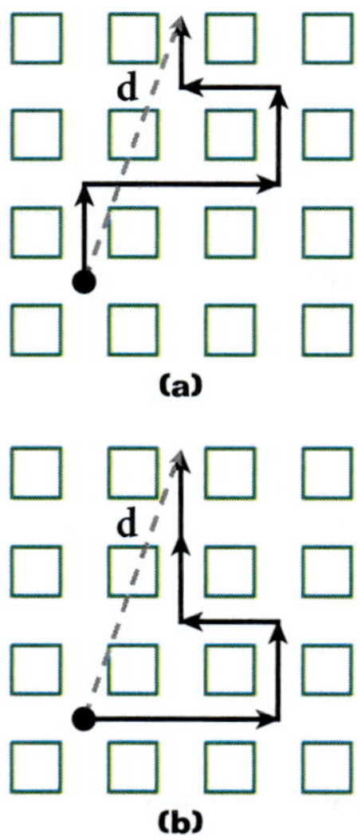

Figure 4
A marathon runner's displacement, **d,** will be the same regardless of whether the runner takes path **(a)** or **(b)** because the vectors can be added in any order.

Vectors can be added in any order

When two or more vectors are added, the sum is independent of the order of the addition. This idea is demonstrated by a runner practicing for a marathon along city streets, as represented in **Figure 4.** The runner executes the same four displacements in each case, but the order is different. Regardless of which path the runner takes, the runner will have the same total displacement, expressed as **d.** Similarly, the vector sum of two or more vectors is the same regardless of the order in which the vectors are added, provided that the magnitude and direction of each vector remain the same.

To subtract a vector, add its opposite

Vector subtraction makes use of the definition of the negative of a vector. The negative of a vector is defined as a vector with the same magnitude as the original vector but opposite in direction. For instance, the negative of the velocity of a car traveling 30 m/s to the west is −30 m/s to the west, or +30 m/s to the east. Thus, adding a vector to its negative vector gives zero.

When subtracting vectors in two dimensions, first draw the negative of the vector to be subtracted. Then add that negative vector to the other vector by using the triangle method of addition.

SCILINKS®
www.scilinks.org
Topic: Vectors
Code: HF61597

Multiplying or dividing vectors by scalars results in vectors

There are mathematical operations in which vectors can multiply other vectors, but they are not needed in this book. This book does, however, make use of vectors multiplied by scalars, with a vector as the result. For example, if a cab driver obeys a customer who tells him to go twice as fast, that cab's original velocity vector, $\mathbf{v}_{cab}$, is multiplied by the scalar number 2. The result, written $2\mathbf{v}_{cab}$, is a vector with a magnitude twice that of the original vector and pointing in the same direction.

On the other hand, if another cab driver is told to go twice as fast in the opposite direction, this is the same as multiplying by the scalar number –2. The result is a vector with a magnitude two times the initial velocity but pointing in the opposite direction, written as $-2\mathbf{v}_{cab}$.

SECTION REVIEW

1. Which of the following quantities are scalars, and which are vectors?
- **a.** the acceleration of a plane as it takes off
- **b.** the number of passengers on the plane
- **c.** the duration of the flight
- **d.** the displacement of the flight
- **e.** the amount of fuel required for the flight

2. A roller coaster moves 85 m horizontally, then travels 45 m at an angle of 30.0° above the horizontal. What is its displacement from its starting point? Use graphical techniques.

3. A novice pilot sets a plane's controls, thinking the plane will fly at 2.50×10^2 km/h to the north. If the wind blows at 75 km/h toward the southeast, what is the plane's resultant velocity? Use graphical techniques.

4. While flying over the Grand Canyon, the pilot slows the plane down to one-half the velocity in item 3. If the wind's velocity is still 75 km/h toward the southeast, what will the plane's new resultant velocity be? Use graphical techniques.

5. Critical Thinking The water used in many fountains is recycled. For instance, a single water particle in a fountain travels through an 85 m system and then returns to the same point. What is the displacement of this water particle during one cycle?

SECTION 1

Teaching Tip — BASIC

Some students may need further explanation and visual examples of the negative of a vector. Give several examples of vectors in different directions, and then show their negative vectors.

SECTION REVIEW ANSWERS

1. **a.** vector
b. scalar
c. scalar
d. vector
e. scalar

2. 126 m at $(1.0 \times 10^1)°$ above the horizontal

3. 204 km/h at 75° north of east

4. 89 km/h at 54° north of east

5. zero

SECTION 2
General Level

Teaching Tip BASIC

Review the sign conventions for coordinate systems that were established in the chapter "Motion in One Dimension." Movements to the right along the *x*-axis and upward along the *y*-axis are considered positive, and movements to the left along the *x*-axis and downward along the *y*-axis are considered negative.

Visual Strategy GENERAL

Figure 6

Point out the two very different choices for coordinate axes.

Q Which set of axes will give the correct answer?

A ***Either set of axes must give the same results.***

SECTION 2

Vector Operations

SECTION OBJECTIVES

- **Identify appropriate coordinate systems for solving problems with vectors.**
- **Apply the Pythagorean theorem and tangent function to calculate the magnitude and direction of a resultant vector.**
- **Resolve vectors into components using the sine and cosine functions.**
- **Add vectors that are not perpendicular.**

Figure 5
A gecko's displacement while climbing a tree can be represented by an arrow pointing along the *y*-axis.

COORDINATE SYSTEMS IN TWO DIMENSIONS

In the chapter "Motion in One Dimension," the motion of a gecko climbing a tree was described as motion along the *y*-axis. The direction of the displacement of the gecko was denoted by a positive or negative sign. The displacement of the gecko can now be described by an arrow pointing along the *y*-axis, as shown in **Figure 5.** A more versatile system for diagraming the motion of an object, however, employs vectors and the use of both the *x*- and *y*-axes simultaneously.

The addition of another axis not only helps describe motion in two dimensions but also simplifies analysis of motion in one dimension. For example, two methods can be used to describe the motion of a jet moving at 300 m/s to the northeast. In one approach, the coordinate system can be turned so that the plane is depicted as moving along the *y*-axis, as in **Figure 6(a).** The jet's motion also can be depicted on a two-dimensional coordinate system whose axes point north and east, as shown in **Figure 6(b).**

One problem with the first method is that the axis must be turned again if the direction of the plane changes. Another problem is that the first method provides no way to deal with a second airplane that is not traveling in the same direction as the first airplane. Thus, axes are often designated using fixed directions. For example, in **Figure 6(b),** the positive *y*-axis points north and the positive *x*-axis points east.

Similarly, when you analyze the motion of objects thrown into the air, orienting the *y*-axis parallel to the vertical direction simplifies problem solving.

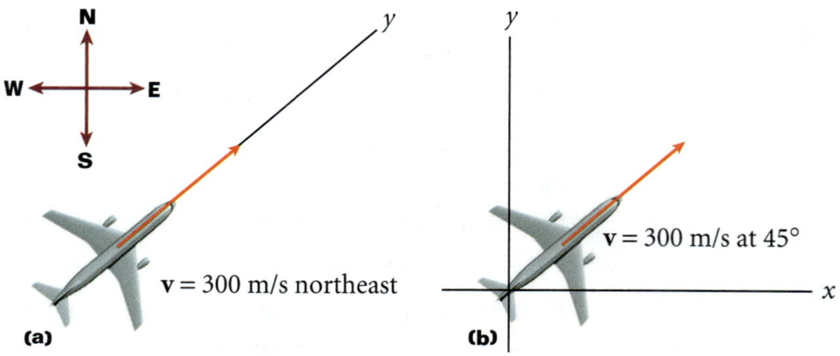

Figure 6
A plane traveling northeast at a velocity of 300 m/s can be represented as either **(a)** moving along a *y*-axis chosen to point to the northeast or **(b)** moving at an angle of 45° to both the *x*- and *y*-axes, which line up with west-east and south-north, respectively.

> **TIP** *There are no firm rules for applying coordinate systems to situations involving vectors. As long as you are consistent, the final answer will be correct regardless of the system you choose. Perhaps your best choice for orienting axes is the approach that makes solving the problem easiest for you.*

DETERMINING RESULTANT MAGNITUDE AND DIRECTION

In Section 1, the magnitude and direction of a resultant were found graphically. However, this approach is time consuming, and the accuracy of the answer depends on how carefully the diagram is drawn and measured. A simpler method uses the Pythagorean theorem and the tangent function.

Use the Pythagorean theorem to find the magnitude of the resultant

Imagine a tourist climbing a pyramid in Egypt. The tourist knows the height and width of the pyramid and would like to know the distance covered in a climb from the bottom to the top of the pyramid. Assume that the tourist climbs directly up the middle of one face.

As can be seen in **Figure 7,** the magnitude of the tourist's vertical displacement, Δy, is the height of the pyramid. The magnitude of the horizontal displacement, Δx, equals the distance from one edge of the pyramid to the middle, or half the pyramid's width. Notice that these two vectors are perpendicular and form a right triangle with the displacement, **d.**

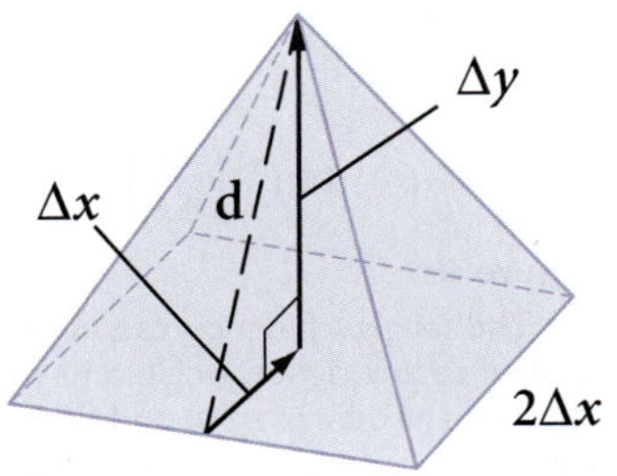

Figure 7
Because the base and height of a pyramid are perpendicular, we can find a tourist's total displacement, **d,** if we know the height, Δy, and width, $2\Delta x$, of the pyramid.

As shown in **Figure 8(a),** the Pythagorean theorem states that for any right triangle, the square of the hypotenuse—the side opposite the right angle—equals the sum of the squares of the other two sides, or legs.

PYTHAGOREAN THEOREM FOR RIGHT TRIANGLES

$$c^2 = a^2 + b^2$$

$$\text{(length of hypotenuse)}^2 = \text{(length of one leg)}^2 + \text{(length of other leg)}^2$$

In **Figure 8(b),** the Pythagorean theorem is applied to find the tourist's displacement. The square of the displacement is equal to the sum of the square of the horizontal displacement and the square of the vertical displacement. In this way, you can find out the magnitude of the displacement, d.

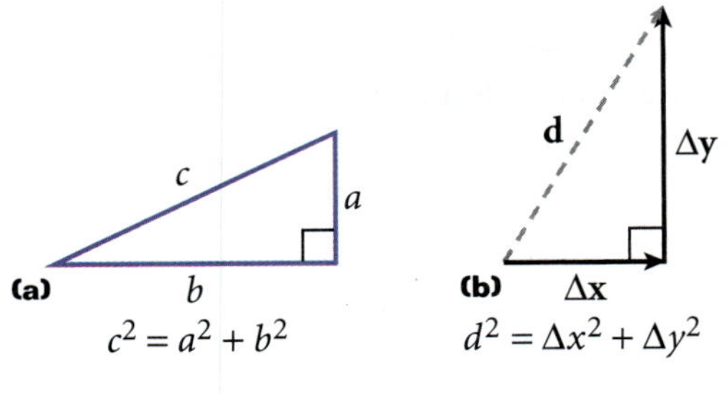

Figure 8
(a) The Pythagorean theorem can be applied to any right triangle. **(b)** It can also be applied to find the magnitude of a resultant displacement.

Misconception Alert — ADVANCED

Students often try to apply the Pythagorean theorem to triangles that do not contain a right angle. Point out that the Pythagorean theorem can be used only with a right triangle. Some students may know the Law of Cosines, which applies to all triangles. This law states that $c^2 = a^2 + b^2 - 2ab\cos\theta$. The Law of Cosines can be used to calculate one side of any triangle when the opposite angle and the lengths of the other two sides are known. In this expression, c is the unknown side, θ is the angle opposite c, and a and b are the two known sides. Some students may attempt to use the Law of Cosines to add nonperpendicular vectors. This approach will give the correct answer, but it entails more computation and is more prone to student error when more than two vectors are to be added.

Teaching Tip — GENERAL

Point out that finding the resultant for the pyramid is fairly simple because the height, half-width, and hypotenuse form a right triangle. It is important to mention at this point that right triangles will also allow students to find the x and y components that are important for vector addition.

Classroom Practice

Finding Resultant Magnitude and Direction
A plane travels from Houston, Texas, to Washington, D.C., which is 1540 km east and 1160 km north of Houston. What is the total displacement of the plane?

Answer
1930 km at 37.0° north of east

A camper travels 4.5 km northeast and 4.5 km northwest. What is the camper's total displacement?

Answer
6.4 km north

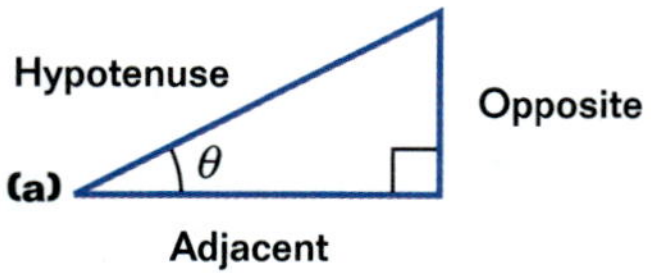

$$\tan\theta = \frac{\text{opp}}{\text{adj}}$$

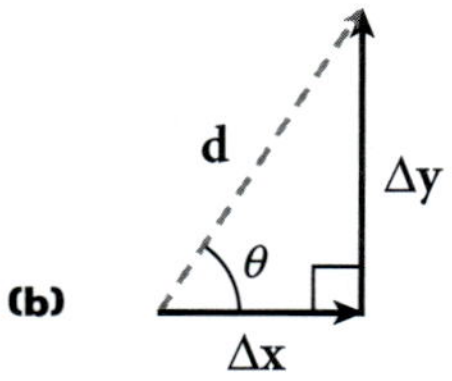

$$\tan\theta = \frac{\Delta y}{\Delta x}$$

$$\theta = \tan^{-1}\left(\frac{\Delta y}{\Delta x}\right)$$

Figure 9
(a) The tangent function can be applied to any right triangle, and **(b)** it can also be used to find the direction of a resultant displacement.

Use the tangent function to find the direction of the resultant

In order to completely describe the tourist's displacement, you must also know the direction of the tourist's motion. Because **Δx, Δy,** and **d** form a right triangle, as shown in **Figure 9(b),** the inverse tangent function can be used to find the angle θ, which denotes the direction of the tourist's displacement.

For any right triangle, the tangent of an angle is defined as the ratio of the opposite and adjacent legs with respect to a specified acute angle of a right triangle, as shown in **Figure 9(a).**

As shown below, the magnitude of the opposite leg divided by the magnitude of the adjacent leg equals the tangent of the angle.

DEFINITION OF THE TANGENT FUNCTION FOR RIGHT TRIANGLES

$$\tan\theta = \frac{\text{opp}}{\text{adj}} \qquad \textbf{tangent of angle} = \frac{\textbf{opposite leg}}{\textbf{adjacent leg}}$$

The inverse of the tangent function, which is shown below, gives the angle.

$$\theta = \tan^{-1}\left(\frac{\text{opp}}{\text{adj}}\right)$$

SAMPLE PROBLEM A

Finding Resultant Magnitude and Direction

PROBLEM

An archaeologist climbs the Great Pyramid in Giza, Egypt. The pyramid's height is 136 m and its width is 2.30×10^2 m. What is the magnitude and the direction of the displacement of the archaeologist after she has climbed from the bottom of the pyramid to the top?

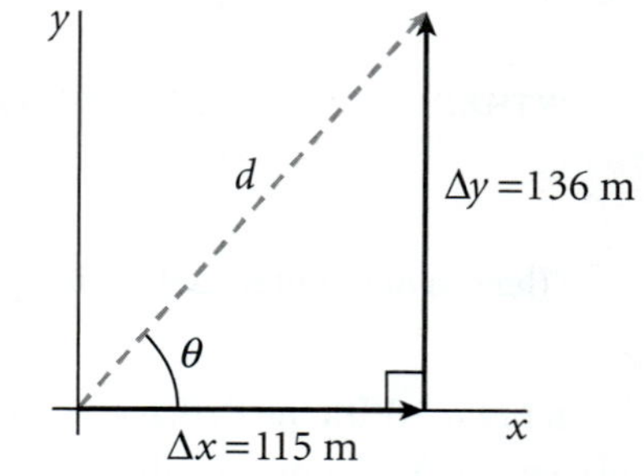

SOLUTION

1. DEFINE **Given:** $\Delta y = 136$ m $\Delta x = \frac{1}{2}(\text{width}) = 115$ m

Unknown: $d = ?$ $\theta = ?$

Diagram: Choose the archaeologist's starting position as the origin of the coordinate system.

2. PLAN **Choose an equation or situation:**
The Pythagorean theorem can be used to find the magnitude of the archaeologist's displacement. The direction of the displacement can be found by using the tangent function.

$$d^2 = \Delta x^2 + \Delta y^2$$

$$\tan\theta = \frac{\Delta y}{\Delta x}$$

Rearrange the equations to isolate the unknowns:

$$d = \sqrt{\Delta x^2 + \Delta y^2}$$

$$\theta = \tan^{-1}\left(\frac{\Delta y}{\Delta x}\right)$$

3. CALCULATE **Substitute the values into the equations and solve:**

$$d = \sqrt{(115 \text{ m})^2 + (136 \text{ m})^2}$$

$$\boxed{d = 178 \text{ m}}$$

$$\theta = \tan^{-1}\left(\frac{136 \text{ m}}{115 \text{ m}}\right)$$

$$\boxed{\theta = 49.8°}$$

TIP *Be sure your calculator is set to calculate angles measured in degrees. Some calculators have a button labeled "DRG" that, when pressed, toggles between degrees, radians, and grads.*

4. EVALUATE Because d is the hypotenuse, the archaeologist's displacement should be less than the sum of the height and half of the width. The angle is expected to be more than 45° because the height is greater than half of the width.

PRACTICE A

Finding Resultant Magnitude and Direction

1. A truck driver is attempting to deliver some furniture. First, he travels 8 km east, and then he turns around and travels 3 km west. Finally, he turns again and travels 12 km east to his destination.
   - **a.** What distance has the driver traveled?
   - **b.** What is the driver's total displacement?
2. While following the directions on a treasure map, a pirate walks 45.0 m north and then turns and walks 7.5 m east. What single straight-line displacement could the pirate have taken to reach the treasure?
3. Emily passes a soccer ball 6.0 m directly across the field to Kara. Kara then kicks the ball 14.5 m directly down the field to Luisa. What is the ball's total displacement as it travels between Emily and Luisa?
4. A hummingbird, 3.4 m above the ground, flies 1.2 m along a straight path. Upon spotting a flower below, the hummingbird drops directly downward 1.4 m to hover in front of the flower. What is the hummingbird's total displacement?

Alternative Problem-Solving Approach

The angle may be calculated using any trigonometric function, such as the following.

$$\theta = \sin^{-1}\left(\frac{\Delta y}{d}\right) = 49.8°$$

PROBLEM GUIDE A

Use this guide to assign problems.
SE = Student Edition Textbook
PW = Problem Workbook
PB = Problem Bank on the One-Stop Planner (OSP)

Solving for:

resultant	**SE**	Sample, 1–4; Ch. Rvw. 21–22, 23*
	PW	2, 4–5, 7*
	PB	Sample, 1–5
component	**PW**	Sample, 1, 3*, 6*
	PB	6–10

***Challenging Problem**
Consult the printed Solutions Manual or the OSP for detailed solutions.

ANSWERS

Practice A

1. a. 23 km
   b. 17 km to the east
2. 45.6 m at 9.5° east of north
3. 15.7 m at 22° to the side of downfield
4. 1.8 m at 49° below the horizontal

Teaching Tip

In mathematics, the components of a vector are called projections. The *x* component is the projection of the vector along the *x*-axis, and the *y* component is the projection of the vector along the *y*-axis.

Misconception Alert — GENERAL

Because of the prominence of angles measured from the *x*-axis, students may develop the misconception that the *x* component of a vector is always calculated using the cosine function. This misconception may be corrected by using examples on the board in which the angles are measured from the *y*-axis.

Teaching Tip — BASIC

Students who need a refresher on trigonometry should be directed to Appendix A, which includes a more detailed discussion of the sine, cosine, and tangent functions, as well as the Pythagorean theorem.

components of a vector

the projections of a vector along the axes of a coordinate system

RESOLVING VECTORS INTO COMPONENTS

In the pyramid example, the horizontal and vertical parts that add up to give the tourist's actual displacement are called **components.** The *x* component is parallel to the *x*-axis. The *y* component is parallel to the *y*-axis. Any vector can be completely described by a set of perpendicular components.

In this textbook, components of vectors are shown as outlined, open arrows. Components have arrowheads to indicate their direction. Components are scalars (numbers), but they are signed numbers, and the direction is important to determine their sign in a particular coordinate system.

You can often describe an object's motion more conveniently by breaking a single vector into two components, or *resolving* the vector. Resolving a vector allows you to analyze the motion in each direction.

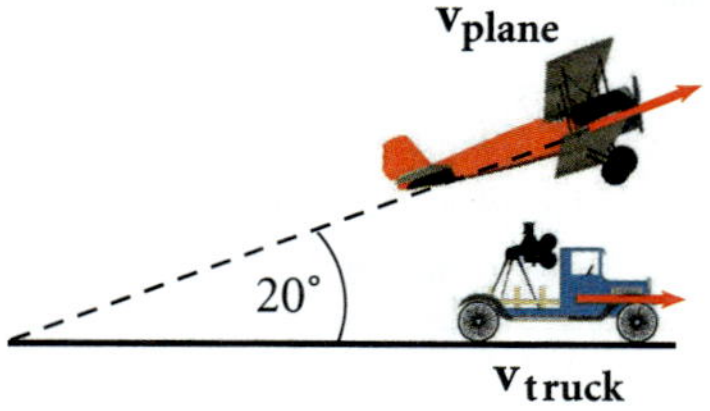

Figure 10
A truck carrying a film crew must be driven at the correct velocity to enable the crew to film the underside of a biplane. The plane flies at 95 km/h at an angle of 20° relative to the ground.

This point may be illustrated by examining a scene on the set of a new action movie. For this scene, a biplane travels at 95 km/h at an angle of 20° relative to the ground. Attempting to film the plane from below, a camera team travels in a truck that is directly beneath the plane at all times, as shown in **Figure 10.**

To find the velocity that the truck must maintain to stay beneath the plane, we must know the horizontal component of the plane's velocity. Once more, the key to solving the problem is to recognize that a right triangle can be drawn using the plane's velocity and its *x* and *y* components. The situation can then be analyzed using trigonometry.

The sine and cosine functions are defined in terms of the lengths of the sides of such right triangles. The sine of an angle is the ratio of the leg opposite that angle to the hypotenuse.

DEFINITION OF THE SINE FUNCTION FOR RIGHT TRIANGLES

$$\sin\theta = \frac{opp}{hyp} \qquad \text{sine of an angle} = \frac{\text{opposite leg}}{\text{hypotenuse}}$$

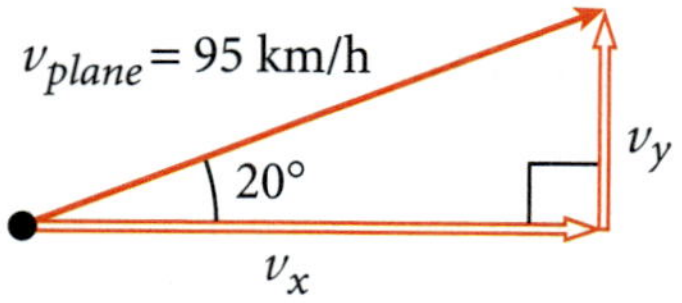

Figure 11
To stay beneath the biplane, the truck must be driven with a velocity equal to the *x* component (v_x) of the biplane's velocity.

In **Figure 11,** the leg opposite the 20° angle represents the *y* component, v_y, which describes the vertical speed of the airplane. The hypotenuse, $\mathbf{v}_{plane}$, is the resultant vector that describes the airplane's total velocity.

The cosine of an angle is the ratio between the leg adjacent to that angle and the hypotenuse.

DEFINITION OF THE COSINE FUNCTION FOR RIGHT TRIANGLES

$$\cos\theta = \frac{adj}{hyp} \qquad \text{cosine of an angle} = \frac{\text{adjacent leg}}{\text{hypotenuse}}$$

In **Figure 11,** the adjacent leg represents the *x* component, v_x, which describes the airplane's horizontal speed. This *x* component equals the speed that the truck must maintain to stay beneath the plane. Thus, the truck must maintain a speed of $v_x = (\cos 20°)(95 \text{ km/h}) = 90$ km/h.

SAMPLE PROBLEM B

Resolving Vectors

PROBLEM

Find the components of the velocity of a helicopter traveling 95 km/h at an angle of 35° to the ground.

SOLUTION

1. DEFINE

Given: $v = 95$ km/h $\quad \theta = 35°$

Unknown: $v_x = ?\quad v_y = ?$

Diagram: The most convenient coordinate system is one with the x-axis directed along the ground and the y-axis directed vertically.

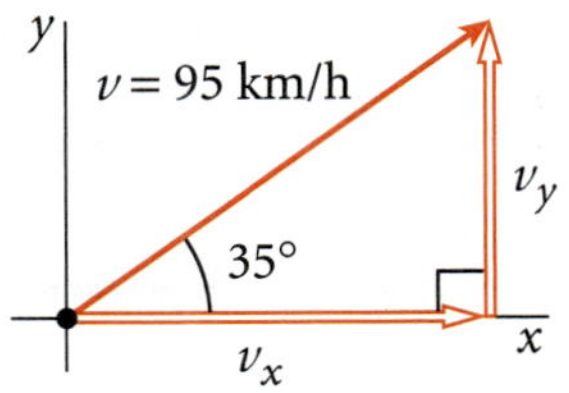

2. PLAN

Choose an equation or situation:
Because the axes are perpendicular, the sine and cosine functions can be used to find the components.

$$\sin\theta = \frac{v_y}{v}$$

$$\cos\theta = \frac{v_x}{v}$$

Rearrange the equations to isolate the unknowns:

$$v_y = v\sin\theta$$
$$v_x = v\cos\theta$$

TIP: *Don't assume that the cosine function can always be used for the x-component and the sine function can always be used for the y-component. The correct choice of function depends on where the given angle is located. Instead, always check to see which component is adjacent and which component is opposite to the given angle.*

3. CALCULATE

Substitute the values into the equations and solve:

$$v_y = (95 \text{ km/h})(\sin 35°)$$

$$\boxed{v_y = 54 \text{ km/h}}$$

$$v_x = (95 \text{ km/h})(\cos 35°)$$

$$\boxed{v_x = 78 \text{ km/h}}$$

4. EVALUATE

Because the components of the velocity form a right triangle with the helicopter's actual velocity, the components must satisfy the Pythagorean theorem.

$$v^2 = v_x^2 + v_y^2$$
$$(95)^2 = (78)^2 + (54)^2$$
$$9025 \approx 9000$$

The slight difference is due to rounding.

Classroom Practice

Resolving Vectors
An arrow is shot from a bow at an angle of 25° above the horizontal with an initial speed of 45 m/s. Find the horizontal and vertical components of the arrow's initial velocity.

Answer
41 m/s, 19 m/s

The arrow strikes the target with a speed of 45 m/s at an angle of −25° with respect to the horizontal. Calculate the horizontal and vertical components of the arrow's final velocity.

Answer
41 m/s, −19 m/s

PROBLEM GUIDE B

Use this guide to assign problems.
SE = Student Edition Textbook
PW = Problem Workbook
PB = Problem Bank on the One-Stop Planner (OSP)

Solving for:

one component	**SE** 1–2 **PW** Sample, 1, 3 **PB** 1–4
both components	**SE** Sample, 3–4; Ch. Rvw. 24–25 **PW** 2–8 **PB** Sample, 5–10

***Challenging Problem**
Consult the printed Solutions Manual or the OSP for detailed solutions.

ANSWERS

Practice B

1. 95 km/h
2. 44 km/h
3. 21 m/s, 5.7 m/s
4. 0 m, 5 m

Teaching Tip — GENERAL

Problems involving vectors that are not perpendicular use both vector addition and vector resolution. Because they act as a nice summary to the concepts of this section, you may want to do several examples involving this type of problem. These problems require a methodical approach to problem solving, which should prove helpful to students while studying more-difficult subjects, such as inclined-plane problems and equilibrium problems.

Teaching Tip — BASIC

Be sure students distinguish between the components of $\mathbf{d_1}$ and $\mathbf{d_2}$ in **Figure 12.** Ask students to draw each vector and its components separately. Then, have them sum the components in each direction and add the summed components together to find **d.** Repeat with additional examples to help prepare students for Sample Problem C on the next page.

PRACTICE B

Resolving Vectors

1. How fast must a truck travel to stay beneath an airplane that is moving 105 km/h at an angle of 25° to the ground?
2. What is the magnitude of the vertical component of the velocity of the plane in item 1?
3. A truck drives up a hill with a 15° incline. If the truck has a constant speed of 22 m/s, what are the horizontal and vertical components of the truck's velocity?
4. What are the horizontal and vertical components of a cat's displacement when the cat has climbed 5 m directly up a tree?

ADDING VECTORS THAT ARE NOT PERPENDICULAR

Until this point, the vector-addition problems concerned vectors that are perpendicular to one another. However, many objects move in one direction and then turn at an angle before continuing their motion.

Suppose that a plane initially travels 5 km at an angle of 35° to the ground, then climbs at only 10° relative to the ground for 22 km. How can you determine the magnitude and direction for the vector denoting the total displacement of the plane?

Because the original displacement vectors do not form a right triangle, you can not apply the tangent function or the Pythagorean theorem when adding the original two vectors.

Determining the magnitude and the direction of the resultant can be achieved by resolving each of the plane's displacement vectors into its x and y components. Then the components along each axis can be added together. As shown in **Figure 12,** these sums will be the two perpendicular components of the resultant, **d.** The resultant's magnitude can then be found by using the Pythagorean theorem, and its direction can be found by using the inverse tangent function.

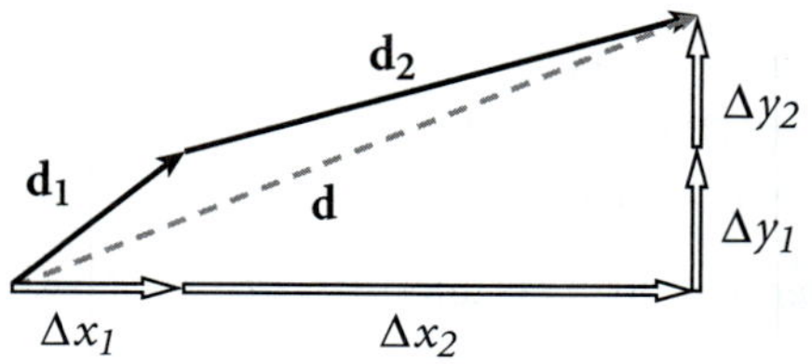

Figure 12
Add the components of the original displacement vectors to find two components that form a right triangle with the resultant vector.

92

SAMPLE PROBLEM C

STRATEGY Adding Vectors Algebraically

PROBLEM

A hiker walks 27.0 km from her base camp at 35° south of east. The next day, she walks 41.0 km in a direction 65° north of east and discovers a forest ranger's tower. Find the magnitude and direction of her resultant displacement between the base camp and the tower.

SOLUTION

1. Select a coordinate system. Then sketch and label each vector.

Given: $d_1 = 27.0$ km $\quad \theta_1 = -35°$

$d_2 = 41.0$ km $\quad \theta_2 = 65°$

Unknown: $d = ?\quad \theta = ?$

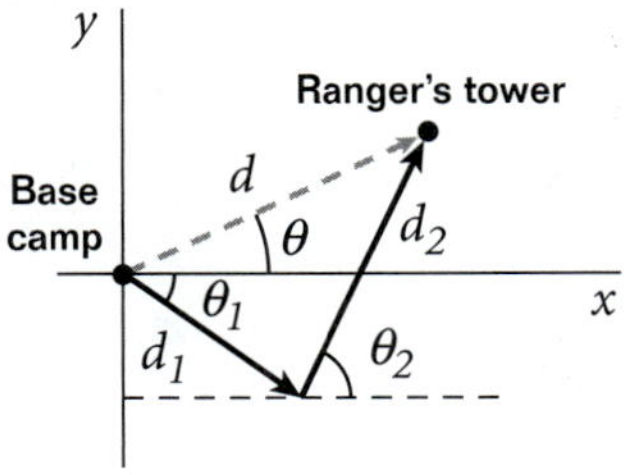

TIP: *θ_1 is negative, because clockwise angles from the positive x-axis are conventionally considered to be negative.*

2. Find the x and y components of all vectors.

Make a separate sketch of the displacements for each day. Use the cosine and sine functions to find the displacement components.

$$\cos\theta = \frac{\Delta x}{d} \qquad \sin\theta = \frac{\Delta y}{d}$$

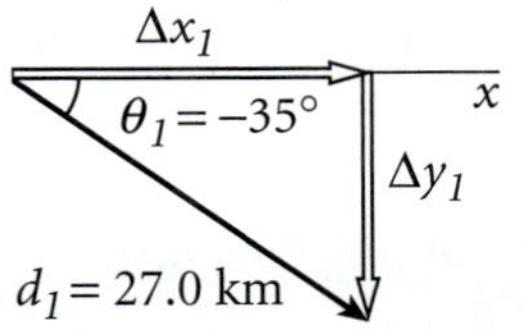

(a)

(a) For day 1: $\Delta x_1 = d_1 \cos\theta_1 = (27.0\text{ km})\,[\cos(-35°)] = 22\text{ km}$

$\Delta y_1 = d_1 \sin\theta_1 = (27.0\text{ km})\,[\sin(-35°)] = -15\text{ km}$

(b) For day 2: $\Delta x_2 = d_2 \cos\theta_2 = (41.0\text{ km})\,(\cos 65°) = 17\text{ km}$

$\Delta y_2 = d_2 \sin\theta_2 = (41.0\text{ km})\,(\sin 65°) = 37\text{ km}$

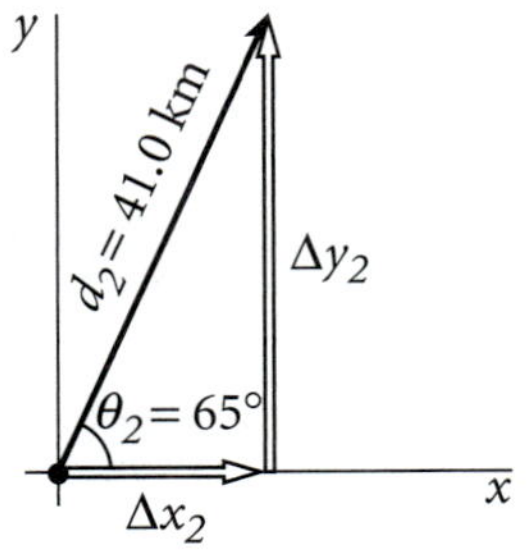

(b)

3. Find the x and y components of the total displacement.

$$\Delta x_{tot} = \Delta x_1 + \Delta x_2 = 22\text{ km} + 17\text{ km} = 39\text{ km}$$

$$\Delta y_{tot} = \Delta y_1 + \Delta y_2 = -15\text{ km} + 37\text{ km} = 22\text{ km}$$

4. Use the Pythagorean theorem to find the magnitude of the resultant vector.

$$d^2 = (\Delta x_{tot})^2 + (\Delta y_{tot})^2$$

$$d = \sqrt{(\Delta x_{tot})^2 + (\Delta y_{tot})^2} = \sqrt{(39\text{ km})^2 + (22\text{ km})^2} = \boxed{45\text{ km}}$$

5. Use a suitable trigonometric function to find the angle.

$$\theta = \tan^{-1}\left(\frac{\Delta y_{tot}}{\Delta x_{tot}}\right) = \tan^{-1}\left(\frac{22\text{ km}}{39\text{ km}}\right) = \boxed{29°\text{ north of east}}$$

Classroom Practice

Adding Vectors Algebraically

A camper walks 4.5 km at 45° north of east then 4.5 km due south. Find the camper's total displacement.

Answer

3.4 km at 22° south of east

A plane flies 118 km at 15.0° south of east and then flies 118 km at 35.0° west of north. Find the magnitude and direction of the total displacement of the plane.

Answer

81 km at 55° north of east

PROBLEM GUIDE C

Use this guide to assign problems.
SE = Student Edition Textbook
PW = Problem Workbook
PB = Problem Bank on the One-Stop Planner (OSP)

Solving for:

vector sum	**SE** Sample, 1–3, 4*; Ch. Rvw. 26, 53* **PW** Sample, 1–5 **PB** Sample, 1–10

***Challenging Problem**
Consult the printed Solutions Manual or the OSP for detailed solutions.

SECTION 2

ANSWERS

Practice C

1. 49 m at 7.3° to the right of downfield
2. 7.5 km at 26° above the horizontal
3. 13.0 m at 57° north of east
4. 171 km at 34° east of north

SECTION REVIEW ANSWERS

1. a. *x*-axis: forward and backward on sidewalk
   y-axis: left and right on sidewalk
   b. *x*-axis: forward and backward on rope
   y-axis: up and down
   c. *x*-axis: horizontal at water level
   y-axis: up and down
2. a. 5.8 m/s at 59° downriver from its intended path
   b. 6.1 m/s at 9.5° from the direction the wave is traveling
3. a. 7.07 km north, 7.07 km east
   b. 1.6 m/s^2 horizontal, 1.1 m/s^2 vertical
4. because the Pythagorean theorem and the tangent function can be applied only to right triangles

PRACTICE C

Adding Vectors Algebraically

1. A football player runs directly down the field for 35 m before turning to the right at an angle of 25° from his original direction and running an additional 15 m before getting tackled. What is the magnitude and direction of the runner's total displacement?
2. A plane travels 2.5 km at an angle of 35° to the ground and then changes direction and travels 5.2 km at an angle of 22° to the ground. What is the magnitude and direction of the plane's total displacement?
3. During a rodeo, a clown runs 8.0 m north, turns 55° north of east, and runs 3.5 m. Then, after waiting for the bull to come near, the clown turns due east and runs 5.0 m to exit the arena. What is the clown's total displacement?
4. An airplane flying parallel to the ground undergoes two consecutive displacements. The first is 75 km 30.0° west of north, and the second is 155 km 60.0° east of north. What is the total displacement of the airplane?

SECTION REVIEW

1. Identify a convenient coordinate system for analyzing each of the following situations:
   a. a dog walking along a sidewalk
   b. an acrobat walking along a high wire
   c. a submarine submerging at an angle of 30° to the horizontal
2. Find the magnitude and direction of the resultant velocity vector for the following perpendicular velocities:
   a. a fish swimming at 3.0 m/s relative to the water across a river that moves at 5.0 m/s
   b. a surfer traveling at 1.0 m/s relative to the water across a wave that is traveling at 6.0 m/s
3. Find the vector components along the directions noted in parentheses.
   a. a car displaced 45° north of east by 10.0 km (north and east)
   b. a duck accelerating away from a hunter at 2.0 m/s^2 at an angle of 35° to the ground (horizontal and vertical)
4. **Critical Thinking** Why do nonperpendicular vectors need to be resolved into components before you can add the vectors together?

Projectile Motion

SECTION OBJECTIVES

- Recognize examples of projectile motion.
- Describe the path of a projectile as a parabola.
- Resolve vectors into their components and apply the kinematic equations to solve problems involving projectile motion.

TWO-DIMENSIONAL MOTION

In the last section, quantities such as displacement and velocity were shown to be vectors that can be resolved into components. In this section, these components will be used to understand and predict the motion of objects thrown into the air.

Use of components avoids vector multiplication

How can you know the displacement, velocity, and acceleration of a ball at any point in time during its flight? All of the kinematic equations could be rewritten in terms of vector quantities. However, when an object is propelled into the air in a direction other than straight up or down, the velocity, acceleration, and displacement of the object do not all point in the same direction. This makes the vector forms of the equations difficult to solve.

One way to deal with these situations is to avoid using the complicated vector forms of the equations altogether. Instead, apply the technique of resolving vectors into components. Then you can apply the simpler one-dimensional forms of the equations for each component. Finally, you can recombine the components to determine the resultant.

Components simplify projectile motion

When a long jumper approaches his jump, he runs along a straight line, which can be called the x-axis. When he jumps, as shown in **Figure 13,** his velocity has both horizontal and vertical components. Movement in this plane can be depicted by using both the x- and y-axes.

Note that in **Figure 14(b),** a jumper's velocity vector is resolved into its two vector components. This way, the jumper's motion can be analyzed using the kinematic equations applied to one direction at a time.

Figure 13
When the long jumper is in the air, his velocity has both a horizontal and a vertical component.

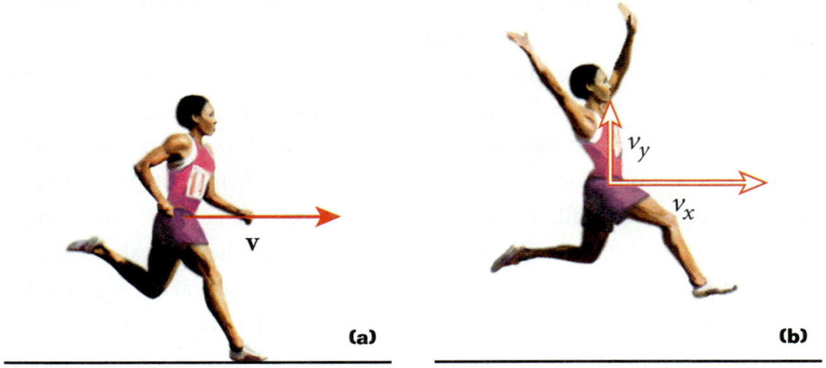

Figure 14
(a) A long jumper's velocity while sprinting along the runway can be represented by a horizontal vector. **(b)** Once the jumper is airborne, the jumper's velocity at any instant can be described by the components of the velocity.

Teaching Tip — GENERAL

On the chalkboard, show examples of vector components and the kinematic equations. Show the simplification of the x-direction equations when the x component of acceleration is zero.

Visual Strategy ADVANCED

Figure 13

Tell students that the long jumper builds up speed in the x direction and jumps, so there is also a component of speed in the y direction.

Q Does the angle of takeoff matter to the jumper? Consider the difference between a very small angle (near 0°) and a larger angle (near 45°).

A ***The angle matters because it affects how long the jumper stays off the ground and how far he goes horizontally while in the air.***

SECTION 3

Demonstration

Air Resistance — GENERAL

Purpose Show the effects of air resistance on the flight of a projectile.

Materials rubber stopper, table-tennis ball

Procedure Toss the stopper at an angle of 45°. Have students note the trajectory. Sketch the parabolic path on the chalkboard. Throw the table-tennis ball at an angle of 45°. Have students note the trajectory. Sketch its path on the chalkboard. Have students compare the two trajectories. Students should note that the path of the table-tennis ball is not symmetrical but has a steeper descent, demonstrating the effect of air resistance.

Teaching Tip

Review the general equation for a parabola ($y = ax^2 + bx + c$), and illustrate a parabola on the board for clarity. You may wish to point out that for low launches, such as the example shown in **Figure 15,** the path of a projectile is nearly the same as a circular arc. The parabolic shape becomes more apparent for higher launches. The path of the yellow ball shown in **Figure 16** is clearly a parabola rather than a circular arc.

projectile motion

the curved path that an object follows when thrown, launched, or otherwise projected near the surface of Earth

In this section, we will focus on the form of two-dimensional motion called **projectile motion.** Objects that are thrown or launched into the air and are subject to gravity are called *projectiles.* Some examples of projectiles are softballs, footballs, and arrows when they are projected through the air. Even a long jumper can be considered a projectile.

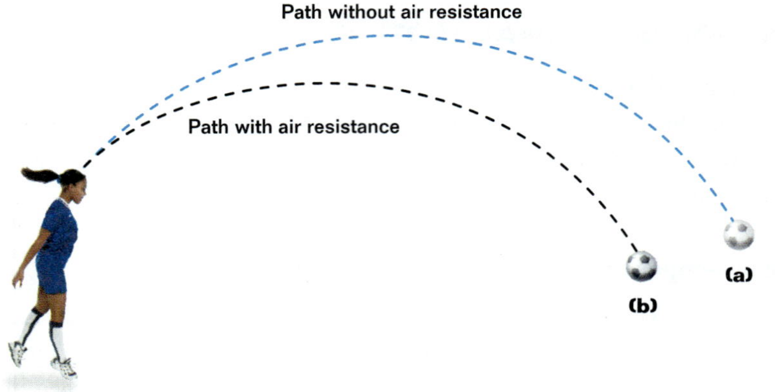

Figure 15
(a) Without air resistance, the soccer ball would travel along a parabola. **(b)** With air resistance, the soccer ball would travel along a shorter path.

Projectiles follow parabolic trajectories

The path of a projectile is a curve called a *parabola,* as shown in **Figure 15(a).** Many people mistakenly believe that projectiles eventually fall straight down in much the same way that a cartoon character does after running off a cliff. But if an object has an initial horizontal velocity in any given time interval, there will be horizontal motion throughout the flight of the projectile. *Note that for the purposes of samples and exercises in this book, the horizontal velocity of the projectile will be considered constant.* This velocity would not be constant if we accounted for air resistance. With air resistance, a projectile slows down as it collides with air particles, as shown in **Figure 15(b).**

Projectile motion is free fall with an initial horizontal velocity

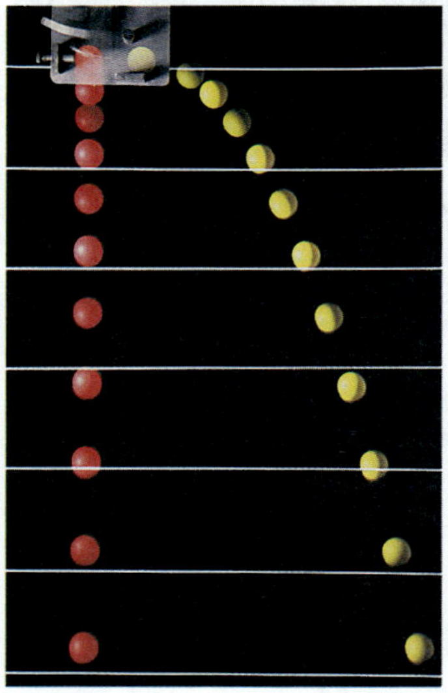

Figure 16
This is a strobe photograph of two table-tennis balls released at the same time. Even though the yellow ball is given an initial horizontal velocity and the red ball is simply dropped, both balls fall at the same rate.

To understand the motion a projectile undergoes, first examine **Figure 16.** The red ball was dropped at the same instant the yellow ball was launched horizontally. If air resistance is disregarded, both balls hit the ground at the same time.

By examining each ball's position in relation to the horizontal lines and to one another, we see that the two balls fall at the same rate. This may seem impossible because one is given an initial velocity and the other begins from rest. But if the motion is analyzed one component at a time, it makes sense.

First, consider the red ball that falls straight down. It has no motion in the horizontal direction. In the vertical direction, it starts from rest ($v_{y,i} = 0$ m/s) and proceeds in free fall. Thus, the kinematic equations from the chapter "Motion in One Dimension" can be applied to analyze the vertical motion of the falling ball, as shown on the next page. Note that on Earth's surface the acceleration (a_y) will equal $-g$ (-9.81 m/s^2) because the only vertical component of acceleration is free-fall acceleration. Note also that Δy is negative.

VERTICAL MOTION OF A PROJECTILE THAT FALLS FROM REST

$$v_{y,f} = a_y \Delta t$$

$$v_{y,f}^2 = 2a_y \Delta y$$

$$\Delta y = \frac{1}{2}a_y(\Delta t)^2$$

Now consider the components of motion of the yellow ball that is launched in **Figure 16.** This ball undergoes the same horizontal displacement during each time interval. This means that the ball's horizontal velocity remains constant (if air resistance is assumed to be negligible). Thus, when the kinematic equations are used to analyze the horizontal motion of a projectile, the initial horizontal velocity is equal to the horizontal velocity throughout the projectile's flight. A projectile's horizontal motion is described by the following equation.

HORIZONTAL MOTION OF A PROJECTILE

$$v_x = v_{x,i} = \text{constant}$$

$$\Delta x = v_x \Delta t$$

Next consider the initial motion of the launched yellow ball in **Figure 16.** Despite having an initial horizontal velocity, the launched ball has no initial velocity in the vertical direction. Just like the red ball that falls straight down, the launched yellow ball is in free fall. The vertical motion of the launched yellow ball is described by the same free-fall equations. In any time interval, the launched ball undergoes the same vertical displacement as the ball that falls straight down. For this reason, both balls reach the ground at the same time.

To find the velocity of a projectile at any point during its flight, find the vector that has the known components. Specifically, use the Pythagorean theorem to find the magnitude of the velocity, and use the tangent function to find the direction of the velocity.

Did you know?

The greatest distance a regulation-size baseball has ever been thrown is 135.9 m, by Glen Gorbous in 1957.

SCLINKS®
www.scilinks.org
Topic: Projectile Motion
Code: HF61223

Quick Lab

Projectile Motion

MATERIALS LIST

- 2 identical balls
- slope or ramp

SAFETY

Perform this experiment away from walls and furniture that can be damaged.

Roll a ball off a table. At the instant the rolling ball leaves the table, drop a second ball from the same height above the floor. Do the two balls hit the floor at the same time? Try varying the speed at which you roll the first ball off the table. Does varying the speed affect whether the two balls strike the ground at the same time? Next roll one of the balls down a slope. Drop the other ball from the base of the slope at the instant the first ball leaves the slope. Which of the balls hits the ground first in this situation?

Quick Lab

TEACHER'S NOTES

Dropping the second ball as the first leaves the table is tricky. Students should try this several times in order to get the timing right. Holding the second ball just past the edge of the table and near the path of the first ball works well. You should point out the limitations of this Quick Lab because of human reaction time.

Classroom Practice

Projectiles Launched Horizontally
People in movies often jump from buildings into pools. If a person jumps horizontally from the 10th floor (30.0 m) to a pool that is 5.0 m away from the building, with what initial speed must the person jump?

Answer
2.0 m/s

PROBLEM GUIDE D

Use this guide to assign problems.
SE = Student Edition Textbook
PW = Problem Workbook
PB = Problem Bank on the One-Stop Planner (OSP)

Solving for:

v_x	**SE** Sample, 1–3; Ch. Rvw. 31–32, 51a **PW** 6, 7*, 8* **PB** 9
Δx	**SE** 4; Ch. Rvw. 33 **PW** Sample, 1–2 **PB** 8, 10
Δy	**SE** Ch. Rvw. 51b **PW** 3–4, 5* **PB** Sample, 1–7

***Challenging Problem**
Consult the printed Solutions Manual or the OSP for detailed solutions.

SAMPLE PROBLEM D

Projectiles Launched Horizontally

PROBLEM

The Royal Gorge Bridge in Colorado rises 321 m above the Arkansas River. Suppose you kick a rock horizontally off the bridge. The magnitude of the rock's horizontal displacement is 45.0 m. Find the speed at which the rock was kicked.

SOLUTION

1. DEFINE

Given: $\Delta y = -321$ m $\Delta x = 45.0$ m $a_y = -g = -9.81$ m/s^2

Unknown: $v_i = v_x = ?$

Diagram: The initial velocity vector of the rock has only a horizontal component. Choose the coordinate system oriented so that the positive y direction points upward and the positive x direction points to the right.

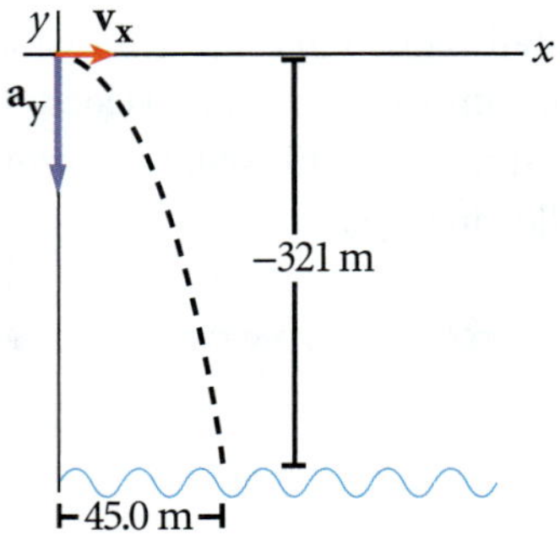

2. PLAN

Choose an equation or situation:
Because air resistance can be neglected, the rock's horizontal velocity remains constant.

$$\Delta x = v_x \Delta t$$

Because there is no initial vertical velocity, the following equation applies.

$$\Delta y = \tfrac{1}{2} a_y (\Delta t)^2$$

Rearrange the equations to isolate the unknowns:
Note that the time interval is the same for the vertical and horizontal displacements, so the second equation can be rearranged to solve for Δt.

$$\Delta t = \sqrt{\frac{2\Delta y}{a_y}}$$

Next rearrange the first equation for v_x, and substitute the above value of Δt into the new equation.

$$v_x = \frac{\Delta x}{\Delta t} = \left(\sqrt{\frac{a_y}{2\Delta y}}\right)\Delta x$$

TIP *The value for v_x can be either positive or negative because of the square root. Because the direction was not asked for, use the positive root.*

3. CALCULATE

Substitute the values into the equation and solve:

$$v_x = \sqrt{\frac{-9.81 \text{ m/s}^2}{(2)(-321 \text{ m})}}(45.0 \text{ m}) = \boxed{5.56 \text{ m/s}}$$

4. EVALUATE

To check your work, estimate the value of the time interval for Δx and solve for Δy. If v_x is about 5.5 m/s and $\Delta x = 45$ m, $\Delta t \approx 8$ s. If you use an approximate value of 10 m/s^2 for g, $\Delta y \approx -320$ m, almost identical to the given value.

PRACTICE D

Projectiles Launched Horizontally

1. A baseball rolls off a 0.70 m high desk and strikes the floor 0.25 m away from the base of the desk. How fast was the ball rolling?
2. A cat chases a mouse across a 1.0 m high table. The mouse steps out of the way, and the cat slides off the table and strikes the floor 2.2 m from the edge of the table. When the cat slid off the table, what was its speed?
3. A pelican flying along a horizontal path drops a fish from a height of 5.4 m. The fish travels 8.0 m horizontally before it hits the water below. What is the pelican's speed?
4. If the pelican in item 3 was traveling at the same speed but was only 2.7 m above the water, how far would the fish travel horizontally before hitting the water below?

Use components to analyze objects launched at an angle

Let us examine a case in which a projectile is launched at an angle to the horizontal, as shown in **Figure 17.** The projectile has an initial vertical component of velocity as well as a horizontal component of velocity.

Suppose the initial velocity vector makes an angle θ with the horizontal. Again, to analyze the motion of such a projectile, you must resolve the initial velocity vector into its components. The sine and cosine functions can be used to find the horizontal and vertical components of the initial velocity.

$$v_{x,i} = v_i \cos\theta \quad \text{and} \quad v_{y,i} = v_i \sin\theta$$

We can substitute these values for $v_{x,i}$ and $v_{y,i}$ into the kinematic equations to obtain a set of equations that can be used to analyze the motion of a projectile launched at an angle.

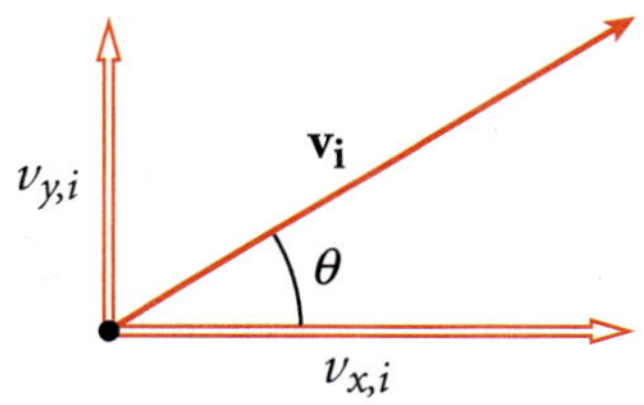

Figure 17
An object is projected with an initial velocity, $\mathbf{v_i}$, at an angle of θ. Resolve the initial velocity into its x and y components. Then, the kinematic equations can be applied to describe the motion of the projectile throughout its flight.

PROJECTILES LAUNCHED AT AN ANGLE

$$v_x = v_{x,i} = v_i \cos\theta = \text{constant}$$
$$\Delta x = (v_i \cos\theta)\Delta t$$
$$v_{y,f} = v_i \sin\theta + a_y \Delta t$$
$$v_{y,f}^2 = v_i^2 (\sin\theta)^2 + 2a_y \Delta y$$
$$\Delta y = (v_i \sin\theta)\Delta t + \frac{1}{2}a_y(\Delta t)^2$$

As we have seen, the velocity of a projectile launched at an angle to the ground has both horizontal and vertical components. The vertical motion is similar to that of an object that is thrown straight up with an initial velocity.

SECTION 3

ANSWERS

Practice D

1. 0.66 m/s
2. 4.9 m/s
3. 7.6 m/s
4. 5.6 m

Visual Strategy GENERAL

Figure 17

Point out to students that from the time immediately after firing until it hits the ground, the projectile follows a parabolic path. Emphasize that the vector $\mathbf{v_i}$ does not represent any part of the path of the projectile but only the direction and magnitude of its initial velocity.

Q What is the acceleration of a projectile just before it hits the ground?

A *-9.81 m/s^2, the same as at any other time during the flight*

Teaching Tip

Students often memorize equations without taking the time to understand them. Discourage students from memorizing the equations shown on this page. Help them see how each of these equations is a form of one of the kinematic equations they learned in the chapter "Motion in One Dimension." Explain that solving projectile problems is done by applying the kinematic equations separately in each direction.

Classroom Practice

Projectiles Launched at an Angle
A golfer practices driving balls off a cliff and into the water below. The edge of the cliff is 15 m above the water. If the golf ball is launched at 51 m/s at an angle of 15°, how far does the ball travel horizontally before hitting the water? (See Appendix A for hints on solving quadratic equations.)

Answer
1.7×10^2 m

PROBLEM GUIDE E

Use this guide to assign problems.
SE = Student Edition Textbook
PW = Problem Workbook
PB = Problem Bank on the One-Stop Planner (OSP)

Solving for:

$\Delta y/\Delta x$	**SE** Sample, 1–3; Ch. Rvw. 34a, 35–36, 55a, 56b, 59–60, 62a, 62c **PW** 4b, 5, 7–8 **PB** 6
v_i	**SE** 4; Ch. Rvw. 47a*, 48a, 49*, 56a, 61 **PW** Sample, 1–3, 4a **PB** 8, 10
Δt	**SE** Ch. Rvw. 34b, 47b*, 55b, 62b **PB** 7, 9
v_f	**SE** Ch. Rvw. 47c*, 48b
θ	**PW** 6 **PB** Sample, 1–5

***Challenging Problem**
Consult the printed Solutions Manual or the OSP for detailed solutions.

SAMPLE PROBLEM E

STRATEGY Projectiles Launched at an Angle

PROBLEM

A zookeeper finds an escaped monkey hanging from a light pole. Aiming her tranquilizer gun at the monkey, she kneels 10.0 m from the light pole, which is 5.00 m high. The tip of her gun is 1.00 m above the ground. At the same moment that the monkey drops a banana, the zookeeper shoots. If the dart travels at 50.0 m/s, will the dart hit the monkey, the banana, or neither one?

SOLUTION

1. Select a coordinate system.
The positive y-axis points up, and the positive x-axis points along the ground toward the pole. Because the dart leaves the gun at a height of 1.00 m, the vertical distance is 4.00 m.

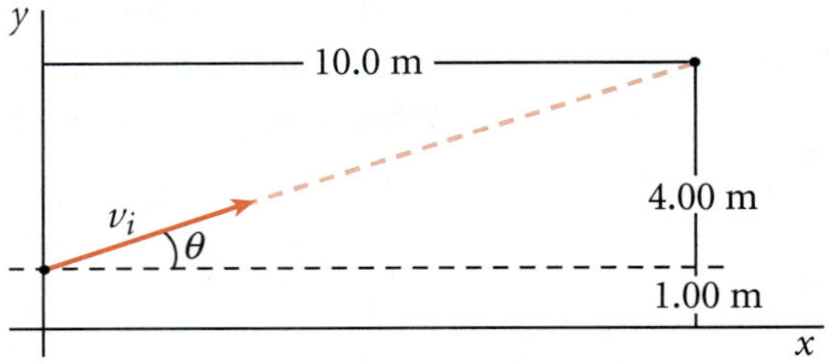

2. Use the inverse tangent function to find the angle that the initial velocity makes with the x-axis.

$$\theta = \tan^{-1}\left(\frac{\Delta y}{\Delta x}\right) = \tan^{-1}\left(\frac{4.00 \text{ m}}{10.0 \text{ m}}\right) = 21.8°$$

3. Choose a kinematic equation to solve for time.
Rearrange the equation for motion along the x-axis to isolate the unknown, Δt, which is the time the dart takes to travel the horizontal distance.

$$\Delta x = (v_i \cos \theta)\Delta t$$

$$\Delta t = \frac{\Delta x}{v_i \cos \theta} = \frac{10.0 \text{ m}}{(50.0 \text{ m/s})(\cos 21.8°)} = 0.215 \text{ s}$$

4. Find out how far each object will fall during this time.
Use the free-fall kinematic equation in both cases. For the banana, $v_i = 0$. Thus:

$$\Delta y_b = \tfrac{1}{2}a_y(\Delta t)^2 = \tfrac{1}{2}(-9.81 \text{ m/s}^2)(0.215 \text{ s})^2 = -0.227 \text{ m}$$

The dart has an initial vertical component of velocity equal to $v_i \sin\theta$, so:

$$\Delta y_d = (v_i \sin \theta)\Delta t + \tfrac{1}{2}a_y(\Delta t)^2$$
$$\Delta y_d = (50.0 \text{ m/s})(\sin 21.8°)(0.215 \text{ s}) + \tfrac{1}{2}(-9.81 \text{ m/s}^2)(0.215 \text{ s})^2$$
$$\Delta y_d = 3.99 \text{ m} - 0.227 \text{ m} = 3.76 \text{ m}$$

5. Analyze the results.
Find the final height of both the banana and the dart.

$$y_{banana,f} = y_{b,i} + \Delta y_b = 5.00 \text{ m} + (-0.227 \text{ m}) = \boxed{4.77 \text{ m above the ground}}$$

$$y_{dart,f} = y_{d,i} + \Delta y_d = 1.00 \text{ m} + 3.76 \text{ m} = \boxed{4.76 \text{ m above the ground}}$$

The dart hits the banana. The slight difference is due to rounding.

PRACTICE E

Projectiles Launched at an Angle

1. In a scene in an action movie, a stuntman jumps from the top of one building to the top of another building 4.0 m away. After a running start, he leaps at a velocity of 5.0 m/s at an angle of 15° with respect to the flat roof. Will he make it to the other roof, which is 2.5 m lower than the building he jumps from?

2. A golfer hits a golf ball at an angle of 25.0° to the ground. If the golf ball covers a horizontal distance of 301.5 m, what is the ball's maximum height? (Hint: At the top of its flight, the ball's vertical velocity component will be zero.)

3. A baseball is thrown at an angle of 25° relative to the ground at a speed of 23.0 m/s. If the ball was caught 42.0 m from the thrower, how long was it in the air? How high above the thrower did the ball travel?

4. Salmon often jump waterfalls to reach their breeding grounds. One salmon starts 2.00 m from a waterfall that is 0.55 m tall and jumps at an angle of 32.0°. What must be the salmon's minimum speed to reach the waterfall?

SECTION REVIEW

1. Which of the following exhibit parabolic motion?

a. a flat rock skipping across the surface of a lake
b. a three-point shot in basketball
c. the space shuttle while orbiting Earth
d. a ball bouncing across a room
e. a life preserver dropped from a stationary helicopter

2. During a thunderstorm, a tornado lifts a car to a height of 125 m above the ground. Increasing in strength, the tornado flings the car horizontally with a speed of 90.0 m/s. How long does the car take to reach the ground? How far horizontally does the car travel before hitting the ground?

3. Interpreting Graphics An Alaskan rescue plane drops a package of emergency rations to a stranded party of explorers, as illustrated in **Figure 18.** The plane is traveling horizontally at 30.0 m/s at a height of 200.0 m above the ground.

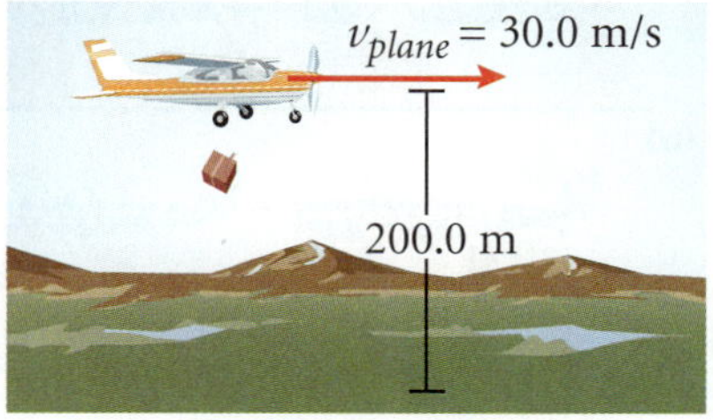

Figure 18

a. What horizontal distance does the package fall before landing?
b. Find the velocity of the package just before it hits the ground.

SECTION 3

ANSWERS

Practice E

1. yes, $\Delta y = -2.3$ m
2. 35.1 m
3. 2.0 s; 4.8 m
4. 6.2 m/s

Teaching Tip

Point out that there are different ways to solve comparison questions such as item 1 in Practice E. In this case, one option is to solve for Δy and then compare the calculated value of Δy with the vertical distance given (2.5 m). Another option is to solve for Δx and compare the calculated value of Δx with the horizontal distance given (4.0 m).

SECTION REVIEW ANSWERS

1. a, b, d
2. 5.05 s; 454 m
3. **a.** 192 m
   b. 69.4 m/s at 64.4° below the horizontal

Visual Strategy ADVANCED

Figure 19

Tell students that this diagram shows what would happen if there were no air resistance.

Q How would the two diagrams change if air resistance were included?

A *In the first diagram, the object would appear to fall down and backward as viewed by the pilot. In the second diagram, the object would follow a shortened path as viewed by the observer on the ground. (At any instant after release, both the x-component and the y-component would be smaller, but the suppression of horizontal motion has the dominant effect on the trajectory.)*

SECTION 4

Relative Motion

SECTION OBJECTIVES

- **Describe situations in terms of frame of reference.**
- **Solve problems involving relative velocity.**

FRAMES OF REFERENCE

If you are moving at 80 km/h north and a car passes you going 90 km/h, to you the faster car seems to be moving north at 10 km/h. Someone standing on the side of the road would measure the velocity of the faster car as 90 km/h toward the north. This simple example demonstrates that velocity measurements depend on the frame of reference of the observer.

Velocity measurements differ in different frames of reference

Observers using different frames of reference may measure different displacements or velocities for an object in motion. That is, two observers moving with respect to each other would generally not agree on some features of the motion.

Consider a stunt dummy that is dropped from an airplane flying horizontally over Earth with a constant velocity. As shown in **Figure 19(a),** a passenger on the airplane would describe the motion of the dummy as a straight line toward Earth. An observer on the ground would view the trajectory of the dummy as that of a projectile, as shown in **Figure 19(b).** Relative to the ground, the dummy would have a vertical component of velocity (resulting from free-fall acceleration and equal to the velocity measured by the observer in the airplane) *and* a horizontal component of velocity given to it by the airplane's motion. If the airplane continued to move horizontally with the same velocity, the dummy would enter the swimming pool directly beneath the airplane (assuming negligible air resistance).

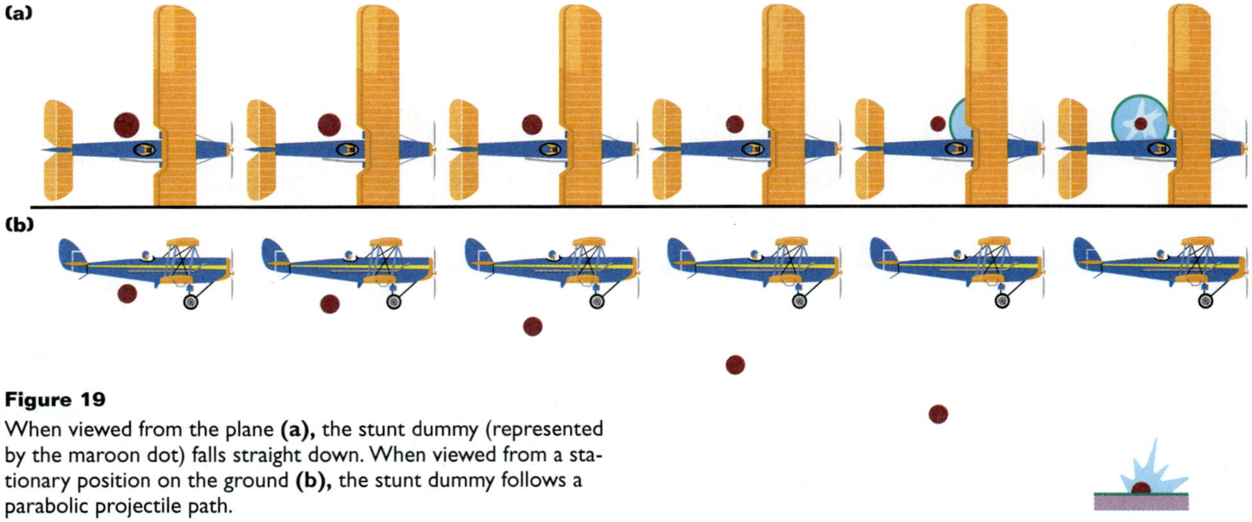

Figure 19
When viewed from the plane **(a),** the stunt dummy (represented by the maroon dot) falls straight down. When viewed from a stationary position on the ground **(b),** the stunt dummy follows a parabolic projectile path.

RELATIVE VELOCITY

The case of the faster car overtaking your car was easy to solve with a minimum of thought and effort, but you will encounter many situations in which a more systematic method of solving such problems is beneficial. To develop this method, write down all the information that is given and that you want to know in the form of velocities with subscripts appended.

$\mathbf{v_{se}} = +80$ km/h north (Here the subscript *se* means the velocity of the *slower* car with respect to *Earth.*)

$\mathbf{v_{fe}} = +90$ km/h north (The subscript *fe* means the velocity of the *fast* car with respect to *Earth.*)

We want to know $\mathbf{v_{fs}}$, which is the velocity of the fast car with respect to the slower car. To find this, we write an equation for $\mathbf{v_{fs}}$ in terms of the other velocities, so on the right side of the equation the subscripts start with *f* and eventually end with *s*. Also, each velocity subscript starts with the letter that ended the preceding velocity subscript.

$$\mathbf{v_{fs}} = \mathbf{v_{fe}} + \mathbf{v_{es}}$$

The boldface notation indicates that velocity is a vector quantity. This approach to adding and monitoring subscripts is similar to vector addition, in which vector arrows are placed head to tail to find a resultant.

We know that $\mathbf{v_{es}} = -\mathbf{v_{se}}$ because an observer in the slow car perceives Earth as moving south at a velocity of 80 km/h while a stationary observer on the ground (Earth) views the car as moving north at a velocity of 80 km/h. Thus, this problem can be solved as follows:

$$\mathbf{v_{fs}} = \mathbf{v_{fe}} + \mathbf{v_{es}} = \mathbf{v_{fe}} - \mathbf{v_{se}}$$

$$\mathbf{v_{fs}} = (+90 \text{ km/h north}) - (+80 \text{ km/h north}) = +10 \text{ km/h north}$$

When solving relative velocity problems, follow the above technique for writing subscripts. The particular subscripts will vary depending on the problem, but the method for ordering the subscripts does not change. A general form of the relative velocity equation is $\mathbf{v_{ac}} = \mathbf{v_{ab}} + \mathbf{v_{bc}}$. This general form may help you remember the technique for writing subscripts.

Did you know?

Like velocity, displacement and acceleration depend on the frame in which they are measured. In some cases, it is instructive to visualize gravity as the ground accelerating toward a projectile rather than the projectile accelerating toward the ground.

ADVANCED TOPICS

See "Special Relativity and Velocities" in **Appendix J: Advanced Topics** to learn about how velocities are added in Einstein's special theory of relativity.

Why it Matters

Conceptual Challenge

1. Elevator Acceleration A boy bounces a rubber ball in an elevator that is going down. If the boy drops the ball as the elevator is slowing down, is the magnitude of the ball's acceleration relative to the elevator less than or greater than the magnitude of its acceleration relative to the ground?

2. Aircraft Carrier Is the velocity of a plane relative to an aircraft carrier slower when it approaches from the stern (rear) or from the bow (front)?

SECTION 4

Teaching Tip

Relative velocity can also be shown as the difference of two vectors.

$$\mathbf{v_{fs}} = \mathbf{v_{fe}} - \mathbf{v_{se}}$$

Another way of stating this equation is that the relative velocity of one moving object to another is the difference between their velocities relative to some common reference point.

You may want to demonstrate on the board that this equation works for noncollinear velocities, as in Sample Problem F on the next page.

ANSWERS

Conceptual Challenge

1. greater than, because the elevator is accelerating upward toward the ball as it falls
2. The plane's velocity is slower relative to the moving carrier when it approaches from the stern.

Classroom Practice

Relative Velocity

A plane flies northeast at an airspeed of 563.0 km/h. (*Airspeed* is the speed of an aircraft relative to the air.) A 48.0 km/h wind is blowing to the southeast. What is the plane's velocity relative to the ground?

Answer

565.0 km/h at 40.1° north of east

The wind shifts to blow 63.0 km/h toward the southwest. What is the plane's velocity relative to the ground if the plane's airspeed remains constant?

Answer

500.0 km/h northeast

SAMPLE PROBLEM F

Relative Velocity

PROBLEM

A boat heading north crosses a wide river with a velocity of 10.00 km/h relative to the water. The river has a uniform velocity of 5.00 km/h due east. Determine the boat's velocity with respect to an observer on shore.

SOLUTION

1. DEFINE

Given: $\mathbf{v_{bw}}$ = 10.00 km/h due north (velocity of the *boat, b,* with respect to the *water, w*)

$\mathbf{v_{we}}$ = 5.00 km/h due east (velocity of the *water, w,* with respect to *Earth, e*)

Unknown: $\mathbf{v_{be}} = ?$

Diagram: See the diagram on the right.

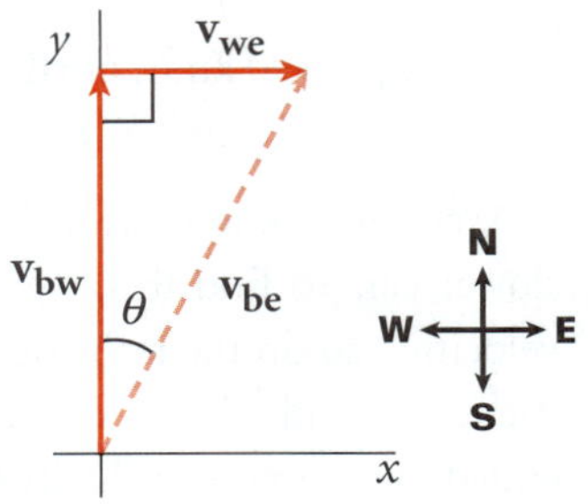

2. PLAN

Choose an equation or situation:

To find $\mathbf{v_{be}}$, write the equation so that the subscripts on the right start with *b* and end with *e*.

$$\mathbf{v_{be}} = \mathbf{v_{bw}} + \mathbf{v_{we}}$$

As in Section 2, we use the Pythagorean theorem to calculate the magnitude of the resultant velocity and the tangent function to find the direction.

$$(v_{be})^2 = (v_{bw})^2 + (v_{we})^2$$

$$\tan \theta = \frac{v_{we}}{v_{bw}}$$

Rearrange the equations to isolate the unknowns:

$$v_{be} = \sqrt{(v_{bw})^2 + (v_{we})^2}$$

$$\theta = \tan^{-1}\left(\frac{v_{we}}{v_{bw}}\right)$$

3. CALCULATE

Substitute the known values into the equations and solve:

$$v_{be} = \sqrt{(10.00 \text{ km/h})^2 + (5.00 \text{ km/h})^2}$$

$$\boxed{v_{be} = 11.18 \text{ km/h}}$$

$$\theta = \tan^{-1}\left(\frac{5.00 \text{ km/h}}{10.0 \text{ km/h}}\right)$$

$$\boxed{\theta = 26.6°}$$

4. EVALUATE

The boat travels at a speed of 11.18 km/h in the direction 26.6° east of north with respect to Earth.

PRACTICE F

Relative Velocity

1. A passenger at the rear of a train traveling at 15 m/s relative to Earth throws a baseball with a speed of 15 m/s in the direction opposite the motion of the train. What is the velocity of the baseball relative to Earth as it leaves the thrower's hand?
2. A spy runs from the front to the back of an aircraft carrier at a velocity of 3.5 m/s. If the aircraft carrier is moving forward at 18.0 m/s, how fast does the spy appear to be running when viewed by an observer on a nearby stationary submarine?
3. A ferry is crossing a river. If the ferry is headed due north with a speed of 2.5 m/s relative to the water and the river's velocity is 3.0 m/s to the east, what will the boat's velocity relative to Earth be? (Hint: Remember to include the direction in describing the velocity.)
4. A pet-store supply truck moves at 25.0 m/s north along a highway. Inside, a dog moves at 1.75 m/s at an angle of 35.0° east of north. What is the velocity of the dog relative to the road?

SECTION REVIEW

1. A woman on a 10-speed bicycle travels at 9 m/s relative to the ground as she passes a little boy on a tricycle going in the opposite direction. If the boy is traveling at 1 m/s relative to the ground, how fast does the boy appear to be moving relative to the woman?
2. A girl at an airport rolls a ball north on a moving walkway that moves east. If the ball's speed with respect to the walkway is 0.15 m/s and the walkway moves at a speed of 1.50 m/s, what is the velocity of the ball relative to the ground?
3. **Critical Thinking** Describe the motion of the following objects if they are observed from the stated frames of reference:
   - **a.** a person standing on a platform viewed from a train traveling north
   - **b.** a train traveling north viewed by a person standing on a platform
   - **c.** a ball dropped by a boy walking at a speed of 1 m/s viewed by the boy
   - **d.** a ball dropped by a boy walking 1 m/s as seen by a nearby viewer who is stationary

SECTION 4

PROBLEM GUIDE F

Use this guide to assign problems.
SE = Student Edition Textbook
PW = Problem Workbook
PB = Problem Bank on the One-Stop Planner (OSP)

Solving for:

v, θ	**SE** Sample, 1–4; Ch. Rvw. 43a, 44a–b, 46a–b, 57a–b **PW** Sample, 1 **PB** 7, 10
Δt	**SE** Ch. Rvw. 45, 50a–b, 52, 54, 58 **PW** 2, 4, 6 **PB** Sample, 1–5
$\Delta y/\Delta x$	**SE** Ch. Rvw. 43b **PW** 5 **PB** 6, 8–9

***Challenging Problem**
Consult the printed Solutions Manual or the OSP for detailed solutions.

ANSWERS

Practice F

1. 0 m/s
2. 14.5 m/s (in the direction that the aircraft carrier is moving)
3. 3.90 m/s at (4.0×10^1)° north of east
4. 26.4 m/s at 2.17° east of north

SECTION REVIEW ANSWERS

1. 10 m/s (in the opposite direction)
2. 1.51 m/s at 5.7° north of east
3. **a.** south with a speed equal to the train's speed
   b. moves north
   c. appears to fall straight down
   d. moves in a parabola

Kinesiology

Lisa Griffin was born in Toronto, Canada, and grew up in a small town in Ontario. Griffin says that she has always been interested in "the nature of reality and the relationship between the mind and the body." As a high school student visiting her public library, she first learned that the link between the mind and the body was the central nervous system (CNS). As an undergraduate at the University of Guelph, she was inspired by Dr. John Brooke, who taught that the spinal cord is also part of the brain and has its own forms of "thinking" and memory.

PHYSICS CAREERS

Kinesiology

How does the body move? This question is just one of the many that kinesiology continually asks. To learn more about kinesiology as a career, read the interview with Lisa Griffin, who teaches in the Department of Kinesiology and Health Education at the University of Texas at Austin.

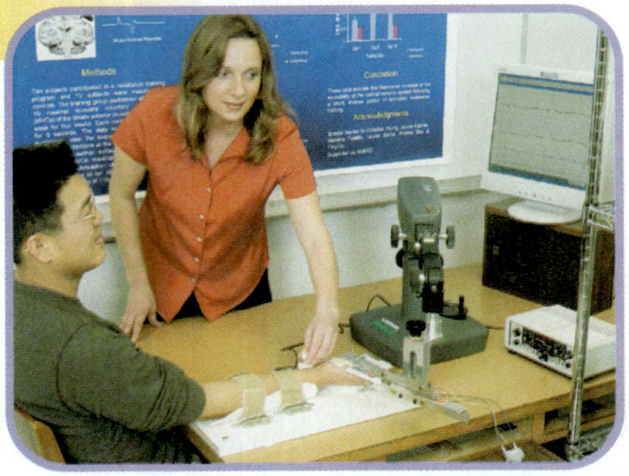

Lisa Griffin applies an electrical stimulus to a nerve in a patient's wrist. This experiment tested the best patterns of stimulation to recreate movement in paralyzed hands.

What training did you receive in order to become a kinesiologist?

I received a B.Sc. degree in human kinetics with a minor in biochemistry and M.Sc. and Ph.D. degrees in neuroscience. Kinesiology typically covers motor control, biomechanics, and exercise physiology. People who work in these branches are known as neuroscientists, biomechanists, and physiologists, respectively.

What makes kinesiology interesting to you?

The field of kinesiology allows me to explore how the central nervous system (CNS) controls human movement. Thus we work with people, and the findings of our work can be used to help others.

What is the nature of your research?

We record force output and single motor unit firing patterns from the muscles of human participants during fatigue and training. We then use these frequency patterns to stimulate their hands artificially with electrical stimulation. We are working toward developing an electrical stimulation system that people with paralysis could use to generate limb movement. This could help many who have spinal cord injuries from accidents or brain damage from stroke.

How does your work address two-dimensional motion and vectors?

I investigate motor unit firing frequencies required to generate force output from muscle over time. Thus we record muscle contraction with strain gauge force transducers, bridge amplifiers, an analog to digital converter, and a computer data acquisition and analysis program. For example, the muscles of the thumb produce force in both *x* and *y* directions. We record the *x* and *y* forces on two different channels, and then we calculate the resultant force online so that we can view the net output during contraction.

What are your most and least favorite things about your work?

My favorite thing is coming up with new ideas and working with students who are excited about their work. The thing I would most like to change is the amount of time it takes to get the results of the experiments after you think of the ideas.

What advice would you offer to students who are interested in this field?

Do not underestimate the depth of the questions that can be addressed with human participants.

CHAPTER 3

Highlights

KEY IDEAS

Section 1 Introduction to Vectors

- A scalar is a quantity completely specified by only a number with appropriate units, whereas a vector is a quantity that has magnitude and direction.
- Vectors can be added graphically using the triangle method of addition, in which the tail of one vector is placed at the head of the other. The resultant is the vector drawn from the tail of the first vector to the head of the last vector.

Section 2 Vector Operations

- The Pythagorean theorem and the inverse tangent function can be used to find the magnitude and direction of a resultant vector.
- Any vector can be resolved into its component vectors by using the sine and cosine functions.

Section 3 Projectile Motion

- Neglecting air resistance, a projectile has a constant horizontal velocity and a constant downward free-fall acceleration.
- In the absence of air resistance, projectiles follow a parabolic path.

Section 4 Relative Motion

- If the frame of reference is denoted with subscripts ($\mathbf{v}_{ab}$ is the velocity of object or frame *a* with respect to object or frame *b*), then the velocity of an object with respect to a different frame of reference can be found by adding the known velocities so that the subscript starts with the letter that ends the preceding velocity subscript: $\mathbf{v}_{ac} = \mathbf{v}_{ab} + \mathbf{v}_{bc}$.
- If the order of the subscripts is reversed, there is a change in sign; for example, $\mathbf{v}_{cd} = -\mathbf{v}_{dc}$.

KEY TERMS

scalar (p. 82)

vector (p. 82)

resultant (p. 83)

components of a vector (p. 90)

projectile motion (p. 96)

PROBLEM SOLVING

See **Appendix D: Equations** for a summary of the equations introduced in this chapter. If you need more problem-solving practice, see **Appendix I: Additional Problems.**

Variable Symbols

Quantities		Units	
d (vector)	displacement	m	meters
v (vector)	velocity	m/s	meters/second
a (vector)	acceleration	m/s^2	meters/second^2
Δx (scalar)	horizontal component	m	meters
Δy (scalar)	vertical component	m	meters

Diagram Symbols

Symbol	Meaning
black arrow	displacement vector
orange arrow	velocity vector
thick blue arrow	acceleration vector
dashed arrow	resultant vector
double-line arrow	component

CHAPTER 3

Highlights

Teaching Tip

Ask students to prepare a concept map for the chapter. The concept map should include most of the vocabulary terms, along with other integral terms or concepts.

ANSWERS

1. A scalar represents the magnitude of a physical quantity.
2. no, must be equal and opposite
3. Speed is the magnitude of velocity.
4. 30 m/s east
5. no, because the scalar has no direction
6. a. 5.00 units at 53.1° below the positive x-axis
   b. 5.00 units at 53.1° above the positive x-axis
   c. 8.54 units at 69.4° below the positive x-axis
   d. 5.00 units at 127° clockwise from the positive x-axis
7. a. 5.20 m at 60.0° above the positive x-axis
   b. 3.00 m at 30.0° below the positive x-axis
   c. 3.00 m at 150° counterclockwise from the positive x-axis
   d. 5.20 m at 60.0° below the positive x-axis
8. 7.9 m at 4.3° north of west
9. 15.3 m at 58.4° south of east
10. when the vectors point in the same direction
11. a car moving straight and speeding up, a car moving straight and slowing down
12. 55; the maximum value of the vector sum is 80 units, the minimum value is 30 units
13. Yes, the distance from the tail of the first vector to the head of the last vector is zero.

CHAPTER 3 *Review*

VECTORS AND THE GRAPHICAL METHOD

Review Questions

1. The magnitude of a vector is a scalar. Explain this statement.

2. If two vectors have unequal magnitudes, can their sum be zero? Explain.

3. What is the relationship between instantaneous speed and instantaneous velocity?

4. What is another way of saying –30 m/s west?

5. Is it possible to add a vector quantity to a scalar quantity? Explain.

6. Vector **A** is 3.00 units in length and points along the positive x-axis. Vector **B** is 4.00 units in length and points along the negative y-axis. Use graphical methods to find the magnitude and direction of the following vectors:
   a. **A** + **B**
   b. **A** – **B**
   c. **A** + 2**B**
   d. **B** – **A**

7. Each of the displacement vectors **A** and **B** shown in the figure below has a magnitude of 3.00 m. Graphically find the following:
   a. **A** + **B**
   b. **A** – **B**
   c. **B** – **A**
   d. **A** – 2**B**

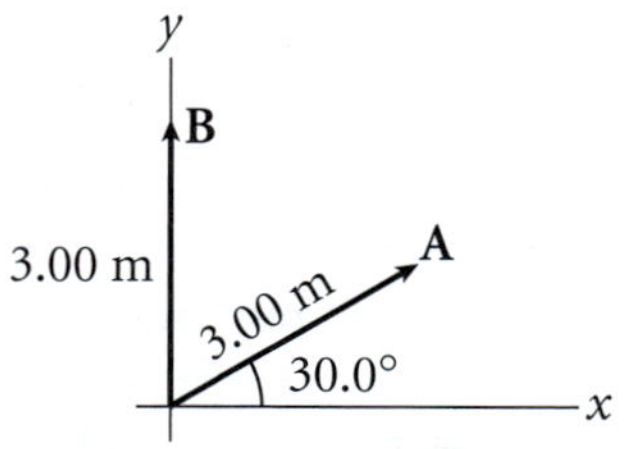

8. A dog searching for a bone walks 3.50 m south, then 8.20 m at an angle of 30.0° north of east, and finally 15.0 m west. Use graphical techniques to find the dog's resultant displacement vector.

9. A man lost in a maze makes three consecutive displacements so that at the end of the walk he is back where he started, as shown below. The first displacement is 8.00 m westward, and the second is 13.0 m northward. Use the graphical method to find the third displacement.

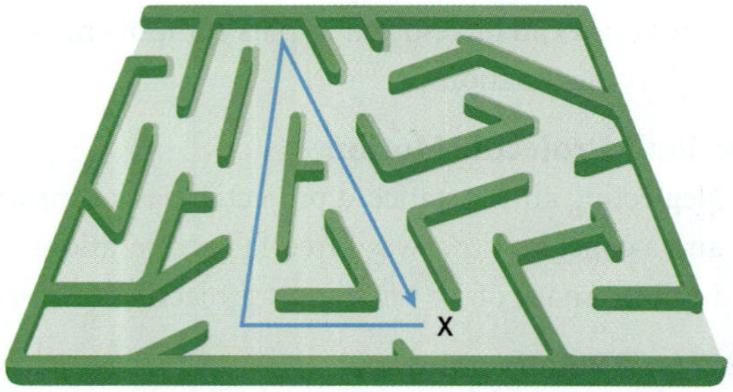

Conceptual Questions

10. If **B** is added to **A**, under what conditions does the resultant have the magnitude equal to $A + B$?

11. Give an example of a moving object that has a velocity vector and an acceleration vector in the same direction and an example of one that has velocity and acceleration vectors in opposite directions.

12. A student accurately uses the method for combining vectors. The two vectors she combines have magnitudes of 55 and 25 units. The answer that she gets is either 85, 20, or 55. Pick the correct answer, and explain why it is the only one of the three that can be correct.

13. If a set of vectors laid head to tail forms a closed polygon, the resultant is zero. Is this statement true? Explain your reasoning.

VECTOR OPERATIONS

Review Questions

14. Can a vector have a component equal to zero and still have a nonzero magnitude?

15. Can a vector have a component greater than its magnitude?

16. Explain the difference between vector addition and vector resolution.

17. How would you add two vectors that are not perpendicular or parallel?

Conceptual Questions

18. If **A** + **B** equals 0, what can you say about the components of the two vectors?

19. Under what circumstances would a vector have components that are equal in magnitude?

20. The vector sum of three vectors gives a resultant equal to zero. What can you say about the vectors?

Practice Problems

For problems 21–23, see Sample Problem A.

21. A girl delivering newspapers travels three blocks west, four blocks north, and then six blocks east.

a. What is her resultant displacement?

b. What is the total distance she travels?

22. A quarterback takes the ball from the line of scrimmage, runs backward for 10.0 yards, and then runs sideways parallel to the line of scrimmage for 15.0 yards. At this point, he throws a 50.0-yard forward pass straight down the field. What is the magnitude of the football's resultant displacement?

23. A shopper pushes a cart 40.0 m south down one aisle and then turns 90.0° and moves 15.0 m. He then makes another 90.0° turn and moves 20.0 m. Find the shopper's total displacement. (There could be more than one correct answer.)

For problems 24–25, see Sample Problem B.

24. A submarine dives 110.0 m at an angle of 10.0° below the horizontal. What are the two components?

25. A person walks 25.0° north of east for 3.10 km. How far would another person walk due north and due east to arrive at the same location?

For problem 26, see Sample Problem C.

26. A person walks the path shown below. The total trip consists of four straight-line paths. At the end of the walk, what is the person's resultant displacement measured from the starting point?

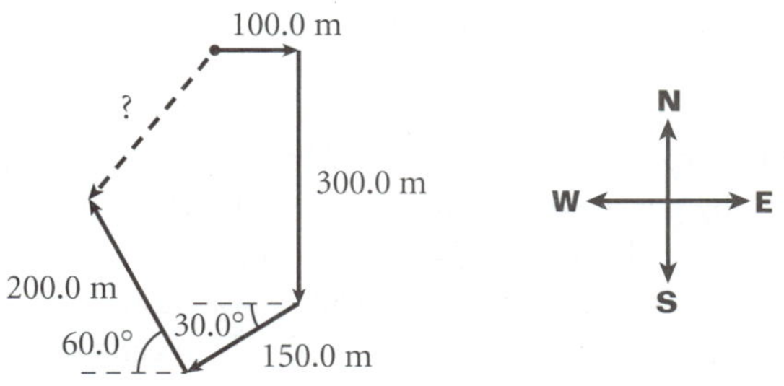

PROJECTILE MOTION

Review Questions

27. A dart is fired horizontally from a dart gun, and another dart is dropped simultaneously from the same height. If air resistance can be neglected, which dart hits the ground first?

28. If a rock is dropped from the top of a sailboat's mast, will it hit the deck at the same point whether the boat is at rest or in motion at constant velocity?

29. Does a ball dropped out of the window of a moving car take longer to reach the ground than one dropped at the same height from a car at rest?

30. A rock is dropped at the same instant that a ball at the same elevation is thrown horizontally. Which will have the greater speed when it reaches ground level?

Practice Problems

For problems 31–33, see Sample Problem D.

31. The fastest recorded pitch in Major League Baseball was thrown by Nolan Ryan in 1974. If this pitch were thrown horizontally, the ball would fall 0.809 m (2.65 ft) by the time it reached home plate, 18.3 m (60 ft) away. How fast was Ryan's pitch?

3 REVIEW

14. Yes, the second component could be nonzero.

15. No, the hypotenuse is always greater than the legs.

16. Vector addition is combining vectors to find a resultant vector. Vector resolution is breaking a vector into its component vectors.

17. by resolving both vectors into their vector components, adding the corresponding components together, and finding the vector sum of the summed components

18. They are equal and opposite.

19. if the vector is oriented at 45° from the axes

20. They form a closed triangle when laid head to tail.

21. **a.** 5 blocks at 53° north of east
b. 13 blocks

22. 42.7 yards

23. 61.8 m at 76.0° S of E (or S of W), 25.0 m at 53.1° S of E (or S of W)

24. 108 m, −19.1 m

25. 2.81 km east, 1.31 km north

26. 2.40×10^2 m at 57.2° south of west

27. Both hit at the same time.

28. yes, neglecting air resistance

29. no, neglecting air resistance

30. The vertical components of each velocity vector will be the same, but the thrown ball will also have a horizontal component of velocity. As a result, the thrown ball will have a greater speed.

31. 45.1 m/s

32. 3.3 s; 36 m/s
33. 11 m
34. a. 2.77×10^5 m
b. 284 s
35. a. clears the goal by 1 m
b. falling
36. 4.11 m
37. 80 m; 210 m
38. Displacement and velocity depend on the frame of reference in which they are measured.
39. the coordinate system used to describe the motion
40. Earth
41. a. 70 m/s east
b. 20 m/s
42. a. To the passenger, the ball appears to move in a straight line. To an outside observer, the ball moves along a parabolic trajectory.
b. The passenger would see the ball move backward, while the stationary observer would see no change from part (a).

32. A person standing at the edge of a seaside cliff kicks a stone over the edge with a speed of 18 m/s. The cliff is 52 m above the water's surface, as shown at right. How long does it take for the stone to fall to the water? With what speed does it strike the water?

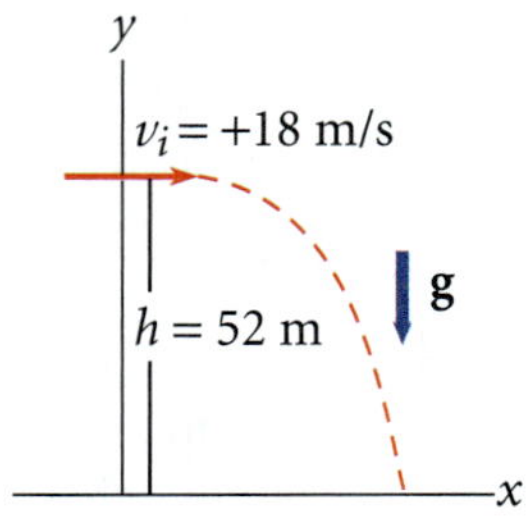

33. A spy in a speed boat is being chased down a river by government officials in a faster craft. Just as the officials' boat pulls up next to the spy's boat, both boats reach the edge of a 5.0 m waterfall. If the spy's speed is 15 m/s and the officials' speed is 26 m/s, how far apart will the two vessels be when they land below the waterfall?

For problems 34–37, see Sample Problem E.

34. A shell is fired from the ground with an initial speed of 1.70×10^3 m/s (approximately five times the speed of sound) at an initial angle of 55.0° to the horizontal. Neglecting air resistance, find

a. the shell's horizontal range

b. the amount of time the shell is in motion

35. A place kicker must kick a football from a point 36.0 m (about 40.0 yd) from the goal. As a result of the kick, the ball must clear the crossbar, which is 3.05 m high. When kicked, the ball leaves the ground with a speed of 20.0 m/s at an angle of 53° to the horizontal.

a. By how much does the ball clear or fall short of clearing the crossbar?

b. Does the ball approach the crossbar while still rising or while falling?

36. When a water gun is fired while being held horizontally at a height of 1.00 m above ground level, the water travels a horizontal distance of 5.00 m. A child, who is holding the same gun in a horizontal position, is also sliding down a 45.0° incline at a constant speed of 2.00 m/s. If the child fires the gun when it is 1.00 m above the ground and the water takes 0.329 s to reach the ground, how far will the water travel horizontally?

37. A ship maneuvers to within 2.50×10^3 m of an island's 1.80×10^3 m high mountain peak and fires a projectile at an enemy ship 6.10×10^2 m on the other side of the peak, as illustrated below. If the ship shoots the projectile with an initial velocity of 2.50×10^2 m/s at an angle of 75.0°, how close to the enemy ship does the projectile land? How close (vertically) does the projectile come to the peak?

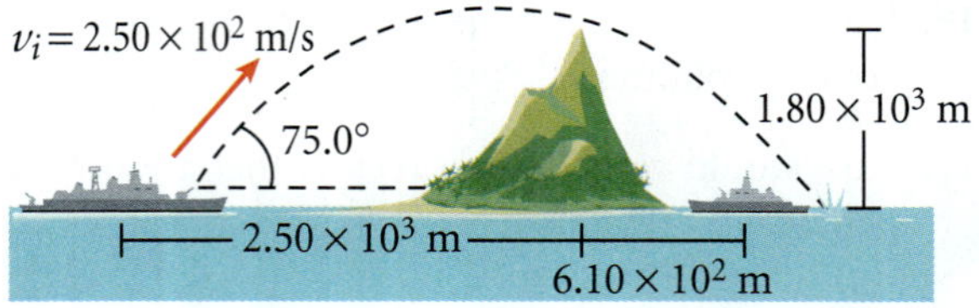

RELATIVE MOTION

Review Questions

38. Explain the statement "All motion is relative."

39. What is a frame of reference?

40. When we describe motion, what is a common frame of reference?

41. A small airplane is flying at 50 m/s toward the east. A wind of 20 m/s toward the east suddenly begins to blow and gives the plane a velocity of 70 m/s east.

a. Which vector is the resultant vector?

b. What is the magnitude of the wind velocity?

42. A ball is thrown upward in the air by a passenger on a train that is moving with constant velocity.

a. Describe the path of the ball as seen by the passenger. Describe the path as seen by a stationary observer outside the train.

b. How would these observations change if the train were accelerating along the track?

Practice Problems

For problems 43–46, see Sample Problem F.

43. A river flows due east at 1.50 m/s. A boat crosses the river from the south shore to the north shore by maintaining a constant velocity of 10.0 m/s due north relative to the water.

a. What is the velocity of the boat as viewed by an observer on shore?

b. If the river is 325 m wide, how far downstream is the boat when it reaches the north shore?

44. The pilot of an aircraft wishes to fly due west in a 50.0 km/h wind blowing toward the south. The speed of the aircraft in the absence of a wind is 205 km/h.

a. In what direction should the aircraft head?
b. What should its speed relative to the ground be?

45. A hunter wishes to cross a river that is 1.5 km wide and that flows with a speed of 5.0 km/h. The hunter uses a small powerboat that moves at a maximum speed of 12 km/h with respect to the water. What is the minimum time necessary for crossing?

46. A swimmer can swim in still water at a speed of 9.50 m/s. He intends to swim directly across a river that has a downstream current of 3.75 m/s.

a. What must the swimmer's direction be?
b. What is his velocity relative to the bank?

MIXED REVIEW

47. A ball player hits a home run, and the baseball just clears a wall 21.0 m high located 130.0 m from home plate. The ball is hit at an angle of 35.0° to the horizontal, and air resistance is negligible. Assume the ball is hit at a height of 1.0 m above the ground.

a. What is the initial speed of the ball?
b. How much time does it take for the ball to reach the wall?
c. Find the components of the velocity and the speed of the ball when it reaches the wall.

48. A daredevil jumps a canyon 12 m wide. To do so, he drives a car up a 15° incline.

a. What minimum speed must he achieve to clear the canyon?
b. If the daredevil jumps at this minimum speed, what will his speed be when he reaches the other side?

49. A 2.00 m tall basketball player attempts a goal 10.00 m from the basket (3.05 m high). If he shoots the ball at a 45.0° angle, at what initial speed must he throw the basketball so that it goes through the hoop without striking the backboard?

50. An escalator is 20.0 m long. If a person stands on the escalator, it takes 50.0 s to ride to the top.

a. If a person walks up the moving escalator with a speed of 0.500 m/s relative to the escalator, how long does it take the person to get to the top?
b. If a person walks down the "up" escalator with the same relative speed as in item (a), how long does it take to reach the bottom?

51. A ball is projected horizontally from the edge of a table that is 1.00 m high, and it strikes the floor at a point 1.20 m from the base of the table.

a. What is the initial speed of the ball?
b. How high is the ball above the floor when its velocity vector makes a 45.0° angle with the horizontal?

52. How long does it take an automobile traveling 60.0 km/h to become even with a car that is traveling in another lane at 40.0 km/h if the cars' front bumpers are initially 125 m apart?

53. The eye of a hurricane passes over Grand Bahama Island. It is moving in a direction 60.0° north of west with a speed of 41.0 km/h. Exactly three hours later, the course of the hurricane shifts due north, and its speed slows to 25.0 km/h, as shown below. How far from Grand Bahama is the hurricane 4.50 h after it passes over the island?

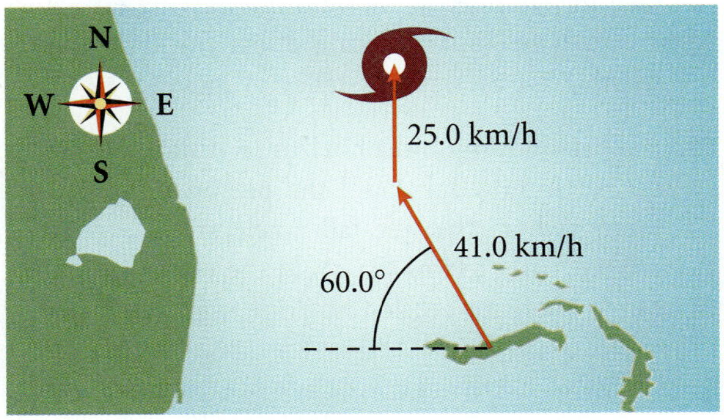

54. A boat moves through a river at 7.5 m/s relative to the water, regardless of the boat's direction. If the water in the river is flowing at 1.5 m/s, how long does it take the boat to make a round trip consisting of a 250 m displacement downstream followed by a 250 m displacement upstream?

3 REVIEW

43. **a.** 10.1 m/s at 8.53° east of north
b. 48.8 m

44. **a.** 14.1° north of west
b. 199 km/h

45. 7.5 min

46. **a.** 23.2° upstream from straight across
b. 8.72 m/s across the river

47. **a.** 41.7 m/s
b. 3.81 s
c. $v_{y,f} = -13.5$ m/s, $v_{x,f} = 34.2$ m/s, $v_f = 36.7$ m/s

48. **a.** 15 m/s
b. 15 m/s

49. 10.5 m/s

50. **a.** 22.2 s
b. 2.00×10^2 s

51. **a.** 2.66 m/s
b. 0.64 m

52. 22.5 s

53. 157 km

54. 7.0×10^1 s

55. a. 32.5 m
    b. 1.78 s
56. a. 5.0×10^1 m/s
    b. 4.0×10^1 m
57. a. 57.7 km/h at 60.0° west of the vertical
    b. 28.8 km/h straight down
58. 12.0 s
59. 18 m; 7.9 m
60. 15.3 m
61. 6.19 m/s downfield
62. a. 2.4×10^4 m
    b. 152 s
    c. 6.0×10^4 m

55. A car is parked on a cliff overlooking the ocean on an incline that makes an angle of 24.0° below the horizontal. The negligent driver leaves the car in neutral, and the emergency brakes are defective. The car rolls from rest down the incline with a constant acceleration of 4.00 m/s^2 and travels 50.0 m to the edge of the cliff. The cliff is 30.0 m above the ocean.

a. What is the car's position relative to the base of the cliff when the car lands in the ocean?

b. How long is the car in the air?

56. A golf ball with an initial angle of 34° lands exactly 240 m down the range on a level course.

a. Neglecting air friction, what initial speed would achieve this result?

b. Using the speed determined in item (a), find the maximum height reached by the ball.

57. A car travels due east with a speed of 50.0 km/h. Rain is falling vertically with respect to Earth. The traces of the rain on the side windows of the car make an angle of 60.0° with the vertical. Find the velocity of the rain with respect to the following:

a. the car

b. Earth

58. A shopper in a department store can walk up a stationary (stalled) escalator in 30.0 s. If the normally functioning escalator can carry the standing shopper to the next floor in 20.0 s, how long would it take the shopper to walk up the moving escalator? Assume the same walking effort for the shopper whether the escalator is stalled or moving.

59. If a person can jump a horizontal distance of 3.0 m on Earth, how far could the person jump on the moon, where the free-fall acceleration is $g/6$ and $g = 9.81$ m/s^2? How far could the person jump on Mars, where the acceleration due to gravity is $0.38g$?

60. A science student riding on a flatcar of a train moving at a constant speed of 10.0 m/s throws a ball toward the caboose along a path that the student judges as making an initial angle of 60.0° with the horizontal. The teacher, who is standing on the ground nearby, observes the ball rising vertically. How high does the ball rise?

61. A football is thrown directly toward a receiver with an initial speed of 18.0 m/s at an angle of 35.0° above the horizontal. At that instant, the receiver is 18.0 m from the quarterback. In what direction and with what constant speed should the receiver run to catch the football at the level at which it was thrown?

62. A rocket is launched at an angle of 53° above the horizontal with an initial speed of 75 m/s, as shown below. It moves for 25 s along its initial line of motion with an acceleration of 25 m/s^2. At this time, its engines fail and the rocket proceeds to move as a free body.

a. What is the rocket's maximum altitude?

b. What is the rocket's total time of flight?

c. What is the rocket's horizontal range?

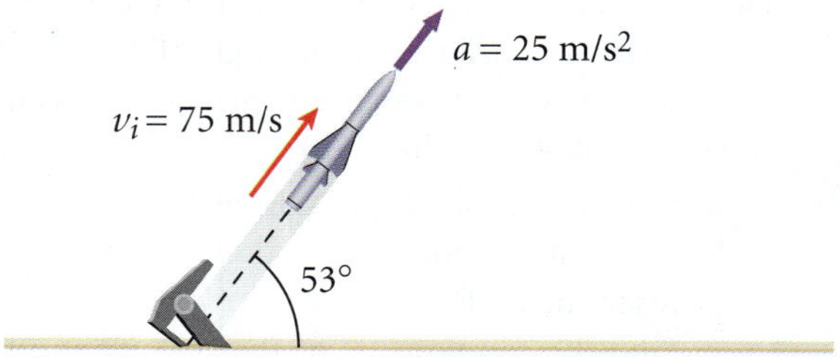

Alternative Assessment

1. Work in cooperative groups to analyze a game of chess in terms of displacement vectors. Make a model chessboard, and draw arrows showing all the possible moves for each piece as vectors made of horizontal and vertical components. Then have two members of your group play the game while the others keep track of each piece's moves. Be prepared to demonstrate how vector addition can be used to explain where a piece would be after several moves.
2. Use a garden hose to investigate the laws of projectile motion. Design experiments to investigate how the angle of the hose affects the range of the water stream. (Assume that the initial speed of water is constant and is determined by the pressure indicated by the faucet's setting.) What quantities will you measure, and how will you measure them? What variables do you need to control? What is the shape of the water stream? How can you reach the maximum range? How can you reach the highest point? Present your results to the rest of the class and discuss the conclusions.
3. You are helping NASA engineers design a basketball court for a colony on the moon. How do you anticipate the ball's motion compared with its motion on Earth? What changes will there be for the players—how they move and how they throw the ball? What changes would you recommend for the size of the court, the basket height, and other regulations in order to adapt the sport to the moon's low gravity? Create a presentation or a report presenting your suggestions, and include the physics concepts behind your recommendations.
4. There is conflicting testimony in a court case. A police officer claims that his radar monitor indicated that a car was traveling at 176 km/h (110 mi/h). The driver argues that the radar must have recorded the relative velocity because he was only going 88 km/h (55 mi/h). Is it possible that both are telling the truth? Could one be lying? Prepare scripts for expert witnesses, for both the prosecution and the defense, that use physics to justify their positions before the jury. Create visual aids to be used as evidence to support the different arguments.

Graphing Calculator Practice

Two Dimensional Motion

Recall the following equation from your studies of projectiles launched at an angle.

$$\Delta y = (v_i \sin \theta)\Delta t + \tfrac{1}{2}a_y(\Delta t)^2$$

Consider a baseball that is thrown straight up in the air. The equation for projectile motion can be entered as Y_1 on a graphing calculator.

$$Y_1 = VX - 4.9X^2$$

Given the initial velocity (V), your graphing calculator can calculate the height (Y1) of the baseball versus the time interval (X) that the ball remains in the air. Why is the factor $\sin \theta$ missing from the equation for Y_1?

In this activity, you will determine the maximum height and flight time of a baseball thrown vertically at various initial velocities.

Visit go.hrw.com and type in the keyword **HF6TDMX** to find this graphing calculator activity. Refer to **Appendix B** for instructions on downloading the program for this activity.

Alternative Assessment ANSWERS

1. Chess pieces can move in ways that require more than one component. Student reports should show how several moves can be reported as a vector sum.
2. Student plans should be safe and include measurements of angle and range. They should find that 45° is the best angle for maximum range and 90° is the best angle for maximum height.
3. Students should recognize that balls will stay in the air longer and that players can jump higher. Players will need to shoot lower and more gently, or the court should be longer and the basket higher.
4. Students should recognize that if the cars were driving toward each other, both could be telling the truth.

extension

Graphing Calculator Practice

Visit go.hrw.com for answers to this Graphing Calculator activity.

Keyword HF6TDMXT

Standardized Test Prep

ANSWERS

1. B
2. H
3. A
4. J
5. C
6. G
7. B
8. J

Standardized Test Prep

MULTIPLE CHOICE

1. Vector **A** has a magnitude of 30 units. Vector **B** is perpendicular to vector **A** and has a magnitude of 40 units. What would the magnitude of the resultant vector **A** + **B** be?

A. 10 units
B. 50 units
C. 70 units
D. zero

2. What term represents the magnitude of a velocity vector?

F. acceleration
G. momentum
H. speed
J. velocity

Use the diagram below to answer questions 3–4.

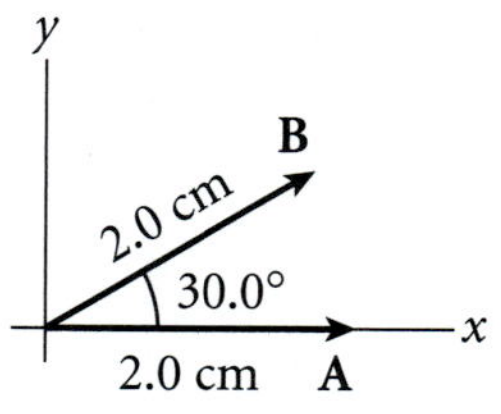

3. What is the direction of the resultant vector **A** + **B**?

A. 15° above the x-axis
B. 75° above the x-axis
C. 15° below the x-axis
D. 75° below the x-axis

4. What is the direction of the resultant vector **A** – **B**?

F. 15° above the x-axis
G. 75° above the x-axis
H. 15° below the x-axis
J. 75° below the x-axis

Use the passage below to answer questions 5–6.

A motorboat heads due east at 5.0 m/s across a river that flows toward the south at a speed of 5.0 m/s.

5. What is the resultant velocity relative to an observer on the shore?

A. 3.2 m/s to the southeast
B. 5.0 m/s to the southeast
C. 7.1 m/s to the southeast
D. 10.0 m/s to the southeast

6. If the river is 125 m wide, how long does the boat take to cross the river?

F. 39 s
G. 25 s
H. 17 s
J. 12 s

7. The pilot of a plane measures an air velocity of 165 km/h south relative to the plane. An observer on the ground sees the plane pass overhead at a velocity of 145 km/h toward the north. What is the velocity of the wind that is affecting the plane relative to the observer?

A. 20 km/h to the north
B. 20 km/h to the south
C. 165 km/h to the north
D. 310 km/h to the south

8. A golfer takes two putts to sink his ball in the hole once he is on the green. The first putt displaces the ball 6.00 m east, and the second putt displaces the ball 5.40 m south. What displacement would put the ball in the hole in one putt?

F. 11.40 m southeast
G. 8.07 m at 48.0° south of east
H. 3.32 m at 42.0° south of east
J. 8.07 m at 42.0° south of east

Use the information below to answer questions 9–12.

A girl riding a bicycle at 2.0 m/s throws a tennis ball horizontally forward at a speed of 1.0 m/s from a height of 1.5 m. At the same moment, a boy standing on the sidewalk drops a tennis ball straight down from a height of 1.5 m.

9. What is the initial speed of the girl's ball relative to the boy?

A. 1.0 m/s
B. 1.5 m/s
C. 2.0 m/s
D. 3.0 m/s

10. If air resistance is disregarded, which ball will hit the ground first?

F. the boy's ball
G. the girl's ball
H. neither
J. The answer cannot be determined from the given information.

11. If air resistance is disregarded, which ball will have a greater speed (relative to the ground) when it hits the ground?

A. the boy's ball
B. the girl's ball
C. neither
D. The answer cannot be determined from the given information.

12. What is the speed of the girl's ball when it hits the ground?

F. 1.0 m/s
G. 3.0 m/s
H. 6.2 m/s
J. 8.4 m/s

SHORT RESPONSE

13. If one of the components of one vector along the direction of another vector is zero, what can you conclude about these two vectors?

14. A roller coaster travels 41.1 m at an angle of 40.0° above the horizontal. How far does it move horizontally and vertically?

15. A ball is thrown straight upward and returns to the thrower's hand after 3.00 s in the air. A second ball is thrown at an angle of 30.0° with the horizontal. At what speed must the second ball be thrown to reach the same height as the one thrown vertically?

EXTENDED RESPONSE

16. A human cannonball is shot out of a cannon at 45.0° to the horizontal with an initial speed of 25.0 m/s. A net is positioned at a horizontal distance of 50.0 m from the cannon. At what height above the cannon should the net be placed in order to catch the human cannonball? Show your work.

Read the following passage to answer question 17.

Three airline executives are discussing ideas for developing flights that are more energy efficient.

Executive A: Because the Earth rotates from west to east, we could operate "static flights"—a helicopter or airship could begin by rising straight up from New York City and then descend straight down four hours later when San Francisco arrives below.

Executive B: This approach could work for one-way flights, but the return trip would take 20 hours.

Executive C: That approach will never work. Think about it. When you throw a ball straight up in the air, it comes straight back down to the same point.

Executive A: The ball returns to the same point because Earth's motion is not significant during such a short time.

17. In a paragraph, state which of the executives is correct, and explain why.

Test TIP If you get stuck answering a question, move on. You can return to the question later if you have time.

9. D
10. H
11. B
12. H
13. They are perpendicular.
14. 31.5 m horizontally, 26.4 m vertically
15. 29.4 m/s
16. 10.8 m (See the Solutions Manual or the One-Stop Planner for a full solution.)
17. Executive C is correct. Student explanations should include the concept of relative velocity—when a helicopter or airship lifts off straight up from the ground, it is already moving horizontally with Earth's horizontal velocity. (We assume that Earth's motion is constant for the purposes of this scenario and does not depend on time.)

Lab Planning

Beginning on page T34 are preparation notes and teaching tips to assist you in planning.

Blank data tables (as well as some sample data) appear on the **One-Stop Planner.**

No Books in the Lab?

See the ***Datasheets for In-Text Labs*** workbook for a reproducible master copy of this experiment.

The same workbook also contains a version of this experiment with explicit procedural steps if you prefer a more directed approach.

Safety Caution

Remind students to be careful; balls on the floor represent a safety hazard.

Tips and Tricks

- With correct alignment, the box should catch the ball every time. Copy-paper boxes work well.
- Carbon paper may be taped down along one side. Handle carbon paper carefully and do not allow it to touch clothing.
- Show students how to use the washer on a cord to find the baseline from which to measure the distance.
- Some students may need help determining where to measure the height and length of the inclined plane.

CHAPTER 3

Inquiry Lab — Velocity of a Projectile

Design Your Own

OBJECTIVES

- **Measure** the velocity of projectiles in terms of the horizontal displacement during free fall.
- **Compare** the velocity and acceleration of projectiles accelerated down different inclined planes.

MATERIALS LIST

- aluminum sheet, edges covered with heavy tape
- C-clamp
- cardboard box
- cord
- inclined plane
- masking tape
- meterstick
- packing tape
- several large sheets white paper
- several sheets carbon paper
- small metal ball
- small metal washer
- support stand and clamp
- towel or cloth

When a ball rolls down an inclined plane, then rolls off the edge of a table, the ball becomes a projectile with some positive horizontal velocity and an initial vertical velocity of zero. However, the length of time that the projectile stays in the air depends not on the horizontal velocity but on the height of the table above the ground. The horizontal velocity determines how far the projectile travels during the time it is in the air.

In this lab, you will roll a ball down an inclined plane, off the edge of a table, and onto a piece of carbon paper on the floor. You will design your own experiment by deciding the details of the setup and the procedure, including how many trials to perform, the angles of the inclined plane, and how high up the plane you will release the ball. Your experiment should include trials with the plane inclined to different angles and multiple trials at different heights along the plane at each angle of inclination. Your procedure should include steps to measure the height from which the ball is released, the length of the ball's travel along the plane, and the horizontal displacement of the ball after it leaves the table.

SAFETY

- **Tie back long hair, secure loose clothing, and remove loose jewelry to prevent its getting caught in moving or rotating parts. Put on goggles.**
- **Perform this experiment in a clear area. Falling or dropped masses can cause serious injury.**

PROCEDURE

1. Study the materials provided, and read the Analysis and Conclusions questions. Design an experiment using the provided materials to meet the goals stated above and to allow you to answer the questions.
2. Write out your lab procedure, including a detailed description of the measurements to take during each step and the number of trials to perform. Create a data table to record your measurements for each trial. You may use **Figure 1** as a guide to one possible setup. Your setup should include a box to catch the ball at the end of each trial.
3. Ask your teacher to approve your procedure.
4. Follow all steps of your procedure.
5. Clean up your work area as directed by your teacher.

ANALYSIS

1. **Organizing Data** Find the time interval for the ball's motion from the edge of the table to the floor using the equation for the vertical motion of a projectile. In those equations, Δy is the vertical displacement of the ball after it leaves the table. The result is the time interval for each trial.

2. **Organizing Data** Using the time interval from item 1 and the value for *Displacement* Δx, calculate the average horizontal velocity for each trial during the ball's motion from the edge of the table to the floor.

3. **Constructing Graphs** Plot a graph of average horizontal velocity versus height of release. You may use graph paper, a computer, or a graphing calculator.

4. **Constructing Graphs** Plot a graph of average horizontal velocity versus length of travel along the plane. You may use graph paper, a computer, or a graphing calculator.

CONCLUSIONS

5. **Drawing Conclusions** What is the relationship between the height of the inclined plane and the horizontal velocity of the ball? Explain.

6. **Drawing Conclusions** What is the relationship between the length of the inclined plane and the horizontal velocity of the ball? Explain.

7. **Evaluating Methods** Why might using the vertical displacement to calculate the time interval be more reliable than using a stopwatch for each trial?

8. **Applying Conclusions** In which trials would the total velocity of the ball when it hits the ground be the greatest?

EXTENSION

9. **Designing Experiments** Design an experiment to test the assumption that the time the ball is in the air is independent of the horizontal velocity of the ball. If you have time and your teacher approves your plan, carry out the experiment.

Figure 1

- Use tape to cover the sharp edges of the aluminum sheet before taping it to the end of the plane. The aluminum keeps the ball from bouncing as it rolls onto the table.
- Use a washer hanging from a string to find the zero-displacement point directly under the edge of the table.
- Use a box lined with a soft cloth to catch the ball after it lands.

ANSWERS

Analysis

1. $\Delta t = 0.39$ s (for $\Delta y = -0.75$ m)

2. Trials 1 & 2: $v = 2.22$ m/s
Trials 3 & 4: $v = 1.54$ m/s
Trials 5 & 6: $v = 1.64$ m/s
Trials 7 & 8: $v = 1.17$ m/s

3. Graphs should show a rough correlation between average horizontal velocity and height of release.

4. Graphs should show no definite correlation between average horizontal velocity and length of travel along the plane.

Conclusions

5. The greater the height of release along the plane is, the greater the average horizontal velocity of the projectile is (because potential energy turns into kinetic energy as the ball rolls down the plane).

6. There is no definite correlation between length of travel along the plane and the average horizontal velocity. At a given angle, longer travel along the plane gives a greater velocity because the height of release is greater.

7. The height can be measured with more precision than the time can be. Also, human reaction time introduces error.

8. The total velocity at impact would be greatest in trials that have the greatest average horizontal velocity.

Extension

9. Plans should be complete and should involve measuring time intervals of projectiles launched at different horizontal velocities.

CHAPTER 4 Forces and the Laws of Motion
Planning Guide

Compression Guide
To shorten instruction because of time limitations, omit the opener and abbreviate the review.

OBJECTIVES	LABS, DEMONSTRATIONS, AND ACTIVITIES	TECHNOLOGY RESOURCES
PACING • 45 min pp. 118–119 **Chapter Opener**	ANC **Discovery Lab** Discovering Newton's Laws*◆ BASIC	CD **Visual Concepts,** Chapter 4 BASIC
PACING • 45 min pp. 120–124 **Section 1 Changes in Motion** • Describe how force affects the motion of an object. • Interpret and construct free-body diagrams.	SE **Quick Lab** Force and Changes in Motion, p. 122 GENERAL	OSP **Lesson Plans** TR 9 Force Diagrams and Free-Body Diagrams TR 10 Free-Body Diagram of a Sled Being Pulled
PACING • 45 min pp. 125–129 **Section 2 Newton's First Law** • Explain the relationship between the motion of an object and the net external force acting on the object. • Determine the net external force on an object. • Calculate the force required to bring an object into equilibrium.	TE **Demonstration** Inertia, p. 125 BASIC SE **Quick Lab** Inertia, p. 126 GENERAL	OSP **Lesson Plans** TR 11 Determining Net Force TR 12 Inertia and the Operation of a Seat Belt
PACING • 90 min pp. 130–134 **Section 3 Newton's Second and Third Laws** • Describe an object's acceleration in terms of its mass and the net force acting on it. • Predict the direction and magnitude of the acceleration caused by a known net force. • Identify action-reaction pairs.	SE **Skills Practice Lab** Force and Acceleration, pp. 152–155◆ GENERAL ANC **Datasheet** Force and Acceleration* GENERAL SE **CBL™ Lab** Force and Acceleration, pp. 934–935◆ GENERAL ANC **CBL™ Experiment** Force and Acceleration*◆ GENERAL	OSP **Lesson Plans** EXT **Integrating Technology** Car Seat Safety BASIC
PACING • 45 min pp. 135–143 **Section 4 Everyday Forces** • Explain the difference between mass and weight. • Find the direction and magnitude of normal forces. • Describe air resistance as a form of friction. • Use coefficients of friction to calculate frictional force.	TE **Demonstration** Static vs. Kinetic Friction, p. 137 GENERAL TE **Demonstration** Friction of Different Surfaces, p. 137 ADVANCED TE **Demonstration** Friction and Surface Area, p. 138 GENERAL ANC **Invention Lab** Friction: Testing Materials*◆ ADVANCED ANC **CBL™ Experiment** Static and Kinetic Friction*◆ ADVANCED ANC **CBL™ Experiment** Air Resistance*◆ ADVANCED	OSP **Lesson Plans** TR 13 Static and Kinetic Friction TR 14 Friction Depends on the Surfaces and the Applied Force TR 18A Coefficients of Friction

PACING • 90 min

CHAPTER REVIEW, ASSESSMENT, AND STANDARDIZED TEST PREPARATION

- SE **Chapter Highlights,** p. 144
- SE **Chapter Review,** pp. 145–148
- SE **Alternative Assessment,** p. 149 ADVANCED
- SE **Graphing Calculator Practice,** p. 149 GENERAL
- SE **Standardized Test Prep,** pp. 150–151 GENERAL
- SE **Appendix D: Equations,** p. 855
- SE **Appendix I: Additional Problems,** pp. 882–883
- ANC **Study Guide Worksheet** Mixed Review* GENERAL
- ANC **Chapter Test A*** GENERAL
- ANC **Chapter Test B*** ADVANCED
- OSP **Test Generator**

Online and Technology Resources

go.hrw.com Holt Online Learning

Visit go.hrw.com to find a variety of online resources. To access this chapter's extensions, enter the keyword HF6FORXT and click the "go" button. Click Holt Online Learning for an online edition of this textbook, and other interactive resources.

HOLT Teacher's One-Stop Planner®

This DVD package includes:
- Holt Calendar Planner
- Customizable Lesson Plans
- Editable Worksheets
- ExamView® Version 6 Assessment Suite
- Interactive Teacher's Edition
- Holt PuzzlePro®
- Holt PowerNotes® Presentations
- MindPoint® Quiz Show

KEY

SE Student Edition
TE Teacher Edition
ANC Ancillary Worksheet
OSP One-Stop Planner
CD CD or CD-ROM
TR Teaching Transparencies
EXT Online Extension
* Also on One-Stop Planner
◆ Requires advance prep

SKILLS DEVELOPMENT RESOURCES	REVIEW AND ASSESSMENT	CORRELATIONS
		National Science Education Standards
SE **Sample Set A** Drawing Free-Body Diagrams, pp. 123–124 GENERAL ANC **Problem Workbook** Sample Set A* GENERAL OSP **Problem Bank** Sample Set A GENERAL	SE **Section Review,** p. 124 GENERAL ANC **Study Guide Worksheet** Section 1* GENERAL ANC **Quiz** Section 1* BASIC	UCP 1, 2, 3 SAI 1, 2 ST 1, 2 SPSP 1 PS 4a
SE **Sample Set B** Determining Net Force, pp. 127–128 GENERAL TE **Classroom Practice,** p. 127 GENERAL ANC **Problem Workbook** Sample Set B* GENERAL OSP **Problem Bank** Sample Set B GENERAL	SE **Section Review,** p. 129 GENERAL ANC **Study Guide Worksheet** Section 2* GENERAL ANC **Quiz** Section 2* BASIC	UCP 1, 2, 3, 5 SAI 1, 2 ST 1, 2 HNS 3 PS 4a
SE **Sample Set C** Newton's Second Law, pp. 131–132 BASIC TE **Classroom Practice,** p. 131 GENERAL ANC **Problem Workbook** Sample Set C* BASIC OSP **Problem Bank** Sample Set C BASIC SE **Conceptual Challenge,** p. 132 GENERAL	SE **Section Review,** p. 134 GENERAL ANC **Study Guide Worksheet** Section 3* GENERAL ANC **Quiz** Section 3* BASIC	UCP 1, 2, 3 SAI 1, 2 HNS 1, 2, 3 SPSP 5 PS 4a
SE **Sample Set D** Coefficients of Friction, p. 139 BASIC TE **Classroom Practice,** p. 139 BASIC ANC **Problem Workbook** Sample Set D* BASIC OSP **Problem Bank** Sample Set D BASIC SE **Sample Set E** Overcoming Friction, pp. 140–141 ADVANCED TE **Classroom Practice,** p. 140 ADVANCED ANC **Problem Workbook** Sample Set E* ADVANCED OSP **Problem Bank** Sample Set E ADVANCED	SE **Section Review,** p. 143 GENERAL ANC **Study Guide Worksheet** Section 4* GENERAL ANC **Quiz** Section 4* BASIC	UCP 1, 2, 3, 5 SAI 1, 2 ST 1, 2 HNS 3 SPSP 3, 4, 5 PS 2d, 4a, 4b

SCILINKS. NSTA
www.scilinks.org

Maintained by the **National Science Teachers Association.**

Topic: Forces
SciLinks Code: HF60604

Topic: Friction
SciLinks Code: HF60622

Topic: Newton's Laws
SciLinks Code: HF61028

Classroom CD-ROMs

- Guided Reading Audio Program
- Student One Stop
- Virtual Investigations
- Visual Concepts

Holt Lab Generator CD-ROM

Search for any lab by topic, standard, difficulty level, or time. Edit any lab to fit your needs, or create your own labs. Use the Lab Materials QuickList software to customize your lab materials list.

CHAPTER 4

Overview

Section 1 defines *force* and introduces free-body diagrams.

Section 2 discusses Newton's first law and the relationship between mass and inertia.

Section 3 introduces the relationships between net force, mass, and acceleration and discusses action-reaction pairs.

Section 4 examines the familiar forces of weight, normal force, and friction.

About the Illustration

Crash-test dummies are equipped with up to 48 sensors: accelerometers and force meters are placed at different positions and depths to record the force applied to the head, the bones, the organs, and the skin. Dummies are designed to resemble people of different shapes and sizes, from a six-month-old baby to a pregnant woman to a 223 lb, 7 ft tall man.

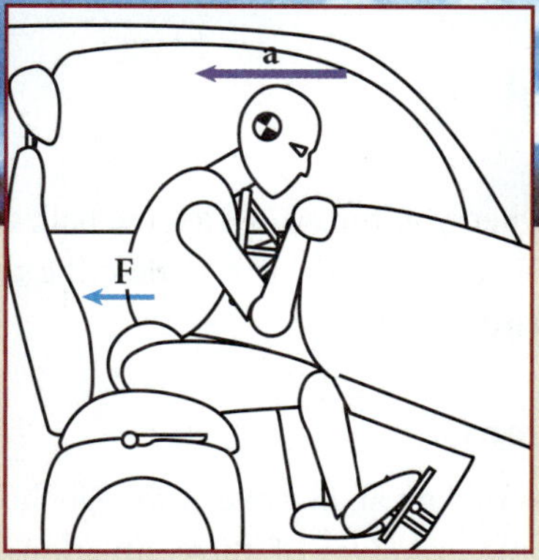

CHAPTER 4

Forces and the Laws of Motion

At General Motors' Milford Proving Grounds in Michigan, technicians place a crash-test dummy behind the steering wheel of a new car. When the car crashes, the dummy continues moving forward and hits the dashboard. The dashboard then exerts a force on the dummy that accelerates the dummy backward, as shown in the illustration. Sensors in the dummy record the forces and accelerations involved in the collision.

WHAT TO EXPECT

In this chapter, you will learn to analyze interactions by identifying the forces involved. Then, you can predict and understand many types of motion.

Why it Matters

Forces play an important role in engineering. For example, technicians study the accelerations and forces involved in car crashes in order to design safer cars and more-effective restraint systems.

CHAPTER PREVIEW

Tapping Prior Knowledge

Knowledge to Review

- ✔ Acceleration is the time rate of change of velocity. Because velocity is a vector quantity, acceleration is also a vector quantity.
- ✔ Kinematics describes the motion of an object without using the concept of force. Kinematic equations for the special case of constant acceleration were discussed in the chapter "Two-Dimensional Motion and Vectors."
- ✔ Vectors are quantities that have both magnitude and direction; the direction and magnitude of vectors can be represented by arrows drawn in the appropriate direction at the appropriate length.

Items to Probe

- ✔ Vector addition: Have students practice resolving vectors into components, adding the components, and finding the resultant of the vector addition.

119

SECTION 1
General Level

Teaching Tip

Now that students have studied motion as complex as projectile motion, explore their understanding of force. Ask them what mechanism causes motion and why some objects accelerate at higher rates than others do. Point out that *force* is attributed to any mechanism that causes or may cause a change in an object's velocity with respect to time.

Visual Strategy - BASIC

Figure 1

Point out to students that the ball is experiencing force in all three pictures.

Q How can you tell that the ball experiences at least one force in each picture?

A ***by changes in the ball's speed or direction***

Teaching Tip

Point out that still photographs such as those in **Figure 1** cannot actually show objects "experiencing force" because forces cause changes in velocity with respect to time.

SECTION 1

Changes in Motion

SECTION OBJECTIVES

- **Describe how force affects the motion of an object.**
- **Interpret and construct free-body diagrams.**

FORCE

You exert a **force** on a ball when you throw or kick the ball, and you exert a force on a chair when you sit in the chair. Forces describe the interactions between an object and its environment.

force

an action exerted on an object which may change the object's state of rest or motion

Forces can cause accelerations

In many situations, a force exerted on an object can change the object's velocity with respect to time. Some examples of these situations are shown in **Figure 1.** A force can cause a stationary object to move, as when you throw a ball. Force also causes moving objects to stop, as when you catch a ball. A force can also cause a moving object to change direction, such as when a baseball collides with a bat and flies off in another direction. Notice that in each of these cases, the force is responsible for a change in velocity with respect to time—an acceleration.

(a)

(b)

(c)

Figure 1
Force can cause objects to **(a)** start moving, **(b)** stop moving, and/or **(c)** change direction.

The SI unit of force is the newton

The SI unit of force is the newton, named after Sir Isaac Newton (1642–1727), whose work contributed much to the modern understanding of force and motion. The newton (N) is defined as the amount of force that, when acting on a 1 kg mass, produces an acceleration of 1 m/s^2. Therefore, 1 N = 1 kg × 1 m/s^2.

The weight of an object is a measure of the magnitude of the gravitational force exerted on the object. It is the result of the interaction of an object's mass with the gravitational field of another object, such as Earth. Many of the

Table 1	Units of Mass, Acceleration, and Force		
System	**Mass**	**Acceleration**	**Force**
SI	kg	m/s^2	$N = kg \bullet m/s^2$
cgs	g	cm/s^2	$dyne = g \bullet cm/s^2$
Avoirdupois	slug	ft/s^2	$lb = slug \bullet ft/s^2$

Did you know?

The symbol for the pound, lb, comes from *libra,* the Latin word for "pound," a unit of measure that has been used since medieval times to measure weight.

terms and units you use every day to talk about weight are really units of force that can be converted to newtons. For example, a $\frac{1}{4}$ lb stick of margarine has a weight equivalent to a force of about 1 N, as shown in the following conversions:

$$1 \text{ lb} = 4.448 \text{ N}$$
$$1 \text{ N} = 0.225 \text{ lb}$$

Forces can act through contact or at a distance

If you pull on a spring, the spring stretches. If you pull on a wagon, the wagon moves. When a football is caught, its motion is stopped. These pushes and pulls are examples of *contact forces,* which are so named because they result from physical contact between two objects. Contact forces are usually easy to identify when you analyze a situation.

Another class of forces—called *field forces*—does not involve physical contact between two objects. One example of this kind of force is gravitational force. Whenever an object falls to Earth, the object is accelerated by Earth's gravity. In other words, Earth exerts a force on the object even when Earth is not in immediate physical contact with the object.

Another common example of a field force is the attraction or repulsion between electric charges. You can observe this force by rubbing a balloon against your hair and then observing how little pieces of paper appear to jump up and cling to the balloon's surface, as shown in **Figure 2.** The paper is pulled by the balloon's electric field.

The theory of fields was developed as a tool to explain how objects could exert force on each other without touching. According to this theory, masses create gravitational fields in the space around them. An object falls to Earth because of the interaction between the object's mass and Earth's gravitational field. Similarly, charged objects create electromagnetic fields.

The distinction between contact forces and field forces is useful when dealing with forces that we observe at the macroscopic level. (*Macroscopic* refers to the realm of phenomena that are visible to the naked eye.) As we will see later, all macroscopic contact forces are actually due to microscopic field forces. For instance, contact forces in a collision are due to electric fields between atoms and molecules. In fact, every force can be categorized as one of four fundamental field forces.

Figure 2
The electric field around the rubbed balloon exerts an attractive electric force on the pieces of paper.

Visual Strategy GENERAL

Figure 2
Tell students that both contact and field forces are being demonstrated in this picture.

Q Identify the contact and field force examples in the picture.

A ***contact: the table supporting the pieces of paper, a person supporting the balloon***
field: gravitational force pulling down on the paper and balloon, electric force pulling up on the paper

Teaching Tip

There is a more detailed discussion of the four fundamental forces at the end of Section 4, and this topic is covered further in the chapter "Subatomic Physics."

Quick Lab

TEACHER'S NOTES

If the toy car is rolled an appreciable distance before the collision, students may observe the car slowing down because of friction. Give students a brief explanation of friction (a complete explanation follows in Section 4).

Teaching Tip — ADVANCED

Explain to students that several forces acting on a body at different points can produce translational movement of the body without rotation, rotation without translational movement, or translational movement and rotation together, depending on exactly where the forces act on the body. Discuss examples of each situation. Then explain that the examples in this chapter are limited to translational movement without rotation, so the sum of forces is all that is required. For this reason, the forces can be drawn as if they act on the body at a common point.

The concept of torque is discussed in the chapter "Circular Motion and Gravitation." Rotational equilibrium and dynamics are covered in the feature "Rotational Dynamics" in **Appendix J: Advanced Topics.**

SCILINKS®
www.scilinks.org
Topic: Forces
Code: HF60604

FORCE DIAGRAMS

When you push a toy car, it accelerates. If you push the car harder, the acceleration will be greater. In other words, the acceleration of the car depends on the force's *magnitude.* The direction in which the car moves depends on the *direction* of the force. For example, if you push the toy car from the front, the car will move in a different direction than if you push it from behind.

Force is a vector

Because the effect of a force depends on both magnitude and direction, force is a vector quantity. Diagrams that show force vectors as arrows, such as **Figure 3(a),** are called *force diagrams.* In this book, the arrows used to represent forces are blue. The tail of an arrow is attached to the object on which the force is acting. A force vector points in the direction of the force, and its length is proportional to the magnitude of the force.

At this point, we will disregard the size and shape of objects and assume that all forces act at the center of an object. In force diagrams, all forces are drawn as if they act at that point, no matter where the force is applied.

A free-body diagram helps analyze a situation

After engineers analyzing a test-car crash have identified all of the forces involved, they isolate the car from the other objects in its environment. One of their goals is to determine which forces affect the car and its passengers. **Figure 3(b)** is a free-body diagram. This diagram represents the same collision that the force diagram **(a)** does but shows only the car and the forces acting on the car. The forces exerted *by* the car on other objects are not included in the free-body diagram because they do not affect the motion of the car.

A free-body diagram is used to analyze only the forces affecting the motion of a single object. Free-body diagrams are constructed and analyzed just like other vector diagrams. In Sample Problem A, you will learn to draw free-body diagrams for some situations described in this book. In Section 2, you will learn to use free-body diagrams to find component and resultant forces.

Quick Lab

Force and Changes in Motion

MATERIALS LIST

- 1 toy car
- 1 book

Use a toy car and a book to model a car colliding with a brick wall. Observe the motion of the car before and after the crash. Identify as many changes in its motion as you can, such as changes in speed or direction. Make a list of all of the changes, and try to identify the forces that caused them. Make a force diagram of the collision.

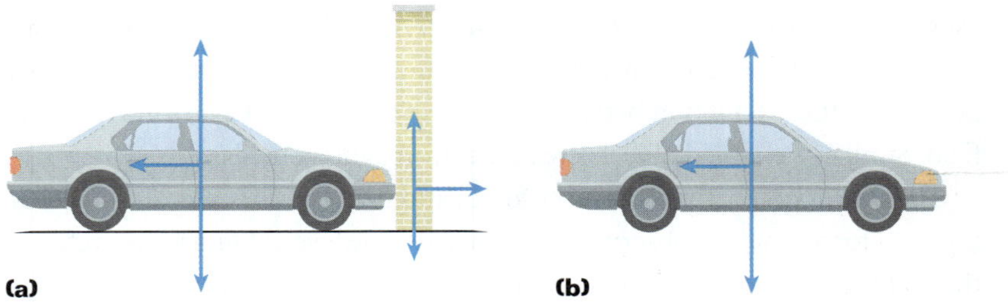

Figure 3
(a) In a force diagram, vector arrows represent all the forces acting in a situation. **(b)** A free-body diagram shows only the forces acting on the object of interest—in this case, the car.

SAMPLE PROBLEM A

STRATEGY Drawing Free-Body Diagrams

PROBLEM

The photograph at right shows a person pulling a sled. Draw a free-body diagram for this sled. The magnitudes of the forces acting on the sled are 60 N by the string, 130 N by the Earth (gravitational force), and 90 N upward by the ground.

SOLUTION

1. Identify the forces acting on the object and the directions of the forces.

- The string exerts 60 N on the sled in the direction that the string pulls.
- The Earth exerts a downward force of 130 N on the sled.
- The ground exerts an upward force of 90 N on the sled.

> **TIP** *In a free-body diagram, only include forces acting on the object. Do not include forces that the object exerts on other objects. In this problem, the forces are given, but later in the chapter, you will need to identify the forces when drawing a free-body diagram.*

(a)
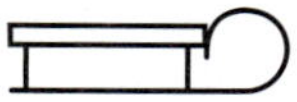

2. Draw a diagram to represent the isolated object.
It is often helpful to draw a very simple shape with some distinguishing characteristics that will help you visualize the object, as shown in **(a).** Free-body diagrams are often drawn using simple squares, circles, or even points to represent the object.

(b)
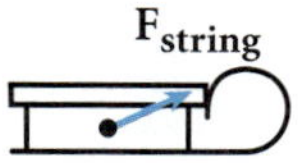

3. Draw and label vector arrows for all external forces acting on the object.
A free-body diagram of the sled will show all the forces acting on the sled as if the forces are acting on the center of the sled. First, draw and label an arrow that represents the force exerted by the string attached to the sled. The arrow should point in the same direction as the force that the string exerts on the sled, as in **(b).**

(c)
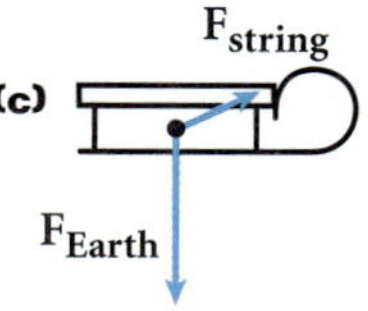

> **TIP** *When you draw an arrow representing a force, it is important to label the arrow with either the magnitude of the force or a name that will distinguish it from the other forces acting on the object. Also, be sure that the length of the arrow approximately represents the magnitude of the force.*

Next, draw and label the gravitational force, which is directed toward the center of Earth, as shown in **(c).** Finally, draw and label the upward force exerted by the ground, as shown in **(d).** Diagram **(d)** is the completed free-body diagram of the sled being pulled.

(d)
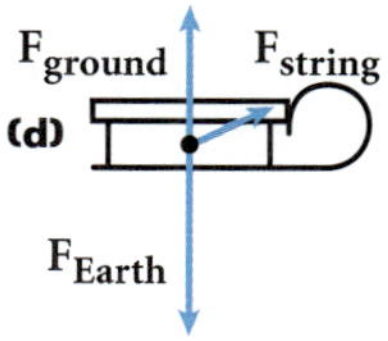

Misconception Alert

It is important to emphasize early and consistently that a free-body diagram shows only the forces acting *on* the object. A separate free-body diagram for the person pulling the sled in Sample Problem A can be used to emphasize this point and to introduce Newton's third law.

Teaching Tip

A good understanding of free-body diagrams is essential to strong physics problem-solving skills. Take this time to make sure students can properly dissect a situation involving several forces. You may wish to do several examples on the board to further emphasize the importance of the diagram step of problem solving.

This Sample Problem focuses on drawing free-body diagrams for given forces. Return to this skill in Section 4, after students have studied Newton's laws and have learned about everyday forces. At that point, ask students to build on this skill by drawing free-body diagrams for given situations where they must identify each force involved.

PROBLEM GUIDE A

Use this guide to assign problems.
SE = Student Edition Textbook
PW = Problem Workbook
PB = Problem Bank on the One-Stop Planner (OSP)

Solving for:

free-body diagrams	**SE**	Sample, 1–2; Ch. Rvw. 7–9
	PW	Sample, 1–3
	PB	Sample, 1–3

***Challenging Problem**
Consult the printed Solutions Manual or the OSP for detailed solutions.

SECTION 1

ANSWERS

Practice A

1. Each diagram should include all forces acting on the object, pointing in the correct directions and with the lengths roughly proportional to the magnitudes of the forces. Be sure each vector is labeled.
2. Diagrams should include a downward gravitational force and an upward force of the desk on the book; both vectors should have the same length and should be labeled.

SECTION REVIEW ANSWERS

1. Answers will vary.
2. gravity and electric force, answers will vary; because they can cause a change in motion
3. the newton; $1\ N = 1\ kg \bullet 1\ m/s^2$
4. because force has both magnitude and direction
5. $\mathbf{F_g}$ points down, and $\mathbf{F_{kicker}}$ points in the direction of the kick.
6. Each arrow should have a label identifying the object exerting the force and the object acted on by the force.

PRACTICE A

Drawing Free-Body Diagrams

1. A truck pulls a trailer on a flat stretch of road. The forces acting on the trailer are the force due to gravity (250 000 N downward), the force exerted by the road (250 000 N upward), and the force exerted by the cable connecting the trailer to the truck (20 000 N to the right). The forces acting on the truck are the force due to gravity (80 000 N downward), the force exerted by the road (80 000 N upward), the force exerted by the cable (20 000 N to the left), and the force causing the truck to move forward (26 400 N to the right).
   a. Draw and label a free-body diagram of the trailer.
   b. Draw and label a free-body diagram of the truck.
2. A physics book is at rest on a desk. Gravitational force pulls the book down. The desk exerts an upward force on the book that is equal in magnitude to the gravitational force. Draw a free-body diagram of the book.

SECTION REVIEW

1. List three examples of each of the following:
   a. a force causing an object to start moving
   b. a force causing an object to stop moving
   c. a force causing an object to change its direction of motion
2. Give two examples of field forces described in this section and two examples of contact forces you observe in everyday life. Explain why you think that these are forces.
3. What is the SI unit of force? What is this unit equivalent to in terms of fundamental units?
4. Why is force a vector quantity?
5. Draw a free-body diagram of a football being kicked. Assume that the only forces acting on the ball are the force due to gravity and the force exerted by the kicker.
6. **Interpreting Graphics** Study the force diagram shown in **Figure 3(a).** Redraw the diagram, and label each vector arrow with a description of the force. In each description, include the object exerting the force and the object on which the force is acting.

Newton's First Law

SECTION OBJECTIVES

- **Explain the relationship between the motion of an object and the net external force acting on the object.**
- **Determine the net external force on an object.**
- **Calculate the force required to bring an object into equilibrium.**

INERTIA

A hovercraft, such as the one in **Figure 4,** glides along the surface of the water on a cushion of air. A common misconception is that an object on which no force is acting will always be at rest. This situation is not always the case. If the hovercraft shown in **Figure 4** is moving at a constant velocity, then there is no net force acting on it. To see why this is the case, consider how a block will slide on different surfaces.

First, imagine a block on a deep, thick carpet. If you apply a force by pushing the block, the block will begin sliding, but soon after you remove the force, the block will come to rest. Next, imagine pushing the same block across a smooth, waxed floor. When you push with the same force, the block will slide much farther before coming to rest. In fact, a block sliding on a perfectly smooth surface would slide forever in the absence of an applied force.

In the 1630s, Galileo concluded correctly that it is an object's nature to *maintain its state of motion or rest.* Note that an object on which no force is acting is not necessarily at rest; the object could also be moving with a constant velocity. This concept was further developed by Newton in 1687 and has come to be known as **Newton's first law of motion.**

Figure 4
A hovercraft floats on a cushion of air above the water. Air provides less resistance to motion than water does.

NEWTON'S FIRST LAW

An object at rest remains at rest, and an object in motion continues in motion with constant velocity (that is, constant speed in a straight line) unless the object experiences a net external force.

Inertia is the tendency of an object not to accelerate. Newton's first law is often referred to as the *law of inertia* because it states that in the absence of a net force, a body will preserve its state of motion. In other words, Newton's first law says that *when the net external force on an object is zero, the object's acceleration (or the change in the object's velocity) is zero.*

inertia

the tendency of an object to resist being moved or, if the object is moving, to resist a change in speed or direction

Teaching Tip — BASIC

It takes force to make an object start moving or change direction. The more massive an object is, the larger the force that is required for a given change.

Demonstration

Inertia — BASIC

Purpose Help students develop a kinesthetic sense of inertia.

Materials physics book, calculator

Procedure Tell students that they will be able to feel the effects of inertia. First, tell them to hold the physics book upright between their hands, palms facing inward. Have them move the book from side to side (oscillating a distance of 30 cm) at regular time intervals. Tell the students to note the effort involved in changing the motion of the book. Repeat the demonstration with the calculator, and have students note the much smaller effort required.

Quick Lab

TEACHER'S NOTES

If students have trouble keeping the ball in place while accelerating the skateboard, they can tape a wooden block onto the skateboard to keep the ball from rolling off the back. Students should recognize that when the skateboard hits the wall, the ball continues moving forward as a result of its inertia.

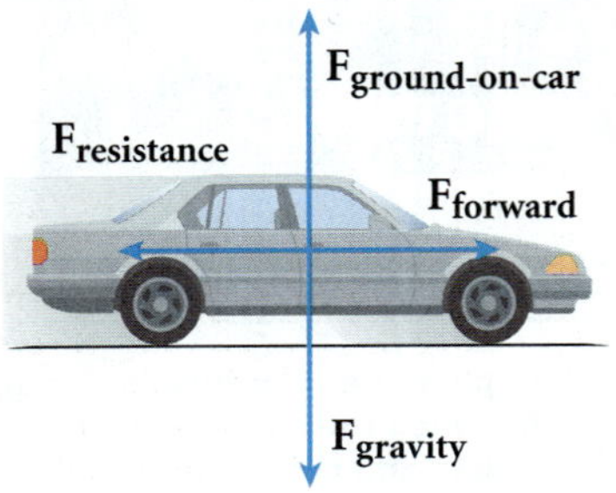

Figure 5
Although several forces are acting on this car, the vector sum of the forces is zero, so the car moves at a constant velocity.

net force

a single force whose external effects on a rigid body are the same as the effects of several actual forces acting on the body

The sum of forces acting on an object is the net force

Consider a car traveling at a constant velocity. Newton's first law tells us that the net external force on the car must be equal to zero. However, **Figure 5** shows that many forces act on a car in motion. The vector $\mathbf{F_{forward}}$ represents the forward force of the road on the tires. The vector $\mathbf{F_{resistance}}$, which acts in the opposite direction, is due partly to friction between the road surface and tires and is due partly to air resistance. The vector $\mathbf{F_{gravity}}$ represents the downward gravitational force on the car, and the vector $\mathbf{F_{ground\text{-}on\text{-}car}}$ represents the upward force that the road exerts on the car.

To understand how a car under the influence of so many forces can maintain a constant velocity, you must understand the distinction between external force and net external force. An *external force* is a single force that acts on an object as a result of the interaction between the object and its environment. All four forces in **Figure 5** are external forces acting on the car. The **net force** is the vector sum of all forces acting on an object.

When all external forces acting on an object are known, the net force can be found by using the methods for finding resultant vectors. The net force is equivalent to the one force that would produce the same effect on the object that all of the external forces combined would. Although four forces are acting on the car in **Figure 5,** the car will maintain its constant velocity as long as the vector sum of these forces is equal to zero.

Mass is a measure of inertia

Imagine a basketball and a bowling ball at rest side by side on the ground. Newton's first law states that both balls remain at rest as long as no net external force acts on them. Now, imagine supplying a net force by pushing each ball. If the two are pushed with equal force, the basketball will accelerate much more than the bowling ball. The bowling ball experiences a smaller acceleration because it has more inertia than the basketball does.

As the example of the bowling ball and the basketball shows, the inertia of an object is proportional to the object's mass. The greater the mass of a body, the less the body accelerates under an applied force. Similarly, a light object undergoes a larger acceleration than does a heavy object under the same force. Therefore, mass, which is a measure of the amount of matter in an object, is also a measure of the inertia of an object.

Quick Lab

Inertia

MATERIALS LIST

- **skateboard or cart**
- **toy balls with various masses**

SAFETY

Perform this experiment away from walls and furniture that can be damaged.

Place a small ball on the rear end of a skateboard or cart. Push the skateboard across the floor and into a wall. You may need to hold the ball in place while pushing the skateboard up to speed or accelerate the skateboard slowly so that friction holds the ball in place. Observe what happens to the ball when the skateboard hits the wall. Can you explain your observation in terms of inertia? Repeat the procedure using balls with different masses, and compare the results.

SAMPLE PROBLEM B

STRATEGY Determining Net Force

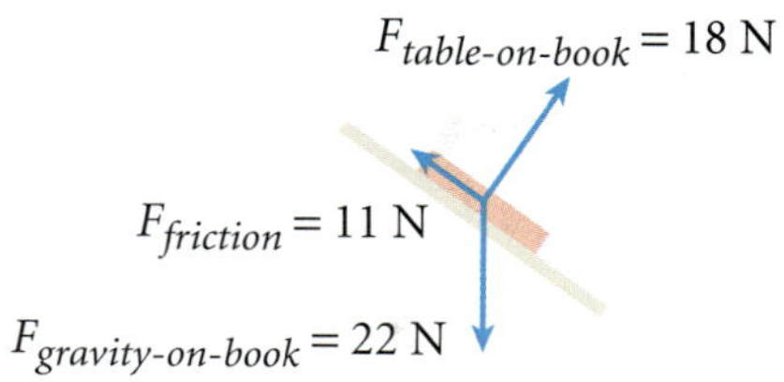

PROBLEM

Derek leaves his physics book on top of a drafting table that is inclined at a 35° angle. The free-body diagram at right shows the forces acting on the book. Find the net force acting on the book.

SOLUTION

1. Define the problem, and identify the variables.

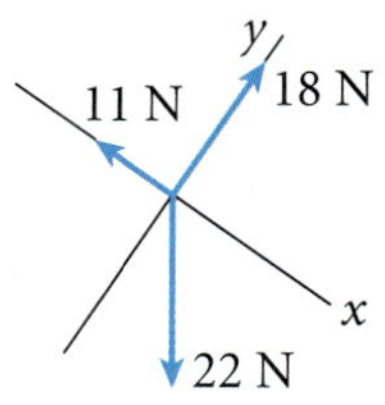

(a)

Given: $F_{gravity\text{-}on\text{-}book} = F_g = 22$ N

$F_{friction} = F_f = 11$ N

$F_{table\text{-}on\text{-}book} = F_t = 18$ N

Unknown: $F_{net} = ?$

2. Select a coordinate system, and apply it to the free-body diagram.

Choose the x-axis parallel to and the y-axis perpendicular to the incline of the table, as shown in **(a).** This coordinate system is the most convenient because only one force needs to be resolved into x and y components.

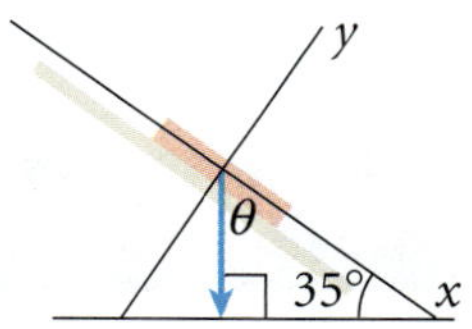

(b)

TIP *To simplify the problem, always choose the coordinate system in which as many forces as possible lie on the x- and y-axes.*

3. Find the x and y components of all vectors.

Draw a sketch, as shown in **(b),** to help find the components of the vector $\mathbf{F_g}$. The angle θ is equal to 180°– 90° – 35° = 55°.

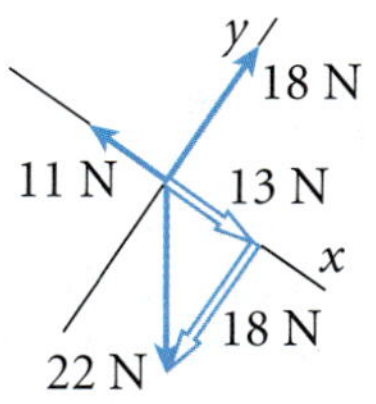

(c)

$$\cos\theta = \frac{F_{g,x}}{F_g} \qquad \sin\theta = \frac{F_{g,y}}{F_g}$$

$$F_{g,x} = F_g \cos\theta \qquad F_{g,y} = F_g \sin\theta$$

$$F_{g,x} = (22 \text{ N})(\cos 55°) = 13 \text{ N} \qquad F_{g,y} = (22 \text{ N})(\sin 55°) = 18 \text{ N}$$

Add both components to the free-body diagram, as shown in **(c).**

4. Find the net force in both the x and y directions.

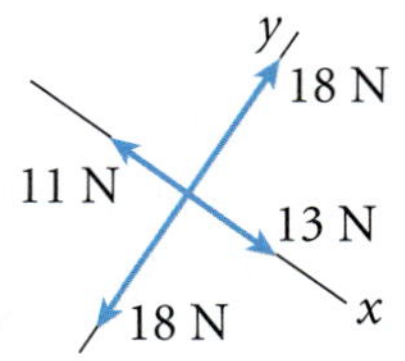

(d)

Diagram **(d)** shows another free-body diagram of the book, now with forces acting only along the x- and y-axes.

For the x direction:

$$\Sigma F_x = F_{g,x} - F_f$$

$$\Sigma F_x = 13 \text{ N} - 11 \text{ N} = 2 \text{ N}$$

For the y direction:

$$\Sigma F_y = F_t - F_{g,y}$$

$$\Sigma F_y = 18 \text{ N} - 18 \text{ N} = 0 \text{ N}$$

5. Find the net force.

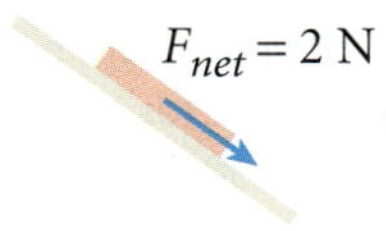

(e)

Add the net forces in the x and y directions together as vectors to find the total net force. In this case, $\mathbf{F_{net}} = 2$ N in the $+x$ direction, as shown in **(e).** Thus, the book accelerates down the incline.

Classroom Practice

Determining Net Force

An agriculture student is designing a support to keep a tree upright. Two wires have been attached to the tree and placed at right angles to each other. One wire exerts a force of 30.0 N on the tree; the other wire exerts a 40.0 N force. Determine where to place a third wire and how much force it should exert so that the net force acting on the tree is equal to zero.

Answer

50.0 N at 143° from the 40.0 N force and at 127° from the 30.0 N force

A flying, stationary kite is acted on by a force of 9.8 N downward. The wind exerts a force of 45 N at an angle of 50.0° above the horizontal. Find the force that the string exerts on the kite.

Answer

38 N, 40° below the horizontal

Teaching Tip

Point out that rather than calculate the magnitude of the y component of gravitational force, you could simply deduce that this force must be equal and opposite to the force of the table on the book because there is no acceleration in the y direction (the book stays on the table).

SECTION 2

ANSWERS

Practice B

1. $F_x = 60.6$ N; $F_y = 35.0$ N
2. 2.48 N at 25.0° counterclockwise from straight down
3. 557 N at 35.7° west of north

PROBLEM GUIDE B

Use this guide to assign problems.
SE = Student Edition Textbook
PW = Problem Workbook
PB = Problem Bank on the One-Stop Planner (OSP)

Solving for:

F_x, F_y	SE	Sample, 1; Ch. Rvw. 11–12
	PW	3, 4*, 5*
	PB	7–10
F_{net}	SE	Sample, 2–3; Ch. Rvw. 10, 22a*
	PW	Sample, 1– 2
	PB	1–6

***Challenging Problem**
Consult the printed Solutions Manual or the OSP for detailed solutions.

Why it Matters

Seat Belts

Many students have difficulty visualizing how mechanical devices operate. You may want to further describe how the seat belt mechanism works so that students fully benefit from this illustration. Ask students: Which way will the rod turn (clockwise or counterclockwise) if the car comes to an abrupt stop? *(clockwise)*

PRACTICE B

Determining Net Force

1. A man is pulling on his dog with a force of 70.0 N directed at an angle of +30.0° to the horizontal. Find the x and y components of this force.
2. A gust of wind blows an apple from a tree. As the apple falls, the gravitational force on the apple is 2.25 N downward, and the force of the wind on the apple is 1.05 N to the right. Find the magnitude and direction of the net force on the apple.
3. The wind exerts a force of 452 N north on a sailboat, while the water exerts a force of 325 N west on the sailboat. Find the magnitude and direction of the net force on the sailboat.

TIP *If there is a net force in both the x and y directions, use vector addition to find the total net force.*

Why it Matters

Seat Belts

The purpose of a seat belt is to prevent serious injury by holding a passenger firmly in place in the event of a collision. A seat belt may also lock when a car rapidly slows down or turns a corner. While inertia causes passengers in a car to continue moving forward as the car slows down, inertia also causes seat belts to lock into place.

The illustration shows how one type of shoulder harness operates. Under normal conditions, the ratchet turns freely to allow the harness to wind or unwind along the pulley. In a collision, the car undergoes a large acceleration and rapidly comes to rest. Because of its inertia, the large mass under the seat continues to slide forward along the tracks, in the direction indicated by the arrow. The pin connection between the mass and the rod causes the rod to pivot and lock the ratchet wheel in place. At this point, the harness no longer unwinds, and the seat belt holds the passenger firmly in place.

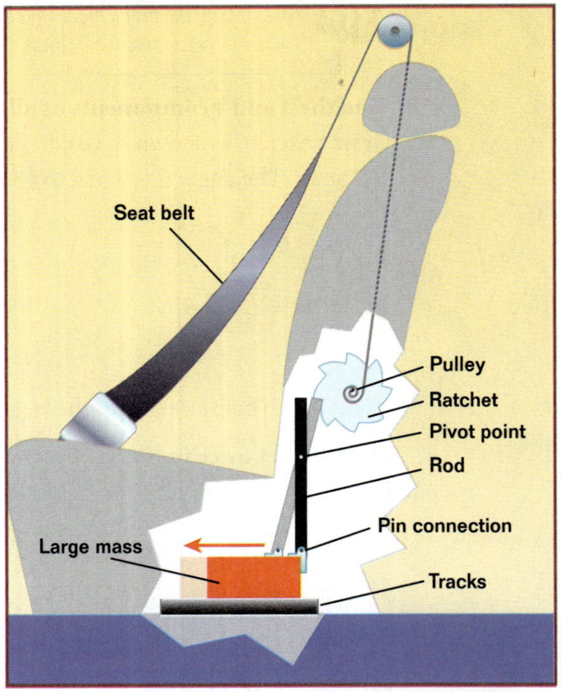

When the car suddenly slows down, inertia causes the large mass under the seat to continue moving, which activates the lock on the safety belt.

EQUILIBRIUM

Objects that are either at rest or moving with constant velocity are said to be in **equilibrium.** Newton's first law describes objects in equilibrium, whether they are at rest or moving with a constant velocity. Newton's first law states one condition that must be true for equilibrium: the net force acting on a body in equilibrium must be equal to zero.

equilibrium

the state in which the net force on an object is zero

The net force on the fishing bob in **Figure 6(a)** is equal to zero because the bob is at rest. Imagine that a fish bites the bait, as shown in **Figure 6(b).** Because a net force is acting on the line, the bob accelerates toward the hooked fish.

Now, consider a different scenario. Suppose that at the instant the fish begins pulling on the line, the person reacts by applying a force to the bob that is equal and opposite to the force exerted by the fish. In this case, the net force on the bob remains zero, as shown in **Figure 6(c),** and the bob remains at rest. In this example, the bob is at rest while in equilibrium, but an object can also be in equilibrium while moving at a constant velocity.

An object is in equilibrium when the vector sum of the forces acting on the object is equal to zero. To determine whether a body is in equilibrium, find the net force, as shown in Sample Problem B. If the net force is zero, the body is in equilibrium. If there is a net force, a second force equal and opposite to this net force will put the body in equilibrium.

(a)

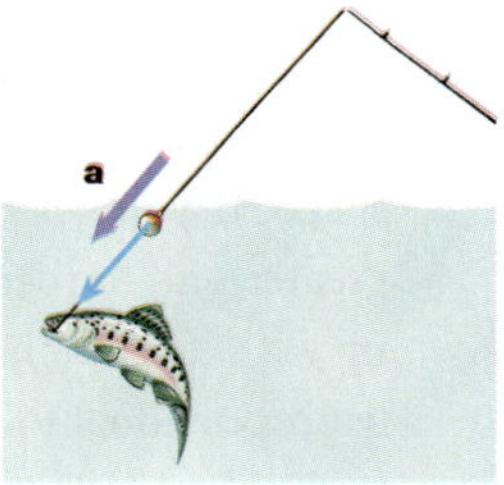

(b)

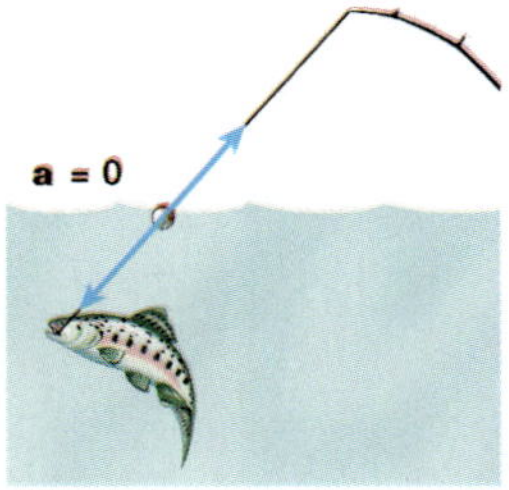

(c)

Figure 6
(a) The bob on this fishing line is at rest. **(b)** When the bob is acted on by a net force, it accelerates. **(c)** If an equal and opposite force is applied, the net force remains zero.

SECTION REVIEW

1. If a car is traveling westward with a constant velocity of 20 m/s, what is the net force acting on the car?
2. If a car is accelerating downhill under a net force of 3674 N, what additional force would cause the car to have a constant velocity?
3. The sensor in the torso of a crash-test dummy records the magnitude and direction of the net force acting on the dummy. If the dummy is thrown forward with a force of 130.0 N while simultaneously being hit from the side with a force of 4500.0 N, what force will the sensor report?
4. What force will the seat belt have to exert on the dummy in item 3 to hold the dummy in the seat?
5. **Critical Thinking** Can an object be in equilibrium if only one force acts on the object?

SECTION 2

Visual Strategy

Figure 6
Point out that in order for the bob to be in equilibrium, all the forces must cancel. You may want to diagram this situation on the board and include the force of the water on the bob (buoyant force).

Q Other than the forces applied by the person and the fish, do any other forces act on the bob?

A ***yes, the upward (buoyant) force of the water on the bob and the downward gravitational force***

SECTION REVIEW ANSWERS

1. zero
2. –3674 N
3. 4502 N at 1.655° forward of the side
4. the same magnitude as the net force in item 3 but in the opposite direction
5. No, either no force or two or more forces are required for equilibrium.

SECTION 3

Newton's Second and Third Laws

SECTION OBJECTIVES

- **Describe an object's acceleration in terms of its mass and the net force acting on it.**
- **Predict the direction and magnitude of the acceleration caused by a known net force.**
- **Identify action-reaction pairs.**

NEWTON'S SECOND LAW

From Newton's first law, we know that an object with no net force acting on it is in a state of equilibrium. We also know that an object experiencing a net force undergoes a change in its velocity. But exactly how much does a known force affect the motion of an object?

Force is proportional to mass and acceleration

Imagine pushing a stalled car through a level intersection, as shown in **Figure 7.** Because a net force causes an object to accelerate, the speed of the car will increase. When you push the car by yourself, however, the acceleration will be so small that it will take a long time for you to notice an increase in the car's speed. If you get several friends to help you, the net force on the car is much greater, and the car will soon be moving so fast that you will have to run to keep up with it. This change happens because the acceleration of an object is directly proportional to the net force acting on the object. (Note that this is an idealized example that disregards any friction forces that would hinder the motion. In reality, the car accelerates when the push is greater than the frictional force. However, when the force exerted by the pushers equals the frictional force, the net force becomes zero, and the car moves at a constant velocity.)

Experience reveals that the mass of an object also affects the object's acceleration. A lightweight car accelerates more than a heavy truck if the same force is applied to both. Thus, it requires less force to accelerate a low-mass object than it does to accelerate a high-mass object at the same rate.

Figure 7
(a) A small force on an object causes a small acceleration, but **(b)** a larger force causes a larger acceleration.

Newton's second law relates force, mass, and acceleration

The relationships between mass, force, and acceleration are quantified in **Newton's second law.**

> **NEWTON'S SECOND LAW**
>
> **The acceleration of an object is directly proportional to the net force acting on the object and inversely proportional to the object's mass.**

According to Newton's second law, if equal forces are applied to two objects of different masses, the object with greater mass will experience a smaller acceleration, and the object with less mass will experience a greater acceleration.

In equation form, we can state Newton's law as follows:

> **NEWTON'S SECOND LAW**
>
> $$\Sigma \mathbf{F} = m\mathbf{a}$$
>
> net force = mass × acceleration

In this equation, **a** is the acceleration of the object and m is the object's mass. Note that Σ is the Greek capital letter *sigma,* which represents the sum of the quantities that come after it. In this case, **ΣF** represents the *vector sum of all external forces acting on the object,* or the net force.

SAMPLE PROBLEM C

Newton's Second Law

PROBLEM

Roberto and Laura are studying across from each other at a wide table. Laura slides a 2.2 kg book toward Roberto. If the net force acting on the book is 1.6 N to the right, what is the book's acceleration?

SOLUTION

Given: $m = 2.2$ kg

$\mathbf{F}_{net} = \Sigma\mathbf{F} = 1.6$ N to the right

Unknown: $\mathbf{a} = ?$

Use Newton's second law, and solve for **a.**

$$\Sigma\mathbf{F} = m\mathbf{a}, \text{ so } \mathbf{a} = \frac{\Sigma\mathbf{F}}{m}$$

$$a = \frac{1.6 \text{ N}}{2.2 \text{ kg}} = 0.73 \text{ m/s}^2$$

$$\boxed{\mathbf{a} = 0.73 \text{ m/s}^2 \text{ to the right}}$$

TIP *If more than one force is acting on an object, you must find the net force as shown in Sample Problem B before applying Newton's second law. The acceleration will be in the direction of the net force.*

SECTION 3

Classroom Practice

Newton's Second Law

Space-shuttle astronauts experience accelerations of about 35 m/s^2 during takeoff. What force does a 75 kg astronaut experience during an acceleration of this magnitude?

Answer

2600 N

An 8.5 kg bowling ball initially at rest is dropped from the top of an 11 m building. The ball hits the ground 1.5 s later. Find the net force on the falling ball.

Answer

83 N

PROBLEM GUIDE C

Use this guide to assign problems.
SE = Student Edition Textbook
PW = Problem Workbook
PB = Problem Bank on the One-Stop Planner (OSP)

Solving for:

a	SE	Sample, 1–3; Ch. Rvw. 20, 22b*, 42a, 44a, 45a*, 48a*
	PW	2b, 9–11, 12*, 13b*, 14b
	PB	7–10
F_{net}	SE	5*; Ch. Rvw. 19, 21, 40*, 41, 42b*, 43, 44b, 45a*, 50*, 51
	PW	Sample, 1–2a, 3–5, 6*, 7*, 8*, 13a*
	PB	4–6
m	SE	4
	PW	14a
	PB	Sample, 1–3

***Challenging Problem**
Consult the printed Solutions Manual or the OSP for detailed solutions.

ANSWERS

Practice C

1. 2.2 m/s^2 forward
2. 1.4 m/s^2 north
3. 4.50 m/s^2 to the east
4. 2.1 kg
5. 14 N

PRACTICE C

Newton's Second Law

1. The net force on the propeller of a 3.2 kg model airplane is 7.0 N forward. What is the acceleration of the airplane?
2. The net force on a golf cart is 390 N north. If the cart has a total mass of 270 kg, what are the magnitude and direction of the cart's acceleration?
3. A car has a mass of 1.50×10^3 kg. If the force acting on the car is 6.75×10^3 N to the east, what is the car's acceleration?
4. A soccer ball kicked with a force of 13.5 N accelerates at 6.5 m/s^2 to the right. What is the mass of the ball?
5. A 2.0 kg otter starts from rest at the top of a muddy incline 85 cm long and slides down to the bottom in 0.50 s. What net force acts on the otter along the incline?

TIP *For some problems, it may be easier to use the equation for Newton's second law twice: once for all of the forces acting in the x direction ($\Sigma F_x = ma_x$) and once for all of the forces acting in the y direction ($\Sigma F_y = ma_y$). If the net force in both directions is zero, then **a** = 0, which corresponds to the equilibrium situation in which **v** is either constant or zero.*

Why it Matters

Conceptual Challenge

1. Gravity and Rocks

The force due to gravity is twice as great on a 2 kg rock as it is on a 1 kg rock. Why doesn't the 2 kg rock have a greater free-fall acceleration?

2. Leaking Truck

A truck loaded with sand accelerates at 0.5 m/s^2 on the highway. If the driving force on the truck remains constant, what happens to the truck's acceleration if sand leaks at a constant rate from a hole in the truck bed?

ANSWERS

Conceptual Challenge

1. A greater force acts on the heavier rock, but the heavier rock also has greater mass, so the acceleration is the same. Free-fall acceleration is independent of mass.
2. The acceleration will increase as the mass decreases.

NEWTON'S THIRD LAW

A force is exerted on an object when that object interacts with another object in its environment. Consider a moving car colliding with a concrete barrier. The car exerts a force on the barrier at the moment of collision. Furthermore, the barrier exerts a force on the car so that the car rapidly slows down after coming into contact with the barrier. Similarly, when your hand applies a force to a door to push it open, the door simultaneously exerts a force back on your hand.

Forces always exist in pairs

From examples like those discussed in the previous paragraph, Newton recognized that a single isolated force cannot exist. Instead, *forces always exist in pairs.* The car exerts a force on the barrier, and at the same time, the barrier exerts a force on the car. Newton described this type of situation with his **third law of motion.**

NEWTON'S THIRD LAW

If two objects interact, the magnitude of the force exerted on object 1 by object 2 is equal to the magnitude of the force simultaneously exerted on object 2 by object 1, and these two forces are opposite in direction.

SCILINKS®
www.scilinks.org
Topic: Newton's Laws
Code: HF61028

An alternative statement of this law is that *for every action, there is an equal and opposite reaction.* When two objects interact with one another, the forces that the objects exert on each other are called an *action-reaction pair.* The force that object 1 exerts on object 2 is sometimes called the *action force,* while the force that object 2 exerts on object 1 is called the *reaction force.* The action force is equal in magnitude and opposite in direction to the reaction force. The terms *action* and *reaction* sometimes cause confusion because they are used a little differently in physics than they are in everyday speech. In everyday speech, the word *reaction* is used to refer to something that happens *after* and *in response to* an event. In physics, however, the reaction force occurs at exactly the same time as the action force.

Because the action and reaction forces coexist, either force can be called the action or the reaction. For example, you could call the force that the car exerts on the barrier the action and the force that the barrier exerts on the car the reaction. Likewise, you could choose to call the force that the barrier exerts on the car the action and the force that the car exerts on the barrier the reaction.

Action and reaction forces each act on different objects

One important thing to remember about action-reaction pairs is that each force acts on a different object. Consider the task of driving a nail into wood, as illustrated in **Figure 8.** To accelerate the nail and drive it into the wood, the hammer exerts a force on the nail. According to Newton's third law, the nail exerts a force on the hammer that is equal to the magnitude of the force that the hammer exerts on the nail.

Figure 8
The force that the nail exerts on the hammer is equal and opposite to the force that the hammer exerts on the nail.

The concept of action-reaction pairs is a common source of confusion because some people assume incorrectly that the equal and opposite forces balance one another and make any change in the state of motion impossible. If the force that the nail exerts on the hammer is equal to the force the hammer exerts on the nail, why doesn't the nail remain at rest?

The motion of the nail is affected only by the forces acting on the nail. To determine whether the nail will accelerate, draw a free-body diagram to isolate the forces acting on the nail, as shown in **Figure 9.** The force of the nail on the hammer is not included in the diagram because it does not act on the nail. According to the diagram, the nail will be driven into the wood because there is a net force acting on the nail. Thus, action-reaction pairs do not imply that the net force on either object is zero. The action-reaction forces are equal and opposite, but either object may still have a net force acting on it.

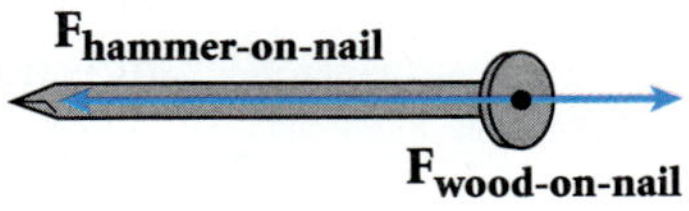

Figure 9
The net force acting on the nail drives the nail into the wood.

SECTION 3

Misconception Alert — GENERAL

It is important to clear up any misconception that action and reaction forces cancel each other. One way to reinforce the true nature of Newton's third law is to use free-body diagrams. On the board, draw separate free-body diagrams for two or more interacting objects, such as a book on a table. Identify the third-law pairs, and point out that the force arrows are on separate bodies. The motion of the book is affected only by forces on the book. The motion of the table is affected only by forces on the table. Have students practice drawing free-body diagrams for multiple objects, building up levels of complexity with each new diagram (for example, a book on an inclined plane on a table on Earth).

Field forces also exist in pairs

Newton's third law also applies to field forces. For example, consider the gravitational force exerted by Earth on an object. During calibration at the crash-test site, engineers calibrate the sensors in the heads of crash-test dummies by removing the heads and dropping them from a known height.

The force that Earth exerts on a dummy's head is $\mathbf{F_g}$. Let's call this force the action. What is the reaction? Because $\mathbf{F_g}$ is the force exerted on the falling head by Earth, the reaction to $\mathbf{F_g}$ is the force exerted on Earth by the falling head.

According to Newton's third law, the force of the dummy on Earth is equal to the force of Earth on the dummy. Thus, as a falling object accelerates toward Earth, Earth also accelerates toward the object.

The thought that Earth accelerates toward the dummy's head may seem to contradict our experience. One way to make sense of this idea is to refer to Newton's second law. The mass of Earth is much greater than that of the dummy's head. Therefore, while the dummy's head undergoes a large acceleration due to the force of Earth, the acceleration of Earth due to this reaction force is negligibly small because of Earth's enormous mass.

extension

Integrating Technology

Visit go.hrw.com for the activity "Car Seat Safety."

Keyword HF6FORX

SECTION REVIEW

1. A 6.0 kg object undergoes an acceleration of 2.0 m/s^2.

a. What is the magnitude of the net force acting on the object?

b. If this same force is applied to a 4.0 kg object, what acceleration is produced?

2. A child causes a wagon to accelerate by pulling it with a horizontal force. Newton's third law says that the wagon exerts an equal and opposite force on the child. How can the wagon accelerate? (Hint: Draw a free-body diagram for each object.)

3. Identify the action-reaction pairs in the following situations:

a. A person takes a step.

b. A snowball hits someone in the back.

c. A baseball player catches a ball.

d. A gust of wind strikes a window.

4. The forces acting on a sailboat are 390 N north and 180 N east. If the boat (including crew) has a mass of 270 kg, what are the magnitude and direction of the boat's acceleration?

5. Critical Thinking If a small sports car collides head-on with a massive truck, which vehicle experiences the greater impact force? Which vehicle experiences the greater acceleration? Explain your answers.

SECTION REVIEW ANSWERS

1. **a.** 12 N
b. 3.0 m/s^2

2. The reaction force acts on the child, not on the wagon itself, so there is still a net force on the wagon.

3. **a.** person pushes on ground; ground pushes on person
b. snowball exerts force on back; back exerts force on snowball
c. ball exerts force on glove; glove exerts force on ball
d. wind exerts force on window; window exerts force on wind

4. 1.6 m/s^2 at an angle of 65° north of east

5. Each impact force has the same magnitude. The sports car experiences the larger acceleration because it has a smaller mass, and acceleration is inversely proportional to mass.

Everyday Forces

SECTION OBJECTIVES

- **Explain the difference between mass and weight.**
- **Find the direction and magnitude of normal forces.**
- **Describe air resistance as a form of friction.**
- **Use coefficients of friction to calculate frictional force.**

WEIGHT

How do you know that a bowling ball weighs more than a tennis ball? If you imagine holding one ball in each hand, you can imagine the downward forces acting on your hands. Because the bowling ball has more mass than the tennis ball does, gravitational force pulls more strongly on the bowling ball. Thus, the bowling ball pushes your hand down with more force than the tennis ball does.

The gravitational force exerted on the ball by Earth, $\mathbf{F_g}$, is a vector quantity, directed toward the center of Earth. The magnitude of this force, F_g, is a scalar quantity called **weight.** The weight of an object can be calculated using the equation $F_g = ma_g$, where a_g is the magnitude of the acceleration due to gravity, or free-fall acceleration. On the surface of Earth, $a_g = g$, and $F_g = mg$. In this book, $g = 9.81\ \text{m/s}^2$ unless otherwise specified.

weight

a measure of the gravitational force exerted on an object; its value can change with the location of the object in the universe

Weight, unlike mass, is not an inherent property of an object. Because it is equal to the magnitude of the force due to gravity, weight depends on location. For example, if the astronaut in **Figure 10** weighs 800 N (180 lb) on Earth, he would weigh only about 130 N (30 lb) on the moon. As you will see in the chapter "Circular Motion and Gravitation," the value of a_g on the surface of a planet depends on the planet's mass and radius. On the moon, a_g is about $1.6\ \text{m/s}^2$—much smaller than $9.81\ \text{m/s}^2$.

Even on Earth, an object's weight may vary with location. Objects weigh less at higher altitudes than they do at sea level because the value of a_g decreases as distance from the surface of Earth increases. The value of a_g also varies slightly with changes in latitude.

Figure 10
On the moon, astronauts weigh much less than they do on Earth.

THE NORMAL FORCE

Imagine a television set at rest on a table. We know that the gravitational force is acting on the television. How can we use Newton's laws to explain why the television does not continue to fall toward the center of Earth?

An analysis of the forces acting on the television will reveal the forces that are in equilibrium. First, we know that the gravitational force of Earth, $\mathbf{F_g}$, is acting downward. Because the television is in equilibrium, we know that another force, equal in magnitude to $\mathbf{F_g}$ but in the opposite direction, must be acting on it. This force is the force exerted on the television by the table. This force is called the **normal force,** $\mathbf{F_n}$.

normal force

a force that acts on a surface in a direction perpendicular to the surface

Visual Strategy ADVANCED

Figure 10
Point out that it is easier to lift a massive object on the moon than on Earth because the object weighs less on the moon, even though its mass remains the same. Also, point out that an object's inertia is the same regardless of the magnitude of free-fall acceleration.

Q Will a dart shot from a dart gun go farther horizontally on Earth or on the moon? Disregard air resistance.

A ***The dart will travel farther on the moon. Because the dart is accelerated downward more slowly on the moon than on Earth, it is in motion for a longer time on the moon. The horizontal velocity will be the same in each case.***

Teaching Tip

For most practical purposes, the gravitational field near the surface of Earth is constant. For example, a person who weighs 180 lb at sea level would weigh 179.5 lb at an altitude of 9 km above sea level. In this example, the difference in weight is only 0.3 percent.

SECTION 4

Visual Strategy GENERAL

Figure 11

Tell students that the television is in equilibrium, so the normal force from the table must be equal in magnitude and opposite in direction to the gravitational force exerted on the television.

Instruct students to draw free-body diagrams for the television, table, and Earth and to identify the third-law pairs.

Q Do the forces in **Figure 11** constitute an action-reaction pair (Newton's third law)?

A ***No, both forces act on the television and therefore cannot be an action-reaction pair.***

Teaching Tip

Be sure students understand why the two angles labeled θ in **Figure 12** are equal. Recognizing equal angles in free-body diagrams is an important problem-solving skill for this chapter.

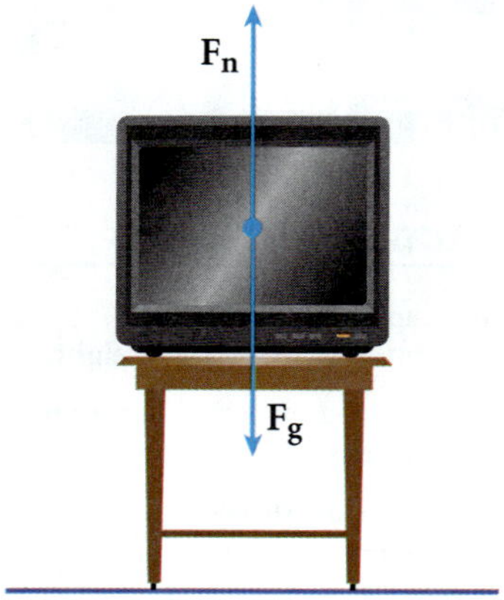

Figure 11
In this example, the normal force, $\mathbf{F_n}$, is equal and opposite to the force due to gravity, $\mathbf{F_g}$.

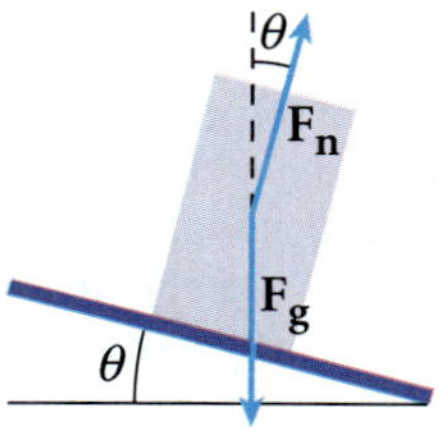

Figure 12
The normal force is not always opposite the force due to gravity, as shown by this example of a refrigerator on a loading ramp.

static friction

the force that resists the initiation of sliding motion between two surfaces that are in contact and at rest

The word *normal* is used because the direction of the contact force is perpendicular to the table surface and one meaning of the word *normal* is "perpendicular." **Figure 11** shows the forces acting on the television.

The normal force is always perpendicular to the contact surface but is not always opposite in direction to the force due to gravity. **Figure 12** shows a free-body diagram of a refrigerator on a loading ramp. The normal force is perpendicular to the ramp, not directly opposite the force due to gravity. In the absence of other forces, the normal force, $\mathbf{F_n}$, is equal and opposite to the component of $\mathbf{F_g}$ that is perpendicular to the contact surface. The magnitude of the normal force can be calculated as $F_n = mg \cos \theta$. The angle θ is the angle between the normal force and a vertical line and is also the angle between the contact surface and a horizontal line.

THE FORCE OF FRICTION

Consider a jug of juice at rest (in equilibrium) on a table, as in **Figure 13(a).** We know from Newton's first law that the net force acting on the jug is zero. Newton's second law tells us that any additional unbalanced force applied to the jug will cause the jug to accelerate and to remain in motion unless acted on by another force. But experience tells us that the jug will not move at all if we apply a very small horizontal force. Even when we apply a force large enough to move the jug, the jug will stop moving almost as soon as we remove this applied force.

Friction opposes the applied force

When the jug is at rest, the only forces acting on it are the force due to gravity and the normal force exerted by the table. These forces are equal and opposite, so the jug is in equilibrium. When you push the jug with a small horizontal force **F**, as shown in **Figure 13(b),** the table exerts an equal force in the opposite direction. As a result, the jug remains in equilibrium and therefore also remains at rest. The resistive force that keeps the jug from moving is called the force of **static friction,** abbreviated as $\mathbf{F_s}$.

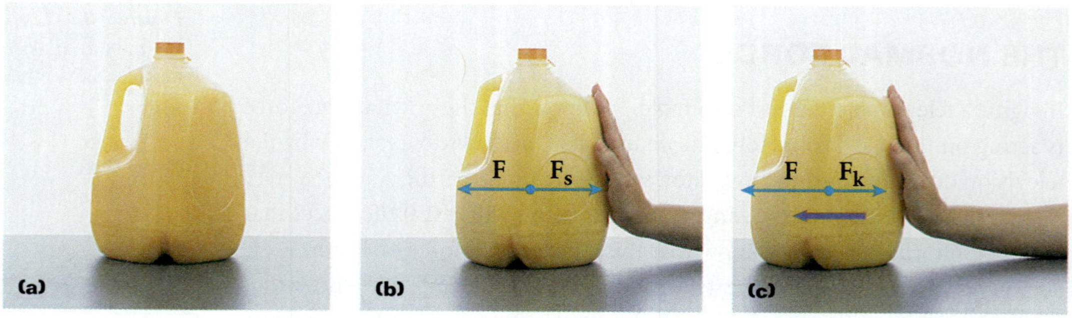

Figure 13 **(a)** Because this jug of juice is in equilibrium, any unbalanced horizontal force applied to it will cause the jug to accelerate. **(b)** When a small force is applied, the jug remains in equilibrium because the static-friction force is equal but opposite to the applied force. **(c)** The jug begins to accelerate as soon as the applied force exceeds the opposing static-friction force.

As long as the jug does not move, the force of static friction is always equal to and opposite in direction to the component of the applied force that is parallel to the surface ($F_s = -F_{applied}$). As the applied force increases, the force of static friction also increases; if the applied force decreases, the force of static friction also decreases. When the applied force is as great as it can be without causing the jug to move, the force of static friction reaches its maximum value, $\mathbf{F_{s,max}}$.

Kinetic friction is less than static friction

When the applied force on the jug exceeds $\mathbf{F_{s,max}}$, the jug begins to move with an acceleration to the left, as shown in **Figure 13(c).** A frictional force is still acting on the jug as the jug moves, but that force is actually less than $\mathbf{F_{s,max}}$. The retarding frictional force on an object in motion is called the force of **kinetic friction** **($\mathbf{F_k}$).** The magnitude of the net force acting on the object is equal to the difference between the applied force and the force of kinetic friction ($F_{applied} - F_k$).

kinetic friction

the force that opposes the movement of two surfaces that are in contact and are sliding over each other

At the microscopic level, frictional forces arise from complex interactions between contacting surfaces. Most surfaces, even those that seem very smooth to the touch, are actually quite rough at the microscopic level, as illustrated in **Figure 14.** Notice that the surfaces are in contact at only a few points. When two surfaces are stationary with respect to each other, the surfaces stick together somewhat at the contact points. This *adhesion* is caused by electrostatic forces between molecules of the two surfaces.

TIP *In free-body diagrams, the force of friction is always parallel to the surface of contact. The force of kinetic friction is always opposite the direction of motion. To determine the direction of the force of static friction, use the principle of equilibrium. For an object in equilibrium, the frictional force must point in the direction that results in a net force of zero.*

The force of friction is proportional to the normal force

It is easier to push a chair across the floor at a constant speed than to push a heavy desk across the floor at the same speed. Experimental observations show that the magnitude of the force of friction is approximately proportional to the magnitude of the normal force that a surface exerts on an object. Because the desk is heavier than the chair, the desk also experiences a greater normal force and therefore greater friction.

Friction can be calculated approximately

Keep in mind that the force of friction is really a macroscopic effect caused by a complex combination of forces at a microscopic level. However, we can approximately calculate the force of friction with certain assumptions. The relationship between normal force and the force of friction is one factor that affects friction. For instance, it is easier to slide a light textbook across a desk than it is to slide a heavier textbook. The relationship between the normal force and the force of friction provides a good approximation for the friction between dry, flat surfaces that are at rest or sliding past one another.

Figure 14
On the microscopic level, even very smooth surfaces make contact at only a few points.

Demonstration

Static vs. Kinetic Friction — GENERAL

Purpose Show that kinetic friction is less than static friction.

Materials rectangular block, hook, spring scale

Procedure Use the spring scale to measure the force required to start the rectangular block moving. Then, use the spring scale to measure the frictional force for constant velocity. Perform several trials. Have students record all data and find the average for each. Point out that the normal force and the surfaces remain the same, so the only difference in the two average values is due to motion.

Demonstration

Friction of Different Surfaces — ADVANCED

Purpose Show students that the force of friction depends on the surface.

Materials large cube with different materials (such as glass, carpeting, and sandpaper) covering each of four sides, with two sides left uncovered; hook; spring scale

Procedure Attach the hook to one of the two uncovered sides of the block. Pull the block across the table with the spring scale. Repeat the demonstration with a new surface of the cube exposed to the table. Repeat the demonstration for the two remaining covered sides. Have students summarize the results and reach a conclusion concerning the nature of the surfaces in contact and the frictional force.

Demonstration

Friction and Surface Area ADVANCED

Purpose Show the relation between surface area and frictional forces.

Materials rectangular block, hook, spring scale

Procedure Attach the hook to the block. Pull the block across the table with the spring scale. Have students note the force required to pull the block at a constant velocity. Repeat the demonstration for another surface area in contact with the table, and have students note the force. Ask students to summarize the results and to reach a conclusion concerning the areas in contact and frictional forces.

Visual Strategy ADVANCED

Figure 15
Remind students that frictional force depends on the coefficient of friction and the normal force.

Q What changes in environment might cause a change in the frictional force experienced by the snowboarder on the way down the hill?

A *Answers will vary but could include the following: surface conditions (such as wet or dry snow, ice, and dirt), whether the snowboarder is moving or not moving, and the angle of the hill.*

The force of friction also depends on the composition and qualities of the surfaces in contact. For example, it is easier to push a desk across a tile floor than across a floor covered with carpet. Although the normal force on the desk is the same in both cases, the force of friction between the desk and the carpet is higher than the force of friction between the desk and the tile. The quantity that expresses the dependence of frictional forces on the particular surfaces in contact is called the **coefficient of friction.** The coefficient of friction between a waxed snowboard and the snow will affect the acceleration of the snowboarder shown in **Figure 15.** The coefficient of friction is represented by the symbol μ, the lowercase Greek letter *mu.*

coefficient of friction

the ratio of the magnitude of the force of friction between two objects in contact to the magnitude of the normal force with which the objects press against each other

The coefficient of friction is a ratio of forces

The coefficient of friction is defined as the ratio of the force of friction to the normal force between two surfaces. The *coefficient of kinetic friction* is the ratio of the force of kinetic friction to the normal force.

$$\mu_k = \frac{F_k}{F_n}$$

The *coefficient of static friction* is the ratio of the maximum value of the force of static friction to the normal force.

$$\mu_s = \frac{F_{s,max}}{F_n}$$

If the value of μ and the normal force on the object are known, then the magnitude of the force of friction can be calculated directly.

$$F_f = \mu F_n$$

Figure 15
Snowboarders wax their boards to minimize the coefficient of friction between the boards and the snow.

Table 2 shows some experimental values of μ_s and μ_k for different materials. Because kinetic friction is less than or equal to the maximum static friction, the coefficient of kinetic friction is always less than or equal to the coefficient of static friction.

Table 2 Coefficients of Friction (Approximate Values)

	μ_s	μ_k		μ_s	μ_k
steel on steel	0.74	0.57	waxed wood on wet snow	0.14	0.1
aluminum on steel	0.61	0.47	waxed wood on dry snow	—	0.04
rubber on dry concrete	1.0	0.8	metal on metal (lubricated)	0.15	0.06
rubber on wet concrete	—	0.5	ice on ice	0.1	0.03
wood on wood	0.4	0.2	Teflon on Teflon	0.04	0.04
glass on glass	0.9	0.4	synovial joints in humans	0.01	0.003

SAMPLE PROBLEM D

Coefficients of Friction

PROBLEM

A 24 kg crate initially at rest on a horizontal floor requires a 75 N horizontal force to set it in motion. Find the coefficient of static friction between the crate and the floor.

SOLUTION

Given: $F_{s,max} = F_{applied} = 75 \text{ N}$ $m = 24 \text{ kg}$

Unknown: $\mu_s = ?$

Use the equation for the coefficient of static friction.

$$\mu_s = \frac{F_{s,max}}{F_n} = \frac{F_{s,max}}{mg}$$

$$\mu_s = \frac{75 \text{ N}}{24 \text{ kg} \times 9.81 \text{ m/s}^2}$$

$$\boxed{\mu_s = 0.32}$$

TIP *Because the crate is on a horizontal surface, the magnitude of the normal force (F_n) equals the crate's weight (mg).*

PRACTICE D

Coefficients of Friction

1. Once the crate in Sample Problem D is in motion, a horizontal force of 53 N keeps the crate moving with a constant velocity. Find μ_k, the coefficient of kinetic friction, between the crate and the floor.

2. A 25 kg chair initially at rest on a horizontal floor requires a 165 N horizontal force to set it in motion. Once the chair is in motion, a 127 N horizontal force keeps it moving at a constant velocity.

a. Find the coefficient of static friction between the chair and the floor.
b. Find the coefficient of kinetic friction between the chair and the floor.

3. A museum curator moves artifacts into place on various different display surfaces. Use the values in **Table 2** to find $F_{s,max}$ and F_k for the following situations:

a. moving a 145 kg aluminum sculpture across a horizontal steel platform
b. pulling a 15 kg steel sword across a horizontal steel shield
c. pushing a 250 kg wood bed on a horizontal wood floor
d. sliding a 0.55 kg glass amulet on a horizontal glass display case

Classroom Practice

Coefficients of Friction
A refrigerator is placed on a ramp. The refrigerator begins to slide when the ramp is raised to an angle of 34°. What is the coefficient of static friction?

Answer 0.67

PROBLEM GUIDE D

Use this guide to assign problems.
SE = Student Edition Textbook
PW = Problem Workbook
PB = Problem Bank on the One-Stop Planner (OSP)

Solving for:

μ	**SE** Sample, 1–2; Ch. Rvw. 35, 36*, 37*, 49 **PW** 4–7, 10* **PB** 8–10
F_f	**SE** 3 **PW** Sample, 1–3, 7, 10* **PB** 5–7
F_n, m	**PW** 8–9 **PB** Sample, 1–4

***Challenging Problem**
Consult the printed Solutions Manual or the OSP for detailed solutions.

ANSWERS

Practice D

1. 0.23
2. **a.** 0.67
b. 0.52
3. **a.** 8.7×10^2 N, 6.7×10^2 N
b. 1.1×10^2 N, 84 N
c. 1×10^3 N, 5×10^2 N
d. 5 N, 2 N

Classroom Practice

Overcoming Friction
Two students are sliding a 225 kg sofa at constant speed across a wood floor. One student pulls with a force of 225 N at an angle of 13° above the horizontal. The other student pushes with a force of 250 N at an angle of 23° below the horizontal. What is the coefficient of kinetic friction between the sofa and the floor?

Answer
0.22

How could the students make moving the sofa easier?

Answer
They could change the angles, put the sofa on rollers, or wax the floors.

SAMPLE PROBLEM E

Overcoming Friction

PROBLEM

A student attaches a rope to a 20.0 kg box of books. He pulls with a force of 90.0 N at an angle of 30.0° with the horizontal. The coefficient of kinetic friction between the box and the sidewalk is 0.500. Find the acceleration of the box.

SOLUTION

1. DEFINE

Given: $m = 20.0$ kg $\quad \mu_k = 0.500$

$\mathbf{F_{applied}} = 90.0$ N at $\theta = 30.0°$

Unknown: $\mathbf{a} = ?$

Diagram:

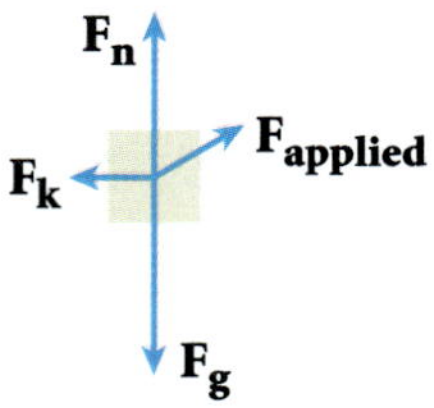

2. PLAN

Choose a convenient coordinate system, and find the *x* and *y* components of all forces.

The diagram at right shows the most convenient coordinate system, because the only force to resolve into components is $\mathbf{F_{applied}}$.

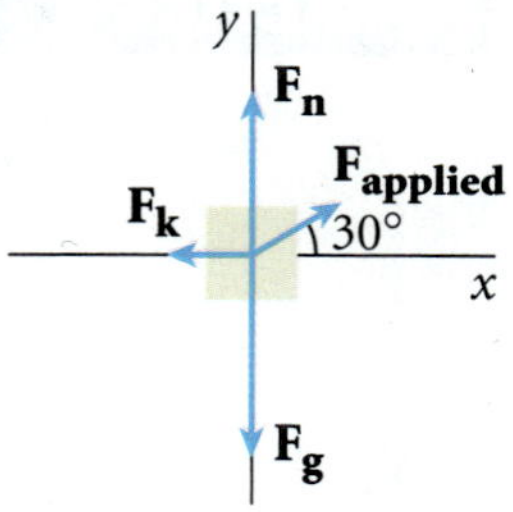

$F_{applied,y} = (90.0 \text{ N})(\sin 30.0°) = 45.0$ N (upward)

$F_{applied,x} = (90.0 \text{ N})(\cos 30.0°) = 77.9$ N (to the right)

Choose an equation or situation:

A. Find the normal force, F_n, by applying the condition of equilibrium in the vertical direction: $\Sigma F_y = 0$.

B. Calculate the force of kinetic friction on the box: $F_k = \mu_k F_n$.

C. Apply Newton's second law along the horizontal direction to find the acceleration of the box: $\Sigma F_x = ma_x$.

3. CALCULATE

Substitute the values into the equations and solve:

A. To apply the condition of equilibrium in the vertical direction, you need to account for all of the forces in the *y* direction: F_g, F_n, and $F_{applied,y}$. You know $F_{applied,y}$ and can use the box's mass to find F_g.

$F_{applied,y} = 45.0$ N

$F_g = (20.0 \text{ kg})(9.81 \text{ m/s}^2) = 196$ N

Next, apply the equilibrium condition, $\Sigma F_y = 0$, and solve for F_n.

$$\Sigma F_y = F_n + F_{applied,y} - F_g = 0$$

$$F_n + 45.0\ \text{N} - 196\ \text{N} = 0$$

$$F_n = -45.0\ \text{N} + 196\ \text{N} = 151\ \text{N}$$

> **TIP** *Remember to pay attention to the direction of forces. Here, F_g is subtracted from F_n and $F_{applied,y}$ because $\mathbf{F_g}$ is directed downward.*

B. Use the normal force to find the force of kinetic friction.

$$F_k = \mu_k F_n = (0.500)(151\ \text{N}) = 75.5\ \text{N}$$

C. Use Newton's second law to determine the horizontal acceleration.

> **TIP** *$\mathbf{F_k}$ is directed toward the left, opposite the direction of $\mathbf{F_{applied,x}}$. As a result, when you find the sum of the forces in the x direction, you need to subtract F_k from $F_{applied,x}$.*

$$\Sigma F_x = F_{applied,x} - F_k = ma_x$$

$$a_x = \frac{F_{applied,x} - F_k}{m} = \frac{77.9\ \text{N} - 75.5\ \text{N}}{20.0\ \text{kg}} = \frac{2.4\ \text{N}}{20.0\ \text{kg}} = \frac{2.4\ \text{kg}\bullet\text{m/s}^2}{20.0\ \text{kg}}$$

$$\boxed{\mathbf{a} = 0.12\ \text{m/s}^2 \text{ to the right}}$$

4. EVALUATE The normal force is not equal in magnitude to the weight because the *y* component of the student's pull on the rope helps support the box.

PRACTICE E

Overcoming Friction

1. A student pulls on a rope attached to a box of books and moves the box down the hall. The student pulls with a force of 185 N at an angle of 25.0° above the horizontal. The box has a mass of 35.0 kg, and μ_k between the box and the floor is 0.27. Find the acceleration of the box.
2. The student in item 1 moves the box up a ramp inclined at 12° with the horizontal. If the box starts from rest at the bottom of the ramp and is pulled at an angle of 25.0° with respect to the incline and with the same 185 N force, what is the acceleration up the ramp? Assume that $\mu_k = 0.27$.
3. A 75 kg box slides down a 25.0° ramp with an acceleration of 3.60 m/s^2.
   a. Find μ_k between the box and the ramp.
   b. What acceleration would a 175 kg box have on this ramp?
4. A box of books weighing 325 N moves at a constant velocity across the floor when the box is pushed with a force of 425 N exerted downward at an angle of 35.2° below the horizontal. Find μ_k between the box and the floor.

PROBLEM GUIDE E

Use this guide to assign problems.
SE = Student Edition Textbook
PW = Problem Workbook
PB = Problem Bank on the One-Stop Planner (OSP)

Solving for:

F_f, a	**SE**	Sample, 1–3; Ch. Rvw. 38*, 39*, 47a–b*, 48c
	PW	5–7
	PB	4–7
F_n, m	**SE**	Ch. Rvw. 21, 29, 41, 50, 52*
	PW	Sample, 1–3
	PB	8–10
μ	**SE**	3, 4; Ch. Rvw. 36–37, 48b*
	PW	4
	PB	Sample, 1–3

***Challenging Problem**
Consult the printed Solutions Manual or the OSP for detailed solutions.

ANSWERS

Practice E

1. 2.7 m/s^2 in the positive *x* direction
2. 0.77 m/s^2 up the ramp
3. **a.** 0.061
   b. 3.61 m/s^2 down the ramp (Note that *m* cancels in the solution, and *a* is the same in both cases; the slight difference is due to rounding.)
4. 0.609

Key Models and Analogies

Objects moving in outer space do not experience air resistance. Thus, Earth continually orbits the sun without slowing down. (Earth's speed is actually decreasing because of frequent collisions with small masses such as meteoroids, but this effect is minor.)

Why it Matters

Driving and Friction

The coefficient of friction between the ground and the tires of a car is smaller when rain or snow is on the ground. Snow and rain tires are excellent examples of ways that we have adapted tires to regain some of the necessary frictional forces.

Point out to students that the friction between a tire and pavement is more complex than the simple sliding friction between dry surfaces, which they have been studying. The force of friction on a car tire is not necessarily simply proportional to the normal force.

The fact that there is not a simple proportion between the frictional and normal forces is due in part to the fact that the tires are rolling, so they peel vertically away from the surface rather than continuously slide across it. Also, when the road is covered with water or snow, other factors such as viscosity come into play.

SCILINKS.
www.scilinks.org
Topic: Friction
Code: HF60622

Air resistance is a form of friction

Another type of friction, the retarding force produced by air resistance, is important in the analysis of motion. Whenever an object moves through a fluid medium, such as air or water, the fluid provides a resistance to the object's motion.

For example, the force of air resistance, $\mathbf{F_R}$, on a moving car acts in the direction opposite the direction of the car's motion. At low speeds, the magnitude of $\mathbf{F_R}$ is roughly proportional to the car's speed. At higher speeds, $\mathbf{F_R}$ is roughly proportional to the square of the car's speed. When the magnitude of $\mathbf{F_R}$ equals the magnitude of the force moving the car forward, the net force is zero and the car moves at a constant speed.

A similar situation occurs when an object falls through air. As a free-falling body accelerates, its velocity increases. As the velocity increases, the resistance of the air to the object's motion also constantly increases. When the upward force of air resistance balances the downward gravitational force, the net force on the object is zero and the object continues to move downward with a constant maximum speed, called the *terminal speed.*

Why it Matters

Driving and Friction

Accelerating a car seems simple to the driver. It is just a matter of pressing on a pedal or turning a wheel. But what are the forces involved?

A car moves because as its wheels turn, they push back against the road. It is actually the reaction force of the road pushing on the car that causes the car to accelerate. Without the friction between the tires and the road, the wheels would not be able to exert this force and the car would not experience a reaction force. Thus, acceleration requires this friction. Water and snow provide less friction and therefore reduce the amount of control the driver has over the direction and speed of the car.

As a car moves slowly over an area of water on the road, the water is squeezed out from under the tires. If the car moves too quickly, there is not enough time for the weight of the car to squeeze the water out from under the tires. The water trapped between the tires and the road will lift the tires and car off the road, a phenomenon called *hydroplaning.* When this situation occurs, there is very little friction between the tires and the water, and the car becomes difficult to control. To prevent hydroplaning, rain tires, such as the ones shown above, keep water from accumulating between the tire and the road. Deep channels down the center of the tire provide a place for the water to accumulate, and curved grooves in the tread channel the water outward.

Because snow moves even less easily than water, snow tires have several deep grooves in their tread, enabling the tire to cut through the snow and make contact with the pavement. These deep grooves push against the snow and, like the paddle blades of a riverboat, use the snow's inertia to provide resistance.

There are four fundamental forces

At the microscopic level, friction results from interactions between the protons and electrons in atoms and molecules. Magnetic force also results from atomic phenomena. These forces are classified as *electromagnetic forces.* The electromagnetic force is one of four fundamental forces in nature. The other three fundamental forces are gravitational force, the strong nuclear force, and the weak nuclear force. All four fundamental forces are field forces.

The strong and weak nuclear forces have very small ranges, so their effects are not directly observable. The electromagnetic and gravitational forces act over long ranges. Thus, any force you can observe at the macroscopic level is either due to gravitational or electromagnetic forces.

The strong nuclear force is the strongest of all four fundamental forces. Gravity is the weakest. Although the force due to gravity holds the planets, stars, and galaxies together, its effect on subatomic particles is negligible. This explains why electric and magnetic effects can easily overcome gravity. For example, a bar magnet has the ability to lift another magnet off a desk.

SECTION REVIEW

1. Draw a free-body diagram for each of the following objects:

a. a projectile accelerating downward in the presence of air resistance

b. a crate being pushed across a flat surface at a constant speed

2. A bag of sugar has a mass of 2.26 kg.

a. What is its weight in newtons on the moon, where the acceleration due to gravity is one-sixth that on Earth?

b. What is its weight on Jupiter, where the acceleration due to gravity is 2.64 times that on Earth?

3. A 2.0 kg block on an incline at a 60.0° angle is held in equilibrium by a horizontal force.

a. Determine the magnitude of this horizontal force. (Disregard friction.)

b. Determine the magnitude of the normal force on the block.

4. A 55 kg ice skater is at rest on a flat skating rink. A 198 N horizontal force is needed to set the skater in motion. However, after the skater is in motion, a horizontal force of 175 N keeps the skater moving at a constant velocity. Find the coefficients of static and kinetic friction between the skates and the ice.

5. Critical Thinking The force of air resistance acting on a certain falling object is roughly proportional to the square of the object's velocity and is directed upward. If the object falls fast enough, will the force of air resistance eventually exceed the weight of the object and cause the object to move upward? Explain.

SECTION REVIEW ANSWERS

1. a. An arrow labeled $\mathbf{F_g}$ should point down, and an arrow labeled $\mathbf{F_{air}}$ should point opposite the direction of motion. The arrow $\mathbf{F_g}$ should be longer than the arrow $\mathbf{F_{air}}$.

b. $\mathbf{F_g}$ points down, $\mathbf{F_n}$ points up, $\mathbf{F_{applied}}$ is horizontal, and $\mathbf{F_{friction}}$ points in the opposite direction. The two vertical arrows are equal in length, as are the two horizontal arrows.

2. a. 3.70 N

b. 58.5 N

3. a. 34 N

b. 39 N

4. 0.37, 0.32

5. no; Once at equilibrium, the velocity will not increase, so the force of air resistance will not increase.

Highlights

Teaching Tip

Ask students to prepare a concept map for the chapter. The concept map should include most of the vocabulary terms, along with other integral terms or concepts.

CHAPTER 4

Highlights

KEY TERMS

force (p. 120)

inertia (p. 125)

net force (p. 126)

equilibrium (p. 129)

weight (p. 135)

normal force (p. 135)

static friction (p. 136)

kinetic friction (p. 137)

coefficient of friction (p. 138)

KEY IDEAS

Section 1 Changes in Motion

- Force is a vector quantity that causes acceleration (when unbalanced).
- Force can act either through the physical contact of two objects (contact force) or at a distance (field force).
- A free-body diagram shows only the forces that act on one object. These forces are the only ones that affect the motion of that object.

Section 2 Newton's First Law

- The tendency of an object not to accelerate is called *inertia*. Mass is the physical quantity used to measure inertia.
- The net force acting on an object is the vector sum of all external forces acting on the object. An object is in a state of equilibrium when the net force acting on the object is zero.

Section 3 Newton's Second and Third Laws

- The net force acting on an object is equal to the product of the object's mass and the object's acceleration.
- When two bodies exert force on each other, the forces are equal in magnitude and opposite in direction. These forces are called an action-reaction pair. Forces always exist in such pairs.

Section 4 Everyday Forces

- The weight of an object is the magnitude of the gravitational force on the object and is equal to the object's mass times the acceleration due to gravity.
- A normal force is a force that acts on an object in a direction perpendicular to the surface of contact.
- Friction is a resistive force that acts in a direction opposite to the direction of the relative motion of two contacting surfaces. The force of friction between two surfaces is proportional to the normal force.

PROBLEM SOLVING

See **Appendix D: Equations** for a summary of the equations introduced in this chapter. If you need more problem-solving practice, see **Appendix I: Additional Problems.**

Variable Symbols

Quantities		Units		Conversions
$\mathbf{F}$ (vector)	force	N	newtons	$= kg \bullet m/s^2$
F (scalar)				
μ	coefficient of friction	(no units)		

FORCES AND NEWTON'S FIRST LAW

Review Questions

1. Is it possible for an object to be in motion if no net force is acting on it? Explain.

2. If an object is at rest, can we conclude that no external forces are acting on it?

3. An object thrown into the air stops at the highest point in its path. Is it in equilibrium at this point? Explain.

4. What physical quantity is a measure of the amount of inertia an object has?

Conceptual Questions

5. A beach ball is left in the bed of a pickup truck. Describe what happens to the ball when the truck accelerates forward.

6. A large crate is placed on the bed of a truck but is not tied down.

a. As the truck accelerates forward, the crate slides across the bed until it hits the tailgate. Explain what causes this.

b. If the driver slammed on the brakes, what could happen to the crate?

Practice Problems

For problems 7–9, see Sample Problem A.

7. Earth exerts a downward gravitational force of 8.9 N on a cake that is resting on a plate. The plate exerts a force of 11.0 N upward on the cake, and a knife exerts a downward force of 2.1 N on the cake. Draw a free-body diagram of the cake.

8. A chair is pushed forward with a force of 185 N. The gravitational force of Earth on the chair is 155 N downward, and the floor exerts a force of 155 N upward on the chair. Draw a free-body diagram showing the forces acting on the chair.

9. Draw a free-body diagram representing each of the following objects:

a. a ball falling in the presence of air resistance
b. a helicopter lifting off a landing pad
c. an athlete running along a horizontal track

For problems 10–12, see Sample Problem B.

10. Four forces act on a hot-air balloon, shown from the side in the figure below. Find the magnitude and direction of the resultant force on the balloon.

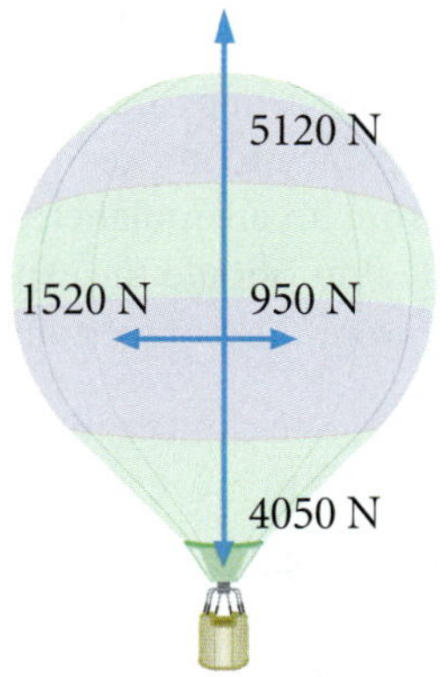

11. Two lifeguards pull on ropes attached to a raft. If they pull in the same direction, the raft experiences a net force of 334 N to the right. If they pull in opposite directions, the raft experiences a net force of 106 N to the left.

a. Draw a free-body diagram representing the raft for each situation.
b. Find the force exerted by each lifeguard on the raft for each situation. (Disregard any other forces acting on the raft.)

12. A dog pulls on a pillow with a force of 5 N at an angle of 37° above the horizontal. Find the x and y components of this force.

Review

ANSWERS

1. yes; The object could move at a constant velocity.

2. no, just that the net force equals zero

3. no; It has a net force downward (gravitational force).

4. mass

5. The ball moves toward the back of the truck because inertia keeps it in place relative to the ground.

6. **a.** the inertia of the crate
b. It could continue forward by inertia and hit the cab.

7. $\mathbf{F_g}$ (8.9 N) and $\mathbf{F_{applied}}$ (2.1 N) point downward, and $\mathbf{F_n}$ (11.0 N) points upward.

8. $\mathbf{F_{applied}}$ (185 N) points forward, $\mathbf{F_g}$ (155 N) points downward, and $\mathbf{F_n}$ (155 N) points upward. The diagram may also include $\mathbf{F_{friction}}$ backward.

9. **a.** $\mathbf{F_g}$ points down, and $\mathbf{F_R}$ points up.
b. $\mathbf{F_{rotors}}$ points up, and $\mathbf{F_g}$ points down.
c. $\mathbf{F_g}$ points down, $\mathbf{F_{track}}$ points in the direction of motion, and $\mathbf{F_n}$ points up.

10. 1210 N at 62° above the 1520 N force

11. **a.** $\mathbf{F_1}$ (220 N) and $\mathbf{F_2}$ (114 N) both point right; $\mathbf{F_1}$ (220 N) points left, and $\mathbf{F_2}$ (114 N) points right.
b. first situation: 220 N to the right, 114 N to the right; second situation: 220 N to the left, 114 N to the right

12. 4 N; 3 N

13. because Earth has a very large mass

14. An object with greater mass requires a larger force for a given acceleration.

15. One-sixth of the force needed to lift an object on Earth is needed on the moon.

16. on the horse: the force of the cart, F_g down, F_n up, a reaction force of the ground on the hooves; on the cart: the force of the horse, F_g down, F_n up, kinetic friction

17. push it gently; With a smaller force, the astronaut will experience a smaller reaction force.

18. As the climber exerts a force downward, the rope supplies a reaction force that is directed upward. When this reaction force is greater than the climber's weight, the climber accelerates upward.

19. **a.** zero
b. zero

20. 3.52 m/s^2

21. 55 N to the right

22. **a.** 770 N at 8.1° to the right of forward
b. 0.24 m/s^2 at 8.1° to the right of forward

23. Mass is the inertial property of matter. Weight is the gravitational force acting on an object. Weight is equal to mass times the free-fall acceleration.

24. **a.** −1.47 N
b. −1.47 N

25. **a.** $\mathbf{F_g}$ points down, and $\mathbf{F_n}$ points up.
b. $\mathbf{F_g}$ points down, and $\mathbf{F_n}$ points up perpendicular to the surface of the ramp.
c. same as (b)
d. same as (b)

26. **a.** 54 N
b. 53 N

NEWTON'S SECOND AND THIRD LAWS

Review Questions

13. The force that attracts Earth to an object is equal to and opposite the force that Earth exerts on the object. Explain why Earth's acceleration is not equal to and opposite the object's acceleration.

14. State Newton's second law in your own words.

15. An astronaut on the moon has a 110 kg crate and a 230 kg crate. How do the forces required to lift the crates straight up on the moon compare with the forces required to lift them on Earth? (Assume that the astronaut lifts with constant velocity in both cases.)

16. Draw a force diagram to identify all the action-reaction pairs that exist for a horse pulling a cart.

Conceptual Questions

17. A space explorer is moving through space far from any planet or star and notices a large rock, taken as a specimen from an alien planet, floating around the cabin of the ship. Should the explorer push it gently or kick it toward the storage compartment? Why?

18. Explain why a rope climber must pull downward on the rope in order to move upward. Discuss the force exerted by the climber's arms in relation to the weight of the climber during the various stages of each "step" up the rope.

19. An 1850 kg car is moving to the right at a constant speed of 1.44 m/s.

a. What is the net force on the car?

b. What would be the net force on the car if it were moving to the left?

Practice Problems

For problems 20–22, see Sample Problem C.

20. What acceleration will you give to a 24.3 kg box if you push it horizontally with a net force of 85.5 N?

21. What net force is required to give a 25 kg suitcase an acceleration of 2.2 m/s^2 to the right?

22. Two forces are applied to a car in an effort to accelerate it, as shown below.

a. What is the resultant of these two forces?

b. If the car has a mass of 3200 kg, what acceleration does it have? (Disregard friction.)

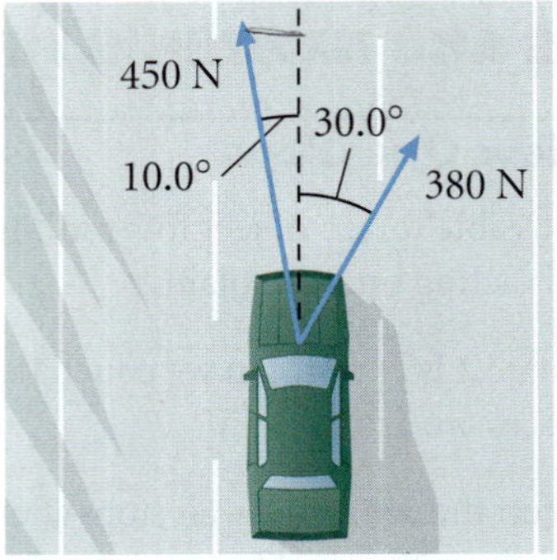

WEIGHT, FRICTION, AND NORMAL FORCE

Review Questions

23. Explain the relationship between mass and weight.

24. A 0.150 kg baseball is thrown upward with an initial speed of 20.0 m/s.

a. What is the force on the ball when it reaches half of its maximum height? (Disregard air resistance.)

b. What is the force on the ball when it reaches its peak?

25. Draw free-body diagrams showing the weight and normal forces on a laundry basket in each of the following situations:

a. at rest on a horizontal surface

b. at rest on a ramp inclined 12° above the horizontal

c. at rest on a ramp inclined 25° above the horizontal

d. at rest on a ramp inclined 45° above the horizontal

26. If the basket in item 25 has a mass of 5.5 kg, find the magnitude of the normal force for the situations described in (a) through (d).

27. A teapot is initially at rest on a horizontal tabletop, then one end of the table is lifted slightly. Does the normal force increase or decrease? Does the force of static friction increase or decrease?

28. Which is usually greater, the maximum force of static friction or the force of kinetic friction?

29. A 5.4 kg bag of groceries is in equilibrium on an incline of angle $\theta = 15°$. Find the magnitude of the normal force on the bag.

Conceptual Questions

30. Imagine an astronaut in space at the midpoint between two stars of equal mass. If all other objects are infinitely far away, what is the weight of the astronaut? Explain your answer.

31. A ball is held in a person's hand.

a. Identify all the external forces acting on the ball and the reaction force to each.

b. If the ball is dropped, what force is exerted on it while it is falling? Identify the reaction force in this case. (Disregard air resistance.)

32. Explain why pushing downward on a book as you push it across a table increases the force of friction between the table and the book.

33. Analyze the motion of a rock dropped in water in terms of its speed and acceleration. Assume that a resistive force acting on the rock increases as the speed increases.

34. A sky diver falls through the air. As the speed of the sky diver increases, what happens to the sky diver's acceleration? What is the acceleration when the sky diver reaches terminal speed?

Practice Problems

For problems 35–37, see Sample Problem D.

35. A 95 kg clock initially at rest on a horizontal floor requires a 650 N horizontal force to set it in motion. After the clock is in motion, a horizontal force of 560 N keeps it moving with a constant velocity. Find μ_s and μ_k between the clock and the floor.

36. A box slides down a 30.0° ramp with an acceleration of 1.20 m/s^2. Determine the coefficient of kinetic friction between the box and the ramp.

37. A 4.00 kg block is pushed along the ceiling with a constant applied force of 85.0 N that acts at an angle of 55.0° with the horizontal, as in the figure. The block accelerates to the right at 6.00 m/s^2. Determine the coefficient of kinetic friction between the block and the ceiling.

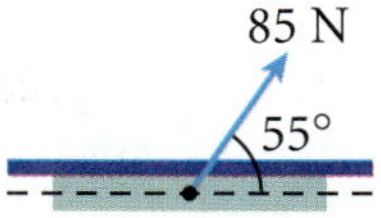

For problems 38–39, see Sample Problem E.

38. A clerk moves a box of cans down an aisle by pulling on a strap attached to the box. The clerk pulls with a force of 185.0 N at an angle of 25.0° with the horizontal. The box has a mass of 35.0 kg, and the coefficient of kinetic friction between box and floor is 0.450. Find the acceleration of the box.

39. A 925 N crate is being pulled across a level floor by a force **F** of 325 N at an angle of 25° above the horizontal. The coefficient of kinetic friction between the crate and floor is 0.25. Find the magnitude of the acceleration of the crate.

MIXED REVIEW

40. A block with a mass of 6.0 kg is held in equilibrium on an incline of angle $\theta = 30.0°$ by a horizontal force, **F,** as shown in the figure. Find the magnitudes of the normal force on the block and of **F.** (Ignore friction.)

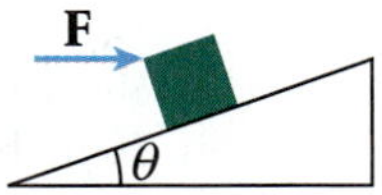

41. A 2.0 kg mass starts from rest and slides down an inclined plane 8.0×10^{-1} m long in 0.50 s. What net force is acting on the mass along the incline?

42. A 2.26 kg book is dropped from a height of 1.5 m.

a. What is its acceleration?

b. What is its weight in newtons?

4 REVIEW

c. 49 N
d. 38 N

27. The normal force decreases; The force of static friction increases to counteract the component of the weight along the table.

28. $\mathbf{F_{s,max}}$

29. 51 N

30. 0 N; The forces exerted by each star cancel.

31. **a.** the weight of the ball and an equal reaction force of the ball on Earth; the force of the person's hand on the ball and an equal reaction force of the ball on the hand

b. F_g; the force of the ball on Earth

32. Pushing down on the book increases the normal force and therefore also increases the friction.

33. The rock will accelerate until the magnitude of the resistive force equals the net downward force on the rock. (This downward force is the rock's weight minus the buoyant force of the water.) Then the rock's speed will be constant.

34. As the sky diver's speed increases, the acceleration decreases because the resistive force increases with increasing speed; zero

35. 0.70, 0.60

36. 0.436

37. 0.816

38. 1.4 m/s^2 down the aisle

39. 1.0 m/s^2

40. 68 N; 34 N

41. 13 N down the incline

42. **a.** 9.81 m/s^2 downward

b. 22.2 N

43. 64 N upward

44. **a.** zero
b. 33.9 N

45. **a.** 0.25 m/s^2 forward
b. 18 m
c. 3.0 m/s

46. 5.0×10^1 m

47. **a.** 2 s
b. The box will never move. The force exerted is not enough to overcome friction.

48. **a.** 1.78 m/s^2
b. 0.37
c. 9.4 N
d. 2.67 m/s

49. -1.2 m/s^2; 0.12

50. -5.0×10^2 N

51. **a.** 2690 N forward
b. 699 N forward

52. 32.2 N

53. 13 N, 13 N, 0 N, −26 N

54. 1.41°

43. A 5.0 kg bucket of water is raised from a well by a rope. If the upward acceleration of the bucket is 3.0 m/s^2, find the force exerted by the rope on the bucket of water.

44. A 3.46 kg briefcase is sitting at rest on a level floor.

a. What is the briefcases's acceleration?
b. What is its weight in newtons?

45. A boat moves through the water with two forces acting on it. One is a 2.10×10^3 N forward push by the motor, and the other is a 1.80×10^3 N resistive force due to the water.

a. What is the acceleration of the 1200 kg boat?
b. If it starts from rest, how far will it move in 12 s?
c. What will its speed be at the end of this time interval?

46. A girl on a sled coasts down a hill. Her speed is 7.0 m/s when she reaches level ground at the bottom. The coefficient of kinetic friction between the sled's runners and the hard, icy snow is 0.050, and the girl and sled together weigh 645 N. How far does the sled travel on the level ground before coming to rest?

47. A box of books weighing 319 N is shoved across the floor by a force of 485 N exerted downward at an angle of 35° below the horizontal.

a. If μ_k between the box and the floor is 0.57, how long does it take to move the box 4.00 m, starting from rest?
b. If μ_k between the box and the floor is 0.75, how long does it take to move the box 4.00 m, starting from rest?

48. A 3.00 kg block starts from rest at the top of a 30.0° incline and accelerates uniformly down the incline, moving 2.00 m in 1.50 s.

a. Find the magnitude of the acceleration of the block.
b. Find the coefficient of kinetic friction between the block and the incline.
c. Find the magnitude of the frictional force acting on the block.
d. Find the speed of the block after it has slid a distance of 2.00 m.

49. A hockey puck is hit on a frozen lake and starts moving with a speed of 12.0 m/s. Exactly 5.0 s later, its speed is 6.0 m/s. What is the puck's average acceleration? What is the coefficient of kinetic friction between the puck and the ice?

50. The parachute on a race car that weighs 8820 N opens at the end of a quarter-mile run when the car is traveling 35 m/s. What net retarding force must be supplied by the parachute to stop the car in a distance of 1100 m?

51. A 1250 kg car is pulling a 325 kg trailer. Together, the car and trailer have an acceleration of 2.15 m/s^2 directly forward.

a. Determine the net force on the car.
b. Determine the net force on the trailer.

52. The coefficient of static friction between the 3.00 kg crate and the 35.0° incline shown here is 0.300. What is the magnitude of the minimum force, F, that must be applied to the crate perpendicularly to the incline to prevent the crate from sliding down the incline?

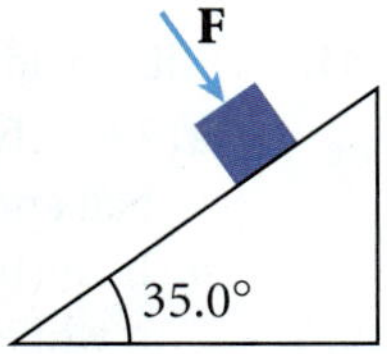

53. The graph below shows a plot of the speed of a person's body during a chin-up. All motion is vertical and the mass of the person (excluding the arms) is 64.0 kg. Find the magnitude of the net force exerted on the body at 0.50 s intervals.

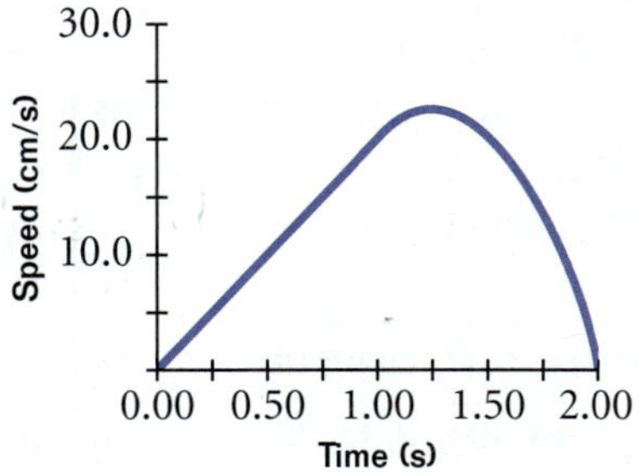

54. A machine in an ice factory is capable of exerting 3.00×10^2 N of force to pull a large block of ice up a slope. The block weighs 1.22×10^4 N. Assuming there is no friction, what is the maximum angle that the slope can make with the horizontal if the machine is to be able to complete the task?

Alternative Assessment

1. Predict what will happen in the following test of the laws of motion. You and a partner face each other, each holding a bathroom scale. Place the scales back to back, and slowly begin pushing on them. Record the measurements of both scales at the same time. Perform the experiment. Which of Newton's laws have you verified?

2. Research how the work of scientists Antoine Lavoisier, Isaac Newton, and Albert Einstein related to the study of mass. Which of these scientists might have said the following?

   a. The mass of a body is a measure of the quantity of matter in the body.

   b. The mass of a body is the body's resistance to a change in motion.

   c. The mass of a body depends on the body's velocity.

   To what extent are these statements compatible or contradictory? Present your findings to the class for review and discussion.

3. Imagine an airplane with a series of special instruments anchored to its walls: a pendulum, a 100 kg mass on a spring balance, and a sealed half-full aquarium. What will happen to each instrument when the plane takes off, makes turns, slows down, lands, etc.? If possible, test your predictions by simulating airplane motion in elevators, car rides, and other situations. Use instruments similar to those described above, and also observe your body sensations. Write a report comparing your predictions with your experiences.

4. With a small group, determine which of the following statements is correct. Use a diagram to explain your answer.

   a. Rockets cannot travel in space because there is nothing for the gas exiting the rocket to push against.

   b. Rockets can travel because gas exerts an unbalanced force on the rocket.

   c. The action and reaction forces are equal and opposite. Therefore, they balance each other, and no movement is possible.

Graphing Calculator Practice

Static Friction

The force of static friction depends on two factors: the coefficient of static friction for the two surfaces in contact, and the normal force between the two surfaces. The relationship can be represented on a graphing calculator by the following equation:

$$Y_1 = SX$$

Given a value for the coefficient of static friction (S), the graphing calculator can calculate and graph the force of static friction (Y_1) as a function of normal force (X).

In this activity, you will use a graphing calculator program to compare the force of static friction of wood boxes on a wood surface with that of steel boxes on a steel surface.

Visit go.hrw.com and type in the keyword **HF6FORX** to find this graphing calculator activity. Refer to **Appendix B** for instructions on downloading the program for this activity.

Alternative Assessment ANSWERS

1. Scales will have identical readings because of Newton's third law.
2. Lavoisier: (a), because Lavoisier is credited with establishing the fact that mass is conserved in chemical reactions; Newton: (a) and (b), because Newton established a definition of force by relating it to mass and acceleration; Einstein: (a), (b), and (c), because Einstein showed that at high speeds, Newton's second law of motion requires an additional correction term that is speed dependent
3. Students may construct accelerometers and/or anchor helium-filled balloons to the elevator or car floor. Students' reports should compare their predictions to their experiences.
4. Statement b is true. The gas pushes in all directions. Some gas pushes forward on the rocket, and some exits through the back. The net force in the forward direction causes the acceleration.

extension

Graphing Calculator Practice

Visit go.hrw.com for answers to this Graphing Calculator activity.

Keyword HF6FORXT

Standardized Test Prep

ANSWERS

1. C
2. G
3. C
4. G
5. A
6. G

Standardized Test Prep

MULTIPLE CHOICE

Use the passage below to answer questions 1–2.

Two blocks of masses m_1 and m_2 are placed in contact with each other on a smooth, horizontal surface. Block m_1 is on the left of block m_2. A constant horizontal force F to the right is applied to m_1.

1. What is the acceleration of the two blocks?

A. $a = \dfrac{F}{m_1}$

B. $a = \dfrac{F}{m_2}$

C. $a = \dfrac{F}{m_1 + m_2}$

D. $a = \dfrac{F}{(m_1)(m_2)}$

2. What is the horizontal force acting on m_2?

F. $m_1 a$
G. $m_2 a$
H. $(m_1 + m_2)a$
J. $m_1 m_2 a$

3. A crate is pulled to the right (positive *x*-axis) with a force of 82.0 N, to the left with a force of 115 N, upward with a force of 565 N, and downward with a force of 236 N. Find the magnitude and direction of the net force on the crate.

A. 3.30 N at 96° counterclockwise from the positive *x*-axis
B. 3.30 N at 6° counterclockwise from the positive *x*-axis
C. 3.30×10^2 N at 96° counterclockwise from the positive *x*-axis
D. 3.30×10^2 N at 6° counterclockwise from the positive *x*-axis

4. A ball with a mass of m is thrown into the air, as shown in the figure below. What is the force exerted on Earth by the ball?

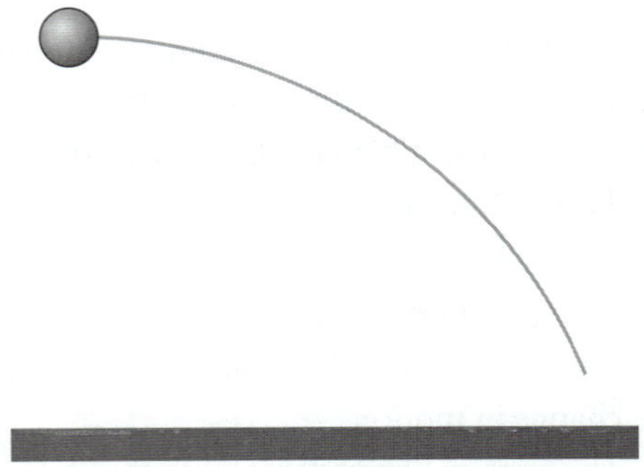

F. $m_{ball}g$, directed down
G. $m_{ball}g$, directed up
H. $m_{Earth}g$, directed down
J. $m_{Earth}g$, directed up

5. A freight train has a mass of 1.5×10^7 kg. If the locomotive can exert a constant pull of 7.5×10^5 N, how long would it take to increase the speed of the train from rest to 85 km/h? (Disregard friction.)

A. 4.7×10^2 s
B. 4.7 s
C. 5.0×10^{-2} s
D. 5.0×10^4 s

Use the passage below to answer questions 6–7.

A truck driver slams on the brakes and skids to a stop through a displacement Δx.

6. If the truck's mass doubles, find the truck's skidding distance in terms of Δx. (Hint: Increasing the mass increases the normal force.)

F. $\Delta x/4$
G. Δx
H. $2\Delta x$
J. $4\Delta x$

7. If the truck's initial velocity were halved, what would be the truck's skidding distance?

A. $\Delta x/4$
B. Δx
C. $2\Delta x$
D. $4\Delta x$

Use the graph below to answer questions 8–9. The graph shows the relationship between the applied force and the force of friction.

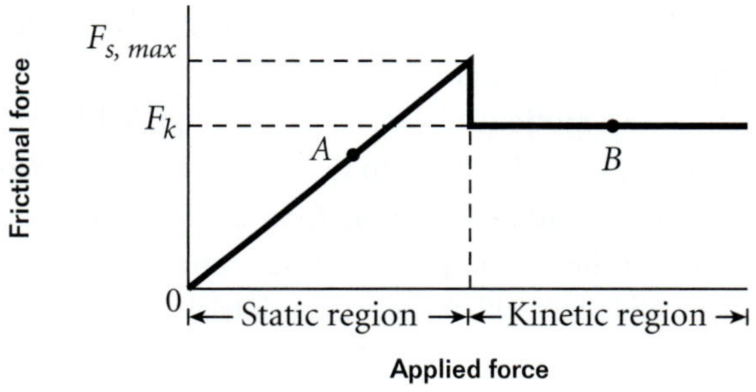

8. What is the relationship between the forces at point *A*?

F. $F_s = F_{applied}$
G. $F_k = F_{applied}$
H. $F_s < F_{applied}$
J. $F_k > F_{applied}$

9. What is the relationship between the forces at point *B*?

A. $F_{s,\,max} = F_k$
B. $F_k > F_{s,\,max}$
C. $F_k > F_{applied}$
D. $F_k < F_{applied}$

SHORT RESPONSE

Base your answers to questions 10–12 on the information below.

A 3.00 kg ball is dropped from rest from the roof of a building 176.4 m high. While the ball is falling, a horizontal wind exerts a constant force of 12.0 N on the ball.

10. How long does the ball take to hit the ground?

11. How far from the building does the ball hit the ground?

12. When the ball hits the ground, what is its speed?

Base your answers to questions 13–15 on the information below.

A crate rests on the horizontal bed of a pickup truck. For each situation described below, indicate the motion of the crate relative to the ground, the motion of the crate relative to the truck, and whether the crate will hit the front wall of the truck bed, the back wall, or neither. Disregard friction.

13. Starting at rest, the truck accelerates to the right.

14. The crate is at rest relative to the truck while the truck moves with a constant velocity to the right.

15. The truck in item 14 slows down.

EXTENDED RESPONSE

16. A student pulls a rope attached to a 10.0 kg wooden sled and moves the sled across dry snow. The student pulls with a force of 15.0 N at an angle of 45.0°. If μ_k between the sled and the snow is 0.040, what is the sled's acceleration? Show your work.

17. You can keep a 3 kg book from dropping by pushing it horizontally against a wall. Draw force diagrams, and identify all the forces involved. How do they combine to result in a zero net force? Will the force you must supply to hold the book up be different for different types of walls? Design a series of experiments to test your answer. Identify exactly which measurements will be necessary and what equipment you will need.

Test TIP For a question involving experimental data, determine the constants, variables, and control *before* answering the question.

7. A

8. F

9. D

10. 6.00 s

11. 72.0 m

12. 63.6 m/s

13. at rest, moves to the left, hits back wall

14. moves to the right (with velocity v), at rest, neither

15. moves to the right, moves to the right, hits front wall

16. 0.71 m/s^2 (See the Solutions Manual or the One-Stop Planner for the full solution.)

17. Student plans should be safe and should involve measuring forces such as weight, applied force, normal force, and frictional force.

Lab Planning

Beginning on page T34 are preparation notes and teaching tips to assist you in planning.

Blank data tables (as well as some sample data) appear on the **One-Stop Planner.**

No Books in the Lab?

See the ***Datasheets for In-Text Labs*** workbook for a reproducible master copy of this experiment.

CBL™ Option

A **CBL™** version of this lab appears in **Appendix K** and in the ***CBL™ Experiments*** workbook.

Safety Caution

Remind students to fasten masses securely, to keep the area clear during the experiment, and to prevent the cart from falling off the table.

Tips and Tricks

- If 110 V ac timers are used, calibration may be omitted: the average period of these timers is

  0.017 s ($\frac{1}{60}$ s).
- If possible, mount timers on stand rods to adjust the height of the timer. The tape should be level from the cart to the timer. If the timer must be table-mounted, leave at least 0.5 m between the timer and the cart to prevent the difference in height from affecting the results.

CHAPTER 4

Skills Practice Lab — Force and Acceleration

OBJECTIVES

- **Compare** the accelerations of a mass acted on by different forces.
- **Compare** the accelerations of different masses acted on by the same force.
- **Examine** the relationships between mass, force, acceleration, and Newton's laws of motion.

MATERIALS LIST

- **balance**
- **C-clamp**
- **calibrated masses and holder**
- **cord**
- **dynamics cart**
- **hooked mass, 1000 g**
- **mass hanger**
- **masking tape**
- **meterstick**
- **pulley with table clamp**
- **recording timer and tape**
- **stopwatch**

Newton's second law states that any net external force applied to a mass causes the mass to accelerate according to the equation $\mathbf{F} = m\mathbf{a}$. Because of frictional forces, experience does not always seem to support this. For example, when you are driving a car, you must apply a constant force to keep the car moving with a constant velocity. In the absence of friction, the car would continue to move with a constant velocity after the force was removed. The continued application of force would cause the car to accelerate.

In this lab, you will study the motion of a dynamics cart pulled by the weight of masses falling from a table to the floor. In the first part of the experiment, the total mass will remain constant while the force acting on the cart will be different for each trial. In the second part, the force acting on the cart will remain constant, but the total mass will change for each trial.

SAFETY

- **Tie back long hair, secure loose clothing, and remove loose jewelry to prevent its getting caught in moving or rotating parts. Put on goggles.**
- **Attach masses securely. Falling or dropped masses can cause serious injury.**

PROCEDURE

Preparation

1. Read the entire lab procedure, and plan the steps you will take.
2. If you are not using a datasheet provided by your teacher, prepare a data table in your lab notebook with six columns and six rows. In the first row, label the first through sixth columns *Trial, Total Mass (kg), Accelerating Mass (kg), Accelerating Force (N), Time Interval (s),* and *Distance (m)*. In the first column, label the second through sixth rows *1, 2, 3, 4,* and *5*.
3. Choose a location where the cart will be able to move a considerable distance without any obstacles and where you will be able to clamp the pulley to a table edge.

Apparatus Setup

4. Set up the apparatus as shown in **Figure 1.** Clamp the pulley to the edge of the table so that it is level with the top of the cart. Clamp the recording timer to a ring stand or to the edge of the table to hold it in place. If the timer is clamped to the table, leave 0.5 m between the timer and the initial position of the cart. Insert the carbon disk into the timer, and thread the tape through the guides under the disk. When your teacher approves your setup, plug the timer into a wall outlet.

5. If you have not used the recording timer before, refer to the lab in the chapter "Motion in One Dimension" for instructions. Calibrate the recording timer with the stopwatch or use the previously determined value for the timer's period.

6. Record the value for the timer's period at the top of the data table.

7. Fasten the timing tape to one end of the cart.

Constant Mass with Varying Force

8. Carefully measure the mass of the cart assembly on the platform balance, making sure that the cart does not roll or fall off the balance. Then load the cart with masses equal to 0.60 kg. Lightly tape the masses to the cart to hold them in place. Add these masses to the mass of the cart and record the total.

9. Attach one end of the cord to a small mass hanger and the other end of the cord to the cart. Pass the cord over the pulley and fasten a small mass to the end to offset the frictional force on the cart. You have chosen the correct mass if the cart moves forward with a constant velocity when you give it a push. *This counterweight should stay on the string throughout the entire experiment.* Add the mass of the counterweight to the mass of the cart and masses, and record the sum as *Total Mass* in your data table.

10. For the first trial, remove a 0.10 kg mass from the cart, and securely fasten it to the end of the string along with the counterweight. Record 0.10 kg as the *Accelerating Mass* in the data table.

11. Hold the cart by holding the tape behind the timer. Make sure the area under the falling mass is clear of obstacles. Start the timer and release the tape simultaneously.

Figure 1

Step 4: Make sure the clamp protrudes as little as possible from the edge of the table.

Step 7: Fold the end of the recording tape over the edge of the rail on the cart and tape it down.

Step 10: Always hold the cart when you are removing and adding masses. When you are ready, release the cart from the same position each time.

CHAPTER 4 LAB

- Show students how to thread the paper tape through the recording timer guides under the carbon disk.
- Students may need practice releasing the cart and starting the timer at the same time. The best method is to hold the cart by holding the tape straight out behind the timer.
- The timing tape should be fastened to the cart by folding the end of the tape over, hooking the fold over the flat rail on the cart, and securing the paper tape with masking tape.

✓ Checkpoints

Step 4: Check each setup before the timer is plugged in. Make sure the tape is level and inserted properly and that all clamps are tight and positioned where they will not protrude and cause injury.

Step 9: Make sure masses are securely attached. Students should be able to demonstrate that the counterweight allows the cart to move at a constant velocity when given a small push.

Step 15: Students should be able to explain how the dots on the tape represent the motion of the cart. They should be able to explain in what order the dots were made.

CHAPTER 4 LAB

ANSWERS

Analysis

1. Trial 1: $F = 0.981$ N, Trial 2: $F = 1.962$ N, Trial 3: $F = 2.943$ N, Trial 4: $F = 2.943$ N, Trial 5: $F = 2.943$ N

2. Trial 1: $a = 0.635$ m/s^2, Trial 2: $a = 1.25$ m/s^2, Trial 3: $a = 1.89$ m/s^2, Trial 4: $a = 1.46$ m/s^2, Trial 5: $a = 0.927$ m/s^2

3. Student graphs should show a straight line beginning at the origin and pointing up and to the right.

12. Carefully stop the cart when the 0.10 kg mass hits the floor, and then stop the timer. ***Do not let the cart fall off the table.***
13. Remove the tape and label it with the trial number.
14. Use a meterstick to measure the distance the weights fell. Record the *Distance* in your data table.
15. On the tape, measure this distance starting from the first clear dot. Mark the end of this distance. Count the number of dots between the first dot and this mark.
16. Calculate and record the *Time Interval* represented by the number of dots. Fasten a new timing tape to the end of the cart.
17. Replace the 0.10 kg mass in the cart. Remove 0.20 kg from the cart and attach it securely to the end of the cord. Repeat the procedure, label the tape, and record the results in your data table as *Trial 2.*
18. Leave the 0.20 kg mass on the end of the cord and attach the 0.10 kg mass from the cart securely to the end of the cord. Repeat the procedure, label the tape, and record the results in your data table as *Trial 3.*

Constant Force with Varying Mass

19. For the two trials in this part of the experiment, keep 0.30 kg and the counterweight on the string. Be sure to include this mass when recording the total mass for these three trials.
20. Add 0.50 kg to the cart. Tape the mass to the cart to keep it in place. Run the experiment and record the total mass, accelerating mass, accelerating force, distance, and time under *Trial 4* in your data table.
21. Tape 1.00 kg to the cart and repeat the procedure. Record the data under *Trial 5* in your data table.
22. Clean up your work area. Put equipment away as instructed.

ANALYSIS

1. **Analyzing Data** Calculate the *Accelerating Force* for each trial. Use Newton's second law equation, $\mathbf{F} = m\mathbf{a}$, where m = *Accelerating Mass* and $a = a_g$. Enter these values in your data table.
2. **Organizing Data** Use your values for the distance and time to find the acceleration of the cart for each trial, using the equation $\Delta x = \frac{1}{2}a\Delta t^2$ for constantly accelerated motion.
3. **Constructing Graphs** Using the data from *Trials 1–3,* plot a graph of the acceleration of the cart versus the accelerating force. Use a graphing calculator, computer, or graph paper.

4. **Analyzing Graphs** Based on your graph from item 3, what is the relationship between the acceleration of the cart and the accelerating force? Explain how your graph supports your answer.

5. **Constructing Graphs** Using the data from *Trials 3–5,* plot a graph of the total mass versus the acceleration. Use a graphing calculator, computer, or graph paper.

6. **Interpreting Graphs** Based on your graph from item 5, what is the relationship between the total mass and the acceleration? Explain how your graph supports your answer.

CONCLUSIONS

7. **Evaluating Methods** Why does the mass in *Trials 1–3* remain constant even though masses are removed from the cart during the trials?

8. **Evaluating Methods** Do the carts move with the same velocity and acceleration as the accelerating masses that are dropped? If not, why not?

9. **Drawing Conclusions** Do your data support Newton's second law? Use your data and your analysis of your graphs to support your conclusions.

10. **Applying Conclusions** A team of automobile safety engineers developed a new type of car and performed some test crashes to find out whether the car is safe. The engineers tested the new car by involving it in a series of different types of accidents. For each test, the engineers applied a known force to the car and measured the acceleration of the car after the crash. The graph in **Figure 2** shows the acceleration of the car plotted against the applied force. Compare this with the data you collected and the graphs you made for this experiment to answer the following questions.

    a. Based on the graph, what is the relationship between the acceleration of the new car and the force of the collision?

    b. Does this graph support Newton's second law? Use your analysis of this graph to support your conclusions.

    c. Do the data from the crash tests meet your expectations based on this lab? Explain what you think may have happened to affect the results. If you were on the engineering team, how would you find out whether your results were in error?

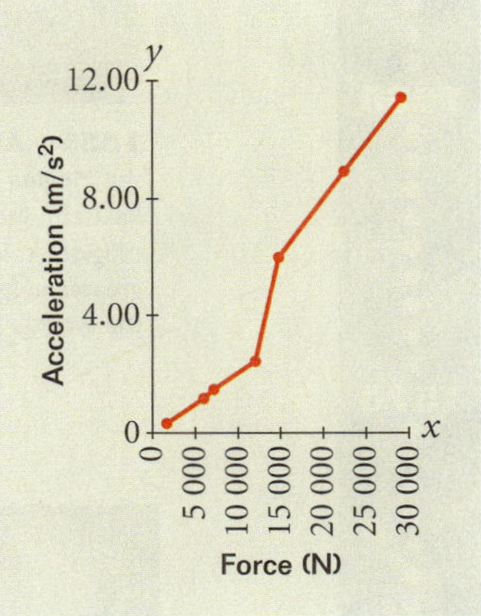

Figure 2

EXTENSION

11. **Designing Experiments** How would your results be affected if you used the mass of the cart and its contents instead of the total mass? Predict what would happen if you performed *Trials 1–3* again, keeping the mass of the cart and its contents constant while varying the accelerating mass. If there is time and your teacher approves your plan, go into the lab and try it. Plot your data using a graphing calculator, computer, or graph paper.

4. There is a direct relationship between the acceleration and the force.

5. Students' graphs should show a straight line pointing down and to the right.

6. There is an inverse relationship between the mass and the acceleration.

Conclusions

7. The masses are moved from the cart to the string; all the mass in the system is part of the accelerated mass.

8. Each cart and mass have the same value for velocity and acceleration. However, the directions of the vectors are different.

9. Data should support Newton's second law, a direct relationship between a and F, and an inverse relationship between m and a.

10. a. As F increases, a increases. The proportion between F and a is different above and below 15 000 N.

b. The graph, which shows a proportional relationship between acceleration and force, supports Newton's second law. Deviation of the curve could be due to experimental error.

c. Student answers should reflect that the data are not what should be expected, possibly due to a change in mass or an error in measurement. To verify, the experiment should be repeated several times.

Extension

11. Student answers should reflect an understanding that the acceleration of the cart depends on force and mass. Plans should be safe and complete.

Physics and Its World *Timeline 1540–1690*

1540 · 1550 · 1560 · 1570 · 1580 · 1590 · 1600 · 1610

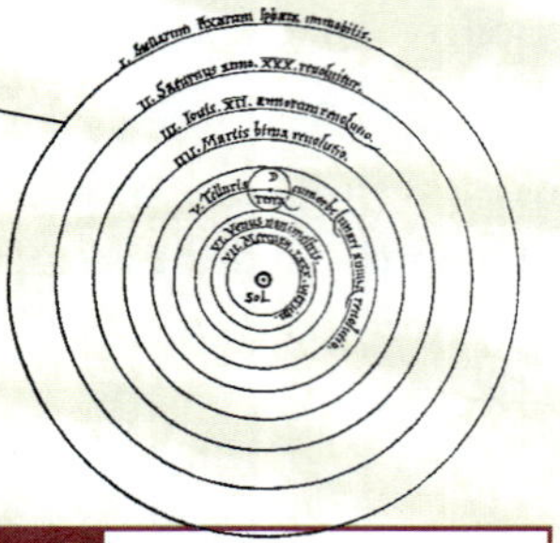

1543 **Nicholas Copernicus'** *On the Revolutions of the Heavenly Bodies* is published. It is the first work on astronomy to provide an analytical basis for the motion of the planets, including Earth, around the sun.

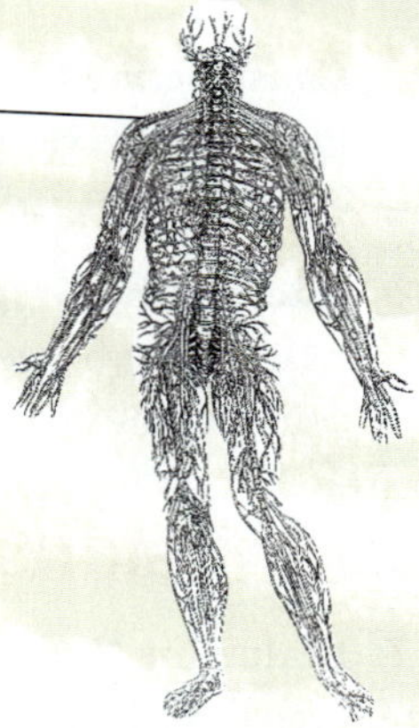

1543 – **Andries van Wesel,** better known as Andreas Vesalius, completes his *Seven Books on the Structure of the Human Body.* It is the first work on anatomy to be based on the dissection of human bodies.

1556 – **Akbar** becomes ruler of the Moghul Empire in North India, Pakistan, and Afghanistan. By ensuring religious tolerance, he establishes greater unity in India, making it one of the world's great powers.

1564 – English writers **Christopher Marlowe** and **William Shakespeare** are born.

1588 – **Queen Elizabeth I** of England sends the English fleet to repel the invasion by the Spanish Armada. The success of the English navy marks the beginning of Great Britain's status as a major naval power.

1592 $\Delta x = v_i \Delta t + \frac{1}{2}a(\Delta t)^2$

Galileo Galilei is appointed professor of mathematics at the University of Padua. While there, he performs experiments on the motions of bodies.

1603 – Kabuki theater achieved broad popularity in Japan.

1605 – The first part of **Miguel de Cervantes's** *Don Quixote* is published.

1608 – The first telescopes are constructed in the Netherlands. Using these instruments as models, **Galileo** constructs his first telescope the following year.

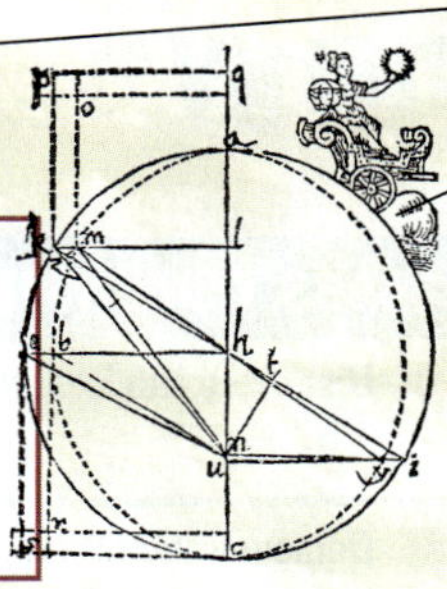

1609

$$T^2 \propto a^3$$

New Astronomy, by **Johannes Kepler**, is published. In it, Kepler demonstrates that the orbit of Mars is elliptical rather than circular.

1637 – René **Descartes's** *Discourse on Method* is published. According to Descartes's philosophy of rationalism, the laws of nature can be deduced by reason.

1644 – The Ch'ing, or Manchu, Dynasty is established in China. China becomes the most prosperous nation in the world, then declines until the Ch'ing Dynasty is replaced by the Chinese Republic in 1911.

1655 – The first paintings of Dutch artist **Jan Vermeer** are produced around this time. Vermeer's paintings portray middle- and working-class people in everyday situations.

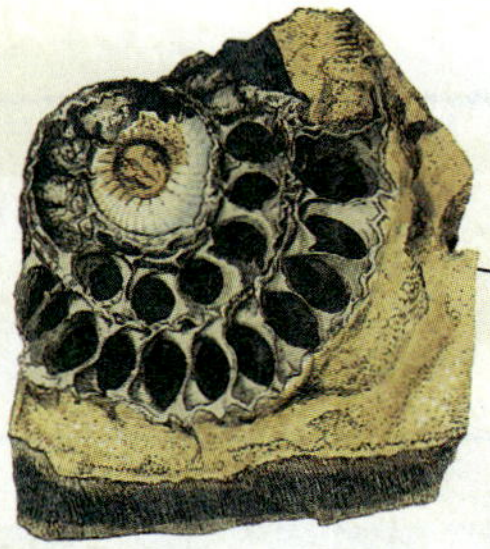

1669 – Danish geologist **Niclaus Steno** correctly determines the structure of crystals and identifies fossils as organic remains.

1678

$$c = f\lambda$$

Christiaan Huygens completes the bulk of his *Treatise on Light*, in which he presents his model of secondary wavelets, known today as Huygens' principle. The completed book is published 12 years later.

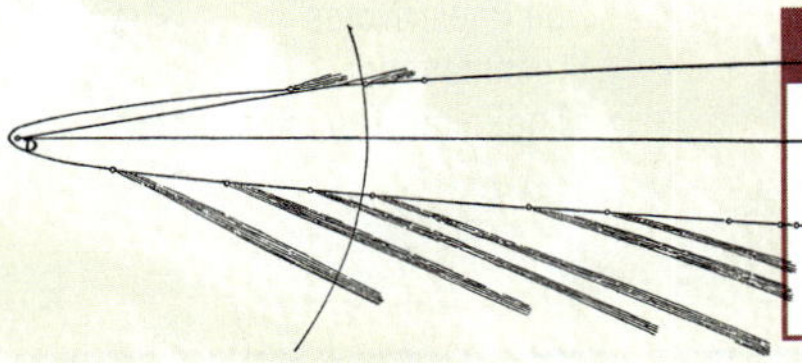

1687

$$F = ma$$

Issac Newton's masterpiece, *Mathematical Principles of Natural Philosophy*, is published. In this extensive work, Newton systematically presents a unified model of mechanics.

1610
1620
1630
1640
1650
1660
1670
1680
1690

CHAPTER 5

Work and Energy

Planning Guide

Compression Guide

To shorten instruction because of time limitations, omit the opener and abbreviate the review.

OBJECTIVES	LABS, DEMONSTRATIONS, AND ACTIVITIES	TECHNOLOGY RESOURCES
PACING • 45 min pp. 158–160 **Chapter Opener**	**ANC Discovery Lab** Exploring Work and Energy*◆ BASIC	**CD Visual Concepts,** Chapter 5 BASIC
PACING • 45 min pp. 160–163 **Section 1 Work** • Recognize the difference between the scientific and ordinary definitions of work. • Define work by relating it to force and displacement. • Identify where work is being performed in a variety of situations. • Calculate the net work done when many forces are applied to an object.	**TE Demonstration** Work, p.160 BASIC **TE Demonstration** Quantifying Work, p.162 GENERAL	**OSP Lesson Plans** **EXT Integrating Biology** Muscles and Work BASIC **TR** 15 Definition of Work **TR** 16 Positive and Negative Values of Work
PACING • 90 min pp. 164–172 **Section 2 Energy** • Identify several forms of energy. • Calculate kinetic energy for an object. • Apply the work–kinetic energy theorem to solve problems. • Distinguish between kinetic and potential energy. • Classify different types of potential energy. • Calculate the potential energy associated with an object's position.	**TE Demonstration** Potential Energy, p.169 GENERAL **ANC Invention Lab** Bungee Jumping: Energy*◆ ADVANCED	**OSP Lesson Plans** **EXT Integrating Health** Energy Costs of Walking and Running BASIC **TR** 17 Defining Potential Energy **TR** 18 Elastic Potential Energy
PACING • 90 min pp. 173–178 **Section 3 Conservation of Energy** • Identify situations in which conservation of mechanical energy is valid. • Recognize the forms that conserved energy can take. • Solve problems using conservation of mechanical energy.	**SE Quick Lab Mechanical Energy,** p. 175 GENERAL **SE Skills Practice Lab** Conservation of Mechanical Energy pp. 192–195◆ GENERAL **ANC Datasheet** Conservation of Mechanical Energy* GENERAL **TE Demonstration** Mechanical Energy, p.173 ADVANCED **TE Demonstration** Conservation of Energy, p.174 GENERAL **ANC CBL™ Experiment** Conserv. Mechanical Energy*◆ GENERAL	**OSP Lesson Plans** **TR** 19 Friction and the Non-conservation of Mechanical Energy **TR** 19A Classification of Energy **TR** 20A Energy of a Falling 75 g Egg
PACING • 45 min pp. 179–181 **Section 4 Power** • Relate the concepts of energy, time, and power. • Calculate power in two different ways. • Explain the effect of machines on work and power.	**ANC CBL™ Experiment** Loss of Mechanical Energy*◆ ADVANCED	**OSP Lesson Plans** **EXT Integrating Chemistry** Chemical Reactions BASIC

PACING • 90 min

CHAPTER REVIEW, ASSESSMENT, AND STANDARDIZED TEST PREPARATION

- **SE Chapter Highlights,** p. 183
- **SE Chapter Review,** pp. 184–189
- **SE Graphing Calculator Practice,** p. 188 GENERAL
- **SE Alternative Assessment,** p. 189 ADVANCED
- **SE Standardized Test Prep,** pp. 190–191 GENERAL
- **SE Appendix D: Equations,** p. 856
- **SE Appendix I: Additional Problems,** pp. 883–884
- **ANC Study Guide Worksheet** Mixed Review* GENERAL
- **ANC Chapter Test A*** GENERAL
- **ANC Chapter Test B*** ADVANCED
- **OSP Test Generator**

Online and Technology Resources

go.hrw.com Holt Online Learning

Visit **go.hrw.com** to find a variety of online resources. To access this chapter's extensions, enter the keyword **HF6WRKXT** and click the "go" button. Click **Holt Online Learning** for an online edition of this textbook, and other interactive resources.

HOLT Teacher's One-Stop Planner®

This DVD package includes:
- Holt Calendar Planner
- Customizable Lesson Plans
- Editable Worksheets
- ExamView® Version 6 Assessment Suite
- Interactive Teacher's Edition
- Holt PuzzlePro®
- Holt PowerNotes® Presentations
- MindPoint® Quiz Show

KEY

SE Student Edition	**OSP** One-Stop Planner	**EXT** Online Extension
TE Teacher Edition	**CD** CD or CD-ROM	* Also on One-Stop Planner
ANC Ancillary Worksheet	**TR** Teaching Transparencies	◆ Requires advance prep

SKILLS DEVELOPMENT RESOURCES	REVIEW AND ASSESSMENT	CORRELATIONS
		National Science Education Standards
SE Sample Set A Work, pp. 161–162 BASIC **TE Classroom Practice,** p. 161 BASIC **ANC Problem Workbook*** and **OSP Problem Bank** Sample Set A BASIC	**SE Section Review,** p. 163 GENERAL **ANC Study Guide Worksheet** Section 1* GENERAL **ANC Quiz** Section 1* BASIC	UCP 1, 2, 3 SAI 1, 2
SE Sample Set B Kinetic Energy, pp. 165–166 BASIC **TE Classroom Practice,** p. 165 BASIC **ANC Problem Workbook*** and **OSP Problem Bank** Sample Set B BASIC **SE Sample Set C** Work–Kinetic Energy Theorem, pp. 167–168 GENERAL **ANC Problem Workbook*** and **OSP Problem Bank** Sample Set C GENERAL **SE Sample Set D** Potential Energy, pp. 171–172 GENERAL **TE Classroom Practice,** p. 171 GENERAL **ANC Problem Workbook*** and **OSP Problem Bank** Sample Set D GENERAL	**SE Section Review,** p. 172 GENERAL **ANC Study Guide Worksheet** Section 2* GENERAL **ANC Quiz** Section 2* BASIC	UCP 1, 2, 3 HNS 1, 3 PS 5b
SE Sample Set E Conservation of Mechanical Energy, pp. 176–177 GENERAL **TE Classroom Practice,** p. 176 GENERAL **ANC Problem Workbook*** and **OSP Problem Bank** Sample Set E GENERAL **SE Appendix J: Advanced Topics** The Equivalence of Mass and Energy, pp. 918–919 ADVANCED	**SE Section Review,** p. 178 GENERAL **ANC Study Guide Worksheet** Section 3* GENERAL **ANC Quiz** Section 3* BASIC	UCP 1, 2, 3 SAI 1, 2 PS 5a, 5b
SE Sample Set F Power, pp. 180–181 BASIC **TE Classroom Practice,** p. 180 BASIC **ANC Problem Workbook*** and **OSP Problem Bank** Sample Set F BASIC **SE Conceptual Challenge,** p. 179	**SE Section Review,** p. 181 GENERAL **ANC Study Guide Worksheet** Section 4* GENERAL **ANC Quiz** Section 4* BASIC	UCP 1, 2, 3, 5 ST 1, 2 SPSP 5 PS 5b

SCILINKS. NSTA

www.scilinks.org

Maintained by the **National Science Teachers Association.**

Topic: Work
SciLinks Code: HF61674

Topic: Potential and Kinetic Energy
SciLinks Code: HF61196

Topic: Conservation of Energy
SciLinks Code: HF60345

Classroom CD-ROMs

- Guided Reading Audio Program
- Student One Stop
- Virtual Investigations
- Visual Concepts

Holt Lab Generator CD-ROM

Search for any lab by topic, standard, difficulty level, or time. Edit any lab to fit your needs, or create your own labs. Use the Lab Materials QuickList software to customize your lab materials list.

CHAPTER 5
Overview

Section 1 introduces work and shows calculations of the work done in a variety of situations.

Section 2 identifies and shows calculations using kinetic energy, the work–kinetic energy theorem, and different types of potential energy.

Section 3 explores the conditions necessary for conservation of mechanical energy and applies this principle to problem solving.

Section 4 introduces the relationships among work, time, power, force, and speed.

About the Illustration

This audiokinetic sculpture was created by George Rhoads, whose sculptures can be seen at the Boston Museum of Science, at the Port Authority Bus Terminal in New York City, and in various shopping centers. After completing the chapter, have students return to this photograph and apply the concepts of work and the conservation of energy to describe which balls probably have mostly potential energy and which have mostly kinetic energy.

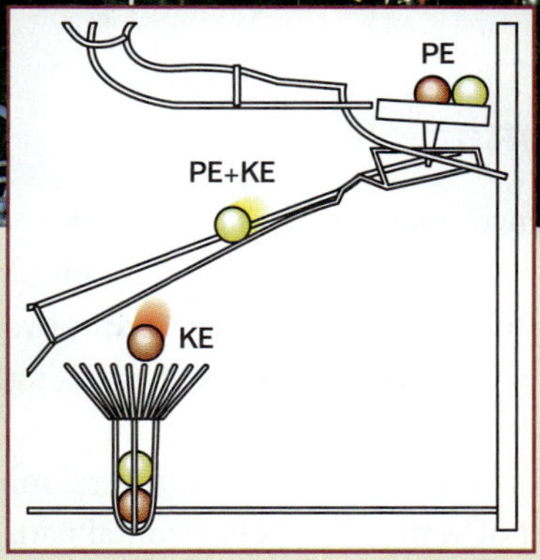

CHAPTER 5

Work and Energy

This whimsical piece of art is called an *audiokinetic sculpture*. Balls are raised to a high point on the curved blue track. As the balls move down the track, they turn levers, spin rotors, and bounce off elastic membranes. The energy that each ball has—whether associated with the ball's motion, the ball's position above the ground, or the ball's loss of mechanical energy due to friction—varies in a way that keeps the total energy of the system constant.

WHAT TO EXPECT

In this chapter, you will learn about work and different types of energy that are relevant to mechanics. Kinetic energy, which is associated with motion, and potential energy, which is related to an object's position, are two forms of energy that you will study.

Why it Matters

Work, energy, and power are related to one another. Everyday machines such as motors are usually described by the amount of work that they are capable of doing or by the amount of power that they produce.

CHAPTER PREVIEW

Tapping Prior Knowledge

Knowledge to Review

- ✔ Review the kinematic equations.
- ✔ Newton's second law states that force = mass × acceleration ($\mathbf{F} = m\mathbf{a}$).
- ✔ Kinetic friction is a resistive force exerted on a moving body by its environment.

Items to Probe

- ✔ Familiarity with phenomena of energy transformation: Ask students to describe the action of jumping up and down on a trampoline in terms of energy.
- ✔ Preconceptions about energy dissipation: Ask students if energy is ever lost in a process.

SECTION 1
General Level

Demonstration

Work BASIC

Purpose Determine whether work is done in various situations.

Materials teacher's text, spring scale, string

Procedure Hang the textbook from the scale with the string. Hold the book stationary, have the students note the scale reading, and record the weight of the book *(mg)*. Ask the students if the spring is exerting a force on the book. *(Yes, the spring exerts a force on the book that is equal to and opposite the book's weight.)* Ask students if the spring is doing work on the book, which is being held in a fixed position. *(No, because the displacement of the book is zero.)*

Now lift the book by pulling up on the spring at a constant velocity and have the students note the scale reading. Again ask the students if the spring is doing work on the book. *(Yes, the force of the lifting, equal in magnitude to the weight of the book, is upward, and the displacement is upward.)* Have students calculate the amount of work $(m \times g \times h)$.

Hold the book at shoulder height and carry it across the room at a constant speed. Ask students if work is being done on the book. *(No, because the upward force is perpendicular to the horizontal displacement.)*

SECTION 1

Work

SECTION OBJECTIVES

- **Recognize the difference between the scientific and ordinary definitions of *work*.**
- **Define *work* by relating it to force and displacement.**
- **Identify where work is being performed in a variety of situations.**
- **Calculate the net work done when many forces are applied to an object.**

DEFINITION OF WORK

Many of the terms you have encountered so far in this book have meanings in physics that are similar to their meanings in everyday life. In its everyday sense, the term *work* means to do something that takes physical or mental effort. But in physics, work has a distinctly different meaning. Consider the following situations:

- A student holds a heavy chair at arm's length for several minutes.
- A student carries a bucket of water along a horizontal path while walking at constant velocity.

It might surprise you to know that as the term work is used in physics, there is no work done on the chair or the bucket, even though effort is required in both cases. We will return to these examples later.

Work is done on an object when a force causes a displacement of the object

Imagine that your car, like the car shown in **Figure 1,** has run out of gas and you have to push it down the road to the gas station. If you push the car with a constant horizontal force, the **work** you do on the car is equal to the magnitude of the force, F, times the magnitude of the displacement of the car. Using the symbol d instead of Δx for displacement, we define work for a constant force as:

work
the product of the component of a force along the direction of displacement and the magnitude of the displacement

$$W = Fd$$

Work is not done on an object unless the object is moved with the action of a force. The application of a force alone does not constitute work. For this reason, no work is done on the chair when a student holds the chair at arm's length. Even though the student exerts a force to support the chair, the chair does not move. The student's tired arms suggest that work is being done, which is indeed true. The quivering muscles in the student's arms go through many small displacements and do work within the student's body. However, work is not done on the chair.

Figure 1
This person exerts a constant force on the car and displaces it to the left. The work done on the car by the person is equal to the force the person exerts times the displacement of the car.

Work is done only when components of a force are parallel to a displacement

When the force on an object and the object's displacement are in different directions, only the component of the force that is parallel to the object's displacement does work. Components of the force perpendicular to a displacement do not do work.

For example, imagine pushing a crate along the ground. If the force you exert is horizontal, all of your effort moves the crate. If your force is at an angle, only the horizontal component of your applied force causes a displacement and contributes to the work. If the angle between the force and the direction of the displacement is θ, as in **Figure 2,** work can be expressed as follows:

$$W = Fd\cos\theta$$

If $\theta = 0°$, then $\cos 0° = 1$ and $W = Fd$, which is the definition of work given earlier. If $\theta = 90°$, however, then $\cos 90° = 0$ and $W = 0$. So, no work is done on a bucket of water being carried by a student walking horizontally. The upward force exerted by the student to support the bucket is perpendicular to the displacement of the bucket, which results in no work done on the bucket.

Finally, if many constant forces are acting on an object, you can find the *net* work done on the object by first finding the net force on the object.

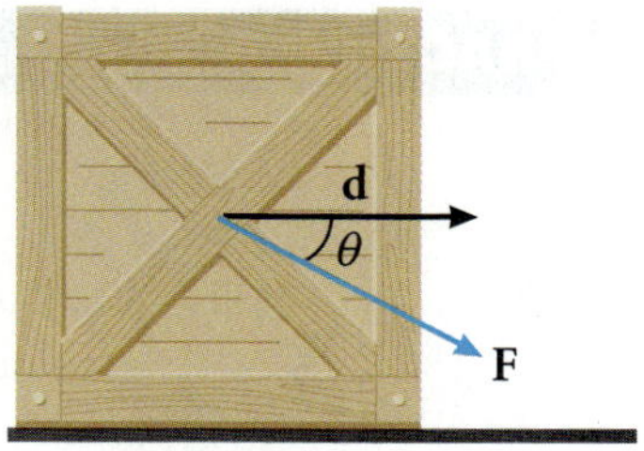

$W = Fd\cos\theta$

Figure 2
The work done on this crate is equal to the force times the displacement times the cosine of the angle between them.

NET WORK DONE BY A CONSTANT NET FORCE

$$W_{net} = F_{net}d\cos\theta$$

net work = net force × displacement × cosine of the angle between them

Work has dimensions of force times length. In the SI system, work has a unit of newtons times meters (N•m), or joules (J). To give you an idea of how large a joule is, consider that the work done in lifting an apple from your waist to the top of your head is about 1 J.

Did you know?

The joule is named for the British physicist James Prescott Joule (1818–1889). Joule made major contributions to the understanding of energy, heat, and electricity.

SAMPLE PROBLEM A

Work

PROBLEM

How much work is done on a vacuum cleaner pulled 3.0 m by a force of 50.0 N at an angle of 30.0° above the horizontal?

SOLUTION

Given: $F = 50.0$ N $\quad \theta = 30.0°$ $\quad d = 3.0$ m

Unknown: $W = ?$

Use the equation for net work done by a constant force:

$$W = Fd\cos\theta$$

Only the horizontal component of the applied force is doing work on the vacuum cleaner.

$$W = (50.0 \text{ N})(3.0 \text{ m})(\cos 30.0°)$$

$$\boxed{W = 130 \text{ J}}$$

SECTION 1

Teaching Tip — GENERAL

Point out that although the crate in **Figure 2** is a large object, we think of its mass as reduced to a point at its center to simplify the situation.

Classroom Practice

Work
A 20.0 kg suitcase is raised 3.0 m above a platform by a conveyor belt. How much work is done on the suitcase?

Answer
5.9×10^2 J

PROBLEM GUIDE A

Use this guide to assign problems.
SE = Student Edition Textbook
PW = Problem Workbook
PB = Problem Bank on the One-Stop Planner (OSP)

Solving for:

W	**SE** Sample, 1–3; Ch. Rvw. 7–10, 45–46, 50 **PW** 8–11 **PB** 5–7
F	**SE** Ch. Rvw. 50 **PW** 5–7 **PB** Sample, 1–4
d	**SE** 4 **PW** Sample, 1–4 **PB** 8–10

***Challenging Problem**
Consult the printed Solutions Manual or the OSP for detailed solutions.

SECTION 1

ANSWERS

Practice A

1. 1.50×10^7 J
2. 7.0×10^2 J
3. 1.6×10^3 J
4. 1.1 m

Demonstration

Quantifying Work — GENERAL

Purpose Demonstrate the relationship between the direction and the magnitude of a force.

Materials plastic sled or piece of cardboard, 3 m and 1 m lengths of rope

Procedure Attach both ropes to the sled, and ask a student volunteer to sit on the sled. Ask students whether it will require more force to pull the sled across the floor with the 1 m rope or with the 3 m rope. *(The 3 m rope will require less force.)* Have a student try to pull the sled with each rope and report to the class which way is easier. Sketch both situations on the board, emphasizing that the horizontal component of the force is smaller with the short rope because it is held at a greater angle above the horizontal.

PRACTICE A

Work

1. A tugboat pulls a ship with a constant net horizontal force of 5.00×10^3 N and causes the ship to move through a harbor. How much work is done on the ship if it moves a distance of 3.00 km?
2. A weight lifter lifts a set of weights a vertical distance of 2.00 m. If a constant net force of 350 N is exerted on the weights, what is the net work done on the weights?
3. A shopper in a supermarket pushes a cart with a force of 35 N directed at an angle of 25° downward from the horizontal. Find the work done by the shopper on the cart as the shopper moves along a 50.0 m length of aisle.
4. If 2.0 J of work is done in raising a 180 g apple, how far is it lifted?

The sign of work is important

Work is a scalar quantity and can be positive or negative, as shown in **Figure 3.** Work is positive when the component of force is in the same direction as the displacement. For example, when you lift a box, the work done by the force you exert on the box is positive because that force is upward, in the same direction as the displacement. Work is negative when the force is in the direction

extension

Integrating Biology

Visit go.hrw.com for the activity "Muscles and Work."

Keyword HF6WRKX

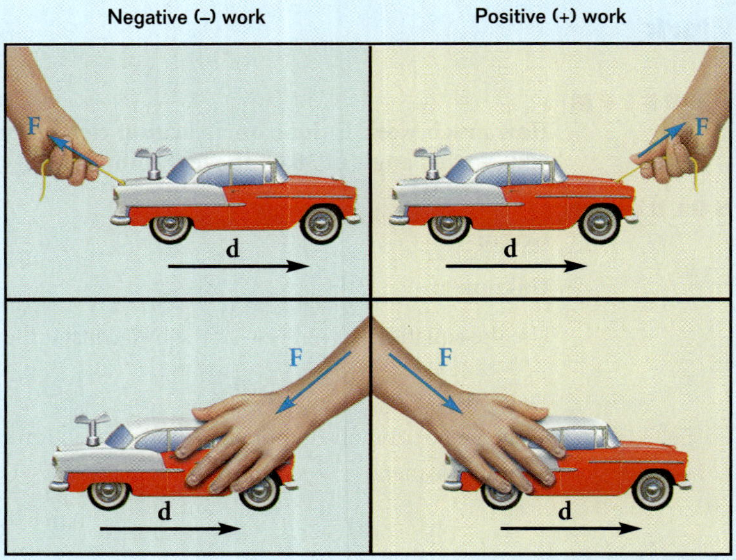

Figure 3
Depending on the angle, an applied force can either cause a moving car to slow down (left), which results in negative work done on the car, or speed up (right), which results in positive work done on the car.

opposite the displacement. For example, the force of kinetic friction between a sliding box and the floor is opposite to the displacement of the box, so the work done by the force of friction on the box is negative. If you are very careful in applying the equation for work, your answer will have the correct sign: $\cos \theta$ is negative for angles greater than 90° but less than 270°.

If the work done on an object results only in a change in the object's speed, the sign of the net work on the object tells you whether the object's speed is increasing or decreasing. If the net work is positive, the object speeds up and work is done *on* the object. If the net work is negative, the object slows down and work is done *by* the object on something else.

SCILINKS.
www.scilinks.org
Topic: Work
Code: HF61674

SECTION REVIEW

1. For each of the following cases, indicate whether the work done on the second object in each example will have a positive or a negative value.

a. The road exerts a friction force on a speeding car skidding to a stop.

b. A rope exerts a force on a bucket as the bucket is raised up a well.

c. Air exerts a force on a parachute as the parachutist falls to Earth.

2. If a neighbor pushes a lawnmower four times as far as you do but exerts only half the force, which one of you does more work and by how much?

3. A worker pushes a 1.50×10^3 N crate with a horizontal force of 345 N a distance of 24.0 m. Assume the coefficient of kinetic friction between the crate and the floor is 0.220.

a. How much work is done by the worker on the crate?

b. How much work is done by the floor on the crate?

c. What is the net work done on the crate?

4. A 0.075 kg ball in a kinetic sculpture moves at a constant speed along a motorized vertical conveyor belt. The ball rises 1.32 m above the ground. A constant frictional force of 0.350 N acts in the direction opposite the conveyor belt's motion. What is the net work done on the ball?

5. Critical Thinking For each of the following statements, identify whether the everyday or the scientific meaning of work is intended.

a. Jack had to work against time as the deadline neared.

b. Jill had to work on her homework before she went to bed.

c. Jack did work carrying the pail of water up the hill.

6. Critical Thinking Determine whether work is being done in each of the following examples:

a. a train engine pulling a loaded boxcar initially at rest

b. a tug of war that is evenly matched

c. a crane lifting a car

SECTION 1

Teaching Tip — BASIC

Write the following table on the board to help students remember the various situations that affect the sign of work.

Force is in direction of motion.	positive work
Force opposes motion.	negative work
Force is 90° to motion.	no work
Object is not in motion.	no work

SECTION REVIEW ANSWERS

1. a. negative
b. positive
c. negative

2. the neighbor; twice as much

3. a. 8.28×10^3 J
b. -7.92×10^3 J
c. 3.6×10^2 J

4. 0.00 J

5. a. everyday sense
b. everyday sense
c. scientific sense

6. a. yes
b. no
c. yes

SECTION 2
General Level

Misconception Alert — BASIC

Students may think that kinetic energy depends on the direction of motion. Ask them to compare the kinetic energy of identical cars traveling at the same speed in each of the following situations: one driving north, one driving south, one driving uphill, and one driving downhill. *(The kinetic energy is the same in each case because kinetic energy depends only on mass and speed, which are the same in each case.)*

The Language of Physics

The symbol for kinetic energy, *KE*, may look like the product of two variables (*K* and *E*) to some students. Point out that the two letters together designate kinetic energy. This symbol for kinetic energy is not universal. Some books use the letter *K* alone; others use *E* alone and specify the kind of energy in context.

SECTION 2

Energy

SECTION OBJECTIVES

- **Identify several forms of energy.**
- **Calculate kinetic energy for an object.**
- **Apply the work–kinetic energy theorem to solve problems.**
- **Distinguish between kinetic and potential energy.**
- **Classify different types of potential energy.**
- **Calculate the potential energy associated with an object's position.**

KINETIC ENERGY

Kinetic energy is energy associated with an object in motion. **Figure 4** shows a cart of mass m moving to the right on a frictionless air track under the action of a constant net force, **F**, acting to the right. Because the force is constant, we know from Newton's second law that the cart moves with a constant acceleration, **a.** While the force is applied, the cart accelerates from an initial velocity v_i to a final velocity v_f. If the cart is displaced a distance of Δx, the work done by **F** during this displacement is

$$W_{net} = F\Delta x = ma\Delta x$$

When you studied one-dimensional motion, you learned that the following relationship holds when an object undergoes constant acceleration:

$$v_f^2 = v_i^2 + 2a\Delta x$$

$$a\Delta x = \frac{v_f^2 - v_i^2}{2}$$

Substituting this result into the equation $W_{net} = ma\Delta x$ gives

$$W_{net} = m\left(\frac{v_f^2 - v_i^2}{2}\right)$$

$$W_{net} = \frac{1}{2}mv_f^2 - \frac{1}{2}mv_i^2$$

kinetic energy

the energy of an object that is due to the object's motion

Kinetic energy depends on speed and mass

The quantity $\frac{1}{2}mv^2$ has a special name in physics: **kinetic energy.** The kinetic energy of an object with mass m and speed v, when treated as a particle, is given by the expression shown on the next page.

extension

Integrating Health
Visit go.hrw.com for the activity "Energy Costs of Walking and Running."

Keyword HF6WRKX

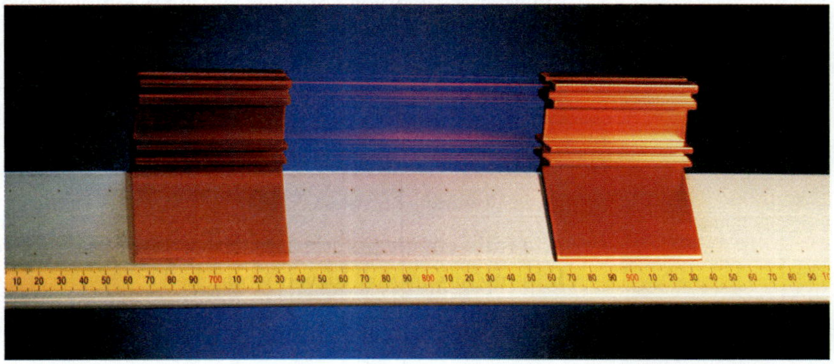

Figure 4
The work done on an object by a constant force equals the object's mass times its acceleration times its displacement.

KINETIC ENERGY

$$KE = \frac{1}{2}mv^2$$

$$\textbf{kinetic energy} = \frac{1}{2} \times \textbf{mass} \times (\textbf{speed})^2$$

Kinetic energy is a scalar quantity, and the SI unit for kinetic energy (and all other forms of energy) is the joule. Recall that a joule is also used as the basic unit for work.

Kinetic energy depends on both an object's speed and its mass. If a bowling ball and a volleyball are traveling at the same speed, which do you think has more kinetic energy? You may think that because they are moving with identical speeds they have exactly the same kinetic energy. However, the bowling ball has more kinetic energy than the volleyball traveling at the same speed because the bowling ball has more mass than the volleyball.

SAMPLE PROBLEM B

Kinetic Energy

PROBLEM

A 7.00 kg bowling ball moves at 3.00 m/s. How fast must a 2.45 g table-tennis ball move in order to have the same kinetic energy as the bowling ball? Is this speed reasonable for a table-tennis ball in play?

SOLUTION

Given: The subscripts b and t indicate the bowling ball and the table-tennis ball, respectively.

$m_b = 7.00$ kg $m_t = 2.45$ g $v_b = 3.00$ m/s

Unknown: $v_t = ?$

First, calculate the kinetic energy of the bowling ball.

$$KE_b = \frac{1}{2}m_b v_b^2 = \frac{1}{2}(7.00 \text{ kg})(3.00 \text{ m/s})^2 = 31.5 \text{ J}$$

Then, solve for the speed of the table-tennis ball having the same kinetic energy as the bowling ball.

$$KE_t = \frac{1}{2}m_t v_t^2 = KE_b = 31.5 \text{ J}$$

$$v_t = \sqrt{\frac{2KE_b}{m_t}} = \sqrt{\frac{(2)(31.5 \text{ J})}{2.45 \times 10^{-3} \text{ kg}}}$$

$$\boxed{v_t = 1.60 \times 10^2 \text{ m/s}}$$

This speed would be very fast for a table-tennis ball.

Teaching Tip — GENERAL

A joule is defined as a kilogram-meter squared per second squared. Point out to students how this equation for kinetic energy has the units of energy. Be sure that they know that mass must be in kg and speed must be in m/s for the units of energy to work out properly.

Classroom Practice

Kinetic Energy

A 6.0 kg cat runs after a mouse at 10.0 m/s. What is the cat's kinetic energy?

Answer

3.0×10^2 J

PROBLEM GUIDE B

Use this guide to assign problems.
SE = Student Edition Textbook
PW = Problem Workbook
PB = Problem Bank on the One-Stop Planner (OSP)

Solving for:

KE	**SE** 3–4; Ch. Rvw. 14, 19, 44 **PW** 6, 8–9 **PB** Sample, 1–4
v	**SE** Sample, 1–2; Ch. Rvw. 20, 37, 44, 48 **PW** 5–7 **PB** 8–10
m	**SE** 5 **PW** Sample, 1–4 **PB** 5–7

***Challenging Problem**
Consult the printed Solutions Manual or the OSP for detailed solutions.

ANSWERS

Practice B

1. 1.7×10^2 m/s
2. 38.8 m/s
3. the bullet with the greater mass; 2 to 1
4. 2.4 J, 9.6 J; the bullet with the greater speed; 1 to 4
5. 1.6×10^3 kg

The Language of Physics — BASIC

The symbol Δ (the Greek letter *delta*) is used to denote change. Students should be familiar with this symbol from earlier chapters. Point out that although the context is different, the symbol means the same thing; namely, a difference between two quantities. The subscripts i and f used with ME stand for initial and final amounts, respectively, of mechanical energy. Thus, ΔME is the difference between ME_f and ME_i, or $\Delta ME = ME_f - ME_i$.

PRACTICE B

Kinetic Energy

1. Calculate the speed of an 8.0×10^4 kg airliner with a kinetic energy of 1.1×10^9 J.
2. What is the speed of a 0.145 kg baseball if its kinetic energy is 109 J?
3. Two bullets have masses of 3.0 g and 6.0 g, respectively. Both are fired with a speed of 40.0 m/s. Which bullet has more kinetic energy? What is the ratio of their kinetic energies?
4. Two 3.0 g bullets are fired with speeds of 40.0 m/s and 80.0 m/s, respectively. What are their kinetic energies? Which bullet has more kinetic energy? What is the ratio of their kinetic energies?
5. A car has a kinetic energy of 4.32×10^5 J when traveling at a speed of 23 m/s. What is its mass?

work–kinetic energy theorem

the net work done by all the forces acting on an object is equal to the change in the object's kinetic energy

Figure 5
The moving hammer has kinetic energy and can do work on the puck, which can rise against gravity and ring the bell.

The net work done on a body equals its change in kinetic energy

The equation $W_{net} = \frac{1}{2}mv_f^2 - \frac{1}{2}mv_i^2$ derived at the beginning of this section says that the net work done by a *net* force acting on an object is equal to the *change* in the kinetic energy of the object. This important relationship, known as the **work–kinetic energy theorem,** is often written as follows:

WORK–KINETIC ENERGY THEOREM

$$W_{net} = \Delta KE$$

net work = change in kinetic energy

When you use this theorem, you must include all the forces that do work on the object in calculating the net work done. From this theorem, we see that the speed of the object increases if the net work done on it is positive, because the final kinetic energy is greater than the initial kinetic energy. The object's speed decreases if the net work is negative, because the final kinetic energy is less than the initial kinetic energy.

The work–kinetic energy theorem allows us to think of kinetic energy as the work that an object can do while the object changes speed or as the amount of energy stored in the motion of an object. For example, the moving hammer in the ring-the-bell game in **Figure 5** has kinetic energy and can therefore do work on the puck. The puck can do work against gravity by moving up and striking the bell. When the bell is struck, part of the energy is converted into sound.

SAMPLE PROBLEM C

Work–Kinetic Energy Theorem

PROBLEM

On a frozen pond, a person kicks a 10.0 kg sled, giving it an initial speed of 2.2 m/s. How far does the sled move if the coefficient of kinetic friction between the sled and the ice is 0.10?

SOLUTION

1. DEFINE

Given: $m = 10.0$ kg $\quad v_i = 2.2$ m/s $\quad v_f = 0$ m/s $\quad \mu_k = 0.10$

Unknown: $d = ?$

Diagram:

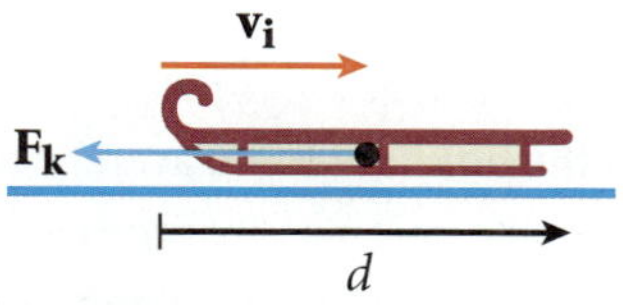

2. PLAN

Choose an equation or situation:
This problem can be solved using the definition of work and the work–kinetic energy theorem.

$$W_{net} = F_{net} d \cos\theta$$

The net work done on the sled is provided by the force of kinetic friction.

$$W_{net} = F_k d \cos\theta = \mu_k mgd \cos\theta$$

The force of kinetic friction is in the direction opposite d, so $\theta = 180°$. Because the sled comes to rest, the final kinetic energy is zero.

$$W_{net} = \Delta KE = KE_f - KE_i = -\tfrac{1}{2} m v_i^2$$

Use the work-kinetic energy theorem, and solve for d.

$$-\tfrac{1}{2} m v_i^2 = \mu_k mgd \cos\theta$$

$$d = \frac{-v_i^2}{2\mu_k g \cos\theta}$$

3. CALCULATE

Substitute values into the equation:

$$d = \frac{-(2.2 \text{ m/s})^2}{2(0.10)(9.81 \text{ m/s}^2)(\cos 180°)}$$

$$\boxed{d = 2.5 \text{ m}}$$

4. EVALUATE

According to Newton's second law, the acceleration of the sled is about -1 m/s^2 and the time it takes the sled to stop is about 2 s. Thus, the distance the sled traveled in the given amount of time should be less than the distance it would have traveled in the absence of friction.

$$2.5 \text{ m} < (2.2 \text{ m/s})(2 \text{ s}) = 4.4 \text{ m}$$

PROBLEM GUIDE C

Use this guide to assign problems.
SE = Student Edition Textbook
PW = Problem Workbook
PB = Problem Bank on the One-Stop Planner (OSP)

Solving for:

d	**SE** Sample, 1–3; Ch. Rvw. 21–22 **PW** 7 **PB** 5
F	**PW** 3 **PB** 3–4
W	**SE** 5; Ch. Rvw. 44, 46 **PW** 5 **PB** 6
KE	**SE** 5; Ch. Rvw. 40, 44 **PW** Sample, 1–2 **PB** 7–8
v	**SE** 5; Ch. Rvw. 38, 44 **PW** 4 **PB** 9–10
m	**PW** 6 **PB** Sample, 1–2

***Challenging Problem**
Consult the printed Solutions Manual or the OSP for detailed solutions.

167

ANSWERS

Practice C

1. 7.8 m
2. 21 m
3. 5.1 m
4. 3.0×10^2 N

Why it Matters

The Energy in Food

This feature introduces students to another form of potential energy: chemical energy. Chemical energy, like gravitational potential energy and elastic potential energy, is a latent, stored form of energy. However, chemical energy is not simply or directly dependent on relative position. Instead, chemical energy depends on the molecular structure and the strength of chemical bonds, and this strength depends on the relative affinities of different atoms in molecules.

PRACTICE C

Work–Kinetic Energy Theorem

1. A student wearing frictionless in-line skates on a horizontal surface is pushed by a friend with a constant force of 45 N. How far must the student be pushed, starting from rest, so that her final kinetic energy is 352 J?

2. A 2.0×10^3 kg car accelerates from rest under the actions of two forces. One is a forward force of 1140 N provided by traction between the wheels and the road. The other is a 950 N resistive force due to various frictional forces. Use the work–kinetic energy theorem to determine how far the car must travel for its speed to reach 2.0 m/s.

3. A 2.1×10^3 kg car starts from rest at the top of a driveway that is sloped at an angle of 20.0° with the horizontal. An average friction force of 4.0×10^3 N impedes the car's motion so that the car's speed at the bottom of the driveway is 3.8 m/s. What is the length of the driveway?

4. A 75 kg bobsled is pushed along a horizontal surface by two athletes. After the bobsled is pushed a distance of 4.5 m starting from rest, its speed is 6.0 m/s. Find the magnitude of the net force on the bobsled.

Why it Matters

The Energy in Food

The food that you eat provides your body with energy. Your body needs this energy to move your muscles, to maintain a steady internal temperature, and to carry out many other bodily processes. The energy in food is stored as a kind of *potential energy* in the chemical bonds within sugars and other organic molecules.

When you digest food, some of this energy is released. The energy is then stored again in sugar molecules, usually as glucose. When cells in your body need energy to carry out cellular processes, the cells break down the glucose molecules through a process called *cellular respiration.* The primary product of cellular respiration is a high-energy molecule called *adenosine triphosphate* (ATP), which has a significant role in many chemical reactions in cells.

Nutritionists and food scientists use units of Calories to quantify the energy in food. A standard calorie (cal) is defined as the amount of energy required to increase the temperature of 1 mL of water by 1°C, which equals 4.186 joules (J). A *food* Calorie is actually 1 kilocalorie, or 4186 J.

People who are trying to lose weight often monitor the number of Calories that they eat each day. These people count Calories because the body stores unused energy as fat. Most food labels show the number of Calories in each serving of food. The amount of energy that your body needs each day depends on many factors, including your age, your weight, and the amount of exercise that you get. A typically healthy and active person requires about 1500 to 2000 Calories per day.

POTENTIAL ENERGY

Consider the balanced boulder shown in **Figure 6.** As long as the boulder remains balanced, it has no kinetic energy. If it becomes unbalanced, it will fall vertically to the desert floor and will gain kinetic energy as it falls. What is the origin of this kinetic energy?

Potential energy is stored energy

Potential energy is associated with an object that has the potential to move because of its position relative to some other location. Unlike kinetic energy, potential energy depends not only on the properties of an object but also on the object's interaction with its environment.

Gravitational potential energy depends on height from a zero level

You learned earlier how gravitational forces influence the motion of a projectile. If an object is thrown up in the air, the force of gravity will eventually cause the object to fall back down. Similarly, the force of gravity will cause the unbalanced boulder in the previous example to fall. The energy associated with an object due to the object's position relative to a gravitational source is called **gravitational potential energy.**

Imagine an egg falling off a table. As it falls, it gains kinetic energy. But where does the egg's kinetic energy come from? It comes from the gravitational potential energy that is associated with the egg's initial position on the table relative to the floor. Gravitational potential energy can be determined using the following equation:

GRAVITATIONAL POTENTIAL ENERGY

$$PE_g = mgh$$

gravitational potential energy = mass × free-fall acceleration × height

The SI unit for gravitational potential energy, like for kinetic energy, is the joule. Note that the definition for gravitational potential energy in this chapter is valid only when the free-fall acceleration is constant over the entire height, such as at any point near the Earth's surface. Furthermore, gravitational potential energy depends on both the height and the free-fall acceleration, neither of which is a property of an object.

Also note that the height, h, is measured from an arbitrary zero level. In the example of the egg, if the floor is the zero level, then h is the height of the table, and mgh is the gravitational potential energy relative to the floor. Alternatively, if the table is the zero level, then h is zero. Thus, the potential energy associated with the egg relative to the table is zero.

Figure 6
Energy is present in this example, but it is not kinetic energy because there is no motion. What kind of energy is it?

potential energy

the energy associated with an object because of the position, shape, or condition of the object

gravitational potential energy

the potential energy stored in the gravitational fields of interacting bodies

SCILINKS.
www.scilinks.org
Topic: Potential and Kinetic Energy
Code: HF61196

SECTION 2

The Language of Physics — GENERAL

In the symbol PE_g, PE stands for potential energy, and the subscript g specifies that the source of this potential energy is gravity. Some texts use U rather than PE to represent potential energy.

Demonstration

Potential Energy — GENERAL

Purpose Show that potential energy is stored energy.

Materials a racquetball cut in half (note that you may have to trim the cut ball slightly more for it to work properly)

Caution *Do not face the area where you drop the ball, because it may rise up high enough to hit you.*

Procedure Pop the hollow hemisphere of the ball inside out and hold it hollow-side up. Before dropping it from a low height, ask students to predict whether it will bounce back and, if so, approximately how high.

Release the ball. The half ball will pop out on impact with the surface and will bounce up to a greater height with its hollow side facing down. Ask students where the additional gravitational potential energy came from. *(Elastic potential energy was stored in the half ball when it was inverted inside out.)*

SECTION 2

Misconception Alert — BASIC

Some students do not realize that the potential energy of an object is relative. Point out that the zero-level for measuring height is arbitrarily defined in each problem. The potential energy is calculated relative to that level. Ask students how they would calculate the potential energy of a book on their desk relative to the desk, to the classroom floor, and to the roof.

Teaching Tip — ADVANCED

Ask students whether it is possible to have a negative potential energy. *(Yes, a negative potential energy means that work must be done to bring an object to the zero-level.)* Then ask whether an object can have a positive potential energy relative to one reference point and a negative potential energy relative to another reference point. Have students give examples to support their answer. *(Yes. For example, a book that is 0.5 m below a table and 0.5 m above the ground has a positive potential energy relative to the ground but a negative potential energy relative to the table.)*

Visual Strategy — GENERAL

Figure 8

Point out that the spring's potential energy depends on the difference between the spring's relaxed and compressed lengths.

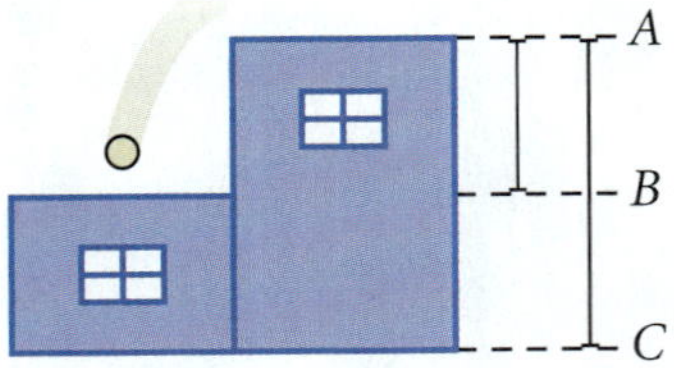

Figure 7
If *B* is the zero level, then all the gravitational potential energy is converted to kinetic energy as the ball falls from *A* to *B*. If *C* is the zero level, then only part of the total gravitational potential energy is converted to kinetic energy during the fall from *A* to *B*.

Suppose you drop a volleyball from a second-floor roof and it lands on the first-floor roof of an adjacent building (see **Figure 7**). If the height is measured from the ground, the gravitational potential energy is not zero because the ball is still above the ground. But if the height is measured from the first-floor roof, the potential energy is zero when the ball lands on the roof.

Gravitational potential energy is a result of an object's position, so it must be measured relative to some *zero level*. The zero level is the vertical coordinate at which gravitational potential energy is defined to be zero. This zero level is arbitrary, and it is chosen to make a specific problem easier to solve. In many cases, the statement of the problem suggests what to use as a zero level.

Elastic potential energy depends on distance compressed or stretched

elastic potential energy

the energy available for use when a deformed elastic object returns to its original configuration

Imagine you are playing with a spring on a tabletop. You push a block into the spring, compressing the spring, and then release the block. The block slides across the tabletop. The kinetic energy of the block came from the stored energy in the compressed spring. This potential energy is called **elastic potential energy.** Elastic potential energy is stored in any compressed or stretched object, such as a spring or the stretched strings of a tennis racket or guitar.

The length of a spring when no external forces are acting on it is called the *relaxed length* of the spring. When an external force compresses or stretches the spring, elastic potential energy is stored in the spring. The amount of energy depends on the distance the spring is compressed or stretched from its relaxed length, as shown in **Figure 8.** Elastic potential energy can be determined using the following equation:

ELASTIC POTENTIAL ENERGY

$$PE_{elastic} = \frac{1}{2}kx^2$$

$$\textbf{elastic potential energy} = \frac{1}{2} \times \textbf{spring constant} \times \left(\begin{matrix}\textbf{distance compressed} \\ \textbf{or stretched}\end{matrix}\right)^2$$

spring constant

a parameter that is a measure of a spring's resistance to being compressed or stretched

The symbol k is called the **spring constant,** or force constant. For a flexible spring, the spring constant is small, whereas for a stiff spring, the spring constant is large. Spring constants have units of newtons divided by meters (N/m).

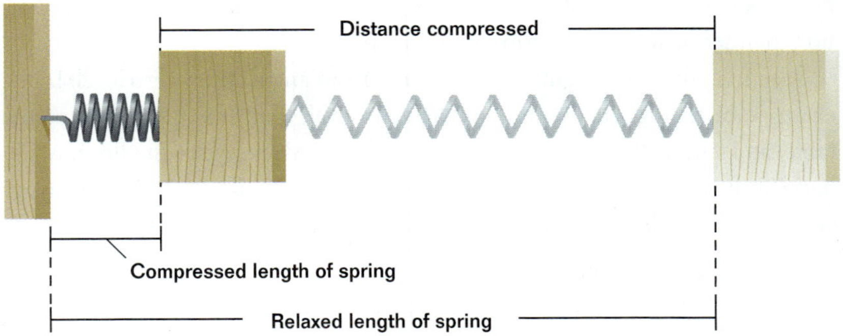

Figure 8
The distance to use in the equation for elastic potential energy is the distance the spring is compressed or stretched from its relaxed length.

SAMPLE PROBLEM D

Potential Energy

PROBLEM

A 70.0 kg stuntman is attached to a bungee cord with an unstretched length of 15.0 m. He jumps off a bridge spanning a river from a height of 50.0 m. When he finally stops, the cord has a stretched length of 44.0 m. Treat the stuntman as a point mass, and disregard the weight of the bungee cord. Assuming the spring constant of the bungee cord is 71.8 N/m, what is the total potential energy relative to the water when the man stops falling?

SOLUTION

1. DEFINE

Given:

$m = 70.0$ kg $k = 71.8$ N/m $g = 9.81$ m/s^2

$h = 50.0$ m $-$ 44.0 m = 6.0 m

$x = 44.0$ m $-$ 15.0 m = 29.0 m

$PE = 0$ J at river level

Unknown: $PE_{tot} = ?$

Diagram:

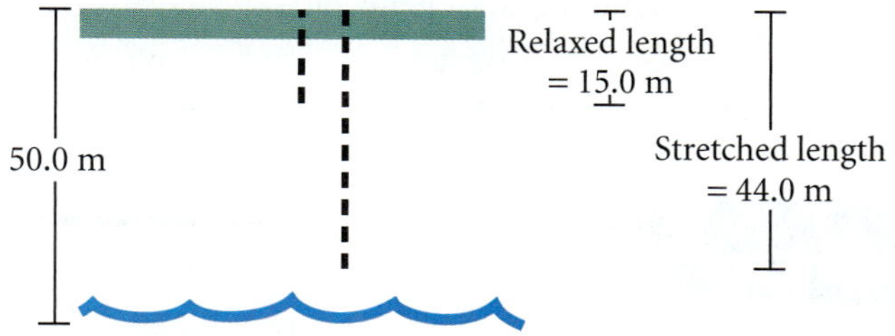

2. PLAN

Choose an equation or situation:

The zero level for gravitational potential energy is chosen to be at the surface of the water. The total potential energy is the sum of the gravitational and elastic potential energy.

$$PE_{tot} = PE_g + PE_{elastic}$$

$$PE_g = mgh$$

$$PE_{elastic} = \frac{1}{2}kx^2$$

> **TIP** *Choose the zero potential energy location that makes the problem easiest to solve.*

3. CALCULATE

Substitute the values into the equations and solve:

$$PE_g = (70.0 \text{ kg})(9.81 \text{ m/s}^2)(6.0 \text{ m}) = 4.1 \times 10^3 \text{ J}$$

$$PE_{elastic} = \frac{1}{2}(71.8 \text{ N/m})(29.0 \text{ m})^2 = 3.02 \times 10^4 \text{ J}$$

$$PE_{tot} = 4.1 \times 10^3 \text{ J} + 3.02 \times 10^4 \text{ J}$$

$$\boxed{PE_{tot} = 3.43 \times 10^4 \text{ J}}$$

4. EVALUATE

One way to evaluate the answer is to make an order-of-magnitude estimate. The gravitational potential energy is on the order of 10^2 kg $\times$ 10 m/s^2 $\times$ 10 m = 10^4 J. The elastic potential energy is on the order of 1×10^2 N/m $\times$ 10^2 m^2 = 10^4 J. Thus, the total potential energy should be on the order of 2×10^4 J. This number is close to the actual answer.

Classroom Practice

Potential Energy

When a 2.00 kg mass is attached to a vertical spring, the spring is stretched 10.0 cm such that the mass is 50.0 cm above the table.

a. What is the gravitational potential energy associated with this mass relative to the table?

b. What is the spring's elastic potential energy if the spring constant is 400.0 N/m?

c. What is the total potential energy of this system?

Answers

a. 9.81 J

b. 2.00 J

c. 11.81 J

PROBLEM GUIDE D

Use this guide to assign problems.
SE = Student Edition Textbook
PW = Problem Workbook
PB = Problem Bank on the One-Stop Planner (OSP)

Solving for:

PE	**SE** Sample, 1–3; Ch. Rvw. 23–25, 37 **PW** 7–9 **PB** 9–10
k	**PW** 10 **PB** Sample, 1–3
h or *d*	**PW** 4–6, 10 **PB** 4–6
m	**PW** Sample, 1–3 **PB** 7–8

***Challenging Problem**
Consult the printed Solutions Manual or the OSP for detailed solutions.

SECTION 2

ANSWERS

Practice D

1. 3.3 J
2. 3.1×10^{-2} J
3. a. 785 J
   b. 105 J
   c. 0.00 J

PRACTICE D

Potential Energy

1. A spring with a force constant of 5.2 N/m has a relaxed length of 2.45 m. When a mass is attached to the end of the spring and allowed to come to rest, the vertical length of the spring is 3.57 m. Calculate the elastic potential energy stored in the spring.
2. The staples inside a stapler are kept in place by a spring with a relaxed length of 0.115 m. If the spring constant is 51.0 N/m, how much elastic potential energy is stored in the spring when its length is 0.150 m?
3. A 40.0 kg child is in a swing that is attached to ropes 2.00 m long. Find the gravitational potential energy associated with the child relative to the child's lowest position under the following conditions:
   a. when the ropes are horizontal
   b. when the ropes make a 30.0° angle with the vertical
   c. at the bottom of the circular arc

SECTION REVIEW

1. A pinball bangs against a bumper, giving the ball a speed of 42 cm/s. If the ball has a mass of 50.0 g, what is the ball's kinetic energy in joules?
2. A student slides a 0.75 kg textbook across a table, and it comes to rest after traveling 1.2 m. Given that the coefficient of kinetic friction between the book and the table is 0.34, use the work–kinetic energy theorem to find the book's initial speed.
3. A spoon is raised 21.0 cm above a table. If the spoon and its contents have a mass of 30.0 g, what is the gravitational potential energy associated with the spoon at that height relative to the surface of the table?
4. **Critical Thinking** What forms of energy are involved in the following situations?
   a. a bicycle coasting along a level road
   b. heating water
   c. throwing a football
   d. winding the mainspring of a clock
5. **Critical Thinking** How do the forms of energy in item 4 differ from one another? Be sure to discuss mechanical versus nonmechanical energy, kinetic versus potential energy, and gravitational versus elastic potential energy.

SECTION REVIEW ANSWERS

1. 4.4×10^{-3} J
2. 2.8 m/s
3. 6.18×10^{-2} J
4. a. kinetic energy
   b. nonmechanical energy
   c. kinetic energy, gravitational potential energy
   d. elastic potential energy
5. The heated water is an instance of nonmechanical energy, because its mass is not displaced with a velocity or with respect to a zero position, as would be the case for the various types of mechanical energy. The bicycle and football both have masses in motion, so they have kinetic energy. The wound spring has been displaced from its relaxed position and so has elastic potential energy, while the football is above the ground and therefore has a gravitational potential energy associated with it.

Conservation of Energy

SECTION OBJECTIVES

- **Identify situations in which conservation of mechanical energy is valid.**
- **Recognize the forms that conserved energy can take.**
- **Solve problems using conservation of mechanical energy.**

ADVANCED TOPICS

See "The Equivalence of Mass and Energy" in **Appendix J: Advanced Topics** to learn about Einstein's theory of relativity.

CONSERVED QUANTITIES

When we say that something is *conserved,* we mean that it remains constant. If we have a certain amount of a conserved quantity at some instant of time, we will have the same amount of that quantity at a later time. This does not mean that the quantity cannot change form during that time, but if we consider all the forms that the quantity can take, we will find that we always have the same amount.

For example, the amount of money you now have is not a conserved quantity because it is likely to change over time. For the moment, however, let us assume that you do not spend the money you have, so your money is conserved. This means that if you have a dollar in your pocket, you will always have that same amount, although it may change form. One day it may be in the form of a bill. The next day you may have a hundred pennies, and the next day you may have an assortment of dimes and nickels. But when you total the change, you always have the equivalent of a dollar. It would be nice if money were like this, but of course it isn't. Because money is often acquired and spent, it is not a conserved quantity.

An example of a conserved quantity that you are already familiar with is mass. For instance, imagine that a light bulb is dropped on the floor and shatters into many pieces. No matter how the bulb shatters, the total mass of all of the pieces together is the same as the mass of the intact light bulb because mass is conserved.

MECHANICAL ENERGY

We have seen examples of objects that have either kinetic or potential energy. The description of the motion of many objects, however, often involves a combination of kinetic and potential energy as well as different forms of potential energy. Situations involving a combination of these different forms of energy can often be analyzed simply. For example, consider the motion of the different parts of a pendulum clock. The pendulum swings back and forth. At the highest point of its swing, there is only gravitational potential energy associated with its position. At other points in its swing, the pendulum is in motion, so it has kinetic energy as well. Elastic potential energy is also present in the many springs that are part of the inner workings of the clock. The motion of the pendulum in a clock is shown in **Figure 9.**

Figure 9
Total potential and kinetic energy must be taken into account in order to describe the total energy of the pendulum in a clock.

The Language of Physics

The symbol ΣPE stands for "sum of the potential energies." Just as the Greek letter Δ (*delta*) is used to denote difference, the Greek letter Σ (*sigma*) is used to denote sum.

Demonstration

Mechanical Energy — ADVANCED

Purpose Show two kinds of energy in a mechanical system.

Materials pendulum attached to a ring stand

Procedure As the pendulum swings to and fro, have students describe the motion in terms of gravitational potential energy and kinetic energy when the bob is at different positions along its path. *(At maximum displacement, the gravitational potential energy is maximum and the bob's kinetic energy is zero. The potential energy is gradually converted into kinetic energy. At the equilibrium position, the kinetic energy is maximum and the gravitational potential energy is zero.)*

Demonstration

Conservation of Energy — GENERAL

Purpose Demonstrate the conservation of mechanical energy.

Materials steel ball, spring balance, meterstick

Procedure Measure the weight of the steel ball, and record this value on the chalkboard. Drop the ball from shoulder height, and ask the students to describe the motion in terms of potential energy and kinetic energy when the ball is at different positions along its path.

Now measure the initial and final heights of the ball, and record these values on the chalkboard. Have students calculate the corresponding potential energy. Ask them to estimate how much kinetic energy the ball should have at its lowest point *(same as its initial potential energy)* and at its midway point, disregarding friction *(half of its initial potential energy)*. Explain that if friction can be disregarded, the ball's potential energy is converted into kinetic energy, while the total amount of mechanical energy remains constant.

Analyzing situations involving kinetic, gravitational potential, and elastic potential energy is relatively simple. Unfortunately, analyzing situations involving other forms of energy—such as chemical potential energy—is not as easy.

We can ignore these other forms of energy if their influence is negligible or if they are not relevant to the situation being analyzed. In most situations that we are concerned with, these forms of energy are not involved in the motion of objects. In ignoring these other forms of energy, we will find it useful to define a quantity called **mechanical energy.** The mechanical energy is the sum of kinetic energy and all forms of potential energy associated with an object or group of objects.

mechanical energy

the sum of kinetic energy and all forms of potential energy

$$ME = KE + \Sigma PE$$

All energy, such as nuclear, chemical, internal, and electrical, that is not mechanical energy is classified as *nonmechanical energy.* Do not be confused by the term *mechanical energy.* It is not a unique form of energy. It is merely a way of classifying energy, as shown in **Figure 10.** As you learn about new forms of energy in this book, you will be able to add them to this chart.

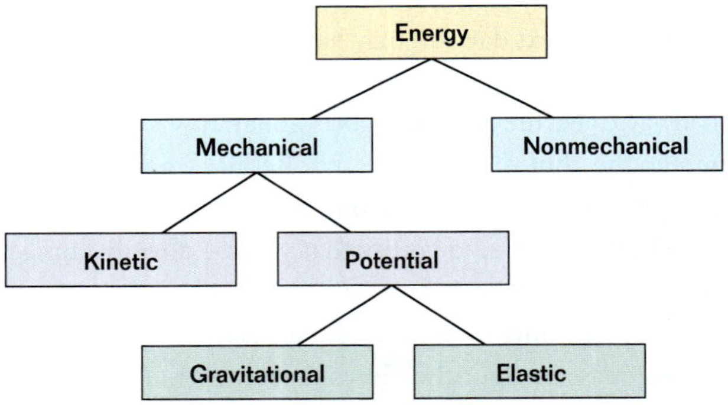

Figure 10
Energy can be classified in a number of ways.

Mechanical energy is often conserved

Imagine a 75 g egg located on a countertop 1.0 m above the ground, as shown in **Figure 11.** The egg is knocked off the edge and falls to the ground. Because the acceleration of the egg is constant as it falls, you can use the kinematic formulas to determine the speed of the egg and the distance the egg has fallen at any subsequent time. The distance fallen can then be subtracted from the initial height to find the height of the egg above the ground at any subsequent time. For example, after 0.10 s, the egg has a speed of 0.98 m/s and has fallen a distance of 0.05 m, corresponding to a height above the ground of 0.95 m. Once the egg's speed and its height above the ground are known as a function of time, you can use what you have learned in this chapter to calculate both the kinetic energy of the egg and the gravitational potential energy associated with the position of the egg at any subsequent time. Adding the kinetic and potential energy gives the total mechanical energy at each position.

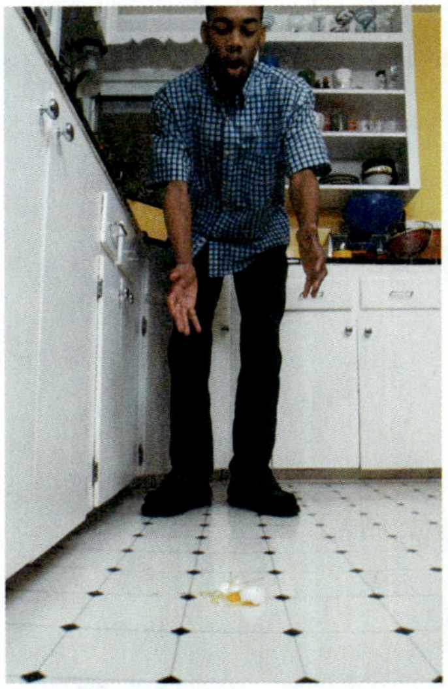

Figure 11
The total mechanical energy, potential energy plus kinetic energy, is conserved as the egg falls.

Table 1 Energy of a Falling 75 g Egg

Time (s)	Height (m)	Speed (m/s)	PE_g (J)	KE (J)	ME (J)
0.00	1.0	0.00	0.74	0.00	0.74
0.10	0.95	0.98	0.70	0.036	0.74
0.20	0.80	2.0	0.59	0.15	0.74
0.30	0.56	2.9	0.41	0.33	0.74
0.40	0.22	3.9	0.16	0.58	0.74

In the absence of friction, the total mechanical energy remains the same. This principle is called *conservation of mechanical energy.* Although the amount of mechanical energy is constant, mechanical energy itself can change form. For instance, consider the forms of energy for the falling egg, as shown in **Table 1.** As the egg falls, the potential energy is continuously converted into kinetic energy. If the egg were thrown up in the air, kinetic energy would be converted into gravitational potential energy. In either case, mechanical energy is conserved. The conservation of mechanical energy can be written symbolically as follows:

CONSERVATION OF MECHANICAL ENERGY

$$ME_i = ME_f$$

initial mechanical energy = final mechanical energy (in the absence of friction)

The mathematical expression for the conservation of mechanical energy depends on the forms of potential energy in a given problem. For instance, if the only force acting on an object is the force of gravity, as in the egg example, the conservation law can be written as follows:

$$\frac{1}{2}mv_i^2 + mgh_i = \frac{1}{2}mv_f^2 + mgh_f$$

If other forces (except friction) are present, simply add the appropriate potential energy terms associated with each force. For instance, if the egg happened to compress or stretch a spring as it fell, the conservation law would also include an elastic potential energy term on each side of the equation.

In situations in which frictional forces are present, the principle of mechanical energy conservation no longer holds because kinetic energy is not simply converted to a form of potential energy. This special situation will be discussed more thoroughly later in this section.

Quick Lab

Mechanical Energy

MATERIALS LIST

- **medium-sized spring (spring balance)**
- **assortment of small balls, each having a different mass**
- **ruler**
- **tape**
- **scale or balance**

SAFETY

Students should wear goggles to perform this lab.

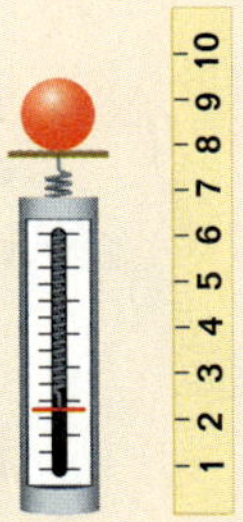

First, determine the mass of each of the balls. Then, tape the ruler to the side of a tabletop so that the ruler is vertical. Place the spring vertically on the tabletop near the ruler, and compress the spring by pressing down on one of the balls. Release the ball, and measure the maximum height it achieves in the air. Repeat this process five times, and be sure to compress the spring by the same amount each time. Average the results. From the data, can you predict how high each of the other balls will rise? Test your predictions. (Hint: Assume mechanical energy is conserved.)

Quick Lab

TEACHER'S NOTES

This activity is meant to demonstrate energy transfer (from the spring to the ball) and the conservation of mechanical energy.

The lab is most effective when the balls have significantly different masses and when the spring is compressed the same amount in each case.

Because the system for all cases has the same $ME_i = \frac{1}{2}kx_i^2$, which is converted into $ME_f = mgh_f$, balls with a larger mass will achieve a lower height.

Point out that if the measurements are reliable, they can be used to determine the spring constant.

Classroom Practice

Conservation of Mechanical Energy
A small 10.0 g ball is held to a slingshot that is stretched 6.0 cm. The spring constant is 2.0×10^2 N/m.

a. What is the elastic potential energy of the slingshot before it is released?

b. What is the kinetic energy of the ball just after the slingshot is released?

c. What is the ball's speed at that instant?

d. How high does the ball rise if it is shot directly upward?

Answers

a. 0.36 J
b. 0.36 J
c. 8.5 m/s
d. 3.7 m

PROBLEM GUIDE E

Use this guide to assign problems.
SE = Student Edition Textbook
PW = Problem Workbook
PB = Problem Bank on the One-Stop Planner (OSP)

Solving for:

v	**SE** Sample, 1–3; Ch. Rvw. 33–34, 47, 51–52 **PW** 4–5 **PB** 8–10
h	**SE** 4–5; Ch. Rvw. 39, 41–42, 47, 52 **PW** Sample, 1–3 **PB** 5–7
E	**SE** Ch. Rvw. 37, 45, 48, 51 **PW** Sample, 6–7 **PB** Sample, 1–4

*Challenging Problem
Consult the printed Solutions Manual or the OSP for detailed solutions.

SAMPLE PROBLEM E

Conservation of Mechanical Energy

PROBLEM

Starting from rest, a child zooms down a frictionless slide from an initial height of 3.00 m. What is her speed at the bottom of the slide? Assume she has a mass of 25.0 kg.

SOLUTION

1. DEFINE

Given: $h = h_i = 3.00$ m $m = 25.0$ kg $v_i = 0.0$ m/s
$h_f = 0$ m

Unknown: $v_f = ?$

2. PLAN

Choose an equation or situation:
The slide is frictionless, so mechanical energy is conserved. Kinetic energy and gravitational potential energy are the only forms of energy present.

$$KE = \tfrac{1}{2}mv^2 \qquad PE = mgh$$

The zero level chosen for gravitational potential energy is the bottom of the slide. Because the child ends at the zero level, the final gravitational potential energy is zero.

$$PE_{g,f} = 0$$

The initial gravitational potential energy at the top of the slide is

$$PE_{g,i} = mgh_i = mgh$$

Because the child starts at rest, the initial kinetic energy at the top is zero.

$$KE_i = 0$$

Therefore, the final kinetic energy is as follows:

$$KE_f = \tfrac{1}{2}mv_f^2$$

3. CALCULATE

Substitute values into the equations:

$$PE_{g,i} = (25.0 \text{ kg})(9.81 \text{ m/s}^2)(3.00 \text{ m}) = 736 \text{ J}$$
$$KE_f = (\tfrac{1}{2})(25.0 \text{ kg})v_f^2$$

Now use the calculated quantities to evaluate the final velocity.

$$ME_i = ME_f$$
$$PE_i + KE_i = PE_f + KE_f$$
$$736 \text{ J} + 0 \text{ J} = 0 \text{ J} + (0.500)(25.0 \text{ kg})v_f^2$$

$$\boxed{v_f = 7.67 \text{ m/s}}$$

CALCULATOR SOLUTION

Your calculator should give an answer of 7.67333, but because the answer is limited to three significant figures, it should be rounded to 7.67.

4. EVALUATE The expression for the square of the final speed can be written as follows:

$$v_f^2 = \frac{2\cancel{m}gh}{\cancel{m}} = 2gh$$

Notice that the masses cancel, so the final speed does not depend on the mass of the child. This result makes sense because the acceleration of an object due to gravity does not depend on the mass of the object.

PRACTICE E

Conservation of Mechanical Energy

1. A bird is flying with a speed of 18.0 m/s over water when it accidentally drops a 2.00 kg fish. If the altitude of the bird is 5.40 m and friction is disregarded, what is the speed of the fish when it hits the water?
2. A 755 N diver drops from a board 10.0 m above the water's surface. Find the diver's speed 5.00 m above the water's surface. Then find the diver's speed just before striking the water.
3. If the diver in item 2 leaves the board with an initial upward speed of 2.00 m/s, find the diver's speed when striking the water.
4. An Olympic runner leaps over a hurdle. If the runner's initial vertical speed is 2.2 m/s, how much will the runner's center of mass be raised during the jump?
5. A pendulum bob is released from some initial height such that the speed of the bob at the bottom of the swing is 1.9 m/s. What is the initial height of the bob?

Energy conservation occurs even when acceleration varies

If the slope of the slide in Sample Problem E was constant, the acceleration along the slide would also be constant and the one-dimensional kinematic formulas could have been used to solve the problem. However, you do not know the shape of the slide. Thus, the acceleration may not be constant, and the kinematic formulas could not be used.

But now we can apply a new method to solve such a problem. Because the slide is frictionless, mechanical energy is conserved. We simply equate the initial mechanical energy to the final mechanical energy and ignore all the details in the middle. The shape of the slide is not a contributing factor to the system's mechanical energy as long as friction can be ignored.

SCILINKS.
www.scilinks.org
Topic: Conservation of Energy
Code: HF60345

Alternative Problem-Solving Approach ADVANCED

The process shown in Sample Problem E can be reversed. Rather than calculating each type of energy separately, begin with the conservation of mechanical energy:

$$ME_i = ME_f$$

Next determine what types of energy are involved and substitute the formulas for each type of energy into the equation.

In this case,

$$PE_i = KE_f$$

$$mgh_i = \tfrac{1}{2}mv_f^2$$

Solve for v in terms of the other variables, and then substitute the given values into this equation.

$$v_f^2 = 2gh_i$$

$$v_f = \sqrt{2gh_i}$$

$$v_f = \sqrt{2(9.81\ \text{m/s}^2)(3.00\ \text{m})}$$

$$v_f = 7.67\ \text{m/s}$$

ANSWERS

Practice E

1. 20.7 m/s
2. 9.9 m/s; 14.0 m/s
3. 14.1 m/s
4. 0.25 m
5. 0.18 m

SECTION 3

Misconception Alert — ADVANCED

Some students may confuse the conservation of mechanical energy with the general energy conservation law. Point out that although mechanical energy is not always conserved, the total energy is always conserved. For example, as the sanding block's kinetic energy decreases, energy is transferred to the rough surface in the form of internal energy (this topic will be discussed in the chapter on heat and temperature). As a result, the temperatures of the block and surface increase slightly. The total energy in the system remains constant, although the mechanical energy decreases.

SECTION REVIEW ANSWERS

1. 2.93 m/s
2. No, the roller coaster will not reach the top of the second hill. If the total mechanical energy is constant, the roller coaster will reach its initial height and then begin rolling back down the hill.
3. **a.** yes
   b. no
   c. yes, if air resistance is disregarded
4. Answers may vary. The downward-sloping track converts potential energy to kinetic energy. Levers employ kinetic energy to increase potential energy. Springs and elastic membranes convert kinetic energy to elastic potential energy and back again. Mechanical energy is not conserved; some energy is lost because of kinetic friction.

(a)

(b)

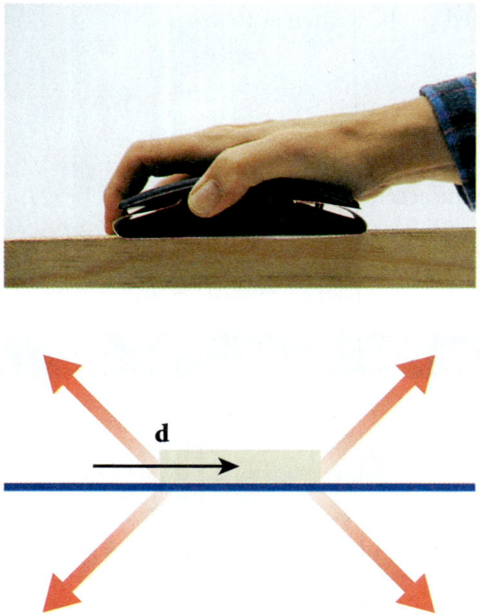

Figure 12
(a) As the block slides, its kinetic energy tends to decrease because of friction. The force from the hand keeps it moving. **(b)** Kinetic energy is dissipated into the block and surface.

Mechanical energy is not conserved in the presence of friction

If you have ever used a sanding block to sand a rough surface, such as in **Figure 12,** you may have noticed that you had to keep applying a force to keep the block moving. The reason is that kinetic friction between the moving block and the surface causes the kinetic energy of the block to be converted into a nonmechanical form of energy. As you continue to exert a force on the block, you are replacing the kinetic energy that is lost because of kinetic friction. The observable result of this energy dissipation is that the sanding block and the tabletop become warmer.

In the presence of kinetic friction, nonmechanical energy is no longer negligible and mechanical energy is no longer conserved. This does not mean that energy in general is not conserved—total energy is *always* conserved. However, the mechanical energy is converted into forms of energy that are much more difficult to account for, and the mechanical energy is therefore considered to be "lost."

SECTION REVIEW

1. If the spring of a jack-in-the-box is compressed a distance of 8.00 cm from its relaxed length and then released, what is the speed of the toy head when the spring returns to its natural length? Assume the mass of the toy head is 50.0 g, the spring constant is 80.0 N/m, and the toy head moves only in the vertical direction. Also disregard the mass of the spring. (Hint: Remember that there are two forms of potential energy in the problem.)
2. You are designing a roller coaster in which a car will be pulled to the top of a hill of height h and then, starting from a momentary rest, will be released to roll freely down the hill and toward the peak of the next hill, which is 1.1 times as high. Will your design be successful? Explain your answer.
3. Is conservation of mechanical energy likely to hold in these situations?
   a. a hockey puck sliding on a frictionless surface of ice
   b. a toy car rolling on a carpeted floor
   c. a baseball being thrown into the air
4. **Critical Thinking** What parts of the kinetic sculpture on the opening pages of this chapter involve the conversion of one form of energy to another? Is mechanical energy conserved in these processes?

Power

SECTION OBJECTIVES

- **Relate the concepts of energy, time, and power.**
- **Calculate power in two different ways.**
- **Explain the effect of machines on work and power.**

RATE OF ENERGY TRANSFER

The rate at which work is done is called **power.** More generally, power is the rate of energy transfer by any method. Like the concepts of energy and work, power has a specific meaning in science that differs from its everyday meaning.

Imagine you are producing a play and you need to raise and lower the curtain between scenes in a specific amount of time. You decide to use a motor that will pull on a rope connected to the top of the curtain rod. Your assistant finds three motors but doesn't know which one to use. One way to decide is to consider the power output of each motor.

If the work done on an object is W in a time interval Δt, then the average power delivered to the object over this time interval is written as follows:

power

a quantity that measures the rate at which work is done or energy is transformed

POWER

$$P = \frac{W}{\Delta t}$$

power = work ÷ time interval

It is sometimes useful to rewrite this equation in an alternative form by substituting the definition of work into the definition of power.

$$W = Fd$$

$$P = \frac{W}{\Delta t} = F\frac{d}{\Delta t}$$

The distance moved per unit time is just the speed of the object.

extension

Integrating Chemistry

Visit go.hrw.com for the activity "Chemical Reactions."

Keyword HF6WRKX

Why it Matters

Conceptual Challenge

1. Mountain Roads Many mountain roads are built so that they zigzag up the mountain rather than go straight up toward the peak. Discuss the advantages of such a design from the viewpoint of energy conservation and power.

2. Light Bulbs A light bulb is described as *having* 60 watts. What's wrong with this statement?

Teaching Tip — GENERAL

To help students understand the relationship $P = Fv$, have them calculate power both ways in a simple example. For example, ask students to calculate the work done, the power using the first equation, the speed, and the power using the second equation when a 100.0 N force moves an object 20.0 m in 5.0 s at constant speed (*2.00×10^3 J, 4.0×10^2 W, 4.0 m/s, 4.0×10^2 W*).

ANSWERS

Conceptual Challenge

1. Assuming mechanical energy is conserved, the same amount of energy is needed to reach the top in both cases. Because the same amount of work must be done, the path with a longer distance takes more time and hence requires less power.
2. Light bulbs don't *have* the energy stored within them; energy is transferred to them in the form of electricity at a rate of 60 J/s.

Classroom Practice

Power
Two horses pull a cart. Each exerts a force of 250.0 N at a speed of 2.0 m/s for 10.0 min.

a. Calculate the power delivered by the horses.

b. How much work is done by the two horses?

Answers

a. 1.0×10^3 W

b. 6.0×10^5 J

Alternative Problem-Solving Approach — GENERAL

Calculate the time it would take each motor to do the same work:

$W = Fd = mgd = 14 \times 10^3$ J

$\Delta t = W/P$

$\Delta t_1 = 14 \times 10^3 \text{ J}/1.0 \times 10^3 \text{ W} = 14$ s

$\Delta t_2 = 14 \times 10^3 \text{ J}/3.5 \times 10^3 \text{ W} = 4.0$ s

$\Delta t_3 = 14 \times 10^3 \text{ J}/5.5 \times 10^3 \text{ W} = 2.5$ s

This approach shows that the second motor comes closest to 5.0 s.

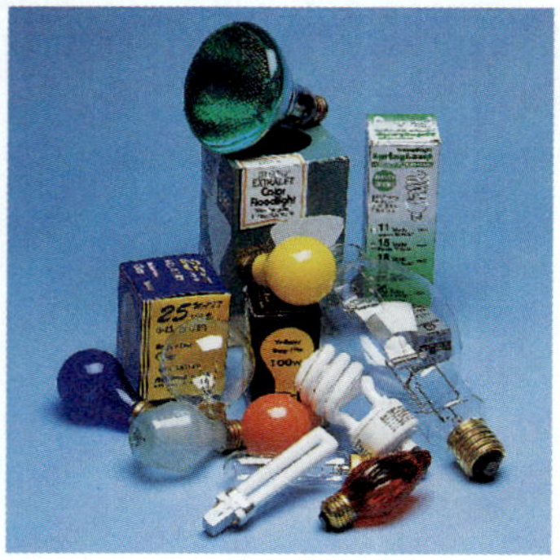

Figure 13
The power of each of these bulbs tells you the rate at which energy is converted by the bulb. The bulbs in this photo have power ratings that range from 0.7 W to 200 W.

POWER (ALTERNATIVE FORM)

$$P = Fv$$

power = force × speed

The SI unit of power is the *watt,* W, which is defined to be one joule per second. The *horsepower,* hp, is another unit of power that is sometimes used. One horsepower is equal to 746 watts.

The watt is perhaps most familiar to you from your everyday experience with light bulbs (see **Figure 13**). A dim light bulb uses about 40 W of power, while a bright bulb can use up to 500 W. Decorative lights use about 0.7 W each for indoor lights and 7.0 W each for outdoor lights.

In Sample Problem F, the three motors would lift the curtain at different rates because the power output for each motor is different. So each motor would do work on the curtain at different rates and would thus transfer energy to the curtain at different rates.

SAMPLE PROBLEM F

Power

PROBLEM

A 193 kg curtain needs to be raised 7.5 m, at constant speed, in as close to 5.0 s as possible. The power ratings for three motors are listed as 1.0 kW, 3.5 kW, and 5.5 kW. Which motor is best for the job?

SOLUTION

Given: $m = 193$ kg $\quad \Delta t = 5.0$ s $\quad d = 7.5$ m

Unknown: $P = ?$

Use the definition of power. Substitute the equation for work.

$$P = \frac{W}{\Delta t} = \frac{Fd}{\Delta t} = \frac{mgd}{\Delta t}$$

$$= \frac{(193 \text{ kg})(9.81 \text{ m/s}^2)(7.5 \text{ m})}{5.0 \text{ s}}$$

$$P = 2.8 \times 10^3 \text{ W} = 2.8 \text{ kW}$$

The best motor to use is the 3.5 kW motor. The 1.0 kW motor will not lift the curtain fast enough, and the 5.5 kW motor will lift the curtain too fast.

PRACTICE F

Power

1. A 1.0×10^3 kg elevator carries a maximum load of 800.0 kg. A constant frictional force of 4.0×10^3 N retards the elevator's motion upward. What minimum power, in kilowatts, must the motor deliver to lift the fully loaded elevator at a constant speed of 3.00 m/s?
2. A car with a mass of 1.50×10^3 kg starts from rest and accelerates to a speed of 18.0 m/s in 12.0 s. Assume that the force of resistance remains constant at 400.0 N during this time. What is the average power developed by the car's engine?
3. A rain cloud contains 2.66×10^7 kg of water vapor. How long would it take for a 2.00 kW pump to raise the same amount of water to the cloud's altitude, 2.00 km?
4. How long does it take a 19 kW steam engine to do 6.8×10^7 J of work?
5. A 1.50×10^3 kg car accelerates uniformly from rest to 10.0 m/s in 3.00 s.
   a. What is the work done on the car in this time interval?
   b. What is the power delivered by the engine in this time interval?

SECTION REVIEW

1. A 50.0 kg student climbs 5.00 m up a rope at a constant speed. If the student's power output is 200.0 W, how long does it take the student to climb the rope? How much work does the student do?
2. A motor-driven winch pulls the 50.0 kg student in the previous item 5.00 m up the rope at a constant speed of 1.25 m/s. How much power does the motor use in raising the student? How much work does the motor do on the student?
3. **Critical Thinking** How are energy, time, and power related?
4. **Critical Thinking** People often use the word *powerful* to describe the engines in some automobiles. In this context, how does the word relate to the definition of *power*? How does this word relate to the alternative definition of *power*?

PROBLEM GUIDE F

Use this guide to assign problems.
SE = Student Edition Textbook
PW = Problem Workbook
PB = Problem Bank on the One-Stop Planner (OSP)

Solving for:

P	**SE**	Sample, 1–2, 5; Ch. Rvw. 36
	PW	5–6
	PB	8–10
Δt	**SE**	3–4; Ch. Rvw. 35
	PW	3–4
	PB	Sample, 1–3
W	**SE**	5
	PW	Sample, 1–2
	PB	4–7

***Challenging Problem**
Consult the printed Solutions Manual or the OSP for detailed solutions.

ANSWERS

Practice F

1. 66 kW
2. 2.38×10^4 W (23.8 kW)
3. 2.61×10^8 s (8.27 years)
4. 3.6×10^3 s (1.0 h)
5. **a.** 7.50×10^4 J
   b. 2.50×10^4 W

SECTION REVIEW ANSWERS

1. 12.3 s; 2.45×10^3 J
2. 613 W; 2.45×10^3 J
3. Power equals energy transferred divided by time of transfer.
4. A powerful engine is capable of doing more work in a given time. The force and speed delivered by a powerful engine is large relative to less powerful engines.

Roller Coaster Designer

Two of Steve's first roller coasters are the Ninjas at Six Flags Over Mid-America and at Six Flags Magic Mountain. His West Coaster, built on Santa Monica Pier, towers five stories above the Pacific Ocean. The cars on the Steel Force at Dorney Park in Pennsylvania reach speeds of over 75 mi/h and drop more than 200 ft to disappear into a 120 ft tunnel. The Mamba at Worlds of Fun in Missouri features two giant back-to-back hills, a fast spiral, and five camelback humps. The camelbacks are designed to pull your seat out from under you, so that you feel like you're floating. Roller coaster fans call this feeling *airtime.*

PHYSICS CAREERS

Roller Coaster Designer

As the name states, the cars of a roller coaster really do coast along the tracks. A motor pulls the cars up a high hill at the beginning of the ride. After the hill, however, the motion of the car is a result of gravity and inertia. As the cars roll down the hill, they must pick up the speed that they need to whiz through the rest of the curves, loops, twists, and bumps in the track. To learn more about designing roller coasters, read the interview with Steve Okamoto.

The roller coaster pictured here is named Wild Thing and is located in Minnesota. The highest point on the track is 63 m off the ground and the cars' maximum speed is 118 km/h.

How did you become a roller coaster designer?

I have been fascinated with roller coasters ever since my first ride on one. I remember going to Disneyland as a kid. My mother was always upset with me because I kept looking over the sides of the rides, trying to figure out how they worked. My interest in finding out how things worked led me to study mechanical engineering.

What sort of training do you have?

I earned a degree in product design. For this degree, I studied mechanical engineering and studio art. Product designers consider an object's form as well as its function. They also take into account the interests and abilities of the product's consumer. Most rides and parks have some kind of theme, so I must consider marketing goals and concerns in my designs.

What is the nature of your work?

To design a roller coaster, I study site maps of the location. Then, I go to the amusement park to look at the actual site. Because most rides I design are for older parks (few parks are built from scratch), fitting a coaster around, above, and in between existing rides and buildings is one of my biggest challenges. I also have to design how the parts of the ride will work together. The towers and structures that support the ride have to be strong enough to hold up a track and speeding cars that are full of people. The cars themselves need special wheels to keep them locked onto the track and seat belts or bars to keep the passengers safely inside. It's like putting together a puzzle, except the pieces haven't been cut out yet.

What advice do you have for a student who is interested in designing roller coasters?

Studying math and science is very important. To design a successful coaster, I have to understand how energy is converted from one form to another as the cars move along the track. I have to calculate speeds and accelerations of the cars on each part of the track. They have to go fast enough to make it up the next hill! I rely on my knowledge of geometry and physics to create the roller coaster's curves, loops, and dips.

Highlights

KEY IDEAS

Section 1 Work

- Work is done on an object only when a net force acts on the object to displace it in the direction of a component of the net force.
- The amount of work done on an object by a force is equal to the component of the force along the direction of motion times the distance the object moves.

Section 2 Energy

- Objects in motion have kinetic energy because of their mass and speed.
- The net work done on or by an object is equal to the change in the kinetic energy of the object.
- Potential energy is energy associated with an object's position. Two forms of potential energy discussed in this chapter are gravitational potential energy and elastic potential energy.

Section 3 Conservation of Energy

- Energy can change form but can never be created or destroyed.
- Mechanical energy is the total kinetic and potential energy present in a given situation.
- In the absence of friction, mechanical energy is conserved, so the amount of mechanical energy remains constant.

Section 4 Power

- Power is the rate at which work is done or the rate of energy transfer.
- Machines with different power ratings do the same amount of work in different time intervals.

Variable Symbols

Quantities		Units		Conversions
W	work	J	joule	$= N \bullet m$ $= kg \bullet m^2/s^2$
KE	kinetic energy	J	joule	
PE_g	gravitational potential energy	J	joule	
$PE_{elastic}$	elastic potential energy	J	joule	
P	power	W	watt	$= J/s$

KEY TERMS

work (p. 160)

kinetic energy (p. 164)

work–kinetic energy theorem (p. 166)

potential energy (p. 169)

gravitational potential energy (p. 169)

elastic potential energy (p. 170)

spring constant (p. 170)

mechanical energy (p. 174)

power (p. 179)

PROBLEM SOLVING

See **Appendix D: Equations** for a summary of the equations introduced in this chapter. If you need more problem-solving practice, see **Appendix I: Additional Problems.**

Highlights

Teaching Tip

Explaining concepts in written form helps solidify students' understanding of difficult concepts and helps enforce good communication skills. Have students summarize the differences between mechanical and nonmechanical energy and between kinetic energy, gravitational potential energy, and elastic potential energy. Essays should also include a thorough discussion of work and its link to kinetic and potential energy. Be sure students explain concepts clearly and correctly and use good sentence structure.

Review

ANSWERS

1. No, a change in speed corresponds to a change in kinetic energy, which cannot occur without work (either positive or negative) being done on the object.
2. **a.** yes, positive
   b. no
   c. yes, positive
   d. yes, negative
3. No, force would decrease, but distance would increase, which would keep work constant.
4. The tension is perpendicular to the bob's motion, so it does not do work on the bob. The component of the bob's weight that is perpendicular to the bob's motion does not do work on the bob, but the component that is in the direction of its motion does.
5. The car leaving longer skid marks was moving faster.
6. yes; no; yes, the ball's weight and air resistance
7. 53 J, −53 J
8. 2.4×10^5 J
9. 47.5 J
10. **a.** 6230 J
    b. −6230 J
    c. 0.640
11. **a.** no
    b. yes
    c. yes

Review

WORK

Review Questions

1. Can the speed of an object change if the net work done on it is zero?

2. Discuss whether any work is being done by each of the following agents and, if so, whether the work is positive or negative.
   a. a chicken scratching the ground
   b. a person reading a sign
   c. a crane lifting a bucket of concrete
   d. the force of gravity on the bucket in **(c)**

3. Furniture movers wish to load a truck using a ramp from the ground to the rear of the truck. One of the movers claims that less work would be required if the ramp's length were increased, reducing its angle with the horizontal. Is this claim valid? Explain.

Conceptual Questions

4. A pendulum swings back and forth, as shown at right. Does the tension force in the string do work on the pendulum bob? Does the force of gravity do work on the bob? Explain your answers.

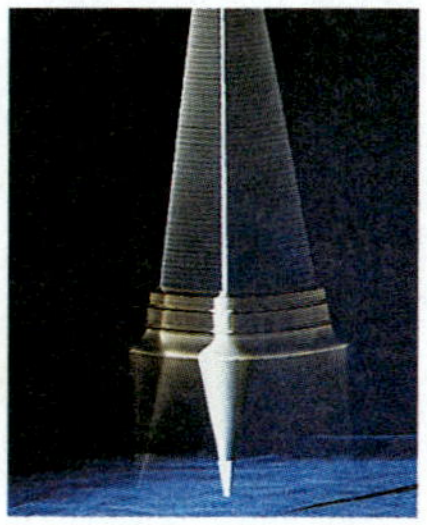

5. The drivers of two identical cars heading toward each other apply the brakes at the same instant. The skid marks of one of the cars are twice as long as the skid marks of the other vehicle. Assuming that the brakes of both cars apply the same force, what conclusions can you draw about the motion of the cars?

6. When a punter kicks a football, is he doing work on the ball while his toe is in contact with it? Is he doing work on the ball after the ball loses contact with his toe? Are any forces doing work on the ball while the ball is in flight?

Practice Problems

For problems 7–10, see Sample Problem A.

7. A person lifts a 4.5 kg cement block a vertical distance of 1.2 m and then carries the block horizontally a distance of 7.3 m. Determine the work done by the person and by the force of gravity in this process.

8. A plane designed for vertical takeoff has a mass of 8.0×10^3 kg. Find the net work done by all forces on the plane as it accelerates upward at 1.0 m/s^2 through a distance of 30.0 m after starting from rest.

9. When catching a baseball, a catcher's glove moves by 10 cm along the line of motion of the ball. If the baseball exerts a force of 475 N on the glove, how much work is done by the ball?

10. A flight attendant pulls her 70.0 N flight bag a distance of 253 m along a level airport floor at a constant velocity. The force she exerts is 40.0 N at an angle of 52.0° above the horizontal. Find the following:
    a. the work she does on the flight bag
    b. the work done by the force of friction on the flight bag
    c. the coefficient of kinetic friction between the flight bag and the floor

ENERGY

Review Questions

11. A person drops a ball from the top of a building while another person on the ground observes the ball's motion. Each observer chooses his or her own location as the level for zero potential energy. Will they calculate the same values for:
    a. the potential energy associated with the ball?
    b. the change in potential energy associated with the ball?
    c. the ball's kinetic energy?

12. Can the kinetic energy of an object be negative? Explain your answer.

13. Can the gravitational potential energy associated with an object be negative? Explain your answer.

14. Two identical objects move with speeds of 5.0 m/s and 25.0 m/s. What is the ratio of their kinetic energies?

Conceptual Questions

15. A satellite is in a circular orbit above Earth's surface. Why is the work done on the satellite by the gravitational force zero? What does the work–kinetic energy theorem predict about the satellite's speed?

16. A car traveling at 50.0 km/h skids a distance of 35 m after its brakes lock. Estimate how far it will skid if its brakes lock when its initial speed is 100.0 km/h. What happens to the car's kinetic energy as it comes to rest?

17. Explain why more energy is needed to walk up stairs than to walk horizontally at the same speed.

18. How can the work–kinetic energy theorem explain why the force of sliding friction reduces the kinetic energy of a particle?

Practice Problems

For problems 19–20, see Sample Problem B.

19. What is the kinetic energy of an automobile with a mass of 1250 kg traveling at a speed of 11 m/s?

20. What speed would a fly with a mass of 0.55 g need in order to have the same kinetic energy as the automobile in item 19?

For problems 21–22, see Sample Problem C.

21. A 50.0 kg diver steps off a diving board and drops straight down into the water. The water provides an upward average net force of 1500 N. If the diver comes to rest 5.0 m below the water's surface, what is the total distance between the diving board and the diver's stopping point underwater?

22. In a circus performance, a monkey on a sled is given an initial speed of 4.0 m/s up a 25° incline. The combined mass of the monkey and the sled is 20.0 kg, and the coefficient of kinetic friction between the sled and the incline is 0.20. How far up the incline does the sled move?

For problems 23–25, see Sample Problem D.

23. A 55 kg skier is at the top of a slope, as shown in the illustration below. At the initial point **A,** the skier is 10.0 m vertically above the final point **B.**
    - **a.** Set the zero level for gravitational potential energy at **B,** and find the gravitational potential energy associated with the skier at **A** and at **B.** Then find the difference in potential energy between these two points.
    - **b.** Repeat this problem with the zero level at point **A.**
    - **c.** Repeat this problem with the zero level midway down the slope, at a height of 5.0 m.

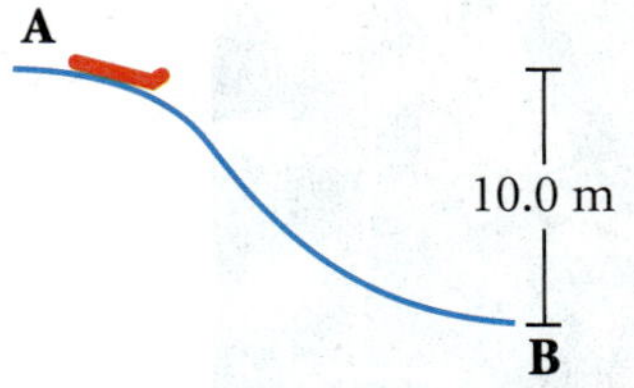

24. A 2.00 kg ball is attached to a ceiling by a string. The distance from the ceiling to the center of the ball is 1.00 m, and the height of the room is 3.00 m. What is the gravitational potential energy associated with the ball relative to each of the following?
    - **a.** the ceiling
    - **b.** the floor
    - **c.** a point at the same elevation as the ball

25. A spring has a force constant of 500.0 N/m. Show that the potential energy stored in the spring is as follows:
    - **a.** 0.400 J when the spring is stretched 4.00 cm from equilibrium
    - **b.** 0.225 J when the spring is compressed 3.00 cm from equilibrium
    - **c.** zero when the spring is unstretched

5 REVIEW

12. No, kinetic energy cannot be negative because mass is always positive and the speed term of the equation is squared.
13. yes, because potential energy depends on the distance to an arbitrary zero level, which can be above or below the object
14. 1 to 25
15. The gravitational force does not do work on the satellite because the force of gravity is always perpendicular to the path of the motion. The satellite's speed must be constant.
16. The work required to stop the car equals the car's initial kinetic energy. If speed is doubled, work is quadrupled. Thus, the car will travel 140 m. Its kinetic energy is changed into internal energy.
17. Work must be done against gravity in order to climb a staircase at a constant speed. Walking on a horizontal surface does not require work to be done against gravity.
18. The work done by friction equals the change in mechanical energy, so the particle's speed decreases.
19. 7.6×10^4 J
20. 1.7×10^4 m/s
21. 2.0×10^1 m
22. 1.4 m
23. **a.** 5400 J, 0 J; 5400 J
    b. 0 J, –5400 J; 5400 J
    c. 2700 J, –2700 J; 5400 J
24. **a.** –19.6 J
    b. 39.2 J
    c. 0 J
25. **a.** $(0.5)(500.0\ \text{N/m})(4.00 \times 10^{-2}\ \text{m})^2 = 0.400\ \text{J}$
    b. $(\frac{1}{2})(500.0\ \text{N/m})(-3.00 \times 10^{-2}\ \text{m})^2 = 0.225\ \text{J}$
    c. $(0.5)(500.0\ \text{N/m})(0\ \text{m})^2 = 0\ \text{J}$

5 REVIEW

26. **a.** nonmechanical
b. mechanical
c. mechanical
d. mechanical
e. both

27. As the athlete runs faster, KE increases. As he is lifted above the ground, KE decreases as PE_g and $PE_{elastic}$ increase ($PE_{elastic}$ comes from the bent pole). At the highest point, $KE = 0$ and PE_g is at its maximum value. As the athlete falls, KE increases and PE_g decreases. When the athlete lands, KE is at its maximum value and $PE_g = 0$.

28. The ball will not hit the lecturer because, according to the principle of energy conservation, it would need an input of energy to reach a height greater than its initial height. If the ball were given a push, the lecturer would be in danger.

29. **a.** Athlete does work on the weight. PE_g increases.
b. No work done on the weight. PE_g is constant.
c. Athlete does negative work on the weight. PE_g decreases.

30. at the ball's lowest height; at its maximum height

31. no, because energy wouldn't be conserved

32. two, gravitational potential energy and elastic potential energy; yes, because total mechanical energy is conserved if there is no dissipation of energy

33. 12.0 m/s

34. **a.** 10.9 m/s
b. 11.6 m/s

35. 17.2 s

36. 5.9×10^8 W

CONSERVATION OF MECHANICAL ENERGY

Review Questions

26. Each of the following objects possesses energy. Which forms of energy are mechanical, which are nonmechanical, and which are a combination?

a. glowing embers in a campfire
b. a strong wind
c. a swinging pendulum
d. a person sitting on a mattress
e. a rocket being launched into space

27. Discuss the energy transformations that occur during the pole-vault event shown in the photograph below. Disregard rotational motion and air resistance.

28. A strong cord suspends a bowling ball from the center of a lecture hall's ceiling, forming a pendulum. The ball is pulled to the tip of a lecturer's nose at the front of the room and is then released. If the lecturer remains stationary, explain why the lecturer is not struck by the ball on its return swing. Would this person be safe if the ball were given a slight push from its starting position at the person's nose?

Conceptual Questions

29. Discuss the work done and change in mechanical energy as an athlete does the following:

a. lifts a weight
b. holds the weight up in a fixed position
c. lowers the weight slowly

30. A ball is thrown straight up. At what position is its kinetic energy at its maximum? At what position is gravitational potential energy at its maximum?

31. Advertisements for a toy ball once stated that it would rebound to a height greater than the height from which it was dropped. Is this possible?

32. A weight is connected to a spring that is suspended vertically from the ceiling. If the weight is displaced downward from its equilibrium position and released, it will oscillate up and down. How many forms of potential energy are involved? If air resistance and friction are disregarded, will the total mechanical energy be conserved? Explain.

Practice Problems

For problems 33–34, see Sample Problem E.

33. A child and sled with a combined mass of 50.0 kg slide down a frictionless hill that is 7.34 m high. If the sled starts from rest, what is its speed at the bottom of the hill?

34. Tarzan swings on a 30.0 m long vine initially inclined at an angle of 37.0° with the vertical. What is his speed at the bottom of the swing if he does the following?

a. starts from rest
b. starts with an initial speed of 4.00 m/s

POWER

Practice Problems

For problems 35–36, see Sample Problem F.

35. If an automobile engine delivers 50.0 hp of power, how much time will it take for the engine to do 6.40×10^5 J of work? (Hint: Note that one horsepower, 1 hp, is equal to 746 watts.)

36. Water flows over a section of Niagara Falls at the rate of 1.2×10^6 kg/s and falls 50.0 m. How much power is generated by the falling water?

MIXED REVIEW

37. A 215 g particle is released from rest at point **A** inside a smooth hemispherical bowl of radius 30.0 cm, as shown at right. Calculate the following:

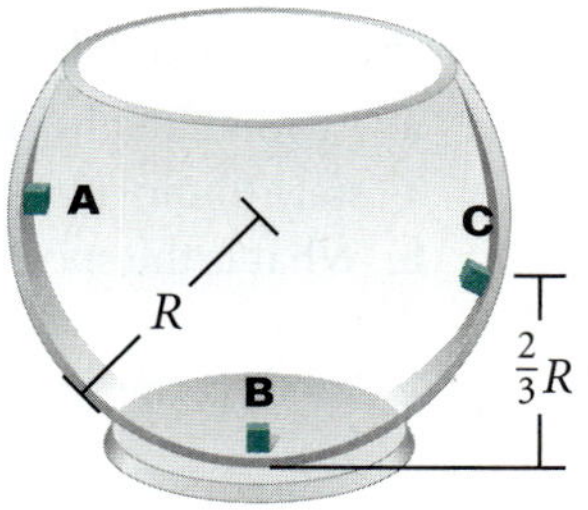

a. the gravitational potential energy at **A** relative to **B**

b. the particle's kinetic energy at **B**

c. the particle's speed at **B**

d. the potential energy and kinetic energy at **C**

38. A person doing a chin-up weighs 700.0 N, disregarding the weight of the arms. During the first 25.0 cm of the lift, each arm exerts an upward force of 355 N on the torso. If the upward movement starts from rest, what is the person's speed at this point?

39. A 50.0 kg pole vaulter running at 10.0 m/s vaults over the bar. If the vaulter's horizontal component of velocity over the bar is 1.0 m/s and air resistance is disregarded, how high was the jump?

40. An 80.0 N box of clothes is pulled 20.0 m up a 30.0° ramp by a force of 115 N that points along the ramp. If the coefficient of kinetic friction between the box and ramp is 0.22, calculate the change in the box's kinetic energy.

41. Tarzan and Jane, whose total mass is 130.0 kg, start their swing on a 5.0 m long vine when the vine is at an angle of 30.0° with the horizontal. At the bottom of the arc, Jane, whose mass is 50.0 kg, releases the vine. What is the maximum height at which Tarzan can land on a branch after his swing continues? (Hint: Treat Tarzan's and Jane's energies as separate quantities.)

42. A 0.250 kg block on a vertical spring with a spring constant of 5.00×10^3 N/m is pushed downward, compressing the spring 0.100 m. When released, the block leaves the spring and travels upward vertically. How high does it rise above the point of release?

43. Three identical balls, all with the same initial speed, are thrown by a juggling clown on a tightrope. The first ball is thrown horizontally, the second is thrown at some angle above the horizontal, and the third is thrown at some angle below the horizontal. Disregarding air resistance, describe the motions of the three balls, and compare the speeds of the balls as they reach the ground.

44. A 0.60 kg rubber ball has a speed of 2.0 m/s at point A and kinetic energy of 7.5 J at point B. Determine the following:

a. the ball's kinetic energy at A

b. the ball's speed at B

c. the total work done on the ball from A to B

45. Starting from rest, a 5.0 kg block slides 2.5 m down a rough 30.0° incline in 2.0 s. Determine the following:

a. the work done by the force of gravity

b. the mechanical energy lost due to friction

c. the work done by the normal force between the block and the incline

46. A skier of mass 70.0 kg is pulled up a slope by a motor-driven cable. How much work is required to pull the skier 60.0 m up a 35° slope (assumed to be frictionless) at a constant speed of 2.0 m/s?

47. An acrobat on skis starts from rest 50.0 m above the ground on a frictionless track and flies off the track at a 45.0° angle above the horizontal and at a height of 10.0 m. Disregard air resistance.

a. What is the skier's speed when leaving the track?

b. What is the maximum height attained?

48. Starting from rest, a 10.0 kg suitcase slides 3.00 m down a frictionless ramp inclined at 30.0° from the floor. The suitcase then slides an additional 5.00 m along the floor before coming to a stop. Determine the following:

a. the suitcase's speed at the bottom of the ramp

b. the coefficient of kinetic friction between the suitcase and the floor

c. the change in mechanical energy due to friction

49. A light horizontal spring has a spring constant of 105 N/m. A 2.00 kg block is pressed against one end of the spring, compressing the spring 0.100 m. After the block is released, the block moves 0.250 m to the right before coming to rest. What is the coefficient of kinetic friction between the horizontal surface and the block?

5 REVIEW

37. **a.** 0.633 J
b. 0.633 J
c. 2.43 m/s
d. 0.422 J, 0.211 J

38. 0.265 m/s

39. 5.0 m

40. 1.2×10^3 J

41. 2.5 m

42. 10.2 m

43. Although the total distance traveled by each ball is different, the displacements are the same, so the change in potential energy for each ball is the same. Also, each ball has the same initial kinetic energy, so the final kinetic energy of each ball (and thus the speed of each) will be the same.

44. **a.** 1.2 J
b. 5.0 m/s
c. 6.3 J

45. **a.** 61 J
b. −45 J
c. 0 J

46. 2.4×10^4 J

47. **a.** 28.0 m/s
b. 30.0 m above the ground

48. **a.** 5.42 m/s
b. 0.300
c. −147 J

49. 0.107

50. a. 310 J
b. −150 J
c. 180 N

51. a. 66 J
b. 2.3 m/s
c. 66 J
d. −16 J

52. a. 1.45 m
b. 1.98 m/s
c. 5.33 m/s

extension

Graphing Calculator Practice

Visit go.hrw.com for answers to this Graphing Calculator activity.

Keyword HF6WRKXT

50. A 5.0 kg block is pushed 3.0 m at a constant velocity up a vertical wall by a constant force applied at an angle of 30.0° with the horizontal, as shown at right. If the coefficient of kinetic friction between the block and the wall is 0.30, determine the following:

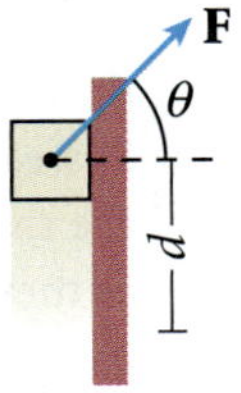

a. the work done by the force on the block

b. the work done by gravity on the block

c. the magnitude of the normal force between the block and the wall

51. A 25 kg child on a 2.0 m long swing is released from rest when the swing supports make an angle of 30.0° with the vertical.

a. What is the maximum potential energy associated with the child?

b. Disregarding friction, find the child's speed at the lowest position.

c. What is the child's total mechanical energy?

d. If the speed of the child at the lowest position is 2.00 m/s, what is the change in mechanical energy due to friction?

52. A ball of mass 522 g starts at rest and slides down a frictionless track, as shown in the diagram. It leaves the track horizontally, striking the ground.

a. At what height above the ground does the ball start to move?

b. What is the speed of the ball when it leaves the track?

c. What is the speed of the ball when it hits the ground?

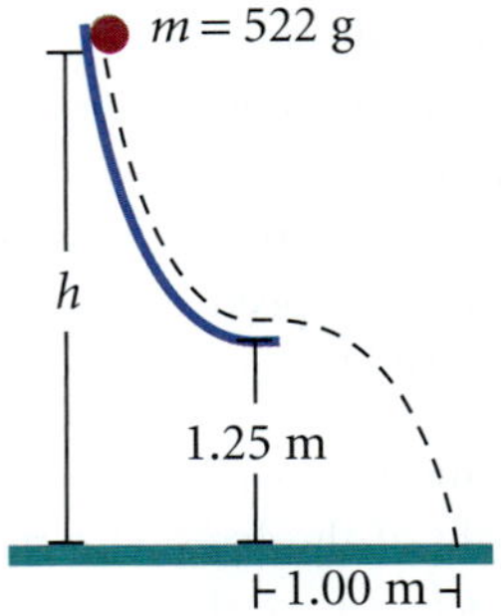

Graphing Calculator *Practice*

Work of Displacement

Work done, as you learned earlier in this chapter, is a result of the net applied force, the distance of the displacement, and the angle of the applied force relative to the direction of displacement. Work done is described by in the following equation:

$$W_{net} = F_{net} d \cos \theta$$

The equation for work done can be represented on a graphing calculator as follows:

$$Y_1 = FXCOS(\theta)$$

In this activity, you will use this equation and your graphing calculator to produce a table of results for various values of θ. Column one of the table will be the displacement (X) in meters, and column two will be the work done (Y_1) in joules.

Visit go.hrw.com and enter the keyword **HF6WRKX** to find this graphing calculator activity. Refer to **Appendix B** for instructions on downloading the program for this activity.

Alternative Assessment

1. Design experiments for measuring your power output when doing push-ups, running up a flight of stairs, pushing a car, loading boxes onto a truck, throwing a baseball, or performing other energy-transferring activities. What data do you need to measure or calculate? Form groups to present and discuss your plans. If your teacher approves your plans, perform the experiments.

2. Investigate the amount of kinetic energy involved when your car's speed is 60 km/h, 50 km/h, 40 km/h, 30 km/h, 20 km/h, and 10 km/h. (Hint: Find your car's mass in the owner's manual.) How much work does the brake system have to do to stop the car at each speed?

   If the owner's manual includes a table of braking distances at different speeds, determine the force the braking system must exert. Organize your findings in charts and graphs to study the questions and to present your conclusions.

3. Investigate the energy transformations of your body as you swing on a swing set. Working with a partner, measure the height of the swing at the high and low points of your motion. What points involve a maximum gravitational potential energy? What points involve a maximum kinetic energy? For three other points in the path of the swing, calculate the gravitational potential energy, the kinetic energy, and the velocity. Organize your findings in bar graphs.

4. Design an experiment to test the conservation of mechanical energy for a toy car rolling down a ramp. Use a board propped up on a stack of books as the ramp. To find the final speed of the car, use the equation final speed = 2(average speed) = 2(length of ramp/time). Before beginning the experiment, make predictions about what to expect. Will the kinetic energy at the bottom equal the potential energy at the top? If not, which might be greater? Test your predictions with various ramp heights, and write a report describing your experiment and your results.

5. In order to save fuel, an airline executive recommended the following changes in the airlines' largest jet flights:
   a. restrict the weight of personal luggage
   b. remove pillows, blankets, and magazines from the cabin
   c. lower flight altitudes by 5 percent
   d. reduce flying speeds by 5 percent

   Research the information necessary to calculate the approximate kinetic and potential energy of a large passenger aircraft. Which of the measures described above would result in significant savings? What might be their other consequences? Summarize your conclusions in a presentation or report.

6. Make a chart of the kinetic energies your body can have. First, measure your mass. Then, measure your speed when walking, running, sprinting, riding a bicycle, and driving a car. Make a poster graphically comparing these findings.

7. You are trying to find a way to bring electricity to a remote village in order to run a water-purifying device. A donor is willing to provide battery chargers that connect to bicycles. Assuming the water-purification device requires 18.6 kW•h daily, how many bicycles would a village need if a person can average 100 W while riding a bicycle? Is this a useful way to help the village? Evaluate your findings for strengths and weaknesses. Summarize your comments and suggestions in a letter to the donor.

8. Many scientific units are named after famous scientists or inventors. The SI unit of power, the watt, was name for the Scottish scientist, James Watt. The SI unit of energy, the joule, was named for the English scientist James Prescott Joule. Use the Internet or library resources to learn about the contributions of one of these two scientists. Write a short report with your findings, and then present your report to the class.

Alternative Assessment ANSWERS

1. Student plans should be safe and should include measuring work and the time intervals.
2. Students should recognize that all of the car's KE must be brought to zero, because $v_f = 0$ m/s. Therefore, the brake system must do as much work as the car's KE (if air resistance and friction are neglected).
3. Student plans should be safe and should include measurements of height, mass, and speed. Kinetic energy is highest at the bottom of the swing.
4. Students should find that mechanical energy is nearly conserved. Kinetic energy may be slightly less than potential energy due to small frictional losses.
5. Students will need to research information about altitude, friction, speed, and masses involved to evaluate the plans.
6. Student posters should indicate that increasing speed causes their *KE* to increase.
7. Students' letters will vary but should acknowledge that 186 h of bicycling are needed for a day of use. Thus, at least eight bicycles would be required.
8. Watt (1736–1819) played an important role in the development of the steam engine. Joule (1818–1889) explored the connection between thermal energy and other forms of energy.

Standardized Test Prep

ANSWERS

1. D
2. H
3. C
4. F
5. D
6. J
7. B
8. J

Standardized Test Prep

MULTIPLE CHOICE

1. In which of the following situations is work *not* being done?

A. A chair is lifted vertically with respect to the floor.
B. A bookcase is slid across carpeting.
C. A table is dropped onto the ground.
D. A stack of books is carried at waist level across a room.

2. Which of the following equations correctly describes the relation between power, work, and time?

F. $W = \frac{P}{t}$
G. $W = \frac{t}{P}$
H. $P = \frac{W}{t}$
J. $P = \frac{t}{W}$

Use the graph below to answer questions 3–5. The graph shows the energy of a 75 g yo-yo at different times as the yo-yo moves up and down on its string.

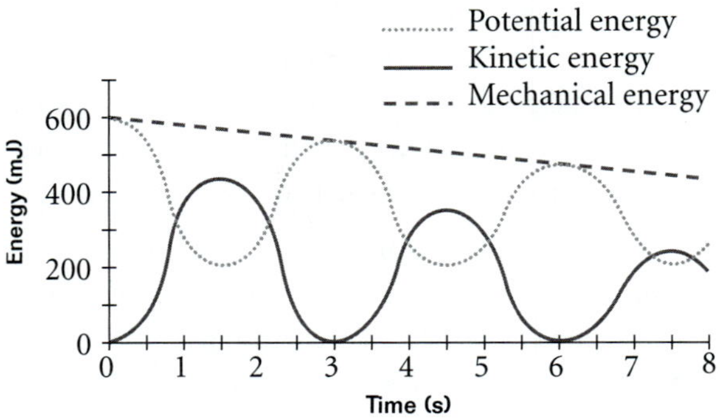

3. By what amount does the mechanical energy of the yo-yo change after 6.0 s?

A. 500 mJ
B. 0 mJ
C. −100 mJ
D. −600 mJ

4. What is the speed of the yo-yo after 4.5 s?

F. 3.1 m/s
G. 2.3 m/s
H. 3.6 m/s
J. 1.6 m/s

5. What is the maximum height of the yo-yo?

A. 0.27 m
B. 0.54 m
C. 0.75 m
D. 0.82 m

6. A car with mass m requires 5.0 kJ of work to move from rest to a final speed v. If this same amount of work is performed during the same amount of time on a car with a mass of $2m$, what is the final speed of the second car?

F. $2v$
G. $\sqrt{2}v$
H. $\frac{v}{2}$
J. $\frac{v}{\sqrt{2}}$

Use the passage below to answer questions 7–8.

A 70.0 kg base runner moving at a speed of 4.0 m/s begins his slide into second base. The coefficient of friction between his clothes and Earth is 0.70. His slide lowers his speed to zero just as he reaches the base.

7. How much mechanical energy is lost because of friction acting on the runner?

A. 1100 J
B. 560 J
C. 140 J
D. 0 J

8. How far does the runner slide?

F. 0.29 m
G. 0.57 m
H. 0.86 m
J. 1.2 m

Use the passage below to answer questions 9–10.

A spring scale has a spring with a force constant of 250 N/m and a weighing pan with a mass of 0.075 kg. During one weighing, the spring is stretched a distance of 12 cm from equilibrium. During a second weighing, the spring is stretched a distance of 18 cm.

9. How much greater is the elastic potential energy of the stretched spring during the second weighing than during the first weighing?

A. $\frac{9}{4}$

B. $\frac{3}{2}$

C. $\frac{2}{3}$

D. $\frac{4}{9}$

10. If the spring is suddenly released after each weighing, the weighing pan moves back and forth through the equilibrium position. What is the ratio of the pan's maximum speed after the second weighing to the pan's maximum speed after the first weighing? Consider the force of gravity on the pan to be negligible.

F. $\frac{9}{4}$ **H.** $\frac{2}{3}$

G. $\frac{3}{2}$ **J.** $\frac{4}{9}$

SHORT RESPONSE

11. A student with a mass of 66.0 kg climbs a staircase in 44.0 s. If the distance between the base and the top of the staircase is 14.0 m, how much power will the student deliver by climbing the stairs?

Base your answers to questions 12–13 on the information below.

A 75.0 kg man jumps from a window that is 1.00 m above a sidewalk.

12. Write the equation for the man's speed when he strikes the ground.

13. Calculate the man's speed when he strikes the ground.

EXTENDED RESPONSE

Base your answers to questions 14–16 on the information below.

A projectile with a mass of 5.0 kg is shot horizontally from a height of 25.0 m above a flat desert surface. The projectile's initial speed is 17 m/s. Calculate the following for the instant before the projectile hits the surface:

14. The work done on the projectile by gravity.

15. The change in kinetic energy since the projectile was fired.

16. The final kinetic energy of the projectile.

17. A skier starts from rest at the top of a hill that is inclined at 10.5° with the horizontal. The hillside is 200.0 m long, and the coefficient of friction between the snow and the skis is 0.075. At the bottom of the hill, the snow is level and the coefficient of friction is unchanged. How far does the skier move along the horizontal portion of the snow before coming to rest? Show all of your work.

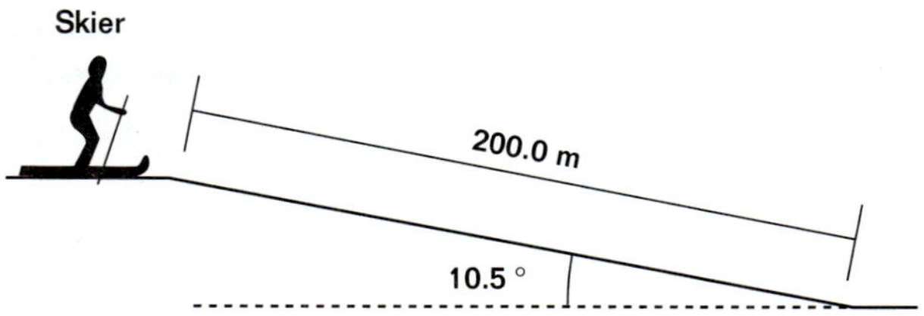

Test TIP When solving a mathematical problem, you must first decide which equation or equations you need to answer the question.

9. A
10. G
11. 206 W
12. $v = \sqrt{2gh}$
13. 4.4 m/s
14. 1200 J
15. 1200 J
16. 1900 J
17. 290 m

Lab Planning

Beginning on page T34 are preparation notes and teaching tips to assist you in planning.

Blank data tables (as well as some sample data) appear on the **One-Stop Planner.**

No Books in the Lab?

See the ***Datasheets for In-Text Labs*** workbook for a reproducible master copy of this experiment.

CBL™ Option

A **CBL™** version of this lab appears in the ***CBL™ Experiments*** workbook.

Safety Caution

Remind students to attach masses securely and to make sure the area is clear before allowing masses to oscillate. Remind students not to pull too hard on the spring because it will not return to the correct equilibrium position. Also, do not add too much mass (which will stretch the spring to the point of deforming it).

CHAPTER 5

Skills Practice Lab — Conservation of Mechanical Energy

OBJECTIVES

- **Determine** the spring constant of a spring.
- **Calculate** elastic potential energy.
- **Calculate** gravitational potential energy.
- **Determine** whether mechanical energy is conserved in an oscillating spring.

MATERIALS LIST

- **Hooke's law apparatus**
- **meterstick**
- **rubber bands**
- **set of masses**
- **support stand and clamp**

A mass on a spring will oscillate vertically when it is lifted to the length of the relaxed spring and released. The gravitational potential energy increases from a minimum at the lowest point to a maximum at the highest point. The elastic potential energy in the spring increases from a minimum at the highest point, where the spring is relaxed, to a maximum at the lowest point, where the spring is stretched. Because the mass is temporarily at rest, the kinetic energy of the mass is zero at the highest and lowest points. Thus, the total mechanical energy at those points is the sum of the elastic potential energy and the gravitational potential energy.

A Hooke's law apparatus combines a stand for mounting a hanging spring and a vertical ruler for measuring the displacement of a mass attached to the spring. In this lab, you will use a Hooke's law apparatus to determine the spring constant of a spring. You will also collect data during the oscillation of a mass on the spring and use your data to calculate gravitational potential energy and elastic potential energy at different points in the oscillation.

SAFETY

- **Tie back long hair, secure loose clothing, and remove loose jewelry to prevent their getting caught in moving or rotating parts. Put on goggles.**
- **Attach masses securely. Perform this experiment in a clear area. Swinging or dropped masses can cause serious injury.**

PROCEDURE

Preparation

1. Read the entire lab procedure, and plan the steps you will take.
2. If you are not using a datasheet provided by your teacher, prepare a data table in your lab notebook with four columns and seven rows. In the first row, label the first through fourth columns *Trial, Mass (kg), Stretched Spring (m),* and *Force (N).* In the first column, label the second through seventh rows *1, 2, 3, 4, 5,* and *6.* Above or below the data table, make a space to enter the value for *Initial Spring (m).*
3. If you are not using a datasheet provided by your teacher, prepare a second data table in your lab notebook with three columns and seven rows. In the first row, label the first through third columns *Trial, Highest Point*

(m), and *Lowest Point (m)*. In the first column, label the second through seventh rows *1, 2, 3, 4, 5,* and *6*. Above or below the data table, make a space to enter the value for *Initial Distance (m)*.

Spring Constant

4. Set up the Hooke's law apparatus as shown in **Figure 1.**

5. Place a rubber band around the scale at the initial resting position of the pointer, or adjust the scale or pan to read 0.0 cm. Record this position of the pointer as *Initial Spring (m)*. If you have set the scale at 0.0 cm, record 0.00 m as the initial spring position.

6. Measure the distance from the floor to the rubber band on the scale. Record this measurement in the second data table under *Initial Distance (m)*. This distance must remain constant throughout the lab.

7. Find a mass that will stretch the spring so that the pointer moves approximately one-quarter of the way down the scale.

8. Record the value of the mass. Also record the position of the pointer under *Stretched Spring* in the data table.

9. Perform several trials with increasing masses until the spring stretches to the bottom of the scale. Record the mass and the position of the pointer for each trial.

Conservation of Mechanical Energy

10. Find a mass that will stretch the spring to about twice its original length. Record the mass in the second data table. Leave the mass in place on the pan.

Figure 1

Step 5: If the scale is adjusted to read 0.0 cm, record 0.00 m as the initial spring length in your data table.

Step 7: In this part of the lab, you will collect data to find the spring constant of the spring.

Step 10: In this part of the lab, you will oscillate a mass on the spring to find out whether mechanical energy is conserved.

CHAPTER 5 LAB

Tips and Tricks

- For best results, use weights of less than 1.0 N for steps 10–14.
- Show students how to read the scale on the Hooke's law apparatus.
- Demonstrate releasing the mass hanger so it will oscillate vertically without twisting.
- Draw a diagram of the apparatus on the chalkboard, and label the distances students will be measuring in the lab: *Initial Distance, Initial Spring, Stretched Spring, Highest Point,* and *Lowest Point.* Show students how to refer to the diagram to find the elongation of the spring and the height of the mass at each point.

✓ Checkpoints

Step 5: Students should adjust the scale to zero at the initial position of the spring if possible.

Step 6: Make sure students are measuring the vertical distance from the floor to the initial position of the spring.

Step 13: Students may need to practice a technique to identify the highest and lowest points while the mass is oscillating. Without disturbing the apparatus, they might use pencils as pointers to mark the place until they can place their rubber bands.

ANSWERS

Analysis

1. For sample data, values range from 0.021 m to 0.118 m.

2. For sample data, values range from $F = 0.050$ N to $F = 2.02$ N.

3. For sample data, $k_{avg} =$ 19.6 N/m.

4. For sample data, values range from 0.000 m to 0.006 m.

5. For sample data, values range from 0.021 m to 0.117 m.

6. For sample data, values range from 0.00 J to 3.52×10^{-4} J.

7. For sample data, values range from 4.31×10^{-3} J to 1.34×10^{-1} J.

8. $PE_{elastic}$ is greatest at the lowest point and least at the highest point because the elongation is greatest at the lowest point and because $PE_{elastic}$ depends on the elongation squared.

9. For sample data, values range from 0.234 m to 0.240 m.

10. For sample data, values range from 0.123 m to 0.219 m.

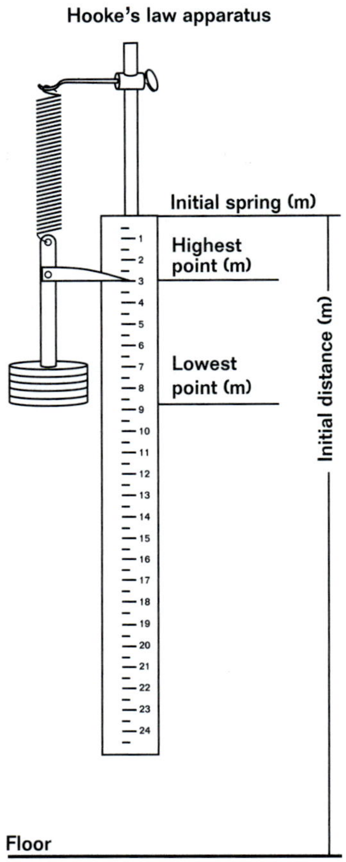

Figure 2

11. Raise the pan until the pointer is at the zero position, the position where you measured the *Initial Spring* measurement.

12. Gently release the pan to let the pan drop. Watch closely to identify the high and low points of the oscillation.

13. Use a rubber band to mark the lowest position to which the pan falls, as indicated by the pointer. This point is the lowest point of the oscillation. Record the values as *Highest Point* and *Lowest Point* in your data table.

14. Perform several more trials, using a different mass for each trial. Record all data in your data table.

15. Clean up your work area. Put equipment away safely so that it is ready to be used again.

ANALYSIS

1. Organizing Data Use your data from the first data table to calculate the elongation of the spring. Use the equation *elongation = stretched spring − initial spring.*

2. Organizing Data For each trial, convert the masses used to measure the spring constant to their force equivalents. Use the equation $F_g = ma_g$.

3. Organizing Data For each trial, calculate the spring constant using the equation $k = \frac{force}{elongation}$. Take the average of all trials, and use this value as the spring constant.

4. Organizing Data Using your data from the second data table, calculate the elongation of the spring at the highest point of each trial. Use the equation *elongation = highest point − initial spring.* Refer to **Figure 2.**

5. Organizing Data Calculate the elongation of the spring at the lowest point of each trial. Use the equation *elongation = lowest point − initial spring.* Refer to **Figure 2.**

6. Organizing Data For each trial, calculate the elastic potential energy, $PE_{elastic} = \frac{1}{2}kx^2$, at the highest point of the oscillation.

7. Organizing Data For each trial, calculate the elastic potential energy at the lowest point of the oscillation.

8. Analyzing Results Based on your calculations in items 6 and 7, where is the elastic potential energy greatest? Where is it the least? Explain these results in terms of the energy stored in the spring.

9. Organizing Data Calculate the height of the mass at the highest point of each trial. Use the equation *highest = initial distance − elongation.*

10. Organizing Data Calculate the height of the mass at the lowest point of each trial. Use the equation *lowest* = *initial distance* – *elongation.*

11. Organizing Data For each trial, calculate the gravitational potential energy, $PE_g = ma_gh$, at the highest point of the oscillation.

12. Organizing Data For each trial, calculate the gravitational potential energy at the lowest point of the oscillation.

13. Analyzing Results According to your calculations in items 11 and 12, where is the gravitational potential energy the greatest? Where is it the least? Explain these results in terms of gravity and the height of the mass and the spring.

14. Organizing Data Find the total potential energy at the top of the oscillation and at the bottom of the oscillation.

CONCLUSIONS

15. Drawing Conclusions Based on your data, is mechanical energy conserved in the oscillating mass on the spring? Explain how your data support your answers.

16. Making Predictions How would using a stiffer spring affect the value for the spring constant? How would this change affect the values for the elastic and gravitational potential energies?

EXTENSIONS

17. Extending Ideas Use your data to find the midpoint of the oscillation for each trial. Calculate the gravitational potential energy and the elastic potential energy at the midpoint. Use the principle of the conservation of mechanical energy to find the kinetic energy and the speed of the mass at the midpoint.

18. Designing Experiments Based on what you have learned in this lab, design an experiment to measure the spring constants of springs and other elastic materials in common products, such as the springs inside ball point pens, rubber bands, or even elastic waistbands. Include in your plan a way to determine how well each spring or elastic material conserves mechanical energy. If you have time and your teacher approves your plan, carry out the experiment on several items, and make a table comparing your results for the various items.

11. For sample data, values range from 0.119 J to 0.475 J.

12. For sample data, values range from 0.073 J to 0.389 J.

13. Gravitational *PE* is greatest at the highest point because it is proportional to the height of the mass.

14. Make sure students use the relationship $PE_{total} = PE_g + PE_{elastic}$. For sample data, values range from 0.119 J to 0.475 J at the high point and 0.161 J to 0.393 J at the low point.

Conclusions

15. Mechanical energy is conserved; the sum of the elastic and gravitational potential energies is the same at the top and bottom of the oscillation.

16. A stiffer spring would give greater values for the spring constant. The elastic potential energy would be greater, but the gravitational potential energy would not change.

Extensions

17. For sample data, PE_g at the midpoint ranges from 0.096 J to 0.408 J, and $PE_{elastic}$ at the midpoint ranges from 1.08×10^{-3} J to 3.70×10^{-2} J. Values for *KE* at the midpoint range from 0.001 J to 0.030 J, and values for the speed at the midpoint range from 0.11 m/s to 0.90 m/s.

18. Student designs and results will vary. Students should recognize that some elastic materials act as a spring only when stretched, not when compressed. If you wish students to carry out their designed experiments, you may provide them with various springs and elastic items, or have them find items at home.

CHAPTER 6

Momentum and Collisions

Planning Guide

Compression Guide

To shorten instruction because of time limitations, omit the opener and Section 3 and abbreviate the review.

OBJECTIVES	LABS, DEMONSTRATIONS, AND ACTIVITIES	TECHNOLOGY RESOURCES
PACING • 45 min pp. 196–197 **Chapter Opener**		CD **Visual Concepts,** Chapter 6 BASIC
PACING • 45 min pp. 198–204 **Section 1 Momentum and Impulse** • Compare the momentum of different moving objects. • Compare the momentum of the same object moving with different velocities. • Identify examples of change in the momentum of an object. • Describe changes in momentum in terms of force and time.	TE **Demonstration** Impulse, p.200 BASIC ANC **CBL™ Experiment** Impulse and Momentum*◆ GENERAL	OSP **Lesson Plans** TR 20 Impulse-Momentum Theorem TR 21A Stopping Distances
PACING • 90 min pp. 205–211 **Section 2 Conservation of Momentum** • Describe the interaction between two objects in terms of the change in momentum of each object. • Compare the total momentum of two objects before and after they interact. • State the law of conservation of momentum. • Predict the final velocities of objects after collisions, given the initial velocities.	SE **Inquiry Lab** Conservation of Momentum, pp. 230–231◆ GENERAL ANC **Datasheet** Inquiry Lab, Conservation of Momentum* GENERAL ANC **Datasheet** Skills Practice Lab, Conservation of Momentum* GENERAL ANC **CBL™ Experiment** Conservation of Momentum*◆ GENERAL	OSP **Lesson Plans** TR 21 Force and Change in Momentum TR 22A Momentum in a Collision
PACING • 45 min pp. 212–220 *Advanced Level* **Section 3 Elastic and Inelastic Conditions** • Identify different types of collisions. • Determine the changes in kinetic energy during perfectly inelastic collisions. • Compare conservation of momentum and conservation of kinetic energy in perfectly inelastic and elastic collisions. • Find the final velocity of an object in perfectly inelastic and elastic collisions.	SE **Quick Lab** Elastic and Inelastic Collisions, p. 217 GENERAL TE **Demonstration** Inelastic Collisions, p.212 GENERAL	OSP **Lesson Plans** TR 22 Types of Collisions TR 23A Inelastic Collision TR 24A Elastic Collision

PACING • 90 min

CHAPTER REVIEW, ASSESSMENT, AND STANDARDIZED TEST PREPARATION

SE **Chapter Highlights,** p. 222
SE **Chapter Review,** pp. 223–227
SE **Graphing Calculator Practice,** p. 226 GENERAL
SE **Alternative Assessment,** p. 227 ADVANCED
SE **Standardized Test Prep,** pp. 228–229 GENERAL
SE **Appendix D: Equations,** pp. 856–857
SE **Appendix I: Additional Problems,** pp. 884–886
ANC **Study Guide Worksheet** Mixed Review* GENERAL
ANC **Chapter Test A*** GENERAL
ANC **Chapter Test B*** ADVANCED
OSP **Test Generator**

Online and Technology Resources

go.hrw.com Holt Online Learning

Visit go.hrw.com to find a variety of online resources. To access this chapter's extensions, enter the keyword **HF6MOMXT** and click the "go" button. Click **Holt Online Learning** for an online edition of this textbook, and other interactive resources.

HOLT Teacher's One-Stop Planner®

This DVD package includes:
- Holt Calendar Planner
- Customizable Lesson Plans
- Editable Worksheets
- ExamView® Version 6 Assessment Suite
- Interactive Teacher's Edition
- Holt PuzzlePro®
- Holt PowerNotes® Presentations
- MindPoint® Quiz Show

KEY

- **SE** Student Edition
- **TE** Teacher Edition
- **ANC** Ancillary Worksheet
- **OSP** One-Stop Planner
- **CD** CD or CD-ROM
- **TR** Teaching Transparencies
- **EXT** Online Extension
- * Also on One-Stop Planner
- ◆ Requires advance prep

SKILLS DEVELOPMENT RESOURCES	REVIEW AND ASSESSMENT	CORRELATIONS
		National Science Education Standards
SE Sample Set A Momentum, pg. 199 BASIC **ANC Problem Workbook*** and **OSP Problem Bank** Sample Set A BASIC **SE Sample Set B** Force and Impulse, pg. 201 BASIC **TE Classroom Practice,** p. 201 BASIC **ANC Problem Workbook*** and **OSP Problem Bank** Sample Set B BASIC **SE Sample Set C** Stopping Distance, pp. 202–203 **TE Classroom Practice,** p. 202 BASIC **ANC Problem Workbook*** and **OSP Problem Bank** Sample Set C BASIC	**SE Section Review,** p. 204 GENERAL **ANC Study Guide Worksheet** Section 1* GENERAL **ANC Quiz** Section 1* BASIC	UCP 1,2,3 HNS 3
SE Sample Set D Conservation of Momentum, pp. 208–209 GENERAL **TE Classroom Practice,** p. 208 GENERAL **ANC Problem Workbook*** and **OSP Problem Bank** Sample Set D GENERAL **SE Conceptual Challenge,** p. 206	**SE Section Review,** p. 211 GENERAL **ANC Study Guide Worksheet** Section 2* GENERAL **ANC Quiz** Section 2* BASIC	UCP 1,2,3,5 SAI 1,2 ST 1,2 SPSP 1,4,5 PS 5a
SE Sample Set E Perfectly Inelastic Collisions, pp. 213–214 GENERAL **TE Classroom Practice,** p. 213 GENERAL **ANC Problem Workbook*** and **OSP Problem Bank** Sample Set E GENERAL **SE Sample Set F** Kinetic Energy in Perfectly Inelastic Collisions, pp. 215–216 GENERAL **TE Classroom Practice,** p. 215 GENERAL **ANC Problem Workbook*** and **OSP Problem Bank** Sample Set F GENERAL **SE Sample Set G** Elastic Collisions, pp. 218–219 ADVANCED **TE Classroom Practice,** p. 218 ADVANCED **ANC Problem Workbook*** and **OSP Problem Bank** Sample Set G ADVANCED	**SE Section Review,** p. 220 ADVANCED **ANC Study Guide Worksheet** Section 3* ADVANCED **ANC Quiz** Section 3* GENERAL	UCP 1,2,3 SAI 1,2 PS 5a

SCILINKS. NSTA
www.scilinks.org

Maintained by the **National Science Teachers Association.**

Topic: Momentum
SciLinks Code: HF60988

Topic: Collisions
SciLinks Code: HF60311

Topic: Rocketry
SciLinks Code: HF61324

Classroom CD-ROMs

- Guided Reading Audio Program
- Student One Stop
- Virtual Investigations
- Visual Concepts

Holt Lab Generator CD-ROM

Search for any lab by topic, standard, difficulty level, or time. Edit any lab to fit your needs, or create your own labs. Use the Lab Materials QuickList software to customize your lab materials list.

CHAPTER 6
Overview

Section 1 defines momentum in terms of mass and velocity, introduces the concept of impulse, and relates impulse and momentum.

Section 2 explores the law of conservation of momentum and uses this law to predict the final velocity of an object after a collision.

Section 3 distinguishes between elastic, perfectly inelastic, and inelastic collisions and discusses whether kinetic energy is conserved in each type of collision.

About the Illustration

Soccer is a good example to help students understand the concept of momentum and distinguish it from force, velocity, and kinetic energy. This photograph is a dramatic example of a player colliding with a ball and changing the momentum of the ball. Use this example to illustrate the vector nature of momentum; the photograph can open a discussion about how the direction as well as the magnitude of momentum is affected by the collision.

CHAPTER 6

Momentum and Collisions

Soccer players must consider much information about the ball and their own bodies in order to play effectively. The player in the photograph determines what force to exert on the ball in order to send the ball where he wants it to go.

WHAT TO EXPECT

In this chapter, you will analyze momentum and collisions between two or more objects. You will consider the mass and velocity of one or more objects and the conservation of momentum and energy.

Why It Matters

Collisions and other transfers of momentum occur frequently in everyday life. Examples in sports include the motion of balls against rackets in tennis and the motion of human bodies against each other in football.

CHAPTER PREVIEW

Tapping Prior Knowledge

Knowledge to Review

- ✔ A force on an object is a push or pull that tends to cause a change in motion. Forces can be field or contact forces.
- ✔ Newton's laws of motion describe the effects of forces on objects and the idea that forces always exist in pairs.
- ✔ Energy of motion, called *kinetic energy,* depends on mass and speed: $KE = \frac{1}{2}mv^2$.
- ✔ Energy is neither created nor destroyed, but it can be converted from one form to another.

Items to Probe

- ✔ Newton's third law: Have students draw free-body diagrams for interacting objects and identify the third law pairs.

SECTION 1
General Level

The Language of Physics — BASIC

As seen in previous chapters, words used in our everyday language, such as *work* and *energy*, often have precise definitions in physics. *Momentum* is another example of such a word. As with work and energy, there is a relationship between the everyday use and the scientific use of *momentum.* For example, popular ideas or trends are sometimes said to be "gaining momentum." In this case, the phrase "gaining momentum" means that an idea is gaining popularity. Ask students how this meaning compares with the meaning of *momentum* in physics. Ask students to provide other examples and compare the different meanings.

SECTION 1

Momentum and Impulse

SECTION OBJECTIVES

- **Compare the momentum of different moving objects.**
- **Compare the momentum of the same object moving with different velocities.**
- **Identify examples of change in the momentum of an object.**
- **Describe changes in momentum in terms of force and time.**

LINEAR MOMENTUM

When a soccer player heads a moving ball during a game, the ball's velocity changes rapidly. After the ball is struck, the ball's speed and the direction of the ball's motion change. The ball moves across the soccer field with a different speed than it had and in a different direction than it was traveling before the collision.

The quantities and kinematic equations describing one-dimensional motion predict the motion of the ball before and after the ball is struck. The concept of force and Newton's laws can be used to calculate how the motion of the ball changes when the ball is struck. In this chapter, we will examine how the force and the duration of the collision between the ball and the soccer player affect the motion of the ball.

Momentum is mass times velocity

momentum

a quantity defined as the product of the mass and velocity of an object

To address such issues, we need a new concept, **momentum.** *Momentum* is a word we use every day in a variety of situations. In physics this word has a specific meaning. The linear momentum of an object of mass m moving with a velocity $\mathbf{v}$ is defined as the product of the mass and the velocity. Momentum is represented by the symbol $\mathbf{p}$.

MOMENTUM

$$\mathbf{p} = m\mathbf{v}$$

$$\text{momentum} = \text{mass} \times \text{velocity}$$

Figure 1
A bicycle rolling downhill has momentum. An increase in either mass or speed will increase the momentum.

As its definition shows, momentum is a vector quantity, with its direction matching that of the velocity. Momentum has dimensions mass × length/time, and its SI units are kilogram-meters per second (kg•m/s).

If you think about some examples of the way the word *momentum* is used in everyday speech, you will see that the physics definition conveys a similar meaning. Imagine coasting down a hill of uniform slope on your bike without pedaling or using the brakes. Because of the force of gravity, you will accelerate; that is, your velocity will increase with time. This idea is often expressed by saying that you are "picking up speed" or "gathering momentum." The faster you move, the more momentum you have and the more difficult it is to come to a stop.

Imagine rolling a bowling ball down one lane at a bowling alley and rolling a playground ball down another lane at the same speed. The more massive bowling ball exerts more force on the pins than the playground ball exerts because the bowling ball has more momentum than the playground ball does. When we think of a massive object moving at a high velocity, we often say that the object has a large momentum. A less massive object with the same velocity has a smaller momentum.

On the other hand, a small object moving with a very high velocity may have a larger momentum than a more massive object that is moving slowly does. For example, small hailstones falling from very high clouds can have enough momentum to hurt you or cause serious damage to cars and buildings.

Did you know?

Momentum is so fundamental in Newton's mechanics that Newton called it simply "quantity of motion." The symbol for momentum, **p**, comes from German mathematician Gottfried Leibniz. Leibniz used the term *progress* to mean "the quantity of motion with which a body proceeds in a certain direction."

SAMPLE PROBLEM A

Momentum

PROBLEM

A 2250 kg pickup truck has a velocity of 25 m/s to the east. What is the momentum of the truck?

SOLUTION

Given: $m = 2250$ kg $\quad \mathbf{v} = 25$ m/s to the east

Unknown: $\mathbf{p} = ?$

Use the definition of momentum.

$$\mathbf{p} = m\mathbf{v} = (2250 \text{ kg})(25 \text{ m/s east})$$

$$\boxed{\mathbf{p} = 5.6 \times 10^4 \text{ kg}\bullet\text{m/s to the east}}$$

TIP *Momentum is a vector quantity, so you must specify both its size and direction.*

PRACTICE A

Momentum

1. A deer with a mass of 146 kg is running head-on toward you with a speed of 17 m/s. You are going north. Find the momentum of the deer.
2. A 21 kg child on a 5.9 kg bike is riding with a velocity of 4.5 m/s to the northwest.
   a. What is the total momentum of the child and the bike together?
   b. What is the momentum of the child?
   c. What is the momentum of the bike?
3. What velocity must a 1210 kg car have in order to have the same momentum as the pickup truck in Sample Problem A?

SECTION 1

PROBLEM GUIDE A

Use this guide to assign problems.
SE = Student Edition Textbook
PW = Problem Workbook
PB = Problem Bank on the One-Stop Planner (OSP)

Solving for:

p	**SE** Sample, 1–2; Ch. Rvw. 11, 37* **PW** 5–6 **PB** 5–7
m	**SE** Ch. Rvw. 36* **PW** Sample, 1–2 **PB** 8–10
v	**SE** 3; Ch. Rvw. 35, 36* **PW** 3–4 **PB** Sample, 1–4

***Challenging Problem**
Consult the printed Solutions Manual or the OSP for detailed solutions.

ANSWERS

Practice A

1. 2.5×10^3 kg•m/s to the south
2. a. 1.2×10^2 kg•m/s to the northwest
   b. 94 kg•m/s to the northwest
   c. 27 kg•m/s to the northwest
3. 46 m/s to the east

Demonstration

Impulse BASIC

Purpose Show that changes in momentum are caused by forces.

Materials dynamics cart

Procedure Have the students observe the cart at rest. Ask the students the value of the momentum of the cart when it is at rest (*zero*). Now push on the cart, and ask what has happened to the momentum of the cart. *(Its momentum has increased.)* How was the cart's momentum changed? *(An external force was applied.)* Stop the cart as it moves across the table. Again, ask the students how the momentum of the cart was changed. *(An external force was applied.)*

Figure 2
When the ball is moving very fast, the player must exert a large force over a short time to change the ball's momentum and quickly bring the ball to a stop.

SCILINKS.
www.scilinks.org
Topic: Momentum
Code: HF60988

A change in momentum takes force and time

Figure 2 shows a player stopping a moving soccer ball. In a given time interval, he must exert more force to stop a fast ball than to stop a ball that is moving more slowly. Now imagine a toy truck and a real dump truck rolling across a smooth surface with the same velocity. It would take much more force to stop the massive dump truck than to stop the toy truck in the same time interval. You have probably also noticed that a ball moving very fast stings your hands when you catch it, while a slow-moving ball causes no discomfort when you catch it. The fast ball stings because it exerts more force on your hand than the slow-moving ball does.

From examples like these, we see that a change in momentum is closely related to force. In fact, when Newton first expressed his second law mathematically, he wrote it not as $\mathbf{F} = m\mathbf{a}$, but in the following form.

$$\mathbf{F} = \frac{\Delta \mathbf{p}}{\Delta t}$$

$$\text{force} = \frac{\text{change in momentum}}{\text{time interval}}$$

We can rearrange this equation to find the change in momentum in terms of the net external force and the time interval required to make this change.

IMPULSE-MOMENTUM THEOREM

$$\mathbf{F}\Delta t = \Delta \mathbf{p} \quad \text{or} \quad \mathbf{F}\Delta t = \Delta \mathbf{p} = m\mathbf{v_f} - m\mathbf{v_i}$$

force × time interval = change in momentum

This equation states that a net external force, **F,** applied to an object for a certain time interval, Δt, will cause a change in the object's momentum equal to the product of the force and the time interval. In simple terms, a small force acting for a long time can produce the same change in momentum as a large force acting for a short time. In this book, all forces exerted on an object are assumed to be constant unless otherwise stated.

The expression $\mathbf{F}\Delta t = \Delta \mathbf{p}$ is called the impulse-momentum theorem. The term on the left side of the equation, $\mathbf{F}\Delta t$, is called the **impulse** of the force **F** for the time interval Δt.

impulse
the product of the force and the time over which the force acts on an object

The equation $\mathbf{F}\Delta t = \Delta \mathbf{p}$ explains why proper technique is important in so many sports, from karate and billiards to softball and croquet. For example, when a batter hits a ball, the ball will experience a greater change in momentum if the batter keeps the bat in contact with the ball for a longer time. Extending the time interval over which a constant force is applied allows a smaller force to cause a greater change in momentum than would result if the force were applied for a very short time. You may have noticed this fact when pushing a full shopping cart or moving furniture.

SAMPLE PROBLEM B

Force and Impulse

PROBLEM

A 1400 kg car moving westward with a velocity of 15 m/s collides with a utility pole and is brought to rest in 0.30 s. Find the force exerted on the car during the collision.

SOLUTION

Given: $m = 1400$ kg $\quad \mathbf{v_i} = 15$ m/s to the west, $v_i = -15$ m/s

$\Delta t = 0.30$ s $\quad \mathbf{v_f} = 0$ m/s

Unknown: $\mathbf{F} = ?$

Use the impulse-momentum theorem.

$$\mathbf{F}\Delta t = \Delta \mathbf{p} = m\mathbf{v_f} - m\mathbf{v_i}$$

$$\mathbf{F} = \frac{m\mathbf{v_f} - m\mathbf{v_i}}{\Delta t}$$

$$F = \frac{(1400 \text{ kg})(0 \text{ m/s}) - (1400 \text{ kg})(-15 \text{ m/s})}{0.30 \text{ s}} = \frac{21\ 000 \text{ kg}\bullet\text{m/s}}{0.30 \text{ s}}$$

$$\mathbf{F} = 7.0 \times 10^4 \text{ N to the east}$$

TIP *Create a simple convention for describing the direction of vectors. For example, always use a negative speed for objects moving west or south and a positive speed for objects moving east or north.*

PRACTICE B

Force and Impulse

1. A 0.50 kg football is thrown with a velocity of 15 m/s to the right. A stationary receiver catches the ball and brings it to rest in 0.020 s. What is the force exerted on the ball by the receiver?

2. An 82 kg man drops from rest on a diving board 3.0 m above the surface of the water and comes to rest 0.55 s after reaching the water. What is the net force on the diver as he is brought to rest?

3. A 0.40 kg soccer ball approaches a player horizontally with a velocity of 18 m/s to the north. The player strikes the ball and causes it to move in the opposite direction with a velocity of 22 m/s. What impulse was delivered to the ball by the player?

4. A 0.50 kg object is at rest. A 3.00 N force to the right acts on the object during a time interval of 1.50 s.

a. What is the velocity of the object at the end of this interval?

b. At the end of this interval, a constant force of 4.00 N to the left is applied for 3.00 s. What is the velocity at the end of the 3.00 s?

Classroom Practice

Force and Impulse

Air bags are designed to protect passengers during collisions. Compare the magnitude of the force required to stop a moving passenger in 0.75 s (by a deployed air bag) with the magnitude of the force required to stop the same passenger at the same speed in 0.026 s (by the dashboard).

Answer

$$F_{air\ bag} = \frac{1}{29} F_{dashboard}$$

PROBLEM GUIDE B

Use this guide to assign problems.
SE = Student Edition Textbook
PW = Problem Workbook
PB = Problem Bank on the One-Stop Planner (OSP)

Solving for:

F	**SE** Sample, 1–2; Ch. Rvw. 12–13, 41, 47* **PW** 7–9 **PB** 5–7
Δt	**SE** 3* **PW** Sample, 1–3 **PB** 8–10
$\Delta p, v$	**SE** 4; Ch. Rvw. 46, 47* **PW** 4–6 **PB** Sample, 1–4

***Challenging Problem**
Consult the printed Solutions Manual or the OSP for detailed solutions.

ANSWERS

Practice B

1. 3.8×10^2 N to the left
2. 1.1×10^3 N upward
3. 16 kg•m/s to the south
4. a. 9.0 m/s to the right
   b. 15 m/s to the left

Visual Strategy BASIC

Figure 3
Be sure students understand the relationship between stopping time and momentum.

Q Why is the loaded truck's stopping time twice as much as the empty truck's when acted on by the same force?

A *The loaded truck's momentum must be twice as large as the unloaded truck, so its change in momentum is also twice as large. Assuming that the applied forces are the same, the time period must be twice as large because $\Delta p = F\Delta t$.*

Q How do the stopping distances of the trucks compare?

A *The loaded truck's time period is twice as large while its acceleration is half as much ($F = ma$). Because $x = v_i\Delta t + \frac{1}{2}a\Delta t^2$, the loaded truck's stopping distance is two times as large as the empty truck's. (The braking force is assumed to be the same in both cases.)*

Classroom Practice

Stopping Distance
If the maximum coefficient of kinetic friction between a 2300 kg car and a road is 0.50, what is the minimum stopping distance for a car entering a skid at 29 m/s?

Answer
86 m

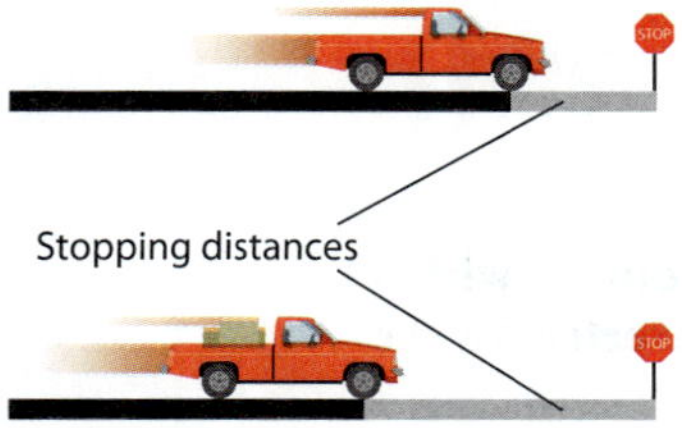

Figure 3
The loaded truck must undergo a greater change in momentum in order to stop than the truck without a load.

Stopping times and distances depend on the impulse-momentum theorem

Highway safety engineers use the impulse-momentum theorem to determine stopping distances and safe following distances for cars and trucks. For example, the truck hauling a load of bricks in **Figure 3** has twice the mass of the other truck, which has no load. Therefore, if both are traveling at 48 km/h, the loaded truck has twice as much momentum as the unloaded truck. If we assume that the brakes on each truck exert about the same force, we find that the stopping time is two times longer for the loaded truck than for the unloaded truck, and the stopping distance for the loaded truck is two times greater than the stopping distance for the truck without a load.

SAMPLE PROBLEM C

Stopping Distance

PROBLEM

A 2240 kg car traveling to the west slows down uniformly from 20.0 m/s to 5.00 m/s. How long does it take the car to decelerate if the force on the car is 8410 N to the east? How far does the car travel during the deceleration?

SOLUTION

Given: $m = 2240$ kg $\quad \mathbf{v_i} = 20.0$ m/s to the west, $v_i = -20.0$ m/s

$\mathbf{v_f} = 5.00$ m/s to the west, $v_f = -5.00$ m/s

$\mathbf{F} = 8410$ N to the east, $F = +8410$ N

Unknown: $\Delta t = ?$ $\quad \Delta \mathbf{x} = ?$

Use the impulse-momentum theorem.

$$\mathbf{F}\Delta t = \Delta\mathbf{p}$$

$$\Delta t = \frac{\Delta \mathbf{p}}{\mathbf{F}} = \frac{m\mathbf{v_f} - m\mathbf{v_i}}{\mathbf{F}}$$

$$\Delta t = \frac{(2240\ \text{kg})(-5.00\ \text{m/s}) - (2240\ \text{kg})(-20.0\ \text{m/s})}{8410\ \text{kg}\bullet\text{m/s}^2}$$

$$\boxed{\Delta t = 4.00\ \text{s}}$$

$$\Delta x = \tfrac{1}{2}(v_i + v_f)\Delta t$$

$$\Delta x = \tfrac{1}{2}(-20.0\ \text{m/s} - 5.00\ \text{m/s})(4.00\ \text{s})$$

$$\boxed{\Delta \mathbf{x} = -50.0\ \text{m} = 50.0\ \text{m to the west}}$$

TIP *For motion in one dimension, take special care to set up the sign of the speed. You can then treat the vectors in the equations of motion as scalars and add direction at the end.*

PRACTICE C

Stopping Distance

1. How long would the car in Sample Problem C take to come to a stop from its initial velocity of 20.0 m/s to the west? How far would the car move before stopping? Assume a constant acceleration.
2. A 2500 kg car traveling to the north is slowed down uniformly from an initial velocity of 20.0 m/s by a 6250 N braking force acting opposite the car's motion. Use the impulse-momentum theorem to answer the following questions:
   - **a.** What is the car's velocity after 2.50 s?
   - **b.** How far does the car move during 2.50 s?
   - **c.** How long does it take the car to come to a complete stop?
3. Assume that the car in Sample Problem C has a mass of 3250 kg.
   - **a.** How much force would be required to cause the same acceleration as in item 1? Use the impulse-momentum theorem.
   - **b.** How far would the car move before stopping? (Use the force found in **a.**)

Force is reduced when the time interval of an impact is increased

The impulse-momentum theorem is used to design safety equipment that reduces the force exerted on the human body during collisions. Examples of this are the nets and giant air mattresses firefighters use to catch people who must jump out of tall burning buildings. The relationship is also used to design sports equipment and games.

Figure 4 shows an Inupiat family playing a traditional game. Common sense tells us that it is much better for the girl to fall onto the outstretched blanket than onto the hard ground. In both cases, however, the change in momentum of the falling girl is exactly the same. The difference is that the blanket "gives way" and extends the time of collision so that the change in the girl's momentum occurs over a longer time interval. A longer time interval requires a smaller force to achieve the same change in the girl's momentum. Therefore, the force exerted on the girl when she lands on the outstretched blanket is less than the force would be if she were to land on the ground.

Figure 4
In this game, the girl is protected from injury because the blanket reduces the force of the collision by allowing it to take place over a longer time interval.

PROBLEM GUIDE C

Use this guide to assign problems.
SE = Student Edition Textbook
PW = Problem Workbook
PB = Problem Bank on the One-Stop Planner (OSP)

Solving for:

Δx	**SE** Sample, 1–3; Ch. Rvw. 14 **PW** 7–9 **PB** 5–7
Δt	**SE** Sample, 1–2; Ch. Rvw. 14 **PW** Sample, 1–3 **PB** 8–10
Δp	**PW** 4–6 **PB** Sample, 1–4
F	**SE** 3; Ch. Rvw. 41 **PW** Sample, 1–3 **PB** 5–7

***Challenging Problem**
Consult the printed Solutions Manual or the OSP for detailed solutions.

ANSWERS

Practice C

1. 5.33 s; 53.3 m to the west
2. **a.** 14 m/s to the north
   b. 42 m to the north
   c. 8.0 s
3. **a.** 1.22×10^4 N to the east
   b. 53.3 m to the west

SECTION 1

Teaching Tip — GENERAL

Make sure students understand that the correspondence between the time interval and the force is a result of the impulse-momentum theorem. When the egg hits the plate, as in **Figure 5(a),** the time period is shorter and the force is greater. When the egg hits the pillow, as in **(b),** the time increases and the force decreases. As the time period increases, the force continues to decrease.

SECTION REVIEW ANSWERS

1. a. momentum increases by a factor of two
   b. kinetic energy increases by a factor of four
2. a. 31.0 m/s
   b. the bullet
3. a. 2.6 kg•m/s downfield
   b. 1.3×10^2 N downfield
4. no; Because $\Delta \mathbf{p} = \mathbf{F}\Delta t$, it is possible for a large force applied over a very short time interval to change the momentum less than a smaller force applied over a longer time period.
5. Impulse is equal to the change in momentum.

Figure 5
A large force exerted over a short time **(a)** causes the same change in the egg's momentum as a small force exerted over a longer time **(b).**

Now consider a falling egg. When the egg hits a hard surface, like the plate in **Figure 5(a),** the egg comes to rest in a very short time interval. The force the hard plate exerts on the egg due to the collision is large. When the egg hits a floor covered with a pillow, as in **Figure 5(b),** the egg undergoes the same change in momentum, but over a much longer time interval. In this case, the force required to accelerate the egg to rest is much smaller. By applying a small force to the egg over a longer time interval, the pillow causes the same change in the egg's momentum as the hard plate, which applies a large force over a short time interval. Because the force in the second situation is smaller, the egg can withstand it without breaking.

SECTION REVIEW

1. The speed of a particle is doubled.
   a. By what factor is its momentum changed?
   b. What happens to its kinetic energy?
2. A pitcher claims he can throw a 0.145 kg baseball with as much momentum as a speeding bullet. Assume that a 3.00 g bullet moves at a speed of 1.50×10^3 m/s.
   a. What must the baseball's speed be if the pitcher's claim is valid?
   b. Which has greater kinetic energy, the ball or the bullet?
3. A 0.42 kg soccer ball is moving downfield with a velocity of 12 m/s. A player kicks the ball so that it has a final velocity of 18 m/s downfield.
   a. What is the change in the ball's momentum?
   b. Find the constant force exerted by the player's foot on the ball if the two are in contact for 0.020 s.
4. **Critical Thinking** When a force is exerted on an object, does a large force always produce a larger change in the object's momentum than a smaller force does? Explain.
5. **Critical Thinking** What is the relationship between impulse and momentum?

Conservation of Momentum

SECTION OBJECTIVES

- **Describe the interaction between two objects in terms of the change in momentum of each object.**
- **Compare the total momentum of two objects before and after they interact.**
- **State the law of conservation of momentum.**
- **Predict the final velocities of objects after collisions, given the initial velocities.**

MOMENTUM IS CONSERVED

So far in this chapter, we have considered the momentum of only one object at a time. Now we will consider the momentum of two or more objects interacting with each other. **Figure 6** shows a stationary billiard ball set into motion by a collision with a moving billiard ball. Assume that both balls are on a smooth table and that neither ball rotates before or after the collision. Before the collision, the momentum of ball B is equal to zero because the ball is stationary. During the collision, ball B gains momentum while ball A loses momentum. The momentum that ball A loses is exactly equal to the momentum that ball B gains.

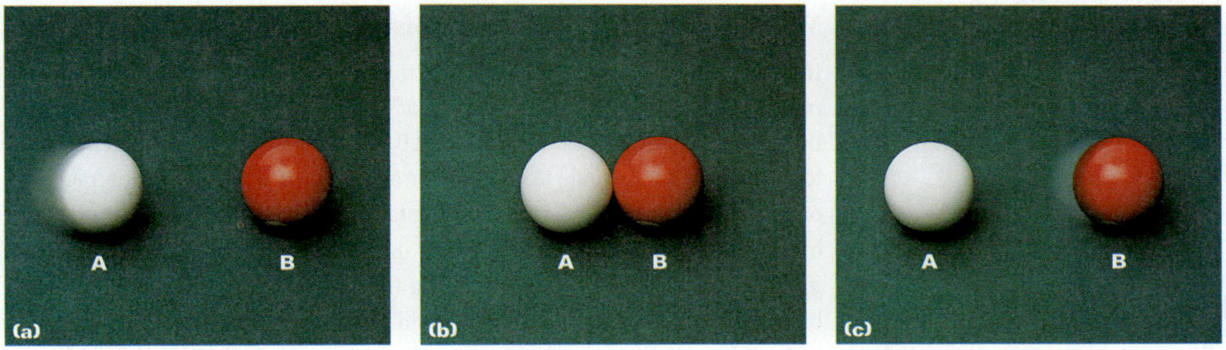

Figure 6
(a) Before the collision, the momentum of ball A is $\mathbf{p}_{A,i}$ and of ball B is zero. **(b)** During the collision, ball A loses momentum, and ball B gains momentum. **(c)** After the collision, ball B has momentum $\mathbf{p}_{B,f}$.

Table 1 shows the velocity and momentum of each billiard ball both before and after the collision. The momentum of each ball changes due to the collision, but the *total* momentum of the two balls together remains constant. In

Table 1 Momentum in a Collision

	Ball A			Ball B		
	Mass	Velocity	Momentum	Mass	Velocity	Momentum
before collision	0.16 kg	4.50 m/s	0.72 kg•m/s	0.16 kg	0 m/s	0 kg•m/s
after collision	0.16 kg	0.11 m/s	0.018 kg•m/s	0.16 kg	4.39 m/s	0.70 kg•m/s

Visual Strategy GENERAL

Figure 6
Point out to students that the two billiard balls interact by physically colliding.

Q How do the force exerted on ball A and the time interval over which it is exerted compare with the force exerted on ball B and its corresponding time interval?

A ***The forces are equal in magnitude and opposite in direction (Newton's third law), and the time intervals are also equal.***

Q Using your answer to the previous question, compare the changes in momentum of the two balls.

A ***The change in momentum of ball A must be equal in magnitude but opposite in direction to the change in momentum of ball B. This relationship is because of Newton's third law and the Impulse-momentum theorem, $\Delta \mathbf{p} = \mathbf{F}\Delta t$.***

ANSWERS

Conceptual Challenge

1. No, the only possible way for their final total momentum to be zero is if the initial total momentum is also zero. This could happen only if both skaters initially have the same magnitude of momentum but opposite directions.
2. The principle of conservation of momentum tells us that the momentum of the spacecraft and its fuel before the rockets are fired must equal the momentum of the two after the rockets are fired. Both begin at rest, so the total initial momentum is zero. When the rockets are fired, the combustion of the fuel gives the exhaust gases momentum. The spacecraft will gain a momentum equal in magnitude but opposite in direction to the exhaust gases. Thus, the total momentum will be kept at zero.

Teaching Tip — ADVANCED

Explain to the students that the cumulative effects of frictional forces during the collision are very small if we consider the system immediately before and immediately after the collision. With this assumption, we can consider momentum to be conserved. If longer periods of time are considered, frictional forces do become significant.

SCILINKS.
www.scilinks.org
Topic: Rocketry
Code: HF61324

other words, the momentum of ball A plus the momentum of ball B before the collision is equal to the momentum of ball A plus the momentum of ball B after the collision.

$$\mathbf{p}_{A,i} + \mathbf{p}_{B,i} = \mathbf{p}_{A,f} + \mathbf{p}_{B,f}$$

This relationship is true for all interactions between isolated objects and is known as the *law of conservation of momentum.*

CONSERVATION OF MOMENTUM

$$m_1\mathbf{v}_{1,i} + m_2\mathbf{v}_{2,i} = m_1\mathbf{v}_{1,f} + m_2\mathbf{v}_{2,f}$$

total initial momentum = total final momentum

For an isolated system, the law of conservation of momentum can be stated as follows:

The total momentum of all objects interacting with one another remains constant regardless of the nature of the forces between the objects.

Momentum is conserved in collisions

In the billiard ball example, we found that the momentum of ball A does not remain constant and the momentum of ball B does not remain constant, but the total momentum of ball A and ball B does remain constant. In general, the total momentum remains constant for a system of objects that interact with one another. In this case, in which the table is assumed to be frictionless, the billiard balls are the only two objects interacting. If a third object exerted a force on either ball A or ball B during the collision, the total momentum of ball A, ball B, and the third object would remain constant.

In this book, most conservation-of-momentum problems deal with only two isolated objects. However, when you use conservation of momentum to solve a problem or investigate a situation, it is important to include all objects that are involved in the interaction. Frictional forces—such as the frictional force between the billiard balls and the table—will be disregarded in most conservation-of-momentum problems in this book.

Momentum is conserved for objects pushing away from each other

Another example of conservation of momentum occurs when two or more interacting objects that initially have no momentum begin moving away from each other. Imagine that you initially stand at rest and then jump up, leaving the ground with a velocity $\mathbf{v}$. Obviously, *your* momentum is not conserved; before the jump, it was zero, and it became $m\mathbf{v}$ as you began to rise. However, the total momentum remains constant if you include Earth in your analysis. The total momentum for you and Earth remains constant.

If your momentum after you jump is 60 kg•m/s upward, then Earth must have a corresponding momentum of 60 kg•m/s downward, because total

Why it Matters

Conceptual Challenge

1. Ice Skating

If a reckless ice skater collides with another skater who is standing on the ice, is it possible for both skaters to be at rest after the collision?

2. Space Travel

A spacecraft undergoes a change of velocity when its rockets are fired. How does the spacecraft change velocity in empty space, where there is nothing for the gases emitted by the rockets to push against?

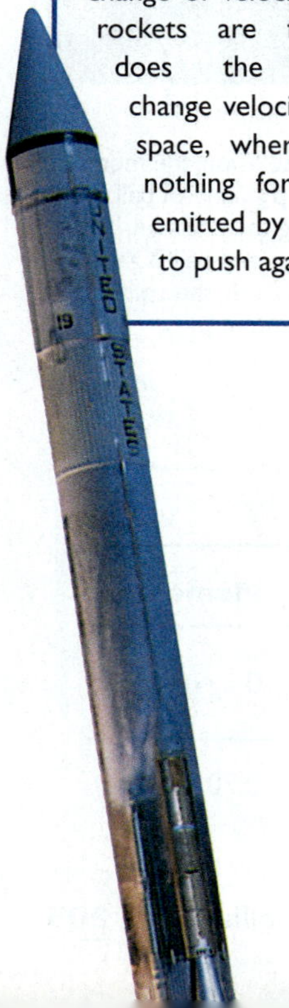

momentum is conserved. However, because Earth has an enormous mass (6×10^{24} kg), its momentum corresponds to a tiny velocity (1×10^{-23} m/s).

Imagine two skaters pushing away from each other, as shown in **Figure 7.** The skaters are both initially at rest with a momentum of $\mathbf{p_{1,i}} = \mathbf{p_{2,i}} = 0$. When they push away from each other, they move in opposite directions with equal but opposite momentum so that the total final momentum is also zero ($\mathbf{p_{1,f}} + \mathbf{p_{2,f}} = 0$).

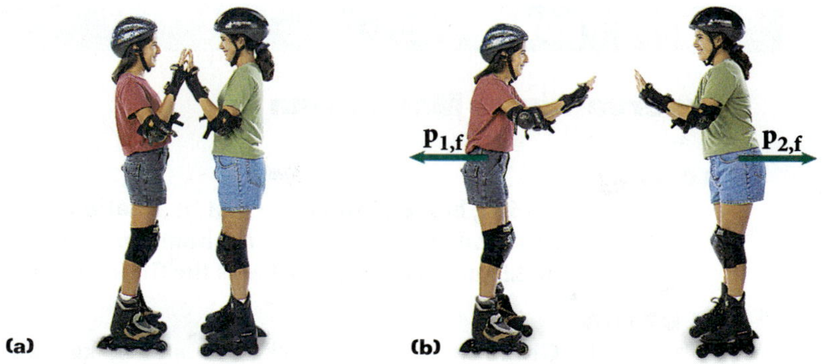

Figure 7
(a) When the skaters stand facing each other, both skaters have zero momentum, so the total momentum of both skaters is zero.
(b) When the skaters push away from each other, their momentum is equal but opposite, so the total momentum is still zero.

Why it Matters

Surviving a Collision

Pucks and carts collide in physics labs all the time with little damage. But when cars collide on a freeway, the resulting rapid change in speed can cause injury or death to the drivers and any passengers.

Many types of collisions are dangerous, but head-on collisions involve the greatest accelerations and thus the greatest forces. When two cars going 100 km/h (62 mi/h) collide head-on, each car dissipates the same amount of kinetic energy that it would dissipate if it hit the ground after being dropped from the roof of a 12-story building.

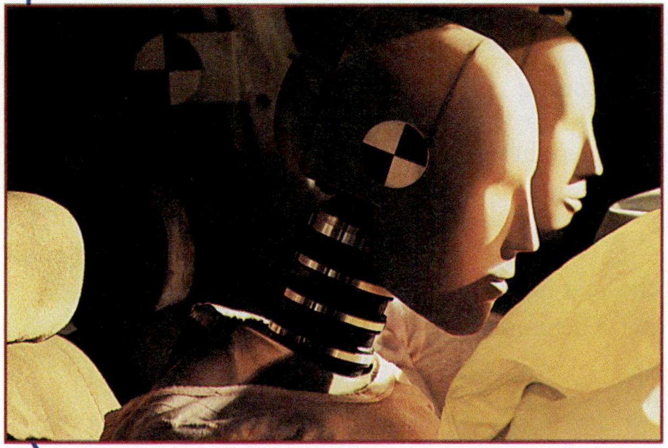

The key to many automobile-safety features is the concept of impulse. One way today's cars make use of the concept of impulse is by crumpling during impact. Pliable sheet metal and frame structures absorb energy until the force reaches the passenger compartment, which is built of rigid metal for protection. Because the crumpling slows the car gradually, it is an important factor in keeping the driver alive.

Even taking into account this built-in safety feature, the National Safety Council estimates that high-speed collisions involve accelerations of 20 times the free-fall acceleration. In other words, an 89 N (20 lb) infant could experience a force of 1780 N (400 lb) in a high-speed collision.

Seat belts are necessary to protect a body from forces of such large magnitudes. They stretch and extend the time it takes a passenger's body to stop, thereby reducing the force on the person. Air bags further extend the time over which the momentum of a passenger changes, decreasing the force even more. As of 1998, all new cars have air bags on both the driver and passenger sides. Seat belts also prevent passengers from hitting the inside frame of the car. During a collision, a person not wearing a seat belt is likely to hit the windshield, the steering wheel, or the dashboard—often with traumatic results.

SECTION 2

Key Models and Analogies — BASIC

Compare the principle of conservation of momentum with conservation of energy. Energy can be transferred from one object to another, but the *total* amount of energy in an isolated system remains constant. In a similar way, momentum is transferred during a collision, but the *total* momentum in an isolated system remains constant.

Misconception Alert

Some students may think that the principle of conservation of momentum applies only to collisions. Use the example of two skaters in **Figure 7** to show that the law holds even when the initial momentum is zero.

Why it Matters

Surviving a Collision

This feature applies the concepts in this chapter to an example most students can understand.

Extension — ADVANCED

Give students values for the mass and speed of two cars, and have them calculate the changes in momentum with the assumption that the cars come to rest after the collision. Estimate a time interval for the collision, and have them calculate the forces experienced by the drivers.

Have students research safety devices and designs that help protect drivers in a collision. Students can give oral reports, presenting their recommendation for a specific car or safety device.

SECTION 2

Classroom Practice

Conservation of Momentum
A 0.40 kg ball approaches a wall perpendicularly at 15 m/s. It collides with the wall and rebounds with an equal speed in the opposite direction. Calculate the impulse exerted on the wall.

Answer
12 kg•m/s in the original direction of motion of the ball

SAMPLE PROBLEM D

Conservation of Momentum

PROBLEM

A 76 kg boater, initially at rest in a stationary 45 kg boat, steps out of the boat and onto the dock. If the boater moves out of the boat with a velocity of 2.5 m/s to the right, what is the final velocity of the boat?

SOLUTION

1. DEFINE

Given: $m_1 = 76$ kg $\quad m_2 = 45$ kg

$\mathbf{v_{1,i}} = 0 \quad \mathbf{v_{2,i}} = 0$

$\mathbf{v_{1,f}} = 2.5$ m/s to the right

Unknown: $\mathbf{v_{2,f}} = ?$

Diagram:

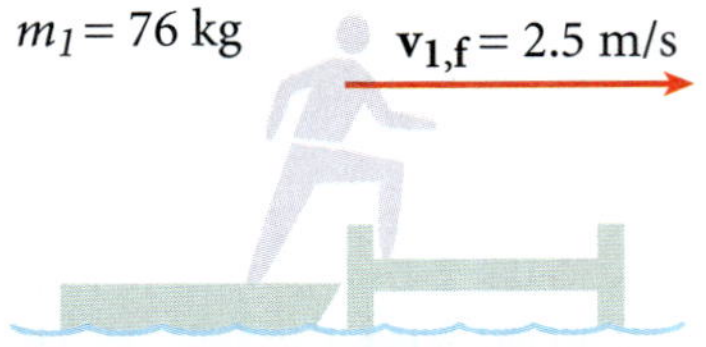

2. PLAN

Choose an equation or situation: Because the total momentum of an isolated system remains constant, the total initial momentum of the boater and the boat will be equal to the total final momentum of the boater and the boat.

$$m_1\mathbf{v_{1,i}} + m_2\mathbf{v_{2,i}} = m_1\mathbf{v_{1,f}} + m_2\mathbf{v_{2,f}}$$

Because the boater and the boat are initially at rest, the total initial momentum of the system is equal to zero. Therefore, the final momentum of the system must also be equal to zero.

$$m_1\mathbf{v_{1,f}} + m_2\mathbf{v_{2,f}} = 0$$

Rearrange the equation to solve for the final velocity of the boat.

$$m_2\mathbf{v_{2,f}} = -m_1\mathbf{v_{1,f}}$$

$$\mathbf{v_{2,f}} = -\frac{m_1}{m_2}\mathbf{v_{1,f}}$$

3. CALCULATE

Substitute the values into the equation and solve:

$$\mathbf{v_{2,f}} = -\frac{76 \text{ kg}}{45 \text{ kg}}(2.5 \text{ m/s to the right})$$

$$\mathbf{v_{2,f}} = -4.2 \text{ m/s to the right}$$

4. EVALUATE

The negative sign for $\mathbf{v_{2,f}}$ indicates that the boat is moving to the left, in the direction *opposite* the motion of the boater. Therefore,

$$\boxed{\mathbf{v_{2,f}} = 4.2 \text{ m/s to the left}}$$

PRACTICE D

Conservation of Momentum

1. A 63.0 kg astronaut is on a spacewalk when the tether line to the shuttle breaks. The astronaut is able to throw a spare 10.0 kg oxygen tank in a direction away from the shuttle with a speed of 12.0 m/s, propelling the astronaut back to the shuttle. Assuming that the astronaut starts from rest with respect to the shuttle, find the astronaut's final speed with respect to the shuttle after the tank is thrown.

2. An 85.0 kg fisherman jumps from a dock into a 135.0 kg rowboat at rest on the west side of the dock. If the velocity of the fisherman is 4.30 m/s to the west as he leaves the dock, what is the final velocity of the fisherman and the boat?

3. Each croquet ball in a set has a mass of 0.50 kg. The green ball, traveling at 12.0 m/s, strikes the blue ball, which is at rest. Assuming that the balls slide on a frictionless surface and all collisions are head-on, find the final speed of the blue ball in each of the following situations:

a. The green ball stops moving after it strikes the blue ball.

b. The green ball continues moving after the collision at 2.4 m/s in the same direction.

4. A boy on a 2.0 kg skateboard initially at rest tosses an 8.0 kg jug of water in the forward direction. If the jug has a speed of 3.0 m/s relative to the ground and the boy and skateboard move in the opposite direction at 0.60 m/s, find the boy's mass.

Newton's third law leads to conservation of momentum

Consider two isolated bumper cars, m_1 and m_2, before and after they collide. Before the collision, the velocities of the two bumper cars are $\mathbf{v_{1,i}}$ and $\mathbf{v_{2,i}}$, respectively. After the collision, their velocities are $\mathbf{v_{1,f}}$ and $\mathbf{v_{2,f}}$, respectively. The impulse-momentum theorem, $\mathbf{F}\Delta t = \Delta \mathbf{p}$, describes the change in momentum of one of the bumper cars. Applied to m_1, the impulse-momentum theorem gives the following:

$$\mathbf{F_1}\Delta t = m_1\mathbf{v_{1,f}} - m_1\mathbf{v_{1,i}}$$

Likewise, for m_2 it gives the following:

$$\mathbf{F_2}\Delta t = m_2\mathbf{v_{2,f}} - m_2\mathbf{v_{2,i}}$$

SECTION 2

PROBLEM GUIDE D

Use this guide to assign problems.
SE = Student Edition Textbook
PW = Problem Workbook
PB = Problem Bank on the One-Stop Planner (OSP)

Solving for:

v_f	**SE** Sample, 1–3; Ch. Rvw. 22–23, 40*, 43*, 44*, 48*, 50* **PW** 5–7 **PB** 5–7
v_i	**PW** 3–4 **PB** Sample, 1–4
m	**SE** 4 **PW** Sample, 1–2 **PB** 8–10

***Challenging Problem**
Consult the printed Solutions Manual or the OSP for detailed solutions.

ANSWERS

Practice D

1. 1.90 m/s
2. 1.66 m/s to the west
3. **a.** 12.0 m/s
   b. 9.6 m/s
4. 38 kg

Teaching Tip — GENERAL

A quick review of Newton's third law may help students better follow the derivation of the conservation of momentum in this section. Remind students that, according to Newton's third law, the force exerted by one body on another is equal in magnitude and opposite in direction to the force exerted on the first body by the second body.

Teaching Tip — GENERAL

Remind students that conservation laws are valid only for a closed system. In the example of two bumper cars colliding, the system consists of the two cars. Most cases considered in this chapter involve just two objects in a collision, but a system can include any number of objects interacting with one another. All examples discussed in this chapter assume an isolated system unless stated otherwise.

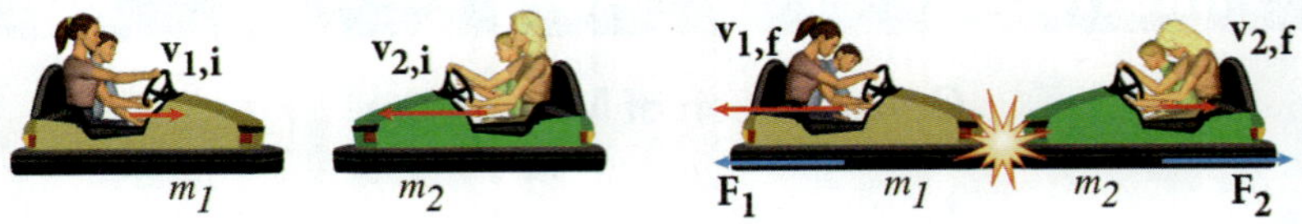

Figure 8
During the collision, the force exerted on each bumper car causes a change in momentum for each car. The total momentum is the same before and after the collision.

$\mathbf{F_1}$ is the force that m_2 exerts on m_1 during the collision, and $\mathbf{F_2}$ is the force that m_1 exerts on m_2 during the collision, as shown in **Figure 8.** Because the only forces acting in the collision are the forces the two bumper cars exert on each other, Newton's third law tells us that the force on m_1 is equal to and opposite the force on m_2 ($\mathbf{F_1} = -\mathbf{F_2}$). Additionally, the two forces act over the same time interval, Δt. Therefore, the force m_2 exerts on m_1 multiplied by the time interval is equal to the force m_1 exerts on m_2 multiplied by the time interval, or $\mathbf{F_1}\Delta t = -\mathbf{F_2}\Delta t$. That is, the impulse on m_1 is equal to and opposite the impulse on m_2. This relationship is true in every collision or interaction between two isolated objects.

Because impulse is equal to the change in momentum, and the impulse on m_1 is equal to and opposite the impulse on m_2, the change in momentum of m_1 is equal to and opposite the change in momentum of m_2. This means that in every interaction between two isolated objects, the change in momentum of the first object is equal to and opposite the change in momentum of the second object. In equation form, this is expressed by the following equation.

$$m_1\mathbf{v_{1,f}} - m_1\mathbf{v_{1,i}} = -(m_2\mathbf{v_{2,f}} - m_2\mathbf{v_{2,i}})$$

This equation means that if the momentum of one object increases after a collision, then the momentum of the other object in the situation must decrease by an equal amount. Rearranging this equation gives the following equation for the conservation of momentum.

$$m_1\mathbf{v_{1,i}} + m_2\mathbf{v_{2,i}} = m_1\mathbf{v_{1,f}} + m_2\mathbf{v_{2,f}}$$

Forces in real collisions are not constant during the collisions

As mentioned in Section 1, the forces involved in a collision are treated as though they are constant. In a real collision, however, the forces may vary in time in a complicated way. **Figure 9** shows the forces acting during the collision of the two bumper cars. At all times during the collision, the forces on the two cars at any instant during the collision are equal in magnitude and opposite in direction. However, the magnitudes of the forces change throughout the collision—increasing, reaching a maximum, and then decreasing.

When solving impulse problems, you should use the average force over the time of the collision as the value for force. Recall that the average velocity of an object undergoing a constant acceleration is equal to the constant velocity required for the object to travel the same displacement in the same time interval. The time-averaged force during a collision is equal to the constant force required to cause the same change in momentum as the real, changing force.

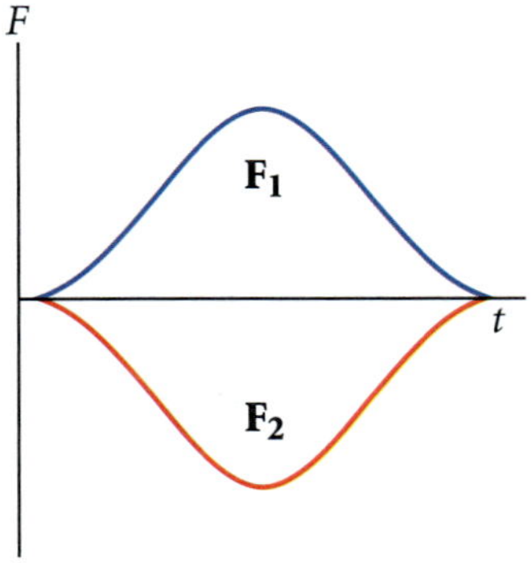

Figure 9
This graph shows the force on each bumper car during the collision. Although both forces vary with time, $\mathbf{F_1}$ and $\mathbf{F_2}$ are always equal in magnitude and opposite in direction.

SECTION REVIEW

1. A 44 kg student on in-line skates is playing with a 22 kg exercise ball. Disregarding friction, explain what happens during the following situations.

a. The student is holding the ball, and both are at rest. The student then throws the ball horizontally, causing the student to glide back at 3.5 m/s.

b. Explain what happens to the ball in part (a) in terms of the momentum of the student and the momentum of the ball.

c. The student is initially at rest. The student then catches the ball, which is initially moving to the right at 4.6 m/s.

d. Explain what happens in part (c) in terms of the momentum of the student and the momentum of the ball.

2. A boy stands at one end of a floating raft that is stationary relative to the shore. He then walks in a straight line to the opposite end of the raft, away from the shore.

a. Does the raft move? Explain.

b. What is the total momentum of the boy and the raft before the boy walks across the raft?

c. What is the total momentum of the boy and the raft after the boy walks across the raft?

3. High-speed stroboscopic photographs show the head of a 215 g golf club traveling at 55.0 m/s just before it strikes a 46 g golf ball at rest on a tee. After the collision, the club travels (in the same direction) at 42.0 m/s. Use the law of conservation of momentum to find the speed of the golf ball just after impact.

4. Critical Thinking Two isolated objects have a head-on collision. For each of the following questions, explain your answer.

a. If you know the change in momentum of one object, can you find the change in momentum of the other object?

b. If you know the initial and final velocity of one object and the mass of the other object, do you have enough information to find the final velocity of the second object?

c. If you know the masses of both objects and the final velocities of both objects, do you have enough information to find the initial velocities of both objects?

d. If you know the masses and initial velocities of both objects and the final velocity of one object, do you have enough information to find the final velocity of the other object?

e. If you know the change in momentum of one object and the initial and final velocities of the other object, do you have enough information to find the mass of either object?

SECTION REVIEW ANSWERS

1. a. The ball will move away at 7.0 m/s.

b. The momentum gained by the ball must be equal to and opposite the momentum gained by the student.

c. The student and the ball will move to the right at 1.5 m/s.

d. The student's initial momentum is zero. When the student catches the ball, some of the ball's momentum is transferred to the student.

2. a. yes; The total initial momentum is zero, so the boy and the raft must move in opposite directions to conserve momentum.

b. zero

c. zero

3. 61 m/s

4. a. Yes, the momentum lost by one object must equal the momentum gained by the other object.

b. No, $\mathbf{v}_{2,f}$ also depends on $\mathbf{v}_{2,i}$ and m_1.

c. No, using the conservation of momentum, you could only find a relationship between $\mathbf{v}_{1,i}$ and $\mathbf{v}_{2,i}$.

d. Yes, using the conservation of momentum, you could substitute the given values and solve for $\mathbf{v}_f$.

e. Using the conservation of momentum, you could find m_1 if $\mathbf{v}_{1,i}$ and $\mathbf{v}_{1,f}$ are given, but you would need $\mathbf{v}_{2,i}$ and $\mathbf{v}_{2,f}$ to find m_2.

SECTION 3
Advanced Level

Demonstration

Inelastic Collisions

Purpose Show the conservation of momentum in an inelastic collision.

Materials two balls with the same mass, string, tape, small piece of modeling clay, meterstick, paper or chalkboard

Procedure Tie a piece of string around each ball, using tape if necessary. Hold the two strings so that the balls hang at the same height in front of either the chalkboard or a length of paper taped to the wall. Place the clay on one of the balls so that the clay will hold the balls together when they collide. Hold up one of the balls, and have a student mark its displacement on the paper or chalkboard.

Release the ball. It should stick to the second ball; both balls should move together. Have a student mark the displacement of the two balls after the collision on the paper or chalkboard. Measure the two displacements with the meterstick. If momentum is conserved, the height of the two balls together will be $\frac{1}{4}$ the original height. Explain to the students that according to the conservation of momentum, $m_1\mathbf{v_{1,i}} + m_2\mathbf{v_{2,i}} = (m_1 + m_2)\mathbf{v_f}$ for a perfectly inelastic collision. Thus, since the second ball starts at rest, the final velocity of the two balls will be half the initial velocity of the first ball. Because the kinetic energy at the bottom of the swing equals the potential energy at the top ($mgh = \frac{1}{2}mv^2$), the two balls should reach $\frac{1}{4}$ the initial height of the first ball.

SECTION 3

Elastic and Inelastic Collisions

SECTION OBJECTIVES

- **Identify different types of collisions.**
- **Determine the changes in kinetic energy during perfectly inelastic collisions.**
- **Compare conservation of momentum and conservation of kinetic energy in perfectly inelastic and elastic collisions.**
- **Find the final velocity of an object in perfectly inelastic and elastic collisions.**

COLLISIONS

As you go about your day-to-day activities, you probably witness many collisions without really thinking about them. In some collisions, two objects collide and stick together so that they travel together after the impact. An example of this action is a collision between football players during a tackle, as shown in **Figure 10.** In an isolated system, the two football players would both move together after the collision with a momentum equal to the sum of their *momenta* (plural of *momentum*) before the collision. In other collisions, such as a collision between a tennis racquet and a tennis ball, two objects collide and bounce so that they move away with two different velocities.

The total momentum remains constant in any type of collision. However, the total kinetic energy is generally not conserved in a collision because some kinetic energy is converted to internal energy when the objects deform. In this section, we will examine different types of collisions and determine whether kinetic energy is conserved in each type. We will primarily explore two extreme types of collisions: elastic and perfectly inelastic collisions.

Perfectly inelastic collisions can be analyzed in terms of momentum

When two objects, such as the two football players, collide and move together as one mass, the collision is called a **perfectly inelastic collision.** Likewise, if a meteorite collides head on with Earth, it becomes buried in Earth and the collision is perfectly inelastic.

perfectly inelastic collision

a collision in which two objects stick together after colliding

Figure 10
When one football player tackles another, they both continue to fall together. This is one familiar example of a perfectly inelastic collision.

Perfectly inelastic collisions are easy to analyze in terms of momentum because the objects become essentially one object after the collision. The final mass is equal to the combined masses of the colliding objects. The combination moves with a predictable velocity after the collision.

Consider two cars of masses m_1 and m_2 moving with initial velocities of $\mathbf{v_{1,i}}$ and $\mathbf{v_{2,i}}$ along a straight line, as shown in **Figure 11.** The two cars stick together and move with some common velocity, $\mathbf{v_f}$, along the same line of motion after the collision. The total momentum of the two cars before the collision is equal to the total momentum of the two cars after the collision.

PERFECTLY INELASTIC COLLISION

$$m_1\mathbf{v_{1,i}} + m_2\mathbf{v_{2,i}} = (m_1 + m_2)\,\mathbf{v_f}$$

This simplified version of the equation for conservation of momentum is useful in analyzing perfectly inelastic collisions. When using this equation, it is important to pay attention to signs that indicate direction. In **Figure 11,** $\mathbf{v_{1,i}}$ has a positive value (m_1 moving to the right), while $\mathbf{v_{2,i}}$ has a negative value (m_2 moving to the left).

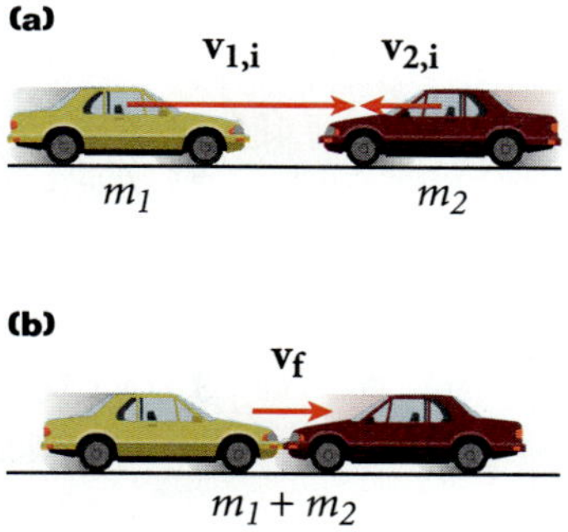

Figure 11
The total momentum of the two cars before the collision **(a)** is the same as the total momentum of the two cars after the inelastic collision **(b).**

SAMPLE PROBLEM E

Perfectly Inelastic Collisions

PROBLEM

A 1850 kg luxury sedan stopped at a traffic light is struck from the rear by a compact car with a mass of 975 kg. The two cars become entangled as a result of the collision. If the compact car was moving at a velocity of 22.0 m/s to the north before the collision, what is the velocity of the entangled mass after the collision?

SOLUTION

Given: $m_1 = 1850$ kg $\quad m_2 = 975$ kg $\quad \mathbf{v_{1,i}} = 0$ m/s
$\mathbf{v_{2,i}} = 22.0$ m/s to the north

Unknown: $\mathbf{v_f} = ?$

Use the equation for a perfectly inelastic collision.

$$m_1\mathbf{v_{1,i}} + m_2\mathbf{v_{2,i}} = (m_1 + m_2)\,\mathbf{v_f}$$

$$\mathbf{v_f} = \frac{m_1\mathbf{v_{1,i}} + m_2\mathbf{v_{2,i}}}{m_1 + m_2}$$

$$\mathbf{v_f} = \frac{(1850\text{ kg})(0\text{ m/s}) + (975\text{ kg})(22.0\text{ m/s north})}{1850\text{ kg} + 975\text{ kg}}$$

$$\boxed{\mathbf{v_f} = 7.59\text{ m/s to the north}}$$

Classroom Practice

Perfectly Inelastic Collisions

An empty train car moving east at 21 m/s collides with a loaded train car initially at rest that has twice the mass of the empty car. The two cars stick together.

a. Find the velocity of the two cars after the collision.

b. Find the final speed if the loaded car moving at 17 m/s had hit the empty car initially at rest.

Answer

a. 7.0 m/s to the east

b. 11 m/s

An empty train car moving at 15 m/s collides with a loaded car of three times the mass moving in the same direction at one-third the speed of the empty car. The cars stick together. Find the speed of the cars after the collision.

Answer

7.5 m/s

SECTION 3

PROBLEM GUIDE E

Use this guide to assign problems.
SE = Student Edition Textbook
PW = Problem Workbook
PB = Problem Bank on the One-Stop Planner (OSP)

Solving for:

v_f	**SE** Sample, 1–3; Ch. Rvw. 28–32 **PW** 7–9 **PB** 5–7
v_i	**SE** 4, 5*; Ch. Rvw. 39, 42 **PW** 4–6 **PB** Sample, 1–4
m	**SE** 5*; Ch. Rvw. 38* **PW** Sample, 1–3 **PB** 8–10

***Challenging Problem**
Consult the printed Solutions Manual or the OSP for detailed solutions.

ANSWERS

Practice E

1. 3.8 m/s to the south
2. 1.8 m/s
3. 4.25 m/s to the north
4. 4.2 m/s to the right
5. a. 3.0 kg
   b. 5.32 m/s

Misconception Alert — ADVANCED

Students may think that elastic materials can undergo only elastic collisions. Consider a large, brass bell with a clapper. The material, brass, is very elastic. After the collision, the bell continues to vibrate and give off sound (energy!) for a long time afterwards: the collision *isn't* elastic even though the materials are. Inelastic materials undergo *only* inelastic collisions. Elastic materials may undergo either elastic or inelastic collisions.

PRACTICE E

Perfectly Inelastic Collisions

1. A 1500 kg car traveling at 15.0 m/s to the south collides with a 4500 kg truck that is initially at rest at a stoplight. The car and truck stick together and move together after the collision. What is the final velocity of the two-vehicle mass?
2. A grocery shopper tosses a 9.0 kg bag of rice into a stationary 18.0 kg grocery cart. The bag hits the cart with a horizontal speed of 5.5 m/s toward the front of the cart. What is the final speed of the cart and bag?
3. A 1.50×10^4 kg railroad car moving at 7.00 m/s to the north collides with and sticks to another railroad car of the same mass that is moving in the same direction at 1.50 m/s. What is the velocity of the joined cars after the collision?
4. A dry cleaner throws a 22 kg bag of laundry onto a stationary 9.0 kg cart. The cart and laundry bag begin moving at 3.0 m/s to the right. Find the velocity of the laundry bag before the collision.
5. A 47.4 kg student runs down the sidewalk and jumps with a horizontal speed of 4.20 m/s onto a stationary skateboard. The student and skateboard move down the sidewalk with a speed of 3.95 m/s. Find the following:
   a. the mass of the skateboard
   b. how fast the student would have to jump to have a final speed of 5.00 m/s

Kinetic energy is not conserved in inelastic collisions

In an inelastic collision, the total kinetic energy does not remain constant when the objects collide and stick together. Some of the kinetic energy is converted to sound energy and internal energy as the objects deform during the collision.

This phenomenon helps make sense of the special use of the words *elastic* and *inelastic* in physics. We normally think of *elastic* as referring to something that always returns to, or keeps, its original shape. In physics, an elastic material is one in which the work done to deform the material during a collision is equal to the work the material does to return to its original shape. During a collision, some of the work done on an *inelastic* material is converted to other forms of energy, such as heat and sound.

The decrease in the total kinetic energy during an inelastic collision can be calculated by using the formula for kinetic energy, as shown in Sample Problem F. It is important to remember that not all of the initial kinetic energy is necessarily lost in a perfectly inelastic collision.

SAMPLE PROBLEM F

Kinetic Energy in Perfectly Inelastic Collisions

PROBLEM

Two clay balls collide head-on in a perfectly inelastic collision. The first ball has a mass of 0.500 kg and an initial velocity of 4.00 m/s to the right. The second ball has a mass of 0.250 kg and an initial velocity of 3.00 m/s to the left. What is the decrease in kinetic energy during the collision?

SOLUTION

1. DEFINE **Given:** $m_1 = 0.500$ kg $\quad m_2 = 0.250$ kg

$\mathbf{v_{1,i}} = 4.00$ m/s to the right, $v_{1,i} = +4.00$ m/s

$\mathbf{v_{2,i}} = 3.00$ m/s to the left, $v_{2,i} = -3.00$ m/s

Unknown: $\Delta KE = ?$

2. PLAN **Choose an equation or situation:** The change in kinetic energy is simply the initial kinetic energy subtracted from the final kinetic energy.

$$\Delta KE = KE_f - KE_i$$

Determine both the initial and final kinetic energy.

Initial: $KE_i = KE_{1,i} + KE_{2,i} = \frac{1}{2}m_1v_{1,i}^2 + \frac{1}{2}m_2v_{2,i}^2$

Final: $KE_f = KE_{1,f} + KE_{2,f} = \frac{1}{2}(m_1 + m_2)v_f^2$

As you did in Sample Problem E, use the equation for a perfectly inelastic collision to calculate the final velocity.

$$\mathbf{v_f} = \frac{m_1\mathbf{v_{1,i}} + m_2\mathbf{v_{2,i}}}{m_1 + m_2}$$

3. CALCULATE **Substitute the values into the equation and solve:** First, calculate the final velocity, which will be used in the final kinetic energy equation.

$$v_f = \frac{(0.500 \text{ kg})(4.00 \text{ m/s}) + (0.250 \text{ kg})(-3.00 \text{ m/s})}{0.500 \text{ kg} + 0.250 \text{ kg}}$$

$\mathbf{v_f} = 1.67$ m/s to the right

Next calculate the initial and final kinetic energy.

$$KE_i = \frac{1}{2}(0.500 \text{ kg})(4.00 \text{ m/s})^2 + \frac{1}{2}(0.250 \text{ kg})(-3.00 \text{ m/s})^2 = 5.12 \text{ J}$$

$$KE_f = \frac{1}{2}(0.500 \text{ kg} + 0.250 \text{ kg})(1.67 \text{ m/s})^2 = 1.05 \text{ J}$$

Finally, calculate the change in kinetic energy.

$$\Delta KE = KE_f - KE_i = 1.05 \text{ J} - 5.12 \text{ J}$$

$$\boxed{\Delta KE = -4.07 \text{ J}}$$

4. EVALUATE The negative sign indicates that kinetic energy is lost.

Classroom Practice

Kinetic Energy in Perfectly Inelastic Collisions

A clay ball with a mass of 0.35 kg hits another 0.35 kg ball at rest, and the two stick together. The first ball has an initial speed of 4.2 m/s.

a. What is the final speed of the balls?

b. Calculate the decrease in kinetic energy that occurs during the collision.

c. What percentage of the initial kinetic energy is converted to other forms of energy?

Answers

a. 2.1 m/s

b. 1.6 J

c. 52 percent (This is actually 50 percent. The difference is due to rounding.)

A 0.75 kg ball moving at 3.8 m/s to the right strikes an identical ball moving at 3.8 m/s to the left. The balls stick together after the collision and stop. What percentage of the initial kinetic energy is converted to other forms?

Answer

100 percent

PROBLEM GUIDE F

Use this guide to assign problems.
SE = Student Edition Textbook
PW = Problem Workbook
PB = Problem Bank on the One-Stop Planner (OSP)

Solving for:

ΔKE	**SE**	Sample, 1–3; Ch. Rvw. 30–31
	PW	Sample, 1–7
	PB	Sample, 1–10

***Challenging Problem**
Consult the printed Solutions Manual or the OSP for detailed solutions.

ANSWERS

Practice F

1. a. 0.43 m/s to the west
   b. 17 J
2. a. 6.2 m/s to the south
   b. 3.9 J
3. a. 4.6 m/s to the south
   b. 3.9×10^3 J

Key Models and Analogies — GENERAL

Just as friction is often disregarded to simplify situations, the decrease in kinetic energy in a nearly elastic collision can be disregarded to create an ideal case. This ideal case can then be used to obtain a very close approximation to the observed result.

Teaching Tip — ADVANCED

Discuss a variety of examples of collisions with students. For each example, ask whether the collision is closer to an elastic collision or to a perfectly inelastic collision. Also ask students where kinetic energy is converted to other forms of energy in each of the different examples.

PRACTICE F

Kinetic Energy in Perfectly Inelastic Collisions

1. A 0.25 kg arrow with a velocity of 12 m/s to the west strikes and pierces the center of a 6.8 kg target.
   a. What is the final velocity of the combined mass?
   b. What is the decrease in kinetic energy during the collision?
2. During practice, a student kicks a 0.40 kg soccer ball with a velocity of 8.5 m/s to the south into a 0.15 kg bucket lying on its side. The bucket travels with the ball after the collision.
   a. What is the final velocity of the combined mass?
   b. What is the decrease in kinetic energy during the collision?
3. A 56 kg ice skater traveling at 4.0 m/s to the north meets and joins hands with a 65 kg skater traveling at 12.0 m/s in the opposite direction. Without rotating, the two skaters continue skating together with joined hands.
   a. What is the final velocity of the two skaters?
   b. What is the decrease in kinetic energy during the collision?

ELASTIC COLLISIONS

When a player kicks a soccer ball, the collision between the ball and the player's foot is much closer to elastic than the collisions we have studied so far. In this case, *elastic* means that the ball and the player's foot remain separate after the collision.

In an **elastic collision,** two objects collide and return to their original shapes with no loss of total kinetic energy. After the collision, the two objects move separately. In an elastic collision, both the total momentum and the total kinetic energy are conserved.

elastic collision
a collision in which the total momentum and the total kinetic energy are conserved

SCILINKS.
www.scilinks.org
Topic: Collisions
Code: HF60311

Most collisions are neither elastic nor perfectly inelastic

In the everyday world, most collisions are not perfectly inelastic. That is, colliding objects do not usually stick together and continue to move as one object. Most collisions are not elastic, either. Even *nearly* elastic collisions, such as those between billiard balls or between a football player's foot and the ball, result in some decrease in kinetic energy. For example, a football deforms when it is kicked. During this deformation, some of the kinetic energy is converted to internal elastic potential energy. In most collisions, some of the kinetic energy is also converted into sound, such as the click of billiard balls colliding. In fact, any collision that produces sound is not elastic; the sound signifies a decrease in kinetic energy.

Elastic and perfectly inelastic collisions are limiting cases; most collisions actually fall into a category between these two extremes. In this third category of collisions, called *inelastic collisions,* the colliding objects bounce and move separately after the collision, but the total kinetic energy decreases in the collision. *For the problems in this book, we will consider all collisions in which the objects do not stick together to be elastic collisions.* Therefore, we will assume that the total momentum and the total kinetic energy each will stay the same before and after a collision in all collisions that are not perfectly inelastic.

Kinetic energy is conserved in elastic collisions

Figure 12 shows an elastic head-on collision between two soccer balls of equal mass. Assume, as in earlier examples, that the balls are isolated on a frictionless surface and that they do not rotate. The first ball is moving to the right when it collides with the second ball, which is moving to the left. When considered as a whole, the entire system has momentum to the left.

After the elastic collision, the first ball moves to the left and the second ball moves to the right. The magnitude of the momentum of the first ball, which is now moving to the left, is greater than the magnitude of the momentum of the second ball, which is now moving to the right. The entire system still has momentum to the left, just as before the collision.

Another example of a nearly elastic collision is the collision between a golf ball and a club. After a golf club strikes a stationary golf ball, the golf ball moves at a very high speed in the same direction as the golf club. The golf club continues to move in the same direction, but its velocity decreases so that the momentum lost by the golf club is equal to and opposite the momentum gained by the golf ball. *The total momentum is always constant throughout the collision. In addition, if the collision is perfectly elastic, the value of the total kinetic energy after the collision is equal to the value before the collision.*

MOMENTUM AND KINETIC ENERGY ARE CONSERVED IN AN ELASTIC COLLISION

$$m_1\mathbf{v}_{1,i} + m_2\mathbf{v}_{2,i} = m_1\mathbf{v}_{1,f} + m_2\mathbf{v}_{2,f}$$

$$\frac{1}{2}m_1v_{1,i}^2 + \frac{1}{2}m_2v_{2,i}^2 = \frac{1}{2}m_1v_{1,f}^2 + \frac{1}{2}m_2v_{2,f}^2$$

Remember that v is positive if an object moves to the right and negative if it moves to the left.

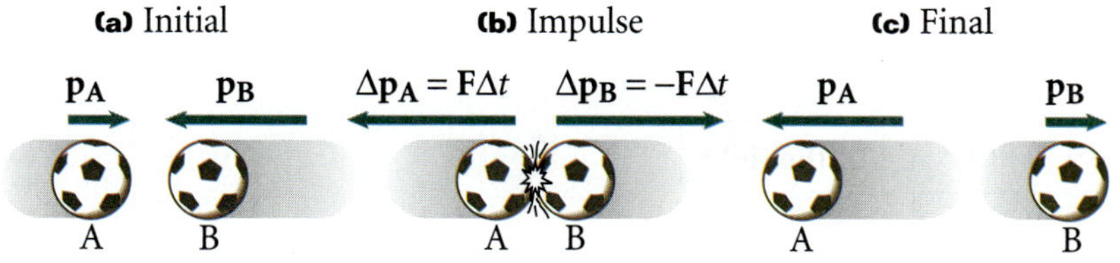

Figure 12
In an elastic collision like this one **(b)**, both objects return to their original shapes and move separately after the collision **(c)**.

Quick Lab

Elastic and Inelastic Collisions

MATERIALS LIST

- **2 or 3 small balls of different types**

SAFETY

Perform this lab in an open space, preferably outdoors, away from furniture and other people.

Drop one of the balls from shoulder height onto a hard-surfaced floor or sidewalk. Observe the motion of the ball before and after it collides with the ground. Next, throw the ball down from the same height. Perform several trials, giving the ball a different velocity each time. Repeat with the other balls.

During each trial, observe the height to which the ball bounces. Rate the collisions from most nearly elastic to most inelastic. Describe what evidence you have for or against conservation of kinetic energy and conservation of momentum for each collision. Based on your observations, do you think the equation for elastic collisions is useful to make predictions?

Quick Lab

TEACHER'S NOTES

The purpose of this lab is to show that in any collision, the elasticity of the materials involved affects the changes in kinetic energy. Test the balls before the lab in order to ensure a noticeable difference in elasticity. An interesting contrast can be observed by comparing new tennis balls with older ones.

Homework Options
This QuickLab can easily be performed outside of the physics lab room.

Teaching Tip GENERAL

Point out to students that they should recognize the first equation in the box. This equation, which expresses the principle of conservation of momentum, holds for both types of collisions. The conservation of kinetic energy, on the other hand, which is expressed by the second equation in the box, is valid only for elastic collisions.

Classroom Practice

Elastic Collisions
Two billiard balls, each with a mass of 0.35 kg, strike each other head-on. One ball is initially moving left at 4.1 m/s and ends up moving right at 3.5 m/s. The second ball is initially moving to the right at 3.5 m/s. Assume that neither ball rotates before or after the collision and that both balls are moving on a frictionless surface. Predict the final velocity of the second ball.

Answer
4.1 m/s to the left

Two nonrotating balls on a frictionless surface collide elastically head on. The first ball has a mass of 15 g and an initial velocity of 3.5 m/s to the right, while the second ball has a mass of 22 g and an initial velocity of 4.0 m/s to the left. The final velocity of the 15 g ball is 5.4 m/s to the left. What is the final velocity of the 22 g ball?

Answer
2.0 m/s to the right

SAMPLE PROBLEM G

Elastic Collisions

PROBLEM

A 0.015 kg marble moving to the right at 0.225 m/s makes an elastic head-on collision with a 0.030 kg shooter marble moving to the left at 0.180 m/s. After the collision, the smaller marble moves to the left at 0.315 m/s. Assume that neither marble rotates before or after the collision and that both marbles are moving on a frictionless surface. What is the velocity of the 0.030 kg marble after the collision?

SOLUTION

1. DEFINE **Given:** $m_1 = 0.015$ kg $\quad m_2 = 0.030$ kg

$\mathbf{v_{1,i}} = 0.225$ m/s to the right, $v_{1,i} = +0.225$ m/s

$\mathbf{v_{2,i}} = 0.180$ m/s to the left, $v_{2,i} = -0.180$ m/s

$\mathbf{v_{1,f}} = 0.315$ m/s to the left, $v_{1,f} = -0.315$ m/s

Unknown: $\mathbf{v_{2,f}} = ?$

Diagram:

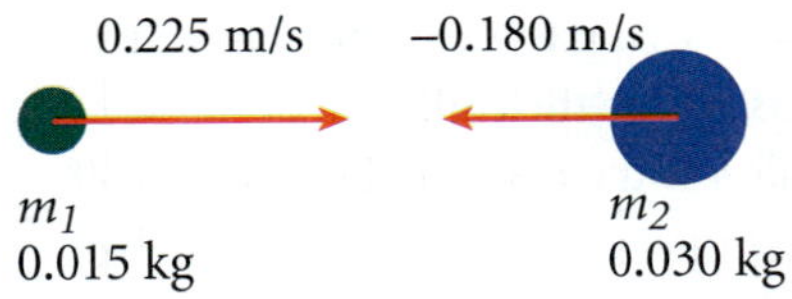

2. PLAN **Choose an equation or situation:** Use the equation for the conservation of momentum to find the final velocity of m_2, the 0.030 kg marble.

$$m_1\mathbf{v_{1,i}} + m_2\mathbf{v_{2,i}} = m_1\mathbf{v_{1,f}} + m_2\mathbf{v_{2,f}}$$

Rearrange the equation to isolate the final velocity of m_2.

$$m_2\mathbf{v_{2,f}} = m_1\mathbf{v_{1,i}} + m_2\mathbf{v_{2,i}} - m_1\mathbf{v_{1,f}}$$

$$\mathbf{v_{2,f}} = \frac{m_1\mathbf{v_{1,i}} + m_2\mathbf{v_{2,i}} - m_1\mathbf{v_{1,f}}}{m_2}$$

3. CALCULATE **Substitute the values into the equation and solve:** The rearranged conservation-of-momentum equation will allow you to isolate and solve for the final velocity.

$$v_{2,f} = \frac{(0.015\text{ kg})(0.225\text{ m/s}) + (0.030\text{ kg})(-0.180\text{ m/s}) - (0.015\text{ kg})(-0.315\text{ m/s})}{0.030\text{ kg}}$$

$$v_{2,f} = \frac{(3.4 \times 10^{-3}\text{ kg}\bullet\text{m/s}) + (-5.4 \times 10^{-3}\text{ kg}\bullet\text{m/s}) - (-4.7 \times 10^{-3}\text{ kg}\bullet\text{m/s})}{0.030\text{ kg}}$$

$$v_{2,f} = \frac{2.7 \times 10^{-3}\text{ kg}\bullet\text{m/s}}{3.0 \times 10^{-2}\text{ kg}}$$

$$\boxed{\mathbf{v_{2,f}} = 9.0 \times 10^{-2}\text{ m/s to the right}}$$

4. EVALUATE Confirm your answer by making sure kinetic energy is also conserved using these values.

Conservation of kinetic energy

$$\frac{1}{2}m_1v_{1,i}^2 + \frac{1}{2}m_2v_{2,i}^2 = \frac{1}{2}m_1v_{1,f}^2 + \frac{1}{2}m_2v_{2,f}^2$$

$$KE_i = \frac{1}{2}(0.015\text{ kg})(0.225\text{ m/s})^2 + \frac{1}{2}(0.030\text{ kg})(-0.180\text{ m/s})^2 = 8.7 \times 10^{-4}\text{ kg}\bullet\text{m}^2/\text{s}^2 = 8.7 \times 10^{-4}\text{ J}$$

$$KE_f = \frac{1}{2}(0.015\text{ kg})(0.315\text{ m/s})^2 + \frac{1}{2}(0.030\text{ kg})(0.090\text{ m/s})^2 = 8.7 \times 10^{-4}\text{ kg}\bullet\text{m}^2/\text{s}^2 = 8.7 \times 10^{-4}\text{ J}$$

Kinetic energy is conserved.

PRACTICE G

Elastic Collisions

1. A 0.015 kg marble sliding to the right at 22.5 cm/s on a frictionless surface makes an elastic head-on collision with a 0.015 kg marble moving to the left at 18.0 cm/s. After the collision, the first marble moves to the left at 18.0 cm/s.

a. Find the velocity of the second marble after the collision.

b. Verify your answer by calculating the total kinetic energy before and after the collision.

2. A 16.0 kg canoe moving to the left at 12.5 m/s makes an elastic head-on collision with a 14.0 kg raft moving to the right at 16.0 m/s. After the collision, the raft moves to the left at 14.4 m/s. Disregard any effects of the water.

a. Find the velocity of the canoe after the collision.

b. Verify your answer by calculating the total kinetic energy before and after the collision.

3. A 4.0 kg bowling ball sliding to the right at 8.0 m/s has an elastic head-on collision with another 4.0 kg bowling ball initially at rest. The first ball stops after the collision.

a. Find the velocity of the second ball after the collision.

b. Verify your answer by calculating the total kinetic energy before and after the collision.

4. A 25.0 kg bumper car moving to the right at 5.00 m/s overtakes and collides elastically with a 35.0 kg bumper car moving to the right. After the collision, the 25.0 kg bumper car slows to 1.50 m/s to the right, and the 35.0 kg car moves at 4.50 m/s to the right.

a. Find the velocity of the 35 kg bumper car before the collision.

b. Verify your answer by calculating the total kinetic energy before and after the collision.

SECTION 3

PROBLEM GUIDE G

Use this guide to assign problems.
SE = Student Edition Textbook
PW = Problem Workbook
PB = Problem Bank on the One-Stop Planner (OSP)

Solving for:

v_f	**SE** Sample, 1–3; Ch. Rvw. 32–34, 46* **PW** Sample, 6–7 **PB** 7–10
v_i	**SE** 4 **PW** Sample, 1–3 **PB** 3–6
m	**PW** 4–5 **PB** Sample, 1–2

***Challenging Problem**
Consult the printed Solutions Manual or the OSP for detailed solutions.

ANSWERS

Practice G

1. a. 22.5 cm/s to the right
   b. $KE_i = 6.2 \times 10^{-4}\text{ J} = KE_f$
2. a. 14.1 m/s to the right
   b. $KE_i = 3.04 \times 10^3$ J, $KE_f = 3.04 \times 10^3$ J, so $KE_i = KE_f$
3. a. 8.0 m/s to the right
   b. $KE_i = 1.3 \times 10^2\text{ J} = KE_f$
4. a. 2.0 m/s to the right
   b. $KE_i = 382\text{ J} = KE_f$

SECTION 3

Visual Strategy GENERAL

Table 2

Point out that the third case (inelastic) contains elements of both ideal cases. Total *KE* is not conserved, as in perfectly inelastic collisions, but the two objects do separate from one another after the collision, as in perfectly elastic collisions.

Q What is common to all cases?

A ***Momentum is conserved in each case.***

SECTION REVIEW ANSWERS

1. For elastic collisions, answers may include billiard balls colliding, a soccer ball hitting a player's foot, or a tennis ball hitting a wall. For inelastic collisions, answers may include a person catching a ball, a meteorite hitting Earth, or two clay balls colliding.
2. **a.** 1.1 m/s to the south
   b. 1.4×10^3 J
3. **a.** 3.5 m/s
   b. 0 J
   c. 0 J
4. No, some *KE* is converted to sound energy and some is converted to internal elastic potential energy as the cars deform, so the collision cannot be elastic.
5. **a.** no; If the collision is perfectly elastic, total *KE* is conserved, but each object can gain or lose *KE*.
   b. no; Total **p** is conserved, but each object can gain or lose **p**.

Table 2 **Types of Collisions**

Type of collision	Diagram	What happens	Conserved quantity
perfectly inelastic	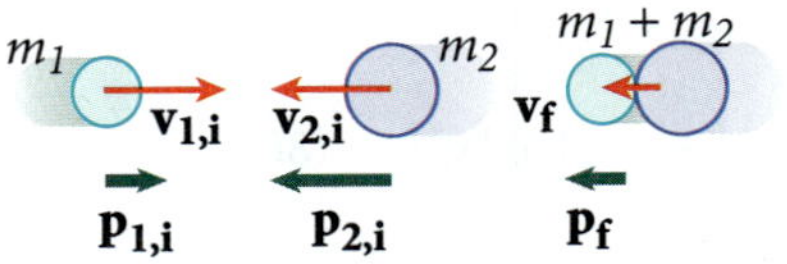	The two objects stick together after the collision so that their final velocities are the same.	momentum
elastic	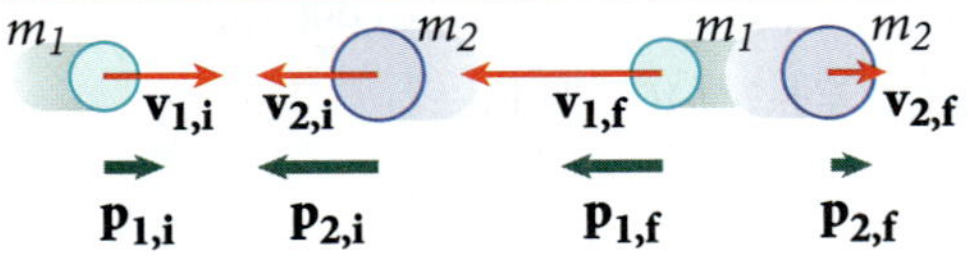	The two objects bounce after the collision so that they move separately.	momentum kinetic energy
inelastic	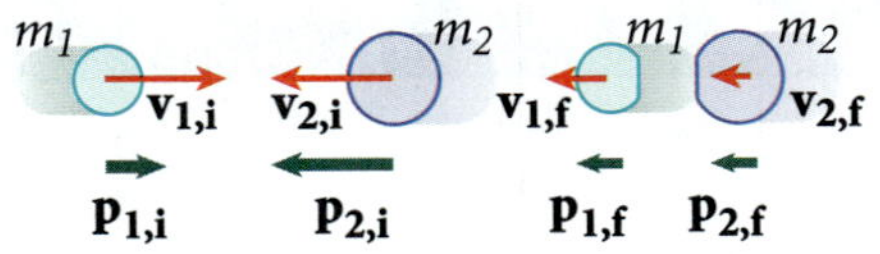	The two objects deform during the collision so that the total kinetic energy decreases, but the objects move separately after the collision.	momentum

SECTION REVIEW

1. Give two examples of elastic collisions and two examples of perfectly inelastic collisions.
2. A 95.0 kg fullback moving south with a speed of 5.0 m/s has a perfectly inelastic collision with a 90.0 kg opponent running north at 3.0 m/s.
   a. Calculate the velocity of the players just after the tackle.
   b. Calculate the decrease in total kinetic energy as a result of the collision.
3. Two 0.40 kg soccer balls collide elastically in a head-on collision. The first ball starts at rest, and the second ball has a speed of 3.5 m/s. After the collision, the second ball is at rest.
   a. What is the final speed of the first ball?
   b. What is the kinetic energy of the first ball before the collision?
   c. What is the kinetic energy of the second ball after the collision?
4. **Critical Thinking** If two automobiles collide, they usually do not stick together. Does this mean the collision is elastic?
5. **Critical Thinking** A rubber ball collides elastically with the sidewalk.
   a. Does each object have the same kinetic energy after the collision as it had before the collision? Explain.
   b. Does each object have the same momentum after the collision as it had before the collision? Explain.

PHYSICS CAREERS

High School Physics Teacher

Physics teachers help students understand this branch of science both in the classroom and in the so-called real world. To learn more about teaching physics as a career, read this interview with Linda Rush, who teaches high school physics at Southside High School in Fort Smith, Arkansas.

Linda Rush enjoys working with students, particularly with hands-on activities.

What does a physics teacher do every day?

I teach anywhere from 100 to 130 students a day. I also take care of the lab and equipment, which is sometimes difficult but necessary. In addition, physics teachers have to attend training sessions to stay current in the field.

What schooling did you take in order to become a physics teacher?

I have two college degrees: a bachelor's in physical science education and a master's in secondary education.

At first, I planned to go into the medical field but changed my mind and decided to become a teacher. I started out as a math teacher, but I changed to science because I enjoy the practical applications.

Did your family influence your career choice?

Neither of my parents went to college, but they both liked to tinker. They built an experimental solar house back in the 1970s. My dad rebuilt antique cars. My mom was a computer programmer. When we moved from the city to the country, my parents were determined that my sister and I wouldn't be helpless, so we learned how to do and fix everything.

What is your favorite thing about your job?

I like to watch my students learn—seeing that light bulb of understanding go on. Students can learn so much from one another. I hope that more students will take physics classes. So many students are afraid to try and don't have confidence in themselves.

What are your students surprised to learn about you?

My students are often surprised to learn that I am a kayaker, a hiker, and the mother of five daughters. Sometimes they forget that teachers are real people.

What advice do you have for students who are interested in teaching physics?

Take as many lab classes in college as possible. Learn as many hands-on activities as you can to use in the classroom. Also, get a broad background in other sciences. Don't be limited to only one field. I think what has helped me is that I'm not *just* a physics person. I have a well-rounded background, having taught all kinds of science and math classes.

High School Physics Teacher

Linda Rush is well known at Southside for conducting skateboard experiments in the hallway, supporting a successful robotics program, and doing similar activities of interest to students. On a more somber note, she recently helped her students cope with the shock of losing a classmate in a traffic accident. But even an unexpected tragedy can bring home the importance of the topics that she teaches.

"When you can relate science to current events—the shuttle disaster, for example—students pay more attention," said Rush. "Science has real-world relevance."

Within the limits of state requirements, Rush tries to allow some flexibility in terms of what is being taught.

"Sometimes it helps to let your students direct you," she said. "It's funny what they find interesting and want to share with the class. It really varies."

CHAPTER 6

Highlights

Teaching Tip

Ask students to prepare a concept map for the chapter. The concept map should include most of the vocabulary terms, along with other integral terms and concepts.

CHAPTER 6

Highlights

KEY TERMS

momentum (p. 198)

impulse (p. 200)

perfectly inelastic collision (p. 212)

elastic collision (p. 216)

PROBLEM SOLVING

See **Appendix D: Equations** for a summary of the equations introduced in this chapter. If you need more problem-solving practice, see **Appendix I: Additional Problems.**

KEY IDEAS

Section 1 Momentum and Impulse

- Momentum is a vector quantity defined as the product of an object's mass and velocity.
- A net external force applied constantly to an object for a certain time interval will cause a change in the object's momentum equal to the product of the force and the time interval during which the force acts.
- The product of the constant applied force and the time interval during which the force is applied is called the impulse of the force for the time interval.

Section 2 Conservation of Momentum

- In all interactions between isolated objects, momentum is conserved.
- In every interaction between two isolated objects, the change in momentum of the first object is equal to and opposite the change in momentum of the second object.

Section 3 Elastic and Inelastic Collisions

- In a perfectly inelastic collision, two objects stick together and move as one mass after the collision.
- Momentum is conserved but kinetic energy is not conserved in a perfectly inelastic collision.
- In an inelastic collision, kinetic energy is converted to internal elastic potential energy when the objects deform. Some kinetic energy is also converted to sound energy and internal energy.
- In an elastic collision, two objects return to their original shapes and move away from the collision separately.
- Both momentum and kinetic energy are conserved in an elastic collision.
- Few collisions are elastic or perfectly inelastic.

Variable Symbols

Quantities		Units
p	momentum	kg•m/s kilogram-meters per second
$\mathbf{F}\Delta t$	impulse	N•s Newton-seconds = kilogram-meters per second

Review

MOMENTUM AND IMPULSE

Review Questions

1. If an object is not moving, what is its momentum?

2. If two particles have equal kinetic energies, must they have the same momentum? Explain.

3. Show that $\mathbf{F} = m\mathbf{a}$ and $\mathbf{F} = \frac{\Delta \mathbf{p}}{\Delta t}$ are equivalent.

Conceptual Questions

4. A truck loaded with sand is moving down the highway in a straight path.

- **a.** What happens to the momentum of the truck if the truck's velocity is increasing?
- **b.** What happens to the momentum of the truck if sand leaks at a constant rate through a hole in the truck bed while the truck maintains a constant velocity?

5. Gymnasts always perform on padded mats. Use the impulse-momentum theorem to discuss how these mats protect the athletes.

6. When a car collision occurs, an air bag is inflated, protecting the passenger from serious injury. How does the air bag soften the blow? Discuss the physics involved in terms of momentum and impulse.

7. If you jump from a table onto the floor, are you more likely to be hurt if your knees are bent or if your legs are stiff and your knees are locked? Explain.

8. Consider a field of insects, all of which have essentially the same mass.

- **a.** If the total momentum of the insects is zero, what does this imply about their motion?
- **b.** If the total kinetic energy of the insects is zero, what does this imply about their motion?

9. Two students hold an open bed sheet loosely by its corners to form a "catching net." The instructor asks a third student to throw an egg into the middle of the sheet as hard as possible. Why doesn't the egg's shell break?

10. How do car bumpers that collapse on impact help protect a driver?

Practice Problems

For problem 11, see Sample Problem A.

11. Calculate the linear momentum for each of the following cases:

- **a.** a proton with mass 1.67×10^{-27} kg moving with a velocity of 5.00×10^6 m/s straight up
- **b.** a 15.0 g bullet moving with a velocity of 325 m/s to the right
- **c.** a 75.0 kg sprinter running with a velocity of 10.0 m/s southwest
- **d.** Earth ($m = 5.98 \times 10^{24}$ kg) moving in its orbit with a velocity equal to 2.98×10^4 m/s forward

For problems 12–13, see Sample Problem B.

12. A 2.5 kg ball strikes a wall with a velocity of 8.5 m/s to the left. The ball bounces off with a velocity of 7.5 m/s to the right. If the ball is in contact with the wall for 0.25 s, what is the constant force exerted on the ball by the wall?

13. A football punter accelerates a 0.55 kg football from rest to a speed of 8.0 m/s in 0.25 s. What constant force does the punter exert on the ball?

For problem 14, see Sample Problem C.

14. A 0.15 kg baseball moving at +26 m/s is slowed to a stop by a catcher who exerts a constant force of −390 N. How long does it take this force to stop the ball? How far does the ball travel before stopping?

Review

ANSWERS

1. zero (because $v = 0$)

2. no; KE is related to the magnitude of $\mathbf{p}$ by $p = \sqrt{2mKE}$. Objects that have the same KE must also have the same mass and direction to have the same $\mathbf{p}$.

3. $\mathbf{F} = \frac{\Delta \mathbf{p}}{\Delta t} = \frac{m\mathbf{v_f} - m\mathbf{v_i}}{\Delta t} = \frac{m(\mathbf{v_f} - \mathbf{v_i})}{\Delta t} = m\frac{\Delta \mathbf{v}}{\Delta t} = m\mathbf{a}$

4. **a.** Momentum increases.
b. Momentum decreases.

5. A mat decreases the average force on the gymnast by increasing the time interval in which the gymnast is brought to rest.

6. The air bag increases the time interval in which the passenger comes to rest, which decreases the average force on the passenger.

7. When your legs are stiff and your knees are locked, the time interval of the collision is short and the average force exerted by the floor is large, which may result in bone fracture.

8. **a.** The net velocity of all insects must equal zero (although each insect could be moving).
b. The velocity of each insect must be zero.

9. The average force on the egg is small because of the large time interval in which the egg is in contact with the sheet.

10. Car bumpers increase the time interval over which a collision occurs, which decreases the force.
11. a. 8.35×10^{-21} kg•m/s upward
b. 4.88 kg•m/s to the right
c. 7.50×10^{2} kg•m/s to the southwest
d. 1.78×10^{29} kg•m/s forward
12. 160 N to the right
13. 18 N
14. 0.010 s; 0.13 m
15. Before they push, the total momentum of the system is zero. So, after they push, the total momentum of the system must remain zero.
16. no; Momentum can be transferred between balls.
17. Part of the ball's momentum is transferred to the ground; Earth's mass is so large that the resulting change in Earth's velocity is imperceptible.
18. As the ball accelerates toward Earth, Earth also accelerates toward the ball. Therefore, Earth is also gaining momentum in the direction opposite the ball's momentum.
19. The gun was pushed with a momentum equal in magnitude but opposite in direction to the momentum of the gases.
20. She should throw the camera in the direction away from the shuttle to cause her to move back toward the shuttle.
21. The gun recoils with a backward momentum equal to the forward momentum of the bullet. Because the gun's mass is so much greater than the bullet's, the gun's velocity will be smaller than the bullet's.
22. a. 2.43 m/s to the right
b. 7.97×10^{-2} m/s to the right

CONSERVATION OF MOMENTUM

Review Questions

15. Two skaters initially at rest push against each other so that they move in opposite directions. What is the total momentum of the two skaters when they begin moving? Explain.

16. In a collision between two soccer balls, momentum is conserved. Is momentum conserved for each soccer ball? Explain.

17. Explain how momentum is conserved when a ball bounces against a floor.

Conceptual Questions

18. As a ball falls toward Earth, the momentum of the ball increases. How would you reconcile this observation with the law of conservation of momentum?

19. In the early 1900s, Robert Goddard proposed sending a rocket to the moon. Critics took the position that in a vacuum such as exists between Earth and the moon, the gases emitted by the rocket would have nothing to push against to propel the rocket. To settle the debate, Goddard placed a gun in a vacuum and fired a blank cartridge from it. (A blank cartridge fires only the hot gases of the burning gunpowder.) What happened when the gun was fired? Explain your answer.

20. An astronaut carrying a camera in space finds herself drifting away from a space shuttle after her tether becomes unfastened. If she has no propulsion device, what should she do to move back to the shuttle?

21. When a bullet is fired from a gun, what happens to the gun? Explain your answer using the principles of momentum discussed in this chapter.

Practice Problems

For problems 22–23, see Sample Problem D.

22. A 65.0 kg ice skater moving to the right with a velocity of 2.50 m/s throws a 0.150 kg snowball to the right with a velocity of 32.0 m/s relative to the ground.
    a. What is the velocity of the ice skater after throwing the snowball? Disregard the friction between the skates and the ice.
    b. A second skater initially at rest with a mass of 60.0 kg catches the snowball. What is the velocity of the second skater after catching the snowball in a perfectly inelastic collision?

23. A tennis player places a 55 kg ball machine on a frictionless surface, as shown below. The machine fires a 0.057 kg tennis ball horizontally with a velocity of 36 m/s toward the north. What is the final velocity of the machine?

ELASTIC AND INELASTIC COLLISIONS

Review Questions

24. Consider a perfectly inelastic head-on collision between a small car and a large truck traveling at the same speed. Which vehicle has a greater change in kinetic energy as a result of the collision?

25. Given the masses of two objects and their velocities before and after a head-on collision, how could you determine whether the collision was elastic, inelastic, or perfectly inelastic? Explain.

26. In an elastic collision between two objects, do both objects have the same kinetic energy after the collision as before? Explain.

27. If two objects collide and one is initially at rest, is it possible for both to be at rest after the collision? Is it possible for one to be at rest after the collision? Explain.

Practice Problems

For problems 28–29, see Sample Problem E.

28. Two carts with masses of 4.0 kg and 3.0 kg move toward each other on a frictionless track with speeds

of 5.0 m/s and 4.0 m/s respectively. The carts stick together after colliding head-on. Find the final speed.

29. A 1.20 kg skateboard is coasting along the pavement at a speed of 5.00 m/s when a 0.800 kg cat drops from a tree vertically downward onto the skateboard. What is the speed of the skateboard-cat combination?

For problems 30–31, see Sample Problem F.

30. A railroad car with a mass of 2.00×10^4 kg moving at 3.00 m/s collides and joins with two railroad cars already joined together, each with the same mass as the single car and initially moving in the same direction at 1.20 m/s.
    a. What is the speed of the three joined cars after the collision?
    b. What is the decrease in kinetic energy during the collision?

31. An 88 kg fullback moving east with a speed of 5.0 m/s is tackled by a 97 kg opponent running west at 3.0 m/s, and the collision is perfectly inelastic. Calculate the following:
    a. the velocity of the players just after the tackle
    b. the decrease in kinetic energy during the collision

For problems 32–34, see Sample Problem G.

32. A 5.0 g coin sliding to the right at 25.0 cm/s makes an elastic head-on collision with a 15.0 g coin that is initially at rest. After the collision, the 5.0 g coin moves to the left at 12.5 cm/s.
    a. Find the final velocity of the other coin.
    b. Find the amount of kinetic energy transferred to the 15.0 g coin.

33. A billiard ball traveling at 4.0 m/s has an elastic head-on collision with a billiard ball of equal mass that is initially at rest. The first ball is at rest after the collision. What is the speed of the second ball after the collision?

34. A 25.0 g marble sliding to the right at 20.0 cm/s overtakes and collides elastically with a 10.0 g marble moving in the same direction at 15.0 cm/s. After the collision, the 10.0 g marble moves to the right at 22.1 cm/s. Find the velocity of the 25.0 g marble after the collision.

MIXED REVIEW

35. If a 0.147 kg baseball has a momentum of $\mathbf{p}$ = 6.17 kg•m/s as it is thrown from home to second base, what is its velocity?

36. A moving object has a kinetic energy of 150 J and a momentum with a magnitude of 30.0 kg•m/s. Determine the mass and speed of the object.

37. A 0.10 kg ball of dough is thrown straight up into the air with an initial speed of 15 m/s.
    a. Find the momentum of the ball of dough at its maximum height.
    b. Find the momentum of the ball of dough halfway to its maximum height on the way up.

38. A 3.00 kg mud ball has a perfectly inelastic collision with a second mud ball that is initially at rest. The composite system moves with a speed equal to one-third the original speed of the 3.00 kg mud ball. What is the mass of the second mud ball?

39. A 5.5 g dart is fired into a block of wood with a mass of 22.6 g. The wood block is initially at rest on a 1.5 m tall post. After the collision, the wood block and dart land 2.5 m from the base of the post. Find the initial speed of the dart.

40. A 730 N student stands in the middle of a frozen pond having a radius of 5.0 m. He is unable to get to the other side because of a lack of friction between his shoes and the ice. To overcome this difficulty, he throws his 2.6 kg physics textbook horizontally toward the north shore at a speed of 5.0 m/s. How long does it take him to reach the south shore?

41. A 0.025 kg golf ball moving at 18.0 m/s crashes through the window of a house in 5.0×10^{-4} s. After the crash, the ball continues in the same direction with a speed of 10.0 m/s. Assuming the force exerted on the ball by the window was constant, what was the magnitude of this force?

42. A 1550 kg car moving south at 10.0 m/s collides with a 2550 kg car moving north. The cars stick together and move as a unit after the collision at a velocity of 5.22 m/s to the north. Find the velocity of the 2550 kg car before the collision.

6 REVIEW

23. 0.037 m/s to the south
24. Because the initial velocities of the truck and the car are the same and the final velocity is the same, the change in *KE* depends only on the mass. The truck has a greater mass, so the change in its *KE* is greater.
25. by calculating the kinetic energy before and after the collision; If *KE* is conserved, the collision is elastic. If the collision is not elastic, look at the final velocities to determine if it is perfectly inelastic.
26. no; Total kinetic energy is conserved but kinetic energy can be transferred from one object to the other.
27. Both cannot be at rest after the collision because the total initial momentum was greater than zero; The object initially in motion can be at rest if its momentum is entirely transferred to the other object.
28. 1 m/s
29. 3.00 m/s
30. a. 1.80 m/s
    b. 2.16×10^4 J
31. a. 0.81 m/s to the east
    b. 1.4×10^3 J
32. a. 12 cm/s to the right
    b. 1.1×10^{-4} J
33. 4.0 m/s
34. 17.2 cm/s to the right
35. 42.0 m/s toward second base
36. 3.0 kg; 1.0×10^1 m/s
37. a. 0.0 kg•m/s
    b. 1.1 kg•m/s upward
38. 6.00 kg
39. 23 m/s
40. 29 s
41. 4.0×10^2 N
42. 14.5 m/s to the north

43. 2.36×10^{-2} m

44. 254 s

45. 0.413

46. **a.** 0.83 m/s to the right
b. 1.2 m/s to the left

47. −22 cm/s, 22 cm/s

extension

Graphing Calculator Practice

Visit go.hrw.com for answers to this Graphing Calculator activity.

Keyword HF6MOMXT

43. The bird perched on the swing shown in the diagram has a mass of 52.0 g, and the base of the swing has a mass of 153 g. The swing and bird are originally at rest, and then the bird takes off horizontally at 2.00 m/s. How high will the base of the swing rise above its original level? Disregard friction.

44. An 85.0 kg astronaut is working on the engines of a spaceship that is drifting through space with a constant velocity. The astronaut turns away to look at Earth and several seconds later is 30.0 m behind the ship, at rest relative to the spaceship. The only way to return to the ship without a thruster is to throw a wrench directly away from the ship. If the wrench has a mass of 0.500 kg, and the astronaut throws the wrench with a speed of 20.0 m/s, how long does it take the astronaut to reach the ship?

45. A 2250 kg car traveling at 10.0 m/s collides with a 2750 kg car that is initially at rest at a stoplight. The cars stick together and move 2.50 m before friction causes them to stop. Determine the coefficient of kinetic friction between the cars and the road, assuming that the negative acceleration is constant and that all wheels on both cars lock at the time of impact.

46. A constant force of 2.5 N to the right acts on a 1.5 kg mass for 0.50 s.

a. Find the final velocity of the mass if it is initially at rest.

b. Find the final velocity of the mass if it is initially moving along the *x*-axis with a velocity of 2.0 m/s to the left.

47. Two billiard balls with identical masses and sliding in opposite directions have an elastic head-on collision. Before the collision, each ball has a speed of 22 cm/s. Find the speed of each billiard ball immediately after the collision. (See Appendix A for hints on solving simultaneous equations.)

Graphing Calculator *Practice*

Momentum

As you learned earlier in this chapter, the linear momentum, **p**, of an object of mass *m* moving with a velocity **v** is defined as the product of the mass and the velocity. A change in momentum requires force and time. This fundamental relationship between force, momentum, and time is shown in Newton's second law of motion.

$$\mathbf{F} = \frac{\Delta \mathbf{p}}{\Delta t}, \text{ where } \Delta \mathbf{p} = m\mathbf{v_f} - m\mathbf{v_i}$$

In this graphing calculator activity, you will determine the force that must be exerted to change the momentum of an object in various time intervals. This activity will help you better understand

- the relationship between time and force
- the consequences of the signs of the force and the velocity

Visit go.hrw.com and enter the keyword **HF6MOMX** to find this graphing calculator activity. Refer to **Appendix B** for instructions on downloading the program for this activity.

48. A 7.50 kg laundry bag is dropped from rest at an initial height of 3.00 m.

a. What is the speed of Earth toward the bag just before the bag hits the ground? Use the value 5.98×10^{24} kg as the mass of Earth.

b. Use your answer to part **(a)** to justify disregarding the motion of Earth when dealing with the motion of objects on Earth.

49. A 55 kg pole-vaulter falls from rest from a height of 5.0 m onto a foam-rubber pad. The pole-vaulter comes to rest 0.30 s after landing on the pad.

a. Calculate the athlete's velocity just before reaching the pad.

b. Calculate the constant force exerted on the pole-vaulter due to the collision.

50. An unstable nucleus with a mass of 17.0×10^{-27} kg initially at rest disintegrates into three particles. One of the particles, of mass 5.0×10^{-27} kg, moves along the positive y-axis with a speed of 6.0×10^6 m/s. Another particle, of mass 8.4×10^{-27} kg, moves along the positive x-axis with a speed of 4.0×10^6 m/s. Determine the third particle's speed and direction of motion. (Assume that mass is conserved.)

Alternative Assessment

1. Design an experiment to test the conservation of momentum. You may use dynamics carts, toy cars, coins, or any other suitable objects. Explore different types of collisions, including perfectly inelastic collisions and elastic collisions. If your teacher approves your plan, perform the experiment. Write a report describing your results.

2. Design an experiment that uses a dynamics cart with other easily found equipment to test whether it is safer to crash into a steel railing or into a container filled with sand. How can you measure the forces applied to the cart as it crashes into the barrier? If your teacher approves your plan, perform the experiment.

3. Obtain a videotape of one of your school's sports teams in action. Create a play-by-play description of a short segment of the videotape, explaining how momentum and kinetic energy change during impacts that take place in the segment.

4. Use your knowledge of impulse and momentum to construct a container that will protect an egg dropped from a two-story building. The container should prevent the egg from breaking when it hits the ground. Do not use a device that reduces air resistance, such as a parachute. Also avoid using any packing materials. Test your container. If the egg breaks, modify your design and then try again.

5. An inventor has asked an Olympic biathlon team to test his new rifles during the target-shooting segment of the event. The new 0.75 kg guns shoot 25.0 g bullets at 615 m/s. The team's coach has hired you to advise him about how these guns could affect the biathletes' accuracy. Prepare figures to justify your answer. Be ready to defend your position.

6 REVIEW

48. a. 9.62×10^{-24} m/s upward

b. The velocity of Earth is so small that the Earth's movement can be disregarded.

49. a. 9.9 m/s downward

b. 1.8×10^3 N upward

50. 1.3×10^7 m/s, 41° below the negative x-axis

Alternative Assessment ANSWERS

1. Student reports will vary, but should show how momentum is conserved in their experiment.

2. Student plans should be safe and involve measuring force or calculating force by measuring change in momentum and the time interval. Rigid objects tend to cause more damage.

3. Student answers will vary, but they should indicate whether collisions are elastic or inelastic and should describe which quantities are conserved.

4. Students should try to increase the time of impact to decrease the force on the egg.

5. Student answers should indicate that the rifle's mass alone is very small. The recoil speed would be unreasonably large (21 m/s). However, if the rifle is held firmly against the shoulder, this action effectively increases the mass of the recoiling gun-athlete system. In the case of a 70 kg person, the recoil speed would be 0.22 m/s.

Standardized Test Prep

ANSWERS

1. A
2. J
3. C
4. G
5. D
6. G
7. B

Standardized Test Prep

MULTIPLE CHOICE

1. If a particle's kinetic energy is zero, what is its momentum?

A. zero
B. 1 kg•m/s
C. 15 kg•m/s
D. negative

2. The vector below represents the momentum of a car traveling along a road.

The car strikes another car, which is at rest, and the result is an inelastic collision. Which of the following vectors represents the momentum of the first car after the collision?

F. 
G. 
H. 
J. 

3. What is the momentum of a 0.148 kg baseball thrown with a velocity of 35 m/s toward home plate?

A. 5.1 kg•m/s toward home plate
B. 5.1 kg•m/s away from home plate
C. 5.2 kg•m/s toward home plate
D. 5.2 kg•m/s away from home plate

Use the passage below to answer questions 4–5.

After being struck by a bowling ball, a 1.5 kg bowling pin slides to the right at 3.0 m/s and collides head-on with another 1.5 kg bowling pin initially at rest.

4. What is the final velocity of the second pin if the first pin moves to the right at 0.5 m/s after the collision?

F. 2.5 m/s to the left
G. 2.5 m/s to the right
H. 3.0 m/s to the left
J. 3.0 m/s to the right

5. What is the final velocity of the second pin if the first pin stops moving when it hits the second pin?

A. 2.5 m/s to the left
B. 2.5 m/s to the right
C. 3.0 m/s to the left
D. 3.0 m/s to the right

6. For a given change in momentum, if the net force that is applied to an object increases, what happens to the time interval over which the force is applied?

F. The time interval increases.
G. The time interval decreases.
H. The time interval stays the same.
J. It is impossible to determine the answer from the given information.

7. Which equation expresses the law of conservation of momentum?

A. $\mathbf{p} = m\mathbf{v}$
B. $m_1\mathbf{v}_{1,i} + m_2\mathbf{v}_{2,i} = m_1\mathbf{v}_{1,f} + m_2\mathbf{v}_{2,f}$
C. $\frac{1}{2}m_1 v_{1,i}^2 + m_2 v_{2,i}^2 = \frac{1}{2}(m_1 + m_2)v_f^2$
D. $KE = \mathbf{p}$

8. Two shuffleboard disks of equal mass, one of which is orange and one of which is yellow, are involved in an elastic collision. The yellow disk is initially at rest and is struck by the orange disk, which is moving initially to the right at 5.00 m/s. After the collision, the orange disk is at rest. What is the velocity of the yellow disk after the collision?

F. zero
G. 5.00 m/s to the left
H. 2.50 m/s to the right
J. 5.00 m/s to the right

Use the information below to answer questions 9–10.

A 0.400 kg bead slides on a straight frictionless wire and moves with a velocity of 3.50 cm/s to the right, as shown below. The bead collides elastically with a larger 0.600 kg bead that is initially at rest. After the collision, the smaller bead moves to the left with a velocity of 0.70 cm/s.

9. What is the large bead's velocity after the collision?

A. 1.68 cm/s to the right
B. 1.87 cm/s to the right
C. 2.80 cm/s to the right
D. 3.97 cm/s to the right

10. What is the total kinetic energy of the system of beads after the collision?

F. 1.40×10^{-4} J
G. 2.45×10^{-4} J
H. 4.70×10^{-4} J
J. 4.90×10^{-4} J

SHORT RESPONSE

11. Is momentum conserved when two objects with zero initial momentum push away from each other?

12. In which type of collision is kinetic energy conserved? What is an example of this type of collision?

Base your answers to questions 13–14 on the information below.

An 8.0 g bullet is fired into a 2.5 kg pendulum bob, which is initially at rest and becomes embedded in the bob. The pendulum then rises a vertical distance of 6.0 cm.

13. What was the initial speed of the bullet? Show your work.

14. What will be the kinetic energy of the pendulum when the pendulum swings back to its lowest point? Show your work.

EXTENDED RESPONSE

15. An engineer working on a space mission claims that if momentum concerns are taken into account, a spaceship will need far less fuel for the return trip than for the first half of the mission. Write a paragraph to explain and support this hypothesis.

Test TIP Work out problems on scratch paper even if you are not asked to show your work. If you get an answer that is not one of the choices, go back and check your work.

8. J
9. C
10. G
11. yes
12. elastic collision; Sample: Two billiard balls collide and then move separately after the collision.
13. 340 m/s (See the Solutions Manual or One-Stop Planner for a full solution.)
14. 1.5 J (See the Solutions Manual or One-Stop Planner for a full solution.)
15. Student answers will vary but should recognize that the ship will have used some of the fuel and thus will have less mass on the return trip.

Lab Planning

Beginning on page T34 are preparation notes and teaching tips to assist you in planning.

Blank data tables (as well as some sample data) appear on the **One-Stop Planner.**

No Books in the Lab?

See the ***Datasheets for In-Text Labs*** workbook for a reproducible master copy of this experiment.

The same workbook also contains a version of this experiment with explicit procedural steps if you prefer a more directed approach.

CBL™ Option

A **CBL™** version of this lab appears in the ***CBL™ Experiments*** workbook.

Safety Caution

Remind students to attach masses to carts securely and to make sure the carts do not fall off the table. Books or wooden blocks may be clamped to the ends of the table to serve as bumpers and keep the carts from falling.

Tips and Tricks

- To attach paper tapes to the carts, create "sidearms" by securely attaching rods to the carts. Remind students that the lattice rods must be included in the mass of the carts.
- Students can mount the timer on a support rod to level the tape path with the tops of the lattice rods.

CHAPTER 6

Inquiry Lab

Conservation of Momentum

Design Your Own

OBJECTIVES

- **Measure** the mass and velocity of two carts.
- **Calculate** the momentum of each cart.
- **Verify** the law of conservation of momentum.

MATERIALS LIST

- **2 carts, one with a spring mechanism**
- **balance**
- **metric ruler**
- **paper tape**
- **recording timer**
- **stopwatch**

When a spring-loaded cart pushes off against another cart, the force on the first cart is accompanied by an equal and opposite force on the second cart. Both of these forces act for exactly the same time interval. So, in the absence of other forces, the change in momentum of the first cart is equal and opposite to the change in momentum of the second cart. In this lab, you will design an experiment to study the momentum of two carts having unequal masses. In your experiment, the carts will be placed together so that they will move apart when a compressed spring between them is released. You will collect data from several trials that will allow you to calculate the momentum of each cart and the total momentum of the system before and after the carts move apart.

SAFETY

- **Tie back long hair, secure loose clothing, and remove loose jewelry to prevent their getting caught in moving or rotating parts.**

PROCEDURE

1. Study the materials provided, and design an experiment to meet the goals stated above. If you have not used a recording timer before, refer to the lab in the chapter "Motion in One Dimension" for instructions.
2. Write out your lab procedure, including a detailed description of the measurements to take during each step and the number of trials to perform. You may use **Figure 1** as a guide to one possible setup. You can use one recording timer for both carts at the same time by threading two tapes through the timer and using two carbon disks back to back between the tapes. Remember to calibrate your recording timer or use a known period for the timer.
3. Ask your teacher to approve your procedure.
4. Follow all steps of your procedure.
5. Clean up your work area. Put equipment away safely so that it is ready to be used again.

Figure 1

- The recording timer will mark the tapes for both carts at the same time. Place two carbon disks back to back with one tape above and one tape below.
- If the spring mechanism has more than one notch, choose the first notch. Press straight down to release the spring mechanism so that you do not affect the motion of the carts. Let the carts move at least 1.0 m before you catch them, but do not let the carts fall off the table.

ANALYSIS

1. **Organizing Data** For each trial in your experiment, find the velocities $\mathbf{v_1}$ and $\mathbf{v_2}$. Because the carts are moving in opposite directions, assign one of the carts a negative velocity to indicate direction.
2. **Organizing Data** For each trial, calculate the momentum of each cart by multiplying its mass by its velocity.
3. **Organizing Data** For each trial, find the total momentum of the two carts.
4. **Applying Ideas** For each trial, what is the total momentum of the two carts before they start moving?

CONCLUSIONS

5. **Drawing Conclusions** In this situation, conservation of velocity would mean that the total velocity for both carts is the same after the spring mechanism is released as it was before the release. Is velocity conserved in this experiment? Support your answer with data from the experiment.
6. **Drawing Conclusions** Is momentum conserved in this experiment? Support your answer with data from the experiment.
7. **Evaluating Methods** What source of experimental error might have affected your results?
8. **Evaluating Methods** How would using two carts with identical masses affect your answers to items 5 and 6?

- Show students how to thread both tapes through the timer at the same time. The lower tape should pass under both carbon disks, and the upper tape should pass over both disks.

ANSWERS

Analysis

1. Answers will vary. Make sure students use the relationship $\mathbf{v}_{\mathbf{avg}} = \frac{\Delta \mathbf{x}}{\Delta t}$. Typical values will range from ±0.2 m/s to ±0.9 m/s.

2. Make sure students use the relationship $\mathbf{p} = m\mathbf{v}$. Typical values will range from ±0.4 kg•m/s to ±0.8 kg•m/s.

3. Make sure students use the relationship $\mathbf{p} = \mathbf{p_1} + \mathbf{p_2}$. Typical values will range from –0.004 kg•m/s to 0.004 kg•m/s.

4. For all trials, the total momentum of the two carts before they start moving is zero, because the carts have no velocity.

Conclusions

5. Velocity is not conserved in this experiment.

6. Momentum is conserved. The values for the total final momentum found in item 3 are very close to zero, the total initial momentum.

7. Students should conclude that human reaction time affects results in this experiment.

8. Momentum will always be conserved. If the carts have the same mass, velocity will also be conserved.

CHAPTER 7

Circular Motion and Gravitation

Planning Guide

Compression Guide

To shorten instruction because of time limitations, omit the opener, Section 4, the lab, and the review.

OBJECTIVES	LABS, DEMONSTRATIONS, AND ACTIVITIES	TECHNOLOGY RESOURCES
PACING • 45 min pp. 232–233 **Chapter Opener**	ANC **Discovery Lab** Circular Motion*◆ BASIC ANC **Discovery Lab** Torque and Center of Mass*◆ BASIC	CD **Visual Concepts,** Chapter 7 BASIC
PACING • 45 min pp. 234–239 **Section 1 Circular Motion** • Solve problems involving centripetal acceleration. • Solve problems involving centripetal force. • Explain how the apparent existence of an outward force in circular motion can be explained as inertia resisting the centripetal force.	TE **Demonstration** Centripetal Acceleration, p. 234 BASIC TE **Demonstration** Centripetal Force, p. 237 GENERAL ANC **CBL™ Experiment** Centripetal Acceleration*◆ ADVANCED	OSP **Lesson Plans** TR 23 Centripetal Acceleration TR 24 Centripetal Force
PACING • 45 min pp. 240–247 **Section 2 Newton's Law of Universal Gravitation** • Explain how Newton's law of universal gravitation accounts for various phenomena, including satellite and planetary orbits, falling objects, and the tides. • Apply Newton's law of universal gravitation to solve problems.	SE **Quick Lab** Gravitational Field Strength, p. 245 GENERAL	OSP **Lesson Plans** TR 25 Newton's Law of Universal Gravitation
PACING • 45 min pp. 248–253 **Section 3 Motion in Space** • Describe Kepler's laws of planetary motion. • Relate Newton's mathematical analysis of gravitational force to the elliptical planetary orbits proposed by Kepler. • Solve problems involving orbital speed and period.	SE **Quick Lab** Kepler's Third Law, p. 249 GENERAL SE **Quick Lab** Elevator Acceleration, p. 252 GENERAL	OSP **Lesson Plans** EXT **Integrating Health** Exercise in Space BASIC TR 26 Kepler's Second Law TR 27 Weight and Weightlessness TR 25A Planetary Data
PACING • 90 min pp. 254–261 **Section 4 Torque and Simple Machines** • Distinguish between torque and force. • Calculate the magnitude of a torque on an object. • Identify the six types of simple machines. • Calculate the mechanical advantage of a simple machine.	SE **Quick Lab** Changing the Lever Arm, p. 255 GENERAL SE **Inquiry Lab** Machines and Efficiency, pp. 270–271◆ GENERAL ANC **Datasheet** Inquiry Lab, Machines and Efficiency* GENERAL ANC **Datasheet** Skills Practice Lab, Machines and Efficiency* ANC **Invention Lab** The Rotating Egg Drop*◆ ADVANCED	OSP **Lesson Plans** TR 28 Torque and the Lever Arm TR 29 Lever Arm of a Wrench TR 30 Simple Machines

PACING • 90 min

CHAPTER REVIEW, ASSESSMENT, AND STANDARDIZED TEST PREPARATION

SE **Chapter Highlights,** p. 262
SE **Chapter Review,** pp. 263–267
SE **Graphing Calculator Practice,** p. 266 GENERAL
SE **Alternative Assessment,** p. 267 ADVANCED
SE **Standardized Test Prep,** pp. 268–269 GENERAL
SE **Appendix D: Equations,** pp. 857–858
SE **Appendix I: Additional Problems,** pp. 886–887
ANC **Study Guide Worksheet** Mixed Review* GENERAL
ANC **Chapter Test A*** GENERAL
ANC **Chapter Test B*** ADVANCED
OSP **Test Generator**

Online and Technology Resources

Holt Online Learning

Visit go.hrw.com to find a variety of online resources. To access this chapter's extensions, enter the keyword **HF6CMGXT** and click the "go" button. Click **Holt Online Learning** for an online edition of this textbook, and other interactive resources.

HOLT Teacher's One-Stop Planner®

This DVD package includes:
- Holt Calendar Planner
- Customizable Lesson Plans
- Editable Worksheets
- ExamView® Version 6 Assessment Suite
- Interactive Teacher's Edition
- Holt PuzzlePro®
- Holt PowerNotes® Presentations
- MindPoint® Quiz Show

KEY

SE Student Edition	**OSP** One-Stop Planner	**EXT** Online Extension
TE Teacher Edition	**CD** CD or CD-ROM	* Also on One-Stop Planner
ANC Ancillary Worksheet	**TR** Teaching Transparencies	◆ Requires advance prep

SKILLS DEVELOPMENT RESOURCES	REVIEW AND ASSESSMENT	CORRELATIONS
		National Science Education Standards
SE Sample Set A Centripetal Acceleration, pp. 235–236 BASIC **TE Classroom Practice,** p. 235 BASIC **ANC Problem Workbook*** and **OSP Problem Bank** Sample Set A BASIC **SE Sample Set B** Centripetal Force, pp. 237–238 BASIC **ANC Problem Workbook*** and **OSP Problem Bank** Sample Set B BASIC **SE Conceptual Challenge,** p. 239 GENERAL **SE Appendix J: Advanced Topics** Tangential Speed and Acceleration, pp. 902–903 ADVANCED	**SE Section Review,** p. 239 GENERAL **ANC Study Guide Worksheet** Section 1* GENERAL **ANC Quiz** Section 1* BASIC	UCP 1, 2 SAI 2 PS 4a
SE Sample Set C Gravitational Force, p. 242 BASIC **TE Classroom Practice,** p. 242 BASIC **ANC Problem Workbook*** and **OSP Problem Bank** Sample Set C BASIC **SE Conceptual Challenge,** p. 246 GENERAL **SE Appendix J: Advanced Topics** General Relativity, pp. 920–921 ADVANCED	**SE Section Review,** p. 247 GENERAL **ANC Study Guide Worksheet** Section 2* GENERAL **ANC Quiz** Section 2* BASIC	UCP 1, 2 ST 2 HNS 1, 3 PS 4a, 4b
SE Sample Set D Period and Speed of an Orbiting Object, p. 251 GENERAL **TE Classroom Practice,** p. 251 GENERAL **ANC Problem Workbook*** and **OSP Problem Bank** Sample Set D GENERAL	**SE Section Review,** p. 253 GENERAL **ANC Study Guide Worksheet** Section 3* GENERAL **ANC Quiz** Section 3* BASIC	UCP 1, 2 SAI 1, 2 HNS 1, 2, 3 PS 4b
SE Sample Set E Torque, pp. 257–258 GENERAL **TE Classroom Practice,** p. 257 GENERAL **ANC Problem Workbook*** and **OSP Problem Bank** Sample Set E GENERAL **SE Appendix J: Advanced Topics** Rotation and Inertia, pp. 904–905 ADVANCED **SE Appendix J: Advanced Topics** Rotational Dynamics, pp. 906–907 ADVANCED	**SE Section Review,** p. 261 GENERAL **ANC Study Guide Worksheet** Section 4* GENERAL **ANC Quiz** Section 4* BASIC	UCP 1, 2, 5 SAI 1, 2 ST 2

SCILINKS. NSTA
www.scilinks.org

Maintained by the **National Science Teachers Association.**

Topic: Gravity and Orbiting Objects
SciLinks Code: HF60692

Topic: Torque
SciLinks Code: HF61538

Classroom CD-ROMs

- Guided Reading Audio Program
- Student One Stop
- Virtual Investigations
- Visual Concepts

Holt Lab Generator CD-ROM

Search for any lab by topic, standard, difficulty level, or time. Edit any lab to fit your needs, or create your own labs. Use the Lab Materials QuickList software to customize your lab materials list.

CHAPTER 7
Overview

Section 1 introduces tangential speed, centripetal acceleration, and centripetal force.

Section 2 examines gravitational force, Newton's law of universal gravitation, and gravitational field strength.

Section 3 introduces Kepler's laws of planetary motion, discusses how Kepler's laws relate to Newton's law of gravitation, and distinguishes between true and apparent weightlessness.

Section 4 explains torque and discusses types of simple machines.

About the Illustration

The fifth flight of the *Challenger* space shuttle was launched into orbit in April 1984. Capturing and repairing the *Solar Max* satellite was one of the mission's most important goals. This involved two "spacewalks," including the one shown here. After several unsuccessful attempts, the satellite was eventually captured and brought to the shuttle's cargo bay, where astronauts Nelson and van Hoften performed the needed repairs. The satellite was successfully redeployed the next day.

The astronaut shown in this photograph is walking out onto the cargo bay area of the space shuttle to attempt repair of a satellite. Although the astronaut's initial attempt to capture the satellite was unsuccessful, this task was later accomplished by a robotic arm. The astronauts were then able to repair the satellite.

CHAPTER 7

Circular Motion and Gravitation

WHAT TO EXPECT

In this chapter, you will learn how to describe circular motion and the forces associated with it, including the force due to gravity.

Why it Matters

Circular motion is present all around you—from a rotating Ferris wheel in an amusement park to a space shuttle orbiting Earth to Earth's orbit around the sun.

CHAPTER PREVIEW

Tapping Prior Knowledge

Knowledge to Review

- ✔ Velocity is speed and the direction of travel.
- ✔ Acceleration is the rate of change of velocity.
- ✔ Forces cause the acceleration of motion.
- ✔ Work is done on an object to change the energy of the object.

Items to Probe

- ✔ Net force: Have students calculate the net force on an object experiencing multiple forces.
- ✔ Preconceptions about inertia and motion: Ask students to explain in their own words the relationship between inertia and motion. Students may mistakenly believe that inertia applies only to motionless objects.

233

SECTION 1
General Level

extension

In-Depth Physics Content
Your students can visit go.hrw.com for online chapters that integrate more in-depth development of the concepts covered here.

Keyword HF6CMGX

Misconception Alert

Some students will have difficulty with terminology at this point because of the previous familiarity with the term *centrifugal.* It is important to emphasize the distinction between *centripetal* (center-seeking) and *centrifugal* (center-fleeing). To avoid reinforcing this misconception, avoid using the term *centrifugal.*

Demonstration

Centripetal Acceleration — BASIC

Purpose Show an example of centripetal acceleration.

Materials battery-operated toy car, 50 cm string

Procedure Tie one end of the string to the midpoint of the car. Hold the other end of the string to the table or floor. Start the car. Have students observe the motion. Release the string and allow the car to move in a straight line.

Point out that even though the car was traveling at constant speed, the car was accelerating because the direction of motion continued to change. Show that the change in velocity was always perpendicular to the direction of motion.

SECTION 1

Circular Motion

SECTION OBJECTIVES

- Solve problems involving centripetal acceleration.
- Solve problems involving centripetal force.
- Explain how the apparent existence of an outward force in circular motion can be explained as inertia resisting the centripetal force.

CENTRIPETAL ACCELERATION

Consider a spinning Ferris wheel, as shown in **Figure 1.** The cars on the rotating Ferris wheel are said to be in *circular motion.* Any object that revolves about a single axis undergoes circular motion. The line about which the rotation occurs is called the *axis of rotation.* In this case, it is a line perpendicular to the side of the Ferris wheel and passing through the wheel's center.

Tangential speed depends on distance

Tangential speed (v_t) can be used to describe the speed of an object in circular motion. The tangential speed of a car on the Ferris wheel is the car's speed along an imaginary line drawn tangent to the car's circular path. This definition can be applied to any object moving in circular motion. When the tangential speed is constant, the motion is described as *uniform circular motion.*

The tangential speed depends on the distance from the object to the center of the circular path. For example, consider a pair of horses side-by-side on a carousel. Each completes one full circle in the same time period, but the horse on the outside covers more distance than the inside horse does, so the outside horse has a greater tangential speed.

ADVANCED TOPICS

See "Tangential Speed and Acceleration" in **Appendix J: Advanced Topics** to learn more about tangential speed, and to be introduced to the concept of tangential acceleration.

Centripetal acceleration is due to a change in direction

Suppose a car on a Ferris wheel is moving at a constant speed around the wheel. Even though the tangential speed is constant, the car still has an acceleration. To see why, consider the equation that defines acceleration:

$$\mathbf{a} = \frac{\mathbf{v_f} - \mathbf{v_i}}{t_f - t_i}$$

Acceleration depends on a change in the velocity. Because velocity is a vector, acceleration can be produced by a change in the *magnitude* of the velocity, a change in the *direction* of the velocity, or both.

Figure 1
Any point on a Ferris wheel spinning about a fixed axis undergoes circular motion.

The acceleration of a Ferris wheel car moving in a circular path and at constant speed is due to a change in direction. An acceleration of this nature is called a **centripetal acceleration.** The magnitude of a centripetal acceleration is given by the following equation:

centripetal acceleration

the acceleration directed toward the center of a circular path

CENTRIPETAL ACCELERATION

$$a_c = \frac{v_t^2}{r}$$

$$\text{centripetal acceleration} = \frac{(\text{tangential speed})^2}{\text{radius of circular path}}$$

What is the direction of centripetal acceleration? To answer this question, consider **Figure 2(a).** At time t_i, an object is at point A and has tangential velocity $\mathbf{v_i}$. At time t_f, the object is at point B and has tangential velocity $\mathbf{v_f}$. Assume that $\mathbf{v_i}$ and $\mathbf{v_f}$ differ in direction but have the same magnitudes.

The change in velocity ($\Delta\mathbf{v} = \mathbf{v_f} - \mathbf{v_i}$) can be determined graphically, as shown by the vector triangle in **Figure 2(b).** Note that when Δt is very small, $\mathbf{v_f}$ will be almost parallel to $\mathbf{v_i}$. The vector $\Delta\mathbf{v}$ will be approximately perpendicular to $\mathbf{v_f}$ and $\mathbf{v_i}$ and will be pointing toward the center of the circle. Because the acceleration is in the direction of $\Delta\mathbf{v}$, the acceleration will also be directed toward the center of the circle. Centripetal acceleration is always directed toward the center of a circle. In fact, the word *centripetal* means "center seeking." This is the reason that the acceleration of an object in uniform circular motion is called *centripetal acceleration.*

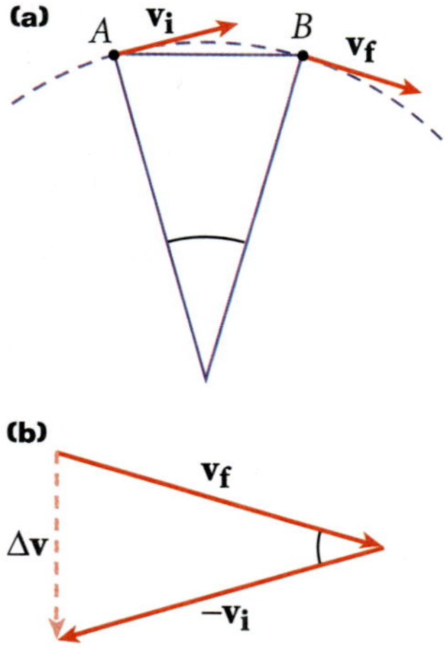

Figure 2
(a) As the particle moves from *A* to *B*, the direction of the particle's velocity vector changes. **(b)** For short time intervals, $\Delta\mathbf{v}$ is directed toward the center of the circle.

SAMPLE PROBLEM A

Centripetal Acceleration

PROBLEM

A test car moves at a constant speed around a circular track. If the car is 48.2 m from the track's center and has a centripetal acceleration of 8.05 m/s², what is the car's tangential speed?

SOLUTION

Given: $r = 48.2$ m $\quad a_c = 8.05$ m/s^2

Unknown: $v_t = ?$

Use the centripetal acceleration equation, and rearrange to solve for v_t.

$$a_c = \frac{v_t^2}{r}$$

$$v_t = \sqrt{a_c r} = \sqrt{(8.05 \text{ m/s}^2)(48.2 \text{ m})}$$

$$\boxed{v_t = 19.7 \text{ m/s}}$$

Classroom Practice

Centripetal Acceleration
The cylindrical tub of a washing machine has a radius of 34 cm. During the spin cycle, the wall of the tub rotates with a tangential speed of 5.5 m/s. Calculate the centripetal acceleration of the clothes sitting against the tub.

Answer
89 m/s^2

PROBLEM GUIDE A

Use this guide to assign problems.
SE = Student Edition Textbook
PW = Problem Workbook
PB = Problem Bank on the One-Stop Planner (OSP)

Solving for:

v_t	**SE** Sample, 1–2; Ch. Rvw. 8–9 **PW** 3–4 **PB** 4–6
a_c	**SE** 3; Ch. Rvw. 50a **PW** 5–6 **PB** Sample, 1–3
r	**SE** 4 **PW** Sample, 1–2 **PB** 7–10

***Challenging Problem**
Consult the printed Solutions Manual or the OSP for detailed solutions.

SECTION 1

ANSWERS

Practice A

1. 2.5 m/s
2. 11 m/s
3. 1.5 m/s^2
4. 58.4 m

Teaching Tip — ADVANCED

Tangential acceleration is covered in more detail in the feature "Tangential Speed and Acceleration" in **Appendix J: Advanced Topics.**

Visual Strategy GENERAL

Figure 3 and Figure 4

Have students examine **Figures 3** and **4.** Point out that the string is not entirely in a horizontal plane. Instead, the string makes an angle of slightly less than 90° with the direction of gravitational force.

Q What provides the force that opposes the ball's weight?

A ***the vertical component of tension in the string***

Q What provides the centripetal force that holds the ball in a circular path?

A ***the horizontal component of tension in the string***

PRACTICE A

Centripetal Acceleration

1. A rope attaches a tire to an overhanging tree limb. A girl swinging on the tire has a centripetal acceleration of 3.0 m/s^2. If the length of the rope is 2.1 m, what is the girl's tangential speed?
2. As a young boy swings a yo-yo parallel to the ground and above his head, the yo-yo has a centripetal acceleration of 250 m/s^2. If the yo-yo's string is 0.50 m long, what is the yo-yo's tangential speed?
3. A dog sits 1.5 m from the center of a merry-go-round. The merry-go-round is set in motion, and the dog's tangential speed is 1.5 m/s. What is the dog's centripetal acceleration?
4. A race car moving along a circular track has a centripetal acceleration of 15.4 m/s^2. If the car has a tangential speed of 30.0 m/s, what is the distance between the car and the center of the track?

Tangential acceleration is due to a change in speed

You have seen that centripetal acceleration results from a change in direction. In circular motion, an acceleration due to a change in speed is called *tangential acceleration.* To understand the difference between centripetal and tangential acceleration, consider a car traveling in a circular track. Because the car is moving in a circle, the car has a centripetal component of acceleration. If the car's speed changes, the car also has a tangential component of acceleration.

Figure 3
When a ball is whirled in a circle, it is acted on by a force directed toward the center of the ball's circular path.

CENTRIPETAL FORCE

Consider a ball of mass m that is tied to a string of length r and that is being whirled in a horizontal circular path, as shown in **Figure 3.** Assume that the ball moves with constant speed. Because the velocity vector, **v,** continuously changes direction during the motion, the ball experiences a centripetal acceleration that is directed toward the center of motion. As seen earlier, the magnitude of this acceleration is given by the following equation:

$$a_c = \frac{v_t^2}{r}$$

The inertia of the ball tends to maintain the ball's motion in a straight path. However, the string exerts a force that overcomes this tendency. The forces acting on the ball are gravitational force and the force exerted by the string, as shown in **Figure 4(a)** on the next page. The force exerted by the string has horizontal and vertical components. The vertical component is equal and opposite to the gravitational force. Thus, the horizontal component is the net force. This net force is directed toward the center of the circle, as shown in **Figure 4(b).** The net force that is directed toward the center of an object's circular path is called *centripetal force.* Newton's second law can be applied to find the magnitude of this force.

$$F_c = ma_c$$

The equation for centripetal acceleration can be combined with Newton's second law to obtain the following equation for centripetal force:

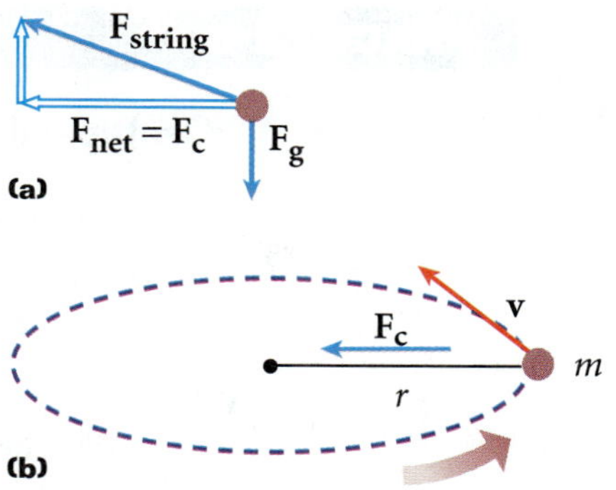

Figure 4
The net force on a ball whirled in a circle **(a)** is directed toward the center of the circle **(b).**

CENTRIPETAL FORCE

$$F_c = \frac{mv_t^2}{r}$$

$$\textbf{centripetal force} = \textbf{mass} \times \frac{\textbf{(tangential speed)}^2}{\textbf{radius of circular path}}$$

Centripetal force is simply the name given to the net force on an object in uniform circular motion. Any type of force or combination of forces can provide this net force. For example, friction between a race car's tires and a circular track is a centripetal force that keeps the car in a circular path. As another example, gravitational force is a centripetal force that keeps the moon in its orbit.

SAMPLE PROBLEM B

Centripetal Force

PROBLEM

A pilot is flying a small plane at 56.6 m/s in a circular path with a radius of 188.5 m. The centripetal force needed to maintain the plane's circular motion is 1.89×10^4 N. What is the plane's mass?

SOLUTION

Given: $v_t = 56.6$ m/s $r = 188.5$ m $F_c = 1.89 \times 10^4$ N

Unknown: $m = ?$

Use the equation for centripetal force. Rearrange to solve for m.

$$F_c = \frac{mv_t^2}{r}$$

$$m = \frac{F_c r}{v_t^2} = \frac{(1.89 \times 10^4 \text{ N})(188.5 \text{ m})}{(56.6 \text{ m/s})^2}$$

$$\boxed{m = 1110 \text{ kg}}$$

SECTION 1

Demonstration

Centripetal Force — GENERAL

Purpose Show that the horizontal component of tension in a rope provides the centripetal force for a ball moving in a horizontal circle.

Materials 100 g mass (with hook), 30 cm string, spring scale, protractor

Procedure Attach one end of the string to the hook on the mass, and attach the other end to the end of the spring scale. Use the protractor to measure the position at which the string makes an angle of 30° with the vertical. Then, rotate the string with a constant speed so that the mass moves in a circular path and the string remains at the 30° angle with the vertical. While maintaining this motion, have a volunteer attempt to read the spring scale. (Maintaining a constant angle can be difficult; you may wish to practice this before demonstrating to the class.)

Point out that the scale reading equals the tension in the string. The vertical component balances the weight of the mass, while the horizontal component of the tension holds the mass in a circular path. Ask students to calculate the vertical tension component and to compare it with the weight ($F_{scale} \cos \theta = mg$) so that they can confirm whether the scale reading and angle are correct. Then, have them calculate the centripetal force from the horizontal component of tension ($F_{scale} \sin \theta = F_c$). Repeat the demonstration at angles equal to 45° and 60°. Ask students to predict what change will occur in the force as measured on the spring scale.

SECTION 1

PROBLEM GUIDE B

Use this guide to assign problems.
SE = Student Edition Textbook
PW = Problem Workbook
PB = Problem Bank on the One-Stop Planner (OSP)

Solving for:

m	**SE** Sample, 1; Ch. Rvw. 10a*, 11* **PW** 3 **PB** 7–10
v_t	**SE** 4; Ch. Rvw. 10b, 39* **PW** Sample, 1–2 **PB** 5–7
r	**SE** 2 **PW** 4 **PB** 6–7
F_c	**SE** 3; Ch. Rvw. 50b, 53 **PW** 5 **PB** Sample, 1–3

***Challenging Problem**
Consult the printed Solutions Manual or the OSP for detailed solutions.

ANSWERS

Practice B

1. 29.6 kg
2. 40.0 m
3. 40.0 N
4. 35.0 m/s

PRACTICE B

Centripetal Force

1. A 2.10 m rope attaches a tire to an overhanging tree limb. A girl swinging on the tire has a tangential speed of 2.50 m/s. If the magnitude of the centripetal force is 88.0 N, what is the girl's mass?
2. A bicyclist is riding at a tangential speed of 13.2 m/s around a circular track. The magnitude of the centripetal force is 377 N, and the combined mass of the bicycle and rider is 86.5 kg. What is the track's radius?
3. A dog sits 1.50 m from the center of a merry-go-round and revolves at a tangential speed of 1.80 m/s. If the dog's mass is 18.5 kg, what is the magnitude of the centripetal force on the dog?
4. A 905 kg car travels around a circular track with a circumference of 3.25 km. If the magnitude of the centripetal force is 2140 N, what is the car's tangential speed?

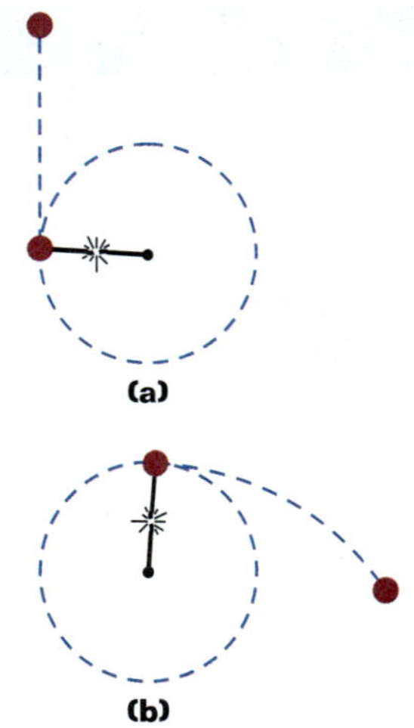

Figure 5
A ball that is on the end of a string is whirled in a vertical circular path. If the string breaks at the position shown in **(a),** the ball will move vertically upward in free fall. **(b)** If the string breaks at the top of the ball's path, the ball will move along a parabolic path.

Centripetal force is necessary for circular motion

Because centripetal force acts at right angles to an object's circular motion, the force changes the direction of the object's velocity. If this force vanishes, the object stops moving in a circular path. Instead, the object moves along a straight path that is tangent to the circle.

For example, consider a ball that is attached to a string and that is whirled in a vertical circle, as shown in **Figure 5.** If the string breaks when the ball is at the position shown in **Figure 5(a),** the centripetal force will vanish. Thus, the ball will move vertically upward, as if it has been thrown straight up in the air. If the string breaks when the ball is at the top of its circular path, as shown in **Figure 5(b),** the ball will fly off horizontally in a direction tangent to the path. The ball will then move in the parabolic path of a projectile.

DESCRIBING A ROTATING SYSTEM

To better understand the motion of a rotating system, consider a car traveling at high speed and approaching an exit ramp that curves to the left. As the driver makes the sharp left turn, the passenger slides to the right and hits the door. At that point, the force of the door keeps the passenger from being ejected from the car. What causes the passenger to move toward the door? A popular explanation is that a force must push the passenger outward. This force is sometimes called the *centrifugal force,* but that term often creates confusion, so it is not used in this textbook.

Inertia is often misinterpreted as a force

The phenomenon is correctly explained as follows: Before the car enters the ramp, the passenger is moving in a straight path. As the car enters the ramp and travels along a curved path, the passenger, because of inertia, tends to move along the original straight path. This movement is in accordance with Newton's first law, which states that the natural tendency of a body is to continue moving in a straight line.

However, if a sufficiently large centripetal force acts on the passenger, the person will move along the same curved path that the car does. The origin of the centripetal force is the force of friction between the passenger and the car seat. If this frictional force is not sufficient, the passenger slides across the seat as the car turns underneath. Eventually, the passenger encounters the door, which provides a large enough force to enable the passenger to follow the same curved path as the car does. The passenger does not slide toward the door because of some mysterious outward force. Instead, the frictional force exerted on the passenger by the seat is not great enough to keep the passenger moving in the same circle as the car.

Why it Matters

Conceptual Challenge

1. Pizza

Pizza makers traditionally form the crust by throwing the dough up in the air and spinning it. Why does this make the pizza crust bigger?

2. Swings

The amusement-park ride pictured below spins riders around on swings attached by cables from above. What causes the swings to move away from the center of the ride when the center column begins to turn?

SECTION REVIEW

1. What are three examples of circular motion?
2. A girl on a spinning amusement park ride is 12 m from the center of the ride and has a centripetal acceleration of 17 m/s^2. What is the girl's tangential speed?
3. Use an example to describe the difference between tangential and centripetal acceleration.
4. Identify the forces that contribute to the centripetal force on the object in each of the following examples:
   a. a *bicyclist* moving around a flat, circular track
   b. a *bicycle* moving around a flat, circular track
   c. a *race car* turning a corner on a steeply banked curve
5. A 90.0 kg person rides a spinning amusement park ride that has a radius of 11.5 m. If the person's tangential speed is 13.2 m/s, what is the magnitude of the centripetal force acting on the person?
6. Explain what makes a passenger in a turning car slide toward the door of the car.
7. **Critical Thinking** A roller coaster's passengers are suspended upside down as it moves at a constant speed through a vertical loop. What is the direction of the force that causes the coaster and its passengers to move in a circle? What provides this force?

SECTION 1

ANSWERS

Conceptual Challenge

1. The forces between each piece of the gooey dough are not large enough to cause the pieces to move in circular paths of small radius.
2. The swings move out until the tension in the chains can produce a large enough centripetal force to make the swings follow a circular path.

SECTION REVIEW ANSWERS

1. Answers may vary.
2. 14 m/s
3. Specific examples may vary, but all should indicate that an object moving in a circular path undergoes centripetal acceleration because the object changes direction. Tangential acceleration occurs when the object changes its speed around the path.
4. **a.** the forces that the bicycle seat, handlebars, and pedals exert on the bicyclist
   b. friction between the tires and the track
   c. components of (1) the normal force from the banked curve and (2) the friction between the tires and the road
5. 1.36×10^3 N
6. When the car turns, the inertia of the passenger keeps the passenger moving in the same direction that the car was initially moving.
7. toward the center of the loop; the track and gravity

SECTION 2
General Level

Visual Strategy GENERAL

Figure 6
Point out that the nature of the cannonball's path depends as much on the cannonball's tangential velocity as on the strength of the gravitational force. Have students compare the path of the cannonball when the ball's tangential speed is low with the paths of the projectiles that were studied in the chapter on two-dimensional motion.

Q How is the motion of an orbiting object similar to the motion of a projectile with two-dimensional motion?

A ***The orbiting object has a constant tangential speed, just as the horizontal component of a projectile's speed is constant.***

Q What happens when the speed of the projectile is so great that the force required to keep the projectile moving in a circle is larger than the gravitational force acting on the projectile?

A ***The projectile continues on a curved path, moving farther away from Earth's center as the force due to gravity weakens.***

SECTION 2

Newton's Law of Universal Gravitation

SECTION OBJECTIVES

- **Explain how Newton's law of universal gravitation accounts for various phenomena, including satellite and planetary orbits, falling objects, and the tides.**
- **Apply Newton's law of universal gravitation to solve problems.**

gravitational force

the mutual force of attraction between particles of matter

GRAVITATIONAL FORCE

Earth and many of the other planets in our solar system travel in nearly circular orbits around the sun. Thus, a centripetal force must keep them in orbit. One of Isaac Newton's great achievements was the realization that the centripetal force that holds the planets in orbit is the very same force that pulls an apple toward the ground—**gravitational force.**

Orbiting objects are in free fall

To see how this idea is true, we can use a thought experiment that Newton developed. Consider a cannon sitting on a high mountaintop, as shown in **Figure 6.** The path of each cannonball is a parabola, and the horizontal distance that each cannonball covers increases as the cannonball's initial speed increases. Newton realized that if an object were projected at just the right speed, the object would fall down toward Earth in just the same way that Earth curved out from under it. In other words, it would orbit Earth. In this case, the gravitational force between the cannonball and Earth is a centripetal force that keeps the cannonball in orbit. Satellites stay in orbit for this same reason. Thus, the force that pulls an apple toward Earth is the same force that keeps the moon and other satellites in orbit around Earth. Similarly, a gravitational attraction between Earth and our sun keeps Earth in its orbit around the sun.

Figure 6
Each successive cannonball has a greater initial speed, so the horizontal distance that the ball travels increases. If the initial speed is great enough, the curvature of Earth will cause the cannonball to continue falling without ever landing.

Gravitational force depends on the masses and the distance

Newton developed the following equation to describe quantitatively the magnitude of the gravitational force if distance r separates masses m_1 and m_2:

NEWTON'S LAW OF UNIVERSAL GRAVITATION

$$F_g = G\frac{m_1 m_2}{r^2}$$

$$\textbf{gravitational force} = \textbf{constant} \times \frac{\textbf{mass 1} \times \textbf{mass 2}}{(\textbf{distance between masses})^2}$$

G is called the *constant of universal gravitation.* The value of G was unknown in Newton's day, but experiments have since determined the value to be as follows:

$$G = 6.673 \times 10^{-11} \frac{\text{N}\bullet\text{m}^2}{\text{kg}^2}$$

Newton demonstrated that the gravitational force that a spherical mass exerts on a particle outside the sphere would be the same if the entire mass of the sphere were concentrated at the sphere's center. When calculating the gravitational force between Earth and our sun, for example, you use the distance between their centers.

SCILINKS.
www.scilinks.org
Topic: Gravity and Orbiting Objects
Code: HF60692

Gravitational force acts between all masses

Gravitational force always attracts objects to one another, as shown in **Figure 7.** The force that the moon exerts on Earth is equal and opposite to the force that Earth exerts on the moon. This relationship is an example of Newton's third law of motion. Also, note that the gravitational forces shown in **Figure 7** are centripetal forces. As a result of these centripetal forces, the moon and Earth each orbit around the center of mass of the Earth-moon system. Because Earth has a much greater mass than the moon, this center of mass lies within Earth.

Gravitational force exists between any two masses, regardless of size. For instance, desks in a classroom have a mutual attraction because of gravitational force. The force between the desks, however, is negligibly small relative to the force between each desk and Earth because of the differences in mass.

If gravitational force acts between all masses, why doesn't Earth accelerate up toward a falling apple? In fact, it does! But, Earth's acceleration is so tiny that you cannot detect it. Because Earth's mass is so large and acceleration is inversely proportional to mass, the Earth's acceleration is negligible. The apple has a much smaller mass and thus a much greater acceleration.

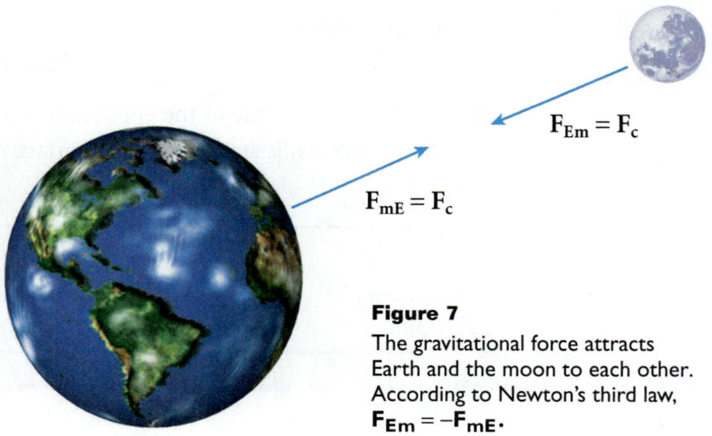

Figure 7
The gravitational force attracts Earth and the moon to each other. According to Newton's third law, $\mathbf{F_{Em}} = -\mathbf{F_{mE}}$.

Teaching Tip — BASIC

To emphasize the concept discussed in the last paragraph of this page, ask students to calculate the acceleration of Earth toward an apple and to use the weight of the apple as the force of attraction. Give the following values for the calculation:

$m_{apple} = 0.15$ kg
$m_{Earth} = 5.97 \times 10^{24}$ kg
$r_{Earth} = 6.38 \times 10^6$ m
(answer: $a_{Earth} = 2.5 \times 10^{-25}$ m/s^2)

Visual Strategy — ADVANCED

Figure 7

Be sure students understand that the moon and Earth each orbit around the center of mass of the Earth-moon system, which is inside Earth.

Q What might Earth's orbit around the moon look like from a point in space above the North Pole?

A ***Because the axis of rotation for Earth's orbit is inside Earth, the orbital motion would look like a slight wobble.***

Classroom Practice

Gravitational Force
Find the gravitational force that Earth (mass = 5.97×10^{24} kg) exerts on the moon (mass = 7.35×10^{22} kg) when the distance between them is 3.84×10^{8} m.

Answer
1.99×10^{20} N

PROBLEM GUIDE C

Use this guide to assign problems.
SE = Student Edition Textbook
PW = Problem Workbook
PB = Problem Bank on the One-Stop Planner (OSP)

Solving for:

r	**SE**	Sample, 1–2; Ch. Rvw. 18–19, 52
	PW	4–5
	PB	7–10
m	**PW**	Sample, 1–3
	PB	4–6
F_g	**SE**	3; Ch. Rvw. 40
	PW	6–7
	PB	Sample, 1–3

*Challenging Problem
Consult the printed Solutions Manual or the OSP for detailed solutions.

ANSWERS

Practice C

1. 0.692 m
2. 9.4×10^{6} m
3. a. 651 N
   b. 246 N
   c. 38.5 N

SAMPLE PROBLEM C

Gravitational Force

PROBLEM

Find the distance between a 0.300 kg billiard ball and a 0.400 kg billiard ball if the magnitude of the gravitational force between them is 8.92×10^{-11} N.

SOLUTION

Given: $m_1 = 0.300$ kg $\quad m_2 = 0.400$ kg $\quad F_g = 8.92 \times 10^{-11}$ N

Unknown: $r = ?$

Use the equation for Newton's law of universal gravitation, and solve for r.

$$F_g = G\frac{m_1 m_2}{r^2}$$

$$r = \sqrt{G\frac{m_1 m_2}{F_g}}$$

$$r = \sqrt{\left(6.673 \times 10^{-11}\ \frac{\text{N}\bullet\text{m}^2}{\text{kg}^2}\right) \times \frac{(0.300\ \text{kg})(0.400\ \text{kg})}{8.92 \times 10^{-11}\ \text{N}}}$$

$$\boxed{r = 3.00 \times 10^{-1}\ \text{m}}$$

PRACTICE C

Gravitational Force

1. What must be the distance between two 0.800 kg balls if the magnitude of the gravitational force between them is equal to that in Sample Problem C?
2. Mars has a mass of about 6.4×10^{23} kg, and its moon Phobos has a mass of about 9.6×10^{15} kg. If the magnitude of the gravitational force between the two bodies is 4.6×10^{15} N, how far apart are Mars and Phobos?
3. Find the magnitude of the gravitational force a 66.5 kg person would experience while standing on the surface of each of the following bodies:

Celestial Body	Mass	Radius
a. Earth	5.97×10^{24} kg	6.38×10^{6} m
b. Mars	6.42×10^{23} kg	3.40×10^{6} m
c. Pluto	1.25×10^{22} kg	1.20×10^{6} m

Why it Matters

Black Holes

This artist's conception shows a disk of material orbiting a black hole. Such disks provide indirect evidence of black holes within our own galaxy.

A black hole is an object that is so massive that nothing, not even light, can escape the pull of its gravity. In 1916, Karl Schwarzschild was the first person to suggest the existence of black holes. He used his solutions to Einstein's general-relativity equations to explain the properties of black holes. In 1967, the physicist John Wheeler coined the term "*black hole*" to describe these objects.

In order for an object to escape the gravitational pull of a planet, such as Earth, the object must be moving away from the planet faster than a certain threshold speed, which is called the *escape velocity.* The escape velocity at the surface of Earth is about 1.1×10^4 m/s, or about 25 000 mi/h.

The escape velocity for a black hole is greater than the speed of light. And, according to Einstein's special theory of relativity, no object can move at a speed equal to or greater than the speed of light. Thus, no object that is within a certain distance of a black hole can move fast enough to escape the gravitational pull of the black hole. That distance, called the *Schwarzschild radius,* defines the edge, or *horizon,* of a black hole.

Newton's laws say that only objects with mass can be subject to forces. How can a black hole trap light if light has no mass? According to Einstein's general theory of relativity, any object with mass bends the fabric of space and time itself. When an object that has mass or even when a ray of light passes near another object, the path of the moving object or ray curves because space-time itself is curved. The curvature is so great inside a black hole that the path of any light that might be emitted from the black hole bends back toward the black hole and remains trapped inside the horizon.

Because black holes trap light, they cannot be observed directly. Instead, astronomers must look for indirect evidence of black holes. For example, astronomers have observed stars orbiting very rapidly around the centers of some galaxies. By measuring the speed of the orbits, astronomers can calculate the mass of the dark object—the black hole—that must be at the galaxy's center. Black holes at the centers of galaxies typically have masses millions or billions of times the mass of the sun.

The figure above shows a disk of material orbiting a black hole. Material that orbits a black hole can move at such high speeds and have so much energy that the material emits X rays. From observations of the X rays coming from such disks, scientists have discovered several black holes within our own galaxy.

This image from NASA's *Chandra X-ray Observatory* is of Sagittarius A*, which is a supermassive black hole at the center of our galaxy. Astronomers are studying the image to learn more about Sagittarius A* and about black holes in the centers of other galaxies.

Why it Matters

Black Holes

A black hole is an object whose escape velocity is greater than the speed of light. Escape velocity is independent of the mass of the escaping object. Understanding how a black hole can trap light, which has no mass, requires an appeal to Einstein's general theory of relativity. You can construct a simple model to help students grasp the difficult concept of gravity as a curvature of space-time. Attach a sheet of flexible rubber to a hoop, and then place a heavy object in the center of the sheet. The sheet will bend downward into a cone. You can then roll a coin in an orbit around the object in the center. Point out to students that this is only a rough three-dimensional model, while the curvature of space-time is four-dimensional.

The extreme bending of space-time that prevents light from escaping the Schwarzschild radius of a black hole occurs in a less extreme form around stars that are less compact than black holes. During a solar eclipse in 1922, astronomers observed light, which came from a distant star, bend as the light passed near the sun. This observation was one of the first pieces of evidence to support Einstein's general theory of relativity. Direct students who wish to learn more about general relativity to the feature "General Relativity" in **Appendix J: Advanced Topics.**

Teaching Tip GENERAL

Tell students that the time at which high and low tides occur at a given location is not the same each day. The times vary because the tidal period is actually 12 hours and 25 minutes. There are two high and low tides every 24 hours and 50 minutes.

Teaching Tip ADVANCED

Expand on the fact that tidal force arises from the difference between the gravitational forces at Earth's near surface, center, and far surface, and mention that this differential force is an inverse-cubed quantity. That is to say, as the distance r between two bodies increases, the tidal force decreases as $\frac{1}{r^3}$. For this reason, the sun, which has a greater mass than the moon does, has less effect on the Earth's ocean tides than the moon does.

Point out that tidal forces are exerted on all substances and that the amount of distortion depends on the elasticity of the body under gravitational influence. The rocky surfaces of the moon and Earth also experience tidal forces, but the bulge is not as great as with Earth's oceans because rock is not as easily extended.

If an orbiting body moves too close to a more massive body, the tidal force on the orbiting body may be large enough to break the body apart. The distance at which tidal forces can become destructive is called *Roche's limit,* after the 19th-century French mathematician Edouard Roche.

APPLYING THE LAW OF GRAVITATION

For about six hours, water slowly rises along the shoreline of many coastal areas and culminates in a high tide. The water level then slowly lowers for about six hours and returns to a low tide. This cycle then repeats. Tides take place in all bodies of water but are most noticeable along seacoasts. In the Bay of Fundy, shown in **Figure 8,** the water rises as much as 16 m from its low point. Because a high tide happens about every 12 hours, there are usually two high tides and two low tides each day. Before Newton developed the law of universal gravitation, no one could explain why tides occur in this pattern.

Did you know?

When the sun and moon are in line, the combined effect produces a greater-than-usual high tide called a *spring tide.* When the sun and moon are at right angles, the result is a lower-than-normal high tide called a *neap tide.* Each revolution of the moon around Earth corresponds to two spring tides and two neap tides.

Newton's law of gravitation accounts for ocean tides

High and low tides are partly due to the gravitational force exerted on Earth by its moon. The tides result from the *difference* between the gravitational force at Earth's surface and at Earth's center. A full explanation is beyond the scope of this text, but we will briefly examine this relationship.

The two high tides take place at locations on Earth that are nearly in line with the moon. On the side of Earth that is nearest to the moon, the moon's gravitational force is *greater* than it is at Earth's center (because gravitational force decreases with distance). The water is pulled toward the moon, creating an outward bulge. On the opposite side of Earth, the gravitational force is *less* than it is at the center. On this side, all mass is still pulled toward the moon, but the water is pulled least. This creates another outward bulge. Two high tides take place each day because when Earth rotates one full time, any given point on Earth will pass through both bulges.

The moon's gravitational force is not the only factor that affects ocean tides. Other influencing factors include the depths of the ocean basins, Earth's tilt and rotation, and friction between the ocean water and the ocean floor. The sun also contributes to Earth's ocean tides, but the sun's effect is not as significant as the moon's is. Although the sun exerts a much greater gravitational force on Earth than the moon does, the *difference* between the force on the far and near sides of Earth is what affects the tides.

Figure 8
Some of the world's highest tides occur at the Bay of Fundy, which is between New Brunswick and Nova Scotia, Canada. These photographs show a river outlet to the Bay of Fundy at low and high tide.

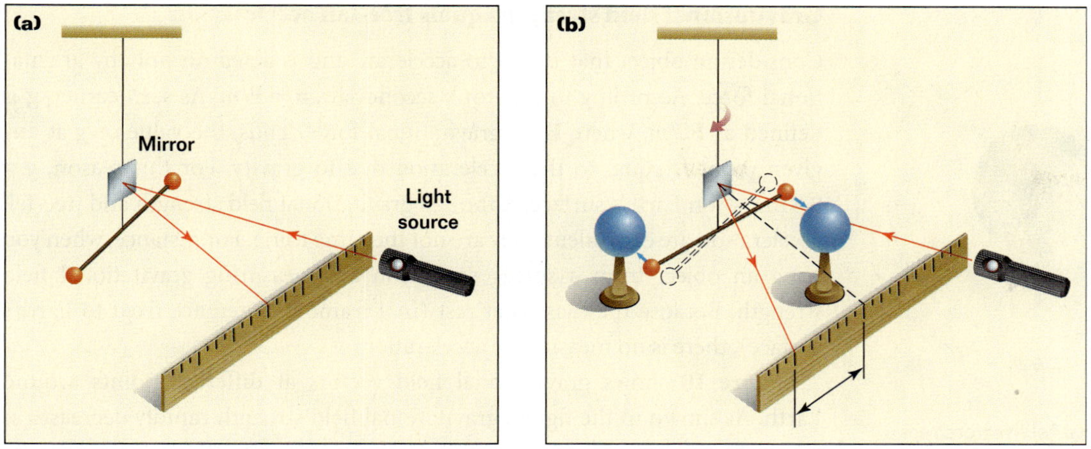

Figure 9
Henry Cavendish used an experiment similar to this one to determine the value of *G*.

Cavendish finds the value of *G* and Earth's mass

In 1798, Henry Cavendish conducted an experiment that determined the value of the constant *G*. This experiment is illustrated in **Figure 9.** As shown in **Figure 9(a),** two small spheres are fixed to the ends of a suspended light rod. These two small spheres are attracted to two larger spheres by the gravitational force, as shown in **Figure 9(b).** The angle of rotation is measured with a light beam and is then used to determine the gravitational force between the spheres. When the masses, the distance between them, and the gravitational force are known, Newton's law of universal gravitation can be used to find *G*. Once the value of *G* is known, the law can be used again to find Earth's mass.

Gravity is a field force

Newton was not able to explain how objects can exert forces on one another without coming into contact. His mathematical theory described gravity, but didn't explain how it worked. Later work also showed that Newton's laws are not accurate for very small objects or for those moving near the speed of light. Scientists later developed a theory of fields to explain how gravity and other field forces operate. According to this theory, masses create a gravitational field in space. A gravitational force is an interaction between a mass and the gravitational field created by other masses.

When you raise a ball to a certain height above Earth, the ball gains potential energy. Where is this potential energy stored? The physical properties of the ball and of Earth have not changed. However, the gravitational field between the ball and Earth *has* changed since the ball has changed position relative to Earth. According to field theory, the gravitational energy is stored in the gravitational field itself.

At any point, Earth's gravitational field can be described by the *gravitational field strength,* abbreviated *g*. The value of *g* is equal to the magnitude of the gravitational force exerted on a unit mass at that point, or $g = F_g/m$. The gravitational field **(g)** is a vector with a magnitude of *g* that points in the direction of the gravitational force.

Quick Lab

Gravitational Field Strength

MATERIALS LIST

- **spring scale**
- **hook (of a known mass)**
- **various masses**

You can attach a mass to a spring scale to find the gravitational force that is acting on that mass. Attach various combinations of masses to the hook, and record the force in each case. Use your data to calculate the gravitational field strength for each trial ($g = F_g/m$). Be sure that your calculations account for the mass of the hook. Average your values to find the gravitational field strength at your location on Earth's surface. Do you notice anything about the value you obtained?

Misconception Alert — GENERAL

Students may not understand why gravitational field strength is equal to the force divided by the mass acted upon, rather than just the gravitational force. Explain that for a given location, the strength associated with the gravitational field must be constant but that the gravitational force exerted on two different masses at the same location will differ as the masses themselves differ.

Quick Lab

TEACHER'S NOTES

Ask students about the significance of their resulting values. If results are not close to free-fall acceleration, you can find a class average. At this point, students should realize that the value of *g* equals the value of free-fall acceleration. Tell students that they will learn more about this later in the section.

The Language of Physics

The symbol *g* represents two different quantities: free-fall acceleration on Earth's surface (9.81 m/s^2) and gravitational field strength. Free-fall acceleration at any location (a_g) is always equal to the gravitational field strength (g) at that location.

ANSWERS

Conceptual Challenge

1. no; The distance from the center of the moon to the moon's surface is less than the distance between Earth's center and Earth's surface. These distances, along with the masses of the two bodies, affect gravitational field strength.
2. The gold should be purchased at a high altitude, where the gold weighs less, and sold at a low altitude, where the gold weighs more. (Even so, this scheme is impractical, in part because of transportation costs.)

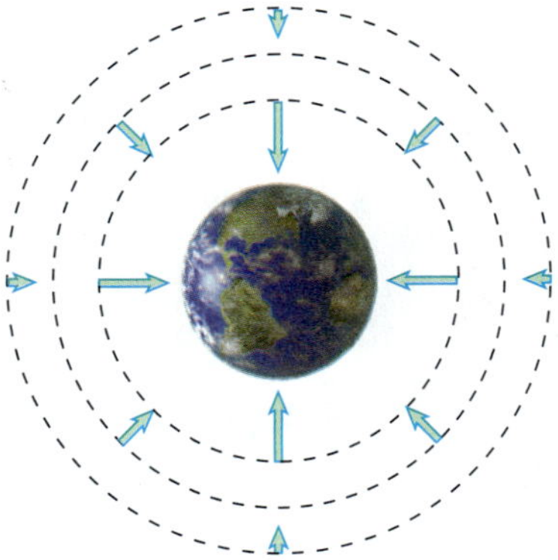

Figure 10
The gravitational field vectors represent Earth's gravitational field at each point. Note that the field has the same strength at equal distances from Earth's center.

Gravitational field strength equals free-fall acceleration

Consider an object that is free to accelerate and is acted on only by gravitational force. According to Newton's second law, $\mathbf{a} = \mathbf{F}/m$. As seen earlier, $\mathbf{g}$ is defined as $\mathbf{F_g}/m$, where $\mathbf{F_g}$ is gravitational force. Thus, the value of g at any given point is equal to the acceleration due to gravity. For this reason, $g = 9.81$ m/s^2 on Earth's surface. Although gravitational field strength and free-fall acceleration are equivalent, they are not the same thing. For instance, when you hang an object from a spring scale, you are measuring gravitational field strength. Because the mass is at rest (in a frame of reference fixed to Earth's surface), there is no measurable acceleration.

Figure 10 shows gravitational field vectors at different points around Earth. As shown in the figure, gravitational field strength rapidly decreases as the distance from Earth increases, as you would expect from the inverse-square nature of Newton's law of universal gravitation.

Weight changes with location

In the chapter about forces, you learned that weight is the magnitude of the force due to gravity, which equals mass times free-fall acceleration. We can now refine our definition of weight as mass times gravitational field strength. The two definitions are mathematically equivalent, but our new definition helps to explain why your weight changes with your location in the universe.

Newton's law of universal gravitation shows that the value of g depends on mass and distance. For example, consider a tennis ball of mass m. The gravitational force between the tennis ball and Earth is as follows:

$$F_g = \frac{Gmm_E}{r^2}$$

Combining this equation with the definition for gravitational field strength yields the following expression for g:

$$g = \frac{F_g}{m} = \frac{Gmm_E}{mr^2} = G\frac{m_E}{r^2}$$

This equation shows that gravitational field strength depends only on mass and distance. Thus, as your distance from Earth's center increases, the value of g decreases, so your weight also decreases. On the surface of any planet, the value of g, as well as your weight, will depend on the planet's mass and radius.

Why it Matters

Conceptual Challenge

1. Gravity on the Moon The magnitude of g on the moon's surface is about $\frac{1}{6}$ of the value of g on Earth's surface. Can you infer from this relationship that the moon's mass is $\frac{1}{6}$ of Earth's mass? Why or why not?

2. Selling Gold A scam artist hopes to make a profit by buying and selling gold at different altitudes for the same price per weight. Should the scam artist buy or sell at the higher altitude? Explain.

Gravitational mass equals inertial mass

Because gravitational field strength equals free-fall acceleration, free-fall acceleration on the surface of Earth likewise depends only on Earth's mass and radius. Free-fall acceleration does not depend on the falling object's mass, because m cancels from each side of the equation, as shown on the previous page.

Although we are assuming that the m in each equation is the same, this assumption was not always an accepted scientific fact. In Newton's second law, m is sometimes called *inertial mass* because this m refers to the property of an object to resist acceleration. In Newton's gravitation equation, m is sometimes called *gravitational mass* because this m relates to how objects attract one another.

How do we know that inertial and gravitational mass are equal? The fact that the acceleration of objects in free fall on Earth's surface is always the same confirms that the two types of masses are equal. A more massive object experiences a greater gravitational force, but the object resists acceleration by just that amount. For this reason, all masses fall with the same acceleration (disregarding air resistance).

There is no obvious reason why the two types of masses should be equal. For instance, the property of electric charges that causes them to be attracted or repelled was originally called *electrical mass.* Even though this term has the word *mass* in it, electrical *mass* has no connection to gravitational or inertial mass. The equality between inertial and gravitational mass has been continually tested and has thus far always held up.

ADVANCED TOPICS

The equality of gravitational and inertial masses puzzled scientists for many years. Einstein's general theory of relativity was the first explanation of this equality. See "General Relativity" in **Appendix J: Advanced Topics** to learn more about this topic.

SECTION REVIEW

1. Explain how the force due to gravity keeps a satellite in orbit.

2. Is there gravitational force between two students sitting in a classroom? If so, explain why you don't observe any effects of this force.

3. Earth has a mass of 5.97×10^{24} kg and a radius of 6.38×10^{6} m, while Saturn has a mass of 5.68×10^{26} kg and a radius of 6.03×10^{7} m. Find the weight of a 65.0 kg person at the following locations:

a. on the surface of Earth
b. 1000 km above the surface of Earth
c. on the surface of Saturn
d. 1000 km above the surface of Saturn

4. What is the magnitude of g at a height above Earth's surface where free-fall acceleration equals 6.5 m/s^2?

5. Critical Thinking Suppose the value of G has just been discovered. Use the value of G and an approximate value for Earth's radius (6.38×10^{6} m) to find an approximation for Earth's mass.

SECTION 2

The Language of Physics

The difference between *inertial mass* and *gravitational mass* can be understood by the ways in which they are measured. Gravitational mass is a measurement of the weight of an object (as weighed on a spring scale) divided by the gravitational field strength at the location of the measurement. The gravitational mass is pulled by a field force but does not accelerate. By contrast, inertial mass is accelerated, as in an Atwood's machine, where the inertial mass accelerates under the force of another falling mass.

SECTION REVIEW ANSWERS

1. A satellite moves tangentially around a planet and would continue to move in a straight line if there were no gravitational force. The combination of the acceleration due to gravity and the tangential speed causes the satellite to follow a path that is parallel to the curvature of the planet.

2. yes; The magnitude of the force is extremely small because the masses of the students are small relative to Earth's mass.

3. **a.** 636 N
b. 475 N
c. 678 N
d. 656 N

4. 6.5 m/s^2

5. 5.98×10^{24} kg

SECTION 3
General Level

Misconception Alert — GENERAL

Students may believe the Copernican heliocentric model solved all of the problems inherent in the geocentric model of the solar system. Point out that even Copernicus's model was not as simple or as accurate as it sounded. He was not able to eliminate the need for epicycles. In an attempt to rectify the problem, Tycho Brahe introduced a modified geocentric model in which all the planets except Earth revolved around the sun, which in turn revolved around Earth. All of these models failed because they were dependent on the ancient belief that orbits must be perfectly circular.

Teaching Tip — ADVANCED

Students may think that Aristarchus's heliocentric model was rejected because it was unorthodox. Explain that his theory, though unconventional, was actually tested. If Earth were moving about the sun, the apparent position of the foreground stars should shift with respect to the apparent position of the background stars. The same shift happens when you are looking at a nearby object with respect to the horizon and then you move to the side. This shifting, called *parallax,* was not observed; however, the failure to observe stellar parallax was due simply to the very great distance of the stars compared to the size of Earth's orbit. The first measurement of stellar parallax was not made until 1838, when telescopes were large enough to detect small stellar motions.

SECTION 3

Motion in Space

SECTION OBJECTIVES

- **Describe Kepler's laws of planetary motion.**
- **Relate Newton's mathematical analysis of gravitational force to the elliptical planetary orbits proposed by Kepler.**
- **Solve problems involving orbital speed and period.**

KEPLER'S LAWS

People have studied the motions of the planets since ancient times. Until the middle of the 16th century, most people believed that Earth was at the center of the universe. Originally, it was believed that the sun and other planets orbited Earth in perfect circles. However, this model did not account for all of the observations of planetary motion.

In the second century CE, Claudius Ptolemy developed an elaborate theory of planetary motion. Ptolemy's theory attempted to reconcile observation with theory and to keep Earth at the center of the universe. In this theory, planets travel in small circles called *epicycles* while simultaneously traveling in larger circular orbits. Even Ptolemy's complex model did not fully agree with observation, although the model did explain more than previous theories.

In 1543, the Polish astronomer Nicolaus Copernicus (1473–1543) published *On the Revolutions of the Heavenly Spheres,* in which he proposed that Earth and other planets orbit the sun in perfect circles. **Figure 11** shows a sun-centered planetary model that is believed to have been made for King George III of England. The idea of a sun-centered universe was not completely new in the 16th century. A Greek named Aristarchus theorized 1700 years before Copernicus did that Earth revolved around the sun, but most other scientists did not accept his theory.

Kepler's three laws describe the motion of the planets

The astronomer Tycho Brahe (1546–1601) made many precise observations of the planets and stars. However, some of Brahe's data did not agree with the Copernican model. The astronomer Johannes Kepler (1571–1630) worked for many years to reconcile Copernican theory with Brahe's data. Kepler's analysis led to three laws of planetary motion, which were developed a generation before Newton's law of universal gravitation. Kepler's three laws can be summarized as shown on the next page.

Figure 11
This elaborate planetary model—called an *orrery*—shows the motions of Mercury, Venus, and Earth around the sun. The model also shows the moon's inclined orbit around Earth.

KEPLER'S LAWS OF PLANETARY MOTION

First Law: Each planet travels in an elliptical orbit around the sun, and the sun is at one of the focal points.

Second Law: An imaginary line drawn from the sun to any planet sweeps out equal areas in equal time intervals.

Third Law: The square of a planet's orbital period (T^2) is proportional to the cube of the average distance (r^3) between the planet and the sun, or $T^2 \propto r^3$.

Kepler's first law states that the planets' orbits are ellipses rather than circles. Kepler discovered this law while working with Brahe's data for the orbit of Mars. While trying to explain the data, Kepler experimented with 70 different circular orbits and generated numerous pages of calculations. He finally realized that if the orbit is an ellipse rather than a circle and the sun is at one focal point of the ellipse, the data fit perfectly.

Kepler's second law states that an imaginary line from the sun to any planet sweeps out equal areas in equal times, as shown in **Figure 12.** In other words, if the time a planet takes to travel the arc on the left (Δt_1) is equal to the time the planet takes to cover the arc on the right (Δt_2), then the area A_1 is equal to the area A_2. Thus, the planets travel faster when they are closer to the sun.

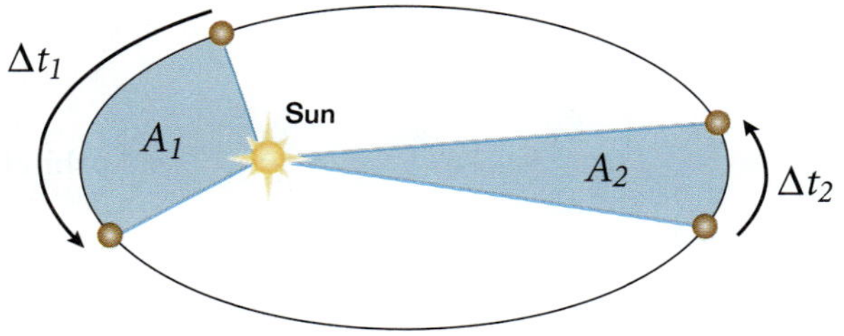

Figure 12
This diagram illustrates a planet moving in an elliptical orbit around the sun. If Δt_1 equals Δt_2, then the two shaded areas are equal. Thus, the planet travels faster when it is closer to the sun and slower when it is farther away.

While Kepler's first two laws describe the motion of each planet individually, his third law relates the orbital periods and distances of one planet to those of another planet. The orbital period (T) is the time a planet takes to finish one full revolution, and the distance (r) is the mean distance between the planet and the sun. Kepler's third law relates the orbital period and mean distance for two orbiting planets as follows:

$$\frac{T_1^2}{T_2^2} = \frac{r_1^3}{r_2^3}, \text{ or } T^2 \propto r^3$$

This law also applies to satellites orbiting Earth, including our moon. In that case, r is the distance between the orbiting satellite and Earth. The proportionality constant depends on the mass of the central object.

Kepler's laws are consistent with Newton's law of gravitation

Newton used Kepler's laws to support his law of gravitation. For example, Newton proved that if force is inversely proportional to distance squared, as stated in the law of universal gravitation, the resulting orbit must be an ellipse or a circle. He also demonstrated that his law of gravitation could be used to derive Kepler's third law. (Try a similar derivation yourself in the Quick Lab at right.) The fact that Kepler's laws closely matched observations gave additional support for Newton's theory of gravitation.

Quick Lab

Kepler's Third Law

You can mathematically show how Kepler's third law can be derived from Newton's law of universal gravitation (assuming circular orbits). To begin, recall that the centripetal force is provided by the gravitational force. Set the equations for gravitational and centripetal force equal to one another, and solve for v_t^2. Because speed equals distance divided by time and because the distance for one period is the circumference ($2\pi r$), $v_t = 2\pi r/T$. Square this value, substitute the squared value into your previous equation, and then isolate T^2. How does your result relate to Kepler's third law?

Quick Lab

TEACHER'S NOTES

Students should find that $T^2 = \left(\frac{4\pi^2}{Gm}\right)r^3$. This equation expresses that T^2 is proportional to r^3, and that $\frac{4\pi^2}{Gm}$ is the proportionality constant. This is a more detailed form of Kepler's third law ($T^2 \propto r^3$). Be sure students understand that m is the mass of the object being orbited.

Homework Options
This QuickLab can easily be performed outside of the physics lab room.

Teaching Tip — ADVANCED

Explain how Kepler's third law provides a fairly simple and powerful means of measuring the mass of stars and planets. In general, the mass used in the third law should equal the sum of the orbiting body's mass and the mass of the body that is orbited. While in most cases the orbiting body's mass is so small that it can be ignored, other cases, such as the two stars in a binary star system, require that one of the masses be known or realistically estimated. The masses of many stars are determined from the stars' colors and levels of brightness.

Kepler's third law is especially useful for binary star systems in which one of the companions is small and thus hard to see but is very massive. Examples of small but massive bodies include white dwarfs and neutron stars. A more dramatic example is a black hole. The mass of the invisible companion can be estimated by observing the orbital motions of a visible companion, whose mass can be estimated from its light.

Teaching Tip — GENERAL

Have students use values in the table to confirm Kepler's third law. (They should get the same ratio of T^2/r^3 for any planet they try.) Students who did the Quick Lab on the previous page can also see how the orbital speed equation is derived. They simply need to take the square root of their expression for v_t^2.

Did you know?

We generally speak of the moon orbiting Earth, but they are actually both in orbit around the center of mass of the Earth-moon system. Because Earth is so much more massive than the moon, their common center of mass lies inside Earth. Thus, the moon appears to orbit Earth. The center of mass does not always lie inside one of the bodies. For example, Pluto and its moon, Charon, orbit a center of mass that lies between them. Also, many binary star systems have two stars that orbit a common center of mass between the stars.

Kepler's third law describes orbital period

According to Kepler's third law, $T^2 \propto r^3$. The constant of proportionality between these two variables turns out to be $4\pi^2/Gm$, where m is the mass of the object being orbited. (To learn why this is the case, try the Quick Lab on the previous page.) Thus, Kepler's third law can also be stated as follows:

$$T^2 = \left(\frac{4\pi^2}{Gm}\right)r^3$$

The square root of the above equation, which is shown below on the left, describes the period of any object that is in a circular orbit. The speed of an object that is in a circular orbit depends on the same factors that the period does, as shown in the equation on the right. The assumption of a circular orbit provides a close approximation for real orbits in our solar system because all planets except Mercury have orbits that are nearly circular.

PERIOD AND SPEED OF AN OBJECT IN CIRCULAR ORBIT

$$T = 2\pi\sqrt{\frac{r^3}{Gm}} \qquad v_t = \sqrt{G\frac{m}{r}}$$

$$\textbf{orbital period} = 2\pi\sqrt{\frac{\textbf{(mean radius)}^3}{\textbf{(constant)(mass of central object)}}}$$

$$\textbf{orbital speed} = \sqrt{\textbf{(constant)}\left(\frac{\textbf{mass of central object}}{\textbf{mean radius}}\right)}$$

Note that m in both equations is the mass of the central object that is being orbited. The mass of the planet or satellite that is in orbit does not affect its speed or period. The mean radius (r) is the distance between the centers of the two bodies. For an artificial satellite orbiting Earth, r is equal to Earth's mean radius plus the satellite's distance from Earth's surface (its "altitude"). **Table 1** gives planetary data that can be used to calculate orbital speeds and periods.

Table 1 Planetary Data

Planet	Mass (kg)	Mean radius (m)	Mean distance from sun (m)	Planet	Mass (kg)	Mean radius (m)	Mean distance from sun (m)
Earth	5.97×10^{24}	6.38×10^6	1.50×10^{11}	Neptune	1.02×10^{26}	2.48×10^7	4.50×10^{12}
Earth's moon	7.35×10^{22}	1.74×10^6	—	Saturn	5.68×10^{26}	6.03×10^7	1.43×10^{12}
Jupiter	1.90×10^{27}	7.15×10^7	7.79×10^{11}	Sun	1.99×10^{30}	6.96×10^8	—
Mars	6.42×10^{23}	3.40×10^6	2.28×10^{11}	Uranus	8.68×10^{25}	2.56×10^7	2.87×10^{12}
Mercury	3.30×10^{23}	2.44×10^6	5.79×10^{10}	Venus	4.87×10^{24}	6.05×10^6	1.08×10^{11}

SAMPLE PROBLEM D

Period and Speed of an Orbiting Object

PROBLEM

The color-enhanced image of Venus shown here was compiled from data taken by *Magellan*, the first planetary spacecraft to be launched from a space shuttle. During the spacecraft's fifth orbit around Venus, *Magellan* traveled at a mean altitude of 361 km. If the orbit had been circular, what would *Magellan*'s period and speed have been?

SOLUTION

1. DEFINE **Given:** $r_1 = 361\ \text{km} = 3.61 \times 10^5\ \text{m}$

Unknown: $T = ?$ $\quad v_t = ?$

2. PLAN **Choose an equation or situation:** Use the equations for the period and speed of an object in a circular orbit.

$$T = 2\pi\sqrt{\frac{r^3}{Gm}} \qquad v_t = \sqrt{G\frac{m}{r}}$$

Use **Table 1** to find the values for the radius (r_2) and mass (m) of Venus.

$$r_2 = 6.05 \times 10^6\ \text{m} \qquad m = 4.87 \times 10^{24}\ \text{kg}$$

Find r by adding the distance between the spacecraft and Venus's surface (r_1) to Venus's radius (r_2).

$$r = r_1 + r_2 = (3.61 \times 10^5\ \text{m}) + (6.05 \times 10^6\ \text{m}) = 6.41 \times 10^6\ \text{m}$$

3. CALCULATE **Substitute the values into the equations and solve:**

$$T = 2\pi\sqrt{\frac{(6.41 \times 10^6\ \text{m})^3}{\left(6.673 \times 10^{-11}\ \frac{\text{N}\bullet\text{m}^2}{\text{kg}^2}\right)(4.87 \times 10^{24}\ \text{kg})}} = \boxed{5.66 \times 10^3\ \text{s}}$$

$$v_t = \sqrt{\left(6.673 \times 10^{-11}\ \frac{\text{N}\bullet\text{m}^2}{\text{kg}^2}\right)\left(\frac{4.87 \times 10^{24}\ \text{kg}}{6.41 \times 10^6\ \text{m}}\right)} = \boxed{7.12 \times 10^3\ \text{m/s}}$$

4. EVALUATE *Magellan* takes $(5.66 \times 10^3\ \text{s})(1\ \text{min}/60\ \text{s}) \approx 94$ min to complete one orbit.

PRACTICE D

Period and Speed of an Orbiting Object

1. Find the orbital speed and period that the *Magellan* satellite from Sample Problem D would have at the same mean altitude above Earth, Jupiter, and Earth's moon.

2. At what distance above Earth would a satellite have a period of 125 min?

Classroom Practice

Period and Speed of an Orbiting Object
A large planet orbiting a distant star is discovered. The planet's orbit is nearly circular and fairly close to the star. The planet's orbital distance is 7.50×10^{10} m, and its period is 105.5 days. Calculate the mass of the star.

Answer
3.00×10^{30} kg

PROBLEM GUIDE D

Use this guide to assign problems.
SE = Student Edition Textbook
PW = Problem Workbook
PB = Problem Bank on the One-Stop Planner (OSP)

Solving for:

v_t	**SE**	Sample, 1; Ch. Rvw. 27, 49b
	PW	3–4
	PB	5–6
T	**SE**	Sample, 1; Ch. Rvw. 27, 49a
	PW	5–6
	PB	7–8
r	**SE**	2; Ch. Rvw. 28
	PW	Sample, 1–2
	PB	3–4
m	**SE**	Ch. Rvw. 29
	PW	7–8
	PB	Sample, 1–2

***Challenging Problem**
Consult the printed Solutions Manual or the OSP for detailed solutions.

ANSWERS

Practice D

1. Earth: 7.69×10^3 m/s, 5.51×10^3 s; Jupiter: 4.20×10^4 m/s, 1.08×10^4 s; moon: 1.53×10^3 m/s, 8.63×10^3 s
2. 1.90×10^6 m

Quick Lab

TEACHER'S NOTES

To perform this lab, select an elevator that accelerates through most of its trip between the top and ground floors. Slow elevators tend to move at a constant speed, so no change in the scale reading will be observed after the moment when the elevator begins to move. The measurements will be easier to make if the building has more than five floors.

Teaching Tip — ADVANCED

A means to overcome apparent weightlessness in space is to have a spacecraft or portion of the spacecraft that is a cylinder rotating about its central axis. Objects or people against the walls experience a normal force that is proportional to the square of the speed of the cylinder's rotation and is inversely proportional to the cylinder's radius. If the cylinder is rotated at the proper speed, the cylinder creates an "artificial gravity" that is equal to the gravitational field strength at Earth's surface.

Quick Lab

Elevator Acceleration

MATERIALS LIST

- elevator
- bathroom scale
- watch or stopwatch

In this activity, you will stand on a bathroom scale while riding an elevator up to the top floor and then back. Stand on the scale in a first-floor elevator, and record your weight. As the elevator moves up, record the scale reading for every two-second interval. Repeat the process as the elevator moves down.

Now, find the net force for each time interval, and then use Newton's second law to calculate the elevator's acceleration for each interval. How does the acceleration change? How does the elevator's maximum acceleration compare with free-fall acceleration?

WEIGHT AND WEIGHTLESSNESS

In the chapter about forces, you learned that weight is the magnitude of the force due to gravity. When you step on a bathroom scale, it does not actually measure your weight. The scale measures the downward force exerted on it. When your weight is the only downward force acting on the scale, the scale reading equals your weight. If a friend pushes down on you while you are standing on the scale, the scale reading will go up. However, your weight has not changed; the scale reading equals your weight plus the extra applied force. Because of Newton's third law, the downward force you exert on the scale equals the upward force exerted on you by the scale (the normal force). Thus, the scale reading is equal to the normal force acting on you.

For example, imagine you are standing in an elevator, as illustrated in **Figure 13.** When the elevator is at rest, as in **Figure 13(a),** the magnitude of the normal force is equal to your weight. A scale in the elevator would record your weight. When the elevator begins accelerating downward, as in **Figure 13(b),** the normal force will be smaller. The scale would now record an amount that is less than your weight. If the elevator's acceleration were equal to free-fall acceleration, as shown in **Figure 13(c),** you would be falling at the same rate as the elevator and would not feel the force of the floor at all. In this case, the scale would read zero. You still have the same weight, but you and the elevator are both falling with free-fall acceleration. In other words, no normal force is acting on you. This situation is called *apparent weightlessness.*

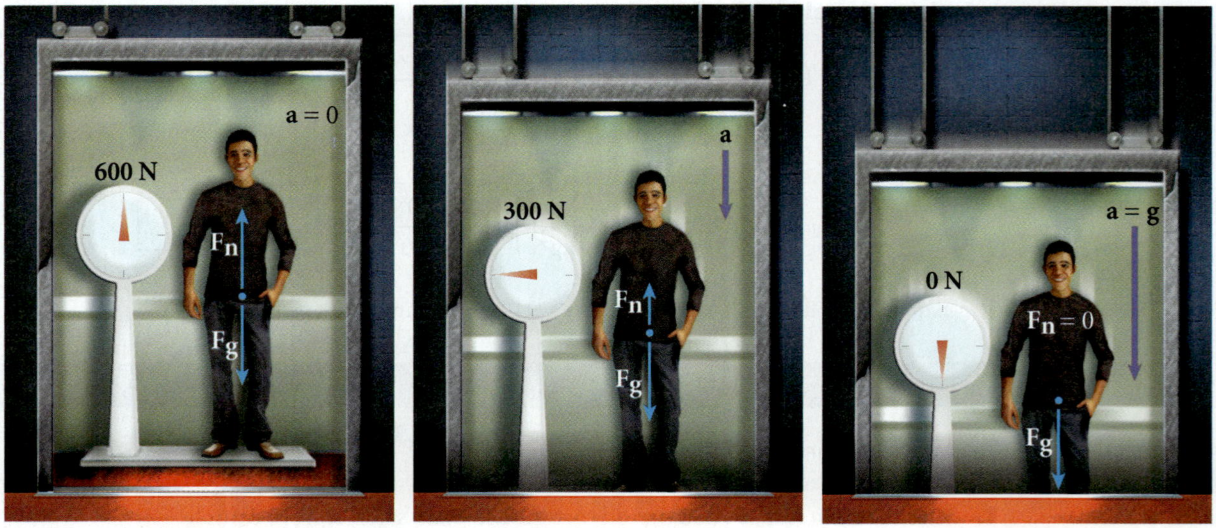

(a) (b) (c)

Figure 13
When this elevator accelerates, the normal force acting on the person changes. If the elevator were in free fall, the normal force would drop to zero and the person would experience a sensation of apparent weightlessness.

Astronauts in orbit experience apparent weightlessness

Astronauts floating in a space shuttle are experiencing apparent weightlessness. Because the shuttle is accelerating at the same rate as the astronauts are, this example is similar to the elevator in **Figure 13(c).** The force due to gravity keeps the astronauts and shuttle in orbit, but the astronauts *feel* weightless because no normal force is acting on them.

The human body relies on gravitational force. For example, this force pulls blood downward so that the blood collects in the veins of your legs when you are standing. Because the body of an astronaut in orbit accelerates along with the space shuttle, gravitational force has no effect on the body. This state can initially cause nausea and dizziness. Over time, it can pose serious health risks, such as weakened muscles and brittle bones. When astronauts return to Earth, their bodies need time to readjust to the effects of the gravitational force.

So far, we have been describing apparent weightlessness. Actual weightlessness occurs only in deep space, far from stars and planets. Gravitational force is never entirely absent, but it can become negligible at distances that are far enough away from any masses. In this case, a star or astronaut would not be pulled into an orbit but would instead drift in a straight line at constant speed.

extension

Integrating Health

Visit go.hrw.com for the activity "Exercise in Space."

Keyword HF6CMGX

SECTION REVIEW

1. Compare Ptolemy's model of the solar system with Copernicus's. How does Kepler's first law of planetary motion refine Copernicus's model?
2. Does a planet in orbit around the sun travel at a constant speed? How do you know?
3. Suppose you know the mean distance between both Mercury and the sun and Venus and the sun. You also know the period of Venus's orbit around the sun. How can you find the period of Mercury's orbit?
4. Explain how Kepler's laws of planetary motion relate to Newton's law of universal gravitation.
5. Find the orbital speed and period of Earth's moon. The average distance between the centers of Earth and of the moon is 3.84×10^8 m.
6. **Critical Thinking** An amusement park ride raises people high into the air, suspends them for a moment, and then drops them at the rate of free-fall acceleration. Is a person in this ride experiencing apparent weightlessness, true weightlessness, or neither? Explain.
7. **Critical Thinking** Suppose you went on the ride described in item 6, held a penny in front of you, and released the penny at the moment the ride started to drop. What would you observe?

SECTION 3

The Language of Physics

Apparent weightlessness is sometimes called *microgravity.*

SECTION REVIEW ANSWERS

1. Both Ptolemy's and Copernicus's models used circular orbits to describe the motions of planets. However, in Ptolemy's model, the planets and the sun moved around a motionless Earth, whereas in Copernicus's model, the planets and Earth moved around the sun. Kepler replaced Copernicus's circular orbits with ellipses.
2. no; a planet moves faster as it comes closer to the sun because, in accordance with Kepler's second law, it sweeps out equal areas in equal time intervals.
3. Use Kepler's third law:
$$\frac{T_M^2}{T_V^2} = \frac{r_M^3}{r_V^3}$$
so
$$T_M = T_V\sqrt{\frac{r_M^3}{r_V^3}}$$
4. Newton demonstrated that an inverse-square force is necessary to produce an elliptical orbit, as predicted by Kepler's first law. It also explains Kepler's third law for circular orbits.
5. $v_t = 1.02 \times 10^3$ m/s; $T = 2.37 \times 10^6$ s $= 27.4$ days
6. apparent weightlessness; The normal force is zero because the person is in free fall.
7. You would see the penny float in front of you.

SECTION 4

General Level

Visual Strategy GENERAL

Figure 14

Point out that the pins were initially at rest and were set in motion by the bowling ball.

Q Is the energy gained by the pins greater than, equal to, or less than the energy lost by the bowling ball?

A ***The energy gained by the pins is equal to the energy lost by the bowling ball (except for a small loss of energy in the form of sound and a slight temperature increase in the pins).***

Misconception Alert BASIC

Many students may think that any force acting on an object will produce a torque. You may want to demonstrate that if you push an object at its center of mass without rotating the object, you are applying a force to the object without producing a torque.

SECTION 4

Torque and Simple Machines

SECTION OBJECTIVES

- **Distinguish between torque and force.**
- **Calculate the magnitude of a torque on an object.**
- **Identify the six types of simple machines.**
- **Calculate the mechanical advantage of a simple machine.**

ROTATIONAL MOTION

Earlier in this chapter, you studied various examples of uniform circular motion, such as a spinning Ferris wheel or an orbiting satellite. During uniform circular motion, an object moves in a circular path and at constant speed. An object that is in circular motion is accelerating because the direction of the object's velocity is constantly changing. This centripetal acceleration is directed toward the center of the circle. The net force causing the acceleration is a centripetal force, which is also directed toward the center of the circle.

In this section, we will examine a related type of motion: the motion of a rotating rigid object. For example, consider a football that is spinning as it flies through the air. If gravity is the only force acting on the football, the football spins around a point called its *center of mass.* As the football moves through the air, its center of mass follows a parabolic path. Note that the center of mass is not always at the center of the object.

ADVANCED TOPICS

See "Rotation and Inertia" and "Rotational Dynamics" in **Appendix J: Advanced Topics** to learn more about rotational motion.

Rotational and translational motion can be separated

Imagine that you roll a strike while bowling. When the bowling ball strikes the pins, as shown in **Figure 14,** the pins spin in the air as they fly backward. Thus, they have both rotational and linear motion. These types of motion can be analyzed separately. In this section, we will isolate rotational motion. In particular, we will explore how to measure the ability of a force to rotate an object.

Figure 14
Pins that are spinning and flying through the air exhibit both rotational and translational motion.

THE MAGNITUDE OF A TORQUE

Imagine a cat trying to leave a house by pushing perpendicularly on a cat-flap door. **Figure 15** shows a cat-flap door hinged at the top. In this configuration, the door is free to rotate around a line that passes through the hinge. This is the door's *axis of rotation.* When the cat pushes at the bottom edge of the door with a force that is perpendicular to the door, the door opens. The ability of a force to rotate an object around some axis is measured by a quantity called **torque.**

torque

a quantity that measures the ability of a force to rotate an object around some axis

Torque depends on the force and the lever arm

If a cat pushed on the door with the same force but at a point closer to the hinge, the door would be more difficult to rotate. How easily an object rotates depends not only on how much force is applied but also on where the force is applied. The farther the force is from the axis of rotation, the easier it is to rotate the object and the more torque is produced. The perpendicular distance from the axis of rotation to a line drawn along the direction of the force is called the **lever arm.**

lever arm

the perpendicular distance from the axis of rotation to a line drawn along the direction of the force

Figure 16 shows a diagram of the force **F** applied by the pet perpendicular to the cat-flap door. If you examine the definition of *lever arm,* you will see that in this case the lever arm is the distance *d* shown in the figure, the distance from the pet's nose to the hinge. That is, *d* is the perpendicular distance from the axis of rotation to the line along which the applied force acts. If the pet pressed on the door at a higher point, the lever arm would be shorter. As a result, the cat would need to exert a greater force to apply the same torque.

Figure 15
The cat-flap door rotates on a hinge, allowing pets to enter and leave a house at will.

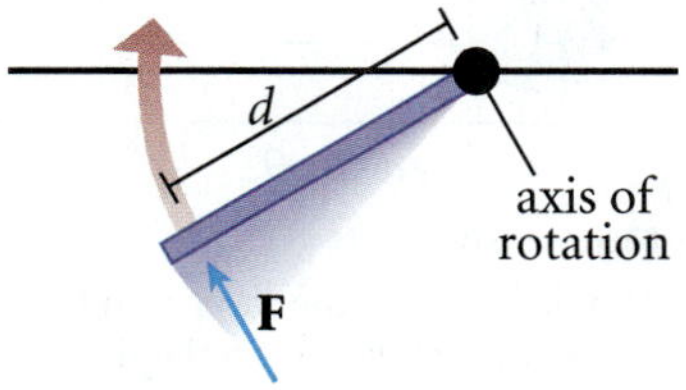

Figure 16
A force applied to an extended object can produce a torque. This torque, in turn, causes the object to rotate.

Quick Lab

Changing the Lever Arm

MATERIALS LIST

- door
- masking tape

In this activity, you will explore how the amount of force required to open a door changes when the lever arm changes. Using only perpendicular forces, open a door several times by applying a force at different distances from the hinge. You may have to tape the latch so that the door will open when you push without turning the knob. Because the angle of the applied force is kept constant, decreasing the distance to the hinge decreases the lever arm. Compare the relative effort required to open the door when pushing near the edge to that required when pushing near the hinged side of the door. Summarize your findings in terms of torque and the lever arm.

Quick Lab

TEACHER'S NOTES

Students should find that the smallest lever arm (closest to the hinge) requires the greatest force to produce the same torque.

Homework Options

This QuickLab can easily be performed outside of the physics lab room.

SECTION 4

Visual Strategy GENERAL

Figure 18

Be certain students recognize how to identify the force, the distance to the axis, the angle between the force and the axis, and the lever arm.

Q Mechanics often use "cheater bars" to loosen stubborn bolts. A cheater bar is usually a length of pipe that fits over the handle of a wrench, which makes the handle longer. How do cheater bars help the mechanic?

A ***Cheater bars increase the lever arm length while keeping the force the same, thus increasing the torque applied to a stubborn bolt.***

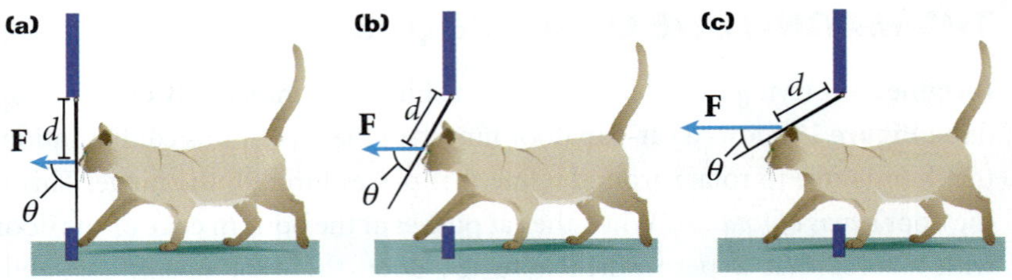

Figure 17
In each example, the cat is pushing on the door at the same distance from the axis. To produce the same torque, the cat must apply greater force for smaller angles.

The lever arm depends on the angle

Forces do not have to be perpendicular to an object to cause the object to rotate. Imagine the cat-flap door again. In **Figure 17(a),** the force exerted by the cat is perpendicular. When the angle is less than 90°, as in **(b)** and **(c),** the door will still rotate, but not as easily. The symbol for torque is the Greek letter *tau* (τ), and the magnitude of the torque is given by the following equation:

TORQUE

$$\tau = Fd \sin \theta$$

torque = force × lever arm

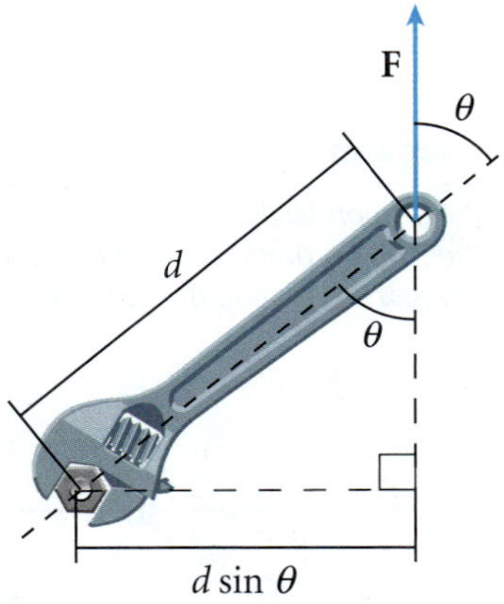

Figure 18
The direction of the lever arm is always perpendicular to the direction of the applied force.

The SI unit of torque is the N•m. Notice that the inclusion of the factor sin θ in this equation takes into account the changes in torque shown in **Figure 17.**

Figure 18 shows a wrench pivoted around a bolt. In this case, the applied force acts at an angle to the wrench. The quantity d is the distance from the axis of rotation to the point where force is applied. The quantity *d sin θ*, however, is the *perpendicular* distance from the axis of rotation to a line drawn along the direction of the force. *Thus, d sin θ is the lever arm.* Note that the perpendicular distance between the door hinge and the point of application of force **F** in **Figure 17** decreases as the cat goes further through the door.

THE SIGN OF A TORQUE

Torque, like displacement and force, is a vector quantity. In this textbook, we will assign each torque a positive or negative sign, depending on the direction the force tends to rotate an object. We will use the convention that the sign of the torque resulting from a force is positive if the rotation is counterclockwise and negative if the rotation is clockwise. In calculations, remember to assign positive and negative values to forces and displacements according to the sign convention established in the chapter "Motion in One Dimension."

TIP *To determine the sign of a torque, imagine that the torque is the only one acting on the object and that the object is free to rotate. Visualize the direction that the object would rotate. If more than one force is acting, treat each force separately. Be careful to associate the correct sign with each torque.*

For example, imagine that you are pulling on a wishbone with a perpendicular force F_1 and that a friend is pulling in the opposite direction with a force F_2. If you pull the wishbone so that it would rotate counterclockwise, then you exert a positive torque of magnitude F_1d_1. Your friend, on the other hand, exerts a negative torque, $-F_2d_2$. To find the net torque acting on the wishbone, simply add up the individual torques.

$$\tau_{net} = \Sigma\tau = \tau_1 + \tau_2 = F_1d_1 + (-F_2d_2)$$

When you properly apply the sign convention, the sign of the net torque will tell you which way the object will rotate, if at all.

SCILINKS.
www.scilinks.org
Topic: Torque
Code: HF61538

SAMPLE PROBLEM E

Torque

PROBLEM

A basketball is being pushed by two players during tip-off. One player exerts an upward force of 15 N at a perpendicular distance of 14 cm from the axis of rotation. The second player applies a downward force of 11 N at a perpendicular distance of 7.0 cm from the axis of rotation. Find the net torque acting on the ball about its center of mass.

SOLUTION

1. DEFINE **Given:** $F_1 = 15$ N $\quad F_2 = 11$ N

$d_1 = 0.14$ m $\quad d_2 = 0.070$ m

Unknown: $\tau_{net} = ?$

Diagram:

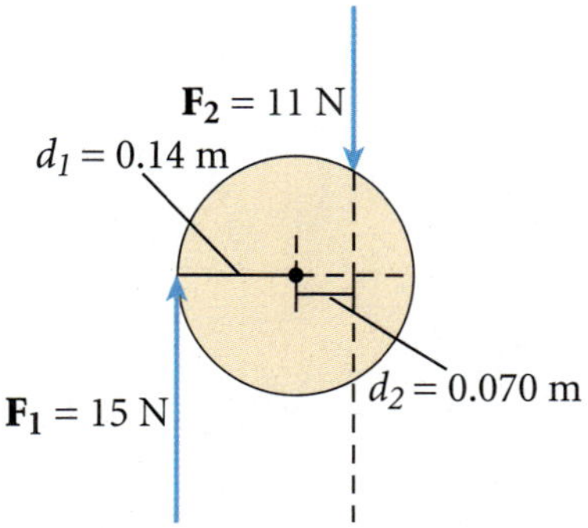

2. PLAN **Choose an equation or situation:** Apply the definition of torque to each force, and add up the individual torques.

$$\tau = Fd$$

$$\tau_{net} = \tau_1 + \tau_2 = F_1d_1 + F_2d_2$$

TIP *The factor sin θ is not included because each given distance is the perpendicular distance from the axis of rotation to a line drawn along the direction of the force.*

3. CALCULATE **Substitute the values into the equations and solve:** First, determine the torque produced by each force. Use the standard convention for signs.

$$\tau_1 = F_1d_1 = (15 \text{ N})(-0.14 \text{ m}) = -2.1 \text{ N}\bullet\text{m}$$

$$\tau_2 = F_2d_2 = (-11 \text{ N})(0.070 \text{ m}) = -0.77 \text{ N}\bullet\text{m}$$

$$\tau_{net} = -2.1 \text{ N}\bullet\text{m} - 0.77 \text{ N}\bullet\text{m}$$

$$\boxed{\tau_{net} = -2.9 \text{ N}\bullet\text{m}}$$

4. EVALUATE The net torque is negative, so the ball rotates in a clockwise direction.

Classroom Practice

Torque
A student pushes with a minimum force of 50.0 N on the middle of a door to open it.
a. What minimum force must be applied at the edge of the door in order for the door to open?
b. What minimum force must be applied to the hinged side of the door in order for the door to open?

Answer
a. 25.0 N
b. The door cannot be opened by a force at the hinge location. It can be *broken* but cannot be opened normally.

PROBLEM GUIDE E

Use this guide to assign problems.
SE = Student Edition Textbook
PW = Problem Workbook
PB = Problem Bank on the One-Stop Planner (OSP)

Solving for:

τ	**SE**	Sample, 1–2; Ch. Rvw. 37–38, 43, 44*, 51a, 54
	PW	6, 7a
	PB	4–6
d	**PW**	Sample, 1–2, 3*
	PB	7–10
F	**SE**	3; Ch. Rvw. 41–42, 51b, 54
	PW	4*, 5*, 7b
	PB	Sample, 1–3

***Challenging Problem**
Consult the printed Solutions Manual or the OSP for detailed solutions.

SECTION 4

ANSWERS

Practice E

1. 0.75 N•m
2. a. 5.1 N•m
   b. 15 N•m
3. 133 N

Teaching Tip — GENERAL

Another way to calculate torque is to use the full distance from the pivot point to the applied force and to use only the perpendicular component of the applied force. The magnitude of this "effective force" is equal to $F \sin \theta$. The resulting torque will be equal to the result that would be obtained using the applied force and lever arm.

Visual Strategy — BASIC

Figure 19

Point out that the reason the bottle opener is so useful is that the length of the handle is longer than the distance between the rim, which pries the bottle cap up, and the cap itself. The amount of force needed to produce a large torque on the cap is small because the lever arm is long.

Q Would the bottle opener still provide a mechanical advantage if the "prying arm" were equal in length to the handle?

A ***No, if both the "prying arm" and the handle were the same length, the force input would equal the force output, and the mechanical advantage would equal 1.***

PRACTICE E

Torque

1. Find the magnitude of the torque produced by a 3.0 N force applied to a door at a perpendicular distance of 0.25 m from the hinge.
2. A simple pendulum consists of a 3.0 kg point mass hanging at the end of a 2.0 m long light string that is connected to a pivot point.
   a. Calculate the magnitude of the torque (due to gravitational force) around this pivot point when the string makes a 5.0° angle with the vertical.
   b. Repeat this calculation for an angle of 15.0°.
3. If the torque required to loosen a nut on the wheel of a car has a magnitude of 40.0 N•m, what *minimum* force must be exerted by a mechanic at the end of a 30.0 cm wrench to loosen the nut?

TYPES OF SIMPLE MACHINES

What do you do when you need to pry a cap off a bottle of soda? You probably use a bottle opener, as shown in **Figure 19.** Similarly, you would probably use scissors to cut paper or a hammer to drive a nail into a board. All of these devices make your task easier. These devices are all examples of *machines.*

The term *machine* may bring to mind intricate systems with multicolored wires and complex gear-and-pulley systems. Compared with internal-combustion engines or airplanes, simple devices such as hammers, scissors, and bottle openers may not seem like machines, but they are.

A machine is any device that transmits or modifies force, usually by changing the force applied to an object. All machines are combinations or modifications of six fundamental types of machines, called *simple machines.* These six simple machines are the lever, pulley, inclined plane, wheel and axle, wedge, and screw, as shown in **Table 2** on the next page.

Figure 19
Because this bottle opener makes work easier, it is an example of a machine.

Using simple machines

Because the purpose of a simple machine is to change the direction or magnitude of an input force, a useful way of characterizing a simple machine is to compare how large the output force is relative to the input force. This ratio, called the machine's *mechanical advantage,* is written as follows:

$$MA = \frac{\text{output force}}{\text{input force}} = \frac{F_{out}}{F_{in}}$$

Table 2 Six Simple Machines

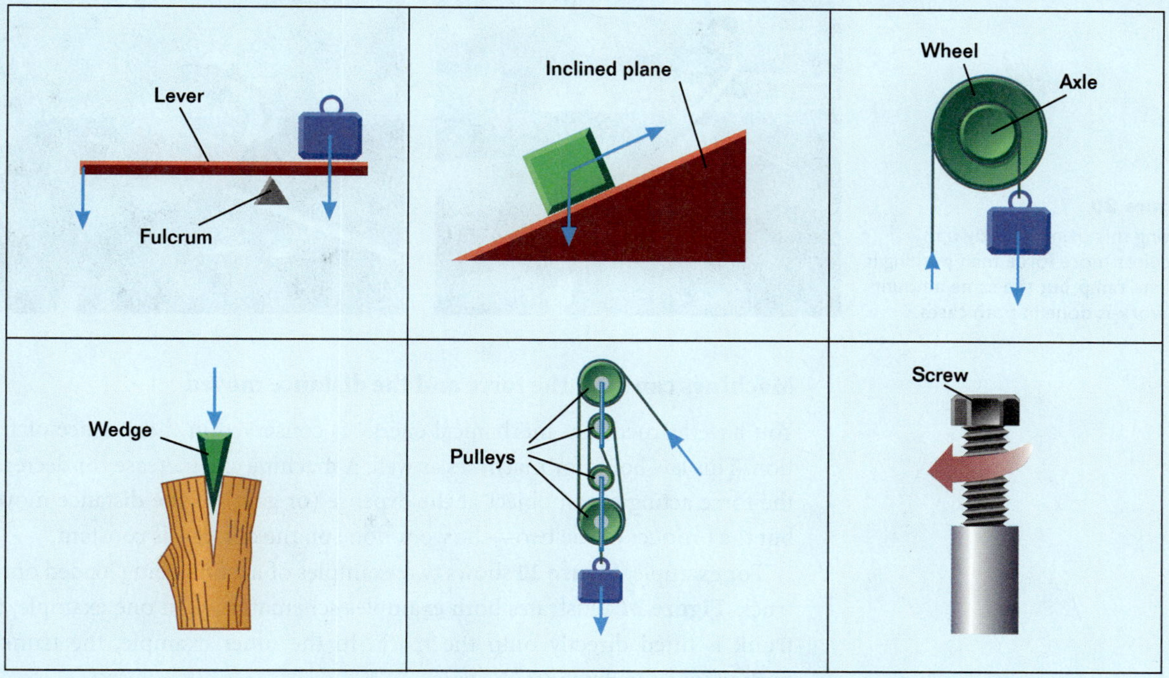

One example of mechanical advantage is the use of the back of a hammer to pry a nail from a board. In this example, the hammer is a type of lever. A person applies an input force to one end of the handle. The handle, in turn, exerts an output force on the head of a nail stuck in a board. If friction is disregarded, the input torque will equal the output torque. This relation can be written as follows:

$$\tau_{in} = \tau_{out}$$
$$F_{in}d_{in} = F_{out}d_{out}$$

Substituting this expression into the definition of mechanical advantage gives the following result:

$$MA = \frac{F_{out}}{F_{in}} = \frac{d_{in}}{d_{out}}$$

TIP *This equation can be used to predict the output force for a given input force if there is no friction. The equation is not valid if friction is taken into account. With friction, the output force will be less than expected, and thus* $\frac{d_{in}}{d_{out}}$ *will not equal* $\frac{F_{out}}{F_{in}}$.

The longer the input lever arm as compared with the output lever arm, the greater the mechanical advantage is. This in turn indicates the factor by which the input force is amplified. If the force of the board on the nail is 99 N and if the mechanical advantage is 10, then an input force of 10 N is enough to pull out the nail. Without a machine, the nail could not be removed unless the input force was greater than 99 N.

Teaching Tip — GENERAL

Building acrostics may help students remember lists, such as the six simple machines. Students should be encouraged to create their own acrostics, because the students will be more likely to recall words of their own choosing. You may want to give them an example like the following: "**L**et **P**eople **Wa**nder **W**hen the **P**arty **S**tarts," for **L**ever, inclined **P**lane, **W**heel and **A**xle, **W**edge, **P**ulley, and **S**crew.

The Language of Physics

The *ideal mechanical advantage (IMA)* refers to the mechanical advantage that would exist if there were no friction. Thus, $IMA = \frac{d_{in}}{d_{out}}$. The *actual mechanical advantage (AMA)* takes friction into account. Because there is always some friction, the ideal mechanical advantage is always greater than the actual mechanical advantage. If F_{out} is the measured output force, then the ratio $\frac{F_{out}}{F_{in}}$ will give the actual mechanical advantage, and $\frac{F_{out}}{F_{in}}$ will not equal $\frac{d_{in}}{d_{out}}$. On the other hand, if the equality on this page is used to calculate F_{out}, then F_{out} is the *predicted* output force in an ideal situation with no friction.

Misconception Alert — BASIC

Reinforce the idea that machines do not create something from nothing. If friction is disregarded, machines use the same amount of energy to achieve the goal. Use numerical examples to illustrate that the work done on the objects is the same. Show the trade-off between force and distance.

Teaching Tip

If you have discussed the difference between ideal and actual mechanical advantage with students (see The Language of Physics on the previous page), point out that efficiency can also be calculated with the ratio $\frac{AMA}{IMA}$ because this ratio is equivalent to the ratio $\frac{W_{out}}{W_{in}}$.

Figure 20
Lifting this trunk directly up requires more force than pushing it up the ramp, but the same amount of work is done in both cases.

Machines can alter the force and the distance moved

You have learned that mechanical energy is conserved in the absence of friction. This law holds for machines as well. A machine can increase (or decrease) the force acting on an object at the expense (or gain) of the distance moved, but the product of the two—the work done on the object—is constant.

For example, **Figure 20** shows two examples of a trunk being loaded onto a truck. **Figure 21** illustrates both examples schematically. In one example, the trunk is lifted directly onto the truck. In the other example, the trunk is pushed up an incline into the truck.

In the first example, a force ($\mathbf{F_1}$) of 360 N is required to lift the trunk, which moves through a distance (d_1) of 1.0 m. This requires 360 N•m of work (360 N × 1 m). In the second example, a lesser force ($\mathbf{F_2}$) of only 120 N would be needed (ignoring friction), but the trunk must be pushed a greater distance (d_2) of 3.0 m. This also requires 360 N•m of work (120 N × 3 m). As a result, the two methods require the same amount of energy.

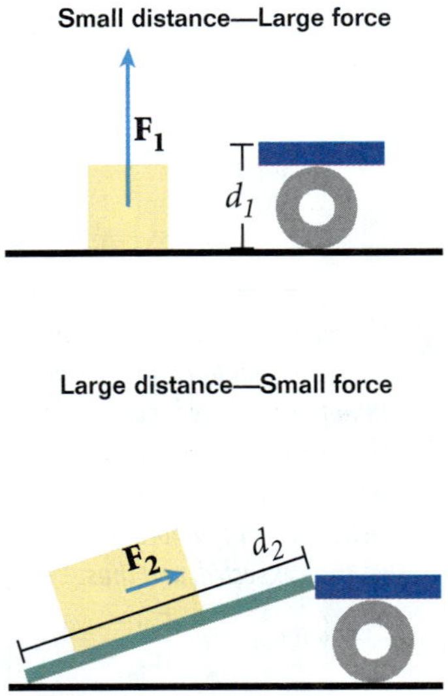

Figure 21
Simple machines can alter both the force needed to perform a task and the distance through which the force acts.

Efficiency is a measure of how well a machine works

The simple machines we have considered so far are ideal, frictionless machines. Real machines, however, are not frictionless. They dissipate energy. When the parts of a machine move and contact other objects, some of the input energy is dissipated as sound or heat. The *efficiency* of a machine is the ratio of useful work output to work input. It is defined by the following equation:

$$eff = \frac{W_{out}}{W_{in}}$$

If a machine is frictionless, then mechanical energy is conserved. This means that the work done on the machine (input work) is equal to the work done by the machine (output work) because work is a measure of energy transfer. Thus, the mechanical efficiency of an ideal machine is 1, or 100 percent. This is the best efficiency a machine can have. Because all real machines have at least a little friction, the efficiency of real machines is always less than 1.

SECTION REVIEW

1. Determine whether each of the following situations involves linear motion, rotational motion, or a combination of the two.

a. a baseball dropped from the roof of a house
b. a baseball rolling toward third base
c. a pinwheel in the wind
d. a door swinging open

2. What quantity describes the ability of a force to rotate an object? How does it differ from a force? On what quantities does it depend?

3. How would the force needed to open a door change if you put the handle in the middle of the door?

4. What are three ways that a cat pushing on a cat-flap door can change the amount of torque applied to the door?

5. The efficiency of a squeaky pulley system is 73 percent. The pulleys are used to raise a mass to a certain height. What force is exerted on the machine if a rope is pulled 18.0 m in order to raise a 58 kg mass a height of 3.0 m?

6. A person lifts a 950 N box by pushing it up an incline. If the person exerts a force of 350 N along the incline, what is the mechanical advantage of the incline?

7. You are attempting to move a large rock by using a long lever. Will the work you do on the lever be greater than, the same as, or less than the work done by the lever on the rock? Explain.

8. Interpreting Graphics Calculate the torque for each force acting on the bar in **Figure 22.** Assume the axis is perpendicular to the page and passes through point *O*. In what direction will the object rotate?

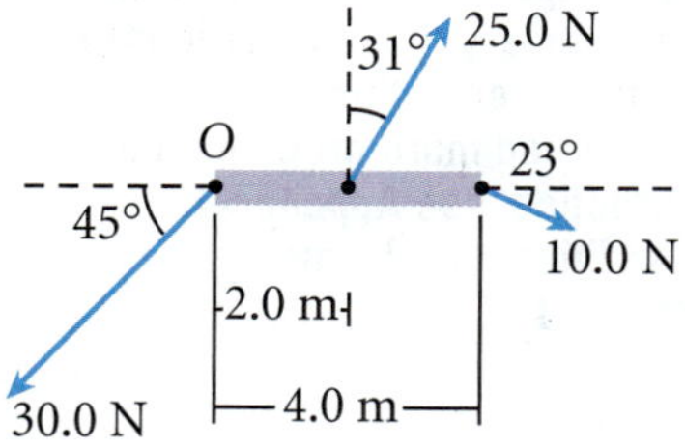

Figure 22

9. Interpreting Graphics **Figure 23** shows an example of a Rube Goldberg machine. Identify two types of simple machines that are included in this compound machine.

10. Critical Thinking A bicycle can be described as a combination of simple machines. Identify two types of simple machines that are used to propel a typical bicycle.

Figure 23

SECTION 4

SECTION REVIEW ANSWERS

1. a. linear motion
b. linear and rotational motion
c. rotational motion
d. rotational motion

2. torque; It is measured in units of N•m, not N. It is a measure of the ability of a force to accelerate an object around an axis; It depends on the force and the lever arm.

3. Twice as much force would be needed to open the door.

4. by changing the point at which the force is applied, the angle at which the force is applied, or the magnitude of the applied force

5. 130 N

6. 2.7

7. Ideally, the amounts of work will be the same, but because no machine is perfectly efficient, the work you do will be greater.

8. $\tau_{30} = 0$ N•m, $\tau_{25} = +43$ N•m, $\tau_{10} = -16$ N•m;
The bar will rotate counterclockwise.

9. Possible answers include lever, wedge, and pulley.

10. the wheels (wheel and axle) and the gear system (pulley)

CHAPTER 7

Highlights

Teaching Tip

This chapter contains several difficult concepts and equations. Students may find it helpful to compile a table of equations with short descriptions of the quantities involved. Students' tables may contain vocabulary terms or other integral terms or concepts. The tables will prove useful to students solving problems on their own.

extension

In-Depth Physics Content

Your students can visit go.hrw.com for online chapters that integrate more in-depth development of the concepts covered here.

Keyword HF6CMGX

CHAPTER 7

Highlights

KEY TERMS

centripetal acceleration (p. 235)

gravitational force (p. 240)

torque (p. 255)

lever arm (p. 255)

KEY IDEAS

Section 1 Circular Motion

- An object that revolves about a single axis undergoes circular motion.
- An object in circular motion has a centripetal acceleration and a centripetal force, which are both directed toward the center of the circular path.

Section 2 Newton's Law of Universal Gravitation

- Every particle in the universe is attracted to every other particle by a force that is directly proportional to the product of the particles' masses and inversely proportional to the square of the distance between the particles.
- Gravitational field strength is the gravitational force that would be exerted on a unit mass at any given point in space and is equal to free-fall acceleration.

Section 3 Motion in Space

- Kepler developed three laws of planetary motion.
- Both the period and speed of an object that is in a circular orbit around another object depend on two quantities: the mass of the central object and the distance between the centers of the objects.

Section 4 Torque and Simple Machines

- Torque is a measure of a force's ability to rotate an object.
- The torque on an object depends on the magnitude of the applied force and on the lever arm.
- Simple machines provide a mechanical advantage.

PROBLEM SOLVING

See **Appendix D: Equations** for a summary of the equations introduced in this chapter. If you need more problem-solving practice, see **Appendix I: Additional Problems.**

Variable Symbols

Quantities		Units	
v_t	tangential speed	m/s	meters/second
a_c	centripetal acceleration	m/s^2	meters/second^2
F_c	centripetal force	N	newtons
F_g	gravitational force	N	newtons
g	gravitational field strength	N/kg	newtons/kilogram
T	orbital period	s	seconds
τ	torque	N•m	newton meter

CHAPTER 7 Review

CIRCULAR MOTION

Review Questions

1. When a solid wheel rotates about a fixed axis, do all of the points of the wheel have the same tangential speed?

2. Correct the following statement: The racing car rounds the turn at a constant velocity of 145 km/h.

3. Describe the path of a moving body whose acceleration is constant in magnitude at all times and is perpendicular to the velocity.

4. Give an example of a situation in which an automobile driver can have a centripetal acceleration but no tangential acceleration.

Conceptual Questions

5. The force exerted by a spring increases as the spring stretches. Imagine that you attach a heavy object to one end of a spring and then, while holding the spring's other end, whirl the spring and object in a horizontal circle. Does the spring stretch? Explain.

6. Can a car move around a circular racetrack so that the car has a tangential acceleration but no centripetal acceleration?

7. Why does mud fly off a rapidly turning wheel?

Practice Problems

For problems 8–9, see Sample Problem A.

8. A building superintendent twirls a set of keys in a circle at the end of a cord. If the keys have a centripetal acceleration of 145 m/s^2 and the cord has a length of 0.34 m, what is the tangential speed of the keys?

9. A sock stuck to the side of a clothes-dryer barrel has a centripetal acceleration of 28 m/s^2. If the dryer barrel has a radius of 27 cm, what is the tangential speed of the sock?

For problems 10–11, see Sample Problem B.

10. A roller-coaster car speeds down a hill past point *A* and then rolls up a hill past point *B*, as shown below.

a. The car has a speed of 20.0 m/s at point *A*. If the track exerts a normal force on the car of 2.06 $\times 10^4$ N at this point, what is the mass of the car? (Be sure to account for gravitational force.)

b. What is the maximum speed the car can have at point *B* for the gravitational force to hold it on the track?

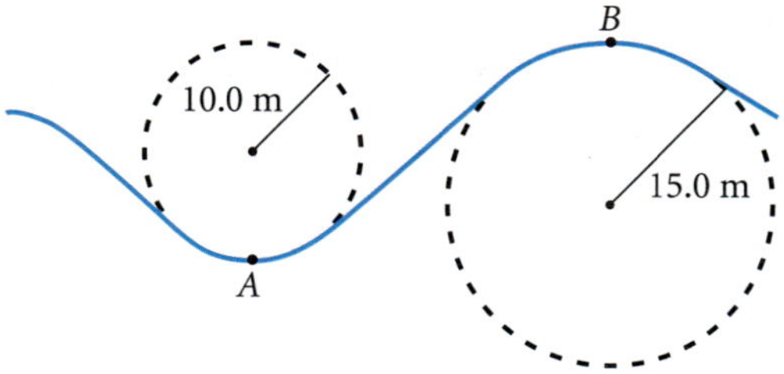

11. Tarzan tries to cross a river by swinging from one bank to the other on a vine that is 10.0 m long. His speed at the bottom of the swing is 8.0 m/s. Tarzan does not know that the vine has a breaking strength of 1.0×10^3 N. What is the largest mass that Tarzan can have and still make it safely across the river?

NEWTON'S LAW OF UNIVERSAL GRAVITATION

Review Questions

12. Identify the influence of mass and distance on gravitational forces.

13. If a satellite orbiting Earth is in free fall, why does the satellite not fall and crash into Earth?

14. How does the gravitational force exerted by Earth on the sun compare with the gravitational force exerted by the sun on Earth?

15. Describe two situations in which Newton's laws are not completely accurate.

CHAPTER 7

Review

ANSWERS

1. no

2. constant *speed*, not *velocity*

3. a circle

4. a car driving in a circle at a constant speed

5. yes; The object moves in a spiral path and because of inertia, the object's distance increases outward until the spring force is great enough to keep the object at a constant radius.

6. No, a_c is necessary for circular motion.

7. The mud's internal cohesion is not able to provide enough force to maintain the circular motion.

8. 7.0 m/s

9. 2.7 m/s

10. a. 414 kg
b. 12.1 m/s

11. 62 kg

12. Gravitational forces are directly related to mass and inversely related to distance squared.

13. The satellite is also moving parallel to Earth and at such a speed that Earth's surface curves away from the orbit at the same rate that the satellite falls.

14. The forces are equal in magnitude and opposite in direction, in accordance with Newton's third law.

15. Newton's laws do not give completely accurate results for objects moving near the speed of light, or for objects so small that quantum effects become significant.

7 REVIEW

16. at the equator; The moon's orbit around Earth is closer to the equator than to the North Pole.

17. You could use the law of universal gravitation $\left(m_E = \frac{gr^2}{G}\right)$.

18. 2.50 m

19. 1.0×10^{-10} m (0.10 nm)

20. In both models, planets orbit the sun. In Kepler's model, the orbits of planets are elliptical, while in Copernicus's model, the orbits are circular.

21. Kepler's laws can be derived from Newton's laws, and Kepler's laws closely match observations.

22. zero

23. The astronauts, the shuttle, and objects in the shuttle are all falling with the same free-fall acceleration.

24. Neither one: they have the same orbital speed.

25. twice as big

26. Strong gravitational fields are present in orbits near planets. Far from any large celestial bodies, the gravitational field may be so small that the force due to gravity is negligible.

27. $v_t = 1630$ m/s; $T = 5.78 \times 10^5$ s

28. 3.58×10^7 m

29. Jupiter ($m = 1.9 \times 10^{27}$ kg)

30. The long-handled screwdriver provides a longer lever arm and greater mechanical advantage.

31. The advantage is in multiplying force applied or changing the direction of the applied force.

32. the perpendicular distance from the axis of rotation to a line in the direction of the force (the lever arm)

Conceptual Questions

16. Would you expect tides to be higher at the equator or at the North Pole? Why?

17. Given Earth's radius, how could you use the value of G to calculate Earth's mass?

Practice Problems

For problems 18–19, see Sample Problem C.

18. The gravitational force of attraction between two students sitting at their desks in physics class is 3.20×10^{-8} N. If one student has a mass of 50.0 kg and the other has a mass of 60.0 kg, how far apart are the students sitting?

19. If the gravitational force between the electron (9.11×10^{-31} kg) and the proton (1.67×10^{-27} kg) in a hydrogen atom is 1.0×10^{-47} N, how far apart are the two particles?

MOTION IN SPACE

Review Questions

20. Compare and contrast Kepler's model of the solar system with Copernicus's model.

21. How do Kepler's laws help support Newton's theory of gravitation?

22. You are standing on a scale in an elevator. For a brief time, the elevator descends with free-fall acceleration. What does the scale show your weight to be during that time interval?

23. Astronauts floating around inside the space shuttle are not actually in a zero-gravity environment. What is the real reason astronauts seem weightless?

Conceptual Questions

24. A tiny alien spaceship (m = 0.25 kg) and the *International Space Station* are both orbiting Earth in circular orbits and at the same distance from Earth. Which one has a greater orbital speed?

25. The planet shown below sweeps out Area 1 in half the time that the planet sweeps out Area 2. How much bigger is Area 2 than Area 1?

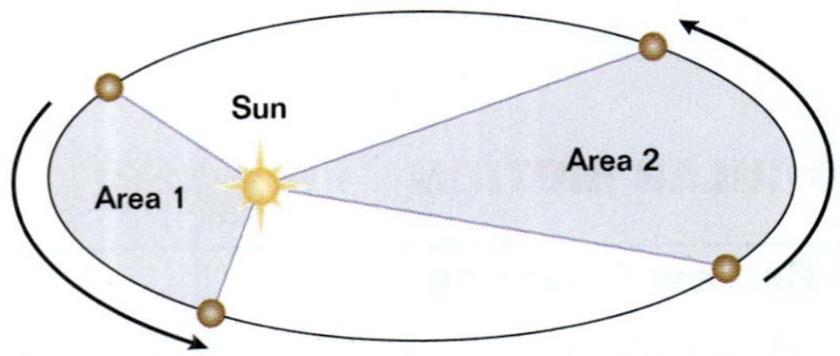

26. Comment on the statement, "There is no gravity in outer space."

Practice Problems

For problems 27–29, see Sample Problem D.

27. What would be the orbital speed and period of a satellite in orbit 1.44×10^8 m above Earth?

28. A satellite with an orbital period of exactly 24.0 h is always positioned over the same spot on Earth. This is known as a *geosynchronous* orbit. Television, communication, and weather satellites use geosynchronous orbits. At what distance would a satellite have to orbit Earth in order to have a geosynchronous orbit?

29. The distance between the centers of a small moon and a planet in our solar system is 2.0×10^8 m. If the moon's orbital period is 5.0×10^4 s, what is the planet? (See **Table 1** of the chapter for planet masses.)

TORQUE AND SIMPLE MACHINES

Review Questions

30. Why is it easier to loosen the lid from the top of a paint can with a long-handled screwdriver than with a short-handled screwdriver?

31. If a machine cannot multiply the amount of work, what is the advantage of using such a machine?

32. In the equation for the magnitude of a torque, what does the quantity $d \sin \theta$ represent?

Conceptual Questions

33. Which of the forces acting on the rod shown below will produce a torque about the axis at the left end of the rod?

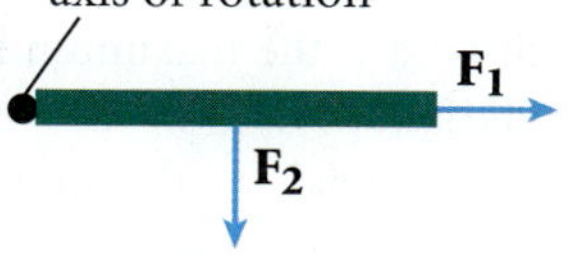

34. Two forces equal in magnitude but opposite in direction act at the same point on an object. Is it possible for there to be a net torque on the object? Explain.

35. You are attempting to move a large rock by using a long lever. Is it more effective to place the lever's axis of rotation nearer to your hands or nearer to the rock? Explain.

36. A perpetual motion machine is a machine that, when set in motion, will never come to a halt. Why is such a machine not possible?

Practice Problems

For problems 37–38, see Sample Problem E.

37. A bucket filled with water has a mass of 54 kg and is hanging from a rope that is wound around a 0.050 m radius stationary cylinder. If the cylinder does not rotate and the bucket hangs straight down, what is the magnitude of the torque the bucket produces around the center of the cylinder?

38. A mechanic jacks up the front of a car to an angle of 8.0° with the horizontal in order to change the front tires. The car is 3.05 m long and has a mass of 1130 kg. Gravitational force acts at the center of mass, which is located 1.12 m from the front end. The rear wheels are 0.40 m from the back end. Calculate the magnitude of the torque exerted by the jack.

MIXED REVIEW

39. A 2.00×10^3 kg car rounds a circular turn of radius 20.0 m. If the road is flat and the coefficient of static friction between the tires and the road is 0.70, how fast can the car go without skidding?

40. During a solar eclipse, the moon, Earth, and sun lie on the same line, with the moon between Earth and the sun. What force is exerted on

a. the moon by the sun?
b. the moon by Earth?
c. Earth by the sun?

(See the table in Appendix F for data on the sun, moon, and Earth.)

41. A wooden bucket filled with water has a mass of 75 kg and is attached to a rope that is wound around a cylinder with a radius of 0.075 m. A crank with a turning radius of 0.25 m is attached to the end of the cylinder. What minimum force directed perpendicularly to the crank handle is required to raise the bucket?

42. If the torque required to loosen a nut that holds a wheel on a car has a magnitude of 58 N•m, what force must be exerted at the end of a 0.35 m lug wrench to loosen the nut when the angle is 56°? (Hint: See **Figure 18** for an example, and assume that θ is 56°.)

43. In a canyon between two mountains, a spherical boulder with a radius of 1.4 m is just set in motion by a force of 1600 N. The force is applied at an angle of 53.5° measured with respect to the vertical radius of the boulder. What is the magnitude of the torque on the boulder?

44. The hands of the clock in the famous Parliament Clock Tower in London are 2.7 m and 4.5 m long and have masses of 60.0 kg and 100.0 kg, respectively. Calculate the torque around the center of the clock due to the weight of these hands at 5:20. The weight of each hand acts at the center of mass (the midpoint of the hand).

45. The efficiency of a pulley system is 64 percent. The pulleys are used to raise a mass of 78 kg to a height of 4.0 m. What force is exerted on the rope of the pulley system if the rope is pulled for 24 m in order to raise the mass to the required height?

46. A crate is pulled 2.0 m at constant velocity along a 15° incline. The coefficient of kinetic friction between the crate and the plane is 0.160. Calculate the efficiency of this procedure.

7 REVIEW

33. F_2
34. no; Because the forces act on the same point, the torques will cancel.
35. Placing the axis of rotation nearer to the rock will increase the mechanical advantage.
36. A perpetual motion machine would require 100 percent efficiency.
37. 26 N•m
38. 1.68×10^4 N•m
39. 12 m/s
40. **a.** 4.34×10^{20} N
b. 1.99×10^{20} N
c. 3.52×10^{22} N
41. 220 N
42. 2.0×10^2 N
43. 1800 N•m
44. –2200 N•m
45. 2.0×10^2 N
46. 63%

47. 72%
48. 6.4 m
49. **a.** 2.25 days
b. 1.60×10^4 m/s
50. **a.** 0.965 m/s^2
b. 1.33×10^3 N
c. 0.0985
51. **a.** 6300 N•m
b. 550 N
52. 6.96×10^8 m
53. 6620 N; no ($F_c = 7880$ N)

47. A pulley system is used to lift a piano 3.0 m. If a force of 2200 N is applied to the rope as the rope is pulled in 14 m, what is the efficiency of the machine? Assume the mass of the piano is 750 kg.

48. A pulley system has an efficiency of 87.5 percent. How much of the rope must be pulled in if a force of 648 N is needed to lift a 150 kg desk 2.46 m? (Disregard friction.)

49. Jupiter's four large moons—Io, Europa, Ganymede, and Callisto—were discovered by Galileo in 1610. Jupiter also has dozens of smaller moons. Jupiter's rocky, volcanically-active moon Io is about the size of Earth's moon. Io has radius of about 1.82×10^6 m, and the mean distance between Io and Jupiter is 4.22×10^8 m.

a. If Io's orbit were circular, how many days would it take for Io to complete one full revolution around Jupiter?

b. If Io's orbit were circular, what would its orbital speed be?

50. A 13 500 N car traveling at 50.0 km/h rounds a curve of radius 2.00×10^2 m. Find the following:

a. the centripetal acceleration of the car

b. the centripetal force

c. the minimum coefficient of static friction between the tires and the road that will allow the car to round the curve safely

51. The arm of a crane at a construction site is 15.0 m long, and it makes an angle of 20.0° with the horizontal. Assume that the maximum load the crane can handle is limited by the amount of torque the load produces around the base of the arm.

a. What is the magnitude of the maximum torque the crane can withstand if the maximum load the crane can handle is 450 N?

b. What is the maximum load for this crane at an angle of 40.0° with the horizontal?

52. At the sun's surface, the gravitational force between the sun and a 5.00 kg mass of hot gas has a magnitude of 1370 N. Assuming that the sun is spherical, what is the sun's mean radius?

53. An automobile with a tangential speed of 55.0 km/h follows a circular road that has a radius of 40.0 m. The automobile has a mass of 1350 kg. The pavement is wet and oily, so the coefficient of kinetic friction between the car's tires and the pavement is only 0.500. How large is the available frictional force? Is this frictional force large enough to maintain the automobile's circular motion?

extension

Graphing Calculator Practice

Visit go.hrw.com for answers to this Graphing Calculator activity.

Keyword HF6CMGXT

Graphing Calculator Practice

Torque

One of the terms introduced in this chapter is *torque.* Torque is a measure of the ability of a force to rotate an object around an axis. As you learned earlier in this chapter, torque is described by the following equation:

$$\tau = Fd \sin \theta$$

In this equation, F is the applied force, d is the distance from the axis of rotation, and θ is the angle at which the force is applied. A mechanic using a long wrench to loosen a "frozen" bolt is a common illustration of this equation.

In this graphing calculator activity, you will determine how torque relates to the angle of the applied force and to the distance of application.

Visit go.hrw.com and type in the keyword **HF6CMGX** to find this graphing calculator activity. Refer to **Appendix B** for instructions on downloading the program for this activity.

54. A force is applied to a door at an angle of 60.0° and 0.35 m from the hinge. The force exerts a torque with a magnitude of 2.0 N•m. What is the magnitude of the force? How large is the maximum torque this force can exert?

55. Imagine a balance with unequal arms. An earring placed in the left basket was balanced by 5.00 g of standard masses on the right. When placed in the right basket, the same earring required 15.00 g on the left to balance. Which was the longer arm? Do you need to know the exact length of each arm to determine the mass of the earring? Explain.

Alternative Assessment

1. Research the historical development of the concept of gravitational force. Find out how scientists' ideas about gravity have changed over time. Identify the contributions of different scientists, such as Galileo, Kepler, Newton, and Einstein. How did each scientist's work build on the work of earlier scientists? Analyze, review, and critique the different scientific explanations of gravity. Focus on each scientist's hypotheses and theories. What are their strengths? What are their weaknesses? What do scientists think about gravity now? Use scientific evidence and other information to support your answers. Write a report or prepare an oral presentation to share your conclusions.

2. In the reduced gravity of space, called *microgravity,* astronauts lose bone and muscle mass, even after a short time. These effects happen more gradually on Earth as people age. Scientists are studying this phenomenon so that they can find ways to counteract it, in space and on Earth. Such studies are essential for future plans that involve astronauts spending significant time on space stations, or for distant missions such as a trip to Mars. Research the causes of this phenomenon and possible methods of prevention, including NASA's current efforts to minimize bone density loss for astronauts on the International Space Station. Create a poster or brochure displaying the results of your research.

3. Research the life and scientific contributions of one of the astronomers discussed in the chapter: Claudius Ptolemy, Nicolaus Copernicus, Tycho Brahe, or Johannes Kepler. On a posterboard, create a visual timeline that summarizes key events in the atsronomer's life and work, including astronomical discoveries and other scientific advances or inventions. Add images to some of the events on the timeline. You may also want to include historical events on the timeline to provide context for the scientific works.

4. Describe exactly which measurements you would need to make in order to identify the torques at work during a ride on a specific bicycle. Your plans should include measurements you can make with equipment available to you. If others in the class analyzed different bicycle models, compare the models for efficiency and mechanical advantage.

5. Prepare a poster or a series of models of simple machines, explaining their use and how they work. Include a schematic diagram next to each sample or picture to identify the fulcrum, lever arm, and resistance. Add your own examples to the following list: nail clipper, wheelbarrow, can opener, nutcracker, electric drill, screwdriver, tweezers, and key in lock.

7 REVIEW

54. 6.6 N; 2.3 N•m
55. The right arm is longer; Lengths do not need to be known; $xr = 5R$ and $xR = 15r$, so $x^2 = 75$ and $x = 8.7$ g.

Alternative Assessment ANSWERS

1. Answers will vary. Newton built on the work of Kepler and Galileo to develop his law of universal gravitation. This law was considered to give a complete description of gravity until Einstein developed the general theory of relativity. Many of Einstein's predictions have been confirmed by observations, but experimenters are still trying to prove the existence of gravity waves and particles called *gravitons.*
2. The loss of Earth's gravity disrupts the process of bone maintenance, which causes calcium loss. NASA scientists are conducting experiments to learn how rapidly this occurs, what the effects are, and how it can be minimized or prevented.
3. Student timelines should visually depict several key events in the life of their chosen astronomer, along with notable scientific achievements.
4. Student plans should be safe and should involve measuring forces and lever arms.
5. Student diagrams should clearly identify force input and force output.

Standardized Test Prep

ANSWERS

1. C
2. H
3. C
4. G
5. D
6. J
7. D
8. F

Standardized Test Prep

MULTIPLE CHOICE

1. An object moves in a circle at a constant speed. Which of the following is *not* true of the object?

A. Its centripetal acceleration points toward the center of the circle.
B. Its tangential speed is constant.
C. Its velocity is constant.
D. A centripetal force acts on the object.

Use the passage below to answer questions 2–3.

A car traveling at 15 m/s on a flat surface turns in a circle with a radius of 25 m.

2. What is the centripetal acceleration of the car?

F. 2.4×10^{-2} m/s^2
G. 0.60 m/s^2
H. 9.0 m/s^2
J. zero

3. What is the most direct cause of the car's centripetal acceleration?

A. the torque on the steering wheel
B. the torque on the tires of the car
C. the force of friction between the tires and the road
D. the normal force between the tires and the road

4. Earth ($m = 5.97 \times 10^{24}$ kg) orbits the sun ($m = 1.99 \times 10^{30}$ kg) at a mean distance of 1.50×10^{11} m. What is the gravitational force of the sun on Earth? ($G = 6.673 \times 10^{-11}$ N•m^2/kg^2)

F. 5.29×10^{32} N
G. 3.52×10^{22} N
H. 5.90×10^{-2} N
J. 1.77×10^{-8} N

5. Which of the following is a correct interpretation of the expression $a_g = g = G\frac{m_E}{r^2}$?

A. Gravitational field strength changes with an object's distance from Earth.
B. Free-fall acceleration changes with an object's distance from Earth.
C. Free-fall acceleration is independent of the falling object's mass.
D. All of the above are correct interpretations.

6. What data do you need to calculate the orbital speed of a satellite?

F. mass of satellite, mass of planet, radius of orbit
G. mass of satellite, radius of planet, area of orbit
H. mass of satellite and radius of orbit only
J. mass of planet and radius of orbit only

7. Which of the following choices correctly describes the orbital relationship between Earth and the sun?

A. The sun orbits Earth in a perfect circle.
B. Earth orbits the sun in a perfect circle.
C. The sun orbits Earth in an ellipse, with Earth at one focus.
D. Earth orbits the sun in an ellipse, with the sun at one focus.

Use the diagram below to answer questions 8–9.

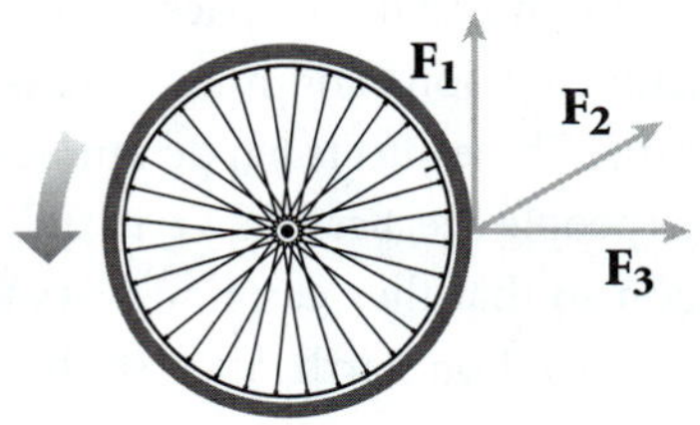

8. The three forces acting on the wheel above have equal magnitudes. Which force will produce the greatest torque on the wheel?

F. F_1
G. F_2
H. F_3
J. Each force will produce the same torque.

9. If each force is 6.0 N, the angle between $\mathbf{F_1}$ and $\mathbf{F_2}$ is 60.0°, and the radius of the wheel is 1.0 m, what is the resultant torque on the wheel?

A. −18 N•m
B. −9.0 N•m
C. 9.0 N•m
D. 18 N•m

10. A force of 75 N is applied to a lever. This force lifts a load weighing 225 N. What is the mechanical advantage of the lever?

F. $\frac{1}{3}$
G. 3
H. 150
J. 300

11. A pulley system has an efficiency of 87.5 percent. How much work must you do to lift a desk weighing 1320 N to a height of 1.50 m?

A. 1510 J
B. 1730 J
C. 1980 J
D. 2260 J

12. Which of the following statements is correct?

F. Mass and weight both vary with location.
G. Mass varies with location, but weight does not.
H. Weight varies with location, but mass does not.
J. Neither mass nor weight varies with location.

13. Which astronomer discovered that planets travel in elliptical rather than circular orbits?

A. Johannes Kepler
B. Nicolaus Copernicus
C. Tycho Brahe
D. Claudius Ptolemy

SHORT RESPONSE

14. Explain how it is possible for all the water to remain in a pail that is whirled in a vertical path, as shown below.

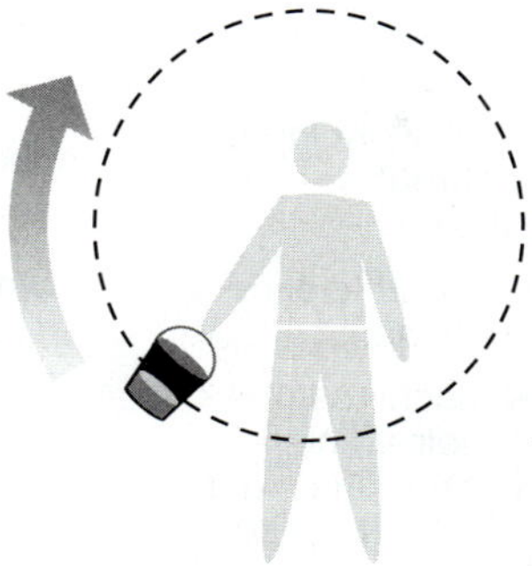

15. Explain why approximately two high tides take place every day at a given location on Earth.

16. If you used a machine to increase the output force, what factor would have to be sacrificed? Give an example.

EXTENDED RESPONSE

17. Mars orbits the sun ($m = 1.99 \times 10^{30}$ kg) at a mean distance of 2.28×10^{11} m. Calculate the length of the Martian year in Earth days. Show all of your work. ($G = 6.673 \times 10^{-11}$ N•m^2/kg^2)

Test TIP If you are solving a quantitative problem, start by writing down the relevant equation(s). Solve the equation(s) to find the variable you need for the answer, and then substitute the given data.

9. C
10. G
11. D
12. H
13. A
14. The water remains in the pail even when the pail is upside down because the water tends to move in a straight path due to inertia.
15. The moon's tidal forces create two bulges on Earth. As Earth rotates on its axis once per day, any given point on Earth passes through both bulges.
16. You would have to apply the input force over a greater distance. Examples may include any machines that increase output force at the expense of input distance.
17. 687 days (See the Solutions Manual or the One-Stop Planner for a full solution.)

Lab Planning

Beginning on page T34 are preparation notes and teaching tips to assist you in planning.

Blank data tables (as well as some sample data) appear on the **One-Stop Planner.**

No Books in the Lab?

See the ***Datasheets for In-Text Labs*** workbook for a reproducible master copy of this experiment.

The same workbook also contains a version of this experiment with explicit procedural steps if you prefer a more directed approach.

Safety Caution

Falling masses can cause injury. Goggles should be worn to shield eyes from hooked masses at eye level.

Tips and Tricks

- Encourage students to place a wooden block or textbook at the bottom of the incline to stop the cart.
- Use single and tandem open-faced pulley sets; multiple-pulley sets are difficult to thread and confusing for students.
- Demonstrate threading pulleys: Hang the upper set (A) from the stand, and hold the lower set (B). Connect B to A by threading the bottom A pulley to the top B pulley. Hang a mass on B to stabilize it, and release. Keep the cord tight and finish threading.

CHAPTER 7

Inquiry Lab

Machines and Efficiency

Design Your Own

OBJECTIVES

- **Measure** the work input and work output of several machines.
- **Calculate** the efficiency of each machine.
- **Compare** machines based on their efficiencies, and determine what factors affect efficiency.

MATERIALS LIST

- balance
- C-clamp
- cord
- dynamics cart
- inclined plane
- mass hanger
- meterstick
- pulleys, single and tandem
- set of hooked masses
- right-angle clamp
- support stand
- suspension clamp

In this lab, you will design an experiment to study the efficiency of two types of simple machines: inclined planes and pulleys. In your experiment, you should use each type of machine to use a smaller mass to lift a larger mass. With each setup, you should collect data that will allow you to calculate the work input and the work output of the system. The ratio of the useful work output to the work input is called the *efficiency* of a machine. By calculating efficiency, you will be able to compare the two types of machines.

SAFETY

- **Tie back long hair, secure loose clothing, and remove loose jewelry to prevent their getting caught in moving parts and pulleys. Put on goggles.**
- **Attach string to masses and objects securely. Falling or dropped masses can cause serious injury.**

PROCEDURE

1. Study the materials provided, and design an experiment to meet the goals stated above.

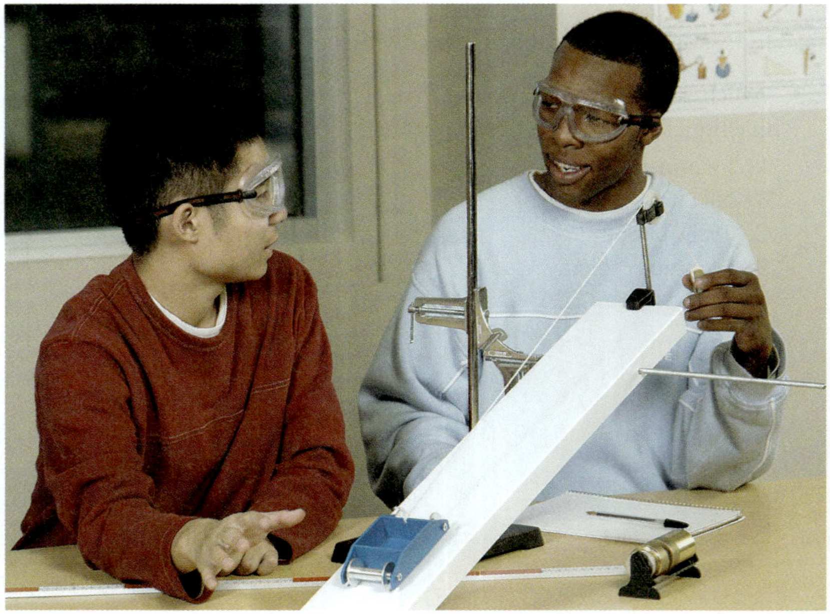

Figure 1
- Choose any angle, but make sure the top of the plane is at least 20 cm above the table.
- Make sure the string is long enough to help prevent the cart from falling off the top of the plane. Attach the mass hanger securely to the end of the string.

2. Write out your lab procedure, including a detailed description of the measurements to take during each step and the number of trials to perform. You may use **Figure 1** as a guide to one possible setup.

3. Ask your teacher to approve your procedure.

4. Follow all steps of your procedure.

5. Clean up your work area. Put equipment away safely so that it is ready to be used again.

ANALYSIS

1. **Organizing Data** For each trial, make the following calculations:
   a. the weight of the mass being raised
   b. the weight of the mass on the string
   c. the work input and the work output

2. **Analyzing Results** In which trial did a machine perform the most work? In which trial did a machine perform the least work?

3. **Organizing Data** Calculate the efficiency for each trial.

4. **Analyzing Results** Is the machine that performed the most work also the most efficient? Is the machine that performed the least work also the least efficient? What is the relationship between work and efficiency?

CONCLUSIONS

5. **Drawing Conclusions** Based on your calculations in item 4, which is more efficient, a pulley system or an inclined plane?

6. **Evaluating Methods** Why is it important to calculate the work input and the work output from measurements made when the object is moving with constant velocity?

EXTENSIONS

7. **Designing Experiments** Design an experiment to measure the efficiency of different lever setups. If there is time and your teacher approves, test your lever setups in the lab. How does the efficiency of a lever compare with the efficiency of the other types of machines you have studied?

8. **Building Models** Compare the trial with the highest efficiency and the trial with the lowest efficiency. Based on their differences, design a more efficient machine than any you built in the lab. If there is time and your teacher approves, test the machine to determine whether it is more efficient.

ANSWERS

Analysis

1. a. **Inclined Plane:** T1: 2.49 N, T2: 4.91 N, T3: 6.87 N; **Pulley:** T1: 5.56 N, T2: 5.62 N, T3: 5.56 N
   b. **Inclined Plane:** T1: 1.12 N, T2: 2.18 N, T3: 3.05 N; **Pulley:** T1: 1.22 N, T2: 1.34 N, T3: 1.50 N
   c. Student answers will vary. Make sure students use the relationship $W = Fd$. Typical values will be within these ranges: 0.271 J–0.774 J (inclined plane), 0.256 J–0.351 J (pulley).

2. The most work was done by the Trial 3 plane; the least work was done by the Trial 3 pulleys.

3. **Inclined Plane:** T1: 91.3%, T2: 93.3%, T3: 93.2%; **Pulley:** T1: 90.4%, T2: 85.0%, T3: 90.7%

4. Efficiency depends on both the work input and the work output.

Conclusions

5. inclined plane

6. Measurements should be made when there is no net external force doing work.

Extensions

7, 8. Student plans should be safe and complete and should include calculations of work input and work output.

CHAPTER 8

Fluid Mechanics
Planning Guide

Compression Guide
To shorten instruction because of time limitations, omit the opener, Advanced sections, and the review.

OBJECTIVES	LABS, DEMONSTRATIONS, AND ACTIVITIES	TECHNOLOGY RESOURCES
PACING • 45 min pp. 272–273 **Chapter Opener**		CD **Visual Concepts,** Chapter 8 BASIC
PACING • 45 min pp. 274–279 *Advanced Level* **Section 1 Fluids and Buoyant Force** • Define a fluid. • Distinguish a gas from a liquid. • Determine the magnitude of the buoyant force exerted on a floating object or a submerged object. • Explain why some objects float and some objects sink.	TE **Demonstration** Volume of Liquids and Gases, p. 274 GENERAL TE **Demonstration** Buoyant Force, p. 275 ADVANCED TE **Demonstration** Float an Egg, p. 277 GENERAL	OSP **Lesson Plans** EXT **Integrating Astronomy** Plasmas GENERAL EXT **Integrating Biology** How Fish Maintain Neutral Buoyancy GENERAL TR 31 Displaced Volume of a Fluid TR 32 Buoyant Force TR 27A Densities of Some Common Substances
PACING • 45 min pp. 280–283 *Advanced Level* **Section 2 Fluid Pressure** • Calculate the pressure exerted by a fluid. • Calculate how pressure varies with depth in a fluid.	TE **Demonstration** Defining Pressure, p. 280 GENERAL	OSP **Lesson Plans** EXT **Integrating Technology** Hydraulic Lift Force GENERAL
PACING • 45 min pp. 284–286 *Advanced Level* **Section 3 Fluids in Motion** • Examine the motion of a fluid using the continuity equation. • Recognize the effects of Bernoulli's principle on fluid motion.	TE **Demonstration** Fluid Flow Around a Table-Tennis Ball, p. 284 GENERAL TE **Demonstration** Fluid Flow Between Two Cans, p. 284 GENERAL	OSP **Lesson Plans** TR 33 The Continuity Equation and Bernoulli's Principle

PACING • 90 min

CHAPTER REVIEW, ASSESSMENT, AND STANDARDIZED TEST PREPARATION

SE **Chapter Highlights,** p. 287
SE **Chapter Review,** pp. 288–291
SE **Graphing Calculator Practice,** p. 291 GENERAL
SE **Alternative Assessment,** p. 291 ADVANCED
SE **Standardized Test Prep,** pp. 292–293 GENERAL
SE **Appendix D: Equations,** p. 858
SE **Appendix I: Additional Problems,** p. 887
ANC **Study Guide Worksheet** Mixed Review* GENERAL
ANC **Chapter Test A*** GENERAL
ANC **Chapter Test B*** ADVANCED
OSP **Test Generator**

Online and Technology Resources

go.hrw.com Holt Online Learning

Visit go.hrw.com to find a variety of online resources. To access this chapter's extensions, enter the keyword **HF6FLUXT** and click the "go" button. Click **Holt Online Learning** for an online edition of this textbook, and other interactive resources.

HOLT Teacher's One-Stop Planner®

This DVD package includes:
- Holt Calendar Planner
- Customizable Lesson Plans
- Editable Worksheets
- ExamView® Version 6 Assessment Suite
- Interactive Teacher's Edition
- Holt PuzzlePro®
- Holt PowerNotes® Presentations
- MindPoint® Quiz Show

KEY

SE	Student Edition	OSP	One-Stop Planner	EXT	Online Extension
TE	Teacher Edition	CD	CD or CD-ROM	*	Also on One-Stop Planner
ANC	Ancillary Worksheet	TR	Teaching Transparencies	◆	Requires advance prep

SKILLS DEVELOPMENT RESOURCES	REVIEW AND ASSESSMENT	CORRELATIONS
		National Science Education Standards
SE Sample Set A Buoyant Force, pp. 278–279 GENERAL **TE Classroom Practice,** p. 278 GENERAL **ANC Problem Workbook** Sample Set A* GENERAL **OSP Problem Bank** Sample Set A GENERAL **SE Appendix J: Advanced Topics** Properties of Gases, pp. 908–909 ADVANCED	**SE Section Review,** p. 279 GENERAL **ANC Study Guide Worksheet** Section 1* GENERAL **ANC Quiz** Section 1* GENERAL	UCP 1, 2, 3, 4, 5 ST 2 HNS 1, 3 PS 2e
SE Sample Set B Pressure, pp. 281–282 GENERAL **TE Classroom Practice,** p. 281 GENERAL **ANC Problem Workbook** Sample Set B* GENERAL **OSP Problem Bank** Sample Set B GENERAL **SE Appendix J: Advanced Topics** Fluid Pressure, pp. 910–911 ADVANCED **EXT Practice Problems** Pressure as a Function of Depth GENERAL	**SE Section Review,** p. 283 GENERAL **ANC Study Guide Worksheet** Section 2* GENERAL **ANC Quiz** Section 2* GENERAL	UCP 1, 2, 3, 4, 5 SAI 1, 2 ST 1, 2 SPSP 5
	SE Section Review, p. 286 GENERAL **ANC Study Guide Worksheet** Section 3* GENERAL **ANC Quiz** Section 3* GENERAL	UCP 1, 2, 3, 4, 5 SAI 1, 2 ST 1, 2 SPSP 1, 5

SCILINKS. NSTA
www.scilinks.org

Maintained by the **National Science Teachers Association.**

Topic: Archimedes
SciLinks Code: HF60093

Topic: Buoyancy
SciLinks Code: HF60201

Topic: Atmospheric Pressure
SciLinks Code: HF60114

Classroom CD-ROMs

- Guided Reading Audio Program
- Student One Stop
- Virtual Investigations
- Visual Concepts

Holt Lab Generator CD-ROM

Search for any lab by topic, standard, difficulty level, or time. Edit any lab to fit your needs, or create your own labs. Use the Lab Materials QuickList software to customize your lab materials list.

CHAPTER 8
Overview

Section 1 defines ideal fluids, calculates buoyant force, and explains why objects float or sink.

Section 2 calculates pressures transferred by a fluid in a hydraulic lift and explains how hydrostatic pressure varies with depth.

Section 3 introduces the equation of continuity and applies Bernoulli's equation to solve problems of fluids in motion.

About the Illustration

This kayaker is hurtling over Rainbow Falls, on the south fork of the Tuolumne River, in northern California. The Tuolumne is a favorite river among advanced kayakers and rafters, prized for its exciting rapids and waterfalls.

extension

In-Depth Physics Content

Your students can visit go.hrw.com for an online chapter that integrates more in-depth development of the concepts covered here.

Keyword HF6FLUX

CHAPTER 8

Fluid Mechanics

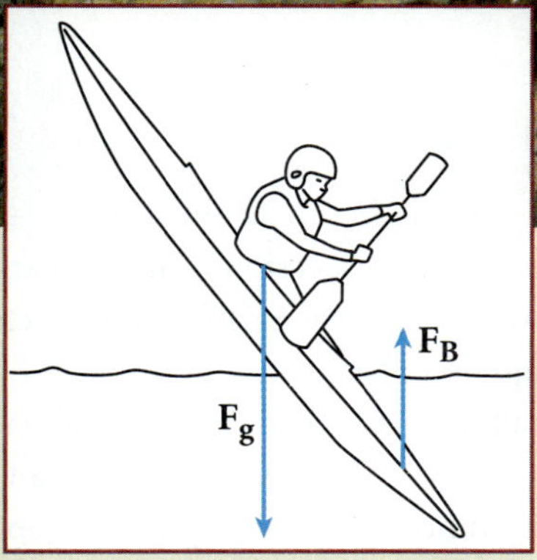

Kayakers know that if their weight (F_g) exceeds the upward, buoyant force (F_B) that causes them to float, they are sunk—literally! For an object, such as a kayak, that is immersed in a fluid, buoyant force equals the weight of the fluid that the object displaces. Buoyant force causes a kayak to pop to the surface after a plunge down a waterfall.

WHAT TO EXPECT

In this chapter, you will learn about buoyant force, fluid pressure, and the basic equations that govern the behavior of fluids. This chapter will also introduce moving fluids and the continuity equation.

Why it Matters

Many kinds of hydraulic devices, such as the brakes in a car and the lifts that move heavy equipment, make use of the properties of fluids. An understanding of the properties of fluids is needed to design such devices.

CHAPTER PREVIEW

Tapping Prior Knowledge

Knowledge to Review

- ✔ Forces can cause changes in an object's motion or in its shape.
- ✔ Energy can be kinetic energy or potential energy.
- ✔ In the absence of friction, the total mechanical energy of a system is constant. The total mass of a closed system is constant.

Items to Probe

- ✔ Operational understanding of the concepts of area and volume: Ask students to compare the volume of containers of different shapes (with approximately the same capacity).
- ✔ Ability to relate density, mass, and volume in a meaningful way: Ask students to calculate m, given ρ and V, using correct units.

273

Demonstration

Volume of Liquids and Gases — GENERAL

Purpose Demonstrate that, unlike gases, liquids have a definite volume and cannot be significantly compressed.

Materials large syringe, large eyedropper or turkey baster

Procedure Pull the piston of the syringe to about half the tube's length, letting air into the syringe. Then, hold your finger tightly over the opening and pull the piston back as far as possible. Read the volume of air inside the syringe. Without removing your finger, push the piston in as far as possible. Have a student record the air volume on the board for each case. Point out that no air has been let into or out of the syringe tube, but the same amount of air occupies all the space that is available. Next, fill the syringe halfway with only water, and repeat the demonstration. Let a student in class try to compress it.

Tightly cover the opening of the eyedropper and squeeze the bulb to show that the volume of air that equals that of the bulb and the pipe can easily be reduced. Fill the eyedropper halfway with water, and note the water level. Hold the end sealed and squeeze the bulb to demonstrate that the air volume can be reduced but the water level will remain the same.

SECTION 1

Fluids and Buoyant Force

SECTION OBJECTIVES

- Define a fluid.
- Distinguish a gas from a liquid.
- Determine the magnitude of the buoyant force exerted on a floating object or a submerged object.
- Explain why some objects float and some objects sink.

fluid

a nonsolid state of matter in which the atoms or molecules are free to move past each other, as in a gas or a liquid

extension

Integrating Astronomy
Visit go.hrw.com for the activity "Plasmas."

Keyword HF6FLUX

ADVANCED TOPICS

See "Properties of Gases" in **Appendix J: Advanced Topics** to learn more about how gases behave.

DEFINING A FLUID

Matter is normally classified as being in one of three states—solid, liquid, or gaseous. Up to this point, this book's discussion of motion and the causes of motion has dealt primarily with the behavior of solid objects. This chapter concerns the mechanics of liquids and gases.

Figure 1(a) is a photo of a liquid; **Figure 1(b)** shows an example of a gas. Pause for a moment and see if you can identify a common trait between them. One property they have in common is the ability to flow and to alter their shape in the process. Materials that exhibit these properties are called **fluids.** Solid objects are not considered to be fluids because they cannot flow and therefore have a definite shape.

Liquids have a definite volume; gases do not

Even though both gases and liquids are fluids, there is a difference between them: one has a definite volume, and the other does not. Liquids, like solids, have a definite volume, but unlike solids, they do not have a definite shape. Imagine filling the tank of a lawn mower with gasoline. The gasoline, a liquid, changes its shape from that of its original container to that of the tank. If there is a gallon of gasoline in the container before you pour, there will be a gallon in the tank after you pour. Gases, on the other hand, have neither a definite volume nor a definite shape. When a gas is poured from a smaller container into a larger one, the gas not only changes its shape to fit the new container but also spreads out and changes its volume within the container.

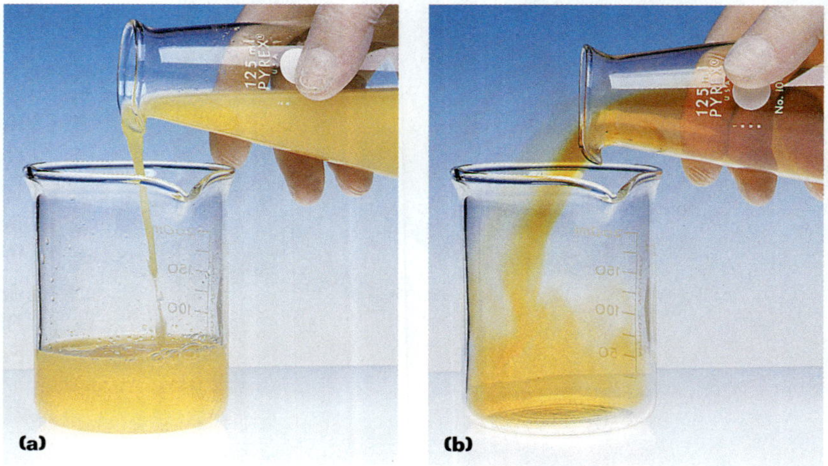

Figure 1
Both **(a)** liquids and **(b)** gases are considered fluids because they can flow and change shape.

DENSITY AND BUOYANT FORCE

Have you ever felt confined in a crowded elevator? You probably felt that way because there were too many people in the elevator for the amount of space available. In other words, the *density* of people was too high. In general, density is a measure of a quantity in a given space. The quantity can be anything from people or trees to mass or energy.

Mass density is mass per unit volume of a substance

When the word *density* is used to describe a fluid, what is really being measured is the fluid's **mass density.** Mass density is the mass per unit volume of a substance. It is often represented by the Greek letter ρ (*rho*).

MASS DENSITY

$$\rho = \frac{m}{V}$$

$$\text{mass density} = \frac{\textbf{mass}}{\textbf{volume}}$$

The SI unit of mass density is kilograms per cubic meter (kg/m^3). In this book we will follow the convention of using the word *density* to refer to *mass density.* **Table 1** lists the densities of some fluids and a few important solids.

Solids and liquids tend to be almost incompressible, meaning that their density changes very little with changes in pressure. Thus, the densities listed in **Table 1** for solids and liquids are approximately independent of pressure. Gases, on the other hand, are compressible and can have densities over a wide range of values. Thus, there is not a standard density for a gas, as there is for solids and liquids. The densities listed for gases in **Table 1** are the values of the density at a stated temperature and pressure. For deviations of temperature and pressure from these values, the density of the gas will vary significantly.

mass density

the concentration of matter of an object, measured as the mass per unit volume of a substance

Table 1
Densities of Some Common Substances*

Substance	ρ (kg/m^3)
Hydrogen	0.0899
Helium	0.179
Steam (100°C)	0.598
Air	1.29
Oxygen	1.43
Carbon dioxide	1.98
Ethanol	0.806×10^3
Ice	0.917×10^3
Fresh water (4°C)	1.00×10^3
Sea water (15°C)	1.025×10^3
Iron	7.86×10^3
Mercury	13.6×10^3
Gold	19.3×10^3

*All densities are measured at 0°C and 1 atm unless otherwise noted.

Buoyant forces can keep objects afloat

Have you ever wondered why things feel lighter underwater than they do in air? The reason is that a fluid exerts an upward force on objects that are partially or completely submerged in it. This upward force is called a **buoyant force.** If you have ever rested on an air mattress in a swimming pool, you have experienced a buoyant force. The buoyant force kept you and the mattress afloat.

Because the buoyant force acts in a direction opposite the force of gravity, the net force acting on an object submerged in a fluid, such as water, is smaller than the object's weight. Thus, the object appears to weigh less in water than it does in air. The weight of an object immersed in a fluid is the object's *apparent weight.* In the case of a heavy object, such as a brick, its apparent weight is less in water than its actual weight is in air, but it may still sink in water because the buoyant force is not enough to keep it afloat.

buoyant force

the upward force exerted by a liquid on an object immersed in or floating on the liquid

Demonstration

Buoyant Force — ADVANCED

Purpose Show the relationship between buoyant force and submerged volume.

Materials large spring scale, several cylinders of the same size but of different materials (preferably of low density), a clear container partially filled with water

Procedure Point out that all of the cylinders have the same volume. Measure this volume using a graduated cylinder or by the overflow method. Hang a cylinder on the scale, read the weight, and slowly lower the cylinder into the water. Students should observe that the scale's reading drops continuously as more of the cylinder is submerged. Explain that the water is exerting an upward force on the cylinder. Have a student record the scale's reading before the cylinder's immersion, midway through its immersion, and when it is completely submerged. Repeat with the other cylinders. Examining all of the data will show that for the same volume submerged, the buoyant force is the same, regardless of the cylinder's weight.

The Language of Physics — GENERAL

Students may need help interpreting all the symbols in the equations and relating them to prior knowledge. It may be helpful to remind students that g is free-fall acceleration, with a value of 9.81 m/s^2, which allows them to find an object's weight in newtons when its mass is known in kilograms.

Misconception Alert

Students may wonder why the buoyant force, F_B, is treated like weight (mg) even though it pushes upward. By carefully reading the statement of Archimedes' principle, they may realize that the *magnitude* of that force equals the weight of the fluid that would otherwise occupy the space taken up by the submerged object.

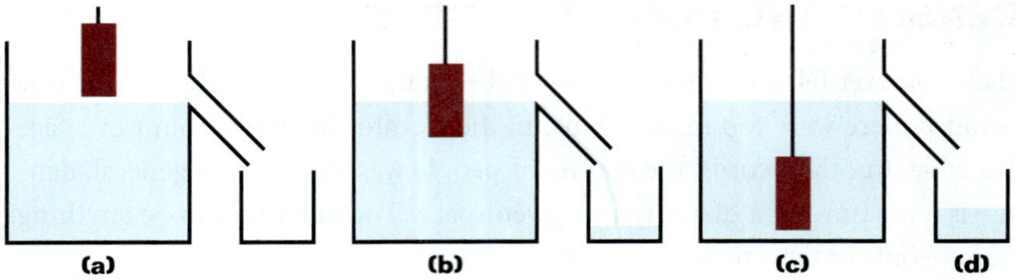

Figure 2
(a) A brick is being lowered into a container of water. **(b)** The brick displaces water, causing the water to flow into a smaller container. **(c)** When the brick is completely submerged, the volume of the displaced water **(d)** is equal to the volume of the brick.

extension
Integrating Biology
Visit go.hrw.com for the activity "How Fish Maintain Neutral Buoyancy."
Keyword HF6FLUX

Archimedes' principle describes the magnitude of a buoyant force

Imagine that you submerge a brick in a container of water, as shown in **Figure 2.** A spout on the side of the container at the water's surface allows water to flow out of the container. As the brick sinks, the water level rises and water flows through the spout into a smaller container. The total volume of water that collects in the smaller container is the *displaced volume* of water from the large container. The displaced volume of water is equal to the volume of the portion of the brick that is underwater.

The magnitude of the buoyant force acting on the brick at any given time can be calculated by using a rule known as *Archimedes' principle.* This principle can be stated as follows: *Any object completely or partially submerged in a fluid experiences an upward buoyant force equal in magnitude to the weight of the fluid displaced by the object.* Everyone has experienced Archimedes' principle. For example, recall that it is relatively easy to lift someone if you are both standing in a swimming pool, even if lifting that same person on dry land would be difficult.

Using m_f to represent the mass of the displaced fluid, Archimedes' principle can be written symbolically as follows:

BUOYANT FORCE

$$F_B = F_g(\text{displaced fluid}) = m_f g$$

magnitude of buoyant force = weight of fluid displaced

Whether an object will float or sink depends on the net force acting on it. This net force is the object's apparent weight and can be calculated as follows:

$$F_{net} = F_B - F_g(\text{object})$$

Now we can apply Archimedes' principle, using m_o to represent the mass of the submerged object.

$$F_{net} = m_f g - m_o g$$

Remember that $m = \rho V$, so the expression can be rewritten as follows:

$$F_{net} = (\rho_f V_f - \rho_o V_o)g$$

Note that in this expression, the fluid quantities refer to the *displaced* fluid.

Did you know?

Archimedes was a Greek mathematician who was born in Syracuse, a city on the island of Sicily. According to legend, the king of Syracuse suspected that a certain golden crown was not pure gold. While bathing, Archimedes figured out how to test the crown's authenticity when he discovered the buoyancy principle. He is reported to have then exclaimed, "Eureka!" meaning "I've found it!"

SCILINKS.
www.scilinks.org
Topic: Archimedes
Code: HF60093

For a floating object, the buoyant force equals the object's weight

Imagine a cargo-filled raft floating on a lake. There are two forces acting on the raft and its cargo: the downward force of gravity and the upward buoyant force of the water. Because the raft is floating in the water, the raft is in equilibrium and the two forces are balanced, as shown in **Figure 3.** For floating objects, the buoyant force and the weight of the object are equal in magnitude.

BUOYANT FORCE ON FLOATING OBJECTS

$$F_B = F_g(object) = m_o g$$

buoyant force = weight of floating object

Notice that Archimedes' principle is not required to find the buoyant force on a floating object if the weight of the object is known.

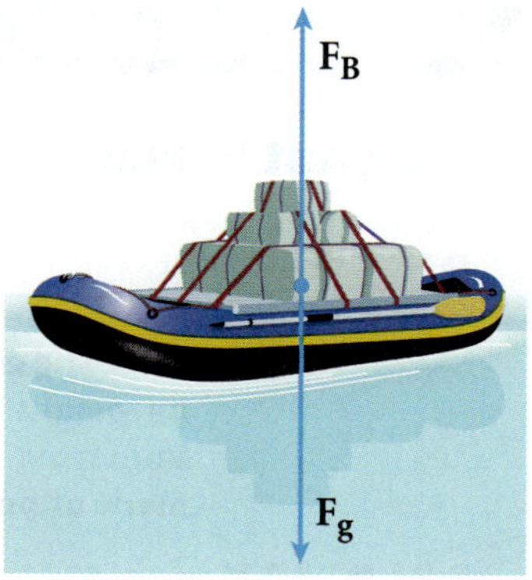

Figure 3
The raft and cargo are floating because their weight and the buoyant force are balanced.

The apparent weight of a submerged object depends on density

Imagine that a hole is accidentally punched in the raft shown in **Figure 3** and that the raft begins to sink. The cargo and raft eventually sink below the water's surface, as shown in **Figure 4.** The net force on the raft and cargo is the vector sum of the buoyant force and the weight of the raft and cargo. As the volume of the raft decreases, the volume of water displaced by the raft and cargo also decreases, as does the magnitude of the buoyant force. This can be written by using the expression for the net force:

$$F_{net} = (\rho_f V_f - \rho_o V_o)g$$

Because the raft and cargo are completely submerged, V_f and V_o are equal:

$$F_{net} = (\rho_f - \rho_o)Vg$$

Notice that both the direction and the magnitude of the net force depend on the difference between the density of the object and the density of the fluid in which it is immersed. If the object's density is greater than the fluid density, the net force is negative (downward) and the object sinks. If the object's density is less than the fluid density, the net force is positive (upward) and the object rises to the surface and floats. If the densities are the same, the object hangs suspended underwater.

A simple relationship between the weight of a submerged object and the buoyant force on the object can be found by considering their ratio as follows:

$$\frac{F_g(object)}{F_B} = \frac{\rho_o \cancel{Vg}}{\rho_f \cancel{Vg}}$$

$$\frac{F_g(object)}{F_B} = \frac{\rho_o}{\rho_f}$$

This last expression is often useful in solving buoyancy problems.

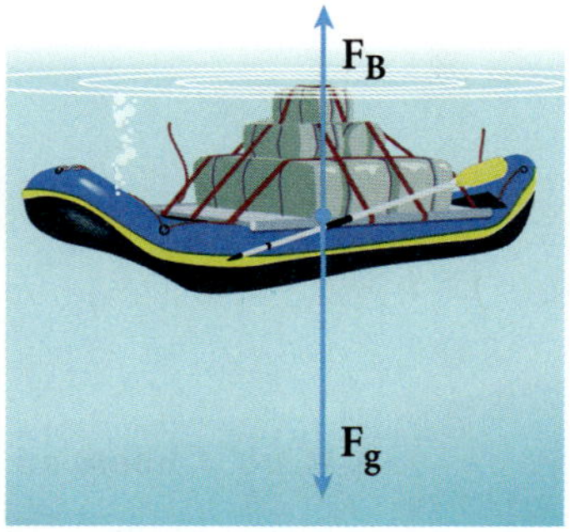

Figure 4
The raft and cargo sink because their density is greater than the density of water.

SCILINKS.
www.scilinks.org
Topic: Buoyancy
Code: HF60201

SECTION 1

Demonstration

Float an Egg — GENERAL

Purpose Demonstrate the effect of liquid density on the buoyant force.

Materials uncooked egg, demonstration scale, glass jar, water, table salt, measuring spoon, stirring stick

Procedure Measure the egg's weight and record it on the board. Fill the glass jar about two-thirds full with water. Demonstrate that the egg sinks in the water. Is a buoyant force acting on it? *(yes, but less than the egg's weight)* Explain that adding salt to the water will increase the density of the water. Have the class watch the egg as you stir one spoonful of salt at a time into the water. The egg will start to float but will remain submerged. Ask students to estimate the buoyant force (*same as the egg's weight*). Keep adding salt until the egg floats partially above water. Again ask students to estimate the buoyant force (*same as the egg's weight*). Explain that the buoyant force balances the egg's weight in both cases, but as the density of the water increases, the volume of the egg that is immersed decreases.

Alternative Problem-Solving Approach

$F_B = F_{actual} - F_{apparent}$

$F_B = 7.84\ N - 6.86\ N = 0.98\ N$

$F_B = \rho_{fluid} V g$

$V = \dfrac{0.98\ N}{(1.00 \times 10^3\ kg/m^3)(9.81\ m/s^2)}$

$V = 0.00010\ m^3$

This volume of gold should weigh 18.9 N ($0.00010\ m^3 \times 9.81\ m/s^2 \times$ density of gold), which is much greater than the 7.84 N weight given in the problem.

Classroom Practice

Buoyant Force
Calculate the actual weight, the buoyant force, and the apparent weight of a $5.00 \times 10^{-5}\ m^3$ iron ball floating at rest in mercury.

Answer
3.86 N, 3.86 N, 0.00 N

How much of the ball's volume is immersed in mercury?

Answer
$2.89 \times 10^{-5}\ m^3$

SAMPLE PROBLEM A

Buoyant Force

PROBLEM

A bargain hunter purchases a "gold" crown at a flea market. After she gets home, she hangs the crown from a scale and finds its weight to be 7.84 N. She then weighs the crown while it is immersed in water, and the scale reads 6.86 N. Is the crown made of pure gold? Explain.

SOLUTION

1. DEFINE

Given: $F_g = 7.84\ N$ apparent weight $= 6.86\ N$

$\rho_f = \rho_{water} = 1.00 \times 10^3\ kg/m^3$

Unknown: $\rho_o = ?$

Diagram:

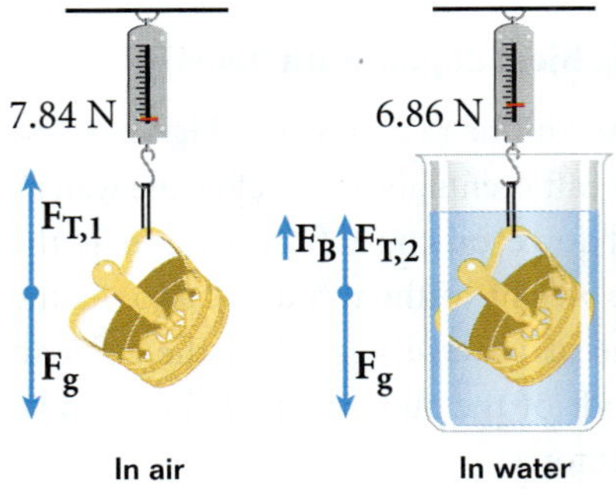

TIP *The use of a diagram can help clarify a problem and the variables involved. In this diagram, $F_{T,1}$ equals the actual weight of the crown, and $F_{T,2}$ is the apparent weight of the crown when immersed in water.*

2. PLAN

Choose an equation or situation:
Because the object is completely submerged, consider the ratio of the weight to the buoyant force.

$$F_g - F_B = \text{apparent weight}$$

$$\frac{F_g}{F_B} = \frac{\rho_o}{\rho_f}$$

Rearrange the equation to isolate the unknown:

$$F_B = F_g - (\text{apparent weight})$$

$$\rho_o = \frac{F_g}{F_B}\rho_f$$

3. CALCULATE

Substitute the values into the equation and solve:

$$F_B = 7.84\ N - 6.86\ N = 0.98\ N$$

$$\rho_o = \frac{F_g}{F_B}\rho_f = \frac{7.84\ N}{0.98\ N}(1.00 \times 10^3\ kg/m^3)$$

$$\boxed{\rho_o = 8.0 \times 10^3\ kg/m^3}$$

4. EVALUATE

From **Table 1,** the density of gold is $19.3 \times 10^3\ kg/m^3$. Because $8.0 \times 10^3\ kg/m^3 < 19.3 \times 10^3\ kg/m^3$, the crown cannot be pure gold.

PRACTICE A

Buoyant Force

1. A piece of metal weighs 50.0 N in air, 36.0 N in water, and 41.0 N in an unknown liquid. Find the densities of the following:
   a. the metal
   b. the unknown liquid
2. A 2.8 kg rectangular air mattress is 2.00 m long, 0.500 m wide, and 0.100 m thick. What mass can it support in water before sinking?
3. A ferry boat is 4.0 m wide and 6.0 m long. When a truck pulls onto it, the boat sinks 4.00 cm in the water. What is the weight of the truck?
4. An empty rubber balloon has a mass of 0.0120 kg. The balloon is filled with helium at 0°C, 1 atm pressure, and a density of 0.179 kg/m^3. The filled balloon has a radius of 0.500 m.
   a. What is the magnitude of the buoyant force acting on the balloon? (Hint: See **Table 1** for the density of air.)
   b. What is the magnitude of the net force acting on the balloon?

SECTION REVIEW

1. What is the difference between a solid and a fluid? What is the difference between a gas and a liquid?
2. Which of the following objects will float in a tub of mercury?
   a. a solid gold bead
   b. an ice cube
   c. an iron bolt
   d. 5 mL of water
3. A 650 kg weather balloon is designed to lift a 4600 kg package. What volume should the balloon have after being inflated with helium at 0°C and 1 atm pressure to lift the total load? (Hint: Use the density values in **Table 1.**)
4. A submerged submarine alters its buoyancy so that it initially accelerates upward at 0.325 m/s^2. What is the submarine's average density at this time? (Hint: the density of sea water is 1.025×10^3 kg/m^3.)
5. **Critical Thinking** Many kayaks are made of plastics and other composite materials that are denser than water. How are such kayaks able to float in water?

SECTION 1

PROBLEM GUIDE A

Use this guide to assign problems.
SE = Student Edition Textbook
PW = Problem Workbook
PB = Problem Bank on the One-Stop Planner (OSP)

Solving for:

ρ	**SE** Sample, 1; Ch. Rvw. 8–9 **PW** 6–8 **PB** 5–7
m	**SE** 2, 3; Ch. Rvw. 23 **PW** Sample, 1–3 **PB** 8–10
F_B	**SE** 4; Ch. Rvw. 24*, 25*, 26*, 29*, 30*, 32*, 34*, 35* **PW** 4–5 **PB** Sample, 1–4

***Challenging Problem**
Consult the printed Solutions Manual or the OSP for detailed solutions.

ANSWERS

Practice A

1. a. 3.57×10^3 kg/m^3
   b. 6.4×10^2 kg/m^3
2. 97 kg
3. 9.4×10^3 N
4. a. 6.63 N
   b. 5.59 N

SECTION REVIEW ANSWERS

1. A solid has a definite shape, while a fluid does not; a liquid has a definite volume, while a gas does not.
2. b, c, and d
3. 4.7×10^3 m^3
4. 9.92×10^2 kg/m^3
5. The kayak's effective density includes the material of and the air within the kayak and is less than water's density.

SECTION 2
Advanced Level

Demonstration

Defining Pressure – GENERAL

Purpose Relate pressure and the area on which a force is exerted.

Materials flat foam pad, brick or other box-shaped object, demonstration scale

Procedure Measure and record the weight of the brick and its dimensions. Place the brick on the pad, large side down. Have students notice the deformation of the pad. Ask how much force the brick exerts on the pad *(same as the brick's weight)*. Over how many square centimeters is this force distributed? *(area of contact)* Now place the brick on the pad with its smaller face down and raise the same questions. Students will observe a deeper deformation. Discuss how this demonstration relates to the definition of pressure as $P = \frac{F}{A}$.

Misconception Alert

Students may confuse the pressure *increase* in Pascal's principle with the pressure of the fluid itself. While an increase in pressure is transmitted equally throughout a fluid, the total pressure at different points in the fluid may vary, for example, with depth.

SECTION 2

Fluid Pressure

SECTION OBJECTIVES

- **Calculate the pressure exerted by a fluid.**
- **Calculate how pressure varies with depth in a fluid.**

PRESSURE

Deep-sea explorers wear atmospheric diving suits like the one shown in **Figure 5** to resist the forces exerted by water in the depths of the ocean. You experience the effects of similar forces on your ears when you dive to the bottom of a swimming pool, drive up a mountain, or ride in an airplane.

Pressure is force per unit area

In the examples above, the fluids exert **pressure** on your eardrums. Pressure is a measure of how much force is applied over a given area. It can be written as follows:

pressure

the magnitude of the force on a surface per unit area

PRESSURE

$$P = \frac{F}{A}$$

$$\text{pressure} = \frac{\text{force}}{\text{area}}$$

ADVANCED TOPICS

See "Fluid Pressure" in **Appendix J: Advanced Topics** to learn more about other properties of fluids.

The SI unit of pressure is the *pascal* (Pa), which is equal to 1 N/m^2. The pascal is a small unit of pressure. The pressure of the atmosphere at sea level is about 1.01×10^5 Pa. This amount of air pressure under normal conditions is the basis for another unit, the *atmosphere* (atm). For the purpose of calculating pressure, 10^5 Pa is about the same as 1 atm. The absolute air pressure inside a typical automobile tire is about 3×10^5 Pa, or 3 atm.

Figure 5
Atmospheric diving suits allow divers to withstand the pressure exerted by the fluid in the ocean at depths of up to 610 m.

Applied pressure is transmitted equally throughout a fluid

When you pump a bicycle tire, you apply a force on the pump that in turn exerts a force on the air inside the tire. The air responds by pushing not only against the pump but also against the walls of the tire. As a result, the pressure increases by an equal amount throughout the tire.

In general, if the pressure in a fluid is increased at any point in a container (such as at the valve of the tire), the pressure increases at all points inside the container by exactly the same amount. Blaise Pascal (1623–1662) noted this fact in what is now called *Pascal's principle* (or *Pascal's law*):

PASCAL'S PRINCIPLE

Pressure applied to a fluid in a closed container is transmitted equally to every point of the fluid and to the walls of the container.

A hydraulic lift, such as the one shown in **Figure 6,** makes use of Pascal's principle. A small force F_1 applied to a small piston of area A_1 causes a pressure increase in a fluid, such as oil. According to Pascal's principle, this increase in pressure, P_{inc}, is transmitted to a larger piston of area A_2 and the fluid exerts a force F_2 on this piston. Applying Pascal's principle and the definition of pressure gives the following equation:

$$P_{inc} = \frac{F_1}{A_1} = \frac{F_2}{A_2}$$

Rearranging this equation to solve for F_2 produces the following:

$$F_2 = \frac{A_2}{A_1}F_1$$

This second equation shows that the output force, F_2, is larger than the input force, F_1, by a factor equal to the ratio of the areas of the two pistons. However, the input force must be applied over a longer distance; the work required to lift the truck is not reduced by the use of a hydraulic lift.

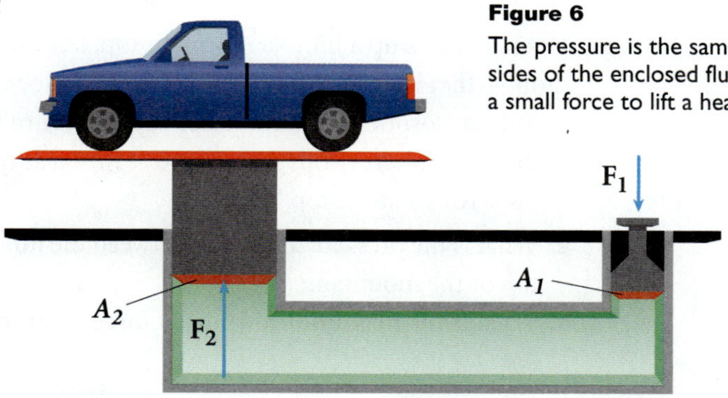

Figure 6
The pressure is the same on both sides of the enclosed fluid, allowing a small force to lift a heavy object.

extension
Integrating Technology
Visit go.hrw.com for the activity "Hydraulic Lift Force."
Keyword HF6FLUX

SAMPLE PROBLEM B

Pressure

PROBLEM

The small piston of a hydraulic lift has an area of 0.20 m^2. A car weighing 1.20 × 10^4 N sits on a rack mounted on the large piston. The large piston has an area of 0.90 m^2. How large a force must be applied to the small piston to support the car?

SOLUTION

Given: $A_1 = 0.20\ \text{m}^2 \qquad A_2 = 0.90\ \text{m}^2$
$F_2 = 1.20 \times 10^4\ \text{N}$

Unknown: $F_1 = ?$

Use the equation for pressure and apply Pascal's principle.

$$\frac{F_1}{A_1} = \frac{F_2}{A_2}$$

$$F_1 = \left(\frac{A_1}{A_2}\right)F_2 = \left(\frac{0.20\ \text{m}^2}{0.90\ \text{m}^2}\right)(1.20 \times 10^4\ \text{N})$$

$$\boxed{F_1 = 2.7 \times 10^3\ \text{N}}$$

Classroom Practice

Pressure
In a hydraulic lift, a 620 N force is exerted on a 0.20 m^2 piston in order to support a weight that is placed on a 2.0 m^2 piston. How much pressure is exerted on the narrow piston? How much weight can the wide piston lift?

Answer
3.1×10^3 Pa; 6.2×10^3 N

Teaching Tip — ADVANCED
A woman wearing snowshoes stands safely in the snow. If she removes her snowshoes, she quickly begins to sink. Explain what happens in terms of force and pressure.
With snowshoes, her weight is applied over a larger area, so the pressure is small. Without snowshoes, her weight is applied over a smaller area, so the pressure is large.

PROBLEM GUIDE B

Use this guide to assign problems.
SE = Student Edition Textbook
PW = Problem Workbook
PB = Problem Bank on the One-Stop Planner (OSP)

Solving for:

F	**SE** Sample, 1a, 3b; Ch. Rvw. 14–16, 21 **PW** Sample, 1–2 **PB** 5–7
P	**SE** 1b, 2, 3a; Ch. Rvw. 22, 27, 33* **PW** 6–7 **PB** 8–10
A	**PW** 3–5 **PB** Sample, 1–4

*Challenging Problem
Consult the printed Solutions Manual or the OSP for detailed solutions.

ANSWERS

Practice B

1. a. 1.48×10^3 N
   b. 1.88×10^5 Pa
2. 2.7×10^2 Pa
3. a. 1.2×10^3 Pa
   b. 6.0×10^{-2} N

Teaching Tip — ADVANCED

Ask students why the roof of a building does not collapse under the tremendous pressure exerted by our atmosphere.
The pressure inside the building is approximately equal to the approximately equal to the pressure outside the building.

PRACTICE B

Pressure

1. In a car lift, compressed air exerts a force on a piston with a radius of 5.00 cm. This pressure is transmitted to a second piston with a radius of 15.0 cm.
   a. How large a force must the compressed air exert to lift a 1.33×10^4 N car?
   b. What pressure produces this force? Neglect the weight of the pistons.
2. A 1.5 m wide by 2.5 m long water bed weighs 1025 N. Find the pressure that the water bed exerts on the floor. Assume that the entire lower surface of the bed makes contact with the floor.
3. A person rides up a lift to a mountaintop, but the person's ears fail to "pop"—that is, the pressure of the inner ear does not equalize with the outside atmosphere. The radius of each eardrum is 0.40 cm. The pressure of the atmosphere drops from 1.010×10^5 Pa at the bottom of the lift to 0.998×10^5 Pa at the top.
   a. What is the pressure difference between the inner and outer ear at the top of the mountain?
   b. What is the magnitude of the net force on each eardrum?

SCILINKS.
www.scilinks.org
Topic: Atmospheric Pressure
Code: HF60114

Pressure varies with depth in a fluid

As a submarine dives deeper in the water, the pressure of the water against the hull of the submarine increases, so the hull must be strong enough to withstand large pressures. Water pressure increases with depth because the water at a given depth must support the weight of the water above it.

Imagine a small area on the hull of a submarine. The weight of the entire column of water above that area exerts a force on the area. The column of water has a volume equal to Ah, where A is the cross-sectional area of the column and h is its height. Hence the mass of this column of water is $m = \rho V = \rho Ah$. Using the definitions of density and pressure, the pressure at this depth due to the weight of the column of water can be calculated as follows:

$$P = \frac{F}{A} = \frac{mg}{A} = \frac{\rho V g}{A} = \frac{\rho A h g}{A} = \rho h g$$

This equation is valid only if the density is the same throughout the fluid.

The pressure in the equation above is referred to as *gauge pressure.* It is not the total pressure at this depth because the atmosphere itself also exerts a pressure at the surface. Thus, the gauge pressure is actually the total pressure minus the atmospheric pressure. By using the symbol P_0 for the atmospheric pressure at the surface, we can express the total pressure, or *absolute pressure,* at a given depth in a fluid of uniform density ρ as follows:

FLUID PRESSURE AS A FUNCTION OF DEPTH

$$P = P_0 + \rho g h$$

absolute pressure =
atmospheric pressure + (density × free-fall acceleration × depth)

This expression for pressure in a fluid can be used to help understand buoyant forces. Consider a rectangular box submerged in a container of water, as shown in **Figure 7.** The water pressure at the top of the box is $-(P_0 + \rho g h_1)$, and the water pressure at the bottom of the box is $P_0 + \rho g h_2$. From the definition of pressure, we know that the downward force on the box is $-A(P_0 + \rho g h_1)$, where A is the area of the top of the box. The upward force on the box is $A(P_0 + \rho g h_2)$. The net force on the box is the sum of these two forces.

$$F_{net} = A(P_0 + \rho g h_2) - A(P_0 + \rho g h_1) = \rho g(h_2 - h_1)A = \rho g V = m_f g$$

Note that this is an expression of Archimedes' principle. In general, we can say that buoyant forces arise from the differences in fluid pressure between the top and the bottom of an immersed object.

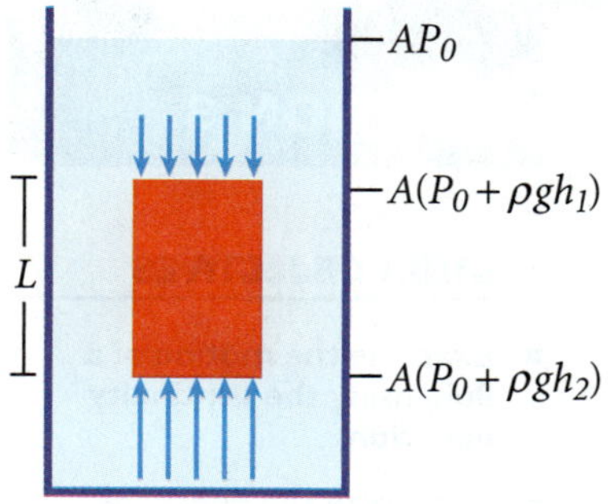

Figure 7
The fluid pressure at the bottom of the box is greater than the fluid pressure at the top of the box.

extension

Practice Problems

Visit go.hrw.com to find a sample and practice problems for pressure as a function of depth.

Keyword HF6FLUX

SECTION REVIEW

1. Which of the following exerts the most pressure while resting on a floor?
   a. a 25 N cube with 1.5 m sides
   b. a 15 N cylinder with a base radius of 1.0 m
   c. a 25 N cube with 2.0 m sides
   d. a 25 N cylinder with a base radius of 1.0 m

2. Water is to be pumped to the top of the Empire State Building, which is 366 m high. What gauge pressure is needed in the water line at the base of the building to raise the water to this height? (Hint: See **Table 1** for the density of water.)

3. When a submarine dives to a depth of 5.0×10^2 m, how much pressure, in Pa, must its hull be able to withstand? How many times larger is this pressure than the pressure at the surface? (Hint: See **Table 1** for the density of sea water.)

4. **Critical Thinking** Calculate the depth in the ocean at which the pressure is three times atmospheric pressure. (Hint: Use the value for the density of sea water given in **Table 1.**)

Visual Strategy ADVANCED

Figure 7

Point out that the fluid exerts force in all directions because of the pressure. Pressure against the sides of the box also increases with depth.

Q Suppose the box in the diagram is immersed in water and $L = 2.0$ m. How does pressure on the box's sides 1.0 m from its top compare with pressure on the sides near the top and near the bottom?

A ***$\Delta P = \rho g \Delta h$ = $(1.00 \times 10^3$ kg/m$^3)(9.81$ m/s$^2)$ $(1.0$ m$)$ = 9800 Pa. This means that the pressure on the middle area of the sides of the box is 9800 Pa higher than the pressure near the top and 9800 Pa lower than the pressure near the bottom of the box.***

SECTION REVIEW ANSWERS

1. a
2. 3.59×10^6 Pa
3. $5. \times 10^6$ Pa; 5.0×10^1
4. 20.1 m

SECTION 3
Advanced Level

Demonstration

Fluid Flow Around a Table-Tennis Ball

Purpose Introduce students to some of the interesting effects of fluid flow and Bernoulli's principle.

Materials table-tennis ball glued to the end of a piece of lightweight string, water faucet

Procedure Tell the class that this section covers the surprising phenomena of fluids in motion "sucking in" things around them.

Hold the end of the string so that the ball is suspended a few centimeters below the faucet but away from the stream path. When you turn on the faucet, the ball moves toward the water.

Demonstration

Fluid Flow Between Two Cans

Purpose Further demonstrate the effects of fluid flow and Bernoulli's principle.

Materials two empty soda cans, flat surface

Procedure Lay the soda cans on their sides approximately 2 cm apart on a table or other flat surface. Ask students what would happen if you were to blow air between the two cans. Ask a student volunteer to blow between the cans. The cans will move toward each other. Explain that the air moves faster between the cans, lowering the pressure between them and drawing them together.

SECTION 3

Fluids in Motion

SECTION OBJECTIVES

- **Examine the motion of a fluid using the continuity equation.**
- **Recognize the effects of Bernoulli's principle on fluid motion.**

FLUID FLOW

Have you ever gone canoeing or rafting down a river? If so, you may have noticed that part of the river flowed smoothly, allowing you to float calmly or to simply paddle along. At other places in the river, there may have been rocks or dramatic bends that created foamy whitewater rapids.

When a fluid, such as river water, is in motion, the flow can be characterized in one of two ways. The flow is said to be *laminar* if every particle that passes a particular point moves along the same smooth path traveled by the particles that passed that point earlier. The smooth stretches of a river are regions of laminar flow.

In contrast, the flow of a fluid becomes irregular, or *turbulent,* above a certain velocity or under conditions that can cause abrupt changes in velocity, such as where there are obstacles or sharp turns in a river. Irregular motions of the fluid, called *eddy currents,* are characteristic of turbulent flow.

Figure 8 shows a photograph of water flowing past a cylinder. Hydrogen bubbles were added to the water to make the streamlines and the eddy currents visible. Notice the dramatic difference in flow patterns between the laminar flow and the turbulent flow. Laminar flow is much easier to model because it is predictable. Turbulent flow is extremely chaotic and unpredictable.

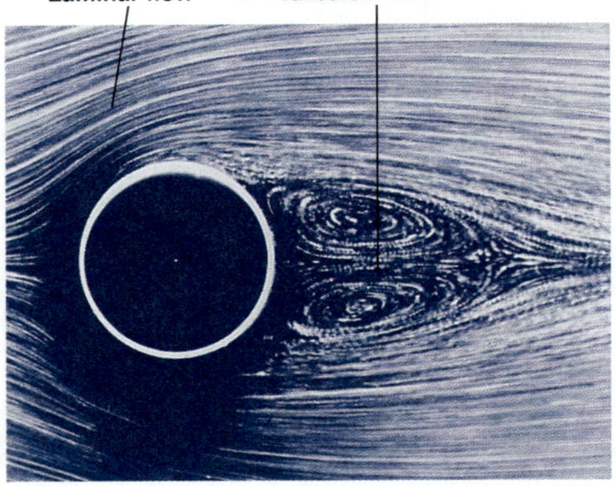

Figure 8
The water flowing around this cylinder exhibits laminar flow and turbulent flow.

The ideal fluid model simplifies fluid-flow analysis

Many features of fluid motion can be understood by considering the behavior of an **ideal fluid.** Although no real fluid has all the properties of an ideal fluid, the ideal fluid model does help explain many properties of real fluids, so the model is a useful tool for analysis. While discussing density and buoyancy, we assumed all of the fluids used in problems were practically incompressible. A fluid is incompressible if the density of the fluid always remains constant.

ideal fluid

a fluid that has no internal friction or viscosity and is incompressible

The term *viscosity* refers to the amount of internal friction within a fluid. A fluid with a high viscosity flows more slowly than does a fluid with a low viscosity. As a viscous fluid flows, part of the kinetic energy of the fluid is transformed into internal energy due to the friction of the fluid particles sliding past each other. Ideal fluids are considered *nonviscous,* so they lose no kinetic energy due to friction as they flow. Ideal fluids are also characterized by a *steady flow.* In other words, the velocity, density, and pressure at each point in the fluid are constant. Ideal flow of an ideal fluid is also *nonturbulent,* which means that there are no eddy currents in the moving fluid.

PRINCIPLES OF FLUID FLOW

Fluid behavior is often very complex. Several general principles describing the flow of fluids can be derived relatively easily from basic physical laws.

The continuity equation results from mass conservation

Imagine that an ideal fluid flows into one end of a pipe and out the other end, as shown in **Figure 9.** The diameter of the pipe is different at each end. How does the speed of fluid flow change as the fluid passes through the pipe?

Because mass is conserved and because the fluid is incompressible, we know that the mass flowing into the bottom of the pipe, m_1, must equal the mass flowing out of the top of the pipe, m_2, during any given time interval:

$$m_1 = m_2$$

This simple equation can be expanded by recalling that $m = \rho V$ and by using the formula for the volume of a cylinder, $V = A\Delta x$.

$$\rho_1 V_1 = \rho_2 V_2$$
$$\rho_1 A_1 \Delta x_1 = \rho_2 A_2 \Delta x_2$$

The length of the cylinder, Δx, is also the distance the fluid travels, which is equal to the speed of flow multiplied by the time interval ($\Delta x = v\Delta t$).

$$\rho_1 A_1 v_1 \Delta t = \rho_2 A_2 v_2 \Delta t$$

The time interval and, for an ideal fluid, the density are the same on each side of the equation, so they cancel each other out. The resulting equation is called the continuity equation:

CONTINUITY EQUATION

$$A_1 v_1 = A_2 v_2$$

area × speed in region 1 = area × speed in region 2

The speed of fluid flow depends on cross-sectional area

Note in the continuity equation that A_1 and A_2 can represent any two different cross-sectional areas of the pipe, not just the ends. This equation implies that the fluid speed is faster where the pipe is narrow and slower where the pipe is wide. The product Av, which has units of volume per unit time, is called the *flow rate.* The flow rate is constant throughout the pipe.

The continuity equation explains an effect you may have observed as water flows slowly from a faucet, as shown in **Figure 10.** Because the water speeds up due to gravity as it falls, the stream narrows, satisfying the continuity equation. The continuity equation also explains why a river tends to flow more rapidly in places where the river is shallow or narrow than in places where the river is deep and wide.

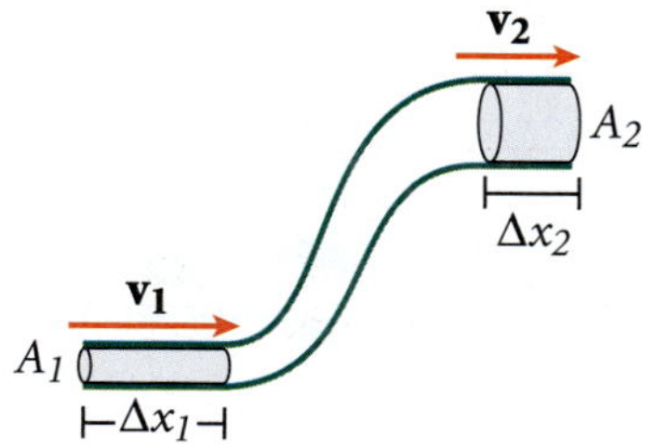

Figure 9
The mass flowing into the pipe must equal the mass flowing out of the pipe in the same time interval.

Figure 10
The width of a stream of water narrows as the water falls and speed up.

Teaching Tip — GENERAL

The viscosity index on motor oil cans consists of a Society of Automotive Engineers (SAE) number (typically from 5 to 50) followed by the letter *W.* The higher the SAE number, the more viscous the oil is. Low-viscosity oils are meant for use in severe winter climates because low-viscosity oils flow more easily in cold temperatures. For hot or high-speed driving conditions, a high-viscosity oil can be used because the excessive heat effectively thins the oil.

Visual Strategy ADVANCED

Figure 9

Point out that the volumes of the two shaded sections of the pipe must be equal in order for the continuity equation to apply.

Q Assume A_1 is smaller than pictured in **Figure 9.** Would Δx_1 need to be longer or shorter for the mass of liquid in each section to still be equal?

A *longer*

SECTION 3

Misconception Alert

One might think that the water coming out of the hose is at a higher pressure than the water in the hose, but the opposite is true. The pressure outside the hose is atmospheric pressure, while the pressure inside the hose is higher than atmospheric pressure. In fact, the water flows out of the hose because the greater pressure in the hose results in a net force on the water at the end of the hose, which pushes the water out.

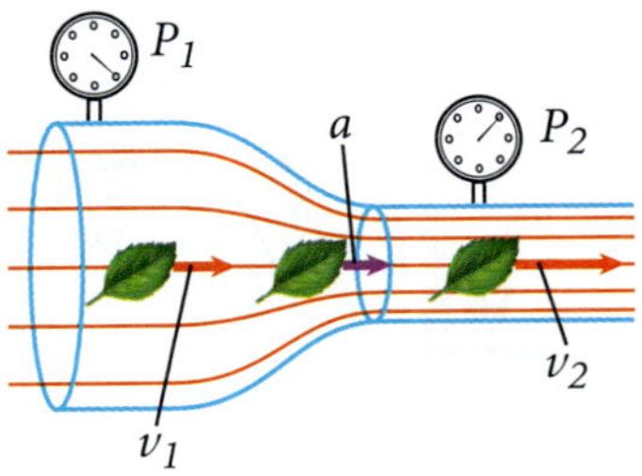

Figure 11
A leaf speeds up as it passes into a constriction in a drainage pipe. The water pressure on the right is less than the pressure on the left.

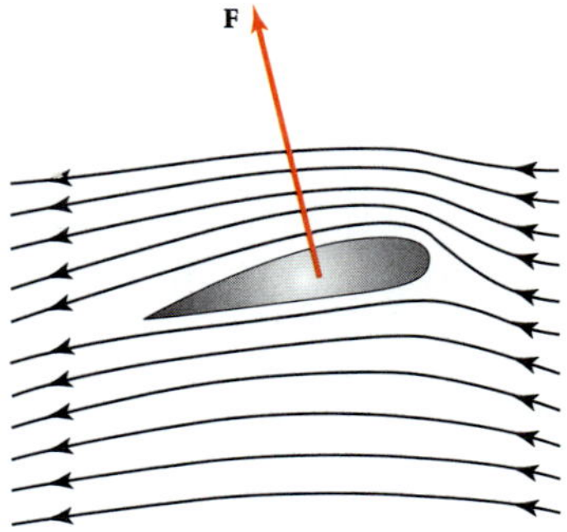

Figure 12
As air flows around an airplane wing, the air above the wing moves faster than the air below, producing lift.

The pressure in a fluid is related to the speed of flow

Suppose there is a water-logged leaf carried along by the water in a drainage pipe, as shown in **Figure 11.** The continuity equation shows that the water moves faster through the narrow part of the tube than through the wider part of the tube. Therefore, as the water carries the leaf into the constriction, the leaf speeds up.

If the water and the leaf are accelerating as they enter the constriction, an unbalanced force must be causing the acceleration, according to Newton's second law. This unbalanced force is a result of the fact that the water pressure in front of the leaf is less than the water pressure behind the leaf. The pressure difference causes the leaf and the water around it to accelerate as it enters the narrow part of the tube. This behavior illustrates a general principle known as *Bernoulli's principle,* which can be stated as follows:

BERNOULLI'S PRINCIPLE

The pressure in a fluid decreases as the fluid's velocity increases.

The lift on an airplane wing can be explained, in part, with Bernoulli's principle. As an airplane flies, air flows around the wings and body of the plane, as shown in **Figure 12.** Airplane wings are designed to direct the flow of air so that the air speed above the wing is greater than the air speed below the wing. Therefore, the air pressure above the wing is less than the pressure below, and there is a net upward force on the wing, called *lift.* The tilt of an airplane wing also adds to the lift on the plane. The front of the wing is tilted upward so that air striking the bottom of the wing is deflected downward.

SECTION REVIEW ANSWERS

1. 16.0 m/s
2. 2.7 m/s
3. 30.0 s, because the volume rate of flow is constant
4. shorter

SECTION REVIEW

1. Water at a pressure of 3.00×10^5 Pa flows through a horizontal pipe at a speed of 1.00 m/s. The pipe narrows to one-fourth its original diameter. What is the speed of the flow in the narrow section?
2. A 2.0 cm diameter faucet tap fills a 2.5×10^{-2} m^3 container in 30.0 s. What is the speed at which the water leaves the faucet?
3. **Critical Thinking** The time required to fill a glass with water from a large container with a spigot is 30.0 s. If you replace the spigot with a smaller one so that the speed of the water leaving the nozzle doubles, how long does it take to fill the glass?
4. **Interpreting Graphics** For this problem, refer back to **Figure 9.** Assume that the cross-sectional area, A_2, in the tube is increased. Would the length, Δx_2, need to be longer or shorter for the mass of liquid in both sections to still be equal?

Highlights

KEY IDEAS

Section 1 Fluids and Buoyant Force
- Force is a vector quantity that causes changes in motion.
- A fluid is a material that can flow, and thus it has no definite shape. Both gases and liquids are fluids.
- Buoyant force is an upward force exerted by a fluid on an object floating on or submerged in the fluid.
- The magnitude of a buoyant force for a submerged object is determined by Archimedes' principle and is equal to the weight of the displaced fluid.
- The magnitude of a buoyant force for a floating object is equal to the weight of the object because the object is in equilibrium.

Section 2 Fluid Pressure
- Pressure is a measure of how much force is exerted over a given area.
- According to Pascal's principle, pressure applied to a fluid in a closed container is transmitted equally to every point of the fluid and to the walls of the container.
- The pressure in a fluid increases with depth.

Section 3 Fluids in Motion
- Moving fluids can exhibit laminar (smooth) flow or turbulent flow.
- An ideal fluid is incompressible, nonviscous, and, when undergoing ideal flow, nonturbulent.
- The continuity equation is derived from the fact that the amount of fluid leaving a pipe during some time interval equals the amount entering the pipe during that same time interval.
- According to Bernoulli's principle, swift-moving fluids exert less pressure than slower-moving fluids.

KEY TERMS

fluid (p. 274)

mass density (p. 275)

buoyant force (p. 275)

pressure (p. 280)

ideal fluid (p. 284)

PROBLEM SOLVING

See **Appendix D: Equations** for a summary of the equations introduced in this chapter. If you need more problem-solving practice, see **Appendix I: Additional Problems.**

Variable Symbols

Quantities		Units		Conversions
ρ	density	kg/m^3	kilogram per meter^3	$= 10^{-3}\ g/cm^3$
P	pressure	Pa	pascal	$= N/m^2$ $= 10^{-5}$ atm

Highlights

Teaching Tip — GENERAL

Ask students to prepare a concept map of the chapter. The concept map should include most of the vocabulary terms, along with other integral terms or concepts.

extension

In-Depth Physics Content

Your students can visit go.hrw.com for an online chapter that integrates more in-depth development of the concepts covered here.

Keyword HF6FLUX

Review

ANSWERS

1. Buoyant force opposes and so reduces the effect of weight.
2. the object's weight
3. The weight of the water displaced is greater than the weight of the ball.
4. The level falls because $\rho_{ice} < \rho_{water}$.
5. in the ocean; Because $\rho_{sea\ water} > \rho_{fresh\ water}$, less sea water must be displaced.
6. because the average density of the boat, including air inside the hollow hull, is less than the density of the water
7. The amount of the block that is submerged will decrease. The volume of water displaced remains the same when the block is inverted. When the block is inverted, the steel occupies some of the displaced volume. Thus, the amount of wood submerged is less than half the block.
8. **a.** 6.3×10^3 kg/m^3
   b. 9.2×10^2 kg/m^3
9. 2.1×10^3 kg/m^3
10. no; A force on a small area can produce a large pressure.
11. the pascal; 1 pascal (1 Pa) = 1 N/m^2
12. The force opposing F_g is spread out over a large number of nails, so no single nail exerts very much pressure.
13. No, there would be no way to reduce the pressure in your mouth below the zero atmospheric pressure outside the liquid.
14. 1.9×10^4 N

CHAPTER 8 Review

DENSITY AND BUOYANCY

Review Questions

1. How is weight affected by buoyant force?

2. Buoyant force equals what for any floating object?

Conceptual Questions

3. If an inflated beach ball is placed beneath the surface of a pool of water and released, the ball shoots upward. Why?

4. An ice cube is submerged in a glass of water. What happens to the level of the water as the ice melts?

5. Will a ship ride higher in an inland freshwater lake or in the ocean? Why?

6. Steel is much denser than water. How, then, do steel boats float?

7. A small piece of steel is tied to a block of wood. When the wood is placed in a tub of water with the steel on top, half of the block is submerged. If the block is inverted so that the steel is underwater, will the amount of the wooden block that is submerged increase, decrease, or remain the same?

Practice Problems

For problems 8–9, see Sample Problem A.

8. An object weighs 315 N in air. When tied to a string, connected to a balance, and immersed in water, it weighs 265 N. When it is immersed in oil, it weighs 269 N. Find the following:

a. the density of the object
b. the density of the oil

9. A sample of an unknown material weighs 300.0 N in air and 200.0 N when submerged in an alcohol solution with a density of 0.70×10^3 kg/m^3. What is the density of the material?

PRESSURE

Review Questions

10. Is a large amount of pressure always caused by a large force? Explain your answer.

11. What is the SI unit of pressure? What is it equal to, in terms of other SI units?

Conceptual Questions

12. After a long class, a physics teacher stretches out for a nap on a bed of nails. How is this possible?

13. When drinking through a straw, you reduce the pressure in your mouth and the atmosphere moves the liquid. Could you use a straw to drink on the moon?

Practice Problems

For problems 14–16, see Sample Problem B.

14. The four tires of an automobile are inflated to an absolute pressure of 2.0×10^5 Pa. Each tire has an area of 0.024 m^2 in contact with the ground. Determine the weight of the automobile.

15. A pipe contains water at 5.00×10^5 Pa above atmospheric pressure. If you patch a 4.00 mm diameter hole in the pipe with a piece of bubble gum, how much force must the gum be able to withstand?

16. A piston, *A*, as shown at right, has a diameter of 0.64 cm. A second piston, *B*, has a diameter of 3.8 cm. Determine the force, **F**, necessary to support the 500.0 N weight in the absence of friction.

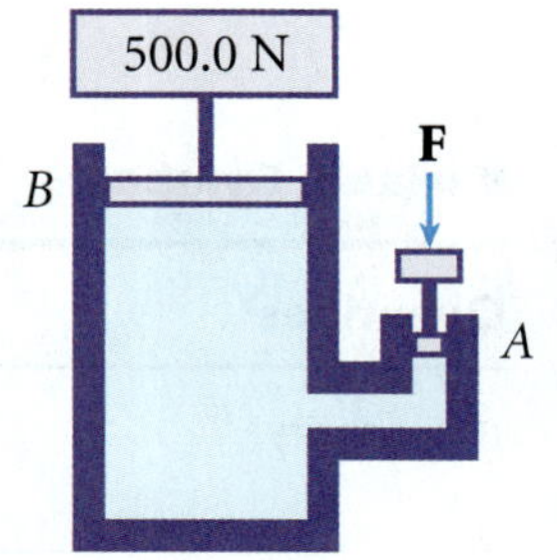

FLUID FLOW

Conceptual Questions

17. Prairie dogs live in underground burrows with at least two entrances. They ventilate their burrows by building a mound around one entrance, which is open to a stream of air. A second entrance at ground level is open to almost stagnant air. Use Bernoulli's principle to explain how this construction creates air flow through the burrow.

18. Municipal water supplies are often provided by reservoirs built on high ground. Why does water from such a reservoir flow more rapidly out of a faucet on the ground floor of a building than out of an identical faucet on a higher floor?

19. If air from a hair dryer is blown over the top of a table-tennis ball, the ball can be suspended in air. Explain how this suspension is possible.

MIXED REVIEW

20. An engineer weighs a sample of mercury ($\rho = 13.6 \times 10^3$ kg/m^3) and finds that the weight of the sample is 4.5 N. What is the sample's volume?

21. About how much force is exerted by the atmosphere on 1.00 km^2 of land at sea level?

22. A 70.0 kg man sits in a 5.0 kg chair so that his weight is evenly distributed on the legs of the chair. Assume that each leg makes contact with the floor over a circular area with a radius of 1.0 cm. What is the pressure exerted on the floor by each leg?

23. A frog in a hemispherical bowl, as shown below, just floats in a fluid with a density of 1.35×10^3 kg/m^3. If the bowl has a radius of 6.00 cm and negligible mass, what is the mass of the frog?

24. When a load of 1.0×10^6 N is placed on a battleship, the ship sinks only 2.5 cm in the water. Estimate the cross-sectional area of the ship at water level. (Hint: See **Table 1** for the density of sea water.)

25. A 1.0 kg beaker containing 2.0 kg of oil with a density of 916 kg/m^3 rests on a scale. A 2.0 kg block of iron is suspended from a spring scale and completely submerged in the oil, as shown at right. Find the equilibrium readings of both scales. (Hint: See **Table 1** for the density of iron.)

26. A raft is constructed of wood having a density of 600.0 kg/m^3. The surface area of the bottom of the raft is 5.7 m^2, and the volume of the raft is 0.60 m^3. When the raft is placed in fresh water having a density of 1.0×10^3 kg/m^3, how deep is the bottom of the raft below water level?

27. A physics book has a height of 26 cm, a width of 21 cm, and a thickness of 3.5 cm.

- **a.** What is the density of the physics book if it weighs 19 N?
- **b.** Find the pressure that the physics book exerts on a desktop when the book lies face up.
- **c.** Find the pressure that the physics book exerts on the surface of a desktop when the book is balanced on its spine.

28. A natural-gas pipeline with a diameter of 0.250 m delivers 1.55 m^3 of gas per second. What is the flow speed of the gas?

29. A 2.0 cm thick bar of soap is floating in water, with 1.5 cm of the bar underwater. Bath oil with a density of 900.0 kg/m^3 is added and floats on top of the water. How high on the side of the bar will the oil reach when the soap is floating in only the oil?

30. Which dam must be stronger, one that holds back 1.0×10^5 m^3 of water 10 m deep or one that holds back 1.0×10^3 m^3 of water 20 m deep?

8 REVIEW

15. 6.28 N
16. 14 N downward
17. The air flow causes the pressure over the entrance in the mound to be lower than the pressure over the other entrance. Thus, air is pushed through the mound by the higher-pressure area.
18. The water at the ground floor has less potential energy relative to the reservoir than the water at the higher floor does. Because energy is conserved, the water flowing to the lower faucet has greater kinetic energy and thus greater velocity than the water flowing to the higher faucet does.
19. The moving air above the ball creates a low pressure area so that the air below the ball exerts a force that is equal and opposite F_g.
20. 3.4×10^{-5} m^3
21. 1.01×10^{11} N
22. 5.9×10^5 Pa
23. 6.11×10^{-1} kg
24. 4.0×10^3 m^2
25. 17 N, 31 N
26. 6.3×10^{-2} m
27. **a.** 1.0×10^3 kg/m^3
b. 3.5×10^2 Pa
c. 2.1×10^3 Pa
28. 31.6 m/s
29. 1.7×10^{-2} m
30. the one that holds back water 20 m deep, because pressure increases with increasing depth

8 REVIEW

31. 0.605 m
32. **a.** 84 g/s
b. 2.8×10^{-2} cm/s
33. 6.3 m
34. 21 Pa
35. **a.** 0.48 m/s^2
b. 4.0 s
36. disagree; because the buoyant force is also proportional to g
37. 1.7×10^{-3} m
38. The buoyant force causes the net force on the astronaut to be close to zero. In space, the astronauts accelerate at the same rate as the craft, so they feel as if they have no net force acting on them.
39. Helium is less dense than air and therefore floats in air.

31. A light spring with a spring constant of 90.0 N/m rests vertically on a table, as shown in **(a)** below. A 2.00 g balloon is filled with helium (0°C and 1 atm pressure) to a volume of 5.00 m^3 and connected to the spring, causing the spring to stretch, as shown in **(b)**. How much does the spring stretch when the system is in equilibrium? (Hint: See **Table 1** for the density of helium. The magnitude of the spring force equals $k\Delta x$.)

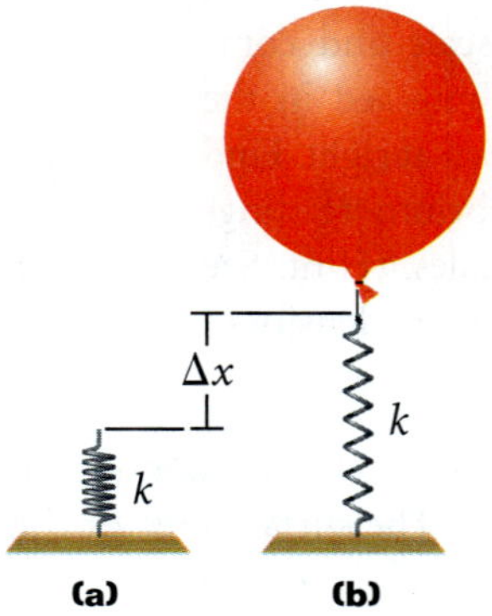

32. The aorta in an average adult has a cross-sectional area of 2.0 cm^2.

a. Calculate the flow rate (in grams per second) of blood ($\rho = 1.0$ g/cm^3) in the aorta if the flow speed is 42 cm/s.

b. Assume that the aorta branches to form a large number of capillaries with a combined cross-sectional area of 3.0×10^3 cm^2. What is the flow speed in the capillaries?

33. A 1.0 kg hollow ball with a radius of 0.10 m is filled with air and is released from rest at the bottom of a 2.0 m deep pool of water. How high above the surface of the water does the ball rise? Disregard friction and the ball's motion when the ball is only partially submerged.

34. In testing a new material for shielding spacecraft, 150 ball bearings each moving at a supersonic speed of 400.0 m/s collide head-on and elastically with the material during a 1.00 min interval. If the ball bearings each have a mass of 8.0 g and the area of the tested material is 0.75 m^2, what is the pressure exerted on the material?

35. A thin, rigid, spherical shell with a mass of 4.00 kg and diameter of 0.200 m is filled with helium (adding negligible mass) at 0°C and 1 atm pressure. It is then released from rest on the bottom of a pool of water that is 4.00 m deep.

a. Determine the upward acceleration of the shell.

b. How long will it take for the top of the shell to reach the surface? Disregard frictional effects.

36. A student claims that if the strength of Earth's gravity doubled, people would be unable to float on water. Do you agree or disagree with this statement? Why?

37. A light spring with a spring constant of 16.0 N/m rests vertically on the bottom of a large beaker of water, as shown in **(a)** below. A 5.00×10^{-3} kg block of wood with a density of 650.0 kg/m^3 is connected to the spring, and the mass-spring system is allowed to come to static equilibrium, as shown in **(b)** below. How much does the spring stretch?

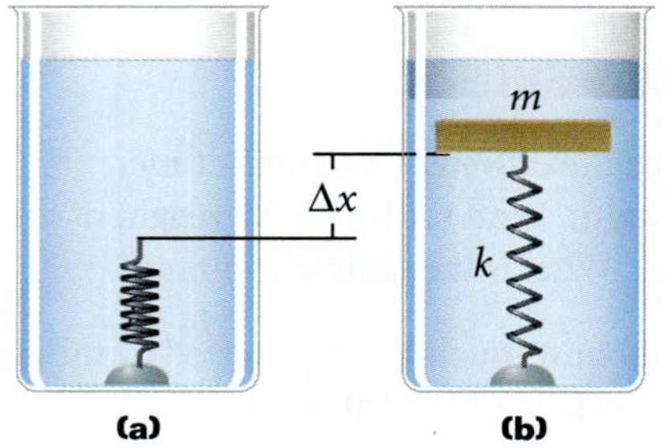

38. Astronauts sometimes train underwater to simulate conditions in space. Explain why.

39. Explain why balloonists use helium instead of air in balloons.

Alternative Assessment

1. Build a hydrometer from a long test tube with some sand at the bottom and a stopper. Adjust the amount of sand as needed so that the tube floats in most liquids. Calibrate it, and place a label with markings on the tube. Measure the densities of the following liquid foods: skim milk, whole milk, vegetable oil, pancake syrup, and molasses. Summarize your findings in a chart or table.

2. The owner of a fleet of tractor-trailers has contacted you after a series of accidents involving tractor-trailers passing each other on the highway. The owner wants to know how drivers can minimize the pull exerted as one tractor-trailer passes another going in the same direction. Should the passing tractor-trailer try to pass as quickly as possible or as slowly as possible? Design experiments to determine the answer by using model motor boats in a swimming pool. Indicate exactly what you will measure and how. If your teacher approves your plan and you are able to locate the necessary equipment, perform the experiment.

3. Record any examples of pumps in the tools, machines, and appliances you encounter in one week, and briefly describe the appearance and function of each pump. Research how one of these pumps works, and evaluate the explanation of the pump's operation for strengths and weaknesses. Share your findings in a group meeting and create a presentation, model, or diagram that summarizes the group's findings.

Graphing Calculator *Practice*

Flow Rates

Flow rate, as you learned earlier in this chapter, is described by the following equation:

$$\text{flow rate} = Av$$

Flow rate is a measure of the volume of a fluid that passes through a tube per unit time. *A* is the cross-sectional area of the tube, and *v* is the flow speed of the fluid. If *A* has units of centimeters squared and *v* has units of centimeters per second, flow rate will have units of cubic centimeters per second.

The graphing calculator will use the following equation to determine flow rate.

$$Y_1 = \pi * V(X/2)^2$$

You will use this equation to study the flow rates (Y_1) for various hose diameters (X) and flow speeds (V). The calculator will produce a table of flow rates in cubic centimeters per second versus hose diameters in centimeters.

In this graphing calculator activity, you will learn how to read a table on the calculator and to use that table to make predictions about flow rates.

Visit go.hrw.com and enter the keyword **HF6FLUX** to find this graphing calculator activity. Refer to **Appendix B** for instructions on downloading the program for this activity.

Alternative Assessment ANSWERS

1. Students should realize that the higher the tube floats (more buoyant force), the greater the fluid's density.
2. Student plans should be safe and should measure force or acceleration perpendicular to the motion. When boats or trucks pass, the speed of the fluid between them increases (as in a narrow pipe), and pressure drops.
3. Student answers should recognize that pumps use pressure differences to move fluids. Possible answers include air pump for tires, gas pump, or vacuum cleaner.

extension

Graphing Calculator Practice

Visit go.hrw.com for answers to this Graphing Calculator activity.

Keyword HF6FLUXT

Standardized Test Prep

ANSWERS

1. C
2. G
3. B
4. J
5. A
6. J

Standardized Test Prep

MULTIPLE CHOICE

1. Which of the following is the correct equation for the net force acting on a submerged object?

A. $F_{net} = 0$
B. $F_{net} = (\rho_{object} - \rho_{fluid})gV_{object}$
C. $F_{net} = (\rho_{fluid} - \rho_{object})gV_{object}$
D. $F_{net} = (\rho_{fluid} + \rho_{object})gV_{object}$

2. How many times greater than the lifting force must the force applied to a hydraulic lift be if the ratio of the area where pressure is applied to the lifted area is $\frac{1}{7}$?

F. $\frac{1}{49}$
G. $\frac{1}{7}$
H. 7
J. 49

3. A typical silo on a farm has many bands wrapped around its perimeter, as shown in the figure below. Why is the spacing between successive bands smaller toward the bottom?

A. to provide support for the silo's sides above them
B. to resist the increasing pressure that the grains exert with increasing depth
C. to resist the increasing pressure that the atmosphere exerts with increasing depth
D. to make access to smaller quantities of grain near the ground possible

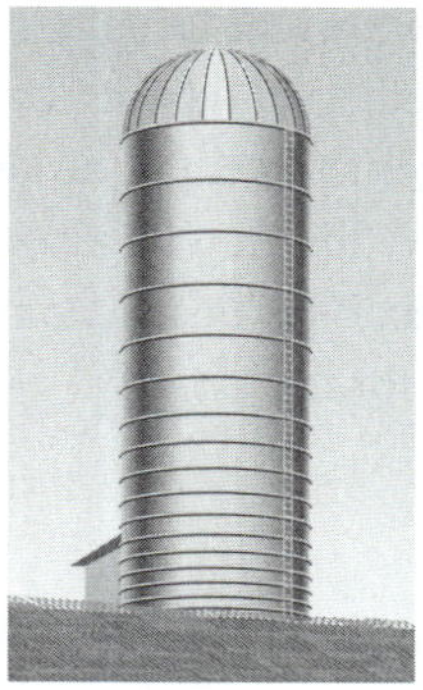

4. A fish rests on the bottom of a bucket of water while the bucket is being weighed. When the fish begins to swim around in the bucket, how does the reading on the scale change?

F. The motion of the fish causes the scale reading to increase.
G. The motion of the fish causes the scale reading to decrease.
H. The buoyant force on the fish is exerted downward on the bucket, causing the scale reading to increase.
J. The mass of the system, and so the scale reading, will remain unchanged.

Use the passage below to answer questions 5–6.

A metal block (ρ = 7900 kg/m^3) is connected to a spring scale by a string 5 cm in length. The block's weight in air is recorded. A second reading is recorded when the block is placed in a tank of fluid and the surface of the fluid is 3 cm below the scale.

5. If the fluid is oil (ρ < 1000 kg/m^3), which of the following must be true?

A. The first scale reading is larger than the second reading.
B. The second scale reading is larger than the first reading.
C. The two scale readings are identical.
D. The second scale reading is zero.

6. If the fluid is mercury (ρ = 13 600 kg/m^3), which of the following must be true?

F. The first scale reading is larger than the second reading.
G. The second scale reading is larger than the first reading.
H. The two scale readings are identical.
J. The second scale reading is zero.

Use the passage below to answer questions 7–8.

Water flows through a pipe of varying width at a constant mass flow rate. At point A the diameter of the pipe is d_A and at point B the diameter of the pipe is d_B.

7. Which of the following equations describes the relationship between the water speed at point A, v_A, and the water speed at point B, v_A?

A. $d_A v_A = d_B v_B$

B. $d_A^2 v_A = d_B^2 v_B$

C. $d_A d_B = v_A v_B$

D. $\frac{1}{2} d_A v_A^2 = \frac{1}{2} d_B v_B^2$

8. If the cross-sectional area of point A is $2.5\ m^2$ and the cross-sectional area of point B is $5.0\ m^2$, how many times faster does the water flow at point A than at point B?

F. $\frac{1}{4}$

G. $\frac{1}{2}$

H. 2

J. 4

SHORT RESPONSE

9. Will an ice cube float higher in water or in mercury? Explain your answer.

10. The approximate inside diameter of the aorta is 1.6 cm, and that of a capillary is 1.0×10^{-6} m. The average flow speed is about 1.0 m/s in the aorta and 1.0 cm/s in the capillaries. If all the blood in the aorta eventually flows through the capillaries, estimate the number of capillaries.

11. A hydraulic brake system is shown below. The area of the piston in the master cylinder is $6.40\ cm^2$, and the area of the piston in the brake cylinder is $1.75\ cm^2$. The coefficient of friction between the brake shoe and wheel drum is 0.50. What is the frictional force between the brake shoe and wheel drum when a force of 44 N is exerted on the pedal?

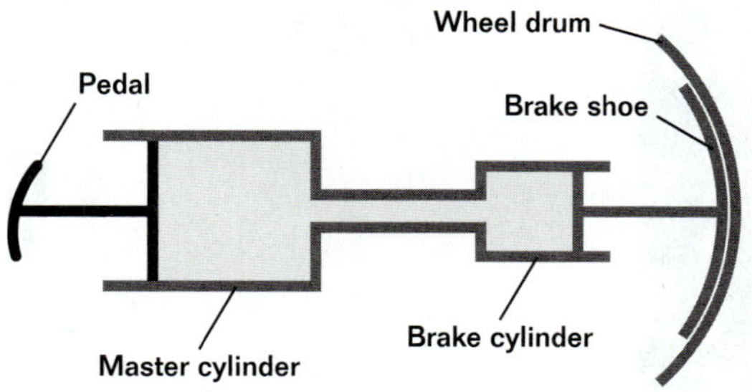

EXTENDED RESPONSE

Base your answers to questions 12–14 on the information below.

Oil, which has a density of $930.0\ kg/m^3$, floats on water. A rectangular block of wood with a height, h, of 4.00 cm and a density of $960.0\ kg/m^3$ floats partly in the water, and the rest floats completely under the oil layer.

12. What is the balanced force equation for this situation?

13. What is the equation that describes y, the thickness of the part of the block that is submerged in water?

14. What is the value for y?

Test TIP For problems involving several forces, write down equations showing how the forces interact.

7. B

8. H

9. mercury; because the density of mercury is greater than that of water

10. 2.5×10^{10} capillaries

11. 6.0 N

12. $F_{B,oil} + F_{B,water} = F_{g,block}$

13. $y = \frac{(\rho_{block} - \rho_{oil})h}{(\rho_{water} - \rho_{oil})}$

14. 1.71×10^{-2} m

Physics and Its World *Timeline 1690–1785*

1690

1698 – The Ashanti empire, the last of the major African kingdoms, emerges in what is now Ghana. The Ashanti's strong centralized government and effective bureaucracy enable them to control the region for nearly two centuries.

1700

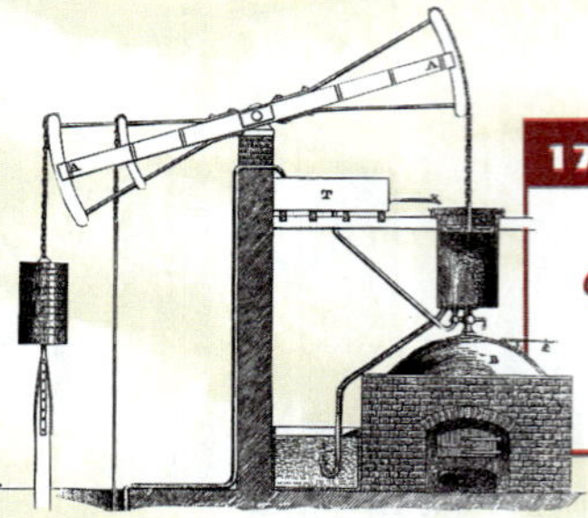

1712

$$eff = \frac{W_{net}}{Q_h}$$

Thomas Newcomen invents the first practical steam engine. Over 50 years later, **James Watt** makes significant improvements to the Newcomen engine.

1710

1715 (approx.) – Chinese writer **Ts'ao Hsüeh-ch'in** is born. The book *The Dream of the Red Chamber,* attributed to him and another writer, is widely regarded today as the greatest Chinese novel.

1721 – **Johann Sebastian Bach** completes the six *Brandenburg Concertos.*

1720

1735 – **John Harrison** constructs the first of four chronometers that will allow navigators to accurately determine a ship's longitude.

1738 – Under the leadership of **Nadir Shah,** the Persian Empire expands into India as the Moghul Empire enters a stage of decline.

1730

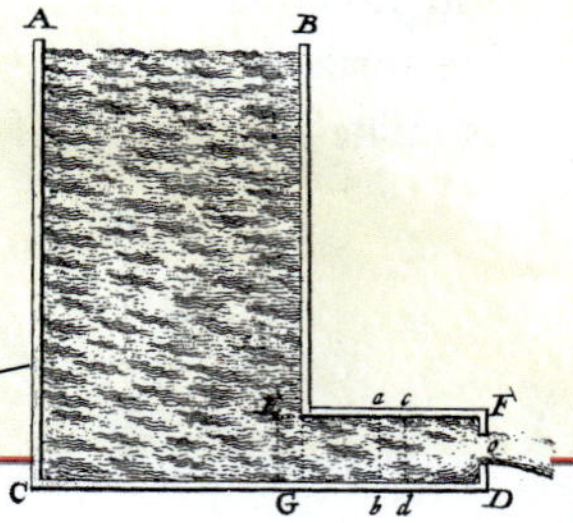

1738

$$P + \frac{1}{2}\rho v^2 + \rho g h = \text{constant}$$

Daniel Bernoulli's *Hydrodynamics,* which includes his research on the mechanical behavior of fluids, is published.

1740

1752

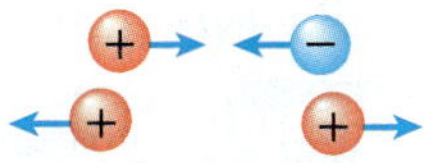

Benjamin Franklin performs the dangerous "kite experiment," in which he demonstrates that lightning consists of electric charge. He would build on the first studies of electricity performed earlier in the century by describing electricity as having positive and negative charge.

1747 – Contrary to the favored idea that heat is a fluid, Russian chemist **Mikhail V. Lomonosov** publishes his hypothesis that heat is the result of motion. Several years later, Lomonosov formulates conservation laws for mass and energy.

1756 – The Seven Year's War begins. British general **James Wolfe** leads the capture of Fort Louisburg, in Canada, in 1758.

1757 – German musician **William Herschel** emigrates to England to avoid fighting in the Seven Year's War. Over the next 60 years, he pursues astronomy, constructing the largest reflecting telescopes of the era and discovering new objects, such as binary stars and the planet Uranus.

1770 – **Antoine Laurent Lavoisier** begins his research on chemical reactions, notably oxidation and combustion.

1772 – **Caroline Herschel,** sister of astronomer **William Herschel,** joins her brother in England. She compiles the most comprehensive star catalog of the era and discovers several nebulae—regions of glowing gas—within our galaxy.

1775 – The American Revolution begins.

1785

$$F_{electric} = k_C\left(\frac{q_1q_2}{r^2}\right)$$

Charles Augustin de Coulomb publishes the results of experiments that will systematically and conclusively prove the inverse-square law for electric force. The law has been suggested for over 30 years by other scientists, such as **Daniel Bernoulli, Joseph Priestly,** and **Henry Cavendish.**

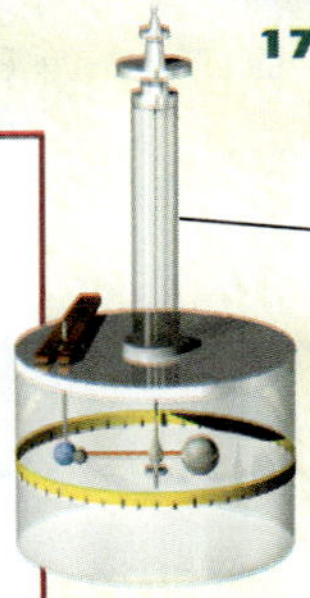

1740

1750

1760

1770

1780

1790

CHAPTER 9

Heat
Planning Guide

Compression Guide

To shorten instruction because of time limitations, omit the opener, Section 3, the lab, and the review.

OBJECTIVES	LABS, DEMONSTRATIONS, AND ACTIVITIES	TECHNOLOGY RESOURCES
PACING • 45 min pp. 296–297 **Chapter Opener**	ANC **Discovery Lab** Temperature and Internal Energy*◆ BASIC	CD **Visual Concepts,** Chapter 9 BASIC
PACING • 90 min pp. 298–304 **Section 1 Temperature and Thermal Equilibrium** • Relate temperature to the kinetic energy of atoms and molecules. • Describe the changes in the temperatures of two objects reaching thermal equilibrium. • Identify the various temperature scales, and convert from one scale to another.	SE **Quick Lab** Sensing Temperature, p. 298 GENERAL TE **Demonstration** Temperature and Internal Energy, p. 299 BASIC	OSP **Lesson Plans** EXT **Integrating Health** Skin Temperature BASIC TR 34 Linear Expansion of Mercury in a Thermometer TR 28A Examples of Different Forms of Energy TR 29A Determining Absolute Zero for an Ideal Gas TR 30A Temperature Scales and Their Uses
PACING • 45 min pp. 305–312 **Section 2 Defining Heat** • Explain heat as the energy transferred between substances that are at different temperatures. • Relate heat and temperature change on the macroscopic level to particle motion on the microscopic level. • Apply the principle of energy conservation to calculate changes in potential, kinetic, and internal energy.	TE **Demonstration** Conduction, p. 308 BASIC SE **Quick Lab** Work and Heat, p. 309 GENERAL TE **Demonstration** Internal Energy, p. 309 GENERAL ANC **Invention Lab** Thermal Conduction*◆ ADVANCED ANC **CBL™ Experiment** Newton's Law of Cooling*◆ ADVANCED	OSP **Lesson Plans** EXT **Integrating Environmental Science** Understanding the Conservation of Energy BASIC TR 35 Transfer of Particles' Kinetic Energy as Heat TR 31A Thermal Units and Their Values in Joules
PACING • 90 min pp. 313–319 *Advanced Level* **Section 3 Changes in Temperature and Phase** • Perform calculations with specific heat capacity. • Interpret the various sections of a heating curve.	SE **Skills Practice Lab** Specific Heat Capacity, pp. 328–331◆ GENERAL ANC **Datasheet** Specific Heat Capacity * GENERAL SE **CBL™ Lab** Specific Heat Capacity, pp. 936–937◆ GENERAL ANC **CBL™ Experiment** Specific Heat Capacity*◆ GENERAL	OSP **Lesson Plans** EXT **Integrating Earth Science** Land and Sea Breezes GENERAL TR 36 A Simple Calorimeter TR 37 Temperature Change of Ice, Water, and Steam with Added Energy TR 32A Specific Heat Capacities TR 33A Latent Heats of Fusion and Vaporization

PACING • 90 min

CHAPTER REVIEW, ASSESSMENT, AND STANDARDIZED TEST PREPARATION

SE **Chapter Highlights,** p. 321
SE **Chapter Review,** pp. 322–325
SE **Graphing Calculator Practice,** p. 324 GENERAL
SE **Alternative Assessment,** p. 325 ADVANCED
SE **Standardized Test Prep,** pp. 326–327 GENERAL
SE **Appendix D: Equations,** pp. 858–859
SE **Appendix I: Additional Problems,** pp. 887–888
ANC **Study Guide Worksheet** Mixed Review* GENERAL
ANC **Chapter Test A*** GENERAL
ANC **Chapter Test B*** ADVANCED
OSP **Test Generator**

Online and Technology Resources

go.hrw.com Holt Online Learning

Visit **go.hrw.com** to find a variety of online resources. To access this chapter's extensions, enter the keyword **HF6HATXT** and click the "go" button. Click **Holt Online Learning** for an online edition of this textbook, and other interactive resources.

HOLT Teacher's One-Stop Planner®

This DVD package includes:

- Holt Calendar Planner
- Customizable Lesson Plans
- Editable Worksheets
- ExamView® Version 6 Assessment Suite
- Interactive Teacher's Edition
- Holt PuzzlePro®
- Holt PowerNotes® Presentations
- MindPoint® Quiz Show

KEY

SE Student Edition	**OSP** One-Stop Planner	**EXT** Online Extension
TE Teacher Edition	**CD** CD or CD-ROM	* Also on One-Stop Planner
ANC Ancillary Worksheet	**TR** Teaching Transparencies	◆ Requires advance prep

SKILLS DEVELOPMENT RESOURCES	REVIEW AND ASSESSMENT	CORRELATIONS
		National Science Education Standards
SE Conceptual Challenge, p. 300 GENERAL **SE Sample Set A** Temperature Conversion, p. 303 BASIC **TE Classroom Practice,** p. 303 BASIC **ANC Problem Workbook** Sample Set A* BASIC **OSP Problem Bank** Sample Set A BASIC	**SE Section Review,** p. 304 GENERAL **ANC Study Guide Worksheet** Section 1* GENERAL **ANC Quiz** Section 1* BASIC	UCP 1, 2, 3, 4, 5 SAI 1, 2 ST 1, 2 SPSP 5 PS 5c
SE Sample Set B Conservation of Energy, pp. 310–311 GENERAL **TE Classroom Practice,** p. 310 GENERAL **ANC Problem Workbook** Sample Set B* GENERAL **OSP Problem Bank** Sample Set B GENERAL	**SE Section Review,** p. 311 GENERAL **ANC Study Guide Worksheet** Section 2* GENERAL **ANC Quiz** Section 2* BASIC	UCP 1, 2, 3, 4 SAI 1, 2 PS 5a, 5b, 5c
SE Sample Set C Calorimetry, pp. 315–316 GENERAL **TE Classroom Practice,** p. 315 GENERAL **ANC Problem Workbook** Sample Set C* GENERAL **OSP Problem Bank** Sample Set C GENERAL **EXT Practice Problems** Latent Heat ADVANCED	**SE Section Review,** p. 319 ADVANCED **ANC Study Guide Worksheet** Section 3* GENERAL **ANC Quiz** Section 3* GENERAL	UCP 1, 2, 3, 4, 5 ST 1, 2 HNS 3 SPSP 2, 5 PS 5a, 5b, 5c

SciLinks NSTA

www.scilinks.org

Maintained by the **National Science Teachers Association.**

Topic: Temperature Scales
SciLinks Code: HF61506

Topic: James Prescott Joule
SciLinks Code: HF60824

Topic: Conduction and Convection
SciLinks Code: HF60338

Topic: Specific Heat Capacity
SciLinks Code: HF61694

Topic: Heat Pumps
SciLinks Code: HF60730

Topic: Greenhouse Gases
SciLinks Code: HF60697

Classroom CD-ROMs

- Guided Reading Audio Program
- Student One Stop
- Virtual Investigations
- Visual Concepts

Holt Lab Generator CD-ROM

Search for any lab by topic, standard, difficulty level, or time. Edit any lab to fit your needs, or create your own labs. Use the Lab Materials QuickList software to customize your lab materials list.

CHAPTER 9
Overview

Section 1 introduces the concepts of temperature, internal energy, and thermal equilibrium and identifies the Fahrenheit, Celsius, and Kelvin temperature scales.

Section 2 relates heat and temperature change to molecular motion and internal energy, discusses thermal conduction, and demonstrates how the conservation of energy can be used to calculate internal energy.

Section 3 introduces specific heat capacity and latent heat and describes the relationship between the amount of energy transferred as heat and either temperature change or phase change.

About the Illustration

Cooking popcorn is a familiar example of the transfer of energy as heat as well as a dramatic example of what happens when water rapidly undergoes a phase change to become steam. The high temperature of the steam and the increase in volume that water undergoes during vaporization cause the popcorn kernels to suddenly explode.

CHAPTER 9

Heat

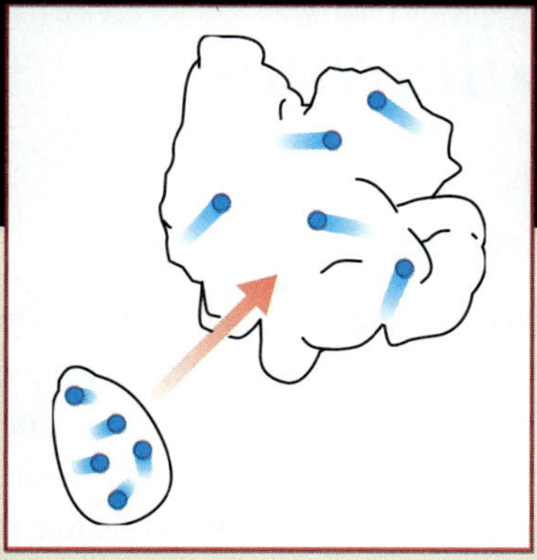

Whether you make popcorn in a pan of hot oil or in a microwave oven, water molecules inside the hard kernels will absorb energy, as shown in the diagram. When the kernels reach a high enough temperature, they rupture. At this point, superheated water suddenly turns into steam and rushes outward, and the kernels burst open to form the fluffy, edible puffs of starch.

WHAT TO EXPECT

In this chapter, you will learn the difference between temperature and heat. You will also learn how different substances change temperature or phase when energy is added to or removed from the substances.

Why it Matters

This type of energy transfer affects many things in the world around you, including making popcorn, turning water into ice cubes, swimming in a sun-warmed pool, and keeping warm in a sleeping bag while camping.

CHAPTER PREVIEW

Tapping Prior Knowledge

Knowledge to Review

- ✔ Work is the product of displacement and the component of force that acts in the same direction as the displacement.
- ✔ Kinetic energy is the energy of an object due to its motion. Potential energy is the energy of an object due to its position, shape, or condition.
- ✔ In the absence of work, the total mechanical energy of a closed system is conserved.

Items to Probe

- ✔ Familiarity with heat and friction: Ask what two metallic objects feel like after they have been rubbed together.
- ✔ Preconceptions about heat and temperature: Ask students what happens when warm things are combined with cooler ones.

297

SECTION 1
General Level

extension

In-Depth Physics Content

Your students can visit go.hrw.com for an online chapter that integrates more in-depth development of the concepts covered here.

Keyword HF6HATX

Misconception Alert

Some students may confuse their perceptions of hot and cold with the temperature of an object; these students think that objects that feel hot have high temperatures and objects that feel cool have low temperatures. The discussion and Quick Lab on this page address this misconception.

Quick Lab

TEACHER'S NOTES

This experiment is meant to demonstrate that whether an object feels hot or cold is not a reliable indicator of the object's temperature. After the experiment, ask students to explain why this is the case *(perceived temperature depends on the temperature difference between the water and your hand)*.

The experiment works best if the ice cubes have just melted and the mixture is stirred to ensure that all parts of the water have the same temperature.

Homework Options

This QuickLab can easily be performed outside of the physics lab room.

SECTION 1

Temperature and Thermal Equilibrium

SECTION OBJECTIVES

- Relate temperature to the kinetic energy of atoms and molecules.
- Describe the changes in the temperatures of two objects reaching thermal equilibrium.
- Identify the various temperature scales, and convert from one scale to another.

DEFINING TEMPERATURE

When you hold a glass of lemonade with ice, such as that shown in **Figure 1,** you feel a sharp sensation in your hand that we describe as "cold." Likewise, you experience a "hot" feeling when you touch a cup of hot chocolate. We often associate temperature with how hot or cold an object feels when we touch it. Our sense of touch serves as a qualitative indicator of temperature. However, this sensation of hot or cold also depends on the temperature of the skin and therefore can be misleading. The same object may feel warm or cool, depending on the properties of the object and on the conditions of your body.

Determining an object's temperature with precision requires a standard definition of temperature and a procedure for making measurements that establish how "hot" or "cold" objects are.

Figure 1

Objects at low temperatures feel cold to the touch, while objects at high temperatures feel hot. However, the sensation of hot and cold can be misleading.

Adding or removing energy usually changes temperature

Consider what happens when you use an electric range to cook food. By turning the dial that controls the electric current delivered to the heating element, you can adjust the element's temperature. As the current is increased, the temperature of the element increases. Similarly, as the current is reduced, the temperature of the element decreases. In general, energy must be either added to or removed from a substance to change its temperature.

Quick Lab

Sensing Temperature

MATERIALS LIST

- 3 identical basins
- hot and cold tap water
- ice

SAFETY

Use only hot tap water. The temperature of the hot water must not exceed 50°C (122°F).

Fill one basin with hot tap water. Fill another with cold tap water, and add ice until about one-third of the mixture is ice. Fill the third basin with an equal mixture of hot and cold tap water.

Place your left hand in the hot water and your right hand in the cold water for 15 s. Then place both hands in the basin of lukewarm water for 15 s. Describe whether the water feels hot or cold to either of your hands.

Temperature is proportional to the kinetic energy of atoms and molecules

The **temperature** of a substance is proportional to the average kinetic energy of particles in the substance. A substance's temperature increases as a direct result of added energy being distributed among the particles of the substance, as shown in **Figure 2.**

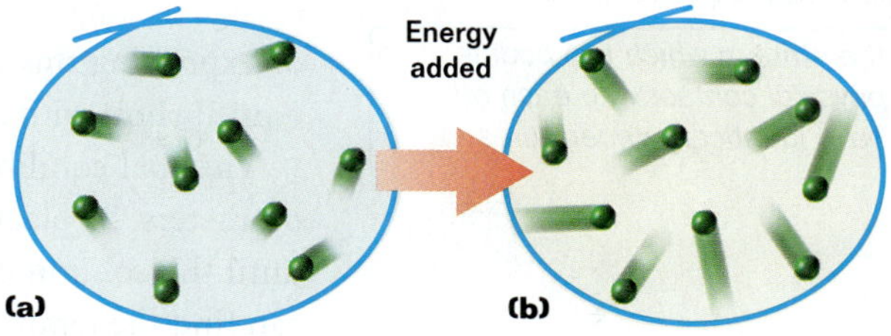

Figure 2
The low average kinetic energy of the particles **(a),** and thus the temperature of the gas, increases when energy is added to the gas **(b).**

A *monatomic* gas contains only one type of atom. For a monatomic gas, temperature can be understood in terms of the translational kinetic energy of the atoms in the gas. For other kinds of substances, molecules can rotate or vibrate, so other types of energy are also present, as shown in **Table 1.**

The energies associated with atomic motion are referred to as **internal energy,** which is proportional to the substance's temperature (assuming no phase change). For an ideal gas, the internal energy depends only on the temperature of the gas. (See "Properties of Gases" in **Appendix J: Advanced Topics** to learn about ideal gases.) For nonideal gases, as well as for liquids and solids, other properties contribute to the internal energy. The symbol U stands for internal energy, and ΔU stands for a change in internal energy.

temperature

a measure of the average kinetic energy of the particles in a substance

internal energy

the energy of a substance due to both the random motions of its particles and to the potential energy that results from the distances and alignments between the particles

Temperature is meaningful only when it is stable

Imagine a can of warm fruit juice immersed in a large beaker of cold water. After about 15 minutes, the can of fruit juice will be cooler and the water

Table 1 Examples of Different Forms of Energy

Form of energy	Macroscopic examples	Microscopic examples	Energy type
Translational	airplane in flight, roller coaster at bottom of rise	CO_2 molecule in linear motion	kinetic energy 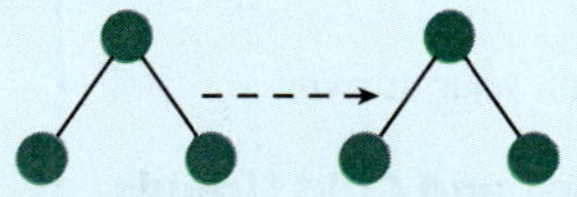
Rotational	spinning top	CO_2 molecule spinning about its center of mass	kinetic energy 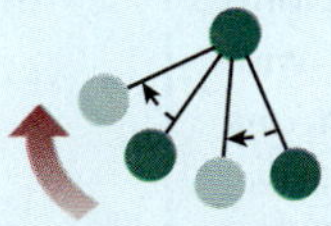
Vibrational	plucked guitar string	bending and stretching of bonds between atoms in a CO_2 molecule	kinetic and potential energy 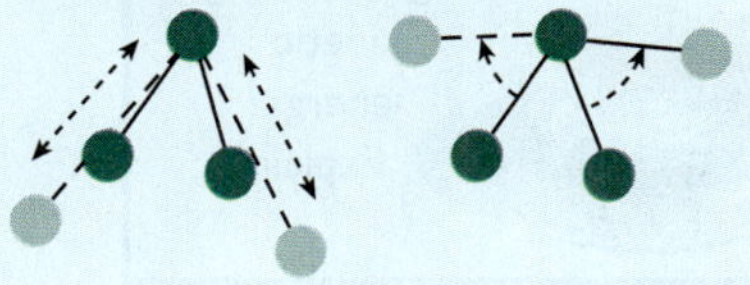

SECTION 1

Teaching Tip — GENERAL

To distinguish between temperature and internal energy, explain that temperature, like color or density, relates to any sample of the object, while internal energy depends on the size (and state) of the sample. Ask students to compare the density, color, temperature, and internal energy of a glass of milk at 20°C with these same properties of half a glass of milk. *(Temperature, like density and color, doesn't change; internal energy, on the other hand, is half as much.)*

Demonstration

Temperature and Internal Energy — BASIC

Purpose Show the relationship between temperature and internal energy.

Materials clear glass of cold water, clear glass of hot water, food coloring

Procedure Place a drop of food coloring in each glass. Let students observe the coloring as it diffuses in each glass without stirring the liquids. Have students note that the coloring diffuses more quickly in the hot water. Ask why this occurs. *(The hot water has more internal energy. So, the molecules of the hot water have a higher kinetic energy and therefore disperse the coloring more quickly.)*

Misconception Alert — GENERAL

Some students do not recognize the universal tendency toward temperature equalization. Because metal sometimes feels colder than wood, students tend to believe that different materials in the same surroundings have different temperatures. Ask them to describe the direction of energy transfer between a variety of objects made of different materials and the air surrounding them. Ask them when this transfer of energy would stop.

ANSWERS

Conceptual Challenge

1. b; If the final temperature were less than 50°C (a) or greater than 60°C (c), energy would not be conserved. For energy to be conserved, the equilibrium temperature must be between the initial temperatures of the substances.
2. The water in the swimming pool has more internal energy. The much larger volume, and therefore the much larger number of particles, more than makes up for the lower temperature; The hot tea has a higher average kinetic energy, because temperature is proportional to average kinetic energy.

thermal equilibrium

the state in which two bodies in physical contact with each other have identical temperatures

surrounding it will be slightly warmer. Eventually, both the can of fruit juice and the water will be at the same temperature. That temperature will not change as long as conditions remain unchanged in the beaker. Another way of expressing this is to say that the water and can of juice are in **thermal equilibrium** with each other.

Thermal equilibrium is the basis for measuring temperature with thermometers. By placing a thermometer in contact with an object and waiting until the column of liquid in the thermometer stops rising or falling, you can find the temperature of the object. The reason is that the thermometer is in thermal equilibrium with the object. Just as in the case of the can of fruit juice in the cold water, the temperature of any two objects in thermal equilibrium always lies between their initial temperatures.

Matter expands as its temperature increases

Increasing the temperature of a gas at constant pressure causes the volume of the gas to increase. This increase occurs not only for gases but also for liquids and solids. In general, if the temperature of a substance increases, so does its volume. This phenomenon is known as *thermal expansion.*

You may have noticed that the concrete roadway segments of a bridge are separated by gaps. This is necessary because concrete expands with increasing temperature. Without these gaps, thermal expansion would cause the segments to push against each other, and they would eventually buckle and break apart.

Different substances undergo different amounts of expansion for a given temperature change. The thermal expansion characteristics of a material are indicated by a quantity called the *coefficient of volume expansion.* Gases have the largest values for this coefficient. Liquids have much smaller values.

In general, the volume of a liquid tends to decrease with decreasing temperature. But, the volume of water increases with decreasing temperature in the range between 0°C and 4°C. Also, as the water freezes, it forms a crystal that has more empty space between the molecules than does liquid water. This explains why ice floats in liquid water. It also explains why a pond freezes from the top down instead of from the bottom up. If this did not happen, fish would likely not survive in freezing temperatures.

Solids typically have the smallest coefficient of volume expansion values. For this reason, liquids in solid containers expand more than the container. This property allows some liquids to be used to measure changes in temperature.

Why it Matters

Conceptual Challenge

1. Hot Chocolate

If two cups of hot chocolate, one at 50°C and the other at 60°C, are poured together in a large container, will the final temperature of the double batch be

a. less than 50°C?
b. between 50°C and 60°C?
c. greater than 60°C?

Explain your answer.

2. Hot and Cold Liquids

A cup of hot tea is poured from a teapot, and a swimming pool is filled with cold water. Which one has a higher total internal energy? Which has a higher average kinetic energy? Explain.

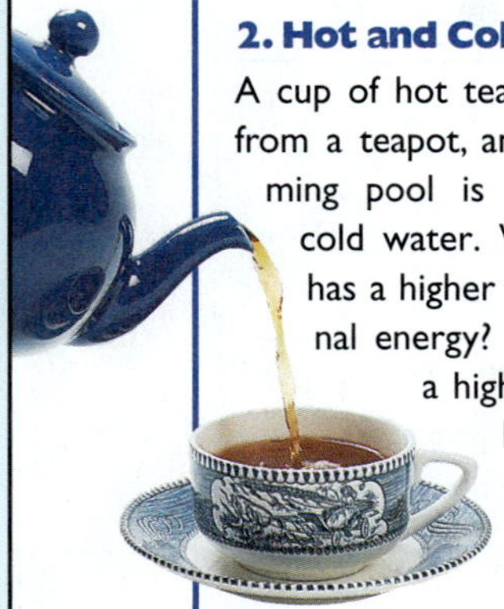

MEASURING TEMPERATURE

In order for a device to be used as a thermometer, it must make use of a change in some physical property that corresponds to changing temperature, such as the volume of a gas or liquid, or the pressure of a gas at constant volume. The most common thermometers use a glass tube containing a thin column of mercury, colored alcohol, or colored mineral spirits. When the thermometer is heated, the volume of the liquid expands. (The cross-sectional area of the tube remains nearly constant during temperature changes.) The change in length of the liquid column is proportional to the temperature change, as shown in **Figure 3.**

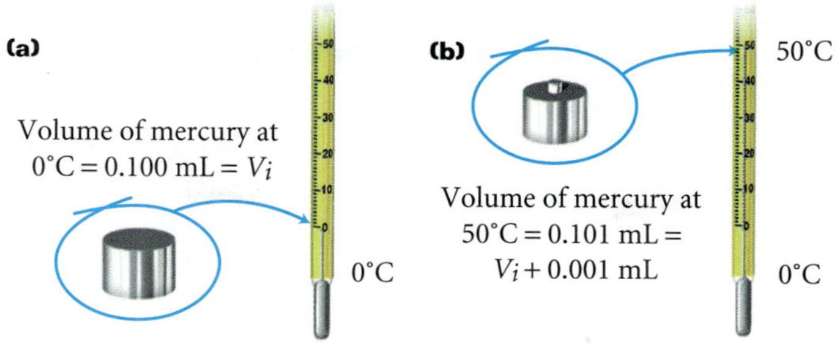

Figure 3
The volume of mercury in this thermometer increases slightly when the mercury's temperature increases from 0°C **(a)** to 50°C **(b)**.

Calibrating thermometers requires fixed temperatures

A thermometer must be more than an unmarked, thin glass tube of liquid; the length of the liquid column at different temperatures must be known. One reference point is etched on the tube and refers to when the thermometer is in thermal equilibrium with a mixture of water and ice at one atmosphere of pressure. This temperature is called the *ice point* or *melting point* of water and is defined as zero degrees Celsius, or 0°C. A second reference mark is made at the point when the thermometer is in thermal equilibrium with a mixture of steam and water at one atmosphere of pressure. This temperature is called the *steam point* or *boiling point* of water and is defined as 100°C.

A temperature scale can be made by dividing the distance between the reference marks into equally spaced units, called *degrees.* This process is based on the assumption that the expansion of the mercury is linear (proportional to the temperature difference), which is a very good approximation.

SCILINKS.
www.scilinks.org
Topic: Temperature Scales
Code: HF61506

Temperature units depend on the scale used

The temperature scales most widely used today are the Fahrenheit, Celsius, and Kelvin scales. The Fahrenheit scale is commonly used in the United States. The Celsius scale is used in countries that have adopted the metric system and by the scientific community worldwide. Celsius and Fahrenheit temperature measurements can be converted to each other using this equation.

CELSIUS-FAHRENHEIT TEMPERATURE CONVERSION

$$T_F = \frac{9}{5}T_C + 32.0$$

$$\textbf{Fahrenheit temperature} = \left(\frac{9}{5} \times \textbf{Celsius temperature}\right) + 32.0$$

The number 32.0 in the equation indicates the difference between the ice point value in each scale. The point at which water freezes is 0.0 degrees on the Celsius scale and 32.0 degrees on the Fahrenheit scale.

Temperature values in the Celsius and Fahrenheit scales can have positive, negative, or zero values. But because the kinetic energy of the atoms in a substance must be positive, the absolute temperature that is proportional to that energy should be positive also. A temperature scale with only positive values is

Did you know?

As a thermometer comes into thermal equilibrium with an object, the object's temperature changes slightly. In most cases the object is so massive compared with the thermometer that the object's temperature change is insignificant.

SECTION 1

Visual Strategy – BASIC

Figure 3
Point out that the tube containing mercury is sealed and that the air above the mercury has been evacuated.

Q Is there more mercury in the tube at 50°C than there is at 0°C?

A ***No, because of thermal expansion, the same amount of mercury occupies a larger volume at 50°C.***

Teaching Tip

Students can learn more about atmospheric pressure in the feature "Fluid Pressure" in **Appendix J: Advanced Topics.**

Teaching Tip — GENERAL

Point out that the first half of the Celsius-Fahrenheit conversion equation ($\frac{9}{5}T_C$) accounts for the difference in the size of a degree in each scale and that the second half of the equation (+32.0) adjusts the zero point of the Fahrenheit scale so that the total number of degrees is measured from the same reference point (in this case, the freezing point of water, 0°C and 32°F). To convert a range of degrees rather than a specific temperature, only the first half of the equation should be used:

$$\Delta T_F = \frac{9}{5}(\Delta T_C)$$

Visual Strategy – BASIC

Figure 4
Point out that, unlike the ideal gas represented in this graph, real gases turn to liquids at low temperatures. Make sure that students interpret the labels and graph properly.

Q What is absolute zero on the Celsius scale?

A *–273.15°C*

Teaching Tip – GENERAL

Compare the Celsius-Kelvin conversion equation with the Celsius-Fahrenheit conversion equation on the previous page. Ask students why T_C is multiplied by $\frac{9}{5}$ in one equation but not in the other *(the size of a degree differs between the Celsius and Fahrenheit scales but not between the Celsius and Kelvin scales)* and why one equation has 32.0 while the other has 273.15 *(the scales have different zero points)*.

The Language of Physics

In common speech, we speak of *hot temperatures* and *cold temperatures.* In physics, we say that objects are *hot* or *cold* relative to our body or senses, but their temperature is *high* or *low* relative to a temperature scale.

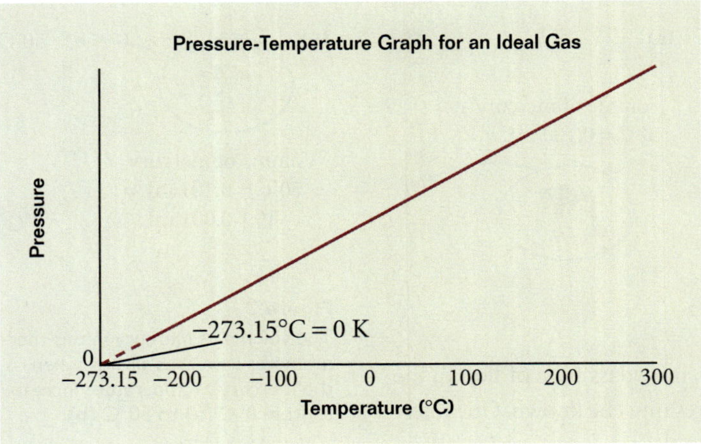

Figure 4
This graph suggests that if the gas's temperature could be lowered to −273.15°C, or 0 K, the gas's pressure would be zero.

suggested in the graph of pressure versus temperature for an ideal gas at constant volume, shown in **Figure 4.** As the gas's temperature decreases, so does its pressure. The graph suggests that if the temperature could be lowered to −273.15°C, the pressure of the sample would be zero. This temperature is designated in the Kelvin scale as 0.00 K, where K is the symbol for the temperature unit called the *kelvin.* Temperatures in this scale are indicated by the symbol T.

A temperature difference of one degree is the same on the Celsius and Kelvin scales. The two scales differ only in the choice of zero point. Thus, the ice point (0.00°C) equals 273.15 K, and the steam point (100.00°C) equals 373.15 K (see **Table 2**). The Celsius temperature can therefore be converted to the Kelvin temperature by adding 273.15.

CELSIUS-KELVIN TEMPERATURE CONVERSION

$$T = T_C + 273.15$$

Kelvin temperature = Celsius temperature + 273.15

Kelvin temperatures for various physical processes can range from around 1 000 000 000 K (10^9 K), which is the temperature of the interiors of the most massive stars, to less than 1 K, which is slightly cooler than the boiling point of liquid helium. The temperature 0 K is often referred to as *absolute zero.* Absolute zero has never been reached, although laboratory experiments have reached temperatures of just a half-billionth of a degree above absolute zero.

Table 2 Temperature Scales and Their Uses

Scale	Ice point	Steam point	Applications
Fahrenheit	32°F	212°F	meteorology, medicine, and non-scientific uses (U.S.)
Celsius	0°C	100°C	meteorology, medicine, and non-scientific uses (outside U.S.); other sciences (international)
Kelvin (absolute)	273.15 K	373.15 K	physical chemistry, gas laws, astrophysics, thermodynamics, low-temperature physics

SAMPLE PROBLEM A

Temperature Conversion

PROBLEM

What are the equivalent Celsius and Kelvin temperatures of 50.0°F?

SOLUTION

Given: $T_F = 50.0°F$

Unknown: $T_C = ?$ $T = ?$

Use the Celsius-Fahrenheit equation to convert Fahrenheit into Celsius.

$$T_F = \frac{9}{5}T_C + 32.0$$

$$T_C = \frac{5}{9}(T_F - 32.0)$$

$$T_C = \frac{5}{9}(50.0 - 32.0)°C = 10.0°C$$

Use the Celsius-Kelvin equation to convert Celsius into Kelvin.

$$T = T_C + 273.15$$

$$T = (10.0 + 273.15)\text{K} = 283.2 \text{ K}$$

$$\boxed{T_C = 10.0°C \quad T = 283.2 \text{ K}}$$

PRACTICE A

Temperature Conversion

1. The lowest outdoor temperature ever recorded on Earth is −128.6°F, recorded at Vostok Station, Antarctica, in 1983. What is this temperature on the Celsius and Kelvin scales?
2. The temperatures of one northeastern state range from 105°F in the summer to −25°F in winter. Express this temperature range in degrees Celsius and in kelvins.
3. The normal human body temperature is 98.6°F. A person with a fever may record 102°F. Express these temperatures in degrees Celsius.
4. A pan of water is heated from 23°C to 78°C. What is the change in its temperature on the Kelvin and Fahrenheit scales?
5. Liquid nitrogen is used to cool substances to very low temperatures. Express the boiling point of liquid nitrogen (77.34 K at 1 atm of pressure) in degrees Celsius and in degrees Fahrenheit.

Classroom Practice

Temperature Conversion
One day it was −40°C at the top of Mont Blanc and −40°F at the top of Mount Whitney. Which place was colder?

Answer
neither (−40°C = −40°F)

PROBLEM GUIDE A

Use this guide to assign problems.
SE = Student Edition Textbook
PW = Problem Workbook
PB = Problem Bank on the One-Stop Planner (OSP)

Solving for:

T_C	**SE**	Sample, 1–3, 5; Ch. Rvw. 9–10
	PW	4–5
	PB	Sample, 1–3, 8–10
T_F	**SE**	4–5; Ch. Rvw. 38
	PW	Sample, 1–3, 5
	PB	Sample, 1–3
T	**SE**	Sample, 1–2, 4; Ch. Rvw. 9–10, 38
	PW	Sample, 1–2, 6–7
	PB	4–7

***Challenging Problem**
Consult the printed Solutions Manual or the OSP for detailed solutions.

ANSWERS

Practice A

1. −89.22°C, 183.93 K
2. 41°C to −32°C, 314 K to 241K
3. 37.0°C, 39°C
4. 55 K, 99°F
5. −195.81°C, −320.5°F

SECTION 1

SECTION REVIEW ANSWERS

1. The pan's temperature decreases if the water's temperature increases; The water and pan have reached thermal equilibrium when their temperatures are the same.
2. −183.0°C, −297.4°F
3. a. 119.0°C
   b. 246.2°F, 832.3°F
   c. 392.2 K, 717.8 K
4. a, c
5. c, a, b; The blue particles in (c) have the greatest average kinetic energy and therefore have the highest temperature.
6. Popcorn kernels pop when moisture inside the kernel turns into steam, creating pressure that ruptures the shell. Old kernels that have dried out may not have enough moisture to pop; Adding moisture to the kernels will make more of them pop.

SECTION REVIEW

extension

Integrating Health

Visit go.hrw.com for the activity "Skin Temperature."

Keyword HF6HATX

1. A hot copper pan is dropped into a tub of water. If the water's temperature rises, what happens to the temperature of the pan? How will you know when the water and copper pan reach thermal equilibrium?
2. Oxygen condenses into a liquid at approximately 90.2 K. To what temperature does this correspond on both the Celsius and Fahrenheit temperature scales?
3. The boiling point of sulfur is 444.6°C. Sulfur's melting point is 586.1°F lower than its boiling point.
   a. Determine the melting point of sulfur in degrees Celsius.
   b. Find the melting and boiling points in degrees Fahrenheit.
   c. Find the melting and boiling points in kelvins.
4. Which of the following is true for popcorn kernels and the water molecules inside them during popping?
   a. The temperature of the kernels increases.
   b. The water molecules are destroyed.
   c. The kinetic energy of the water molecules increases.
   d. The mass of the water molecules changes.
5. **Interpreting Graphics** Two gases that are in physical contact with each other consist of particles of identical mass. In what order should the images shown in **Figure 5** be placed to correctly describe the changing distribution of kinetic energy among the gas particles? Which group of particles has the highest temperature at any time? Explain.

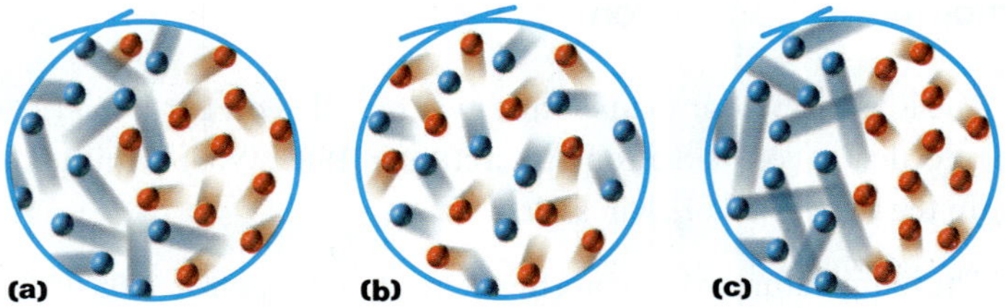

Figure 5

6. **Critical Thinking** Have you ever tried to make popcorn and found that most of the kernels did not pop, as shown in **Figure 6**? What might be the reason that they did not pop? What could you do to try to make more of the kernels pop?

Figure 6

SECTION 2

Defining Heat

SECTION OBJECTIVES

- **Explain heat as the energy transferred between substances that are at different temperatures.**
- **Relate heat and temperature change on the macroscopic level to particle motion on the microscopic level.**
- **Apply the principle of energy conservation to calculate changes in potential, kinetic, and internal energy.**

HEAT AND ENERGY

Thermal physics often appears mysterious at the macroscopic level. Hot objects become cool without any obvious cause. To understand thermal processes, it is helpful to shift attention to the behavior of atoms and molecules. Mechanics can be used to explain much of what is happening at the molecular, or microscopic, level. This in turn accounts for what you observe at the macroscopic level. Throughout this chapter, the focus will shift between these two viewpoints.

What happens when you immerse a warm fruit juice bottle in a container of ice water, as shown in **Figure 7**? As the temperatures of the bottle and of the juice decrease, the water's temperature increases slightly until both final temperatures are the same. Energy is transferred from the bottle of juice to the water because the two objects are at different temperatures. This energy that is transferred is defined as **heat.**

heat

the energy transferred between objects because of a difference in their temperatures

The word *heat* is sometimes used to refer to the *process* by which energy is transferred between objects because of a difference in their temperatures. This textbook will use *heat* to refer only to the energy itself.

Energy is transferred between substances as heat

From a macroscopic viewpoint, energy transferred as heat tends to move from an object at higher temperature to an object at lower temperature. This is similar to the mechanical behavior of objects moving from a higher gravitational potential energy to a lower gravitational potential energy. Just as a pencil will drop from your desk to the floor but will not jump from the floor to your desk, so energy will travel spontaneously from an object at higher temperature to one at lower temperature and not the other way around.

Figure 7
Energy is transferred as heat from objects with higher temperatures (the fruit juice and bottle) to those with lower temperatures (the cold water).

SECTION 2
General Level

The Language of Physics

Some texts refer to internal energy as *heat energy* and the transfer of internal energy as *heat flow.* This language results from earlier theories of heat, which assumed that heat is a substance that flows from one object to another. To avoid confusion, this terminology is not used in this text.

Misconception Alert

Many students think of "cold" and "heat" as substances that flow from one object to another. Point out that in all cases, *energy* is transferred from one object to another. Heat and cold do not flow between objects, but the energy transferred does change the temperature of the objects involved because the distribution of internal energy changes. Also, be sure students understand that heat is not *in* a body; it is the amount of energy that is transferred between two bodies. In this respect, heat is similar to work. Work is not *in* a body but represents the amount of energy transferred from one body to another.

Teaching Tip — GENERAL

Point out that in **Figure 7,** the ice in the water must melt before the water's temperature can increase.

The Language of Physics

The *internal energy, U,* also called the *thermal energy* of an object, is the total energy of all of the molecules in an object, including both their kinetic energy and their potential energy. However, in this chapter we are primarily concerned with changes in the kinetic energy of the molecules.

Misconception Alert — GENERAL

Many students do not clearly distinguish between temperature, heat, and internal energy. Temperature measures the average kinetic energy of the molecules in an object. Heat is the energy transferred from one object to another because of a difference in their temperatures. Internal energy is the sum of the energies of the molecules. Drawing schematic representations of these statements using a specific example will help students understand these relationships.

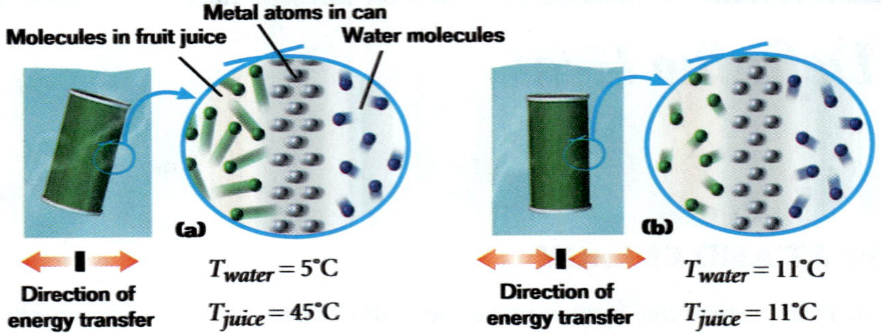

Figure 8
Energy is transferred as heat from the higher-energy particles to lower-energy particles **(a)**. The net energy transferred is zero when thermal equilibrium is reached **(b)**.

SCILINKS.
www.scilinks.org
Topic: James Prescott Joule
Code: HF60824

The direction in which energy travels as heat can be explained at the atomic level. Consider a warm can of fruit juice in ice water. At first, the molecules in the fruit juice have a higher average kinetic energy than do the water molecules that surround the can, as shown in **Figure 8(a).** This energy is transferred from the juice to the can by the juice molecules colliding with the metal atoms of the can. The atoms vibrate more because of their increased energy. This energy is then transferred to the surrounding water molecules, as shown in **Figure 8(b).**

As the energy of the water molecules gradually increases, the energy of the fruit juice's molecules and of the can's atoms decreases until all of the particles have, on the average, equal kinetic energies. In individual collisions, energy may be transferred from the lower-energy water molecules to the higher-energy metal atoms and fruit juice particles. That is, energy can be transferred in either direction. However, because the average kinetic energy of particles is higher in the object at higher temperature, more energy moves out of the object as heat than moves into it. Thus, the net transfer of energy as heat is in only one direction.

The transfer of energy as heat alters an object's temperature

Thermal equilibrium may be understood in terms of energy exchange between two objects at equal temperature. When the can of fruit juice and the surrounding water are at the same temperature, as depicted in **Figure 9,** the quantity of energy transferred from the can of fruit juice to the water is the same as the energy transferred from the water to the can of juice. The net energy transferred between the two objects is zero.

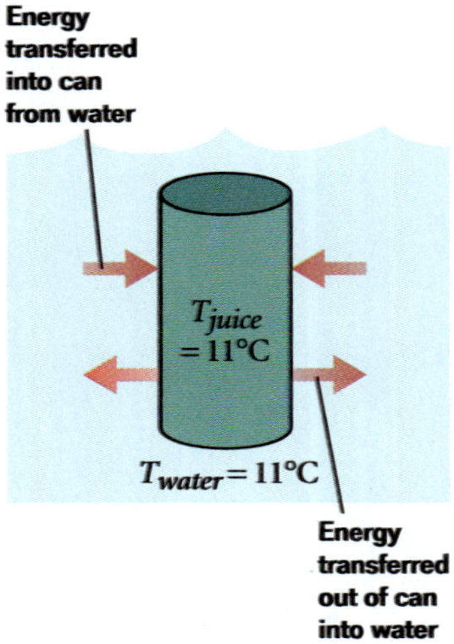

Figure 9
At thermal equilibrium, the net energy exchanged between two objects equals zero.

This reveals the difference between temperature and heat. The atoms of all objects are in continuous motion, so all objects have some internal energy. Because temperature is a measure of that energy, all objects have some temperature. Heat, on the other hand, is the energy transferred from one object to another because of the temperature difference between them. When there is no temperature difference between a substance and its surroundings, no net energy is transferred as heat.

Energy transfer as heat depends on the difference of the temperatures of the two objects. The greater the temperature difference is between two objects, the greater the rate of energy transfer between them as heat (other factors being the same).

For example, in winter, energy is transferred as heat from a car's surface at 30°C to a cold raindrop at 5°C. In the summer, energy is transferred as heat from a car's surface at 45°C to a warm raindrop at 20°C. In each case, the amount of energy transferred each second is the same, because the substances and the temperature difference (25°C) are the same. See **Figure 10.**

The concepts of heat and temperature help to explain why hands held in separate bowls containing hot and cold water subsequently sense the temperature of lukewarm water differently. The nerves in the outer skin of your hand detect energy passing through the skin from objects with temperatures different from your body temperature. If one hand is at thermal equilibrium with cold water, more energy is transferred from the outer layers of your hand than can be replaced by the blood, which has a temperature of about 37.0°C (98.6°F). When the hand is immediately placed in water that is at a higher temperature, energy is transferred from the water to the cooler hand. The energy transferred into the skin causes the water to feel warm. Likewise, the hand that has been in hot water temporarily gains energy from the water. The loss of this energy to the lukewarm water makes that water feel cool.

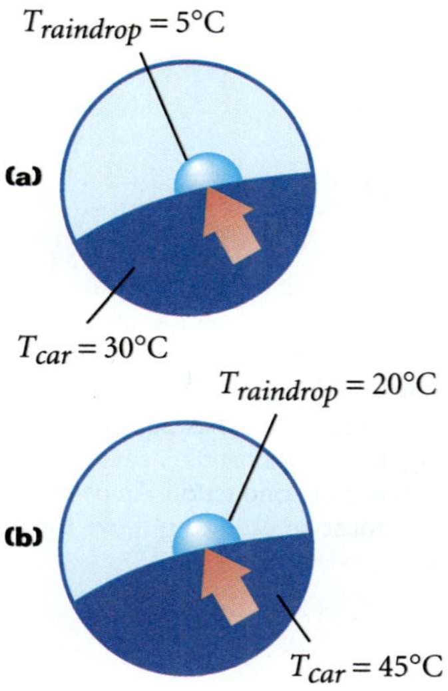

Figure 10
The energy transferred each second as heat from the car's surface to the raindrop is the same for low temperatures **(a)** as for high temperatures **(b),** provided the temperature differences are the same.

Heat has the units of energy

Before scientists arrived at the modern model for heat, several different units for measuring heat had already been developed. These units are still widely used in many applications and therefore are listed in **Table 3.** Because heat, like work, is energy in transit, all heat units can be converted to joules, the SI unit for energy.

Just as other forms of energy have a symbol that identifies them (*PE* for potential energy, *KE* for kinetic energy, *U* for internal energy, *W* for work), heat is indicated by the symbol *Q*.

Table 3 Thermal Units and Their Values in Joules

Heat unit	Equivalent value	Uses
joule (J)	equal to $1\ kg \bullet \left(\frac{m^2}{s^2}\right)$	SI unit of energy
calorie (cal)	4.186 J	non-SI unit of heat; found especially in older works of physics and chemistry
kilocalorie (kcal)	4.186×10^3 J	non-SI unit of heat
Calorie, or dietary Calorie	4.186×10^3 J = 1 kcal	food and nutritional science
British thermal unit (Btu)	1.055×10^3 J	English unit of heat; used in engineering, air-conditioning, and refrigeration
therm	1.055×10^8 J	equal to 100 000 Btu; used to measure natural-gas usage

Key Models and Analogies — GENERAL

Just as the work, *W*, done on an object equals the change in kinetic energy of the object ($W = \Delta KE$ in the work–kinetic energy theorem), the heat, *Q*, transferred to or from an object equals the increase or decrease of the internal energy in the object through temperature change ($Q = \Delta U$) if no work is done.

The Language of Physics

Texts that use the term *heat energy* (rather than internal energy) and *heat flow* (rather than heat) likewise refer to *Q* as heat flow rather than heat.

Teaching Tip — BASIC

Make sure students understand the meaning of the symbols in the table. Remind students that SI stands for *Système International d'Unités*, or International System of Units.

Demonstration

Conduction — Basic

Purpose Show that the rate of conduction depends on the materials involved.

Materials spoons made of different materials, cups of hot water, thermometer

Procedure Fill each of the cups with the same amount of hot water. Verify that the water in each of the cups is at the same temperature. Place a different spoon in each cup. After several minutes, have students feel the handle of each spoon and rank the materials as thermal conductors from best to poorest. Silver spoons will rank at the top. Point out that conduction occurs both between the water and the spoons and between the spoons and students' hands. The spoon that feels hottest is the best conductor, although you cannot tell which spoon actually has a greater temperature.

Figure 11
After this burner has been turned on, the skillet's handle heats up because of conduction. An oven mitt must be used to remove the skillet safely.

THERMAL CONDUCTION

When you first place an iron skillet on a stove, the metal handle feels comfortable to the touch. After a few minutes, the handle becomes too hot to touch without a cooking mitt, as shown in **Figure 11.** The handle is hot because energy was transferred from the high-temperature burner to the skillet. The added energy increased the temperature of the skillet and its contents. This type of energy transfer is called *thermal conduction.*

The rate of thermal conduction depends on the substance

Thermal conduction can be understood by the behavior of atoms in a metal. As the skillet is heated, the atoms nearest to the burner vibrate with greater energy. These vibrating atoms jostle their less energetic neighbors and transfer some of their energy in the process. Gradually, iron atoms farther away from the element gain more energy.

The rate of thermal conduction depends on the properties of the substance being heated. A metal ice tray and a cardboard package of frozen food removed from the freezer are at the same temperature. However, the metal tray feels colder than the package because metal conducts energy more easily and more rapidly than cardboard does. Substances that rapidly transfer energy as heat are called *thermal conductors.* Substances that slowly transfer energy as heat are called *thermal insulators.* In general, metals are good thermal conductors. Materials such as asbestos, cork, ceramic, cardboard, and fiberglass are poor thermal conductors (and therefore good thermal insulators).

Did you know?

Although cooking oil is no better a thermal conductor than most nonmetals are, it is useful for transferring energy uniformly around the surface of the food being cooked. When popping popcorn, for instance, coating the kernels with oil improves the energy transfer to each kernel, so a higher percentage of them pop.

Convection and radiation also transfer energy

There are two other mechanisms for transferring energy between places or objects at different temperatures. *Convection* involves the movement of cold and hot matter, such as hot air rising upward over a flame. This mechanism does not involve heat alone. Instead, it uses the combined effects of pressure differences, conduction, and buoyancy. In the case of air over a flame, the air is heated through particle collisions (conduction), causing it to expand and its density to decrease. The warm air is then displaced by denser, colder air. Thus, the flame heats the air faster than by conduction alone.

The other principal energy transfer mechanism is *electromagnetic radiation.* Unlike convection, energy in this form does not involve the transfer of matter. Instead, objects reduce their internal energy by giving off electromagnetic radiation of particular wavelengths or are heated by electromagnetic radiation like a car is heated by the absorption of sunlight.

SCILINKS.
www.scilinks.org
Topic: Conduction and Convection
Code: HF60338

HEAT AND WORK

Hammer a nail into a block of wood. After several minutes, pry the nail loose from the block and touch the side of the nail. It feels warm to the touch, indicating that energy is being transferred from the nail to your hand. Work is done in pulling the nail out of the wood. The nail encounters friction with the wood, and most of the energy required to overcome this friction is transformed into internal energy. The increase in the internal energy of the nail raises the nail's temperature, and the temperature difference between the nail and your hand results in the transfer of energy to your hand as heat.

Friction is just one way of increasing a substance's internal energy. In the case of solids, internal energy can be increased by deforming their structure. Common examples of this deformation are stretching a rubber band or bending a piece of metal.

Total energy is conserved

When the concept of mechanical energy was introduced, you discovered that whenever friction between two objects exists, not all of the work done appears as mechanical energy. Similarly, when objects collide inelastically, not all of their initial kinetic energy remains as kinetic energy after the collision. Some of the energy is absorbed as internal energy by the objects. For this reason, in the case of the nail pulled from the wood, the nail (and if you could touch it, the wood inside the hole) feels warm. If changes in internal energy are taken into account along with changes in mechanical energy, the total energy is a universally conserved property. In other words, the sum of the changes in potential, kinetic, and internal energy is equal to zero.

CONSERVATION OF ENERGY

$$\Delta PE + \Delta KE + \Delta U = 0$$

the change in potential energy + the change in kinetic energy + the change in internal energy = 0

extension

Integrating Environmental Science

Visit go.hrw.com for the activity "Understanding the Conservation of Energy."

Keyword HF6HATX

Quick Lab

Work and Heat

MATERIALS LIST

- **1 large rubber band about 7–10 mm wide**

SAFETY

To avoid breaking the rubber band, do not stretch it more than a few inches. Do not point a stretched rubber band at another person.

Hold the rubber band between your thumbs. Touch the middle section of the rubber band to your lip and note how it feels. Rapidly stretch the rubber band and keep it stretched. Touch the middle section of the rubber band to your lip again. Notice whether the rubber band's temperature has changed. (You may have to repeat this procedure several times before you can clearly distinguish the temperature difference.)

SECTION 2

Demonstration

Internal Energy — GENERAL

Purpose Show the conversion of work into internal energy.

Materials small mixing bowl, maple syrup, thermometer, electric egg beater

Procedure Fill about $\frac{1}{4}$ of the mixing bowl with maple syrup. Measure and record the temperature. Run the electric beater in the bowl for about 10 min, and record the final temperature. Have students describe the chain of energy transfers that took place *(work done by the beater converted to kinetic energy of the syrup molecules, or the internal energy of the syrup)*. The experiment can be quantified by recording the temperature of the syrup at regular intervals of time throughout the experiment.

Quick Lab

TEACHER'S NOTES

This exercise is meant to demonstrate how work increases an object's internal energy.

Some students may assert that the increase in temperature is caused by the transfer of energy from their hand to the rubber band. This should be recognized as an astute observation. Address it by comparing results reported by students with warm hands with those reported by students with cold hands.

Homework Options

This QuickLab can easily be performed outside of the physics lab room.

PROBLEM GUIDE B

Use this guide to assign problems.
SE = Student Edition Textbook
PW = Problem Workbook
PB = Problem Bank on the One-Stop Planner (OSP)

Solving for:

ΔU	SE Sample, 1–3; Ch. Rvw. 24–25, 34* PW 5–6, 7* PB 8–10
m	PW Sample, 1–2 PB 5–7
PE	PW 3–4 PB Sample, 1–4

*Challenging Problem
Consult the printed Solutions Manual or the OSP for detailed solutions.

Classroom Practice

Conservation of Energy
A 0.10 kg ball falls 10.0 m onto a hard floor and then bounces back up to 9.0 m. How much of its mechanical energy is transformed to the internal energy of the ball and the floor?

Answer
0.98 J

SAMPLE PROBLEM B

Conservation of Energy

PROBLEM

An arrangement similar to the one used to demonstrate energy conservation is shown at right. A vessel contains water. Paddles that are propelled by falling masses turn in the water. This agitation warms the water and increases its internal energy. The temperature of the water is then measured, giving an indication of the water's internal-energy increase. If a total mass of 11.5 kg falls 1.3 m and all of the mechanical energy is converted to internal energy, by how much will the internal energy of the water increase? (Assume no energy is transferred as heat out of the vessel to the surroundings or from the surroundings to the vessel's interior.)

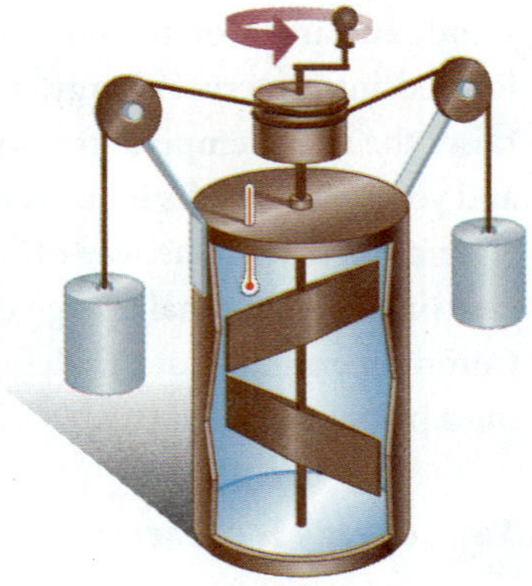

Joule's Apparatus

SOLUTION

1. DEFINE **Given:** $m = 11.5$ kg $\quad h = 1.3$ m $\quad g = 9.81$ m/s^2

Unknown: $\Delta U = ?$

2. PLAN **Choose an equation or situation:**
Use the conservation of energy equation, and solve for ΔU.

$$\Delta PE + \Delta KE + \Delta U = 0$$

$$(PE_f - PE_i) + (KE_f - KE_i) + \Delta U = 0$$

$$\Delta U = -PE_f + PE_i - KE_f + KE_i$$

TIP *Don't forget that a change in any quantity, indicated by the symbol Δ, equals the final value minus the initial value.*

Because the masses begin at rest, KE_i equals zero. If we assume that KE_f is small compared to the loss of PE, we can set KE_f equal to zero also.

$$KE_f = 0 \qquad KE_i = 0$$

Because all of the potential energy is assumed to be converted to internal energy, PE_i can be set equal to mgh if PE_f is set equal to zero.

$$PE_i = mgh \qquad PE_f = 0$$

Substitute each quantity into the equation for ΔU:

$$\Delta U = 0 + mgh + 0 + 0 = mgh$$

3. CALCULATE **Substitute the values into the equation and solve:**

$$\Delta U = (11.5 \text{ kg})(9.81 \text{ m/s}^2)(1.3 \text{ m})$$

$$\boxed{\Delta U = 1.5 \times 10^2 \text{ J}}$$

CALCULATOR SOLUTION

Because the minimum number of significant figures in the data is two, the calculator answer, 146.6595 J, should be rounded to two digits.

4. EVALUATE The answer can be estimated using rounded values for m and g. If $m \approx 10$ kg and $g \approx 10$ m/s^2, then $\Delta U \approx 130$ J, which is close to the actual value calculated.

PRACTICE B

Conservation of Energy

1. In the arrangement described in Sample Problem B, how much would the water's internal energy increase if the mass fell 6.69 m?
2. A worker drives a 0.500 kg spike into a rail tie with a 2.50 kg sledgehammer. The hammer hits the spike with a speed of 65.0 m/s. If one-third of the hammer's kinetic energy is converted to the internal energy of the hammer and spike, how much does the total internal energy increase?
3. A 3.0×10^{-3} kg copper penny drops a distance of 50.0 m to the ground. If 65 percent of the initial potential energy goes into increasing the internal energy of the penny, determine the magnitude of that increase.
4. The amount of internal energy needed to raise the temperature of 0.25 kg of water by 0.2°C is 209.3 J. How fast must a 0.25 kg baseball travel in order for its kinetic energy to equal this internal energy?

SECTION REVIEW

1. Use the microscopic interpretations of temperature and heat to explain how you can blow on your hands to warm them and also blow on a bowl of hot soup to cool it.
2. If a bottle of water is shaken vigorously, will the internal energy of the water change? Why or why not?
3. At Niagara Falls, if 505 kg of water fall a distance of 50.0 m, what is the increase in the internal energy of the water at the bottom of the falls? Assume that all of the initial potential energy goes into increasing the water's internal energy and that the final kinetic energy is zero.
4. **Critical Thinking** A bottle of water at room temperature is placed in a freezer for a short time. An identical bottle of water that has been lying in the sunlight is placed in a refrigerator for the same amount of time. What must you know to determine which situation involves more energy transfer?
5. **Critical Thinking** On a camping trip, your friend tells you that fluffing up a down sleeping bag before you go to bed will keep you warmer than sleeping in the same bag when it is still crushed from being in its storage sack. Explain why this happens.

SECTION 2

ANSWERS

Practice B

1. 755 J
2. 1.76×10^3 J
3. 0.96 J
4. 41 m/s

SECTION REVIEW ANSWERS

1. In the first case, molecules in the exhaled air have a greater average kinetic energy than the air surrounding your cold hands. Energy is transferred to the hands, causing their temperature to increase. In the second case, the molecules in the soup have a greater average kinetic energy than the exhaled air passing over the soup's surface. Energy is therefore transferred from the soup to the relatively cooler air.
2. yes; Shaking the bottle adds kinetic energy to the system, and this kinetic energy is converted into the internal energy of the water molecules.
3. 2.48×10^5 J
4. You would need to know the comparative temperature changes for each bottle of water (assuming the water doesn't freeze in either case).
5. Because air is a good insulator, you will not feel as warm if the insulating layer of air is squeezed out of the sleeping bag.

Why it Matters

Climate and Clothing

Extension ADVANCED

Have students conduct research to find additional examples of specific clothing types that are designed for particular climates. Encourage them to search for unusual examples. Ask each student to choose one example and to create a brochure that "sells" the clothing by explaining how the clothing suits a particular climate. Students should include photographs or illustrations in their brochures.

Why it Matters

Climate and Clothing

To remain healthy, the human body must maintain a temperature of about 37.0°C (98.6°F), which becomes increasingly difficult as the surrounding air becomes hotter or colder than body temperature.

Unless the body is properly insulated, its temperature will drop in its attempt to reach thermal equilibrium with very cold surroundings. If this situation is not corrected in time, the body will enter a state of hypothermia, which lowers pulse, blood pressure, and respiration. Once body temperature reaches 32.2°C (90.0°F), a person can lose consciousness. When body temperature reaches 25.6°C (78.0°F), hypothermia is almost always fatal.

To prevent hypothermia, the transfer of energy from the human body to the surrounding air must be hindered, which is done by surrounding the body with heat-insulating material. An extremely effective and common thermal insulator is air. Like most gases, air is a very poor thermal conductor, so even a thin layer of air near the skin provides a barrier to energy transfer.

The Inupiat parka, called an *atigi*, consists today of a canvas shell over sheepskin. The wool provides layers of insulating air between the wearer and the cold.

The Inupiat people of northern Alaska have designed clothing to protect them from the severe Arctic climate, where average air temperatures range from 10°C (50°F) to −37°C (−35°F). The Inupiat clothing is made from animal skins that make use of air's insulating properties. Until recently, the traditional parka (*atigi*) was made from caribou skins. Two separate parkas are worn in layers, with the fur lining the inside of the inner parka and the outside of the outer parka. Insulation is provided by air that is trapped between the short inner hairs and within the long, hollow hairs of the fur. Today, inner parkas are made from sheepskin, as shown on the left.

At the other extreme, the Bedouins of the Arabian Desert have developed clothing that permits them to survive another of the harshest environments on Earth. Bedouin garments cover most of the body, which protects the wearer from direct sunlight and prevents excessive loss of body water from evaporation. These clothes are also designed to cool the wearer. The Bedouins must keep their body temperatures from becoming too high in desert temperatures, which often are in excess of 38°C (100°F). Heat exhaustion or heatstroke will result if the body's temperature becomes too high.

The Bedouin headcloth, called a *kefiyah,* employs evaporation to remove energy from the air close to the head, which cools the wearer.

Although members of different tribes, as well as men and women within the same tribes, wear different types of clothing, a few basic garments are common to all Bedouins. One such garment is the *kefiyah,* a headcloth worn by Bedouin men, as shown in the photograph above. A similar garment made of two separate cloths, which are called a *mandil* and a *hatta,* is worn by Bedouin women. Firmly wrapped around the head of the wearer, the cloth absorbs perspiration and cools the wearer during evaporation. This same garment is also useful during cold periods in the desert. The garment, wound snugly around the head, has folds that trap air and provide an insulating layer to keep the head warm.

Changes in Temperature and Phase

SECTION 3

SECTION OBJECTIVES

- **Perform calculations with specific heat capacity.**
- **Interpret the various sections of a heating curve.**

SPECIFIC HEAT CAPACITY

On a hot day, the water in a swimming pool, such as the one shown **Figure 12,** may be cool, even if the air around it is hot. This may seem odd, because both the air and water receive energy from sunlight. One reason that the water may be cooler than the air is evaporation, which is a cooling process.

However, evaporation is not the only reason for the difference. Experiments have shown that the change in temperature due to adding or removing a given amount of energy depends on the particular substance. In other words, the same change in energy will cause a different temperature change in equal masses of different substances. This fact is due to differences in the motion of atoms and molecules at the microscopic level.

The **specific heat capacity** of a substance is defined as the energy required to change the temperature of 1 kg of that substance by 1°C. (This quantity is also sometimes known as just *specific heat.*) Every substance has a unique specific heat capacity. This value tells you how much the temperature of a given mass of that substance will increase or decrease, based on how much energy is added or removed as heat. This relationship is expressed mathematically as follows:

SPECIFIC HEAT CAPACITY

$$c_p = \frac{Q}{m\Delta T}$$

$$\textbf{specific heat capacity} = \frac{\textbf{energy transferred as heat}}{\textbf{mass} \times \textbf{change in temperature}}$$

The subscript p indicates that the specific heat capacity is measured at constant pressure. Maintaining constant pressure is an important detail when determining certain thermal properties of gases, which are much more affected by changes in pressure than are solids or liquids. Note that a temperature change of 1°C is equal in magnitude to a temperature change of 1 K, so ΔT gives the temperature change in either scale.

The equation for specific heat capacity applies to both substances that absorb energy from their surroundings and those that transfer energy to their surroundings. When the temperature increases, ΔT and Q are taken to be positive, which corresponds to energy transferred into the substance. Likewise, when the temperature decreases, ΔT and Q are negative and energy is

Figure 12
The air around the pool and the water in the pool receive energy from sunlight. However, the increase in temperature is greater for the air than for the water.

specific heat capacity

the quantity of heat required to raise a unit mass of homogeneous material 1 K or 1°C in a specified way given constant pressure and volume

SECTION 3
Advanced Level

The Language of Physics

The subscript p in c_p indicates the specific heat capacity of a substance when that substance is heated or cooled at a constant pressure. For gases, the specific heat capacity is significantly different at constant pressure than at constant volume. For liquids and solids, the difference is very little, but for the sake of simplicity, the symbol c_p is used for all substances in this text.

Teaching Tip — GENERAL

Ask students to consider equal masses of air and water to which equal amounts of energy are added. If the temperature increase for the air is greater than it is for the water, which substance must have the greater specific heat capacity? *(the water, because it absorbs more energy for each incremental change in temperature)*

The Language of Physics

In physics, the word *capacity* usually indicates that a quantity is divided by a measure of the amount or mass; it does not refer to "how much something will hold." *Specific heat* is an abbreviated form of the term *specific heat capacity.* Be sure students realize that specific heat is not a kind of heat. Reemphasize that heat is not contained in objects. Rather, heat is the amount of energy that is transferred between objects due to a difference in their temperatures.

Teaching Tip

Point out that the specific heat capacities of ice, water, and steam do not have the same value; in other words, the specific heat capacity of a substance depends on its phase.

Visual Strategy GENERAL

Figure 13

Be sure students understand this schematic representation of a calorimeter.

Q Why is water used in a calorimeter?

A ***Water's specific heat capacity has a well-known value, and thus the change in the water's temperature can be used to determine the specific heat capacity of the sample.***

SCILINKS.
www.scilinks.org
Topic: Specific Heat Capacity
Code: HF61694

Table 4 Specific Heat Capacities

Substance	c_p (J/kg•°C)	Substance	c_p (J/kg•°C)
aluminum	8.99×10^2	lead	1.28×10^2
copper	3.87×10^2	mercury	1.38×10^2
glass	8.37×10^2	silver	2.34×10^2
gold	1.29×10^2	steam	2.01×10^3
ice	2.09×10^3	water	4.186×10^3
iron	4.48×10^2		

transferred from the substance. **Table 4** lists specific heat capacities that have been determined for several substances.

Calorimetry is used to determine specific heat capacity

To measure the specific heat capacity of a substance, it is necessary to measure mass, temperature change, and energy transferred as heat. Mass and temperature change are directly measurable, but the direct measurement of heat is difficult. However, the specific heat capacity of water (4.186 kJ/kg•°C) is well known, so the energy transferred as heat between an object of unknown specific heat capacity and a known quantity of water can be measured.

If a hot substance is placed in an insulated container of cool water, energy conservation requires that the energy the substance gives up must equal the energy absorbed by the water. Although some energy is transferred to the surrounding container, this effect is small and will be ignored in this discussion. Energy conservation can be used to calculate the specific heat capacity, $c_{p,x}$, of the substance (indicated by the subscript x), as follows:

$$\text{energy absorbed by water} = \text{energy released by the substance}$$

$$Q_w = -Q_x$$

$$c_{p,w}m_w\Delta T_w = -c_{p,x}m_x\Delta T_x$$

For simplicity, a subscript w will always stand for "water" in problems involving specific heat capacities. As discussed earlier, the energy gained by a substance is expressed as a positive quantity, and the energy released is expressed as a negative quantity. The first equation above can be rewritten as $Q_w + Q_x = 0$, which shows that the net change in energy transferred as heat equals zero. Note that ΔT equals the final temperature minus the initial temperature.

calorimetry

an experimental procedure used to measure the energy transferred from one substance to another as heat

This approach to determining a substance's specific heat capacity is called **calorimetry**, and devices that are used for making this measurement are called *calorimeters*. A calorimeter also contains both a thermometer to measure the final temperature of substances at thermal equilibrium and a stirrer to ensure the uniform mixture of energy throughout the water. See **Figure 13.**

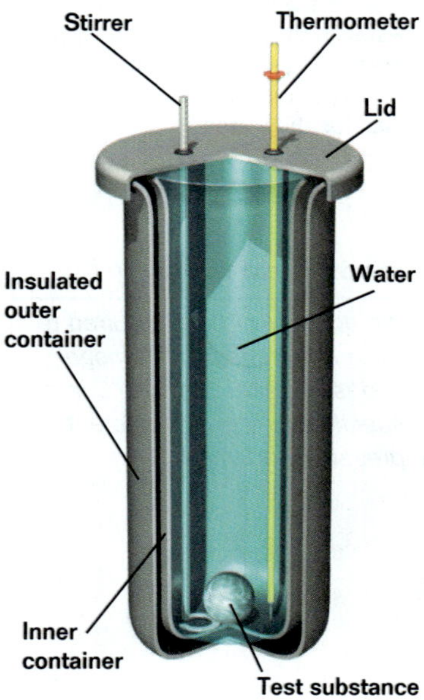

Figure 13
A simple calorimeter allows the specific heat capacity of a substance to be determined.

SAMPLE PROBLEM C

Calorimetry

PROBLEM

A 0.050 kg metal bolt is heated to an unknown initial temperature. It is then dropped into a calorimeter containing 0.15 kg of water with an initial temperature of 21.0°C. The bolt and the water then reach a final temperature of 25.0°C. If the metal has a specific heat capacity of 899 J/kg•°C, find the initial temperature of the metal.

SOLUTION

1. DEFINE **Given:**

$m_{metal} = m_m = 0.050$ kg $c_{p,m} = 899$ J/kg•°C

$m_{water} = m_w = 0.15$ kg $c_{p,w} = 4186$ J/kg•°C

$T_{water} = T_w = 21.0$°C $T_{final} = T_f = 25.0$°C

Unknown: $T_{metal} = T_m = ?$

Diagram:

Before placing hot sample in calorimeter

After thermal equilibrium has been reached

$m_m = 0.050$ kg

$m_w = 0.15$ kg
$T_w = 21.0$°C

$T_f = 25.0$°C

2. PLAN **Choose an equation or situation:**

The energy absorbed by the water equals the energy removed from the bolt.

$$Q_w = -Q_m$$

$$c_{p,w} m_w \Delta T_w = -c_{p,m} m_m \Delta T_m$$

$$c_{p,w} m_w (T_f - T_w) = -c_{p,m} m_m (T_f - T_m)$$

Rearrange the equation to isolate the unknown:

$$T_m = \frac{c_{p,w} m_w (T_f - T_w)}{c_{p,m} m_m} + T_f$$

TIP *Because T_w is less than T_f, you know that T_m must be greater than T_f.*

3. CALCULATE **Substitute the values into the equation and solve:**

$$T_m = \frac{(4186 \text{ J/kg•°C})(0.15 \text{ kg})(25.0\text{°C} - 21.0\text{°C})}{(899 \text{ J/kg•°C})(0.050 \text{ kg})} + 25.0\text{°C}$$

$$\boxed{T_m = 81\text{°C}}$$

4. EVALUATE T_m is greater than T_f, as expected.

Classroom Practice

Calorimetry

You are preparing to take a bath. The cold-water faucet supplies water at 20.0°C, and the water from the hot-water faucet is 60.0°C. Each faucet has poured 25.0 kg of water into the tub. What is the temperature of the bath?

Answer

40.0°C

You prefer your bath at 30.0°C. The hot-water faucet has already poured 20.0 kg of water at 60.0°C into the tub. How much cold water (20.0°C) should you add?

Answer

60.0 kg

PROBLEM GUIDE C

Use this guide to assign problems.
SE = Student Edition Textbook
PW = Problem Workbook
PB = Problem Bank on the One-Stop Planner (OSP)

Solving for:

T	**SE** Sample, 1–2; Ch. Rvw. 31–32 **PW** 3–5 **PB** 8–10
c_p	**SE** 3 **PW** 6–7 **PB** Sample, 1–4
m	**SE** 4; Ch. Rvw. 37, 40 **PW** Sample, 1–2 **PB** 5–7

*Challenging Problem
Consult the printed Solutions Manual or the OSP for detailed solutions.

ANSWERS

Practice C

1. 47°C
2. 18°C
3. 390 J/kg•°C
4. 135 g

Why it Matters

Earth-Coupled Heat Pumps

The earth-coupled heat pump demonstrates an excellent use of the transfer of energy as heat. These pumps are also known as *geothermal heat pumps, ground-water heat pumps, ground-source heat pumps,* or *geoexchange systems.* Today, over 650,000 geoexchange units have been installed in the United States, and new installations have been increasing by about 20% each year.

Extension ADVANCED

Have students conduct research to learn more about the advantages and disadvantages of earth-coupled heat pumps. Ask them to compare earth-coupled heat pumps with traditional heating and cooling systems in terms of cost (installation, maintenance, and operation), efficiency, environmental impact, and consumer satisfaction. Ask students to write a position paper supporting one type of system over the other.

PRACTICE C

Calorimetry

1. What is the final temperature when a 3.0 kg gold bar at 99°C is dropped into 0.22 kg of water at 25°C?
2. A 0.225 kg sample of tin initially at 97.5°C is dropped into 0.115 kg of water. The initial temperature of the water is 10.0°C. If the specific heat capacity of tin is 230 J/kg•°C, what is the final equilibrium temperature of the tin-water mixture?
3. Brass is an alloy made from copper and zinc. A 0.59 kg brass sample at 98.0°C is dropped into 2.80 kg of water at 5.0°C. If the equilibrium temperature is 6.8°C, what is the specific heat capacity of brass?
4. A hot, just-minted copper coin is placed in 101 g of water to cool. The water temperature changes by 8.39°C, and the temperature of the coin changes by 68.0°C. What is the mass of the coin?

Why it Matters

Earth-Coupled Heat Pumps

As the earliest cave dwellers knew, a good way to stay warm in the winter and cool in the summer is to go underground. Now, scientists and engineers are using the same premise—and using existing technology in a new, more efficient way—to heat and cool above-ground homes for a fraction of the cost of conventional systems.

The average specific heat capacity of earth is smaller than the average specific heat capacity of air. However, earth has a greater density than air does, which means that near a house, there are more kilograms of earth than of air. So, a 1°C change in temperature involves transferring more energy to or from the ground than to or from the air. Thus, the temperature of the ground in the winter will probably be higher than the temperature of the air above it. In the summer, the temperature of the ground will likely be lower than the temperature of the air.

An earth-coupled heat pump enables homeowners to tap the temperature just below the ground to heat their homes in the winter or cool them in the summer. The system includes a network of plastic pipes placed in trenches or inserted in holes drilled 2 to 3 m (6 to 10 ft) beneath the ground's surface. To heat a home, a fluid circulates through the pipe, absorbs energy from the surrounding earth, and transfers this energy to a heat pump inside the house. Although the system can function anywhere on Earth's surface, it is most appropriate in severe climates, where dramatic temperature swings may not be ideal for air-based systems.

LATENT HEAT

Suppose you place an ice cube with a temperature of –25°C in a glass, and then you place the glass in a room. The ice cube slowly warms, and the temperature of the ice will increase until the ice begins to melt at 0°C. The graph in **Figure 14** and data in **Table 5** show how the temperature of 10.0 g of ice changes as energy is added.

You can see that temperature steadily increases from –25°C to 0°C (segment **A** of the graph). You could use the mass and the specific heat capacity of ice to calculate how much energy is added to the ice during this segment.

At 0°C, the temperature stops increasing. Instead, the ice begins to melt and to change into water (segment **B**). The ice-and-water mixture remains at this temperature until all of the ice melts. Suppose that you now heat the water in a pan on a stovetop. From 0°C to 100°C, the water's temperature steadily increases (segment **C**). At 100°C, however, the temperature stops rising, and the water turns into steam (segment **D**). Once the water has completely vaporized, the temperature of the steam increases (segment **E**).

SCILINKS.
www.scilinks.org
Topic: Heat Pumps
Code: HF60730

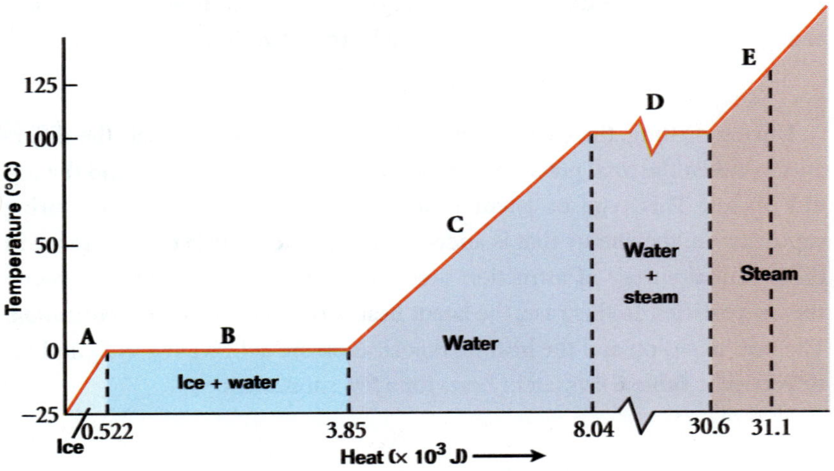

Figure 14
This idealized graph shows the temperature change of 10.0 g of ice as it is heated from –25°C in the ice phase to steam above 125°C at atmospheric pressure. (Note that the horizontal scale of the graph is not uniform.)

Table 5 Changes Occurring During the Heating of 10.0 g of Ice

Segment of graph	Type of change	Amount of energy transferred as heat	Temperature range of segment
A	temperature of ice increases	522 J	–25°C to 0°C
B	ice melts; becomes water	3.33×10^3 J	0°C
C	temperature of water increases	4.19×10^3 J	0°C to 100°C
D	water boils; becomes steam	2.26×10^4 J	100°C
E	temperature of steam increases	500 J	100°C to 125°C

Visual Strategy GENERAL

Figure 14
Make sure students understand the meaning of symbols in the graph. Explain that the —/\/— indicates a break in the scale on the *x*-axis.

Q What do the horizontal segments of the graph (**B** and **D**) indicate?

A ***They represent the times when the temperature was constant and the substance was undergoing a phase change. In B the ice was melting, and in D the water was turning into steam.***

Q Did energy transfer happen continuously throughout the process? Did the temperature increase at the same rate throughout the process?

A ***The water absorbed energy continuously, but the temperature increased only when all of the water was in one phase (segments A, C, and E); The rate of temperature increase varied for each phase because ice, water, and steam have different specific heat capacities.***

Misconception Alert

Students know from lessons in chemistry that heating and cooling can cause changes in the chemical properties of materials. To ensure they distinguish these phenomena from the physical process of phase change, ask if a substance chemically changes in the course of melting or boiling.

The Language of Physics — ADVANCED

The term *fusion* has more than one use in physics. Some students may be familiar with fusion as a nuclear process. In the present context, the term *fusion* refers not to a nuclear process but to the phase change from a solid to a liquid. Point out that the reverse process of fusion is freezing and that it occurs at the same temperature, called the *melting point* (or *freezing point*). Likewise, the reverse of vaporization is condensation. Vaporization and condensation occur at the boiling point, which is different for each substance. Discuss whether the term *boiling* always means "high-temperature."

Teaching Tip — GENERAL

Be sure that students understand the relationship between heat of fusion, heat of vaporization, and latent heat. Latent heat can be either the heat of fusion or the heat of vaporization. The heat of fusion corresponds to melting or solidification, and the heat of vaporization corresponds to vaporization or condensation.

phase change

the physical change of a substance from one state (solid, liquid, or gas) to another at constant temperature and pressure

latent heat

the energy per unit mass that is transferred during a phase change of a substance

When substances melt, freeze, boil, condense, or sublime (change from a solid to vapor or from vapor to a solid), the energy added or removed changes the internal energy of the substance without changing the substance's temperature. These changes in matter are called **phase changes.**

Latent heat is energy transferred during phase changes

To understand the behavior of a substance undergoing a phase change, you need to consider the changes in potential energy. Potential energy is present among a collection of particles in a solid or in a liquid in the form of attractive bonds. These bonds result from the charges within atoms and molecules. Potential energy is associated with the electric forces between these charges.

Phase changes result from a change in the potential energy between particles of a substance. When energy is added to or removed from a substance that is undergoing a phase change, the particles of the substance rearrange themselves to make up for their change of energy. This rearrangement occurs without a change in the average kinetic energy of the particles. The energy that is added or removed per unit mass is called **latent heat,** abbreviated as L. Note that according to this definition, the energy transferred as heat during a phase change simply equals the mass multiplied by the latent heat, as follows:

$$Q = mL$$

During melting, the energy that is added to a substance equals the difference between the total potential energies for particles in the solid and the liquid phases. This type of latent heat is called the *heat of fusion.* During vaporization, the energy that is added to a substance equals the difference in the potential energy of attraction between the liquid particles and between the gas particles. In this case, the latent heat is called the *heat of vaporization.* The heat of fusion and the heat of vaporization are abbreviated as L_f and L_v, respectively. **Table 6** lists latent heats for a few substances.

Table 6 Latent Heats of Fusion and Vaporization at Standard Pressure

Substance	Melting point (°C)	L_f (J/kg)	Boiling point (°C)	L_v (J/kg)
nitrogen	−209.97	2.55×10^4	−195.81	2.01×10^5
oxygen	−218.79	1.38×10^4	−182.97	2.13×10^5
ethyl alcohol	−114	1.04×10^5	78	8.54×10^5
water	0.00	3.33×10^5	100.00	2.26×10^6
lead	327.3	2.45×10^4	1745	8.70×10^5
aluminum	660.4	3.97×10^5	2467	1.14×10^7

extension

Practice Problems

Visit go.hrw.com for a sample problem and practice problems covering latent heat.

Keyword HF6HATX

SECTION REVIEW

1. A jeweler working with a heated 47 g gold ring must lower the ring's temperature to make it safe to handle. If the ring is initially at 99°C, what mass of water at 25°C is needed to lower the ring's temperature to 38°C?

2. How much energy must be added to a bowl of 125 popcorn kernels in order for them to reach a popping temperature of 175°C? Assume that their initial temperature is 21°C, that the specific heat capacity of popcorn is 1650 J/kg•°C, and that each kernel has a mass of 0.105 g.

3. Because of the pressure inside a popcorn kernel, water does not vaporize at 100°C. Instead, it stays liquid until its temperature is about 175°C, at which point the kernel ruptures and the superheated water turns into steam. How much energy is needed to pop 95.0 g of corn if 14 percent of a kernel's mass consists of water? Assume that the latent heat of vaporization for water at 175°C is 0.90 times its value at 100°C and that the kernels have an initial temperature of 175°C.

4. **Critical Thinking** Using the concepts of latent heat and internal energy, explain why it is difficult to build a fire with damp wood.

5. **Critical Thinking** Why does steam at 100°C cause more severe burns than does liquid water at 100°C?

6. **Interpreting Graphics** From the heating curve for a 15 g sample, as shown in **Figure 15,** estimate the following properties of the substance.
   - **a.** the specific heat capacity of the liquid
   - **b.** the latent heat of fusion
   - **c.** the specific heat capacity of the solid
   - **d.** the specific heat capacity of the vapor
   - **e.** the latent heat of vaporization

extension

Integrating Earth Science

Visit go.hrw.com for the activity "Land and Sea Breezes."

Keyword HF6HATX

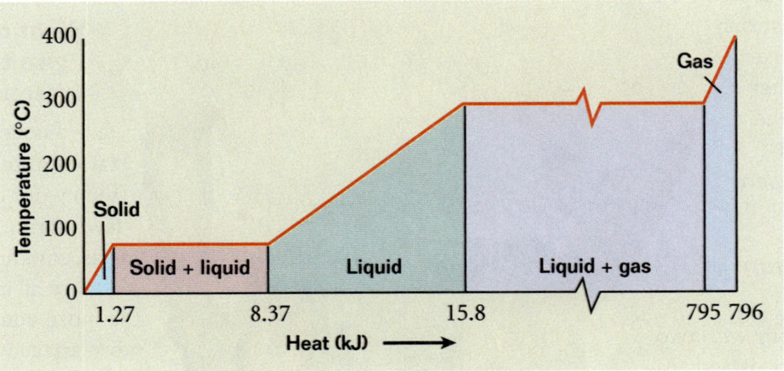

Figure 15

SECTION 3

SECTION REVIEW ANSWERS

1. 6.8 g
2. 3340 J
3. 2.7×10^4 J
4. When firewood is damp, a large amount of energy is used to increase the water's temperature and then to vaporize the water (because water has both a high specific heat capacity and a high latent heat of vaporization). After this has been accomplished, the remaining energy is used to burn the wood. Thus, much more energy is required when the wood is damp.
5. The steam has more internal energy in the form of latent heat and thus will transfer a great deal of energy at 100°C before its temperature decreases. The temperature of water that is initially at 100°C will immediately begin decreasing as energy is transferred to the body.
6. (Estimated values may vary.)
   - **a.** 2×10^3 J/kg•°C
   - **b.** 4.7×10^5 J/kg
   - **c.** 1×10^3 J/kg•°C
   - **d.** 7×10^2 J/kg•°C
   - **e.** 5.2×10^7 J/kg

HVAC Technician

According to Doug Garner, HVAC technicians do more than merely keep buildings at a comfortable temperature. They also help clients such as the food service industry by keeping food refrigerated and bacteria in check. "Heat and temperature affect us more than you might think at first," says Garner.

Although someone interested in becoming an HVAC technician can begin to train while still in high school, Garner stresses that a traditional education is still important. "Our technicians must be able to communicate well and have a good mechanical aptitude." According to Garner, such background won't be wasted; the HVAC business has a place for just about any skill that a student brings to the field. "Our industry needs good, qualified people, and there are a lot of companies looking for them."

PHYSICS CAREERS

HVAC Technician

HVAC stands for *heating, ventilation, and air conditioning.* An HVAC technician knows what it takes to keep buildings warm in winter and cool in summer. To learn more about working with HVAC as a career, read the interview with contractor and business owner Doug Garner.

Doug Garner is checking a *potential relay.* This relay is connected to a capacitor that starts the compressor.

What does an HVAC technician do?

Basically, we sell, replace, and repair air-conditioning and heating equipment. We replace obsolete A/C and heating units in older homes and buildings, we install new units in new homes and buildings, and we repair units when they break down.

How did you become an HVAC technician?

There are numerous ways to get into the business. When I was about 17 years old, I was given an opportunity to work for a man with whom I went to church. I worked as an apprentice for three years after high school, and I learned from him and a couple of very good technicians. I also took some business courses at a local community college to help with the business end.

What about HVAC made it more interesting than other fields?

There were other things that I was interested in doing, but realistically HVAC was more practical. In other words, that's where the money and opportunities were for me.

What is the nature of your work?

I have a company with two service technicians and an apprentice. Most of my duties involve getting jobs secured, bidding on and designing the different systems to suit the needs of the customer. I have to have a basic understanding of advertising, marketing, and sales as well as of the technical areas as they apply to this field. Our technicians must be able to communicate well and have a good mechanical aptitude.

What do you like most about your job?

You get to work in a lot of different places and situations. It is never boring, and you meet a lot of people. You can make as much money as you are willing to work for.

What advice would you give to students who are interested in your field?

Take a course in HVAC at a technical institute or trade school, and then work as an apprentice for a few years. Mechanical engineering, sales, communication, and people skills are all important in this field; the more education you have, the more attractive you can be to a company.

CHAPTER 9 Highlights

KEY IDEAS

Section 1 Temperature and Thermal Equilibrium
- Temperature can be changed by transferring energy to or from a substance.
- Thermal equilibrium is the condition in which the temperature of two objects in physical contact with each other is the same.
- The most common temperature scales are the Fahrenheit, Celsius, and Kelvin (or absolute) scales.

Section 2 Defining Heat
- Heat is energy that is transferred from objects at higher temperatures to objects at lower temperatures.
- Energy is transferred by thermal conduction through particle collisions.
- Energy is conserved when mechanical energy and internal energy are taken into account. Thus, for a closed system, the sum of the changes in kinetic energy, potential energy, and internal energy must equal zero.

Section 3 Changes in Temperature and Phase
- Specific heat capacity is a measure of the energy needed to change a substance's temperature.
- By convention, the energy that is gained by a substance is positive, and the energy that is released by a substance is negative.
- Latent heat is the energy required to change the phase of a substance.

KEY TERMS

temperature (p. 299)

internal energy (p. 299)

thermal equilibrium (p. 300)

heat (p. 305)

specific heat capacity (p. 313)

calorimetry (p. 314)

phase change (p. 318)

latent heat (p. 318)

Variable Symbols

Quantities		Units	
T	temperature (Kelvin)	K	kelvins
T_C	temperature (Celsius)	°C	degrees Celsius
T_F	temperature (Fahrenheit)	°F	degrees Fahrenheit
ΔU	change in internal energy	J	joules
Q	heat	J	joules
c_p	specific heat capacity at constant pressure	$\frac{J}{kg \bullet °C}$	
L	latent heat	$\frac{J}{kg}$	

PROBLEM SOLVING

See **Appendix D: Equations** for a summary of the equations introduced in this chapter. If you need more problem-solving practice, see **Appendix I: Additional Problems.**

CHAPTER 9

Highlights

Teaching Tip — GENERAL

Because students often confuse many of the terms introduced in this chapter with one another, have students discuss the relationships between the various key terms as part of their vocabulary review.

extension

In-Depth Physics Content

Your students can visit go.hrw.com for an online chapter that integrates more in-depth development of the concepts covered here.

Keyword HF6HATX

Review

ANSWERS

1. Temperature increases as internal energy increases, except during a phase change.
2. Their temperatures are the same.
3. Answers may vary but should include anything that changes linearly with temperature.
4. constant temperature points, such as water's melting and boiling points at 1 atm
5. a. the hot griddle
   b. the 1 kg block of ice
   For equal temperatures, the object with larger mass has a greater internal energy.
6. a. the metal knife
   b. the seven 12 g ice cubes
   For equal internal energies, the object with smaller mass has a greater temperature.
7. Water's steam and ice points are fixed for a given atmospheric pressure, while human body temperature varies.
8. The water's temperature will decrease during the measurement; If the thermometer is much less massive than the water, this change is negligible.
9. 57.8°C, 331.0 K
10. 1064°C, 1337 K
11. b; Both have the same temperature, so no energy is transferred.
12. when air temperature is 35°C
13. none; Toast and oven are in thermal equilibrium.
14. Atoms in one end transfer energy to neighboring particles; this process continues until the other end is heated.

CHAPTER 9 Review

TEMPERATURE AND THERMAL EQUILIBRIUM

Review Questions

1. What is the relationship between temperature and internal energy?

2. What must be true of two objects if the objects are in a state of thermal equilibrium?

3. What are some physical properties that could be used in developing a temperature scale?

Conceptual Questions

4. What property must a substance have in order to be used for calibrating a thermometer?

5. Which object in each of the following pairs has greater total internal energy, assuming that the two objects in each pair are in thermal equilibrium? Explain your reasoning in each case.
   a. a metal knife in thermal equilibrium with a hot griddle
   b. a 1 kg block of ice at −25°C or seven 12 g ice cubes at −25°C

6. Assume that each pair of objects in item 5 has the same internal energy instead of the same temperature. Which item in each pair will have the higher temperature?

7. Why are the steam and ice points of water better fixed points for a thermometer than the temperature of a human body?

8. How does the temperature of a tub of hot water as measured by a thermometer differ from the water's temperature before the measurement is made? What property of a thermometer is necessary for the difference between these two temperatures to be minimized?

Practice Problems

For problems 9–10, see Sample Problem A.

9. The highest recorded temperature on Earth was 136°F, at Azizia, Libya, in 1922. Express this temperature in degrees Celsius and in kelvins.

10. The melting point of gold is 1947°F. Express this temperature in degrees Celsius and in kelvins.

DEFINING HEAT

Review Questions

11. Which drawing below shows the direction in which net energy is transferred as heat between an ice cube and the freezer walls when the temperature of both is −10°C? Explain your answer.

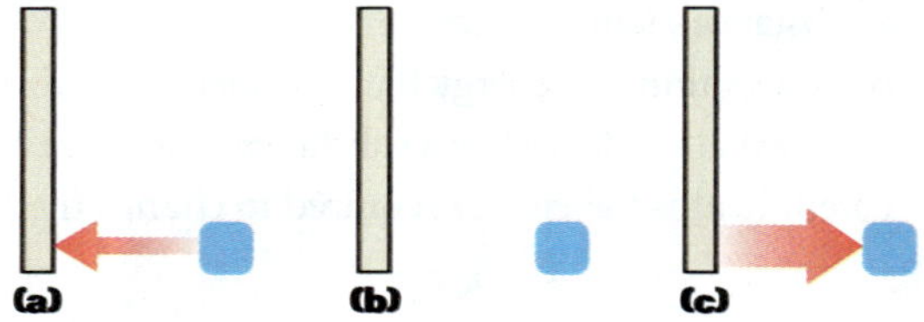

12. A glass of water has an initial temperature of 8°C. In which situation will the rate of energy transfer be greater, when the air's temperature is 25°C or 35°C?

13. How much energy is transferred between a piece of toast and an oven when both are at a temperature of 55°C? Explain.

14. How does a metal rod conduct energy from one end, which has been placed in a fire, to the other end, which is at room temperature?

15. How does air within winter clothing keep you warm on cold winter days?

Conceptual Questions

16. If water in a sealed, insulated container is stirred, is its temperature likely to increase slightly, decrease slightly, or stay the same? Explain your answer.

17. Given your answer to item 16, why does stirring a hot cup of coffee cool it down?

18. Given any two bodies, the one with the higher temperature contains more heat. What is wrong with this statement?

19. Explain how conduction causes water on the surface of a bridge to freeze sooner than water on the road surface on either side of the bridge.

20. A tile floor may feel uncomfortably cold to your bare feet, but a carpeted floor in an adjoining room at the same temperature feels warm. Why?

21. Why is it recommended that several items of clothing be worn in layers on cold days?

22. Why does a fan make you feel cooler on a hot day?

23. A paper cup is filled with water and then placed over an open flame, as shown at right. Explain why the cup does not catch fire and burn.

Practice Problems

For problems 24–25, see Sample Problem B.

24. A force of 315 N is applied horizontally to a crate in order to displace the crate 35.0 m across a level floor at a constant velocity. As a result of this work, the crate's internal energy is increased by an amount equal to 14 percent of the crate's initial internal energy. Calculate the initial internal energy of the crate. (Disregard the work done on the floor, and assume that all work goes into the crate.)

25. A 0.75 kg spike is hammered into a railroad tie. The initial speed of the spike is equal to 3.0 m/s.

a. If the tie and spike together absorb 85 percent of the spike's initial kinetic energy as internal energy, calculate the increase in internal energy of the tie and spike.

b. What happens to the remaining energy?

CHANGES IN TEMPERATURE AND PHASE

Review Questions

26. What principle permits calorimetry to be used to determine the specific heat capacity of a substance? Explain.

27. Why does the temperature of melting ice not change even though energy is being transferred as heat to the ice?

Conceptual Questions

28. Why does the evaporation of water cool the air near the water's surface?

29. Until refrigerators were invented, many people stored fruits and vegetables in underground cellars. Why was this more effective than keeping them in the open air?

30. During the winter, the people mentioned in item 29 would often place an open barrel of water in the cellar alongside their produce. Explain why this was done and why it would be effective.

Practice Problems

For problems 31–32, see Sample Problem C.

31. A 25.5 g silver ring ($c_p = 234$ J/kg•°C) is heated to a temperature of 84.0°C and then placed in a calorimeter containing 5.00×10^{-2} kg of water at 24.0°C. The calorimeter is not perfectly insulated, however, and 0.140 kJ of energy is transferred to the surroundings before a final temperature is reached. What is the final temperature?

9 REVIEW

15. Air is an insulator, so the internal energy of your body is not as easily transferred to the environment.

16. Temperature increases slightly because stirring adds energy.

17. Hotter portions of liquid are in contact with the air, so energy is transferred from coffee to air at a faster rate.

18. Heat is energy in transit, not a substance. Temperature is proportional to internal energy, not to heat.

19. The bridge's surface is directly exposed to cold air both above and below, whereas the road's surface is exposed only on one side. In addition, the stable temperature of the ground below the road helps to keep the road's temperature stable.

20. The tile floor is a better thermal conductor than the carpet. Thus, the energy transfer from your feet to the tile happens faster.

21. The air between each of the layers acts as a thermal insulator.

22. Evaporation requires energy and thus cools the surrounding air. A fan continuously supplies new, dry air, so perspiration is continually evaporated and the body is cooled.

23. The paper nearest to the flame maintains a temperature equal to that of the boiling water as long as the water has not entirely boiled away. The cup doesn't burn, because the heat input at the bottom of the cup is rapidly conducted to the boiling water.

24. 7.9×10^4 J

25. **a.** 2.9 J

b. It goes into the air, the ground, and the hammer.

26. conservation of energy; Energy removed from sample equals energy added to water.
27. Energy is used to break bonds between molecules in ice. (There is a change in *PE*, not *KE*.)
28. Evaporation requires energy, which comes from the air.
29. A large mass of cool earth around the cellar does not undergo as large a temperature change during the year as does the outside air.
30. Water has a large latent heat of fusion. While the water is freezing, energy is transferred as heat to the surrounding air. (Once the water is frozen, this technique no longer works.)
31. 25.0°C
32. 120°C
33. **a.** $T_R = T_F + 459.7$, or $T_F = T_R - 459.7$
b. $T = \frac{5}{9} T_R$, or $T_R = \frac{9}{5} T$
34. 14 m
35. **a.** $T_{TH} = \frac{3}{2} T_C + 50$, or $T_C = \frac{2}{3}(T_{TH} - 50)$
b. −360° TH
36. 5.1×10^4 s, or 14 h

extension
Graphing Calculator Practice
Visit go.hrw.com for answers to this Graphing Calculator activity.
Keyword HF6HATXT

32. When a driver brakes an automobile, friction between the brake disks and the brake pads converts part of the car's translational kinetic energy to internal energy. If a 1500 kg automobile traveling at 32 m/s comes to a halt after its brakes are applied, how much can the temperature rise in each of the four 3.5 kg steel brake disks? Assume the disks are made of iron ($c_p = 448$ J/kg•°C) and that all of the kinetic energy is distributed in equal parts to the internal energy of the brakes.

MIXED REVIEW

33. Absolute zero on a temperature scale called the *Rankine* scale is $T_R = 0$°R, and the scale's unit is the same size as the Fahrenheit degree.
  a. Write a formula that relates the Rankine scale to the Fahrenheit scale.
  b. Write a formula that relates the Rankine scale to the Kelvin scale.

34. A 3.0 kg rock is initially at rest at the top of a cliff. Assuming the rock falls into the sea at the foot of the cliff and that its kinetic energy is transferred entirely to the water, how high is the cliff if the temperature of 1.0 kg of water is raised 0.10°C? (Neglect the heat capacity of the rock.)

35. The freezing and boiling points of water on the imaginary "Too Hot" temperature scale are selected to be exactly 50 and 200 degrees TH.
  a. Derive an equation relating the Too Hot scale to the Celsius scale. (Hint: Make a graph of one temperature scale versus the other, and solve for the equation of the line.)
  b. Calculate absolute zero in degrees TH.

36. A hot-water heater is operated by solar power. If the solar collector has an area of 6.0 m^2 and the power delivered by sunlight is 550 W/m^2, how long will it take to increase the temperature of 1.0 m^3 of water from 21°C to 61°C?

Graphing Calculator Practice

Specific Heat Capacity
Specific heat capacity (c_p), as you learned earlier in this chapter, is equal to the amount of energy required to change the temperature of 1 kg of a substance by 1°C. This relationship is expressed by the following equation:

$$\Delta T = \frac{Q}{mc_p}$$

In this equation, ΔT is the change in temperature, Q is the amount of energy absorbed by the substance as heat, c_p is the specific heat capacity of the substance, and m is the mass of the substance.

This equation can be represented on a graphing calculator as follows:

$$Y_1 = T + (X/(MC))$$

A graph of this equation will illustrate the relationship between energy absorbed as heat and temperature.

In this graphing calculator activity, you will enter various values for the energy absorbed and will determine the resulting temperature. Then, you can explore how changing the specific heat capacity, mass, and initial temperature changes your results.

Visit go.hrw.com and type in the keyword **HF6HATX** to find this graphing calculator activity. Refer to **Appendix B** for instructions on downloading the program for this activity.

37. A student drops two metallic objects into a 120 g steel container holding 150 g of water at 25°C. One object is a 253 g cube of copper that is initially at 85°C, and the other is a chunk of aluminum that is initially at 5°C. To the surprise of the student, the water reaches a final temperature of 25°C, its initial temperature. What is the mass of the aluminum chunk?

38. At what Fahrenheit temperature are the Kelvin and Fahrenheit temperatures numerically equal?

39. A 250 g aluminum cup holds and is in thermal equilibrium with 850 g of water at 83°C. The combination of cup and water is cooled uniformly so that the temperature decreases by 1.5°C per minute. At what rate is energy being removed?

40. A jar of tea is placed in sunlight until it reaches an equilibrium temperature of 32°C. In an attempt to cool the liquid, which has a mass of 180 g, 112 g of ice at 0°C is added. At the time at which the temperature of the tea (and melted ice) is 15°C, determine the mass of the remaining ice in the jar. Assume the specific heat capacity of the tea to be that of pure liquid water.

Alternative Assessment

1. According to legend, Archimedes determined whether the king's crown was pure gold by comparing its water displacement with the displacement of a piece of pure gold of equal mass. But this procedure is difficult to apply to very small objects. Use the concept of specific heat capacity to design a method for determining whether a ring is pure gold. Present your plan to the class, and ask others to suggest improvements to your design. Discuss each suggestion's advantages and disadvantages.

2. The host of a cooking show on television claims that you can greatly reduce the baking time for potatoes by inserting a nail through each potato. Explain whether this advice has a scientific basis. Would this approach be more efficient than wrapping the potatoes in aluminum foil? List all arguments and discuss their strengths and weaknesses.

3. The graph of decreasing temperature versus time of a hot object is called its cooling curve. Design and perform an experiment to determine the cooling curve of water in containers of various materials and shapes. Draw cooling curves for each one. Which trends represent good insulation? Use your findings and graphs to design a lunch box that keeps food warm or cold.

4. Research the life and work of James Prescott Joule, who is best known for his apparatus demonstrating the equivalence of work and heat and the conservation of energy. Many scientists initially did not accept Joule's conclusions. Research the reasoning behind their objections. Prepare a presentation for a class discussion either supporting the objections of Joule's critics or defending Joule's conclusion before England's Royal Academy of Sciences.

5. Research how scientists measure the temperature of the following: the sun, a flame, a volcano, outer space, liquid hydrogen, mice, and insects. Find out what instruments are used in each case and how they are calibrated to known temperatures. Using what you learn, prepare a chart or other presentation on the tools used to measure temperature and the limitations on their ranges.

6. Get information on solar water heaters that are available where you live. How does each type work? Compare prices and operating expenses for solar water heaters versus gas water heaters. What are some of the other advantages and limitations of solar water heaters? Prepare an informative brochure for homeowners who are interested in this technology.

9 REVIEW

37. 330 g

38. 574.6 K = 574.6°F

39. 5.7×10^3 J/min = 95 J/s

40. 8.0×10^1 g

Alternative Assessment ANSWERS

1. Student plans should be safe and complete. They should include a list of equipment, measurements, and calculations. One technique is to measure the ring's mass and to then measure the temperature change of the ring and warm water when they are placed in a calorimeter.

2. Student answers should indicate that metal is a good conductor and that the faster way to transfer energy to the potato's interior is with the nail.

3. Student plans should be safe and complete and should include a list of equipment, measurements, and calculations. Graphs for insulators should show slow energy loss.

4. Student research will vary. Joule's work was accepted after Helmholtz confirmed its theoretical foundations.

5. Answers will vary. For temperatures below mercury's freezing point, organic liquids are used in thermometers. Temperatures of distant objects are measured indirectly.

6. Student analyses should include considerations of economics, environment, and convenience.

Standardized Test Prep

ANSWERS

1. D
2. F
3. C
4. G
5. C
6. J
7. B

Standardized Test Prep

MULTIPLE CHOICE

1. What must be true about two given objects for energy to be transferred as heat between them?

A. The objects must be large.
B. The objects must be hot.
C. The objects must contain a large amount of energy.
D. The objects must have different temperatures.

2. A metal spoon is placed in one of two identical cups of hot coffee. Why does the cup with the spoon have a lower temperature after a few minutes?

F. Energy is removed from the coffee mostly by conduction through the spoon.
G. Energy is removed from the coffee mostly by convection through the spoon.
H. Energy is removed from the coffee mostly by radiation through the spoon.
J. The metal in the spoon has an extremely large specific heat capacity.

Use the passage below to answer questions 3–4.

The boiling point of liquid hydrogen is −252.87°C.

3. What is the value of this temperature on the Fahrenheit scale?

A. 20.28°F
B. −220.87°F
C. −423.2°F
D. 0°F

4. What is the value of this temperature in kelvins?

F. 273 K
G. 20.28 K
H. −423.2 K
J. 0 K

5. A cup of hot chocolate with a temperature of 40°C is placed inside a refrigerator at 5°C. An identical cup of hot chocolate at 90°C is placed on a table in a room at 25°C. A third identical cup of hot chocolate at 80°C is placed on an outdoor table, where the surrounding air has a temperature of 0°C. For which of the three cups has the most energy been transferred as heat when equilibrium has been reached?

A. The first cup has the largest energy transfer.
B. The second cup has the largest energy transfer.
C. The third cup has the largest energy transfer.
D. The same amount of energy is transferred as heat for all three cups.

6. What data are required in order to determine the specific heat capacity of an unknown substance by means of calorimetry?

F. $c_{p,water}$, T_{water}, $T_{substance}$, T_{final}, V_{water}, $V_{substance}$
G. $c_{p,substance}$, T_{water}, $T_{substance}$, T_{final}, m_{water}, $m_{substance}$
H. $c_{p,water}$, $T_{substance}$, m_{water}, $m_{substance}$
J. $c_{p,water}$, T_{water}, $T_{substance}$, T_{final}, m_{water}, $m_{substance}$

7. During a cold spell, Florida orange growers often spray a mist of water over their trees during the night. Why is this done?

A. The large latent heat of vaporization for water keeps the trees from freezing.
B. The large latent heat of fusion for water prevents it and thus the trees from freezing.
C. The small latent heat of fusion for water prevents the water and thus the trees from freezing.
D. The small heat capacity of water makes the water a good insulator.

Use the heating curve below to answer questions 8–10. The graph shows the change in temperature of a 23 g sample of a substance as energy is added to the substance as heat.

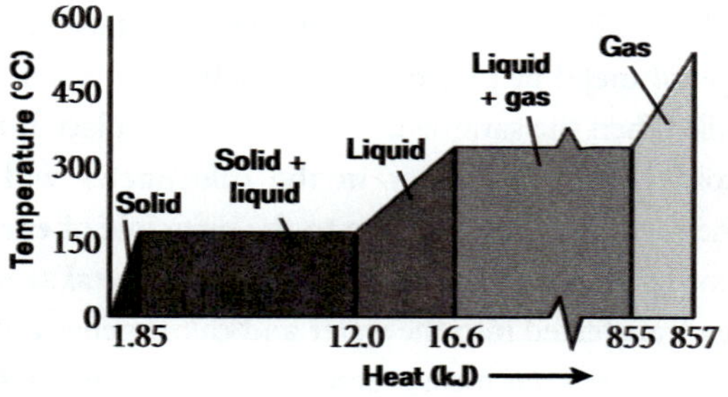

8. What is the specific heat capacity of the liquid?
   F. 4.4×10^5 J/kg•°C
   G. 4.0×10^2 J/kg•°C
   H. 5.0×10^2 J/kg•°C
   J. 1.1×10^3 J/kg•°C

9. What is the latent heat of fusion?
   A. 4.4×10^5 J/kg
   B. 4.0×10^2 J/kg•°C
   C. 10.15×10^3 J
   D. 3.6×10^7 J/kg

10. What is the specific heat capacity of the solid?
   F. 1.85×10^3 J/kg•°C
   G. 4.0×10^2 J/kg•°C
   H. 5.0×10^2 J/kg•°C
   J. 1.1×10^3 J/kg•°C

SHORT RESPONSE

Base your answers to questions 11–12 on the information below.

The largest of the Great Lakes, Lake Superior, contains 1.20×10^{16} kg of fresh water, which has a specific heat capacity of 4186 J/kg•°C and a latent heat of fusion of 3.33×10^5 J/kg.

11. How much energy would be needed to increase the temperature of Lake Superior by 1.0°C?

12. If Lake Superior were still liquid at 0°C, how much energy would need to be removed from the lake for it to become completely frozen?

13. Ethyl alcohol has about one-half the specific heat capacity of water. If equal masses of alcohol and water in separate beakers at the same temperature are supplied with the same amount of energy, which will have the higher final temperature?

14. A 0.200 kg glass holds 0.300 kg of hot water, as shown below. The glass and water are set on a table to cool. After the temperature has decreased by 2.0°C, how much energy has been removed from the water and glass? (The specific heat capacity of glass is 837 J/kg•°C, and that of water is 4186 J/kg•°C.)

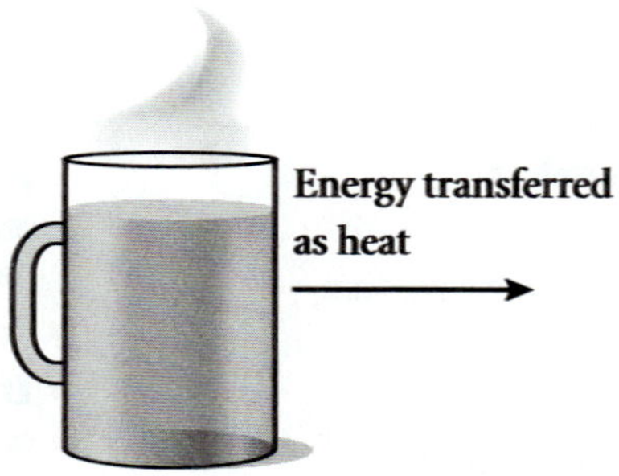

EXTENDED RESPONSE

15. How is thermal energy transferred by the process of convection?

16. Show that the temperature −40.0° is unique in that it has the same numerical value on the Celsius and Fahrenheit scales. Show all of your work.

Test TIP Use dimensional analysis to check your work when solving mathematical problems. Include units in each step of your calculation. If you do not end up with the correct unit in your answer, check each step of your calculation for errors.

8. J

9. A

10. H

11. 5.0×10^{19} J

12. 4.00×10^{21} J

13. the ethyl alcohol

14. 2900 J

15. The increasing temperature of a liquid or gas causes it to become less dense, so it rises above colder liquid or gas, transferring thermal energy with it.

16. $T_F = \frac{9}{5}(-40.0°C) + 32.0 =$

−40.0°F

CHAPTER 9

Skills Practice Lab Specific Heat Capacity

Lab Planning

Beginning on page T34 are preparation notes and teaching tips to assist you in planning.

Blank data tables (as well as some sample data) appear on the **One-Stop Planner.**

No Books in the Lab?

See the ***Datasheets for In-Text Labs*** workbook for a reproducible master copy of this experiment.

CBL™ Option

A **CBL™** version of this lab appears in **Appendix K** and in the ***CBL™ Experiments*** workbook.

Safety Caution

This lab presents several safety hazards.

- With larger groups, make sure all students are involved and paying attention to safety.
- Remind students never to leave a hot plate unattended while it is on.
- Make sure all students wear the appropriate safety gear at all times.
- Students must take extra care not to break thermometers in this lab.
- Students must keep their hands out of the steam.
- Demonstrate how to carefully pour the shot into the metal heating dipper around the thermometer. The bulb must be surrounded by shot. The thermometer must be removed very slowly and carefully, and it must ***never*** be inserted into the dipper full of metal shot.

OBJECTIVES

- **Measure** temperature.
- **Apply** the specific heat capacity equation for calorimetry to calculate the specific heat capacity of a metal.
- **Identify** unknown metals by comparing their specific heat capacities with accepted values for specific heat capacities.

MATERIALS LIST

- **2 beakers**
- **2 thermometers**
- **balance**
- **hand-held magnifying lens**
- **hot plate**
- **ice**
- **metal calorimeter and stirring rod**
- **metal heating vessel with metal heating dipper**
- **samples of beads or shot formed from various metals**
- **small plastic dish**

In this experiment, you will use calorimetry to identify various metals. In each trial, you will heat a sample of metal by placing it above a bath of water and bringing the water to a boil. When the sample is heated, you will place it in a calorimeter containing cold water. The water in the calorimeter will be warmed by the metal as the metal cools. According to the principle of energy conservation, the total amount of energy transferred out of the metal sample as it cools equals the energy transferred into the water and calorimeter as they are warmed. In this lab, you will use your measurements to determine the specific heat capacity and identity of each metal.

SAFETY

- **When using a burner or hot plate, always wear goggles and an apron to protect your eyes and clothing. Tie back long hair, secure loose clothing, and remove loose jewelry. If your clothing catches on fire, walk to the emergency lab shower and use the shower to put out the fire.**
- **Never leave a hot plate unattended while it is turned on.**
- **If a thermometer breaks, notify the teacher immediately.**
- **Do not heat glassware that is broken, chipped, or cracked. Use tongs or a mitt to handle heated glassware and other equipment because it does not always look hot when it is hot. Allow all equipment to cool before storing it.**
- **Never put broken glass or ceramics in a regular waste container. Use a dustpan, brush, and heavy gloves to carefully pick up broken pieces and dispose of them in a container specifically provided for this purpose.**

PROCEDURE

Preparation

1. Read the entire lab procedure, and plan the steps you will take. Determine which steps can be performed while you are waiting for the water to heat.
2. If you are not using a datasheet provided by your teacher, prepare a data table in your lab notebook with four columns and eight rows. In the first row, label the second through fourth columns *Trial 1, Trial 2,* and *Trial 3.* In the first column, label the second through eighth rows *Sample Number, Mass of Metal, Mass of Calorimeter Cup and Stirrer, Mass of Water, Initial Temperature of Metal, Initial Temperature of Water and Calorimeter,* and *Final Temperature of Metal, Water, and Calorimeter.*

3. In the appendix of this book, look up the specific heat capacity of the material the calorimeter is made of and record the information in the top left corner of your data table.

Finding the Specific Heat Capacity of a Metal

4. Choose a location where you can set up the experiment away from the edge of the table and from other groups. Make sure the switch of the hot plate is in the "off" position before you plug it in.
5. Fill a metal heating vessel with 200 mL of water and place it on the hot plate, as shown in **Figure 1.** Turn on the hot plate and adjust the heating control to heat the water.
6. Measure out about 100 g of the metal sample. Record the number of the metal sample (1, 2, and so on) in your data table. Hold the thermometer in the metal heating dipper, and very carefully pour the sample into the metal heating dipper. Make sure the bulb of the thermometer is surrounded by the metal. Place the dipper with metal contents into the heating vessel. Hold the thermometer while the sample is heating.
7. While the sample is heating, determine the mass of the stirring rod and empty inner cup of the calorimeter. Record the mass in your data table. Do not leave the hot plate unattended.
8. Use the second thermometer to measure room temperature. For the water in the calorimeter, you will need about 100 g of water that is a little colder than room temperature. Put the water in a beaker. Place the thermometer in the water to check the temperature of the water. (Do not use water colder than 5°C below room temperature. You may need to use ice to get the initial temperature low enough, but make sure all the ice has melted before pouring the water into the calorimeter.)

Figure 1

Step 5: Start heating the water before you begin the rest of the lab. Never leave a hot plate unattended when it is turned on.

Step 6: Be very careful when pouring the metal sample in the dipper around the thermometer. Make sure the thermometer bulb is surrounded by the metal sample.

Step 12: Begin taking temperature readings a few seconds before adding the sample to the calorimeter.

Step 15: Record the *highest* temperature reached by the water, sample, and calorimeter combination.

Tips and Tricks

- Using two thermometers improves the speed and accuracy of the lab: use a −20°C to 110°C thermometer to measure the temperature of the sample and a −5°C to 50°C thermometer to measure the temperature of the water in the calorimeter.

✓ Checkpoints

Step 4: All hot plates and liquids must be kept away from the edge of the table. Keep metal shot in a container at all times.

Step 5: The hot plate must be turned up to the highest level.

Step 6: Students must be careful when pouring metal shot into the dipper. The thermometer must be held upright until it is removed from the heating dipper.

Step 7: For the next few steps, do not leave the hot plate unattended.

Step 8: Only a small amount of ice is required; it should be removed before students measure the mass of the water.

Step 10: If not doing more trials, make sure hot plates are turned off. If they remain on, make sure that heating vessels have enough water in them and that hot plates are attended to.

Step 13: Students should use mitts to handle the hot dipper.

Step 14: Stir gently; violent motion of the sample in the water could break the thermometer.

Step 16: Because the water is still hot, subsequent trials will take less time.

ANSWERS

Analysis

1. Student answers will vary. Make sure students use the relationship $\Delta T = T_f - T_i$. Typical values range from 0.9°C to 1.3°C.

2. Student answers will vary. Make sure students use the relationship $c_p = \frac{Q}{m\Delta T}$.

a. Typical values range from 43.7 J to 63.1 J.

b. Typical values range from 558 J to 850 J.

3. Make sure students use the relationship $Q_{total} = Q_1 + Q_2$. Typical values range from 602 J to 913 J.

4. Make sure students use the relationship $c_p = \frac{Q}{m\Delta T}$. Typical values for the specific heat capacity range from 255 J/kg•°C to 383 J/kg•°C.

Figure 2
Step 10: Use only the stirring rod—not the thermometer—to stir the water in the calorimeter.

9. Place the calorimeter and stirrer on the balance, and carefully add 100 g of the water. Record the mass of the water in your data table. Replace the cup in its insulating shell, and cover.

10. Use the thermometer to measure the temperature of the sample when the water is boiling and the sample reaches a constant temperature. Record this temperature as the initial temperature of the metal sample. (Note: When making temperature readings, take care not to touch the hot plate and the water.) Use the hand-held magnifying lens to measure to the nearest 0.5°C. Make sure that the thermometer bulb is completely surrounded by the metal sample, and keep your line of sight at a right angle to the stem of the thermometer. Carefully remove the thermometer and set it aside in a secure place.

11. Use the stirring rod to gently stir the water in the calorimeter, as shown in **Figure 2. Do not use the thermometer to stir the water.**

12. Place the second thermometer in the covered calorimeter. Measure the temperature of the water in the calorimeter to the nearest 0.1°C. Record this temperature in your data table as the initial temperature of the water and calorimeter.

13. Quickly transfer the sample to the cold water in the calorimeter and replace the cover. Use a mitt when handling the metal heating dipper. If you are not doing any more trials, make sure the hot plate is turned off. Otherwise, make sure there is plenty of water in the heating vessel, and do not leave the hot plate unattended.

14. Use the stirring rod to gently agitate the sample and stir the water in the calorimeter. **Do not use the thermometer to stir the water.**

15. Take readings every 5.0 s until five consecutive readings are the same. Record the highest reading in your data table.

16. If time permits, make additional trials with other metals. Record the data for all trials in your data table.

17. Clean up your work area. Put equipment away safely so that it is ready to be used again.

ANALYSIS

1. **Organizing Data** For each trial, calculate the temperature change of the water and calorimeter.

2. **Organizing Data** Use your data for each trial.

   a. Calculate the energy transferred to the calorimeter cup and stirring rod as heat, using the value for the specific heat capacity you found in step 3.

   b. Calculate the energy transferred to the water as heat.

3. **Organizing Data** Calculate the total energy transferred as heat into the water and the calorimeter.

4. **Organizing Data** For each trial, find the temperature change of the sample and calculate the specific heat capacity of the sample.

CONCLUSIONS

5. **Drawing Conclusions** Use the accepted values for the specific heat capacities of various metals in this chapter and in the appendix to determine what metal makes up each sample.

6. **Evaluating Results** Calculate the absolute and relative errors of the experimental values. Check with your teacher to see if you have correctly identified the metals.

7. **Evaluating Methods** Explain why the energy transferred as heat into the calorimeter and the water is equal to the energy transferred as heat from the metal sample.

8. **Evaluating Methods** Explain why it is important to calculate the temperature change using the highest temperature as the final temperature, rather than the last temperature recorded.

9. **Evaluating Methods** Why should the water be a few degrees colder than room temperature when the initial temperature is taken?

10. **Making Predictions** How would your results be affected if the initial temperature of the water in the calorimeter were 50°C instead of slightly cooler than room temperature?

11. **Drawing Conclusions** How is the temperature change of the calorimeter and the water within the calorimeter affected by the specific heat capacity of the metal? Did a metal with a high specific heat capacity raise the temperature of the water and the calorimeter more or less than a metal with a low specific heat capacity?

12. **Applying Conclusions** An environmentally conscious engineering team wants to design tea kettles out of a metal that will allow the water to reach its boiling point using the least possible amount of energy from a range or other heating source. Using the values for specific heat capacity in this chapter, choose a material that would work well, considering only the implications of transfer of energy as heat. Explain how the specific heat capacity of water will affect the operation of the tea kettle.

EXTENSION

13. **Evaluating Methods** What is the purpose of the outer shell of the calorimeter and the insulating ring in this experiment?

Conclusions

5. Student answers should correctly identify the metal used in each trial.

6. For sample data, absolute error values range from 2.5 J/kg•°C to 3.5 J/kg•°C. Typical values of relative error range from 0.007 to 0.151.

7. Energy is conserved in a closed system.

8. The highest temperature represents the total energy transferred as heat.

9. A greater temperature difference between the water and the sample allows more accurate results to be obtained; the smaller the difference, the higher the precision that is required.

10. The temperature difference would be smaller, so the error would be greater.

11. The higher specific heat capacities caused a greater change in the temperature of the water.

12. Students should choose a metal with a low specific heat capacity.

Extensions

13. The shell and insulated ring decrease energy transfer to the environment.

Science • Technology • Society

Global Warming

Data recorded from various locations around the world over the past century indicate that the average atmospheric temperature is currently 0.6°C higher than it was 100 years ago. However, historical studies indicate that some short-term fluctuations in climate are natural, such as the Little Ice Age of the 17th century. Does this recent increase represent a trend toward global warming or is it simply part of a natural cyclic variation in climate? Although the answer cannot be determined with certainty, most scientists now believe that global warming is a significant issue that requires worldwide attention.

The Greenhouse Effect

Global warming may be due, in part, to the greenhouse effect. The glass of a greenhouse traps sunlight inside the greenhouse, and thus a warm environment is created, even in the winter. Earth's atmosphere functions in a similar way, as the diagram shows. Molecules of "greenhouse gases," primarily carbon dioxide and methane, absorb energy that is radiated from Earth's surface. These molecules then release energy as heat, causing the atmosphere to be warmer than it would be without these gases. The greenhouse effect is beneficial—without it, Earth would be far too cold to support life. However, an "increased" greenhouse effect, caused by increased levels of greenhouse gases in the atmosphere, could contribute to global warming.

Carbon dioxide and methane are natural components of our atmosphere. However, the levels of atmospheric carbon dioxide and methane have increased rapidly during the last 100 years. This increase has been determined by analyzing air trapped in the ice layers of Greenland. Deeper sections of the ice contain air from earlier times. During the last ice age, our atmosphere contained about 185 ppm (parts per million) of carbon dioxide, CO_2. The levels 130 years ago were about 300 ppm. Today, the levels are about 380 ppm. This increase can be attributed to the increase in combustion reactions, due primarily to coal and petroleum burning, and to deforestation, which has decreased the number of trees that consume CO_2.

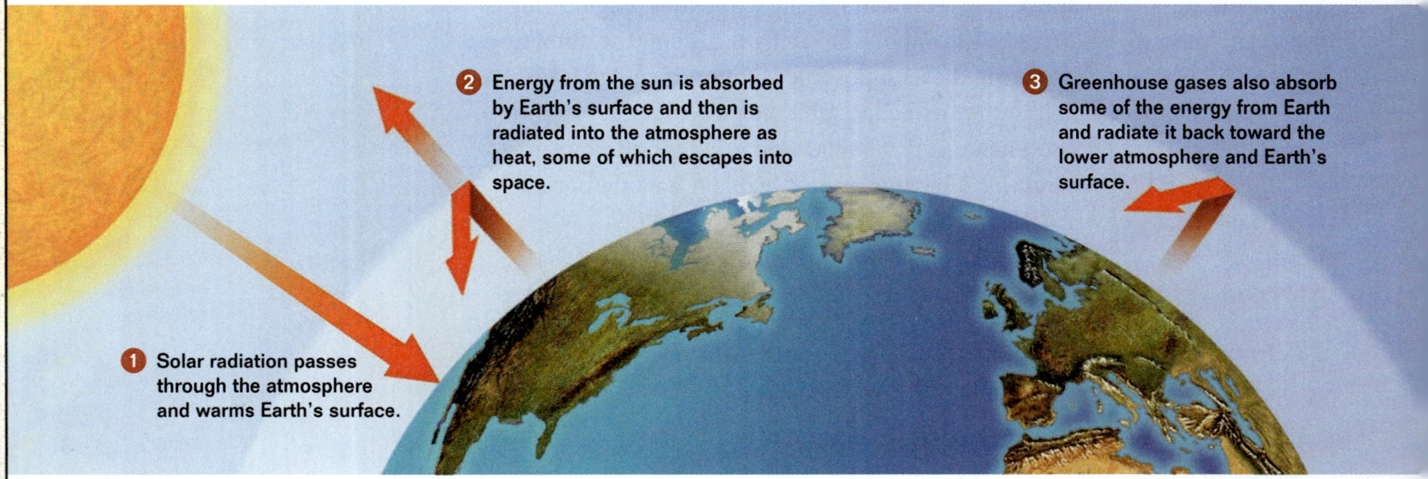

332

Modeling Climate Changes

Does the well-documented increase in greenhouse gas levels enable detailed predictions? Atmospheric physicists have greatly improved their models in recent years. The models are able to correctly predict past ice ages and to account for the energy-absorbing qualities of oceans. But such models remain oversimplified, partly because of a lack of detailed long-term data. In addition, the effects of many variables, such as fluctuations in solar energy output and volcanic processes, are poorly understood and cannot be factored into predictions.

Effects of Global Warming

Although an increase in atmospheric temperature of 0.6°C over 100 years sounds small, small increases in temperature can have pronounced effects if they continue to occur. Increased temperatures can eventually cause the ice in polar regions to melt. Then, ocean levels will rise, which may flood some coastal areas. Such disasters depend on whether global temperatures continue to increase. Although this issue is being debated, most scientists believe that action should be taken to reduce the levels of greenhouse gases in the environment. Opponents argue that economic costs must also be taken into account and weighed together with the environmental concerns to make the best overall decision.

Taking Action

In February 2005, the Kyoto Protocol went into effect. The Kyoto Protocol, ratified by more than 160 countries, commits developed nations to reducing their emissions of carbon dioxide and other greenhouse gases. Although the United States has not ratified the Kyoto Protocol, in 2002 President Bush announced a commitment to reduce greenhouse gas levels in the United States by 18% over the next decade. This reduction will require a combination of voluntary, incentive-based, and mandatory measures. Proponents of this plan claim that it will provide reductions similar to those mandated by the Kyoto Protocol.

This false-color image shows the energy radiating from Earth's upper atmosphere. The blue areas are the coldest. The American southwest is in the upper right-hand corner.

Researching the Issue

1. Carbon dioxide levels in the atmosphere have varied during Earth's history. Research the roles of volcanoes, plants, and limestone formation, and determine whether these processes have any bearing on the current increase in CO_2 concentrations. Can you think of any practical means of using these processes to reduce CO_2 concentrations? What would be the advantages and disadvantages?

2. Find out what technological developments have been suggested for slowing climatic warming. Can they be easily implemented? What are the drawbacks of these methods?

SCiLINKS.
www.scilinks.org
Topic: Greenhouse Gases
Code: HF60697

CHAPTER 10 Thermodynamics
Planning Guide

Compression Guide

To shorten instruction because of time limitations, omit the opener, Advanced sections, and the review.

OBJECTIVES	LABS, DEMONSTRATIONS, AND ACTIVITIES	TECHNOLOGY RESOURCES
PACING • 45 min pp. 334–335 **Chapter Opener**		CD **Visual Concepts,** Chapter 10 BASIC
PACING • 45 min pp. 336–341 **Section 1 Relationships Between Heat and Work** • Recognize that a system can absorb or release energy as heat in order for work to be done on or by the system and that work done on or by a system can result in the transfer of energy as heat. • Compute the amount of work done during a thermodynamic process. • Distinguish between isovolumetric, isothermal, and adiabatic thermodynamic processes.	TE **Demonstration** Work from Heat, p. 336 GENERAL	OSP **Lesson Plans** TR 38 A Bomb Calorimeter TR 39 An Isothermal Process
PACING • 45 min pp. 342–351 *Advanced Level* **Section 2 The First Law of Thermodynamics** • Illustrate how the first law of thermodynamics is a statement of energy conservation. • Calculate heat, work, and the change in internal energy by applying the first law of thermodynamics. • Apply the first law of thermodynamics to describe cyclic processes.		OSP **Lesson Plans** TR 40 Conservation of Total Energy TR 41 The Steps of a Gasoline Engine Cycle TR 42 The Steps of a Refrigeration Cycle TR 34A Signs of Q and W for a System TR 35A The First Law of Thermodynamics for Special Processes TR 36A Thermodynamics of a Refrigerator
PACING • 45 min pp. 352–358 *Advanced Level* **Section 3 The Second Law of Thermodynamics** • Recognize why the second law of thermodynamics requires two bodies at different temperatures for work to be done. • Calculate the efficiency of a heat engine. • Relate the disorder of a system to its ability to do work or transfer energy as heat.	SE **Quick Lab** Entropy and Probability, p. 357 ADVANCED	OSP **Lesson Plans** EXT **Integrating Environmental Science** Thermal Pollution GENERAL TR 43 Low and High Entropy Systems TR 44 Entropy Changes Produced by a Refrigerator Freezing Water TR 37A Typical Efficiencies for Engines

PACING • 90 min

CHAPTER REVIEW, ASSESSMENT, AND STANDARDIZED TEST PREPARATION

- SE **Chapter Highlights,** p. 359
- SE **Chapter Review,** pp. 360–363
- SE **Graphing Calculator Practice,** p. 362 GENERAL
- SE **Alternative Assessment,** p. 363 ADVANCED
- SE **Standardized Test Prep,** pp. 364–365 GENERAL
- SE **Appendix D: Equations,** p. 859
- SE **Appendix I: Additional Problems,** p. 888
- ANC **Study Guide Worksheet** Mixed Review* GENERAL
- ANC **Chapter Test A*** GENERAL
- ANC **Chapter Test B*** ADVANCED
- OSP **Test Generator**

Online and Technology Resources

go.hrw.com Holt Online Learning

Visit go.hrw.com to find a variety of online resources. To access this chapter's extensions, enter the keyword **HF6TDYXT** and click the "go" button. Click **Holt Online Learning** for an online edition of this textbook, and other interactive resources.

HOLT Teacher's One-Stop Planner®

This DVD package includes:

- Holt Calendar Planner
- Customizable Lesson Plans
- Editable Worksheets
- ExamView® Version 6 Assessment Suite
- Interactive Teacher's Edition
- Holt PuzzlePro®
- Holt PowerNotes® Presentations
- MindPoint® Quiz Show

KEY

SE Student Edition	OSP One-Stop Planner	EXT Online Extension
TE Teacher Edition	CD CD or CD-ROM	* Also on One-Stop Planner
ANC Ancillary Worksheet	TR Teaching Transparencies	◆ Requires advance prep

SKILLS DEVELOPMENT RESOURCES	REVIEW AND ASSESSMENT	CORRELATIONS
		National Science Education Standards
SE **Sample Set A** Work Done on or by a Gas, p. 338 BASIC TE **Classroom Practice,** p. 338 BASIC ANC **Problem Workbook** Sample Set A* BASIC OSP **Problem Bank** Sample Set A BASIC	SE **Section Review,** p. 341 GENERAL ANC **Study Guide Worksheet** Section 1* GENERAL ANC **Quiz** Section 1* BASIC	UCP 1, 2, 3, 4, 5 ST 1, 2 SPSP 5
SE **Sample Set B** The First Law of Thermodynamics, pp. 345–346 GENERAL TE **Classroom Practice,** p. 345 GENERAL ANC **Problem Workbook** Sample Set B* GENERAL OSP **Problem Bank** Sample Set B GENERAL	SE **Section Review,** p. 349 ADVANCED ANC **Study Guide Worksheet** Section 2* GENERAL ANC **Quiz** Section 2* GENERAL	UCP 1, 2, 3, 4, 5 ST 1, 2 SPSP 2, 5 PS 5a, 5b, 5c
SE **Conceptual Challenge,** p. 353 ADVANCED SE **Sample Set C** Heat-Engine Efficiency, pp. 354–355 GENERAL TE **Classroom Practice,** p. 354 GENERAL ANC **Problem Workbook** Sample Set C* GENERAL OSP **Problem Bank** Sample Set C GENERAL	SE **Section Review,** p. 357 ADVANCED ANC **Study Guide Worksheet** Section 3* GENERAL ANC **Quiz** Section 3* GENERAL	UCP 1, 2, 3, 4, 5 ST 1, 2 HNS 1 SPSP 2, 5 PS 5a, 5d

SCILINKS. NSTA
www.scilinks.org

Maintained by the **National Science Teachers Association.**

Topic: Energy Transfer
SciLinks Code: HF60516

Topic: Thermodynamics
SciLinks Code: HF61514

Topic: Heat Engines
SciLinks Code: HF60728

Topic: Stirling Engines
SciLinks Code: HF61455

Topic: Entropy
SciLinks Code: HF60523

Classroom CD-ROMs

- Guided Reading Audio Program
- Student One Stop
- Virtual Investigations
- Visual Concepts

Holt Lab Generator CD-ROM

Search for any lab by topic, standard, difficulty level, or time. Edit any lab to fit your needs, or create your own labs. Use the Lab Materials QuickList software to customize your lab materials list.

CHAPTER 10
Overview

Section 1 explains that a system can absorb energy by heat or work and then transfer energy to its surroundings as work or heat, and it distinguishes between isovolumetric, isothermal, and adiabatic processes.

Section 2 introduces the first law of thermodynamics and the relationships between heat, work, and internal energy, and it applies the first law of thermodynamics to cyclic processes in refrigeration, heat engines, and combustion engines.

Section 3 introduces the second law of thermodynamics, shows how to calculate the efficiency of heat engines, and discusses entropy with respect to the second law.

About the Illustration

For many students, the concepts of thermodynamics are difficult and abstract. Using examples that students are familiar with and that can be represented by fairly simple models, such as the balloon shown in this photograph, helps students see how thermodynamics applies to the world around them. Return to this example throughout the chapter as new concepts are introduced.

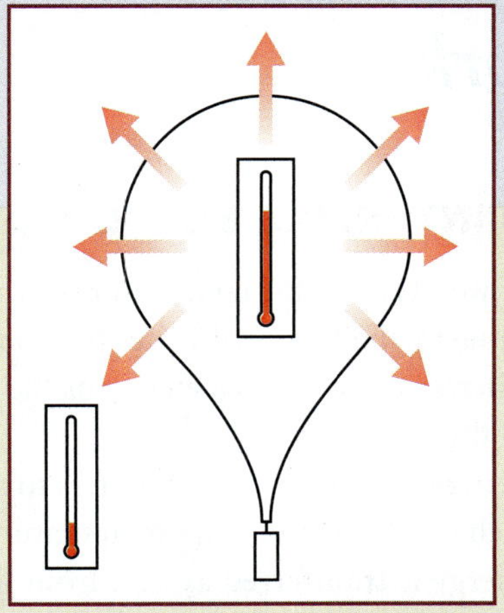

CHAPTER 10

Thermodynamics

The balloon shown in this photograph is used to lift scientific instruments into the upper atmosphere. The balloon can be modeled as a simple thermodynamic system. For instance, changes in temperature outside the balloon may cause energy transfers between the gas in the balloon and the outside air. This transfer of energy as heat changes the balloon's internal energy.

WHAT TO EXPECT

In this chapter, you will learn how two types of energy transfer—work and heat—serve to change a system's internal energy. You will also learn a new form of the law of energy conservation and will see how machine efficiency is limited.

Why it Matters

The principles of thermodynamics explain how many cyclic processes work, from refrigerators to the internal-combustion engines of automobiles.

CHAPTER PREVIEW

Tapping Prior Knowledge

Knowledge to Review

- ✔ Work is the product of force and displacement.
- ✔ Internal energy is the sum of the kinetic and potential energies of all the molecules in an object.
- ✔ Heat is the energy transferred from an area of higher temperature to an area of lower temperature until thermal equilibrium is achieved.
- ✔ Energy can be transformed from one form to another, but the total energy of an isolated system is constant.

Items to Probe

- ✔ Interactions between parts of a system: Ask students to describe the thermal processes that occur when a metal container with ice in it is placed in hot water.

335

SECTION 1
General Level

extension

In-Depth Physics Content

Your students can visit go.hrw.com for an online chapter that integrates more in-depth development of the concepts covered here.

Keyword HF6TDYX

Demonstration

Work from Heat — GENERAL

Purpose Show that internal energy can be converted to kinetic energy.

Materials a thermoelectric converter with fan, four plastic-foam cups, ice water, hot water, thermometer

Procedure Fill two cups with cold water and two with hot water. Measure the temperature of the water in each cup. Keep the cups covered. Show students that no electric source is connected to the fan. Ask students if these cups of water could make the fan work with (a) two legs in ice water, (b) two legs in hot water, or (c) one leg in ice water and one leg in hot water. Try each of the options. The fan will turn only in the last case. Let the fan operate until it slows down or stops. Measure the final temperature of the water in each cup. The temperature difference should be less than it was originally. Ask students what conclusions they can draw from their observations. *(Internal energy from the water has been used to make the fan turn. The difference in temperature allows the conversion of internal energy transferred as heat to electrical energy. The electrical energy sets the fan in motion by means of the motor, so it is converted to kinetic energy.)*

SECTION 1

Relationships Between Heat and Work

SECTION OBJECTIVES

- **Recognize that a system can absorb or release energy as heat in order for work to be done on or by the system and that work done on or by a system can result in the transfer of energy as heat.**
- **Compute the amount of work done during a thermodynamic process.**
- **Distinguish between isovolumetric, isothermal, and adiabatic thermodynamic processes.**

HEAT, WORK, AND INTERNAL ENERGY

Pulling a nail from a piece of wood causes the temperature of the nail and the wood to increase. Work is done by the frictional forces between the nail and the wood fibers. This work increases the internal energy of the iron atoms in the nail and the molecules in the wood.

The increase in internal energy of the nail corresponds to an increase in the nail's temperature, which is higher than the temperature of the surrounding air. As a result, energy is transferred as heat from the nail to the surrounding air. When the nail and surrounding air are at the same temperature, this energy transfer ceases.

Internal energy can be used to do work

The example of the hammer and nail illustrates that work can increase the internal energy of a substance. This internal energy can then decrease through the transfer of energy as heat. The reverse is also possible. Energy can be transferred to a substance as heat, and this internal energy can then be used to do work.

Consider a flask of water. A balloon is placed over the mouth of the flask, and the flask is heated until the water boils. Energy transferred as heat from the flame of the gas burner to the water increases the internal energy of the water. When the water's temperature reaches the boiling point, the water changes phase and becomes steam. At this constant temperature, the volume of the steam increases. This expansion provides a force that pushes the balloon outward and does work on the atmosphere, as shown in **Figure 1.** Thus, the steam does work, and the steam's internal energy decreases as predicted by the principle of energy conservation.

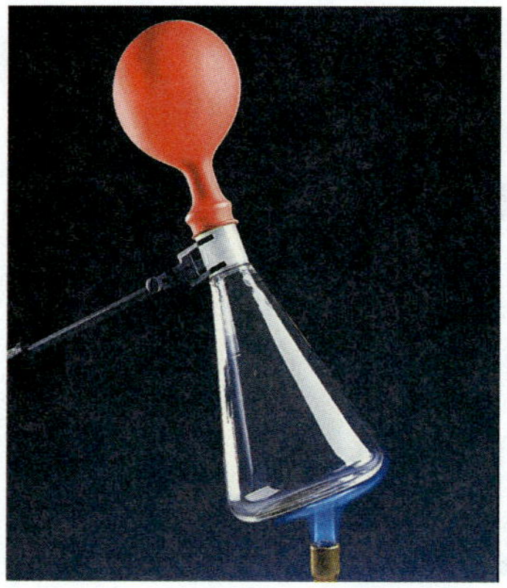

Figure 1
Energy transferred as heat turns water into steam. Energy from the steam does work on the air outside the balloon.

system

a set of particles or interacting components considered to be a distinct physical entity for the purpose of study

Heat and work are energy transferred to or from a system

On a microscopic scale, heat and work are similar. In this textbook, both are defined as energy that is transferred to or from a substance. This changes the substance's internal energy (and thus its temperature or phase). In other words, the terms *heat* and *work* always refer to energy in transit. An object never has "heat" or "work" in it; it has only internal energy.

In the previous examples, the internal energy of a substance or combination of substances has been treated as a single quantity to which energy is added or from which energy is taken away. Such a substance or combination of substances is called a **system.**

An example of a system would be the flask, balloon, water, and steam that were heated over the burner. As the burner transferred energy as heat to the system, the system's internal energy increased. When the expanding steam did work on the air outside the balloon by pushing it back (as the balloon expanded), the system's internal energy decreased. Some of the energy transferred to the system as heat was transferred out of the system as work done on the air.

A system is rarely completely isolated from its surroundings. In the example above, a heat interaction occurs between the burner and the system, and work is done by the system on the surroundings (the balloon moves the outside air outward). Energy is also transferred as heat to the air surrounding the flask because of the temperature difference between the flask and the surrounding air. In such cases, we must account for all of the interactions between the system and its **environment** that could affect the system's internal energy.

SCILINKS.
www.scilinks.org
Topic: Energy Transfer
Code: HF60516

environment

the combination of conditions and influences outside a system that affect the behavior of the system

Work done on or by a gas is pressure multiplied by volume change

In thermodynamic systems, work is defined in terms of pressure and volume change. Pressure is a measure of how much force is applied over a given area ($P = F/A$). Change in volume is equal to area multiplied by displacement ($\Delta V = Ad$). These expressions can be substituted into the definition of work introduced in the chapter "Work and Energy" to derive a new definition for the work done on or by a gas, as follows:

$$W = Fd$$

$$W = Fd\left(\frac{A}{A}\right) = \left(\frac{F}{A}\right)(Ad) = P\Delta V$$

WORK DONE BY A GAS

$$W = P\Delta V$$

work = pressure × volume change

This chapter will use only this new definition of work. Note that this definition assumes that P is constant.

If the gas expands, as shown in **Figure 2,** ΔV is positive, and the work done by the gas on the piston is positive. If the gas is compressed, ΔV is negative, and the work done by the gas on the piston is negative. (In other words, the piston does work on the gas.) When the gas volume remains constant, there is no displacement and no work is done on or by the system.

Although the pressure can change during a process, work is done only if the volume changes. A situation in which pressure increases and volume remains constant is comparable to one in which a force does not displace a mass even as the force is increased. Work is not done in either situation.

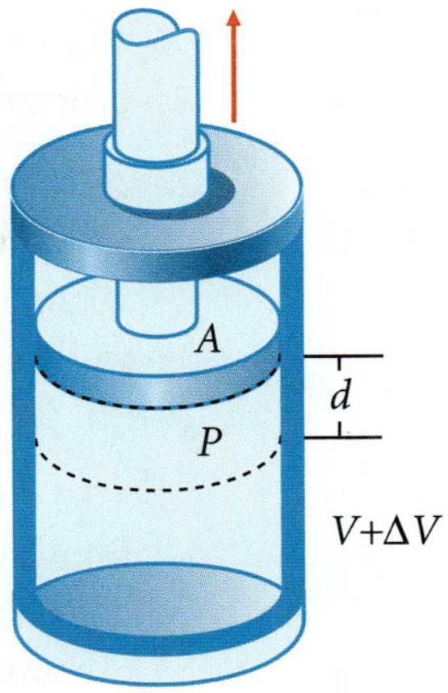

Figure 2
Work done on or by the gas is the product of the volume change (area A multiplied by the displacement d) and the pressure of the gas.

SECTION 1

Teaching Tip

Some of the first experiments demonstrating the equivalence between heat and work were performed by James Prescott Joule. In perhaps the most well known experiment, a paddle wheel was turned by falling weights. When the paddle wheel was placed in water, work was done on the water by the friction between the wheel and the water. As a result, the water's temperature increased. Joule found that the increase in temperature was proportional to the energy expended. In this experiment, the work done on a system was used to increase the system's internal energy by raising its temperature. In the example given on the previous page (in which steam is used to inflate a balloon), the reverse process occurs. The steam does work, and its internal energy decreases.

Misconception Alert — BASIC

Students are often confused by the concepts in this chapter. Small group discussions will help students review and organize their concepts of heat, work, pressure, temperature, and volume of a gas. Ask how they could use different containers (soft or rigid, thermally isolated or not) to control these variables.

SECTION 1

Classroom Practice

Work Done on or by a Gas
The cross-sectional area of the piston in **Figure 2** on the previous page is $0.20\ m^2$. A 400.0 N weight pushes the piston down 0.15 m and compresses the gas in the cylinder. How much pressure is exerted on the gas? By how much did the gas volume decrease?

Answer
2.0×10^3 Pa; $0.030\ m^3$

PROBLEM GUIDE A

Use this guide to assign problems.
SE = Student Edition Textbook
PW = Problem Workbook
PB = Problem Bank on the One-Stop Planner (OSP)

Solving for:

W	**SE** Sample, 1–3; Ch. Rvw. 9–10 **PW** 5–6 **PB** 8–10
P	**SE** 4 **PW** 3–4 **PB** Sample, 1–3
ΔV	**PW** Sample, 1–2 **PB** 4–7

***Challenging Problem**
Consult the printed Solutions Manual or the OSP for detailed solutions.

ANSWERS

Practice A

1. a. 6.4×10^5 J
   b. -4.8×10^5 J
2. −167.5 J; Work is done on the gas because the volume change is negative.
3. 3.3×10^2 J
4. 2.0×10^5 Pa

SAMPLE PROBLEM A

Work Done on or by a Gas

PROBLEM

An engine cylinder has a cross-sectional area of $0.010\ m^2$. How much work can be done by a gas in the cylinder if the gas exerts a constant pressure of 7.5×10^5 Pa on the piston and moves the piston a distance of 0.040 m?

SOLUTION

Given: $A = 0.010\ m^2 \qquad d = 0.040\ m$

$P = 7.5 \times 10^5\ Pa = 7.5 \times 10^5\ N/m^2$

Unknown: $W = ?$

Use the equation for the work done on or by a gas.

$$W = P\Delta V = PAd$$

$$W = (7.5 \times 10^5\ N/m^2)(0.010\ m^2)(0.040\ m)$$

$$\boxed{W = 3.0 \times 10^2\ J}$$

TIP *Because W is positive, we can conclude that the work is done by the gas rather than on the gas.*

PRACTICE A

Work Done on or by a Gas

1. Gas in a container is at a pressure of 1.6×10^5 Pa and a volume of $4.0\ m^3$. What is the work done by the gas if

a. it expands at constant pressure to twice its initial volume?

b. it is compressed at constant pressure to one-quarter of its initial volume?

2. A gas is enclosed in a container fitted with a piston. The applied pressure is maintained at 599.5 kPa as the piston moves inward, which changes the volume of the gas from $5.317 \times 10^{-4}\ m^3$ to $2.523 \times 10^{-4}\ m^3$. How much work is done? Is the work done *on* or *by* the gas? Explain your answer.

3. A balloon is inflated with helium at a constant pressure that is 4.3×10^5 Pa in excess of atmospheric pressure. If the balloon inflates from a volume of $1.8 \times 10^{-4}\ m^3$ to $9.5 \times 10^{-4}\ m^3$, how much work is done on the surrounding air by the helium-filled balloon during this expansion?

4. Steam moves into the cylinder of a steam engine at a constant pressure and does 0.84 J of work on a piston. The diameter of the piston is 1.6 cm, and the piston travels 2.1 cm. What is the pressure of the steam?

THERMODYNAMIC PROCESSES

In this section, three distinct quantities have been related to each other: internal energy (U), heat (Q), and work (W). Processes that involve only work or only heat are rare. In most cases, energy is transferred as both heat and work. However, in many processes, one type of energy transfer is dominant, and the other type is negligible. In these cases, the real process can be approximated with an ideal process. For example, if the dominant form of energy transfer is work and the energy transferred as heat is extremely small, we can neglect the heat transfer and still obtain an accurate model. In this way, many real processes can be approximated by one of three ideal processes.

The rest of this chapter deals with ideal processes in gases. All objects have internal energy, which is the sum of the kinetic and potential energies of their molecules. However, monatomic gases present a simpler situation to study because all of their internal energy is kinetic. (The reason is that the molecules of a gas are too far apart to interact with each other significantly.)

No work is done in a constant-volume process

In general, when a gas undergoes a change in temperature but no change in volume, no work is done on or by the system. Such a process is called a constant-volume process, or **isovolumetric process.**

isovolumetric process

a thermodynamic process that takes place at constant volume so that no work is done on or by the system

One example of an isovolumetric process takes place inside a *bomb calorimeter,* shown in **Figure 3.** This device is a thick container in which a small quantity of a substance undergoes a combustion reaction. The energy released by the reaction increases the pressure and temperature of the gaseous reaction products. Because the container's walls are thick, there is no change in the volume of the gas. Energy can be transferred to or from the container only as heat. As in the case of the simple calorimeter discussed in the chapter "Heat," the increase in the temperature of water surrounding the bomb calorimeter provides information for calculating the total amount of energy produced by the reaction.

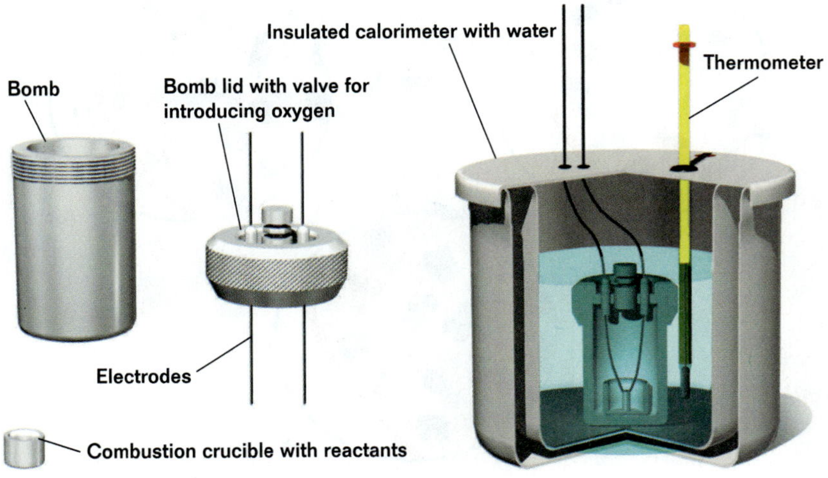

Figure 3
The volume inside the bomb calorimeter is nearly constant, so most of the energy is transferred to or from the calorimeter as heat.

Visual Strategy GENERAL

Figure 3

Q What happens to the temperature and internal energy of the gas and water after the combustion reaction?

A ***The water's temperature and internal energy increase as the gas's temperature and internal energy decrease.***

Teaching Tip — BASIC

Point out to students that the Greek word *isos* means "the same." This tip can be used to remember the words *isovolumetric* (same volume) and *isothermal* (same temperature).

Misconception Alert — ADVANCED

Students may wonder why energy should flow into the balloon described on this page because the isothermal expansion restores equilibrium between the inside and outside pressures. Point out that a loss of internal energy is a decrease in molecular kinetic energy, so the temperature inside the balloon decreases. Because of the temperature difference, energy is transferred from the environment to restore thermal equilibrium.

Teaching Tip — GENERAL

Tell students that the absence of a change in the internal energy of an isothermal process is true only for systems in which there is no phase change. During a phase change, such as when water becomes steam, the temperature remains constant, but internal energy increases.

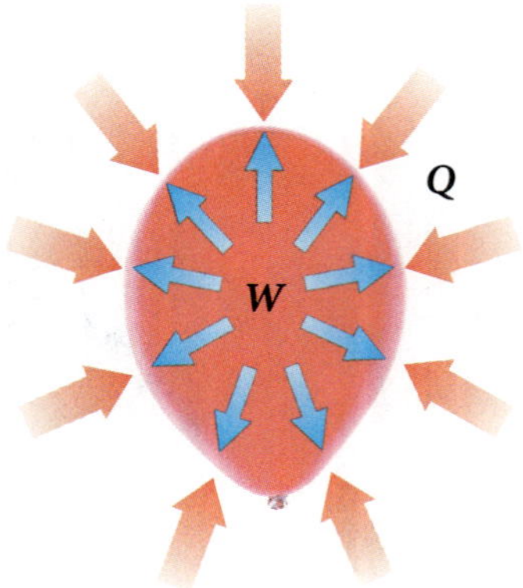

Figure 4
An isothermal process can be approximated if energy is slowly removed from a system as work while an equivalent amount of energy is added as heat.

isothermal process

a thermodynamic process that takes place at constant temperature

Internal energy is constant in a constant-temperature process

Although you may think of a toy balloon that has been inflated and sealed as a static system, it is subject to continuous thermodynamic effects. Consider what happens to such a balloon during an approaching storm. (To simplify this example, we will assume that the balloon is only partially inflated and thus does not store elastic energy.) During the few hours before the storm arrives, the barometric pressure of the atmosphere steadily decreases by about 2000 Pa. If you are indoors and the temperature of the building is controlled, any change in outside temperature will not take place indoors. But because no building is perfectly sealed, changes in the pressure of the air outside also take place inside.

As the atmospheric pressure inside the building slowly decreases, the balloon expands and slowly does work on the air outside the balloon. At the same time, energy is slowly transferred into the balloon as heat. The net result of these two processes is that the air inside the balloon stays at the same temperature as the air outside the balloon. Thus, the internal energy of the balloon's air does not change. The energy transferred out of the balloon as work is matched by the energy transferred into the balloon as heat. This process is illustrated in **Figure 4.**

This example is a close approximation of an **isothermal process.** In an isothermal process, the system's temperature remains constant and internal energy does not change when energy is transferred to or from the system as heat or work.

You may wonder how energy can be transferred as heat from the air outside the balloon to the air inside when both gases are at the same constant temperature. The reason that energy can be transferred as heat in an isothermal process may be seen if you consider the process as consisting of a large number of very gradual, very small, sequential changes, as shown in **Figure 5.**

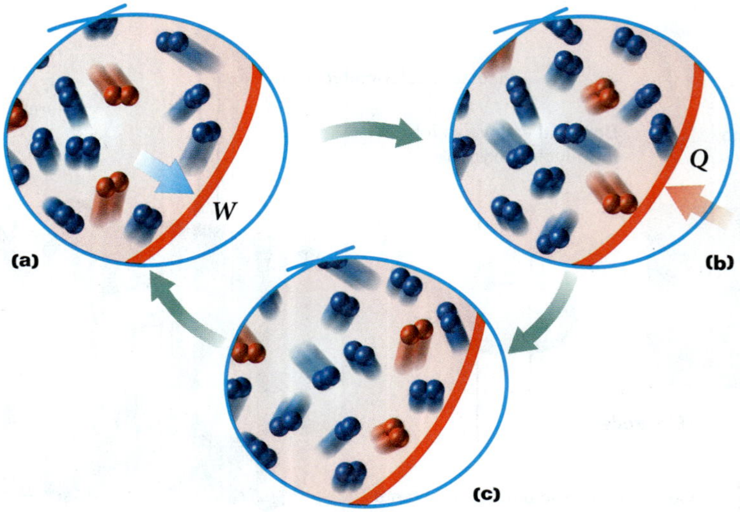

Figure 5
In an isothermal process in a partially inflated balloon, **(a)** small amounts of energy are removed as work. **(b)** Energy is added to the gas within the balloon's interior as heat so that **(c)** thermal equilibrium is quickly restored.

Energy is not transferred as heat in an adiabatic process

When a tank of compressed gas is opened to fill a toy balloon, the process of inflation occurs rapidly. The internal energy of the gas does not remain constant. Instead, as the pressure of the gas in the tank decreases, so do the gas's internal energy and temperature.

If the balloon and the tank are thermally insulated, no energy can be transferred from the expanding gas as heat. A process in which changes occur but no energy is transferred to or from a system as heat is called an **adiabatic process.** The decrease in internal energy must therefore be equal to the energy transferred from the gas as work. This work is done by the confined gas as it pushes the wall of the balloon outward, overcoming the pressure exerted by the air outside the balloon. As a result, the balloon inflates, as shown in **Figure 6.** Note that unlike an isothermal process, which must happen slowly, an adiabatic process must happen rapidly.

As mentioned earlier, the three processes described here rarely occur ideally, but many situations can be approximated by one of the three processes. This allows you to make predictions. For example, both refrigerators and internal-combustion engines require that gases be compressed or expanded rapidly. By making the approximation that these processes are adiabatic, one can make quite good predictions about how these machines will operate.

adiabatic process

a thermodynamic process during which no energy is transferred to or from the system as heat

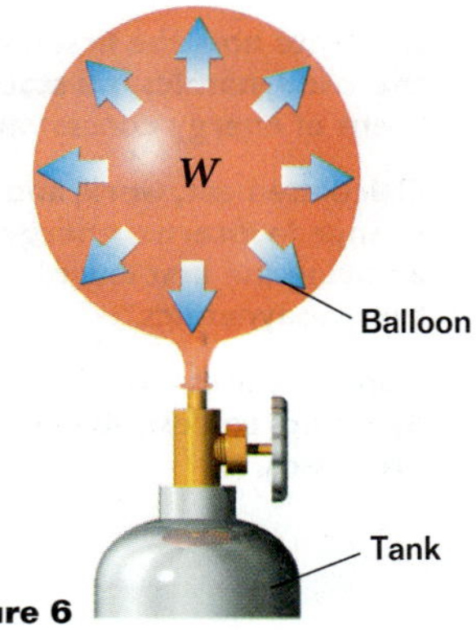

Figure 6
As the gas inside the tank and balloon rapidly expands, its internal energy decreases. This energy leaves the system by means of work done against the outside air.

SECTION REVIEW

1. In which of the situations listed below is energy being transferred as heat to the system in order for the system to do work?
   a. Two sticks are rubbed together to start a fire.
   b. A firecracker explodes.
   c. A red-hot iron bar is set aside to cool.
2. A mixture of gasoline vapor and air is placed in an engine cylinder. The piston has an area of 7.4×10^{-3} m^2 and is displaced inward by 7.2×10^{-2} m. If 9.5×10^5 Pa of pressure is placed on the piston, how much work is done during this process? Is work being done *on* or *by* the gas mixture?
3. A weather balloon slowly expands as energy is transferred as heat from the outside air. If the average net pressure is 1.5×10^3 Pa and the balloon's volume increases by 5.4×10^{-5} m^3, how much work is done by the expanding gas?
4. **Critical Thinking** Identify the following processes as isothermal, isovolumetric, or adiabatic:
   a. a tire being rapidly inflated
   b. a tire expanding gradually at a constant temperature
   c. a steel tank of gas being heated

Teaching Tip — ADVANCED

Point out that the rapid inflation of the balloon described in the student text is a process that is only approximately adiabatic. Some transfer of energy as heat actually does take place because neither the balloon nor the tank is perfectly insulated. The decrease in the internal energy and the temperature of the rapidly expanding gas accounts for the sudden drop in temperature of the outside surface of the tank when a compressed gas is being released. Once the adiabatic expansion is complete, the temperature of the gas gradually increases as energy from the outside air is transferred into the tank as heat.

SECTION REVIEW ANSWERS

1. (b) (energy is transferred as heat to the firecracker, which does work when it explodes)
2. -5.1×10^2 J; on the gas mixture
3. 8.1×10^{-2} J
4. a. adiabatic
   b. isothermal
   c. isovolumetric

SECTION 2
Advanced Level

Visual Strategy GENERAL

Figure 7
Point out that the *KE* bars represent the energy of the car alone, the *PE* bars represent the energy of the car-Earth system, and the *U* bars represent the sum of the internal energies of the car and the track.

Q How does the potential energy vary as the car rolls up and down the track? How is this reflected in the energy bars?

A ***The potential energy depends only on the car's elevation. Accordingly, the PE bar is highest at (b), second highest at (e), second lowest at (c), and lowest at (d).***

Q Draw a fourth bar representing the *mechanical* energy, *ME*, at locations **(b)**, **(c)**, **(d)**, and **(e)**. How do these *ME* bars relate to the *U* bars?

A ***The new ME bar should equal KE + PE in each case. Thus, it will get shorter from (b) to (e) as the U bar gets taller by the same amount.***

SECTION 2

The First Law of Thermodynamics

SECTION OBJECTIVES

- **Illustrate how the first law of thermodynamics is a statement of energy conservation.**
- **Calculate heat, work, and the change in internal energy by applying the first law of thermodynamics.**
- **Apply the first law of thermodynamics to describe cyclic processes.**

ENERGY CONSERVATION

Imagine a roller coaster that operates without friction. The car is raised against gravitational force by work. Once the car is freely moving, it will have a certain kinetic energy (*KE*) and a certain potential energy (*PE*). Because there is no friction, the mechanical energy (*KE* + *PE*) remains constant throughout the ride's duration. Thus, when the car is at the top of the rise, it moves relatively slowly (larger *PE* + smaller *KE*). At lower points in the track, the car has less potential energy and so moves more quickly (smaller *PE* + larger *KE*).

If friction is taken into account, mechanical energy is no longer conserved, as shown in **Figure 7.** A steady decrease in the car's total mechanical energy occurs because of work being done against the friction between the car's axles and its bearings and between the car's wheels and the coaster track. Mechanical energy is transferred to the atoms and molecules throughout the entire roller coaster (both the car and the track). Thus, the roller coaster's internal energy increases by an amount equal to the decrease in the mechanical energy. Most of this energy is then gradually dissipated to the air surrounding the roller coaster as heat. If the internal energy for the roller coaster (the system) and the energy dissipated to the surrounding air (the environment) are taken into account, then the total energy will be constant.

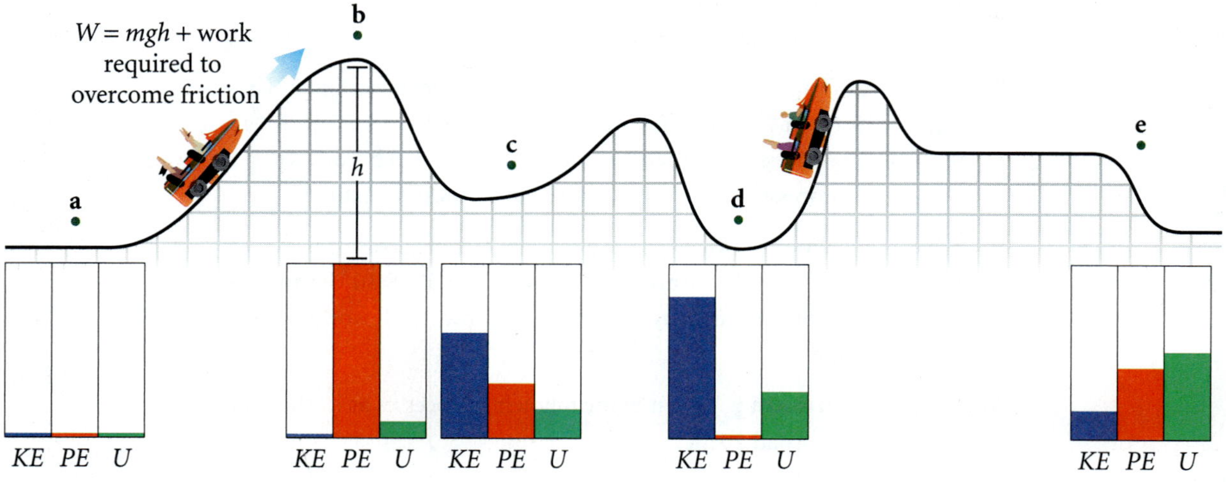

Figure 7
In the presence of friction, the internal energy (*U*) of the roller coaster increases as *KE* + *PE* decreases.

The principle of energy conservation that takes into account a system's internal energy as well as work and heat is called the *first law of thermodynamics.*

Imagine that the isothermally expanding toy balloon in the previous section is squeezed rapidly. The process is no longer isothermal. Instead, it is a combination of two processes. On the one hand, work (W) is done on the system. The balloon and the air inside it (the system) are compressed, so the air's internal energy and temperature increase. Work is being done on the system, so W is a negative quantity. The rapid squeezing of the balloon can be treated as an adiabatic process, so $Q = 0$ and, therefore, $\Delta U = -W$.

After the compression step, energy is transferred from the system as heat (Q). Some of the internal energy of the air inside the balloon is transferred to the air outside the balloon. During this step, the internal energy of the gas decreases, so ΔU has a negative value. Similarly, because energy is removed from the system, Q has a negative value. The change in internal energy for this step can be expressed as $-\Delta U = -Q$, or $\Delta U = Q$.

The signs for heat and work for a system are summarized in **Table 1.** To remember whether a system's internal energy increases or decreases, you may find it helpful to visualize the system as a circle. When work is done on the system or energy is transferred as heat into the system, an arrow points into the circle. This shows that internal energy increases. When work is done by the system or energy is transferred as heat out of the system, the arrow points out of the circle. This shows that internal energy decreases.

Table 1 Signs of *Q* and *W* for a System

$Q > 0$	energy added to system as heat
$Q < 0$	energy removed from system as heat
$Q = 0$	no transfer of energy as heat
$W > 0$	work done by system (expansion of gas)
$W < 0$	work done on system (compression of gas)
$W = 0$	no work done

The first law of thermodynamics can be expressed mathematically

In all the thermodynamic processes described so far, energy has been conserved. To describe the overall change in the system's internal energy, one must account for the transfer of energy to or from the system as heat and work. The total change in the internal energy is the difference between the final internal energy value (U_f) and the initial internal energy value (U_i). That is, $\Delta U = U_f - U_i$. Energy conservation requires that the total change in internal energy from its initial to its final equilibrium conditions be equal to the net transfer of energy as both heat and work. This statement of total energy conservation, shown mathematically on the next page, is the first law of thermodynamics.

Did you know?

Not all ways of transferring energy can be classified simply by work or by heat. Other processes that can change the internal energy of a substance include changes in the chemical and magnetic properties of the substance.

SCiLINKS.
www.scilinks.org
Topic: Thermodynamics
Code: HF61514

Teaching Tip — ADVANCED

Point out to students that this chapter assumes that internal energy depends only on temperature. As discussed in the chapter "Heat," this is true for ideal gases. However, for nonideal gases and for liquids and solids, other properties in addition to temperature contribute to internal energy.

Key Models and Analogies — GENERAL

Show these simple examples to model the signs of Q and W in typical situations. Students can reduce problems to one case or to a combination of two cases.

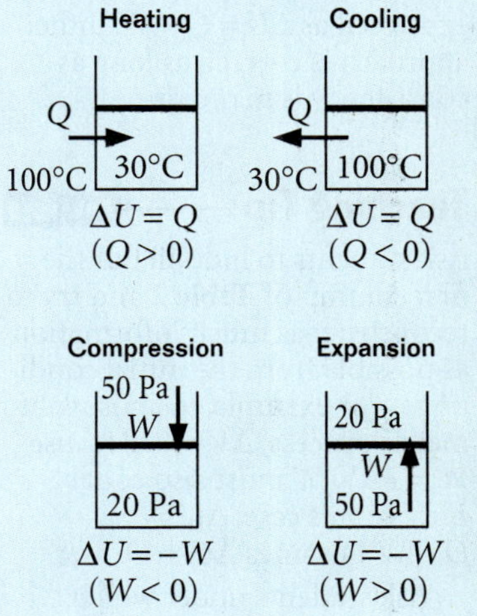

The Language of Physics

The first law is a statement of conservation of energy. Because the Q term is positive, it represents the *energy added to* the system as heat. Because the W term is negative, it represents the *work done by* the system. In other words, because energy added to the system as heat *increases* the internal energy and work done by the system *decreases* the system's internal energy, the first law is expressed as $\Delta U = Q - W$.

In this text, positive W is defined as the work done *by* the system. Alternatively, some texts define positive W as the work done *on* the system. With this definition, the first law is expressed as $\Delta U = Q + W$. Either approach is correct, as long as consistency is maintained.

Teaching Tip — ADVANCED

Ask students to hide all but the first column of **Table 2** and try to reconstruct as much information as possible from the initial conditions. For example, in an isovolumetric process, $\Delta V = 0$. Because $W = P\Delta V$, W must also equal zero. In this case, $\Delta U = Q - W$ becomes $\Delta U = Q$.

Ask students under which conditions work can be done on a gas without changing its internal energy *(in an isothermal process, where $\Delta U = 0$).*

THE FIRST LAW OF THERMODYNAMICS

$$\Delta U = Q - W$$

Change in system's internal energy = energy transferred to or from system as heat – energy transferred to or from system as work

When this equation is used, all quantities must have the same energy units. Throughout this chapter, the SI unit for energy, the joule, will be used.

According to the first law of thermodynamics, a system's internal energy can be changed by transferring energy as either work, heat, or a combination of the two. The thermodynamic processes discussed in Section 1 can therefore be expressed using the equation for the first law of thermodynamics, as shown in **Table 2.**

Table 2 First Law of Thermodynamics for Special Processes

Process	Conditions	First law of thermodynamics	Interpretation
Isovolumetric	no work done	$\Delta V = 0$, so $P\Delta V = 0$ and $W = 0$; therefore, $\Delta U = Q$	Energy added to the system as heat ($Q > 0$) increases the system's internal energy. Energy removed from the system as heat ($Q < 0$) decreases the system's internal energy.
Isothermal	no change in temperature or internal energy	$\Delta T = 0$, so $\Delta U = 0$; therefore, $\Delta U = Q - W = 0$, or $Q = W$	Energy added to the system as heat is removed from the system as work done by the system. Energy added to the system by work done on it is removed from the system as heat.
Adiabatic	no energy transferred as heat	$Q = 0$, so $\Delta U = -W$	Work done on the system ($W < 0$) increases the system's internal energy. Work done by the system ($W > 0$) decreases the system's internal energy.
Isolated system	no energy transferred as heat and no work done on or by the system	$Q = 0$ and $W = 0$, so $\Delta U = 0$ and $U_i = U_f$	There is no change in the system's internal energy.

SAMPLE PROBLEM B

The First Law of Thermodynamics

PROBLEM

A total of 135 J of work is done on a gaseous refrigerant as it undergoes compression. If the internal energy of the gas increases by 114 J during the process, what is the total amount of energy transferred as heat? Has energy been added to or removed from the refrigerant as heat?

SOLUTION

1. DEFINE **Given:** $W = -135\ \text{J}$ $\quad \Delta U = 114\ \text{J}$

Unknown: $Q = ?$

> **TIP** *Work is done on the gas, so work (W) has a negative value. The internal energy increases during the process, so the change in internal energy (ΔU) has a positive value.*

Diagram:

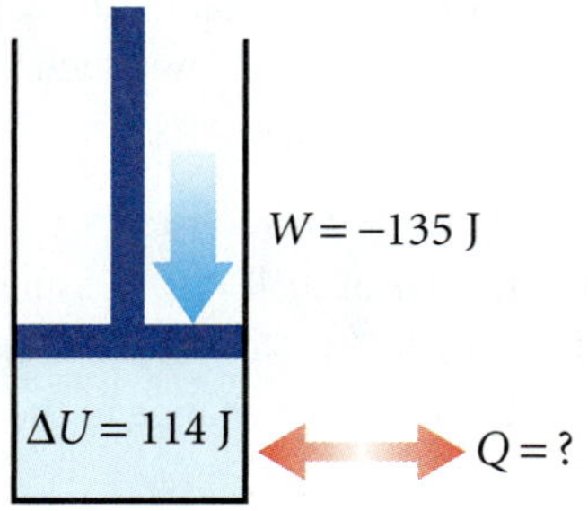

2. PLAN **Choose an equation or situation:**
Apply the first law of thermodynamics using the values for ΔU and W in order to find the value for Q.

$$\Delta U = Q - W$$

Rearrange the equation to isolate the unknown:

$$Q = \Delta U + W$$

3. CALCULATE **Substitute the values into the equation and solve:**

$$Q = 114\ \text{J} + (-135\ \text{J}) = -21\ \text{J}$$

$$\boxed{Q = -21\ \text{J}}$$

> **TIP** *The sign for the value of Q is negative. From **Table 1,** Q < 0 indicates that energy is transferred as heat from the refrigerant.*

4. EVALUATE Although the internal energy of the refrigerant increases under compression, more energy is added as work than can be accounted for by the increase in the internal energy. This energy is removed from the gas as heat, as indicated by the minus sign preceding the value for Q.

Classroom Practice

The First Law of Thermodynamics
A gas is trapped in a small metal cylinder with a movable piston and is submerged in a large amount of ice water so that the initial temperature of the gas is 0°C. A total of 1200 J of work is done by a force that slowly pushes the piston inward.

a. Is this process isothermal, adiabatic, or isovolumetric?

b. How much energy is transferred as heat between the gas and the ice water?

Answers

a. isothermal (the large amount of ice water and the slow process maintain the gas at 0°C)

b. Q from the gas to the ice water = 1200 J

PROBLEM GUIDE B

Use this guide to assign problems.
SE = Student Edition Textbook
PW = Problem Workbook
PB = Problem Bank on the One-Stop Planner (OSP)

Solving for:

Q	**SE** Sample, 1–3; Ch. Rvw. 16–17, 39 **PW** 3, 5 **PB** 7–10
W	**SE** 4; Ch. Rvw. 17 **PW** 2, 6–7 **PB** Sample, 1–3
ΔU	**SE** 5; Ch. Rvw. 17, 29 **PW** Sample, 1, 2–3, 4 **PB** 4–6

***Challenging Problem**
Consult the printed Solutions Manual or the OSP for detailed solutions.

SECTION 2

ANSWERS

Practice B

1. 33 J
2. −143 J; removed as heat
3. 1.00×10^4 J
4. 0 J; 344 J done by gas
5. 1.74×10^8 J

PRACTICE B

The First Law of Thermodynamics

1. Heat is added to a system, and the system does 26 J of work. If the internal energy increases by 7 J, how much heat was added to the system?
2. The internal energy of the gas in a gasoline engine's cylinder decreases by 195 J. If 52.0 J of work is done by the gas, how much energy is transferred as heat? Is this energy added to or removed from the gas?
3. A 2.0 kg quantity of water is held at constant volume in a pressure cooker and heated by a range element. The system's internal energy increases by 8.0×10^3 J. However, the pressure cooker is not well insulated, and as a result, 2.0×10^3 J of energy is transferred to the surrounding air. How much energy is transferred from the range element to the pressure cooker as heat?
4. The internal energy of a gas decreases by 344 J. If the process is adiabatic, how much energy is transferred as heat? How much work is done on or by the gas?
5. A steam engine's boiler completely converts 155 kg of water to steam. This process involves the transfer of 3.50×10^8 J as heat. If steam escaping through a safety valve does 1.76×10^8 J of work expanding against the outside atmosphere, what is the net change in the internal energy of the water-steam system?

CYCLIC PROCESSES

cyclic process

a thermodynamic process in which a system returns to the same conditions under which it started

A refrigerator performs mechanical work to create temperature differences between its closed interior and its environment (the air in the room). This process leads to the transfer of energy as heat. A heat engine does the opposite: it uses heat to do mechanical work. Both of these processes have something in common: they are examples of **cyclic processes.**

In a cyclic process, the system's properties at the end of the process are identical to the system's properties before the process took place. The final and initial values of internal energy are the same, and the change in internal energy is zero.

$$\Delta U_{net} = 0 \text{ and } Q_{net} = W_{net}$$

A cyclic process resembles an isothermal process in that all energy is transferred as work and heat. But now the process is repeated with no net change in the system's internal energy.

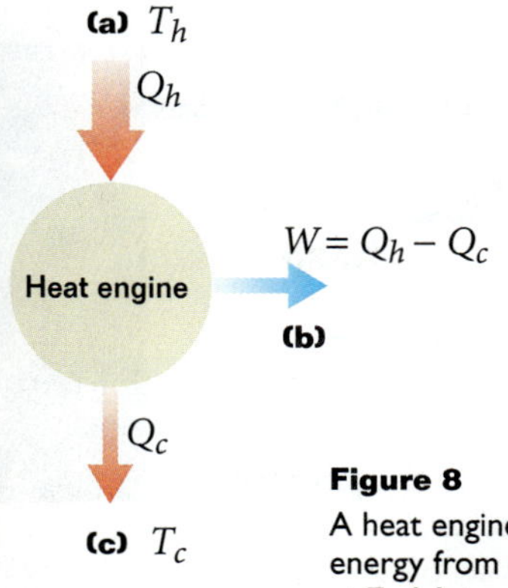

Figure 8
A heat engine is able to do work **(b)** by transferring energy from a high-temperature substance (the boiler) at T_h **(a)** to a substance at a lower temperature (the air surrounding the engine) at T_c **(c).**

Heat engines use heat to do work

A heat engine is a device that uses heat to do mechanical work. A heat engine is similar to a water wheel, which uses a difference in potential energy to do work. A water wheel uses the energy of water falling from one level above Earth's surface to another. The change in potential energy increases the water's kinetic energy so that the water can do work on one side of the wheel and thus turn it.

Instead of using the difference in potential energy to do work, heat engines do work by transferring energy from a high-temperature substance to a lower-temperature substance, as indicated for the steam engine shown in **Figure 8.** For each complete cycle of the heat engine, the net work done will equal the difference between the energy transferred as heat from a high-temperature substance to the engine (Q_h) and the energy transferred as heat from the engine to a lower-temperature substance (Q_c).

$$W_{net} = Q_h - Q_c$$

The larger the difference between the energy transferred as heat into the engine and out of the engine, the more work the engine can do in each cycle.

The internal-combustion engine found in most vehicles is an example of a heat engine. Internal-combustion engines burn fuel within a closed chamber (the cylinder). The potential energy of the chemical bonds in the reactant gases is converted to kinetic energy of the particle products of the reaction. These gaseous products push against a piston and thus do work on the environment (in this case, a crankshaft that transforms the linear motion of the piston to the rotational motion of the axle and wheels).

Although the basic operation of any internal-combustion engine resembles that of an ideal cyclic heat engine, certain steps do not fit the idealized model. When gas is taken in or removed from the cylinder, matter enters or leaves the system so that the matter in the system is not isolated. No heat engine operates perfectly. Only part of the available internal energy leaves the engine as work done on the environment; most of the energy is removed as heat.

SCILINKS.
www.scilinks.org
Topic: Heat Engines
Code: HF60728

Key Models and Analogies

The model of a system interacting with its environment is useful when dealing with thermodynamic cycles. Point out that the steam in the cylinder is the thermodynamic system, which exchanges energy with its surroundings. Remind students that energy is transferred as heat when the temperature of the system and that of the environment are different. The air is the environment for the cooling phase, and the boiler is the environment in the heating phase.

Teaching Tip — GENERAL

Have students describe the four steps in the cycle in terms of energy transfer, work done, and changes in internal energy of the steam.

Heater-boiler: Q_{in} into steam
W_{out} done by steam on piston
Air: Q_{out} out of steam
W_{in} done by piston on steam

STOP Misconception Alert — GENERAL

Students tend to believe that $W_{out} = W_{in}$ because volume increases and then decreases by the same amount as the piston returns to its original position. Remind them that $W = P\Delta V$ applies only when pressure is constant, which is not the case in this example.

Why it Matters

Gasoline Engines

Point out that another difference between an ideal heat engine and the internal-combustion engine is that in the heat engine, the energy source is external and energy is transferred to the engine as heat. In the internal-combustion engine, the energy source is internal (because it comes from a chemical reaction in the cylinder), so energy is not transferred into the engine as heat.

Key Models and Analogies — ADVANCED

A graph of pressure (P) versus volume (V) is often used for depicting the steps of thermodynamic cycles. Draw a simple graph, as shown below, on the chalkboard. Ask students which segment of the graph corresponds to heating *(ab)*, cooling *(cd)*, compression *(da)*, and expansion *(bc)*. Use this simple graph to tell students that the net work done in the cycle equals the area enclosed by the graph connecting points **a, b, c,** and **d.**

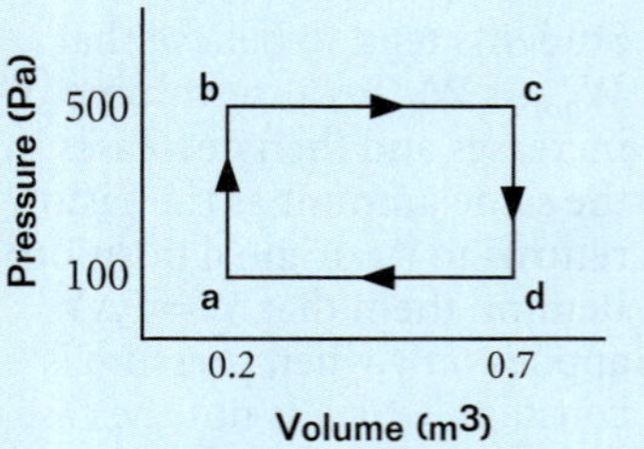

Why it Matters

Gasoline Engines

A gasoline engine is one type of internal-combustion engine. The diagram below illustrates the steps in one cycle of operation for a gasoline engine. During compression, shown in **(a),** work is done by the piston as it adiabatically compresses the fuel-and-air mixture in the cylinder. Once maximum compression of the gas is reached, combustion takes place. The chemical potential energy released during combustion increases the internal energy of the gas, as shown in **(b).** The hot, high-pressure gases from the combustion reaction expand in volume, pushing the piston and turning the crankshaft, as shown in **(c).** Once all of the work is done by the piston, some energy is transferred as heat through the walls of the cylinder. Even more energy is transferred by the physical removal of the hot exhaust gases from the cylinder, as shown in **(d).** A new fuel-air mixture is then drawn through the intake valve into the cylinder by the downward-moving piston, as shown in **(e).**

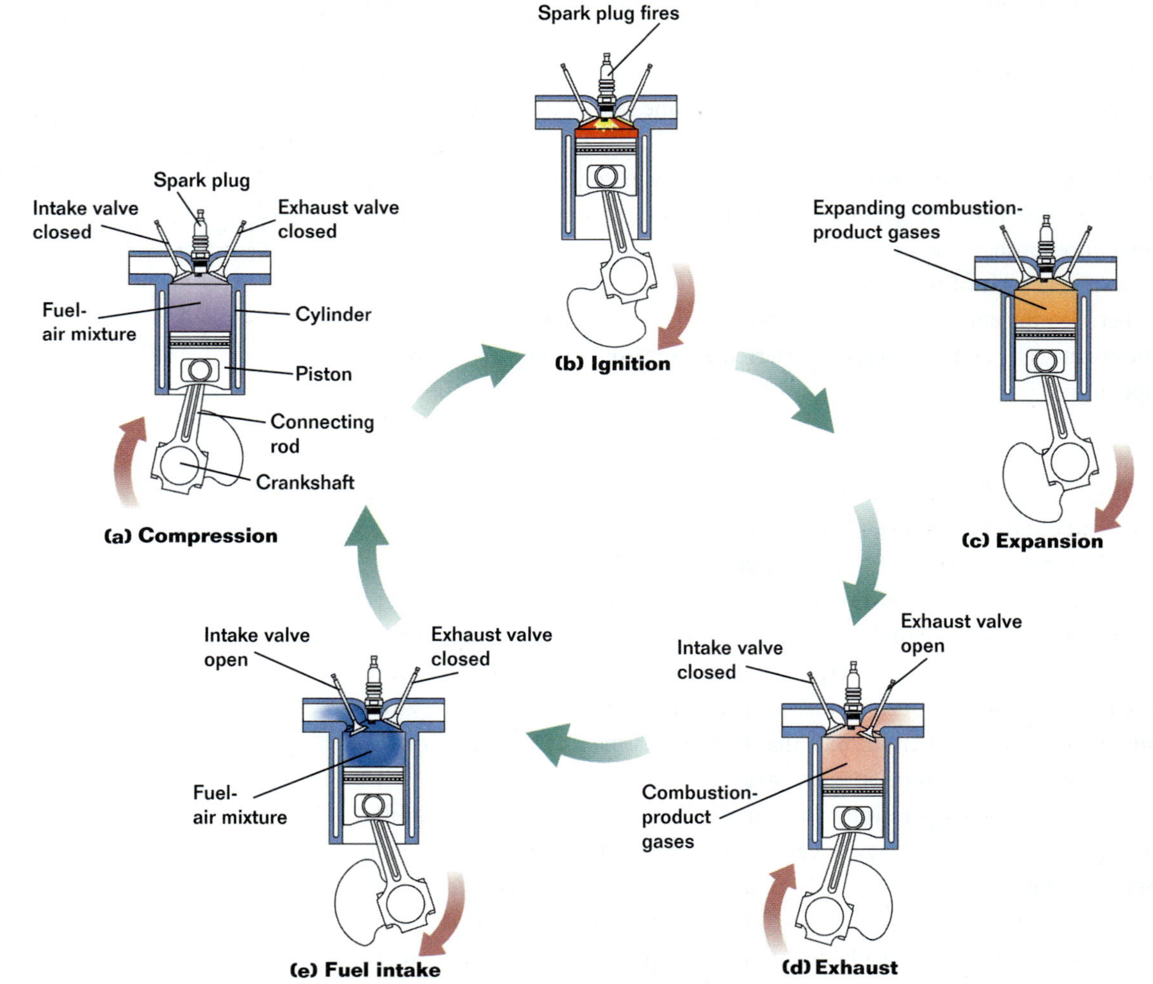

SECTION REVIEW

1. Use the first law of thermodynamics to show that the internal energy of an isolated system is always conserved.

2. In the systems listed below, identify where energy is transferred as heat and work and where changes in internal energy occur. Is energy conserved in each case?
   - **a.** the steam in a steam engine consisting of a boiler, a firebox, a cylinder, a piston, and a flywheel
   - **b.** the drill bit of a power drill and a metal block into which a hole is being drilled

3. Express the first law of thermodynamics for the following processes:
   - **a.** isothermal
   - **b.** adiabatic
   - **c.** isovolumetric

4. A compressor for a jackhammer expands the air in the hammer's cylinder at a constant pressure of 8.6×10^5 Pa. The increase in the cylinder's volume is 4.05×10^{-4} m^3. During the process, 9.5 J of energy is transferred out of the cylinder as heat.
   - **a.** What is the work done by the air?
   - **b.** What is the change in the air's internal energy?
   - **c.** What type of ideal thermodynamic process does this approximate?

5. A mixture of fuel and air is enclosed in an engine cylinder fitted with a piston. The gas pressure is maintained at 7.07×10^5 Pa as the piston moves slowly inward. If the gas volume decreases by 1.1×10^{-4} m^3 and the internal energy of the gas increases by 62 J, how much energy is added to or removed from the system as heat?

6. Over several cycles, a refrigerator does 1.51×10^4 J of work on the refrigerant. The refrigerant in turn removes 7.55×10^4 J as heat from the air inside the refrigerator.
   - **a.** How much energy is transferred as heat to the outside air?
   - **b.** What is the net change in the internal energy of the refrigerant?
   - **c.** What is the amount of work done on the air inside the refrigerator?
   - **d.** What is the net change in the internal energy of the air inside the refrigerator?

7. If a weather balloon in flight gives up 15 J of energy as heat and the gas within it does 13 J of work on the outside air, by how much does its internal energy change?

8. **Critical Thinking** After reading the feature on the next page, explain why opening the refrigerator door on a hot day does not cause your kitchen to become cooler.

SECTION REVIEW ANSWERS

1. By definition, no energy is transferred to or from isolated systems, so $Q = W = 0$ and $\Delta U = Q - W = 0$.
2. **a.** The firebox and boiler transfer energy to the steam via the cylinder. The steam then does work (decreasing its internal energy) on the piston. The piston communicates this energy to the flywheel by doing work on it.
   b. Work is done by the drill on the bit and by the bit on the block, increasing U for each.
   Energy is conserved for both cases when all forms of energy are taken into account. (In part b, this includes the electrical energy supplied to the power drill.)
3. **a.** $Q = W$
   b. $\Delta U = -W$
   c. $\Delta U = Q$
4. **a.** $W = 3.5 \times 10^2$ J
   b. $\Delta U = -3.6 \times 10^2$ J
   c. adiabatic
5. $Q = -16$ J (removed as heat)
6. **a.** $Q = 9.06 \times 10^4$ J
   b. $\Delta U = 0$ J (cyclic process)
   c. $W = 0$ J ($\Delta V = 0$)
   d. $\Delta U = -7.55 \times 10^4$ J
7. $\Delta U = -28$ J
8. Energy removed from the refrigerator is added to the same air that is being cooled by the refrigerator, so there is no decrease in temperature.

Why it Matters

Refrigerators

Point out that the refrigerant as a system is in thermal contact with two environments (the inside of the refrigerator and the outside air) at two separate moments in time. Between these steps of the cycle, the refrigerant does work and work is done on it adiabatically. The end result of drawing energy as heat from inside (Q_c) to outside (Q_h) occurs when the electric motor in the refrigerator does work to compress the refrigerant.

Why it Matters

Refrigerators

As shown in the photograph below, a refrigerator can be represented schematically as a system that transfers energy from a body at a low temperature **(c)** to one at a high temperature **(a).** The refrigerator uses work performed by an electric motor to compress the *refrigerant,* which is a substance that evaporates at a very low temperature. In the past, ammonia was used as a refrigerant in home refrigerators. However, ammonia leaks pose a risk because pure ammonia is highly toxic to people. In the 1930s, home refrigerators began using a newly developed, nontoxic class of refrigerants called *CFCs* (*chlorofluorocarbons*). Today, it is known that CFCs damage the ozone layer. Since the 1990s, home refrigerators have used refrigerants that are less harmful to the ozone layer.

The process by which a refrigerator operates consists of four basic steps, as illustrated in the diagram on the next page. The system to and from which energy is transferred is defined here as the refrigerant contained within the inner surface of the tubing. Initially, the liquid refrigerant is at a low temperature and pressure so that it is colder than the air inside the refrigerator. The refrigerant absorbs energy from inside the refrigerator and lowers the refrigerator's interior temperature. This transfer of energy as heat increases the temperature of the liquid refrigerant until it begins to boil, as shown in **(a).** The refrigerant continues to absorb energy until it has completely vaporized.

Once it is in the vapor phase, the refrigerant is passed through a *compressor.* The compressor does work on the gas by decreasing its volume without transferring energy as heat, as shown in **(b).** This adiabatic process increases the pressure and internal energy (and thus the temperature) of the gaseous refrigerant.

In the next step, the refrigerant is moved to the outer parts of the refrigerator, where thermal contact is made with the air in the room. The refrigerant gives up energy to the environment, which is at a lower temperature, as shown in **(c).** The gaseous refrigerant at high pressure then condenses at a constant temperature to a liquid.

The liquefied refrigerant is then brought back into the refrigerator. Just outside the low-temperature interior of the refrigerator, the refrigerant goes through an *expansion valve* and expands without absorbing energy as heat. The liquid then does work as it moves from a high-pressure region to a low-pressure region, and its volume increases, as shown in **(d).**

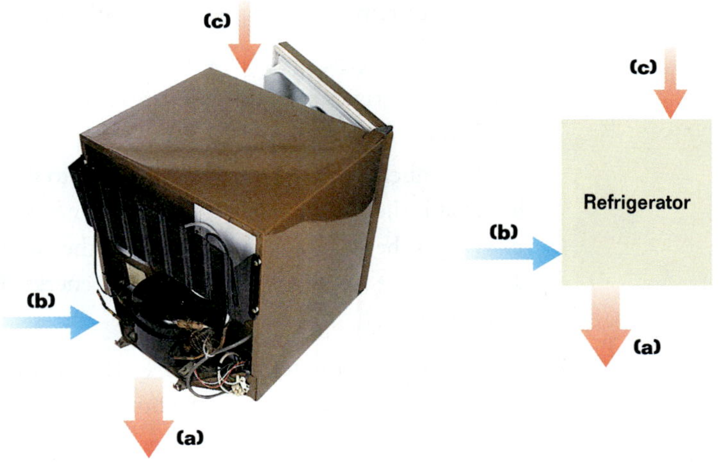

A refrigerator does work **(b)** in order to transfer energy as heat from the inside of the refrigerator **(c)** to the air outside the refrigerator **(a).**

In doing so, the gas expands and cools.

The refrigerant now has the same internal energy and phase as it did at the start of the process. If the temperature of the refrigerant is still lower than the temperature of the air inside the refrigerator, the cycle will repeat. Because the final internal energy is equal to the initial internal energy, this process is cyclic. The first law of thermodynamics can be used to describe the signs of each thermodynamic quantity in the four steps listed, as shown in the table.

Thermodynamics of a Refrigerant

Step	Q	W	ΔU
A	+	0	+
B	0	−	+
C	−	0	−
D	0	+	−

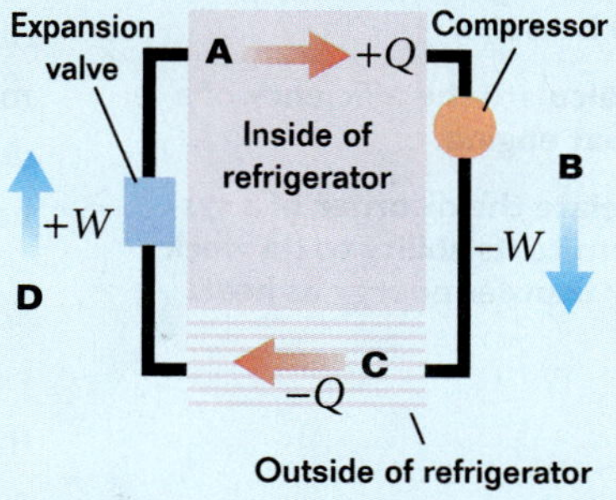

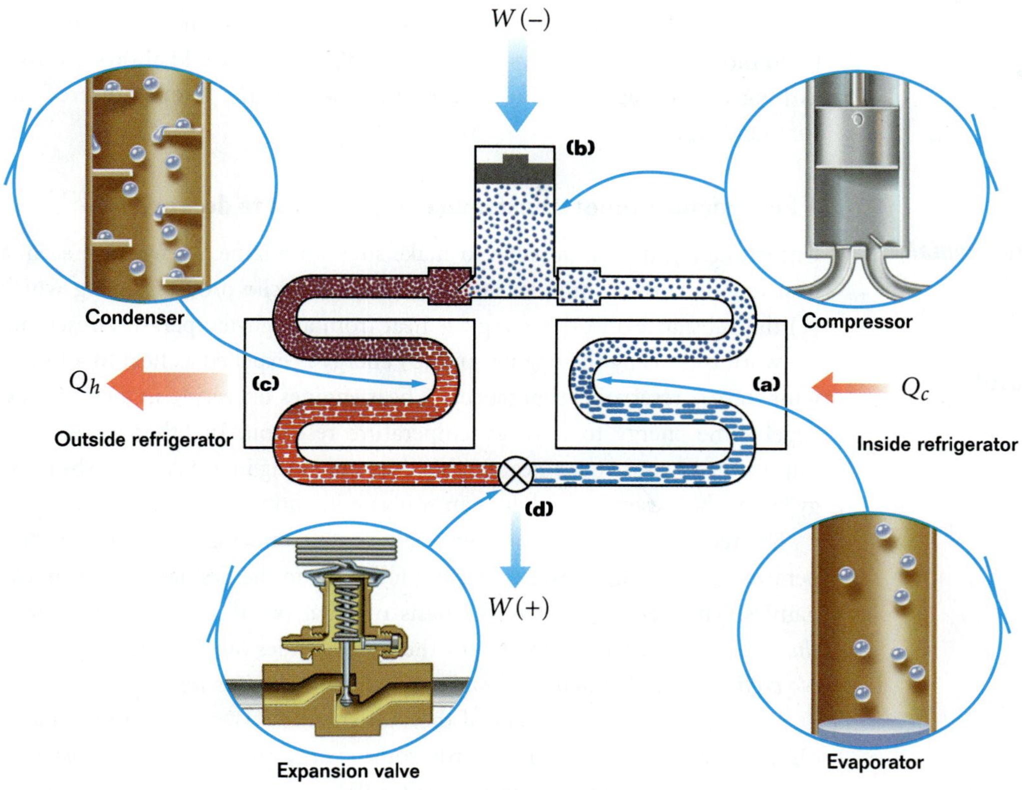

In each of the four steps of a refrigeration cycle, energy is transferred to or from the refrigerant either by heat or by work.

Visual Strategy GENERAL

Be sure that students recognize areas of low pressure (the refrigerant inside the refrigerator) and high pressure (the refrigerant outside the refrigerator) in the fluid shown in the diagram.

Q What happens to the refrigerant's volume, internal energy, and temperature when the refrigerant passes through the expansion valve?

A ***During the adiabatic expansion of liquid refrigerant, the refrigerant's volume increases, which causes its internal energy to decrease ($\Delta U < 0$). This internal energy decrease results in a lowering of the refrigerant's temperature.***

Q As the refrigerant goes from **(d)** to **(b)** through the coiled pipe **(a),** which area will be at a lower temperature?

A ***The lowest temperature zone will be near (d)—where the freezer compartment would be.***

The Second Law of Thermodynamics

SECTION OBJECTIVES

- **Recognize why the second law of thermodynamics requires two bodies at different temperatures for work to be done.**
- **Calculate the efficiency of a heat engine.**
- **Relate the disorder of a system to its ability to do work or transfer energy as heat.**

EFFICIENCY OF HEAT ENGINES

In the previous section, you learned how a heat engine absorbs a quantity of energy from a high-temperature body as heat, does work on the environment, and then gives up energy to a low-temperature body as heat. The work derived from each cycle of a heat engine equals the difference between the heat input and heat output during the cycle, as follows:

$$W_{net} = Q_{net} = Q_h - Q_c$$

This equation, obtained from the first law of thermodynamics, indicates that all energy entering and leaving the system is accounted for and is thus conserved. The equation also suggests that more work is gained by taking more energy at a higher temperature and giving up less energy at a lower temperature. If no energy is given up at the lower temperature ($Q_c = 0$), then it seems that work could be obtained from energy transferred as heat from any body, such as the air around the engine. Such an engine would be able to do more work on hot days than on cold days, but it would always do work as long as the engine's temperature was less than the temperature of the surrounding air.

extension

Integrating Environmental Science

Visit go.hrw.com for the activity "Thermal Pollution."

Keyword HF6TDYX

A heat engine cannot transfer all energy as heat to do work

Unfortunately, it is impossible to make such an engine. As we have seen, a heat engine carries some substance through a cyclic process during which (1) the substance absorbs energy as heat from a high-temperature reservoir, (2) work is done by the engine, and (3) energy is expelled as heat to a lower-temperature reservoir. In practice, all heat engines operating in a cycle must expel some energy to a lower-temperature reservoir. In other words, it is impossible to construct a heat engine that, operating in a cycle, absorbs energy from a hot reservoir and does an equivalent amount of work.

The requirement that a heat engine give up some energy at a lower temperature in order to do work does not follow from the first law of thermodynamics. This requirement is the basis of what is called the *second law of thermodynamics.* The second law of thermodynamics can be stated as follows: *No cyclic process that converts heat entirely into work is possible.*

According to the second law of thermodynamics, W can never be equal to Q_h in a cyclic process. In other words, some energy must always be transferred as heat to the system's surroundings ($Q_c > 0$).

352

Efficiency measures how well an engine operates

A cyclic process cannot completely convert energy transferred as heat into work, nor can it transfer energy as heat from a low-temperature body to a high-temperature body without work being done in the process. However, we can measure how closely a cyclic process approaches these ideal situations. A measure of how well an engine operates is given by the engine's *efficiency* (*eff*). In general, efficiency is a measure of the useful energy taken out of a process relative to the total energy that is put into the process. Efficiencies for different types of engines are listed in **Table 3.**

Recall from the first law of thermodynamics that the work done on the environment by the engine is equal to the difference between the energy transferred to and from the system as heat. For a heat engine, the efficiency is the ratio of work done by the engine to the energy added to the system as heat during one cycle.

EQUATION FOR THE EFFICIENCY OF A HEAT ENGINE

$$eff = \frac{W_{net}}{Q_h} = \frac{Q_h - Q_c}{Q_h} = 1 - \frac{Q_c}{Q_h}$$

$$\text{efficiency} = \frac{\text{net work done by engine}}{\text{energy added to engine as heat}}$$

$$= \frac{\text{energy added as heat} - \text{energy removed as heat}}{\text{energy added as heat}}$$

$$= 1 - \frac{\text{energy removed as heat}}{\text{energy added as heat}}$$

Notice that efficiency is a unitless quantity that can be calculated using only the *magnitudes* for the energies added to and taken away from the engine.

This equation confirms that a heat engine has 100 percent efficiency ($eff = 1$) only if there is no energy transferred away from the engine as heat ($Q_c = 0$).

Table 3
Typical Efficiencies for Engines

Engine type	*eff* (calculated maximum values)
steam engine	0.29
steam turbine	0.40
gasoline engine	0.60
diesel engine	0.56

Engine type	*eff* (measured values)
steam engine	0.17
steam turbine	0.30
gasoline engine	0.25
diesel engine	0.35

SCILINKS.
www.scilinks.org
Topic: Stirling Engines
Code: HF61455

Why it Matters

Conceptual Challenge

1. Cooling Engines Use the second law of thermodynamics to explain why an automobile engine requires a cooling system to operate.

2. Power Plants Why are many coal-burning and nuclear power plants located near rivers?

SECTION 3

The Language of Physics

The term *efficiency* is used to compare the amount of work output to the amount of energy invested in a process. Efficiency is usually represented either by *eff* or simply by *e*. In this text, the abbreviation *eff* is used. Efficiency may be expressed as a decimal number or as a percentage.

Teaching Tip — GENERAL

Students may need to be reminded that W_{net} denotes the difference between work output and work input in a cycle. $W_{net} = Q_h - Q_c$ because the first law of thermodynamics requires that $W_{out} - W_{in} = Q_h - Q_c$ over a cycle. Working through a simple numerical example may help students understand the meaning of efficiency. For example, if $W_{out} = 1000$ J, $W_{in} = 600$ J, and $Q_h = 2000$ J, they can calculate $W_{net} = 400$ J and $eff = 0.2$.

ANSWERS

Conceptual Challenge

1. According to the second law of thermodynamics, some energy must be lost to the environment. For the engine to be efficient, it must be able to give up energy as heat to a substance at a much lower temperature. This is provided by passing water, coolant, or air around the engine's cylinders.
2. Exhaust energy must be removed as heat to a low-temperature substance, such as water. A river provides a continuously replenished source of cool water.

Classroom Practice

Heat-Engine Efficiency
A steam engine takes in 198×10^3 J and exhausts 149×10^3 J as heat per cycle. What is its efficiency?

Answer
0.247

A turbine takes in 67 500 J as heat and does 18 100 J of work during each cycle. Calculate its efficiency.

Answer
0.268

PROBLEM GUIDE C

Use this guide to assign problems.
SE = Student Edition Textbook
PW = Problem Workbook
PB = Problem Bank on the One-Stop Planner (OSP)

Solving for:

eff	**SE** Sample, 1–3; Ch. Rvw. 25–27 **PW** 4–6 **PB** 8–10
W	**SE** 4 **PW** Sample, 1 **PB** Sample, 1–4
Q	**SE** 5–6 **PW** Sample, 1–4 **PB** 5–7

***Challenging Problem**
Consult the printed Solutions Manual or the OSP for detailed solutions.

Unfortunately, there can be no such heat engine, so the efficiencies of all engines are less than 1.0. The smaller the fraction of usable energy that an engine can provide, the lower its efficiency is.

The equation also provides some important information for increasing engine efficiency. If the amount of energy added to the system as heat is increased or the amount of energy given up by the system is reduced, the ratio of Q_c/Q_h becomes much smaller and the engine's efficiency comes closer to 1.0.

The efficiency equation gives only a maximum value for an engine's efficiency. Friction, thermal conduction, and the inertia of moving parts in the engine hinder the engine's performance, and experimentally measured efficiencies are significantly lower than the calculated efficiencies (see **Table 3**).

SAMPLE PROBLEM C

Heat-Engine Efficiency

PROBLEM

Find the efficiency of a gasoline engine that, during one cycle, receives 204 J of energy from combustion and loses 153 J as heat to the exhaust.

SOLUTION

1. DEFINE

Given: $Q_h = 204$ J $\quad Q_c = 153$ J

Unknown: $eff = ?$

Diagram:

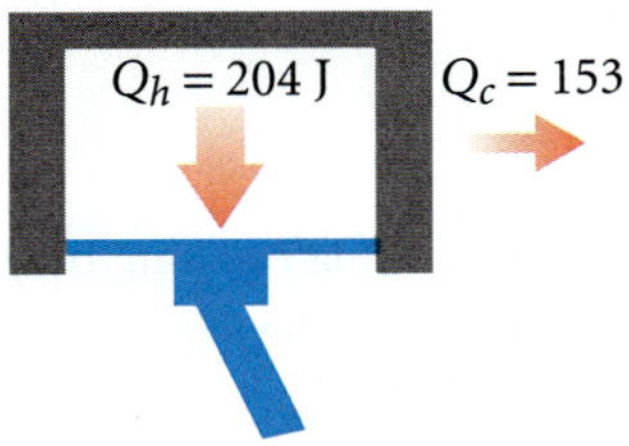

2. PLAN

Choose an equation or situation:
The efficiency of a heat engine is the ratio of the work done by the engine to the energy transferred to it as heat.

$$eff = \frac{W_{net}}{Q_h} = 1 - \frac{Q_c}{Q_h}$$

3. CALCULATE

Substitute the values into the equation and solve:

$$eff = 1 - \frac{153 \text{ J}}{204 \text{ J}} = 0.250$$

$$\boxed{eff = 0.250}$$

4. EVALUATE

Only 25 percent of the energy added as heat is used by the engine to do work. As expected, the efficiency is less than 1.0.

PRACTICE C

Heat-Engine Efficiency

1. If a steam engine takes in 2.254×10^4 kJ from the boiler and gives up 1.915×10^4 kJ in exhaust during one cycle, what is the engine's efficiency?

2. A test model for an experimental gasoline engine does 45 J of work in one cycle and gives up 31 J as heat. What is the engine's efficiency?

3. A steam engine absorbs 1.98×10^5 J and expels 1.49×10^5 J in each cycle. Assume that all of the remaining energy is used to do work.
   a. What is the engine's efficiency?
   b. How much work is done in each cycle?

4. If a gasoline engine has an efficiency of 21 percent and loses 780 J to the cooling system and exhaust during each cycle, how much work is done by the engine?

5. A certain diesel engine performs 372 J of work in each cycle with an efficiency of 33.0 percent. How much energy is transferred from the engine to the exhaust and cooling system as heat?

6. If the energy removed from an engine as heat during one cycle is 6.0×10^2 J, how much energy must be added to the engine during one cycle in order for it to operate at 31 percent efficiency?

ENTROPY

When you shuffle a deck of cards, it is highly improbable that the cards would end up separated by suit and in numerical sequence. Such a highly ordered arrangement can be formed in only a few ways, but there are more than 8×10^{67} ways to arrange 52 cards (because $52! = 8 \times 10^{67}$).

In thermodynamics, a system left to itself tends to go from a state with a very ordered set of energies (one that has only a small probability of being randomly formed) to one in which there is less order (or that has a high probability of being randomly formed). The measure of a system's disorder is called the **entropy** of the system. The greater the entropy of a system is, the greater the system's disorder.

The greater probability of a disordered arrangement indicates that an ordered system is likely to become disordered. Put another way, the entropy of a system tends to increase. This greater probability also reduces the chance that a disordered system will become ordered at random. Thus, once a system has reached a state of greatest disorder, it will tend to remain in that state and have *maximum entropy.*

SCINKS.
www.scilinks.org
Topic: Entropy
Code: HF60523

entropy

a measure of the randomness or disorder of a system

SECTION 3

ANSWERS

Practice C

1. 0.1504
2. 0.59
3. **a.** 0.247
   b. 4.9×10^4 J
4. 210 J
5. 755 J
6. 8.7×10^2 J

SECTION 3

The Language of Physics

The term *entropy* refers to the measure of the disorder in a system. This measure is related to the ability of a system to do useful work. It depends on the temperature and other system characteristics. The term *entropy* is also used in reference to disorder in other areas. In the field of information theory, well-organized information systems have a lower entropy than random ones. For example, when the key terms in this text are indexed, the information in the book is more highly organized. The entropy of the book (as an information system, not as a physical system) is lower. Similarly, a library without a catalog has a higher entropy of information.

Quick Lab

TEACHER'S NOTES

This experiment (on the next page) is intended to illustrate that a system can be in any one of many states but that some states are formed more frequently than others.

Their results should be as follows:

2–1 way	8–5 ways
3–2 ways	9–4 ways
4–3 ways	10–3 ways
5–4 ways	11–2 ways
6–5 ways	12–1 way
7–6 ways	

Homework Options
This QuickLab can easily be performed outside of the physics lab room.

Figure 9
If all gas particles moved toward the piston, all of the internal energy could be used to do work. This extremely well ordered situation is highly improbable.

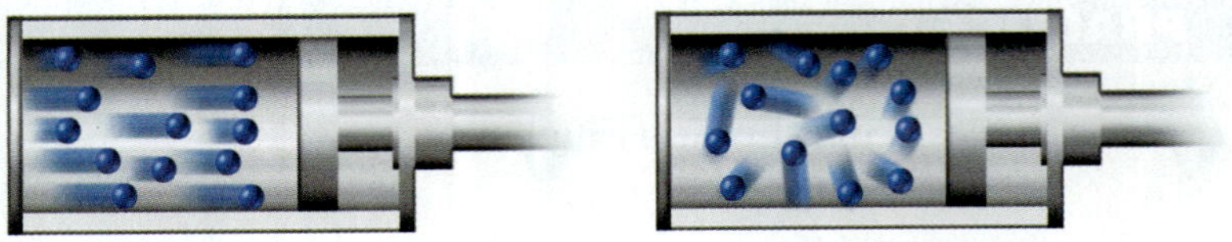

Greater disorder means there is less energy to do work

Heat engines are limited in that only some of the energy added as heat can be used to do work. Not all of the gas particles move in an orderly fashion toward the piston and give up all of their energy in collision with the piston, as shown on the left in **Figure 9.** Instead, they move in all available directions, as shown on the right in **Figure 9.** They transfer energy through collisions with the walls of the engine cylinder as well as with each other. Although energy is conserved, not all of it is available to do useful work. The motion of the particles of a system (in this case, the gas in the cylinder) is not well ordered and therefore is less useful for doing work.

Because of the connection between a system's entropy, its ability to do work, and the direction of energy transfer, the second law of thermodynamics can also be expressed in terms of entropy change. This law applies to the entire universe, not only to a system that interacts with its environment. So, the second law can be stated as follows: *The entropy of the universe increases in all natural processes.*

Note that entropy can decrease for parts of systems, such as the water in the freezer shown in **Figure 10,** provided this decrease is offset by a greater increase in entropy elsewhere in the universe. The water's entropy decreases as it becomes ice, but the entropy of the air in the room is increased by a greater amount as energy is transferred by heat from the refrigerator. The result is that the total entropy of the refrigerator and the room together has increased.

Did you know?

Entropy decreases in many systems on Earth. For example, atoms and molecules become incorporated into complex and orderly biological structures such as cells and tissues. These appear to be spontaneous because we think of the Earth itself as a closed system. So much energy comes from the sun that the disorder in chemical and biological systems is reduced, while the total entropy of the Earth, sun, and intervening space increases.

Figure 10
Because of the refrigerator's less-than-perfect efficiency, the entropy of the outside air molecules increases more than the entropy of the freezing water decreases.

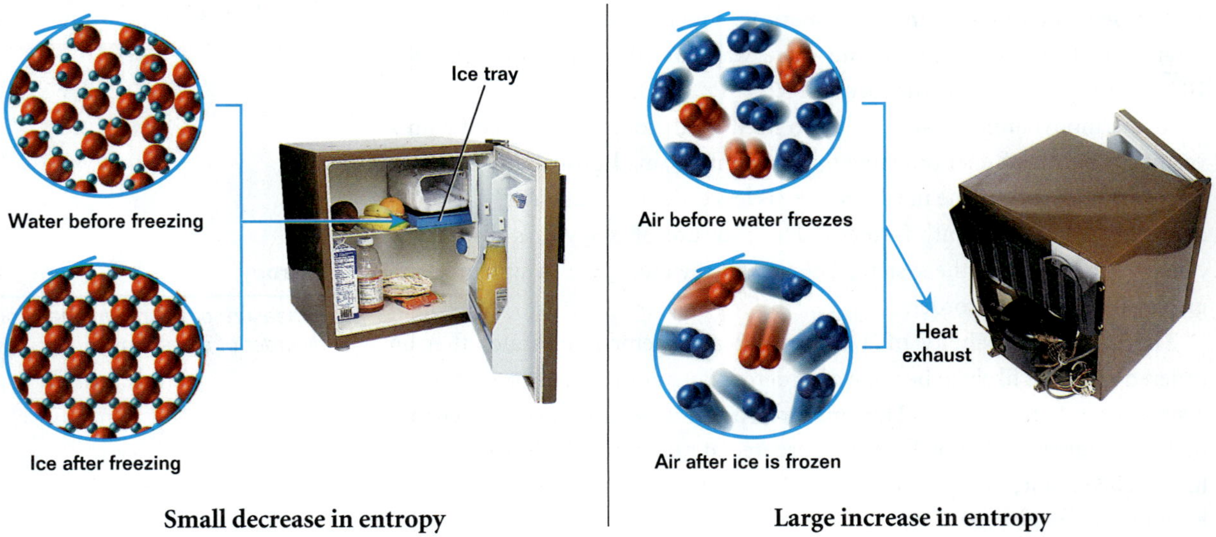

Quick Lab

Entropy and Probability

MATERIALS LIST

- 3 dice
- a sheet of paper
- a pencil

Take two dice from a board game. Record all the possible ways to obtain the numbers 2 through 12 on the sheet of paper. How many possible dice combinations can be rolled? How many combinations of both dice will produce the number 5? the number 8? the number 11? Which number(s) from 2 through 12 is most probable? How many ways out of the total number of ways can this number(s) be rolled? Which number(s) from 2 through 12 is least probable? How many ways out of the total number of ways can this number(s) be rolled?

Repeat the experiment with three dice. Write down all of the possible combinations that will produce the numbers 3 through 18. What number is most probable?

SECTION REVIEW

1. Is it possible to construct a heat engine that doesn't transfer energy to its surroundings? Explain.

2. An engineer claims to have built an engine that takes in 7.5×10^4 J and expels 3.5×10^4 J.

a. How much energy can the engine provide by doing work?
b. What is the efficiency of the engine?
c. Is this efficiency possible? Explain your answer.

3. Of the items listed below, which ones have high entropy?

a. papers scattered randomly across a desk
b. papers organized in a report
c. a freshly opened pack of cards
d. a mixed deck of cards
e. a room after a party
f. a room before a party

4. Some compounds have been observed to form spontaneously, even though they are more ordered than their components. Explain how this is consistent with the second law of thermodynamics.

5. Discuss three common examples of natural processes that involve decreases in entropy. Identify the corresponding entropy increases in the environments of these processes.

6. Critical Thinking A steam-driven turbine is one major component of an electric power plant. Why is it advantageous to increase the steam's temperature as much as possible?

7. Critical Thinking Show that three purple marbles and three light blue marbles in two groups of three marbles each can be arranged in four combinations: two with only one possible arrangement each and two with nine possible arrangements each.

SECTION 3

SECTION REVIEW ANSWERS

1. no; In order for a heat engine to do work in a thermodynamic cycle, some energy must be transferred as heat to surroundings at a temperature lower than that of the engine.

2. **a.** 4.0×10^4 J
b. 0.53
c. yes; because $eff < 1$

3. a, d, and e

4. Individual compounds can decrease entropy spontaneously if there is an exchange of energy between the molecules and the environment such that the entropy of the environment increases by more than the molecules' entropy decreases.

5. Answers may vary. One example is water freezing in winter (energy transfer from water to air increases air's entropy).

6. This increases the amount of energy transferred to the engine as heat (Q_h) and thus raises the engine's efficiency.

7. The 2 ordered states are PPP/BBB and BBB/PPP. There are 9 mixed states with 2 purple marbles on the left (PPB/BBP, PPB/BPB, PPB/PBB, PBP/BBP, PBP/BPB, PBP/PBB, BPP/BBP, BPP/BPB, BPP/PBB) and 9 with 1 purple marble on the left (opposite of the 9 previous cases).

Why it Matters

Deep-Sea Air Conditioning

Deep-sea air conditioning is also known as *sea-water air conditioning,* or *SWAC.* Makai's air-conditioning system takes advantage of cold, deep-sea water as a natural refrigerant. Pipes for the system will be placed at a depth of about 600 m (or deeper). At this depth, the water's temperature is equal to or less than 7°C.

Extension — ADVANCED

As discussed in the feature, Makai Ocean Engineering plans to make this technology ecologically safe. Have students read recent periodicals about local businesses to research the effects of business on ecological systems and the measures businesses have taken to avoid a negative impact on the environment. Students should write a letter to one of these local businesses to give support to or render criticism of its present practices.

Why it Matters

Deep-Sea Air Conditioning

Deep beneath the ocean, about half a mile down, sunlight barely penetrates the still waters. Scientists at Makai Ocean Engineering in Hawaii are now tapping into that pitch-dark region as a resource for air conditioning.

In tropical locations where buildings are cooled year-round, air-conditioning systems operate with cold water. Refrigeration systems cool the water, and pumps circulate it throughout the walls of a building, where the water absorbs heat from the rooms. Unfortunately, powering these compressors is neither cheap nor efficient.

Instead of cooling the water in their operating systems, the systems designed by Makai use frigid water from the ocean's depths. First, engineers install a pipeline that reaches deep into the ocean, where the water is nearly freezing. Then, powerful pumps on the shoreline move the water directly into a building's air-conditioning system. There, a system of heat exchangers uses the sea water to cool the fresh water in the air-conditioning system.

One complicating factor is that the water must also be returned to the ocean in a manner that will not disrupt the local ecosystem. It must be either piped to a depth of a few hundred feet, where its temperature is close to that of the ocean at that level, or poured into onshore pits, where it eventually seeps through the land and comes to an acceptable temperature by the time it reaches the ocean.

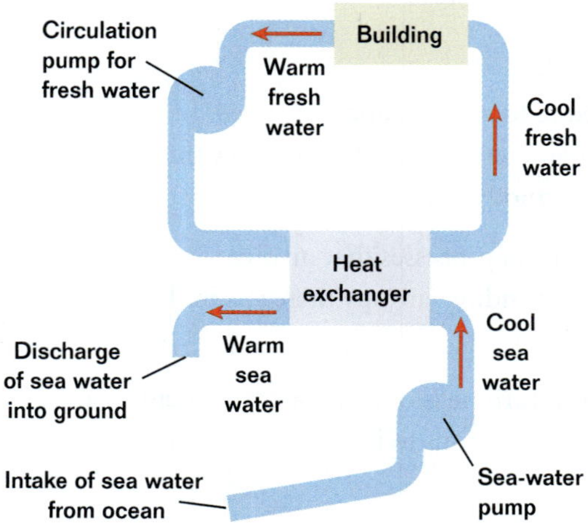

"This deep-sea air conditioning benefits the environment by operating with a renewable resource instead of freon," said Dr. Van Ryzin, the president of Makai. "Because the system eliminates the need for compressors, it uses only about 10 percent of the electricity of current methods, saving fossil fuels and a lot of money." However, deep-sea air-conditioning technology works only for buildings within a few kilometers of the shore and carries a hefty installation cost of several million dollars. For this reason, Dr. Van Ryzin thinks this type of system is most appropriate for large central air-conditioning systems, such as those necessary to cool resorts or large manufacturing plants, where the electricity savings can eventually make up for the installation costs. Under the right circumstances, air conditioning with sea water can be provided at one-third to one-half the cost of conventional air conditioning.

Highlights

KEY IDEAS

Section 1 Relationships Between Heat and Work

- A thermodynamic system is an object or set of objects considered to be a distinct physical entity to or from which energy is added or removed. The surroundings make up the system's environment.
- Energy can be transferred to or from a system as heat and/or work, changing the system's internal energy in the process.
- For gases at constant pressure, work is defined as the product of gas pressure and the change in the volume of the gas.

Section 2 The First Law of Thermodynamics

- Energy is conserved for any system and its environment and is described by the first law of thermodynamics.
- A cyclic process returns a system to conditions identical to those it had before the process began, so its internal energy is unchanged.

Section 3 The Second Law of Thermodynamics

- The second law of thermodynamics states that no machine can transfer all of its absorbed energy as work.
- The efficiency of a heat engine depends on the amount of energy transferred as heat to and from the engine.
- Entropy is a measure of the disorder of a system. As a system becomes more disordered, less of its energy is available to do work.
- The entropy of a system can increase or decrease, but the total entropy of the universe is always increasing.

KEY TERMS

system (p. 336)

environment (p. 337)

isovolumetric process (p. 339)

isothermal process (p. 340)

adiabatic process (p. 341)

cyclic process (p. 346)

entropy (p. 355)

PROBLEM SOLVING

See **Appendix D: Equations** for a summary of the equations introduced in this chapter. If you need more problem-solving practice, see **Appendix I: Additional Problems.**

Variable Symbols

Quantities		Units	
ΔU	change in internal energy	J	joules
Q	heat	J	joules
W	work	J	joules
eff	efficiency	(unitless)	

Diagram Symbols

Energy transferred as heat

Energy transferred as work

Thermodynamic cycle

CHAPTER 10

Highlights

Teaching Tip

Have students create concept maps of the following: compression, expansion, heating, cooling, work, heat, and internal energy. Let students add terms such as *volume, temperature,* or *entropy* to link related concepts. There are many ways to represent these concepts and their relationships. This exercise will help students synthesize the information presented in the chapter.

extension

In-Depth Physics Content

Your students can visit go.hrw.com for an online chapter that integrates more in-depth development of the concepts covered here.

Keyword HF6TDYX

Review

ANSWERS

1. see glossary definitions
2. energy transfers to the system as heat or as work
3. b, c, d, e
4. a. ΔU
   b. Q
   c. W
5. Work done by gas causes a decrease in U and T (temperature) of gas and an increase in U of surroundings.
6. a. work done by forces due to friction between hands, which raises hands' internal energy, energy transferred as heat from warmed hands to cold air
   b. work done by forces due to friction, which increases U of drill and block, with water, U of drill and block decreases as energy is transferred as heat to water, U of water increases until large enough for phase change
7. adiabatic
8. isovolumetric
9. 1.08×10^3 J; done by the gas
10. 3.50×10^2 J
11. $\Delta U = Q - W$; change in a system's internal energy equals energy transferred as heat or work to or from a system
12. a. $\Delta U = 0, Q = W$
    b. $\Delta U = -W, Q = 0$
    c. $\Delta U = Q, W = 0$
13. The energy source used to power the refrigerator is the source of additional energy.

CHAPTER 10 Review

HEAT, WORK, AND INTERNAL ENERGY

Review Questions

1. Define a thermodynamic system and its environment.

2. In what two ways can the internal energy of a system be increased?

3. Which of the following expressions have units that are equivalent to the units of work?
   a. mg
   b. $\frac{1}{2}mv^2$
   c. mgh
   d. Fd
   e. $P\Delta V$
   f. $V\Delta T$

4. For each of the following, which thermodynamic quantities (ΔU, Q, and W) have values equal to zero?
   a. an isothermal process
   b. an adiabatic process
   c. an isovolumetric process

Conceptual Questions

5. When an ideal gas expands adiabatically, it does work on its surroundings. Describe the various transfers of energy that take place.

6. In each of the following cases, trace the chain of energy transfers (as heat or as work) as well as changes in internal energy.
   a. You rub your hands together to warm them on a cold day, and they soon become cold again.
   b. A hole is drilled into a block of metal. When a small amount of water is placed in the drilled hole, steam rises from the hole.

7. Paint from an aerosol can is sprayed continuously for 30 s. The can was initially at room temperature, but now it feels cold to the touch. What type of thermodynamic process occurs for a small sample of gas as it leaves the high-pressure interior of the can and moves to the outside atmosphere?

8. The can of spray paint in item 7 is set aside for an hour. During this time the contents of the can return to room temperature. What type of thermodynamic process takes place in the can during the time the can is not in use?

Practice Problems

For problems 9–10, see Sample Problem A.

9. How much work is done when a tire's volume increases from 35.25×10^{-3} m^3 to 39.47×10^{-3} m^3 at a pressure of 2.55×10^5 Pa in excess of atmospheric pressure? Is work done on or by the gas?

10. Helium in a toy balloon does work on its surroundings as it expands with a constant pressure of 2.52×10^5 Pa in excess of atmospheric pressure. The balloon's initial volume is 1.1×10^{-4} m^3, and its final volume is 1.50×10^{-3} m^3. Determine the amount of work done by the gas in the balloon.

ENERGY CONSERVATION AND CYCLIC PROCESSES

Review Questions

11. Write the equation for the first law of thermodynamics, and explain why it is an expression of energy conservation.

12. Rewrite the equation for the first law of thermodynamics for each of the following special thermodynamic processes:
    a. an isothermal process
    b. an adiabatic process
    c. an isovolumetric process

13. How is energy conserved if more energy is transferred as heat from a refrigerator to the outside air than is removed from the inside air of the refrigerator?

Conceptual Questions

14. A bomb calorimeter is placed in a water bath, and a mixture of fuel and oxygen is burned inside it. The temperature of the water is observed to rise during the combustion reaction. The calorimeter and the water remain at constant volume.

a. If the reaction products are the system, which thermodynamic quantities—ΔU, Q, or W—are positive and which are negative?

b. If the water bath is the system, which thermodynamic quantities—ΔU, Q, or W—are positive and which are negative?

15. Which of the thermodynamic values (ΔU, Q, or W) would be negative for the following systems?

a. a steel rail (system) undergoing slow thermal expansion on a hot day displaces the spikes and ties that hold the rail in place

b. the interior of a closed refrigerator (system)

c. the helium in a thermally insulated weather balloon (system) expands during inflation

Practice Problems

For problems 16–17, see Sample Problem B.

16. Heat is added to an open pan of water at 100.0°C, vaporizing the water. The expanding steam that results does 43.0 kJ of work, and the internal energy of the system increases by 604 kJ. How much energy is transferred to the system as heat?

17. A 150 kg steel rod in a building under construction supports a load of 6050 kg. During the day, the rod's temperature increases from 22°C to 47°C. This temperature increase causes the rod to thermally expand and raise the load 5.5 mm.

a. Find the energy transferred as heat to or from the rod. (Hint: Assume the specific heat capacity of steel is the same as for iron.)

b. Find the work done in this process. Is work done on or by the rod?

c. How great is the change in the rod's internal energy? Does the rod's internal energy increase or decrease?

EFFICIENCY AND ENTROPY

Review Questions

18. The first law of thermodynamics states that you cannot obtain more energy from a process than you originally put in. The second law states that you cannot obtain as much usable energy from a system as you put into it. Explain why these two statements do not contradict each other.

19. What conditions are necessary for a heat engine to have an efficiency of 1.0?

20. In which of the following systems is entropy increasing? (Do not include the surroundings as part of the system.)

a. An egg is broken and scrambled.

b. A cluttered room is cleaned and organized.

c. A thin stick is placed in a glass of sugar-saturated water, and sugar crystals form on the stick.

21. Why is it not possible for all of the energy transferred as heat from a high-temperature source to be expelled from an engine by work?

Conceptual Questions

22. If a cup of very hot water is used as an energy source and a cup of cold water is used as an energy "sink," the cups can, in principle, be used to do work, as shown below. If the contents are mixed together and the resulting lukewarm contents are separated into two cups, no work can be done. Use the second law of thermodynamics to explain this. Has the first law of thermodynamics been violated by mixing and separating the contents of the two cups?

10 REVIEW

14. a. $\Delta U < 0$, $Q < 0$, $W = 0$
b. $\Delta U > 0$, $Q > 0$, $W = 0$

15. a. none (Q, W, and $\Delta U > 0$)
b. $\Delta U < 0$, $Q < 0$ for refrigerator interior ($W = 0$)
c. $\Delta U < 0$ ($Q = 0$, $W > 0$)

16. 647 kJ

17. a. 1.7×10^6 J, to the rod
b. 3.3×10^2 J; by the rod
c. 1.7×10^6 J; it increases

18. Energy is always conserved (first law), but not all of the energy transferred into a system can be used to do work (second law). Energy that must be wasted, according to the second law, is still accounted for by the first law.

19. All energy transferred into the engine must be used to do work ($Q_h = W$). No energy is wasted ($Q_c = 0$). This condition cannot be met by real cyclic heat engines.

20. (a)

21. If all the energy from a high-temperature source were used by the engine to do work, the engine would not be able to expel energy to a lower-temperature body.

22. Energy must be transferred as heat for work to be done. This cannot occur if the water in both cups has same temperature; No, the total energy in the water is unchanged, although usable energy has decreased.

23. The plant's efficiency would increase, but the advantage gained would be more than offset by the use of energy needed to refrigerate the water.

24. no; Entropy of water and air during the water's evaporation increases by more than the entropy of the sodium and chloride ions decreases.

25. In inelastic collisions, some kinetic energy is converted to the internal energy of the colliding objects, so the system's total entropy increases. Kinetic energy is not conserved, but total energy is conserved.

26. 0.210

27. 0.32

28. 0.41

29. **a.** 188 J
b. 1.400×10^3 J

extension

Graphing Calculator Practice
Visit go.hrw.com for answers to this Graphing Calculator activity.
Keyword HF6TDYXT

23. Suppose the waste heat at a power plant is exhausted to a pond of water. Could the efficiency of the plant be increased by refrigerating the water in the pond?

24. A salt solution is placed in a bowl and set in sunlight. The salt crystals that remain after the water has evaporated are more highly ordered than the randomly dispersed sodium and chloride ions in the solution. Has the requirement that total entropy increase been violated? Explain your answer.

25. Use a discussion of internal energy and entropy to explain why the statement, "Energy is not conserved in an inelastic collision," is not true.

Practice Problems

For problems 26–28, see Sample Problem C.

26. In one cycle, an engine burning a mixture of air and methanol (methyl alcohol) absorbs 525 J and expels 415 J. What is the engine's efficiency?

27. The energy provided each hour as heat to the turbine in an electric power plant is 9.5×10^{12} J. If 6.5×10^{12} J of energy is exhausted each hour from the engine as heat, what is the efficiency of this heat engine?

28. A heat engine absorbs 850 J of energy per cycle from a high-temperature source. The engine does 3.5×10^2 J of work during each cycle, expelling 5.0×10^2 J as heat. What is the engine's efficiency?

MIXED REVIEW

29. A gas expands when 606 J of energy is added to it as heat. The expanding gas does 418 J of work on its surroundings.

a. What is the overall change in the internal energy of the gas?

b. If the work done by the gas were equal to 1212 J (rather than 418 J), how much energy would need to be added as heat in order for the change in internal energy to equal the change in internal energy in part **a**?

Graphing Calculator *Practice*

Carnot Efficiency

Sadi Carnot (1796-1832), a French engineer, studied the efficiencies of heat engines. He described an ideal engine—now called the *Carnot engine*—that consists of an ideal gas inside a thermally nonconductive cylinder that has a piston and a replaceable base.

In the Carnot engine, the piston moves upward as the cylinder's conductive base is brought in contact with a heat reservoir, T_h. The piston continues to rise when the base is replaced by a nonconductive base. Then, the energy is transferred to a cooler reservoir at a temperature, T_c, followed by further compression when the base is again replaced. Carnot discovered that the efficiency of such an engine can be determined by the following equation:

$$\textit{highest theoretical efficiency} = 1 - \frac{T_c}{T_h}$$

In this graphing calculator activity, you will enter various values for T_h and T_c to calculate the highest theoretical efficiency of a heat engine. Because of friction and other problems, the actual efficiency of a heat engine will be lower than the calculated efficiency.

Visit go.hrw.com and type in the keyword **HF6TDYX** to find this graphing calculator activity. Refer to **Appendix B** for instructions on downloading the program for this activity.

30. The lid of a pressure cooker forms a nearly airtight seal. Steam builds up pressure and increases temperature within the pressure cooker so that food cooks faster than it does in an ordinary pot. The system is defined as the pressure cooker and the water and steam within it. Suppose that 2.0 g of water is sealed in a pressure cooker and then vaporized by heating.

a. What happens to the water's internal energy?
b. Is energy transferred as heat to or from the system?
c. Is energy transferred as work to or from the system?
d. If 5175 J must be added as heat to completely vaporize the water, what is the change in the water's internal energy?

Alternative Assessment

1. Imagine that an inventor is asking you to invest your savings in the development of a new turbine that will produce cheap electricity. The turbine will take in 1000 J of energy from fuel to supply 650 J of work, which can then be used to power a generator. The energy removed as heat to a cooling system will raise the temperature of 0.10 kg of water by 1.2°C. Are these figures consistent with the first and second laws of thermodynamics? Would you consider investing in this project? Write a business letter to the inventor explaining how your analysis affected your decision.
2. Talk to someone who works on air conditioners or refrigerators to find out what fluids are used in these systems. What properties should refrigerant fluids have? Research the use of freon and freon substitutes. Why is using freon forbidden by international treaty? What fluids are now used in refrigerators and car air conditioners? For what temperature ranges are these fluids appropriate? What are the advantages and disadvantages of each fluid? Summarize your research in the form of a presentation or report.
3. Research how an internal-combustion engine operates. Describe the four steps of a combustion cycle. What materials go in and out of the engine during each step? How many cylinders are involved in one cycle? What energy processes take place during each stroke? In which steps is work done? Summarize your findings with diagrams or in a report. Contact an expert auto mechanic, and ask the mechanic to review your report for accuracy.
4. The law of entropy can also be called the law of increasing disorder, but this law seems to contradict the existence of living organisms that are able to organize chemicals into organic molecules. Prepare for a class debate on the validity of the following arguments:
   a. Living things are not subject to the laws of thermodynamics.
   b. The increase in the universe's entropy due to life processes is greater than the decrease in entropy within a living organism.
5. Work in groups to create a classroom presentation on the life, times, and work of James Watt, inventor of the first commercially successful steam engine in the early nineteenth century. Include material about how this machine affected transportation and industry in the United States.
6. Most major appliances are required by law to have an *EnergyGuide* label attached to them. The label indicates the average amount of energy used by the appliance in a year, and gives the average cost of using the appliance based on a national average of cost per energy unit. In a store, look at the *EnergyGuide* labels attached to three different models of one brand of a major appliance. Create a graph showing the total yearly cost of each appliance over ten years (including the initial cost of the appliance with year one). Determine which model you would purchase, and write a paragraph defending your choice.

30. **a.** it increases ($\Delta U > 0$)
b. to the system ($Q > 0$)
c. no ($\Delta V = 0$; therefore, $W = 0$)
d. 5175 J

Alternative Assessment ANSWERS

1. The proposal is invalid because the energy provided by work is greater than the differences in the energy transferred as heat to and from the engine.
2. Students' answers will vary. Freon is no longer used in new systems because of harm it may do to the ozone layer. Factors in choosing coolants include latent heat, boiling point, cost, and safety.
3. Air and fuel go in; exhaust goes out. The number of cylinders depends on the type of engine. Work is done as the gas expands and as the piston expels the exhaust.
4. Students should realize that (b) best describes the second law.
5. Students' answers will vary. Watt's engine marked the start of the mechanization of the Industrial Revolution.
6. Students may include both environmental and economic factors in their decisions. They may find that more expensive models are cheaper in the long run due to higher efficiencies.

Standardized Test Prep

ANSWERS

1. B
2. H
3. A
4. G
5. C
6. J
7. B

Standardized Test Prep

MULTIPLE CHOICE

1. If there is no change in the internal energy of a gas, even though energy is transferred to the gas as heat and work, what is the thermodynamic process that the gas undergoes called?

A. adiabatic
B. isothermal
C. isovolumetric
D. isobaric

2. To calculate the efficiency of a heat engine, which thermodynamic property does *not* need to be known?

F. the energy transferred as heat to the engine
G. the energy transferred as heat from the engine
H. the change in the internal energy of the engine
J. the work done by the engine

3. In which of the following processes is no work done?

A. Water is boiled in a pressure cooker.
B. A refrigerator is used to freeze water.
C. An automobile engine operates for several minutes.
D. A tire is inflated with an air pump.

4. A thermodynamic process occurs in which the entropy of a system decreases. From the second law of thermodynamics, what can you conclude about the entropy change of the environment?

F. The entropy of the environment decreases.
G. The entropy of the environment increases.
H. The entropy of the environment remains unchanged.
J. There is not enough information to state what happens to the environment's entropy.

Use the passage and diagrams below to answer questions 5–8.

A system consists of steam within the confines of a steam engine, whose cylinder and piston are shown in the figures below.

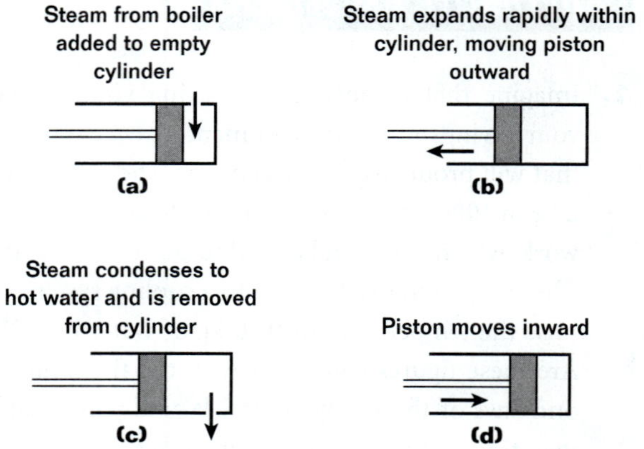

5. Which of the figures describes a situation in which $\Delta U < 0$, $Q < 0$, and $W = 0$?

A. (a)
B. (b)
C. (c)
D. (d)

6. Which of the figures describes a situation in which $\Delta U > 0$, $Q = 0$, and $W < 0$?

F. (a)
G. (b)
H. (c)
J. (d)

7. Which of the figures describes a situation in which $\Delta U < 0$, $Q = 0$, and $W > 0$?

A. (a)
B. (b)
C. (c)
D. (d)

8. Which of the figures describes a situation in which $\Delta U > 0$, $Q > 0$, and $W = 0$?
   F. (a)
   G. (b)
   H. (c)
   J. (d)

9. A power plant has a power output of 1055 MW and operates with an efficiency of 0.330. Excess energy is carried away as heat from the plant to a nearby river. How much energy is transferred away from the power plant as heat?
   A. 0.348×10^9 J/s
   B. 0.520×10^9 J/s
   C. 0.707×10^9 J/s
   D. 2.14×10^9 J/s

10. How much work must be done by air pumped into a tire if the tire's volume increases from 0.031 m^3 to 0.041 m^3 and the net, constant pressure of the air is 300.0 kPa?
   F. 3.0×10^2 J
   G. 3.0×10^3 J
   H. 3.0×10^4 J
   J. 3.0×10^5 J

SHORT RESPONSE

Use the passage below to answer questions 11–12.

An air conditioner is left running on a table in the middle of the room, so none of the air that passes through the air conditioner is transferred to outside the room.

11. Does passing air through the air conditioner affect the temperature of the room? (Ignore the thermal effects of the motor running the compressor.)

12. Taking the compressor motor into account, what would happen to the temperature of the room?

13. If 1600 J of energy are transferred as heat to an engine and 1200 J are transferred as heat away from the engine to the surrounding air, what is the efficiency of the engine?

EXTENDED RESPONSE

14. How do the temperature of combustion and the temperatures of coolant and exhaust affect the efficiency of automobile engines?

Base your answers to questions 15–18 on the information below. In each problem, show all of your work.

A steam shovel raises 450.0 kg of dirt a vertical distance of 8.6 m. The steam shovel's engine provides 2.00×10^5 J of energy as heat for the steam shovel to lift the dirt.

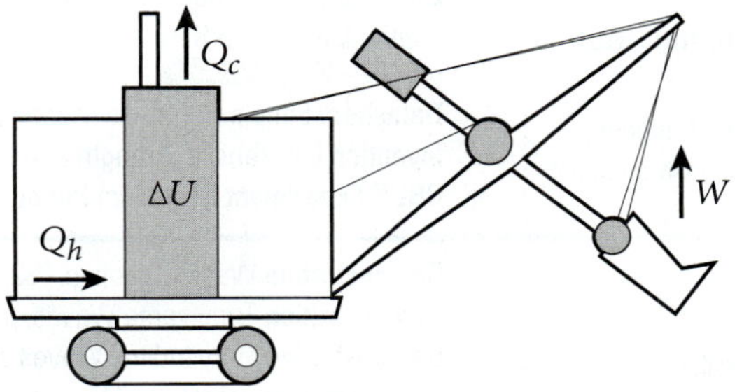

15. How much work is done by the steam shovel in lifting the dirt?

16. What is the efficiency of the steam shovel?

17. Assuming there is no change in the internal energy of the steam shovel's engine, how much energy is given up by the shovel as waste heat?

18. Suppose the internal energy of the steam shovel's engine increases by 5.0×10^3 J. How much energy is given up now as waste heat?

19. One way to look at heat and work is to think of energy transferred as heat as a "disorganized" form of energy and energy transferred as work as an "organized" form. Use this interpretation to show that the increased order obtained by freezing water is less than the total disorder that results from the freezer used to form the ice.

Test TIP Identify each of the quantities given in each problem; then write down the necessary equations for solving the problem, making sure that you have values for each term in each equation.

8. F

9. D

10. G

11. No, because the energy removed from the cooled air is returned to the room.

12. The temperature increases.

13. 0.25

14. The greater the temperature difference is the greater is the amount of energy transferred as heat. For efficiency to increase, the heat transferred between the combustion reaction and the engine (Q_h) should be made to increase, whereas the energy given up as waste heat to the coolant and exhaust (Q_c) should be made to decrease.

15. 3.8×10^4 J

16. 0.19

17. 1.62×10^5 J

18. 1.57×10^5 J

19. Disorganized energy is removed from water to form ice, but a greater amount of organized energy must become disorganized in order to operate the freezer.

CHAPTER 11

Vibrations and Waves

Planning Guide

Compression Guide

To shorten instruction because of time limitations, omit the opener and abbreviate the review.

OBJECTIVES	LABS, DEMONSTRATIONS, AND ACTIVITIES	TECHNOLOGY RESOURCES
PACING • 45 min pp. 366–367 **Chapter Opener**	ANC **Discovery Lab** Pendulums and Spring Waves*◆ BASIC	CD **Visual Concepts,** Chapter 11 BASIC
PACING • 45 min pp. 368–375 **Section 1 Simple Harmonic Motion** • Identify the conditions of simple harmonic motion. • Explain how force, velocity, and acceleration change as an object vibrates with simple harmonic motion. • Calculate the spring force using Hooke's law.	TE **Demonstration** A Vibrating Spring, p. 368 GENERAL TE **Demonstration** An Oscillating Pendulum, p. 368 GENERAL TE **Demonstration** Hooke's Law, p. 369 GENERAL SE **Quick Lab** Energy of a Pendulum, p. 374 GENERAL	OSP **Lesson Plans** TR 45 A Mass-Spring System TR 46 A Simple Pendulum TR 47 Simple Harmonic Motion
PACING • -90 min pp. 376–381 **Section 2 Measuring Simple Harmonic Motion** • Identify the amplitude of vibration. • Recognize the relationship between period and frequency. • Calculate the period and frequency of an object vibrating with simple harmonic motion.	TE **Demonstration** Period and Frequency, p. 376 GENERAL SE **Inquiry Lab** Simple Harmonic Motion of a Pendulum, pp. 402–403◆ GENERAL ANC **Datasheet** Simple Harmonic Motion of a Pendulum* ANC **Invention Lab** Tensile Strength and Hooke's Law*◆ ANC **CBL™ Experiment** Pendulum Periods*◆ ADVANCED	OSP **Lesson Plans** EXT **Integrating Technology** Bicycle Design and Shock Absorption BASIC TR 39A Measures of Simple Harmonic Motion
PACING • 90 min pp. 382–388 **Section 3 Properties of Waves** • Distinguish local particle vibrations from overall wave motion. • Differentiate between pulse waves and periodic waves. • Interpret waveforms of transverse and longitudinal waves. • Apply the relationship among wave speed, frequency, and wavelength to solve problems. Relate energy and amplitude.	TE **Demonstration** Wave Motion, p. 382 BASIC TE **Demonstration** Transverse Waves, p. 383 GENERAL TE **Demonstration** Longitudinal Waves, p. 385 GENERAL TE **Demonstration** Amplitude, Wavelength, and Wave Speed, p. 386 GENERAL	OSP **Lesson Plans** EXT **Integrating Earth Science** Earthquake Waves TR 48 The Relationship Between SHM and Wave Motion
PACING • 45 min pp. 389–394 **Section 4 Wave Interactions** • Apply the superposition principle. Differentiate between constructive and destructive interference. • Predict when a reflected wave will be inverted. • Predict whether specific traveling waves will produce a standing wave. Identify nodes and antinodes of a standing wave.	TE **Demonstration** Wave Superposition, p. 389 GENERAL TE **Demonstration** Waves Passing Each Other, p. 390 BASIC TE **Demonstration** Wave Reflection, p. 392 BASIC TE **Demonstration** Standing Waves, p. 394 GENERAL	OSP **Lesson Plans** TR 50 Constructive Interference TR 51 Destructive Interference TR 52 Reflection of a Pulse Wave TR 53 Standing Waves

PACING • 90 min

CHAPTER REVIEW, ASSESSMENT, AND STANDARDIZED TEST PREPARATION

- SE **Chapter Highlights,** p. 395
- SE **Chapter Review,** pp. 396–398
- SE **Graphing Calculator Practice,** p. 399 GENERAL
- SE **Alternative Assessment,** p. 399 ADVANCED
- SE **Standardized Test Prep,** pp. 400–401 GENERAL
- SE **Appendix D: Equations,** p. 859
- SE **Appendix I: Additional Problems,** pp. 888–889
- ANC **Study Guide Worksheet** Mixed Review* GENERAL
- ANC **Chapter Test A*** GENERAL
- ANC **Chapter Test B*** ADVANCED
- OSP **Test Generator**

Online and Technology Resources

go.hrw.com Holt Online Learning

Visit go.hrw.com to find a variety of online resources. To access this chapter's extensions, enter the keyword **HF6VIBXT** and click the "go" button. Click **Holt Online Learning** for an online edition of this textbook, and other interactive resources.

HOLT Teacher's One-Stop Planner®

This DVD package includes:
- Holt Calendar Planner
- Customizable Lesson Plans
- Editable Worksheets
- ExamView® Version 6 Assessment Suite
- Interactive Teacher's Edition
- Holt PuzzlePro®
- Holt PowerNotes® Presentations
- MindPoint® Quiz Show

KEY

SE	Student Edition	OSP	One-Stop Planner	EXT	Online Extension
TE	Teacher Edition	CD	CD or CD-ROM	*	Also on One-Stop Planner
ANC	Ancillary Worksheet	TR	Teaching Transparencies	◆	Requires advance prep

SKILLS DEVELOPMENT RESOURCES	REVIEW AND ASSESSMENT	CORRELATIONS
		National Science Education Standards
SE **Conceptual Challenge,** p. 369 GENERAL SE **Sample Set A** Hooke's Law, pp. 370–371 GENERAL TE **Classroom Practice,** p. 370 GENERAL ANC **Problem Workbook** Sample Set A* GENERAL OSP **Problem Bank** Sample Set A GENERAL	SE **Section Review,** p. 375 GENERAL ANC **Study Guide Worksheet** Section 1* GENERAL ANC **Quiz** Section 1* BASIC	UCP 1, 2, 3, 4, 5 SAI 1, 2 ST 1, 2 HNS 3 SPSP 5 PS 4a
SE **Sample Set B** SHM of a Simple Pendulum pp. 378–379 BASIC TE **Classroom Practice,** p. 378 BASIC ANC **Problem Workbook*** and OSP **Problem Bank** Sample Set B BASIC SE **Conceptual Challenge,** p. 379 GENERAL SE **Sample Set C** SHM of a Mass-Spring System, pp. 380–381 BASIC ANC **Problem Workbook*** and OSP **Problem Bank** Sample Set C BASIC	SE **Section Review,** p. 381 GENERAL ANC **Study Guide Worksheet** Section 2* GENERAL ANC **Quiz** Section 2* BASIC	UCP 1, 2, 3, 4 SAI 1, 2 ST 2 HNS 1, 2 SPSP 2, 5
SE **Sample Set D** Wave Speed, p. 387 BASIC ANC **Problem Workbook*** and OSP **Problem Bank** Sample Set D BASIC	SE **Section Review,** p. 388 GENERAL ANC **Study Guide Worksheet** Section 3* GENERAL ANC **Quiz** Section 3* BASIC	UCP 1, 2, 3 PS 6a, 6b
SE **Appendix J: Advanced Topics** DeBroglie Waves, pp. 922–923 ADVANCED	SE **Section Review,** p. 394 GENERAL ANC **Study Guide Worksheet** Section 4* GENERAL ANC **Quiz** Section 4* BASIC	UCP 1, 2, 3, 4, 5 ST 1, 2 HNS 1, 3 SPSP 1, 5 PS 6a

SCILINKS. NSTA
www.scilinks.org

Maintained by the **National Science Teachers Association.**

Topic: Hooke's Law
SciLinks Code: HF60756

Topic: Pendulums
SciLinks Code: HF61121

Topic: Wave Motion
SciLinks Code: HF61639

Classroom CD-ROMs

- Guided Reading Audio Program
- Student One Stop
- Virtual Investigations
- Visual Concepts

Holt Lab Generator CD-ROM

Search for any lab by topic, standard, difficulty level, or time. Edit any lab to fit your needs, or create your own labs. Use the Lab Materials QuickList software to customize your lab materials list.

CHAPTER 11
Overview

Section 1 introduces restoring force, the conditions of simple harmonic motion, Hooke's law, and the relationship between force, velocity, and acceleration in simple harmonic motion.

Section 2 identifies the variables affecting amplitude, period, and frequency in a simple pendulum and in a mass-spring system.

Section 3 introduces concepts of wave motion, including wave speed, frequency, wavelength, amplitude, and energy, and discusses their relationships.

Section 4 explores how to use the superposition principle to predict patterns of interference and to identify the conditions for standing waves.

About the Illustration

The mechanical metronome was invented by Dietrich Winkel (c. 1776–1826) but was patented by Johann N. Maelzel in 1816. Today, electronic digital metronomes, which typically include both a flashing light and a ticking sound, are often used.

CHAPTER 11

Vibrations and Waves

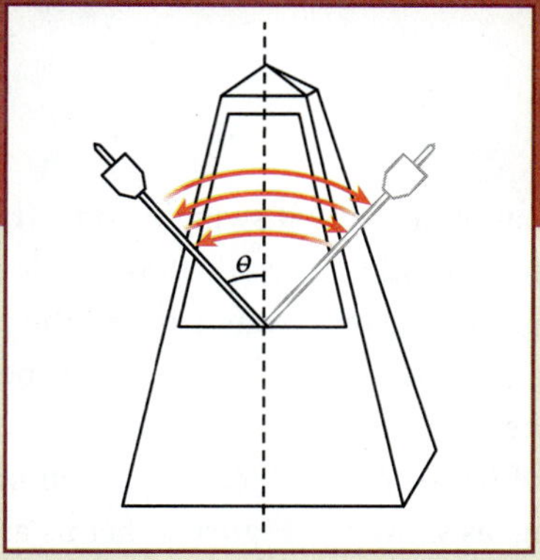

A mechanical metronome consists of an inverted pendulum and a counterweight on opposite sides of a pivot. A sliding weight above the pivot is used to change the rate of vibration. As the pendulum vibrates, the metronome ticks, and musicians use the sound to keep a steady tempo. This vibration is an example of a *periodic motion.*

WHAT TO EXPECT

In this chapter, you will study a kind of periodic motion called *simple harmonic motion* and will learn about the relationship between simple harmonic vibrations and waves.

Why it Matters

Waves can carry information, such as conversations and television broadcasts. Much of your perception of the physical world is dependent on waves. You could not hear or see anything without sound waves and light waves.

CHAPTER PREVIEW

Tapping Prior Knowledge

Knowledge to Review

- ✔ Elastic potential energy is the energy stored in a stretched or compressed elastic object.
- ✔ A spring constant is a parameter that expresses how resistant a spring is to being compressed or stretched.
- ✔ Gravitational potential energy is the energy associated with an object due to its position relative to Earth.

Items to Probe

- ✔ Familiarity with periodic motion: Ask students to describe the motion of objects that move in a closed path.
- ✔ Preconceptions about waves: Ask students to identify the source and the propagating medium in various examples of wave phenomena.

367

SECTION 1
General Level

Demonstration

A Vibrating Spring GENERAL

Purpose Show the changes in velocity and restoring force for a vibrating mass-spring system.

Materials spring, ring stand, ring weight holder, weight

Procedure Hang the spring from the ring stand, suspend a weight holder, and add a weight to the holder. Stretch the weight down, and release it to start a harmonic motion. Have the students sketch diagrams representing the position and direction of motion of the weight at various points of the cycle. Ask them when the magnitude of the velocity increases or decreases, and have them draw the vector of the net force that may cause this motion in each case. Point out that the net force always pulls the weight toward its equilibrium position.

Demonstration

An Oscillating Pendulum GENERAL

Purpose Show the changes in velocity and restoring force for an oscillating pendulum.

Materials pendulum bob, string, ring stand, ring

Procedure Attach the bob to the string, and suspend the string from the ring stand. Start the pendulum swinging, and then repeat the process described in the demonstration above. Point out that the two cases are analogous and that the net force always pulls the pendulum bob toward its equilibrium position.

SECTION 1

Simple Harmonic Motion

SECTION OBJECTIVES

- **Identify the conditions of simple harmonic motion.**
- **Explain how force, velocity, and acceleration change as an object vibrates with simple harmonic motion.**
- **Calculate the spring force using Hooke's law.**

HOOKE'S LAW

A repeated motion, such as that of an acrobat swinging on a trapeze, is called a periodic motion. Other periodic motions include those made by a child on a playground swing, a wrecking ball swaying to and fro, and the pendulum of a grandfather clock or a metronome. In each of these cases, the periodic motion is back and forth over the same path.

One of the simplest types of back-and-forth periodic motion is the motion of a mass attached to a spring, as shown in **Figure 1.** Let us assume that the mass moves on a frictionless horizontal surface. When the spring is stretched or compressed and then released, it vibrates back and forth about its unstretched position. We will begin by considering this example, and then we will apply our conclusions to the swinging motion of a trapeze acrobat.

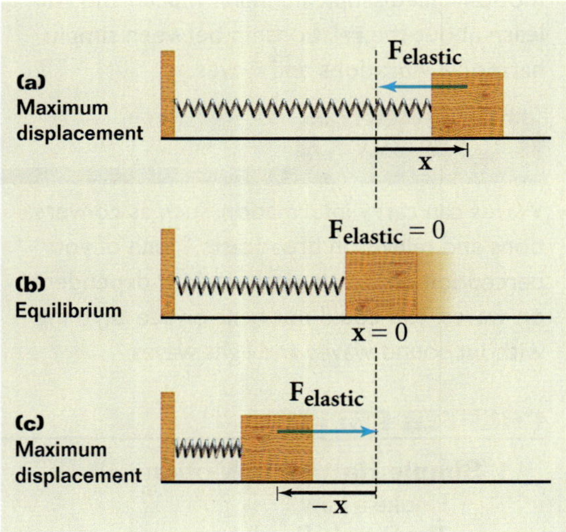

Figure 1
The direction of the force acting on the mass ($\mathbf{F}_{elastic}$) is always opposite the direction of the mass's displacement from equilibrium ($\mathbf{x} = 0$). **(a)** When the spring is stretched to the right, the spring force pulls the mass to the left. **(b)** When the spring is unstretched, the spring force is zero. **(c)** When the spring is compressed to the left, the spring force is directed to the right.

At the equilibrium position, speed reaches a maximum

In **Figure 1(a),** the spring is stretched away from its unstretched, or equilibrium, position ($x = 0$). In this stretched position, the spring exerts a force on the mass toward the equilibrium position. This spring force decreases as the spring moves toward the equilibrium position, and it reaches zero at equilibrium, as illustrated in **Figure 1(b).** The mass's acceleration also becomes zero at equilibrium.

Though the spring force and acceleration decrease as the mass moves toward the equilibrium position, the speed of the mass increases. At the equilibrium position, when acceleration reaches zero, the speed reaches a maximum. At that point, although the spring force is zero, the mass's momentum causes it to overshoot the equilibrium position and compress the spring.

At maximum displacement, spring force and acceleration reach a maximum

As the mass moves beyond equilibrium, the spring force and the acceleration increase. But the direction of the spring force and of the acceleration (toward equilibrium) is opposite the mass's direction of motion (away from equilibrium), and the mass begins to slow down.

When the spring's compression is equal to the distance the spring was originally stretched away from the equilibrium position (x), as shown in **Figure 1(c),** the mass is at maximum displacement, and the spring force and acceleration of the mass reach a maximum. At this point, the speed of the mass becomes zero. The spring force acting to the right causes the mass to change its direction, and the mass begins moving back toward the equilibrium position. Then the entire process begins again, and the mass continues to oscillate back and forth over the same path.

In an ideal system, the mass-spring system would oscillate indefinitely. But in the physical world, friction retards the motion of the vibrating mass, and the mass-spring system eventually comes to rest. This effect is called *damping.* In most cases, the effect of damping is minimal over a short period of time, so the ideal mass-spring system provides an approximation for the motion of a physical mass-spring system.

In simple harmonic motion, restoring force is proportional to displacement

As you have seen, the spring force always pushes or pulls the mass toward its original equilibrium position. For this reason, it is sometimes called a *restoring force.* Measurements show that the restoring force is directly proportional to the displacement of the mass. This relationship was determined in 1678 by Robert Hooke and is known as *Hooke's Law.* The following equation mathematically describes Hooke's Law:

HOOKE'S LAW

$$F_{elastic} = -kx$$

spring force = −(spring constant × displacement)

The negative sign in the equation signifies that the direction of the spring force is always opposite the direction of the mass's displacement from equilibrium. In other words, the negative sign shows that the spring force will tend to move the object back to its equilibrium position.

As mentioned in the chapter "Work and Energy," the quantity k is a positive constant called the *spring constant.* The value of the spring constant is a measure of the stiffness of the spring. A greater value of k means a stiffer spring because a greater force is needed to stretch or compress that spring a given amount. The SI units of k are N/m. As a result, N is the unit of the spring force when the spring constant (N/m) is multiplied by the displacement (m). The motion of a vibrating mass-spring system is an example of **simple harmonic motion.** Simple harmonic motion describes any periodic motion that is the result of a restoring force that is proportional to displacement. Because simple harmonic motion involves a restoring force, every simple harmonic motion is a back-and-forth motion over the same path.

SCI LINKS.
www.scilinks.org
Topic: Hooke's Law
Code: HF60756

simple harmonic motion

vibration about an equilibrium position in which a restoring force is proportional to the displacement from equilibrium

Why it Matters

Conceptual Challenge

1. Earth's Orbit

The motion of Earth orbiting the sun is periodic. Is this motion simple harmonic? Why or why not?

2. Pinball

In pinball games, the force exerted by a compressed spring is used to release a ball. If the distance the spring is compressed is doubled, how will the force exerted on the ball change? If the spring is replaced with one that is half as stiff, how will the force acting on the ball change?

SECTION 1

Demonstration

Hooke's Law — GENERAL

Purpose Verify Hooke's law experimentally.

Materials 2 springs with different spring constants, 2 ring stands, 2 rings, 2 weight holders, incremental weights, ruler

Procedure Hang the springs from the ring stands, and suspend a weight holder from each spring. Add incremental weights to the holders, measure the resulting displacements, and record these values on the board.

From the data for each spring, sketch a graph of force versus displacement on the board. Show that the relationship between force and displacement is linear, and calculate the slope (which equals the spring constant). Ask students why a negative sign appears in the equation. *(The elastic force on the weight is opposite the weight's displacement from equilibrium.)*

ANSWERS

Conceptual Challenge

1. no; because Earth does not oscillate about an equilibrium position
2. The force will double; The force will be half as large.

Classroom Practice

Hooke's Law
A 76 N crate is hung from a spring ($k = 450$ N/m). How much displacement is caused by the weight of this crate?

Answer
−0.17 m

A spring of $k = 1962$ N/m loses its elasticity if stretched more than 50.0 cm. What is the mass of the heaviest object the spring can support without being damaged?

Answer
1.00×10^2 kg

Alternative Problem-Solving Approach

The weight of the object pulls downward.

$F_g = -mg = -5.4$ N

The spring stretches until its restoring force ($F_{elastic} = -kx$) balances the −5.4 N. This occurs when $x = -0.020$ m. Thus, 5.4 N = $-k(-0.020$ m) and $k = 270$ N/m.

SAMPLE PROBLEM A

Hooke's Law

PROBLEM

If a mass of 0.55 kg attached to a vertical spring stretches the spring 2.0 cm from its original equilibrium position, what is the spring constant?

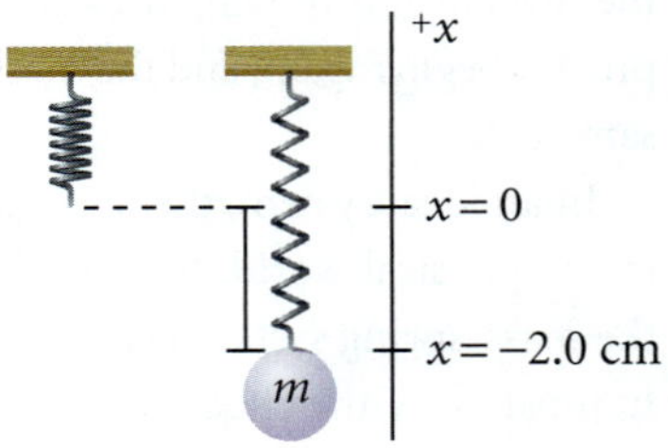

SOLUTION

1. DEFINE

Given: $m = 0.55$ kg $x = -2.0$ cm $= -0.020$ m
$g = 9.81$ m/s^2

Unknown: $k = ?$

Diagram:

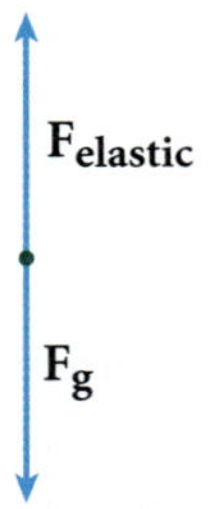

2. PLAN

Choose an equation or situation: When the mass is attached to the spring, the equilibrium position changes. At the new equilibrium position, the net force acting on the mass is zero. So the spring force (given by Hooke's law) must be equal and opposite to the weight of the mass.

$$\mathbf{F_{net}} = 0 = \mathbf{F_{elastic}} + \mathbf{F_g}$$

$$F_{elastic} = -kx$$

$$F_g = -mg$$

$$-kx - mg = 0$$

Rearrange the equation to isolate the unknown:

$$kx = -mg$$

$$k = \frac{-mg}{x}$$

3. CALCULATE

Substitute the values into the equation and solve:

$$k = \frac{-(0.55 \text{ kg})(9.81 \text{ m/s}^2)}{-0.020 \text{ m}}$$

$$\boxed{k = 270 \text{ N/m}}$$

CALCULATOR SOLUTION

The calculator answer for k is 269.775. This answer is rounded to two significant figures, 270 N/m.

4. EVALUATE

The value of k implies that 270 N of force is required to displace the spring 1 m.

PRACTICE A

Hooke's Law

1. Suppose the spring in Sample Problem A is replaced with a spring that stretches 36 cm from its equilibrium position.
   a. What is the spring constant in this case?
   b. Is this spring stiffer or less stiff than the one in Sample Problem A?
2. A load of 45 N attached to a spring that is hanging vertically stretches the spring 0.14 m. What is the spring constant?
3. A slingshot consists of a light leather cup attached between two rubber bands. If it takes a force of 32 N to stretch the bands 1.2 cm, what is the equivalent spring constant of the two rubber bands?
4. How much force is required to pull a spring 3.0 cm from its equilibrium position if the spring constant is 2.7×10^3 N/m?

A stretched or compressed spring has elastic potential energy

As you saw in the chapter "Work and Energy," a stretched or compressed spring stores elastic potential energy. To see how mechanical energy is conserved in an ideal mass-spring system, consider an archer shooting an arrow from a bow, as shown in **Figure 2.** Bending the bow by pulling back the bowstring is analogous to stretching a spring. To simplify this situation, we will disregard friction and internal energy.

Once the bowstring has been pulled back, the bow stores elastic potential energy. Because the bow, arrow, and bowstring (the system) are now at rest, the kinetic energy of the system is zero, and the mechanical energy of the system is solely elastic potential energy.

When the bowstring is released, the bow's elastic potential energy is converted to the kinetic energy of the arrow. At the moment the arrow leaves the bowstring, it gains most of the elastic potential energy originally stored in the bow. (The rest of the elastic potential energy is converted to the kinetic energy of the bow and the bowstring.) Thus, once the arrow leaves the bowstring, the mechanical energy of the bow-and-arrow system is solely kinetic. Because mechanical energy must be conserved, the total kinetic energy of the bow, arrow, and bowstring is equal to the elastic potential energy originally stored in the bow.

Figure 2
The elastic potential energy stored in this stretched bow is converted into the kinetic energy of the arrow.

PROBLEM GUIDE A

Use this guide to assign problems.
SE = Student Edition Textbook
PW = Problem Workbook
PB = Problem Bank on the One-Stop Planner (OSP)

Solving for:

k	**SE** Sample, 1–3; Ch. Rvw. 8–9 **PW** 3, 4*, 5* **PB** 5–7
F	**SE** 4; Ch. Rvw. 44–45, 50 **PW** 6*, 7, 8* **PB** Sample, 1–4
x	**PW** Sample, 1–5 **PB** 8–10

***Challenging Problem**
Consult the printed Solutions Manual or the OSP for detailed solutions.

ANSWERS

Practice A

1. a. 15 N/m
   b. less stiff
2. 3.2×10^2 N/m
3. 2.7×10^3 N/m
4. 81 N

Why it Matters

Shock Absorbers

The spring–shock absorber system on modern cars is an excellent example of damped harmonic oscillation.

A shock absorber consists of a piston moving up and down in a chamber filled with oil. As the piston moves, the oil is squeezed through the channels between the piston and the tube, causing the piston to decelerate.

Why it Matters

Shock Absorbers

Bumps in the road are certainly a nuisance, but without strategic use of damping devices, they could also prove deadly. To control a car going 110 km/h (70 mi/h), a driver needs all the wheels on the ground. Bumps in the road lift the wheels off the ground and rob the driver of control. A good solution is to fit the car with springs at each wheel. The springs absorb energy as the wheels rise over the bumps and push the wheels back to the pavement to keep the wheels on the road. However, once set in motion, springs tend to continue to go up and down in simple harmonic motion. This affects the driver's control of the car and can also be uncomfortable.

One way to cut down on unwanted vibrations is to use stiff springs that compress only a few centimeters under thousands of newtons of force. Such springs have very high spring constants and thus do not vibrate as freely as softer springs with lower constants. However, this solution reduces the driver's ability to keep the car's wheels on the road.

To completely solve the problem, energy-absorbing devices known as *shock absorbers* are placed parallel to the springs in some automobiles, as shown in part (a) of the illustration below. Shock absorbers are fluid-filled tubes that turn the simple harmonic motion of the springs into damped harmonic motion. In damped harmonic motion, each cycle of stretch and compression of the spring is much smaller than the previous cycle. Modern auto suspensions are set up so that all of a spring's energy is absorbed by the shock absorbers, eliminating vibrations in just one up-and-down cycle. This keeps the car from continually bouncing without sacrificing the spring's ability to keep the wheels on the road.

Different spring constants and shock absorber damping are combined to give a wide variety of road responses. For example, larger vehicles have heavy-duty leaf springs made of stacks of steel strips, which have a larger spring constant than coil springs do. In this type of suspension system, the shock absorber is perpendicular to the spring, as shown in part (b) of the illustration. The stiffness of the spring can affect steering response time, traction, and the general feel of the car.

As a result of the variety of combinations that are possible, your driving experiences can range from the luxurious floating of a limousine to the bone-rattling road feel of a sports car.

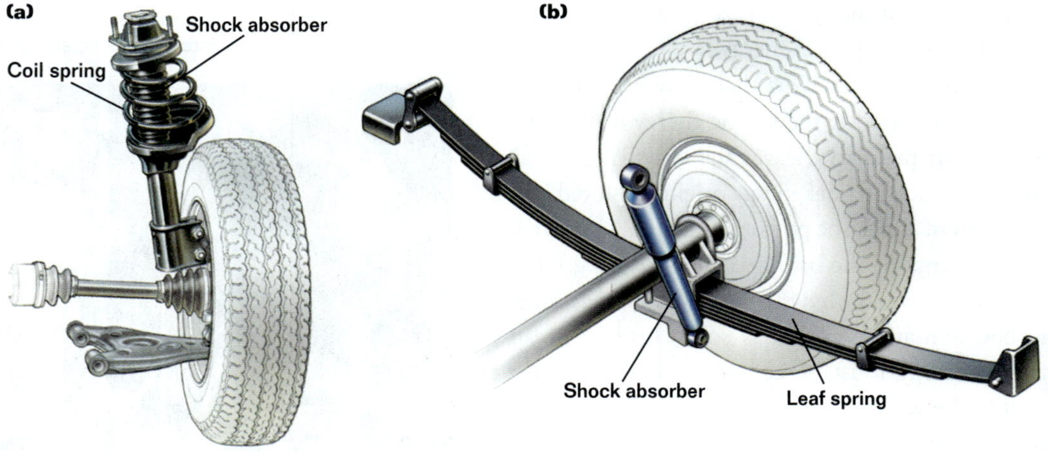

THE SIMPLE PENDULUM

As you have seen, the periodic motion of a mass-spring system is one example of simple harmonic motion. Now consider the trapeze acrobats shown in **Figure 3(a).** Like the vibrating mass-spring system, the swinging motion of a trapeze acrobat is a periodic vibration. Is a trapeze acrobat's motion an example of simple harmonic motion?

To answer this question, we will use a simple pendulum as a model of the acrobat's motion, which is a physical pendulum. A simple pendulum consists of a mass called a *bob,* which is attached to a fixed string, as shown in **Figure 3(b).** When working with a simple pendulum, we assume that the mass of the bob is concentrated at a point and that the mass of the string is negligible. Furthermore, we disregard the effects of friction and air resistance. For a physical pendulum, on the other hand, the distribution of the mass must be considered, and friction and air resistance also must be taken into account. To simplify our analysis, we will disregard these complications and use a simple pendulum to approximate a physical pendulum in all of our examples.

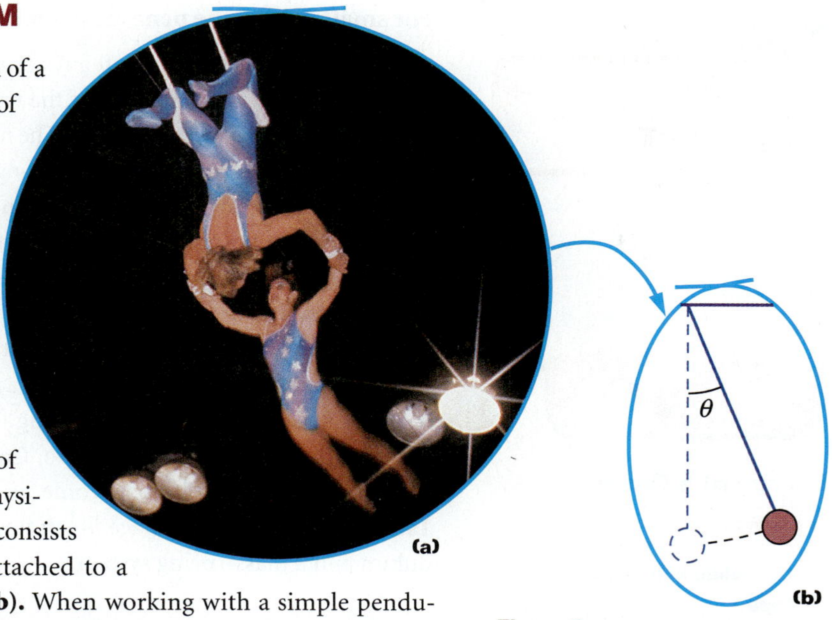

Figure 3
(a) The motion of these trapeze acrobats is modeled by **(b)** a simple pendulum.

The restoring force of a pendulum is a component of the bob's weight

To see whether the pendulum's motion is simple harmonic, we must first examine the forces exerted on the pendulum's bob to determine which force acts as the restoring force. If the restoring force is proportional to the displacement, then the pendulum's motion is simple harmonic. Let us select a coordinate system in which the *x*-axis is tangent to the direction of motion and the *y*-axis is perpendicular to the direction of motion. Because the bob is always changing its position, these axes will change at each point of the bob's motion.

The forces acting on the bob at any point include the force exerted by the string and the gravitational force. The force exerted by the string always acts along the *y*-axis, which is along the string. The gravitational force can be resolved into two components along the chosen axes, as shown in **Figure 4.** Because both the force exerted by the string and the *y* component of the gravitational force are perpendicular to the bob's motion, the *x* component of the gravitational force is the net force acting on the bob in the direction of its motion. In this case, the *x* component of the gravitational force always pulls the bob toward its equilibrium position and hence is the restoring force. Note that the restoring force ($F_{g,x} = F_g \sin \theta$) is zero at equilibrium because θ equals zero at this point.

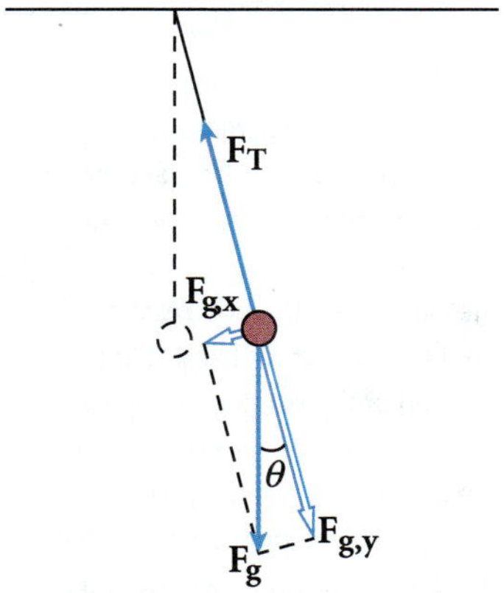

Figure 4
At any displacement from equilibrium, the weight of the bob (**F_g**) can be resolved into two components. The *x* component (**$F_{g,x}$**), which is perpendicular to the string, is the only force acting on the bob in the direction of its motion.

The Language of Physics

Point out that the term *simple pendulum* is used because we have simplified our analysis by disregarding complications such as friction and air resistance. The term *ideal mass-spring system,* seen earlier in this section, was used to express the same concept for a mass-spring system.

Visual Strategy GENERAL

Figure 4

Be certain students make a distinction between force diagrams and schematic diagrams representing physical objects. In particular, string *length* and string *force* should not be confused.

Q Draw the vectors representing the forces on the bob at the equilibrium position when the pendulum is at rest. Which component of $\mathbf{F_g}$ is 0? Which component of $\mathbf{F_g}$ is equal to $\mathbf{F_g}$?

A ***Students' vector diagrams should show that F_g is equal and opposite to F_T (both are vertical); $F_{g,x} = 0$; $F_{g,y} = F_g$.***

Quick Lab

TEACHER'S NOTES

This activity is meant to demonstrate that the kinetic energy of the pendulum at the equilibrium position increases as the pendulum's maximum displacement from equilibrium increases.

For this lab to be effective, it is best to arrange the pendulum so that it transfers all of its energy to the toy car and comes to rest after striking the car. This is best achieved when the collision between the car and the bob is head-on and when the mass of the bob is nearly equal to the mass of the car.

Of course, the energy transferred to the car will be quickly dissipated, in part because of the inelasticity of the collision and in part because of friction on the wheels of the car. As a result, the displacement of the car is only a very rough indication of the energy of the pendulum.

Visual Strategy – BASIC

Figure 5

Point out to students that the total mechanical energy of the system is represented by a horizontal line because the sum of the kinetic energy and the potential energy is always constant. This means that as one increases, the other decreases by the same amount, and vice versa.

Q Does this graph apply to a vibrating mass-spring system as well?

A *yes, because the energy changes in a mass-spring system are analogous to those in a simple pendulum*

SCILINKS®
www.scilinks.org
Topic: Pendulums
Code: HF61121

Quick Lab

Energy of a Pendulum

MATERIALS LIST

- pendulum bob and string
- tape
- toy car
- protractor
- meterstick or tape measure

Tie one end of a piece of string around the pendulum bob, and use tape to secure it in place. Set the toy car on a smooth surface, and hold the string of the pendulum directly above the car so that the bob rests on the car. Use your other hand to pull back the bob of the pendulum, and have your partner measure the angle of the pendulum with the protractor.

Release the pendulum so that the bob strikes the car. Measure the displacement of the car. What happened to the pendulum's potential energy after you released the bob? Repeat the process using different angles. How can you account for your results?

For small angles, the pendulum's motion is simple harmonic

As with a mass-spring system, the restoring force of a simple pendulum is not constant. Instead, the magnitude of the restoring force varies with the bob's distance from the equilibrium position. The magnitude of the restoring force is proportional to $\sin\theta$. When the maximum angle of displacement θ is relatively small ($<15°$), $\sin\theta$ is approximately equal to θ in radians. As a result, the restoring force is very nearly proportional to the displacement and the pendulum's motion is an excellent approximation of simple harmonic motion. We will assume small angles of displacement unless otherwise noted.

Because a simple pendulum vibrates with simple harmonic motion, many of our earlier conclusions for a mass-spring system apply here. At maximum displacement, the restoring force and acceleration reach a maximum while the speed becomes zero. Conversely, at equilibrium, the restoring force and acceleration become zero and speed reaches a maximum. **Table 1** on the following page illustrates the analogy between a simple pendulum and a mass-spring system.

Gravitational potential increases as a pendulum's displacement increases

As with the mass-spring system, the mechanical energy of a simple pendulum is conserved in an ideal (frictionless) system. However, the spring's potential energy is elastic, while the pendulum's potential energy is gravitational. We define the gravitational potential energy of a pendulum to be zero when it is at the lowest point of its swing.

Figure 5 illustrates how a pendulum's mechanical energy changes as the pendulum oscillates. At maximum displacement from equilibrium, a pendulum's energy is entirely gravitational potential energy. As the pendulum swings toward equilibrium, it gains kinetic energy and loses potential energy. At the equilibrium position, its energy becomes solely kinetic.

As the pendulum swings past its equilibrium position, the kinetic energy decreases while the gravitational potential energy increases. At maximum displacement from equilibrium, the pendulum's energy is once again entirely gravitational potential energy.

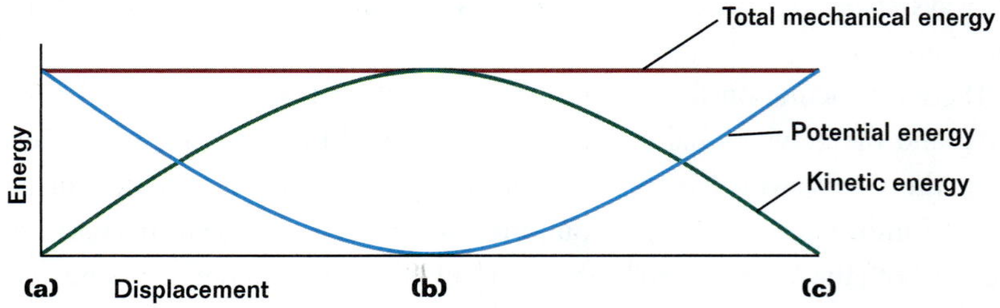

Figure 5
Whether at maximum displacement **(a)**, equilibrium **(b)**, or maximum displacement in the other direction **(c)**, the pendulum's total mechanical energy remains the same. However, as the graph shows, the pendulum's kinetic energy and potential energy are constantly changing.

Table 1 Simple Harmonic Motion

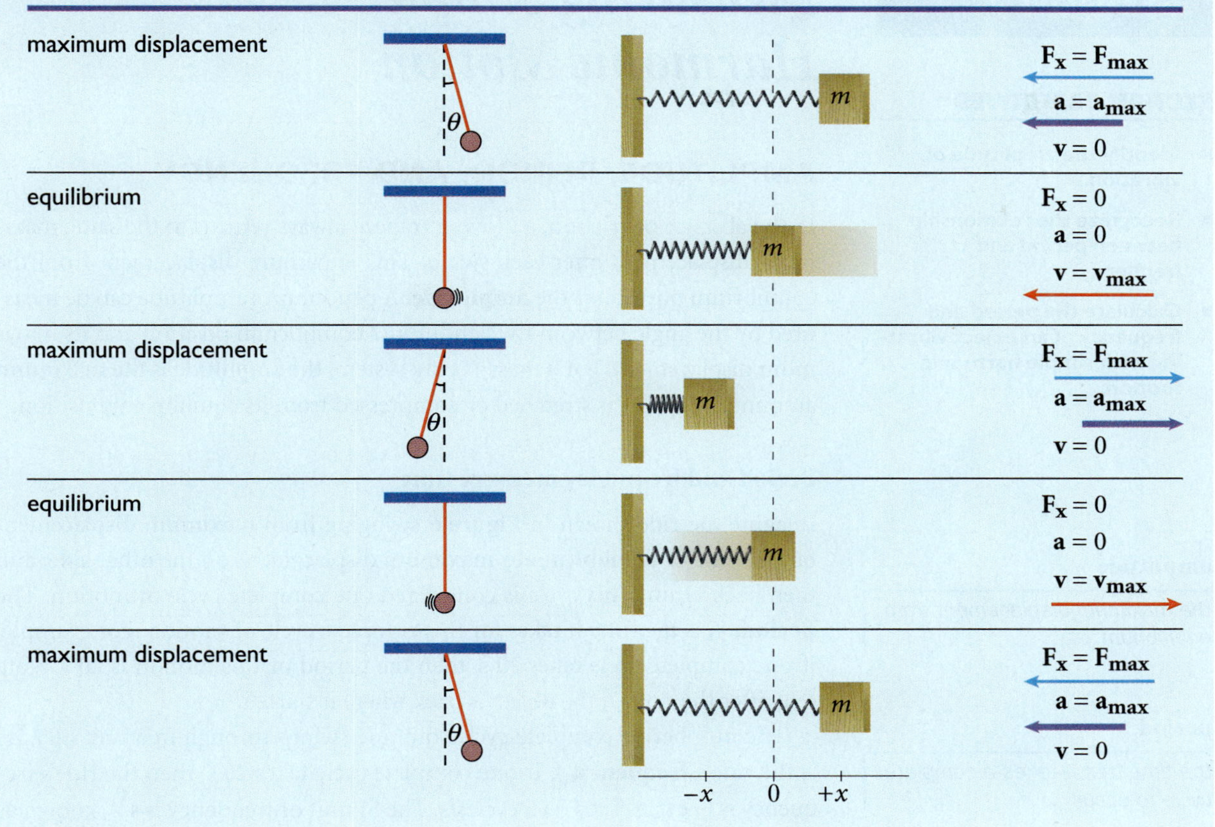

SECTION REVIEW

1. Which of these periodic motions are simple harmonic?
   - **a.** a child swinging on a playground swing ($\theta = 45°$)
   - **b.** a CD rotating in a player
   - **c.** an oscillating clock pendulum ($\theta = 10°$)
2. A pinball machine uses a spring that is compressed 4.0 cm to launch a ball. If the spring constant is 13 N/m, what is the force on the ball at the moment the spring is released?
3. How does the restoring force acting on a pendulum bob change as the bob swings toward the equilibrium position? How do the bob's acceleration (along the direction of motion) and velocity change?
4. **Critical Thinking** When an acrobat reaches the equilibrium position, the net force acting along the direction of motion is zero. Why does the acrobat swing past the equilibrium position?

SECTION 1

Visual Strategy – BASIC

Table 1

Table 1 can be used to review the concepts discussed in this section. Comparing these two different types of simple harmonic motion will help students grasp the essential aspects of simple harmonic motion.

Q Does the fourth column of the table refer to the simple pendulum or to the mass-spring system?

A ***both; The displacements, velocities, and restoring forces in these two cases are analogous.***

SECTION REVIEW ANSWERS

1. c
2. 0.52 N
3. force decreases; acceleration decreases; velocity increases
4. because the acrobat's momentum carries him or her through the equilibrium position

SECTION 2
General Level

Demonstration

Period and Frequency — GENERAL

Purpose Show period and frequency empirically and verify their inverse relationship.

Materials pendulum bob, string, ring stand, clock

Procedure Attach the pendulum bob to the string, and suspend the string from the ring stand. Set the pendulum in motion. Have a student record the time required to complete 20 oscillations. Meanwhile, have another student determine how many oscillations take place each second.

Have students use the first measurement to find the pendulum's period (T = number of seconds/20), and ask the students what the second measurement indicates *(frequency)*. Compare the values for the period and frequency of the pendulum. *(The two should be inversely related.)*

The Language of Physics

Students may be confused by the transition from cycles/s as a measure of frequency to the SI unit s^{-1} (hertz). Point out that the term *cycle* is not part of the SI unit because this term refers to an event rather than a unit of measure. The same holds true for the period, which can be considered as s/cycle but whose SI unit is simply s. This should help clarify the inverse relationship between frequency (cycles/s or s^{-1}) and period (s/cycle or s).

SECTION 2

Measuring Simple Harmonic Motion

SECTION OBJECTIVES

- Identify the amplitude of vibration.
- Recognize the relationship between period and frequency.
- Calculate the period and frequency of an object vibrating with simple harmonic motion.

AMPLITUDE, PERIOD, AND FREQUENCY

In the absence of friction, a moving trapeze always returns to the same maximum displacement after each swing. This maximum displacement from the equilibrium position is the **amplitude.** A pendulum's amplitude can be measured by the angle between the pendulum's equilibrium position and its maximum displacement. For a mass-spring system, the amplitude is the maximum amount the spring is stretched or compressed from its equilibrium position.

amplitude

the maximum displacement from equilibrium

Period and frequency measure time

Imagine the ride shown in **Figure 6** swinging from maximum displacement on one side of equilibrium to maximum displacement on the other side, and then back again. This cycle is considered one complete cycle of motion. The **period,** T, is the time it takes for this complete cycle of motion. For example, if one complete cycle takes 20 s, then the period of this motion is 20 s. Note that after the time T, the object is back where it started.

period

the time that it takes a complete cycle to occur

The number of complete cycles the ride swings through in a unit of time is the ride's **frequency,** f. If one complete cycle takes 20 s, then the ride's frequency is $\frac{1}{20}$ cycles/s, or 0.05 cycles/s. The SI unit of frequency is s^{-1}, known as hertz (Hz). In this case, the ride's frequency is 0.05 Hz.

frequency

the number of cycles or vibrations per unit of time

Period and frequency can be confusing because both are concepts involving time in simple harmonic motion. Notice that the period is the time per cycle and that the frequency is the number of cycles per unit time, so they are inversely related.

$$f = \frac{1}{T} \text{ or } T = \frac{1}{f}$$

This relationship was used to determine the frequency of the ride.

$$f = \frac{1}{T} = \frac{1}{20\ \text{s}} = 0.05\ \text{Hz}$$

In any problem where you have a value for period or frequency, you can calculate the other value. These terms are summarized in **Table 2** on the next page.

Figure 6
For any periodic motion—such as the motion of this amusement park ride in Helsinki, Finland—period and frequency are inversely related.

Table 2 Measures of Simple Harmonic Motion

Term	Example	Definition	SI unit
amplitude	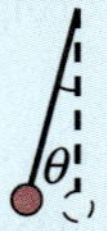	maximum displacement from equilibrium	radian, rad meter, m
period, T		time that it takes to complete a full cycle	second, s
frequency, f		number of cycles or vibrations per unit of time	hertz, Hz ($Hz = s^{-1}$)

extension

Integrating Technology

Visit go.hrw.com for the activity "Bicycle Design and Shock Absorption."

Keyword HF6VIBX

The period of a simple pendulum depends on pendulum length and free-fall acceleration

Although both a simple pendulum and a mass-spring system vibrate with simple harmonic motion, calculating the period and frequency of each requires a separate equation. This is because in each, the period and frequency depend on different physical factors.

Consider an experimental setup of two pendulums of the same length but with bobs of different masses. The length of a pendulum is measured from the pivot point to the center of mass of the pendulum bob. If you were to pull each bob aside the same small distance and then release them at the same time, each pendulum would complete one vibration in the same amount of time. If you then changed the amplitude of one of the pendulums, you would find that they would still have the same period. Thus, for small amplitudes, the period of a pendulum does not depend on the mass or on the amplitude.

However, changing the length of a pendulum *does* affect its period. A change in the free-fall acceleration also affects the period of a pendulum. The exact relationship between these variables can be derived mathematically or found experimentally.

PERIOD OF A SIMPLE PENDULUM IN SIMPLE HARMONIC MOTION

$$T = 2\pi\sqrt{\frac{L}{a_g}}$$

period = 2π × square root of (length divided by free-fall acceleration)

Did you know?

Galileo is credited as the first person to notice that the motion of a pendulum depends on its length and is independent of its amplitude (for small angles). He supposedly observed this while attending church services at a cathedral in Pisa. The pendulum he studied was a swinging chandelier that was set in motion when someone bumped it while lighting the candles. Galileo is said to have measured its frequency, and hence its period, by timing the swings with his pulse.

Demonstration

Relationship Between the Length and the Period of a Pendulum GENERAL

Purpose Verify the equation for a pendulum's period experimentally.

Materials pendulum bob, string, ring stand, clock, meterstick

Procedure Repeat the demonstration "Period and Frequency" (on the previous page) with a variety of lengths. Record each length and its corresponding period. (Frequency does not need to be measured in this demonstration.) Verify that the results are consistent with the following equation:

$$T = 2\pi\sqrt{\frac{L}{a_g}}$$

Next ask the students to calculate the length required for a pendulum to have a period of 1.0 s. Have the students construct such a pendulum to test their prediction.

Misconception Alert

Remind students that, as seen in Section 1, a pendulum's amplitude must be less than about 15° in order for its motion to approximate simple harmonic motion. For greater amplitudes, the pendulum's amplitude *does* affect its period; in those cases, this equation for the period of a simple pendulum would not apply.

Classroom Practice

SHM of a Simple Pendulum
What is the period of a 3.98 m long pendulum? What is the period of a 99.4 cm long pendulum?

Answer
4.00 s; 2.00 s

A desktop toy pendulum swings back and forth once every 1.0 s. How long is this pendulum?

Answer
0.25 m

What is the free-fall acceleration at a location where a 6.00 m long pendulum swings through exactly 100 cycles in 492 s?

Answer
9.79 m/s^2

PROBLEM GUIDE B

Use this guide to assign problems.
SE = Student Edition Textbook
PW = Problem Workbook
PB = Problem Bank on the One-Stop Planner (OSP)

Solving for:

L	**SE** Sample, 1–3; Ch. Rvw. 19 **PW** 4, 5 **PB** 5–7
T, f	**SE** 4; Ch. Rvw. 20, 27a **PW** Sample, 1–3 **PB** 8–10
a_g	**PW** 6 **PB** Sample, 1–4

*Challenging Problem
Consult the printed Solutions Manual or the OSP for detailed solutions.

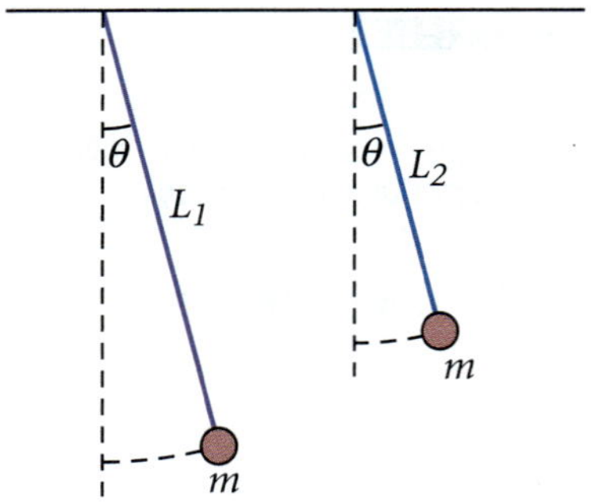

Figure 7
When the length of one pendulum is decreased, the distance that the pendulum travels to equilibrium is also decreased. Because the accelerations of the two pendulums are equal, the shorter pendulum will have a smaller period.

Why does the period of a pendulum depend on pendulum length and free-fall acceleration? When two pendulums have different lengths but the same amplitude, the shorter pendulum will have a smaller arc to travel through, as shown in **Figure 7.** Because the distance from maximum displacement to equilibrium is less while the acceleration caused by the restoring force remains the same, the shorter pendulum will have a shorter period.

Why don't mass and amplitude affect the period of a pendulum? When the bobs of two pendulums differ in mass, the heavier mass provides a larger restoring force, but it also needs a larger force to achieve the same acceleration. This is similar to the situation for objects in free fall, which all have the same acceleration regardless of their mass. Because the acceleration of both pendulums is the same, the period for both is also the same.

For small angles (<15°), when the amplitude of a pendulum increases, the restoring force also increases proportionally. Because force is proportional to acceleration, the initial acceleration will be greater. However, the distance this pendulum must cover is also greater. For small angles, the effects of the two increasing quantities cancel and the pendulum's period remains the same.

SAMPLE PROBLEM B

Simple Harmonic Motion of a Simple Pendulum

PROBLEM

You need to know the height of a tower, but darkness obscures the ceiling. You note that a pendulum extending from the ceiling almost touches the floor and that its period is 12 s. How tall is the tower?

SOLUTION

Given: $T = 12 \text{ s}$ $\quad a_g = g = 9.81 \text{ m/s}^2$

Unknown: $L = ?$

Use the equation for the period of a simple pendulum, and solve for L.

$$T = 2\pi\sqrt{\frac{L}{a_g}}$$

$$\frac{T\sqrt{a_g}}{2\pi} = \sqrt{L}$$

$$\frac{T^2 a_g}{4\pi^2} = L$$

$$L = \frac{(12 \text{ s})^2(9.81 \text{ m/s}^2)}{4\pi^2}$$

$$\boxed{L = 36 \text{ m}}$$

TIP *Remember that on Earth's surface, $a_g = g = 9.81 \text{ m/s}^2$. Use this value in the equation for the period of a pendulum if a problem does not specify otherwise. At higher altitudes or on different planets, use the given value of a_g instead.*

PRACTICE B

Simple Harmonic Motion of a Simple Pendulum

1. If the period of the pendulum in the preceding sample problem were 24 s, how tall would the tower be?
2. You are designing a pendulum clock to have a period of 1.0 s. How long should the pendulum be?
3. A trapeze artist swings in simple harmonic motion with a period of 3.8 s. Calculate the length of the cables supporting the trapeze.
4. Calculate the period and frequency of a 3.500 m long pendulum at the following locations:
   a. the North Pole, where $a_g = 9.832\ m/s^2$
   b. Chicago, where $a_g = 9.803\ m/s^2$
   c. Jakarta, Indonesia, where $a_g = 9.782\ m/s^2$

Period of a mass-spring system depends on mass and spring constant

Now consider the period of a mass-spring system. In this case, according to Hooke's law, the restoring force acting on the mass is determined by the displacement of the mass and by the spring constant ($F_{elastic} = -kx$). The magnitude of the mass does not affect the restoring force. So, unlike in the case of the pendulum, in which a heavier mass increased both the force on the bob and the bob's inertia, a heavier mass attached to a spring increases inertia without providing a compensating increase in force. Because of this increase in inertia, a heavy mass has a smaller acceleration than a light mass has. Thus, a heavy mass will take more time to complete one cycle of motion. In other words, the heavy mass has a greater period. Thus, as mass increases, the period of vibration increases when there is no compensating increase in force.

Why it Matters

Conceptual Challenge

1. Pendulum on the Moon

The free-fall acceleration on the surface of the moon is approximately one-sixth of the free-fall acceleration on the surface of Earth. Compare the period of a pendulum on the moon with that of an identical pendulum set in motion on Earth.

2. Pendulum Clocks

Why is a wound mainspring often used to provide energy to a pendulum clock in order to prevent the amplitude of the pendulum from decreasing?

ANSWERS

Practice B

1. 1.4×10^2 m
2. 25 cm
3. 3.6 m
4. **a.** 3.749 s, 0.2667 Hz
   b. 3.754 s, 0.2664 Hz
   c. 3.758 s, 0.2661 Hz

Teaching Tip

At the end of Section 1, students compared a simple pendulum with a mass-spring system to find their similarities (see **Table 1** in Section 1). The discussion of mass on this page can be used to demonstrate their differences. Mass does not affect the period of a pendulum, but it does affect the period of a mass-spring system. This difference exists because the restoring forces are different.

ANSWERS

Conceptual Challenge

1. The period of the pendulum on the moon would be a little less than 2.5 times as long as the period on Earth.
2. Because a pendulum's motion only approximates simple harmonic motion, if the amplitude decreased, the clock would not keep accurate time.

Classroom Practice

SHM of a Mass-Spring System
A 1.0 kg mass attached to one end of a spring completes one oscillation every 2.0 s. Find the spring constant.

Answer
9.9 N/m

Alternative Problem-Solving Approach

Think of the total mass (1275 kg + 153 kg = 1428 kg) as being supported by one spring with four times the strength of each spring:

$$k = \frac{4\pi^2 m}{T^2}$$

$$k = \frac{4\pi^2(1428 \text{ kg})}{(0.840 \text{ s})^2}$$

$$k = 7.99 \times 10^4 \text{ N/m}$$

The spring constant of a single spring is one-fourth of this value.

$$k = 0.25(7.99 \times 10^4 \text{ N/m})$$
$$k = 2.00 \times 10^4 \text{ N/m}$$

The greater the spring constant (k), the stiffer the spring; hence a greater force is required to stretch or compress the spring. When force is greater, acceleration is greater and the amount of time required for a single cycle should decrease (assuming that the amplitude remains constant). Thus, for a given amplitude, a stiffer spring will take less time to complete one cycle of motion than one that is less stiff.

As with the pendulum, the equation for the period of a mass-spring system can be derived mathematically or found experimentally.

PERIOD OF A MASS-SPRING SYSTEM IN SIMPLE HARMONIC MOTION

$$T = 2\pi\sqrt{\frac{m}{k}}$$

period = 2π × square root of (mass divided by spring constant)

Note that changing the amplitude of the vibration does not affect the period. This statement is true only for systems and circumstances in which the spring obeys Hooke's law.

SAMPLE PROBLEM C

Simple Harmonic Motion of a Mass-Spring System

PROBLEM

The body of a 1275 kg car is supported on a frame by four springs. Two people riding in the car have a combined mass of 153 kg. When driven over a pothole in the road, the frame vibrates with a period of 0.840 s. For the first few seconds, the vibration approximates simple harmonic motion. Find the spring constant of a single spring.

SOLUTION

Given: $m = \frac{(1275 \text{ kg} + 153 \text{ kg})}{4} = 357 \text{ kg}$ $T = 0.840 \text{ s}$

Unknown: $k = ?$

Use the equation for the period of a mass-spring system, and solve for k.

$$T = 2\pi\sqrt{\frac{m}{k}}$$

$$T^2 = 4\pi^2\left(\frac{m}{k}\right)$$

$$k = \frac{4\pi^2 m}{T^2} = \frac{4\pi^2(357 \text{ kg})}{(0.840 \text{ s})^2}$$

$$\boxed{k = 2.00 \times 10^4 \text{ N/m}}$$

PRACTICE C

Simple Harmonic Motion of a Mass-Spring System

1. A mass of 0.30 kg is attached to a spring and is set into vibration with a period of 0.24 s. What is the spring constant of the spring?
2. When a mass of 25 g is attached to a certain spring, it makes 20 complete vibrations in 4.0 s. What is the spring constant of the spring?
3. A 125 N object vibrates with a period of 3.56 s when hanging from a spring. What is the spring constant of the spring?
4. When two more people get into the car described in Sample Problem C, the total mass of all four occupants of the car becomes 255 kg. Now what is the period of vibration of the car when it is driven over a pothole in the road?
5. A spring of spring constant 30.0 N/m is attached to different masses, and the system is set in motion. Find the period and frequency of vibration for masses of the following magnitudes:
   a. 2.3 kg
   b. 15 g
   c. 1.9 kg

SECTION REVIEW

1. The reading on a metronome indicates the number of oscillations per minute. What are the frequency and period of the metronome's vibration when the metronome is set at 180?
2. A child swings on a playground swing with a 2.5 m long chain.
   a. What is the period of the child's motion?
   b. What is the frequency of vibration?
3. A 0.75 kg mass attached to a vertical spring stretches the spring 0.30 m.
   a. What is the spring constant?
   b. The mass-spring system is now placed on a horizontal surface and set vibrating. What is the period of the vibration?
4. **Critical Thinking** Two mass-spring systems vibrate with simple harmonic motion. If the spring constants are equal and the mass of one system is twice that of the other, which system has a greater period?

SECTION 2

PROBLEM GUIDE C

Use this guide to assign problems.
SE = Student Edition Textbook
PW = Problem Workbook
PB = Problem Bank on the One-Stop Planner (OSP)

Solving for:

k	**SE** Sample, 1–3 **PW** 4 **PB** 4–6
T, f	**SE** 4, 5; Ch. Rvw. 21 **PW** Sample, 1, 2, 3* **PB** 7–10
m	**PW** 5, 6 **PB** Sample, 1–3

***Challenging Problem**
Consult the printed Solutions Manual or the OSP for detailed solutions.

ANSWERS

Practice C

1. 2.1×10^2 N/m
2. 25 N/m
3. 39.7 N/m
4. 0.869 s
5. **a.** 1.7 s, 0.59 Hz
   b. 0.14 s, 7.1 Hz
   c. 1.6 s, 0.62 Hz

SECTION REVIEW ANSWERS

1. 3.0 Hz, 0.33 s
2. **a.** 3.2 s
   b. 0.31 Hz
3. **a.** 25 N/m
   b. 1.1 s
4. The system with the larger mass has a greater period.

SECTION 3
General Level

Demonstration

Wave Motion — BASIC

Purpose Distinguish between wave motion and particle vibration.

Materials long, coiled spring

Procedure Stretch the spring about 5–10 m. Have a student hold one end of the spring securely. As you hold the other end, send a single pulse along the spring by quickly jerking the spring sideways and then back to its equilibrium position. This will produce a transverse pulse. Have students note that the wave pulse is moving from you toward the student and that no part of the spring is being carried from you to the student. Emphasize this by having a student attach a small piece of paper to one of the coils of the spring.

Then send another pulse along the spring and show that while the wave moves along the spring, the coil returns to its original position. Tell students that this is an example of a mechanical wave whose medium is the spring.

Teaching Tip

Electromagnetic waves are discussed in Section 1 of the chapter "Light and Reflection." They are covered in greater detail in Section 4 of the chapter "Electromagnetic Induction."

SECTION 3

Properties of Waves

SECTION OBJECTIVES

- **Distinguish local particle vibrations from overall wave motion.**
- **Differentiate between pulse waves and periodic waves.**
- **Interpret waveforms of transverse and longitudinal waves.**
- **Apply the relationship among wave speed, frequency, and wavelength to solve problems.**
- **Relate energy and amplitude.**

WAVE MOTION

Consider what happens to the surface of a pond when you drop a pebble into the water. The disturbance created by the pebble generates water waves that travel away from the disturbance, as seen in **Figure 8.** If you examined the motion of a leaf floating near the disturbance, you would see that the leaf moves up and down and back and forth about its original position. However, the leaf does not undergo any net displacement from the motion of the waves.

The leaf's motion indicates the motion of the particles in the water. The water molecules move locally, like the leaf does, but they do not travel across the pond. That is, the water wave moves from one place to another, but the water itself is not carried with it.

Figure 8
A pebble dropped into a pond creates ripple waves similar to those shown here.

A wave is the motion of a disturbance

Ripple waves in a pond start with a disturbance at some point in the water. This disturbance causes water on the surface near that point to move, which in turn causes points farther away to move. In this way, the waves travel outward in a circular pattern away from the original disturbance.

In this example, the water in the pond is the **medium** through which the disturbance travels. Particles in the medium—in this case, water molecules—move in vertical circles as waves pass. Note that the medium does not actually travel with the waves. After the waves have passed, the water returns to its original position.

medium
a physical environment through which a disturbance can travel

Waves of almost every kind require a material medium in which to travel. Sound waves, for example, cannot travel through outer space, because space is very nearly a vacuum. In order for sound waves to travel, they must have a medium such as air or water. Waves that require a material medium are called **mechanical waves.**

mechanical wave
a wave that requires a medium through which to travel

Not all wave propagation requires a medium. Electromagnetic waves, such as visible light, radio waves, microwaves, and X rays, can travel through a vacuum. You will study electromagnetic waves in later chapters.

WAVE TYPES

One of the simplest ways to demonstrate wave motion is to flip one end of a taut rope whose opposite end is fixed, as shown in **Figure 9.** The flip of your wrist creates a pulse that travels to the fixed end with a definite speed. A wave that consists of a single traveling pulse is called a *pulse wave.*

Figure 9
A single flip of the wrist creates a pulse wave on a taut rope.

Now imagine that you continue to generate pulses at one end of the rope. Together, these pulses form what is called a *periodic wave.* Whenever the source of a wave's motion is a periodic motion, such as the motion of your hand moving up and down repeatedly, a periodic wave is produced.

Sine waves describe particles vibrating with simple harmonic motion

Figure 10 depicts a blade that vibrates with simple harmonic motion and thus makes a periodic wave on a string. As the wave travels to the right, any single point on the string vibrates up and down. Because the blade is vibrating with simple harmonic motion, the vibration of each point of the string is also simple harmonic. A wave whose source vibrates with simple harmonic motion is called a *sine wave.* Thus, a sine wave is a special case of a periodic wave in which the periodic motion is simple harmonic. The wave in **Figure 10** is called a sine wave because a graph of the trigonometric function $y = \sin x$ produces this curve when plotted.

A close look at a single point on the string illustrated in **Figure 10** shows that its motion resembles the motion of a mass hanging from a vibrating spring. As the wave travels to the right, the point vibrates around its equilibrium position with simple harmonic motion. This relationship between simple harmonic motion and wave motion enables us to use some of the terms and concepts from simple harmonic motion in our study of wave motion.

SCILINKS.
www.scilinks.org
Topic: Wave Motion
Code: HF61639

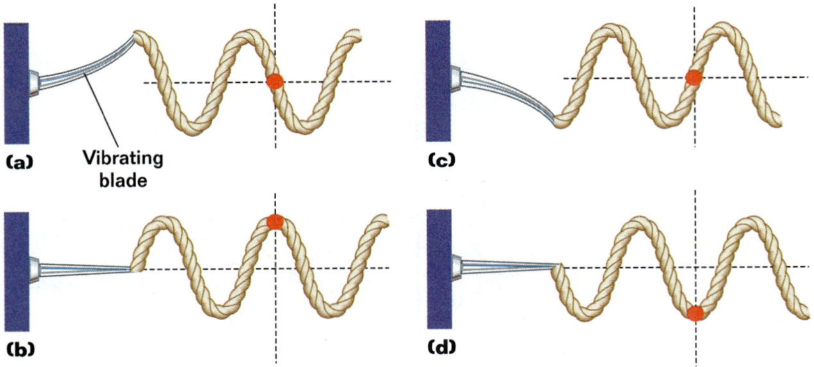

Figure 10
As the sine wave created by this vibrating blade travels to the right, a single point on the string vibrates up and down with simple harmonic motion.

Demonstration

Transverse Waves GENERAL

Purpose Demonstrate that in a transverse pulse, particle vibration and wave motion are perpendicular to each other.

Materials long, coiled spring

Procedure Generate a pulse as you did in the demonstration "Wave Motion" (on the previous page). Have the students point in the direction in which the spring is displaced *(perpendicular to the spring).* Then, have them point in the direction in which the wave moves along the spring *(parallel to the spring).* The students should see that the disturbance of the spring is perpendicular to the direction of the motion of the disturbance along the spring. In other words, the medium is displaced perpendicular to the direction of the motion of the pulse. Tell students this is an example of a transverse pulse.

Visual Strategy BASIC

Figure 10

Ask the students to visualize a particle attached to the red point on the wave. As the wave travels, the particle would, like a mass on a spring, vibrate with simple harmonic motion.

Q For each component of the figure, determine in what part of its cycle the vibrating particle would be.

A (a) *equilibrium*
(b) *maximum displacement*
(c) *equilibrium*
(d) *maximum displacement*

Visual Strategy GENERAL

Figure 11

Make sure students understand the meanings of displacement, amplitude, and wavelength.

Q Ask students to sketch graphs describing the waves that will be produced if a hand shaking the rope makes the following changes without changing the frequency of the waves:

(a) makes motions that are twice as high

(b) starts shaking the rope in the opposite direction

Compare the displacements, amplitudes, and wavelengths of each graph with those in the original case (**Figure 11**).

A **(a)** *The wavelength doesn't change, the amplitude is twice the original, and the displacement is twice the original at each point. Thus, the crests of this graph are twice as high and the troughs are twice as low, but they occur at the same positions along the x-axis.*

(b) *The amplitude and wavelength are the same as the original, but the displacements have opposite signs. This graph is a mirror image of the original one.*

Vibrations of a transverse wave are perpendicular to the wave motion

Figure 11(a) is a representation of the wave shown in **Figure 10** (on the previous page) at a specific instant of time, *t*. This wave travels to the right as the particles of the rope vibrate up and down. Thus, the vibrations are perpendicular to the direction of the wave's motion. A wave such as this, in which the particles of the disturbed medium move perpendicularly to the wave motion, is called a **transverse wave.**

transverse wave

a wave whose particles vibrate perpendicularly to the direction the wave is traveling

The wave shown in **Figure 11(a)** can be represented on a coordinate system, as shown in **Figure 11(b).** A picture of a wave like the one in **Figure 11(b)** is sometimes called a *waveform.* A waveform can represent either the displacements of each point of the wave at a single moment in time or the displacements of a single particle as time passes.

In this case, the waveform depicts the displacements at a single instant. The x-axis represents the equilibrium position of the string, and the y coordinates of the curve represent the displacement of each point of the string at time t. For example, points where the curve crosses the x-axis (where $y = 0$) have zero displacement. Conversely, at the highest and lowest points of the curve, where displacement is greatest, the absolute values of y are greatest.

Wave measures include crest, trough, amplitude, and wavelength

A wave can be measured in terms of its displacement from equilibrium. The highest point above the equilibrium position is called the wave **crest.** The lowest point below the equilibrium position is the **trough** of the wave. As in simple harmonic motion, amplitude is a measure of maximum displacement from equilibrium. The amplitude of a wave is the distance from the equilibrium position to a crest or to a trough, as shown in **Figure 11(b).**

crest

the highest point above the equilibrium position

trough

the lowest point below the equilibrium position

As a wave passes a given point along its path, that point undergoes cyclical motion. The point is displaced first in one direction and then in the other direction. Finally, the point returns to its original equilibrium position, thereby completing one cycle. The distance the wave travels along its path during one cycle is called the **wavelength,** λ (the Greek letter *lambda*). A simple way to find the wavelength is to measure the distance between two adjacent similar points of the wave, such as from crest to crest or from trough to trough. Notice in **Figure 11(b)** that the distances between adjacent crests or troughs in the waveform are equal.

wavelength

the distance between two adjacent similar points of a wave, such as from crest to crest or from trough to trough

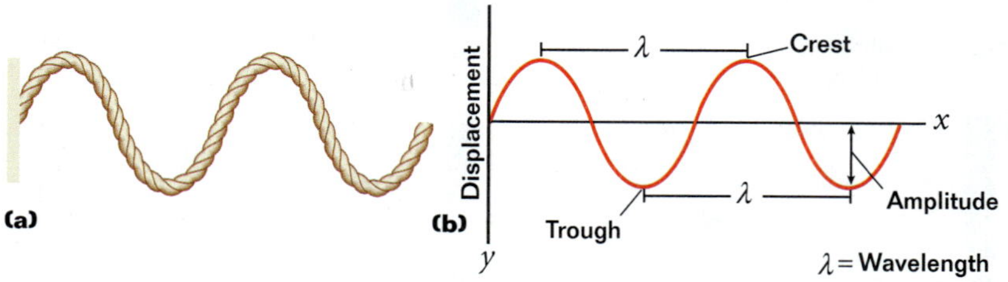

Figure 11
(a) A picture of a transverse wave at some instant *t* can be turned into **(b)** a graph. The *x*-axis represents the equilibrium position of the string. The curve shows the displacements of the string at time *t*.

Vibrations of a longitudinal wave are parallel to the wave motion

You can create another type of wave with a spring. Suppose that one end of the spring is fixed and that the free end is pumped back and forth along the length of the spring, as shown in **Figure 12.** This action produces compressed and stretched regions of the coil that travel along the spring. The displacement of the coils is in the direction of wave motion. In other words, the vibrations are parallel to the motion of the wave.

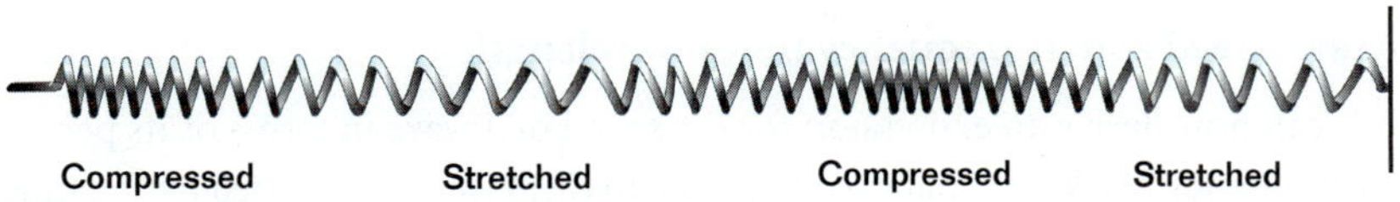

Figure 12
As this wave travels to the right, the coils of the spring are tighter in some regions and looser in others. The displacement of the coils is parallel to the direction of wave motion, so this wave is longitudinal.

When the particles of the medium vibrate parallel to the direction of wave motion, the wave is called a **longitudinal wave.** Sound waves in the air are longitudinal waves because air particles vibrate back and forth in a direction parallel to the direction of wave motion.

longitudinal wave

a wave whose particles vibrate parallel to the direction the wave is traveling

A longitudinal wave can also be described by a sine curve. Consider a longitudinal wave traveling on a spring. **Figure 13(a)** is a snapshot of the longitudinal wave at some instant *t*, and **Figure 13(b)** shows the sine curve representing the wave. The compressed regions correspond to the crests of the waveform, and the stretched regions correspond to troughs.

The type of wave represented by the curve in **Figure 13(b)** is often called a *density wave* or a *pressure wave.* The crests, where the spring coils are compressed, are regions of high density and pressure (relative to the equilibrium density or pressure of the medium). Conversely, the troughs, where the coils are stretched, are regions of low density and pressure.

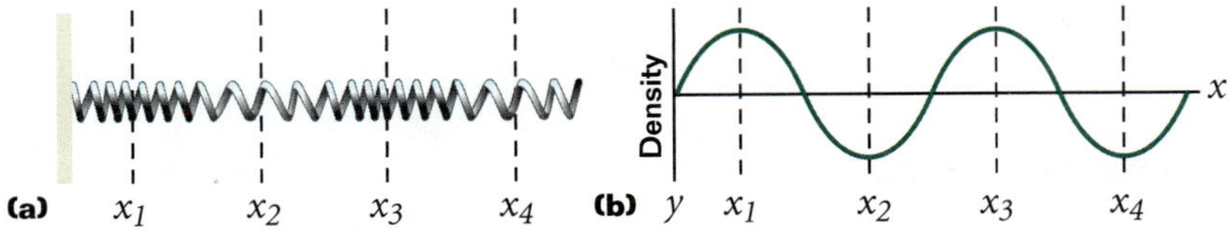

Figure 13
(a) A longitudinal wave at some instant *t* can also be represented by **(b)** a graph. The crests of this waveform correspond to compressed regions, and the troughs correspond to stretched regions.

PERIOD, FREQUENCY, AND WAVE SPEED

Sound waves may begin with the vibrations of your vocal cords, a guitar string, or a taut drumhead. In each of these cases, the source of wave motion is a vibrating object. The vibrating object that causes a sine wave always has a characteristic frequency. Because this motion is transferred to the particles of the medium, the frequency of vibration of the particles is equal to the frequency of the source. When the vibrating particles of the medium complete one full cycle, one complete wavelength passes any given point. Thus, wave frequency describes the number of waves that pass a given point in a unit of time.

Demonstration

Longitudinal Waves GENERAL

Purpose Demonstrate that in a longitudinal pulse, particle vibration and wave motion are parallel.

Materials long, coiled spring

Procedure With the spring lying flat on the floor, compress approximately 10 cm of the coil. Instruct the students to observe the spring and to listen carefully as you release the pulse. Ask them to indicate the direction of the displacement of the spring and the direction in which the disturbance moved along the spring.

The students should see and hear that the displacement of the spring and the motion of the displacement along the spring are in the same direction. In this case, the direction in which the medium is disturbed is the same as the direction in which the disturbance moves through the medium. This is an example of a longitudinal pulse.

Misconception Alert

Some students may confuse the graph of a transverse pulse, **Figure 11(b),** with the graph of a longitudinal pulse, **Figure 13(b).** Point out that in the first case, the *y*-axis represents *displacement,* while in the latter case, the *y*-axis represents *density.* Although the graphs look similar, this difference must be kept in mind when interpreting the two different kinds of graphs.

Demonstration

Amplitude, Wavelength, and Wave Speed — GENERAL

Purpose Show that wave speed is independent of amplitude and wavelength.

Materials long, coiled spring and a clock or stopwatch

Procedure Generate a transverse pulse with a small amplitude, and have a student record the time it takes for the pulse to travel the length of the spring. Repeat this process for a pulse with a larger amplitude. Have students compare the two times *(they should be approximately equal)*, and ask what conclusion they can draw from this observation *(that wave speed is independent of amplitude)*. Repeat the process, but vary the wavelength rather than the amplitude by varying the frequency. Have students compare the times *(they should be approximately equal)*, and ask what conclusion they can draw from this observation *(that wave speed is independent of wavelength)*.

Teaching Tip

Although this chapter primarily deals with mechanical waves, the wave-speed equation holds true for mechanical and electromagnetic waves, and both are included in practice problems.

Did you know?

The frequencies of sound waves that are audible to humans range from 20 Hz to 20 000 Hz. Electromagnetic waves, which include visible light, radio waves, and microwaves, have an even broader range of frequencies—from about 10^4 Hz and lower to 10^{25} Hz and higher.

The period of a wave is the time required for one complete cycle of vibration of the medium's particles. As the particles of the medium complete one full cycle of vibration at any point of the wave, one wavelength passes by that point. Thus, the period of a wave describes the time it takes for a complete wavelength to pass a given point. The relationship between period and frequency seen earlier in this chapter holds true for waves as well; the period of a wave is inversely related to its frequency.

Wave speed equals frequency times wavelength

We can now derive an expression for the speed of a wave in terms of its period or frequency. We know that speed is equal to displacement divided by the time it takes to undergo that displacement.

$$v = \frac{\Delta x}{\Delta t}$$

For waves, a displacement of one wavelength (λ) occurs in a time interval equal to one period of the vibration (T).

$$v = \frac{\lambda}{T}$$

As you saw earlier in this chapter, frequency and period are inversely related.

$$f = \frac{1}{T}$$

Substituting this frequency relationship into the previous equation for speed gives a new equation for the speed of a wave.

$$v = \frac{\lambda}{T} = f\lambda$$

SPEED OF A WAVE

$$v = f\lambda$$

speed of a wave = frequency × wavelength

The speed of a mechanical wave is constant for any given medium. For example, at a concert, sound waves from different instruments reach your ears at the same moment, even when the frequencies of the sound waves are different. Thus, although the frequencies and wavelengths of the sounds produced by each instrument may be different, the product of the frequency and wavelength is always the same at the same temperature. As a result, when a mechanical wave's frequency is increased, its wavelength must decrease in order for its speed to remain constant. The speed of a wave changes only when the wave moves from one medium to another or when certain properties of the medium (such as temperature) are varied.

SAMPLE PROBLEM D

Wave Speed

PROBLEM

A piano string tuned to middle C vibrates with a frequency of 262 Hz. Assuming the speed of sound in air is 343 m/s, find the wavelength of the sound waves produced by the string.

SOLUTION

Given: $v = 343$ m/s $f = 262$ Hz

Unknown: $\lambda = ?$

Use the equation relating speed, wavelength, and frequency for a wave.

$$v = f\lambda$$

$$\lambda = \frac{v}{f} = \frac{343 \text{ m/s}}{262 \text{ Hz}} = \frac{343 \text{ m} \bullet \text{s}^{-1}}{262 \text{ s}^{-1}}$$

$$\boxed{\lambda = 1.31 \text{ m}}$$

PRACTICE D

Wave Speed

1. A piano emits frequencies that range from a low of about 28 Hz to a high of about 4200 Hz. Find the range of wavelengths in air attained by this instrument when the speed of sound in air is 340 m/s.
2. The speed of all electromagnetic waves in empty space is 3.00×10^8 m/s. Calculate the wavelength of electromagnetic waves emitted at the following frequencies:
   a. radio waves at 88.0 MHz
   b. visible light at 6.0×10^8 MHz
   c. X rays at 3.0×10^{12} MHz
3. The red light emitted by a He-Ne laser has a wavelength of 633 nm in air and travels at 3.00×10^8 m/s. Find the frequency of the laser light.
4. A tuning fork produces a sound with a frequency of 256 Hz and a wavelength in air of 1.35 m.
   a. What value does this give for the speed of sound in air?
   b. What would be the wavelength of this same sound in water in which sound travels at 1500 m/s?

PROBLEM GUIDE D

Use this guide to assign problems.
SE = Student Edition Textbook
PW = Problem Workbook
PB = Problem Bank on the One-Stop Planner (OSP)

Solving for:

λ	**SE** Sample, 1–2, 4b*; Ch. Rvw. 35, 48 **PW** 4 **PB** 6, 7
f	**SE** 3 **PW** Sample, 1–3 **PB** 8–10
v	**SE** 4a **PW** 5, 6 **PB** Sample, 1–5

***Challenging Problem**
Consult the printed Solutions Manual or the OSP for detailed solutions.

ANSWERS

Practice D

1. $0.081 \text{ m} \le \lambda \le 12 \text{ m}$
2. **a.** 3.41 m
   b. 5.0×10^{-7} m
   c. 1.0×10^{-10} m
3. 4.74×10^{14} Hz
4. **a.** 346 m/s
   b. 5.86 m

SECTION 3

The Language of Physics

Students may be familiar with the term *damping* in a musical context, such as damping the sound from a guitar string or a drumhead. Point out that this is essentially the same usage of the word in physics. In these musical examples, when a sound wave is damped, the amplitude of the sound wave decreases. As a result, the sound becomes softer.

SECTION REVIEW ANSWERS

1. The disturbance moves, not the medium.
2. **a.** One portion of the spring should have a single compressed region and a single stretched region.
   b. The spring should have several compressed regions and several stretched regions.
   c. The spring should contain a single hump either above or below its equilibrium position.
   d. The spring should contain several humps above and below its equilibrium position.
3. The graph for **(b)** should look like **Figure 11(b)** but shoud have a *y*-axis labeled *density.* The graph for **(d)** should resemble **Figure 11(b).**
4. The energy will be 16 times as great.
5. 6.0×10^4 Hz

extension

Integrating Earth Science

Visit go.hrw.com for the activity "Earthquake Waves."

Keyword HF6VIBX

Waves transfer energy

When a pebble is dropped into a pond, the water wave that is produced carries a certain amount of energy. As the wave spreads to other parts of the pond, the energy likewise moves across the pond. Thus, the wave transfers energy from one place in the pond to another while the water remains in essentially the same place. In other words, waves transfer energy by the vibration of matter rather than by the transfer of matter itself. For this reason, waves are often able to transport energy efficiently.

The rate at which a wave transfers energy depends on the amplitude at which the particles of the medium are vibrating. The greater the amplitude, the more energy a wave carries in a given time interval. For a mechanical wave, the energy transferred is proportional to the square of the wave's amplitude. When the amplitude of a mechanical wave is doubled, the energy it carries in a given time interval increases by a factor of four. Conversely, when the amplitude is halved, the energy decreases by a factor of four.

As with a mass-spring system or a simple pendulum, the amplitude of a wave gradually diminishes over time as its energy is dissipated. This effect, called *damping,* is usually minimal over relatively short distances. For simplicity, we have disregarded damping in our analysis of wave motions.

SECTION REVIEW

1. As waves pass by a duck floating on a lake, the duck bobs up and down but remains in essentially one place. Explain why the duck is not carried along by the wave motion.
2. Sketch each of the following waves that are on a spring that is attached to a wall at one end:
   a. a pulse wave that is longitudinal
   b. a periodic wave that is longitudinal
   c. a pulse wave that is transverse
   d. a periodic wave that is transverse
3. Draw a graph for each of the waves described in items **(b)** and **(d)** above, and label the *y*-axis of each graph with the appropriate variable. Label the following on each graph: crest, trough, wavelength, and amplitude.
4. If the amplitude of a sound wave is increased by a factor of four, how does the energy carried by the sound wave in a given time interval change?
5. The smallest insects that a bat can detect are approximately the size of one wavelength of the sound the bat makes. What is the minimum frequency of sound waves required for the bat to detect an insect that is 0.57 cm long? (Assume the speed of sound is 340 m/s.)

Wave Interactions

SECTION OBJECTIVES

- **Apply the superposition principle.**
- **Differentiate between constructive and destructive interference.**
- **Predict when a reflected wave will be inverted.**
- **Predict whether specific traveling waves will produce a standing wave.**
- **Identify nodes and antinodes of a standing wave.**

WAVE INTERFERENCE

When two bumper boats collide, as shown in **Figure 14,** each bounces back in another direction. The two bumper boats cannot occupy the same space, and so they are forced to change the direction of their motion. This is true not just of bumper boats but of all matter. Two different material objects can never occupy the same space at the same time.

Figure 14
Two of these bumper boats cannot be in the same place at one time. Waves, on the other hand, can pass through one another.

When two waves come together, they do not bounce back as bumper boats do. If you listen carefully at a concert, you can distinguish the sounds of different instruments. Trumpet sounds are different from flute sounds, even when the two instruments are played at the same time. The sound waves of each instrument are unaffected by the other waves that are passing through the same space at the same moment. Because mechanical waves are not matter but rather are displacements of matter, two waves can occupy the same space at the same time. The combination of two overlapping waves is called *superposition.*

Figure 15 shows two sets of water waves in a ripple tank. As the waves move outward from their respective sources, they pass through one another. As they pass through one another, the waves interact to form an *interference pattern* of light and dark bands. Although this superposition of mechanical waves is fairly easy to observe, these are not the only kind of waves that can pass through the same space at the same time. Visible light and other forms of electromagnetic radiation also undergo superposition, and they can interact to form interference patterns.

Figure 15
This ripple tank demonstrates the interference of water waves.

Demonstration

Wave Superposition — GENERAL

Purpose Show that the amplitudes of traveling waves add as the waves cross one another.

Materials long, coiled spring

Procedure Hold one end of the spring, and have a student hold the opposite end. Generate a transverse pulse, and have the students observe its motion along the spring. After the pulse has dissipated, tell the student at the opposite end to generate an identical pulse. Ask students to predict what will happen when you and the student generate pulses simultaneously, and then do so. Point out that the pulses create a larger disturbance at the point where they cross each other along the spring. How can you tell that the pulses are passing through each other when they collide, not bouncing off each other? To confirm that pulses pass through one another, send two pulses of visibly different amplitudes toward each other.

Demonstration

Waves Passing Each Other

BASIC

Purpose Show that wave pulses are unaffected after they pass through one another.

Materials long, coiled spring

Procedure Repeat the demonstration "Wave Superposition" (on the previous page), but in this case, the two pulses should have opposite displacements. Have the students observe the pulse that reaches your hand after the two pulses have passed through each other; repeat this demonstration several times. Next repeat the process for two pulses of different amplitudes, and also repeat the process for pulses with displacements on the same side. Guide the students to conclude that in all cases the two pulses passed through each other and were unaffected by the presence of the other pulse. Also have students determine which examples were constructive *(displacements on the same side)* and which were destructive *(displacements on opposite sides)*.

Displacements in the same direction produce constructive interference

In **Figure 16(a),** two wave pulses are traveling toward each other on a stretched rope. The larger pulse is moving to the right, while the smaller pulse moves toward the left. At the moment the two wave pulses meet, a resultant wave is formed, as shown in **Figure 16(b).**

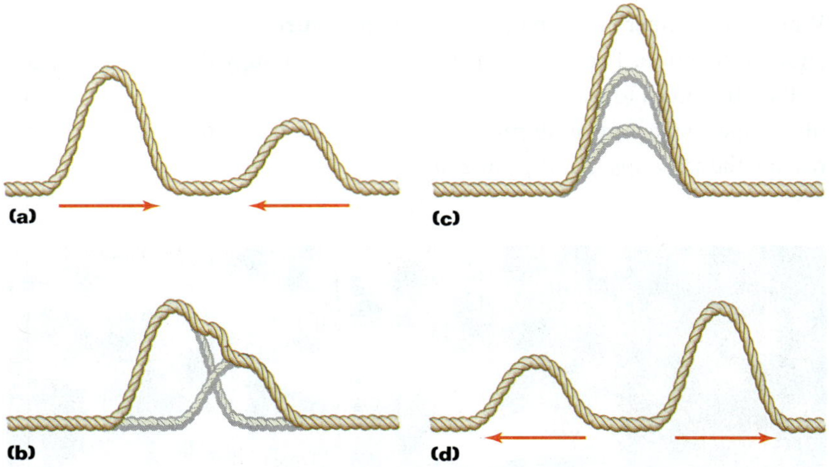

Figure 16
When these two wave pulses meet, the displacements at each point add up to form a resultant wave. This is an example of constructive interference.

At each point along the rope, the displacements due to the two pulses are added together, and the result is the displacement of the resultant wave. For example, when the two pulses exactly coincide, as they do in **Figure 16(c),** the amplitude of the resultant wave is equal to the sum of the amplitudes of each pulse. This method of summing the displacements of waves is known as the *superposition principle.* According to this principle, when two or more waves travel through a medium at the same time, the resultant wave is the sum of the displacements of the individual waves at each point. Ideally, the superposition principle holds true for all types of waves, both mechanical and electromagnetic. However, experiments show that in reality the superposition principle is valid only when the individual waves have small amplitudes—an assumption we make in all our examples.

Notice that after the two pulses pass through each other, each pulse has the same shape it had before the waves met and each is still traveling in the same direction, as shown in **Figure 16(d).** This is true for sound waves at a concert, water waves in a pond, light waves, and other types of waves. Each wave maintains its own characteristics after interference, just as the two pulses do in our example above.

You have seen that when more than one wave travels through the same space at the same time, the resultant wave is equal to the sum of the individual displacements. If the displacements are on the same side of equilibrium, as in **Figure 16,** they have the same sign. When added together, the resultant wave is larger than the individual displacements. This is called **constructive interference.**

constructive interference

a superposition of two or more waves in which individual displacements on the same side of the equilibrium position are added together to form the resultant wave

Displacements in opposite directions produce destructive interference

What happens if the pulses are on opposite sides of the equilibrium position, as they are in **Figure 17(a)**? In this case, the displacements have different signs, one positive and one negative. When the positive and negative displacements are added, as shown in **Figure 17(b)** and **(c),** the resultant wave is the difference between the pulses. This is called **destructive interference.** After the pulses separate, their shapes are unchanged, as seen in **Figure 17(d).**

destructive interference

a superposition of two or more waves in which individual displacements on opposite sides of the equilibrium position are added together to form the resultant wave

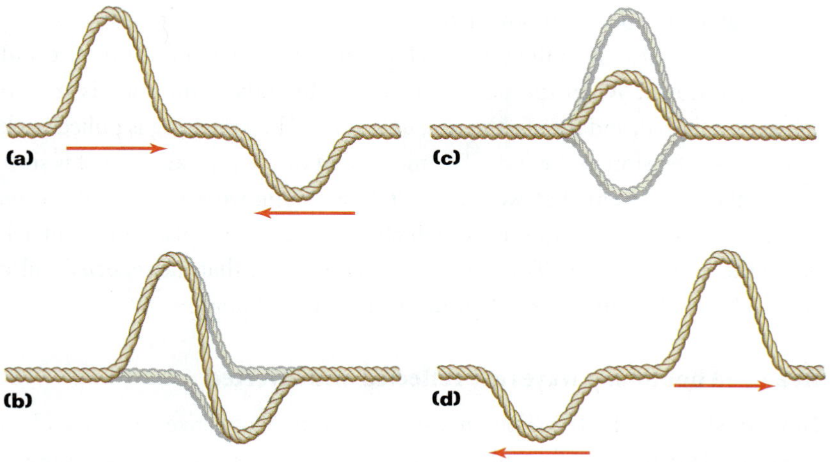

Figure 17
In this case, known as destructive interference, the displacement of one pulse is subtracted from the displacement of the other.

Figure 18 shows two pulses of equal amplitude but with displacements of opposite signs. When the two pulses coincide and the displacements are added, the resultant wave has a displacement of zero. In other words, at the instant the two pulses overlap, they completely cancel each other; it is as if there were no disturbance at all. This situation is known as *complete destructive interference.*

If these waves were water waves coming together, one of the waves would be acting to pull an individual drop of water upward at the same instant and with the same force that another wave would be acting to pull it downward. The result would be no net force on the drop, and there would be no net displacement of the water at that moment.

Thus far, we have considered the interference produced by two transverse pulse waves. The superposition principle is valid for longitudinal waves as well. In a *compression,* particles are moved closer together, while in a *rarefaction,* particles are spread farther apart. So, when a compression and a rarefaction interfere, there is destructive interference.

In our examples, we have considered constructive and destructive interference separately, and we have dealt only with pulse waves. With periodic waves, complicated patterns arise that involve regions of constructive and destructive interference. The locations of these regions may remain fixed or may vary with time as the individual waves travel.

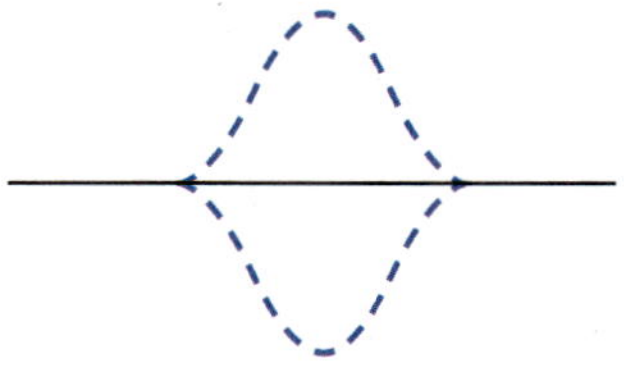

Figure 18
The resultant displacement at each point of the string is zero, so the two pulses cancel one another. This is complete destructive interference.

Misconception Alert

Students may believe that a new wave is created that replaces the original two wave pulses. Common sense and daily experience with collisions that *do* affect the colliding objects support this preconception.

Nevertheless, the *resultant wave* occurs only at the time and place that the waves meet each other. Each wave keeps its characteristics and continues with its original speed and wavelength after the encounter.

Demonstration

Wave Reflection — BASIC

Purpose Demonstrate that a wave reflected by a fixed boundary is inverted.

Materials long, coiled spring

Procedure Fix one end of the spring, and hold the other end in your hand. Send a transverse pulse along the spring, and ask students to observe the reflected pulse. Students should note that the reflected pulse is inverted.

ADVANCED TOPICS

See "De Broglie Waves" in **Appendix J: Advanced Topics** to learn about the wave characteristics that all matter exhibits at the microscopic level.

REFLECTION

In our discussion of waves so far, we have assumed that the waves being analyzed could travel indefinitely without striking anything that would stop them or otherwise change their motion. But what happens to the motion of a wave when it reaches a boundary?

At a free boundary, waves are reflected

Consider a pulse wave traveling on a stretched rope whose end forms a ring around a post, as shown in **Figure 19(a).** We will assume that the ring is free to slide along the post without friction.

As the pulse travels to the right, each point of the rope moves up once and then back down. When the pulse reaches the boundary, the rope is free to move up as usual, and it pulls the ring up with it. Then, the ring is pulled back down by the tension in the rope. The movement of the rope at the post is similar to the movement that would result if someone were to whip the rope upward to send a pulse to the left, which would cause a pulse to travel back along the rope to the left. This is called *reflection.* Note that the reflected pulse is upright and has the same amplitude as the incident pulse.

At a fixed boundary, waves are reflected and inverted

Now consider a pulse traveling on a stretched rope that is fixed at one end, as in **Figure 19(b).** When the pulse reaches the wall, the rope exerts an upward force on the wall, and the wall in turn exerts an equal and opposite reaction force on the rope. This downward force on the rope causes a displacement in the direction opposite the displacement of the original pulse. As a result, the pulse is inverted after reflection.

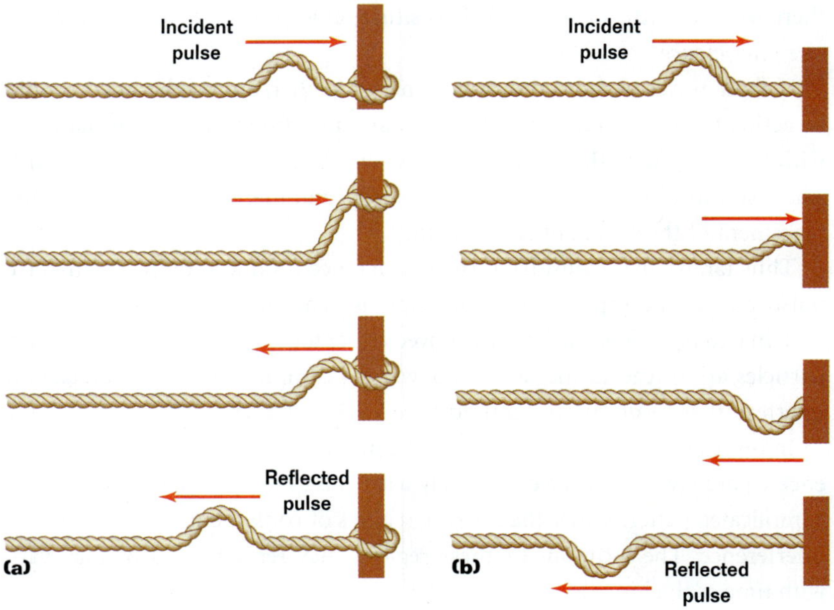

Figure 19
(a) When a pulse travels down a rope whose end is free to slide up the post, the pulse is reflected from the free end. **(b)** When a pulse travels down a rope that is fixed at one end, the reflected pulse is inverted.

STANDING WAVES

Consider a string that is attached on one end to a rigid support and that is shaken up and down in a regular motion at the other end. The regular motion produces waves of a certain frequency, wavelength, and amplitude traveling down the string. When the waves reach the other end, they are reflected back toward the oncoming waves. If the string is vibrated at exactly the right frequency, a **standing wave**—a resultant wave pattern that appears to be stationary on the string—is produced. The standing wave consists of alternating regions of constructive and destructive interference.

standing wave

a wave pattern that results when two waves of the same frequency, wavelength, and amplitude travel in opposite directions and interfere

Standing waves have nodes and antinodes

Figure 20(a) shows four possible standing waves for a given string length. The points at which complete destructive interference happens are called **nodes.** There is no motion in the string at the nodes. But midway between two adjacent nodes, the string vibrates with the largest amplitude. These points are called **antinodes.**

node

a point in a standing wave that maintains zero displacement

antinode

a point in a standing wave, halfway between two nodes, at which the largest displacement occurs

Figure 20(b) shows the oscillation of the second case shown in **Figure 20(a)** during half a cycle. All points on the string oscillate vertically with the same frequency, except for the nodes, which are stationary. In this case, there are three nodes (N) and two antinodes (A), as illustrated in the figure.

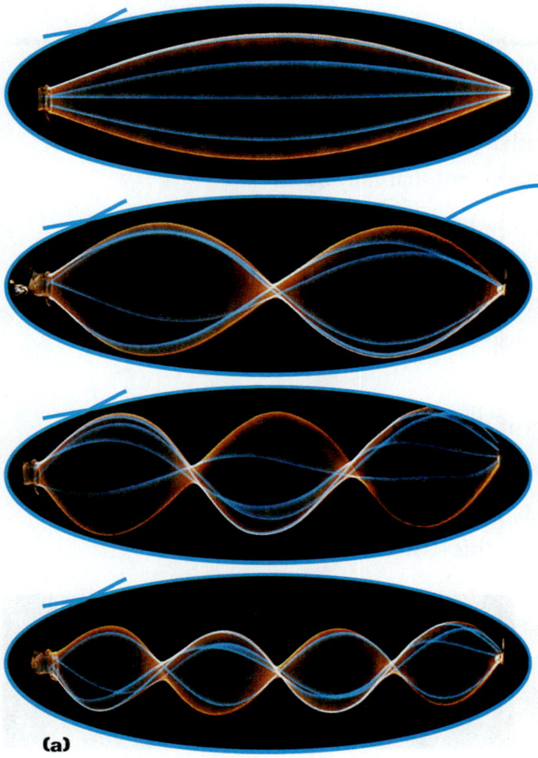
(a)

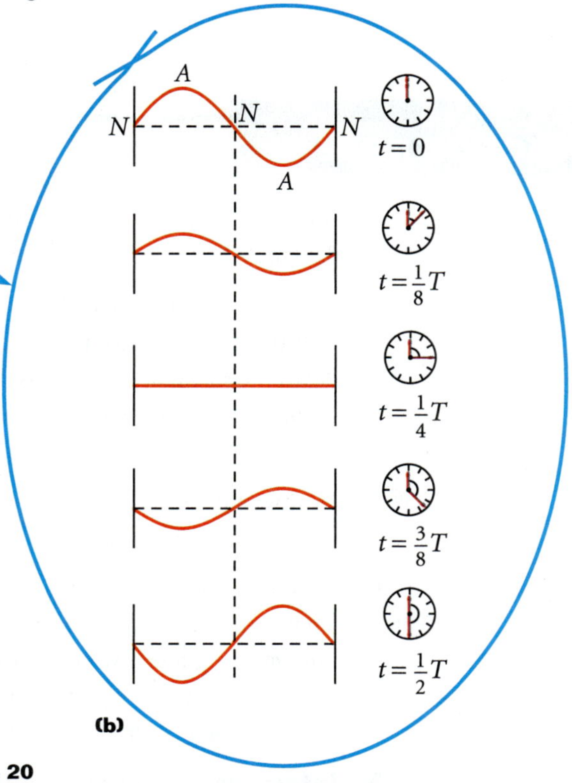

(b)

Figure 20
(a) This photograph shows four possible standing waves that can exist on a given string. **(b)** The diagram shows the progression of the second standing wave for one-half of a cycle.

SECTION 4

The Language of Physics

The term *standing wave* may mislead students. Point out that the individual waves that compose standing waves are actually *traveling* waves. It is only the resultant wave pattern, which is the superposition of various individual traveling waves, that appears to stand still.

Visual Strategy GENERAL

Figure 20

Make sure students understand that the waves in **Figure 20(a)** are examples of different possible standing waves for a given string length, while the diagrams in **Figure 20(b)** represent the vibrations of just one of these standing waves. Specifically, the diagram corresponds to the standing wave shown in the second photograph from the top.

Q Have students draw a schematic diagram, like the one shown in **Figure 20(b),** for the wave shown in the top photograph, in which the wavelength equals twice the string length.

A ***Students' diagrams should look like the left half (left of the dotted vertical line) of Figure 20(b).***

Demonstration

Standing Waves — GENERAL

Purpose Demonstrate longitudinal standing waves around a circle, and verify that only certain frequencies produce standing waves.

Materials small toy spring, 2 in. diameter cylinder, pen, overhead projector

Procedure Connect the toy spring end to end with a wire, and place it around the cylinder on the overhead projector. Brush the tip of the pen against the spring's coils. Have students measure the time it takes for the wave to travel around the spring, and calculate the frequency. Now brush the tip of the pen once back and forth with that frequency. Repeat with frequencies that are integral multiples of the original frequency. Guide the students to conclude that only certain frequencies of vibrations produce standing waves.

SECTION REVIEW ANSWERS

1. 0.50 m
2. 0.00 m; destructive
3. 0.30 m; constructive
4. Wavelengths that will produce standing waves include 4.0 m, 2.0 m, and 1.3 m; Any value that does not allow both ends of the string to be nodes is acceptable.
5. 4 nodes; 3 antinodes

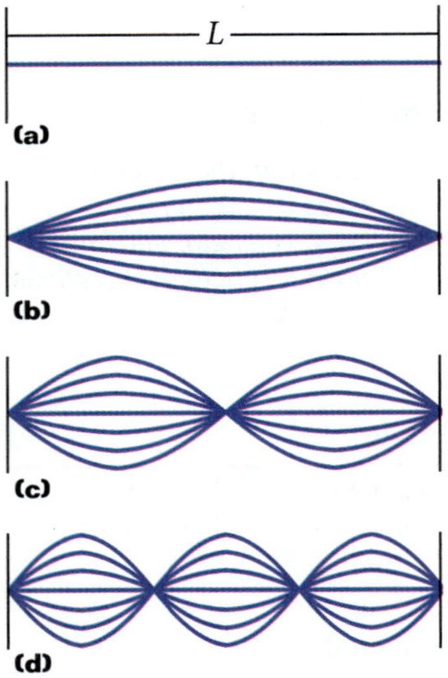

Figure 21
Only certain frequencies produce standing waves on this fixed string. The wavelength of these standing waves depends on the string length. Possible wavelengths include $2L$ **(b)**, L **(c)**, and $\frac{2}{3}L$ **(d)**.

Only certain frequencies, and therefore wavelengths, produce standing wave patterns. **Figure 21** shows standing waves for a given string length. In each case, the curves represent the position of the string at different instants of time. If the string were vibrating rapidly, the several positions would blur together and give the appearance of loops, like those shown in the diagram. A single loop corresponds to either a crest or trough alone, while two loops correspond to a crest and a trough together, or one wavelength.

The ends of the string must be nodes because these points cannot vibrate. As you can see in **Figure 21,** a standing wave can be produced for any wavelength that allows both ends of the string to be nodes. For example, in **Figure 21(b),** each end is a node, and there are no nodes in between. Because a single loop corresponds to either a crest or trough alone, this standing wave corresponds to one-half of a wavelength. Thus, the wavelength in this case is equal to twice the string length ($2L$).

The next possible standing wave, shown in **Figure 21(c),** has three nodes: one at either end and one in the middle. In this case, there are two loops, which correspond to a crest and a trough. Thus, this standing wave has a wavelength equal to the string length (L). The next case, shown in **Figure 21(d),** has a wavelength equal to two-thirds of the string length $\left(\frac{2}{3}L\right)$, and the pattern continues. Wavelengths between the values shown here do not produce standing waves because they allow only one end of the string to be a node.

SECTION REVIEW

1. A wave of amplitude 0.30 m interferes with a second wave of amplitude 0.20 m. What is the largest resultant displacement that may occur?
2. A string is rigidly attached to a post at one end. Several pulses of amplitude 0.15 m sent down the string are reflected at the post and travel back down the string without a loss of amplitude. What is the amplitude at a point on the string where the maximum displacement points of two pulses cross? What type of interference is this?
3. How would your answer to item 2 change if the same pulses were sent down a string whose end is free? What type of interference is this?
4. A stretched string fixed at both ends is 2.0 m long. What are three wavelengths that will produce standing waves on this string? Name at least one wavelength that would not produce a standing wave pattern, and explain your answer.
5. **Interpreting Graphics** Look at the standing wave shown in **Figure 22.** How many nodes does this wave have? How many antinodes?

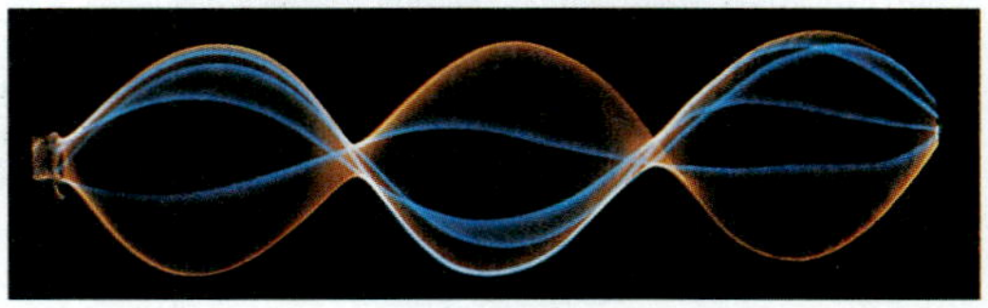

Figure 22

CHAPTER 11 Highlights

KEY IDEAS

Section 1 Simple Harmonic Motion

- In simple harmonic motion, restoring force is proportional to displacement.
- A mass-spring system vibrates with simple harmonic motion, and the spring force is given by Hooke's law.
- For small angles of displacement (<15°), a simple pendulum swings with simple harmonic motion.
- In simple harmonic motion, restoring force and acceleration are maximum at maximum displacement and speed is maximum at equilibrium.

Section 2 Measuring Simple Harmonic Motion

- The period of a mass-spring system depends only on the mass and the spring constant. The period of a simple pendulum depends only on the string length and the free-fall acceleration.
- Frequency is the inverse of period.

Section 3 Properties of Waves

- As a wave travels, the particles of the medium vibrate around an equilibrium position.
- In a transverse wave, vibrations are *perpendicular* to the direction of wave motion. In a longitudinal wave, vibrations are *parallel* to the direction of wave motion.
- Wave speed equals frequency times wavelength.

Section 4 Wave Interactions

- If two or more waves are moving through a medium, the resultant wave is found by adding the individual displacements together point by point.
- Standing waves are formed when two waves that have the same frequency, amplitude, and wavelength travel in opposite directions and interfere.

Variable Symbols

Quantities		Units	
$F_{elastic}$	spring force	N	newtons
k	spring constant	N/m	newtons/meter
T	period	s	seconds
f	frequency	Hz	hertz = s^{-1}
λ	wavelength	m	meters

KEY TERMS

simple harmonic motion (p. 369)

amplitude (p. 376)

period (p. 376)

frequency (p. 376)

medium (p. 382)

mechanical wave (p. 382)

transverse wave (p. 384)

crest (p. 384)

trough (p. 384)

wavelength (p. 384)

longitudinal wave (p. 385)

constructive interference (p. 390)

destructive interference (p. 391)

standing wave (p. 393)

node (p. 393)

antinode (p. 393)

PROBLEM SOLVING

See **Appendix D: Equations** for a summary of the equations introduced in this chapter. If you need more problem-solving practice, see **Appendix I: Additional Problems.**

CHAPTER 11

Highlights

Teaching Tip

Ask students to prepare a concept map of the chapter. The concept map should include most of the vocabulary terms, along with other integral terms or concepts.

CHAPTER 11

Review

ANSWERS

1. oscillation about an equilibrium position in which a restoring force is proportional to displacement
2. mass-spring system, child on a swing, pendulum of a grandfather clock, metronome
3. No, acceleration changes throughout the oscillator's motion. It is zero at the equilibrium position and greatest at maximum displacement.
4. No, a pendulum's displacement is approximately proportional to its restoring force only at angles smaller than 15°.
5. gravitational potential energy; When April lets go of the bob, $PE = \text{max}$ and $KE = 0$. At the bottom of its swing, $KE = \text{max}$ and $PE = 0$.
6. because frictional forces are neglected in an ideal mass-spring system
7. the tangent component; because it always pulls the bob toward the equilibrium position
8. 130 N/m
9. 580 N/m
10. twice
11. $4A$
12. They are inversely related.
13. becomes $\sqrt{2}$ times as long; remains the same because mass does not affect period

CHAPTER 11 Review

SIMPLE HARMONIC MOTION

Review Questions

1. What characterizes an object's motion as simple harmonic?
2. List four examples of simple harmonic motion.
3. Does the acceleration of a simple harmonic oscillator remain constant during its motion? Is the acceleration ever zero? Explain.
4. A pendulum is released 40° from its resting position. Is its motion simple harmonic?
5. April is about to release the bob of a pendulum. Before she lets go, what sort of potential energy does the bob have? How does the energy of the bob change as it swings through one full cycle of motion?

Conceptual Questions

6. An ideal mass-spring system vibrating with simple harmonic motion would oscillate indefinitely. Explain why.
7. In a simple pendulum, the weight of the bob can be divided into two components, one tangent to the bob's direction of motion and the other perpendicular to the bob's direction of motion. Which of these is the restoring force, and why?

Practice Problems

For problems 8–9, see Sample Problem A.

8. Janet wants to find the spring constant of a given spring, so she hangs the spring vertically and attaches a 0.40 kg mass to the spring's other end. If the spring stretches 3.0 cm from its equilibrium position, what is the spring constant?
9. In preparing to shoot an arrow, an archer pulls a bowstring back 0.40 m by exerting a force that increases uniformly from 0 to 230 N. What is the equivalent spring constant of the bow?

PERIOD AND FREQUENCY

Review Questions

10. A child swings on a playground swing. How many times does the child swing through the swing's equilibrium position during the course of a single period of motion?
11. What is the total distance traveled by an object moving back and forth in simple harmonic motion in a time interval equal to its period when its amplitude is equal to A?
12. How is the period of a simple harmonic vibration related to its frequency?

Conceptual Questions

13. What happens to the period of a simple pendulum when the pendulum's length is doubled? What happens when the suspended mass is doubled?
14. A pendulum bob is made with a ball filled with water. What would happen to the frequency of vibration of this pendulum if a hole in the ball allowed water to slowly leak out? (Treat the pendulum as a simple pendulum.)
15. If a pendulum clock keeps perfect time at the base of a mountain, will it also keep perfect time when moved to the top of the mountain? Explain.
16. If a grandfather clock is running slow, how can you adjust the length of the pendulum to correct the time?
17. A simple pendulum can be used as an altimeter on a plane. How will the period of the pendulum vary as the plane rises from the ground to its final cruising altitude?
18. Will the period of a vibrating mass-spring system on Earth be different from the period of an identical mass-spring system on the moon? Why or why not?

Practice Problems

For problems 19–20, see Sample Problem B.

19. Find the length of a pendulum that oscillates with a frequency of 0.16 Hz.

20. A pendulum that moves through its equilibrium position once every 1.000 s is sometimes called a *seconds pendulum.*

- **a.** What is the period of any seconds pendulum?
- **b.** In Cambridge, England, a seconds pendulum is 0.9942 m long. What is the free-fall acceleration in Cambridge?
- **c.** In Tokyo, Japan, a seconds pendulum is 0.9927 m long. What is the free-fall acceleration in Tokyo?

For problem 21, see Sample Problem C.

21. A spring with a spring constant of 1.8×10^2 N/m is attached to a 1.5 kg mass and then set in motion.

- **a.** What is the period of the mass-spring system?
- **b.** What is the frequency of the vibration?

PROPERTIES OF WAVES

Review Questions

22. What is common to all waves?

23. How do transverse and longitudinal waves differ?

24. The figure below depicts a pulse wave traveling on a spring.

- **a.** In which direction are the particles of the medium vibrating?
- **b.** Is this wave transverse or longitudinal?

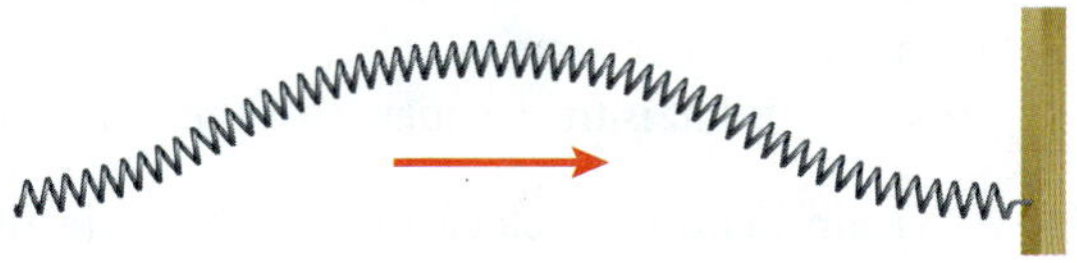

25. In a stretched spring, several coils are pinched together and others are spread farther apart than usual. What sort of wave is this?

26. How far does a wave travel in one period?

27. If you shook the end of a rope up and down three times each second, what would be the period of the waves set up in the rope? What would be the frequency?

28. Give three examples of mechanical waves. How are these different from electromagnetic waves, such as light waves?

Conceptual Questions

29. How does a single point on a string move as a transverse wave passes by that point?

30. What happens to the wavelength of a wave on a string when the frequency is doubled? What happens to the speed of the wave?

31. Why do sound waves need a medium through which to travel?

32. Two tuning forks with frequencies of 256 Hz and 512 Hz are struck. Which of the sounds will move faster through the air?

33. What is one advantage of transferring energy by electromagnetic waves?

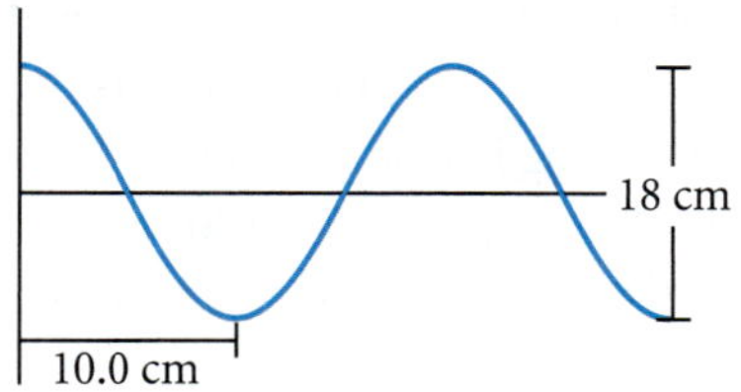

34. A wave traveling in the positive x direction with a frequency of 25.0 Hz is shown in the figure above. Find the following values for this wave:

- **a.** amplitude
- **b.** wavelength
- **c.** period
- **d.** speed

Practice Problems

For problem 35, see Sample Problem D.

35. Microwaves travel at the speed of light, 3.00×10^8 m/s. When the frequency of microwaves is 9.00×10^9 Hz, what is their wavelength?

11 REVIEW

14. The frequency would not change, because the frequency of a simple pendulum does not depend on mass.

15. no; a_g would change slightly, so T would also change.

16. Make the pendulum shorter to decrease the period.

17. The period will increase as the altitude increases.

18. They will be the same because the period is independent of free-fall acceleration.

19. 9.7 m

20. **a.** 2.000 s
b. 9.812 m/s^2
c. 9.798 m/s^2

21. **a.** 0.57 s
b. 1.8 Hz

22. movement of a disturbance

23. Transverse wave particles vibrate perpendicular to wave motion. Longitudinal wave particles vibrate parallel to wave motion.

24. **a.** vertically, perpendicular to wave motion
b. transverse

25. longitudinal

26. one wavelength

27. 1/3 s; 3 Hz

28. sound waves, water waves, and waves on a spring; Light waves do not need a medium to move through, but mechanical waves do.

29. up and down, no horizontal movement

30. It becomes half as long; It stays the same.

31. because sound waves are vibrations of particles; Without particles, no propagation occurs.

32. neither, because the speed of sound is constant in air

33. They can transport energy through space (a vacuum).

397

34. **a.** 9.0 cm
b. 20.0 cm
c. 0.0400 s
d. 5.00 m/s

35. 0.0333 m

36. **a.** a sine wave with twice the amplitude
b. a straight line (the waves cancel each other completely)

37. In constructive interference, individual displacements are on the same side of the equilibrium position. In destructive interference, the individual displacements are on opposite sides of the equilibrium position.

38. y_3

39. **a.** 0.0 cm
b. 48 cm

40. yes, because waves do not collide like other matter; They add to form a resultant wave.

41. zero

42. yes; when constructive interference occurs

43. a, b, and d ($\lambda = 0.5L$, L, and $2L$, respectively)

44. 14 N

45. 1.7 N

46. 2.0 Hz, 0.50 s, 0.30 m/s

47. 446 m

48. 0.129 m $\leq \lambda \leq$ 1.73 m

49. 9.70 m/s^2

50. 5.17×10^{14} Hz

51. 9:48 A.M.

WAVE INTERACTIONS

Review Questions

36. Using the superposition principle, draw the resultant waves for each of the examples below.

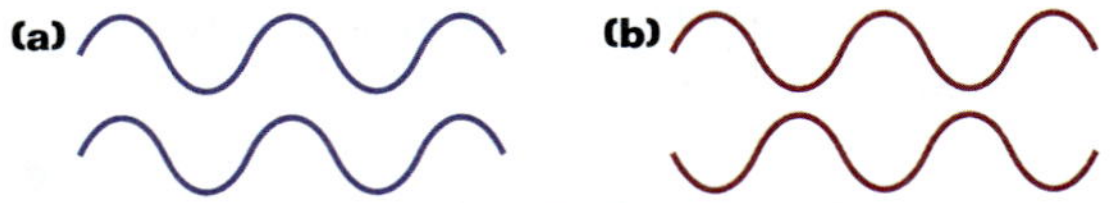

37. What is the difference between constructive interference and destructive interference?

38. Which one of the waveforms shown below is the resultant waveform?

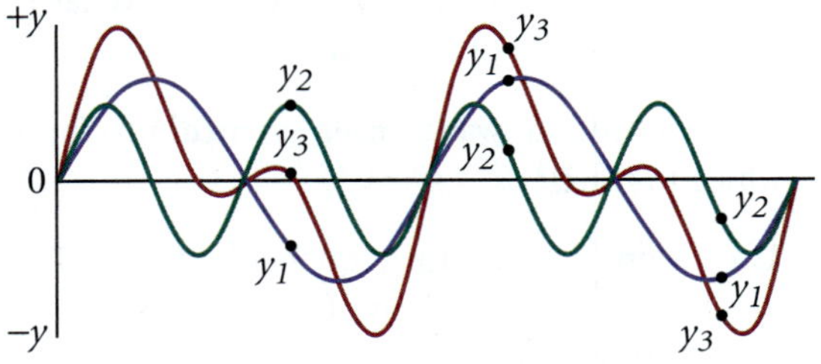

39. Anthony sends a series of pulses of amplitude 24 cm down a string that is attached to a post at one end. Assuming the pulses are reflected with no loss of amplitude, what is the amplitude at a point on the string where two pulses are crossing if

a. the string is rigidly attached to the post?
b. the end at which reflection occurs is free to slide up and down?

Conceptual Questions

40. Can more than two waves interfere in a given medium?

41. What is the resultant displacement at a position where destructive interference is complete?

42. When two waves interfere, can the resultant wave be larger than either of the two original waves? If so, under what conditions?

43. Which of the following wavelengths will produce standing waves on a string that is 3.5 m long?

a. 1.75 m
b. 3.5 m
c. 5.0 m
d. 7.0 m

MIXED REVIEW

44. In an arcade game, a 0.12 kg disk is shot across a frictionless horizontal surface by being compressed against a spring and then released. If the spring has a spring constant of 230 N/m and is compressed from its equilibrium position by 6.0 cm, what is the magnitude of the spring force on the disk at the moment it is released?

45. A child's toy consists of a piece of plastic attached to a spring, as shown at right. The spring is compressed against the floor a distance of 2.0 cm and released. If the spring constant is 85 N/m, what is the magnitude of the spring force acting on the toy at the moment it is released?

46. You dip your finger into a pan of water twice each second, producing waves with crests that are separated by 0.15 m. Determine the frequency, period, and speed of these water waves.

47. A sound wave traveling at 343 m/s is emitted by the foghorn of a tugboat. An echo is heard 2.60 s later. How far away is the reflecting object?

48. The notes produced by a violin range in frequency from approximately 196 Hz to 2637 Hz. Find the possible range of wavelengths in air produced by this instrument when the speed of sound in air is 340 m/s.

49. What is the free-fall acceleration in a location where the period of a 0.850 m long pendulum is 1.86 s?

50. Yellow light travels through a certain glass block at a speed of 1.97×10^8 m/s. The wavelength of the light in this particular type of glass is 3.81×10^{-7} m (381 nm). What is the frequency of the yellow light?

51. A certain pendulum clock that works perfectly on Earth is taken to the moon, where $a_g = 1.63$ m/s^2. If the clock is started at 12:00 A.M., what will it read after 24.0 h have passed on Earth?

Alternative Assessment

1. Design an experiment to compare the spring constant and period of oscillation of a system built with two (or more) springs connected in two ways: in series (attached end to end) and in parallel (one end of each spring anchored to a common point). If your teacher approves your plan, obtain the necessary equipment and perform the experiment.

2. The rule that the period of a pendulum is determined by its length is a good approximation for amplitudes below 15°. Design an experiment to investigate how amplitudes of oscillation greater than 15° affect the motion of a pendulum.

   List what equipment you would need, what measurements you would perform, what data you would record, and what you would calculate. If your teacher approves your plan, obtain the necessary equipment and perform the experiment.

3. Research earthquakes and different kinds of seismic waves. Create a presentation about earthquakes that includes answers to the following questions as well as additional information: Do earthquakes travel through oceans? What is transferred from place to place as seismic waves propagate? What determines their speed?

4. Identify examples of periodic motion in nature. Create a chart describing the objects involved, their path of motion, their periods, and the forces involved. Which of the periodic motions are harmonic and which are not?

5. Research the active noise reduction (ANR) technology used in noise-cancelling headphones. How does it work? What are some other applications that use ANR technology? Choose one application, and create a brochure to explain how it works.

Graphing Calculator *Practice*

Pendulum

Would a pendulum have the same period of oscillation on Mars, Venus, or Neptune? A pendulum's period, as you learned earlier in this chapter, is described by the following equation:

$$T = 2\pi\sqrt{\frac{L}{a_g}}$$

In this equation, T is the period, L is the length of the pendulum, and a_g is the free-fall acceleration (9.81 m/s^2 on Earth's surface). This equation can be rearranged to solve for L if T is known.

$$L = \frac{a_g T^2}{4\pi^2}$$

In this graphing calculator activity, you will enter the period of a pendulum on Earth's surface. The calculator will use the previous equation to determine L, the length of the pendulum. The calculator will then use this length to display a graph showing how the period of this pendulum changes as free-fall acceleration changes. You will use this graph to find the period of a pendulum on various planets.

Visit go.hrw.com and type in the keyword **HF6VIBX** to find this graphing calculator activity. Refer to **Appendix B** for instructions on downloading the program for this activity.

Alternative Assessment ANSWERS

1. Student plans should be safe and complete and should include a list of equipment, measurements, and calculations. For springs in series, $1/k = 1/k_1 + 1/k_2$; for springs in parallel, $k = k_1 + k_2$.
2. Student plans should be safe and complete and should include a list of equipment, measurements, and calculations.
3. Answers should indicate that wave speed depends on the medium. Earthquakes involve longitudinal P waves, transverse S waves, and Rayleigh waves (circular motion).
4. Examples should involve repetitive motion. Vibrations often are harmonic, but many other motions, including circular motion, typically are not harmonic.
5. Noise-cancelling headphones and other ANR applications generate "antinoise" sound waves that interfere destructively with existing sound waves.

extension

Graphing Calculator Practice

Visit go.hrw.com for answers to this Graphing Calculator activity.

Keyword HF6VIBXT

Standardized Test Prep

ANSWERS

1. C

2. J

3. C

4. F

5. C

6. G

7. B

8. G

Standardized Test Prep

MULTIPLE CHOICE

Base your answers to questions 1–6 on the information below.

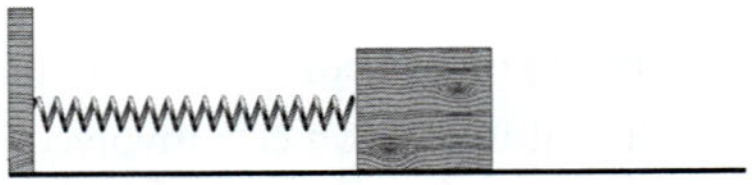

A mass is attached to a spring and moves with simple harmonic motion on a frictionless horizontal surface, as shown above.

1. In what direction does the restoring force act?

A. to the left
B. to the right
C. to the left or to the right depending on whether the spring is stretched or compressed
D. perpendicular to the motion of the mass

2. If the mass is displaced −0.35 m from its equilibrium position, the restoring force is 7.0 N. What is the spring constant?

F. -5.0×10^{-2} N/m
G. -2.0×10^{1} N/m
H. 5.0×10^{-2} N/m
J. 2.0×10^{1} N/m

3. In what form is the energy in the system when the mass passes through the equilibrium position?

A. elastic potential energy
B. gravitational potential energy
C. kinetic energy
D. a combination of two or more of the above

4. In what form is the energy in the system when the mass is at maximum displacement?

F. elastic potential energy
G. gravitational potential energy
H. kinetic energy
J. a combination of two or more of the above

5. Which of the following does *not* affect the period of the mass-spring system?

A. mass
B. spring constant
C. amplitude of vibration
D. All of the above affect the period.

6. If the mass is 48 kg and the spring constant is 12 N/m, what is the period of the oscillation?

F. 8π s
G. 4π s
H. π s
J. $\frac{\pi}{2}$ s

Base your answers to questions 7–10 on the information below.

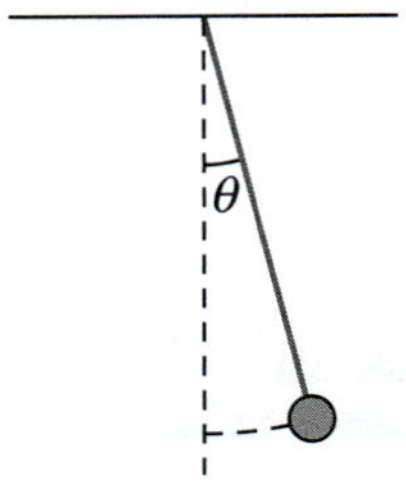

A pendulum bob hangs from a string and moves with simple harmonic motion, as shown above.

7. What is the restoring force in the pendulum?

A. the total weight of the bob
B. the component of the bob's weight tangent to the motion of the bob
C. the component of the bob's weight perpendicular to the motion of the bob
D. the elastic force of the stretched string

8. Which of the following does *not* affect the period of the pendulum?

F. the length of the string
G. the mass of the pendulum bob
H. the free-fall acceleration at the pendulum's location
J. All of the above affect the period.

9. If the pendulum completes exactly 12 cycles in 2.0 min, what is the frequency of the pendulum?

A. 0.10 Hz **C.** 6.0 Hz
B. 0.17 Hz **D.** 10 Hz

10. If the pendulum's length is 2.00 m and a_g = 9.80 m/s^2, how many complete oscillations does the pendulum make in 5.00 min?

F. 1.76 **H.** 106
G. 21.6 **J.** 239

Base your answers to questions 11–13 on the graph below.

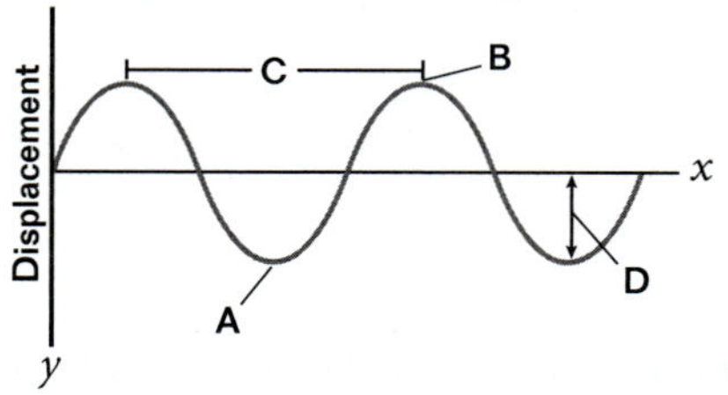

11. What kind of wave does this graph represent?

A. transverse wave
B. longitudinal wave
C. electromagnetic wave
D. pulse wave

12. Which letter on the graph is used for the wavelength?

F. A **H.** C
G. B **J.** D

13. Which letter on the graph is used for a trough?

A. A **C.** C
B. B **D.** D

Base your answers to questions 14–15 on the passage below.

A wave with an amplitude of 0.75 m has the same wavelength as a second wave with an amplitude of 0.53 m. The two waves interfere.

14. What is the amplitude of the resultant wave if the interference is constructive?

F. 0.22 m **H.** 0.75 m
G. 0.53 m **J.** 1.28 m

15. What is the amplitude of the resultant wave if the interference is destructive?

A. 0.22 m **C.** 0.75 m
B. 0.53 m **D.** 1.28 m

16. Two successive crests of a transverse wave are 1.20 m apart. Eight crests pass a given point every 12.0 s. What is the wave speed?

F. 0.667 m/s **H.** 1.80 m/s
G. 0.800 m/s **J.** 9.60 m/s

SHORT RESPONSE

17. Green light has a wavelength of 5.20×10^{-7} m and a speed in air of 3.00×10^{8} m/s. Calculate the frequency and the period of the light.

18. What kind of wave does not need a medium through which to travel?

19. List three wavelengths that could form standing waves on a 2.0 m string that is fixed at both ends.

EXTENDED RESPONSE

20. A visitor to a lighthouse wishes to find out the height of the tower. The visitor ties a spool of thread to a small rock to make a simple pendulum. Then, the visitor hangs the pendulum down a spiral staircase in the center of the tower. The period of oscillation is 9.49 s. What is the height of the tower? Show all of your work.

21. A harmonic wave is traveling along a rope. The oscillator that generates the wave completes 40.0 vibrations in 30.0 s. A given crest of the wave travels 425 cm along the rope in a period of 10.0 s. What is the wavelength? Show all of your work.

Test TIP Take a little time to look over a test before you start. Look for questions that may be easy for you to answer, and answer those first. Then, move on to the harder questions.

9. A
10. H
11. A
12. H
13. A
14. J
15. A
16. G
17. 5.77×10^{14} Hz, 1.73×10^{-15} s
18. electromagnetic waves
19. Possible correct answers include 4.0 m, 2.0 m, 1.3 m, 1.0 m, or other wavelengths such that $n\lambda = 4.0$ m (where n is a positive integer).
20. 22.4 m (See the Solutions Manual or the One-Stop Planner for a full solution.)
21. 0.319 m (See the Solutions Manual or the One-Stop Planner for a full solution.)

CHAPTER 11
Inquiry Lab

Lab Planning

Beginning on page T34 are preparation notes and teaching tips to assist you in planning.

Blank data tables (as well as some sample data) appear on the **One-Stop Planner.**

No Books in the Lab?

See the ***Datasheets for In-Text Labs*** workbook for a reproducible master copy of this experiment.

The same workbook also contains a version of this experiment with explicit procedural steps if you prefer a more directed approach.

Safety Caution

Falling masses can cause injury. Students should wear goggles to shield eyes from clamps and swinging masses at eye level.

Tips and Tricks

- Use tall support stands and pendulum clamps for best results and ease of adjusting the length of the cord.
- The pendulum must be kept at a small angle. (When θ is small enough, cosine θ can be approximated by 1.)

CHAPTER 11

Inquiry Lab

Simple Harmonic Motion of a Pendulum

Design Your Own

OBJECTIVES

- **Construct** simple pendulums, and find their periods.
- **Calculate** the value for a_g, the free-fall acceleration.
- **Examine** the relationships between length, mass, and period for different pendulums.

MATERIALS LIST

- **balance**
- **cord**
- **meterstick**
- **pendulum bobs**
- **pendulum clamp**
- **protractor**
- **stopwatch**
- **support stand**

The period of a pendulum is the time required for the pendulum to complete one cycle. In this lab, you will construct models of a simple pendulum using different masses and lengths of cord. You will design an experiment to measure the period of each model and to determine how the period depends on length and mass. Your experiment should include several trials at a constant mass but at different lengths, and several more trials at constant length but with different masses. In the Analysis, you will also use the period and the length of the cord for each trial to calculate the free-fall acceleration, a_g, at your geographical location.

SAFETY

- **Tie back long hair, secure loose clothing, and remove loose jewelry to prevent their getting caught in moving parts or pulleys. Put on goggles.**
- **Attach masses to the thread and the thread to clamps securely. Swing masses in areas free of people and obstacles. Swinging or dropped masses can cause serious injury.**

PROCEDURE

1. Study the materials provided, and design an experiment to meet the goals stated above.
2. Write out your lab procedure, including a detailed description of the measurements to take during each step and the number of trials to perform. You may use **Figure 1** as a guide to a possible setup. You should keep the amplitude of the swing less than 15° in each trial.
3. Ask your teacher to approve your procedure.
4. Follow all steps of your procedure.
5. Clean up your work area. Put equipment away safely so that it is ready to be used again.

ANALYSIS

1. **Organizing Data** For each trial, calculate the period of the pendulum.

2. **Organizing Data** Calculate the value for the free-fall acceleration, a_g, for each trial. Use the equation for the period of a simple pendulum, rearranged to solve for a_g.

3. **Constructing Graphs** Plot the following graphs:

   a. the period vs. the length (for constant-mass trials)

   b. the period vs. the mass of the bob (for constant-length trials)

   c. the period vs. the square root of the length (for constant-mass trials)

CONCLUSIONS

4. **Evaluating Results** Use 9.81 m/s^2 as the accepted value for a_g.

   a. Compute the absolute error for each trial using the following equation:

   $$absolute\ error = |experimental - accepted|$$

   b. Compute the relative error for each trial using the following equation:

   $$relative\ error = \frac{(experimental - accepted)}{accepted}$$

5. **Drawing Conclusions** Based on your data and your graphs, how does the mass of the pendulum bob affect the period of vibration?

6. **Drawing Conclusions** Based on your data and your graphs, how does the length of the pendulum affect the period of vibration?

Figure 1

- Hold the bob so that the cord is perfectly straight while you measure the angle.
- Release the bob gently so that it swings smoothly. Practice counting and timing cycles to get good results.

ANSWERS

Analysis

1. Constant mass: Trial 1: 1.77 s; Trial 2: 1.41 s; Trial 3: 1.06 s. **Constant length:** Trial 4: 1.77 s; Trial 5: 1.77 s; Trial 6: 1.76 s.

2. Student answers will vary. Make sure students use the relationship $T = 2\pi\sqrt{\frac{L}{a_g}}$. Typical values will range from 9.50 m/s^2 to 9.82 m/s^2.

3. a. The graph should show a transformed square-root function, a parabolic curve that opens to the right.

b. Graphs should show a straight line parallel to the x-axis.

c. Graphs should show a straight line pointing up and to the right.

Conclusions

4. a. For sample data, values range from 0.01 m/s^2 to 0.29 m/s^2.

b. For sample data, values range from 0.001 to 0.030.

5. The mass has no effect.

6. The longer the pendulum is, the longer the period.

Physics and Its World *Timeline 1785–1830*

1780

1789 – The storming of the Bastille marks the climax of the French Revolution.

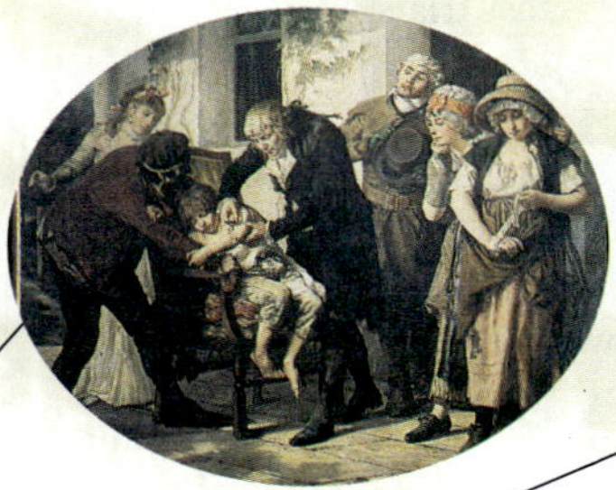

1790

1796 – **Edward Jenner** develops the smallpox vaccine.

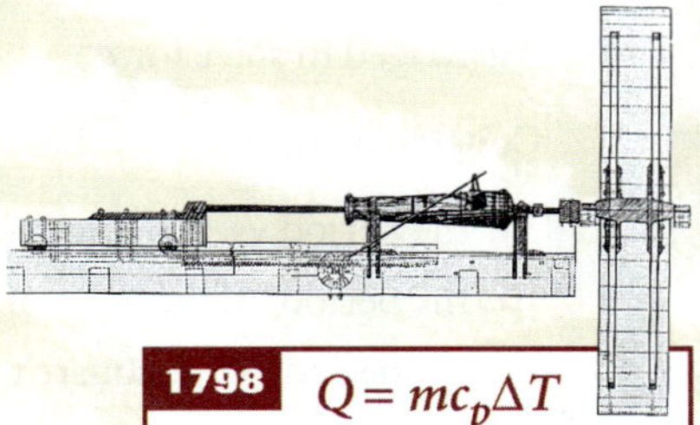

1798 $Q = mc_p\Delta T$

Benjamin Thompson (Count Rumford) demonstrates that energy transferred as heat results from mechanical processes, rather than the release of caloric, the heat fluid that has been widely believed to exist in all substances.

1800 $\Delta V = \frac{\Delta PE_{electric}}{q}$

Alessandro Volta develops the first current-electricity cell using alternating plates of silver and zinc.

1800

1801 $m\lambda = d(\sin\theta)$

Thomas Young demonstrates that light rays interfere, providing the first substantial support for a wave theory of light.

1810

1804 – Saint-Domingue, under the control of the French-African majority led by **Toussaint-Louverture,** becomes the independent Republic of Haiti. Over the next two decades most of Europe's western colonies become independent.

1804 – **Richard Trevithick** builds and tests the first steam locomotive. It pulls 10 tons along a distance of 15 km at a speed of 8 km/h.

1811 – Mathematician **Sophie Germain** writes the first of three papers on the mathematics of vibrating surfaces. She later addresses one of the most famous problems in mathematics—Fermat's last theorem—proving it to be true for a wide range of conditions.

1810 – **Kamehameha I** unites the Hawaiian islands under a monarchy.

1814

$$\sin\theta = \frac{m\lambda}{a}$$

Augustin Fresnel begins his research in optics, the results of which will confirm and explain **Thomas Young's** discovery of interference and will firmly establish the wave model of light first suggested by **Christiaan Huygens** over a century earlier.

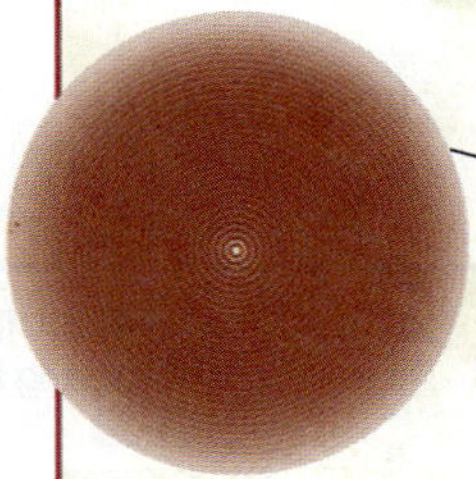

1818 – **Mary Shelley** writes *Frankenstein, or the Modern Prometheus.* Primarily thought of as a horror novel, the book's emphasis on science and its moral consequences also qualifies it as the first "science fiction" novel.

1820

$$F_{magnetic} = BI\ell$$

Hans Christian Oersted demonstrates that an electric current produces a magnetic field. (**Gian Dominico Romagnosi,** an amateur scientist, discovered the effect 18 years earlier, but at the time attracted no attention.) **André-Marie Ampére** repeats Oersted's experiment and formulates the law of electromagnetism that today bears his name.

1826 – **Katsushika Hokusai** begins his series of prints *Thirty-Six Views of Mount Fuji.*

1830 – **Hector Berlioz** composes his *Symphonie Fantastique*, one of the first Romantic works for large orchestra that tells a story with music.

405

CHAPTER 12

Sound
Planning Guide

Compression Guide

To shorten instruction because of time limitations, omit the opener and Section 3 and abbreviate the review.

OBJECTIVES	LABS, DEMONSTRATIONS, AND ACTIVITIES	TECHNOLOGY RESOURCES
PACING • 45 min pp. 406–407 **Chapter Opener**	**ANC Discovery Lab** Resonance and the Nature of Sound*◆ BASIC	**CD Visual Concepts,** Chapter 12 BASIC
PACING • 90 min pp. 408–413 **Section 1 Sound Waves** • Explain how sound waves are produced. • Relate frequency to pitch. • Compare the speed of sound in various media. • Relate plane waves to spherical waves. • Recognize the Doppler effect, and determine the direction of a frequency shift when there is relative motion between a source and an observer.	**TE Demonstration** Sound Waves in a Solid, p. 410 BASIC **TE Demonstration** The Doppler Effect, p. 412 GENERAL **SE Skills Practice Lab** Speed of Sound, pp. 440–441◆ GENERAL **ANC Datasheet** Speed of Sound* GENERAL **SE CBL™ Lab** Speed of Sound, pp. 938–939◆ GENERAL **ANC CBL™ Experiment** Speed of Sound*◆ GENERAL	**OSP Lesson Plans** TR 54 Production of a Sound Wave TR 55 Graph of a Sound Wave TR 56 Spherical Waves TR 57 The Doppler Effect TR 40A Speed of Sound in Various Media
PACING • 45 min pp. 414–421 **Section 2 Sound Intensity and Resonance** • Calculate the intensity of sound waves. • Relate intensity, decibel level, and perceived loudness. • Explain why resonance occurs.	**SE Quick Lab** Resonance, p. 418 GENERAL **TE Demonstration** Resonance, p. 418 GENERAL	**OSP Lesson Plans** **EXT Integrating Health** Why Your Ears Pop BASIC TR 58 Diagram of the Human Ear TR 41A Range of Audibility of the Average Human Ear TR 42A Conversion of Intensity to Decibel Level
PACING • 45 min pp. 422–431 *Advanced Level* **Section 3 Harmonics** • Differentiate between the harmonic series of open and closed pipes. • Calculate the harmonics of a vibrating string and of open and closed pipes. • Relate harmonics and timbre. • Relate the frequency difference between two waves to the number of beats heard per second.	**TE Demonstration** Seeing Sounds, p. 422 GENERAL **SE Quick Lab** A Pipe Closed at One End, p. 425 GENERAL **ANC Invention Lab** Building a Musical Instrument*◆ ADVANCED **ANC CBL™ Experiment** Sound Waves and Beats*◆ ADVANCED	**OSP Lesson Plans** TR 59 Harmonics of Open and Closed Pipes TR 60 Harmonics of Musical Instruments TR 61 Beats TR 43A The Harmonic Series

PACING • 90 min

CHAPTER REVIEW, ASSESSMENT, AND STANDARDIZED TEST PREPARATION

- SE **Chapter Highlights,** p. 433
- SE **Chapter Review,** pp. 434–437
- SE **Graphing Calculator Practice,** p. 436 GENERAL
- SE **Alternative Assessment,** p. 437 ADVANCED
- SE **Standardized Test Prep,** pp. 438–439 GENERAL
- SE **Appendix D: Equations,** p. 860
- SE **Appendix I: Additional Problems,** p. 889
- ANC **Study Guide Worksheet** Mixed Review* GENERAL
- ANC **Chapter Test A*** GENERAL
- ANC **Chapter Test B*** ADVANCED
- OSP **Test Generator**

Online and Technology Resources

go.hrw.com Holt Online Learning

Visit go.hrw.com to find a variety of online resources. To access this chapter's extensions, enter the keyword **HF6SNDXT** and click the "go" button. Click **Holt Online Learning** for an online edition of this textbook, and other interactive resources.

HOLT Teacher's One-Stop Planner®

This DVD package includes:
- Holt Calendar Planner
- Customizable Lesson Plans
- Editable Worksheets
- ExamView® Version 6 Assessment Suite
- Interactive Teacher's Edition
- Holt PuzzlePro®
- Holt PowerNotes® Presentations
- MindPoint® Quiz Show

KEY

SE Student Edition	**OSP** One-Stop Planner	**EXT** Online Extension
TE Teacher Edition	**CD** CD or CD-ROM	* Also on One-Stop Planner
ANC Ancillary Worksheet	**TR** Teaching Transparencies	◆ Requires advance prep

SKILLS DEVELOPMENT RESOURCES	REVIEW AND ASSESSMENT	CORRELATIONS
		National Science Education Standards
SE Conceptual Challenge, p. 411 GENERAL **SE Appendix J: Advanced Topics** The Doppler Effect and the Big Bang, pp. 912–913 ADVANCED	**SE Section Review,** p. 413 GENERAL **ANC Study Guide Worksheet** Section 1* GENERAL **ANC Quiz** Section 1* BASIC	UCP 1, 2, 3, 5 SAI 1, 2 ST 1, 2 HNS 1, 3 SPSP 2, 5 PS 6a
SE Sample Set A Intensity of Sound Waves, p. 415 BASIC **TE Classroom Practice,** p. 415 BASIC **ANC Problem Workbook** Sample Set A* BASIC **OSP Problem Bank** Sample Set A BASIC **SE Conceptual Challenge,** p. 419 GENERAL	**SE Section Review,** p. 420 GENERAL **ANC Study Guide Worksheet** Section 2* GENERAL **ANC Quiz** Section 2* BASIC	UCP 1, 2, 3, 5 SAI 1, 2 ST 1, 2 HNS 3 SPSP 1, 2, 3, 4 PS 6a
SE Sample Set B Harmonics, pp. 426–427 GENERAL **TE Classroom Practice,** p. 426 GENERAL **ANC Problem Workbook** Sample Set B* GENERAL **OSP Problem Bank** Sample Set B GENERAL **SE Conceptual Challenge,** p. 430 ADVANCED	**SE Section Review,** p. 431 ADVANCED **ANC Study Guide Worksheet** Section 3* GENERAL **ANC Quiz** Section 3* GENERAL	UCP 1, 2, 3, 4, 5 SAI 1, 2 ST 1, 2 HNS 1, 2 SPSP 2, 5

SCILINKS. NSTA

www.scilinks.org

Maintained by the **National Science Teachers Association.**

Topic: Sound
SciLinks Code: HF61426

Topic: Resonance
SciLinks Code: HF61303

Topic: Harmonics
SciLinks Code: HF60715

Topic: Acoustics
SciLinks Code: HF60015

Classroom CD-ROMs

- Guided Reading Audio Program
- Student One Stop
- Virtual Investigations
- Visual Concepts

Holt Lab Generator CD-ROM

Search for any lab by topic, standard, difficulty level, or time. Edit any lab to fit your needs, or create your own labs. Use the Lab Materials QuickList software to customize your lab materials list.

CHAPTER 12

Overview

Section 1 explains how sound waves are produced, explores the basic characteristics of sound waves, and introduces the Doppler effect.

Section 2 explains how to calculate intensity; relates intensity, decibel level, and perceived loudness; and explores the phenomenon of resonance.

Section 3 introduces standing waves on a vibrating string and in open and closed pipes, calculates harmonics, relates harmonics and timbre, and discusses how beats occur.

About the Illustration

The dolphins shown in this photograph are bottlenose dolphins in captivity in Hawaii. Dolphins use sounds for navigation, communication, and echolocation. A variety of other marine mammals and most bats also use sound waves to echolocate.

Some marine mammals, such as dolphins, use sound waves to locate distant objects. In this process, called *echo-location,* a dolphin produces a rapid train of short sound pulses that travel through the water, bounce off distant objects, and reflect back to the dolphin. From these echoes, dolphins can determine the size, shape, speed, and distance of their potential prey.

CHAPTER 12

Sound

WHAT TO EXPECT

In this chapter, you will study many physical aspects of sound, including the nature of sound waves, frequency, intensity, resonance, and harmonics.

Why it Matters

Some animals, including dolphins and bats, use sound waves to learn about their prey. Musical instruments create a variety of pleasing sounds through different harmonics.

CHAPTER PREVIEW

Tapping Prior Knowledge

Knowledge to Review

- ✔ Longitudinal waves are waves in which the particles vibrate parallel to the direction of wave motion.
- ✔ Wave speed is equal to the product of wavelength and frequency.
- ✔ Standing waves are formed when two waves of the same frequency, amplitude, and wavelength travel in opposite directions through a medium and interfere with each other.

Items to Probe

- ✔ Familiarity with wave-related phenomena: Ask students to describe how sound travels between a speaker and a listener.
- ✔ Preconceptions about pitch and loudness: Ask students what determines the pitch and loudness associated with a sound.

SECTION 1
General Level

Visual Strategy GENERAL

Figure 1
Point out that the vibrations of the prongs cause the air molecules to move back and forth.

Q How is the position of the air molecules related to the motion of the prongs over time?

A ***The air molecules follow the prongs' motion. There is an increase of air pressure when a prong "pushes," leaving fewer molecules and lower pressure behind. When the prong returns, the pressure pattern shifts back.***

SECTION 1

Sound Waves

SECTION OBJECTIVES

- **Explain how sound waves are produced.**
- **Relate frequency to pitch.**
- **Compare the speed of sound in various media.**
- **Relate plane waves to spherical waves.**
- **Recognize the Doppler effect, and determine the direction of a frequency shift when there is relative motion between a source and an observer.**

THE PRODUCTION OF SOUND WAVES

Whether a sound wave conveys the shrill whine of a jet engine or the melodic whistling of a bird, it begins with a vibrating object. We will explore how sound waves are produced by considering a vibrating tuning fork, as shown in **Figure 1(a).**

The vibrating prong of a tuning fork, shown in **Figure 1(b),** sets the air molecules near it in motion. As the prong swings to the right, as in **Figure 1(c),** the air molecules in front of the movement are forced closer together. (This situation is exaggerated in the figure for clarity.) Such a region of high molecular density and high air pressure is called a **compression.** As the prong moves to the left, as in **Figure 1(d),** the molecules to the right spread apart, and the density and air pressure in this region become lower than normal. This region of lower density and pressure is called a **rarefaction.**

As the tuning fork continues to vibrate, a series of compressions and rarefactions forms and spreads away from each prong. These compressions and rarefactions spread out in all directions, like ripple waves on a pond. When the tuning fork vibrates with simple harmonic motion, the air molecules also vibrate back and forth with simple harmonic motion.

compression

the region of a longitudinal wave in which the density and pressure are at a maximum

rarefaction

the region of a longitudinal wave in which the density and pressure are at a minimum

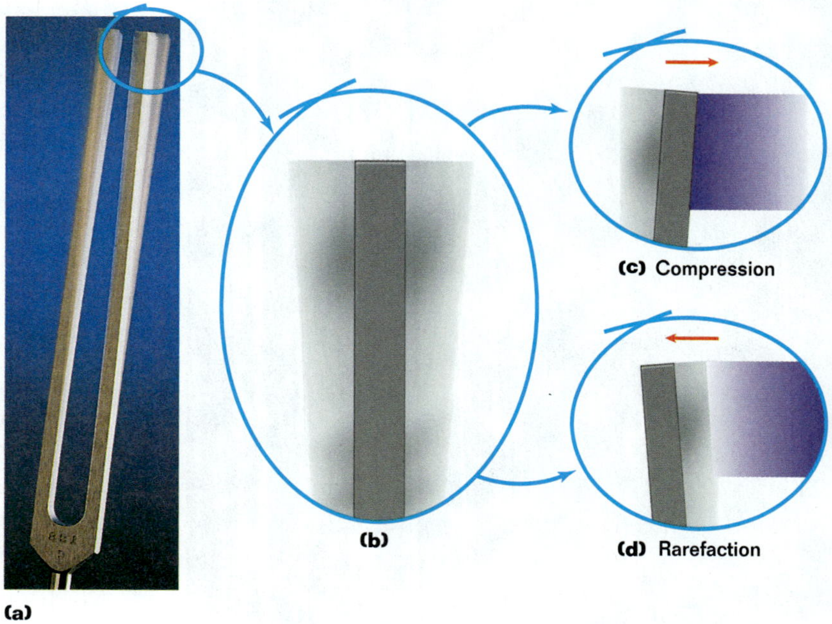

Figure 1
(a) The sound from a tuning fork is produced by **(b)** the vibrations of each of its prongs. **(c)** When a prong swings to the right, there is a region of high density and pressure. **(d)** When the prong swings back to the left, a region of lower density and pressure exists.

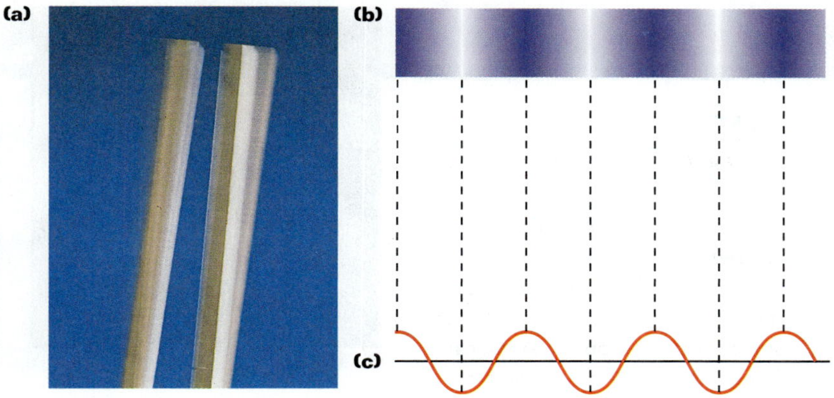

Figure 2
(a) As this tuning fork vibrates, **(b)** a series of compressions and rarefactions moves away from each prong. **(c)** The crests of this sine wave correspond to compressions, and the troughs correspond to rarefactions.

Sound waves are longitudinal

In sound waves, the vibrations of air molecules are parallel to the direction of wave motion. Thus, sound waves are longitudinal. The simplest longitudinal wave produced by a vibrating object can be represented by a sine curve. In **Figure 2,** the crests correspond to compressions (regions of higher pressure), and the troughs correspond to rarefactions (regions of lower pressure). Thus, the sine curve represents the changes in air pressure due to the propagation of the sound waves. Note that **Figure 2** shows an idealized case. This example disregards energy losses that would decrease the wave amplitude.

SCILINKS®
www.scilinks.org
Topic: Sound
Code: HF61426

CHARACTERISTICS OF SOUND WAVES

As discussed earlier, *frequency* is defined as the number of cycles per unit of time. Sound waves that the average human ear can hear, called *audible* sound waves, have frequencies between 20 and 20 000 Hz. (An individual's hearing depends on a variety of factors, including age and experiences with loud noises.) Sound waves with frequencies less than 20 Hz are called *infrasonic* waves, and those above 20 000 Hz are called *ultrasonic* waves.

It may seem confusing to use the term *sound waves* for infrasonic or ultrasonic waves because humans cannot hear these sounds, but these waves consist of the same types of vibrations as the sounds that we can hear. The range of audible sound waves depends on the ability of the average human ear to detect their vibrations. Dogs can hear ultrasonic waves that humans cannot.

Did you know?

Elephants use infrasonic sound waves to communicate with one another. Their large ears enable them to detect these low-frequency sound waves, which have relatively long wavelengths. Elephants can effectively communicate in this way, even when they are separated by many kilometers.

Frequency determines pitch

The frequency of an audible sound wave determines how high or low we perceive the sound to be, which is known as **pitch.** As the frequency of a sound wave increases, the pitch rises. The frequency of a wave is an objective quantity that can be measured, while pitch refers to how different frequencies are perceived by the human ear. Pitch depends not only on frequency but also on other factors, such as background noise and loudness.

pitch

a measure of how high or low a sound is perceived to be, depending on the frequency of the sound wave

SECTION 1

Teaching Tip — ADVANCED

Figure 2 uses a sine curve to represent the compressions and rarefactions of a longitudinal wave produced by a vibrating object. Compressions correspond to crests, and rarefactions correspond to troughs. Sometimes a sine curve is used to represent *displacement* rather than pressure and density. For any given longitudinal wave, the sine curve representing pressure and the sine curve representing displacement are 90° out of phase.

Misconception Alert — BASIC

Point out that some individuals may be able to hear sounds slightly below 20 Hz or above 20 000 Hz because the range of frequencies defined as audible is based on the ability of the *average* human ear.

Why it Matters

Ultrasound Images

Students may be familiar with X rays as a means of examining organs within the body. Point out that ultrasound waves are much safer than X rays, a form of electromagnetic radiation. Ultrasound waves are used to observe fetuses because X rays can produce birth defects and because ultrasound provides more detail for soft tissue.

Demonstration

Sound Waves in a Solid — BASIC

Purpose Show sound waves traveling through a solid.

Materials coat hanger, two strings

Procedure Open the coat hanger, and tie a string at each end. Ask a volunteer to hold the ends of the strings tautly next to his or her ears while you hit the coat hanger with a pen. Ask the volunteer to describe the sounds he or she heard (*sounds like bells*). Have other students in the class hold the strings to their ears as you continue to hit the coat hanger. Explain that the vibrations of the coat hanger traveled through the strings and produced the ringing sounds that students observed.

Why it Matters

Ultrasound Images

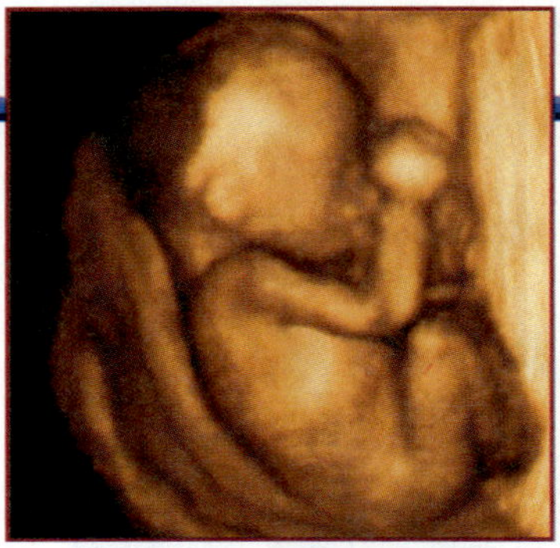

Ultrasonic waves can be used to produce images of objects inside the body. Such imaging is possible because sound waves are partially reflected when they reach a boundary between two materials of different densities. The images produced by ultrasonic waves are clearer and more detailed than those that can be produced by lower-frequency sound waves because the short wavelengths of ultrasonic waves are easily reflected off small objects. Audible and infrasonic sound waves are not as effective because their longer wavelengths pass around small objects.

In order for ultrasonic waves to "see" an object inside the body, the wavelength of the waves used must be about the same size as or smaller than the object. A typical frequency used in an ultrasonic device is about 10 MHz. The speed of an ultrasonic wave in human tissue is about 1500 m/s, so the wavelength of 10 MHz waves is $\lambda = v/f = 0.15$ mm. A 10 MHz ultrasonic device will not detect objects smaller than this size.

Physicians commonly use ultrasonic waves to observe fetuses. In this process, a crystal emits ultrasonic pulses. The same crystal acts as a receiver and detects the reflected sound waves. These reflected sound waves are converted to an electrical signal, which forms an image on a fluorescent screen. By repeating this process for different portions of the mother's abdomen, a physician can obtain a complete picture of the fetus, as shown above. These images allow doctors to detect some types of fetal abnormalities.

Table 1 Speed of Sound in Various Media

Medium	v (m/s)
Gases	
air (0°C)	331
air (25°C)	346
air (100°C)	366
helium (0°C)	972
hydrogen (0°C)	1290
oxygen (0°C)	317
Liquids at 25°C	
methyl alcohol	1140
sea water	1530
water	1490
Solids	
aluminum	5100
copper	3560
iron	5130
lead	1320
vulcanized rubber	54

Speed of sound depends on the medium

Sound waves can travel through solids, liquids, and gases. Because waves consist of particle vibrations, the speed of a wave depends on how quickly one particle can transfer its motion to another particle. For example, solid particles respond more rapidly to a disturbance than gas particles do because the molecules of a solid are closer together than those of a gas are. As a result, sound waves generally travel faster through solids than through gases. **Table 1** shows the speed of sound waves in various media.

The speed of sound also depends on the temperature of the medium. As temperature rises, the particles of a gas collide more frequently. Thus, in a gas, the disturbance can spread faster at higher temperatures than at lower temperatures. In liquids and solids, the particles are close enough together that the difference due to temperature changes is less noticeable.

Sound waves propagate in three dimensions

In the chapter "Vibrations and Waves," waves were shown as traveling in a single direction. But sound waves actually travel away from a vibrating source in all three dimensions. When a musician plays a saxophone in the middle of a room, the resulting sound can be heard throughout the room because the sound waves spread out in all directions. The wave fronts of sound waves spreading in three dimensions are approximately spherical. To simplify, we shall assume that the wave fronts are exactly spherical unless stated otherwise.

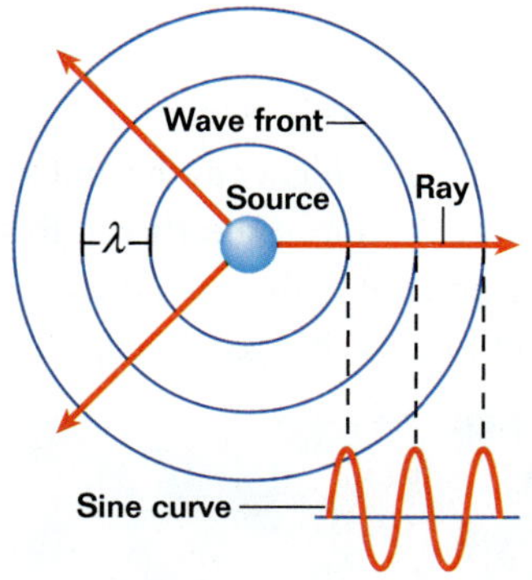

Figure 3
In this representation of a spherical wave, the wave fronts represent compressions, and the rays show the direction of wave motion. Each wave front corresponds to a crest of the sine curve. In turn, the sine curve corresponds to a single ray.

Spherical waves can be represented graphically in two dimensions with a series of circles surrounding the source, as shown in **Figure 3.** The circles represent the centers of compressions, called *wave fronts.* Because we are considering a three-dimensional phenomenon in two dimensions, each circle represents a spherical area.

Because each wave front locates the center of a compression, the distance between adjacent wave fronts is equal to one wavelength, λ. The radial lines perpendicular to the wave fronts are called *rays.* Rays indicate the direction of the wave motion. The sine curve used in our previous representation of sound waves, also shown in **Figure 3,** corresponds to a single ray. Because crests of the sine curve represent compressions, each wave front crossed by this ray corresponds to a crest of the sine curve.

Now, consider a small portion of a spherical wave front that is many wavelengths away from the source, as shown in **Figure 4.** In this case, the rays are nearly parallel lines, and the wave fronts are nearly parallel planes. Thus, at distances from the source that are great relative to the wavelength, we can approximate spherical wave fronts with parallel planes. Such waves are called *plane waves.* Any small portion of a spherical wave that is far from the source can be considered a plane wave. Plane waves can be treated as one-dimensional waves all traveling in the same direction, as in the chapter "Vibrations and Waves."

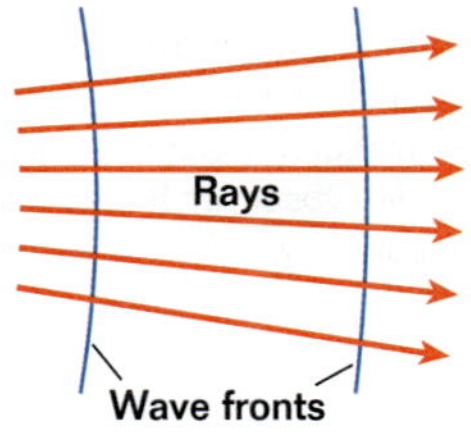

Figure 4
Spherical wave fronts that are a great distance from the source can be approximated with parallel planes known as *plane waves.*

Why it Matters

Conceptual Challenge

1. Music from a Trumpet Suppose you hear music being played from a trumpet that is across the room from you. Compressions and rarefactions from the sound wave reach your ear, and you interpret these vibrations as sound. Were the air particles that are vibrating near your ear carried across the room by the sound wave? How do you know?

2. Lightning and Thunder Light waves travel nearly 1 million times faster than sound waves in air. With this in mind, explain how the distance to a lightning bolt can be determined by counting the seconds between the flash and the sound of the thunder.

ANSWERS

Conceptual Challenge

1. no; As with all waves, the *disturbance* travels, not the material itself. Individual air molecules do not move across the room with the sound wave. Instead, each molecule vibrates in place, and the vibrations are transferred from one particle to the next.
2. Because light travels so much faster than sound, the speed of light can be considered to be effectively infinite in this case, and the flash is considered to have occurred at the same instant we see it. Thus, the time it takes the sound wave to reach the listener multiplied by the speed of sound in air gives the approximate distance between the observer and the lightning bolt.

Demonstration

The Doppler Effect GENERAL

Purpose Show that the observed frequency of sound waves depends on the relative motion between the source of the sound waves and the observer.

Materials battery-operated high-volume oscillator (available at most local electronics stores), appropriate batteries for the oscillator, foam ball large enough to hold the oscillator and batteries

Procedure Carefully cut into the foam ball and remove enough material so that the oscillator and the batteries will snugly fit inside the ball. Connect the batteries to the oscillator, and place the oscillator and batteries securely inside the ball. Allow the students to toss the ball about the classroom. Have the students note the differences in the observed frequency of the sound of the oscillator when the ball is traveling toward them, when the ball is traveling away from them, and when the ball is at rest.

Misconception Alert BASIC

Some students may think that the observed frequency *rises* as the source of sound approaches an observer and *decreases* as the source moves away. Stress the fact that the observed frequency is *higher* or *lower* and that it changes only when the source passes the observer. This concept is illustrated in **Figure 5,** which shows that the distance between wave fronts is constant for each observer.

SCILINKS.
www.scilinks.org
Topic: Doppler Effect
Code: HF60424

THE DOPPLER EFFECT

If you stand on the street while an ambulance speeds by with its siren on, you will notice the pitch of the siren change. The pitch will be higher as the ambulance approaches and will be lower as it moves away. As you read earlier in this section, the pitch of a sound depends on its frequency. But in this case, the siren is not changing its frequency. How can we account for this change in pitch?

Figure 5
As this ambulance moves to the left, Observer *A* hears the siren at a higher frequency than the driver does, while Observer *B* hears a lower frequency.

Relative motion creates a change in frequency

If a siren sounds in a parked ambulance, an observer standing on the street hears the same frequency that the driver hears, as you would expect. When an ambulance is moving, as shown in **Figure 5,** there is relative motion between the moving ambulance and a stationary observer. This relative motion affects the way the wave fronts of the sound waves produced by the siren are perceived by an observer. (For simplicity's sake, we will assume that the sound waves produced by the siren are spherical.)

Although the frequency of the siren remains constant, the wave fronts reach an observer in front of the ambulance (Observer *A*) more often than they would if the ambulance were stationary. The reason is that the source of the sound waves is moving toward the observer. The speed of sound in the air does not change, because the speed depends only on the temperature of the air. Thus, the product of wavelength and frequency remains constant. Because the wavelength is less, the frequency heard by Observer *A* is *greater* than the source frequency.

For the same reason, the wave fronts reach an observer behind the ambulance (Observer *B*) less often than they would if the ambulance were stationary. As a result, the frequency heard by Observer *B* is *less* than the source frequency. This frequency shift is known as the **Doppler effect.** The Doppler effect is named for the Austrian physicist Christian Doppler (1803–1853), who first described it.

Doppler effect
an observed change in frequency when there is relative motion between the source of waves and an observer

We have considered a moving source with respect to a stationary observer, but the Doppler effect also occurs when the observer is moving with respect to a stationary source or when both are moving at different velocities. In other words, the Doppler effect occurs whenever there is *relative motion* between the source of waves and an observer. (If the observer is moving instead of the source, the wavelength in air does not change, but the frequency at which waves arrive at the ear is altered by the motion of the ear relative to the medium.) Although the Doppler effect is most commonly experienced with sound waves, it is a phenomenon common to all waves, including electromagnetic waves, such as visible light.

ADVANCED TOPICS

See "The Doppler Effect and the Big Bang" in **Appendix J: Advanced Topics** to learn how observations of the Doppler effect with light waves have provided evidence for the expansion of the universe.

SECTION REVIEW

1. What is the relationship between frequency and pitch?
2. Dolphin echolocation is similar to ultrasound. Reflected sound waves allow a dolphin to form an image of the object that reflected the waves. Dolphins can produce sound waves with frequencies ranging from 0.25 kHz to 220 kHz, but only those at the upper end of this spectrum are used in echolocation. Explain why high-frequency waves work better than low-frequency waves.
3. Sound pulses emitted by a dolphin travel through 20°C ocean water at a rate of 1450 m/s. In 20°C air, these pulses would travel 342.9 m/s. How can you account for this difference in speed?
4. **Interpreting Graphics** Could a portion of the innermost wave front shown in **Figure 6** be approximated by a plane wave? Why or why not?

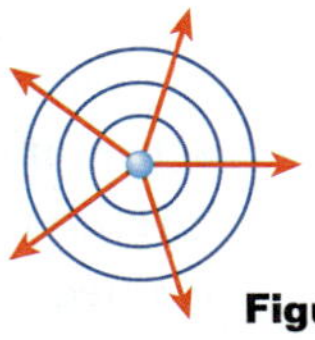

Figure 6

5. **Interpreting Graphics** **Figure 7** is a diagram of the Doppler effect in a ripple tank. In which direction is the source of these ripple waves moving?
6. **Interpreting Graphics** If the source of the waves in **Figure 7** is stationary, which way must the ripple tank be moving?

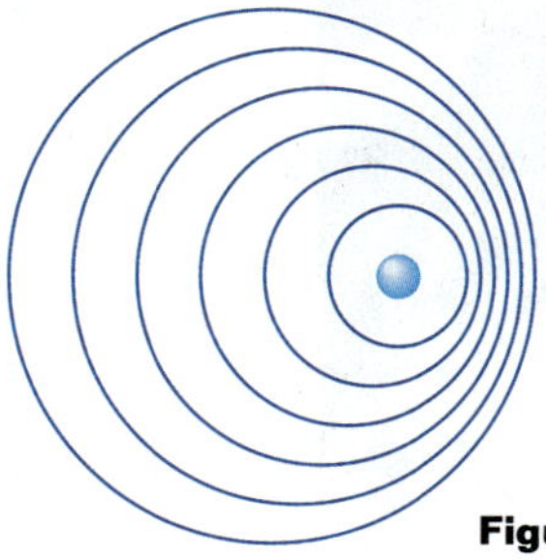

Figure 7

7. **Critical Thinking** As a dolphin swims toward a fish, the dolphin sends out sound waves to determine the direction the fish is moving. If the frequency of the reflected waves is higher than that of the emitted waves, is the dolphin catching up to the fish or falling behind?

SECTION REVIEW ANSWERS

1. A greater frequency is perceived as a higher pitch.
2. Higher frequencies function well in echolocation because their relatively short wavelengths are able to detect smaller objects. (Longer wavelengths would disperse around small objects.)
3. Sound waves travel faster through water than through air because the molecules of water are closer together and, as a result, can spread vibrations more quickly.
4. no; The wave front must be far from the source (relative to the wavelength) to be approximated by plane waves.
5. to the right
6. to the left
7. The dolphin is catching up to the fish.

SECTION 2
General Level

Misconception Alert — BASIC

Some students may have trouble distinguishing between energy and the *rate* of energy transfer, or power. Briefly review the idea of power (introduced in the chapter "Work and Energy") to make sure this distinction is clear to students before intensity is introduced.

Key Models and Analogies — GENERAL

Ask students to imagine a balloon that is being inflated. The surface of the expanding balloon is analogous to a spherical sound wave. Point out that the same amount of material is spread over a larger area as the balloon is inflated. In a similar way, the same amount of energy is spread over a larger area at greater distances from the source of the sound waves. Thus, intensity and distance are inversely related, as seen in the equation for the intensity of a spherical wave.

SECTION 2

Sound Intensity and Resonance

SECTION OBJECTIVES

- Calculate the intensity of sound waves.
- Relate intensity, decibel level, and perceived loudness.
- Explain why resonance occurs.

SOUND INTENSITY

When a piano player strikes a piano key, a hammer inside the piano strikes a wire and causes it to vibrate, as shown in **Figure 8.** The wire's vibrations are then transferred to the piano's soundboard. As the soundboard vibrates, it exerts a force on air molecules around it, causing the air molecules to move. Because this force is exerted through displacement of the soundboard, the soundboard does work on the air. Thus, as the soundboard vibrates back and forth, its kinetic energy is converted into sound waves. This is one reason that the vibration of the soundboard gradually dies out.

Intensity is the rate of energy flow through a given area

As described in Section 1, sound waves traveling in air are longitudinal waves. As the sound waves travel outward from the source, energy is transferred from one air molecule to the next. The rate at which this energy is transferred through a unit area of the plane wave is called the **intensity** of the wave. Because power, P, is defined as the rate of energy transfer, intensity can also be described in terms of power.

intensity
the rate at which energy flows through a unit area perpendicular to the direction of wave motion

$$\text{intensity} = \frac{\Delta E/\Delta t}{\text{area}} = \frac{P}{\text{area}}$$

The SI unit for power is the watt. Thus, intensity has units of watts per square meter (W/m^2). In a spherical wave, energy propagates equally in all directions; no one direction is preferred over any other. In this case, the power emitted by the source (P) is distributed over a spherical surface (area $= 4\pi r^2$), assuming that there is no absorption in the medium.

INTENSITY OF A SPHERICAL WAVE

$$\text{intensity} = \frac{P}{4\pi r^2}$$

$$\textbf{intensity} = \frac{\textbf{(power)}}{(4\pi)\textbf{(distance from the source)}^2}$$

Figure 8
As a piano wire vibrates, it transfers energy to the piano's soundboard, which in turn transfers energy into the air in the form of sound.

This equation shows that the intensity of a sound wave decreases as the distance from the source (r) increases. This occurs because the same amount of energy is spread over a larger area.

SAMPLE PROBLEM A

Intensity of Sound Waves

PROBLEM

What is the intensity of the sound waves produced by a trumpet at a distance of 3.2 m when the power output of the trumpet is 0.20 W? Assume that the sound waves are spherical.

SOLUTION

Given: $P = 0.20$ W $\quad r = 3.2$ m

Unknown: Intensity = ?

Use the equation for the intensity of a spherical wave.

$$\text{Intensity} = \frac{P}{4\pi r^2}$$

$$\text{Intensity} = \frac{0.20 \text{ W}}{4\pi(3.2 \text{ m})^2}$$

$$\boxed{\text{Intensity} = 1.6 \times 10^{-3} \text{ W/m}^2}$$

CALCULATOR SOLUTION

The calculator answer for intensity is 0.0015542. This is rounded to 1.6×10^{-3} because each of the given quantities has two significant figures.

PRACTICE A

Intensity of Sound Waves

1. Calculate the intensity of the sound waves from an electric guitar's amplifier at a distance of 5.0 m when its power output is equal to each of the following values:

a. 0.25 W

b. 0.50 W

c. 2.0 W

2. At a maximum level of loudness, the power output of a 75-piece orchestra radiated as sound is 70.0 W. What is the intensity of these sound waves to a listener who is sitting 25.0 m from the orchestra?

3. If the intensity of a person's voice is 4.6×10^{-7} W/m^2 at a distance of 2.0 m, how much sound power does that person generate?

4. How much power is radiated as sound from a band whose intensity is 1.6×10^{-3} W/m^2 at a distance of 15 m?

5. The power output of a tuba is 0.35 W. At what distance is the sound intensity of the tuba 1.2×10^{-3} W/m^2?

Classroom Practice

Intensity of Sound Waves
The intensity of the sound from an explosion is 0.10 W/m^2 at a distance of 1.0×10^3 m. Find the intensity of the sound at distances of 5.0×10^2 m, 1.0×10^2 m, and 10.0 m.

Answer

0.41 W/m^2, 1.0×10^1 W/m^2, 1.0×10^3 W/m^2

PROBLEM GUIDE A

Use this guide to assign problems.
SE = Student Edition Textbook
PW = Problem Workbook
PB = Problem Bank on the One-Stop Planner (OSP)

Solving for:

I	**SE** Sample, 1–2; Ch. Rvw. 22–23 **PW** 7–9 **PB** 5–7
P	**SE** 3–4; Ch. Rvw. 43* **PW** Sample, 1–3 **PB** 8–10
r	**SE** 5 **PW** 4–6 **PB** Sample, 1–4

***Challenging Problem**
Consult the printed Solutions Manual or the OSP for detailed solutions.

ANSWERS

Practice A

1. a. 8.0×10^{-4} W/m^2
   b. 1.6×10^{-3} W/m^2
   c. 6.4×10^{-3} W/m^2
2. 8.91×10^{-3} W/m^2
3. 2.3×10^{-5} W
4. 4.5 W
5. 4.8 m

Visual Strategy GENERAL

Figure 9

Be certain that students understand the information contained in the different regions of the graph. Also point out that the scale of the y-axis of this graph is logarithmic. Thus, the intensity represented by each horizontal line is 100 times greater than the intensity represented by the line immediately below that line.

Q Are there musical sounds of 1000 Hz and 1.0×10^{-6} W/m^2?

A ***yes, because the speech region is a subset of the music region***

Q Does this graph describe an individual's hearing exactly?

A ***No, the graph is based on the average human ear. Each individual's hearing may vary.***

Misconception Alert

The relationship between frequency, intensity, and audibility is complex and often confusing to students. Stress that neither frequency nor intensity alone can determine which sounds are audible; both factors must be taken into account.

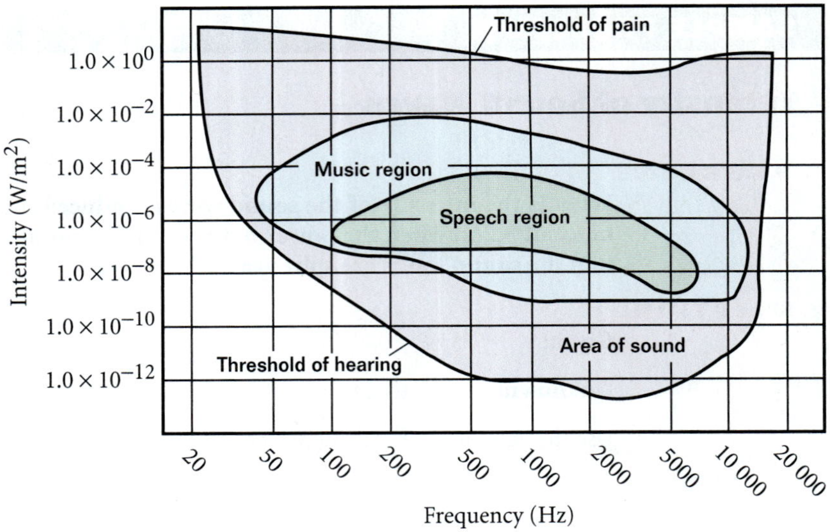

Figure 9
Human hearing depends on both the frequency and the intensity of sound waves. Sounds in the middle of the spectrum of frequencies can be heard more easily (at lower intensities) than those at lower and higher frequencies.

Intensity and frequency determine which sounds are audible

As you saw in Section 1, the frequency of sound waves heard by the average human ranges from 20 to 20 000 Hz. Intensity is also a factor in determining which sound waves are audible. **Figure 9** shows how the range of audibility of the average human ear depends on both frequency and intensity. As you can see in this graph, sounds at low frequencies (those below 50 Hz) or high frequencies (those above 12 000 Hz) must be relatively intense to be heard, whereas sounds in the middle of the spectrum are audible at lower intensities.

The softest sounds that can be heard by the average human ear occur at a frequency of about 1000 Hz and an intensity of 1.0×10^{-12} W/m^2. Such a sound is said to be at the *threshold of hearing*. (Note that some humans can hear slightly softer sounds, at a frequency of about 3300 Hz.) The threshold of hearing at each frequency is represented by the lowest curve in **Figure 9.**

For frequencies near 1000 Hz and at the threshold of hearing, the changes in pressure due to compressions and rarefactions are about three ten-billionths of atmospheric pressure. The maximum displacement of an air molecule at the threshold of hearing is approximately 1×10^{-11} m. Comparing this number to the diameter of a typical air molecule (about 1×10^{-10} m) reveals that the ear is an extremely sensitive detector of sound waves.

The loudest sounds that the human ear can tolerate have an intensity of about 1.0 W/m^2. This is known as the *threshold of pain* because sounds with greater intensities can produce pain in addition to hearing. The highest curve in **Figure 9** represents the threshold of pain at each frequency. Exposure to sounds above the threshold of pain can cause immediate damage to the ear, even if no pain is felt. Prolonged exposure to sounds of lower intensities can also damage the ear. For this reason, many musicians wear earplugs during their performances. Note that the threshold of hearing and the threshold of pain merge at both high and low ends of the spectrum.

Did you know?

A 75-piece orchestra produces about 75 W at its loudest. This is comparable to the power required to keep one medium-sized electric light bulb burning. Speech has even less power. It would take the conversation of about 2 million people to provide the amount of power required to keep a 50 W light bulb burning.

Relative intensity is measured in decibels

Just as the frequency of a sound wave determines its pitch, the intensity of a wave approximately determines its perceived loudness. However, loudness is not directly proportional to intensity. The reason is that the sensation of loudness is approximately logarithmic in the human ear.

Relative intensity is the ratio of the intensity of a given sound wave to the intensity at the threshold of hearing. Because of the logarithmic dependence of perceived loudness on intensity, using a number equal to 10 times the logarithm of the relative intensity provides a good indicator for human perceptions of loudness. This measure of loudness is referred to as the *decibel level.* The decibel level is dimensionless because it is proportional to the logarithm of a ratio. A dimensionless unit called the **decibel** (dB) is used for values on this scale.

The conversion of intensity to decibel level is shown in **Table 2.** Notice in **Table 2** that when the intensity is multiplied by 10, 10 dB are added to the decibel level. A given difference in decibels corresponds to a fixed difference in perceived loudness. Although much more intensity (0.9 W/m^2) is added between 110 and 120 dB than between 10 and 20 dB (9×10^{-11} W/m^2), in each case the perceived loudness increases by the same amount.

decibel

a dimensionless unit that describes the ratio of two intensities of sound; the threshold of hearing is commonly used as the reference intensity

Table 2 Conversion of Intensity to Decibel Level

Intensity (W/m^2)	Decibel level (dB)	Examples
1.0×10^{-12}	0	threshold of hearing
1.0×10^{-11}	10	rustling leaves
1.0×10^{-10}	20	quiet whisper
1.0×10^{-9}	30	whisper
1.0×10^{-8}	40	mosquito buzzing
1.0×10^{-7}	50	normal conversation
1.0×10^{-6}	60	air conditioning at 6 m
1.0×10^{-5}	70	vacuum cleaner
1.0×10^{-4}	80	busy traffic, alarm clock
1.0×10^{-3}	90	lawn mower
1.0×10^{-2}	100	subway, power motor
1.0×10^{-1}	110	auto horn at 1 m
1.0×10^{0}	120	threshold of pain
1.0×10^{1}	130	thunderclap, machine gun
1.0×10^{3}	150	nearby jet airplane

Did you know?

The original unit of decibel level is the *bel,* named in honor of Alexander Graham Bell, the inventor of the telephone. The decibel is equivalent to 0.1 bel.

extension

Integrating Health

Visit go.hrw.com for the activity "Why Your Ears Pop."

Keyword HF6SNDX

SECTION 2

Misconception Alert

Some students may confuse decibel level with intensity. Point out that the ratio of one decibel level to another does *not* give the ratio between the intensities of these sounds because the decibel scale is logarithmic. Work with examples in **Table 2** to correct this misconception.

Key Models and Analogies

The decibel scale of loudness is designed similarly to the Richter scale, which measures earthquake intensity, and the pH scale, which measures acidity levels. The values on these scales correspond to the order of magnitude (power of 10) of the original quantities because these quantities span a very large range of values.

Visual Strategy GENERAL

Table 2

Point out that zero on the decibel scale does not mean zero intensity or that there is no sound but instead that the sound is inaudible.

Q How does the sound intensity of a subway compare with that of a conversation?

A ***At equal distances, noise from a subway brings to our ears 100 000 ($10^{-2}/10^{-7}$) times more energy than a conversation. (Some students may think the answer is twice as much, or the ratio of 100 to 50. Explain why this is not the case.)***

SECTION 2

Demonstration

Resonance GENERAL

Purpose Show resonance with tuning forks.

Materials several tuning forks (including two of the same frequency), two identical resonance boxes, a rubber mallet

Procedure Place the resonance boxes 15 cm apart with the open ends facing each other. Using the two forks that have the same frequency, mount a tuning fork on each box. Strike one fork with the mallet. Students who are listening carefully should begin to hear a faint sound from the second tuning fork when you stop the vibrations of the first tuning fork. Explain that because energy is transferred through air and wood, the second fork picked up these vibrations because its natural frequency matched that of the first fork. Repeat the experiment with forks of different frequencies to show that resonance does not occur in those cases.

Quick Lab

TEACHER'S NOTES

Students should recognize that resonance occurs when the frequency of the applied force (the pumping or pushing) matches the natural frequency of the system (the swing).

Homework Options This QuickLab can easily be performed outside of the physics lab room.

FORCED VIBRATIONS AND RESONANCE

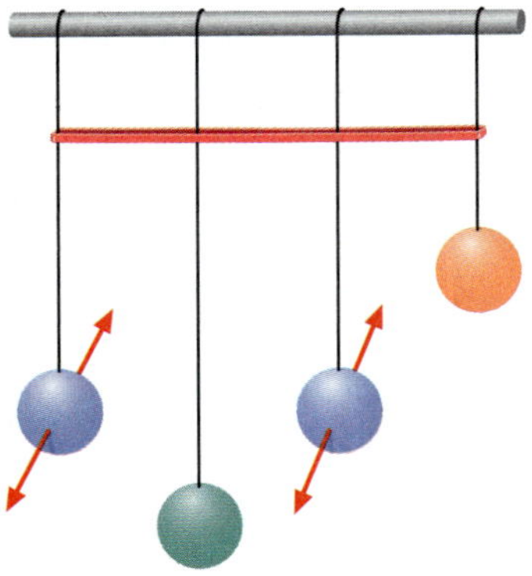

Figure 10
If one blue pendulum is set in motion, only the other blue pendulum, whose length is the same, will eventually oscillate with a large amplitude, or resonate.

When an isolated guitar string is held taut and plucked, hardly any sound is heard. When the same string is placed on a guitar and plucked, the intensity of the sound increases dramatically. What is responsible for this difference? To find the answer to this question, consider a set of pendulums suspended from a beam and bound by a loose rubber band, as shown in **Figure 10.** If one of the pendulums is set in motion, its vibrations are transferred by the rubber band to the other pendulums, which will also begin vibrating. This is called a *forced vibration.*

The vibrating strings of a guitar force the bridge of the guitar to vibrate, and the bridge in turn transfers its vibrations to the guitar body. These forced vibrations are called *sympathetic vibrations.* Because the guitar body has a larger area than the strings do, it enables the strings' vibrations to be transferred to the air more efficiently. As a result, the intensity of the sound is increased, and the strings' vibrations die out faster than they would if they were not attached to the body of the guitar. In other words, the guitar body allows the energy exchange between the strings and the air to happen more efficiently, thereby increasing the intensity of the sound produced.

In an electric guitar, string vibrations are translated into electrical impulses, which can be amplified as much as desired. An electric guitar can produce sounds that are much more intense than those of an unamplified acoustic guitar, which uses only the forced vibrations of the guitar's body to increase the intensity of the sound from the vibrating strings.

SCILINKS.
www.scilinks.org
Topic: Resonance
Code: HF61303

Vibration at the natural frequency produces resonance

As you saw in the chapter on waves, the frequency of a pendulum depends on its string length. Thus, every pendulum will vibrate at a certain frequency, known as its *natural frequency.* In **Figure 10,** the two blue pendulums have the same natural frequency, while the red and green pendulums have different natural frequencies. When the first blue pendulum is set in motion, the red and green pendulums will vibrate only slightly, but the second blue pendulum will oscillate with a much larger amplitude because its natural frequency

Quick Lab

Resonance

MATERIALS LIST

- swing set

Go to a playground, and swing on one of the swings. Try pumping (or being pushed) at different rates—faster than, slower than, and equal to the natural frequency of the swing. Observe whether the rate at which you pump (or are pushed) affects how easily the amplitude of the vibration increases. Are some rates more effective at building your amplitude than others? You should find that the pushes are most effective when they match the swing's natural frequency. Explain how your results support the statement that resonance works best when the frequency of the applied force matches the system's natural frequency.

matches the frequency of the pendulum that was initially set in motion. This system is said to be in **resonance.** Because energy is transferred from one pendulum to the other, the amplitude of vibration of the first blue pendulum will decrease as the second blue pendulum's amplitude increases.

resonance

a phenomenon that occurs when the frequency of a force applied to a system matches the natural frequency of vibration of the system, resulting in a large amplitude of vibration

A striking example of structural resonance occurred in 1940, when the Tacoma Narrows bridge, in Washington, shown in **Figure 11,** was set in motion by the wind. High winds set up standing waves in the bridge, causing the bridge to oscillate at one of its natural frequencies. The amplitude of the vibrations increased until the bridge collapsed. A more recent example of structural resonance occurred during the Loma Prieta earthquake near Oakland, California, in 1989, when part of the upper deck of a freeway collapsed. The collapse of this particular section of roadway has been traced to the fact that the earthquake waves had a frequency of 1.5 Hz, very close to the natural frequency of that section of the roadway.

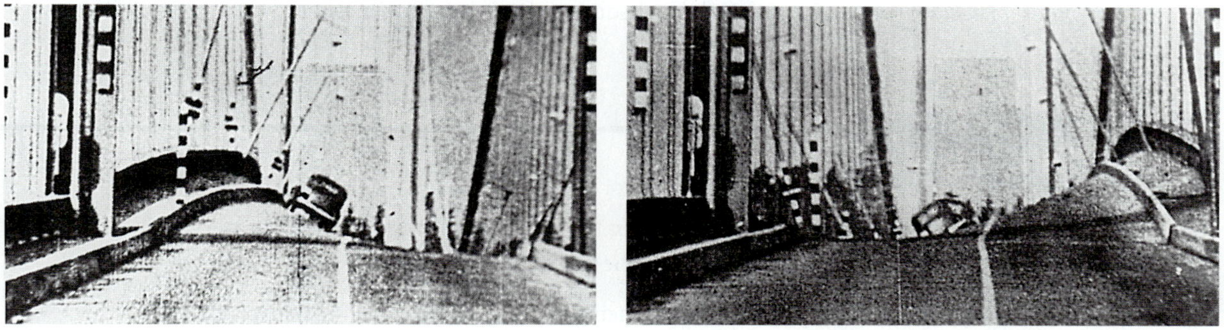

Figure 11
On November 7, 1940, the Tacoma Narrows suspension bridge collapsed, just four months after it opened. Standing waves caused by strong winds set the bridge in motion and led to its collapse.

Why it Matters

Conceptual Challenge

1. Concert If a 15-person musical ensemble gains 15 new members, so that its size doubles, will a listener perceive the music created by the ensemble to be twice as loud? Why or why not?

2. A Noisy Factory Federal regulations require that no office or factory worker be exposed to noise levels that average above 90 dB over an 8 h day. Thus, a factory that currently averages 100 dB must reduce its noise level by 10 dB. Assuming that each piece of machinery produces the same amount of noise, what percentage of equipment must be removed? Explain your answer.

3. Broken Crystal Opera singers have been known to set crystal goblets in vibration with their powerful voices. In fact, an amplified human voice can shatter the glass, but only at certain fundamental frequencies. Speculate about why only certain fundamental frequencies will break the glass.

4. Electric Guitars Electric guitars, which use electric amplifiers to magnify their sound, can have a variety of shapes, but acoustic guitars all have the same basic shape. Explain why.

ANSWERS

Conceptual Challenge

1. no; The intensity doubles, but loudness is not directly proportional to intensity.
2. Reducing the decibel level by 10 requires reducing the intensity by a factor of 10. Thus, 90 percent of the equipment must be removed.
3. Only frequencies that match one of the natural frequencies of the glass can establish a resonance condition. Only then can the vibrations become large enough to shatter the goblet.
4. The body of an acoustic guitar is designed so that it can transmit the strings' vibrations to the air efficiently. (The guitar's particular shape also determines the characteristic sound that distinguishes the guitar from similar instruments, such as banjos and lutes.) Because the vibrations of an electric guitar's strings are converted into electrical signals, there are fewer restrictions on the electric guitar's shape.

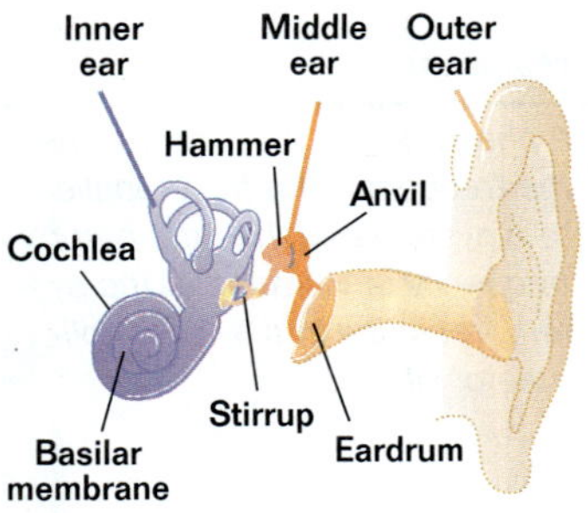

Figure 12
Sound waves travel through the three regions of the ear and are then transmitted to the brain as impulses through nerve endings on the basilar membrane.

The human ear transmits vibrations that cause nerve impulses

The human ear is divided into three sections—outer, middle, and inner—as shown in **Figure 12.** Sound waves travel down the ear canal of the outer ear. The ear canal terminates at a thin, flat piece of tissue called the *eardrum.*

The eardrum vibrates with the sound waves and transfers these vibrations to the three small bones of the middle ear, known as the *hammer,* the *anvil,* and the *stirrup.* These bones in turn transmit the vibrations to the inner ear, which contains a snail-shaped tube about 2 cm long called the *cochlea.*

The *basilar membrane* runs through the coiled cochlea, dividing it roughly in half. The basilar membrane has different natural frequencies at different positions along its length, according to the width and thickness of the membrane at that point. Sound waves of varying frequencies resonate at different spots along the basilar membrane, creating impulses in hair cells—specialized nerve cells—embedded in the membrane. These impulses are then sent to the brain, which interprets them as sounds of varying frequencies.

SECTION REVIEW

1. When the decibel level of traffic in the street goes from 40 to 60 dB, how much greater is the intensity of the noise?
2. If two flutists play their instruments together at the same intensity, is the sound twice as loud as that of either flutist playing alone at that intensity? Why or why not?
3. A tuning fork consists of two metal prongs that vibrate at a single frequency when struck lightly. What will happen if a vibrating tuning fork is placed near another tuning fork of the same frequency? Explain.
4. A certain microphone placed in the ocean is sensitive to sounds emitted by dolphins. To produce a usable signal, sound waves striking the microphone must have a decibel level of 10 dB. If dolphins emit sound waves with a power of 0.050 W, how far can a dolphin be from the microphone and still be heard? (Assume the sound waves propagate spherically, and disregard absorption of the sound waves.)
5. **Critical Thinking** Which of the following factors change when a sound gets louder? Which change when a pitch gets higher?
   - **a.** intensity
   - **b.** speed of the sound waves
   - **c.** frequency
   - **d.** decibel level
   - **e.** wavelength
   - **f.** amplitude

SECTION REVIEW ANSWERS

1. The intensity has increased by a factor of 100 (10^2).
2. no; because the sensation of loudness is approximately logarithmic in the human ear
3. The second tuning fork will pick up the vibrations of the first tuning fork, and a faint sound will be heard from the second fork. This occurs because the two forks have the same natural frequency, which is the condition required for resonance.
4. 2.0×10^4 m
5. a, d, f; c, e

Why it Matters

Hearing Loss

To prevent damage to their ears, people should wear ear protection when working with power tools.

About 10 percent of all Americans have some degree of hearing loss. There are three basic types of hearing loss. *Conductive hearing loss* is an impairment of the transmission of sound waves in the outer ear or transmission of vibrations in the middle ear. Conductive hearing loss is most often caused by improper development of the parts of the outer or middle ear or by damage to these parts of the ear by physical trauma or disease. Conductive hearing loss can often be corrected with medicine or surgery. *Neural hearing loss* is caused by problems with the auditory nerve, which carries signals from the inner ear to the brain. One common cause of neural hearing loss is a tumor pressing against the auditory nerve. *Sensory hearing loss* is caused by damage to the inner ear, particularly the microscopic hair cells in the cochlea.

Sensory hearing loss can be present at birth and may be genetic or due to disease or developmental disorders. However, the most common source of damage to hair cells is exposure to loud noise. Short-term exposure to loud noise can cause ringing in the ears and temporary hearing impairment. Frequent or long-term exposure to noise above 70 dB—including noise from familiar sources such as hair dryers or lawn mowers—can damage the hair cells permanently.

The hair cells in the cochlea are not like the hair on your head or skin. They are highly specialized nerve cells that cannot be repaired or replaced by the body when they are severely damaged or destroyed. Cochlear hair cells can recover from minor damage, but if the source of the damage recurs frequently, even if it is only moderately loud noise, the hair cells may not have time to recover and can become permanently damaged. It is therefore important to protect yourself from sensory hearing loss by reducing your exposure to loud noise or by using a noise-dampening headset or earplugs that fully block the ear canal when you must be exposed to loud noise.

Permanent sensory hearing loss usually occurs gradually, sometimes over 20 years or more. Because the hair cells that respond to higher-pitched sounds are smaller and more delicate, sensitivity to sounds with frequencies around 20 kHz is usually the first to be lost. Loss of sensitivity to sounds with frequencies around 4 kHz is often the first to be noticed because these frequencies are in the upper range of human speech. People who are starting to lose their hearing often have trouble hearing higher-pitched voices or hearing consonant sounds such as *s, t, p, d,* and *f*. As the hearing loss advances, loss of sensitivity to a wider range of sounds follows.

Although there is currently no true "cure" for sensory hearing loss, some remedies are available. *Hearing aids* act like tiny amplifiers, making any sounds that reach the ear louder. *Assistive listening devices* serve to amplify a specific small range of frequencies for people who have only partial hearing loss in that range. *Cochlear implants* use an electrode that is surgically implanted into the cochlea through a hole behind the outer ear. Electrical signals to the electrode stimulate the auditory nerve directly, in effect bypassing the hair cells altogether.

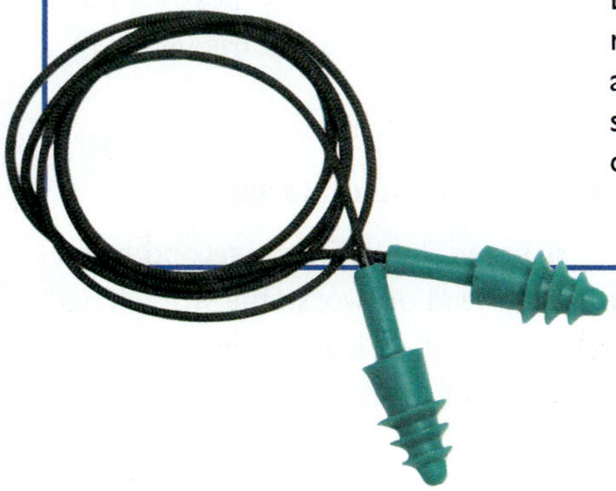

Why it Matters

Hearing Loss

Understanding sensory hearing loss due to noise is important because it can be prevented. It is especially important for younger people because the effects of exposure to noise accumulate over time. Damage to cochlear hair cells may have no obvious effect at a younger age, but if exposure to noise continues, over time it can lead to significant or even total hearing loss. Musicians (even classical musicians) and people who live or work in locations with persistent loud noise are at high risk of sensory hearing loss. Even occasional exposure to high noise levels, such as at music concerts, can cause significant hearing loss over time. Remind students that listening to headphones frequently, especially at high volumes, can also damage hair cells. Because the ear is so close to the source of sound from headphone speakers, the sound intensity can be very high at the ear.

Extension

Use a sound-level meter to measure the sound intensities or decibel levels of sounds around the classroom or school. If you have more than one meter, you can split the class into groups to measure sound levels. You may wish to loan sound-level meters out to students who show a particular interest so that they can measure sound levels in other environments.

Demonstration

Seeing Sounds — GENERAL

Purpose Observe sound waves from a variety of sources.

Materials oscilloscope, microphone, small amplifier (if needed), assorted sound sources

CAUTION *Consult the oscilloscope's user's manual for instructions on the proper use of the oscilloscope.*

Procedure Connect the microphone or the output from the microphone and amplifier to the input of the oscilloscope. Display the sound pattern of the human voice by having several students speak or sing into the microphone. Have the students note the patterns on the screen of the oscilloscope. Show the characteristic of a single frequency sound by inputting the sound from an oscillator, sine wave generator, tuning fork, or a student's voice at a single pitch.

Then have several students play the same note on various instruments one at a time. Have the students note the characteristic harmonics of each instrument as they are displayed on the oscilloscope screen.

SECTION 3

Harmonics

SECTION OBJECTIVES

- **Differentiate between the harmonic series of open and closed pipes.**
- **Calculate the harmonics of a vibrating string and of open and closed pipes.**
- **Relate harmonics and timbre.**
- **Relate the frequency difference between two waves to the number of beats heard per second.**

STANDING WAVES ON A VIBRATING STRING

As discussed in the chapter "Vibrations and Waves," a variety of standing waves can occur when a string is fixed at both ends and set into vibration. The vibrations on the string of a musical instrument, such as the violin in **Figure 13,** usually consist of many standing waves together at the same time, each of which has a different wavelength and frequency. So, the sounds you hear from a stringed instrument, even those that sound like a single pitch, actually consist of multiple frequencies.

Table 3, on the next page, shows several possible vibrations on an idealized string. The ends of the string, which cannot vibrate, must always be nodes (N). The simplest vibration that can occur is shown in the first row of **Table 3.** In this case, the center of the string experiences the most displacement, and so it is an antinode (A). Because the distance from one node to the next is always half a wavelength, the string length (L) must equal $\lambda_1/2$. Thus, the wavelength is twice the string length ($\lambda_1 = 2L$).

As described in the chapter on waves, the speed of a wave equals the frequency times the wavelength, which can be rearranged as shown.

$$v = f\lambda, \text{ so } f = \frac{v}{\lambda}$$

By substituting the value for wavelength found above into this equation for frequency, we see that the frequency of this vibration is equal to the speed of the wave divided by twice the string length.

$$\text{fundamental frequency} = f_1 = \frac{v}{\lambda_1} = \frac{v}{2L}$$

Figure 13
The vibrating strings of a violin produce standing waves whose frequencies depend on the string lengths.

This frequency of vibration is called the **fundamental frequency** of the vibrating string. Because frequency is inversely proportional to wavelength and because we are considering the greatest possible wavelength, the fundamental frequency is the lowest possible frequency of a standing wave on this string.

fundamental frequency
the lowest frequency of vibration of a standing wave

Harmonics are integral multiples of the fundamental frequency

The next possible standing wave for a string is shown in the second row of **Table 3.** In this case, there are three nodes instead of two, so the string length is equal to one wavelength. Because this wavelength is half the previous wavelength, the frequency of this wave is twice that of the fundamental frequency.

$$f_2 = 2f_1$$

Table 3 The Harmonic Series

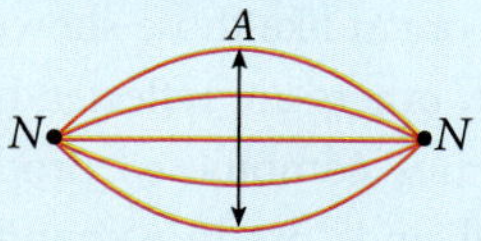	$\lambda_1 = 2L$	f_1	fundamental frequency, or first harmonic
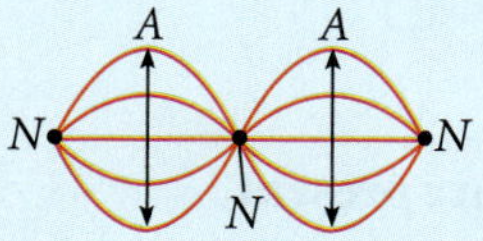	$\lambda_2 = L$	$f_2 = 2f_1$	second harmonic
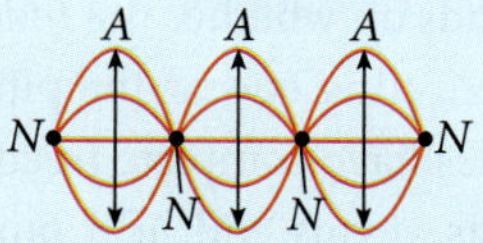	$\lambda_3 = \frac{2}{3}L$	$f_3 = 3f_1$	third harmonic
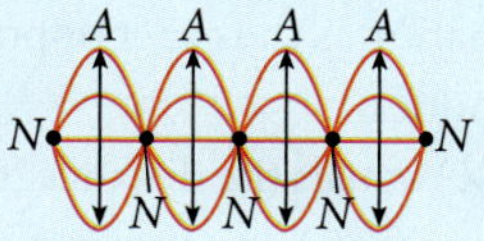	$\lambda_4 = \frac{1}{2}L$	$f_4 = 4f_1$	fourth harmonic

This pattern continues, and the frequency of the standing wave shown in the third row of **Table 3** is three times the fundamental frequency. More generally, the frequencies of the standing wave patterns are all integral multiples of the fundamental frequency. These frequencies form what is called a **harmonic series.** The fundamental frequency (f_1) corresponds to the first harmonic, the next frequency (f_2) corresponds to the second harmonic, and so on.

Because each harmonic is an integral multiple of the fundamental frequency, the equation for the fundamental frequency can be generalized to include the entire harmonic series. Thus, $f_n = nf_1$, where f_1 is the fundamental frequency ($f_1 = \frac{v}{2L}$) and f_n is the frequency of the nth harmonic. The general form of the equation is written as follows:

HARMONIC SERIES OF STANDING WAVES ON A VIBRATING STRING

$$f_n = n\frac{v}{2L} \quad n = 1, 2, 3, \ldots$$

$$\textbf{frequency} = \textbf{harmonic number} \times \frac{\textbf{(speed of waves on the string)}}{\textbf{(2)(length of vibrating string)}}$$

Note that v in this equation is the speed of waves on the vibrating string and not the speed of the resultant sound waves in air. If the string vibrates at one of these frequencies, the sound waves produced in the surrounding air will have the same frequency. However, the speed of these waves will be the speed of sound waves in air, and the wavelength of these waves will be that speed divided by the frequency.

SC*i*LINKS.
www.scilinks.org
Topic: Harmonics
Code: HF60715

harmonic series

a series of frequencies that includes the fundamental frequency and integral multiples of the fundamental frequency

Did you know?

When a guitar player presses down on a guitar string at any point, that point becomes a node and only a portion of the string vibrates. As a result, a single string can be used to create a variety of fundamental frequencies. In the equation on this page, L refers to the portion of the string that is vibrating.

Teaching Tip — BASIC

Have students visualize the harmonics to help them recall the appropriate equations. Both ends of a string fixed at each end must be nodes. For the first harmonic (one loop), the length L of the string must equal $\frac{1}{2}\lambda$. Thus, $\frac{1}{2}\lambda = L$ and $\lambda = 2L$. For other harmonics, students can use this same technique with the appropriate multiples of $\frac{1}{2}\lambda$.

Visual Strategy GENERAL

Table 3

Be sure students understand that each loop corresponds to half a wavelength.

Q Find the wavelength (λ_5) and frequency (f_5) for the next possible case in the harmonic series.

A $\lambda_5 = \frac{2}{5}L, f_5 = 5f_1$

Teaching Tip — ADVANCED

Point out that frequency depends on both string length and wave speed, as shown by the equation for the harmonic series. Thus, two strings of the same length will not necessarily have the same fundamental frequency. The string's tension and mass per unit length affect the speed of waves on the string, so the fundamental frequency can be changed by varying either of these factors. For example, a guitar player tunes each string of the guitar by adjusting the tension. This adjustment changes the speed of the waves on the string and thus also changes the string's fundamental frequency.

Teaching Tip

The memory device used for a vibrating string can be used for open and closed pipes as well. Ask students to visualize the standing wave, remembering that closed ends must be nodes and open ends must be antinodes. Sketching diagrams for each case will show that pipes closed at one end must have half a loop ($\frac{1}{4}\lambda$) more than pipes open at both ends.

Figure 14
The harmonic series present in each of these organ pipes depends on whether the end of the pipe is open or closed.

STANDING WAVES IN AN AIR COLUMN

Standing waves can also be set up in a tube of air, such as the inside of a trumpet, the column of a saxophone, or the pipes of an organ like those shown in **Figure 14.** While some waves travel down the tube, others are reflected back upward. These waves traveling in opposite directions combine to produce standing waves. Many brass instruments and woodwinds produce sound by means of these vibrating air columns.

If both ends of a pipe are open, all harmonics are present

The harmonic series present in an organ pipe depends on whether the reflecting end of the pipe is open or closed. When the reflecting end of the pipe is open, as is illustrated in **Figure 15,** the air molecules have complete freedom of motion, so an antinode (of displacement) exists at this end. If a pipe is open at both ends, each end is an antinode. This situation is the exact opposite of a string fixed at both ends, where both ends are nodes.

Because the distance from one node to the next ($\frac{1}{2}\lambda$) equals the distance from one antinode to the next, the pattern of standing waves that can occur in a pipe open at both ends is the same as that of a vibrating string. Thus, the entire harmonic series is present in this case, as shown in **Figure 15,** and our earlier equation for the harmonic series of a vibrating string can be used.

HARMONIC SERIES OF A PIPE OPEN AT BOTH ENDS

$$f_n = n\frac{v}{2L} \quad n = 1, 2, 3, \ldots$$

$$\textbf{frequency} = \textbf{harmonic number} \times \frac{\textbf{(speed of sound in the pipe)}}{\textbf{(2)(length of vibrating air column)}}$$

In this equation, L represents the length of the vibrating air column. Just as the fundamental frequency of a string instrument can be varied by changing the string length, the fundamental frequency of many woodwind and brass instruments can be varied by changing the length of the vibrating air column.

Did you know?

A flute is similar to a pipe open at both ends. When all keys of a flute are closed, the length of the vibrating air column is approximately equal to the length of the flute. As the keys are opened one by one, the length of the vibrating air column decreases, and the fundamental frequency increases.

Harmonics in an open-ended pipe

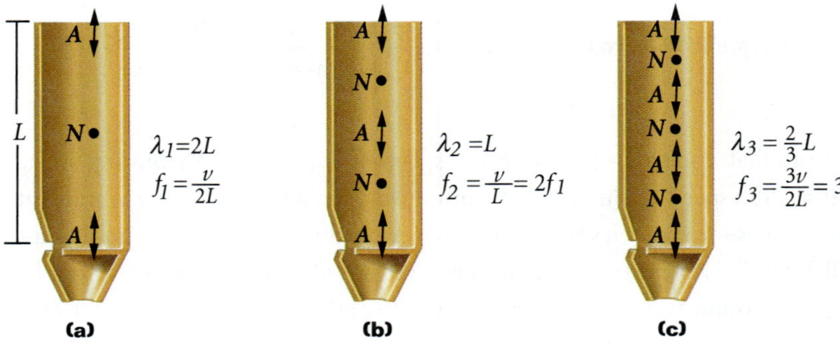

Figure 15
In a pipe open at both ends, each end is an antinode of displacement, and all harmonics are present. Shown here are the **(a)** first, **(b)** second, and **(c)** third harmonics.

Harmonics in a pipe closed at one end

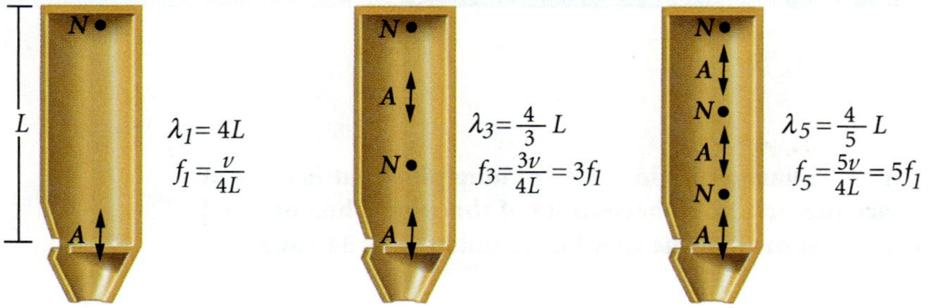

Figure 16
In a pipe closed at one end, the closed end is a node of displacement and the open end is an antinode of displacement. In this case, only the odd harmonics are present. The **(a)** first, **(b)** third, and **(c)** fifth harmonics are shown here.

If one end of a pipe is closed, only odd harmonics are present

When one end of an organ pipe is closed, as is illustrated in **Figure 16,** the movement of air molecules is restricted at this end, making this end a node. In this case, one end of the pipe is a node and the other is an antinode. As a result, a different set of standing waves can occur.

As shown in **Figure 16(a),** the simplest possible standing wave that can exist in this pipe is one for which the length of the pipe is equal to one-fourth of a wavelength. Hence, the wavelength of this standing wave equals four times the length of the pipe. Thus, in this case, the fundamental frequency equals the velocity divided by four times the pipe length.

$$f_1 = \frac{v}{\lambda_1} = \frac{v}{4L}$$

For the case shown in **Figure 16(b),** the length of the pipe is equal to three-fourths of a wavelength, so the wavelength is four-thirds the length of the pipe ($\lambda_3 = \frac{4}{3}L$). Substituting this value into the equation for frequency gives the frequency of this harmonic.

$$f_3 = \frac{v}{\lambda_3} = \frac{v}{\frac{4}{3}L} = \frac{3v}{4L} = 3f_1$$

The frequency of this harmonic is *three* times the fundamental frequency. Repeating this calculation for the case shown in **Figure 16(c)** gives a frequency equal to *five* times the fundamental frequency. Thus, only the odd-numbered harmonics vibrate in a pipe closed at one end. We can generalize the equation for the harmonic series of a pipe closed at one end as follows:

HARMONIC SERIES OF A PIPE CLOSED AT ONE END

$$f_n = n\frac{v}{4L} \qquad n = 1, 3, 5, \ldots$$

$$\textbf{frequency} = \textbf{harmonic number} \times \frac{\textbf{(speed of sound in the pipe)}}{\textbf{(4)(length of vibrating air column)}}$$

Quick Lab

A Pipe Closed at One End

MATERIALS LIST

- **straw**
- **scissors**

SAFETY

Always use caution when working with scissors.

Snip off the corners of one end of the straw so that the end tapers to a point, as shown above. Chew on this end to flatten it, and you create a double-reed instrument! Put your lips around the tapered end of the straw, press them together tightly, and blow through the straw. When you hear a steady tone, slowly snip off pieces of the straw at the other end. Be careful to keep about the same amount of pressure with your lips. How does the pitch change as the straw becomes shorter? How can you account for this change in pitch? You may be able to produce more than one tone for any given length of the straw. How is this possible?

Quick Lab

TEACHER'S NOTES

This lab demonstrates the effect of pipe length on pitch. The experiment works best with narrow paper straws.

Students should find that the shorter the straw is, the higher the fundamental frequency is. One length of straw can produce more than one possible tone because there are several possible wavelengths of standing waves for a given length of straw.

The sound frequencies can be demonstrated visually by blowing into a microphone connected to an oscilloscope and watching the results on the oscilloscope screen.

Homework Options
This QuickLab can easily be performed outside of the physics lab room.

Classroom Practice

Harmonics

One string on a toy guitar is 34.5 cm long.

a. What is the wavelength of its first harmonic?

b. The string is plucked, and the speed of waves on the string is 410 m/s. What are the frequencies of the first three harmonics?

Answers

a. 69.0 cm

b. 590 Hz, 1200 Hz, 1800 Hz

Alternative Problem-Solving Approach

For the open pipe, both open ends must be antinodes; thus $L = \frac{1}{2}\lambda_1$, or $\lambda_1 = 2L = 2(2.45 \text{ m}) = 4.90$ m. The first harmonic can now be found with the equation $f = \frac{v}{\lambda}$, as follows:

$$f_1 = \frac{v}{\lambda_1} = \frac{345 \text{ m/s}}{4.90 \text{ m}} = 70.4 \text{ Hz}$$

For the closed pipe, one end is a node and one is an antinode; hence, $L = \frac{1}{4}\lambda_1$, or $\lambda_1 = 4L = 4(2.45\text{m}) = 9.80$ m. Thus, the first harmonic is as follows:

$$f_1 = \frac{v}{\lambda_1} = \frac{345 \text{ m/s}}{9.80 \text{ m}} = 35.2 \text{ Hz}$$

The other harmonics can be found by multiplying the harmonic number by the fundamental frequency, as in Sample Problem B.

SAMPLE PROBLEM B

Harmonics

PROBLEM

What are the first three harmonics in a 2.45 m long pipe that is open at both ends? What are the first three harmonics of this pipe when one end of the pipe is closed? Assume that the speed of sound in air is 345 m/s.

SOLUTION

1. DEFINE

Given: $L = 2.45$ m $\quad v = 345$ m/s

Unknown: Pipe open at both ends: $f_1 \quad f_2 \quad f_3$

Pipe closed at one end: $f_1 \quad f_3 \quad f_5$

2. PLAN

Choose an equation or situation:

When the pipe is open at both ends, the fundamental frequency can be found by using the equation for the entire harmonic series:

$$f_n = n\frac{v}{2L}, \; n = 1, 2, 3, \ldots$$

When the pipe is closed at one end, use the following equation:

$$f_n = n\frac{v}{4L}, \; n = 1, 3, 5, \ldots$$

In both cases, the second two harmonics can be found by multiplying the harmonic numbers by the fundamental frequency.

3. CALCULATE

Substitute the values into the equations and solve:

For a pipe open at both ends:

$$f_1 = n\frac{v}{2L} = (1)\left(\frac{345 \text{ m/s}}{(2)(2.45 \text{ m})}\right) = \boxed{70.4 \text{ Hz}}$$

The next two harmonics are the second and the third:

$$f_2 = 2f_1 = (2)(70.4 \text{ Hz}) = \boxed{141 \text{ Hz}}$$

$$f_3 = 3f_1 = (3)(70.4 \text{ Hz}) = \boxed{211 \text{ Hz}}$$

For a pipe closed at one end:

$$f_1 = n\frac{v}{4L} = (1)\left(\frac{345 \text{ m/s}}{(4)(2.45 \text{ m})}\right) = \boxed{35.2 \text{ Hz}}$$

The next possible harmonics are the third and the fifth:

$$f_3 = 3f_1 = (3)(35.2 \text{ Hz}) = \boxed{106 \text{ Hz}}$$

$$f_5 = 5f_1 = (5)(35.2 \text{ Hz}) = \boxed{176 \text{ Hz}}$$

TIP *Be sure to use the correct harmonic numbers for each situation. For a pipe open at both ends, n = 1, 2, 3, etc. For a pipe closed at one end, only odd harmonics are present, so n = 1, 3, 5, etc.*

4. EVALUATE In a pipe open at both ends, the first possible wavelength is $2L$; in a pipe closed at one end, the first possible wavelength is $4L$. Because frequency and wavelength are inversely proportional, the fundamental frequency of the open pipe should be twice that of the closed pipe, that is, $70.4 = (2)(35.2)$.

PRACTICE B

Harmonics

1. What is the fundamental frequency of a 0.20 m long organ pipe that is closed at one end, when the speed of sound in the pipe is 352 m/s?
2. A flute is essentially a pipe open at both ends. The length of a flute is approximately 66.0 cm. What are the first three harmonics of a flute when all keys are closed, making the vibrating air column approximately equal to the length of the flute? The speed of sound in the flute is 340 m/s.
3. What is the fundamental frequency of a guitar string when the speed of waves on the string is 115 m/s and the effective string lengths are as follows?
   a. 70.0 cm **b.** 50.0 cm **c.** 40.0 cm
4. A violin string that is 50.0 cm long has a fundamental frequency of 440 Hz. What is the speed of the waves on this string?

Trumpets, saxophones, and clarinets are similar to a pipe closed at one end. For example, although the trumpet shown in **Figure 17** has two open ends, the player's mouth effectively closes one end of the instrument. In a saxophone or a clarinet, the reed closes one end.

Despite the similarity between these instruments and a pipe closed at one end, our equation for the harmonic series of pipes does not directly apply to such instruments. One reason the equation does not apply is that any deviation from the cylindrical shape of a pipe affects the harmonic series of an instrument. Another reason is that the open holes in many instruments affect the harmonics. For example, a clarinet is primarily cylindrical, but there are some even harmonics in a clarinet's tone at relatively small intensities. The shape of a saxophone is such that the harmonic series in a saxophone is similar to that in a cylindrical pipe open at both ends even though only one end of the saxophone is open. These deviations are in part responsible for the variety of sounds that can be produced by different instruments.

Figure 17
Variations in shape give each instrument a different harmonic series.

SECTION 3

PROBLEM GUIDE B

Use this guide to assign problems.
SE = Student Edition Textbook
PW = Problem Workbook
PB = Problem Bank on the One-Stop Planner (OSP)

Solving for:

f_n	**SE** Sample, 1–3; Ch. Rvw. 34–35, 36b, 40 **PW** 7–9 **PB** 5–7
v	**SE** 4 **PW** Sample, 1–3 **PB** 8–10
L	**SE** Ch. Rvw. 36a, 41* **PW** 4–6 **PB** Sample, 1–4

***Challenging Problem**
Consult the printed Solutions Manual or the OSP for detailed solutions.

ANSWERS

Practice B

1. 440 Hz
2. 260 Hz, 520 Hz, 780 Hz
3. **a.** 82.1 Hz
   b. 115 Hz
   c. 144 Hz
4. 440 m/s

Teaching Tip — GENERAL

Explain the distinction sometimes made between musical sounds and noise. When individual waveforms are regular, the resultant waveform, although complex, is composed of repeating patterns. On the other hand, when the individual waveforms are not periodic, the resultant waveform is nonrepeating. Sounds that are considered to be musical typically have waveforms with repeating patterns, while those that are considered to be noise do not.

Visual Strategy ADVANCED

Table 4

Point out that the second column in **Table 4** represents the spectrum of wave amplitudes of harmonics produced by each instrument.

Q The harmonics shown in **Table 4** are for the musical note A-natural ($f_1 = 220$ Hz). List the frequencies produced by the clarinet in order of intensity, from greatest to least.

A *660 Hz, 220 Hz, 440 Hz, 1540 Hz, 880 Hz, 1100 Hz*

SCILINKS.
www.scilinks.org
Topic: Acoustics
Code: HF60015

timbre

the musical quality of a tone resulting from the combination of harmonics present at different intensities

Harmonics account for sound quality, or timbre

Table 4 shows the harmonics present in a tuning fork, a clarinet, and a viola when each sounds the musical note A-natural. Each instrument has its own characteristic mixture of harmonics at varying intensities.

The harmonics shown in the second column of **Table 4** add together according to the principle of superposition to give the resultant waveform shown in the third column. Since a tuning fork vibrates at only its fundamental frequency, its waveform is simply a sine wave. (Some tuning forks also vibrate at higher frequencies when they are struck hard enough.) The waveforms of the other instruments are more complex because they consist of many harmonics, each at different intensities. Each individual harmonic waveform is a sine wave, but the resultant wave is more complex than a sine wave because each individual waveform has a different frequency.

In music, the mixture of harmonics that produces the characteristic sound of an instrument is referred to as the *spectrum of the sound.* From the perspective of the listener, this spectrum results in *sound quality,* or **timbre.** A clarinet sounds different from a viola because of differences in timbre, even when both instruments are sounding the same note at the same volume. The rich harmonics of most instruments provide a much fuller sound than that of a tuning fork.

Table 4 Harmonics of a Tuning Fork, a Clarinet, and a Viola at the Same Pitch

Tuning fork		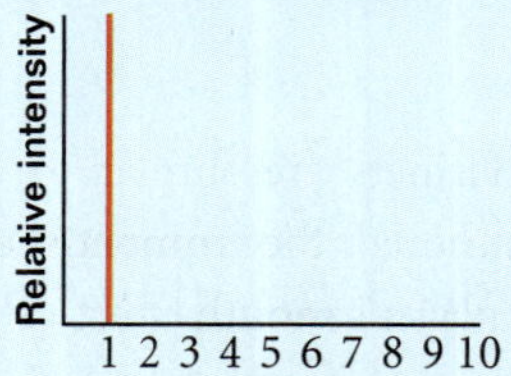	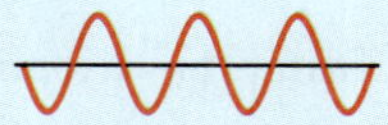
		Harmonics	Resultant waveform
Clarinet	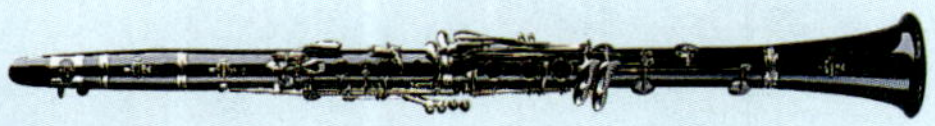	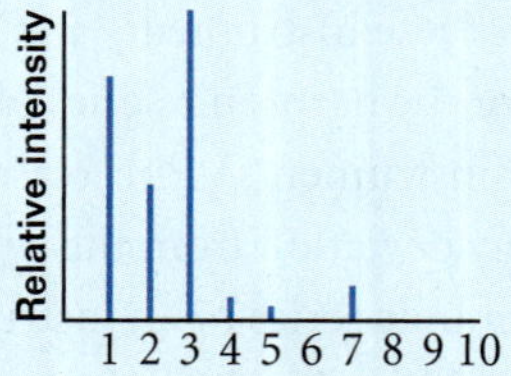	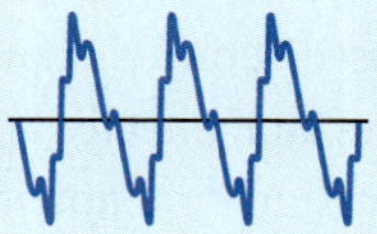
		Harmonics	Resultant waveform
Viola		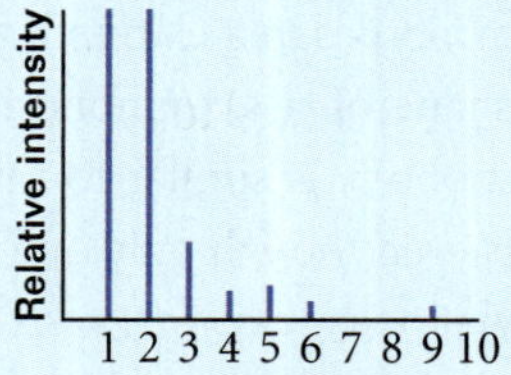	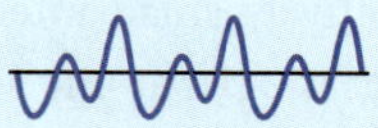
		Harmonics	Resultant waveform

Why it Matters

Reverberation

Auditoriums, churches, concert halls, libraries, and music rooms are designed with specific functions in mind. One auditorium may be made for rock concerts, while another is constructed for use as a lecture hall. Your school's auditorium, for instance, may allow you to hear a speaker well but make a band sound damped and muffled.

Rooms are often constructed so that sounds made by a speaker or a musical instrument bounce back and forth against the ceiling, walls, floor, and other surfaces. This repetitive echo is called *reverberation*. The reverberation time is the amount of time it takes for a sound's intensity to decrease by 60 dB.

For speech, the auditorium should be designed so that the reverberation time is relatively short. A repeated echo of each word could become confusing to listeners.

Music halls may differ in construction depending on the type of music usually played there. For example, rock music is generally less pleasing with a large amount of reverberation, but more reverberation is sometimes desired for orchestral and choral music.

For these reasons, you may notice a difference in the way ceilings, walls, and furnishings are designed in different rooms. Ceilings designed for a lot of reverberation are flat and hard. Ceilings in libraries and other quiet places are often made of soft or textured material to muffle sounds. Padded furnishings and plants can also be strategically arranged to absorb sound. All of these different factors are considered and combined to accommodate the auditory function of a room.

The intensity of each harmonic varies within a particular instrument, depending on frequency, amplitude of vibration, and a variety of other factors. With a violin, for example, the intensity of each harmonic depends on where the string is bowed, the speed of the bow on the string, and the force the bow exerts on the string. Because there are so many factors involved, most instruments can produce a wide variety of tones.

Even though the waveforms of a clarinet and a viola are more complex than those of a tuning fork, note that each consists of repeating patterns. Such waveforms are said to be *periodic*. These repeating patterns occur because each frequency is an integral multiple of the fundamental frequency.

Fundamental frequency determines pitch

As you saw in Section 1, the frequency of a sound determines its pitch. In musical instruments, the fundamental frequency of a vibration typically determines pitch. Other harmonics are sometimes referred to as *overtones*. In the chromatic (half-step) musical scale, there are 12 notes, each of which has a characteristic frequency. The frequency of the thirteenth note is exactly twice that of the first note, and together the 13 notes constitute an *octave*. For stringed instruments and open-ended wind instruments, the frequency of the second harmonic of a note corresponds to the frequency of the octave above that note.

Why it Matters

Reverberation

The discussion of harmonics in this section considers the interference of the sound waves from a single instrument. This feature discusses the interference that occurs when echoes interfere with the original sound waves and with one another.

Extension — GENERAL

Students may be surprised to learn that reverberation is an important consideration when buildings are being designed. This opportunity can be used to discuss acoustical engineering as a possible career choice for students who are interested in acoustics.

Visual Strategy GENERAL

Figure 18
Be sure students understand that this figure depicts two waves at a particular point in space as time passes (rather than an expanse of space at an instant of time, as in previous wave representations).

Q At what time(s) are the two waves exactly out of phase? At what time(s) are the two waves exactly in phase?

A *t_1 and t_3; t_2*

ANSWERS

Conceptual Challenge

1. The fundamental frequencies are getting closer together because if the number of beats heard each second is decreasing, the two waves are closer to being completely in phase at all points.
2. When the two flutists play the same note, the number of beats heard each second is the frequency difference between the two flutes. If one of the flutes is adjusted until no beats are heard, the two flutes will be in tune with each other.
3. The speed of the sound waves in air will not be the same as the speed of waves on the string. There is no frequency change because the vibrations are still occurring at the same rate, so the wavelength must change with the wave speed (because $v = \lambda f$).

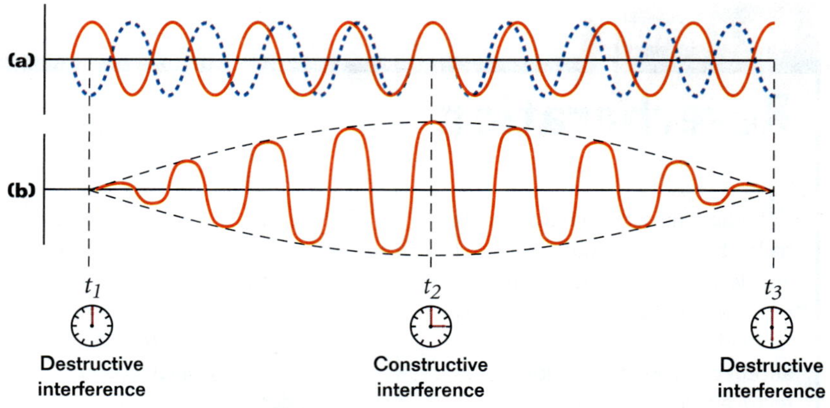

Figure 18
Beats are formed by the interference of two waves of slightly different frequencies traveling in the same direction. In this case, constructive interference is greatest at t_2, when the two waves are in phase.

BEATS

So far, we have considered the superposition of waves in a harmonic series, where each frequency is an integral multiple of the fundamental frequency. When two waves of *slightly* different frequencies interfere, the interference pattern varies in such a way that a listener hears an alternation between loudness and softness. The variation from soft to loud and back to soft is called a **beat.**

beat
the periodic variation in the amplitude of a wave that is the superposition of two waves of slightly different frequencies

Sound waves at slightly different frequencies produce beats

Figure 18 shows how beats occur. In **Figure 18(a),** the waves produced by two tuning forks of different frequencies start exactly opposite one another. These waves combine according to the superposition principle, as shown in **Figure 18(b).** When the two waves are exactly opposite one another, they are said to be *out of phase,* and complete destructive interference occurs. For this reason, no sound is heard at t_1.

Because these waves have different frequencies, after a few more cycles, the crest of the blue wave matches up with the crest of the red wave, as at t_2. At this

Why it Matters

Conceptual Challenge

1. Concert Violins Before a performance, musicians tune their instruments to match their fundamental frequencies. If a conductor hears the number of beats decreasing as two violin players are tuning, are the fundamental frequencies of these violins becoming closer together or farther apart? Explain.

2. Tuning Flutes How could two flute players use beats to ensure that their instruments are in tune with each other?

3. Sounds from a Guitar Will the speed of waves on a vibrating guitar string be the same as the speed of the sound waves in the air that are generated by this vibration? How will the frequency and wavelength of the waves on the string compare with the frequency and wavelength of the sound waves in the air?

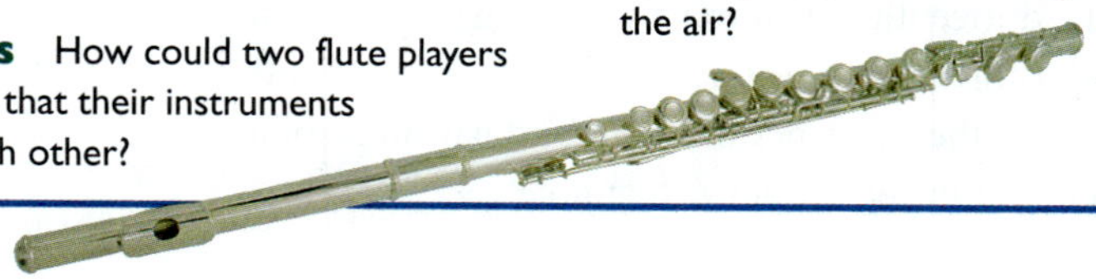

point, the waves are said to be *in phase.* Now constructive interference occurs, and the sound is louder. Because the blue wave has a higher frequency than the red wave, the waves are out of phase again at t_3, and no sound is heard.

As time passes, the waves continue to be in and out of phase, the interference constantly shifts between constructive interference and destructive interference, and the listener hears the sound getting softer and louder and then softer again. You may have noticed a similar phenomenon on a playground swing set. If two people are swinging next to one another at different frequencies, the two swings may alternate between being in phase and being out of phase.

The number of beats per second corresponds to the difference between frequencies

In our previous example, there is one beat, which occurs at t_2. One beat corresponds to the blue wave gaining one entire cycle on the red wave. This is because to go from one destructive interference to the next, the red wave must lag one entire cycle behind the blue wave. If the time that lapses from t_1 to t_3 is one second, then the blue wave completes one more cycle per second than the red wave. In other words, its frequency is greater by 1 Hz. By generalizing this, you can see that the frequency difference between two sounds can be found by the number of beats heard per second.

SECTION REVIEW

1. On a piano, the note middle C has a fundamental frequency of 262 Hz. What is the second harmonic of this note?
2. If the piano wire in item 1 is 66.0 cm long, what is the speed of waves on this wire?
3. A piano tuner using a 392 Hz tuning fork to tune the wire for G-natural hears four beats per second. What are the two possible frequencies of vibration of this piano wire?
4. In a clarinet, the reed end of the instrument acts as a node and the first open hole acts as an antinode. Because the shape of the clarinet is nearly cylindrical, its harmonic series approximately follows that of a pipe closed at one end. What harmonic series is predominant in a clarinet?
5. **Critical Thinking** Which of the following are different for a trumpet and a banjo when both play notes at the same fundamental frequency?
   a. wavelength in air of the first harmonic
   b. which harmonics are present
   c. intensity of each harmonic
   d. speed of sound in air

Teaching Tip

Point out that the number of beats is always the *absolute value* of the difference between two frequencies.

Measuring the number of beats does not tell you which of two instruments has a higher frequency. If two waves are exactly in phase, there are no beats and the sounds are exactly in tune.

SECTION REVIEW ANSWERS

1. 524 Hz
2. 348 m/s
3. 388 Hz and 396 Hz
4. the odd harmonics
5. b, c

Piano Tuner

The piano was invented in the early 1700s by an Italian harpsichord maker named Bartolomeo Christofori. The piano differs from the harpsichord in that piano strings are struck by felt hammers rather than plucked. The piano's full name was originally *gravicembalo con piano e forte* (now shortened to *pianoforte*), which emphasizes the piano's primary advantage over the harpsichord: a player can control the piano's volume. By 1800, the piano had supplanted the harpsichord and other instruments in popularity, and important composers such as C.P.E. Bach, Haydn, and Mozart had begun composing music for it. But the influence of the harpsichord can still be seen today: the grand piano is constructed in the shape of its predecessor.

Harpsichords and early pianos required frequent tuning but were relatively easy to tune. Therefore, the performers tuned their instruments themselves. Structural changes in the 19th century allowed pianos to remain in tune longer, but they also became more difficult to tune. This resulted in the need for a piano tuner, a specialist who could devote the time and effort required to develop the more-demanding tuning skills.

PHYSICS CAREERS

Piano Tuner

Piano tuners apply their knowledge of one aspect of physics—sound—to their everyday work. To learn more about piano tuning as a career, read the interview with Ramón Ramírez.

Each piano string is wrapped around a tuning pin. Rotating the pins with a tuning hammer alters the string tension, which changes the pitch.

What schooling did you receive in order to become a registered piano technician (RPT)?

I started off as a music education major and completed that degree. Then, I became the first person in the United States to receive a master's degree in applied music with piano technology as the major.

Did you receive encouragement from a teacher or some other person?

Yes. First, my parents and siblings, who helped me decide to major in music, encouraged me. Later, I was instructed by acoustician Owen Jorgensen, who authored three of the most influential books on historical tuning. Also an accomplished performer, Owen Jorgensen is a phenomenal piano tuner and technician. His work with experimental tuning has opened a new direction for music of the future.

What sort of equipment do you use?

The three most basic tools are a type of wrench called a *tuning hammer,* mutes to silence strings that should not be sounding at a given moment, and a tuning fork, which is used to establish precise pitch. Additionally, a metronome and a watch or clock are useful for timing beats. A calculator can be used for operations such as converting beats per second to beats per minute.

What is your favorite thing about your job?

This question is difficult because there are so many details about my work that I like. Possibly, it is the people I work with on a daily basis. Piano owners tend to be interesting and often enjoyable people.

How does physics influence your work?

Physics is the vehicle by which the complex mathematics of tuning moves from theory to audible reality. The harmonic series might seem only theoretical on paper, but modern tuners have to clearly hear individual pitches up to the sixth harmonic, and historical systems required a working ability to hear to the seventh. (A few tuners can hear to the twelfth harmonic.)

What advice would you give to students who are interested in piano tuning?

Obtain a used piano, some basic tools, and a copy of Owen Jorgensen's 1992 book, *Tuning.* Begin by tuning the simplest historical systems and gradually work your way through more complex systems. Because attending one of the very few schools that teach piano technology will probably require relocating, you can find out if you have talent and/or interest before making a larger investment.

CHAPTER 12 Highlights

KEY IDEAS

Section 1 Sound Waves

- The frequency of a sound wave determines its pitch.
- The speed of sound depends on the medium.
- The relative motion between the source of waves and an observer creates an apparent frequency shift known as the Doppler effect.

Section 2 Sound Intensity and Resonance

- The sound intensity of a spherical wave is the power per area.
- Sound intensity is inversely proportional to the square of the distance from the source because the same energy is spread over a larger area.
- Intensity and frequency determine which sounds are audible.
- Decibel level is a measure of relative intensity on a logarithmic scale.
- A given difference in decibels corresponds to a fixed difference in perceived loudness.
- A forced vibration at the natural frequency produces resonance.
- The human ear transmits vibrations that cause nerve impulses. The brain interprets these impulses as sounds of varying frequencies.

Section 3 Harmonics

- Harmonics are integral multiples of the fundamental frequency.
- A vibrating string or a pipe open at both ends produces all harmonics.
- A pipe closed at one end produces only odd harmonics.
- The number and intensity of harmonics account for the sound quality of an instrument, also known as timbre.

Variable Symbols

	Quantities	Units	
	sound intensity	W/m^2	watts/meters squared
	decibel level	dB	decibels
f_n	frequency of the *n*th harmonic	Hz	Hertz = s^{-1}
L	length of a vibrating string or an air column	m	meters

KEY TERMS

compression (p. 408)

rarefaction (p. 408)

pitch (p. 409)

Doppler effect (p. 412)

intensity (p. 414)

decibel (p. 417)

resonance (p. 419)

fundamental frequency (p. 422)

harmonic series (p. 423)

timbre (p. 428)

beat (p. 430)

PROBLEM SOLVING

See **Appendix D: Equations** for a summary of the equations introduced in this chapter. If you need more problem-solving practice, see **Appendix I: Additional Problems.**

CHAPTER 12

Highlights

Teaching Tip

Ask students to prepare a concept map for the chapter. The concept map should include all of the vocabulary terms, along with other integral terms or concepts.

Review

ANSWERS

1. because air molecules vibrate in a direction parallel to the direction of wave motion
2. Diagrams should depict a sine curve that begins and ends at its lowest point and that has three crests and two troughs.
3. Frequency is an objective measure of the rate of particle vibration. Pitch is a subjective quality that depends on the listener.
4. Infrasonic waves are below 20 Hz, audible waves are between 20 and 20 000 Hz, and ultrasonic waves are greater than 20 000 Hz.
5. Molecules that have more motion (higher temperature) can transfer their vibrations more easily. This is less noticeable in liquids and solids because the particles are closer together.
6. The siren's pitch will drop.
7. because their short wavelengths can image small objects
8. Frequency doubles; Speed remains constant.
9. Sound travels faster through the ground.
10. Notes that are played at the same time reach your ears at the same time.
11. the driver of the van
12. greater than 40 kHz
13. Intensity is power per area; decibel level is a measure of *relative* intensity.
14. 90 dB, 30 dB, 20 dB, 60 dB (Answers may vary slightly.)

CHAPTER 12 Review

SOUND WAVES

Review Questions

1. Why are sound waves in air characterized as longitudinal?
2. Draw the sine curve that corresponds to the sound wave depicted below.

3. What is the difference between frequency and pitch?
4. What are the differences between infrasonic, audible, and ultrasonic sound waves?
5. Explain why the speed of sound depends on the temperature of the medium. Why is this temperature dependence more noticeable in a gas than in a solid or a liquid?
6. You are at a street corner and hear an ambulance siren. Without looking, how can you tell when the ambulance passes by?
7. Why do ultrasound waves produce images of objects inside the body more effectively than audible sound waves do?

Conceptual Questions

8. If the wavelength of a sound source is reduced by a factor of 2, what happens to the wave's frequency? What happens to its speed?
9. As a result of a distant explosion, an observer first senses a ground tremor, then hears the explosion. What accounts for this time lag?
10. By listening to a band or an orchestra, how can you determine that the speed of sound is the same for all frequencies?
11. A fire engine is moving at 40 m/s and sounding its horn. A car in front of the fire engine is moving at 30 m/s, and a van in front of the car is stationary. Which observer hears the fire engine's horn at a higher pitch, the driver of the car or the driver of the van?
12. A bat flying toward a wall emits a chirp at 40 kHz. Is the frequency of the echo received by the bat greater than, less than, or equal to 40 kHz?

SOUND INTENSITY AND RESONANCE

Review Questions

13. What is the difference between intensity and decibel level?
14. Using **Table 2** (Section 2) as a guide, estimate the decibel levels of the following sounds: a cheering crowd at a football game, background noise in a church, the pages of this textbook being turned, and light traffic.
15. Why is the threshold of hearing represented as a curve in **Figure 9** (Section 2) rather than as a single point?
16. Under what conditions does resonance occur?

Conceptual Questions

17. The decibel level of an orchestra is 90 dB, and a single violin achieves a level of 70 dB. How does the sound intensity from the full orchestra compare with that from the violin alone?
18. A noisy machine in a factory produces a decibel rating of 80 dB. How many identical machines could you add to the factory without exceeding the 90 dB limit set by federal regulations?
19. Why is the intensity of an echo less than that of the original sound?

20. Why are pushes given to a playground swing more effective if they are given at certain, regular intervals than if they are given at random positions in the swing's cycle?

21. Although soldiers are usually required to march together in step, they must break their march when crossing a bridge. Explain the possible danger of crossing a rickety bridge without taking this precaution.

Practice Problems

For problems 22–23, see Sample Problem A.

22. A baseball coach shouts loudly at an umpire standing 5.0 meters away. If the sound power produced by the coach is 3.1×10^{-3} W, what is the decibel level of the sound when it reaches the umpire? (Hint: Use **Table 2** in this chapter.)

23. A stereo speaker represented by P in the figure on the right emits sound waves with a power output of 100.0 W. What is the intensity of the sound waves at point x when $r = 10.0$ m?

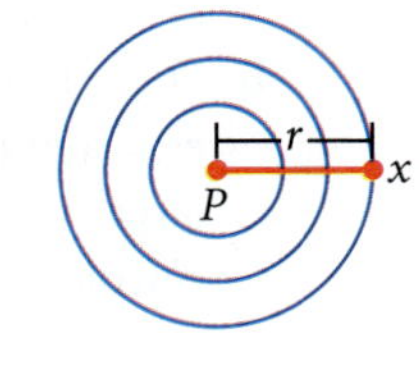

HARMONICS

Review Questions

24. What is fundamental frequency? How are harmonics related to the fundamental frequency?

25. The figures below show a stretched string vibrating in several of its modes. If the length of the string is 2.0 m, what is the wavelength of the wave on the string in **(a)**, **(b)**, **(c)**, and **(d)**?

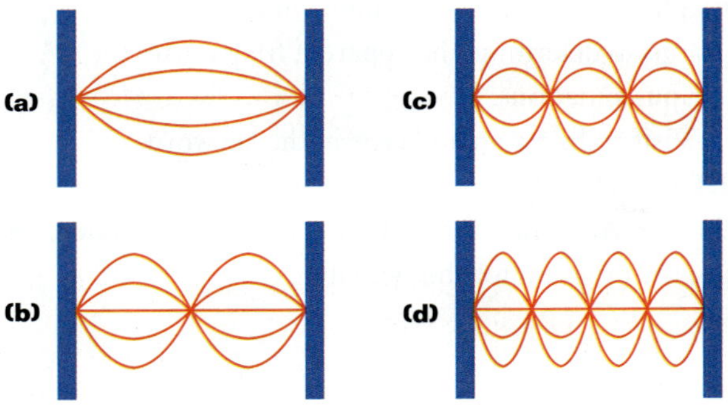

26. Why does a pipe closed at one end have a different harmonic series than an open pipe?

27. Explain why a saxophone sounds different from a clarinet, even when they sound the same fundamental frequency at the same decibel level.

Conceptual Questions

28. Why does a vibrating guitar string sound louder when it is on the instrument than it does when it is stretched on a work bench?

29. Two violin players tuning their instruments together hear six beats in 2 s. What is the frequency difference between the two violins?

30. What is the purpose of the slide on a trombone and the valves on a trumpet?

31. A student records the first 10 harmonics for a pipe. Is it possible to determine whether the pipe is open or closed by comparing the difference in frequencies between the adjacent harmonics with the fundamental frequency? Explain.

32. A flute is similar to a pipe open at both ends, while a clarinet is similar to a pipe closed at one end. Explain why the fundamental frequency of a flute is about twice that of the clarinet, even though the length of these two instruments is approximately the same.

33. The fundamental frequency of any note produced by a flute will vary slightly with temperature changes in the air. For any given note, will an increase in temperature produce a slightly higher fundamental frequency or a slightly lower one?

Practice Problems

For problems 34–35, see Sample Problem B.

34. What are the first three harmonics of a note produced on a 31.0 cm long violin string if waves on this string have a speed of 274.4 m/s?

35. The human ear canal is about 2.8 cm long and can be regarded as a tube open at one end and closed at the eardrum. What is the frequency around which we would expect hearing to be best when the speed of sound in air is 340 m/s? (Hint: Find the fundamental frequency for the ear canal.)

12 REVIEW

15. because the threshold of hearing depends on both frequency and intensity
16. when a forced vibration is the same as the natural frequency of a vibrating system
17. The sound intensity from the orchestra is 100 times that from the violin.
18. 9 machines (for a total of 10)
19. because intensity decreases with distance and the sound has traveled from the source to a reflecting surface and back (and because of imperfect reflection)
20. The swing's amplitude is maximized when the pushes match the swing's natural frequency.
21. Vibrations could set the bridge in motion if they match one of the bridge's natural frequencies.
22. 70 dB
23. 7.96×10^{-2} W/m^2
24. the lowest possible frequency of a vibrating system; They are integral multiples of the fundamental frequency.
25. **a.** 4.0 m
    b. 2.0 m
    c. 1.3 m
    d. 1.0 m
26. because a closed end is a node, while an open end is an antinode
27. The instruments have different harmonics present at various intensity levels.
28. The guitar's body transfers the string's vibrations to the air more efficiently, which increases the intensity of the sound.
29. 3 Hz
30. to change the length of the air column, thereby changing the fundamental frequency

31. yes; This difference will equal the fundamental frequency if the pipe is open at both ends but will equal twice the fundamental frequency if the pipe is closed at one end.
32. The first possible wavelength is $2L$ for the flute and $4L$ for the clarinet. Because the speed of sound is the same in each and $v = \lambda f$, the flute's fundamental frequency is twice the clarinet's.
33. As temperature increases, the speed of sound in air increases. Because f_1 is proportional to v, fundamental frequency likewise increases.
34. 443 Hz, 886 Hz, 1330 Hz
35. 3.0×10^3 Hz
36. **a.** 52 cm
b. 640 Hz, 960 Hz
37. 5 beats per second
38. 20 m, 2×10^{-2} m
39. 0.20 s
40. **a.** 70.1 Hz
b. 285
41. $L_{closed} = 1.5\ (L_{open})$
42. 1.9×10^{-2} m
43. **a.** 5.0×10^4 W
b. 2.8×10^{-3} W

extension

Graphing Calculator Practice

Visit go.hrw.com for answers to this Graphing Calculator activity.

Keyword HF6SNDXT

MIXED REVIEW

36. A pipe that is open at both ends has a fundamental frequency of 320 Hz when the speed of sound in air is 331 m/s.

a. What is the length of this pipe?

b. What are the next two harmonics?

37. When two tuning forks of 132 Hz and 137 Hz, respectively, are sounded simultaneously, how many beats per second are heard?

38. The range of human hearing extends from approximately 20 Hz to 20 000 Hz. Find the wavelengths of these extremes when the speed of sound in air is equal to 343 m/s.

39. A dolphin in 25°C sea water emits a sound directed toward the bottom of the ocean 150 m below. How much time passes before it hears an echo? (See **Table 1** in this chapter for the speed of the sound.)

40. An open organ pipe is 2.46 m long, and the speed of the air in the pipe is 345 m/s.

a. What is the fundamental frequency of this pipe?

b. How many harmonics are possible in the normal hearing range, 20 Hz to 20 000 Hz?

41. The fundamental frequency of an open organ pipe corresponds to the note middle C ($f = 261.6$ Hz on the chromatic musical scale). The third harmonic (f_3) of another organ pipe that is closed at one end has the same frequency. Compare the lengths of these two pipes.

42. Some studies indicate that the upper frequency limit of hearing is determined by the diameter of the eardrum. The wavelength of the sound wave and the diameter of the eardrum are approximately equal at this upper limit. If this is so, what is the diameter of the eardrum of a person capable of hearing 2.0×10^4 Hz? Assume 378 m/s is the speed of sound in the ear.

43. The decibel level of the noise from a jet aircraft is 130 dB when measured 20.0 m from the aircraft.

a. How much sound power does the jet aircraft emit?

b. How much sound power would strike the eardrum of an airport worker 20.0 m from the aircraft? (Use the diameter found in item 42 to calculate the area of the eardrum.)

Graphing Calculator Practice

Doppler Effect

As you learned earlier in this chapter, relative motion between a source of sound and an observer can create changes in the observed frequency. This frequency shift is known as the *Doppler effect.* The frequencies heard by the observer can be described by the following two equations, where f' represents the apparent frequency and f represents the actual frequency.

$$f' = f\left(\frac{v_{sound}}{v_{sound} - v_{source}}\right)$$

$$f' = f\left(\frac{v_{sound}}{v_{sound} + v_{source}}\right)$$

The first equation applies when the source of sound is approaching the observer, and the second equation applies when the source of sound is moving away from the observer.

In this graphing calculator activity, you will graph these two equations and will analyze the graphs to determine the apparent frequencies for various situations.

Visit go.hrw.com and type in the keyword **HF6SNDX** to find this graphing calculator activity. Refer to **Appendix B** for instructions on downloading the program for this activity.

Alternative Assessment

1. A new airport is being built 750 m from your school. The noise level 50 m from planes that will land at the airport is 130 dB. In open spaces, such as the fields between the school and the airport, the level decreases by 20 dB each time the distance increases tenfold. Work in a cooperative group to research the options for keeping the noise level tolerable at the school. How far away would the school have to be moved to make the sound manageable? Research the cost of land near your school. What options are available for soundproofing the school's buildings? How expensive are these options? Have each member in the group present the advantages and disadvantages of such options.
2. Use soft-drink bottles and water to make a musical instrument. Adjust the amount of water in different bottles to create musical notes. Play them as percussion instruments (by tapping the bottles) or as wind instruments (by blowing over the mouths of individual bottles). What media are vibrating in each case? What affects the fundamental frequency? Use a microphone and an oscilloscope to analyze your performance and to demonstrate the effects of tuning your instrument.
3. Interview members of the medical profession to learn about human hearing. What are some types of hearing disabilities? How are hearing disabilities related to disease, age, and occupational or environmental hazards? What procedures and instruments are used to test hearing? How do hearing aids help? What are the limitations of hearing aids? Present your findings to the class.
4. Do research on the types of architectural acoustics that would affect a restaurant. What are some of the acoustics problems in places where many people gather? How do odd-shaped ceilings, decorative panels, draperies, and glass windows affect echo and noise? Find the shortest wavelengths of sounds that should be absorbed, considering that conversation sounds range from 500 to 5000 Hz. Prepare a plan or a model of your school cafeteria, and show what approaches you would use to keep the level of noise to a minimum.
5. Doppler radar systems use the Doppler effect to identify the speed of objects such as aircraft, ships, automobiles, and weather systems. For example, meteorologists use Doppler radar to track the movement of storm systems. Police use Doppler radar to determine whether a motorist is speeding. Doppler radar systems use electromagnetic waves, rather than sound waves. Choose an application of Doppler radar to research. Create a poster showing how the application works.
6. How does a piano produce sound? Why do grand pianos sound different than upright pianos? How are harpsichords and early pianos different from modern pianos? What types of tuning systems were used in the past, and which are used today? Use library and/or Internet sources to answer these questions. If possible, try playing notes on different pianos, and compare the resulting sounds. Create a presentation to share your results with the class.
7. Research the speed of sound in different media (including solids, liquids, and gases), and at different temperatures. Also investigate the concept of *supersonic* speed and find some examples of objects that can move at supersonic speeds. Create a bar chart to compare your results.
8. Bats rely on echolocation to find and track prey. Conduct research to find out how this works. Which species of bats use echolocation? What type of sounds do bats emit? What can a bat learn from reflected sounds, and how do bats process the information? Write a paper with the results of your research.

Alternative Assessment ANSWERS

1. Plans should consider the threshold of pain (120 dB). To achieve levels of 90 to 70 dB, the school must be moved 5 to 50 km.
2. When the bottle is tapped, the water vibrates. More water corresponds to a lower pitch. When air is blown over the bottle, the air vibrates. More water (less air) corresponds to a higher pitch.
3. Student reports should relate the physics of sound to the physiology of the ear and the workings of hearing aids and testing devices.
4. Soft irregular surfaces decrease noise because they do not reflect sound waves. Plans to reduce echo effects could include using grooves or ridges to break up smooth surfaces or using insulating materials.
5. Student posters should illustrate how Doppler radar can be used to determine the speed of a moving object.
6. Student presentations should describe several differences between early and modern pianos, and between uprights and grand pianos.
7. Bar charts should compare the speed of sound in various media. A supersonic speed is greater than the speed at which sound travels in a particular medium.
8. Bats echolocate by interpreting reflected sound waves to determine the location, size, and speed of moving prey.

Standardized Test Prep

ANSWERS

1. B
2. J
3. D
4. H
5. C
6. F
7. B
8. G

Standardized Test Prep

MULTIPLE CHOICE

1. When a part of a sound wave travels from air into water, which property of the wave remains unchanged?

A. speed
B. frequency
C. wavelength
D. amplitude

2. What is the wavelength of the sound wave shown in the figure below?

F. 1.00 m
G. 0.75 m
H. 0.50 m
J. 0.25 m

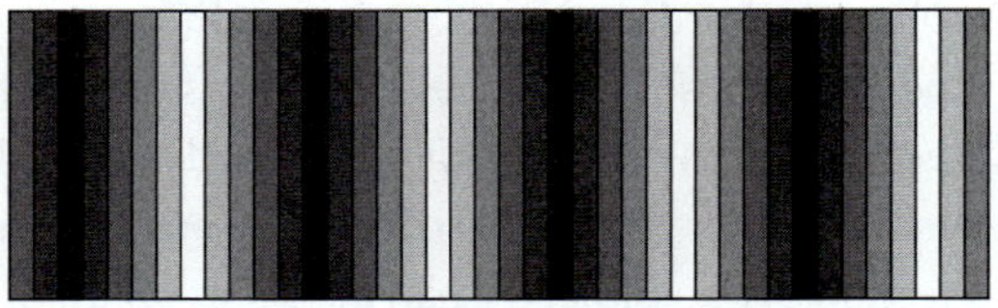

3. If a sound seems to be getting louder, which of the following is probably increasing?

A. speed of sound
B. frequency
C. wavelength
D. intensity

4. The intensity of a sound wave increases by 1000 W/m^2. What is this increase equal to in decibels?

F. 10
G. 20
H. 30
J. 40

5. The Doppler effect occurs in all but which of the following situations?

A. A source of sound moves toward a listener.
B. A listener moves toward a source of sound.
C. A listener and a source of sound remain at rest with respect to each other.
D. A listener and a source of sound move toward or away from each other.

6. If the distance from a point source of sound is tripled, by what factor is the sound intensity changed?

F. $\frac{1}{9}$
G. $\frac{1}{3}$
H. 3
J. 9

7. Why can a dog hear a sound produced by a dog whistle, but its owner cannot?

A. Dogs detect sounds of less intensity than do humans.
B. Dogs detect sounds of higher frequency than do humans.
C. Dogs detect sounds of lower frequency than do humans.
D. Dogs detect sounds of higher speed than do humans.

8. The greatest value ever achieved for the speed of sound in air is about 1.0×10^4 m/s, and the highest frequency ever produced is about 2.0×10^{10} Hz. If a single sound wave with this speed and frequency were produced, what would its wavelength be?

F. 5.0×10^{-6} m
G. 5.0×10^{-7} m
H. 2.0×10^{6} m
J. 2.0×10^{14} m

9. The horn of a parked automobile is stuck. If you are in a vehicle that passes the automobile, as shown below, what is the nature of the sound that you hear?

A. The original sound of the horn rises in pitch.
B. The original sound of the horn drops in pitch.
C. A lower pitch is heard rising to a higher pitch.
D. A higher pitch is heard dropping to a lower pitch.

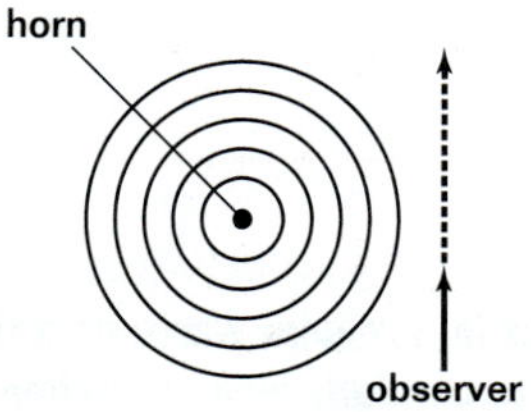

10. The second harmonic of a guitar string has a frequency of 165 Hz. If the speed of waves on the string is 120 m/s, what is the string's length?

F. 0.36 m
G. 0.73 m
H. 1.1 m
J. 1.4 m

SHORT RESPONSE

11. Two wind instruments produce sound waves with frequencies of 440 Hz and 447 Hz, respectively. How many beats per second are heard from the superposition of the two waves?

12. If you blow across the open end of a soda bottle and produce a tone of 250 Hz, what will be the frequency of the next harmonic heard if you blow much harder?

13. The figure below shows a string vibrating in the sixth harmonic. The length of the string is 1.0 m. What is the wavelength of the wave on the string?

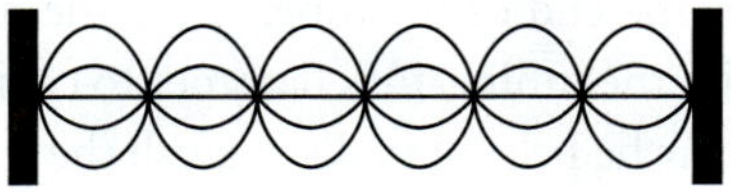

14. The power output of a certain loudspeaker is 250.0 W. If a person listening to the sound produced by the speaker is sitting 6.5 m away, what is the intensity of the sound?

EXTENDED RESPONSE

Use the following information to solve problems 15–16. Be sure to show all of your work.

The area of a typical eardrum is approximately equal to 5.0×10^{-5} m^2.

15. What is the sound power (the energy per second) incident on the eardrum at the threshold of pain (1.0 W/m^2)?

16. What is the sound power (the energy per second) incident on the eardrum at the threshold of hearing (1.0×10^{-12} W/m^2)?

Use the following information to solve problems 17–19. Be sure to show all of your work.

A pipe that is open at both ends has a fundamental frequency of 456 Hz when the speed of sound in air is 331 m/s.

17. How long is the pipe?

18. What is the frequency of the pipe's second harmonic?

19. What is the fundamental frequency of this pipe when the speed of sound in air is increased to 367 m/s as a result of a rise in the temperature of the air?

Test TIP Be certain that the equations used in harmonic calculations are for the right kind of sound source (vibrating string, pipe open at both ends, or pipe closed at one end).

9. D
10. G
11. 7 beats per second (7 Hz)
12. 750 Hz
13. 0.33 m
14. 0.47 W/m^2
15. 5.0×10^{-5} W
16. 5.0×10^{-17} W
17. 0.363 m
18. 912 Hz
19. 506 Hz

Lab Planning

Beginning on page T34 are preparation notes and teaching tips to assist you in planning.

Blank data tables (as well as some sample data) appear on the **One-Stop Planner.**

No Books in the Lab?

See the ***Datasheets for In-Text Labs*** workbook for a reproducible master copy of this experiment.

CBL™ Option

A **CBL™** version of this lab appears in **Appendix K** and in the ***CBL™ Experiments*** workbook.

Safety Caution

Remind students that broken glass must be disposed of in a separate container.

Tips and Tricks

- Adding food coloring to the water makes the water levels in the tube easier to see.
- Students may need practice striking the tuning fork. Emphasize that students should listen very carefully to hear the resonance. They should also practice moving the reservoir up and down *very slowly.*
- Values for λ can be improved by using the end correction: effective tube length $= L + 0.4d$, where d is the internal diameter of the tube. This takes into account the air above the tube end that also vibrates. If you want to use this correction, have students measure the internal diameter of the tube.

CHAPTER 12

Skills Practice Lab — Speed of Sound

OBJECTIVES

- **Measure** the speed of sound in air using a resonance apparatus.

MATERIALS LIST

- **4 tuning forks of different frequencies**
- **Erlenmeyer flask, 1000 mL**
- **resonance apparatus with clamp**
- **thermometer**
- **tuning-fork hammer**
- **water**

The speed of sound can be determined using a tuning fork to produce resonance in a tube that is closed at the bottom but open on top. The wavelength of the sound may be calculated from the resonant length of the tube. In this experiment, you will use a resonance apparatus to measure the speed of sound.

SAFETY

- **Put on goggles.**
- **Never put broken glass or ceramics in a regular waste container. Use a dustpan, brush, and heavy gloves to carefully pick up broken pieces and dispose of them in a container specifically provided for this purpose.**
- **If a thermometer breaks, notify the teacher immediately.**

PROCEDURE

Preparation

1. Read the entire lab, and plan what steps you will take.
2. If you are not using a datasheet provided by your teacher, prepare a data table in your lab notebook with four columns and five rows. In the first row, label the first through fourth columns *Trial, Length of Tube (m), Frequency (Hz),* and *Temperature (°C).* In the first column, label the second through fifth rows *1, 2, 3,* and *4.*

Finding the Speed of Sound

3. Set up the resonance apparatus as shown in **Figure 1.**
4. Raise the reservoir so that the top is level with the top of the tube. Fill the reservoir with water until the level in the tube is at the 5 cm mark.
5. Measure and record the temperature of the air inside the tube. Select a tuning fork, and record the frequency of the fork in your data table.
6. Securely clamp the tuning fork in place as shown in the figure, with the lower tine about 1 cm above the end of the tube. Strike the tuning fork sharply, but not too hard, with the tuning-fork hammer to create a vibration. A few practice strikes may be helpful to distinguish the tonal sound of the tuning fork from the unwanted metallic "ringing" sound that may result from striking the fork too hard. ***Do not strike the fork with anything other than a hard rubber mallet.***

7. While the tuning fork is vibrating directly above the tube, slowly lower the reservoir about 20 cm or until you locate the position of the reservoir where the resonance is loudest. (Note: To locate the exact position of the resonance, you may need to strike the tuning fork again while the water level is falling.) Raise the reservoir to about 2 cm above the approximate level where you think the resonance is loudest. Strike the tuning fork with the tuning fork hammer and carefully lower the reservoir about 5 cm until you find the exact position of resonance.
8. Using the scale marked on the tube, record the level of the water in the tube when the resonance is loudest. Record this level to the nearest millimeter in your data table.
9. Repeat the procedure for several trials, using tuning forks of different frequencies.
10. Clean up your work area. Put equipment away safely so that it is ready to be used again. Recycle or dispose of used materials as directed by your teacher.

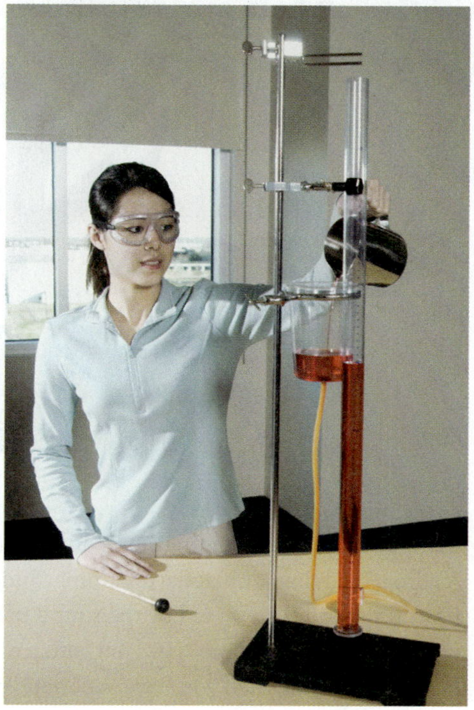

Figure 1
Step 7: From the position of greatest resonance, move the reservoir up 2 cm and down again until you find the exact position.

ANALYSIS

1. **Organizing Data** For each trial, calculate the wavelength of the sound by using the equation for the fundamental wavelength, $\lambda = 4L$, where L is the length of the tube.
2. **Organizing Data** For each trial, find the speed of sound. Use the equation $v = f\lambda$, where f is the frequency of the tuning fork.

CONCLUSIONS

3. **Evaluating Results** Find the accepted value for the speed of sound in air at room temperature (see **Appendix F**). Find the average of your results for the speed of sound, and use the average as the experimental value.
   a. Compute the absolute error using the following equation:

   $$\text{absolute error} = |\text{experimental} - \text{accepted}|$$

   b. Compute the relative error using the following equation:

   $$\text{relative error} = \frac{(\text{experimental} - \text{accepted})}{\text{accepted}}$$

4. **Analyzing Results** Based on your results, is the speed of sound in air at a given temperature the same for all sounds, or do some sounds move more quickly or more slowly than other sounds? Explain.
5. **Applying Ideas** How could you find the speed of sound in air at different temperatures?

✓ Checkpoints

Step 3: Make sure the apparatus is secure. The apparatus should be away from the edge of the table; otherwise, use a C-clamp to secure the base of the support stand to the edge of the table. Make sure students wear goggles to protect themselves from clamps and support rods at eye level.

Step 6: The lower tine of the tuning fork should be 1 cm away from the top of the tube.

Step 7: Remind students to work quietly and to move the reservoir very slowly and steadily. They should demonstrate that they have found the greatest resonance.

ANSWERS

Analysis

1. Trial 1: $\lambda = 1.030$ m
Trial 2: $\lambda = 0.848$ m
Trial 3: $\lambda = 0.632$ m
Trial 4: $\lambda = 0.292$ m

2. Trial 1: $v = 329$ m/s
Trial 2: $v = 326$ m/s
Trial 3: $v = 324$ m/s
Trial 4: $v = 299$ m/s

Conclusions

3. Accepted value: 346 m/s

4. Students should realize that all sound waves travel at the same speed.

5. Answers will vary. Students could look up the values in a table or perform the experiment in different places or under different conditions.

Science • Technology • Society

NOISE POLLUTION

Suppose you are spending some quiet time alone–reading, studying, or just daydreaming. Suddenly your peaceful mood is shattered by the sound of a lawn mower, loud music, or an airplane taking off. If this has happened to you, then you have experienced noise pollution.

Noise is defined as any loud, discordant, or disagreeable sound, so classifying sounds as noise is often a matter of personal opinion. When you are at a party, you might enjoy listening to loud music, but when you are at home trying to sleep, you may find the same music very disturbing.

There are two kinds of noise pollution, both of which can result in long-term hearing problems and even physical damage to the ear. The chapter "Sound" explains how we receive and interpret sound.

How Can Noise Damage Hearing?

The small bones and hairlike cells of the inner ear are delicate and very sensitive to the compression waves we interpret as sounds. The first type of noise pollution involves noises that are so loud they endanger the sensitive parts of the ear. Prolonged exposure to sounds of about 85 dB can begin to damage hearing irreversibly. Certain sounds above 120 dB can cause immediate damage. The sound level produced by a food blender or by diesel truck traffic is about 85 dB. A jet engine heard from a few meters away is about 140 dB.

Have you ever noticed the "headphones" worn by ground crew at an airport or by workers using chain saws or jackhammers? In most cases, these are ear protectors worn to prevent the hearing loss brought on by damage to the inner ear.

Whose Noise Annoys?

The second kind of noise pollution is more controversial because it involves noises that are considered annoyances. No one knows for sure how to measure levels of annoyance, but sometimes annoying noise becomes intolerable. Lack of

sleep due to noise causes people to have slow reaction times and poor judgment, which can result in mistakes at work or school and accidents on the job or on the road. Scientists have found that continuous, irritating noise can raise blood pressure, which leads to other health problems.

A major debate involves noise made by aircraft. Airport traffic in the United States nearly doubled from 1980 to 1990 and continues to grow at a rapid pace. People who live near airports once found aircraft noise an occasional annoyance, but because of increased traffic and runways added to accommodate growth, they now suffer sleep disruptions and other health effects.

Many people have organized groups to oppose airport expansion. Their primary concerns are the increase in noise and the decrease in property values associated with airport expansion.

But, city governments argue that an airport benefits the entire community both socially and economically and that airports must expand to meet the needs of increased populations. Officials have also argued that people knew they were taking chances by building or buying near an airport and that the community cannot compensate for their losses. Airlines contend that attempts to reduce noise by using less power during takeoffs or by veering away from populated areas can pose a serious threat to passenger safety.

Other Annoyances

Besides airports, people currently complain most about noise pollution from nearby construction sites, personal watercraft, loud stereos in homes and cars, all-terrain vehicles, snowmobiles, and power lawn equipment, such as mowers and leaf blowers. Many people want to control such noise by passing laws to limit the use of this equipment to certain times of the day or by requiring that sound-muffling devices be used.

Opponents to these measures argue that much of this activity takes place on private property and that, in the case of building sites and industries, noise limitation would increase costs. Some public officials would like to control annoying noise but point out that laws to do so fall under the category of nuisance laws, which are notoriously difficult to enforce.

Noise pollution is also a problem in areas where few or no people live. Unwanted noise in wilderness areas can affect animal behavior and reproduction. Sometimes animals are simply scared away from their habitats. For this reason, the government has taken action in some national parks to reduce sightseeing flights, get rid of noisy campers, and limit or eliminate certain noisy vehicles. Some parks have drastically limited the number of people who can be in a park at any one time.

Researching the Issue

1. Obtain a sound-level meter, and measure the noise level at places where you and your friends might be during an average week. Also make some measurements at locations where sound is annoyingly loud. Be sure to hold the meter at head level and read the meter for 30 seconds to obtain an average. Present your findings to the class in a graphic display.

2. Measure the sound levels at increasing distances from two sources of steady, loud noise. Record all of your locations and measurements. Graph your data, and write an interpretation describing how sound level varies with distance from the source.

3. Is there a source of noise in your community that most people recognize to be a problem? If so, find out what causes the noise and what people want to do to relieve the problem. Hold a panel discussion to analyze the opinions of each side, and propose your own solution.

443

CHAPTER 13 Light and Reflection
Planning Guide

Compression Guide
To shorten instruction because of time limitations, omit the opener and abbreviate the review.

OBJECTIVES	LABS, DEMONSTRATIONS, AND ACTIVITIES	TECHNOLOGY RESOURCES
PACING • 90 min pp. 444–445 **Chapter Opener**	**ANC** **Discovery Lab** Light and Mirrors*◆ BASIC	**CD** **Visual Concepts,** Chapter 13 BASIC
PACING • 90 min pp. 446–450 **Section 1 Characteristics of Light** • Identify the components of the electromagnetic spectrum. • Calculate the frequency or wavelength of electromagnetic radiation. Recognize that light has a finite speed. • Describe how the brightness of a light source is affected by distance.	**SE** **Skills Practice** Brightness of Light, pp. 484–485◆ GENERAL **ANC** **Datasheet** Brightness of Light* GENERAL **TE** **Demonstration** Infrared Light, p. 446 GENERAL **TE** **Demonstration** Radio Waves, p. 447 GENERAL **TE** **Demonstration** How Light Travels, p. 448 GENERAL **ANC** **CBL™ Experiment** Brightness of Light*◆ GENERAL	**OSP** **Lesson Plans** **EXT** **Integrating Astronomy** Starlight, Star Heat BASIC **TR** 62 Components of an Electromagnetic Wave **TR** 44A The Electromagnetic Spectrum **TR** 45A Predicting Wave Front Position Using Huygens' Principle
PACING • 45 min pp. 451–454 **Section 2 Flat Mirrors** • Distinguish between specular and diffuse reflection of light. • Apply the law of reflection for flat mirrors. • Describe the nature of images formed by flat mirrors.	**TE** **Demonstration** Diffuse Reflection, p. 451 BASIC **TE** **Demonstration** Specular Reflection, p. 452 GENERAL **TE** **Demonstration** Flat Mirror Images, p. 453 GENERAL **ANC** **Invention Lab** Designing a Device to Trace Drawings*◆ ADVANCED	**OSP** **Lesson Plans** **TR** 63 Image Formation by a Flat Mirror
PACING • 90 min pp. 455–468 **Section 3 Curved Mirrors** • Calculate distances and focal lengths using the mirror equation for concave and convex spherical mirrors. • Draw ray diagrams to find the image distance and magnification for concave and convex spherical mirrors. • Distinguish between real and virtual images. • Describe how parabolic mirrors differ from spherical mirrors.	**SE** **Quick Lab** Curved Mirrors, p. 457 GENERAL **TE** **Demonstration** Image Formed by a Concave Mirror, p. 455 GENERAL **TE** **Demonstration** Focal Point of a Concave Mirror, p. 457 GENERAL **TE** **Demonstration** Beams Reflected from a Concave Mirror, p. 457 BASIC **TE** **Demonstration** Convex Mirror, p. 463 BASIC	**OSP** **Lesson Plans** **TR** 64 Concave Spherical Mirror **TR** 65 Images Created by Concave Mirrors **TR** 66 Convex Spherical Mirror **TR** 67 Spherical Aberration and Parabolic Mirrors **TR** 46A Rules for Drawing Reference Rays
PACING • 45 min pp. 469–474 **Section 4 Color and Polarization** • Recognize how additive colors affect the color of light. • Recognize how pigments affect the color of reflected light. • Explain how linearly polarized light is formed and detected.	**SE** **Quick Lab** Polarization of Sunlight, p. 473 GENERAL **TE** **Demonstration** Reflection and Absorption of Color, p. 469 BASIC **TE** **Demonstration** Polarizing by Transmission, p. 472 GENERAL **TE** **Demonstration** Crossed Polarizers, p. 473 ADVANCED **TE** **Demonstration** Polarizing by Reflection, p. 474 GENERAL **ANC** **CBL™ Experiment** Polarization of Light*◆ ADVANCED	**OSP** **Lesson Plans** **EXT** **Integrating Biology** How Does Sunscreen Work? BASIC **TR** 68 Additive and Subtractive Primary Colors **TR** 69 Aligned and Crossed Polarizing Filters **TR** 48A Polarization of Light by Transmission, Reflection, and Scattering

PACING • 90 min

CHAPTER REVIEW, ASSESSMENT, AND STANDARDIZED TEST PREPARATION

- **SE** **Chapter Highlights,** p. 475
- **SE** **Chapter Review,** pp. 476–480
- **SE** **Alternative Assessment,** p. 480 ADVANCED
- **SE** **Graphing Calculator Practice,** p. 481 GENERAL
- **SE** **Standardized Test Prep,** pp. 482–483 GENERAL
- **SE** **Appendix D: Equations,** p. 860
- **SE** **Appendix I: Additional Problems,** pp. 889–890
- **ANC** **Study Guide Worksheet** Mixed Review* GENERAL
- **ANC** **Chapter Test A*** GENERAL
- **ANC** **Chapter Test B*** ADVANCED
- **OSP** **Test Generator**

Online and Technology Resources

go.hrw.com — Holt Online Learning

Visit go.hrw.com to find a variety of online resources. To access this chapter's extensions, enter the keyword **HF6LGTXT** and click the "go" button. Click **Holt Online Learning** for an online edition of this textbook, and other interactive resources.

HOLT Teacher's One-Stop Planner®

This DVD package includes:
- Holt Calendar Planner
- Customizable Lesson Plans
- Editable Worksheets
- ExamView® Version 6 Assessment Suite
- Interactive Teacher's Edition
- Holt PuzzlePro®
- Holt PowerNotes® Presentations
- MindPoint® Quiz Show

KEY

- **SE** Student Edition
- **TE** Teacher Edition
- **ANC** Ancillary Worksheet
- **OSP** One-Stop Planner
- **CD** CD or CD-ROM
- **TR** Teaching Transparencies
- **EXT** Online Extension
- * Also on One-Stop Planner
- ◆ Requires advance prep

SKILLS DEVELOPMENT RESOURCES	REVIEW AND ASSESSMENT	CORRELATIONS
		National Science Education Standards
SE **Sample Set A** Electromagnetic Waves, pp. 448–449 BASIC **ANC** **Problem Workbook** Sample Set A* BASIC **OSP** **Problem Bank** Sample Set A BASIC	**SE** **Section Review,** p. 450 GENERAL **ANC** **Study Guide Worksheet** Section 1* GENERAL **ANC** **Quiz** Section 1* BASIC	UCP 1, 2, 3, 4, 5 SAI 1, 2 ST 1, 2 HNS 1, 3 SPSP 5 PS 6b
	SE **Section Review,** p. 454 GENERAL **ANC** **Study Guide Worksheet** Section 2* GENERAL **ANC** **Quiz** Section 2* BASIC	UCP 1, 2, 3, 5
SE **Sample Set B** Imaging with Concave Mirrors, pp. 461–462 ADVANCED **TE** **Classroom Practice,** p. 461 GENERAL **ANC** **Problem Workbook** Sample Set B* GENERAL **OSP** **Problem Bank** Sample Set B GENERAL **SE** **Sample Set C** Convex Mirrors, pp. 465–466 GENERAL **TE** **Classroom Practice,** p. 465 GENERAL **ANC** **Problem Workbook** Sample Set C* GENERAL **OSP** **Problem Bank** Sample Set C GENERAL	**SE** **Section Review,** p. 468 GENERAL **ANC** **Study Guide Worksheet** Section 3* GENERAL **ANC** **Quiz** Section 3* BASIC	UCP 1, 2, 3, 4, 5 SAI 1, 2 ST 1, 2 HNS 1
SE **Conceptual Challenge,** p. 471 GENERAL	**SE** **Section Review,** p. 474 GENERAL **ANC** **Study Guide Worksheet** Section 4* GENERAL **ANC** **Quiz** Section 4* BASIC	UCP 1, 2, 5 SAI 1, 2 ST 1, 2 SPSP 5

SCILINKS. NSTA

www.scilinks.org

Maintained by the **National Science Teachers Association.**

Topic: Electromagnetic Spectrum
SciLinks Code: HF60482

Topic: Light Bulbs
SciLinks Code: HF60879

Topic: Mirrors
SciLinks Code: HF60970

Topic: Telescopes
SciLinks Code: HF61500

Topic: Color
SciLinks Code: HF60313

Classroom CD-ROMs

- Guided Reading Audio Program
- Student One Stop
- Virtual Investigations
- Visual Concepts

Holt Lab Generator CD-ROM

Search for any lab by topic, standard, difficulty level, or time. Edit any lab to fit your needs, or create your own labs. Use the Lab Materials QuickList software to customize your lab materials list.

Section 1 identifies the components of the electromagnetic spectrum, relates their frequency and wavelength to the speed of light, and introduces the relationship between brightness and distance for a light source.

Section 2 applies the laws of reflection to plane mirrors and uses ray diagrams to determine image location.

Section 3 shows how image location and magnification are calculated for concave and convex mirrors, uses ray diagrams to confirm calculated results, and explains spherical aberration.

Section 4 investigates additive and subtractive colors and explores the phenomenon of polarization.

About the Illustration

The Very Large Array (VLA) is part of a project called the National Radio Astronomy Observatory (NRAO). Radio astronomy is used to produce images of celestial bodies. A number of celestial objects emit more strongly at radio wavelengths than at those of visible light. The VLA is used to study both distant celestial objects as well as our sun.

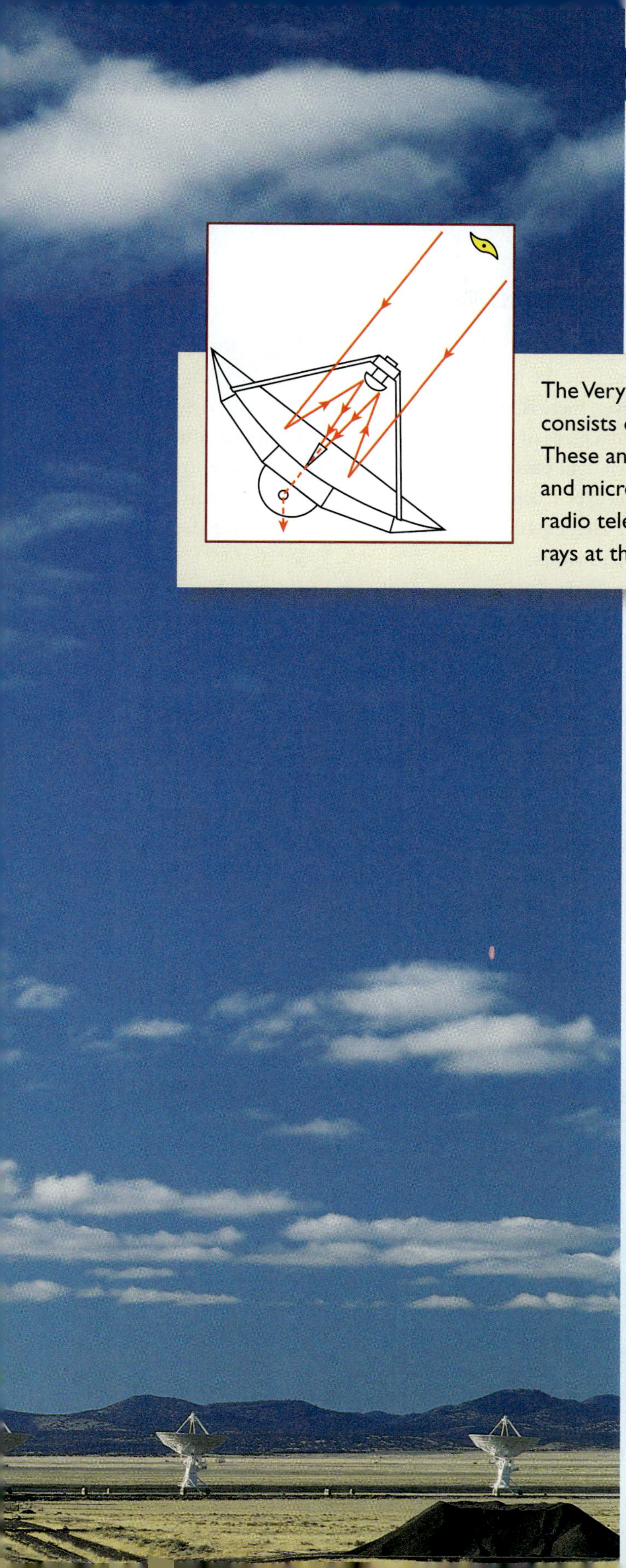

The Very Large Array, located near Socorro, New Mexico, consists of 27 radio antennas, each 25 meters in diameter. These antennas detect *electromagnetic radiation* in the radio and microwave regions of the spectrum. The dish of a radio telescope reflects the radio waves and focuses the rays at the receiver poised above the dish.

CHAPTER 13

Light and Reflection

WHAT TO EXPECT

In this chapter, you will learn about the characteristics of light and other forms of electromagnetic radiation. You will learn how flat and curved mirrors can be used to reflect light and create real and virtual images of objects.

Why it Matters

Mirrors have many applications both for scientists and in everyday life. For example, a reflector telescope uses two mirrors to gather, focus, and reflect light onto the eyepiece. The reflector telescope remains one of the most popular designs used by amateur astronomers, even though it was invented over 300 years ago.

CHAPTER PREVIEW

Tapping Prior Knowledge

Knowledge to Review

- ✔ Waves transport energy. They can be transverse or longitudinal. Waves have amplitude, frequency, wavelength, and velocity.

Items to Probe

- ✔ The ability to describe spatial relationships in geometric terms: Make sure students understand terms such as *perpendicular* and *parallel* and can solve equations of the form $\left(\frac{1}{A}\right)+\left(\frac{1}{B}\right)=\left(\frac{1}{C}\right)$.
- ✔ Preconceptions about light: Ask students to describe the path of sound waves when we hear something and the path of light rays when we see an object.

Demonstration

Infrared Light — GENERAL

Purpose Demonstrate one form of invisible electromagnetic radiation.

Materials incandescent light source, black box, prism, white paper, thermometer or temperature sensor with probeware system

Procedure Make a slit in one side of the black box, and place the light source inside the box. Set the prism in the path of the beam of light emerging from the slit. Lower the classroom lights, and place the white paper on the other side of the prism so that the light beam's spectrum is cast on the paper. Tape the thermometer on the paper so that the red light of the spectrum shines on the thermometer bulb. Have a student record the initial temperature and the temperature after 5 min.

Repeat this procedure, placing the thermometer bulb in the dark region to the left of the red part of the spectrum. As an optional step, repeat the procedure for other colors in the spectrum and for the region just past the violet light. Explain that infrared (IR) radiation, which is one example of nonvisible electromagnetic radiation, increases a substance's temperature. Energy transferred away from the substance as heat can be photographically detected with infrared-sensitive film. (Note: If using a probeware system for this demonstration, place the temperature sensor in the same positions as the thermometer bulb.)

Characteristics of Light

SECTION OBJECTIVES

- **Identify the components of the electromagnetic spectrum.**
- **Calculate the frequency or wavelength of electromagnetic radiation.**
- **Recognize that light has a finite speed.**
- **Describe how the brightness of a light source is affected by distance.**

ELECTROMAGNETIC WAVES

When most people think of light, they think of the light that they can see. Some examples include the bright, white light that is produced by a light bulb or the sun. However, there is more to light than these examples.

When you hold a piece of green glass or plastic in front of a source of white light, you see green light pass through. This phenomenon is also true for other colors. What your eyes recognize as "white" light is actually light that can be separated into six elementary colors of the visible *spectrum:* red, orange, yellow, green, blue, and violet.

If you examine a glass prism, such as the one shown in **Figure 1,** or any thick, triangular-shaped piece of glass, you will find that sunlight passes through the glass and emerges as a rainbowlike band of colors.

The spectrum includes more than visible light

Not all light is visible to the human eye. If you were to use certain types of photographic film to examine the light dispersed through a prism, you would find that the film records a much wider spectrum than the one you see. A variety of forms of radiation—including X rays, microwaves, and radio waves—have many of the same properties as visible light. The reason is that they are all examples of **electromagnetic waves.**

electromagnetic wave

a wave that consists of oscillating electric and magnetic fields, which radiate outward from the source at the speed of light

Light has been described as a particle, as a wave, and even as a combination of the two. Although the current model incorporates aspects of both particle and wave theories, the wave model is best suited for an introductory discussion of light, and it is the one that will be used in this section.

Figure 1
A prism separates light into its component colors.

SCILINKS
www.scilinks.org
Topic: Electromagnetic Spectrum
Code: HF60482

Electromagnetic waves vary depending on frequency and wavelength

In classical electromagnetic wave theory, light is considered to be a wave composed of oscillating electric and magnetic fields. These fields are perpendicular to the direction in which the wave moves, as shown in **Figure 2.** Therefore, electromagnetic waves are transverse waves. The electric and magnetic fields are also at right angles to each other.

Electromagnetic waves are distinguished by their different frequencies and wavelengths. In visible light, these differences in frequency and wavelength account for different colors. The difference in frequencies and wavelengths also distinguishes visible light from invisible electromagnetic radiation, such as X rays.

Types of electromagnetic waves are listed in **Table 1.** Note the wide range of wavelengths and frequencies. Although specific ranges are indicated in the table, the electromagnetic spectrum is, in reality, continuous. There is no sharp division between one kind of wave and the next. Some types of waves even have overlapping ranges.

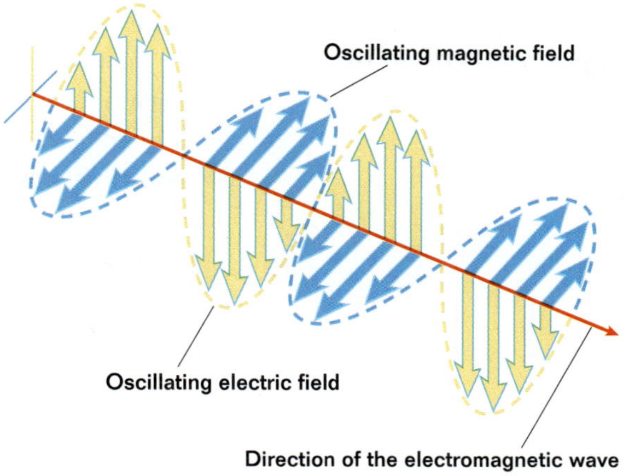

Figure 2
An electromagnetic wave consists of electric and magnetic field waves at right angles to each other.

Table 1 The Electromagnetic Spectrum

Classification	Range	Applications
radio waves	$\lambda > 30$ cm $f < 1.0 \times 10^9$ Hz	AM and FM radio; television
microwaves	30 cm $> \lambda >$ 1 mm 1.0×10^9 Hz $< f < 3.0 \times 10^{11}$ Hz	radar; atomic and molecular research; aircraft navigation; microwave ovens
infrared (IR) waves	1 mm $> \lambda >$ 700 nm 3.0×10^{11} Hz $< f < 4.3 \times 10^{14}$ Hz	molecular vibrational spectra; infrared photography; physical therapy
visible light	700 nm (red) $> \lambda >$ 400 nm (violet) 4.3×10^{14} Hz $< f < 7.5 \times 10^{14}$ Hz	visible-light photography; optical microscopy; optical astronomy
ultraviolet (UV) light	400 nm $> \lambda >$ 60 nm 7.5×10^{14} Hz $< f < 5.0 \times 10^{15}$ Hz	sterilization of medical instruments; identification of fluorescent minerals
X rays	60 nm $> \lambda > 10^{-4}$ nm 5.0×10^{15} Hz $< f < 3.0 \times 10^{21}$ Hz	medical examination of bones, teeth, and vital organs; treatment for types of cancer
gamma rays	0.1 nm $> \lambda > 10^{-5}$ nm 3.0×10^{18} Hz $< f < 3.0 \times 10^{22}$ Hz	examination of thick materials for structural flaws; treatment of types of cancer; food irradiation

Demonstration

Radio Waves

Purpose Demonstrate another example of radiation in the electromagnetic spectrum.

Materials pocket-size transistor radio, metal can with metal lid, glass jar, aluminum foil, paper, plastic wrap

Procedure Tell students that radio waves have a longer wavelength than infrared waves and that these waves can be detected with a radio. Turn on the radio, and tune it to a station. Have students notice changes in reception as you place the radio in the metal can and then in the glass jar. Wrap the radio in the paper, in the aluminum foil, and in the plastic wrap. Have students notice that paper is "transparent" to radio waves. Point out that different types of radiation can penetrate different materials.

Teaching Tip — BASIC

Point out that all of the frequencies in **Table 1** are expressed in hertz but that the wavelengths are in centimeters, millimeters, and nanometers, which stand for 10^{-2}, 10^{-3}, and 10^{-9} m, respectively. Ask students if the wavelengths increase or decrease as the frequencies decrease (*increase*).

Demonstration

How Light Travels GENERAL

Purpose Demonstrate that light waves can be approximated as rays.

Materials laser, index card, two dusty chalkboard erasers, plane mirror

CAUTION *Avoid pointing the laser beam near students' eyes; retinal damage may occur.*

Procedure Direct the laser beam across the room. Point out that in order for students to see the beam, it is necessary to have an object in the path of the beam that will reflect some light.

Place the index card in the path of the beam, and slowly walk across the room while keeping the beam centered on the card. Students should see that the beam travels in a straight line.

Stand beside the beam, and tap the erasers together above the beam. As the chalk dust falls from the erasers, the beam will become visible. Quickly walk along the length of the beam, tapping the erasers until the entire beam becomes visible.

extension

Integrating Astronomy

Visit go.hrw.com for the activity "Starlight, Star Heat."

Keyword HF6LGTX

All electromagnetic waves move at the speed of light

All forms of electromagnetic radiation travel at a single high speed in a vacuum. Early experimental attempts to determine the speed of light failed because this speed is so great. As experimental techniques improved, especially during the nineteenth and early twentieth centuries, the speed of light was determined with increasing accuracy and precision. By the mid-twentieth century, the experimental error was less than 0.001 percent. The currently accepted value for light traveling in a vacuum is $2.997\,924\,58 \times 10^8$ m/s. Light travels slightly slower in air, with a speed of $2.997\,09 \times 10^8$ m/s. For calculations in this book, the value used for both situations will be 3.00×10^8 m/s.

The relationship between frequency, wavelength, and speed described in the chapter on vibrations and waves also holds true for light waves.

WAVE SPEED EQUATION

$$c = f\lambda$$

speed of light = frequency × wavelength

SAMPLE PROBLEM A

Electromagnetic Waves

PROBLEM

The AM radio band extends from 5.4×10^5 Hz to 1.7×10^6 Hz. What are the longest and shortest wavelengths in this frequency range?

SOLUTION

Given: $f_1 = 5.4 \times 10^5$ Hz $\quad f_2 = 1.7 \times 10^6$ Hz $\quad c = 3.00 \times 10^8$ m/s

Unknown: $\lambda_1 = ?$ $\quad \lambda_2 = ?$

Use the wave speed equation on this page to find the wavelengths:

$$c = f\lambda \quad \lambda = \frac{c}{f}$$

$$\lambda_1 = \frac{3.00 \times 10^8 \text{ m/s}}{5.4 \times 10^5 \text{ Hz}}$$

$$\boxed{\lambda_1 = 5.6 \times 10^2 \text{ m}}$$

$$\lambda_2 = \frac{3.00 \times 10^8 \text{ m/s}}{1.7 \times 10^6 \text{ Hz}}$$

$$\boxed{\lambda_2 = 1.8 \times 10^2 \text{ m}}$$

CALCULATOR SOLUTION

Although the calculator solutions are 555.5555556 m and 176.470588 m, both answers must be rounded to two digits because the frequencies have only two significant figures.

PRACTICE A

Electromagnetic Waves

1. Gamma-ray bursters are objects in the universe that emit pulses of gamma rays with high energies. The frequency of the most energetic bursts has been measured at around 3.0×10^{21} Hz. What is the wavelength of these gamma rays?
2. What is the wavelength range for the FM radio band (88 MHz–108 MHz)?
3. Shortwave radio is broadcast between 3.50 and 29.7 MHz. To what range of wavelengths does this correspond? Why do you suppose this part of the spectrum is called shortwave radio?
4. What is the frequency of an electromagnetic wave if it has a wavelength of 1.0 km?
5. The portion of the visible spectrum that appears brightest to the human eye is around 560 nm in wavelength, which corresponds to yellow-green. What is the frequency of 560 nm light?
6. What is the frequency of highly energetic ultraviolet radiation that has a wavelength of 125 nm?

Waves can be approximated as rays

Consider an ocean wave coming toward the shore. The broad crest of the wave that is perpendicular to the wave's motion consists of a line of water particles. Similarly, another line of water particles forms a low-lying trough in the wave, and still another line of particles forms the crest of a second wave. In any type of wave, these lines of particles are called *wave fronts.*

All the points on the wave front of a plane wave can be treated as point sources, that is, coming from a source of neglibible size. A few of these points are shown on the initial wave front in **Figure 3.** Each of these point sources produces a circular or spherical secondary wave, or *wavelet.* The radii of these wavelets are indicated by the blue arrows in **Figure 3.** The line that is tangent to each of these wavelets at some later time determines the new position of the initial wave front (the new wave front in **Figure 3**). This approach to analyzing waves is called *Huygens' principle,* named for the physicist Christian Huygens, who developed it.

Huygens' principle can be used to derive the properties of any wave (including light) that interacts with matter, but the same results can be obtained by treating the propagating wave as a straight line perpendicular to the wave front. This line is called a *ray,* and this simplification is called the *ray approximation.*

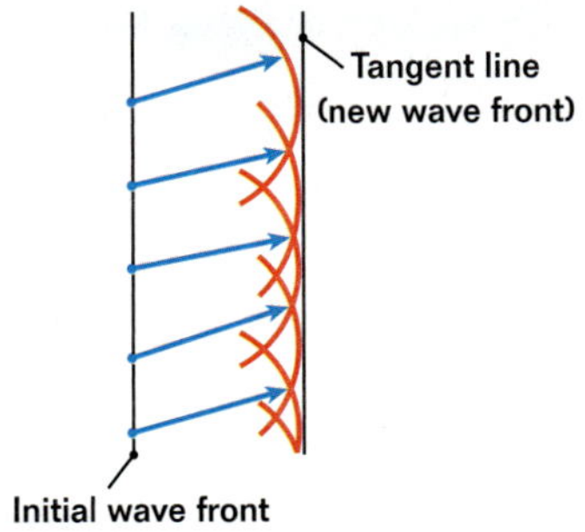

Figure 3
According to Huygens' principle, a wave front can be divided into point sources. The line tangent to the wavelets from these sources marks the wave front's new position.

PROBLEM GUIDE A

Use this guide to assign problems.
SE = Student Edition Textbook
PW = Problem Workbook
PB = Problem Bank on the One-Stop Planner (OSP)

Solving for:

λ	**SE** Sample, 1–3; Ch. Rvw. 10–13 **PW** 5–7 **PB** 7–10
f	**SE** 4–6 **PW** Sample, 1–4 **PB** 3–6
c	**PW** 8 **PB** Sample, 1, 2

***Challenging Problem**
Consult the printed Solutions Manual or the OSP for detailed solutions.

ANSWERS

Practice A

1. 1.0×10^{-13} m
2. 3.4 m–2.78 m
3. 85.7 m–10.1 m; The wavelengths are shorter than those of the AM radio band.
4. 3.0×10^{5} Hz
5. 5.4×10^{14} Hz
6. 2.40×10^{15} Hz

SECTION 1

Teaching Tip — GENERAL

Point out that the illuminance at any location is analogous to the power per unit of area at that location. Ask students to compare the surface areas of two spherical lampshades and the brightness of the light that hits their surface when identical light bulbs are placed at the center of each. Use values for their radii, such as 10 cm and 50 cm (surface area of a sphere = $4\pi r^2$).

Teaching Tip — BASIC

As compact fluorescent light bulbs become more prevalent, the use of luminous flux (lumens) rather than power rating (watts) is becoming more common in rating a bulb's light output. Incandescent bulbs generally produce about 15 lumens/watt, whereas fluorescent bulbs produce 50–100 lumens/watt. Ask students which type of light bulb is more efficient. Have them compare the average power usage over week, month, or year of normal usage for both types of light bulbs.

SECTION REVIEW ANSWERS

1. **a.** microwave
   b. radio waves, visible light
   c. visible light
2. 3.96×10^{-7} m; near ultraviolet
3. The speed of light is too great to be measured over such a short distance. The time of travel for the light in Galileo's experiment was about 1.0×10^{-5} s.
4. $\frac{1}{16}$ of the sun's brightness on Earth's surface

SCILINKS
www.scilinks.org
Topic: Light Bulbs
Code: HF60879

Illuminance decreases as the square of the distance from the source

You have probably noticed that it is easier to read a book beside a lamp using a 100 W bulb rather than a 25 W bulb. It is also easier to read nearer to a lamp than farther from a lamp. These experiences suggest that the intensity of light depends on both the amount of light energy emitted from a source and the distance from the light source.

Light bulbs are rated by their power input (measured in watts) and their light output. The rate at which light is emitted from a source is called the *luminous flux* and is measured in *lumens* (lm). Luminous flux is a measure of power output but is weighted to take into account the response of the human eye to light. The idea of luminous flux helps us understand why the illumination on a book page is reduced as you move away from a light source. Imagine spherical surfaces of different sizes with a point light source at the center of the sphere, shown in **Figure 4.** A point source of light provides light equally in all directions. The principle of conservation of energy tells us that the luminous flux is the same on each sphere. However, the luminous flux divided by the area of the surface, which is called the *illuminance* (measured in lm/m^2, or *lux*), decreases as the radius squared when you move away from a light source.

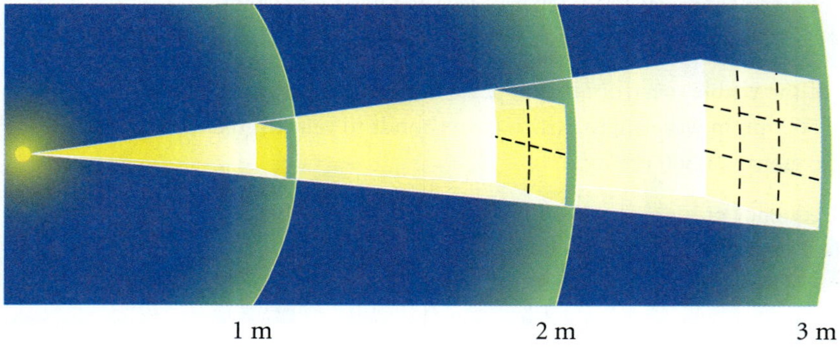

Figure 4
Less light falls on each unit square as the distance from the source increases.

SECTION REVIEW

1. Identify which portions of the electromagnetic spectrum are used in each of the devices listed.
   a. a microwave oven
   b. a television set
   c. a single-lens reflex camera
2. If an electromagnetic wave has a frequency of 7.57×10^{14} Hz, what is its wavelength? To what part of the spectrum does this wave belong?
3. Galileo performed an experiment to measure the speed of light by timing how long it took light to travel from a lamp he was holding to an assistant about 1.5 km away and back again. Why was Galileo unable to conclude that light had a finite speed?
4. **Critical Thinking** How bright would the sun appear to an observer on Earth if the sun were four times farther from Earth than it actually is? Express your answer as a fraction of the sun's brightness on Earth's surface.

Flat Mirrors

SECTION OBJECTIVES

- **Distinguish between specular and diffuse reflection of light.**
- **Apply the law of reflection for flat mirrors.**
- **Describe the nature of images formed by flat mirrors.**

REFLECTION OF LIGHT

Suppose you have just had your hair cut and you want to know what the back of your head looks like. You can do this seemingly impossible task by using two mirrors to direct light from behind your head to your eyes. Redirecting light with mirrors reveals a basic property of light's interaction with matter.

Light traveling through a uniform substance, whether it is air, water, or a vacuum, always travels in a straight line. However, when the light encounters a different substance, its path will change. If a material is opaque to the light, such as the dark, highly polished surface of a wooden table, the light will not pass into the table more than a few wavelengths. Part of the light is absorbed, and the rest of it is deflected at the surface. This change in the direction of the light is called **reflection.** All substances absorb at least some incoming light and reflect the rest. A good mirror can reflect about 90 percent of the incident light, but no surface is a perfect reflector. Notice in **Figure 5** that the images of the person get successively darker.

reflection

the change in direction of an electromagnetic wave at a surface that causes it to move away from the surface

The texture of a surface affects how it reflects light

The manner in which light is reflected from a surface depends on the surface's smoothness. Light that is reflected from a rough, textured surface, such as paper, cloth, or unpolished wood, is reflected in many different directions, as shown in **Figure 6(a).** This type of reflection is called *diffuse reflection.* Diffuse reflection will be discussed further in Section 4.

Light reflected from smooth, shiny surfaces, such as a mirror or water in a pond, is reflected in one direction only, as shown in **Figure 6(b).** This type of reflection is called *specular reflection.* A surface is considered smooth if its surface variations are small compared with the wavelength of the incoming light. For our discussion, reflection will be used to mean only specular reflection.

Figure 5
Mirrors reflect nearly all incoming light, so multiple images of an object between two mirrors are easily formed.

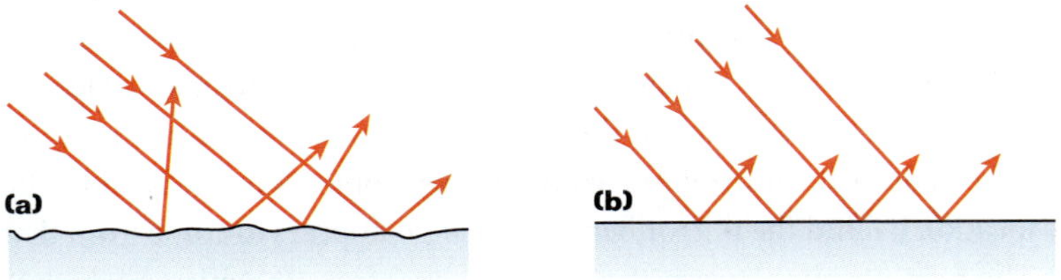

Figure 6
Diffusely reflected light is reflected in many directions **(a),** whereas specularly reflected light is reflected in the same forward direction only **(b).**

Demonstration

Diffuse Reflection

Purpose Demonstrate that light reflected from a rough surface reflects in many directions.

Materials laser, index card

CAUTION *Avoid directing the laser beam near students' eyes.*

Procedure Tape the index card to a wall in the classroom. Direct the laser beam onto the card from across the room. Make sure the beam is not perpendicular to the card's surface. Explain to students that the light strikes the index card as a single beam but that because the card is a diffuse reflector, the beam undergoes reflection in all directions. Only a small part of the reflected light will go to each part of the room, enabling everyone to see the spot where the laser beam strikes the card.

Visual Strategy – BASIC

Figure 6

Point out that the incident light rays are parallel.

Q Why are the reflected rays from the surface in **(a)** not parallel?

A ***Because the surface in (a) has irregularities, it forms a different angle with the incident ray at each point.***

Visual Strategy ADVANCED

Figure 7

In the photo, the light from the flashlight strikes the surface.

Q Why is there a bright spot where it strikes the surface?

A ***Light is scattered off the imperfect or dirty surface.***

Q Below the bright spot, there seem to be rays at the lower left and lower right. What causes these rays?

A ***We are seeing reflections off the mirror surface of the incident and reflected beams above.***

Demonstration

Specular Reflection

Purpose Demonstrate that all parallel rays of light reflected from a smooth surface are reflected in the same direction.

Materials laser, flat mirror, dusty chalkboard erasers

CAUTION *Avoid directing the primary beam of the laser and the reflected beam from the mirror toward the students.*

Procedure Tape the flat mirror to a wall in the classroom. Direct the beam of the laser onto the mirror from across the room. Make sure the beam is not perpendicular to the mirror's surface. Explain to students that the beam is reflected at the mirror's surface and that the angle of incidence is equal to the angle of reflection. This can be shown qualitatively by gently tapping the erasers in front of the mirror so that both the incoming and reflected beams become visible.

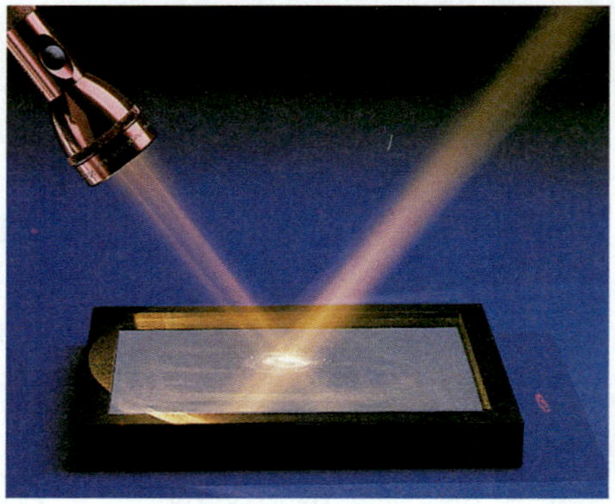

(a)

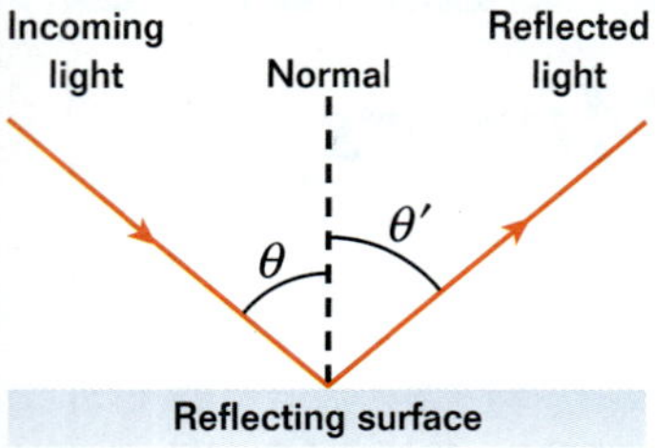

(b)

Figure 7
The symmetry of reflected light **(a)** is described by the law of reflection, which states that the angles of the incoming and reflected rays are equal **(b).**

Incoming and reflected angles are equal

You probably have noticed that when incoming rays of light strike a smooth reflecting surface, such as a polished table or mirror, at an angle close to the surface, the reflected rays are also close to the surface. When the incoming rays are high above the reflecting surface, the reflected rays are also high above the surface. An example of this similarity between incoming and reflected rays is shown in **Figure 7(a).**

If a straight line is drawn perpendicular to the reflecting surface at the point where the incoming ray strikes the surface, the **angle of incidence** and the **angle of reflection** can be defined with respect to the line. Careful measurements of the incident and reflected angles θ and θ', respectively, reveal that the angles are equal, as illustrated in **Figure 7(b).**

angle of incidence

the angle between a ray that strikes a surface and the line perpendicular to that surface at the point of contact

angle of reflection

the angle formed by the line perpendicular to a surface and the direction in which a reflected ray moves

$$\theta = \theta'$$

angle of incoming light ray = angle of reflected light ray

The line perpendicular to the reflecting surface is referred to as the *normal* to the surface. It therefore follows that the angle between the incoming ray and the surface equals $90° - \theta$, and the angle between the reflected ray and the surface equals $90° - \theta'$.

FLAT MIRRORS

The simplest mirror is the *flat mirror*. If an object, such as a pencil, is placed at a distance in front of a flat mirror and light is bounced off the pencil, light rays will spread out from the pencil and reflect from the mirror's surface. To an observer looking at the mirror, these rays appear to come from a location on the other side of the mirror. As a convention, an object's image is said to be at this location behind the mirror because the light appears to come from that point. The relationship between the *object distance* from the mirror, which is represented as p, and the *image distance*, which is represented as q, is such that the object and image distances are equal. Similarly, the image of the object is the same size as the object.

The image formed by rays that appear to come from the image point behind the mirror—but never really do—is called a **virtual image.** As shown in **Figure 8(a),** a flat mirror always forms a virtual image, which always appears as if it is behind the surface of the mirror. For this reason, a virtual image can never be displayed on a physical surface.

virtual image

an image from which light rays appear to diverge, even though they are not actually focused there; a virtual image cannot be projected on a screen

Image location can be predicted with ray diagrams

Ray diagrams, such as the one shown in **Figure 8(b),** are drawings that use simple geometry to locate an image formed by a mirror. Suppose you want to make a ray diagram for a pencil placed in front of a flat mirror. First, sketch the situation. Draw the location and arrangement of the mirror and the position of the pencil with respect to the mirror. Construct the drawing so that the object and the image distances (p and q, respectively) are proportional to their actual sizes. To simplify matters, we will consider only the tip of the pencil.

To pinpoint the location of the pencil tip's image, draw two rays on your diagram. Draw the first ray from the pencil tip perpendicular to the mirror's surface. Because this ray makes an angle of 0° with a line perpendicular (or *normal*) to the mirror, the angle of reflection also equals 0°, causing the ray to reflect back on itself. In **Figure 8(b),** this ray is denoted by the number **1** and is shown with arrows pointing in both directions because the incident ray reflects back on itself.

Draw the second ray from the tip of the pencil to the mirror, but this time place the ray at an angle that is not perpendicular to the surface of the mirror. The second ray is denoted in **Figure 8(b)** by the number **2.** Then, draw the reflected ray, keeping in mind that it will reflect away from the surface of the mirror at an angle, θ', equal to the angle of incidence, θ.

Next, trace both reflected rays back to the point from which they appear to have originated, that is, behind the mirror. Use dotted lines when drawing these rays that appear to emerge from behind the mirror to distinguish them from the actual rays of light (the solid lines) in front of the mirror. The point at which these dotted lines meet is the image point, which in this case is where the image of the pencil's tip forms.

By continuing this process for all of the other parts of the pencil, you can locate the complete virtual image of the pencil. Note that the pencil's image appears as far behind the mirror as the pencil is in front of the mirror ($p = q$). Likewise, the object height, h, equals the image height, h'.

SCILINKS.
www.scilinks.org
Topic: Mirrors
Code: HF60970

(a)

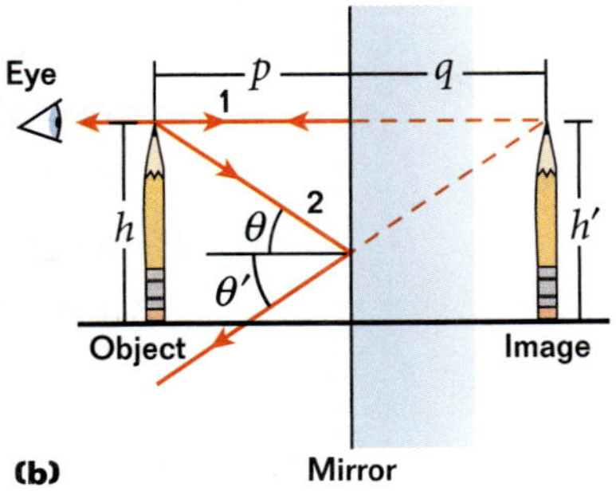

(b)

Figure 8
The position and size of the virtual image that forms in a flat mirror **(a)** can be predicted by constructing a ray diagram **(b).**

Demonstration

Flat Mirror Images GENERAL

Purpose Demonstrate that the image behind a mirror is virtual.

Materials sheet of high-quality plate glass (0.5 m × 0.5 m), two identical candles, black chalkboard or black surface

Procedure Place the sheet of glass vertically in front of the chalkboard. Place one candle about 30 cm in front of the glass, and place the other candle 30 cm behind it.

Have students who might be able to see the candle behind the glass without looking through the glass move to the rear of the classroom. Have students close their eyes while you light the front candle. Discreetly adjust the position of the candle behind the glass so that the image of the front candle as seen in the glass coincides with the position of the back candle. When properly positioned, the back candle will appear lit when it is viewed through the glass sheet.

Hold a match near the back candle and tell students to open their eyes. Raise the match and blow it out, giving the impression that you have just finished lighting the back candle. Ask how many candles are lit. Then lift the glass. Explain that the image of the flame appeared to be at the same distance behind the glass as the rear candle.

SECTION 2

Teaching Tip — GENERAL

Point out that the object and its image are equidistant from the mirror's surface ($p = q$). Students can demonstrate this conclusion using simple geometry to show that the ray-tracing procedure produces congruent triangles.

SECTION REVIEW ANSWERS

1. a. specular
   b. diffuse
   c. specular
   d. diffuse
2. 75°; 75°
3. Diagrams will vary. Verify that the rays' angles of incidence equal the angles of reflection in the students' drawings.
4. The light must be reflected from one mirror to the other in order to form images of images. Therefore, the mirrors must be exactly or very nearly parallel to each other, with their mirrored surfaces facing each other.
5. It will seem twice as large, because the object distance (the distance between the mirror and any point in the room, including the back wall) equals the image distance.
6. An object facing the mirror produces an image that faces the object, and the front of the object corresponds to the back of the image. It appears as if the image is an object located behind the mirror that is left-right reversed.

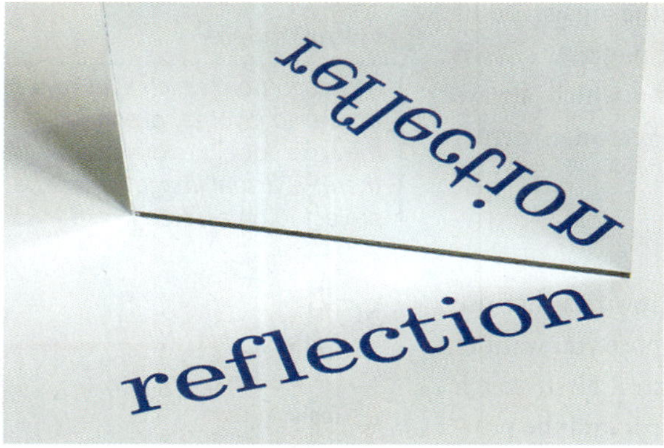

Figure 9
The front of an object becomes the back of its image.

This ray-tracing procedure will work for any object placed in front of a flat mirror. By selecting a single point on the object (usually its uppermost tip or edge), you can use ray tracing to locate the same point on the image. The rest of the image can be added once the image point and image distance have been determined.

The image formed by a flat mirror appears reversed to an observer in front of the mirror. You can easily observe this effect by placing a piece of writing in front of a mirror, as shown in **Figure 9.** In the mirror, each of the letters is reversed. You may also notice that the angle the word and its reflection make with respect to the mirror is the same.

SECTION REVIEW

1. Which of the following are examples of specular reflection, and which are examples of diffuse reflection?
   a. reflection of light from the surface of a lake on a calm day
   b. reflection of light from a plastic trash bag
   c. reflection of light from the lens of eyeglasses
   d. reflection of light from a carpet
2. Suppose you are holding a flat mirror and standing at the center of a giant clock face built into the floor. Someone standing at 12 o'clock shines a beam of light toward you, and you want to use the mirror to reflect the beam toward an observer standing at 5 o'clock. What should the angle of incidence be to achieve this? What should the angle of reflection be?
3. Some department-store windows are slanted inward at the bottom. This is to decrease the glare from brightly illuminated buildings across the street, which would make it difficult for shoppers to see the display inside and near the bottom of the window. Sketch a light ray reflecting from such a window to show how this technique works.
4. **Interpreting Graphics** The photograph in **Figure 5** shows multiple images that were created by multiple reflections between two flat mirrors. What conclusion can you make about the relative orientation of the mirrors? Explain your answer.
5. **Critical Thinking** If one wall of a room consists of a large flat mirror, how much larger will the room appear to be? Explain your answer.
6. **Critical Thinking** Why does a flat mirror appear to reverse the person looking into a mirror left to right, but not up and down?

Curved Mirrors

SECTION OBJECTIVES

- Calculate distances and focal lengths using the mirror equation for concave and convex spherical mirrors.
- Draw ray diagrams to find the image distance and magnification for concave and convex spherical mirrors.
- Distinguish between real and virtual images.
- Describe how parabolic mirrors differ from spherical mirrors.

CONCAVE SPHERICAL MIRRORS

Small, circular mirrors, such as those used on dressing tables, may appear at first glance to be the same as flat mirrors. However, the images they form differ from those formed by flat mirrors. The images for objects close to the mirror are larger than the object, as shown in **Figure 10(a),** whereas the images of objects far from the mirror are smaller and upside down, as shown in **Figure 10(b).** Images such as these are characteristic of curved mirrors. The image in **Figure 10(a)** is a virtual image like those created by flat mirrors. In contrast, the image in **Figure 10(b)** is a *real* image.

Concave mirrors can be used to form real images

One basic type of curved mirror is the spherical mirror. A spherical mirror, as its name implies, has the shape of part of a sphere's surface. A spherical mirror with light reflecting from its silvered, concave surface (that is, the inner surface of a sphere) is called a **concave spherical mirror.** Concave mirrors are used whenever a magnified image of an object is needed, as in the case of the dressing-table mirror.

concave spherical mirror

a mirror whose reflecting surface is a segment of the inside of a sphere

One factor that determines where the image will appear in a concave spherical mirror and how large that image will be is the radius of curvature, R, of the mirror. The radius of curvature is the same as the radius of the spherical shell of which the mirror is a small part; R is therefore the distance from the mirror's surface to the center of curvature, C.

Figure 10
Curved mirrors can be used to form images that are larger **(a)** or smaller **(b)** than the object.

Demonstration

Image Formed by a Concave Mirror

Purpose Demonstrate that changing the curvature of a concave mirror produces different images.

Materials flexible polyester film with reflective coating (about 1 m × 0.5 m), poster board

Procedure Tape the reflective film flat to the poster board, and place a can or a bottle with a large label on it on a table 20 to 50 cm away. Let students observe that the image in the reflective film is like one they would expect to see in a flat mirror. Slowly start bending the sides of the sheet of film to turn it into a concave mirror (until it has a cylindrical shape). Have students describe changes in the image as you roll the reflective film. *(The virtual image grows wider and moves farther away, eventually disappearing when the radius of curvature becomes too small.)* Repeat if necessary.

Explain that by bending the mirror, you changed its radius of curvature. Have a student hold the mirror at one constant curvature as you move the object slowly closer to the mirror and then farther away. Ask students to describe how changes in the object's distance affect the position and size of the object's image.

Visual Strategy GENERAL

Figure 11
This image shows a real image of a light bulb on a glass plate. The bulb itself is off to the left, too far away to fit in the photo frame. There is actually an image of the entire bulb on the plate, but only the image of the filament is bright enough to be seen in the photograph.

Q What represents the size of the object? What represents its distance from the mirror? What represents the size of the image? What represents the image's distance from *C*? What does *f* refer to?

A ***h; p; h'; R – q; distance from focal point to mirror***

Key Models and Analogies ADVANCED

To help students understand the reflections in **Figure 11,** point out that light striking the mirror is reflected according to the law of reflection, as if the curved mirror were made of many small plane mirrors positioned to form a circle. The ray through *C* would be normal to such a mirror, so it is reflected back in the same direction from which it came.

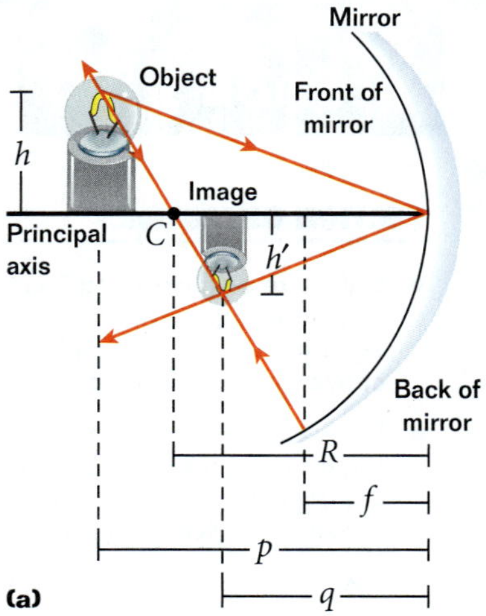

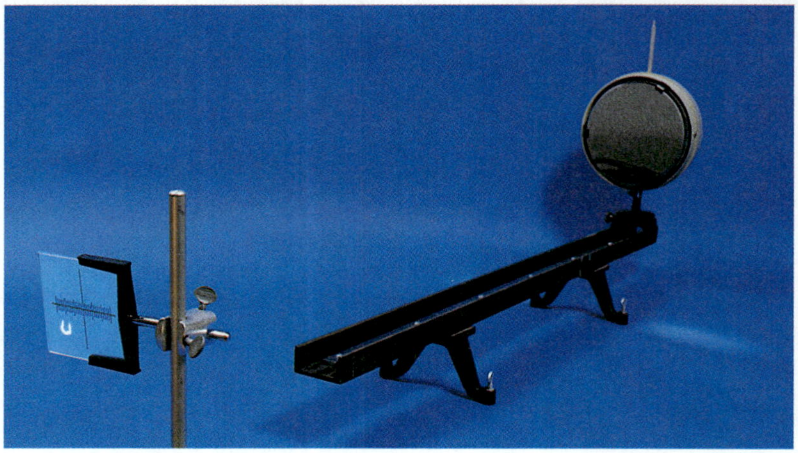

Figure 11
(a) The rays from an object, such as a light bulb, converge to form a real image in front of a concave mirror. **(b)** In this lab setup, the real image of a light-bulb filament appears on a glass plate in front of a concave mirror.

Imagine a light bulb placed upright at a distance *p* from a concave spherical mirror, as shown in **Figure 11(a).** The base of the bulb is along the mirror's principal axis, which is the line that extends infinitely from the center of the mirror's surface through the center of curvature, *C*. Light rays diverge from the light bulb, reflect from the mirror's surface, and converge at some distance, *q*, in front of the mirror. Because the light rays reflected by the mirror actually pass through the image point, which in this case is below the principal axis, the image forms in front of the mirror.

If you place a piece of paper at the image point, you will see on the paper a sharp and clear image of the light bulb. As you move the paper in either direction away from the image point, the rays diverge, and the image becomes unfocused. An image of this type is called a **real image.** Unlike the virtual images that appear behind a flat mirror, real images can be displayed on a surface, like the images on a movie screen. **Figure 11(b)** shows a real image of a light-bulb filament on a glass plate in front of a concave mirror. This light bulb itself is outside the photograph, to the left.

real image

an image that is formed by the intersection of light rays; a real image can be projected on a screen

Images created by spherical mirrors suffer from spherical aberration

As you draw ray diagrams, you may notice that certain rays do not exactly intersect at the image point. This phenomenon is particularly noticeable for rays that are far from the principal axis and for mirrors with a small radius of curvature. This situation, called *spherical aberration,* also occurs with real light rays and real spherical mirrors and will be discussed further at the end of this section when we introduce *parabolic mirrors.*

In the next pages of this section, you will learn about the mirror equation and ray diagrams. Both of these concepts are valid only for *paraxial rays,* but they do provide quite useful approximations. Paraxial rays are those light rays that are very near the principal axis of the mirror. We will assume that all of the rays used in our drawings and calculations with spherical mirrors are paraxial, even though they may not appear to be so in all of the diagrams accompanying the text.

Image location can be predicted with the mirror equation

By looking at **Figure 11(a),** you can see that object distance, image distance, and radius of curvature are interdependent. If the object distance and radius of curvature of the mirror are known, you can predict where the image will appear. Alternatively, the radius of curvature of a mirror can be determined if you know where the image appears for a given object distance. The following equation relates object distance, p, image distance, q, and the radius of curvature, R, is called *the mirror equation.*

$$\frac{1}{p}+\frac{1}{q}=\frac{2}{R}$$

If the light bulb is placed very far from the mirror, the object distance, p, is great enough compared with R that $1/p$ is almost 0. In this case, q is almost $R/2$, so the image forms about halfway between the center of curvature and the center of the mirror's surface. The image point, as shown in **Figure 12(a)** and **(b),** is in this special case called the *focal point* of the mirror and is denoted by the capital letter F. Because the light rays are reversible, the reflected rays from a light source at the focal point will emerge parallel to each other and will not form an image.

For light emerging from a source very far away from a mirror, the light rays are essentially parallel to one another. In this case, an image forms at the focal point, F, and the image distance is called the *focal length,* denoted by the lowercase letter f. For a spherical mirror, the focal length is equal to half the radius of curvature of the mirror. The mirror equation can therefore be expressed in terms of the focal length.

MIRROR EQUATION

$$\frac{1}{p}+\frac{1}{q}=\frac{1}{f}$$

$$\frac{1}{\text{object distance}}+\frac{1}{\text{image distance}}=\frac{1}{\text{focal length}}$$

Quick Lab

Curved Mirrors

MATERIALS LIST

- stainless-steel or silver spoon
- short pencil

Observe the pencil's reflection in the inner portion of the spoon. Slowly move the spoon closer to the pencil. Note any changes in the appearance of the pencil's reflection. Repeat these steps using the other side of the spoon as the mirror.

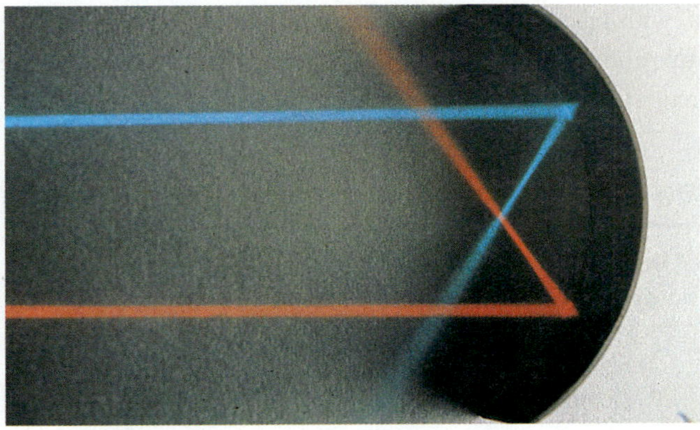
(a)

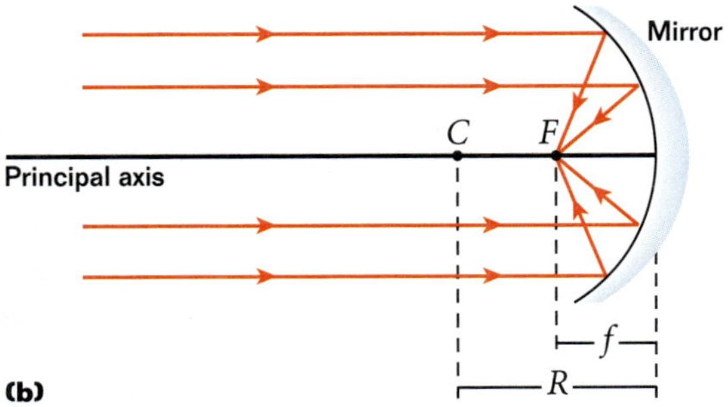

(b)

Figure 12
Light rays that are parallel converge at a single point **(a),** which can be represented in a diagram **(b),** when the rays are assumed to be from a distant object ($p \approx \infty$).

Quick Lab

TEACHER'S NOTES

This activity is intended to explore how an object's distance affects the object's image in concave and convex mirrors. This experiment works best with a very shiny spoon that has a large radius of curvature.

Homework Options
This QuickLab can easily be performed outside of the physics lab room.

Demonstration

Focal Point of a Concave Mirror GENERAL

Purpose Demonstrate that rays parallel to the principal axis are reflected through the focal point, and show that $f=\frac{R}{2}$.

Materials light source, ray filter, concave mirror, white paper

Procedure Use the ray filter to produce five beams. Dim the lights, and hold the sheet of paper in front of the beams to let students observe that the incident rays are parallel. Place the concave mirror 20 to 30 cm from the light source, and let students observe the beams converging. Tell them that the point of convergence is called the focal point. Explain that past that point, the beams diverge. Draw the mirror's curve and the principal axis, and mark the focal point on the chalkboard. Ask students if that point could be the center of the circle from which the mirror was cut (*no*). Have students mark where the approximate center of the circle is. Measure R, and compare it with f.

Misconception Alert — GENERAL

The term *magnification* sometimes leads students to think that the image is larger than the object. Use a numerical example to show that this is not always the case. Ask students to calculate the image height of an 8 cm object when $p = 12$ cm and $q = 3$ cm (-2 *cm*).

Demonstration

Beams Reflected from a Concave Mirror — BASIC

Purpose Demonstrate the formation of a virtual image.

Materials light source, ray filter, concave mirror

Procedure Ask students if reflected beams always converge from concave mirrors. Lower the lights in the room so that the light can be easily seen. Place the mirror as far away as possible from the front of the light source, and move the mirror closer and closer to the light source until the reflected beams diverge. Tell students that the point behind the mirror from which the beam seems to come is a virtual image of that point on the source.

A set of sign conventions for the three variables must be established for use with the mirror equation. The region in which light rays reflect and form real images is called the front side of the mirror. The other side, where light rays do not exist—and where virtual images are formed—is called the back side of the mirror.

Object and image distances have a positive sign when measured from the center of the mirror to any point on the mirror's front side. Distances for images that form on the back side of the mirror always have a negative sign. Because the mirrored surface is on the front side of a concave mirror, its focal length always has a positive sign. The object and image heights are positive when both are above the principal axis and negative when either is below.

Magnification relates image and object sizes

Unlike flat mirrors, curved mirrors form images that are not the same size as the object. The measure of how large or small the image is with respect to the original object's size is called the *magnification* of the image.

If you know where an object's image will form for a given object distance, you can determine the magnification of the image. Magnification, M, is defined as the ratio of the height of the bulb's image to the bulb's actual height. M also equals the negative of the ratio of the image distance to the object distance. If an image is smaller than the object, the magnitude of its magnification is less than 1. If the image is larger than the object, the magnitude of its magnification is greater than 1. Magnification is a unitless quantity.

EQUATION FOR MAGNIFICATION

$$M = \frac{h'}{h} = -\frac{q}{p}$$

$$\text{magnification} = \frac{\text{image height}}{\text{object height}} = -\frac{\text{image distance}}{\text{object distance}}$$

For an image in front of the mirror, M is negative and the image is upside down, or *inverted,* with respect to the object. When the image is behind the mirror, M is positive and the image is *upright* with respect to the object. The conventions for magnification are listed in **Table 2.**

Table 2 Sign Conventions for Magnification

Orientation of image with respect to object	Sign of M	Type of image this applies to
upright	+	virtual
inverted	−	real

Ray diagrams can be used for concave spherical mirrors

Ray diagrams are useful for checking values calculated from the mirror and magnification equations. The techniques for ray diagrams that were used to locate the image for an object in front of a flat mirror can also be used for concave spherical mirrors. When drawing ray diagrams for concave mirrors, follow the basic procedure for a flat mirror, but also measure all distances along the principal axis and mark the center of curvature, *C*, and the focal point, *F*. As with a flat mirror, draw the diagram to scale. For instance, if the object distance is 50 cm, you can draw the object distance as 5 cm.

For spherical mirrors, three reference rays are used to find the image point. The intersection of any *two* rays locates the image. The third ray should intersect at the same point and can be used to check the diagram. These reference rays are described in **Table 3.**

Table 3 Rules for Drawing Reference Rays

Ray	Line drawn from object to mirror	Line drawn from mirror to image after reflection
1	parallel to principal axis	through focal point *F*
2	through focal point *F*	parallel to principal axis
3	through center of curvature *C*	back along itself through *C*

The image distance in the diagram should agree with the value for *q* calculated from the mirror equation. However, the image distance may differ because of inaccuracies that arise from drawing the ray diagrams at a reduced scale and far from the principal axis. Ray diagrams should therefore be used to obtain *approximate* values only; they should not be relied on for the best quantitative results.

Concave mirrors can produce both real and virtual images

When an object is moved toward a concave spherical mirror, its image changes, as shown in **Table 4** on next page. If the object is very far from the mirror, the light rays converge very near the focal point, *F*, of the mirror and form an image there. For objects at a finite distance greater than the radius of curvature, *C*, the image is real, smaller than the object, inverted, and located between *C* and *F*. When the object is at *C*, the image is real, located at *C*, and inverted. For an object at *C*, the image is the same size as the object. If the object is located between *C* and *F*, the image will be real, inverted, larger than the object, and located outside of *C*. When the object is at the focal point, no image is formed. When the object lies between *F* and the mirror surface, the image forms again, but now it becomes virtual, upright, and larger.

459

Visual Strategy GENERAL

Table 4
Point out that from all the possible rays coming from the tip of the pencil, these three rays are selected because they are reflected according to the simple rules listed in **Table 3.** Make sure that students understand the application of these rules in the three diagrams.

Table 4 Images Created by Concave Mirrors

Ray diagrams

1.

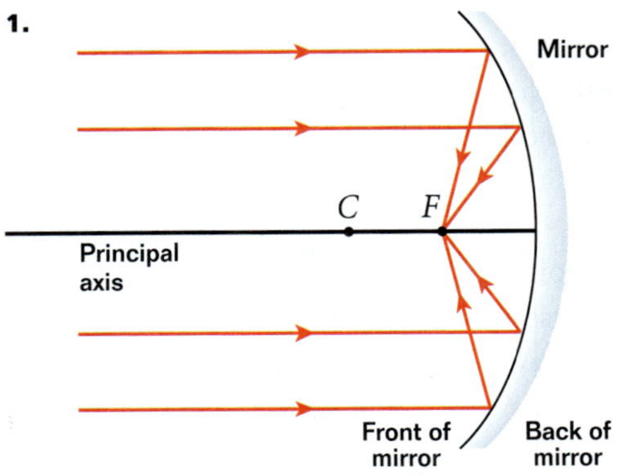

Configuration: object at infinity

Image: real image at *F*

2.

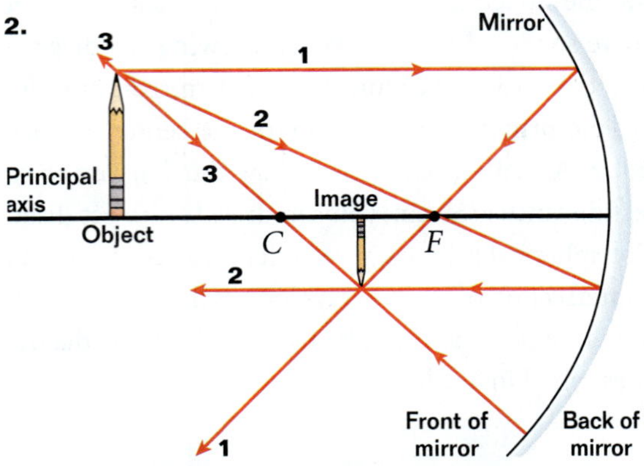

Configuration: object outside *C*

Image: real image between *C* and *F*, inverted with magnification < 1

3.

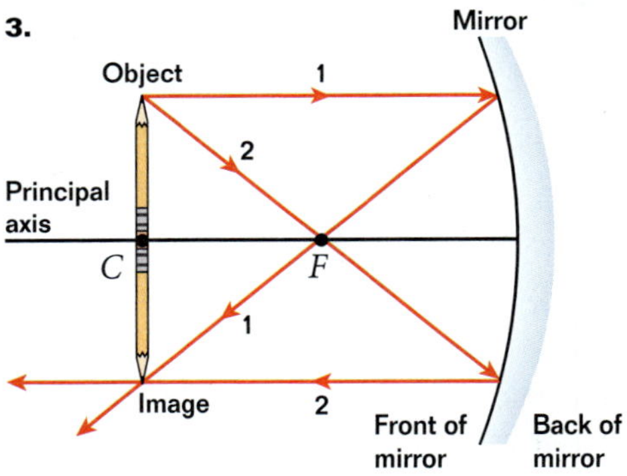

Configuration: object at *C*

Image: real image at *C*, inverted with magnification = 1

4.

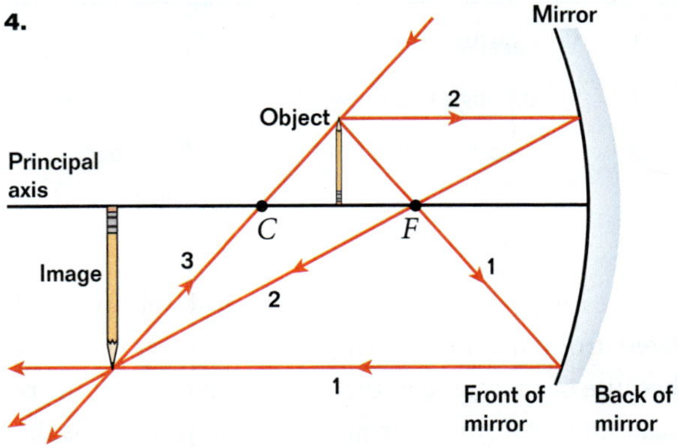

Configuration: object between *C* and *F*

Image: real image at *C*, inverted with magnification > 1

5.

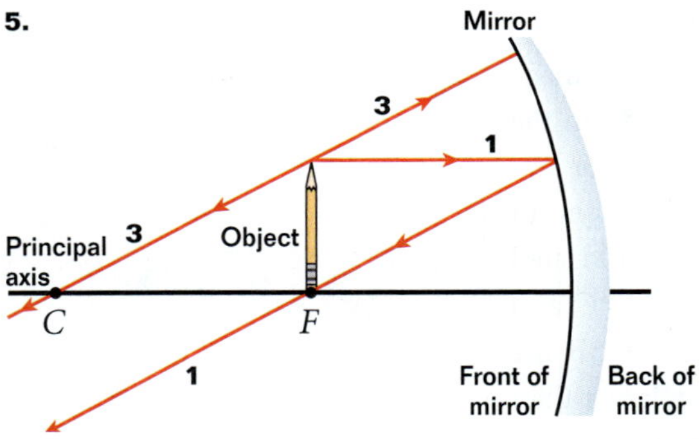

Configuration: object at *F*

Image: image at infinity (no image)

6.

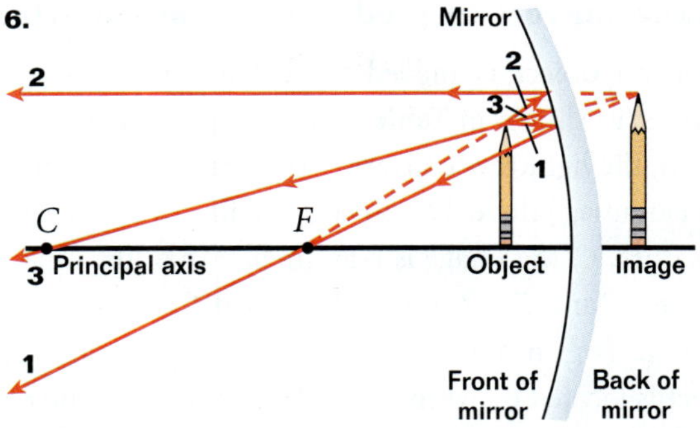

Configuration: object inside *F*

Image: virtual, upright image at *C* with magnification >1

SAMPLE PROBLEM B

STRATEGY Imaging with Concave Mirrors

PROBLEM

A concave spherical mirror has a focal length of 10.0 cm. Locate the image of a pencil that is placed upright 30.0 cm from the mirror. Find the magnification of the image. Draw a ray diagram to confirm your answer.

SOLUTION

1. **Determine the sign and magnitude of the focal length and object size.**

$f = +10.0$ cm $\qquad p = +30.0$ cm

The mirror is concave, so f is positive. The object is in front of the mirror, so p is positive.

2. **Draw a ray diagram using the rules given in Table 3.**

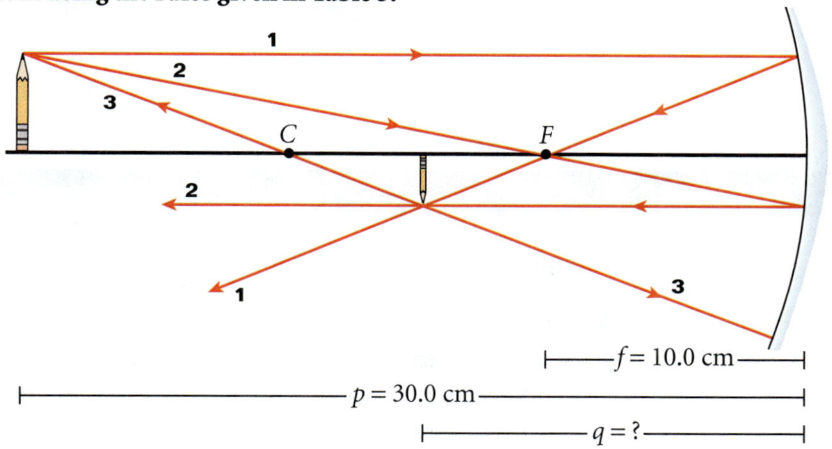

3. **Use the mirror equation to relate the object and image distances to the focal length.**

$$\frac{1}{p} + \frac{1}{q} = \frac{1}{f}$$

4. **Use the magnification equation in terms of object and image distances.**

$$M = -\frac{q}{p}$$

5. **Rearrange the equation to isolate the image distance, and calculate.**

Subtract the reciprocal of the object distance from the reciprocal of the focal length to obtain an expression for the unknown image distance.

$$\frac{1}{q} = \frac{1}{f} - \frac{1}{p}$$

continued on next page

Classroom Practice

Imaging with Concave Mirrors

When an object is placed 30.0 cm in front of a concave mirror, a real image is formed 60.0 cm from the mirror's surface. Find the focal length.

Answer

20.0 cm

A square object is placed 15 cm in front of a concave mirror with a focal length of 25 cm. A round object is placed 45 cm in front of the same mirror. Find the image distance, magnification, and type of image formed for each object. Draw ray diagrams for each object to confirm your answers.

Answer

$q_{square} = -38$ cm, $M_{square} = 2.5$, virtual and upright; $q_{round} = 56$ cm, $M_{round} = -1.2$, real and inverted

PROBLEM GUIDE B

Use this guide to assign problems.
SE = Student Edition Textbook
PW = Problem Workbook
PB = Problem Bank on the One-Stop Planner (OSP)

Solving for:

q	**SE** Sample, 1–2; Ch. Rvw. 35*, 49* **PW** 3–5 **PB** 7–10
p	**SE** Ch. Rvw. 36, 46, 48 **PW** 6–9 **PB** Sample, 1–3
R, f	**SE** 3–4; Ch. Rvw. 46–47 **PW** Sample, 1, 2 **PB** 4–6
M	**SE** Sample 1–4; Ch. Rvw. 34–36, 49* **PW** 3–4, 6–8 **PB** 8–10
h, h'	**PW** 5, 9 **PB** Sample, 1–3, 7

***Challenging Problem**
Consult the printed Solutions Manual or the OSP for detailed solutions.

ANSWERS

Practice B

1. $p = 10.0$ cm: no image (infinite q); $p = 5.00$ cm: $q = -10.0$ cm, $M = 2.00$; virtual, upright image
2. $q = 53$ cm; $M = -0.57$, real, inverted image
3. $R = 1.00 \times 10^2$ cm; $M = 2.00$; virtual image
4. $f = 6.00$ cm; $M = -1.20$; $q = 7.71$ cm; $M = -0.286$; real image

Substitute the values for f and p into the mirror equation and the magnification equation to find the image distance and magnification.

$$\frac{1}{q} = \frac{1}{10.0 \text{ cm}} - \frac{1}{30.0 \text{ cm}} = \frac{0.100}{1 \text{ cm}} - \frac{0.033}{1 \text{ cm}} = \frac{0.067}{1 \text{ cm}}$$

$$\boxed{q = 15 \text{ cm}}$$

$$\boxed{M = -\frac{q}{p} = -\frac{15 \text{ cm}}{30.0 \text{ cm}} = -0.50}$$

6. Evaluate your answer in terms of the image location and size.
The image appears between the focal point (10.0 cm) and the center of curvature (20.0 cm), as confirmed by the ray diagram. The image is smaller than the object and inverted ($-1 < M < 0$), as is also confirmed by the ray diagram. The image is therefore real.

PRACTICE B

Imaging with Concave Mirrors

1. Find the image distance and magnification of the mirror in the sample problem when the object distances are 10.0 cm and 5.00 cm. Are the images real or virtual? Are the images inverted or upright? Draw a ray diagram for each case to confirm your results.
2. A concave shaving mirror has a focal length of 33 cm. Calculate the image position of a cologne bottle placed in front of the mirror at a distance of 93 cm. Calculate the magnification of the image. Is the image real or virtual? Is the image inverted or upright? Draw a ray diagram to show where the image forms and how large it is with respect to the object.
3. A concave makeup mirror is designed so that a person 25.0 cm in front of it sees an upright image at a distance of 50.0 cm behind the mirror. What is the radius of curvature of the mirror? What is the magnification of the image? Is the image real or virtual?
4. A pen placed 11.0 cm from a concave spherical mirror produces a real image 13.2 cm from the mirror. What is the focal length of the mirror? What is the magnification of the image? If the pen is placed 27.0 cm from the mirror, what is the new position of the image? What is the magnification of the new image? Is the new image real or virtual? Draw ray diagrams to confirm your results.

CONVEX SPHERICAL MIRRORS

On recent models of automobiles, there is a side-view mirror on the passenger's side of the car. Unlike the flat mirror on the driver's side, which produces unmagnified images, the passenger's mirror bulges outward at the center. Images in this mirror are distorted near the mirror's edges, and the image is smaller than the object. This type of mirror is called a **convex spherical mirror.**

convex spherical mirror

a mirror whose reflecting surface is an outward-curved segment of a sphere

A convex spherical mirror is a segment of a sphere that is silvered so that light is reflected from the sphere's outer, convex surface. This type of mirror is also called a diverging mirror because the incoming rays diverge after reflection as though they were coming from some point behind the mirror. The resulting image is therefore always virtual, and the image distance is always negative. Because the mirrored surface is on the side opposite the radius of curvature, a convex spherical mirror also has a negative focal length. The sign conventions for all mirrors are summarized in **Table 5.**

The technique for drawing ray diagrams for a convex mirror differs slightly from that for concave mirrors. The focal point and center of curvature are situated behind the mirror's surface. Dotted lines are extended along the reflected reference rays to points behind the mirror, as shown in **Figure 13(a).** A virtual, upright image forms where the three rays apparently intersect. Magnification for convex mirrors is always less than 1, as shown in **Figure 13(b).**

Convex spherical mirrors take the objects in a large field of view and produce a small image, so they are well suited for providing a fixed observer with a complete view of a large area. Convex mirrors are often placed in stores to help employees monitor customers and at the intersections of busy hallways so that people in both hallways can tell when others are approaching.

The side-view mirror on the passenger's side of a car is another application of the convex mirror. This mirror usually carries the warning, "objects are closer than they appear." Without this warning, a driver might think that he or she is looking into a flat mirror, which does not alter the size of the image. The driver could therefore be fooled into believing that a vehicle is farther away than it is because the image is smaller than the actual object.

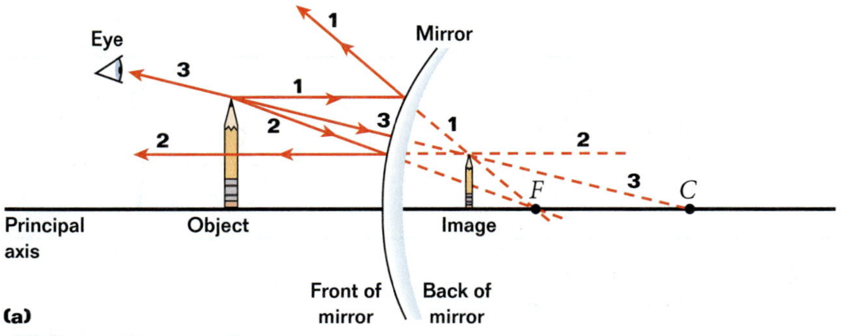

(a)

(b)

Figure 13
Light rays diverge upon reflection from a convex mirror **(a)**, forming a virtual image that is always smaller than the object **(b)**.

Demonstration

Convex Mirror

Purpose Demonstrate that parallel beams reflected by convex mirrors are diverging.

Materials light source, ray filter, convex mirror, white paper

Procedure Use the ray filter to produce five beams. Dim the lights. Place the paper in front of the beams, and let students observe that the incident rays are parallel. Place the convex mirror as far as possible from the front of the light source, and let students observe the beams diverging. Move the mirror closer to the light source. Students will notice that the beam is always diverging. Remind students that concave mirrors produce both real and virtual images from real objects. In contrast, convex mirrors produce only virtual images from real objects.

Visual Strategy – BASIC

Table 5

Make sure that students properly interpret information related to all the cases listed in **Table 5.** Point out that as a general rule, distances in front of the mirror are assigned a positive sign and distances behind the mirror are assigned a negative sign.

Q The image formed by a concave mirror is upright and virtual. What would be the signs of R, f, q, and h'?

A +, +, −, +

Q The image formed by a convex mirror is also upright and virtual. What would be the signs of R, f, q, and h'?

A −, −, −, +

Did you know?

There are certain circumstances in which the object for one mirror is the image that appears behind another mirror. In these cases, the object is virtual and has a negative object distance. Because of the rarity of these situations, virtual object distance ($p < 0$) has not been listed in **Table 5.**

Table 5 Sign Conventions for Mirrors

Symbol	Situation	Sign	
p	object is in front of the mirror (real object)	+	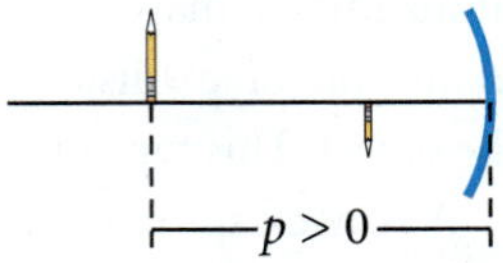
q	image is in front of the mirror (real image)	+	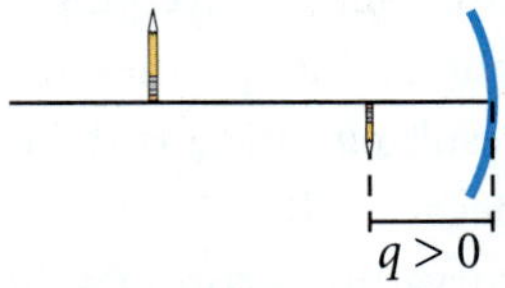
q	image is behind the mirror (virtual image)	−	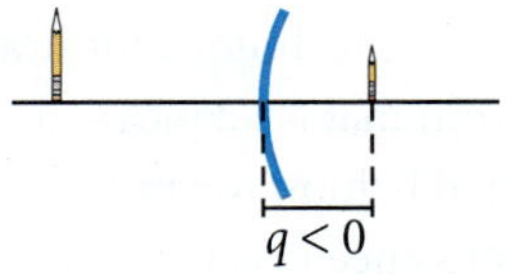
R, f	center of curvature is in front of the mirror (concave spherical mirror)	+	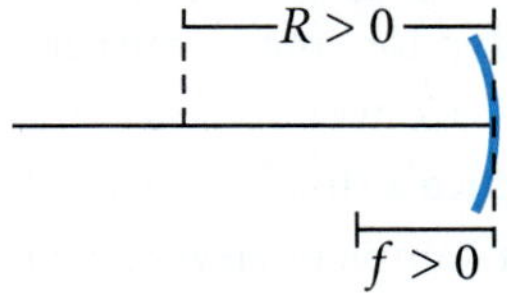
R, f	center of curvature is behind the mirror (convex spherical mirror)	−	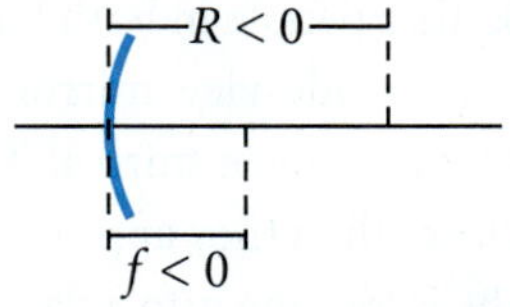
R, f	mirror has no curvature (flat mirror)	∞	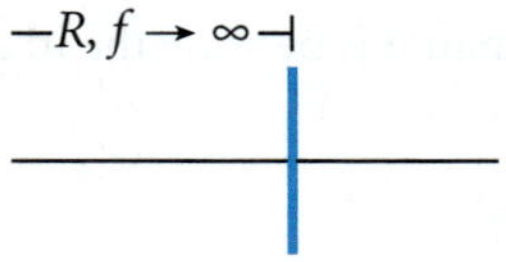
h'	image is above the principal axis	+	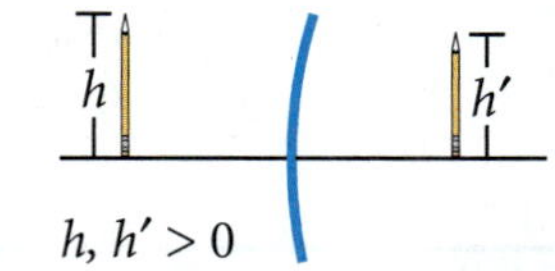
h'	image is below the principal axis	−	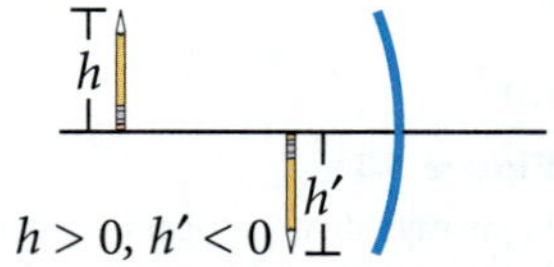

SAMPLE PROBLEM C

Convex Mirrors

PROBLEM

An upright pencil is placed in front of a convex spherical mirror with a focal length of 8.00 cm. An erect image 2.50 cm tall is formed 4.44 cm behind the mirror. Find the position of the object, the magnification of the image, and the height of the pencil.

SOLUTION

1. DEFINE

Given: $f = -8.00$ cm $q = -4.44$ cm $h' = 2.50$ cm

Because the mirror is convex, the focal length is negative.

The image is behind the mirror, so q is also negative.

Unknown: $p = ?$ $h = ?$ $M = ?$

Diagram: Construct a ray diagram.

2. PLAN

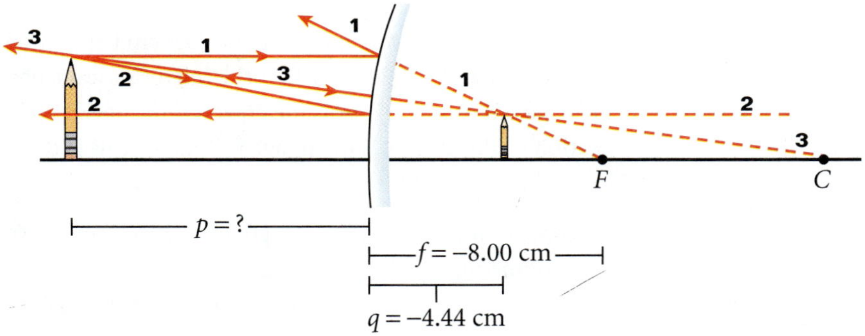

Choose an equation or situation: Use the mirror equation.

$$\frac{1}{p} + \frac{1}{q} = \frac{1}{f}$$

Use the magnification formula.

$$M = \frac{h'}{h} = -\frac{q}{p}$$

Rearrange the equation to isolate the unknown:

$$\frac{1}{p} = \frac{1}{f} - \frac{1}{q} \quad \text{and} \quad h = -\frac{p}{q}h'$$

3. CALCULATE **Substitute the values into the equation and solve:**

$$\frac{1}{p} = \frac{1}{-8.00 \text{ cm}} - \frac{1}{-4.44 \text{ cm}}$$

$$\frac{1}{p} = \frac{-0.125}{1 \text{ cm}} - \frac{-0.225}{1 \text{ cm}} = \frac{0.100}{1 \text{ cm}}$$

$$\boxed{p = 10.0 \text{ cm}}$$

continued on next page

Classroom Practice

Convex Mirrors

The radius of curvature of a convex mirror is 12.0 cm. Where is the focal point located?

Answer

6.00 cm behind the mirror ($f = -6.00$ cm)

Find the position of the image for an object placed at the following distances from the mirror in the previous question: $p = 1.00$ cm, 2.00 cm, 3.00 cm, 6.00 cm, 12.0 cm, 30.0 cm, 50.0 cm

Answer

−0.855 cm, −1.50 cm, −2.00 cm, −2.99 cm, −4.00 cm, −5.00 cm, −5.35 cm

How does the position of the image vary as the object in the previous question moves farther away from the mirror?

Answer

The image is always behind the mirror and between the mirror and the focal point. It moves from $q = 0$ to $q = f$ as the object moves away from the mirror (from $p = 0$ to infinity).

PROBLEM GUIDE C

Use this guide to assign problems.
SE = Student Edition Textbook
PW = Problem Workbook
PB = Problem Bank on the One-Stop Planner (OSP)

Solving for:

p	**SE** Sample, 1–3; Ch. Rvw. 36, 51* **PW** 7–9 **PB** 3–6
q	**SE** 4–6 **PW** Sample, 1–3 **PB** 7–10
R, f	**SE** Ch. Rvw. 36, 48, 50, 55 **PW** 4–6 **PB** Sample, 1, 2
M	**SE** 1–6; Ch. Rvw. 50 **PW** Sample, 1–2, 4–5, 7–8 **PB** Sample, 1, 3–5, 7–8
h, h'	**PW** 3, 6, 9 **PB** 2, 6, 9

***Challenging Problem**
Consult the printed Solutions Manual or the OSP for detailed solutions.

ANSWERS

Practice C

1. $p = 46.0$ cm; $M = 0.500$; virtual, upright image; $h = 3.40$ cm
2. $M = 0.04$; $p = 6$ m; $h = 2$ m; virtual, upright image
3. $p = 45$ cm; $h = 17$ cm; $M = 0.41$; virtual, upright image
4. $q = -0.25$ m; $M = 0.081$; virtual, upright image
5. $q = -1.31$ cm; $M = 0.125$; virtual, upright image
6. $q = -2.0 \times 10^1$ cm; $M = 0.41$; virtual, upright image

Substitute the values for p and q to find the magnification of the image.

$$M = -\frac{q}{p} = -\frac{-4.44 \text{ cm}}{10.0 \text{ cm}}$$

$$\boxed{M = 0.444}$$

Substitute the values for p, q, and h' to find the height of the object.

$$h = -\frac{p}{q}h' = -\frac{10.0 \text{ cm}}{-4.44 \text{ cm}}(2.50 \text{ cm})$$

$$\boxed{h = 5.63 \text{ cm}}$$

PRACTICE C

Convex Mirrors

1. The image of a crayon appears to be 23.0 cm behind the surface of a convex mirror and is 1.70 cm tall. If the mirror's focal length is 46.0 cm, how far in front of the mirror is the crayon positioned? What is the magnification of the image? Is the image virtual or real? Is the image inverted or upright? How tall is the actual crayon?

2. A convex mirror with a focal length of 0.25 m forms a 0.080 m tall image of an automobile at a distance of 0.24 m behind the mirror. What is the magnification of the image? Where is the car located, and what is its height? Is the image real or virtual? Is the image upright or inverted?

3. A convex mirror of focal length 33 cm forms an image of a soda bottle at a distance of 19 cm behind the mirror. If the height of the image is 7.0 cm, where is the object located, and how tall is it? What is the magnification of the image? Is the image virtual or real? Is the image inverted or upright? Draw a ray diagram to confirm your results.

4. A convex mirror with a radius of curvature of 0.550 m is placed above the aisles in a store. Determine the image distance and magnification of a customer lying on the floor 3.1 m below the mirror. Is the image virtual or real? Is the image inverted or upright?

5. A spherical glass ornament is 6.00 cm in diameter. If an object is placed 10.5 cm away from the ornament, where will its image form? What is the magnification? Is the image virtual or real? Is the image inverted or upright?

6. A candle is 49 cm in front of a convex spherical mirror that has a focal length of 35 cm. What are the image distance and magnification? Is the image virtual or real? Is the image inverted or upright? Draw a ray diagram to confirm your results.

PARABOLIC MIRRORS

You have probably noticed that certain rays in ray diagrams do not intersect exactly at the image point. This occurs especially with rays that reflect at the mirror's surface far from the principal axis. The situation also occurs with real light rays and real spherical mirrors.

If light rays from an object are near the principal axis, all of the reflected rays pass through the image point. Rays that reflect at points on the mirror far from the principal axis converge at slightly different points on the principal axis, as shown in **Figure 14.** This produces a blurred image. This effect, called *spherical aberration,* is present to some extent in any spherical mirror.

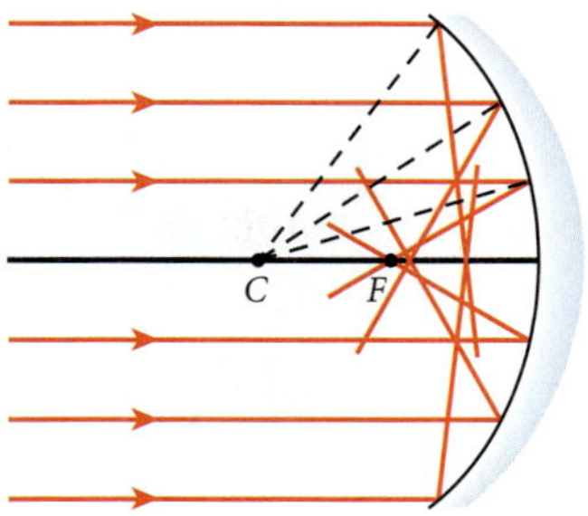

Figure 14
Spherical aberration occurs when parallel rays far from the principal axis converge away from the mirror's focal point.

Parabolic mirrors eliminate spherical aberration

A simple way to reduce the effect of spherical aberration is to use a mirror with a small diameter; that way, the rays are never far from the principal axis. If the mirror is large to begin with, shielding its outer portion will limit how much of the mirror is used and thus will accomplish the same effect. However, many concave mirrors, such as those used in astronomical telescopes, are made large so that they will collect a large amount of light. An alternative approach is to use a mirror that is not a segment of a sphere but still focuses light rays in a manner similar to a small spherical concave mirror. This is accomplished with a parabolic mirror.

Parabolic mirrors are segments of a paraboloid (a three-dimensional parabola) whose inner surface is reflecting. All rays parallel to the principal axis converge at the focal point regardless of where on the mirror's surface the rays reflect. Thus, a real image forms without spherical aberration, as illustrated in **Figure 15.** Similarly, light rays from an object at the focal point of a parabolic mirror will be reflected from the mirror in parallel rays. Parabolic reflectors are ideal for flashlights and automobile headlights. (Spherical mirrors are extensively used because they are easier to manufacture than parabolic mirrors, and thus are less expensive.)

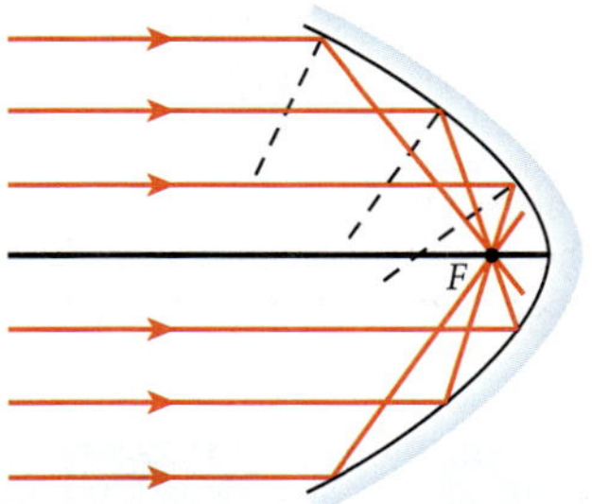

Figure 15
All parallel rays converge at a parabolic mirror's focal point. The curvature in this figure is much greater than it is in real parabolic mirrors.

Reflecting telescopes use parabolic mirrors

A telescope permits you to view distant objects, whether they are buildings a few kilometers away or galaxies that are millions of light-years from Earth. Not all telescopes are intended for visible light. Because all electromagnetic radiation obeys the law of reflection, parabolic surfaces can be constructed to reflect and focus electromagnetic radiation of different wavelengths. For instance, a radio telescope consists of a large metal parabolic surface that reflects radio waves in order to receive radio signals from objects in space.

There are two types of telescopes that use visible light. One type, called a *refracting telescope,* uses a combination of lenses to form an image. The other kind uses a curved mirror and small lenses to form an image. This type of telescope is called a *reflecting telescope.*

SCILINKS®
www.scilinks.org
Topic: Telescopes
Code: HF61500

Visual Strategy ADVANCED

Figure 14
Remind students that for a spherical mirror, the normal to the surface at any point lies along the radius at that point. Point out that each pair of reflected rays at equal distances on opposite sides of the principal axis cross one another on the principal axis.

Q How does the angle of incidence vary when the incoming ray is farther away from the principal axis of a spherical mirror? Mark the points where each pair of reflected rays intersects. Where will the next pair intersect if the mirror's surface is extended?

A ***The angle of incidence increases; The points should be marked on the principal axis, to the right of F; The points of intersection come closer to the mirror's surface, and the focal point becomes more ill-defined.***

Misconception Alert

Students may be unclear about the difference between segments of a circle and of a parabolic curve. Ask them to draw the continuation of the circular and parabolic curves of the mirrors to see that the curves are similar for a short segment. Have students compare the curves of a circle and parabola by using their graphing calculators to compare the curves whose equations are $y = \sqrt{1 - x^2} - 1$ and $y = -x^2$.

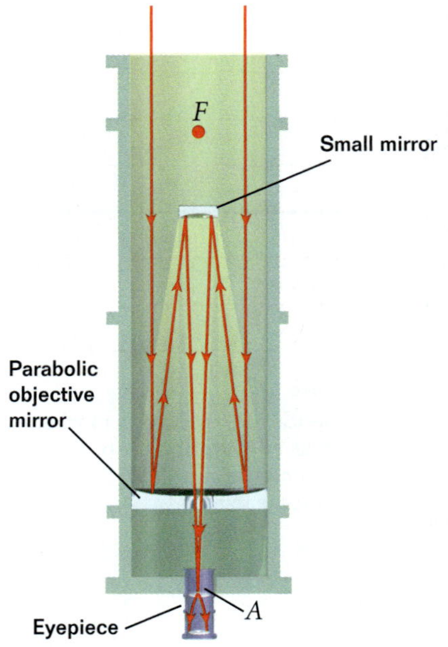

Figure 16
The parabolic objective mirror in a Cassegrain reflector focuses incoming light.

Reflecting telescopes employ a parabolic mirror (called an *objective mirror*) to focus light. One type of reflecting telescope, called a *Cassegrain reflector,* is shown in **Figure 16.** Parallel light rays pass down the barrel of the telescope and are reflected by the parabolic objective mirror at the telescope's base. These rays converge toward the objective mirror's focal point, *F*, where a real image would normally form. However, a small curved mirror that lies in the path of the light rays reflects the light back toward the center of the objective mirror. The light then passes through a small hole in the center of the objective mirror and comes to a focus at point *A*. An eyepiece near point *A* magnifies the image.

You may wonder how a hole can be placed in the objective mirror without affecting the final image formed by the telescope. Each part of the mirror's surface reflects light from distant objects, so a complete image is always formed. The presence of the hole merely reduces the amount of light that is reflected. Even that is not severely affected by the hole because the light-gathering capacity of an objective mirror is dependent on the mirror's area. For instance, a 1 m diameter hole in a mirror that is 4 m in diameter reduces the mirror's reflecting surface by only $\frac{1}{16}$, or 6.25 percent.

SECTION REVIEW

1. A steel ball bearing with a radius of 1.5 cm forms an image of an object that has been placed 1.1 cm away from the bearing's surface. Determine the image distance and magnification. Is the image virtual or real? Is the image inverted or upright? Draw a ray diagram to confirm your results.
2. A spherical mirror is to be used in a motion-picture projector to form an inverted, real image 95 times as tall as the picture in a single frame of film. The image is projected onto a screen 13 m from the mirror. What type of mirror is required, and how far should it be from the film?
3. Which of the following images are real and which are virtual?
   a. the image of a distant illuminated building projected onto a piece of heavy, white cardboard by a small reflecting telescope
   b. the image of an automobile in a flat rearview mirror
   c. the image of shop aisles in a convex observation mirror
4. **Critical Thinking** Why is an image formed by a parabolic mirror sharper than the image of the same object formed by a concave spherical mirror?
5. **Critical Thinking** The reflector of the radio telescope at Arecibo has a radius of curvature of 265.0 m. How far above the reflector must the radio-detecting equipment be placed in order to obtain clear radio images?

SECTION REVIEW ANSWERS

1. $q = -0.45$ cm; $M = 0.41$; virtual, upright image
2. concave mirror; $p = 0.14$ m
3. **a.** real
   b. virtual
   c. virtual
4. The rays reflected by a parabolic mirror all focus at one point, whereas the rays reflected by a concave spherical mirror reflect along a line that includes the mirror's focal point.
5. 132.5 m, which is the telescope's focal length

Color and Polarization

SECTION OBJECTIVES

- **Recognize how additive colors affect the color of light.**
- **Recognize how pigments affect the color of reflected light.**
- **Explain how linearly polarized light is formed and detected.**

COLOR

You have probably noticed that the color of an object can appear different under different lighting conditions. These differences are due to differences in the reflecting and light-absorbing properties of the object being illuminated.

So far, we have assumed that objects are either like mirrors, which reflect almost all light uniformly, or like rough objects, which reflect light diffusely in several directions. But, as mentioned in Section 1, objects absorb certain wavelengths from the light striking them and reflect the rest. The color of an object depends on which wavelengths of light shine on the object and which wavelengths are reflected (see **Figure 17**).

If all wavelengths of incoming light are completely reflected by an object, that object appears to have the same color as the light illuminating it. This gives the object the same appearance as a white object illuminated by the light. An object of a particular color, such as the green leaf in **Figure 17,** absorbs light of all colors except the light whose color is the same as the object's color. By contrast, an object that reflects no light appears black. In truth, leaves appear green only when their primary pigment, chlorophyll, is present. In the autumn, when the green pigment is destroyed, other colors are reflected by the leaves.

Additive primary colors produce white light when combined

Because white light can be dispersed into its elementary colors, it is reasonable to suppose that elementary colors can be combined to form white light. One way of doing this is to use a lens to recombine light that has been dispersed by a prism. Another way is to combine light that has been passed through red, green, and blue filters. These colors are called the *additive primary colors* because when they are added in varying proportions, they can form all of the colors of the spectrum.

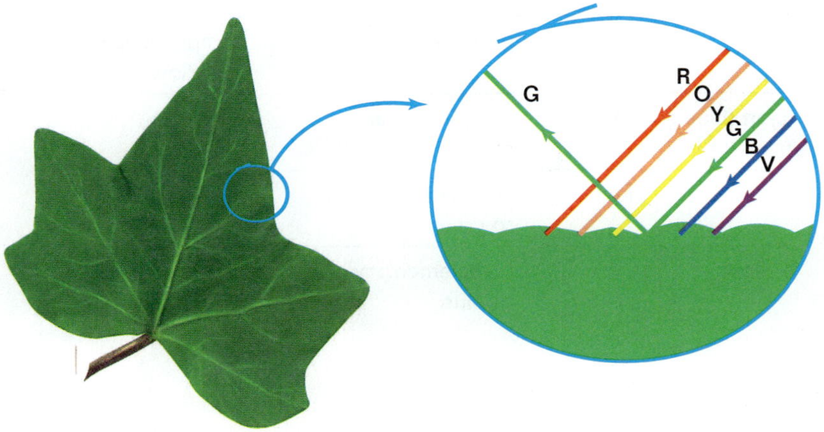

Figure 17
A leaf appears green under white light because the primary pigment in the leaf reflects only green light.

SCILINKS.
www.scilinks.org
Topic: Color
Code: HF60313

extension
Integrating Biology
Visit go.hrw.com for the activity "How Does Sunscreen Work?"
Keyword HF6LGTX

Demonstration

Reflection and Absorption of Color — BASIC

Purpose Help students realize that the colors we see depend on the properties of the light incident on an object and on the colors the object reflects.

Materials white paper; crayons of different colors; projector with red, blue, and green filters or cellophane

Procedure Write a sentence on the paper, alternating crayons for each word. Turn off the lights and challenge students to read the sentence while you cover the projector with one or two filters at a time. Students should have their eyes closed while the filters are being changed. Ask students to write their guesses for the complete text that has been partially revealed under different lights. Finally, have students explain what combination of light color and crayon color made a particular word visible or invisible.

Teaching Tip

Students can use color controls on a personal computer to explore color mixing. Different computers vary, but you might suggest using a simple drawing program to create a filled rectangle. Many programs allow you to choose the color with RGB (red-green-blue) or CMY (cyan-magenta-yellow) percentages.

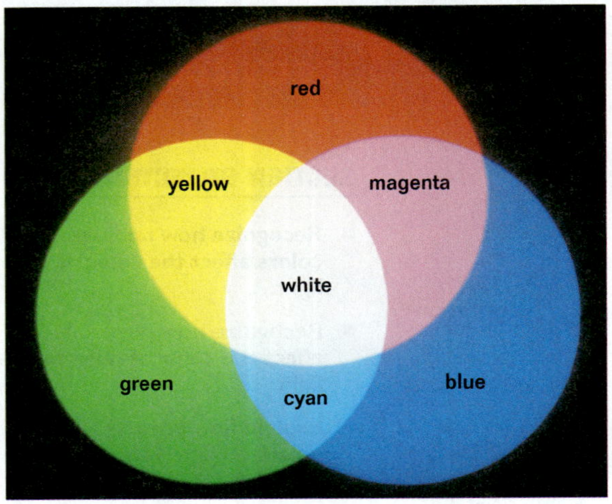

Figure 18
The combination of the additive primary colors in any two circles produces the complementary color of the third additive primary color.

When light passed through a red filter is combined with green light produced with a green filter, a patch of yellow light appears. When two primary colors are combined, a *complementary* color is formed. For example, when red light and green light are combined, they produce yellow light. As noted in **Figure 18** and **Table 6,** the complements of the primary colors red, green, and blue are cyan, magenta, and yellow, respectively.

Certain chemical compounds can be added to glass to give it color. Iron compounds give glass a green color, while manganese compounds give glass a magenta color. Because green and magenta are complementary colors, the correct proportion of iron and manganese compounds added to the glass can cause the glass to appear colorless.

Another example of additive colors is the image produced on a color television screen. A television screen consists of small, luminous dots, or *pixels,* that glow either red, green, or blue when they are struck by electrons (see **Figure 19**). Varying the brightness of different pixels in different parts of the picture produces a picture that appears to have many colors present at the same time.

Humans can see in color because there are three kinds of color receptors in the eye. Each receptor, called a *cone cell,* is sensitive to either red, green, or blue light. Light of different wavelengths stimulates a combination of these receptors so that a wide range of colors can be perceived.

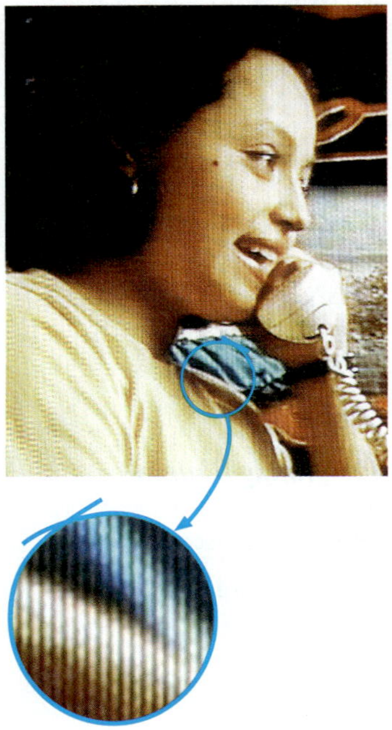

Figure 19
The brightness of the red, green, and blue pixels of a color television screen are adjusted such that, from a distance, all of the colors in a single picture appear.

Table 6 Additive and Subtractive Primary Colors

Colors	Additive (mixing light)	Subtractive (mixing pigments)
red	primary	complementary to cyan
green	primary	complementary to magenta
blue	primary	complementary to yellow
cyan (blue green)	complementary to red	primary
magenta (red blue)	complementary to green	primary
yellow	complementary to blue	primary

Subtractive primary colors filter out all light when combined

When blue light and yellow light are mixed, white light results. However, if you mix a cyan pigment (such as paint or the colored wax of a crayon) with a yellow pigment, the resulting color is green, not white. This difference is due to the fact that pigments rely on colors of light that are absorbed, or subtracted, from the incoming light.

For example, yellow pigment subtracts blue and violet colors from white light and reflects red, orange, yellow, and green light. Cyan pigment subtracts red, orange, and yellow from the light and reflects green, cyan, and violet. When yellow and blue pigments are combined, only green light is reflected.

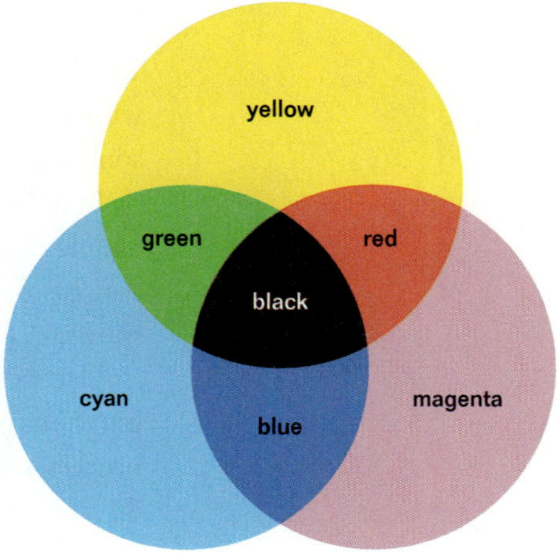

Figure 20
The combination of the subtractive primary colors by any two filters produces the complementary color of the third subtractive primary color.

When pigments are mixed, each one subtracts certain colors from white light, and the resulting color depends on the frequencies that are not absorbed. The primary pigments (or *primary subtractive colors,* as they are sometimes called) are cyan, magenta, and yellow. These are the same colors that are complementary to the additive primary colors (see **Table 6**). When any two primary subtractive colors are combined, they produce either red, green, or blue pigments. When the three primary pigments are mixed together in the proper proportions, all of the colors are subtracted from white light, and the mixture is black, as shown in **Figure 20.**

Combining yellow pigment and its complementary color, blue, should produce a black pigment. Yet earlier, blue and yellow were combined to produce green. The difference between these two situations is explained by the broad use of color names. The "blue" pigment that is added to a "yellow" pigment to produce green is not a pure blue. If it were, only blue light would be reflected from it. Similarly, a pure yellow pigment will reflect only yellow light. Because most pigments found in paints and dyes are combinations of different substances, they reflect light from nearby parts of the visible spectrum. Without knowledge of the light-absorption characteristics of these pigments, it is hard to predict exactly what colors will result from different combinations.

Why it Matters

Conceptual Challenge

1. Colors in a Blanket Brown is a mixture of yellow with small amounts of red and green. If you shine red light on a brown woolen blanket, what color will the blanket appear? Will it appear lighter or darker than it would under white light? Explain your answers.

2. Blueprints If a blueprint (a blue drawing on a white background) is viewed under blue light, will you still be able to perceive the drawing? What will the blueprint look like under yellow light?

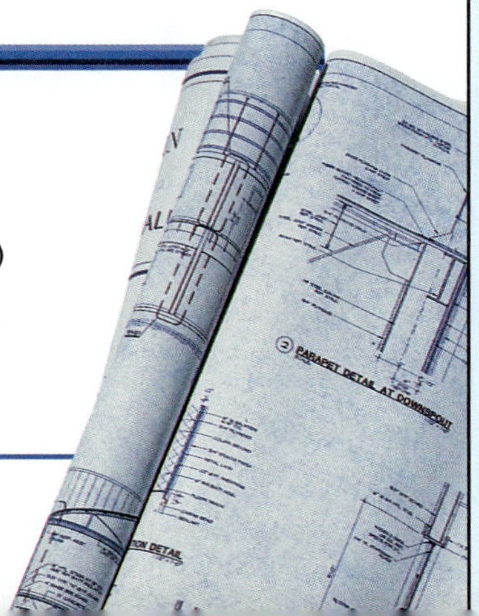

SECTION 4

Key Models and Analogies

The names of the primary additive and subtractive colors can be arranged on a triangle, as shown below.

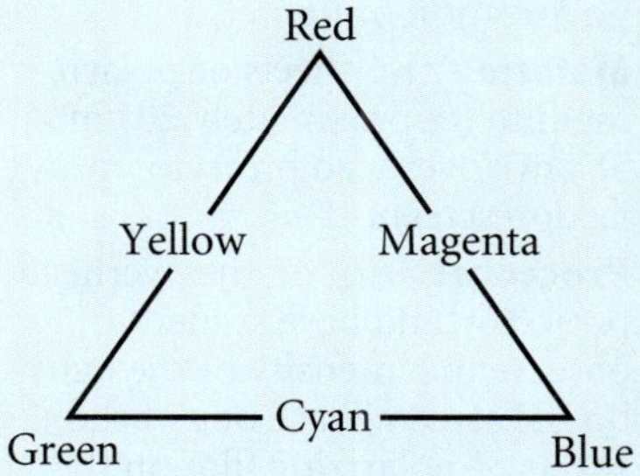

Each side of the triangle is the complementary color of the color at the opposite vertex. This diagram also shows the results of combining pigments: the colors at any two sides of the triangle combine to form the color at the vertex between them. For example, pure magenta pigment mixed with pure cyan pigment will only reflect blue light. The pigment whose name is on one side reflects the light composed of the colors that are indicated at the vertices on either end of that side. This same pigment absorbs the light with the color whose name is on the opposite vertex.

ANSWERS

Conceptual Challenge

1. red; darker; Some of the red light is absorbed by the green pigment (assume that yellow pigment consists of green and red pigments).
2. No, the entire page will look blue (the blueprint may appear darker, depending on the light source); The blue drawing will become black, and the page will be yellow.

Demonstration

Polarizing Light by Transmission

Purpose Demonstrate linear polarization of light.

Materials two sheets of polarizing film (approximately 20 cm × 25 cm), overhead projector, projection screen

Procedure Turn on the overhead projector, and have students observe the intensity of the light on the screen. Place one of the sheets of polarizing film on the overhead projector. Have students note the decreased intensity of the light. Explain to the class that the light from the projector is randomly oriented in all directions. Only the components of light parallel to the polarizer's transmission axis are transmitted, and therefore the intensity of light on the screen is reduced.

Hold the second polarizer in front of the top lens of the overhead projector. Have students note that the intensity of the light remains constant when the second polarizer's transmission axis is parallel to the transmission axis of the original polarizer. Rotate the second polarizer 90° so that its transmission axis is perpendicular to the transmission axis of the original polarizer. Have students note that the intensity of the light is almost zero.

POLARIZATION OF LIGHT WAVES

You have probably seen sunglasses with polarized lenses that reduce glare without blocking the light entirely. There is a property of light that allows some of the light to be filtered by certain materials in the lenses.

In an electromagnetic wave, the electric field is at right angles to both the magnetic field and the direction of propagation. Light from a typical source consists of waves that have electric fields oscillating in random directions, as shown in **Figure 21.** Light of this sort is said to be *unpolarized.*

Electric-field oscillations of unpolarized light waves can be treated as combinations of vertical and horizontal electric-field oscillations. There are certain processes that separate waves with electric-field oscillations in the vertical direction from those in the horizontal direction, producing a beam of light with electric field waves oriented in the same direction, as shown in **Figure 22.** These waves are said to have **linear polarization.**

linear polarization

the alignment of electromagnetic waves in such a way that the vibrations of the electric fields in each of the waves are parallel to each other

Light can be linearly polarized through transmission

Certain transparent crystals cause unpolarized light that passes through them to become linearly polarized. The direction in which the electric fields are polarized is determined by the arrangement of the atoms or molecules in the

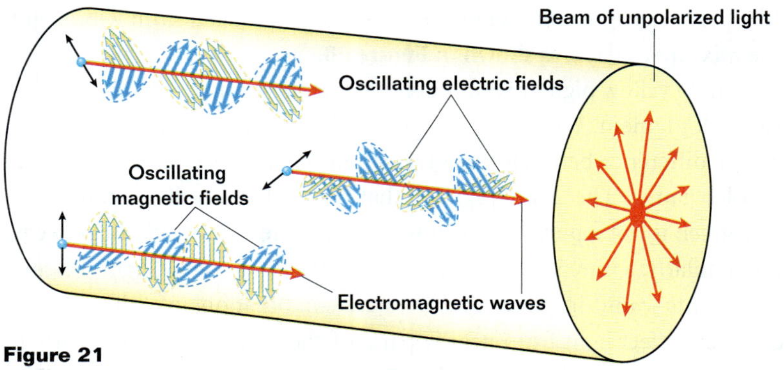

Figure 21
Randomly oscillating electric fields produce unpolarized light.

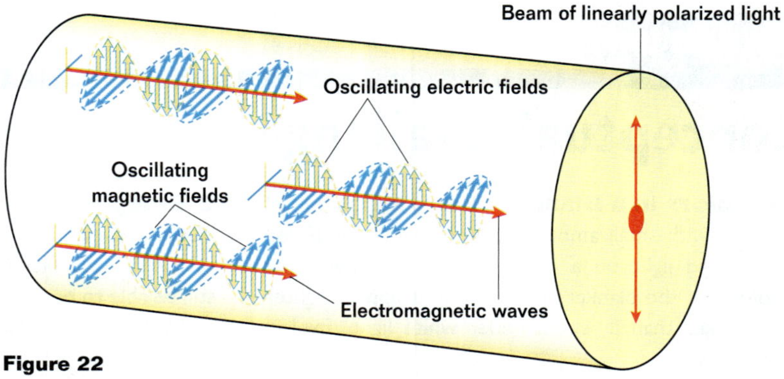

Figure 22
Light waves with aligned electric fields are linearly polarized.

crystal. For substances that polarize light by transmission, the line along which light is polarized is called the *transmission axis* of the substance. Only light waves that are linearly polarized with respect to the transmission axis of the polarizing substance can pass freely through the substance. All light that is polarized at an angle of 90° to the transmission axis does not pass through.

When two polarizing films are held with the transmission axes parallel, light will pass through the films, as shown in **Figure 23(a).** If they are held with the transmission axes perpendicular to each other, as in **Figure 23(b),** no light will pass through the films.

A polarizing substance can be used not only to linearly polarize light but also to determine if and how light is linearly polarized. By rotating a polarizing substance as a beam of polarized light passes through it, a change in the intensity of the light can be seen (see **Figure 24**). The light is brightest when its plane of polarization is parallel to the transmission axis. The larger the angle is between the electric-field waves and the transmission axis, the smaller the component of light that passes through the polarizer will be and the less bright the light will be. When the transmission axis is perpendicular to the plane of polarization for the light, no light passes through.

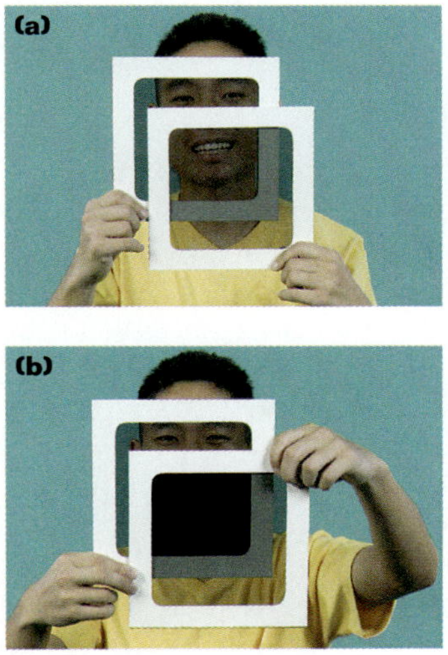

Figure 23
(a) Light will pass through a pair of polarizing films when their polarization axes are aligned in the same direction. **(b)** When the axes are at right angles to one another, light will not get through.

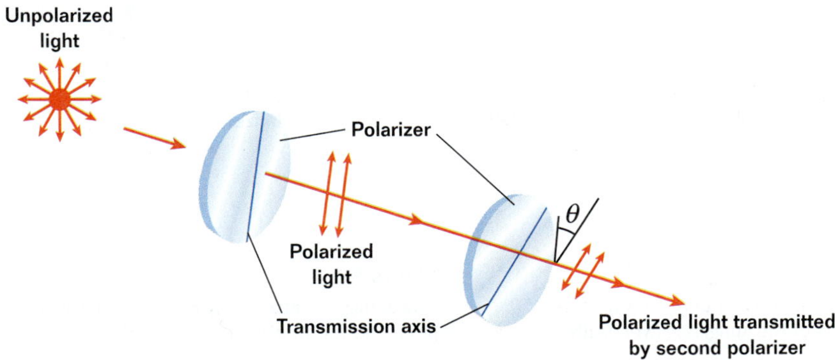

Figure 24
The brightness of the polarized light decreases as the angle, θ, increases between the transmission axis of the second polarizer and the plane of polarization of the light.

Light can be polarized by reflection and scattering

When light is reflected at a certain angle from a surface, the reflected light is completely polarized parallel to the reflecting surface. If the surface is parallel to the ground, the light is polarized horizontally. This is the case with glaring light that reflects at a low angle from roads, bodies of water, and car hoods.

Quick Lab

Polarization of Sunlight

MATERIALS LIST

- **a sheet of polarizing filter or sunglasses with polarizing lenses**

SAFETY

Never look directly at the sun.

During mid-morning or mid-afternoon, when the sun is well above the horizon but not directly overhead, look directly up at the sky through the polarizing filter. Note how the light's intensity is reduced.

Rotate the polarizer. Take note of which orientations of the polarizer make the sky darker and thus best reduce the amount of transmitted light.

Repeat the test with light from other parts of the sky. Test light reflected off a table near a window. Compare the results of these various experiments.

Demonstration

Transmitting Light with Crossed Polarizers

Purpose Demonstrate the vector nature of light.

Materials three sheets of polarizing film, overhead projector, projection screen

Procedure This demonstration is best done following the previous one, "Polarizing Light by Transmission." Set up two of the films with their axes of transmission crossed at 90° to each other. Ask students to predict what will happen if a third film is inserted between the two films. Insert the third sheet of polarizing film between the first two at an angle of 45°. Have students note that some light is now passed through the three films.

What is happening? Light, which consists of oscillating electric and magnetic fields, has a vector nature. The light passing through the first film is polarized in the direction of that film's axis of polarization. When this polarized light reaches the second film, the second film then passes the component of the polarized light parallel to its axis of polarization. The light is now polarized at 45° when it reaches the third film. Again, the third film passes the component of the light parallel to its axis of polarization.

SECTION 4

Demonstration

Polarizing Light by Reflection — GENERAL

Purpose Demonstrate that a reflected beam is polarized.

Materials focusable flashlight or projector, pane of glass (or the glass surface of an overhead projector), sheet of polarizing film, projection screen

Procedure Dim the classroom lights. Shine a flashlight beam directly on the screen. Place the sheet of polarizing film across the beam, and rotate the sheet. Have students observe that the initial intensity is reduced and that rotating the polarizing sheet does not make a difference. Explain that this is because the light is randomly polarized. Now, shine the light beam onto the glass pane. Place the polarizing film across the reflected beam, and rotate the film. The brightness of the light on the screen will vary with the rotation. With a particular angle (about 34° between the beam and the glass plate), complete polarization may be obtained, and the image on the screen will completely fade.

SECTION REVIEW ANSWERS

1. blue
2. magenta; red
3. yellow and cyan (more yellow than cyan); green and red (more green than red)
4. If you turn the polarizer and very little light is transmitted, then the reflected light is polarized.

Because the light that causes glare is in most cases horizontally polarized, it can be filtered out by a polarizing substance whose transmission axis is oriented vertically. This is the case with polarizing sunglasses. As shown in **Figure 25,** the angle between the polarized reflected light and the transmission axis of the polarizer is 90°. Thus, none of the polarized light passes through.

In addition to reflection and absorption, scattering can also polarize light. Scattering, or the absorption and reradiation of light by particles in the atmosphere, causes sunlight to be polarized, as shown in **Figure 26.** When an unpolarized beam of sunlight strikes air molecules, the electrons in the molecules begin vibrating with the electric field of the incoming wave. A horizontally polarized wave is emitted by the electrons as a result of their horizontal motion, and a vertically polarized wave is emitted parallel to Earth as a result of their vertical motion. Thus, an observer with his or her back to the sun will see polarized light when looking up toward the sky.

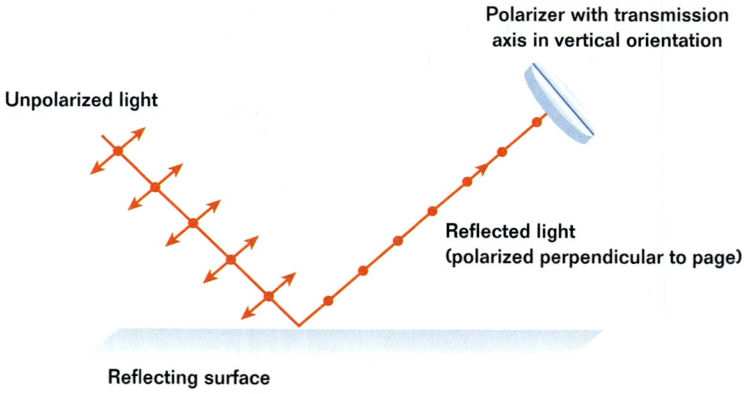

Figure 25
At a particular angle, reflected light is polarized horizontally. This light can be blocked by aligning the transmission axes of the sunglasses vertically.

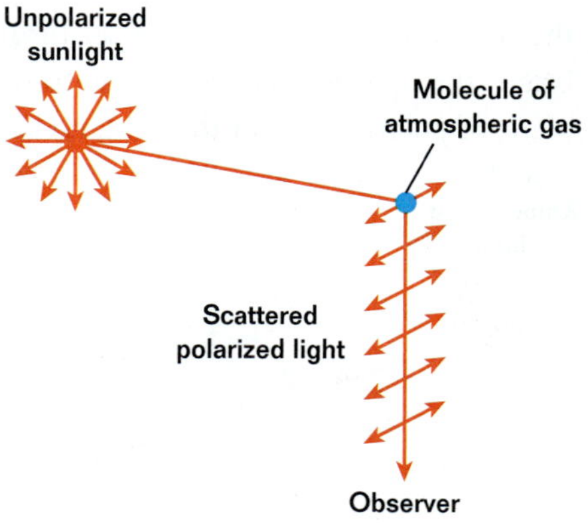

Figure 26
The sunlight scattered by air molecules is polarized for an observer on Earth's surface.

SECTION REVIEW

1. A lens for a spotlight is coated so that it does not transmit yellow light. If the light source is white, what color is the spotlight?
2. A house is painted with pigments that reflect red and blue light but absorb all other colors. What color does the house appear to be when it is illuminated by white light? What color does it appear to be under red light?
3. What primary pigments would an artist need to mix to obtain a pale yellow green color? What primary additive colors would a theater-lighting designer need to mix in order to produce the same color with light?
4. **Critical Thinking** The light reflected from the surface of a pool of water is observed through a polarizer. How can you tell if the reflected light is polarized?

CHAPTER 13 Highlights

KEY IDEAS

Section 1 Characteristics of Light

- Light is electromagnetic radiation that consists of oscillating electric and magnetic fields with different wavelengths.
- The frequency times the wavelength of electromagnetic radiation is equal to *c*, the speed of light.
- The brightness of light is inversely proportional to the square of the distance from the light source.

Section 2 Flat Mirrors

- Light obeys the law of reflection, which states that the incident and reflected angles of light are equal.
- Flat mirrors form virtual images that are the same distance from the mirror's surface as the object is.

Section 3 Curved Mirrors

- The mirror equation relates object distance, image distance, and focal length of a spherical mirror.
- The magnification equation relates image height or distance to object height or distance, respectively.

Section 4 Color and Polarization

- Light of different colors can be produced by adding light consisting of the primary additive colors (red, green, and blue).
- Pigments can be produced by combining subtractive colors (magenta, yellow, and cyan).
- Light can be linearly polarized by transmission, reflection, or scattering.

Variable Symbols

Quantities		Units	
p	object distance	m	meters
q	image distance	m	meters
R	radius of curvature	m	meters
f	focal length	m	meters
M	magnification	(unitless)	

KEY TERMS

electromagnetic wave (p. 446)

reflection (p. 451)

angle of incidence (p. 452)

angle of reflection (p. 452)

virtual image (p. 453)

concave spherical mirror (p. 455)

real image (p. 456)

convex spherical mirror (p. 463)

linear polarization (p. 472)

PROBLEM SOLVING

See **Appendix D: Equations** for a summary of the equations introduced in this chapter. If you need more problem-solving practice, see **Appendix I: Additional Problems.**

Diagram Symbols

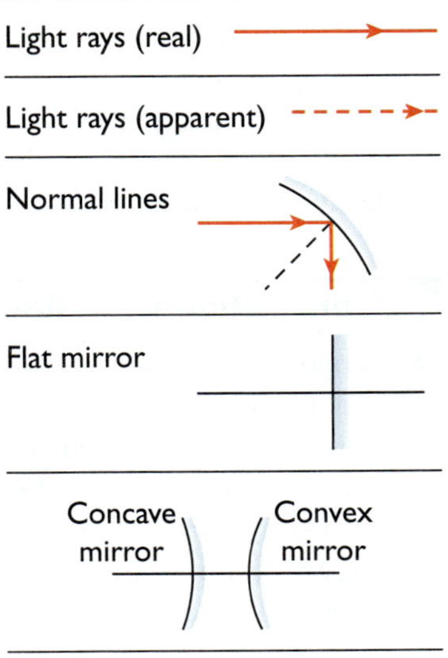

CHAPTER 13

Highlights

Teaching Tip

Writing down difficult concepts can help students better understand them and can enhance students' communication skills. Have students summarize the differences between images formed by convex mirrors and images formed by concave mirrors. Their writings should include a thorough explanation of the mirror equation, sign conventions, and ray diagrams for each case. Be sure students explain concepts clearly and correctly and use good sentence structure.

CHAPTER 13

Review

ANSWERS

1. a. radio waves
   b. gamma rays
2. b
3. Its speed is accurately known. Measuring the time it takes light to travel a distance allows the distance to be determined. (Alternatively, if the source's brightness is known, its apparent brightness can be measured and its distance calculated.)
4. The wave front at *B* would be an arc of a large circle. The rays would point radially outward from *A* to *B*.
5. Apparent brightness equals the actual brightness divided by the square of the distance between observer and source.
6. $1999 + 2(95) = 2189$
7. 3.00×10^8 m/s
8. The light from galaxies was emitted millions of years ago.
9. no; Those stars may be closer and so appear brighter.
10. 4.0×10^{-7} m, 3.0×10^{-7} m
11. 1×10^{-6} m
12. 3.02 m
13. 9.1×10^{-3} m (9.1 mm)
14. a. diffusely
    b. specularly
    c. specularly
    d. diffusely
    e. specularly

CHAPTER 13 Review

CHARACTERISTICS OF LIGHT

Review Questions

1. Which band of the electromagnetic spectrum has
   a. the lowest frequency?
   b. the shortest wavelength?

2. Which of the following electromagnetic waves has the highest frequency?
   a. radio
   b. ultraviolet radiation
   c. blue light
   d. infrared radiation

3. Why can light be used to measure distances accurately? What must be known in order to make distance measurements?

4. For the diagram below, use Huygens' principle to show what the wave front at point *A* will look like at point *B*. How would you represent this wave front in the ray approximation?

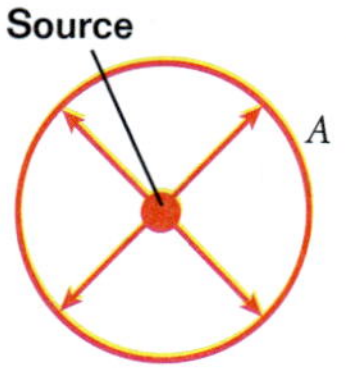

5. What is the relationship between the actual brightness of a light source and its apparent brightness from where you see it?

Conceptual Questions

6. Suppose an intelligent society capable of receiving and transmitting radio signals lives on a planet orbiting Procyon, a star 95 light-years away from Earth. If a signal were sent toward Procyon in 1999, what is the earliest year that Earth could expect to receive a return message? (Hint: A light-year is the distance a ray of light travels in one year.)

7. How fast do X rays travel in a vacuum?

8. Why do astronomers observing distant galaxies talk about looking backward in time?

9. Do the brightest stars that you see in the night sky necessarily give off more light than dimmer stars? Explain your answer.

Practice Problems

For problems 10–13, see Sample Problem A.

10. The compound eyes of bees and other insects are highly sensitive to light in the ultraviolet portion of the spectrum, particularly light with frequencies between 7.5×10^{14} Hz and 1.0×10^{15} Hz. To what wavelengths do these frequencies correspond?

11. The brightest light detected from the star Antares has a frequency of about 3×10^{14} Hz. What is the wavelength of this light?

12. What is the wavelength for an FM radio signal if the number on the dial reads 99.5 MHz?

13. What is the wavelength of a radar signal that has a frequency of 33 GHz?

FLAT MIRRORS

Review Questions

14. For each of the objects listed below, identify whether light is reflected diffusely or specularly.
    a. a concrete driveway
    b. an undisturbed pond
    c. a polished silver tray
    d. a sheet of paper
    e. a mercury column in a thermometer

15. If you are stranded on an island, where would you align a mirror to use sunlight to signal a searching aircraft?

16. If you are standing 2 m in front of a flat mirror, how far behind the mirror is your image? What is the magnification of the image?

Conceptual Questions

17. When you shine a flashlight across a room, you see the beam of light on the wall. Why do you not see the light in the air?

18. How can an object be a specular reflector for some electromagnetic waves yet be diffuse for others?

19. A flat mirror that is 0.85 m tall is attached to a wall so that its upper edge is 1.7 m above the floor. Use the law of reflection and a ray diagram to determine if this mirror will show a person who is 1.7 m tall his or her complete reflection.

20. Two flat mirrors make an angle of 90.0° with each other, as diagrammed at right. An incoming ray makes an angle of 35° with the normal of mirror A. Use the law of reflection to determine the angle of reflection from mirror B. What is unusual about the incoming and reflected rays of light for this arrangement of mirrors?

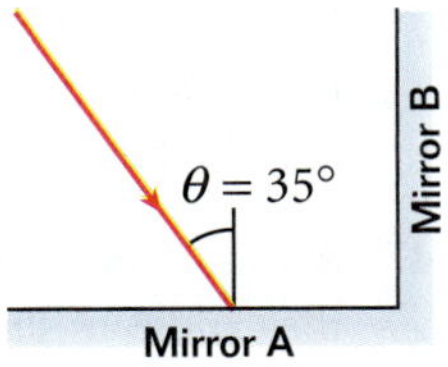

21. If you walk 1.2 m/s toward a flat mirror, how fast does your image move with respect to the mirror? In what direction does your image move with respect to you?

22. Why do the images produced by two opposing flat mirrors appear to be progressively smaller?

CURVED MIRRORS

Review Questions

23. Which type of mirror should be used to project movie images on a large screen?

24. If an object is placed outside the focal length of a concave mirror, which type of image will be formed? Will it appear in front of or behind the mirror?

25. Can you use a convex mirror to burn a hole in paper by focusing light rays from the sun at the mirror's focal point?

26. A convex mirror forms an image from a real object. Can the image ever be larger than the object?

27. Why are parabolic mirrors preferred over spherical concave mirrors for use in reflecting telescopes?

Conceptual Questions

28. Where does a ray of light that is parallel to the principal axis of a concave mirror go after it is reflected at the mirror's surface?

29. What happens to the real image produced by a concave mirror if you move the original object to the location of the image?

30. Consider a concave spherical mirror and a real object. Is the image always inverted? Is the image always real? Give conditions for your answers.

31. Explain why enlarged images seem dimmer than the original objects.

32. What test could you perform to determine if an image is real or virtual?

33. You've been given a concave mirror that may or may not be parabolic. What test could you perform to determine whether it is parabolic?

Practice Problems

For problems 34–35, see Sample Problem B.

34. A concave shaving mirror has a radius of curvature of 25.0 cm. For each of the following cases, find the magnification, and determine whether the image formed is real or virtual and upright or inverted.
    a. an upright pencil placed 45.0 cm from the mirror
    b. an upright pencil placed 25.0 cm from the mirror
    c. an upright pencil placed 5.00 cm from the mirror

13 REVIEW

15. point normal halfway between sun and aircraft
16. 2 m behind; $M = 1$
17. The gas molecules in air do not reflect the light.
18. Reflection is diffuse if λ is smaller than surface irregularities of reflector.
19. no; Diagram should show that the ray from feet reflected at bottom of mirror goes to the top of the head *above* the eyes.
20. $\theta_2' = 55°$; Ray reflected from the second mirror is always parallel to the incoming ray.
21. 1.2 m/s; The image moves toward the mirror's surface.
22. Images serve as objects for more images. Each reflection doubles the apparent distance from "object" to mirror.
23. concave
24. real, inverted image; in front
25. No, rays always diverge from a convex mirror.
26. no, $h' < h$ for convex mirrors
27. no spherical aberration
28. through the focal point
29. A real image appears at the former object position.
30. no; no; image is upright and virtual when $p < f$
31. The light is spread out more in the larger image.
32. try to project image on paper
33. Produce rays parallel to and far from the principal axis. All rays focus at F for a parabolic mirror.
34. a. $M = -0.384$; real, inverted
    b. $M = -1.00$; real, inverted
    c. $M = 1.67$; virtual, upright

35. $q = 26$ cm; real, inverted; $M = -2.0$

36. $p = 52.9$ cm; $h = 5.69$ cm; $M = 0.299$; virtual, upright

37. red, green, blue; They make white light.

38. cyan, magenta, yellow; They make black pigment.

39. The polarized light from the first polarizer is blocked by the second polarizer when the component of the light that is parallel to the second polarizer's transmission axis equals zero; The light must be perpendicular (90°) to the second polarizer's transmission axis.

40. **a.** green pigment
b. white light
c. black pigment
d. yellow light
e. cyan light

41. **a.** magenta
b. red
c. blue
d. black
e. red

42. cyan; blue

43. Rotate the sunglasses while looking at the sky or sunlight reflecting off a horizontal surface. If brightness changes, the glasses have polarizing lenses.

44. Light reflected from a horizontal surface like an auto hood is polarized horizontally and is blocked by the lenses. Light reflected from tall narrow surfaces like the tank will be vertically polarized, and almost all of it will pass through the lenses.

45. yes; Light from the sky is polarized, but light from the clouds is not polarized.

46. $p = 4.1 \times 10^2$ cm; $f = 32$ cm; $R = 64$ cm; real, inverted image

35. A concave spherical mirror can be used to project an image onto a sheet of paper, allowing the magnified image of an illuminated real object to be accurately traced. If you have a concave mirror with a focal length of 8.5 cm, where would you place a sheet of paper so that the image projected onto it is twice as far from the mirror as the object is? Is the image upright or inverted, real or virtual? What would the magnification of the image be?

For problem 36, see Sample Problem C.

36. A convex mirror with a radius of curvature of 45.0 cm forms a 1.70 cm tall image of a pencil at a distance of 15.8 cm behind the mirror. Calculate the object distance for the pencil and its height. Is the image real or virtual? What is the magnification? Is the image inverted or upright?

COLOR AND POLARIZATION

Review Questions

37. What are the three primary additive colors? What happens when you mix them?

38. What are the three primary subtractive colors (or primary pigments)? What happens when you mix them?

39. Explain why a polarizing disk used to analyze light can block light from a beam that has been passed through another polarizer. What is the relative orientation of the two polarizing disks?

Conceptual Questions

40. Explain what could happen when you mix the following:

a. cyan and yellow pigment
b. blue and yellow light
c. pure blue and pure yellow pigment
d. green and red light
e. green and blue light

41. What color would an opaque magenta shirt appear to be under the following colors of light?

a. white
b. red
c. cyan
d. green
e. yellow

42. A substance is known to reflect green and blue light. What color would it appear to be when it is illuminated by white light? by blue light?

43. How can you tell if a pair of sunglasses has polarizing lenses?

44. Why would sunglasses with polarizing lenses remove the glare from your view of the hood of your car or a distant body of water but not from a tall metal tank used for storing liquids?

45. Is light from the sky polarized? Why do clouds seen through polarizing glasses stand out in bold contrast to the sky?

MIXED REVIEW

46. The real image of a tree is magnified −0.085 times by a telescope's primary mirror. If the tree's image forms 35 cm in front of the mirror, what is the distance between the mirror and the tree? What is the focal length of the mirror? What is the value for the mirror's radius of curvature? Is the image virtual or real? Is the image inverted or upright?

47. A candlestick holder has a concave reflector behind the candle, as shown below. The reflector magnifies a candle −0.75 times and forms an image 4.6 cm away from the reflector's surface. Is the image inverted or upright? What are the object distance and the reflector's focal length? Is the image virtual or real?

48. A child holds a candy bar 15.5 cm in front of the convex side-view mirror of an automobile. The image height is reduced by one-half. What is the radius of curvature of the mirror?

49. A glowing electric light bulb placed 15 cm from a concave spherical mirror produces a real image 8.5 cm from the mirror. If the light bulb is moved to a position 25 cm from the mirror, what is the position of the image? Is the final image real or virtual? What are the magnifications of the first and final images? Are the two images inverted or upright?

50. A convex mirror is placed on the ceiling at the intersection of two hallways. If a person stands directly underneath the mirror, the person's shoe is a distance of 195 cm from the mirror. The mirror forms an image of the shoe that appears 12.8 cm behind the mirror's surface. What is the mirror's focal length? What is the magnification of the image? Is the image real or virtual? Is the image upright or inverted?

51. The side-view mirror of an automobile has a radius of curvature of 11.3 cm. The mirror produces a virtual image one-third the size of the object. How far is the object from the mirror?

52. An object is placed 10.0 cm in front of a mirror. What type must the mirror be to form an image of the object on a wall 2.00 m away from the mirror? What is the magnification of the image? Is the image real or virtual? Is the image inverted or upright?

53. The reflecting surfaces of two intersecting flat mirrors are at an angle of θ ($0° < \theta < 90°$), as shown in the figure below. A light ray strikes the horizontal mirror. Use the law of reflection to show that the emerging ray will intersect the incident ray at an angle of $\phi = 180° - 2\theta$.

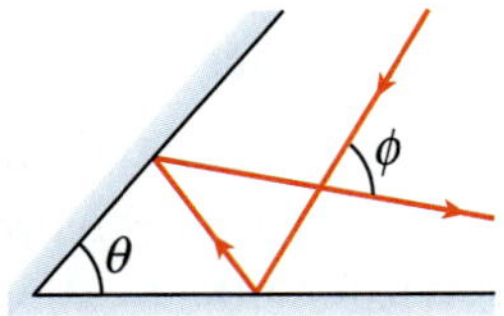

54. Show that if a flat mirror is assumed to have an "infinite" radius of curvature, the mirror equation reduces to $q = -p$.

55. A real object is placed at the zero end of a meterstick. A large concave mirror at the 100.0 cm end of the meterstick forms an image of the object at the 70.0 cm position. A small convex mirror placed at the 20.0 cm position forms a final image at the 10.0 cm point. What is the radius of curvature of the convex mirror? (Hint: The first image created by the concave mirror acts as an object for the convex mirror.)

56. A dedicated sports-car enthusiast polishes the inside and outside surfaces of a hubcap that is a section of a sphere. When he looks into one side of the hubcap, he sees an image of his face 30.0 cm behind the hubcap. He then turns the hubcap over and sees another image of his face 10.0 cm behind the hubcap.

a. How far is his face from the hubcap?
b. What is the radius of curvature of the hubcap?
c. What is the magnification for each image?
d. Are the images real or virtual?
e. Are the images upright or inverted?

57. An object 2.70 cm tall is placed 12.0 cm in front of a mirror. What type of mirror and what radius of curvature are needed to create an upright image that is 5.40 cm in height? What is the magnification of the image? Is the image real or virtual?

58. A "floating coin" illusion consists of two parabolic mirrors, each with a focal length of 7.5 cm, facing each other so that their centers are 7.5 cm apart (see the figure below). If a few coins are placed on the lower mirror, an image of the coins forms in the small opening at the center of the top mirror. Use the mirror equation, and draw a ray diagram to show that the final image forms at that location. Show that the magnification is 1 and that the image is real and upright. (Note: A flashlight beam shined on these images has a very startling effect. Even at a glancing angle, the incoming light beam is seemingly reflected off the images of the coins. Do you understand why?)

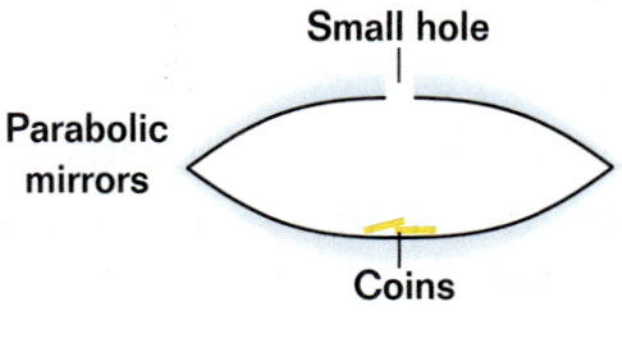

13 REVIEW

47. inverted; $p = 6.1$ cm; $f =$ 2.6 cm; real

48. $R = -31.0$ cm

49. $q_2 = 6.7$ cm; real; $M_1 = -0.57$, $M_2 = -0.27$; inverted

50. $f = -13.7$ cm; $M = 0.0656$; virtual, upright

51. $p = 11.3$ cm

52. concave; $M = -20.0$; real; inverted

53. (See *Teacher's Solution Manual and Answer Key.*)

54. (See *Teacher's Solution Manual and Answer Key.*)

55. $R = -25.0$ cm

56. **a.** 15.0 cm
b. 59.9 cm
c. $M_{convex} = 2.00$, $M_{concave} = 0.667$
d. virtual
e. upright

57. concave, $R = 48.1$ cm; $M = 2.00$; virtual

58. (See *Teacher's Solution Manual and Answer Key.*)

479

59. (See *Teacher's Solution Manual and Answer Key.*)

60. (See *Teacher's Solution Manual and Answer Key.*)

Alternative Assessment ANSWERS

1. Students' answers will vary. Be sure plans are safe and experimental designs reflect a controlled experiment.
2. Alhazen (965–1038) studied light, pinhole cameras, and parabolic mirrors. When he failed to invent a machine to control the Nile's floods, the Caliph sentenced him to death. Alhazen escaped by pretending to be insane.
3. Students' discussions will vary. Students should recognize that convex mirrors will work best. Check that plans take into account the law of reflection.
4. Students' answers will vary. Be sure the class covers the entire spectrum. Students should provide source references.
5. Students' answers will vary but should indicate that when the sun is on one side of Earth and the moon is on the other side, the moon will be bright at night.
6. Students' summaries will vary. Parallel mirrors produce an infinite number of images that are smaller and smaller. The angles for one, two, three, five, and seven images are 180°, 120°, 90°, 60°, and 45°, respectively.

59. Use the mirror equation and the equation for magnification to prove that the image of a real object formed by a convex mirror is always upright, virtual, and smaller than the object. Use the same equations to prove that the image of a real object placed in front of any spherical mirror is always virtual and upright when $p < |f|$.

60. Use trigonometry to derive the mirror and magnification equations. (Hint: Note that the incoming ray between the object and the mirror forms the hypotenuse of a right triangle. The reflected ray between the image point and the mirror is also the hypotenuse of a right triangle.)

Alternative Assessment

1. Suntan lotions include compounds that absorb the ultraviolet radiation in sunlight and therefore prevent the ultraviolet radiation from damaging skin cells. Design experiments to test the properties of varying grades (SPFs) of suntan lotions. Plan to use blueprint paper, film, plants, or other light-sensitive items. Write down the questions that will guide your inquiry, the materials you will need, the procedures you plan to follow, and the measurements you will take. If your teacher approves your plan, perform the experiments and report or demonstrate your findings in class.

2. The Egyptian scholar Alhazen studied lenses, mirrors, rainbows, and other light phenomena early in the Middle Ages. Research his scholarly work, his life, and his relationship with the Caliph al-Hakim. How advanced were Alhazen's inventions and theories? Summarize your findings and report them to the class.

3. Work in cooperative groups to explore the use of corner and ceiling mirrors as low-tech surveillance devices. Make a floor plan of an existing store, or devise a floor plan for an imaginary one. Determine how much of the store could be monitored by a clerk if flat mirrors were placed in the corners. If you could use curved mirrors in such a system, would you use concave or convex mirrors? Where would you place them? Identify which parts of the store could be observed with the curved mirrors in place. Note any disadvantages that your choice of mirrors may have.

4. Research the characteristics, effects, and applications of a specific type of electromagnetic wave in the spectrum. Find information about the range of wavelengths, frequencies, and energies; natural and artificial sources of the waves; and the methods used to detect them. Find out how they were discovered and how they affect matter. Learn about any dangers associated with them and about their uses in technology. Work together with others in the class who are researching other parts of the spectrum to build a group presentation, brochure, chart, or Web page that covers the entire spectrum.

5. The Chinese astronomer Chang Heng (78–139 CE) recognized that moonlight was a reflection of sunlight. He applied this theory to explain lunar eclipses. Make diagrams showing how Heng might have represented the moon's illumination and the path of light when the Earth, moon, and sun were in various positions on ordinary nights and on nights when there were lunar eclipses. Find out more about Heng's other scientific work, and report your findings to the class.

6. Explore how many images are produced when you stand between two flat mirrors whose reflecting surfaces face each other. What are the locations of the images? Are they identical? Investigate these questions with diagrams and calculations. Then test your calculated results with parallel mirrors, perpendicular mirrors, and mirrors at angles in between. Which angles produce one, two, three, five, and seven images? Summarize your results with a chart, diagram, or computer presentation.

Graphing Calculator Practice

Mirrors

Mirrors produce many types of images: virtual or real, enlarged or reduced, and upright or inverted. The mirror equation and the magnification equation can help sort things out. The mirror equation relates the object distance (p), image distance (q), and focal length (f) to one another.

$$\frac{1}{p}+\frac{1}{q}=\frac{1}{f}$$

Image size can be determined from the magnification equation.

$$M=-\frac{q}{p}$$

Magnification values that are greater than 1 or less than –1 indicate that the image of an object is larger than the object itself. Negative magnification values indicate that an image is real and inverted, while positive magnification values indicate that an image is virtual and upright.

In this graphing calculator activity, the calculator will produce a table of image distance and magnification for various object distances for a mirror with a known focal length. You will use this table to determine the characteristics of the images produced by a variety of mirrors and object distances.

Visit go.hrw.com and enter the keyword **HF6LGTX** to find this graphing calculator activity. Refer to **Appendix B** for instructions on downloading the program for this activity.

extension

Graphing Calculator Practice

Visit go.hrw.com for answers to this Graphing Calculator activity.

Keyword HF6LGTXT

Standardized Test Prep

ANSWERS

1. B
2. H
3. C
4. H
5. B
6. F
7. A
8. G

Standardized Test Prep

MULTIPLE CHOICE

1. Which equation is correct for calculating the focal point of a spherical mirror?

A. $1/f = 1/p - 1/q$
B. $1/f = 1/p + 1/q$
C. $1/p = 1/f + 1/q$
D. $1/q = 1/f + 1/p$

2. Which of the following statements is true about the speeds of gamma rays and radio waves in a vacuum?

F. Gamma rays travel faster than radio waves.
G. Radio rays travel faster than gamma rays.
H. Gamma rays and radio waves travel at the same speed in a vacuum.
J. The speed of gamma rays and radio waves in a vacuum depends on their frequencies.

3. Which of the following correctly states the law of reflection?

A. The angle between an incident ray of light and the normal to the mirror's surface equals the angle between the mirror's surface and the reflected light ray.
B. The angle between an incident ray of light and the mirror's surface equals the angle between the normal to the mirror's surface and the reflected light ray.
C. The angle between an incident ray of light and the normal to the mirror's surface equals the angle between the normal and the reflected light ray.
D. The angle between an incident ray of light and the normal to the mirror's surface is complementary to the angle between the normal and the reflected light ray.

4. Which of the following processes does not linearly polarize light?

F. scattering
G. transmission
H. refraction
J. reflection

Use the ray diagram below to answer questions 5–7.

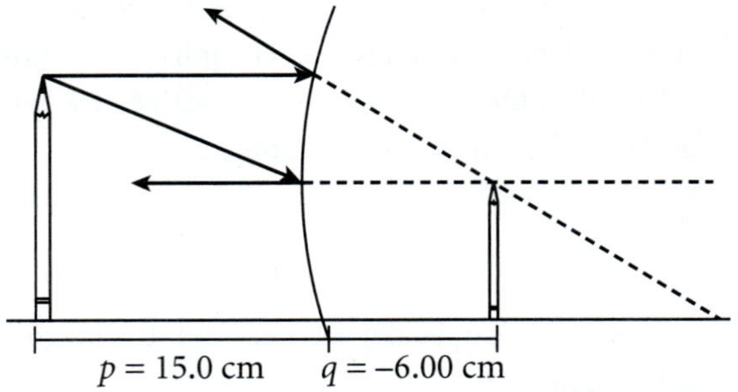

5. Which kind of mirror is shown in the ray diagram?

A. flat
B. convex
C. concave
D. Not enough information is available to draw a conclusion.

6. What is true of the image formed by the mirror?

F. virtual, upright, and diminished
G. real, inverted, and diminished
H. virtual, upright, and enlarged
J. real, inverted, and enlarged

7. What is the focal length of the mirror?

A. −10.0 cm
B. −4.30 cm
C. 4.30 cm
D. 10.0 cm

8. Which combination of primary additive colors will produce magenta-colored light?

F. green and blue
G. red and blue
H. green and red
J. cyan and yellow

9. What is the frequency of an infrared wave that has a vacuum wavelength of 5.5 μm?

A. 165 Hz
B. 5.5×10^{10} Hz
C. 5.5×10^{13} Hz
D. 5.5×10^{16} Hz

10. If the distance from a light source is increased by a factor of 5, by how many times brighter does the light appear?

F. 25
G. 5
H. 1/5
J. 1/25

SHORT RESPONSE

11. White light is passed through a filter that allows only yellow, green, and blue light to pass through it. This light is then shone on a piece of blue fabric and on a piece of red fabric. Which colors do the two pieces of fabric appear to have under this light?

12. The clothing department of a store has a mirror that consists of three flat mirrors, each arranged so that a person standing before the mirrors can see how an article of clothing looks from the side and back. Suppose a ray from a flashlight is shined on the mirror on the left. If the incident ray makes an angle of 65° with respect to the normal to the mirror's surface, what will be the angle θ of the ray reflected from the mirror on the right?

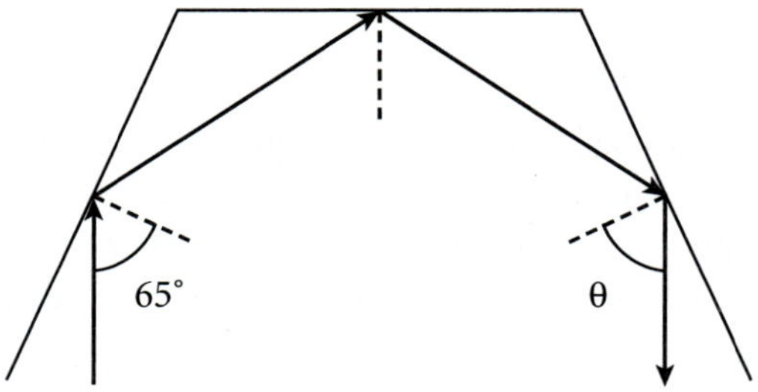

13. X rays emitted from material around compact massive stars, such as neutron stars or black holes, serve to help locate and identify such objects. What would be the wavelength of the X rays emitted from material around such an object if the X rays have a frequency of 5.0×10^{19} Hz?

EXTENDED RESPONSE

14. Explain how you can use a piece of polarizing plastic to determine if light is linearly polarized.

Use the ray diagram below to answer questions 15–19.

A candle is placed 30.0 cm from the reflecting surface of a concave mirror. The radius of curvature of the mirror is 20.0 cm.

15. What is the distance between the surface of the mirror and the image?

16. What is the focal length of the mirror?

17. What is the magnification of the image?

18. If the candle is 12 cm tall, what is the image height?

19. Is the image real or virtual? Is it upright or inverted?

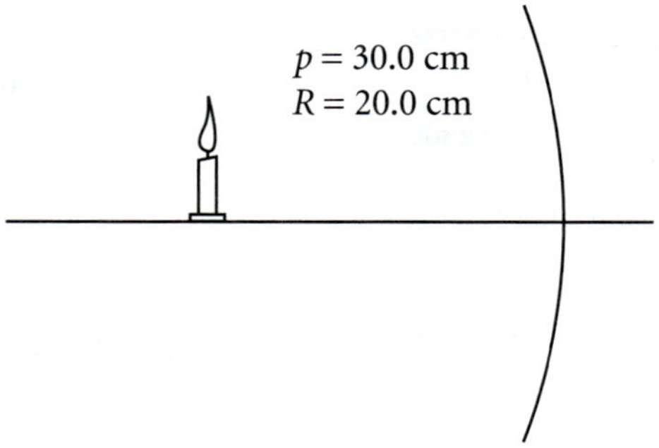

Test TIP Double-check the signs of all values to be used in the mirror and magnification equations.

9. C

10. J

11. The blue fabric appears blue. The red fabric appears black.

12. 65°

13. 6.0×10^{-12} m = 6.0 pm

14. Polarized light will pass through the plastic when the transmission axis of the plastic is parallel with the light's plane of polarization. Rotating the plastic 90° will prevent the polarized light from passing through the plastic, so the plastic appears dark. If light is not linearly polarized, rotating the plastic 90° will have no effect on the light's intensity.

15. 15.0 cm

16. 10.0 cm

17. −0.500

18. −6.0 cm

19. real; inverted

Lab Planning

Beginning on page T34 are preparation notes and teaching tips to assist you in planning.

Blank data tables (as well as some sample data) appear on the **One-Stop Planner.**

No Books in the Lab?

See the ***Datasheets for In-Text Labs*** workbook for a reproducible master copy of this experiment.

CBL™ Option

A CBL™ version of this lab appears in the ***CBL™ Experiments*** workbook.

Safety Caution

Remind students to report all breakage immediately. Students should be instructed not to look directly at a light source.

Tips and Tricks

- Most commercially available light meters measure in the metric unit of lux. Make sure you know what units the light meters in your lab use.
- Show students how to adjust the light meter if the values for the intensity are fluctuating too much for them to read a constant value. Set the light meter to the "fast" mode, and use the "data hold" option to find the correct value.

CHAPTER 13

Skills Practice Lab — Brightness of Light

OBJECTIVES

- **Determine** the relationship between the intensity of the light emitted by a light source and the distance from the source.
- **Explore** the inverse square law in terms of the intensity of light.

MATERIALS LIST

- black aperture tube for light meter
- black paper aperture stop for light meter
- black tube to cover bulb and socket
- black paper square
- clamp for support stand
- light meter
- meterstick
- meterstick-mounted bulb socket
- meterstick supports
- power supply
- small, clear incandescent bulb
- support stand
- masking tape

The measured brightness of a light depends on the distance between the light meter and the light source. In this lab, you will use a light meter to measure the intensity of light at different distances from a light source in order to investigate the relationship between the distance and the brightness of a light source.

SAFETY

- **Use a hot mitt to handle resistors, light sources, and other equipment that may be hot. Allow all equipment to cool before storing it.**
- **If a bulb breaks, notify your teacher immediately. Do not remove broken bulbs from sockets.**
- **Never put broken glass in a regular waste container. Use a dustpan, brush, and heavy gloves to carefully pick up broken pieces, and dispose of them in a container specifically provided for this purpose.**
- **Avoid looking directly at a light source. Looking directly at a light source may cause permanent eye damage. Put on goggles.**

PROCEDURE

Preparation

1. If you are not using a datasheet provided by your teacher, prepare a data table in your lab notebook with two columns and nine rows. In the first row, label the columns *Distance (m)* and *Intensity.* In the first column of your data table, label the second through ninth rows *0.20, 0.25, 0.30, 0.35, 0.40, 0.50, 0.75,* and *1.00.*

Brightness of Light

2. Set up the meterstick, meterstick supports, light source (bulb and socket), power supply, and light meter with aperture as shown in **Figure 1.** Carefully screw the bulb into the socket. Tape the meterstick and supports to the lab table. Set the 0.00 m mark on the meterstick directly below the face of the light meter as shown.
3. Set the bulb socket 0.20 m away from the light meter. Align the clamp and the aperture of the meter so that the aperture is level, parallel to the meterstick, and at the same height as the hole in the tube covering the bulb. Adjust the bulb socket to the 0.20 m mark on the meterstick.

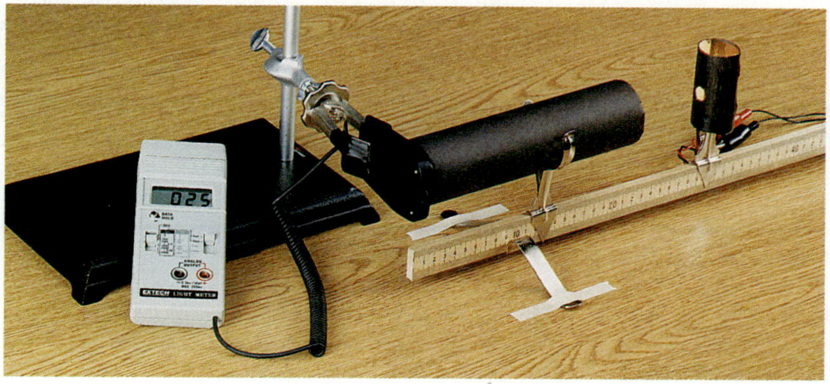

Figure 1
Step 2: Make sure the apparatus is set up securely. The meterstick should be taped to the table so that it cannot move during the experiment. Use a support stand and clamp to hold the light meter in place.
Step 3: Use the meterstick to place the light source at the correct position. Make sure the light source and light meter are lined up.

4. Carefully connect the power supply to the wires from the light socket, and adjust the power supply to 5.0 V. ***Do not plug in the power supply until your teacher has approved your setup.*** When your teacher has approved your setup, carefully plug the power supply into the wall outlet to light the lamp.
5. Use the light meter to read the intensity. Select the "fast" mode, and use the "data hold" option if the intensity values continually fluctuate. Record the intensity value for that distance in your data table.
6. Repeat this procedure for all other bulb distances recorded in your data table.
7. After the last trial turn off the light source. Cover the opening of the phototube with the piece of black paper. Draw a line above or below your data table, and label it *Background.* Use this space to record the intensity reading of the light meter without the bulb illuminated.
8. Clean up your work area. Put equipment away as directed by your teacher.

ANALYSIS

1. **Organizing Data** For each trial, find the real value of the measured light intensity by subtracting the background from the measured value.
2. **Constructing Graphs** Make a graph of the intensity plotted against the distance. Use a graphing calculator, computer, or graph paper.
3. **Organizing Data** For each trial, calculate $1/(Distance^2)$. This value represents the inverse of the distance squared.
4. **Constructing Graphs** Make a graph of the intensity plotted against the inverse of the distance squared.

CONCLUSIONS

5. **Analyzing** Based on your graphs, what is the relationship between the intensity of the light and the distance from the light source? Explain how your graphs support your answer.

✓ Checkpoints

Step 3: The bulb must be securely placed on the meterstick. If necessary, use electrical tape to hold the bulb in position.

Step 4: If the students are using a dc power supply, show them how to adjust the voltage. Power supplies should be set at around 5.0 V for this exercise.

ANSWERS

Analysis

1. Answers will vary. Typical measured values will range from 7 lux to 125 lux for the light meter.
2. The graph should have an inverse square relationship and show that intensity decreases as distance increases.
3. Typical values will range from $1.00\ m^{-2}$ to $100.00\ m^{-2}$.
4. The graph should have a straight line pointing up and to the right.

Conclusions

5. The intensity of a light varies directly with the inverse square of the distance from the light source.

CHAPTER 14

Refraction

Planning Guide

Compression Guide

To shorten instruction because of time limitations, omit the opener and Section 3 and abbreviate the review.

OBJECTIVES	LABS, DEMONSTRATIONS, AND ACTIVITIES	TECHNOLOGY RESOURCES
PACING • 45 min pp. 486–487 **Chapter Opener**	ANC **Discovery Lab** Refraction and Lenses*◆ BASIC	CD **Visual Concepts,** Chapter 14 BASIC
PACING • 45 min pp. 488–493 **Section 1 Refraction** • Recognize situations in which refraction will occur. • Identify which direction light will bend when it passes from one medium to another. • Solve problems using Snell's law.	TE **Demonstration** Refraction from Air to Water, p. 488 GENERAL TE **Demonstration** Refraction in Various Materials, p. 490 GENERAL TE **Demonstration** Underwater Appearance, p. 491 GENERAL	OSP **Lesson Plans** TR 70 Refraction TR 71 Refraction and the Wave Model of Light TR 72 Image Position for Objects in Different Media TR 49A Indices of Refraction for Various Substances
PACING • 135 min pp. 494–505 **Section 2 Thin Lenses** • Use ray diagrams to find the position of an image produced by a converging or diverging lens, and identify the image as real or virtual. • Solve problems using the thin-lens equation. • Calculate the magnification of lenses. • Describe the positioning of lenses in compound microscopes and refracting telescopes.	SE **Quick Lab** Focal Length, p. 496 GENERAL SE **Quick Lab** Prescription Glasses, p. 502 GENERAL SE **Skills Practice Lab** Converging Lenses, pp. 522–523◆ GENERAL ANC **Datasheet** Converging Lenses* GENERAL TE **Demonstration** The Effect of Lenses on Light Beams, p. 494 GENERAL TE **Demonstration** Focal Lengths of Lenses, p. 495 BASIC TE **Demonstration** Microscope, p. 502 BASIC ANC **Invention Lab** Camera Design*◆ ADVANCED	OSP **Lesson Plans** EXT **Integrating Astronomy** The Refracting Telescope at Yerkes BASIC TR 73 Lenses and Focal Length TR 74 Images Created by Converging Lenses TR 75 Image Created by a Diverging Lens TR 76 Nearsighted and Farsighted
PACING • 45 min pp. 506–511 *Advanced Level* **Section 3 Optical Phenomena** • Predict whether light will be refracted or undergo total internal reflection. • Recognize atmospheric conditions that cause refraction. • Explain dispersion and phenomena such as rainbows in terms of the relationship between the index of refraction and the wavelength.	SE **Quick Lab** Periscope, p. 507 GENERAL TE **Demonstration** Critical Angle, p. 506 GENERAL TE **Demonstration** Fiber Optic—Bending Light, p. 509 ADVANCED TE **Demonstration** Dispersion, p. 509 ADVANCED TE **Demonstration** Rainbow, p. 510 GENERAL TE **Demonstration** Chromatic Aberration, p. 511 ADVANCED	OSP **Lesson Plans** TR 77 Rainbows

PACING • 90 min

CHAPTER REVIEW, ASSESSMENT, AND STANDARDIZED TEST PREPARATION

SE **Chapter Highlights,** p. 513
SE **Chapter Review,** pp. 514–519
SE **Graphing Calculator Practice,** p. 518 GENERAL
SE **Alternative Assessment,** p. 519 ADVANCED
SE **Standardized Test Prep,** pp. 520–521 GENERAL
SE **Appendix D: Equations,** p. 860–861
SE **Appendix I: Additional Problems,** pp. 890–891
ANC **Study Guide Worksheet** Mixed Review* GENERAL
ANC **Chapter Test A*** GENERAL
ANC **Chapter Test B*** ADVANCED
OSP **Test Generator**

Online and Technology Resources

go.hrw.com Holt Online Learning

Visit go.hrw.com to find a variety of online resources. To access this chapter's extensions, enter the keyword **HF6REFXT** and click the "go" button. Click **Holt Online Learning** for an online edition of this textbook, and other interactive resources.

HOLT Teacher's One-Stop Planner®

This DVD package includes:
- Holt Calendar Planner
- Customizable Lesson Plans
- Editable Worksheets
- ExamView® Version 6 Assessment Suite
- Interactive Teacher's Edition
- Holt PuzzlePro®
- Holt PowerNotes® Presentations
- MindPoint® Quiz Show

KEY

SE Student Edition	**OSP** One-Stop Planner	**EXT** Online Extension
TE Teacher Edition	**CD** CD or CD-ROM	* Also on One-Stop Planner
ANC Ancillary Worksheet	**TR** Teaching Transparencies	◆ Requires advance prep

SKILLS DEVELOPMENT RESOURCES	REVIEW AND ASSESSMENT	CORRELATIONS
		National Science Education Standards
SE **Sample Set A** Snell's Law, pp. 492–493 BASIC **TE** **Classroom Practice,** p. 492 BASIC **ANC** **Problem Workbook** Sample Set A* BASIC **OSP** **Problem Bank** Sample Set A BASIC **SE** **Conceptual Challenge,** p. 491 ADVANCED	**SE** **Section Review,** p. 493 GENERAL **ANC** **Study Guide Worksheet** Section 1* GENERAL **ANC** **Quiz** Section 1* BASIC	UCP 1, 2 SAI 1, 2
SE **Sample Set B** Lenses, pp. 500–501 GENERAL **TE** **Classroom Practice,** p. 500 GENERAL **ANC** **Problem Workbook** Sample Set B* GENERAL **OSP** **Problem Bank** Sample Set B GENERAL	**SE** **Section Review,** p. 505 GENERAL **ANC** **Study Guide Worksheet** Section 2* GENERAL **ANC** **Quiz** Section 2* BASIC	UCP 1, 2, 3, 5 SAI 1, 2 ST 1, 2 HNS 1 SPSP 1, 5
SE **Sample Set C** Critical Angle, pp. 507–508 GENERAL **TE** **Classroom Practice,** p. 507 GENERAL **ANC** **Problem Workbook** Sample Set C* GENERAL **OSP** **Problem Bank** Sample Set C GENERAL	**SE** **Section Review,** p. 511 ADVANCED **ANC** **Study Guide Worksheet** Section 3* ADVANCED **ANC** **Quiz** Section 3* GENERAL	UCP 1, 2, 3, 5 SAI 1, 2 ST 1, 2 HNS 1 SPSP 5

SCILINKS. NSTA
www.scilinks.org

Maintained by the **National Science Teachers Association.**

Topic: Snell's Law
SciLinks Code: HF61404

Topic: Lenses
SciLinks Code: HF60868

Topic: Fiber Optics
SciLinks Code: HF60572

Topic: Abnormalities of the Eye
SciLinks Code: HF60002

Topic: Dispersion of Light
SciLinks Code: HF60416

Classroom CD-ROMs

- Guided Reading Audio Program
- Student One Stop
- Virtual Investigations
- Visual Concepts

Holt Lab Generator CD-ROM

Search for any lab by topic, standard, difficulty level, or time. Edit any lab to fit your needs, or create your own labs. Use the Lab Materials QuickList software to customize your lab materials list.

CHAPTER 14
Overview

Section 1 investigates which direction light will bend when it enters another medium and uses Snell's law to solve problems.

Section 2 solves problems involving image formation by converging and diverging lenses using ray diagrams and the thin-lens equation, explores eye disorders and eyeglasses, and examines the positioning of lenses in microscopes and refracting telescopes.

Section 3 calculates critical angle; predicts when total internal reflection will occur; explains atmospheric phenomena, including mirages and rainbows; and briefly describes lens aberrations.

About the Illustration

This photo of a group of hikers and a rainbow was taken in Yosemite National Park in California along the "Mist Trail" overlooking Vernal Falls.

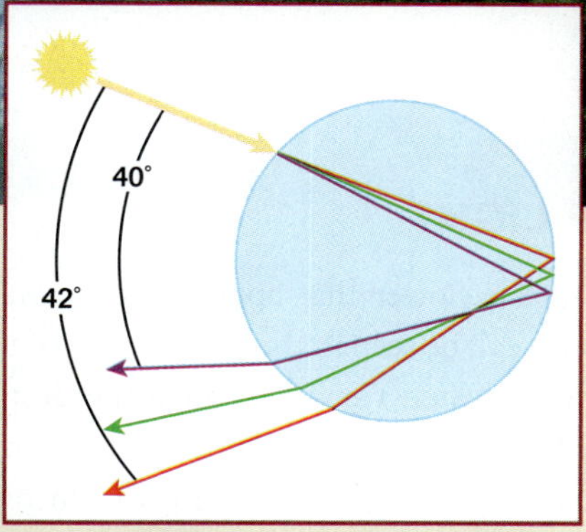

CHAPTER 14

Refraction

Most of us have seen a rainbow when sunlight hits droplets of water in the air. Sunlight is bent, or *refracted,* as it passes through a raindrop. Longer wavelengths of light (red) are bent the least, and shorter wavelengths of light (violet) are bent the most.

WHAT TO EXPECT

In this chapter, you will study optical phenomena associated with the refraction of light as it passes from one transparent medium to another. You will learn how to analyze *converging* and *diverging* lenses. You will then better understand how optical devices work.

Why it Matters

Optical devices, such as cameras, microscopes, and telescopes, use the principles of reflection and refraction to create images that we can then use for many artistic and scientific applications. An understanding of how lenses function is also essential to the practice of optometry.

CHAPTER PREVIEW

1 Refraction
- Refraction of Light
- The Law of Refraction

2 Thin Lenses
- Types of Lenses
- Characteristics of Lenses
- The Thin-Lens Equation and Magnification
- Eyeglasses and Contact Lenses
- Combination of Thin Lenses

3 Optical Phenomena
- Total Internal Reflection
- Atmospheric Refraction
- Dispersion
- Lens Aberrations

Tapping Prior Knowledge

Knowledge to Review

- ✔ Wave speed equals frequency times wavelength.
- ✔ Reflection is the turning back of an electromagnetic wave at the surface of a substance.
- ✔ The focal point is the point at which a beam parallel to the principal axis will converge after reflection from a concave mirror.
- ✔ The focal length is the distance from the focal point to the mirror.
- ✔ Spherical aberration is an effect in which the image produced by a spherical mirror is blurred. It results from light rays converging at different points when the mirror is not parabolic.

Items to Probe

- ✔ Sine function: Ask what the highest value is that the sine of an angle may have and what the angle is at that value (*1, 90°*).

SECTION 1
General Level

Demonstration

Refraction from Air to Water — GENERAL

Purpose Demonstrate the phenomenon of refraction, and explore how the angle of incidence affects the angle of refraction.

Materials laser, dusty chalkboard erasers, aquarium filled with water to which a few drops of whole milk have been added

CAUTION *Lasers can damage the eyes. Avoid directing the laser beam toward the students.*

Procedure Adjust the position of the laser so that it shines into the water at an angle of about 30° from the normal. Gently tap the erasers together above the water so that students can see the path of the beam in air. Have students observe the bending of the light ray as it enters the water. Now, move the laser so that the beam strikes the water at various angles. Have students observe that the path of the light ray in water depends on the angle at which the light strikes the surface of the water.

Visual Strategy — BASIC

Figure 2

Point out that a portion of the incident light is reflected. Make sure students realize that all angles are measured relative to the normal. Remind students that the normal is an imaginary line drawn perpendicular to the surface.

Q What is the angle between the normal line and the boundary between air and water?

A *90°*

SECTION 1

Refraction

SECTION OBJECTIVES

- **Recognize situations in which refraction will occur.**
- **Identify which direction light will bend when it passes from one medium to another.**
- **Solve problems using Snell's law.**

refraction

the bending of a wave front as the wave front passes between two substances in which the speed of the wave differs

Figure 1
The flower looks small when viewed through the water droplet. The light from the flower is bent because of the shape of the water droplet and the change in material as the light passes through the water.

REFRACTION OF LIGHT

Look at the tiny image of the flower that appears in the water droplet in **Figure 1.** The blurred flower can be seen in the background of the photo. Why does the flower look different when viewed through the droplet? This phenomenon occurs because light is bent at the boundary between the water and the air around it. The bending of light as it travels from one medium to another is called **refraction.**

If light travels from one transparent medium to another at any angle other than straight on (normal to the surface), the light ray changes direction when it meets the boundary. As in the case of reflection, the angles of the incoming and refracted rays are measured with respect to the normal. For studying refraction, the normal line is extended into the refracting medium, as shown in **Figure 2.** The angle between the refracted ray and the normal is called the *angle of refraction,* θ_r, and the angle of incidence is designated as θ_i.

Refraction occurs when light's velocity changes

Glass, water, ice, diamonds, and quartz are all examples of transparent media through which light can pass. The speed of light in each of these materials is different. The speed of light in water, for instance, is less than the speed of light in air. And the speed of light in glass is less than the speed of light in water.

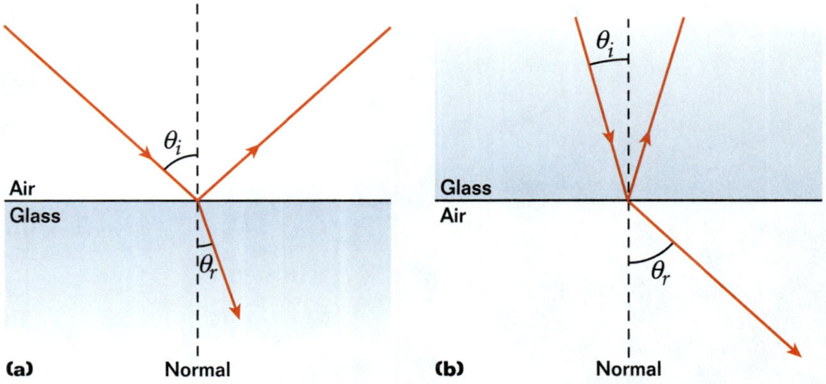

Figure 2
When light moves from one medium to another, part of it is reflected and part is refracted. **(a)** When the light ray moves from air into glass, the refracted portion is bent toward the normal, **(b)** whereas the path of the light ray moving from glass into air is bent away from the normal.

When light moves from a material in which its speed is higher to a material in which its speed is lower, such as from air to glass, the ray is bent toward the normal, as shown in **Figure 2(a).** If the ray moves from a material in which its speed is lower to one in which its speed is higher, as in **Figure 2(b),** the ray is bent away from the normal. If the incident ray of light is parallel to the normal, then no refraction (bending) occurs in either case.

Note that the path of a light ray that crosses a boundary between two different media is reversible. If the ray in **Figure 2(a)** originated inside the glass block, it would follow the same path as shown in the figure, but the reflected ray would be inside the block.

Refraction can be explained in terms of the wave model of light

In the previous chapter on light and refraction, you learned how to use wave fronts and light rays to approximate light waves. This analogy can be extended to light passing from one medium into another. In **Figure 3,** the wave fronts are shown in red and are assumed to be spherical. The combined wave front (dotted line connecting the individual wave fronts) is a superposition of all the spherical wave fronts. The direction of propagation of the wave is perpendicular to the wave front and is what we call the *light ray.*

Consider wave fronts of a plane wave of light traveling at an angle to the surface of a block of glass, as shown in **Figure 3.** As the light enters the glass, the wave fronts slow down, but the wave fronts that have not yet reached the surface of the glass continue traveling at the speed of light in air. During this time, the slower wave fronts travel a smaller distance than do the wave fronts in the air, so the entire plane wave changes directions.

Note the difference in wavelength (the space between the wave fronts) between the plane wave in air and the plane wave in the glass. Because the wave fronts inside the glass are traveling more slowly, in the same time interval they move through a shorter distance than the wave fronts that are still traveling in air. Thus, the wavelength of the light in the glass, λ_{glass}, is shorter than the wavelength of the incoming light, λ_{air}. The frequency of the light does *not* change when the light passes from one medium to another.

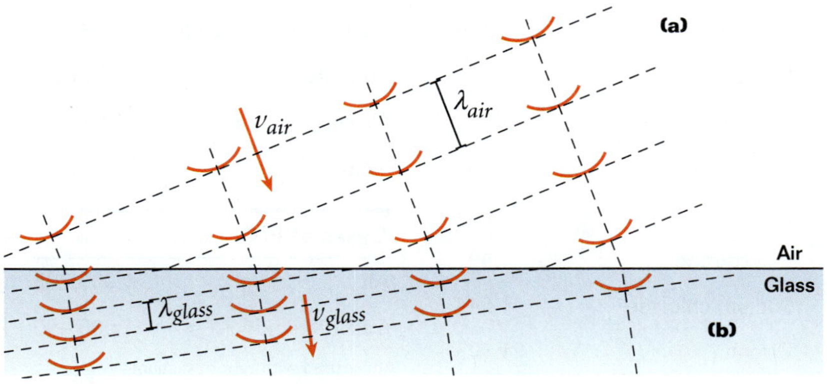

Figure 3
A plane wave traveling in air **(a)** has a wavelength of λ_{air} and velocity of v_{air}. Each wave front turns as it strikes the glass. Because the speed of the wave fronts in the glass **(b),** v_{glass}, is slower, the wavelength of the light becomes shorter, and the wave fronts change direction.

Did you know?

The speed of light in a vacuum, *c*, is an important constant used by physicists. It has been measured to be about 3.00×10^8 m/s. Inside of other mediums, such as air, glass, or water, the speed of light is different and is less than *c*.

SECTION 1

Misconception Alert

Students may think that the frequency of light changes as light enters a different medium. Point out that the frequency cannot change. If the refracted frequency were *less* than the incident frequency, wave crests would have to "pile up" somewhere. If the refracted frequency were *greater* than the incident frequency, wave crests would have to "pop up" from nowhere.

Demonstration

Refraction in Various Materials

Purpose Demonstrate that different materials have different refractive indices.

Materials laser, dusty chalkboard erasers, wide rectangular piece of plastic (acrylic polymer), rectangular piece of glass or other material, aquarium filled with water to which a few drops of milk have been added

CAUTION *Lasers can damage the eyes. Avoid directing the laser beam toward the students.*

Procedure Adjust the position of the laser so that it shines into the side of the plastic, at an angle of about 30° from the normal (choose a point of incidence such that the light inside the plastic will exit the opposite parallel side of the block). Gently tap the erasers together to make the incoming laser beam visible. Have students observe and record the bending of the light rays as they enter and exit the material. Use chalk dust to see where the refracted beam goes.

Repeat this demonstration with the other materials. Point out that the angle of incidence is the same in every experiment. Ask students to compare the angles of refraction in the different materials. Ask in which material the speed of light is lowest (*the most refractive one*). Place the blocks inside the water and repeat. Have students observe that the bending is less dramatic.

index of refraction

the ratio of the speed of light in a vacuum to the speed of light in a given transparent medium

Did you know?

The index of refraction of any medium can also be expressed as the ratio of the wavelength of light in a vacuum, λ_0, to the wavelength of light in that medium, λ_n, as shown in the following relation.

$$n = \frac{\lambda_0}{\lambda_n}$$

THE LAW OF REFRACTION

An important property of transparent substances is the **index of refraction.** The index of refraction for a substance is the ratio of the speed of light in a vacuum to the speed of light in that substance.

INDEX OF REFRACTION

$$n = \frac{c}{v}$$

$$\text{index of refraction} = \frac{\text{speed of light in vacuum}}{\text{speed of light in medium}}$$

From this definition, we see that the index of refraction is a dimensionless number that is always greater than 1 because light always travels slower in a substance than in a vacuum. **Table 1** lists the indices of refraction for different substances. Note that the larger the index of refraction is, the slower light travels in that substance and the more a light ray will bend when it passes from a vacuum into that material.

Imagine, as an example, light passing between air and water. When light begins in the air (high speed of light and low index of refraction) and travels into the water (lower speed of light and higher index of refraction), the light rays are bent toward the normal. Conversely, when light passes from the water to the air, the light rays are bent away from the normal.

Note that the value for the index of refraction of air is nearly that of a vacuum. For simplicity, *use the value n = 1.00 for air when solving problems.*

Table 1 Indices of Refraction for Various Substances*

Solids at 20°C	n	Liquids at 20°C	n
Cubic zirconia	2.20	Benzene	1.501
Diamond	2.419	Carbon disulfide	1.628
Fluorite	1.434	Carbon tetrachloride	1.461
Fused quartz	1.458	Ethyl alcohol	1.361
Glass, crown	1.52	Glycerine	1.473
Glass, flint	1.66	Water	1.333
Ice (at 0°C)	1.309	**Gases at 0°C, 1 atm**	**n**
Polystyrene	1.49	Air	1.000 293
Sodium chloride	1.544	Carbon dioxide	1.000 450
Zircon	1.923		

*measured with light of vacuum wavelength = 589 nm

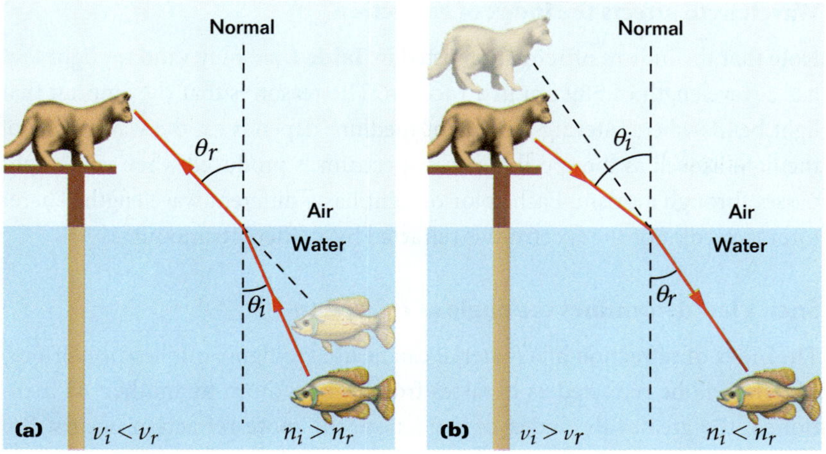

Figure 4
(a) To the cat on the pier, the fish looks closer to the surface than it really is. **(b)** To the fish, the cat seems to be farther from the surface than it actually is.

Objects appear to be in different positions due to refraction

When looking at a fish underwater, a cat sitting on a pier perceives the fish to be closer to the water's surface than it actually is, as shown in **Figure 4(a).** Conversely, the fish perceives the cat on the pier to be farther from the water's surface than it actually is, as shown in **Figure 4(b).**

Because of the reversibility of refraction, both the fish and the cat see along the same path, as shown by the solid lines in both figures. However, the light ray that reaches the fish forms a smaller angle with respect to the normal than does the light ray from the cat to the water's surface. The reason is that light is bent toward the normal when it travels from a medium with a lower index of refraction (the air) to one with a higher index of refraction (the water). Extending this ray along a straight line shows the cat's image to be above the cat's actual position.

On the other hand, the light ray that reaches the cat from the water's surface forms a larger angle with respect to the normal, because the light from the fish travels from a medium with a higher index of refraction to one with a lower index of refraction. Note that the fish's image is closer to the water's surface than the fish actually is. An underwater object seen from the air above appears larger than its actual size because the image, which is the same size as the object, is closer to the observer.

Why it Matters

Conceptual Challenge

1. The Invisible Man H. G. Wells wrote a famous novel about a man who made himself invisible by changing his index of refraction. What would his index of refraction have to be to accomplish this?

2. Visibility for the Invisible Man Would the invisible man be able to see anything?

3. Fishing When trying to catch a fish, should a pelican dive into the water horizontally in front of or behind the image of the fish it sees?

Demonstration

Underwater Appearance

Purpose View the change in apparent position due to refraction.

Materials (for each student pair) beaker wrapped with opaque paper, penny, container of water

Procedure Have one student place the penny in the beaker against the far side of the beaker. Have the other student position his or her eyes so that the penny can just barely be seen over the rim of the beaker.

Have the first student slowly pour water into the beaker. Have the second student note that the penny will appear to rise with the rising water.

ANSWERS — ADVANCED

Conceptual Challenge

1. It would have to be the index of refraction of air. If his index of refraction were any larger, the images behind him would appear distorted, giving away his position.
2. No, he would not, because light rays would not be bent by his cornea or lens. Thus, no image would be formed on his retina.
3. in front of the image, because the light from the fish bends away from the normal as it enters the air

SECTION 1

Teaching Tip

Point out that the spectrum of visible light is between 400 nm (violet) and 700 nm (red). Light with 589 nm wavelength is yellow.

Classroom Practice

Snell's Law

Find the angle of refraction of a light ray entering diamond from the following materials at an angle of 30.00°. (Hint: Use data from **Table 1.**)

a. water

b. cubic zirconia

Answers

a. 15.99°

b. 27.0°

A light ray (589 nm) traveling through air strikes an unknown substance at 60.00° and forms an angle of 41.42° with the normal inside. What material is it?

Answer

ice ($n = 1.309$)

The Language of Physics

$\sin^{-1}$ denotes the inverse function of sine, not $\frac{1}{\sin}$. Just as $\sin 30° = 0.5$, $\sin^{-1}(0.5) = 30°$.

Wavelength affects the index of refraction

Note that the indices of refraction listed in **Table 1** are only valid for light that has a wavelength of 589 nm in a vacuum. The reason is that the amount that light bends when entering a different medium depends on the wavelength of the light as well as the speed. Thus, a spectrum is produced when white light passes through a prism. Each color of light has a different wavelength. Therefore, each color of the spectrum is refracted by a different amount.

SCILINKS.
www.scilinks.org
Topic: Snell's Law
Code: HF61404

Snell's law determines the angle of refraction

The index of refraction of a material can be used to figure out how much a ray of light will be refracted as it passes from one medium to another. As mentioned, the greater the index of refraction, the more refraction occurs. But how can the angle of refraction be found?

In 1621, Willebrord Snell experimented with light passing through different media. He developed a relationship called Snell's law, which can be used to find the angle of refraction for light traveling between any two media.

SNELL'S LAW

$$n_i \sin \theta_i = n_r \sin \theta_r$$

index of refraction of first medium × sine of the angle of incidence = index of refraction of second medium × sine of the angle of refraction

SAMPLE PROBLEM A

Snell's Law

PROBLEM

A light ray of wavelength 589 nm (produced by a sodium lamp) traveling through air strikes a smooth, flat slab of crown glass at an angle of 30.0° to the normal. Find the angle of refraction, θ_r.

SOLUTION

Given: $\theta_i = 30.0°$ $n_i = 1.00$ $n_r = 1.52$

Unknown: $\theta_r = ?$

Use the equation for Snell's law.

$$n_i \sin \theta_i = n_r \sin \theta_r$$

$$\theta_r = \sin^{-1}\left[\frac{n_i}{n_r}(\sin \theta_i)\right] = \sin^{-1}\left[\frac{1.00}{1.52}(\sin 30.0°)\right]$$

$$\boxed{\theta_r = 19.2°}$$

PRACTICE A

Snell's Law

1. Find the angle of refraction for a ray of light that enters a bucket of water from air at an angle of 25.0° to the normal. (Hint: Use **Table 1.**)

2. For an incoming ray of light of vacuum wavelength 589 nm, fill in the unknown values in the following table. (Hint: Use **Table 1.**)

	from (medium)	to (medium)	θ_i	θ_r
a.	flint glass	crown glass	25.0°	?
b.	air	?	14.5°	9.80°
c.	air	diamond	31.6°	?

3. A ray of light of vacuum wavelength 550 nm traveling in air enters a slab of transparent material. The incoming ray makes an angle of 40.0° with the normal, and the refracted ray makes an angle of 26.0° with the normal. Find the index of refraction of the transparent material. (Assume that the index of refraction of air for light of wavelength 550 nm is 1.00.)

SECTION REVIEW

1. Sunlight passes into a raindrop at an angle of 22.5° from the normal at one point on the droplet. What is the angle of refraction?

2. For each of the following cases, will light rays be bent toward or away from the normal?
   a. $n_i > n_r$, where $\theta_i = 20°$
   b. $n_i < n_r$, where $\theta_i = 20°$
   c. from air to glass with an angle of incidence of 30°
   d. from glass to air with an angle of incidence of 30°

3. Find the angle of refraction of a ray of light that enters a diamond from air at an angle of 15.0° to the normal. (Hint: Use **Table 1.**)

4. **Critical Thinking** In which of the following situations will light from a laser be refracted?
   a. traveling from air into a diamond at an angle of 30° to the normal
   b. traveling from water into ice along the normal
   c. upon striking a metal surface
   d. traveling from air into a glass of iced tea at an angle of 25° to the normal

PROBLEM GUIDE A

Use this guide to assign problems.
SE = Student Edition Textbook
PW = Problem Workbook
PB = Problem Bank on the One-Stop Planner (OSP)

Solving for:

θ_r	**SE** Sample, 1–2; Ch. Rvw. 10–14, 42a, 49 **PW** 5 **PB** 7–10
θ_i	**SE** Ch. Rvw. 40, 42b **PW** 3 **PB** Sample, 1–3
n	**SE** 3; Ch. Rvw. 39, 41 **PW** Sample, 1–2, 4 **PB** 4–6

***Challenging Problem**
Consult the printed Solutions Manual or the OSP for detailed solutions.

ANSWERS

Practice A

1. 18.5°
2. a. 27.5°
   b. glycerine ($n = 1.47$)
   c. 12.5°
3. 1.47

SECTION REVIEW ANSWERS

1. 16.7°
2. a. away
   b. toward
   c. toward
   d. away
3. 6.14°
4. a, d

SECTION 2
General Level

Demonstration

The Effect of Lenses on Light Beams — GENERAL

Purpose Demonstrate the effect of converging and diverging lenses on parallel and non-parallel beams.

Materials light source, white sheet of paper, ray filter, converging lens, diverging lens

Procedure Use the ray filter to produce five beams of light. Place the converging lens in front of the light source. Let students observe the beams converging on the white paper. Point out that past the focal point the beams diverge. Turn the light source off, and place it on the other side of the lens. Ask students to predict the pattern of the beam in this case and to sketch a diagram of it. Shine the light, and let them conclude that the lens has two focal points. Repeat the process with the diverging lens. Have students formulate the rules about parallel beams and rays passing through the focal point. Repeat with a combination of both lenses, first close to each other and then farther and farther apart, to show that the *convex* lens causes beams to converge and the *concave* lens causes beams to diverge.

SECTION 2

Thin Lenses

SECTION OBJECTIVES

- **Use ray diagrams to find the position of an image produced by a converging or diverging lens, and identify the image as real or virtual.**
- **Solve problems using the thin-lens equation.**
- **Calculate the magnification of lenses.**
- **Describe the positioning of lenses in compound microscopes and refracting telescopes.**

TYPES OF LENSES

When light traveling in air enters a pane of glass, it is bent toward the normal. As the light exits the pane of glass, it is bent again. When the light exits, however, its speed increases as it enters the air, so the light bends away from the normal. Because the amount of refraction is the same regardless of whether light is entering or exiting a medium, the light rays are bent as much on exiting the pane of glass as they were on entering.

Curved surfaces change the direction of light

When the surfaces of a medium are curved, the direction of the normal line differs for each spot on the surface of the medium. Thus, when light passes through a medium that has one or more curved surfaces, the change in the direction of the light rays varies from point to point. This principle is applied in media called **lenses.** Like mirrors, lenses form images, but lenses do so by refraction rather than by reflection. The images formed can be either real or virtual, depending on the type of lens and on the placement of the object. Recall that a real image is formed when rays of light actually intersect to form the image. Virtual images form at a point from which light rays appear to come but do not actually come. Real images can be projected onto a screen; virtual images cannot.

lens

a transparent object that refracts light rays such that they converge or diverge to create an image

Lenses are commonly used to form images in optical instruments, such as cameras, telescopes, and microscopes. In fact, transparent tissue in the front of the human eye acts as a lens, converging light toward the light-sensitive retina, which lines the back of the eye.

A typical lens consists of a piece of glass or plastic ground so that each of its two refracting surfaces is a segment of either a sphere or a plane. **Figure 5** shows examples of lenses. Notice that the lenses are shaped differently. The lens that is thicker at the middle than it is at the rim, shown in **Figure 5(a),** is an example of a *converging* lens. The lens that is thinner at the middle than it

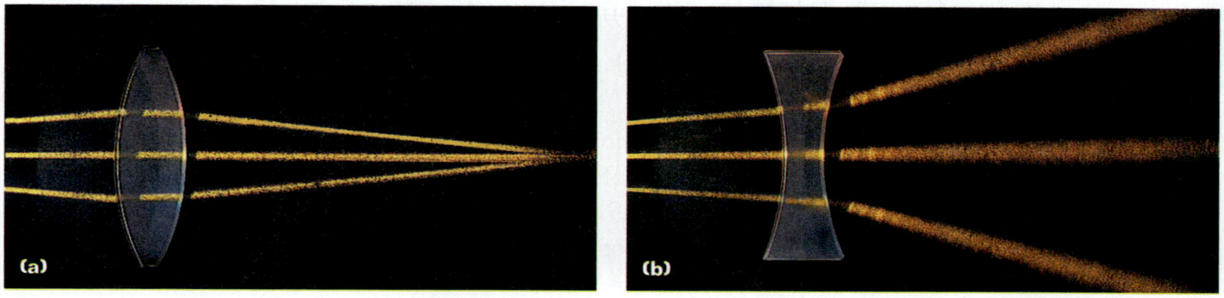

Figure 5
When rays of light pass through **(a)** a converging lens (thicker at the middle), they are bent inward. When they pass through **(b)** a diverging lens (thicker at the edge), they are bent outward.

is at the rim, shown in **Figure 5(b),** is an example of a *diverging* lens. The light rays show why the names *converging* and *diverging* are applied to these lenses.

Focal length is the image distance for an infinite object distance

As with mirrors, it is convenient to define a point called the *focal point* for a lens. Note that light rays from an object far away are nearly parallel. The focal point of a converging lens is the location where the image of an object at an infinite distance from the lens is focused. For example, in **Figure 6(a)** a group of rays parallel to the principal axis passes through a focal point, F, after being bent inward by the lens. Unlike mirrors, every lens has a focal point on each side of the lens because light can pass through the lens from either side, as illustrated in **Figure 6.** The distance from the focal point to the center of the lens is called the *focal length, f.* The focal length is the image distance that corresponds to an infinite object distance.

Rays parallel to the principal axis diverge after passing through a diverging lens, as shown in **Figure 6(b).** In this case, the focal point is defined as the point from which the diverged rays appear to originate. Again, the focal length is defined as the distance from the center of the lens to the focal point.

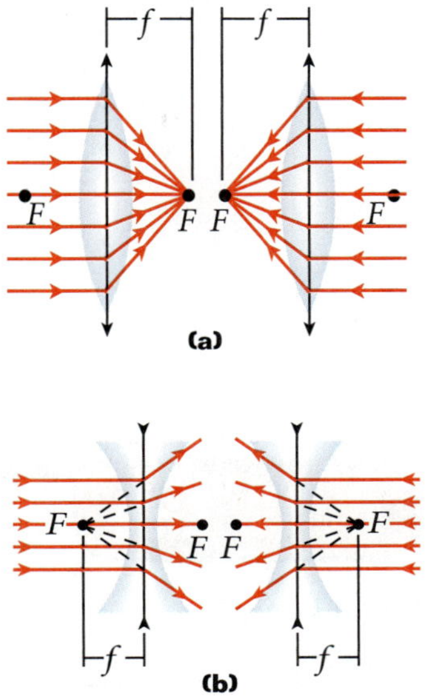

Figure 6
Both **(a)** converging lenses and **(b)** diverging lenses have two focal points but only one focal length.

Ray diagrams of thin-lens systems help identify image height and location

In the chapter on light and reflection, we used a set of standard rays and a ray diagram to predict the characteristics of images formed by spherical mirrors. A similar approach can be used for lenses.

We know, as shown in **Figure 5,** that refraction occurs at a boundary between two materials with different indexes of refraction. However, for *thin lenses* (lenses for which the thickness of the lens is small compared to the radius of curvature of the lens or the distance of the object from the lens), we can represent the front and back boundaries of the lens as a line segment passing through the center of the lens. To draw ray diagrams in the thin-lens approximation, we will use a line segment with arrow ends to indicate a converging lens, as in **Figure 6(a).** To show a diverging lens, we will draw a line segment with "upside-down" arrow ends, as illustrated in **Figure 6(b).** We can then draw ray diagrams using the set of rules outlined in **Table 2.**

Table 2 Rules for Drawing Reference Rays

Ray	From object to lens	From *converging* lens to image	From *diverging* lens to image
Parallel ray	parallel to principal axis	passes through focal point, F	directed away from focal point, F
Central ray	to the center of the lens	from the center of the lens	from the center of the lens
Focal ray	passes through focal point, F	parallel to principal axis	parallel to principal axis

Demonstration

Focal Lengths of Lenses

Purpose Locate the focal point of a converging lens.

Materials optical bench, light source, ray filter, converging lens, screen

Procedure Use the ray filter to produce five beams. Place the converging lens in front of the light source. Move the screen back and forth on the other side of the lens until the image is sharp. This is the focal point of the converging lens. Have students record the focal length.

SECTION 2

Key Models and Analogies — GENERAL

Converging lenses can be compared to concave mirrors, and diverging lenses can be compared to convex mirrors. Light rays pass through lenses but behave the same way as reflected rays do in mirrors. This analogy allows students to apply the rules for mirrors when studying lenses.

The equations and sign conventions for distances are also analogous. For example, remembering that the focal length in convex mirrors has a negative sign and that such mirrors form virtual images of real objects allows students to anticipate image formation in diverging lenses.

Quick Lab

TEACHER'S NOTES

This activity is meant to show a simple way to locate the focal point and to measure the focal length of a converging lens.

On a cloudy day, this lab can be done with a strong flashlight or penlight.

SCILINKS® www.scilinks.org
Topic: Lenses
Code: HF60868

Quick Lab

Focal Length

MATERIALS LIST

- **magnifying glass**
- **ruler**

SAFETY

Care should be taken not to focus the sunlight onto a flammable surface or any body parts, such as hands or arms. Also, DO NOT look at the sun through the magnifying glass because serious eye injury can result.

On a sunny day, hold the magnifying glass, which is a converging lens, above a nonflammable surface, such as a sidewalk, so that a round spot of light is formed on the surface. Move the magnifying glass up and down to find the height at which the spot formed by the lens is most distinct, or smallest. Use the ruler to measure the distance between the magnifying glass and the surface. This distance is the approximate focal length of the lens.

The reasons why these rules work relate to concepts already covered in this textbook. From the definition of a focal point, we know that light traveling parallel to the principal axis (parallel ray) will be focused at the focal point. For a converging lens, this means that light will come together at the focal point in back of the lens. (In this book, the *front* of the lens is defined as the side of the lens that the light rays first encounter. The *back* of the lens refers to the side of the lens opposite where the light rays first encounter the lens.) But a similar ray passing through a diverging lens will exit the lens as if it originated from the focal point in front of the lens. Because refraction is reversible, a ray entering a converging lens from either focal point will be refracted so that it is parallel to the principal axis.

For both lenses, a ray passing through the center of the lens will continue in a straight line with no net refraction. This occurs because both sides of a lens are parallel to one another along any path through the center of the lens. As with a pane of glass, the exiting ray will be parallel to the ray that entered the lens. For ray diagrams, the usual assumption is that the lens is negligibly thin, so it is assumed that the ray is not displaced sideways but instead continues in a straight line.

CHARACTERISTICS OF LENSES

Table 3 summarizes the possible relationships between object and image positions for converging lenses. The rules for drawing reference rays were used to create each of these diagrams. Note that applications are listed along with each ray diagram to show the varied uses of the different configurations.

Converging lenses can produce real or virtual images of real objects

An object infinitely far away from a converging lens will create a point image at the focal point, as shown in the first diagram in **Table 3.** This image is real, which means that it can be projected on a screen.

As a distant object approaches the focal point, the image becomes larger and farther away, as shown in the second, third, and fourth diagrams in **Table 3.** When the object is at the focal point, as shown in the fifth diagram, the light rays from the object are refracted so that they exit the lens parallel to each other. (Because the object is at the focal point, it is impossible to draw a third ray that passes through that focal point, the lens, and the tip of the object.)

When the object is between a converging lens and its focal point, the light rays from the object diverge when they pass through the lens, as shown in the sixth diagram in **Table 3.** This image appears to an observer in back of the lens as being on the same side of the lens as the object. In other words, the brain interprets these diverging rays as coming from an object directly along the path of the rays that reach the eye. The ray diagram for this final case is less straightforward than those drawn for the other cases in the table. The first two rays (parallel to the axis and through the center of the lens) are drawn in the usual

Table 3 Images Created by Converging Lenses

Ray diagrams

1.

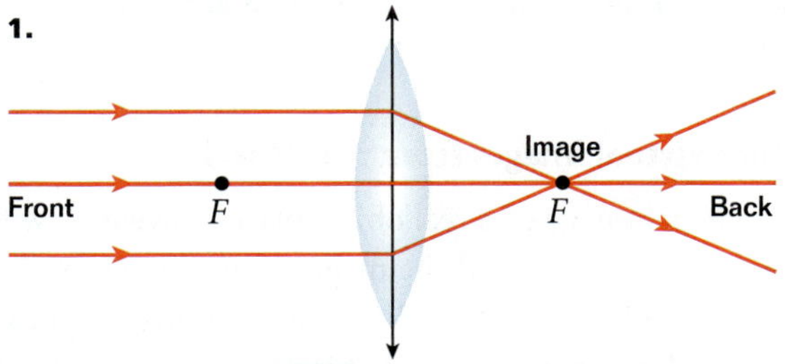

Configuration: object at infinity; point image at *F*

Applications: burning a hole with a magnifying glass

2.

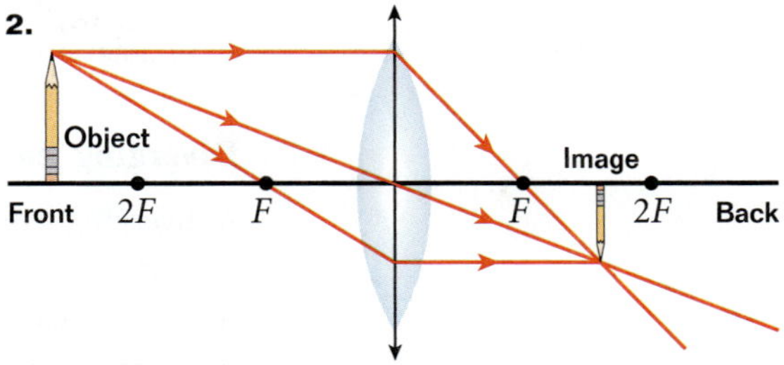

Configuration: object outside 2*F*; real, smaller image between *F* and 2*F*

Applications: lens of a camera, human eyeball lens, and objective lens of a refracting telescope

3.

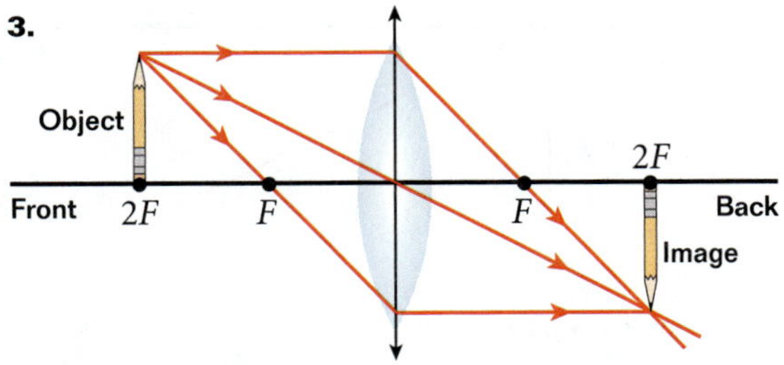

Configuration: object at 2*F*; real image at 2*F* same size as object

Applications: inverting lens of a field telescope

4.

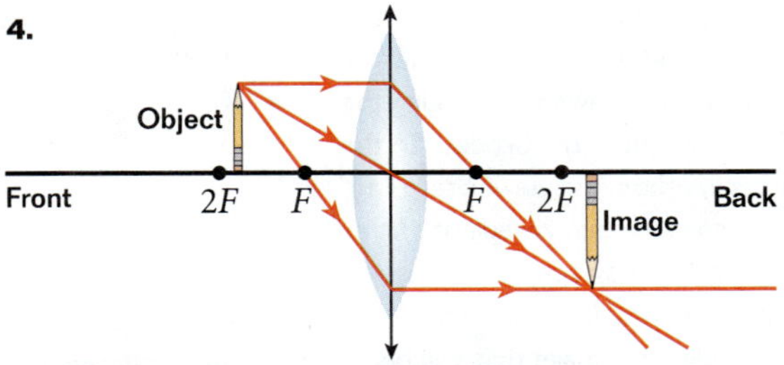

Configuration: object between *F* and 2*F*; magnified real image outside 2*F*

Applications: motion-picture or slide projector and objective lens in a compound microscope

5.

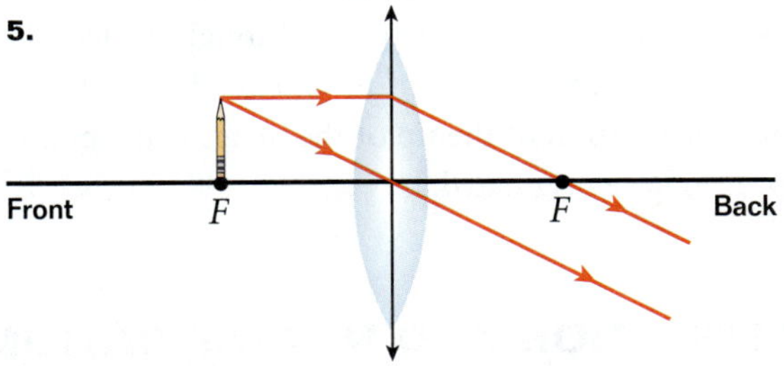

Configuration: object at *F*; image at infinity

Applications: lenses used in lighthouses and searchlights

6.

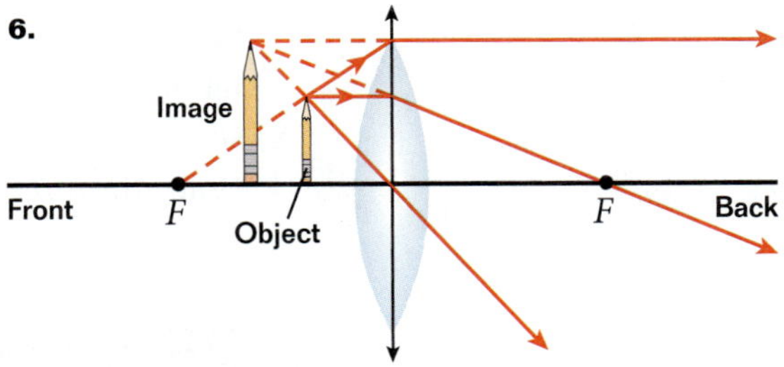

Configuration: object inside *F*; magnified virtual image on the same side of the lens as the object

Applications: magnifying with a magnifying glass; eye-piece lens of microscope, binoculars, and telescope

Visual Strategy – BASIC

Table 3

Point out that in the first case, the object is so far away that the rays from all its points (top to bottom) converge at the focal point on the axis. In the other diagrams, rays are drawn from the object's top only, and their intersection—after passing through the lens—determines the image's top (the bottom is assumed to be on the axis at the same distance from the lens as the top).

Q Suppose that in each case the lens's focal length is 10 cm. What values (or range of values) will the object and image distances have for each case?

A ***case 1:*** *object at* ∞, *image at 10 cm*

case 2: $\infty > p > 20$ *cm*, $20 \text{ cm} > q > 10$ *cm*

case 3: $p = 20$ *cm*, $q = 20$ *cm*

case 4: $20 \text{ cm} > p > 10$ *cm*, $\infty > q > 20$ *cm*

case 5: $p = 10$ *cm*, *image at infinity*

case 6: $p < 10$ *cm*, *negative image distance*

Misconception Alert — ADVANCED

Students might wonder about the need for diverging lenses because virtual images are also formed with converging lenses, as shown in the sixth case in **Table 3.** Ask them to compare the object's location when using each kind of lens for creating a virtual image. A diverging lens creates a virtual image of objects at any distance in front of it. The converging lens can do that only for objects inside *F*, and this virtual image is always magnified. The image produced by a diverging lens is always virtual, smaller, and closer than the object regardless of the object's location. This last characteristic is useful for correcting nearsightedness with eyeglasses.

fashion. The third ray, however, is drawn so that if it were extended, it would connect the focal point in front of the lens, the tip of the object, and the lens in a straight line. To determine where the image is, draw lines extending from the rays exiting the lens back to the point where they would appear to have originated to an observer on the back side of the lens (these lines are dashed in the sixth diagram in **Table 3**).

Diverging lenses produce virtual images from real objects

A diverging lens creates a virtual image of a real object placed anywhere with respect to the lens. The image is upright, and the magnification is always less than one; that is, the image size is reduced. Additionally, the image appears inside the focal point for any placement of the real object.

Did you know?

The lens of a camera forms an inverted image on the film in the back of the camera. Two methods are used to view this image before taking a picture. In one, a system of mirrors and prisms reflects the image to the viewfinder, making the image upright in the process. In the other method, the viewfinder is a diverging lens that is separate from the main lens system. This lens forms an upright virtual image that resembles the image that will be projected onto the film.

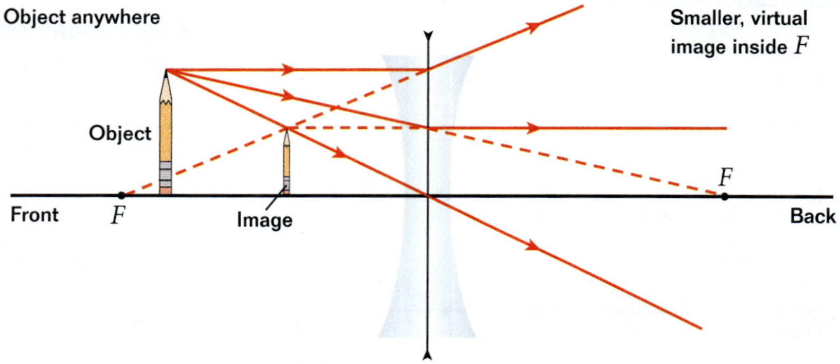

Figure 7
The image created by a diverging lens is always a virtual, smaller image.

The ray diagram shown in **Figure 7** for diverging lenses was created using the rules given in **Table 2.** The first ray, parallel to the axis, appears to come from the focal point on the same side of the lens as the object. This ray is indicated by the oblique dashed line. The second ray passes through the center of the lens and is not refracted. The third ray is drawn as if it were going to the focal point in back of the lens. As this ray passes through the lens, it is refracted parallel to the principal axis and must be extended backward, as shown by the dashed line. The location of the tip of the image is the point at which the three rays appear to have originated.

THE THIN-LENS EQUATION AND MAGNIFICATION

Ray diagrams for lenses give a good estimate of image size and distance, but it is also possible to calculate these values. The equation that relates object and image distances for a lens is called the *thin-lens equation* because it is derived using the assumption that the lens is very thin. In other words, this equation applies when the lens thickness is much smaller than its focal length.

THIN-LENS EQUATION

$$\frac{1}{p}+\frac{1}{q}=\frac{1}{f}$$

$$\frac{1}{\text{distance from object to lens}}+\frac{1}{\text{distance from image to lens}}=\frac{1}{\text{focal length}}$$

When using the thin-lens equation, we often illustrate it using the ray diagram model in which, for clarity, we magnify the vertical axis and show the lens position as a thin line. Always remember that the actual light rays bend at the lens surfaces and that our diagram showing bending at a single central line is an idealized model, which is quite good for thin lenses. But the model, and the equation, must be modified to deal properly with thick lenses, systems of lenses, and object and image points far from the principal axis.

The thin-lens equation can be applied to both converging and diverging lenses if we adhere to a set of sign conventions. **Table 4** gives the sign conventions for lenses. Under this convention, an image in back of the lens (that is, a real image) has a positive image distance, and an image in front of the lens, or a virtual image, has a negative image distance. A converging lens has a positive focal length and a diverging lens has a negative focal length. Therefore, converging lenses are sometimes called *positive lenses* and diverging lenses are sometimes called *negative lenses.*

Table 4
Sign Conventions for Lenses

	+	−
p	real object in front of the lens	real object in back of the lens
q	virtual image in back of the lens	virtual image in front of the lens
f	converging lens	diverging lens

Magnification by a lens depends on object and image distances

Recall that magnification (M) is defined as the ratio of image height to object height. The following equation can be used to calculate the magnification of both converging and diverging lenses.

MAGNIFICATION OF A LENS

$$M=\frac{h'}{h}=-\frac{q}{p}$$

$$\text{magnification}=\frac{\text{image height}}{\text{object height}}=-\frac{\text{distance from image to lens}}{\text{distance from object to lens}}$$

If close attention is given to the sign conventions defined in **Table 4,** then the magnification will describe the image's size and orientation. When the magnitude of the magnification of an object is less than one, the image is smaller than the object. Conversely, when the magnitude of the magnification is greater than one, the image is larger than the object.

Additionally, a negative sign for the magnification indicates that the image is real and inverted. A positive magnification signifies that the image is upright and virtual.

Misconception Alert GENERAL

Students may not be sure about all the cases shown in **Table 4,** particularly about the meaning of a negative distance for the object, which suggests that the object is in back of the lens. Some students may need to be reminded that according to these conventions, light rays always travel from the front to the back of a lens. Ask them if an object could be virtual. Point out that the real image formed by a converging lens may become a virtual object for another lens if the lens is located so that it interrupts the rays before they converge.

Classroom Practice

Lenses
When an object is placed 3.00 cm in front of a converging lens, a real image is formed 6.00 cm in back of the lens. Find the focal distance.

Answer
$f = 2.00$ cm

Where would you place an object in order to produce a virtual image 15.0 cm in front of a converging lens with a focal length of 10.0 cm? How about a diverging lens with the same focal length?

Answer
converging lens: 6.00 cm;
diverging lens: –30.0 cm

Alternative Problem-Solving Approach — ADVANCED

Draw a ray diagram to scale on graph paper. Determine the answer using two rays, and draw the third one to confirm.

Another possibility is to apply this method using a computer art/paint/draw application. The number of pixels can be translated to scale. Working with a grid in the background may be helpful.

SAMPLE PROBLEM B

Lenses

PROBLEM

An object is placed 30.0 cm in front of a converging lens and then 12.5 cm in front of a diverging lens. Both lenses have a focal length of 10.0 cm. For both cases, find the image distance and the magnification. Describe the images.

SOLUTION

1. DEFINE

Given: $f_{converging} = 10.0$ cm $\quad f_{diverging} = -10.0$ cm
$p_{converging} = 30.0$ cm $\quad p_{diverging} = 12.5$ cm

Unknown: $q_{converging} = ?\quad M = ?\quad q_{diverging} = ?\quad M = ?$

Diagrams:

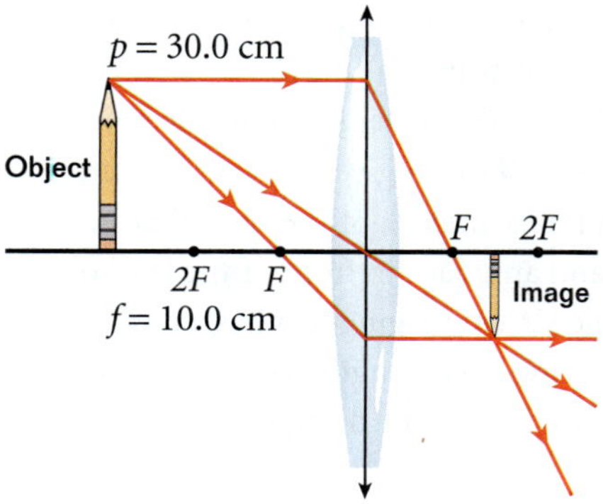

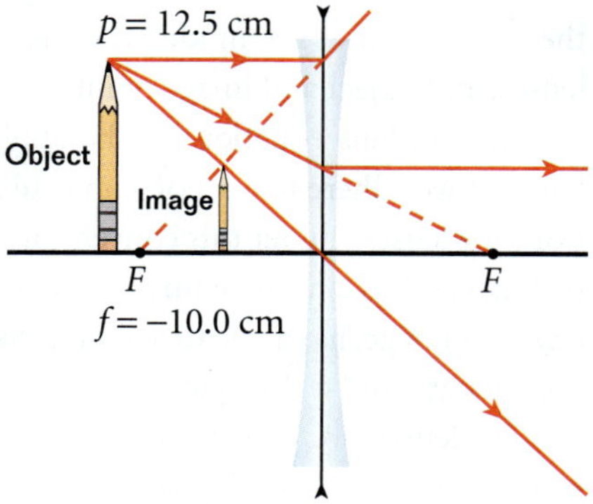

2. PLAN

Choose an equation or situation:
The thin-lens equation can be used to find the image distance, and the equation for magnification will serve to describe the size and orientation of the image.

$$\frac{1}{p} + \frac{1}{q} = \frac{1}{f} \qquad M = -\frac{q}{p}$$

Rearrange the equation to isolate the unknown:

$$\frac{1}{q} = \frac{1}{f} - \frac{1}{p}$$

3. CALCULATE

For the converging lens:

$$\frac{1}{q} = \frac{1}{f} - \frac{1}{p} = \frac{1}{10.0 \text{ cm}} - \frac{1}{30.0 \text{ cm}} = \frac{2}{30.0 \text{ cm}}$$

$$\boxed{q = 15.0 \text{ cm}}$$

$$M = -\frac{q}{p} = -\frac{15.0 \text{ cm}}{30.0 \text{ cm}}$$

$$\boxed{M = -0.500}$$

For the diverging lens:

$$\frac{1}{q} = \frac{1}{f} - \frac{1}{p} = \frac{1}{-10.0 \text{ cm}} - \frac{1}{12.5 \text{ cm}} = -\frac{22.5}{125 \text{ cm}}$$

$$\boxed{q = -5.56 \text{ cm}}$$

$$M = -\frac{q}{p} = -\frac{-5.56 \text{ cm}}{12.5 \text{ cm}}$$

$$\boxed{M = 0.445}$$

4. EVALUATE These values and signs for the converging lens indicate a real, inverted, smaller image. This is expected because the object distance is longer than twice the focal length of the converging lens. The values and signs for the diverging lens indicate a virtual, upright, smaller image formed inside the focal point. This is the only kind of image diverging lenses form.

PRACTICE B

Lenses

1. An object is placed 20.0 cm in front of a converging lens of focal length 10.0 cm. Find the image distance and the magnification. Describe the image.
2. Sherlock Holmes examines a clue by holding his magnifying glass at arm's length and 10.0 cm away from an object. The magnifying glass has a focal length of 15.0 cm. Find the image distance and the magnification. Describe the image that he observes.
3. An object is placed 20.0 cm in front of a diverging lens of focal length 10.0 cm. Find the image distance and the magnification. Describe the image.
4. Fill in the missing values in the following table.

	f	p	q	M
		Converging lens		
a.	6.0 cm	?	−3.0 cm	?
b.	2.9 cm	?	7.0 cm	?
		Diverging lens		
c.	−6.0 cm	4.0 cm	?	?
d.	?	5.0 cm	?	0.50

PROBLEM GUIDE B

Use this guide to assign problems.
SE = Student Edition Textbook
PW = Problem Workbook
PB = Problem Bank on the One-Stop Planner (OSP)

Solving for:

q	**SE** Sample, 1–4; Ch. Rvw. 24, 26, 55* **PW** 3, 6–9, 16 **PB** 4–6
M	**SE** Sample, 1–4; Ch. Rvw. 24–26, 48 **PW** 1–2 **PB** 4–6
p	**SE** 4; Ch. Rvw. 43, 45, 47–48 **PW** Sample, 1–7, 16 **PB** 3–4, 7–10
h	**PW** Sample, 1–2, 4–5, 7–9, 14–15 **PB** Sample, 1–2
f	**SE** 4; Ch. Rvw. 46, 64–65 **PW** 10–15 **PB** Sample, 1–2

***Challenging Problem**
Consult the printed Solutions Manual or the OSP for detailed solutions.

ANSWERS

Practice B

1. 20.0 cm, $M = -1.00$; real, inverted image
2. −30.0 cm, $M = 3.00$; virtual, upright image
3. −6.67 cm, $M = 0.333$; virtual, upright image
4. a. $p = 2.0$ cm, $M = 1.5$
   b. $p = 5.0$ cm, $M = -1.4$
   c. $q = -2.4$ cm, $M = 0.60$
   d. $f = -5.0$ cm, $q = -2.5$ cm

Quick Lab

TEACHER'S NOTES

This activity is meant to let students experience how eyeglasses work to correct hyperopia and myopia.

This lab is more effective when using eyeglasses of 1.5 diopters or higher and when students make systematic observations of the same object at increasing distance or of objects of increasing size at a constant distance. The lines and numbers on a ruler make good objects for this experiment.

Demonstration

Microscope

Purpose Explain the function of lenses in a microscope.

Materials optical bench, light source with small aperture, two converging lenses (one with a very short *f* and one with a long *f*), screen

Procedure Place the lens with the short focal length in front of the light source. Position the screen to show that the image is real and inverted. Tell students that this is what the objective does in a microscope. Point out that a screen is needed in order to see the image. Ask if it is magnified (*barely*).

Place the second lens in back of the screen, and remove the screen. Tell students that this is the eyepiece. Ask students to try to locate the new image. (They may try to place the screen behind the second lens and move the second lens or the screen.) Have them look through the second lens toward the light source and move the second lens until they see the virtual image, upright and magnified.

Quick Lab

Prescription Glasses

MATERIALS LIST

- **several pairs of prescription eyeglasses**

Hold a pair of prescription glasses at various distances from your eye, and look at different objects through the lenses. Try this with different types of glasses, such as those for farsightedness and nearsightedness, and describe what effect the differences have on the image you see. If you have bifocals, how do the images produced by the top and bottom portions of the bifocal lens compare?

SCI*LINKS*.
www.scilinks.org
Topic: Abnormalities of the Eye
Code: HF60002

EYEGLASSES AND CONTACT LENSES

The transparent front of the eye, called the *cornea,* acts like a lens, directing light rays toward the light-sensitive *retina* in the back of the eye. Although most of the refraction of light occurs at the cornea, the eye also contains a small lens, called the *crystalline lens,* that refracts light as well.

When the eye attempts to produce a focused image of a nearby object but the image position is behind the retina, the abnormality is known as *hyperopia,* and the person is said to be *farsighted.* With this defect, distant objects are seen clearly, but near objects are blurred. Either the hyperopic eye is too short or the ciliary muscle that adjusts the shape of the lens cannot adjust enough to properly focus the image. **Table 5** shows how hyperopia can be corrected with a converging lens.

Another condition, known as *myopia,* or *nearsightedness,* occurs either when the eye is longer than normal or when the maximum focal length of the lens is insufficient to produce a clear image on the retina. In this case, light from a distant object is focused in front of the retina. The distinguishing feature of this imperfection is that distant objects are not seen clearly. Nearsightedness can be corrected with a diverging lens, as shown in **Table 5.**

A contact lens is simply a lens worn directly over the cornea of the eye. The lens floats on a thin layer of tears.

Table 5 Farsighted and Nearsighted

Farsighted

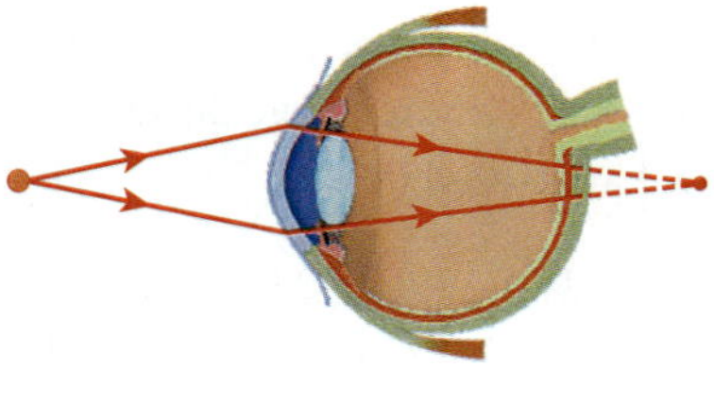

Hyperopia

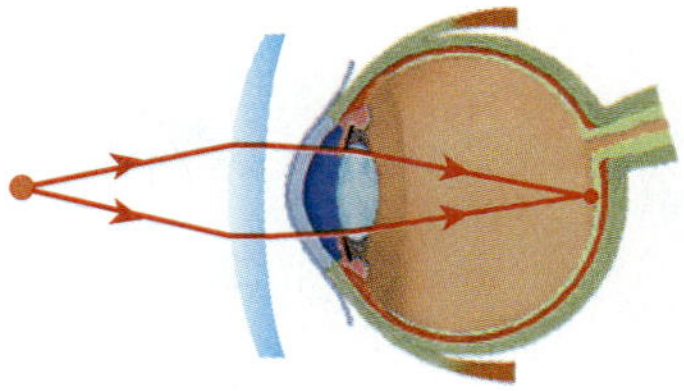

Corrected with a converging lens

Nearsighted

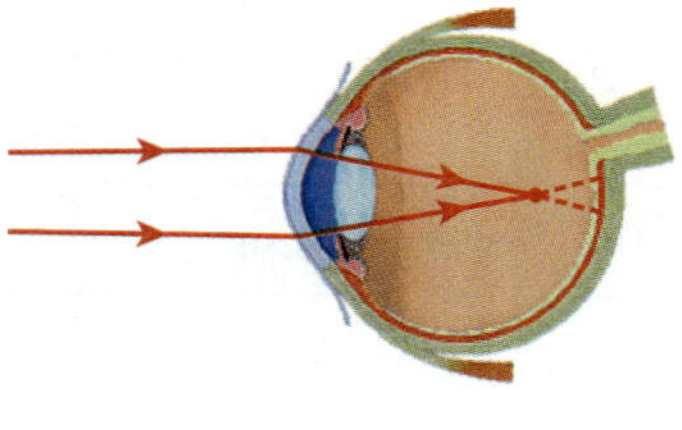

Myopia

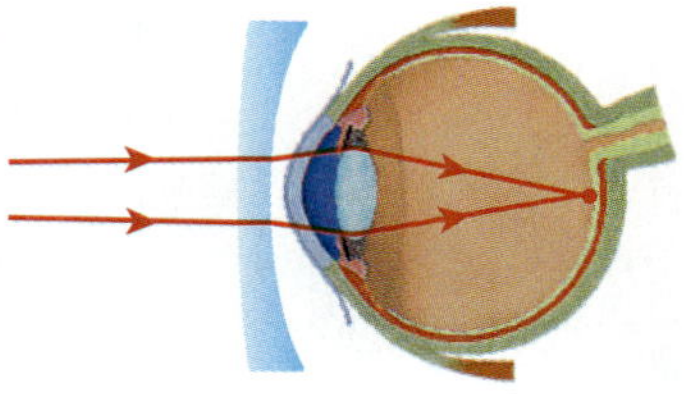

Corrected with a diverging lens

COMBINATION OF THIN LENSES

If two lenses are used to form an image, the system can be treated in the following manner. First, the image of the first lens is calculated as though the second lens were not present. The light then approaches the second lens as if it had come from the image formed by the first lens. Hence, *the image formed by the first lens is treated as the object for the second lens.* The image formed by the second lens is the final image of the system. The overall magnification of a system of lenses is the product of the magnifications of the separate lenses. If the image formed by the first lens is in back of the second lens, then the image is treated as a virtual object for the second lens (that is, p is negative). The same procedure can be extended to a system of three or more lenses.

Compound microscopes use two converging lenses

A simple magnifier, such as a magnifying glass, provides only limited assistance when inspecting the minute details of an object. Greater magnification can be achieved by combining two lenses in a device called a *compound microscope.* It consists of two lenses: an objective lens (near the object) with a focal length of less than 1 cm and an eyepiece with a focal length of a few centimeters. As shown in **Figure 8,** the object placed just outside the focal point of the objective lens forms a real, inverted, and enlarged image that is at or just inside the focal point of the eyepiece. The eyepiece, which serves as a simple magnifier, uses this enlarged image as its object and produces an even more enlarged virtual image. The image viewed through a microscope is upside-down with respect to the actual orientation of the specimen, as shown in **Figure 8.**

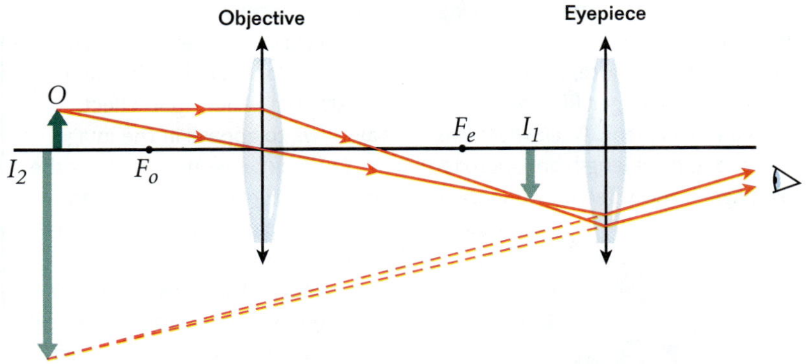

Figure 8
In a compound microscope, the real, inverted image produced by the objective lens is used as the object for the eyepiece lens.

The microscope has extended our vision into the previously unknown realm of incredibly small objects. A question that is often asked about microscopes is, "With extreme patience and care, would it be possible to construct a microscope that would enable us to see an atom?" As long as visible light is used to illuminate the object, the answer is no. In order to be seen, the object under a microscope must be at least as large as a wavelength of light. An atom is many times smaller than a wavelength of visible light, so its mysteries must be probed through other techniques.

Visual Strategy ADVANCED

Figure 8
Make sure that students transfer their understanding of the diagram to that of a real microscope.

Q Normally, a microscope is placed in a vertical position. Redraw the diagram so that it represents a microscope in a vertical position, and indicate where you would place the slide.

A *slide at O*

Key Models and Analogies — GENERAL

The optical functions of different parts of the eye may be compared to those of a camera. Ask: Which part of the eye corresponds to the film in a camera? (*retina*) Which components of the projector or camera might be compared to the eye's ability to accommodate? (*the ability to move the objective lens in a camera back and forth and to add lenses to it*)

Why it Matters

Cameras

More expensive cameras contain special lens elements and groups of lenses designed to minimize distortions and aberrations. Special lens elements, such as *achromatic doublets* and *apochromatic lenses* are designed to minimize chromatic aberration. Achromatic lenses are made with two different types of glass sandwiched together, whereas apochromatic lenses are made from specialty materials, often expensive, that have a similar index of refraction over the visible spectrum. Other lenses, such as *aspherical lenses,* are designed to reduce spherical aberration by careful design of the lens surface in nonspherical shapes.

Extension

Students can make a simple pinhole camera using a box and photographic film. Numerous Web sites and books contain instructions for making pinhole cameras. Have students research how to make a pinhole camera. They may build a pinhole camera and photograph subjects with different exposure times and light levels. Have students photograph both still and moving images.

Why it Matters

Cameras

This cross-sectional view of an SLR camera shows the many optical elements used to form an image on the film.

Cameras come in many types and sizes, from the small and simple "point-and-shoot" camera you might use to snap photos on a vacation to the large and complex video camera used to film a Hollywood motion picture. Most cameras have at least one lens, and more complex cameras may have 30 or more lenses and may even contain mirrors and prisms. However, the simplest camera, called a *pinhole camera,* consists of a closed, light-tight box with a small (about 0.5 mm) hole in it. A surprisingly good image can be made with a pinhole camera! The film is placed on the wall opposite the hole and must be exposed for quite a long time because not much light passes through the hole.

Making the hole a bit larger and adding a single, converging lens and a shutter, which opens and closes quickly to allow light to pass through the lens and expose the film, can make another simple camera called a *fixed-focus camera.* The film is located at the focal length of the lens, and a typical disposable camera is of this kind. This type of camera usually gives good images only for objects far from the camera. For close objects, the focus falls behind the film. Because the film location is fixed, the lens must be able to be moved away from the film and thus be "focused."

There are many types of camera lenses, and they are easily interchangeable on most *single-lens reflex* (SLR) cameras. A normal lens is one that provides about the same field of view as a human eye. Sometimes, however, a photographer wants to photograph distant objects with more detail or capture a larger object without taking multiple shots. A *wide-angle lens* has a very short focal length and can capture a larger field of view than a normal lens. A *telephoto lens* has a long focal length and increases magnification. Telephoto lenses have a narrow angle of view. *Zoom lenses* allow you to change the focal length without changing lenses. These camera lenses contain multiple lenses that can be moved relative to one another.

High-quality cameras contain quite a few lenses, both converging and diverging, to minimize the *distortions* and *aberrations* that are created by a single converging lens. The most prevalent aberration occurs because lenses bend light of different colors by different amounts, causing, in effect, rainbows to appear in the image.

You may be wondering how the optics change for digital cameras. The lenses and shutters are essentially the same as those used in film cameras. However, the film is replaced by a *charge-coupled device* (CCD) array, an array of tiny sensors that produce a current when hit by light from the subject being photographed. Lenses must still focus the light coming from the subject onto the CCD array, as they must on film.

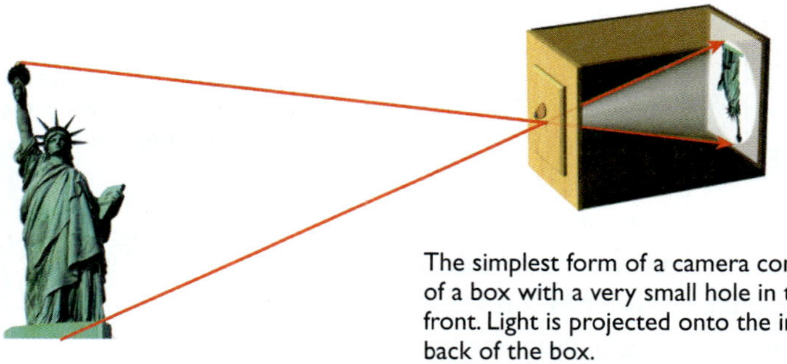

The simplest form of a camera consists of a box with a very small hole in the front. Light is projected onto the inside back of the box.

Refracting telescopes also use two converging lenses

As mentioned in the chapter on light and reflection, there are two types of telescopes, reflecting and refracting. In a refracting telescope, an image is formed at the eye in much the same manner as is done with a microscope. A small, inverted image is formed at the focal point of the objective lens, F_O, because the object is essentially at infinity. The eyepiece is positioned so that its focal point lies very close to the focal point of the objective lens, where the image is formed, as shown in **Figure 9.** Because the image is now just inside the focal point of the eyepiece, F_e, the eyepiece acts like a simple magnifier and allows the viewer to examine the object in detail.

extension

Integrating Astronomy
Visit go.hrw.com for the activity "The Refracting Telescope at Yerkes."

Keyword HF6REFX

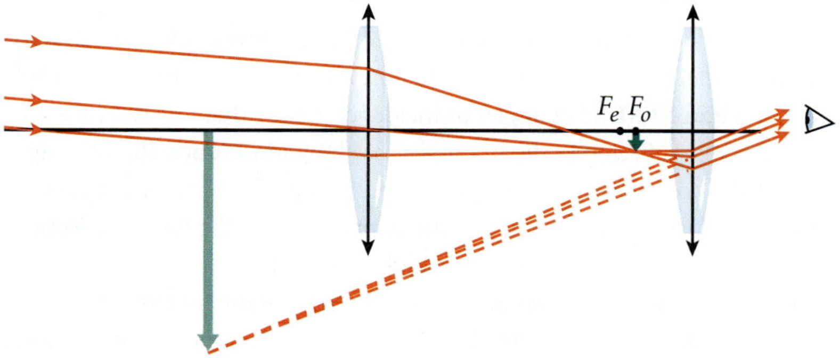

Figure 9
The image produced by the objective lens of a refracting telescope is a real, inverted image that is at its focal point. This inverted image, in turn, is the object from which the eyepiece creates a magnified, virtual image.

SECTION REVIEW

1. What type of image is produced by the cornea and the lens on the retina?
2. What type of image, virtual or real, is produced in the following cases?
   a. an object inside the focal point of a camera lens
   b. an object outside the focal point of a refracting telescope's objective lens
   c. an object outside the focal point of a camera's viewfinder
3. Find the image position for an object placed 3.0 cm outside the focal point of a converging lens with a 4.0 cm focal length.
4. What is the magnification of the object from item 3?
5. **Interpreting Graphics** Using a ray diagram, find the position and height of an image produced by a viewfinder in a camera with a focal length of 5.0 cm if the object is 1.0 cm tall and 10.0 cm in front of the lens. A camera viewfinder is a diverging lens.
6. **Critical Thinking** Compare the length of a refracting telescope with the sum of the focal lengths of its two lenses.

Teaching Tip — GENERAL

Viewing objects on land upside down is troublesome. Discuss ways in which this problem can be solved.

Terrestrial (or field) telescopes contain an extra converging lens to make the images of objects on land appear upright. Galileo, who invented the first refracting telescope, used a diverging lens as the eyepiece to solve this inversion problem.

SECTION REVIEW ANSWERS

1. real, inverted
2. **a.** virtual
   b. real
   c. virtual
3. 9.3 cm
4. −1.3
5. $q = -3.3$ cm, $h' = 0.33$ cm
6. The length of the telescope is slightly shorter than $f_O + f_e$.

Demonstration

Critical Angle — GENERAL

Purpose Demonstrate critical angle and total internal reflection.

Materials 90° prism, laser, chalk dust

CAUTION *Avoid directing the laser beam toward the students.*

Procedure Adjust the position of the laser so that the beam is in position 1, as shown below.

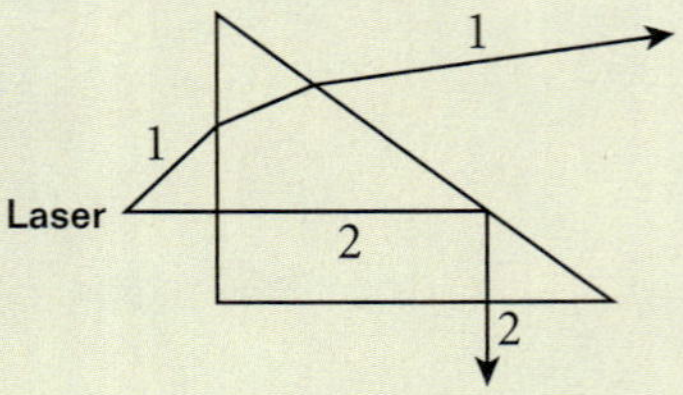

Gently tap the erasers together above the laser beam to make the incoming and the exiting laser beam visible. Let students observe the refracted beam inside and outside the prism.

Slowly rotate the beam from position 1 to position 2. When the exiting beam no longer passes through the hypotenuse of the prism, add chalk dust near the other side of the right angle (where beam 2 would exit). Have students observe the beam inside the prism and observe that it is totally internally reflected.

SECTION 3

Optical Phenomena

SECTION OBJECTIVES

- **Predict whether light will be refracted or undergo total internal reflection.**
- **Recognize atmospheric conditions that cause refraction.**
- **Explain dispersion and phenomena such as rainbows in terms of the relationship between the index of refraction and the wavelength.**

total internal reflection

the complete reflection that takes place within a substance when the angle of incidence of light striking the surface boundary is greater than the critical angle

critical angle

the angle of incidence at which the refracted light makes an angle of 90° with the normal

TOTAL INTERNAL REFLECTION

An interesting effect called **total internal reflection** can occur when light moves along a path from a medium with a *higher* index of refraction to one with a *lower* index of refraction. Consider light rays traveling from water into air, as shown in **Figure 10(a).** Four possible directions of the rays are shown in the figure.

At some particular angle of incidence, called the **critical angle,** the refracted ray moves parallel to the boundary, making the angle of refraction equal to 90°, as shown in **Figure 10(b).** For angles of incidence greater than the critical angle, the ray is entirely reflected at the boundary, as shown in **Figure 10.** This ray is reflected at the boundary as though it had struck a perfectly reflecting surface. Its path and the path of all rays like it can be predicted by the law of reflection; that is, the angle of incidence equals the angle of reflection.

In optical equipment, prisms are arranged so that light entering the prism is totally internally reflected off the back surface of the prism. Prisms are used in place of silvered or aluminized mirrors because they reflect light more efficiently and are more scratch resistant.

Snell's law can be used to find the critical angle. As mentioned above, when the angle of incidence, θ_i, equals the critical angle, θ_c, then the angle of refraction, θ_r, equals 90°. Substituting these values into Snell's law gives the following relation.

$$n_i \sin \theta_c = n_r \sin 90°$$

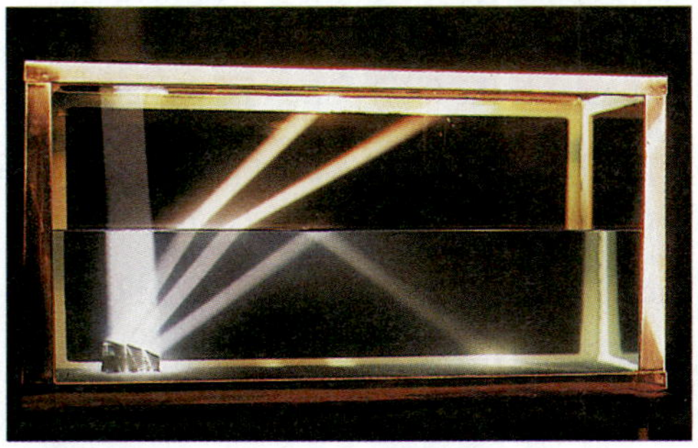

(a)

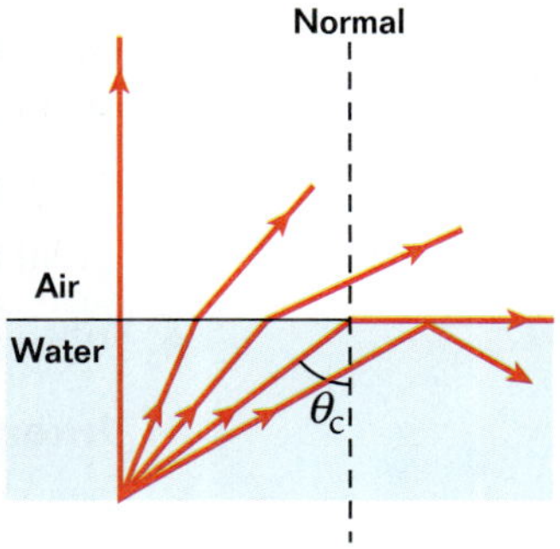

(b)

Figure 10
(a) This photo demonstrates several different paths of light radiated from the bottom of an aquarium. **(b)** At the critical angle, θ_c, a light ray will travel parallel to the boundary. Any rays with an angle of incidence greater than θ_c will be totally internally reflected at the boundary.

Because the sine of 90° equals 1, the following relationship results.

CRITICAL ANGLE

$$\sin \theta_c = \frac{n_r}{n_i} \quad \text{for } n_i > n_r$$

$$\text{sine (critical angle)} = \frac{\textbf{index of refraction of second medium}}{\textbf{index of refraction of first medium}}$$

but only if index of refraction of first medium > index of refraction of second medium

Note that this equation can be used only when n_i is greater than n_r. In other words, *total internal reflection occurs only when light moves along a path from a medium of higher index of refraction to a medium of lower index of refraction.* If n_i were less than n_r, this equation would give $\sin \theta_c > 1$, which is an impossible result because by definition the sine of an angle can never be greater than 1.

When the second substance is air, the critical angle is small for substances with large indices of refraction. Diamonds, which have an index of refraction of 2.419, have a critical angle of 24.4°. By comparison, the critical angle for crown glass, a very clear optical glass, where $n = 1.52$, is 41.0°. Because diamonds have such a small critical angle, most of the light that enters a cut diamond is totally internally reflected. The reflected light eventually exits the diamond from the most visible faces of the diamond. Jewelers cut diamonds so that the maximum light entering the upper surface is reflected back to these faces.

Quick Lab

Periscope

MATERIALS LIST

- **two 90° prisms**

Align the two prisms side by side as shown below.

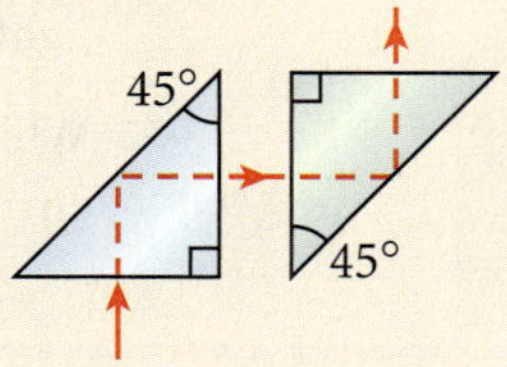

Note that this configuration can be used like a periscope to see an object above your line of sight if the configuration is oriented vertically and to see around a corner if it is oriented horizontally. How would you arrange the prisms to see behind you? Draw your design on paper and test it.

SAMPLE PROBLEM C

Critical Angle

PROBLEM

Find the critical angle for a water-air boundary if the index of refraction of water is 1.333.

SOLUTION

Given: $n_i = 1.333 \quad n_r = 1.000$

Unknown: $\theta_c = ?$

Use the equation for critical angle on this page.

$$\sin \theta_c = \frac{n_r}{n_i}$$

$$\theta_c = \sin^{-1}\left(\frac{n_r}{n_i}\right) = \sin^{-1}\left(\frac{1.00}{1.333}\right)$$

$$\boxed{\theta_c = 48.6°}$$

TIP *Remember that the critical angle equation is valid only if the light is moving from a higher to a lower index of refraction.*

Quick Lab

TEACHER'S NOTES

This activity is meant to demonstrate an application of total internal reflection. Point out that the critical angle for glass is about 41° and that the angle of incidence from glass to air in this arrangement is 45°.

This lab is more effective if students look through each prism separately to experience the path followed by the light rays and realize that the prisms could be replaced by mirrors.

Classroom Practice

Critical Angle

Calculate the critical angle of light traveling from the following substances into air.

a. quartz ($n = 1.46$)

b. acrylic resin ($n = 1.51$)

c. flint glass ($n = 1.66$)

Answers

a. 43.2° **b.** 41.5° **c.** 37.0°

PROBLEM GUIDE C

Use this guide to assign problems.
SE = Student Edition Textbook
PW = Problem Workbook
PB = Problem Bank on the One-Stop Planner (OSP)

Solving for:

θ_c	**SE** Sample, 1–4; Ch. Rvw. 36–37, 38*, 56* **PW** 5 **PB** 7–10
n_r	**SE** Ch. Rvw. 58–59 **PW** Sample, 1–2 **PB** 4–6
n_i	**PW** 3 **PB** Sample, 1–3

***Challenging Problem**
Consult the printed Solutions Manual or the OSP for detailed solutions.

ANSWERS

Practice C

1. 42.8°
2. 64.82°
3. 49.8°
4. diamond (24.4°); cubic zirconia (27.0°)

PRACTICE C

Critical Angle

1. Glycerine is used to make soap and other personal care products. Find the critical angle for light traveling from glycerine ($n = 1.473$) into air.
2. Calculate the critical angle for light traveling from glycerine ($n = 1.473$) into water ($n = 1.333$).
3. Ice has a lower index of refraction than water. Find the critical angle for light traveling from ice ($n = 1.309$) into air.
4. Which has a smaller critical angle in air, diamond ($n = 2.419$) or cubic zirconia ($n = 2.20$)? Show your work.

Why it Matters

Fiber Optics

Another interesting application of total internal reflection is the use of glass or transparent plastic rods, like the ones shown in the photograph, to transfer light from one place to another. As indicated in the illustration below, light is confined to traveling within the rods, even around gentle curves, as a result of successive internal reflections. Such a *light pipe* can be flexible if thin fibers rather than thick rods are used. If a bundle of parallel fibers is used to construct an optical transmission line, images can be transferred from one point to another.

This technique is used in a technology known as *fiber optics.* Very little light intensity is lost in these fibers as a result of reflections on the sides. Any loss of intensity is due essentially to reflections from the two ends and absorption by the fiber material. Fiber-optic devices are particularly useful for viewing images produced at inaccessible locations. For example, a fiber-optic cable can be threaded through the esophagus and into the stomach to look for ulcers.

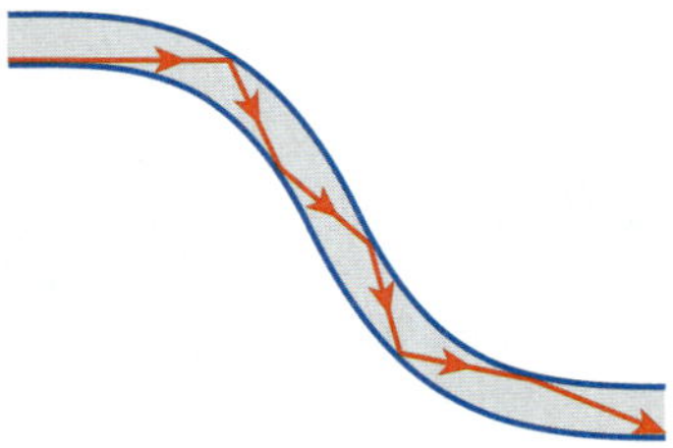

Light is guided along a fiber by multiple internal reflections.

Fiber optic cables are widely used in telecommunications because the fibers can carry much higher volumes of telephone calls and computer signals than can electrical wires.

ATMOSPHERIC REFRACTION

We see an example of refraction every day: the sun can be seen even after it has passed below the horizon. Rays of light from the sun strike Earth's atmosphere and are bent because the atmosphere has an index of refraction different from that of the near-vacuum of space. The bending in this situation is gradual and continuous because the light moves through layers of air that have a continuously changing index of refraction. Our eyes follow them back along the direction from which they appear to have come.

SCILINKS.
www.scilinks.org
Topic: Fiber Optics
Code: HF60572

Refracted light produces mirages

The *mirage* is another phenomenon of nature produced by refraction in the atmosphere. A mirage can be observed when the ground is so hot that the air directly above it is warmer than the air at higher elevations.

These layers of air at different heights above Earth have different densities and different refractive indices. The effect this can have is pictured in **Figure 11.** In this situation, the observer sees a tree in two different ways. One group of light rays reaches the observer by the straight-line path *A*, and the eye traces these rays back to see the tree in the normal fashion. A second group of rays travels along the curved path *B*. These rays are directed toward the ground and are then bent as a result of refraction. Consequently, the observer also sees an inverted image of the tree by tracing these rays back to the point at which they appear to have originated. Because both an upright image and an inverted image are seen when the image of a tree is observed in a reflecting pool of water, the observer subconsciously calls upon this past experience and concludes that a pool of water must be in front of the tree.

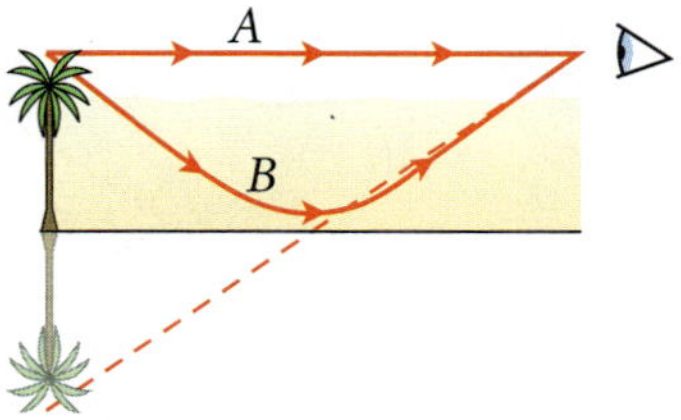

Figure 11
A mirage is produced by the bending of light rays in the atmosphere when there are large temperature differences between the ground and the air.

DISPERSION

An important property of the index of refraction is that its value in anything but a vacuum depends on the wavelength of light. Because the index of refraction is a function of wavelength, Snell's law indicates that incoming light of different wavelengths is bent at different angles as it moves into a refracting material. This phenomenon is called **dispersion.** As mentioned in Section 1, the index of refraction decreases with increasing wavelength. For instance, blue light ($\lambda \approx 470$ nm) bends more than red light ($\lambda \approx 650$ nm) when passing into a refracting material.

dispersion

the process of separating polychromatic light into its component wavelengths

White light passed through a prism produces a visible spectrum

To understand how dispersion can affect light, consider what happens when light strikes a prism, as in **Figure 12.** Because of dispersion, the blue component of the incoming ray is bent more than the red component, and the rays that emerge from the second face of the prism fan out in a series of colors known as a *visible spectrum.* These colors, in order of decreasing wavelength, are red, orange, yellow, green, blue, and violet.

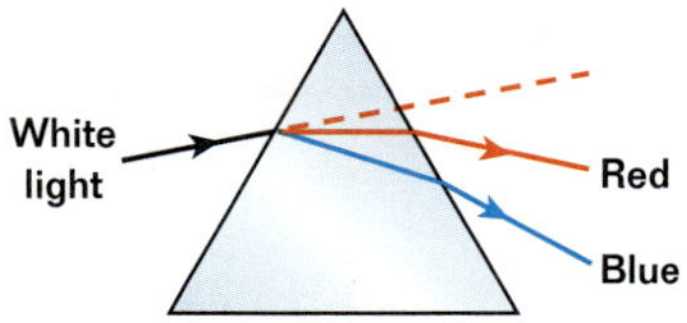

Figure 12
When white light enters a prism, the blue light is bent more than the red, and the prism disperses the white light into its various spectral components.

Demonstration

Fiber Optics—Bending Light

Purpose Demonstrate total internal reflection in fiber optics.

Materials laser, clear plastic bottle with a 3 mm diameter hole in its side near the bottom, piece of optical fiber, white sheet of paper

CAUTION *Avoid directing the laser beam toward the students.*

Procedure Tape over the hole in the bottle; fill the bottle with water. Have one student hold the bottle above a sink while you adjust the position of the laser so it shines through the bottle toward the hole. Have another student pull the tape from the hole and watch the light flowing out of the bottle with the water.

Adjust the laser beam to shine directly into one end of the optical fiber. Place paper in front of the other end. Bend the fiber in loops, and show that the light follows the curve inside.

Demonstration

Dispersion

Purpose Demonstrate the color spectrum formed by refraction.

Materials slide projector, unwanted slide, opaque tape, triangular prism

Procedure Remind students that the index of refraction of each material varies with wavelength. Construct a small slit on the slide with two pieces of opaque tape placed close together, and place the slide in the slide projector. Shine the resulting beam on the triangular prism. Adjust the projector and prism so that the spectrum falls on a screen or a wall. Have students observe that the shorter wavelengths bend more.

Demonstration

Rainbow

Purpose Create a visual, rainbow-like display.

Materials overhead projector; large, clear plastic cup with steeply sloped sides filled with water

Procedure Place the plastic cup filled with water on the overhead projector. Upward light rays from the projector are at an angle of incidence with the steep sides; refraction separates the colors, causing them to form a rainbow on the ceiling. Point out that this demonstration does not model how a true rainbow is formed because the cup is not spherical.

Rainbows are created by dispersion of light in water droplets

The dispersion of light into a spectrum is demonstrated most vividly in nature by a rainbow, often seen by an observer positioned between the sun and a rain shower. When a ray of sunlight strikes a drop of water in the atmosphere, it is first refracted at the front surface of the drop, with the violet light refracting the most and the red light the least. Then, at the back surface of the drop, the light is reflected and returns to the front surface, where it again undergoes refraction as it moves from water into air. The rays leave the drop so that the angle between the incident white light and the returning violet ray is 40° and the angle between the white light and the returning red ray is 42°, as shown in **Figure 13(b).**

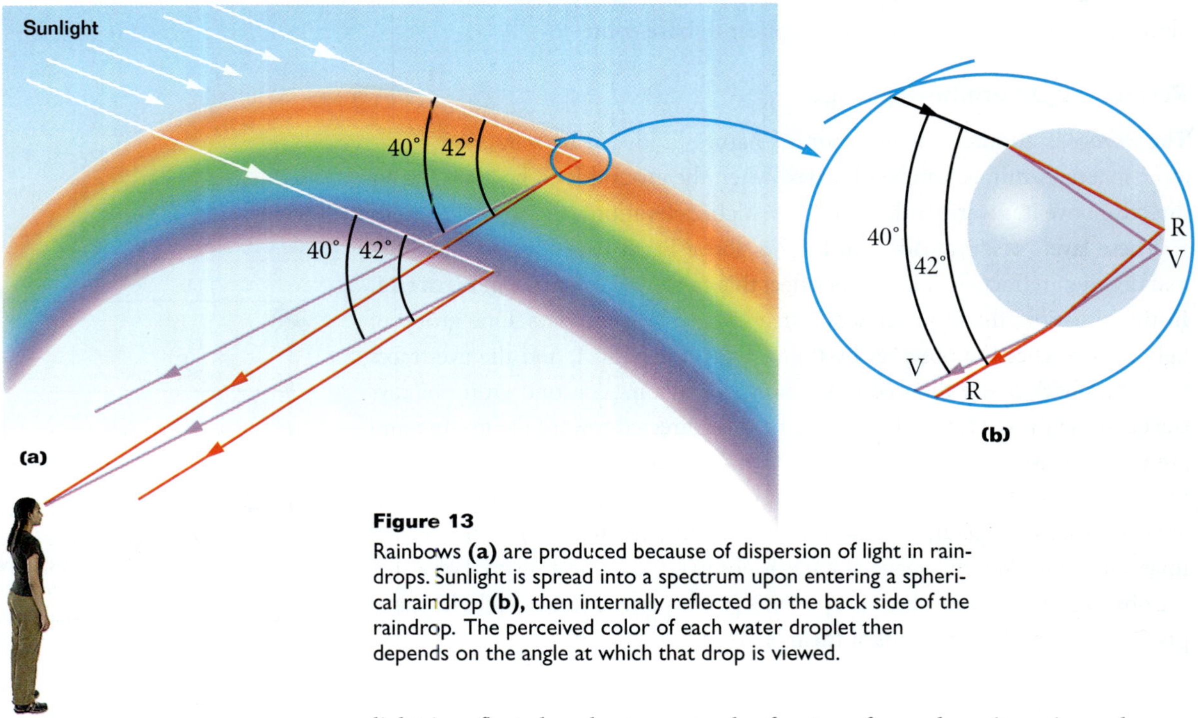

Figure 13
Rainbows **(a)** are produced because of dispersion of light in raindrops. Sunlight is spread into a spectrum upon entering a spherical raindrop **(b),** then internally reflected on the back side of the raindrop. The perceived color of each water droplet then depends on the angle at which that drop is viewed.

Now, consider **Figure 13(a).** When an observer views a raindrop high in the sky, the red light reaches the observer, but the violet light, like the other spectral colors, passes over the observer because it deviates from the path of the white light more than the red light does. Hence, the observer sees this drop as being red. Similarly, a drop lower in the sky would direct violet light toward the observer and appear to be violet. (The red light from this drop would strike the ground and not be seen.) The remaining colors of the spectrum would reach the observer from raindrops lying between these two extreme positions.

Note that rainbows are most commonly seen above the horizon, where the ends of the rainbow disappear into the ground. However, if an observer is at an elevated vantage point, such as on an airplane or at the rim of a canyon, a complete circular rainbow can be seen.

LENS ABERRATIONS

One of the basic problems of lenses and lens systems is the imperfect quality of the images. The simple theory of mirrors and lenses assumes that rays make small angles with the principal axis and that all rays reaching the lens or mirror from a point source are focused at a single point, producing a sharp image. Clearly, this is not always true in the real world. Where the approximations used in this theory do not hold, imperfect images are formed.

As with spherical mirrors, *spherical aberration* occurs for lenses also. It results from the fact that the focal points of light rays far from the principal axis of a spherical lens are different from the focal points of rays with the same wavelength passing near the axis. Rays near the middle of the lens are focused farther from the lens than rays at the edges.

Another type of aberration, called **chromatic aberration,** arises from the wavelength dependence of refraction. Because the index of refraction of a material varies with wavelength, different wavelengths of light are focused at different focal points by a lens. For example, when white light passes through a lens, violet light is refracted more than red light, as shown in **Figure 14;** thus, the focal length for red light is greater than that for violet light. Other colors' wavelengths have intermediate focal points. Because a diverging lens has the opposite shape, the chromatic aberration for a diverging lens is opposite that for a converging lens. Chromatic aberration can be greatly reduced by the use of a combination of converging and diverging lenses made from two different types of glass.

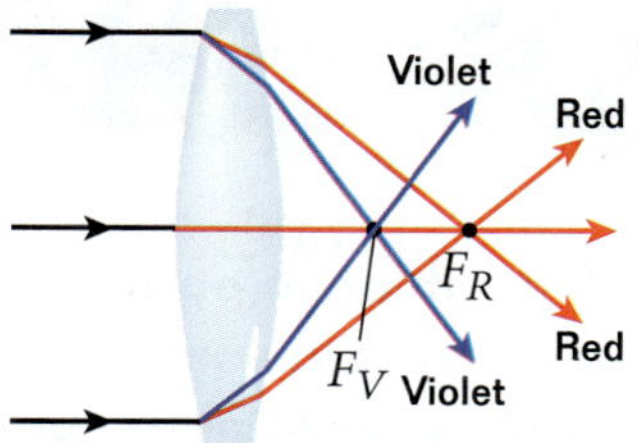

Figure 14
Because of dispersion, white light passing through a converging lens is focused at different focal points for each wavelength of light. (The angles in this figure are exaggerated for clarity.)

chromatic aberration

the focusing of different colors of light at different distances behind a lens

SECTION REVIEW

1. Find the critical angle for light traveling from water ($n = 1.333$) into ice ($n = 1.309$).
2. Which of the following describe places where a mirage is likely to appear?
   a. above a warm lake on a warm day
   b. above an asphalt road on a hot day
   c. above a ski slope on a cold day
   d. above the sand on a beach on a hot day
   e. above a black car on a sunny day
3. When white light passes through a prism, which will be bent more, the red or green light?
4. **Critical Thinking** After a storm, a man walks out onto his porch. Looking to the east, he sees a rainbow that has formed above his neighbor's house. What time of day is it, morning or evening?

Demonstration

Chromatic Aberration

Purpose Demonstrate that the focal point of a lens depends on the wavelength.

Materials lens with a small focal length, light source or projector, screen

Procedure Place the lens in front of the light source and adjust the screen so that the beam appears to focus. Move the screen forward and back slightly to show that each color forms a sharp image at a different distance.

SECTION REVIEW ANSWERS

1. 79.11°
2. b, d, e
3. green
4. evening, because the sun must be behind him to see the rainbow

Optometrist

Dr. Dewey Handy practices optometry in Jackson, Mississippi, and is a board member of the Mississippi Optometric Association. He also has served as an assistant professor of physics at Jackson State University, and currently he is an adjunct faculty member at the Southern College of Optometry. In 2003, the National Optometric Association named Handy Optometrist of the Year.

An optometrist named David White catalyzed Handy's interest in optometry in college. Handy realized that optometry offered to him the perfect combination of physics and a health career, allowing him to directly serve people.

At present, Handy serves in a similar mentoring capacity, providing opportunities for interns and externs. According to Handy, the benefits are mutual: "They get practical experience, and I get to learn newer techniques."

PHYSICS CAREERS

Optometrist

The job of an optometrist is to correct imperfect vision using optical devices such as eyeglasses or contact lenses. Optometrists also treat diseases of the eye such as glaucoma. To learn more about optometry as a career, read the interview with Dewey Handy, O.D.

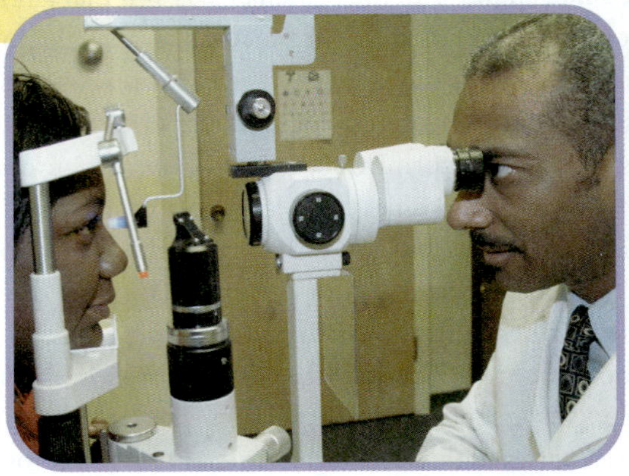

Dr. Dewey Handy uses optical devices to test the vision of a patient.

How did you decide to become an optometrist?

For a while, I didn't know what career I was going to choose. In high school, I had a great love for geometry and an interest in science and anatomy. In college, I was looking for a challenge, so I ended up majoring in physics—almost by accident.

In college, I decided to apply my abilities in science to directly help people. I wasn't excited about dentistry or general medicine, but I was looking for something in a health career that would allow me to use physics.

What education is required to become an optometrist?

I have a bachelor of science in physics, and I attended optometry school for four years.

What sort of work does an optometrist do?

After taking a complete eye and medical history, the doctor may use prisms and/or lenses to determine the proper prescription for the patient. Then, a series of neurological, health, and binocular vision tests are done. After the history and data have been collected, a diagnosis and treatment plan are developed. This treatment may include glasses, contact lenses, low-vision aids, vision training, or medication for treatment of eye disease.

What do you enjoy most about your job?

I like the problem-solving nature of the work, putting the data together to come up with solutions. We read the problem, compile data, develop a formula, and solve the problem—just as in physics, but with people instead of abstract problems. I also like helping people.

What advice do you have for students who are interested in optometry?

You definitely need to have a good background in basic science: chemistry, biology, and physics. Even if you don't major in science, you need to have a good grasp of it by the time you get to optometry school.

Being well rounded will help you get into optometry school—and get out, too. You have to be comfortable doing the science, but you also have to be comfortable dealing with people.

CHAPTER 14

Highlights

KEY IDEAS

Section 1 Refraction

- According to Snell's law, as a light ray travels from one medium into another medium where its speed is different, the light ray will change its direction unless it travels along the normal.
- When light passes from a medium with a smaller index of refraction to one with a larger index of refraction, the ray bends toward the normal. For the opposite situation, the ray bends away from the normal.

Section 2 Thin Lenses

- The image produced by a converging lens is real and inverted when the object is outside the focal point and virtual and upright when the object is inside the focal point. Diverging lenses always produce upright, virtual images.
- The location of an image created by a lens can be found using either a ray diagram or the thin-lens equation.

Section 3 Optical Phenomena

- Total internal reflection can occur when light attempts to move from a material with a higher index of refraction to one with a lower index of refraction. If the angle of incidence of a ray is greater than the critical angle, the ray is totally reflected at the boundary.
- Mirages and the visibility of the sun after it has physically set are natural phenomena that can be attributed to refraction of light in Earth's atmosphere.

Variable Symbols

	Quantities		Units
θ_i	angle of incidence	°	degrees
θ_r	angle of refraction	°	degrees
n	index of refraction		
p	distance from object to lens	m	meters
q	distance from image to lens	m	meters
h'	image height	m	meters
h	object height	m	meters
θ_c	critical angle	°	degrees

KEY TERMS

refraction (p. 488)

index of refraction (p. 490)

lens (p. 494)

total internal reflection (p. 506)

critical angle (p. 506)

dispersion (p. 509)

chromatic aberration (p. 511)

PROBLEM SOLVING

See **Appendix D: Equations** for a summary of the equations introduced in this chapter. If you need more problem-solving practice, see **Appendix I: Additional Problems.**

CHAPTER 14

Highlights

Teaching Tip

Explaining concepts in written form helps to solidify students' understanding of difficult concepts and helps to enforce good communication skills. Have students write an essay summarizing the differences between images formed by converging lenses and those formed by diverging lenses. Essays should include a thorough explanation of the thin-lens equation, sign conventions, and ray diagrams for each case.

Review

ANSWERS

1. no, not when $n_i > n_r$
2. Yes, its wavelength gets shorter; Yes, its velocity gets slower; no
3. $n = \frac{c}{v}$, where c is the speed of light in a vacuum
4. Light rays from the bottom bend away from the normal so that the image is closer to the observer.
5. $\theta_i \neq 0$, $n_i \neq n_r$, both media must be transparent
6. X
7. The image of the oar underwater is closer to the observer.
8. behind—when you are outside the water, the image of the coin appears farther away than the actual coin
9. n of liquid helium is approximately equal to n of air, so light is bent very little at the interface
10. 30.3°
11. 26°
12. 25.5°
13. 30.0°, 19.5°, 19.5°, 30.0°
14. $\theta_1 = 30.4°$, $\theta_2 = 22.3°$
15. converging
16. Rays are refracted parallel to one another.

CHAPTER 14 Review

REFRACTION AND SNELL'S LAW

Review Questions

1. Does a light ray traveling from one medium into another always bend toward the normal?

2. As light travels from a vacuum ($n = 1$) to a medium such as glass ($n > 1$), does its wavelength change? Does its speed change? Does its frequency change?

3. What is the relationship between the speed of light and the index of refraction of a transparent substance?

4. Why does a clear stream always appear to be shallower than it actually is?

5. What are the three conditions that must be met for refraction to occur?

Conceptual Questions

6. Two colors of light (X and Y) are sent through a glass prism, and X is bent more than Y. Which color travels more slowly in the prism?

7. Why does an oar appear to be bent when part of it is in the water?

8. A friend throws a coin into a pool. You close your eyes and dive toward the spot where you saw it from the edge of the pool. When you reach the bottom, will the coin be in front of you or behind you?

9. The level of water ($n = 1.33$) in a clear glass container is easily observed with the naked eye. The level of liquid helium ($n = 1.03$) in a clear glass container is extremely difficult to see with the naked eye. Explain why.

Practice Problems

For problems 10–14, see Sample Problem A.

10. Light passes from air into water at an angle of incidence of 42.3°. Determine the angle of refraction in the water.

11. A ray of light enters the top of a glass of water at an angle of 36° with the vertical. What is the angle between the refracted ray and the vertical?

12. A narrow ray of yellow light from glowing sodium ($\lambda_0 = 589$ nm) traveling in air strikes a smooth surface of water at an angle of $\theta_i = 35.0°$. Determine the angle of refraction, θ_r.

13. A ray of light traveling in air strikes a flat 2.00 cm thick block of glass ($n = 1.50$) at an angle of 30.0° with the normal. Trace the light ray through the glass, and find the angles of incidence and refraction at each surface.

14. The light ray shown in the figure below makes an angle of 20.0° with the normal line at the boundary of linseed oil and water. Determine the angles θ_1 and θ_2. Note that $n = 1.48$ for linseed oil.

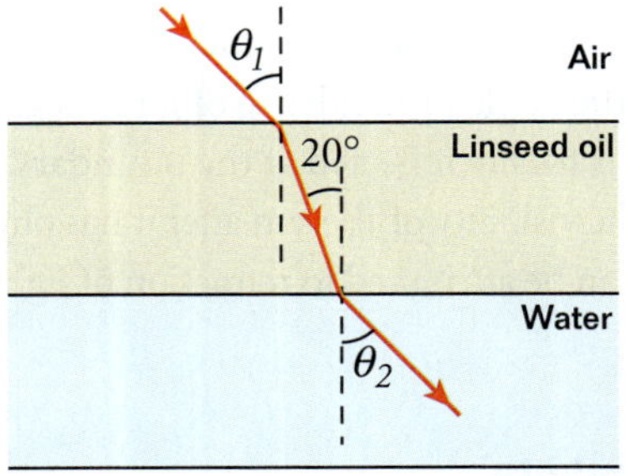

RAY DIAGRAMS AND THIN LENSES

Review Questions

15. Which type of lens can focus the sun's rays?

16. Why is no image formed when an object is at the focal point of a converging lens?

17. Consider the image formed by a thin converging lens. Under what conditions will the image be
    a. inverted?
    b. upright?
    c. real?
    d. virtual?
    e. larger than the object?
    f. smaller than the object?

18. Repeat a–f of item 17 for a thin diverging lens.

19. Explain this statement: The focal point of a converging lens is the location of an image of a point object at infinity. Based on this statement, can you think of a quick method for determining the focal length of a positive lens?

Conceptual Questions

20. If a glass converging lens is submerged in water, will its focal length be longer or shorter than when the lens is in air?

21. In order to get an upright image, slides must be placed upside down in a slide projector. What type of lens must the slide projector have? Is the slide inside or outside the focal point of the lens?

22. If there are two converging lenses in a compound microscope, why is the image still inverted?

23. In a Jules Verne novel, a piece of ice is shaped into the form of a magnifying lens to focus sunlight and thereby start a fire. Is this possible?

Practice Problems

For problems 24–26, see Sample Problem B.

24. An object is placed in front of a diverging lens with a focal length of 20.0 cm. For each object distance, find the image distance and the magnification. Describe each image.
    a. 40.0 cm
    b. 20.0 cm
    c. 10.0 cm

25. A person looks at a gem using a converging lens with a focal length of 12.5 cm. The lens forms a virtual image 30.0 cm from the lens. Determine the magnification. Is the image upright or inverted?

26. An object is placed in front of a converging lens with a focal length of 20.0 cm. For each object distance, find the image distance and the magnification. Describe each image.
    a. 40.0 cm
    b. 10.0 cm

TOTAL INTERNAL REFLECTION, ATMOSPHERIC REFRACTION, AND ABERRATIONS

Review Questions

27. Is it possible to have total internal reflection for light incident from air on water? Explain.

28. What are the conditions necessary for the occurrence of a mirage?

29. On a hot day, what is it that we are seeing when we observe a “water on the road” mirage?

30. Why does the arc of a rainbow appear with red colors on top and violet colors on the bottom?

31. What type of aberration is involved in each of the following situations?
    a. The edges of the image appear reddish.
    b. The central portion of the image cannot be clearly focused.
    c. The outer portion of the image cannot be clearly focused.
    d. The central portion of the image is enlarged relative to the outer portions.

Conceptual Questions

32. A laser beam passing through a nonhomogeneous sugar solution follows a curved path. Explain.

33. On a warm day, the image of a boat floating on cold water appears above the boat. Explain.

34. Explain why a mirror cannot give rise to chromatic aberration.

35. Why does a diamond show flashes of color when observed under ordinary white light?

14 REVIEW

17. a. object outside F
    b. object inside F
    c. object outside F
    d. object inside F
    e. object inside $2F$
    f. object outside $2F$
18. a. never b. always c. never
    d. always e. never f. always
19. Light from a point image at ∞ will enter the lens parallel to the principal axis and converge at F; Focus sunlight on the ground and measure f.
20. longer
21. converging lens; outside
22. The image produced by the objective lens is inside F_e. Therefore, the eyepiece produces a virtual image of a real, inverted image.
23. yes, because $n_{ice} > n_{air}$
24. a. −13.3 cm, $M = 0.332$; virtual, upright
    b. −10.0 cm, $M = 0.500$; virtual, upright
    c. −6.67 cm, $M = 0.667$; virtual, upright
25. 3.40; upright
26. a. 40.0 cm, $M = -1.00$; real, inverted
    b. −20.0 cm, $M = 2.00$; virtual, upright
27. no; $n_{air} < n_{water}$
28. The air next to the ground must be hotter than the air above it.
29. Light from the blue sky is refracted upward as it passes close to the ground.
30. Violet light is deviated more than red light as it passes through the drops of water.
31. a. chromatic
    b. spherical
    c. spherical
    d. spherical

32. As ρ changes, the speed of the light through it changes. Thus, the light is continually refracted as n changes.

33. Rays initially moving upward are bent because T increases with height.

34. θ' does not depend on the wavelength of the light.

35. Light entering the diamond is dispersed. Each color is totally internally reflected until $\theta_i < \theta_c$.

36. 42.8°

37. **a.** 31.3°
b. 44.2°
c. 49.8°

38. It will be totally internally reflected because θ_i (45°) > θ_c(41.1°).

39. 1.31

40. 51.9°

41. 1.62; carbon disulfide

42. **a.** 36.7°
b. 53.4°

43. 7.50 cm

44. 8.55 cm

45. **a.** 6.00 cm
b. A diverging lens cannot form an image larger than the object.

46. −80.0 cm

47. **a.** 3.01 cm
b. 2.05 cm

48. 1.20×10^2 cm; 0.250

49. blue: 47.8°, red: 48.2°

Practice Problems

For problems 36–38, see Sample Problem C.

36. Calculate the critical angle for light going from glycerine into air.

37. Assuming that $\lambda = 589$ nm, calculate the critical angles for the following materials when they are surrounded by air:

a. zircon
b. fluorite
c. ice

38. Light traveling in air enters the flat side of a prism made of crown glass (n = 1.52), as shown at right. Will the light pass through the other side of the prism or will it be totally internally reflected? Be sure to show your work.

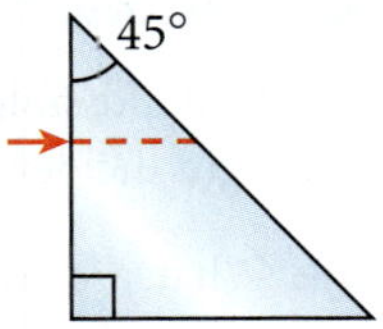

MIXED REVIEW

39. The angle of incidence and the angle of refraction for light going from air into a material with a higher index of refraction are 63.5° and 42.9°, respectively. What is the index of refraction of this material?

40. A person shines a light at a friend who is swimming underwater. If the ray in the water makes an angle of 36.2° with the normal, what is the angle of incidence?

41. What is the index of refraction of a material in which the speed of light is 1.85×10^8 m/s? Look at the indices of refraction in **Table 1** to identify this material.

42. Light moves from flint glass into water at an angle of incidence of 28.7°.

a. What is the angle of refraction?
b. At what angle would the light have to be incident to give an angle of refraction of 90.0°?

43. A magnifying glass has a converging lens of focal length 15.0 cm. At what distance from a nickel should you hold this lens to get an image with a magnification of +2.00?

44. The image of the United States postage stamps in the figure above is 1.50 times the size of the actual stamps in front of the lens. Determine the focal length of the lens if the distance from the lens to the stamps is 2.84 cm.

45. Where must an object be placed to have a magnification of 2.00 in each of the following cases? Show your work.

a. a converging lens of focal length 12.0 cm
b. a diverging lens of focal length 12.0 cm

46. A diverging lens is used to form a virtual image of an object. The object is 80.0 cm in front of the lens, and the image is 40.0 cm in front of the lens. Determine the focal length of the lens.

47. A microscope slide is placed in front of a converging lens with a focal length of 2.44 cm. The lens forms an image of the slide 12.9 cm from the slide.

a. How far is the lens from the slide if the image is real?
b. How far is the lens from the slide if the image is virtual?

48. Where must an object be placed to form an image 30.0 cm from a diverging lens with a focal length of 40.0 cm? Determine the magnification of the image.

49. The index of refraction for red light in water is 1.331, and that for blue light is 1.340. If a ray of white light traveling in air enters the water at an angle of incidence of 83.0°, what are the angles of refraction for the red and blue components of the light?

50. A ray of light traveling in air strikes the surface of mineral oil at an angle of 23.1° with the normal to the surface. If the light travels at 2.17×10^8 m/s through the oil, what is the angle of refraction? (Hint: Remember the definition of the index of refraction.)

51. A ray of light traveling in air strikes the surface of a liquid. If the angle of incidence is 30.0° and the angle of refraction is 22.0°, find the critical angle for light traveling from the liquid back into the air.

52. The laws of refraction and reflection are the same for sound and for light. The speed of sound is 340 m/s in air and 1510 m/s in water. If a sound wave that is traveling in air approaches a flat water surface with an angle of incidence of 12.0°, what is the angle of refraction?

53. A jewel thief decides to hide a stolen diamond by placing it at the bottom of a crystal-clear fountain. He places a circular piece of wood on the surface of the water and anchors it directly above the diamond at the bottom of the fountain, as shown below. If the fountain is 2.00 m deep, find the minimum diameter of the piece of wood that would prevent the diamond from being seen from outside the water.

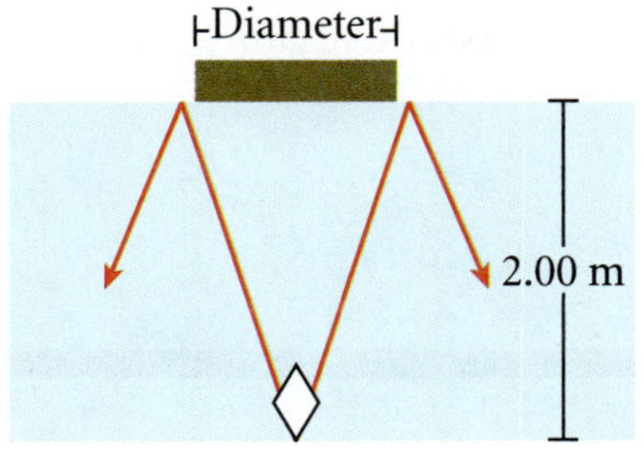

54. A ray of light traveling in air strikes the surface of a block of clear ice at an angle of 40.0° with the normal. Part of the light is reflected, and part is refracted. Find the angle between the reflected and refracted light.

55. An object's distance from a converging lens is 10 times the focal length. How far is the image from the lens? Express the answer as a fraction of the focal length.

56. A fiber-optic cable used for telecommunications has an index of refraction of 1.53. For total internal reflection of light inside the cable, what is the minimum angle of incidence to the inside wall of the cable if the cable is in the following:

a. air

b. water

57. A ray of light traveling in air strikes the midpoint of one face of an equiangular glass prism ($n = 1.50$) at an angle of exactly 30.0°, as shown below.

a. Trace the path of the light ray through the glass, and find the angle of incidence of the ray at the bottom of the prism.

b. Will the ray pass through the bottom surface of the prism, or will it be totally internally reflected?

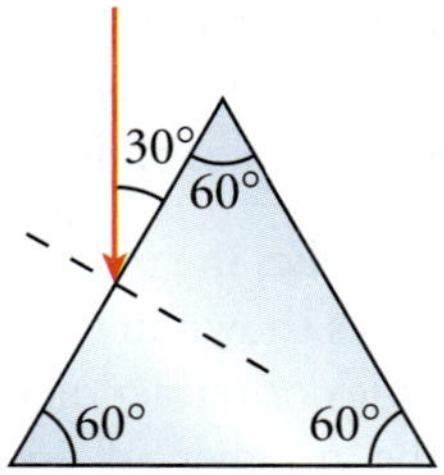

58. Light strikes the surface of a prism, $n = 1.8$, as shown in the figure below. If the prism is surrounded by a fluid, what is the maximum index of refraction of the fluid that will still cause total internal reflection within the prism?

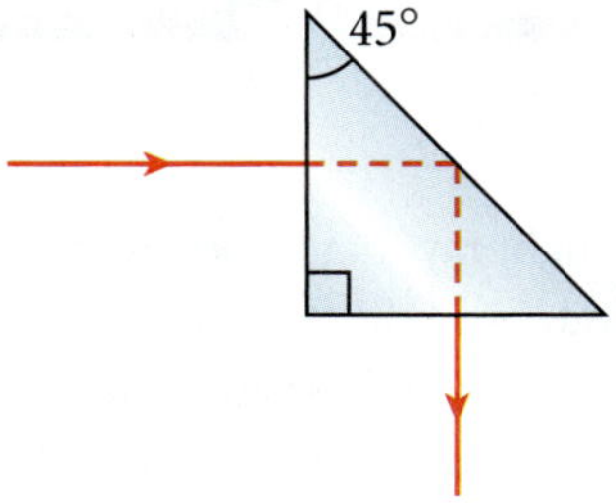

59. A fiber-optic rod consists of a central strand of material surrounded by an outer coating. The interior portion of the rod has an index of refraction of 1.60. If all rays striking the interior walls of the rod with incident angles greater than 59.5° are subject to total internal reflection, what is the index of refraction of the coating?

14 REVIEW

50. 16.5°

51. 48.8°

52. 67°

53. 4.54 m

54. 110.6°

55. $\frac{10}{9}f$

56. **a.** 40.8°
b. 60.6°

57. **a.** 24.7°
b. It will pass through the bottom surface because $\theta_i < \theta_c$ ($\theta_c = 41.8°$).

58. 1.3

59. 1.38

60. 53.4°
61. 58.0 m
62. 82 internal reflections
63. **a.** 4.83 cm
b. The lens must be moved 0.12 cm.
64. 1.80 cm
65. 1.90 cm

extension

Graphing Calculator Practice
Visit go.hrw.com for answers to this Graphing Calculator activity.
Keyword HF6REFXT

60. A flashlight on the bottom of a 4.00 m deep swimming pool sends a ray upward and at an angle so that the ray strikes the surface of the water 2.00 m from the point directly above the flashlight. What angle (in air) does the emerging ray make with the water's surface? (Hint: To determine the angle of incidence, consider the right triangle formed by the light ray, the pool bottom, and the imaginary line straight down from where the ray strikes the surface of the water.)

61. A submarine is 325 m horizontally out from the shore and 115 m beneath the surface of the water. A laser beam is sent from the submarine so that it strikes the surface of the water at a point 205 m from the shore. If the beam strikes the top of a building standing directly at the water's edge, find the height of the building. (Hint: To determine the angle of incidence, consider the right triangle formed by the light beam, the horizontal line drawn at the depth of the submarine, and the imaginary line straight down from where the beam strikes the surface of the water.)

62. A laser beam traveling in air strikes the midpoint of one end of a slab of material, as shown in the figure (in the next column). The index of refraction of the slab is 1.48. Determine the number of internal reflections of the laser beam before it finally emerges from the opposite end of the slab.

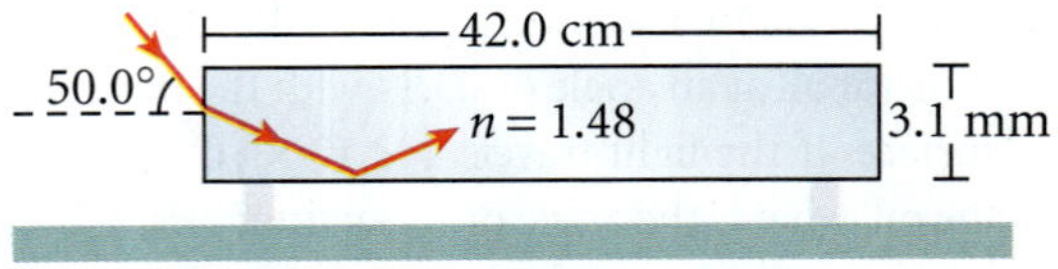

63. A nature photographer is using a camera that has a lens with a focal length of 4.80 cm. The photographer is taking pictures of ancient trees in a forest and wants the lens to be focused on a very old tree that is 10.0 m away.

a. How far must the lens be from the film in order for the resulting picture to be clearly focused?

b. How much would the lens have to be moved to take a picture of another tree that is only 1.75 m away?

64. The distance from the front to the back of your eye is approximately 1.90 cm. If you can see a clear image of a book when it is 35.0 cm from your eye, what is the focal length of the lens/cornea system?

65. Suppose you look out the window and see your friend, who is standing 15.0 m away. To what focal length must your eye muscles adjust the lens of your eye so that you may see your friend clearly? Remember that the distance from the front to the back of your eye is about 1.90 cm.

Graphing Calculator *Practice*

Snell's Law

What happens to a light ray that passes from air into a medium whose index of refraction differs from that of air? Snell's law, as you learned earlier in this chapter, describes the relationship between the angle of refraction and the index of refraction.

$$n_i \sin \theta_i = n_r \sin \theta_r$$

In this equation, n_i is the index of refraction of the medium of the incident light ray, and θ_i is the angle of incidence; n_r is the index of refraction of the medium of the refracted light, and θ_r is the angle of refraction.

In this graphing calculator activity, you will enter the angle of incidence and will view a graph of the index of refraction versus the angle of refraction. You can use this graph to better understand the relationship between the index of refraction and the angle of refraction.

Visit go.hrw.com and enter the keyword **HF6REFX** to find this graphing calculator activity. Refer to **Appendix B** for instructions on downloading the program for this activity.

Alternative Assessment

1. Interview an optometrist, optician, or ophthalmologist. Find out what equipment and tools each uses. What kinds of eye problems is each able to correct? What training is necessary for each career?
2. Obtain permission to use a microscope and slides from your school's biology teacher. Identify the optical components (lenses, mirror, object, and light source) and knobs. Find out how they function at different magnifications and what adjustments must be made to obtain a clear image. Sketch a ray diagram for the microscope's image formation. Estimate the size of the images you see, and calculate the approximate size of the actual cells or microorganisms you observe. How closely do your estimates match the magnification indicated on the microscope?
3. Construct your own telescope with mailing tubes (one small enough to slide inside the other), two lenses, cardboard disks for mounting the lenses, glue, and masking tape. Test your instrument at night. Try to combine different lenses and explore ways to improve your telescope's performance. Keep records of your results to make a brochure documenting the development of your telescope.
4. Study the history of the camera. Possible topics include the following. How did the camera obscura work? What discovery made the first permanent photograph possible? How do instant cameras work? How do modern digital cameras differ from film cameras? Give a short presentation to the class to share the information.
5. Create a pinhole camera with simple household materials. Find instructions for constructing a pinhole camera on the Internet, and follow them to make your own pinhole camera. Partner with a photography student to develop the pictures in your school's darkroom. Create a visual presentation to share your photographs with the class.
6. Research how phone, television, and radio signals are transmitted over long distances through fiber-optic devices. Obtain information from companies that provide telephone or cable television service. What materials are fiber-optic cables made of? What are their most important properties? Are there limits on the kind of light that travels in these cables? What are the advantages of fiber-optic technology over broadcast transmission? Produce a brochure or informational video to explain this technology to consumers.
7. When the Indian physicist Venkata Raman first saw the Mediterranean Sea, he proposed that its blue color was due to the structure of water molecules rather than to the scattering of light from suspended particles. Later, he won the Nobel Prize for work relating to the implications of this hypothesis. Research Raman's life and work. Find out about his background and the challenges and opportunities he met on his way to becoming a physicist. Create a presentation about him in the form of a report, poster, short video, or computer presentation.
8. Choose a radio telescope to research. Possibilities include the Very Large Array in New Mexico, the Arecibo telescope in Puerto Rico, or the Green Bank Telescope in West Virginia. Use the Internet to learn about observations that have been made with the telescope. How long has the telescope been operating? How large is the telescope? What discoveries have been made with it? Has the telescope been used for any SETI *(Search for Extra-Terrestrial Intelligence)* investigations? After your research is complete, write a list of questions that you still have about the telescope. If possible, call the observatory and interview a member of the staff. Write a magazine article with the results of your research.

Alternative Assessment ANSWERS

1. Interview questions should demonstrate an understanding that refraction of light underlies the physiology of vision and corrective procedures.
2. Students' diagrams should indicate that microscopes have two lenses, an objective lens and an eyepiece. The magnification of the lenses multiplied together equals the net magnification.
3. Students should recognize that having one tube sliding within another will allow for easy adjustments to the distance between lenses when focusing.
4. Student presentations should cover some aspect of the history of the camera.
5. Encourage students to bring their pinhole cameras to class, and to show the class how they were constructed.
6. Students should recognize that one advantage of fiber-optic transmission over broadcast technology is that absorption and dispersion of the signal are limited.
7. Raman (1888–1970) was the first Asian to win the Nobel Prize in one of the sciences. At that time in India, there were no opportunities for becoming a career physicist. Like Einstein, he took another job and studied science in his spare time. He later obtained a position at the University of Calcutta.
8. Student articles should be in a magazine format and should discuss the radio telescope of their choice.

Standardized Test Prep

ANSWERS

1. D
2. H
3. A
4. J
5. B
6. J
7. A
8. F

Standardized Test Prep

MULTIPLE CHOICE

1. How is light affected by an increase in the index of refraction?

A. Its frequency increases.
B. Its frequency decreases.
C. Its speed increases.
D. Its speed decreases.

2. Which of the following conditions is *not* necessary for refraction to occur?

F. Both the incident and refracting substances must be transparent.
G. Both substances must have different indices of refraction.
H. The light must have only one wavelength.
J. The light must enter at an angle greater than 0° with respect to the normal.

Use the ray diagram below to answer questions 3–4.

3. What is the focal length of the lens?

A. −12.5 cm
B. −8.33 cm
C. 8.33 cm
D. 12.5 cm

4. What is true of the image formed by the lens?

F. real, inverted, and enlarged
G. real, inverted, and diminished
H. virtual, upright, and enlarged
J. virtual, upright, and diminished

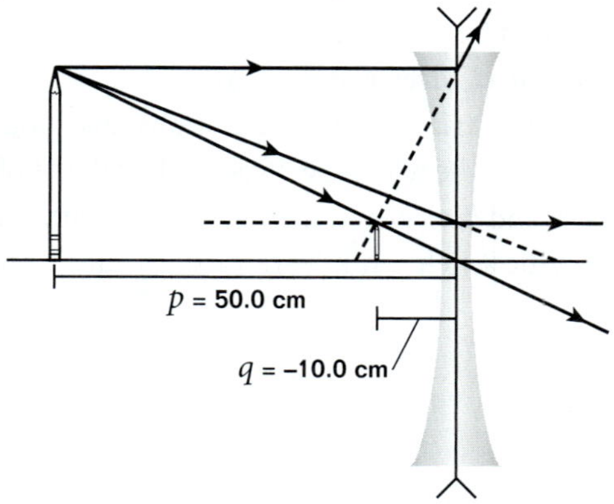

5. A block of flint glass with an index of refraction of 1.66 is immersed in oil with an index of refraction of 1.33. How does the critical angle for a refracted light ray in the glass vary from when the glass is surrounded by air?

A. It remains unchanged.
B. It increases.
C. It decreases.
D. No total internal reflection takes place when the glass is placed in the oil.

6. Which color of light is most refracted during dispersion by a prism?

F. red
G. yellow
H. green
J. violet

7. If an object in air is viewed from beneath the surface of water below, where does the object appear to be?

A. The object appears above its true position.
B. The object appears exactly at its true position.
C. The object appears below its true position.
D. The object cannot be viewed from beneath the water's surface.

8. The phenomenon called "looming" is similar to a mirage, except that the inverted image appears above the object instead of below it. What must be true if looming is to occur?

F. The temperature of the air must increase with distance above the surface.
G. The temperature of the air must decrease with distance above the surface.
H. The mass of the air must increase with distance above the surface.
J. The mass of the air must increase with distance above the surface.

9. Light with a vacuum wavelength of 500.0 nm passes into benzene, which has an index of refraction of 1.5. What is the wavelength of the light within the benzene?
   A. 0.0013 nm
   B. 0.0030 nm
   C. 330 nm
   D. 750 nm

10. Which of the following is *not* a necessary condition for seeing a magnified image with a lens?
   F. The object and image are on the same side of the lens.
   G. The lens must be converging.
   H. The observer must be placed within the focal length of the lens.
   J. The object must be placed within the focal length of the lens.

SHORT RESPONSE

11. In telescopes, at least two converging lenses are used: one for the objective and one for the eyepiece. These lenses must be positioned in such a way that the final image is virtual and very much enlarged. In terms of the focal points of the two lenses, how must the lenses be positioned?

12. A beam of light passes from the fused quartz of a bottle ($n = 1.46$) into the ethyl alcohol ($n = 1.36$) that is contained inside the bottle. If the beam of the light inside the quartz makes an angle of 25.0° with respect to the normal of both substances, at what angle to the normal will the light enter the alcohol?

13. A layer of glycerine ($n = 1.47$) covers a zircon slab ($n = 1.92$). At what angle to the normal must a beam of light pass through the zircon toward the glycerine so that the light undergoes total internal reflection?

EXTENDED RESPONSE

14. Explain how light passing through raindrops is reflected and dispersed so that a rainbow is produced. Include in your explanation why the lower band of the rainbow is violet and the outer band is red.

Use the ray diagram below to answer questions 15–18.

A collector wishes to observe a coin in detail and so places it 5.00 cm in front of a converging lens. An image forms 7.50 cm in front of the lens, as shown in the figure below.

15. What is the focal length of the lens?

16. What is the magnification of the coin's image?

17. If the coin has a diameter of 2.8 cm, what is the diameter of the coin's image?

18. Is the coin's image virtual or real? upright or inverted?

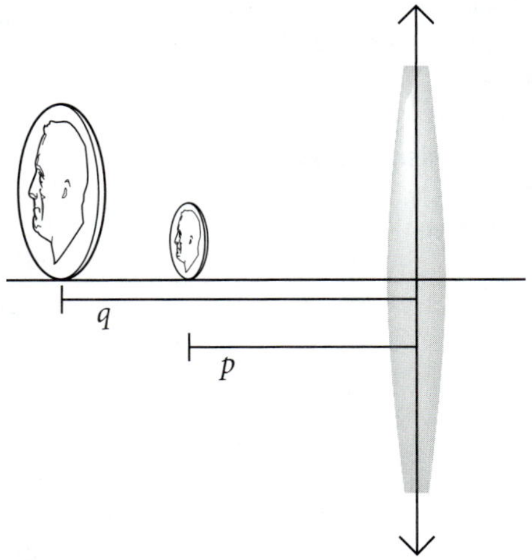

Test TIP When calculating the value of an angle by taking the arcsine of a quantity, recall that the quantity must be positive and no greater than 1.

9. C

10. H

11. The focal point of the objective must lie within the focal point of the eyepiece.

12. 27.0°

13. 50.0°

14. There are three effects—a refraction, a reflection, and then a final refraction. The light of each wavelength in the visible spectrum is refracted by a different amount: the red light undergoes the least amount of refraction, and the violet light undergoes the most. At the far side of the raindrop, the light is internally reflected and undergoes refraction again when it leaves the front side of the raindrop. Because of the internal reflection, the final dispersion of the light is such that the violet light makes an angle of 40° with the incident ray, and the red light makes an angle of 42° with the incident ray. For an observer, the upper edge of the rainbow has the color of the light that bends farthest from the incident light, so the outer band of the rainbow is red. Similarly, the lower edge has the color of the light that bends least from the incident light, so the inner band is violet. The net effect is that the ray that is refracted the most ends up closest to the incident light, that is, the smallest angular displacement.

15. 15 cm

16. 1.5

17. 4.2 cm

18. virtual; upright

Lab Planning

Beginning on page T34 are preparation notes and teaching tips to assist you in planning.

Blank data tables (as well as some sample data) appear on the **One-Stop Planner.**

No Books in the Lab?

See the ***Datasheets for In-Text Labs*** workbook for a reproducible master copy of this experiment.

Safety Caution

Remind students to report all breakage immediately. Students should be instructed not to look directly at a light source.

Tips and Tricks

- If power supplies are used that can supply more than the rated voltage of the bulb, make sure students do not burn out bulbs by using excessive voltage.

CHAPTER 14

Skills Practice Lab Converging Lenses

OBJECTIVES

- **Investigate** the relationships between the positions of the lens and object, and the position and size of the image.

MATERIALS LIST

- 2 screen support riders
- cardboard image screen with metric scale
- converging lens
- dc power supply
- insulated copper wire, 2 lengths
- lens support rider
- meterstick and meterstick supports
- metric ruler
- miniature lamp and base on rider
- object screen

Converging lenses can produce both real and virtual images, and they can produce images that are smaller, the same size as, or larger than the object. In this experiment, you will study image formation using a converging lens.

SAFETY

- **Use a hot mitt to handle resistors, light sources, and other equipment that may be hot. Allow all equipment to cool before storing it.**
- **Never put broken glass or ceramics in a regular waste container. Use a dustpan, brush, and heavy gloves to carefully pick up broken pieces, and dispose of them in a container specifically provided for this purpose.**

PROCEDURE

Preparation

1. Read the entire lab, and plan what measurements you will take.
2. If you are not using a datasheet provided by your teacher, prepare a data table in your lab notebook with eight columns and six rows. In the first row, label the columns *Trial, Position of Lens (cm), Position of Object (cm), Position of Image (cm), q (cm), p (cm),* h_o *(cm),* and h_i *(cm).* In the first column, label the second through sixth rows *1, 2, 3, 4,* and *5.*

Formation of Images

3. Set up the meterstick, meterstick supports, image screen, and lens as shown in **Figure 1.** Locate and mark the point on the mounted screen where it intercepts the principal axis of the mounted lens.
4. Place the illuminated object screen at one end of the meterstick. Make adjustments so that the center of the object screen coincides with the principal axis of the lens.
5. Place the lens far enough from the object screen to give an object distance greater than twice the focal length of the lens. Move the image screen along the meterstick until the image is as well defined as possible. Read and record in your data table the positions of the object, lens, and image to the nearest millimeter on the meterstick. Also record the object distance, p, the image distance, q, the height of the object, h_o, and the height of the image, h_i.

6. Repeat step 5 four times with the lens at a different position each time. These positions should give the following object distances:

   a. exactly twice the focal length
   b. between one and two focal lengths
   c. exactly one focal length
   d. less than one focal length

   Record all measurements as in step 5. If you do not see an image, place *X*s in your data table for that trial.

7. Set the object distance less than one focal length, remove the image screen, and place your eye close to the lens. Look through the lens at the object, and record your observations.

8. Clean up your work area. Put equipment away safely. Recycle or dispose of used materials as directed by your teacher.

Figure 1

Step 3: Make sure the image screen is securely held in the screen support rider to prevent its moving during the experiment.

Step 4: Use the illuminated object screen as the object for this part of the lab.

ANALYSIS

1. **Organizing Data** For each trial recorded in the data table, perform the following calculations:

   a. Find the reciprocal of the object distance, p.
   b. Find the reciprocal of the image distance, q.
   c. Add the reciprocals found in **(a)** and **(b)**.
   d. Find the inverse of your answer in **(c)**.

2. **Organizing Data** For each trial, perform the following calculations:

   a. Find the ratio between q and p.
   b. Find the ratio between h_i and h_o.

CONCLUSIONS

3. **Recognizing Patterns** Compare the inverse of the sum of the reciprocals for each trial with the focal length of the lens. What is the relationship? Is this true for all trials? Explain.

4. **Recognizing Patterns** For each trial, compare the ratios found in item 2.

   a. Based on your results, what physical quantity is expressed by each of the ratios found in item 2?
   b. What is the relationship between the two ratios for each trial? Is this true for all trials? Explain.

CHAPTER 14 LAB

✓ Checkpoints

Step 3: Students may think that the center of the screen must coincide exactly with the principal axis. Actually, as long as the principal axis is near the center of the screen, the procedure will work.

Step 5: Some students may be confused by the description of the distance in terms of the focal length of the lens. Help them find the appropriate position.

Step 6: Students will not see an image on the screen with the lens at distances from the object screen less than or equal to the focal length.

ANSWERS

Analysis

1. Answers will vary. Typical values for (a) and (b) will range from 0.012 cm^{-1} to 0.050 cm^{-1}. For (c), students' answers should have nearly the same value for all trials.

2. For sample data, values range from 0.733 to 4.05.

Conclusions

3. The sum of these reciprocals is the reciprocal of the focal length; The relationship is constant for all trials.

4. a. Both ratios represent the magnification.
b. For each trial, the ratios are equal.

CHAPTER 15

Interference and Diffraction

Planning Guide

Compression Guide

To shorten instruction because of time limitations, omit the opener and Section 3 and abbreviate the review.

OBJECTIVES	LABS, DEMONSTRATIONS, AND ACTIVITIES	TECHNOLOGY RESOURCES
PACING • 45 min pp. 524–525 **Chapter Opener**		**CD Visual Concepts,** Chapter 15 BASIC
PACING • 45 min pp. 526–531 **Section 1 Interference** • Describe how light waves interfere with each other to produce bright and dark fringes. • Identify the conditions required for interference to occur. • Predict the location of interference fringes using the equation for double-slit interference.	TE **Demonstration** Interference in Sound Waves, p. 526 BASIC TE **Demonstration** Interference in a Ripple Tank p. 527 BASIC TE **Demonstration** How Distance Traveled Affects Interference, p. 528 ADVANCED TE **Demonstration** Thin-Film Interference, p. 529 BASIC	OSP **Lesson Plans** TR 78 Interference Between Transverse Waves TR 79 Conditions for Interference of Light Waves TR 50A Comparison of Waves in Phase and 180° out of Phase TR 51A Path Difference for Light Waves from Two Slits TR 52A Position of Higher-Order Interference Fringes
PACING • 135 min pp. 532–540 **Section 2 Diffraction** • Describe how light waves bend around obstacles and produce bright and dark fringes. • Calculate the positions of fringes for a diffraction grating. • Describe how diffraction determines an optical instrument's ability to resolve images.	SE **Skills Practice Lab** Diffraction, pp. 554–555◆ GENERAL ANC **Datasheet** Diffraction* GENERAL TE **Demonstration** Waves Bending Around Corners, p. 532 GENERAL TE **Demonstration** Diffraction and Interference by a Single Slit, p. 533 GENERAL TE **Demonstration** Light Diffraction by an Obstacle: Poisson Spot, p. 534 ADVANCED TE **Demonstration** Effect of Slit Size on Diffraction Patterns, p. 535 GENERAL TE **Demonstration** Multiple-Slit Diffraction p. 536 GENERAL	OSP **Lesson Plans** TR 80 Diffraction of Light with Decreasing Slit Width TR 81 Constructive Interference by a Diffraction Grating TR 82 Function and Use of a Diffraction Grating in a Spectrometer TR 83 Resolution of Two Light Sources TR 53A Destructive Interference in Single-Slit Diffraction
PACING • 45 min pp. 541–545 *Advanced Level* **Section 3 Lasers** • Describe the properties of laser light. • Explain how laser light has particular advantages in certain applications.	TE **Demonstration** Dancing Light, p. 541 ADVANCED TE **Demonstration** Interference in Laser Light, p. 542 GENERAL	OSP **Lesson Plans** TR 84 Operation of a Laser TR 85 Components of a Compact Disc Player TR 54A Wave Fronts from Noncoherent and Coherent Light Sources

PACING • 90 min

CHAPTER REVIEW, ASSESSMENT, AND STANDARDIZED TEST PREPARATION

- SE **Chapter Highlights,** p. 547
- SE **Chapter Review,** pp. 548–550
- SE **Alternative Assessment,** p. 550 ADVANCED
- SE **Graphing Calculator Practice,** p. 551 GENERAL
- SE **Standardized Test Prep,** pp. 552–553 GENERAL
- SE **Appendix D: Equations,** p. 861
- SE **Appendix I: Additional Problems,** pp. 891
- ANC **Study Guide Worksheet** Mixed Review* GENERAL
- ANC **Chapter Test A*** GENERAL
- ANC **Chapter Test B*** ADVANCED
- OSP **Test Generator**

Online and Technology Resources

go.hrw.com Holt Online Learning

Visit go.hrw.com to find a variety of online resources. To access this chapter's extensions, enter the keyword **HF6INFXT** and click the "go" button. Click **Holt Online Learning** for an online edition of this textbook, and other interactive resources.

HOLT Teacher's One-Stop Planner®

This DVD package includes:
- Holt Calendar Planner
- Customizable Lesson Plans
- Editable Worksheets
- ExamView® Version 6 Assessment Suite
- Interactive Teacher's Edition
- Holt PuzzlePro®
- Holt PowerNotes® Presentations
- MindPoint® Quiz Show

KEY

SE Student Edition	**OSP** One-Stop Planner	**EXT** Online Extension
TE Teacher Edition	**CD** CD or CD-ROM	* Also on One-Stop Planner
ANC Ancillary Worksheet	**TR** Teaching Transparencies	◆ Requires advance prep

SKILLS DEVELOPMENT RESOURCES	REVIEW AND ASSESSMENT	CORRELATIONS
		National Science Education Standards
SE Sample Set A Interference, pp. 530–531 GENERAL **TE Classroom Practice,** p. 530 GENERAL **ANC Problem Workbook** Sample Set A* GENERAL **OSP Problem Bank** Sample Set A GENERAL	**SE Section Review,** p. 531 GENERAL **ANC Study Guide Worksheet** Section 1* GENERAL **ANC Quiz** Section 1* BASIC	UCP 1, 2, 3
SE Sample Set B Diffraction Gratings, pp. 537–538 GENERAL **TE Classroom Practice,** p. 537 GENERAL **ANC Problem Workbook** Sample Set B* GENERAL **OSP Problem Bank** Sample Set B GENERAL **SE Conceptual Challenge,** p. 535 ADVANCED	**SE Section Review,** p. 540 GENERAL **ANC Study Guide Worksheet** Section 2* GENERAL **ANC Quiz** Section 2* BASIC	UCP 1, 2, 3, 5 SAI 1, 2 ST 1, 2 HNS 1 SPSP 5
	SE Section Review, p. 545 ADVANCED **ANC Study Guide Worksheet** Section 3* ADVANCED **ANC Quiz** Section 3* GENERAL	UCP 1, 2, 3, 5 ST 1, 2 HNS 1 SPSP 5

SCILINKS. NSTA
www.scilinks.org

Maintained by the **National Science Teachers Association.**

Topic: Interference
SciLinks Code: HF60806

Topic: Diffraction
SciLinks Code: HF60405

Topic: Lasers
SciLinks Code: HF60853

Topic: Bar Codes
SciLinks Code: HF60135

Classroom CD-ROMs

- Guided Reading Audio Program
- Student One Stop
- Virtual Investigations
- Visual Concepts

Holt Lab Generator CD-ROM

Search for any lab by topic, standard, difficulty level, or time. Edit any lab to fit your needs, or create your own labs. Use the Lab Materials QuickList software to customize your lab materials list.

CHAPTER 15
Overview

Section 1 identifies the conditions required for interference to occur and shows how to calculate the location of bright and dark fringes in double-slit interference.

Section 2 describes how diffracted light waves interfere, shows how to calculate the position of fringes produced by a diffraction grating, and discusses the resolving power of optical instruments.

Section 3 describes how a laser produces coherent light and explores applications of lasers.

About the Illustration

The colors visible on the discs' surfaces are characteristic of the light used in the photograph. Have students verify the ability of a compact disc to separate light into its particular spectrum by reflecting light from different sources off a compact disc's surface. For example, sunlight is separated into all of the visible colors, while most fluorescent lighting produces only a few colors.

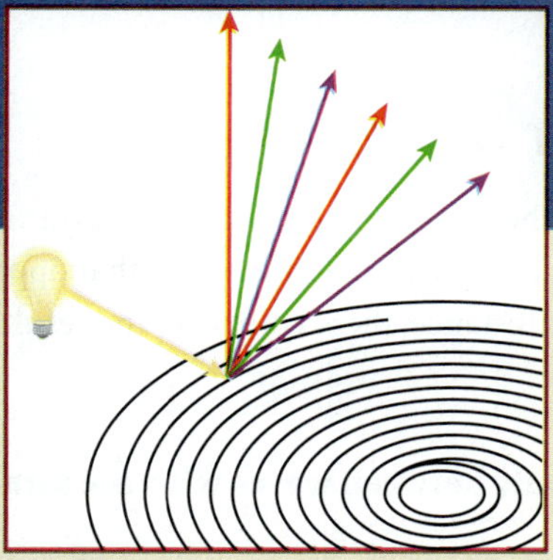

CHAPTER 15

Interference and Diffraction

The streaks of colored light you see coming from a compact disc resemble the colors that appear when white light passes through a prism. However, the compact disc does not separate light by means of refraction. Instead, the light waves undergo interference.

WHAT TO EXPECT

In this chapter, you will learn about interference of light. In interference, light waves combine to produce resultant waves that are either brighter or less bright than the component waves.

Why it Matters

Devices called *diffraction gratings* use the principle of interference to separate light into its component wavelengths. Diffraction gratings are used in instruments called *spectrometers,* which are used to study the chemical composition and temperature of stars.

CHAPTER PREVIEW

Tapping Prior Knowledge

Knowledge to Review

- ✔ The superposition principle: When two mechanical waves pass through the same space at the same time, their displacements at each point add.
- ✔ When two waves with the same frequency and amplitude overlap, the resulting wave has the same frequency as the individual waves. If the waves are in phase, the resultant wave has twice their amplitude. If they are 180° out of phase, the amplitudes cancel.

Items to Probe

- ✔ Preconceptions about waves: Ask students what happens when two waves (A and B) travel toward each other on a string and meet at a point where A's displacement is 4 cm up and B's displacement is 3 cm up. *(The displacement at that point is 7 cm up.)* What if A is 4 cm up and B is 3 cm down? *(The displacement is 1 cm up.)*

525

SECTION 1
General Level

Demonstration

Interference in Sound Waves — GENERAL

Purpose Introduce students to interference patterns using sound waves from two coherent sources.

Materials sine-wave generator, amplifier, two speakers, tape measures, overhead projector with grid transparency

Procedure Connect the generator and the speakers to the amplifier. Place the speakers about 3 m apart so that both face the class. Have students stand in rows perpendicular to a line joining the two speakers. To reduce the effect of echoes from the walls, have students cover the ear that is opposite the speakers.

Set the generator to a frequency of about 440 Hz. Turn on the generator and amplifier, and adjust the two speakers to equal intensity. Have students slowly walk forward in their rows and listen to the intensity of the sound. Tell them to stand still when they find a location where the sound is at a minimum. Have students measure the distance between each point where sound intensity is at a minimum. Using these data, have students draw the *destructive interference* fringes on the transparency. When all of the fringe positions are recorded, turn the overhead projector on and have students note that they were standing in places where the superposition of waves resulted in destructive interference.

SECTION 1

Interference

SECTION OBJECTIVES

- Describe how light waves interfere with each other to produce bright and dark fringes.
- Identify the conditions required for interference to occur.
- Predict the location of interference fringes using the equation for double-slit interference.

COMBINING LIGHT WAVES

You have probably noticed the bands of color that form on the surface of a soap bubble, as shown in **Figure 1.** Unlike the colors that appear when light passes through a refracting substance, these colors are the result of light waves combining with each other.

Figure 1
Light waves interfere to form bands of color on a soap bubble's surface.

Interference takes place only between waves with the same wavelength

To understand how light waves combine with each other, let us review how other kinds of waves combine. If two waves with identical wavelengths interact, they combine to form a resultant wave. This resultant wave has the same wavelength as the component waves, but according to the superposition principle, its displacement at any instant equals the sum of the displacements of the component waves. The resultant wave is the consequence of the *interference* between the two waves.

Figure 2 can be used to describe pairs of mechanical waves or electromagnetic waves with the same wavelength. A light source that has a single wavelength is called *monochromatic,* which means single colored. In the case of *constructive interference,* the component waves combine to form a resultant wave with the same wavelength but with an amplitude that is greater than the amplitude of either of the individual component waves. For light, the result of constructive interference is light that is brighter than the light from the contributing waves. In the case of *destructive interference,* the resultant amplitude is less than the amplitude of the larger component wave. For light, the result of destructive interference is dimmer light or dark spots.

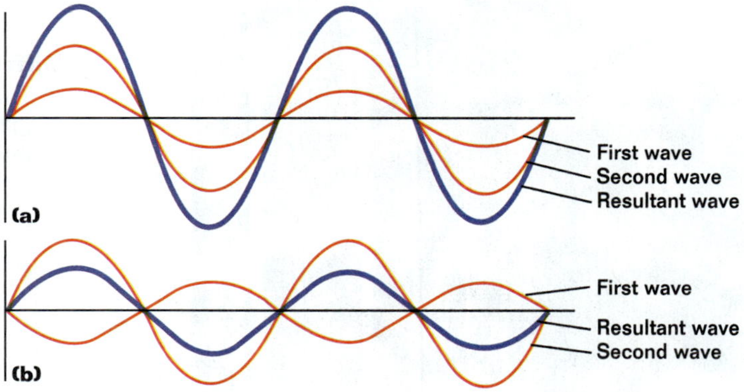

Figure 2
Two waves can interfere **(a)** constructively or **(b)** destructively. In interference, energy is not lost but is instead redistributed.

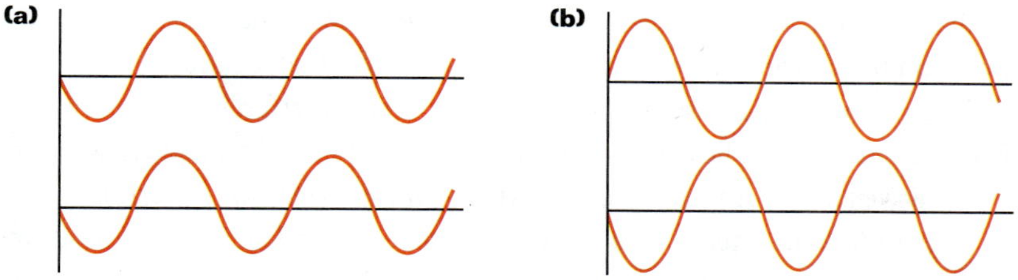

Figure 3
(a) The features of two waves in phase completely match, whereas **(b)** they are opposite each other in waves that are 180° out of phase.

SCILINKS.
www.scilinks.org
Topic: Interference
Code: HF60806

Waves must have a constant phase difference for interference to be observed

For two waves to produce a stable interference pattern, the phases of the individual waves must remain unchanged relative to one another. If the crest of one wave overlaps the crest of another wave, as in **Figure 3(a),** the two have a phase difference of 0° and are said to be *in phase.* If the crest of one wave overlaps the trough of the other wave, as in **Figure 3(b),** the two waves have a phase difference of 180° and are said to be *out of phase.*

Waves are said to have **coherence** when the phase difference between two waves is constant and the waves do not shift relative to each other as time passes. Sources of such waves are said to be *coherent.*

coherence
the correlation between the phases of two or more waves

When two light bulbs are placed side by side, no interference is observed, even if the lights are the same color. The reason is that the light waves from one bulb are emitted independently of the waves from the other bulb. Random changes occurring in the light from one bulb do not necessarily occur in the light from the other bulb. Thus, the phase difference between the light waves from the two bulbs is not constant. The light waves still interfere, but the conditions for the interference change with each phase change, and therefore, no single interference pattern is observed. Light sources of this type are said to be *incoherent.*

DEMONSTRATING INTERFERENCE

Interference in light waves from two sources can be demonstrated in the following way. Light from a single source is passed through a narrow slit and then through two narrow parallel slits. The slits serve as a pair of coherent light sources because the waves emerging from them come from the same source. Any random change in the light emitted by the source will occur in the two separate beams at the same time.

If monochromatic light is used, the light from the two slits produces a series of bright and dark parallel bands, or *fringes,* on a distant viewing screen, as shown in **Figure 4.** When the light from the two slits arrives at a point on the viewing screen where constructive interference occurs, a bright fringe appears

Figure 4
An interference pattern consists of alternating light and dark fringes.

Teaching Tip GENERAL

Point out to students that waves can also interfere in an intermediate way in which the relative phase between the waves is not exactly 0° or 180° as it is in **Figure 2** and **Figure 3.** Ask students to draw waves of the same wavelength that are 45°, 90°, and 135° out of phase. What do the resultant waves look like for each of these three cases?

Demonstration

Interference in a Ripple Tank

Purpose Demonstrate interference patterns using two types of point sources.

Materials ripple tank, ripple generator, two pencils, overhead projector or light source, screen

Procedure Touch the water in the tank with a pencil point several times to generate a wave. Let students observe the wave pattern. Repeat the procedure with two pencils 20 cm apart so that the two waves vary in frequency and phase. Explain that the pencils act as point sources but that they generate waves of different phase or wavelength, or both. Use the ripple generator to create two coherent sources. Point out that the speakers in the previous Demonstration were also coherent sources. Explain that when the sources are coherent, they produce a stable interference pattern.

Demonstration

How Distance Traveled Affects Interference

Purpose Demonstrate that for constructive interference the difference between the distances traveled by two coherent wave fronts is equal to a whole number times the wavelength.

Materials sine-wave generator, amplifier, two speakers, tape measures

Procedure Set up the equipment as in the first Demonstration, and ask students to return to their positions in line. Set the generator to a frequency of about 440 Hz, and ask students to step forward and back until they find a place where the sound is at a maximum. Have them measure the distance from their location to each of the speakers and calculate the difference in distance (path difference). Given the sound's frequency and speed (330 m/s), ask students to calculate its wavelength (*0.75 m*), compare it to the path difference, and record the results on the chalkboard.

Increase the frequency to 660 Hz ($\lambda = 0.50$ m), and repeat the demonstration. Have students notice that they are now standing closer to each other.

Have students examine all the results and note that although each of them has detected a different wave crest, the path difference between the speakers and the wave crest equals a whole number times the wavelength in all cases.

Figure 5
When waves of white light from two coherent sources interfere, the pattern is indistinct because different colors interfere constructively and destructively at different positions.

at that location. When the light from the two slits combines destructively at a point on the viewing screen, a dark fringe appears at that location.

When a white-light source is used to observe interference, the situation becomes more complicated. The reason is that white light includes waves of many wavelengths. An example of a white-light interference pattern is shown in **Figure 5.** The interference pattern is stable or well defined at positions where there is constructive interference between light waves of the same wavelength. This explains the color bands on either side of the center band of white light. This effect also accounts for the bands of color seen on soap bubbles.

Figure 6 shows some of the ways that two coherent waves leaving the slits can combine at the viewing screen. When the waves arrive at the central point of the screen, as in **Figure 6(a),** they have traveled equal distances. Thus, they arrive in phase at the center of the screen, constructive interference occurs, and a bright fringe forms at that location.

When the two light waves combine at a specific point off the center of the screen, as in **Figure 6(b),** the wave from the more distant slit must travel one wavelength farther than the wave from the nearer slit. Because the second wave has traveled exactly one wavelength farther than the first wave, the two waves are in phase when they combine at the screen. Constructive interference therefore occurs, and a second bright fringe appears on the screen.

If the waves meet midway between the locations of the two bright fringes, as in **Figure 6(c),** the first wave travels half a wavelength farther than the second wave. In this case, the trough of the first wave overlaps the crest of the second wave, giving rise to destructive interference. Consequently, a dark fringe appears on the viewing screen between the bright fringes.

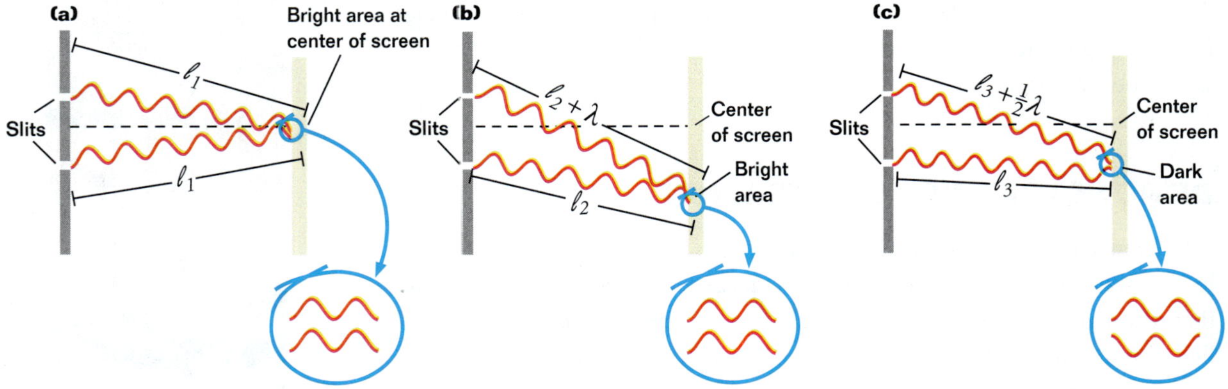

Figure 6
(a) When both waves of light travel the same distance (ℓ_1), they arrive at the screen in phase and interfere constructively. **(b)** If the difference between the distances traveled by the light from each source equals a whole wavelength (λ), the waves still interfere constructively. **(c)** If the distances traveled by the light differ by a half wavelength, the waves interfere destructively.

Predicting the location of interference fringes

Consider two narrow slits that are separated by a distance d, as shown in **Figure 7,** and through which two coherent, monochromatic light waves, l_1 and l_2, pass and are projected onto a screen. If the distance from the slits to the viewing screen is very large compared with the distance between the slits, then l_1 and l_2 are nearly parallel. As a result of this approximation, l_1 and l_2 make the same angle, θ, with the horizontal dotted lines that are perpendicular to the slits. The angle θ also indicates the position at which the waves combine with respect to the central point of the viewing screen.

The difference in the distance traveled by the two waves is called their **path difference.** Study the right triangle shown in **Figure 7,** and note that the path difference between the two waves is equal to $d\sin\theta$. Note carefully that the value for the path difference varies with angle θ and that each value of θ defines a specific position on the screen.

The value of the path difference determines whether the two waves are in or out of phase when they arrive at the viewing screen. If the path difference is either zero or some whole-number multiple of the wavelength, the two waves are in phase, and constructive interference results. The condition for bright fringes (constructive interference) is given by:

EQUATION FOR CONSTRUCTIVE INTERFERENCE

$$d\sin\theta = \pm m\lambda \quad m = 0, 1, 2, 3, \ldots$$

the path difference between two waves = an integer multiple of the wavelength

In this equation, m is the **order number** of the fringe. The central bright fringe at $\theta = 0$ ($m = 0$) is called the *zeroth-order maximum,* or the *central maximum;* the first maximum on either side of the central maximum, which occurs when $m = 1$, is called the *first-order maximum,* and so forth.

Similarly, when the path difference is an odd multiple of $\frac{1}{2}\lambda$, the two waves arriving at the screen are 180° out of phase, giving rise to destructive interference. The condition for dark fringes, or destructive interference, is given by the following equation:

EQUATION FOR DESTRUCTIVE INTERFERENCE

$$d\sin\theta = \pm(m + \tfrac{1}{2})\lambda \quad m = 0, 1, 2, 3, \ldots$$

the path difference between two waves = an odd number of half wavelengths

If $m = 0$ in this equation, the path difference is $\pm\frac{1}{2}\lambda$, which is the condition required for the first dark fringe on either side of the bright central maximum.

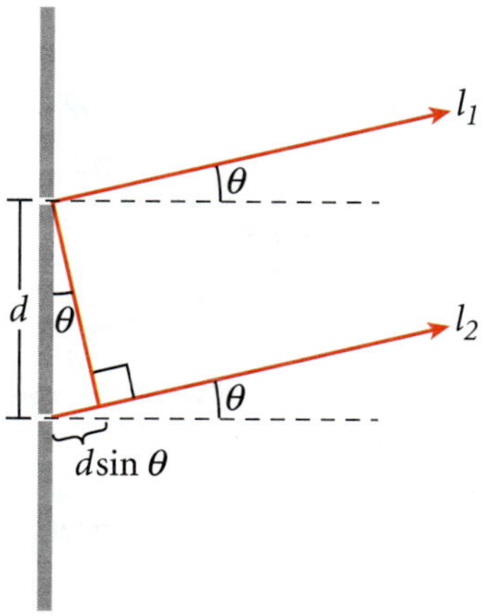

Figure 7
The path difference for two light waves equals $d\sin\theta$. In order to emphasize the path difference, the figure is not drawn to scale.

path difference

the difference in the distance traveled by two beams when they are scattered in the same direction from different points

order number

the number assigned to interference fringes with respect to the central bright fringe

SECTION 1

Visual Strategy GENERAL

Figure 7
Point out that lines l_1 and l_2 should intersect for interference to occur. They are represented as parallel only as an approximation so that the path difference can be found mathematically.

Q If $d = 0.50$ mm and $\theta = 0.30°$, how large is the path difference? How many wavelengths does this equal for green light with a wavelength of 520 nm?

A *$d\sin\theta = 2600$ nm, or 5 wavelengths of the light*

Demonstration

Thin-Film Interference

Purpose Demonstrate that wavelength affects the position of interference fringes.

Materials bottles of soap solution; overhead or slide projector; red, green, and blue cellophane

Procedure Use the soap solution to blow bubbles for the students to observe. Explain that the colors seen on the soap bubble result from interference between two reflected waves. One is reflected from the bubble's outer surface and the other from its inner surface. Because the soap bubble is very thin, the path difference for these waves is so small that interference can be observed. Have students note how the bands of color swirl over the bubbles' surfaces. Cover the light source with one of the pieces of cellophane, and have students repeat the demonstration. This time, have students note the widths of the alternating monochromatic and black bands. After using all three cellophane covers, point out that the fringe width in each pattern varies with color (wavelength).

SECTION 1

Visual Strategy – BASIC

Figure 8
Have students notice that the dark and bright fringes form a symmetrical pattern.

Q How many zeroth-order bright lines are there? How many dark ones?

A *one; two*

Classroom Practice

Interference
The distance between two slits is 0.0050 mm. Find the angles of the zeroth-, first-, second-, third-, and fourth-order bright fringes of interference produced with light with a wavelength of 550 nm.

Answer
0°, 6.3°, 13°, 19°, 26°

When monochromatic light falls on two slits with a separation of 0.010 mm, the zeroth-order dark fringes are observed at a 2.0° angle. Find the wavelength.

Answer
7.0×10^{-7} m $= 7.0 \times 10^2$ nm

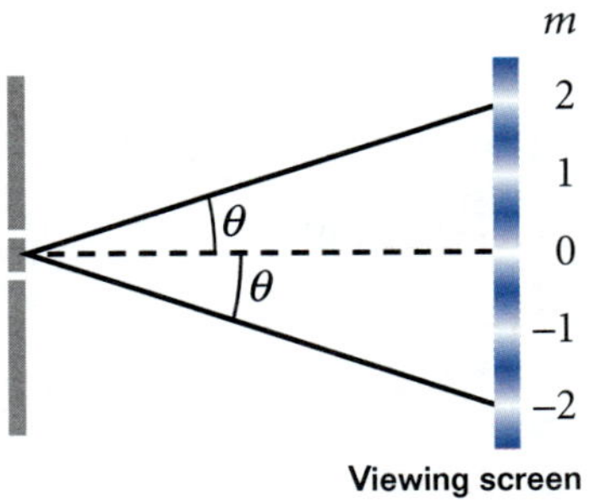

Figure 8
The higher-order ($m = 1, 2$) maxima appear on either side of the central maximum ($m = 0$).

Likewise, if $m = 1$, the path difference is $\pm\frac{3}{2}\lambda$, which is the condition for the second dark fringe on each side of the central maximum, and so forth.

A representation of the interference pattern formed by double-slit interference is shown in **Figure 8.** The numbers indicate the two *maxima* (the plural of *maximum*) that form on either side of the central (zeroth-order) maximum. The darkest areas indicate the positions of the dark fringes, or *minima* (the plural of *minimum*), that also appear in the pattern.

Because the separation between interference fringes varies for light of different wavelengths, double-slit interference provides a method of measuring the wavelength of light. In fact, this technique was used to make the first measurement of the wavelength of light.

SAMPLE PROBLEM A

Interference

PROBLEM

The distance between the two slits is 0.030 mm. The second-order bright fringe ($m = 2$) is measured on a viewing screen at an angle of 2.15° from the central maximum. Determine the wavelength of the light.

SOLUTION

1. DEFINE

Given: $d = 3.0 \times 10^{-5}$ m $m = 2$ $\theta = 2.15°$

Unknown: $\lambda = ?$

Diagram:

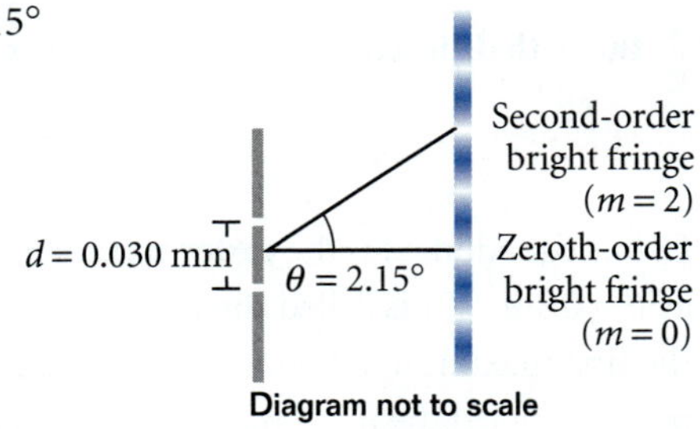

2. PLAN

Choose an equation or situation: Use the equation for constructive interference.

$$d \sin\theta = m\lambda$$

Rearrange the equation to isolate the unknown:

$$\lambda = \frac{d \sin\theta}{m}$$

3. CALCULATE

Substitute the values into the equation and solve:

$$\lambda = \frac{(3.0 \times 10^{-5}\ \text{m})(\sin 2.15°)}{2}$$

$$\lambda = 5.6 \times 10^{-7}\ \text{m} = 5.6 \times 10^2\ \text{nm}$$

$$\boxed{\lambda = 5.6 \times 10^2\ \text{nm}}$$

CALCULATOR SOLUTION

Because the minimum number of significant figures for the data is two, the calculator answer 5.627366×10^{-7} should be rounded to two significant figures.

4. EVALUATE

This wavelength of light is in the visible spectrum. The wavelength corresponds to light of a yellow-green color.

PRACTICE A

Interference

1. A double-slit interference experiment is performed with blue-green light from an argon-gas laser (lasers will be discussed further in Section 3). The separation between the slits is 0.50 mm, and the first-order maximum of the interference pattern is at an angle of 0.059° from the center of the pattern. What is the wavelength of argon laser light?
2. Light falls on a double slit with slit separation of 2.02×10^{-6} m, and the first bright fringe is seen at an angle of 16.5° relative to the central maximum. Find the wavelength of the light.
3. A pair of narrow parallel slits separated by a distance of 0.250 mm is illuminated by the green component from a mercury vapor lamp ($\lambda = 546.1$ nm). Calculate the angle from the central maximum to the first bright fringe on either side of the central maximum.
4. Using the data from item 2, determine the angle between the central maximum and the second dark fringe in the interference pattern.

SECTION REVIEW

1. What is the necessary condition for a path length difference between two waves that interfere constructively? destructively?
2. If white light is used instead of monochromatic light to demonstrate interference, how does the interference pattern change?
3. If the distance between two slits is 0.0550 mm, find the angle between the first-order and second-order bright fringes for yellow light with a wavelength of 605 nm.
4. **Interpreting Graphics** Two radio antennas simultaneously transmit identical signals with a wavelength of 3.35 m, as shown in **Figure 9.** A radio several miles away in a car traveling parallel to the straight line between the antennas receives the signals. If the second maximum is located at an angle of 1.28° north of the central maximum for the interfering signals, what is the distance, *d*, between the two antennas?

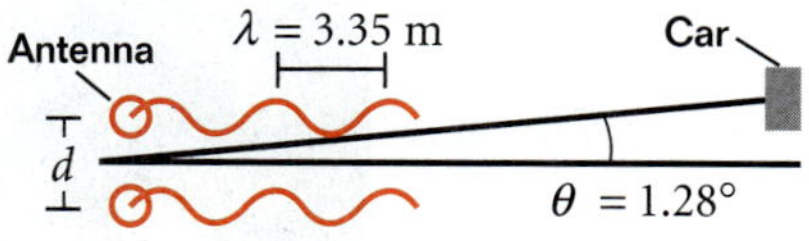

Figure 9

SECTION 1

PROBLEM GUIDE A

Use this guide to assign problems.
SE = Student Edition Textbook
PW = Problem Workbook
PB = Problem Bank on the One-Stop Planner (OSP)

Solving for:

λ	**SE** Sample, 1–2; Ch. Rvw. 9–11, 29 **PW** 3–4 **PB** 4–6
θ	**SE** 3–4 **PW** Sample, 1–2 **PB** 7–10
d	**SE** Ch. Rvw. 28 **PW** 5–7 **PB** Sample, 1–3

***Challenging Problem**
Consult the printed Solutions Manual or the OSP for detailed solutions.

ANSWERS

Practice A

1. 5.1×10^{-7} m $= 5.1 \times 10^{2}$ nm
2. 574 nm
3. 0.125°
4. 25.2°

SECTION REVIEW ANSWERS

1. a difference of an integral number of wavelengths; a difference of an odd integral number of half wavelengths
2. It becomes blurred, and the bright fringes are made up of narrow, colored bands.
3. 0.63°
4. 3.00×10^{2} m

Demonstration

Waves Bending Around Corners — GENERAL

Purpose Demonstrate wave diffraction in a ripple tank.

Materials ripple tank, straight-wave generator, barrier, overhead projector or light source, screen

Procedure Place the barrier in the tank, and turn on the straight-wave generator. Let students examine the edges of the "shadow" of quiet water extending beyond the barrier. Explain that the waves that appear to start at the corners illustrate Huygens' principle. Ask students to sketch the patterns that they observe and to describe areas of light and shadow that would be formed if this were a light wave. Point out that the divergence of a wave from its initial path by an obstacle is called diffraction.

Diffraction

SECTION OBJECTIVES

- **Describe how light waves bend around obstacles and produce bright and dark fringes.**
- **Calculate the positions of fringes for a diffraction grating.**
- **Describe how diffraction determines an optical instrument's ability to resolve images.**

THE BENDING OF LIGHT WAVES

If you stand near the corner of a building, you can hear someone who is talking around the corner, but you cannot see the person. The reason is that sound waves are able to bend around the corner. In a similar fashion, water waves bend around obstacles, such as the barriers shown in **Figure 10.** Light waves can also bend around obstacles, but because of their short wavelengths, the amount they bend is too small to be easily observed.

If light traveled only in straight lines, you would not be able to observe an interference pattern in the double-slit demonstration. Instead, you would see two thin strips of light where each slit and the source were lined up perfectly. The rest of the screen would be completely dark. The edges of the slits would appear on the screen as sharply defined shadows. But this does not happen. Some of the light bends to the right and to the left as it passes through each slit.

The bending of light as it passes through each of the two slits can be understood using Huygens' principle, which states that any point on a wave front can be treated as a point source of waves. Because each slit serves as a point source of light, the waves spread out from the slits. The result is that light deviates from a straight-line path and enters the region that would otherwise be shadowed. This divergence of light from its initial direction of travel is called **diffraction.**

diffraction

a change in the direction of a wave when the wave encounters an obstacle, an opening, or an edge

In general, diffraction occurs when waves pass through small openings, around obstacles, or by sharp edges. When a wide slit (1 mm or more) is placed between a distant light source and a screen, the light produces a bright rectangle with clearly marked edges on the screen. But if the slit is gradually

Figure 10
A property of all waves is that they bend, or *diffract,* around objects.

narrowed, the light eventually begins to spread out and produce a *diffraction pattern,* such as that shown in **Figure 11.** Like the interference fringes in the double-slit demonstration, this pattern of light and dark bands arises from the combination of light waves.

Wavelets in a wave front interfere with each other

Diffraction patterns resemble interference patterns because they also result from constructive and destructive interference. In the case of interference, it is assumed that the slits behave as point sources of light. For diffraction, the actual width of a single slit is considered.

According to Huygens' principle, each portion of a slit acts as a source of waves. Hence, light from one portion of the slit can interfere with light from another portion. The resultant intensity of the diffracted light on the screen depends on the angle, θ, through which the light is diffracted.

To understand the single-slit diffraction pattern, consider **Figure 12(a),** which shows an incoming plane wave passing through a slit of width a. Each point (or, more accurately, each infinitely thin slit) within the wide slit is a source of Huygens wavelets. The figure is simplified by showing only five among this infinite number of sources. As with double-slit interference, the viewing screen is assumed to be so far from the slit that the rays emerging from the slit are nearly parallel. At the viewing screen's midpoint, all rays from the slit travel the same distance, so a bright fringe appears.

The wavelets from the five sources can also interfere destructively when they arrive at the screen, as shown in **Figure 12(b).** When the extra distance traveled by the wave originating at point 3 is half a wavelength longer than the wave from point 1, these two waves interfere destructively at the screen. At this same time, the wave from point 5 travels half a wavelength farther than the wave from point 3, so these waves also interfere destructively. With all pairs of points interfering destructively, this point on the screen is dark.

For angles other than those at which destructive interference completely occurs, some of the light waves remain uncanceled. At these angles light appears on the screen as part of a bright band. The brightest band appears in the pattern's center, while the bands to either side are much dimmer.

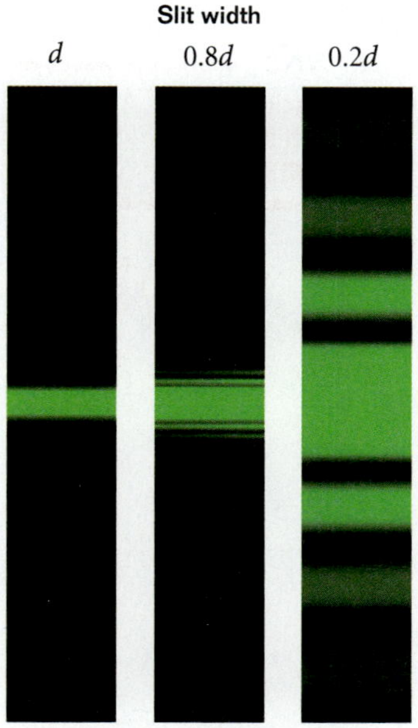

Figure 11
Diffraction becomes more evident as the width of the slit is narrowed. (Note: The wavelength of this light is 510 nm.)

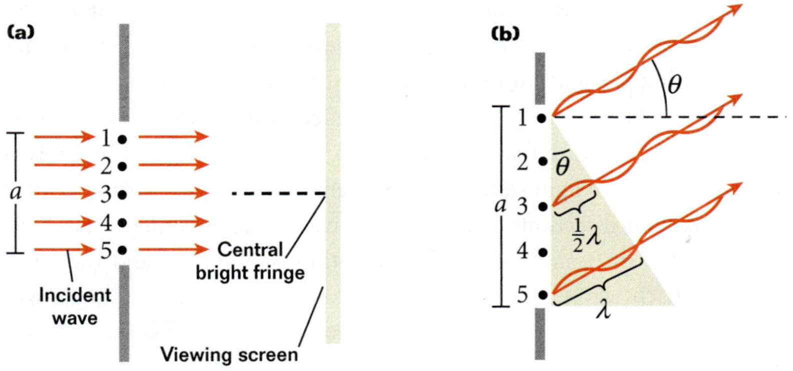

Figure 12
(a) By treating the light coming through the slit as a line of infinitely thin sources along the slit's width, one can determine **(b)** the conditions at which destructive interference occurs between the waves from the upper half of the slit and the waves from the lower half.

Demonstration

Diffraction and Interference by a Single Slit — GENERAL

Purpose Demonstrate diffraction and single-slit interference in a ripple tank.

Materials ripple tank, straight-wave generator, two straight barriers, overhead projector or light source, screen

Procedure Place two barriers about 10 cm from the straight-wave generator, and leave a wide opening between them. Ask students to predict how these obstacles will affect the wave; then, generate the wave to confirm their prediction. Ask what will change when the width of the opening decreases. Have students note that diffraction becomes more evident as the width of the slit narrows.

Tell students that you will increase the wavelength by decreasing the wave generator's frequency until interference patterns appear. (Increase the slit's width, if necessary.) Explain that each portion of the slit acts as a point source. Ask students under what conditions they might expect light to produce similar diffraction and interference patterns (*when the slit is wider—but not vastly wider—than the wavelength*).

SECTION 2

Demonstration

Light Diffraction by an Obstacle: Poisson Spot — ADVANCED

Purpose Demonstrate the bright spot of light produced by interference of diffracted light around the edge of an obstacle.

Materials laser, pin with a round head, clay, screen

CAUTION *Direct the laser beam away from the students.*

Procedure Place the pin tip in the clay, and place the pinhead in the path of the laser beam (far enough away that the beam is a little larger than the pinhead). Ask students to describe the pattern they expect to see on the screen. (Some students may expect a dark shadow of the pinhead; others may correctly expect an interference pattern in the shadow due to light bending around the edges of the head.) Have students note the bright spot at the center of the shadow. Explain that this experiment was crucial in confirming the wave theory of light. If light travels in straight-line paths with no bending around obstacles, as it would if light were composed of a stream of particles, the center of the shadow would be dark. The wave theory of light, however, predicts constructive interference will occur at this point producing a bright spot.

Figure 13
In a diffraction pattern, the central maximum is twice as wide as the secondary maxima.

Figure 14
A diffraction pattern forms in the penny's shadow when light is diffracted at the penny's edge. Note the bright spot that is formed at the center of the shadow.

Light diffracted by an obstacle also produces a pattern

The diffraction pattern that results from monochromatic light passing through a single slit consists of a broad, intense central band—the *central maximum*—flanked by a series of narrower, less intense secondary bands (called *secondary maxima*) and a series of dark bands, or *minima.* An example of such a pattern is shown in **Figure 13.** The points at which maximum constructive interference occurs lie approximately halfway between the dark fringes. Note that the central bright fringe is quite a bit brighter and about twice as wide as the next brightest maximum.

Diffraction occurs around the edges of all objects. **Figure 14** shows the diffraction pattern that appears in the shadow of a penny. The pattern consists of the shadow, with a bright spot at its center, and a series of bright and dark bands of light that continue to the shadow's edge. The penny is large compared with the wavelength of the light, and a magnifying glass is required to observe the pattern.

Figure 15
Compact discs disperse light into its component colors in a manner similar to that of a diffraction grating.

DIFFRACTION GRATINGS

You have probably noticed that if white light is incident on a compact disc, streaks of color are visible. These streaks appear because the digital information (alternating pits and smooth reflecting surfaces) on the disc forms closely spaced rows. These rows of data do not reflect nearly as much light as the thin portions of the disc that separate them. These areas consist entirely of reflecting material, so light reflected from them undergoes constructive interference in certain directions. This constructive interference depends on the direction of the incoming light, the orientation of the disc, and the light's wavelength. Each wavelength of light can be seen at a particular angle with respect to the disc's surface, causing you to see a "rainbow" of color, as shown in **Figure 15.**

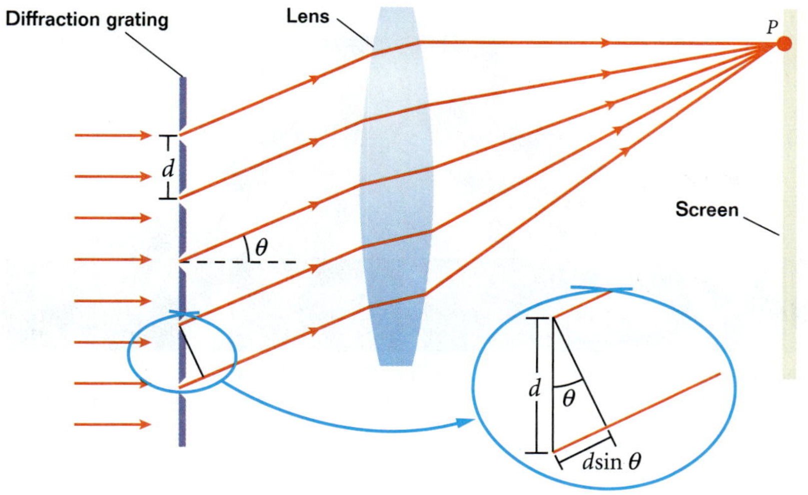

Figure 16
Light of a single wavelength passes through each of the slits of a diffraction grating to constructively interfere at a particular angle θ.

This phenomenon has been put to practical use in a device called a *diffraction grating.* A diffraction grating, which can be constructed to either transmit or reflect light, uses diffraction and interference to disperse light into its component colors with an effect similar to that of a glass prism. A transmission grating consists of many equally spaced parallel slits. Gratings are made by ruling equally spaced lines on a piece of glass using a diamond cutting point driven by an elaborate machine called a *ruling engine.* Replicas are then made by pouring liquid plastic on the grating and then peeling it off once it has set. This plastic grating is then fastened to a flat piece of glass or plastic for support.

Figure 16 shows a schematic diagram of a section of a diffraction grating. A monochromatic plane wave is incoming from the left, normal to the plane of the grating. The waves that emerge nearly parallel from the grating are brought together at a point P on the screen by the lens. The intensity of the pattern on the screen is the result of the combined effects of interference and diffraction. Each slit produces diffraction, and the diffracted beams in turn interfere with one another to produce the pattern.

For some arbitrary angle, θ, measured from the original direction of travel of the wave, the waves must travel *different* path lengths before reaching point P on the screen. Note that the path difference between waves from any two adjacent slits is $d\sin\theta$. If this path difference equals one wavelength or some integral multiple of a wavelength, waves from all slits will be in phase at P, and a bright line will be observed. The condition for bright line formation at angle θ is therefore given by the equation for constructive interference:

$$d\sin\theta = \pm m\lambda \qquad m = 0, 1, 2, 3, \ldots$$

This equation can be used to calculate the wavelength of light if you know the grating spacing and the angle of deviation. The integer m is the order number for the bright lines of a given wavelength. If the incident radiation contains several wavelengths, each wavelength deviates by a specific angle, which can be determined from the equation.

Why it Matters

Conceptual Challenge

1. Spiked Stars

Photographs of stars always show spikes extending from the stars. Given that the aperture of a camera's rectangular shutter has straight edges, explain how diffraction accounts for the spikes.

2. Radio Diffraction

Visible light waves are not observed diffracting around buildings or other obstacles. However, radio waves can be detected around buildings or mountains, even when the transmitter is not visible. Explain why diffraction is more evident for radio waves than for visible light.

Demonstration

Effect of Slit Size on Diffraction Patterns — GENERAL

Purpose Demonstrate that light diffraction is more evident in narrow slits.

Materials two razor blades, clear glass or plastic plate, unfrosted colored light bulb, adhesive tape

Procedure Tape the two razor blades flat on the plate with one blade's edge nearly touching the edge of the other blade. In a dark room, have students stand 3 to 6 m away from the light and look through the slit to observe the diffraction pattern. Ask how the distance from the first dark line to the bright central fringe would vary as the slit's width decreases (*the distance increases*). Point out that the central fringe is wider and much brighter than the others. Explain that diffraction gratings have many very narrow slits, allowing the observer to see the first bright fringe from each slit. (The other ones are less evident.)

ANSWERS — ADVANCED

Conceptual Challenge

1. The edges of the camera shutter diffract light, causing secondary maxima to appear on the sides of the star's image. These maxima give the appearance of light spikes around the star.
2. The wavelength of a radio wave is much longer than that of visible light.

SECTION 2

The Language of Physics

As in double-slit interference, d represents the distance between two adjacent slits. Bright lines occur for special values of θ that exist when the path difference from adjacent slits is a whole number of wavelengths ($m = 0, 1, 2, 3$, and so forth).

Demonstration

Multiple-Slit Diffraction

Purpose Demonstrate patterns formed by a diffraction grating and the effect of different grating line separations.

Materials laser, two optical gratings with different grating constants, screen

CAUTION *Avoid directing the laser beam toward the students.*

Procedure Shine the laser beam onto the screen, and place the optical grating directly in front of the beam. Have students note the bright central fringe (the zeroth-order maximum) and the first-order maxima, one on each side of the zeroth-order maximum.

Replace the first optical grating with the second. Ask students why there are differences in the separation between the zeroth-order and first-order maxima produced by each grating. *(Smaller line spacing produces greater separation between the zeroth-order and first-order maxima.)*

Remind students that diffraction is greatest when the size of the openings and the wavelength of waves are of the same order of magnitude.

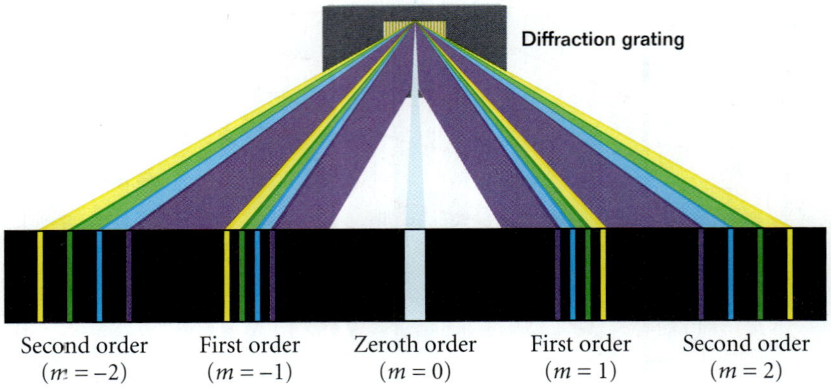

Figure 17
Light is dispersed by a diffraction grating. The angle of deviation for the first-order maximum is smaller for blue light than for yellow light.

Note in **Figure 17** that all wavelengths combine at $\theta = 0$, which corresponds to $m = 0$. This is called the *zeroth-order maximum.* The *first-order maximum,* corresponding to $m = 1$, is observed at an angle that satisfies the relationship $\sin \theta = \lambda/d$. The *second-order maximum*, corresponding to $m = 2$, is observed at an angle where $\sin \theta = 2\lambda/d$.

The sharpness of the principal maxima and the broad range of the dark areas depend on the number of lines in a grating. The number of lines per unit length in a grating is the inverse of the line separation d. For example, a grating ruled with 5000 lines/cm has a slit spacing, d, equal to the inverse of this number; hence, $d = (1/5000)$ cm $= 2 \times 10^{-4}$ cm. The greater the number of lines per unit length in a grating, the less separation between the slits and the farther spread apart the individual wavelengths of light are.

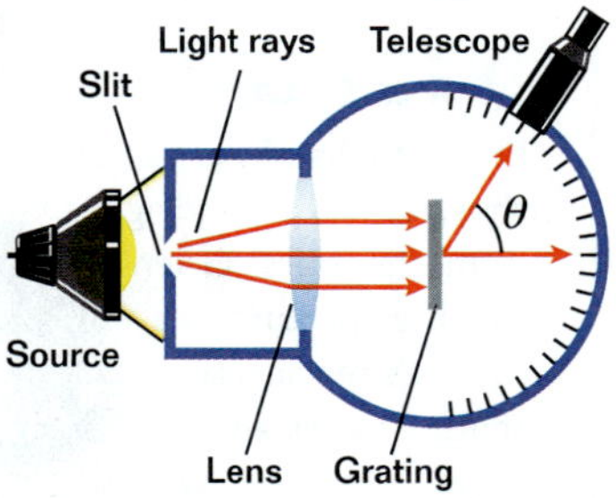

Figure 18
The spectrometer uses a grating to disperse the light from a source.

Diffraction gratings are frequently used in devices called *spectrometers,* which separate the light from a source into its monochromatic components. A diagram of the basic components of a spectrometer is shown in **Figure 18.** The light to be analyzed passes through a slit and is formed into a parallel beam by a lens. The light then passes through the grating. The diffracted light leaves the grating at angles that satisfy the diffraction grating equation. A telescope with a calibrated scale is used to observe the first-order maxima and to measure the angles at which they appear. From these measurements, the wavelengths of the light can be determined and the chemical composition of the light source can be identified. An example of a spectrum produced by a spectrometer is shown in **Figure 19.** Spectrometers are used extensively in astronomy to study the chemical compositions and temperatures of stars, interstellar gas clouds, and galaxies.

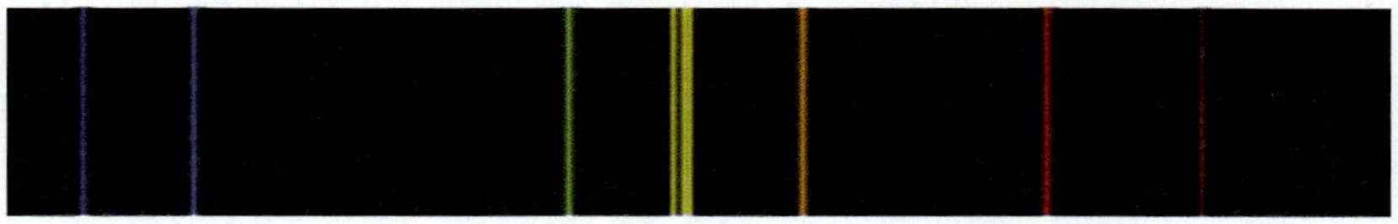

Figure 19
The light from mercury vapor is passed through a diffraction grating, producing the spectrum shown.

SAMPLE PROBLEM B

Diffraction Gratings

PROBLEM

Monochromatic light from a helium-neon laser ($\lambda = 632.8$ nm) shines at a right angle to the surface of a diffraction grating that contains 150 500 lines/m. Find the angles at which one would observe the first-order and second-order maxima.

SOLUTION

1. DEFINE **Given:** $\lambda = 632.8 \text{ nm} = 6.328 \times 10^{-7} \text{ m}$ $\quad m = 1 \text{ and } 2$

$$d = \frac{1}{150\,500 \frac{\text{lines}}{\text{m}}} = \frac{1}{150\,500} \text{ m}$$

Unknown: $\theta_1 = ?$ $\quad \theta_2 = ?$

Diagram:

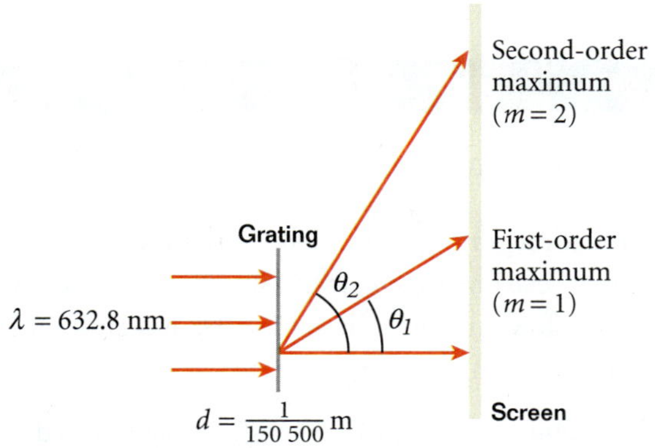

2. PLAN **Choose an equation or situation:** Use the equation for a diffraction grating.

$$d \sin \theta = \pm m\lambda$$

Rearrange the equation to isolate the unknown:

$$\theta = \sin^{-1}\left(\frac{m\lambda}{d}\right)$$

3. CALCULATE **Substitute the values into the equation and solve:**

For the first-order maximum, $m = 1$:

$$\theta_1 = \sin^{-1}\left(\frac{\lambda}{d}\right) = \sin^{-1}\left(\frac{6.328 \times 10^{-7} \text{ m}}{\frac{1}{150\,500} \text{ m}}\right)$$

$$\boxed{\theta_1 = 5.465^\circ}$$

continued on next page

Classroom Practice

Diffraction Gratings
Monochromatic light shines at the surface of a diffraction grating with 5.0×10^3 lines/cm. The first-order maximum is observed at a 15° angle. Find the wavelength.

Answer
5.2×10^2 nm

Find the first-order and the second-order angles of diffraction observed through a 1.00×10^4 lines/cm diffraction grating with light of wavelengths 400.0 nm, 500.0 nm, and 600.0 nm.

Answers
400.0 nm: $\theta_1 = 23.6^\circ$, $\theta_2 = 53.1^\circ$
500.0 nm: $\theta_1 = 30.0^\circ$, $\theta_2 = 90.0^\circ$ (does not occur)
600.0 nm: $\theta_1 = 36.9^\circ$, $\sin\theta_2 = 1.2$ (does not occur)

Light of 400.0 nm wavelength is shined on a 5.0×10^3 lines/cm grating. How many diffraction lines can be observed?

Answer
five ($m = 0, 1, 2, 3, 4$; The $m = 5$ fringe cannot be seen because it is at an angle of 90°.)

537

PROBLEM GUIDE B

Use this guide to assign problems.
SE = Student Edition Textbook
PW = Problem Workbook
PB = Problem Bank on the One-Stop Planner (OSP)

Solving for:

θ	**SE** Sample, 1–2; Ch. Rvw. 19–21 **PW** 4–5 **PB** 4–6
λ	**PW** Sample, 1–2 **PB** 7–10
d	**SE** 5*; Ch. Rvw. 28, 30 **PW** 3 **PB** 3–4
m	**SE** 3–4 **PW** 6–7 **PB** Sample, 1–2

***Challenging Problem**
Consult the printed Solutions Manual or the OSP for detailed solutions.

ANSWERS

Practice B

1. 0.02°, 0.04°, 0.11°
2. a. 11.0°
   b. 17.2°
3. 11
4. 1
5. 6.62×10^3 lines/cm

For $m = 2$:

$$\theta_2 = \sin^{-1}\left(\frac{2\lambda}{d}\right)$$

$$\theta_2 = \sin^{-1}\left(\frac{2(6.328 \times 10^{-7}\ \text{m})}{\frac{1}{150\ 500}\ \text{m}}\right)$$

$$\boxed{\theta_2 = 10.98°}$$

CALCULATOR SOLUTION

Because the minimum number of significant figures for the data is four, the calculator answers 5.464926226 and 10.98037754 should be rounded to four significant figures.

4. EVALUATE The second-order maximum is spread slightly more than twice as far from the center as the first-order maximum. This diffraction grating does not have high dispersion, and it can produce spectral lines up to the tenth-order maxima (where $\sin\theta = 0.9524$).

PRACTICE B

Diffraction Gratings

1. A diffraction grating with 5.000×10^3 lines/cm is used to examine the sodium spectrum. Calculate the angular separation of the two closely spaced yellow lines of sodium (588.995 nm and 589.592 nm) in each of the first three orders.

2. A diffraction grating with 4525 lines/cm is illuminated by direct sunlight. The first-order solar spectrum is spread out on a white screen hanging on a wall opposite the grating.
   a. At what angle does the first-order maximum for blue light with a wavelength of 422 nm appear?
   b. At what angle does the first-order maximum for red light with a wavelength of 655 nm appear?

3. A grating with 1555 lines/cm is illuminated with light of wavelength 565 nm. What is the highest-order number that can be observed with this grating? (Hint: Remember that $\sin\theta$ can never be greater than 1 for a diffraction grating.)

4. Repeat item 3 for a diffraction grating with 15 550 lines/cm that is illuminated with light of wavelength 565 nm.

5. A diffraction grating is calibrated by using the 546.1 nm line of mercury vapor. The first-order maximum is found at an angle of 21.2°. Calculate the number of lines per centimeter on this grating.

DIFFRACTION AND INSTRUMENT RESOLUTION

The ability of an optical system, such as a microscope or a telescope, to distinguish between closely spaced objects is limited by the wave nature of light. To understand this limitation, consider **Figure 20,** which shows two light sources far from a narrow slit. The sources can be taken as two point sources that are not coherent. For example, they could be two distant stars that appear close to each other in the night sky.

If no diffraction occurred, you would observe two distinct bright spots (or images) on the screen at the far right. However, because of diffraction, each source is shown to have a bright central region flanked by weaker bright and dark rings. What is observed on the screen is the resultant from the superposition of two diffraction patterns, one from each source.

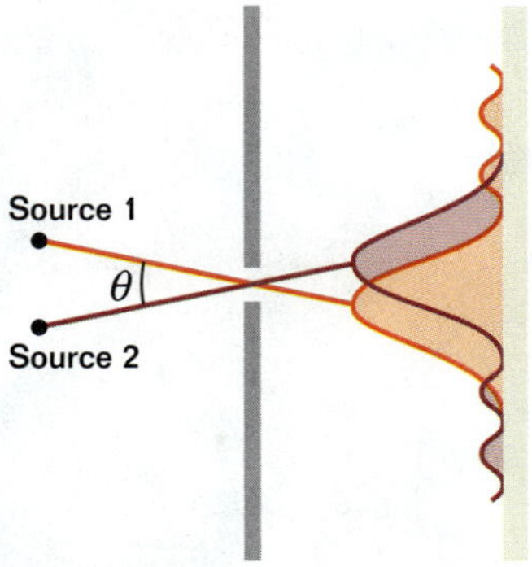

Figure 20
Each of two distant point sources produces a diffraction pattern.

Resolution depends on wavelength and aperture width

If the two sources are separated so that their central maxima do not overlap, as in **Figure 21,** their images can just be distinguished and are said to be barely *resolved.* To achieve high resolution or **resolving power,** the angle between the resolved objects, θ, should be as small as possible as shown in **Figure 20.** The shorter the wavelength of the incoming light or the wider the opening, or *aperture,* through which the light passes, the smaller the angle of resolution, θ, will be and the greater the resolving power will be. For visible-light telescopes, the aperture width, D, is approximately equal to the diameter of the mirror or lens. The equation to determine the limiting angle of resolution *in radians* for an optical instrument with a circular aperture is as follows:

resolving power

the ability of an optical instrument to form separate images of two objects that are close together

$$\theta = 1.22\frac{\lambda}{D}$$

The constant 1.22 comes from the derivation of the equation for circular apertures and is absent for long slits. Note that one radian equals $(180/\pi)°$, as discussed in the Appendix J feature "Angular Kinematics." The equation indicates that for light with a short wavelength, such as an X ray, a small aperture is sufficient for high resolution. On the other hand, if the wavelength of the light is long, as in the case of a radio wave, the aperture must be large in order to resolve distant objects. This is one reason why radio telescopes have large dish-like antennas.

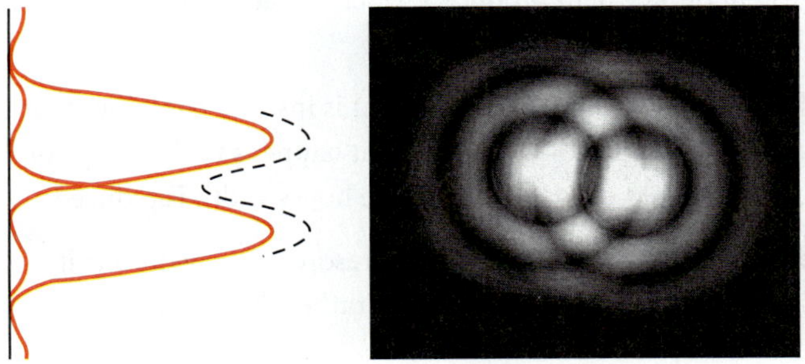

Figure 21
Two point sources are barely resolved if the central maxima of their diffraction patterns do not overlap.

Teaching Tip — BASIC

Explain that the trade-off for seeing more clearly is seeing a smaller portion of an object. Ask students if, with the unaided eye, they can see all of the moon (*yes*) and any details (*few*). Ask them if, when they look through a telescope, they see more details, and ask why that is so (*yes, because of a higher resolving power*). Ask if they see all of the moon (*no, because of a more limited field of vision*). Suggest that they try reading very small print with a magnifying glass. When the spaces between words are very large (because of a high resolving power), fewer words can be read.

The Language of Physics — ADVANCED

Point out that the equation for resolving power is known as the Rayleigh criterion. This equation gives the angle θ in *radians.* One rad equals $(180/\pi)°$. For example, for light with a wavelength of 500.0 nm, a lens that is 0.10 m in diameter has a limiting angle of 6.1×10^{-6} rad, or $(3.5 \times 10^{-4})°$. This may seem like a very small angle, but for viewing the surface of Mars (nearly 75×10^6 km away), it corresponds to a distance of about 460 km. This means a telescope with such a lens would show two mountains separated by a distance of less than 460 km as one blurred mountain.

SECTION 2

Teaching Tip

Ask which color, blue or red, gives a better resolution with a microscope (*blue, because it has a shorter wavelength*). Point out that electron microscopes allow us to distinguish smaller details in cells than is possible with visible light because the effective wavelength of electrons used in electron microscopes typically is measured in hundredths of a nanometer. This is about 10 000 times shorter than the wavelength of red light.

SECTION REVIEW ANSWERS

1. **a.** 5.89×10^{-7} m (589 nm)
   b. 24.7°
2. The width of the central maximum increases as the width of the slit decreases.
3. a human hair; Its size is closest to the size of the wavelength of visible light.
4. orange light; Longer wavelengths are diffracted more.
5. The image of the point source casts a sharp shadow when the hole is much wider than the light's wavelength. As the hole narrows, the light waves bend more around the hole's edges, causing the image to widen. Light waves from the hole destructively interfere, producing a regular pattern of dark fringes on either side of the bright central fringe.
6. yes; Ultraviolet light has a shorter wavelength than visible light, and resolving power is greater for short wavelengths.

Figure 22
The 27 antennas at the Very Large Array in New Mexico are used together to provide improved resolution for observing distant radio sources. The antennas can be arranged to have the resolving power of a 36 km wide radio telescope.

Yet, even with their large sizes, radio telescopes cannot resolve sources as easily as visible-light telescopes resolve visible-light sources. At the shortest radio wavelength (1 mm), the largest single antenna for a radio telescope—the 305 m dish at Arecibo, Puerto Rico—has a resolution angle of 4×10^{-6} rad. The same resolution angle can be obtained for the longest visible light waves (700 nm) by an optical telescope with a 21 cm mirror.

To compensate for the poor resolution of radio waves, one can combine several radio telescopes so that they will function like a much larger telescope. An example of this is shown in **Figure 22.** If the radio antennas are arranged in a line and computers are used to process the signals that each antenna receives, the resolution of the radio "images" is the same as it would be if the radio telescope had a diameter of several kilometers.

It should be noted that the resolving power for optical telescopes on Earth is limited by the constantly moving layers of air in the atmosphere, which blur the light from objects in space. The images from the *Hubble Space Telescope* are of superior quality largely because the telescope operates in the vacuum of space. Under these conditions, the actual resolving power of the telescope is close to the telescope's theoretical resolving power.

SECTION REVIEW

1. Light passes through a diffraction grating with 3550 lines/cm and forms a first-order maximum at an angle of 12.07°.
   a. What is the wavelength of the light?
   b. At what angle will the second maximum appear?
2. Describe the change in width of the central maximum of the single-slit diffraction pattern as the width of the slit is made smaller.
3. Which object would produce the most distinct diffraction pattern: an apple, a pencil lead, or a human hair? Explain your answer.
4. Would orange light or blue light produce a wider diffraction pattern? Explain why.
5. **Critical Thinking** A point source of light is inside a container that is opaque except for a single hole. Discuss what happens to the image of the point source projected onto a screen as the hole's width is reduced.
6. **Critical Thinking** Would it be easier to resolve nearby objects if you detected them using ultraviolet radiation rather than visible light? Explain.

Lasers

SECTION OBJECTIVES

- **Describe the properties of laser light.**
- **Explain how laser light has particular advantages in certain applications.**

LASERS AND COHERENCE

At this point, you are familiar with electromagnetic radiation that is produced by glowing, or *incandescent,* light sources. This includes light from light bulbs, candle flames, or the sun. You may have seen another form of light that is very different from the light produced by incandescent sources. The light produced by a **laser** has unique properties that make it very useful for many applications.

To understand how laser light is different from conventional light, consider the light produced by an incandescent light bulb, as shown in **Figure 23.** When electric charges move through the filament, electromagnetic waves are emitted in the form of visible light. In a typical light bulb, there are variations in the structure of the filament and in the way charges move through it. As a result, electromagnetic waves are emitted at different times from different parts of the filament. These waves have different intensities and move in different directions. The light also covers a wide range of the electromagnetic spectrum because it includes light of different wavelengths. Because so many different wavelengths exist, and because the light is changing almost constantly, the light produced is incoherent. That is, the component waves do not maintain a constant phase difference at all times. The wave fronts of incoherent light are like the wave fronts that result when rain falls on the surface of a pond. No two wave fronts are caused by the same event, and they therefore do not produce a stable interference pattern.

laser

a device that produces coherent light at a single wavelength

Did you know?

The light from an ordinary electric lamp undergoes about 100 million (10^8) random changes every second.

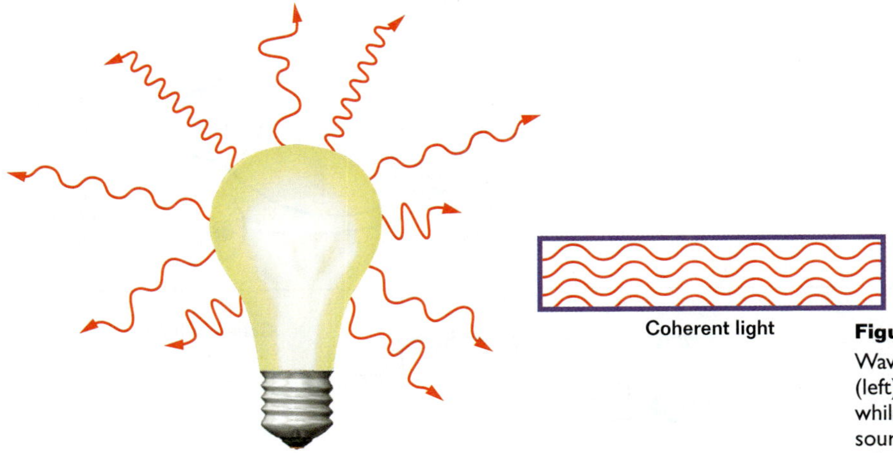

Figure 23
Waves from an incoherent light source (left) have changing phase relationships, while waves from a coherent light source (right) have constant phase relationships.

Demonstration

Dancing Light — ADVANCED

Purpose Demonstrate characteristics of a laser beam.

Materials cassette-tape or CD player with an open speaker, clear plastic wrap, reflective polyester film, laser, light pen

CAUTION *Avoid directing the laser beam toward students.*

Procedure Before students arrive, pull the clear wrap tightly around the speaker, and secure it with tape. Tape the reflective film to the clear wrap in front of the speaker. In class, direct the laser beam so that it is reflected off the reflective film attached to the speaker and onto the ceiling. Play the music, and darken the room. Replace the laser with a light pen, and have students compare the two beams. Have them note that laser light is emitted in one narrow beam in one direction and is almost monochromatic.

Demonstration

Interference in Laser Light — GENERAL

Purpose Demonstrate interference of light waves with the same wavelength.

Materials laser, light pen, projection screen, opaque slide with two slits (to make one, spray a layer of paint on a glass plate and use a pin to etch two parallel slits about 1 mm apart through the paint)

CAUTION *Avoid directing the light from the laser at the students.*

Procedure Shine the laser through two slits and project the beam toward the screen. Point out that each slit acts as a point source of light. Have students examine the bands of brightness and darkness. Explain that the areas of darkness correspond to the nodal lines observed in the experiments with sound and water waves in Section 1, in the first two Demonstrations of this chapter.

Repeat the experiment with a light pen. Explain that although the light from the two slits is nearly coherent, white light contains many wavelengths. Although these interfere with each other, the pattern is blurred.

SCILINKS.
www.scilinks.org
Topic: Lasers
Code: HF60853

Lasers, on the other hand, typically produce a narrow beam of coherent light. The waves emitted by a laser are in phase, and they do not shift relative to each other as time progresses. Because all the waves are in phase, they interfere constructively at all points. The individual waves effectively behave like a single wave with a very large amplitude. In addition, the light produced by a laser is monochromatic, so all the waves have exactly the same wavelength. As a result of these properties, the intensity, or brightness, of laser light can be made much greater than that of incoherent light. For light, intensity is a measure of the energy transferred per unit time over a given area.

Lasers transform energy into coherent light

A laser is a device that converts light, electrical energy, or chemical energy into coherent light. There are a variety of different types of lasers, but they all have some common features. They all use a substance called the *active medium* to which energy is added to produce coherent light. The active medium can be a solid, liquid, or gas. The composition of the active medium determines the wavelength of the light produced by the laser.

The basic operation of a laser is shown in **Figure 24.** When high-energy light or electrical or chemical energy is added to the active medium, as in **Figure 24(a),** the atoms in the active medium absorb some of the energy. You will learn that atoms exist at different *energy states* in the chapter "Atomic

Did you know?

The word *laser* is an acronym (a word made from the first letters of several words) that stands for "*l*ight *a*mplification by *s*timulated *e*mission of *r*adiation."

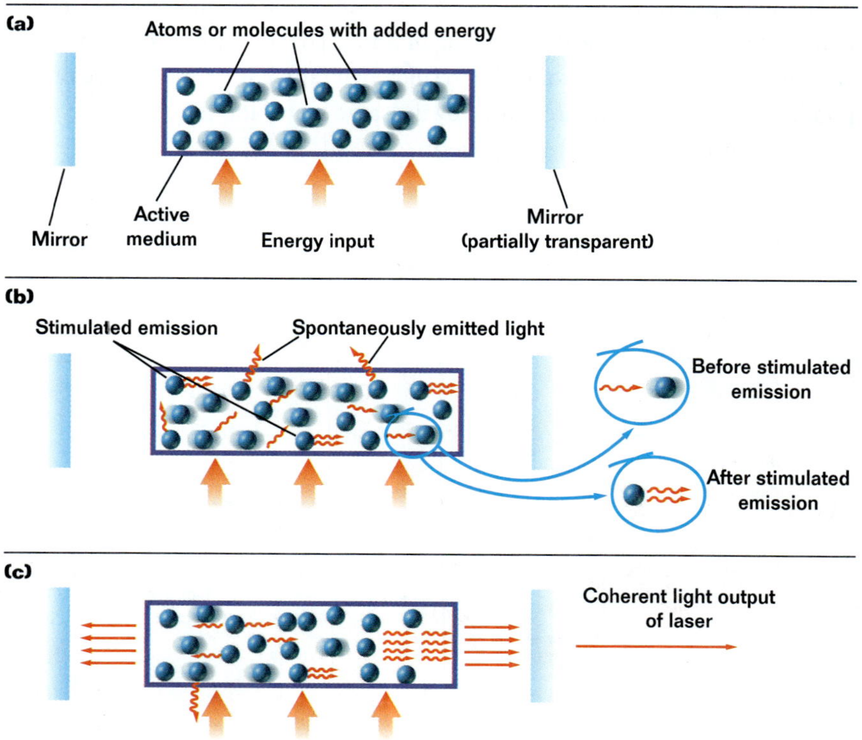

Figure 24
(a) Atoms or molecules in the active medium of a laser absorb energy from an external source. **(b)** When a spontaneously emitted light wave interacts with an atom, it may cause the atom to emit an identical light wave. **(c)** Stimulated emission increases the amount of coherent light in the active medium, and the coherent waves behave as a single wave.

Physics." When energy is added to an atom that is at a lower energy state, the atom can be excited to a higher energy state. These *excited* atoms then release their excess energy in the form of electromagnetic radiation when they return to their original, lower energy states.

When light of a certain wavelength is applied to excited atoms, the atoms can be induced to release light waves that have the same properties. After one atom spontaneously releases its energy in the form of a light wave, this initial wave can cause other energized atoms to release their excess energy as light waves with the same wavelength, phase, and direction as the initial wave, as shown in **Figure 24(b).** This process is called *stimulated emission.*

Most of the light produced by stimulated emission escapes out the sides of the glass tube. However, some of the light moves along the length of the tube, producing more stimulated emission as it goes. Mirrors on the ends of the material return these coherent light waves into the active medium, where they stimulate the emission of more coherent light waves, as shown in **Figure 24(c).** As the light passes back and forth through the active medium, it becomes more and more intense. One of the mirrors is slightly transparent, which allows the intense coherent light to be emitted by the laser.

APPLICATIONS OF LASERS

There are a wide variety of laser types, with wavelengths ranging from the far infrared to the X-ray region of the spectrum. Scientists have also created *masers,* devices similar to lasers but operate in the microwave region of the spectrum. Lasers are used in many ways, from common household uses to a wide variety of industrial uses and very specialized medical applications.

Lasers are used to measure distances with great precision

Of the properties of laser light, the one that is most evident is that it emerges from the laser as a narrow beam. Unlike the light from a light bulb or even the light that is focused by a parabolic reflector, the light from a laser undergoes very little spreading with distance. One reason is that all the light waves emitted by the laser have the same direction. As a result, a laser can be used to measure large distances, because it can be pointed at distant reflectors and the reflected light can be detected.

As shown in **Figure 25,** astronomers direct laser light at particular points on the moon's surface to determine the Earth-to-moon distance. A pulse of light is directed toward one of several 0.25 m^2 reflectors that were placed on the moon's surface by astronauts during the *Apollo* missions. By knowing the speed of light and measuring the time the light takes to travel to the moon and back, scientists have measured the Earth-to-moon distance to be about 3.84×10^5 km. Geologists use repeated measurements to record changes in the height of Earth's crust from geological processes. Lasers can be used for these measurements even when the height changes by only a few centimeters.

Figure 25
A laser beam is fired at reflectors on the moon, which is more than 380 000 km away.

SECTION 3

Key Models and Analogies — GENERAL

Have students think of a busy shopping mall, where people walk in all directions and with different strides. Tell students that this scene can represent incoherent light emitted by a light bulb, with each person being analogous to a light wave with its own direction, wavelength, and phase. Then, have students think of a marching band, with all members of the band lifting their feet at precisely the same time and marching in the same direction with steps that are the same size. Explain that this is analogous to laser light.

STOP Misconception Alert — ADVANCED

Students may think that fluorescence and phosphorescence are related to the way lasers work. Make sure they realize that in fluorescent lamps, any atoms that become excited (by absorbing energy) immediately lose the energy by emitting light. In phosphorescent paint, the excited atoms are able to remain at a higher level of energy for a longer period of time, but they emit light spontaneously. In lasers, most of the atoms remain in an excited state until something triggers the excited atoms to emit light.

Why it Matters

Compact Disc Players

CDs, CD-ROMs, CD-Rs, and DVDs all fall into a class of digital media called *optical storage devices.*

The laser typically used for a CD player has a wavelength of 780 nm, light in the near-infrared region. A DVD player laser has a wavelength of 650 nm, visible as red light. Both of these lasers are diode lasers. These lasers are semiconductor devices that produce low-intensity, coherent light when they carry current.

Why it Matters

Compact Disc Players

An interesting application of the laser is the compact disc (CD) player. In a CD player, light from a laser is directed through a series of optics toward a compact disc on which the music or data have been digitally recorded. The CD player "reads" the data in the way the laser light is reflected from the compact disc.

In digital recording, a sound signal is sampled at regular intervals of time. Each sampling is converted to an electrical signal, which in turn is converted into a series of binary numbers. Binary numbers consist only of zeros and ones. The binary numbers are coded to contain information about the signal, including the frequencies and harmonics that are present, the volume for the left and right channels, and the speed of the motor that rotates the disc. This process is called *analog-to-digital* (a-d) conversion.

These binary, digital data in a CD are stored as a series of pits and smooth areas (called *lands*) on the surface of the disc. The series of pits and lands is recorded starting at the center of the disc and spiraling outward along *tracks* in the CD. These tracks are just 500 nm wide and spaced 1600 nm apart. If you could stretch the data track of a CD out, it would be almost 5 km long!

When a CD is played, the laser light is reflected off this series of pits and lands into a detector. In fact, the depth of the pit is chosen so that destructive interference occurs when the laser transitions from a pit to a land or from a land to a pit. The detector records the changes in light reflection between the pits and lands as ones and smooth areas as zeros—binary data that are then converted back to the analog signal you hear as music. This step is called *digital-to-analog* (d-a) conversion, and the analog signal can then be amplified to the speaker system.

A CD read-only memory (CD-ROM) drive on your computer works in much the same way. Data from a computer are already in a digital format, so no a-d or d-a conversion is needed.

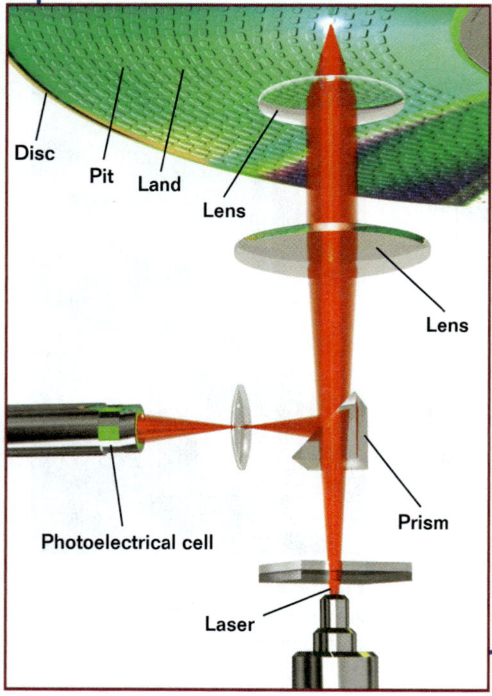

Light from a laser is directed toward the surface of the compact disc. Smooth parts of the disc reflect the light back to the photoelectrical cell.

You may wonder how a CD-recordable (CD-R) disc is different. These discs don't have any pits and lands at all. Instead, they have a layer of light-sensitive dye sandwiched between a smooth reflective metal, usually aluminum, and clear plastic. A CD-R drive has an additional laser, about 10 times more powerful than a CD reading laser, that writes the digital data along the tracks of the CD-R disc. When the writing laser shines on the light-sensitive dye, the dye turns dark and creates nonreflecting areas along the track. This process creates the digital pattern that behaves like the pits and lands, which a standard CD player can read.

A digital versatile disc (DVD) player operates on the same principle. However, the laser in a DVD player has a shorter wavelength than the laser in a CD player. This shorter wavelength allows the DVD player to read data that are spaced closer together than data on a CD. Additionally, some DVDs contain two layers of data and may even be written on both sides! The lower layer of a two-layer DVD has a thinner coating of reflective material, usually gold, that allows some of the light to pass through it so that the upper level of the DVD can be read.

Lasers have many applications in medicine

Lasers are also used for many medical procedures by making use of the fact that specific body tissues absorb different wavelengths of laser light. For example, lasers can be used to lighten or remove scars and certain types of birthmarks without affecting surrounding tissues. The scar tissue responds to the wavelength of light used in the laser, but other body tissues are protected.

Many medical applications of lasers take advantage of the fact that water can be vaporized by high-intensity infrared light produced by carbon dioxide lasers having a wavelength of 10 μm. Carbon dioxide lasers can cut through muscle tissue by heating and evaporating the water contained in the cells. One advantage of a laser is that the energy from the laser also coagulates blood in the newly opened blood vessels, thereby reducing blood loss and decreasing the risk of infection. A laser beam can also be trapped in an optical fiber endoscope, which can be inserted through an orifice and directed to internal body structures. As a result, surgeons can stop internal bleeding or remove tumors without performing massive surgery.

Lasers can also be used to treat tissues that cannot be reached by conventional surgical methods. For example, some very specific wavelengths of lasers can pass through certain structures at the front of the eye—the cornea and lens—without damaging them. Therefore, lasers can be effective at treating lesions of the retina, inside the eye. Lasers are used for other eye surgeries, including surgery to correct *glaucoma,* a condition in which the fluid pressure within the eye is too great. Left untreated, glaucoma can lead to damage of the optic nerve and eventual blindness. Focusing a laser at the clogged drainage port allows a tiny hole to be burned in the tissue, which relieves the pressure. Lasers can also be used to correct nearsightedness by focusing the beam on the central portion of the cornea to cause it to become flatter.

Did you know?

The principle behind reading the information stored on a compact disc is also the basis for the reading of bar codes found on many products. When these products are scanned, laser light reflected from the bars and spaces of the bar code reproduces the binary codes that represent the product's inventory number. This information is transmitted to the store's computer system, which returns the product's name and price to the cash register.

SCILINKS.
www.scilinks.org
Topic: Bar Codes
Code: HF60135

SECTION REVIEW

1. How does light from a laser differ from light whose waves all have the same wavelength but are not coherent?

2. The process of stimulated emission involves producing a second wave that is identical to the first. Does this gaining of a second wave violate the principle of energy conservation? Explain your answer.

3. **Critical Thinking** Fiber-optic systems transmit light by means of internal reflection within thin strands of extremely pure glass. In these fiber-optic systems, laser light is used instead of white light to transmit the signal. Apply your knowledge of refraction to explain why.

Teaching Tip — ADVANCED

Point out that there are many types of lasers, including semiconductor (diode) lasers, CO_2 lasers, ruby lasers, microwave lasers, X-ray lasers, and tunable lasers. Lasers now have medical, military, industrial, and scientific applications. They are used as bloodless scalpels, saws, and moonquake detectors. Astronomers have even found laser action in stars. In 1996, the *Hubble Space Telescope* discovered an ultraviolet laser star. Have students find more information about how lasers are used, and have them report on one of these topics.

SECTION REVIEW ANSWERS

1. Waves emitted by a laser do not shift relative to each other as time progresses (they are coherent and continuously in phase).
2. no; The second light wave is obtained from the energy that is added to an atom in the active medium by an external energy source.
3. Laser light is nearly monochromatic, so it does not spread out very much into different components with different wavelengths as it passes between the fiber and transmission and receiving equipment (that is, dispersion is nearly absent over short distances).

Laser Surgeon

Dr. Shawn Wong practices ophthalmology in Austin, Texas, and specializes in LASIK (*l*aser-*a*ssisted *i*n *s*itu *k*eratomileusis) surgery to correct vision problems. In his spare time, Dr. Wong enjoys bicycle racing and flying model airplanes.

Dr. Wong particularly likes the problem-solving aspects of his work. He also enjoys "not only making a living, but helping to solve people's problems along the way. Laser eye surgery is a very dynamic field—what we can't do now is just around the corner."

PHYSICS CAREERS

Laser Surgeon

Laser surgery combines two fields—eye care and high-tech engineering—to give perfect vision to people who otherwise would need glasses or contacts. To learn more about this career, read the interview with ophthalmologist Dr. L. Shawn Wong, who runs a laser center in Austin, Texas.

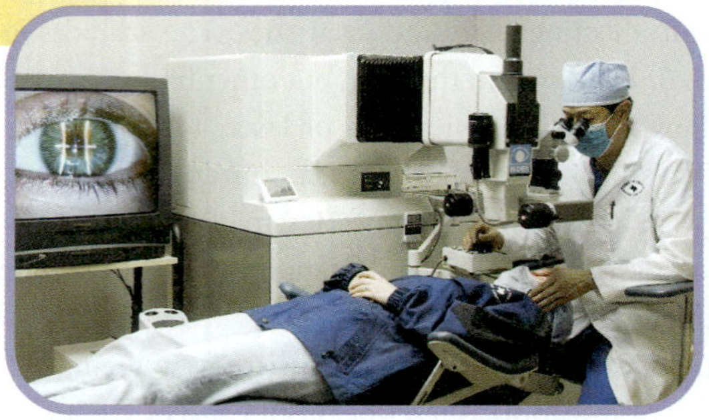

Dr. Wong makes measurements of the eye in preparation for laser surgery.

What sort of education helped you become a laser surgeon?

Besides using my medical school training, I use a lot of engineering in my work; physics and math courses are very helpful. In high school, even in junior high, having a love of math and science is extremely helpful.

Who helped you find your career path?

Of all my teachers, my junior high earth science teacher made the biggest impression on me. What I learned in those classes I actually still use today: problem solving. Interestingly, I work in the town where I grew up; a lot of my former teachers are my patients today.

What makes laser surgery interesting to you?

It's nice to be able to help people. Unlike glasses and contacts, laser surgery is not a correction; it's a cure. When you are improving people's vision, everybody in the room gets to see the results. I don't need to tell patients they're doing well—they can tell.

What is the nature of your work?

A typical patient is somebody born with poor vision. We make these patients undergo a lot of formal diagnostic testing and informal screening to be sure they are good candidates. Lasers are used for diagnosing as well as treating. Laser tolerances are extremely small—we're talking in terms of submicrons, the individual cells of the eye.

What is your favorite thing about your job? What would you most like to change about it?

My favorite thing is making people visually free. I would like to be able to solve an even wider range of problems. We can't solve everything.

How does your work relate to the physics of interference and diffraction?

Measuring diffraction and interference is part of every aspect of what we do. The approach is based on doing many small things correctly. Applying small physics principles in the right order can solve very big problems.

What advice would you give to somebody who is considering a career in laser surgery?

My education didn't start in medical school; it started by asking questions as a kid. You need a genuine love of taking on complex problems. A background in physics and math is extremely helpful. Technology in medicine is based on engineering.

Being well rounded will help you get into medical school—and get out, too. You have to be comfortable doing the science, but you also have to be comfortable dealing with people.

CHAPTER 15 Highlights

KEY IDEAS

Section 1 Interference

- Light waves with the same wavelength and constant phase differences interfere with each other to produce light and dark interference patterns.
- In double-slit interference, the position of a bright fringe requires that the path difference between two interfering point sources be equal to a whole number of wavelengths.
- In double-slit interference, the position of a dark fringe requires that the path difference between two interfering point sources be equal to an odd number of half wavelengths.

Section 2 Diffraction

- Light waves form a diffraction pattern by passing around an obstacle or bending through a slit and interfering with each other.
- The position of a maximum in a pattern created by a diffraction grating depends on the separation of the slits in the grating, the order of the maximum, and the wavelength of the light.

Section 3 Lasers

- A laser is a device that transforms energy into a beam of coherent monochromatic light.

Variable Symbols

Quantities		Units	
λ	wavelength	m	meters
θ	angle from the center of an interference pattern	°	degrees
d	slit separation	m	meters
m	order number	(unitless)	

KEY TERMS

coherence (p. 527)

path difference (p. 529)

order number (p. 529)

diffraction (p. 532)

resolving power (p. 539)

laser (p. 541)

PROBLEM SOLVING

See **Appendix D: Equations** for a summary of the equations introduced in this chapter. If you need more problem-solving practice, see **Appendix I: Additional Problems.**

CHAPTER 15

Highlights

Teaching Tip

Because double-slit interference can be a difficult subject to understand, students may find it helpful to write an expository essay with diagrams, explaining how an interference pattern is created. Essays should include a discussion of diffraction as it relates to double-slit interference.

Review

ANSWERS

1. The amplitude of the resultant wave is twice the amplitude of either interfering wave, so a bright fringe forms; The amplitude of the resultant wave is zero, so a dark fringe forms.
2. by differences in brightness
3. $\lambda_{blue\ light} < \lambda_{red\ light}$, so interference fringes form at smaller angles for blue light.
4. d, m, θ
5. θ would decrease because λ is shorter in water.
6. no; because the light from the stars is not coherent
7. No interference is observed because the two waves have different wavelengths.
8. The separation of the fringes decreases because d and $\sin\theta$ are inversely proportional.
9. 630 nm
10. a. 589 nm
    b. 0.327°
    c. 0.436°
11. 160 μm
12. At certain angles the diffracted waves from different parts of the slit destructively interfere with each other, forming dark fringes like those observed in an interference pattern.
13. It increases as the light with longer wavelength is diffracted more.
14. Light of a particular wavelength interferes constructively at a particular angle.
15. The larger an orbiting telescope is, the more expensive it is to place in orbit. In order to

CHAPTER 15 Review

INTERFERENCE

Review Questions

1. What happens if two light waves with the same amplitude interfere constructively? What happens if they interfere destructively?
2. Interference in sound is recognized by differences in volume; how is interference in light recognized?
3. A double-slit interference experiment is performed with red light and then again with blue light. In what ways do the two interference patterns differ? (Hint: Consider the difference in wavelength for the two colors of light.)
4. What data would you need to collect to correctly calculate the wavelength of light in a double-slit interference experiment?

Conceptual Questions

5. If a double-slit experiment were performed underwater, how would the observed interference pattern be affected? (Hint: Consider how light changes in a medium with a higher index of refraction.)
6. Because of their great distance from us, stars are essentially point sources of light. If two stars were near each other in the sky, would the light from them produce an interference pattern? Explain your answer.
7. Assume that white light is provided by a single source in a double-slit experiment. Describe the interference pattern if one slit is covered with a red filter and the other slit is covered with a blue filter.
8. An interference pattern is formed by using green light and an apparatus in which the two slits can move. If the slits are moved farther apart, will the separation of the bright fringes in the pattern decrease, increase, or remain unchanged? Why?

Practice Problems

For problems 9–11, see Sample Problem A.

9. Light falls on two slits spaced 0.33 mm apart. If the angle between the first dark fringe and the central maximum is 0.055°, what is the wavelength of the light?
10. A sodium-vapor street lamp produces light that is nearly monochromatic. If the light shines on a wooden door in which there are two straight, parallel cracks, an interference pattern will form on a distant wall behind the door. The slits have a separation of 0.3096 mm, and the second-order maximum occurs at an angle of 0.218° from the central maximum. Determine the following quantities:
    a. the wavelength of the light
    b. the angle of the third-order maximum
    c. the angle of the fourth-order maximum
11. All but two gaps within a set of venetian blinds have been blocked off to create a double-slit system. These gaps are separated by a distance of 3.2 cm. Infrared radiation is then passed through the two gaps in the blinds. If the angle between the central and the second-order maxima in the interference pattern is 0.56°, what is the wavelength of the radiation?

DIFFRACTION

Review Questions

12. Why does light produce a pattern similar to an interference pattern when it passes through a single slit?
13. How does the width of the central region of a single-slit diffraction pattern change as the wavelength of the light increases?
14. Why is white light separated into a spectrum of colors when it is passed through a diffraction grating?

15. Why might orbiting telescopes be problematic for the radio portion of the electromagnetic spectrum?

Conceptual Questions

16. Monochromatic light shines through two different diffraction gratings. The second grating produces a pattern in which the first-order and second-order maxima are more widely spread apart. Use this information to tell if there are more or fewer lines per centimeter in the second grating than in the first.

17. Why is the resolving power of your eye better at night than during the day?

18. Globular clusters, such as the one shown below, are spherical groupings of stars that form a ring around the Milky Way galaxy. Because there can be millions of stars in a single cluster and because they are distant, resolving individual stars within the cluster is a challenge. Of the following conditions, which would make it easier to resolve the component stars? Which would make it more difficult?

   a. The number of stars per unit volume is half as great.
   b. The cluster is twice as far away.
   c. The cluster is observed in the ultraviolet portion instead of in the visible region of the electromagnetic spectrum.
   d. The telescope's mirror or lens is twice as wide.

Practice Problems

For problems 19–21, see Sample Problem B.

19. Light with a wavelength of 707 nm is passed through a diffraction grating with 795 slits/cm. Find the angle at which one would observe the first-order maximum.

20. If light with a wavelength of 353 nm is passed through the diffraction grating with 795 slits/cm, find the angle at which one would observe the second-order maximum.

21. By attaching a diffraction-grating spectroscope to an astronomical telescope, one can measure the spectral lines from a star and determine the star's chemical composition. Assume the grating has 3661 lines/cm.
   a. If the wavelengths of the star's light are 478.5 nm, 647.4 nm, and 696.4 nm, what are the angles at which the first-order spectral lines occur?
   b. At what angles are these lines found in the second-order spectrum?

LASERS

Review Questions

22. What properties does laser light have that are not found in the light used to light your home?

23. Laser light is commonly used to demonstrate double-slit interference. Explain why laser light is preferable to light from other sources for observing interference.

24. Give two examples in which the uniform direction of laser light is advantageous. Give two examples in which the high intensity of laser light is advantageous.

25. Laser light is often linearly polarized. How would you show that this statement is true?

MIXED REVIEW

26. The 546.1 nm line in mercury is measured at an angle of 81.0° in the third-order spectrum of a diffraction grating. Calculate the number of lines per centimeter for the grating.

27. Recall from your study of heat and entropy that the entropy of a system is a measure of that system's disorder. Why is it appropriate to describe a laser as an entropy-reducing device?

have adequate resolution, an orbiting radio telescope would have to be very large.

16. The sin θ is inversely proportional to d, which is the reciprocal of the number of lines per centimeter. The grating that spreads the pattern the most has the most lines per centimeter.
17. The pupil is larger at night, so the angle of resolution is smaller and resolving power is greater.
18. easier: a, c, d
    harder: b
19. 3.22°
20. 3.22°
21. a. 10.09°, 13.71°, 14.77°
    b. 20.51°, 28.30°, 30.66°
22. Laser light is coherent and monochromatic.
23. Light must be coherent for an interference pattern to form. An interference pattern is well defined with monochromatic light.
24. Answers may include distance measurements, compact disc players, and fiber-optic communications; Answers may include laser surgery and fiber-optic communications.
25. by rotating a polarizing sheet in front of the laser light—if the intensity of the transmitted light varies, the light is polarized
26. 6030 lines/cm
27. Energy (light, for example) added to the active medium's atoms is not highly ordered. This light travels in all directions, is incoherent, and is not monochromatic. The laser converts a portion of this energy into a beam of more coherent, monochromatic light.

15 REVIEW

28. 2.41×10^{-4} m
29. 432.0 nm
30. 8.000×10^{-7} m
31. 1.93×10^{-3} mm = 3 λ; a maximum

Alternative Assessment ANSWERS

1. To represent constant wavelength, the distances between lines should not exceed the lines' thickness. Students can create the same effects by overlapping drawings on computer screens.
2. One hole should produce concentric diffraction rings; two holes should produce a striped pattern. Suggested improvements will vary.
3. Students should use path difference and refraction to predict wavelengths.
4. Presentations will vary but should indicate the source material used. Young (1773–1829) was a child prodigy. Fresnel (1788–1827) was eight years old before he could read.
5. Student presentations will vary. Findings should be explained clearly.

28. A double-slit interference experiment is performed using blue light from a hydrogen discharge tube (λ = 486 nm). The fifth-order bright fringe in the interference pattern is 0.578° from the central maximum. How far apart are the two slits separated?

29. A beam containing light of wavelengths λ_1 and λ_2 passes through a set of parallel slits. In the interference pattern, the fourth bright line of the λ_1 light occurs at the same position as the fifth bright line of the λ_2 light. If λ_1 is known to be 540.0 nm, what is the value of λ_2?

30. Visible light from an incandescent light bulb ranges from 400.0 nm to 700.0 nm. When this light is focused on a diffraction grating, the entire first-order spectrum is seen, but none of the second-order spectrum is seen. What is the maximum spacing between lines on this grating?

31. In an arrangement to demonstrate double-slit interference, λ = 643 nm, θ = 0.737°, and d = 0.150 mm. For light from the two slits interfering at this angle, what is the path difference both in millimeters and in terms of the number of wavelengths? Will the interference correspond to a maximum, a minimum, or an intermediate condition?

Alternative Assessment

1. Design simulations of interference patterns. Use a computer to draw many concentric circles at regular distances to represent waves traveling from a point source. Photocopy the page onto two transparencies, and lay them on an overhead projector. Vary the distances between "source points," and observe how these variations affect interference patterns. Design transparencies with thicker lines with larger separations to explore the effect of wavelength on interference.

2. Investigate the effect of slit separation on interference patterns. Wrap a flashlight or a pen light tightly with tin foil and make pinholes in the foil. First, record the pattern you see on a screen a few inches away with one hole; then, do the same with two holes. How does the distance between the holes affect the distance between the bright parts of the pattern? Draw schematic diagrams of your observations, and compare them with the results of double-slit interference. How would you improve your equipment?

3. Soap bubbles exhibit different colors because light that is reflected from the outer layer of the soap film interferes with light that is refracted and then reflected from the inner layer of the soap film. Given a refractive index of n = 1.35 and thicknesses ranging from 600 nm to 1000 nm for a soap film, can you predict the colors of a bubble? Test your answer by making soap bubbles and observing the order in which the different colors appear. Can you tell the thickness of a soap bubble from its colors? Organize your findings into a chart, or create a computer program to predict the thicknesses of a bubble based on the wavelengths of light it appears to reflect.

4. Thomas Young's 1803 experiment provided crucial evidence for the wave nature of light, but it was met with strong opposition in England until Augustin Fresnel presented his wave theory of light to the French Academy of Sciences in 1819. Research the lives and careers of these two scientists. Create a presentation about one of them. The presentation can be in the form of a report, poster, short video, or computer presentation.

5. Research waves that surround you, including those used in commercial, medicinal, and industrial applications. Interpret how the waves' characteristics and behaviors make them useful. For example, investigate what kinds of waves are used in medical procedures such as MRI and ultrasound. What are their wavelengths? Research how lasers are used in medicine. How are they used in industry? Prepare a poster or chart describing your findings, and present it to the class.

Graphing Calculator Practice

Double-Slit Experiment

One of the classic experiments that demonstrate the wave nature of light is the double-slit experiment. In this experiment, light from a single source is passed through a narrow slit and then through two narrow parallel slits. When the light appears on a viewing screen behind the slits, you see a pattern of alternating bright and dark fringes corresponding to constructive and destructive interference of the light.

As you studied earlier in the chapter, the bright fringes are described by the following equation.

$$d \sin \theta = \pm m\lambda$$

In this equation, d is the slit separation, θ is the fringe angle, m is the order number, and λ is the wavelength of the incident wave. Typically, only the first few fringes ($m = 0, 1, 2, 3$) are bright enough to see.

In this graphing calculator activity, you will calculate a table of fringe angles. By analyzing this table, you will gain a better understanding of the relationship between fringe angles, wavelength, and slit separation.

Visit go.hrw.com and enter the keyword **HF6INFX** to find this graphing calculator activity. Refer to **Appendix B** for instructions on downloading the program for this activity.

extension

Graphing Calculator Practice

Visit go.hrw.com for answers to this Graphing Calculator activity.

Keyword HF6INFXT

CHAPTER 15

Standardized Test Prep

ANSWERS

1. B
2. H
3. C
4. G
5. C
6. H
7. B

Standardized Test Prep

MULTIPLE CHOICE

1. In the equations for interference, what does the term *d* represent?

A. the distance from the midpoint between the two slits to the viewing screen
B. the distance between the two slits through which a light wave passes
C. the distance between two bright interference fringes
D. the distance between two dark interference fringes

2. Which of the following must be true for two waves with identical amplitudes and wavelengths to undergo complete destructive interference?

F. The waves must be in phase at all times.
G. The waves must be 90° out of phase at all times.
H. The waves must be 180° out of phase at all times.
J. The waves must be 270° out of phase at all times.

3. Which equation correctly describes the condition for observing the third dark fringe in an interference pattern?

A. $d \sin \theta = \lambda/2$
B. $d \sin \theta = 3\lambda/2$
C. $d \sin \theta = 5\lambda/2$
D. $d \sin \theta = 3\lambda$

4. Why is the diffraction of sound easier to observe than the diffraction of visible light?

F. Sound waves are easier to detect than visible light waves.
G. Sound waves have longer wavelengths than visible light waves and so bend more around barriers.
H. Sound waves are longitudinal waves, which diffract more than transverse waves.
J. Sound waves have greater amplitude than visible light waves.

5. Monochromatic infrared waves with a wavelength of 750 nm pass through two narrow slits. If the slits are 25 μm apart, at what angle will the fourth-order bright fringe appear on a viewing screen?

A. 4.3°
B. 6.0°
C. 6.9°
D. 7.8°

6. Monochromatic light with a wavelength of 640 nm passes through a diffraction grating that has 5.0×10^4 lines/m. A bright line on a screen appears at an angle of 11.1° from the central bright fringe. What is the order of this bright line?

F. $m = 2$
G. $m = 4$
H. $m = 6$
J. $m = 8$

7. For observing the same object, how many times better is the resolution of the telescope shown on the left in the figure below than that of the telescope shown on the right?

A. 4
B. 2
C. $\frac{1}{2}$
D. $\frac{1}{4}$

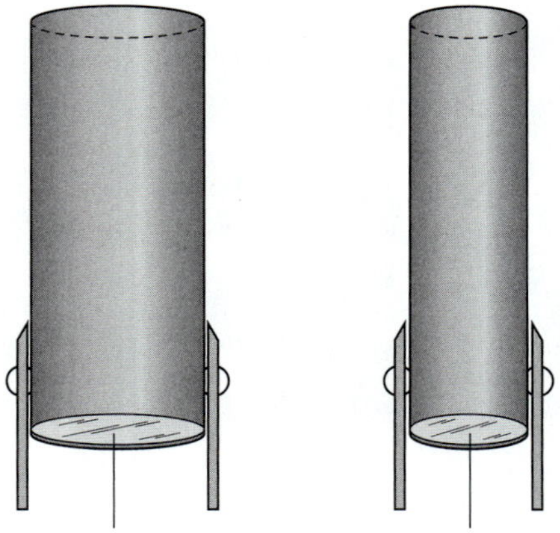

Area of mirror = 80 m^2 Area of mirror = 20 m^2

8. What steps should you employ to design a telescope with a high degree of resolution?
   F. Widen the aperture, or design the telescope to detect light of short wavelength.
   G. Narrow the aperture, or design the telescope to detect light of short wavelength.
   H. Widen the aperture, or design the telescope to detect light of long wavelength.
   J. Narrow the aperture, or design the telescope to detect light of long wavelength.

9. What is the property of a laser called that causes coherent light to be emitted?
   A. different intensities
   B. light amplification
   C. monochromaticity
   D. stimulated emission

10. Which of the following is *not* an essential component of a laser?
   F. a partially transparent mirror
   G. a fully reflecting mirror
   H. a converging lens
   J. an active medium

SHORT RESPONSE

11. Why is laser light useful for the purposes of making astronomical measurements and surveying?

12. A diffraction grating used in a spectrometer causes the third-order maximum of blue light with a wavelength of 490 nm to form at an angle of 6.33° from the central maximum ($m = 0$). What is the ruling of the grating in lines/cm?

13. Telescopes that orbit Earth provide better images of distant objects because orbiting telescopes are more able to operate near their theoretical resolution than telescopes on Earth. The orbiting telescopes needed to provide high resolution in the visible part of the spectrum are much larger than the orbiting telescopes that provide similar images in the ultraviolet and X-ray portion of the spectrum. Explain why the sizes must vary.

EXTENDED RESPONSE

14. Radio signals often reflect from objects and recombine at a distance. Suppose you are moving in a direction perpendicular to a radio signal source and its reflected signal. How would interference between these two signals sound on a radio receiver?

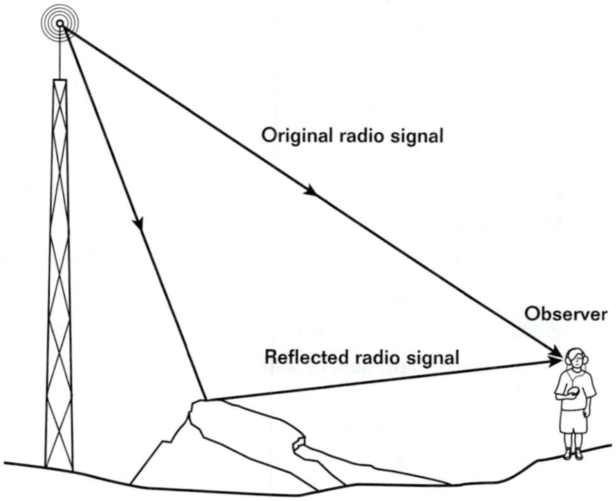

Base your answers to questions 15–17 on the information below. In each problem, show all of your work.

A double-slit apparatus for demonstrating interference is constructed so that the slits are separated by 15.0 μm. A first-order fringe for constructive interference appears at an angle of 2.25° from the zeroth-order (central) fringe.

15. What is the wavelength of the light?

16. At what angle would the third-order ($m = 3$) bright fringe appear?

17. At what angle would the third-order ($m = 3$) dark fringe appear?

Test TIP Be sure that angles in all calculations involving trigonometric functions are computed in the proper units (degrees or radians).

8. F
9. D
10. H
11. The beam does not spread out much or lose intensity over long distances.
12. 7.5×10^4 lines/m = 750 lines/cm
13. The resolving power of a telescope depends on the ratio of the wavelength to the diameter of the aperture. Telescopes using longer wavelength radiation (visible light) must be larger than those using shorter wavelengths (ultraviolet, X ray) to achieve the same resolving power.
14. The interference pattern for radio signals would "appear" on a radio receiver as an alternating increase in signal intensity followed by a loss of intensity (heard as static or "white noise").
15. 589 nm
16. 6.77°
17. 7.90°

553

Lab Planning

Beginning on page T34 are preparation notes and teaching tips to assist you in planning.

Blank data tables (as well as some sample data) appear on the **One-Stop Planner.**

No Books in the Lab?

See the ***Datasheets for In-Text Labs*** workbook for a reproducible master copy of this experiment.

Safety Caution

Make sure all students know the proper procedure for cleaning up broken glass. Remind students not to attempt to remove broken bulbs from sockets.

Tips and Tricks

- Before the lab, demonstrate diffraction using an excited gas source. Set up the optical bench as described in step 3 of the lab, and place the light source about 5 cm away from the slit. Adjust the grating and the slit scale distance and angle so a full first-order and part of a second-order spectrum appear centered about the slit. Show students how to recognize the images and how to bend paper clips to form riders for marking each similar first-order image and each similar second-order image.

CHAPTER 15

Skills Practice Lab Diffraction

OBJECTIVES

- **Discover** wavelengths of diffracted light.

MATERIALS LIST

- **bent paper-clip riders for meterstick**
- **black cardboard**
- **cellophane tape**
- **diffraction grating**
- **grating holder and support**
- **incandescent light source and power supply**
- **meterstick and 2 supports**
- **metric scale and slit**

In this experiment, you will pass white light through a diffraction grating and make measurements to determine wavelengths of the light's components.

SAFETY

- **Avoid looking directly at a light source. Looking directly at a light source can cause permanent eye damage. Put on goggles.**
- **Use a hot mitt to handle resistors, light sources, and other equipment that may be hot. Allow all equipment to cool before storing it.**
- **If a bulb breaks, notify your teacher immediately. Do not remove broken bulbs from sockets.**
- **Never put broken glass or ceramics in a regular waste container. Use a dustpan, brush, and heavy gloves to carefully pick up broken pieces and dispose of them in a container specifically provided for this purpose.**

PROCEDURE

Preparation

1. Read the entire lab, and plan the steps you will take.
2. If you are not using a datasheet provided by your teacher, prepare a data table in your lab notebook with six columns and four rows. In the first row, label the columns *Light Source, Image Color, Order, Image 1 (m), Image 2 (m),* and *Slit (m).* In the first column, label the second through fourth rows *White.* Above or below the data table, prepare a space to record the slit spacing, *d,* of the grating.

Wavelengths of White Light

3. Set up the optical bench as shown in **Figure 1.** Mount the scale and slit on one end of the optical bench, and place a piece of tape over the slit. Place a cardboard shield around the light source to direct all the light through the slit. Illuminate the slit with white light. Mount the grating near the opposite end of the optical bench.
4. Adjust the apparatus so that the white-light source is centered on the slit and the slit scale is perpendicular to the optical bench. Tape the optical bench and the white-light source securely in place.

5. With your eye close to the grating, observe the first-order spectra. Move the grating forward or backward as required so that the entire spectrum appears on each side of the scale. Place a bent paper-clip rider on the scale at the point in each first-order spectrum where the yellow light is the purest. Adjust the grating and slit scale by rotating the grating around its vertical axis so that the two yellow points end up equidistant from the source slit. Reposition the riders if necessary.

Figure 1

Step 3: Use the cardboard to make a shield around the light source. Make sure the light is directed through the slit.

Step 4: From above, the slit scale should form right angles with the meterstick. Measurements can be strongly affected if the equipment is moved even slightly during the procedure.

Step 5: Place a bent paper-clip rider to mark the position of the images on the scale.

6. Use the scale to measure the distance from the slit to each rider to the nearest millimeter. Record these distances in your data table as *Image 1* and *Image 2*. Also measure the distance from the slit to the grating. Record this distance in your data table as the *Slit (m)*. Record the order number and the image color.

7. Next, adjust the grating and slit to find the clearest first-order continuous spectrum. Measure and record the distance from the slit to the grating. Place a rider on the scale at the point in each first-order spectrum where you see the extreme end of the violet spectrum. Measure and record the distance from the slit to each rider.

8. Repeat step 7 for the extreme red end of the spectrum. Record all data.

9. Clean up your work area. Put equipment away safely so that it is ready to be used again. Recycle or dispose of used materials as directed by your teacher.

ANALYSIS

1. **Organizing Data** Use your data for each trial.
   a. For each trial, find the average image position.
   b. Use the average image position and the distance from the slit to the grating to find the distance from the grating to the image for each trial. (Hint: Use the Pythagorean theorem.)
   c. To find sin θ for each trial, divide the average image position by the distance found in (b).

CONCLUSIONS

2. **Drawing Conclusions** For each trial, find the wavelength of the light using the equation $\lambda = \frac{d(\sin\theta)}{m}$, where λ is the wavelength of light (in meters, m), d is the diffraction-grating spacing (1/[number of lines/m]), and m is the order number of the spectrum containing the image.

CHAPTER 15 LAB

✓ Checkpoints

Step 4: Make sure the apparatus is firmly attached to the table. Large errors can occur if the equipment is moved even slightly during the procedure.

Step 5: Students may need help finding the correct position of the grating. Students should be able to demonstrate that they have found the point in the spectrum where the color is purest.

Step 6: Students should be able to demonstrate how they used the scale to measure the distances.

ANSWERS

Analysis

1. **a.** yellow: 0.096 m
   violet: 0.067 m
   red: 0.118 m
   b. Student answers will vary. For sample data, values range from 0.302 m to 0.318 m.
   c. For sample data, values range from 0.22 to 0.37.

Conclusions

2. yellow: $\lambda = 5.68 \times 10^{-7}$ m
violet: $\lambda = 4.04 \times 10^{-7}$ m
red: $\lambda = 6.80 \times 10^{-7}$ m

CHAPTER 16

Electric Forces and Fields

Planning Guide

Compression Guide

To shorten instruction because of time limitations, omit the opener and Section 3 and abbreviate the review.

OBJECTIVES	LABS, DEMONSTRATIONS, AND ACTIVITIES	TECHNOLOGY RESOURCES
PACING • 45 min pp. 556–557 **Chapter Opener**	ANC **Discovery Lab** Charges and Electrostatics*◆ BASIC	CD **Visual Concepts,** Chapter 16 BASIC
PACING • 90 min pp. 558–563 **Section 1 Electric Charge** • Understand the basic properties of electric charge. • Differentiate between conductors and insulators. • Distinguish between charging by contact, charging by induction, and charging by polarization.	SE **Quick Lab** Polarization, p. 562 GENERAL SE **Skills Practice Lab** Electrostatics, pp. 588–591◆ GENERAL ANC **Datasheet** Electrostatics* GENERAL TE **Demonstration** Effects of Charge, p. 559 BASIC TE **Demonstration** Jumping Spices, p. 561 GENERAL TE **Demonstration** Polarization, p. 563 GENERAL	OSP **Lesson Plans** TR 86 The Millikan Experiment TR 87 Charging by Induction TR 88 Electrical Polarization
PACING • 90 min pp. 564–571 **Section 2 Electric Force** • Calculate electric force using Coulomb's law. • Compare electric force with gravitational force. • Apply the superposition principle to find the resultant force on a charge and to find the position at which the net force on a charge is zero.	TE **Demonstration** Electric Force, p. 564 GENERAL	OSP **Lesson Plans**
PACING • 45 min pp. 572–579 *Advanced Level* **Section 3 The Electric Field** • Calculate electric field strength. • Draw and interpret electric field lines. • Identify the four properties associated with a conductor in electrostatic equilibrium.	TE **Demonstration** Electric Field Strength, p. 573 ADVANCED TE **Demonstration** Charge Accumulation, p. 578 ADVANCED ANC **Invention Lab** Levitating Toys*◆ ADVANCED	OSP **Lesson Plans** EXT **Integrating Astronomy** Ion Propulsion in *Deep Space 1* GENERAL TR 55A Electric Field Lines for a Single Charge TR 56A Electric Field Lines for Opposite Charges TR 57A Electric Field Lines for Two Like Charges

PACING • 90 min

CHAPTER REVIEW, ASSESSMENT, AND STANDARDIZED TEST PREPARATION

SE **Chapter Highlights,** p. 580
SE **Chapter Review,** pp. 581–585
SE **Graphing Calculator Practice,** p. 584 GENERAL
SE **Alternative Assessment,** p. 585 ADVANCED
SE **Standardized Test Prep,** pp. 586–587 GENERAL
SE **Appendix D: Equations,** p. 861
SE **Appendix I: Additional Problems,** pp. 891–892
ANC **Study Guide Worksheet** Mixed Review* GENERAL
ANC **Chapter Test A*** GENERAL
ANC **Chapter Test B*** ADVANCED
OSP **Test Generator**

Online and Technology Resources

go.hrw.com — Holt Online Learning

Visit **go.hrw.com** to find a variety of online resources. To access this chapter's extensions, enter the keyword **HF6ELFXT** and click the "go" button. Click **Holt Online Learning** for an online edition of this textbook, and other interactive resources.

HOLT Teacher's One-Stop Planner®

This DVD package includes:
- Holt Calendar Planner
- Customizable Lesson Plans
- Editable Worksheets
- ExamView® Version 6 Assessment Suite
- Interactive Teacher's Edition
- Holt PuzzlePro®
- Holt PowerNotes® Presentations
- MindPoint® Quiz Show

KEY

SE Student Edition	**OSP** One-Stop Planner	**EXT** Online Extension
TE Teacher Edition	**CD** CD or CD-ROM	* Also on One-Stop Planner
ANC Ancillary Worksheet	**TR** Teaching Transparencies	◆ Requires advance prep

SKILLS DEVELOPMENT RESOURCES	REVIEW AND ASSESSMENT	CORRELATIONS
		National Science Education Standards
SE **Conceptual Challenge,** p. 561	SE **Section Review,** p. 563 GENERAL ANC **Study Guide Worksheet** Section 1* GENERAL ANC **Quiz** Section 1* BASIC	UCP 1, 2, 3, 5 SAI 1, 2 ST 1, 2 HNS 1, 2, 3 SPSP 5 PS 6d
SE **Sample Set A** Coulomb's Law, pp. 565–566 GENERAL TE **Classroom Practice,** p. 565 GENERAL ANC **Problem Workbook** Sample Set A* GENERAL OSP **Problem Bank** Sample Set A GENERAL SE **Sample Set B** The Superposition Principle, pp. 567–568 ADVANCED TE **Classroom Practice,** p. 567 ADVANCED ANC **Problem Workbook** Sample Set B* ADVANCED OSP **Problem Bank** Sample Set B ADVANCED SE **Sample Set C** Equilibrium, pp. 569–570 GENERAL TE **Classroom Practice,** p. 569 GENERAL ANC **Problem Workbook** Sample Set C* GENERAL OSP **Problem Bank** Sample Set C GENERAL SE **Conceptual Challenge,** p. 566	SE **Section Review,** p. 571 GENERAL ANC **Study Guide Worksheet** Section 2* GENERAL ANC **Quiz** Section 2* BASIC	UCP 1, 2, 3, 4, 5 ST 2 HNS 1, 3 SPSP 5 PS 4c
SE **Sample Set D** Electric Field Strength, pp. 574–575 ADVANCED TE **Classroom Practice,** p. 575 ADVANCED ANC **Problem Workbook** Sample Set D* ADVANCED OSP **Problem Bank** Sample Set D ADVANCED	SE **Section Review,** p. 579 ADVANCED ANC **Study Guide Worksheet** Section 3* ADVANCED ANC **Quiz** Section 3* GENERAL	UCP 1, 2, 3, 4, 5 SAI 1, 2 ST 1, 2 SPSP 5

SCILINKS. NSTA
www.scilinks.org

Maintained by the **National Science Teachers Association.**

Topic: Electric Charge
SciLinks Code: HF60470

Topic: Coulomb's Law
SciLinks Code: HF60361

Topic: Microwaves
SciLinks Code: HF60959

Topic: Conductors and Insulators
SciLinks Code: HF60341

Topic: Van de Graaff Generator
SciLinks Code: HF61592

Classroom CD-ROMs

- Guided Reading Audio Program
- Student One Stop
- Virtual Investigations
- Visual Concepts

Holt Lab Generator CD-ROM

Search for any lab by topic, standard, difficulty level, or time. Edit any lab to fit your needs, or create your own labs. Use the Lab Materials QuickList software to customize your lab materials list.

CHAPTER 16

Overview

Section 1 introduces positive and negative electric charge, the conservation of charge, and the quantization of charge and discusses conductors, insulators, and methods of charging.

Section 2 examines Coulomb's law and calculations of net electric forces using the superposition principle.

Section 3 introduces the electric field, electric field lines, and electric field strength; explores the electric fields around various charged objects; and discusses the properties of conductors in electrostatic equilibrium.

About the Illustration

Electrostatic spray painting is used in a variety of industries. Automobile bodies, furniture, toys, refrigerators, and various other mass-produced items are often painted electrostatically. Electrostatic spray painting can be used with metal and some types of wood. Plastic, rubber, and glass cannot be painted electrostatically.

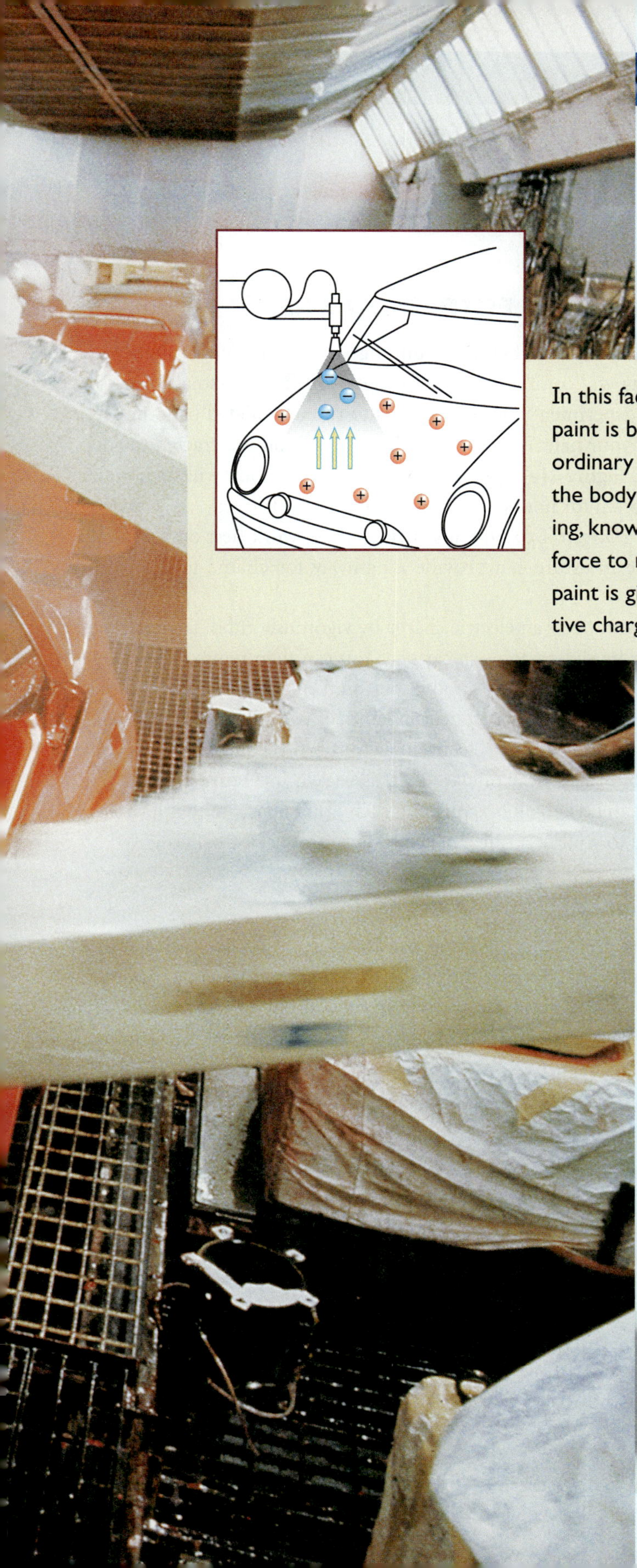

CHAPTER 16

Electric Forces and Fields

In this factory in Bowling Green, Kentucky, a fresh coat of paint is being applied to an automobile by spray guns. With ordinary spray guns, any paint that does not happen to hit the body of the car is wasted. A special type of spray painting, known as *electrostatic spray painting,* utilizes electric force to minimize the amount of paint that is wasted. The paint is given a negative charge and the car is given a positive charge. Thus, the paint is attracted to the car.

WHAT TO EXPECT

In this chapter, you will learn about the basic properties of electric charges. You will learn to calculate the electric force produced by point charges and will learn to interpret electric field lines.

Why it Matters

According to one estimate, electrostatic spray painting saves industries in the United States as much as $50 million each year. You will study how electric force is used in electrostatic spray painting.

CHAPTER PREVIEW

1 Electric Charge
- Properties of Electric Charge
- Transfer of Electric Charge

2 Electric Force
- Coulomb's Law

3 The Electric Field
- Electric Field Strength
- Electric Field Lines
- Conductors in Electrostatic Equilibrium

Tapping Prior Knowledge

Knowledge to Review

- ✔ Vector addition gives a resultant vector that is equivalent to the added effects of each of the individual vectors. Vector addition may be accomplished graphically or mathematically.
- ✔ Field forces act on objects without physical contact. Field forces are often said to act at a distance.

Items to Probe

- ✔ Vector addition and the superposition principle: Have students calculate the resultant force on an object experiencing two or three individual forces.

557

SECTION 1
General Level

Teaching Tip — GENERAL

Encourage students to experiment with electrostatics and to observe the attractive and repulsive effects that charged objects have on one another. Experiments students can easily perform include charging balloons, running a comb through hair and attracting small pieces of paper with the comb, and attaching two pieces of transparent tape to a desktop and then quickly removing them and holding them together to demonstrate repulsion. Further, glass can be charged with silk, and rubber can be charged with wool or fur. The charged glass and rubber will attract one another, as will the glass and silk or the rubber and wool. However, the glass and wool will repel one another as will the silk and rubber.

Visual Strategy — BASIC

Figure 1

Have students consider all forces that may be acting on the balloons in each case.

Q Why do we infer that some sort of electric force is acting on the balloons in (**b**)?

A ***The balloons' strings are at an angle to the vertical. Thus, the tensions in the strings have a horizontal component, which would pull the balloons toward each other. An additional horizontal force in the opposite direction must be preventing this from happening. This force is the repulsive electric force between the balloons.***

SECTION 1

Electric Charge

SECTION OBJECTIVES

- **Understand the basic properties of electric charge.**
- **Differentiate between conductors and insulators.**
- **Distinguish between charging by contact, charging by induction, and charging by polarization.**

PROPERTIES OF ELECTRIC CHARGE

You have probably noticed that after running a plastic comb through your hair on a dry day, the comb attracts strands of your hair or small pieces of paper. A simple experiment you might try is to rub an inflated balloon back and forth across your hair. You may find that the balloon is attracted to your hair, as shown in **Figure 1(a).** On a dry day, a rubbed balloon will stick to the wall of a room, often for hours. When materials behave this way, they are said to be *electrically charged.* Experiments such as these work best on a dry day because excessive moisture can provide a pathway for charge to leak off a charged object.

You can give your body an electric charge by vigorously rubbing your shoes on a wool rug or by sliding across a car seat. You can then remove the charge on your body by lightly touching another person. Under the right conditions, you will see a spark just before you touch, and both of you will feel a slight tingle.

Another way to observe static electricity is to rub two balloons across your hair and then hold them near one another, as shown in **Figure 1(b).** In this case, you will see the two balloons pushing each other apart. Why is a rubbed balloon attracted to your hair but repelled by another rubbed balloon?

There are two kinds of electric charge

The two balloons must have the same kind of charge because each became charged in the same way. Because the two charged balloons repel one another, we see that *like charges repel.* Conversely, a rubbed balloon and your hair, which do not have the same kind of charge, are attracted to one another. Thus, *unlike charges attract.*

Table 1 Conventions for Representing Charges and Electric Field Vectors

Positive charge	⊕ $+q$
Negative charge	⊖ $-q$
Electric field vector	→ $\mathbf{E}$
Electric field lines	

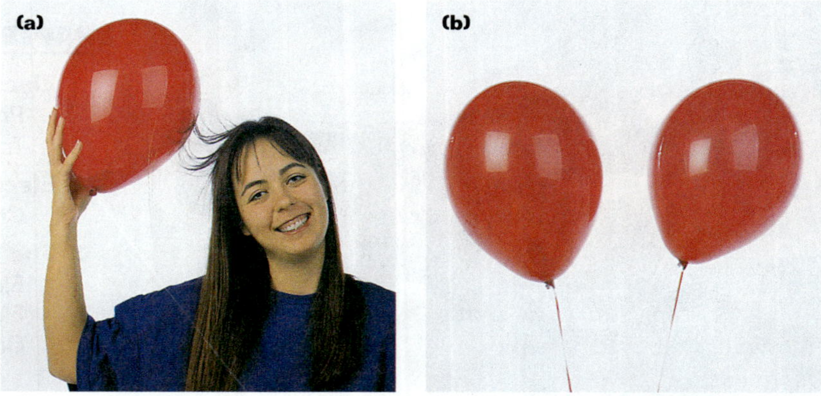

Figure 1
(a) If you rub a balloon across your hair on a dry day, the balloon and your hair become charged and attract each other. **(b)** Two charged balloons, on the other hand, repel each other.

Benjamin Franklin (1706–1790) named the two different kinds of charge *positive* and *negative.* By convention, when you rub a balloon across your hair, the charge on your hair is referred to as *positive* and that on the balloon is referred to as *negative,* as shown in **Figure 2.** Positive and negative charges are said to be *opposite* because an object with an equal amount of positive and negative charge has no net charge.

Electrostatic spray painting utilizes the principle of attraction between unlike charges. Paint droplets are given a negative charge, and the object to be painted is given a positive charge. In ordinary spray painting, many paint droplets drift past the object being painted. But in electrostatic spray painting, the negatively charged paint droplets are attracted to the positively charged target object, so more of the paint droplets hit the object being painted and less paint is wasted.

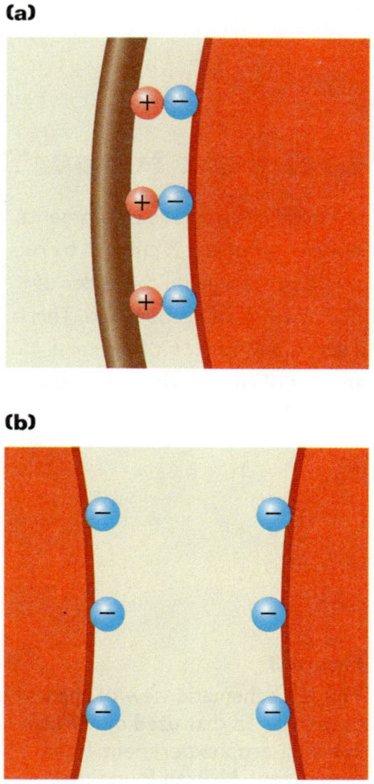

Figure 2
(a) This negatively charged balloon is attracted to positively charged hair because the two have opposite charges. **(b)** Two negatively charged balloons repel one another because they have the same charge.

Electric charge is conserved

When you rub a balloon across your hair, how do the balloon and your hair become electrically charged? To answer this question, you'll need to know a little about the atoms that make up the matter around you. Every atom contains even smaller particles. Positively charged particles, called *protons,* and uncharged particles, called *neutrons,* are located in the center of the atom, called the *nucleus.* Negatively charged particles, known as *electrons,* are located outside the nucleus and move around it. (You will study the structure of the atom and the particles within the atom in greater detail in later chapters on atomic and subatomic physics in this book.)

Protons and neutrons are relatively fixed in the nucleus of the atom, but electrons are easily transferred from one atom to another. When the electrons in an atom are balanced by an equal number of protons, the atom has no net charge. If an electron is transferred from one neutral atom to another, the second atom gains a negative charge and the first atom loses a negative charge, thereby becoming positive. Atoms that are positively or negatively charged are called *ions.*

Both a balloon and your hair contain a very large number of neutral atoms. Charge has a natural tendency to be transferred between unlike materials. Rubbing the two materials together serves to increase the area of contact and thus enhance the charge-transfer process. When a balloon is rubbed against your hair, some of your hair's electrons are transferred to the balloon. Thus, the balloon gains a certain amount of negative charge while your hair loses an equal amount of negative charge and hence is left with a positive charge. In this and similar experiments, only a small portion of the total available charge is transferred from one object to another.

The positive charge on your hair is equal in magnitude to the negative charge on the balloon. Electric charge is conserved in this process; no charge is created or destroyed. This principle of conservation of charge is one of the fundamental laws of nature.

Did you know?

Some cosmetic products contain an organic compound called *chitin,* which is found in crabs and lobsters, and in butterflies and other insects. Chitin is positively charged, so it helps cosmetic products stick to human hair and skin, which are usually slightly negatively charged.

SECTION 1

Demonstration

Effects of Charge

Purpose Show two kinds of charge by use of a pith ball.

Materials pith ball suspended by thread; rubber rod and fur (or balloon and hair); glass rod and silk (or plastic wrap)

Procedure Charge the rubber rod with the fur (or the balloon with hair) by friction. Briefly touch the pith ball with the rubber rod and withdraw the rod. Have students observe the reaction of the pith ball, which is now negatively charged, as you bring the rubber rod near, and ask students to explain this reaction.

Now, ground the pith ball with your finger, and explain the grounding effect. *(Excess electrons leave the pith ball.)*

Repeat the above procedure with the glass rod, which becomes positively charged, and have students observe the same results. Without grounding the pith ball, bring a negatively charged rubber rod near the pith ball. Have students explain the resulting attraction. Explain that at least two kinds of charges are demonstrated here: one that repels the charged pith ball and one that attracts it.

Visual Strategy GENERAL

Figure 3
Point out to students that Millikan used a positively charged metal plate to cause a negatively charged drop to rise.

Q Will all of the oil drops rise toward the top plate when it is given a positive charge?

A ***No, those that are positively charged will continue moving downward.***

Teaching Tip — ADVANCED

Advanced students may want more information about how Millikan found that charge is quantized. Millikan measured the time intervals of a drop falling due to the gravitational force and rising due to electrical attraction and the corresponding distances. He then calculated the drop's velocities with these data. Through a relationship between the drop's upward velocity and the electric field strength, Millikan determined the charge on the drop. By repeating this process for thousands of drops, Millikan found that the charge on each drop was an integral multiple of a fundamental unit of charge. Encourage interested students to research these details after electric field strength has been introduced. Students can present their findings to the class.

Did you know?

In typical electrostatic experiments, in which an object is charged by rubbing, a net charge on the order of 10^{-6} C (= 1 μC) is obtained. This is a very small fraction of the total amount of charge within each object.

Electric charge is quantized

In 1909, Robert Millikan (1886–1953) performed an experiment at the University of Chicago in which he observed the motion of tiny oil droplets between two parallel metal plates, as shown in **Figure 3.** The oil droplets were charged by friction in an atomizer and allowed to pass through a hole in the top plate. Initially, the droplets fell due to their weight. The top plate was given a positive charge as the droplets fell, and the droplets with a negative charge were attracted back upward toward the positively charged plate. By turning the charge on this plate on and off, Millikan was able to watch a single oil droplet for many hours as it alternately rose and fell.

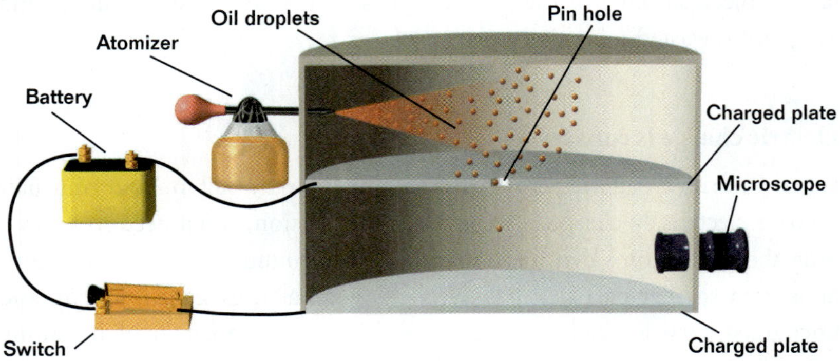

Figure 3
This is a schematic view of apparatus similar to that used by Millikan in his oil-drop experiment. In his experiment, Millikan found that there is a fundamental unit of charge.

After repeating this process for thousands of drops, Millikan found that when an object is charged, its charge is always a multiple of a fundamental unit of charge, symbolized by the letter *e*. In modern terms, charge is said to be *quantized*. This means that charge occurs as integer multiples of *e* in nature. Thus, an object may have a charge of $\pm e$, or $\pm 2e$, or $\pm 3e$, and so on.

Other experiments in Millikan's time demonstrated that the electron has a charge of $-e$ and the proton has an equal and opposite charge, $+e$. The value of *e* has since been determined to be $1.602\,176 \times 10^{-19}$ C, where the coulomb (C) is the SI unit of electric charge. For calculations, this book will use the approximate value given in **Table 2.** A total charge of −1.0 C contains 6.2×10^{18} electrons. Comparing this with the number of free electrons in 1 cm^3 of copper, which is on the order of 10^{23}, shows that 1.0 C is a substantial amount of charge.

Table 2 Charge and Mass of Atomic Particles

Particle	Charge (C)	Mass (kg)
electron	-1.60×10^{-19}	9.109×10^{-31}
proton	$+1.60 \times 10^{-19}$	1.673×10^{-27}
neutron	0	1.675×10^{-27}

SC*LINKS*.
www.scilinks.org
Topic: Electric Charge
Code: HF60470

TRANSFER OF ELECTRIC CHARGE

When a balloon and your hair are charged by rubbing, only the rubbed areas become charged, and there is no tendency for the charge to move into other regions of the material. In contrast, when materials such as copper, aluminum, and silver are charged in some small region, the charge readily distributes itself over the entire surface of the material. For this reason, it is convenient to classify substances in terms of their ability to transfer electric charge.

Materials in which electric charges move freely, such as copper and aluminum, are called **electrical conductors.** Most metals are conductors. Materials in which electric charges do not move freely, such as glass, rubber, silk, and plastic, are called **electrical insulators.**

Semiconductors are a third class of materials characterized by electrical properties that are somewhere between those of insulators and conductors. In their pure state, semiconductors are insulators. But the carefully controlled addition of specific atoms as impurities can dramatically increase a semiconductor's ability to conduct electric charge. Silicon and germanium are two well-known semiconductors that are used in a variety of electronic devices.

Certain metals and compounds belong to a fourth class of materials, called *superconductors.* Superconductors have zero electrical resistance when they are at or below a certain temperature. Thus, superconductors can conduct electricity indefinitely without heating.

electrical conductor

a material in which charges can move freely

electrical insulator

a material in which charges cannot move freely

Insulators and conductors can be charged by contact

In the experiments discussed above, a balloon and hair become charged when they are rubbed together. This process is known as *charging by contact.* Another example of charging by contact is a common experiment in which a glass rod is rubbed with silk and a rubber rod is rubbed with wool or fur. The two rods become oppositely charged and attract one another, as a balloon and your hair do. If two glass rods are charged, the rods have the same charge and repel each other, just as two charged balloons do. Likewise, two charged rubber rods repel one another. All of the materials used in these experiments—glass, rubber, silk, wool, and fur—are insulators. Can conductors also be charged by contact?

If you try a similar experiment with a copper rod, the rod does not attract or repel another charged rod. This result might suggest that a metal cannot be charged by contact. However, if you hold the copper rod with an insulating handle and then rub it with wool or fur, the rod attracts a charged glass rod and repels a charged rubber rod.

In the first case, the electric charges produced by rubbing readily move from the copper through your body and finally to Earth because copper and the human body are both conductors. The copper rod does become charged, but it soon becomes neutral again. In the second case, the insulating handle prevents the flow of charge to Earth, and the copper rod remains charged. Thus, both insulators and conductors can become charged by contact.

Why it Matters

Conceptual Challenge

1. Plastic Wrap

Plastic wrap becomes electrically charged as it is pulled from its container, and, as a result, it is attracted to objects such as food containers. Explain why plastic is a good material for this purpose.

2. Charge Transfer

If a glass rod is rubbed with silk, the glass becomes positively charged and the silk becomes negatively charged. Compare the mass of the glass rod before and after it is charged.

3. Electrons

Many objects in the large-scale world have no net charge, even though they contain an extremely large number of electrons. How is this possible?

Demonstration

Jumping Spices

Purpose Show charging by contact and charging by induction.

Materials sheet of Plexiglas or plastic (plastic picture holder works well), cloth (wool or silk), dried spices (parsley, sage, etc.), overhead projector

Procedure *Note:* Try both wool and silk cloth ahead of time to see which creates a greater charge on the plastic.

Sprinkle spices on the projector. Use cloth to charge the plastic. Hold the charged plastic $\frac{1}{2}$ in. to 1 in. above the spices. Some pieces will stick to the plastic, but many will jump up and down repeatedly. Challenge students to explain the jumping of the initially neutral spices. This will work even in high humidity with vigorous rubbing for approximately 45 s.

ANSWERS — GENERAL

Conceptual Challenge

1. Plastic is an insulator, which can hold electric charges.
2. The glass rod's mass is slightly less after the rubbing because some of its electrons are transferred to the silk.
3. Materials with no net charge have an equal number of protons and neutrons.

Quick Lab

TEACHER'S NOTES

In this experiment, a polarized stream of water is deflected by a charged comb. The lab works extremely well, even on humid days. Make sure the stream is small. Have students draw a schematic diagram representing the charges of the water and the comb.

By changing the height and horizontal distance of the comb, students can drastically change the path of the stream. Challenge the students to explain these results in terms of Newton's laws and projectile motion.

Homework Options
This QuickLab can easily be performed outside of the physics lab room.

Visual Strategy BASIC

Figure 4

Be sure students understand that the positive charge on the sphere does not move.

Q Why do electrons leave the sphere in **(b)**?

A ***The electrons are repelled by the negatively charged rod.***

Quick Lab

Polarization

MATERIALS LIST

- **plastic comb**
- **water faucet**

Turn on a water faucet, and adjust the flow of water so that you have a small but steady stream. The stream should be as slow as possible without producing individual droplets. Comb your hair vigorously. Hold the charged end of the comb near the stream without letting the comb get wet. What happens to the stream of water? What might be causing this to happen?

Conductors can be charged by induction

When a conductor is connected to Earth by means of a conducting wire or copper pipe, the conductor is said to be *grounded.* The Earth can be considered to be an infinite reservoir for electrons because it can accept an unlimited number of electrons. This fact is the key to understanding another method of charging a conductor.

Consider a negatively charged rubber rod brought near a neutral (uncharged) conducting sphere that is insulated so that there is no conducting path to ground. The repulsive force between the electrons in the rod and those in the sphere causes a redistribution of negative charge on the sphere, as shown in **Figure 4(a).** As a result, the region of the sphere nearest the negatively charged rod has an excess of positive charge.

If a grounded conducting wire is then connected to the sphere, as shown in **Figure 4(b),** some of the electrons leave the sphere and travel to Earth. If the wire to ground is then removed while the negatively charged rod is held in place, as shown in **Figure 4(c),** the conducting sphere is left with an excess of induced positive charge. Finally, when the rubber rod is removed from the vicinity of the sphere, as in **Figure 4(d),** the induced net positive charge remains on the ungrounded sphere. The motion of negative charges on the sphere causes the charge to become uniformly distributed over the outside surface of the ungrounded sphere. This process is known as **induction,** and the charge is said to be *induced* on the sphere.

induction

the process of charging a conductor by bringing it near another charged object and grounding the conductor

Notice that charging an object by induction requires no contact with the object inducing the charge but does require contact with a third object, which serves as either a *source* or a *sink* of electrons. A sink is a system which can absorb a large number of charges, such as Earth, without becoming locally charged itself. In the process of inducing a charge on the sphere, the charged rubber rod did not come in contact with the sphere and thus did not lose any of its negative charge. This is in contrast to charging an object by contact, in which charges are transferred directly from one object to another.

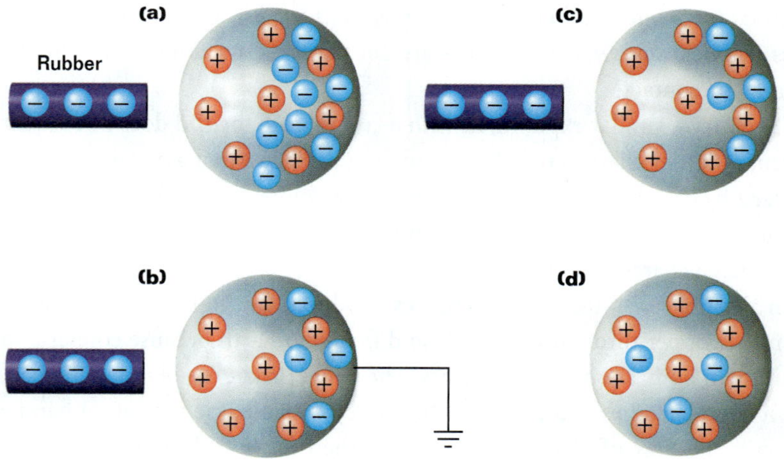

Figure 4
(a) When a charged rubber rod is brought near a metal sphere, the electrons move away from the rod, and the charge on the sphere becomes redistributed. **(b)** If the sphere is grounded, some of the electrons travel through the wire to the ground. **(c)** When this wire is removed, the sphere has an excess of positive charge. **(d)** The electrons become evenly distributed on the surface of the sphere when the rod is removed.

A surface charge can be induced on insulators by polarization

A process very similar to charging by induction in conductors takes place in insulators. In most neutral atoms or molecules, the center of positive charge coincides with the center of negative charge. In the presence of a charged object, these centers may shift slightly, resulting in more positive charge on one side of a molecule than on the other. This is known as *polarization.*

This realignment of charge within individual molecules produces an induced charge on the surface of the insulator, as shown in **Figure 5(a).** When an object becomes polarized, it has no net charge but is still able to attract or repel objects due to this realignment of charge. This explains why a plastic comb can attract small pieces of paper that have no net charge, as shown in **Figure 5(b).** As with induction, in polarization one object induces a charge on the surface of another object with no physical contact.

SCILINKS
www.scilinks.org
Topic: Conductors and Insulators
Code: HF60341

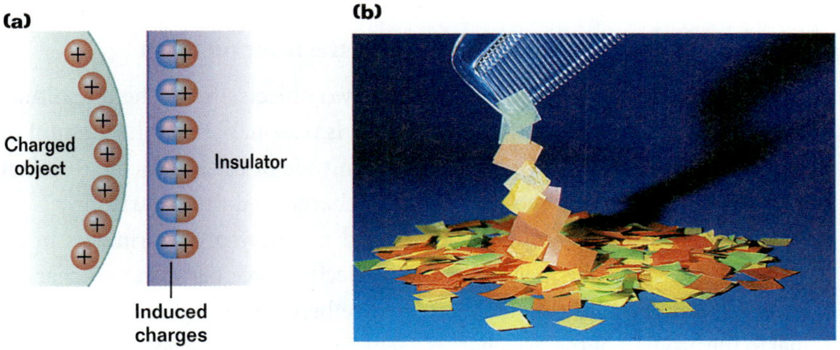

Figure 5
(a) The charged object on the left induces charges on the surface of an insulator, which is said to be *polarized.* **(b)** This charged comb induces a charge on the surface of small pieces of paper that have no net charge.

SECTION REVIEW

1. When a rubber rod is rubbed with wool, the rod becomes negatively charged. What can you conclude about the magnitude of the wool's charge after the rubbing process? Why?
2. What did Millikan's oil-drop experiment reveal about the nature of electric charge?
3. A typical lightning bolt has about 10.0 C of charge. How many excess electrons are in a typical lightning bolt?
4. If you stick a piece of transparent tape on your desk and then quickly pull it off, you will find that the tape is attracted to other areas of your desk that are not charged. Why does this happen?
5. **Critical Thinking** Metals, such as copper and silver, can become charged by induction, while plastic materials cannot. Explain why.
6. **Critical Thinking** Why is an electrostatic spray gun more efficient than an ordinary spray gun?

Demonstration

Polarization GENERAL

Purpose Show the effects of polarization.

Materials rubber rod and fur, small piece of paper (approximately 5 mm × 5 mm), watch glass, meterstick

Procedure Charge the rubber rod with the fur, touch the rod to the paper, and lift the rod. The paper will be lifted up by the rod. Explain that the negatively charged rod induced a positive charge on the surface of the paper, and the two are attracted to one another.

Next, balance the meterstick on the inverted watch glass. Charge the rod with the wool and bring it close to the end of the meterstick, which slowly begins to rotate. Point out that again there was a redistribution of electrons, producing areas of induced positive and negative charge on the end of the meterstick.

SECTION REVIEW ANSWERS

1. It is equal to the magnitude of the rod's charge; Charge is conserved.
2. Charge is quantized.
3. 6.25×10^{19} electrons
4. The tape becomes charged when it is pulled off the desk, thus inducing an opposite surface charge on the desk.
5. because plastic, an insulator, does not easily conduct charge
6. More paint hits the object being painted because of an electrical attraction between the charged droplets and the oppositely charged object.

Demonstration

Electric Force — GENERAL

Purpose Qualitatively illustrate the dependence of electric force on distance. (*Optional:* Qualitatively illustrate the dependence of electric force on charge.)

Materials balloon, water faucet, Leyden jar (optional)

Procedure Repeat the Quick Lab "Polarization" in Section 1 of this chapter a number of times, varying the distance between the balloon and the water stream in each trial. Have students observe the displacement of the water stream for different distances, and ask them to describe the relationship between distance and the force exerted on the stream.

Optional: Using a constant distance, show the effects of varying the charge of a Leyden jar on the water stream. Have students describe the relationship between the amount of charge and the force on the stream.

The Language of Physics

Some texts simply use k for the Coulomb constant. This book uses k_C instead so that students do not confuse the Coulomb constant with the spring constant (k).

SECTION 2

Electric Force

SECTION OBJECTIVES

- **Calculate electric force using Coulomb's law.**
- **Compare electric force with gravitational force.**
- **Apply the superposition principle to find the resultant force on a charge and to find the position at which the net force on a charge is zero.**

COULOMB'S LAW

Two charged objects near one another may experience acceleration either toward or away from each other because each object exerts a force on the other object. This force is called the *electric force.* The two balloon experiments described in the first section demonstrate that the electric force is attractive between opposite charges and repulsive between like charges. What determines how small or large the electric force will be?

The closer two charges are, the greater is the force on them

It seems obvious that the distance between two objects affects the magnitude of the electric force between them. Further, it is reasonable that the amount of charge on the objects will also affect the magnitude of the electric force. What is the precise relationship between distance, charge, and the electric force?

In the 1780s, Charles Coulomb conducted a variety of experiments in an attempt to determine the magnitude of the electric force between two charged objects. Coulomb found that the electric force between two charges is proportional to the product of the two charges. Hence, if one charge is doubled, the electric force likewise doubles, and if both charges are doubled, the electric force increases by a factor of four. Coulomb also found that the electric force is inversely proportional to the square of the distance between the charges. Thus, when the distance between two charges is halved, the force between them increases by a factor of four. The following equation, known as Coulomb's law, expresses these conclusions mathematically for two charges separated by a distance, r.

COULOMB'S LAW

$$F_{electric} = k_C\left(\frac{q_1q_2}{r^2}\right)$$

$$\text{electric force} = \text{Coulomb constant} \times \frac{(\text{charge 1})(\text{charge 2})}{(\text{distance})^2}$$

When dealing with Coulomb's law, remember that force is a vector quantity and must be treated accordingly. The electric force between two objects always acts along the line that connects their centers of charge. Also, note that Coulomb's law applies exactly only to point charges or particles and to spherical distributions of charge. When applying Coulomb's law to spherical distributions of charge, use the distance between the centers of the spheres as r.

Did you know?

The symbol k_C, called the *Coulomb constant*, has SI units of $N\bullet m^2/C^2$ because this gives N as the unit of electric force. The value of k_C depends on the choice of units. Experiments have determined that in SI units, k_C has the value $8.9875 \times 10^9\ N\bullet m^2/C^2$.

SCILINKS.
www.scilinks.org
Topic: Coulomb's Law
Code: HF60361

SAMPLE PROBLEM A

Coulomb's Law

PROBLEM

The electron and proton of a hydrogen atom are separated, on average, by a distance of about 5.3×10^{-11} m. Find the magnitudes of the electric force and the gravitational force that each particle exerts on the other.

SOLUTION

1. DEFINE **Given:**

$r = 5.3 \times 10^{-11}$ m $\quad q_e = -1.60 \times 10^{-19}$ C

$k_C = 8.99 \times 10^9$ N•m²/C² $\quad q_p = +1.60 \times 10^{-19}$ C

$m_e = 9.109 \times 10^{-31}$ kg $\quad G = 6.673 \times 10^{-11}$ N•m²/kg²

$m_p = 1.673 \times 10^{-27}$ kg

Unknown: $F_{electric} = ?$ $\quad F_g = ?$

2. PLAN **Choose an equation or situation:**

Find the magnitude of the electric force using Coulomb's law and the magnitude of the gravitational force using Newton's law of gravitation (introduced in the chapter "Circular Motion and Gravitation" in this book).

$$F_{electric} = k_C \frac{q_1 q_2}{r^2} \qquad F_g = G\frac{m_e m_p}{r^2}$$

3. CALCULATE **Substitute the values into the equations and solve:**

Because we are finding the magnitude of the electric force, which is a scalar, we can disregard the sign of each charge in our calculation.

$$F_{electric} = k_C \frac{q_e q_p}{r^2} = \left(8.99 \times 10^9 \frac{\text{N}\bullet\text{m}^2}{\text{C}^2}\right)\left(\frac{(1.60 \times 10^{-19}\ \text{C})^2}{(5.3 \times 10^{-11}\ \text{m})^2}\right)$$

$$\boxed{F_{electric} = 8.2 \times 10^{-8}\ \text{N}}$$

$$F_g = G\frac{m_e m_p}{r^2} =$$

$$\left(6.673 \times 10^{-11} \frac{\text{N}\bullet\text{m}^2}{\text{kg}^2}\right)\left(\frac{(9.109 \times 10^{-31}\ \text{kg})(1.673 \times 10^{-27}\ \text{kg})}{(5.3 \times 10^{-11}\ \text{m})^2}\right)$$

$$\boxed{F_g = 3.6 \times 10^{-47}\ \text{N}}$$

4. EVALUATE The electron and the proton have opposite signs, so the electric force between the two particles is attractive. The ratio $F_{electric}/F_g \approx 2 \times 10^{39}$; hence, the gravitational force between the particles is negligible compared with the electric force between them. Because each force is inversely proportional to distance squared, their ratio is independent of the distance between the two particles.

Classroom Practice

Coulomb's Law

Consider the forces $F_1 = F$ and $F_2 = -F$ acting on two charged particles separated by a distance, d. Explain the change in the forces exerted on each particle under the following conditions:

a. the distance between the two particles doubles

b. the charge on one particle doubles

c. the charge on each particle doubles

d. the charge on each particle and the distance between the two particles double

Answer

a. $F_1 = \frac{1}{4}F$, $F_2 = -\frac{1}{4}F$

b. $F_1 = 2F$, $F_2 = -2F$

c. $F_1 = 4F$, $F_2 = -4F$

d. $F_1 = F$, $F_2 = -F$

PROBLEM GUIDE A

Use this guide to assign problems.
SE = Student Edition Textbook
PW = Problem Workbook
PB = Problem Bank on the One-Stop Planner (OSP)

Solving for:

$F_{electric}$	**SE** Sample, 1–2, 3a; Ch. Rvw. 15–17 **PW** 5–7 **PB** 4–6
q	**PW** Sample, 1–4 **PB** 7–10
r	**SE** 4 **PW** 8–10 **PB** Sample, 1–3

***Challenging Problem**
Consult the printed Solutions Manual or the OSP for detailed solutions.

ANSWERS

Practice A

1. 230 N (attractive)
2. a. 2.2×10^{-5} N (attractive)
   b. 9.0×10^{-7} N (repulsive)
3. 0.393 m

ANSWERS — BASIC

Conceptual Challenge

1. Earth's gravitational effects are significant because Earth has such a large mass. Electric effects are not usually felt because most objects are electrically neutral, that is, they have the same number of electrons and protons.
2. Their attraction to the protons in the nickel overcomes their repulsion.
3. The force decreases by a factor of four.

PRACTICE A

Coulomb's Law

1. A balloon rubbed against denim gains a charge of −8.0 μC. What is the electric force between the balloon and the denim when the two are separated by a distance of 5.0 cm? (Assume that the charges are located at a point.)
2. Two identical conducting spheres are placed with their centers 0.30 m apart. One is given a charge of $+12 \times 10^{-9}$ C and the other is given a charge of -18×10^{-9} C.
   a. Find the electric force exerted on one sphere by the other.
   b. The spheres are connected by a conducting wire. After equilibrium has occurred, find the electric force between the two spheres.
3. Two electrostatic point charges of +60.0 μC and +50.0 μC exert a repulsive force on each other of 175 N. What is the distance between the two charges?

Resultant force on a charge is the vector sum of the individual forces on that charge

Frequently, more than two charges are present, and it is necessary to find the net electric force on one of them. As demonstrated in Sample Problem A, Coulomb's law gives the electric force between any pair of charges. Coulomb's law also applies when more than two charges are present. Thus, the resultant force on any single charge equals the vector sum of the individual forces exerted on that charge by all of the other individual charges that are present. This is an example of the *principle of superposition.* Once the magnitudes of the individual electric forces are found, the vectors are added together exactly as you learned earlier. This process is demonstrated in Sample Problem B.

Why it Matters

Conceptual Challenge

1. Electric Force
The electric force is significantly stronger than the gravitational force. However, although we feel our attraction to Earth by gravity, we do not usually feel the effects of the electric force. Explain why.

2. Electrons in a Coin
An ordinary nickel contains about 10^{24} electrons, all repelling one another. Why don't these electrons fly off the nickel?

3. Charged Balloons
When the distance between two negatively charged balloons is doubled, by what factor does the repulsive force between them change?

SAMPLE PROBLEM B

STRATEGY The Superposition Principle

PROBLEM

Consider three point charges at the corners of a triangle, as shown at right, where $q_1 = 6.00 \times 10^{-9}$ C, $q_2 = -2.00 \times 10^{-9}$ C, and $q_3 = 5.00 \times 10^{-9}$ C. Find the magnitude and direction of the resultant force on q_3.

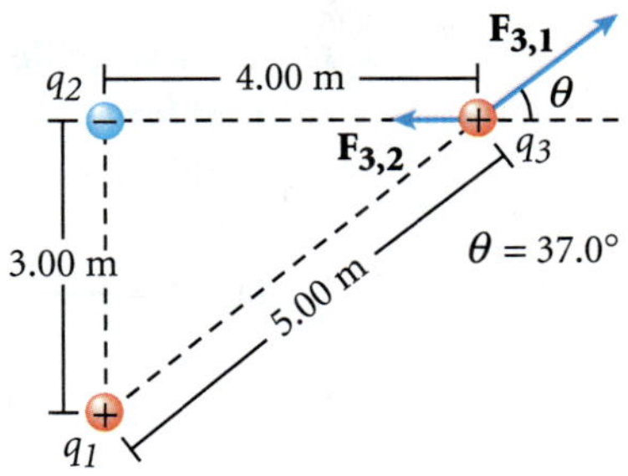

SOLUTION

1. Define the problem, and identify the known variables.

Given:

$q_1 = +6.00 \times 10^{-9}$ C $r_{2,1} = 3.00$ m

$q_2 = -2.00 \times 10^{-9}$ C $r_{3,2} = 4.00$ m

$q_3 = +5.00 \times 10^{-9}$ C $r_{3,1} = 5.00$ m

$\theta = 37.0°$

Unknown: $\mathbf{F_{3,tot}} = ?$

Diagram:

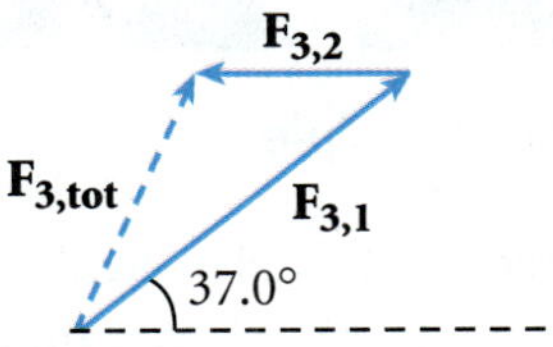

TIP *According to the superposition principle, the resultant force on the charge q_3 is the vector sum of the forces exerted by q_1 and q_2 on q_3. First, find the force exerted on q_3 by each, and then add these two forces together vectorially to get the resultant force on q_3.*

2. Determine the direction of the forces by analyzing the charges.

The force $\mathbf{F_{3,1}}$ is repulsive because q_1 and q_3 have the same sign.
The force $\mathbf{F_{3,2}}$ is attractive because q_2 and q_3 have opposite signs.

3. Calculate the magnitude of the forces with Coulomb's law.

$$F_{3,1} = k_C \frac{q_3 q_1}{(r_{3,1})^2} = (8.99 \times 10^9 \text{ N}\bullet\text{m}^2/\text{C}^2)\left(\frac{(5.00 \times 10^{-9}\text{ C})(6.00 \times 10^{-9}\text{ C})}{(5.00\text{ m})^2}\right)$$

$$F_{3,1} = 1.08 \times 10^{-8}\text{ N}$$

$$F_{3,2} = k_C \frac{q_3 q_2}{(r_{3,2})^2} = (8.99 \times 10^9 \text{ N}\bullet\text{m}^2/\text{C}^2)\left(\frac{(5.00 \times 10^{-9}\text{ C})(2.00 \times 10^{-9}\text{ C})}{(4.00\text{ m})^2}\right)$$

$$F_{3,2} = 5.62 \times 10^{-9}\text{ N}$$

4. Find the *x* and *y* components of each force.

At this point, the direction of each component must be taken into account.

For $\mathbf{F_{3,1}}$: $F_x = (F_{3,1})(\cos 37.0°) = (1.08 \times 10^{-8}\text{ N})(\cos 37.0°) = 8.63 \times 10^{-9}$ N

$F_y = (F_{3,1})(\sin 37.0°) = (1.08 \times 10^{-8}\text{ N})(\sin 37.0°) = 6.50 \times 10^{-9}$ N

For $\mathbf{F_{3,2}}$: $F_x = -F_{3,2} = -5.62 \times 10^{-9}$ N

$F_y = 0$ N

5. Calculate the magnitude of the total force acting in both directions.

$$F_{x,tot} = 8.63 \times 10^{-9}\text{ N} - 5.62 \times 10^{-9}\text{ N} = 3.01 \times 10^{-9}\text{ N}$$

$$F_{y,tot} = 6.50 \times 10^{-9}\text{ N} + 0\text{ N} = 6.50 \times 10^{-9}\text{ N}$$

continued on next page

Classroom Practice

The Superposition Principle
Four equal charges of 1.5 μC are placed at the corners of a square with 5.0 cm sides. Find the net force on a fifth charge placed in the center of the square if the new charge is

a. −1.5 μC
b. +3.0 μC

Answer
a. 0.0 N
b. 0.0 N

Three charges are located on the *x*-axis. A 5.0 μC charge is located at $x = 0.0$ cm, a 1.5 μC charge is located at $x = 3.0$ cm, and a −3.0 μC charge is located at $x = 5.0$ cm. Find the magnitude and direction of the resultant force on the 5.0 μC charge.

Answer
21 N, along the negative *x*-axis

PROBLEM GUIDE B

Use this guide to assign problems.
SE = Student Edition Textbook
PW = Problem Workbook
PB = Problem Bank on the One-Stop Planner (OSP)

Solving for:

F	**SE** Sample, 1–2; Ch. Rvw. 18–19 **PW** Sample, 1–6 **PB** Sample, 1–10

*Challenging Problem
Consult the printed Solutions Manual or the OSP for detailed solutions.

ANSWERS

Practice B

1. 47 N, along the negative x-axis; 157 N, along the positive x-axis; 11.0×10^1 N, along the negative x-axis
2. a. 13.0 N, 31° below the positive x-axis
   b. 25 N, 78° above the negative x-axis
   c. 18 N, 75° below the positive x-axis

6. Use the Pythagorean theorem to find the magnitude of the resultant force.

$$F_{3,tot} = \sqrt{(F_{x,tot})^2 + (F_{y,tot})^2} = \sqrt{(3.01 \times 10^{-9}\ \text{N})^2 + (6.50 \times 10^{-9}\ \text{N})^2}$$

$$\boxed{F_{3,tot} = 7.16 \times 10^{-9}\ \text{N}}$$

7. Use a suitable trigonometric function to find the direction of the resultant force.

In this case, you can use the inverse tangent function:

$$\tan \varphi = \frac{F_{y,tot}}{F_{x,tot}} = \frac{6.50 \times 10^{-9}\ \text{N}}{3.01 \times 10^{-9}\ \text{N}}$$

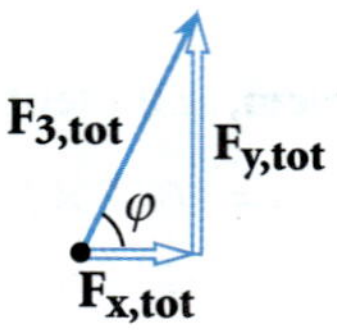

$$\boxed{\varphi = 65.2°}$$

PRACTICE B

The Superposition Principle

1. Three point charges, q_1, q_2, and q_3, lie along the x-axis at $x = 0$, $x = 3.0$ cm, and $x = 5.0$ cm, respectively. Calculate the magnitude and direction of the electric force on each of the three point charges when $q_1 = +6.0\ \mu C$, $q_2 = +1.5\ \mu C$, and $q_3 = -2.0\ \mu C$.
2. Four charged particles are placed so that each particle is at the corner of a square. The sides of the square are 15 cm. The charge at the upper left corner is $+3.0\ \mu C$, the charge at the upper right corner is $-6.0\ \mu C$, the charge at the lower left corner is $-2.4\ \mu C$, and the charge at the lower right corner is $-9.0\ \mu C$.
   a. What is the net electric force on the $+3.0\ \mu C$ charge?
   b. What is the net electric force on the $-6.0\ \mu C$ charge?
   c. What is the net electric force on the $-9.0\ \mu C$ charge?

Consider an object that is in equilibrium. According to Newton's first law, the net external force acting on a body in equilibrium must equal zero. In electrostatic situations, the equilibrium position of a charge is the location at which the net electric force on the charge is zero. To find this location, you must find the position at which the electric force from one charge is equal and opposite the electric force from another charge. This can be done by setting the forces (found by Coulomb's law) equal and then solving for the distance between either charge and the equilibrium position. This is demonstrated in Sample Problem C.

SAMPLE PROBLEM C

Equilibrium

PROBLEM

Three charges lie along the *x*-axis. One positive charge, q_1 = 15 µC, is at x = 2.0 m, and another positive charge, q_2 = 6.0 µC, is at the origin. At what point on the *x*-axis must a negative charge, q_3, be placed so that the resultant force on it is zero?

SOLUTION

Given: $q_1 = 15\ \mu C$ $\quad r_{3,1} = 2.0\ m - d$

$q_2 = 6.0\ \mu C$ $\quad r_{3,2} = d$

Unknown: the distance (d) between the negative charge q_3 and the positive charge q_2 such that the resultant force on q_3 is zero

Diagram:

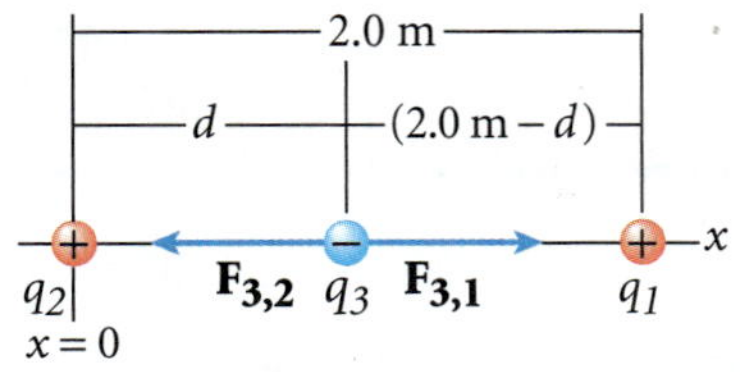

Because we require that the resultant force on q_3 be zero, $F_{3,1}$ must equal $F_{3,2}$. Each force can be found by using Coulomb's law.

$$F_{3,1} = F_{3,2}$$

$$k_C\left(\frac{q_3q_1}{(r_{3,1})^2}\right) = k_C\left(\frac{q_3q_2}{(r_{3,2})^2}\right)$$

$$\frac{q_1}{(2.0\ m - d)^2} = \frac{q_2}{d^2}$$

TIP *Because k_C and q_3 are common terms, they can be canceled from both sides of the equation.*

Now, solve for d to find the location of q_3.

$$(d^2)(q_1) = (2.0\ m - d)^2(q_2)$$

Take the square root of both sides, and then isolate d.

$$d\sqrt{q_1} = (2.0\ m - d)\sqrt{q_2}$$

$$d\left(\sqrt{q_1} + \sqrt{q_2}\right) = \sqrt{q_2}(2.0\ m)$$

$$d = \frac{\sqrt{q_2}(2.0\ m)}{\sqrt{q_1} + \sqrt{q_2}} = \frac{\sqrt{6.0\ \mu C}\ (2.0\ m)}{\sqrt{15\ \mu C} + \sqrt{6.0\ \mu C}} = 0.77\ m$$

$$\boxed{d = 0.77\ m}$$

Classroom Practice

Equilibrium

Two charges, q_1 and q_2, lie on the x-axis. The first charge is at the origin and the second is at x = 1.0 m. Determine the equilibrium position for a third charge, q_3, with respect to q_1 and q_2 for each of the following cases:

a. $q_1 = +10.0\ \mu C$, $q_2 = +7.5\ \mu C$

b. $q_1 = +3.7\ nC$, $q_2 = +5.2\ nC$

c. $q_1 = -3.7\ nC$, $q_2 = -5.2\ nC$

Answer

a. 0.54 m from q_1

b. 0.45 m from q_1

c. 0.45 m from q_1

PROBLEM GUIDE C

Use this guide to assign problems.
SE = Student Edition Textbook
PW = Problem Workbook
PB = Problem Bank on the One-Stop Planner (OSP)

Solving for:

P	**SE** Sample, 1–2; Ch. Rvw. 20–21 **PW** 7–10 **PB** 4–6
q	**SE** Ch. Rvw. 40*, 41 **PW** Sample, 1–6 **PB** 7–10
F	**PW** 11–12, 13* **PB** Sample, 1–3

***Challenging Problem**
Consult the printed Solutions Manual or the OSP for detailed solutions.

SECTION 2

ANSWERS

Practice C

1. $x = 0.62$ m
2. 24.5 cm from q_1 (15.5 cm from q_2)
3. 5.07 m

Misconception Alert — BASIC

The gravitational force acts on a very large scale, and the effects of the gravitational force are more apparent than the effects of the electric force in our typical experiences. These observations may cause some students to think that the gravitational force is stronger than the electric force. Remind students that the electric force between a proton and an electron is much larger than the gravitational force between the two particles (shown in Sample Problem A).

PRACTICE C

Equilibrium

1. A charge of $+2.00 \times 10^{-9}$ C is placed at the origin, and another charge of $+4.00 \times 10^{-9}$ C is placed at $x = 1.5$ m. Find the point between these two charges where a charge of $+3.00 \times 10^{-9}$ C should be placed so that the net electric force on it is zero.
2. A charge q_1 of -5.00×10^{-9} C and a charge q_2 of -2.00×10^{-9} C are separated by a distance of 40.0 cm. Find the equilibrium position for a third charge of $+15.0 \times 10^{-9}$ C.
3. An electron is released above the Earth's surface. A second electron directly below it exerts just enough of an electric force on the first electron to cancel the gravitational force on it. Find the distance between the two electrons.

SCILINKS.
www.scilinks.org
Topic: Van de Graaff Generator
Code: HF61592

Electric force is a field force

The Coulomb force is the second example we have studied of a force that is exerted by one object on another even though there is no physical contact between the two objects. Such a force is known as a *field force.* Recall that another example of a field force is gravitational attraction. Notice that the mathematical form of the Coulomb force is very similar to that of the gravitational force. Both forces are inversely proportional to the square of the distance of separation.

However, there are some important differences between electric and gravitational forces. First of all, as you have seen, electric forces can be either attractive or repulsive. Gravitational forces, on the other hand, are always attractive. The reason is that charge comes in two types—positive and negative—but mass comes in only one type, which results in an attractive gravitational force.

Another difference between the gravitational force and the electric force is their relative strength. As shown in Sample Problem A, the electric force is significantly stronger than the gravitational force. As a result, the electric force between charged atomic particles is much stronger than their gravitational attraction to Earth and between each other.

In the large-scale world, the relative strength of these two forces can be seen by noting that the amount of charge required to overcome the gravitational force is relatively small. For example, if you rub a balloon against your hair and hold the balloon directly above your hair, your hair will stand on end because it is attracted toward the balloon. Although only a small amount of charge is transferred from your hair to the balloon, the electric force between the two is nonetheless stronger than the gravitational force that pulls your hair toward the ground.

Coulomb quantified electric force with a torsion balance

Earlier in this chapter, you learned that Charles Coulomb was the first person to quantify the electric force and establish the inverse square law for electric charges. Coulomb measured electric forces between charged objects with a torsion balance, as shown in **Figure 6.** A torsion balance consists of two small spheres fixed to the ends of a light horizontal rod. The rod is made of an insulating material and is suspended by a silk thread.

In this experiment, one of the spheres is given a charge and another charged object is brought near the charged sphere. The attractive or repulsive force between the two causes the rod to rotate and to twist the suspension. The angle through which the rod rotates is measured by the deflection of a light beam reflected from a mirror attached to the suspension. The rod rotates through some angle against the restoring force of the twisted thread before reaching equilibrium. The value of the angle of rotation increases as the charge increases, thereby providing a quantitative measure of the electric force. With this experiment, Coulomb established the equation for electric force introduced at the beginning of this section. More recent experiments have verified these results to within a very small uncertainty.

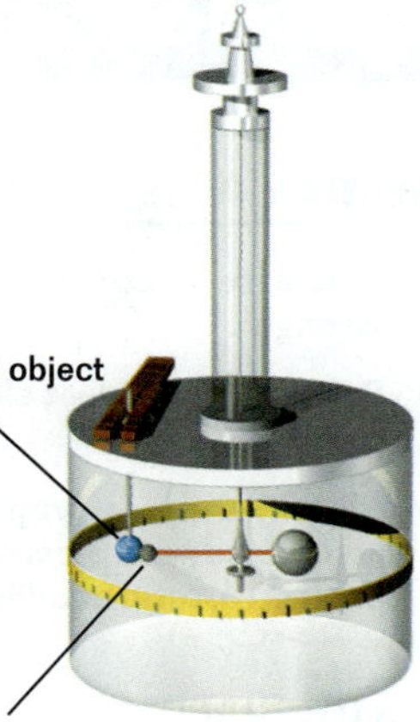

Figure 6
Coulomb's torsion balance was used to establish the inverse square law for the electric force between two charges.

SECTION REVIEW

1. A small glass ball rubbed with silk gains a charge of +2.0 μC. The glass ball is placed 12 cm from a small charged rubber ball that carries a charge of −3.5 μC.
   a. What is the magnitude of the electric force between the two balls?
   b. Is this force attractive or repulsive?
   c. How many electrons has the glass ball lost in the rubbing process?
2. The electric force between a negatively charged paint droplet and a positively charged automobile body is increased by a factor of two, but the charges on each remain constant. How has the distance between the two changed? (Assume that the charge on the automobile is located at a single point.)
3. A $+2.2 \times 10^{-9}$ C charge is on the x-axis at $x = 1.5$ m, a $+5.4 \times 10^{-9}$ C charge is on the x-axis at $x = 2.0$ m, and a $+3.5 \times 10^{-9}$ C charge is at the origin. Find the net force on the charge at the origin.
4. A charge q_1 of -6.00×10^{-9} C and a charge q_2 of -3.00×10^{-9} C are separated by a distance of 60.0 cm. Where could a third charge be placed so that the net electric force on it is zero?
5. **Critical Thinking** What are some similarities between the electric force and the gravitational force? What are some differences between the two forces?

Teaching Tip — ADVANCED

Charles Coulomb (1738–1806) invented the torsion balance (**Figure 6**), which he used to show the relationships between charge, distance, and electric force. By varying the distance between the two charged objects and observing the corresponding angles of rotation, Coulomb found that electric force is inversely proportional to distance squared. Coulomb also developed a method for placing controlled amounts of charge on each object. By varying the charge on each object, he found that electric force is proportional to the charge on each object.

SECTION REVIEW ANSWERS

1. **a.** 4.4 N
   b. attractive
   c. 1.2×10^{13} electrons
2. new distance = $\frac{1}{\sqrt{2}}$ (old distance)
3. 7.3×10^{-8} N, along the negative x-axis
4. 35.2 cm from q_1 (24.8 cm from q_2)
5. both are field forces, and both follow inverse square laws; electric forces are attractive or repulsive while gravitational forces are always attractive, and electric force is significantly stronger than gravitational force

SECTION 3
Advanced Level

Visual Strategy BASIC

Figure 7
Be sure students understand that by convention the direction of the electric field at any point is the direction of the electric force experienced by a positive test charge at that point.

Q Suppose a small negative test charge were placed at the location q_0. Would this change the direction of the electric field vector at that point? Explain.

A ***no; By definition, the direction of the electric field at any point corresponds to the direction in which a positive charge would experience an electric force, regardless of what kind of charge is actually located at that point.***

Q Imagine a similar diagram for the gravitational field of Earth with the moon as the test charge. Which diagram, **(a)** or **(b),** would the gravitational field look like?

A ***Diagram (b) is like a gravitational field because the gravitational force is attractive.***

SECTION 3

The Electric Field

SECTION OBJECTIVES

- **Calculate electric field strength.**
- **Draw and interpret electric field lines.**
- **Identify the four properties associated with a conductor in electrostatic equilibrium.**

electric field

a region where an electric force on a test charge can be detected

ELECTRIC FIELD STRENGTH

As discussed earlier in this chapter, electric force, like gravitational force, is a field force. Unlike contact forces, which require physical contact between objects, field forces are capable of acting through space, producing an effect even when there is no physical contact between the objects involved. The concept of a field is a model that is frequently used to understand how two objects can exert forces on each other at a distance. For example, a charged object sets up an **electric field** in the space around it. When a second charged object enters this field, forces of an electrical nature arise. In other words, the second object interacts with the field of the first particle.

To define an electric field more precisely, consider **Figure 7(a),** which shows an object with a small positive charge, q_0, placed near a second object with a larger positive charge, Q. The strength of the electric field, E, at the location of q_0 is defined as the magnitude of the electric force acting on q_0 divided by the charge of q_0:

$$E = \frac{F_{electric}}{q_0}$$

Note that this is the electric field at the location of q_0 produced by the charge Q, and *not* the field produced by q_0.

Because electric field strength is a ratio of force to charge, the SI units of E are newtons per coulomb (N/C). The electric field is a vector quantity. By convention, the direction of **E** at a point is defined as the direction of the electric force that would be exerted on a small *positive* charge (called a test charge) placed at that point. Thus, in **Figure 7(a),** the direction of the electric field is horizontal and away from the sphere because a positive charge would be repelled by the positive sphere. In **Figure 7(b),** the direction of the electric field is toward the sphere because a positive charge would be attracted toward the negatively charged sphere. In other words, the direction of **E** depends on the sign of the charge producing the field.

Figure 7
(a) A small object with a positive charge q_0 placed in the field, **E,** of an object with a larger positive charge experiences an electric force away from the object. **(b)** A small object with a positive charge q_0 placed in the field, **E,** of a negatively charged object experiences an electric force toward the object.

Now, consider the positively charged conducting sphere in **Figure 8(a).** The field in the region surrounding the sphere could be explored by placing a positive test charge, q_0, in a variety of places near the sphere. To find the electric field at each point, you would first find the electric force on this charge, then divide this force by the magnitude of the test charge.

However, when the magnitude of the test charge is great enough to influence the charge on the conducting sphere, a difficulty with our definition arises. According to Coulomb's law, a strong test charge will cause a rearrangement of the charges on the sphere, as shown in **Figure 8(b).** As a result, the force exerted on the test charge is different from what the force would be if the movement of charge on the sphere had not taken place. Furthermore, the strength of the measured electric field is different from what it would be in the absence of the test charge. To eliminate this problem, we assume that the test charge is small enough to have a negligible effect on the location of the charges on the sphere, the situation shown in **Figure 8(a).**

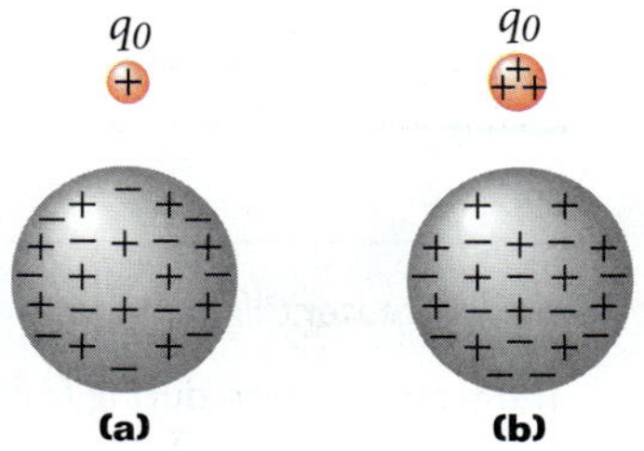

Figure 8
We must assume a small test charge, as in **(a),** because a larger test charge, as in **(b),** can cause a redistribution of the charge on the sphere, which changes the electric field strength.

Electric field strength depends on charge and distance

To reformulate our equation for electric field strength from a point charge, consider a small test charge, q_0, located a distance, r, from a charge, q. According to Coulomb's law, the magnitude of the force on the test charge is given by the following equation:

$$F_{electric} = k_C \frac{qq_0}{r^2}$$

We can find the magnitude of the electric field due to the point charge q at the position of q_0 by substituting this value into our previous equation for electric field strength.

$$E = \frac{F_{electric}}{q_0} = k_C \frac{q\cancel{q_0}}{r^2\cancel{q_0}}$$

Notice that q_0 cancels, and we have a new equation for electric field strength due to a point charge.

ELECTRIC FIELD STRENGTH DUE TO A POINT CHARGE

$$E = k_C \frac{q}{r^2}$$

electric field strength = Coulomb constant × $\frac{\textbf{charge producing the field}}{\textbf{(distance)}^2}$

As stated above, electric field, **E,** is a vector. If q is positive, the field due to this charge is directed outward radially from q. If q is negative, the field is directed toward q. As with electric force, the electric field due to more than one charge is calculated by applying the principle of superposition. A strategy for solving superposition problems is given in Sample Problem D.

extension

Integrating Astronomy
Visit go.hrw.com for the activity "Ion Propulsion in *Deep Space 1*."
Keyword HF6ELFX

SCILINKS.
www.scilinks.org
Topic: Microwaves
Code: HF60959

SECTION 3

Demonstration

Electric Field Strength

Purpose Show the dependence of field strength on charge.

Materials rubber rod and fur, demonstration Leyden jar (available from most science education supply houses), electroscope

Procedure Charge the rubber rod with the fur. Hold the Leyden jar in one hand and touch the charged rod to the center knob of the jar. Avoid discharging the Leyden jar—do not touch the knob with your hand or other conducting material. Bring the knob of the jar close to the knob of the electroscope, and have students note that the leaves of the electroscope diverge.

Add more charge to the Leyden jar by again charging the rubber rod and touching it to the Leyden jar. Bring the knob of the Leyden jar close to the electroscope as before, and have students note that the leaves diverge more than before. Ask the students what the increased divergence of the leaves indicates. *(The electric field is stronger because the charge on the Leyden jar has increased.)* Continue charging the Leyden jar and showing its increased charge as evidenced by the increased divergence of the electroscope leaves.

SECTION 3

Teaching Tip — GENERAL

Students may ask how we can say there is an electric field without a test charge present. Point out that this conclusion is drawn from the fact that q_0 (the test charge) cancels from the equation for electric field strength. This conclusion cannot be demonstrated experimentally because the only way to observe a field is to observe its effects on an object. This could lead to a discussion of the fact that we do not *know* the field exists without a test charge, but since it is there whenever we test the field, we assume the field is there at all times.

PROBLEM GUIDE D

Use this guide to assign problems.
SE = Student Edition Textbook
PW = Problem Workbook
PB = Problem Bank on the One-Stop Planner (OSP)

Solving for:

E	**SE** Sample, 1–2; Ch. Rvw. 32–33, 35b, 37a, 38b, 48, 55 **PW** 5–6 **PB** 4–5
q	**PW** Sample, 1–2 **PB** 6–7
F	**SE** 3 **PB** Sample, 1–3
r	**PW** 3–4 **PB** 8–10

***Challenging Problem**
Consult the printed Solutions Manual or the OSP for detailed solutions.

Table 3 Electric Fields

Examples	E, N/C
in a fluorescent lighting tube	10
in the atmosphere during fair weather	100
under a thundercloud or in a lightning bolt	10 000
at the electron in a hydrogen atom	5.1×10^{11}

Our new equation for electric field strength points out an important property of electric fields. As the equation indicates, an electric field at a given point depends only on the charge, q, of the object setting up the field and on the distance, r, from that object to a specific point in space. As a result, we can say that an electric field exists at any point near a charged body even when there is no test charge at that point. The examples in **Table 3** show the magnitudes of various electric fields.

SAMPLE PROBLEM D

STRATEGY Electric Field Strength

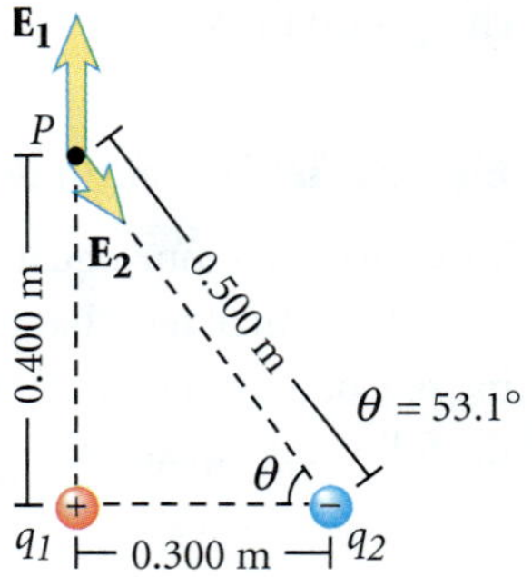

PROBLEM

A charge q_1 = +7.00 μC is at the origin, and a charge q_2 = −5.00 μC is on the x-axis 0.300 m from the origin, as shown at right. Find the electric field strength at point P, which is on the y-axis 0.400 m from the origin.

SOLUTION

1. Define the problem, and identify the known variables.

Given: $q_1 = +7.00\ \mu C = 7.00 \times 10^{-6}$ C $r_1 = 0.400$ m

$q_2 = -5.00\ \mu C = -5.00 \times 10^{-6}$ C $r_2 = 0.500$ m

$\theta = 53.1°$

Unknown: **E** at P ($y = 0.400$ m)

> **TIP** *Apply the principle of superposition. You must first calculate the electric field produced by each charge individually at point P and then add these fields together as vectors.*

2. Calculate the electric field strength produced by each charge.

Because we are finding the magnitude of the electric field, we can neglect the sign of each charge.

$$E_1 = k_C\frac{q_1}{r_1^2} = (8.99 \times 10^9\ \text{N}\bullet\text{m}^2/\text{C}^2)\left(\frac{7.00 \times 10^{-6}\ \text{C}}{(0.400\ \text{m})^2}\right) = 3.93 \times 10^5\ \text{N/C}$$

$$E_2 = k_C\frac{q_2}{r_2^2} = (8.99 \times 10^9\ \text{N}\bullet\text{m}^2/\text{C}^2)\left(\frac{5.00 \times 10^{-6}\ \text{C}}{(0.500\ \text{m})^2}\right) = 1.80 \times 10^5\ \text{N/C}$$

3. Analyze the signs of the charges.

The field vector $\mathbf{E_1}$ at P due to q_1 is directed vertically upward, as shown in the figure above, because q_1 is positive. Likewise, the field vector $\mathbf{E_2}$ at P due to q_2 is directed toward q_2 because q_2 is negative.

4. Find the *x* and *y* components of each electric field vector.

For $\mathbf{E_1}$: $E_{x,1} = 0$ N/C

$E_{y,1} = 3.93 \times 10^5$ N/C

For $\mathbf{E_2}$: $E_{x,2} = (E_2)(\cos 53.1°) = (1.80 \times 10^5 \text{ N/C})(\cos 53.1°) = 1.08 \times 10^5$ N/C

$E_{y,2} = -(E_2)(\sin 53.1°) = -(1.80 \times 10^5 \text{ N/C})(\sin 53.1°) = -1.44 \times 10^5$ N/C

5. Calculate the total electric field strength in both directions.

$$E_{x,tot} = E_{x,1} + E_{x,2} = 0 \text{ N/C} + 1.08 \times 10^5 \text{ N/C} = 1.08 \times 10^5 \text{ N/C}$$

$$E_{y,tot} = E_{y,1} + E_{y,2} = 3.93 \times 10^5 \text{ N/C} - 1.44 \times 10^5 \text{ N/C} = 2.49 \times 10^5 \text{ N/C}$$

6. Use the Pythagorean theorem to find the magnitude of the resultant electric field strength vector.

$$E_{tot} = \sqrt{(E_{x,tot})^2 + (E_{y,tot})^2} = \sqrt{(1.08 \times 10^5 \text{ N/C})^2 + (2.49 \times 10^5 \text{ N/C})^2}$$

$$\boxed{E_{tot} = 2.71 \times 10^5 \text{ N/C}}$$

7. Use a suitable trigonometric function to find the direction of the resultant electric field strength vector.

In this case, you can use the inverse tangent function:

$$\tan \varphi = \frac{E_{y,tot}}{E_{x,tot}} = \frac{2.49 \times 10^5 \text{ N/C}}{1.08 \times 10^5 \text{ N/C}}$$

$$\boxed{\varphi = 66.6°}$$

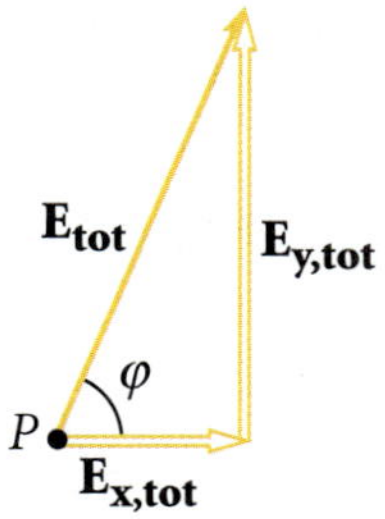

8. Evaluate your answer.

The electric field at point *P* is pointing away from the charge q_1, as expected, because q_1 is a positive charge and is larger than the negative charge q_2.

PRACTICE D

Electric Field Strength

1. A charge, $q_1 = 5.00\ \mu$C, is at the origin, and a second charge, $q_2 = -3.00\ \mu$C, is on the *x*-axis 0.800 m from the origin. Find the electric field at a point on the *y*-axis 0.500 m from the origin.
2. A proton and an electron in a hydrogen atom are separated on the average by about 5.3×10^{-11} m. What is the magnitude and direction of the electric field set up by the proton at the position of the electron?
3. An electric field of 2.0×10^4 N/C is directed along the positive *x*-axis.
   a. What is the electric force on an electron in this field?
   b. What is the electric force on a proton in this field?

SECTION 3

Classroom Practice

Electric Field Strength
An electric field around a charged object is 5.95×10^6 N/C at a distance of 0.100 m. Find the charge on the object.

Answer
6.62×10^{-6} C = 6.62 μC

A charge $q_1 = 4.50\ \mu$C experiences an attractive force of 1.35 N at a distance of 0.150 m from a charged object, q_2.

a. Find the strength of the electric field due to q_2 at a distance of 0.150 m.
b. Find the charge, q_2.

Answer
a. 3.00×10^5 N/C
b. 0.751 μC

ANSWERS

Practice D

1. 1.66×10^5 N/C, 81.1° above the positive *x*-axis
2. 5.1×10^{11} N/C, away from the proton
3. **a.** 3.2×10^{-15} N, along the negative *x*-axis
   b. 3.2×10^{-15} N, along the positive *x*-axis

Misconception Alert

Some students may think that electric field lines are a physical phenomenon. Much like the magnetic field lines with which students may be familiar, electric field lines do not actually exist. Students should be told this explicitly. The lines are a visual representation of the field that would be experienced by a test charge. Explain to the students that an electric field is much like a gravitational field in that it cannot be directly observed. We can only observe the effects of the field.

Visual Strategy GENERAL

Figure 9

Be sure students understand the similarities and differences between the two cases shown in this figure.

Q What can you conclude about the charges in **(a)** and **(b)** by comparing the electric field lines for each case?

A ***Because the number of field lines leaving the charge in (a) is equal to the number of field lines approaching the charge in (b), the charges must be equal in magnitude. The field lines also show that the charge in (a) is positive (because the lines are pointing away from the charge) and that the charge in (b) is negative (because the lines are pointing toward the charge).***

ELECTRIC FIELD LINES

A convenient aid for visualizing electric field patterns is to draw lines pointing in the direction of the electric field, called *electric field lines.* Although electric field lines do not really exist, they offer a useful means of analyzing fields by representing both the strength and the direction of the field at different points in space. This is useful because the field at each point is often the result of more than one charge, as seen in Sample Problem D. Field lines make it easier to visualize the net field at each point.

The number of field lines is proportional to the electric field strength

By convention, electric field lines are drawn so that the electric field vector, **E,** is tangent to the lines at each point. Further, the number of lines per unit area through a surface perpendicular to the lines is proportional to the strength of the electric field in a given region. Thus, E is stronger where the field lines are close together and weaker where they are far apart.

Figure 9(a) shows some representative electric field lines for a positive point charge. Note that this two-dimensional drawing contains only the field lines that lie in the plane containing the point charge. The lines are actually directed outward radially from the charge in all directions, somewhat like quills radiate from the body of a porcupine. Because a positive test charge placed in this field would be repelled by the positive charge q, the lines are directed away from the positive charge, extending to infinity. Similarly, the electric field lines for a single negative point charge, which begin at infinity, are directed inward toward the charge, as shown in **Figure 9(b).** Note that the lines are closer together as they get near the charge, indicating that the strength of the field is increasing. This is consistent with our equation for electric field strength, which is inversely proportional to distance squared. **Figure 9(c)** shows grass seeds in an insulating liquid. When a small charged conductor is placed in the center, these seeds align with the electric field produced by the charged body.

The rules for drawing electric field lines are summarized in **Table 4.** Note that no two field lines from the same field can cross one another. The reason is that at every point in space, the electric field vector points in a single direction and any field line at that point must also point in that direction.

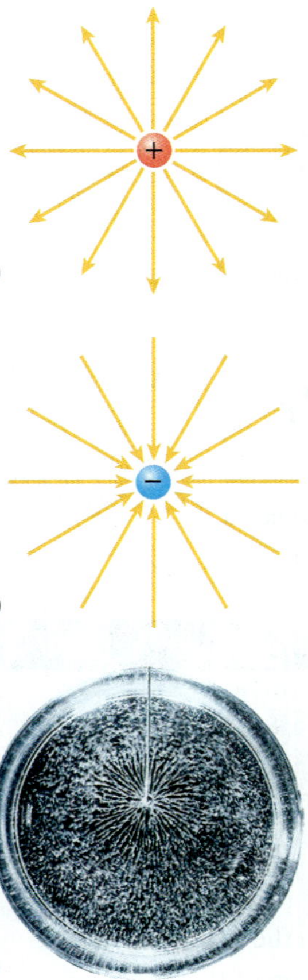

(a)

(b)

(c)

Figure 9
The diagrams **(a)** and **(b)** show some representative electric field lines for a positive and a negative point charge. In **(c),** grass seeds align with a similar field produced by a charged body.

Table 4 Rules for Drawing Electric Field Lines

The lines must begin on positive charges or at infinity and must terminate on negative charges or at infinity.
The number of lines drawn leaving a positive charge or approaching a negative charge is proportional to the magnitude of the charge.
No two field lines from the same field can cross each other.

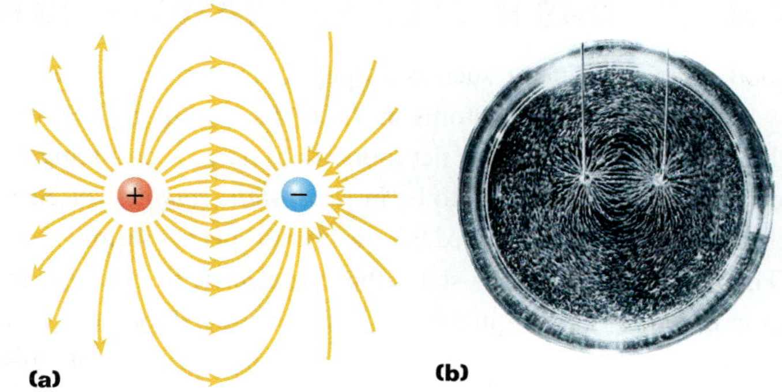

Figure 10
(a) This diagram shows the electric field lines for two equal and opposite point charges. Note that the number of lines leaving the positive charge equals the number of lines terminating on the negative charge. **(b)** In this photograph, grass seeds in an insulating liquid align with a similar electric field produced by two oppositely charged conductors.

Figure 10 shows the electric field lines for two point charges of equal magnitudes but opposite signs. This charge configuration is called an *electric dipole.* In this case, the number of lines that begin at the positive charge must equal the number of lines that terminate on the negative charge. At points very near the charges, the lines are nearly radial. The high density of lines between the charges indicates a strong electric field in this region.

In electrostatic spray painting, field lines between a negatively charged spray gun and a positively charged target object are similar to those shown in **Figure 10.** As you can see, the field lines suggest that paint droplets that narrowly miss the target object still experience a force directed toward the object, sometimes causing them to wrap around from behind and hit it. This does happen and increases the efficiency of an electrostatic spray gun.

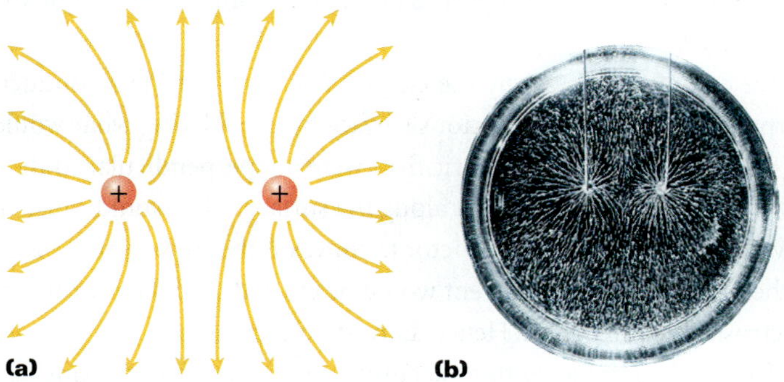

Figure 11
(a) This diagram shows the electric field lines for two positive point charges. **(b)** The photograph shows the analogous case for grass seeds in an insulating liquid around two conductors with the same charge.

Figure 11 shows the electric field lines in the vicinity of two equal positive point charges. Again, close to either charge, the lines are nearly radial. The same number of lines emerges from each charge because the charges are equal in magnitude. At great distances from the charges, the field approximately equals that of a single point charge of magnitude $2q$.

Finally, **Figure 12** is a sketch of the electric field lines associated with a positive charge $+2q$ and a negative charge $-q$. In this case, the number of lines leaving the charge $+2q$ is twice the number terminating on the charge $-q$. Hence, only half the lines that leave the positive charge end at the negative charge. The remaining half terminate at infinity. At distances that are great compared with the separation between the charges, the pattern of electric field lines is equivalent to that of a single charge, $+q$.

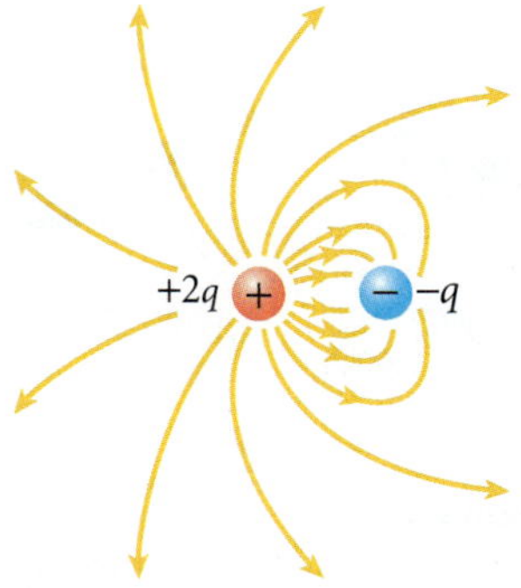

Figure 12
In this case, only half the lines originating from the positive charge terminate on the negative charge because the positive charge is twice as great as the negative charge.

Visual Strategy GENERAL

Figures 10 and 11

Have students examine these two figures and think about the spacing of the field lines.

Q How is the space between field lines related to the strength of the field?

A ***The electric field is strongest where the lines are closest together and weakest where the lines are farthest apart.***

Q Where would an object in either of these fields experience the greatest force?

A ***at the location where the field strength is greatest (where the lines are closest together)***

SECTION 3

Demonstration

Charge Accumulation

Purpose Provide a dramatic visual demonstration of charge accumulation by a Van de Graaff generator.

Materials Van de Graaff generator, tape, aluminum pie pans, puffed rice (or small pieces of paper), glass beaker

Procedure Stack 10 to 12 pie pans upside down on top of the Van de Graaff generator. Turn the generator on. Have students explain the flying pans.

Next, tape one pie pan upright to the top of the generator, and pour in some of the puffed rice or paper pieces. Turn the generator on. Once again, have students explain the results.

Finally, tape the glass beaker upright to the top of the generator. Pour in some puffed rice. Turn the generator on. Have students explain the results.

Teaching Tip — BASIC

Point out that the properties discussed for conductors in equilibrium do not hold true for insulators because charges do not move readily in insulators.

CONDUCTORS IN ELECTROSTATIC EQUILIBRIUM

A good electric conductor, such as copper, contains charges (electrons) that are only weakly bound to the atoms in the material and are free to move about within the material. When no net motion of charge is occurring within a conductor, the conductor is said to be in *electrostatic equilibrium.* As we shall see, such a conductor that is isolated has the four properties summarized in **Table 5.**

The first property, which states that the electric field is zero inside a conductor in electrostatic equilibrium, can be understood by examining what would happen if this were not true. If there were an electric field inside a conductor, the free charges would move and a flow of charge, or current, would be created. However, if there were a net movement of charge, the conductor would no longer be in electrostatic equilibrium.

Table 5 Conductors in Electrostatic Equilibrium

The electric field is zero everywhere inside the conductor.
Any excess charge on an isolated conductor resides entirely on the conductor's outer surface.
The electric field just outside a charged conductor is perpendicular to the conductor's surface.
On an irregularly shaped conductor, charge tends to accumulate where the radius of curvature of the surface is smallest, that is, at sharp points.

The fact that any excess charge resides on the outer surface of the conductor is a direct result of the repulsion between like charges described by Coulomb's law. If an excess of charge is placed inside a conductor, the repulsive forces arising between the charges force them as far apart as possible, causing them to quickly migrate to the surface.

We can understand why the electric field just outside a conductor must be perpendicular to the conductor's surface by considering what would happen if this were not true. If the electric field were *not* perpendicular to the surface, the field would have a component along the surface. This would cause the free negative charges within the conductor to move on the surface of the conductor. But if the charges moved, a current would be created, and there would no longer be electrostatic equilibrium. Hence, **E** must be perpendicular to the surface.

To see why charge tends to accumulate at sharp points, consider a conductor that is fairly flat at one end and relatively pointed at the other. Any excess charge placed on the object moves to its surface. **Figure 13** shows the forces between two charges at each end of such an object. At the flatter end, these forces are predominantly directed parallel to the surface. Thus, the charges move apart until repulsive forces from other nearby charges create a state of equilibrium.

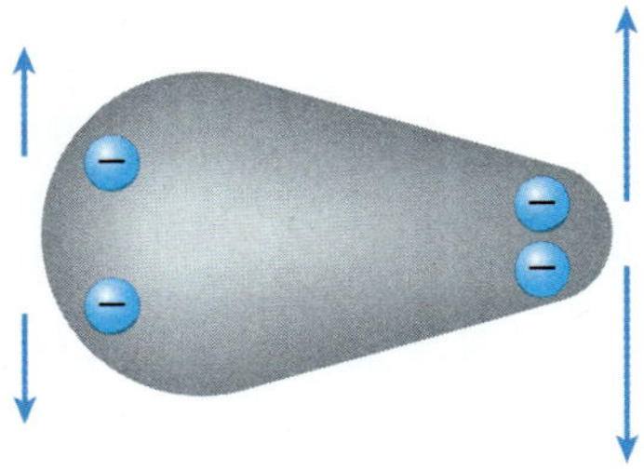

Figure 13
When one end of a conductor is more pointed than the other, excess charge tends to accumulate at the sharper end, resulting in a larger charge per unit area and therefore a larger repulsive electric force between charges at this end.

At the sharp end, however, the forces of repulsion between two charges are directed predominantly perpendicular to the surface. As a result, there is less tendency for the charges to move apart along the surface and the amount of charge per unit area is greater than at the flat end. The cumulative effect of many such outward forces from nearby charges at the sharp end produces a large electric field directed away from the surface.

Why it Matters

Microwave Ovens

It would be hard to find a town in America that does not have a microwave oven. Most homes, convenience stores, and restaurants have this marvelous invention that somehow heats only the soft parts of the food and leaves the inorganic and hard materials, like ceramic and the surfaces of bone, at approximately the same temperature. A neat trick, indeed, but how is it done?

Microwave ovens take advantage of a property of water molecules called *bipolarity.* Water molecules are considered bipolar because each molecule has a positive and a negative end. In other words, more of the electrons in these molecules are at one end of the molecule than the other.

Because microwaves are a high-frequency form of electromagnetic radiation, they supply an electric field that changes polarity billions of times a second. As this electric field passes a bipolar molecule, the positive side of the molecule experiences a force in one direction, and the negative side of the molecule is pushed or pulled in the other direction. When the field changes polarity, the directions of these forces are reversed. Instead of tearing apart, the molecules swing around and line up with the electric field.

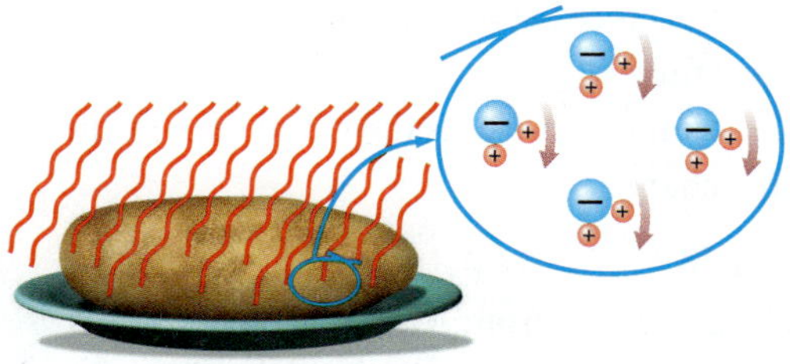

As the bipolar molecules swing around, they rub against one another, producing friction. This friction in turn increases the internal energy of the food. Energy is transferred to the food by radiation (the microwaves) as opposed to conduction from hot air, as in a conventional oven.

Depending on the microwave oven's power and design, this rotational motion can generate up to about 3 J of internal energy each second in 1 g of water. At this rate, a top-power microwave oven can boil a cup (250 mL) of water in 2 min using about 0.033 kW•h of electricity.

Items such as dry plates and the air in the oven are unaffected by the fluctuating electric field because they are not polarized. Because energy is not wasted on heating these nonpolar items, the microwave oven cooks food faster and more efficiently than other ovens.

SECTION REVIEW

1. Find the electric field at a point midway between two charges of $+40.0 \times 10^{-9}$ C and $+60.0 \times 10^{-9}$ C separated by a distance of 30.0 cm.

2. Two point charges are a small distance apart.

a. Sketch the electric field lines for the two if one has a charge four times that of the other and if both charges are positive.

b. Repeat **(a),** but assume both charges are negative.

3. Interpreting Graphics **Figure 14** shows the electric field lines for two point charges separated by a small distance.

a. Determine the ratio q_1/q_2.

b. What are the signs of q_1 and q_2?

4. Critical Thinking Explain why you're more likely to get a shock from static electricity by touching a metal object with your finger instead of with your entire hand.

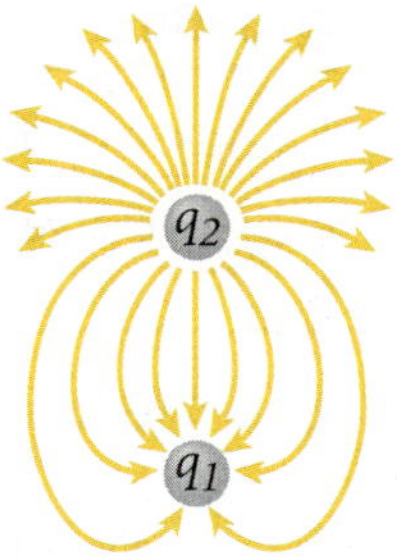

Figure 14

SECTION 3

Why it Matters

Microwave Ovens

Students should be familiar with the changing polarity of electric fields. With this background knowledge, this feature will serve as an introduction to the working principles behind microwave ovens.

Extension

Controversy over the safety of microwave ovens in the home has surrounded this technology since its introduction. Have students research microwave-oven safety tests and present their results to the class.

SECTION REVIEW ANSWERS

1. 8.0×10^3 N/C, directed toward the 40.0×10^{-9} C charge

2. **a.** All lines should point away from the two charges, and one charge should have four times as many lines as the other.

b. All lines should point toward the two charges, and one charge should have four times as many lines as the other.

3. **a.** $-\frac{3}{8}$

b. q_1 is negative; q_2 is positive

4. because charge accumulates at sharp points

CHAPTER 16

Highlights

Teaching Tip

Explaining concepts in written form helps to solidify students' understanding of difficult concepts and helps to enforce good communication skills. Have students summarize the main topics of this chapter in an essay, including methods of charging, the conservation of charge, Coulomb's law, and the electric field. Be sure students explain concepts clearly and correctly and use good sentence structure.

CHAPTER 16

Highlights

KEY TERMS

electrical conductor (p. 561)

electrical insulator (p. 561)

induction (p. 562)

electric field (p. 572)

PROBLEM SOLVING

See **Appendix D: Equations** for a summary of the equations introduced in this chapter. If you need more problem-solving practice, see **Appendix I: Additional Problems.**

Diagram Symbols

Positive charge	$+q$
Negative charge	$-q$
Electric field vector	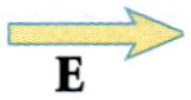 **E**
Electric field lines	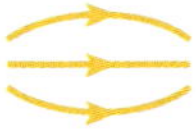

KEY IDEAS

Section 1 Electric Charge

- There are two kinds of electric charge: positive and negative. Like charges repel, and unlike charges attract.
- Electric charge is conserved.
- The fundamental unit of charge, *e*, is the magnitude of the charge of a single electron or proton.
- Conductors and insulators can be charged by contact. Conductors can also be charged by induction. A surface charge can be induced on an insulator by polarization.

Section 2 Electric Force

- According to Coulomb's law, the electric force between two charges is proportional to the magnitude of each of the charges and inversely proportional to the square of the distance between them.
- The electric force is a field force.
- The resultant electric force on any charge is the vector sum of the individual electric forces on that charge.

Section 3 The Electric Field

- An electric field exists in the region around a charged object.
- Electric field strength depends on the magnitude of the charge producing the field and the distance between that charge and a point in the field.
- The direction of the electric field vector, **E,** is the direction in which an electric force would act on a positive test charge.
- Field lines are tangent to the electric field vector at any point, and the number of lines is proportional to the magnitude of the field strength.

Variable Symbols

Quantities		Units		Conversions
$F_{electric}$	electric force	N	newtons	$= kg \bullet m/s^2$
q	charge	C e	coulomb (SI unit of charge) fundamental unit of charge	$= 6.3 \times 10^{18}\ e$ $= 1.60 \times 10^{-19}$ C
k_C	Coulomb constant	$N \bullet \frac{m^2}{C^2}$	newtons $\times \frac{\text{meters}^2}{\text{coulombs}^2}$	$= 8.99 \times 10^9\ N \bullet m^2$
E	electric field strength	N/C	newtons/coulomb	

CHAPTER 16 Review

ELECTRIC CHARGE

Review Questions

1. How are conductors different from insulators?

2. When a conductor is charged by induction, is the induced surface charge on the conductor the same or opposite the charge of the object inducing the surface charge?

3. A negatively charged balloon has 3.5 μC of charge. How many excess electrons are on this balloon?

Conceptual Questions

4. Would life be different if the electron were positively charged and the proton were negatively charged? Explain your answer.

5. Explain from an atomic viewpoint why charge is usually transferred by electrons.

6. Because of a higher moisture content, air is a better conductor of charge in the summer than in the winter. Would you expect the shocks from static electricity to be more severe in summer or winter? Explain your answer.

7. A balloon is negatively charged by rubbing and then clings to a wall. Does this mean that the wall is positively charged?

8. Which effect proves more conclusively that an object is charged, attraction to or repulsion from another object? Explain.

ELECTRIC FORCE

Review Questions

9. What determines the direction of the electric force between two charges?

10. In which direction will the electric force from the two equal positive charges move the negative charge shown below?

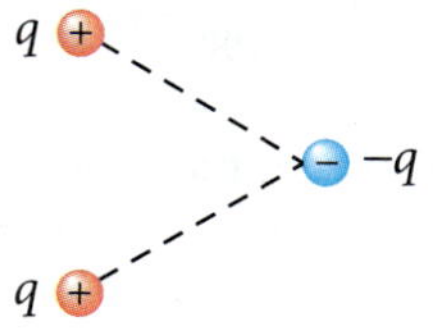

11. The gravitational force is always attractive, while the electric force is both attractive and repulsive. What accounts for this difference?

12. When more than one charged object is present in an area, how can the total electric force on one of the charged objects be predicted?

13. Identify examples of electric forces in everyday life.

Conceptual Questions

14. According to Newton's third law, every action has an equal and opposite reaction. When a comb is charged and held near small pieces of paper, the comb exerts an electric force on the paper pieces and pulls them toward it. Why don't you observe the comb moving toward the paper pieces as well?

Practice Problems

For problems 15–17, see Sample Problem A.

15. At the point of fission, a nucleus of ^{235}U that has 92 protons is divided into two smaller spheres, each of which has 46 protons and a radius of 5.90×10^{-15} m. What is the magnitude of the repulsive force pushing these two spheres apart?

16. What is the electric force between a glass ball that has +2.5 μC of charge and a rubber ball that has −5.0 μC of charge when they are separated by a distance of 5.0 cm?

CHAPTER 16

Review

ANSWERS

1. conductors transfer charge easily, insulators do not
2. opposite
3. 2.2×10^{13} electrons
4. no; *Positive* and *negative* are arbitrary designations.
5. Protons are relatively fixed in the nucleus, whereas the surrounding electrons can be transferred from one atom to another.
6. winter; because more charge can accumulate before electric discharge occurs
7. The balloon induces a local surface charge on the wall, but the wall as a whole is not charged.
8. repulsion; because attraction can be the result of an induced surface charge, but repulsion occurs only when two objects each have a net charge
9. the signs of the charges
10. to the left
11. mass is positive, charges are positive or negative
12. Each force exerted on an object is found, and then the forces are added together vectorially.
13. Answers will vary but may include the force between hair and a comb or the force that acts when people receive a "shock" by touching an object.
14. The comb has a significantly greater mass, so its acceleration is much less.
15. 3.50×10^{3} N
16. 45 N (attractive)

16 REVIEW

17. 91 N (repulsive)
18. 1.00×10^{-7} N, 12° below the positive x-axis
19. 1.48×10^{-7} N, along the $+x$ direction
20. $y = 0.8$ m
21. 18 cm from the 3.5 nC charge
22. See the definition in Section 3 of this chapter.
23. $E = \frac{F_{electric}}{q_0} = \frac{k_C q q_0}{r^2 q_0} = \frac{k_C q}{r^2}$
24. Answers should agree with the discussion in Section 3 of this chapter.
25. Lines should originate from the $+q$ charge and end on the $-3q$ charge, and the ratio of lines from $+q$ to lines ending on $-3q$ should be $\frac{1}{3}$.
26. magnitude of the charge; sign of the charge
27. **a.** where curvature is greatest; where curvature is least
b. where curvature is greatest
28. No, they are a tool used to visualize and analyze the electric field.
29. so the test charge does not significantly affect the charges responsible for the field
30. because at any point in space, **E** points in a single direction
31. Electric forces are equal and opposite; The proton's acceleration is less because it has a greater mass ($F = ma$).
32. 12.0×10^3 N/C, toward the 30.0×10^{-9} C charge
33. 5.7×10^3 N/C, 75° above the positive x-axis

17. An alpha particle (charge = $+2.0e$) is sent at high speed toward a gold nucleus (charge = $+79e$). What is the electric force acting on the alpha particle when the alpha particle is 2.0×10^{-14} m from the gold nucleus?

For problems 18–19, see Sample Problem B.

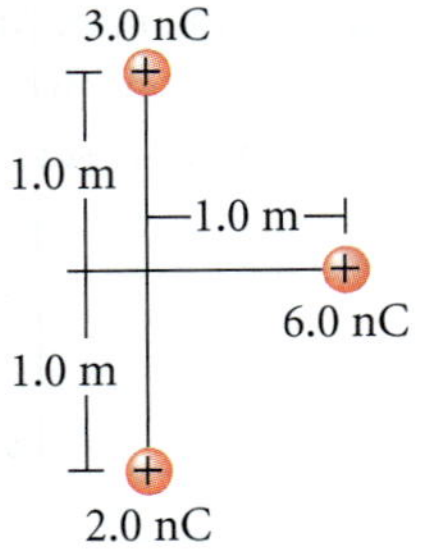

18. Three positive point charges of 3.0 nC, 6.0 nC, and 2.0 nC, respectively, are arranged in a triangular pattern, as shown at right. Find the magnitude and direction of the electric force acting on the 6.0 nC charge.

19. Two positive point charges, each of which has a charge of 2.5×10^{-9} C, are located at $y = +0.50$ m and $y = -0.50$ m. Find the magnitude and direction of the resultant electric force acting on a charge of 3.0×10^{-9} C located at $x = 0.70$ m.

For problems 20–21, see Sample Problem C.

20. Three point charges lie in a straight line along the y-axis. A charge of $q_1 = -9.0$ μC is at $y = 6.0$ m, and a charge of $q_2 = -8.0$ μC is at $y = -4.0$ m. The net electric force on the third point charge is zero. Where is this charge located?

21. A charge of +3.5 nC and a charge of +5.0 nC are separated by 40.0 cm. Find the equilibrium position for a −6.0 nC charge.

THE ELECTRIC FIELD

Review Questions

22. What is an electric field?

23. Show that the definition of electric field strength ($E = F_{electric}/q_0$) is equivalent to the equation $E = k_C q/r^2$ for point charges.

24. As you increase the potential on an irregularly shaped conductor, a bluish purple glow called a *corona* forms around a sharp end sooner than around a smoother end. Explain why.

25. Draw some representative electric field lines for two charges of $+q$ and $-3q$ separated by a small distance.

26. When electric field lines are being drawn, what determines the number of lines originating from a charge? What determines whether the lines originate from or terminate on a charge?

27. Consider the electric field lines in the figure below.
a. Where is charge density the highest? Where is it the lowest?
b. If an opposite charge were brought into the vicinity, where would charge on the pear-shaped object "leak off" most readily?

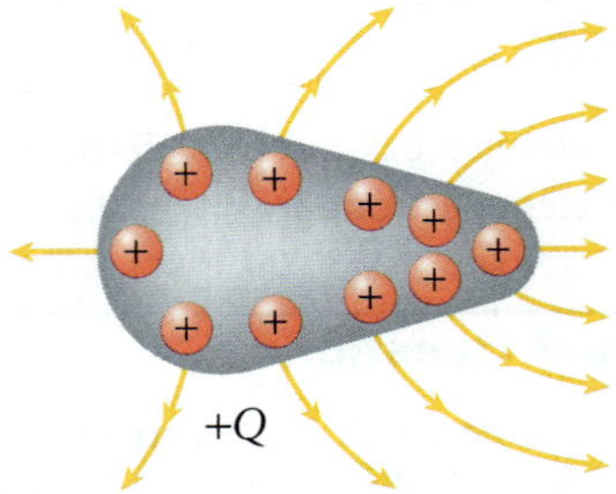

28. Do electric field lines actually exist?

Conceptual Questions

29. When defining the electric field, why must the magnitude of the test charge be very small?

30. Why can't two field lines from the same field cross one another?

31. A "free" electron and "free" proton are placed in an identical electric field. Compare the electric force on each particle. How do their accelerations compare?

Practice Problems

For problems 32–33, see Sample Problem D.

32. Find the electric field at a point midway between two charges of $+30.0 \times 10^{-9}$ C and $+60.0 \times 10^{-9}$ C separated by a distance of 30.0 cm.

33. A +5.7 μC point charge is on the x-axis at $x = -3.0$ m, and a +2.0 μC point charge is on the x-axis at $x = +1.0$ m. Determine the net electric field (magnitude and direction) on the y-axis at $y = +2.0$ m.

MIXED REVIEW

34. Calculate the net charge on a substance consisting of a combination of 7.0×10^{13} protons and 4.0×10^{13} electrons.

35. An electron moving through an electric field experiences an acceleration of 6.3×10^3 m/s^2.

a. Find the electric force acting on the electron.
b. What is the strength of the electric field?

36. One gram of copper has 9.48×10^{21} atoms, and each copper atom has 29 electrons.

a. How many electrons are contained in 1.00 g of copper?
b. What is the total charge of these electrons?

37. Consider three charges arranged as shown below.

a. What is the electric field strength at a point 1.0 cm to the left of the middle charge?
b. What is the magnitude of the force on a −2.0 μC charge placed at this point?

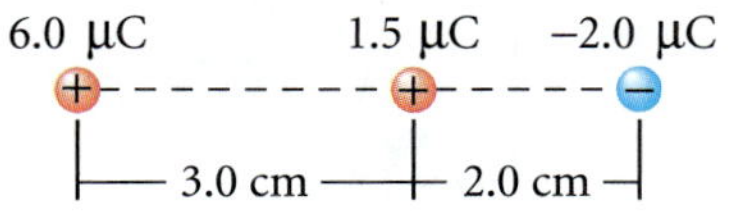

38. Consider three charges arranged in a triangle as shown below.

a. What is the net electric force acting on the charge at the origin?
b. What is the net electric field at the position of the charge at the origin?

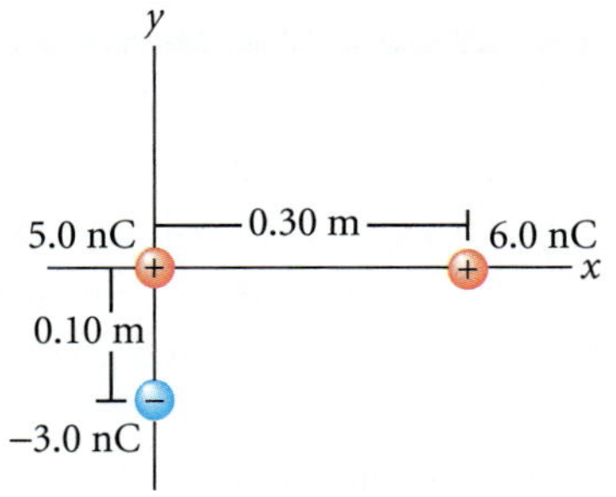

39. Sketch the electric field pattern set up by a positively charged hollow conducting sphere. Include regions both inside and outside the sphere.

40. The moon ($m = 7.36 \times 10^{22}$ kg) is bound to Earth ($m = 5.98 \times 10^{24}$ kg) by gravity. If, instead, the force of attraction were the result of each having a charge of the same magnitude but opposite in sign, find the quantity of charge that would have to be placed on each to produce the required force.

41. Two small metallic spheres, each with a mass of 0.20 g, are suspended as pendulums by light strings from a common point. They are given the same electric charge, and the two come to equilibrium when each string is at an angle of 5.0° with the vertical. If the string is 30.0 cm long, what is the magnitude of the charge on each sphere?

42. What are the magnitude and the direction of the electric field that will balance the weight of an electron? What are the magnitude and direction of the electric field that will balance the weight of a proton?

43. An electron and a proton are each placed at rest in an external uniform electric field of magnitude 520 N/C. Calculate the speed of each particle after 48 ns.

44. A Van de Graaff generator is charged so that the magnitude of the electric field at its surface is 3.0×10^4 N/C.

a. What is the magnitude of the electric force on a proton released at the surface of the generator?
b. Find the proton's acceleration at this instant.

45. Thunderstorms can have an electric field of up to 3.4×10^5 N/C. What is the magnitude of the electric force on an electron in such a field?

46. An object with a net charge of 24 μC is placed in a uniform electric field of 610 N/C, directed vertically. What is the mass of this object if it floats in this electric field?

47. Three identical point charges, with mass $m = 0.10$ kg, hang from three strings, as shown below. If $L = 30.0$ cm and $\theta = 45°$, what is the value of q?

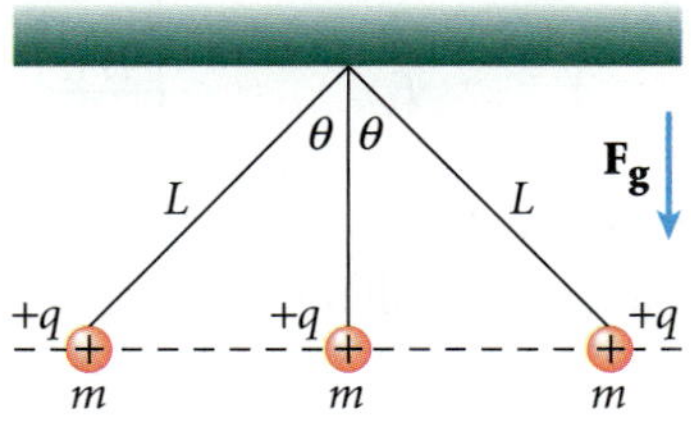

16 REVIEW

34. 4.8×10^{-6} C
35. **a.** 5.7×10^{-27} N, in a direction opposite **E**
b. 3.6×10^{-8} N/C
36. **a.** 2.75×10^{23} electrons
b. 4.40×10^4 C
37. **a.** 2.0×10^7 N/C, along the positive x-axis
b. 4.0×10^1 N
38. **a.** 1.3×10^{-5} N, 77° below the negative x-axis
b. 2.6×10^3 N/C, 77° below the negative x-axis
39. There are no lines inside the sphere. All lines outside the sphere are evenly spaced and are directed away from the sphere radially.
40. 5.72×10^{13} C
41. 7.2×10^{-9} C
42. 5.58×10^{-11} N/C, downward; 1.03×10^{-7} N/C, upward
43. $v_{electron} = 4.4 \times 10^6$ m/s; $v_{proton} = 2.4 \times 10^3$ m/s
44. **a.** 4.8×10^{-15} N
b. 2.9×10^{12} m/s^2
45. 5.4×10^{-14} N
46. 1.5×10^{-3} kg
47. 2.0×10^{-6} C

48. 0 N/C
49. 32.5 m
50. a. 5.27×10^{13} m/s^2
b. 5.27×10^{5} m/s
51. a. 5.3×10^{17} m/s^2
b. 8.5×10^{-4} m
c. 2.9×10^{14} m/s^2
52. 1.62×10^{4} N/C, opposite the proton's velocity
53. a. positive
b. 5.3×10^{-7} C

48. In a laboratory experiment, five equal negative point charges are placed symmetrically around the circumference of a circle of radius *r*. Calculate the electric field at the center of the circle.

49. An electron and a proton both start from rest and from the same point in a uniform electric field of 370.0 N/C. How far apart are they 1.00 μs after they are released? Ignore the attraction between the electron and the proton. (Hint: Imagine the experiment performed with the proton only, and then repeat with the electron only.)

50. An electron is accelerated by a constant electric field of magnitude 300.0 N/C.

a. Find the acceleration of the electron.
b. Find the electron's speed after 1.00×10^{-8} s, assuming it starts from rest.

51. If the electric field strength is increased to about 3.0×10^{6} N/C, air "breaks down" and loses its insulating quality. Under these conditions, sparking results.

a. What acceleration does an electron experience when the electron is placed in such an electric field?
b. If the electron starts from rest when it is placed in an electric field under these conditions, in what distance does it acquire a speed equal to 10.0 percent of the speed of light?
c. What acceleration does a proton experience when the proton is placed in such an electric field?

52. Each of the protons in a particle beam has a kinetic energy of 3.25×10^{-15} J. What are the magnitude and direction of the electric field that will stop these protons in a distance of 1.25 m?

53. A small 2.0 g plastic ball is suspended by a 20.0 cm string in a uniform electric field of 1.0×10^{4} N/C, as shown below.

a. Is the ball's charge positive or negative?
b. If the ball is in equilibrium when the string makes a 15° angle with the vertical as indicated, what is the net charge on the ball?

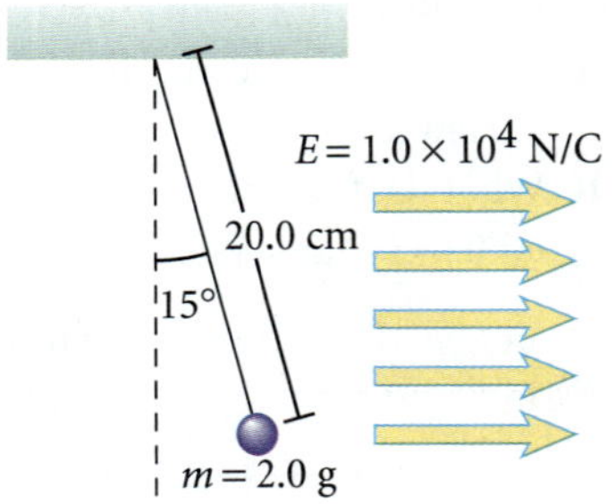

extension

Graphing Calculator Practice

Visit go.hrw.com for answers to this Graphing Calculator activity.

Keyword HF6ELFXT

Graphing Calculator *Practice*

Coulomb's Law

One of the most important and fundamental laws of physics—and of all science—is Coulomb's law. As you learned earlier in this chapter, this law states that the electric force, $F_{electric}$, between two charges, q_1 and q_2, which are separated by a distance, *r*, is given by the following equation.

$$F_{electric} = k_C\left(\frac{q_1 q_2}{r^2}\right)$$

In this graphing calculator activity, you will enter the charges and will observe a graph of electric force versus distance. By analyzing graphs for various sets of charges (positive with positive, negative with negative, and positive with negative), you will better understand Coulomb's law and how charge and distance affect electric force.

Visit go.hrw.com and enter the keyword **HF6ELFX** to find this graphing calculator activity. Refer to **Appendix B** for instructions on downloading the program for this activity.

54. A constant electric field directed along the positive x-axis has a strength of 2.0×10^3 N/C.

   a. Find the electric force exerted on a proton by the field.
   b. Find the acceleration of the proton.
   c. Find the time required for the proton to reach a speed of 1.00×10^6 m/s, assuming it starts from rest.

55. Consider an electron that is released from rest in a uniform electric field.

   a. If the electron is accelerated to 1.0 percent of the speed of light after traveling 2.0 mm, what is the strength of the electric field?
   b. What speed does the electron have after traveling 4.0 mm from rest?

56. A DNA molecule (deoxyribonucleic acid) is 2.17 μm long. The ends of the molecule become singly ionized so that there is -1.60×10^{-19} C on one end and $+1.60 \times 10^{-19}$ C on the other. The helical molecule acts as a spring and compresses 1.00 percent upon becoming charged. Find the effective spring constant of the molecule.

Alternative Assessment

1. A metal can is placed on a wooden table. If a positively charged ball suspended by a thread is brought close to the can, the ball will swing toward the can, make contact, then move away. Explain why this happens and predict whether the ball is likely to make contact a second time. Sketch diagrams showing the charges on the ball and on the can at each phase. How can you test whether your explanation is correct? If your teacher approves of your plan, try testing your explanation.

2. The common copying machine was designed in the 1960s, after the American inventor Chester Carlson developed a practical device for attracting carbon-black to paper using localized electrostatic action. Research how this process works and determine why the last copy made when several hundred copies are made can be noticeably less sharp than the first copy. Create a report, poster, or brochure for office workers containing tips for using copiers.

3. The *triboelectric series* is an ordered list of materials that can be charged by friction. Use the Internet to find a copy of the triboelectric series, and to learn about how it works. Design a series of demonstrations to illustrate charging by friction, and use the triboelectric series to determine the resulting charges for each material. If your teacher approves of your plan, conduct your demonstrations for the class. Explain to the class how the triboelectric series works, and discuss whether it is always completely accurate.

4. Research how an electrostatic precipitator works to remove smoke and dust particles from the polluting emissions of fuel-burning industries. Find out what industries use precipitators. What are their advantages and costs? What alternatives are available? Summarize your findings in a brochure.

5. Imagine you are a member of a research team interested in lightning and you are preparing a grant proposal. Research information about the frequency, location, and effects of thunderstorms. Write a proposal that includes background information, research questions, a description of necessary equipment, and recommended locations for data collection.

6. Electric force is also known as the *Coulomb force*. Research the historical development of the concept of electric force. Describe the work of Coulomb and other scientists such as Priestley, Cavendish, and Faraday.

7. Benjamin Franklin (1706-1790) first suggested the terms *positive* and *negative* for the two different types of electric charge. Franklin was the first person to realize that lightning is a huge electric discharge. He demonstrated this with a dangerous experiment in which he used a kite to gather charges during a thunderstorm. Franklin also invented the first lightning rod. Conduct research to find out more about one of these discoveries, or about another one of Franklin's famous inventions. Create a poster showing how the invention works, or how the discovery was made.

16 REVIEW

54. a. 3.2×10^{-16} N, along the positive x-axis
    b. 1.9×10^{11} m/s^2, along the positive x-axis
    c. 5.3×10^{-6} s

55. a. 1.3×10^4 N/C
    b. 4.2×10^6 m/s

56. 2.25×10^{-9} N/m

Alternative Assessment ANSWERS

1. The ball induces an opposite charge, some of which is transferred upon contact, repelling the ball. Student plans should include methods for measuring charge polarity.
2. Repeated charging of the cylinder makes the pattern spread, causing blurry copies.
3. The triboelectric series ranks materials by their tendency to gain or lose electrons. Student demonstrations should be safe, and should demonstrate several different examples of charging by friction.
4. Advantages include pollution reduction. Disadvantages include costs.
5. Student plans will vary. Be certain plans correspond to the research questions asked.
6. Coulomb's work built on the work of Priestley, Franklin, and others. Faraday developed a theory of field lines to explain electric forces.
7. Franklin's other inventions include the Franklin stove and bifocals. Student posters should illustrate one of Franklin's inventions or discoveries.

CHAPTER 16

Standardized Test Prep

ANSWERS

1. D
2. G
3. C
4. J
5. B
6. H
7. B

Standardized Test Prep

MULTIPLE CHOICE

1. In which way is the electric force similar to the gravitational force?

A. Electric force is proportional to the mass of the object.
B. Electric force is similar in strength to gravitational force.
C. Electric force is both attractive and repulsive.
D. Electric force decreases in strength as the distance between the charges increases.

2. What must the charges be for A and B in the figure below so that they produce the electric field lines shown?

F. A and B must both be positive.
G. A and B must both be negative.
H. A must be negative, and B must be positive.
J. A must be positive, and B must be negative.

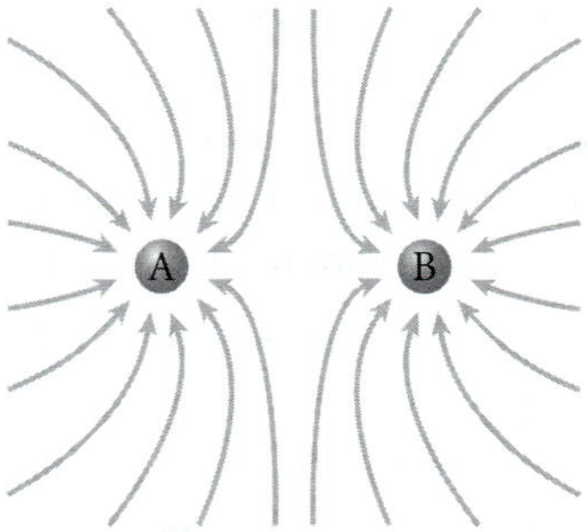

3. Which activity does not produce the same results as the other three?

A. sliding over a plastic-covered automobile seat
B. walking across a woolen carpet
C. scraping food from a metal bowl with a metal spoon
D. brushing dry hair with a plastic comb

4. By how much does the electric force between two charges change when the distance between them is doubled?

F. 4
G. 2
H. $\frac{1}{2}$
J. $\frac{1}{4}$

Use the passage below to answer questions 5–6.

A negatively charged object is brought close to the surface of a conductor, whose opposite side is then grounded.

5. What is this process of charging called?

A. charging by contact
B. charging by induction
C. charging by conduction
D. charging by polarization

6. What kind of charge is left on the conductor's surface?

F. neutral
G. negative
H. positive
J. both positive and negative

Use the graph on the next page to answer questions 7–10. The graph shows the electric field strength at different distances from the center of the charged conducting sphere of a Van de Graaff generator.

7. What is the electric field strength 2.0 m from the center of the conducting sphere?

A. 0 N/C
B. 5.0×10^2 N/C
C. 5.0×10^3 N/C
D. 7.2×10^3 N/C

8. What is the strength of the electric field at the surface of the conducting sphere?
   F. 0 N/C
   G. 1.5×10^2 N/C
   H. 2.0×10^2 N/C
   J. 7.2×10^3 N/C

9. What is the strength of the electric field inside the conducting sphere?
   A. 0 N/C
   B. 1.5×10^2 N/C
   C. 2.0×10^2 N/C
   D. 7.2×10^3 N/C

10. What is the radius of the conducting sphere?
   F. 0.5 m
   G. 1.0 m
   H. 1.5 m
   J. 2.0 m

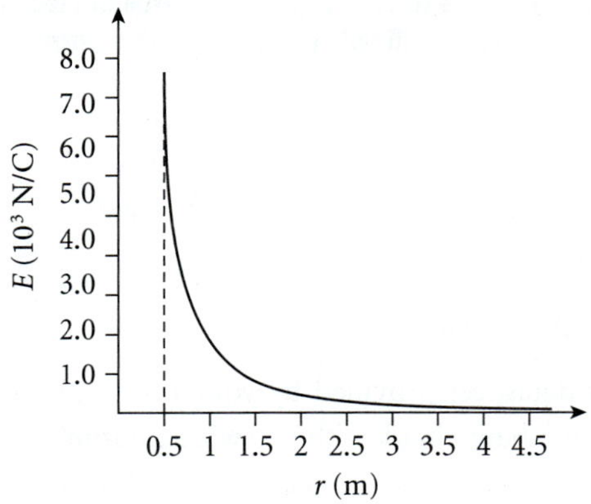

SHORT RESPONSE

11. Three identical charges ($q = +5.0$ mC) are along a circle with a radius of 2.0 m at angles of 30°, 150°, and 270°, as shown in the figure below. What is the resultant electric field at the center?

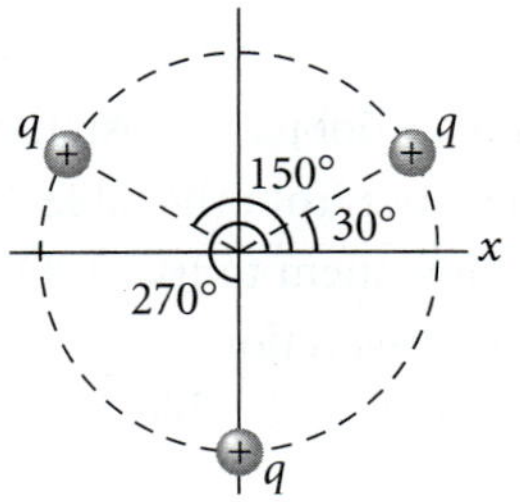

12. If a suspended object is attracted to another object that is charged, can you conclude that the suspended object is charged? Briefly explain your answer.

13. One gram of hydrogen contains 6.02×10^{23} atoms, each with one electron and one proton. Suppose that 1.00 g of hydrogen is separated into protons and electrons, that the protons are placed at Earth's north pole, and that the electrons are placed at Earth's south pole. Assuming the radius of Earth to be 6.38×10^6 m, what is the magnitude of the resulting compressional force on Earth?

14. Air becomes a conductor when the electric field strength exceeds 3.0×10^6 N/C. Determine the maximum amount of charge that can be carried by a metal sphere 2.0 m in radius.

EXTENDED RESPONSE

Use the information below to answer questions 15–18.
A proton, which has a mass of 1.673×10^{-27} kg, accelerates from rest in a uniform electric field of 640 N/C. At some time later, its speed is 1.2×10^6 m/s.

15. What is the magnitude of the acceleration of the proton?

16. How long does it take the proton to reach this speed?

17. How far has it moved in this time interval?

18. What is its kinetic energy at the later time?

19. A student standing on a piece of insulating material places her hand on a Van de Graaff generator. She then turns on the generator. Shortly thereafter, her hairs stand on end. Explain how charge is or is not transferred in this situation, why the student is not shocked, and what causes her hairs to stand up after the generator is started.

Test TIP In problems for which resultant forces are asked, the solution can be made much easier by drawing a sketch of the situation described and seeing if a symmetrical arrangement of components, and thus a canceling of forces, exists.

8. J
9. A
10. F
11. 0.0 N/C
12. not necessarily; The suspended object might have a charge induced on it, but its overall charge could be neutral.
13. 5.12×10^5 N
14. 1.3×10^{-3} C
15. 6.1×10^{10} m/s^2
16. 2.0×10^{-5} s
17. 12 m
18. 1.2×10^{-15} J
19. The charge on the sphere of the Van de Graaff generator is transferred to the student by means of conduction. This charge remains on the student because she is insulated from the ground. As there is no path between the student and the generator and the student and the ground by which charge can escape, the student is not shocked. The accumulation of charges of the same sign on the strands of the student's hair causes the strands to repel each other and so stand on end.

Lab Planning

Beginning on page T34 are preparation notes and teaching tips to assist you in planning.

Blank data tables (as well as some sample data) appear on the **One-Stop Planner.**

No Books in the Lab?

See the ***Datasheets for In-Text Labs*** workbook for a reproducible master copy of this experiment.

Safety Caution

Remind students to make sure rods are secure when not in use. Glass rods may break, and all rods present a safety hazard if they roll onto the floor.

CHAPTER 16

Skills Practice Lab Electrostatics

OBJECTIVES

- **Investigate** the use of an electroscope.
- **Use** an electroscope and other materials to analyze properties of static electricity.
- **Determine** the number of the kinds of electric charge.

MATERIALS LIST

- roll of cellophane tape
- wool pad
- 2 polystyrene rods
- 2 PVC rods
- demonstration capacitor
- electroscope
- flint glass rod
- insulated copper wire
- insulated wire with 2 alligator clips
- meterstick
- nylon cord
- silk cloth
- silk thread
- support stand with clamp
- suspension support for rod

When objects made of two different materials are rubbed together, electron movement causes a surplus or deficiency of electrons on each object, and hence a net negative or positive charge. When an object has an electric charge, it attracts some things and repels others. In this experiment, you will develop charges on different objects and distinguish between the types of charges. You will also use an electroscope to examine the transfer of charges and the conductivity of different materials.

SAFETY

- **Put on goggles.**
- **Never put broken glass or ceramics in a regular waste container. Use a dustpan, brush, and heavy gloves to carefully pick up broken pieces and dispose of them in a container specifically provided for this purpose.**

PROCEDURE

Preparation

1. Read the entire lab, and plan what steps you will take.
2. If you are not using a datasheet provided by your teacher, prepare an observation table in your lab notebook with two wide columns. Label the columns *Experiment* and *Observation.* For each part of the lab, you will write a brief description of what you do in each step under *Experiment.* In the *Observation* column, record your observations of what happens.

Electric Charge

3. Cut four strips of cellophane tape 20 cm long. Fold over a tab at the end of each tape. Tape strips to the lab table, and label the strips *A, B, C,* and *D* with a pencil.
4. Vigorously rub tapes A and B with a wool pad. Grasp the tabbed ends of A and B and carefully remove the tapes from the table. Slowly bring the tapes close together, but do not allow them to touch. Observe how they affect one another. Record your observations in your lab notebook. Carefully place tapes A and B back on the lab table.

5. Carefully remove tape C from the lab table. Tape it firmly down on top of tape D. Carefully remove tapes C and D together from the lab table. Quickly separate them, being careful not to tangle the tapes. Bring the tapes close together—but not touching—and observe how they affect one another. Record your observations in your lab notebook. Place tape D back on the lab table, and place tape C down on top of tape D.

6. Vigorously rub tape A with a wool pad again. Grasp tape A by the tab and carefully remove it from the table. Remove C and D together from the table. Quickly pull C and D apart. Bring C close to tape A, but do not let them touch. Observe how they affect each other. Move C away, and bring D close to tape A. Record your observations in your lab notebook. Throw the four tapes away.

7. Tape a meterstick flat on the surface of the table so that the end of the meterstick extends over the edge of the table. Take another 20 cm long piece of tape, fold a tab on one end, and tape it down on the table. Vigorously rub the tape with a wool pad. Grasp the tab, and carefully remove the tape. Attach the tape to the end of the meterstick so that the tape freely hangs straight down.

8. Rub a polystyrene rod with the wool pad. Bring the rod near the end of the tape that is hanging down, and observe the effect on the tape. Record your observations in your lab notebook. Throw away the tape, and remove the meterstick from the tabletop.

9. Tie the suspension support securely to a string, and suspend the string from the support stand and clamp. Attach a polystyrene rod to the support, and rub the rod with the wool pad. Rub a second polystyrene rod with wool, and bring this rod near one end of the suspended rod. Observe what happens, and record your observations in your lab notebook.

10. Rub the PVC rod with the wool pad, and bring the rod near one end of the suspended polystyrene rod, as you see in **Figure 1.** Observe what happens, and record your observations.

11. Suspend a glass rod on the support, and rub the rod with silk. Rub the PVC rod with the wool pad, and bring the rod near one end of the glass rod. Observe what happens, and record your observations.

12. Suspend a PVC rod on the support, and rub the rod with the wool pad. Rub another PVC rod with wool, and bring the rod near one end of the suspended PVC rod. Observe what happens, and record your observations.

Figure 1

Step 9: Suspend the rod securely so that it hangs freely.

Step 10: Charge the rod by rubbing it vigorously with the wool pad. Bring the charged rod near one end of the suspended rod.

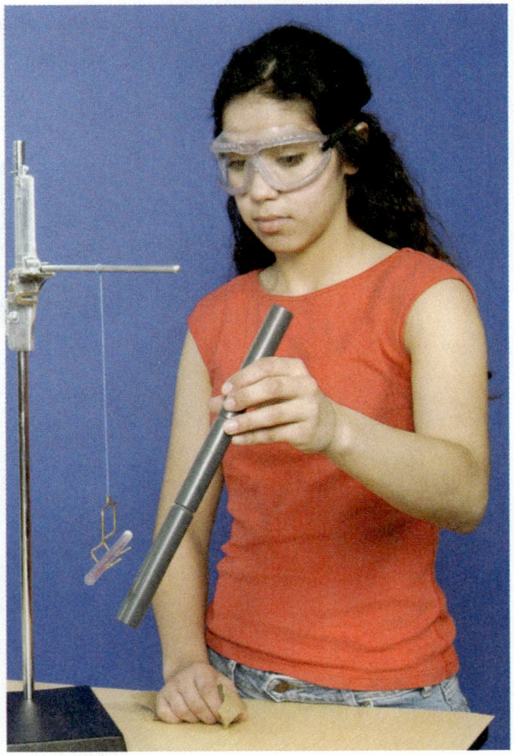

Tips and Tricks

- Before the lab, remind students that they will need to determine how many types of charge there are, based on their observations.
- Before beginning the lab, give a brief demonstration using two strips of adhesive tape. Prepare the tape as in step 3 of the lab. Rub the strips with a wool pad, and pull them up. Hold them near each other, as in step 4 of the lab. Touch one of the strips of tape with your hand to remove the charge. Remind students that touching items in this lab will affect their results.

✓ Checkpoints

Step 7: Make sure metersticks are firmly attached to the tabletop and that they do not extend too far over the edge. Remind students to be careful and to keep their work areas clear of people and other obstacles.

Step 9: Make sure all setups are secure. Students should be starting to develop a hypothesis about the number of types of charge there are. They should be able to explain how their results will help them decide.

Step 14: Encourage students to determine what type of charge is on the electroscope.

Step 17: Students may discharge the electroscope by connecting the wire to any object at ground potential. A good choice is a water faucet or pipe.

Step 23: Based on their observation when the second capacitor plate is removed, students should begin to formulate an idea of how a capacitor works.

Charging an Electroscope by Conduction and Induction

13. Charge a polystyrene rod with a wool cloth, and touch it briefly against the knob of the electroscope. Record your observations.
14. Touch the knob of the electroscope with your hand. Observe what happens.
15. Bring a charged polystyrene rod near, but not touching, the electroscope knob. Observe what happens.
16. Continue holding the polystyrene rod near the electroscope knob. Briefly touch the knob of the electroscope with your finger. Observe what happens. Remove the rod, and observe what happens.
17. Discharge the electroscope by connecting a wire from it to a grounded object. Your teacher will tell you where to connect the wire.
18. Charge a glass rod with a silk cloth, and repeat the procedure in steps 13–17 using the glass rod instead of the polystyrene rod.

Conductors, Insulators, and Capacitors

19. Set up the apparatus as shown in **Figure 2.** (You will not use the second capacitor plate until step 22.) The insulated wire has an alligator clip at each end. One alligator clip connects to the rod beneath the ball on the electroscope. Run the insulated copper wire from the other alligator clip to one of the plates of the demonstration capacitor.
20. Follow the procedure in steps 13–17, but this time bring the rod near the capacitor plate. Observe what happens to the leaves of the electroscope when the rod is brought near the capacitor plate. Record your observations.

Figure 2
Step 20: Use insulated wire to connect the electroscope to one plate of the capacitor. Touch the charged rod to the plate.

21. Replace the copper wire with a piece of thread. Follow the procedure in steps 13–17, but this time bring the rod near the capacitor plate. Observe what happens to the leaves of the electroscope when the rod is brought near the capacitor plate. Record your observations.
22. Connect the rod beneath the knob of the electroscope to one plate of the demonstration capacitor with a short piece of copper wire. Touch the charged polystyrene rod to the plate, and observe what happens to the leaves of the electroscope. Bring the second plate of the capacitor near, but not touching, the first. Observe what happens, and record your observations in your notebook.
23. Remove the second capacitor plate. Observe what happens. Record your observations in your notebook.

24. Bring the second plate near the first again. Using both capacitor plates, try to cause the same result as you obtained using only one plate.

25. Clean up your work area. Put equipment away safely so that it is ready to be used again. Recycle or dispose of used materials as directed by your teacher.

ANALYSIS

1. **Classifying** Use examples from your observations to explain your answers to the following questions. Assume that the polystyrene rod takes a negative charge when it is rubbed with wool. To answer these questions, assume that like charges repel one another and unlike charges attract one another.
   a. What type of charge is on tape A?
   b. What type of charge is on tape B?
   c. Are the charges on C and D the same?
   d. Are the charges on C or D the same as the charges on A or B?
   e. What type of charge is on the charged suspended glass rod? Is the charge on the suspended glass rod the same as or different from the charge on tape A?
   f. What type of charge is on the charged suspended polystyrene rod?
   g. What type of charge is on the second charged polystyrene rod?
   h. What type of charge is on the charged suspended PVC rod?

2. **Describing Events** Use your observations to answer the following:
   a. After you touch the knob of the electroscope with your hand, what type of charge is on the electroscope? Explain how your observations support this conclusion.
   b. When the charged polystyrene rod is used to charge an electroscope by induction, what type of charge is on the electroscope?
   c. What type of charge is on the electroscope when it is charged by induction using the charged glass rod?

3. **Classifying** Is copper a conductor or an insulator? Is silk a conductor or an insulator? Use your observations to support your answers.

CONCLUSIONS

4. **Drawing Conclusions** Based on your observations, how many types of charge are there? Explain how your observations support this conclusion.

5. **Applying Conclusions** Use your results to explain what a capacitor does.

ANSWERS

Analysis

1. a. Tape A is negatively charged.
   b. Tape B is negatively charged.
   c. C and D do not have the same charge.
   d. Either C has the same charge as A and B or D has the same charge as A and B.
   e. The glass rod is positively charged; it does not have the same charge as A.
   f. The polystyrene rod is negatively charged.
   g. The second polystyrene rod is negatively charged.
   h. The PVC rod is negatively charged.

2. a. no charge
   b. positive
   c. negative

3. Copper is a conductor; silk is an insulator.

Conclusions

4. Students should recognize that there are two types of charge.

5. Capacitors store charge.

CHAPTER 17 Electrical Energy and Current

Planning Guide

Compression Guide

To shorten instruction because of time limitations, omit the opener and abbreviate the review.

OBJECTIVES	LABS, DEMONSTRATIONS, AND ACTIVITIES	TECHNOLOGY RESOURCES
PACING • 45 min pp. 592–593 **Chapter Opener**	**ANC** **Discovery Lab** Resistors and Current*◆ BASIC	**CD** **Visual Concepts,** Chapter 17 BASIC
PACING • 45 min pp. 594–601 **Section 1 Electric Potential** • Distinguish between electrical potential energy, electric potential, and potential difference. • Solve problems involving electrical energy and potential difference. • Describe the energy conversions that occur in a battery.	**SE** **Quick Lab** A Voltaic Pile, p. 600 GENERAL **TE** **Demonstration** Potential Energy, p. 594 BASIC	**OSP** **Lesson Plans** **EXT** **Integrating Biology** Electric Eels BASIC **TR** 58A Charge Moved in a Uniform Electric Field
PACING • 45 min pp. 602–607 **Section 2 Capacitance** • Relate capacitance to the storage of electrical potential energy in the form of separated charges. • Calculate the capacitance of various devices. • Calculate the energy stored in a capacitor.	**TE** **Demonstration** Capacitor Discharge, p. 603 ADVANCED **TE** **Demonstration** Functions of a Capacitor, p. 605 GENERAL **ANC** **CBL™ Experiment** Capacitors*◆ ADVANCED	**OSP** **Lesson Plans** **EXT** **Integrating Biology** Electric Shock: Caution! BASIC **TR** 89 Charging a Capacitor **TR** 90 A Capacitor with a Dielectric **TR** 91 Capacitors in Keyboards
PACING • 135 min pp. 608–616 **Section 3 Current and Resistance** • Describe the basic properties of electric current, and solve problems relating current, charge, and time. • Distinguish between the drift speed of a charge carrier and the average speed of the charge carrier between collisions. • Calculate resistance, current, and potential difference by using the definition of resistance. • Distinguish between ohmic and non-ohmic materials.	**SE** **Quick Lab** A Lemon Battery, p. 610 GENERAL **SE** **Skills Practice Lab** Current and Resistance, pp. 634–635◆ GENERAL **ANC** **Datasheet** Current and Resistance* GENERAL **TE** **Demonstration** Drift Speed, p. 611 BASIC **TE** **Demonstration** Non-Ohmic Resistance, p. 613 GENERAL **TE** **Demonstration** Resistance Factors, p. 613 GENERAL **ANC** **CBL™ Experiment** Current and Resistance*◆ GENERAL	**OSP** **Lesson Plans** **EXT** **Integrating Biology** The Brain's Signals BASIC **TR** 92 Conventional Current **TR** 93 Factors that Affect Resistance **TR** 59A Graphs for Ohmic and Non-Ohmic Materials
PACING • 45 min pp. 618–623 **Section 4 Electric Power** • Differentiate between direct current and alternating current. • Relate electric power to the rate at which electrical energy is converted to other forms of energy. Calculate electric power and the cost of running electrical appliances.	**SE** **Quick Lab** Energy Use in Appliances, p. 620 GENERAL **TE** **Demonstration** Potential Difference as a Source of Current, p. 618 BASIC **ANC** **Invention Lab** Battery-Operated Heater*◆ ADVANCED **ANC** **CBL™ Experiment** Electrical Energy*◆ ADVANCED	**OSP** **Lesson Plans** **EXT** **Integrating Chemistry** Rechargeable Ni-Cd Batteries BASIC

PACING • 90 min

CHAPTER REVIEW, ASSESSMENT, AND STANDARDIZED TEST PREPARATION

SE **Chapter Highlights,** p. 625
SE **Chapter Review,** pp. 626–631
SE **Graphing Calculator Practice,** p. 630 GENERAL
SE **Alternative Assessment,** p. 631 ADVANCED
SE **Standardized Test Prep,** pp. 632–633 GENERAL
SE **Appendix D: Equations,** p. 862
SE **Appendix I: Additional Problems,** pp. 892–893
ANC **Study Guide Worksheet** Mixed Review* GENERAL
ANC **Chapter Test A*** GENERAL
ANC **Chapter Test B*** ADVANCED
OSP **Test Generator**

Online and Technology Resources

go.hrw.com Holt Online Learning

Visit go.hrw.com to find a variety of online resources. To access this chapter's extensions, enter the keyword **HF6ELCXT** and click the "go" button. Click **Holt Online Learning** for an online edition of this textbook, and other interactive resources.

HOLT Teacher's One-Stop Planner®

This DVD package includes:
- Holt Calendar Planner
- Customizable Lesson Plans
- Editable Worksheets
- ExamView® Version 6 Assessment Suite
- Interactive Teacher's Edition
- Holt PuzzlePro®
- Holt PowerNotes® Presentations
- MindPoint® Quiz Show

KEY

SE Student Edition	**OSP** One-Stop Planner	**EXT** Online Extension
TE Teacher Edition	**CD** CD or CD-ROM	* Also on One-Stop Planner
ANC Ancillary Worksheet	**TR** Teaching Transparencies	◆ Requires advance prep

SKILLS DEVELOPMENT RESOURCES	REVIEW AND ASSESSMENT	CORRELATIONS
		National Science Education Standards
SE Sample Set A Potential Energy and Potential Difference, p. 599 BASIC **TE Classroom Practice,** p. 598 BASIC **ANC Problem Workbook** Sample Set A* BASIC **OSP Problem Bank** Sample Set A BASIC **EXT Practice Problems** Potential Difference GENERAL	**SE Section Review,** p. 601 GENERAL **ANC Study Guide Worksheet** Section 1* GENERAL **ANC Quiz** Section 1* BASIC	UCP 1, 2, 3, 5 ST 1, 2 SPSP 4, 5
SE Sample Set B Capacitance, pp. 606–607 BASIC **ANC Problem Workbook** Sample Set B* BASIC **OSP Problem Bank** Sample Set B BASIC **SE Conceptual Challenge,** p. 604 GENERAL	**SE Section Review,** p. 607 GENERAL **ANC Study Guide Worksheet** Section 2* GENERAL **ANC Quiz** Section 2* BASIC	UCP 1, 2, 3, 4, 5 SAI 1, 2 ST 1, 2 SPSP 2, 4, 5
SE Sample Set C Current, pp. 609 BASIC **TE Classroom Practice,** p. 609 BASIC **ANC Problem Workbook** and **OSP Problem Bank** Sample Set C BASIC **SE Sample Set D** Resistance, pp. 614–615 BASIC **TE Classroom Practice,** p. 514 BASIC **ANC Problem Workbook** and **OSP Problem Bank** Sample Set D BASIC **SE Conceptual Challenge,** p. 611 GENERAL **SE Appendix J: Advanced Topics** Superconductors pp. 928–929 ADVANCED	**SE Section Review,** p. 616 GENERAL **ANC Study Guide Worksheet** Section 3* GENERAL **ANC Quiz** Section 3* BASIC	UCP 1, 2, 3, 4, 5 SAI 1, 2 ST 1, 2 HNS 1, 3 SPSP 1, 2, 4, 5
SE Sample Set E Electric Power, pp. 621 BASIC **ANC Problem Workbook** Sample Set E* BASIC **OSP Problem Bank** Sample Set E BASIC **SE Appendix J: Advanced Topics** Electron Tunneling, pp. 924–925 ADVANCED **EXT Practice Problems** Electrical Energy BASIC	**SE Section Review,** p. 623 GENERAL **ANC Study Guide Worksheet** Section 4* GENERAL **ANC Quiz** Section 4* BASIC	UCP 1, 2, 3, 4 SAI 1, 2 ST 2 HNS 1, 2, 3 SPSP 2, 5

SCILINKS. NSTA

www.scilinks.org

Maintained by the **National Science Teachers Association.**

Topic: Electrical Energy
SciLinks Code: HF60475

Topic: Batteries
SciLinks Code: HF60139

Topic: Michael Faraday
SciLinks Code: HF60955

Topic: Capacitors
SciLinks Code: HF60211

Topic: Electric Current
SciLinks Code: HF60472

Topic: Ohm's Law
SciLinks Code: HF61071

Topic: Superconductors
SciLinks Code: HF61478

Classroom CD-ROMs

- Guided Reading Audio Program
- Student One Stop
- Virtual Investigations
- Visual Concepts

Holt Lab Generator CD-ROM

Search for any lab by topic, standard, difficulty level, or time. Edit any lab to fit your needs, or create your own labs. Use the Lab Materials QuickList software to customize your lab materials list.

CHAPTER 17
Overview

Section 1 introduces electrical potential energy and defines *potential difference.* The potential energy associated with a charge in a uniform electric field is defined. This section also applies the concepts of electric potential and potential difference to a battery in a circuit.

Section 2 relates capacitance to the storage of electrical potential energy in the form of separated charges, discusses the dependence of capacitance on the shape of the capacitor and the material between the plates of the capacitor, and calculates the energy stored by a capacitor.

Section 3 establishes the concept of current as the rate of charge movement; discusses conventional current, drift velocity, and sources of current; and distinguishes between direct and alternating currents. This section also explores resistance in terms of both Ohm's law and the factors that affect resistance and distinguishes between ohmic and nonohmic materials.

Section 4 introduces electric power as the rate at which electrical energy is transferred, solves problems involving electric power and the cost of electrical energy, and discusses why electrical energy is transported at high potential differences.

About the Illustration

This photograph of a lightning storm was taken in southern Arizona. Lightning, a familiar sight to all students, can be used to clarify the concepts of potential difference, capacitance, and electrical breakdown.

CHAPTER 17

Electrical Energy and Current

During a thunderstorm, different charges accumulate in different parts of a cloud to create an electric field between the cloud and the ground. Eventually, a critical *breakdown voltage* is reached. At this point, the molecules in the air are broken down into charged particles, forming a state of matter called a *plasma*. Because a plasma conducts electricity, an electric charge flows between the cloud and the ground, an event known as lightning.

WHAT TO EXPECT

In this chapter, you will learn about electric potential and electrical energy and will learn about how capacitors can be used to store electrical energy. You will be introduced to electric current and resistance.

Why it Matters

The use of electrical energy is universal in our modern society. An understanding of electrical energy and the factors that affect its rate of use can help us use electric power more wisely.

CHAPTER PREVIEW

Tapping Prior Knowledge

Knowledge to Review

- ✔ Vector addition gives a resultant vector that is equivalent to the added effects of each of the individual vectors.
- ✔ Field forces are forces that act on objects without coming into physical contact with other objects.
- ✔ Potential energy is the energy associated with an object due to its position relative to another.
- ✔ Power is the rate at which energy is transferred or the rate at which work is done.

Items to Probe

- ✔ Vector addition and the superposition principle: Ask students to calculate the resultant force on an object experiencing two or three individual forces.

SECTION 1
General Level

Demonstration

Potential Energy — BASIC

Purpose Show the relationship between electric field and electrical potential energy by comparing gravitational field and gravitational potential energy.

Materials lump of clay

Procedure Place the clay on the floor, and ask students to estimate the gravitational potential energy of the clay relative to the floor. *(The clay-Earth system's gravitational potential energy is zero because $h = 0$.)*

Next hold the clay about 1 m above the floor, and have students estimate the new gravitational potential energy (mgh, $h = 1$ m). Ask students how the clay-Earth system acquired gravitational potential energy (*work was done on the system as the clay was lifted in a direction opposite the direction of the field*). Explain that this is analogous to moving a negative charge in an electric field in the direction of **E** or moving a positive charge in an electric field in a direction opposite to **E.** Work is required to move the charge, just as work was required to lift the clay.

Next ask students to describe what happens if you let go of the clay. *(The gravitational field of the Earth exerts a force on the clay, which accelerates toward the Earth.)* Point out that the results would be the same in an electric field if the Earth and the clay were particles having opposite charge.

SECTION 1

Electric Potential

SECTION OBJECTIVES

- **Distinguish between electrical potential energy, electric potential, and potential difference.**
- **Solve problems involving electrical energy and potential difference.**
- **Describe the energy conversions that occur in a battery.**

ELECTRICAL POTENTIAL ENERGY

You have learned that when two charges interact, there is an electric force between them. As with the gravitational force associated with an object's position relative to Earth, there is a potential energy associated with this force. This kind of potential energy is called **electrical potential energy.** Unlike gravitational potential energy, electrical potential energy results from the interaction of two objects' charges, not their masses.

electrical potential energy

potential energy associated with a charge due to its position in an electric field

Electrical potential energy is a component of mechanical energy

Mechanical energy is conserved as long as friction and radiation are not present. As with gravitational and elastic potential energy, electrical potential energy can be included in the expression for mechanical energy. If a gravitational force, an elastic force, and an electric force are all acting on an object, the mechanical energy can be written as follows:

$$ME = KE + PE_{grav} + PE_{elastic} + PE_{electric}$$

To account for the forces (except friction) that may also be present in a problem, the appropriate potential-energy terms associated with each force are added to the expression for mechanical energy.

Figure 1
As the charges in these sparks move, the electrical potential energy decreases, just as gravitational potential energy decreases as an object falls.

Recall from your study of work and energy that any time a force is used to move an object, work is done on that object. This statement is also true for charges moved by an electric force. Whenever a charge moves—because of the electric field produced by another charge or group of charges—work is done on that charge.

For example, negative electric charges build up on the plate in the center of the device, called a *Tesla coil,* shown in **Figure 1.** The electrical potential energy associated with each charge decreases as the charge moves from the central plate to the walls (and through the walls to the ground).

Electrical potential energy can be associated with a charge in a uniform field

Consider a positive charge in a uniform electric field. (A uniform field is a field that has the same value and direction at all points.) Assume the charge is displaced at a constant velocity *in the same direction as the electric field,* as shown in **Figure 2.**

There is a change in the electrical potential energy associated with the charge's new position in the electric field. The change in the electrical potential energy depends on the charge, q, as well as the strength of the electric field, E, and the displacement, d. It can be written as follows:

$$\Delta PE_{electric} = -qEd$$

The negative sign indicates that the electrical potential energy will increase if the charge is negative and decrease if the charge is positive.

As with other forms of potential energy, it is the *difference* in electrical potential energy that is physically important. If the displacement in the expression above is chosen so that it is the distance in the direction of the field from the reference point, or zero level, then the initial electrical potential energy is zero and the expression can be rewritten as shown below. As with other forms of energy, the SI unit for electrical potential energy is the joule (J).

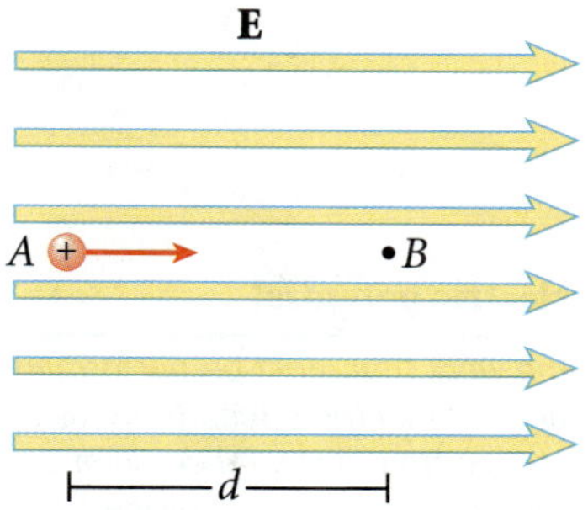

Figure 2
A positive charge moves from point *A* to point *B* in a uniform electric field, and the potential energy changes as a result.

ELECTRICAL POTENTIAL ENERGY IN A UNIFORM ELECTRIC FIELD

$$PE_{electric} = -qEd$$

electrical potential energy =
−(charge × electric field strength × displacement from the reference point in the direction of the field)

This equation is valid only for a uniform electric field, such as that between two oppositely charged parallel plates. In contrast, the electric field lines for a point charge are farther apart as the distance from the charge increases. Thus, the electric field of a point charge is an example of a nonuniform electric field.

Electrical potential energy is similar to gravitational potential energy

When electrical potential energy is calculated, d is the magnitude of the displacement's component *in the direction of the electric field.* The electric field does work on a positive charge by moving the charge in the direction of $\mathbf{E}$ (just as Earth's gravitational field does work on a mass by moving the mass toward Earth). After such a movement, the system's final potential energy is less than its initial potential energy. A negative charge behaves in the opposite manner, because a negative charge undergoes a force in the opposite direction. Moving a charge in a direction that is perpendicular to $\mathbf{E}$ is analogous to moving an object horizontally in a gravitational field: no work is done, and the potential energy of the system remains constant.

SCILINKS.
www.scilinks.org
Topic: Electrical Energy
Code: HF60475

Teaching Tip — BASIC

Remind students that the vector for the electric field points in the direction in which a *positive* test charge experiences an electric force. The following table can be used as a memory aid:

	+ charge	− charge
Along **E**	loses $PE_{electric}$	gains $PE_{electric}$
Opposite **E**	gains $PE_{electric}$	loses $PE_{electric}$

Key Models and Analogies — GENERAL

Drawing a parallel between gravitational and electric potential difference can be a useful tool. First, explain to students that gravitational potential is the gravitational potential energy per unit mass, while electric potential is the electrical potential energy per unit charge. Then ask students to imagine carrying a book up to the second floor of a building. The change in the gravitational potential energy of the book is *mgh*. Thus, the change in gravitational potential (gravitational potential difference) is $\frac{mgh}{m}$, or *gh*.

Point out that this gravitational potential is independent of the mass of the book; it depends only on the field and on the change in height. Likewise, in a uniform electric field, electric potential position depends only on the field and on the change in position, as seen by the equation for potential difference in a uniform electric field ($\Delta V = -Ed$).

The Language of Physics

Make sure students distinguish between the symbols for electric potential (V) and potential difference (ΔV) with the abbreviation for the unit of potential difference, the volt (V). There is a relationship between the two. Because it is measured in volts, the symbol V is used for electric potential. The symbol ΔV is used for a *change* in electric potential, or potential difference, which is sometimes referred to as *voltage.*

POTENTIAL DIFFERENCE

The concept of electrical potential energy is useful in solving problems, particularly those involving charged particles. But at any point in an electric field, as the magnitude of the charge increases, the magnitude of the associated electrical potential energy increases. It is more convenient to express the potential in a manner independent of the charge at that point, a concept called **electric potential.**

electric potential

the work that must be performed against electric forces to move a charge from a reference point to the point in question, divided by the charge

The electric potential at some point is defined as the electrical potential energy associated with a charged particle in an electric field divided by the charge of the particle.

$$V = \frac{PE_{electric}}{q}$$

The potential at a point is the result of the fields due to all *other* charges near enough and large enough to contribute force on a charge at that point. In other words, the electric potential at a point *is independent of the charge at that point.* The force that a test charge at the point in question experiences is proportional to the magnitude of the charge.

Potential difference is a change in electric potential

The **potential difference** between two points can be expressed as follows:

potential difference

the work that must be performed against electric forces to move a charge between the two points in question, divided by the charge

POTENTIAL DIFFERENCE

$$\Delta V = \frac{\Delta PE_{electric}}{q}$$

$$\textbf{potential difference} = \frac{\textbf{change in electrical potential energy}}{\textbf{electric charge}}$$

Figure 3
For a typical car battery, there is a potential difference of 13.2 V between the negative (black) and the positive (red) terminals.

Potential difference is a measure of the difference in the electrical potential energy between two positions in space divided by the charge. The SI unit for potential difference (and for electric potential) is the *volt,* V, and is equivalent to one joule per coulomb. As a 1 C charge moves through a potential difference of 1 V, the charge gains 1 J of energy. The potential difference between the two terminals of a battery can range from about 1.5 V for a small battery to about 13.2 V for a car battery like the one the student is looking at in **Figure 3.**

Because the reference point for measuring electrical potential energy is arbitrary, the reference point for measuring electric potential is also arbitrary. Thus, only changes in electric potential are significant.

Remember that electrical potential energy is a quantity of energy, with units in joules. However, electric potential and potential difference are both measures of energy per unit charge (measured in units of volts), and potential difference describes a change in energy per unit charge.

The potential difference in a uniform field varies with the displacement from a reference point

The expression for potential difference can be combined with the expressions for electrical potential energy. The resulting equations are often simpler to apply in certain situations. For example, consider the electrical potential energy of a charge in a uniform electric field.

$$PE_{electric} = -qEd$$

This expression can be substituted into the equation for potential difference.

$$\Delta V = \frac{\Delta(-qEd)}{q}$$

As the charge moves in a uniform electric field, the quantity in the parentheses does not change from the reference point. Thus, the potential difference in this case can be rewritten as follows:

POTENTIAL DIFFERENCE IN A UNIFORM ELECTRIC FIELD

$$\Delta V = -Ed$$

potential difference =
−(magnitude of the electric field × displacement)

Keep in mind that d is the displacement *parallel* to the field and that motion perpendicular to the field does not change the electrical potential energy.

The reference point for potential difference near a point charge is often at infinity

To determine the potential difference between two points in the field of a point charge, first calculate the electric potential associated with each point. Imagine a point charge q_2 at point A in the electric field of a point charge q_1 at point B some distance, r, away as shown in **Figure 4.** The electric potential at point A due to q_1 can be expressed as follows:

$$V_A = \frac{PE_{electric}}{q_2} = k_C\frac{q_1q_2}{rq_2} = k_C\frac{q_1}{r}$$

Do not confuse the two charges in this example. The charge q_1 is responsible for the electric potential at point A. Therefore, *an electric potential exists at some point in an electric field regardless of whether there is a charge at that point.* In this case, the electric potential at a point depends on only two quantities: the charge responsible for the electric potential (in this case q_1) and the distance r from this charge to the point in question.

Did you know?

A unit of energy commonly used in atomic and nuclear physics that is convenient because of its small size is the *electron volt,* eV. It is defined as the energy that an electron (or proton) gains when accelerated through a potential difference of 1 V. One electron volt is equal to 1.60×10^{-19} J.

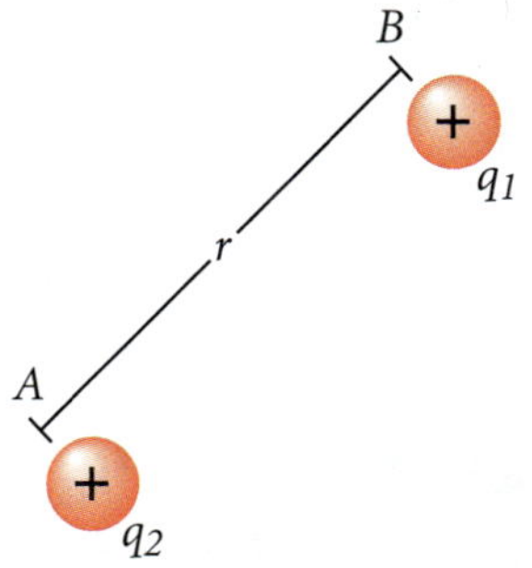

Figure 4
The electric potential at point A depends on the charge at point B and the distance r.

SECTION 1

Teaching Tip ADVANCED

Remind students that $F_{electric} = qE$. Thus, the first equation on this page can be derived as follows:
$PE_{electric} = -W$
$PE_{electric} = -Fd$
$PE_{electric} = -(qE)d$

Visual Strategy GENERAL

Figure 4
Point out to students that the potential energy of the system increases as the distance r decreases.

Q How would you modify the figure to make this case similar to gravitational potential?

A ***Change q_2 to a negative charge. When q_2 is negative and q_1 is positive, the potential energy of the system increases as the distance r increases.***

Misconception Alert

Some students may confuse electric potential ($V = \frac{k_C q_1}{r}$) and potential difference ($\Delta V = \frac{k_C q_1}{r}$) because of the similarity between their equations. Remind them that this similarity is the result of choosing zero potential at infinite distance as a reference point. To reinforce the individual nature of the two concepts, work out an example on the chalkboard or overhead projector using a different reference point.

Classroom Practice

Potential Difference

Between a point some distance r_i from a point charge q_i and an infinite distance, there exists a potential difference of 1.0 V. Determine the resulting potential difference for the following cases:

a. $r = r_i, q = 2q_i$

b. $r = \frac{1}{2}r_i, q = q_i$

c. $r = 2r_i, q = q_i$

d. $r = 2r_i, q = 2q_i$

Answer

a. $\Delta V = 2.0$ V

b. $\Delta V = 2.0$ V

c. $\Delta V = 0.50$ V

d. $\Delta V = 1.0$ V

Did you know?

The volt is named after the Italian physicist Alessandro Volta (1745–1827), who developed the first practical electric battery, known as a voltaic pile. Because potential difference is measured in units of volts, it is sometimes referred to as *voltage*.

extension

Practice Problems

Visit go.hrw.com to find a sample and practice problems covering potential difference.

Keyword HF6ELCX

extension

Integrating Biology

Visit go.hrw.com for the activity "Electric Eels."

Keyword HF6ELCX

To determine the potential difference between any two points near the point charge q_1, first note that the electric potential at each point depends only on the distance from each point to the charge q_1. If the two distances are r_1 and r_2, then the potential difference between these two points can be written as follows:

$$\Delta V = k_C \frac{q_1}{r_2} - k_C \frac{q_1}{r_1} = k_C q_1 \left(\frac{1}{r_2} - \frac{1}{r_1} \right)$$

If the distance r_1 between the point and q_1 is large enough, it is assumed to be infinitely far from the charge q_1. In that case, the quantity $1/r_1$ is zero. The expression then simplifies to the following (dropping the subscripts):

POTENTIAL DIFFERENCE BETWEEN A POINT AT INFINITY AND A POINT NEAR A POINT CHARGE

$$\Delta V = k_C \frac{q}{r}$$

$$\textbf{potential difference} = \textbf{Coulomb constant} \times \frac{\textbf{value of the point charge}}{\textbf{distance to the point charge}}$$

This result for the potential difference associated with a point charge appears identical to the electric potential associated with a point charge. The two expressions look the same only because we have chosen a special reference point from which to measure the potential difference.

One common application of the concept of potential difference is in the operation of electric circuits. Recall that the reference point for determining the electric potential at some point is arbitrary and must be defined. Earth is frequently designated to have an electric potential of zero and makes a convenient reference point. Thus, *grounding* an electrical device (connecting it to Earth) creates a possible reference point, which is commonly used to measure the electric potential in an electric circuit.

The superposition principle can be used to calculate the electric potential for a group of charges

The electric potential at a point near two or more charges is obtained by applying a rule called the *superposition principle.* This rule states that the total electric potential at some point near several point charges is the algebraic sum of the electric potentials resulting from each of the individual charges. While this is similar to the method used previously to find the resultant electric field at a point in space, here the summation is much easier to evaluate because the electric potentials are scalar quantities, not vector quantities. There are no vector components to consider.

To evaluate the electric potential at a point near a group of point charges, you simply take the algebraic sum of the potentials resulting from all charges. Remember, you must keep track of signs. The electric potential at some point near a positive charge is positive, and the potential near a negative charge is negative.

SAMPLE PROBLEM A

Potential Energy and Potential Difference

PROBLEM

A charge moves a distance of 2.0 cm in the direction of a uniform electric field whose magnitude is 215 N/C. As the charge moves, its electrical potential energy decreases by 6.9×10^{-19} J. Find the charge on the moving particle. What is the potential difference between the two locations?

SOLUTION

Given: $\Delta PE_{electric} = -6.9 \times 10^{-19}$ J $\quad d = 0.020$ m

$E = 215$ N/C

Unknown: $q = ?$ $\quad \Delta V = ?$

Use the equation for the change in electrical potential energy.

$$\Delta PE_{electric} = -qEd$$

Rearrange to solve for q, and insert values.

$$q = -\frac{\Delta PE_{electric}}{Ed} = -\frac{(-6.9 \times 10^{-19}\ \text{J})}{(215\ \text{N/C})(0.020\ \text{m})}$$

$$\boxed{q = 1.6 \times 10^{-19}\ \text{C}}$$

The potential difference is the magnitude of E times the displacement.

$$\Delta V = -Ed = -(215\ \text{N/C})(0.020\ \text{m})$$

$$\boxed{\Delta V = -4.3\ \text{V}}$$

TIP *Remember that a newton·meter is equal to a joule and that a joule per coulomb is a volt. Thus, potential difference is expressed in volts.*

PRACTICE A

Potential Energy and Potential Difference

1. As a particle moves 10.0 m along an electric field of strength 75 N/C, its electrical potential energy decreases by 4.8×10^{-16} J. What is the particle's charge?
2. What is the potential difference between the initial and final locations of the particle in Problem 1?
3. An electron moves 4.5 m in the direction of an electric field of strength 325 N/C. Determine the change in electrical potential energy.

PROBLEM GUIDE A

Use this guide to assign problems.
SE = Student Edition Textbook
PW = Problem Workbook
PB = Problem Bank on the One-Stop Planner (OSP)

Solving for:

ΔV	**SE** Sample, 2–3; Ch. Rvw. 8, 60 **PW** 4–5 **PB** 5–7
q	**SE** Sample 1, Ch. Rvw. 66 **PW** 3–5 **PB** Sample, 1–2
V	**SE** Ch. Rvw. 9, 64 **PW** 5–8 **PB** 8–10

***Challenging Problem**
Consult the printed Solutions Manual or the OSP for detailed solutions.

ANSWERS

Practice A

1. 6.4×10^{-19} C
2. −750 V
3. 2.3×10^{-16} J

Misconception Alert — GENERAL

A battery does not supply the charges that move all the way around a circuit. Rather, the battery *does work* on charges to increase their electrical potential energy. Batteries provide a constant potential difference, which supplies charges with kinetic energy.

Quick Lab

TEACHER'S NOTES

The first batteries were made by Volta in 1800 and consisted of alternating layers of zinc, blotting paper soaked in brine, and silver. In this lab, students simulate this type of battery by using coins and brine-soaked paper towel. The pile can be stacked as high as you like, with each layer increasing the voltage by a fixed amount.

SciLINKS
www.scilinks.org
Topic: Batteries
Code: HF60139

A battery does work to move charges

A good illustration of the concepts of electric potential and potential difference is the way in which a battery powers an electrical apparatus, such as a flashlight, a motor, or a clock. A battery is an energy-storage device that provides a constant potential difference between two locations, called *terminals*, inside the battery.

Recall that the reference point for determining the electric potential at a location is arbitrary. For example, consider a typical 1.5 V alkaline battery. This type of battery maintains a potential difference across its terminals such that the positive terminal has an electric potential that is 1.5 V higher than the electric potential of the negative terminal. If we designate that the negative terminal of the battery is at zero potential, the positive terminal would have a potential of 1.5 V. We could just as correctly choose the potential of the negative terminal to be −0.75 V and the positive terminal to be +0.75 V.

Inside a battery, a chemical reaction produces electrons (negative charges) that collect on the negative terminal of the battery. Negative charges move inside the battery from the positive terminal to the negative terminal, through a potential difference of $\Delta V = -1.5$ V. The chemical reaction inside the battery does work on—that is, provides energy to—the charges when moving them from the positive terminal to the negative terminal. This transit increases the magnitude of the electrical potential energy associated with the charges. The result of this motion is that every coulomb of charge that leaves the positive terminal of the battery is associated with a total of 1.5 J of electrical potential energy.

Now, consider the movement of electrons in an electrical device that is connected to a battery. As 1 C of charge moves through the device toward the positive terminal of the battery, the charge gives up its 1.5 J of electrical energy to the device. When the charge reaches the positive terminal, the charge's electrical potential energy is again zero. Electrons must travel to the positive terminal for the chemical reaction in a battery to occur. For this reason, a battery can be unused for a period of time and still have power available.

Quick Lab

A Voltaic Pile

MATERIALS LIST

- salt
- water
- paper towel
- pennies
- nickels
- voltmeter (1 V range)

Dissolve as much salt as possible in the water. Soak the paper towel in the salt water and then tear it into small circles that are slightly bigger than a nickel. Make a stack alternating one penny, a piece of paper towel and then one nickel. Repeat this stack by placing the second penny on top of the first nickel. Measure the voltage between the first penny and the last nickel by placing the leads of the voltmeter at each end of the stack. Be sure to have your voltmeter on the lowest dc voltage setting. Try stacking additional layers of penny–paper towel–nickel, and measure the voltage again. What happens if you replace the nickels or pennies with dimes or quarters?

SECTION REVIEW

1. What is the difference between $\Delta PE_{electric}$ and $PE_{electric}$?
2. In a uniform electric field, what factors does the electrical potential energy depend on?
3. Describe the conditions that are necessary for mechanical energy to be a conserved quantity.
4. Is there a single correct reference point from which all electrical potential energy measurements must be taken?
5. A uniform electric field with a magnitude of 250 N/C is directed in the positive x direction. A 12 μC charge moves from the origin to the point (20.0 cm, 50.0 cm). What is the change in the electrical potential energy of the system as a result of the change in position of this charge?
6. What is the change in the electrical potential energy in a lightning bolt if 35 C of charge travel to the ground from a cloud 2.0 km above the ground in the direction of the field? Assume the electric field is uniform and has a magnitude of 1.0×10^6 N/C.
7. The gap between electrodes in a spark plug is 0.060 cm. Producing an electric spark in a gasoline-air mixture requires an electric field of 3.0×10^6 V/m. What minimum potential difference must be supplied by the ignition circuit to start a car?
8. A proton is released from rest in a uniform electric field with a magnitude of 8.0×10^4 V/m. The proton is displaced 0.50 m as a result.
   a. Find the potential difference between the proton's initial and final positions.
   b. Find the change in electrical potential energy of the proton as a result of this displacement.
9. In a thunderstorm, the air must be ionized by a high voltage before a conducting path for a lightning bolt can be created. An electric field of about 1.0×10^6 V/m is required to ionize dry air. What would the breakdown voltage in air be if a thundercloud were 1.60 km above ground? Assume that the electric field between the cloud and the ground is uniform.
10. Explain how electric potential and potential difference are related. What units are used for each one?
11. **Critical Thinking** Given the electrical potential energy, how do you calculate electric potential?
12. **Critical Thinking** Why is electric potential a more useful quantity for most calculations than electrical potential energy is?

SECTION 1

SECTION REVIEW ANSWERS

1. $\Delta PE_{electric} = -qE\Delta d = PE_f - PE_i$ is the change in electrical potential energy between two points. If the initial position is considered to be the zero level, $PE_i = 0$, so $PE_{electric} = PE_f = -qEd$.
2. charge, electric field strength, and displacement in the direction of the field
3. Mechanical energy is conserved in the absence of friction and radiation.
4. No, any reference point can be used, but the initial position is typically used as the zero level to simplify calculations.
5. -6.0×10^{-4} J
6. -7.0×10^{10} J
7. 1.8×10^3 V
8. **a.** -4.0×10^4 V
   b. -6.4×10^{-15} J
9. 1.6×10^9 V
10. Potential difference is the change in electric potential over a distance. Both have the same SI unit, called the volt (V).
11. $V = \frac{PE_{electric}}{q}$
12. Electrical potential energy at a point depends on the charge located at that point, while the electric potential at any point is independent of the charge at that point.

SECTION 2
General Level

Key Models and Analogies — GENERAL

Explain to students that a capacitor stores electrical potential energy much in the way a stretched or a compressed spring stores elastic potential energy. Work must be done to charge a capacitor, just as work must be done to stretch or compress a spring. In both cases, the potential energy acquired by doing work is stored and used at a later time.

Visual Strategy — BASIC

Figure 5

Point out to the students that **Figure 5** represents the net charge accumulation on a charged capacitor.

Q Is the net charge of a charged capacitor greater than, less than, or equal to the net charge of the same capacitor when it is uncharged?

A ***The net charges of the two capacitors are equal.***

The Language of Physics

In everyday language, *capacity* means "the amount something can hold." *Capacitance* in physics does *not* express how much charge a capacitor can hold. It expresses how much charge the capacitor plates have relative to the potential difference between the plates. The maximum charge that the capacitor can have depends on the breakdown potential.

SECTION 2

Capacitance

SECTION OBJECTIVES

- **Relate capacitance to the storage of electrical potential energy in the form of separated charges.**
- **Calculate the capacitance of various devices.**
- **Calculate the energy stored in a capacitor.**

CAPACITORS AND CHARGE STORAGE

A *capacitor* is a device that is used to store electrical potential energy. It has many uses, including tuning the frequency of radio receivers, eliminating sparking in automobile ignition systems, and storing energy in electronic flash units.

An *energized* (or charged) capacitor is useful because energy can be reclaimed from the capacitor when needed for a specific application. A typical design for a capacitor consists of two parallel metal plates separated by a small distance. This type of capacitor is called a *parallel-plate capacitor.* When we speak of *the charge on a capacitor,* we mean the magnitude of the charge on either plate.

The capacitor is energized by connecting the plates to the two terminals of a battery or other sources of potential difference, as **Figure 5** shows. When this connection is made, charges are removed from one of the plates, leaving the plate with a net charge. An equal and opposite amount of charge accumulates on the other plate. Charge transfer between the plates stops when the potential difference between the plates is equal to the potential difference between the terminals of the battery. This charging process is shown in **Figure 5(b).**

Capacitance is the ratio of charge to potential difference

capacitance

the ability of a conductor to store energy in the form of electrically separated charges

The ability of a conductor to store energy in the form of electrically separated charges is measured by the **capacitance** of the conductor. Capacitance is defined as the ratio of the net charge on each plate to the potential difference created by the separated charges.

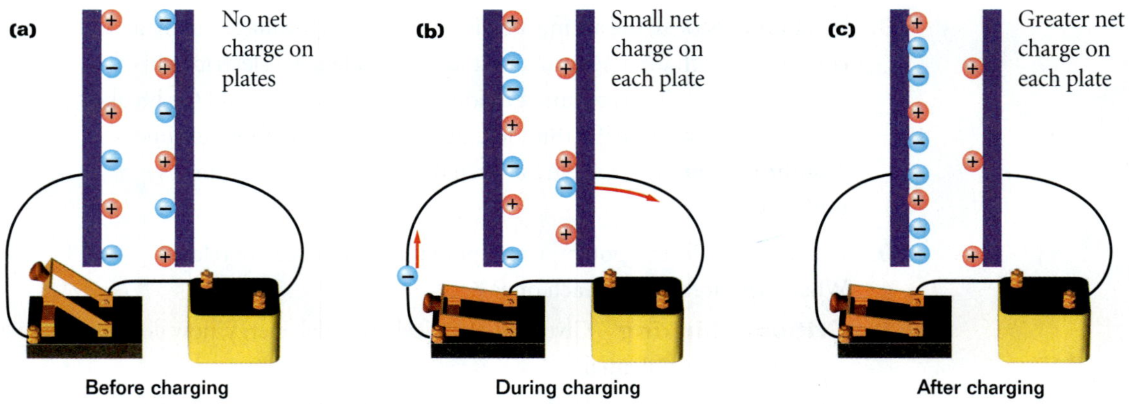

Figure 5
When connected to a battery, the plates of a parallel-plate capacitor become oppositely charged.

CAPACITANCE

$$C = \frac{Q}{\Delta V}$$

$$\text{capacitance} = \frac{\textbf{magnitude of charge on each plate}}{\textbf{potential difference}}$$

The SI unit for capacitance is the *farad,* F, which is equivalent to a coulomb per volt (C/V). In practice, most typical capacitors have capacitances ranging from microfarads ($1\ \mu F = 1 \times 10^{-6}$ F) to picofarads ($1\ pF = 1 \times 10^{-12}$ F).

Capacitance depends on the size and shape of the capacitor

The capacitance of a parallel-plate capacitor with no material between its plates is given by the following expression:

CAPACITANCE FOR A PARALLEL-PLATE CAPACITOR IN A VACUUM

$$C = \varepsilon_0 \frac{A}{d}$$

$$\textbf{capacitance} = \textbf{permittivity of a vacuum} \times \frac{\textbf{area of one of the plates}}{\textbf{distance between the plates}}$$

In this expression, the Greek letter ε (epsilon) represents a constant called the *permittivity* of the medium. When it is followed by a subscripted zero, it refers to a vacuum. It has a magnitude of $8.85 \times 10^{-12}\ C^2/N \bullet m^2$.

We can combine the two equations for capacitance to find an expression for the charge stored on a parallel-plate capacitor.

$$Q = \frac{\varepsilon_0 A}{d} \Delta V$$

This equation tells us that for a given potential difference, ΔV, the charge on a plate is proportional to the area of the plates and inversely proportional to the separation of the plates.

Suppose an isolated conducting sphere has a radius R and a charge Q. The potential difference between the surface of the sphere and infinity is the same as it would be for an equal point charge at the center of the sphere.

$$\Delta V = k_C \frac{Q}{R}$$

Substituting this expression into the definition of capacitance results in the following expression:

$$C_{sphere} = \frac{Q}{\Delta V} = \frac{R}{k_C}$$

Did you know?

The farad is named after Michael Faraday (1791–1867), a prominent nineteenth-century English chemist and physicist. Faraday made many contributions to our understanding of electromagnetic phenomena.

SCILINKS.
www.scilinks.org
Topic: Michael Faraday
Code: HF60955

Demonstration

Capacitor Discharge — ADVANCED

Purpose Show a capacitor being charged and discharged.

Materials 9 V battery, 10 μF capacitor, two switches, ammeter, light bulb and socket, insulated wire

Procedure Construct the circuit so that the battery charges the capacitor when switch 1 is closed and the capacitor powers the bulb when switch 2 is closed. Explain to students that a circuit must be complete for charges to move. Charge the capacitor by closing switch 1 and opening switch 2. Ask what will happen if switch 1 is opened (*nothing*), and then open the switch.

Next ask what will happen if switch 1 is kept open and switch 2 is closed. Have the students observe the bulb as you close switch 2. To convince students that the battery is not affecting the bulb, remove the battery from the circuit, and repeat the demonstration. Repeat the demonstration a third time with an ammeter placed in series with the light bulb. Point out to students that a charged capacitor would work well in a flashlight if only a flash of light were needed. However, in most cases a continuous current is needed.

Teaching Tip — ADVANCED

When a dielectric completely fills the region between two plates of a capacitor, the capacitance is multiplied by a factor known as the dielectric constant, κ (Greek letter *kappa*). The dielectric constant of air is very close to that of a vacuum, so calculations assuming a vacuum provide a good approximation for air. A few dielectric constants are given in the following table:

Material	Dielectric constant (κ)
Vacuum	1.000 00
Air	1.000 59
Nylon	3.4
Pyrex glass	5.6
Paper	3.7
Water	80

ANSWERS

Conceptual Challenge

1. yes, because charge is conserved
2. electrical potential energy

SCILINKS.
www.scilinks.org
Topic: Capacitors
Code: HF60211

This equation indicates that the capacitance of a sphere increases as the size of the sphere increases. Because Earth is so large, it has an extremely large capacitance. Thus, Earth can provide or accept a large amount of charge without its electric potential changing too much. This is the reason why Earth is often used as a reference point for measuring potential differences in electric circuits.

The material between a capacitor's plates can change its capacitance

So far, we have assumed that the space between the plates of a parallel-plate capacitor is a vacuum. However, in many parallel-plate capacitors, the space is filled with a material called a *dielectric.* A dielectric is an insulating material, such as air, rubber, glass, or waxed paper. When a dielectric is inserted between the plates of a capacitor, the capacitance increases. The capacitance increases because the molecules in a dielectric can align with the applied electric field, causing an excess negative charge near the surface of the dielectric at the positive plate and an excess positive charge near the surface of the dielectric at the negative plate. The surface charge on the dielectric effectively reduces the charge on the capacitor plates, as shown in **Figure 6.** Thus, the plates can store more charge for a given potential difference. According to the expression $Q = C\Delta V$, if the charge increases and the potential difference is constant, the capacitance must increase. A capacitor with a dielectric can store more charge and energy for a given potential difference than can the same capacitor without a dielectric. In this book, problems will assume that capacitors are in a vacuum, with no dielectrics.

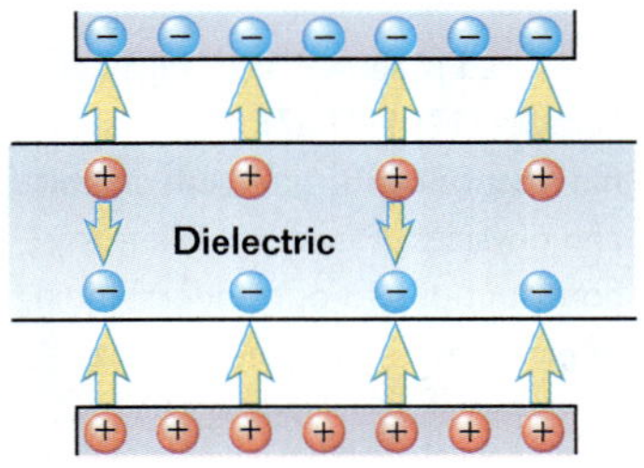

Figure 6
The effect of a dielectric is to reduce the strength of the electric field in a capacitor.

Discharging a capacitor releases its charge

Once a capacitor is charged, the battery or other source of potential difference that charged it can be removed from the circuit. The two plates of the capacitor will remain charged unless they are connected with a material that conducts. Once the plates are connected, the capacitor will *discharge.* This process

Why it Matters

Conceptual Challenge

1. Charge on a Capacitor Plate
A certain capacitor is designed so that one plate is large and the other is small. Do the plates have the same magnitude of charge when connected to a battery?

2. Capacitor Storage
What does a capacitor store, given that the net charge in a parallel-plate capacitor is always zero?

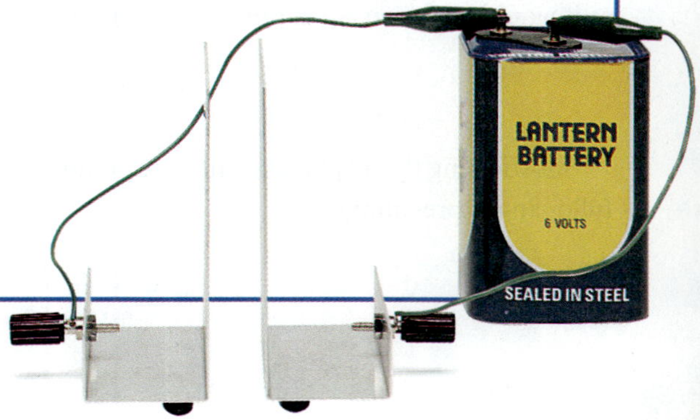

is the opposite of charging. The charges move back from one plate to another until both plates are uncharged again because this is the state of lowest potential energy.

One device that uses a capacitor is the flash attachment of a camera. A battery is used to charge the capacitor, and this stored charge is then released when the shutter-release button is pressed to take a picture. One advantage of using a discharging capacitor instead of a battery to power a flash is that with a capacitor, the stored charge can be delivered to a flash tube much faster, illuminating the subject at the instant more light is needed.

Computers make use of capacitors in many ways. For example, one type of computer keyboard has capacitors at the base of its keys, as shown in **Figure 7.** Each key is connected to a movable plate, which represents one side of the capacitor. The fixed plate on the bottom of the keyboard represents the other side of the capacitor. When a key is pressed, the capacitor spacing decreases, causing an increase in capacitance. External electronic circuits recognize that a key has been pressed when its capacitance changes.

Because the area of the plates and the distance between the plates can be controlled, the capacitance, and thus the electric field strength, can also be easily controlled.

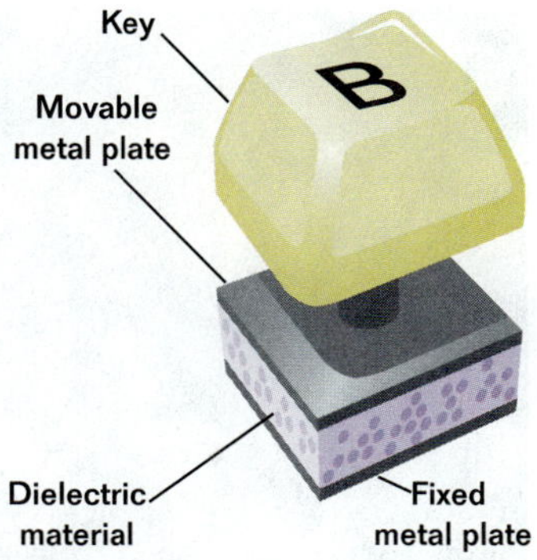

Figure 7
A parallel-plate capacitor is often used in keyboards.

extension

Integrating Biology
Visit go.hrw.com for the activity "Electric Shock: Caution!"

Keyword HF6ELCX

ENERGY AND CAPACITORS

A charged capacitor stores electrical potential energy because it requires work to move charges through a circuit to the opposite plates of a capacitor. The work done on these charges is a measure of the transfer of energy.

For example, if a capacitor is initially uncharged so that the plates are at the same electric potential, that is, if both plates are neutral, then almost no work is required to transfer a small amount of charge from one plate to the other. However, once a charge has been transferred, a small potential difference appears between the plates. As additional charge is transferred through this potential difference, the electrical potential energy of the system increases. This increase in energy is the result of work done on the charge. The electrical potential energy stored in a capacitor that is charged from zero to some charge, Q, is given by the following expression:

ELECTRICAL POTENTIAL ENERGY STORED IN A CHARGED CAPACITOR

$$PE_{electric} = \frac{1}{2}Q\Delta V$$

electrical potential energy =
$\frac{1}{2}$(charge on one plate)(final potential difference)

Note that this equation is also an expression for the work required to charge the capacitor.

SECTION 2

Teaching Tip — GENERAL

Living cells have characteristics similar to capacitors. Different concentrations of ions inside a cell and in the fluid surrounding the cell set up a charge distribution across the cell membrane. The cell is equivalent to a small capacitor separated by a dielectric (the cell membrane wall). The potential difference across some cell membranes is typically on the order of 100 mV.

Demonstration

Functions of a Capacitor — GENERAL

Purpose Show the energy storage and delivery functions of a capacitor.

Materials hand-crank generator and large (1 F) capacitor

Procedure Before connecting the hand-crank generator to the capacitor, crank the generator and show the students how quickly the handle stops when it is released.

Connect the generator to the capacitor. Crank the generator, and release the handle. Have students observe the motor effect of the crank handle being driven by the stored energy in the capacitor.

SECTION 2

Teaching Tip

The maximum electric field that can be produced before a dielectric breaks down and begins to conduct is known as *dielectric strength.* Most insulating materials have dielectric strengths greater than that of air.

PROBLEM GUIDE B

Use this guide to assign problems.
SE = Student Edition Textbook
PW = Problem Workbook
PB = Problem Bank on the One-Stop Planner (OSP)

Solving for:

Q	**SE** Sample, 1, 3; Ch. Rvw. 18 **PW** 3–4 **PB** 5–7
$PE_{electric}$	**SE** 1–2; Ch. Rvw. 19 **PW** 6 **PB** 5–6
C	**SE** 2 **PW** Sample, 1–2, 7–8 **PB** 3–4
ΔV	**SE** 3 **PW** 5 **PB** 8–10
d	**PW** 7, 9 **PB** Sample, 1–3

*Challenging Problem
Consult the printed Solutions Manual or the OSP for detailed solutions.

Figure 8
The markings caused by electrical breakdown in this material look similar to the lightning bolts produced when air undergoes electrical breakdown to form a plasma of charged particles.

By substituting the definition of capacitance ($C = Q/\Delta V$), we can see that these alternative forms are also valid:

$$PE_{electric} = \frac{1}{2}C(\Delta V)^2$$

$$PE_{electric} = \frac{Q^2}{2C}$$

These results apply to any capacitor. In practice, there is a limit to the maximum energy (or charge) that can be stored because electrical breakdown ultimately occurs between the plates of the capacitor for a sufficiently large potential difference. So, capacitors are usually labeled with a maximum operating potential difference. Electrical breakdown in a capacitor is like a lightning discharge in the atmosphere. **Figure 8** shows a pattern created in a block of plastic resin that has undergone electrical breakdown. This book's problems assume that all potential differences are below the maximum.

SAMPLE PROBLEM B

Capacitance

PROBLEM

A capacitor, connected to a 12 V battery, holds 36 µC of charge on each plate. What is the capacitance of the capacitor? How much electrical potential energy is stored in the capacitor?

SOLUTION

Given: $Q = 36\ \mu C = 3.6 \times 10^{-5}$ C $\quad \Delta V = 12$ V

Unknown: $C = ?$ $\quad PE_{electric} = ?$

To determine the capacitance, use the definition of capacitance.

$$C = \frac{Q}{\Delta V} = \frac{3.6 \times 10^{-5}\ \text{C}}{12\ \text{V}}$$

$$\boxed{C = 3.0 \times 10^{-6}\ \text{F} = 3.0\ \mu\text{F}}$$

To determine the potential energy, use the alternative form of the equation for the potential energy of a charged capacitor shown on this page:

$$PE_{electric} = \frac{1}{2}C(\Delta V)^2$$

$$PE_{electric} = (0.5)(3.0 \times 10^{-6}\ \text{F})(12\ \text{V})^2$$

$$\boxed{PE_{electric} = 2.2 \times 10^{-4}\ \text{J}}$$

PRACTICE B

Capacitance

1. A 4.00 μF capacitor is connected to a 12.0 V battery.
   a. What is the charge on each plate of the capacitor?
   b. If this same capacitor is connected to a 1.50 V battery, how much electrical potential energy is stored?
2. A parallel-plate capacitor has a charge of 6.0 μC when charged by a potential difference of 1.25 V.
   a. Find its capacitance.
   b. How much electrical potential energy is stored when this capacitor is connected to a 1.50 V battery?
3. A capacitor has a capacitance of 2.00 pF.
   a. What potential difference would be required to store 18.0 pC?
   b. How much charge is stored when the potential difference is 2.5 V?
4. You are asked to design a parallel-plate capacitor having a capacitance of 1.00 F and a plate separation of 1.00 mm. Calculate the required surface area of each plate. Is this a realistic size for a capacitor?

SECTION REVIEW

1. Assume Earth and a cloud layer 800.0 m above the Earth can be treated as plates of a parallel-plate capacitor.
   a. If the cloud layer has an area of 1.00×10^6 m^2, what is the capacitance?
   b. If an electric field strength of 2.0×10^6 N/C causes the air to conduct charge (lightning), what charge can the cloud hold?
   c. Describe what must happen to its molecules for air to conduct electricity.
2. A parallel-plate capacitor has an area of 2.0 cm^2, and the plates are separated by 2.0 mm.
   a. What is the capacitance?
   b. How much charge does this capacitor store when connected to a 6.0 V battery?
3. A parallel-plate capacitor has a capacitance of 1.35 pF. If a 12.0 V battery is connected to this capacitor, how much electrical potential energy would it store?
4. **Critical Thinking** Explain why two metal plates near each other will not become charged until connected to a source of potential difference.

ANSWERS

Practice B

1. **a.** 4.80×10^{-5} C
   b. 4.50×10^{-6} J
2. **a.** 4.8×10^{-6} F
   b. 5.4×10^{-6} J
3. **a.** 9.00 V
   b. 5.0×10^{-12} C
4. 1.13×10^8 m^2; no

SECTION REVIEW ANSWERS

1. **a.** 1.11×10^{-8} F
   b. ±18 C
   c. The molecules and atoms must be broken down into charged particles (a plasma) that can conduct electricity.
2. **a.** 8.8×10^{-13} F
   b. 5.3×10^{-12} C
3. 9.72×10^{-11} J
4. If there is no potential difference, there is no electric force to set charges in motion.

SECTION 3
General Level

Misconception Alert

Some students may think that electric current is the flow of electrical energy. Explain that current refers to the movement of matter, not the movement of energy.

Visual Strategy BASIC

Figure 9

Use this diagram to reinforce the definition of *current* as the rate of charge movement.

Q How does the current change if the number of charge carriers increases?

A *The current increases.*

Q How does the current change if the time interval during which a given number of charge carriers pass the cross-sectional area increases?

A *The current decreases.*

SECTION 3

Current and Resistance

SECTION OBJECTIVES

- **Describe the basic properties of electric current, and solve problems relating current, charge, and time.**
- **Distinguish between the drift speed of a charge carrier and the average speed of the charge carrier between collisions.**
- **Calculate resistance, current, and potential difference by using the definition of *resistance*.**
- **Distinguish between ohmic and non-ohmic materials, and learn what factors affect resistance.**

CURRENT AND CHARGE MOVEMENT

Although many practical applications and devices are based on the principles of static electricity, electricity did not become an integral part of our daily lives until scientists learned to control the movement of electric charge, known as *current.* Electric currents power our lights, radios, television sets, air conditioners, and refrigerators. Currents are also used in automobile engines, travel through miniature components that make up the chips of computers, and perform countless other invaluable tasks.

Electric currents are even part of the human body. This connection between physics and biology was discovered by Luigi Galvani (1737–1798). While conducting electrical experiments near a frog he had recently dissected, Galvani noticed that electrical sparks caused the frog's legs to twitch and even convulse. After further research, Galvani concluded that electricity was present in the frog. Today, we know that electric currents are responsible for transmitting messages between body muscles and the brain. In fact, every function involving the nervous system is initiated by electrical activity.

Current is the rate of charge movement

A current exists whenever there is a net movement of electric charge through a medium. To define *current* more precisely, suppose electrons are moving through a wire, as shown in **Figure 9.** The **electric current** is the rate at which these charges move through the cross section of the wire. If ΔQ is the amount of charge that passes through this area in a time interval, Δt, then the current, I, is the ratio of the amount of charge to the time interval. Note that the direction of current is *opposite* the movement of the negative charges. We will further discuss this detail later in this section.

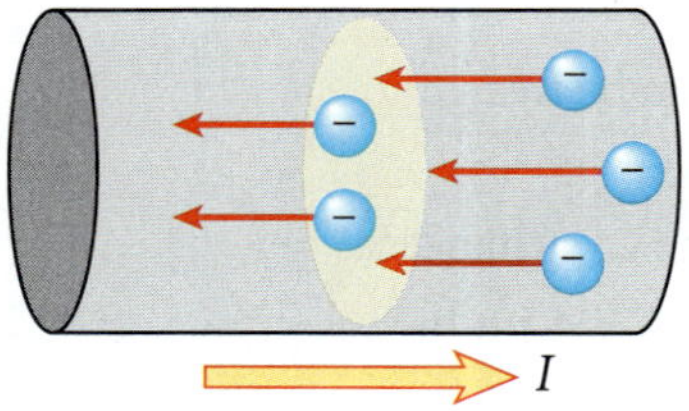

Figure 9
The current in this wire is defined as the rate at which electric charges pass through a cross-sectional area of the wire.

electric current

the rate at which electric charges pass through a given area

ELECTRIC CURRENT

$$I = \frac{\Delta Q}{\Delta t}$$

$$\text{electric current} = \frac{\text{charge passing through a given area}}{\text{time interval}}$$

The SI unit for current is the *ampere*, A. One ampere is equivalent to one coulomb of charge passing through a cross-sectional area in a time interval of one second (1 A = 1 C/s).

SAMPLE PROBLEM C

Current

PROBLEM

The current in a light bulb is 0.835 A. How long does it take for a total charge of 1.67 C to pass through the filament of the bulb?

SOLUTION

Given: $\Delta Q = 1.67$ C $\quad I = 0.835$ A

Unknown: $\Delta t = ?$

Use the definition of electric current. Rearrange to solve for the time interval.

$$I = \frac{\Delta Q}{\Delta t}$$

$$\Delta t = \frac{\Delta Q}{I}$$

$$\boxed{\Delta t = \frac{1.67\ \text{C}}{0.835\ \text{A}} = 2.00\ \text{s}}$$

PRACTICE C

Current

1. If the current in a wire of a CD player is 5.00 mA, how long would it take for 2.00 C of charge to pass through a cross-sectional area of this wire?
2. In a particular television tube, the beam current is 60.0 µA. How long does it take for 3.75×10^{14} electrons to strike the screen? (Hint: Recall that an electron has a charge of -1.60×10^{-19} C.)
3. If a metal wire carries a current of 80.0 mA, how long does it take for 3.00×10^{20} electrons to pass a given cross-sectional area of the wire?
4. The compressor on an air conditioner draws 40.0 A when it starts up. If the start-up time is 0.50 s, how much charge passes a cross-sectional area of the circuit in this time?
5. A total charge of 9.0 mC passes through a cross-sectional area of a nichrome wire in 3.5 s.
   a. What is the current in the wire?
   b. How many electrons pass through the cross-sectional area in 10.0 s?
   c. If the number of charges that pass through the cross-sectional area during the given time interval doubles, what is the resulting current?

SECTION 3

Classroom Practice

Current

A 100.0 W light bulb draws 0.83 A of current. How long does it take for 1.9×10^{22} electrons to pass a given cross-sectional area of the filament?

Answer

1.0 h

PROBLEM GUIDE C

Use this guide to assign problems.
SE = Student Edition Textbook
PW = Problem Workbook
PB = Problem Bank on the One-Stop Planner (OSP)

Solving for:

Δt	**SE** Sample, 1–3; Ch. Rvw. 32–33 **PW** 6 **PB** 4–6
ΔQ	**SE** 4; Ch. Rvw. 70, 78a* **PW** Sample, 1–2, 3* **PB** 7–10
I	**SE** 5; Ch. Rvw. 69, 78b **PW** 4–5 **PB** Sample, 1–3

***Challenging Problem**
Consult the printed Solutions Manual or the OSP for detailed solutions.

ANSWERS

Practice C

1. 4.00×10^2 s
2. 1.00 s
3. 6.00×10^2 s
4. 2.0×10^1 C
5. a. 2.6×10^{-3} A
   b. 1.6×10^{17} electrons
   c. 5.1×10^{-3} A

Misconception Alert — GENERAL

Some students may think that the charges moving in a circuit are always positive. Stress that charge carriers can be positive, negative, or a combination of the two.

Teaching Tip — ADVANCED

Currents that consist of both positive and negative charge carriers in motion include those that exist in batteries, in the human body, in the ocean, and in the ground. Electric currents in the brain and nerves of the human body consist of moving sodium and potassium ions.

Quick Lab

TEACHER'S NOTES

This lab is meant to give an example of charge movement through an electrolytic solution. Be sure that the ends of the copper wire are not sharp. Use sandpaper to smooth any sharp ends.

Misconception Alert — BASIC

Many students have the misconception that charge carriers move at the speed of light. When discussing the concept of drift velocity, address this misconception directly with students.

SCILINKS.
www.scilinks.org
Topic: Electric Current
Code: HF60472

Quick Lab

A Lemon Battery

MATERIALS LIST

- lemon
- copper wire
- paper clip

Straighten the paper clip, and insert it and the copper wire into the lemon to construct a chemical cell. Touch the ends of both wires with your tongue. Because a potential difference exists across the two metals and because your saliva provides an electrolytic solution that conducts electric current, you should feel a slight tingling sensation on your tongue. CAUTION: Do not share battery set-ups with other students. Dispose of your materials according to your teacher's instructions.

Conventional current is defined in terms of positive charge movement

The moving charges that make up a current can be positive, negative, or a combination of the two. In a common conductor, such as copper, current is due to the motion of negatively charged electrons, because the atomic structure of solid conductors allows the electrons to be transferred easily from one atom to the next. In contrast, the protons are relatively fixed inside the nucleus of the atom. In certain particle accelerators, a current exists when positively charged protons are set in motion. In some cases—in gases and dissolved salts, for example—current is the result of positive charges moving in one direction and negative charges moving in the opposite direction.

Positive and negative charges in motion are sometimes called *charge carriers. Conventional current* is defined in terms of the flow of positive charges. Thus, negative charge carriers, such as electrons, would have a conventional current in the direction opposite their physical motion. The three possible cases of charge flow are shown in **Table 1.** We will use conventional current in this book unless stated otherwise.

Table 1 Conventional Current

	First case	Second case	Third case
Motion of charge carriers			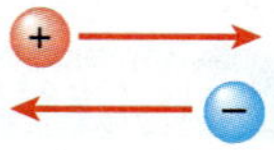
Equivalent conventional current			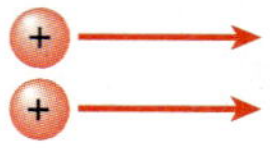

As you learned in Section 1, an electric field in a material sets charges in motion. For a material to be a good conductor, charge carriers in the material must be able to move easily through the material. Many metals are good conductors because metals usually contain a large number of free electrons. Body fluids and salt water are able to conduct electric charge because they contain charged atoms called *ions.* Because dissolved ions can move through a solution easily, they can be charge carriers. A solute that dissolves in water to give a solution that conducts electric current is called an *electrolyte.*

DRIFT VELOCITY

When you turn on a light switch, the light comes on almost immediately. For this reason, many people think that electrons flow very rapidly from the switch to the light bulb. However, this is not the case. When you turn on the switch, electron motion near the switch changes the electric field there, and the change propagates throughout the wire very quickly. Such changes travel through the wire at nearly the speed of light. The charges themselves, however, travel much more slowly.

Drift velocity is the net velocity of charge carriers

To see how the electrons move, consider a solid conductor in which the charge carriers are free electrons. When the conductor is in electrostatic equilibrium, the electrons move randomly, similar to the movement of molecules in a gas. When a potential difference is applied across the conductor, an electric field is set up inside the conductor. The force due to that field sets the electrons in motion, thereby creating a current.

These electrons do not move in straight lines along the conductor in a direction opposite the electric field. Instead, they undergo repeated collisions with the vibrating metal atoms of the conductor. If these collisions were charted, the result would be a complicated zigzag pattern like the one shown in **Figure 10.** The energy transferred from the electrons to the metal atoms during the collisions increases the vibrational energy of the atoms, and the conductor's temperature increases.

The electrons gain kinetic energy as they are accelerated by the electric field in the conductor. They also lose kinetic energy because of the collisions described above. However, despite the internal collisions, the individual electrons move slowly along the conductor in a direction opposite the electric field, **E,** with a velocity known as the **drift velocity,** $\mathbf{v}_{drift}$.

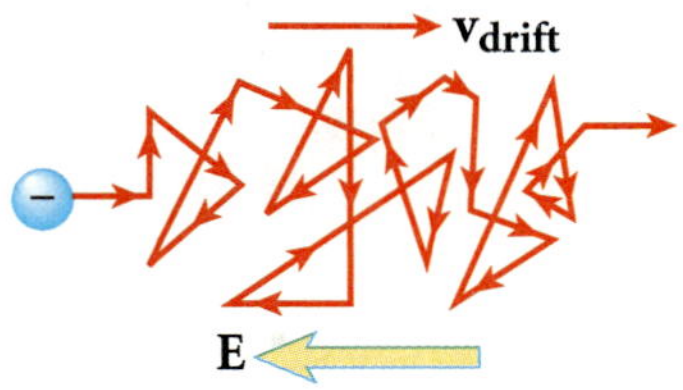

Figure 10
When an electron moves through a conductor, collisions with the vibrating metal atoms of the conductor force the electron to change its direction constantly.

drift velocity
the net velocity of a charge carrier moving in an electric field

Drift speeds are relatively small

The magnitudes of drift velocities, or drift speeds, are typically very small. In fact, the drift speed is much less than the average speed between collisions. For example, in a copper wire that has a current of 10.0 A, the drift speed of electrons is only 2.46×10^{-4} m/s. These electrons would take about 68 min to travel 1 m! The electric field, on the other hand, reaches electrons throughout the wire at a speed approximately equal to the speed of light.

Why it Matters

Conceptual Challenge

1. Electric Field Inside a Conductor
We concluded in our study of electrostatics that the field inside a conductor is zero, yet we have seen that an electric field exists inside a conductor that carries a current. How is this zero electric field possible?

2. Turning on a Light
If charges travel very slowly through a metal (approximately 10^{-4} m/s), why doesn't it take several hours for a light to come on after you flip a switch?

3. Particle Accelerator
The positively charged dome of a Van de Graaff generator can be used to accelerate positively charged protons. A current exists due to the motion of these protons. In this case, how does the direction of conventional current compare with the direction in which the charge carriers move?

Demonstration

Drift Speed — BASIC

Purpose Illustrate the difference between drift speed and the speed of a signal in a circuit.

Materials 30 cm plastic ruler with center groove, about 35 marbles

Procedure Place enough marbles in the center groove of the ruler to fill the ruler from end to end. Tell the students that the marbles represent the free electrons in a conductor. Add one more marble to the ruler by sliding it horizontally onto the end of the ruler. The marble at the other end of the ruler should roll off the ruler almost immediately.

Point out that each individual marble moved only a small distance (drift speed of about 10^{-4} m/s in a circuit) but the information traveled to the other end of the ruler at a high speed (nearly the speed of light in a circuit). Repeat the demonstration several times while students observe the motion of an individual marble.

ANSWERS

Conceptual Challenge

1. The electric field in a conductor must be zero when the conductor is in electrostatic equilibrium, but this conclusion does not apply if charges are in motion.
2. because the signal for charges to begin moving (the electric field) travels at nearly the speed of light
3. They are the same.

Misconception Alert BASIC

Some students may develop the idea that resistance is a variable that can change, like force or acceleration. Emphasize that the resistance of an object is more like mass (assuming temperature remains constant). Once a resistor is built, it usually has a fixed resistance. (Exceptions are non-ohmic resistors.) The resistance of a circuit can be changed by adding or changing resistors, just as adding mass to an object changes its total mass.

Teaching Tip GENERAL

Many books refer to the equation $\Delta V = IR$ as Ohm's law. However, this equation can be used with both ohmic and non-ohmic materials. Ohm's law refers to the empirical observation that for many materials, the resistance is constant over a wide range of potential differences. Thus, the equation for resistance ($\Delta V = IR$) is an expression of Ohm's law only for cases in which R is independent of ΔV. To avoid misconceptions, this book introduces the equation for resistance first, then Ohm's law.

RESISTANCE TO CURRENT

When a light bulb is connected to a battery, the current in the bulb depends on the potential difference across the battery. For example, a 9.0 V battery connected to a light bulb generates a greater current than a 6.0 V battery connected to the same bulb. But potential difference is not the only factor that determines the current in the light bulb. The materials that make up the connecting wires and the bulb's filament also affect the current in the bulb. Even though most materials can be classified as conductors or insulators, some conductors allow charges to move through them more easily than others. The opposition to the motion of charge through a conductor is the conductor's **resistance.** Quantitatively, resistance is defined as the ratio of potential difference to current, as follows:

resistance

the opposition presented to electric current by a material or device

RESISTANCE

$$R = \frac{\Delta V}{I}$$

$$\text{resistance} = \frac{\textbf{potential difference}}{\textbf{current}}$$

The SI unit for resistance, the *ohm,* is equal to one volt per ampere and is represented by the Greek letter Ω (*omega*).

Resistance is constant over a range of potential differences

For many materials, including most metals, experiments show that *the resistance is constant over a wide range of applied potential differences.* This statement, known as Ohm's law, is named for Georg Simon Ohm (1789–1854), who was the first to conduct a systematic study of electrical resistance. Mathematically, Ohm's law is stated as follows:

$$\frac{\Delta V}{I} = \text{constant}$$

As can be seen by comparing the definition of resistance with Ohm's law, the constant of proportionality in the Ohm's law equation is resistance. It is common practice to express Ohm's law as $\Delta V = IR$.

Ohm's law does not hold for all materials

Ohm's law is not a fundamental law of nature like the conservation of energy or the universal law of gravitation. Instead, it is a behavior that is valid only for certain materials. Materials that have a constant resistance over a wide range of potential differences are said to be *ohmic.* A graph of current versus potential difference for an ohmic material is linear, as shown in **Figure 11(a).** This is because the slope of such a graph ($I/\Delta V$) is inversely proportional to resistance. When resistance is constant, the current is proportional to the potential difference and the resulting graph is a straight line.

SCILINKS®
www.scilinks.org
Topic: Ohm's Law
Code: HF61071

Materials that do not function according to Ohm's law are said to be *non-ohmic.* **Figure 11(b)** shows a graph of current versus potential difference for a non-ohmic material. In this case, the slope is not constant because resistance varies. Hence, the resulting graph is nonlinear. One common semiconducting device that is non-ohmic is the *diode.* Its resistance is small for currents in one direction and large for currents in the reverse direction. Diodes are used in circuits to control the direction of current. This book assumes that all resistors function according to Ohm's law unless stated otherwise.

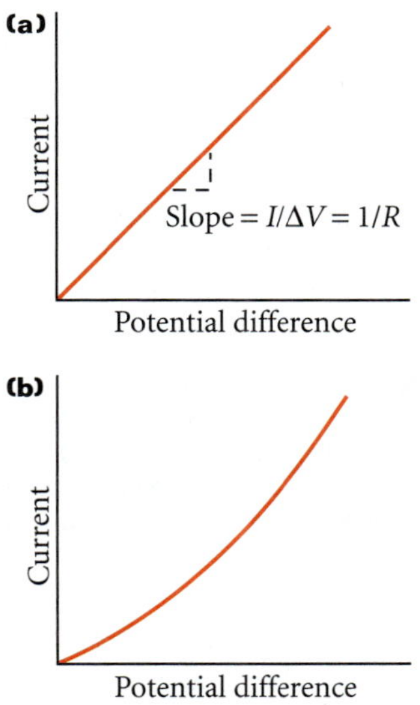

Figure 11
(a) The current–potential difference curve of an ohmic material is linear, and the slope is the inverse of the material's resistance. **(b)** The current–potential difference curve of a non-ohmic material is nonlinear.

Resistance depends on length, area, material, and temperature

Earlier in this section, you learned that electrons do not move in straight-line paths through a conductor. Instead, they undergo repeated collisions with the metal atoms. These collisions affect the motion of charges somewhat as a force of internal friction would. This is the origin of a material's resistance. Thus, any factors that affect the number of collisions will also affect a material's resistance. Some of these factors are shown in **Table 2.**

Two of these factors—length and cross-sectional area—are purely geometrical. It is intuitive that a longer length of wire provides more resistance than a shorter length of wire does. Similarly, a wider wire allows charges to flow more easily than a thinner wire does, much as a larger pipe allows water to flow more easily than a smaller pipe does. The material effects have to do with the structure of the atoms making up the material. Finally, for most materials, resistance increases as the temperature of the metal increases. When a material is hot, its atoms vibrate fast, and it is more difficult for an electron to flow through the material.

Table 2 Factors That Affect Resistance

Factor	Less resistance	Greater resistance
Length	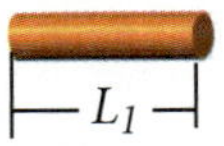 L_1	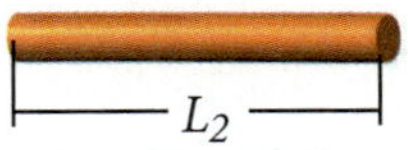 L_2
Cross-sectional area	 A_1	 A_2
Material	 Copper	 Iron
Temperature	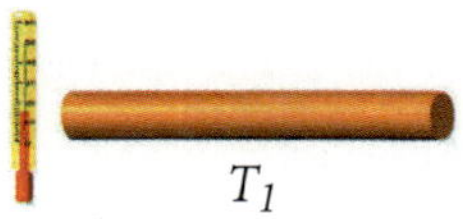 T_1	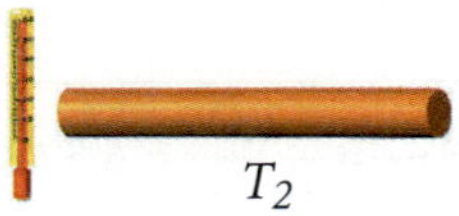 T_2

SECTION 3

Demonstration

Non-Ohmic Resistance GENERAL

Purpose Show the non-ohmic response of a resistor.

Materials flashlight bulb, 12 V power supply, ammeter, ohmmeter

Procedure Measure the resistance of the bulb. Plot ΔV versus I on the chalkboard. Measure I in 2 V increments up to 10 V. Do not exceed 12 V, which will burn out the bulb. As you plot the curve, you will see that the curve is nonlinear. Why? (*The bulb is non-ohmic. Resistance in the filament varies as a function of temperature.*)

Demonstration

Factors That Affect Resistance GENERAL

Purpose Show the dependence of resistance on cross-sectional area and length.

Materials two 50.0 Ω resistors, ohmmeter, connecting wires

Procedure Tell the students you have two equal resistors with the same length, cross-sectional area, and temperature. Ask students to predict whether the two resistors attached side by side will allow easier flow (lower R) or make the flow more difficult (higher R). Connect the two resistors side by side (in parallel), measure the resistance, and have students explain the results. (*Area is doubled, so resistance decreases.*) Repeat the demonstration for two resistors attached end to end (in series). (*Length is doubled, so resistance increases.*)

Teaching Tip — ADVANCED

The first color band of a typical four-band resistor represents the first digit of the resistor's resistance, the second band represents the second digit, and the third band represents the power of ten by which these digits are multiplied. The fourth band, which is gold or silver, indicates the accuracy of the resistor. Color codes are as follows:

black	0
brown	1
red	2
orange	3
yellow	4
green	5
blue	6
violet	7
gray	8
white	9

Classroom Practice

Resistance

A potential difference ΔV_1 is applied across a resistance R_1, resulting in a current I_1. Determine the new current, I, when the following changes are made:

a. $\Delta V = 2\Delta V_1$, $R = R_1$

b. $\Delta V = \Delta V_1$, $R = 2R_1$

c. $\Delta V = 2\Delta V_1$, $R = 2R_1$

Answers

a. $I = 2I_1$

b. $I = \frac{1}{2}I_1$

c. $I = I_1$

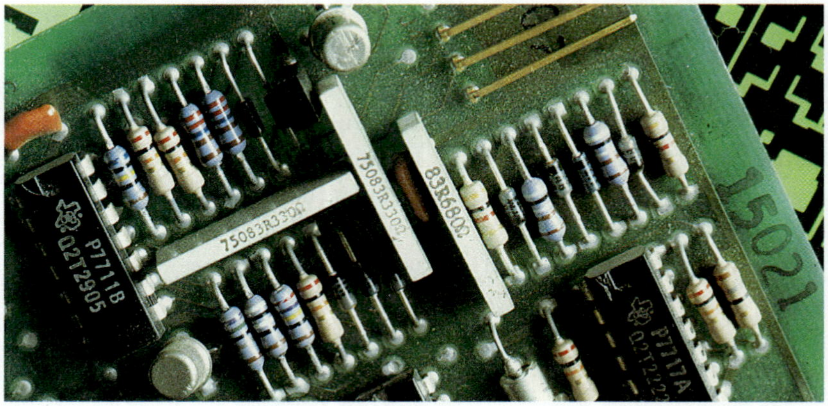

Figure 12
Resistors, such as those shown here, are used to control current. The colors of the bands represent a code for the values of the resistances.

Resistors can be used to control the amount of current in a conductor

One way to change the current in a conductor is to change the potential difference across the ends of the conductor. But in many cases, such as in household circuits, the potential difference does not change. How can the current in a certain wire be changed if the potential difference remains constant?

According to the definition of resistance, if ΔV remains constant, current decreases when resistance increases. Thus, the current in a wire can be decreased by replacing the wire with one of higher resistance. The same effect can be accomplished by making the wire longer or by connecting a *resistor* to the wire. A resistor is a simple electrical element that provides a specified resistance. **Figure 12** shows a group of resistors in a circuit board. Resistors are sometimes used to control the current in an attached conductor because this is often more practical than changing the potential difference or the properties of the conductor.

SCILINKS.
www.scilinks.org
Topic: Superconductors
Code: HF61478

SAMPLE PROBLEM D

Resistance

PROBLEM

The resistance of a steam iron is 19.0 Ω. What is the current in the iron when it is connected across a potential difference of 120 V?

SOLUTION

Given: $R = 19.0\ \Omega$ $\quad \Delta V = 120\ \text{V}$

Unknown: $I = ?$

Use Ohm's law to relate resistance to potential difference and current.

$$R = \frac{\Delta V}{I}$$

$$\boxed{I = \frac{\Delta V}{R} = \frac{120\ \text{V}}{19.0\ \Omega} = 6.32\ \text{A}}$$

PRACTICE D

Resistance

1. A 1.5 V battery is connected to a small light bulb with a resistance of 3.5 Ω. What is the current in the bulb?
2. A stereo with a resistance of 65 Ω is connected across a potential difference of 120 V. What is the current in this device?
3. Find the current in the following devices when they are connected across a potential difference of 120 V.
   a. a hot plate with a resistance of 48 Ω
   b. a microwave oven with a resistance of 20 Ω
4. The current in a microwave oven is 6.25 A. If the resistance of the oven's circuitry is 17.6 Ω, what is the potential difference across the oven?
5. A typical color television draws 2.5 A of current when connected across a potential difference of 115 V. What is the effective resistance of the television set?
6. The current in a certain resistor is 0.50 A when it is connected to a potential difference of 110 V. What is the current in this same resistor if
   a. the operating potential difference is 90.0 V?
   b. the operating potential difference is 130 V?

Salt water and perspiration lower the body's resistance

The human body's resistance to current is on the order of 500 000 Ω when the skin is dry. However, the body's resistance decreases when the skin is wet. If the body is soaked with salt water, its resistance can be as low as 100 Ω. This is because ions in salt water readily conduct electric charge. Such low resistances can be dangerous if a large potential difference is applied between parts of the body because current increases as resistance decreases. Currents in the body that are less than 0.01 A either are imperceptible or generate a slight tingling feeling. Greater currents are painful and can disturb breathing, and currents above 0.15 A disrupt the electrical activity of the heart and can be fatal.

Perspiration also contains ions that conduct electric charge. In a *galvanic skin response* (GSR) test, commonly used as a stress test and as part of some so-called lie detectors, a very small potential difference is set up across the body. Perspiration increases when a person is nervous or stressed, thereby decreasing the resistance of the body. In GSR tests, a state of low stress and high resistance, or "normal" state, is used as a control, and a state of higher stress is reflected as a decreased resistance compared with the normal state.

extension

Integrating Biology
Visit go.hrw.com for the activity "The Brain's Signals."
Keyword HF6ELCX

PROBLEM GUIDE D

Use this guide to assign problems.
SE = Student Edition Textbook
PW = Problem Workbook
PB = Problem Bank on the One-Stop Planner (OSP)

Solving for:

I	**SE** Sample, 1–3, 6*; Ch. Rvw. 40–42 **PW** 6–7 **PB** 4–6
ΔV	**SE** 4; Ch. Rvw. 71 **PW** 4–5 **PB** Sample, 1–3
R	**SE** 5 **PW** Sample, 1–3 **PB** 7–10

***Challenging Problem**
Consult the printed Solutions Manual or the OSP for detailed solutions.

ANSWERS

Practice D

1. 0.43 A
2. 1.8 A
3. **a.** 2.5 A
   b. 6.0 A
4. 1.10×10^2 V
5. 46 Ω
6. **a.** 0.41 A
   b. 0.59 A

Teaching Tip

Students may better understand the discussion of potentiometers if you provide a potentiometer with its top pried open to illustrate the resistor's operation.

SECTION REVIEW ANSWERS

1. yes; when charge carriers are negative
2. **a.** 0.60 A
   b. 2.2×10^{20} electrons
3. 12 A
4. 3.6 Ω
5. The diode is non-ohmic.
6. Resistors regulate the magnitude of the current. Diodes regulate the direction of current.
7. 1.5 A; 2.4 A
8. Although electrons undergo a force that moves them across the wire, collisions with atoms continually randomize their motion. As a result, the drift speed is much less than the average speed between collisions.
9. You could decrease current by making the wire as long as possible, thereby increasing its resistance.

ADVANCED TOPICS

See "Superconductors and BCS Theory" in **Appendix J: Advanced Topics** to learn more about superconducting materials.

Potentiometers have variable resistance

A *potentiometer* is a special type of resistor that has a fixed contact on one end and an adjustable, sliding contact that allows the user to tap off different potential differences. The sliding contact is frequently mounted on a rotating shaft, and the resistance is adjusted by rotating a knob. Potentiometers (frequently called *pots* for short) have many applications. In fact, most of the knobs on everyday items, such as volume controller on a stereo, are potentiometers. Potentiometers may also be mounted linearly. One example is a dimmer switch to control the light output of a light fixture. The joystick on your video game controller uses two potentiometers, one for motion in the x direction and one for motion in the y direction, to tell the computer the movements that you make when playing a game.

SECTION REVIEW

1. Can the direction of conventional current ever be opposite the direction of charge movement? If so, when?
2. The charge that passes through the filament of a certain light bulb in 5.00 s is 3.0 C.
   a. What is the current in the light bulb?
   b. How many electrons pass through the filament of the light bulb in a time interval of 1.0 min?
3. How much current would a 10.2 Ω toaster oven draw when connected to a 120 V outlet?
4. An ammeter registers 2.5 A of current in a wire that is connected to a 9.0 V battery. What is the wire's resistance?
5. In a particular diode, the current triples when the applied potential difference is doubled. What can you conclude about the diode?
6. What is the function of resistors in a circuit board? What is the function of diodes in a circuit board?
7. Calculate the current in a 75 Ω resistor when a potential difference of 115 V is placed across it. What will the current be if the resistor is replaced with a 47 Ω resistor?
8. **Critical Thinking** In a conductor that carries a current, which is less, the drift speed of an electron or the average speed of the electron between collisions? Explain your answer.
9. **Critical Thinking** You have only one type of wire. If you are connecting a battery to a light bulb with this wire, how could you decrease the current in the wire?

Why it Matters

Superconductors

Take a moment to imagine the many things that could be created with materials that conduct electricity with *zero* resistance. There would be no heating or reduction in the current when conducting electricity with such a material. These materials exist and are called *superconductors*.

Superconductors have zero resistance below a certain temperature, called the *critical temperature*. The graph of resistance as a function of temperature for a superconductor resembles that of a normal metal at temperatures well above the critical temperature. But when the temperature is near or below the critical temperature, the resistance suddenly drops to zero, as the graph below shows. This graph shows the resistance of mercury just above and below its critical temperature of 4.15 K.

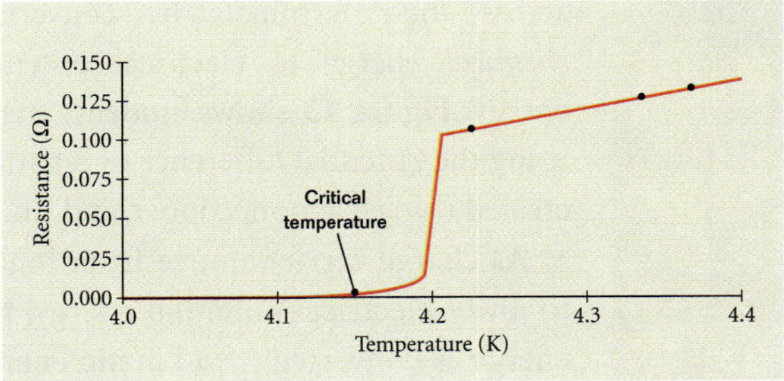

Today, there are thousands of known superconductors, including common metals such as aluminum, tin, lead, and zinc. However, for common metals that exhibit superconductivity, the critical temperature is extremely low—near absolute zero. For example, aluminum reaches superconductivity at 1.19 K, just a little more than one degree above absolute zero. Temperatures near absolute zero are difficult to achieve and maintain. Interestingly, copper, silver, and gold, which are excellent conductors at room temperature, do not exhibit superconductivity.

An important recent development in physics is the discovery of high-temperature superconductors. The excitement began with a 1986 publication by scientists at the IBM Zurich Research Laboratory in Switzerland. In this publication, scientists reported evidence for superconductivity at a temperature near 30 K. More recently, scientists have found superconductivity at temperatures as high as 150 K. However, 150 K is still −123°C, which is much colder than room temperature. The search continues for a material that has superconducting qualities at room temperature. This important search has both scientific and practical applications.

This express train in Shanghai, China, which utilizes the Meissner effect, levitates above the track and travels at up to 430 km/h in normal operations.

One of the truly remarkable features of superconductors is that once a current is established in them, the current continues even if the applied potential difference is removed. In fact, steady currents have been observed to persist for many years in superconducting loops with no apparent decay. This feature makes superconducting materials attractive for a wide variety of applications.

Because electric currents produce magnetic effects, current in a superconductor can be used to float a magnet in the air over a superconductor. This effect, known as the Meissner effect, is used with high-speed express trains, such as the one shown in the figure above. This type of train levitates a few inches above the track.

One useful application of superconductivity is superconducting magnets. Such magnets are being considered for storing energy. The idea of using superconducting power lines to transmit power more efficiently is also being researched. Modern superconducting electronic devices that consist of two thin-film superconductors separated by a thin insulator have been constructed. They include magnetometers (magnetic-field measuring devices) and various microwave devices.

Why it Matters

Superconductors

Magnetic levitation (MAGLEV) is an application ideally suited to high-temperature superconductors. Passenger and cargo transport trains float above the track, such that friction is nearly eliminated between the train and its tracks. The lack of friction allows the train to travel at very high speeds. MAGLEV technology has now been proven, but commercial application has been slow because of the high cost and logistical concerns.

Extension

Point out to students that magnetic levitation has major advantages (it reduces friction) but has disadvantages, too.

Q Why are superconductors not used widely for trains or floating cars?

A ***Answers may vary but could include high construction costs, high cooling costs, and temperature-control problems, which could lead to crashes.***

SECTION 4
General Level

Demonstration

Potential Difference as a Source of Current — BASIC

Purpose Show the function of a battery or generator and the conservation of charge.

Materials inclined plane, marbles

Procedure Set up the inclined plane, and roll the marbles down the plane. As the marbles reach the bottom of the incline, return them to the top and let them roll down again. Have the students explain the parallels between this demonstration and a battery supplying energy to a circuit. *(The hand provides gravitational potential energy just as the battery provides electrical potential energy. The marbles lose energy on the ramp but are not "used up" in the process, just as charge carriers are not "used up" in a circuit.)*

SECTION 4

Electric Power

SECTION OBJECTIVES

- **Differentiate between direct current and alternating current.**
- **Relate electric power to the rate at which electrical energy is converted to other forms of energy.**
- **Calculate electric power and the cost of running electrical appliances.**

SOURCES AND TYPES OF CURRENT

When you drop a ball, it falls to the ground, moving from a place of higher gravitational potential energy to one of lower gravitational potential energy. As discussed in Section 1, charges behave in similar ways. For example, free electrons in a conductor move randomly when all points in the conductor are at the same potential. But when a potential difference is applied across the conductor, they will move from a position of higher electric potential to a position of lower electric potential. Thus, a potential difference maintains current in a circuit.

Figure 13
Batteries maintain electric current by converting chemical energy into electrical energy.

Batteries and generators supply energy to charge carriers

Batteries maintain a potential difference across their terminals by converting *chemical* energy to electrical potential energy. **Figure 13** shows students measuring the potential difference of a battery created using a lemon, copper, and tin.

As charge carriers move from higher to lower electrical potential energy, this energy is converted into kinetic energy. This motion allows collisions to occur between the moving charges and the remaining material in the circuit elements. These collisions transfer energy (in the form of heat) back to the circuit.

A battery stores energy in the form of chemical energy, and its energy is released through a chemical reaction that occurs inside the battery. The battery continues to supply electrical energy to the charge carriers until its chemical energy is depleted. At this point, the battery must be replaced or recharged.

Because batteries must often be replaced or recharged, generators are sometimes preferable. Generators convert *mechanical* energy into electrical energy. For example, a hydroelectric power plant converts the kinetic energy of falling water into electrical potential energy. Generators are the source of the current to a wall outlet in your home and supply the electrical energy to operate your appliances. When you plug an appliance into an outlet, an effective potential difference of 120 V is applied to the device.

extension
Integrating Chemistry
Visit go.hrw.com for the activity "Rechargeable Ni-Cd Batteries."
Keyword HF6ELCX

Current can be direct or alternating

There are two different types of current: *direct current* (dc) and *alternating current* (ac). In direct current, charges move in only one direction with negative charges moving from a lower to higher electric potential. Hence, the conventional current is directed from the positive terminal to the negative terminal of a battery. Note, however, that the electrons actually move in the opposite direction.

Consider a light bulb connected to a battery. The potential difference between the terminals of a battery is fixed, so batteries always generate a direct current.

In alternating current, the terminals of the source of potential difference are constantly changing sign. Hence, there is no net motion of the charge carriers in alternating current; they simply vibrate back and forth. If this vibration were slow enough, you would notice flickering in lights and similar effects in other appliances. To eliminate this problem, alternating current is made to change direction rapidly. In the United States, alternating current oscillates 60 times every second. Thus, its frequency is 60 Hz. The graphs in **Figure 14** compare direct and alternating current. Alternating current has advantages that make it more practical for use in transferring electrical energy. For this reason, the current supplied to your home by power companies is alternating current rather than direct current.

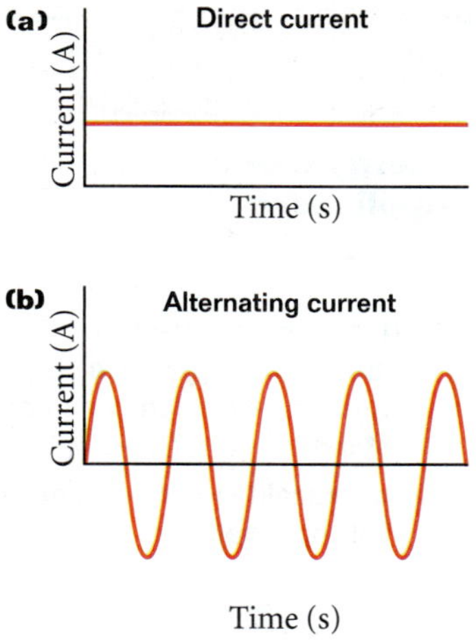

Figure 14
(a) The direction of direct current does not change, while **(b)** the direction of alternating current continually changes.

ENERGY TRANSFER

When a battery is used to maintain an electric current in a conductor, chemical energy stored in the battery is continuously converted to the electrical energy of the charge carriers. As the charge carriers move through the conductor, this electrical energy is converted to internal energy due to collisions between the charge carriers and other particles in the conductor.

For example, consider a light bulb connected to a battery, as shown in **Figure 15(a).** Imagine a charge Q moving from the battery's terminal to the light bulb and then back to the other terminal. The changes in electrical potential energy are shown in **Figure 15(b).** If we disregard the resistance of the connecting wire, no loss in energy occurs as the charge moves through the wire (A to B). But when the charge moves through the filament of the light bulb (B to C), which has a higher resistance than the wire has, it loses electrical potential energy due to collisions. This electrical energy is converted into internal energy, and the filament warms up and glows.

When the charge first returns to the battery's terminal (D), its potential energy is, by convention, zero, and the battery must do work on the charge. As the charge moves between the terminals of the battery (D to A), its electrical potential energy increases by $Q\Delta V$ (where ΔV is the potential difference across the two terminals). The battery's chemical energy must decrease by the same amount.

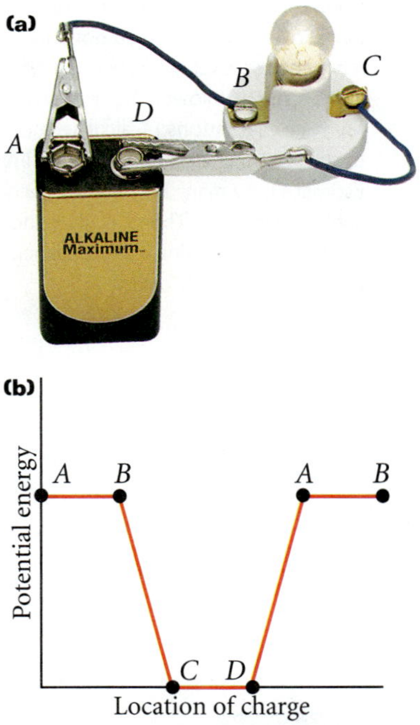

Figure 15
A charge leaves the battery at *A* with a certain amount of electrical potential energy. The charge loses this energy while moving from *B* to *C*, and then regains the energy as it moves through the battery from *D* to *A*.

Teaching Tip — ADVANCED

The decision to supply homes and businesses with alternating current rather than direct current came after a long debate between George Westinghouse's company (Westinghouse Electric Company), which supplied alternating current, and Thomas Edison's company (Edison Electric Light Company), which supplied direct current. Have students research this debate and the advantages and disadvantages of transporting each type of current.

Visual Strategy GENERAL

Figure 15

Use an analogy with mechanical energy to help students understand the changes in electrical potential energy.

Q Compare the electrical energy changes shown in **Figure 15** to the mechanical energy changes involved in carrying tennis balls up a flight of stairs and then dropping the balls back down to the ground.

A ***The tennis balls are analogous to charges, and the bulb uses energy just as dropping the balls does. Charges moving through the battery are analogous to the balls being carried up the stairs; the battery increases electrical potential energy just as raising the balls increases gravitational potential energy. The movement of charge through the wire is analogous to the balls being carried horizontally because the potential energies do not change in either case.***

Quick Lab

TEACHER'S NOTES

The energy a lamp uses depends on the type of bulb in it. You may want to discuss the advantages and disadvantages of lower-wattage bulbs with students.

To extend this activity, have students compare their family's electrical usage over a given billing period with that of other students' families and with the class average.

Quick Lab

Energy Use in Home Appliances

MATERIALS LIST

- **three small household appliances, such as a toaster, television, lamp, or stereo**
- **household electric-company bill (optional)**

SAFETY

Unplug appliances before examination. Use extreme caution when handling electrical equipment.

Look for a label on the back or bottom of each appliance. Record the power rating, which is given in units of watts (W). Use the billing statement to find the cost of energy per kilowatt-hour. (If you don't have a bill, choose a value between $0.05 and $0.20 per kilowatt-hour to use for your calculations.) Calculate the cost of running each appliance for 1 h. Estimate how many hours a day each appliance is used. Then calculate the monthly cost of using each appliance based on your daily estimate.

extension

Practice Problems

Visit go.hrw.com to find a sample and practice problems on the cost of electrical energy.

Keyword HF6ELCX

Electric power is the rate of conversion of electrical energy

Earlier in the text, power was described as the rate at which work is done. *Electric power,* then, is the rate at which charge carriers do work. Put another way, electric power is the rate at which charge carriers convert electrical potential energy to nonelectrical forms of energy.

$$P = \frac{W}{\Delta t} = \frac{\Delta PE}{\Delta t}$$

Potential difference is the change in potential energy per unit of charge.

$$\Delta V = \frac{\Delta PE}{q}$$

This equation can be rewritten in terms of potential energy.

$$\Delta PE = q\Delta V$$

We can then substitute this expression for potential energy into the equation for power.

$$P = \frac{\Delta PE}{\Delta t} = \frac{q\Delta V}{\Delta t}$$

Because current, I, is defined as the rate of charge movement ($q/\Delta t$), we can express electric power as current multiplied by potential difference.

ELECTRIC POWER

$$P = I\Delta V$$

electric power = current × potential difference

This equation describes the rate at which charge carriers lose electrical potential energy. In other words, power is the rate of conversion of electrical energy. Recall that the SI unit of power is the *watt,* W. In terms of the dissipation of electrical energy, 1 W is equivalent to 1 J of electrical energy being converted to other forms of energy per second.

Most light bulbs are labeled with their power ratings. The amount of heat and light given off by a bulb is related to the power rating, also known as *wattage,* of the bulb.

Because $\Delta V = IR$ for ohmic resistors, we can express the power dissipated by a resistor in the following alternative forms:

$$P = I\Delta V = I(IR) = I^2R$$

$$P = I\Delta V = \left(\frac{\Delta V}{R}\right)\Delta V = \frac{(\Delta V)^2}{R}$$

The conversion of electrical energy to internal energy in a resistant material is called *joule heating,* also often referred to as an I^2R loss.

SAMPLE PROBLEM E

Electric Power

PROBLEM

An electric space heater is connected across a 120 V outlet. The heater dissipates 1320 W of power in the form of electromagnetic radiation and heat. Calculate the resistance of the heater.

SOLUTION

Given: $\Delta V = 120\ \text{V}$ $\quad P = 1320\ \text{W}$

Unknown: $R = ?$

Because power and potential difference are given but resistance is unknown, use the form of the power equation that relates power to the other two variables.

$$P = \frac{(\Delta V)^2}{R}$$

Rearrange the equation to solve for resistance.

$$R = \frac{(\Delta V)^2}{P} = \frac{(120\ \text{V})^2}{1320\ \text{W}} = \frac{(120)^2\ \text{J}^2/\text{C}^2}{1320\ \text{J/s}}$$

$$R = \frac{(120)^2\ \text{J/C}}{1320\ \text{C/s}} = 10.9\ \text{V/A}$$

$$\boxed{R = 10.9\ \Omega}$$

PRACTICE E

Electric Power

1. A 1050 W electric toaster operates on a household circuit of 120 V. What is the resistance of the wire that makes up the heating element of the toaster?
2. A small electronic device is rated at 0.25 W when connected to 120 V. What is the resistance of this device?
3. A calculator is rated at 0.10 W and has an internal resistance of 22 Ω. What battery potential difference is required for this device?
4. An electric heater is operated by applying a potential difference of 50.0 V across a wire of total resistance 8.00 Ω. Find the current in the wire and the power rating of the heater.
5. What would the current in the heater in Problem 4 be if the wire developed a short and the resistance was reduced to 0.100 Ω?

PROBLEM GUIDE E

Use this guide to assign problems.
SE = Student Edition Textbook
PW = Problem Workbook
PB = Problem Bank on the One-Stop Planner (OSP)

Solving for:

R	**SE** Sample, 1–3; Ch. Rvw. 55–56 **PW** 5 **PB** 4–6
I	**SE** 4; Ch. Rvw. 56, 72 **PW** Sample, 1–3 **PB** 7–8
ΔV	**PW** 4 **PB** 9–10
P	**SE** Ch. Rvw. 73 **PW** 6 **PB** Sample, 1–3

***Challenging Problem**
Consult the printed Solutions Manual or the OSP for detailed solutions.

ANSWERS

Practice E

1. 14 Ω
2. $5.8 \times 10^4\ \Omega$
3. 1.5 V
4. 6.25 A; 312 W
5. 5.00×10^2 A

Teaching Tip — BASIC

To give students practice converting from one unit to another, list a variety of values for energy on the chalkboard. Include some in joules and others in kilowatt-hours. Ask students to convert those in kilowatt-hours to joules, and vice versa.

Why it Matters

Household Appliance Power Usage

The power ratings for appliances must be displayed on an Underwriters Laboratory label somewhere on the appliance. Hair dryers generally have a power rating of 1200 to 1875 W. A clothes dryer has a much higher power rating and may consume up to 5000 W.

Extension — GENERAL

Ask students to research the power consumption of small household appliances, such as hair dryers, coffee makers, toasters, and space heaters, in their homes. Ask students to prepare a report estimating how much electrical energy they use during the course of a week.

ADVANCED TOPICS

See "Electron Tunneling" in **Appendix J: Advanced Topics** to learn how electrons behave in the theory of quantum mechanics.

Electric companies measure energy consumed in kilowatt-hours

Electric power, as discussed previously, is the rate of energy transfer. Power companies charge for energy, not power. However, the unit of energy used by electric companies to calculate consumption, the *kilowatt-hour,* is defined in terms of power. One kilowatt-hour (kW•h) is the energy delivered in 1 h at the constant rate of 1 kW. The following equation shows the relationship between the kilowatt-hour and the SI unit of energy, the joule:

$$1\ \text{kW}\bullet\text{h} \times \frac{10^3\ \text{W}}{1\ \text{kW}} \times \frac{60\ \text{min}}{1\ \text{h}} \times \frac{60\ \text{s}}{1\ \text{min}} = 3.6 \times 10^6\ \text{W}\bullet\text{s} = 3.6 \times 10^6\ \text{J}$$

On an electric bill, the electrical energy used in a given period is usually stated in multiples of kilowatt-hours. An electric meter, such as the one outside your home, is used by the electric company to determine how much energy is consumed over some period of time. So, the electric company does not charge for the amount of power delivered but instead charges for the amount of energy used.

Why it Matters

Household Appliance Power Usage

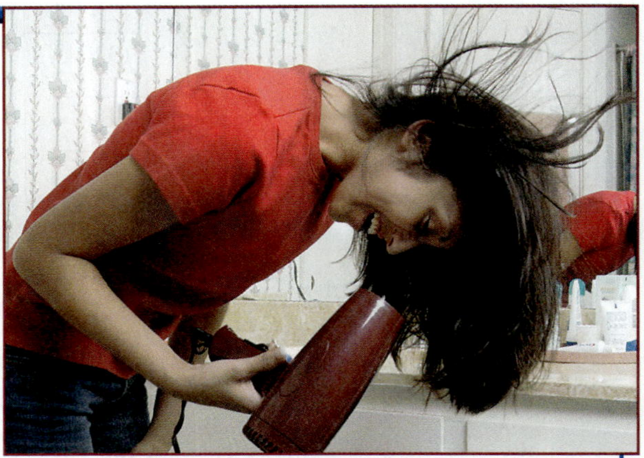

Hair dryers contain a resistive coil that becomes hot when there is an electric current in the coil.

The electrical energy supplied by power companies is used to generate electric currents. These currents are used to operate household appliances. When the charge carriers that make up an electric current encounter resistance, some of the electrical energy is converted to internal energy by collisions and the conductor warms up. This effect is used in many appliances, such as hair dryers, electric heaters, electric clothes dryers, steam irons, and toasters.

Hair dryers contain a long, thin heating coil that becomes very hot when there is an electric current in the coil. This coil is commonly made of an alloy of the two metals nickel and chromium. This nickel chromium alloy conducts electricity poorly.

In a hair dryer, a fan behind the heating coil blows air through the hot coils. The air is then heated and blown out of the hair dryer. The same principle is also used in clothes dryers and electric heaters.

In a steam iron, a heating coil warms the bottom of the iron and also turns water into steam. An electric toaster has heating elements around the edges and in the center. When bread is loaded into the toaster, the heating coils turn on and a timer controls how long the elements remain on before the bread is popped out of the toaster.

Appliances that use resistive heater coils consume a relatively large amount of electric energy. This energy consumption occurs because a large amount of current is required to heat the coils to a useful level. Because power is proportional to the current squared times the resistance, energy consumption is high.

Electrical energy is transferred at high potential differences to minimize energy loss

When transporting electrical energy by power lines, such as those shown in **Figure 16,** power companies want to minimize the I^2R loss and maximize the energy delivered to a consumer. This can be done by decreasing either current or resistance. Although wires have little resistance, recall that resistance is proportional to length. Hence, resistance becomes a factor when power is transported over long distances. Even though power lines are designed to minimize resistance, some energy will be lost due to the length of the power lines.

As expressed by the equation $P = I^2R$, energy loss is proportional to the *square* of the current in the wire. For this reason, decreasing current is even more important than decreasing resistance. Because $P = I\Delta V$, the same amount of power can be transported either at high currents and low potential differences or at low currents and high potential differences. Thus, transferring electrical energy at low currents, thereby minimizing the I^2R loss, requires that electrical energy be transported at very high potential differences. Power plants transport electrical energy at potential differences of up to 765 000 V. Locally, this potential difference is reduced by a transformer to about 4000 V. At your home, this potential difference is reduced again to about 120 V by another transformer.

Figure 16
Power companies transfer electrical energy at high potential differences in order to minimize the I^2R loss.

SECTION REVIEW

1. What does the power rating on a light bulb describe?
2. If the resistance of a light bulb is increased, how will the electrical energy used by the light bulb over the same time period change?
3. The potential difference across a resting neuron in the human body is about 70 mV, and the current in it is approximately 200 μA. How much power does the neuron release?
4. How much does it cost to watch an entire World Series (21 h) on a 90.0 W black-and-white television set? Assume that electrical energy costs $0.070/kW•h.
5. Explain why it is more efficient to transport electrical energy at high potential differences and low currents rather than at low potential differences and high currents.

Classroom Practice

Cost of Electrical Energy
A person replaces a 100.0 W bulb with a 75.0 W bulb. The light is used about 3.0 h per day. Approximately how much money can the person expect to save on the electric bill each month (30 days) if the cost of electrical energy is $0.10/kW•h?

Answer
$0.22 per month

SECTION REVIEW ANSWERS

1. the rate at which electrical energy is delivered to the bulb
2. It will decrease.
3. 1×10^{-5} W
4. $0.13
5. A low current minimizes the I^2R loss, making power transfer more efficient. Because $P = I\Delta V$, for a given amount of power, a low current corresponds to a high potential difference.

Electrician

According to David Ellison, becoming a master electrician can take as much time and require as much learning as going to graduate school. A beginner, known as an *apprentice,* receives a license and works for four to five years. The next level, in which the individual is called a *journeyman,* can last for two to three years. After that, according to Ellison, "you are qualified to take a master test," but "a lot of people don't want the master. A lead journeyman does earn some good money."

PHYSICS CAREERS

Electrician

Electricity enables us to see at night, to cook, to have heat and hot water, to communicate, to be entertained, and to do many other things. Without electricity, our lives would be unimaginably different. To learn more about being an electrician, read the interview with master electrician David Ellison.

David Ellison teaches electrician skills to students at a local community college.

How did you become an electrician?

I went to junior college to learn electronics—everything from TVs and radios to radio towers and television stations. But I didn't particularly like that sort of work. While working in a furniture factory, I got to know the master electrician for the factory, and I began working with him. Eventually he got me a job with a master electrician in town.

Most of my experience has been on the job—very little schooling. But back then, there wasn't a lot of schooling. Now they have some good classes.

What about electrical work made it more interesting than other fields?

I enjoy working with something you can't see or smell—but if you do touch it, it'll let you know. And if you flip a light switch, there it is. I also enjoy wiring up the switches and safeties, and solving problems when they don't work.

Where do you currently work?

I have been self-employed since 1989. About three years ago, I was invited to teach at the community college. I enjoy it. My students seem to relate better to the fact that I'm still working in the field. When I explain something to them, I can talk from recent experience. Teaching helps me stay on top of the field, too.

Are there any drawbacks to your work?

Electricity is dangerous. I've been burned twice over 30 percent of my body. Also, the hours can be bad. Owning my own business, I go from 6 in the morning until 9 or 10 at night. I am on call at the local hospital—I was there on Thanksgiving day. But that's the nature of my relationship with my customers.

What advice do you have for a student who is interested in becoming an electrician?

If you know a local electrical contractor, go talk or visit for the day. Or take a class at the local community college to see if it interests you. Some companies have their own classes, usually one night a week. Going to school gives you some technical knowledge, but getting out and doing it is still the best way to learn.

Highlights

KEY IDEAS

Section 1 Electric Potential

- Electrical potential energy is energy that a charged object has because of its shape and its position in an electric field.
- Electric potential is electrical potential energy divided by charge.
- Only differences in electric potential (potential differences) from one position to another are useful in calculations.

Section 2 Capacitance

- The capacitance, C, of an object is the magnitude of the charge, Q, on each of a capacitor's plates divided by the potential difference, ΔV, between the plates.
- A capacitor is a device that is used to store electrical potential energy. The potential energy stored in a charged capacitor depends on the charge and the potential difference between the capacitor's two plates.

Section 3 Current and Resistance

- Current is the rate of charge movement.
- Resistance equals potential difference divided by current.
- Resistance depends on length, cross-sectional area, temperature, and material.

Section 4 Electric Power

- In direct current, charges move in a single direction; in alternating current, the direction of charge movement continually alternates.
- Electric power is the rate of conversion of electrical energy.
- The power dissipated by a resistor equals current squared times resistance.
- Electric companies measure energy consumed in kilowatt-hours.

Variable Symbols

Quantities		Units		Conversions
$PE_{electric}$	electrical potential energy	J	joule	$= \text{N}\bullet\text{m} = \text{kg}\bullet\text{m}^2/\text{s}^2$
ΔV	potential difference	V	volt	$= \text{J/C}$
C	capacitance	F	farad	$= \text{C/V}$
I	current	A	ampere	$= \text{C/s}$
R	resistance	Ω	ohm	$= \text{V/A}$
P	electric power	W	watt	$= \text{J/s}$

KEY TERMS

electrical potential energy (p. 594)

electric potential (p. 596)

potential difference (p. 596)

capacitance (p. 602)

electric current (p. 608)

drift velocity (p. 611)

resistance (p. 612)

PROBLEM SOLVING

See **Appendix D: Equations** for a summary of the equations introduced in this chapter. If you need more problem-solving practice, see **Appendix I: Additional Problems.**

Diagram Symbols

Electric field	E (arrow)
Current	I (arrow)
Positive charge	+
Negative charge	−

Highlights

Teaching Tip

Written explanations help to solidify students' understanding of difficult concepts and to enforce good communication skills. Have students write essays in which they summarize the differences between electrical potential energy, electric potential, and potential difference, and also have them summarize how these quantities are related to one another. Essays should also include a thorough discussion of the factors that affect capacitance. Be sure students explain concepts clearly and correctly, and use good sentence structure.

extension

In-Depth Physics Content

Your students can visit go.hrw.com for online chapters that integrate more in-depth development of the concepts covered here.

Keyword HF6ELCX

Review

ANSWERS

1. The positive charge moves in the direction of the electric field. As the field does work on the charge, it converts the electrical potential energy associated with the charge to kinetic energy.
2. b
3. Electric potential is the electrical potential energy associated with a charge divided by the magnitude of the charge. Thus, electric potential is a characteristic of a point in space and is independent of the charge at that point.
4. Potential difference is the change in electric potential between two points in space.
5. at infinity
6. no; The point of zero electric potential can be chosen to be anywhere.
7. decrease; decrease
8. 2.6×10^4 V
9. -4.2×10^5 V
10. The charge on each plate doubles.
11. Use a dielectric between the capacitor plates.
12. The Earth is large enough that it can accept or supply an unlimited number of charges without its electric potential changing significantly; Any object with this ability can act as a ground.
13. $PE_{electric}$ is 4 times as great.
14. yes; The plates' capacitance depends on the area of the plates and the distance between them.

CHAPTER 17 Review

ELECTRICAL POTENTIAL ENERGY AND POTENTIAL DIFFERENCE

Review Questions

1. Describe the motion and explain the energy conversions that are involved when a positive charge is placed in a uniform electric field. Be sure your discussion includes the following terms: *electrical potential energy, work,* and *kinetic energy.*

2. If a point charge is displaced perpendicular to a uniform electric field, which of the following expressions is likely to be equal to the change in electrical potential energy?
   a. $-qEd$
   b. 0
   c. $-k_C\left(\frac{q^2}{r^2}\right)$

3. Differentiate between electrical potential energy and electric potential.

4. Differentiate between electric potential and potential difference.

5. At what location in relationship to a point charge is the electric potential considered by convention to be zero?

Conceptual Questions

6. If the electric field in some region is zero, must the electric potential in that same region also be zero? Explain your answer.

7. If a proton is released from rest in a uniform electric field, does the corresponding electric potential at the proton's changing locations increase or decrease? What about the electrical potential energy?

Practice Problems

For problems 8–9, see Sample Problem A.

8. The magnitude of a uniform electric field between two plates is about 1.7×10^6 N/C. If the distance between these plates is 1.5 cm, find the potential difference between the plates.

9. In the figure below, find the electric potential at point *P* due to the grouping of charges at the other corners of the rectangle.

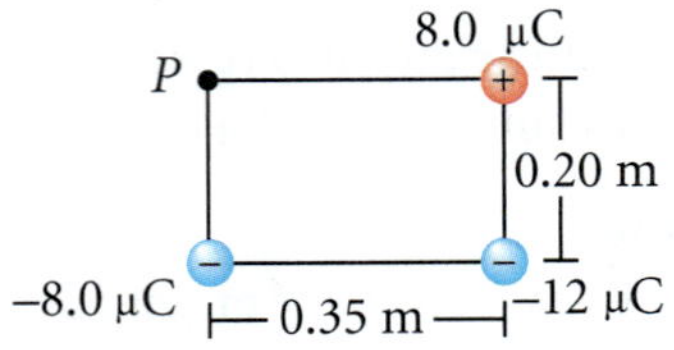

CAPACITANCE

Review Questions

10. What happens to the charge on a parallel-plate capacitor if the potential difference doubles?

11. You want to increase the maximum potential difference of a parallel-plate capacitor. Describe how you can do this for a fixed plate separation.

12. Why is the Earth considered a "ground" in electric terms? Can any other object act as a ground?

Conceptual Questions

13. If the potential difference across a capacitor is doubled, by what factor is the electrical potential energy stored in the capacitor multiplied?

14. Two parallel plates are uncharged. Does the set of plates have a capacitance? Explain.

15. If you were asked to design a small capacitor with high capacitance, what factors would be important in your design?

16. A parallel-plate capacitor is charged and then disconnected from a battery. How much does the stored energy change when the plate separation is doubled?

17. Why is it dangerous to touch the terminals of a high-voltage capacitor even after the potential difference has been removed? What can be done to make the capacitor safe to handle?

Practice Problems

For problems 18–19, see Sample Problem B.

18. A 12.0 V battery is connected to a 6.0 pF parallel-plate capacitor. What is the charge on each plate?

19. Two devices with capacitances of 25 μF and 5.0 μF are each charged with separate 120 V power supplies. Calculate the total energy stored in the two capacitors.

ELECTRIC CURRENT

Review Questions

20. What is electric current? What is the SI unit for electric current?

21. In a metal conductor, current is the result of moving electrons. Can charge carriers ever be positive?

22. What is meant by the term *conventional current*?

23. What is the difference between the drift speed of an electron in a metal wire and the average speed of the electron between collisions with the atoms of the metal wire?

24. There is a current in a metal wire due to the motion of electrons. Sketch a possible path for the motion of a single electron in this wire, the direction of the electric field vector, and the direction of conventional current.

25. What is an electrolyte?

26. What is the direction of conventional current in each case shown below?

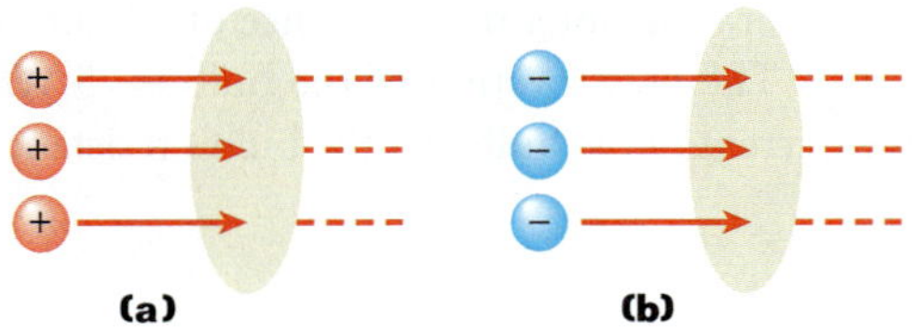

Conceptual Questions

27. In an analogy between traffic flow and electric current, what would correspond to the charge, Q? What would correspond to the current, I?

28. Is current ever "used up"? Explain your answer.

29. Why do wires usually warm up when an electric current is in them?

30. When a light bulb is connected to a battery, charges begin moving almost immediately, although each electron travels very slowly across the wire. Explain why the bulb lights up so quickly.

31. What is the net drift velocity of an electron in a wire that has alternating current in it?

Practice Problems

For problems 32–33, see Sample Problem C.

32. How long does it take a total charge of 10.0 C to pass through a cross-sectional area of a copper wire that carries a current of 5.0 A?

33. A hair dryer draws a current of 9.1 A.

a. How long does it take for 1.9×10^3 C of charge to pass through the hair dryer?

b. How many electrons does this amount of charge represent?

RESISTANCE

Review Questions

34 What factors affect the resistance of a conductor?

35. Each of the wires shown below is made of copper. Assuming each piece of wire is at the same temperature, which has the greatest resistance? Which has the least resistance?

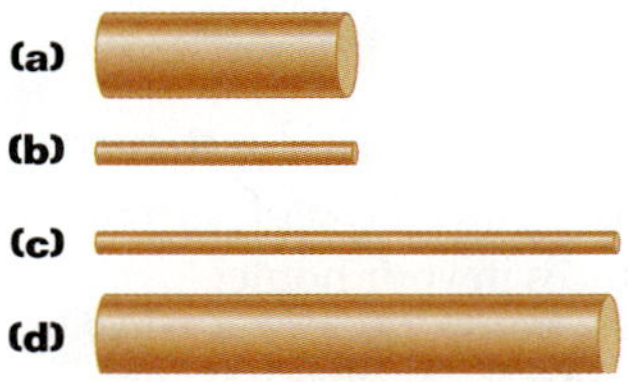

36. Why are resistors used in circuit boards?

15. The plates should be close together, have a large surface area, and have a dielectric between them.

16. $PE_{electric}$ is doubled when d is doubled ($PE_{electric} = \frac{1Q^2}{2C} = \frac{1}{2}Q^2\frac{d}{\varepsilon_0 A}$).

17. The capacitor stores electrical potential energy that could be discharged through the body, a conductor. This could be prevented by first discharging the capacitor so that the potential across it is zero.

18. $\pm 7.2 \times 10^{-11}$ C

19. 0.22 J

20. rate of charge movement; ampere

21. yes

22. the current made of positive charge that has the same effect as the actual current

23. $v_{avg} >> v_{drift}$

24. The electron changes direction but slowly moves opposite **E** and conventional current.

25. a solution that conducts an electric current

26. **a.** to the right
b. to the left

27. cars; the rate at which cars pass a given point

28. no; Electrical energy, not current, is "used up."

29. because of collisions between charge carriers and atoms

30. Electrons in the wire move when the electric field reaches them.

31. zero

32. 2.0 s

33. **a.** 3.5 min
b. 1.2×10^{22} electrons

34. length, cross-sectional area, temperature, material

35. c (greatest); a (least)

36. to regulate current

17 REVIEW

37. They are proportional.
38. They are inversely proportional.
39. At a higher temperature, atoms vibrate with greater amplitudes, which makes it more difficult for electrons to move through the material.
40. 0.20 A
41. 3.4 A
42. a. 1.8 A
    b. 4.5 A
    c. 0.45 A
43. $PE_{electric}$ of charges is converted to KE of charges in the device.
44. Batteries convert chemical energy, while generators convert mechanical energy. Batteries produce dc, while generators can produce dc or ac.
45. In dc, charges move in one direction, and in ac, charges oscillate; ac
46. Electric power is the rate at which electrical energy is converted. Mechanical power is the rate at which mechanical energy is converted. Electrical energy is a form of mechanical energy.
47. electrical energy; power
48. because resistance increases as length increases
49. 3.6×10^6 J
50. Batteries supply energy to free electrons in the conductor rather than supplying the charge.
51. the 75 W bulb
52. the conductor with the greater resistance
53. 2.0×10^{16} J
54. The wires have much less resistance than the filament of the light bulb.
55. 93 Ω
56. 0.62 A; 190 Ω

Conceptual Questions

37. For a constant resistance, how are potential difference and current related?

38. If the potential difference across a conductor is constant, how is current dependent on resistance?

39. Using the atomic theory of matter, explain why the resistance of a material should increase as its temperature increases.

Practice Problems

For problems 40–42, see Sample Problem D.

40. A nichrome wire with a resistance of 15 Ω is connected across the terminals of a 3.0 V flashlight battery. How much current is in the wire?

41. How much current is drawn by a television with a resistance of 35 Ω that is connected across a potential difference of 120 V?

42. Calculate the current that each resistor shown below would draw when connected to a 9.0 V battery.

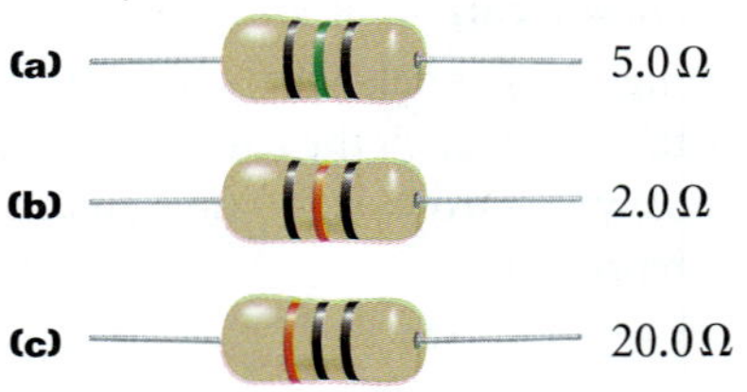

ELECTRIC POWER

Review Questions

43. Why must energy be continuously pumped into a circuit by a battery or a generator to maintain an electric current?

44. Name at least two differences between batteries and generators.

45. What is the difference between direct current and alternating current? Which type of current is supplied to the appliances in your home?

46. Compare and contrast mechanical power with electric power.

47. What quantity is measured in kilowatt-hours? What quantity is measured in kilowatts?

48. If electrical energy is transmitted over long distances, the resistance of the wires becomes significant. Why?

49. How many joules are in a kilowatt-hour?

Conceptual Questions

50. A student in your class claims that batteries work by supplying the charges that move in a conductor, generating a current. What is wrong with this reasoning?

51. A 60 W light bulb and a 75 W light bulb operate from 120 V. Which bulb has a greater current in it?

52. Two conductors of the same length and radius are connected across the same potential difference. One conductor has twice as much resistance as the other. Which conductor dissipates more power?

53. It is estimated that in the United States (population 250 million) there is one electric clock per person, with each clock using energy at a rate of 2.5 W. Using this estimate, how much energy is consumed by all of the electric clocks in the United States in a year?

54. When a small lamp is connected to a battery, the filament becomes hot enough to emit electromagnetic radiation in the form of visible light, while the wires do not. What does this tell you about their relative resistances of the filament and the wires?

Practice Problems

For problems 55–56, see Sample Problem E.

55. A computer is connected across a 110 V power supply. The computer dissipates 130 W of power in the form of electromagnetic radiation and heat. Calculate the resistance of the computer.

56. The operating potential difference of a light bulb is 120 V. The power rating of the bulb is 75 W. Find the current in the bulb and the bulb's resistance.

MIXED REVIEW

57. At some distance from a point charge, the electric potential is 600.0 V and the magnitude of the electric field is 200.0 N/C. Determine the distance from the charge and the charge.

58. A circular parallel-plate capacitor with a spacing of 3.0 mm is charged to produce a uniform electric field with a strength of 3.0×10^6 N/C. What plate radius is required if the stored charge is -1.0 μC?

59. A 12 V battery is connected across two parallel metal plates separated by 0.30 cm. Find the magnitude of the electric field.

60. A parallel-plate capacitor has an area of 5.00 cm^2, and the plates are separated by 1.00 mm. The capacitor stores a charge of 400.0 pC.

a. What is the potential difference across the plates of the capacitor?

b. What is the magnitude of the uniform electric field in the region that is located between the plates?

61. A proton is accelerated from rest through a potential difference of 25 700 V.

a. What is the kinetic energy of this proton in joules after this acceleration?

b. What is the speed of the proton after this acceleration?

62. A proton is accelerated from rest through a potential difference of 120 V. Calculate the final speed of this proton.

63. A pair of oppositely charged parallel plates are separated by 5.33 mm. A potential difference of 600.0 V exists between the plates.

a. What is the magnitude of the electric field strength in the region that is located between the plates?

b. What is the magnitude of the force on an electron that is in the region between the plates at a point that is exactly 2.90 mm from the positive plate?

c. The electron is moved to the negative plate from an initial position 2.90 mm from the positive plate. What is the change in electrical potential energy due to the movement of this electron?

64. The three charges shown at right are located at the vertices of an isosceles triangle. Calculate the electric potential at the midpoint of the base if each one of the charges at the corners has a magnitude of 5.0×10^{-9} C.

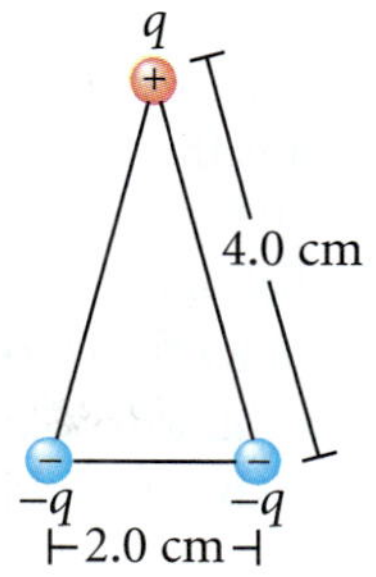

65. A charge of -3.00×10^{-9} C is at the origin of a coordinate system, and a charge of 8.00×10^{-9} C is on the x-axis at 2.00 m. At what two locations on the x-axis is the electric potential zero? (Hint: One location is between the charges, and the other is to the left of the y-axis.)

66. An ion is displaced through a potential difference of 60.0 V and experiences an increase of electrical potential energy of 1.92×10^{-17} J. Calculate the charge on the ion.

67. A proton is accelerated through a potential difference of 4.5×10^6 V.

a. How much kinetic energy has the proton acquired?

b. If the proton started at rest, how fast is it moving?

68. Each plate on a 3750 pF capacitor carries a charge with a magnitude of 1.75×10^{-8} C.

a. What is the potential difference across the plates when the capacitor has been fully charged?

b. If the plates are 6.50×10^{-4} m apart, what is the magnitude of the electric field between the two plates?

69. A net charge of 45 mC passes through the cross-sectional area of a wire in 15 s.

a. What is the current in the wire?

b. How many electrons pass the cross-sectional area in 1.0 min?

70. The current in a lightning bolt is 2.0×10^5 A. How many coulombs of charge pass through a cross-sectional area of the lightning bolt in 0.50 s?

17 REVIEW

57. 3.000 m; 2.00×10^{-7} C
58. 0.11 m
59. 4.0×10^3 V/m
60. **a.** 90.4 V
b. 9.04×10^4 V/m
61. **a.** 4.11×10^{-15} J
b. 2.22×10^6 m/s
62. 1.5×10^5 m/s
63. **a.** 1.13×10^5 V/m
b. 1.81×10^{-14} N
c. 4.39×10^{-17} J
64. −7800 V
65. 0.545 m, −1.20 m
66. 3.20×10^{-19} C
67. **a.** 7.2×10^{-13} J
b. 2.9×10^7 m/s
68. **a.** 4.67 V
b. 7180 V/m
69. **a.** 3.0×10^{-3} A
b. 1.1×10^{18} electrons
70. 1.0×10^5 C

71. a. 32 V
b. 0.16 V
72. 2.7 A
73. 1.0×10^5 W
74. 61.1 A
75. 3.2×10^5 J
76. 1.1×10^3 s (18 min)
77. 13.5 h

71. A person notices a mild shock if the current along a path through the thumb and index finger exceeds 80.0 μA. Determine the maximum allowable potential difference without shock across the thumb and index finger for the following:

a. a dry-skin resistance of $4.0 \times 10^5\ \Omega$
b. a wet-skin resistance of $2.0 \times 10^3\ \Omega$

72. A color television has a power rating of 325 W. How much current does this set draw from a potential difference of 120 V?

73. An X-ray tube used for cancer therapy operates at 4.0 MV with a beam current of 25 mA striking a metal target. Calculate the power of this beam.

74. The mass of a gold atom is 3.27×10^{-25} kg. If 1.25 kg of gold is deposited on the negative electrode of an electrolytic cell in a period of 2.78 h, what is the current in the cell in this period? Assume that each gold ion carries one elementary unit of positive charge.

75. The power supplied to a typical black-and-white television is 90.0 W when the set is connected across a potential difference of 120 V. How much electrical energy does this set consume in 1.0 h?

76. A color television set draws about 2.5 A of current when connected to a potential difference of 120 V. How much time is required for it to consume the same energy that the black-and-white model described in item 75 consumes in 1.0 h?

77. The headlights on a car are rated at 80.0 W. If they are connected to a fully charged 90.0 A•h, 12.0 V battery, how long does it take the battery to completely discharge?

extension

Graphing Calculator Practice

Visit go.hrw.com for answers to this Graphing Calculator activity.

Keyword HF6ELCXT

Graphing Calculator *Practice*

Resistance and Current

When you install a 100 W light bulb, what is the resistance of and current passing through this light bulb? The answer to this question and similar questions is found in two equations that you learned earlier in this chapter:

$$P = \frac{(\Delta V)^2}{R} \text{ and } P = I\Delta V$$

These equations describe the power dissipated by a resistor. In these equations, P is the power in watts, ΔV is the potential difference in volts, R is the resistance in ohms, and I is the current in amperes.

In this graphing calculator activity, you will calculate a series of tables of resistance and current versus potential difference for various values of dissipated power. By analyzing these tables, you will better understand the relationships between power, potential difference, resistance, and current. (You will also be able to answer the question about the 100 W light bulb.)

Visit go.hrw.com and type in the keyword **HF6ELCX** to find this graphing calculator activity. Refer to **Appendix B** for instructions on downloading the program for this activity.

78. The current in a conductor varies over time as shown in the graph below.

a. How many coulombs of charge pass through a cross section of the conductor in the time interval $t = 0$ to $t = 5.0$ s?

b. What constant current would transport the same total charge during the 5.0 s interval as does the actual current?

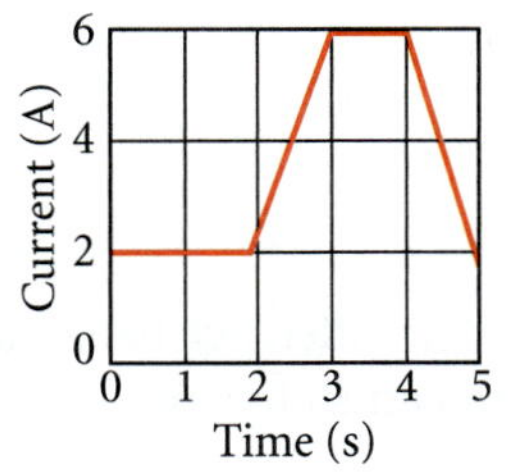

79. Birds resting on high-voltage power lines are a common sight. A certain copper power line carries a current of 50.0 A, and its resistance per unit length is 1.12×10^{-5} Ω/m. If a bird is standing on this line with its feet 4.0 cm apart, what is the potential difference across the bird's feet?

80. An electric car is designed to run on a bank of batteries with a total potential difference of 12 V and a total energy storage of 2.0×10^7 J.

a. If the electric motor draws 8.0 kW, what is the current delivered to the motor?

b. If the electric motor draws 8.0 kW as the car moves at a steady speed of 20.0 m/s, how far will the car travel before it is "out of juice"?

Alternative Assessment

1. Imagine that you are assisting nuclear scientists who need to accelerate electrons between electrically charged plates. Design and sketch a piece of equipment that could accelerate electrons to 10^7 m/s. What should the potential difference be between the plates? How would protons move inside this device? What would you change in order to accelerate the electrons to 100 m/s?

2. Tantalum is an element widely used in electrolytic capacitors. Research tantulum and its properties. Where on Earth is it found? In what form is it found? How expensive is it? Present your findings to the class in the form of a report, poster, or computer presentation.

3. Research an operational maglev train, such as the commercially operating train in Shangai, China, or the demonstration trains in Japan or Germany. Alternatively, research a maglev system that is under construction or being proposed for development. Investigate the cost of development, major hurdles that had to be overcome or will need to be overcome, and the advantages and disadvantages of the train. Suppose that there is a proposal for a maglev train in your area. Develop an argument for or against the proposed train, based on your research. Write a paper to convince other citizens of your position.

4. Visit an electric parts or electronic parts store or consult a print or on-line catalog to learn about different kinds of resistors. Find out what the different resistors look like, what they are made of, what their resistance is, how they are labeled, and what they are used for. Summarize your findings in a poster or a brochure entitled *A Consumer's Guide to Resistors.*

5. The units of measurement you learned about in this chapter were named after four famous scientists: Andre-Marie Ampere, Michael Faraday, Georg Simon Ohm, and Alessandro Volta. Research their lives, works, discoveries, and contributions. Create a presentation about one of these scientists. The presentation can be in the form of a report, poster, short video, or computer presentation.

6. A *thermistor* is a device that changes its resistance as its temperature changes. Thermistors are often used in digital thermometers. Another common temperature sensor is the *thermocouple,* which generates a potential difference that depends on its temperature. Many thermostats use thermistors or thermocouples to regulate temperature. Research how thermistors or thermocouples work, and how they are used in one of the applications mentioned above. Create a slideshow or a poster with the results of your research.

17 REVIEW

78. a. 18 C
b. 3.6 A

79. 2.2×10^{-5} V

80. a. 670 A
b. 5.0×10^4 m

Alternative Assessment ANSWERS

1. Student answers should include the need for potential differences of 300 V and 3×10^{-8} V to accelerate electrons to each speed.
2. Student answers will vary. Be sure students include sources. Thin films of Ta_2O_5 are used as dielectrics in some capacitors. Powder price for tantalum ranges from $5/g to $10/g.
3. Advantages include less maintenance due to the lack of friction and the ability to achieve very high speeds. The main disadvantage is high construction costs.
4. Presentations should describe various types and sizes of resistors. Students may include the color codes for resistors.
5. Ampere (1775–1836) founded the field of electrodynamics. Faraday (1791–1867) discovered electromagnetic induction and was the first to use lines of force to describe fields. Ohm (1787–1854) discovered the relationship between resistance and the length and cross section of a wire. Volta (1745–1827) invented the first battery, the "Voltaic pile."
6. The resistance of a thermistor is used to determine temperature. In a thermocouple, a junction between two different metals produces a temperature-dependent potential difference.

Standardized Test Prep

ANSWERS

1. A

2. J

3. A

4. G

5. D

6. G

7. D

Standardized Test Prep

MULTIPLE CHOICE

Use the diagram below to answer questions 1–2.

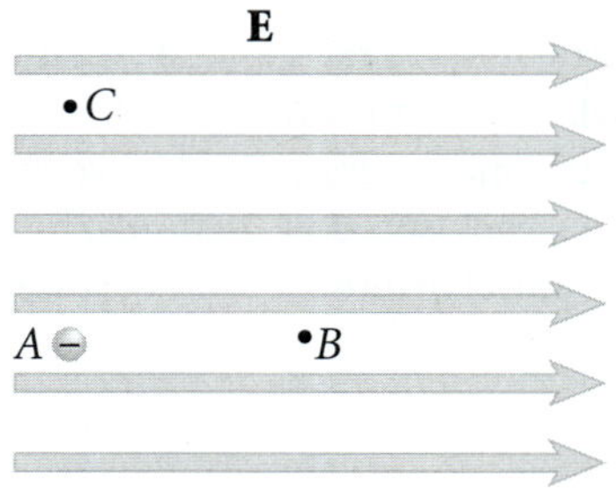

1. What changes would take place if the electron moved from point A to point B in the uniform electric field?

A. The electron's electrical potential energy would increase; its electric potential would increase.
B. The electron's electrical potential energy would increase; its electric potential would decrease.
C. The electron's electrical potential energy would decrease; its electric potential would decrease.
D. Neither the electron's electrical potential energy nor its electric potential would change.

2. What changes would take place if the electron moved from point A to point C in the uniform electric field?

F. The electron's electrical potential energy would increase; its electric potential would increase.
G. The electron's electrical potential energy would increase; its electric potential would decrease.
H. The electron's electrical potential energy would decrease; its electric potential would decrease.
J. Neither the electron's electrical potential energy nor its electric potential would change.

Use the following passage to answer questions 3–4.

A proton ($q = 1.6 \times 10^{-19}$ C) moves 2.0×10^{-6} m in the direction of an electric field that has a magnitude of 2.0 N/C.

3. What is the change in the electrical potential energy associated with the proton?

A. -6.4×10^{-25} J
B. -4.0×10^{-6} V
C. $+6.4 \times 10^{-25}$ J
D. $+4.0 \times 10^{-6}$ V

4. What is the potential difference between the proton's starting point and ending point?

F. -6.4×10^{-25} J
G. -4.0×10^{-6} V
H. $+6.4 \times 10^{-25}$ J
J. $+4.0 \times 10^{-6}$ V

5. If the negative terminal of a 12 V battery is grounded, what is the potential of the positive terminal?

A. -12 V
B. $+0$ V
C. $+6$ V
D. $+12$ V

6. If the area of the plates of a parallel-plate capacitor is doubled while the spacing between the plates is halved, how is the capacitance affected?

F. C is doubled
G. C is increased by four times
H. C is decreased by 1/4
J. C does not change

Use the following passage to answer questions 7–8.

A potential difference of 10.0 V exists across the plates of a capacitor when the charge on each plate is 40.0 μC.

7. What is the capacitance of the capacitor?

A. 2.00×10^{-4} F
B. 4.00×10^{-4} F
C. 2.00×10^{-6} F
D. 4.00×10^{-6} F

8. How much electrical potential energy is stored in the capacitor?

F. 2.00×10^{-4} J
G. 4.00×10^{-4} J
H. 2.00×10^{-6} J
J. 4.00×10^{-6} J

9. How long does it take 5.0 C of charge to pass through a given cross section of a copper wire if $I = 5.0$ A?

A. 0.20 s
B. 1.0 s
C. 5.0 s
D. 25 s

10. A potential difference of 12 V produces a current of 0.40 A in a piece of copper wire. What is the resistance of the wire?

F. 4.8 Ω
G. 12 Ω
H. 30 Ω
J. 36 Ω

11. How many joules of energy are dissipated by a 50.0 W light bulb in 2.00 s?

A. 25.0 J
B. 50.0 J
C. 100 J
D. 200 J

12. How much power is needed to operate a radio that draws 7.0 A of current when a potential difference of 115 V is applied across it?

F. 6.1×10^{-2} W
G. 2.3×10^{0} W
H. 1.6×10^{1} W
J. 8.0×10^{2} W

SHORT RESPONSE

13. Electrons are moving from left to right in a wire. No other charged particles are moving in the wire. In what direction is the conventional current?

14. What is drift velocity, and how does it compare with the speed at which an electric field travels through a wire?

15. List four factors that can affect the resistance of a wire.

EXTENDED RESPONSE

16. A parallel-plate capacitor is made of two circular plates, each of which has a diameter of 2.50×10^{-3} m. The plates of the capacitor are separated by a space of 1.40×10^{-4} m.

a. Assuming that the capacitor is operating in a vacuum and that the permittivity of a vacuum ($\varepsilon_0 = 8.85 \times 10^{-12}$ $C^2/N\bullet m^2$) can be used, determine the capacitance of the capacitor.

b. How much charge will be stored on each plate of the capacitor when the capacitor's plates are connected across a potential difference of 0.12 V?

c. What is the electrical potential energy stored in the capacitor when fully charged by the potential difference of 0.12 V?

d. What is the potential difference between a point midway between the plates and a point that is 1.10×10^{-4} m from one of the plates?

e. If the potential difference of 0.12 V is removed from the circuit and the circuit is allowed to discharge until the charge on the plates has decreased to 70.7 percent of its fully charged value, what will the potential difference across the capacitor be?

Test TIP If at any point while taking a test you do not clearly understand the directions or the wording of a question, raise your hand and ask for help.

8. F
9. B
10. H
11. C
12. J
13. right to left
14. Drift velocity is the net velocity of a charge carrier moving in an electric field. Drift velocities in a wire are typically much smaller than the speeds at which changes in the electric field propagate through the wire.
15. length, cross-sectional area (thickness), temperature, and material
16. **a.** 3.10×10^{-13} F
b. 3.7×10^{-14} C
c. 2.2×10^{-15} J
d. 3.4×10^{-2} V
e. 8.5×10^{-2} V
(See the Solutions Manual or One-Stop Planner for a full solution.)

Lab Planning

Beginning on page T34 are preparation notes and teaching tips to assist you in planning.

Blank data tables (as well as some sample data) appear on the **One-Stop Planner.**

No Books in the Lab?

See the ***Datasheets for In-Text Labs*** workbook for a reproducible master copy of this experiment.

CBL™ Option

A **CBL™** version of this lab appears in the ***CBL™ Experiments*** workbook.

Safety Caution

Emphasize the dangers of working with electricity. For the safety of the students and of the equipment, remind students to have you check their circuits before turning on the power supply or closing the switch.

Tips and Tricks

- See the preparation notes for instructions on setting up multimeters to measure potential difference and current.
- Make sure students understand how to wire current meters (in series) and voltage meters (in parallel) in a circuit.
- Show students how to wire the circuit to include a resistance coil and how to move the meters to the next coil.
- Demonstrate how to follow the circuit with your finger to make sure all the connections are correct.

CHAPTER 17

Skills Practice Lab Current and Resistance

OBJECTIVES

- **Determine** the resistance of conductors, using the definition of resistance.
- **Explore** the relationships between length, diameter, material, and the resistance of a conductor.

MATERIALS LIST

- **2 multimeters or 1 dc ammeter and 1 voltmeter**
- **insulated connecting wire**
- **momentary contact switch**
- **mounted resistance coils**
- **power supply**

In this experiment, you will study the effects of length, cross-sectional area, and material on the resistance of conductors. You will use a set of mounted resistance coils, which will provide wire coils of different lengths, diameters, and metals. You will measure the potential difference across the resistance coil, and you will find the current in the conductor. Then you will use these values to calculate the resistance of each resistance coil using the definition of resistance.

SAFETY

- **Never close a circuit until it has been approved by your teacher. Never rewire or adjust any element of a closed circuit. Never work with electricity near water; be sure the floor and all work surfaces are dry.**
- **If the pointer on any kind of meter moves off scale, open the circuit immediately by opening the switch.**
- **Do not attempt this exercise with any batteries or electrical devices other than those provided by your teacher for this purpose.**
- **Use a hot mitt to handle resistors, light sources, and other equipment that may be hot. Allow all equipment to cool before storing it.**

PROCEDURE

Preparation

1. Read the entire lab, and plan what steps you will take.
2. If you are not using a datasheet provided by your teacher, prepare a data table in your lab notebook with seven columns and six rows. Label the first through seventh columns *Trial, Metal, Gauge Number, Length (cm), Cross-sectional Area* (cm^2), ΔV_x *(V)*, and *I (A)*. In the first column, label the second through sixth rows *1, 2, 3, 4,* and *5.*

Current at Varied Resistances

3. Set up the apparatus as shown in **Figure 1.** Construct a circuit that includes a power supply, a switch, a current meter, a voltmeter, and the mounted resistance coils. ***Do not turn on the power supply. Do not close the switch until your teacher has approved your circuit.***

4. With the switch open, connect the current meter in a straight line in series with the mounted resistance coils. Make sure the black lead on the meter is connected to the black pin on the power supply. Connect the black lead on the voltmeter to the side of the first resistance coil that is connected to the black pin on the power supply, and connect the red lead to the other side of the coil in parallel. ***Do not close the switch until your teacher approves your circuit.***

5. When your teacher has approved your circuit, make sure the power supply dial is turned completely counterclockwise. Turn on the power supply, and slowly turn the dial clockwise. Periodically close the switch briefly and read the current value on the current meter. Adjust the dial until the current is approximately 0.15 A.

6. Close the switch. Quickly record the current in and the potential difference across the resistance coil in your data table. Open the switch immediately. Turn off the power supply by turning the dial completely counterclockwise. Your teacher will tell you the length and cross-sectional area of the wire on the coil. Record these values in your data table.

7. Repeat steps 3–6 with different coils until five coils have been studied.

8. Clean up your work area. Put equipment away safely.

ANALYSIS

1. **Organizing Data** Use the measurements for current and potential difference to calculate the resistance, R_C, for each resistance coil you tested. Use the definition of resistance, $R = \frac{\Delta V}{I}$.

CONCLUSIONS

2. **Drawing Conclusions** Rate the coils from lowest to highest resistance. Record your ratings.
   a. According to your results for this experiment, how does the length of the wire affect the resistance of the coil?
   b. According to your results for this experiment, how does the cross-sectional area affect the resistance of the coil?

3. **Drawing Conclusions** Based on your results for the metals used in this experiment, which metal has the greatest resistance? Which metal has the least resistance? Explain how you arrived at these conclusions.

Figure 1

Step 3: The set of mounted resistance coils shown includes five different resistance coils. In this lab, you will measure the current and potential difference for each coil in turn.

Step 4: Use your finger to trace the circuit from the black pin on the power supply through the circuit to the red pin on the power supply to check for proper connections.

Step 6: Close the switch only long enough to take readings. Open the switch as soon as you have taken the readings.

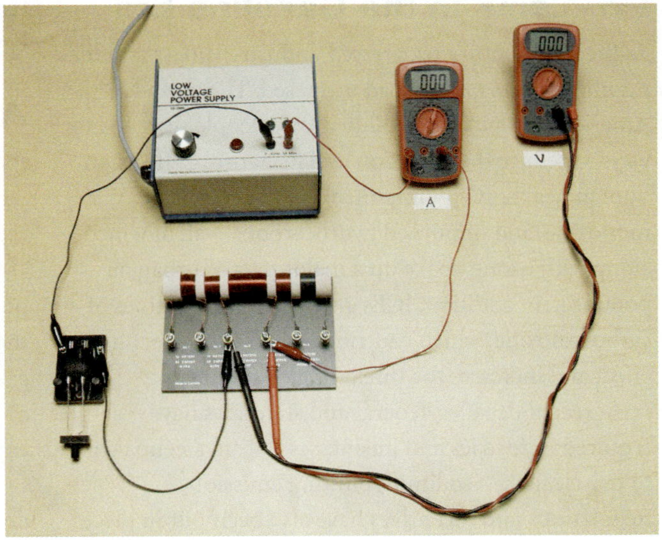

✓ Checkpoints

Step 4: Check each circuit to make sure all connections are made properly and power supplies are set at appropriate levels. Students should be able to demonstrate that they can trace the circuit to check the connections.

Step 7: For each trial, make sure students connect all circuits properly. At the beginning of each trial, the knob on the power supply should be turned completely counterclockwise.

ANSWERS

Analysis

1. Student answers will vary. Make sure students use the relationship $R_C = \Delta V_C / I$. For sample data, values for the potential difference range from 0.08 V to 2.42 V. Values for R_C range from 0.53 Ω to 16 Ω.

Conclusions

2. a. Longer wires provide more resistance.
   b. Thinner wires provide more resistance.

3. For resistance coils of Cu and Cu/Ni, Cu/Ni provides much greater resistance than Cu. Students should compare resistors of the same length and cross-sectional area.

Science • Technology • Society

Hybrid Electric Vehicles

At the start of the 20th century, electric-powered vehicles and gasoline-powered vehicles were competing for dominance in the emerging automobile industry. Electric cars were considered more reliable, and certainly quieter and less polluting, than gasoline-powered cars. However, they could go only a few miles before they needed recharging, so they were only suitable for use over short distances. A few inventors developed hybrid cars that used both electricity and gasoline engines for power, but these never caught on. As more roads were paved and as more people wanted to travel farther, electric cars were abandoned in favor of cars that burned gasoline in *internal combustion engines* (ICEs).

Problems in the Gasoline Era

As the 20th century progressed, industry spread and the number of cars on the road increased. The air in North America became more polluted, and people searched for ways to reduce the pollution and its harmful effects on human health. ICEs emit nitrogen oxides, carbon monoxide, and unburned hydrocarbons—all of which, along with ozone, make up a major part of urban air pollution. In addition, ICEs give off large quantities of carbon dioxide, which contributes to Earth's greenhouse effect and increases the threat of global warming.

In recent decades, federal and state laws have required industries and businesses—from steelmakers to dry cleaners—to limit polluting emissions. Regulations and incentives have also been put in place to increase the fuel-efficiency and reduce the emissions of passenger cars. Although overall air quality has improved as a result of these efforts, air pollution is still a serious problem, largely due to emissions from vehicles with ICEs.

In the 1970s, a global energy crisis emerged as several oil-exporting countries cut off their oil exports for political and economic reasons. Oil and gas prices rose dramatically, and many people suddenly had no access to gasoline or could no longer afford it. Although the crisis subsided, worldwide economic and political instability, and a growing awareness that global oil supplies are finite, has kept oil and gas prices uncertain ever since. The United States currently imports more than half of the oil it uses. As a result, access to oil resources plays a key role in U.S. foreign policy.

A Return to Electric Cars?

As the problems with air pollution and rising oil prices have become more apparent, people have started to reexamine alternatives to gasoline-powered ICEs. In the 1990s, several *electric vehicles* (EVs), which run solely on electricity, were developed for passenger use. While the performance of these EVs was comparable to gasoline-powered cars, they typically had driving ranges of only 80–240 km (50–150 miles), and were more expensive than gasoline-powered models.

In the mid to late 1990's, several automakers designed and developed *hybrid electric vehicles* (HEVs),

636

which use electricity in combination with a gasoline engine. HEVs have been more commercially successful than pure EVs. Today, several HEV models are available to consumers, and more are appearing on the road every day.

Advantages of Hybrid Electric Vehicles

All HEVs combine the power of a battery-driven electric motor with the power of an ICE. However, different models do this in different ways. In a *series design,* an electric motor powers the car directly, while the ICE serves only to power a generator that recharges the battery for the electric motor. With this design, the ICE is very efficient because it is always recharging the battery. However, series-design HEVs have less power-on-demand for acceleration.

In a *parallel design,* both the electric motor and the ICE attach directly to the drive train to power the wheels. With this design, the electric motor provides the primary power when driving in stop-and-go traffic, while the ICE kicks in at higher speeds, when the ICE is more efficient. Unlike conventional gasoline-powered vehicles, parallel-design HEVs get better gas mileage and produce fewer emissions in town than they do on the highway.

Both series and parallel HEVs have longer driving ranges than their pure EV counterparts. Some HEVs can travel 700 miles before they need refueling. Furthermore, because the ICE charges the battery, an HEV never needs to be plugged in. In addition to maximizing the efficiency of the electric motor and the engine, many HEVs also have *regenerative braking* systems that recapture some of the power lost during braking and use it to recharge the battery. The result is a more efficient car that produces fewer emissions and gets better gas mileage than a comparable car powered solely by gasoline. HEVs are currently more expensive than conventional gasoline-powered cars, but the cost savings over time due to increased fuel economy may be enough to compensate for the greater initial price.

HEVs are a step toward solving the problems with air pollution and the politics of oil. As research continues, HEVs will likely become even more efficient. New technologies, such as hydrogen fuel cells, may replace the ICE component in HEVs, resulting in an even better car for the future.

Researching the Issue

1. Go to a local car dealer and ask about hybrid electric vehicles. Do they have any HEV models available? Are they going to offer any new HEV models in the future? Do these models use a series design, a parallel design, or another type of design?

2. Unlike pure electric vehicles, HEVs still burn gasoline in combustion engines. Do you think the increase in fuel economy and the reduced emissions of HEVs go far enough to address the problems at hand? What alternative or additional solutions can you recommend?

3. The federal government and some states offer tax deductions and other incentives for people who own HEVs or other alternative-fuel vehicles. Hold a discussion or debate on the question, "Should the government spend taxpayers' money to subsidize the purchase of alternative-fuel vehicles that people might otherwise not buy?"

4. Research hydrogen fuel cell technology. How do fuel cells work? Do they produce any harmful emissions? In what form is the hydrogen stored? What are some possible sources of hydrogen fuel? What problems must be solved before hydrogen fuel cells are ready for widespread use?

SCILINKS.
www.scilinks.org
Topic: Electric Vehicles
Code: HF60474

637

CHAPTER 18

Circuits and Circuit Elements

Planning Guide

Compression Guide

To shorten instruction because of time limitations, omit the opener and Section 3 and abbreviate the review.

OBJECTIVES	LABS, DEMONSTRATIONS, AND ACTIVITIES	TECHNOLOGY RESOURCES
PACING • 45 min pp. 638–639 **Chapter Opener**	ANC **Discovery Lab** Exploring Circuit Elements*◆ BASIC	CD **Visual Concepts**, Chapter 18 BASIC
PACING • 45 min pp. 640–646 **Section 1 Schematic Diagrams and Circuits** • Interpret and construct circuit diagrams. • Identify circuits as open or closed. • Deduce the potential difference across the circuit load, given the potential difference across the battery's terminals.	SE **Quick Lab** Simple Circuits, p. 644 GENERAL	OSP **Lesson Plans** EXT **Integrating Technology** Incandescent Light Bulbs BASIC TR 94 Schematic Diagram Symbols TR 95 Light Bulb
PACING • 135 min pp. 647–656 **Section 2 Resistors in Series or in Parallel** • Calculate the equivalent resistance for a circuit of resistors in series, and find the current in and potential difference across each resistor in the circuit. • Calculate the equivalent resistance for a circuit of resistors in parallel, and find the current in and potential difference across each resistor in the circuit.	TE **Demonstration** Resistors in Series, p. 647 BASIC SE **Quick Lab** Series and Parallel Circuits, p. 652 GENERAL TE **Demonstration** Resistors in Parallel, p. 652 BASIC SE **Inquiry Lab** Resistors in Series and in Parallel, pp. 674–675◆ GENERAL ANC **Datasheet** Inquiry Lab, Resistors in Series and in Parallel* GENERAL ANC **Datasheet** Skills Practice Lab, Resistors in Series and in Parallel* GENERAL ANC **CBL™ Experiment** Resistors in Series and in Parallel*◆ GENERAL ANC **CBL™ Experiment** Series and Parallel Circuits *◆ ADVANCED	OSP **Lesson Plans** TR 60A Resistors in Series and Parallel TR 61A Series and Parallel Decorative Lights
PACING • 45 min pp. 657–664 *Advanced Level* **Section 3 Complex Resistor Combinations** • Calculate the equivalent resistance for a complex circuit involving both series and parallel portions. • Calculate the current in and potential difference across individual elements within a complex circuit.	ANC **Invention Lab** Designing a Dimmer Switch*◆ ADVANCED	OSP **Lesson Plans** EXT **Integrating Health** Recording Electricity in the Brain GENERAL TR 96 Finding Equivalent Resistance TR 97 Components of a Decorative Light Bulb

PACING • 90 min

CHAPTER REVIEW, ASSESSMENT, AND STANDARDIZED TEST PREPARATION

- SE **Chapter Highlights**, p. 665
- SE **Chapter Review**, pp. 666–671
- SE **Graphing Calculator Practice**, p. 670 GENERAL
- SE **Alternative Assessment**, p. 671 ADVANCED
- SE **Standardized Test Prep**, pp. 672–673 GENERAL
- SE **Appendix D: Equations**, p. 863
- SE **Appendix I: Additional Problems**, pp. 893–894
- ANC **Study Guide Worksheet** Mixed Review* GENERAL
- ANC **Chapter Test A*** GENERAL
- ANC **Chapter Test B*** ADVANCED
- OSP **Test Generator**

Online and Technology Resources

go.hrw.com Holt Online Learning

Visit go.hrw.com to find a variety of online resources. To access this chapter's extensions, enter the keyword **HF6CIRXT** and click the "go" button. Click **Holt Online Learning** for an online edition of this textbook, and other interactive resources.

HOLT Teacher's One-Stop Planner®

This DVD package includes:
- Holt Calendar Planner
- Customizable Lesson Plans
- Editable Worksheets
- ExamView® Version 6 Assessment Suite
- Interactive Teacher's Edition
- Holt PuzzlePro®
- Holt PowerNotes® Presentations
- MindPoint® Quiz Show

KEY		
SE Student Edition	**OSP** One-Stop Planner	**EXT** Online Extension
TE Teacher Edition	**CD** CD or CD-ROM	***** Also on One-Stop Planner
ANC Ancillary Worksheet	**TR** Teaching Transparencies	**◆** Requires advance prep

SKILLS DEVELOPMENT RESOURCES	REVIEW AND ASSESSMENT	CORRELATIONS
		National Science Education Standards
SE Conceptual Challenge, p. 642 GENERAL	**SE Section Review,** p. 645 GENERAL **ANC Study Guide Worksheet** Section 1* GENERAL **ANC Quiz** Section 1* BASIC	UCP 1, 2, 3, 4, 5 SAI 1, 2 ST 1, 2 SPSP 4, 5
SE Sample Set A Resistors in Series, pp. 649–650 GENERAL **TE Classroom Practice,** p. 649 GENERAL **ANC Problem Workbook** Sample Set A* GENERAL **OSP Problem Bank** Sample Set A GENERAL **SE Conceptual Challenge,** p. 653 GENERAL **SE Sample Set B** Resistors in Parallel, pp. 654–655 GENERAL **TE Classroom Practice,** p. 654 GENERAL **ANC Problem Workbook** Sample Set B* GENERAL **OSP Problem Bank** Sample Set B GENERAL	**SE Section Review,** p. 656 GENERAL **ANC Study Guide Worksheet** Section 2* GENERAL **ANC Quiz** Section 2* BASIC	UCP 1, 2, 3, 5 SAI 1, 2 ST 1, 2 SPSP 5
SE Sample Set C Equivalent Resistance, pp. 658–659 ADVANCED **TE Classroom Practice,** p. 658 ADVANCED **ANC Problem Workbook** Sample Set C* ADVANCED **OSP Problem Bank** Sample Set C ADVANCED **SE Sample Set D** Current in and Potential Difference Across a Resistor, pp. 660–662 ADVANCED **TE Classroom Practice,** p. 660 ADVANCED **ANC Problem Workbook** Sample Set D* ADVANCED **OSP Problem Bank** Sample Set D ADVANCED	**SE Section Review,** p. 663 GENERAL **ANC Study Guide Worksheet** Section 3* GENERAL **ANC Quiz** Section 3* GENERAL	UCP 1, 2, 3, 5 SAI 1, 2 ST 1 SPSP 5

SCILINKS. NSTA

www.scilinks.org

Maintained by the **National Science Teachers Association.**

Topic: Electric Circuits
SciLinks Code: HF60471

Topic: Resistors
SciLinks Code: HF61302

Classroom CD-ROMs

- Guided Reading Audio Program
- Student One Stop
- Virtual Investigations
- Visual Concepts

Holt Lab Generator CD-ROM

Search for any lab by topic, standard, difficulty level, or time. Edit any lab to fit your needs, or create your own labs. Use the Lab Materials QuickList software to customize your lab materials list.

CHAPTER 18
Overview

Section 1 introduces the concept of an electric circuit, distinguishes between open and closed circuits, and describes the concept of a short circuit.

Section 2 describes the relationships between equivalent resistance, current, and potential difference for series circuits and parallel circuits.

Section 3 explores complicated circuits containing portions in series and portions in parallel.

About the Illustration

The Riverwalk in San Antonio, Texas, is a downtown shopping and entertainment district built on the banks of the San Antonio River. The Riverwalk began as a Works Progress Administration project in the Great Depression of the 1930s. In the 1970s and 1980s, redevelopment and expansion of the Riverwalk sparked an economic revival of downtown San Antonio.

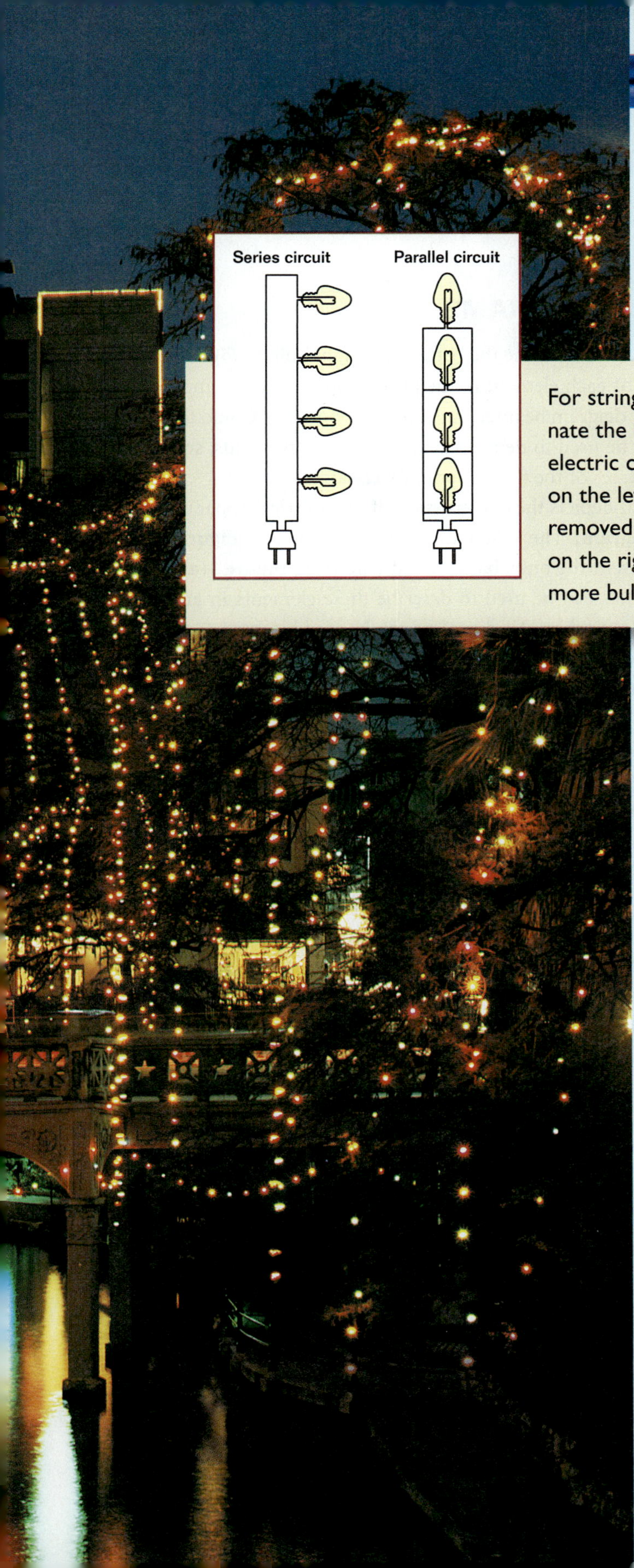

CHAPTER 18

Circuits and Circuit Elements

For strings of decorative lights—such as these that illuminate the Riverwalk in San Antonio, Texas—two types of electric circuits can be used. In a series circuit, illustrated on the left, the entire set goes dark when one bulb is removed from the circuit. In a parallel circuit, illustrated on the right, other bulbs remain lighted even when one or more bulbs are removed.

WHAT TO EXPECT

In this chapter, you will explore the basic properties of series and parallel circuits.

Why it Matters

All electric circuits are wired in series, parallel, or a combination. The type of circuit affects the current and potential difference of elements connected to the circuit, such as decorative light bulbs on strands or appliances in your home.

CHAPTER PREVIEW

Tapping Prior Knowledge

Knowledge to Review

- ✔ Potential difference is the change in electrical potential energy per unit charge from one point to another.
- ✔ Current is the rate at which electric charges move through a given area.
- ✔ $\Delta V = IR$ can be used to relate current and potential difference for specific electrical devices.

Items to Probe

- ✔ Preconceptions about wiring: Ask students to trace the path of charges moving in a string of decorative lights.
- ✔ Familiarity with schematic diagrams: Ask students to draw their own picture representing the wiring of decorative lights.

639

SECTION 1
General Level

Visual Strategy GENERAL

Figure 1
Students should be encouraged to create alternative representations of the circuit shown in **(a).** Students should discuss what their symbols stand for, how convenient their symbols would be for others to use, and in what way each symbol reflects *relevant information.*

Q Identify information about the group of elements that is *not* relevant to its function and is unnecessary in a schematic.

A ***The colors and sizes of the items shown and whether the wires are coiled, bent, or straight are irrelevant to the function of the group of elements.***

Visual Strategy GENERAL

Figure 1
Students should recognize that the straight-line symbols connecting the battery symbol with the bulb symbol in **(b)** represent not only the wire but also all parts of the conducting connection between the bulb and battery.

Q Identify the parts of the photo symbolized by the black straight lines in the diagrams.

A ***The black lines symbolize the conducting path provided by the wires, clips, and socket.***

SECTION 1

Schematic Diagrams and Circuits

SECTION OBJECTIVES

- **Interpret and construct circuit diagrams.**
- **Identify circuits as open or closed.**
- **Deduce the potential difference across the circuit load, given the potential difference across the battery's terminals.**

schematic diagram

a representation of a circuit that uses lines to represent wires and different symbols to represent components

SCHEMATIC DIAGRAMS

Take a few minutes to examine the battery and light bulb in **Figure 1(a);** then draw a diagram of each element in the photograph and its connection. How easily could your diagram be interpreted by someone else? Could the elements in your diagram be used to depict a string of decorative lights, such as those draped over the trees of the San Antonio Riverwalk?

A diagram that depicts the construction of an electrical apparatus is called a **schematic diagram.** The schematic diagram shown in **Figure 1(b)** uses symbols to represent the bulb, battery, and wire from **Figure 1(a).** Note that these same symbols can be used to describe these elements in any electrical apparatus. This way, schematic diagrams can be read by anyone familiar with the standard set of symbols.

Reading schematic diagrams allows us to determine how the parts in an electrical device are arranged. In this chapter, you will see how the arrangement of resistors in an electrical device can affect the current in and potential difference across the other elements in the device. The ability to interpret schematic diagrams for complicated electrical equipment is an essential skill for solving problems involving electricity.

As shown in **Table 1,** each element used in a piece of electrical equipment is represented by a symbol in schematic diagrams that reflects the element's construction or function. For example, the schematic-diagram symbol that represents an open switch resembles the open knife switch that is shown in the corresponding photograph. Note that **Table 1** also includes other forms of schematic-diagram symbols; these alternative symbols will not be used in this book.

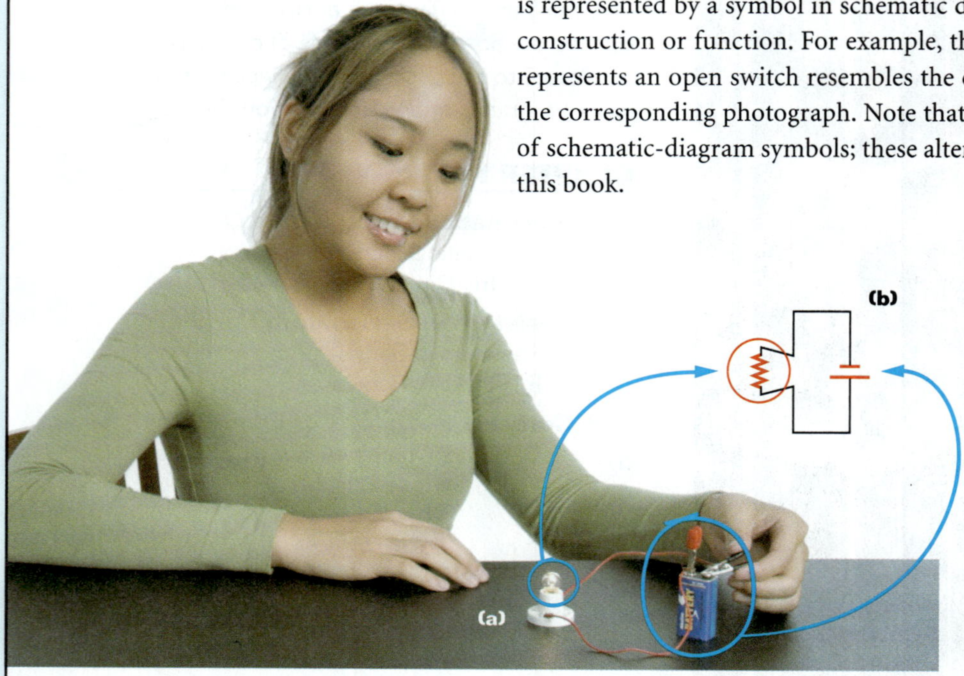

Figure 1
(a) When this battery is connected to a light bulb, the potential difference across the battery generates a current that illuminates the bulb. **(b)** The connections between the light bulb and battery can be represented in a schematic diagram.

Table 1 Schematic-Diagram Symbols

Component	Symbol used in this book	Other forms of this symbol	Explanation
Wire or conductor 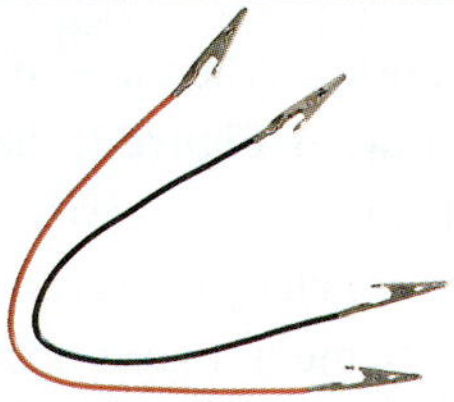			• Wires that connect elements are conductors. • Because wires offer negligible resistance, they are represented by straight lines.
Resistor or circuit load 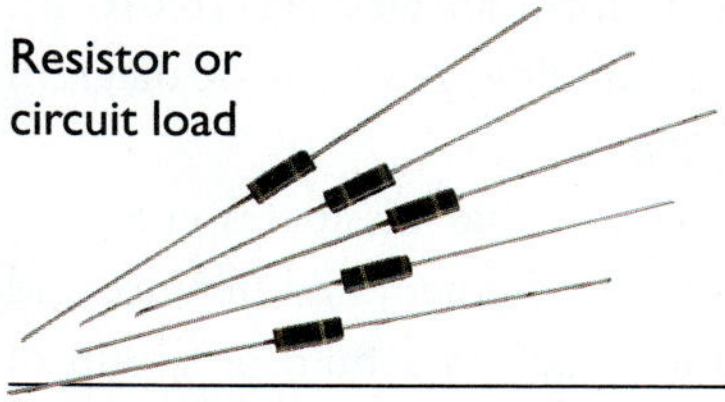			• Resistors are shown having multiple bends, illustrating resistance to the movement of charges.
Bulb or lamp 			• The multiple bends of the filament indicate that the light bulb behaves as a resistor. • The symbol for the filament of the bulb is often enclosed in a circle to emphasize the enclosure of a resistor in a bulb.
Plug 			• The plug symbol looks like a container for two prongs. • The emf between the two prongs of a plug is symbolized by lines of unequal length.
Battery 	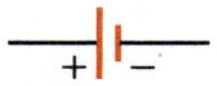	 Multiple cells	• Differences in line length indicate a potential difference between positive and negative terminals of the battery. • The longer line represents the positive terminal of the battery.
Switch  Open  Closed	 Open  Closed		• The small circles indicate the two places where the switch makes contact with the wires. Most switches work by breaking only one of the contacts, not both.
Capacitor  	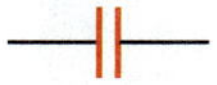		• The two parallel plates of a capacitor are symbolized by two parallel lines of equal length. • One curved line indicates that the capacitor can be used with only direct current sources with the polarity as shown.

Visual Strategy ADVANCED

Table 1

Be sure students recognize that the different symbols represent devices with different functions.

Q Challenge students to identify which devices have the following functions: storing energy, transforming energy, and conducting current.

A ***batteries and capacitors store energy; resistors, bulbs, and batteries transform energy; wires, resistors, bulbs, plugs, closed switches, and batteries conduct current***

The Language of Physics

Although **Table 1** contains several schematic-diagram symbols, several stylistic variations exist. For example, some other symbols for light bulbs are shown below.

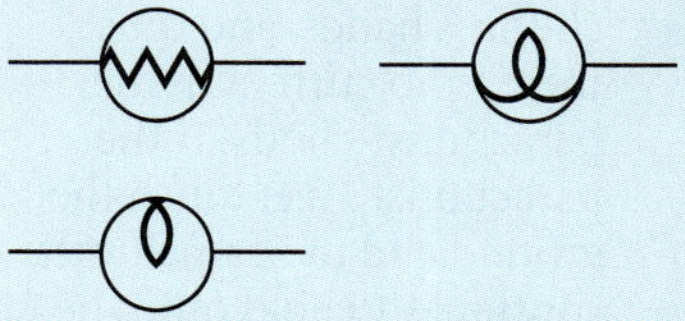

Because light bulbs behave as resistors for small changes in voltage, the symbols for resistors are often used for light bulbs.

Key Models and Analogies

Many teachers use a fluid model of electric current. In this model, charges moving due to potential difference are analogous to water moving to a level of lower gravitational potential energy. Wires are analogous to horizontal pipes, and resistors are analogous to water wheels, which transform the energy to another form. Batteries and generators act like pumps in that they lift water up, increasing its potential energy.

ANSWERS

Conceptual Challenge

1. Because there is no potential difference between the bird's feet, there is no current in the bird's body.
2. At first there is no potential difference between the parachutist's hands, and thus there is no current in the parachutist's body. If the parachutist's feet touch the ground and the parachutist continues to hold onto the wire, however, there will be current in the parachutist's body because of the potential difference between the wire in the parachutist's hands and the ground.

ELECTRIC CIRCUITS

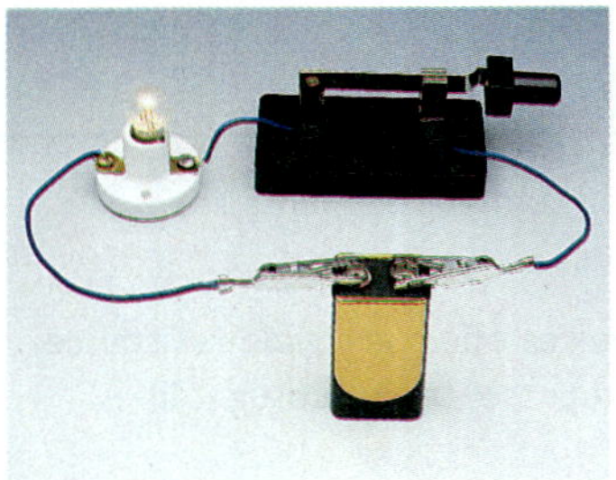

Figure 2
When all electrical components are connected, charges can move freely in a circuit. The movement of charges in a circuit can be halted by opening the switch.

electric circuit

a set of electrical components connected such that they provide one or more complete paths for the movement of charges

Think about how you get the bulb in **Figure 2** to light up. Will the bulb stay lit if the switch is opened? Is there any way to light the bulb without connecting the wires to the battery?

The filament of the light bulb acts as a resistor. When a wire connects the terminals of the battery to the light bulb, as shown in **Figure 2,** charges built up on one terminal of the battery have a path to follow to reach the opposite charges on the other terminal. Because there are charges moving through the wire, a current exists. This current causes the filament to heat up and glow.

Together, the bulb, battery, switch, and wire form an **electric circuit.** An electric circuit is a path through which charges can flow. A schematic diagram for a circuit is sometimes called a *circuit diagram.*

Any element or group of elements in a circuit that dissipates energy is called a *load.* A simple circuit consists of a source of potential difference and electrical energy, such as a battery, and a load, such as a bulb or group of bulbs. Because the connecting wire and switch have negligible resistance, we will not consider these elements as part of the load.

In **Figure 2,** the path from one battery terminal to the other is complete, a potential difference exists, and electrons move from one terminal to the other. In other words, there is a closed-loop path for electrons to follow. This is called a *closed circuit.* The switch in the circuit in **Figure 2** must be closed in order for a steady current to exist.

Without a complete path, there is no charge flow and therefore no current. This situation is an *open circuit.* If the switch in **Figure 2** were open, as shown in **Table 1,** the circuit would be open, the current would be zero, and the bulb would not light up.

Why it Matters

Conceptual Challenge

1. Bird on a Wire Why is it possible for a bird to be perched on a high-voltage wire without being electrocuted? (Hint: Consider the potential difference between the bird's two feet.)

2. Parachutist on a Wire Suppose a parachutist lands on a high-voltage wire and grabs the wire in preparation to be rescued. Will the parachutist be electrocuted? If the wire breaks, why should the parachutist let go of the wire as it falls to the ground? (Hint: First consider the potential difference between the parachutist's two hands holding the wire. Then consider the potential difference between the wire and the ground.)

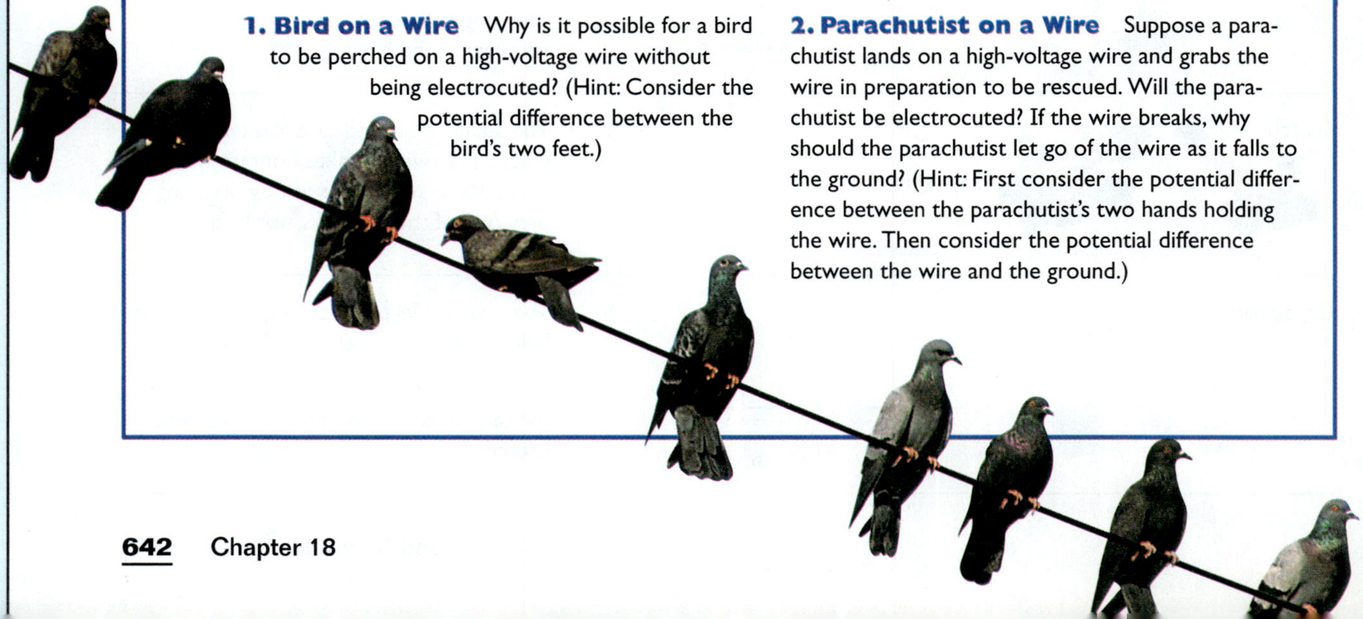

Why it Matters

Light Bulbs

How does a light bulb contain a complete conducting path? When you look at a clear light bulb, you can see the twisted filament inside that provides a portion of the conducting path for the circuit. However, the bulb screws into a single socket; it seems to have only a single contact, the rounded part at the bulb's base.

Closer examination of the socket reveals that it has two contacts inside. One contact, in the bottom of the socket, is connected to the wire going to one side of the filament. The other contact is in the side of the socket, and it is connected to the wire going to the other side of the filament.

The placement of the contacts within the socket indicates how the bulb completes the circuit, as shown on the right. Within the bulb, one side of the filament is connected with wires to the contact at the light bulb's base, **(a).** The other side of the filament is connected to the side of the metal base, **(c).** Insulating material between the side of the base and the contact on the bottom prevents the wires from being connected to each other with a conducting material. In this way, charges have only one path to follow when passing through a light bulb—through the filament, **(b).**

When a light bulb is screwed in, the contact on one side of the socket touches the threads on the side of the bulb's base. The contact on the bottom of the socket touches the contact on the bottom of the bulb's base. Charges then enter through the bulb's base, move through the bulb to the filament, and exit the bulb through the threads. For most light bulbs, the bulb will glow regardless of which direction the charges move. Thus, the positive terminal of a battery can be connected to either the base of the bulb or the threads of the bulb, as long as the negative terminal is connected to the threads or base, respectively. All that matters is that there is a complete conducting path for the charges to move through the circuit.

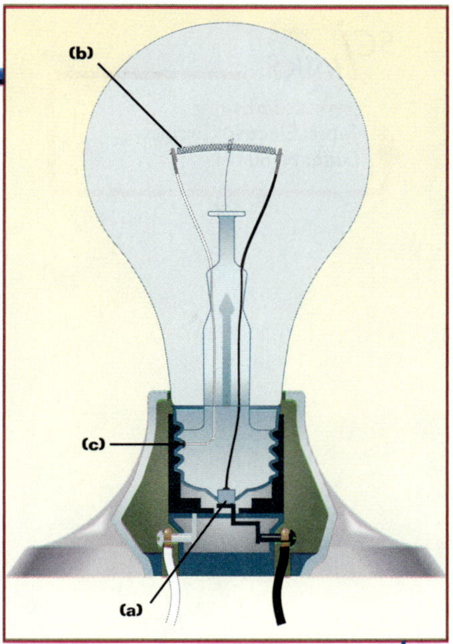

Light bulbs contain a complete conducting path. When a light bulb is screwed in, charges can enter through the base **(a),** move along the wire to the filament **(b),** and exit the bulb through the threads **(c).**

Short circuits can be hazardous

Without a load, such as a bulb or other resistor, the circuit contains little resistance to the movement of charges. This situation is called a *short circuit.* For example, a short circuit occurs when a wire is connected from one terminal of a battery to the other by a wire with little resistance. This commonly occurs when uninsulated wires connected to different terminals come into contact with each other.

When short circuits occur in the wiring of your home, the increase in current can become unsafe. Most wires cannot withstand the increased current, and they begin to overheat. The wire's insulation may even melt or cause a fire.

extension

Integrating Technology

Visit go.hrw.com for the activity "Incandescent Light Bulbs."

Keyword HF6CIRX

Why it Matters

Light Bulbs

Many electrical products, such as decorative lights, extension cords, and appliances, have a prominent tag labeled "UL." This mark, from Underwriters Laboratories, indicates that the product has been tested by UL engineers for electrical, fire, and other hazards.

Misconception Alert — BASIC

Because batteries are said to *run down,* many students believe that current is consumed by a circuit.

To check for this misconception, ask students to draw arrows representing the current in a simple circuit.

Some may believe that current is used up in the resistor. Their diagrams will show charges moving only from the battery to the bulb.

Others may think that the current comes back to the battery but has decreased in magnitude. Arrows representing current in their diagrams may get smaller after the resistor.

Point out that the number of charges entering a part of the circuit in some time interval equals the number of charges leaving it in the same time interval.

The Language of Physics

The term *emf* originally stood for *electromotive force.* This term may be misleading because emf is not a force. Rather, it refers to a potential difference measured in volts. The voltage value on a battery label denotes its emf.

In this text, internal resistance will be disregarded unless specifically noted. The value of the terminal voltage, ΔV, can be found from the emf, ε, the total current, I, and the internal resistance, r, with the following equation:

$$\Delta V = \varepsilon - Ir$$

Quick Lab

TEACHER'S NOTES

To light the bulb, students should connect the bottom of the bulb to one terminal of the battery and the side of the bulb's base to the other terminal. The bulb can be lit with one wire by holding the base of the bulb to one of the battery's terminals and using the wire to connect the side of the bulb's base to the other terminal.

SCILINKS®
www.scilinks.org
Topic: Electric Circuits
Code: HF60471

The source of potential difference and electrical energy is the circuit's emf

Will a bulb in a circuit light up if you remove the battery? Without a potential difference, there is no charge flow and no current. The battery is necessary because the battery is the source of potential difference and electrical energy for the circuit. So, the bulb must be connected to the battery to be lit.

Any device that increases the potential energy of charges circulating in a circuit is a source of *emf*. The emf is the energy per unit charge supplied by a source of electric current. Think of such a source as a "charge pump" that forces electrons to move in a certain direction. Batteries and generators are examples of emf sources.

For conventional current, the terminal voltage is less than the emf

Look at the battery attached to the light bulb in the circuit shown in **Figure 3.** As shown in the inset, instead of behaving only like a source of emf, the battery behaves as if it contains both an emf source and a resistor. The battery's internal resistance to current is the result of moving charges colliding with atoms inside the battery while the charges are traveling from one terminal to the other. Thus, when charges move conventionally in a battery, the potential difference across the battery's terminals, the *terminal voltage,* is actually slightly less than the emf.

Unless otherwise stated, any reference in this book to the potential difference across a battery should be thought of as the potential difference measured across the battery's terminals rather than as the emf of the battery. In other words, all examples and end-of-chapter problems will disregard the internal resistance of the battery.

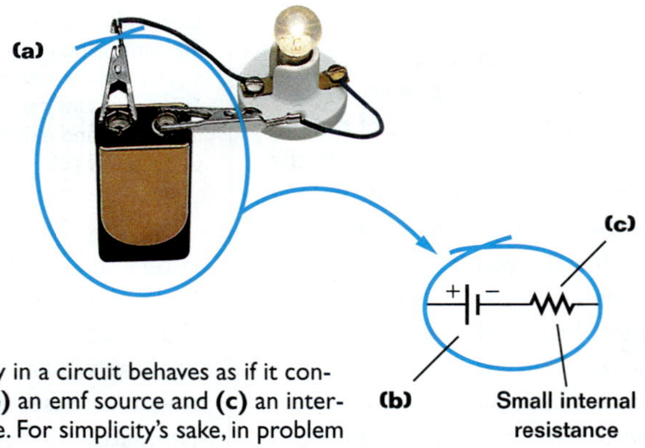

Figure 3
(a) A battery in a circuit behaves as if it contains both **(b)** an emf source and **(c)** an internal resistance. For simplicity's sake, in problem solving it will be assumed that this internal resistance is insignificant.

Quick Lab

Simple Circuits

MATERIALS LIST

- **1 miniature light bulb**
- **1 D-cell battery**
- **wires**
- **rubber band or tape**

SAFETY

Do not perform this lab with any batteries or electrical devices other than those listed here.

Never work with electricity near water. Be sure the floor and all work surfaces are dry.

Connect the bulb to the battery using two wires, using a rubber band or tape to hold the wire to the battery. Once you have gotten the bulb to light, try different arrangements to see whether there is more than one way to get the bulb to light. Can you make the bulb light using just one wire? Diagram each arrangement that you try, and note whether it produces light.

Explain exactly which parts of the bulb, battery, and wire must be connected for the light bulb to produce light.

Potential difference across a load equals the terminal voltage

When charges move within a battery from one terminal to the other, the chemical energy of the battery is converted to the electrical potential energy of the charges. As charges move through the circuit, their electrical potential energy is converted to other forms of energy. For instance, when the load is a resistor, the electrical potential energy of the charges is converted to the internal energy of the resistor and dissipated as thermal energy and light energy.

Because energy is conserved, the energy gained and the energy lost must be equal for one complete trip around the circuit (starting and ending at the same place). Thus, the electrical potential energy gained in the battery must equal the energy dissipated by the load. Because the potential difference is the measurement of potential energy per amount of charge, the potential increase across the battery must equal the potential decrease across the load.

SECTION REVIEW

1. Identify the types of elements in the schematic diagram illustrated in **Figure 4** and the number of each type.

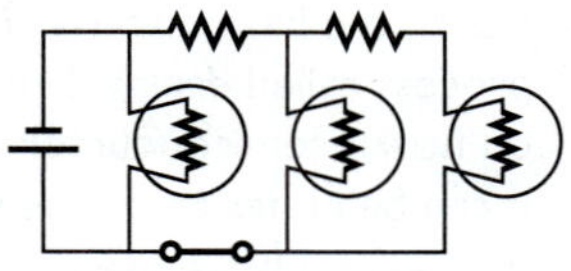

Figure 4

2. Using the symbols listed in **Table 1,** draw a schematic diagram of a working circuit that contains two resistors, an emf source, and a closed switch.

3. In which of the circuits pictured below will there be no current?

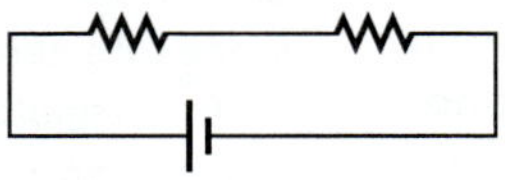

Figure 5

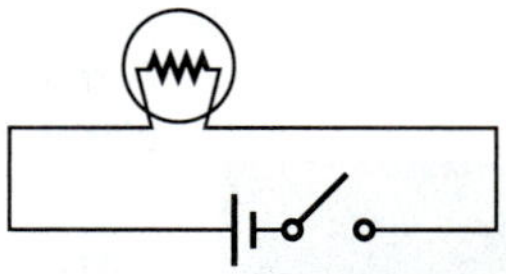

Figure 6

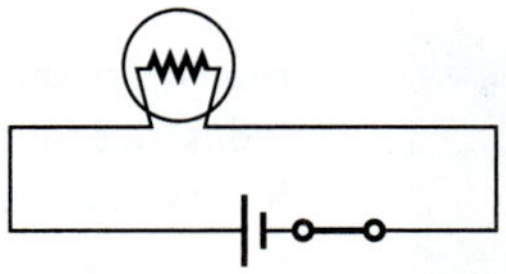

Figure 7

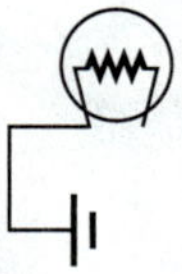

Figure 8

4. If the potential difference across the bulb in a certain flashlight is 3.0 V, what is the potential difference across the combination of batteries used to power it?

5. **Critical Thinking** In what forms is the electrical energy that is supplied to a string of decorative lights dissipated?

SECTION 1

Key Models and Analogies

From an energy-transformation perspective, think of batteries as electrical-energy-supply devices and of resistors and light bulbs as electrical-energy-consuming devices. The electric current conveys this energy from the battery to the resistor.

SECTION REVIEW ANSWERS

1. one battery, one closed switch, two resistors, and three bulbs
2. The circuit should include the circuit elements as they appear in **Table 1.**
3. **Figure 6** and **Figure 8** will have no current in them.
4. 3.0 V
5. It is converted to thermal energy and light energy.

Why it Matters

Transistors and Integrated Circuits

The branch of physics that studies the properties of semiconductors and related technologies is called *solid-state physics.*

The transistor was invented at Bell Labs in 1947. The integrated circuit was invented by Jack Kilby of Texas Instruments in 1958. In 1959, Robert Noyce received a patent for the silicon-based integrated circuit. Noyce later founded Intel, the company responsible for the creation of the microprocessor. The invention of the integrated circuit brought about an enormous boom in technology, as large, complex circuits could be contained in a small area. The microprocessor chip in a typical personal computer contains tens of millions of transistors.

Why it Matters

Transistors and Integrated Circuits

A computer motherboard is a multi-layered collection of integrated circuits containing millions of transistors and other circuit elements.

In the chapter "Electrical Energy and Current," you learned about a class of materials called *semiconductors,* which have properties between those of insulators and conductors. Semiconductors play many important roles in today's world, as they are the foundation of circuits found in virtually every electronic device.

Most commercial semiconductors are made primarily of either silicon or germanium. The conductive properties of semiconductors can be enhanced by adding impurities to the base material in a process called *doping.* Depending on how a semiconductor is doped, it can be either an n-type semiconductor or a p-type semiconductor. N-type semiconductors carry negative charges (in the form of electrons), and p-type semiconductors carry positive charges. The positive charges in a p-type semiconductor are not actually positively charged particles. They are "holes" created by the absence of electrons.

The most interesting and useful properties of semiconductors emerge when more than one type of semiconductor is used in a device. One such device is a *diode,* which is made by placing a p-type semiconductor next to an n-type semiconductor. The junction where the two types meet is called a *p-n junction.* A diode has almost infinite resistance in one direction and nearly zero resistance in the other direction. One useful application of diodes is the conversion of alternating current to direct current.

A *transistor* is a device that contains three layers of semiconductors. Transistors can be either *pnp transistors* or *npn transistors,* depending on the order of the layers. A transistor is like two diodes placed back-to-back. You might think this would mean that no current exists in a transistor, as there is infinite resistance at one or the other of the p-n junctions. However, if a small voltage is applied to the middle layer of the transistor, the p-n junctions are altered in such a way that a large amount of current can be in the transistor. As a result, transistors can be used as switches, allowing a small current to turn a larger current on or off. Transistor-based switches are the building blocks of computers. A single switch turned on or off can represent a binary digit, or *bit,* which is always either a one or a zero.

An *integrated circuit* is a collection of transistors, diodes, capacitors, and resistors embedded in a single piece of silicon, known as a *chip.* Much of the rapid progress in the computer and electronics industries in the past few decades has been a result of improvements in semiconductor technologies. These improvements allow smaller and smaller transistors and other circuit elements to be placed on chips. A typical computer motherboard, such as the one shown here, contains several integrated circuits, each one containing several million transistors.

Resistors in Series or in Parallel

SECTION 2

SECTION OBJECTIVES

- Calculate the equivalent resistance for a circuit of resistors in series, and find the current in and potential difference across each resistor in the circuit.
- Calculate the equivalent resistance for a circuit of resistors in parallel, and find the current in and potential difference across each resistor in the circuit.

RESISTORS IN SERIES

In a circuit that consists of a single bulb and a battery, the potential difference across the bulb equals the terminal voltage. The total current in the circuit can be found using the equation $\Delta V = IR$.

What happens when a second bulb is added to such a circuit, as shown in **Figure 9**? When moving through this circuit, charges that pass through one bulb must also move through the second bulb. Because all charges in the circuit must follow the same conducting path, these bulbs are said to be connected in **series.**

series

describes two or more components of a circuit that provide a single path for current

Resistors in series carry the same current

Light-bulb filaments are resistors; thus, **Figure 9(b)** represents the two bulbs in **Figure 9(a)** as resistors. Because charge is conserved, charges cannot build up or disappear at a point. For this reason, the amount of charge that enters one bulb in a given time interval equals the amount of charge that exits that bulb in the same amount of time. Because there is only one path for a charge to follow, the amount of charge entering and exiting the first bulb must equal the amount of charge that enters and exits the second bulb in the same time interval.

Because the current is the amount of charge moving past a point per unit of time, the current in the first bulb must equal the current in the second bulb. This is true for any number of resistors arranged in series. *When many resistors are connected in series, the current in each resistor is the same.*

The total current in a series circuit depends on how many resistors are present and on how much resistance each offers. Thus, to find the total current, first use the individual resistance values to find the total resistance of the circuit, called the *equivalent resistance.* Then the equivalent resistance can be used to find the current.

Figure 9
These two light bulbs are connected in series. Because light-bulb filaments are resistors, **(a)** the two bulbs in this series circuit can be represented by **(b)** two resistors in the schematic diagram shown on the right.

SECTION 2
General Level

Demonstration

Resistors in Series BASIC

Purpose Demonstrate that series circuits require all elements to conduct.

Materials 2 flashlight bulbs, bulb holders, battery, battery holder, 3 short pieces of wire

Procedure Wire the bulbs in series with the battery, and point out the lit bulbs. Trace the path for the movement of charges. Ask students to predict what will happen if you unscrew the second bulb. Unscrew it. Point out that the charges no longer have a complete path.

Visual Strategy GENERAL

Figure 9

Point out that even though the resistors are different and must be labeled R_1 and R_2, there is only one value for current.

Q Is the current within the battery less than, equal to, or greater than the circuit current?

A ***The current within the battery is the same as the circuit current.***

Visual Strategy GENERAL

Figure 10

Be certain students understand what is meant by the idea that the resistor labeled R_{eq} can replace the other two resistors. The current in and potential difference across the equivalent resistor is the same as if the two resistors are taken together.

Q Explain why it was not necessary to label the current in **Figure 10(b)** as I_{eq}.

A ***The current will be the same in this equivalent resistor as in the original circuit.***

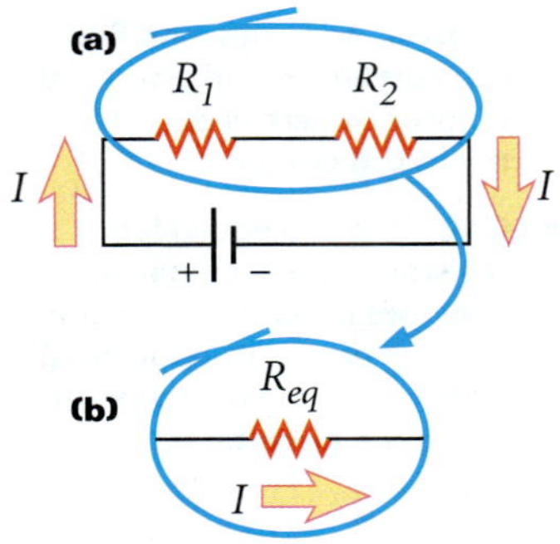

Figure 10
(a) The two resistors in the actual circuit have the same effect on the current in the circuit as **(b)** the equivalent resistor.

The equivalent resistance in a series circuit is the sum of the circuit's resistances

As described in Section 1, the potential difference across the battery, ΔV, must equal the potential difference across the load, $\Delta V_1 + \Delta V_2$, where ΔV_1 is the potential difference across R_1 and ΔV_2 is the potential difference across R_2.

$$\Delta V = \Delta V_1 + \Delta V_2$$

According to $\Delta V = IR$, the potential difference across each resistor is equal to the current in that resistor multiplied by the resistance.

$$\Delta V = I_1R_1 + I_2R_2$$

Because the resistors are in series, the current in each is the same. For this reason, I_1 and I_2 can be replaced with a single variable for the current, I.

$$\Delta V = I(R_1 + R_2)$$

Finding a value for the equivalent resistance of the circuit is now possible. If you imagine the equivalent resistance replacing the original two resistors, as shown in **Figure 10,** you can treat the circuit as if it contains only one resistor and use $\Delta V = IR$ to relate the total potential difference, current, and equivalent resistance.

$$\Delta V = I(R_{eq})$$

Now set the last two equations for ΔV equal to each other, and divide by the current.

$$\Delta V = I(R_{eq}) = I(R_1 + R_2)$$

$$R_{eq} = R_1 + R_2$$

Thus, the equivalent resistance of the series combination is the sum of the individual resistances. An extension of this analysis shows that the equivalent resistance of two or more resistors connected in series can be calculated using the following equation.

RESISTORS IN SERIES

$$R_{eq} = R_1 + R_2 + R_3 \ldots$$

Equivalent resistance equals the total of individual resistances in series.

Because R_{eq} represents the sum of the individual resistances that have been connected in series, *the equivalent resistance of a series combination of resistors is always greater than any individual resistance.*

To find the total current in a series circuit, first simplify the circuit to a single equivalent resistance using the boxed equation above; then use $\Delta V = IR$ to calculate the current.

$$I = \frac{\Delta V}{R_{eq}}$$

Because the current in each bulb is equal to the total current, you can also use $\Delta V = IR$ to calculate the potential difference across each resistor.

$$\Delta V_1 = IR_1 \quad \text{and} \quad \Delta V_2 = IR_2$$

The method described above can be used to find the potential difference across resistors in a series circuit containing any number of resistors.

SAMPLE PROBLEM A

Resistors in Series

PROBLEM

A 9.0 V battery is connected to four light bulbs, as shown at right. Find the equivalent resistance for the circuit and the current in the circuit.

SOLUTION

1. DEFINE **Given:** $\Delta V = 9.0$ V $\quad R_1 = 2.0\ \Omega$
$R_2 = 4.0\ \Omega \quad R_3 = 5.0\ \Omega$
$R_4 = 7.0\ \Omega$

Unknown: $R_{eq} = ? \quad I = ?$

Diagram:

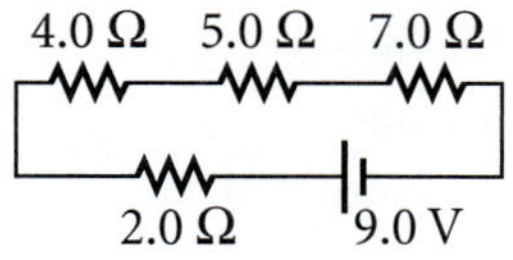

2. PLAN **Choose an equation or situation:**
Because the resistors are connected end to end, they are in series. Thus, the equivalent resistance can be calculated with the equation for resistors in series.

$$R_{eq} = R_1 + R_2 + R_3 \ldots$$

The following equation can be used to calculate the current.

$$\Delta V = IR_{eq}$$

Rearrange the equation to isolate the unknown:
No rearrangement is necessary to calculate R_{eq}, but $\Delta V = IR_{eq}$ must be rearranged to calculate current.

$$I = \frac{\Delta V}{R_{eq}}$$

3. CALCULATE **Substitute the values into the equation and solve:**

$$R_{eq} = 2.0\ \Omega + 4.0\ \Omega + 5.0\ \Omega + 7.0\ \Omega$$

$$\boxed{R_{eq} = 18.0\ \Omega}$$

continued on next page

Classroom Practice

Resistors in Series
Calculate the equivalent resistance, the current in each resistor, and the potential difference across each resistor if a 24.0 V battery is connected in series to the following:

a. five 2.0 Ω resistors
b. 50 2.0 Ω resistors
c. 500 2.0 Ω resistors

Answers
a. $1.0 \times 10^1\ \Omega$, 2.4 A, 4.8 V
b. $1.0 \times 10^2\ \Omega$, 0.24 A, 0.48 V
c. $1.0 \times 10^3\ \Omega$, 0.024 A, 0.048 V

Alternative Problem-Solving Approach

Because the resistors are in series, the current is the same in each resistor. If I is the value of current, the problem can also be solved by first applying $\Delta V = IR$ to each resistor and then using the sum of potential differences to calculate I:

$$\Delta V_1 = R_1 I = 2.0I$$
$$\Delta V_2 = R_2 I = 4.0I$$
$$\Delta V_3 = R_3 I = 5.0I$$
$$\Delta V_4 = R_4 I = 7.0I$$
$$\Delta V_1 + \Delta V_2 + \Delta V_3 + \Delta V_4 = \Delta V$$
$$\Delta V = 18.0I$$

Now substitute the given value for ΔV:

$$18.0I = 9.0\text{ V}$$
$$I = 0.50\text{ A}$$

SECTION 2

PROBLEM GUIDE A

Use this guide to assign problems.
SE = Student Edition Textbook
PW = Problem Workbook
PB = Problem Bank on the One-Stop Planner (OSP)

Solving for:

R_{eq}	**SE** Sample, 1–2, 4, 6; Ch. Rvw. 16–17 **PW** Sample, 1–2, 5 **PB** 4–6
I	**SE** Sample, 1–2, 4; Ch. Rvw. 17 **PW** Sample, 4 **PB** 7–10
R	**SE** 5 **PW** 3, 6 **PB** Sample, 1–3
ΔV	**SE** 3, 4 **PW** 4
P	**PW** 7

*Challenging Problem
Consult the printed Solutions Manual or the OSP for detailed solutions.

ANSWERS

Practice A

1. a. 43.6 Ω
   b. 0.275 A
2. a. 24.0 Ω
   b. 1.00 A
   c. 1.00 A
3. 1.0 V, 2.0 V, 2.5 V, 3.5 V
4. a. 11.28 Ω, 0.798 A
   b. 5.79 V, 3.22 V
5. 0.5 Ω
6. a. 67.6 Ω
   b. 45 bulbs

Substitute the equivalent resistance value into the equation for current.

$$I = \frac{\Delta V}{R_{eq}} = \frac{9.0\ \text{V}}{18.0\ \Omega}$$

$$\boxed{I = 0.50\ \text{A}}$$

4. EVALUATE For resistors connected in series, the equivalent resistance should be greater than the largest resistance in the circuit.

$$18.0\ \Omega > 7.0\ \Omega$$

PRACTICE A

Resistors in Series

1. A 12.0 V storage battery is connected to three resistors, 6.75 Ω, 15.3 Ω, and 21.6 Ω, respectively. The resistors are joined in series.
   a. Calculate the equivalent resistance.
   b. What is the current in the circuit?
2. A 4.0 Ω resistor, an 8.0 Ω resistor, and a 12.0 Ω resistor are connected in series with a 24.0 V battery.
   a. Calculate the equivalent resistance.
   b. Calculate the current in the circuit.
   c. What is the current in each resistor?
3. Because the current in the equivalent resistor of Sample Problem A is 0.50 A, it must also be the current in each resistor of the original circuit. Find the potential difference across each resistor.
4. A series combination of two resistors, 7.25 Ω and 4.03 Ω, is connected to a 9.00 V battery.
   a. Calculate the equivalent resistance of the circuit and the current.
   b. What is the potential difference across each resistor?
5. A 7.0 Ω resistor is connected in series with another resistor and a 4.5 V battery. The current in the circuit is 0.60 A. Calculate the value of the unknown resistance.
6. Several light bulbs are connected in series across a 115 V source of emf.
   a. What is the equivalent resistance if the current in the circuit is 1.70 A?
   b. If each light bulb has a resistance of 1.50 Ω, how many light bulbs are in the circuit?

Series circuits require all elements to conduct

What happens to a series circuit when a single bulb burns out? Consider what a circuit diagram for a string of lights with one broken filament would look like. As the schematic diagram in **Figure 11** shows, the broken filament means that there is a gap in the conducting pathway used to make up the circuit. Because the circuit is no longer closed, there is no current in it and all of the bulbs go dark.

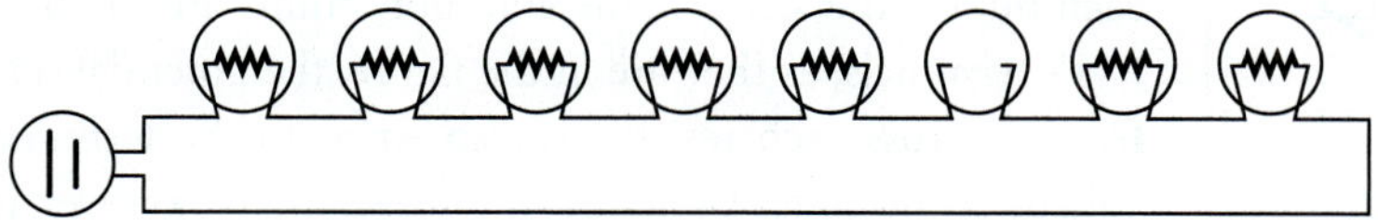

Figure 11
A burned-out filament in a bulb has the same effect as an open switch. Because this series circuit is no longer complete, there is no current in the circuit.

SCILINKS.
www.scilinks.org
Topic: Resistors
Code: HF61302

Why, then, would anyone arrange resistors in series? Resistors can be placed in series with a device in order to regulate the current in that device. In the case of decorative lights, adding an additional bulb will decrease the current in each bulb. Thus, the filament of each bulb need not withstand such a high current. Another advantage to placing resistors in series is that several lesser resistances can be used to add up to a single greater resistance that is unavailable. Finally, in some cases, it is important to have a circuit that will have no current if any one of its component parts fails. This technique is used in a variety of contexts, including some burglar alarm systems.

RESISTORS IN PARALLEL

As discussed above, when a single bulb in a series light set burns out, the entire string of lights goes dark because the circuit is no longer closed. What would happen if there were alternative pathways for the movement of charge, as shown in **Figure 12**?

A wiring arrangement that provides alternative pathways for the movement of a charge is a **parallel** arrangement. The bulbs of the decorative light set shown in the schematic diagram in **Figure 12** are arranged in parallel with each other.

parallel

describes two or more components of a circuit that provide separate conducting paths for current because the components are connected across common points or junctions

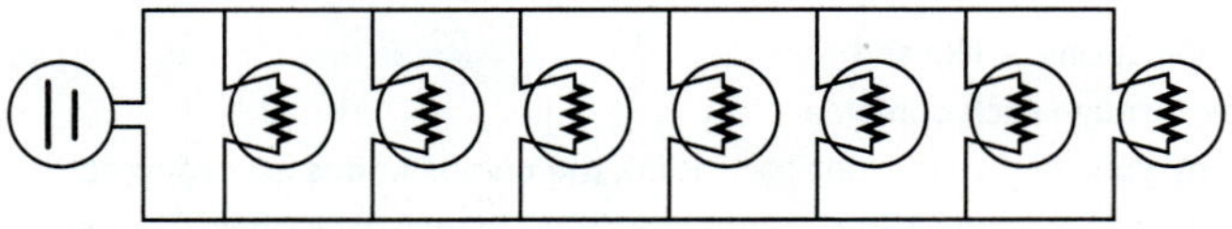

Figure 12
These decorative lights are wired in parallel. Notice that in a parallel arrangement there is more than one path for current.

SECTION 2

Visual Strategy GENERAL

Figure 11
Have students examine a burned-out bulb to see the broken filament. Tell students that a bulb is said to *burn out* when its filament breaks. Remind students how a bulb works, referring them to the "Light Bulbs" feature in Section 1 for clarification. Point out that when the filament is broken, charges no longer have a complete pathway from the base of the bulb to the threads.

Q Would the other bulbs light if a wire were attached from the base of the burned-out bulb to its threads?

A ***Yes, charges would have a complete path to follow.***

Visual Strategy BASIC

Figure 12
Have students trace each of the alternative pathways in their book. Point out that if any of these pathways remains intact, there will be current in the circuit.

Q What parts of the circuit would have current if all of the bulbs except the last one had broken filaments?

A ***There would be current in the intact bulb and in the wires that connect the bulb across the potential difference.***

Visual Strategy GENERAL

Figure 13
Working through a diagram like **Figure 13(b)** with numerical examples may help students understand the relationships for current in a parallel circuit.

Q Assume that I from the battery = 5 A and I_1 = 2 A. What must I_2 be?

A *3 A*

Demonstration

Resistors in Parallel BASIC

Purpose Demonstrate that parallel circuits do not require all elements to conduct.

Materials two flashlight bulbs, bulb holders, battery, battery holder, four short pieces of wire

Procedure Connect the bulbs in parallel with the battery as shown in **Figure 13.** Trace the path for the movement of the charges. Ask students to predict what will happen if you unscrew the second bulb. Unscrew it. Point out that the charges still have a complete path in the other bulb.

Quick Lab

TEACHER'S NOTES

For this lab to be effective, it is very important that the straws be taped together. Crimping one end of a straw and stuffing it into another straw will not work well.

Homework Options
This QuickLab can easily be performed outside of the physics lab room.

Resistors in parallel have the same potential differences across them

To explore the consequences of arranging resistors in parallel, consider the two bulbs connected to a battery in **Figure 13(a).** In this arrangement, the left side of each bulb is connected to the positive terminal of the battery, and the right side of each bulb is connected to the negative terminal. Because the sides of each bulb are connected to common points, the potential difference across each bulb is the same. If the common points are the battery's terminals, as they are in the figure, the potential difference across each resistor is also equal to the terminal voltage of the battery. The current in each bulb, however, is not always the same.

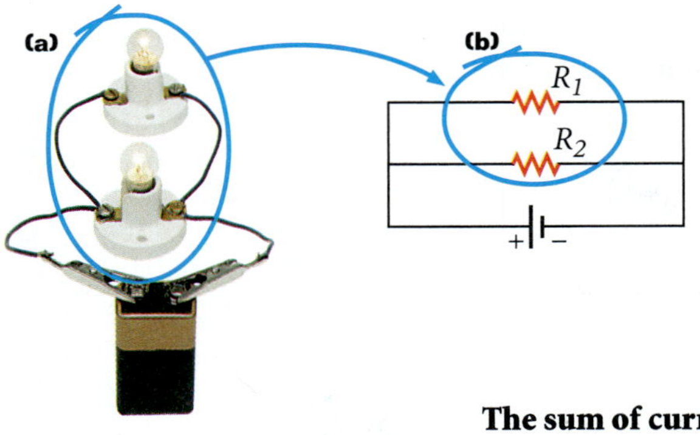

Figure 13
(a) This simple parallel circuit with two bulbs connected to a battery can be represented by **(b)** the schematic diagram shown on the right.

The sum of currents in parallel resistors equals the total current

In **Figure 13,** when a certain amount of charge leaves the positive terminal and reaches the branch on the left side of the circuit, some of the charge moves through the top bulb and some moves through the bottom bulb. If one of the bulbs has less resistance, more charge moves through that bulb because the bulb offers less opposition to the flow of charges.

Because charge is conserved, the sum of the currents in each bulb equals the current I delivered by the battery. This is true for all resistors in parallel.

$$I = I_1 + I_2 + I_3 \ldots$$

The parallel circuit shown in **Figure 13** can be simplified to an equivalent resistance with a method similar to the one used for series circuits. To do this, first show the relationship among the currents.

$$I = I_1 + I_2$$

Then substitute the equivalents for current according to $\Delta V = IR$.

$$\frac{\Delta V}{R_{eq}} = \frac{\Delta V_1}{R_1} + \frac{\Delta V_2}{R_2}$$

Quick Lab

Series and Parallel Circuits

MATERIALS LIST

- 4 regular drinking straws
- 4 stirring straws or coffee stirrers
- tape

Cut the regular drinking straws and thin stirring straws into equal lengths. Tape them end to end in long tubes to form series combinations. Form parallel combinations by taping the straws together side by side.

Try several combinations of like and unlike straws. Blow through each combination of tubes, holding your fingers in front of the opening(s) to compare the airflow (or current) that you achieve with each combination.

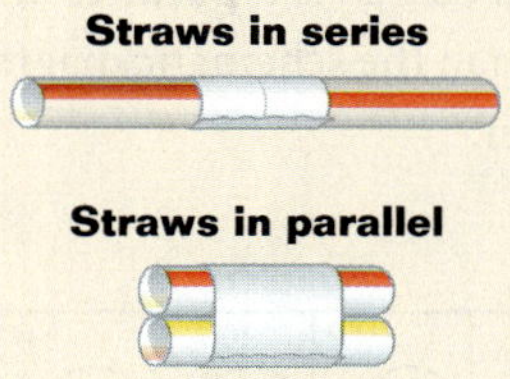

Rank the combinations according to how much resistance they offer. Classify them according to the amount of current created in each.

Because the potential difference across each bulb in a parallel arrangement equals the terminal voltage ($\Delta V = \Delta V_1 = \Delta V_2$), you can divide each side of the equation by ΔV to get the following equation.

$$\frac{1}{R_{eq}} = \frac{1}{R_1} + \frac{1}{R_2}$$

An extension of this analysis shows that the equivalent resistance of two or more resistors connected in parallel can be calculated using the following equation.

RESISTORS IN PARALLEL

$$\frac{1}{R_{eq}} = \frac{1}{R_1} + \frac{1}{R_2} + \frac{1}{R_3} \ldots$$

The equivalent resistance of resistors in parallel can be calculated using a reciprocal relationship.

Notice that this equation does not give the value of the equivalent resistance directly. You must take the reciprocal of your answer to obtain the value of the equivalent resistance.

Because of the reciprocal relationship, *the equivalent resistance for a parallel arrangement of resistors must always be less than the smallest resistance in the group of resistors.*

The conclusions made about both series and parallel circuits are summarized in **Table 2.**

Why it Matters

Conceptual Challenge

1. Car Headlights

How can you tell that the headlights on a car are wired in parallel rather than in series? How would the brightness of the bulbs differ if they were wired in series across the same 12 V battery instead of in parallel?

2. Simple Circuits

Sketch as many different circuits as you can using three light bulbs—each of which has the same resistance—and a battery.

Table 2 Resistors in Series or in Parallel

	Series	Parallel
schematic diagram	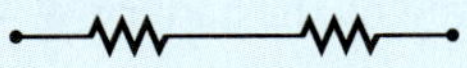	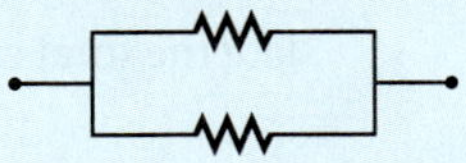
current	$I = I_1 = I_2 = I_3 \ldots$ = same for each resistor	$I = I_1 + I_2 + I_3 \ldots$ = sum of currents
potential difference	$\Delta V = \Delta V_1 + \Delta V_2 + \Delta V_3 \ldots$ = sum of potential differences	$\Delta V = \Delta V_1 = \Delta V_2 = \Delta V_3 \ldots$ = same for each resistor
equivalent resistance	$R_{eq} = R_1 + R_2 + R_3 \ldots$ = sum of individual resistances	$\frac{1}{R_{eq}} = \frac{1}{R_1} + \frac{1}{R_2} + \frac{1}{R_3} \ldots$ = reciprocal sum of resistances

ANSWERS

Conceptual Challenge

1. Car headlights must be wired in parallel so that if one burns out, the other will stay lit. If they were wired in series, they would be less bright.
2. There are four possible circuits: all resistors in series, all resistors in parallel, one resistor in series with two others in parallel, and one resistor in parallel with two others in series.

Misconception Alert — BASIC

Use a simple numerical example to demonstrate that mathematically adding the inverses is not the same as taking the inverse of the sum. The example below uses resistances that have values of 2 and 3 in parallel.

Correct: $\frac{1}{2} + \frac{1}{3} = \frac{5}{6}$, $R_{eq} = \frac{6}{5}$

Incorrect: $2 + 3 = 5$, $R_{eq} \neq \frac{1}{5}$

Classroom Practice

Resistors in Parallel
Find the equivalent resistance, the current in each resistor, and the current drawn by the circuit load for a 9.0 V battery connected in parallel to three 30.0 Ω resistors.

Answer
10.0 Ω, 0.30 A, 0.90 A

PROBLEM GUIDE B

Use this guide to assign problems.
SE = Student Edition Textbook
PW = Problem Workbook
PB = Problem Bank on the One-Stop Planner (OSP)

Solving for:

R_{eq}	**SE** Sample, 2–4; Ch. Rvw. 18–19 **PW** Sample, 1–2, 4–6 **PB** 4–6
I	**SE** Sample, 1, 3–4; Ch. Rvw. 18–19 **PW** Sample, 6–7 **PB** 7–10
R	**PW** 3 **PB** Sample, 1–3
ΔV	**SE** 4b

***Challenging Problem**
Consult the printed Solutions Manual or the OSP for detailed solutions.

SAMPLE PROBLEM B

Resistors in Parallel

PROBLEM

A 9.0 V battery is connected to four resistors, as shown at right. Find the equivalent resistance for the circuit and the total current in the circuit.

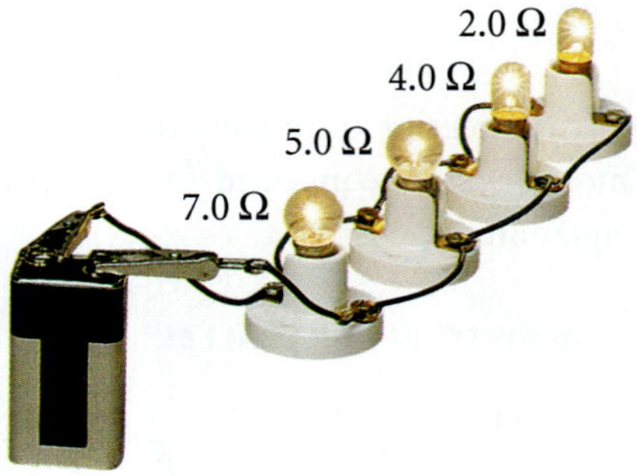

SOLUTION

1. DEFINE

Given: $\Delta V = 9.0$ V $\quad R_1 = 2.0\ \Omega$ $\quad R_2 = 4.0\ \Omega$ $\quad R_3 = 5.0\ \Omega$ $\quad R_4 = 7.0\ \Omega$

Unknown: $R_{eq} = ?\quad I = ?$

Diagram:

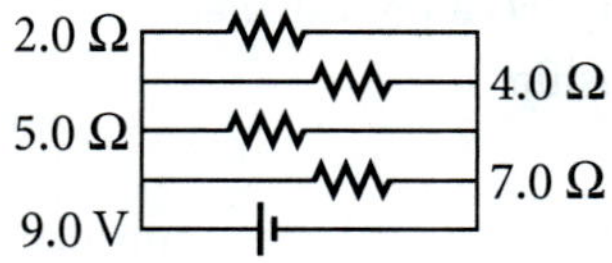

2. PLAN

Choose an equation or situation:
Because both sides of each resistor are connected to common points, they are in parallel. Thus, the equivalent resistance can be calculated with the equation for resistors in parallel.

$$\frac{1}{R_{eq}} = \frac{1}{R_1} + \frac{1}{R_2} + \frac{1}{R_3} \ldots \text{ for parallel}$$

The following equation can be used to calculate the total current.

$$\Delta V = IR_{eq}$$

Rearrange the equation to isolate the unknown:
No rearrangement is necessary to calculate R_{eq}; rearrange $\Delta V = IR_{eq}$ to calculate the total current delivered by the battery.

$$I = \frac{\Delta V}{R_{eq}}$$

3. CALCULATE

Substitute the values into the equation and solve:

$$\frac{1}{R_{eq}} = \frac{1}{2.0\ \Omega} + \frac{1}{4.0\ \Omega} + \frac{1}{5.0\ \Omega} + \frac{1}{7.0\ \Omega}$$

$$\frac{1}{R_{eq}} = \frac{0.50}{1\ \Omega} + \frac{0.25}{1\ \Omega} + \frac{0.20}{1\ \Omega} + \frac{0.14}{1\ \Omega} = \frac{1.09}{1\ \Omega}$$

$$R_{eq} = \frac{1\ \Omega}{1.09}$$

$$\boxed{R_{eq} = 0.917\ \Omega}$$

TIP *The equation for resistors in parallel gives you the reciprocal of the equivalent resistance. Be sure to take the reciprocal of this value in the final step to find the equivalent resistance.*

Substitute that equivalent resistance value in the equation for current.

$$I = \frac{\Delta V_{tot}}{R_{eq}} = \frac{9.0\ \text{V}}{0.917\ \Omega}$$

$$\boxed{I = 9.8\ \text{A}}$$

CALCULATOR SOLUTION

The calculator answer is 9.814612868, but because the potential difference, 9.0 V, has only two significant digits, the answer is reported as 9.8 A.

4. EVALUATE For resistors connected in parallel, the equivalent resistance should be less than the smallest resistance.

$$0.917\ \Omega < 2.0\ \Omega$$

PRACTICE B

Resistors in Parallel

1. The potential difference across the equivalent resistance in Sample Problem B equals the potential difference across each of the individual parallel resistors. Calculate the value for the current in each resistor.

2. A length of wire is cut into five equal pieces. The five pieces are then connected in parallel, with the resulting resistance being 2.00 Ω. What was the resistance of the original length of wire before it was cut up?

3. A 4.0 Ω resistor, an 8.0 Ω resistor, and a 12.0 Ω resistor are connected in parallel across a 24.0 V battery.
   a. What is the equivalent resistance of the circuit?
   b. What is the current in each resistor?

4. An 18.0 Ω, 9.00 Ω, and 6.00 Ω resistor are connected in parallel to an emf source. A current of 4.00 A is in the 9.00 Ω resistor.
   a. Calculate the equivalent resistance of the circuit.
   b. What is the potential difference across the source?
   c. Calculate the current in the other resistors.

Parallel circuits do not require all elements to conduct

What happens when a bulb burns out in a string of decorative lights that is wired in parallel? There is no current in that branch of the circuit, but each of the parallel branches provides a separate alternative pathway for current. Thus, the potential difference supplied to the other branches and the current in these branches remain the same, and the bulbs in these branches remain lit.

Alternative Problem-Solving Approach

The problem can also be solved by applying $\Delta V = IR$ to each resistor to find its current, then adding these to get the total current. Finally, use $R_{eq} = \frac{\Delta V}{I_{tot}}$ to find R_{eq}.

$$I_1 = \frac{\Delta V}{R_1} = \frac{9.0\ \text{V}}{2.0\ \Omega} = 4.5\ \text{A}$$

$$I_2 = \frac{\Delta V}{R_2} = \frac{9.0\ \text{V}}{4.0\ \Omega} = 2.2\ \text{A}$$

$$I_3 = \frac{\Delta V}{R_3} = \frac{9.0\ \text{V}}{5.0\ \Omega} = 1.8\ \text{A}$$

$$I_4 = \frac{\Delta V}{R_4} = \frac{9.0\ \text{V}}{7.0\ \Omega} = 1.3\ \text{A}$$

$$I_{tot} = I_1 + I_2 + I_3 + I_4$$

$$I_{tot} = 9.8\ \text{A}$$

$$R_{eq} = \frac{9.0\ \text{V}}{9.8\ \text{A}} = 0.92\ \Omega$$

The slight difference in the answer obtained this way is due to rounding.

ANSWERS

Practice B

1. 4.5 A, 2.2 A, 1.8 A, 1.3 A
2. 50.0 Ω
3. **a.** 2.2 Ω
b. 6.0 A, 3.0 A, 2.00 A
4. **a.** 2.99 Ω
b. 36.0 V
c. 2.00 A, 6.00 A

SECTION REVIEW ANSWERS

1. in the series circuit
2. **a.** All have equal I.
   b. 15 Ω
3. **a.** 5 Ω
   b. All have equal ΔV.
4. **a.** 2.0 Ω: 2.0 A, 4.0 V
   4.0 Ω: 2.0 A, 8.0 V
   b. 2.0 Ω: 6.0 A, 12 V
   4.0 Ω: 3.0 A, 12 V
5. Because the resistance of each bulb is the same, the brightness depends only on the potential difference. Bulbs (a), (d), and (e) have an equal potential difference across them and thus equal brightnesses. Because bulbs (b) and (c) have the same resistance and are in series, they have equal but lesser potential differences and thus are equally bright but less bright than bulbs (a), (d), and (e).

Did you know?

Because the potential difference provided by a wall outlet in a home in North America is not the same as the potential difference that is standard on other continents, appliances made in North America are not always compatible with wall outlets in homes on other continents.

When resistors are wired in parallel with an emf source, the potential difference across each resistor always equals the potential difference across the source. Because household circuits are arranged in parallel, appliance manufacturers are able to standardize their design, producing devices that all operate at the same potential difference. As a result, manufacturers can choose the resistance to ensure that the current will be neither too high nor too low for the internal wiring and other components that make up the device.

Additionally, the equivalent resistance of several parallel resistors is less than the resistance of any of the individual resistors. Thus, a low equivalent resistance can be created with a group of resistors of higher resistances.

SECTION REVIEW

1. Two resistors are wired in series. In another circuit, the same two resistors are wired in parallel. In which circuit is the equivalent resistance greater?
2. A 5 Ω, a 10 Ω, and a 15 Ω resistor are connected in series.
   a. Which resistor has the most current in it?
   b. Which resistor has the largest potential difference across it?
3. A 5 Ω, a 10 Ω, and a 15 Ω resistor are connected in parallel.
   a. Which resistor has the most current in it?
   b. Which resistor has the largest potential difference across it?
4. Find the current in and potential difference across each of the resistors in the following circuits:
   a. a 2.0 Ω and a 4.0 Ω resistor wired in series with a 12 V source
   b. a 2.0 Ω and a 4.0 Ω resistor wired in parallel with a 12 V source
5. **Interpreting Graphics** The brightness of a bulb depends only on the bulb's resistance and on the potential difference across it. A bulb with a greater potential difference dissipates more power and thus is brighter. The five bulbs shown in **Figure 14** are identical, and so are the three batteries. Rank the bulbs in order of brightness from greatest to least, indicating if any are equal. Explain your reasoning. (Disregard the resistance of the wires.)

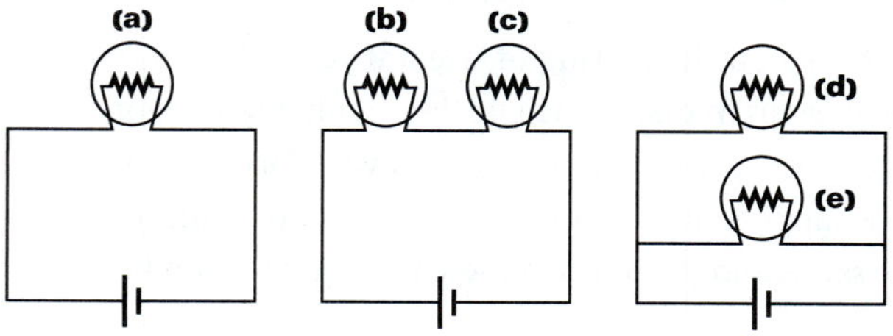

Figure 14

Complex Resistor Combinations

SECTION OBJECTIVES

- **Calculate the equivalent resistance for a complex circuit involving both series and parallel portions.**
- **Calculate the current in and potential difference across individual elements within a complex circuit.**

RESISTORS COMBINED BOTH IN PARALLEL AND IN SERIES

Series and parallel circuits are not often encountered independent of one another. Most circuits today employ both series and parallel wiring to utilize the advantages of each type.

A common example of a complex circuit is the electrical wiring typical in a home. In a home, a fuse or circuit breaker is connected in series to numerous outlets, which are wired to one another in parallel. An example of a typical household circuit is shown in **Figure 15.**

As a result of the outlets being wired in parallel, all the appliances operate independently; if one is switched off, any others remain on. Wiring the outlets in parallel ensures that an identical potential difference exists across any appliance. This way, appliance manufacturers can produce appliances that all use the same standard potential difference.

To prevent excessive current, a fuse or circuit breaker must be placed in series with all of the outlets. Fuses and circuit breakers open the circuit when the current becomes too high. A fuse is a small metallic strip that melts if the current exceeds a certain value. After a fuse has melted, it must be replaced. A circuit breaker, a more modern device, triggers a switch when current reaches a certain value. The switch must be reset, rather than replaced, after the circuit overload has been removed. Both fuses and circuit breakers must be in series with the entire load to prevent excessive current from reaching any appliance. In fact, if all the devices in **Figure 15** were used at once, the circuit would be overloaded. The circuit breaker would interrupt the current.

Fuses and circuit breakers are carefully selected to meet the demands of a circuit. If the circuit is to carry currents as large as 30 A, an appropriate fuse or circuit breaker must be used. Because the fuse or circuit breaker is placed in series with the rest of the circuit, the current in the fuse or circuit breaker is the same as the total current in the circuit. To find this current, one must determine the equivalent resistance.

When determining the equivalent resistance for a complex circuit, you must simplify the circuit into groups of series and parallel resistors and then find the equivalent resistance for each group by using the rules for finding the equivalent resistance of series and parallel resistors.

(a)

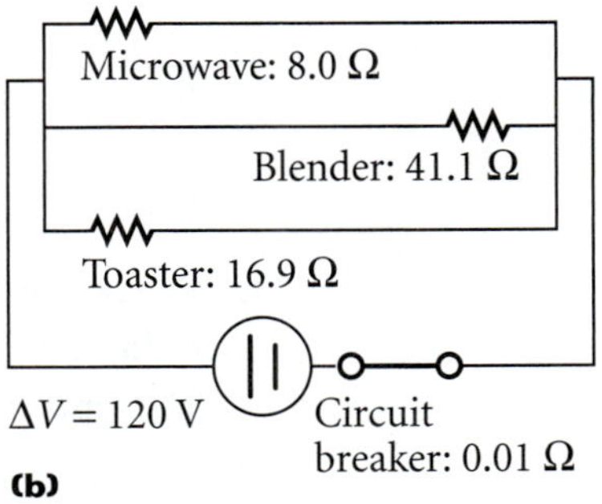

(b)

Figure 15
(a) When all of these devices are plugged into the same household circuit, **(b)** the result is a parallel combination of resistors in series with a circuit breaker.

Visual Strategy GENERAL

Figure 15

Explain that the circuit breaker is shown as a switch because it contains a switch that opens when the current in the circuit becomes too large.

Q How much of the total current is in the circuit breaker when it is in series with the parallel combination of devices shown?

A ***All of the total current is in the circuit breaker when it is in series.***

Teaching Tip ADVANCED

Point out that each device added in parallel to a circuit draws more current from the emf source.

Mathematically, because of the following equation, the more resistors that are added in parallel, the more current there will be in the main wires of the circuit.

$$I_{tot} = \frac{\Delta V}{R_1} + \frac{\Delta V}{R_2} \ldots$$

If the current is too great, the main wires, plugs, and outlet connections will heat up. Excessive current can damage equipment and can even cause fires.

Classroom Practice

Equivalent Resistance
Use the following values with the circuit in **Figure 16** on the following page. What is the equivalent resistance for each circuit?

a. $R_a = 5.0\ \Omega$, $R_b = 3.0\ \Omega$, $R_c = 6.0\ \Omega$

b. $R_a = 6.0\ \Omega$, $R_b = 8.0\ \Omega$, $R_c = 2.0\ \Omega$

Answers

a. 7.0 Ω

b. 7.6 Ω

PROBLEM GUIDE C

Use this guide to assign problems.
SE = Student Edition Textbook
PW = Problem Workbook
PB = Problem Bank on the One-Stop Planner (OSP)

Solving for:

R_{eq}	**SE** Sample, 1–2; Ch. Rvw. 23–24 **PW** Sample, 1, 4–5 **PB** 4–6
I	**PW** 2–3 **PB** 7–10
R	**PW** 3 **PB** Sample, 1–3
P	**PB** 4, 6

***Challenging Problem**
Consult the printed Solutions Manual or the OSP for detailed solutions.

SAMPLE PROBLEM C

STRATEGY Equivalent Resistance

PROBLEM

Determine the equivalent resistance of the complex circuit shown below.

REASONING

The best approach is to divide the circuit into groups of series and parallel resistors. This way, the methods presented in Sample Problems A and B can be used to calculate the equivalent resistance for each group.

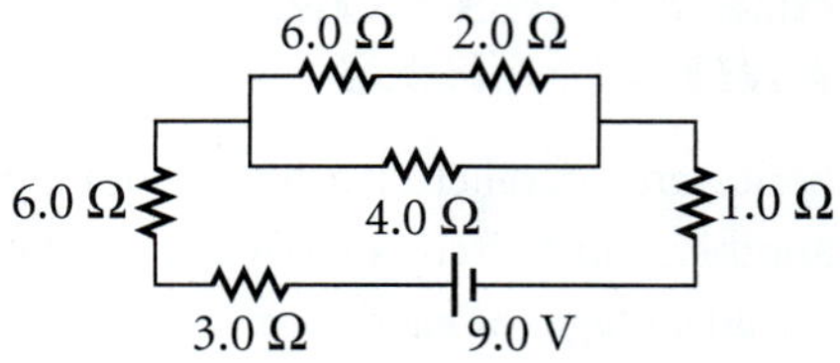

SOLUTION

1. Redraw the circuit as a group of resistors along one side of the circuit.
Because bends in a wire do not affect the circuit, they do not need to be represented in a schematic diagram. Redraw the circuit without the corners, keeping the arrangement of the circuit elements the same, as shown at right.

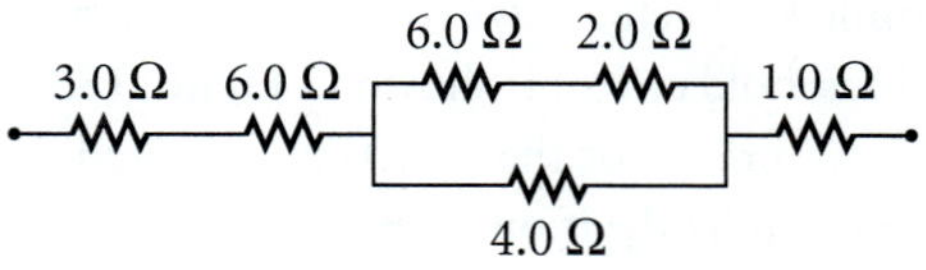

TIP *For now, disregard the emf source, and work only with the resistances.*

2. Identify components in series, and calculate their equivalent resistance.
Resistors in groups **(a)** and **(b)** are in series.
For group **(a):** $R_{eq} = 3.0\ \Omega + 6.0\ \Omega = 9.0\ \Omega$
For group **(b):** $R_{eq} = 6.0\ \Omega + 2.0\ \Omega = 8.0\ \Omega$

3. Identify components in parallel, and calculate their equivalent resistance.
Resistors in group **(c)** are in parallel.

For group **(c):**

$$\frac{1}{R_{eq}} = \frac{1}{8.0\ \Omega} + \frac{1}{4.0\ \Omega} = \frac{0.12}{1\ \Omega} + \frac{0.25}{1\ \Omega} = \frac{0.37}{1\ \Omega}$$

$$R_{eq} = 2.7\ \Omega$$

4. Repeat steps 2 and 3 until the resistors in the circuit are reduced to a single equivalent resistance.
The remainder of the resistors, group **(d),** are in series.

For group **(d):** $R_{eq} = 9.0\ \Omega + 2.7\ \Omega + 1.0\ \Omega$

$$\boxed{R_{eq} = 12.7\ \Omega}$$

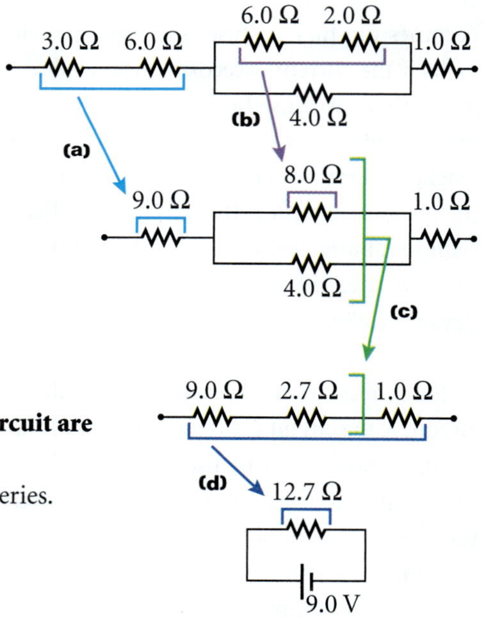

TIP *It doesn't matter in what order the operations of simplifying the circuit are done, as long as the simpler equivalent circuits still have the same current in and potential difference across the load.*

PRACTICE C

Equivalent Resistance

1. For each of the following sets of values, determine the equivalent resistance for the circuit shown in **Figure 16.**

a. $R_a = 25.0\ \Omega$ $R_b = 3.0\ \Omega$ $R_c = 40.0\ \Omega$
b. $R_a = 12.0\ \Omega$ $R_b = 35.0\ \Omega$ $R_c = 25.0\ \Omega$
c. $R_a = 15.0\ \Omega$ $R_b = 28.0\ \Omega$ $R_c = 12.0\ \Omega$

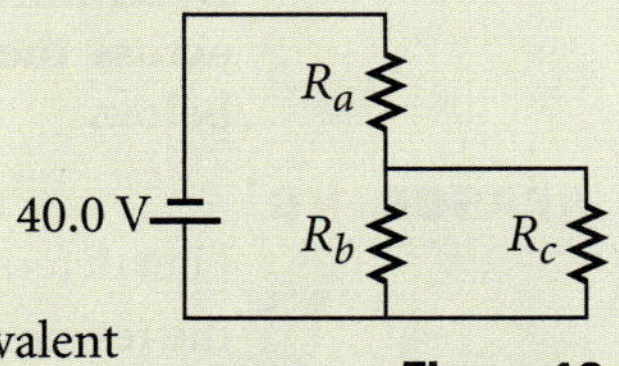

Figure 16

2. For each of the following sets of values, determine the equivalent resistance for the circuit shown in **Figure 17.**

a. $R_a = 25.0\ \Omega$ $R_b = 3.0\ \Omega$ $R_c = 40.0\ \Omega$
$R_d = 15.0\ \Omega$ $R_e = 18.0\ \Omega$

b. $R_a = 12.0\ \Omega$ $R_b = 35.0\ \Omega$ $R_c = 25.0\ \Omega$
$R_d = 50.0\ \Omega$ $R_e = 45.0\ \Omega$

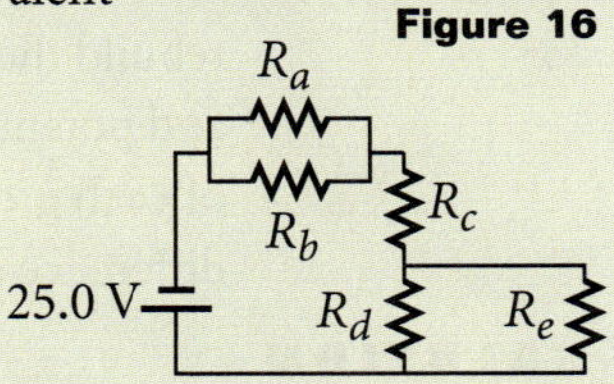

Figure 17

Work backward to find the current in and potential difference across a part of a circuit

Now that the equivalent resistance for a complex circuit has been determined, you can work backward to find the current in and potential difference across any resistor in that circuit. In the household example, substitute potential difference and equivalent resistance in $\Delta V = IR$ to find the total current in the circuit. Because the fuse or circuit breaker is in series with the load, the current in it is equal to the total current. Once this total current is determined, $\Delta V = IR$ can again be used to find the potential difference across the fuse or circuit breaker.

There is no single formula for finding the current in and potential difference across a resistor buried inside a complex circuit. Instead, $\Delta V = IR$ and the rules reviewed in **Table 3** must be applied to smaller pieces of the circuit until the desired values are found.

Table 3 Series and Parallel Resistors

	Series	Parallel
current	same as total	add to find total
potential difference	add to find total	same as total

ANSWERS

Practice C

1. **a.** 27.8 Ω
b. 26.6 Ω
c. 23.4 Ω

2. **a.** 50.9 Ω
b. 57.6 Ω

Alternative Problem-Solving Approach

Students should be encouraged to suggest and examine alternative ways and sequences for grouping the resistors.

For example, you could first find the equivalent resistance of the 6.0 Ω and 2.0 Ω resistors shown as group **(b)** in Sample Problem C, then find the equivalent resistance of the 8.0 Ω and 4.0 Ω resistors shown as group **(c).** The result would be four resistors in series: 6.0 Ω, 3.0 Ω, 2.7 Ω, and 1.0 Ω. The equivalent resistance of these four resistors is 12.7 Ω.

Classroom Practice

Current in and Potential Difference Across a Resistor
Use the following values with the circuit in **Figure 18** in Practice D. What is the current in and potential difference across each of the resistors?

$R_a = 8.0\ \Omega$, $R_b = 4.0\ \Omega$, $R_c = 6.0\ \Omega$, $R_d = 3.0\ \Omega$, $R_e = 9.0\ \Omega$, $R_f = 7.0\ \Omega$

Answers

$I_a = 0.35$ A, $\Delta V_a = 2.8$ V
$I_b = 0.35$ A, $\Delta V_b = 1.4$ V
$I_c = 0.70$ A, $\Delta V_c = 4.2$ V
$I_d = 0.80$ A, $\Delta V_d = 2.4$ V
$I_e = 0.27$ A, $\Delta V_e = 2.4$ V
$I_f = 1.05$ A, $\Delta V_f = 7.4$ V

SAMPLE PROBLEM D

STRATEGY Current in and Potential Difference Across a Resistor

PROBLEM

Determine the current in and potential difference across the 2.0 Ω resistor highlighted in the figure below.

REASONING

First determine the total circuit current by reducing the resistors to a single equivalent resistance. Then rebuild the circuit in steps, calculating the current and potential difference for the equivalent resistance of each group until the current in and potential difference across the 2.0 Ω resistor are known.

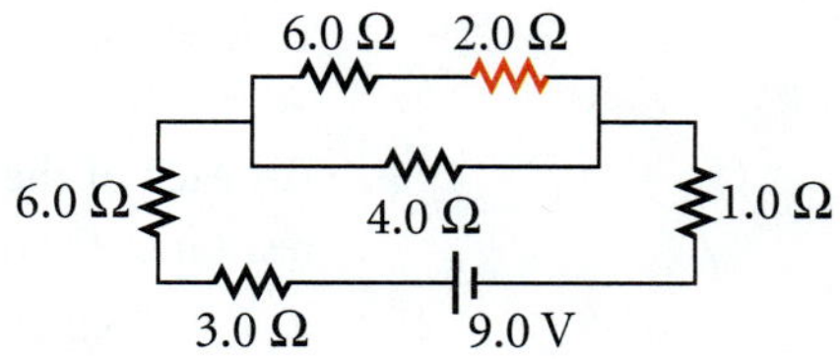

SOLUTION

1. Determine the equivalent resistance of the circuit.
The equivalent resistance of the circuit is 12.7 Ω; this value is calculated in Sample Problem C.

2. Calculate the total current in the circuit.
Substitute the potential difference and equivalent resistance in $\Delta V = IR$, and rearrange the equation to find the current delivered by the battery.

$$I = \frac{\Delta V}{R_{eq}} = \frac{9.0\ \text{V}}{12.7\ \Omega} = 0.71\ \text{A}$$

3. Determine a path from the equivalent resistance found in step 1 to the 2.0 Ω resistor.
Review the path taken to find the equivalent resistance in the figure at right, and work backward through this path. The equivalent resistance for the entire circuit is the same as the equivalent resistance for group **(d).** The center resistor in group **(d)** in turn is the equivalent resistance for group **(c).** The top resistor in group **(c)** is the equivalent resistance for group **(b),** and the right resistor in group **(b)** is the 2.0 Ω resistor.

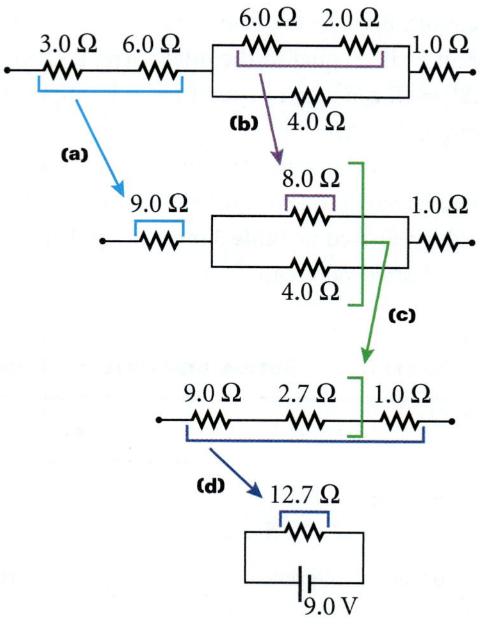

TIP *It is not necessary to solve for R_{eq} first and then work backward to find current in or potential difference across a particular resistor, as shown in this Sample Problem, but working through these steps keeps the mathematical operations at each step simpler.*

4. Follow the path determined in step 3, and calculate the current in and potential difference across each equivalent resistance. Repeat this process until the desired values are found.

A. Regroup, evaluate, and calculate.

Replace the circuit's equivalent resistance with group **(d).** The resistors in group **(d)** are in series; therefore, the current in each resistor is the same as the current in the equivalent resistance, which equals 0.71 A. The potential difference across the 2.7 Ω resistor in group **(d)** can be calculated using $\Delta V = IR$.

Given: $I = 0.71$ A $\quad R = 2.7\ \Omega$

Unknown: $\Delta V = ?$

$$\Delta V = IR = (0.71\ \text{A})(2.7\ \Omega) = 1.9\ \text{V}$$

B. Regroup, evaluate, and calculate.

Replace the center resistor with group **(c).**

The resistors in group **(c)** are in parallel; therefore, the potential difference across each resistor is the same as the potential difference across the 2.7 Ω equivalent resistance, which equals 1.9 V. The current in the 8.0 Ω resistor in group **(c)** can be calculated using $\Delta V = IR$.

Given: $\Delta V = 1.9$ V $\quad R = 8.0\ \Omega$

Unknown: $I = ?$

$$I = \frac{\Delta V}{R} = \frac{1.9\ \text{V}}{8.0\ \Omega} = 0.24\ \text{A}$$

C. Regroup, evaluate, and calculate.

Replace the 8.0 Ω resistor with group **(b).**

The resistors in group **(b)** are in series; therefore, the current in each resistor is the same as the current in the 8.0 Ω equivalent resistance, which equals 0.24 A.

$$\boxed{I = 0.24\ \text{A}}$$

The potential difference across the 2.0 Ω resistor can be calculated using $\Delta V = IR$.

Given: $I = 0.24$ A $\quad R = 2.0\ \Omega$

Unknown: $\Delta V = ?$

$$\Delta V = IR = (0.24\ \text{A})\ (2.0\ \Omega) = 0.48\ \text{V}$$

$$\boxed{\Delta V = 0.48\ \text{V}}$$

TIP *You can check each step in problems like Sample Problem D by using $\Delta V = IR$ for each resistor in a set. You can also check the sum of ΔV for series circuits and the sum of I for parallel circuits.*

Alternative Problem-Solving Approach

Remind students that they can check each step by using $\Delta V = IR$ for each resistor in a set, as discussed in the Tip on this student page. They can also check the sum of ΔV for series circuits and the sum of I for parallel circuits.

For **A,** the potential difference across the 2.7 Ω resistor is 1.9 V. For the other two resistors in series in group **(d):**

$$\Delta V = (0.71\ \text{A})(9.0\ \Omega) = 6.4\ \text{V}$$
$$\Delta V = (0.71\ \text{A})(1.0\ \Omega) = 0.71\ \text{V}$$

The total ΔV across group **(d)** matches the terminal voltage.

$$1.9\ \text{V} + 6.4\ \text{V} + 0.71\ \text{V} = 9.0\ \text{V}$$

For **B,** the current across the 8.0 Ω resistor is 0.24 A. For the other resistor in group **(c):**

$$I = \frac{1.9\ \text{V}}{4.0\ \Omega} = 0.48\ \text{A}$$

The total of these currents is 0.72 A, which differs from 0.71 A because of rounding.

For **C,** the potential difference across the 2.0 Ω resistor is 0.48 V. For the other resistor:

$$\Delta V = (0.24\ \text{A})(6.0\ \Omega) = 1.4\ \text{V}$$

The total of these potential differences is 1.9 V, which was given in the previous step.

PROBLEM GUIDE D

Use this guide to assign problems.
SE = Student Edition Textbook
PW = Problem Workbook
PB = Problem Bank on the One-Stop Planner (OSP)

Solving for:

I	**SE**	Sample, Practice; Ch. Rvw. 25–26
	PW	Sample, 3
	PB	Sample, 1–10
ΔV	**SE**	Sample, Practice; Ch. Rvw. 25–26
	PW	Sample, 1–3
	PB	Sample, 1–10

***Challenging Problem**
Consult the printed Solutions Manual or the OSP for detailed solutions.

ANSWERS

Practice D

R_a: 0.50 A, 2.5 V
R_b: 0.50 A, 3.5 V
R_c: 1.5 A, 6.0 V
R_d: 1.0 A, 4.0 V
R_e: 1.0 A, 4.0 V
R_f: 2.0 A, 4.0 V

Why it Matters

Decorative Lights and Bulbs

Although decorative lights are an excellent topic during classroom discussion of series and parallel circuits, many decorative light sets use the jumpers described in this feature to avoid the pitfalls of each type of circuit. In effect, the jumper functions like a switch that remains open while the filament conducts and closes to connect the wires when the filament burns out.

A more in-depth discussion of this mechanism can be found in the December 1992 edition of *The Physics Teacher.*

PRACTICE D

Current in and Potential Difference Across a Resistor

Calculate the current in and potential difference across each of the resistors shown in the schematic diagram in **Figure 18.**

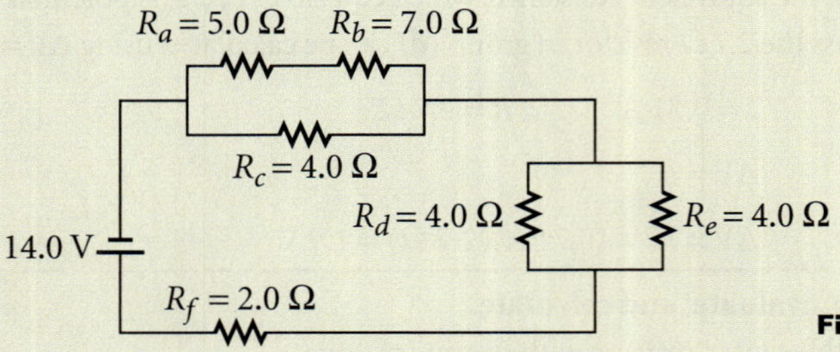

Figure 18

Why it Matters

Decorative Lights and Bulbs

Light sets arranged in series cannot remain lit if a bulb burns out. Wiring in parallel can eliminate this problem, but each bulb must then be able to withstand 120 V. To eliminate the drawbacks of either approach, modern light sets typically contain two or three sections connected to each other in parallel, each of which contains bulbs in series.

When one bulb is removed from a modern light set, half or one-third of the lights in the set go dark because the bulbs in that section are wired in series. When a bulb *burns out,* however, all of the other bulbs in the set remain lit. How is this possible?

Modern decorative bulbs have a short loop of insulated wire, called the *jumper,* that is wrapped around the wires connected to the filament, as shown at left. There is no current in the insulated wire when the bulb is functioning properly. When the filament breaks, however, the current in the section is zero and the potential difference across the two wires connected to the broken filament is then 120 V. This large potential difference creates a spark across the two wires that burns the insulation off the small loop of wire. Once that occurs, the small loop closes the circuit, and the other bulbs in the section remain lit.

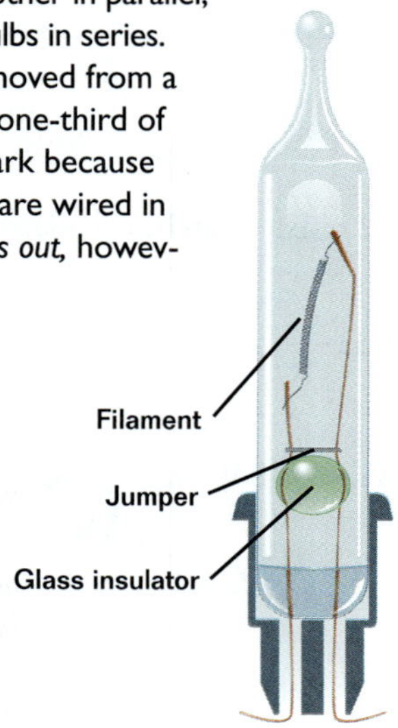

Because the small loop in the burned-out bulb has very little resistance, the equivalent resistance of that portion of the light set decreases; its current increases. This increased current results in a slight increase in each bulb's brightness. As more bulbs burn out, the temperature in each bulb increases and can become a fire hazard; thus, bulbs should be replaced soon after burning out.

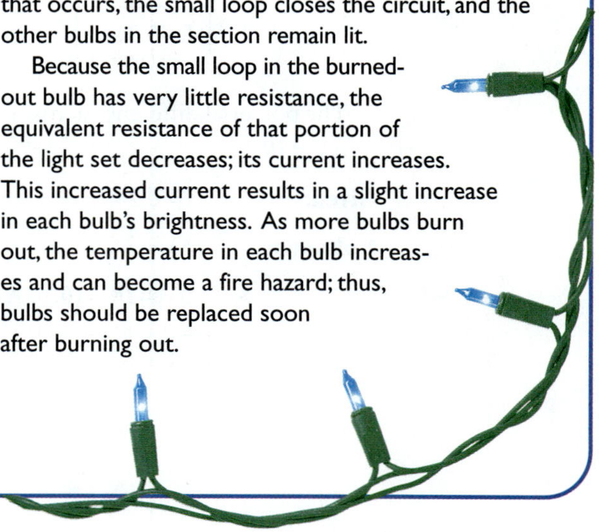

SECTION REVIEW

1. Find the equivalent resistance of the complex circuit shown in **Figure 19.**

2. What is the current in the 1.5 Ω resistor in the complex circuit shown in **Figure 19**?

3. What is the potential difference across the 1.5 Ω resistor in the circuit shown in **Figure 19**?

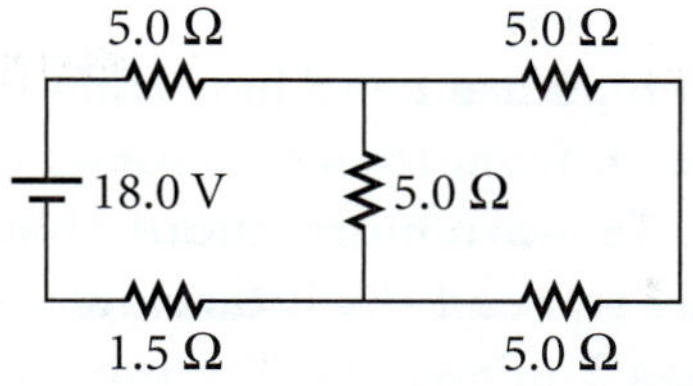

Figure 19

4. A certain strand of miniature lights contains 35 bulbs wired in series, with each bulb having a resistance of 15.0 Ω. What is the equivalent resistance when three such strands are connected in parallel across a potential difference of 120.0 V?

5. What is the current in and potential difference across each of the bulbs in the strands of lights described in item 4?

6. If one of the bulbs in one of the three strands of lights in item 4 goes out while the other bulbs in that strand remain lit, what is the current in and potential difference across each of the lit bulbs in that strand?

7. **Interpreting Graphics** **Figure 20** depicts a household circuit containing several appliances and a circuit breaker attached to a 120 V source of potential difference.
   a. Is the current in the toaster equal to the current in the microwave?
   b. Is the potential difference across the microwave equal to the potential difference across the popcorn popper?
   c. Is the current in the circuit breaker equal to the total current in all of the appliances combined?
   d. Determine the equivalent resistance for the circuit.
   e. Determine how much current is in the toaster.

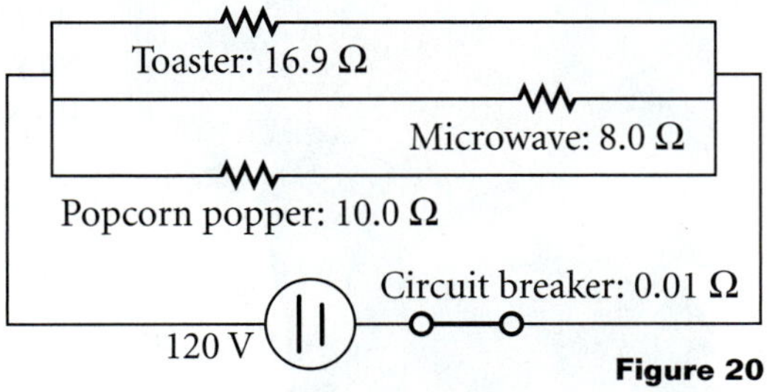

Figure 20

extension

Integrating Health

Visit go.hrw.com for the activity "Recording Electricity in the Brain."

Keyword HF6CIRX

SECTION 3

SECTION REVIEW ANSWERS

1. 9.8 Ω
2. 1.8 A
3. 2.7 V
4. 175 Ω
5. 0.229 A, 3.44 V
6. 0.235 A, 3.52 V
7. **a.** No, the current in the toaster is less than the current in the microwave.
   b. Yes, the potential differences are equal because they are in parallel.
   c. yes, because it is in series with the rest of the circuit
   d. 3.6 Ω
   e. 7.1 A

Semiconductor Technician

Brad Baker credits his father's example with helping guide him into a career in technology. "My father was a systems analyst and into computers," says Baker. "That piqued my interest growing up."

When Baker uses a cooking analogy to describe his work, it is more than mere wordplay. After receiving guidance from the engineer, Baker and his team begin a tinkering process that tweaks the different variables to achieve the desired result. The variables include power, wattage, chemistry gas flow, temperature, and the actual time that the wafer is in the process.

Baker works with another process called *photo engineering,* which helps devices do more work while getting smaller. In Baker's words, engineers "print the design on a wafer, and then we etch away the parts that aren't needed." After the etch process, another team adds material to connect different layers. Then, the process is repeated.

PHYSICS CAREERS

Semiconductor Technician

Brad Baker is creating a recipe on the plasma etch tool to test a new process.

Electronic chips are used in a wide variety of devices, from toys to phones to computers. To learn more about chip making as a career, read the interview with Etch Process Engineering Technician Brad Baker, who works for Motorola.

What training did you receive in order to become a semiconductor technician?

My experience is fairly unique. My degree is in psychology. You have to have an associate's degree in some sort of electrical or engineering field or an undergraduate degree in any field.

What about semiconductor manufacturing made it more interesting than other fields?

While attending college, I worked at an airline. There was not a lot of opportunity to advance, which helped point me in other directions. Circuitry has a lot of parallels to the biological aspects of the brain, which is what I studied in school. We use the scientific method a lot.

What is the nature of your work?

I work on the etch process team. Device engineers design the actual semiconductor. Our job is to figure out how to make what they have requested. It's sort of like being a chef. Once you have experience, you know which ingredient to add.

What is your favorite thing about your job?

I feel like a scientist. My company gives us the freedom to try new things and develop new processes.

Has your job changed since you started it?

Each generation of device is smaller, so we have to do more in less space. As the devices get smaller, it becomes more challenging to get a design process that is powerful enough but doesn't etch too much or too little.

What advice do you have for students who are interested in semiconductor engineering?

The field is very science oriented, so choose chemical engineering, electrical engineering, or material science as majors. Other strengths are the ability to understand and meet challenges, knowledge of trouble-shooting techniques, patience, and analytical skills. Also, everything is computer automated, so you have to know how to use computers.

CHAPTER 18 Highlights

KEY IDEAS

Section 1 Schematic Diagrams and Circuits

- Schematic diagrams use standardized symbols to summarize the contents of electric circuits.
- A circuit is a set of electrical components connected so that they provide one or more complete paths for the movement of charges.
- Any device that transforms nonelectrical energy into electrical energy, such as a battery or a generator, is a source of emf.
- If the internal resistance of a battery is neglected, the emf can be considered equal to the terminal voltage, the potential difference across the source's two terminals.

Section 2 Resistors in Series or in Parallel

- Resistors in series have the same current.
- The equivalent resistance of a set of resistors connected in series is the sum of the individual resistances.
- The sum of currents in parallel resistors equals the total current.
- The equivalent resistance of a set of resistors connected in parallel is calculated using an inverse relationship.

Section 3 Complex Resistor Combinations

- Many complex circuits can be understood by isolating segments that are in series or in parallel and simplifying them to their equivalent resistances.

Variable Symbols

Quantities	Units	Conversions
I current	A amperes	= C/s = coulombs of charge per second
R resistance	Ω ohms	= V/A = volts per ampere of current
ΔV potential difference	V volts	= J/C = joules of energy per coulomb of charge

KEY TERMS

schematic diagram (p. 640)

electric circuit (p. 642)

series (p. 647)

parallel (p. 651)

PROBLEM SOLVING

See **Appendix D: Equations** for a summary of the equations introduced in this chapter. If you need more problem-solving practice, see **Appendix I: Additional Problems.**

Diagram Symbols

Wire or conductor	
Resistor or circuit load	
Bulb or lamp	
Plug	
Battery/ direct-current emf source	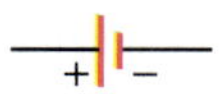
Switch	
Capacitor	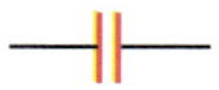

CHAPTER 18

Highlights

Teaching Tip

Ask students to prepare a concept map for the chapter. The concept map should include most of the vocabulary terms, along with other integral terms or concepts.

Review

ANSWERS

1. Schematic diagrams are useful because they summarize the contents of an electric circuit.
2. Accept any schematic diagram that contains three resistors, a battery, and a switch.
3. *B, C*
4. 12.0 V
5. b
6. Charges move through both the emf source and the load.
7. When the circuit is open, there is no complete path for charge flow and hence no current.
8. Some of the electrical energy is dissipated by heat, and the remainder is converted into light energy.
9. A potential difference across the body from contact with a faulty wire can generate a current in the body.
10. all; *D*
11. b
12. a
13. Because the resistance is very low, the current in a short circuit is high ($I = \frac{\Delta V}{R}$). Such high currents can cause wires to overheat, causing a fire.
14. A fuse will not work in parallel because there is an alternative path for the current.

SCHEMATIC DIAGRAMS AND CIRCUITS

Review Questions

1. Why are schematic diagrams useful?

2. Draw a circuit diagram for a circuit containing three 5.0 Ω resistors, a 6.0 V battery, and a switch.

3. The switch in the circuit shown at right can be set to connect to points *A, B,* or *C.* Which of these connections will provide a complete circuit?

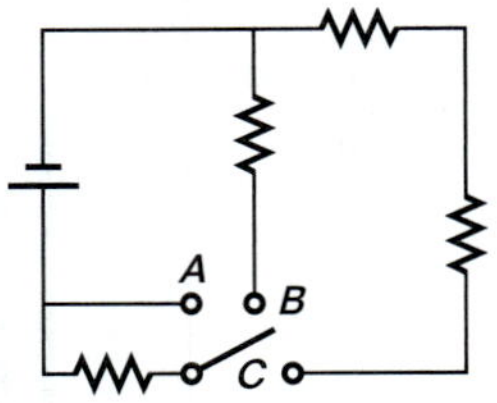

4. If the batteries in a cassette recorder provide a terminal voltage of 12.0 V, what is the potential difference across the entire recorder?

5. In a case in which the internal resistance of a battery is significant, which is greater?
   a. the terminal voltage
   b. the emf of the battery

Conceptual Questions

6. Do charges move from a source of potential difference into a load or through both the source and the load?

7. Assuming that you want to create a circuit that has current in it, why should there be no openings in the circuit?

8. Suppose a 9 V battery is connected across a light bulb. In what form is the electrical energy supplied by the battery dissipated by the light bulb?

9. Why is it dangerous to use an electrical appliance when you are in the bathtub?

10. Which of the switches in the circuit below will complete a circuit when closed? Which will cause a short circuit?

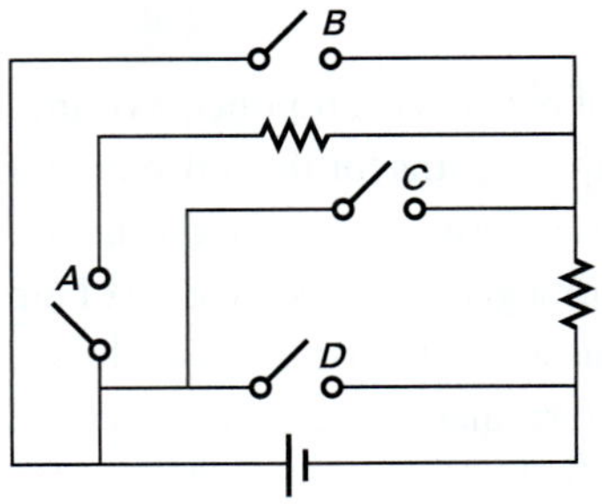

RESISTORS IN SERIES OR IN PARALLEL

Review Questions

11. If four resistors in a circuit are connected in series, which of the following is the same for the resistors in the circuit?
    a. potential difference across the resistors
    b. current in the resistors

12. If four resistors in a circuit are in parallel, which of the following is the same for the resistors in the circuit?
    a. potential difference across the resistors
    b. current in the resistors

Conceptual Questions

13. A short circuit is a circuit containing a path of very low resistance in parallel with some other part of the circuit. Discuss the effect of a short circuit on the current within the portion of the circuit that has very low resistance.

14. Fuses protect electrical devices by opening a circuit if the current in the circuit is too high. Would a fuse work successfully if it were connected in parallel with the device that it is supposed to protect?

15. What might be an advantage of using two identical resistors in parallel that are connected in series with another identical parallel pair, as shown below, instead of using a single resistor?

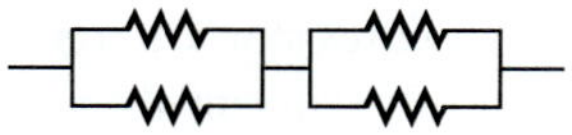

Practice Problems

For problems 16–17, see Sample Problem A.

16. A length of wire is cut into five equal pieces. If each piece has a resistance of 0.15 Ω, what was the resistance of the original length of wire?

17. A 4.0 Ω resistor, an 8.0 Ω resistor, and a 12 Ω resistor are connected in series with a 24 V battery. Determine the following:

a. the equivalent resistance for the circuit

b. the current in the circuit

For problems 18–19, see Sample Problem B.

18. The resistors in item 17 are connected in parallel across a 24 V battery. Determine the following:

a. the equivalent resistance for the circuit

b. the current delivered by the battery

19. An 18.0 Ω resistor, 9.00 Ω resistor, and 6.00 Ω resistor are connected in parallel across a 12 V battery. Determine the following:

a. the equivalent resistance for the circuit

b. the current delivered by the battery

COMPLEX RESISTOR COMBINATIONS

Conceptual Questions

20. A technician has two resistors, each of which has the same resistance, R.

a. How many different resistances can the technician achieve?

b. Express the effective resistance of each possibility in terms of R.

21. The technician in item 20 finds another resistor, so now there are three resistors with the same resistance.

a. How many different resistances can the technician achieve?

b. Express the effective resistance of each possibility in terms of R.

22. Three identical light bulbs are connected in circuit to a battery, as shown below. Compare the level of brightness of each bulb when all the bulbs are illuminated. What happens to the brightness of each bulb if the following changes are made to the circuit?

a. Bulb *A* is removed from its socket.

b. Bulb *C* is removed from its socket.

c. A wire is connected directly between points *D* and *E*.

d. A wire is connected directly between points *D* and *F*.

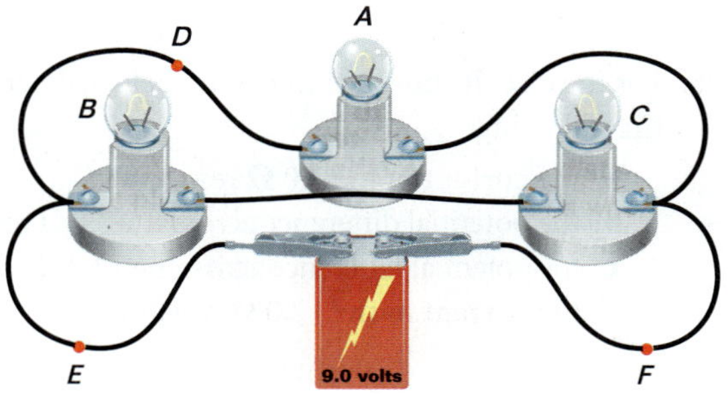

Practice Problems

For problems 23–24, see Sample Problem C.

23. Find the equivalent resistance of the circuit shown in the figure below.

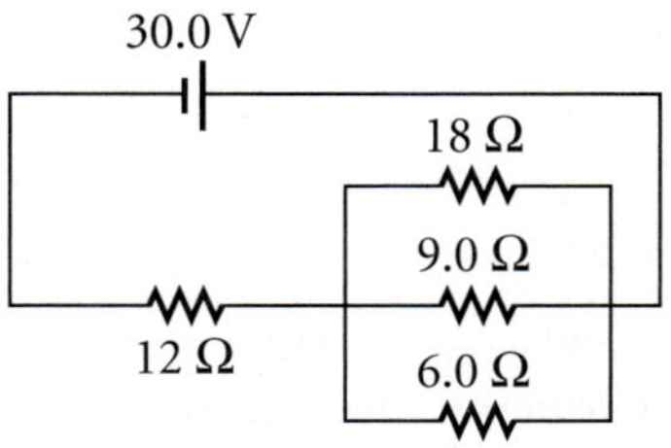

18 REVIEW

15. Because the resistors in each group are in parallel, a broken resistor does not open the circuit.

16. 0.75 Ω

17. **a.** 24 Ω
b. 1.0 A

18. **a.** 2.2 Ω
b. 11 A

19. **a.** 2.99 Ω
b. 4.0 A

20. **a.** three combinations
b. $R, 2R, \frac{R}{2}$

21. **a.** seven combinations
b. $R, 2R, 3R, \frac{R}{2}, \frac{R}{3}, \frac{2R}{3}, \frac{3R}{2}$

22. Bulb *A* is brighter than bulbs *B* and *C*. Bulbs *B* and *C* have the same brightness.

a. Bulbs *B* and *C* stay the same.

b. Bulb *A* stays the same, and bulb *B* goes dark because the circuit is open at *C*.

c. There is no change in any of the bulbs.

d. No bulbs light because there is a short circuit across the battery.

23. 15 Ω

24. 13.3 Ω

25. 3.0 Ω: 1.8 A, 5.4 V
6.0 Ω: 1.1 A, 6.5 V
9.0 Ω: 0.72 A, 6.5 V

26. **a.** 1.7 A
b. 3.4 V
c. 5.1 V
d. 0.42 A

27. 28 V

28. 2.2 V

29. 3.8 V

30. 3.0×10^1 V

31. **a.** 33.0 Ω
b. 132 V
c. 4.00 A, 4.00 A

32. **a.** Place one 20 Ω resistor in series with two parallel 50 Ω resistors.
b. Place two parallel 50 Ω resistors in series with two parallel 20 Ω resistors, or place two circuits, each composed of a 20 Ω resistor in series with a 50 Ω resistor, in parallel.

33. 10.0 Ω

34. 1875 Ω

24. Find the equivalent resistance of the circuit shown in the figure below.

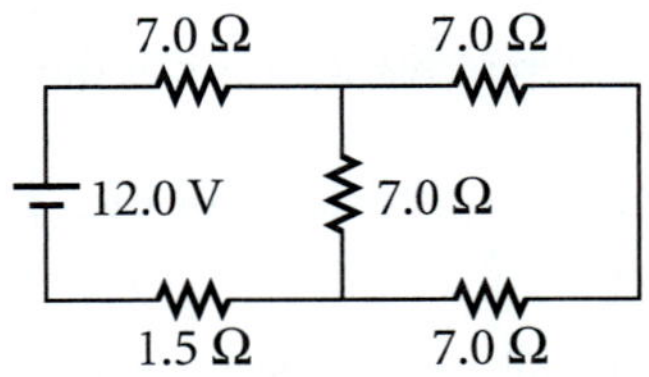

For problems 25–26, see Sample Problem D.

25. For the circuit shown below, determine the current in each resistor and the potential difference across each resistor.

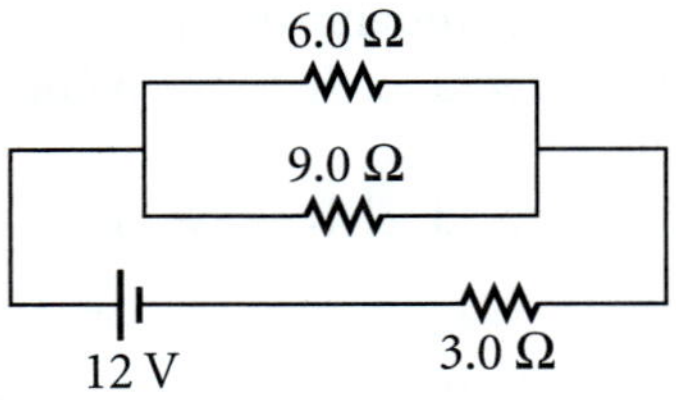

26. For the circuit shown in the figure below, determine the following:

a. the current in the 2.0 Ω resistor
b. the potential difference across the 2.0 Ω resistor
c. the potential difference across the 12.0 Ω resistor
d. the current in the 12.0 Ω resistor

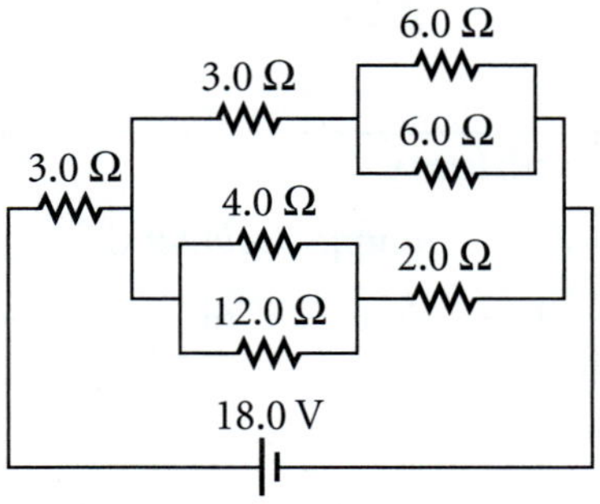

MIXED REVIEW

27. An 8.0 Ω resistor and a 6.0 Ω resistor are connected in series with a battery. The potential difference across the 6.0 Ω resistor is measured as 12 V. Find the potential difference across the battery.

28. A 9.0 Ω resistor and a 6.0 Ω resistor are connected in parallel to a battery, and the current in the 9.0 Ω resistor is found to be 0.25 A. Find the potential difference across the battery.

29. A 9.0 Ω resistor and a 6.0 Ω resistor are connected in series to a battery, and the current through the 9.0 Ω resistor is 0.25 A. What is the potential difference across the battery?

30. A 9.0 Ω resistor and a 6.0 Ω resistor are connected in series with an emf source. The potential difference across the 6.0 Ω resistor is measured with a voltmeter to be 12 V. Find the potential difference across the emf source.

31. An 18.0 Ω, 9.00 Ω, and 6.00 Ω resistor are connected in series with an emf source. The current in the 9.00 Ω resistor is measured to be 4.00 A.

a. Calculate the equivalent resistance of the three resistors in the circuit.
b. Find the potential difference across the emf source.
c. Find the current in the other resistors.

32. The stockroom has only 20 Ω and 50 Ω resistors.

a. You need a resistance of 45 Ω. How can this resistance be achieved using three resistors?
b. Describe two ways to achieve a resistance of 35 Ω using four resistors.

33. The equivalent resistance of the circuit shown below is 60.0 Ω. Use the diagram to determine the value of R.

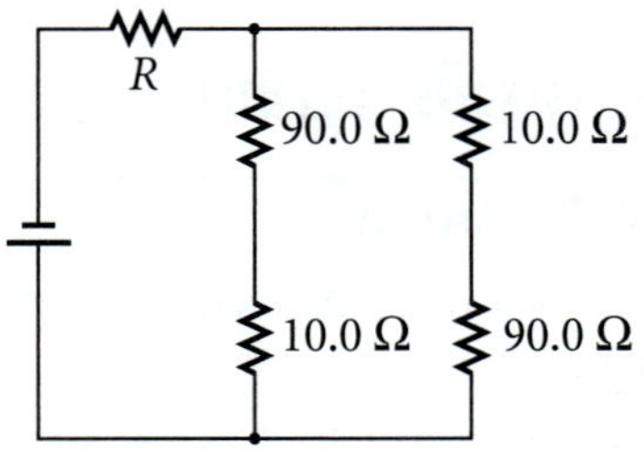

34. Two identical parallel-wired strings of 25 bulbs are connected to each other in series. If the equivalent resistance of the combination is 150.0 Ω and it is connected across a potential difference of 120.0 V, what is the resistance of each individual bulb?

35. The figures **(a)–(e)** below depict five resistance diagrams. Each individual resistance is 6.0 Ω.

a. Which resistance combination has the largest equivalent resistance?

b. Which resistance combination has the smallest equivalent resistance?

c. Which resistance combination has an equivalent resistance of 4.0 Ω?

d. Which resistance combination has an equivalent resistance of 9.0 Ω?

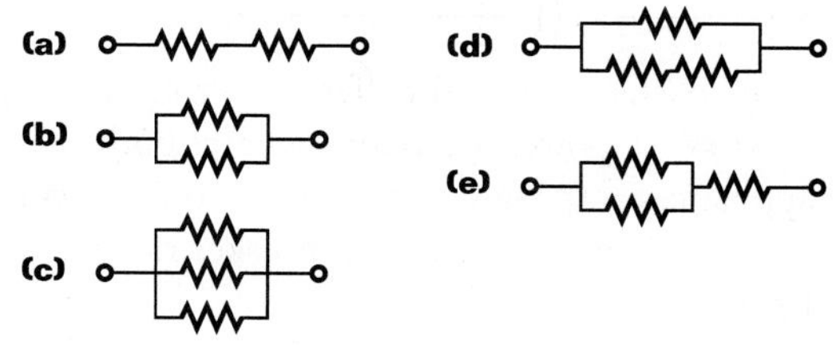

36. Three small lamps are connected to a 9.0 V battery, as shown below.

a. What is the equivalent resistance of this circuit?

b. What is the current in the battery?

c. What is the current in each bulb?

d. What is the potential difference across each bulb?

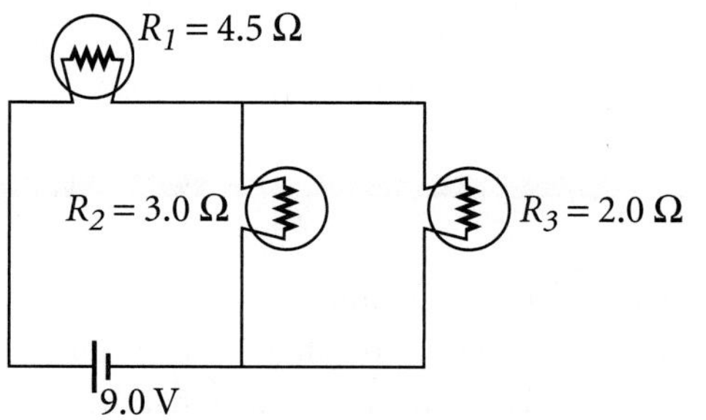

37. An 18.0 Ω resistor and a 6.0 Ω resistor are connected in series to an 18.0 V battery. Find the current in and the potential difference across each resistor.

38. A 30.0 Ω resistor is connected in parallel to a 15.0 Ω resistor. These are joined in series to a 5.00 Ω resistor and a source with a potential difference of 30.0 V.

a. Draw a schematic diagram for this circuit.

b. Calculate the equivalent resistance.

c. Calculate the current in each resistor.

d. Calculate the potential difference across each resistor.

39. A resistor with an unknown resistance is connected in parallel to a 12 Ω resistor. When both resistors are connected to an emf source of 12 V, the current in the unknown resistor is measured with an ammeter to be 3.0 A. What is the resistance of the unknown resistor?

40. The resistors described in item 37 are reconnected in parallel to the same 18.0 V battery. Find the current in each resistor and the potential difference across each resistor.

41. The equivalent resistance for the circuit shown below drops to one-half its original value when the switch, *S*, is closed. Determine the value of *R*.

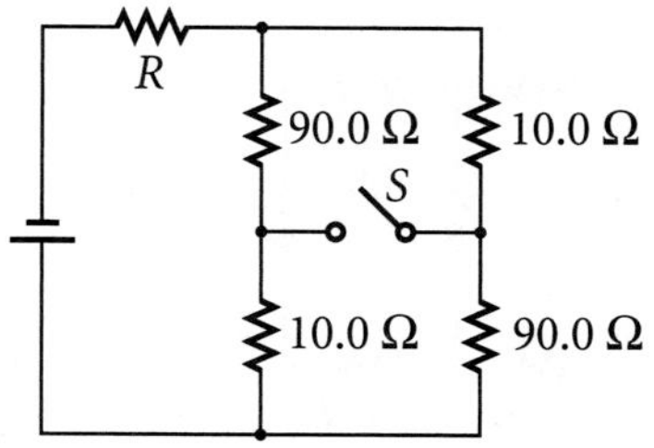

42. You can obtain only four 20.0 Ω resistors from the stockroom.

a. How can you achieve a resistance of 50.0 Ω under these circumstances?

b. What can you do if you need a 5.0 Ω resistor?

43. Four resistors are connected to a battery with a terminal voltage of 12.0 V, as shown below. Determine the following:

a. the equivalent resistance for the circuit

b. the current in the battery

c. the current in the 30.0 Ω resistor

d. the power dissipated by the 50.0 Ω resistor

e. the power dissipated by the 20.0 Ω resistor

(Hint: Remember that $P = \frac{(\Delta V)^2}{R} = I\Delta V$.)

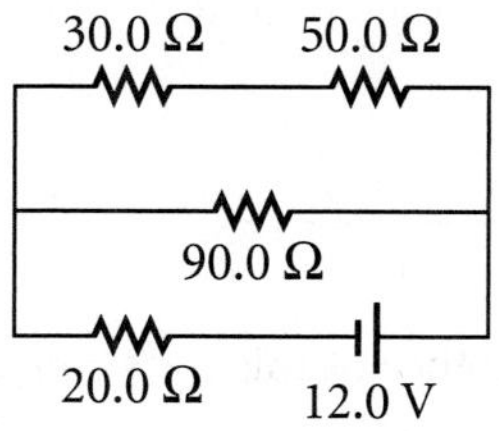

18 REVIEW

35. **a.** a
b. c
c. d
d. e

36. **a.** 5.7 Ω
b. 1.6 A
c. 1.6 A (R_1), 0.63 A (R_2), 0.95 A (R_3)
d. 7.2 V (R_1), 1.9 V (R_2), 1.9 V (R_3)

37. 18.0 Ω: 0.750 A, 13.5 V
6.0 Ω: 0.750 A, 4.5 V

38. **a.**

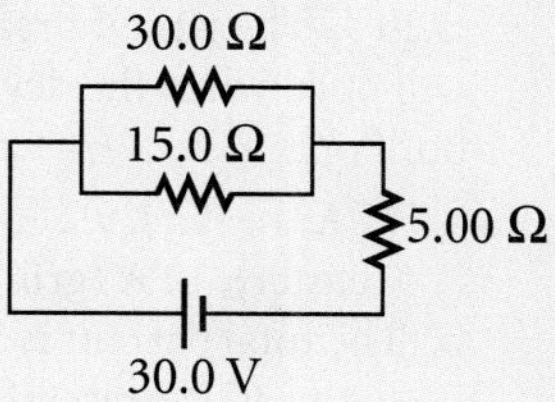

b. 15.00 Ω
c. 5.00 Ω: 2.00 A;
15.0 Ω: 1.33 A;
30.0 Ω: 0.667 A
d. 5.00 Ω: 10.0 V;
15.0 Ω: 20.0 V;
30.0 Ω: 20.0 V

39. 4.0 Ω

40. 18.0 Ω: 1.00 A, 18.0 V
6.0 Ω: 3.0 A, 18.0 V

41. 13.96 Ω

42. **a.** two resistors in series with two parallel resistors
b. four parallel resistors

43. **a.** 62.4 Ω
b. 0.192 A
c. 0.102 A
d. 0.520 W
e. 0.737 W

18 REVIEW

44. 6.0 Ω (*A*), 3.0 Ω (*B*)

45. The circuit must contain three groups of resistors, each containing three resistors in parallel, that are connected to one another in series.

46. a. 14.0 Ω
 b. 2.0 A

47. a. 5.1 Ω
 b. 4.5 V

48. no; Assuming devices are wired in parallel, total current is 20 A. The circuit breaker will open when the devices are both on.

49. a. 11 A (heater), 9.2 A (toaster), 12 A (grill)
 b. The total current is 32.2 A, so the 30.0 A circuit breaker will open the circuit if these appliances are all on.

extension

Graphing Calculator Practice

Visit go.hrw.com for answers to this Graphing Calculator activity.

Keyword HF6CIRXT

44. Two resistors, *A* and *B*, are connected in series to a 6.0 V battery. A voltmeter connected across resistor *A* measures a potential difference of 4.0 V. When the two resistors are connected in parallel across the 6.0 V battery, the current in *B* is found to be 2.0 A. Find the resistances of *A* and *B*.

45. Draw a schematic diagram of nine 100 Ω resistors arranged in a series-parallel network so that the total resistance of the network is also 100 Ω. All nine resistors must be used.

46. For the circuit below, find the following:

a. the equivalent resistance of the circuit

b. the current in the 5.0 Ω resistor

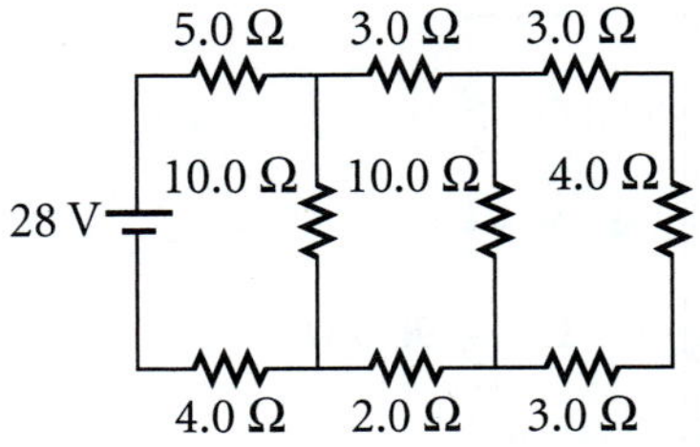

47. The power supplied to the circuit shown below is 4.00 W. Determine the following:

a. the equivalent resistance of the circuit

b. the potential difference across the battery

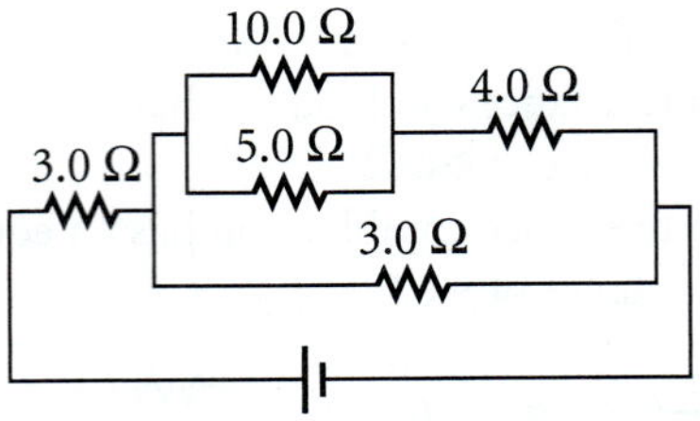

48. Your toaster oven and coffee maker each dissipate 1200 W of power. Can you operate both of these appliances at the same time if the 120 V line you use in your kitchen has a circuit breaker rated at 15 A? Explain.

49. An electric heater is rated at 1300 W, a toaster is rated at 1100 W, and an electric grill is rated at 1500 W. The three appliances are connected in parallel across a 120 V emf source.

a. Find the current in each appliance.

b. Is a 30.0 A circuit breaker sufficient in this situation? Explain.

Graphing Calculator *Practice*

Parallel Resistors

Electric circuits are often composed of combinations of series and parallel circuits. The overall resistance of a circuit is determined by dividing the circuit into groups of series and parallel resistors and determining the equivalent resistance of each group. As you learned earlier in this chapter, the equivalent resistance of parallel resistors is given by the following equation:

$$\frac{1}{R_{eq}} = \frac{1}{R_1} + \frac{1}{R_2} + \frac{1}{R_3} + \cdots$$

One interesting consequence of this equation is that the equivalent resistance for resistors in parallel will always be less than the smallest resistor in the group.

In this graphing calculator activity, you will determine the equivalent resistance for various resistors in parallel. You will confirm that the equivalent resistance is always less than the smallest resistor, and you will relate the number of resistors and changes in resistance to the equivalent resistance.

Visit go.hrw.com and type in the keyword **HF6CIRX** to find this graphing calculator activity. Refer to **Appendix B** for instructions on downloading the program for this activity.

Alternative Assessment

1. How many ways can two or more batteries be connected in a circuit with a light bulb? How will the current change depending on the arrangement? First draw diagrams of the circuits you want to test. Then identify the measurements you need to make to answer the question. If your teacher approves your plan, obtain the necessary equipment and perform the experiment.

2. Research the career of an electrical engineer or technician. Prepare materials for people interested in this career field. Include information on where people in this career field work, which tools and equipment they use, and the challenges of their field. Indicate what training is typically necessary to enter the field.

3. The manager of an automotive repair shop has been contacted by two competing firms that are selling ammeters to be used in testing automobile electrical systems. One firm has published claims that its ammeter is better because it has high internal resistance. The other firm has published claims that its ammeter is better because it has low resistance. Write a report with your recommendation to the manager of the automotive repair shop. Include diagrams and calculations that explain how you reached your conclusion.

4. You and your friend want to start a business exporting small electrical appliances. You have found people willing to be your partners to distribute these appliances in Germany. Write a letter to these potential partners that describes your product line and that asks for the information you will need about the electric power, sources, consumption, and distribution in Germany.

5. Contact an electrician, builder, or contractor, and ask to see a house electrical plan. Study the diagram to identify the circuit breakers, their connections to different appliances in the home, and the limitations they impose on the circuit's design. Find out how much current, on average, is in each appliance in the house. Draw a diagram of the house, showing which circuit breakers control which appliances. Your diagram should also keep the current in each of these appliances under the performance and safety limits.

Alternative Assessment ANSWERS

1. Students' plans should be safe and should test series and parallel combinations of batteries.
2. Students should recognize that the principles of circuits are applied by electrical engineers and technicians.
3. An ammeter is connected in series in a circuit, so it must have low internal resistance in order to measure all parts of the circuit without interfering with it.
4. Students' letters should formulate clear direct questions and request meaningful information.
5. Students should draw several parallel sets of appliances, each of which shows the appliances wired in series to a circuit breaker.

Standardized Test Prep

ANSWERS

1. C

2. J

3. B

4. F

5. B

6. J

7. D

8. G

9. C

Standardized Test Prep

MULTIPLE CHOICE

1. Which of the following is the correct term for a circuit that does not have a closed-loop path for electron flow?

A. closed circuit
B. dead circuit
C. open circuit
D. short circuit

2. Which of the following is the correct term for a circuit in which the load has been unintentionally bypassed?

F. closed circuit
G. dead circuit
H. open circuit
J. short circuit

Use the diagram below to answer questions 3–5.

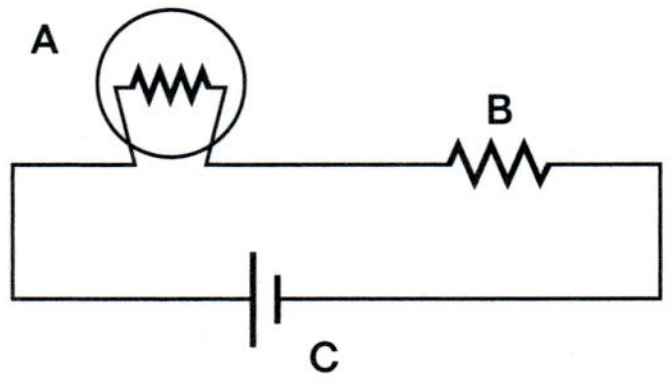

3. Which of the circuit elements contribute to the load of the circuit?

A. Only A
B. A and B, but not C
C. Only C
D. A, B, and C

4. Which of the following is the correct equation for the equivalent resistance of the circuit?

F. $R_{eq} = R_A + R_B$

G. $\frac{1}{R_{eq}} = \frac{1}{R_A} + \frac{1}{R_B}$

H. $R_{eq} = I\Delta V$

J. $\frac{1}{R_{eq}} = \frac{1}{R_A} + \frac{1}{R_B} + \frac{1}{R_C}$

5. Which of the following is the correct equation for the current in the resistor?

A. $I = I_A + I_B + I_C$

B. $I_B = \frac{\Delta V}{R_{eq}}$

C. $I_B = I_{total} + I_A$

D. $I_B = \frac{\Delta V}{R_B}$

Use the diagram below to answer questions 6–7.

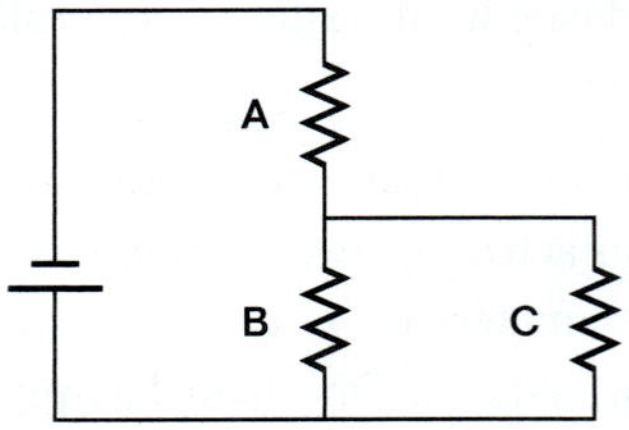

6. Which of the following is the correct equation for the equivalent resistance of the circuit?

F. $R_{eq} = R_A + R_B + R_C$

G. $\frac{1}{R_{eq}} = \frac{1}{R_A} + \frac{1}{R_B} + \frac{1}{R_C}$

H. $R_{eq} = I\Delta V$

J. $R_{eq} = R_A + \left(\frac{1}{R_B} + \frac{1}{R_C}\right)^{-1}$

7. Which of the following is the correct equation for the current in resistor B?

A. $I = I_A + I_B + I_C$

B. $I_B = \frac{\Delta V}{R_{eq}}$

C. $I_B = I_{total} + I_A$

D. $I_B = \frac{\Delta V_B}{R_B}$

8. Three 2.0 Ω resistors are connected in series to a 12 V battery. What is the potential difference across each resistor?
   F. 2.0 V
   G. 4.0 V
   H. 12 V
   J. 36 V

Use the following passage to answer questions 9–11.
Six light bulbs are connected in parallel to a 9.0 V battery. Each bulb has a resistance of 3.0 Ω.

9. What is the potential difference across each bulb?
   A. 1.5 V
   B. 3.0 V
   C. 9.0 V
   D. 27 V

10. What is the current in each bulb?
   F. 0.5 A
   G. 3.0 A
   H. 4.5 A
   J. 18 A

11. What is the total current in the circuit?
   A. 0.5 A
   B. 3.0 A
   C. 4.5 A
   D. 18 A

SHORT RESPONSE

12. Which is greater, a battery's terminal voltage or the same battery's emf? Explain why these two quantities are not equal.

13. Describe how a short circuit could lead to a fire.

14. Explain the advantage of wiring the bulbs in a string of decorative lights in parallel rather than in series.

EXTENDED RESPONSE

15. Using standard symbols for circuit elements, draw a diagram of a circuit that contains a battery, an open switch, and a light bulb in parallel with a resistor. Add an arrow to indicate the direction of current if the switch were closed.

Use the diagram below to answer questions 16–17.

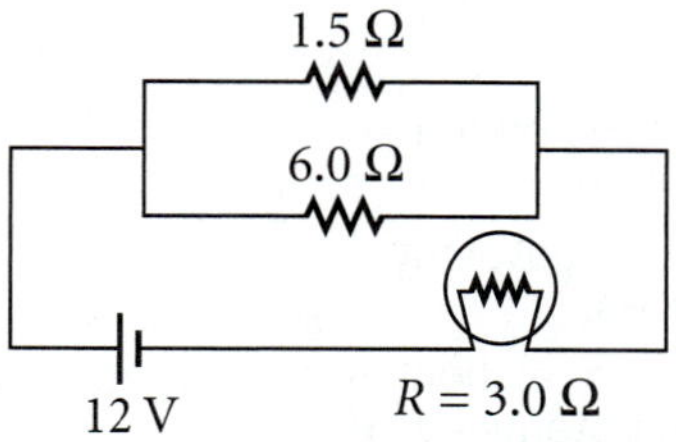

16. For the circuit shown, calculate the following:
   a. the equivalent resistance of the circuit
   b. the current in the light bulb.
   Show all your work for both calculations.

17. After a period of time, the 6.0 Ω resistor fails and breaks. Describe what happens to the brightness of the bulb. Support your answer.

18. Find the current in and potential difference across each of the resistors in the following circuits:
   a. a 4.0 Ω and a 12.0 Ω resistor wired in series with a 4.0 V source.
   b. a 4.0 Ω and a 12.0 Ω resistor wired in parallel with a 4.0 V source.
   Show all your work for each calculation.

19. Find the current in and potential difference across each of the resistors in the following circuits:
   a. a 150 Ω and a 180 Ω resistor wired in series with a 12 V source.
   b. a 150 Ω and a 180 Ω resistor wired in parallel with a 12 V source.
   Show all your work for each calculation.

Test TIP Prepare yourself for taking an important test by getting plenty of sleep the night before and by eating a healthy breakfast on the day of the test.

10. G

11. D

12. A battery's emf is slightly greater than its terminal voltage. The difference is due to the battery's internal resistance.

13. In a short circuit, the equivalent resistance of the circuit drops very low, causing the current to be very high. The higher current can cause wires still in the circuit to overheat, which may in turn cause a fire in materials contacting the wires.

14. If one bulb is removed, the other bulbs will still carry current.

15.

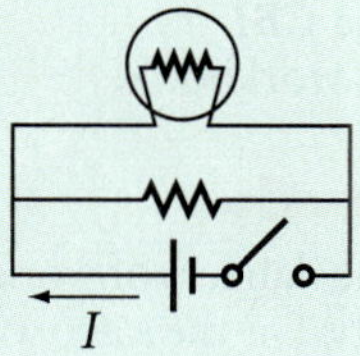

16. a. 4.2 Ω **b.** 2.9 A (See the Solutions Manual or One-Stop Planner for the full solution.)

17. The bulb will grow dim. The loss of the 6.0 Ω resistor causes the equivalent resistance of the circuit to increase to 4.5 Ω. As a result, the current in the bulb drops to 2.7 A, and the brightness of the bulb decreases.

18. a. 4.0 Ω: 0.25 A, 1.0 V
12.0 Ω: 0.25 A, 3.0 V
b. 4.0 Ω: 1.0 A, 4.0 V
12.0 Ω: 0.33 A, 4.0 V

19. a. 150 Ω: 0.036 A, 5.4 V
180 Ω: 0.036 A, 6.5 V
b. 150 Ω: 0.080 A, 12 V
180 Ω: 0.067 A, 12 V

Lab Planning

Beginning on page T34 are preparation notes and teaching tips to assist you in planning.

Blank data tables (as well as some sample data) appear on the **One-Stop Planner.**

No Books in the Lab?

See the ***Datasheets for In-Text Labs*** workbook for a reproducible master copy of this experiment.

The same workbook also contains a version of this experiment with explicit procedural steps if you prefer a more directed approach.

CBL™ Option

A CBL™ version of this lab appears in the ***CBL™ Experiments*** workbook.

Safety Caution

Emphasize the dangers of working with electricity. For the safety of the students and of the equipment, remind students to have you check their circuits before turning on the power supply or closing the switch.

Tips and Tricks

- See the preparation notes and teaching tips beginning on page T34 for instructions on setting up multimeters to measure potential difference and current.
- Make sure students understand how to wire current meters (in series) and voltage meters (in parallel) in a circuit.
- If the students are using a dc power supply, show them how to adjust the voltage.

CHAPTER 18

Inquiry Lab

Resistors in Series and in Parallel

Design Your Own

OBJECTIVES

- **Measure** current in and potential difference across resistors in series and in parallel.
- **Find** the unknown resistances of two resistors.
- **Calculate** equivalent resistances.
- **Analyze** the relationships between potential difference, current, and resistance in a circuit.

MATERIALS LIST

- **2 multimeters, or 1 dc ammeter and 1 voltmeter**
- **2 resistors**
- **insulated connecting wire**
- **power supply**
- **switch**

In this lab, you will design an experiment to compare the circuits created by wiring two unknown resistors first in series and then in parallel. By taking measurements of the current in and the potential difference (voltage) across the resistors, and the potential difference across the whole circuit, you will find the value of the resistance of each resistor and the equivalent resistance for both resistors to compare the total current in each of the two circuits.

SAFETY

- **Never close a circuit until it has been approved by your teacher.**
- **Never rewire or adjust any element of a closed circuit.**
- **Never work with electricity near water; be sure the floor and all work surfaces are dry.**
- **If the pointer on any kind of meter moves off scale, open the circuit immediately by opening the switch.**
- **Do not attempt this exercise with any batteries or electrical devices other than those provided by your teacher for this purpose.**
- **Use a hot mitt to handle resistors, light sources, and other equipment that may be hot. Allow all equipment to cool before storing it.**

PROCEDURE

1. Study the materials provided, and design an experiment to meet the goals stated above. If you are not certain how to use the power supply, the meters, or any of the other materials, ask your teacher for help.
2. Write out your lab procedure, including a detailed description of the measurements to take during each step. You may use **Figure 1** and **Figure 2** as guides to possible setups. The dc power supplies should be set to about 5.0 V when you are taking measurements.
3. Ask your teacher to approve your procedure.
4. Follow all steps of your procedure. **IMPORTANT: Your teacher must approve your circuit before you turn on the power supply in any step. Always open the switch immediately after you have taken measurements. Do not change the circuit or any of the meter connections**

while the switch is closed. Any time you change your circuit, including the points of connection for any of the meters, your teacher must approve the circuit again before you close the switch.

5. Clean up your work area. Put equipment away safely so that it is ready to be used again.

Figure 1
• Use the voltage meter to measure potential difference across the resistors.

ANALYSIS

1. **Organizing Data** Using your measurements for potential difference and current, compute the resistance values of R_1 and R_2 in each circuit.

2. **Analyzing Results** Compare your results from item 1 for the different circuits.
   a. Do R_1 and R_2 have the same values in each circuit?
   b. Did you expect R_1 and R_2 to have the same values? Explain. If the results are different, suggest a possible reason.

3. **Organizing Data** Compute the equivalent resistance R_{eq} using the values found in item 1 for each circuit.

4. **Analyzing Results** Based on your calculations in item 3, did the two resistors provide the same equivalent resistance in both circuits? If not, which combination had the greater resistance? Explain how the combination of resistors affects the total resistance in the circuit.

5. **Organizing Data** Compute the total current in each circuit using the calculated value for R_{eq} and the measured value for ΔV_T.

6. **Analyzing Results** Do both circuits have the same total current? If not, which circuit has the greater current? Explain how the combination of resistors affects the total current in the circuit.

CONCLUSIONS

7. **Drawing Conclusions** Compare the total current in each circuit with the current in each resistor. What is the relationship between the current in an individual resistor and the total current in the circuit?

8. **Drawing Conclusions** For each circuit, compare the potential difference across each resistor with ΔV_T. What is the relationship?

• Explain the black (negative) and red (positive) convention, and show students how to connect the battery or power supply and the meters or probes correctly.

ANSWERS

Analysis

1. Answers must use the relationship $R = \frac{\Delta V}{I}$. Typical values for sample data range from 61 Ω to 142 Ω.

2. **a, b.** The resistors are the same. If not, answers will vary but should include an analysis of error in the lab.

3. Make sure students use the relationships $R_{eq} = R_1 + R_2$ and $\frac{1}{R_{eq}} = \frac{1}{R_1} + \frac{1}{R_2}$. Answers will vary, but typical values for sample data range from 46 Ω to 194 Ω.

4. The resistors have a higher equivalent resistance in the series circuit.

5. $I_{total} = \frac{\Delta V_T}{R_{eq}}$. Typical values will range from 0.0077 A to 0.031 A.

6. The parallel circuit has a higher current.

Conclusions

7. *series:* the current in each resistor is equal to the total current in the circuit; *parallel:* the total current in the circuit is equal to the sum of the currents in each resistor

8. *series:* ΔV_T is approximately equal to the sum of the voltage across each resistor; *parallel:* ΔV_T is equal to the voltage across each individual resistor

CHAPTER 19

Magnetism
Planning Guide

Compression Guide
To shorten instruction because of time limitations, omit the opener and abbreviate the review.

OBJECTIVES	LABS, DEMONSTRATIONS, AND ACTIVITIES	TECHNOLOGY RESOURCES
PACING • 45 min pp. 676–677 **Chapter Opener**	ANC **Discovery Lab** Magnetism*◆ BASIC	CD **Visual Concepts,** Chapter 19 BASIC
PACING • 45 min pp. 678–682 **Section 1 Magnets and Magnetic Fields** • For given situations, predict whether magnets will repel or attract each other. • Describe the magnetic field around a permanent magnet. • Describe the orientation of Earth's magnetic field.	SE **Quick Lab** Magnetic Field of a File Cabinet, p. 681 GENERAL TE **Demonstration** Magnetic Poles, p. 678 BASIC TE **Demonstration** Magnetic Domains, p. 679 GENERAL TE **Demonstration** Magnetic Fields, p. 680 GENERAL	OSP **Lesson Plans** EXT **Integrating Chemistry** Molecular Magnetism BASIC EXT **Integrating Technology** Magnetic Resonance Imaging BASIC TR 98 Magnetic Field of a Bar Magnet TR 99 Earth's Magnetic Field
PACING • 90 min pp. 684–686 **Section 2 Magnetism from Electricity** • Describe the magnetic field produced by current in a straight conductor and in a solenoid. • Use the right-hand rule to determine the direction of the magnetic field in a current-carrying wire.	SE **Quick Lab** Electromagnetism, p. 685 GENERAL SE **Skills Practice Lab** Magnetic Field of a Conducting Wire, pp. 702–703◆ GENERAL ANC **Datasheet** Magnetic Field of a Conducting Wire* GENERAL SE **CBL™ Lab** Magnetic Field of a Conducting Wire, pp. 940–941◆ GENERAL ANC **CBL™ Experiments** Magnetic Field of a Conducting Wire*◆ GENERAL TE **Demonstration** Current-Carrying Wire, p. 684 BASIC	OSP **Lesson Plans** TR 100 Magnetic Field of a Current-Carrying Wire TR 101 The Right-Hand Rule TR 102 Magnetic Field of a Current Loop and of a Solenoid
PACING • 90 min pp. 687–693 **Section 3 Magnetic Force** • Given the force on a charge in a magnetic field, determine the strength of the magnetic field. • Use the right-hand rule to find the direction of the force on a charge moving through a magnetic field. • Determine the magnitude and direction of the force on a wire carrying current in a magnetic field.	TE **Demonstration** Electromagnetic Force, p. 687 GENERAL TE **Demonstration** Force Between Parallel Conductors, p. 691 ADVANCED ANC **Invention Lab** Designing a Magnetic Spring*◆ ADVANCED ANC **CBL™ Experiments** Magnetic Field Strength*◆ ADVANCED	OSP **Lesson Plans** TR 103 Force on a Moving Charge TR 104 Charge Moving in a Magnetic Field TR 105 Force on a Current-Carrying Wire TR 106 Force Between Parallel Wires TR 62A Cathode Ray Tube TR 63A Loudspeaker

PACING • 90 min

CHAPTER REVIEW, ASSESSMENT, AND STANDARDIZED TEST PREPARATION

SE **Chapter Highlights,** p. 694
SE **Chapter Review,** pp. 695–699
SE **Graphing Calculator Practice,** p. 698 GENERAL
SE **Alternative Assessment,** p. 699 ADVANCED
SE **Standardized Test Prep,** pp. 700–701 GENERAL
SE **Appendix D: Equations,** p. 863
SE **Appendix I: Additional Problems,** pp. 894–895
ANC **Study Guide Worksheet** Mixed Review* GENERAL
ANC **Chapter Test A*** GENERAL
ANC **Chapter Test B*** ADVANCED
OSP **Test Generator**

Online and Technology Resources

Holt Online Learning

Visit go.hrw.com to find a variety of online resources. To access this chapter's extensions, enter the keyword HF6MAGXT and click the "go" button. Click Holt Online Learning for an online edition of this textbook, and other interactive resources.

HOLT Teacher's One-Stop Planner®

This DVD package includes:
- Holt Calendar Planner
- Customizable Lesson Plans
- Editable Worksheets
- ExamView® Version 6 Assessment Suite
- Interactive Teacher's Edition
- Holt PuzzlePro®
- Holt PowerNotes® Presentations
- MindPoint® Quiz Show

KEY

SE	Student Edition	OSP	One-Stop Planner	EXT	Online Extension
TE	Teacher Edition	CD	CD or CD-ROM	*	Also on One-Stop Planner
ANC	Ancillary Worksheet	TR	Teaching Transparencies	◆	Requires advance prep

SKILLS DEVELOPMENT RESOURCES	REVIEW AND ASSESSMENT	CORRELATIONS
		National Science Education Standards
	SE **Section Review,** p. 682 GENERAL ANC **Study Guide Worksheet** Section 1* GENERAL ANC **Quiz** Section 1* BASIC	UCP 1, 2, 3, 4, 5 SAI 1, 2 SPSP 3, 5 PS 5b
	SE **Section Review,** p. 686 GENERAL ANC **Study Guide Worksheet** Section 2* GENERAL ANC **Quiz** Section 2* BASIC	UCP 1, 2, 3, 5 SAI 1, 2 ST 1, 2 SPSP 1, 2, 4, 5 PS 2d, 4e, 5b
SE **Sample Set A** Particle in a Magnetic Field, p. 689 BASIC TE **Classroom Practice,** p. 688 BASIC ANC **Problem Workbook** Sample Set A* BASIC OSP **Problem Bank** Sample Set A BASIC SE **Sample Set B** Force on a Current-Carrying conductor, p. 692 BASIC TE **Classroom Practice,** p. 692 BASIC ANC **Problem Workbook** Sample Set B* BASIC OSP **Problem Bank** Sample Set B BASIC	SE **Section Review,** p. 693 GENERAL ANC **Study Guide Worksheet** Section 3* GENERAL ANC **Quiz** Section 3* BASIC	UCP 1, 2, 3, 5 SAI 1, 2 ST 1, 2 HNS 1 SPSP 5 PS 4a, 4c, 5b

SCILINKS. NSTA
www.scilinks.org

Maintained by the **National Science Teachers Association.**

Topic: Magnets
SciLinks Code: HF60901

Topic: Magnetic Fields
SciLinks Code: HF60898

Topic: Electromagnets
SciLinks Code: HF60484

Classroom CD-ROMs

- Guided Reading Audio Program
- Student One Stop
- Virtual Investigations
- Visual Concepts

Holt Lab Generator CD-ROM

Search for any lab by topic, standard, difficulty level, or time. Edit any lab to fit your needs, or create your own labs. Use the Lab Materials QuickList software to customize your lab materials list.

CHAPTER 19
Overview

Section 1 introduces magnets and magnetic fields and discusses magnetization.

Section 2 applies the right-hand rule for magnetism, explores electromagnetism and solenoids, and introduces magnetic domains.

Section 3 concentrates on calculations of magnetic fields and magnetic forces.

About the Illustration

Astronauts Dale A. Gardner and Joseph P. Allen IV work together to bring the *Westar VI* telecommunications satellite into the *Discovery* space shuttle's payload bay. Allen is on a mobile foot restraint, which is attached to the *Discovery*'s Remote Manipulator System. The satellite had to be recovered because a propulsion systems defect prevented it from reaching a sufficient orbital radius (altitude) for telecommunications purposes.

CHAPTER 19

Magnetism

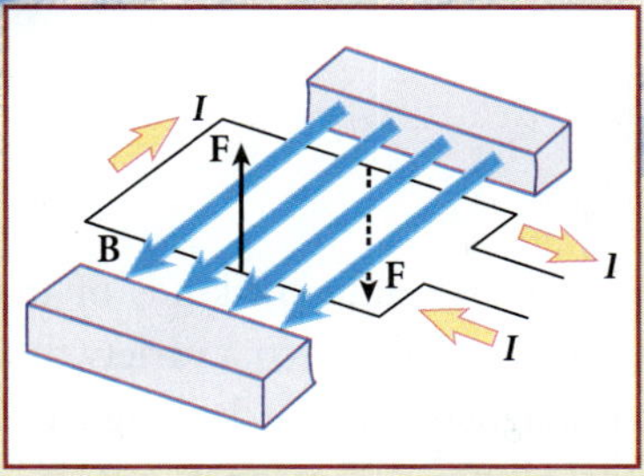

Satellites sometimes contain loops of wire called *magnetic torque coils* that a satellite operator on Earth can activate. When current is in the coil, the magnetic field of Earth exerts a torque on the loop of wire. Torque coils are used to align a satellite in the orientation needed for its instruments to work.

WHAT TO EXPECT

In this chapter, you will learn that a current-carrying coil of wire behaves like a magnet. You will also study the forces exerted on charged particles that are moving in a magnetic field.

Why it Matters

Permanent magnets and electromagnets are used in many everyday and scientific applications. Huge electromagnets are used to pick up and move heavy loads, such as scrap iron at a recycling plant.

CHAPTER PREVIEW

Tapping Prior Knowledge

Knowledge to Review

- ✔ A net force causes a change in the motion of an object.
- ✔ Torque is the cause of changes in rotation. The magnitude of a torque equals the product of the force and the lever arm.
- ✔ Electric fields surround charged objects and exert forces on other charged objects.
- ✔ Electric current is the rate at which electric charges move through a cross-sectional area.

Items to Probe

- ✔ Electric fields: Have students relate electric-field strength to the force exerted on a charged particle in the field.

677

Demonstration

Magnetic Poles

Purpose Show that all magnets have north and south poles and that there are attractive and repulsive forces between two magnets.

Materials two bar magnets, ring stand, string, various types of magnets, such as those shown in **Figure 1**

Procedure Use the string to suspend one bar magnet horizontally from the ring stand. Have students note that the magnet points north.

Bring the north pole of the other bar magnet near the north pole of the suspended magnet. Have students observe the reaction. Ask students to predict what will happen when you bring the south pole of the unattached magnet near the north pole of the suspended magnet. Then ask them what will happen when you bring the south pole of the unattached magnet near the south pole of the suspended magnet. Demonstrate both cases.

Perform similar demonstrations with the other types of magnets.

Magnets and Magnetic Fields

SECTION OBJECTIVES

- **For given situations, predict whether magnets will repel or attract each other.**
- **Describe the magnetic field around a permanent magnet.**
- **Describe the orientation of Earth's magnetic field.**

SCILINKS
www.scilinks.org
Topic: Magnets
Code: HF60901

MAGNETS

Most people have had experience with different kinds of magnets, such as those shown in **Figure 1.** You have probably seen a variety of magnet shapes, such as horseshoe magnets, bar magnets, and the flat magnets frequently used to attach items to a refrigerator. All types of magnets attract iron-containing objects such as paper clips and nails. In the following discussion, we will assume that the magnet has the shape of a bar. Iron objects are most strongly attracted to the ends of such a magnet. These ends are called *poles;* one is called the *north pole,* and the other is called the *south pole.* The names derive from the behavior of a magnet on Earth. If a bar magnet is suspended from its midpoint so that it can swing freely in a horizontal plane, it will rotate until its north pole points north and its south pole points south. In fact, a compass is just a magnetized needle that swings freely on a pivot.

The list of important technological applications of magnetism is very long. For instance, large electromagnets are used to pick up heavy loads. Magnets are also used in meters, motors, generators, and loudspeakers. Magnetic tapes are routinely used in sound- and video-recording equipment, and magnetic recording material is used on computer disks. Superconducting magnets are currently being used to contain extremely high-temperature plasmas that are used in controlled nuclear fusion research. Superconducting magnets are also used to levitate modern trains. These *maglev* trains are faster and provide a smoother ride than the ordinary track system because of the absence of friction between the train and the track.

Figure 1
Magnets come in a variety of shapes and sizes, but like poles of two magnets always repel one another.

Like poles repel each other, and unlike poles attract each other

The magnetic force between two magnets can be likened to the electric force between charged objects in that unlike poles of two magnets attract one another and like poles repel one another. Thus, the north pole of a magnet is attracted to the south pole of another magnet, and two north poles (or two south poles) brought close together repel each other. Electric charges differ from magnetic poles in that they can be isolated, whereas magnetic poles cannot. In fact, no matter how many times a permanent magnet is cut, each piece always has a north pole and a south pole. Thus, magnetic poles always occur in pairs.

MAGNETIC DOMAINS

The magnetic properties of many materials are explained in terms of a model in which an electron is said to spin on its axis much like a top does. (This classical description should not be taken literally. The property of electron spin can be understood only with the methods of quantum mechanics.) The spinning electron represents a charge that is in motion. As you will learn in the next section of this chapter, moving charges create magnetic fields.

In atoms containing many electrons, the electrons usually pair up with their spins opposite each other causing their fields to cancel each other. For this reason, most substances, such as wood and plastic, are not magnetic. However, in materials such as iron, cobalt, and nickel, the magnetic fields produced by the electron spins do not cancel completely. Such materials are said to be *ferromagnetic.*

In ferromagnetic materials, strong coupling occurs between neighboring atoms to form large groups of atoms whose net spins are aligned; these groups are called **magnetic domains.** Domains typically range in size from about 10^{-4} cm to 10^{-1} cm. In an unmagnetized substance, the domains are randomly oriented, as shown in **Figure 2.** When an external magnetic field is applied, the orientation of the magnetic fields of each domain may change slightly to more closely align with the external magnetic field, or the domains that are already aligned with the external field may grow at the expense of the other domains. This alignment enhances the applied magnetic field.

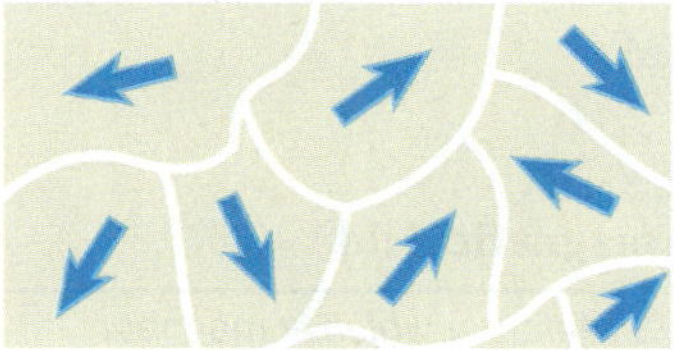

Figure 2
When a substance is unmagnetized, its domains are randomly oriented.

magnetic domain

a region composed of a group of atoms whose magnetic fields are aligned in the same direction

Some materials can be made into permanent magnets

Just as two materials, such as rubber and wool, can become charged after they are rubbed together, an unmagnetized piece of iron can become a permanent magnet by being stroked with a permanent magnet. Magnetism can be induced by other means as well. For example, if a piece of unmagnetized iron is placed near a strong permanent magnet, the piece of iron will eventually become magnetized. The process can be reversed either by heating and cooling the iron or by hammering the iron, because these actions cause the magnetic domains to jiggle and lose their alignment.

A magnetic piece of material is classified as magnetically *hard* or *soft,* depending on the extent to which it retains its magnetism. Soft magnetic materials, such as iron, are easily magnetized but also tend to lose their magnetism easily. In hard magnetic materials, domain alignment persists after the external magnetic field is removed; the result is a permanent magnet. In contrast, hard magnetic materials, such as cobalt and nickel, are difficult to magnetize, but once they are magnetized, they tend to retain their magnetism. In soft magnetic materials, once the external field is removed, the random motion of the particles in the material changes the orientation of the domains and the material returns to an unmagnetized state.

extension

Integrating Chemistry
Visit go.hrw.com for the activity "Molecular Magnetism."

Keyword HF6MAGX

SECTION 1

Demonstration

Magnetic Domains

Purpose Show the effects of impact on a magnetized ferromagnetic material.

Materials two small paper clips, bar magnet

Procedure Pick up one paper clip with the magnet. Touch the second paper clip to the bottom of the first paper clip so that both are suspended from the magnet. Remove the first paper clip from the bar magnet, and have students note that the second paper clip remains suspended from the first. Have students conclude that the second paper clip has also become magnetized. Drop both paper clips onto the table top. Try to pick up one paper clip with the other, and have students note that the clips are no longer magnetic. Ask students to explain. *(The impact of the clips hitting the table caused the domains to once again return to random orientations.)*

Demonstration

Magnetic Fields

Purpose Show the interaction of magnets and magnetic fields.

Materials two bar magnets, one horseshoe magnet, one blank transparency, iron filings, overhead projector

Procedure Set one of the bar magnets on the overhead projector, and lay the blank transparency over the magnet. Sprinkle the iron filings onto the transparency, and have students observe the behavior of the filings. Repeat the demonstration using the following:

- **a.** two bar magnets about 4 cm apart, aligned with opposite poles facing each other
- **b.** two bar magnets about 4 cm apart, aligned with like poles facing each other
- **c.** horseshoe magnet

MAGNETIC FIELDS

You know that the interaction between charged objects can be described using the concept of an electric field. A similar approach can be used to describe the **magnetic field** that surrounds any magnetized material. As with an electric field, a magnetic field, **B,** is a vector quantity that has both magnitude and direction.

magnetic field

a region in which a magnetic force can be detected

Magnetic field lines can be drawn with the aid of a compass

The magnetic field of a bar magnet can be explored using a compass, as illustrated in **Figure 3.** If a small, freely suspended bar magnet, such as the needle of a compass, is brought near a magnetic field, the compass needle will align with the magnetic field lines. The direction of the magnetic field, **B,** at any location is defined as the direction that the north pole of a compass needle points to at that location.

Magnetic field lines appear to begin at the north pole of a magnet and to end at the south pole of a magnet. However, magnetic field lines have no beginning or end. Rather, they always form a closed loop. In a permanent magnet, the field lines actually continue within the magnet itself to form a closed loop. (These lines are not shown in the illustration.)

This text will follow a simple convention to indicate the direction of **B.** An arrow will be used to show a magnetic field that is in the same plane as the page, as shown in **Table 1.** When the field is directed into the page, we will use a series of blue crosses to represent the tails of arrows. If the field is directed out of the page, we will use a series of blue dots to represent the tips of arrows.

Table 1
Conventions for Representing the Direction of a Magnetic Field

In the plane of the page	↑
Into the page	✕
Out of the page	●

Magnetic flux relates to the strength of a magnetic field

One useful way to model magnetic field strength is to define a quantity called *magnetic flux,* Φ_M. It is defined as the number of field lines that cross a certain area at right angles to that area. Magnetic flux can be calculated by the following equation.

MAGNETIC FLUX

$$\Phi_M = AB\cos\theta$$

magnetic flux = (surface area) × (magnetic field component normal to the plane of surface)

Now look again at **Figure 3.** Imagine two circles of the same size that are perpendicular to the axis of the magnet. One circle is located near one pole of the magnet, and the other circle is alongside the magnet. More magnetic field lines cross the circle that is near the pole of the magnet. This greater flux indicates that the magnetic field is strongest at the magnet's poles.

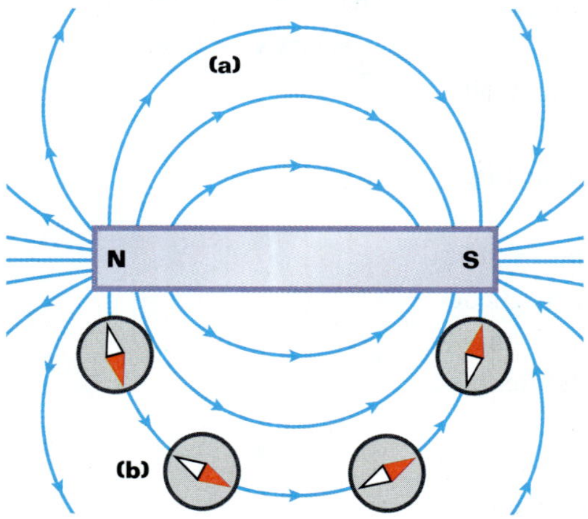

Figure 3
The magnetic field **(a)** of a bar magnet can be traced with a compass **(b).** Note that the north poles of the compasses point in the direction of the field lines from the magnet's north pole to its south pole.

Earth has a magnetic field similar to that of a bar magnet

The north and south poles of a small bar magnet are correctly described as the "north-seeking" and "south-seeking" poles. This description means that if a magnet is used as a compass, the north pole of the magnet will seek, or point to, a location near the geographic North Pole of Earth. Because unlike poles attract, we can deduce that the geographic North Pole of Earth corresponds to the magnetic south pole and the geographic South Pole of Earth corresponds to the magnetic north pole. Note that the configuration of Earth's magnetic field, pictured in **Figure 4,** resembles the field that would be produced if a bar magnet were buried within Earth.

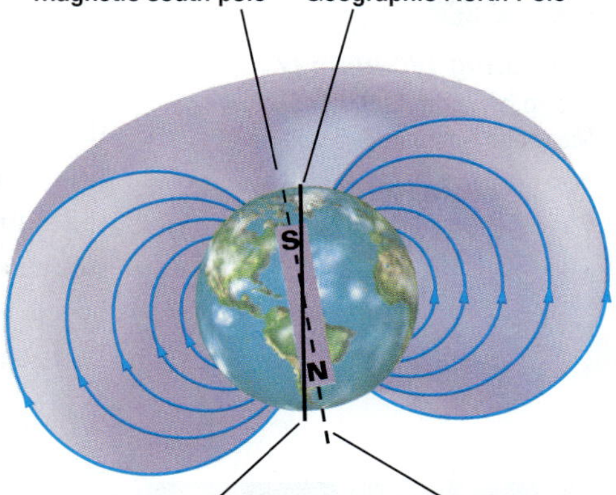

Figure 4
Earth's magnetic field has a configuration similar to a bar magnet's. Note that the magnetic south pole is near the geographic North Pole and that the magnetic north pole is near the geographic South Pole.

If a compass needle is allowed to rotate both perpendicular to and parallel to the surface of Earth, the needle will be exactly parallel with respect to Earth's surface only near the equator. As the compass is moved northward, the needle will rotate so that it points more toward the surface of Earth. Finally, at a point just north of Hudson Bay, in Canada, the north pole of the needle will point perpendicular to Earth's surface. This site is considered to be the location of the magnetic south pole of Earth. It is approximately 1500 km from Earth's geographic North Pole. Similarly, the magnetic north pole of Earth is roughly the same distance from the geographic South Pole.

The difference between true north, which is defined by the axis of rotation of Earth, and north indicated by a compass, varies from point to point on Earth. This difference is referred to as *magnetic declination.* An imaginary line running roughly north-south near the center of North America currently has zero declination. Along the line a compass will indicate true north. However, in the state of Washington, a compass aligns about 20° east of true north. To further complicate matters, geological evidence indicates that Earth's magnetic field has changed—and even reversed—throughout Earth's history.

Although Earth has large deposits of iron ore deep beneath its surface, the high temperatures in Earth's liquid core prevent the iron from retaining any permanent magnetization. It is considered more likely that the source of Earth's magnetic field is the movement of charges in *convection currents* in

Did you know?

By convention, the north pole of a magnet is frequently painted red. This practice comes from the longstanding use of magnets, in the form of compasses, as navigational aids. Long before Global Positioning System (GPS) satellites, the compass gave humans an easy way to orient themselves.

Quick Lab

Magnetic Field of a File Cabinet

MATERIALS LIST

- compass
- metal file cabinet

Stand in front of the file cabinet, and hold the compass face up and parallel to the ground. Now move the compass from the top of the file cabinet to the bottom. Making sure that the compass is parallel to the ground, check to see if the direction of the compass needle changes as it moves from the top of the cabinet to the bottom. If the compass needle changes direction, the file cabinet is magnetized. Can you explain what might have caused the file cabinet to become magnetized? Remember that Earth's magnetic field has a vertical component as well as a horizontal component.

Try tracing the field around some large metal objects around your house. Can you find an object that has been magnetized by the horizontal component of Earth's magnetic field?

Teaching Tip — GENERAL

Point out that the direction of Earth's magnetic field has reversed several times during the last million years. Evidence for this is provided by basalt (an iron-containing rock) that is sometimes spewed forth by volcanic activity on the ocean floor. When the lava is molten, the domains of the ferromagnetic material align with Earth's magnetic field. As the lava cools, it solidifies and retains a picture of Earth's magnetic field direction. Dated basalt deposits provide evidence that Earth's magnetic field has reversed periodically over time.

Quick Lab

TEACHER'S NOTES

If performing this lab in class, test the file cabinet before the lab.

Possible substitutes include iron flagpoles and iron fence posts, such as those around tennis courts.

SECTION 1

SECTION REVIEW ANSWERS

1. **a.** repel
   b. attract
   c. attract
2. two
3. a, b
4. parallel to Earth's surface, pointing from approximately the geographic South Pole (magnetic north pole) to approximately the geographic North Pole (magnetic south pole)
5. hard magnetic material, because it is less easily magnetized

extension

Integrating Technology
Visit go.hrw.com for the activity "Magnetic Resonance Imaging."
Keyword HF6MAGX

Earth's core. These currents occur because the temperature in Earth's liquid core is not evenly distributed. Charged ions or electrons circling in the liquid interior of Earth could produce a magnetic field. There is also evidence that the strength of a planet's magnetic field is related to the planet's rate of rotation. For example, Jupiter rotates at a faster rate than Earth does, and recent space probes indicate that Jupiter's magnetic field is stronger than Earth's is. Conversely, Venus rotates more slowly than Earth does and has been found to have a weaker magnetic field than Earth does. Investigation into the cause of Earth's magnetism continues.

SECTION REVIEW

1. For each of the cases in the figure below, identify whether the magnets will attract or repel one another.

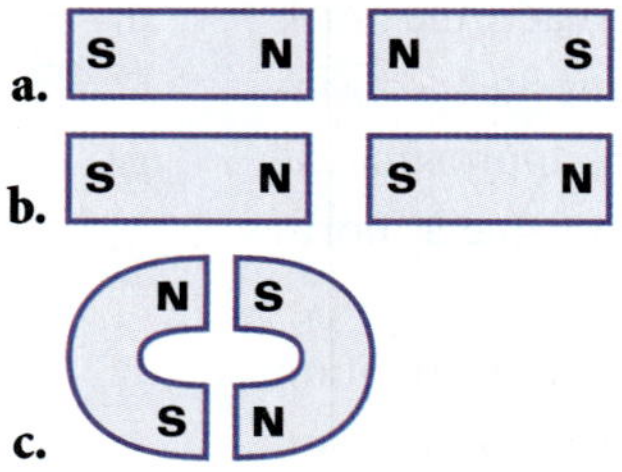

2. When you break a bar magnet in half, how many poles does each piece have?

3. **Interpreting Graphics** Which of the compass-needle orientations in the figure below might correctly describe the magnet's field at that point?

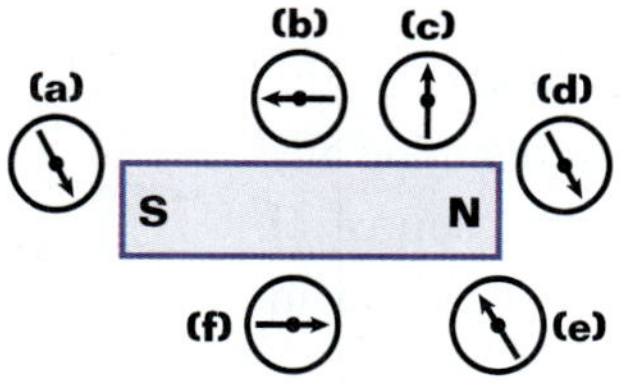

4. **Critical Thinking** Satellite ground operators use the feedback from a device called a magnetometer, which senses the direction of Earth's magnetic field, to decide which torque coil to activate. What direction will the magnetometer read for Earth's magnetic field when the satellite passes over Earth's equator?

5. **Critical Thinking** In order to protect other equipment, the body of a satellite must remain unmagnetized, even when the torque coils have been activated. Would hard or soft magnetic materials be best for building the rest of the satellite?

Why it Matters

Magnetic Resonance Imaging

Magnetic resonance imaging, or MRI, is an imaging technique that has been used in clinical medicine since the early 1980s. MRI allows doctors to make two-dimensional images of or three-dimensional models of parts of the human body. The use of MRI in medicine has grown rapidly. MRI produces high-resolution images that can be tailored to study different types of tissues, depending on the application. Also, MRI procedures are generally much safer than *computerized axial tomography (CAT)* scans, which flood the body with X rays.

A typical MRI machine looks like a giant cube, 2–3 meters on each side, with a cylindrical hollow in the center to accommodate the patient as shown in the illustration. The MRI machine uses electromagnets to create magnetic fields ranging in strength from 0.5–2.0 T. These fields are strong enough to erase credit cards and to pull pens out of pockets, even across the MRI exam room. Because resistance would cause normal electromagnets to dissipate a huge amount of heat when creating fields this strong, the electromagnets in most MRI machines contain superconducting wires that have zero resistance.

The creation of an image with MRI depends on the behavior of atomic nuclei within a magnetic field. In a strong magnetic field, the nucleus of an atom tends to line up along the direction of the field. This behavior is particularly true for hydrogen atoms, which are the most common atoms in the body.

The primary magnet in an MRI system creates a strong, uniform magnetic field centered on the part of the patient that is being examined. The field causes hydrogen nuclei in the body to line up in the direction of the field. Smaller magnets, called *gradient magnets,* are then turned on and off to create small variations, or pulses, in the overall magnetic field. Each pulse causes the hydrogen nuclei to shift away from their alignment. After the pulse, the nuclei return to alignment, and as they do so, they emit radio frequency electromagnetic waves. Scanners within the MRI machine detect these radio waves, and a computer processes the waves into images.

Different types of tissues can be seen with MRI, depending on the frequency and duration of the pulses. MRI is particularly good for imaging the brain and spinal tissues and can be used to study brain function, brain tumors, multiple sclerosis, and other neurological disorders. MRI can also be used to create images of blood vessels without the surrounding tissue, which can be very useful for studying the circulatory system. The main drawbacks of MRI are that MRI systems are very expensive and that MRI cannot be used on some patients, such as those with pacemakers or certain types of metal implants.

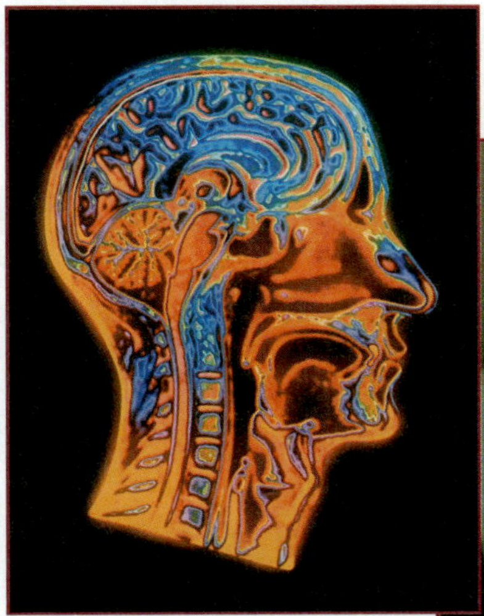

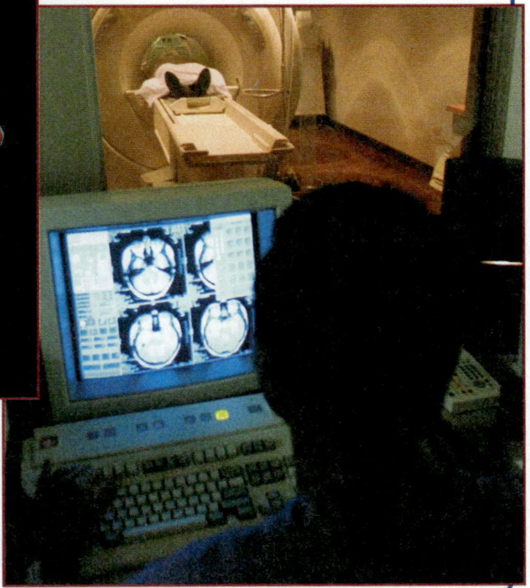

The imaging magnet in most MRI machines is of the superconducting type. The magnet is the most expensive component of the MRI system.

Why it Matters

Magnetic Resonance Imaging

Magnetic resonance imaging (MRI) evolved from nuclear magnetic resonance (NMR) spectroscopy, a technique that has been used in physics research since the middle of the 20th century. The 1952 Nobel Prize in physics was awarded to Felix Bloch and Edward Purcell for their use of NMR spectroscopy to study the behavior of atomic nuclei in the presence of strong magnetic fields.

Raymond Damadian produced the first two-dimensional MRI images in 1973. The first use of MRI on a human was in 1977, and MRI became a regular part of clinical medicine in the 1980s. The 2003 Nobel Prize in physiology or medicine was awarded to Paul C. Lauterbur and Sir Peter Mansfield for their contributions to the development of MRI techniques for diagnostic and research applications in medicine.

SECTION 2
General Level

Demonstration

Current-Carrying Wire — BASIC

Purpose Show that a long, straight, current-carrying wire has a magnetic field.

Materials wire, dc power supply, small compasses, cardboard, ring stand, two clamps

Procedure Cut a small hole in the center of the cardboard. Use the clamp and ring stand to hold the cardboard parallel to the desktop. Thread the wire through the hole in the cardboard, and clamp the wire to the top of the ring stand so that it is perpendicular to the cardboard. Leave at least 10 cm of wire above and below the cardboard. Connect the wire to the dc power supply. Place the compasses on the cardboard in a circular pattern around the wire. Turn on the power supply momentarily, and have students note the deflection of the compass needles. Ask students to describe the magnetic field around the wire *(concentric circles)*. Ask students to predict what will happen if the leads of the wires are reversed *(the compasses will reverse)*. Demonstrate.

SECTION 2

Magnetism from Electricity

SECTION OBJECTIVES

- **Describe the magnetic field produced by current in a straight conductor and in a solenoid.**
- **Use the right-hand rule to determine the direction of the magnetic field in a current-carrying wire.**

MAGNETIC FIELD OF A CURRENT-CARRYING WIRE

Scientists in the late 1700s suspected that there was a relationship between electricity and magnetism, but no theory had been developed to guide their experiments. In 1820, Danish physicist Hans Christian Oersted devised a method to study this relationship. Following a lecture to his advanced class, Oersted demonstrated that when brought near a current-carrying wire, a compass needle is deflected from its usual north-south orientation. He published an account of this discovery in July 1820, and his work stimulated other scientists all over Europe to repeat the experiment.

A long, straight, current-carrying wire has a cylindrical magnetic field

The experiment shown in **Figure 5(a)** uses iron filings to show that a current-carrying conductor produces a magnetic field. In a similar experiment, several compass needles are placed in a horizontal plane near a long vertical wire, as illustrated in **Figure 5(b).** When no current is in the wire, all needles point in the same direction (that of Earth's magnetic field). However, when the wire carries a strong, steady current, all the needles deflect in directions tangent to concentric circles around the wire. This result points out the direction of **B,** the magnetic field *induced* by the current. When the current is reversed, the needles reverse direction.

Figure 5
(a) When the wire carries a strong current, the alignments of the iron filings show that the magnetic field induced by the current forms concentric circles around the wire. **(b)** Compasses can be used to show the direction of the magnetic field induced by the wire.

(a) (b)

The right-hand rule can be used to determine the direction of the magnetic field

These observations show that the direction of **B** is consistent with a simple rule for conventional current, known as *the right-hand rule:* If the wire is grasped in the right hand with the thumb in the direction of the current, as shown in **Figure 6,** the four fingers will curl in the direction of **B.**

As shown in **Figure 5(a),** the lines of **B** form concentric circles about the wire. By symmetry, the magnitude of **B** is the same everywhere on a circular path centered on the wire and lying in a plane perpendicular to the wire. Experiments show that **B** is proportional to the current in the wire and inversely proportional to the distance from the wire.

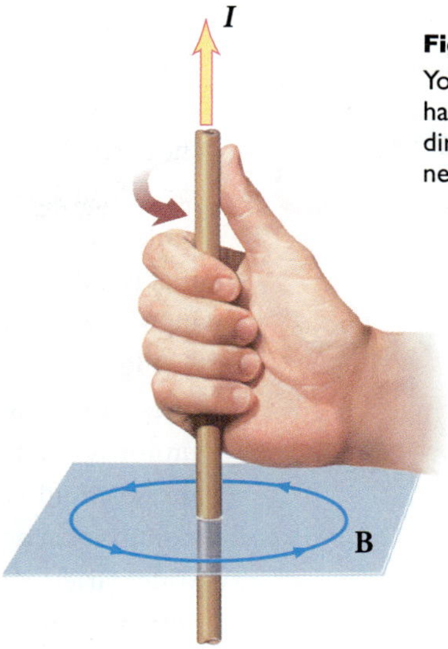

Figure 6
You can use the right-hand rule to find the direction of this magnetic field.

MAGNETIC FIELD OF A CURRENT LOOP

The right-hand rule can also be applied to find the direction of the magnetic field of a current-carrying loop, such as the loop represented in **Figure 7(a).** Regardless of where on the loop you apply the right-hand rule, the field within the loop points in the same direction—upward. Note that the field lines of the current-carrying loop resemble those of a bar magnet, as shown in **Figure 7(b).** If a long, straight wire is bent into a coil of several closely spaced loops, as shown on the next page in **Figure 8,** the resulting device is called a **solenoid.**

solenoid

a long, helically wound coil of insulated wire

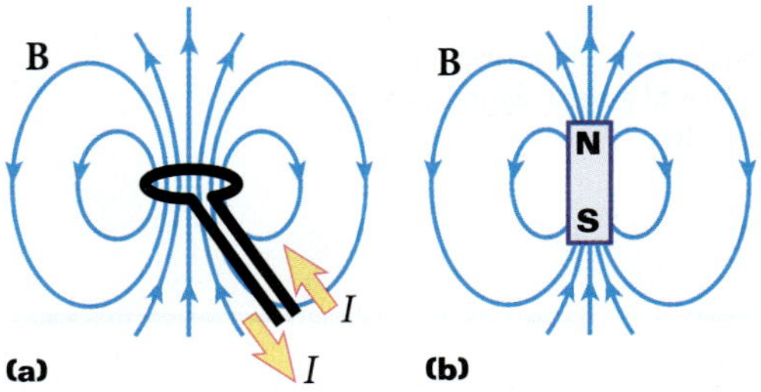

Figure 7
(a) The magnetic field of a current loop is similar to **(b)** that of a bar magnet.

Quick Lab

Electromagnetism

MATERIALS LIST

- **D-cell battery**
- **1 m length of insulated wire**
- **large nail**
- **compass**
- **metal paper clips**

Wind the wire around the nail, as shown below. Remove the insulation from the ends of the wire, and hold these ends against the metal terminals of the battery.

Use the compass to determine whether the nail is magnetized. Next, flip the battery so that the direction of the current is reversed. Again, bring the compass toward the same part of the nail. Can you explain why the compass needle now points in a different direction?

Bring paper clips near the nail while connected to the battery. What happens to the paper clips? How many can you pick up?

Quick Lab

TEACHER'S NOTES

Tell students that a short circuit is being created and that the potential difference across the battery goes to zero quickly. Be sure students disconnect the battery after making their observations.

Have students compare the number of paper clips they can pick up with respect to the number of windings around the nail. More windings result in a larger field, so students should be able to pick up more paper clips with more windings.

SECTION 2

Visual Strategy BASIC

Figure 8
Have students practice using the right-hand rule by confirming that the direction of the magnetic field shown in **Figure 8** is correctly drawn at all points around the solenoid. Point out to students that the fingers curl along the turns of the solenoid.

Q If the current in the solenoid is reversed, which end of the solenoid is the north pole?

A *the right side*

SECTION REVIEW ANSWERS

1. concentric circles around the wire
2. The fields produced by the top and bottom of every loop all point in the same direction and are confined to a small region of space.
3. Electrons usually pair up with their spins opposite each other, and their fields cancel each other.
4. The magnetic field that is induced (by domain variation) in the rod adds to the magnetic field of the solenoid.

SCiLINKS®
www.scilinks.org
Topic: Electromagnets
Code: HF60484

Solenoids produce a strong magnetic field by combining several loops
A solenoid is important in many applications because it acts as a magnet when it carries a current. The magnetic field strength inside a solenoid increases with the current and is proportional to the number of coils per unit length. The magnetic field of a solenoid can be increased by inserting an iron rod through the center of the coil; this device is often called an *electromagnet.* The magnetic field that is induced in the rod adds to the magnetic field of the solenoid, often creating a powerful magnet.

Figure 8 shows the magnetic field lines of a solenoid. Note that the field lines inside the solenoid point in the same direction, are nearly parallel, are uniformly spaced, and are close together. This indicates that the field inside the solenoid is strong and nearly uniform. The field outside the solenoid is nonuniform and much weaker than the interior field. Solenoids are used in a wide variety of applications, from most of the appliances in your home to very high-precision medical equipment.

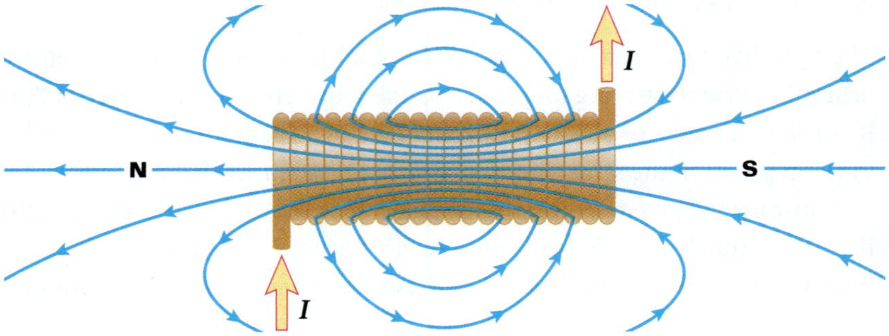

Figure 8
The magnetic field inside a solenoid is strong and nearly uniform. Note that the field lines resemble those of a bar magnet, so a solenoid effectively has north and south poles.

SECTION REVIEW

1. What is the shape of the magnetic field produced by a straight current-carrying wire?
2. Why is the magnetic field inside a solenoid stronger than the magnetic field outside?
3. If electrons behave like magnets, then why aren't all atoms magnets?
4. **Critical Thinking** In some satellites, torque coils are replaced by devices called *torque rods.* In torque rods, a ferromagnetic material is inserted inside the coil. Why does a torque rod have a stronger magnetic field than a torque coil?

Magnetic Force

SECTION OBJECTIVES

- **Given the force on a charge in a magnetic field, determine the strength of the magnetic field.**
- **Use the right-hand rule to find the direction of the force on a charge moving through a magnetic field.**
- **Determine the magnitude and direction of the force on a wire carrying current in a magnetic field.**

CHARGED PARTICLES IN A MAGNETIC FIELD

Although experiments show that a constant magnetic field does not exert a net force on a stationary charged particle, charges moving through a magnetic field do experience a magnetic force. This force has its maximum value when the charge moves perpendicular to the magnetic field, decreases in value at other angles, and becomes zero when the particle moves along the field lines. To keep the math simple in this book, we will limit our discussion to situations in which charges move parallel or perpendicular to the magnetic field lines.

A charge moving through a magnetic field experiences a force

Recall that the electric field at a point in space is defined as the electric force per unit charge acting on some test charge placed at that point. In a similar manner, we can describe the properties of the magnetic field, **B,** in terms of the magnetic force exerted on a test charge at a given point. Our test object is assumed to be a positive charge, q, moving with velocity **v** perpendicular to **B.** It has been found experimentally that the strength of the magnetic force on the particle moving perpendicular to the field is equal to the product of the magnitude of the charge, q, the magnitude of the velocity, v, and the strength of the external magnetic field, B, as shown by the following relationship.

$$F_{magnetic} = qvB$$

This expression can be rearranged as follows:

MAGNITUDE OF A MAGNETIC FIELD

$$B = \frac{F_{magnetic}}{qv}$$

$$\textbf{magnetic field} = \frac{\textbf{magnetic force on a charged particle}}{\textbf{(magnitude of charge)(speed of charge)}}$$

If the force is in newtons, the charge is in coulombs, and the speed is in meters per second, the unit of magnetic field strength is the tesla (T). Thus, if a 1 C charge moving at 1 m/s perpendicular to a magnetic field experiences a magnetic force of 1 N, the magnitude of the magnetic field is equal to 1 T. Most magnetic fields are much smaller than 1 T. We can express the units of the magnetic field as follows:

$$T = \frac{N}{C \bullet m/s} = \frac{N}{A \bullet m} = \frac{V \bullet s}{m^2}$$

SC*i*LINKS.
www.scilinks.org
Topic: Magnetic Fields
Code: HF60898

Demonstration

Electromagnetic Force

Purpose Show that moving charges in a magnetic field experience a force.

Materials strong horseshoe magnet, wire, two ring stands and supports, variable dc power supply

Procedure Lay the horseshoe magnet on its side so that one pole is above the other. Connect the wire to the power supply, and support the wire so that it passes through the center of the poles of the horseshoe magnet. Remind students that current is made up of moving charges and that they will observe the magnetic field exerting a force on the charges. Turn on the power supply, and gradually increase the current in the wire until the wire is forced to one side or the other.

Ask students what will happen if the direction of the current is reversed. *(The wire will move in the opposite direction.)* Turn off the power supply, detach the wires, and attach them to the opposite terminals. Turn on the power supply, and increase the current to confirm the students' hypotheses.

Have students sketch the magnetic fields of the magnet and the wire.

Classroom Practice

Particle in a Magnetic Field
An electron moving north at 4.5×10^4 m/s enters a 1.0 mT magnetic field pointed upward.

a. What is the magnitude and direction of the force exerted on the electron?

b. What would the force be if the particle were a proton?

c. What would the force be if the particle were a neutron?

Answers

a. 7.2×10^{-18} N west

b. 7.2×10^{-18} N east

c. 0.0 N

Why it Matters

Television Screens

The electron gun is the heart of a CRT. A small, hot filament (called the *heater*) heats up a negatively charged cathode, which emits a cloud of electrons. Two positively charged anodes, one for accelerating the electrons and one for focusing, form the electrons into a beam, which is then directed toward the phosphor-coated screen. The phosphors glow as the electrons strike them.

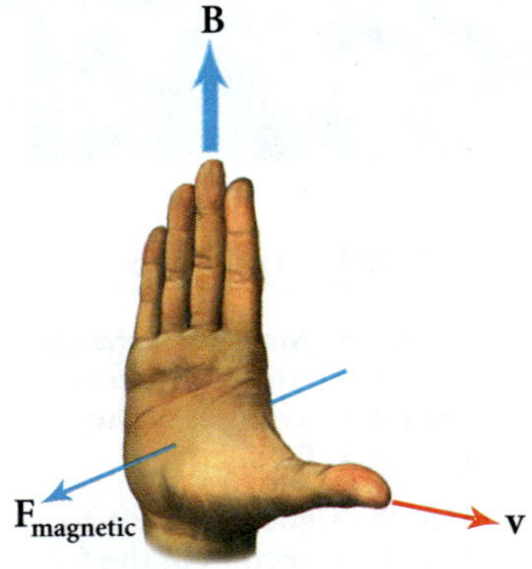

Figure 9
Use this alternative right-hand rule to find the direction of the magnetic force on a positive charge.

Conventional laboratory magnets can produce magnetic fields up to about 1.5 T. Superconducting magnets that can generate magnetic fields as great as 30 T have been constructed. For comparison, Earth's magnetic field near its surface is about 50 μT (5×10^{-5} T).

An alternative right-hand rule can be used to find the direction of the magnetic force

Experiments show that the direction of the magnetic force is always perpendicular to both the velocity, **v**, and the magnetic field, **B**. To determine the direction of the force, use the right-hand rule. As before, place your fingers in the direction of **B** with your thumb pointing in the direction of **v**, as illustrated in **Figure 9.** The magnetic force, $\mathbf{F_{magnetic}}$, on a positive charge is directed *out* of the palm of your hand.

If the charge is negative rather than positive, the force is directed *opposite* that shown in **Figure 9.** That is, if q is negative, simply use the right-hand rule to find the direction of $\mathbf{F_{magnetic}}$ for positive q, and then reverse this direction for the negative charge.

Why it Matters

Television Screens

The force on a moving charge due to a magnetic field is used to create pictures on a television screen. The main component of a television is the *cathode ray tube,* which is essentially a vacuum tube in which electric fields are used to form a beam of electrons. Phosphor on the television screen glows when it is struck by the electrons in the beam. Without magnetism, however, only the center of the screen would be illuminated by the beam. The direction of the beam is changed by two electromagnets, one deflecting the beam horizontally, the other deflecting the beam vertically. The direction of the beam can be changed by changing the direction of the current in each electromagnet. In this way, the beam illuminates the entire screen.

In a color television, three different colors of phosphor—red, green, and blue—make up the screen. Three electron beams, one for each color, scan over the screen to produce a color picture.

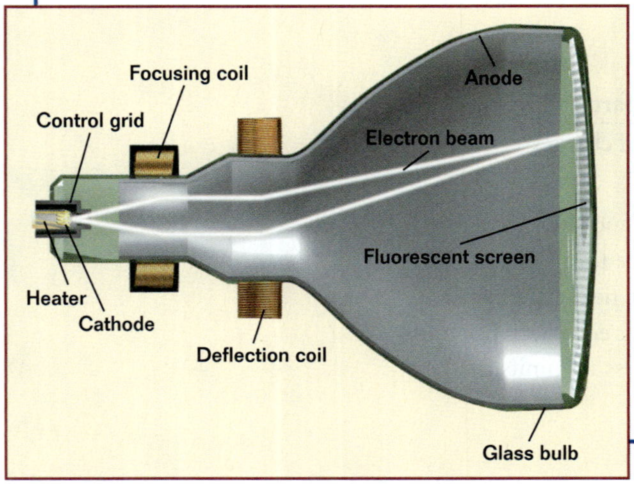

In a cathode ray tube, the *cathode* is a heated filament inside a vacuum tube, similar to the filament in a light bulb. The *ray* is a stream of electrons that come off the heated filament into a vacuum.

SAMPLE PROBLEM A

Particle in a Magnetic Field

PROBLEM

A proton moving east experiences a force of 8.8×10^{-19} N upward due to the Earth's magnetic field. At this location, the field has a magnitude of 5.5×10^{-5} T to the north. Find the speed of the particle.

SOLUTION

Given: $q = 1.60 \times 10^{-19}$ C $\quad B = 5.5 \times 10^{-5}$ T

$F_{magnetic} = 8.8 \times 10^{-19}$ N

Unknown: $v = ?$

Use the definition of magnetic field strength. Rearrange to solve for v.

$$B = \frac{F_{magnetic}}{qv}$$

$$v = \frac{F_{magnetic}}{qB}$$

$$v = \frac{8.8 \times 10^{-19}\ \text{N}}{(1.60 \times 10^{-19}\ \text{C})(5.5 \times 10^{-5}\ \text{T})} = 1.0 \times 10^{5}\ \text{m/s}$$

TIP *The directions given can be used to verify the right-hand rule. Imagine standing at this location and facing north. Turn the palm of your right hand upward (the direction of the force) with your thumb pointing east (the direction of the velocity). If your palm and thumb point in these directions, your fingers point directly north in the direction of the magnetic field, as they should.*

PRACTICE A

Particle in a Magnetic Field

1. A proton moves perpendicularly to a magnetic field that has a magnitude of 4.20×10^{-2} T. What is the speed of the particle if the magnitude of the magnetic force on it is 2.40×10^{-14} N?
2. If an electron in an electron beam experiences a downward force of 2.0×10^{-14} N while traveling in a magnetic field of 8.3×10^{-2} T west, what is the direction and magnitude of the velocity?
3. A uniform 1.5 T magnetic field points north. If an electron moves vertically downward (toward the ground) with a speed of 2.5×10^{7} m/s through this field, what force (magnitude and direction) will act on it?

PROBLEM GUIDE A

Use this guide to assign problems.
SE = Student Edition Textbook
PW = Problem Workbook
PB = Problem Bank on the One-Stop Planner (OSP)

Solving for:

v	**SE** Sample, 1–3; Ch. Rvw. 30–31 **PW** 6–7 **PB** 4–6
$F_{magnetic}$	**SE** 4–5; Ch. Rvw. 35–37, 40* **PW** Sample, 1–3 **PB** 7–10
B	**SE** 6; Ch. Rvw. 34, 39 **PW** 4–5 **PB** Sample, 1–3

***Challenging Problem**
Consult the printed Solutions Manual or the OSP for detailed solutions.

ANSWERS

Practice A

1. 3.57×10^{6} m/s
2. 1.5×10^{6} m/s north
3. 6.0×10^{-12} N west

Visual Strategy GENERAL

Figure 10
Students should be encouraged to apply the right-hand rule to describe the force on the charge at several points on the circle.

Q What direction would the force on a moving charged particle be if the particle were on the left side of the circle?

A *to the right, toward the center*

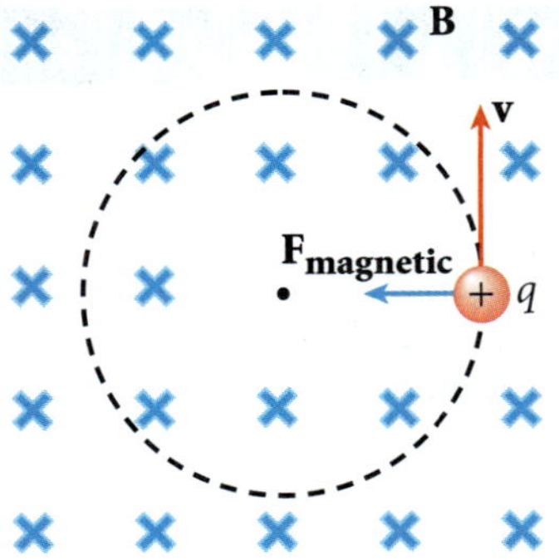

Figure 10
When the velocity, **v**, of a charged particle is perpendicular to a uniform magnetic field, the particle moves in a circle whose plane is perpendicular to **B.**

A charge moving through a magnetic field follows a circular path

Consider a positively charged particle moving in a uniform magnetic field. Suppose the direction of the particle's initial velocity is exactly perpendicular to the field, as in **Figure 10.** Application of the right-hand rule for the charge q shows that the direction of the magnetic force, $\mathbf{F_{magnetic}}$, at the charge's location is to the left. Furthermore, application of the right-hand rule at any point shows that the magnetic force is always directed toward the center of the circular path. Therefore, the magnetic force is, in effect, a force that maintains circular motion and changes only the direction of **v,** not its magnitude.

Now consider a charged particle traveling with its initial velocity at some angle to a uniform magnetic field. A component of the particle's initial velocity is parallel to the magnetic field. This parallel part is not affected by the magnetic field, and that part of the motion will remain the same. The perpendicular part results in a circular motion, as described above. The particle will follow a helical path, like the red stripes on a candy cane, whose axis is parallel to the magnetic field.

MAGNETIC FORCE ON A CURRENT-CARRYING CONDUCTOR

Recall that current consists of many charged particles in motion. If a force is exerted on a single charged particle when the particle moves through a magnetic field, it should be no surprise that a current-carrying wire also experiences a force when it is placed in a magnetic field. The resultant force on the wire is the sum of the individual magnetic forces on the charged particles. The force on the particles is transmitted to the bulk of the wire through collisions with the atoms making up the wire.

Consider a straight segment of wire of length ℓ carrying current, I, in a uniform external magnetic field, **B,** as in **Figure 11.** When the current and magnetic field are perpendicular, the magnitude of the total magnetic force on the wire is given by the following relationship.

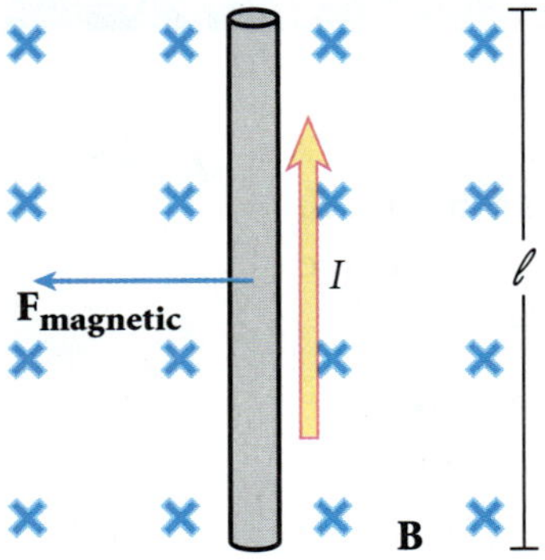

Figure 11
A current-carrying conductor in a magnetic field experiences a force that is perpendicular to the direction of the current.

FORCE ON A CURRENT-CARRYING CONDUCTOR PERPENDICULAR TO A MAGNETIC FIELD

$$F_{magnetic} = BI\ell$$

magnitude of magnetic force = (magnitude of magnetic field) (current)(length of conductor within B)

The direction of the magnetic force on a wire can be obtained by using the right-hand rule. However, in this case, you must place your thumb in the direction of the current rather than in the direction of the velocity, **v.** In **Figure 11,** the direction of the magnetic force on the wire is to the left. When the current is either in the direction of the field or opposite the direction of the field, the magnetic force on the wire is zero.

Two parallel conducting wires exert a force on one another

Because a current in a conductor creates its own magnetic field, it is easy to understand that two current-carrying wires placed close together exert magnetic forces on each other. When the two conductors are parallel to each other, the direction of the magnetic field created by one is perpendicular to the direction of the current of the other, and vice versa. In this way, a force of $F_{magnetic} = BI\ell$ acts on each wire, where B is the magnitude of the magnetic field created by the other wire.

Consider the two long, straight, parallel wires shown in **Figure 12.** When the current in each is in the same direction, the two wires attract one another. Confirm this by using the right-hand rule. Point your thumb in the direction of current in one wire, and point your fingers in the direction of the field produced by the other wire. By doing this, you find that the direction of the force (pointing out from the palm of your hand) is toward the other wire. When the currents in each wire are in opposite directions, the wires repel one another.

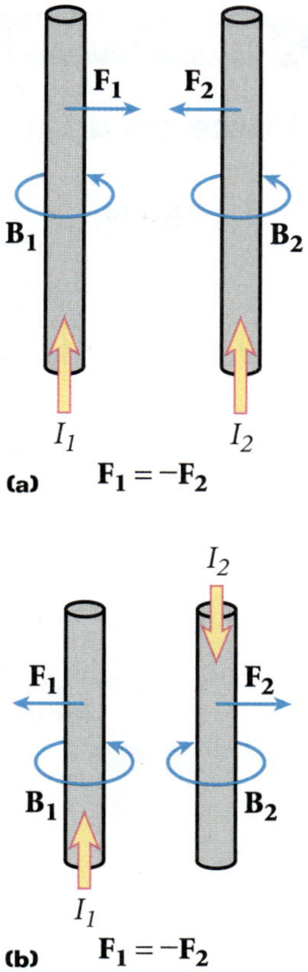

Figure 12
Two parallel wires, each carrying a steady current, exert magnetic forces on each other. The force is **(a)** attractive if the currents have the same direction and **(b)** repulsive if the two currents have opposite directions.

Loudspeakers use magnetic force to produce sound

The loudspeakers in most sound systems use a magnetic force acting on a current-carrying wire in a magnetic field to produce sound waves. One speaker design, shown in **Figure 13,** consists of a coil of wire, a flexible paper cone attached to the coil that acts as the speaker, and a permanent magnet. In a speaker system, a sound signal is converted to a varying electric signal by the microphone. This electrical signal is amplified and sent to the loudspeaker. At the loudspeaker, this varying electrical current causes a varying magnetic force on the coil. This alternating force on the coil results in vibrations of the attached cone, which produce variations in the density of the air in front of it. In this way, an electric signal is converted to a sound wave that closely resembles the sound wave produced by the source.

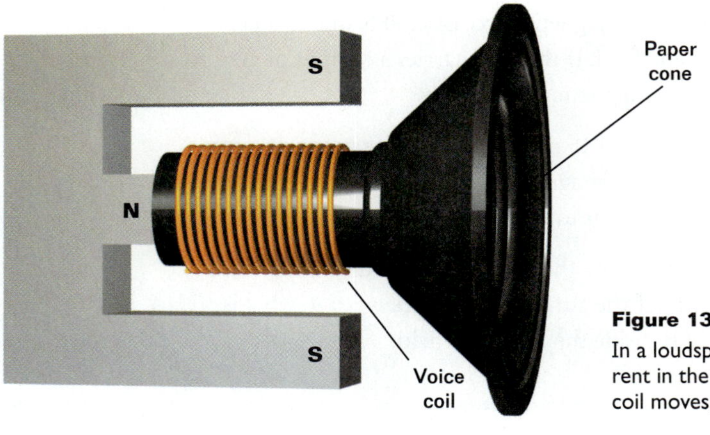

Figure 13
In a loudspeaker, when the direction and magnitude of the current in the coil of wire change, the paper cone attached to the coil moves, producing sound waves.

Demonstration

Force Between Parallel Conductors

Purpose Show that two current-carrying parallel wires exert a force on each other.

Materials two 10 cm × 1 cm strips of aluminum foil, two 9.0 V NiCad batteries, clay, connecting wire with alligator clips, two switches, overhead projector

Procedure Arrange the strips of aluminum foil so that they are parallel to one another, and secure the ends of the foil to the overhead projector with the clay. Make sure the two pieces of foil can be seen clearly when projected onto a screen. Using the wire and alligator clips, connect each piece of foil to a battery and a switch so that the direction of the current in each piece of foil is the same. Leave the switches open until you begin the demonstration.

Point out to students that the direction of the current in each strip will be the same when the switch is closed. Momentarily close the switch, and have students observe that the strips move together.

Ask what effect the wires will have on each other if the direction of the current in one strip is reversed. Rearrange one of the circuits so that the direction of the current is reversed, and have students note that the two strips of aluminum repel one another.

Classroom Practice

Force on a Current-Carrying Conductor
A 4.5 m wire carries a current of 12.5 A from north to south. If the magnetic force on the wire due to a uniform magnetic field is 1.1×10^3 N downward, what is the magnitude and direction of the magnetic field?

Answer
2.0×10^1 T, to the west

PROBLEM GUIDE B

Use this guide to assign problems.
SE = Student Edition Textbook
PW = Problem Workbook
PB = Problem Bank on the One-Stop Planner (OSP)

Solving for:

B	**SE** Sample, 1–3; Ch. Rvw. 32–33, 41–42, 44 **PW** 5–6 **PB** 3–5
F	**PW** 9–10 **PB** 8–10
ℓ	**SE** 4 **PW** 7–8 **PB** Sample, 1–2
I	**PW** Sample, 1–4 **PB** 6–7

*Challenging Problem
Consult the printed Solutions Manual or the OSP for detailed solutions.

ANSWERS

Practice B

1. 1.7×10^{-7} T in $+z$ direction
2. 0.050 T
3. 1.5 T
4. 0.59 m

SAMPLE PROBLEM B

Force on a Current-Carrying Conductor

PROBLEM

A wire 36 m long carries a current of 22 A from east to west. If the magnetic force on the wire due to Earth's magnetic field is downward (toward Earth) and has a magnitude of 4.0×10^{-2} N, find the magnitude and direction of the magnetic field at this location.

SOLUTION

Given: $\ell = 36$ m $I = 22$ A $F_{magnetic} = 4.0 \times 10^{-2}$ N

Unknown: $B = ?$

Use the equation for the force on a current-carrying conductor perpendicular to a magnetic field.

$$F_{magnetic} = BI\ell$$

Rearrange to solve for B.

$$B = \frac{F_{magnetic}}{I\ell} = \frac{4.0 \times 10^{-2}\ \text{N}}{(22\ \text{A})(36\ \text{m})} = 5.0 \times 10^{-5}\ \text{T}$$

Using the right-hand rule to find the direction of **B,** face north with your thumb pointing to the west (in the direction of the current) and the palm of your hand down (in the direction of the force). Your fingers point north. Thus, Earth's magnetic field is from south to north.

PRACTICE B

Force on a Current-Carrying Conductor

1. A 6.0 m wire carries a current of 7.0 A toward the $+x$ direction. A magnetic force of 7.0×10^{-6} N acts on the wire in the $-y$ direction. Find the magnitude and direction of the magnetic field producing the force.
2. A wire 1.0 m long experiences a magnetic force of 0.50 N due to a perpendicular uniform magnetic field. If the wire carries a current of 10.0 A, what is the magnitude of the magnetic field?
3. The magnetic force on a straight 0.15 m segment of wire carrying a current of 4.5 A is 1.0 N. What is the magnitude of the component of the magnetic field that is perpendicular to the wire?
4. The magnetic force acting on a wire that is perpendicular to a 1.5 T uniform magnetic field is 4.4 N. If the current in the wire is 5.0 A, what is the length of the wire that is inside the magnetic field?

GALVANOMETERS

A *galvanometer* is a device used in the construction of both ammeters and voltmeters. Its operation is based on the fact that a torque acts on a current loop in the presence of a magnetic field. **Figure 14** shows a simplified arrangement of the main components of a galvanometer. It consists of a coil of wire wrapped around a soft iron core mounted so that it is free to pivot in the magnetic field provided by the permanent magnet. The torque experienced by the coil is proportional to the current in the coil. This means that the larger the current, the greater the torque and the more the coil will rotate before the spring tightens enough to stop the movement. Hence, the amount of deflection of the needle is proportional to the current in the coil. When there is no current in the coil, the spring returns the needle to zero. Once the instrument is properly calibrated, it can be used in conjunction with other circuit elements as an ammeter (to measure currents) or as a voltmeter (to measure potential differences).

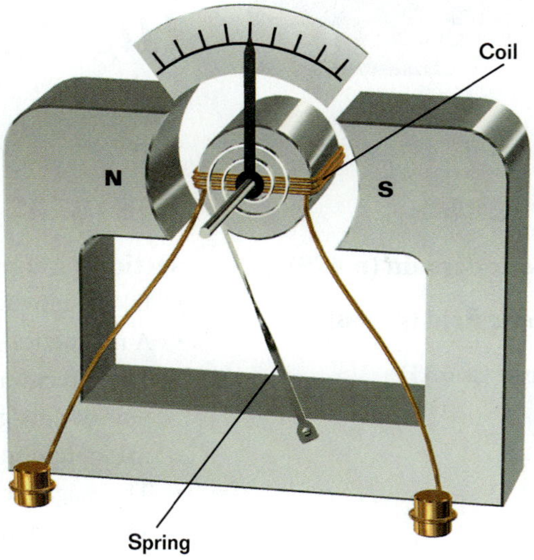

Figure 14
In a galvanometer, when current enters the coil, which is in a magnetic field, the magnetic force causes the coil to twist.

SECTION REVIEW

1. A particle with a charge of 0.030 C experiences a magnetic force of 1.5 N while moving at right angles to a uniform magnetic field. If the speed of the charge is 620 m/s, what is the magnitude of the magnetic field the particle passes through?

2. An electron moving north encounters a uniform magnetic field. If the magnetic field points east, what is the direction of the magnetic force on the electron?

3. A straight segment of wire has a length of 25 cm and carries a current of 5.0 A. If the wire is perpendicular to a magnetic field of 0.60 T, then what is the magnitude of the magnetic force on this segment of the wire?

4. Two parallel wires have charges moving in the same direction. Is the force between them attractive or repulsive?

5. **Interpreting Graphics** Find the direction of the magnetic force on the current-carrying wire in **Figure 15.**

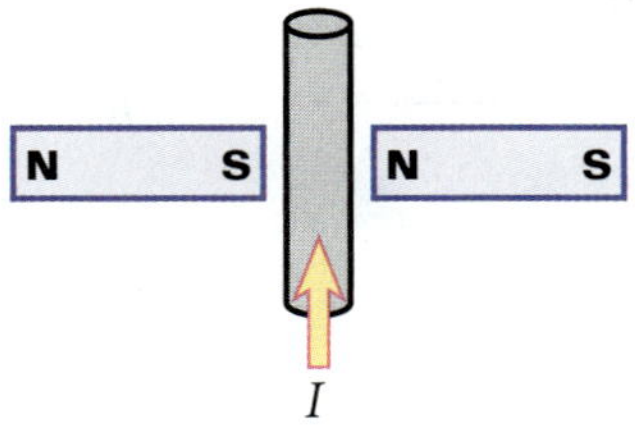

Figure 15

SECTION 3

Visual Strategy ADVANCED

Figure 14
Encourage students to use the right-hand rule to understand how a galvanometer works.

Q If charges move through the coil from the left terminal to the right terminal, which direction will the needle rotate?

A *clockwise*

SECTION REVIEW ANSWERS

1. 0.081 T
2. upward
3. 0.75 N
4. attractive
5. out of the page

CHAPTER 19

Highlights

Teaching Tip

Have students write an essay summarizing the various cases of current-carrying wires discussed in this chapter. Essays should include a thorough explanation of the magnetic field produced by a current-carrying wire, the magnetic force on a current-carrying wire that is in a magnetic field, and the force between two parallel current-carrying wires. Remind students to include the right-hand rule for each case.

CHAPTER 19

Highlights

KEY TERMS

magnetic domain (p. 679)

magnetic field (p. 680)

solenoid (p. 685)

KEY IDEAS

Section 1 Magnets and Magnetic Fields

- Like magnetic poles repel, and unlike poles attract.
- A magnetic domain is a group of atoms whose magnetic fields are aligned.
- The direction of any magnetic field is defined as the direction the north pole of a magnet would point if placed in the field. The magnetic field of a magnet points from the north pole of the magnet to the south pole.
- The magnetic north pole of Earth corresponds to the geographic South Pole, and the magnetic south pole corresponds to the geographic North Pole.

Section 2 Magnetism from Electricity

- A magnetic field exists around any current-carrying wire; the direction of the magnetic field follows a circular path around the wire.
- The magnetic field created by a solenoid or coil is similar to the magnetic field of a permanent magnet.

Section 3 Magnetic Force

- The direction of the force on a positive charge moving through a magnetic field can be found by using the alternate right-hand rule.
- A current-carrying wire in an external magnetic field undergoes a magnetic force. The direction of the magnetic force on the wire can be found by using the alternate right-hand rule.
- Two parallel current-carrying wires exert on one another forces that are equal in magnitude and opposite in direction. If the currents are in the same direction, the two wires attract one another. If the currents are in opposite directions, the wires repel one another.

PROBLEM SOLVING

See **Appendix D: Equations** for a summary of the equations introduced in this chapter. If you need more problem-solving practice, see **Appendix I: Additional Problems.**

Diagram Symbols

Magnetic field vector	↑
Magnetic field pointing into the page	✕
Magnetic field pointing out of the page	●

Variable Symbols

Quantities		Units		Conversions
B	magnetic field	T	tesla	$= \frac{N}{C \bullet m/s} = \frac{N}{A \bullet m}$
$F_{magnetic}$	magnetic force	N	newtons	$= \frac{kg \bullet m}{s^2}$
ℓ	length of conductor in field	m	meters	

CHAPTER 19 Review

MAGNETS AND MAGNETIC FIELDS

Review Questions

1. What is the minimum number of poles for a magnet?
2. When you break a magnet in half, how many poles does each piece have?
3. The north pole of a magnet is attracted to the geographic North Pole of Earth, yet like poles repel. Can you explain this?
4. Which way would a compass needle point if you were at the magnetic north pole?
5. What is a magnetic domain?
6. Why are iron atoms so strongly affected by magnetic fields?
7. When a magnetized steel needle is strongly heated in a Bunsen burner flame, it becomes demagnetized. Explain why.
8. If an unmagnetized piece of iron is attracted to one pole of a magnet, will it be repelled by the opposite pole?

Conceptual Questions

9. In the figure below, two permanent magnets with holes bored through their centers are placed one over the other. Because the poles of the upper magnet are the reverse of those of the lower, the upper magnet levitates above the lower magnet. If the upper magnet were displaced slightly, either up or down, what would be the resulting motion? Explain. What would happen if the upper magnet were inverted?

10. You have two iron bars and a ball of string in your possession; one iron bar is magnetized, and one iron bar is not. How can you determine which iron bar is magnetized?
11. Why does a very strong magnet attract both poles of a weak magnet?
12. A magnet attracts a piece of iron. The iron can then attract another piece of iron. Explain, on the basis of alignment of domains, what happens in each piece of iron.
13. When a small magnet is repeatedly dropped, it becomes demagnetized. Explain what happens to the magnet at the atomic level.

MAGNETISM FROM ELECTRICITY

Review Questions

14. A conductor carrying a current is arranged so that electrons flow in one segment from east to west. If a compass is held over this segment of the wire, in what direction is the needle deflected? (Hint: Recall that current is defined as the motion of *positive* charges.)
15. What factors does the strength of the magnetic field of a solenoid depend on?

Conceptual Questions

16. A solenoid with ends marked *A* and *B* is suspended by a thread so that the core can rotate in the horizontal plane. A current is maintained in the coil so that the electrons move clockwise when viewed from end *A* toward end *B*. How will the coil align itself in Earth's magnetic field?
17. Is it possible to orient a current-carrying loop of wire in a uniform magnetic field so that the loop will not tend to rotate?

Review

ANSWERS

1. two
2. two
3. The geographic North Pole is near the magnetic south pole.
4. perpendicular to Earth's surface
5. a group of atoms whose magnetic fields are aligned in a common direction
6. They have unpaired electron spins.
7. The added energy causes the domains to become less aligned.
8. No, unmagnetized iron is attracted to either pole of a magnet.
9. a damped periodic oscillation; Gravitational force and the repulsive force along with a displacement from equilibrium result in periodic motion. When inverted, the two magnets would attract each other.
10. Hang each of the bars by the string. The magnetized bar will align itself with Earth's magnetic field.
11. It realigns the domains of the weaker magnet.
12. first piece: the domains align with the magnetic field of the magnet; second piece: the domains align with the magnetic field of the first piece
13. The energy absorbed disturbs the alignment of the domains.
14. south
15. number of coils per unit length, amount of current
16. End *A* will point toward the geographic South Pole.

17. yes, by aligning the plane of the loop perpendicular to the magnetic field
18. Yes, the north pole of the solenoid would point to Earth's geographic North Pole; No, the solenoid would oscillate back and forth as its poles continually reversed.
19. They have opposite charge.
20. The proton would go left, and the electron would go right.
21. The magnetic field of the magnet exerts a force on the moving electrons in the electron beam.
22. The proton would move up in a half circle and exit above its point of entry. The electron would move down in a half circle and exit below its point of entry.
23. The magnetic field from one wire is perpendicular to the second wire (and thus the current in it) at the second wire's location. The magnetic force on the second wire is away from the first wire.
24. no; Magnetic fields only exert a net force on moving charges.
25. positive y direction; no; It moves in circles in the x-y plane.
26. a. into the page
    b. to the right
    c. down the page
27. a. The stream moves away from the wire.
    b. The stream moves toward the wire.
28. The stream moves toward the observer.
29. Because the wires are twisted together, the region where the magnetic field is non-zero is very small.
30. 15 m/s

18. If a solenoid were suspended by a string so that it could rotate freely, could it be used as a compass when it carried a direct current? Could it also be used if the current were alternating in direction?

MAGNETIC FORCE

Review Questions

19. Two charged particles are projected into a region where there is a magnetic field perpendicular to their velocities. If the particles are deflected in opposite directions, what can you say about them?

20. Suppose an electron is chasing a proton up this page when suddenly a magnetic field pointing into the page is applied. What would happen to the particles?

21. Why does the picture on a television screen become distorted when a magnet is brought near the screen?

22. A proton moving horizontally enters a region where there is a uniform magnetic field perpendicular to the proton's velocity, as shown below. Describe the proton's subsequent motion. How would an electron behave under the same circumstances?

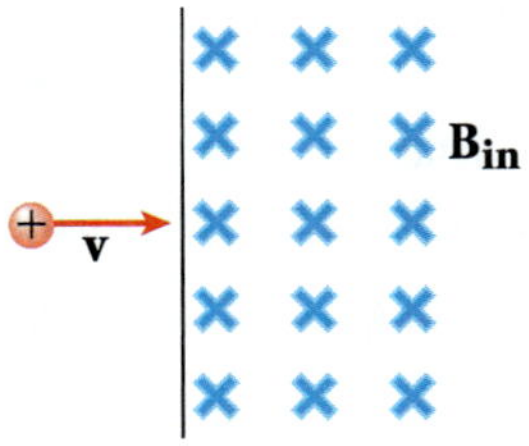

23. Explain why two parallel wires carrying currents in opposite directions repel each other.

24. Can a stationary magnetic field set a resting electron in motion? Explain.

25. At a given instant, a proton moves in the positive x direction in a region where there is a magnetic field in the negative z direction. What is the direction of the magnetic force? Does the proton continue to move along the x-axis? Explain.

26. For each situation below, use the movement of the positively charged particle and the direction of the magnetic force acting on it to find the direction of the magnetic field.

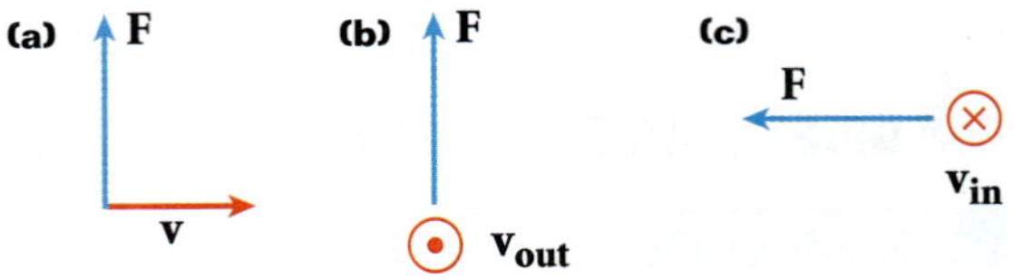

Conceptual Questions

27. A stream of electrons is projected horizontally to the right. A straight conductor carrying a current is supported parallel to and above the electron stream.

a. What is the effect on the electron stream if the current in the conductor is left to right?

b. What is the effect if the current is reversed?

28. If the conductor in item 27 is replaced by a magnet with a downward magnetic field, what is the effect on the electron stream?

29. Two wires carrying equal but opposite currents are twisted together in the construction of a circuit. Why does this technique reduce stray magnetic fields?

Practice Problems

For problems 30–31, see Sample Problem A.

30. A duck flying due east passes over Atlanta, where the magnetic field of Earth is 5.0×10^{-5} T directed north. The duck has a positive charge of 4.0×10^{-8} C. If the magnetic force acting on the duck is 3.0×10^{-11} N upward, what is the duck's velocity?

31. A proton moves eastward in the plane of Earth's magnetic equator, where Earth's magnetic field points north and has a magnitude of 5.0×10^{-5} T. What velocity must the proton have for the magnetic force to just cancel the gravitational force?

For problems 32–33, see Sample Problem B.

32. A wire carries a 10.0 A current at an angle 90.0° from the direction of a magnetic field. If the magnitude of the magnetic force on a 5.00 m length of the wire is 15.0 N, what is the strength of the magnetic field?

33. A thin 1.00 m long copper rod in a uniform magnetic field has a mass of 50.0 g. When the rod carries a current of 0.245 A, it floats in the magnetic field. What is the field strength of the magnetic field?

MIXED REVIEW

34. A proton moves at 2.50×10^6 m/s horizontally at a right angle to a magnetic field.
   a. What is the strength of the magnetic field required to exactly balance the weight of the proton and keep it moving horizontally?
   b. Should the direction of the magnetic field be in a horizontal or a vertical plane?

35. Find the direction of the force on a proton moving through each magnetic field in the four figures below.

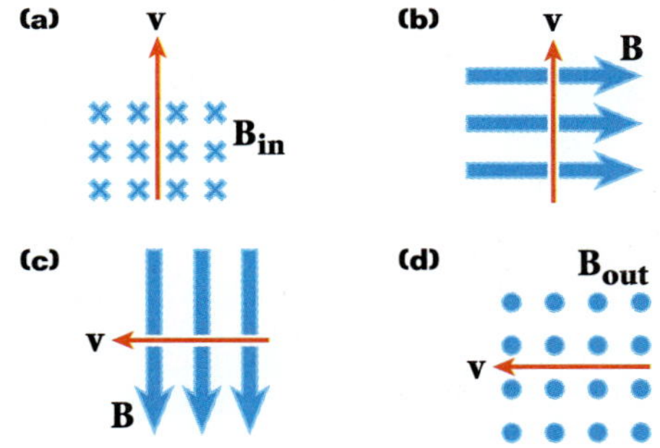

36. Find the direction of the force on an electron moving through each magnetic field in the four figures in item 35 above.

37. In the four figures in item 35, assume that in each case the velocity vector shown is replaced with a wire carrying a current in the direction of the velocity vector. Find the direction of the magnetic force acting on each wire.

38. A proton moves at a speed of 2.0×10^7 m/s at right angles to a magnetic field with a magnitude of 0.10 T. Find the magnitude of the acceleration of the proton.

39. A proton moves perpendicularly to a uniform magnetic field, **B,** with a speed of 1.0×10^7 m/s and experiences an acceleration of 2.0×10^{13} m/s^2 in the positive *x* direction when its velocity is in the positive *z* direction. Determine the magnitude and direction of the field.

40. A proton travels with a speed of 3.0×10^6 m/s at an angle of 37° west of north. A magnetic field of 0.30 T points to the north. Determine the following:
   a. the magnitude of the magnetic force on the proton
   b. the direction of the magnetic force on the proton
   c. the proton's acceleration as it moves through the magnetic field

   (Hint: The magnetic force experienced by the proton in the magnetic field is proportional to the component of the proton's velocity that is perpendicular to the magnetic field.)

41. In the figure below, a 15 cm length of conducting wire that is free to move is held in place between two thin conducting wires. All the wires are in a magnetic field. When a 5.0 A current is in the wire, as shown in the figure, the wire segment moves upward at a constant velocity. Assuming the wire slides without friction on the two vertical conductors and has a mass of 0.15 kg, find the magnitude and direction of the minimum magnetic field that is required to move the wire.

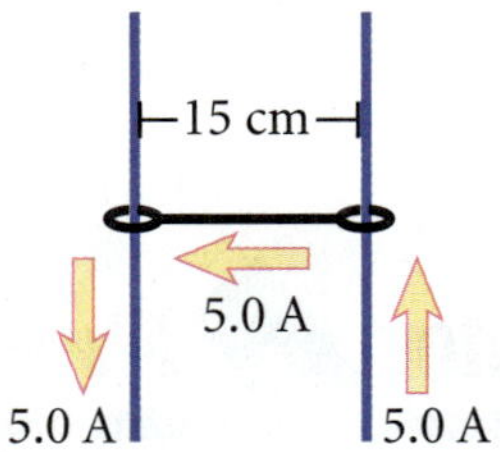

42. A current, $I = 15$ A, is directed along the positive *x*-axis and perpendicular to a uniform magnetic field. The conductor experiences a magnetic force per unit length of 0.12 N/m in the negative *y* direction. Calculate the magnitude and direction of the magnetic field in the region through which the current passes.

43. A proton moving perpendicular to a magnetic field of strength 3.5 mT experiences a force due to the field of 4.5×10^{-21} N. Calculate the following:
   a. the speed of the proton
   b. the kinetic energy of the proton

   Recall that a proton has a charge of 1.60×10^{-19} C and a mass of 1.67×10^{-27} kg.

19 REVIEW

31. 2.1×10^{-3} m/s
32. 0.300 T
33. 2.00 T
34. a. 4.10×10^{-14} T
b. horizontal
35. a. to the left
b. into the page
c. out of the page
d. up the page
36. a. to the right
b. out of the page
c. into the page
d. down the page
37. a. to the left
b. into the page
c. out of the page
d. up the page
38. 1.9×10^{14} m/s^2
39. 2.1×10^{-2} T, in the negative *y* direction
40. a. 8.7×10^{-14} N
b. downward
c. 5.2×10^{13} m/s^2
41. 2.0 T, out of the page
42. 8.0×10^{-3} T, in the positive *z* direction
43. a. 8.0 m/s
b. 5.4×10^{-26} J

44. 1.39×10^{-2} T, toward the observer

45. 2.82×10^{7} m/s

46. **a.** -2.9×10^{-12} N

b. -2.6×10^{-17} N

c. 8.9×10^{-30} N

extension

Graphing Calculator Practice

Visit go.hrw.com for answers to this Graphing Calculator activity.

Keyword HF6MAGXT

44. A singly charged positive ion that has a mass of 6.68×10^{-27} kg moves clockwise with a speed of 1.00×10^{4} m/s. The positively charged ion moves in a circular path that has a radius of 3.00 cm. Find the direction and strength of the uniform magnetic field through which the charge is moving. (Hint: The magnetic force exerted on the positive ion is the centripetal force, and the speed given for the positive ion is its tangential speed.)

45. What speed would a proton need to achieve in order to circle Earth 1000.0 km above the magnetic equator? Assume that Earth's magnetic field is everywhere perpendicular to the path of the proton and that Earth's magnetic field has an intensity of 4.00×10^{-8} T. (Hint: The magnetic force exerted on the proton is equal to the centripetal force, and the speed needed by the proton is its tangential speed. Remember that the radius of the circular orbit should also include the radius of Earth. Ignore relativistic effects.)

46. Calculate the force on an electron in each of the following situations:

a. moving at 2.0 percent the speed of light and perpendicular to a 3.0 T magnetic field

b. 3.0×10^{-6} m from a proton

c. in Earth's gravitational field at the surface of Earth

Use the following: $q_e = -1.6 \times 10^{-19}$ C; $m_e = 9.1 \times 10^{-31}$ kg; $q_p = 1.6 \times 10^{-19}$ C; $c = 3.0 \times 10^{8}$ m/s; $k_C = 9.0 \times 10^{9}$ N•m²/C²

Graphing Calculator *Practice*

Solenoids

A solenoid consists of a long, helically wound coil of insulated wire. When it carries a current, a solenoid acts as a magnet. The magnetic field strength (B) increases linearly with the current (I) and with the number of coils per unit length. Because there is a direct relation between B and I, the following equation applies to any solenoid:

$$B = aI + b$$

In this equation, the parameters a and b are different for different solenoids. The a and b parameters can be determined if the magnetic field strength of the solenoid is known at two different currents.

Once you determine a and b, you can predict the magnetic field strength of a solenoid for various currents.

The graphing calculator program that accompanies this activity uses this procedure. You will be given the magnetic field and current data for various solenoids. You will then use this information and the program to predict the magnetic field strength of each solenoid.

Visit go.hrw.com and type in the keyword **HF6MAGX** to find this graphing calculator activity. Refer to **Appendix B** for instructions on downloading the program for this activity.

Alternative Assessment

1. During a field investigation with your class, you find a roundish chunk of metal that attracts iron objects. Design a procedure to determine whether the object is magnetic and, if so, to locate its poles. Describe the limitations of your method. What materials would you need? How would you draw your conclusions? List all the possible results you can anticipate and the conclusions you could draw from each result.
2. Imagine you have been hired by a manufacturer interested in making kitchen magnets. The manufacturer wants you to determine how to combine several magnets to get a very strong magnet. He also wants to know what protective material to use to cover the magnets. Develop a method for measuring the strength of different magnets by recording the maximum number of paper clips they can hold under various conditions. First open a paper clip to use as a hook. Test the strength of different magnets and combinations of magnets by holding up the magnet, placing the open clip on the magnet, and hooking the rest of the paper clips so that they hang below the magnet. Examine the effect of layering different materials between the magnet and the clips. Organize your data in tables and graphs to present your conclusions.
3. Research phenomena related to one of the following topics, and prepare a report or presentation with pictures and data.
   - **a.** How does Earth's magnetic field vary with latitude, with longitude, with the distance from Earth, and in time?
   - **b.** How do people who rely on compasses account for these differences in Earth's magnetic field?
   - **c.** What is the Van Allen belt?
   - **d.** How do solar flares occur?
   - **e.** How do solar flares affect Earth?
4. Obtain old buzzers, bells, telephone receivers, speakers, motors from power or kitchen tools, and so on to take apart. Identify the mechanical and electromagnetic components. Examine their connections. How do they produce magnetic fields? Work in a cooperative group to describe and organize your findings about several devices for a display entitled "Anatomy of Electromagnetic Devices."
5. Magnetic force was first described by the ancient Greeks, who mined a magnetic mineral called magnetite. Magnetite was used in early experiments on magnetic force. Research the historical development of the concept of magnetic force. Describe the work of Peregrinus, William Gilbert, Oersted, Faraday, and other scientists.

Alternative Assessment ANSWERS

1. Students' plans should be safe and logical. A compass could indicate the identity and location of the two poles.
2. Students' plans should include safe and complete plans for testing the strength of each magnet based on how many paper clips the magnet can lift.
3. Earth's magnetic field is not constant through time or space; People who rely on compasses keep tables of corrections to apply, depending on their approximate location; The Van Allen belt is a cloud of charged particles around Earth; Solar flares are large outflows of charged particles that can affect Earth's magnetic field.
4. Students' answers should clearly indicate which devices are likely to produce a magnetic field (*solenoids and coils*). Students should indicate how such magnetic forces are converted into mechanical motions.
5. Peregrinus is credited with the first experiments with magnetism, using a thin piece of iron to map the magnetic field of magnetite. William Gilbert, Oersted, and Faraday studied both magnetism and electricity.

Standardized Test Prep

ANSWERS

1. C
2. G
3. B
4. J
5. C
6. F
7. A

Standardized Test Prep

MULTIPLE CHOICE

1. Which of the following statements best describes the domains in unmagnetized iron?

A. There are no domains.
B. There are domains, but the domains are smaller than in magnetized iron.
C. There are domains, but the domains are oriented randomly.
D. There are domains, but the domains are not magnetized.

2. Which of the following statements is most correct?

F. The north pole of a freely rotating magnet points north because the magnetic pole near the geographic North Pole is like the north pole of a magnet.
G. The north pole of a freely rotating magnet points north because the magnetic pole near the geographic North Pole is like the south pole of a magnet.
H. The north pole of a freely rotating magnet points south because the magnetic pole near the geographic South Pole is like the north pole of a magnet.
J. The north pole of a freely rotating magnet points south because the magnetic pole near the geographic South Pole is like the south pole of a magnet.

3. If you are standing at Earth's magnetic north pole and holding a bar magnet that is free to rotate in three dimensions, which direction will the south pole of the magnet point?

A. straight up
B. straight down
C. parallel to the ground, toward the north
D. parallel to the ground, toward the south

4. How can you increase the strength of a magnetic field inside a solenoid?

F. increase the number of coils per unit length
G. increase the current
H. place an iron rod inside the solenoid
J. all of the above

Use the diagram below to answer questions 5–6.

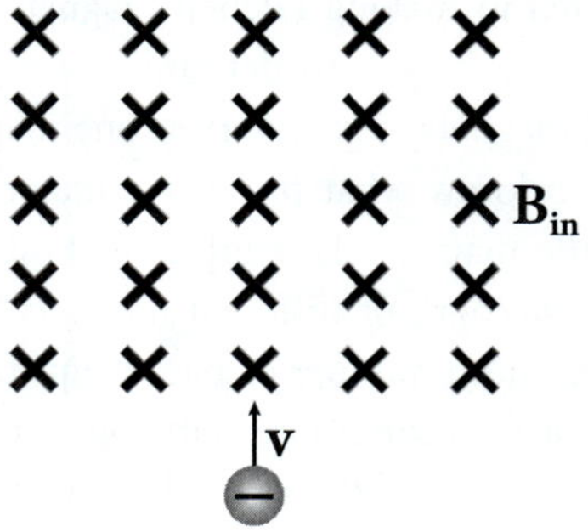

5. How will the electron move once it passes into the magnetic field?

A. It will curve to the right and then continue moving in a straight line to the right.
B. It will curve to the left and then continue moving in a straight line to the left.
C. It will move in a clockwise circle.
D. It will move in a counterclockwise circle.

6. What will be the magnitude of the force on the electron once it passes into the magnetic field?

F. qvB
G. $-qvB$
H. $\frac{qv}{B}$
J. $BI\ell$

7. An alpha particle ($q = 3.2 \times 10^{-19}$ C) moves at a speed of 2.5×10^{6} m/s perpendicular to a magnetic field of strength 2.0×10^{-4} T. What is the magnitude of the magnetic force on the particle?

A. 1.6×10^{-16} N
B. -1.6×10^{-16} N
C. 4.0×10^{-9} N
D. zero

Use the passage below to answer questions 8–9.

A wire 25 cm long carries a 12 A current from east to west. Earth's magnetic field at the wire's location has a magnitude of 4.8×10^{-5} T and is directed from south to north.

8. What is the magnitude of the magnetic force on the wire?

F. 2.3×10^{-5} N
G. 1.4×10^{-4} N
H. 2.3×10^{-3} N
J. 1.4×10^{-2} N

9. What is the direction of the magnetic force on the wire?

A. north
B. south
C. up, away from Earth
D. down, toward Earth

Use the diagram below to answer questions 10–12.

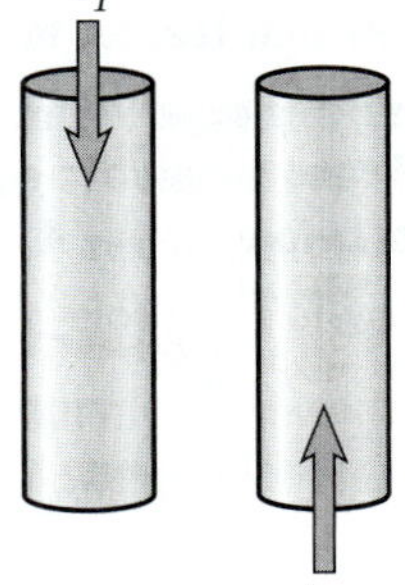

Wire 1 carries current I_1 and creates magnetic field B_1. Wire 2 carries current I_2 and creates magnetic field B_2.

10. What is the direction of the magnetic field B_1 at the location of wire 2?

F. to the left
G. to the right
H. into the page
J. out of the page

11. What is the direction of the force on wire 2 as a result of B_1?

A. to the left
B. to the right
C. into the page
D. out of the page

12. What is the magnitude of the magnetic force on wire 2?

F. $B_1 I_1 \ell_1$
G. $B_1 I_1 \ell_2$
H. $B_1 I_2 \ell_2$
J. $B_2 I_2 \ell_2$

SHORT RESPONSE

13. Sketch the magnetic field lines around a bar magnet.

14. Describe how to use the right-hand rule to determine the direction of a magnetic field around a current-carrying wire.

15. Draw a diagram showing the path of a positively charged particle moving in the plane of a piece of paper if a uniform magnetic field is coming out of the page.

EXTENDED RESPONSE

16. A proton ($q = 1.6 \times 10^{-19}$ C; $m = 1.7 \times 10^{-27}$ kg) is in a uniform 0.25 T magnetic field. The proton moves in a clockwise circle with a tangential speed of 2.8×10^5 m/s.

a. What is the direction of the magnetic field? Explain how you determined this.

b. What is the radius of the circle? Show your work.

Test TIP If you are asked to write out an answer, to show your calculations, or to draw a diagram, be sure to write clearly, to show all steps of your work, and to add clear labels to your diagrams. You may receive some credit for using the right approach to a problem, even if you do not arrive at the correct final answer.

8. G

9. D

10. J

11. B

12. H

13. Sketches should show a bar magnet with N and S poles labeled. Field lines should be directed away from the north end and toward the south end. See diagram in chapter for an example.

14. Imagine wrapping the fingers of your right hand around the wire and pointing your thumb in the direction of the current. The magnetic field lines form concentric circles that are centered on the wire and curve in the same direction as your fingers.

15. Diagrams should represent the magnetic field coming out of the page as an evenly spaced area of dots. The path of the particle is a clockwise circle.

16. a. The field is directed out of the page. Using the right-hand rule, the thumb points in the direction of the tangential motion of the proton. The force is directed toward the center of the circle and comes out of the palm of the hand. For this result to be true, the fingertips must point upward, indicating that the magnetic field points up and out of the page.

b. 1.2×10^{-2} m (See the Solutions Manual or One-Stop Planner for a full solution.)

Lab Planning

Beginning on page T34 are preparation notes and teaching tips to assist you in planning.

Blank data tables (as well as some sample data) appear on the **One-Stop Planner.**

No Books in the Lab?

See the ***Datasheets for In-Text Labs*** workbook for a reproducible master copy of this experiment.

CBL™ Option

A **CBL™** version of this lab appears in **Appendix K** and in the ***CBL™ Experiments*** workbook.

Safety Caution

Emphasize the dangers of working with electricity. Remind students to have you check their circuits before turning on the power supply or closing the switch.

CHAPTER 19

Skills Practice Lab

Magnetic Field of a Conducting Wire

OBJECTIVES

- **Use** a compass to explore the direction of the magnetic field of a current-carrying wire.
- **Analyze** the relationship between the direction of the magnetic field of a conducting wire and the direction of the current in the wire.

MATERIALS LIST

- **1 Ω resistor**
- **compass**
- **galvanometer**
- **insulated connecting wires and bare copper wire**
- **multimeter or dc ammeter**
- **power supply**
- **switch**

In this lab, you will study the magnetic field that occurs around a current-carrying wire. You will construct a circuit with a current-carrying wire and use a magnetic compass needle to investigate the relationship between the magnetic field and the current in the wire. You will be able to determine the magnitude and direction of the magnetic field surrounding the wire.

SAFETY

- **Never close a circuit until it has been approved by your teacher. Never rewire or adjust any element of a closed circuit. Never work with electricity near water; be sure the floor and all work surfaces are dry.**
- **If the pointer on any kind of meter moves off scale, open the circuit immediately by opening the switch.**
- **Do not attempt this exercise with any batteries, electrical devices, or magnets other than those provided by your teacher for this purpose.**
- **Wire coils may heat up rapidly during this experiment. If heating occurs, open the switch immediately and handle the equipment with a hot mitt. Allow all equipment to cool before storing it.**

PROCEDURE

Preparation

1. Read the entire lab procedure, and plan the steps you will take.
2. If you are not using a datasheet provided by your teacher, prepare a data table in your lab notebook with four columns and nine rows. In the first row, label the columns *Turns, Current (A), Current Direction,* and *Compass Reading.* In the first column, label the second through ninth rows *One, One, Two, Two, Three, Three, Four,* and *Four.*

Magnetic Field of a Current-Carrying Wire

3. Wrap the wire once around the galvanometer. Place the large compass on the stand of the galvanometer so that the compass needle is parallel to and directly below the wire, as shown in **Figure 1.** (**Figure 1** shows multiple wire windings, not only one.) Turn the galvanometer until the turn of wire is in the north-to-south plane, as indicated by the compass needle.

4. Construct a circuit that contains the power supply, a current meter, a 1 Ω resistor, and a switch, all wired in series with the galvanometer. Connect the galvanometer so that the direction of the current will be from south to north through the segment of the loop above the compass needle. ***Do not close the switch until your teacher has approved your circuit.***

5. Set the power supply to its lowest output. When your teacher has approved your circuit, close the switch briefly. Using the potentiometer on the power supply, adjust the current in the circuit to 1.5 A. Use the potentiometer to maintain a current of 1.5 A throughout the lab. Record the current, the current direction, and the compass reading in your data table. *Open the switch as soon as you have made your measurements so that the power supply and wires don't overheat.*

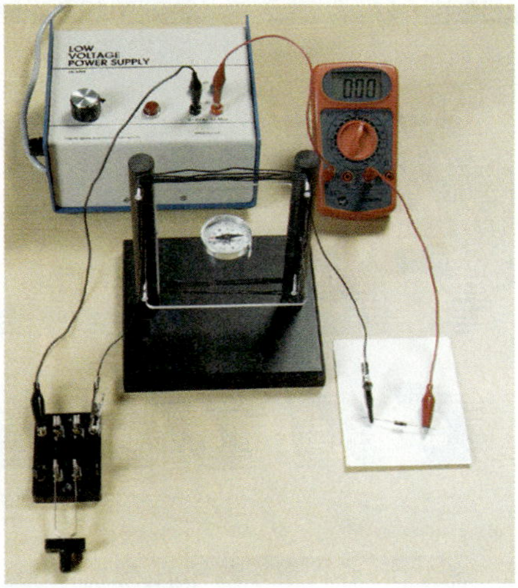

Figure 1
Step 3: Use the support pins on the galvanometer to wrap the wire into a loop. Adjust the apparatus so that the needle and wire are in the north-to-south plane.

6. Reverse the direction of the current in the segment of the loop above the needle by reversing the wires connecting to the power supply.

7. Close the switch. Adjust the power supply to 1.5 A. Record your observations in your data table, and then immediately open the switch.

8. Remove the galvanometer from the circuit. Add a second turn of wire, and reconnect the galvanometer to the circuit so that the current direction will be south to north.

9. Close the switch. Adjust the power supply to 1.5 A. Record your observations in your data table. Open the switch immediately.

10. Repeat Steps 6 and 7 for two turns of wire.

11. Repeat the experiment for three turns and then four turns of wire. For each, connect the circuit so that the direction of the current is from south to north and then north to south. Record all information.

12. Clean up your work area. Put equipment away safely.

ANALYSIS

1. **Organizing Data** For each trial, find the tangent of the angle of the compass needle's deflection.

2. **Constructing Graphs** Use a computer, graphing calculator, or graph paper to plot the tangents (item 1) against the number of turns in the wire.

CONCLUSIONS

3. **Drawing Conclusions** What is the relationship between the tangent of the angle and the number of turns? Explain.

4. **Drawing Conclusions** What is the relationship between the direction of current in the wire and the direction of the magnetic field? Explain.

Tips and Tricks

- Make sure students know how to read a compass and record the readings.
- See the teaching tips for instructions on setting up multimeters to measure current.
- Make sure students understand how to wire current meters (in series) in a circuit.
- Show students how to set up the wire coil on the galvanometer apparatus and how to determine the direction of the current.

✓ Checkpoints

Step 4: Make sure the power supply, resistor, current meter, and switch are connected correctly and are at the proper settings.

Step 6: Make sure students reconnect the wires properly to reverse the current.

Step 8: Make sure students have two loops of wire and that all connections are correct.

ANSWERS

Analysis

1. Student answers will vary. For sample data, values for $\tan\theta$ range from 0.577 to 2.747.

2. Student graphs should show that $\tan\theta$ increases as the number of turns increases.

Conclusions

3. Because $\tan\theta$ increases as the number of turns increases, students should realize that $\tan\theta$ is proportional to the magnetic field strength.

4. When the current direction changes, the direction of the magnetic field also changes.

Science• Technology• Society

CAN CELL PHONES CAUSE CANCER?

In 2005, lawsuits were filed against cell phone manufacturers in Georgia, Louisiana, Maryland, New York, and Pennsylvania. These lawsuits alleged that there may be a link between cell phone use and brain cancer and accused the manufacturers of failing to adequately protect users from radiation. Is there any truth to this claim?

Nonionizing Radiation

Cell phones transfer messages by sending and receiving electromagnetic waves. The electromagnetic spectrum includes low-energy waves, such as radio waves, and high-energy waves, such as X rays and gamma rays. High-energy electromagnetic waves are *ionizing*, which means they have enough energy to remove an electron from its orbit. Ionizing electromagnetic radiation can damage living tissue and cause DNA mutations, which is why exposure to X rays should be limited.

Cell phones use radio frequencies (RFs) ranging from about 1800 to 2200 MHz. These *nonionizing* waves do not alter the molecular structure of living tissue. They can cause the atoms in a molecule to vibrate but do not have enough energy to remove electrons from their orbits. At high enough levels, however, nonionizing radiation can cause biological damage by heating living tissue. But the amount of heat that a cell phone's radiation generates is very small, much smaller than the energy generated in a microwave oven.

What Do the Studies Say?

The effects of nonionizing radiation on the human body are not fully known. Several studies have been conducted to determine whether there is any possible link between cell phone use and brain cancer. Scientists conducting these studies have attempted to determine whether the risk of brain cancer is greater for cell phone users than for nonusers. Even if a link is found, it is not necessarily a cause-and-effect link. In other words, even if cell phone users *do* have a higher risk of cancer, cell phone use is not necessarily the cause.

Further, several issues complicate the research. In the past, scientists had to estimate RF exposure based on interviews with patients. (Scientists are now using RF meters to measure the amount of exposure, so newer studies will be more accurate in this regard.) Also, cell

704

phones were not widely available until the 1990s, and brain tumors develop over many years. Hence, the early studies did not cover long-term cell phone use.

A study funded by the National Cancer Institute and published in 2001 focused on 782 patients who had brain cancer and 799 patients who did not have brain tumors. This study did not find any increased risk of brain cancer for the patients who had used cell phones. Many other studies have found similar results. However, a few studies have found a possible link, including one Swedish study that found an increased risk of acoustic neuroma (a benign tumor) in long-term cell phone users. In 2006, the FDA announced that it would revisit the issue. The FDA stated that the findings of two Swedish studies are inconsistent with earlier studies and acknowledged that these studies are difficult to interpret.

Reducing RF Exposure

Although they have not demonstrated a link between cell phone use and brain cancer, scientists have not concluded that there is no risk. More research is needed, especially on the long-term effects of RF exposure. In the meantime, concerned cell phone users can take measures to limit their exposure to RF. Exposure depends on a number of factors, including the amount of time spent using the phone, the amount of cell phone traffic in the area, and the distance between the antenna and the user's head. One way to reduce exposure is to minimize the time spent on cellular calls. Another option is to use a hands-free device that puts the antenna farther from the head.

Researching the Issue

1. Cell phone makers are now required to report the *specific absorption rate* (SAR), the amount of RF energy absorbed by the user. The maximum allowed SAR is 1.6 watts per kilogram. Conduct research to find the SAR of several top models of phones. If you own a cell phone, see if you can determine the SAR of your phone.

2. Research the effects of ionizing radiation on the human body. What are some sources of ionizing radiation? How does ionizing radiation affect living tissue?

3. Use the Internet to research one of the recent epidemiological studies done on cell phone use and brain cancer. Write a short report describing the study, including the subjects and control group, the method of obtaining data, and the conclusions reached by the researchers. Share your report with the class.

4. As cell phones have grown in popularity, concerns have arisen about the safety of driving while using a cell phone. Several countries have banned the use of cell phones while driving. Conduct research to find out about studies conducted on this issue. Is it hazardous? Should we pass laws to prevent it? Choose a position on the issue, and write a paper defending your position.

705

CHAPTER 20

Electromagnetic Induction
Planning Guide

Compression Guide
To shorten instruction because of time limitations, omit the opener, Advanced sections, and the review.

OBJECTIVES	LABS, DEMONSTRATIONS, AND ACTIVITIES	TECHNOLOGY RESOURCES
PACING • 45 min pp. 706–707 **Chapter Opener**	**ANC** **Discovery Lab** Electricity and Magnetism*◆ BASIC	**CD** **Visual Concepts,** Chapter 20 BASIC
PACING • 135 min pp. 708–715 **Section 1 Electricity and Magnetism** • Recognize that relative motion between a conductor and a magnetic field induces an emf in the conductor. • Describe how the change in the number of magnetic field lines through a circuit loop affects the induced electric current. • Apply Lenz's law and Faraday's law of induction to solve problems involving induced emf and current.	**SE** **Skills Practice Lab** Electromagnetic Induction, pp. 746–747◆ GENERAL **ANC** **Datasheet** Electromagnetic Induction* GENERAL **TE** **Demonstration** Induced Current, p. 710 GENERAL **TE** **Demonstration** Lenz's Law, p. 712 GENERAL	**OSP** **Lesson Plans** **TR** 107 Ways of Inducing a Current in a Circuit **TR** 108 Direction of Induced Current and Their Magnetic Fields **TR** 64A Magnetic Field of a Conducting Loop
PACING • 45 min pp. 712–722 **Section 2 Generators, Motors, and Mutual Inductance** • Describe how generators and motors operate. Explain the energy conversions that take place in generators and motors. • Describe how mutual induction occurs in circuits.	**TE** **Demonstration** Electric Motor, p. 720 GENERAL **TE** **Demonstration** ac and dc, p. 721 ADVANCED	**OSP** **Lesson Plans** **TR** 109 Induction of an emf in an ac Generator **TR** 110 Components of a DC Motor
PACING • 45 min pp. 723–730 *Advanced Level* **Section 3 AC Circuits and Transformers** • Distinguish between rms values and maximum values of current and potential difference. Solve problems involving rms and maximum values of current and emf for ac circuits. • Apply the transformer equation to solve problems involving step-up and step-down transformers.	**TE** **Demonstration** Effects of Alternating Current, p. 724 ADVANCED **TE** **Demonstration** Induced emf, p. 725 GENERAL **TE** **Demonstration** Transformers, p. 727 GENERAL **ANC** **Invention Lab** Building a Circuit Breaker*◆ ADVANCED	**OSP** **Lesson Plans** **TR** 111 Induction by a Fluctuating Current **TR** 112 A Step-Up Transformer
PACING • 45 min pp. 731–737 *Advanced Level* **Section 4 Electromagnetic Waves** • Describe what electromagnetic waves are and how they are produced. Recognize that electricity and magnetism are two aspects of a single electromagnetic force. • Explain how electromagnetic waves transfer energy. Describe various applications of electromagnetic waves.	**TE** **Demonstration** Van de Graaff Generator, p. 735 ADVANCED	**OSP** **Lesson Plans** **EXT** **Integrating Technology** Radio Waves GENERAL **TR** 113 The Sun at Different Wavelengths **TR** 114 The Electromagnetic Spectrum

PACING • 90 min

CHAPTER REVIEW, ASSESSMENT, AND STANDARDIZED TEST PREPARATION

- **SE** **Chapter Highlights,** p. 738
- **SE** **Chapter Review,** pp. 739–743
- **SE** **Graphing Calculator Practice,** p. 742 GENERAL
- **SE** **Alternative Assessment,** p. 743 ADVANCED
- **SE** **Standardized Test Prep,** pp. 744–745 GENERAL
- **SE** **Appendix D: Equations,** pp. 863–864
- **SE** **Appendix I: Additional Problems,** p. 895
- **ANC** **Study Guide Worksheet** Mixed Review* GENERAL
- **ANC** **Chapter Test A*** GENERAL
- **ANC** **Chapter Test B*** ADVANCED
- **OSP** **Test Generator**

Online and Technology Resources

go.hrw.com Holt Online Learning

Visit go.hrw.com to find a variety of online resources. To access this chapter's extensions, enter the keyword **HF6EMIXT** and click the "go" button. Click **Holt Online Learning** for an online edition of this textbook, and other interactive resources.

HOLT Teacher's One-Stop Planner®

This DVD package includes:
- Holt Calendar Planner
- Customizable Lesson Plans
- Editable Worksheets
- ExamView® Version 6 Assessment Suite
- Interactive Teacher's Edition
- Holt PuzzlePro®
- Holt PowerNotes® Presentations
- MindPoint® Quiz Show

KEY

SE	Student Edition	OSP	One-Stop Planner	EXT	Online Extension
TE	Teacher Edition	CD	CD or CD-ROM	*	Also on One-Stop Planner
ANC	Ancillary Worksheet	TR	Teaching Transparencies	◆	Requires advance prep

SKILLS DEVELOPMENT RESOURCES	REVIEW AND ASSESSMENT	CORRELATIONS
		National Science Education Standards
SE **Sample Set A** Induced emf and Current, pp. 713–714 GENERAL TE **Classroom Practice,** p. 713 GENERAL ANC **Problem Workbook** Sample Set A* GENERAL OSP **Problem Bank** Sample Set A GENERAL SE **Conceptual Challenge,** p. 711 ADVANCED	SE **Section Review,** p. 715 GENERAL ANC **Study Guide Worksheet** Section 1* GENERAL ANC **Quiz** Section 1* BASIC	UCP 1, 2, 3, 4, 5 SAI 1, 2 ST 1, 2 SPSP 5
SE **Appendix J: Advanced Topics** Angular Kinematics, pp. 898–901 ADVANCED EXT **Practice Problems** Induction in Generators ADVANCED	SE **Section Review,** p. 722 GENERAL ANC **Study Guide Worksheet** Section 2* GENERAL ANC **Quiz** Section 2* BASIC	UCP 1, 2, 3, 4, 5 ST 1, 2 SPSP 5
SE **Sample Set B** rms Current and emf, pp. 725–726 GENERAL TE **Classroom Practice,** p. 725 GENERAL ANC **Problem Workbook*** and OSP **Problem Bank** Sample Set B GENERAL SE **Sample Set C** Transformers, pp. 728–729 GENERAL TE **Classroom Practice,** p. 728 GENERAL ANC **Problem Workbook** and OSP **Problem Bank** Sample Set C* GENERAL	SE **Section Review,** p. 730 ADVANCED ANC **Study Guide Worksheet** Section 3* ADVANCED ANC **Quiz** Section 3* GENERAL	UCP 1, 2, 3, 4, 5 SAI 1, 2 ST 1, 2 HNS 3 SPSP 5
	SE **Section Review,** p. 737 ADVANCED ANC **Study Guide Worksheet** Section 4* ADVANCED ANC **Quiz** Section 4* GENERAL	UCP 1, 2, 3, 4, 5 SAI 1, 2 HNS 1, 3 SPSP 5 PS 6b

SCILINKS. NSTA

www.scilinks.org

Maintained by the **National Science Teachers Association.**

Topic: Electromagnetic Induction
SciLinks Code: HF60481

Topic: Alternating Current
SciLinks Code: HF60050

Topic: Electrical Safety
SciLinks Code: HF60476

Topic: Transformers
SciLinks Code: HF61549

Classroom CD-ROMs

- Guided Reading Audio Program
- Student One Stop
- Virtual Investigations
- Visual Concepts

Holt Lab Generator CD-ROM

Search for any lab by topic, standard, difficulty level, or time. Edit any lab to fit your needs, or create your own labs. Use the Lab Materials QuickList software to customize your lab materials list.

CHAPTER 20

Overview

Section 1 introduces induced current, discusses Lenz's law, and applies Faraday's law of induction to calculate induced emf and induced current.

Section 2 introduces generators and motors—devices that convert energy from one form to another—and examines mutual inductance.

Section 3 shows how to calculate the rms current for ac circuits, discusses transformers, and shows how to calculate the emf for a step-up or step-down transformer.

Section 4 further explores electromagnetic waves and the electromagnetic spectrum, first introduced in the chapter "Light and Reflection."

About the Illustration

The strings of an electric guitar are magnetized by a permanent magnet located inside coils of wire beneath the strings. Two sets of coils—one at the base of the neck and the other just above the bridge—can be seen in each of the guitars in the photograph. As a string vibrates, the changing magnetic field induces in the coil an alternating current, whose frequency depends on the string's frequency.

The vibrations of the strings in an electric guitar change the magnetic field near a coil of wire called the pickup. In turn, this induces an electric current in the coil, which is then amplified to create the unique sound of an electric guitar.

CHAPTER 20

Electromagnetic Induction

WHAT TO EXPECT

In this chapter, you will learn how induction produces and changes alternating currents. You will also explore electromagnetic waves and the electromagnetic spectrum.

Why it Matters

Electric guitars have many different types of pickups, but all generate electric current by the process of induction. An understanding of the induction of electromagnetic fields is essential to the good design of an electric guitar.

CHAPTER PREVIEW

Tapping Prior Knowledge

Knowledge to Review

- ✔ The definition of resistance states that the current in a circuit is proportional to the applied potential difference and inversely proportional to the resistance of the circuit.
- ✔ Emf produces an electric field that causes a force on free charges in a conductor. This force causes the free charges to move through the conductor. These moving charges are what we call an electric current.
- ✔ A charged particle moving in a magnetic field experiences a magnetic force. When the particle moves with a component perpendicular to the field, $F_{magnetic} = qvB$.

Items to Probe

- ✔ Resistance: Have students calculate current for various cases involving changing potential differences.

SECTION 1
General Level

Teaching Tip — GENERAL

Remind students that as discussed in the chapter on magnetism, electric charges that move through a magnetic field experience a magnetic force. The maximum value of this force is qvB. When charges are at rest with respect to the magnetic field ($v = 0$ m/s), they do not experience a magnetic force.

Visual Strategy GENERAL

Figure 1

Be sure students understand the conditions required to induce a current in a circuit.

Q Why is a current generated in a moving conductor in a magnetic field but not in a moving insulator in the same magnetic field?

A ***A force acts on the charges in both cases. In conductors, the electrons are free to move in response to the magnetic force. A force also acts on the electrons in an insulator, but these electrons are not free to move.***

Q What is the condition required for current to exist in the circuit shown in this figure?

A ***There must be relative motion between the circuit and the magnetic field; when the two are at rest relative to one another, charges in the conducting wires do not experience a magnetic force, so there is no induced current.***

SECTION 1

Electricity from Magnetism

SECTION OBJECTIVES

- **Recognize that relative motion between a conductor and a magnetic field induces an emf in the conductor.**
- **Describe how the change in the number of magnetic field lines through a circuit loop affects the magnitude and direction of the induced electric current.**
- **Apply Lenz's law and Faraday's law of induction to solve problems involving induced emf and current.**

ELECTROMAGNETIC INDUCTION

Recall that when you were studying circuits, you were asked if it was possible to produce an electric current using only wires and no battery. So far, all electric circuits that you have studied have used a battery or an electrical power supply to create a potential difference within a circuit. The electric field associated with that potential difference causes charges to move through the circuit and to create a current.

It is also possible to *induce* a current in a circuit without the use of a battery or an electrical power supply. You have learned that a current in a circuit is the source of a magnetic field. Conversely, a current results when a closed electric circuit moves with respect to a magnetic field, as shown in **Figure 1.** The process of inducing a current in a circuit by changing the magnetic field that passes through the circuit is called **electromagnetic induction.**

electromagnetic induction

the process of creating a current in a circuit loop by changing the magnetic flux in the loop

Consider a closed circuit consisting of only a resistor that is in the vicinity of a magnet. There is no battery to supply a current. If neither the magnet nor the circuit is moving with respect to the other, no current will be present in the circuit. But, if the circuit moves toward or away from the magnet or the magnet moves toward or away from the circuit, a current is induced. As long as there is relative motion between the two, a current is created in the circuit.

The separation of charges by the magnetic force induces an emf

It may seem strange that there can be an induced emf and a corresponding induced current without a battery or similar source of electrical energy. Recall from the previous chapter that a moving charge can be deflected by a magnetic field. This deflection can be used to explain how an emf occurs in a wire that moves through a magnetic field.

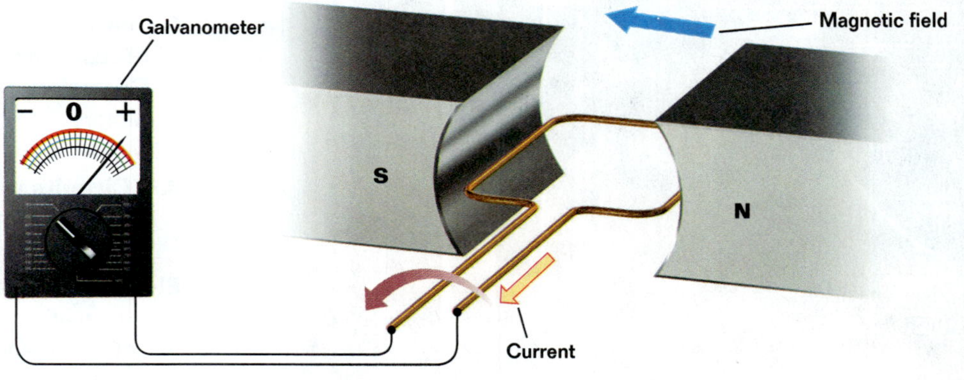

Figure 1
When the circuit loop crosses the lines of the magnetic field, a current is induced in the circuit, as indicated by the movement of the galvanometer needle.

Consider a conducting wire pulled through a magnetic field, as shown on the left in **Figure 2.** You learned when studying magnetism that charged particles moving with a velocity at an angle to the magnetic field will experience a magnetic force. According to the right-hand rule, this force will be perpendicular to both the magnetic field and the motion of the charges. For free positive charges in the wire, the force is directed downward along the wire. For negative charges, the force is upward. This effect is equivalent to replacing the segment of wire and the magnetic field with a battery that has a potential difference, or emf, between its terminals, as shown on the right in **Figure 2.** As long as the conducting wire moves through the magnetic field, the emf will be maintained.

The polarity of the induced emf depends on the direction in which the wire is moved through the magnetic field. For instance, if the wire in **Figure 2** is moved to the right, the right-hand rule predicts that the negative charges will be pushed upward. If the wire is moved to the left, the negative charges will be pushed downward. The magnitude of the induced emf depends on the velocity with which the wire is moving through the magnetic field, on the length of the wire in the field, and on the strength of the magnetic field.

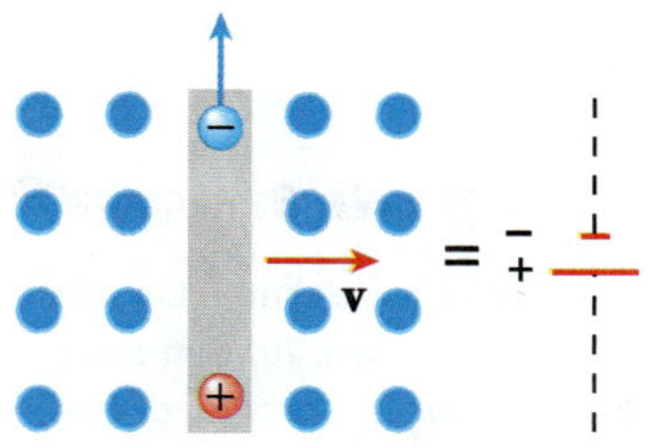

Figure 2
The separation of positive and negative moving charges by the magnetic force creates a potential difference (emf) between the ends of the conductor.

The angle between a magnetic field and a circuit affects induction

One way to induce an emf in a closed loop of wire is to move all or part of the loop into or out of a constant magnetic field. No emf is induced if the loop is static and the magnetic field is constant.

The magnitude of the induced emf and current depend partly on how the loop is oriented to the magnetic field, as shown in **Figure 3.** The induced current is largest if the plane of the loop is perpendicular to the magnetic field, as in **(a)**; it is smaller if the plane is tilted into the field, as in **(b)**; and it is zero if the plane is parallel to the field, as in **(c).**

The role that the orientation of the loop plays in inducing the current can be explained by the force that the magnetic field exerts on the charges in the moving loop. Only the component of the magnetic field *perpendicular* to both the plane and the motion of the loop exerts a magnetic force on the charges in the loop. If the area of the loop is moved *parallel* to the magnetic field, there is no magnetic field component perpendicular to the plane of the loop and therefore no induced emf to move the charges around the circuit.

SCILINKS.
www.scilinks.org
Topic: Electromagnetic Induction
Code: HF60481

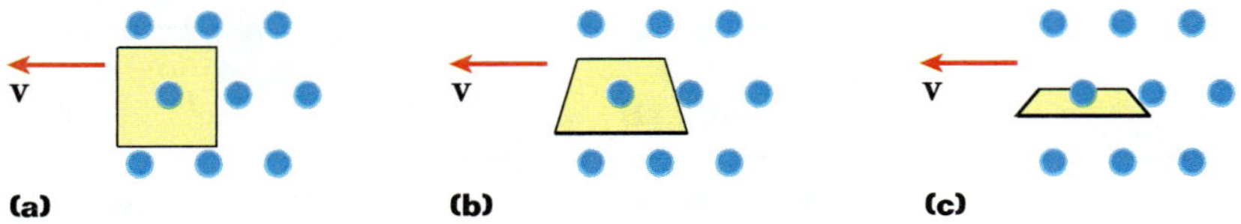

Figure 3
These three loops of wire are moving out of a region that has a constant magnetic field. The induced emf and current are largest when the plane of the loop is perpendicular to the magnetic field **(a),** smaller when the plane of the loop is tilted **(b),** and zero when the plane of the loop and the magnetic field are parallel **(c).**

Teaching Tip

Remind students that, as seen in the chapter on magnetism, a current creates a magnetic field whose direction can be found with the right-hand rule. If the thumb points in the direction of conventional current (moving positive charges), the right hand wraps around the wire in the direction of the magnetic field.

Visual Strategy GENERAL

Figure 3

Point out that any charge with a velocity component perpendicular to a magnetic field will experience a force that is perpendicular both to the magnetic field and to that velocity component.

Q Why does a loop that is parallel to the field and that moves as shown in **Figure 3(c)** experience no current around the loop, even though the loop is moving in the magnetic field?

A ***The electrons do experience a magnetic force perpendicular to the field, but they cannot move in that direction because the loop is parallel to the field.***

Demonstration

Induced Current

Purpose Show students an example of an induced current.

Materials flashlight bulb in holder, coil of wire, bar magnet (two may be required), connecting wires

Procedure Connect the flashlight bulb to the coil of wire with the connecting wires. Tell students to observe the demonstration with the intent of explaining the energy conversions. Move the bar magnet into and out of the coil several times in rapid succession. Have students explain the energy conversions on the board or in their notebooks. The following energy conversions should be discussed: kinetic energy is converted to electrical energy (moving magnet generates a current) and electrical energy is converted to light (the current heats the light bulb's filament).

Did you know?

In 1996, the space shuttle *Columbia* attempted to use a 20.7 km conducting tether to study Earth's magnetic field in space. The plan was to drag the tether through the magnetic field, inducing an emf in the tether. The magnitude of the emf would directly vary with the strength of the magnetic field. Unfortunately, the tether broke before it was fully extended, so the experiment was abandoned.

Change in the number of magnetic field lines induces a current

So far, you have learned that moving a circuit loop into or out of a magnetic field can induce an emf and a current in the circuit. Changing the size of the loop or the strength of the magnetic field also will induce an emf in the circuit.

One way to predict whether a current will be induced in a given situation is to consider how many magnetic field lines cut through the loop. For example, moving the circuit into the magnetic field causes some lines to move into the loop. Changing the size of the circuit loop or rotating the loop changes the number of field lines passing through the loop, as does changing the magnetic field's strength or direction. **Table 1** summarizes these three ways of inducing a current.

CHARACTERISTICS OF INDUCED CURRENT

Suppose a bar magnet is pushed into a coil of wire. As the magnet moves into the coil, the strength of the magnetic field within the coil increases, and a current is induced in the circuit. This induced current in turn produces its own magnetic field, whose direction can be found by using the right-hand rule. If you were to apply this rule for several cases, you would notice that the induced magnetic field direction depends on the change in the applied field.

As the magnet approaches, the magnetic field passing through the coil increases in strength. The induced current in the coil is in a direction that produces a magnetic field that opposes the increasing strength of the approaching field. So, the induced magnetic field is in the opposite direction of the increasing magnetic field.

Table 1 Ways of Inducing a Current in a Circuit

Description	Before	After
Circuit is moved into or out of magnetic field (either circuit or magnet moving).		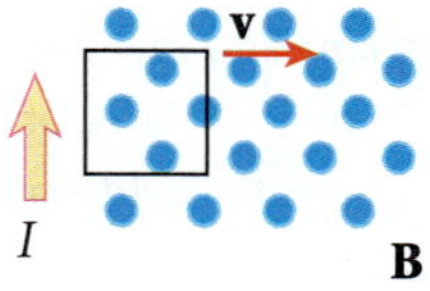
Circuit is rotated in the magnetic field (angle between area of circuit and magnetic field changes).	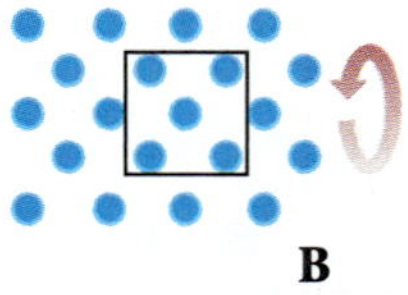	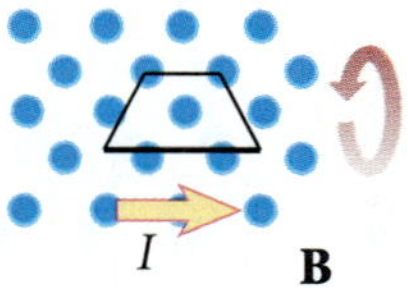
Intensity and/or direction of magnetic field is varied.	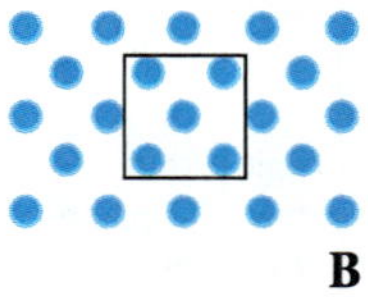	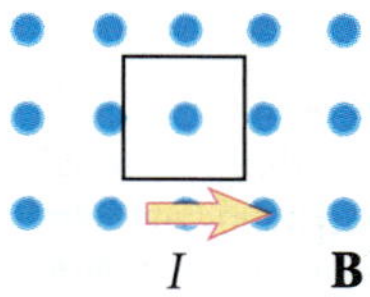

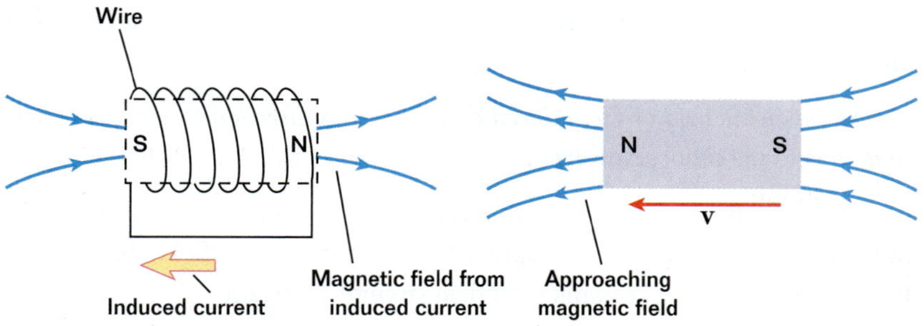

Figure 4
When a bar magnet is moved toward a coil, the induced magnetic field is similar to the field of a bar magnet with the orientation shown.

The induced magnetic field is similar to the field of a bar magnet that is oriented as shown in **Figure 4.** The coil and the approaching magnet create a pair of forces that repel each other.

If the magnet is moved away from the coil, the magnetic field passing through the coil decreases in strength. Again, the current induced in the coil produces a magnetic field that opposes the decreasing strength of the receding field. This means that the magnetic field that the coil sets up is in the same direction as the receding magnetic field.

The induced magnetic field is similar to the field of a bar magnet oriented as shown in **Figure 5.** In this case the coil and magnet attract each other.

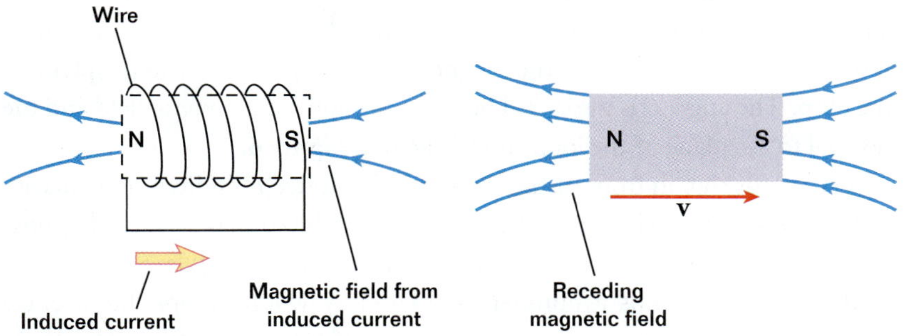

Figure 5
When a bar magnet is moved away from a coil, the induced magnetic field is similar to the field of a bar magnet with the orientation shown.

Why it Matters

Conceptual Challenge

1. Falling Magnet A bar magnet is dropped toward the floor, on which lies a large ring of conducting metal. The magnet's length—and thus the poles of the magnet—is parallel to the direction of motion. Disregarding air resistance, does the magnet fall toward the ring with the constant acceleration of a freely falling body? Explain your answer.

2. Induction in a Bracelet Suppose you are wearing a bracelet that is an unbroken ring of copper. If you walk briskly into a strong magnetic field while wearing the bracelet, how would you hold your wrist with respect to the magnetic field in order to avoid inducing a current in the bracelet?

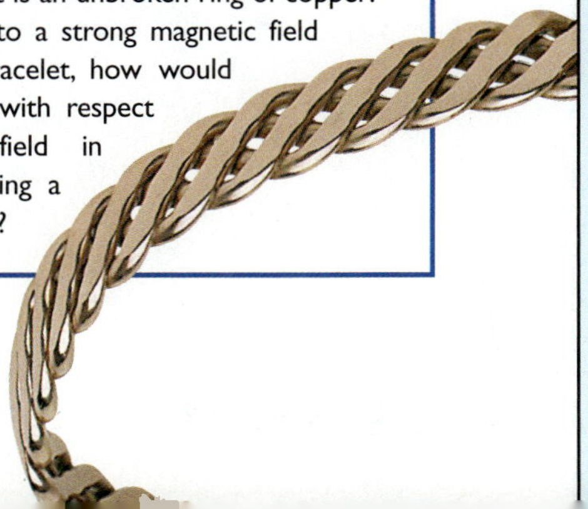

The Language of Physics — ADVANCED

The term *emf* originally stood for electromotive force. Today, emf is considered to be analogous to a potential difference instead of a force. Specifically, emf makes charges in a circuit move, just as a potential difference does. To avoid misconceptions, the term electromotive force is not used in this text.

Misconception Alert — GENERAL

Some students may not distinguish between the external magnetic field that induces a current and the magnetic field that is set up by the induced current. Use the two examples discussed on this page to clarify the difference between these two magnetic fields.

ANSWERS

Conceptual Challenge

1. no; The magnet's acceleration is slightly smaller. The magnet induces a current in the conducting ring, and the magnetic field of this current opposes the field of the falling magnet. This induced field exerts an upward force on the magnet, reducing the magnet's net acceleration downward.
2. Upon entering and leaving the field, the plane of the bracelet must be parallel to the direction of the field.

Demonstration

Lenz's Law GENERAL

Purpose Illustrate Lenz's law experimentally.

Materials flashlight bulb in holder, coil of wire, diode, connecting wires, bar magnet (or two)

Procedure Repeat the first demonstration, but include a diode in the circuit. Tell students that the diode allows charges to move in only one direction.

Explain to students that you will be testing Lenz's law experimentally. Point out that inserting the magnet one way (with the north pole first) will generate a current in one direction. Reversing the magnet will generate a current in the opposite direction, which the diode will stop. The light bulb will serve as an indication of the direction of current. Thus, the bulb should light up in one case but not in the other (because the diode blocks current in one direction). Perform the demonstration and verify these conclusions.

The rule for finding the direction of the induced current is called *Lenz's law* and is expressed as follows:

The magnetic field of the induced current is in a direction to produce a field that opposes the change causing it.

Note that the field of the induced current does not oppose the applied field but rather the change in the applied field. If the applied field changes, the induced field tends to keep the total field strength constant.

Faraday's law of induction predicts the magnitude of the induced emf

Lenz's law allows you to determine the direction of an induced current in a circuit. Lenz's law does not provide information on the magnitude of the induced current or the induced emf. To calculate the magnitude of the induced emf, you must use *Faraday's law of magnetic induction.* For a single loop of a circuit, this may be expressed as follows:

$$\text{emf} = -\frac{\Delta\Phi_M}{\Delta t}$$

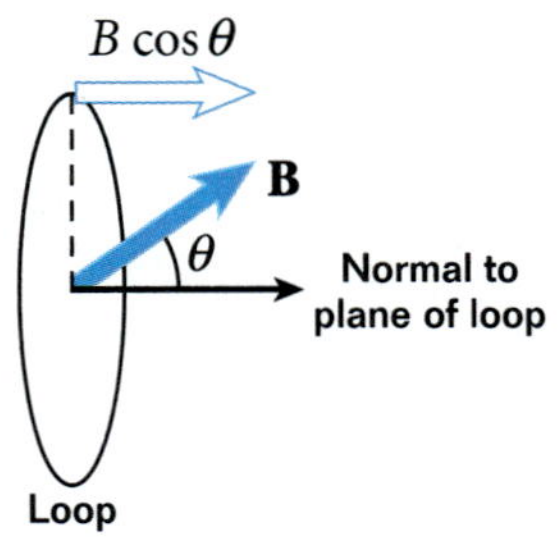

Figure 6
The angle θ is defined as the angle between the magnetic field and the normal to the plane of the loop. $B\cos\theta$ equals the strength of the magnetic field perpendicular to the plane of the loop.

Recall from the chapter on magnetism that the magnetic flux, Φ_M, can be written as $AB\cos\theta$. This equation means that a change with time of any of the three variables—applied magnetic field strength, B; circuit area, A; or angle of orientation, θ—can give rise to an induced emf. The term $B\cos\theta$ represents the component of the magnetic field perpendicular to the plane of the loop. The angle θ is measured between the applied magnetic field and the normal to the plane of the loop, as indicated in **Figure 6.**

The minus sign in front of the equation is included to indicate the polarity of the induced emf. The sign indicates that the induced magnetic field opposes the change in the applied magnetic field as stated by Lenz's law.

If a circuit contains a number, N, of tightly wound loops, the average induced emf is simply N times the induced emf for a single loop. The equation thus takes the general form of Faraday's law of magnetic induction.

FARADAY'S LAW OF MAGNETIC INDUCTION

$$\text{emf} = -N\frac{\Delta\Phi_M}{\Delta t}$$

average induced emf = –the number of loops in the circuit × the time rate of change of the magnetic flux

In this chapter, N is always assumed to be a whole number.

Recall that the SI unit for magnetic field strength is the tesla (T), which equals one newton per ampere-meter, or N/(A•m). The tesla can also be expressed in the equivalent units of one volt-second per meter squared, or $(V\bullet s)/m^2$. Thus, the unit for emf, as for electric potential, is the volt.

SAMPLE PROBLEM A

Induced emf and Current

PROBLEM

A coil with 25 turns of wire is wrapped around a hollow tube with an area of 1.8 m^2. Each turn has the same area as the tube. A uniform magnetic field is applied at a right angle to the plane of the coil. If the field increases uniformly from 0.00 T to 0.55 T in 0.85 s, find the magnitude of the induced emf in the coil. If the resistance in the coil is 2.5 Ω, find the magnitude of the induced current in the coil.

SOLUTION

1. DEFINE

Given: $\Delta t = 0.85$ s $\quad A = 1.8$ m^2 $\quad \theta = 0.0°$ $\quad N = 25$ turns

$B_i = 0.00$ T $= 0.00$ V•s/m^2 $\quad B_f = 0.55$ T $= 0.55$ V•s/m^2

$R = 2.5\ \Omega$

Unknown: emf = ? $\quad I = ?$

Diagram: Show the coil before and after the change in the magnetic field.

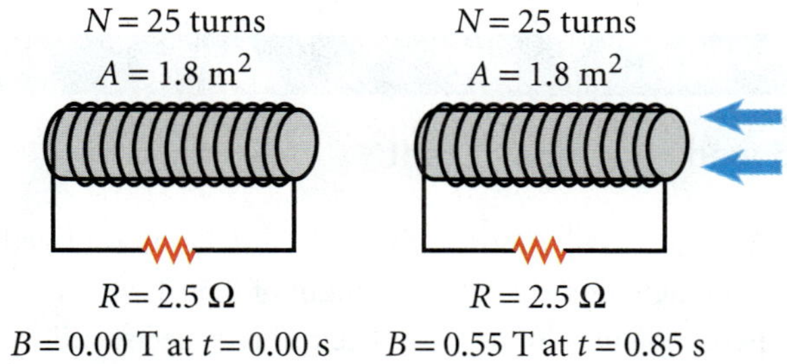

2. PLAN

Choose an equation or situation: Use Faraday's law of magnetic induction to find the induced emf in the coil.

$$\text{emf} = -N\frac{\Delta\Phi_M}{\Delta t} = -N\frac{\Delta[AB\cos\theta]}{\Delta t}$$

Substitute the induced emf into the definition of resistance to determine the induced current in the coil.

$$I = \frac{\text{emf}}{R}$$

Rearrange the equation to isolate the unknown: In this example, only the magnetic field strength changes with time. The other components (the coil area and the angle between the magnetic field and the coil) remain constant.

$$\text{emf} = -NA\cos\theta\frac{\Delta B}{\Delta t}$$

continued on next page

The Language of Physics — GENERAL

Both emf and potential difference have similar current-driving abilities, and both are measurable in the same units (volts). In terms of emf, Ohm's law takes the following form:

$$I = \frac{\text{emf}}{R}$$

This form of the definition of resistance is used in Sample Problem A.

Classroom Practice

Induced emf and Current

A coil with 25 turns of wire is moving in a uniform magnetic field of 1.5 T. The magnetic field is perpendicular to the plane of the coil. The coil has a cross-sectional area of 0.80 m^2. The coil exits the field in 1.0 s.

a. Find the induced emf.

b. Determine the induced current if the coil's resistance is 1.5 Ω.

Answers

a. 3.0×10^1 V

b. 2.0×10^1 A

Teaching Tip

Point out to students that Lenz's law is used in the fourth step of Sample Problem A to find the direction of the induced current.

PROBLEM GUIDE A

Use this guide to assign problems.
SE = Student Edition Textbook
PW = Problem Workbook
PB = Problem Bank on the One-Stop Planner (OSP)

Solving for:

emf	**SE** Sample, 1, 3; Ch. Rvw. 10, 12, 42 **PW** 4 **PB** 4–6
I	**SE** Sample, 2; Ch. Rvw. 11 **PB** 7–8
Δt	**SE** Ch. Rvw. 37 **PW** Sample, 1–3 **PB** 9–10
B	**SE** 4; Ch. Rvw. 39 **PB** Sample, 1–3
N	**SE** Ch. Rvw. 38 **PW** 5
A	**PW** 6

***Challenging Problem**
Consult the printed Solutions Manual or the OSP for detailed solutions.

ANSWERS

Practice A

1. 0.30 V
2. 14 A
3. 0.14 V
4. 4.83×10^{-5} T

3. CALCULATE **Substitute the values into the equation and solve:**

$$emf = -(25)(1.8\ \cancel{m^2})(\cos 0.0°)\frac{\left((0.55-0.00)\dfrac{V\bullet \cancel{s}}{\cancel{m^2}}\right)}{(0.85\ \cancel{s})} = -29\ V$$

$$I = \frac{-29\ V}{2.5\ \Omega} = -12\ A$$

$$emf = -29\ V$$
$$I = -12\ A$$

TIP *Because the minimum number of significant figures for the data is two, the calculator answer, 29.11764706, should be rounded to two digits.*

4. EVALUATE The induced emf, and therefore the induced current, is directed through the coil so that the magnetic field produced by the induced current opposes the change in the applied magnetic field. For the diagram shown on the previous page, the induced magnetic field is directed to the right and the current that produces it is directed from left to right through the resistor.

PRACTICE A

Induced emf and Current

1. A single circular loop with a radius of 22 cm is placed in a uniform external magnetic field with a strength of 0.50 T so that the plane of the coil is perpendicular to the field. The coil is pulled steadily out of the field in 0.25 s. Find the average induced emf during this interval.
2. A coil with 205 turns of wire, a total resistance of 23 Ω, and a cross-sectional area of 0.25 m^2 is positioned with its plane perpendicular to the field of a powerful electromagnet. What average current is induced in the coil during the 0.25 s that the magnetic field drops from 1.6 T to 0.0 T?
3. A circular wire loop with a radius of 0.33 m is located in an external magnetic field of strength +0.35 T that is perpendicular to the plane of the loop. The field strength changes to −0.25 T in 1.5 s. (The plus and minus signs for a magnetic field refer to opposite directions through the coil.) Find the magnitude of the average induced emf during this interval.
4. A 505-turn circular-loop coil with a diameter of 15.5 cm is initially aligned so that its plane is perpendicular to Earth's magnetic field. In 2.77 ms the coil is rotated 90.0° so that its plane is parallel to Earth's magnetic field. If an average emf of 0.166 V is induced in the coil, what is the value of Earth's magnetic field?

Why it Matters

Electric Guitar Pickups

The word *pickup* refers to a device that "picks up" the sound of an instrument and turns the sound into an electrical signal. The most common type of electric guitar pickup uses electromagnetic induction to convert string vibrations into electrical energy.

In their most basic form, magnetic pickups consist simply of a permanent magnet and a coil of copper wire. A pole piece under each guitar string concentrates and shapes the magnetic field. Because guitar strings are made from magnetic materials (steel and/or nickel), a vibrating guitar string causes a change in the magnetic field above the pickup. This changing magnetic field induces a current in the pickup coil.

Many turns of very fine gauge wire—finer than the hair on your head—are wound around each pole piece. The number of turns determines the current that the pickup produces, with more windings resulting in a larger current.

Electric guitar pickups come in many different styles, and a single electric guitar can have two or three different types of pickups in it. One such style, the humbucking pickup, is designed to reduce the noise, or hum, that single-coil pickups generate from ac electricity. The windings, magnets, and location of the pickup all affect the sound that the guitar and pickup produce.

SECTION REVIEW

1. A circular current loop made of flexible wire is located in a magnetic field. Describe three ways an emf can be induced in the loop.

2. A bar magnet is positioned near a coil of wire, as shown to the right. What is the direction of the current in the resistor when the magnet is moved to the left, as in **(a)**? to the right, as in **(b)**?

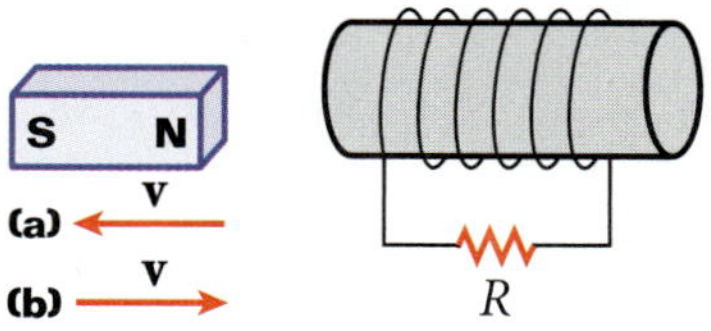

3. A 256-turn coil with a cross-sectional area of 0.0025 m^2 is placed in a uniform external magnetic field of strength 0.25 T so that the plane of the coil is perpendicular to the field. The coil is pulled steadily out of the field in 0.75 s. Find the average induced emf during this interval.

4. **Critical Thinking** Electric guitar strings are made of ferromagnetic materials that can be magnetized. The strings lie closely over and perpendicular to a coil of wire. Inside the coil are permanent magnets that magnetize the segments of the strings overhead. Using this arrangement, explain how the vibrations of a plucked string produce an electrical signal at the same frequency as the vibration of the string.

Why it Matters

Electric Guitar Pickups

The Rickenbacker Company made the first magnetic electric-guitar pickup in 1931. This first magnetic pickup consisted of two U-shaped magnets, pole pieces, and a single coil of wire.

A simple bobbin of wire around a magnet will create a pickup, but also acts as an antenna and picks up stray electrical noise like the 60 Hz hum from ac electricity. In 1956, Seth Lover designed the humbucking pickup to cancel out stray electrical noise. The humbucking pickup has two coils that are wired such that any noise that the coils detect is cancelled and any signal created by magnetic induction (from the string vibrations) is accepted.

SECTION REVIEW ANSWERS

1. Answers should include any three of the following: moving the loop into or out of the magnetic field; rotating the loop within the magnetic field; changing the strength of the magnetic field through the static loop; altering the loop's shape.
2. from left to right; from right to left
3. 0.21 V
4. When the magnetized strings vibrate, the strength of their magnetic field changes periodically with respect to the wire loops in the coil, inducing an emf in the coil. The fluctuations in the induced current match the frequencies and amplitudes of the sounds produced on the vibrating strings.

Misconception Alert — BASIC

Some students may think that no work is required to generate electricity. Stress that this is not the case; work is required to rotate a loop in a magnetic field. In a hydroelectric power plant, this work is generated by falling water. The water loses gravitational potential energy as it gains kinetic energy that changes as it does work.

Misconception Alert — GENERAL

When trying to determine the direction of current in **Figure 8,** students might try using the right-hand rule for a current-carrying wire in a magnetic field instead of the right-hand rule for charges moving in a magnetic field. Explain that the latter rule must be used because charges are not initially moving along the wire; rather, they move with the wire, in a direction perpendicular to the length of the wire. Thus, when applying the right-hand rule to find the direction of induced current, the thumb points along the direction in which the wire is moving, not along the wire, and the fingers point along the direction of the magnetic field. The resulting force on the charges, and thus the current, points out of the palm of the hand. Have students apply this form of the right-hand rule for segments *a, b, c,* and *d* in each case shown in **Figure 8** to determine the direction of the current in each segment.

Generators, Motors, and Mutual Inductance

SECTION OBJECTIVES

- **Describe how generators and motors operate.**
- **Explain the energy conversions that take place in generators and motors.**
- **Describe how mutual induction occurs in circuits.**

GENERATORS AND ALTERNATING CURRENT

In the previous section, you learned that a current can be induced in a circuit either by changing the magnetic field strength or by moving the circuit loop in or out of the magnetic field. Another way to induce a current is to change the orientation of the loop with respect to the magnetic field.

This second approach to inducing a current represents a practical means of generating electrical energy. In effect, the mechanical energy used to turn the loop is converted to electrical energy. A device that does this conversion is called an electric **generator.**

generator

a machine that converts mechanical energy into electrical energy

In most commercial power plants, mechanical energy is provided in the form of rotational motion. For example, in a hydroelectric plant, falling water directed against the blades of a turbine causes the turbine to turn. In a coal or natural-gas-burning plant, energy produced by burning fuel is used to convert water to steam, and this steam is directed against the turbine blades to turn the turbine.

Basically, a generator uses the turbine's rotary motion to turn a wire loop in a magnetic field. A simple generator is shown in **Figure 7.** As the loop rotates, the effective area of the loop changes with time, inducing an emf and a current in an external circuit connected to the ends of the loop.

A generator produces a continuously changing emf

Consider a single loop of wire that is rotated with a constant angular frequency in a uniform magnetic field. The loop can be thought of as four conducting wires. In this example, the loop is rotating counterclockwise within a magnetic field directed to the left.

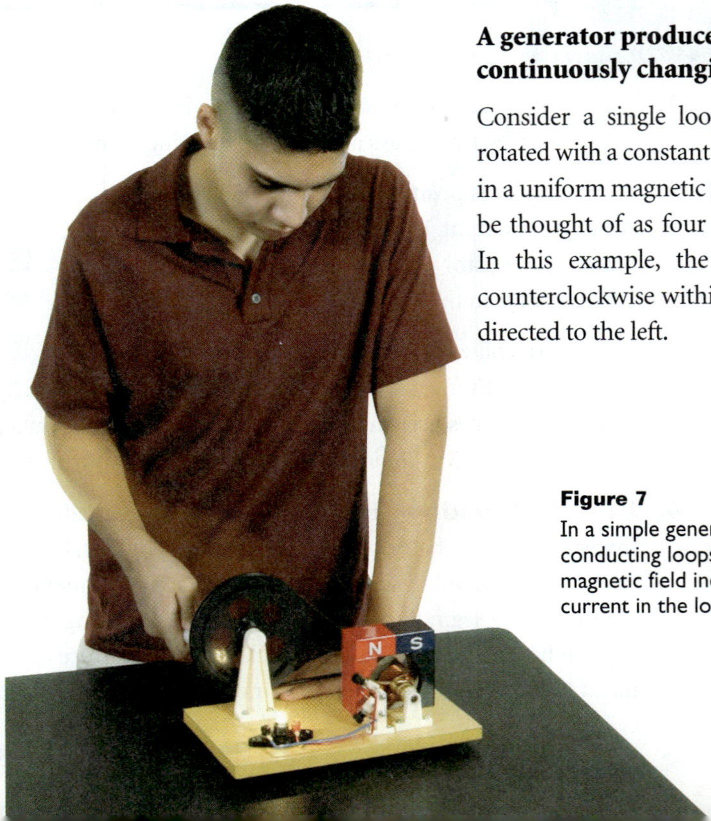

Figure 7
In a simple generator, the rotation of conducting loops through a constant magnetic field induces an alternating current in the loops.

When the area of the loop is perpendicular to the magnetic field lines, as shown in **Figure 8(a),** every segment of wire in the loop is moving parallel to the magnetic field lines. At this instant, the magnetic field does not exert force on the charges in any part of the wire, so the induced emf in each segment is therefore zero.

As the loop rotates away from this position, segments *a* and *c* cross magnetic field lines, so the magnetic force on the charges in these segments, and thus the induced emf, increases. The magnetic force on the charges in segments *b* and *d* cancel each other, so the motion of these segments does not contribute to the emf or the current. The greatest magnetic force on the charges and the greatest induced emf occur at the instant when segments *a* and *c* move perpendicularly to the magnetic field lines, as in **Figure 8(b).** This occurs when the plane of the loop is parallel to the field lines.

Because segment *a* moves downward through the field while segment *c* moves upward, their emfs are in opposite directions, but both produce a counterclockwise current. As the loop continues to rotate, segments *a* and *c* cross fewer lines, and the emf decreases. When the plane of the loop is perpendicular to the magnetic field, the motion of segments *a* and *c* is again parallel to the magnetic lines and the induced emf is again zero, as shown in **Figure 8(c).** Segments *a* and *c* now move in directions opposite those in which they moved from their positions in **(a)** to those in **(b).** As a result, the polarity of the induced emf and the direction of the current are reversed, as shown in **Figure 8(d).**

SCILINKS.
www.scilinks.org
Topic: Alternating Current
Code: HF60050

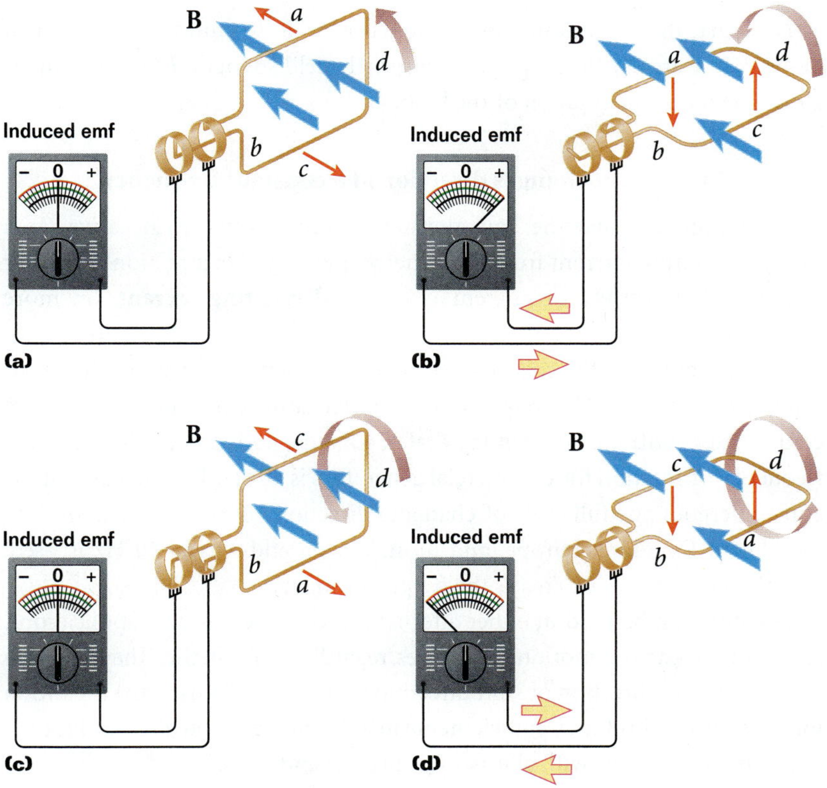

Figure 8
For a rotating loop in a magnetic field, the induced emf is zero when the loop is perpendicular to the magnetic field, as in **(a)** and **(c),** and is at a maximum when the loop is parallel to the field, as in **(b)** and **(d).**

Visual Strategy ADVANCED

Figure 8

Have students compare **Figure 8** with the analogy introduced in the demonstration on alternating current. After asking students the following question, plot graphs of $y = \sin x$ and $y = |\sin x|$ on the board, and use the graphs to compare the two cases.

Misconception Alert BASIC

Some students may think that the equation for the maximum emf of a generator holds for all cases. Be sure they understand that this equation only holds when the plane of the rotating loop and the magnetic field vectors are parallel. When the plane of the loop and the magnetic field vectors are not parallel, emf is not at a maximum, and the first equation on this page must be used.

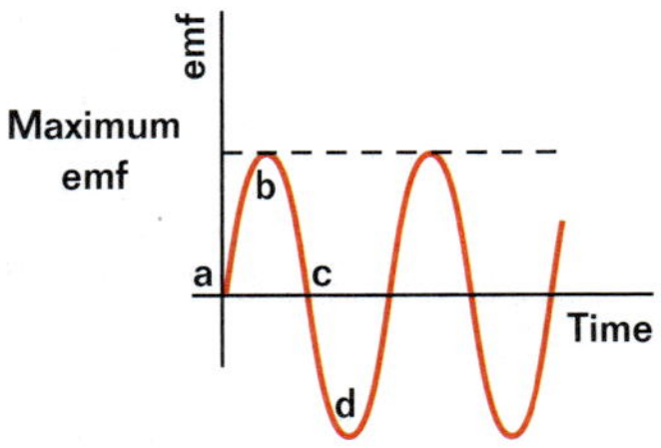

Figure 9
The change with time of the induced emf in a rotating loop is depicted by a sine wave. The letters on the plot correspond to the coil locations in **Figure 8.**

ADVANCED TOPICS

See "Angular Kinematics" in **Appendix J: Advanced Topics** to learn more about angular frequency.

alternating current

an electric current that changes direction at regular intervals

extension

Practice Problems

Visit go.hrw.com to find a sample and practice problems for induction in generators.

Keyword HF6EMIX

A graph of the change in emf versus time as the loop rotates is shown in **Figure 9.** Note the similarities between this graph and a sine curve. The four locations marked on the curve correspond to the orientation of the loop with respect to the magnetic field in **Figure 8.** At locations a and c, the emf is zero. These locations correspond to the instants when the plane of the loop is parallel to the direction of the magnetic field. At locations b and d, the emf is at its maximum and minimum, respectively. These locations correspond to the instants when the plane of the loop is perpendicular to the magnetic field.

The induced emf is the result of the steady change in the angle θ between the magnetic field lines and the normal to the loop. The following equation for the emf produced by a generator can be derived from Faraday's law of induction. The derivation is not shown here because it requires the use of calculus. In this equation, the angle of orientation, θ, has been replaced with the equivalent expression ωt, where ω is the angular frequency of rotation ($2\pi f$).

$$\text{emf} = NAB\omega \sin \omega t$$

The equation describes the sinusoidal variation of emf with time, as graphed in **Figure 9.**

The maximum emf strength can be easily calculated for a sinusoidal function. The emf has a maximum value when the plane of a loop is parallel to a magnetic field; that is, when $\sin \omega t = 1$, which occurs when $\omega t = \theta = 90°$. In this case, the expression above reduces to the following:

$$\text{maximum emf} = NAB\omega$$

Note that the maximum emf is a function of four things: the number of loops, N; the area of the loop, A; the magnetic field strength, B; and the angular frequency of the rotation of the loop, ω.

Alternating current changes direction at a constant frequency

Note in **Figure 9** that the emf alternates from positive to negative. As a result, the output current from the generator changes its direction at regular intervals. This variety of current is called **alternating current,** or, more commonly, *ac.*

The rate at which the coil in an ac generator rotates determines the maximum generated emf. The frequency of the alternating current can differ from country to country. In the United States, Canada, and Central America, the frequency of rotation for commercial generators is 60 Hz. This means that the emf undergoes one full cycle of changing direction 60 times each second. In the United Kingdom, Europe, and most of Asia and Africa, 50 Hz is used. (Recall that $\omega = 2\pi f$, where f is the frequency in Hz.)

Resistors can be used in either alternating- or direct-current applications. A resistor resists the motion of charges regardless of whether they move in one continuous direction or shift direction periodically. Thus, if the definition for resistance holds for circuit elements in a dc circuit, it will also hold for the same circuit elements with alternating currents and emfs.

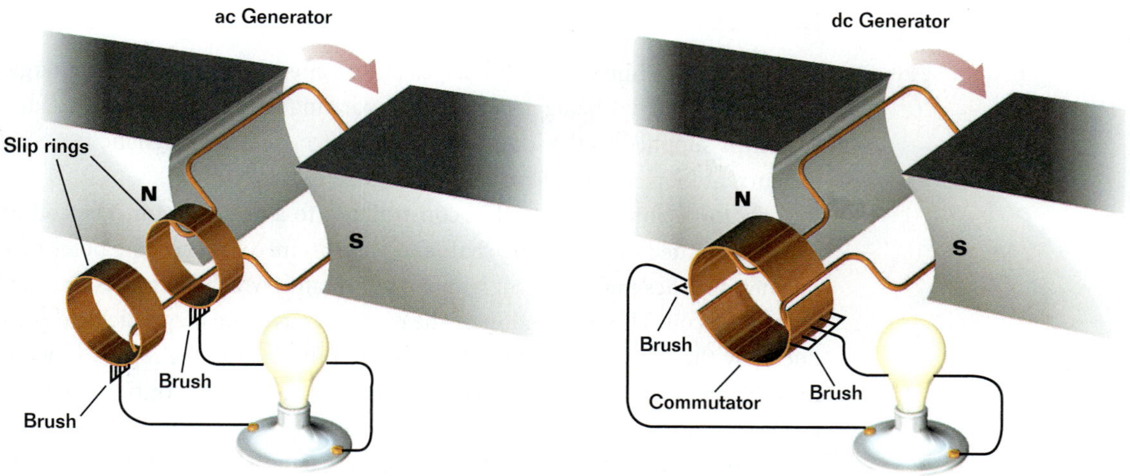

Figure 10
A simple dc generator (shown on the right) employs the same design as an ac generator (shown on the left). A split slip ring converts alternating current to direct current.

Alternating current can be converted to direct current

The conducting loop in an ac generator must be free to rotate through the magnetic field. Yet it must also be part of an electric circuit at all times. To accomplish this, the ends of the loop are connected to conducting rings, called *slip rings,* that rotate with the loop. Connections to the external circuit are made by stationary graphite strips, called *brushes,* that make continuous contact with the slip rings. Because the current changes direction in the loop, the output current through the brushes alternates direction as well.

By varying this arrangement slightly, an ac generator can be converted to a dc generator. Note in **Figure 10** that the components of a dc generator are essentially the same as those of the ac generator except that the contacts to the rotating loop are made by a single split slip ring, called a *commutator.*

At the point in the loop's rotation when the current has dropped to zero and is about to change direction, each half of the commutator comes into contact with the brush that was previously in contact with the other half of the commutator. The reversed current in the loop changes directions again so that the output current has the same direction as it originally had, although it still changes from a maximum value to zero. A plot of this pulsating direct current is shown in **Figure 11.**

A steady direct current can be produced by using many loops and commutators distributed around the rotation axis of the dc generator. This generator uses slip rings to continually switch the output of the generator to the commutator that is producing its maximum emf. This switching produces an output that has a slight ripple but is nearly constant.

SCILINKS.
www.scilinks.org
Topic: Electrical Safety
Code: HF60476

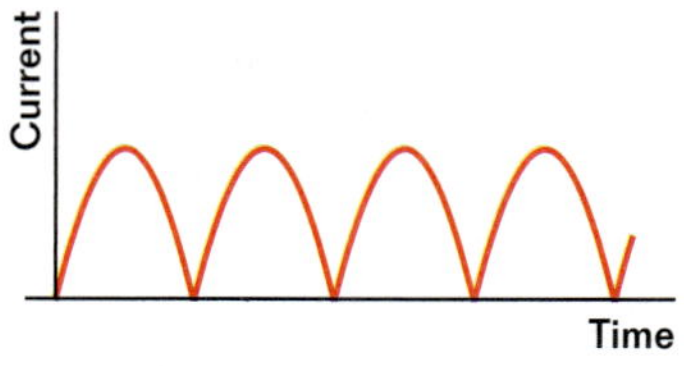

Figure 11
The output current for a dc generator with a single loop is a sine wave with the negative parts of the curve made positive.

Teaching Tip — GENERAL
Inform students that converting from ac to dc may also be accomplished electronically. This is done, for example, by ac-to-dc power converters, such as those used for portable electronic games or CD players.

Teaching Tip — ADVANCED
Point out that, as shown by the graph of direct current in **Figure 11,** the direct current produced by a generator alternates over time but does not change polarity, as does alternating current. This kind of current is called *half-wave rectified.* The direct current generated by a battery, on the other hand, is steady and does not fluctuate. Generators can produce a steady direct current similar to that produced by batteries if many loops and commutators are used.

SECTION 2

Teaching Tip — BASIC

Point out that generators and motors have opposite functions. Generators convert mechanical energy to electrical energy, while motors convert electrical energy to mechanical energy.

Demonstration

Electric Motor — GENERAL

Purpose Illustrate the similarity between generators and motors.

Materials hand-operated generator, capacitor (1 F), 100 Ω resistor, connecting wires

Procedure Connect the capacitor to the hand-operated generator. Charge the capacitor with the generator. Do not apply a large potential difference. Release the handle, and watch the generator become a motor.

Optional: An interesting comparison can be made between the responses of the generator to different loads. Repeat the demonstration with and without the resistor attached.

MOTORS

Motors are machines that convert electrical energy to mechanical energy. Instead of a current being generated by a rotating loop in a magnetic field, a current is supplied to the loop by an emf source, and the magnetic force on the current loop causes it to rotate (see **Figure 12**).

A motor is almost identical in construction to a dc generator. The coil of wire is mounted on a rotating shaft and is positioned between the poles of a magnet. Brushes make contact with a commutator, which alternates the current in the coil. This alternation of the current causes the magnetic field produced by the current to regularly reverse and thus always be repelled by the fixed magnetic field. Thus, the coil and the shaft are kept in continuous rotational motion.

A motor can perform mechanical work when a shaft connected to its rotating coil is attached to some external device. As the coil in the motor rotates, however, the changing normal component of the magnetic field through it induces an emf that acts to reduce the current in the coil. If this were not the case, Lenz's law would be violated. This induced emf is called the **back emf.**

back emf

the emf induced in a motor's coil that tends to reduce the current in the coil of the motor

The back emf increases in magnitude as the magnetic field changes at a higher rate. In other words, the faster the coil rotates, the greater the back emf becomes. The potential difference available to supply current to the motor equals the difference between the applied potential difference and the back emf. Consequently, the current in the coil is also reduced because of the presence of back emf. As the motor turns faster, both the net emf across the motor and the net current in the coil become smaller.

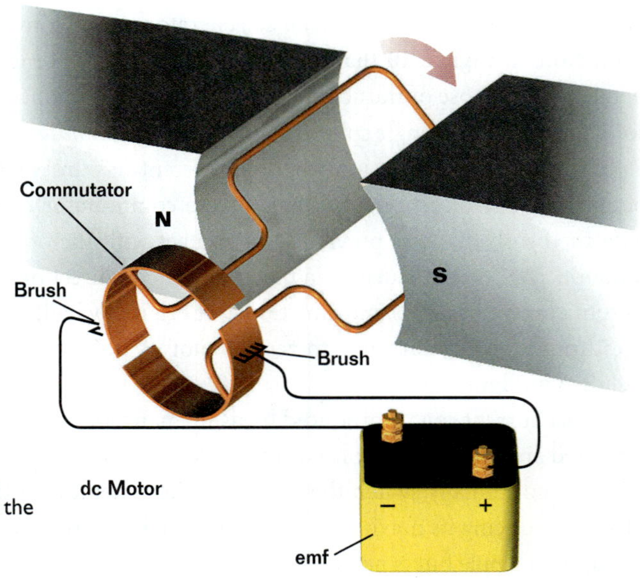

Figure 12
In a motor, the current in the coil interacts with the magnetic field, causing the coil and the shaft on which the coil is mounted to turn.

MUTUAL INDUCTANCE

The basic principle of electromagnetic induction was first demonstrated by Michael Faraday. His experimental apparatus, which resembled the arrangement shown in **Figure 13,** used a coil connected to a switch and a battery instead of a magnet to produce a magnetic field. This coil is called the *primary coil,* and its circuit is called the *primary circuit.* The magnetic field is strengthened by the magnetic properties of the iron ring around which the primary coil is wrapped.

A second coil is wrapped around another part of the iron ring and is connected to a galvanometer. An emf is induced in this coil, called the *secondary coil,* when the magnetic field of the primary coil is changed. When the switch in the primary circuit is closed, the galvanometer in the secondary circuit deflects in one direction and then returns to zero. When the switch is opened, the galvanometer deflects in the opposite direction and again returns to zero. When there is a steady current in the primary circuit, the galvanometer reads zero.

The magnitude of this emf is predicted by Faraday's law of induction. However, Faraday's law can be rewritten so that the induced emf is proportional to the changing current in the primary coil. This can be done because of the direct proportionality between the magnetic field produced by a current in a coil, or solenoid, and the current itself. The form of Faraday's law in terms of changing primary current is as follows:

$$\text{emf} = -N\frac{\Delta\Phi_M}{\Delta t} = -M\frac{\Delta I}{\Delta t}$$

The constant, M, is called the **mutual inductance** of the two-coil system. The mutual inductance depends on the geometrical properties of the coils and their orientation to each other. A changing current in the secondary coil can also induce an emf in the primary circuit. In fact, when the current through the second coil varies, the induced emf in the first coil is governed by an analogous equation *with the same value of M.*

mutual inductance

the ability of one circuit to induce an emf in a nearby circuit in the presence of a changing current

The induced emf in the secondary circuit can be changed by changing the number of turns of wire in the secondary coil. This arrangement is the basis of an extremely useful electrical device: the transformer.

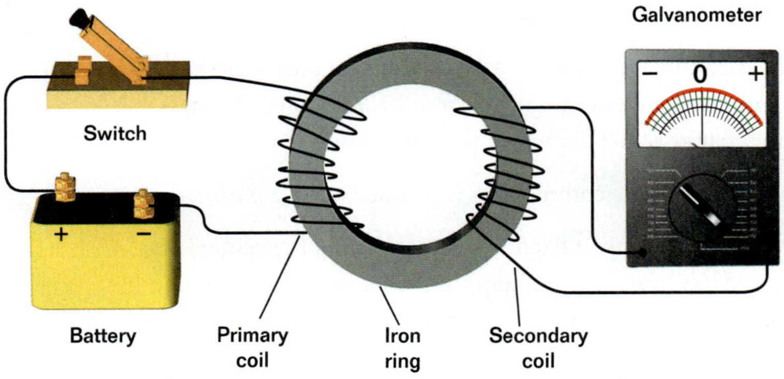

Figure 13
Faraday's electromagnetic-induction experiment used a changing current in one circuit to induce a current in another circuit.

Demonstration

ac and dc ADVANCED

Purpose Show the difference between alternating current and direct current.

Materials hand-operated generator, galvanometer, wires, oscilloscope, variable autotransformer, dc source

Procedure Show students that the current measured by the galvanometer varies from positive to negative as the ac generator is working. Hook up the dc source to the oscilloscope to show students that the oscilloscope is a type of voltmeter. Explain that the sweep allows you to see changes in the potential difference that may be too quick to see with a galvanometer.

Replace the dc source with the autotransformer. Tell students what the sweep frequency is, and have them determine the frequency of the alternating current. Repeat the process with a different sweep frequency. Show students what happens to the trace as you increase the potential difference. If time permits, you may wish to quantify the discussion by showing that the heating effects of the alternating current will be identical to those of the direct current if the effective (rather than the peak-to-peak) ac potential difference is used for the comparison.

Why it Matters

Avoiding Electrocution

Ground fault interrupters make use of Faraday's law. Both the wire that leads from the wall outlet to the appliance and the wire that leads from the appliance back to the wall pass through an iron ring. Part of the iron ring is wrapped in a coil called a *sensing coil*. When the current to the appliance equals the current from the appliance, the net magnetic field through the sensing coil is zero.

If a short circuit occurs, the net magnetic field through the coil is no longer zero. Because the current in the wire is alternating, an ac potential difference is induced in the sensing coil. This induced potential difference in the coil is used to trigger a circuit breaker, stopping the current before it reaches a level that might be harmful to the person using the appliance.

SECTION REVIEW ANSWERS

1. 120 V
2. 8.3×10^{-2} T
3. The commutator reverses the direction of the current in the coil so that the magnetic field that the current produces always opposes the external field. Without the commutator, the coil would stop turning once the coil's magnetic field was aligned with the external field.
4. Maximum mutual inductance takes place when the planes of the two loops are parallel. Minimum mutual inductance takes place when the planes of the loops are perpendicular to each other.

Why it Matters

Avoiding Electrocution

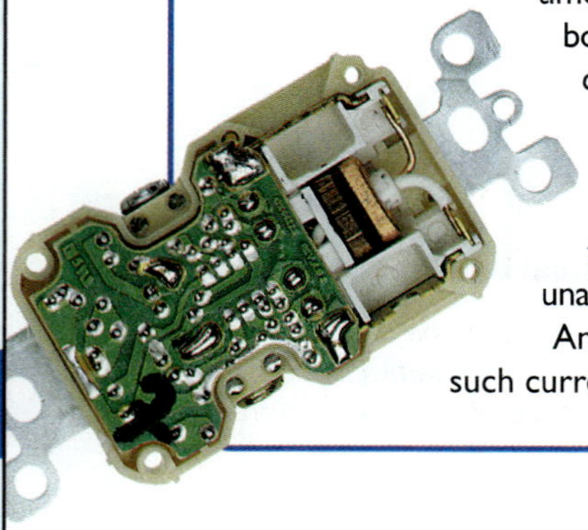

A person can receive an electric shock by touching something that is at a different electric potential than your body. For example, you might touch a high electric potential object while in contact with a cold-water pipe (normally at zero potential) or while standing on the floor with wet feet (because impure water is a good conductor).

Electric shock can result in fatal burns or can cause the muscles of vital organs, such as the heart, to malfunction. The degree of damage to the body depends on the magnitude of the current, the length of time it acts, and the part of the body through which it passes. A current of 100 milliamps (mA) can be fatal. If the current is larger than about 10 mA, the hand muscles contract and the person may be unable to let go of the wire.

Any wires designed to have such currents in them are wrapped in insulation, usually plastic or rubber, to prevent electrocution. However, with frequent use, electrical cords can fray, exposing some of the conductors. In these and other situations in which electrical contact can be made, devices called a ground fault circuit interrupter (GFCI) and a ground fault interrupter (GFI) are mounted in electrical outlets and individual appliances to prevent further electrocution.

GFCIs and GFIs provide protection by comparing the current in one side of the electrical outlet socket to the current in the other socket. The two currents are compared by induction in a device called a *differential transformer*. If there is even a 5 mA difference, the interrupter opens the circuit in a few milliseconds (thousandths of a second). The quick motion needed to open the circuit is again provided by induction, with the use of a solenoid switch.

Despite these safety devices, you can still be electrocuted. Never use electrical appliances near water or with wet hands. Use a battery-powered radio near water because batteries cannot supply enough current to harm you. It is also a good idea to replace old outlets with GFCI-equipped units or to install GFI-equipped circuit breakers.

SECTION REVIEW

1. A loop with 37 turns and an area of 0.33 m^2 is rotating at 281 rad/s. The loop's axis of rotation is perpendicular to a uniform magnetic field with a strength of 0.035 T. What is the maximum emf induced?
2. A generator coil has 25 turns of wire and a cross-sectional area of 36 cm^2. The maximum emf developed in the generator is 2.8 V at 60 Hz. What is the strength of the magnetic field in which the coil rotates?
3. Explain what would happen if a commutator were not used in a motor.
4. **Critical Thinking** Suppose a fixed distance separates the centers of two circular loops. What relative orientation of the loops will give the maximum mutual inductance? What orientation will give the minimum mutual inductance?

AC Circuits and Transformers

SECTION OBJECTIVES

- **Distinguish between rms values and maximum values of current and potential difference.**
- **Solve problems involving rms and maximum values of current and emf for ac circuits.**
- **Apply the transformer equation to solve problems involving step-up and step-down transformers.**

EFFECTIVE CURRENT

In the previous section, you learned that an electrical generator could produce an alternating current that varies as a sine wave with respect to time. Commercial power plants use generators to provide electrical energy to power the many electrical devices in our homes and businesses. In this section, we will investigate the characteristics of simple ac circuits.

As with the discussion about direct-current circuits, the resistance, the current, and the potential difference in a circuit are all relevant to a discussion about alternating-current circuits. The emf in ac circuits is analogous to the potential difference in dc circuits. One way to measure these three important circuit parameters is with a digital multimeter, as shown in **Figure 14.** The resistance, current, or emf can be measured by choosing the proper settings on the multimeter and locations in the circuit.

Figure 14
The effective current and emf of an electric circuit can be measured using a digital multimeter.

Effective current and effective emf are measured in ac circuits

An ac circuit consists of combinations of circuit elements and an ac generator or an ac power supply, which provides the alternating current. As shown earlier, the emf produced by a typical ac generator is sinusoidal and varies with time. The induced emf as a function of time (Δv) can be written in terms of the maximum emf (ΔV_{max}), and the emf produced by a generator can be expressed as follows:

$$\Delta v = \Delta V_{max} \sin \omega t$$

A simple ac circuit can be treated as an equivalent resistance and an ac source. In a circuit diagram, the ac source is represented by the symbol ⊙, as shown in **Figure 15.**

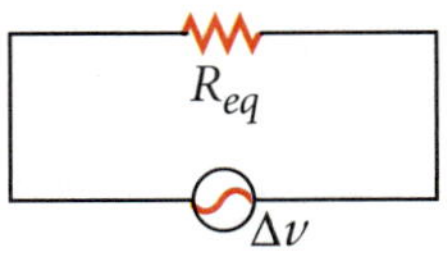

Figure 15
An ac circuit represented schematically consists of an ac source and an equivalent resistance.

The instantaneous current that changes with the potential difference can be determined using the definition for resistance. The instantaneous current, i, is related to maximum current by the following expression:

$$i = I_{max} \sin \omega t$$

The rate at which electrical energy is converted to internal energy in the resistor (the power, P) has the same form as in the case of direct current. The electrical energy converted to internal energy at some point in time in a resistor is proportional to the *square* of the instantaneous current and is independent of the direction of the current. However, the energy produced by an alternating current with a maximum value of I_{max} is not the same as that produced by a direct current of the same value. The energies are different because during a cycle, the alternating current is at its maximum value for only an instant.

Teaching Tip — ADVANCED

Because both emf and potential difference are measured in volts, the term *voltage* is often used interchangeably when referring to either quantity. This usage can lead to some confusion between the two quantities. Strictly speaking, an induced emf does not have a potential difference. According to the definition of potential difference, the value of potential difference must be obtained independently of the path chosen to calculate the value. This definition is not true of induced emf. Thus, potential difference has no meaning when discussing magnetic induction.

Demonstration

Effects of Alternating Current

Purpose Illustrate the effects of alternating current.

Materials bicolored LED (available at many electronics stores), step-down transformer, resistor (100–250 Ω), 2 m of flexible wire

Procedure Connect the wire, the resistor, and the LED in series with the transformer. Explain to students that the diode is red and green, respectively, for currents with opposite directions. Turn the lights down, and show the class that the LED appears yellow. The reason is that the alternating polarity of the current switches the color of the LED from red to green 60 times each second.

Hold the wire about halfway between the transformer and the diode, and twirl the wire in a vertical circle so that the diode moves in a circular path. Students should see red and green bars at equally spaced intervals along the diode's path. Have students count the number of green bars they see in the circle, and then have them measure the time it takes for the diode to travel 10 times around the circular path. Ask students how much time it takes for the diode to change from green to red to green $\left(\frac{1}{60}\,s\right)$.

rms current

the value of alternating current that gives the same heating effect that the corresponding value of direct current does

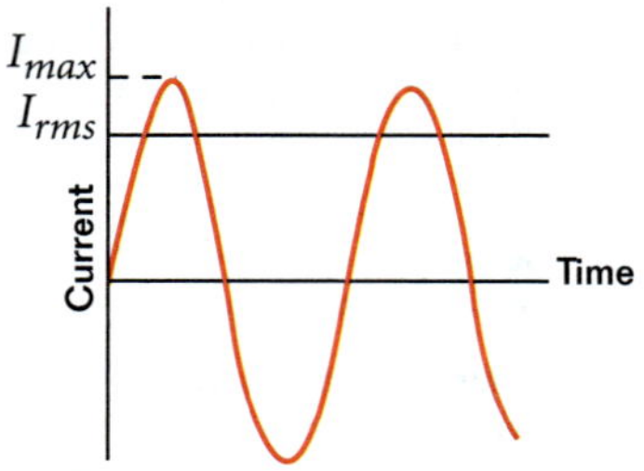

Figure 16
The rms current is a little more than two-thirds as large as the maximum current.

An important measure of the current in an ac circuit is the **rms current.** The rms (or *root-mean-square*) current is the same as the amount of direct current that would dissipate the same energy in a resistor as is dissipated by the instantaneous alternating current over a complete cycle.

Figure 16 shows a graph in which instantaneous and rms currents are compared. **Table 2** summarizes the notations used in this chapter for these and other ac quantities.

The equation for the average power dissipated in an ac circuit has the same form as the equation for power dissipated in a dc circuit except that the dc current I is replaced by the rms current (I_{rms}).

$$P = (I_{rms})^2 R$$

This equation is identical in form to the one for direct current. However, the power dissipated in the ac circuit equals half the power dissipated in a dc circuit when the dc current equals I_{max}.

$$P = (I_{rms})^2 R = \tfrac{1}{2}(I_{max})^2 R$$

From this equation you may note that the rms current is related to the maximum value of the alternating current by the following equation:

$$(I_{rms})^2 = \frac{(I_{max})^2}{2}$$

$$I_{rms} = \frac{I_{max}}{\sqrt{2}} = 0.707\ I_{max}$$

This equation says that an alternating current with a maximum value of 5 A produces the same heating effect in a resistor as a direct current of $(5/\sqrt{2})$ A, or about 3.5 A.

Alternating emfs are also best discussed in terms of their rms values, with the relationship between rms and maximum values analogous to the one for currents. The rms and maximum values are related as follows:

$$\Delta V_{rms} = \frac{\Delta V_{max}}{\sqrt{2}} = 0.707\ V_{max}$$

Table 2 Notation Used for ac Circuits

	Induced or Applied emf	Current
Instantaneous values	Δv	i
Maximum values	ΔV_{max}	I_{max}
rms values	$\Delta V_{rms} = \frac{\Delta V_{max}}{\sqrt{2}}$	$I_{rms} = \frac{I_{max}}{\sqrt{2}}$

SAMPLE PROBLEM B

rms Current and emf

PROBLEM

A generator with a maximum output emf of 205 V is connected to a 115 Ω resistor. Calculate the rms potential difference. Find the rms current through the resistor. Find the maximum ac current in the circuit.

SOLUTION

1. DEFINE **Given:** $\Delta V_{max} = 205$ V $R = 115\ \Omega$

Unknown: $\Delta V_{rms} = ?$ $I_{rms} = ?$ $I_{max} = ?$

Diagram:

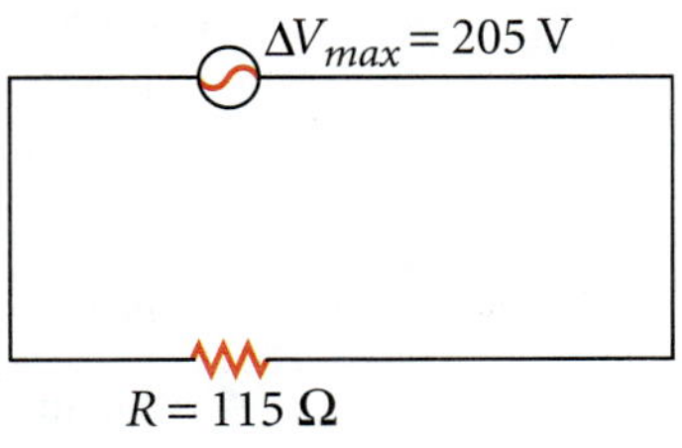

2. PLAN **Choose an equation or situation:** Use the equation for the rms potential difference to find ΔV_{rms}.

$$\Delta V_{rms} = 0.707\ \Delta V_{max}$$

Rearrange the definition for resistance to calculate I_{rms}.

$$I_{rms} = \frac{\Delta V_{rms}}{R}$$

Use the equation for rms current to find I_{max}.

$$I_{rms} = 0.707\ I_{max}$$

Rearrange the equation to isolate the unknown:
Rearrange the equation relating rms current to maximum current so that maximum current is calculated.

$$I_{max} = \frac{I_{rms}}{0.707}$$

TIP *Because emf is measured in volts, maximum emf is frequently abbreviated as ΔV_{max}, and rms emf can be abbreviated as ΔV_{rms}.*

3. CALCULATE **Substitute the values into the equation and solve:**

$$\Delta V_{rms} = (0.707)(205\text{ V}) = 145\text{ V}$$

$$I_{rms} = \frac{145\text{ V}}{115\ \Omega} = 1.26\text{ A}$$

$$I_{max} = \frac{1.26\text{ A}}{0.707} = 1.78\text{ A}$$

$$\boxed{\begin{aligned}\Delta V_{rms} &= 145\text{ V}\\ I_{rms} &= 1.26\text{ A}\\ I_{max} &= 1.78\text{ A}\end{aligned}}$$

4. EVALUATE The rms values for the emf and current are a little more than two-thirds the maximum values, as expected.

Classroom Practice

rms Current and emf
A generator supplies 110 V rms to a 25 Ω circuit. What is the maximum current supplied to the circuit?

Answer
6.2 A

Demonstration

Induced emf — GENERAL

Purpose Show that induced emf depends on the number of turns of wire.

Materials 9 V battery, knife switch, 30 cm (or longer) iron bar, two (or more) flashlight bulbs in holders, three connecting wires (each about 1 m long)

Procedure Set up the primary coil before class by connecting the battery and switch in series with one of the wires coiled around the iron bar. Set up two secondary coils, one with 25 turns of wire and one with 50 turns of wire, each in series with a flashlight bulb. Close the switch, and have students compare the brightness of the two bulbs. You may wish to open and close the switch several times for repeated observations.

SECTION 3

PROBLEM GUIDE B

Use this guide to assign problems.
SE = Student Edition Textbook
PW = Problem Workbook
PB = Problem Bank on the One-Stop Planner (OSP)

Solving for:

ΔV	**SE**	Sample, 1, 4, 5a, 6; Ch. Rvw. 25–28
	PW	Sample, 4–6
	PB	Sample, 1, 4–6, 9
I	**SE**	Sample, 1–2, 3a, 4, 5b; Ch. Rvw. 26–28
	PW	Sample, 1–5, 7
	PB	Sample, 1–5, 7, 8–10
R	**SE**	3b
	PW	3, 5
	PB	3, 5
P	**SE**	Ch. Rvw. 27
	PW	Sample, 1
	PB	8–10

***Challenging Problem**
Consult the printed Solutions Manual or the OSP for detailed solutions.

ANSWERS

Practice B

1. 4.8 A; 6.8 A, 170 V
2. 7.8 A
3. a. 7.42 A
   b. 14.8 Ω
4. 1.44 A; 2.04 A, 21.2 V
5. a. 1.10×10^2 V
   b. 2.1 A
6. 319 V

PRACTICE B

rms Current and emf

1. What is the rms current in a light bulb that has a resistance of 25 Ω and an rms emf of 120 V? What are the maximum values for current and emf?

2. The current in an ac circuit is measured with an ammeter. The meter gives a reading of 5.5 A. Calculate the maximum ac current.

3. A toaster is plugged into a source of alternating emf with an rms value of 110 V. The heating element is designed to convey a current with a peak value of 10.5 A. Find the following:
   a. the rms current in the heating element
   b. the resistance of the heating element

4. An audio amplifier provides an alternating rms emf of 15.0 V. A loudspeaker connected to the amplifier has a resistance of 10.4 Ω. What is the rms current in the speaker? What are the maximum values of the current and the emf?

5. An ac generator has a maximum emf output of 155 V.
   a. Find the rms emf output.
   b. Find the rms current in the circuit when the generator is connected to a 53 Ω resistor.

6. The largest emf that can be placed across a certain capacitor at any instant is 451 V. What is the largest rms emf that can be placed across the capacitor without damaging it?

Resistance influences current in an ac circuit

The ac potential difference (commonly called the *voltage*) of 120 V measured from an electrical outlet is actually an rms emf of 120 V. (This, too, is a simplification that assumes that the voltmeter has infinite resistance.) A quick calculation shows that such an emf has a maximum value of about 170 V.

The resistance of a circuit modifies the current in an ac circuit just as it does in a dc circuit. If the definition of resistance is valid for an ac circuit, the rms emf across a resistor equals the rms current multiplied by the resistance. Thus, all maximum and rms values can be calculated if only one current or emf value and the circuit resistance are known.

Ammeters and voltmeters that measure alternating current are calibrated to measure rms values. In this chapter, all values of alternating current and emf will be given as rms values unless otherwise noted. The equations for ac circuits have the same form as those for dc circuits when rms values are used.

TRANSFORMERS

It is often desirable or necessary to change a small ac applied emf to a larger one or to change a large applied emf to a smaller one. The device that makes these conversions possible is the **transformer.**

transformer

a device that increases or decreases the emf of alternating current

In its simplest form, an ac transformer consists of two coils of wire wound around a core of soft iron, like the apparatus for the Faraday experiment. The coil on the left in **Figure 17** has N_1 turns and is connected to the input ac potential difference source. This coil is called the primary winding, or the *primary*. The coil on the right, which is connected to a resistor R and consists of N_2 turns, is the *secondary*. As in Faraday's experiment, the iron core "guides" the magnetic field lines so that nearly all of the field lines pass through both of the coils.

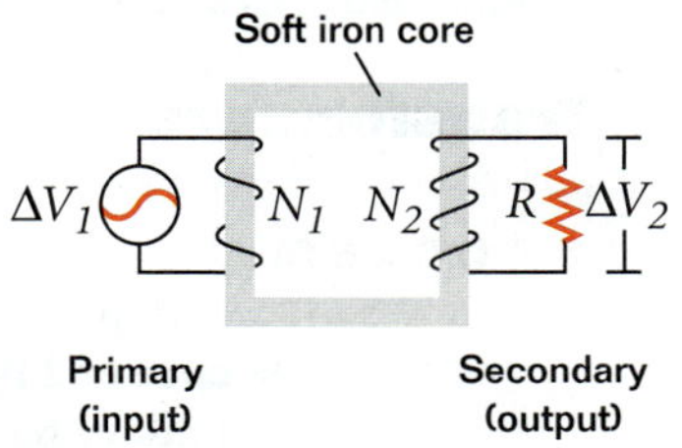

Figure 17
A transformer uses the alternating current in the primary circuit to induce an alternating current in the secondary circuit.

Because the strength of the magnetic field in the iron core and the cross-sectional area of the core are the same for both the primary and secondary windings, the measured ac potential differences across the two windings differ only because of the different number of turns of wire for each. The applied emf that gives rise to the changing magnetic field in the primary is related to that changing field by Faraday's law of induction.

$$\Delta V_1 = -N_1 \frac{\Delta \Phi_M}{\Delta t}$$

Similarly, the induced emf across the secondary coil is

$$\Delta V_2 = -N_2 \frac{\Delta \Phi_M}{\Delta t}$$

Taking the ratio of ΔV_1 to ΔV_2 causes all terms on the right side of both equations except for N_1 and N_2 to cancel. This result is the transformer equation.

TRANSFORMER EQUATION

$$\Delta V_2 = \frac{N_2}{N_1} \Delta V_1$$

induced emf in secondary =
$$\left(\frac{\textbf{number of turns in secondary}}{\textbf{number of turns in primary}}\right) \textbf{applied emf in primary}$$

Another way to express this equation is to equate the ratio of the potential differences to the ratio of the number of turns.

$$\frac{\Delta V_2}{\Delta V_1} = \frac{N_2}{N_1}$$

When N_2 is greater than N_1, the secondary emf is greater than that of the primary, and the transformer is called a *step-up transformer*. When N_2 is less than

SCILINKS.
www.scilinks.org
Topic: Transformers
Code: HF61549

SECTION 3

Demonstration

Transformers GENERAL

Purpose Show students how a transformer works, and reinforce the idea that a changing current is required.

Materials iron bar, 9 V battery, knife switch, flashlight bulb in holder, two 1 m long wires

Procedure Set up the primary side of the transformer before class by connecting the battery and switch in series with one of the wires coiled around the iron bar. The primary coil should have 50 turns. Set up the secondary coil with 25 turns, and connect the flashlight bulb to the secondary coil.

In class, demonstrate this step-down transformer by momentarily closing the switch and then opening it again. Have students discuss the transfer of energy that occurs in this situation.

Close the switch, and keep it closed. Have students note the behavior of the bulb. Discuss the brief illumination and fading of the bulb with students. Lead students to consider the concept of changing current. Open the switch, and note the illumination. Discuss this effect.

Teaching Tip GENERAL

Some students may wonder whether dc transformers are possible. The dc produced by a battery would not work with a transformer. The reason is that a changing current is required, and a battery generates a steady direct current. However, a generator can produce a fluctuating direct current. Such a current could be used by a transformer.

Classroom Practice

Transformers
A transformer has 75 turns on the primary and 1500 turns on the secondary.

a. If the potential difference across the primary is 120 V, what is the potential difference across the secondary?

b. If the transformer has 1625 turns on the secondary instead of 1500, what is the potential difference across the secondary?

Answers

a. 2400 V

b. 2600 V

N_1, the secondary emf is less than that of the primary, and the transformer is called a *step-down transformer.*

It may seem that a transformer provides something for nothing. For example, a step-up transformer can change an applied emf from 10 V to 100 V. However, the power output at the secondary is, at best, equal to the power input at the primary. In reality, energy is lost to heating and radiation, so the output power will be *less* than the input power. Thus, an increase in induced emf at the secondary means that there must be a proportional decrease in current.

SAMPLE PROBLEM C

Transformers

PROBLEM

A step-up transformer is used on a 120 V line to provide a potential difference of 2400 V. If the primary has 75 turns, how many turns must the secondary have?

SOLUTION

1. DEFINE

Given: $\Delta V_1 = 120$ V $\quad \Delta V_2 = 2400$ V $\quad N_1 = 75$ turns

Unknown: $N_2 = ?$

Diagram:

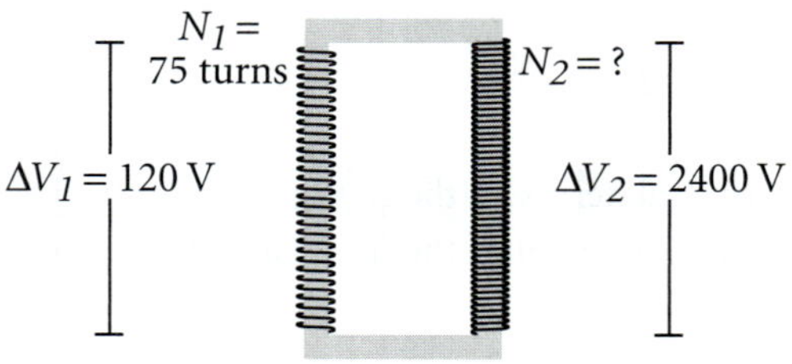

2. PLAN

Choose an equation or situation: Use the transformer equation.

$$\Delta V_2 = \frac{N_2}{N_1}\Delta V_1$$

Rearrange the equation to isolate the unknown:

$$N_2 = \frac{\Delta V_2}{\Delta V_1} N_1$$

3. CALCULATE

Substitute the values into the equation and solve:

$$N_2 = \left(\frac{2400 \text{ V}}{120 \text{ V}}\right) 75 \text{ turns} = 1500 \text{ turns}$$

$$\boxed{N_2 = 1500 \text{ turns}}$$

4. EVALUATE

The greater number of turns in the secondary accounts for the increase in the emf in the secondary. The step-up factor for the transformer is 20:1.

PRACTICE C

Transformers

1. A step-down transformer providing electricity for a residential neighborhood has exactly 2680 turns in its primary. When the potential difference across the primary is 5850 V, the potential difference at the secondary is 120 V. How many turns are in the secondary?

2. A step-up transformer used in an automobile has a potential difference across the primary of 12 V and a potential difference across the secondary of 2.0×10^4 V. If the number of turns in the primary is 21, what is the number of turns in the secondary?

3. A step-up transformer for long-range transmission of electric power is used to create a potential difference of 119 340 V across the secondary. If the potential difference across the primary is 117 V and the number of turns in the secondary is 25 500, what is the number of turns in the primary?

4. A potential difference of 0.750 V is needed to provide a large current for arc welding. If the potential difference across the primary of a step-down transformer is 117 V, what is the ratio of the number of turns of wire on the primary to the number of turns on the secondary?

5. A step-down transformer has 525 turns in its secondary and 12 500 turns in its primary. If the potential difference across the primary is 3510 V, what is the potential difference across the secondary?

Real transformers are not perfectly efficient

The transformer equation assumes that no power is lost between the transformer's primary and secondary coils. Real transformers typically have efficiencies ranging from 90 percent to 99 percent. Power is lost because of the small currents induced by changing magnetic fields in the transformer's iron core and because of resistance in the wires of the windings.

The power lost to resistive heating in transmission lines varies as I^2R. To minimize I^2R loss and maximize the deliverable energy, power companies use a high emf and a low current when transmitting power over long distances. By reducing the current by a factor of 10, the power loss is reduced by a factor of 100. In practice, the emf is stepped up to around 230 000 V at the generating station, is stepped down to 20 000 V at a regional distribution station, and is finally stepped down to 120 V at the customer's utility pole. The high emf in long-distance transmission lines makes the lines especially dangerous when high winds knock them down.

PROBLEM GUIDE C

Use this guide to assign problems.
SE = Student Edition Textbook
PW = Problem Workbook
PB = Problem Bank on the One-Stop Planner (OSP)

Solving for:

N	**SE** Sample, 1–3; Ch. Rvw. 35–36 **PW** 4–5, 7 **PB** 6, 8–10
$\frac{N_1}{N_2}$	**SE** 4; Ch. Rvw. 41 **PB** 7
ΔV	**SE** 5–6; Ch. Rvw. 40 **PW** Sample, 1–3, 6b **PB** Sample, 1–5
I	**PW** 6a, 7 **PB** 10

***Challenging Problem**
Consult the printed Solutions Manual or the OSP for detailed solutions.

ANSWERS

Practice C

1. 55 turns
2. 3.5×10^4 turns
3. 25 turns
4. 156:1
5. 147 V

SECTION 3

Teaching Tip — GENERAL

Point out to students that spark plugs create a spark because the potential difference across the gap increases by up to 10 000 V. At such high potential differences, air is ionized and becomes a conductor.

SECTION REVIEW ANSWERS

1. 0.035 mA; 0.11 V; 0.15 V
2. 1.7×10^4 V
3. 2.6×10^4 V
4. zero; The current is positive for exactly the same amount of time that it is negative. The direction of current (positive or negative) does not affect the heating of the resistor. That is, the resistor is being heated throughout the cycle whether the current is positive or negative.

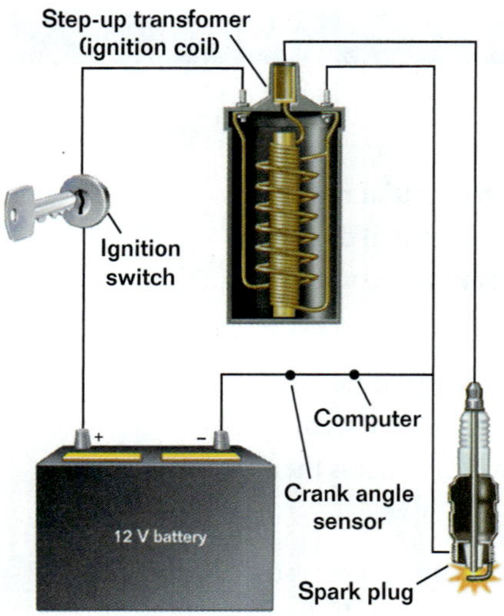

Figure 18
The transformer in an automobile engine raises the potential difference across the gap in a spark plug so that sparking occurs.

The ignition coil in a gasoline engine is a transformer

An automobile battery provides a constant emf of 12 dc volts to power various systems in your automobile. The ignition system uses a transformer, called the *ignition coil,* to convert the car battery's 12 dc volts to a potential difference that is large enough to cause sparking between the gaps of the spark plugs. The diagram in **Figure 18** shows a type of ignition system that has been used in automobiles since about 1990. In this arrangement, called an *electronic ignition,* each cylinder has its own transformer coil.

The ignition system on your car has to work in perfect concert with the rest of the engine. The goal is to ignite the fuel at the exact moment when the expanding gases can do the maximum amount of work. A photoelectric detector, called a *crank angle sensor,* uses the crankshaft's position to determine when the cylinder's contents are near maximum compression.

The sensor then sends a signal to the automobile's computer. Upon receiving this signal, the computer closes the primary circuit to the cylinder's coil, causing the current in the primary to rapidly increase. As we learned earlier in this chapter, the increase in current induces a rapid change in the magnetic field of the transformer. Because the change in magnetic field on the primary side is so quick, the change induces a very large emf, from 40 000 to 100 000 V. The emf is applied across the spark plug and creates a spark that ignites and burns the fuel that powers your automobile.

SECTION REVIEW

1. The rms current that a single coil of an electric guitar produces is 0.025 mA. The coil's resistance is 4.3 kΩ. What is the maximum instantaneous current? What is the rms emf produced by the coil? What is the maximum emf produced by the coil?
2. A step-up transformer has exactly 50 turns in its primary and exactly 7000 turns in its secondary. If the applied emf in the primary is 120 V, what emf is induced in the secondary?
3. A television picture tube requires a high potential difference, which a step-up transformer provides in older models. The transformer has 12 turns in its primary and 2550 turns in its secondary. If 120 V is applied across the primary, what is the output emf?
4. **Critical Thinking** What is the average value of current over one cycle of an ac signal? Why, then, is a resistor heated by an ac current?

Electromagnetic Waves

SECTION 4

SECTION OBJECTIVES

- **Describe what electromagnetic waves are and how they are produced.**
- **Recognize that electricity and magnetism are two aspects of a single electromagnetic force.**
- **Explain how electromagnetic waves transfer energy.**
- **Describe various applications of electromagnetic waves.**

PROPAGATION OF ELECTROMAGNETIC WAVES

Light is a phenomenon known as an *electromagnetic wave.* As the name implies, oscillating electric and magnetic fields create electromagnetic waves. In this section, you will learn more about the nature and the discovery of electromagnetic waves.

The wavelength and frequency of electromagnetic waves vary widely, from radio waves with very long wavelengths to gamma rays with extremely short wavelengths. The visible light that our eyes can detect occupies an intermediate range of wavelengths. Familiar objects "look" quite different at different wavelengths. **Figure 19** shows how a person might appear to us if we could see beyond the red end of the visible spectrum.

In this chapter, you have learned that a changing magnetic field can induce a current in a circuit (Faraday's law of induction). From Coulomb's law, which describes the electrostatic force between two charges, you know that electric field lines start on positive charges and end at negative charges. On the other hand, magnetic field lines always form closed loops and have no beginning or end. Finally, you learned in the chapter on magnetism that a magnetic field is created around a current-carrying wire, as stated by Ampere's law.

Electromagnetic waves consist of changing electric and magnetic fields

In the mid-1800s, Scottish physicist James Clerk Maxwell created a simple but sophisticated set of equations to describe the relationship between electric and magnetic fields. Maxwell's equations summarized the known phenomena of his time: the observations that were described by Coulomb, Faraday, Ampere, and other scientists of his era. Maxwell believed that nature is symmetric, and he hypothesized that a changing electric field should produce a magnetic field in a manner analogous to Faraday's law of induction.

Maxwell's equations described many of the phenomena, such as magnetic induction, that had already been observed. However, other phenomena that had not been observed could be derived from the equations. For example, Maxwell's equations predicted that a changing magnetic field would create a changing electric field, which would, in turn, create a changing magnetic field, and so on. The predicted result of those changing fields is a wave that moves through space at the speed of light.

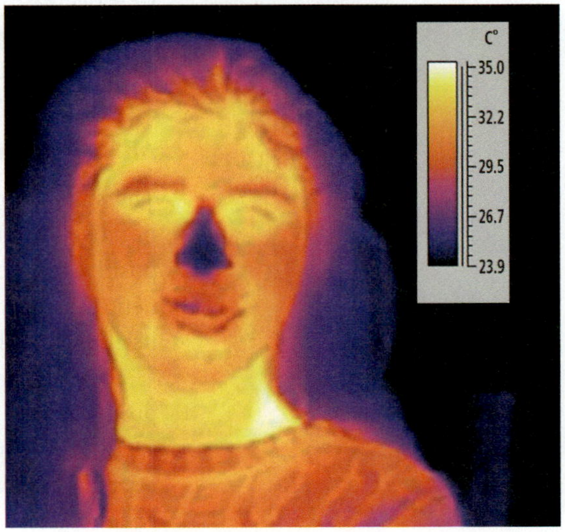

Figure 19
At normal body temperature, humans radiate most strongly in the infrared, at a wavelength of about 10 microns (10^{-5} m). The wavelength of the infrared radiation can be correlated to temperature.

SECTION 4
Advanced Level

Visual Strategy GENERAL

Figure 19
Explain to students that the colors in the infrared photograph are added to provide contrast between the different parts of the image so that the parts can be easily seen in the visible spectrum. Point out that all images that are obtained using nonvisible electromagnetic radiation show only light and dark areas, as in a black-and-white photograph. The stronger that radiation is in a particular part of the image, the greater the exposure is in that part of the image. False color is often added later to these images.

Q From which parts of the person is the most infrared radiation emitted?

A ***The neck, eyes, and forehead emit the most infrared radiation.***

Q From which parts of the person is the least infrared radiation emitted?

A ***The nose and hair emit the least infrared radiation.***

Teaching Tip — ADVANCED

Explain to students that one can take the product of the electric and magnetic field strengths to determine the power that electromagnetic radiation transmits. Because the electromagnetic radiation moves in a direction perpendicular to both the electric and magnetic fields, the power radiated is a vector quantity. This vector is called the *Poynting vector,* after the English physicist John Henry Poynting.

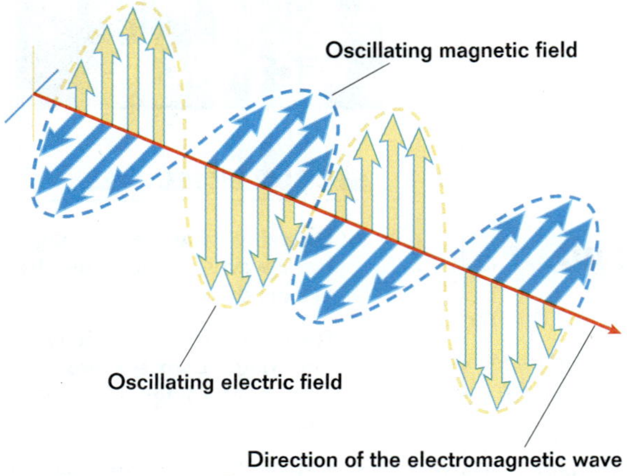

Figure 20
An electromagnetic wave consists of electric and magnetic field waves at right angles to each other. The wave moves in the direction perpendicular to both oscillating waves.

Maxwell predicted that light was electromagnetic in nature. The scientific community did not immediately accept Maxwell's equations. However, in 1887, a German physicist named Heinrich Hertz generated and detected electromagnetic waves in his laboratory. Hertz's experimental confirmation of Maxwell's work convinced the scientific community to accept the work.

Electromagnetic waves are simply oscillating electric and magnetic fields. The electric and magnetic fields are at right angles to each other and also at right angles to the direction that the wave is moving. **Figure 20** is a simple illustration of an electromagnetic wave at a single point in time. The electric field oscillates back and forth in one plane while the magnetic field oscillates back and forth in a perpendicular plane. The wave travels in the direction that is perpendicular to both of the oscillating fields. In the chapter on vibrations and waves you learned that this kind of wave is called a *transverse wave.*

Electric and magnetic forces are aspects of a single force

Although magnetism and electricity seem like very different things, we know that both electric and magnetic fields can produce forces on charged particles. These forces are aspects of one and the same force, called the *electromagnetic force.* Physicists have identified four *fundamental forces* in the universe: the strong force, which holds together the nucleus of an atom; the electromagnetic force, which is discussed here; the weak force, which is involved in nuclear decay; and the gravitational force, discussed in the chapter "Circular Motion and Gravitation". In the 1970s, physicists came to regard the electromagnetic and the weak force as two aspects of a single *electroweak interaction.*

The electromagnetic force obeys the *inverse-square law.* The force's magnitude decreases as one over the distance from the source squared. The inverse-square law applies to phenomena—such as gravity, light, and sound—that spread their influence equally in all directions and with an infinite range.

All electromagnetic waves are produced by accelerating charges

The simplest radiation source is an oscillating charged particle. Consider a negatively charged particle (electron) moving back and forth beside a fixed positive charge (proton). Recall that the changing electric field induces a magnetic field perpendicular to the electric field. In this way, the wave *propagates* itself as each changing field induces the other.

The frequency of oscillation determines the frequency of the wave that is produced. In an antenna, two metal rods are connected to an alternating voltage source that is changed from positive to negative voltage at the desired frequency. The wavelength λ of the wave is related to the frequency f by the equation $\lambda = c/f$, in which c is the speed of light.

Electromagnetic waves transfer energy

All types of waves, whether they are mechanical or electromagnetic or are longitudinal or transverse, have an energy associated with their motion. In the case of electromagnetic waves, that energy is stored in the oscillating electric and magnetic fields.

The simplest definition of energy is the capacity to do work. When work is performed on a body, a force moves the body in the direction of the force. The force that electromagnetic fields exert on a charged particle is proportional to the electric field strength, E, and the magnetic field strength, B. So, we can say that energy is stored in electric and magnetic fields in much the same way that energy is stored in gravitational fields.

The energy transported by electromagnetic waves is called **electromagnetic radiation.** The energy carried by electromagnetic waves can be transferred to objects in the path of the waves or converted to other forms, such as heat. An everyday example is the use of the energy from microwave radiation to warm food. Energy from the sun reaches Earth via electromagnetic radiation across a variety of wavelengths. Some of these wavelengths are illustrated in **Figure 21.**

electromagnetic radiation

the transfer of energy associated with an electric and magnetic field; it varies periodically and travels at the speed of light

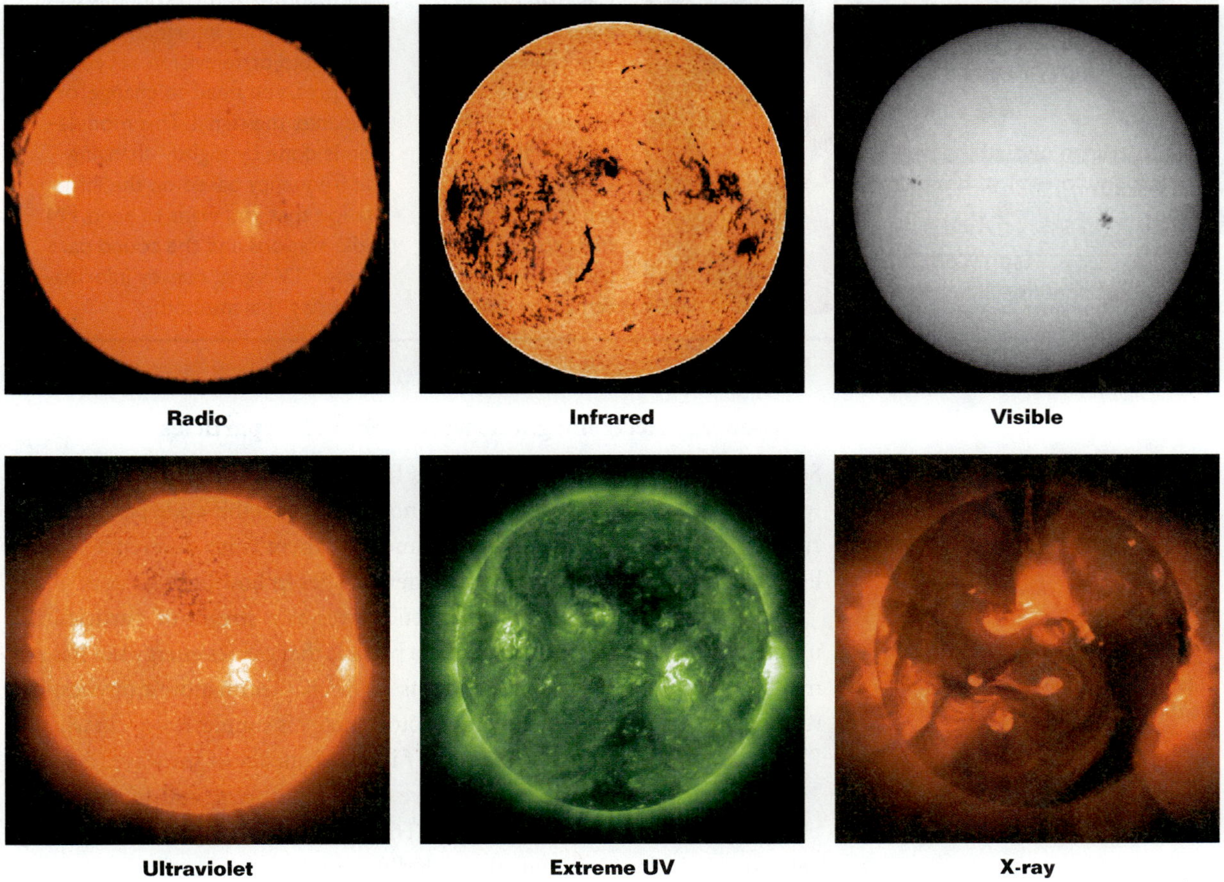

Figure 21
The sun radiates in all parts of the electromagnetic spectrum, not just in the visible light that we are accustomed to observing. These images show what the sun would look like if we could "see" at different wavelengths of electromagnetic radiation.

SECTION 4

Visual Strategy ADVANCED

Figure 21
Point out to students that an area of the sun that emits strong electromagnetic radiation in one part of the spectrum may not emit as strongly in another part.

Q The sun's corona is a thin region of the solar atmosphere that emits high-frequency electromagnetic radiation. In which of the photographs does the corona appear?

A ***The photographs showing ultraviolet and X-ray radiation reveal bright regions in the solar atmosphere that correspond to the corona.***

Q In what part of the spectrum is the corona brightest?

A ***X-ray***

Why it Matters

Radio and TV Broadcasts

In 1920, Reginald Fessenden accomplished the first modulation of radio waves by sound signals. The basic principle of transmitting sound by radio waves, or *radiotelephony,* involved imposing the waveform of a sound signal onto a radio carrier wave. Because sound waves have a much lower frequency than radio waves do, the amplitude or frequency of the carrier wave can be compressed or expanded (modulated) to the general shape of the sound signal.

Amplitude modulation (AM), which uses lower-frequency radio waves, often takes advantage of the ionosphere. This region of the atmosphere reflects the lower-frequency radio waves, giving them a greater broadcast range. Some high-power AM stations can be received in distant parts of the world.

Frequency modulation (FM) has the advantage of being less susceptible to interference from other stations at the same frequency or from static. However, FM's shorter wavelengths do not reflect off the atmosphere, so they cannot be picked up over the horizon.

The Language of Physics

In 1926, the American physical chemist Gilbert N. Lewis introduced the word *photon.*

Why it Matters

Radio and TV Broadcasts

"You are listening to 97.7 WKID, student-run radio from Central High School." What does the radio announcer mean by this greeting? Where do those numbers and letters come from? The numbers mean that the radio station is broadcasting a frequency modulated (FM) radio signal of 97.7 megahertz (MHz). In other words, the electric and magnetic fields of the radio wave are changing back and forth between their minimum and maximum values 97,700,000 times per second. That's a lot of oscillations in a three-minute long song!

The Federal Communications Commission (FCC) assigns the call letters, such as WKID, and the frequencies that the various stations will use. All FM radio stations are located in the band of frequencies that range from 88 to 108 MHz. Similarly, amplitude modulated (AM) radio stations are all in the 535 to 1,700 kHz band. A kilohertz (kHz) is 1,000 cycles per second, so the AM band is broadcast at lower frequencies than the FM band is. The television channels 2 to 6 broadcast between 54 MHz and 88 MHz. Channels 7 to 13 are in the 174 MHz to 220 MHz band, and the remaining channels occupy even higher frequency bands in the spectrum.

How are these radio waves transmitted? To create a simple radio transmitter, you need to create a rapidly changing electric current in a wire. The easiest form of a changing current is a sine wave. A sine wave can be created with a few simple circuit components, such as a capacitor and an inductor. The wave is amplified, sent to an antenna, and transmitted into space.

If you have a sine wave generator and a transmitter, you have a radio station. The only problem is that a sine wave contains very little information! To turn sound waves or pictures into information that your radio or television set can interpret, you need to change, or *modulate,* the signal. This modulation is done by slightly changing the frequency based on the information that you want to send. FM radio stations and the sound part of your TV signal convey information using this method.

High-energy electromagnetic waves behave like particles

Sometimes, an electromagnetic wave's frequency (or wavelength) makes the wave behave more like a particle. This notion is called the *wave-particle duality* of light. It is important to understand that there is no difference in what light *is* at different frequencies. The difference lies in how light *behaves.*

When thinking about electromagnetic waves as a stream of particles, it is useful to define a **photon.** A photon is a particle that carries energy but has zero mass. You will learn more about photons in the chapter on atomic physics. The relationship between frequency and photon energy is simple: $E = hf$, in which h, Plank's constant, is a fixed number and f is the frequency of the wave.

photon

a unit or quantum of light; a particle of electromagnetic radiation that has zero mass and carries a quantum of energy

Low-energy photons tend to behave more like waves, and higher energy photons behave more like particles. This distinction helps scientist design detectors and telescopes to distinguish different frequencies of radiation.

THE ELECTROMAGNETIC SPECTRUM

At first glance, radio waves seem completely different from visible light and gamma rays. They are produced and detected in very different ways. A large antenna is needed to detect radio waves, your eye can see visible light, and sophisticated scientific equipment must be used to observe gamma rays. Even though they appear quite different, all the different parts of the *electromagnetic spectrum* are fundamentally the same thing. They are all electromagnetic waves.

The electromagnetic spectrum can be expressed in terms of wavelength, frequency, or energy. The electromagnetic spectrum is illustrated in **Figure 22.** Longer wavelengths, such as radio waves and microwaves, are usually described in terms of frequency. If your favorite FM radio station is 90.5, the frequency is 90.5 MHz (9.05×10^7 Hz). Infrared, visible, and ultraviolet light are usually described in terms of their wavelength. We see the wavelength 670 nm (6.70×10^{-7} m) as red light. The shortest wavelength radiation is generally described in terms of the energy of one photon. For example, the element cesium-137 emits gamma rays with energy of 662 keV (10^{-13} J). (A keV is a *kilo-electron volt,* equal to 1000 eV or 1.60×10^{-16} J.)

extension

Integrating Technology

Visit go.hrw.com for the activity "Radio Waves."

Keyword HF6EMIX

Radio waves

Radio waves have the longest wavelengths in the spectrum. The wavelengths range in size from the diameter of a soccer ball to the length of a soccer field and beyond. Because long wavelengths can easily travel around objects, they work well for transmitting information across long distances. In the United States, the FCC regulates the radio spectrum, assigning the bands that certain stations can use for radio and television broadcasting.

Objects that are far away in deep space also emit radio waves. Because these waves can pass through Earth's atmosphere, scientists can use huge antennas on land to collect the waves, which can help the scientists to understand the nature of the universe.

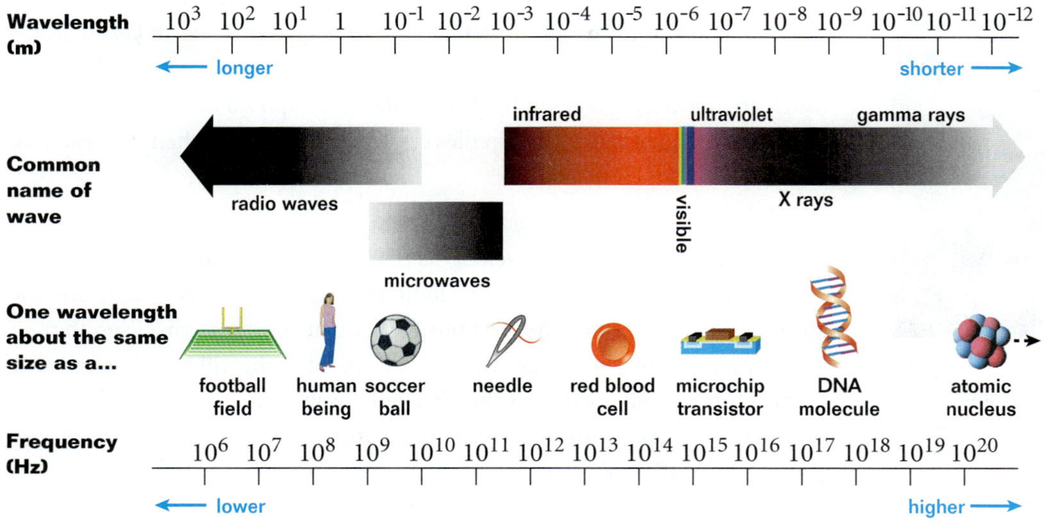

Figure 22

The electromagnetic spectrum ranges from very long radio waves, with wavelengths equal to the height of a tall building, to very short-wavelength gamma rays, with wavelengths as short as the diameter of the nucleus of an atom.

Misconception Alert

Students may think that radio waves are always associated with sound. Remind them that modulated radio waves in the FM part of the spectrum provide the picture component of a broadcast television signal. Mention also that radio astronomy involves "seeing" rather than "listening" at very long electromagnetic wavelengths.

Demonstration

Van de Graaff Generator

Purpose To show how an electric charge that is moving between surfaces produces electromagnetic radiation in the radio part of the spectrum.

Materials Van de Graaff generator, metal rod with insulated handle, portable radio

Procedure Turn the radio on and set it to a part of the AM band where there is no channel and very little noise. Turn up the volume and place the radio several meters away from the Van de Graaff generator. Turn on the generator, hold the insulated end of the metal rod, and bring the opposite end of the rod close to the sphere of the generator. Have students notice that the radio clicks each time that a spark jumps between the generator and the rod. Explain that the sparks consist of changing electric currents, which produce changing electric and magnetic fields and thus electromagnetic waves.

Teaching Tip — Advanced

Explain to students that one of the most common forms of electromagnetic radiation in the universe is in the microwave region of the spectrum. Hydrogen is the most abundant element in the universe. When a hydrogen electron undergoes a transition from its higher-energy spin state to its lower-energy spin state, the atom emits a photon. The photon has a specific wavelength associated with the energy change of the transition. This low-energy microwave photon, with a wavelength of 21 cm, occurs in regions of hydrogen clouds where no stars have formed. Because interstellar dust does not readily absorb radio waves, this radiation is easily observed throughout the Milky Way galaxy.

Misconception Alert — General

Students may be perplexed about the bright colors that appear under low-frequency ultraviolet light, or black light. Explain that the materials used in black-light posters absorb ultraviolet radiation and then reemit the energy at longer wavelengths of light. This process, called *fluorescence,* produces the glowing colors that are seen under the invisible ultraviolet light.

Microwaves

The wavelengths of microwaves range from 30 cm to 1 mm in length. These waves are considered to be part of the radio spectrum and are also regulated by the FCC. Microwaves are used to study the stars, to talk with satellites in orbit, and to heat up your after-school snack.

Microwave ovens use the longer-wavelength microwaves to cook your popcorn quickly. Microwaves are also useful for transmitting information because they can penetrate mist, clouds, smoke, and haze. Microwave towers throughout the world convey telephone calls and computer data from city to city. Shorter-wavelength microwaves are used for radar. Radar works by sending out bursts of microwaves and detecting the reflections off of objects the waves hit.

Infrared

Infrared light lies between the microwave and the visible parts of the electromagnetic spectrum. The *far-infrared* wavelengths, which are close to the microwave end of the spectrum, are about the size of the head of a pin. Short, *near-infrared* wavelengths are microscopic. They are about the size of a cell.

You experience far-infrared radiation every day as heat given off by anything warm: sunlight, a warm sidewalk, a flame, and even your own body! Television remote controls and some burglar alarm systems use near-infrared radiation. Night-vision goggles show the world as it looks in the infrared, which helps police officers and rescue workers to locate people, animals, and other warm objects in the dark. Mosquitoes can also "see" in the infrared, which is one of the tools in their arsenal for finding dinner.

Visible light

The wavelengths that the human eye can see range from about 700 nm (red light) to 400 nm (violet light). This range is a very small part of the electromagnetic spectrum! We see the visible spectrum as a rainbow, as shown in **Figure 23.**

Visible light is produced in many ways. An incandescent light bulb gives off light—and heat—from a glowing filament. In neon lights and in lasers, atoms emit light directly. Televisions and fluorescent lights make use of *phosphors,* which are materials that emit light when they are exposed to high-energy electrons or ultraviolet radiation. Fireflies create light through a chemical reaction.

Figure 23
When white light shines through a prism or through water, such as in this rainbow, you can see the colors of the visible light spectrum.

Ultraviolet

Ultraviolet (UV) light has wavelengths that are shorter than visible light, just beyond the violet. Our sun emits light throughout the spectrum, but the ultraviolet waves are the ones responsible for causing sunburns. Even though you cannot see ultraviolet light with your eyes, this light will also damage your retina. Only a small portion of the ultraviolet waves that the sun emits actually penetrates Earth's atmosphere. Various atmospheric gases, such as ozone, block most of the UV waves.

Ultraviolet light is often used as a disinfectant to kill bacteria in city water supplies or to sterilize equipment in hospitals. Scientists use ultraviolet light to determine the chemical makeup of atoms and molecules and also the nature of stars and other celestial bodies. Ultraviolet light is also used to harden some kinds of dental fillings.

X rays

As the wavelengths of electromagnetic waves decrease, the associated photons increase in energy. X rays have very short wavelengths, about the size of atoms, and are usually thought of in terms of their energy instead of their wavelength.

While the German scientist Wilhelm Conrad Roentgen was experimenting with vacuum tubes, he accidentally discovered X rays. A week later, he took an X-ray photograph of his wife's hand, which clearly revealed her wedding ring and her bones. This first X ray is shown in **Figure 24.** Roentgen called the phenomenon *X ray* to indicate that it was an unknown type of radiation, and the name remains in use today.

You are probably familiar with the use of X rays in medicine and dentistry. Airport security also uses X rays to see inside luggage. Emission of X rays from otherwise dark areas of space suggests the existence of black holes.

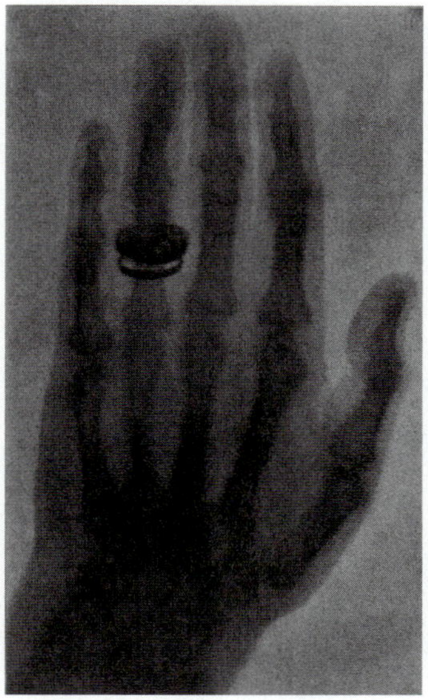

Figure 24
Wilhelm Roentgen took this X-ray image of Bertha Roentgen's hand one week after his discovery of this new type of electromagnetic radiation.

Gamma rays

The shortest wavelength electromagnetic waves are called *gamma rays.* As with X rays, gamma rays are usually described by their energy. The highest-energy gamma rays observed by scientists come from the hottest regions of the universe.

Radioactive atoms and nuclear explosions produce gamma rays. Gamma rays can kill living cells and are used in medicine to destroy cancer cells. The universe is a huge generator of gamma rays. Because gamma rays do not fully pierce Earth's atmosphere, astronomers frequently mount gamma-ray detectors on satellites.

SECTION REVIEW

1. What concepts did Maxwell use to help create his theory of electricity and magnetism? What phenomenon did Maxwell's equations predict?
2. What do electric and magnetic forces have in common?
3. The parts of the electromagnetic spectrum are commonly described in one of three ways. What are these ways?
4. **Critical Thinking** Where is the energy of an electromagnetic wave stored? Describe how this energy can be used.

SECTION 4

The Language of Physics

Gamma rays, which were first observed by the French physicist P. Villard in 1900, were classified as such by Ernest Rutherford, who had already discovered alpha and beta rays the previous year.

SECTION REVIEW ANSWERS

1. Maxwell combined the following observations: static electric and magnetic fields, the magnetic field of a current-carrying wire and electromagnetic induction. He also hypothesized that a changing electric field would produce a magnetic field. Maxwell's equations predicted electromagnetic radiation.
2. Both act on charged particles.
3. frequency, wavelength, and energy of a photon
4. The energy of electromagnetic waves is stored in the electric and magnetic fields that makeup the wave. The energy can be added to materials to increase their energy, such as the heating of food.

Highlights

Teaching Tip

Ask students to prepare a concept map for the chapter. The concept map should include most of the vocabulary terms, along with other integral terms or concepts.

extension

In-Depth Physics Content

Your students can visit go.hrw.com for an online chapter that integrates more in-depth development of the concepts covered here.

Keyword HF6EMIX

Highlights

KEY TERMS

electromagnetic induction (p. 708)

generator (p. 716)

alternating current (p. 718)

back emf (p. 720)

mutual inductance (p. 721)

rms current (p. 724)

transformer (p. 727)

electromagnetic radiation (p. 733)

photon (p. 734)

PROBLEM SOLVING

See **Appendix D: Equations** for a summary of the equations introduced in this chapter. If you need more problem-solving practice, see **Appendix I: Additional Problems.**

KEY IDEAS

Section 1 Electricity from Magnetism

- A change in the magnetic flux through a conducting coil induces an electric current in the coil. This concept is called *electromagnetic induction.*
- Lenz's law states that the magnetic field of an induced current opposes the change that caused it.
- The magnitude of the induced emf can be calculated using Faraday's law of induction.

Section 2 Generators, Motors, and Mutual Inductance

- Generators use induction to convert mechanical energy into electrical energy.
- Motors use an arrangement similar to that of generators to convert electrical energy into mechanical energy.
- *Mutual inductance* is the process by which an emf is induced in one circuit as a result of a changing current in another nearby circuit.

Section 3 AC Circuits and Transformers

- The root-mean-square (rms) current and rms emf in an ac circuit are important measures of the characteristics of an ac circuit.
- Transformers change the emf of an alternating current in an ac circuit.

Section 4 Electromagnetic Waves

- Electromagnetic waves are transverse waves that are traveling at the speed of light and are associated with oscillating electric and magnetic fields.
- Electromagnetic waves transfer energy. The energy of electromagnetic waves is stored in the waves' electric and magnetic fields.
- The electromagnetic spectrum has a wide variety of applications and characteristics that cover a broad range of wavelengths and frequencies.

Variable Symbols

Quantities		Units	
N	number of turns	(unitless)	
ΔV_{max}	maximum emf	V	volt
ΔV_{rms}	rms emf	V	volt
I_{max}	maximum current	A	ampere
I_{rms}	rms current	A	ampere
M	mutual inductance	H	henry = V•s/A

ELECTRICITY FROM MAGNETISM

Review Questions

1. Suppose you have two circuits. One consists of an electromagnet, a dc emf source, and a variable resistor that permits you to control the strength of the magnetic field. In the second circuit, you have a coil of wire and a galvanometer. List three ways that you can induce a current in the second circuit.

2. Explain how Lenz's law allows you to determine the direction of an induced current.

3. What four factors affect the magnitude of the induced emf in a coil of wire?

4. If you have a fixed magnetic field and a length of wire, how can you increase the induced emf across the ends of the wire?

Conceptual Questions

5. Rapidly inserting the north pole of a bar magnet into a coil of wire connected to a galvanometer causes the needle of the galvanometer to deflect to the right. What will happen to the needle if you do the following?

- **a.** pull the magnet out of the coil
- **b.** let the magnet sit at rest in the coil
- **c.** thrust the south end of the magnet into the coil

6. Explain how Lenz's law illustrates the principle of energy conservation.

7. Does dropping a strong magnet down a long copper tube induce a current in the tube? If so, what effect will the induced current have on the motion of the magnet?

8. Two bar magnets are placed side by side so that the north pole of one magnet is next to the south pole of the other magnet. If these magnets are then pushed toward a coil of wire, would you expect an emf to be induced in the coil? Explain your answer.

9. An electromagnet is placed next to a coil of wire in the arrangement shown below. According to Lenz's law, what will be the direction of the induced current in the resistor R in the following cases?

- **a.** The magnetic field suddenly decreases after the switch is opened.
- **b.** The coil is moved closer to the electromagnet.

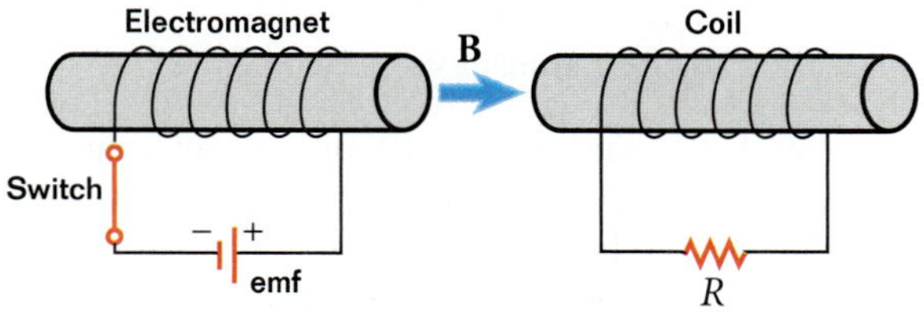

Practice Problems

For problems 10–12, see Sample Problem A.

10. A flexible loop of conducting wire has a radius of 0.12 m and is perpendicular to a uniform magnetic field with a strength of 0.15 T, as in figure **(a)** below. The loop is grasped at opposite ends and stretched until it closes to an area of 3×10^{-3} m^2, as in figure **(b)** below. If it takes 0.20 s to close the loop, find the magnitude of the average emf induced in the loop during this time.

11. A rectangular coil 0.055 m by 0.085 m is positioned so that its cross-sectional area is perpendicular to the direction of a magnetic field, B. If the coil has 75 turns and a total resistance of 8.7 Ω and the field decreases at a rate of 3.0 T/s, what is the magnitude of the induced current in the coil?

Review

ANSWERS

1. Place plane of coil perpendicular to magnetic field and move it into or out of field, rotate coil about axis perpendicular to magnetic field, and change strength of magnetic field with variable resistor.

2. The direction of the induced current is such that its magnetic field will oppose the change in the external magnetic field.

3. magnetic field component perpendicular to plane of coil, area of coil, time in which changes occur, number of turns of wire in coil

4. Wrap the wire into a coil that has many turns (large N), or move it in and out of the B field quickly (small Δt).

5. **a.** needle deflects to the left
b. no deflection of needle
c. needle deflects to the left

6. By opposing changes in the external field, the induced B field prevents the system's energy from increasing or decreasing.

7. yes; The induced field opposes the magnetic field of the magnet. The resulting force slows the magnet's speed through the tube.

8. no; Effects of one magnet cancel those of other magnet.

9. **a.** from left to right
b. from right to left

10. 3.2×10^{-2} V

11. 0.12 A

12. −0.63 V

13. B field (induces emf in turning coil), wire coil (conducts induced current), slip rings (maintain contact with rest of circuit by means of conducting brushes)

14. turn the handle faster

15. *Frequency* indicates how often each second the current goes from a maximum value in one direction to a maximum value in the other direction and back.

16. Replace the slip rings with a commutator, which prevents the reversal of the current every half-cycle.

17. an emf with polarity opposite that of the emf powering the motor; The coil's rotation in the B field induces a back emf that reduces the net potential difference across the motor.

18. The changing B field produced by a changing current in one circuit induces an emf and current in a nearby circuit.

19. A step-up transformer uses the B field of an alternating current to induce an increased emf in the secondary. A step-down transformer uses the same principle to induce a smaller emf in the secondary.

20. no; The change in potential difference in a transformer is accompanied by an inverse change in the current. In an ideal transformer, power is unchanged, as expected from energy conservation.

21. The magnetic forces are greatest on charges in the sides of a loop that move perpendicular to the B field (that is, when the plane of the loop is parallel to the field lines).

22. a step-down transformer; I is larger in the secondary, so wire with a lower R is needed to reduce energy dissipation.

12. A 52-turn coil with an area of 5.5×10^{-3} m^2 is dropped from a position where $B = 0.00$ T to a new position where $B = 0.55$ T. If the displacement occurs in 0.25 s and the area of the coil is perpendicular to the magnetic field lines, what is the resulting average emf induced in the coil?

GENERATORS, MOTORS, AND MUTUAL INDUCTANCE

Review Questions

13. List the essential components of an electric generator, and explain the role of each component in generating an alternating emf.

14. A student turns the handle of a small generator attached to a lamp socket containing a 15 W bulb. The bulb barely glows. What should the student do to make the bulb glow more brightly?

15. What is meant by the term *frequency* in reference to an alternating current?

16. How can an ac generator be converted to a dc generator? Explain your answer.

17. What is meant by back emf? How is it induced in an electric motor?

18. Describe how mutual induction occurs.

19. What is the difference between a step-up transformer and a step-down transformer?

20. Does a step-up transformer increase power? Explain your answer.

Conceptual Questions

21. When the plane of a rotating loop of wire is parallel to the magnetic field lines, the number of lines passing through the loop is zero. Why is the current at a maximum at this point in the loop's rotation?

22. In many transformers, the wire around one winding is thicker, and therefore has lower resistance, than the wire around the other winding. If the thicker wire is wrapped around the secondary winding, is the device a step-up or a step-down transformer? Explain.

23. A bar magnet is attached perpendicular to a rotating shaft. The magnet is then placed in the center of a coil of wire. In which of the arrangements shown below could this device be used as an electric generator? Explain your choice.

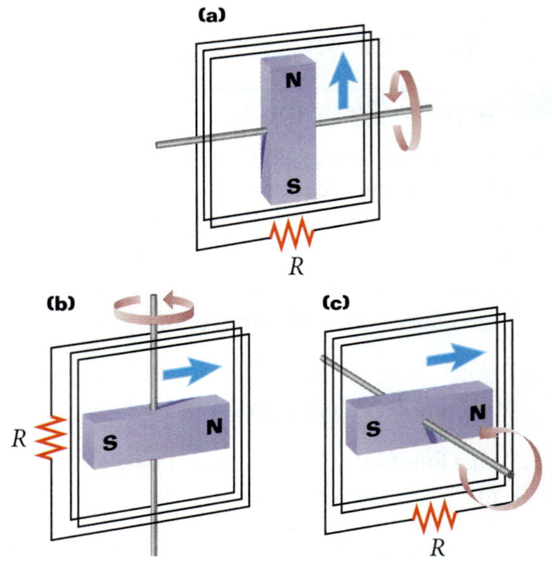

24. Would a transformer work with pulsating direct current? Explain your answer.

25. The faster the coil of loops, or *armature,* of an ac generator rotates, the harder it is to turn the armature. Use Lenz's law to explain why this happens.

Practice Problems

For problems 26–29, see Sample Problem B.

26. The rms applied emf across high-voltage transmission lines in Great Britain is 220 000 V. What is the maximum emf?

27. The maximum applied emf across certain heavy-duty appliances is 340 V. If the total resistance of an appliance is 120 Ω, calculate the following:

a. the rms applied emf

b. the rms current

28. The maximum current that can pass through a light bulb filament is 0.909 A when its resistance is 182 Ω.

a. What is the rms current conducted by the filament of the bulb?

b. What is the rms emf across the bulb's filament?

c. How much power does the light bulb use?

29. A 996 W hair dryer is designed to carry a peak current of 11.8 A.

   a. How large is the rms current in the hair dryer?
   b. What is the rms emf across the hair dryer?

AC CIRCUITS AND TRANSFORMERS

Review Questions

30. Which quantities remain constant when alternating currents are generated?

31. How does the power dissipated in a resistor by an alternating current relate to the power dissipated by a direct current that has potential difference and current values that are equal to the maximum values of the alternating current?

Conceptual Questions

32. In a Ground Fault Interrupter, would the difference in current across an outlet be measured in terms of the rms value of current or the actual current at a given moment? Explain your answer.

33. Voltmeters and ammeters that measure ac quantities are calibrated to measure the rms values of emf and current, respectively. Why would this be preferred to measuring the maximum emf or current?

Practice Problems

For problems 34–37, see Sample Problem C.

34. A transformer is used to convert 120 V to 9.0 V for use in a portable CD player. If the primary, which is connected to the outlet, has 640 turns, how many turns does the secondary have?

35. Suppose a 9.00 V CD player has a transformer for converting current in Great Britain. If the ratio of the turns of wire on the primary to the secondary coils is 24.6 to 1, what is the outlet potential difference?

36. A transformer is used to convert 120 V to 6.3 V in order to power a toy electric train. If there are 210 turns in the primary, how many turns should there be in the secondary?

37. The transformer shown in the figure below is constructed so that the coil on the left has five times as many turns of wire as the coil on the right does.

   a. If the input potential difference is across the coil on the left, what type of transformer is this?
   b. If the input potential difference is 24 000 V, what is the output potential difference?

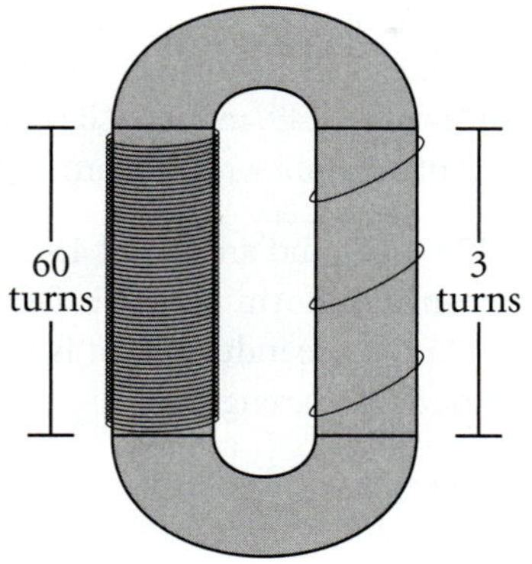

ELECTROMAGNETIC WAVES

Review Questions

38. How are electric and magnetic fields oriented to each other in an electromagnetic wave?

39. How does the behavior of low-energy electromagnetic radiation differ from that of high-energy electromagnetic radiation?

Conceptual Questions

40. Why does electromagnetic radiation obey the inverse-square law?

41. Why is a longer antenna needed to produce a low-frequency radio wave than to produce a high-frequency radio wave?

MIXED REVIEW PROBLEMS

42. A student attempts to make a simple generator by passing a single loop of wire between the poles of a horseshoe magnet with a 2.5×10^{-2} T field. The area of the loop is 7.54×10^{-3} m^2 and is moved perpendicular to the magnetic field lines. In what time interval will the student have to move the loop out

20 REVIEW

23. a, b; The *B* field lines in these cases are in a plane perpendicular to the plane of the loop, so the loop crosses the field lines.
24. yes; Current changes continually, so its changing *B* field can induce an emf in a transformer's secondary.
25. The *B* field of an induced current opposes the change (due to coil rotation) in the external *B* field. Faster rotation of the coil increases this induced current and thus the opposing field.
26. 3.1×10^5 V
27. a. 2.4×10^2 V
    b. 2.0 A
28. a. 0.643 A
    b. 117 V
    c. 75.2 W
29. a. 8.34 A
    b. 119 V
30. frequency, maximum current, and maximum voltage
31. The ac power is half as large as the dc power.
32. rms value; The difference between different phases of an alternating current may be large enough to cause the GIF to break the circuit. Rms current is a measure of an effective current with constant value so that changes in the current can be reliably measured.
33. The rms values are convenient because they give the same heating effect as the same value of direct current.
34. 48 turns
35. 221 V
36. 11 turns
37. a. a step-down transformer
    b. 1.2×10^3 V

38. Changing electric and magnetic fields are at right angles to each other and to the direction in which the electromagnetic wave moves.

39. Low-energy electromagnetic radiation behaves more like a wave, while high-energy radiation behaves more like a particle.

40. Electromagnetic waves move radially outward from the source. The surface area of the sphere around the wave source increases with the square of the distance from the source, so the amount of radiation on any given part of the spherical surface decreases with the square of the radius.

41. The wave's frequency is proportional to the inverse of the time that an electron takes to travel the length of the antenna. Because an electron takes longer to travel the length of a long antenna than a short one, the wave's frequency will be smaller.

42. 1.3×10^{-4} s; no

43. 790 turns

44. 4.2×10^{-2} T

45. **a.** a step-up transformer
b. 440 V

46. 1.03×10^{5} V

extension

Graphing Calculator Practice

Visit go.hrw.com for answers to this Graphing Calculator activity.

Keyword HF6EMIXT

of the magnetic field in order to induce an emf of 1.5 V? Is this a practical generator?

43. The same student in item 42 modifies the simple generator by wrapping a much longer piece of wire around a cylinder with about one-fourth the area of the original loop (1.886×10^{-3} m^2). Again using a uniform magnetic field with a strength of 2.5×10^{-2} T, the student finds that by removing the coil perpendicular to the magnetic field lines during 0.25 s, an emf of 149 mV can be induced. How many turns of wire are wrapped around the coil?

44. A coil of 325 turns and an area of 19.5×10^{-4} m^2 is removed from a uniform magnetic field at an angle of 45° in 1.25 s. If the induced emf is 15 mV, what is the magnetic field's strength?

45. A transformer has 22 turns of wire in its primary and 88 turns in its secondary.

a. Is this a step-up or step-down transformer?

b. If 110 V ac is applied to the primary, what is the output potential difference?

46. A bolt of lightning, such as the one shown on the left side of the figure below, behaves like a vertical wire conducting electric current. As a result, it produces a magnetic field whose strength varies with the distance from the lightning. A 105-turn circular coil is oriented perpendicular to the magnetic field, as shown on the right side of the figure below. The coil has a radius of 0.833 m. If the magnetic field at the coil drops from 4.72×10^{-3} T to 0.00 T in 10.5 μs, what is the average emf induced in the coil?

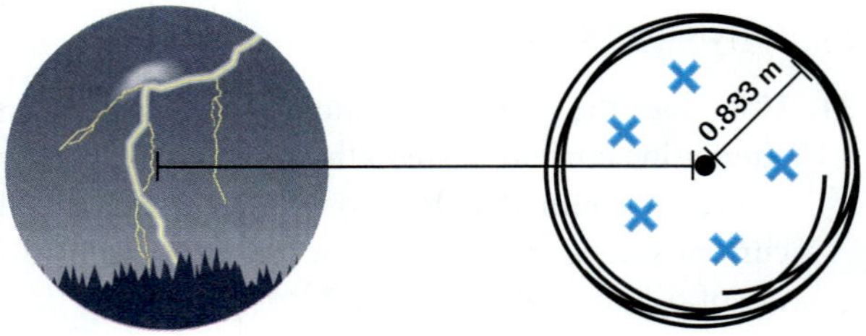

Graphing Calculator *Practice*

Alternating Current

In alternating current (ac), the emf alternates from positive to negative. The current responds to the changes in emf by oscillating with the same frequency that the emf does. This relationship is shown in the following equation for the instantaneous current:

$$i = I_{max} \sin \omega t$$

In this equation, ω is the ac frequency, and I_{max} is the maximum current. The effective current of an ac circuit is the root-mean-square current (rms current), I_{rms}. The rms current is related to the maximum current by the following equation:

$$I_{rms} = \frac{I_{max}}{\sqrt{2}}$$

In this graphing calculator activity, the calculator will use these two equations to make graphs of instantaneous current and rms current versus time. By analyzing these graphs, you will be able to determine what the values of the instantaneous current and the rms current are at any point in time. The graphs will give you a better understanding of current in ac circuits.

Visit go.hrw.com and type in the keyword **HF6EMIX** to find this graphing calculator activity. Refer to **Appendix B** for instructions on downloading the program for this activity.

47. The potential difference in the lines that carry electric power to homes is typically 20.0 kV. What is the ratio of the turns in the primary to the turns in the secondary of the transformer if the output potential difference is 117 V?

48. The alternating emf of a generator is represented by the equation emf = (245 V) sin 560*t*, in which emf is in volts and *t* is in seconds. Use these values to find the frequency of the emf and the maximum emf output of the source.

49. A pair of adjacent coils has a mutual inductance of 1.06 H. Determine the average emf induced in the secondary circuit when the current in the primary circuit changes from 0 A to 9.50 A in a time interval of 0.0336 s.

50. A generator supplies 5.0×10^3 kW of power. The output emf is 4500 V before it is stepped up to 510 kV. The electricity travels 410 mi (6.44×10^5 m) through a transmission line that has a resistance per unit length of 4.5×10^{-4} Ω/m.

a. How much power is lost through transmission of the electrical energy along the line?

b. How much power would be lost through transmission if the generator's output emf were not stepped up? What does this answer tell you about the role of large emfs (voltages) in power transmission?

Alternative Assessment

1. Two identical magnets are dropped simultaneously from the same point. One of them passes through a coil of wire in a closed circuit. Predict whether the two magnets will hit the ground at the same time. Explain your reasoning. Then, plan an experiment to test which of the following variables measurably affect how long each magnet takes to fall: magnetic strength, coil cross-sectional area, and the number of loops the coil has. What measurements will you make? What are the limits of precision in your measurements? If your teacher approves your plan, obtain the necessary materials and perform the experiments. Report your results to the class, describing how you made your measurements, what you concluded, and what additional questions need to be investigated.

2. What do adapters do to potential difference, current, frequency, and power? Examine the input/output information on several adapters to find out. Do they contain step-up or step-down transformers? How does the output current compare to the input? What happens to the frequency? What percentage of the energy do they transfer? What are they used for?

3. Research the debate between the proponents of alternating current and those who favored direct current in the 1880–1890s. How were Thomas Edison and George Westinghouse involved in the controversy? What advantages and disadvantages did each side claim? What uses of electricity were anticipated? What kind of current was finally generated in the Niagara Falls hydroelectric plant? Had you been in a position to fund these projects at that time, which projects would you have funded? Prepare your arguments to re-enact a meeting of businesspeople in Buffalo in 1887.

4. Research the history of telecommunication. Who invented the telegraph? Who patented it in England? Who patented it in the United States? Research the contributions of Charles Wheatstone, Joseph Henry, and Samuel Morse. How did each of these men deal with issues of fame, wealth, and credit to other people's ideas? Write a summary of your findings, and prepare a class discussion about the effect patents and copyrights have had on modern technology.

20 REVIEW

47. 171 : 1
48. $f = 89$ Hz, 245 V
49. 300 V
50. **a.** 28 kW
b. 3.6×10^5 kW; The power dissipated by the alternating current whose emf has not been stepped up is more than the power generated. This indicates that without stepping up its emf, electricity cannot be conveyed very far along a transmission line.

Alternative Assessment ANSWERS

1. Students' plans will vary. Be sure proposed tests are safe, measure time accurately, and adjust one variable at a time.

2. Most adapters use step-down transformers. Students should note that power output is always less than input.

3. Students' answers will vary. Edison claimed that ac was dangerous and unreliable. Westinghouse noted that dc could not be stepped up and that long-distance transmission was inefficient. Batteries can store dc for peak use, but new industries that used electricity at all hours favored ac.

4. Wheatstone (1802–1875) took out the first telegraph patent in England in 1837. In the United States, Henry (1797–1878) invented the telegraph 10 years before Morse (1791–1872), who claimed in court that he invented it by himself.

CHAPTER 20

Standardized Test Prep

ANSWERS

1. C
2. G
3. D
4. H
5. B
6. J
7. A
8. F
9. C
10. G

Standardized Test Prep

MULTIPLE CHOICE

1. Which of the following equations correctly describes Faraday's law of induction?

A. $\text{emf} = -N\dfrac{\Delta(AB\tan\theta)}{\Delta t}$

B. $\text{emf} = N\dfrac{\Delta(AB\cos\theta)}{\Delta t}$

C. $\text{emf} = -N\dfrac{\Delta(AB\cos\theta)}{\Delta t}$

D. $\text{emf} = M\dfrac{\Delta(AB\cos\theta)}{\Delta t}$

2. For the coil shown in the figure below, what must be done to induce a clockwise current?

F. Either move the north pole of a magnet down into the coil, or move the south pole of the magnet up and out of the coil.
G. Either move the south pole of a magnet down into the coil, or move the north pole of the magnet up and out of the coil.
H. Move either pole of the magnet down into the coil.
J. Move either pole of the magnet up and out of the coil.

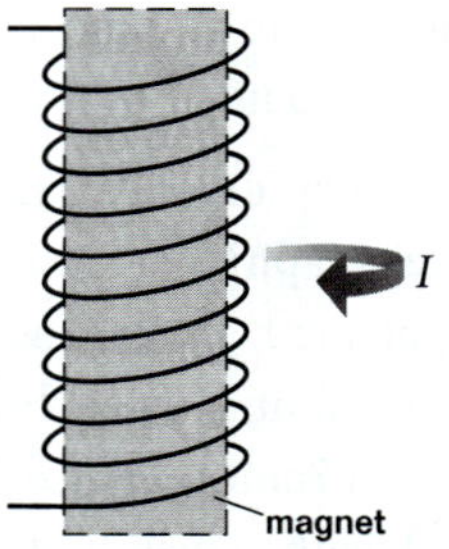

3. Which of the following would *not* increase the emf produced by a generator?

A. rotating the generator coil faster
B. increasing the strength of the generator magnets
C. increasing the number of turns of wire in the coil
D. reducing the cross-sectional area of the coil

4. By what factor do you multiply the maximum emf to calculate the rms emf for an alternating current?

F. 2
G. $\sqrt{2}$
H. $\dfrac{1}{\sqrt{2}}$
J. $\dfrac{1}{2}$

5. Which of the following correctly describes the composition of an electromagnetic wave?

A. a transverse electric wave and a magnetic transverse wave that are parallel and are moving in the same direction
B. a transverse electric wave and a magnetic transverse wave that are perpendicular and are moving in the same direction
C. a transverse electric wave and a magnetic transverse wave that are parallel and are moving at right angles to each other
D. a transverse electric wave and a magnetic transverse wave that are perpendicular and are moving at right angles to each other

6. A coil is moved out of a magnetic field in order to induce an emf. The wire of the coil is then rewound so that the area of the coil is increased by 1.5 times. Extra wire is used in the coil so that the number of turns is doubled. If the time in which the coil is removed from the field is reduced by half and the magnetic field strength remains unchanged, how many times greater is the new induced emf than the original induced emf?

F. 1.5 times
G. 2 times
H. 3 times
J. 6 times

Use the passage below to answer questions 7–8.

A pair of transformers is connected in series, as shown in the figure below.

7. From left to right, what are the types of the two transformers?

A. Both are step-down transformers.
B. Both are step-up transformers.
C. One is a step-down transformer; and one is a step-up transformer.
D. One is a step-up transformer; and one is a step-down transformer.

8. What is the output potential difference from the secondary coil of the transformer on the right?

F. 400 V
G. 12 000 V
H. 160 000 V
J. 360 000 V

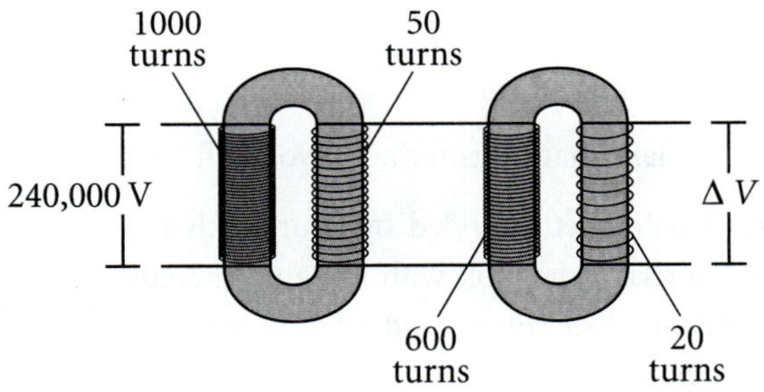

9. What are the particles that can be used to describe electromagnetic radiation called?

A. electrons
B. magnetons
C. photons
D. protons

10. The maximum values for the current and potential difference in an ac circuit are 3.5 A and 340 V, respectively. How much power is dissipated in this circuit?

F. 300 W
G. 600 W
H. 1200 W
J. 2400 W

SHORT RESPONSE

11. The alternating current through an electric toaster has a maximum value of 12.0 A. What is the rms value of this current?

12. What is the purpose of a commutator in an ac generator?

13. How does the energy of one photon of an electromagnetic wave relate to the wave's frequency?

14. A transformer has 150 turns of wire on the primary coil and 75 000 turns on the secondary coil. If the input potential difference across the primary is 120 V, what is the output potential difference across the secondary?

EXTENDED RESPONSE

15. Why is alternating current used for power transmission instead of direct current? Be sure to include power dissipation and electrical safety considerations in your answer.

Base your answers to questions 16–18 on the information below.

A device at a carnival's haunted house involves a metal ring that flies upward from a table when a patron passes near the table's edge. The device consists of a photoelectric switch that activates the circuit when anyone walks in front of the switch and of a coil of wire into which a current is suddenly introduced when the switch is triggered.

16. Why must the current enter the coil just as someone comes up to the table?

17. Using Lenz's law, explain why the ring flies upward when there is an increasing current in the coil?

18. Suppose the change in the magnetic field is 0.10 T/s. If the radius of the ring is 2.4 cm and the ring is assumed to consist of one turn of wire, what is the emf induced in the ring?

Test TIP Be sure to convert all units of given quantities to proper SI units.

11. 8.48 A

12. It converts ac to a changing current in one direction only.

13. The energy is directly proportional to the wave's frequency ($E = hf$).

14. 6.0×10^4 V

15. For electric power to be transferred over long distances without a large amount of power dissipation, the electric power must have a high potential difference and low current. However, to be safely used in homes, the potential difference must be lower than that used for long-distance power transmission. Because of induction, the potential difference and current of electricity can be transformed to higher or lower values, but the current must change continuously (alternate) for this to happen.

16. The change in current in the coil will produce a changing magnetic field, which will induce a current in the ring. The induced current produces a magnetic field that interacts with the magnetic field from the coil, causing the ring to rise from the table.

17. According to Lenz's law, the magnetic field induced in the ring must oppose the magnetic field that induces the current in the ring. The opposing fields cause the ring, which can move freely, to rise upward from the coil under the table's surface.

18. 1.8×10^{-4} V

Lab Planning

Beginning on page T34 are preparation notes and teaching tips to assist you in planning.

Blank data tables (as well as some sample data) appear on the **One-Stop Planner.**

No Books in the Lab?

See the ***Datasheets for In-Text Labs*** workbook for a reproducible master copy of this experiment.

Safety Caution

Emphasize the dangers of working with electricity. For the safety of the students and of the equipment, remind students to have you check their circuits before turning on the power supply or closing the switch.

Tips and Tricks

- Show students how to set up the circuit for the second part of the lab. Make sure they know how to use the rheostat to adjust the current in the circuit.

✓ Checkpoints

Step 3: Because this lab requires no measurements, students may be unsure what to record in their notebooks. Guide students to come up with a format for recording their observations.

Step 7: Make sure the power supply or battery, rheostat, galvanometer, coil, and switch are connected properly and set at the proper settings. Students should be able to demonstrate that all connections are properly made.

Step 12: Students should be able to demonstrate that they reversed the current in the circuit.

CHAPTER 20

Skills Practice Lab Electromagnetic Induction

OBJECTIVES

- **Determine** the relationship between the magnetic field of a magnet and the current induced in a conductor.
- **Detect** an induced current using a galvanometer.
- **Determine** what factors affect the direction and magnitude of an induced current.

MATERIALS LIST

- galvanometer
- insulated connecting wires
- momentary contact switch
- pair of bar magnets or a large horseshoe magnet
- power supply
- rheostat (10 Ω) or potentiometer
- student primary and secondary coil set with iron core

In this laboratory, you will use a magnet, a conductor, and a galvanometer to explore electromagnets and the principle of self-induction.

SAFETY

- **Never close a circuit until it has been approved by your teacher. Never rewire or adjust any element of a closed circuit. Never work with electricity near water; be sure the floor and all work surfaces are dry.**
- **If the pointer on any kind of meter moves off scale, open the circuit immediately by opening the switch.**
- **Do not attempt this exercise with any batteries, electrical devices, or magnets other than those provided by your teacher for this purpose.**

PROCEDURE

Preparation

1. Read the entire lab, and plan what measurements you will take.
2. If you are not using a datasheet provided by your teacher, prepare an observation table in your lab notebook with three wide columns. Label the columns *Sketch of Setup, Experiment,* and *Observations.* For each part of the lab, you will sketch the apparatus and label the poles of the magnet, write a brief description, and record your observations.

Induction with a Permanent Magnet

3. Connect the ends of the smaller coil to the galvanometer. Hold the magnet still and move the coil quickly over the north pole of the magnet, as shown in **Figure 1.** Remove the coil quickly. Observe the galvanometer.
4. Repeat, moving the coil more slowly. Observe the galvanometer.
5. Turn the magnet over, and repeat steps 3 and 4, moving the coil over the south pole of the magnet. Observe the galvanometer.
6. Hold the coil stationary, and quickly move the north pole of the magnet in and out of the coil. Repeat slowly. Turn the magnet, and repeat this step using the south pole. Observe the galvanometer.

Induction with an Electromagnet

7. Rewire the galvanometer and connect it to the larger coil. Slip the smaller coil inside the larger coil. Connect the small coil in series with a switch, battery, and rheostat, so that the arrangement resembles that shown in **Figure 2.** Close the switch. Adjust the rheostat so that the galvanometer reading registers on the scale. Observe the galvanometer.

8. Open the switch to interrupt the current in the small coil. Observe the galvanometer.

9. Close the switch again, and open it after a few seconds. Observe the galvanometer.

10. Adjust the rheostat to increase the current in the small coil. Close the switch, and observe the galvanometer.

11. Decrease the current in the circuit, and observe the galvanometer. Open the switch.

12. Reverse the direction of the current by reversing the battery connections. Close the switch, and observe the galvanometer.

13. Place an iron rod inside the small coil. Open and close the switch while observing the galvanometer. Record all observations in your notebook.

14. Clean up your work area. Put equipment away safely so that it is ready to be used again.

Figure 1

Step 3: Connect the coil to the galvanometer. Holding the magnet still, move the coil over the magnet quickly.

Step 4: Holding the magnet still, move the coil over the magnet slowly.

Step 6: Repeat the procedure, but hold the coil still while moving the magnet.

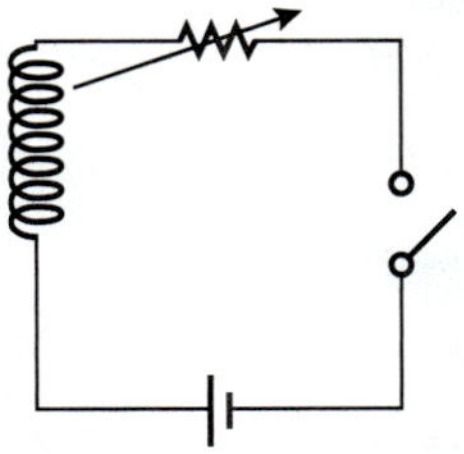

Figure 2

Step 7: Connect the larger coil to the galvanometer. Place the smaller coil inside the larger coil. Connect the smaller coil in series with the battery, switch, and rheostat.

ANALYSIS

1. **Describing Events** Based on your observations from the first part of the lab, did the speed of the motion have any effect on the galvanometer?

2. **Explaining Events** In the first part of the lab, did it make any difference whether the coil or the magnet moved? Explain why or why not.

CONCLUSIONS

3. **Drawing Conclusions** Explain what the galvanometer readings revealed to you about the magnet and the wire coil.

4. **Drawing Conclusions** Based on your observations, what conditions are required to induce a current in a circuit?

5. **Drawing Conclusions** Based on your observations, what factors influence the direction and magnitude of the induced current?

ANSWERS

Analysis

1. Yes, the faster the motion, the greater the deflection of the galvanometer needle.

2. It did not matter which one moved. The important thing was the motion between them, or relative motion.

Conclusions

3. The galvanometer showed that there was a current in the wire coil. The direction of the deflection indicated the direction of the current.

4. A changing magnetic field is needed to induce a current in a circuit.

5. The direction of the current is influenced by the direction the magnet is moved and by whether the field is increasing or decreasing. The magnitude is influenced by the speed of the change.

1820

Physics and Its World *Timeline 1830–1890*

1831

$$\text{emf} = -N\Delta \frac{[AB(\cos\theta)]}{\Delta t}$$

Michael Faraday begins experiments demonstrating electromagnetic induction. Similar experiments are conducted around the same time by **Joseph Henry** in the United States, but he doesn't publish the results of his work at this time.

1830

1831 – **Charles Darwin** sets sail on the H.M.S. *Beagle* to begin studies of life-forms in South America, New Zealand, and Australia. His discoveries form the foundation for the theory of evolution by natural selection.

1837 – **Queen Victoria** ascends the British throne at the age of 18. Her reign continues for 64 years, setting the tone for the Victorian era.

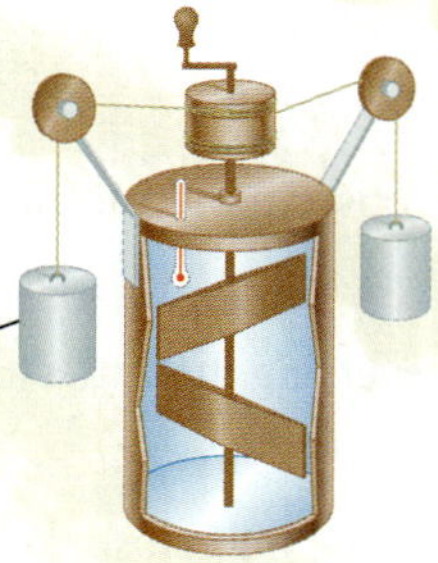

1843

$$\Delta U = Q - W$$

James Prescott Joule determines that mechanical energy is equivalent to energy transferred as heat, laying the foundation for the principle of energy conservation.

1840

1843 – **Richard Wagner's** first major operatic success, *The Flying Dutchman*, premieres in Dresden, Germany.

1844

Samuel Morse sends the first telegraph message from Washington, D. C. to Baltimore.

1850

1850 – **Harriet Tubman,** an ex-slave from Maryland, becomes a "conductor" on the Underground Railroad. Over the next decade, she helps more than 300 slaves escape to northern "free" states.

1850

$$W = Q_h - Q_c$$

Rudolph Clausius formulates the second law of thermodynamics, the first step in the transformation of thermo-dynamics into an exact science.

1860

1861 – **Benito Juárez** is elected president of Mexico. During his administration, the invasion by France is repelled and basic social reforms are implemented.

Scala/Art Resource, NY

1861 – The American Civil War begins at Fort Sumter in Charleston, South Carolina.

1873

$$c = \frac{1}{\sqrt{\mu_0 \varepsilon_0}}$$

James Clerk Maxwell completes his *Treatise on Electricity and Magnetism.* In this work, Maxwell gives **Michael Faraday's** discoveries a mathematical framework.

1874 – The first exhibition of impressionist paintings, including works by **Claude Monet, Camille Pissarro,** and **Pierre-Auguste Renoir,** takes place in Paris.

Giraudon/Art Resource, NY

1878 – The first commercial telephone exchange in the United States begins operation in New Haven, Connecticut.

1884 – *Adventures of Huckleberry Finn,* by **Samuel L. Clemens** (better known as Mark Twain), is published.

1888

$$\lambda = \frac{c}{f}$$

Heinrich Hertz experimentally demonstrates the existence of electromagnetic waves, which were predicted by **James Clerk Maxwell. Oliver Lodge** makes the same discovery independently.

1860

1870

1880

1890

1900

749

CHAPTER 21

Atomic Physics
Planning Guide

Compression Guide

To shorten instruction because of time limitations, omit the opener and Section 3 and abbreviate the review.

OBJECTIVES	LABS, DEMONSTRATIONS, AND ACTIVITIES	TECHNOLOGY RESOURCES
PACING • 45 min pp. 750–751 **Chapter Opener**		CD **Visual Concepts,** Chapter 21 BASIC
PACING • 90 min pp. 752–761 **Section 1 Quantization of Energy** • Explain how Planck resolved the ultraviolet catastrophe in blackbody radiation. • Calculate energy of quanta using Planck's equation. • Solve problems involving maximum kinetic energy, work function, and threshold frequency in the photoelectric effect.	TE **Demonstration** A Blackbody, p. 752 GENERAL TE **Demonstration** Color of Thermal Sources, p. 753 GENERAL TE **Demonstration** Photoelectric Effect, p. 756 GENERAL SE **Skills Practice Lab** The Photoelectric Effect, pp. 784–785◆ GENERAL ANC **Datasheet** The Photoelectric Effect* GENERAL	OSP **Lesson Plans** TR 115 Blackbody Radiation TR 116 Motion Picture Sound and the Photoelectric Effect TR 68A Absorption of Light by a Blackbody TR 69A The Photoelectric Effect TR 70A Compton Shift
PACING • 45 min pp. 762–770 **Section 2 Models of the Atom** • Explain the strengths and weaknesses of Rutherford's model of the atom. • Recognize that each element has a unique emission and absorption spectrum. • Explain atomic spectra using Bohr's model of the atom. • Interpret energy-level diagrams.	TE **Demonstration** Particle Scattering, p. 762 GENERAL TE **Demonstration** Spectral Lines, p. 764 GENERAL SE **Quick Lab** Atomic Spectra, p. 765 GENERAL	OSP **Lesson Plans** EXT **Integrating Technology** Seeing Atoms: The STM BASIC TR 117 The Production of an Emission Spectrum TR 118 Emission and Absorption Spectra of Hydrogen TR 119 Electron Transitions and Spectral Lines TR 71A Rutherford's Scattering Experiment and Atomic Model TR 72A Absorption, Electron Transition, and Emission
PACING • 45 min pp. 771–777 *Advanced Level* **Section 3 Quantum Mechanics** • Recognize the dual nature of light and matter. • Calculate the de Broglie wavelength of matter waves. • Distinguish between classical ideas of measurement and Heisenberg's uncertainty principle. • Describe the quantum-mechanical picture of the atom, including the electron cloud and probability waves.	TE **Demonstration** Probability Distribution, p. 776 GENERAL	OSP **Lesson Plans**

PACING • 90 min

CHAPTER REVIEW, ASSESSMENT, AND STANDARDIZED TEST PREPARATION

SE **Chapter Highlights,** p. 778
SE **Chapter Review,** pp. 779–781
SE **Graphing Calculator Practice,** p. 780 GENERAL
SE **Alternative Assessment,** p. 781 ADVANCED
SE **Standardized Test Prep,** pp. 782–783 GENERAL
SE **Appendix D: Equations,** p. 864
SE **Appendix I: Additional Problems,** pp. 895–896
ANC **Study Guide Worksheet** Mixed Review* GENERAL
ANC **Chapter Test A*** GENERAL
ANC **Chapter Test B*** ADVANCED
OSP **Test Generator**

Online and Technology Resources

go.hrw.com Holt Online Learning

Visit go.hrw.com to find a variety of online resources. To access this chapter's extensions, enter the keyword **HF6ATMXT** and click the "go" button. Click **Holt Online Learning** for an online edition of this textbook, and other interactive resources.

HOLT Teacher's One-Stop Planner®

This DVD package includes:
- Holt Calendar Planner
- Customizable Lesson Plans
- Editable Worksheets
- ExamView® Version 6 Assessment Suite
- Interactive Teacher's Edition
- Holt PuzzlePro®
- Holt PowerNotes® Presentations
- MindPoint® Quiz Show

KEY

SE	Student Edition	OSP	One-Stop Planner	EXT	Online Extension
TE	Teacher Edition	CD	CD or CD-ROM	*	Also on One-Stop Planner
ANC	Ancillary Worksheet	TR	Teaching Transparencies	◆	Requires advance prep

SKILLS DEVELOPMENT RESOURCES	REVIEW AND ASSESSMENT	CORRELATIONS
		National Science Education Standards
SE **Sample Set A** Quantum Energy, p. 755 BASIC ANC **Problem Workbook** Sample Set A* BASIC OSP **Problem Bank** Sample Set A BASIC SE **Sample Set B** The Photoelectric Effect, p. 758 GENERAL ANC **Problem Workbook** Sample Set B* GENERAL OSP **Problem Bank** Sample Set B GENERAL SE **Conceptual Challenge,** p. 759 GENERAL	SE **Section Review,** p. 760 GENERAL ANC **Study Guide Worksheet** Section 1* GENERAL ANC **Quiz** Section 1* BASIC	UCP 1, 2, 3, 5 SAI 1, 2 ST 1, 2 HNS 1, 2, 3 SPSP 5
SE **Conceptual Challenge,** p. 767 GENERAL SE **Sample Set C** Interpreting Energy-Level Diagrams, pp. 768–769 ADVANCED ANC **Problem Workbook** Sample Set C* ADVANCED OSP **Problem Bank** Sample Set C ADVANCED SE **Appendix J: Advanced Topics** Semiconductor Doping, pp. 926–927 ADVANCED	SE **Section Review,** p. 770 GENERAL ANC **Study Guide Worksheet** Section 2* GENERAL ANC **Quiz** Section 2* BASIC	UCP 1, 2, 3, 5 SAI 1, 2 ST 1, 2 HNS 1, 3 SPSP 1, 4, 5 PS 1a, 6c
SE **Sample Set D** De Broglie Waves, pp. 773–774 GENERAL TE **Classroom Practice,** p. 773 GENERAL ANC **Problem Workbook** Sample Set D* GENERAL OSP **Problem Bank** Sample Set D GENERAL SE **Appendix J: Advanced Topics** De Broglie Waves, pp. 922–923 ADVANCED	SE **Section Review,** p. 777 GENERAL ANC **Study Guide Worksheet** Section 3* GENERAL ANC **Quiz** Section 3* GENERAL	UCP 1, 2, 3 HNS 1, 3

SCILINKS. NSTA
www.scilinks.org

Maintained by the **National Science Teachers Association.**

Topic: Max Planck
SciLinks Code: HF60923

Topic: Photoelectric Effect
SciLinks Code: HF61138

Topic: Early Atomic Theory
SciLinks Code: HF60442

Topic: Modern Atomic Theory
SciLinks Code: HF60978

Classroom CD-ROMs

- Guided Reading Audio Program
- Student One Stop
- Virtual Investigations
- Visual Concepts

Holt Lab Generator CD-ROM

Search for any lab by topic, standard, difficulty level, or time. Edit any lab to fit your needs, or create your own labs. Use the Lab Materials QuickList software to customize your lab materials list.

CHAPTER 21
Overview

Section 1 introduces the quantization of energy in blackbody radiation and the photoelectric effect; solves problems involving energy quanta, threshold frequency, and work function; and discusses the Compton shift as it pertains to the particle theory of light.

Section 2 explores Rutherford's model of the atom, introduces emission and absorption spectra, explains atomic spectra in terms of Bohr's model of hydrogen, and evaluates the strengths and weaknesses of Bohr's model.

Section 3 discusses the wave-particle duality of light and matter, shows how to calculate de Broglie wavelengths, introduces the uncertainty principle, and describes the quantum-mechanical picture of the atom.

About the Illustration

This photograph of the aurora borealis was taken in Denali National Park, Alaska. In the background are the Alaska Range Mountains. Auroras occur when electrically charged particles, mainly from the sun, become trapped in Earth's atmosphere over the magnetic poles and collide with other atoms, resulting in the emission of visible light.

CHAPTER 21

Atomic Physics

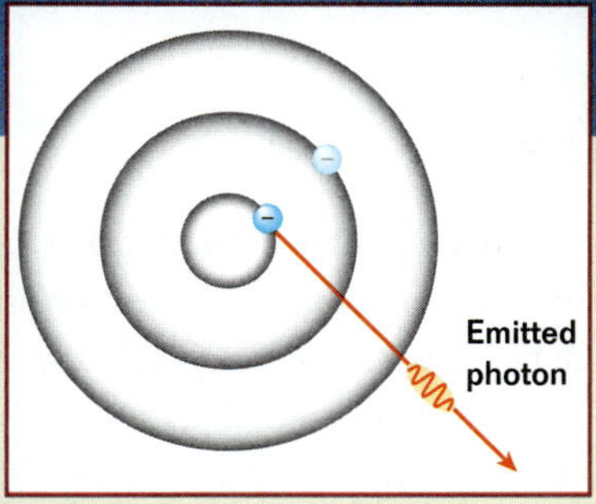

Colorful lights similar to these in Denali National Park, Alaska, are commonly seen in the sky in northern latitudes. They are known as the *aurora borealis,* or the *northern lights.* The northern lights—which can extend thousands of kilometers—appear as arcs, bands, or streaks of color, sometimes flickering or pulsating. This is caused by billions of atomic "jumps," as shown in the diagram.

WHAT TO EXPECT

In this chapter, you will learn about the development of the field known as *quantum mechanics,* which is required to describe phenomena at the atomic level, where classical physics fails.

Why it Matters

The theory of quantum mechanics has led to many technological advances. For instance, it has enabled us to use nuclear energy and to perform life-saving MRIs. It also explains how the stars shine, and why the northern lights occur.

CHAPTER PREVIEW

1 Quantization of Energy
- Blackbody Radiation
- The Photoelectric Effect

2 Models of the Atom
- Early Models of the Atom
- Atomic Spectra
- The Bohr Model of the Hydrogen Atom

3 Quantum Mechanics
- The Dual Nature of Light
- Matter Waves
- The Uncertainty Principle
- The Electron Cloud

Tapping Prior Knowledge

Knowledge to Review

- ✔ Standing waves occur when two waves of the same frequency, amplitude, and wavelength interfere as they travel through a medium in opposite directions.
- ✔ Light in the wave model consists of electromagnetic waves.
- ✔ Electromagnetic waves are transverse waves that consist of oscillating electric and magnetic fields at right angles to each other.

Items to Probe

- ✔ Familiarity with atomic structure: Ask students to draw a schematic diagram of the atom.
- ✔ Preconceptions about the nature of light: Ask students to list arguments for and against the wave and particle theories of light.

Demonstration

A Blackbody — GENERAL

Purpose Show an example of a blackbody.

Materials two closed cardboard boxes thick enough to prevent light from passing through, one with a white interior and one with a black interior; index cards with holes of different sizes

Procedure Cut a hole slightly smaller than the size of an index card in the front of each box. Have students hold the index cards up to the hole in each box one at a time, and have them look through the different index cards into each box. If the hole in the index card is not too large, they will find that even the box with the white interior appears dark. Ask them if light enters the box (*yes*) and if the light comes back out of the box (*most of it does not*). Point out that each time light hits a wall inside the box, some of the light is absorbed. If the box is reasonably insulated, the energy absorbed by the walls will cause their temperature to rise, and energy will be radiated inside the box. Tell students that both boxes are examples of blackbodies.

SECTION 1

Quantization of Energy

SECTION OBJECTIVES

- **Explain how Planck resolved the ultraviolet catastrophe in blackbody radiation.**
- **Calculate energy of quanta using Planck's equation.**
- **Solve problems involving maximum kinetic energy, work function, and threshold frequency in the photoelectric effect.**

BLACKBODY RADIATION

By the end of the nineteenth century, scientists thought that classical physics was nearly complete. One of the few remaining questions to be solved involved electromagnetic radiation and thermodynamics. Specifically, scientists were concerned with the glow of objects when they reach a high temperature.

All objects emit electromagnetic radiation. This radiation, which depends on the temperature and other properties of an object, typically consists of a continuous distribution of wavelengths from the infrared, visible, and ultraviolet portions of the spectrum. The distribution of the intensity of the different wavelengths varies with temperature.

At low temperatures, radiation wavelengths are mainly in the infrared region. So, they cannot be seen by the human eye. As the temperature of an object increases, the range of wavelengths given off shifts into the visible region of the electromagnetic spectrum. For example, the molten metal shown in **Figure 1** seems to have a yellow glow. At even higher temperatures, the object appears to have a white glow, as in the hot tungsten filament of a light bulb, and then a bluish glow.

Figure 1
This molten metal has a bright yellow glow because of its high temperature.

Classical physics cannot account for blackbody radiation

One problem at the end of the 1800s was understanding the distribution of wavelengths given off by a blackbody. Most objects absorb some incoming radiation and reflect the rest. An ideal system that absorbs all incoming radiation is called a *blackbody.* Physicists study **blackbody radiation** by observing a hollow object with a small opening, as shown in **Figure 2.** The system is a good example of how a blackbody works; it traps radiation. The light given off by the opening is in equlibrium with light from the walls of the object, because the light has been given off and reabsorbed many times.

blackbody radiation

the radiation emitted by a blackbody, which is a perfect radiator and absorber and emits radiation based only on its temperature

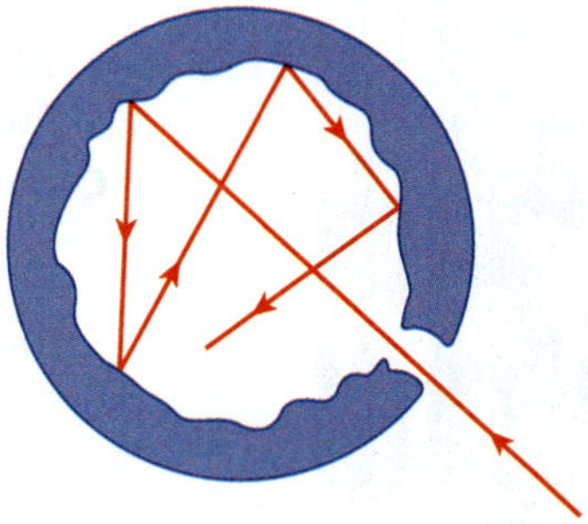

Figure 2
Light enters this hollow object through the small opening and strikes the interior wall. Some of the energy is absorbed by the wall, but some is reflected at a random angle. After each reflection, part of the light is absorbed by the wall. After many reflections, essentially all of the incoming energy is absorbed by the cavity wall. Only a small fraction of the incident energy escapes through the opening.

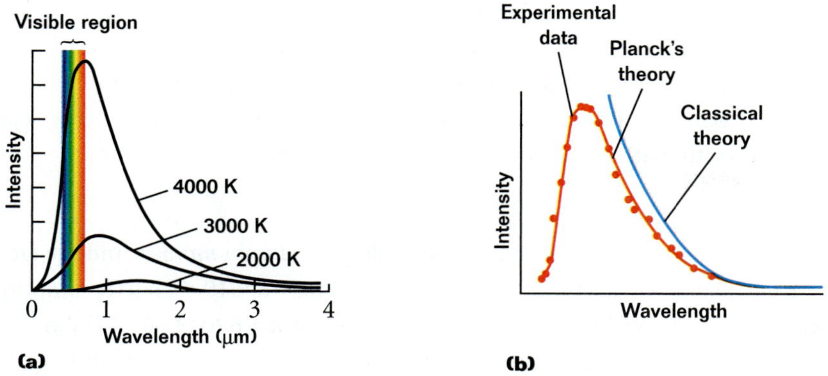

Figure 3
(a) This graph shows the intensity of blackbody radiation at three different temperatures. **(b)** Classical theory's prediction for blackbody radiation (the blue curve) did not correspond to the experimental data (the red data points) at all wavelengths, whereas Planck's theory (the red curve) did.

Experimental data for the radiation given off by an object at three different temperatures are shown in **Figure 3(a).** Note that as the temperature increases, the total energy given off by the body (the area under the curve) also increases. In addition, as the temperature increases, the peak of the distribution shifts to shorter wavelengths.

Scientists could not account for these experimental results with classical physics. **Figure 3(b)** compares an experimental plot of the blackbody radiation spectrum (the red data points) with the theoretical picture of what this curve should look like based on classical theories (the blue curve). Classical theory predicts that as the wavelength approaches zero, the amount of energy being radiated should become infinite. This prediction is contrary to the experimental data, which show that as the wavelength approaches zero, the amount of energy being radiated also approaches zero. This contradiction is often called the **ultraviolet catastrophe** because the disagreement occurs at the ultraviolet end of the spectrum.

ultraviolet catastrophe

the failed prediction of classical physics that the energy radiated by a blackbody at extremely short wavelengths is extremely large and that the total energy radiated is infinite

Experimental data for blackbody radiation support the quantization of energy

In 1900, Max Planck (1858–1947) developed a formula for blackbody radiation that was in complete agreement with experimental data at all wavelengths. Planck's original theoretical approach is rather abstract in that it involves arguments based on entropy and thermodynamics. The arguments presented in this book are easier to visualize, and they convey the spirit and revolutionary impact of Planck's original work.

Planck proposed that blackbody radiation was produced by submicroscopic electric oscillators, which he called *resonators.* He assumed that the walls of a glowing cavity were composed of billions of these resonators, all vibrating at different frequencies. Although most scientists naturally assumed that the energy of these resonators was continuous, Planck made the radical assumption that these resonators could only absorb and then give off certain discrete amounts of energy.

When he first discovered this idea, Planck was using a mathematical technique in which quantities that are known to be continuous are temporarily

SCILINKS.
www.scilinks.org
Topic: Max Planck
Code: HF60923

Demonstration

Color of Thermal Sources

GENERAL

Purpose Show that different thermal sources emit different spectra.

Materials matches, candle, Bunsen burner, nail, incandescent light bulb, tongs

CAUTION *Wear goggles and an apron when using the burner.*

Procedure Ask students to rank the objects by temperature based on past experience. Tell students to carefully observe and record the colors of the different flames (match, candle, Bunsen burner) and of the hot objects (nail, light bulb) that you are about to demonstrate. Then, turn on the light bulb, strike a match, light the candle, light the Bunsen burner, and hold the nail in the flame of the Bunsen burner with tongs until the nail glows. Point out that brightness does not necessarily mean a higher temperature. Have students rank the dominant color of each object by wavelengths and compare their results with their original ranking of temperatures.

Teaching Tip — BASIC

Point out that $1\ C \times 1\ V = 1\ J$. The magnitude of charge on 1 electron is 1.60×10^{-19} C. Therefore, an energy of 1 eV is the magnitude of charge on 1 electron × 1 V. Hence, $1\ eV = 1.60 \times 10^{-19}$ J.

considered to be discrete. After the calculations are made, the discrete units are taken to be infinitesimally small. Planck found that the calculations worked if he omitted this step and considered energy to come in discrete units throughout his calculations. With this method, Planck found that the total energy (E_n) of a resonator with frequency f is an integral multiple of hf, as follows:

$$E_n = nhf$$

In this equation, n is a positive integer called a *quantum number,* and the factor h is Planck's constant, which equals $6.626\ 068\ 96 \times 10^{-34}$ J•s. To simplify calculations, we will use the approximate value of $h = 6.63 \times 10^{-34}$ J•s in this textbook. Because the energy of each resonator comes in discrete units, it is said to be *quantized,* and the allowed energy states are called *quantum states* or *energy levels.* With the assumption that energy is quantized, Planck was able to derive the red curve shown in **Figure 3(b)** on the previous page.

According to Planck's theory, the resonators absorb or give off energy in discrete multiples of hf. Einstein later applied the concept of quantized energy to light. The units of light energy called *quanta* (now called *photons*) are absorbed or given off as a result of electrons "jumping" from one quantum state to another. As seen by the equation above, if the quantum number (n) changes by one unit, the amount of energy radiated changes by hf. For this reason, the energy of a light quantum, which corresponds to the energy difference between two adjacent levels, is given by the following equation:

ENERGY OF A LIGHT QUANTUM

$$E = hf$$

energy of a quantum (n = 1) = Planck's constant × frequency

Did you know?

Max Planck became president of the Kaiser Wilhelm Institute of Berlin in 1930. Although Planck remained in Germany during the Hitler regime, he openly protested the Nazis' treatment of his Jewish colleagues and consequently was forced to resign his presidency in 1937. Following World War II, he was reinstated as president, and the institute was renamed the Max Planck Institute in his honor.

A resonator will radiate or absorb energy only when it changes quantum states. The idea that energy comes in discrete units marked the birth of a new theory known as *quantum mechanics.*

If Planck's constant is expressed in units of J•s, the equation $E = hf$ gives the energy in joules. However, when dealing with the parts of atoms, energy is often expressed in units of the electron volt, eV. An *electron volt* is defined as the energy that an electron or proton gains when it is accelerated through a potential difference of 1 V. Because 1 V = 1 J/C, the relation between the electron volt and the joule is as follows:

$$1\ eV = 1.60 \times 10^{-19}\ C{\bullet}V = 1.60 \times 10^{-19}\ C{\bullet}J/C = 1.60 \times 10^{-19}\ J$$

Planck's idea that energy is quantized was so radical that most scientists, including Planck himself, did not consider the quantization of energy to be realistic. Planck thought of his assumption as a mathematical approach to be used in calculations rather than a physical explanation. Therefore, he and other scientists continued to search for a different explanation of blackbody radiation that was consistent with classical physics.

SAMPLE PROBLEM A

Quantum Energy

PROBLEM

At the peak of the sun's radiation spectrum, each photon carries an energy of about 2.7 eV. What is the frequency of this light?

SOLUTION

Given: $E = 2.7$ eV $\qquad h = 6.63 \times 10^{-34}$ J•s

Unknown: $f = ?$

Use the equation for the energy of a light quantum, and isolate frequency.

$$E = hf \quad \text{or} \quad f = \frac{E}{h}$$

$$f = \frac{E}{h} = \frac{(2.7 \text{ eV})(1.60 \times 10^{-19} \text{ J/eV})}{6.63 \times 10^{-34} \text{ J•s}}$$

$$\boxed{f = 6.5 \times 10^{14} \text{ Hz}}$$

TIP *Always be sure that your units cancel properly. In this problem, you need to convert energy from electron volts to joules. For this reason, 2.7 eV is multiplied by the conversion factor of 1.60×10^{-19} J/eV.*

PRACTICE A

Quantum Energy

1. Assume that the pendulum of a grandfather clock acts as one of Planck's resonators. If it carries away an energy of 8.1×10^{-15} eV in a one-quantum change, what is the frequency of the pendulum? (Note that an energy this small would not be measurable. For this reason, we do not notice quantum effects in the large-scale world.)
2. A vibrating mass-spring system has a frequency of 0.56 Hz. How much energy of this vibration is carried away in a one-quantum change?
3. A photon in a laboratory experiment has an energy of 5.0 eV. What is the frequency of this photon?
4. Radiation emitted from human skin reaches its peak at $\lambda = 940$ µm.
   a. What is the frequency of this radiation?
   b. What type of electromagnetic waves are these?
   c. How much energy (in electron volts) is carried by one quantum of this radiation?

PROBLEM GUIDE A

Use this guide to assign problems.
SE = Student Edition Textbook
PW = Problem Workbook
PB = Problem Bank on the One-Stop Planner (OSP)

Solving for:

E	**SE** 2, 4*; Ch. Rvw. 12 **PW** 6*, 7, 8* **PB** Sample, 1–2
f	**SE** Sample, 1, 3, 4*; Ch. Rvw. 11 **PW** 4*, 5*, 7, 8* **PB** 3–5
λ	**PW** Sample, 1–3 **PB** 6–8

***Challenging Problem**
Consult the printed Solutions Manual or the OSP for detailed solutions.

ANSWERS

Practice A

1. 2.0 Hz
2. 3.7×10^{-34} J
3. 1.2×10^{15} Hz
4. **a.** 3.19×10^{11} Hz
   b. infrared
   c. 1.32×10^{-3} eV

Visual Strategy BASIC

Figure 4

Q Can one directly observe electrons being emitted from a metal surface?

A ***No, electrons cannot be directly observed, but the effect of electrons leaving the metal surface can be detected because the metal becomes charged when it loses electrons.***

Demonstration

Photoelectric Effect GENERAL

Purpose Show that light can release electrons from a metal.

Materials electroscope, rubber rod, wool, thin rectangular piece of zinc (3 cm × 8 cm), incandescent light, ultraviolet light

Procedure Attach the zinc plate to the knob of the electroscope, charge the rubber rod with the wool, and charge the zinc plate with the rubber rod. Turn on the incandescent light, and shine it on the zinc plate. Have students note that there is no change in the separation of the leaves. Turn off the incandescent light. Turn on the ultraviolet light, and shine it on the zinc plate. Have students observe that the leaves of the electroscope collapse (discharge). Explain that this occurs because as electrons are ejected from the zinc, the magnitude of the net (negative) charge decreases.

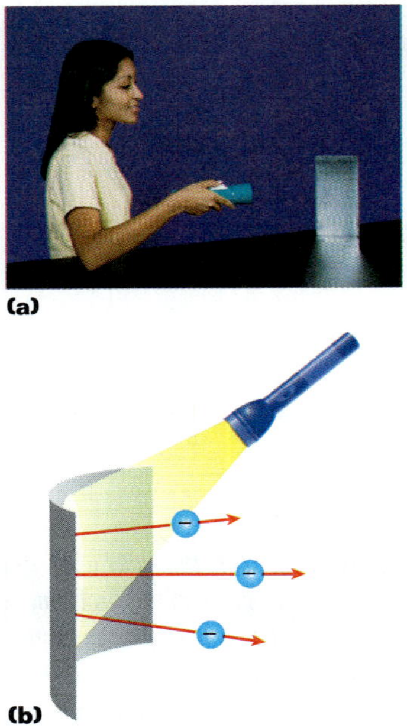

(a)

(b)

Figure 4
A light beam shining on a metal **(a)** may eject electrons from the metal **(b).** Because this interaction involves both light and electrons, it is called the photoelectric effect.

photoelectric effect

the emission of electrons from a material surface that occurs when light of certain frequencies shines on the surface of the material

THE PHOTOELECTRIC EFFECT

As you learned in the chapter "Electromagnetic Induction," James Maxwell discovered in 1873 that light was a form of electromagnetic waves. Experiments by Heinrich Hertz provided experimental evidence of Maxwell's theories. However, the results of some later experiments by Hertz could not be explained by the wave model of the nature of light. One of these was the **photoelectric effect.** When light strikes a metal surface, the surface may emit electrons, as **Figure 4** illustrates. Scientists call this effect the photoelectric effect. They refer to the electrons that are emitted as *photoelectrons.*

Classical physics cannot explain the photoelectric effect

The fact that light waves can eject electrons from a metal surface does not contradict the principles of classical physics. Light waves have energy, and if that energy is great enough, an electron could be stripped from its atom and have enough energy to escape the metal. However, the details of the photoelectric effect cannot be explained by classical theories. In order to see where the conflict arises, we must consider what should happen according to classical theory and then compare these predictions with experimental observations.

As was stated in the chapter on waves, the energy of a wave increases as its intensity increases. Thus, according to classical physics, light waves of any frequency should have sufficient energy to eject electrons from the metal if the intensity of the light is high enough. Moreover, at lower intensities, electrons should be ejected if light shines on the metal for a sufficient time period. (Electrons would take time to absorb the incoming energy before acquiring enough kinetic energy to escape from the metal.) Furthermore, increasing the intensity of the light waves should increase the kinetic energy of the photoelectrons, and the maximum kinetic energy of any electron should be determined by the light's intensity. These classical predictions are summarized in the second column of **Table 1.**

Table 1 The Photoelectric Effect

	Classical predictions	Experimental evidence
Whether electrons are ejected depends on . . .	the intensity of the light.	the frequency of the light.
The kinetic energy of ejected electrons depends on . . .	the intensity of the light.	the frequency of the light.
At low intensities, electron ejection . . .	takes time.	occurs almost instantaneously above a certain frequency.

SCILINKS.
www.scilinks.org
Topic: Photoelectric Effect
Code: HF61138

Scientists found that *none* of these classical predictions are observed experimentally. No electrons are emitted if the frequency of the incoming light falls below a certain frequency, even if the intensity is very high. This frequency, known as the *threshold frequency* (f_t), differs from metal to metal.

If the light frequency exceeds the threshold frequency, the photoelectric effect is observed. The number of photoelectrons emitted is proportional to the light intensity, but the maximum kinetic energy of the photoelectrons is independent of the light intensity. Instead, the maximum kinetic energy of the photoelectrons increases with increasing frequency. Furthermore, electrons are emitted from the surface almost instantaneously, even at low intensities. See **Table 1.**

Einstein proposed that all electromagnetic waves are quantized

Albert Einstein resolved this conflict in his 1905 paper on the photoelectric effect, for which he received the Nobel Prize in 1921, by extending Planck's concept of quantization to electromagnetic waves. Einstein assumed that an electromagnetic wave can be viewed as a stream of particles, now called **photons.** Each photon has an energy, E, given by Planck's equation ($E = hf$). In this theory, each photon is absorbed as a unit by an electron. When a photon's energy is transferred to an electron in a metal, the energy acquired by the electron is equal to hf.

photon

a unit or quantum of light; a particle of electromagnetic radiation that has zero mass and carries a quantum of energy

Threshold frequency depends on the work function of the surface

In order to be ejected from a metal, an electron must overcome the force that binds it to the metal. The smallest amount of energy the electron must have to escape the surface of a metal is the **work function** of the metal. The work function is equal to hf_t, where f_t is the threshold frequency for the metal. Photons with energy greater than hf_t eject electrons from the surface of and from within the metal. Because energy must be conserved, the maximum kinetic energy (of photoelectrons ejected from the surface) is the difference between the photon energy and the work function of the metal. This relationship is expressed mathematically by the following equation:

work function

the minimum energy needed to remove an electron from a metal atom

MAXIMUM KINETIC ENERGY OF A PHOTOELECTRON

$$KE_{max} = hf - hf_t$$

maximum kinetic energy =
(Planck's constant × frequency of incoming photon) – work function

According to this equation, there should be a linear relationship between f and KE_{max} because h is a constant and the work function, hf_t, is constant for any given metal. Experiments have verified that this is indeed the case, as shown in **Figure 5,** and the slope of such a curve ($\Delta KE/\Delta f$) gives a value for h that corresponds to Planck's value.

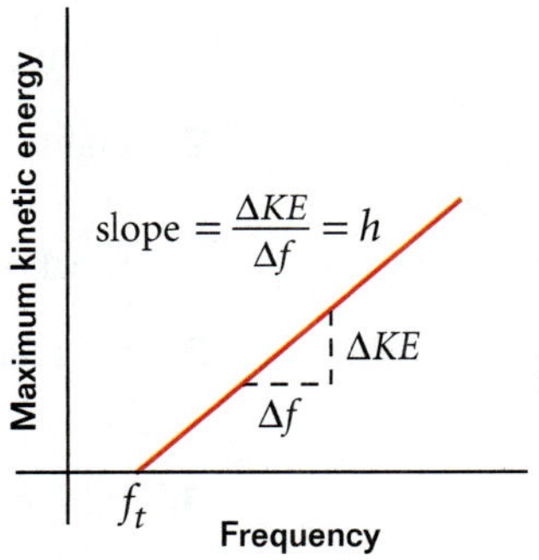

Figure 5
This graph shows a linear relationship between the maximum kinetic energy of emitted electrons and the frequency of incoming light. The intercept with the horizontal axis is the threshold frequency.

The Language of Physics

Some texts use the Greek letter *phi* (ϕ) to represent work function; such texts express the maximum kinetic energy of a photoelectron as $KE_{max} = hf - \phi$. You may also see frequency represented by the Greek letter *nu* (ν), leading to the expression $KE_{max} = h\nu - \phi$.

Teaching Tip

When light has a frequency equal to the threshold frequency of a surface ($f = f_t$), $KE_{max} = 0$. In other words, all of the photons' energy is used to dislodge electrons from the surface. As a result, the electrons have no kinetic energy.

Teaching Tip — ADVANCED

Some students may wish to know why the equation on this page gives the *maximum* kinetic energy of the ejected photoelectrons. Explain that the work required to eject an electron depends on how much force binds the electron to the metal. The work function is the *minimum* amount of energy required for an electron to escape from the surface of a metal. In other words, the work function is the amount of energy required for electrons that are the least strongly bound to the metal and that would therefore have the greatest amount of energy remaining as kinetic energy. Electrons that are bound more strongly will have less kinetic energy when ejected.

SECTION 1

PROBLEM GUIDE B

Use this guide to assign problems.
SE = Student Edition Textbook
PW = Problem Workbook
PB = Problem Bank on the One-Stop Planner (OSP)

Solving for:

KE_{max}	**PW** 6* **PB** Sample, 1–2
f_t	**SE** Sample, 1, 2*, 4*; Ch. Rvw. 13 **PW** 2*, 3*, 4*, 6* **PB** 3–4
hf_t	**SE** 2*, 3, 4*; Ch. Rvw. 14 **PW** 1, 5*, 7*, 9* **PB** 5–6
λ	**PW** Sample, 8* **PB** 7–8
v	**PW** 3*, 4* **PB** 9–10

***Challenging Problem**
Consult the printed Solutions Manual or the OSP for detailed solutions.

ANSWERS

Practice B

1. 4.83×10^{14} Hz
2. 2.3 eV; 5.6×10^{14} Hz
3. 2.36 eV
4. lithium, cesium

SAMPLE PROBLEM B

The Photoelectric Effect

PROBLEM

Light of frequency of 1.00×10^{15} Hz illuminates a sodium surface. The ejected photoelectrons are found to have a maximum kinetic energy of 1.78 eV. Find the threshold frequency for this metal.

SOLUTION

Given: $KE_{max} = (1.78 \text{ eV})(1.60 \times 10^{-19} \text{ J/eV})$

$KE_{max} = 2.85 \times 10^{-19}$ J $\qquad f = 1.00 \times 10^{15}$ Hz

Unknown: $f_t = ?$

Use the expression for maximum kinetic energy, and solve for f_t.

$$KE_{max} = hf - hf_t$$

$$f_t = \frac{hf - KE_{max}}{h}$$

$$f_t = \frac{(6.63 \times 10^{-34} \text{ J}\bullet\text{s})(1.00 \times 10^{15} \text{ Hz}) - (2.85 \times 10^{-19} \text{ J})}{6.63 \times 10^{-34} \text{ J}\bullet\text{s}}$$

$$\boxed{f_t = 5.70 \times 10^{14} \text{ Hz}}$$

PRACTICE B

The Photoelectric Effect

1. In the photoelectric effect, it is found that incident photons with energy 5.00 eV will produce electrons with a maximum kinetic energy 3.00 eV. What is the threshold frequency of this material?
2. Light of wavelength 350 nm falls on a potassium surface, and the photoelectrons have a maximum kinetic energy of 1.3 eV. What is the work function of potassium? What is the threshold frequency for potassium?
3. Calculate the work function of sodium using the information given in Sample Problem B.
4. Which of the following metals will exhibit the photoelectric effect when light of 7.0×10^{14} Hz frequency is shined on it?
   a. lithium, $hf_t = 2.3$ eV
   b. silver, $hf_t = 4.7$ eV
   c. cesium, $hf_t = 2.14$ eV

Photon theory accounts for observations of the photoelectric effect

The photon theory of light explains features of the photoelectric effect that cannot be understood using classical concepts. The photoelectric effect is not observed below a certain threshold frequency because the energy of the photon must be greater than or equal to the work function of the material. If the energy of each incoming photon is not equal to or greater than the work function, electrons will never be ejected from the surface, regardless of how many photons are present (how great the intensity is). Because the energy of each photon depends on the frequency of the incoming light ($E = hf$), the photoelectric effect is not observed when the incoming light is below a certain frequency (f_t).

Above the threshold frequency, if the light intensity is doubled, the number of photons is doubled. This in turn doubles the number of electrons ejected from the metal. However, the equation for the maximum kinetic energy of an electron shows that the kinetic energy depends only on the light frequency and the work function, not on the light intensity. Thus, even though there are more electrons ejected, the maximum kinetic energy of individual electrons remains the same.

Finally, the fact that the electrons are emitted almost instantaneously is consistent with the particle theory of light, in which energy appears in small packets. Because each photon affects a single electron, there is no significant time delay between shining light on the metal and observing electrons being ejected.

Einstein's success in explaining the photoelectric effect by assuming that electromagnetic waves are quantized led scientists to realize that the quantization of energy must be considered a real description of the physical world rather than a mathematical contrivance, as most had initially supposed. The discreteness of energy had not been considered a viable possibility because the energy quantum is not detected in our everyday experiences. However, scientists began to believe that the true nature of energy is seen in the submicroscopic level of atoms and molecules, where quantum effects become important and measurable.

Did you know?

Einstein published his paper on the photoelectric effect in 1905 while working in a patent office in Bern, Switzerland. In that same magical year, he published three other well-known papers, including the theory of special relativity.

Why it Matters

Conceptual Challenge

1. Photoelectric Effect

Even though bright red light delivers more total energy per second than dim violet light, the red light cannot eject electrons from a certain metallic surface, while the dimmer violet light can. How does Einstein's photon theory explain this observation?

2. Photographs

Suppose a photograph were made of a person's face using only a few photons. According to Einstein's photon theory, would the result be simply a very faint image of the entire face? Why or why not?

3. Glowing Objects

The color of a hot object depends on the object's temperature. As temperature increases, the color turns from red to orange to yellow to white and finally to blue. Classical physics cannot explain this color change, while quantum mechanics can. What explanation is given by quantum mechanics?

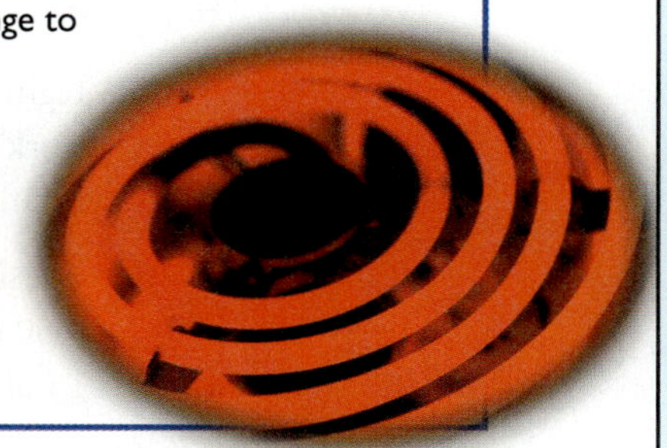

Teaching Tip

When the light intensity is very low, the average length of exposure before the first photoelectron is ejected can be substantial. But even at low intensity, in some experimental trials the time lag is essentially zero.

ANSWERS

Conceptual Challenge

1. Although there are more photons in the red light, each photon has a lower frequency than those that make up the violet light. Energy is proportional to frequency, so the photons of red light have less energy than the photons of violet light, and thus they cannot eject electrons from the surface.
2. No; The resulting photograph would show the image only at places where photons hit the film. As more photons are used, the dots become closer together, so the image eventually appears continuous.
3. Quantum mechanics predicts that as the energy increases, the frequency also increases, and the color of the glow changes accordingly.

SECTION 1

SECTION REVIEW ANSWERS

1. Classical physics predicts that radiation should increase as wavelength decreases, while experimental data shows that instead, radiation increases, reaches a peak, and then decreases. Planck resolved this conflict by assuming that energy is quantized. In classical physics, energy is considered to be continuous.
2. 2.8 eV
3. A high-intensity source delivers energy to the surface more rapidly than a low-intensity source. Thus electrons should receive the energy necessary to escape the surface sooner using the high-intensity source. However, electrons are ejected almost immediately for either source. A high-intensity light should also produce photoelectrons with greater kinetic energy. But observations show that the maximum kinetic energy of photoelectrons depends on frequency rather than intensity.
4. If energy is delivered to electrons in packets (photons) whose energy depends on the frequency, then the intensity has no effect on whether photoelectrons are emitted.
5. yes; 7.8 eV
6. There are more photons in 1 J of red light. A longer wavelength corresponds to a smaller frequency, and energy is proportional to frequency. Thus, there is less energy in each quantum of red light, and more quanta are needed to make up 1 J.

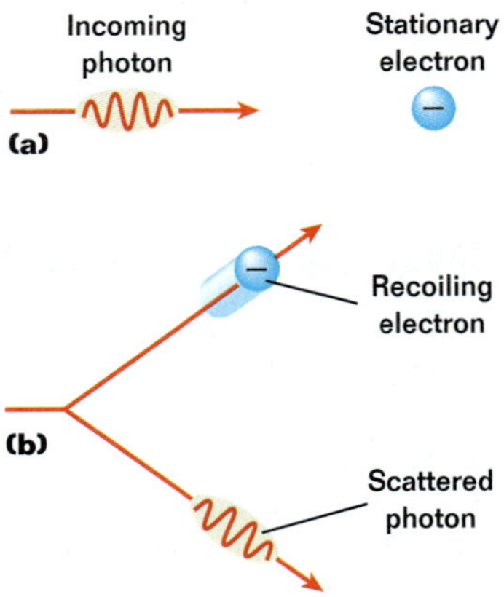

Figure 6
(a) When a photon collides with an electron, **(b)** the scattered photon has less energy and a longer wavelength than the incoming photon.

Compton shift

an increase in the wavelength of the photon scattered by an electron relative to the wavelength of the incident photon

Compton shift supports the photon theory of light

The American physicist Arthur Compton (1892–1962) realized that if light behaves like a particle, then a collision between an electron and a photon should be similar to a collision between two billiard balls. Photons should have momentum as well as energy; both quantities should be conserved in elastic collisions. So, when a photon collides with an electron initially at rest, as in **Figure 6,** the photon transfers some of its energy and momentum to the electron. As a result, the energy and frequency of the scattered photon are lowered; its wavelength should increase.

In 1923, to test this theory, Compton directed electromagnetic waves (X rays) toward a block of graphite. He found that the scattered waves had less energy and longer wavelengths than the incoming waves, just as he had predicted. This change in wavelength, known as the **Compton shift,** provides support for Einstein's photon theory of light.

The amount that the wavelength shifts depends on the angle through which the photon is scattered. Note that even the largest change in wavelength is very small in relation to the wavelengths of visible light. For this reason, the Compton shift is difficult to detect using visible light, but it can be observed using electromagnetic waves with much shorter wavelengths, such as X rays.

SECTION REVIEW

1. Describe the conflict known as the ultraviolet catastrophe. How did Planck resolve this conflict? How does Planck's assumption depart from classical physics?
2. What is the energy (in eV units) carried by one photon of violet light that has a wavelength of 4.5×10^{-7} m?
3. What effects did scientists originally think that the intensity of light shining on a photosensitive surface would have on electrons ejected from that surface? How did these predictions differ from observations?
4. How does Einstein's theory that electromagnetic waves are quantized explain the fact that the frequency of light (rather than the intensity) determines whether electrons are ejected from a photosensitive surface?
5. Light with a wavelength of 1.00×10^{-7} m shines on tungsten, which has a work function of 4.6 eV. Are electrons ejected from the tungsten? If so, what is their maximum kinetic energy?
6. **Critical Thinking** Is the number of photons in 1 J of red light (650 nm) greater than, equal to, or less than the number of photons in 1 J of blue light (450 nm)? Explain.

Why it Matters

Movie Theater Sound

Think about the last movie you watched in a theater. What would the movie have been like without sound? The soundtrack of a movie contributes a great deal to the mood of each scene. For instance, a scary scene often becomes much less frightening if the sound is muted.

The soundtrack of a movie is located along the side of the film in the form of an optical pattern of light and dark lines. A beam of light in the projector is directed through the soundtrack toward a *phototube,* as shown in the diagram below. The phototube contains a cathode and an anode. The cathode, which is made of a material that exhibits the photoelectric effect, is connected to the negative terminal of a power source. The anode is connected to the positive terminal. Thus, a potential difference exists between the cathode and the anode. When a beam of light shines on the cathode, electrons are ejected from it. Because of the potential difference, the photoelectrons are pulled toward the anode, producing a current.

The variation in shading on the soundtrack varies the light intensity falling on the plate of the phototube. When the light intensity increases, the number of photoelectrons also increases because there are more photons striking the surface. Likewise, when the intensity decreases, the number of photoelectrons decreases. Thus, the current is constantly changing as the intensity changes. This changing current electrically simulates the original sound wave. The sound wave is then reproduced by speakers for all moviegoers in the theater to enjoy.

Because the shading on an optical soundtrack varies through a range of values, this type of signal is an example of an *analog* signal. *Digital* signals, on the other hand, are discrete values of "on" and "off." Digital sound is clearer than analog sound—for instance, a compact disc sounds clearer than a cassette tape.

Some newer movies also contain a digital soundtrack. The first commercial movie to use digital sound was *Jurassic Park,* in 1993. Because a theater must have special equipment to play a digital soundtrack, the analog optical soundtrack is still included. The analog soundtrack is used in theaters that do not have digital capabilities, or in situations where there are problems with the digital sound.

The digital soundtrack is compressed onto compact discs. A time code is placed directly on the film, between the images and the optical soundtrack, as a series of dots and dashes. A device mounted on the projector reads the code in a manner similar to the one described above for optical soundtracks. The reader translates the code into pulses of current, which are sent to a computer. The computer controls the CD players that play the soundtrack, using the time code to ensure that the sound remains synchronized with the movie.

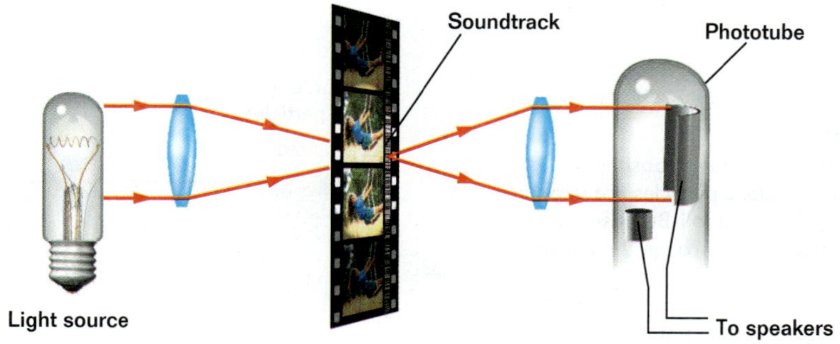

The shading on the soundtrack varies the light intensity reaching the phototube, which varies the current sent to the speaker. This changing current reproduces the original sound waves in the speaker.

SECTION 1

Why it Matters

Movie Theater Sound

Extension — ADVANCED

Encourage students who are interested in this feature to find out about the advances in speaker technologies in recent years. The first optical soundtracks contained just a single channel of sound. Eventually, a process called *matrixing* was developed, which allowed two optical lines to create four channels of sound (left, right, center, and rear). Many of today's digital formats offer six to eight distinct channels of sound.

SECTION 2
General Level

Demonstration

Particle Scattering GENERAL

Purpose Demonstrate the research method used in Rutherford's gold-foil experiment.

Materials two pounds of clay, a dozen marbles, a square piece of cardboard (about 1 m × 1 m), a few books

Procedure Build a small mound with the clay, and set it on the floor. Place the cardboard over the mound, and support it with books at the corners if necessary. Explain to the class that you would like them to determine what is beneath the cardboard without looking. Then roll the marbles beneath the cardboard one at a time. Have students observe where each marble emerges. (This should be done at all locations on your side of the cardboard.)

The students should conclude that most of the space beneath the cardboard is clear because most of the marbles continue under the cardboard without being deflected. Have students try to determine the location, size, and shape of the mound of clay from the deflection of the marbles.

The frustration that the students feel when forbidden to look under the cardboard is a good analogy for what drives scientists to explore the unknown. Point out that Rutherford never got to "look beneath the cardboard" to see if his theory was correct.

SECTION 2

Models of the Atom

SECTION OBJECTIVES

- **Explain the strengths and weaknesses of Rutherford's model of the atom.**
- **Recognize that each element has a unique emission and absorption spectrum.**
- **Explain atomic spectra using Bohr's model of the atom.**
- **Interpret energy-level diagrams.**

EARLY MODELS OF THE ATOM

The model of the atom in the days of Newton was that of a tiny, hard, indestructible sphere. This model was a good basis for the kinetic theory of gases. However, new models had to be devised when experiments revealed the electrical nature of atoms. The discovery of the electron in 1897 prompted J. J. Thomson (1856–1940) to suggest a new model of the atom. In Thomson's model, electrons are embedded in a spherical volume of positive charge like seeds in a watermelon, as shown in **Figure 7.**

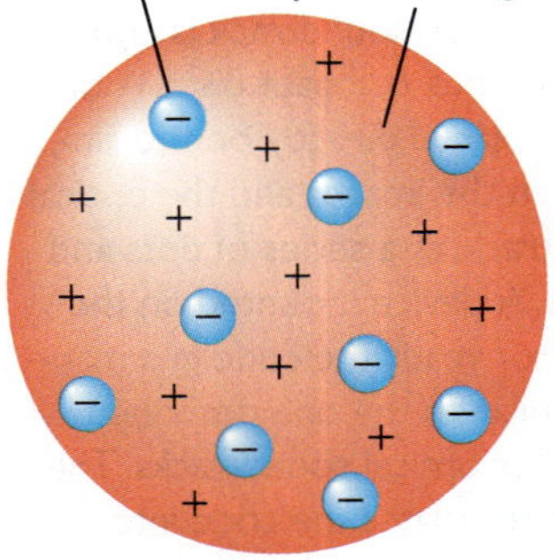

Figure 7
In Thomson's model of the atom, electrons are embedded inside a larger region of positive charge like seeds in a watermelon.

Rutherford proposed a planetary model of the atom

In 1911, Hans Geiger and Ernest Marsden, under the supervision of Ernest Rutherford (1871–1937), performed an important experiment showing that Thomson's model could not be correct. In this experiment, a beam of positively charged *alpha particles*—which consist of two protons and two neutrons—was projected against a thin metal foil, as shown in **Figure 8.** Most of the alpha particles passed through the foil as if it were empty space. Some of the alpha particles were deflected from their original direction through very large angles. Some particles were even deflected backward. Such deflections were completely unexpected on the basis of the Thomson model. Rutherford wrote, "It was quite the most incredible event that has ever happened to me in my life. It was almost as incredible as if you fired a 15-inch shell at a piece of tissue paper and it came back and hit you."

Such large deflections could not occur on the basis of Thomson's model, in which positive charge is evenly distributed throughout the atom, because the positively charged alpha particles would never come close to a positive charge concentrated enough to cause such large-angle deflections.

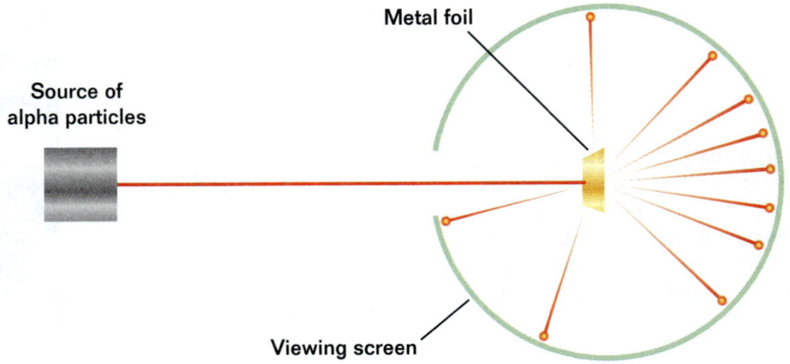

Figure 8
In this experiment, positively charged alpha particles are directed at a thin metal foil. Because many particles pass through the foil and only a few are deflected, Rutherford concluded that the atom's positive charge is concentrated at the center of the atom.

On the basis of his observations, Rutherford concluded that all of the positive charge in an atom and most of the atom's mass are found in a region that is small compared to the size of the atom. He called this concentration of positive charge and mass the *nucleus* of the atom. Any electrons in the atom were assumed to be in the relatively large volume outside the nucleus. So, according to Rutherford's theory, most alpha particles missed the nuclei of the metal atoms entirely and passed through the foil, while only a few came close enough to the nuclei to be deflected.

extension

Integrating Technology

Visit go.hrw.com for the activity "Seeing Atoms: The STM."

Keyword HF6ATMX

Rutherford's model predicts that atoms are unstable

To explain why electrons in this outer region of the atom were not pulled into the nucleus, Rutherford viewed the electrons as moving in orbits about the nucleus, much like the planets orbit the sun, as shown in **Figure 9.**

However, this assumption posed a serious difficulty. If electrons orbited the nucleus, they would undergo a centripetal acceleration. According to Maxwell's theory of electromagnetism, accelerated charges should radiate electromagnetic waves, losing energy. So, the radius of an atom's orbit would steadily decrease. This would lead to an ever-increasing frequency of emitted radiation and a rapid collapse of the atom as the electrons plunged into the nucleus. In fact, calculations show that according to this model, the atom would collapse in about one-billionth of a second. This difficulty with Rutherford's model led scientists to continue searching for a new model of the atom.

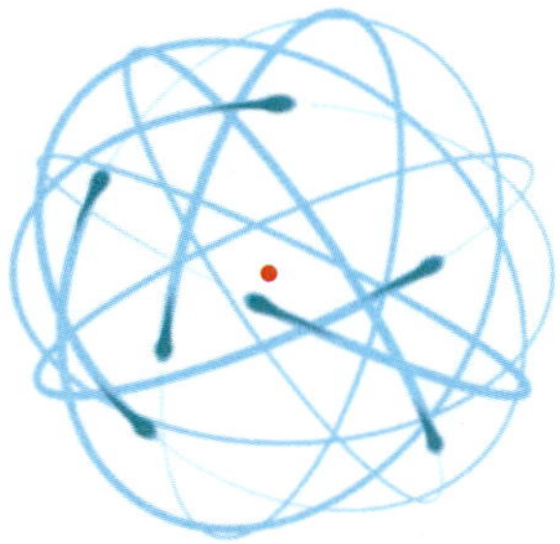

Figure 9

In Rutherford's model of the atom, electrons orbit the nucleus in a manner similar to planets orbiting the sun.

ATOMIC SPECTRA

In addition to solving the problems with Rutherford's planetary model, scientists hoped that a new model of the atom would explain another mysterious fact about gases. When an evacuated glass tube is filled with a pure atomic gas and a sufficiently high potential difference is applied between metal electrodes in the tube, a current is produced in the gas. As a result, the tube gives off light, as shown in **Figure 10.** The light's color is characteristic of the gas in the tube. This is how a neon sign works. The variety of colors seen in neon signs is the result of the light given off by different gases in the tubes.

SCILINKS.

www.scilinks.org

Topic: Early Atomic Theory

Code: HF60442

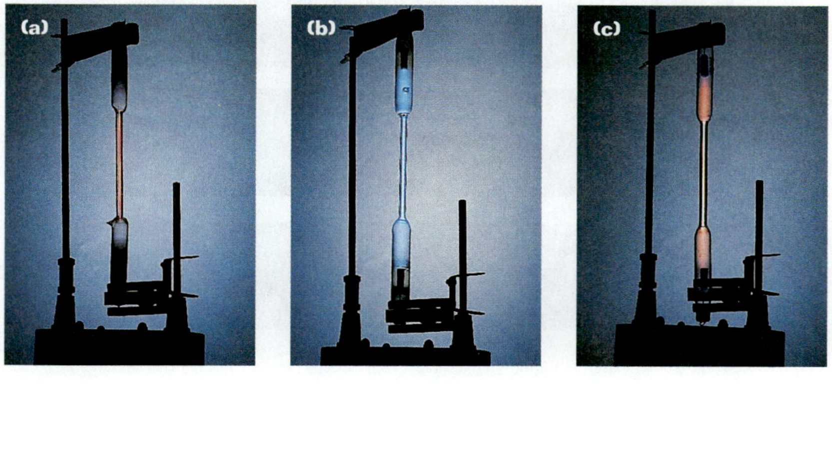

Figure 10

When a potential difference is applied across an atomic gas in a tube—here, hydrogen **(a),** mercury **(b),** and nitrogen **(c)**—the gas glows. The color of the glow depends on the type of gas.

SECTION 2

Key Models and Analogies

Point out that the positively charged nucleus attracts the negatively charged electrons, which end up in orbits, similar to those of satellites around Earth. But Maxwell's theory of electromagnetism predicts that charges moving on a curve (accelerating) should radiate energy.

Demonstration

Spectral Lines — GENERAL

Purpose Show line spectra of different gases.

Materials sodium and mercury vapor light sources, spectral tubes of different gases (argon, helium, nitrogen, hydrogen, etc.), power supply, diffraction gratings, dark room

Procedure Have students look at each light source through a diffraction grating and draw a diagrammatic representation of the spectra they observe. Explain that spectral lines of emission are considered to be the "fingerprints" of a gas. After they discuss why this is the case, explain that each gas emits radiation of certain characteristic wavelengths. Point out that they are observing only the spectral lines of radiation emitted in the visible spectrum. Other equipment may reveal radiation in the infrared and ultraviolet ranges or farther.

Teaching Tip

The light is first passed through slits so that the beams projected onto the screen have a well-defined, linear shape. Without any slits at all, images of the bulb itself would appear on the screen.

Did you know?

When the solar spectrum was first being studied, a set of spectral lines was found that did not correspond to any known element. A new element had been discovered. Because the Greek word for sun is *helios,* this new element was named helium. Helium was later found on Earth.

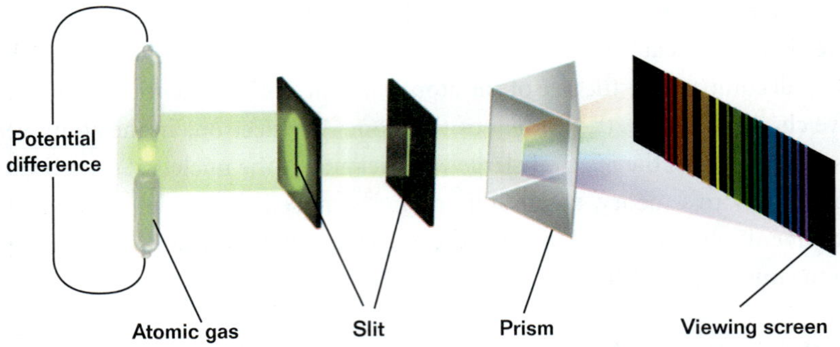

Figure 11
When the light from an atomic gas is passed through a prism or a diffraction grating, the dispersed light appears as a series of distinct, bright spectral lines.

Each gas has a unique emission and absorption spectrum

When the light given off (emitted) by an atomic gas is passed through a prism, as shown in **Figure 11,** a series of distinct bright lines is seen. Each line corresponds to a different wavelength, or color, of light. Such a series of spectral lines is commonly referred to as an **emission spectrum.**

emission spectrum
a diagram or graph that indicates the wavelengths of radiant energy that a substance emits

As shown in **Figure 12,** the emission spectra for hydrogen, mercury, and helium are each unique. Further analysis of other substances reveals that every element has a distinct emission spectrum. In other words, the wavelengths contained in a given spectrum are characteristic of the element giving off the light. Because no two elements give off the same line spectrum, it is possible to use spectroscopy to identify elements in a mixture.

In addition to giving off light at specific wavelengths, an element can also absorb light at specific wavelengths. The spectral lines corresponding to this process form what is known as an **absorption spectrum.** An absorption spectrum can be seen by passing light containing all wavelengths through a

absorption spectrum
a diagram or graph that indicates the wavelengths of radiant energy that a substance absorbs

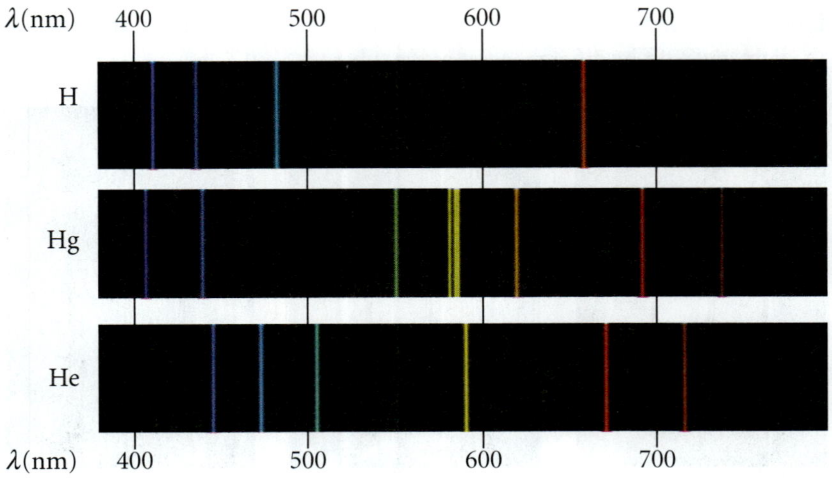

Figure 12
Each of these gases—hydrogen, mercury, and helium—has a unique emission spectrum.

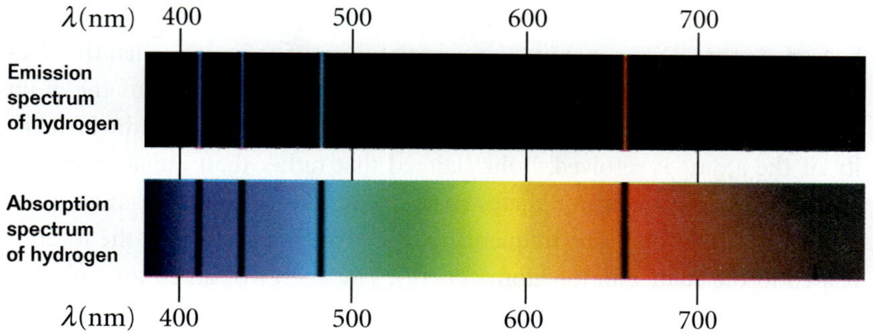

Figure 13
Hydrogen's dark absorption lines occur at the same wavelengths as its bright emission lines.

vapor of the element being analyzed. The absorption spectrum consists of a series of dark lines placed over the otherwise continuous spectrum.

Each line in the absorption spectrum of a given element coincides with a line in the emission spectrum of that element, as shown in **Figure 13** for hydrogen. In everyday experience, more emission lines are usually seen than absorption lines. The reason for this will be discussed shortly.

The absorption spectrum of an element has many practical applications. For example, the continuous spectrum of radiation emitted by the sun must pass through the cooler gases of the solar atmosphere and then through Earth's atmosphere. The various absorption lines seen in the solar spectrum have been used to identify elements in the solar atmosphere. Scientists are also able to examine the light from stars other than our sun in this fashion. With careful observation and analysis, astronomers have determined the proportions of various elements present in individual stars.

Historically, the occurrence of atomic spectra was of great importance to scientists attempting to find a new model of the atom. Long after atomic spectra had been discovered, their cause remained unexplained. There was nothing in Rutherford's planetary model to account for the fact that each element has a unique series of spectral lines. Scientists hoped that a new model of the atom would explain this phenomenon.

THE BOHR MODEL OF THE HYDROGEN ATOM

In 1913, the Danish physicist Niels Bohr (1885–1962) proposed a new model of the hydrogen atom that explained atomic spectra. Bohr's model of hydrogen contains some classical features and some revolutionary principles that could not be explained by classical physics.

Bohr's model is similar to Rutherford's in that the electron moves in circular orbits about the nucleus. The electric force between the positively charged proton inside the nucleus and the negatively charged electron is the force that holds the electron in orbit. However, in Bohr's model, only certain orbits are allowed. The electron is never found between these orbits; instead, it is said to "jump" instantly from one orbit to another without ever being between orbits.

Quick Lab

Atomic Spectra

MATERIALS LIST

- **a diffraction grating**
- **a variety of light sources, such as:**
  - ✓ **a fluorescent light**
  - ✓ **an incandescent light**
  - ✓ **a clear aquarium bulb**
  - ✓ **a sodium-vapor street light**
  - ✓ **a gym light**
  - ✓ **a neon sign**

SAFETY

Be careful of high potential differences that may be present near some of these light sources.

Certain types of light sources produce a continuous spectrum when viewed through a diffraction grating, while others produce discrete lines. Observe a variety of different light sources through a diffraction grating, and compare your results. Try to find at least one example of a continuous spectrum and a few examples of discrete lines.

SECTION 2

Visual Strategy – BASIC

Figure 13
Use this example to compare emission and absorption spectra.

Q How was the absorption spectrum produced?

A ***A light source with a continuous spectrum was passed through hydrogen.***

Q Why is most of the emission spectrum dark and most of the absorption spectrum bright?

A ***The emission spectrum is primarily dark because it consists of only those wavelengths that hydrogen emits. The absorption spectrum, on the other hand, is primarily bright because it contains light from a continuous source; only the wavelengths that hydrogen absorbs are dim or even black.***

Quick Lab

TEACHER'S NOTES

This activity is intended to have students observe and compare a variety of atomic and thermal sources of light. For this lab to be effective, it is best to separate light sources that are close together with dark partitions.

SECTION 2

Teaching Tip

Point out that in his model, Bohr made the assumption that the radii of permitted orbits for electrons may have only discrete values ($r_1, r_2, r_3, \ldots$), which correspond to the energies ($E_1, E_2, E_3, \ldots$) of stationary states. Thus, the difference in energy between two states is $E_2 - E_1 = hf$, where f is the frequency of the corresponding spectral line.

Misconception Alert

Students may think that electrons jump only between upper states and the lowest state. Make sure they understand that transitions between any two levels can occur and that each transition involves the emission or absorption of radiation of the frequency that corresponds to the difference in energy. Ask students which jump would emit radiation of higher frequency, E_6 to E_5 or E_6 to E_4 (*E_6 to E_4*). Explain that the spectral lines we observe are produced by millions of electrons, each of which may be found at any level and may jump to any level.

ADVANCED TOPICS

See "Semiconductor Doping" in **Appendix J: Advanced Topics** to learn how energy levels are used to explain electrical conduction in solids, including semiconductors.

Bohr's model further departs from classical physics by assuming that the hydrogen atom does not emit energy in the form of radiation when the electron is in any of these allowed orbits. Hence, the total energy of the atom remains constant, and one difficulty with the Rutherford model (the instability of the atom) is resolved. Bohr claimed that rather than radiating energy continuously, the electron radiates energy only when it jumps from an outer orbit to an inner one. The frequency of the radiation emitted in the jump is related to the change in the atom's energy. The energy of an emitted photon (E) is equal to the energy decrease of the atom ($-\Delta E_{atom}$). Because $\Delta E_{atom} = E_{final} - E_{initial}$, $E = -\Delta E_{atom} = E_{initial} - E_{final}$. Planck's equation can then be used to find the frequency of the emitted radiation: $E = E_{initial} - E_{final} = hf$.

In Bohr's model, transitions between stable orbits with different energy levels account for the discrete spectral lines

The lowest energy state in the Bohr model, which corresponds to the smallest possible radius, is often called the *ground state* of the atom, and the radius of this orbit is called the *Bohr radius.* At ordinary temperatures, most electrons are in the ground state, with the electron relatively close to the nucleus. When light of a continuous spectrum shines on the atom, only the photons whose energy (hf) matches the energy separation between two levels can be absorbed by the atom. When this occurs, an electron jumps from a lower energy state to a higher energy state, which corresponds to an orbit farther from the nucleus, as shown in **Figure 14(a).** This is called an *excited state.* The absorbed photons account for the dark lines in the absorption spectrum.

Once an electron is in an excited state, there is a certain probability that it will jump back to a lower energy level by emitting a photon, as shown in **Figure 14(b).** This process is known as *spontaneous emission.* The emitted photons are responsible for the bright lines in the emission spectrum.

In both cases, there is a correlation between the "size" of an electron's jump and the energy of the photon. For example, an electron in the fourth energy level could jump to the third level, the second level, or the ground state. Because Planck's equation gives the energy from one level to the next level, a greater jump means that more energy is emitted. Thus, jumps between different levels correspond to the various spectral lines that are observed. The

SCILINKS®
www.scilinks.org
Topic: Modern Atomic Theory
Code: HF60978

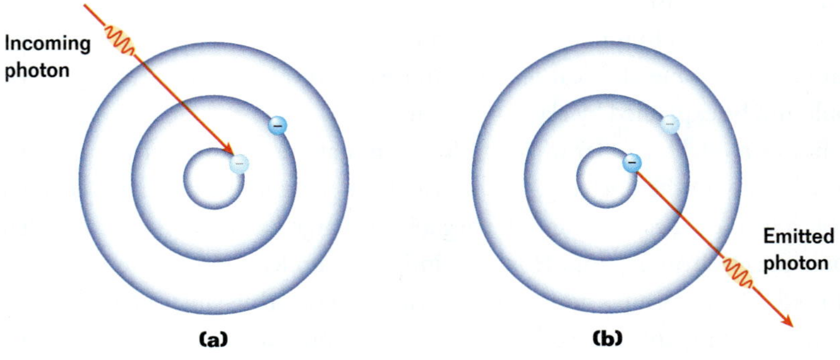

Figure 14
(a) When a photon is absorbed by an atom, an electron jumps to a higher energy level. **(b)** When the electron falls back to a lower energy level, the atom releases a photon.

jumps that correspond to the four spectral lines in the visible spectrum of hydrogen are shown in **Figure 15.** Bohr's calculations successfully account for the wavelengths of all the spectral lines of hydrogen.

As noted earlier, fewer absorption lines than emission lines are typically observed. The reason is that absorption spectra are usually observed when a gas is at room temperature. Thus, most electrons are in the ground state, so all transitions observed are from a single level (E_1) to higher levels. Emission spectra, on the other hand, are seen by raising a gas to a high temperature and viewing downward transitions between any two levels. In this case, all transitions are possible, so more spectral lines are observed.

Bohr's idea of the quantum jump between energy levels provides an explanation for the aurora borealis, or northern lights. Charged particles from the sun sometimes become trapped in Earth's magnetic field and are deposited around the northern and southern magnetic poles. (Light shows in southern latitudes are called *aurora australis,* or *southern lights.*) When deposited, these charged particles from the sun collide with the electrons of the atoms in our atmosphere and transfer energy to these electrons, causing them to jump to higher energy levels. When an electron returns to a lower orbit, some of the energy is released as a photon. The northern lights are the result of billions of these quantum jumps happening at the same time.

The colors of the northern lights are determined by the type of gases in the atmosphere. The charged particles from the sun are most commonly released from Earth's magnetic field into a part of the atmosphere that contains oxygen, which releases green light. Red lights are the result of collisions with nitrogen atoms. Because each type of gas releases a unique color, the northern lights contain only a few distinct colors rather than a continuous spectrum.

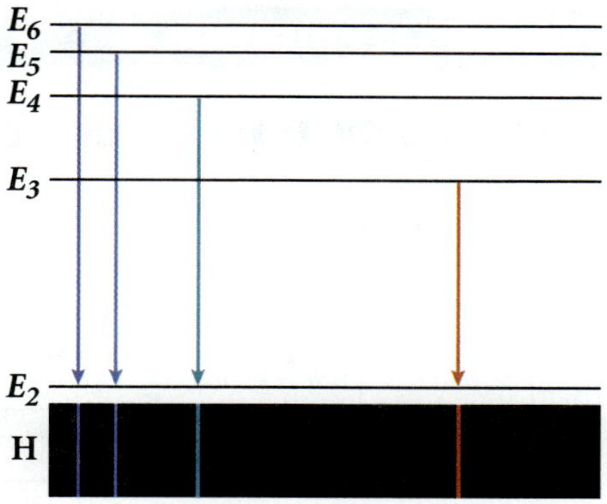

Figure 15
Every jump from one energy level to another corresponds to a specific spectral line. This example shows the transitions that result in the visible spectral lines of hydrogen. The lowest energy level, E_1, is not shown in this diagram.

Why it Matters

Conceptual Challenge

1. Neon Signs When a potential difference is placed across electrodes at the ends of a tube that contains neon, such as a neon sign, the neon glows. Is the light emitted by a neon sign composed of a continuous spectrum or only a few lines? Defend your answer.

2. Energy Levels If a certain atom has four possible energy levels and an electron can jump between any two energy levels of the atom, how many different spectral lines could be emitted?

3. Identifying Gases Neon is not the only type of gas used in neon signs. As you have seen, a variety of gases exhibit similar effects when there is a potential difference across them. While the colors observed are sometimes different, certain gases do glow with the same color. How could you distinguish two such gases?

SECTION 2

Visual Strategy GENERAL

Figure 15
Explain that all of the arrows showing transitions between energy levels are represented relative to the E_2 level. Also point out that this is a representation of the emission spectrum of hydrogen in the *visible* range only. (This part of the hydrogen spectrum—the visible lines—is known as the Balmer series.) Jumps from all levels to E_1 emit ultraviolet radiation, which is not visible.

ANSWERS

Conceptual Challenge

1. The light is composed of only a few lines because the spectrum contains only the lines corresponding to the electron transitions for a neon atom.
2. 6 ($E_4 - E_1$, $E_3 - E_1$, $E_2 - E_1$, $E_4 - E_2$, $E_3 - E_2$, $E_4 - E_3$)
3. Although two gases may glow with the same color, each atomic gas has a unique spectrum that distinguishes it from any other element.

PROBLEM GUIDE C

Use this guide to assign problems.
SE = Student Edition Textbook
PW = Problem Workbook
PB = Problem Bank on the One-Stop Planner (OSP)

SE 1–3, 4*, 5*; Ch. Rvw. 23, 24
PW Sample, 1–3
PB Sample, 1–3

***Challenging Problem**
Consult the printed Solutions Manual or the OSP for detailed solutions.

SAMPLE PROBLEM C

STRATEGY Interpreting Energy-Level Diagrams

PROBLEM

An electron in a hydrogen atom drops from energy level E_4 to energy level E_2. What is the frequency of the emitted photon, and which line in the emission spectrum corresponds to this event?

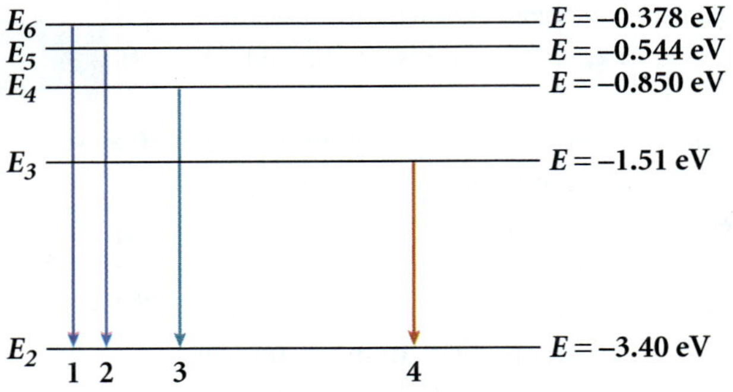

SOLUTION

1. Find the energy of the photon.

The energy of the photon is equal to the change in the energy of the electron. The electron's initial energy level was E_4, and the electron's final energy level was E_2. Using the values from the energy-level diagram gives the following:

$$E = E_{initial} - E_{final}$$

$$E = (-0.850 \text{ eV}) - (-3.40 \text{ eV}) = 2.55 \text{ eV}$$

TIP *Note that the energies for each energy level are negative. The reason is that the energy of an electron in an atom is defined with respect to the amount of work required to remove the electron from the atom. In some energy-level diagrams, the energy of E_1 is defined as zero, and the higher energy levels are positive. In either case, the difference between a higher energy level and a lower one is always positive, indicating that the electron loses energy when it drops to a lower level.*

2. Use Planck's equation to find the frequency.

$$E = hf$$

$$f = \frac{E}{h}$$

$$f = \frac{(2.55 \text{ eV})(1.60 \times 10^{-19} \text{ J/eV})}{6.63 \times 10^{-34} \text{ J}\bullet\text{s}}$$

$$\boxed{f = 6.15 \times 10^{14} \text{ Hz}}$$

TIP *Note that electron volts were converted to joules so that the units cancel properly.*

3. **Find the corresponding line in the emission spectrum.**
Examination of the diagram shows that the electron's jump from energy level E_4 to energy level E_2 corresponds to *Line 3* in the emission spectrum.

4. **Evaluate your answer.**
Line 3 is in the visible part of the electromagnetic spectrum and appears to be blue. The frequency $f = 6.15 \times 10^{14}$ Hz lies within the range of the visible spectrum and is toward the violet end, so it is reasonable that light of this frequency would be visible blue light.

PRACTICE C

Interpreting Energy-Level Diagrams

1. An electron in a hydrogen atom drops from energy level E_3 to E_2. What is the frequency of the emitted photon, and which line in the emission spectrum shown in Sample Problem C corresponds to this event?

2. An electron in a hydrogen atom drops from energy level E_6 to energy level E_3. What is the frequency of the emitted photon, and in which range of the electromagnetic spectrum is this photon? (See **Table 1** in the chapter "Light and Reflection" for ranges in the electromagnetic spectrum.)

3. The energy-level diagram in **Figure 16** shows the first five energy levels for mercury vapor. The energy of E_1 is defined as zero. What is the frequency of the photon emitted when an electron drops from energy level E_5 to E_1 in a mercury atom?

Figure 16

4. How many different spectral lines *could* be emitted if mercury vapor were excited by photons with 6.67 eV of energy? (Hint: An electron could move, for example, from energy level E_5 to E_3, then from E_3 to E_2, and then from E_2 to E_1.)

5. The emission spectrum of hydrogen has one emission line at a frequency of 7.29×10^{14} Hz. Calculate which two energy levels electrons must jump between to produce this line, and identify the line in the energy-level diagram in Sample Problem C. (Hint: First, find the energy of the photons, and then use the energy-level diagram.)

ANSWERS

Practice C

1. 4.56×10^{14} Hz; *Line 4*
2. 2.73×10^{14} Hz; infrared
3. 1.61×10^{15} Hz
4. 10
5. E_6 to E_2; *Line 1*

SECTION 2

SECTION REVIEW ANSWERS

1. If positive charges were spread throughout the atom, as in Thomson's model, the alpha particles would be deflected slightly at most.
2. because most particles were not deflected at all, while a few were sharply deflected
3. It predicted an unstable atom, and it couldn't explain spectral lines.
4. Because spectral lines are unique to each element, an analysis of the spectra would reveal what elements are present.
5. classical physics: electron orbits the nucleus, electron radiates energy; quantum mechanics: only certain orbits are allowed, the electron radiates energy only when it jumps between levels, energy radiated depends on the difference between energy levels
6. Each line corresponds to a jump from one energy level to another.
7. Electrons can jump only from one allowed orbit to another; only lines that correspond to these jumps are found in the spectral lines.

Bohr's model is incomplete

The Bohr model of hydrogen was a tremendous success in some respects because it explained several features of the spectra of hydrogen that had previously defied explanation. Bohr's model gave an expression for the radius of the atom and predicted the energy levels of hydrogen. This model was also successful when applied to hydrogen-like atoms, that is, atoms that contain only one electron. But while many attempts were made to extend the Bohr model to multi-electron atoms, the results were unsuccessful.

Bohr's model of the atom also raised new questions. For example, Bohr assumed that electrons do not radiate energy when they are in a stable orbit, but his model offered no explanation for this. Another problem with Bohr's model was that it could not explain why electrons always have certain stable orbits, while other orbits do not occur. Finally, the model followed classical physics in certain respects but radically departed from classical physics in other respects. For all of these reasons, Bohr's model was not considered to be a complete picture of the structure of the atom, and scientists continued to search for a new model that would resolve these difficulties.

SECTION REVIEW

1. Based on the Thomson model of the atom, what did Rutherford expect to happen when he projected positively charged alpha particles against a metal foil?
2. Why did Rutherford conclude that an atom's positive charge and most of its mass are concentrated in the center of the atom?
3. What are two problems with Rutherford's model of the atom?
4. How could the atomic spectra of gases be used to identify the elements present in distant stars?
5. Bohr's model of the atom follows classical physics in some respects and quantum mechanics in others. Which assumptions of the Bohr model correspond to classical physics? Which correspond to quantum mechanics?
6. How does Bohr's model of the atom account for the emission and absorption spectra of an element?
7. **Critical Thinking** A Norwegian scientist, Lars Vegard, determined the different wavelengths that are part of the northern lights. He found that only a few wavelengths of light, rather than a continuous spectrum, are present in the lights. How does Bohr's model of the atom account for this observation?

Quantum Mechanics

SECTION 3

SECTION 3
Advanced Level

THE DUAL NATURE OF LIGHT

As discussed in Section 1, there is considerable evidence for the photon theory of light. In this theory, all electromagnetic waves consist of photons, particle-like pulses that have energy and momentum. On the other hand, light and other electromagnetic waves exhibit interference and diffraction effects that are considered to be wave behaviors. So, which model is correct? We will see that each is correct and that a specific phenomenon often exhibits only one or the other of these natures of light.

Light is both a wave and a particle

Some experiments can be better explained or only explained by the photon concept, whereas others require a wave model. Most physicists accept both models and believe that the true nature of light is not describable in terms of a single classical picture.

For an example of how photons can be compatible with electromagnetic waves, consider radio waves at a frequency of 2.5 MHz. The energy of a photon having this frequency can be found using Planck's equation, as follows:

$$E = hf = (6.63 \times 10^{-34}\ \text{J}\bullet\text{s})(2.5 \times 10^{6}\ \text{Hz}) = 1.7 \times 10^{-27}\ \text{J}$$

From a practical viewpoint, this energy is too small to be detected as a single photon. A sensitive radio receiver might need as many as 10^{10} of these photons to produce a detectable signal. With such a large number of photons reaching the detector every second, we would not be able to detect the individual photons striking the antenna. Thus, the signal would appear as a continuous wave.

Now consider what happens as we go to higher frequencies and hence shorter wavelengths. In the visible region, it is possible to observe both the photon and the wave characteristics of light. As we mentioned earlier, a light beam can show interference phenomena and produce photoelectrons. The interference phenomena are best explained by the wave model of light, while the photoelectrons are best explained by the particle theory of light.

At even higher frequencies and correspondingly shorter wavelengths, the momentum and energy of the photons increase. Consequently, the photon nature of light becomes very evident. In addition, as the wavelength decreases, wave effects, such as interference and diffraction, become more difficult to observe. Very indirect methods are required to detect the wave nature of very high frequency radiation, such as gamma rays.

SECTION OBJECTIVES

- **Recognize the dual nature of light and matter.**
- **Calculate the de Broglie wavelength of matter waves.**
- **Distinguish between classical ideas of measurement and Heisenberg's uncertainty principle.**
- **Describe the quantum-mechanical picture of the atom, including the electron cloud and probability waves.**

The Language of Physics

Modern physics has introduced ideas that may conflict with intuition and common sense. The dual nature of light is one example. Although the words *particle* and *wave* are familiar, students need to think about them in a new way when dealing with atomic and subatomic phenomena.

Misconception Alert

The concept of matter waves is probably new and unusual to most students. Because students have no experience with matter waves, they may not understand the nature of such waves. The meaning of de Broglie's hypothesis is that we may picture the electron in two ways: as a particle in motion or as a wave occupying a certain region in space. Encourage students to discuss their mental images of electron waves to prepare them for the probability interpretation of matter waves. Also assure them that these ideas remain puzzling to many people.

Teaching Tip — ADVANCED

Some students might ask why the wavelength of a photon is equal to Planck's constant divided by momentum ($\lambda = h/p$). If students have read the feature "The Equivalence of Mass and Energy" in **Appendix J: Advanced Topics,** you can present the following derivation. First, point out that relativistic equations must be used because photons travel at the speed of light. Tell students to recall that $E_R = mc^2$. After rearranging to solve for mass, we can substitute the equivalent of $m = E_R/c^2$ into the momentum equation. Thus, $p = mv = (E_R/c^2)v$. Because $v = c$, this equation reduces to $p = E_R/c$. Finally, because $E_R = hf$ (Planck's equation), $p = hf/c = h/\lambda$.

Thus, all forms of electromagnetic radiation can be described from two points of view. At one extreme, the electromagnetic wave description suits the overall interference pattern formed by a large number of photons. At the other extreme, the particle description is more suitable for dealing with highly energetic photons of very short wavelengths.

MATTER WAVES

In the world around us, we are accustomed to regarding things such as thrown baseballs solely as particles, and things such as sound waves solely as forms of wave motion. As already noted, this rigid distinction cannot be made with light, which has both wave and particle characteristics. In 1924, the French physicist Louis de Broglie (1892–1987) extended the wave-particle duality. In his doctoral dissertation, de Broglie proposed that all forms of matter may have both wave properties and particle properties. At that time, this was a highly revolutionary idea with no experimental support. Now, however, scientists accept the concept of matter's dual nature.

Did you know?

Louis de Broglie's doctoral thesis about the wave nature of matter was so radical and speculative that his professors were uncertain about whether they should accept it. They resolved the issue by asking Einstein to read the paper. Einstein gave his approval, and de Broglie's paper was accepted. Five years after his thesis was accepted, de Broglie won the Nobel Prize for his theory.

The wavelength of a photon is equal to Planck's constant (h) divided by the photon's momentum (p). De Broglie speculated that this relationship might also hold for matter waves, as follows:

WAVELENGTH OF MATTER WAVES

$$\lambda = \frac{h}{p} = \frac{h}{mv}$$

$$\textbf{de Broglie wavelength} = \frac{\textbf{Planck's constant}}{\textbf{momentum}}$$

As seen by this equation, the larger the momentum of an object, the smaller its wavelength. In an analogy with photons, de Broglie postulated that the frequency of a matter wave can be found with Planck's equation as illustrated below:

FREQUENCY OF MATTER WAVES

$$f = \frac{E}{h}$$

$$\textbf{de Broglie frequency} = \frac{\textbf{energy}}{\textbf{Planck's constant}}$$

ADVANCED TOPICS

See "De Broglie Waves" in **Appendix J: Advanced Topics** to learn more about matter waves.

The dual nature of matter suggested by de Broglie is quite apparent in these two equations, both of which contain particle concepts (E and mv) and wave concepts (λ and f).

(a) (b)

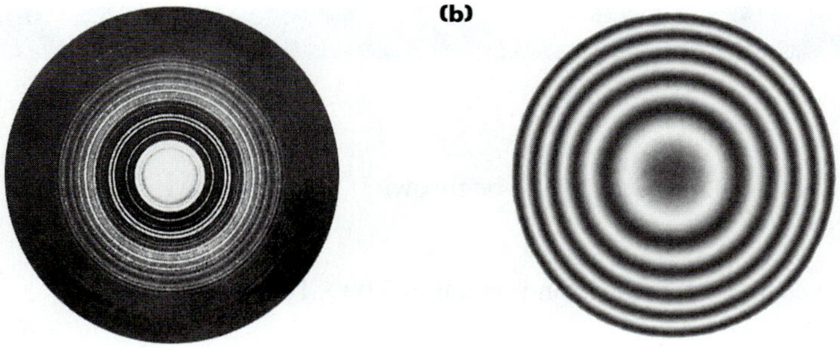

Figure 17
(a) Electrons show interference patterns similar to those of **(b)** light waves. This demonstrates that electrons sometimes behave like waves.

At first, de Broglie's proposal that all particles also exhibit wave properties was regarded as pure speculation. If particles such as electrons had wave properties, then under certain conditions they should exhibit interference phenomena. Three years after de Broglie's proposal, C. J. Davisson and L. Germer, of the United States, discovered that electrons can be diffracted by a single crystal of nickel. This important discovery provided the first experimental confirmation of de Broglie's theory. An example of electron diffraction compared with light diffraction is shown in **Figure 17.**

Electron diffraction by a crystal is possible because the de Broglie wavelength of a low-energy electron is approximately equal to the distance between atoms in a crystal. In principle, diffraction effects should be observable even for objects in our large-scale world. However, the wavelengths of material objects in our everyday world are much smaller than any possible aperture through which the object could pass.

SAMPLE PROBLEM D

De Broglie Waves

PROBLEM

With what speed would an electron with a mass of 9.109×10^{-31} kg have to move if it had a de Broglie wavelength of 7.28×10^{-11} m?

SOLUTION

Given: $m = 9.109 \times 10^{-31}$ kg $\quad \lambda = 7.28 \times 10^{-11}$ m
$h = 6.63 \times 10^{-34}$ J•s

Unknown: $v = ?$

Use the equation for the de Broglie wavelength, and isolate v.

$$\lambda = \frac{h}{mv} \quad \text{or} \quad v = \frac{h}{\lambda m}$$

$$v = \frac{6.63 \times 10^{-34}\ \text{J•s}}{(7.28 \times 10^{-11}\ \text{m})(9.109 \times 10^{-31}\ \text{kg})} = 1.00 \times 10^{7}\ \text{m/s}$$

$$\boxed{v = 1.00 \times 10^{7}\ \text{m/s}}$$

SECTION 3

Classroom Practice

De Broglie Waves
Calculate the de Broglie wavelength of an electron moving at the following speeds:

a. 8.0×10^{4} m/s
b. 8.0×10^{5} m/s
c. 8.0×10^{6} m/s

Answers
a. 9.1×10^{-9} m
b. 9.1×10^{-10} m
c. 9.1×10^{-11} m

PROBLEM GUIDE D

Use this guide to assign problems.
SE = Student Edition Textbook
PW = Problem Workbook
PB = Problem Bank on the One-Stop Planner (OSP)

Solving for:

m	**SE** 5 **PW** 2 **PB** Sample, 1–4
λ	**SE** 4; Ch. Rvw. 34 **PW** Sample, 3, 5, 7* **PB** 5–8
v	**SE** Sample, 1–3; Ch. Rvw. 33 **PW** 1, 4, 6* **PB** 9–10

***Challenging Problem**
Consult the printed Solutions Manual or the OSP for detailed solutions.

SECTION 3

ANSWERS

Practice D

1. 39.9 m/s
2. 1.46×10^3 m/s
3. 8.84×10^{-27} m/s
4. 4.0×10^{-38} m
5. 1.0×10^{-15} kg

Visual Strategy GENERAL

Figure 18

Point out that the electron has a different wavelength in each orbit. Also, just as for a satellite around Earth, a smaller radius involves a faster speed.

Q The radius of the hydrogen atom at the ground state (lowest energy level) is $r_1 = 5.3 \times 10^{-11}$ m. What is the electron's wavelength in the first orbit?

A ***The electron's wavelength equals the circumference of the circle of radius r_1, that is, $\lambda = 2\pi r_1 = 3.3 \times 10^{-10}$ m.***

PRACTICE D

De Broglie Waves

1. With what speed would a 50.0 g rock have to be thrown if it were to have a wavelength of 3.32×10^{-34} m?
2. If the de Broglie wavelength of an electron is equal to 5.00×10^{-7} m, how fast is the electron moving?
3. How fast would one have to throw a 0.15 kg baseball if it were to have a wavelength equal to 5.00×10^{-7} m (the same wavelength as the electron in problem 2)?
4. What is the de Broglie wavelength of a 1375 kg car traveling at 43 km/h?
5. A bacterium moving across a Petri dish at 3.5 μm/s has a de Broglie wavelength of 1.9×10^{-13} m. What is the bacterium's mass?

De Broglie waves account for the allowed orbits of Bohr's model

At first, no one could explain why only some orbits were stable. Then, de Broglie saw a connection between his theory of the wave character of matter and the stable orbits in the Bohr model. De Broglie assumed that an electron orbit would be stable only if it contained an integral (whole) number of electron wavelengths, as shown in **Figure 18.** The first orbit contains one wavelength, the second orbit contains two wavelengths, and so on.

De Broglie's hypothesis compares with the example of standing waves on a vibrating string of a given length, as discussed in the chapter "Vibrations and Waves." In this analogy, the circumference of the electron's orbit corresponds to the string's length. So, the condition for an electron orbit is that the circumference must contain an integral multiple of electron wavelengths.

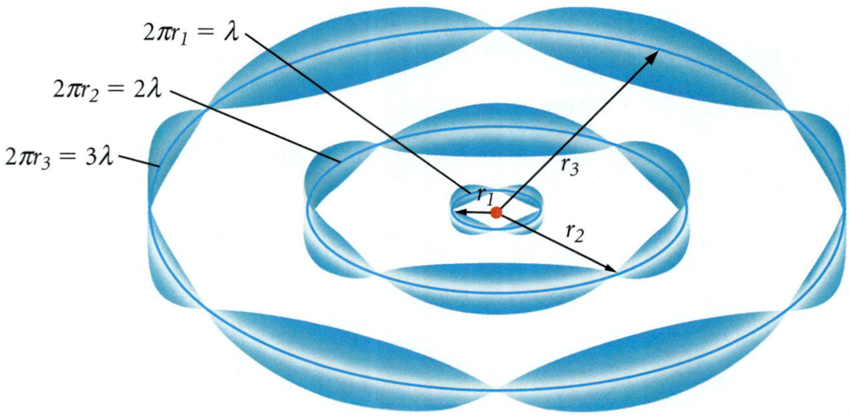

Figure 18
De Broglie's hypothesis that there is always an integral number of electron wavelengths around each circumference explains why only certain orbits are stable.

THE UNCERTAINTY PRINCIPLE

In classical mechanics, there is no limitation to the accuracy of our measurements in experiments. In principle, we could always make a more precise measurement using a more finely detailed meterstick or a stronger magnifier. This unlimited precision does not hold true in quantum mechanics. The absence of such precision is not due to the limitations of our instruments or to our perturbation of the system when we make measurements. It is a fundamental limitation inherent in nature due to the wave nature of particles.

Simultaneous measurements of position and momentum cannot be completely certain

In 1927, Werner Heisenberg argued that *it is fundamentally impossible to make simultaneous measurements of a particle's position and momentum with infinite accuracy.* In fact, the more we learn about a particle's momentum, the less we know of its position, and the reverse is also true. This principle is known as Heisenberg's **uncertainty principle.**

To understand the uncertainty principle, consider the following thought experiment. Suppose you wish to measure the position and momentum of an electron as accurately as possible. You might be able to do this by viewing the electron with a powerful microscope. In order for you to see the electron and thus determine its location, at least one photon of light must bounce off the electron and pass through the microscope into your eye. This incident photon is shown moving toward the electron in **Figure 19(a).** When the photon strikes the electron as in **Figure 19(b),** it transfers some of its energy and momentum to the electron. So, in the process of attempting to locate the electron very accurately, we become less certain of its momentum. The measurement procedure limits the accuracy to which we can determine position and momentum simultaneously.

uncertainty principle

the principle that states that it is impossible to simultaneously determine a particle's position and momentum with infinite accuracy

A thought experiment for viewing an electron with a powerful microscope

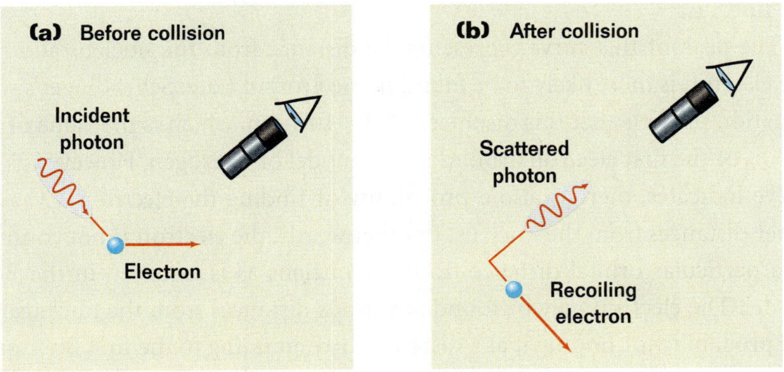

Figure 19
(a) The electron is viewed before colliding with the photon. **(b)** The electron recoils (is disturbed) as the result of the collision with the photon.

Teaching Tip — ADVANCED

Position and momentum are not the only uncertain quantities at the atomic level. Energy and time are also uncertain in a similar way. The longer we take to measure the energy of a particle, the more certain we can be about its energy. Conversely, the shorter the time period, the less certain we can be of the particle's energy. The mathematical form of this version of the uncertainty principle is stated as follows: $\Delta E \Delta t \geq \frac{h}{4\pi}$. In this form of the equation, ΔE is the uncertainty in a measurement of the energy and Δt is the time it takes to make the measurement.

Demonstration

Probability Distribution — GENERAL

Purpose Illustrate the meaning of the wave function as a probability function.

Materials 100 index cards, jar, overhead projector, transparency

Procedure Write a number on each of the index cards as follows: the number *1* on 4 cards, the number *2* on 6 cards, the number *3* on 12 cards, the number *4* on 40 cards, the number *5* on 14 cards, the number *6* on 10 cards, the number *7* on 8 cards, the number *8* on 4 cards, and the number *9* on 2 cards. Place all of the cards in a jar. Draw nine concentric circles on a transparency, and label the circles 1 through 9.

Have each student pick four or five cards from the jar. Point out that each of the numbered index cards corresponds to a numbered circle on the transparency. Tell students that for every index card they pick, they should draw a dot on the corresponding circle on the transparency. A cluster of dots will form around circle 4. Tell students that this is analogous to the way electrons move in the atom. An electron could be anywhere a dot is seen, but in this example it spends most of its time on orbit 4.

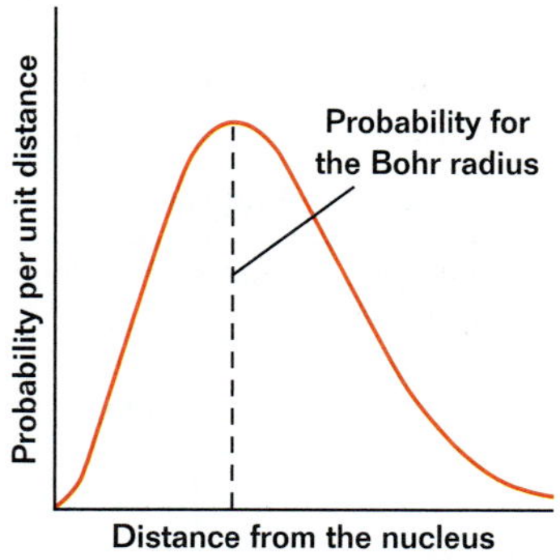

Figure 20
The height of this curve is proportional to the probability of finding the electron at different distances from the nucleus in the ground state of hydrogen.

The mathematical form of the uncertainty principle states that the product of the uncertainties in position and momentum will always be larger than some minimum value. Arguments similar to those given here show that this minimum value is Planck's constant (h) divided by 4π. Thus, $\Delta x \Delta p \geq \frac{h}{4\pi}$. In this equation, Δx and Δp represent the uncertainty in the measured values of a particle's position and momentum, respectively, at some instant. This equation shows that if Δx is made very small, Δp will be large, and vice versa.

THE ELECTRON CLOUD

In 1926, Erwin Schrödinger proposed a wave equation that described the manner in which de Broglie's matter waves change in space and time. Although this equation and its derivation are beyond the scope of this book, we will consider Schrödinger's equation qualitatively. Solving Schrödinger's equation yields a quantity called the *wave function*, represented by ψ (Greek letter *psi*). A particle is represented by a wave function, ψ, that depends on the position of the particle and time.

An electron's location is described by a probability distribution

As discussed earlier, simultaneous measurements of position and momentum cannot be completely certain. Because the electron's location cannot be precisely determined, it is useful to discuss the *probability* of finding the electron at different locations. It turns out that the quantity $|\psi|^2$ is proportional to the probability of finding the electron at a given position. This interpretation of Schrödinger's wave function was first proposed by the German physicist Max Born in 1926.

Figure 20 shows the probability per unit distance of finding the electron at various distances from the nucleus in the ground state of hydrogen. The height of the curve at each point is proportional to the probability of finding the electron, and the x coordinate represents the electron's distance from the nucleus. Note that there is a near-zero probability of finding the electron in the nucleus.

The peak of this curve represents the distance from the nucleus at which the electron is most likely to be found in the ground state. Schrödinger's wave equation predicts that this distance is 5.3×10^{-11} m, which is the value of the radius of the first electron orbit in Bohr's model of hydrogen. However, as the curve indicates, there is also a probability of finding the electron at various other distances from the nucleus. In other words, the electron is not confined to a particular orbital distance from the nucleus as is assumed in the Bohr model. The electron may be found at various distances from the nucleus, but the probability of finding it at a distance corresponding to the first Bohr orbit is greater than that of finding it at any other distance. This new model of the atom is consistent with Heisenberg's uncertainty principle, which states that we cannot know the electron's location with complete certainty.

Did you know?

Although Einstein was one of the founders of quantum theory, he did not believe that it could be a final description of nature. His convictions in this matter led to his famous statement, "In any case, I am convinced that He [God] does not play dice."

Quantum mechanics also predicts that the wave function for the hydrogen atom in the ground state is spherically symmetrical; hence, the electron can be found in a spherical region surrounding the nucleus. This is in contrast to the Bohr theory, which confines the position of the electron to points in a plane. This result is often interpreted by viewing the electron as a cloud surrounding the nucleus, called an *electron cloud.* The density of the cloud at each location is related to the probability of finding the electron at that location. The most probable distance for the electron's location in the ground state is equal to the first Bohr radius.

Analysis of each of the energy levels of hydrogen reveals that the most probable electron location in each case is in agreement with each of the radii predicted by the Bohr theory. The discrete energy levels that could not be explained by Bohr's theory can be derived from Schrödinger's wave equation. In addition, the de Broglie wavelengths account for the allowed orbits that were unexplainable in Bohr's theory. Thus, the new quantum-mechanical model explains certain aspects of the structure of the atom that Bohr's model could not account for.

Although probability waves and electron clouds cannot be simply visualized as Bohr's planetary model could, they offer a mathematical picture of the atom that is more accurate than Bohr's model. The material presented in this chapter is only an introduction to quantum theory. Although we have focused on the simplest example—the hydrogen atom—quantum mechanics has been successfully applied to multi-electron atomic structures. In fact, it forms the basis for understanding the structure of all known atoms and the existence of all molecules. Although most scientists believe that quantum mechanics may be nearly the final picture of the deepest levels of nature, a few continue to search for other explanations, and debates about the implications of quantum mechanics continue.

SECTION REVIEW

1. Is light considered to be a wave or a particle? Explain your answer.
2. How did de Broglie account for the fact that the electrons in Bohr's model are always found at certain distinct distances from the nucleus?
3. Calculate the de Broglie wavelength of a proton moving at 1.00×10^4 m/s.
4. What is the physical significance of the square of the Schrödinger wave function, $|\psi|^2$?
5. Why is the electron sometimes viewed as an electron cloud?
6. **Critical Thinking** In classical physics, the accuracy of measurements has always been limited by the measuring instruments used, and no instrument is perfect. How is this limitation different from that formulated by Heisenberg in the uncertainty principle?

SECTION 3

SECTION REVIEW ANSWERS

1. Light exhibits some aspects that are characteristic of classical waves and others that are characteristic of classical particles.
2. The allowed orbits are those that contain an integral number of electron wavelengths.
3. 3.96×10^{-11} m
4. It is proportional to the probability of finding the electron at a given position.
5. because the electron is found in a spherical region surrounding the nucleus, and its location is not known with complete certainty
6. Even with perfect measuring instruments, according to Heisenberg's uncertainty principle, simultaneous measurements of position and momentum cannot be completely certain. In fact, the smaller the uncertainty is in one of these quantities, the larger is the minimum uncertainty in the other quantity. This inherent limitation is a result of the wave nature of matter.

CHAPTER 21

Highlights

Teaching Tip

Ask students to prepare a concept map for the chapter. The concept map should include most of the vocabulary terms, along with other integral terms or concepts.

CHAPTER 21

Highlights

KEY TERMS

blackbody radiation (p. 752)

ultraviolet catastrophe (p. 753)

photoelectric effect (p. 756)

photon (p. 757)

work function (p. 757)

Compton shift (p. 760)

emission spectrum (p. 764)

absorption spectrum (p. 764)

uncertainty principle (p. 775)

KEY IDEAS

Section 1 Quantization of Energy

- Blackbody radiation and the photoelectric effect contradict classical physics, but they can be explained with the assumption that energy comes in discrete units, or is quantized.
- The energy of a light quantum, or photon, depends on the frequency of the light. Specifically, the energy of a photon is equal to frequency multiplied by Planck's constant.
- Planck's constant (h) is approximately equal to 6.63×10^{-34} J•s.
- The relation between the electron volt and the joule is as follows: $1 \text{ eV} = 1.60 \times 10^{-19}$ J.
- The minimum energy required for an electron to escape from a metal depends on the threshold frequency of the metal.
- The maximum kinetic energy of photoelectrons depends on the work function and the frequency of the light shining on the metal.

Section 2 Models of the Atom

- Rutherford's scattering experiment revealed that all of an atom's positive charge and most of an atom's mass are concentrated at its center.
- Each gas has a unique emission and absorption spectrum.
- Atomic spectra are explained by Bohr's model of the atom, in which electrons move from one energy level to another when they absorb or emit photons.

Section 3 Quantum Mechanics

- Light has both wave and particle characteristics.
- De Broglie proposed that matter has both wave and particle characteristics.
- Simultaneous measurements of position and momentum cannot be made with infinite accuracy.

PROBLEM SOLVING

See **Appendix D: Equations** for a summary of the equations introduced in this chapter. If you need more problem-solving practice, see **Appendix I: Additional Problems.**

Variable Symbols

Quantities		Units	
E	photon energy	J eV	joules electron volts
f_t	threshold frequency	Hz	hertz
hf_t	work function	eV	electron volts
KE_{max}	maximum kinetic energy	eV	electron volts

Review

QUANTIZATION OF ENERGY

Review Questions

1. Why is the term *ultraviolet catastrophe* used to describe the discrepancy between the predictions of classical physics and the experimental data for blackbody radiation?
2. What is meant by the term *quantum*?
3. What did Planck assume in order to explain the experimental data for blackbody radiation? How did Planck's assumption contradict classical physics?
4. What is the relationship between a joule and an electron volt?
5. How do observations of the photoelectric effect conflict with the predictions of classical physics?
6. What does Compton scattering demonstrate?

Conceptual Questions

7. Which has more energy, a photon of ultraviolet radiation or a photon of yellow light?
8. If the photoelectric effect is observed for one metal using light of a certain wavelength, can you conclude that the effect will also be observed for another metal under the same conditions?
9. What effect, if any, would you expect the temperature of a material to have on the ease with which electrons can be ejected from the metal in the photoelectric effect?
10. A photon is deflected by a collision with a moving electron. Can the photon's wavelength ever become shorter as a result of the collision? Explain your answer.

Practice Problems

For problems 11–12, see Sample Problem A.

11. A quantum of electromagnetic radiation has an energy of 2.0 keV. What is its frequency?
12. Calculate the energy in electron volts of a photon having a wavelength in the following ranges:
    a. the microwave range, 5.00 cm
    b. the visible light range, 5.00×10^{-7} m
    c. the X-ray range, 5.00×10^{-8} m

For problems 13–14, see Sample Problem B.

13. Light of frequency 1.5×10^{15} Hz illuminates a piece of tin, and the tin emits photoelectrons of maximum kinetic energy 1.2 eV. What is the threshold frequency of the metal?
14. The threshold frequency of silver is 1.14×10^{15} Hz. What is the work function of silver?

MODELS OF THE ATOM

Review Questions

15. What did Rutherford's foil experiment reveal?
16. If Rutherford's planetary model were correct, atoms would be extremely unstable. Explain why.
17. How can the absorption spectrum of a gas be used to identify the gas?
18. What restriction does the Bohr model place on the movement of an electron in an atom?
19. How is Bohr's model of the hydrogen atom similar to Rutherford's planetary model? How are the two models different?
20. How does Bohr's model account for atomic spectra?

Review

ANSWERS

1. because the discrepancy happens at the ultraviolet end of the spectrum
2. a discrete quantity
3. Planck assumed that the energy radiated by a blackbody is quantized; in classical physics, energy is continuous.
4. $1 \text{ eV} = 1.60 \times 10^{-19}$ J
5. Frequency rather than intensity determines whether electrons are ejected; KE_{max} depends on frequency, not intensity.
6. Photons have momentum.
7. photon of ultraviolet radiation
8. No, the second surface could have a higher work function.
9. At a higher temperature, electrons would be easier to eject.
10. Yes, this could occur if a moving electron transferred kinetic energy to a photon during a collision.
11. 4.8×10^{17} Hz
12. **a.** 2.49×10^{-5} eV
    b. 2.49 eV
    c. 24.9 eV
13. 1.2×10^{15} Hz
14. 4.72 eV
15. Charge and mass are concentrated at the atom's center.
16. because orbiting electrons would radiate energy
17. by comparing observed lines to known spectra of elements
18. The electron is limited to certain orbits.
19. The electron orbits the nucleus in both models. Rutherford: any orbits are possible; Bohr: only certain orbits are stable.

20. Spectral lines correspond to jumps between stable orbits.
21. In both cases, each line corresponds to a transition between energy levels.
22. Absorption spectra are usually observed at room temperature and therefore involve upward jumps by electrons from the ground level. Emission spectra are seen at a high temperature and involve downward jumps between any two levels.
23. a. 2.46×10^{15} Hz
b. 2.92×10^{15} Hz
c. 3.09×10^{15} Hz
d. 3.16×10^{15} Hz
24. Distances between emission lines in sketches should be proportional to the differences between frequencies. The lines are in the ultraviolet range.
25. diffraction, polarization; photoelectric effect, blackbody radiation, Compton shift
26. It is impossible to simultaneously measure position and momentum with complete certainty.
27. Allowed orbits are those that contain an integral number of wavelengths.
28. The electron's location is proportional to $|\psi|^2$. Electrons can be anywhere, but the most probable radial distances are Bohr radii.
29. Bohr's model predicts the electron's location exactly.
30. λ is inversely proportional to m; when m is large, λ is too small to detect.

extension

Graphing Calculator Practice
Visit go.hrw.com for answers to this Graphing Calculator activity.
Keyword HF6ATMXT

Conceptual Questions

21. Explain why all of the wavelengths in an element's absorption spectrum are also found in that element's emission spectrum.

22. More emission lines than absorption lines are usually observed in the atomic spectra of most elements. Explain why this occurs.

Practice Problems

For problems 23–24, see Sample Problem C.

23. Electrons in the ground state of hydrogen (energy level E_1) have an energy of −13.6 eV. Use this value and the energy-level diagram in Sample Problem C to calculate the frequencies of photons emitted when electrons drop to the ground state from the following energy levels:

a. E_2
b. E_3
c. E_4
d. E_5

24. Sketch an emission spectrum showing the relative positions of the emission lines produced by the photons in problem 23. In what part of the electromagnetic spectrum are these lines?

QUANTUM MECHANICS

Review Questions

25. Name two situations in which light behaves like a wave and two situations in which light behaves like a particle.

26. What does Heisenberg's uncertainty principle claim?

27. How do de Broglie's matter waves account for the "allowed" electron orbits?

28. Describe the quantum-mechanical model of the atom. How is this model similar to Bohr's model? How are the two different?

Conceptual Questions

29. How does Heisenberg's uncertainty principle conflict with the Bohr model of hydrogen?

30. Why can the wave properties of an electron be observed, while those of a speeding car cannot?

31. An electron and a proton are accelerated from rest through the same potential difference. Which particle has the longer wavelength? (Hint: Note that $\Delta PE = q\Delta V = \Delta KE$.)

32. Discuss why the term *electron cloud* is used to describe the arrangement of electrons in the quantum-mechanical view of the atom.

Graphing Calculator Practice

De Broglie Wavelength
In 1924, Louis de Broglie proposed the radical new idea that all forms of matter have both wave and particle properties. As you learned earlier in this chapter, this idea is demonstrated in the de Broglie equation.

$$\lambda = \frac{h}{mv}$$

In this equation, mass (m) and velocity (v) are particle properties, and wavelength (λ) is a wave property.

In this graphing calculator activity, you will use this equation to study the de Broglie wavelengths associated with moving particles of various masses and various speeds. You will discover why this equation has very different consequences for subatomic particles, such as electrons and neutrons, than for macroscopic particles, such as baseballs.

Visit go.hrw.com and type in the keyword **HF6ATMX** to find this graphing calculator activity. Refer to **Appendix B** for instructions on downloading the program for this activity.

Practice Problems

For problems 33–34, see Sample Problem D.

33. How fast must an electron move if it is to have a de Broglie wavelength of 5.2×10^{-11} m?

34. Calculate the de Broglie wavelength of a 0.15 kg baseball moving at 45 m/s.

MIXED REVIEW

35. A light source of wavelength λ illuminates a metal and ejects photoelectrons with a maximum kinetic energy of 1.00 eV. A second light source of wavelength $\frac{1}{2}\lambda$ ejects photoelectrons with a maximum kinetic energy of 4.00 eV. What is the work function of the metal?

36. A 0.50 kg mass falls from a height of 3.0 m. If all of the energy of this mass could be converted to visible light of wavelength 5.0×10^{-7} m, how many photons would be produced?

37. Red light (λ = 670.0 nm) produces photoelectrons from a certain material. Green light (λ = 520.0 nm) produces photoelectrons from the same material with 1.50 times the previous maximum kinetic energy. What is the material's work function?

38. Find the de Broglie wavelength of a ball with a mass of 0.200 kg just before it strikes the Earth after it has been dropped from a building 50.0 m tall.

Alternative Assessment

1. Calculate the de Broglie wavelength for an electron, a neutron, a baseball, and your body, at speeds varying from 1.0 m/s to 3.0×10^{7} m/s. Organize your findings in a table. The distance between atoms in a crystal is approximately 10^{-10} m. Which wavelengths could produce diffraction patterns using crystal as a diffraction grating? What can you infer about the wave characteristics of large objects? Explain your conclusions.

2. Bohr, Einstein, Planck, and Heisenberg each received the Nobel Prize for their contributions to twentieth-century physics. Their lives were also affected by the extraordinary events of World War II. Research their stories and the ways the war affected their work. What were their opinions about science and politics during and after the war? Write a report about your findings and about the opinions in your groups regarding the involvement and responsibility of scientists in politics.

3. Conduct research on the history of atomic theory. Create a timeline that shows the development of modern atomic theory, beginning with John Dalton's contributions in 1808. Include the discoveries of J.J. Thomson, Ernest Rutherford, Niels Bohr, and Erwin Schrodinger. You may also include other significant discoveries in the history of atomic theory. In addition, add historical events to the timeline to provide context for the scientific discoveries, and include illustrations with key entries.

4. Choose a simple element, and then create three-dimensional models of an atom of this element. Create at least three different models, corresponding to different versions of atomic theory throughout history. Include information about which historical theories you are representing in each model, and which parts of those theories are no longer accepted today. Also include information about the limitations of your models.

21 REVIEW

31. the electron

32. because the electron's location cannot be determined exactly

33. 1.4×10^{7} m/s

34. 9.8×10^{-35} m

35. 2.00 eV

36. 3.7×10^{19} photons

37. 0.80 eV

38. 1.06×10^{-34} m

Alternative Assessment ANSWERS

1. Students' tables should show that wavelength decreases as an object's mass or speed increases. Wavelengths should be near 10^{-10} m for diffraction in a crystal.

2. Student answers will vary. A good resource is *Heisenberg Probably Slept Here: The Lives, Times, and Ideas of the Great Physicists of the 20th Century*, by Richard Brennan (NY: Wiley, 1997).

3. Student timelines should include significant discoveries in atomic theory, along with notable historical events during roughly the same time periods.

4. Student atomic models should accurately represent their chosen elements (in terms of protons, neutrons, and electrons), and each model should also reflect the atomic theory it corresponds to.

Standardized Test Prep

ANSWERS

1. D
2. G
3. A
4. G
5. B
6. J
7. C
8. G

Standardized Test Prep

MULTIPLE CHOICE

1. What is another word for "quantum of light"?

A. blackbody radiation
B. energy level
C. frequency
D. photon

2. According to classical physics, when a light illuminates a photosensitive surface, what should determine how long it takes before electrons are ejected from the surface?

F. frequency
G. intensity
H. photon energy
J. wavelength

3. According to Einstein's photon theory of light, what does the intensity of light shining on a metal determine?

A. the number of photons hitting the metal in a given time interval
B. the energy of photons hitting the metal
C. whether or not photoelectrons will be emitted
D. KE_{max} of emitted photoelectrons

4. An X-ray photon is scattered by a stationary electron. How does the frequency of this scattered photon compare to its frequency before being scattered?

F. The new frequency is higher.
G. The new frequency is lower.
H. The frequency stays the same.
J. The scattered photon has no frequency.

5. Which of the following summarizes Thomson's model of the atom?

A. Atoms are hard, uniform, indestructible spheres.
B. Electrons are embedded in a sphere of positive charge.
C. Electrons orbit the nucleus in the same way that planets orbit the sun.
D. Electrons exist only at discrete energy levels.

6. What happens when an electron moves from a higher energy level to a lower energy level in an atom?

F. Energy is absorbed from a source outside the atom.
G. The energy contained in the electromagnetic field inside the atom increases.
H. Energy is released across a continuous range of values.
J. A photon is emitted with energy equal to the difference in energy between the two levels.

The diagram below is an energy-level diagram for hydrogen. Use the diagram to answer questions 7–8.

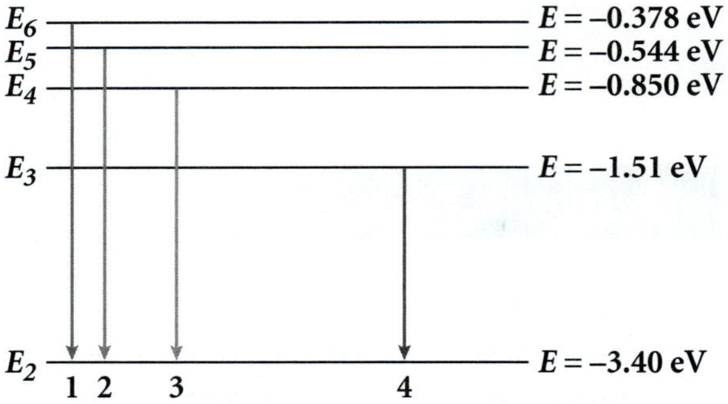

7. What is the frequency of the photon emitted when an electron jumps from E_5 to E_2?

A. 2.86 eV
B. 6.15×10^{14} Hz
C. 6.90×10^{14} Hz
D. 4.31×10^{33} Hz

8. What frequency of photon would be absorbed when an electron jumps from E_2 to E_3?

F. 1.89 eV
G. 4.56×10^{14} Hz
H. 6.89×10^{14} Hz
J. 2.85×10^{33} Hz

9. What type of spectrum is created by applying a high potential difference to a pure atomic gas?

A. an emission spectrum
B. an absorption spectrum
C. a continuous spectrum
D. a visible spectrum

10. What type of spectrum is used to identify elements in the atmospheres of stars?

F. an emission spectrum
G. an absorption spectrum
H. a continuous spectrum
J. a visible spectrum

11. What is the speed of a proton ($m = 1.67 \times 10^{-27}$ kg) with a de Broglie wavelength of 4.00×10^{-14} m?

A. 1.59×10^{-30} m/s
B. 1.01×10^{-7} m/s
C. 9.93×10^{6} m/s
D. 1.01×10^{7} m/s

12. What does Heisenberg's uncertainty principle state?

F. It is impossible to simultaneously measure a particle's position and momentum with infinite accuracy.
G. It is impossible to measure both a particle's position and its momentum.
H. The more accurately we know a particle's position, the more accurately we know the particle's momentum.
J. All measurements are uncertain.

SHORT RESPONSE

13. What is the energy of a photon of light with frequency $f = 2.80 \times 10^{14}$ Hz? Give your answer in both J and eV.

14. Light of wavelength 3.0×10^{-7} m shines on the metals lithium, iron, and mercury, which have work functions of 2.3 eV, 3.9 eV, and 4.5 eV, respectively. Which of these metals will exhibit the photoelectric effect? For each metal that does exhibit the photoelectric effect, what is the maximum kinetic energy of the photoelectrons?

15. Identify the behavior of an electron as primarily like a wave or like a particle in each of the following situations:

a. traversing a circular orbit in a magnetic field
b. absorbing a photon and being ejected from the surface of a metal
c. forming an interference pattern

EXTENDED RESPONSE

16. Describe Bohr's model of the atom. Identify the assumptions that Bohr made that were a departure from those of classical physics. Explain how Bohr's model accounts for atomic spectra.

17. Electrons are ejected from a surface with speeds ranging up to 4.6×10^{5} m/s when light with a wavelength of 625 nm is used.

a. What is the work function of this surface?
b. What is the threshold frequency for this surface?

Show all your work.

18. The wave nature of electrons makes an electron microscope, which uses electrons rather than light, possible. The resolving power of any microscope is approximately equal to the wavelength used. A resolution of approximately 1.0×10^{-11} m would be required in order to "see" an atom.

a. If electrons were used, what minimum kinetic energy of the electrons (in eV) would be required to obtain this degree of resolution?
b. If photons were used, what minimum photon energy would be required?

Test TIP When answering multiple-choice questions, read each answer carefully. Do not be misled by wrong answers that seem right at first glance.

9. A
10. G
11. C
12. F
13. 1.86×10^{-19} J, 1.16 eV
14. lithium and iron; lithium: 1.8 eV, iron: 0.2 eV
15. **a.** particle; **b.** particle; **c.** wave
16. Answers should describe electrons orbiting a nucleus only in discrete energy levels. The model departs from classical physics in that the electrons are only allowed to have certain energies and they do not lose energy simply by moving in an electromagnetic field. Atomic spectra are the result of photons being emitted or absorbed when electrons jump between energy levels.
17. **a.** 1.39 eV
b. 3.35×10^{14} Hz
18. **a.** 1.5×10^{4} eV (15 keV)
b. 1.2×10^{5} eV (120 keV)

Lab Planning

Beginning on page T34 are preparation notes and teaching tips to assist you in planning.

Blank data tables (as well as some sample data) appear on the **One-Stop Planner.**

No Books in the Lab?

See the ***Datasheets for In-Text Labs*** workbook for a reproducible master copy of this experiment.

Safety Caution

Emphasize the dangers of working with electricity. Remind students to have you check their circuits before turning on the power supply or closing the switch.

Tips and Tricks

- Give students a brief demonstration of the apparatus. Point out the parts, and explain how each item works.
- If possible to do safely, use low light levels in the room during this laboratory. Stress to students the importance of preventing stray light from entering the device.
- Inform students that slowly adjusting the potential difference and stopping just as the current reaches 0.0 mA will yield the best results.
- Give students the wavelengths of light for the filters used.

CHAPTER 21

Skills Practice Lab — The Photoelectric Effect

OBJECTIVES

- **Observe** and **measure** the photoelectric effect using a phototube and photoelectric-effect device.
- **Calculate** the maximum kinetic energy of the emitted electrons using the measured value for the stopping voltage.

MATERIALS LIST

- **adhesive tape**
- **black construction paper**
- **insulated connecting wire**
- **lens supports**
- **light sources and power supply**
- **meterstick and meterstick supports**
- **multimeter or dc voltmeter**
- **photoelectric-effect device with amplifier and filters**
- **wood block**

In this laboratory, you will explore the photoelectric effect.

SAFETY

- **Avoid looking directly at a light source. Looking directly at a light may cause permanent eye damage. Put on goggles.**
- **Never close a circuit until it has been approved by your teacher. Never rewire or adjust any element of a closed circuit. Never work with electricity near water; be sure the floor and all work surfaces are dry.**
- **If the pointer on any kind of meter moves off scale, open the circuit immediately by opening the switch.**

PROCEDURE

Preparation

1. Read the entire lab, and plan what measurements you will take. In this laboratory, use black paper to shield the photoelectric-effect device from room lights and block all light when zeroing and when inserting filters.
2. If you are not using a datasheet provided by your teacher, prepare a data table with thirteen columns and four rows. In the first row, label the columns *Filter*, I_1 *(mA)*, ΔV_1 *(V)*, I_2 *(mA)*, ΔV_2 *(V)*, I_3 *(mA)*, ΔV_3 *(V)*, I_4 *(mA)*, ΔV_4 *(V)*, I_5 *(mA)*, ΔV_5 *(V)*, I_6 *(mA)*, and ΔV_{stop} *(V)*. In the first column, label the second through fourth rows *Blue, Green,* and *Red.*

Stopping Voltage

3. Connect the voltmeter across the photoelectric-effect device. ***Do not turn on the power supply or light source until your teacher approves your setup.*** With approval, read the current meter on the device. Turn on the power supply and adjust it until the current meter reads 15 mA. Set the zero point by turning the voltage-adjustment knob on the device until the applied potential difference stops the photocurrent. (The current meter should show the initial reading.) Adjust the "Zero adjust" (or similar) knob until the current meter reads zero. Turn off the device.
4. Make a tube from a sheet of black paper. Place the blue filter into the clips. Use the xenon bulb if available. Set up the apparatus as shown in **Figure 1,** and have your teacher approve your setup. Turn on the light and the device.

5. Turn the voltage-adjustment knob counterclockwise until the voltage meter reads zero. Adjust the power supply until the current meter on the photoelectric-effect device reads about 10 mA.

6. Slowly increase the potential difference by turning the voltage-adjustment knob. Turn the knob as slowly as possible as you approach 0.0 mA and stop just as the meter reaches 0.0 mA.

7. When the current meter reads 0.0 mA, read the voltage meter. Record the values in your data table as I_1 and ΔV_1. Repeat steps 5 through 7 to find ΔV when the current goes to zero four more times. Record the average current as I_6 and the average potential difference as ΔV_{stop} in your data table. The lowest potential difference for which the current is zero is the *stopping voltage.*

8. Prevent light from entering the device, and replace the blue filter with the green filter. Repeat steps 5 through 7 for the green filter. Record all data in your data table.

9. Turn the light source off. Prevent light from entering the device, and replace the green filter with the red filter. Replace the light source with the second light source if available. Place the light source close to the red filter. Turn on the light source, and repeat steps 5 through 7.

10. Clean up your work area. Put equipment away safely so that it is ready to be used again.

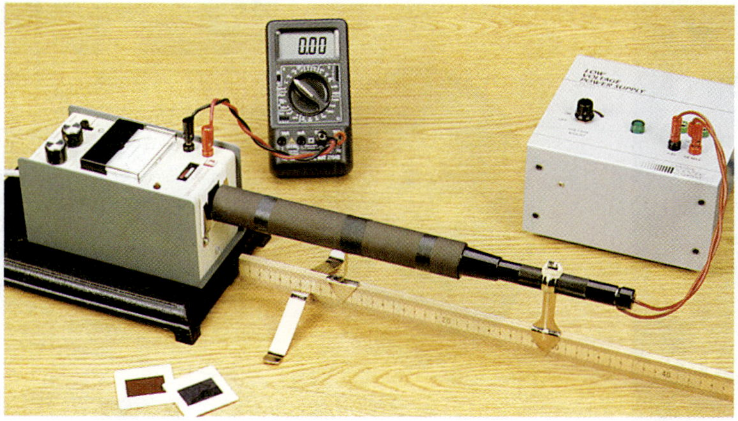

Figure 1

Step 3: When the photoelectric-effect device is operating, the current meter will probably not read zero even though there is no photocurrent. Set the meter to zero by starting a photocurrent and applying a potential difference to stop it. When the current meter returns to its original setting, set the meter to zero.

Step 5: Make sure the device is shielded from light in the room. Start a photocurrent in the device.

Step 7: Adjust the potential difference very slowly. Stop just as the meter reaches 0.0 mA.

ANALYSIS

1. **Organizing Data** For each trial, calculate the maximum kinetic energy (in eV) of the emitted electrons using the equation $KE_{max} = \Delta V_{stop} \times e$, where ΔV_{stop} is the stopping voltage and $e = 1.60 \times 10^{-19}$ C.

2. **Constructing Graphs** Graph the maximum kinetic energy of the emitted electrons versus the wavelength of the light source.

CONCLUSIONS

3. **Analyzing Graphs** Use your graph to answer the following questions.
   a. What is the relationship between the maximum kinetic energy of the emitted electrons and the wavelength of the light?
   b. What is the wavelength of light at which the kinetic energy is zero?

4. **Drawing Conclusions** What is the relationship between the stopping voltage and the wavelength of light?

✓ Checkpoints

Step 3: A light source that needs time to warm up should be turned on before students zero the device. Make sure that light cannot enter the device before approving the setup. Make sure the voltage-adjustment knob is turned until the current meter returns to its initial value. At this point, the current has been stopped and the current meter can be adjusted to read zero.

Steps 5–7: Make sure the device is shielded from light in the room. Repeat five times, and find the average value for the stopping voltage.

Step 9: The light source must be placed very close to the red filter for best results.

ANSWERS

Analysis

1. blue: 2.88×10^{-20} J; green: 1.76×10^{-20} J; red: 8.0×10^{-21} J

2. Student graphs should show a straight line pointing down and to the right.

Conclusions

3. a. The maximum kinetic energy of the electrons decreases as the wavelength increases.
b. Answers will vary. Students should find the answer by extending the line of the graph.

4. The stopping voltage decreases as the wavelength of light increases.

Physics and Its World *Timeline 1890–1950*

1890

1895 – In Paris, the brothers **Auguste** and **Louis Lumière** show a motion picture to the public for the first time.

1898 Po, Ra

Marie and **Pierre Curie** are the first to isolate the radioactive elements polonium and radium.

1900

1903 – **Wilbur** and **Orville Wright** fly the first successful heavier-than-air craft.

1905 $E_0 = mc^2$

Vol. 17 of *Annalen der Physik* contains three extraordinarily original and important papers by **Albert Einstein**. In one paper he introduces his special theory of relativity. In another he presents the quantum theory of light.

1910

1912 – **Henrietta Leavitt** discovers the period-luminosity relation for variable stars, making them among the most accurate and useful objects for determining astronomical distances.

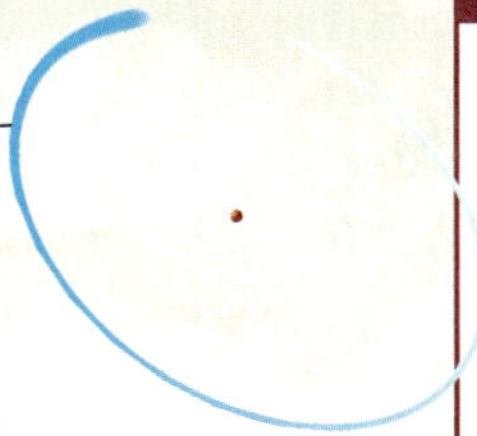

1913 $E_n = \frac{13.6}{n^2}$ eV

Niels Bohr—building on the discoveries of **Ernest Rutherford** and **J. J. Thomson,** and the quantum theories of **Max Planck** and **Albert Einstein**—develops a model of atomic structure based on energy levels that accounts for emission spectra.

1914 – World War I begins.

1920

1922 – **James Joyce's** *Ulysses* is published.

1920

1930

1940

1950

1926 $p = \frac{h}{\lambda}$

Erwin Schrödinger uses the wave-particle model for light and matter to develop the theory of wave mechanics, which describes atomic systems. About the same time, **Werner Heisenberg** develops a mathematically equivalent theory called *quantum mechanics,* by which the probability that matter has certain properties is determined.

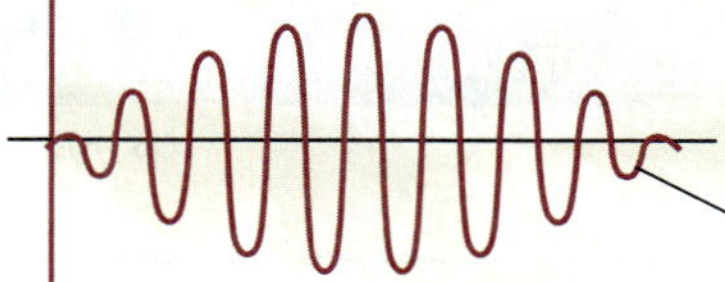

1929 – The New York Stock Exchange collapses, ushering in a global economic crisis known in the United States as The Great Depression.

Museo Nacional Centro de Arte Reina Sofia, Madrid. © 2007 Estate of Pablo Picasso/Artists Rights Society (ARS), New York.

1937 – **Pablo Picasso** paints *Guernica* in outraged response to the Nazi bombing of that town during the Spanish Civil War.

1938 $^{1}_{0}n + ^{235}_{92}\text{U} \rightarrow ^{141}_{56}\text{Ba} + ^{92}_{36}\text{Kr} + 3\,^{1}_{0}n$

Otto Hahn and **Fritz Strassman** achieve nuclear fission. Early the next year, **Lise Meitner** and her nephew **Otto Frisch** explain the process and introduce the term *fission* to describe the division of a nucleus into lighter nuclei.

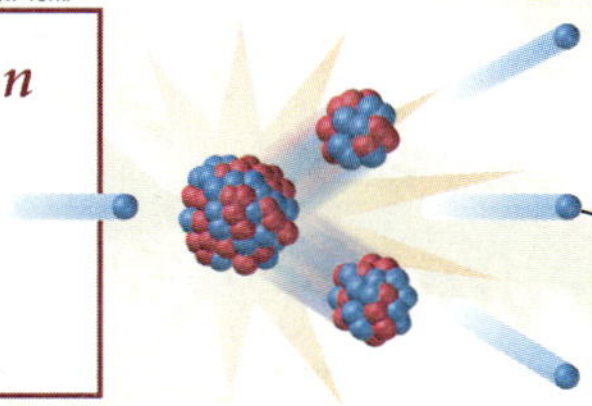

1939 – World War II begins with the Nazi invasion of Poland.

1942

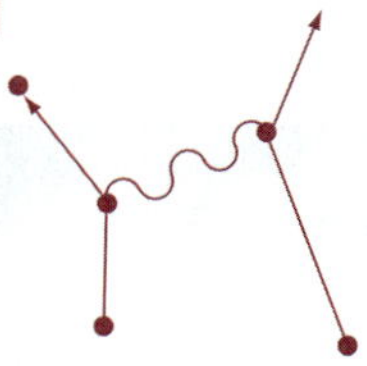

Shin'ichiro Tomonaga proposes an important tenet of quantum electrodynamics, which describes the interactions between charged particles and light at the quantum level. The theory is later independently developed by **Richard Feynman** and **Julian Schwinger.**

1948 – **Martin Luther King, Jr.** graduates from Morehouse College and enters Crozer Theological Seminary where he becomes acquainted with the principles of **Mohandas Gandhi**. During the next two decades he becomes one of the most forceful and articulate voices in the US civil rights movement.

787

CHAPTER 22

Subatomic Physics
Planning Guide

Compression Guide

To shorten instruction because of time limitations, omit the opener, Advanced sections, and the review.

OBJECTIVES	LABS, DEMONSTRATIONS, AND ACTIVITIES	TECHNOLOGY RESOURCES
PACING • 45 min pp. 788–789 **Chapter Opener**		CD **Visual Concepts,** Chapter 22 BASIC
PACING • 45 min pp. 790–796 **Section 1 The Nucleus** • Identify the properties of the nucleus of an atom. • Explain why some nuclei are unstable. • Calculate the binding energy of various nuclei.	TE **Demonstration** Nuclear Stability, p. 793 GENERAL	OSP **Lesson Plans** TR 73A Symbols for Nuclear Quantities TR 74A Mass and Rest Energy of Atomic Particles TR 75A Proton-Neutron Ratio for Stable Nuclei
PACING • 90 min pp. 797–806 *Advanced Level* **Section 2 Nuclear Decay** • Describe the three modes of nuclear decay. • Predict the products of nuclear decay. • Calculate the decay constant and the half-life of a radioactive substance.	TE **Demonstration** Electron Beam Deflection, p. 799 ADVANCED SE **Skills Practice Lab** Half-Life, pp. 826–827◆ GENERAL ANC **Datasheet** Half-Life* GENERAL	OSP **Lesson Plans** EXT **Integrating Earth Science** Radioactivity Within the Earth GENERAL EXT **Integrating Environmental Science** Environmental Radiation GENERAL TR 120 Nuclear Stability and Nuclear Decay TR 121 Half-Life of Carbon-14 TR 76A Types of Radiation and Rules for Decay
PACING • 45 min pp. 807–810 **Section 3 Nuclear Reactions** • Distinguish between nuclear fission and nuclear fusion. • Explain how a chain reaction is utilized by nuclear reactors. • Compare fission and fusion reactors.		OSP **Lesson Plans** EXT **Integrating Astronomy** The Life Cycle of a Star BASIC TR 122 Nuclear Chain Reaction TR 77A Binding Energy per Nucleon
PACING • 45 min pp. 811–817 *Advanced Level* **Section 4 Particle Physics** • Define the four fundamental interactions of nature. • Identify the elementary particles that make up matter. • Describe the standard model of the universe.		OSP **Lesson Plans** EXT **Integrating Astronomy** Our Changing Universe GENERAL TR 78A Electrons Exchanging a Photon TR 79A The Fundamental Interactions of Nature TR 80A Particle Classifications TR 81A Charges of Quarks and Antiquarks TR 82A The Standard Model TR 83A Evolution of the Universe

PACING • 90 min

CHAPTER REVIEW, ASSESSMENT, AND STANDARDIZED TEST PREPARATION

SE **Chapter Highlights,** p. 819
SE **Chapter Review,** pp. 820–823
SE **Graphing Calculator Practice,** p. 822 GENERAL
SE **Alternative Assessment,** p. 823 ADVANCED
SE **Standardized Test Prep,** pp. 824–825 GENERAL
SE **Appendix D: Equations,** p. 864
SE **Appendix I: Additional Problems,** p. 896
ANC **Study Guide Worksheet** Mixed Review* GENERAL
ANC **Chapter Test A*** GENERAL
ANC **Chapter Test B*** ADVANCED
OSP **Test Generator**

Online and Technology Resources

go.hrw.com Holt Online Learning

Visit **go.hrw.com** to find a variety of online resources. To access this chapter's extensions, enter the keyword **HF6SUBXT** and click the "go" button. Click **Holt Online Learning** for an online edition of this textbook, and other interactive resources.

HOLT Teacher's One-Stop Planner®

This DVD package includes:

- Holt Calendar Planner
- Customizable Lesson Plans
- Editable Worksheets
- ExamView® Version 6 Assessment Suite
- Interactive Teacher's Edition
- Holt PuzzlePro®
- Holt PowerNotes® Presentations
- MindPoint® Quiz Show

KEY

SE Student Edition	**OSP** One-Stop Planner	**EXT** Online Extension
TE Teacher Edition	**CD** CD or CD-ROM	* Also on One-Stop Planner
ANC Ancillary Worksheet	**TR** Teaching Transparencies	◆ Requires advance prep

SKILLS DEVELOPMENT RESOURCES	REVIEW AND ASSESSMENT	CORRELATIONS
		National Science Education Standards
SE **Sample Set A** Binding Energy, pp. 795–796 GENERAL TE **Classroom Practice,** p. 795 GENERAL ANC **Problem Workbook*** and OSP **Problem Bank** Sample Set A GENERAL SE **Appendix J: Advanced Topics** The Equivalence of Mass and Energy, pp. 918–919 ADVANCED	SE **Section Review,** p. 796 GENERAL ANC **Study Guide Worksheet** Section 1* GENERAL ANC **Quiz** Section 1* BASIC	UCP 1, 2, 3, 4 HNS 1, 3 PS 1a, 1b, 1c
SE **Sample Set B** Nuclear Decay, p. 802 GENERAL TE **Classroom Practice,** p. 802 GENERAL ANC **Problem Workbook*** and OSP **Problem Bank** Sample Set B GENERAL SE **Conceptual Challenge,** p. 803 ADVANCED SE **Sample Set C** Measuring Nuclear Decay, pp. 804–805 GENERAL TE **Classroom Practice,** p. 804 GENERAL ANC **Problem Workbook*** and OSP **Problem Bank** Sample Set C GENERAL SE **Appendix J: Advanced Topics** Antimatter, pp. 930–931 ADVANCED	SE **Section Review,** p. 806 ADVANCED ANC **Study Guide Worksheet** Section 2* GENERAL ANC **Quiz** Section 2* GENERAL	UCP 1, 2, 3, 4 SAI 1, 2 ST 1, 2 HNS 1, 2, 3 SPSP 1, 2, 3, 4, 5 PS 1c, 1d
	SE **Section Review,** p. 810 GENERAL ANC **Study Guide Worksheet** Section 3* GENERAL ANC **Quiz** Section 3* BASIC	UCP 2, 3, 4 ST 1, 2 HNS 1, 3 SPSP 2, 4, 5 PS 1c
SE **Conceptual Challenge,** p. 814 ADVANCED	SE **Section Review,** p. 817 ADVANCED ANC **Study Guide Worksheet** Section 4* GENERAL ANC **Quiz** Section 4* GENERAL	UCP 1, 2, 3, 4 SAI 3 HNS 1, 2, 3 SPSP 5

SCILINKS. NSTA
www.scilinks.org

Maintained by the **National Science Teachers Association.**

Topic: Atomic Nucleus
SciLinks Code: HF60117

Topic: Fission/Fusion
SciLinks Code: HF60581

Topic: Radioactive Decay
SciLinks Code: HF61254

Classroom CD-ROMs

- Guided Reading Audio Program
- Student One Stop
- Virtual Investigations
- Visual Concepts

Holt Lab Generator CD-ROM

Search for any lab by topic, standard, difficulty level, or time. Edit any lab to fit your needs, or create your own labs. Use the Lab Materials QuickList software to customize your lab materials list.

CHAPTER 22
Overview

Section 1 discusses properties of the atomic nucleus; identifies mass number, atomic number, and neutron number; explains why some nuclei are unstable; and shows how to calculate binding energy.

Section 2 describes the three types of radioactive decay, shows how to predict the outcome of nuclear decay, defines the decay constant, and shows how to calculate the half-life of a radioactive substance.

Section 3 distinguishes between nuclear fission and nuclear fusion and explores nuclear chain reactions.

Section 4 discusses the four fundamental interactions of nature and their mediating particles, classifies the elementary particles that make up matter, and describes the standard model of the universe.

About the Illustration

The tusks shown in this photograph are from a 20 000-year-old woolly mammoth that was discovered in the Taimyr Peninsula of Siberia, Russia. The mammoth was named *Jarkov* after the indigenous Dolgan tribesmen who first discovered the tusks sticking out of the snow. A 23-ton block of ice containing the mammoth was chiseled from the peninsula and flown to an ice tunnel for study. It was originally believed that the mammoth was preserved fully intact. However, as scientists thawed the ice block with hair dryers, they learned that the ice contained only scattered remains of the mammoth.

CHAPTER 22

Subatomic Physics

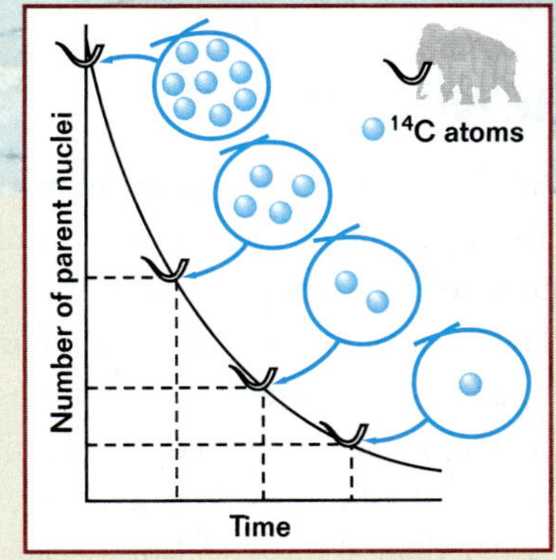

This reindeer sled carries the tusks of a 20 000-year-old woolly mammoth found in Russia. Scientists used a process called *carbon dating* to estimate the mammoth's age. All living organisms have the same ratio of carbon-14 atoms to carbon-12 atoms. The carbon-14 atoms decay into other atoms when an organism dies. Thus, the ratio of carbon-14 to carbon-12 can be used to date the organism.

WHAT TO EXPECT

In this chapter, you will study the atomic nucleus, radioactive decay, and the processes of fission and fusion. You will also learn about the standard model of the universe.

Why it Matters

Radioactive decay makes it possible to date organic materials that are 1000 to 25 000 years old. Nuclear fission is an important energy source today, and nuclear fusion may be an important energy source for the future.

CHAPTER PREVIEW

Tapping Prior Knowledge

Knowledge to Review

- ✔ The force between two charges is proportional to the magnitudes of the charges and inversely proportional to the distance between them squared: $F_{electric} = k_C \frac{q_1 q_2}{r^2}$.
- ✔ Atoms consist of a positive nucleus surrounded by negatively charged electrons.

SECTION 1
General Level

Teaching Tip

For the first three sections of this chapter, students will need to refer to the "Periodic Table of the Elements" (Appendix G) and the "Abbreviated Table of Isotopes and Atomic Masses" (Appendix H).

Misconception Alert — GENERAL

Students might not distinguish between the order of magnitudes involved at the atomic and nuclear levels. Point out that the size of the nuclear radius is about 10^{-14} m. This is approximately 10 000 times smaller than the radius of the atom. Although the nucleus is much smaller than the atom, the energies involved in nuclear processes, such as radioactivity, are a million times higher than the energies exchanged in atomic processes, such as the release of photoelectrons.

SECTION 1

The Nucleus

SECTION OBJECTIVES

- **Identify the properties of the nucleus of an atom.**
- **Explain why some nuclei are unstable.**
- **Calculate the binding energy of various nuclei.**

PROPERTIES OF THE NUCLEUS

In the chapter "Electric Forces and Fields," you learned that atoms are composed of electrons, protons, and neutrons. Except for the ordinary hydrogen nucleus, which consists of a single proton, both protons and neutrons are found in the nucleus. Together, protons and neutrons are referred to as *nucleons.*

As seen in the chapter "Atomic Physics," Rutherford's scattering experiment led to the conclusion that all of an atom's positive charge and most of its mass are concentrated in the nucleus. Rutherford's calculations revealed that the nucleus has a radius of no greater than about 10^{-14} m. Because such small lengths are common in nuclear physics, a convenient unit of length is the *femtometer* (fm). Sometimes called the *fermi,* this unit is equal to 10^{-15} m.

A nucleus can be specified by an atomic number and a mass number

There are a few important quantities that are used to describe the charge and mass of the nucleus. **Table 1** lists these quantities and the symbols commonly used to represent them. The mass number (A) represents the total number of protons and neutrons—or nucleons—in the nucleus. The atomic number (Z) represents the number of protons in the nucleus, and the neutron number (N) represents the number of neutrons in the nucleus. Note that A, Z, and N are always integers.

Table 1 Symbols for Nuclear Quantities

Symbol	Name	Explanation
A	mass number	the number of nucleons (protons and neutrons) in the nucleus
Z	atomic number	the number of protons in the nucleus
N	neutron number	the number of neutrons in the nucleus

As an example, a typical atom of aluminum has a mass number (A) of 27 and an atomic number (Z) of 13. Therefore, it has 13 protons and 14 neutrons ($27 - 13 = 14$). A periodic table of the elements usually includes the atomic number of each element above or near the element's chemical symbol.

Frequently, the mass number and the atomic number of the nucleus of an atom are written before the atom's chemical symbol, as shown in **Figure 1.**

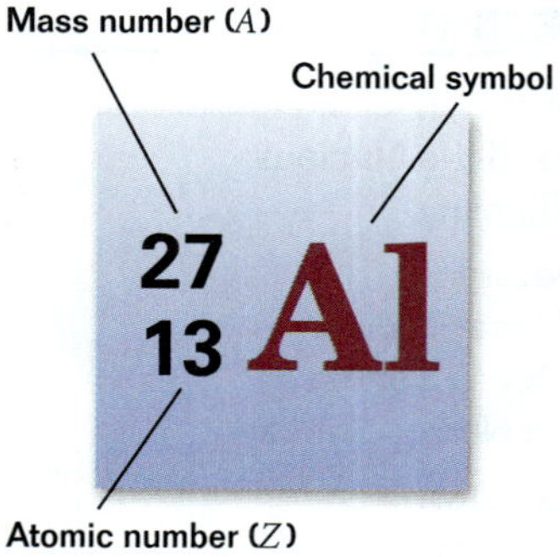

Figure 1
The chemical symbol of an element is often written with its mass number and atomic number, as shown here.

The chemical symbol for aluminum is Al. The superscript refers to the mass number A (27 in the case of aluminum), and the subscript refers to the atomic number Z (13 in the case of aluminum).

An element can be identified by its atomic number, Z. Because the number of protons determines the element, the atomic number of any given element does not change. Thus, the chemical symbol, such as Al, or the name of the element, such as aluminum, can always be used to determine the atomic number. For this reason, the atomic number is sometimes omitted.

Although atomic number does not change within an element, atoms of the same element can have different mass numbers. This is because the number of neutrons in a particular element can vary. Atoms that have the same atomic number but different neutron numbers (and thus different mass numbers) are called **isotopes.** The neutron number for an isotope can be found from the following relationship:

$$A = Z + N$$

This expression says that the mass number of an atom (A) equals the number of protons (Z) plus the number of neutrons (N) in the nucleus of the atom.

The natural abundance of isotopes can vary greatly. For example, $^{11}_{6}C$, $^{12}_{6}C$, $^{13}_{6}C$, and $^{14}_{6}C$ are four isotopes of carbon. The natural abundance of the $^{12}_{6}C$ isotope is about 98.9 percent, while that of the $^{13}_{6}C$ isotope is only about 1.1 percent. Some isotopes do not occur naturally but can be produced in the laboratory. Even the simplest element, hydrogen, has isotopes: $^{1}_{1}H$, called hydrogen; $^{2}_{1}H$, called *deuterium* (or *heavy hydrogen*); and $^{3}_{1}H$, called *tritium* (or *heavy heavy hydrogen*).

A nucleus is very dense

Experiments have shown that most nuclei are approximately spherical and that the volume of a nucleus is proportional to the total number of nucleons, and thus to the mass of the nucleus. This suggests that *all nuclei have nearly the same density,* which is about 2.3×10^{17} kg/m^3, which is 2.3×10^{14} times greater than the density of water (1.0×10^3 kg/m^3). Nucleons combine to form a nucleus as though they were tightly packed spheres, as shown in **Figure 2.**

The unified mass unit and rest energy are used to express the mass of a nucleus

Because the mass of a nucleus is extremely small, the *unified mass unit,* u, is often used for atomic masses. This unit is sometimes referred to as the *atomic mass unit.* 1 u is defined so that 12 u is equal to the mass of one atom of carbon-12. That is, the mass of a nucleus (or atom) is measured relative to the mass of an atom of the neutral carbon-12 isotope (the nucleus plus six electrons). Based on this definition, 1 u = $1.660\,538\,86 \times 10^{-27}$ kg. The proton and neutron each have a mass of about 1 u, and the electron has a mass that is only a small fraction of a unified mass unit—about 5×10^{-4} u.

SCILINKS.
www.scilinks.org
Topic: Atomic Nucleus
Code: HF60117

isotope

an atom that has the same number of protons (or the same atomic number) as other atoms of the same element do but that has a different number of neutrons (and thus a different atomic mass)

Figure 2
A nucleus can be visualized as a cluster of tightly packed spherical protons and neutrons. This illustration is just a representation; nucleons actually fill very little of the volume of the nucleus and are in rapid motion.

Misconception Alert — GENERAL

To help students differentiate between *atomic mass* and *mass number,* explain that the periodic table gives the atomic number and the atomic mass (rather than the mass number) of each element. Ask students if both numbers given on the periodic table are integers. *(Atomic numbers are; atomic masses are not.)* Explain that the atomic mass reported on the periodic table is an average of many nuclei with the same atomic numbers *(Z)* but different mass numbers *(A)*. Mass number, on the other hand, relates to an individual nucleus and is always an integer.

The Language of Physics

The symbol u represents the *unified mass unit,* also known as the *atomic mass unit.* Some texts use the symbol amu to represent this unit.

Teaching Tip — BASIC

Have students draw schematic diagrams like the one shown in **Figure 2** for the four isotopes of carbon discussed on this page ($^{11}_{6}C$, $^{12}_{6}C$, $^{13}_{6}C$, and $^{14}_{6}C$). This will help students visualize the relationship between atomic number, mass number, and the structure of the nucleus.

The Language of Physics

As seen in the chapter "Atomic Physics," the electron volt, eV, is a unit of energy typically used to describe atomic processes ($1 \text{ eV} = 1.60 \times 10^{-19}$ J). Because energies at the nuclear level are much greater than those at the atomic level, units of MeV ($M = 1 \times 10^6$) are often more convenient. $1 \text{ MeV} = (1 \times 10^6)(1.60 \times 10^{-19} \text{ J}) = 1.60 \times 10^{-13}$ J.

Teaching Tip — GENERAL

Table 2 shows the masses of the proton, neutron, and electron in various units. Each unit has advantages; the choice of unit depends on the situation. For example, kg can be convenient when working with other compound SI units. When working only with atomic masses, though, u is usually more convenient. Units of MeV are advantageous because they describe the rest energy of a particle.

Misconception Alert — BASIC

Explain to students that no particle has exactly 1 u of rest energy; the masses of the proton and neutron are only approximately equal to 1 u. Emphasize that $(1 \text{ u})(c^2) = 931.49$ MeV. Have students compare this value to the rest energies of the proton and neutron.

ADVANCED TOPICS

See "The Equivalence of Mass and Energy" in **Appendix J: Advanced Topics** to learn more about rest energy.

Did you know?

The equivalence between mass and energy is predicted by Einstein's special theory of relativity. Another aspect of this theory is that time and length are relative. That is, they depend on an observer's frame of reference, while the speed of light is absolute.

strong force

the interaction that binds nucleons together in a nucleus

Alternatively, the mass of the nucleus is often expressed in terms of rest energy. A particle has a certain amount of energy, called *rest energy,* associated with its mass. The following equation expresses the relationship between mass and rest energy mathematically:

RELATIONSHIP BETWEEN REST ENERGY AND MASS

$$E_R = mc^2$$

rest energy = (mass)(speed of light)^2

This expression is often used because mass is not conserved in many nuclear processes, as we will see. Because the rest energy of a particle is given by $E_R = mc^2$, it is convenient to express a particle's mass in terms of its energy equivalent. The equation that follows is for the rest energy of a particle with a mass of exactly 1 u.

$$E_R = mc^2 = \frac{(1.660\ 538\ 782 \times 10^{-27}\ \text{kg})(299\ 792\ 458\ \text{m/s})^2}{1.602\ 176\ 53 \times 10^{-19}\ \text{J/eV}} \approx 931.49\ \text{MeV}$$

Thus, the conversion of 1 u of mass into energy would produce about 931.49 MeV. This book will use the value 931.49 MeV for calculations. (Recall that M is an abbreviation for the SI prefix *mega-*, which indicates 10^6.)

The masses and energy equivalent of the proton, neutron, and electron are summarized in **Table 2.** Notice that in order to distinguish between the mass of the proton and the mass of the neutron, you must know their masses to at least four significant figures. The masses and some other properties of selected isotopes are given in Appendix H.

Table 2 Mass and Rest Energy of Atomic Particles

Particle	m (kg)	m (u)	E_R (MeV)
proton	1.673×10^{-27}	1.007 276	938.3
neutron	1.675×10^{-27}	1.008 665	939.6
electron	9.109×10^{-31}	0.000 549	0.5110

NUCLEAR STABILITY

Given that the nucleus consists of a closely packed collection of protons and neutrons, you might be surprised that it can exist. It seems that the Coulomb repulsion between protons would cause a nucleus to fly apart. There must be some attractive force to overcome this repulsive force. This force is called the *nuclear force,* or the **strong force.**

Number of Protons Versus Number of Neutrons for Stable Nuclei

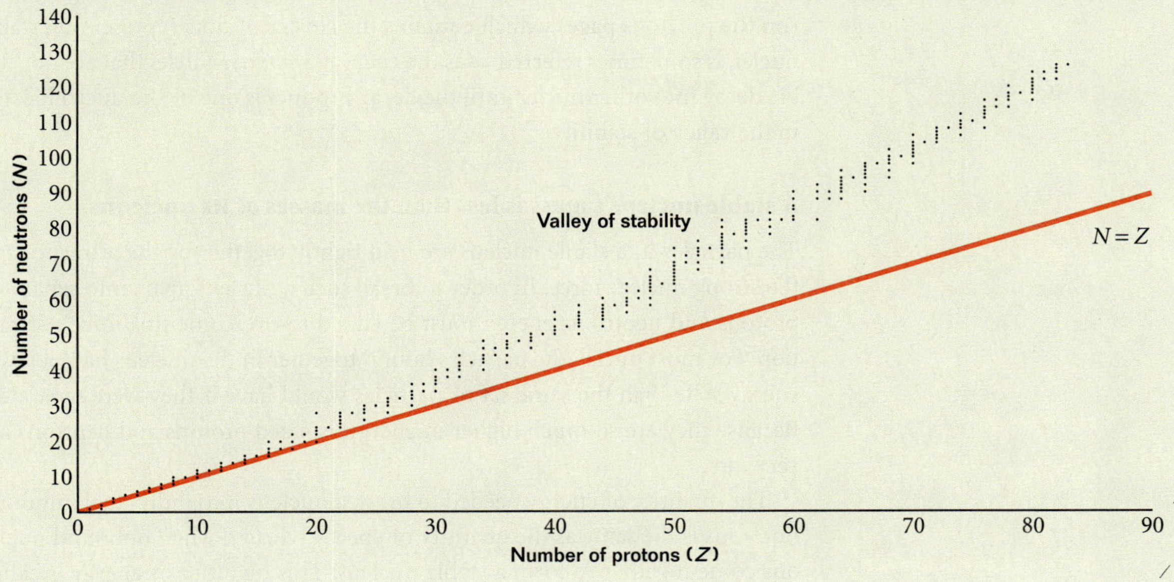

Figure 3
Each data point in this graph represents a stable nucleus. Note that as the number of protons increases, the ratio of neutrons to protons also increases. In other words, heavy nuclei have more neutrons per proton than lighter nuclei.

The strong force has some properties that make it very much unlike other types of force. The strong force is almost completely independent of electric charge. For a given separation, the force of attraction between two protons, two neutrons, or a proton and a neutron has the same magnitude.

Another unusual property of the strong force is its very short range, only about 10^{-15} m. For longer distances, the strong force is virtually zero.

Neutrons help to stabilize a nucleus

A plot of neutron number versus atomic number (the number of protons) for stable nuclei is shown in **Figure 3.** The solid line in the plot shows the location of nuclei that have an equal number of protons and neutrons ($N = Z$). Notice that only light nuclei are on this line, while all heavier nuclei fall above this line. This means that heavy nuclei are stable only when they have more neutrons than protons. This can be understood in terms of the characteristics of the strong force.

For a nucleus to be stable, the repulsion between positively-charged protons must be balanced by the strong nuclear force's attraction between all the particles in the nucleus. The repulsive force exists between all protons in a nucleus because the electrostatic force is long range. But a proton or a neutron attracts only its nearest neighbors because of the nuclear force's short range. So, as the number of protons increases, the number of neutrons has to increase even more to add enough attractive forces to maintain stability.

Visual Strategy GENERAL

Figure 3
Point out that the solid red line is only a reference line. It shows that instead of having equal numbers of protons and neutrons, most stable nuclei have more neutrons than protons. The stable nuclei are represented as clusters of dots rather than as a line because all nuclei have an integral number of protons (Z). For example, there are no stable nuclei between 50 and 51 because it is impossible to have an element with 50.5 protons. Also have students notice that more than one isotope may be stable.

Q Some dots on the empirical graph form vertical lines. What do these dots represent?

A ***Multiple dots on a vertical pattern represent isotopes of nuclei of the same element. For example, with $Z = 48$ (cadmium), the nucleus may contain 58, 60, 62, 63, 64, 65, 66, or 68 neutrons.***

Demonstration

Nuclear Stability —GENERAL

Purpose Simulate forces and stability in the nucleus.

Materials bowl, large number of steel marbles and magnetic marbles of about the same size

Procedure Tell students that the steel marbles represent neutrons and the magnetic marbles represent protons. Although this is not an exact analogy, it does show the relationship between atomic number and stability. Challenge students to build the largest and most stable nucleus, and have them explain how the steel marbles affect stability.

Misconception Alert — ADVANCED

Students may wonder whether elements with Z greater than 83 exist, because these elements are unstable and do not appear in **Figure 3,** on the previous page. After the concept of half-life has been introduced in Section 2, point out that the half-lives of unstable elements vary from fractions of seconds to thousands of years. Thus, some unstable elements exist for very long periods of time.

The Language of Physics — GENERAL

Make sure that students interpret the symbols and subscripts in the equations correctly. Ask how the total mass changes when a helium nucleus is broken down into its components of two protons and two neutrons. *(The total mass increases by the mass defect, Δm.)*

Misconception Alert — GENERAL

Some students may think that binding energy is present in the nucleus. Emphasize that binding energy is the amount of energy required to break the nucleus into its component parts, which destroys its structure. Explain that because very light and very heavy elements have the least binding energy, they are the only elements from which we can extract energy. (This concept is discussed further in Section 3.)

For Z greater than 83, the repulsive forces between protons cannot be compensated by the addition of more neutrons. That is, *elements that contain more than 83 protons do not have stable nuclei.* The long, narrow region in **Figure 3** (on the previous page), which contains the cluster of dots representing stable nuclei, is sometimes referred to as the *valley of stability.* Nuclei that are not stable decay into other nuclei until the decay product is one of the nuclei located in the valley of stability.

A stable nucleus's mass is less than the masses of its nucleons

The particles in a stable nucleus are held tightly together by the attractions of the strong nuclear force. In order to break such a nucleus apart into separated protons and neutrons, energy must be added to overcome this force's attraction. For most nuclei, the particles bound together in the nucleus have a lower energy state than the same set of particles would have if they were separated. Because they are so much higher in energy, isolated protons and neutrons are very rare.

binding energy

the energy released when unbound nucleons come together to form a stable nucleus, which is equivalent to the energy required to break the nucleus into individual nucleons

The quantity of energy needed to break a nucleus into individual unbound nucleons is the same as the quantity of energy released when unbound nucleons come together to form a stable nucleus. This quantity of energy is called the **binding energy** of the nucleus. It is equal to the difference in energy between the nucleons when bound and the same nucleons when unbound. (Note that except for very small values of A, unbound nucleons do not simply combine into a full-grown nucleus.) Binding energy can be calculated from the rest energies of the particles making up a nucleus as follows:

$$E_{bind} = E_{R,unbound} - E_{R,bound}$$

Using the equation for rest energy, we can rewrite this as follows:

$$E_{bind} = m_{unbound}c^2 - m_{bound}c^2 = (m_{unbound} - m_{bound})c^2$$

The mass of the nucleons when unbound minus the mass of the nucleons when bound is called the *mass defect* and is expressed as Δm. Thus, the previous equation for binding energy can be expressed as follows:

BINDING ENERGY OF A NUCLEUS

$$E_{bind} = \Delta mc^2$$

binding energy = mass defect × (speed of light)^2

Note that the total mass of a stable nucleus (m_{bound}) is always less than the sum of the masses of its individual nucleons ($m_{unbound}$). It is often useful to find the mass defect in terms of u so that it can be converted to energy as described earlier in this chapter (1 u = 931.49 MeV).

The mass of the unbound nucleus is the sum of the individual nucleon masses, and the mass of the bound nucleus is about equal to the atomic mass minus the mass of the electrons. Thus, $\Delta m = (Zm_p + Nm_n) - (\text{atomic mass} - Zm_e)$.

One way to rearrange this equation is $\Delta m = (Zm_p + Zm_e) + Nm_n -$ (atomic mass). Because a hydrogen atom contains one proton and one electron, the first term is equal to Z(atomic mass of H). Thus, the equation for mass defect can be rewritten as follows:

$$\Delta m = Z(\text{atomic mass of H}) + Nm_n - \text{atomic mass}$$

Use this equation and the atomic masses given in Appendix H to calculate mass defect when solving problems involving binding energy. (In this discussion, we have disregarded the binding energies of the electrons. This is reasonable because nuclear binding energies are many tens of thousands of times greater than electronic binding energies.)

The binding energy per nucleon for light nuclei ($A < 20$) is much smaller than the binding energy per nucleon for heavier nuclei. In other words, particles in lighter nuclei are less tightly bound on average than particles in heavier nuclei. Except for the lighter nuclei, the average binding energy per nucleon is about 8 MeV. Of all nuclei, iron-58 has the greatest binding energy per nucleon.

SAMPLE PROBLEM A

Binding Energy

PROBLEM

The nucleus of the deuterium atom, called the *deuteron,* consists of a proton and a neutron. Given that the atomic mass of deuterium is 2.014 102 u, calculate the deuteron's binding energy in MeV.

SOLUTION

1. DEFINE **Given:** $Z = 1$ atomic mass of deuterium = 2.014 102 u

$N = 1$ atomic mass of H = 1.007 825 u

$m_n = 1.008\ 665$ u

Unknown: $E_{bind} = ?$

2. PLAN **Choose an equation or situation:**

First, find the mass defect with the following relationship:

$$\Delta m = Z\,(\text{atomic mass of H}) + Nm_n - \text{atomic mass}$$

Then, find the binding energy by converting the mass defect to rest energy.

3. CALCULATE **Substitute the values into the equation and solve:**

$$\Delta m = 1(1.007\ 825\ \text{u}) + 1(1.008\ 665\ \text{u}) - 2.014\ 102\ \text{u}$$

$$\Delta m = 0.002\ 388\ \text{u}$$

$$E_{bind} = (0.002\ 388\ \text{u})\ (931.49\ \text{MeV/u})$$

$$\boxed{E_{bind} = 2.224\ \text{MeV}}$$

4. EVALUATE In order for a deuteron to be separated into its constituents—a proton and a neutron—2.224 MeV of energy must be added.

Teaching Tip — GENERAL

Point out that the *binding energy per nucleon* is the binding energy (E_{bind}) of the entire nucleus divided by the mass number (A). Ask students to find the binding energy per nucleon of $^{56}_{26}Fe$ given $E_{bind} = 492.4$ MeV (*8.793 MeV*). Also explain that when they are solving binding-energy problems, students must find the mass number ($A = Z + N$) to determine the correct atomic mass in Appendix H.

Classroom Practice

Binding Energy

Calculate the binding energy of the following, referring to Appendix H for the atomic masses:

a. tritium (1 proton, 2 neutrons)

b. helium (2 protons, 2 neutrons)

Answers

a. 8.482 MeV

b. 28.297 MeV

PROBLEM GUIDE A

Use this guide to assign problems.
SE = Student Edition Textbook
PW = Problem Workbook
PB = Problem Bank on the One-Stop Planner (OSP)

Solving for:

E_{bind}	**SE** Sample, 1–4; Ch. Rvw. 7–9, 35 **PW** 1–4 **PB** Sample, 1–6, 9
m	**PW** 5–7 **PB** 7–8, 10

***Challenging Problem**
Consult the printed Solutions Manual or the OSP for detailed solutions.

SECTION 1

ANSWERS

Practice A

1. 160.65 MeV; 342.05 MeV
2. 0.764 MeV
3. 7.933 MeV
4. 7.5701 MeV/nucleon

SECTION REVIEW ANSWERS

1. the charge of the nucleus (the atomic number)
2. They have the same number of protons and electrons. They have a different number of neutrons and hence a different number of nucleons.
3. Z is the same for each isotope; A and N are different.
4. the strong force
5. For light nuclei, $N \approx Z$. As the number of protons (Z) increases, more neutrons per proton are needed to compensate for the Coulomb repulsion, so $N > Z$.
6. **a.** 805.77 MeV
   b. 1559.4 MeV
   c. 224.96 MeV
7. six; eight; six
8. 0.2106 MeV; Because of the extra proton in $^{23}_{12}Mg$, the Coulomb repulsion in the Mg nucleus is greater than that in the Na nucleus. Thus, breaking the Mg nucleus into its components requires less energy; in other words, its binding energy is less.

PRACTICE A

Binding Energy

1. Calculate the total binding energy of $^{20}_{10}Ne$ and $^{40}_{20}Ca$. (Refer to Appendix H for this and the following problems.)
2. Determine the difference in the binding energy of $^{3}_{1}H$ and $^{3}_{2}He$.
3. Calculate the binding energy of the last neutron in the $^{43}_{20}Ca$ nucleus. (Hint: Compare the mass of $^{43}_{20}Ca$ with the mass of $^{42}_{20}Ca$ plus the mass of a neutron.)
4. Find the binding energy per nucleon of $^{238}_{92}U$ in MeV.

SECTION REVIEW

1. Does the nuclear mass or the charge of the nucleus determine what element an atom is?
2. Oxygen has several isotopes. What do these isotopes have in common? How do they differ?
3. Of atomic number, mass number, and neutron number, which are the same for each isotope of an element, and which are different?
4. The protons in a nucleus repel one another with the Coulomb force. What holds these protons together?
5. Describe the relationship between the number of protons, the number of neutrons, and the stability of a nucleus.
6. Calculate the total binding energy of the following:
   a. $^{93}_{41}Nb$
   b. $^{197}_{79}Au$
   c. $^{27}_{13}Al$
   (Refer to Appendix H.)
7. How many protons are there in the nucleus $^{14}_{6}C$? How many neutrons? How many electrons are there in the neutral atom?
8. **Critical Thinking** Two isotopes having the same mass number are known as *isobars.* Calculate the difference in binding energy per nucleon for the isobars $^{23}_{11}Na$ and $^{23}_{12}Mg$. How do you account for this difference?

Nuclear Decay

SECTION OBJECTIVES

- **Describe the three modes of nuclear decay.**
- **Predict the products of nuclear decay.**
- **Calculate the decay constant and the half-life of a radioactive substance.**

NUCLEAR DECAY MODES

So far, we have considered what happens when nucleons are bound together to form stable nuclei. However, not all nuclei are stable. There are about 400 stable nuclei; hundreds of others are unstable and tend to break apart into other particles. This process is called *nuclear decay.*

The nuclear decay process can be a natural event or can be induced artificially. In either case, when a nucleus decays, radiation is emitted in the form of particles, photons, or both. The emission of particles and photons is called *radiation,* and the process is called *radioactivity.* For example, the hands and numbers of the watch shown in **Figure 4** contain small amounts of radium salts. The nuclei within these salts decay, releasing light energy that causes the watch to glow in the dark. The nucleus before decay is called the *parent nucleus,* and the nucleus remaining after decay is called the *daughter nucleus.* In all nuclear reactions, the energy released is found by the equation $E = \Delta mc^2$.

Figure 4
The radioactive decay of radium nuclei causes the hands and numbers of this watch to glow in the dark.

A radioactive material can emit three types of radiation

Three types of radiation can be emitted by a nucleus as it undergoes radioactive decay: alpha (α) particles, in which the emitted particles are ${}^4_2$He nuclei; beta (β) particles, in which the emitted particles are either electrons or positrons (positively charged particles with a mass equal to that of the electron); and gamma (γ) rays, in which the emitted "rays" are high-energy photons. These three types of radiation are summarized in **Table 3.**

Table 3 Alpha, Beta, and Gamma Radiation

Particle	Symbols	Composition	Charge	Effect on parent nucleus
alpha	α (${}^4_2$He)	2 protons, 2 neutrons	+2	mass loss; new element produced
beta	β^- (${}^{\ 0}_{-1}$e) β^+ (${}^0_1$e)	electron positron	−1 +1	no change in mass number; new element produced
gamma	γ	photon	0	energy loss

Misconception Alert — GENERAL

Students may not understand why there are only 109 known elements but 400 stable nuclei have been identified. To clarify their understanding, have students review the meaning of isotopes (introduced in Section 1) and list examples.

Teaching Tip — ADVANCED

Point out that all three types of radiation listed in **Table 3** are emitted from the nucleus. Have students consider possible sources for beta radiation (*protons and neutrons may decay*).

SECTION 2

Key Models and Analogies — GENERAL

Relate balancing nuclear reactions to students' knowledge of chemical equations. Point out that chemical reactions involve atomic masses and electrons, while nuclear processes involve nuclear particles. Students should realize that in both cases the procedures for balancing equations are based on the conservation of nucleon number and charge. Have students verify this in the example of the decay of uranium-238. *(The conservation of nucleon number leads to balancing the A numbers: 238 = 234 + 4; conservation of charge requires balancing the Z numbers: 92 = 90 + 2.)*

Misconception Alert — GENERAL

In the equation for beta decay, the symbol ${}_{-1}^{0}e$ suggests that the mass of the electron is zero. This may seem to contradict students' prior understanding of atomic physics. Explain that the zero means that the electron's mass is negligible compared with the other masses involved in nuclear decay. Also emphasize that in beta decay, the daughter nucleus has the same *mass number* that the parent does. In other words, the daughter contains the same number of nucleons that the parent does.

SCILINKS®
www.scilinks.org
Topic: Radioactive Decay
Code: HF61254

The ability of radiation to pass through a material depends on the type of radiation. Alpha particles can usually be stopped by a piece of paper, beta particles can penetrate a few millimeters of aluminum, and gamma rays can penetrate several centimeters of lead.

Helium nuclei are emitted in alpha decay

When a nucleus undergoes alpha decay, it emits an alpha particle (${}_{2}^{4}He$). Thus, the nucleus loses two protons and two neutrons. This makes the nucleus lighter and decreases its positive charge. (Because the electrons around the nuclei do not participate in nuclear reactions, they are ignored.)

For example, the nucleus of uranium-238 (${}_{92}^{238}U$) can decay by alpha emission to a thorium-234 nucleus and an alpha particle, as follows:

$${}_{92}^{238}U \rightarrow {}_{90}^{234}Th + {}_{2}^{4}He$$

This expression says that a parent nucleus, ${}_{92}^{238}U$, emits an alpha particle, ${}_{2}^{4}He$, and thereby changes to a daughter nucleus, ${}_{90}^{234}Th$ (thorium-234). This nuclear reaction and all others follow the rules summarized in **Table 4.** These two rules can be used to determine the unknown daughter atom when a parent atom undergoes alpha decay.

Table 4 Rules for Nuclear Decay

The total of the atomic numbers on the left is the same as the total on the right because charge must be conserved.
The total of the mass numbers on the left is the same as the total on the right because nucleon number must be conserved.

ADVANCED TOPICS

A positron is the *antiparticle* of the electron. See "Antimatter" in **Appendix J: Advanced Topics** to learn more about antiparticles.

Electrons or positrons are emitted in beta decay

When a radioactive nucleus undergoes beta decay, the nucleus emits either an electron or a positron. (A positron has the same mass as the electron but is positively charged.) The atomic number is increased or decreased by one, with an opposite change in the neutron number. Because the daughter nucleus contains the same number of nucleons as the parent nucleus, the mass number does not change. Thus, beta decay does little to change the mass of a nucleus. Instead, the ratio of neutrons to protons in a nucleus is changed. This ratio affects the stability of the nucleus, as seen in Section 1.

A typical beta decay event involves carbon-14, as follows:

$${}_{6}^{14}C \rightarrow {}_{7}^{14}N + {}_{-1}^{0}e \quad \text{(partial equation)}$$

This decay produces an electron, written as ${}_{-1}^{0}e$. In this decay, the atomic number of the daughter nucleus is increased by 1.

Another beta decay event involves nitrogen-12, as follows:

$$^{12}_{7}\text{N} \rightarrow {}^{12}_{6}\text{C} + {}^{0}_{1}e \quad \text{(partial equation)}$$

This decay produces a positron, written as ${}^{0}_{1}e$. In this decay, the atomic number of the daughter nucleus is decreased by 1.

The superscripts and subscripts on the carbon and nitrogen nuclei follow our usual conventions, but those on the electron and the positron may need some explanation. The −1 indicates that the electron has a charge whose magnitude is equal to that of the proton but is negative. Similarly, the 1 indicates that the positron has a charge that is equal to that of the proton in magnitude and sign. Thus, the subscript can be thought of as the charge of the particle. The 0 used for the mass number of the electron and the positron reflects the fact that electrons and positrons are not nucleons; thus, their emission does not change the mass number. Notice that both subscripts and superscripts must balance in the equations for beta decay, just as in alpha decay.

Figure 5
The spiral in this bubble-chamber image is the track left by a positron. This reaction took place in a magnetic field, which caused the positron to spiral as it lost energy.

Beta decay transforms neutrons and protons

A bubble-chamber image of a positron is shown in **Figure 5.** The emission of electrons or positrons from a nucleus is surprising because the nucleus is made of only protons and neutrons. This apparent discrepancy can be explained by noting that in beta decay, either a neutron is transformed into a proton, creating an electron in the process, or a proton is transformed into a neutron, creating a positron in the process. These two beta decays can be written as follows:

$$\begin{array}{l} {}^{1}_{0}n \rightarrow {}^{1}_{1}p + {}^{0}_{-1}e \\ {}^{1}_{1}p \rightarrow {}^{1}_{0}n + {}^{0}_{1}e \end{array} \quad \text{(partial equations)}$$

Decay events can be written in this way because other particles in the nucleus, much like the electrons around the nucleus, do not directly participate in the beta decay. The electrons and positrons involved in beta decay, on the other hand, are produced in the nuclear-decay process. Because they do not come from the shells around the nucleus, they cannot be ignored.

Neutrinos and antineutrinos are emitted in beta decay

Before we conclude our discussion of beta decay, there is one problem that must be resolved. In analyzing the experimental results of beta decay reactions, scientists noticed a disturbing fact. If carbon-14 beta decay actually occurred as described on the previous page, energy, linear momentum, and angular momentum would not be conserved. In 1930, to solve this problem, Wolfgang Pauli proposed that a third particle must be missing from the equation. He reasoned that this new particle, called a *neutrino,* is necessary to conserve energy and momentum. Experimental evidence confirmed the existence of such a particle in 1956.

extension

Integrating Earth Science
Visit go.hrw.com for the activity "Radioactivity Within the Earth."

Keyword HF6SUBX

Demonstration

Electron Beam Deflection ADVANCED

Purpose Explain the curved tracks observed in a bubble chamber.

Materials CRT, high-voltage source, horseshoe magnet

CAUTION *The deflection of the electron beam causes low-energy X rays.*

Procedure Explain to students that a CRT contains a cathode that emits electrons when it is heated. These electrons are propelled from the anode to the cathode and beyond by a high potential difference. When the electrons hit the fluorescent screen, they produce a small amount of light. Turn off the lights in the classroom, and show the students the electron beam on the fluorescent screen. Then place the horseshoe magnet around the CRT so that the electron beam passes through the magnetic field formed at the ends of the magnet. Have students use the appropriate hand rule to predict the direction of the electron beam's deflection, and then test their predictions. Reverse the polarity of the magnet so that the beam is deflected in the opposite direction.

Explain that tracks in a bubble chamber are formed by a bubbling around the ions created by the passage of charged beta or alpha particles. Point out that the direction of a particle's deflection in a magnetic field can be used to determine the particle's charge. Ask students if they could tell which tracks were created by alpha, beta, or gamma radiation, given the magnetic field. Discuss the effect of mass on curvature.

Key Models and Analogies

To help students understand the discovery of the neutrino, have them think of the C-14 decay as an explosion. If you traced the fragments' masses and velocities but a fragment got lost, the final momentum would not appear to equal the initial momentum. Similarly, by adding the directions and speeds that correspond to the tracks from N-14 and beta particles, it was revealed that a fragment of the reaction (the neutrino) appeared to be missing.

Teaching Tip

The equations for nuclear reactions apply the principles of conservation of nucleon number and charge. Other universal conservation laws, including the conservation of momentum and the conservation of energy, must also be satisfied in nuclear processes.

Misconception Alert

Students may not understand how the antineutrino is able to be the neutrino's antiparticle, because both have charges of zero. The neutron also has an antiparticle, called an *antineutron.* Explain that in addition to charge, antiparticles are opposite their corresponding particle in quantum angular momentum, also called *spin.*

The Language of Physics

Because the gamma decay follows the beta decay immediately, the two processes are sometimes expressed by a single equation:
$^{12}_{5}\mathrm{B} \rightarrow {}^{12}_{6}\mathrm{C} + \gamma + {}^{0}_{-1}e + \overline{\nu}$

Table 5
Properties of the Neutrino

The neutrino has zero electric charge.
The neutrino's mass was once believed to be zero; recent experiments suggest a very small nonzero mass (much smaller than the mass of the electron).
The neutrino interacts very weakly with matter and is therefore very difficult to detect.

Did you know?

The word *neutrino* means "little neutral one." It was suggested by the physicist Enrico Fermi because the neutrino had to have zero electric charge and little or no mass.

The Greek letter *nu* (ν) is used to represent a neutrino. When a bar is drawn above the nu ($\overline{\nu}$), the particle is an antineutrino, or the antiparticle of a neutrino. The properties of the neutrino are summarized in **Table 5.** Note that the neutrino has no electric charge and that its mass is very small, perhaps even zero. As a result, the neutrino is difficult to detect experimentally.

With the neutrino, we can now describe the beta decay process of carbon-14 in a form that takes energy and momentum conservation into account, as follows:

$$^{14}_{6}\mathrm{C} \rightarrow {}^{14}_{7}\mathrm{N} + {}^{0}_{-1}e + \overline{\nu}$$

According to this expression, carbon-14 decays into a nitrogen nucleus, releasing an electron and an antineutrino in the process.

The decay of nitrogen-12 can also be rewritten, as follows:

$$^{12}_{7}\mathrm{N} \rightarrow {}^{12}_{6}\mathrm{C} + {}^{0}_{1}e + \nu$$

Here we see that when $^{12}_{7}\mathrm{N}$ decays into $^{12}_{6}\mathrm{C}$, a positron and a neutrino are produced. To avoid confusing these two types of beta decay, keep in mind this simple rule: *In beta decay, an electron is always accompanied by an antineutrino and a positron is always accompanied by a neutrino.*

High-energy photons are emitted in gamma decay

Very often, a nucleus that undergoes radioactive decay, either alpha or beta, is left in an excited energy state. The nucleus can then undergo a gamma decay in which one or more nucleons make transitions from a higher energy level to a lower energy level. In the process, one or more photons are emitted. Such photons, or *gamma rays,* have very high energy relative to the energy of visible light. The process of nuclear de-excitation, or gamma decay, is very similar to the emission of light by an atom, in which an electron makes a transition from a state of higher energy to a state of lower energy (as discussed in the chapter "Atomic Physics"). Note that in gamma decay, energy is emitted but the parts of the nucleus are left unchanged. Thus, both the atomic number and the mass number stay the same. Nonetheless, gamma decay is still considered to be a form of nuclear decay because it involves protons or neutrons in the nucleus.

Two common reasons for a nucleus being in an excited state are alpha and beta decay. The following sequence of events represents a typical situation in which gamma decay occurs:

$$^{12}_{5}\mathrm{B} \rightarrow {}^{12}_{6}\mathrm{C}^{*} + {}^{0}_{-1}e + \overline{\nu}$$
$$^{12}_{6}\mathrm{C}^{*} \rightarrow {}^{12}_{6}\mathrm{C} + \gamma$$

The first step is a beta decay in which $^{12}_{5}\mathrm{B}$ decays to $^{12}_{6}\mathrm{C}^{*}$. The asterisk indicates that the carbon nucleus is left in an excited state following the decay. The excited carbon nucleus then decays in the second step to the ground state by emitting a gamma ray.

NUCLEAR DECAY SERIES

If the product of a nuclear decay is stable, the decay process ends. In other cases, the decay product—the daughter nucleus—is itself unstable. The daughter nucleus then becomes the parent nucleus for an additional decay process. Such a sequence is called a *decay series.*

Figure 6(a) depicts the number of protons versus neutrons for all stable nuclei. A small portion of this graph is enlarged in **Figure 6(b),** which shows a naturally occurring decay series. This decay series begins with thorium, Th, and ends with lead, Pb.

Each square in **Figure 6(b)** corresponds to a possible nucleus. The black dots represent stable nuclei, and the red dots represent unstable nuclei. Thus, each black dot in **Figure 6(b)** corresponds to a data point in the circled portion of **Figure 6(a).** The decay series continues until a stable nucleus is reached, in this case ^{208}Pb. Notice that there is a branch in the decay path; there are actually two ways that thorium can decay into lead.

The entire series in **Figure 6(b)** consists of 10 decays: 6 alpha decays and 4 beta decays. When α decay occurs, the nucleus moves down two squares and to the left two squares because it loses two protons and two neutrons. When β^- decay occurs, the nucleus moves down one square and to the right one square because it loses one neutron and gains one proton. Gamma decays are not represented in this series because they do not alter the ratio of protons to neutrons. In other words, gamma decays do not change the atomic number (Z) or the neutron number (N). Note that the result of the decay series is to lighten the nucleus.

extension

Integrating Environmental Science

Visit go.hrw.com for the activity "Environmental Radiation."

Keyword HF6SUBX

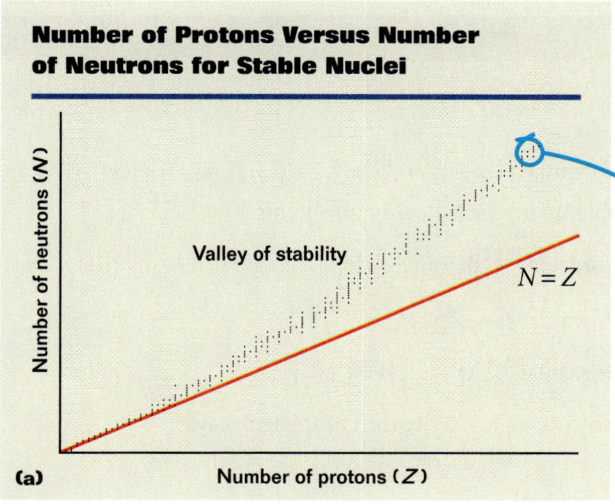

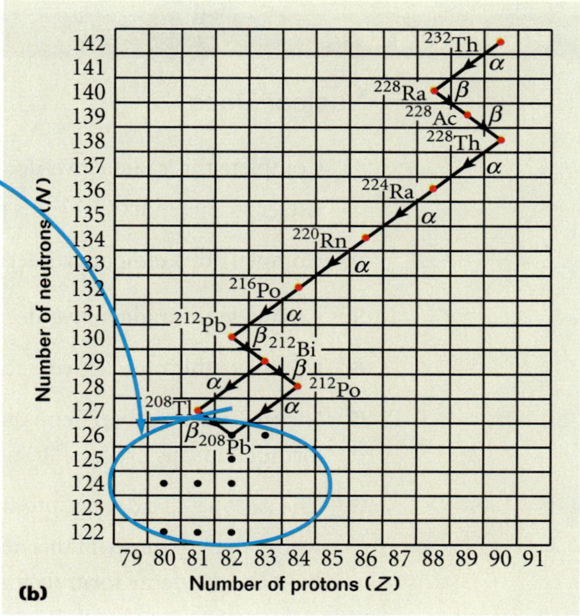

Figure 6

The heaviest nuclei in the graph of all stable nuclei **(a)** are represented by the black dots in the enlarged view **(b).** Note that an unstable nucleus (represented by the red dots) will continue to decay until the daughter nucleus is stable.

SECTION 2

Visual Strategy GENERAL

Figure 6

Be sure students understand the relationship between the two parts of this figure. Point out that the numbers shown next to each chemical symbol on the graph in **(b)** are mass numbers A $(= Z + N)$. For example, with Pb, $82 + 126 = 208$.

Q Why does alpha decay move the plot of the number of particles in the nucleus two squares to the left and two squares down?

A ***When an alpha particle is emitted, the nucleus loses two protons and two neutrons. That gives $Z - 2$ (two squares to the left) and $N - 2$ (two squares down).***

Q Write the nuclear reaction showing the decay of Ra-228 to Ac-228.

A $^{228}_{88}Ra \rightarrow ^{228}_{89}Ac + ^{0}_{-1}e + \overline{v}$

Q Can you tell from the graph in **(b)** whether gamma radiation was emitted?

A ***No; the graph only shows changes in mass and charge. Gamma rays have neither, so the energy released by gamma radiation is not shown on this representation.***

SECTION 2

Classroom Practice

Nuclear Decay

Complete the following reactions:

a. ${}^{228}_{88}\text{Ra} \rightarrow \text{X} + {}^{0}_{-1}e + \overline{\nu}$

b. ${}^{220}_{86}\text{Rn} \rightarrow {}^{216}_{84}\text{Po} + \text{X}$

c. ${}^{212}_{82}\text{Pb} \rightarrow \text{X} + {}^{0}_{-1}e + \overline{\nu}$

Answers

a. ${}^{228}_{89}\text{Ac}$

b. ${}^{4}_{2}\text{He}$

c. ${}^{212}_{83}\text{Bi}$

PROBLEM GUIDE B

Use this guide to assign problems.
SE = Student Edition Textbook
PW = Problem Workbook
PB = Problem Bank on the One-Stop Planner (OSP)

Solving for:

nuclear decay	**SE**	Sample, 1–6; Ch. Rvw. 20–21, 37*, 38–43
	PW	1–7
	PB	Sample, 1–10

***Challenging Problem**
Consult the printed Solutions Manual or the OSP for detailed solutions.

ANSWERS

Practice B

1. ${}^{12}_{6}\text{C}$
2. ${}^{208}_{81}\text{Tl}$
3. ${}^{14}_{6}\text{C}$
4. ${}^{4}_{2}\text{He}$
5. ${}^{63}_{28}\text{Ni} \rightarrow {}^{63}_{29}\text{Cu} + {}^{0}_{-1}e + \overline{\nu}$
6. **a.** β^{-} decay
   b. ${}^{56}_{26}\text{Fe} \rightarrow {}^{56}_{27}\text{Co} + {}^{0}_{-1}e + \overline{\nu}$

SAMPLE PROBLEM B

Nuclear Decay

PROBLEM

The element radium was discovered by Marie and Pierre Curie in 1898. One of the isotopes of radium, ${}^{226}_{88}\text{Ra}$, decays by alpha emission. What is the resulting daughter element?

SOLUTION

Given: The decay can be written symbolically as follows:

$${}^{226}_{88}\text{Ra} \rightarrow \text{X} + {}^{4}_{2}\text{He}$$

Unknown: the daughter element (X)

The mass numbers and atomic numbers on the two sides of the expression must be the same so that both charge and nucleon number are conserved during the course of this particular decay.

$$\text{Mass number of X} = 226 - 4 = 222$$

$$\text{Atomic number of X} = 88 - 2 = 86$$

$${}^{226}_{88}\text{Ra} \rightarrow {}^{222}_{86}\text{X} + {}^{4}_{2}\text{He}$$

The periodic table (Appendix G) shows that the nucleus with an atomic number of 86 is radon, Rn. Thus, the decay process is as follows:

$$\boxed{{}^{226}_{88}\text{Ra} \rightarrow {}^{222}_{86}\text{Rn} + {}^{4}_{2}\text{He}}$$

PRACTICE B

Nuclear Decay

1. Complete this radioactive-decay formula: ${}^{12}_{5}\text{B} \rightarrow ? + {}^{0}_{-1}e + \overline{\nu}$
   (Refer to Appendix G for this problem and the following problems.)
2. Complete this radioactive-decay formula: ${}^{212}_{83}\text{Bi} \rightarrow ? + {}^{4}_{2}\text{He}$
3. Complete this radioactive-decay formula: $? \rightarrow {}^{14}_{7}\text{N} + {}^{0}_{-1}e + \overline{\nu}$
4. Complete this radioactive-decay formula: ${}^{225}_{89}\text{Ac} \rightarrow {}^{221}_{87}\text{Fr} + ?$
5. Nickel-63 decays by β^{-} emission to copper-63. Write the complete decay formula for this process.
6. The isotope ${}^{56}_{26}\text{Fe}$ decays into the isotope ${}^{56}_{27}\text{Co}$.
   a. By what process will this decay occur?
   b. Write the decay formula for this process.

MEASURING NUCLEAR DECAY

Imagine that you are studying a sample of radioactive material. You know that the atoms in the material are decaying into other types of atoms. How many of the unstable parent atoms remain after a certain amount of time?

The decay constant indicates the rate of radioactive decay

If the sample contains N radioactive parent nuclei at some instant, the number of parent nuclei that decay into daughter nuclei (ΔN) in a small time interval (Δt) is proportional to N, as follows:

$$\Delta N = -\lambda N \Delta t$$

The negative sign signifies that N decreases with time; that is, ΔN is negative. The quantity λ is called the *decay constant.* The value of λ for any isotope indicates the rate at which that isotope decays. Isotopes with a large decay constant decay quickly, and those with a small decay constant decay slowly. The number of decays per unit time, $-\Delta N/\Delta t$, is called the *decay rate,* or *activity,* of the sample. Note that the activity of a sample equals the decay constant times the number of radioactive nuclei in the sample, as follows:

$$\text{activity} = \frac{-\Delta N}{\Delta t} = \lambda N$$

The SI unit of activity is the *becquerel* (Bq). One becquerel is equal to 1 decay/s. The *curie* (Ci), which was the original unit of activity, is the approximate activity of 1 g of radium. One curie is equal to 3.7×10^{10} Bq.

Did you know?

In 1898, Marie and Pierre Curie discovered two previously unknown elements, polonium and radium, both of which were radioactive. They were awarded the Nobel Prize in physics in 1903 for their studies of radioactive substances.

Half-life measures how long it takes half a sample to decay

Another quantity that is useful for characterizing radioactive decay is the **half-life,** written as $T_{1/2}$. The half-life of a radioactive substance is the time it takes for half of the radioactive nuclei in a sample to decay. The half-life of any substance is inversely proportional to the decay constant of the substance.

half-life

the time needed for half of the original nuclei of a sample of a radioactive substance to undergo radioactive decay

Why it Matters

Conceptual Challenge

1. Decay Series

Suppose a radioactive parent substance with a very long half-life has a daughter with a very short half-life. Describe what happens to a freshly purified sample of the parent substance.

2. Probability of Decay

"The more probable the decay, the shorter the half-life." Explain this statement.

3. Decay of Radium

The radioactive nucleus $^{226}_{88}$Ra (radium-226) has a half-life of about 1.6×10^3 years. Although the solar system is approximately 5 billion years old, we still find this radium nucleus in nature. Explain how this is possible.

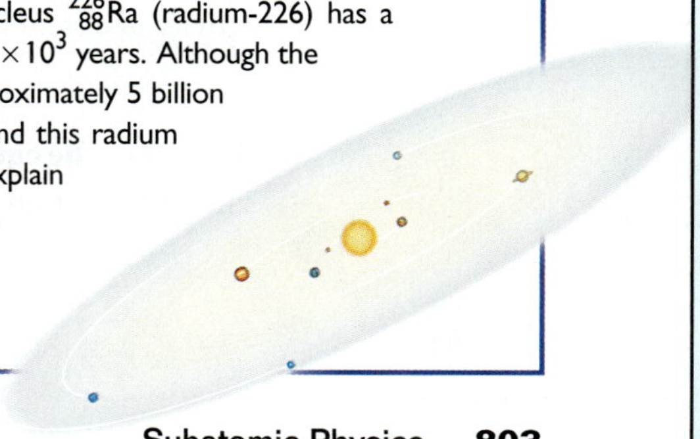

Misconception Alert — GENERAL

Students might not understand that every substance has a decay constant, λ, but that the decay rate, or activity, is not constant. A numerical example can help them see that the activity depends on the number of radioactive nuclei in the sample at some instant. Ask students to compare the activity for two samples of radium ($\lambda = 1.4 \times 10^{-11}\ s^{-1}$). Sample A contains 5.0×10^{15} radioactive nuclei, and sample B contains 2.0×10^{18} radioactive nuclei. Using λN, students will find that A's activity is 7.0×10^4 Bq, while B's activity is 2.8×10^7 Bq.

Also emphasize that half-life does not depend on the size of a sample. For example, any sample of carbon-14 has a half-life of 5715 years.

ANSWERS

Conceptual Challenge

1. The parent substance will decay very slowly. Many of its nuclei will remain undecayed for a long period of time. As soon as some of its nuclei have decayed, the daughter nuclei will decay rapidly.
2. As the probability of decay increases, the amount of nuclei that decay per unit time increases, so it takes less time for the sample to decay.
3. Other elements with longer half-lives decay to $^{226}_{88}$Ra.

Classroom Practice

Measuring Nuclear Decay
The half-life of $^{144}_{56}Ba$ is about 12 s. What is the decay constant of barium-144?

Answer
$5.8 \times 10^{-2}\ s^{-1}$

A sample of barium contains 5.0×10^9 undecayed nuclei of barium-144. How many radioactive nuclei remain after the following time periods:

a. 12 s
b. 24 s
c. 36 s

Answers
a. $N = 2.5 \times 10^9$
b. $N = 1.2 \times 10^9$
c. $N = 6.2 \times 10^8$

Determine the activity in becquerels and in curies after each time period given above.

Answers
a. 1.4×10^8 Bq; 3.9×10^{-3} Ci
b. 7.0×10^7 Bq; 1.9×10^{-3} Ci
c. 3.6×10^7 Bq; 9.7×10^{-4} Ci

Teaching Tip — GENERAL

Point out that, as seen in the Classroom Practice, the activity of a sample decreases over time because the number of radioactive nuclei (N) decreases.

Substances with large decay constants have short half-lives. The relationship between half-life and decay constant is given in the equation below. A derivation of this equation is beyond the scope of this book, but it involves the natural logarithm of 2. Because ln 2 = 0.693, this factor occurs in the final equation.

HALF-LIFE

$$T_{1/2} = \frac{0.693}{\lambda}$$

$$\textbf{half-life} = \frac{\textbf{0.693}}{\textbf{decay constant}}$$

Consider a sample that begins with N radioactive nuclei. By definition, after one half-life, $\frac{1}{2}N$ radioactive nuclei remain. After two half-lives, half of these will have decayed, so $\frac{1}{4}N$ radioactive nuclei remain. After three half-lives, $\frac{1}{8}N$ will remain, and so on.

SAMPLE PROBLEM C

Measuring Nuclear Decay

PROBLEM

The half-life of the radioactive radium (^{226}Ra) nucleus is 5.0×10^{10} s. A sample contains 3.0×10^{16} nuclei. What is the decay constant for this decay? How many radium nuclei, in curies, will decay per second?

SOLUTION

1. DEFINE

Given: $T_{1/2} = 5.0 \times 10^{10}$ s $N = 3.0 \times 10^{16}$

Unknown: $\lambda = ?$ activity = ? Ci

2. PLAN

Choose an equation or situation:
To find the decay constant, use the equation for half-life.

$$T_{1/2} = \frac{0.693}{\lambda}$$

The number of nuclei that decay per second is given by the equation for the activity of a sample.

$$\text{activity} = \lambda N$$

Rearrange the equation to isolate the unknown:
The first equation must be rearranged to isolate the decay constant, λ.

$$\lambda = \frac{0.693}{T_{1/2}}$$

3. CALCULATE **Substitute the values into the equations and solve:**

$$\lambda = \frac{0.693}{T_{1/2}} = \frac{0.693}{5.0 \times 10^{10}\ \text{s}}$$

$$\boxed{\lambda = 1.4 \times 10^{-11}\text{s}^{-1}}$$

$$\text{activity} = \lambda N = \frac{(1.4 \times 10^{-11}\ \text{s}^{-1})(3.0 \times 10^{16})}{3.7 \times 10^{10}\ \text{s}^{-1}/\text{Ci}}$$

TIP *Always pay attention to units. Here, the activity is divided by the conversion factor $3.7 \times 10^{10}\ s^{-1}$/Ci to convert the answer from becquerels to curies, as specified in the problem statement.*

$$\boxed{\text{activity} = 1.1 \times 10^{-5}\ \text{Ci}}$$

4. EVALUATE Because the half-life is on the order of 10^{10} s, the decay constant, which approximately equals 0.7 divided by the half-life, should equal a little less than $10^{-10}\ \text{s}^{-1}$. Thus, $1.4 \times 10^{-11}\ \text{s}^{-1}$ is a reasonable answer for the decay constant.

PRACTICE C

Measuring Nuclear Decay

1. The half-life of $^{214}_{84}$Po is 164 μs. A polonium-214 sample contains 2.0×10^6 nuclei. What is the decay constant for the decay? How many polonium nuclei, in curies, will decay per second?

2. The half-life of $^{214}_{83}$Bi is 19.7 min. A bismuth-214 sample contains 2.0×10^9 nuclei. What is the decay constant for the decay? How many bismuth nuclei, in curies, will decay per second?

3. The half-life of $^{131}_{53}$I is 8.07 days. Calculate the decay constant for this isotope. What is the activity in Ci for a sample that contains 2.5×10^{10} iodine-131 nuclei?

4. Suppose that you start with 1.00×10^{-3} g of a pure radioactive substance and determine 2.0 h later that only 0.25×10^{-3} g of the substance is left undecayed. What is the half-life of this substance?

5. Radon-222 ($^{222}_{86}$Rn) is a radioactive gas with a half-life of 3.82 days. A gas sample contains 4.0×10^8 radon atoms initially.

a. Estimate how many radon atoms will remain after 12 days.
b. Estimate how many radon nuclei will have decayed by this time.

SECTION 2

PROBLEM GUIDE C

Use this guide to assign problems.
SE = Student Edition Textbook
PW = Problem Workbook
PB = Problem Bank on the One-Stop Planner (OSP)

Solving for:

λ	**SE** Sample, 1–3; Ch. Rvw. 22 **PW** 5–6 **PB** 6, 9
$T_{1/2}$	**SE** 4; Ch. Rvw. 45 **PW** 7 **PB** Sample, 1–3
N	**SE** 5; Ch. Rvw. 44*, 47–48 **PW** 1, 4 **PB** 7, 10
Δt	**SE** Ch. Rvw. 23–24 **PW** Sample, 2–3 **PB** 4–5
activity	**SE** Ch. Rvw. 22, 36 **PB** 8

***Challenging Problem**
Consult the printed Solutions Manual or the OSP for detailed solutions.

ANSWERS

Practice C

1. $4.23 \times 10^3\ \text{s}^{-1}$, 0.23 Ci
2. $5.86 \times 10^{-4}\ \text{s}^{-1}$, 3.2×10^{-5} Ci
3. $9.94 \times 10^{-7}\ \text{s}^{-1}$, 6.7×10^{-7} Ci
4. 1.0 h
5. **a.** about 5.0×10^7 atoms
   b. about 3.5×10^8 atoms

SECTION 2

Misconception Alert

Some students may think that the half-life time is half of the time it takes for a sample to be used up. Ask students how many radioactive nuclei out of 600 are left after 2.0 h if $T_{1/2}$ is 1.0 h (*150*).

Visual Strategy GENERAL

Figure 7

Be sure students interpret the graph properly.

Q Given that the half-life of C-14 is 5715 years, what years are shown on the graph points?

A *0; 5715; 11 430; 17 145*

Q If the initial sample contained 80 000 radioactive nuclei, when would it fall below 20 000 nuclei?

A *after 11 430 years (two half-lives)*

SECTION REVIEW ANSWERS

1. In α, ${}^4_2$He particles are emitted; in β, electrons or positrons are emitted; in γ, photons are emitted. In α and β, a new element is produced; in γ, energy is lost.
2. a. ${}^{228}_{88}$Ra
   b. ${}^{12}_{6}$C
   c. ${}^{149}_{62}$Sm
3. a. 1.2×10^{-10} s^{-1}
   b. 5.8×10^{9} s (180 years)
4. 17 140 years

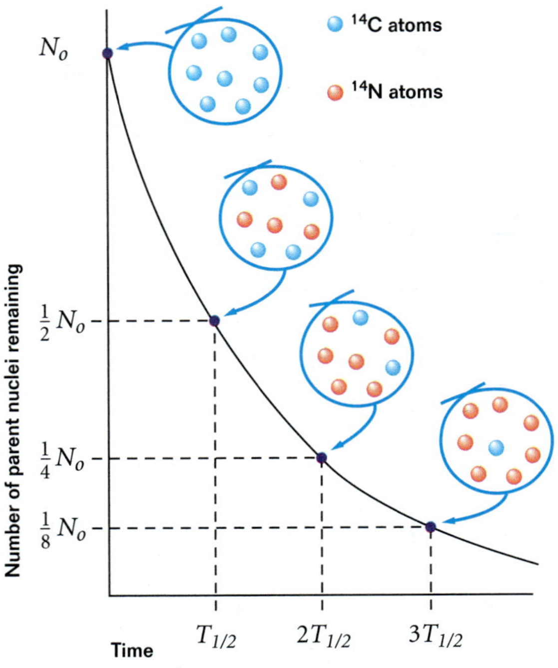

Figure 7
The radioactive isotope carbon-14 has a half-life of 5715 years. In each successive 5715-year period, half the remaining carbon-14 nuclei decay to nitrogen-14.

A *decay curve* is a plot of the number of radioactive parent nuclei remaining in a sample as a function of time. A typical decay curve for a radioactive sample is shown in **Figure 7.** After each half-life, half the remaining parent nuclei have decayed. This is represented in the circles to the right of the decay curve. The blue spheres are the parent nuclei (carbon-14), and the red spheres are daughter nuclei (nitrogen-14). Notice that the total number of nuclei remains constant, while the number of carbon atoms continually decreases over time.

For example, the initial sample contains 8 carbon-14 atoms. After one half-life, there are 4 carbon-14 atoms and 4 nitrogen-14 atoms. By the next half-life, the number of carbon-14 atoms is reduced to 2, and the process continues. As the number of carbon-14 atoms decreases, the number of nitrogen-14 atoms increases.

Living organisms have a constant ratio of carbon-14 to carbon-12 because they continuously exchange carbon dioxide with their surroundings. When an organism dies, this ratio changes due to the decay of carbon-14. Measuring the ratio between carbon-14, which decays as shown in **Figure 7,** and carbon-12, which does not decay, provides an approximate date as to when the organism was alive.

SECTION REVIEW

1. Explain the main differences between alpha, beta, and gamma decays.
2. Complete the following radioactive decay formulas:
   a. ${}^{232}_{90}\mathrm{Th} \rightarrow ? + {}^{4}_{2}\mathrm{He}$
   b. ${}^{12}_{5}\mathrm{B} \rightarrow ? + {}^{0}_{-1}e + \overline{v}$
   c. $? \rightarrow {}^{4}_{2}\mathrm{He} + {}^{145}_{60}\mathrm{Nd}$
3. A radioactive sample consists of 5.3×10^5 nuclei. There is one decay every 4.2 h.
   a. What is the decay constant for the sample?
   b. What is the half-life for the sample?
4. **Critical Thinking** The ^{14}C content decreases after the death of a living system with a half-life of 5715 years. If an archaeologist finds an ancient fire pit containing partially consumed firewood and if the ^{14}C content of the wood is only 12.5 percent that of an equal carbon sample from a present-day tree, what is the age of the ancient site?

Nuclear Reactions

SECTION 3

SECTION 3
General Level

SECTION OBJECTIVES

- Distinguish between nuclear fission and nuclear fusion.
- Explain how a chain reaction is utilized by nuclear reactors.
- Compare fission and fusion reactors.

FISSION AND FUSION

Any process that involves a change in the nucleus of an atom is called a *nuclear reaction.* Nuclear reactions include *fission,* in which a nucleus splits into two or more nuclei, and *fusion,* in which two or more nuclei combine.

Stable nuclei can be converted to unstable nuclei

When a nucleus is bombarded with energetic particles, it may capture a particle, such as a neutron. As a result, the nucleus will no longer be stable and will disintegrate. For example, protons can be released when alpha particles collide with nitrogen atoms, as follows:

$$^{4}_{2}\mathrm{He} + {}^{14}_{7}\mathrm{N} \rightarrow \mathrm{X} + {}^{1}_{1}\mathrm{H}$$

According to this expression, an alpha particle ($^{4}_{2}$He) strikes a nitrogen nucleus ($^{14}_{7}$N) and produces an unknown product nucleus (X) and a proton ($^{1}_{1}$H). By balancing atomic numbers and mass numbers, we can conclude that the unknown product has a mass number of 17 and an atomic number of 8. Because the element with an atomic number of 8 is oxygen, the product can be written symbolically as $^{17}_{8}$O, and the reaction can be written as follows:

$$^{4}_{2}\mathrm{He} + {}^{14}_{7}\mathrm{N} \rightarrow {}^{18}_{9}\mathrm{F} \rightarrow {}^{17}_{8}\mathrm{O} + {}^{1}_{1}\mathrm{H}$$

This nuclear reaction starts with two stable isotopes—helium and nitrogen—that form an unstable intermediate nucleus, fluorine. The intermediate nucleus then disintegrates into two different stable isotopes, hydrogen and oxygen. This reaction, which was the first nuclear reaction to be observed, was detected by Rutherford in 1919.

SCILINKS.
www.scilinks.org
Topic: Fission/Fusion
Code: HF60581

Heavy nuclei can undergo nuclear fission

Nuclear fission occurs when a heavy nucleus splits into two lighter nuclei. For fission to occur naturally, the nucleus must release energy. This means that the nucleons in the daughter nuclei must be more tightly bound and therefore have less mass than the nucleons in the parent nucleus. This decrease in mass per nucleon appears as released energy when fission occurs, often in forms such as photons or kinetic energy of the fission products. Because fission produces lighter nuclei, the binding energy per nucleon must *increase* with decreasing atomic number. **Figure 8** shows that this is possible only for atoms in which $A > 58$. Thus, *fission occurs naturally only for heavy atoms.*

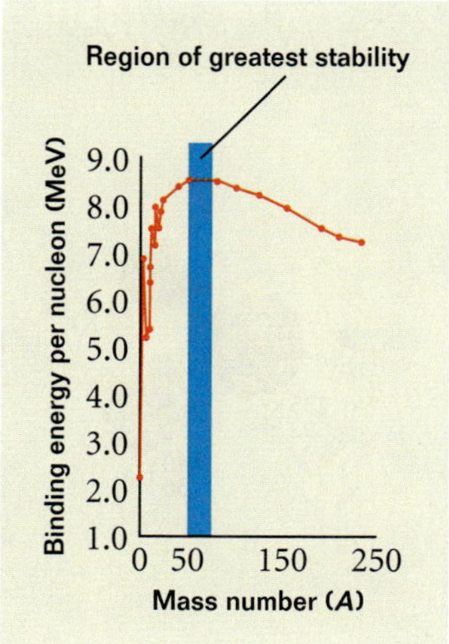

Figure 8
Light nuclei are very loosely bound. The binding energy of heavy nuclei is roughly the same for all nuclei.

Visual Strategy GENERAL

Figure 8
Point out that the graph represents empirical information about energy of all known nuclei, including isotopes of mass numbers $A = 1$ to $A \approx 230$.

Q According to the graph, what is the average binding energy per nucleon of the most stable nucleus?

A ***about 8.5 MeV per nucleon***

Q Suppose a nucleus of mass number $A = 24$ splits into two nuclei of equal masses. Determine the approximate binding energy per nucleon for each daughter nucleus.

A ***about 7 MeV; This is less than the binding energy per nucleon of the parent, so the daughter nuclei will be less stable. This example illustrates why nuclei with $A < 58$ do not undergo fission.***

Q Repeat the last exercise with a parent nucleus of $A = 200$.

A ***about 8.5 MeV; In this case, the binding energy per nucleon increases after fission.***

SECTION 3

The Language of Physics — BASIC

Remind students that the upper number next to the element symbol is the mass number (A) and that the lower number indicates the number of protons (Z). Also remind them that these equations show the conservation of nucleon number and charge in a nuclear reaction. Ask how many protons are found in the parents and in the daughters in the nuclear fission of U-235 *(0 + 92 = 56 + 36 + 0)*. How many nucleons are found in each? *(1 + 235 = 140 + 93 + (3 × 1))*

Misconception Alert — ADVANCED

Students may wonder about the meaning of "an average of about 2.5 neutrons" being emitted when ^{235}U undergoes fission, as described in the last paragraph on this student page. Point out that the products of U-235 fission may be different daughter nuclei than those shown in the diagram. Some reactions release three neutrons, while others release two. Have students find an example of each in **Figure 9.** *(three neutrons:* $^{1}_{0}n + ^{235}_{92}U \rightarrow ^{87}_{35}Br + ^{146}_{57}La + 3\,^{1}_{0}n$*; two neutrons:* $^{1}_{0}n + ^{235}_{92}U \rightarrow ^{144}_{55}Cs + ^{90}_{37}Rb + 2\,^{1}_{0}n$*)*

Teaching Tip

Point out that **Figure 9** is a schematic diagram. It does not represent the actual location of U-235 nuclei. There are about 2.6×10^{21} nuclei in 1 g of uranium. Most of these nuclei are U-238, which are more stable. Only a small portion are U-235, which will decay when hit by a neutron.

extension

Integrating Astronomy
Visit go.hrw.com for the activity "The Life Cycle of a Star."

Keyword HF6SUBX

One example of this process is the fission of uranium-235. First, the nucleus is bombarded with neutrons. When the nucleus absorbs a neutron, it becomes unstable and decays. The fission of ^{235}U can be represented as follows:

$$^{1}_{0}n + ^{235}_{92}U \rightarrow ^{236}_{92}U^{*} \rightarrow X + Y + \text{neutrons}$$

The isotope $^{236}_{92}U^{*}$ is an intermediate state that lasts only for about 10^{-12} s before splitting into X and Y. Many combinations of X and Y are possible. In the fission of uranium, about 90 different daughter nuclei can be formed. The process also results in the production of about two or three neutrons per fission event.

A typical reaction of this type is as follows:

$$^{1}_{0}n + ^{235}_{92}U \rightarrow ^{140}_{56}Ba + ^{93}_{36}Kr + 3\,^{1}_{0}n$$

To estimate the energy released in a typical fission process, note that the binding energy per nucleon is about 7.6 MeV for heavy nuclei (those having a mass number of approximately 240) and about 8.5 MeV for nuclei of intermediate mass (see **Figure 8** on the previous page). The amount of energy released in a fission event is the difference in these binding energies (8.5 MeV − 7.6 MeV, or about 0.9 MeV per nucleon). Assuming a total of 240 nucleons, this is about 220 MeV. This is a very large amount of energy relative to the energy released in typical chemical reactions. For example, the energy released in burning one molecule of the octane used in gasoline engines is about one hundred-millionth the energy released in a single fission event.

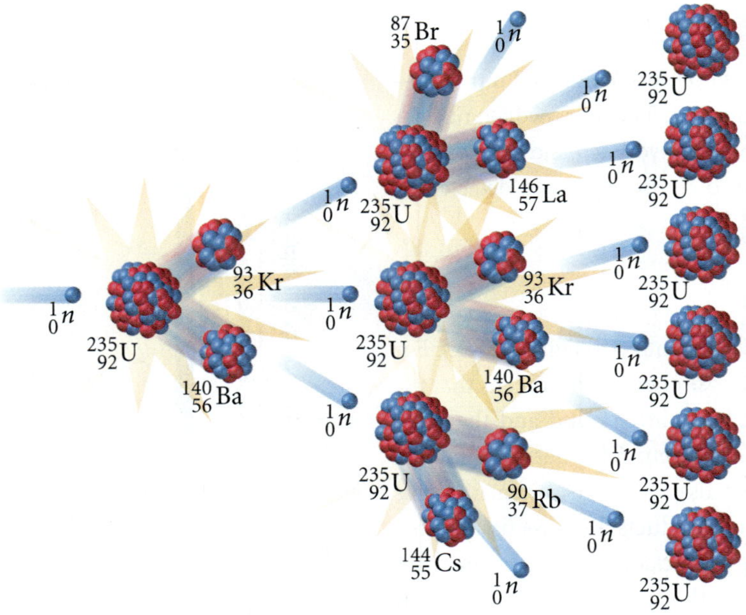

Figure 9
A nuclear chain reaction can be initiated by the capture of a neutron.

Neutrons released in fission can trigger a chain reaction

When ^{235}U undergoes fission, an average of about 2.5 neutrons are emitted per event. The released neutrons can be captured by other nuclei, making these nuclei unstable. This triggers additional fission events, which lead to the possibility of a *chain reaction,* as shown in **Figure 9.** Calculations show that if the chain reaction is not controlled—that is, if it does not proceed slowly—it could result in the release of an enormous amount of energy and a violent explosion. If the energy in 1 kg of ^{235}U were released, it would equal the energy released by the detonation of about 20 000 tons of TNT. This is the principle behind the first nuclear bomb, shown in **Figure 10,** which was essentially an uncontrolled fission reaction.

A *nuclear reactor* is a system designed to maintain a controlled, self-sustained chain reaction. Such a system was first achieved with uranium as the fuel in 1942 by Enrico Fermi, at the University of Chicago. Primarily, it is the uranium-235 isotope that releases energy through nuclear fission. Uranium from ore typically contains only about 0.7 percent of ^{235}U, with the remaining 99.3 percent being the ^{238}U isotope. Because uranium-238 tends to absorb neutrons without fissioning, reactor fuels must be processed to increase the proportion of ^{235}U so that the reaction can sustain itself. This process is called *enrichment.*

At this time, all nuclear reactors operate through fission. One difficulty associated with fission reactors is the safe disposal of radioactive materials when the core is replaced. Transportation of reactor fuel and reactor wastes poses safety risks. As with all energy sources, the risks must be weighed against the benefits and the availability of the energy source.

Figure 10
The first nuclear fission bomb, often called the *atomic bomb,* was tested in New Mexico in 1945.

Light nuclei can undergo nuclear fusion

Nuclear fusion occurs when two light nuclei combine to form a heavier nucleus. As with fission, the product of a fusion event must have a greater binding energy than the original nuclei for energy to be released in the reaction. Because fusion reactions produce heavier nuclei, the binding energy per nucleon must increase as atomic number increases. As shown in **Figure 8** (on the first page of this section), this is possible only for atoms with $A < 58$. Hence, *fusion occurs naturally only for light atoms.*

One example of this process is the fusion reactions that occur in stars. All stars generate energy through fusion. About 90 percent of the stars, including our sun, fuse hydrogen and possibly helium. Some other stars fuse helium or other heavier elements. The *proton-proton cycle* is a series of three nuclear-fusion reactions that are believed to be stages in the liberation of energy in our sun and other stars rich in hydrogen. In the proton-proton cycle, four protons combine to form an alpha particle and two positrons, releasing 25 MeV of energy in the process. The first two steps in this cycle are as follows:

$$^1_1\text{H} + ^1_1\text{H} \rightarrow ^2_1\text{H} + ^0_1e + \nu$$

$$^1_1\text{H} + ^2_1\text{H} \rightarrow ^3_2\text{He} + \gamma$$

This is followed by either of the following processes:

$$^1_1\text{H} + ^3_2\text{He} \rightarrow ^4_2\text{He} + ^0_1e + \nu$$

$$^3_2\text{He} + ^3_2\text{He} \rightarrow ^4_2\text{He} + ^1_1\text{H} + ^1_1\text{H}$$

The released energy is carried primarily by gamma rays, positrons, and neutrinos. These energy-liberating fusion reactions are called *thermonuclear fusion reactions.* The hydrogen (fusion) bomb, first detonated in 1952, is an example of an uncontrolled thermonuclear fusion reaction.

Did you know?

What has been called the *atomic bomb* since 1945 is actually a tremendous *nuclear fission* reaction. Likewise, the so-called *hydrogen bomb* is an uncontrolled *nuclear fusion* reaction in which hydrogen nuclei merge to form helium nuclei.

Teaching Tip — BASIC

Have students use **Figure 8** (on the first page of this section) to answer the following questions: If two nuclei of 90 nucleons each were to fuse, would the binding energy per nucleon in the 180-nucleon nucleus be higher or lower? *(The binding energy per nucleon would decrease from 8.2 MeV to 7.5 MeV.)* What can you infer about the stability of the product? *(The product would be less stable.)* What if two nuclei with five nucleons each were fused? *(The binding energy per nucleon would increase from 7.1 MeV to 8.5 MeV; the product would be more stable.)*

The Language of Physics

The terms *atomic bomb* and *hydrogen bomb* are misleading because terms associated with atoms and elements are used to denote *nuclear* reactions. Students should be reminded that the order of magnitude of energies at the nuclear level is a million times higher than it is at the atomic level.

Teaching Tip

Tell students that stars do not normally fuse elements beyond iron-58, because it is the most stable element. Fusing elements beyond iron-58 would require inputting energy rather than producing energy.

SECTION 3

Teaching Tip — ADVANCED

Have students choose a question about fusion to research and report to class. Example questions may include:

What will happen when the sun runs out of hydrogen, which currently produces helium and energy?

Why don't power plants use fusion to produce electricity?

SECTION REVIEW ANSWERS

1. Both are nuclear reactions that release energy. In fission, a heavy nucleus splits; in fusion, two light nuclei combine.
2. The excess neutrons produced by each reaction are used to induce additional reactions.
3. Enrichment is the process of increasing the amount of ^{235}U in reactor fuels. This is necessary because uranium from ore contains mostly ^{238}U.
4. **a.** $^{1}_{0}n + ^{235}_{92}U \rightarrow ^{141}_{56}Ba + ^{92}_{36}Kr + 3\,^{1}_{0}n$

   b. three neutrons
5. Fusion reactors produce less radioactive waste, and the fuel (water) is plentiful. Difficulties include the high temperatures and densities necessary to achieve controlled fusion reactions.
6. A fusion reactor would produce less radioactive waste than a fission reactor does because fusion produces more-stable nuclei than fission does.

Fusion reactors are being developed

The enormous amount of energy released in fusion reactions suggests the possibility of harnessing this energy for useful purposes on Earth. Efforts are under way to create controlled thermonuclear reactions in the form of a *fusion reactor.* Because of the ready availability of its fuel source—water—controlled fusion is often called the ultimate energy source.

For example, if deuterium ($^{2}_{1}H$) were used as the fuel, 0.16 g of deuterium could be extracted from just 1 L of water at a cost of about one cent. Such rates would make the fuel costs of even an inefficient reactor almost insignificant. An additional advantage of fusion reactors is that few radioactive byproducts are formed. The proton-proton cycle shows that the end product of the fusion of hydrogen nuclei is safe, nonradioactive helium. Unfortunately, a thermonuclear reactor that can deliver a net power output for an extended time is not yet a reality. Many difficulties must be resolved before a successful device is constructed.

For example, the energy released in a gas undergoing nuclear fusion depends on the number of fusion reactions that can occur in a given amount of time. This varies with the density of the gas because collisions are more frequent in a denser gas. It also depends on the amount of time the gas is confined.

In addition, the Coulomb repulsion force between two charged nuclei must be overcome before they can fuse. The fundamental challenge is to give the nuclei enough kinetic energy to overcome this repulsive force. This can be accomplished by heating the fuel to extremely high temperatures (about 10^8 K, or about 10 times greater than the interior temperature of the sun). Such high temperatures are difficult and expensive to obtain in a laboratory or a power plant.

SECTION REVIEW

1. What are the similarities and differences between fission and fusion?
2. Explain how nuclear reactors utilize chain reactions.
3. What is enrichment? Why is enrichment necessary when uranium is used as a reactor fuel?
4. A fission reaction leads to the formation of ^{141}Ba and ^{92}Kr when ^{235}U absorbs a neutron.

   a. How is this reaction expressed symbolically?

   b. How many neutrons are released in this reaction?
5. What are some advantages to fusion reactors (as opposed to fission reactors)? What are some difficulties in the development of a fusion reactor?
6. **Critical Thinking** Why would a fusion reactor produce less radioactive waste material than a fission reactor does?

Particle Physics

SECTION 4

SECTION OBJECTIVES

- **Define the four fundamental interactions of nature.**
- **Identify the elementary particles that make up matter.**
- **Describe the standard model of the universe.**

THE PARTICLE VIEW OF NATURE

Particle physics seeks to discover the ultimate structure of matter: *elementary particles.* Elementary particles, which are the fundamental units that compose matter, do not appear to be divisible and have neither size nor structure.

Many new particles have been produced in accelerators

Until 1932, scientists thought protons and electrons were elementary particles because these particles were stable. However, beginning in 1945, experiments at particle accelerators, such as the Stanford Linear Accelerator shown in **Figure 11,** have demonstrated that new particles are often formed in high-energy collisions between known particles. These new particles tend to be very unstable and have very short half-lives, ranging from 10^{-6}s to 10^{-23}s. So far, more than 300 new particles have been catalogued.

Figure 11
The Stanford Linear Accelerator, in California, creates high-energy particle collisions that provide evidence of new particles.

There are four fundamental interactions in nature

The key to understanding the properties of elementary particles is to be able to describe the interactions between them. All particles in nature are subject to four fundamental interactions: *strong, electromagnetic, weak,* and *gravitational.*

The strong interaction is responsible for the binding of neutrons and protons into nuclei, as we have seen. This interaction, which represents the "glue" that holds the nucleons together, is the strongest of all the fundamental interactions. It is very short-ranged and is negligible for separations greater than about 10^{-15} m (the approximate size of a nucleus).

The electromagnetic interaction, which is about 10^{-2} times the strength of the strong interaction at nuclear distances, is responsible for the attraction of unlike charges and the repulsion of like charges. This interaction is responsible for the binding of atoms and molecules. It is a long-range interaction that decreases in strength as the inverse square of the separation between interacting particles, as described in the chapter "Electric Forces and Fields."

The weak interaction is a short-range nuclear interaction that is involved in beta decay. Its strength is only about 10^{-13} times that of the strong interaction. However, because the strength of an interaction depends on the distance through which it acts, the relative strengths of two interactions differ depending on what separation distance is used. The strength of the weak interaction, for example, is sometimes cited to be as large as 10^{-6} times that of the strong interaction. Keep in mind that these relative strengths are merely estimates and they depend on the assumed separation distance.

SECTION 4
Advanced Level

Teaching Tip — GENERAL

Remind students that until the end of the nineteenth century, atoms were thought to be the most fundamental particles with no substructure. Scientific advances such as the discovery of the electron, the nucleus (with its protons and neutrons), and—more recently—the substructure of particles within the nucleus have changed our view of matter. It is possible that when we reach higher energy densities and measure shorter periods of time, we will be able to detect smaller and smaller particles; so far, this has produced a "subatomic zoo" with over 300 new particles. Encourage students to discuss whether there will be an end to breaking down particles into more-elementary components.

STOP **Misconception Alert** — BASIC

Students will be familiar with electric and gravitational forces, but they may not have a clear idea of how the two compare with one another. Point out that these two forces differ in order of magnitude.

The Language of Physics

A simple diagram representing the interaction between two particles is called a *Feynman diagram.* Such diagrams are named after the American scientist Richard P. Feynman (1918–1988), who developed them. **Figure 12** is an example of a Feynman diagram.

Teaching Tip — BASIC

Make sure students understand the information contained in **Table 6.** Remind them that fm stands for fermi (10^{-15} m), the unit of distance that is used at the nuclear scale. Ask students the following: Which force has the shortest range? *(the weak force)* What are the mediating particles between two protons repelling each other electrostatically? *(photons)* What are the mediating particles between an apple and Earth? *(gravitons)*

Finally, the gravitational interaction is a long-range interaction with a strength of only about 10^{-38} times that of the strong interaction. Although this familiar interaction is what holds the planets, stars, and galaxies together, its effect on elementary particles is negligible. The gravitational interaction is the weakest of all four fundamental interactions.

A force can be thought of as mediated by an exchange of particles

Notice that in this section, we have referred to a force as an *interaction.* This is because in particle physics the interaction of matter is usually described not in terms of forces but in terms of the exchange of special particles. In the case of the electromagnetic interaction, the particles are photons. Thus, it is said that the electromagnetic force is *mediated* by photons.

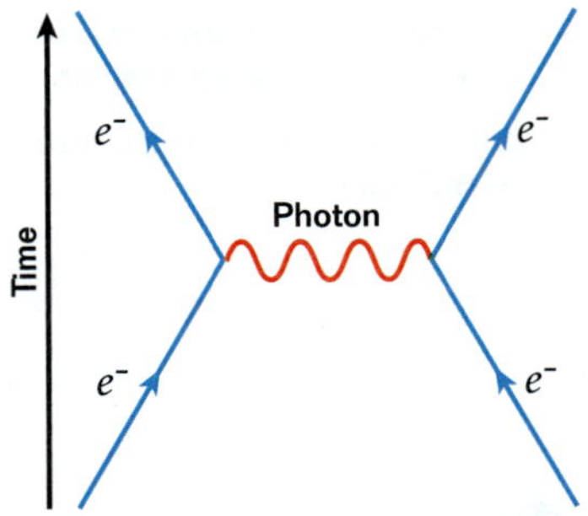

Figure 12
In particle physics, the electromagnetic interaction is modeled as an exchange of photons. The wavy red line represents a photon, and the blue lines represent electrons.

Figure 12 shows how two electrons might repel each other through the exchange of a photon. Because momentum is conserved, the electron emitting a photon changes direction slightly. As the photon is absorbed, the other electron's direction must also change. The net effect is that the two electrons change direction and move away from each other.

Likewise, the strong interaction is mediated by particles called *gluons,* the weak interaction is mediated by particles called the *W* and *Z bosons,* and the gravitational interaction is mediated by *gravitons.* All of these except gravitons have been detected. The four fundamental interactions of nature and their mediating field particles are summarized in **Table 6.**

Table 6 The Fundamental Interactions of Nature

Interaction (force)	Relative strength	Range of force	Mediating field particle
strong	1	≈ 1 fm	gluon
electromagnetic	10^{-2}	proportional to $1/r^2$	photon
weak	10^{-13}	$< 10^{-3}$ fm	$W^{\pm}$ and Z bosons
gravitational	10^{-38}	proportional to $1/r^2$	graviton

Did you know?

The interaction of charged particles by the exchange of photons is described by a theory called *quantum electrodynamics,* or *QED.*

CLASSIFICATION OF PARTICLES

All particles other than the mediating field particles can be classified into two broad categories: *leptons* and *hadrons.* The difference between the two is whether they interact through the strong interaction. Leptons are a group of particles that participate in the weak, gravitational, and electromagnetic interactions but not in the strong interaction. Hadrons are particles that interact through all four fundamental interactions, including the strong interaction.

Leptons are thought to be elementary particles

Electrons and neutrinos are both leptons. Like all leptons, they have no measurable size or internal structure and do not seem to break down into smaller units. Because of this, *leptons appear to be truly elementary.*

The number of known leptons is small. Currently, scientists believe there are only six leptons: the electron, the muon, the tau, and a neutrino associated with each. Each of these six leptons also has an antiparticle.

Hadrons include mesons and baryons

Hadrons, the strongly interacting particles, can be further divided into two classes: *mesons* and *baryons.* Originally, mesons and baryons were classified according to their masses. Baryons were heavier than mesons, and both were heavier than leptons. However, this distinction no longer holds. Today, mesons and baryons are distinguished by their internal structure.

All mesons are unstable. Because of this, they are not constituents of normal, everyday matter. Baryons have masses equal to or greater than the proton mass. The most common examples of baryons are protons and neutrons, which are constituents of normal, everyday matter. A summary of this classification of particles is given in **Figure 13.**

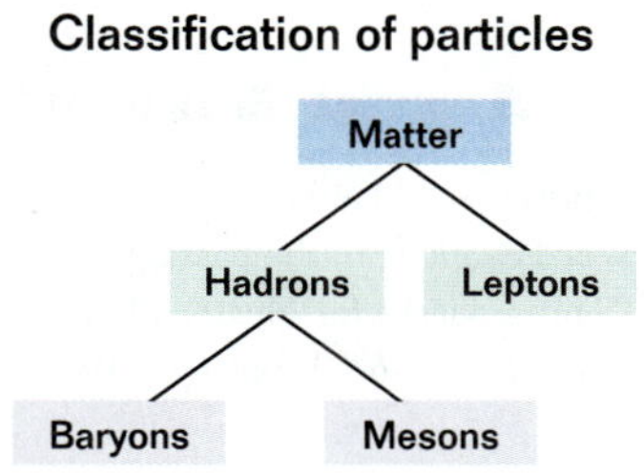

Figure 13
Leptons appear to be elementary, while hadrons consist of smaller particles called *quarks.* As a result, hadrons can be further subdivided into baryons and mesons, based on their internal composition.

Hadrons are thought to be made of quarks

Particle-collision experiments involving hadrons seem to involve many short-lived particles, implying that hadrons are made up of more-fundamental particles. Furthermore, there are numerous hadrons, and many of them are known to decay into other hadrons. These facts strongly suggest that hadrons, unlike leptons, cannot be truly elementary.

In 1963, Murray Gell-Mann and George Zweig independently proposed that hadrons have a more elementary substructure. According to their model, all hadrons are composed of two or three fundamental particles, which came to be called *quarks.* In the original model, there were three types of quarks, designated by the symbols *u, d,* and *s.* These were given the arbitrary names *up, down,* and *sideways* (now more commonly referred to as *strange*). Associated with each quark is an antiquark of opposite charge.

The difference between mesons and baryons is due to the number of quarks that compose them. The compositions of all hadrons known when Gell-Mann and Zweig presented their models could be completely specified by three simple rules, which are summarized in **Table 7.**

Table 7 Hadrons

Particle	Composition
meson	one quark and one antiquark
baryon	three quarks
antibaryon	three antiquarks

Later evidence from collision experiments encouraged theorists to propose the existence of three more quarks, now known as *charm, top,* and *bottom.* These six quarks seem to fit together in pairs: up and down, charm and strange, and top and bottom.

All quarks have a charge associated with them. The charge of a hadron is equal to the sum of the charges of its constituent quarks and is either zero or a multiple of *e,* the fundamental unit of charge. This implies that quarks have a

Teaching Tip — BASIC

This section contains an extensive amount of new information, including classification schemes, names, and concepts. Many of the new terms can easily be confused with each other. Students will need help making connections between the new bits of information and identifying which details are most important to remember. You may choose to use the section summary to prepare an overview of the material. Also, use a board or an overhead projector to clarify which properties or criteria are used for grouping the particles in different categories and subgroups.

STOP Misconception Alert — GENERAL

Students might not differentiate clearly between the mediating field particles and the classification of other particles. To build on their prior knowledge, suggest that they classify the particles they already know into two groups using the following system:

Building blocks or particles of matter interact with, attract, or repel other particles of matter. They may or may not be broken into smaller particles that will in turn attract and repel each other.

Mediating field particles serve as messengers of the forces between the building blocks of matter.

SECTION 4

Teaching Tip — GENERAL

Have students revise their classification systems (see the Misconception Alert on the previous page) to include more details. Ask them to place examples in each category. An example follows:

Mediating particles (field messengers): photons, bosons, gluons, gravitons

Matter particles (building blocks):

- *Leptons (indivisible):* electrons, muons, taus, neutrinos of each
- *Hadrons (made of quarks):*
  - *Mesons (1 quark, 1 antiquark):* π^+, K^-
  - *Baryons (3 quarks):* proton, neutron

Visual Strategy GENERAL

Figure 14

Explain that although many of the quarks have the same charge, they have special properties called *flavor* and *color* that make them different.

Q What is the charge of a π^+ meson, shown in **Figure 14?**

A $\left(+\frac{2}{3}e\right)+\left(+\frac{1}{3}e\right)=1e$

ANSWERS

Conceptual Challenge

1. no; An antibaryon is made of three antiquarks, and a meson is made of one quark and one antiquark. Thus, a baryon, which consists of three quarks, could not be produced in this reaction.
2. yes, although the strong interaction has a much greater effect

Did you know?

Murray Gell-Mann borrowed the word *quark* from the passage "Three quarks for Muster Mark" in James Joyce's book *Finnegans Wake.*

Why it Matters

Conceptual Challenge

1. Particle-Antiparticle Interactions

An antibaryon interacts with a meson. Can a baryon be produced in such an interaction? Explain.

2. Strong and Weak Interactions

Two protons in a nucleus interact via the strong interaction. Are they also subject to the weak interaction?

very unusual property—fractional electric charge. In other words, the charge of the electron is no longer thought to be the smallest possible nonzero charge that a particle can have. The charges for all six quarks that have been discovered and their corresponding antiquarks are summarized in **Table 8.**

Table 8 Quarks and Their Charges

Quark	Charge	Antiquark	Charge
up (u)	$+\frac{2}{3}e$	$\bar{u}$	$-\frac{2}{3}e$
down (d)	$-\frac{1}{3}e$	$\bar{d}$	$+\frac{1}{3}e$
charm (c)	$+\frac{2}{3}e$	$\bar{c}$	$-\frac{2}{3}e$
strange (s)	$-\frac{1}{3}e$	$\bar{s}$	$+\frac{1}{3}e$
top (t)	$+\frac{2}{3}e$	$\bar{t}$	$-\frac{2}{3}e$
bottom (b)	$-\frac{1}{3}e$	$\bar{b}$	$+\frac{1}{3}e$

Figure 14 represents the quark compositions of several hadrons, both baryons and mesons. Just two of the quarks, u and d, are needed to construct the hadrons encountered in ordinary matter (protons and neutrons). The other quarks are needed only to construct rare forms of matter that are typically found only in high-energy situations, such as particle collisions.

The charges of the quarks that make up each hadron in **Figure 14** add up to zero or a multiple of e. For example, the proton contains three quarks (u, u, and d) having charges of $+\frac{2}{3}e$, $+\frac{2}{3}e$, and $-\frac{1}{3}e$. The total charge of the proton is $+e$, as you would expect. Likewise, the total charge of quarks in a neutron is zero ($+\frac{2}{3}e$, $-\frac{1}{3}e$, and $-\frac{1}{3}e$).

You may be wondering whether such discoveries will ever end. At present, physicists believe that six quarks and six leptons (and their antiparticles) are the fundamental particles.

Figure 14

Baryons contain three quarks, while mesons contain a quark and an antiquark. The baryons represented are a proton (p^+) and a neutron (n). The mesons shown are a pion (π^+) and a kaon (K^-), both rare particles.

Despite many extensive efforts, no isolated quark has ever been observed. Physicists now believe that quarks are permanently confined inside ordinary particles by the strong force. This force is often called the *color force* for quarks. Of course, quarks are not really colored. Color is merely a name given to the property of quarks that allows them to attract one another and form composite particles. The attractive force between nucleons is a byproduct of the strong force between quarks.

THE STANDARD MODEL

The current model used in particle physics to understand matter is called the *standard model.* This model was developed over many years by a variety of people. Although the details of the standard model are complex, the model's essential elements can be summarized by using **Figure 15.**

According to the standard model, the strong force is mediated by gluons. This force holds quarks together to form composite particles, such as protons, neutrons, and mesons. Leptons participate only in the electromagnetic, gravitational, and weak interactions. The combination of composite particles, such as protons and neutrons, with leptons, such as electrons, makes the constituents of all matter, which are atoms.

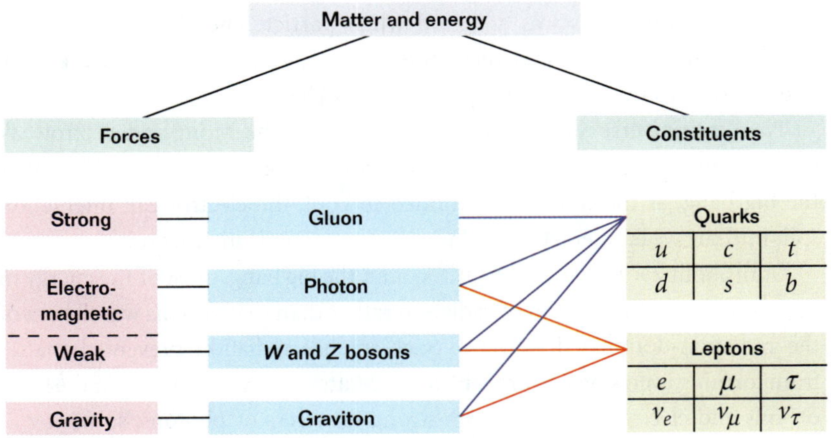

Figure 15
This schematic diagram summarizes the main elements of the standard model, including the fundamental forces, the mediating field particles, and the constituents of matter.

The standard model can help explain the early universe

Particle physics helps us understand the evolution of the universe. If we extrapolate our knowledge of the history of the universe, we find that time itself goes back only about 13 billion to 15 billion years. At that time, the universe was inconceivably dense. In the brief instant after this singular moment, the universe expanded rapidly in an event called the *big bang.* Immediately afterward, there were such extremes in the density of matter and energy that all four fundamental interactions of physics operated in a single, unified way. The temperatures and energy present reduced everything into an undifferentiated "quark soup."

Did you know?

The word *atom* is from the Greek word *atomos,* meaning "indivisible." At one time, atoms were thought to be the indivisible constituents of matter; that is, they were regarded as elementary particles. Today, quarks and leptons are considered to be elementary particles.

The Language of Physics

The Greek letter μ (mu) represents muon particles, and the Greek letter ν (nu) designates neutrino particles. The symbols ν_e, ν_μ, and ν_τ represent the neutrinos associated with the electron, the muon, and the tau, respectively.

Visual Strategy GENERAL

Figure 15

Use the following questions to make sure students interpret the information in **Figure 15** properly:

Q Which forces hold quarks together?

A ***gravitational force, the weak force, the electromagnetic force, and the strong force (the most significant force)***

Q Which field particles mediate the interaction between neutrinos?

A ***photons, W and Z bosons, and gravitons***

Q Which particles are held together by gluons?

A ***quarks***

The Language of Physics

For students, the term *big bang* may conjure up images of the vast explosion of a bomb. Emphasize that the big bang was extremely rapid but was not destructive or violent.

Teaching Tip

The theory of the big bang is also discussed in the feature "The Doppler Effect and the Big Bang" in **Appendix J: Advanced Topics.** Students may benefit from reviewing this feature at this time.

Visual Strategy GENERAL

Figure 16

Be sure students correctly interpret the information presented in **Figure 16.**

Q When did the strong force and the electroweak force separate?

A *at approximately 10^{-32} s*

Q Which of the colored sections shown in the figure occurred during the first second after the big bang?

A *the red section, the yellow section, the green section, and part of the blue section*

Q According to this figure, what is the present age of the universe?

A *approximately 10^{18} s*

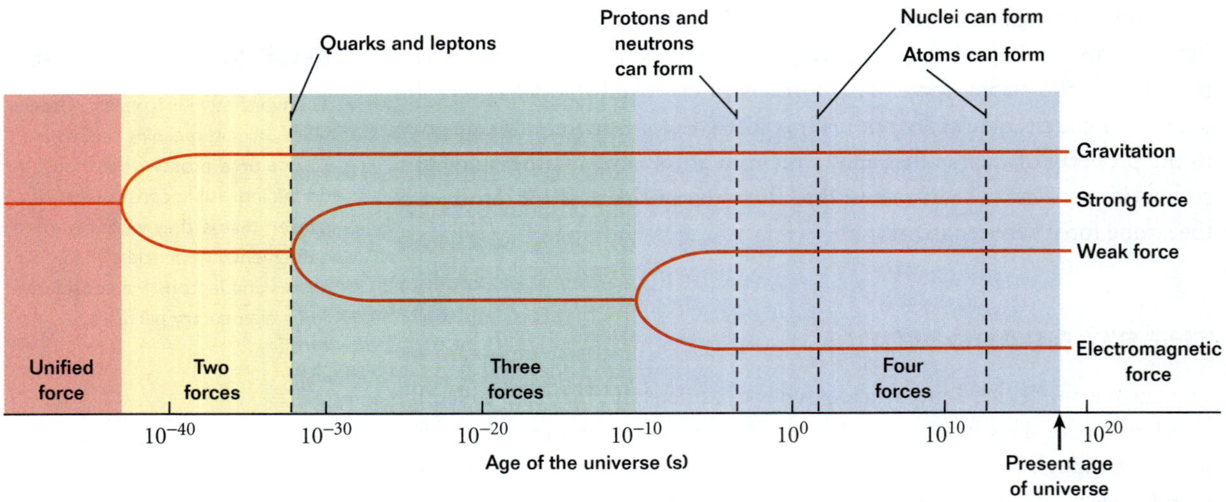

Figure 16
The four fundamental interactions of nature were indistinguishable during the early moments of the big bang.

The evolution of the four fundamental interactions from the big bang to the present is shown in **Figure 16.** During the first 10^{-43} s, it is presumed that the strong, electroweak (electromagnetic and weak), and gravitational interactions were joined together. From 10^{-43} s to 10^{-32} s after the big bang, gravity broke free of this unification while the strong and electroweak interactions remained as one. This was a period when particle energies were so great (greater than 10^{16} GeV) that very massive particles that are now rare, as well as quarks, leptons, and their antiparticles, existed.

Then the universe rapidly expanded and cooled, the strong and electroweak interactions parted, and the grand unification was broken. About 10^{-10} s after the big bang, as the universe continued to cool, the electroweak interaction split into the weak interaction and the electromagnetic interaction.

Until about 7×10^5 years (2×10^{13} s) after the big bang, most of the energy in the universe was in the form of radiation rather than matter. This was the era of the radiation-dominated universe. Such intense radiation prevented matter from forming even single hydrogen atoms. Matter did exist, but only in the form of ions and electrons. Electrons are strong scatterers of photons, so matter at this time was opaque to radiation. Matter continuously absorbed and reemitted photons, thereby ensuring thermal equilibrium of radiation and matter.

By the time the universe was about 380,000 years (1×10^{13} s) old, it had expanded and cooled to about 3000 K. At this temperature, protons could bind to electrons to form hydrogen atoms. Without free electrons to scatter photons, the universe suddenly became transparent. Matter and radiation no longer interacted as strongly, and each evolved separately. By this time, most of the energy in the universe was in the form of matter. Clumps of neutral matter steadily grew: first atoms, followed by molecules, gas clouds, stars, and finally galaxies. This period, referred to as the matter-dominated universe, continues to this day.

extension

Integrating Astronomy
Visit go.hrw.com for the activity "Our Changing Universe."

Keyword HF6SUBX

The standard model is still incomplete

While particle physicists have been exploring the realm of the very small, cosmologists have been exploring cosmic history back to the first microsecond of the big bang. Observation of the events that occur when two particles collide in an accelerator is essential to reconstructing the early moments in cosmic history. Perhaps the key to understanding the early universe is to first understand the world of elementary particles. Cosmologists and particle physicists find that they have many common goals, and they are working together to attempt to study the physical world at its most fundamental level.

Our understanding of physics at short distances is far from complete. Particle physics still faces many questions. For example, why does the photon have no mass, while the W and Z bosons do? Because of this mass difference, the electromagnetic and weak forces are quite distinct at low energies, such as those in everyday life, but they behave in similar ways at very high energies.

To account for these changes, the standard model proposes the existence of a particle called the *Higgs boson,* which exists only at the high energies at which the electromagnetic and weak forces begin to merge. The Higgs boson has not yet been found. According to the standard model, its mass should be less than 1 TeV (10^{12} eV). International efforts are under way to build a device capable of reaching energies close to 1 TeV to search for the Higgs boson.

There are still other questions that the standard model has yet to answer. Is it possible to unify the strong and electroweak theories in a logical and consistent manner? Why do quarks and leptons form three similar but distinct families? Are muons the same as electrons (apart from their different masses), or do they have other subtle differences that have not been detected? Why are some particles charged and others neutral? Why do quarks carry a fractional charge? What determines the masses of the fundamental constituents? Can isolated quarks exist? The questions go on and on. Because of the rapid advances and new discoveries in the field of particle physics, by the time you read this book, some of these questions may have been resolved, while new questions may have emerged.

SECTION REVIEW

1. Name the four fundamental interactions and the particles that mediate each interaction.
2. What are the differences between hadrons and leptons? What are the differences between baryons and mesons?
3. Describe the main stages of the evolution of the universe according to the big bang theory.

Teaching Tip — ADVANCED

Encourage students to formulate their own questions about the standard model and about the age of the universe.

SECTION REVIEW ANSWERS

1. The strong force is mediated by gluons, the electromagnetic force is mediated by photons, the weak force is mediated by W and Z bosons, and gravity is mediated by gravitons.
2. Hadrons consist of quarks, while leptons are thought to be indivisible. Hadrons interact with gluons (the strong interaction), but leptons do not. Mesons consist of one quark and one antiquark, and baryons consist of three quarks.
3. Up to 10^{-43} s, all forces were unified. At 10^{-43} s, gravity broke free; at 10^{-32} s, the strong force broke free; and at 10^{-10} s, the weak force broke free. From 10^{-43} s to 10^{-32} s, quarks, leptons, and their antiparticles existed. Until 7×10^5 years (2×10^{13} s), most energy was in the form of radiation. At 380,000 years (1×10^{13} s), atoms began forming, and soon most energy was in the form of matter, as it is today.

Radiologist

As Katherine Maturen explains, radiologists work with many different types of medical images and processes, including the following:

- **Conventional radiography:** X rays are transmitted through the patient's body to make the kind of picture most people think of when they imagine a chest X ray or an X ray of a broken leg.
- **Computed tomography, or CT scans:** Computers take X rays from many different angles and then combine the data. The resulting image shows what the patient's body would look like if it were sliced like a loaf of bread.
- **Angiography:** Radiopaque material is injected into blood vessels to make sure they are not blocked or narrowed.
- **Fluoroscopy:** Patients swallow barium so that their intestines can be seen in real time.
- **Nuclear medicine:** Tiny amounts of radioactive material are injected into patients (or ingested), and parts of their body become radioactive for a short time. In this kind of imaging, the rays come from inside the patient instead of being projected through the patient, and the rays are gamma rays instead of X rays.
- **Ultrasound:** Sound waves bounce off objects within the body and are then received by a transducer.
- **Magnetic resonance imaging, or MRI:** Magnetic fields alter the orientation of protons within tissue, and a radio antenna receives the resulting signals.

PHYSICS CAREERS

Radiologist

A radiologist's job is to interpret many different kinds of medical images, including those from X rays, CT scans, fluoroscopy, and angiography. To learn more about radiology as a career, read the interview with Katherine Maturen, who works in a large, university-based hospital in Michigan.

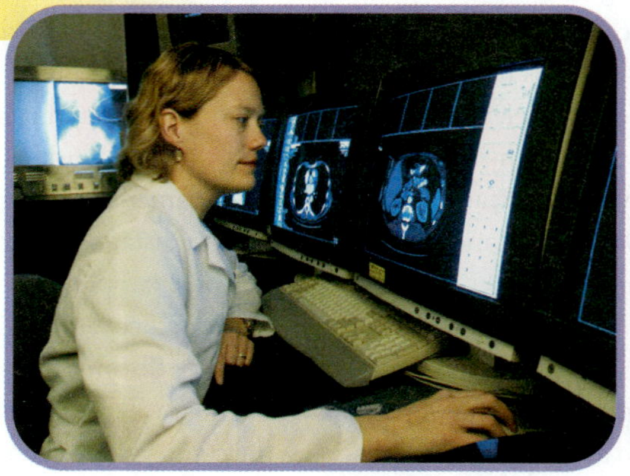

Katherine Maturen reviews a CT scan of the abdomen on a digital workstation.

What schooling did you receive in order to become a radiologist?

I attended four years of college, four years of medical school, and five years of specialty training in radiology.

What influenced your decision to become a radiologist?

During college, I studied a lot of different things and did not decide to go to medical school until my senior year. In medical school, I considered several different specialties before deciding on radiology. The fact that many radiologists seem to really enjoy their work was certainly influential.

What about radiology makes it interesting to you?

Radiology is primarily concerned with diagnosis, which can be like a fun puzzle in really challenging cases. I like the problem-solving aspect. I love anatomy, and I find the many different kinds of radiological images aesthetically pleasing. Another attraction of radiology is to see and understand disease processes.

What kinds of skills are important for a radiologist?

Good observation skills, attention to detail, and a strong knowledge of normal anatomy are essential for a radiologist. The best radiologists also have a thorough understanding of disease processes and the physics of imaging. Finally, the ability to develop rapport both with patients and medical colleagues is very important.

What is your favorite thing about your job?

My favorite thing is reading studies and making diagnoses. The only thing I don't like is long hours and working all night, for obvious reasons!

What advice do you have for students who are interested in radiology?

You should pursue what interests you, regardless of other people's expectations of you or what you have done in the past. Try new things and follow your intellectual curiosity. Remember, you want a job that is actually interesting to you, not just a way to pay the bills.

If you think medical school sounds like too much school, consider radiological technologist programs, where you will be the one working with patients and actually taking most of the pictures. If you decide to go to medical school, don't spend all of your time in college taking science classes. Broaden your horizons and learn about things outside of medicine. And in medical school, pay attention in anatomy lab!

CHAPTER 22 Highlights

KEY IDEAS

Section 1 The Nucleus

- The nucleus, which consists of protons and neutrons, is the small, dense core of an atom.
- A nucleus can be characterized by a mass number, A, an atomic number, Z, and a neutron number, N.
- The binding energy of a nucleus is the difference in energy between its nucleons when bound and its nucleons when unbound.

Section 2 Nuclear Decay

- An unstable nucleus can decay in three ways: alpha (α) decay, beta (β) decay, or gamma (γ) decay.
- The decay constant, λ, indicates the rate of radioactive decay.
- The half-life, $T_{1/2}$, is the time required for half the original nuclei of a radioactive substance to undergo radioactive decay.

Section 3 Nuclear Reactions

- Nuclear reactions involve a change in the nucleus of an atom.
- In fission, a heavy nucleus splits into two lighter nuclei. In fusion, two light nuclei combine to form a heavier nucleus.

Section 4 Particle Physics

- There are four fundamental interactions in nature: strong, weak, gravitational, and electromagnetic.
- The constituents of matter can be classified as leptons or hadrons, and hadrons can be further divided into mesons and baryons. Electrons and neutrinos are leptons. Protons and neutrons are baryons.
- Mesons consist of a quark-antiquark pair; baryons consist of three quarks.

Variable Symbols

Quantities		Units		Conversions
m	mass	u	unified mass unit or atomic mass unit	$= 1.660\ 539 \times 10^{-27}$ kg $= 931.49$ MeV/c^2
λN	activity or decay rate	Bq Ci	becquerel curie	$= 1$ decay/s $= 3.7 \times 10^{10}$ Bq
$T_{1/2}$	half-life	s	seconds	

KEY TERMS

isotope (p. 791)

strong force (p. 792)

binding energy (p. 794)

half-life (p. 803)

PROBLEM SOLVING

See **Appendix D: Equations** for a summary of the equations introduced in this chapter. If you need more problem-solving practice, see **Appendix I: Additional Problems.**

Particle Symbols

Particle name	Symbol
alpha particle (helium nucleus)	α ($^4_2$He)
beta particle (electron)	β^- ($^{0}_{-1}$e)
beta particle (positron)	β^+ ($^{0}_{+1}$e)
gamma ray	γ
neutron	n (^1_0n)
proton	p (^1_1p)
neutrino	v
antineutrino	$\overline{v}$

Highlights

Teaching Tip

Explaining concepts in written form helps to solidify students' understanding of difficult concepts and to enforce good communication skills. Have students choose one major concept from the chapter, such as binding energy, nuclear reactions, or particle physics, and summarize the topic in an essay. Essays should explain the concept accurately and thoroughly.

CHAPTER 22

Review

ANSWERS

1. 79; 118; 79
2. Isotopes are atoms with the same Z but different N and A.
3. the strong force
4. no; because an atom's mass also depends on its neutrons
5. The nucleons would spontaneously break apart.
6. because of a greater Coulomb repulsion between the protons
7. 92.162 MeV
8. 8.482 MeV; 7.718 MeV
9. 8.2607 MeV/nucleon; 8.6974 MeV/nucleon
10. See **Table 3** of this chapter.
11. For each α, A decreases by four and Z decreases by two. For each β^-, A stays the same and Z increases by one.
12. the high temperatures and densities required
13. α particles cannot penetrate the box, but β particles can.
14. α has more mass and thus more momentum than β.
15. This cannot be predicted because half-life applies to a large sample of nuclei, not to a particular nucleus.
16. The amount of C-14 left undecayed would be too small to be measured accurately.
17. No; $m_n > m_p$, so the decay of a free proton into a neutron requires an input of energy.
18. no; The three α particles have a mass of 12.007 806 u. Thus, $^{12}_{6}C$ does not have enough energy to break into three α particles.

CHAPTER 22 *Review*

THE NUCLEUS

Review Questions

1. How many protons are there in the nucleus $^{197}_{79}Au$? How many neutrons? How many electrons are there in the neutral atom?
2. What are isotopes?
3. What holds the nucleons in a nucleus together?

Conceptual Questions

4. Is it possible to accurately predict an atom's mass from its atomic number? Explain.
5. What would happen if the binding energy of a nucleus was zero?
6. Why do heavier elements require more neutrons to maintain stability?

Practice Problems

For problems 7–9, see Sample Problem A and refer to Appendix H.

7. Calculate the total binding energy of $^{12}_{6}C$.
8. Calculate the total binding energy of tritium ($^{3}_{1}H$) and helium-3 ($^{3}_{2}He$).
9. Calculate the average binding energy per nucleon of $^{24}_{12}Mg$ and $^{85}_{37}Rb$.

NUCLEAR DECAY AND REACTIONS

Review Questions

10. Explain the main differences between alpha, beta, and gamma emissions.
11. The figure below shows the steps by which $^{235}_{92}U$ decays to $^{207}_{82}Pb$. Draw this diagram, and enter the correct isotope symbol in each square.

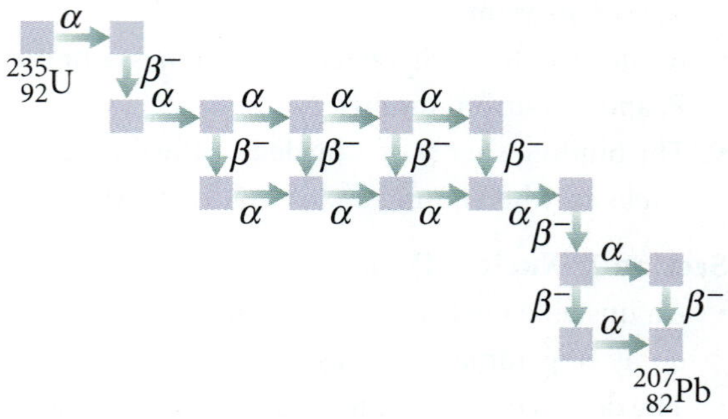

12. What factors make fusion difficult to achieve?

Conceptual Questions

13. If a film is kept in a box, alpha particles from a radioactive source outside the box cannot expose the film, but beta particles can. Explain.
14. An alpha particle has twice the charge of a beta particle. Why does the beta particle deflect more when both pass between electrically charged plates, assuming they both have the same speed?
15. Suppose you have a single atom of a radioactive material whose half-life is one year. Can you be certain that the nucleus will have decayed after two years? Explain.
16. Why is carbon dating unable to provide accurate estimates of very old materials?
17. A free neutron undergoes beta decay with a half-life of about 15 min. Can a free proton undergo a similar decay? (Hint: Compare the masses of the proton and the neutron.)
18. Is it possible for a $^{12}_{6}C$ (12.000 000 u) nucleus to spontaneously decay into three alpha particles? Explain.

19. Why is the temperature required for deuterium-tritium fusion lower than that needed for deuterium-deuterium fusion? (Hint: Consider the Coulomb repulsion and nuclear attraction for each case.)

Practice Problems

For problems 20–21, see Sample Problem B.

20. Determine the product of the following reaction:
$^7_3Li + ^4_2He \rightarrow ? + ^1_0n$

21. Complete the following nuclear reactions:
 a. $? + ^{14}_7N \rightarrow ^1_1H + ^{17}_8O$
 b. $^7_3Li + ^1_1H \rightarrow ^4_2He + ?$

For problems 22–24, see Sample Problem C.

22. A radioactive sample contains 1.67×10^{11} atoms of $^{108}_{47}Ag$ (half-life = 2.42 min) at some instant. Calculate the decay constant and the activity of the sample in mCi.

23. How long will it take a sample of polonium-210 with a half-life of 140 days to decay to one-sixteenth its original strength?

24. The amount of carbon-14 ($^{14}_6C$) in a wooden artifact is measured to be 6.25 percent the amount in a fresh sample of wood from the same region. The half-life of carbon-14 is 5715 years. Assuming the same amount of carbon-14 was initially present in the artifact, determine the age of the artifact.

PARTICLE PHYSICS

Review Questions

25. Describe the properties of quarks.

26. What is the electric charge of the particles with the following quark compositions?
 a. *udd*
 b. *uud*
 c. $u\overline{d}$

27. What is the electric charge of the baryons with the following quark compositions?
 a. $\overline{u}\overline{u}\overline{d}$
 b. $\overline{u}\overline{d}\overline{d}$

28. What are each of the baryons in item 27 called?

29. How many quarks or antiquarks are there in the following particles?
 a. a baryon
 b. an antibaryon
 c. a meson
 d. an antimeson

Conceptual Questions

30. Compare a neutrino with a photon.

31. Consider the statement, "All mesons are hadrons, but not all hadrons are mesons." Is this statement true? Explain.

MIXED REVIEW

32. Complete the following nuclear reaction:
$^{27}_{13}Al + ^4_2He \rightarrow ? + ^{30}_{15}P?$

33. Consider the hydrogen atom to be a sphere with a radius equal to the Bohr radius, 0.53×10^{-10} m, and calculate the approximate value of the ratio of atomic density to nuclear density.

34. Certain stars are thought to collapse at the end of their lives, combining their protons and electrons to form a neutron star. Such a star could be thought of as a giant atomic nucleus. If a star with a mass equal to that of the sun (1.99×10^{30} kg) were to collapse into neutrons, what would be the radius of the star?

35. Calculate the difference in binding energy for the two nuclei $^{15}_8O$ and $^{15}_7N$.

36. A piece of charcoal known to be approximately 25 000 years old contains 7.96×10^{10} C-14 atoms.
 a. Determine the number of decays per minute expected from this sample. (The half-life of C-14 is 5715 years.)
 b. If the radioactive background in the counter without a sample is 20.0 counts per minute and we assume 100.0 percent efficiency in counting, explain why 25 000 is close to the limit of dating with this technique.

22 REVIEW

19. Although the Coulomb repulsion is the same in each, the extra neutron in tritium provides a stronger nuclear force, so it is easier to bring the two nuclei together.
20. $^{10}_5B$
21. **a.** ^4_2He
 b. ^4_2He
22. $4.77 \times 10^{-3}\ s^{-1}$, 22 mCi
23. 560 days
24. 22 860 years old
25. Quarks make up hadrons; they have fractional charge, have never been isolated, and are attracted to one another by the strong force.
26. **a.** 0
 b. *e*
 c. *e*
27. **a.** $-e$
 b. 0
28. **a.** anti-proton
 b. anti-neutron
29. **a.** three quarks
 b. three antiquarks
 c. one quark, one antiquark
 d. one quark, one antiquark
30. The photon is a mediating field particle for the electromagnetic force, while the neutrino is a particle emitted in beta decay.
31. yes; Hadrons consist of mesons and baryons. Thus, all mesons are hadrons—but not all hadrons are mesons; some are baryons.
32. ^1_0n
33. 1.2×10^{-14}
34. 1.3×10^4 m
35. 3.53 MeV
36. **a.** 18.4 decays per minute
 b. The observed count rate is slightly less than the average background and thus would be difficult to measure accurately.

22 REVIEW

37. a. ${}^{1}_{0}n + {}^{197}_{79}Au \rightarrow {}^{198}_{80}Hg + {}^{0}_{-1}e + \overline{v}$

b. 7.885 MeV

38. a. $2\,{}^{1}_{0}n$

b. $3\,{}^{1}_{0}n$

39. ${}^{3}_{2}He$

40. ${}^{13}_{6}C$

41. 2.6×10^{21} atoms

42. ${}^{1}_{0}n$

43. a. ${}^{8}_{4}Be$

b. ${}^{12}_{6}C$

44. 1.6×10^{16} nuclei

extension

Graphing Calculator Practice

Visit go.hrw.com for answers to this Graphing Calculator activity.

Keyword HF6SUBXT

37. Natural gold has only one stable isotope, ${}^{197}_{79}Au$. If gold is bombarded with slow neutrons, β^- particles are emitted.

a. Write the appropriate reaction equation.

b. Calculate the maximum energy of the emitted beta particles.

38. Two ways ^{235}U can undergo fission when bombarded with a neutron are described below. In each case, neutrons are also released. Find the number of neutrons released in each of the following:

a. ^{140}Xe and ^{94}Sr released as fission fragments

b. ^{132}Sn and ^{101}Mo released as fission fragments

39. When a ${}^{6}_{3}Li$ nucleus is struck by a proton, an alpha particle and a product nucleus are released. What is the product nucleus?

40. Suppose ${}^{10}_{5}B$ is struck by an alpha particle, releasing a proton and a product nucleus in the reaction. What is the product nucleus?

41. An all-electric home uses about 2.0×10^3 kW•h of electrical energy per month. How many ^{235}U atoms would be required to provide this house with its energy needs for one year? Assume 100.0 percent conversion efficiency and 208 MeV released per fission.

42. When ^{18}O is struck by a proton, ^{18}F and another particle are produced. What is the other particle?

43. When a star has exhausted its hydrogen fuel, it may fuse other nuclear fuels, such as helium. At temperatures above 1.0×10^8 K, helium fusion can occur.

a. Two alpha particles fuse to produce a nucleus, A, and a gamma ray. What is nucleus A?

b. Nucleus A absorbs an alpha particle to produce a nucleus, B, and a gamma ray. What is nucleus B?

44. A sample of a radioactive isotope is measured to have an activity of 240.0 mCi. If the sample has a half-life of 14 days, how many nuclei of the isotope are there at this time?

Graphing Calculator *Practice*

Nuclear Decay

In nuclear decay, a radioactive substance is transformed into another substance that may or may not be radioactive. The amount of radioactive material remaining is given by the following equation:

$$m = m_0 e^{-\lambda t}$$

In this nuclear decay equation, m_0 is the initial mass and λ is the decay constant. As you learned earlier in this chapter, the decay constant is related to the half-life by the following equation:

$$T_{1/2} = \frac{0.693}{\lambda}$$

One of the interesting aspects of nuclear decay is that radioactive substances have a wide range of half-lives—from femtoseconds to billions of years. And all of these radioactive substances obey both of these equations.

In this graphing calculator activity, the calculator will use these equations to make graphs of the amount of remaining mass versus time. By analyzing these graphs, you will be able to make predictions about radioactive substances that have various initial masses and various half-lives.

Visit go.hrw.com and type in the keyword **HF6SUBX** to find this graphing calculator activity. Refer to **Appendix B** for instructions on downloading the program for this activity.

45. At some instant of time, the activity of a sample of radioactive material is 5.0 μCi. If the sample contains 1.0×10^9 radioactive nuclei, what is the half-life of the material?

46. It has been estimated that Earth has 9.1×10^{11} kg of natural uranium that can be economically mined. Of this total, 0.70 percent is ^{235}U. If all the world's energy needs (7.0×10^{12} J/s) were supplied by ^{235}U fission, how long would this supply last? Assume that 208 MeV of energy is released per fission event and that the mass of ^{235}U is about 3.9×10^{-25} kg.

47. If the average energy released in a fission event is 208 MeV, find the total number of fission events required to provide enough energy to keep a 100.0 W light bulb burning for 1.0 h.

48. How many atoms of ^{235}U must undergo fission to operate a 1.0×10^3 MW power plant for one day if the conversion efficiency is 30.0 percent? Assume 208 MeV released per fission event.

Alternative Assessment

1. You are designing a nuclear power plant for a space station to be established on Mars. Material A is radioactive and has a half-life of two years. Material B is also radioactive and has a half-life of one year. Atoms of material B have one half the mass of atoms of material A. Discuss the benefits and drawbacks involved with each of these fuels.

2. Design a questionnaire to investigate what people in your community know about nuclear power and how they feel about it. Give the questionnaire to your classmates for their comments, and if your teacher approves, conduct a study with people in your community. Present your results in the form of a class presentation and discussion.

3. Investigate careers in nuclear medicine. Interview people who work with radiation or with isotopic tracers in a hospital. Find out what kind of patients they treat or test and the technology they use. What training is necessary for this type of career?

4. Research the lives and careers of female nuclear physicists such as Marie Curie, Lise Meitner, Ida Tacke Noddack, and Maria Goeppert-Mayer. Create a presentation about one of these scientists. The presentation can be in the form of a report, poster, short video, or computer presentation.

5. Research how radioactive decay is used to date archaeological remains and fossils. What nuclear reactions are involved in the carbon-14 dating technique? What assumptions are made when the carbon-14 dating technique is used? What time scale is the carbon-14 technique suitable for? Is the carbon-14 technique appropriate to determine the age of a painting suspected to be 375 years old? Summarize your findings in a brochure or poster for visitors to a science museum.

6. Research the problem of nuclear waste in the United States. How much is there? What kinds of radioactive waste are there? Where are they produced? What are the costs and hazards associated with different techniques for disposal of radioactive waste? How do other countries deal with the problem? Choose the disposal option you think is most appropriate, and write a position paper. Include information about all options and the reasons for your choice.

7. Some modern physicists have developed *string theory* in an attempt to unify the four fundamental forces. Conduct research to learn about this theory. What are the main principles of string theory? Why do some scientists oppose it? Share your results with the class in a short lecture presentation.

45. 3.8×10^3 s
46. 7.8×10^{10} s (2500 years)
47. 1.1×10^{16} fission events
48. 8.7×10^{24} atoms

Alternative Assessment ANSWERS

1. Students should consider how long the supply will last. In two years, the rates of decay will be the same, but later, B will decrease more.
2. Students' questionnaires will vary. Questionnaires should gauge people's actual knowledge as well as their opinions.
3. Students' answers will vary. Typically, isotopic tracers use lower levels of radiation than radiation therapy.
4. Student answers will vary. Curie and Noddack discovered several radioactive elements. Meitner and Goeppert-Mayer explored the phenomenon of nuclear fission. Each faced prejudice during her career.
5. ^{14}C undergoes beta decay to ^{14}N. Carbon dating can be used with organic materials only and works best for artifacts between 1000 and 25 000 years old.
6. Students' answers will vary. Be certain pros and cons are identified.
7. According to string theory, fundamental particles such as electrons and quarks consist of tiny, oscillating loops called *strings*. Some scientists oppose the theory because it is highly theoretical; string theorists have not yet made predictions that could be confirmed with experiments.

Standardized Test Prep

ANSWERS

1. C
2. H
3. D
4. F
5. C
6. H
7. D
8. F

Standardized Test Prep

MULTIPLE CHOICE

1. Which of the following statements correctly describes a nucleus with the symbol $^{14}_{6}C$?

A. It is the nucleus of a cobalt atom with eight protons and six neutrons.
B. It is the nucleus of a carbon atom with eight protons and six neutrons.
C. It is the nucleus of a carbon atom with six protons and eight neutrons.
D. It is the nucleus of a carbon atom with six protons and fourteen neutrons.

2. One unified mass unit (u) is equivalent to a mass of 1.66×10^{-27} kg. What is the equivalent rest energy in joules?

F. 8.27×10^{-46} J
G. 4.98×10^{-19} J
H. 1.49×10^{-10} J
J. 9.31×10^{8} J

3. What kind of force holds protons and neutrons together in a nucleus?

A. electric force
B. gravitational force
C. binding force
D. strong force

4. What type of nuclear decay most often produces the greatest mass loss?

F. alpha decay
G. beta decay
H. gamma decay
J. All of the above produce the same mass loss.

5. A nuclear reaction of major historical note took place in 1932, when a beryllium target was bombarded with alpha particles. Analysis of the experiment showed that the following reaction took place: $^{4}_{2}He + ^{9}_{4}Be \longrightarrow ^{12}_{6}C + X$. What is X in this reaction?

A. $^{0}_{1}e$
B. $^{0}_{-1}p$
C. $^{1}_{0}n$
D. $^{1}_{1}p$

6. What fraction of a radioactive sample has decayed after two half-lives have elapsed?

F. $\frac{1}{4}$
G. $\frac{1}{2}$
H. $\frac{3}{4}$
J. The whole sample has decayed.

7. A sample of organic material is found to contain 18 g of carbon-14. Based on samples of pottery found at a dig, investigators believe the material is about 23 000 years old. The half-life of carbon-14 is 5715 years. Estimate what percentage of the material's carbon-14 has decayed.

A. 4.0%
B. 25%
C. 75%
D. 94%

8. The half-life of radium-228 is 5.76 years. At some instant, a sample contains 2.0×10^{9} nuclei. Calculate the decay constant and the activity of the sample.

F. $\lambda = 3.81 \times 10^{-9}\ s^{-1}$; activity $= 2.1 \times 10^{-10}$ Ci
G. $\lambda = 3.81 \times 10^{-9}\ s^{-1}$; activity $= 7.8$ Ci
H. $\lambda = 0.120\ s^{-1}$; activity $= 6.5 \times 10^{-3}$ Ci
J. $\lambda = 2.6 \times 10^{8}\ s^{-1}$; activity $= 1.4 \times 10^{7}$ Ci

9. What must be true in order for a nuclear reaction to happen naturally?
   A. The nucleus must release energy in the reaction.
   B. The binding energy per nucleon must decrease in the reaction.
   C. The binding energy per nucleon must increase in the reaction.
   D. There must be an input of energy to cause the reaction.

10. Which is the weakest of the four fundamental interactions?
   F. electromagnetic
   G. gravitational
   H. strong
   J. weak

11. Which of the following choices does *not* correctly match a fundamental interaction with its mediating particles?
   A. strong: gluons
   B. electromagnetic: electrons
   C. weak: *W* and *Z* bosons
   D. gravitational: gravitons

12. What is the charge of a baryon containing one up quark (u) and two down quarks (d)?
   F. −1
   G. 0
   H. +1
   J. +2

SHORT RESPONSE

13. Suppose it could be shown that the ratio of carbon-14 to carbon-12 in living organisms was much greater thousands of years ago than it is today. How would this affect the ages we assign to ancient samples of once-living matter?

14. A fission reactor produces energy to drive a generator. Describe briefly how this energy is produced.

15. Balance the following nuclear reaction:
$^{1}_{0}n + ? \longrightarrow {}^{4}_{2}He + {}^{7}_{3}Li$

16. Smoke detectors use the isotope ^{241}Am in their operation. The half-life of Am is 432 years. If the smoke detector is improperly discarded in a landfill, estimate how long its activity will take to decrease to a relatively safe level of 0.1 percent of its original activity. (Hint: The estimation process that you should use notes that the activity decreases to 50% in one half-life, to 25% in two half-lives, and so on.)

EXTENDED RESPONSE

17. Iron-56 ($^{56}_{26}Fe$) has an atomic mass of 55.934 940 u. The atomic mass of hydrogen is 1.007 825 u, and m_n = 1.008 665 u. Show your work for the following calculations:
   a. Find the mass defect in the iron-56 nucleus.
   b. Calculate the binding energy in the iron-56 nucleus.
   c. How much energy would be needed to dissociate all the particles in an iron-56 nucleus?

18. Use the table below to calculate the energy released in the alpha decay of $^{238}_{92}U$. Show your work.

Nucleus	Mass
$^{238}_{92}U$	238.050 784 u
$^{234}_{90}Th$	234.043 593 u
$^{4}_{2}He$	4.002 602 u

Test TIP If you finish a test early, go back and check your work before turning in the test.

9. A
10. G
11. B
12. G
13. More carbon-14 would have decayed, so the samples would be older than predicted.
14. Energy is released when heavy nuclei break down into lighter nuclei.
15. $^{10}_{5}B$
16. 4320 years
17. **a.** 0.528 460 u
   b. 492.26 MeV
   c. 492.26 MeV (same as **b**)
   (See the Solutions Manual or the One-Stop Planner for full solutions.)
18. 4.275 MeV (See the Solutions Manual or the One-Stop Planner for a full solution.)

Lab Planning

Beginning on page T34 are preparation notes and teaching tips to assist you in planning.

Blank data tables (as well as some sample data) appear on the **One-Stop Planner.**

No Books in the Lab?

See the ***Datasheets for In-Text Labs*** workbook for a reproducible master copy of this experiment.

Safety Caution

Remind students of laboratory procedures for the use of chemicals and radioactive materials. Make sure students know the appropriate disposal methods.

Tips and Tricks

- Show students how to use the nuclear-lab station. Point out the parts of the apparatus, and explain how each item works.

✓ Checkpoints

Step 3: To get more students involved in getting the background count, take the count over two 5 min intervals or three 3 min intervals, and find the average. Have different students perform different tasks during each time period.

CHAPTER 22

Skills Practice Lab Half-Life

The half-life of a radioactive element is the time it takes for half a given number of atoms of the element to decay. In this experiment, you will measure the activity of a radioisotope with a short half-life, using a nuclear-lab station over a period of 10 s. You will plot the resulting data on a graph, and you will calculate the half-life from points on the graph.

OBJECTIVES

- **Measure** background radiation using a nuclear-lab station.
- **Measure** the activity of a radioactive substance using a nuclear-lab station.
- **Determine** the half-life of a radioisotope.

MATERIALS LIST

- **cesium-137/barium-137 isogenerator set**
- **dropper**
- **liquid soap**
- **nuclear-lab station, including counter, timer, variable high voltage, absorber set, and sources**
- **small disposable culture dish**
- **support base and rod**
- **symmetrical clamp**

SAFETY

- **Do not eat or drink anything in the laboratory. Never taste chemicals or touch them with your bare hands.**
- **Tie back long hair, secure loose clothing, and remove loose jewelry. Put on a lab apron and goggles.**
- **Do not allow radioactive materials to come into contact with your skin, hair, clothing, or personal belongings. Although the materials used in this lab are not hazardous when used properly, radioactive materials can cause serious illness and may have permanent effects.**
- **Dispose of chemicals as instructed by your teacher. Never pour hazardous chemicals into a regular waste container. Never pour radioactive materials down the drain.**
- **Wash your hands thoroughly when you finish your work and before you leave the laboratory.**

PROCEDURE

Preparation

1. Read the entire lab, and plan the steps you will take.
2. If you are not using a datasheet provided by your teacher, prepare a data table in your lab notebook with 2 columns and 31 rows. Label this table *Data Table.* Label the columns *Time (s)* and *Total Count.* In the first column, label the 2nd through 31st rows in increments of 10 from *10* to *300.* Above the table, make a space to record the *Background Count* (s^{-1}).
3. Turn on the nuclear-lab station. Even when no radioactive substances are near the station, counts will register. This is caused by background radiation. Take the radiation count for 10 min, and divide the total by 600 s (10 min × 60 s/min). This calculation will give you the background count per second. Record this value in your data table.

4. Wearing goggles, gloves, and a protective apron, set up the isogenerator vertically on a stand and clamp it securely in place. Carefully loosen the cap on the isogenerator, and turn the stopcock. Release 3–4 mL of the barium solution into a small culture dish. The dish now contains the radioisotope Ba-137m, a metastable isomer.

5. Add a small drop of soap to the solution so that it spreads evenly over the bottom of the dish.

6. Immediately insert the dish into the nuclear-lab station, and start the counter. Read the total count at the end of each 10 s interval, and record the count in your data table.

7. Clean up your work area. Put equipment away safely so that it is ready to be used again. Recycle or dispose of used materials as directed by your teacher.

ANALYSIS

1. **Organizing Data** For each 10 s time interval, calculate the count per second by dividing the count by 10 s. This will give you the average count per second.

2. **Organizing Data** For each value found in item 1, calculate the actual count per second due to the source by subtracting the background count from the average count.

3. **Constructing Graphs** Plot a graph of the actual count versus the total elapsed time. Use a graphing calculator, computer, or graph paper.

CONCLUSIONS

4. **Analyzing Graphs** Select a value for the actual count near the left end of the graph and read from the graph the amount of time it took for the activity to decrease to one-half that value. Record this value as the half-life.

5. **Analyzing Graphs** Repeat the procedure in step 4 three more times, each time starting at a different place on the graph. Find the average of the four values for the half-life.

6. **Evaluating Methods** Why is using the graph a better way to find the half-life than simply using the data table?

ANSWERS

Analysis

1. Student answers will vary. For sample data, values range from 30 counts/s to 55 counts/s.

2. For sample data, the background count is 1.24 counts/s.

3. Student graphs should show a gradual curve down and to the right, an example of exponential decay.

Conclusions

4. Student answers will vary. Typical values will range from 2.08 min to 2.52 min.

5. For sample data, the average half-life value is 2.25 min.

6. Student answers will vary, but they should recognize that using the graph makes it easier to find the value at exact intervals.

Science•
Technology•
Society

What Can We Do With Nuclear Waste?

For about the past 50 years, people have been arguing about what to do with radioactive waste. Since the waste is harmful to humans—as well as to the environment—deciding where to put it is a serious problem.

Protection for 10 000 Years

As radioactive isotopes decay, nuclear waste emits all common forms of radioactivity–α-particles, β-particles, γ-radiation, and X rays. When this radiation penetrates living cells, it knocks electrons away from atoms, causing them to become electrically charged ions. As a result, vital biological molecules break apart or form abnormal chemical bonds with other molecules. Often, a cell can repair this damage, but if too many molecules are disrupted, the cell will die. This ionizing radiation can also damage a cell's genetic material (DNA and RNA), causing the cell to divide again and again, out of control. This condition is called cancer.

Because of these hazards, nuclear waste must be sealed and stored until the radioactive isotopes in the waste decay to the point at which radiation reaches a safe level. Some kinds of radioactive waste will require safe storage for at least 10 000 years.

Questions of Disposal

Low-level waste includes materials from the nuclear medicine departments at hospitals, where radioactive isotopes are used to diagnose and treat diseases. The greatest disposal problem involves high-level waste, or HLW. Nearly all HLW consists of used fuel rods from reactors at nuclear power plants; about a third of these rods are replaced every year or two because their supply of fissionable uranium-235 becomes depleted, or spent.

When nuclear power plants in the United States began operating in 1957, engineers had planned to reprocess spent fuel to reclaim fissionable isotopes of uranium and plutonium to make new fuel rods. But people feared that the plutonium made available by reprocessing might be used to build bombs, so that plan was abandoned. Since that time, HLW has continued to accumulate at power plant sites in "temporary" storage facilities that are now nearly full. When there is no more storage space, plants will have to cease operation. Consequently, states and utility

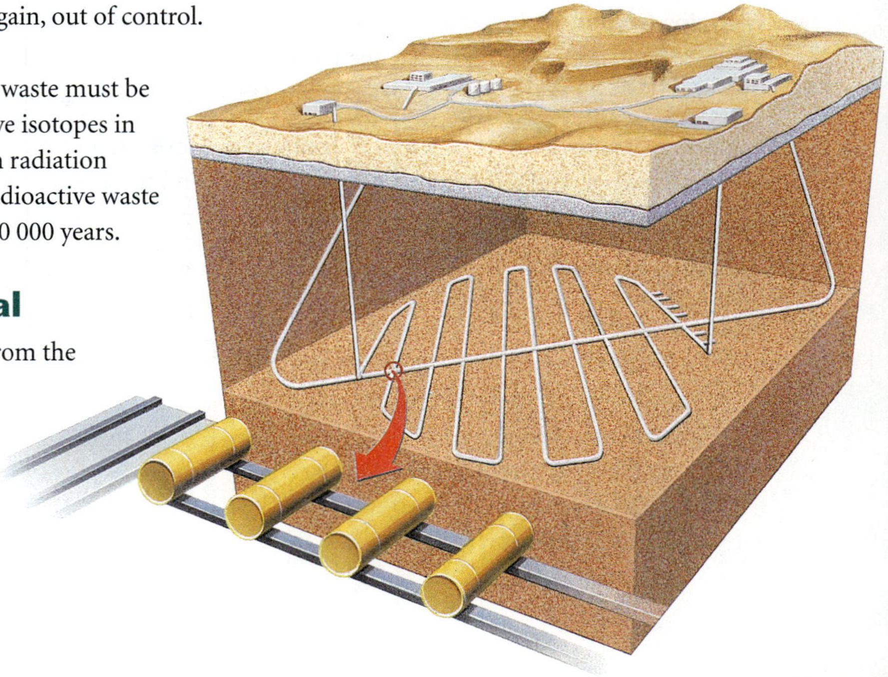

828

companies are demanding that the federal government honor the Nuclear Policy Act of 1982 in which the federal government agreed to provide permanent storage sites.

Disposal Possibilities

Any site used for disposal of HLW must be far away from population centers and likely to remain geologically stable for thousands of years.

One possibility lies at certain deep spots of the oceans, where, some scientists claim, the seabed is geologically stable as well as devoid of life. Sealed stainless-steel canisters of waste could be packed into rocket-shaped carriers which would bury themselves deep into sediments when they hit the ocean bottom. Opponents say that the canisters have not been proven safe and that, if released, the radioactivity could kill off photosynthetic marine algae that replenish much of the world's oxygen. Proponents claim that the ocean bottom already contains many radioactive minerals and that the radioactivity from all HLWs in existence would not harm marine algae.

Scientists in the United States have considered other proposals as well, but since the Nuclear Policy Act of 1982, most of the attention has focused on the development of a disposal site beneath Yucca Mountain in Nevada. The design of this site includes sloping shafts that lead to a 570 hectare (1400 acre) storage area 300 m deep in the mountain's interior. (See cross-section on previous page.)

The U.S. Department of Energy is committed to developing Yucca Mountain. Engineers believe that it will be 2017 before the site is ready to receive waste. Until that time, there is a plan to begin moving HLW from power plants to a remote interim site in a western state. HLW would be transported to the storage site by truck or rail in sealed, steel canisters placed inside reinforced shipping casks.

Objections to Yucca Mountain

There are two main sources of opposition to the Yucca Mountain plan. One source maintains that Yucca Mountain has not been proven to be geologically secure, citing evidence that gases emitted at the lowest depth were able to reach the outside air. In addition, the group is concerned about the possibility of collisions or other accidents that might break the casks open while in transport.

The Department of Energy claims it has proven the casks safe. A variety of tests have been performed, subjecting the casks to forces they would undergo in typical truck or train collisions.

The other source of opposition to Yucca Mountain are people who maintain that the costs of overcoming legal challenges will ultimately make the plan financially infeasible. These people believe that the government should stop spending money on the Yucca Mountain project and resume the plan to reprocess waste to make new nuclear fuel.

Researching the Issue

1. Visit your local library to find sources that will help you learn about the policies of your city, county, or state concerning the transport and/or storage of nuclear waste. Write a report describing these policies and include your opinion of them. Give reasons for the stance you take.

2. Scientists have considered the possibility of using powerful rockets to blast waste into the sun. Organize a classroom discussion of the benefits and hazards of this method.

3. Whenever public officials must decide on a location for a facility that people perceive as hazardous, they must find a way to overcome the NIMBY syndrome. What does *NIMBY* mean? How would you, as a public official, try to overcome this problem?

Contents

831

APPENDIX A Mathematical Review

Mathematical Review

Scientific Notation

Positive exponents Many quantities that scientists deal with often have very large or very small values. For example, the speed of light is about 300 000 000 m/s, and the ink required to make the dot over an *i* in this textbook has a mass of about 0.000 000 001 kg. Obviously, it is cumbersome to work with numbers such as these. We avoid this problem by using a method based on powers of the number 10.

$$10^0 = 1$$
$$10^1 = 10$$
$$10^2 = 10 \times 10 = 100$$
$$10^3 = 10 \times 10 \times 10 = 1000$$
$$10^4 = 10 \times 10 \times 10 \times 10 = 10\,000$$
$$10^5 = 10 \times 10 \times 10 \times 10 \times 10 = 100\,000$$

The number of zeros determines the power to which 10 is raised, or the *exponent* of 10. For example, the speed of light, 300 000 000 m/s, can be expressed as 3×10^8 m/s. In this case, the exponent of 10 is 8.

Negative exponents For numbers less than one, we note the following:

$$10^{-1} = \frac{1}{10} = 0.1$$
$$10^{-2} = \frac{1}{10 \times 10} = 0.01$$
$$10^{-3} = \frac{1}{10 \times 10 \times 10} = 0.001$$
$$10^{-4} = \frac{1}{10 \times 10 \times 10 \times 10} = 0.0001$$
$$10^{-5} = \frac{1}{10 \times 10 \times 10 \times 10 \times 10} = 0.000\,01$$

The value of the negative exponent equals the number of places the decimal point must be moved to be to the right of the first nonzero digit (in these cases, the digit 1). Numbers that are expressed as a number between 1 and 10 multiplied by a power of 10 are said to be in *scientific notation.* For example, 5 943 000 000 is 5.943×10^9 when expressed in scientific notation, and 0.000 083 2 is 8.32×10^{-5} when expressed in scientific notation.

Multiplication and division in scientific notation When numbers expressed in scientific notation are being multiplied, the following general rule is very useful:

$$10^n \times 10^m = 10^{(n+m)}$$

Note that n and m can be any numbers; they are not necessarily integers. For example, $10^2 \times 10^5 = 10^7$, and $10^{1/4} \times 10^{1/2} = 10^{3/4}$. The rule also applies to negative exponents. For example, $10^3 \times 10^{-8} = 10^{-5}$. When dividing numbers expressed in scientific notation, note the following:

$$\frac{10^n}{10^m} = 10^n \times 10^{-m} = 10^{(n-m)}$$

For example, $\frac{10^3}{10^2} = 10^{(3-2)} = 10^1$.

Fractions

The rules for multiplying, dividing, adding, and subtracting fractions are summarized in **Table 1,** where a, b, c, and d are four numbers.

Table 1 Basic Operations for Fractions

Operation	Rule	Example
Multiplication	$\left(\frac{a}{b}\right)\left(\frac{c}{d}\right) = \frac{ac}{bd}$	$\left(\frac{2}{3}\right)\left(\frac{4}{5}\right) = \frac{(2)(4)}{(3)(5)} = \frac{8}{15}$
Division	$\frac{\left(\frac{a}{b}\right)}{\left(\frac{c}{d}\right)} = \frac{ad}{bc}$	$\frac{\left(\frac{2}{3}\right)}{\left(\frac{4}{5}\right)} = \frac{(2)(5)}{(3)(4)} = \frac{10}{12} = \frac{5}{6}$
Addition and subtraction	$\frac{a}{b} \pm \frac{c}{d} = \frac{ad \pm bc}{bd}$	$\frac{2}{3} - \frac{4}{5} = \frac{(2)(5) - (3)(4)}{(3)(5)} = -\frac{2}{15}$

Powers

Rules of exponents When powers of a given quantity, x, are multiplied, the rule used for scientific notation applies:

$$(x^n)(x^m) = x^{(n+m)}$$

For example, $(x^2)(x^4) = x^{(2+4)} = x^6$.

When dividing the powers of a given quantity, note the following:

$$\frac{x^n}{x^m} = x^{(n-m)}$$

For example, $\frac{x^8}{x^2} = x^{(8-2)} = x^6$.

833

A power that is a fraction, such as $\frac{1}{3}$, corresponds to a root as follows:

$$x^{1/n} = \sqrt[n]{x}$$

For example, $4^{1/3} = \sqrt[3]{4} = 1.5874$. (A scientific calculator is useful for such calculations.)

Finally, any quantity, x^n, that is raised to the mth power is as follows:

$$(x^n)^m = x^{nm}$$

For example, $(x^2)^3 = x^{(2)(3)} = x^6$.

The basic rules of exponents are summarized in **Table 2.**

Table 2 Rules of Exponents

$x^0 = 1$	$x^1 = x$	$(x^n)(x^m) = x^{(n+m)}$
$\frac{x^n}{x^m} = x^{(n-m)}$	$x^{(1/n)} = \sqrt[n]{x}$	$(x^n)^m = x^{(nm)}$

Algebra

Solving for unknowns When algebraic operations are performed, the laws of arithmetic apply. Symbols such as x, y, and z are usually used to represent quantities that are not specified. Such unspecified quantities are called *unknowns.*

First, consider the following equation:

$$8x = 32$$

If we wish to solve for x, we can divide each side of the equation by the same factor without disturbing the equality. In this case, if we divide both sides by 8, we have the following:

$$\frac{8x}{8} = \frac{32}{8}$$

$$x = 4$$

Next, consider the following equation:

$$x + 2 = 8$$

In this type of expression, we can add or subtract the same quantity from each side. If we subtract 2 from each side, we get the following:

$$x + 2 - 2 = 8 - 2$$

$$x = 6$$

In general, if $x + a = b$, then $x = b - a$.

Now, consider the following equation:

$$\frac{x}{5} = 9$$

If we multiply each side by 5, we are left with x isolated on the left and a value of 45 on the right.

$$(5)\left(\frac{x}{5}\right) = (9)(5)$$
$$x = 45$$

In all cases, *whatever operation is performed on the left side of the equation must also be performed on the right side.*

Factoring

Some useful formulas for factoring an equation are given in **Table 3.** As an example of a common factor, consider the equation $5x + 5y + 5z = 0$. This equation can be expressed as $5(x + y + z) = 0$. The expression $a^2 + 2ab + b^2$, which is an example of a perfect square, is equivalent to the expression $(a + b)^2$. For example, if $a = 2$ and $b = 3$, then $2^2 + (2)(2)(3) + 3^2 = (2 + 3)^2$, or $(4 + 12 + 9) = 5^2 = 25$. Finally, for an example of the difference of two squares, let $a = 6$ and $b = 3$. In this case, $(6^2 - 3^2) = (6 + 3)(6 - 3)$, or $(36 - 9) = (9)(3) = 27$.

Table 3 Factoring Equations

$ax + ay + az = a(x + y + z)$	common factor
$a^2 + 2ab + b^2 = (a + b)^2$	perfect square
$a^2 - b^2 = (a + b)(a - b)$	difference of two squares

Quadratic Equations

The general form of a quadratic equation is as follows:

$$ax^2 + bx + c = 0$$

In this equation, x is the unknown quantity and a, b, and c are numerical factors known as *coefficients.* This equation has two roots, given by the following:

$$x = \frac{-b \pm \sqrt{b^2 - 4ac}}{2a}$$

If $b^2 \geq 4ac$, the value inside the square-root symbol will be positive or zero and the roots will be real. If $b^2 < 4ac$, the value inside the square-root symbol will be negative and the roots will be imaginary numbers. In problems in this physics book, imaginary roots should not occur.

835

Example
Find the solutions for the equation $x^2 + 5x + 4 = 0$.

Solution
The given equation can be expressed as $(1)x^2 + (5)x + (4) = 0$. In other words, $a = 1$, $b = 5$, and $c = 4$. The two roots of this equation can be found by substituting these values into the quadratic equation, as follows:

$$x = \frac{-b \pm \sqrt{b^2 - 4ac}}{2a} = \frac{-5 \pm \sqrt{5^2 - (4)(1)(4)}}{(2)(1)} = \frac{-5 \pm \sqrt{9}}{2} = \frac{-5 \pm 3}{2}$$

The two roots are $x = \frac{-5 + 3}{2} = -1$ and $x = \frac{-5 - 3}{2} = -4$.

$x = -1$ and $x = -4$

We can evaluate these answers by substituting them into the given equation and verifying that the result is zero.

$$x^2 + 5x + 4 = 0$$

For $x = -1$, $(-1)^2 + 5(-1) + 4 = 1 - 5 + 4 = 0$.

For $x = -4$, $(-4)^2 + 5(-4) + 4 = 16 - 20 + 4 = 0$.

Example
Factor the equation $2x^2 - 3x - 4 = 0$.

Solution
The given equation can be expressed as $(2)x^2 + (-3)x + (-4) = 0$. Thus, $a = 2$, $b = -3$, and $c = -4$. Substitute these values into the quadratic equation to factor the given equation.

$$x = \frac{-b \pm \sqrt{b^2 - 4ac}}{2a} = \frac{3 \pm \sqrt{(-3)^2 - (4)(2)(-4)}}{(2)(2)} = \frac{3 \pm \sqrt{41}}{4} = \frac{3 \pm 6.403}{4}$$

The two roots are $x = \frac{3 + 6.403}{4} = 2.351$ and $x = \frac{3 - 6.403}{4} = -0.851$.

$x = 2.351$ and $x = -0.851$

Again, evaluate these answers by substituting them into the given equation.

$$2x^2 - 3x - 4 = 0$$

For $x = 2.351$, $2(2.351)^2 - 3(2.351) - 4 = 11.054 - 7.053 - 4 \approx 0$.

For $x = -0.851$, $2(-0.851)^2 - 3(-0.851) - 4 = 1.448 + 2.553 - 4 \approx 0$.

Linear Equations

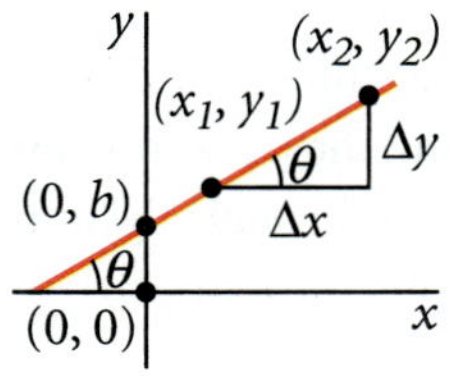

Figure 1

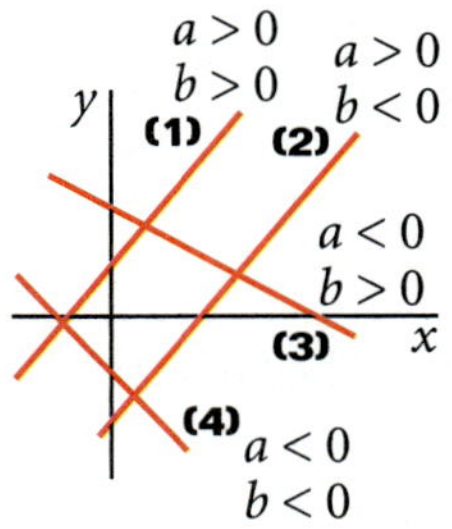

Figure 2

A linear equation has the following general form:

$$y = ax + b$$

In this equation, a and b are constants. This equation is called linear because the graph of y versus x is a straight line, as shown in **Figure 1.** The constant b, called the *intercept,* represents the value of y where the straight line intersects the y-axis. The constant a is equal to the *slope* of the straight line and is also equal to the tangent of the angle that the line makes with the x-axis (θ). If any two points on the straight line are specified by the coordinates (x_1, y_1) and (x_2, y_2), as in **Figure 1,** then the slope of the straight line can be expressed as follows:

$$\text{slope} = \frac{y_2 - y_1}{x_2 - x_1} = \frac{\Delta y}{\Delta x}$$

For example, if the two points shown in **Figure 1** are (2, 4) and (6, 9), then the slope of the line is as follows:

$$\text{slope} = \frac{(9-4)}{(6-2)} = \frac{5}{4}$$

Note that a and b can have either positive or negative values. If $a > 0$, the straight line has a *positive* slope, as in **Figure 1.** If $a < 0$, the straight line has a *negative* slope. Furthermore, if $b > 0$, the y intercept is positive (above the x-axis), while if $b < 0$, the y intercept is negative (below the x-axis). **Figure 2** gives an example of each of these four possible cases, which are summarized in **Table 4.**

Table 4 Linear Equations

Constants	Slope	y intercept
$a > 0, b > 0$	positive slope	positive y intercept
$a > 0, b < 0$	positive slope	negative y intercept
$a < 0, b > 0$	negative slope	positive y intercept
$a < 0, b < 0$	negative slope	negative y intercept

Solving Simultaneous Linear Equations

Consider the following equation:

$$3x + 5y = 15$$

This equation has two unknowns, x and y. Such an equation does not have a unique solution. That is, $(x = 0, y = 3)$, $(x = 5, y = 0)$, and $\left(x = 2, y = \frac{9}{5}\right)$ are all solutions to this equation.

If a problem has two unknowns, a unique solution is possible only if there are two independent equations. In general, if a problem has n unknowns, its solution requires n independent equations. There are three basic methods that can be used to solve simultaneous equations. Each of these methods is discussed below, and an example is given for each.

837

First method: substitution One way to solve two simultaneous equations involving two unknowns, x and y, is to solve one of the equations for one of the unknown values in terms of the other unknown value. In other words, either solve one equation for x in terms of y or solve one equation for y in terms of x. Once you have an expression for either x or y, substitute this expression into the other original equation. At this point, the equation has only one unknown quantity. This unknown can be found through algebraic manipulations and then can be used to determine the other unknown.

Example

Solve the following two simultaneous equations:

1. $5x + y = -8$
2. $2x - 2y = 4$

Solution

First solve for either x or y in one of the equations. We'll begin by solving equation 2 for x.

2. $2x - 2y = 4$

$$2x = 4 + 2y$$

$$x = \frac{4 + 2y}{2} = 2 + y$$

Next, we substitute this equation for x into equation 1 and solve for y.

1. $5x + y = -8$

$$5(2 + y) + y = -8$$

$$10 + 5y + y = -8$$

$$6y = -18$$

$$\boxed{y = -3}$$

To find x, substitute this value for y into the equation for x derived from equation 2.

$$x = 2 + y = 2 + -3$$

$$\boxed{x = -1}$$

There is always more than one way to solve simultaneous equations by substitution. In this example, we first solved equation 2 for x. However, we could have begun by solving equation 2 for y or equation 1 for x or y. Any of these processes would result in the same answer.

838

Second method: canceling one term Simultaneous equations can also be solved by multiplying both sides of one of the equations by a value that will make either the x value or the y value in that equation equal to and opposite the corresponding value in the second equation. When the two equations are added together, that unknown value drops out and only one of the unknown values remains. This unknown can be found through algebraic manipulations and then can be used to determine the other unknown.

Example

Solve the following two simultaneous equations:

1. $3x + y = -6$
2. $-4x - 2y = 6$

Solution

First, multiply each term of one of the equations by a factor that will make either the x or the y values cancel when the two equations are added together. In this case, we can multiply each term in equation 1 by the factor 2. The positive $2y$ in equation 1 will then cancel the negative $2y$ in equation 2.

1. $3x + y = -6$
$(2)(3x) + (2)(y) = -(2)(6)$
$6x + 2y = -12$

Next, add the two equations together and solve for x.

2. $-4x - 2y = 6$

1. $6x + 2y = -12$
$2x = -6$

$\boxed{x = -3}$

Then, substitute this value of x into either equation to find y.

1. $3x + y = -6$
$y = -6 - 3x = -6 - (3)(-3) = -6 + 9$

$\boxed{y = 3}$

In this example, we multiplied both sides of equation 1 by 2 so that the y terms would cancel when the two equations were added together. As with substitution, this is only one of many possible ways to solve the equations. For example, we could have multiplied both sides of equation 2 by $\frac{3}{4}$ so that the x terms would cancel when the two equations were added together.

839

Third method: graphing the equations Two linear equations with two unknowns can also be solved by a graphical method. If the straight lines corresponding to the two equations are plotted in a conventional coordinate system, the intersection of the two lines represents the solution.

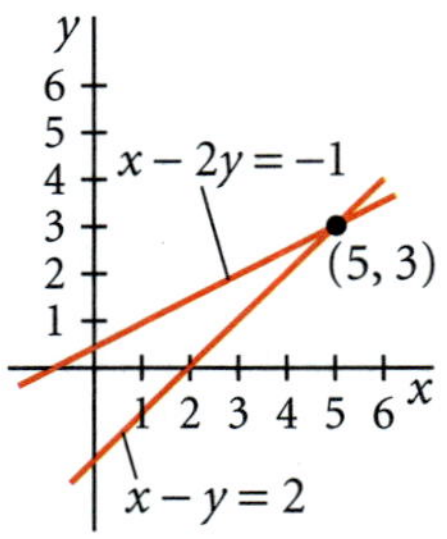

Figure 3

Example

Solve the following two simultaneous equations:

1. $x - y = 2$

2. $x - 2y = -1$

Solution

These two equations are plotted in **Figure 3.** To plot an equation, rewrite the equation in the form $y = ax + b$, where a is the slope and b is the y intercept. In this example, the equations can be rewritten as follows:

$$y = x - 2$$
$$y = \frac{1}{2}x + \frac{1}{2}$$

Once one point of a line is known, any other point on that line can be found with the slope of the line. For example, the slope of the first line is 1, and we know that (0, −2) is a point on this line. If we choose the point $x = 2$, we have $(2, y_2)$. The coordinate y_2 can be found as follows:

$$\text{slope} = \frac{y_2 - y_1}{x_2 - x_1} = \frac{y_2 - (-2)}{2 - 0} = 1$$
$$y_2 = 0$$

Connecting the two known coordinates, (0, −2) and (2, 0), results in a graph of the line. The second line can be plotted with the same method.

As shown in **Figure 3,** the intersection of the two lines has the coordinates $x = 5$, $y = 3$. This intersection represents the solution to the equations. You should check this solution using either of the analytical techniques discussed above.

Logarithms

Suppose that a quantity, x, is expressed as a power of another quantity, a.

$$x = a^y$$

The number a is called the *base number.* The *logarithm* of x with respect to the base, a, is equal to the exponent to which a must be raised in order to satisfy the expression $x = a^y$.

$$y = \log_a x$$

Conversely, the *antilogarithm* of y is the number x.

$$x = \text{antilog}_a y$$

840

Common and natural bases In practice, the two bases most often used are base 10, called the *common* logarithm base, and base $e = 2.718...$, called the *natural* logarithm base. When common logarithms are used, y and x are related as follows:

$$y = \log_{10} x, \text{ or } x = 10^y$$

When natural logarithms are used, the symbol ln is used to signify that the logarithm has a base of e; in other words, $\log_e x = \ln x$.

$$y = \ln x, \text{ or } x = e^y$$

For example, $\log_{10} 52 = 1.716$, so $\text{antilog}_{10} 1.716 = 10^{1.716} = 52$. Likewise, $\ln 52 = 3.951$, so $\text{antiln } 3.951 = e^{3.951} = 52$.

Note that you can convert between base 10 and base e with the equality

$$\ln x = (2.302\ 585)\log_{10} x.$$

Some useful properties of logarithms are summarized in **Table 5.**

Table 5 Properties of Logarithms

Rule	Example
$\log(ab) = \log a + \log b$	$\log(2)(5) = \log 2 + \log 5$
$\log\left(\frac{a}{b}\right) = \log a - \log b$	$\log\frac{3}{4} = \log 3 - \log 4$
$\log(a^n) = n \log a$	$\log 7^3 = 3 \log 7$
$\ln e = 1$	
$\ln e^a = a$	$\ln e^5 = 5$
$\ln\left(\frac{1}{a}\right) = -\ln a$	$\ln\frac{1}{8} = -\ln 8$

Conversions Between Fractions, Decimals, and Percentages

The rules for converting numbers from fractions to decimals and percentages and from percentages to decimals are summarized in **Table 6.**

Table 6 Conversions

Conversion	Rule	Example
Fraction to decimal	divide numerator by denominator	$\frac{31}{45} = 0.69$
Fraction to percentage	convert to decimal, then multiply by 100%	$\frac{31}{45} = (0.69)(100\%) = 69\%$
Percentage to decimal	move decimal point two places to the left, and remove the percent sign	$69\% = 0.69$

841

Geometry

Table 7 provides equations for the area and volume of several geometrical shapes used throughout this text.

Table 7 Geometrical Areas and Volumes

Geometrical shape	Useful equations
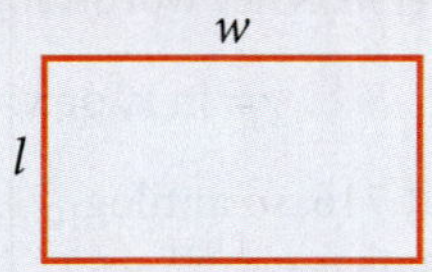 rectangle	area = lw perimeter = $2(l + w)$
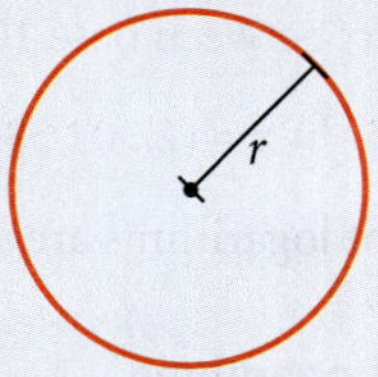 circle	area = πr^2 circumference = $2\pi r$
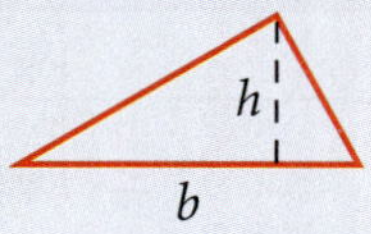 triangle	area = $\frac{1}{2}bh$
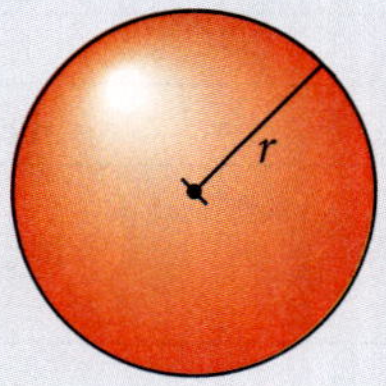 sphere	surface area = $4\pi r^2$ volume = $\frac{4}{3}\pi r^3$
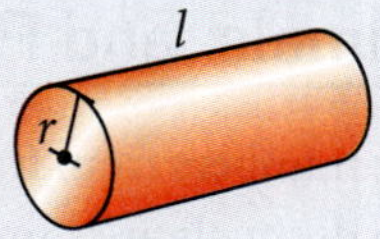 cylinder	surface area = $2\pi r^2 + 2\pi rl$ volume = $\pi r^2 l$
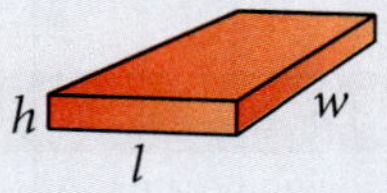 rectangular box	surface area = $2(lh + lw + hw)$ volume = lwh

Trigonometry and the Pythagorean Theorem

$\sin\theta = \frac{a}{c}$

$\cos\theta = \frac{b}{c}$

$\tan\theta = \frac{a}{b}$

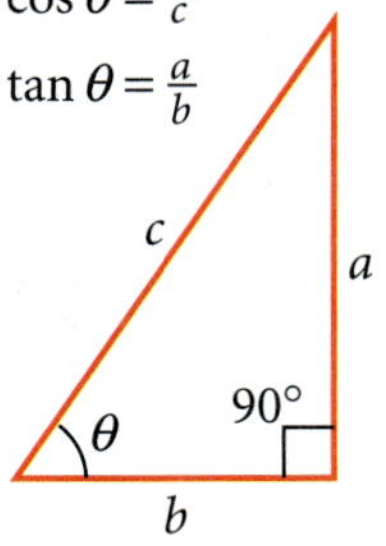

Figure 4

The portion of mathematics that is based on the relationships between the sides and angles of triangles is called *trigonometry.* Many of the concepts of this branch of mathematics are of great importance in the study of physics. To review some of the basic concepts of trigonometry, consider the right triangle shown in **Figure 4,** where side a is opposite the angle θ, side b is adjacent to the angle θ, and side c is the hypotenuse of the triangle (the side opposite the right angle). The most common trigonometry functions are summarized in **Table 8,** using this figure as an example.

Table 8 Trigonometry Functions

sine (sin)	$\sin\theta = \dfrac{\text{side opposite }\theta}{\text{hypotenuse}} = \dfrac{a}{c}$
cosine (cos)	$\cos\theta = \dfrac{\text{side adjacent to }\theta}{\text{hypotenuse}} = \dfrac{b}{c}$
tangent (tan)	$\tan\theta = \dfrac{\text{side opposite }\theta}{\text{side adjacent to }\theta} = \dfrac{a}{b}$
inverse sine ($\sin^{-1}$)	$\theta = \sin^{-1}\left(\dfrac{\text{side opposite }\theta}{\text{hypotenuse}}\right) = \sin^{-1}\left(\dfrac{a}{c}\right)$
inverse cosine ($\cos^{-1}$)	$\theta = \cos^{-1}\left(\dfrac{\text{side adjacent to }\theta}{\text{hypotenuse}}\right) = \cos^{-1}\left(\dfrac{b}{c}\right)$
inverse tangent ($\tan^{-1}$)	$\theta = \tan^{-1}\left(\dfrac{\text{side opposite }\theta}{\text{side adjacent to }\theta}\right) = \tan^{-1}\left(\dfrac{a}{b}\right)$

When $\theta = 30°$, for example, the ratio of a to c is always 0.50. In other words, $\sin 30° = 0.50$. Sine, cosine, and tangent are quantities without units because each represents the ratio of two lengths. Furthermore, note the following trigonometry identity:

$$\frac{\sin\theta}{\cos\theta} = \frac{\dfrac{\text{side opposite }\theta}{\text{hypotenuse}}}{\dfrac{\text{side adjacent to }\theta}{\text{hypotenuse}}} = \frac{\text{side opposite }\theta}{\text{side adjacent to }\theta} = \tan\theta$$

Some additional trigonometry identities are as follows:

$$\sin^2\theta + \cos^2\theta = 1$$
$$\sin\theta = \cos(90° - \theta)$$
$$\cos\theta = \sin(90° - \theta)$$

843

Determining an unknown side The first three functions given in **Table 8** can be used to determine any unknown side of a right triangle when one side and one of the non-right angles are known. For example, if $\theta = 30°$ and $a = 1.0$ m, the other two sides of the triangle can be found as follows:

$$\sin\theta = \frac{a}{c}$$

$$c = \frac{a}{\sin\theta} = \frac{1.0\ \text{m}}{\sin 30°}$$

$$\boxed{c = 2.0\ \text{m}}$$

$$\tan\theta = \frac{a}{b}$$

$$b = \frac{a}{\tan\theta} = \frac{1.0\ \text{m}}{\tan 30°}$$

$$\boxed{b = 1.7\ \text{m}}$$

Determining an unknown angle In some cases, you might know the value of the sine, cosine, or tangent of an angle and need to know the value of the angle itself. The inverse sine, cosine, and tangent functions given in **Table 8** can be used for this purpose. For example, in **Figure 4,** suppose you know that side $a = 1.0$ m and side $c = 2.0$ m. To find the angle θ, you could use the inverse sine function, $\sin^{-1}$, as follows:

$$\theta = \sin^{-1}\left(\frac{a}{c}\right) = \sin^{-1}\left(\frac{1.0\ \text{m}}{2.0\ \text{m}}\right) = \sin^{-1}(0.50)$$

$$\boxed{\theta = 30°}$$

Converting from degrees to radians The two most common units used to measure angles are degrees and radians. A full circle is represented by 360 degrees (360°) or by 2π radians (2π rad). As such, the following conversions can be used:

$$[\text{angle } (°)] = \frac{180}{\pi}[\text{angle (rad)}]$$

$$[\text{angle (rad)}] = \frac{\pi}{180}[\text{angle } (°)]$$

Pythagorean theorem Another useful equation when working with right triangles is the Pythagorean theorem. If a and b are the two legs of a right triangle and c is the hypotenuse, as in **Figure 5,** the Pythagorean theorem can be expressed as follows:

$$c^2 = a^2 + b^2$$

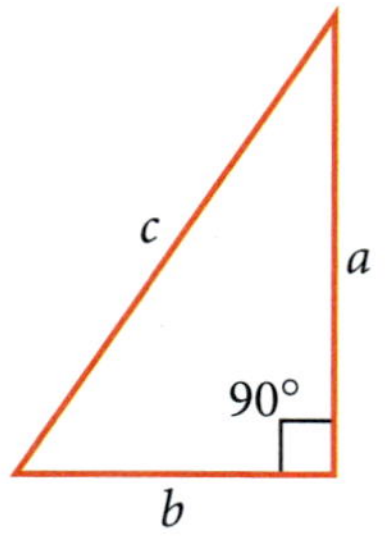

Figure 5

In other words, the square of the hypotenuse of a right triangle equals the sum of the squares of the other two legs of the triangle. The Pythagorean

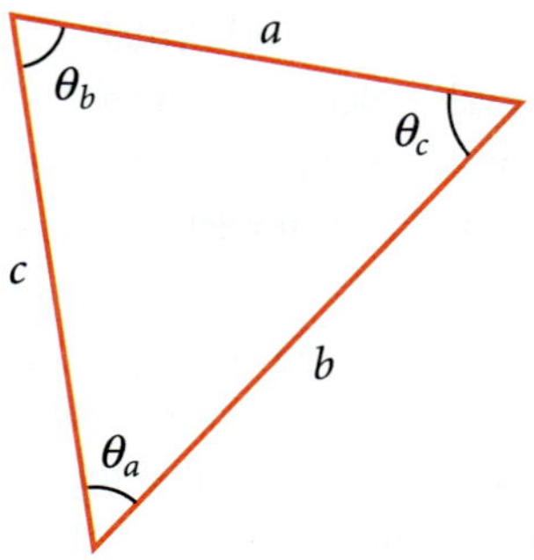

Figure 6

theorem is useful when two sides of a right triangle are known but the third side is not. For example, if $c = 2.0$ m and $a = 1.0$ m, you could find b using the Pythagorean theorem as follows:

$$b = \sqrt{c^2 - a^2} = \sqrt{(2.0 \text{ m})^2 - (1.0 \text{ m})^2}$$
$$b = \sqrt{4.0 \text{ m}^2 - 1.0 \text{ m}^2} = \sqrt{3.0 \text{ m}^2}$$

$$\boxed{b = 1.7 \text{ m}}$$

Law of sines and law of cosines The law of sines may be used to find angles of any general triangle. The law of cosines is used for calculating one side of a triangle when the angle opposite and the other two sides are known. If *a*, *b*, and *c* are the three sides of the triangle and θ_a, θ_b, and θ_c are the three angles opposite those sides, as shown in **Figure 6,** the following relationships hold true:

$$\frac{a}{\sin \theta_a} = \frac{b}{\sin \theta_b} = \frac{c}{\sin \theta_c}$$
$$c^2 = a^2 + b^2 - 2ab \cos \theta_c$$

Accuracy in Laboratory Calculations

Absolute error Some of the laboratory experiments in this book involve finding a value that is already known, such as free-fall acceleration. In this type of experiment, the accuracy of your measurements can be determined by comparing your results with the accepted value. The absolute value of the difference between your experimental or calculated result and the accepted value is called the *absolute error.* Thus, absolute error can be found with the following equation:

$$\text{absolute error} = |\text{experimental} - \text{accepted}|$$

Be sure not to confuse accuracy with precision. The *accuracy* of a measurement refers to how close that measurement is to the accepted value for the quantity being measured. *Precision* depends on the instruments used to measure a quantity. A meterstick that includes millimeters, for example, will give a more precise result than a meterstick whose smallest unit of measure is a centimeter. Thus, a measurement of 9.61 m/s^2 for free-fall acceleration is more precise than a measurement of 9.8 m/s^2, but 9.8 m/s^2 is more accurate than 9.61 m/s^2.

Relative error Note that a measurement that has a relatively large absolute error may be more accurate than a measurement that has a smaller absolute error if the first measurement involved much larger quantities. For this reason, the percentage error, or *relative error,* is often more meaningful than the absolute error. The relative error of a measured value can be found with the following equation:

$$\text{relative error} = \frac{(\text{experimental} - \text{accepted})}{\text{accepted}}$$

845

In other words, the relative error is the difference between the experimental value and the accepted value divided by the accepted value. Because relative error takes the size of the quantity being measured into account, the accuracy of two different measurements can be compared by comparing their relative errors.

For example, consider two laboratory experiments in which you are determining values that are fairly well known. In the first, you determine that free-fall acceleration at Earth's surface is 10.31 m/s^2. In the second, you find that the speed of sound in air at 25°C is 355 m/s. The accepted values for these quantities are 9.81 m/s^2 and 346 m/s, respectively. Now we'll find the absolute and relative errors for each experiment.

For the first experiment, the absolute and relative errors can be calculated as follows:

$$\text{absolute error} = |\text{experimental} - \text{accepted}| = |10.31 \text{ m/s}^2 - 9.81 \text{ m/s}^2|$$

$$\boxed{\text{absolute error} = 0.50 \text{ m/s}^2}$$

$$\text{relative error} = \frac{(\text{experimental} - \text{accepted})}{\text{accepted}} = \frac{(10.31 \text{ m/s}^2 - 9.81 \text{ m/s}^2)}{9.81 \text{ m/s}^2}$$

$$\boxed{\text{relative error} = 0.051 = 5.1\%}$$

For the second experiment, the absolute and relative errors can be calculated as follows:

$$\text{absolute error} = |\text{experimental} - \text{accepted}| = |355 \text{ m/s} - 346 \text{ m/s}|$$

$$\boxed{\text{absolute error} = 9 \text{ m/s}}$$

$$\text{relative error} = \frac{(\text{experimental} - \text{accepted})}{\text{accepted}} = \frac{(355 \text{ m/s} - 346 \text{ m/s})}{346 \text{ m/s}}$$

$$\boxed{\text{relative error} = 0.026 = 2.6\%}$$

Note that the *absolute* error is less in the first experiment, while the *relative* error is less in the second experiment. The absolute error is less in the first experiment because typical values for free-fall acceleration are much smaller than typical values for the speed of sound in air. The relative errors take this difference into account. Thus, comparing the relative errors shows that the speed of sound is measured with greater accuracy than is the free-fall acceleration.

846

Downloading Graphing Calculator Programs

internet connect

go.hrw.com

Topic: Graphing Calculator
Go To: go.hrw.com
Keyword: HF6 CALC

Visit the HRW Web site for links for downloading graphing calculator programs.

CBL2 Labs

The following steps describe how to transfer the DataMate data collection program from the CBL2 or LabPro interface to the calculator.

1. Prepare the calculator to receive the program.
- For the TI-73, TI-73 Explorer, TI-82, TI-83, TI-83 Plus, TI-83 Plus Silver Edition, TI-84 Plus, and TI-84 Plus Silver Edition, turn on your calculator and press [2nd] [LINK]. (On the TI-73, press [APPS] and then select "Link…".) Press [▶] to RECEIVE and then press [ENTER]. ["Waiting…" appears on your screen.]
- For the TI-86, turn on your calculator, press [2nd] [LINK], and then press [F2]. ["Waiting…" appears on your screen.]
- For the TI-89, TI-89 Titanium, TI-92, TI-92 Plus, and Voyage 200, turn on your calculator and make sure that the calculator is on the Home screen. (You do not need to put the calculator into the Receiving mode as is required with the other calculators.)

2. Press the Transfer button on CBL2 or LabPro interface. ["Receiving…" followed by a list of the loaded programs (or application) is displayed on the calculator.]

3. A "Done" message on the calculator and two beeps from the interface will indicate that the transfer is complete. Press [2nd] [QUIT].

4. Verify that the programs have been successfully loaded.
- On the TI-82 and TI-83, press [PRGM] to see DATAMATE and its subprograms.
- On the TI-73, TI-73 Explorer, TI-83 Plus, TI-83 Plus Silver Edition, TI-84 Plus, and TI-84 Plus Silver Edition, press [APPS] to see DATAMATE.
- On the TI-86, press [PRGM] and then [F1] to see DATAMATE and its subprograms.
- On the TI-89, TI-89 Titanium, TI-92, TI-92 Plus, and Voyage 200, press [2nd] [VAR-LINK] to see DATAMATE and its subprograms.

Graphing Calculator Exercises

If you will be doing the Technology and Learning exercises, click Download Graphing Calculator Programs. The file GRAPHING.ZIP will be loaded onto your computer. Once the file is downloaded, double-click the icon and the file will be extracted into a file called Graphing.8xg. Follow the instructions for your TI-Graph Link to load Graphing.8xg onto your TI calculator. When the file is sent to the calculator, it should expand into 22 programs (one for each chapter). These programs should appear in the PRGM menu.

Troubleshooting

- Calculator and CBL instructions in the *Holt Physics* program are written for the TI graphing calculators, for the CBL2 interface from TI or the LabPro interface from Vernier Software & Technology, and for probes from Vernier. You may use other hardware, but some of the programs and instructions may not work exactly as described.
- If you have problems loading programs onto your calculator, you may need to clear programs or other data from your calculator's memory.
- If you need additional help, both TI and Vernier can provide technical support.

847

Symbols

Diagram Symbols

Mechanics

Symbol	Meaning
	displacement vector, displacement component
	velocity vector, velocity component
	acceleration vector
	force vector, force component
	momentum vector
	gravitational field vector
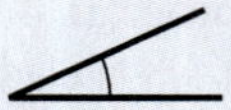	angle marking
	rotational motion

Thermodynamics

Symbol	Meaning
	energy transferred as heat
	energy transferred as work
	cycle or process

Waves and Electromagnetism

Symbol	Meaning
	ray (light or sound)
	positive charge
	negative charge
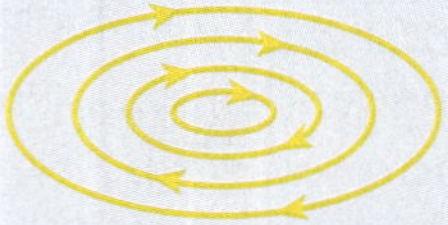	electric field lines
	electric field vector
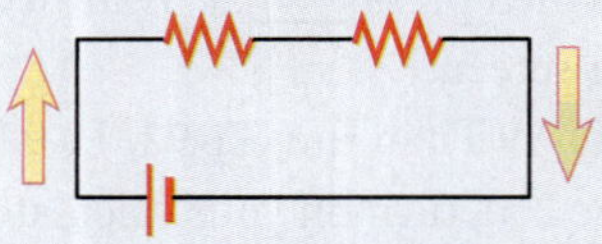	electric current
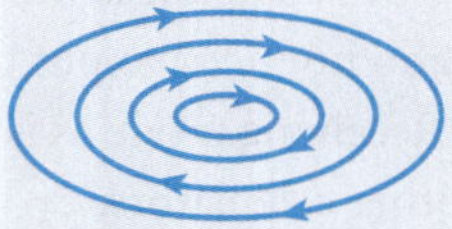	magnetic field lines
	magnetic field vector (into page, out of page)

Mathematical Symbols

Symbol	Meaning	Symbol	Meaning
Δ	(Greek *delta*) change in some quantity	$\leq$	less than or equal to
Σ	(Greek *sigma*) sum of quantities	$\propto$	is proportional to
θ	(Greek *theta*) any angle	$\approx$	is approximately equal to
$=$	equal to	$\vert n \vert$	absolute value or magnitude of
$>$	greater than	sin	sine
$\geq$	greater than or equal to	cos	cosine
$<$	less than	tan	tangent

Quantity Symbols Used Throughout

Symbols that are **boldfaced** refer to vector quantities that have both a magnitude and a direction. Symbols that are *italicized* refer to quantities with only a magnitude. Symbols that are neither are usually units.

Symbol	Quantity
A	area
D	diameter
$\mathbf{F}$, F	force
m	mass
M	total mass
R	radius (of a spherical body, a curved mirror, or a curved lens)
r	radius (of sphere, shell, or disk)
t	time
V	volume

849

Translational Mechanics Symbols Used in This Book

Symbols that are **boldfaced** refer to vector quantities that have both a magnitude and a direction. Symbols that are *italicized* refer to quantities with only a magnitude. Symbols that are neither are usually units.

Symbol	Quantity
$\mathbf{a}$, a	acceleration
$\mathbf{a_g}$	free-fall acceleration (acceleration due to gravity)
$\mathbf{d}$, d	displacement
$\mathbf{F}\Delta t$	impulse
$\mathbf{F_g}$, F_g	gravitational force (weight)
$\mathbf{F_k}$, F_k	force of kinetic friction
$\mathbf{F_n}$, F_n	normal force
$\mathbf{F_{net}}$, F_{net}	net force
$\mathbf{F_R}$, F_R	force of air resistance
$\mathbf{F_s}$, F_s	force of static friction
$\mathbf{F_{s,max}}$, $F_{s,max}$	maximum force of static friction
h	height
k	spring constant
KE	kinetic energy
KE_{trans}	translational kinetic energy
MA	mechanical advantage
ME	mechanical energy (sum of all kinetic and potential energies)
μ_k	(Greek *mu*) coefficient of kinetic friction
μ_s	(Greek *mu*) coefficient of static friction
P	power
$\mathbf{p}$, p	momentum
PE	potential energy
$PE_{elastic}$	elastic potential energy
PE_g	gravitational potential energy
r	separation between point masses
$\mathbf{v}$, v	velocity or speed
W	work
$W_{friction}$	work done by a frictional force (or work required to overcome a frictional force)
W_{net}	net work done
$\Delta\mathbf{x}$, Δx	displacement in the x direction
$\Delta\mathbf{y}$, Δy	displacement in the y direction

Rotational Mechanics Symbols Used in This Book

Symbols that are **boldfaced** refer to vector quantities that have both a magnitude and a direction. Symbols that are *italicized* refer to quantities with only a magnitude. Symbols that are neither are usually units.

Symbol	Quantity
a_t	tangential acceleration
a_c	centripetal acceleration
α	(Greek *alpha*) angular acceleration
$d \sin \theta$	lever arm (for torque calculations)
$\mathbf{F_c}$, F_c	centripetal force
I	moment of inertia
KE_{rot}	rotational kinetic energy
L	angular momentum
ℓ	length of a rotating rod
s	arc length
τ	(Greek *tau*) torque
τ_{net}	(Greek *tau*) net torque
θ	(Greek *theta*) angle of rotation
$\Delta\theta$	(Greek *delta* and *theta*) angular displacement (in radians)
v_t	tangential speed
ω	(Greek *omega*) angular speed

Fluid Dynamics and Thermodynamics Symbols Used in This Book

Symbols that are **boldfaced** refer to vector quantities that have both a magnitude and a direction. Symbols that are *italicized* refer to quantities with only a magnitude. Symbols that are neither are usually units.

Symbol	Quantity
c_p	specific heat capacity
eff	efficiency of a simple machine, thermal efficiency of a heat engine
$\mathbf{F_B}$, F_B	buoyant force
L	latent heat
L_f	latent heat of fusion
L_v	latent heat of vaporization
N	number of gas particles or nuclei
P	pressure
P_0	initial pressure, atmospheric pressure
P_{net}	net pressure
ρ	(Greek *rho*) mass density
Q	heat
Q_c	energy transferred as heat to or from a low-temperature (cold) substance
Q_h	energy transferred as heat to or from a high-temperature (hot) substance
Q_{net}	net amount of energy transferred as heat to or from a system
T	temperature (absolute)
T_C	temperature in degrees Celsius
T_c	temperature of a low-temperature (cool) substance
T_F	temperature in degrees Fahrenheit
T_h	temperature of a high-temperature (hot) substance
U	internal energy

851

Vibrations, Waves, and Optics Symbols Used in This Book

Symbols that are **boldfaced** refer to vector quantities that have both a magnitude and a direction. Symbols that are *italicized* refer to quantities with only a magnitude. Symbols that are neither are usually units.

Symbol	Quantity
C	center of curvature for spherical mirror
d	slit separation in double-slit interference of light
$d \sin \theta$	path difference for interfering light waves
$\mathbf{F}_{\mathbf{elastic}}$, $F_{elastic}$	spring force
F	focal point
f	focal length
f	frequency
f_n	*n*th harmonic frequency
h	object height
h'	image height
k	spring constant
L	length of a pendulum, vibrating string, or vibrating column of air
ℓ	path length of light wave
λ	(Greek *lambda*) wavelength
m	order number for interference fringes
M	magnification of image
n	harmonic number (sound)
n	index of refraction
p	object distance
q	image distance
T	period of a pendulum (simple harmonic motion)
θ	(Greek *theta*) angle of incidence of a beam of light (reflection)
θ	(Greek *theta*) angle of fringe separation from center of interference pattern
θ'	(Greek *theta*) angle of reflection
θ_c	(Greek *theta*) critical angle of refraction
θ_i	(Greek *theta*) angle of incidence of a beam of light (refraction)
θ_r	(Greek *theta*) angle of refraction

Electromagnetism Symbols Used in This Book

Symbols that are **boldfaced** refer to vector quantities that have both a magnitude and a direction. Symbols that are *italicized* refer to quantities with only a magnitude. Symbols that are neither are usually units.

Symbol	Quantity
$\mathbf{B}$, B	magnetic field
C	capacitance
d	separation of plates in a capacitor
$\mathbf{E}$, E	electric field
emf	emf (potential difference) produced by a battery or electromagnetic induction
$\mathbf{F}_{\mathbf{electric}}$, $F_{electric}$	electric force
$\mathbf{F}_{\mathbf{magnetic}}$, $F_{magnetic}$	magnetic force
I	electric current
i	instantaneous current (ac circuit)
I_{max}	maximum current (ac circuit)
I_{rms}	root-mean-square current (ac circuit)
L	self-inductance
ℓ	length of an electrical conductor in a magnetic field
M	mutual inductance
N	number of turns in a current-carrying loop or a transformer coil
$PE_{electric}$	electrical potential energy
Q	large charge or charge on a fully charged capacitor
q	charge
R	resistance
r	separation between charges
R_{eq}	equivalent resistance
V	electric potential
ΔV	potential difference
Δv	instantaneous potential difference (ac circuit)
ΔV_{max}	maximum potential difference (ac circuit)
ΔV_{rms}	root-mean-square potential difference (ac circuit)
ω	(Greek *omega*) angular frequency

852

Particle and Electronic Symbols Used in This Book

For this part of the book, two tables are given because some symbols refer to quantities and others refer to specific particles. The symbol's context should make clear which table should be consulted.

Symbol	Quantity
A	mass number
β	(Greek *beta*) current or potential difference gain of an amplifier
E	photon energy
E_R	rest energy
f_t	threshold frequency (photoelectric effect)
hf_t	work function (photoelectric effect)
KE_{max}	maximum energy of ejected photoelectron
λ	(Greek *lambda*) decay constant
λN	decay rate (activity)
N	neutron number, number of decayed particles
n	energy quantum number
$T_{1/2}$	half-life
Z	atomic number

Symbol	Particle
α	alpha particle
$b, \bar{b}$	bottom quark, antiquark
β^+	(Greek *beta*) positron (beta particle)
β^-	(Greek *beta*) electron (beta particle)
$c, \bar{c}$	charmed quark, antiquark
$d, \bar{d}$	down quark, antiquark
e^+, ${}^{0}_{+1}e$	positron
e^-, ${}^{0}_{-1}e$	electron
γ	(Greek *gamma*) photon (gamma rays)
${}^{4}_{2}\mathrm{He}$	alpha particle (helium-4 nucleus)
μ	(Greek *mu*) muon
${}^{1}_{0}n$	neutron
${}^{1}_{1}p$	proton
$s, \bar{s}$	strange quark, antiquark
$t, \bar{t}$	top quark, antiquark
$u, \bar{u}$	up quark, antiquark
τ	(Greek *tau*) tauon
$\nu, \bar{\nu}$	(Greek *nu*) neutrino, antineutrino
$\mathrm{W}^+, \mathrm{W}^-$	boson (weak force)
Z	boson (weak force)

853

APPENDIX D Equations

Equations

Chapter 2 Motion in One Dimension

DISPLACEMENT	$\Delta x = x_f - x_i$
AVERAGE VELOCITY	$v_{avg} = \frac{\Delta x}{\Delta t} = \frac{x_f - x_i}{t_f - t_i}$
AVERAGE SPEED	$\text{average speed} = \frac{\text{distance traveled}}{\text{time of travel}}$
AVERAGE ACCELERATION	$a_{avg} = \frac{\Delta v}{\Delta t} = \frac{v_f - v_i}{t_f - t_i}$
DISPLACEMENT *These equations are valid only for constantly accelerated, straight-line motion.*	$\Delta x = \frac{1}{2}(v_i + v_f)\Delta t$ $\Delta x = v_i \Delta t + \frac{1}{2}a(\Delta t)^2$
FINAL VELOCITY *These equations are valid only for constantly accelerated, straight-line motion.*	$v_f = v_i + a\Delta t$ $v_f^2 = v_i^2 + 2a\Delta x$

Chapter 3 Two-Dimensional Motion and Vectors

PYTHAGOREAN THEOREM *This equation is valid only for right triangles.*	$c^2 = a^2 + b^2$
TANGENT, SINE, AND COSINE FUNCTIONS *These equations are valid only for right triangles.*	$\tan\theta = \frac{\text{opp}}{\text{adj}}$ $\sin\theta = \frac{\text{opp}}{\text{hyp}}$ $\cos\theta = \frac{\text{adj}}{\text{hyp}}$
VERTICAL MOTION OF A PROJECTILE THAT FALLS FROM REST *These equations assume that air resistance is negligible, and apply only when the initial vertical velocity is zero. On Earth's surface, $a_y = -g = -9.81$ m/s^2.*	$v_{y,f} = a_y\Delta t$ $v_{y,f}^2 = 2a_y\Delta y$ $\Delta y = \frac{1}{2}a_y(\Delta t)^2$
HORIZONTAL MOTION OF A PROJECTILE *These equations assume that air resistance is negligible.*	$v_x = v_{x,i} = \text{constant}$ $\Delta x = v_x \Delta t$

PROJECTILES LAUNCHED AT AN ANGLE *These equations assume that air resistance is negligible. On Earth's surface, $a_y = -g = -9.81$ m/s^2.*	$v_x = v_i \cos\theta = \text{constant}$ $\Delta x = (v_i \cos\theta)\Delta t$ $v_{y,f} = v_i \sin\theta + a_y \Delta t$ $v_{y,f}{}^2 = v_i^2(\sin\theta)^2 + 2a_y \Delta y$ $\Delta y = (v_i \sin\theta)\Delta t + \frac{1}{2}a_y(\Delta t)^2$
RELATIVE VELOCITY	$\mathbf{v_{ac}} = \mathbf{v_{ab}} + \mathbf{v_{bc}}$

Chapter 4 Forces and the Laws of Motion

NEWTON'S FIRST LAW	An object at rest remains at rest, and an object in motion continues in motion with constant velocity (that is, constant speed in a straight line) unless the object experiences a net external force.
NEWTON'S SECOND LAW *$\Sigma\mathbf{F}$ is the vector sum of all external forces acting on the object.*	$\Sigma\mathbf{F} = m\mathbf{a}$
NEWTON'S THIRD LAW	If two objects interact, the magnitude of the force exerted on object 1 by object 2 is equal to the magnitude of the force exerted on object 2 by object 1, and these two forces are opposite in direction.
WEIGHT *On Earth's surface, $a_g = g = 9.81$ m/s^2.*	$F_g = ma_g$
COEFFICIENT OF STATIC FRICTION	$\mu_s = \frac{F_{s,max}}{F_n}$
COEFFICIENT OF KINETIC FRICTION *The coefficient of kinetic friction varies with speed, but we neglect any such variations here.*	$\mu_k = \frac{F_k}{F_n}$
FORCE OF FRICTION	$F_f = \mu F_n$

855

Chapter 5 Work and Energy

NET WORK *This equation applies only when the force is constant.*	$W_{net} = F_{net}d \cos \theta$
KINETIC ENERGY	$KE = \frac{1}{2} mv^2$
WORK-KINETIC ENERGY THEOREM	$W_{net} = \Delta KE$
GRAVITATIONAL POTENTIAL ENERGY	$PE_g = mgh$
ELASTIC POTENTIAL ENERGY	$PE_{elastic} = \frac{1}{2}kx^2$
MECHANICAL ENERGY	$ME = KE + \Sigma PE$
CONSERVATION OF MECHANICAL ENERGY *This equation is valid only if nonmechanical forms of energy (such as friction) are disregarded.*	$ME_i = ME_f$
POWER	$P = \frac{W}{\Delta t} = Fv$

Chapter 6 Momentum and Collisions

MOMENTUM	$\mathbf{p} = m\mathbf{v}$
IMPULSE-MOMENTUM THEOREM *This equation is valid only when the force is constant.*	$\mathbf{F}\Delta t = \Delta\mathbf{p} = m\mathbf{v_f} - m\mathbf{v_i}$
CONSERVATION OF MOMENTUM *These equations are valid for a closed system, that is, when no external forces act on the system during the collision. When such external forces are either negligibly small or act for too short a time to make a significant change in the momentum, these equations represent a good approximation. The second equation is valid for two-body collisions.*	$\mathbf{p_i} = \mathbf{p_f}$ $m_1\mathbf{v_{1,i}} + m_2\mathbf{v_{2,i}} = m_1\mathbf{v_{1,f}} + m_2\mathbf{v_{2,f}}$

CONSERVATION OF MOMENTUM FOR A PERFECTLY INELASTIC COLLISION *This is a simplified version of the conservation of momentum equation valid only for perfectly inelastic collisions between two bodies.*	$m_1\mathbf{v_{1,i}} + m_2\mathbf{v_{2,i}} = (m_1 + m_2)\,\mathbf{v_f}$
CONSERVATION OF KINETIC ENERGY FOR AN ELASTIC COLLISION *No collision is perfectly elastic; some kinetic energy is always converted to other forms of energy. But if these losses are minimal, this equation can provide a good approximation.*	$\frac{1}{2}m_1v_{1,i}^2 + \frac{1}{2}m_2v_{2,i}^2 = \frac{1}{2}m_1v_{1,f}^2 + \frac{1}{2}m_2v_{2,f}^2$

Chapter 7 Circular Motion and Gravitation

CENTRIPETAL ACCELERATION	$a_c = \frac{v_t^2}{r}$
CENTRIPETAL FORCE	$F_c = \frac{mv_t^2}{r}$
NEWTON'S LAW OF UNIVERSAL GRAVITATION *The constant of universal gravitation (G) equals 6.673×10^{-11} N•m²/kg².*	$F_g = G\frac{m_1m_2}{r^2}$
KEPLER'S LAWS OF PLANETARY MOTION	**First Law:** Each planet travels in an elliptical orbit around the sun, and the sun is at one of the focal points. **Second Law:** An imaginary line drawn from the sun to any planet sweeps out equal areas in equal time intervals. **Third Law:** The square of a planet's orbital period (T^2) is proportional to the cube of the average distance (r^3) between the planet and the sun, or $T^2 \propto r^3$.
PERIOD AND SPEED OF AN OBJECT IN CIRCULAR ORBIT *The constant of universal gravitation (G) equals 6.673×10^{-11} N•m²/kg².*	$T = 2\pi\sqrt{\frac{r^3}{Gm}}$ $v_t = \sqrt{G\frac{m}{r}}$
TORQUE	$\tau = Fd\sin\theta$

857

MECHANICAL ADVANTAGE *This equation disregards friction.*	$MA = \frac{F_{out}}{F_{in}} = \frac{d_{in}}{d_{out}}$
EFFICIENCY *This equation accounts for friction.*	$eff = \frac{W_{out}}{W_{in}}$

Chapter 8 Fluid Mechanics

MASS DENSITY	$\rho = \frac{m}{V}$
BUOYANT FORCE *The first equation is for an object that is completely or partially submerged. The second equation is for a floating object.*	$F_B = F_g(displaced\ fluid) = m_f g$ $F_B = F_g(object) = mg$
PRESSURE	$P = \frac{F}{A}$
PASCAL'S PRINCIPLE	Pressure applied to a fluid in a closed container is transmitted equally to every point of the fluid and to the walls of the container.
HYDRAULIC LIFT EQUATION	$F_2 = \frac{A_2}{A_1}F_1$
FLUID PRESSURE AS A FUNCTION OF DEPTH	$P = P_0 + \rho gh$
CONTINUITY EQUATION	$A_1 v_1 = A_2 v_2$
BERNOULLI'S PRINCIPLE	The pressure in a fluid decreases as the fluid's velocity increases.

Chapter 9 Heat

TEMPERATURE CONVERSIONS	$T_F = \frac{9}{5}T_C + 32.0$ $T = T_C + 273.15$

CONSERVATION OF ENERGY	$\Delta PE + \Delta KE + \Delta U = 0$
SPECIFIC HEAT CAPACITY	$c_p = \frac{Q}{m\Delta T}$
CALORIMETRY *These equations assume that the energy transferred to the surrounding container is negligible.*	$Q_w = -Q_x$ $c_{p,w} m_w \Delta T_w = -c_{p,x} m_x \Delta T_x$
LATENT HEAT	$Q = mL$

Chapter 10 Thermodynamics

WORK DONE BY A GAS *This equation is valid only when the pressure is constant. When the work done by the gas (W) is negative, positive work is done on the gas.*	$W = PAd = P\Delta V$
THE FIRST LAW OF THERMODYNAMICS *Q represents the energy added to the system as heat and W represents the work done by the system.*	$\Delta U = Q - W$
CYCLIC PROCESSES	$\Delta U_{net} = 0$ and $Q_{net} = W_{net}$
EFFICIENCY OF A HEAT ENGINE	$eff = \frac{W_{net}}{Q_h} = \frac{Q_h - Q_c}{Q_h} = 1 - \frac{Q_c}{Q_h}$

Chapter 11 Vibrations and Waves

HOOKE'S LAW	$F_{elastic} = -kx$
PERIOD OF A SIMPLE PENDULUM IN SIMPLE HARMONIC MOTION *This equation is valid only when the amplitude is small (less than about 15°).*	$T = 2\pi\sqrt{\frac{L}{a_g}}$
PERIOD OF A MASS-SPRING SYSTEM IN SIMPLE HARMONIC MOTION	$T = 2\pi\sqrt{\frac{m}{k}}$
SPEED OF A WAVE	$v = f\lambda$

859

Chapter 12 Sound

INTENSITY OF A SPHERICAL WAVE *This equation assumes that there is no absorption in the medium.*	$\text{intensity} = \frac{P}{4\pi r^2}$
HARMONIC SERIES OF A VIBRATING STRING OR A PIPE OPEN AT BOTH ENDS	$f_n = n\frac{v}{2L} \quad n = 1, 2, 3, \ldots$
HARMONIC SERIES OF A PIPE CLOSED AT ONE END	$f_n = n\frac{v}{4L} \quad n = 1, 3, 5, \ldots$
BEATS	frequency difference = number of beats per second

Chapter 13 Light and Reflection

SPEED OF ELECTROMAGNETIC WAVES *This book uses the value $c = 3.00 \times 10^8$ m/s for the speed of EM waves in a vacuum or in air.*	$c = f\lambda$
LAW OF REFLECTION	angle of incidence (θ) = angle of reflection (θ')
MIRROR EQUATION *This equation is derived assuming that the rays incident on the mirror are very close to the principal axis of the mirror.*	$\frac{1}{p} + \frac{1}{q} = \frac{1}{f}$
MAGNIFICATION OF A CURVED MIRROR	$M = \frac{h'}{h} = -\frac{q}{p}$

Chapter 14 Refraction

INDEX OF REFRACTION *For any material other than a vacuum, the index of refraction varies with the wavelength of light.*	$n = \frac{c}{v}$
SNELL'S LAW	$n_i \sin \theta_i = n_r \sin \theta_r$

THIN-LENS EQUATION *This equation is derived assuming that the thickness of the lens is much less than the focal length of the lens.*	$\frac{1}{p}+\frac{1}{q}=\frac{1}{f}$
MAGNIFICATION OF A LENS *This equation can be used only when the index of refraction of the first medium (n_i) is greater than the index of refraction of the second medium (n_r).*	$M=\frac{h'}{h}=-\frac{q}{p}$ (for $n_i > n_r$)
CRITICAL ANGLE *This equation can be used only when the index of refraction of the first medium (n_i) is greater than the index of refraction of the second medium (n_r).*	$\sin\theta_c=\frac{n_r}{n_i}$ (for $n_i > n_r$)

Chapter 15 Interference and Diffraction

CONSTRUCTIVE AND DESTRUCTIVE INTERFERENCE *The grating spacing multiplied by the sine of the angle of deviation is the path difference between two waves. To observe interference effects, the sources must be coherent and have identical wavelengths.*	Constructive Interference: $d\sin\theta=\pm m\lambda$ $m=0, 1, 2, 3, \ldots$ Destructive Interference: $d\sin\theta=\pm(m+\frac{1}{2})\lambda$ $m=0, 1, 2, 3, \ldots$
DIFFRACTION GRATING	*See the equation above for constructive interference.*
LIMITING ANGLE OF RESOLUTION *This equation gives the angle θ in radians and applies only to circular apertures.*	$\theta=1.22\frac{\lambda}{D}$

Chapter 16 Electric Forces and Fields

COULOMB'S LAW *This equation assumes either point charges or spherical distributions of charge.*	$F_{electric}=k_C\left(\frac{q_1q_2}{r^2}\right)$
ELECTRIC FIELD STRENGTH DUE TO A POINT CHARGE	$E=k_C\frac{q}{r^2}$

Chapter 17 Electrical Energy and Current

ELECTRICAL POTENTIAL ENERGY

The displacement, d, is from the reference point and is parallel to the field. This equation is valid only for a uniform electric field.

$$PE_{electric} = -qEd$$

POTENTIAL DIFFERENCE

The second half of this equation is valid only for a uniform electric field, and Δd is parallel to the field.

$$\Delta V = \frac{\Delta PE_{electric}}{q} = -E\Delta d$$

POTENTIAL DIFFERENCE BETWEEN A POINT AT INFINITY AND A POINT NEAR A POINT CHARGE

$$\Delta V = k_C \frac{q}{r}$$

CAPACITANCE

$$C = \frac{Q}{\Delta V}$$

CAPACITANCE FOR A PARALLEL-PLATE CAPACITOR IN A VACUUM

The permittivity in a vacuum (ε_0) equals $8.85 \times 10^{-12}\ C^2/(N\bullet m^2)$.

$$C = \varepsilon_0 \frac{A}{d}$$

ELECTRICAL POTENTIAL ENERGY STORED IN A CHARGED CAPACITOR

There is a limit to the maximum energy (or charge) that can be stored in a capacitor because electrical breakdown ultimately occurs between the plates of the capacitor for a sufficiently large potential difference.

$$PE_{electric} = \tfrac{1}{2}Q\Delta V = \tfrac{1}{2}C(\Delta V)^2 = \frac{Q^2}{2C}$$

ELECTRIC CURRENT

$$I = \frac{\Delta Q}{\Delta t}$$

RESISTANCE

$$R = \frac{\Delta V}{I}$$

OHM'S LAW

Ohm's law is not universal, but it does apply to many materials over a wide range of applied potential differences.

$$\frac{\Delta V}{I} = \text{constant}$$

ELECTRIC POWER

$$P = I\Delta V = I^2R = \frac{(\Delta V)^2}{R}$$

Chapter 18 Circuits and Circuit Elements

RESISTORS IN SERIES: EQUIVALENT RESISTANCE AND CURRENT	$R_{eq} = R_1 + R_2 + R_3 \ldots$ The current in each resistor is the same and is equal to the total current.
RESISTORS IN PARALLEL: EQUIVALENT RESISTANCE AND CURRENT	$\frac{1}{R_{eq}} = \frac{1}{R_1} + \frac{1}{R_2} + \frac{1}{R_3} \ldots$ The sum of the current in each resistor equals the total current.

Chapter 19 Magnetism

MAGNETIC FLUX	$\Phi_M = AB\cos\theta$
MAGNITUDE OF A MAGNETIC FIELD *The direction of $F_{magnetic}$ is always perpendicular to both B and v, and can be found with the right-hand rule.*	$B = \frac{F_{magnetic}}{qv}$
FORCE ON A CURRENT-CARRYING CONDUCTOR PERPENDICULAR TO A MAGNETIC FIELD *This equation can be used only when the current and the magnetic field are at right angles to each other.*	$F_{magnetic} = BI\ell$

Chapter 20 Electromagnetic Induction

FARADAY'S LAW OF MAGNETIC INDUCTION *N is assumed to be a whole number.*	$\text{emf} = -N\frac{\Delta\Phi_M}{\Delta t}$
EMF PRODUCED BY A GENERATOR *N is assumed to be a whole number.*	$\text{emf} = NAB\omega \sin \omega t$ $\text{maximum emf} = NAB\omega$
FARADAY'S LAW FOR MUTUAL INDUCTANCE	$\text{emf} = -M\frac{\Delta I}{\Delta t}$

863

RMS CURRENT AND POTENTIAL DIFFERENCE	$I_{rms} = \frac{I_{max}}{\sqrt{2}} = 0.707\ I_{max}$ $\Delta V_{rms} = \frac{\Delta V_{max}}{\sqrt{2}} = 0.707\ \Delta V$
TRANSFORMERS *N is assumed to be a whole number.*	$\Delta V_2 = \frac{N_2}{N_1} \Delta V_1$

Chapter 21 Atomic Physics

ENERGY OF A LIGHT QUANTUM	$E = hf$
MAXIMUM KINETIC ENERGY OF A PHOTOELECTRON	$KE_{max} = hf - hf_t$
WAVELENGTH AND FREQUENCY OF MATTER WAVES *Planck's constant (h) equals 6.63×10^{-34} J•s.*	$\lambda = \frac{h}{p} = \frac{h}{mv}$ $f = \frac{E}{h}$

Chapter 22 Subatomic Physics

RELATIONSHIP BETWEEN REST ENERGY AND MASS	$E_R = mc^2$
BINDING ENERGY OF A NUCLEUS	$E_{bind} = \Delta mc^2$
MASS DEFECT	$\Delta m = Z(\text{atomic mass of H}) + Nm_n - \text{atomic mass}$
ACTIVITY (DECAY RATE)	$\text{activity} = -\frac{\Delta N}{\Delta t} = \lambda N$
HALF-LIFE	$T_{1/2} = \frac{0.693}{\lambda}$

Appendix J Advanced Topics

CONVERSION BETWEEN RADIANS AND DEGREES	$\theta(\text{rad}) = \frac{\pi}{180°}\theta(\text{deg})$
ANGULAR DISPLACEMENT *This equation gives $\Delta\theta$ in radians.*	$\Delta\theta = \frac{\Delta s}{r}$
AVERAGE ANGULAR VELOCITY	$\omega_{avg} = \frac{\Delta\theta}{\Delta t}$
AVERAGE ANGULAR ACCELERATION	$\alpha_{avg} = \frac{\Delta\omega}{\Delta t}$
ROTATIONAL KINEMATICS *These equations apply only when the angular acceleration is constant. The symbol ω represents instantaneous rather than average angular velocity.*	$\omega_f = \omega_i + \alpha\Delta t$ $\Delta\theta = \omega_i\Delta t + \frac{1}{2}\alpha(\Delta t)^2$ $\omega_f^2 = \omega_i^2 + 2\alpha(\Delta\theta)$ $\Delta\theta = \frac{1}{2}(\omega_i + \omega_f)\Delta t$
TANGENTIAL SPEED *For this equation to be valid, ω must be in rad/s.*	$v_t = r\omega$
TANGENTIAL ACCELERATION *For this equation to be valid, α must be in rad/s^2.*	$a_t = r\alpha$
NEWTON'S SECOND LAW FOR ROTATING OBJECTS	$\tau = I\alpha$
ANGULAR MOMENTUM	$L = I\omega$
ROTATIONAL KINETIC ENERGY	$KE_{rot} = \frac{1}{2}I\omega^2$
IDEAL GAS LAW *Boltzmann's constant (k_B) equals 1.38×10^{-23} J/K.*	$PV = Nk_BT$
BERNOULLI'S EQUATION	$P + \frac{1}{2}\rho v^2 + \rho gh = \text{constant}$

APPENDIX E SI Units

SI Units

SI Base Units Used in This Book

Symbol	Name	Quantity
A	ampere	current
K	kelvin	absolute temperature
kg	kilogram	mass
m	meter	length
s	second	time

SI Prefixes

Symbol	Name	Numerical equivalent
a	atto	10^{-18}
f	femto	10^{-15}
p	pico	10^{-12}
n	nano	10^{-9}
μ	micro	10^{-6}
m	milli	10^{-3}
c	centi	10^{-2}
d	deci	10^{-1}
k	kilo	10^{3}
M	mega	10^{6}
G	giga	10^{9}
T	tera	10^{12}
P	peta	10^{15}
E	exa	10^{18}

Other Commonly Used Units

Symbol	Name	Quantity	Conversions
atm	standard atmosphere	pressure	$1.013\,250 \times 10^5$ Pa
Btu	British thermal unit	energy	1.055×10^3 J
Cal	food calorie	energy	= 1 kcal = 4.186×10^3 J
cal	calorie	energy	4.186 J
Ci	curie	decay rate or activity	$3.7 \times 10^{10}\ s^{-1}$
°F	degree Fahrenheit	temperature	0.5556°C
ft	foot	length	0.3048 m
ft•lb	foot-pound	work and energy	1.356 J
g	gram	mass	0.001 kg
gal	gallon	volume	$3.785 \times 10^{-3}\ m^3$
hp	horsepower	power	746 W
in	inch	length	2.54×10^{-2} m
kcal	kilocalorie	energy	4.186×10^3 J
lb	pound	force	4.45 N
mi	mile	length	1.609×10^3 m
rev	revolution	angular displacement	2π rad
°	degrees	angular displacement	$= \left(\frac{2\pi}{360}\right)$ rad $= 1.745 \times 10^{-2}$ rad

Other Units Acceptable with SI

Symbol	Name	Quantity	Conversion
Bq	becquerel	decay rate or activity	$\frac{1}{s}$
C	coulomb	electric charge	1 A•s
°C	degree Celsius	temperature	1 K
dB	decibel	relative intensity (sound)	(unitless)
eV	electron volt	energy	1.60×10^{-19} J
F	farad	capacitance	$1 \frac{A^2 \bullet s^4}{kg \bullet m^2} = 1 \frac{C}{V}$
H	henry	inductance	$1 \frac{kg \bullet m^2}{A^2 \bullet s^2} = 1 \frac{J}{A^2}$
h	hour	time	3.600×10^3 s
Hz	hertz	frequency	$\frac{1}{s}$
J	joule	work and energy	$1 \frac{kg \bullet m^2}{s^2} = 1$ N•m
kW•h	kilowatt-hour	energy	3.60×10^6 J
L	liter	volume	10^{-3} m^3
min	minute	time	6.0×10^1 s
N	newton	force	$1 \frac{kg \bullet m}{s^2}$
Pa	pascal	pressure	$1 \frac{kg}{m \bullet s^2} = 1 \frac{N}{m^2}$
rad	radian	angular displacement	(unitless)
T	tesla	magnetic field strength	$1 \frac{kg}{A \bullet s^2} = 1 \frac{N}{A \bullet m} = 1 \frac{V \bullet s}{m^2}$
u	unified mass unit	mass (atomic masses)	$1.660\ 538\ 782 \times 10^{-27}$ kg
V	volt	electric potential difference	$1 \frac{kg \bullet m^2}{A \bullet s^3} = 1 \frac{J}{C}$
W	watt	power	$1 \frac{kg \bullet m^2}{s^3} = 1 \frac{J}{s}$
Ω	ohm	resistance	$1 \frac{kg \bullet m^2}{A^2 \bullet s^3} = 1 \frac{V}{A}$

867

Useful Tables

Fundamental Constants

Symbol	Quantity	Established value	Value used for calculations in this book
c	speed of light in a vacuum	299 792 458 m/s	3.00×10^8 m/s
e^-	elementary charge	$1.602\ 176\ 487 \times 10^{-19}$ C	1.60×10^{-19} C
e[1]	base of natural logarithms	2.718 2818 28	2.72
ε_0	(Greek *epsilon*) permittivity of a vacuum	$8.854\ 187\ 817 \times 10^{-12}$ $C^2/(N•m^2)$	8.85×10^{-12} $C^2/(N•m^2)$
G	constant of universal gravitation	$6.672\ 59 \times 10^{-11}$ $N•m^2/kg^2$	6.673×10^{-11} $N•m^2/kg^2$
g	free-fall acceleration at Earth's surface	9.806 65 m/s^2	9.81 m/s^2
h	Planck's constant	$6.626\ 068\ 96 \times 10^{-34}$ J•s	6.63×10^{-34} J•s
k_B	Boltzmann's constant (R/N_A)	$1.380\ 6504 \times 10^{-23}$ J/K	1.38×10^{-23} J/K
k_C	Coulomb constant	$8.987\ 551\ 787 \times 10^9$ $N•m^2/C^2$	8.99×10^9 $N•m^2/C^2$
R	molar (universal) gas constant	8.314 472 J/(mol•K)	8.31 J/(mol•K)
π	(Greek *pi*) ratio of the circumference to the diameter of a circle	3.141 592 654	calculator value

Coefficients of Friction (Approximate Values)

	μ_s	μ_k		μ_s	μ_k
steel on steel	0.74	0.57	waxed wood on wet snow	0.14	0.1
aluminum on steel	0.61	0.47	waxed wood on dry snow	—	0.04
rubber on dry concrete	1.0	0.8	metal on metal (lubricated)	0.15	0.06
rubber on wet concrete	—	0.5	ice on ice	0.1	0.03
wood on wood	0.4	0.2	Teflon on Teflon	0.04	0.04
glass on glass	0.9	0.4	synovial joints in humans	0.01	0.003

Useful Astronomical Data

Symbol	Quantity	Value used for calculations in this book
I_E	moment of inertia of Earth	8.03×10^{37} $kg•m^2$
M_E	mass of Earth	5.97×10^{24} kg
R_E	radius of Earth	6.38×10^6 m
	Average Earth–moon distance	3.84×10^8 m
	Average Earth–sun distance	1.50×10^{11} m
	mass of the moon	7.35×10^{22} kg
	mass of the sun	1.99×10^{30} kg
yr	period of Earth's orbit	3.16×10^7 s

The Moment of Inertia for a Few Shapes

Shape		Moment of inertia	Shape		Moment of inertia
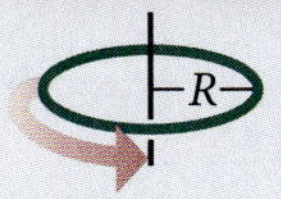	thin hoop about symmetry axis	MR^2	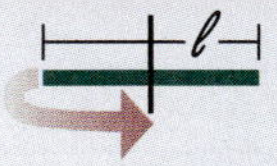	thin rod about perpendicular axis through center	$\frac{1}{12}M\ell^2$
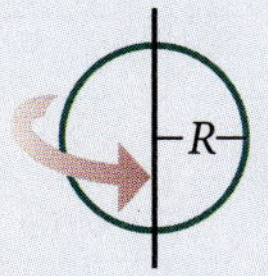	thin hoop about diameter	$\frac{1}{2}MR^2$	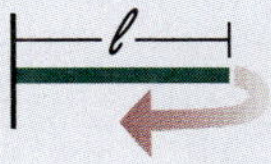	thin rod about perpendicular axis through end	$\frac{1}{3}M\ell^2$
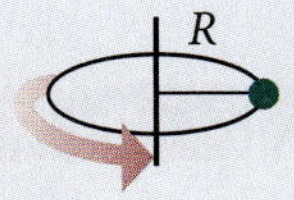	point mass about axis	MR^2	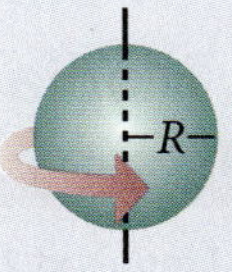	solid sphere about diameter	$\frac{2}{5}MR^2$
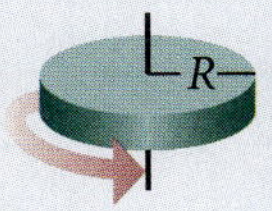	disk or cylinder about symmetry axis	$\frac{1}{2}MR^2$	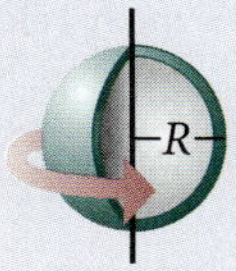	thin spherical shell about diameter	$\frac{2}{3}MR^2$

Densities of Some Common Substances*

Substance	ρ (kg/m^3)
hydrogen	0.0899
helium	0.179
steam (100°C)	0.598
air	1.29
oxygen	1.43
carbon dioxide	1.98
ethanol	0.806×10^3
ice	0.917×10^3
fresh water (4°C)	1.00×10^3
sea water (15°C)	1.025×10^3
glycerine	1.26×10^3
aluminum	2.70×10^3
iron	7.86×10^3
copper	8.92×10^3
silver	10.5×10^3
lead	11.3×10^3
mercury	13.6×10^3
gold	19.3×10^3

*All densities are measured at 0°C and 1 atm unless otherwise noted.

Specific Heat Capacities

Substance	c_p (J/kg•°C)
aluminum	8.99×10^2
copper	3.87×10^2
glass	8.37×10^2
gold	1.29×10^2
ice	2.09×10^3
iron	4.48×10^2
lead	1.28×10^2
mercury	1.38×10^2
silver	2.34×10^2
steam	2.01×10^3
water	4.186×10^3

869

Latent Heats of Fusion and Vaporization at Standard Pressure

Substance	Melting point (°C)	L_f (J/kg)	Boiling point (°C)	L_v (J/kg)
nitrogen	−209.97	2.55×10^4	−195.81	2.01×10^5
oxygen	−218.79	1.38×10^4	−182.97	2.13×10^5
ethyl alcohol	−114	1.04×10^5	78	8.54×10^5
water	0.00	3.33×10^5	100.00	2.26×10^6
lead	327.3	2.45×10^4	1745	8.70×10^5
aluminum	660.4	3.97×10^5	2467	1.14×10^7

Speed of Sound in Various Media

Medium	v (m/s)	Medium	v (m/s)	Medium	v (m/s)
Gases		**Liquids at 25°C**		**Solids**	
air (0°C)	331	methyl alcohol	1140	aluminum	5100
air (25°C)	346	sea water	1530	copper	3560
air (100°C)	366	water	1490	iron	5130
helium (0°C)	972			lead	1320
hydrogen (0°C)	1290			vulcanized rubber	54
oxygen (0°C)	317				

Conversion of Intensity to Decibel Level

Intensity (W/m^2)	Decibel level (dB)	Examples
1.0×10^{-12}	0	threshold of hearing
1.0×10^{-11}	10	rustling leaves
1.0×10^{-10}	20	quiet whisper
1.0×10^{-9}	30	whisper
1.0×10^{-8}	40	mosquito buzzing
1.0×10^{-7}	50	normal conversation
1.0×10^{-6}	60	air conditioning at 6 m
1.0×10^{-5}	70	vacuum cleaner
1.0×10^{-4}	80	busy traffic, alarm clock
1.0×10^{-3}	90	lawn mower
1.0×10^{-2}	100	subway, power motor
1.0×10^{-1}	110	auto horn at 1 m
1.0×10^{0}	120	threshold of pain
1.0×10^{1}	130	thunderclap, machine gun
1.0×10^{3}	150	nearby jet airplane

Indices of Refraction for Various Substances*

Solids at 20°C	n	Liquids at 20°C	n	Gases at 0°C, 1 atm	n
cubic zirconia	2.20	benzene	1.501	air	1.000 293
diamond	2.419	carbon disulfide	1.628	carbon dioxide	1.000 450
fluorite	1.434	carbon tetrachloride	1.461		
fused quartz	1.458	ethyl alcohol	1.361		
glass, crown	1.52	glycerine	1.473		
glass, flint	1.66	water	1.333		
ice (at 0°C)	1.309				
polystyrene	1.49				
sodium chloride	1.544				
zircon	1.923				

*measured with light of vacuum wavelength = 589 nm

Useful Atomic Data

Symbol	Quantity	Established value	Value used for calculations in this book
m_e	mass of electron	$9.109\,382\,15 \times 10^{-31}$ kg $5.485\,799\,0943 \times 10^{-4}$ u 0.510 998 910 MeV	9.109×10^{-31} kg 5.49×10^{-4} u 5.110×10^{-1} MeV
m_n	mass of neutron	$1.674\,927\,211 \times 10^{-27}$ kg 1.008 664 915 97 u 939.565 346 MeV	1.675×10^{-27} kg 1.008 665 u 9.396×10^{2} MeV
m_p	mass of proton	$1.672\,621\,637 \times 10^{-27}$ kg 1.007 276 466 77 u 938.272 013 MeV	1.673×10^{-27} kg 1.007 276 u 9.383×10^{2} MeV

871

APPENDIX G Periodic Table

Periodic Table of the Elements

Key:

Atomic number	6
Symbol	C
Name	Carbon
Average atomic mass	12.0107
Electron configuration	[He]$2s^22p^2$

Period	Group 1	Group 2	Group 3	Group 4	Group 5	Group 6	Group 7	Group 8	Group 9
1	1 **H** Hydrogen 1.007 94 $1s^1$								
2	3 **Li** Lithium 6.941 [He]$2s^1$	4 **Be** Beryllium 9.012 182 [He]$2s^2$							
3	11 **Na** Sodium 22.989 770 [Ne]$3s^1$	12 **Mg** Magnesium 24.3050 [Ne]$3s^2$							
4	19 **K** Potassium 39.0983 [Ar]$4s^1$	20 **Ca** Calcium 40.078 [Ar]$4s^2$	21 **Sc** Scandium 44.955 910 [Ar]$3d^14s^2$	22 **Ti** Titanium 47.867 [Ar]$3d^24s^2$	23 **V** Vanadium 50.9415 [Ar]$3d^34s^2$	24 **Cr** Chromium 51.9961 [Ar]$3d^54s^1$	25 **Mn** Manganese 54.938 049 [Ar]$3d^54s^2$	26 **Fe** Iron 55.845 [Ar]$3d^64s^2$	27 **Co** Cobalt 58.933 200 [Ar]$3d^74s^2$
5	37 **Rb** Rubidium 85.4678 [Kr]$5s^1$	38 **Sr** Strontium 87.62 [Kr]$5s^2$	39 **Y** Yttrium 88.905 85 [Kr]$4d^15s^2$	40 **Zr** Zirconium 91.224 [Kr]$4d^25s^2$	41 **Nb** Niobium 92.906 38 [Kr]$4d^45s^1$	42 **Mo** Molybdenum 95.94 [Kr]$4d^55s^1$	43 **Tc** Technetium (98) [Kr]$4d^65s^1$	44 **Ru** Ruthenium 101.07 [Kr]$4d^75s^1$	45 **Rh** Rhodium 102.905 50 [Kr]$4d^85s^1$
6	55 **Cs** Cesium 132.905 43 [Xe]$6s^1$	56 **Ba** Barium 137.327 [Xe]$6s^2$	57 **La** Lanthanum 138.9055 [Xe]$5d^16s^2$	72 **Hf** Hafnium 178.49 [Xe]$4f^{14}5d^26s^2$	73 **Ta** Tantalum 180.9479 [Xe]$4f^{14}5d^36s^2$	74 **W** Tungsten 183.84 [Xe]$4f^{14}5d^46s^2$	75 **Re** Rhenium 186.207 [Xe]$4f^{14}5d^56s^2$	76 **Os** Osmium 190.23 [Xe]$4f^{14}5d^66s^2$	77 **Ir** Iridium 192.217 [Xe]$4f^{14}5d^76s^2$
7	87 **Fr** Francium (223) [Rn]$7s^1$	88 **Ra** Radium (226) [Rn]$7s^2$	89 **Ac** Actinium (227) [Rn]$6d^17s^2$	104 **Rf** Rutherfordium (261) [Rn]$5f^{14}6d^27s^2$	105 **Db** Dubnium (262) [Rn]$5f^{14}6d^37s^2$	106 **Sg** Seaborgium (266) [Rn]$5f^{14}6d^47s^2$	107 **Bh** Bohrium (264) [Rn]$5f^{14}6d^57s^2$	108 **Hs** Hassium (277) [Rn]$5f^{14}6d^67s^2$	109 **Mt** Meitnerium (268) [Rn]$5f^{14}6d^77s^2$

* The systematic names and symbols for elements greater than 111 will be used until the approval of trivial names by IUPAC.

58 **Ce** Cerium 140.116 [Xe]$4f^15d^16s^2$	59 **Pr** Praseodymium 140.907 65 [Xe]$4f^36s^2$	60 **Nd** Neodymium 144.24 [Xe]$4f^46s^2$	61 **Pm** Promethium (145) [Xe]$4f^56s^2$	62 **Sm** Samarium 150.36 [Xe]$4f^66s^2$
90 **Th** Thorium 232.0381 [Rn]$6d^27s^2$	91 **Pa** Protactinium 231.035 88 [Rn]$5f^26d^17s^2$	92 **U** Uranium 238.028 91 [Rn]$5f^36d^17s^2$	93 **Np** Neptunium (237) [Rn]$5f^46d^17s^2$	94 **Pu** Plutonium (244) [Rn]$5f^67s^2$

Hydrogen
Semiconductors
(also known as *metalloids*)

Metals
- Alkali metals
- Alkaline-earth metals
- Transition metals
- Other metals

Nonmetals
- Halogens
- Noble gases
- Other nonmetals

Group 10	Group 11	Group 12	Group 13	Group 14	Group 15	Group 16	Group 17	Group 18
								2 **He** Helium 4.002 602 $1s^2$
			5 **B** Boron 10.811 [He]$2s^22p^1$	6 **C** Carbon 12.0107 [He]$2s^22p^2$	7 **N** Nitrogen 14.0067 [He]$2s^22p^3$	8 **O** Oxygen 15.9994 [He]$2s^22p^4$	9 **F** Fluorine 18.998 4032 [He]$2s^22p^5$	10 **Ne** Neon 20.1797 [He]$2s^22p^6$
			13 **Al** Aluminum 26.981 538 [Ne]$3s^23p^1$	14 **Si** Silicon 28.0855 [Ne]$3s^23p^2$	15 **P** Phosphorus 30.973 761 [Ne]$3s^23p^3$	16 **S** Sulfur 32.065 [Ne]$3s^23p^4$	17 **Cl** Chlorine 35.453 [Ne]$3s^23p^5$	18 **Ar** Argon 39.948 [Ne]$3s^23p^6$
28 **Ni** Nickel 58.6934 [Ar]$3d^84s^2$	29 **Cu** Copper 63.546 [Ar]$3d^{10}4s^1$	30 **Zn** Zinc 65.409 [Ar]$3d^{10}4s^2$	31 **Ga** Gallium 69.723 [Ar]$3d^{10}4s^24p^1$	32 **Ge** Germanium 72.64 [Ar]$3d^{10}4s^24p^2$	33 **As** Arsenic 74.921 60 [Ar]$3d^{10}4s^24p^3$	34 **Se** Selenium 78.96 [Ar]$3d^{10}4s^24p^4$	35 **Br** Bromine 79.904 [Ar]$3d^{10}4s^24p^5$	36 **Kr** Krypton 83.798 [Ar]$3d^{10}4s^24p^6$
46 **Pd** Palladium 106.42 [Kr]$4d^{10}5s^0$	47 **Ag** Silver 107.8682 [Kr]$4d^{10}5s^1$	48 **Cd** Cadmium 112.411 [Kr]$4d^{10}5s^2$	49 **In** Indium 114.818 [Kr]$4d^{10}5s^25p^1$	50 **Sn** Tin 118.710 [Kr]$4d^{10}5s^25p^2$	51 **Sb** Antimony 121.760 [Kr]$4d^{10}5s^25p^3$	52 **Te** Tellurium 127.60 [Kr]$4d^{10}5s^25p^4$	53 **I** Iodine 126.904 47 [Kr]$4d^{10}5s^25p^5$	54 **Xe** Xenon 131.293 [Kr]$4d^{10}5s^25p^6$
78 **Pt** Platinum 195.078 [Xe]$4f^{14}5d^96s^1$	79 **Au** Gold 196.966 55 [Xe]$4f^{14}5d^{10}6s^1$	80 **Hg** Mercury 200.59 [Xe]$4f^{14}5d^{10}6s^2$	81 **Tl** Thallium 204.3833 [Xe]$4f^{14}5d^{10}6s^26p^1$	82 **Pb** Lead 207.2 [Xe]$4f^{14}5d^{10}6s^26p^2$	83 **Bi** Bismuth 208.980 38 [Xe]$4f^{14}5d^{10}6s^26p^3$	84 **Po** Polonium (209) [Xe]$4f^{14}5d^{10}6s^26p^4$	85 **At** Astatine (210) [Xe]$4f^{14}5d^{10}6s^26p^5$	86 **Rn** Radon (222) [Xe]$4f^{14}5d^{10}6s^26p^6$
110 **Ds** Darmstadtium (271) [Rn]$5f^{14}6d^97s^1$	111 **Rg** Roentgenium (272) [Rn]$5f^{14}6d^{10}7s^1$	112 **Uub*** Ununbium (285) [Rn]$5f^{14}6d^{10}7s^2$		114 **Uuq*** Ununquadium (289) [Rn]$5f^{14}6d^{10}7s^27p^2$		116 **UUh*** Ununhexium (292) [Rn]$5f^{14}6d^{10}7s^27p^4$		

The discoveries of elements with atomic numbers 112, 114, and 116 have been reported but not fully confirmed.

63 **Eu** Europium 151.964 [Xe]$4f^76s^2$	64 **Gd** Gadolinium 157.25 [Xe]$4f^75d^16s^2$	65 **Tb** Terbium 158.925 34 [Xe]$4f^96s^2$	66 **Dy** Dysprosium 162.500 [Xe]$4f^{10}6s^2$	67 **Ho** Holmium 164.930 32 [Xe]$4f^{11}6s^2$	68 **Er** Erbium 167.259 [Xe]$4f^{12}6s^2$	69 **Tm** Thulium 168.934 21 [Xe]$4f^{13}6s^2$	70 **Yb** Ytterbium 173.04 [Xe]$4f^{14}6s^2$	71 **Lu** Lutetium 174.967 [Xe]$4f^{14}5d^16s^2$
95 **Am** Americium (243) [Rn]$5f^77s^2$	96 **Cm** Curium (247) [Rn]$5f^76d^17s^2$	97 **Bk** Berkelium (247) [Rn]$5f^97s^2$	98 **Cf** Californium (251) [Rn]$5f^{10}7s^2$	99 **Es** Einsteinium (252) [Rn]$5f^{11}7s^2$	100 **Fm** Fermium (257) [Rn]$5f^{12}7s^2$	101 **Md** Mendelevium (258) [Rn]$5f^{13}7s^2$	102 **No** Nobelium (259) [Rn]$5f^{14}7s^2$	103 **Lr** Lawrencium (262) [Rn]$5f^{14}6d^17s^2$

The atomic masses listed in this table reflect the precision of current measurements. (Each value listed in parentheses is the mass number of that radioactive element's most stable or most common isotope.)

873

Abbreviated Table of Isotopes and Atomic Masses

Z	Element	Symbol	Average atomic mass (u)	Mass number (* indicates radioactive) A	Atomic mass (u)	Percent abundance	Half-life (if radioactive) $T_{1/2}$
0	(Neutron)	*n*		1*	1.008 665		10.4 m
1	Hydrogen	H	1.0079	1	1.007 825	99.985	
	Deuterium	D		2	2.014 102	0.015	
	Tritium	T		3*	3.016 049		12.33 y
2	Helium	He	4.002 60	3	3.016 029	0.000 14	
				4	4.002 602	99.999 86	
				6*	6.018 886		0.81 s
3	Lithium	Li	6.941	6	6.015 121	7.5	
				7	7.016 003	92.5	
4	Beryllium	Be	9.0122	7*	7.016 928		53.3 d
				8*	8.005 305		6.7×10^{-17} s
				9	9.012 174	100	
				10*	10.013 584		1.5×10^{6} y
5	Boron	B	10.81	10	10.012 936	19.9	
				11	11.009 305	80.1	
6	Carbon	C	12.011	10*	10.016 854		19.3 s
				11*	11.011 433		20.4 m
				12	12.000 000	98.9	
				13	13.003 355	1.10	
				14*	14. 003 242		5715 y
7	Nitrogen	N	14.0067	13*	13.005 738		996 m
				14	14.003 074	99.63	
				15	15.000 108	0.37	
				16*	16.006 100		7.13 s
8	Oxygen	O	15.9994	15*	15.003 065		122 s
				16	15.994 915	99.761	
				17	16.999 132	0.039	
				18	17.999 160	0.200	
				19*	19.003 577		26.9 s
9	Fluorine	F	18.998 40	18*	18.000 937		109.8 m
				19	18.998 404	100	
				20*	19.999 982		11.0 s
10	Neon	Ne	20.180	19*	19.001 880		17.2 s
				20	19.992 435	90.48	
				21	20.993 841	0.27	
				22	21.991 383	9.25	
11	Sodium	Na	22.989 87	22*	21.994 434		2.61 y
				23	22.989 767	100	
				24*	23.990 961		14.96 h
12	Magnesium	Mg	24.305	23*	22.994 124		11.3 s
				24	23.985 042	78.99	
				25	24.985 838	10.00	
				26	25.982 594	11.01	
13	Aluminum	Al	26.981 54	26*	25.986 892		7.4×10^{5} y
				27	26.981 534	100	

874

Z	Element	Symbol	Average atomic mass (u)	Mass number (* indicates radioactive) A	Atomic mass (u)	Percent abundance	Half-life (if radioactive) $T_{1/2}$
14	Silicon	Si	28.086	28	27.976 927	92.23	
				29	28.976 495	4.67	
				30	29.973 770	3.10	
15	Phosphorus	P	30.973 76	30*	29.978 307		2.50 m
				31	30.973 762	100	
				32*	31.973 907		14.263 d
16	Sulfur	S	32.066	32	31.972 071	95.02	
				33	32.971 459	0.75	
				34	33.967 867	4.21	
				35*	34.969 033		87.5 d
17	Chlorine	Cl	35.453	35	34.968 853	75.77	
				36*	35.968 307		3.0×10^5 y
				37	36.975 893	24.23	
18	Argon	Ar	39.948	36	35.967 547	0.337	
				37*	36.966 776		35.04 d
				38	37.962 732	0.063	
				39*	38.964 314		269 y
				40	39.962 384	99.600	
19	Potassium	K	39.0983	39	38.963 708	93.2581	
				40*	39.964 000	0.0117	1.28×10^9 y
				41	40.961 827	6.7302	
20	Calcium	Ca	40.08	40	39.962 591	96.941	
				41*	40.962 279		1.0×10^5 y
				42	41.958 618	0.647	
				43	42.958 767	0.135	
				44	43.955 481	2.086	
21	Scandium	Sc	44.9559	41*	40.969 250		0.596 s
				45	44.955 911	100	
22	Titanium	Ti	47.88	44*	43.959 691		60 y
				47	46.951 765	7.3	
				48	47.947 947	73.8	
23	Vanadium	V	50.9415	50*	49.947 161	0.25	1.5×10^{17} y
				51	50.943 962	99.75	
24	Chromium	Cr	51.996	48*	47.954 033		21.6 h
				52	51.940 511	83.79	
				53	52.940 652	9.50	
25	Manganese	Mn	54.938 05	54*	53.940 361		312.1 d
				55	54.938 048	100	
26	Iron	Fe	55.847	54	53.939 613	5.9	
				55*	54.938 297		2.7 y
				56	55.934 940	91.72	
27	Cobalt	Co	58.933 20	59	58.933 198	100	
				60*	59.933 820		5.27 y
28	Nickel	Ni	58.793	58	57.935 345	68.077	
				59*	58.934 350		7.5×10^4 y
				60	59.930 789	26.223	
29	Copper	Cu	63.54	63	62.929 599	69.17	
				65	64.927 791	30.83	
30	Zinc	Zn	65.39	64	63.929 144	48.6	
				66	65.926 035	27.9	
				67	66.927 129	4.1	
				68	67.924 845	18.8	

875

Z	Element	Symbol	Average atomic mass (u)	Mass number (* indicates radioactive) A	Atomic mass (u)	Percent abundance	Half-life (if radioactive) $T_{1/2}$
31	Gallium	Ga	69.723	69	68.925 580	60.108	
				71	70.924 703	39.892	
32	Germanium	Ge	72.61	70	69.924 250	21.23	
				72	71.922 079	27.66	
				73	72.923 462	7.73	
				74	73.921 177	35.94	
				76	75.921 402	7.44	
33	Arsenic	As	74.9216	75	74.921 594	100	
34	Selenium	Se	78.96	76	75.919 212	9.36	
				77	76.919 913	7.63	
				78	77.917 397	23.78	
				80	79.916 519	49.61	
				82*	81.916 697	8.73	1.4×10^{20} y
35	Bromine	Br	79.904	79	78.918 336	50.69	
				81	80.916 287	49.31	
36	Krypton	Kr	83.80	81*	80.916 589		2.1×10^{5} y
				82	81.913 481	11.6	
				83	82.914 136	11.4	
				84	83.911 508	57.0	
				85*	84.912 531		10.76 y
				86	85.910 615	17.3	
37	Rubidium	Rb	85.468	85	84.911 793	72.17	
				87*	86.909 186	27.83	4.75×10^{10} y
38	Strontium	Sr	87.62	86	85.909 266	9.86	
				87	86.908 883	7.00	
				88	87.905 618	82.58	
				90*	89.907 737		29.1 y
39	Yttrium	Y	88.9058	89	88.905 847	100	
40	Zirconium	Zr	91.224	90	89.904 702	51.45	
				91	90.905 643	11.22	
				92	91.905 038	17.15	
				93*	92.906 473		1.5×10^{6} y
				94	93.906 314	17.38	
41	Niobium	Nb	92.9064	93	92.906 376	100	
				94*	93.907 280		2×10^{4} y
42	Molybdenum	Mo	95.94	92	91.906 807	14.84	
				93*	92.906 811		3.5×10^{3} y
				94	93.905 085	9.25	
				95	94.905 841	15.92	
				96	95.904 678	16.68	
				97	96.906 020	9.55	
				98	97.905 407	24.13	
				100	99.907 476	9.63	
43	Technetium	Tc		97*	96.906 363		2.6×10^{6} y
				98*	97.907 215		4.2×10^{6} y
				99*	98.906 254		2.1×10^{5} y
44	Ruthenium	Ru	101.07	99	98.905 939	12.7	
				100	99.904 219	12.6	
				101	100.905 558	17.1	
				102	101.904 348	31.6	
				104	103.905 558	18.6	
45	Rhodium	Rh	102.9055	103	102.905 502	100	

Z	Element	Symbol	Average atomic mass (u)	Mass number (* indicates radioactive) A	Atomic mass (u)	Percent abundance	Half-life (if radioactive) $T_{1/2}$
46	Palladium	Pd	106.42	104	103.904 033	11.14	
				105	104.905 082	22.33	
				106	105.903 481	27.33	
				108	107.903 898	26.46	
				110	109.905 158	11.72	
47	Silver	Ag	107.868	107	106.905 091	51.84	
				109	108.904 754	48.16	
48	Cadmium	Cd	112.41	109*	108.904 984		462 d
				110	109.903 004	12.49	
				111	110.904 182	12.80	
				112	111.902 760	24.13	
				113*	112.904 401	12.22	9.3×10^{15} y
				114	113.903 359	28.73	
49	Indium	In	114.82	113	112.904 060	4.3	
				115*	114.903 876	95.7	4.4×10^{14} y
50	Tin	Sn	118.71	116	115.901 743	14.53	
				117	116.902 953	7.58	
				118	117.901 605	24.22	
				119	118.903 308	8.58	
				120	119.902 197	32.59	
				121*	120.904 237		55 y
51	Antimony	Sb	121.76	121	120.903 820	57.36	
				123	122.904 215	42.64	
52	Tellurium	Te	127.60	125	124.904 429	7.12	
				126	125.903 309	18.93	
				128*	127.904 468	31.79	$> 8 \times 10^{24}$ y
				130*	129.906 228	33.87	$< 1.25 \times 10^{21}$ y
53	Iodine	I	126.9045	127	126.904 474	100	
				129*	128.904 984		1.6×10^{7} y
54	Xenon	Xe	131.29	129	128.904 779	26.4	
				131	130.905 069	21.2	
				132	131.904 141	26.9	
				134	133.905 394	10.4	
				136*	135.907 214	8.9	$> 2.36 \times 10^{21}$ y
55	Cesium	Cs	132.9054	133	132.905 436	100	
				135*	134.905 891		2×10^{6} y
				137*	136.907 078		30 y
56	Barium	Ba	137.33	133*	132.905 990		10.5 y
				137	136.905 816	11.23	
				138	137.905 236	71.70	
57	Lanthanum	La	138.905	138*	137.907 105	0.0902	1.05×10^{11} y
				139	138.906 346	99.9098	
58	Cerium	Ce	140.12	138	137.905 986	0.25	
				140	139.905 434	88.43	
				142*	141.909 241	11.13	$> 5 \times 10^{16}$ y
59	Praseodymium	Pr	140.9076	141	140.907 647	100	
60	Neodymium	Nd	144.24	142	141.907 718	27.13	
				143	142.909 809	12.18	
				144*	143.910 082	23.80	2.3×10^{15} y
				145	144.912 568	8.30	
				146	145.913 113	17.19	

877

Z	Element	Symbol	Average atomic mass (u)	Mass number (* indicates radioactive) A	Atomic mass (u)	Percent abundance	Half-life (if radioactive) $T_{1/2}$
61	Promethium	Pm		145*	144.912 745		17.7 y
				146*	145.914 968		5.5 y
62	Samarium	Sm	150.36	147*	146.914 894	15.0	1.06×10^{11} y
				148*	147.914 819	11.3	7×10^{15} y
				149*	148.917 180	13.8	$> 2 \times 10^{15}$ y
				150	149.917 273	7.4	
				152	151.919 728	26.7	
				154	153.922 206	22.7	
63	Europium	Eu	151.96	151	150.919 846	47.8	
				152*	151.921 740		13.5 y
				153	152.921 226	52.2	
64	Gadolinium	Gd	157.25	155	154.922 618	14.80	
				156	155.922 119	20.47	
				157	156.923 957	15.65	
				158	157.924 099	24.84	
				160	159.927 050	21.86	
65	Terbium	Tb	158.9253	159	158.925 345	100	
66	Dysprosium	Dy	162.5	161	160.926 930	18.9	
				162	161.926 796	25.5	
				163	162.928 729	24.9	
				164	163.929 172	28.2	
67	Holmium	Ho	164.9303	165	164.930 316	100	
68	Erbium	Er	167.26	166	165.930 292	33.6	
				167	166.932 047	22.95	
				168	167.932 369	27.8	
				170	169.935 462	14.9	
69	Thulium	Tm	168.9342	169	168.934 213	100	
				171*	170.936 428		1.92 y
70	Ytterbium	Yb	173.04	171	170.936 324	14.3	
				172	171.936 379	21.9	
				173	172.938 209	16.12	
				174	173.938 861	31.8	
				176	175.942 564	12.7	
71	Lutetium	Lu	174.967	175	174.940 772	97.41	
				176*	175.942 679	2.59	3.78×10^{10} y
72	Hafnium	Hf	178.49	177	176.943 218	18.606	
				178	177.943 697	27.297	
				179	178.945 813	13.029	
				180	179.946 547	35.100	
73	Tantalum	Ta	180.9479	181	180.947 993	99.988	
74	Tungsten	W	183.85	182	181.948 202	26.3	
				183	182.950 221	14.28	
				184	183.950 929	30.7	
				186	185.954 358	28.6	
75	Rhenium	Re	186.207	185	184.952 951	37.40	
				187*	186.955 746	62.60	4.4×10^{10} y
76	Osmium	Os	190.2	188	187.955832	13.3	
				189	188.958 139	16.1	
				190	189.958 439	26.4	
				192	191.961 468	41.0	
77	Iridium	Ir	192.2	191	190.960 585	37.3	
				193	192.962 916	62.7	

Z	Element	Symbol	Average atomic mass (u)	Mass number (* indicates radioactive) A	Atomic mass (u)	Percent abundance	Half-life (if radioactive) $T_{1/2}$
78	Platinum	Pt	195.08	194	193.962 655	32.9	
				195	194.964 765	33.8	
				196	195.964 926	25.3	
79	Gold	Au	196.9665	197	196.966 543	100	
80	Mercury	Hg	200.59	198	197.966 743	9.97	
				199	198.968 253	16.87	
				200	199.968 299	23.10	
				201	200.970 276	13.10	
				202	201.970 617	29.86	
81	Thallium	Tl	204.383	203	202.972 320	29.524	
				204*	203.073 839		3.78 y
				205	204.974 400	70.476	
				208*	207.981 992		3.053 m
82	Lead	Pb	207.2	206	205.974 440	24.1	
				207	206.974 871	22.1	
				208	207.976 627	52.4	
				212*	211.991 872		10.64 h
83	Bismuth	Bi	208.9803	209	208.980 374	100	
				212*	211.991 259		60.6 m
84	Polonium	Po		209*	208.982 405		102 y
				212*	211.988 842		0.30 μs
				216*	216.001 889		0.145 s
85	Astatine	At		218*	218.008 685		1.6 s
				219*	219.011 294		0.9 m
86	Radon	Rn		220*	220.011 369		55.6 s
				222*	222.017 571		3.823 d
87	Francium	Fr		223*	223.019 733		22 m
88	Radium	Ra		224*	224.020 187		3.66 d
				226*	226.025 402		1.6×10^3 y
				228*	228.031 064		5.75 y
89	Actinium	Ac		227*	227.027 701		18.72 y
				228*	228.028 716		1.913 y
90	Thorium	Th		232*	232.038 051	100	1.40×10^{10} y
				234*	234.043 593		24.1 d
91	Protactinium	Pa		231*	231.035 880		32.760 y
				234*	234.043 300		6.7 h
92	Uranium	U		234*	234.040 946	0.0055	2.46×10^5 y
				235*	235.043 924	0.720	7.04×10^8 y
				238*	238.050 784	99.2745	4.47×10^9 y
93	Neptunium	Np		236*	236.046 560		1.15×10^5 y
				237*	237.048 168		2.14×10^6 y
94	Plutonium	Pu		239*	239.052 157		2.412×10^5 y
				244*	244.064 200		8.1×10^7 y

879

Answers

1. 11.68 m
2. 4.0469×10^{-3} km^2
3. 6.4×10^{-2} m^3
4. 6.0×10^9 mg
5. 6.7×10^{-5} ps
6. 3.53×10^3 km/h, west
7. 2.80 h = 2 h, 48 min
8. 107 s
9. 4.0×10^1 km/h
10. 0.46 m/s^2
11. 48 m/h
12. 1.74 m/s
13. +25.0 m/s = 25.0 m/s, upward
14. −3.31 m/s
15. 44.8 m/s
16. +6.2 m/s^2 = 6.2 m/s^2, upward
17. −21.5 m/s^2 = 21.5 m/s^2, backward
18. 15.8 m
19. 38.5 m
20. −221 m = 221 m, downward
21. 126 s
22. 1.26×10^3 cm = 12.6 m
23. 1.27 s
24. 15.8 km/s
25. 11 km/h
26. 5.4 m/s^2
27. 2.74 s
28. 6.50 s
29. 10.5 m, forward
30. −8.6 m/s = 8.6 m/s, backward
31. 5.9 s

APPENDIX I Additional Problems

Additional Problems

Chapter 1 The Science of Physics

1. Mt. Waialeale in Hawaii gets 1.168×10^3 cm of rainfall per year. Express this quantity in meters.
2. An acre is equal to about 4.0469×10^3 m^2. Express this area in square kilometers.
3. A group drinks about 6.4×10^4 cm^3 of water per person per year. Express this in cubic meters.
4. The largest stone jar on the Plain of Jars in Laos has a mass of 6.0×10^3 kg. Express this mass in milligrams.
5. Half of a sample of the radioactive isotope beryllium-8 decays in 6.7×10^{-17} s. Express this time in picoseconds.

Chapter 2 Motion in One Dimension

6. The fastest airplane is the Lockheed SR-71. If an SR-71 flies 15.0 km west in 15.3 s, what is its average velocity in kilometers per hour?
7. Except for a 22.0 min rest stop, Emily drives with a constant velocity of 89.5 km/h, north. How long does the trip take if Emily's average velocity is 77.8 km/h, north?
8. A spaceship accelerates uniformly for 1220 km. How much time is required for the spaceship to increase its speed from 11.1 km/s to 11.7 km/s?
9. A polar bear initially running at 4.0 m/s accelerates uniformly for 18 s. If the bear travels 135 m in this time, what is its maximum speed?
10. A walrus accelerates from 7.0 km/h to 34.5 km/h over a distance of 95 m. What is the magnitude of the walrus's acceleration?
11. A snail can move about 4.0 m in 5.0 min. What is the average speed of the snail?
12. A crate is accelerated at 0.035 m/s^2 for 28.0 s along a conveyor belt. If the crate's initial speed is 0.76 m/s, what is its final speed?
13. A person throws a ball vertically and catches it after 5.10 s. What is the ball's initial velocity?
14. A bicyclist accelerates −0.870 m/s^2 during a 3.80 s interval. What is the change in the velocity of the bicyclist and bicycle?
15. A hockey puck slides 55.0 m in 1.25 s with a uniform acceleration. If the puck's final speed is 43.2 m/s, what was its initial speed?
16. A small rocket launched from rest travels 12.4 m upward in 2.0 s. What is the rocket's net acceleration?
17. A jet slows uniformly from 153 km/h to 0 km/h over 42.0 m. What is the jet's acceleration?
18. A softball thrown straight up at 17.5 m/s is caught 3.60 s later. How high does the ball rise?
19. A child, starting from rest, sleds down a snow-covered slope in 5.50 s. If the child's final speed is 14.0 m/s, what the length of the slope?
20. A sky diver opens her parachute and drifts down for 34.0 s with a constant velocity of 6.50 m/s. What is the sky diver's displacement?
21. In a race, a tortoise runs at 10.0 cm/s and a hare runs at 200.0 cm/s. Both start at the same time, but the hare stops to rest for 2.00 min. The tortoise wins by 20.0 cm. At what time does the tortoise cross the finish line?
22. What is the length of the race in problem 21?
23. The cable pulling an elevator upward at 12.5 m/s breaks. How long does it take for the elevator to come to rest?
24. A disk is uniformly accelerated from rest for 0.910 s over 7.19 km. What is its final speed?
25. A tiger accelerates 3.0 m/s^2 for 4.1 s to reach a final speed of 55.0 km/h. What was its initial speed in kilometers per hour?
26. A shark accelerates uniformly from 2.8 km/h to 32.0 km/h in 1.5 s. How large is its acceleration?
27. The 1903 Wright flyer was accelerated at 4.88 m/s^2 along a track that was 18.3 m long. How long did it take to accelerate the flyer from rest?
28. A drag racer starts at rest and reaches a speed of 386.0 km/h with an average acceleration of 16.5 m/s^2. How long does this acceleration take?
29. A hummingbird accelerates at −9.20 m/s^2 such that its velocity changes from +50.0 km/h to 0 km/h. What is its displacement?
30. A train backs up from an initial velocity of −4.0 m/s and an average acceleration of −0.27 m/s^2. What is the train's velocity after 17 s?
31. A cross-country skier skiing with an initial velocity of +4.42 m/s slows uniformly at −0.75 m/s^2. How long does it take the skier to stop?

32. What is the skier's displacement in problem 31?

33. A speedboat uniformly increases its speed from 25 m/s west to 35 m/s west. How long does it take the boat to travel 250 m west?

34. A ship accelerates at -7.6×10^{-2} m/s^2 so that it comes to rest at the dock 255 m away in 82.0 s. What is the ship's initial speed?

35. A student skates downhill with an average acceleration of 0.85 m/s^2. Her initial speed is 4.5 m/s, and her final speed is 10.8 m/s. How long does she take to skate down the hill?

36. A wrench dropped from a tall building is caught in a safety net when the wrench has a velocity of −49.5 m/s. How far did it fall?

37. A rocket sled comes to a complete stop from a speed of 320 km/h in 0.18 s. What is the sled's average acceleration?

38. A racehorse uniformly accelerates 7.56 m/s^2, reaching its final speed after running 19.0 m. If the horse starts at rest, what is its final speed?

39. An arrow is shot upward at a speed of 85.1 m/s. How long does the archer have to move from the launching spot before the arrow returns to Earth?

40. A handball strikes a wall with a forward speed of 13.7 m/s and bounces back with a speed of 11.5 m/s. If the ball changes velocity in 0.021 s, what is the handball's average acceleration?

41. A ball accelerates at 6.1 m/s^2 from 1.8 m/s to 9.4 m/s. How far does the ball travel?

42. A small sandbag is dropped from rest from a hovering hot-air balloon. After 2.0 s, what is the sandbag's displacement below the balloon?

43. A hippopotamus accelerates at 0.678 m/s^2 until it reaches a speed of 8.33 m/s. If the hippopotamus runs 46.3 m, what was its initial speed?

44. A ball is hit upward with a speed of 7.5 m/s. How long does the ball take to reach maximum height?

45. A surface probe on the planet Mercury falls 17.6 m downward from a ledge. If free-fall acceleration near Mercury is −3.70 m/s^2, what is the probe's velocity when it reaches the ground?

Chapter 3 Two-Dimensional Motion and Vectors

46. A plane moves 599 m northeast along a runway. If the northern component of this displacement is 89 m, how large is the eastern component?

47. Find the displacement direction in problem 46.

48. A train travels 478 km southwest along a straight stretch. If the train is displaced south by 42 km, what is the train's displacement to the west?

49. Find the displacement direction in problem 48.

50. A ship's total displacement is 7400 km at 26° south of west. If the ship sails 3200 km south, what is the western component of its journey?

51. The distance from an observer on a plain to the top of a nearby mountain is 5.3 km at 8.4° above the horizontal. How tall is the mountain?

52. A skyrocket travels 113 m at an angle of 82.4° with respect to the ground and toward the south. What is the rocket's horizontal displacement?

53. A hot-air balloon descends with a velocity of 55 km/h at an angle of 37° below the horizontal. What is the vertical velocity of the balloon?

54. A stretch of road extends 55 km at 37° north of east, then continues for 66 km due east. What is a driver's resultant displacement along this road?

55. A driver travels 4.1 km west, 17.3 km north, and finally 1.2 km at an angle of 24.6° west of north. What is the driver's displacement?

56. A tornado picks up a car and hurls it horizontally 125 m with a speed of 90.0 m/s. How long does it take the car to reach the ground?

57. A squirrel knocks a nut horizontally at a speed of 10.0 cm/s. If the nut lands at a horizontal distance of 18.6 cm, how high up is the squirrel?

58. A flare is fired at an angle of 35° to the ground at an initial speed of 250 m/s. How long does it take for the flare to reach its maximum altitude?

59. A football kicked with an initial speed of 23.1 m/s reaches a maximum height of 16.9 m. At what angle was the ball kicked?

60. A bird flies north at 58.0 km/h relative to the wind. The wind is blowing at 55.0 km/h south relative to Earth. How long will it take the bird to fly 1.4 km relative to Earth?

61. A race car moving at 286 km/h is 0.750 km behind a car moving at 252 km/h. How long will it take the faster car to catch up to the slower car?

62. A helicopter flies 165 m horizontally and then moves downward to land 45 m below. What is the helicopter's resultant displacement?

63. A toy parachute floats 13.0 m downward. If the parachute travels 9.0 m horizontally, what is the resultant displacement?

32. 13 m, forward
33. 8.3 s
34. 6.2 m/s
35. 7.4 s
36. −125 m = 125 m, downward
37. −490 m/s^2 = 490 m/s^2, backward
38. 16.9 m/s
39. 17.3 s
40. −1200 m/s^2 = 1200 m/s^2, backward
41. 7.0 m
42. 2.0×10^1 m
43. 2.6 m/s
44. 0.76 s
45. −11.4 m/s = 11.4 m/s, downward
46. 592 m, east
47. 8.5° north of east
48. 475 km, west
49. 5.0° south of west
50. 6700 km, west
51. 770 m
52. 14.9 m, south
53. −33 km/h = 33 km/h, downward
54. 115 km, 17° north of east
55. 18.9 km, 76° north of west
56. 5.05 s
57. 17.0 m
58. 15 s
59. 52.0°
60. 0.47 h = 28 min
61. 79 s
62. 171 m, 15° below the horizontal
63. 15.8 m, 55° below the horizontal

881

64. 2.6 m along table's length; 0.61 m along table's width
65. 0.290 m/s, east; 1.16 m/s, north
66. 12.4 km/h, upward; 53.6 km/h, forward
67. 2.6 km
68. 1.42×10^3 m, 16° to the side of the initial displacement
69. 66 km, 46° south of east
70. 404 m
71. 10.7 m
72. 70.0 m/s, 1.23° from the vertical
73. 3.0 s
74. $+2.0 \times 10^1$ m $= 2.0 \times 10^1$ m, forward; -16 m $= 16$ m, downward
75. 76.9 km/h, 60.1° west of north
76. 9.58 km, west; 16.7 km, north
77. 7.0×10^2 m, 3.8° above the horizontal
78. 1.26×10^3 km, 48° north of west
79. 47.2 m
80. 3.6 m/s
81. 6.36 m/s
82. 11 m, west; 18 m, north
83. 13.6 km/h, 73° south of east
84. 47° north of west
85. 58 N
86. 49 N
87. 14.0 N; 2.0 N
88. -6.12 m/s^2 $= 6.12$ m/s^2, downward
89. 9.5×10^4 kg
90. $+13$ N $= 13$ N, upward
91. 258 N, up the slope
92. 0.087
93. 15.9 N
94. 0.73

64. A billiard ball travels 2.7 m at an angle of 13° with respect to the long side of the table. What are the components of the ball's displacement?

65. A golf ball has a velocity of 1.20 m/s at 14.0° east of north. What are the velocity components?

66. A tiger leaps with an initial velocity of 55.0 km/h at an angle of 13.0° with respect to the horizontal. What are the components of the tiger's velocity?

67. A tramway extends 3.88 km up a mountain from a station 0.8 km above sea level. If the horizontal displacement is 3.45 km, how far above sea level is the mountain peak?

68. A bullet travels 850 m, ricochets, and moves another 640 m at an angle of 36° from its previous forward motion. What is the bullet's resultant displacement?

69. A bird flies 46 km at 15° south of east, then 22 km at 13° east of south, and finally 14 km at 14° west of south. What is the bird's displacement?

70. A ball is kicked with a horizontal speed of 9.37 m/s off the top of a mountain. The ball moves 85.0 m horizontally before hitting the ground. How tall is the mountain?

71. A ball is kicked with a horizontal speed of 1.50 m/s from a height of 2.50×10^2 m. What is its horizontal displacement when it hits the ground?

72. What is the velocity of the ball in problem 71 when it reaches the ground?

73. A shingle slides off a roof at a speed of 2.0 m/s and an angle of 30.0° below the horizontal. How long does it take the shingle to fall 45 m?

74. A ball is thrown with an initial speed of 10.0 m/s and an angle of 37.0° above the horizontal. What are the vertical and horizontal components of the ball's displacement after 2.5 s?

75. A rocket moves north at 55.0 km/h with respect to the air. It encounters a wind from 17.0° north of west at 40.0 km/h with respect to Earth. What is the rocket's velocity with respect to Earth?

76. How far to the north and west does the rocket in problem 75 travel after 15.0 min?

77. A cable car travels 2.00×10^2 m on level ground, then 3.00×10^2 m at an incline of 3.0°, and then 2.00×10^2 m at an incline of 8.8°. What is the final displacement of the cable car?

78. A hurricane moves 790 km at 18° north of west, then due west for 150 km, then north for 470 km, and finally 15° east of north for 240 km. What is the hurricane's resultant displacement?

79. What is the range of an arrow shot horizontally at 85.3 m/s from 1.50 m above the ground?

80. A drop of water in a fountain takes 0.50 s to travel 1.5 m horizontally. The water is projected upward at an angle of 33°. What is the drop's initial speed?

81. A golf ball is hit up a 41.0° ramp to travel 4.46 m horizontally and 0.35 m below the edge of the ramp. What is the ball's initial speed?

82. A flare is fired with a velocity of 87 km/h west from a car traveling 145 km/h north. With respect to Earth, what is the flare's resultant displacement 0.45 s after being launched?

83. A sailboat travels south at 12.0 km/h with respect to the water against a current 15.0° south of east at 4.0 km/h. What is the boat's velocity?

Chapter 4 Forces and the Laws of Motion

84. A boat exerts a 9.5×10^4 N force 15.0° north of west on a barge. Another exerts a 7.5×10^4 N force north. What direction is the barge moved?

85. A shopper exerts a force on a cart of 76 N at an angle of 40.0° below the horizontal. How much force pushes the cart in the forward direction?

86. How much force pushes the cart in problem 85 against the floor?

87. What are the magnitudes of the largest and smallest net forces that can be produced by combining a force of 6.0 N and a force of 8.0 N?

88. A buoyant force of 790 N lifts a 214 kg sinking boat. What is the boat's net acceleration?

89. A house is lifted by a net force of 2850 N and moves from rest to an upward speed of 15 cm/s in 5.0 s. What is the mass of the house?

90. An 8.0 kg bag is lifted 20.0 cm in 0.50 s. If it is initially at rest, what is the net force on the bag?

91. A 90.0 kg skier glides at constant speed down a 17.0° slope. Find the frictional force on the skier.

92. A snowboarder slides down a 5.0° slope at a constant speed. What is the coefficient of kinetic friction between the snow and the board?

93. A 2.00 kg block is in equilibrium on a 36.0° incline. What is the normal force on the block?

94. A 1.8×10^3 kg car is parked on a hill on a 15.0° incline. A 1.25×10^4 N frictional force holds the car in place. Find the coefficient of static friction.

95. The coefficient of kinetic friction between a jar slid across a table and the table is 0.20. What is the magnitude of the jar's acceleration?

96. A force of 5.0 N to the left causes a 1.35 kg book to have a net acceleration of 0.76 m/s^2 to the left. What is the frictional force on the book?

97. A child pulls a toy by exerting a force of 15.0 N at an angle of 55.0° with respect to the floor. What are the components of the force?

98. A car is pulled by three forces: 600.0 N to the north, 750.0 N to the east, and 675 N at 30.0° south of east. What direction does the car move?

99. Suppose a catcher exerts a force of −65.0 N to stop a baseball with a mass of 0.145 kg. What is the ball's net acceleration as it is being caught?

100. A 2.0 kg fish pulled upward by a fisherman rises 1.9 m in 2.4 s, starting from rest. What is the net force on the fish during this interval?

101. An 18.0 N force pulls a cart against a 15.0 N frictional force. The speed of the cart increases 1.0 m/s every 5.0 s. What is the cart's mass?

102. A 47 kg sled carries a 33 kg load. The coefficient of kinetic friction between the sled and snow is 0.075. What is the magnitude of the frictional force on the sled as it moves up a hill with a 15° incline?

103. Ice blocks slide with an acceleration of 1.22 m/s^2 down a chute at an angle of 12.0° below the horizontal. What is the coefficient of kinetic friction between the ice and chute?

104. A 1760 N force pulls a 266 kg load up a 17° incline. What is the coefficient of static friction between the load and the incline?

105. A 4.26×10^7 N force pulls a ship at a constant speed along a dry dock. The coefficient of kinetic friction between the ship and dry dock is 0.25. Find the normal force exerted on the ship.

106. If the incline of the dry dock in problem 105 is 10.0°, what is the ship's mass?

107. A 65.0 kg skier is pulled up an 18.0° slope by a force of 2.50×10^2 N. If the net acceleration uphill is 0.44 m/s^2, what is the frictional force between the skis and the snow?

108. Four forces are acting on a hot-air balloon: $\mathbf{F_1}$ = 2280.0 N up, $\mathbf{F_2}$ = 2250.0 N down, $\mathbf{F_3}$ = 85.0 N west, and $\mathbf{F_4}$ = 12.0 N east. What is the direction of the net external force on the balloon?

109. A traffic signal is supported by two cables, each of which makes an angle of 40.0° with the vertical. If each cable can exert a maximum force of 7.50×10^2 N, what is the largest weight they can support?

110. A certain cable of an elevator is designed to exert a force of 4.5×10^4 N. If the maximum acceleration that a loaded car can withstand is 3.5 m/s^2, what is the combined mass of the car and its contents?

111. A frictional force of 2400 N keeps a crate of machine parts from sliding down a ramp with an incline of 30.0°. The coefficient of static friction between the box and the ramp is 0.20. What is the normal force of the ramp on the box?

112. Find the mass of the crate in problem 111.

113. A 5.1×10^2 kg bundle of bricks is pulled up a ramp at an incline of 14° to a construction site. The force needed to move the bricks up the ramp is 4.1×10^3 N. What is the coefficient of static friction between the bricks and the ramp?

Chapter 5 Work and Energy

114. If 2.13×10^6 J of work must be done on a roller-coaster car to move it 3.00×10^2 m, how large is the net force acting on the car?

115. A force of 715 N is applied to a roller-coaster car to push it horizontally. If 2.72×10^4 J of work is done on the car, how far has it been pushed?

116. In 0.181 s, through a distance of 8.05 m, a test pilot's speed decreases from 88.9 m/s to 0 m/s. If the pilot's mass is 70.0 kg, how much work is done against his body?

117. What is the kinetic energy of a disk with a mass of 0.20 g and a speed of 15.8 km/s?

118. A 9.00×10^2 kg walrus is swimming at a speed of 35.0 km/h. What is its kinetic energy?

119. A golf ball with a mass of 47.0 g has a kinetic energy of 1433 J. What is the ball's speed?

120. A turtle, swimming at 9.78 m/s, has a kinetic energy of 6.08×10^4 J. What is the turtle's mass?

121. A 50.0 kg parachutist is falling at a speed of 47.00 m/s when her parachute opens. Her speed upon landing is 5.00 m/s. How much work is done by the air to reduce the parachutist's speed?

122. An 1100 kg car accelerates from 48.0 km/h to 59.0 km/h over 100.0 m. What was the magnitude of the net force acting on it?

95. 2.0 m/s^2

96. 4.0 N, to the right

97. F_x = 8.60 N; F_y = 12.3 N

98. 11.1° north of east

99. −448 m/s^2 = 448 m/s^2, backward

100. 1.3 N, upward

101. 15 kg

102. 57 N

103. 0.085

104. 0.40

105. 1.7×10^8 N

106. 1.8×10^7 kg

107. 24 N, downhill

108. 22.3° up from west

109. 1.150×10^3 N

110. 3.4×10^3 kg

111. 1.2×10^4 N

112. 1400 kg

113. 0.60

114. 7.10×10^3 N

115. 38.0 m

116. -2.77×10^5 J

117. 2.5×10^4 J

118. 4.25×10^4 J

119. 247 m/s

120. 1.27×10^3 kg

121. -5.46×10^4 J

122. 5.1×10^2 N

123. 3.35×10^6 J
124. 4.0×10^{-2} J
125. 1.23 J
126. -2.05×10^3 J
127. 12 s
128. 1.17×10^{10} J
129. 0.600 m
130. 17.2 N
131. 133 J
132. 9.4×10^9 J
133. 53.3 m/s
134. 0.17 kg
135. 72.2 m
136. 9.6×10^2 kg
137. 0.13 m = 13 cm
138. 33.5 m/s
139. 7.7 m/s
140. 2.7×10^{10} W = 27 GW
141. 8.0 s
142. 3.0×10^7 J
143. 230 J
144. 8.72×10^6 N/m
145. 7.96 m
146. 12 m
147. 6.0×10^1 m/s
148. 15 m
149. 1.58×10^3 kg•m/s, north
150. 6.66 m/s, south
151. 3.38×10^{31} kg
152. −0.897 N = 0.879 N, to the left
153. 18 s

123. What is the gravitational potential energy of a 64.0 kg person at 5334 m above sea level?

124. A spring has a force constant of 550 N/m. What is the elastic potential energy stored in the spring when the spring is compressed 1.2 cm?

125. What is the kinetic energy of a 0.500 g raindrop that falls 0.250 km? Ignore air resistance.

126. A 50.0 g projectile is fired upward at 3.00×10^2 m/s and lands at 89.0 m/s. How much mechanical energy is lost to air resistance?

127. How long does it take for 4.5×10^6 J of work to be done by a 380.3 kW engine?

128. A ship's engine has a power output of 13.0 MW. How much work can it do in 15.0 min?

129. A catcher picks up a baseball from the ground with a net upward force of 7.25×10^{-2} N so that 4.35×10^{-2} J of net work is done. How far is the ball lifted?

130. A crane does 1.31×10^3 J of net work when lifting cement 76.2 m. How large is the net force doing this work?

131. A girl exerts a force of 35.0 N at an angle of 20.0° to the horizontal to move a wagon 15.0 m along a level path. What is the net work done on it if a frictional force of 24.0 N is present?

132. The *Queen Mary* had a mass of 7.5×10^7 kg and a top cruising speed of 57 km/h. What was the kinetic energy of the ship at that speed?

133. How fast is a 55.0 kg sky diver falling when her kinetic energy is 7.81×10^4 J?

134. A hockey puck with an initial speed of 8.0 m/s coasts 45 m to a stop. If the force of friction on the puck is 0.12 N, what is the puck's mass?

135. How far does a 1.30×10^4 kg jet travel if it is slowed from 2.40×10^2 km/h to 0 km/h by an acceleration of -30.8 m/s^2?

136. An automobile is raised 7.0 m, resulting in an increase in gravitational potential energy of 6.6×10^4 J. What is the automobile's mass?

137. A spring in a pogo stick has a force constant of 1.5×10^4 N/m. How far is the spring compressed when its elastic potential energy is 120 J?

138. A 100.0 g arrow is pulled back 30.0 cm against a bowstring. The bowstring's force constant is 1250 N/m. What speed will the arrow leave the bow?

139. A ball falls 3.0 m down a vertical pipe, the end of which bends horizontally. How fast does the ball leave the pipe if no energy is lost to friction?

140. A spacecraft's engines do 1.4×10^{13} J of work in 8.5 min. What is the power output of these engines?

141. A runner exerts a force of 334 N against the ground while using 2100 W of power. How long does it take him to run a distance of 50.0 m?

142. A high-speed boat has four 300.0 kW motors. How much work is done in 25 s by the motors?

143. A 92 N force pushes an 18 kg box of books, initially at rest, 7.6 m across a floor. The coefficient of kinetic friction between the floor and the box is 0.35. What is the final kinetic energy of the box of books?

144. A guardrail can be bent by 5.00 cm and then restore its shape. What is its force constant if struck by a car with 1.09×10^4 J of kinetic energy?

145. A 25.0 kg trunk strikes the ground with a speed of 12.5 m/s. If no energy is lost from air resistance, what is the height from which the trunk fell?

146. Sliding a 5.0 kg stone up a frictionless ramp with a 25.0° incline increases its gravitational potential energy by 2.4×10^2 J. How long is the ramp?

147. A constant 4.00×10^2 N force moves a 2.00×10^2 kg iceboat 0.90 km. Frictional force is negligible, and the boat starts at rest. Find the boat's final speed.

148. A 50.0 kg circus clown jumps from a platform into a net 1.00 m above the ground. The net is stretched 0.65 m and has a force constant of 3.4×10^4 N/m. What is the height of the platform?

Chapter 6 Momentum and Collisions

149. If a 50.0 kg cheetah, initially at rest, runs 274 m north in 8.65 s, what is its momentum?

150. If a 1.46×10^5 kg whale has a momentum of 9.73×10^5 kg•m/s to the south, what is its velocity?

151. A star has a momentum of 8.62×10^{36} kg•m/s and a speed of 255 km/s. What is its mass?

152. A 5.00 g projectile has a velocity of 255 m/s right. Find the force to stop this projectile in 1.45 s.

153. How long does it take a 0.17 kg hockey puck to decrease its speed by 9.0 m/s if the coefficient of kinetic friction is 0.050?

154. A 705 kg race car driven by a 65 kg driver moves with a velocity of 382 km/h right. Find the force to bring the car and driver to a stop in 12.0 s.

155. Find the stopping distance in problem 154.

156. A 50.0 g shell fired from a 3.00 kg rifle has a speed of 400.0 m/s. With what velocity does the rifle recoil in the opposite direction?

157. A twig at rest in a pond moves with a speed of 0.40 cm/s opposite a 2.5 g snail, which has a speed of 1.2 cm/s. What is the mass of the twig?

158. A 25.0 kg sled holding a 42.0 kg child has a speed of 3.50 m/s. They collide with and pick up a snowman, initially at rest. The resulting speed of the snowman, sled, and child is 2.90 m/s. What is the snowman's mass?

159. An 8500 kg railway car moves right at 4.5 m/s, and a 9800 kg railway car moves left at 3.9 m/s. The cars collide and stick together. What is the final velocity of the system?

160. What is the change in kinetic energy for the two railway cars in problem 159?

161. A 55 g clay ball moving at 1.5 m/s collides with a 55 g clay ball at rest. By what percentage does the kinetic energy change after the inelastic collision?

162. A 45 g golf ball collides elastically with an identical ball at rest and stops. If the second ball's final speed is 3.0 m/s, what was the first ball's initial speed?

163. A 5.00×10^2 kg racehorse gallops with a momentum of 8.22×10^3 kg•m/s to the west. What is the horse's velocity?

164. A 3.0×10^7 kg ship collides elastically with a 2.5×10^7 kg ship moving north at 4.0 km/h. After the collision, the first ship moves north at 3.1 km/h and the second ship moves south at 6.9 km/h. Find the unknown velocity.

165. A high-speed train has a mass of 7.10×10^5 kg and moves at a speed of 270.0 km/h. What is the magnitude of the train's momentum?

166. A bird with a speed of 50.0 km/h has a momentum of magnitude of 0.278 kg•m/s. What is the bird's mass?

167. A 75 N force pulls a child and sled initially at rest down a snowy hill. If the combined mass of the sled and child is 55 kg, what is their speed after 7.5 s?

168. A student exerts a net force of −1.5 N over a period of 0.25 s to bring a falling 60.0 g egg to a stop. What is the egg's initial speed?

169. A 1.1×10^3 kg walrus starts swimming east from rest and reaches a velocity of 9.7 m/s in 19 s. What is the net force acting on the walrus?

170. A 12.0 kg wagon at rest is pulled by a 15.0 N force at an angle of 20.0° above the horizontal. If an 11.0 N frictional force resists the forward force, how long will the wagon take to reach a speed of 4.50 m/s?

171. A 42 g meteoroid moving forward at 7.82×10^3 m/s collides with a spacecraft. What force is needed to stop the meteoroid in 1.0×10^{-6} s?

172. A 455 kg polar bear slides for 12.2 s across the ice. If the coefficient of kinetic friction between the bear and the ice is 0.071, what is the change in the bear's momentum as it comes to a stop?

173. How far does the bear in problem 172 slide?

174. How long will it take a -1.26×10^4 N force to stop a 2.30×10^3 kg truck moving at a speed of 22.2 m/s?

175. A 63 kg skater at rest catches a sandbag moving north at 5.4 m/s. The skater and bag then move north at 1.5 m/s. Find the sandbag's mass.

176. A 1.36×10^4 kg barge is loaded with 8.4×10^3 kg of coal. What was the unloaded barge's speed if the loaded barge has a speed of 1.3 m/s?

177. A 1292 kg automobile moves east at 88.0 km/h. If all forces remain constant, what is the car's velocity if its mass is reduced to 1255 kg?

178. A 68 kg student steps into a 68 kg boat at rest, causing both to move west at a speed of 0.85 m/s. What was the student's initial velocity?

179. A 1400 kg automobile, heading north at 45 km/h, collides inelastically with a 2500 kg truck traveling east at 33 km/h. What is the vehicles' final velocity?

180. An artist throws 1.3 kg of paint onto a 4.5 kg canvas at rest. The paint-covered canvas slides backward at 0.83 m/s. What is the change in the kinetic energy of the paint and canvas?

181. Find the change in kinetic energy if a 0.650 kg fish leaping to the right at 15.0 m/s collides inelastically with a 0.950 kg fish leaping to the left at 13.5 m/s.

182. A 10.0 kg cart moving at 6.0 m/s hits a 2.5 kg cart moving at 3.0 m/s in the opposite direction. Find the carts' final speed after an inelastic collision.

183. A ball, thrown right 6.00 m/s, hits a 1.25 kg panel at rest, then bounces back at 4.90 m/s. The panel moves right at 1.09 m/s. Find the ball's mass.

154. -6.81×10^3 N = 6.81×10^3 N, to the left

155. 637 m, to the right

156. −6.67 m/s = 6.67 m/s, backward

157. 7.5 g

158. 14 kg

159. 0.0 m/s

160. -1.61×10^5 J

161. -5.0×10^1 percent

162. 3.0 m/s

163. 16.4 m/s, west

164. 6.0 km/h, south

165. 5.33×10^7 kg•m/s

166. 2.00×10^{-2} kg = 20.0 g

167. 1.0×10^1 m/s

168. 6.2 m/s

169. 560 N, east

170. 17 s

171. -3.3×10^8 N = 3.3×10^8 N, backward

172. -3.9×10^3 kg•m/s = 3.9×10^3 kg•m/s, opposite the polar bear's motion

173. 52 m

174. 4.06 s

175. 24 kg

176. 2.1 m/s

177. 90.6 km/h, east

178. 1.7 m/s, west

179. 26 km/h, 37° north of east

180. −6.9 J

181. −157 J

182. 4.2 m/s

183. 0.125 kg

184. 2.1 m/s, east
185. -4.1×10^4 J
186. 12.8 cm/s, to the right
187. 9.8 kg
188. -6.6×10^{16} J
189. 1.0 m/s, 60° south of east
190. 1.2 kg
191. 4.04×10^3 m/s^2
192. 256 m
193. 42 m/s
194. 6350 N
195. 8.9 kg
196. 4.47×10^{15} m
197. 1.04×10^4 m/s = 10.4 km/s
198. 8.34×10^{-7} N
199. 1.48×10^{23} kg
200. 6.96×10^8 m
201. 1.10×10^{12} m
202. 1.2×10^5 s = 34 h
203. 6.6×10^3 m/s = 6.6 km/s
204. 7.49×10^4 N•m
205. 0.87 m
206. 2.35×10^7 m = 2.35×10^4 km
207. 254 N
208. 3.4×10^{-5} m/s^2
209. 0.42 m = 42 cm
210. 38 m/s
211. 25 N

184. A 2150 kg car, moving east at 10.0 m/s, collides and joins with a 3250 kg car. The cars move east together at 5.22 m/s. What is the 3250 kg car's initial velocity?

185. Find the change in kinetic energy in problem 184.

186. A 15.0 g toy car moving to the right at 20.0 cm/s collides elastically with a 20.0 g toy car moving left at 30.0 cm/s. The 15.0 g car then moves left at 37.1 cm/s. Find the 20.0 g car's final velocity.

187. A remora swimming right at 5.0 m/s attaches to a 150.0 kg shark moving left at 7.00 m/s. Both move left at 6.25 m/s. Find the remora's mass.

188. A 6.5×10^{12} kg comet, moving at 420 m/s, catches up to and collides inelastically with a 1.50×10^{13} kg comet moving at 250 m/s. Find the change in the comets' kinetic energy.

189. A 7.00 kg ball moves east at 2.00 m/s, collides with a 7.00 kg ball at rest, and then moves 30.0° north of east at 1.73 m/s. What is the second ball's final velocity?

190. A 2.0 kg block moving at 8.0 m/s on a frictionless surface collides elastically with a block at rest. The first block moves in the same direction at 2.0 m/s. What is the second block's mass?

Chapter 7 Circular Motion and Gravitation

191. A pebble that is 3.81 m from the eye of a tornado has a tangential speed of 124 m/s. What is the magnitude of the pebble's centripetal acceleration?

192. A race car speeds along a curve with a tangential speed of 75.0 m/s. The centripetal acceleration on the car is 22.0 m/s^2. Find the radius of the curve.

193. A subject in a large centrifuge has a radius of 8.9 m and a centripetal acceleration of 20g ($g = 9.81$ m/s^2). What is the tangential speed of the subject?

194. A 1250 kg automobile with a tangential speed of 48.0 km/h follows a circular road that has a radius of 35.0 m. How large is the centripetal force?

195. A rock in a sling is 0.40 m from the axis of rotation and has a tangential speed of 6.0 m/s. What is the rock's mass if the centripetal force is 8.00×10^2 N?

196. A 7.55×10^{13} kg comet orbits the sun with a speed of 0.173 km/s. If the centripetal force on the comet is 505 N, how far is it from the sun?

197. A 2.05×10^8 kg asteroid has an orbit with a 7378 km radius. The centripetal force on the asteroid is 3.00×10^9 N. Find the asteroid's tangential speed.

198. Find the gravitational force between a 0.500 kg mass and a 2.50×10^{12} kg mountain that is 10.0 km away.

199. The gravitational force between Ganymede and Jupiter is 1.636×10^{22} N. Jupiter's mass is 1.90×10^{27} kg, and the distance between the two bodies is 1.071×10^6 km. What is Ganymede's mass?

200. At the sun's surface, the gravitational force on 1.00 kg is 274 N. The sun's mass is 1.99×10^{30} kg. If the sun is assumed spherical, what is the sun's radius?

201. At the surface of a red giant star, the gravitational force on 1.00 kg is only 2.19×10^{-3} N. If its mass equals 3.98×10^{31} kg, what is the star's radius?

202. Uranus has a mass of 8.6×10^{25} kg. The mean distance between the centers of the planet and its moon Miranda is 1.3×10^5 km. If the orbit is circular, what is Miranda's period in hours?

203. What is the tangential speed in problem 202?

204. The rod connected halfway along the 0.660 m radius of a wheel exerts a 2.27×10^5 N force. How large is the maximum torque?

205. A golfer exerts a torque of 0.46 N•m on a golf club. If the club exerts a force of 0.53 N on a stationary golf ball, what is the length of the club?

206. What is the orbital radius of the Martian moon Deimos if it orbits 6.42×10^{23} kg Mars in 30.3 h?

207. A 4.00×10^2 N•m torque is produced applying a force 1.60 m from the fulcrum and at an angle of 80.0° to the lever. How large is the force?

208. A customer 11 m from the center of a revolving restaurant has a speed of 1.92×10^{-2} m/s. How large a centripetal acceleration acts on the customer?

209. A toy train on a circular track has a tangential speed of 0.35 m/s and a centripetal acceleration of 0.29 m/s^2. What is the radius of the track?

210. A person against the inner wall of a hollow cylinder with a 150 m radius feels a centripetal acceleration of 9.81 m/s^2. Find the cylinder's tangential speed.

211. The tangential speed of 0.20 kg toy carts is 5.6 m/s when they are 0.25 m from a turning shaft. How large is the centripetal force on the carts?

212. A 1250 kg car on a curve with a 35.0 m radius has a centripetal force from friction and gravity of 8.07×10^3 N. What is the car's tangential speed?

213. Two wrestlers, 2.50×10^{-2} m apart, exert a 2.77×10^{-3} N gravitational force on each other. One has a mass of 157 kg. What is the other's mass?

214. A 1.81×10^5 kg blue whale is 1.5 m from a 2.04×10^4 kg whale shark. What is the gravitational force between them?

215. Triton's orbit around Neptune has a radius of 3.56×10^5 km. Neptune's mass is 1.03×10^{26} kg. What is Triton's period?

216. Find the tangential speed in problem 215.

217. A moon orbits a 1.0×10^{26} kg planet in 365 days. What is the radius of the moon's orbit?

218. What force is required to produce a 1.4 N•m torque when applied to a door at a 60.0° angle and 0.40 m from the hinge?

219. What is the maximum torque that the force in problem 218 can exert?

220. A worker hanging 65.0° from the vane of a windmill exerts an 8.25×10^3 N•m torque. If the worker weighs 587 N, what is the vane's length?

Chapter 8 Fluid Mechanics

221. A cube of volume 1.00 m^3 floats in gasoline, which has a density of 675 kg/m^3. How large a buoyant force acts on the cube?

222. A cube 10.0 cm on each side has a density of 2.053×10^4 kg/m^3. Its apparent weight in fresh water is 192 N. Find the buoyant force.

223. A 1.47×10^6 kg steel hull has a base that is 2.50×10^3 m^2 in area. If it is placed in sea water ($\rho = 1.025 \times 10^3$ kg/m^3), how deep does the hull sink?

224. What size force will open a door of area 1.54 m^2 if the net pressure on the door is 1.013×10^3 Pa?

225. Gas at a pressure of 1.50×10^6 Pa exerts a force of 1.22×10^4 N on the upper surface of a piston. What is the piston's upper surface area?

226. In a barometer, the mercury column's weight equals the force from air pressure on the mercury's surface. Mercury's density is 13.6×10^3 kg/m^3. What is the air's pressure if the column is 760 mm high?

227. A cube of osmium with a volume of 166 cm^3 is placed in fresh water. The cube's apparent weight is 35.0 N. What is the density of osmium?

228. A block of ebony with a volume of 2.5×10^{-3} m^3 is placed in fresh water. If the apparent weight of the block is 7.4 N, what is the density of ebony?

229. One piston of a hydraulic lift holds 1.40×10^3 kg. The other holds an ice block ($\rho = 917$ kg/m^3) that is 0.076 m thick. Find the first piston's area.

230. A hydraulic-lift piston raises a 4.45×10^4 N weight by 448 m. How large is the force on the other piston if it is pushed 8.00 m downward?

231. A platinum flute with a density of 21.5 g/cm^3 is submerged in fresh water. If its apparent weight is 40.2 N, what is the flute's mass?

Chapter 9 Heat

232. Surface temperature on Mercury ranges from 463 K during the day to 93 K at night. Express this temperature range in degrees Celsius.

233. Solve problem 233 for degrees Fahrenheit.

234. The temperature in Fort Assiniboine, Montana, went from −5°F to +37°F on January 19, 1892. Calculate this change in temperature in kelvins.

235. An acorn falls 9.5 m, absorbing 0.85 of its initial potential energy. If 1200 J/kg will raise the acorn's temperature 1.0°C, what is its temperature increase?

236. A bicyclist on level ground brakes from 13.4 m/s to 0 m/s. What is the cyclist's and bicycle's mass if the increase in internal energy is 5836 J?

237. A 61.4 kg roller skater on level ground brakes from 20.5 m/s to 0 m/s. What is the total change in the internal energy of the system?

238. A 0.225 kg tin can ($c_p = 2.2 \times 10^3$ J/kg•°C) is cooled in water, to which it transfers 3.9×10^4 J of energy. By how much does the can's temperature change?

239. What mass of bismuth ($c_p = 121$ J/kg•°C) increases temperature by 5.0°C when 25 J are added by heat?

240. Placing a 0.250 kg pot in 1.00 kg of water raises the water's temperature 1.00°C. The pot's temperature drops 17.5°C. Find the pot's specific heat capacity.

241. Lavas at Kilauea in Hawaii have temperatures of 2192°F. Express this quantity in degrees Celsius.

242. The present temperature of the background radiation in the universe is 2.7 K. What is this temperature in degrees Celsius?

212. 15.0 m/s = 54.0 km/h
213. 165 kg
214. 0.11 N
215. 5.09×10^5 s = 141 h
216. 4.39×10^3 m/s = 4.39 km/s
217. 5.5×10^9 m = 5.5×10^6 km
218. 4.0 N
219. 1.6 N•m
220. 15.5 m
221. 6.62×10^3 N
222. 9 N
223. 0.574 m
224. 1.56×10^3 N
225. 8.13×10^{-3} m^2
226. 1.0×10^5 Pa
227. 2.25×10^4 kg/m^3
228. 1.30×10^3 kg/m^3
229. 2.0×10^1 m^2
230. 2.49×10^6 N
231. 4.30 kg
232. 1.90×10^2°C to -1.80×10^2°C
233. 374°F to −292°F
234. 24 K
235. 6.6×10^{-2}°C
236. 65.0 kg
237. 1.29×10^4 J
238. −79°C
239. 4.1×10^{-2} kg
240. 957 J/kg•°C
241. 1.200×10^3°C
242. -2.70×10^2°C

243. 315 K
244. 1.4×10^4 J
245. 1.91×10^{-2} kg = 19.1 g
246. 1.2×10^{-4} kg
247. 530 J/kg•°C
248. 820 kg
249. −930°C
250. 4.70×10^6 J
251. 1.50×10^3 Pa = 1.50 kPa
252. 1.89×10^{-2} m^3
253. 873 J
254. -1.0×10^1 J
255. 244 J
256. 2.0×10^2 J
257. 5.3×10^3 J
258. 0.189
259. 2.4×10^3 Pa = 2.4 kPa
260. 6.00×10^{-4} m^3
261. 5895 J
262. 978 kJ = 9.78×10^5 J
263. 5.30×10^2 kJ = 5.30×10^5 J
264. 2.6×10^8 J
265. 1.0×10^4 J
266. 0.220
267. −18 N
268. 1.0×10^4 N/m
269. −0.11 m = −11 cm
270. 3.57 m/s^2
271. 4.0×10^{-2} m = 4.0 cm
272. 4.993 s
273. 0.2003 Hz
274. 1.6 s
275. 730 N/m

243. The human body cannot survive at a temperature of 42°C for very long. Express this quantity in kelvins.

244. Two sticks rubbed together gain 2.15×10^4 J from kinetic energy and lose 33 percent of it to the air. How much does the sticks' internal energy change?

245. A stone falls 561.7 m. When the stone lands, the internal energy of the ground and the stone increases by 105 J. What is the stone's mass?

246. A 2.5 kg block of ice at 0.0°C slows on a level floor from 5.7 m/s to 0 m/s. If 3.3×10^5 J cause 1.0 kg of ice to melt, how much of the ice melts?

247. Placing a 3.0 kg skillet in 5.0 kg of water raises the water's temperature 2.25°C and lowers the skillet's temperature 29.6°C. Find the skillet's specific heat.

248. Air has a specific heat of 1.0×10^3 J/kg•°C. If air's temperature increases 55°C when 45×10^6 J are added to it by heat, what is the air's mass?

249. A 0.23 kg tantalum part has a specific heat capacity of 140 J/kg•°C. By how much does the part's temperature change if it gives up 3.0×10^4 J as heat?

Chapter 10 Thermodynamics

250. A volume of air increases 0.227 m^3 at a net pressure of 2.07×10^7 Pa. How much work is done on the air?

251. The air in a hot-air balloon does 3.29×10^6 J of work, increasing the balloon's volume by 2190 m^3. What is the net pressure in the balloon?

252. Filling a fire extinguisher with nitrogen gas at a net pressure of 25.0 kPa requires 472.5 J of work on the gas. Find the change in the gas's volume.

253. The internal energy of air in a closed car rises 873 J. How much heat energy is transferred to the air?

254. A system's initial internal energy increases from 39 J to 163 J. If 114 J of heat are added to the system, how much work is done on the system?

255. A gas does 623 J of work on its surroundings when 867 J are added to the gas as heat. What is the change in the internal energy of the gas?

256. An engine with an efficiency of 0.29 takes in 693 J as heat. How much work does the engine do?

257. An engine with an efficiency of 0.19 does 998 J of work. How much energy is taken in by heat?

258. Find the efficiency of an engine that receives 571 J as heat and loses 463 J as heat per cycle.

259. A 5.4×10^{-4} m^3 increase in steam's volume does 1.3 J of work on a piston. What is the pressure?

260. A pressure of 655 kPa does 393 J of work inflating a bike tire. Find the change in volume.

261. An engine's internal energy changes from 8093 J to 2.0920×10^4 J. If 6932 J are added as heat, how much work is done on or by the system?

262. Steam expands from a geyser to do 192 kJ of work. If the system's internal energy increases by 786 kJ, how much energy is transferred as heat?

263. If 632 kJ are added to a boiler and 102 kJ of work are done as steam escapes from a safety valve, what is the net change in the system's internal energy?

264. A power plant with an efficiency of 0.35 percent requires 7.37×10^8 J of energy as heat. How much work is done by the power plant?

265. An engine with an efficiency of 0.11 does 1150 J of work. How much energy is taken in as heat?

266. A test engine performs 128 J of work and receives 581 J of energy as heat. What is the engine's efficiency?

Chapter 11 Vibrations and Waves

267. A scale with a spring constant of 420 N/m is compressed 4.3 cm. What is the spring force?

268. A 669 N weight attached to a giant spring stretches it 6.5 cm. What is the spring constant?

269. An archer applies a force of 52 N on a bowstring with a spring constant of 490 N/m. What is the bowstring's displacement?

270. On Mercury, a pendulum 1.14 m long would have a 3.55 s period. Calculate a_g for Mercury.

271. Find the length of a pendulum that oscillates with a frequency of 2.5 Hz.

272. Calculate the period of a 6.200 m long pendulum in Oslo, Norway, where $a_g = 9.819$ m/s^2.

273. Find the pendulum's frequency in problem 272.

274. A 24 kg child jumps on a trampoline with a spring constant of 364 N/m. What is the oscillation period?

275. A 32 N weight oscillates with a 0.42 s period when on a spring scale. Find the spring constant.

276. Find the mass of a ball that oscillates at a period of 0.079 s on a spring with a constant of 63 N/m.

277. A dolphin hears a 280 kHz sound with a wavelength of 0.51 cm. What is the wave's speed?

278. If a sound wave with a frequency of 20.0 Hz has a speed of 331 m/s, what is its wavelength?

279. A sound wave has a speed of 2.42×10^4 m/s and a wavelength of 1.1 m. Find the wave's frequency.

280. An elastic string with a spring constant of 65 N/m is stretched 15 cm and released. What is the spring force exerted by the string?

281. The spring in a seat compresses 7.2 cm under a 620 N weight. What is the spring constant?

282. A 3.0 kg mass is hung from a spring with a spring constant of 36 N/m. Find the displacement.

283. Calculate the period of a 2.500 m long pendulum in Quito, Ecuador, where $a_g = 9.780$ m/s^2.

284. How long is a pendulum with a frequency of 0.50 Hz?

285. A tractor seat supported by a spring with a spring constant of 2.03×10^3 N/m oscillates at a frequency of 0.79 Hz. What is the mass on the spring?

286. An 87 N tree branch oscillates with a period of 0.64 s. What is the branch's spring constant?

287. What is the oscillation period for an 8.2 kg baby in a seat that has a spring constant of 221 N/m?

288. An organ creates a sound with a speed of 331 m/s and a wavelength of 10.6 m. Find the frequency.

289. What is the speed of an earthquake s-wave with a 2.3×10^4 m wavelength and a 0.065 Hz frequency?

Chapter 12 Sound

290. What is the distance from a sound with 5.88×10^{-5} W power if its intensity is 3.9×10^{-6} W/m^2?

291. Sound waves from a stereo have a power output of 3.5 W at 0.50 m. What is the sound's intensity?

292. What is a vacuum cleaner's power output if the sound's intensity 1.5 m away is 4.5×10^{-4} W/m^2?

293. Waves travel at 499 m/s on a 0.850 m long cello string. Find the string's fundamental frequency.

294. A mandolin string's first harmonic is 392 Hz. How long is the string if the wave speed on it is 329 m/s?

295. A 1.53 m long pipe that is closed on one end has a seventh harmonic frequency of 466.2 Hz. What is the speed of the waves in the pipe?

296. A pipe open at both ends has a fundamental frequency of 125 Hz. If the pipe is 1.32 m long, what is the speed of the waves in the pipe?

297. Traffic has a power output of 1.57×10^{-3} W. At what distance is the intensity 5.20×10^{-3} W/m^2?

298. If a mosquito's buzzing has an intensity of 9.3×10^{-8} W/m^2 at a distance of 0.21 m, how much sound power does the mosquito generate?

299. A note from a flute (a pipe with a closed end) has a first harmonic of 392.0 Hz. How long is the flute if the sound's speed is 331 m/s?

300. An organ pipe open at both ends has a first harmonic of 370.0 Hz when the speed of sound is 331 m/s. What is the length of this pipe?

Chapter 13 Light and Reflection

301. A 7.6270×10^8 Hz radio wave has a wavelength of 39.296 cm. What is this wave's speed?

302. An X ray's wavelength is 3.2 nm. Using the speed of light in a vacuum, calculate the frequency of the X ray.

303. What is the wavelength of ultraviolet light with a frequency of 9.5×10^{14} Hz?

304. A concave mirror has a focal length of 17 cm. Where must a 2.7 cm tall coin be placed for its image to appear 23 cm in front of the mirror's surface?

305. How tall is the coin's image in problem 304?

306. A concave mirror's focal length is 9.50 cm. A 3.0 cm tall pin appears to be 15.5 cm in front of the mirror. How far from the mirror is the pin?

307. How tall is the pin's image in problem 306?

308. A convex mirror's magnification is 0.11. Suppose you are 1.75 m tall. How tall is your image?

309. How far in front of the mirror in problem 308 are you if your image is 42 cm behind the mirror?

310. A mirror's focal length is −12 cm. What is the object distance if an image forms 9.00 cm behind the surface of the mirror?

311. What is the magnification in problem 310?

312. A metal bowl is like a concave spherical mirror. You are 35 cm in front of the bowl and see an image at 42 cm. What is the bowl's focal length?

276. 1.0×10^{-2} kg $= 1.0 \times 10^1$ g
277. 1.4×10^3 m/s
278. 16.6 m
279. 2.2×10^4 Hz
280. 9.8 N
281. 8.6×10^3 N/m
282. −0.82 m = −82 cm
283. 3.177 s
284. 0.99 m = 99 cm
285. 82 kg
286. 850 N/m
287. 1.2 s
288. 31.2 Hz
289. 1.5×10^3 m/s
290. 1.1 m
291. 1.1 W/m^2
292. 1.3×10^{-2} W
293. 294 Hz
294. 0.420 m
295. 408 m/s
296. 3.30×10^2 m/s
297. 0.155 m
298. 5.2×10^{-8} W
299. 0.211 m = 21.1 cm
300. 0.447 m = 44.7 cm
301. 2.9971×10^8 m/s
302. 9.4×10^{16} Hz
303. 3.2×10^{-7} m = 320 nm
304. 65 cm
305. −0.96 cm
306. 24.7 cm
307. −1.9 cm
308. 0.19 m
309. 3.8 m
310. 36 cm
311. 0.25
312. 19 cm

313. 38 cm
314. −84.0 cm
315. 2.40
316. −7.7 cm
317. 0.98 cm
318. 5.67×10^{18} Hz
319. 10.5 cm
320. 17 cm to 15 cm
321. 64.0 cm in front of the mirror
322. 1.0×10^{1} cm
323. 8.3 cm
324. −17 cm
325. 0.40
326. −5.6 cm
327. −11 cm
328. 4.1 cm
329. 2.9979×10^{8} m/s
330. 1.2 cm
331. 33.3 cm
332. 27 cm
333. 0.19
334. 63°
335. 32.2°
336. 1.52
337. −10.4 cm
338. −0.50
339. 18 cm
340. 9.0 cm
341. 1.63
342. 34.49°
343. 39.38°

313. For problem 312, find the bowl's radius of curvature.

314. A concave spherical mirror on a dressing table has a focal length of 60.0 cm. If someone sits 35.0 cm in front of it, where is the image?

315. What is the magnification in problem 314?

316. An image appears 5.2 cm behind the surface of a convex mirror when the object is 17 cm in front of the mirror. What is the mirror's focal length?

317. If the object in problem 316 is 3.2 cm tall, how tall is its image?

318. In order for someone to observe an object, the wavelength of the light must be smaller than the object. The Bohr radius of a hydrogen atom is $5.291\,770 \times 10^{-11}$ m. What is the lowest frequency that can be used to locate a hydrogen atom?

319. Meteorologists use Doppler radar to watch the movement of storms. If a weather station uses electromagnetic waves with a frequency of 2.85×10^{9} Hz, what is the wavelength of the radiation?

320. PCS cellular phones have antennas that use radio frequencies from 1800–2000 MHz. What range of wavelengths corresponds to these frequencies?

321. Suppose you have a mirror with a focal length of 32.0 cm. Where would you place your right hand so that you appear to be shaking hands with yourself?

322. A car's headlamp is made of a light bulb in front of a concave spherical mirror. If the bulb is 5.0 cm in front of the mirror, what is the radius of the mirror?

323. Suppose you are 19 cm in front of the bell of your friend's trumpet and you see your image at 14 cm. If the trumpet's bell is a concave mirror, what would be its focal length?

324. A soup ladle is like a spherical convex mirror with a focal length of 27 cm. If you are 43 cm in front of the ladle, where does the image appear?

325. What is the magnification in problem 324?

326. Just after you dry a spoon, you look into the convex part of the spoon. If the spoon has a focal length of −8.2 cm and you are 18 cm in front of the spoon, where does the image appear?

327. The base of a lamp is made of a convex spherical mirror with a focal length of −39 cm. Where does the image appear when you are 16 cm from the base?

328. Consider the lamp and location in problem 327. If your nose is 6.0 cm long, how long does the image appear?

329. How fast does microwave radiation that has a frequency of $1.173\,06 \times 10^{11}$ Hz and a wavelength of 2.5556 mm travel?

330. Suppose the microwaves in your microwave oven have a frequency of 2.5×10^{10} Hz. What is the wavelength of these microwaves?

331. You place an electric heater 3.00 m in front of a concave spherical mirror that has a focal length of 30.0 cm. Where would your hand feel warmest?

332. You see an image of your hand as you reach for a doorknob with a focal length of 6.3 cm. How far from the doorknob is your hand when the image appears at 5.1 cm behind the doorknob?

333. What is the magnification of the image in problem 332?

Chapter 14 Refraction

334. A ray of light in air enters an amethyst crystal ($n = 1.553$). If the angle of refraction is 35°, what is the angle of incidence?

335. Light passes from air at an angle of incidence of 59.2° into a nephrite jade vase ($n = 1.61$). Determine the angle of refraction in the jade.

336. Light entering a pearl travels at a speed of 1.97×10^{8} m/s. What is the pearl's index of refraction?

337. An object in front of a diverging lens of focal length 13.0 cm forms an image with a magnification of +5.00. How far from the lens is the object placed?

338. An object with a height of 18 cm is placed in front of a converging lens. The image height is −9.0 cm. What is the magnification of the lens?

339. If the focal length of the lens in problem 338 is 6.0 cm, how far in front of the lens is the object?

340. Where does the image appear in problem 339?

341. The critical angle for light traveling from a green tourmaline gemstone into air is 37.8°. What is tourmaline's index of refraction?

342. Find the critical angle for light traveling from ruby ($n = 1.766$) into air.

343. Find the critical angle for light traveling from emerald ($n = 1.576$) into air.

344. Malachite has two indices of refraction: $n_1 = 1.91$ and $n_2 = 1.66$. A ray of light in air enters malachite at an incident angle of 35.2°. Calculate both of the angles of refraction.

345. A ray of light in air enters a serpentine figurine ($n = 1.555$). If the angle of refraction is 33°, what is the angle of incidence?

346. The critical angle for light traveling from an aquamarine gemstone into air is 39.18°. What is the index of refraction for aquamarine?

347. A 15 cm tall object is placed 44 cm in front of a diverging lens. A virtual image appears 14 cm in front of the lens. What is the lens's focal length?

348. What is the image height in problem 347?

349. A lighthouse converging lens has a focal length of 4 m. What is the image distance for an object placed 4 m in front of the lens?

350. What is the magnification in problem 349?

351. Light moves from olivine ($n = 1.670$) into onyx. If the critical angle for olivine is 62.85°, what is the index of refraction for onyx?

352. When light in air enters an opal mounted on a ring, the light travels at a speed of 2.07×10^8 m/s. What is opal's index of refraction?

353. When light in air enters albite, it travels at a velocity of 1.95×10^8 m/s. What is albite's index of refraction?

354. A searchlight is constructed by placing a 500 W bulb 0.5 m in front of a converging lens. The focal length of the lens is 0.5 m. What is the image distance?

355. A microscope slide is placed in front of a converging lens with a focal length of 3.6 cm. The lens forms a real image of the slide 15.2 cm behind the lens. How far is the lens from the slide?

356. Where must an object be placed to form an image 12 cm in front of a diverging lens with a focal length of 44 cm?

357. The critical angle for light traveling from almandine garnet into air ranges from 33.1° to 35.3°. Calculate the range of almandine garnet's index of refraction.

358. Light moves from a clear andalusite ($n = 1.64$) crystal into ivory. If the critical angle for andalusite is 69.9°, what is the index of refraction for ivory?

Chapter 15 Interference and Diffraction

359. Light with a 587.5 nm wavelength passes through two slits. A second-order bright fringe forms 0.130° from the center. Find the slit separation.

360. Light passing through two slits with a separation of 8.04×10^{-6} m forms a third bright fringe 13.1° from the center. Find the wavelength.

361. Two slits are separated by 0.0220 cm. Find the angle at which a first-order bright fringe is observed for light with a wavelength of 527 nm.

362. For 546.1 nm light, the first-order maximum for a diffraction grating forms at 75.76°. How many lines per centimeter are on the grating?

363. Infrared light passes through a diffraction grating of 3600 lines/cm. The angle of the third-order maximum is 76.54°. What is the wavelength?

364. A diffraction grating with 1950 lines/cm is used to examine light with a wavelength of 497.3 nm. Find the angle of the first-order maximum.

365. At what angle does the second-order maximum in problem 364 appear?

366. Light passes through two slits separated by 3.92×10^{-6} m to form a second-order bright fringe at an angle of 13.1°. What is the light's wavelength?

367. Light with a wavelength of 430.8 nm shines on two slits that are 0.163 mm apart. What is the angle at which a second dark fringe is observed?

368. Light of wavelength 656.3 nm passes through two slits. The fourth-order dark fringe is 0.548° from the central maximum. Find the slit separation.

369. The first-order maximum for light with a wavelength of 447.1 nm is found at 40.25°. How many lines per centimeter does the grating have?

370. Light through a diffraction grating of 9550 lines/cm forms a second-order maximum at 54.58°. What is the wavelength of the light?

Chapter 16 Electric Forces and Fields

371. Charges of −5.3 μC and +5.3 μC are separated by 4.2 cm. Find the electric force between them.

372. A dog's fur is combed, and the comb gains a charge of 8.0 nC. Find the electric force between the fur and comb when they are 2.0 cm apart.

373. Two equal charges are separated by 6.5×10^{-11} m. If the magnitude of the electric force between the charges is 9.92×10^{-4} N, what is the value of q?

344. 17.6° and 20.3°
345. 58°
346. 1.583
347. −21 cm
348. 4.8 cm
349. ∞
350. ∞
351. 1.486
352. 1.45
353. 1.54
354. ∞
355. 4.8 cm
356. 17 cm
357. 1.73 to 1.83
358. 1.54
359. 5.18×10^{-4} m = 0.518 mm
360. 6.07×10^{-7} m = 607 nm
361. 0.137°
362. 1.775×10^4 lines/cm
363. 9.0×10^{-7} m = 9.0×10^2 nm
364. 5.56°
365. 11.2°
366. 4.44×10^{-7} m = 444 nm
367. 0.227°
368. 2.40×10^{-4} m = 0.240 mm
369. 1.445×10^4 lines/cm
370. 4.27×10^{-7} m = 427 nm
371. 140 N attractive
372. 1.4×10^{-3} N attractive
373. 2.2×10^{-17} C

374. 0.387 m = 38.7 cm

375. 0.00 N

376. 1.91×10^{-10} N, 45.0° above the horizontal

377. 4.0×10^{-8} N, 9.3° below the negative x-axis

378. 0.16 m = 16 cm

379. 260 N from either charge

380. −54 N (54 N along the negative x-axis)

381. 1.6×10^{-12} C

382. 3.15×10^{5} N/C, upward

383. 0.585 m = 58.5 cm

384. 1.60×10^{-19} C

385. 3.97×10^{-6} N, upward

386. 1.74×10^{-7} N, 21.2° above the negative x-axis

387. 0.073 m = 7.3 cm

388. 9.2 C

389. 7.5×10^{-6} N, along the positive y-axis

390. 5.06×10^{-12} C

391. 4.40×10^{5} N/C, 89.1° above the negative x-axis

392. 1.66×10^{-10} m

393. −7.4 C

394. 1.6×10^{-19} C

395. 1.6×10^{-19} C

396. 4.82×10^{-19} N, 45° above the positive x-axis

397. 36 cm

398. −512 C

399. 4.4×10^{-4} J

400. 56 C

374. Two point charges of −13.0 µC and −16.0 µC exert repulsive forces on each other of 12.5 N. What is the distance between the two charges?

375. Three equal point charges of 4.00 nC lie 4.00 m apart on a line. Calculate the magnitude and direction of the net force on the middle charge.

376. A proton is at each corner of a square with sides 1.52×10^{-9} m long. Calculate the resultant force vector on the proton at the upper right corner.

377. Three 2.0 nC charges are located at coordinates (0 m, 0 m), (1.0 m, 0 m), and (1.0 m, 2.0 m). Find the resultant force on the first charge.

378. Charges of 7.2 nC and 6.7 nC are 32 cm apart. Find the equilibrium position for a −3.0 nC charge.

379. A −12.0 µC charge is between two 6.0 µC charges, 5.0 cm away from each. What electric force keeps the central charge in equilibrium?

380. A 9.0 N/C electric field is directed along the x-axis. Find the electric force vector on a −6.0 C charge.

381. What charge experiences an electric force of 6.43×10^{-9} N in an electric field of 4.0×10^{3} N/C?

382. A 5.00 µC charge is 0.500 m above a 15.0 µC charge. Calculate the electric field at a point 1.00 m above the 15.0 mC charge.

383. Two static point charges of 99.9 µC and 33.3 µC exert repulsive forces on each other of 87.3 N. What is the distance between the two charges?

384. Two particles are separated by 9.30×10^{-11} m. If the magnitude of the electric force between the charges is 2.66×10^{-8} N, what is the value of q?

385. A −23.4 nC charge is 0.500 m below a 4.65 nC charge and 1.00 m below a 0.299 nC charge. Find the resultant force vector on the −23.4 nC charge.

386. Three point charges are on the corners of a triangle: q_1 = −9.00 nC is at the origin; q_2 = −8.00 nC is at x = 2.00 m; and q_3 = 7.00 nC is at y = 3.00 m. Find the magnitude and direction of the resultant force on q_1.

387. Charges of −2.50 nC and −7.50 nC are 20.0 cm apart. Find a 5.0 nC charge's equilibrium position.

388. A −4.6 C charge is in equilibrium with a −2.3 C charge 2.0 m to the right, and an unknown charge 4.0 m to the right. What is the unknown charge?

389. Find the electric force vector on a 5.0 nC charge in a 1500 N/C electric field directed along the y-axis.

390. What electric charge experiences an 8.42×10^{-9} N electric force in an electric field of 1663 N/C?

391. Two 3.00 µC charges lie 2.00 m apart on the x-axis. Find the resultant electric field vector at a point 0.250 m on the y-axis, above the charge on the left.

392. Two electrons are 2.00×10^{-10} m and 3.00×10^{-10} m, respectively, from a point. Where with respect to that point must a proton be placed so that the resultant electric field strength is zero?

393. A −7.0 C charge is in equilibrium with a 49 C charge 18 m to the right and an unknown charge 25 m to the right. What is the unknown charge?

394. Suppose two pions are separated by 8.3×10^{-10} m. If the magnitude of the electric force between the charges is 3.34×10^{-10} N, what is the value of q?

395. Suppose two muons having equal but opposite charge are separated by 6.4×10^{-8} m. If the magnitude of the electric force between the charges is 5.62×10^{-14} N, what is the value of q?

396. Consider four electrons at the corners of a square. Each side of the square is 3.02×10^{-5} m. Find the magnitude and direction of the resultant force on q_3 if it is at the origin.

397. A charge of 5.5 nC and a charge of 11 nC are separated by 88 cm. Find the equilibrium position for a −22 nC charge.

398. Three charges are on the y-axis. At the origin is a charge, q_1 = 72 C; an unknown charge, q_2, is at y = 15 mm. A third charge, q_3 = −8.0 C, is placed at y = −9.0 mm so that it is in electrostatic equilibrium with q_1 and q_2. What is the charge on q_2?

Chapter 17 Electrical Energy and Current

399. A helium-filled balloon with a 14.5 nC charge rises 290 m above Earth's surface. By how much does the electrical potential energy change if Earth's electric field is −105 N/C?

400. A charged airplane rises 7.3 km in a 3.4×10^{5} N/C electric field. The electrical potential energy changes by -1.39×10^{11} J. What is the charge on the plane?

401. Earth's radius is 6.4×10^6 m. What is Earth's capacitance if it is regarded as a conducting sphere?

402. A 0.50 pF capacitor is connected across a 1.5 V battery. How much charge can this capacitor store?

403. A 76 C charge passes through a wire's cross-sectional area in 19 s. Find the current in the wire.

404. The current in a telephone is 1.4 A. How long does 98 C of charge take to pass a point in the wire?

405. What is a television's total resistance if it is plugged into a 120 V outlet and carries 0.75 A of current?

406. A motor with a resistance of 12.2 Ω is plugged into a 120.0 V outlet. What is the current in the motor?

407. The potential difference across a motor with a 0.30 Ω resistance is 720 V. How much power is used?

408. What is a microwave oven's resistance if it uses 1750 W of power at a voltage of 120.0 V?

409. A 64 nC charge moves 0.95 m with an electrical potential energy change of -3.88×10^{-5} J. What is the electric field strength?

410. A −14 nC charge travels through a 156 N/C electric field with a change of 2.1×10^{-6} J in the electrical potential energy. How far does the charge travel?

411. A 5.0×10^{-5} F polyester capacitor stores 6.0×10^{-4} C. Find the potential difference across the capacitor.

412. Some ceramic capacitors can store 3×10^{-2} C with a potential difference of 30 kV across them. What is the capacitance of such a capacitor?

413. The area of the plates in a 4550 pF parallel-plate capacitor is 6.4×10^{-3} m^2. Find the plate separation.

414. A television receiver contains a 14 μF capacitor charged across a potential difference of 1.5×10^4 V. How much charge does this capacitor store?

415. A photocopier uses 9.3 A in 15 s. How much charge passes a point in the copier's circuit in this time?

416. A 114 μC charge passes through a gold wire's cross-sectional area in 0.36 s. What is the current?

417. If the current in a blender is 7.8 A, how long do 56 C of charge take to pass a point in the circuit?

418. A computer uses 3.0 A in 2.0 min. How much charge passes a point in the circuit in this time?

419. A battery-powered lantern has a resistance of 6.4 Ω. What potential difference is provided by the battery if the total current is 0.75 A?

420. The potential difference across an electric eel is 650 V. How much current would an electric eel deliver to a body with a resistance of 1.0×10^2 Ω?

421. If a garbage-disposal motor has a resistance of 25.0 Ω and carries a current of 4.66 A, what is the potential difference across the motor's terminals?

422. A medium-sized oscillating fan draws 545 mA of current when the potential difference across its motor is 120 V. How large is the fan's resistance?

423. A generator produces a 2.5×10^4 V potential difference across power lines that carry 20.0 A of current. How much power is generated?

424. A computer with a resistance of 91.0 Ω uses 230.0 W of power. Find the current in the computer.

425. A laser uses 6.0×10^{13} W of power. What is the potential difference across the laser's circuit if the current in the circuit is 8.0×10^6 A?

426. A blender with a 75 Ω resistance uses 350 W of power. What is the current in the blender's circuit?

Chapter 18 Circuits and Circuit Elements

427. A theater has 25 surround-sound speakers wired in series. Each speaker has a resistance of 12.0 Ω. What is the equivalent resistance?

428. In case of an emergency, a corridor on an airplane has 57 lights wired in series. Each light bulb has a resistance of 2.00 Ω. Find the equivalent resistance.

429. Four resistors with resistances of 39 Ω, 82 Ω, 12 Ω, and 42 Ω are connected in parallel across a 3.0 V potential difference. Find the equivalent resistance.

430. Four resistors with resistances of 33 Ω, 39 Ω, 47 Ω, and 68 Ω are connected in parallel across a 1.5 V potential difference. Find the equivalent resistance.

431. A 16 Ω resistor is connected in series with another resistor across a 12 V battery. The current in the circuit is 0.42 A. Find the unknown resistance.

432. A 24 Ω resistor is connected in series with another resistor across a 3.0 V battery. The current in the circuit is 62 mA. Find the unknown resistance.

401. 7.1×10^{-4} F
402. 7.5×10^{-13} C
403. 4.0 A
404. 7.0×10^1 s
405. 160 Ω
406. 9.84 A
407. 1.7×10^6 W = 1.7 MW
408. 8.23 Ω
409. 6.4×10^2 N/C
410. 0.96 m
411. 12 V
412. 1 μF
413. 1.2×10^{-5} m
414. 0.21 C
415. 1.4×10^2 C
416. 3.2×10^{-4} A = 0.32 mA
417. 7.2 s
418. 3.6×10^2 C
419. 4.8 V
420. 6.5 A
421. 116 V
422. 220 Ω
423. 5.0×10^5 W = 0.50 MW
424. 1.59 A
425. 7.5×10^6 V
426. 2.2 A
427. 3.00×10^2 Ω
428. 114 Ω
429. 6.0 Ω
430. 11 Ω
431. 13 Ω
432. 24 Ω

433. 6.0 Ω
434. 4.0 Ω
435. 0.056 A = 56 mA
436. 0.665 A = 665 mA
437. 1.6 A (refrigerator); 1.3 A (oven)
438. 1.5 A (computer); 5.0 A (printer)
439. 12.6 Ω
440. 0.952 A
441. 2.6 V
442. 0.43 A
443. 9.4 Ω
444. 1.6 A
445. 1.6 A
446. 16.6 Ω
447. 1.45 A
448. 0.97 A
449. 4×10^{-12} N
450. 3.0×10^{-10} T
451. 7.6×10^{6} m/s
452. 1×10^{-2} N
453. 0.70 A
454. 0.63 m
455. 5.1×10^{-4} T
456. 0.30 T

433. A 3.3 Ω resistor and another resistor are connected in parallel across a 3.0 V battery. The current in the circuit is 1.41 A. Find the unknown resistance.

434. A 56 Ω resistor and another resistor are connected in parallel across a 12 V battery. The current in the circuit is 3.21 A. Find the unknown resistance.

435. Three bulbs with resistances of 56 Ω, 82 Ω, and 24 Ω are wired in series. If the voltage across the circuit is 9.0 V, what is the current in the circuit?

436. Three bulbs with resistances of 96 Ω, 48 Ω, and 29 Ω are wired in series. What is the current through the bulbs if the voltage across them is 115 V?

437. A refrigerator ($R_1 = 75$ Ω) wired in parallel with an oven ($R_2 = 91$ Ω) is plugged into a 120 V outlet. What is the current in the circuit of each appliance?

438. A computer ($R_1 = 82$ Ω) and printer ($R_2 = 24$ Ω) are wired in parallel across a 120 V potential difference. Find the current in each machine's circuit.

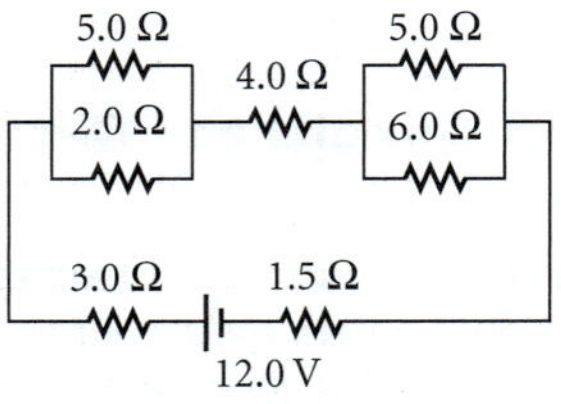

439. For the figure above, what is the equivalent resistance of the circuit?

440. For the figure above, find the current in the circuit.

441. For the figure above, what is the potential difference across the 6.0 Ω resistor?

442. For the figure above, what is the current through the 6.0 Ω resistor?

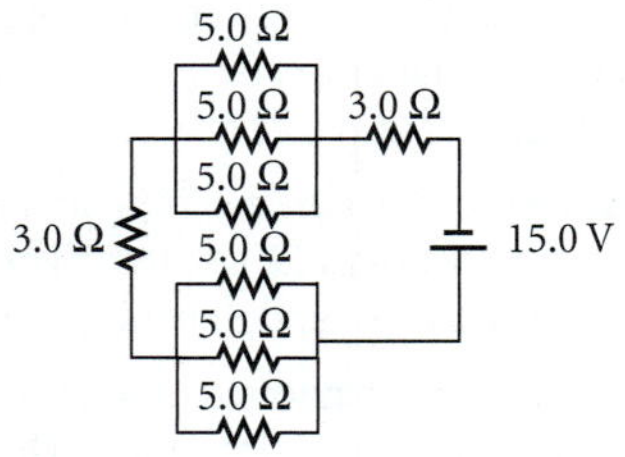

443. For the figure above, calculate the equivalent resistance of the circuit.

444. For the figure above, what is the total current in the circuit?

445. For the figure above, what is the current in the 3.0 Ω resistors?

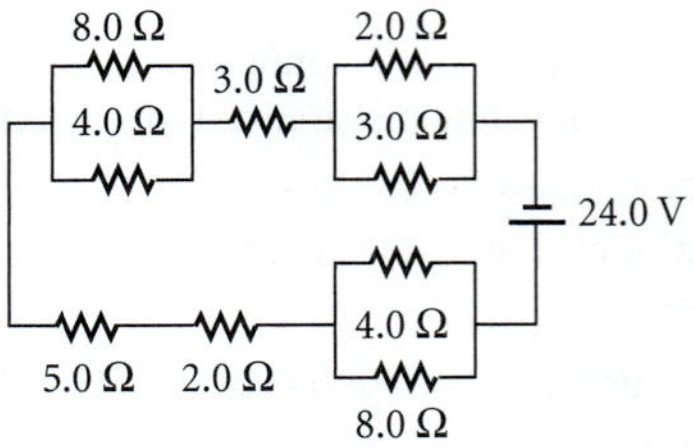

446. For the figure above, calculate the equivalent resistance of the circuit.

447. For the figure above, what is the total current in the circuit?

448. For the figure above, what is the current in either of the 8.0 Ω resistors?

Chapter 19 Magnetism

449. A proton moves at right angles to a magnetic field of 0.8 T. If the proton's speed is 3.0×10^{7} m/s, how large is the magnetic force exerted on the proton?

450. A weak magnetic field exerts a 1.9×10^{-22} N force on an electron moving 3.9×10^{6} m/s perpendicular to the field. What is the magnetic field strength?

451. A 5.0×10^{-5} T magnetic field exerts a 6.1×10^{-17} N force on a 1.60×10^{-19} C charge, which moves at a right angle to the field. What is the charge's speed?

452. A 14 A current passes through a 2 m wire. A 3.6×10^{-4} T magnetic field is at right angles to the wire. What is the magnetic force on the wire?

453. A 1.0 m printer cable is perpendicular to a 1.3×10^{-4} T magnetic field. What current must the cable carry to experience a 9.1×10^{-5} N magnetic force?

454. A wire perpendicular to a 4.6×10^{-4} T magnetic field experiences a 2.9×10^{-3} N magnetic force. How long is the wire if it carries a 10.0 A current?

455. A 12 m wire carries a 12 A current. What magnetic field causes a 7.3×10^{-2} N magnetic force to act on the wire when it is perpendicular to the field?

456. A magnetic force of 3.7×10^{-13} N is exerted on an electron moving at 7.8×10^{6} m/s perpendicular to a sunspot. How large is the sunspot's magnetic field?

457. An electron moves with a speed of 2.2×10^6 m/s at right angles through a 1.1×10^{-2} T magnetic field. How large is the magnetic force on the electron?

458. A pulsar's magnetic field is 1×10^{-8} T. How fast does an electron move perpendicular to this field so that a 3.2×10^{-22} N magnetic force acts on the charge?

459. A levitation device designed to suspend 75 kg uses 10.0 m of wire and a 4.8×10^{-4} T magnetic field, perpendicular to the wire. What current is needed?

460. A power line carries 1.5×10^3 A for 15 km. Earth's magnetic field is 2.3×10^{-5} T at a 45° angle to the power line. What is the magnetic force on the line?

Chapter 20 Electromagnetic Induction

461. A coil with 540 turns and a 0.016 m^2 area is rotated exactly from 0° to 90.0° in 0.050 s. How strong must a magnetic field be to induce an emf of 3.0 V?

462. A 550-turn coil with an area of 5.0×10^{-5} m^2 is in a magnetic field that decreases by 2.5×10^{-4} T in 2.1×10^{-5} s. What is the induced emf in the coil?

463. A 246-turn coil has a 0.40 m^2 area in a magnetic field that increases from 0.237 T to 0.320 T. What time interval is needed to induce an emf of −9.1 V?

464. A 9.5 V emf is induced in a coil that rotates from 0.0° to 90.0° in a 1.25×10^{-2} T magnetic field for 25 ms. The coil's area is 250 cm^2. How many turns of wire are in the coil?

465. A generator provides a rms emf of 320 V across 100 Ω. What is the maximum emf?

466. Find the rms current in the circuit in problem 465.

467. Some wind turbines can provide an rms current of 1.3 A. What is the maximum ac current?

468. A transformer has 1400 turns on the primary and 140 turns on the secondary. What is the voltage across the primary if secondary voltage is 6.9 kV?

469. A transformer has 140 turns on the primary and 840 turns on the secondary. What is the voltage across the secondary if the primary voltage is 5.6 kV?

470. A step-down transformer converts a 3.6 kV voltage to 1.8 kV. If the primary (input) coil has 58 turns, how many turns does the secondary have?

471. A step-up transformer converts a 4.9 kV voltage to 49 kV. If the secondary (output) coil has 480 turns, how many turns does the primary have?

472. A 320-turn coil rotates from 0° to 90.0° in a 0.046 T magnetic field in 0.25 s, which induces an average emf of 4.0 V. What is the area of the coil?

473. A 180-turn coil with a 5.0×10^{-5} m^2 area is in a magnetic field that decreases by 5.2×10^{-4} T in 1.9×10^{-5} s. What is the induced current if the coil's resistance is 1.0×10^2 W?

474. A generator provides a maximum ac current of 1.2 A and a maximum output emf of 211 V. Calculate the rms potential difference.

475. Calculate the rms current for problem 474.

476. A generator can provide a maximum output emf of 170 V. Calculate the rms potential difference.

477. A step-down transformer converts 240 V across the primary to 5.0 V across the secondary. What is the step-down ratio (N_1:N_2)?

Chapter 21 Atomic Physics

478. Determine the energy of a photon of green light with a wavelength of 527 nm.

479. Calculate the de Broglie wavelength of an electron with a velocity of 2.19×10^6 m/s.

480. Calculate the frequency of ultraviolet (UV) light having a photon energy of 20.7 eV.

481. X-ray radiation can have an energy of 12.4 MeV. To what wavelength does this correspond?

482. Light of wavelength 240 nm shines on a potassium surface. Potassium has a work function of 2.3 eV. What is the maximum kinetic energy of the photoelectrons?

483. Manganese has a work function of 4.1 eV. What is the wavelength of the photon that will just have the threshold energy for manganese?

484. What is the speed of a proton with a de Broglie wavelength of 2.64×10^{-14} m?

485. A cheetah can run as fast as 28 m/s. If the cheetah has a de Broglie wavelength of 8.97×10^{-37} m, what is the cheetah's mass?

486. What is the energy of a photon of blue light with a wavelength of 430.8 nm?

457. 3.9×10^{-15} N
458. 2×10^5 m/s
459. 1.5×10^5 A
460. 3.7×10^2 N
461. 1.7×10^{-2} T
462. 0.33 V
463. 0.90 s
464. 7.6×10^2 turns
465. 450 V
466. 3 A
467. 1.8 A
468. 6.9×10^4 V = 69 kV
469. 3.4×10^4 V = 34 kV
470. 29 turns
471. 48 turns
472. 6.8×10^{-2} m^2
473. 2.5×10^{-3} A = 2.5 mA
474. 149 V
475. 0.85 A
476. 120 V
477. 48:1
478. 3.77×10^{-19} J
479. 3.32×10^{-10} m
480. 5.00×10^{15} Hz
481. 1.00×10^{-13} m
482. 2.9 eV
483. 3.0×10^{-7} m
484. 1.50×10^7 m/s
485. 26 kg
486. 4.62×10^{-19} J

487. 4.30×10^{14} Hz
488. 0.40 m
489. 6.0×10^{14} Hz
490. 0.24 eV
491. 4.0×10^{-21} kg
492. 5.64 eV
493. 2.5×10^{-43} m
494. 2.5×10^{-7} m
495. 7.72×10^{14} Hz
496. 7.1×10^{6} m/s
497. 333.73 MeV
498. 363.89 MeV
499. 0.543 705 u
500. $^{208}_{82}Pb$
501. $^{16}_{8}O$
502. $^{4}_{2}He$
503. 15.0 s
504. 1.6×10^{-9} Ci
505. 31.92 h
506. 1020.6 MeV
507. 35.46 MeV
508. 0.600216 u
509. $^{131}_{53}I$
510. $^{4}_{2}He$
511. $^{111}_{54}Xe$
512. 1.25 days
513. 924 days
514. $8.1 \times 10^{-9}\ s^{-1}$

487. Calculate the frequency of infrared (IR) light with a photon energy of 1.78 eV.

488. Calculate the wavelength of a radio wave that has a photon energy of 3.1×10^{-6} eV.

489. Light of frequency 6.5×10^{14} Hz illuminates a lithium surface. The ejected photoelectrons are found to have a maximum kinetic energy of 0.20 eV. Find the threshold frequency of this metal.

490. Light of wavelength 519 nm shines on a rubidium surface. Rubidium has a work function of 2.16 eV. What is the maximum kinetic energy of the photoelectrons?

491. The smallest known virus moves across a Petri dish at 5.6×10^{-6} m/s. If the de Broglie wavelength of the virus is 2.96×10^{-8} m, what is the virus's mass?

492. The threshold frequency of platinum is 1.36×10^{15} Hz. What is the work function of platinum?

493. The ship *Queen Elizabeth II* has a mass of 7.6×10^{7} kg. Calculate the de Broglie wavelength if this ship sails at 35 m/s.

494. Cobalt has a work function of 5.0 eV. What is the wavelength of the photon that will just have the threshold energy for cobalt?

495. Light of frequency 9.89×10^{14} Hz illuminates a calcium surface. The ejected photoelectrons are found to have a maximum kinetic energy of 0.90 eV. Find the threshold frequency of this metal.

496. What is the speed of a neutron with a de Broglie wavelength of 5.6×10^{-14} m?

Chapter 22 Subatomic Physics

497. Calculate the binding energy of $^{39}_{19}K$.

498. Determine the difference in the binding energy of $^{107}_{47}Ag$ and $^{63}_{29}Cu$.

499. Find the mass defect of $^{58}_{28}Ni$.

500. Complete this radioactive-decay formula: $^{212}_{84}Po \longrightarrow ? + ^{4}_{2}He$.

501. Complete this radioactive-decay formula: $^{16}_{7}N \longrightarrow ? + ^{0}_{-1}e + \overline{v}$.

502. Complete this radioactive-decay formula: $^{147}_{62}Sm \longrightarrow ^{143}_{60}Nd + ?$.

503. A 3.29×10^{-3} g sample of a pure radioactive substance is found after 30.0 s to have only 8.22×10^{-4} g left undecayed. What is the half-life of the substance?

504. The half-life of $^{48}_{24}Cr$ is 21.6 h. A chromium-48 sample contains 6.5×10^{6} nuclei. Calculate the activity of the sample in mCi.

505. How long will it take a sample of lead-212 (which has a half-life of 10.64 h) to decay to one-eighth its original strength?

506. Compute the binding energy of $^{120}_{50}Sn$.

507. Calculate the difference in the binding energy of $^{12}_{6}C$ and $^{16}_{8}O$.

508. What is the mass defect of $^{64}_{30}Zn$?

509. Complete this radioactive-decay formula: $? \longrightarrow ^{131}_{54}Xe + ^{0}_{-1}e + \overline{v}$.

510. Complete this radioactive-decay formula: $^{160}_{74}W \longrightarrow ^{156}_{72}Hf + ?$.

511. Complete this radioactive-decay formula: $? \longrightarrow ^{107}_{52}Te + ^{4}_{2}He$.

512. A 4.14×10^{-4} g sample of a pure radioactive substance is found after 1.25 days to have only 2.07×10^{-4} g left undecayed. What is the substance's half-life?

513. How long will it take a sample of cadmium-109 with a half-life of 462 days to decay to one-fourth its original strength?

514. The half-life of $^{55}_{26}Fe$ is 2.7 years. What is the decay constant for the isotope?

Advanced Topics

CONTENTS

extension

Teacher Resources

Visit go.hrw.com for teacher resources related to online extensions in this appendix.

Keyword HF6APJXT

897

The Language of Physics

There is a relationship between rotational motion and circular motion. When a solid object (such as a Ferris wheel) undergoes rotational motion about its fixed axis, a point on the rotating object (such as a light bulb on a Ferris wheel) undergoes circular motion.

Visual Strategy

Figure 1

Point out that the angle through which the light bulb moves is related to the distance around the circle that the light bulb moves (the arc length).

Q If the light bulb moves through an angle twice as large as the one shown in **(b),** how would the new arc length compare with the arc length shown?

A ***The new arc length would be twice as large.***

Visual Strategy

Figure 2

Strengthen students' understanding of radian measurement by having them use the values given in **Figure 2** to estimate the radian measures not shown. The students can then verify their answers by using the conversion equation shown on the next page.

Q What is the radian measure equal to 75°?

A $\frac{5}{12}\pi$

Angular Kinematics

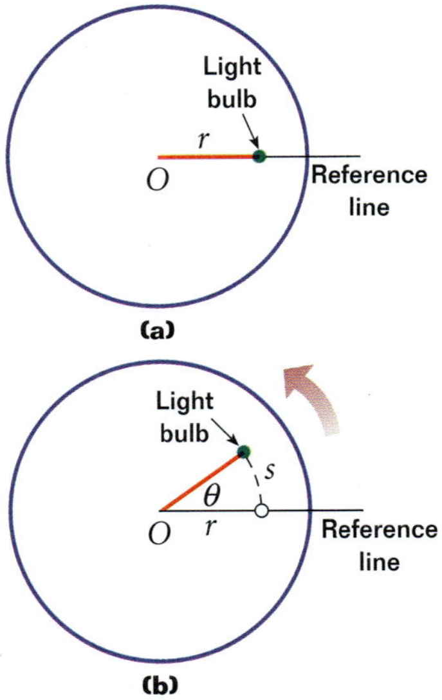

Figure 1
A light bulb on a rotating Ferris wheel **(a)** begins at a point along a reference line and **(b)** moves through an arc length s and therefore through the angle θ.

A point on an object that rotates about a fixed axis undergoes circular motion around that axis. The linear quantities introduced in the chapter "Motion in One Dimension" cannot be used for circular motion because we are considering the rotational motion of an extended object rather than the linear motion of a particle. For this reason, circular motion is described in terms of the change in angular position. All points on a rigid rotating object, except the points on the axis, move through the same angle during any time interval.

Measuring angles with radians

Many of the equations that describe circular motion require that angles be measured in **radians** (rad) rather than in degrees. To see how radians are measured, consider **Figure 1,** which illustrates a light bulb on a rotating Ferris wheel. At $t = 0$, the bulb is on a fixed reference line, as shown in **Figure 1(a).** After a time interval Δt, the bulb advances to a new position, as shown in **Figure 1(b).** In this time interval, the line from the center to the bulb (depicted with a red line in both diagrams) moved through the angle θ with respect to the reference line. Likewise, the bulb moved a distance s, measured along the circumference of the circle; s is the *arc length.*

In general, any angle θ measured in radians is defined by the following equation:

$$\theta = \frac{\text{arc length}}{\text{radius}} = \frac{s}{r}$$

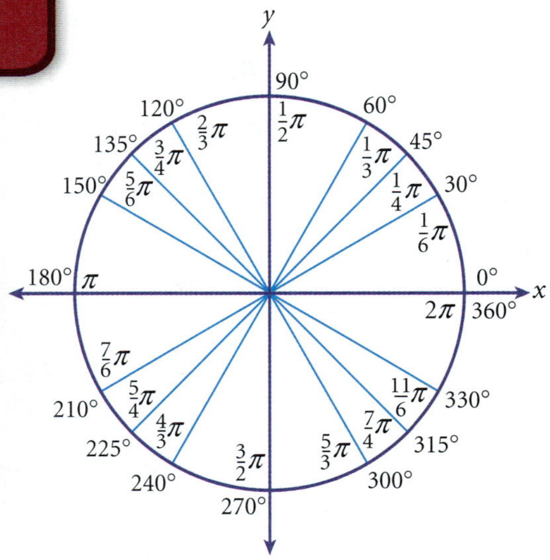

Figure 2
Angular motion is measured in units of radians. Because there are 2π radians in a full circle, radians are often expressed as a multiple of π.

Note that if the arc length, s, is equal to the length of the radius, r, the angle θ swept by r is equal to 1 rad. Because θ is the ratio of an arc length (a distance) to the length of the radius (also a distance), the units cancel and the abbreviation *rad* is substituted in their place. In other words, the radian is a pure number, with no dimensions.

When the bulb on the Ferris wheel moves through an angle of 360° (one revolution of the wheel), the arc length s is equal to the circumference of the circle, or $2\pi r$. Substituting this value for s into the equation above gives the corresponding angle in radians.

$$\theta = \frac{s}{r} = \frac{2\pi r}{r} = 2\pi \text{ rad}$$

Thus, 360° equals 2π rad, or one complete revolution. In other words, one revolution corresponds to an angle of approximately 2(3.14) = 6.28 rad. **Figure 2** on the previous page depicts a circle marked with both radians and degrees.

It follows that any angle in degrees can be converted to an angle in radians by multiplying the angle measured in degrees by $2\pi/360°$. In this way, the degrees cancel out and the measurement is left in radians. The conversion relationship can be simplified as follows:

$$\theta(\text{rad}) = \frac{\pi}{180°}\theta(\text{deg})$$

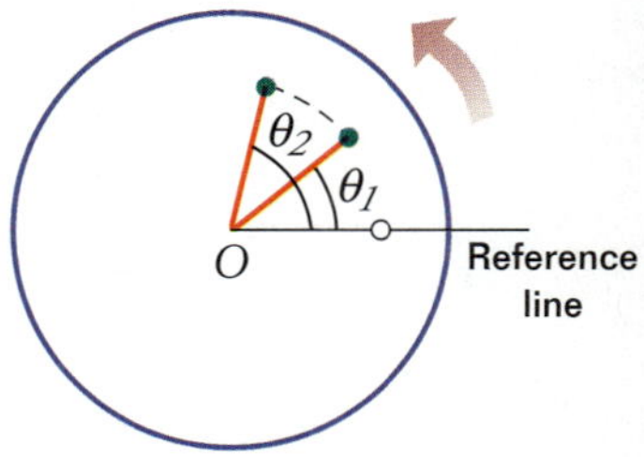

Figure 3
A light bulb on a rotating Ferris wheel rotates through an angular displacement of $\Delta\theta = \theta_2 - \theta_1$.

Angular displacement

Just as an angle in radians is the ratio of the arc length to the radius, the **angular displacement** traveled by the bulb on the Ferris wheel is the change in the arc length, Δs, divided by the distance of the bulb from the axis of rotation. This relationship is depicted in **Figure 3.**

ANGULAR DISPLACEMENT

$$\Delta\theta = \frac{\Delta s}{r}$$

$$\textbf{angular displacement (in radians)} = \frac{\textbf{change in arc length}}{\textbf{distance from axis}}$$

This equation is similar to the equation for linear displacement in that this equation denotes a change in position. The difference is that this equation gives a change in *angular* position rather than a change in *linear* position.

For the purposes of this textbook, when a rotating object is viewed from above, the arc length, s, is considered positive when the point rotates counterclockwise and negative when it rotates clockwise. In other words, $\Delta\theta$ is positive when the object rotates counterclockwise and negative when the object rotates clockwise.

Quick Lab

Radians and Arc Length

MATERIALS LIST

- drawing compass
- paper
- thin wire
- wire cutters or scissors

SAFETY

Cut ends of wire are sharp. Cut and handle wire carefully.

Use the compass to draw a circle on a sheet of paper, and mark the center point of the circle. Measure the radius of the circle, and cut several pieces of wire equal to the length of this radius. Bend the pieces of wire, and lay them along the circle you drew with your compass. Approximately how many pieces of wire do you use to go all the way around the circle? Draw lines from the center of the circle to each end of one of the wires. Note that the angle between these two lines equals 1 rad. How many of these angles are there in this circle? Repeat the experiment with a larger circle, and compare the results of each trial.

Classroom Practice

Angular Displacement
Earth has an equatorial radius of approximately 6380 km and rotates 360° every 24 h.

a. What is the angular displacement (in degrees) of a person standing at the equator for 1.0 h?

b. Convert this angular displacement to radians.

c. What is the arc length traveled by this person?

Answers

a. 15°

b. 0.26 rad

c. approximately 1700 km

Quick Lab

TEACHER'S NOTES

Students should find the same results with both circles. In each case, it takes approximately 6 pieces of wire ($6r$) to go around the circle because circumference = $2\pi r \approx 6r$.

Homework Options
This QuickLab can easily be performed outside of the physics lab room.

Classroom Practice

Angular Velocity
An Indy car can complete 120 laps in 1.5 h. Even though the track is an oval rather than a circle, you can still find the *average* angular speed. Calculate the average angular speed of the Indy car.

Answer
0.14 rad/s

Angular Acceleration
A top that is spinning at 15 rev/s spins for 55 s before coming to a stop. What is the average angular acceleration of the top while it is slowing?

Answer
-1.7 rad/s^2

Angular velocity

Angular velocity is defined in a manner similar to that for linear velocity. The average angular velocity of a rotating rigid object is the ratio of the angular displacement, $\Delta\theta$, to the corresponding time interval, Δt. Thus, angular velocity describes how quickly the rotation occurs. Angular velocity is abbreviated as ω_{avg} (ω is the Greek letter omega).

ANGULAR VELOCITY

$$\omega_{avg} = \frac{\Delta\theta}{\Delta t}$$

$$\text{average angular velocity} = \frac{\text{angular displacement}}{\text{time interval}}$$

Angular velocity is given in units of radians per second (rad/s). Sometimes, angular velocities are given in revolutions per unit time. Recall that 1 rev = 2π rad. The magnitude of angular velocity is called *angular speed.*

Angular acceleration

Figure 4 shows a bicycle turned upside down so that a repairperson can work on the rear wheel. The bicycle pedals are turned so that at time t_1 the wheel has angular velocity ω_1, as shown in **Figure 4(a).** At a later time, t_2, it has angular velocity ω_2, as shown in **Figure 4(b).** Because the angular velocity is changing, there is an **angular acceleration.** The average angular acceleration, α_{avg} (α is the Greek letter *alpha*), of an object is given by the relationship shown below. Angular acceleration has the units radians per second per second (rad/s^2).

ANGULAR ACCELERATION

$$\alpha_{avg} = \frac{\omega_2 - \omega_1}{t_2 - t_1} = \frac{\Delta\omega}{\Delta t}$$

$$\text{average angular acceleration} = \frac{\text{change in angular velocity}}{\text{time interval}}$$

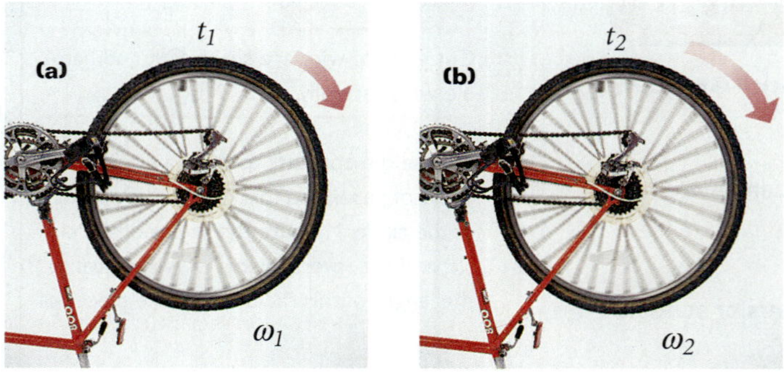

Figure 4
An accelerating bicycle wheel rotates with **(a)** an angular velocity ω_1 at time t_1 and **(b)** an angular velocity ω_2 at time t_2. Thus, the wheel has an angular acceleration.

The relationships between the signs of angular displacement, angular velocity, and angular acceleration are similar to those of the related linear quantities. As discussed earlier, by convention, angular displacement is positive when an object rotates counterclockwise and negative when an object rotates clockwise. Thus, by definition, angular velocity is also positive when an object rotates counterclockwise and negative when an object rotates clockwise. Angular acceleration has the same sign as the angular velocity when it increases the magnitude of the angular velocity, and the opposite sign when it decreases the magnitude.

If a point on the rim of a bicycle wheel had an angular velocity greater than a point nearer the center, the shape of the wheel would be changing. Thus, for a rotating object to remain rigid, as does a bicycle wheel or a Ferris wheel, every portion of the object must have the same angular velocity and the same angular acceleration. This fact is precisely what makes angular velocity and angular acceleration so useful for describing rotational motion.

Kinematic equations for constant angular acceleration

All of the equations for rotational motion defined thus far are analogous to the linear quantities defined in the chapter "Motion in One Dimension." For example, consider the following two equations:

$$\omega_{avg} = \frac{\theta_f - \theta_i}{t_f - t_i} = \frac{\Delta\theta}{\Delta t} \qquad v_{avg} = \frac{x_f - x_i}{t_f - t_i} = \frac{\Delta x}{\Delta t}$$

The equations are similar, with θ replacing x and ω replacing v. The correlations between angular and linear variables are shown in **Table 1.**

In light of the similarities between variables in linear motion and those in rotational motion, it should be no surprise that the kinematic equations of rotational motion are similar to the linear kinematic equations. The equations of rotational kinematics under constant angular acceleration are summarized in **Table 2,** along with the corresponding equations for linear motion under constant acceleration. The rotational motion equations apply only for objects rotating about a fixed axis with constant angular acceleration.

Table 1 Angular Substitutes for Linear Quantities

Angular	Linear
θ	x
ω	v
α	a

Table 2 Rotational and Linear Kinematic Equations

Rotational motion with constant angular acceleration	Linear motion with constant acceleration
$\omega_f = \omega_i + \alpha\Delta t$	$v_f = v_i + a\Delta t$
$\Delta\theta = \omega_i\Delta t + \frac{1}{2}\alpha(\Delta t)^2$	$\Delta x = v_i\Delta t + \frac{1}{2}a(\Delta t)^2$
$\omega_f^2 = \omega_i^2 + 2\alpha\Delta\theta$	$v_f^2 = v_i^2 + 2a\Delta x$
$\Delta\theta = \frac{1}{2}(\omega_i + \omega_f)\Delta t$	$\Delta x = \frac{1}{2}(v_i + v_f)\Delta t$

The quantity ω in these equations represents the *instantaneous angular velocity* of the rotating object rather than the average angular velocity.

extension

Practice Problems

Visit go.hrw.com to find sample and practice problems for angular displacement, angular velocity, angular acceleration, and angular kinematics.

Keyword HF6APJX

Demonstration

Equal Angular Speed

Purpose Illustrate that angular speed is constant at any radius for a rigid extended object.

Materials record player/turntable or bicycle; tape; colored markers

Procedure Use the tape and markers to make two brightly colored flags, and attach them to the turntable so that one flag is near the rim and the other is near, but not at, the center. (Alternatively, if a turntable is not available, you can use a bicycle wheel.) Start the turntable at a moderately slow speed ($33\frac{1}{3}$ rpm) so that the flags are easily observed. Have students note the rotational speed of both flags. Point out that each flag makes a complete rotation in the same amount of time. Change speeds on the turntable, and repeat the observations.

Classroom Practice

Angular Kinematics

A barrel is given a downhill rolling start of 1.5 rad/s at the top of a hill. Assume a constant angular acceleration of 2.9 rad/s^2.

a. If the barrel takes 11.5 s to get to the bottom of the hill, what is the final angular speed of the barrel?

b. What angular displacement does the barrel experience during the 11.5 s ride?

Answers

a. 35 rad/s

b. 2.1×10^2 rad

901

Demonstration

Tangential Speed Versus Angular Speed

Purpose Show that tangential speed depends on radius.

Materials two tennis balls attached to different lengths of string (approx. 1.0 m and 1.5 m)

Procedure Outside on an athletic field, hold the ends of both strings and whirl the tennis balls at constant angular speed over your head. Point out the equal angular speeds of the tennis balls. Ask students to predict the flights of the tennis balls when the strings are released.

Aiming away from students and any breakable items, release the strings and have students observe the flights. Discuss the longer horizontal displacement of the outer ball as a function of its greater tangential speed.

Visual Strategy

Figure 1

Q Which horse would travel farther before hitting the ground if the horses were released from the carousel?

A ***The outer horse would travel farther because it has a greater tangential speed.***

APPENDIX J Advanced Topics

Tangential Speed and Acceleration

The chapter on circular motion introduced the concepts of tangential speed and acceleration. This feature explores these concepts in greater detail. Before reading further, be sure you have read the appendix feature "Angular Kinematics."

Tangential speed

Imagine an amusement-park carousel rotating about its center. Because a carousel is a rigid object, any two horses attached to the carousel have the same angular speed and angular acceleration. However, if the two horses are different distances from the axis of rotation, they have different **tangential speeds.** The tangential speed of a horse on the carousel is its speed along a line drawn tangent to its circular path.

The tangential speeds of two horses at different distances from the center of a carousel are represented in **Figure 1.** Note that the two horses travel the same angular displacement during the same time interval. To achieve this, the horse on the outside must travel a greater distance (Δs) than the horse on the inside. Thus, the outside horse at point B has a greater tangential speed than the inside horse at point A. In general, an object that is farther from the axis of a rigid rotating body must travel at a higher tangential speed to cover the same angular displacement as would an object closer to the axis.

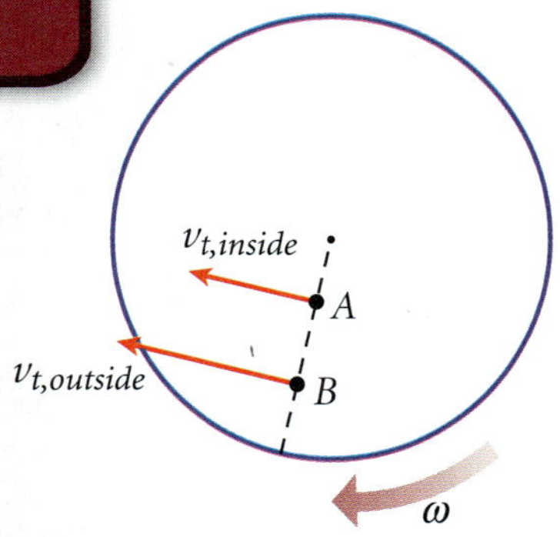

Figure 1
Horses on a carousel move at the same angular speed but different tangential speeds.

If the carousel rotates through an angle θ, a horse rotates through an arc length Δs in the interval Δt. To find the tangential speed, start with the equation for angular displacement:

$$\Delta\theta = \frac{\Delta s}{r}$$

Next, divide both sides of the equation by the time it takes to travel Δs:

$$\frac{\Delta\theta}{\Delta t} = \frac{\Delta s}{r\Delta t}$$

As discussed in the appendix feature "Angular Kinematics," the left side of the equation equals ω_{avg}. Also, Δs is a linear distance, so Δs divided by Δt is a linear speed along an arc length. If Δt is very short, then Δs is so small that it is nearly tangent to the circle; therefore, $\Delta s/\Delta t$ is the tangential speed, v_t.

TANGENTIAL SPEED

$$v_t = r\omega$$

tangential speed = distance from axis × angular speed

In the tangential speed equation, ω is the instantaneous angular speed, rather than the average angular speed, because the time interval is so short.

This equation is valid only when ω is measured in radians per unit of time. Other measures of angular speed must not be used in this equation.

Tangential acceleration

If a carousel speeds up, the horses on it experience an angular acceleration. The linear acceleration related to this angular acceleration is tangent to the circular path and is called the **tangential acceleration.** If an object rotating about a fixed axis changes its angular speed by $\Delta\omega$ in the interval Δt, the tangential speed of a point on the object has changed by the amount Δv_t. Using the equation for tangential speed and dividing by Δt results in the following:

$$\Delta v_t = r\Delta\omega$$

$$\frac{\Delta v_t}{\Delta t} = r\frac{\Delta\omega}{\Delta t}$$

If the time interval Δt is very small, then the left side of this relationship gives the tangential acceleration of the point. The angular speed divided by the time interval on the right side is the angular acceleration. Thus, the tangential acceleration (a_t) of a point on a rotating object is given by the following relationship:

TANGENTIAL ACCELERATION

$$a_t = r\alpha$$

tangential acceleration = distance from axis × angular acceleration

The angular acceleration in this equation refers to the instantaneous angular acceleration. This equation must use the unit radians to be valid. In SI, angular acceleration is expressed as radians per second per second.

Finding total acceleration

Recall that any object moving in a circle has a centripetal acceleration, as discussed in the chapter on circular motion. When both components of acceleration exist simultaneously, the tangential acceleration is tangent to the circular path and the centripetal acceleration points toward the center of the circular path. Because these components of acceleration are perpendicular to each other, the magnitude of the *total acceleration* can be found using the Pythagorean theorem, as follows:

$$a_{total} = \sqrt{a_t^2 + a_c^2}$$

The direction of the total acceleration, as shown in **Figure 2,** depends on the magnitude of each component of acceleration and can be found using the inverse of the tangent function. Note that when there is a tangential acceleration, the tangential speed is changing, and thus this situation is *not* an example of uniform circular motion.

extension

Practice Problems

Visit go.hrw.com to find sample and practice problems for tangential speed and tangential acceleration.

Keyword HF6APJX

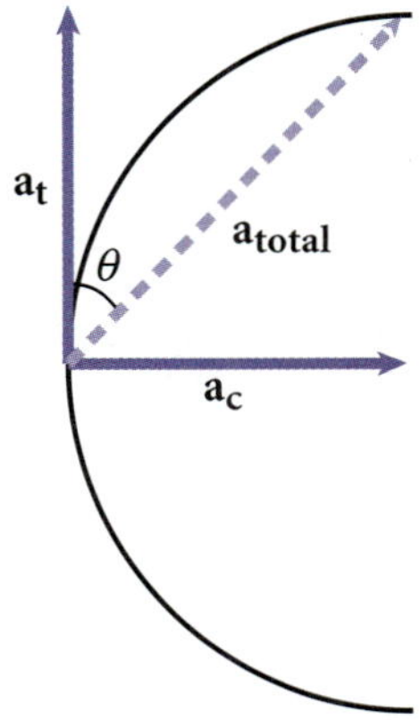

Figure 2

The direction of the total acceleration of a rotating object can be found using the inverse tangent function.

Classroom Practice

Tangential Speed

A golfer has a maximum angular speed of 6.3 rad/s for her swing. She can choose between two drivers, one placing the club head 1.9 m from her axis of rotation and the other placing it 1.7 m from the axis.

a. Find the tangential speed of the club head for each driver.

b. All other factors being equal, which driver is likely to hit the ball farther?

Answers

a. 12 m/s, 11 m/s

b. The longer driver will hit the ball farther because its club head has a higher tangential speed.

Tangential Acceleration

A yo-yo has a tangential acceleration of 0.98 m/s^2 when it is released. The string is wound around a central shaft of radius 0.35 cm. What is the angular acceleration of the yo-yo?

Answer

2.8×10^2 rad/s^2

Teaching Tip

To simulate the large accelerations involved in spaceflight, *Mercury* astronauts rode in the U. S. Navy's centrifuge in Johnsville, Pennsylvania. The astronauts sat in a gondola at the end of a 15.2 m arm that spun around a central axis. During the spin, the astronauts experienced a combination of centripetal and tangential accelerations of the gondola that ranged from 8 to 10 times the acceleration due to gravity.

Misconception Alert

Point out that objects rotate around their center of mass *in the absence of other forces.* For example, a ruler thrown through the air will rotate around its own center of mass because air resistance is evenly distributed and produces zero net torque. However, a ruler with an index card taped to one end will not rotate around its own center of mass when it is thrown through the air because the card provides air resistance, which in turn produces a torque.

Teaching Tip

Let students examine objects that have a center of mass outside the object itself, such as a doughnut, a coat hanger, or a boomerang.

APPENDIX J Advanced Topics

Rotation and Inertia

In Section 4 of the chapter "Circular Motion and Gravitation," you learned that torque measures the ability of a force to rotate an object around some axis, such as a cat-flap door rotating on a hinge. Locating the axis of rotation for a cat-flap door is simple: it rotates on its hinges because the house applies a force that keeps the hinges in place. Now imagine you are playing fetch with your dog, and you throw a stick up into the air for the dog to retrieve. How can you determine the point around which the stick will rotate as it travels through the air?

Center of mass

Unlike the cat-flap door, the stick is not attached to anything. There is a special point around which the stick rotates if gravity is the only force acting on the stick. This point is called the stick's **center of mass.**

SCILINKS®
www.scilinks.org
Topic: Center of Mass
Code: HF60245

The center of mass is also the point at which all the mass of the body can be considered to be concentrated (for translational motion). This means that the complete motion of the stick is a combination of both translational and rotational motion. The stick rotates in the air around its center of mass. The center of mass, in turn, moves as if the stick were a point mass, with all of its mass concentrated at that point for purposes of analyzing its translational motion. For example, the hammer in **Figure 1** rotates about its center of mass as it moves through the air. As the rest of the hammer spins, the center of mass moves along the parabolic path of a projectile.

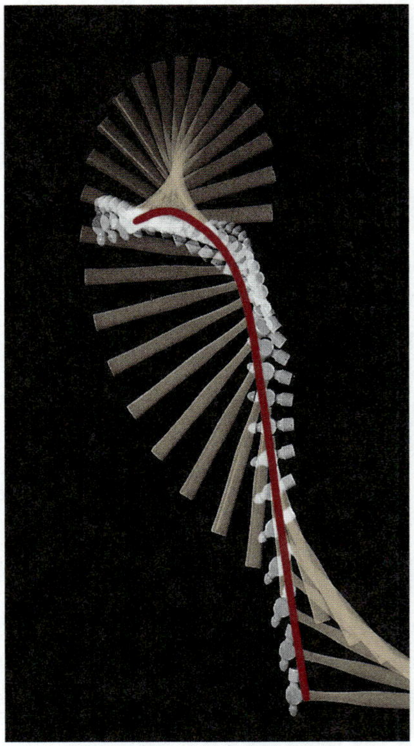

Figure 1
The point around which this hammer rotates is the hammer's center of mass. The center of mass traces out the parabola that is characteristic of projectile motion.

For regularly shaped objects, such as a sphere or a cube, the center of mass is at the geometric center of the object. For more complicated objects, finding the center of mass is more difficult. Although the center of mass is the position at which an extended object's mass can be treated as a point mass, the *center of gravity* is the position at which the gravitational force acts on the extended object as if it were a point mass. For many situations, the center of mass and the center of gravity are equivalent.

Moment of inertia

You may have noticed that it is easier to rotate a baseball bat around some axes than others. The resistance of an object to changes in rotational motion is measured by a quantity called the **moment of inertia.**

The moment of inertia, which is abbreviated as *I*, is similar to mass because they are both forms of inertia. However, there is an important difference between them. Mass is an intrinsic property of an object, and the moment of inertia is not. The moment of inertia depends on the object's mass and the distribution of that mass around the axis of rotation. The farther the mass of an object is, on average, from the axis of rotation, the greater is the object's moment of inertia and the more difficult it is to rotate the object.

According to Newton's second law, when a net force acts on an object, the resulting acceleration of the object depends on the object's mass. Similarly, when a net torque acts on an object, the resulting change in the rotational motion of the object depends on the object's moment of inertia. (This law is covered in more detail in the appendix feature "Rotational Dynamics.")

Some simple formulas for calculating the moment of inertia of common shapes are shown in **Table 1.** The units for moment of inertia are $kg \bullet m^2$. To get an idea of the size of this unit, note that bowling balls typically have moments of inertia about an axis through their centers ranging from about $0.7\ kg \bullet m^2$ to $1.8\ kg \bullet m^2$, depending on the mass and size of the ball.

Did you know?

A baseball bat can be modeled as a rotating thin rod. When a bat is held at its end, its length is greatest with respect to the rotation axis, so its moment of inertia is greatest. Thus, the bat is easier to swing if you hold the bat closer to the center. Baseball players sometimes do this either because a bat is too heavy (large *M*) or too long (large ℓ).

Table 1 The Moment of Inertia for Various Rigid Objects of Mass *M*

Shape		Moment of inertia	Shape		Moment of inertia
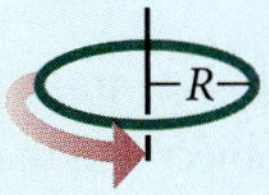	thin hoop about symmetry axis	MR^2	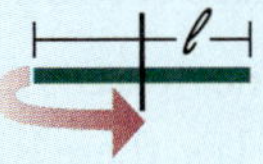	thin rod about perpendicular axis through center	$\frac{1}{12}M\ell^2$
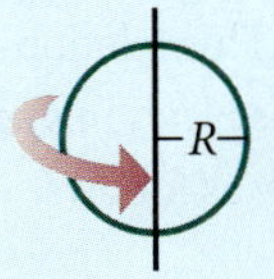	thin hoop about diameter	$\frac{1}{2}MR^2$	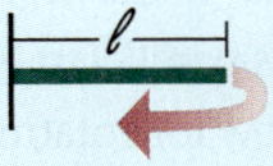	thin rod about perpendicular axis through end	$\frac{1}{3}M\ell^2$
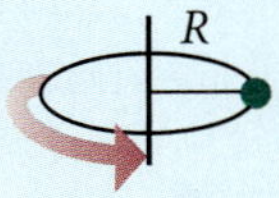	point mass about axis	MR^2	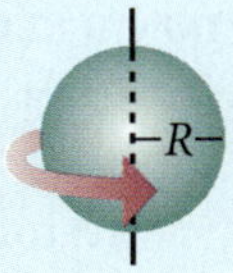	solid sphere about diameter	$\frac{2}{5}MR^2$
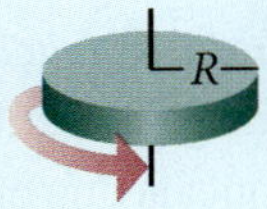	disk or cylinder about symmetry axis	$\frac{1}{2}MR^2$	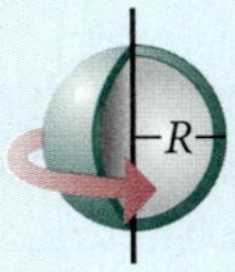	thin spherical shell about diameter	$\frac{2}{3}MR^2$

Demonstration

Moment of Inertia of a Rod

Purpose Give visual examples of the two cases of a thin rod and the case of a cylinder described in **Table 1.**

Materials broomstick or dowel

Procedure If possible, let student volunteers assist by trying the various demonstrations.

Thin rod about center: Hold the rod in the center with one hand. Rotate the rod back and forth through half rotations at regular time intervals. Note the force required to change the motion.

Thin rod about end: Hold rod at one end. Rotate the rod through half circles in the same time interval as previously used. Note the increased force required (corresponding to the larger moment of inertia).

Cylinder: Hold the rod vertically between your palms with your fingers extended. Rotate the cylinder by moving your palms back and forth in the same regular time interval as used previously. Note the much smaller force required in this case.

Misconception Alert

Students will likely confuse zero net torque with zero rotation. State explicitly that an object rotating at constant speed is experiencing zero net torque. This situation is analogous to linear motion: an object moving at a constant velocity has zero net force acting on it.

The Language of Physics

Another way to state the second equilibrium condition is to say that the sum of the clockwise torques must equal the sum of the counterclockwise torques.

Teaching Tip

Students may be confused about which axis of rotation to choose when applying the second condition for equilibrium to an object. Tell students that any axis can be used; the resultant torque acting on an object in rotational equilibrium is independent of where the axis is placed. This fact is useful in solving rotational equilibrium problems because an unknown force that acts along a line passing through this axis of rotation will not produce any torque. Beginning a diagram by arbitrarily setting an axis where a force acts can eliminate an unknown in the problem.

Rotational Dynamics

The appendix feature "Angular Kinematics" developed the kinematic equations for rotational motion. Similarly, the appendix feature "Rotation and Inertia" applied the concept of inertia to rotational motion. In this feature, you will see how torque relates to rotational equilibrium and angular acceleration. You will also learn how momentum and kinetic energy are described in rotational motion.

Figure 1
The two forces exerted on this table are equal and opposite, yet the table moves. How is this possible?

Rotational equilibrium

If you and a friend push on opposite sides of a table, as shown in **Figure 1,** the two forces acting on the table are equal in magnitude and opposite in direction. You might think that the table won't move because the two forces balance each other. But it does; it rotates in place.

The piece of furniture can move even though the net force acting on it is zero because the net torque acting on it is not zero. If the net force on an object is zero, the object is in *translational equilibrium.* If the net torque on an object is zero, the object is in *rotational equilibrium.* For an object to be completely in equilibrium, both rotational and translational, there must be both zero net force and zero net torque. The dependence of equilibrium on the absence of net torque is called the *second condition for equilibrium.*

Newton's second law for rotation

Just as net force is related to translational acceleration according to Newton's second law, there is a relationship between the net torque on an object and the angular acceleration given to the object. Specifically, the net torque on an object is equal to the moment of inertia times the angular acceleration. This relationship is parallel to Newton's second law of motion and is known as Newton's second law for rotating objects. This law is expressed mathematically as follows:

NEWTON'S SECOND LAW FOR ROTATING OBJECTS

$$\tau_{net} = I\alpha$$

net torque = moment of inertia × angular acceleration

This equation shows that a net positive torque corresponds to a positive angular acceleration, and a net negative torque corresponds to a negative angular acceleration. Thus, it is important to keep track of the signs of the torques acting on the object when using this equation to calculate an object's angular acceleration.

Angular momentum

Because a rotating object has inertia, it also possesses momentum associated with its rotation. This momentum is called **angular momentum.** Angular momentum is defined by the following equation:

ANGULAR MOMENTUM

$$L = I\omega$$

angular momentum = moment of inertia × angular speed

The unit of angular momentum is kg•m^2/s. When the net external torque acting on an object or objects is zero, the angular momentum of the object(s) does not change. This is the law of *conservation of angular momentum.* For example, assuming the friction between the skates and the ice is negligible, there is no torque acting on the skater in **Figure 2,** so his angular momentum is conserved. When he brings his hands and feet closer to his body, more of his mass, on average, is nearer to his axis of rotation. As a result, the moment of inertia of his body decreases. Because his angular momentum is constant, his angular speed increases to compensate for his smaller moment of inertia.

Figure 2
When this skater brings his hands and feet closer to his body, his moment of inertia decreases, and his angular speed increases to keep total angular momentum constant.

Angular kinetic energy

Rotating objects possess kinetic energy associated with their angular speed. This is called **rotational kinetic energy** and is expressed by the following equation:

ROTATIONAL KINETIC ENERGY

$$KE_{rot} = \frac{1}{2}I\omega^2$$

$$\textbf{rotational kinetic energy} = \tfrac{1}{2} \times \textbf{moment of inertia} \times (\textbf{angular speed})^2$$

As shown in **Table 1,** rotational kinetic energy is analogous to the translational kinetic energy of a particle, given by the expression $\frac{1}{2}mv^2$. The unit of rotational kinetic energy is the joule, the SI unit for energy.

Table 1 Comparing Translational and Rotational Motion

	Translational motion	Rotational motion
Equilibrium	$\Sigma F = 0$	$\Sigma\tau = 0$
Newton's second law	$\Sigma F = ma$	$\Sigma\tau = I\alpha$
Momentum	$p = mv$	$L = I\omega$
Kinetic Energy	$KE = \frac{1}{2}mv^2$	$KE = \frac{1}{2}I\omega^2$

extension

Practice Problems

Visit go.hrw.com to find sample and practice problems for rotational equilibrium, Newton's second law for rotation, conservation of angular momentum, and angular kinetic energy.

Keyword HF6APJX

Demonstration

Colliding Magnetic Marbles

Purpose This demonstration illustrates conservation of angular momentum and shows that objects moving linearly can also have angular momentum.

Materials 2 magnetic marbles, overhead projector

Procedure Place one magnetic marble in the center of the overhead projector. Send the second magnetic marble toward the first to cause a glancing collision. The marbles will stick together and spin about their center of mass. Point out to students that the linear momentum is *not* converted to angular momentum. Repeat the demonstration and have students observe the slight translation of the two marbles after the collision. This is the linear momentum, which is lost due to friction. The spinning is due to conservation of the angular momentum of the moving marble relative to the stationary marble.

Demonstration

Temperature, Pressure, and Volume

Purpose Demonstrate that the same amount of gas occupies a larger volume at a higher temperature.

Materials heat-resistant flask, balloon, Bunsen burner, goggles

CAUTION Be sure to wear safety goggles at all times when operating the Bunsen burner, and use tongs when handling the heated glassware.

Procedure Cap the flask with the balloon. Emphasize that no gas can flow in or out of the flask or the balloon. Ask students what the air volume is in this closed system (*about equal to that of the flask*). Heat the flask until the balloon expands. Have students estimate and record the balloon's approximate size. Ask them how much air is in the flask now (*same volume, but fewer molecules*) and in the balloon and flask together (*same amount as before, but in a larger volume*). Place the balloon under cold water (not the flask, to avoid breakage) and watch it return to its original size.

Teaching Tip

For an alternative demonstration, place a helium-filled balloon in a refrigerator. After 15 minutes, remove the balloon and have students observe how it behaves as it adjusts to room temperature.

Properties of Gases

SCILINKS. www.scilinks.org
Topic: Gas Laws
Code: HF60637

When the density of a gas is sufficiently low, the pressure, volume, and temperature of the gas tend to be related to one another in a fairly simple way. This relationship is a good approximation for the behavior of many real gases over a wide range of temperatures and pressures, provided their particles are not charged, as in a plasma. These observations have led scientists to develop the concept of an *ideal gas.*

Volume, pressure, and temperature are the three variables that completely describe the macroscopic state of an ideal gas. One of the most important equations in fluid mechanics relates these three quantities to each other.

The ideal gas law

The *ideal gas law* is an expression that relates the volume, pressure, and temperature of a gas. This relationship can be written as follows:

IDEAL GAS LAW

$$PV = Nk_BT$$

pressure × volume =
number of gas particles × Boltzmann's constant × temperature

The symbol k_B represents *Boltzmann's constant.* Its value has been experimentally determined to be approximately 1.38×10^{-23} J/K. Note that when applying the ideal gas law, you must express the temperature in the Kelvin scale. (See the chapter "Heat" to learn about the Kelvin scale.) Also, the ideal gas law makes no mention of the composition of the gas. The gas particles could be oxygen, carbon dioxide, or any other gas. In this sense, the ideal gas law is universally applicable to all gases.

If a gas undergoes a change in volume, pressure, or temperature (or any combination of these), the ideal gas law can be expressed in a particularly useful form. If the number of particles in the gas is constant, the initial and final states of the gas are related as follows:

$$N_1 = N_2$$
$$\frac{P_1V_1}{T_1} = \frac{P_2V_2}{T_2}$$

Figure 1
The balloon is inflated because the volume and pressure of the air inside are both increasing.

This relation is illustrated in the experiment shown in **Figure 1.** In this experiment, a flask filled with air (V_1 equals the volume of the flask) at room temperature (T_1) and atmospheric pressure ($P_1 = P_0$) is placed over a heat source, with a balloon placed over the opening of the flask. As the flask sits over the burner, the temperature of the air inside it increases from T_1 to T_2.

According to the ideal gas law, when the temperature increases, either the pressure or the volume—or both—must also increase. Thus, the air inside the flask exerts a pressure (P_2) on the balloon that serves to inflate the balloon. Because the balloon is expandable, the air expands to a larger volume (V_2) to fill the balloon. When the flask is taken off the burner, the pressure, volume, and temperature of the air inside will slowly return to their initial states.

Another alternative form of the ideal gas law indicates the law's dependence on mass density. Assuming each particle in the gas has a mass m, the total mass of the gas is $N \times m = M$. The ideal gas law can then be written as follows:

$$PV = Nk_BT = \frac{Mk_BT}{m}$$

$$P = \frac{Mk_BT}{mV} = \left(\frac{M}{V}\right)\frac{k_BT}{m} = \frac{\rho k_BT}{m}$$

A real gas

An ideal gas is defined as a gas whose behavior is accurately described by the ideal gas law. Although no real gas obeys the ideal gas law exactly for all temperatures and pressures, the ideal gas law holds for a broad range of physical conditions for all gases. The behavior of real gases departs from the behavior of an ideal gas at high pressures or low temperatures, conditions under which the gas nearly liquefies. However, when a real gas has a relatively high temperature and a relatively low pressure, such as at room temperature and atmospheric pressure, its behavior approximates that of an ideal gas.

For problems involving the motion of fluids, we have assumed that all gases and liquids are ideal fluids. An ideal fluid is a liquid or gas that is assumed to be incompressible. This is usually a good assumption because it is difficult to compress a fluid—even a gas—when it is not confined to a container. A fluid will tend to flow under the action of a force, changing its shape while maintaining a constant volume, rather than compress.

This feature, however, considers confined gases whose pressure, volume, and temperature may change. For example, when a force is applied to a piston, the gas inside the cylinder below the piston is compressed. Even though an ideal gas behaves like an ideal fluid in many situations, it cannot be treated as incompressible when confined to a container.

Did you know?

A third way of writing the ideal gas law may be familiar to you from your study of chemistry:

$$PV = nRT$$

In this equation, n is the number of moles of gas (one mole is equal to 6.02×10^{23} particles). The quantity R is a number called the *molar (universal) gas constant* and has a value of 8.31 J/(mol•K).

extension

Practice Problems

Visit go.hrw.com to find sample and practice problems for the ideal gas law.

Keyword HF6APJX

Quick Lab

Ideal Gas Law

MATERIALS LIST

- **1 plastic 1 L bottle**
- **1 quarter**

Make sure the bottle is empty, and remove the cap. Place the bottle in the freezer for at least 10 min. Wet the quarter with water, and place the quarter over the bottle's opening as you take the bottle out of the freezer. Set the bottle on a nearby tabletop; then observe the bottle and quarter while the air in the bottle warms up. As the air inside the bottle begins to return to room temperature, the quarter begins to jiggle around on top of the bottle. What does this movement tell you about the pressure inside the bottle? What causes this change in pressure? Hypothesize as to why you need to wet the quarter before placing it on top of the bottle.

Key Models and Analogies

Graphs offer a convenient way to represent the relationship between temperature, volume, and pressure for the following special cases of the ideal gas law:

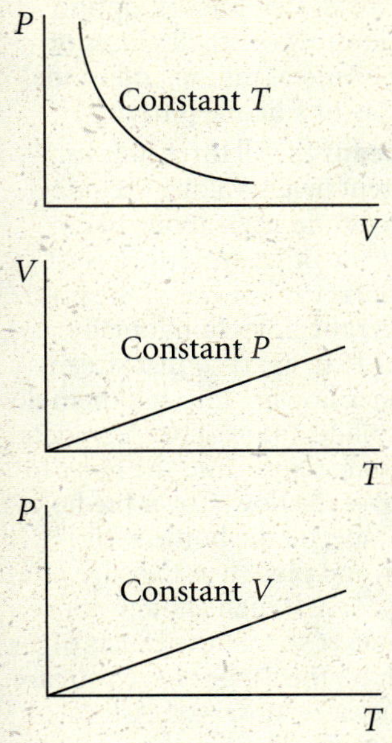

Teaching Tip

If students are not familiar with the Kelvin temperature scale, have them read Section 1 of the chapter "Heat" before attempting the online practice problems that cover the ideal gas law.

Quick Lab

TEACHER'S NOTES

This activity is meant to demonstrate that pressure increases with temperature when the volume of a gas is constant. Eventually, the pressure becomes great enough to overcome the weight of the quarter, and it jumps up slightly, allowing air to escape.

Homework Options

This QuickLab can easily be performed outside of the physics lab room.

Demonstration

Hydrostatic Pressure

Purpose Demonstrate that pressure increases with depth.

Materials 32 oz plastic soda bottle, tape, water, bucket; optional: plastic straw or small pieces of glass or metal tubing and modeling clay or silicone putty

Procedure Drill three holes at different heights along the side of the bottle, each about 10 cm above the other with the lowest hole near the bottom of the bottle. Stagger the holes horizontally about 1 cm apart so that streams of water flowing through them do not collide. Insert short segments of tubing tightly into the holes to improve the flow. Cover the holes with tape, fill the bottle with water, and place it at the edge of a table. Place it high enough above the bucket so that the effects of depth on the range of each stream will be easily observed. Ask students to predict how the water streams will compare. Quickly pull the tape away from all three holes, and have students observe the shape of each stream. Water shooting out from near the bottom of the bottle exits the bottle at a higher speed than water shooting out from near the top does. The reason is that the pressure in the bottle increases with depth.

Fluid Pressure

You learned about fluid pressure and Bernoulli's principle in the chapter "Fluid Mechanics." This feature discusses some additional topics related to fluid pressure, including atmospheric pressure and the kinetic theory of gases. It also covers Bernoulli's equation, which is a more general form of Bernoulli's principle.

SCILINKS.
www.scilinks.org
Topic: Atmospheric Pressure
Code: HF60114

Atmospheric pressure

The weight of the air in the upper portion of Earth's atmosphere exerts pressure on the layers of air below. This pressure is called *atmospheric pressure.* The force that atmospheric pressure exerts on our bodies is extremely large. (Assuming a body area of 2 m^2, this force is on the order of 200 000 N, or 40 000 lb.) How can we exist under such tremendous forces without our bodies collapsing? The answer is that our body cavities and tissues are permeated with fluids and gases that are pushing outward with a pressure equal to that of the atmosphere. Consequently, our bodies are in equilibrium—the force of the atmosphere pushing in equals the internal force pushing out.

An instrument that is commonly used to measure atmospheric pressure is the *mercury barometer.* **Figure 1** shows a very simple mercury barometer. A long tube that is open at one end and closed at the other is filled with mercury and then inverted into a dish of mercury. Once the tube is inverted, the mercury does not empty into the bowl. Instead, the atmosphere exerts a pressure on the mercury in the bowl. This atmospheric pressure pushes the mercury in the tube to some height above the bowl. In this way, the force exerted on the bowl of mercury by the atmosphere is equal to the weight of the column of mercury in the tube. Any change in the height of the column of mercury means that the atmosphere's pressure has changed.

Figure 1
The height of the mercury in the tube of a barometer indicates the atmospheric pressure. (This illustration is not drawn to scale.)

Kinetic theory of gases

Many models of a gas have been developed over the years. Almost all of these models attempt to explain the macroscopic properties of a gas, such as pressure, in terms of events occurring in the gas on a microscopic scale. The most successful model by far is the *kinetic theory of gases.*

In kinetic theory, gas particles are likened to a collection of billiard balls that constantly collide with one another. This simple model is successful in explaining many of the macroscopic properties of a gas. For instance, as these particles strike a wall of a container, they transfer some of their momentum during the collision. The rate of transfer of momentum to the container wall is equal to the force exerted by the gas on the container wall, in accordance with the impulse-momentum theorem. This force per unit area is the gas pressure.

Bernoulli's equation

Imagine a fluid moving through a pipe of varying cross-sectional area and elevation, as shown in **Figure 2.** When the cross-sectional area changes, the pressure and speed of the fluid can change. This change in kinetic energy may be compensated for by a change in gravitational potential energy or by a change in pressure (so energy is still conserved). The expression for the conservation of energy in fluids is called *Bernoulli's equation.* Bernoulli's equation is expressed mathematically as follows:

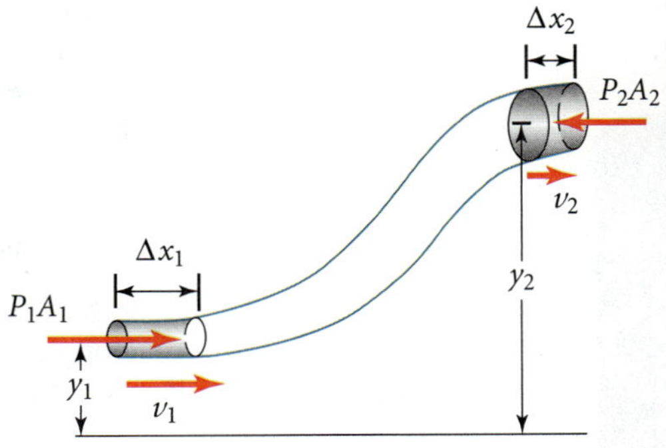

Figure 2
As a fluid flows through this pipe, it may change velocity, pressure, and elevation.

BERNOULLI'S EQUATION

$$P + \frac{1}{2}\rho v^2 + \rho g h = \text{constant}$$

pressure + kinetic energy per unit volume + gravitational potential energy per unit volume = constant along a given streamline

Bernoulli's equation differs slightly from the law of conservation of energy. For example, two of the terms on the left side of the equation look like the terms for kinetic energy and gravitational potential energy, but they contain density, ρ, instead of mass, m. The reason is that the conserved quantity in Bernoulli's equation is energy per unit volume, not just energy. This statement of the conservation of energy in fluids also includes an additional term: pressure, P. If you wish to compare the energy in a given volume of fluid at two different points, Bernoulli's equation takes the following equivalent form:

$$P_1 + \frac{1}{2}\rho v_1^2 + \rho g h_1 = P_2 + \frac{1}{2}\rho v_2^2 + \rho g h_2$$

extension

Practice Problems

Visit go.hrw.com to find a sample and practice problems for Bernoulli's equation.

Keyword HF6APJX

Comparing Bernoulli's principle and Bernoulli's equation

Two special cases of Bernoulli's equation are worth mentioning here. First, if the fluid is at rest, then both speeds are zero. This case is a static situation, such as a column of water in a cylinder. If the height at the top of the column, h_1, is defined as zero and h_2 is the depth, then Bernoulli's equation reduces to the equation for pressure as a function of depth, introduced in the chapter on fluids:

$$P_1 = P_2 + \rho g h_2 \quad \text{(static fluid)}$$

Second, imagine again a fluid flowing through a horizontal pipe with a constriction. Because the height of the fluid is constant, the gravitational potential energy does not change. Bernoulli's equation then reduces to the following:

$$P_1 + \frac{1}{2}\rho v_1^2 = P_2 + \frac{1}{2}\rho v_2^2 \quad \text{(horizontal pipe)}$$

This equation suggests that if v_1 is greater than v_2 at two different points in the flow, then P_1 must be less than P_2. In other words, the pressure decreases as speed increases—Bernoulli's principle. Thus, Bernoulli's principle is a special case of Bernoulli's equation and is strictly true only when elevation is constant.

Classroom Practice

Bernoulli's Equation

A camper creates a shower by attaching a tube to the bottom of a hanging bucket that is open to the atmosphere on top. If the water level in the bucket is 3.15 m above the end of the tube (the shower head), then what is the speed of the water exiting the tube?

Answer

7.86 m/s

A horizontal pipe narrows from a cross section of 2.0 m^2 to 0.30 m^2. If the speed of the water flowing through the wider area of the pipe is 8.0 m/s, what is the speed of the water flowing through the narrow part?

Answer

53 m/s

Assuming incompressible flow, what is the change in pressure as the pipe narrows?

Answer

1.4×10^6 Pa

Teaching Tip

This feature discusses the Doppler effect with light waves. It explores how the idea of an expanding universe originated from observed red shifts and how this in turn suggests that at some point in the past the universe was infinitely small and dense. The current scientific model of the evolution of matter from the moment of the big bang to the present day is discussed at the end of the chapter "Subatomic Physics."

The Doppler Effect and the Big Bang

In the chapter "Sound," you learned that relative motion between the source of sound waves and an observer creates a frequency shift known as the Doppler effect. For visible light, the Doppler effect is observed as a change in color because the frequency of light waves determines color.

Frequency shifts

Of the colors of the visible spectrum, red light has the lowest frequency and violet light has the highest. When a source of light waves is moving toward an observer, the frequency detected is higher than the source frequency. This corresponds to a shift toward the blue end of the spectrum, which is called a *blue shift.* When a source of light waves is moving away from an observer, the observer detects a lower frequency, which corresponds to a shift toward the red end of the spectrum, called a *red shift.* Visible light is one form of electromagnetic radiation. Blue shift and red shift can occur with any type of electromagnetic radiation, not just visible light. **Table 1** illustrates blue shift and red shift.

In astronomy, the light from distant stars or galaxies is analyzed by a process called *spectroscopy.* In this process, starlight is passed through a prism or diffraction grating to produce a spectrum. Dark lines appear in the spectrum at specific frequencies determined by the elements present in the atmospheres of stars. When these lines are shifted toward the blue end of the spectrum, astronomers know the star is moving toward Earth; when the lines are shifted toward the red end, the star is moving away from Earth.

The expansion of the universe

As scientists began to study other galaxies with spectroscopy, the results were astonishing: nearly all of the galaxies that were observed exhibited a red shift,

Table 1 The Doppler Effect for Light

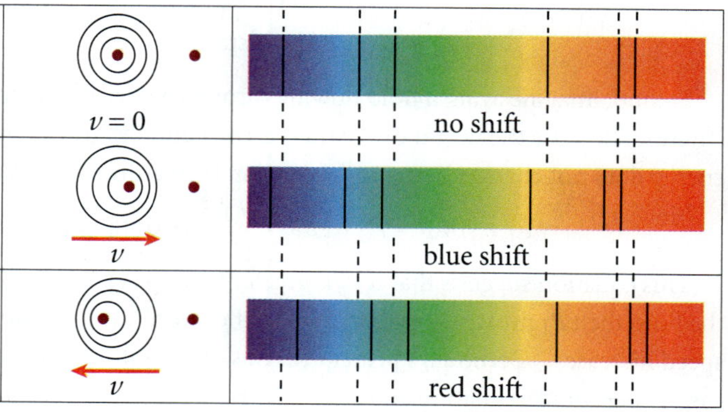

stationary source	$v = 0$	no shift
approaching source	v	blue shift
receding source	v	red shift

which suggested that they were moving away from Earth. If all galaxies are moving away from Earth, the universe must be expanding. This does not imply that Earth is at the center of the expansion; the same phenomenon would be observed from any other point in the universe.

The expansion of the universe suggests that at some point in the past the universe must have had infinite density. The eruption of the universe is often referred to as the *big bang,* which is generally considered to have occurred between about 13 billion and 15 billion years ago. Current models indicate that the big bang involved such great amounts of energy in such a small space that matter could not form clumps or even individual atoms. It took about 380,000 years for the universe to cool from around 10^{32} K to around 3000 K, a temperature cool enough for atoms to begin forming.

Figure 1
Penzias and Wilson detected microwave background radiation, presumably left over from the big bang, with the horn antenna (in the background) at Bell Telephone Laboratories in New Jersey.

Experimental verification

In the 1960s, a group of scientists at Princeton predicted that the explosion of the big bang was so momentous that a small amount of radiation—the leftover glow from the big bang—should still be found in the universe. Around this time, Arno Penzias and Robert Wilson of Bell Labs noticed a faint background hiss interfering with satellite-communications experiments they were conducting. This signal, which was detected in equal amounts in all directions, remained despite all attempts to remove it. Penzias and Wilson learned of the Princeton group's work and realized that the interference they were experiencing matched the characteristics of the radiation expected from the big bang. Subsequent experiments have confirmed the existence of this radiation, known as *cosmic microwave background radiation.* This background radiation is considered to be the most conclusive evidence for the big bang theory.

Figure 2
This image, called the *Hubble Ultra Deep Field,* is a compilation of images taken by two cameras on the *Hubble Space Telescope* between September 2003 and January 2004. It shows the youngest galaxies ever to be seen. These galaxies may have formed as early as 400 million years after the big bang.

The big bang theory is generally accepted by scientists today. Research now focuses on more detailed issues. However, there are certain phenomena that the standard big bang model cannot account for, such as the uniform distribution of matter on a large scale and the large-scale clustering of galaxies. As a result, some scientists are currently working on modifications and refinements to the standard big bang theory.

In March of 2004, astronomers released a new image from the *Hubble Space Telescope.* This image, called the *Hubble Ultra Deep Field* (HUDF), looks further back in time than any previously-recorded images. The image contains an estimated 10,000 galaxies. Scientists will study the HUDF to search for galaxies that existed from 400 million to 800 million years after the big bang. Because galaxies evolved quickly, many important changes happened within a billion years of the big bang. Scientists hope that studies of the HUDF image will resolve some of the current questions regarding the origin and evolution of the universe.

Key Models and Analogies

Ask students to visualize a balloon with dots placed randomly on its surface. From the perspective of any single dot, as the balloon is inflated, all other dots appear to be moving away. This is analogous to the expansion of the universe; the fact that we observe all galaxies as moving away from Earth does not mean that Earth is the center of the expansion. The same phenomenon would be observed from any other point in the universe.

EXTENSION

- Have students research various theories regarding the age of the universe. Their reports should include a discussion of the following: How is the Hubble constant related to the age of the universe? What methods are used to measure the distances to stars? How do current theories on the age of the universe differ?
- Have students research Alan Guth's *inflation theory,* a modification of the standard big bang theory. What difficulties with the big bang theory does the inflation theory resolve?
- Spectroscopy is widely used in astronomy for a variety of purposes. Have students study its uses, along with other practical applications of the Doppler effect.
- Have students explore various possibilities for the future of the universe. Their research should include an investigation into the question of how much dark matter the universe contains and how the answer to this question will affect the fate of the universe.

Teaching Tip

This feature discusses how Einstein's special theory of relativity modifies the notion of time. Although the concept of time dilation is challenging, the basic aspects can be understood conceptually at this level, and the equation for time dilation can be used to see how the theories merge at small speeds but diverge at speeds that are large relative to the speed of light.

Before Einstein published the special theory of relativity in 1905, physicists were struggling with an apparent contradiction in physics. The principle of relativity, which states that the laws of physics are the same in any inertial reference frame, did not allow for the fact that the speed of light is independent of the reference frame from which it is measured. (This contradiction is not immediately obvious to students; a class discussion of the subject will help to bring out the contradiction.)

While most physicists assumed that one of these postulates must be abandoned, Einstein had the insight that both postulates could be true if our ideas of space and time are modified. Time dilation is one of the necessary modifications.

Special Relativity and Time Dilation

In the kinematics chapters, you worked with equations that describe motion in terms of a time interval (Δt). Before Einstein developed the special theory of relativity, everyone assumed that Δt must be the same for any observer, whether that observer is at rest or in motion with respect to the event being measured. This idea is often expressed by the statement that time is *absolute.*

The relativity of time

In 1905, Einstein challenged the assumption that time is absolute in a paper titled "The Electrodynamics of Moving Bodies," which contained his special theory of relativity. The special theory of relativity applies to observers and events that are moving with constant velocity (in uniform motion) with respect to one another. One of the consequences of this theory is that Δt *does* depend on the observer's motion. Consider a passenger in a train that is moving uniformly with respect to an observer standing beside the track, as shown in **Figure 1.** The passenger on the train shines a pulse of light toward a mirror directly above him and measures the amount of time it takes for the pulse to return. Because the passenger is moving along with the train, he sees the pulse of light travel directly up and then directly back down, as in **Figure 1(a).** The observer beside the track, however, sees the pulse hit the mirror at an angle, as in **Figure 1(b),** because the train is moving with respect to the track. Thus, the distance the light travels according to the observer is *greater* than the distance the light travels from the perspective of the passenger.

One of the postulates of Einstein's theory of relativity, which follows from James Clerk Maxwell's equations about light waves, is that the speed of light is the same for *any* observer, even when there is motion between the source of light and the observer. Light is different from all other phenomena in this respect. Although this postulate seems counterintuitive, it was strongly supported by an experiment performed in 1851 by Armand Fizeau. But if the speed of light is the same for both the passenger on the train and the observer beside the track while the distances traveled are different, the time intervals observed by each person must also be different. Thus, the observer beside the track measures a longer time interval than the passenger does. This effect is known as *time dilation.*

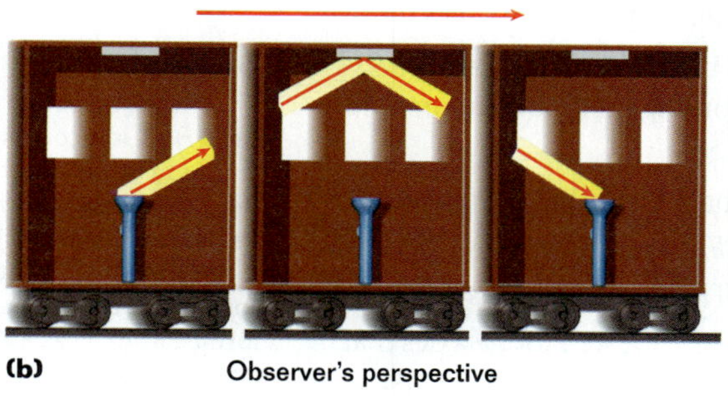

Figure 1
(a) A passenger on a train sends a pulse of light toward a mirror directly above. **(b)** Relative to a stationary observer beside the track, the distance the light travels is greater than that measured by the passenger.

Calculating time dilation

Time dilation is given by the following equation, where $\Delta t'$ represents the time interval measured by the person beside the track, and Δt represents the time interval measured by the person on the train:

$$\Delta t' = \frac{\Delta t}{\sqrt{1 - \frac{v^2}{c^2}}}$$

In this equation, v represents the speed of the train relative to the person beside the track, and c is the speed of light in a vacuum, 3.00×10^8 m/s. At speeds with which we are familiar, where v is much smaller than c, the term $\frac{v^2}{c^2}$ is such a small fraction that $\Delta t'$ is essentially equal to Δt. For this reason, we do not observe the effects of time dilation in our typical experiences. But when speeds are closer to the speed of light, time dilation becomes more noticeable. As seen by this equation, time dilation becomes infinite as v approaches the speed of light.

According to Einstein, the motion between the train and the track is *relative;* that is, either system can be considered to be in motion with respect to the other. For the passenger, the train is stationary and the observer beside the track is in motion. If the light experiment is repeated by the observer beside the track, then the passenger would see the light travel a greater distance than the observer would. So, according to the passenger, it is the observer beside the track whose clock runs more slowly. Observers see their clocks running as if they were not moving. Any clocks in motion relative to the observers will seem to the observers to run slowly. Similarly, by comparing the differences between the time intervals of their own clocks and clocks moving relative to theirs, observers can determine how fast the other clocks are moving with respect to their own.

Experimental verification

The effects we have been considering hold true for all physical processes, including chemical and biological reactions. Scientists have demonstrated time dilation by comparing the lifetime of muons (a type of unstable elementary particle) traveling at $0.9994c$ with the lifetime of stationary muons. In another experiment, atomic clocks on jet planes flying around the world were compared with identical clocks at the U.S. Naval Observatory. In both cases, time dilations were observed that matched the predictions of Einstein's theory of special relativity within the limits of experimental error.

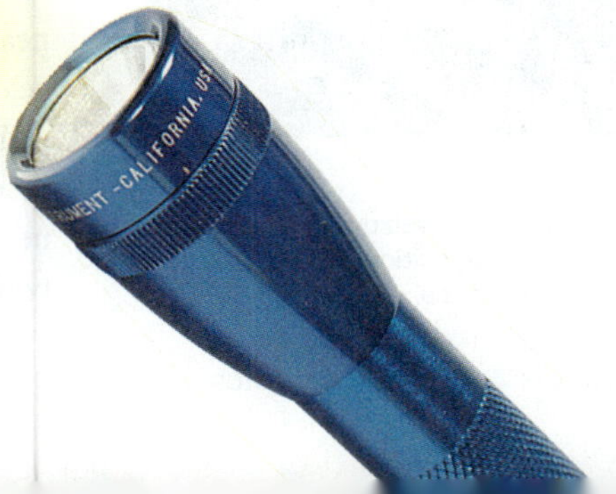

SCILINKS
www.scilinks.org
Topic: Relativity of Time
Code: HF61289

EXTENSION

- Have students research length contraction, another consequence of the theory of special relativity. If L is the length between two points as measured by an observer at rest with respect to the points and v is the speed between two reference frames, then the contracted length L' (measured in the second reference frame) is given by the following equation:

$$L' = L\sqrt{1 - \frac{v^2}{c^2}}$$

  Ask students to try different values of speed in the equations for time dilation and length contraction to determine at what speeds the effects of special relativity become significant.
- Have students research experiments that support the theory of special relativity, including the Michelson-Morley experiment and Armand Fizeau's measurement of the speed of light in water.

Teaching Tip

The theory of special relativity is valid only when the motion between the two reference frames is uniform. When one reference frame is accelerating in relation to the other, the general theory of relativity must be used.

Misconception Alert

Because students have no direct experience with time dilation, many students initially think that time dilation is a theoretical idea rather than a physical effect that can be observed. Use the experiments discussed at the end of this feature to stress the fact that time dilation is a real physical effect.

915

Teaching Tip

This feature builds on the previous appendix feature, "Special Relativity and Time Dilation." In that feature, the assumption that the speed of light is the same for all observers was used to explain why time measurements depend on an observer's frame of reference. In this feature, this assumption and its consequences are discussed in greater detail.

The feature begins by comparing the behavior of light with a typical case to show how the two differ. Next, the need to revise the classical addition of velocities is discussed. Finally, Einstein's relativistic addition of velocities is introduced, along with a discussion of how all cases are covered with this equation.

Today there is overwhelming physical evidence that the speed of light is absolute. Experiments in particle accelerators, in which particles reach speeds very close to *c*, support the relativistic rather than the classical addition of velocities.

Teaching Tip

The constancy of the speed of light is difficult for some students to grasp because we have no direct experience of this phenomenon. Use examples to familiarize your students with Einstein's theory of the constancy of the speed of light. For example, ask students to compare the speed of sound and light waves as viewed by two different observers, one at rest and one moving toward the source of the waves.

Special Relativity and Velocities

In Section 4 of the chapter "Two-Dimensional Motion and Vectors," you learned that velocity measurements are not absolute; every velocity measurement depends on the frame of reference of the observer with respect to the moving object. For example, imagine that someone riding a bike toward you at 25 m/s (v) throws a softball toward you. If the bicyclist measures the softball's speed (u') to be 15 m/s, you would perceive the ball to be moving toward you at 40 m/s (u) because you have a different frame of reference than the bicyclist does. This is expressed mathematically by the equation $u = v + u'$, which is also known as the classical addition of velocities.

The speed of light

As stated in the appendix feature "Special Relativity and Time Dilation," according to Einstein's special theory of relativity, the speed of light is absolute, or independent of all frames of reference. If, instead of a softball, the bicyclist were to shine a beam of light toward you, both you and the bicyclist would measure the light's speed as 3.0×10^8 m/s. This would remain true even if the bicyclist were moving toward you at 99 percent of the speed of light. Thus, Einstein's theory requires a different approach to the addition of velocities. Einstein's modification of the classical formula, which he derived in his 1905 paper on special relativity, covers both the case of the softball and the case of the light beam.

$$u = \frac{v + u'}{1 + (vu'/c^2)}$$

In the equation, u is the velocity of an object in a reference frame, u' is the velocity of the same object in another reference frame, v is the velocity of one reference frame relative to another, and c is the speed of light.

Figure 1
According to Einstein's relativistic equation for the addition of velocities, material particles can never reach the speed of light.

The universality of Einstein's equation

How does Einstein's equation cover both cases? First we shall consider the bicyclist throwing a softball. Because c^2 is such a large number, the vu'/c^2 term in the denominator is very small for velocities typical of our everyday experience. As a result, the denominator of the equation is essentially equal to 1. Hence, for speeds that are small compared with c, the two theories give nearly the same result, $u = v + u'$, and the classical addition of velocities can be used.

However, when speeds approach the speed of light, vu'/c^2 increases, and the denominator becomes greater than 1 but never more than 2. When this occurs, the difference between the two theories becomes significant. For example, if a bicyclist moving toward you at 80 percent of the speed of light were to throw a ball to you at 70 percent of the speed of light, you would observe the ball moving toward you at about 96 percent of the speed of light rather than the 150 percent of the speed of light predicted by classical theory. In this case, the difference between the velocities predicted by each theory cannot be ignored, and the relativistic addition of velocities must be used.

In this last example, it is significant that classical addition predicts a speed greater than the speed of light ($1.5c$), while the relativistic addition predicts a speed less than the speed of light ($0.96c$). In fact, no matter how close the speeds involved are to the speed of light, the relativistic equation yields a result less than the speed of light, as seen in **Table 1.**

How does Einstein's equation cover the second case, in which the bicyclist shines a beam of light toward you? Einstein's equation predicts that any object traveling at the speed of light ($u' = c$) will appear to travel at the speed of light ($u = c$) for an observer in any reference frame:

$$u = \frac{v + u'}{1 + (vu'/c^2)} = \frac{v + c}{1 + (vc/c^2)} = \frac{v + c}{1 + (v/c)} = \frac{v + c}{(c + v)/c} = c$$

This corresponds with our earlier statement that the bicyclist measures the beam of light traveling at the same speed that you do, 3.0×10^8 m/s, even though you have a different reference frame than the bicyclist does. This occurs regardless of how fast the bicycle is moving because v (the bicycle's speed) cancels from the equation. Thus, Einstein's relativistic equation successfully covers both cases. So, Einstein's equation is a more general case of the classical equation, which is simply the limiting case.

SCILINKS.
www.scilinks.org
Topic: Speed of Light
Code: HF61439

Table 1 Classical and Relativistic Addition of Velocities

c = 299 792 458 m/s		**Classical addition**	**Relativistic addition**
Speed between frames (v)	Speed measured in A (u')	Speed measured in B (u)	Speed measured in B (u)
25 m/s	15 m/s	40 m/s	40 m/s
100 000 m/s	100 000 m/s	200 000 m/s	200 000 m/s
50% of c	50% of c	299 792 458 m/s	239 833 966 m/s
90% of c	90% of c	539 626 424 m/s	298 136 146 m/s
99.99% of c	99.99% of c	599 524 958 m/s	299 792 457 m/s

EXTENSION

- Have students research particle-accelerator experiments involving particles traveling at speeds close to c. Their reports should include a discussion of how observations support the relativistic addition of velocities.
- The *ether* was originally conceived as the medium through which light waves traveled. Have students investigate the concept of the ether and the Michelson-Morley experiment, which was intended to detect the ether. Then have a class discussion about why the concept of the ether was originally believed to be necessary and how the Michelson-Morley experiment and the special theory of relativity affected the theory of the ether.

Teaching Tip

The equivalence between mass and energy is a consequence of Einstein's special theory of relativity. Einstein introduced the subject in 1905 in a paper titled "Does the inertia of a body depend on its energy-content?"

Although the derivation of relativistic kinetic energy is beyond the scope of this book, the mass-energy equivalence can be understood conceptually at this level and will enhance students' understanding of energy.

The equivalence between mass and energy in terms of binding energy is explored in the chapter "Subatomic Physics." Fission and fusion are also discussed in greater detail in that chapter, along with applications such as nuclear power and the nuclear bomb.

EXTENSION

- Have students research fission and fusion as energy sources. In their reports, they should include a discussion of the following questions: Why are fusion reactors considered to be safer than fission reactors? What are the difficulties associated with developing a fusion reactor? What are the advantages and disadvantages of using nuclear reactors as sources of energy?
- Einstein was deeply concerned about the possibility of nuclear weapons suggested by his theory. Have students research Einstein's opinions on the subject of nuclear weapons to prepare for a classroom debate on the social responsibility of scientists.

APPENDIX J Advanced Topics

The Equivalence of Mass and Energy

Einstein's $E_R = mc^2$ is one of the most famous equations of the twentieth century. Einstein discovered this equation through his work with relative velocity and kinetic energy.

Relativistic kinetic energy

In the appendix feature "Special Relativity and Velocities," you learned how Einstein's special theory of relativity modifies the classical addition of velocities. The classical equation for kinetic energy ($KE = \frac{1}{2}mv^2$) must also be modified for relativity. In 1905, Einstein derived a new equation for kinetic energy based on the principles of special relativity:

$$KE = \frac{mc^2}{\sqrt{1 - \left(\frac{v^2}{c^2}\right)}} - mc^2$$

In this equation, m is the mass of the object, v is the velocity of the object, and c is the speed of light. Although it isn't immediately obvious, this equation reduces to the classical equation $KE = \frac{1}{2}mv^2$ for speeds that are small relative to the speed of light, as shown in **Figure 1.** The graph also illustrates that velocity can never be greater than 1.0c in the theory of special relativity.

Einstein's relativistic expression for kinetic energy has been confirmed by experiments in which electrons are accelerated to extremely high speeds in particle accelerators. In all cases, the experimental data correspond to Einstein's equation rather than to the classical equation. Nonetheless, the difference between the two theories at low speeds (relative to c) is so minimal that the classical equation can be used in all such cases when the speed is much less than c.

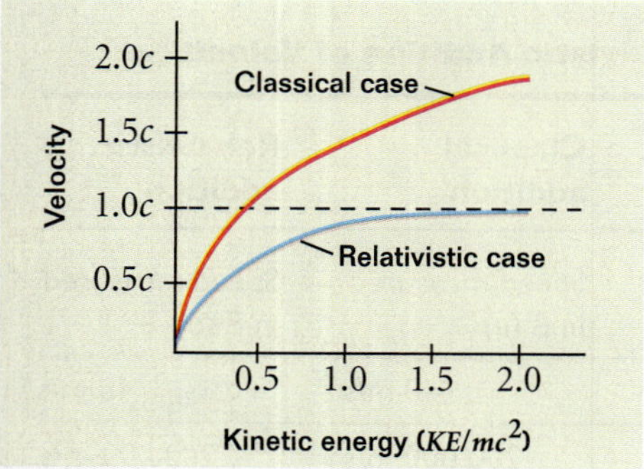

Figure 1
This graph of velocity versus kinetic energy for both the classical and relativistic equations shows that the two theories are in agreement when v is much less than c. Note that v is always less than c in the relativistic case.

Rest energy

The second term of Einstein's equation for kinetic energy, $-mc^2$, is required so that $KE = 0$ when $v = 0$. Note that this term is independent of velocity. This suggests that the *total* energy of an object equals its kinetic energy plus some additional form of energy equal to mc^2. The mathematical expression of this additional energy is the familiar Einstein equation:

$$E_R = mc^2$$

This equation shows that an object has a certain amount of energy (E_R), known as *rest energy*, simply by virtue of its mass. The

Figure 2
Electrons in the Stanford Linear Accelerator in California (SLAC) reach 99.999999967 percent of the speed of light. At such great speeds, the difference between classical and relativistic theories becomes significant.

rest energy of a body is equal to its mass, *m,* multiplied by the speed of light squared, c^2. Thus, the mass of a body is a measure of its rest energy. This equation is significant because rest energy is an aspect of special relativity that was not predicted by classical physics.

Experimental verification

The magnitude of the conversion factor between mass and rest energy ($c^2 = 9 \times 10^{16}\ m^2/s^2$) is so great that even a very small mass has a huge amount of rest energy. Nuclear reactions utilize this relationship by converting mass (rest energy) into other forms of energy. In nuclear fission, which is the energy source of nuclear power plants, the nucleus of an atom is split into two or more nuclei. Taken together, the mass of these nuclei is slightly less than the mass of the original nucleus, and a very large amount of energy is released. In typical nuclear reactions, about one-thousandth of the initial mass is converted from rest energy into other forms of energy. This change in mass, although very small, can be detected experimentally.

SCILINKS.
www.scilinks.org
Topic: Nuclear Reactions
Code: HF61053

Another type of nuclear reaction that converts mass into energy is fusion, which is the source of energy for our sun and other stars. About 4.5 million tons of the sun's mass is converted into other forms of energy every second, by fusing hydrogen into helium. Fortunately, the sun has enough mass to continue to fuse hydrogen into helium for approximately 5 billion more years.

Most of the energy changes encountered in your typical experiences are much smaller than the energy changes that occur in nuclear reactions. Such changes are far too small to be detected experimentally. Thus, for typical cases, the classical equation still holds, and mass and energy can be thought of as separate.

Before Einstein's theory of relativity, conservation of energy and conservation of mass were regarded as two separate laws. The equivalence between mass and energy reveals that in fact these two laws are one. In the words of Einstein, "Prerelativity physics contains two conservation laws of fundamental importance. . . . Through relativity theory, they melt together into *one* principle."

Figure 3
Our sun uses a nuclear reaction called *fusion* to convert mass to energy. About 90 percent of the stars, including our sun, fuse hydrogen, and some older stars fuse helium.

Teaching Tip

Figure 2 shows the PEP-II Collider at SLAC. Electrons and positrons (represented by blue and pink lights) travel along two separate rings in opposite directions; when they collide, both disappear and new particles are created in their place.

The Language of Physics

The notion of mass in relativity has been undergoing a transformation in recent years. In earlier treatments of relativity, which you will still see in some current textbooks, the notion of *relativistic mass* is used. In these treatments, the symbol *m* represents the relativistic mass, which increases as the speed of the object increases. As a result, Einstein's equation is the commonly seen $E = mc^2$, and the energy is the relativistic *total energy*. In these treatments, the notion of *rest mass* is used to represent the mass of an object when its speed is zero.

In modern treatments of relativity, the symbol *m* represents simply the mass of an object, which remains constant. There is no notion of a change of mass with speed. Einstein's equation is $E_R = mc^2$, in which the energy is the *rest energy* of the object. The total energy of the object is then expressed as follows:

$$E = \frac{mc^2}{\sqrt{1 - \frac{v^2}{c^2}}}$$

There is no use of the term *rest mass.*

Teaching Tip

Encourage students who are interested in this feature to read portions of Einstein's popular account of the theories of special and general relativity, titled *Relativity: The Special and the General Theory.* In the first few chapters of Part II of this book, Einstein discusses the equivalence between gravitational and inertial mass and introduces the thought experiment described in this feature.

General Relativity

In the appendix features on Einstein's theory of special relativity, you studied situations involving observers in different reference frames, such as one observer on a moving train and another on the ground. These examples all assumed that the two reference frames were moving uniformly with respect to each other. In other words, neither reference frame was accelerating relative to the other. Special relativity applies only to nonaccelerating reference frames. Einstein expanded his special theory of relativity into the general theory to cover all cases, including accelerating reference frames.

Gravitational attraction and accelerating reference frames

Einstein began with a simple question: "If we pick up a stone and then let it go, why does it fall to the ground?" You might answer that it falls because it is attracted by gravitational force. As usual, Einstein was not satisfied with this typical answer. He was also intrigued by the fact that in a vacuum, all objects in free fall have the same acceleration, regardless of their mass. As you learned in the chapter on gravity, the reason is that gravitational mass is equal to inertial mass. Because the two masses are equivalent, the extra gravitational force from a larger gravitational mass is exactly canceled out by its larger inertial mass, thus producing the same acceleration. Einstein considered this equivalence to be a great puzzle.

To explore these questions, Einstein used a thought experiment similar to the one shown in **Figure 1.** In **Figure 1(a),** a person in an elevator at rest on Earth's surface drops a ball. The ball is in free fall and accelerates downward, as you would expect. In **Figure 1(b),** a similar elevator in space is moving upward with a constant acceleration. If an astronaut in this elevator releases a ball, the floor acceler-

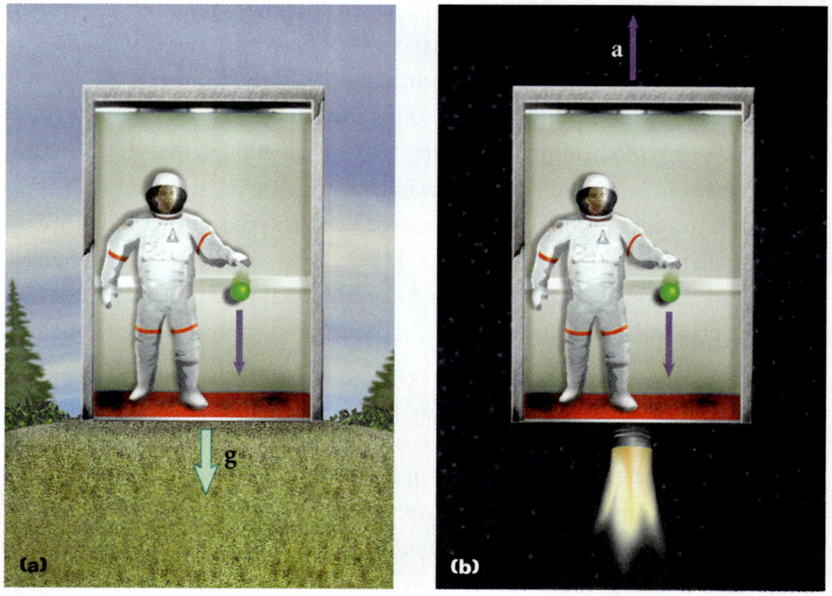

Figure 1
Einstein discovered that there is no way to distinguish between **(a)** a gravitational field and **(b)** an accelerating reference frame.

ates up toward the ball. To the astronaut, the ball appears to be accelerating downward, and this situation is identical to the situation described in **(a).** In other words, the astronaut may *think* that his spaceship is on Earth and that the ball falls because of gravitational attraction. Because gravitational mass equals inertial mass, the astronaut cannot conduct any experiments to distinguish between the two cases. Einstein described this realization as "the happiest thought of my life."

SCILINKS.
www.scilinks.org
Topic: Albert Einstein
Code: HF60097

Gravity and light

Now, imagine a ray of light crossing the accelerating elevator. Suppose that the ray of light enters the elevator from the left side. As the light ray travels across the elevator from left to right, the floor of the elevator accelerates upward. Thus, to an astronaut in the elevator, the light ray would appear to follow a parabolic path.

If Einstein's theory of the equivalence between gravitational fields and accelerating reference frames is correct, then light must also bend this way in a gravitational field. Einstein proposed using the sun's gravitational field to test this idea. The effect is small and difficult to measure, but Einstein predicted that it could be done during a solar eclipse. Einstein published the theory of general relativity in 1916. Just three years later, in 1919, the British astronomer Arthur S. Eddington conducted observations of the light from stars during an eclipse. This experiment provided support for Einstein's theory of general relativity.

Curved spacetime

Although light traveling near a massive object such as the sun appears to bend, is it possible that the light is actually following the straightest path? Einstein theorized that the answer to this question is yes. In general relativity, the three dimensions of space and the one dimension of time are considered together as four-dimensional space-time. When no masses are present, an object moves through "flat" space-time. Einstein proposed that masses change the shape of space-time, as shown in **Figure 2.** A light ray that bends near the sun is following the new shape of the distorted space-time.

For example, imagine rolling a tennis ball across a water bed. If the water bed is flat, the tennis ball will roll straight across. If you place a heavy bowling ball in the center, the bowling ball changes the shape of the water bed. As a result, the tennis ball will then follow a curved path, which, in Newton's theory, is due to the gravitational force between the two. In general relativity, the tennis ball is simply following the curved path of space-time, which is distorted by the bowling ball. Unlike Newton's *mathematical* theory of gravitation, Einstein's theory of curved space-time offers a *physical* explanation for gravitational force.

Figure 2
In the theory of general relativity, masses distort four-dimensional space-time, as illustrated here. This distortion creates the effect we describe as gravitational attraction.

Today, Einstein's theory of general relativity is well accepted. However, scientists have not yet been able to incorporate it with another well-accepted theory that describes things at the microscopic level: quantum mechanics. Many scientists are now working toward a unification of these two theories.

921

Teaching Tip

Louis de Broglie, a French physicist, was awarded the Nobel Prize in 1929 for his prediction of the wave nature of material particles. De Broglie originally proposed this theory in his doctoral thesis in 1924.

The de Broglie family was an aristocratic French family, and Louis de Broglie held the title of Prince. De Broglie came late to the study of theoretical physics because he first studied history. Only after serving as a radio operator in World War I did he begin his study of physics.

De Broglie later found a connection between the wave nature of electrons and the Bohr model of hydrogen, in which only certain orbits of the electron are stable. De Broglie postulated that the stable orbits are those that contain an integral number of wavelengths; this is analogous to the standing waves that can exist on a vibrating string. This connection showed that the stable orbits allowed in Bohr's model are a result of the interference patterns of the electrons. De Broglie Waves are also covered in the chapter "Atomic Physics."

De Broglie Waves

In the chapter on waves, we treated waves and particles as if there were a clear distinction between the two. For most of the history of science, this was believed to be the case. However, in the early twentieth century, scientists were confronted with experimental evidence suggesting that the properties of matter are not always as clear-cut as everyone had assumed.

The dual nature of light

This scientific revolution began in 1900, when Max Planck introduced the possibility that energy could come in discrete units. In 1905, Einstein extended Planck's theory, suggesting that all electromagnetic waves (such as light) sometimes behave like particles. According to this theory, light can behave both like a wave and like a particle; some experiments reveal its wave nature, and other experiments display its particle nature. Although this idea was initially greeted with skepticism, it explained certain phenomena that the wave theory of light could not account for and was soon confirmed empirically in a variety of experiments.

Matter waves

The idea that light has a dual nature led Louis de Broglie to hypothesize that perhaps all matter has wavelike characteristics. De Broglie believed that there should not be two separate branches of physics, one for electromagnetic waves and another for matter. In his doctoral thesis, submitted in 1924, he proposed a theory of matter waves to reconcile this discrepancy. At that time, there was no experimental evidence to support his theory.

De Broglie's calculations suggested that matter waves had a wavelength, λ, often called the de Broglie wavelength, given by the following equation:

$$\lambda = \frac{h}{p} = \frac{h}{mv}$$

The variable h in this equation is called Planck's constant, which is approximately equal to 6.63×10^{-34} J•s. The variable p is the object's momentum, which is equivalent to its mass, m, times its velocity, v. Note that the dual nature of matter suggested by de Broglie is evident in this equation, which includes both a wave concept (λ) and a particle concept (mv).

De Broglie's equation shows that the smaller the momentum of an object, the larger its de Broglie wavelength. But even when the momentum of an object is very small from our perspective, h is so small that the wavelength is still much too small for us to detect. In order to detect a wavelength, one must use an opening comparable in size to the wavelength because waves passing through such an opening

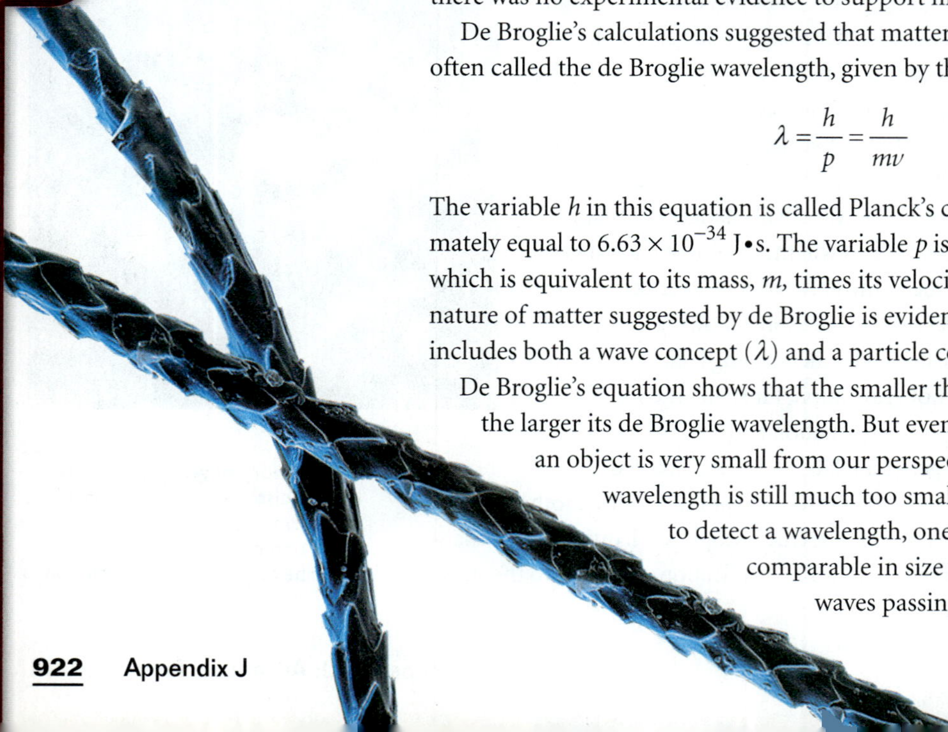

Figure 1
This image of cat hairs, produced by an electron microscope, is magnified 500 times.

will display patterns of constructive and destructive interference. When the opening is much larger than the wavelength, waves travel through it without being affected.

The de Broglie wavelength of a 0.15 kg baseball moving at 30 m/s is about 1.5×10^{-34} m. This is almost a trillion trillion times smaller than the diameter of a typical air molecule—much smaller than any possible opening through which we could observe interference effects. This explains why the de Broglie wavelength of objects cannot be observed in our everyday experience.

However, in the microscopic world, the wave effects of matter can be observed. Electrons ($m = 9.109 \times 10^{-31}$ kg) accelerated to a speed of 1.4×10^{7} m/s have a de Broglie wavelength of about 10^{-10} m, which is approximately equal to the distance between atoms in a crystal. Thus, the atoms in a crystal can act as a three-dimensional grating that should diffract electron waves. Such an experiment was performed three years after de Broglie's thesis by Clinton J. Davisson and Lester H. Germer, and the electrons did create patterns of constructive and destructive interference, such as the pattern in **Figure 2.** This experiment gave confirmation of de Broglie's theory of the dual nature of matter.

Figure 2
In this photograph, electron waves are diffracted by a crystal. Experiments such as this show the wave nature of electrons and thereby provide empirical evidence for de Broglie's theory of the dual nature of matter.

The electron microscope

A practical device that relies on the wave characteristics of matter is the electron microscope. In principle, the electron microscope is similar to an ordinary compound microscope. But while ordinary microscopes use lenses to bend rays of light that are reflected from a small object, electron microscopes use electric and magnetic fields to accelerate and focus a beam of electrons. Rather than examining the image through an eyepiece, as in an ordinary microscope, a magnetic lens forms an image on a fluorescent screen. Without the fluorescent screen, the image would not be visible.

Electron microscopes are able to distinguish details about 100 times smaller than optical microscopes can. Because of their great resolving power, electron microscopes are widely used in many areas of scientific research.

SCILINKS.
www.scilinks.org
Topic: Electron Microscopes
Code: HF60487

Figure 3
These color-enhanced images from a scanning electron microscope show, from left to right, a bean weevil emerging from a bean seed, two strips of velcro fastened together, and pollen grains.

EXTENSION

Have students research the electron microscope and its uses. In their research, ask students to compare the electron microscope with a typical optical microscope. Then have students research the various fields in which electron microscopes are used and what kinds of images they produce. Have students share their results with the class.

Teaching Tip

The equation for the de Broglie wavelength of an electron is derived from the equation for the momentum of a particle of light, also known as a photon ($p = h/\lambda$). De Broglie postulated that this equation might apply to the electron as well ($\lambda = h/p = h/mv$).

Teaching Tip

This feature discusses some basic ideas from quantum mechanics and builds on these ideas to explore a phenomenon called *electron tunneling*. The feature concludes with a discussion of the scanning tunneling microscope, which utilizes electron tunneling to generate highly detailed images that depict the atomic structure on the surface of a material.

Misconception Alert

Some students may think that the potential well shown in **Figure 1** represents a physical well or area. Be sure they understand that the height of the well corresponds to an amount of energy rather than to a physical height. Classically, the electron is confined to the well if its energy is less than *U*.

APPENDIX J Advanced Topics

Electron Tunneling

In the chapter "Electrical Energy and Current," we discussed current as the motion of charge carriers, which we treated as particles. But, as discussed in the "De Broglie Waves" appendix feature, the electron has both particle and wave characteristics. The wave nature of the electron leads to some strange consequences that cannot be explained in terms of classical physics. One example is *tunneling*, a phenomenon whereby electrons can pass into regions that, according to classical physics, they do not have the energy to reach.

Probability waves

To see how tunneling is possible, we must explore matter waves in greater detail. De Broglie's revolutionary idea that particles have a wave nature raised the question of how matter waves behave. In 1926, Erwin Schrödinger proposed a wave equation that described the manner in which de Broglie matter waves change in space and time. Two years later, in an attempt to relate the wave and particle natures of matter, Max Born suggested that the square of the amplitude of a matter wave is proportional to the probability of finding the corresponding particle at that location. This theory is called *quantum mechanics*.

Tunneling

Born's interpretation makes it possible for a particle to be found in a location that is not allowed by classical physics. Consider an electron with a potential energy of zero in the region between 0 and L (region II) of **Figure 1.** We call this region the *potential well.* The electron has a potential energy of some finite value U outside this area (regions I and III). If the energy of the electron is less than U, then according to classical physics, the electron cannot escape the well without first acquiring additional energy.

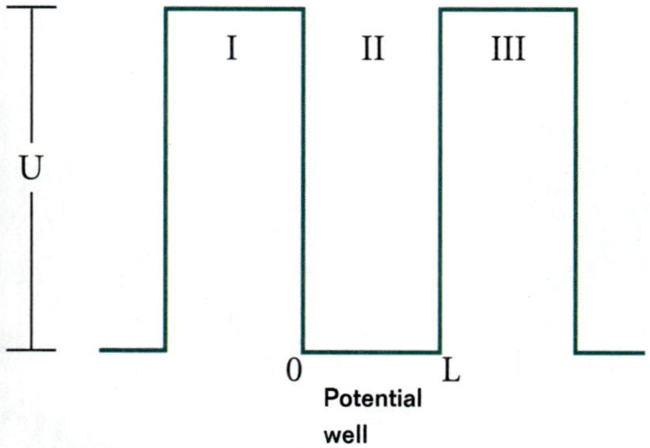

Figure 1
An electron has a potential energy of zero inside the well (region II) and a potential energy of U outside the well. According to classical physics, if the electron's energy is less than U, it cannot escape the well without absorbing energy.

The probability wave for this electron (in its lowest energy state) is shown in **Figure 2** on the next page. Between any two points of this curve, the area under the corresponding part of the curve is proportional to the probability of finding the electron in that region. The highest point of the curve corresponds to the most probable location of the electron, while the lower points correspond to less probable locations. Note that the curve never actually meets the x-axis. This means that the electron has some finite probability of being anywhere in space. Hence, there is a probability that the electron will actually be found outside the potential well. In other words, according to quantum mechanics, the electron is no longer confined to strict boundaries because of its energy. When the electron is found outside the boundaries established by classical physics, it is said to have *tunneled* to its new location.

The scanning tunneling microscope

In 1981, Gerd Binnig and Heinrich Rohrer, at IBM Zurich, discovered a practical application of tunneling current: a powerful microscope called the *scanning tunneling microscope,* or *STM.* The STM can produce highly detailed images with resolution comparable to the size of a single atom. The image of the surface of graphite shown in **Figure 3** demonstrates the power of the STM. Note that individual carbon atoms are recognizable. The smallest detail that can be discerned is about 0.2 nm, or approximately the size of an atom's radius. A typical optical microscope has a resolution no better than 200 nm, or about half the wavelength of visible light, so it could never show the detail seen in **Figure 3.**

In the STM, a conducting probe with a very sharp tip (about the width of an atom) is brought near the surface to be studied. According to classical physics, electrons cannot move between the surface and the tip because they lack the energy to escape either material. But according to quantum theory, electrons can tunnel across the barrier, provided the distance is small enough (about 1 nm). Scientists can apply a potential difference between the surface and the tip to make electrons tunnel preferentially from surface to tip. In this way, the tip samples the distribution of electrons just above the surface.

The STM works because the probability of tunneling decreases exponentially with distance. By monitoring changes in the tunneling current as the tip is scanned over the surface, scientists obtain a sensitive measure of the topography of the electron distribution on the surface. The result is used to make images such as the one in **Figure 3.** The STM can measure the height of surface features to within 0.001 nm, approximately 1/100 of an atomic diameter.

Although the STM was originally designed for imaging atoms, other practical applications are being developed. Engineers have greatly reduced the size of the STM and hope to someday develop a computer in which every piece of data is held by a single atom or by small groups of atoms and then read by an STM.

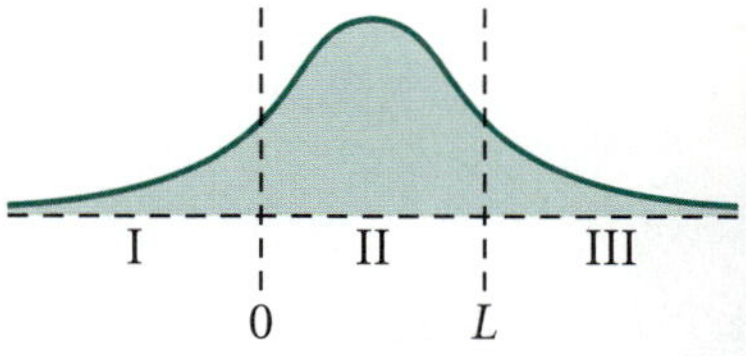

Figure 2
The probability curve for an electron in its lowest energy state shows that there is a certain probability of finding the electron outside the potential well.

Figure 3
A scanning tunneling microscope (STM) was used to produce this image of the surface of graphite, a form of carbon. The contours represent the arrangement of individual carbon atoms on the surface. An STM enables scientists to see small details on surfaces with a lateral resolution of 0.2 nm and a vertical resolution of 0.001 nm.

EXTENSION

Have students form small groups to brainstorm about possible uses of the STM. Then have them research uses of the STM that are currently being developed. Have each group share its ideas and research with the class.

Teaching Tip

Be sure students understand that regions I, II, and III of the potential well (**Figure 1**) correspond to regions I, II, and III of the probability wave (**Figure 2**). Because there is a finite probability that the electron will be found in regions I and III of **Figure 2,** there is a probability that the electron will be found outside the potential well.

Key Models and Analogies

Be sure students understand the difference between the classification of materials into categories based on their ability to conduct charge and the physical characteristics that lead to the different classifications. Remind them that electrical attraction and repulsion can be explained at the atomic level by considering the transfer of electrons between objects. Then tell them that the ability of some materials to conduct charge better than others is explained at the atomic level by band theory.

Visual Strategy

Figure 1

Be sure students understand that the third case, **(c)**, is simply an extension of the second case, **(b)**, in which the energy levels are so close together that they are represented as continuous bands.

Q What causes the splitting of energy levels in each of these cases?

A ***When more than one atom is present, the electric field produced by each atom affects the energy levels of other atoms, causing the splitting shown in the figure.***

Q Could splitting be observed for a single atom?

A ***no***

APPENDIX J Advanced Topics

Semiconductor Doping

Materials can be classified according to their ability to conduct electricity. A good *conductor* has a large number of free charge carriers that can move easily through the material, whereas an *insulator* has a small number of free charge carriers that are relatively immobile. *Semiconductors* exhibit electronic properties between those of insulators and those of conductors. The development of *band theory* uses basic physical principles to explain some of the properties of these three categories of materials.

Electron energy levels

As seen in the chapter "Atomic Physics," the electrons in an atom can possess only certain amounts of energy. For this reason, the electrons are often said to occupy specific *energy levels.* Electrons in a shell sometimes form a set of closely spaced energy levels. Normally, electrons are in the lowest energy level available to them. The specific arrangement of electrons in which all are in the lowest possible energy levels of an atom is called the atom's **ground state.**

If an atom absorbs sufficient energy from the environment, some of the atom's electrons can move to higher energy levels. The atom is then said to be in an **excited state.** If an electron absorbs so much energy that it is no longer bound to the atom, it is then called a *free electron.*

Band theory

Band theory uses the concept of energy levels to explain the mechanisms of conduction in many solids. When identical atoms are far apart, they have identical energy-level diagrams. No two electrons in the same system can occupy the same state. As a result, when two atoms are brought closer together, the energy levels of each atom are altered by the influence of the electric field of the other atom. **Figure 1** shows how two energy levels split when there are two atoms **(a),** four atoms **(b),** and many atoms **(c)** at different separation distances. In the case of two atoms, each energy level splits into two different energy levels, as shown in **Figure 1(a).** Notice that the energy difference between two new energy levels depends on the distance between the atoms.

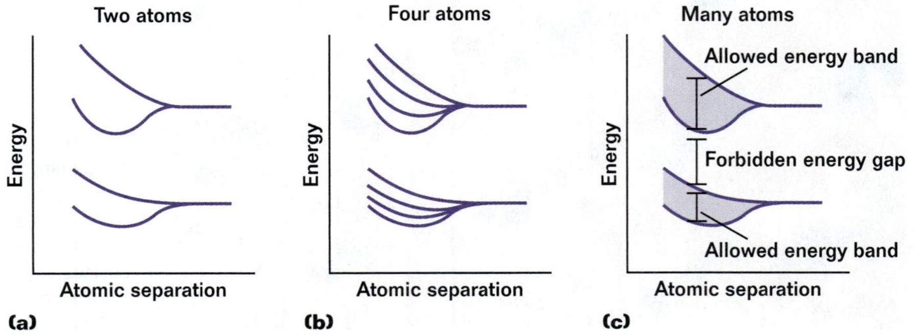

Figure 1
Energy levels split when two atoms are close together **(a).** Adding a few more nearby atoms causes further splitting **(b).** When many atoms interact, the energy levels are so closely spaced that they can be represented as energy bands **(c).**

When more atoms are brought close together, each energy level splits into more levels. If there are many atoms, the energy level splits so many times and the new energy levels are so closely spaced that they may be regarded as a continuous band of energies, as in **Figure 1(c).** The highest band containing occupied energy levels is called the *valence band,* as shown in **Figure 2.** The band immediately above the valence band is called the *conduction band.*

Electron-hole pairs and intrinsic semiconductors

Imagine that a few electrons are excited from the valence band to the conduction band by an electric field, as in **Figure 3.** The electrons in the conduction band are free to move through the material. Normally, electrons in the valence band are unable to move because all nearby energy levels are occupied. But when an electron moves from the valence band into the conduction band, it leaves a vacancy, or **hole,** in an otherwise filled valence band. The hole is positively charged because it results from the removal of an electron from a neutral atom. Whenever another valence electron from this or a nearby atom moves into the hole, a new hole is created at its former location. So, the net effect can be viewed as a positive hole migrating through the material in a direction opposite the motion of the electrons in the conduction band.

In a material containing only one element or compound, there are an equal number of conduction electrons and holes. Such combinations of charges are called *electron-hole pairs,* and a semiconductor that contains such pairs is called an *intrinsic semiconductor.* In the presence of an electric field, the holes move in the direction of the field and the conduction electrons move opposite the field.

Adding impurities to enhance conduction

One way to change the concentration of charge carriers is to add *impurities,* atoms that are different from those of an intrinsic semiconductor. This process is called **doping.** Even a few added impurity atoms (about one part in a million) can have a large effect on a semiconductor's resistance. The semiconductor's conductivity increases as the doping level increases. When impurities dominate conduction, the material is called an *extrinsic semiconductor.*

There are two methods for doping a semiconductor: either add impurities that have extra valence electrons or add impurities that have fewer valence electrons compared with the atoms in the intrinsic semiconductor.

Semiconductors used in commercial devices are usually doped silicon or germanium. These elements have four valence electrons. Semiconductors are doped by replacing an atom of silicon or germanium with one containing either three valence electrons or five valence electrons. Note that a doped semiconductor is electrically neutral because it is made of neutral atoms. The balance of positive and negative charges has not changed, but the number of charges that are free and able to move has. These charges are therefore able to participate in electrical conduction.

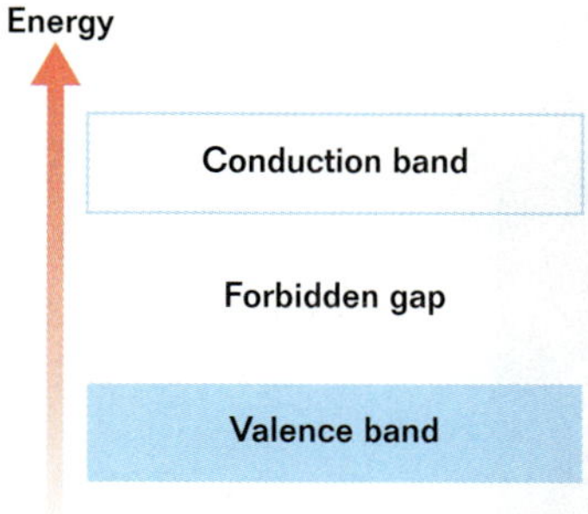

Figure 2
Energy levels of atoms become energy bands in solids. The valence band is the highest occupied band.

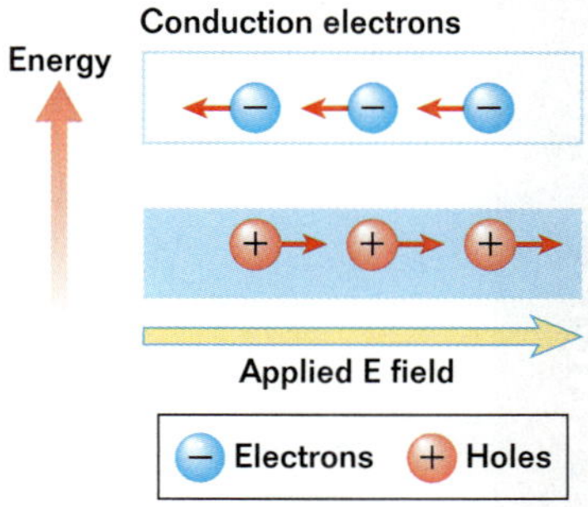

Figure 3
An electric field can excite valence electrons into the conduction band, where they are free to move through the material. Holes in the valence band can then move in the opposite direction.

SCILINKS.
www.scilinks.org
Topic: Semiconductors
Code: HF61376

Demonstration

Hole Flow

Purpose Illustrate hole flow.
Materials nine rubber stoppers
Procedure Have 10 students stand facing the class with their right palms out. Place a stopper in the hand of every person except the student on the far right.

Beginning at the far right, have each student look to their right and place their stopper in their neighbor's palm if that person does not already have a stopper. All of the stoppers will shift one palm to the right, and the person on the far left will be without a stopper. Ask the students to consider the movement of the empty space as an electron "hole." Point out that the hole moves to the left as the stoppers (the electrons) move to the right.

extension

In-Depth Physics Content
Your students can visit go.hrw.com for an online chapter that integrates more in-depth development of the concepts covered here.
Keyword HF6APJX

Demonstration

Lattice Imperfections

Purpose Visually demonstrate the effects of lattice imperfections on electron movement.

Materials board, nails, marble

Procedure

Before class: Prepare a "lattice" by placing nails in a board in a regular pattern (about every 2 cm) with occasional variations. There should be no clear routes from one edge of the board to the opposite edge.

In class: Show students that there is room for a marble to pass unobstructed between the nails. Point out that the occasional misplaced nails prevent a free path through the lattice. Demonstrate this by holding the board at an angle and allowing the marble to roll down. You may want to make an additional lattice with more imperfections to show students the comparison between the two.

Superconductors and BCS Theory

The resistance of many solids (other than semiconductors) increases with increasing temperature. The reason is that at a nonzero temperature, the atoms in a solid are always vibrating, and the higher the temperature, the larger the amplitude of the vibrations. It is more difficult for electrons to move through the solid when the atoms are moving with large amplitudes. This situation is somewhat similar to walking through a crowded room. It is much harder to do so when the people are in motion than when they are standing still.

If the resistance depended only on atomic vibrations, we would expect the resistance of the material that is cooled to absolute zero to go gradually to zero. Experiments have shown, however, that this does not happen. In fact, the resistances of very cold solids behave in two very different ways—either the substance suddenly begins superconducting at temperatures above absolute zero or it never superconducts, no matter how cold it gets.

Resistance from lattice imperfections

The graph in **Figure 1** shows the temperature dependence of the resistance of two similar objects, one made of silver and the other made of tin. The temperature dependence of the resistance of the silver object is similar to that of a typical metal. At higher temperatures, the resistance decreases as the metal is cooled. This decrease in resistance suggests that the amplitude of the lattice vibrations is decreasing, as expected. But at a temperature of about 10 K, the curve levels off and the resistance becomes constant. Cooling the metal further does not appreciably lower the resistance, even though the vibrations of the metal's atoms have been lessened.

Part of the cause of this nonzero resistance, even at absolute zero, is *lattice imperfection.* The regular, geometric pattern of the crystal, or lattice, in a solid is often flawed. A lattice imperfection occurs when some of the atoms do not line up perfectly.

Imagine you are walking through a crowded room in which the people are standing in perfect rows. It would be easy to walk through the room between two rows. Now imagine that occasionally one person stands in the middle of the aisle instead of in the row, making it harder for you to pass. This is similar to the effect of a lattice imperfection. Even in the absence of thermal vibrations, many materials exhibit a *residual resistance* due to the imperfect geometric arrangement of their atoms.

Figure 1 shows that the resistance of tin jumps to zero below a certain temperature that is well above absolute zero. A solid whose resistance is zero below a certain nonzero temperature is called a **superconductor.** The temperature at which the resistance goes to zero is the critical temperature of the superconductor, as described in the chapter "Electrical Energy and Current."

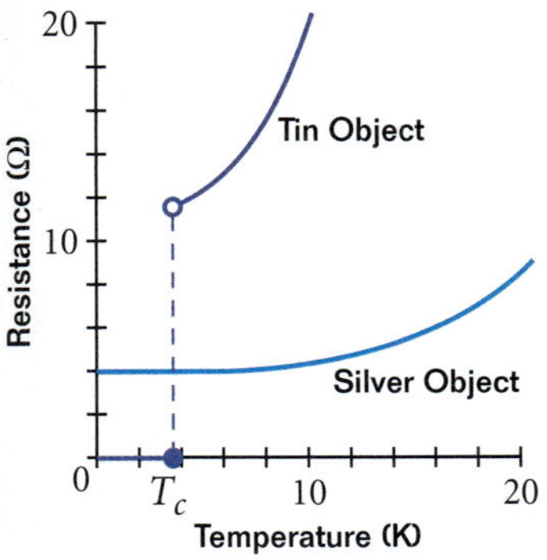

Figure 1
The resistance of silver exhibits the behavior of a normal metal. The resistance of tin goes to zero at temperature T_c, the temperature at which tin becomes a superconductor.

BCS theory

Before the discovery of superconductivity, it was thought that all materials should have some nonzero resistance due to lattice vibrations and lattice imperfections, much like the behavior of the silver in **Figure 1.**

The first complete microscopic theory of superconductivity was not developed until 1957. This theory is called *BCS theory* after the three scientists who first developed it: John Bardeen, Leon Cooper, and Robert Schrieffer. The crucial breakthrough of BCS theory is a new understanding of the special way that electrons traveling in pairs move through the lattice of a superconductor. According to BCS theory, electrons do suffer collisions in a superconductor, just as they do in any other material. However, the collisions do not alter the total momentum of a pair of electrons. The net effect is as if the electrons moved unimpeded through the lattice.

Cooper pairs

Imagine an electron moving through a lattice, such as electron 1 in **Figure 2.** There is an attractive force between the electron and the nearby positively charged atoms in the lattice. As the electron passes by, the attractive force causes the lattice atoms to be pulled toward the electron. The result is a concentration of positive charge near the electron. If a second electron is nearby, it can be attracted to this excess positive charge in the lattice before the lattice has had a chance to return to its equilibrium position.

Through the process of deforming the lattice, the first electron gives up some of its momentum. The deformed region of the lattice attracts the second electron, transferring excess momentum to the second electron. The net effect of this two-step process is a weak, delayed attractive force between the two electrons, resulting from the motion of the lattice as it is deformed by the first electron. The two electrons travel through the lattice acting as if they were a single particle. This particle is called a *Cooper pair.* In BCS theory, Cooper pairs are responsible for superconductivity.

The reason superconductivity has been found at only low temperatures so far is that Cooper pairs are weakly bound. Random thermal motions in the lattice tend to destroy the bonds between Cooper pairs. Even at very low temperatures, Cooper pairs are constantly being formed, destroyed, and reformed in a superconducting material, usually with different pairings of electrons.

Calculations of the properties of a Cooper pair have shown that this peculiar bound state of two electrons has zero total momentum in the absence of an applied electric field. When an external electric field is applied, the Cooper pairs move through the lattice under the influence of the field. However, the center of mass for every Cooper pair has exactly the same momentum. This crucial feature of Cooper pairs explains superconductivity. If one electron scatters, the other electron in a pair also scatters in a way that keeps the total momentum constant. The net result is that scattering due to lattice imperfections and lattice vibrations has no net effect on Cooper pairs.

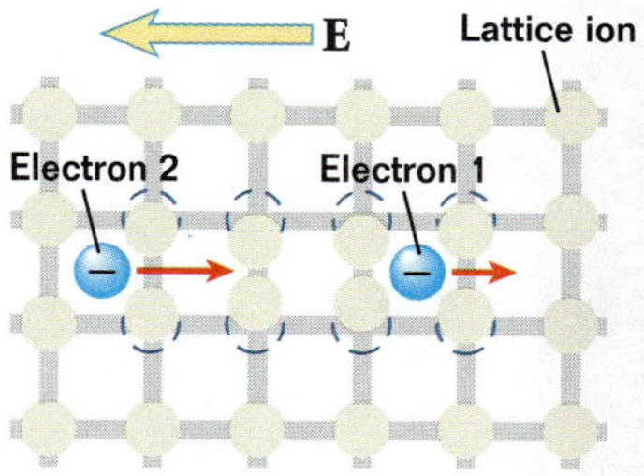

Figure 2
The first electron deforms the lattice, and the deformation affects the second electron. The net result is as if the two electrons were loosely bound together. Such a two-electron bound state is called a *Cooper pair.*

SCILINKS.
www.scilinks.org
Topic: Superconductors
Code: HF61478

Demonstration

Cooper Pairs

Purpose Help students visualize Cooper pairs.

Materials two tennis balls joined by a 20 cm string

CAUTION *Throw the tennis balls away from students.*

Procedure Explain that this demonstration is limited in its ability to show electron-lattice interactions but that it does show conservation of momentum (horizontally) for a pair of objects linked together, such as the electrons in a Cooper pair. Hold one of the tennis balls, whirl the free tennis ball over your head, and then release the pair into the air. The whirling pair shows horizontal conservation of momentum (disregarding gravitational effects). The momentum of an individual ball changes, but the pair acts as one object and travels in a straight line.

extension

In-Depth Physics Content

Your students can visit go.hrw.com for an online chapter that integrates more in-depth development of the concepts covered here.

Keyword HF6APJX

Antimatter

Startling discoveries made in the twentieth century have confirmed that electrons and other particles of matter have *antiparticles.* Antiparticles have the same mass as their corresponding particle but an opposite charge.

The discovery of antiparticles

The discovery of antiparticles began in the 1920s with work by the theoretical physicist Paul Adrien Maurice Dirac (1902–1984), who developed a version of quantum mechanics that incorporated Einstein's theory of special relativity. Dirac's theory was successful in many respects, but it had one major problem: its relativistic wave equation required solutions corresponding to negative energy states. This negative set of solutions suggested the existence of something like an electron but with an opposite charge, just as the negative energy states were opposite to an electron's typical energy states. At the time, there was no experimental evidence of such antiparticles.

In 1932, shortly after Dirac's theory was introduced, evidence of the anti-electron was discovered by the American physicist Carl Anderson. The anti-electron, also known as the *positron,* has the same mass as the electron but is positively charged. Anderson found the positron while examining tracks created by electronlike particles in a cloud chamber placed in a magnetic field. As described in the chapter "Magnetism," such a field will cause moving particles to follow curved paths. The direction in which a particle moves depends on whether its charge is positive or negative. Anderson noted that some of the tracks had deflections typical of an electron's mass, but in the opposite direction, corresponding to a positively charged particle.

Pair production and annihilation

Since Anderson's initial discovery, the positron has been observed in a number of experiments. In perhaps the most common process, a gamma ray with sufficiently high energy collides with a nucleus, creating an electron-positron pair. An example of this process, known as *pair production,* is shown in **Figure 1** on the next page.

During pair production, the energy of the photon is completely converted into the rest energy and kinetic energy of the electron and the positron. Thus, pair production is a striking verification of the equivalence of mass (rest energy) and other forms of energy as predicted by Einstein's special theory of relativity. (This equivalence is discussed in the appendix feature "The Equivalence of Mass and Energy.")

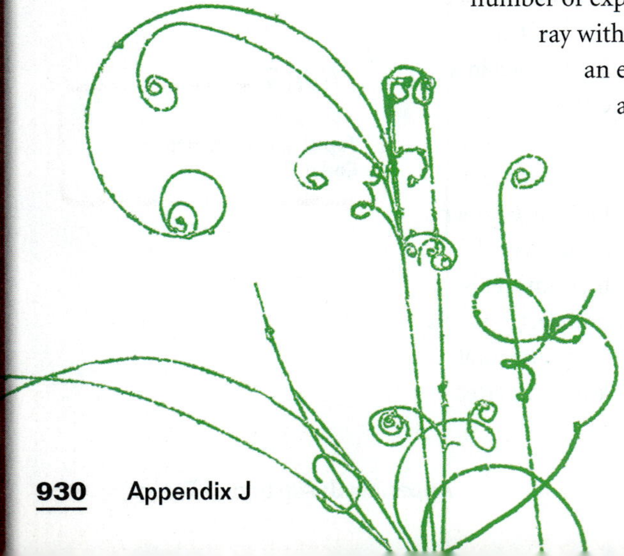

Teaching Tip

This feature discusses the discovery of antimatter particles, the processes of pair production and annihilation, and the production of anti-hydrogen atoms in a particle accelerator. The charge of an anti-electron, also known as a positron, is equal in magnitude but opposite in sign to the charge of an electron. The positron and electron have equal masses and opposite spins. Antimatter was theoretically predicted by Dirac in the late 1920s. The anti-electron was detected experimentally in 1932, and the anti-proton and anti-neutron were detected in 1955 and 1956, respectively. Today, it is believed that every particle has a corresponding antiparticle.

Teaching Tip

Point out that charge must be conserved in pair production and annihilation. For example, in pair production, the initial photon has zero charge, and the total charge of the positron and electron is also zero. It is impossible for a photon to produce a single electron or positron because this would violate the conservation of charge.

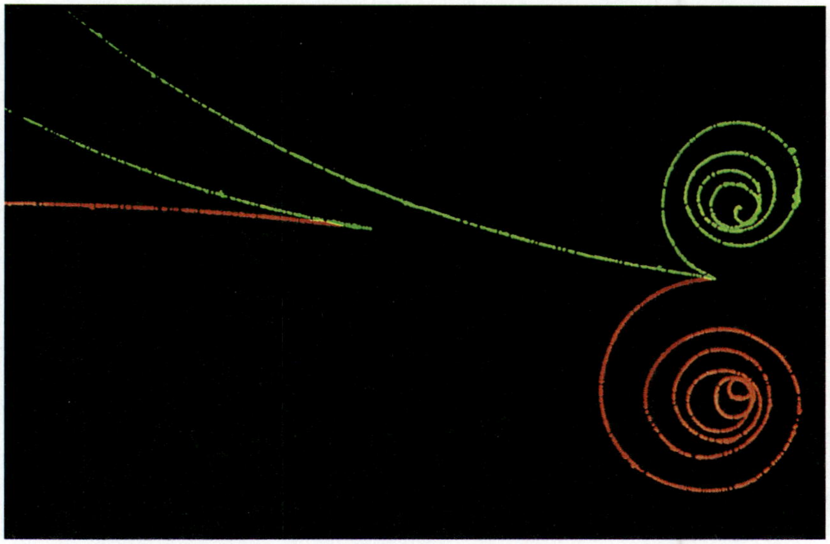

Figure 1
The red and green spirals shown here are the paths of a positron and an electron moving through a magnetic field. Note that these paths have about the same shape but are opposite in direction.

Once formed, a positron will most likely soon collide with an oppositely charged electron in a process known as *pair annihilation.* This process is the opposite of pair production—an electron-positron pair produces two photons. In the simplest example of pair annihilation, an electron and a positron initially at rest combine with each other and disappear, leaving behind two photons. Because the initial momentum of the electron-positron pair is zero, it is impossible to produce a single photon. Momentum can be conserved only if two photons moving in opposite directions, both with the same energy and magnitude of momentum, are produced.

SCILINKS.
www.scilinks.org
Topic: Antiparticles
Code: HF60084

Antimatter produced in a particle accelerator

After the positron was discovered, physicists began to search for the anti-proton and anti-neutron. However, because the proton and neutron are much more massive than the electron, a much greater amount of energy is required to produce their antiparticles. By 1955, technological advances in particle accelerators brought evidence of the anti-proton, and evidence of the anti-neutron was found a year later.

The discovery of other antiparticles leads to the question of whether these antiparticles can be combined to form antimatter and, if so, how that antimatter would behave. In 1995, physicists at the CERN particle accelerator in Geneva, Switzerland, succeeded in producing anti-hydrogen atoms, that is, atoms with a single anti-electron orbiting an anti-proton. Researchers observed nine anti-hydrogen atoms during a three-week period. Unfortunately, the anti-hydrogen atoms had a short lifetime—less than 37 billionths of a second—because as soon as an anti-hydrogen atom encountered ordinary matter, the two annihilated one another. Attempts to produce antimatter for greater time periods are currently under way.

Teaching Tip

Point out that the tracks shown in **Figure 1** are created when charged particles move through a liquid that is just below its boiling point. The chamber containing the liquid is placed in a magnetic field. Charged particles create bubbles as they move through the liquid, and these bubbles are photographed to create images like the one shown. Because the charges are in a magnetic field, the direction in which a particle moves depends on the particle's charge, and the curvature depends on the particle's momentum. The particles shown in **Figure 1** have equal and opposite momenta and charges.

EXTENSION

In April 1997, researchers announced the exciting discovery of evidence of a gigantic stream of positrons at the center of the galaxy. This discovery was made by NASA's Earth-orbiting *Compton Gamma Ray Observatory* satellite. This satellite detected 511 000 eV gamma rays that are thought to have been produced in pair annihilation. The cause and nature of this positron fountain are currently being researched. Have interested students research this discovery and any recent developments related to the discovery, and have them present their findings to the class.

Background
This probeware version of the Skills Practice Lab "Free-Fall Acceleration" from the chapter "Motion in One Dimension" uses a Vernier motion detector instead of a motion timer. The motion detector is easier to use than a motion timer and requires no calibration.

Safety Caution
Remind students to keep the area clear during the experiment and to prevent the blocks from striking their feet.

Tips and Tricks

- Students should have the program DataMate® on their graphing calculators. Refer to Appendix B for instructions. The motion detector must be clamped tightly so that it can't move during the experiment. Students should practice dropping the blocks so that they don't rotate. Remind students to keep their hands out of the way of the motion detector's signal.
- Remind students to keep at least 0.5 m between the block and the sensor to allow the sensor to work properly.
- Show the students how to use the arrow keys to trace the graph in step 12. Explain how to choose the points and how to find the difference between the *y*-values of the points in step 12. Review what the graphs of velocity and acceleration in steps 13 and 14 reveal about the motion of the block. Remind students to make sure the calculator is turned on before they perform each trial.

Free-Fall Acceleration

MATERIALS LIST

- **3 wooden blocks of different masses**
- **balance**
- **C-clamp**
- **LabPro® or CBL 2™ interface**
- **masking tape**
- **meterstick**
- **support stand with V-clamp**
- **thin foam pad**
- **TI graphing calculator with link cable**
- **Vernier motion detector**

You can use a motion detector and a LabPro or CBL 2 interface instead of a recording timer to determine velocity and acceleration. The motion detector measures the position of an object by sending sound waves toward the object and measuring the time that the waves take to echo back to the sensor.

SAFETY

- **Tie back long hair, secure loose clothing, and remove loose jewelry to prevent its getting caught in moving or rotating parts.**
- **Attach masses securely. Falling or dropped masses can cause serious injury.**

PROCEDURE

Preparation

Follow Preparation steps 1–2 for the Skills Practice Lab "Free-Fall Acceleration" in the chapter "Motion in One Dimension."

Apparatus Setup

3. Connect the LabPro or CBL 2 interface to the calculator with the unit-to-unit link cable. Connect the motion detector to either of the two DIG/SONIC ports on the interface.

4. Set up the apparatus as shown in **Figure 1.** Place the ring stand near the edge of the lab table. Use the C-clamp to clamp the base of the ring stand securely to the table. Position the clamp so that it protrudes as little as possible.

5. Using the V-clamp, securely clamp the motion detector to the ring stand so that the detector faces down, over the edge of the table. Make sure the motion detector is far enough away from the edge of the table so the signal will not hit the tabletop, clamp, or table leg. Cover the floor under the motion detector with a foam pad to reduce feedback.

6. Use a meterstick to measure a distance 0.5 m below the motion detector, and mark the point with tape on the ring stand. This point is the starting position from which the blocks will be dropped from rest.

7. Start the program DataMate® on your graphing calculator. Press CLEAR to reset the program.

8. Test to be sure the motion detector is positioned properly.

a. Read the measurement displayed on the calculator screen for the distance between the motion detector and the floor. If the reading seems unusually low, adjust the motion detector to make sure the signal is not hitting the table.

b. Hold the wooden block directly beneath the motion detector, and move the block up and down. Note the distance measurement displayed on the calculator. Make sure the motion detector is not "seeing" other objects, such as the stand base, the tabletop, or the table leg.

Speed and Acceleration of a Falling Object

9. Measure the mass of the first wooden block, and label the block with tape. Record the mass in your data table. Hold the block horizontally between your hands, with your hands flat. Position the block directly below the motion detector and level with the 0.5 m mark.

10. When the area is clear of people and objects, one student should select START to begin data collection. When the motion detector begins to click, the student holding the block should wait two seconds and then release the block by pulling both hands out to the side. Releasing the block this way will prevent the block from twisting as it falls, which may affect your results.

11. When the motion detector has stopped clicking, the graph selection screen will appear on the calculator. Press ENTER to plot a graph of the distance in meters against time in seconds. (Note: If the graph has spikes or black lines, repeat the trial to obtain a smooth graph.)

12. Use the arrow keys to trace along the curve. On the far left and the far right, the curve represents the position of the block before and after its motion. The middle section of the curve represents the motion of the falling block. Sketch this graph in your lab notebook.

a. Choose a point on the curve at the beginning of this middle section. Press the right arrow key once to select the next point. The difference between the x-values for these two readings will be 0.05 s, the time interval between successive readings. Find the difference between the y-values of these two points, and record it as *A–B* for *Trial 1* in your data table.

b. Press the right arrow key twice to move to another point on the curve. Press the right arrow key once more to select the next point. Find the difference between the y-values of these two points, and record it as *C–D* for *Trial 1* in your data table.

c. Choose two more pairs of points along the curve. Select points at even intervals along the curve so that the first pair is at the beginning of the block's motion and the last pair is at the end. Find the difference between the y-values of each pair of points. Record them as *E–F* and *G–H* for *Trial 1*. Press ENTER to return to the graph selection screen.

13. Press the down arrow key once, and then press ENTER to display a graph of the velocity in m/s against time in seconds. Sketch this graph in your lab notebook. Press ENTER to return to the graph selection screen.

14. Press the down arrow key once and then press ENTER to display a graph of the acceleration in m/s^2 against time in seconds. Sketch this graph in your lab notebook. Press ENTER to return to the graph selection screen. Select MAIN SCREEN to return to the main screen of DataMate®.

15. Repeat this procedure using wooden blocks of different masses. Drop each block from the same level in each trial. Record all data for each trial.

16. Clean up your work area. Put equipment away safely so that it is ready to be used again.

ANALYSIS, CONCLUSIONS, AND EXTENSION

Complete the Analysis and Conclusions items for the Skills Practice Lab "Free-Fall Acceleration." Your teacher may also instruct you to complete the Extension exercise. (Note: For Analysis item 1, use 0.02 s for the average period of the timer.)

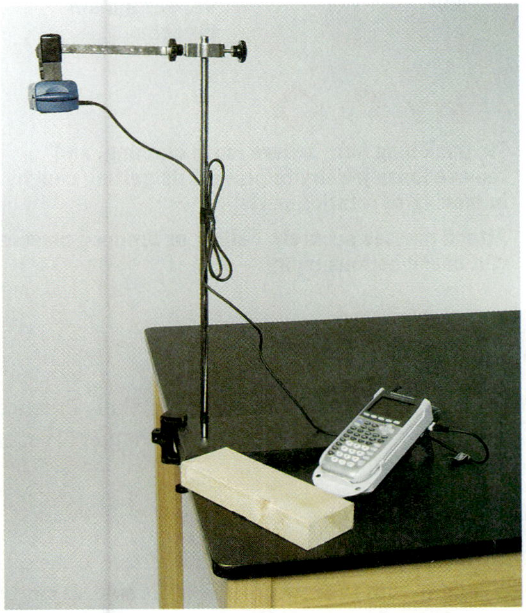

Figure 1

Step 8b: While you read the distance measurements displayed on the calculator, move the wooden block up and down below the motion detector to check the readings.

Step 9: Hold the block flat and parallel to the tape mark.

Step 10: Release the block by pulling hands straight to the sides. It may take some practice to release the block so that it falls straight down without turning. If the block turns while falling, the motion detector will measure the distance to the closest part of the block and introduce error into your results.

- If the CBL 2 yields values that seem out of range, check the calculator batteries.

✓ Checkpoints

Step 6: Make sure students realize that the motion detector is the reference point for all distance measurements. The tape mark must be measured below the detector, not above the ground.

Step 8: Students should be able to demonstrate that the detector is working properly. They should be able to explain why the distance measurement increases as the block moves away and decreases as the block moves toward the detector.

Step 10: Make sure students can drop the block without letting it turn or getting their hands in the way.

Step 11: The graph should be smooth; dense, black lines or spikes signify "drop outs," times when the detector could not detect the block. Check the setup, and have students practice dropping the block before repeating the trial.

Step 12: Students should identify the portion of the curve that represents the block's motion. Guide students to choose four pairs of consecutive points on the graph.

Step 13: Students should be able to interpret the velocity graph on the calculator. They should be able to explain that the velocity of the block increased as the block fell.

Step 14: Students should be able to use the acceleration graph to explain whether the motion of the block meets their expectations. If there are any problems with the setup, students should be alerted by the graphs.

Background

This probeware version of the Skills Practice Lab "Force and Acceleration" from the chapter "Forces and the Laws of Motion" uses a Vernier motion detector instead of a motion timer. The motion detector is easier to use than a motion timer and requires no calibration. This version also adds a Vernier dual-range force sensor, which allows students not only to calculate force from the mass but also to measure force directly.

Safety Caution

Remind students to fasten masses securely, to keep the area clear during the experiment, and to prevent the cart from falling off the table.

Tips and Tricks

- Narrow, smooth drapery cord works better than ordinary string or twine.
- Students should have the program DataMate® on their calculators. Refer to Appendix B for instructions. The motion detector must be stabilized so that it can't move.
- Show the students how to use the arrow keys to trace the graph. Explain how to choose the points in step 14 and how to find the difference between the *x*- and *y*-values in step 15.

✓ Checkpoints

Step 6: Make sure that all clamps are placed so that they protrude as little as possible. If the sensor has multiple settings, set it to 10 N. Warn students not to pull on the force sensor.

Force and Acceleration

MATERIALS LIST

- balance
- calibrated masses and holder
- cord, smooth
- dynamics cart
- hooked mass, 1000 g
- LabPro® or CBL 2™ interface
- mass hanger
- meterstick
- pulley with table clamp
- rod and parallel clamp
- square of poster board, 25 cm × 25 cm
- support stand with V-jaw clamp
- tape
- TI graphing calculator with link cable
- Vernier dual-range force sensor
- Vernier motion detector

SAFETY

- **Tie back long hair, secure loose clothing, and remove loose jewelry to prevent its getting caught in moving or rotating parts.**
- **Attach masses securely. Falling or dropped masses can cause serious injury.**

PROCEDURE

Preparation

Follow Preparation steps 1–3 for the Skills Practice Lab "Force and Acceleration" in the chapter "Forces and the Laws of Motion."

Apparatus Setup

4. Connect the LabPro or CBL 2 interface to the calculator with the unit-to-unit link cable. Connect the Dual-Range Force Sensor to the CH1 port on the interface, and set the switch on the sensor to 10N. Connect the motion detector to the DIG/SONIC 1 port on the interface.

5. Turn on the calculator, and start the DataMate® program. Press CLEAR to reset the program.

6. Set up the apparatus as shown in **Figure 1.** Securely tape the force sensor to the dynamics cart. Tape the poster board to the opposite end of the dynamics cart to make a flat, vertical surface. Clamp the pulley to the table edge using a table clamp, rod, and parallel clamp so that the pulley is level with the force sensor hook. Position the motion detector so that the cart will move away from it in a straight line. Securely clamp the motion detector to the ring stand. Place a piece of tape 0.5 m in front of the motion detector to serve as a starting line for the cart. (**Note: Do not pull on the force sensor.**)

Constant Mass with Varying Force

7. Carefully measure the mass of the cart assembly on the platform balance, making sure that the cart does not roll or fall off the balance. Then, load it with masses equal to 0.60 kg. Lightly tape the masses to the cart to hold them in place.

8. Attach one end of the cord to a small mass hanger and the other end of the cord to the force sensor. Pass the cord over the pulley, and fasten a small mass to the end to offset the frictional force on the cart. The mass is correct when the car moves forward with a constant velocity when you give it a push. The car will have constant velocity for only a short period after it is pushed, then it will accelerate as the counterweight drops. *This counterweight should stay on the cord throughout the entire experiment.* Add the mass of the counterweight to the mass of the cart and masses, and record the sum as *Total Mass* in your data table.

9. For the first trial, remove a 0.10 kg mass from the cart, and securely fasten it to the end of the cord along with the counterweight. Record 0.10 kg as the *Accelerating Mass* in the data table.

10. Place the cart so that the poster-board end is closest to the motion detector and is lined up with the tapeline, 0.5 m in front of the motion detector. Keep the force sensor cord clear so that the cart will be able to move freely.

11. Make sure that the calculator is turned on. Make sure that the area under the falling mass is clear of obstacles. Select START to begin collecting data, and release the cart simultaneously. The motion detector will begin to click as it collects data.

12. Carefully stop the cart when the 0.10 kg mass hits the floor. Do not let the cart fall off the table.

13. When the motion detector has stopped clicking, the graph selection screen will appear on the calculator. Press ENTER to plot a graph of the force sensor reading against time. Use the arrow keys to trace along the curve. The y-value is the force in newtons; it should be fairly constant. Record this value as the *Accelerating Force* in the data table. Press ENTER to return to the graph selection screen.

14. Press the down arrow key once, and then press ENTER to display a graph of the distance in meters against time. Use the arrow keys to trace along the curve. On the far left and the far right, the flat portion of the curve represents the positions of the cart before and after its motion. The middle section of the curve represents the motion of the cart. Choose a point on the curve near the beginning of this middle section (but not the beginning point itself), and choose another point near the end.

15. Find the difference between the y-values of these two points, and record it as the *Distance* for *Trial 1* in your data table. Find the difference between the x-values for these two readings to find the time elapsed between measurements. Record this as the *Time Interval* for *Trial 1* in your data table. Press ENTER to return to the graph selection screen.

a. Press the down arrow key once, and then press ENTER to display a graph of the velocity in m/s against time. Press ENTER to return to the graph selection screen.

b. Press the down arrow key once, and then press ENTER to display a graph of the acceleration in m/s^2 against time. Press ENTER to return to the graph selection screen. Select MAIN SCREEN to return to the main screen.

16. Replace the 0.10 kg mass in the cart. Remove the 0.20 kg mass from the cart, and attach it securely to the end of the cord. Repeat the procedure for *Trial 2*.

17. Leave the 0.20 kg mass on the end of the cord, and attach the 0.10 kg mass from the cart securely to the end of the cord. Repeat the procedure for *Trial 3*.

Constant Force with Varying Mass

18. For the two trials in this part of the experiment, keep 0.30 kg and the counterweight on the cord. Be sure to include this mass when recording the total mass for these trials.

19. Add a 0.50 kg mass to the cart. Tape the mass to the cart to keep it in place. Run the experiment and record the total mass, accelerating mass, accelerating force, distance, and time under *Trial 4* in your data table.

20. Tape a 1.00 kg mass to the cart, and repeat the procedure. Record the data under *Trial 5* in your data table.

21. Clean up your work area. Put equipment away safely so that it is ready to be used again.

ANALYSIS, CONCLUSIONS, AND EXTENSION

Complete the Analysis and Conclusions items for the Skills Practice Lab "Force and Acceleration." Your teacher may also instruct you to complete the Extension exercise. (Note: For Analysis item 1, do not enter new values for *Accelerating Force* in your data table. Instead, compare your calculated values with the values you entered earlier.)

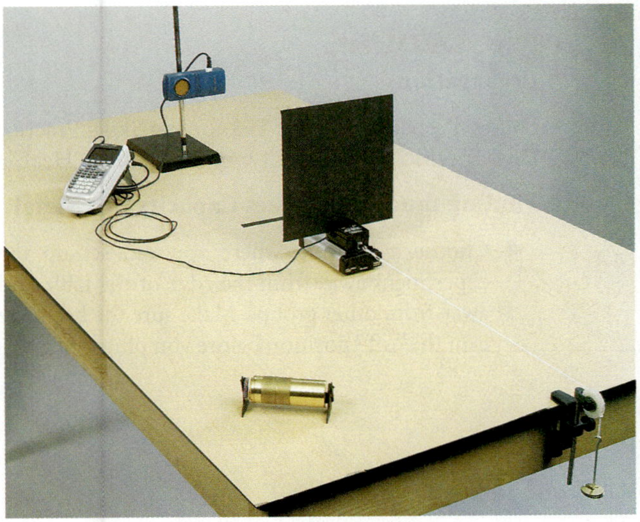

Figure 1

Step 6: Make sure that the motion detector has a clear view of the cart. For all trials, start the cart from the same position.

Step 13: Using the data collection interface with the graphing calculator will allow you to view graphs of the motion after each trial.

Step 8: Make sure that masses are securely attached. Students should be able to demonstrate that the counterweight allows the cart to move at a constant velocity when the cart is given a small push.

Step 10: The force sensor cord must be kept out of the path of the cart; it must also be prevented from dragging behind the cart and slowing it down.

Step 12: It may help to clamp a wood block to the edge of the table to serve as a bumper to keep the cart from rolling off the table.

Step 13: The graph will not be a perfectly straight line. Students may be confused by the values above and below the line. Guide students to select the most frequently appearing value.

Step 14: Students should identify the part of the curve that shows the motion of the cart. If the curve does not show the cart moving away from the detector, make sure that the detector points toward the cart with no interference. If there is interference from the tabletop, tilt the detector slightly upward. Cover the tabletop and any objects on the table, such as gas jets or faucets, with soft cloths.

Step 15: Students should be able to interpret the velocity and acceleration graphs on the graphing calculator. Students should be able to use the graphs to explain whether the motion of the cart meets their expectations. If there are any problems with the setup, the students should be alerted by the graphs.

Step 18: Students should recognize the difference between these two trials and the previous trials. They must add mass to the cart each time without changing the mass on the cord.

Background

This probeware version of the Skills Practice Lab "Specific Heat Capacity" from the chapter "Heat" uses temperature probes instead of standard thermometers. Because temperature probes are not prone to shattering, they are safer than thermometers.

Safety Caution

This lab presents several safety hazards. Remind students never to leave a hot plate unattended while it is on. Make sure all students wear the appropriate safety gear at all times.

- Each group must have at least two students. Groups must perform more than one step at a time, so groups may be larger than usual. With larger groups, make sure all students are involved and paying attention to safety.
- Bringing water to a boil takes 10–20 min; other tasks can be completed as the water heats.
- Each group must have a level work surface large enough so that students are not too close to the hot plates.
- Students should place wet metal shot in designated containers at the end of the lab to be dried and used again.

Tips and Tricks

- Students should have the program DataMate® on their graphing calculators. Refer to Appendix B for instructions. Students should begin collecting data for the calorimeter *before* adding the metal sample to the calorimeter. When adding the metal to the calorimeter, don't let it touch the sensor in the water.

Specific Heat Capacity

MATERIALS LIST

- **beakers, 2**
- **balance**
- **hot plate**
- **ice**
- **LabPro® or CBL 2™ interface**
- **metal calorimeter and stirring rod**
- **metal heating vessel with metal heating dipper**
- **samples of various metals in shot or bead form**
- **small plastic dish**
- **stainless steel temperature probe**
- **TI graphing calculator with link cable**

SAFETY

- **When using a burner or hot plate, always wear goggles and an apron to protect your eyes and clothing. Tie back long hair, secure loose clothing, and remove loose jewelry. If your clothing catches on fire, walk to the emergency lab shower and use the shower to put out the fire.**
- **Never leave a hot plate unattended while it is turned on.**
- **Do not heat glassware that is broken, chipped, or cracked. Use tongs or a mitt to handle heated metal, glassware, and other equipment because it does not always look hot when it is hot. Allow all equipment to cool before storing it.**
- **Never put broken glass or ceramics in a regular waste container. Use a dustpan, brush, and heavy gloves to carefully pick up broken pieces and dispose of them in a container specifically provided for this purpose.**

PROCEDURE

Preparation

Follow Preparation steps 1–3 for the Skills Practice Lab "Specific Heat Capacity" in the chapter "Heat."

Finding the Specific Heat Capacity of a Metal

4. Choose a location where you can set up the experiment away from the edge of the table and away from other groups. Make sure the hot plate is in the "off" position before you plug it in.
5. Fill a metal heating vessel with 200 mL of water, and place it on the hot plate. Turn on the hot plate, and adjust the heating controls to heat the water.
6. Set up the calculator and interface for data collection:
   a. Connect the LabPro or CBL2 interface to the calculator with the unit-to-unit link cable.
   b. Connect the temperature probe to the CH1 port on the interface. Turn on the calculator and start the DataMate® program.
   c. Press CLEAR to reset the program.
   d. Hold the temperature probe in the air with nothing touching it. Observe the temperature readings displayed on the calculator. When the readings are stable, record the displayed temperature in your data table as the room temperature.
7. Obtain about 100 g of the metal shot. First, measure the mass of the small plastic dish. Place the metal shot in the dish, and determine the mass of the shot. Record the number and the mass of the shot in your data table. Place the temperature probe in the metal heating dipper, and carefully pour the shot into the metal heating dipper. Make sure the temperature probe is surrounded by the metal shot.
8. Place the dipper containing the shot into the top of the heating vessel, as shown in **Figure 1.** Make sure the temperature probe cable does not touch the hot plate or any heated surface.
9. While the sample is heating, find the mass of the empty inner cup of the calorimeter and the stirring rod. Record the mass in your data table. *Do not leave the hot plate unattended.*
10. For the water in the calorimeter, you will need about 100 g of cold water. The water must be colder than room temperature. (Do not use water colder than 5°C below room temperature. If ice is used to cool the water, make sure all the ice has melted before pouring the water into the calorimeter.)

11. Place the calorimeter and stirrer on the balance, and carefully add 100 g of the water. Record the mass of the water in your data table. Replace the cup in its insulating shell, and cover.

12. Determine the temperature of the metal shot by observing the temperature readings in degrees Celsius displayed on the calculator. When the readings are stable, record the displayed temperature in your data table as the metal's initial temperature.

13. Carefully remove the temperature probe from the dipper.

14. Use the stirring rod to stir the water in the calorimeter. Place the temperature probe in the calorimeter. Determine the initial temperature of the water in the calorimeter by observing the temperature readings displayed on the calculator. When the readings are stable, record the displayed temperature in your data table as the calorimeter's initial temperature. Leave the probe in the calorimeter.

15. From the DataMate® main screen, select START to begin collecting the temperature readings for the water in the calorimeter.

16. Quickly transfer the metal shot to the cold water in the calorimeter and replace the cover. Use a mitt when handling the metal heating dipper. Use the stirring rod to gently agitate the sample and to stir the water in the calorimeter. If you are not doing any more trials, make sure the hot plate is turned off. Otherwise, make sure there is plenty of water in the heating vessel, and do not leave the hot plate unattended.

17. When data collection is finished, a graph of temperature and time will be displayed. Time in seconds is graphed on the *x*-axis, and the temperature readings are graphed on the *y*-axis. Use the arrow keys to trace along the curve. Record the highest temperature reading displayed on your graph in the data table. Press ENTER to return to the main screen.

18. If time permits, perform additional trials with other samples. Record data for all trials in your data table.

ANALYSIS, CONCLUSIONS, AND EXTENSION

Complete the Analysis and Conclusions items for the Skills Practice Lab "Specific Heat Capacity." Your teacher may also instruct you to complete the Extension exercise.

Figure 1

Step 5: Start heating the water before you set up the calculator and temperature probe. Never leave a hot plate unattended when it is turned on.

Step 7: Be very careful when pouring the metal shot into the dipper around the temperature probe.

Step 15: Begin taking temperature readings a few seconds before adding the shot to the calorimeter.

Step 17: Record the *highest* temperature reached by the water, shot, and calorimeter combination, not the final temperature.

- Demonstrate how to carefully pour the shot into the heating dipper so that it surrounds the temperature probe. The probe must not touch the dipper's sides.

✓ Checkpoints

Step 4: All hot plates and containers of liquids must be kept away from the edge of the table. Metal shot must be kept in a container at all times.

Step 5: The hot plate must be turned up to the highest level.

Step 6: The graphing calculator, interface, and cords must be kept away from the hot plate.

Step 8: Students must exercise care when placing the heating dipper into the heating vessel.

Step 9: For the next few steps, make sure the hot plate is not left unattended.

Step 10: Only a small amount of ice is required; if too much is used, it should be removed before students measure the mass of the water.

Step 16: Students should use mitts when handling the hot metal heating dipper. Remind students not to let the hot metal shot touch the temperature probe in the calorimeter.

Step 17: Students should be able to use the graph to find the temperature at specific times during the experiment.

Step 18: If students are not doing more trials, make sure all hot plates are turned off. If they remain on, make sure that heating vessels have enough water in them and that hot plates are attended to. The vessel, water, and hot plate are still hot; student groups should move away from the hot plate and exercise caution.

Background

This probeware version of the Skills Practice Lab "Speed of Sound" from the chapter "Sound" uses a microphone, which is positioned above an open cardboard tube, instead of a resonance apparatus.

Tips and Tricks

- Students should have the program DataMate® on their graphing calculators. Refer to Appendix B for instructions. The sound used should not be too loud, because a loud sound causes the graph to have too many peaks. The graphs for different types of sounds will look very similar; the tube is dominated by its own low-resonance frequency for most sounds.
- Show students how to identify the second peak on the graph and how to determine whether the sound was too loud.

✓ Checkpoints

Step 5: Make sure students wear goggles if there are clamps and support rods at eye level.

Step 7: Remind students to work quietly; random noises may affect other groups' results. Any quiet, sharp sound will work well. After making the initial sound, students must be silent until the CBL 2 displays DONE.

Step 8: Students should realize that the sound traveled twice the length of the tube during the time interval.

Step 9: Guide students to find the time values for each peak.

Speed of Sound

MATERIALS LIST

- **cardboard tube**
- **LabPro® or CBL 2™ interface**
- **masking tape**
- **meterstick**
- **stainless-steel temperature probe**
- **support stand with clamp**
- **TI graphing calculator with link cable**
- **Vernier microphone**

In this lab, you will determine the speed of sound. You will place a microphone directly above the opening of a large tube, where the microphone will record a short, sharp noise. After the sound travels down the tube and reflects back, the microphone will record the sound again. You can use the time between recordings and the distance that the sound traveled to determine the speed of sound in air.

SAFETY

- **Put on goggles before you install clamps at eye level.**

PREPARATION

Follow Preparation steps 1–2 for the Skills Practice Lab "Speed of Sound" in the chapter "Sound," but for the data table, label the first through fourth columns *Trial, Temperature (°C), Distance from microphone to bottom of tube (m),* and *Time interval(s).*

PROCEDURE

Finding the Speed of Sound

3. Connect the LabPro or CBL 2 interface to the calculator with the unit-to-unit link cable. Connect the microphone to the CH1 port on the interface. Connect the temperature probe to the CH2 port on the interface.
4. Turn on the calculator, and start the DataMate® program. Observe the temperature readings displayed on the calculator. When the readings are stable, record the displayed temperature in your data table. Unplug the temperature probe, and press CLEAR to reset the program.
5. Set up the microphone, ring stand, and tube as shown in **Figure 1.** Tape or clamp the tube securely in place. Clamp the microphone to the edge of the table or to a ring stand so that the microphone points down and is directly above the open end of the tube.
6. Set up the calculator for data collection.
   - **a.** Select SETUP from the main screen.
   - **b.** Press the up arrow key once to select MODE and press ENTER.
   - **c.** Select TIME GRAPH from the SELECT MODE screen.
   - **d.** Select ADVANCED from TIME GRAPH SETTINGS.
   - **e.** Select CHANGE TRIGGERING from ADV. TIME GRAPH SETTINGS.
   - **f.** Select CH1-MICROPHONE from SELECT TRIGGERING.
   - **g.** Select INCREASING from TRIGGER TYPE.
   - **h.** Enter "0.1" for a trigger threshold.
   - **i.** Enter "0" for the pre-store.
   - **j.** Select OK three times to return to the main screen.
7. Select START to prepare for data collection. Make a loud, short noise—such as a snap of the fingers—directly above the tube. This noise will trigger the interface to collect the sound data.
8. Use the meterstick to measure the length from the bottom of the microphone to the bottom of the tube. Record this length to the nearest millimeter in the data table.

9. Look at the graph on the graphing calculator, which shows the sound plotted against time in seconds. There should be two peaks on the graph, one near the beginning and one a little later. The first peak is the sound, and the second peak is the echo of the sound. Use the arrow keys to trace the graph. (Note: If the graph has spikes or black lines, repeat the trial to obtain a smooth graph.)

10. Find the difference between the *x*-values of the two peaks to find the time interval between them. Record the time interval in your data table. Sketch the graph in your lab notebook. Press ENTER on the calculator to return to the main screen.

11. Repeat the procedure for several trials. Try different sounds, such as a soft noise, a loud noise, a high-pitched sound, and a low-pitched sound. Record all data in your data table.

12. Clean up your work area. Put equipment away safely so that it is ready to be used again. Recycle or dispose of used materials as directed by your teacher.

Figure 1

Step 7: The interface will begin collecting sound data as soon as you make a sound, so work quietly until you are ready to begin the experiment. Remain quiet until data collection has finished. Background noise may affect your results.

Step 9: On the graph, the first and second peaks may not be the same height, but they should both be noticeably higher than the other points on the graph. If the sound was too loud, the graph will show many high and low points. Repeat with a softer sound for better results.

ANALYSIS

1. **Organizing Data** For each trial, multiply the measured distance by 2 to find the total distance that the sound traveled.
2. **Organizing Data** Use the values for the distance traveled and the time interval from your data table to find the speed for each trial.

CONCLUSIONS

Complete the Conclusions items for the Skills Practice Lab "Speed of Sound" in the chapter "Sound."

ANSWERS

Analysis

1. Sample data:
distance = 1.962 m

2. Sample data:
Trial 1: v = 327.0 m/s
Trial 2: v = 280.3 m/s
Trial 3: v = 392.4 m/s
Trial 4: v = 327.0 m/s

939

Background
This probeware version of the Skills Practice Lab "Magnetic Field of a Conducting Wire" from the chapter "Magnetism" uses a voltage probe instead of a standard ammeter or multimeter. This version also adds a magnetic field sensor, which allows students to measure the strength of the magnetic field (*B*).

Tips and Tricks

- Students should have the program DataMate® on their graphing calculators. Refer to Appendix B for instructions. Zeroing the magnetic field sensor in step 9 will set the sensor to read Earth's magnetic field strength as zero, so the readings taken during the lab will not include Earth's magnetic field.
- Show students how to set up the wire coil on the galvanometer apparatus and how to use tape to mark the reference circle for the magnetic field sensor. Make sure students know the correct position of the magnetic field sensor.

✓ Checkpoints

Step 5: Make sure that the power supply, resistor, and switch are connected properly and are at the proper settings. Students should be able to demonstrate that all connections are properly made.

Step 8: Make sure that the magnetic field sensor is securely clamped in the right position.

Magnetic Field of a Conducting Wire

MATERIALS LIST

- **alligator clips**
- **galvanometer**
- **LabPro® or CBL 2™ interface**
- **masking tape**
- **power supply**
- **resistor**
- **support stand and buret clamp**
- **double-pole, double-throw switch**
- **TI graphing calculator with link cable**
- **Vernier magnetic field sensor**
- **voltage probe**
- **insulated connecting wires and bare copper wire**

In this lab, you will construct a circuit with a current-carrying wire and will use a magnetic field sensor to investigate the relationship between the magnetic field and the current in the wire. You will be able to determine the magnitude and direction of the magnetic field that surrounds the wire.

SAFETY

- **Never close a circuit until it has been approved by your teacher. Never rewire or adjust any element of a closed circuit. Never work with electricity near water; be sure the floor and all work surfaces are dry.**
- **Do not attempt this exercise with any batteries, electrical devices, or magnets other than those provided by your teacher for this purpose.**
- **Wire coils may heat up rapidly during this experiment. If heating occurs, open the switch immediately and handle the equipment with a hot mitt. Allow all equipment to cool before storing it.**

PROCEDURE

Preparation

1. Read the Safety cautions for the Skills Practice Lab "Magnetic Field of a Conducting Wire" in the chapter "Magnetism."
2. Read the entire lab, and plan what steps you will take. Prepare a table with 4 columns and 7 rows. In the first row, label the columns *Trial, Current Direction,* ΔV_R *(V),* and $B_{Measured}$ *(T).* Label the second through seventh rows as *1* through *6.*

Magnetic Field Strength

3. Set up the apparatus as shown in **Figure 1.** Use 1 m of copper wire to make a square loop around the coil support pins on the galvanometer apparatus. Attach alligator clips to the ends of the wire. Label one clip *A,* and label the other *B.* Place the galvanometer apparatus so that you are facing the plane of the coil.
4. Use masking tape to mark a line on the stand of the galvanometer directly under the top of the coil. Make another tape line perpendicular to the first, as shown in **Figure 1.** The two tape segments should cross in the middle of the apparatus. On the second tape, on the side away from you, mark a point 2 cm from the center. Using this point as the center point, draw a circle with a 1 cm radius.
5. Construct a circuit that contains the power supply and a 1 Ω resistor wired in series through the middle set of posts on the switch. Place the switch so that it moves from left to right. Connect the front right post of the switch to the end of the coil marked *A,* and connect the rear right post of the switch to the end of the coil marked *B.* Now, connect the front left post of the switch to the end of the coil marked *B,* and connect the rear left post of the switch to the end of the coil marked *A.* Do not close the switch or turn on the power supply until your teacher has approved your circuit.
6. Connect the interface to the calculator with the unit-to-unit link cable. Connect the voltage probe to the CH1 port and the magnetic field sensor to the CH2 port on the interface. Set the switch on the magnetic field sensor to HIGH. Connect the voltage probe to measure the voltage across the resistor.
7. Turn on the calculator, and start the DataMate® program. Press CLEAR to reset the program.
8. Set up a support stand with a buret clamp to hold the magnetic field sensor vertically. Position the magnetic sensor securely so that the white dot is facing you and the sensor is directly above the 1 cm circle marked on the tape.

9. Set the magnetic field sensor to read zero.

    a. Select SETUP from the main screen of DataMate®.

    b. Select ZERO from the setup screen.

    c. Select CH2-MAGNET F(MT) from the SELECT CHANNEL menu.

    d. Monitor the reading displayed on the calculator screen. When the reading appears to be stable, press ENTER to zero the sensor.

    e. Keep the magnetic field sensor in this same position for the remainder of the experiment.

10. Make sure the dial on the power supply is turned completely counterclockwise. When your teacher has approved your circuit, turn the dial on the power supply about halfway to its full value.

11. Close the switch briefly. Read the potential difference across the resistor and the strength of the magnetic field. Open the switch as soon as you have made your observations. Record ΔV_R *(V)* and $B_{Measured}$ *(T)* for *Trial 1* in your data table. Determine and record the *Current Direction* (*A* to *B* or *B* to *A*).

12. Reverse the direction of the current by closing the switch in the opposite direction. Read and record the potential difference and the strength of the magnetic field for *Trial 2*. Open the switch as soon as you have made your observations. Determine and record the *Current Direction* (*A* to *B* or *B* to *A*).

13. Increase the setting on the power supply to about two-thirds of the maximum setting on the dial. Repeat the procedure in steps 11 through 12. Record all data in your data table as *Trials 3* and *4*.

14. Repeat step 13 with the power supply set to one-third of the maximum setting. Record all data in your data table as *Trials 5* and *6*.

15. Clean up your work area. Put equipment away safely so that it is ready to be used again.

ANALYSIS

1. **Organizing Data** For each trial, use the equation $\Delta V = IR$ to find the current.

2. **Constructing Graphs** Use the data from *Trials 1, 3,* and *5* to plot a graph of B_{wire} in teslas against the current in amperes in the circuit. Also, plot graphs for *Trials 2, 4,* and *6*. Use a computer, graphing calculator, or graph paper.

CONCLUSIONS

3. **Drawing Conclusions** For each position, what is the relationship between the current in the wire loop and the magnetic field strength?

4. **Drawing Conclusions** What is the relationship between the direction of current in the wire and the direction of the magnetic field? Explain.

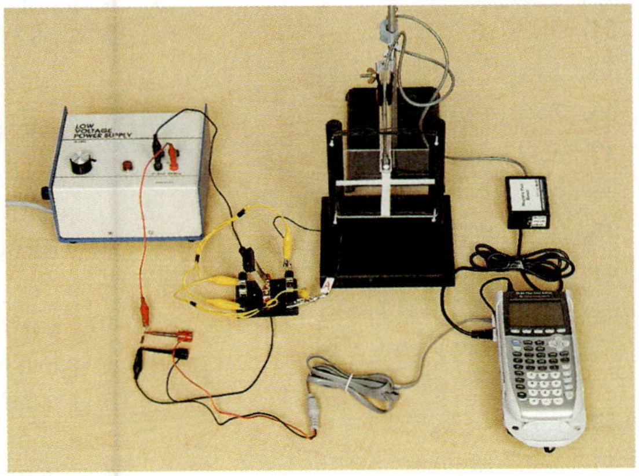

Figure 1

Step 3: Loop the copper wire around the support pins, and attach alligator clips to the ends. Place the galvanometer with one support pin on the left and one on the right.

Step 4: Use two pieces of tape to mark perpendicular lines, and mark a circle to use as a reference for placing the sensor.

Step 5: Place the switch in front of you so that it moves from left to right. Check all connections carefully.

Step 11: Some students may need help determining the *Current Direction*. Students should be able to demonstrate that they have recorded the correct direction.

ANSWERS

Analysis

1. Student answers will vary. For sample data, values range from $I = 1.07$ A to $I = 3.10$ A.

2. Student graphs should show that the magnetic field strength increases as the current increases.

Conclusions

3. For each position, the magnetic field strength increases as the current increases.

4. When the current direction changes, the direction of the magnetic field also changes.

Selected Answers

CHAPTER 1

Practice A, p. 15

1. 5×10^{-5} m

3. a. 1×10^{-8} m
b. 1×10^{-5} mm
c. 1×10^{-2} µm

5. 1.440×10^{3} kg

1 Review, pp. 27–31

11. a. 2×10^{2} mm
b. 7.8×10^{3} s
c. 1.6×10^{7} µg
d. 7.5×10^{4} cm
e. 6.75×10^{-4} g
f. 4.62×10^{-2} cm
g. 9.7 m/s

13. 1.08×10^{9} km

19. a. 3
b. 4
c. 3
d. 2

21. 228.8 cm

23. b, c

29. 4×10^{8} breaths

31. 5.4×10^{8} s

33. 2×10^{3} balls

35. 7×10^{2} tuners

37. a. 22 cm; 38 cm^2
b. 29.2 cm; 67.9 cm^2

39. 9.818×10^{-2} m

41. The ark (6×10^{4} m^3) was about 100 times as large as a typical house (6×10^{2} m^3).

43. 1.0×10^{3} kg

45. a. 0.618 g/cm^3
b. 4.57×10^{16} m^2

CHAPTER 2

Practice A, p. 44

1. 2.0 km to the east

3. 680 m to the north

5. 0.43 h

Practice B, p. 49

1. 2.2 s

3. 5.4 s

5. a. 1.4 m/s
b. 3.1 m/s

Practice C, p. 53

1. 21 m

3. 9.1 s

Practice D, p. 55

1. 9.8 m/s; 29 m

3. −7.5 m/s; 19 m

Practice E, p. 58

1. +2.51 m/s

3. a. 16 m/s
b. 7.0 s

5. +2.3 m/s^2

Practice F, p. 64

1. a. −42 m/s
b. 11 s

3. a. 8.0 m/s
b. 1.63 s

2 Review, pp. 68–73

1. 5.0 m; +5.0 m

3. t_1: negative; t_2: positive; t_3: positive; t_4: negative; t_5: zero

7. 10.1 km to the east

9. a. +70.0 m
b. +140.0 m
c. +14 m/s
d. +28 m/s

11. 0.2 km west of the flagpole

17. 0.0 m/s^2; +1.36 m/s^2; +0.680 m/s^2

19. 110 m

21. a. −15 m/s
b. −38 m

23. 17.5 m

25. 0.99 m/s

31. 3.94 s

33. 1.51 h

35. a. 2.00 min
b. 1.00 min
c. 2.00 min

37. 931 m

39. −26 m/s; 31 m

41. 1.6 s

43. 5 s; 85 s; +60 m/s

45. -1.5×10^{3} m/s^2

47. a. 3.40 s
b. −9.2 m/s
c. −31.4 m/s; −33 m/s

49. a. 4.6 s after stock car starts
b. 38 m
c. +17 m/s (stock car), +21 m/s (race car)

51. 4.44 m/s

CHAPTER 3

Practice A, p. 89

1. a. 23 km
b. 17 km to the east

3. 15.7 m at 22° to the side of downfield

Practice B, p. 92

1. 95 km/h

3. 21 m/s, 5.7 m/s

Practice C, p. 94

1. 49 m at 7.3° to the right of downfield

3. 13.0 m at 57° north of east

Practice D, p. 99

1. 0.66 m/s

3. 7.6 m/s

Practice E, p. 101

1. yes, $\Delta y = -2.3$ m

3. 2.0 s; 4.8 m

Practice F, p. 105

1. 0 m/s
3. 3.90 m/s at $(4.0 \times 10^1)°$ north of east

3 Review, pp. 108–113

7. a. 5.20 m at 60.0° above the positive x-axis
   b. 3.00 m at 30.0° below the positive x-axis
   c. 3.00 m at 150° counter-clockwise from the positive x-axis
   d. 5.20 m at 60.0° below the positive x-axis
9. 15.3 m at 58.4° south of east
19. if the vector is oriented at 45° from the axes
21. a. 5 blocks at 53° north of east
    b. 13 blocks
23. 61.8 m at 76.0° S of E (or S of W), 25.0 m at 53.1° S of E (or S of W)
25. 2.81 km east, 1.31 km north
31. 45.1 m/s
33. 11 m
35. a. clears the goal by 1 m
    b. falling
37. 80 m; 210 m
41. a. 70 m/s east
    b. 20 m/s
43. a. 10.1 m/s at 8.53° east of north
    b. 48.8 m
45. 7.5 min
47. a. 41.7 m/s
    b. 3.81 s
    c. $v_{y,f} = -13.5$ m/s, $v_{x,f} = 34.2$ m/s, $v_f = 36.7$ m/s
49. 10.5 m/s
51. a. 2.66 m/s
    b. 0.64 m
53. 157 km
55. a. 32.5 m
    b. 1.78 s
57. a. 57.7 km/h at 60.0° west of the vertical
    b. 28.8 km/h straight down
59. 18 m; 7.9 m
61. 6.19 m/s downfield

CHAPTER 4

Practice B, p. 128

1. $F_x = 60.6$ N; $F_y = 35.0$ N
3. 557 N at 35.7° west of north

Practice C, p. 132

1. 2.2 m/s^2 forward
3. 4.50 m/s^2 to the east
5. 14 N

Practice D, p. 139

1. 0.23
3. a. 8.7×10^2 N, 6.7×10^2 N
   b. 1.1×10^2 N, 84 N
   c. 1×10^3 N, 5×10^2 N
   d. 5 N, 2 N

Practice E, p. 141

1. 2.7 m/s^2 in the positive x direction
3. a. 0.061
   b. 3.61 m/s^2 down the ramp

4 Review, pp. 145–149

11. a. $\mathbf{F_1}$ (220 N) and $\mathbf{F_2}$ (114 N) both point right; $\mathbf{F_1}$ (220 N) points left, and $\mathbf{F_2}$ (114 N) points right.
    b. first situation: 220 N to the right, 114 N to the right; second situation: 220 N to the left, 114 N to the right
21. 55 N to the right
29. 51 N
35. 0.70, 0.60
37. 0.816
39. 1.0 m/s^2
41. 13 N down the incline
43. 64 N upward
45. a. 0.25 m/s^2 forward
    b. 18 m
    c. 3.0 m/s
47. a. 2 s
    b. The box will never move. The force exerted is not enough to overcome friction.
49. −1.2 m/s^2; 0.12
51. a. 2690 N forward
    b. 699 N forward
53. 13 N, 13 N, 0 N, −26 N

CHAPTER 5

Practice A, p. 162

1. 1.50×10^7 J
3. 1.6×10^3

Practice B, p. 166

1. 1.7×10^2 m/s
3. the bullet with the greater mass; 2 to 1
5. 1.6×10^3 kg

Practice C, p. 168

1. 7.8 m
3. 5.1 m

Practice D, p. 172

1. 3.3 J
3. a. 785 J
   b. 105 J
   c. 0.00 J

Practice E, p. 177

1. 20.7 m/s
3. 14.1 m/s
5. 0.18 m

943

Practice F, p. 181

1. 66 kW
3. 2.61×10^8 s (8.27 years)
5. a. 7.50×10^4 J
   b. 2.50×10^4 W

5 Review, pp. 184–189

7. 53 J, –53 J
9. 47.5 J
19. 7.6×10^4 J
21. 2.0×10^1 m
23. a. 5400 J, 0 J; 5400 J
    b. 0 J, –5400 J; 5400 J
    c. 2700 J, –2700 J; 5400 J
33. 12.0 m/s
35. 17.2 s
37. a. 0.633 J
    b. 0.633 J
    c. 2.43 m/s
    d. 0.422 J, 0.211 J
39. 5.0 m
41. 2.5 m
45. a. 61 J
    b. –45 J
    c. 0 J
47. a. 28.0 m/s
    b. 30.0 m above the ground
49. 0.107
51. a. 66 J
    b. 2.3 m/s
    c. 66 J
    d. –16 J

CHAPTER 6

Practice A, p. 199

1. 2.5×10^3 kg•m/s to the south
3. 46 m/s to the east

Practice B, p. 201

1. 3.8×10^2 N to the left
3. 16 kg•m/s to the south

Practice C, p. 203

1. 5.33 s; 53.3 m to the west
3. a. 1.22×10^4 N to the east
   b. 53.3 m to the west

Practice D, p. 209

1. 1.90 m/s
3. a. 12.0 m/s
   b. 9.6 m/s

Practice E, p. 214

1. 3.8 m/s to the south
3. 4.25 m/s to the north
5. a. 3.0 kg
   b. 5.32 m/s

Practice F, p. 216

1. a. 0.43 m/s to the west
   b. 17 J
3. a. 4.6 m/s to the south
   b. 3.9×10^3 J

Practice G, p. 219

1. a. 22.5 cm/s to the right
   b. $KE_i = 6.2 \times 10^{-4}$ J $= KE_f$
3. a. 8.0 m/s to the right
   b. $KE_i = 1.3 \times 10^2$ J $= KE_f$

6 Review, pp. 223–227

11. a. 8.35×10^{-21} kg•m/s upward
    b. 4.88 kg•m/s to the right
    c. 7.50×10^2 kg•m/s to the southwest
    d. 1.78×10^{29} kg•m/s forward
13. 18 N
23. 0.037 m/s to the south
29. 3.00 m/s
31. a. 0.81 m/s to the east
    b. 1.4×10^3 J
33. 4.0 m/s
35. 42.0 m/s toward second base
37. a. 0.0 kg•m/s
    b. 1.1 kg•m/s upward
39. 23 m/s
41. 4.0×10^2 N
43. 2.36×10^{-2} m
45. 0.413
47. –22 cm/s, 22 cm/s
49. a. 9.9 m/s downward
    b. 1.8×10^3 N upward

CHAPTER 7

Practice A, p. 236

1. 2.5 m/s
3. 1.5 m/s^2

Practice B, p. 238

1. 29.6 kg
3. 40.0 N

Practice C, p. 242

1. 0.692 m
3. a. 651 N
   b. 246 N
   c. 38.5 N

Practice D, p. 251

1. Earth: 7.69×10^3 m/s, 5.51×10^3 s; Jupiter: 4.20×10^4 m/s, 1.08×10^4 s; moon: 1.53×10^3 m/s, 8.63×10^3 s

Practice E, p. 258

1. 0.75 N•m
3. 133 N

7 Review, pp. 263–267

9. 2.7 m/s
11. 62 kg
19. 1.0×10^{-10} m (0.10 nm)

944

27. $v_t = 1630$ m/s; $T = 5.78 \times 10^5$ s
29. Jupiter ($m = 1.9 \times 10^{27}$ kg)
33. F_2
37. 26 N•m
39. 12 m/s
41. 220 N
43. 1800 N•m
45. 2.0×10^2 N
47. 72%
49. a. 2.25 days
  b. 1.60×10^4 m/s
51. a. 6300 N•m
  b. 550 N
53. 6620 N; no ($F_c = 7880$ N)

CHAPTER 8

Practice A, p. 279

1. a. 3.57×10^3 kg/m^3
  b. 6.4×10^2 kg/m^3
3. 9.4×10^3 N

Practice B, p. 282

1. a. 1.48×10^3 N
  b. 1.88×10^5 Pa
3. a. 1.2×10^3 Pa
  b. 6.0×10^{-2} N

8 Review, pp. 288–291

9. 2.1×10^3 kg/m^3
15. 6.28 N
21. 1.01×10^{11} N
23. 6.11×10^{-1} kg
25. 17 N, 31 N
27. a. 1.0×10^3 kg/m^3
  b. 3.5×10^2 Pa
  c. 2.1×10^3 Pa
29. 1.7×10^{-2} m
31. 0.605 m
33. 6.3 m
35. a. 0.48 m/s^2
  b. 4.0 s
37. 1.7×10^{-3} m

CHAPTER 9

Practice A, p. 303

1. −89.22°C, 183.93 K
3. 37.0°C, 39°C
5. −195.81°C, −320.5°F

Practice B, p. 311

1. 755 J
3. 0.96 J

Practice C, p. 316

1. 47°C
3. 390 J/kg•°C

9 Review, pp. 322–325

9. 57.8°C, 331.0 K
25. a. 2.9 J
  b. It goes into the air, the ground, and the hammer.
31. 25.0°C
33. a. $T_R = T_F + 459.7$, or $T_F = T_R - 459.7$
  b. $T = \frac{5}{9} T_R$, or $T_R = \frac{9}{5} T$
35. a. $T_{TH} = \frac{3}{2}T_C + 50$, or $T_C = \frac{2}{3}(T_{TH} - 50)$
  b. −360° TH
37. 330 g
39. 5.7×10^3 J/min = 95 J/s

CHAPTER 10

Practice A, p. 338

1. a. 6.4×10^5 J
  b. -4.8×10^5 J
3. 3.3×10^2 J

Practice B, p. 346

1. 33 J
3. 1.00×10^4 J
5. 1.74×10^8 J

Practice C, p. 355

1. 0.1504
3. a. 0.247
  b. 4.9×10^4 J
5. 755 J

10 Review, pp. 360–363

3. b, c, d, e
9. 1.08×10^3 J; done by the gas
15. a. none (Q, W, and $\Delta U > 0$)
  b. $\Delta U < 0$, $Q < 0$ for refrigerator interior ($W = 0$)
  c. $\Delta U < 0$ ($Q = 0$, $W > 0$)
17. a. 1.7×10^6 J, to the rod
  b. 3.3×10^2 J; by the rod
  c. 1.7×10^6 J; it increases
27. 0.32
29. a. 188 J
  b. 1.400×10^3 J

CHAPTER 11

Practice A, p. 371

1. a. 15 N/m
  b. less stiff
3. 2.7×10^3 N/m

Practice B, p. 379

1. 1.4×10^2 m
3. 3.6 m

Practice C, p. 381

1. 2.1×10^2 N/m
3. 39.7 N/m
5. a. 1.7 s, 0.59 Hz
  b. 0.14 s, 7.1 Hz
  c. 1.6 s, 0.62 Hz

Practice D, p. 387

1. $0.081 \text{ m} \le \lambda \le 12$ m
3. 4.74×10^{14} Hz

945

11 Review, pp. 396–399

9. 580 N/m

11. 4*A*

19. 9.7 m

21. **a.** 0.57 s
b. 1.8 Hz

27. 1/3 s; 3 Hz

35. 0.0333 m

39. **a.** 0.0 cm
b. 48 cm

43. a, b, and d ($\lambda = 0.5L$, L, and $2L$, respectively)

45. 1.7 N

47. 446 m

49. 9.70 m/s^2

51. 9:48 A.M.

CHAPTER 12

Practice A, p. 415

1. **a.** 8.0×10^{-4} W/m^2
b. 1.6×10^{-3} W/m^2
c. 6.4×10^{-3} W/m^2

3. 2.3×10^{-5} W

5. 4.8 m

Practice B, p. 427

1. 440 Hz

3. **a.** 82.1 Hz
b. 115 Hz
c. 144 Hz

12 Review, pp. 434–437

23. 7.96×10^{-2} W/m^2

25. **a.** 4.0 m
b. 2.0 m
c. 1.3 m
d. 1.0 m

29. 3 Hz

35. 3.0×10^3 Hz

37. 5 beats per second

39. 0.20 s

41. $L_{closed} = 1.5\ (L_{open})$

43. **a.** 5.0×10^4 W
b. 2.8×10^{-3} W

CHAPTER 13

Practice A, p. 449

1. 1.0×10^{-13} m

3. 85.7 m–10.1 m; The wavelengths are shorter than those of the AM radio band.

5. 5.4×10^{14} Hz

Practice B, p. 462

1. $p = 10.0$ cm: no image (infinite q); $p = 5.00$ cm: $q = -10.0$ cm, $M = 2.00$; virtual, upright image

3. $R = 1.00 \times 10^2$ cm; $M = 2.00$; virtual image

Practice C, p. 466

1. $p = 46.0$ cm; $M = 0.500$; virtual, upright image; $h = 3.40$ cm

3. $p = 45$ cm; $h = 17$ cm; $M = 0.41$; virtual, upright image

5. $q = -1.31$ cm; $M = 0.125$; virtual, upright image

13 Review, pp. 476–480

7. 3.00×10^8 m/s

11. 1×10^{-6} m

13. 9.1×10^{-3} m (9.1 mm)

21. 1.2 m/s; The image moves toward the mirror's surface.

35. $q = 26$ cm; real, inverted; $M = -2.0$

47. inverted; $p = 6.1$ cm; $f = 2.6$ cm; real

49. $q_2 = 6.7$ cm; real; $M_1 = -0.57$, $M_2 = -0.27$; inverted

51. $p = 11.3$ cm

55. $R = -25.0$ cm

57. concave, $R = 48.1$ cm; $M = 2.00$; virtual

CHAPTER 14

Practice A, p. 493

1. 18.5°

3. 1.47

Practice B, p. 501

1. 20.0 cm, $M = -1.00$; real, inverted image

3. −6.67 cm, $M = 0.333$; virtual, upright image

Practice C, p. 508

1. 42.8°

3. 49.8°

14 Review, pp. 514–519

11. 26°

13. 30.0°, 19.5°, 19.5°, 30.0°

23. yes, because $n_{ice} > n_{air}$

25. 3.40; upright

37. **a.** 31.3°
b. 44.2°
c. 49.8°

39. 1.31

41. 1.62; carbon disulfide

43. 7.50 cm

45. **a.** 6.00 cm
b. A diverging lens cannot form an image larger than the object.

47. **a.** 3.01 cm
b. 2.05 cm

49. blue: 47.8°, red: 48.2°

51. 48.8°

53. 4.54 m

55. $\frac{10}{9}f$

57. a. 24.7°
    b. It will pass through the bottom surface because $\theta_i < \theta_c$ ($\theta_c = 41.8°$).

59. 1.38

61. 58.0 m

63. a. 4.83 cm
    b. The lens must be moved 0.12 cm.

65. 1.90 cm

CHAPTER 15

Practice A, p. 531

1. 5.1×10^{-7} m = 5.1×10^{2} nm
3. 0.125°

Practice B, p. 538

1. 0.02°, 0.04°, 0.11°
3. 11
5. 6.62×10^{3} lines/cm

15 Review, pp. 548–551

5. θ would decrease because λ is shorter in water.
9. 630 nm
11. 160 μm
19. 3.22°
21. a. 10.09°, 13.71°, 14.77°
    b. 20.51°, 28.30°, 30.66°
29. 432.0 nm
31. 1.93×10^{-3} mm = 3 λ; a maximum

CHAPTER 16

Practice A, p. 566

1. 230 N (attractive)
3. 0.393 m

Practice B, p. 568

1. 47 N, along the negative x-axis; 157 N, along the positive x-axis; 11.0×10^{1} N, along the negative x-axis

Practice C, p. 570

1. $x = 0.62$ m
3. 5.07 m

Practice D, p. 575

1. 1.66×10^{5} N/C, 81.1° above the positive x-axis
3. a. 3.2×10^{-15} N, along the negative x-axis
   b. 3.2×10^{-15} N, along the positive x-axis

16 Review, pp. 581–585

15. 3.50×10^{3} N
17. 91 N (repulsive)
19. 1.48×10^{-7} N, along the $+x$ direction
21. 18 cm from the 3.5 nC charge
33. 5.7×10^{3} N/C, 75° above the positive x-axis
35. a. 5.7×10^{-27} N, in a direction opposite **E**
    b. 3.6×10^{-8} N/C
37. a. 2.0×10^{7} N/C, along the positive x-axis
    b. 4.0×10^{1} N
41. 7.2×10^{-9} C
43. $v_{electron} = 4.4 \times 10^{6}$ m/s; $v_{proton} = 2.4 \times 10^{3}$ m/s
45. 5.4×10^{-14} N
47. 2.0×10^{-6} C
49. 32.5 m
51. a. 5.3×10^{17} m/s^2
    b. 8.5×10^{-4} m
    c. 2.9×10^{14} m/s^2
53. a. positive
    b. 5.3×10^{-7} C
55. a. 1.3×10^{4} N/C
    b. 4.2×10^{6} m/s

CHAPTER 17

Practice A, p. 599

1. 6.4×10^{-19} C
3. 2.3×10^{-16} J

Practice B, p. 607

1. a. 4.80×10^{-5} C
   b. 4.50×10^{-6} J
3. a. 9.00 V
   b. 5.0×10^{-12} C

Practice C, p. 609

1. 4.00×10^{2} s
3. 6.00×10^{2} s
5. a. 2.6×10^{-3} A
   b. 1.6×10^{17} electrons
   c. 5.1×10^{-3} A

Practice D, p. 615

1. 0.43 A
3. a. 2.5 A
   b. 6.0 A
5. 46 Ω

Practice E, p. 621

1. 14 Ω
3. 1.5 V
5. 5.00×10^{2} A

17 Review, pp. 626–631

9. -4.2×10^{5} V
19. 0.22 J
23. $v_{avg} >> v_{drift}$
33. a. 3.5 min
    b. 1.2×10^{22} electrons
41. 3.4 A
49. 3.6×10^{6} J
51. the 75 W bulb
53. 2.0×10^{16} J

947

55. 93 Ω
57. 3.000 m; 2.00×10^{-7} C
59. 4.0×10^3 V/m
61. a. 4.11×10^{-15} J
    b. 2.22×10^6 m/s
63. a. 1.13×10^5 V/m
    b. 1.81×10^{-14} N
    c. 4.39×10^{-17} J
65. 0.545 m, −1.20 m
67. a. 7.2×10^{-13} J
    b. 2.9×10^7 m/s
69. a. 3.0×10^{-3} A
    b. 1.1×10^{18} electrons/min
71. a. 32 V
    b. 0.16 V
73. 1.0×10^5 W
75. 3.2×10^5 J
77. 13.5 h
79. 2.2×10^{-5} V

CHAPTER 18

Practice A, p. 650

1. a. 43.6 Ω
   b. 0.275 A
3. 1.0 V, 2.0 V, 2.5 V, 3.5 V
5. 0.5 Ω

Practice B, p. 655

1. 4.5 A, 2.2 A, 1.8 A, 1.3 A
3. a. 2.2 Ω
   b. 6.0 A, 3.0 A, 2.00 A

Practice C, p. 659

1. a. 27.8 Ω
   b. 26.6 Ω
   c. 23.4 Ω

Practice D, p. 662

R_a: 0.50 A, 2.5 V
R_b: 0.50 A, 3.5 V
R_c: 1.5 A, 6.0 V
R_d: 1.0 A, 4.0 V
R_e: 1.0 A, 4.0 V
R_f: 2.0 A, 4.0 V

18 Review, pp. 666–671

17. a. 24 Ω
    b. 1.0 A
19. a. 2.99 Ω
    b. 4.0 A
21. a. seven combinations
    b. $R, 2R, 3R, \frac{R}{2}, \frac{R}{3}, \frac{2R}{3}, \frac{3R}{2}$
23. 15 Ω
25. 3.0 Ω: 1.8 A, 5.4 V
    6.0 Ω: 1.1 A, 6.5 V
    9.0 Ω: 0.72 A, 6.5 V
27. 28 V
29. 3.8 V
31. a. 33.0 Ω
    b. 132 V
    c. 4.00 A, 4.00 A
33. 10.0 Ω
35. a. a
    b. c
    c. d
    d. e
37. 18.0 Ω: 0.750 A, 13.5 V
    6.0 Ω: 0.750 A, 4.5 V
39. 4.0 Ω
41. 13.96 Ω
43. a. 62.4 Ω
    b. 0.192 A
    c. 0.102 A
    d. 0.520 W
    e. 0.737 W
47. a. 5.1 Ω
    b. 4.5 V
49. a. 11 A (heater), 9.2 A (toaster), 12 A (grill)
    b. The total current is 32.2 A, so the 30.0 A circuit breaker will open the circuit if these appliances are all on.

CHAPTER 19

Practice A, p. 689

1. 3.57×10^6 m/s
3. 6.0×10^{-12} N west

Practice B, p. 692

1. 1.7×10^{-7} T in $+z$ direction
3. 1.5 T

19 Review, pp. 695–699

31. 2.1×10^{-3} m/s
33. 2.00 T
39. 2.1×10^{-2} T, in the negative y direction
41. 2.0 T, out of the page
43. a. 8.0 m/s
    b. 5.4×10^{-26} J
45. 2.82×10^7 m/s

CHAPTER 20

Practice A, p. 714

1. 0.30 V
3. 0.14 V

Practice B, p. 726

1. 4.8 A; 6.8 A, 170 V
3. a. 7.42 A
   b. 14.8 Ω
5. a. 1.10×10^2 V
   b. 2.1 A

Practice C, p. 729

1. 55 turns
3. 25 turns
5. 147 V

20 Review, pp. 739–743

11. 0.12 A
27. a. 2.4×10^2 V
    b. 2.0 A

948

29. a. 8.34 A
b. 119 V
35. 221 V
37. a. a step-down transformer
b. 1.2×10^3 *V*
43. 790 turns
45. a. a step-up transformer
b. 440 V
47. 171 : 1
49. 300 V

CHAPTER 21

Practice A, p. 755

1. 2.0 Hz
3. 1.2×10^{15} Hz

Practice B, p. 758

1. 4.83×10^{14} Hz
3. 2.36 eV

Practice C, p. 769

1. 4.56×10^{14} Hz; *Line 4*
3. 1.61×10^{15} Hz
5. E_6 to E_2; *Line 1*

Practice D, p. 774

1. 39.9 m/s
3. 8.84×10^{-27} m/s
5. 1.0×10^{-15} kg

21 Review, pp. 779–781

11. 4.8×10^{17} Hz
13. 1.2×10^{15} Hz
23. a. 2.46×10^{15} Hz
b. 2.92×10^{15} Hz
c. 3.09×10^{15} Hz
d. 3.16×10^{15} Hz
33. 1.4×10^7 m/s
35. 2.00 eV
37. 0.80 eV

CHAPTER 22

Practice A, p. 796

1. 160.65 MeV; 342.05 MeV
3. 7.933 MeV

Practice B, p. 802

1. ${}^{12}_{6}C$
3. ${}^{14}_{6}C$
5. ${}^{63}_{28}Ni \rightarrow {}^{63}_{29}Cu + {}^{0}_{-1}e + \overline{v}$

Practice C, p. 805

1. $4.23 \times 10^3\ s^{-1}$, 0.23 Ci
3. $9.94 \times 10^{-7}\ s^{-1}$, 6.7×10^{-7} Ci
5. a. about 5.0×10^7 atoms
b. about 3.5×10^8 atoms

22 Review, pp. 820–823

1. 79; 118; 79
7. 92.162 MeV
9. 8.2607 MeV/nucleon; 8.6974 MeV/nucleon
21. a. ${}^{4}_{2}He$
b. ${}^{4}_{2}He$
23. 560 days
27. a. $-e$
b. 0
33. 1.2×10^{-14}
35. 3.53 MeV
37. a. ${}^{1}_{0}n + {}^{197}_{79}Au \rightarrow {}^{198}_{80}Hg + {}^{0}_{-1}e + \overline{v}$
b. 7.885 MeV
39. ${}^{3}_{2}He$
41. 2.6×10^{21} atoms
43. a. ${}^{8}_{4}Be$
b. ${}^{12}_{6}C$
45. 3.8×10^3 s
47. 1.1×10^{16} fission events

APPENDIX I

Additional Problems

1. 11.68 m
3. $6.4 \times 10^{-2}\ m^3$
5. 6.7×10^{-5} ps
7. 2.80 h = 2 h, 48 min
9. 4.0×10^1 km/h
11. 48 m/h
13. +25.0 m/s = 25.0 m/s, upward
15. 44.8 m/s
17. $-21.5\ m/s^2 = 21.5\ m/s^2$, backward
19. 38.5 m
21. 126 s
23. 1.27 s
25. 11 km/h
27. 2.74 s
29. 10.5 m, forward
31. 5.9 s
33. 8.3 s
35. 7.4 s
37. $-490\ m/s^2 = 490\ m/s^2$, backward
39. 17.3 s
41. 7.0 m
43. 2.6 m/s
45. −11.4 m/s = 11.4 m/s, downward
47. 8.5° north of east
49. 5.0° south of west
51. 770 m
53. −33 km/h = 33 km/h, downward
55. 18.9 km, 76° north of west
57. 17.0 m
59. 52.0°
61. 79 s
63. 15.8 m, 55° below the horizontal
65. 0.290 m/s, east; 1.16 m/s, north
67. 2.6 km

949

69. 66 km, 46° south of east
71. 10.7 m
73. 3.0 s
75. 76.9 km/h, 60.1° west of north
77. 7.0×10^2 m, 3.8° above the horizontal
79. 47.2 m
81. 6.36 m/s
83. 13.6 km/h, 73° south of east
85. 58 N
87. 14.0 N; 2.0 N
89. 9.5×10^4 kg
91. 258 N, up the slope
93. 15.9 N
95. 2.0 m/s^2
97. $F_x = 8.60$ N; $F_y = 12.3$ N
99. −448 m/s^2 = 448 m/s^2, backward
101. 15 kg
103. 0.085
105. 1.7×10^8 N
107. 24 N, downhill
109. 1.150×10^3 N
111. 1.2×10^4 N
113. 0.60
115. 38.0 m
117. 2.5×10^4 J
119. 247 m/s
121. -5.46×10^4 J
123. 3.35×10^6 J
125. 1.23 J
127. 12 s
129. 0.600 m
131. 133 J
133. 53.3 m/s
135. 72.2 m
137. 0.13 m = 13 cm
139. 7.7 m/s
141. 8.0 s
143. 230 J
145. 7.96 m
147. 6.0×10^1 m/s
149. 1.58×10^3 kg•m/s, north
151. 3.38×10^{31} kg
153. 18 s
155. 637 m, to the right
157. 7.5 g
159. 0.0 m/s
161. -5.0×10^1 percent
163. 16.4 m/s, west
165. 5.33×10^7 kg•m/s
167. 1.0×10^1 m/s
169. 560 N, east
171. -3.3×10^8 N = 3.3×10^8 N, backward
173. 52 m
175. 24 kg
177. 90.6 km/h, east
179. 26 km/h, 37° north of east
181. −157 J
183. 0.125 kg
185. -4.1×10^4 J
187. 9.8 kg
189. 1.0 m/s, 60° south of east
191. 4.04×10^3 m/s^2
193. 42 m/s
195. 8.9 kg
197. 1.04×10^4 m/s = 10.4 km/s
199. 1.48×10^{23} kg
201. 1.10×10^{12} m
203. 6.6×10^3 m/s = 6.6 km/s
205. 0.87 m
207. 254 N
209. 0.42 m = 42 cm
211. 25 N
213. 165 kg
215. 5.09×10^5 s = 141 h
217. 5.5×10^9 m = 5.5×10^6 km
219. 1.6 N•m
221. 6.62×10^3 N
223. 0.574 m
225. 8.13×10^{-3} m^2
227. 2.25×10^4 kg/m^3
229. 2.0×10^1 m^2
231. 4.30 kg
233. 374°F to −292°F
235. 6.6×10^{-2}°C
237. 1.29×10^4 J
239. 4.1×10^{-2} kg
241. 1.200×10^3°C
243. 315 K
245. 1.91×10^{-2} kg = 19.1 g
247. 530 J/kg•°C
249. −930°C
251. 1.50×10^3 Pa = 1.50 kPa
253. 873 J
255. 244 J
257. 5.3×10^3 J
259. 2.4×10^3 Pa = 2.4 kPa
261. 5895 J
263. 5.30×10^2 kJ = 5.30×10^5 J
265. 1.0×10^4 J
267. −18 N
269. −0.11 m = −11 cm
271. 4.0×10^{-2} m = 4.0 cm
273. 0.2003 Hz
275. 730 N/m
277. 1.4×10^3 m/s
279. 2.2×10^4 Hz
281. 8.6×10^3 N/m
283. 3.177 s
285. 82 kg
287. 1.2 s
289. 1.5×10^3 m/s
291. 1.1 W/m^2
293. 294 Hz
295. 408 m/s
297. 0.155 m
299. 0.211 m = 21.1 cm
301. 2.9971×10^8 m/s
303. 3.2×10^{-7} m = 320 nm
305. −0.96 cm
307. −1.9 cm
309. 3.8 m
311. 0.25
313. 38 cm

950

315. 2.40
317. 0.98 cm
319. 10.5 cm
321. 64.0 cm in front of the mirror
323. 8.3 cm
325. 0.40
327. −11 cm
329. 2.9979×10^8 m/s
331. 33.3 cm
333. 0.19
335. 32.2°
337. −10.4 cm
339. 18 cm
341. 1.63
343. 39.38°
345. 58°
347. −21 cm
349. ∞
351. 1.486
353. 1.54
355. 4.8 cm
357. 1.73 to 1.83
359. 5.18×10^{-4} m = 0.518 mm
361. 0.137°
363. 9.0×10^{-7} m = 9.0×10^2 nm
365. 11.2°
367. 0.227°
369. 1.445×10^4 lines/cm
371. 140 N attractive
373. 2.2×10^{-17} C
375. 0.00 N
377. 4.0×10^{-8} N, 9.3° below the negative *x*-axis
379. 260 N from either charge
381. 1.6×10^{-12} C
383. 0.585 m = 58.5 cm
385. 3.97×10^{-6} N, upward
387. 0.073 m = 7.3 cm
389. 7.5×10^{-6} N, along the +*y*-axis
391. 4.40×10^5 N/C, 89.1° above the −*x*-axis
393. −7.4 C
395. 1.6×10^{-19} C
397. 36 cm
399. 4.4×10^{-4} J
401. 7.1×10^{-4} F
403. 4.0 A
405. 160 Ω
407. 1.7×10^6 W = 1.7 MW
409. 6.4×10^2 N/C
411. 12 V
413. 1.2×10^{-5} m
415. 1.4×10^2 C
417. 7.2 s
419. 4.8 V
421. 116 V
423. 5.0×10^5 W = 0.50 MW
425. 7.5×10^6 V
427. 3.00×10^2 Ω
429. 6.0 Ω
431. 13 Ω
433. 6.0 Ω
435. 0.056 A = 56 mA
437. 1.6 A (refrigerator); 1.3 A (oven)
439. 12.6 Ω
441. 2.6 V
443. 9.4 Ω
445. 1.6 A
447. 1.45 A
449. 4×10^{-12} N
451. 7.6×10^6 m/s
453. 0.70 A
455. 5.1×10^{-4} T
457. 3.9×10^{-15} N
459. 1.5×10^5 A
461. 1.7×10^{-2} T
463. 0.90 s
465. 450 V
467. 1.8 A
469. 3.4×10^4 V = 34 kV
471. 48 turns
473. 2.5×10^{-3} A = 2.5 mA
475. 0.85 A
477. 48:1
479. 3.32×10^{-10} m
481. 1.00×10^{-13} m
483. 3.0×10^{-7} m
485. 26 kg
487. 4.30×10^{14} Hz
489. 6.0×10^{14} Hz
491. 4.0×10^{-21} kg
493. 2.5×10^{-43} m
495. 7.72×10^{14} Hz
497. 333.73 MeV
499. 0.543 705 u
501. $^{16}_{8}O$
503. 15.0 s
505. 31.92 h
507. 35.46 MeV
509. $^{131}_{53}I$
511. $^{111}_{54}Xe$
513. 924 days

951

Glossary

Glossary

A

absorption spectrum a diagram or graph that indicates the wavelengths of radiant energy that a substance absorbs (p. 764)

acceleration the rate at which velocity changes over time; an object accelerates if its speed, direction, or both change (p. 48)

accuracy a description of how close a measurement is to the correct or accepted value of the quantity measured (p. 16)

adiabatic process a thermodynamic process in which no energy is transferred to or from the system as heat (p. 341)

alternating current an electric current that changes direction at regular intervals (p. 718)

amplitude the maximum displacement from equilibrium (p. 376)

angle of incidence the angle between a ray that strikes a surface and the line perpendicular to that surface at the point of contact (p. 452)

angle of reflection the angle formed by the line perpendicular to a surface and the direction in which a reflected ray moves (p. 452)

angular acceleration the time rate of change of angular velocity, usually expressed in radians per second per second (p. 900)

angular displacement the angle through which a point, line, or body is rotated in a specified direction and about a specified axis (p. 899)

angular momentum for a rotating object, the product of the object's moment of inertia and angular velocity about the same axis (p. 907)

angular velocity the rate at which a body rotates about an axis, usually expressed in radians per second (p. 900)

antinode a point in a standing wave, halfway between two nodes, at which the largest displacement occurs (p. 393)

average velocity the total displacement divided by the time interval during which the displacement occurred (p. 43)

B

back emf the emf induced in a motor's coil that tends to reduce the current in the coil of the motor (p. 720)

beat the periodic variation in the amplitude of a wave that is the superposition of two waves of slightly different frequencies (p. 430)

binding energy the energy released when unbound nucleons come together to form a stable nucleus, which is equivalent to the energy required to break the nucleus into individual nucleons (p. 794)

blackbody radiation the radiation emitted by a blackbody, which is a perfect radiator and absorber and emits radiation based only on its temperature (p. 752)

buoyant force the upward force exerted by a liquid on an object immersed in or floating on the liquid (p. 275)

C

calorimetry an experimental procedure used to measure the energy transferred from one substance to another as heat (p. 314)

capacitance the ability of a conductor to store energy in the form of electrically separated charges (p. 602)

center of mass the point in a body at which all the mass of the body can be considered to be concentrated when analyzing translational motion (p. 904)

centripetal acceleration the acceleration directed toward the center of a circular path (p. 235)

chromatic aberration the focusing of different colors of light at different distances behind a lens (p. 511)

coefficient of friction the ratio of the magnitude of the force of friction between two objects in contact to the magnitude of the normal force with which the objects press against each other (p. 138)

coherence the correlation between the phases of two or more waves (p. 527)

components of a vector the projections of a vector along the axes of a coordinate system (p. 90)

compression the region of a longitudinal wave in which the density and pressure are at a maximum (p. 408)

Compton shift an increase in the wavelength of the photon scattered by an electron relative to the wavelength of the incident photon (p. 760)

concave spherical mirror a mirror whose reflecting surface is an inward-curved segment of a sphere (p. 455)

constructive interference a superposition of two or more waves in which individual displacements on the same side of the equilibrium position are added together to form the resultant wave (p. 390)

controlled experiment an experiment that tests only one factor at a time by using a comparison of a control group with an experimental group (p. 9)

convex spherical mirror a mirror whose reflecting surface is an outward-curved segment of a sphere (p. 463)

crest the highest point above the equilibrium position (p. 384)

critical angle the minimum angle of incidence for which total internal reflection occurs (p. 506)

cyclic process a thermodynamic process in which a system returns to the same conditions under which it started (p. 346)

D

decibel a dimensionless unit that describes the ratio of two intensities of sound; the threshold of hearing is commonly used as the reference intensity (p. 417)

destructive interference a superposition of two or more waves in which individual displacements on opposite sides of the equilibrium position are added together to form the resultant wave (p. 391)

diffraction a change in the direction of a wave when the wave encounters an obstacle, an opening, or an edge (p. 532)

dispersion the process of separating polychromatic light into its component wavelengths (p. 509)

displacement the change in position of an object (p. 41)

doping the addition of an impurity element to a semiconductor (p. 927)

Doppler effect an observed change in frequency when there is relative motion between the source of waves and an observer (p. 412)

drift velocity the net velocity of a charge carrier moving in an electric field (p. 611)

E

elastic collision a collision in which the total momentum and total kinetic energy remain constant (p. 216)

elastic potential energy the energy available for use when an elastic body returns to its original configuration (p. 170)

electric circuit a set of electrical components connected such that they provide one or more complete paths for the movement of charges (p. 642)

electric current the rate at which charges pass through a given area (p. 608)

electric field a region where an electric force on a test charge can be detected (p. 572)

electric potential the work that must be performed against electric forces to move a charge from a reference point to the point in question divided by the charge (p. 596)

electrical conductor a material in which charges can move freely (p. 561)

electrical insulator a material in which charges cannot move freely (p. 561)

electrical potential energy potential energy associated with a charge due to its position in an electric field (p. 594)

electromagnetic induction the process of creating a current in a circuit by a changing magnetic field (p. 708)

electromagnetic radiation the transfer of energy associated with an electric and magnetic field; it varies periodically and travels at the speed of light (p. 733)

electromagnetic wave a wave that consists of oscillating electric and magnetic fields, which radiate outward from the source at the speed of light (p. 446)

emission spectrum a diagram or graph that indicates the wavelengths of radiant energy that a substance emits (p. 764)

entropy a measure of the randomness or disorder of a system (p. 355)

environment the combination of conditions and influences outside a system that affect the behavior of the system (p. 337)

equilibrium in physics, the state in which the net force on an object is zero (p. 129)

excited state a state in which an atom has more energy than it does at its ground state (p. 926)

F

fluid a nonsolid state of matter in which the atoms or molecules are free to move past each other, as in a gas or liquid (p. 274)

force an action exerted on an object which may change the object's state of rest or motion; force has magnitude and direction (p. 120)

frame of reference a system for specifying the precise location of objects in space and time (p. 40)

free fall the motion of a body when only the force due to gravity is acting on the body (p. 60)

953

frequency the number of cycles or vibrations per unit of time; also the number of waves produced per unit of time (p. 376)

fundamental frequency the lowest frequency of vibration of a standing wave (p. 422)

G

generator a machine that converts mechanical energy into electrical energy (p. 716)

gravitational force the mutual force of attraction between particles of matter (p. 240)

gravitational potential energy the potential energy stored in the gravitational fields of interacting bodies (p. 169)

ground state the lowest energy state of a quantized system (p. 926)

H

half-life the time needed for half of the original nuclei of a sample of a radioactive substance to undergo radioactive decay (p. 803)

harmonic series a series of frequencies that includes the fundamental frequency and integral multiples of the fundamental frequency (p. 423)

heat the energy transferred between objects because of a difference in their temperatures; energy is always transferred from higher-temperature objects to lower-temperature objects until thermal equilibrium is reached (p. 305)

hole an energy level that is not occupied by an electron in a solid (p. 927)

hypothesis an explanation that is based on prior scientific research or observations and that can be tested (p. 8)

I

ideal fluid a fluid that has no internal friction or viscosity and is incompressible (p. 284)

impulse the product of the force and the time over which the force acts on an object (p. 200)

index of refraction the ratio of the speed of light in a vacuum to the speed of light in a given transparent medium (p. 490)

induction the process of charging a conductor by bringing it near another charged object and grounding the conductor (p. 562)

inertia the tendency of an object to resist being moved or, if the object is moving, to resist a change in speed or direction (p. 125)

instantaneous velocity the velocity of an object at some instant or at a specific point in the object's path (p. 46)

intensity the rate at which energy flows through a unit area perpendicular to the direction of wave motion (p. 414)

internal energy the energy of a substance due to both the random motions of its particles and to the potential energy that results from the distances and alignments between the particles (p. 299)

isothermal process a thermodynamic process that takes place at constant temperature (p. 340)

isotope an atom that has the same number of protons (or the same atomic number) as other atoms of the same element do but that has a different number of neutrons (and thus a different atomic mass) (p. 791)

isovolumetric process a thermodynamic process that takes place at constant volume so that no work is done on or by the system (p. 339)

K

kinetic energy the energy of an object that is due to the object's motion (p. 164)

kinetic friction the force that opposes the movement of two surfaces that are in contact and are sliding over each other (p. 137)

L

laser a device that produces coherent light of only one wavelength (p. 541)

latent heat the energy per unit mass that is transferred during a phase change of a substance (p. 318)

lens a transparent object that refracts light waves such that they converge or diverge to create an image (p. 494)

lever arm the perpendicular distance from the axis of rotation to a line drawn along the direction of the force (p. 255)

linear polarization the alignment of electromagnetic waves in such a way that the vibrations of the electric fields in each of the waves are parallel to each other (p. 472)

longitudinal wave a wave whose particles vibrate parallel to the direction the wave is traveling (p. 385)

M

magnetic domain a region composed of a group of atoms whose magnetic fields are aligned in the same direction (p. 679)

magnetic field a region where a magnetic force can be detected (p. 680)

mass density the concentration of matter of an object, measured as the mass per unit volume of a substance (p. 275)

mechanical energy the sum of kinetic energy and all forms of potential energy (p. 174)

mechanical wave a wave that requires a medium through which to travel (p. 382)

medium a physical environment through which a disturbance can travel (p. 382)

model a pattern, plan, representation, or description designed to show the structure or workings of an object, system, or concept (p. 6)

moment of inertia the tendency of a body that is rotating about a fixed axis to resist a change in this rotating motion (p. 905)

momentum a quantity defined as the product of the mass and velocity of an object (p. 198)

mutual inductance the ability of one circuit to induce an emf in a nearby circuit in the presence of a changing current (p. 721)

N

net force a single force whose external effects on a rigid body are the same as the effects of several actual forces acting on the body (p. 126)

node a point in a standing wave that maintains zero displacement (p. 393)

normal force a force that acts on a surface in a direction perpendicular to the surface (p. 135)

O

order number the number assigned to interference fringes relative to the central bright fringe (p. 529)

P

parallel describes two or more components of a circuit that provide separate conducting paths for current because the components are connected across common points or junctions (p. 651)

path difference the difference in the distance traveled by two beams when they are scattered in the same direction from different points (p. 529)

perfectly inelastic collision a collision in which two objects stick together after colliding (p. 212)

period the time that it takes a complete cycle or wave oscillation to occur (p. 376)

phase change the physical change of a substance from one state (solid, liquid, or gas) to another at constant temperature and pressure (p. 318)

photoelectric effect the emission of electrons from a material when light of certain frequencies shines on the surface of the material (p. 756)

photon a unit or quantum of light; a particle of electromagnetic radiation that has zero mass and carries a quantum of energy (pp. 734, 757)

pitch a measure of how high or low a sound is perceived to be, depending on the frequency of the sound wave (p. 409)

potential difference the work that must be performed against electric forces to move a charge between the two points in question divided by the charge (p. 596)

potential energy the energy associated with an object because of the position, shape, or condition of the object (p. 169)

power a quantity that measures the rate at which work is done or energy is transformed (p. 179)

precision the degree of exactness of a measurement (p. 16)

pressure the magnitude of the force on a surface per unit area (p. 280)

projectile motion the curved path that an object follows when thrown, launched, or otherwise projected near the surface of Earth (p. 96)

R

radian an angle whose arc length is equal to the radius of the circle, which is approximately equal to 57.3° (p. 898)

rarefaction the region of a longitudinal wave in which the density and pressure are at a minimum (p. 408)

real image an image that is formed by the intersection of light rays; a real image can be projected on a screen (p. 456)

reflection the turning back of an electromagnetic wave at a surface (p. 451)

refraction the bending of a wavefront as the wavefront passes between two substances in which the speed of the wave differs (p. 488)

resistance the opposition presented to electric current by a material or device (p. 612)

resolving power the ability of an optical instrument to form separate images of two objects that are close together (p. 539)

955

resonance a phenomenon that occurs when the frequency of a force applied to a system matches the natural frequency of vibration of the system, resulting in a large amplitude of vibration (p. 419)

resultant a vector that represents the sum of two or more vectors (p. 83)

rms current the value of alternating current that gives the same heating effect that the corresponding value of direct current does (p. 724)

rotational kinetic energy the energy of an object that is due to the object's rotational motion (p. 907)

S

scalar a physical quantity that has magnitude but no direction (p. 82)

schematic diagram a representation of a circuit that uses lines to represent wires and different symbols to represent components (p. 640)

series describes two or more components of a circuit that provide a single path for current (p. 647)

significant figures those digits in a measurement that are known with certainty plus the first digit that is uncertain (p. 17)

simple harmonic motion vibration about an equilibrium position in which a restoring force is proportional to the displacement from equilibrium (p. 369)

solenoid a long, helically wound coil of insulated wire (p. 685)

specific heat capacity the quantity of heat required to raise a unit mass of homogeneous material 1 K or 1°C in a specified way given constant pressure and volume (p. 313)

spring constant the energy available for use when a deformed elastic object returns to its original configuration (p. 170)

standing wave a wave pattern that results when two waves of the same frequency, wavelength, and amplitude travel in opposite directions and interfere (p. 393)

static friction the force that resists the initiation of sliding motion between two surfaces that are in contact and at rest (p. 136)

strong force the interaction that binds nucleons together in a nucleus (p. 792)

superconductor a material whose resistance is zero at a certain critical temperature, which varies with each material (p. 928)

system a set of particles or interacting components considered to be a distinct physical entity for the purpose of study (pp. 7, 336)

T

tangential acceleration the acceleration of an object that is tangent to the object's circular path (p. 903)

tangential speed the speed of an object that is tangent to the object's circular path (p. 902)

temperature a measure of the average kinetic energy of the particles in an object (p. 299)

thermal equilibrium the state in which two bodies in physical contact with each other have identical temperatures (p. 300)

timbre the musical quality of a tone resulting from the combination of harmonics present at different intensities (p. 428)

torque a quantity that measures the ability of a force to rotate an object around some axis (p. 255)

total internal reflection the complete reflection that takes place within a substance when the angle of incidence of light striking the surface boundary is less than the critical angle (p. 506)

transformer a device that increases or decreases the emf of alternating current (p. 727)

transistor a semiconductor device that can amplify current and that is used in amplifiers, oscillators, and switches (p. 646)

transverse wave a wave whose particles vibrate perpendicularly to the direction the wave is traveling (p. 384)

trough the lowest point below the equilibrium position (p. 384)

U

ultraviolet catastrophe the failed prediction of classical physics that the energy radiated by a blackbody at extremely short wavelengths is extremely large and that the total energy radiated is infinite (p. 753)

uncertainty principle the principle that states that it is impossible to simultaneously determine a particle's position and momentum with infinite accuracy (p. 775)

V

vector a physical quantity that has both magnitude and a direction (p. 82)

virtual image an image from which light rays appear to diverge, even though they are not actually focused there; a virtual image cannot be projected on a screen (p. 453)

W

wavelength the distance between two adjacent similar points of a wave, such as from crest to crest or from trough to trough (p. 384)

weight a measure of the gravitational force exerted on an object; its value can change with the location of the object in the universe (p. 135)

work the product of the component of a force along the direction of displacement and the magnitude of the displacement (p. 160)

work function the minimum energy needed to remove an electron from a metal atom (p. 757)

work-kinetic energy theorem the net work done by all the forces acting on an object is equal to the change in the object's kinetic energy (p. 166)

957

Index

Page references followed by *f* refer to figures. Page references followed by *t* refer to tables.

A

B

C

959

D

E

961

962

963

H

I

965

N

O

P

967

Q

968

969

970

971

Acknowledgments, continued

Staff Credits

The people who contributed to **Holt Physics** are listed below. They represent editorial, design, inventory, multimedia production, manufacturing, marketing, permissions, and production.

Brigitta Arden, Karen Arneson, Wes Bain, Kimberly Barr, Cynthia B. Brooks, Soojinn Choi, Eddie Dawson, Julie Dervin, Michelle Dike, Lydia Doty, James Foster, Jeff Galvez, Leigh Ann García, Mark Grayson, Jessica Gonzalez, Wilonda Ieans, Jevara Jackson, Stephanie Janda, Simon Key, Jane Kirschman, Cathy Kuhles, Denise Mahoney, Michael Mazza, Sean McCormick, Virginia Messler, Richard Metzger, Cathy Murphy, Mercedes Newman, Cathy Paré, Nanda Patel, Jenny Patton, Peter Reid, Mike Roche, Karen Ross, Teresa Ruiz, Kay Selke, Angela Senicz, Chris Smith, Dawn Marie Spinozza, Sherry Sprague, Jeff Streber, Jeannie Taylor, David Trevino, Bob Tucek, Kira Watkins, and Holly Whittaker.

Photo Credits

Abbreviations used: (t) top, (b) bottom, (c) center, (l) left, (r) right, (bkgd) background
CBL Calculator, Andy Christiansen/HRW

Front Cover: (c), Gusto/Photo Researchers, Inc.; (bl) Darren Garnier/Columbia University/Photo Researchers, Inc.; (bc) Mark Garlick/Photo Researchers, Inc.; (br) © Charles Orrico/SuperStock **Title Page:** 1 (c), Gusto/Photo Researchers, Inc. **Table of Contents:** v (tl), © Robert Harding World Imagery/Alamy Photos; v (bl), Courtesy of the Archives, California Institute of Technology; vi (tl), Mark Gallup/Pictor/Image State; vi (bl, bc), Michelle Bridwell/HRW Photo; vii (cl), © David Madison/Getty Images; viii (tl), Rafael Macia/Photo Researchers, Inc.; viii (bl), © RubberBall Productions; ix (cl), © Comstock Images/Alamy Photos; ix (bl, bc), Educational Development Center; x (tl), GmbH & Co./Alamy; x (bl), Sam Dudgeon/HRW; xii (tl), © Laguna Design/Photo Researchers, Inc.; xiii (br), © Zefa Visual Media - Germany/Index Stock Imagery, Inc.; xvi (tr), Victoria Smith/HRW **Chapter One:** 2–3 (all), © Philippe Psaila/Photo Researchers, Inc.; 4 (b), Sam Dudgeon/HRW; 5 (tr), © Roger Ressmeyer/CORBIS ; 6 (b), Michelle Bridwell/HRW Photo; 7 (c), Michelle Bridwell/HRW Photo; 10 (bl), The National Institute of Standards and Technology; 11 (br), Stephen Dalton/Photo Researchers, Inc.; 13 (tr, br), Courtesy NASA/JPL-Caltech; 14 (t), Sam Dudgeon/HRW; 16 (all), Michelle Bridwell/HRW Photo; 17 (all), Michelle Bridwell/HRW Photo; 18 (tl), © Ron Thomas/Getty Images; 21 (br), Richard Megna/Fundamental Photographs, New York; 23 (br), © John P. Kelly/Getty Images; 28 (all), Sam Dudgeon/HRW; 35 (tr), Victoria Smith/HRW; 36 (all), Victoria Smith/HRW **Chapter Two:** 38–39 (c), © Robert Harding World Imagery/Alamy Photos; 40 (bl), Courtesy of New York Transit Museum Archives, Brooklyn.; 41 (br), © VCG/Getty Images; 45 (br), Michelle Bridwell/HRW Photo; 50 (tl), Craig Lovell/CORBIS; 50 (br), Aaron Haupt/Photo Researchers, Inc.; 51 (r), Richard Megna/Fundamental Photographs, New York; 57 (tr), Michelle Bridwell/HRW Photo; 60 (bl), James A. Sugar/Black Star; 61 (tc, tr), Richard Megna/Fundamental Photographs, New York; 63 (tr), Victoria Smith/HRW; 64 (br), © Zefa Visual Media - Germany/Index Stock Imagery, Inc.; 66 (tr, bc), Sam Dudgeon/HRW; 69 (cr), Courtesy of David Rogers; 70 (tr), Educational Development Center; 77 (b), Victoria Smith/HRW; 78 (tl), Mike Fager/HRW Photo **Chapter Three:** 80–81 (all), © Wendell Metzen/Index Stock Imagery; 82 (bl), Michelle Bridwell/HRW Photo; 95 (cr), © Dennis O'Clair/Getty Images; 96 (bl), Richard Megna/Fundamental Photographs, New York; 96 (c), Sam Dudgeon/HRW; 103 (br), © Philip Wallick/CORBIS; 106 (tr, bc), Sam Dudgeon/HRW; 117 (tr, cr), Victoria Smith/HRW **Chapter Four:** 118–119 (all), © Nicholas Pinturas/Getty Images; 120 (all), Michelle Bridwell/HRW Photo; 121 (br), Michelle Bridwell/HRW Photo; 123 (tr), age fotostock/Birgit Koch; 129 (tr), Tony Freeman/PhotoEdit; 130 (l, r), Michelle Bridwell/HRW Photo; 132 (bl), Michelle Bridwell/HRW Photo; 133 (cr), © Mark Richards/PhotoEdit; 135 (cr), George Hunter/SuperStock; 135 (cr), Photo Researchers, Inc.; 136 (all), Michelle Bridwell/HRW Photo; 138 (cl), Mark Gallup/Pictor/Image State; 140 (tr), Michelle Bridwell/HRW Photo; 142 (br), Goodyear Tire and Rubber Company; 153 (br), Victoria Smith/HRW **Timeline (1540–1690):** 156 (tc, bl), Image Select/Ann Ronan Picture Library; 156 (tr, br), CORBIS; 156 (tl), Giraudon/Art Resource, NY; 156 (c), © Kean Collection/Getty Images; 156 (cl), Erich Lessing/Art Resource, NY; 156 (bc), AKG Photo, London; 157 (tl), Scala/Art Resource, NY; 157 (tr), Image Select/Ann Ronan Picture Library; 157 (cl) (1644), Burstein/CORBIS; 157 (cr), AKG Photo, London; 157 (cl) (1669), J. Sowerby/The Art Archive; 157 (bl), University of California Press, Berkeley, California 1934; 157 (tc), © Kean Collection/Getty Images **Chapter Five:** 158–159 (all), Wayne Sorce; 160 (bl), Mavournea Hay/HRW Photo; 164 (b), Richard Megna/Fundamental Photographs, New York; 166 (bl), Pulse Productions/SuperStock/PictureQuest; 168 (br), Peter Van Steen/HRW Photo; 169 (tr), © George Schwartz/Getty Images; 173 (br), Robert Mathena/Fundamental Photographs, New York; 174 (bl), Sam Dudgeon/HRW; 176 (tr), Tony Freeman/PhotoEdit; 178 (tl), Robert Wolf; 179 (br), Werner H. Muller/Peter Arnold, Inc.; 180 (tl), Sam Dudgeon/HRW; 182 (tr), Courtesy Steve Okamoto; 182 (bc), Jeff Singer/HRW Photo; 184 (c), Leonard Lessin/Peter Arnold, Inc.; 186 (cl), © Harold & Esther Edgerton Foundation, 2006, Courtesy of Palm Press, Inc.; 193 (b), Victoria Smith/HRW

973

Chapter Six: 196–197 (all), © Mike Powell/Getty Images; 198 (bl), © Glenn Paulina/Transtock; 200 (tl), © Brian Bahr/Getty Images; 203 (cr), Lawrence Migdale; 204 (tl, tc), Richard Megna/Fundamental Photographs, New York; 205 (all), Michelle Bridwell/Frontera Fotos; 206 (bl), NASA; 207 (tr), Michelle Bridwell/HRW Photo; 207 (b), © Wayne Eastep/Getty Images; 210 (tl), © Tony Anderson/Getty Images; 212 (br), © Nathan Bilow/Getty Images; 221 (bc, tr) Steven Jones/HRW Photo; 231 (t), Victoria Smith/HRW **Chapter Seven:** 232–233 (all), © NASA/Roger Ressmeyer/CORBIS; 234 (br), © Blaine Harrington III/CORBIS; 236 (bl), Sam Dudgeon/HRW; 239 (cr), Werner H. Muller/Peter Arnold, Inc.; 243 (tr), © NASA/Photo Researchers, Inc.; 243 (bl), NASA/CXC/MIT/F.K.Baganoff et al.; 244 (bl, br) © Andrew J. Martinez/Photo Researchers, Inc.; 246 (br), Digital Image copyright © 2006 PhotoDisc; 248 (bl), Adler Planetarium & Astronomy Museum; 251 (tr), Courtesy NASA/JPL-Caltech; 254 (bc), © Rene Sheret/Getty Images; 255 (c), Michelle Bridwell/HRW Photo; 258 (bl), Sam Dudgeon/HRW; 260 (bl, br) Scott Van Osdol/HRW Photo; 261 (cr), Rube Goldberg Inc.; 270 (bc), Victoria Smith/HRW **Chapter Eight:** 272–273 (all), © David Madison/Getty Images; 274 (l, r), Richard Megna/Fundamental Photographs, New York; 280 (bl), Daniel A. Nord; 284 (tl), M. Van Dyke/Parabolic Press; 285 (br), Victoria Smith/HRW **Timeline (1690–1785):** 294 (tl), R. Sheridan/The Ancient Art & Architecture Collection LTD; 294 (tr), CORBIS; 294 (cl), An Introduction to Chinese Literature, by Liu Wu-chi, Indiana University Press, 1966, pg. 238; 294 (cr), Johann Ihle/AKG Photo, London; 294 (bl), National Maritime Museum London; 294 (br), Hydrodynamics by Daniel Bernoulli & Hydraulics by Johann Bernoulli, 1968 Dover Publications, Inc., New York; 294 (bc), SEF/Art Resource, NY; 295 (tr), The Ancient Art & Architecture Collection, Ltd.; 295 (tc), AKG Photo, London; 295 (c, bl) Image Select/Ann Ronan Picture Library; 295 (cr), The Art Archive; 295 (br, bc) CORBIS; 295 (cl), © Bettmann/CORBIS **Chapter Nine:** 296–297 (all), Larry Lefever/Grant Heilman Photography; 298 (cl), Michelle Bridwell/HRW Photo; 300 (bl), Sam Dudgeon/HRW; 304 (br), Michelle Bridwell/HRW Photo; 305 (bl), Victoria Smith/HRW; 308 (tl), Victoria Smith/HRW; 312 (bl), Lawrence Migdale; 312 (tr), © Hugh Sitton/Getty Images; 313 (tr), © James Randkley/Getty Images; 320 (bc, tr) Sam Dudgeon/HRW; 323 (c), Richard Megna/Fundamental Photographs, New York; 329 (bc), Victoria Smith/HRW; 330 (tl), Victoria Smith/HRW **Chapter Ten:** 334–335 (all), Cindy Yamanaka/National Geographic Society; 336 (cl), Richard Megna/Fundamental Photographs, New York; 340 (tl), Michelle Bridwell/HRW Photo; 342 (cl), Rafael Macia/Photo Researchers, Inc.; 347 (tl), Sam Dudgeon/HRW; 348 (tr), Robert Wood/HRW Photo; 350 (bc), Michelle Bridwell/HRW Photo; 350 (tr), John Langford/HRW Photo; 353 (b), © William Holmes/Getty Images; 356 (bl, br) Michelle Bridwell/HRW Photo; 358 (tr), © Douglas Peebles/CORBIS; 361 (l, r), Mike Fager/HRW Photo **Chapter Eleven:** 366–367 (all), Comstock; 369 (br), © Steve Bronstein/Getty Images; 371 (br), Photo Researchers, Inc.; 372 (tl), © Verna Bice/Acclaim Stock Photography; 373 (tr), SuperStock; 376 (bl), Palais Didier/SuperStock; 379 (br), Robert Mathena/Fundamental Photographs, New York; 382 (c), NSPI-IFA/Nawrocki Stock Photo, Inc.; 383 (c), Victoria Smith/HRW; 389 (br), Richard Megna/Fundamental Photographs, New York; 389 (c), © Lou Jacobs Jr./Grant Heilman Photography, Inc.; 393 (all), Richard Megna/Fundamental Photographs, New York; 394 (b), Richard Megna/Fundamental Photographs, New York; 403 (b), Victoria Smith/HRW **Timeline (1785–1830):** 404 (tl, tcl) The Art Archive; 404 (c), Image Select/Ann Ronan Picture Library; 404 (tr, br, bl) CORBIS; 404 (bcr), Courtesy of Saunders Publishing; 405 (c), Dr. Gerald Wheeler/Courtesy Saunders Publishing; 405 (tr), Portrait by Gemalde 1821/AKG Photo, London; 405 (tl), Giraudon/Art Resource, NY; 405 (cr), CORBIS; 405 (cl), Richard Menga/Fundamental Photographs, New York; 405 (bl), Ronald Sheridan/The Ancient Art & Architecture Collection, Ltd.; 405 (br), Image Select/Ann Ronan Picture Library **Chapter Twelve:** 406–407 (all), © Flip Nicklin/Minden Pictures; 408 (bl), Richard Megna/Fundamental Photographs, New York; 409 (tl), Richard Megna/Fundamental Photographs, New York; 410 (tr), Dr. Najeeb Layyous/Photo Researchers, Inc.; 411 (br), Peter Arnold, Inc.; 412 (tc), Peter Van Steen/HRW Photo; 414 (bl), Tony Freeman/PhotoEdit; 419 (cl, cr) CORBIS; 419 (br), Photo Researchers, Inc.; 421 (tr), © Tony Freeman/PhotoEdit; 421 (bl), Sam Dudgeon/HRW; 422 (bl), © RubberBall Productions; 424 (tl), Joseph Barnell/SuperStock; 427 (br), © Robert W. Ginn/PhotoEdit; 428 (t), Sam Dudgeon/HRW; 428 (c), SuperStock; 428 (b), © 2006 photo by David Boehl/ViolinsEtc; 429 (t), Fred Maroon/Photo Researchers, Inc.; 430 (br), © PhotoDisc/gettyimages; 432 (tr, bc), Sam Dudgeon/HRW; 441 (tr), Victoria Smith/HRW **Science, Technology & Society (Noise Pollution):** 442 (br), Scott Barrow, Inc./SuperStock; 442 (tl), © COMSTOCK, Inc.; 442 (tr), ©Stockbyte; 443 (cr), © Royalty Free/CORBIS **Chapter Thirteen:** 444–445 (all), Laurence Parent; 446 (br), Photo Researchers, Inc.; 451 (br), Loren Winters/Visuals Unlimited; 452 (tl), Richard Megna/Fundamental Photographs, New York; 453 (br), Richard Megna/Fundamental Photographs, New York; 454 (tl), Sam Dudgeon/HRW; 455 (l, r), Richard Megna/Fundamental Photographs, New York; 456 (t), Sam Dudgeon/HRW; 457 (bl), Richard Megna/Fundamental Photographs, New York; 463 (br), Richard Megna/Fundamental Photographs, New York; 470 (tl), Leonard Lessin/Peter Arnold, Inc.; 470 (cl, bl) Victoria Smith/HRW; 471 (tr), Sam Dudgeon/HRW; 471 (b), Michelle Bridwell/HRW Photo; 473 (tr, tc) Victoria Smith/HRW; 485 (t), Mike Fager/HRW Photo **Chapter Fourteen:** 486–487 (all), Kerrick James Photography; 488 (cl), © Christopher Burki/Getty Images; 491 (br), © George Lepp/Getty Images; 494 (l, r), Richard Megna/Fundamental Photographs, New York; 506 (b), Ken Kay/Fundamental Photographs, New York; 508 (cr), © Comstock Images/Alamy Photos; 510 (cl), Victoria Smith/HRW; 512 (tr, bc), Jan A. Allinder/HRW Photo; 516 (tr), Victoria Smith/HRW; 523 (tr), Victoria Smith/HRW **Chapter Fifteen:** 524–525 (all), © Charles Thatcher/Getty Images; 526 (cl), © Ian Mckinnell/Getty Images; 527 (br), Kodansha, Inc. Courtesy of Saunders Publishing; 528 (tl), Kodansha, Inc. Courtesy of Saunders Publishing; 532 (b), Fundamental Photographs, New York; 533 (tr), Courtesy of Saunders Publishing; 534 (cl), Dr. Gerald Wheeler/Courtesy of Saunders Publishing; 534 (bl), © VCG/Getty Images; 534 (t), Courtesy of Saunders Publishing; 535 (b), ASTROSTOCK; 539 (br), Courtesy School of Physics & Astronomy University of Birmingham; 540 (tl), © Getty Images; 543 (br), Dr. Randy Ricklefs/University of Texas; 544 (tr), © image100; 546 (tr, bc), Sam Dudgeon/HRW; 549 (cl), Photo Researchers, Inc.; 555 (tr), Mike Fager/HRW Photo **Chapter Sixteen:** 556–557 (all), © Andrew Sacks/Getty Images; 558 (bc, br) Michelle Bridwell/HRW Photo; 561 (cr), Michelle Bridwell/HRW Photo; 563 (c), Fundamental Photographs, New York; 566 (br), Michelle Bridwell/HRW Photo; 576 (bl), Educational Development Center; 577 (t, c) Educational Development Center; 579 (tr), Sam Dudgeon/HRW; 589 (br), Victoria Smith/HRW; 590 (bl), Victoria Smith/HRW **Chapter Seventeen:** 592–593 (all), Warren Faidley/WeatherStock; 594 (bc), © Arthur S. Aubry/PhotoDisc/gettyimages; 596 (bl), © Tom Stewart/CORBIS; 604 (br), Sam Dudgeon/HRW; 606 (tl), Sam Dudgeon/Courtesy High Voltage Lab/MIT; 611 (br), Photo Researchers, Inc.; 614 (t), Roger Allyn Lee/SuperStock; 617 (tr), © Joseph Brignolo/Getty Images; 618 (cl), Sam Dudgeon/HRW; 619 (cr), Sam Dudgeon/HRW; 622 (c), Sam Dudgeon/HRW; 623 (tr), Comstock; 624 (tr, bc), Sam Dudgeon/HRW; 635 (br), Victoria Smith/HRW **Science, Technology & Society (Hybrid Electric Vehicles):** 636 (tr), Courtesy of Honda/NewsCom **Chapter Eighteen:** 638–639 (all), Laurence Parent; 640 (bl), Sam Dudgeon/HRW; 641 (bulb, wire, resistor, switch) Sergio Purtell/Foca/HRW; 641 (plug, battery, capacitor) Sam Dudgeon/HRW; 642 (b), © Ben Simmons/CORBIS; 642 (tl), Sam Dudgeon/HRW; 643 (cl), Sam Dudgeon/HRW; 644 (c), Sam Dudgeon/HRW; 646 (bc), © David Young-Wolff/Photo Edit; 646 (bl), © Walter Hodges/Getty Images; 646 (tr), © Royalty Free/CORBIS; 647 (bl), Sam Dudgeon/HRW; 649 (cr), Sam Dudgeon/HRW; 652 (tl), Sam Dudgeon/HRW; 653 (tr), KRT/NewsCom; 654 (tr), Sam Dudgeon/HRW; 657 (cr), Victoria Smith/HRW; 662 (br), Victoria Smith/HRW; 664 (tr, bc), Sam Dudgeon/HRW; 675 (tr), Victoria Smith/HRW **Chapter Nineteen:** 676–677 (all), © World Perspectives/Getty Images; 678 (bl), Sam Dudgeon/HRW; 683 (bc), © Alfred Pasieka/Photo Researchers, Inc.; 683 (br), iPhoto.ca/Jeff McIntosh/NewsCom; 684 (bl, br) Richard Megna/Fundamental Photographs, New York; 695 (bl), Gamma Photo/Central Scientific Company; 703 (tr), Victoria Smith/HRW **Chapter Twenty:** 706–707 (all), © Aaron Jones Studio/Getty Images; 711 (br), Sam Dudgeon/HRW; 715 (tr), © SW Productions/PhotoDisc/gettyimages; 716 (b), Victoria Smith/HRW; 722 (tl), Mike Fager/HRW Photo; 723 (cr), Sam Dudgeon/HRW; 731 (br), Infrared Processing and Analysis Center, California Institute of Technology; 733 (tl), The Nobeyama Radio Observatory; 733 (tc), National Solar Observatory/AURA/NSF; 733 (tr), Big Bear Solar Observatory/New Jersey Institute of Technology; 733 (bl, bc), Courtesy of SOHO/EIT consortium. SOHO is a project of international cooperation between ESA and NASA.; 733 (br), Yohkoh Solar Observatory; 734 (cl), LINCOLN JOURNAL STAR/Ken BLACKBIRD/AP/Wide World Photos; 736 (bl), B. Glanzmann/Masterfile; 737 (cr), Library of Congress; 747 (tr), Victoria Smith/HRW **Timeline (1830–1890):** 748 (tr), CORBIS; 748 (cr, br), © Kean Collection/Getty Images; 748 (c), UPI/CORBIS; 748 (bl), Courtesy of the Library of Congress; 748 (tl), National Portrait Gallery, London, courtesy AIP Emilio Segrè Visual Archives; 748 (cl), © Franz Xavier Winterhalter/The Bridgeman Art Library/Getty Images; 749 (tr), Scala/Art Resource, NY; 749 (tl, br), CORBIS; 749 (cl), Giraudon/Art Resource, NY; 749 (bl), © Museum of the City of New York/Getty Images; 749 (bcr), © E. M. Kimble/Getty Images; 749 (tcr), © Baldwin H. Ward & Kathryn C. Ward/CORBIS **Chapter Twenty-One:** 750–751 (all), © Johnny Johnson/Getty Images; 752 (cl), © VOLKER STEGER/Photo RESEARCHERS, INC.; 756 (tl), Victoria Smith/HRW; 759 (br), © D. Hurst/Alamy Photos; 761 (tr), © GDT/Getty Images; 763 (all), Richard Megna/Fundamental Photographs, New York; 767 (br), Phil Lauro/Index Stock Imagery, Inc.; 773 (tl), Loria and Klinger/Photo Researchers, Inc.; 773 (tr), Mike Fager/HRW Photo; 785 (cr), Mike Fager/HRW Photo **Timeline (1890–1950):** 786 (tl), Snark/Art Resource, NY; 786 (tr), AIP Emilio Segrè Visual Archives; 786 (c), NASA; 786 (cl), Photo Deutsches Museum München; 786 (cr), AIP Emilio Segre' Visual Archives, Shapley Collection; 786 (bl), Courtesy of U.S. Army; 786 (br), © Hulton Archive/Getty Images; 787 (br), © Bettmann/CORBIS; 787 (tl), Dorothea Lange/Courtesy of the Library of Congress; 787 (tr), Art Resource, NY; 787 (cl), © Pictorial Parade/Getty Images; 787 (bl), University of Tsukuba, Tomonaga Memorial Room courtesy AIP Emilio Segrè Visual Archives **Chapter Twenty-Two:** 788–789 (all), AP Photo/Francis Latreille/Nova Productions; 797 (cr), Richard Megna/Fundamental Photographs, New York; 799 (tr), Courtesy of Brookhaven National Laboratory; 809 (tr), Scott Camazine/Photo Researchers, Inc.; 811 (cr), David Parker/Photo Researchers, Inc.; 814 (cl), P. Loiez Cern/Photo Researchers, Inc.; 818 (tr, bc), Dwight Cendrowski/HRW Photo **Appendices:** 830–831 (all), © Daryl Benson/Masterfile; 897 (tc), © Dr. Dennis Kunkel/Visuals Unlimited ; 897 (bl), Fermilab/Photo Researchers, Inc.; 900 (l, r), Michelle Bridwell/HRW Photo; 904 (bc), © Harold & Esther Edgerton Foundation, 2003, Courtesy of Palm Press, Inc.; 906 (cl), Michelle Bridwell/HRW Photo; 907 (l, r), © David Madison; 910 (bl), Richard Megna/Fundamental Photographs, New York; 915 (tr), Property of AT&T Archives. Printed with permission of AT&T.; 917 (br), Sam Dudgeon/HRW; 917 (b), NASA, ESA, S. Beckwith (STScI) and the HUDF Team; 919 (bl), Courtesy of the Archives, California Institute of Technology; 921 (br), Photo Researchers, Inc.; 921 (tc), © David Parker/Photo Researchers, Inc.; 923 (br), © Laguna Design/Photo Researchers, Inc.; 924 (bl), Oliver Mecker/Photo Researchers, Inc.; 925 (tr), Loria and Klinger/Photo Researchers, Inc.; 925 (bl), © Dr. Dennis Kunkel/Visuals Unlimited ; 925 (bc), © Andrew Syred/Photo Researchers, Inc.; 925 (br), © Susumu Nishinaga/Photo Researchers, Inc.; 927 (b), © Colin Cuthbert/Photo Researchers, Inc.; 932 (bl), Fermilab/Photo Researchers, Inc.; 933 (t), LBNL/Photo Researchers, Inc.; 933 (br), Brookhaven National Laboratory/Photo Researchers, Inc.; 934–935 (bkgd), © Royalty Free/CORBIS; 937 (br), Sam Dudgeon/HRW; 939 (b), Sam Dudgeon/HRW; 941 (bl), Sam Dudgeon/HRW; 943 (br), Sam Dudgeon/HRW; 945 (br), Sam Dudgeon/HRW

974

Fundamental Constants

Symbol	Quantity	Established value	Value used for calculations in this book
c	speed of light in a vacuum	299 792 458 m/s	3.00×10^{8} m/s
e^{-}	elementary charge	$1.602\ 176\ 487 \times 10^{-19}$ C	1.60×10^{-19} C
e^{1}	base of natural logarithms	2.718 2818 28	2.72
ε_0	(Greek *epsilon*) permittivity of a vacuum	$8.854\ 187\ 817 \times 10^{-12}$ $C^2/(N\bullet m^2)$	8.85×10^{-12} $C^2/(N\bullet m^2)$
G	constant of universal gravitation	$6.672\ 59 \times 10^{-11}$ $N\bullet m^2/kg^2$	6.673×10^{-11} $N\bullet m^2/kg^2$
g	free-fall acceleration at Earth's surface	9.806 65 m/s^2	9.81 m/s^2
h	Planck's constant	$6.626\ 068\ 96 \times 10^{-34}$ J•s	6.63×10^{-34} J•s
k_B	Boltzmann's constant (R/N_A)	$1.380\ 6504 \times 10^{-23}$ J/K	1.38×10^{-23} J/K
k_C	Coulomb constant	$8.987\ 551\ 787 \times 10^{9}$ $N\bullet m^2/C^2$	8.99×10^{9} $N\bullet m^2/C^2$
R	molar (universal) gas constant	8.314 472 J/(mol•K)	8.31 J/(mol•K)
π	(Greek *pi*) ratio of the circumference to the diameter of a circle	3.141 592 654	calculator value

Useful Astronomical Data

Symbol	Quantity	Value used for calculations in this book
I_E	moment of inertia of Earth	8.03×10^{37} $kg\bullet m^2$
M_E	mass of Earth	5.97×10^{24} kg
R_E	radius of Earth	6.38×10^{6} m

Useful Atomic Data

Symbol	Quantity	Established value	Value used for calculations in this book
m_e	mass of electron	$9.109\ 382\ 15 \times 10^{-31}$ kg $5.485\ 799\ 0943 \times 10^{-4}$ u 0.510 998 910 MeV	9.109×10^{-31} kg 5.49×10^{-4} u 5.110×10^{-1} MeV
m_n	mass of neutron	$1.674\ 927\ 211 \times 10^{-27}$ kg 1.008 664 915 97 u 939.565 346 MeV	1.675×10^{-27} kg 1.008 665 u 9.396×10^{2} MeV
m_p	mass of proton	$1.672\ 621\ 637 \times 10^{-27}$ kg 1.007 276 466 77 u 938.272 013 MeV	1.673×10^{-27} kg 1.007 276 u 9.383×10^{2} MeV